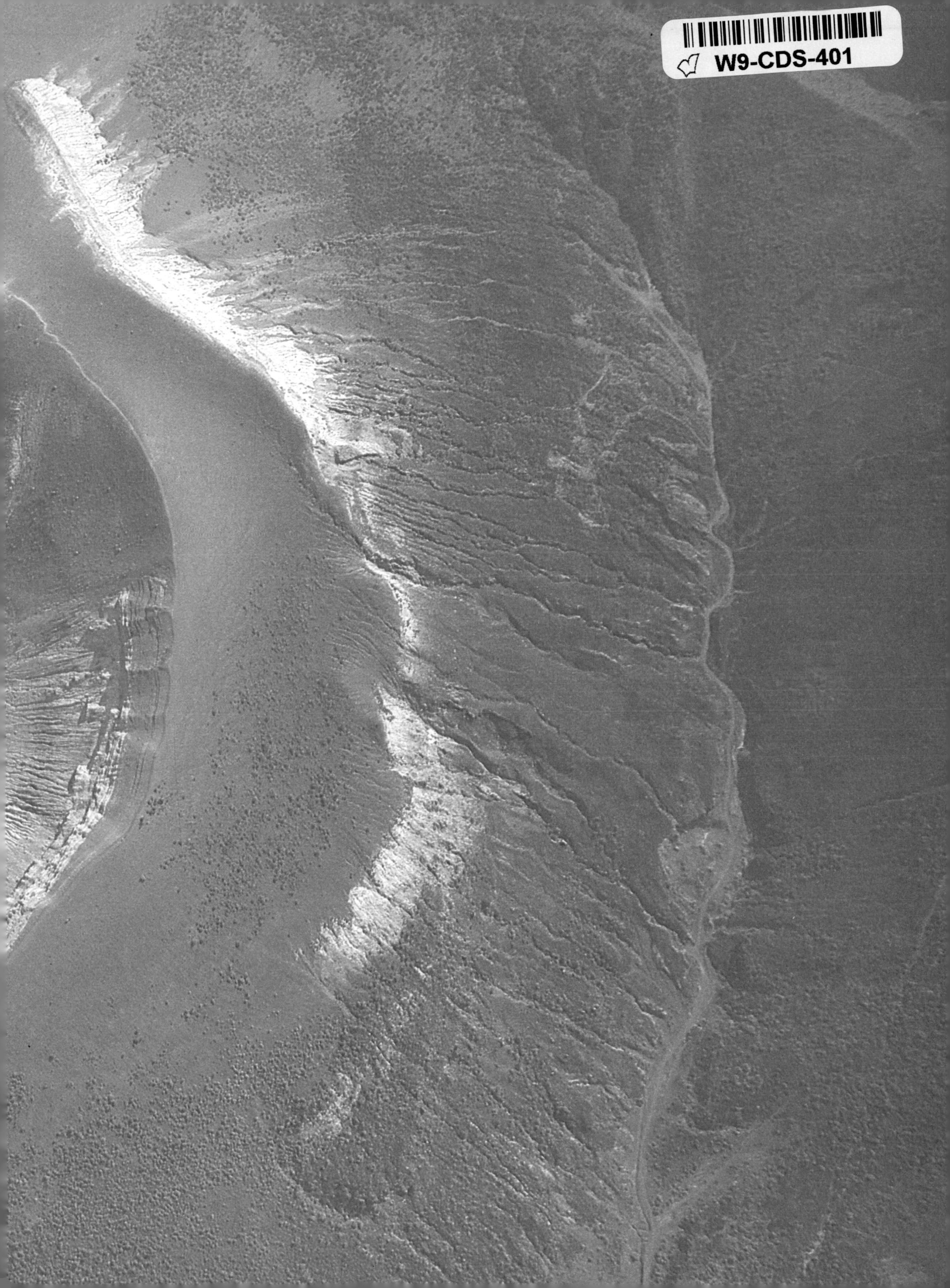

Guide to Places of the World

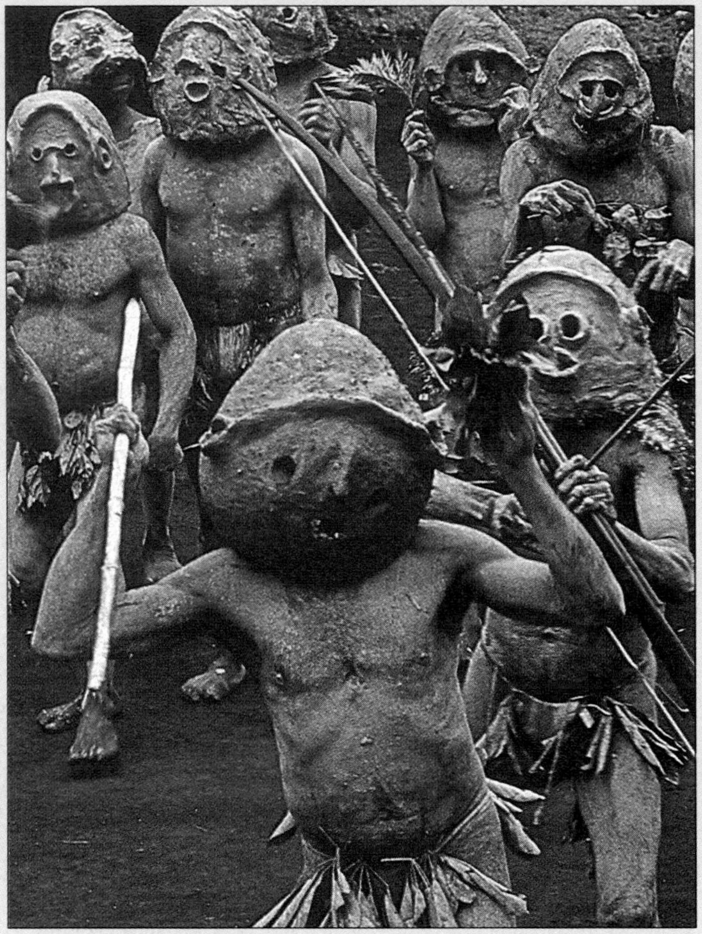

Inside front cover: Crater of Mount Vulcanello, Lipari Islands, Italy.
Inside back cover: Cultivation terraces, Judaea, Israel.
Preceding page: Waterfront scene, Calcutta, India.
Above: Masked tribal dancers, Papua New Guinea.

READER'S DIGEST

Guide to Places of the World

Published by The Reader's Digest Association Limited
London · New York · Sydney · Cape Town · Montreal

Contributors

Guide to Places of the World

was edited and designed by
The Reader's Digest Association Limited
London

Designer: Bob Hook

First Edition

Printed in Great Britain

The typefaces used for the text in this book are Times Roman and Helvetica.

The publishers wish to express their gratitude to the following who, as contributors or consultants, assisted in the preparation of *Guide to Places of the World*.

Consultant editor

Alan B. Mountjoy, MC, MA, FRGS
Reader in Geography
Royal Holloway and Bedford New College,
University of London

Major contributors

Guy Arnold, MA

Kathleen Baker, BSc, PhD, AKC
Lecturer in Geography
School of Oriental and African Studies,
University of London

David Burtenshaw, MA
Principal Lecturer in Geography
Portsmouth Polytechnic

Sylvia Chant, PhD
Lecturer in Geography
Liverpool University

Graham P. Chapman, MA, PhD
Lecturer in South Asian Geography
University of Cambridge

Hugh Clout, BA, MPhil, PhD, D de l'Université
Reader in Geography
University College London

John Davis, BA, PhD
Head of Department of Geography
Birkbeck College, University of London

Harvey Demaine, PhD
Senior Lecturer in Regional Development
Planning, Asian Institute of Technology,
Bangkok, Thailand

F.C. Evans, MA
Senior Master
King's College School, Wimbledon

Alan Gilbert, BSc, PhD
Reader in Geography
University College London

F.E. Ian Hamilton, BSc, PhD
Senior Lecturer in Economics and Social
Studies of Eastern Europe, London School of
Economics and Political Science

Adrian Jansen, BA, MA, Diplome d'Etudes
Approfundis Sciences Politiques
Regional Director South, British Council

Russell King, BA, MSc, PhD
Reader in Geography
University of Leicester

B.J. Knapp, BSc, PhD
Head of Geography Department
Leighton Park School, Easley, Berkshire

David Livingstone, BA, DipEd, PhD
Research Officer, Department of Geography
The Queen's University of Belfast

Keith Lye, BA, FRGS

David McDowall, MA, MLitt

W.R. Mead, BSc, PhD
Emeritus Professor of Geography
University College London

Roy Millward, BA
Late Reader in Historical Geography
University of Leicester

Anthony O'Connor, BA, PhD
Reader in Geography
University College London

John H. Paterson, MA
Professor of Geography
University of Leicester

Deborah Potts, BSc
Lecturer in Geography
School of Oriental and African Studies,
University of London

Stephen A. Royle, MA, PhD, FRGS
Lecturer in Geography
The Queen's University of Belfast

John Sargent, BA, FRGS
Reader in Geography
School of Oriental and African Studies,
University of London

P.T.H. Unwin, MA, PhD
Lecturer in Historical Geography
Royal Holloway and Bedford New College,
University of London

Lawrie Wright, BA, PhD
Lecturer in Geography
Queen Mary College, University of London

Consultants

Robert P. Beckinsale, MA, DPhil
Life Fellow of University College London
(Retired)

Ian Black, BA, PhD
Senior Lecturer in History
University of New South Wales

John Carter

D.J. Dwyer, BA, PhD
Professor of Geography
University of Keele

Valerie Fifer

T.W. Freeman, MA
Emeritus Professor of Geography
University of Manchester

Ivan Jolliffe, BSc, MSc, PhD, CompICE
Lecturer in Geography
Royal Holloway and Bedford New College,
University of London

Hamish Keith, OBE

Richard I. Lawless, BA, PhD, FRAI
Senior Lecturer and Assistant Director
Centre for Middle Eastern and Islamic Studies,
University of Durham

Robert B. Potter, BSc, PhD
Lecturer in Geography
Royal Holloway and Bedford New College,
University of London

Craig J. Reynolds, BA, PhD
Senior Lecturer in History
University of Sydney

A.G. Terry, BSc
Second Master
Latymer Upper School, Hammersmith

Carl Thayer, BA, MA, PhD
Senior Lecturer in Politics
Australian Defence Force Academy

E.M. Yates, PhD, MSc
Reader in Geography
King's College, University of London

Illustrator

Gary Hincks

The publishers also wish to thank the following organisations, which provided assistance:

Africa Institute of South Africa

British Museum, Public Relations Office

Commonwealth Institute

The Institute of Petroleum, London

Royal Geographical Society

United Nations Information Centre, London

Embassies and High Commissions of many nations

Top left: Marketplace, Quercy, France.
Top right: Poncho seller, Ecuador.
Above: Tuaregs and their flocks, Niger.

Entries in this book are alphabetised word-for-word, hyphens being ignored. Thus, for example, entries appear in this order:

New Zealand
Newark
Newcastle *Australia*
Newcastle *South Africa*
Newcastle upon Tyne
Newcastle-under-Lyme

Features on nations and major dependent territories appear in approximately their correct alphabetical position, with a cross-reference inserted at the correct point in the main A to Z sequence, except where this would in any case appear on the same two-page spread as the feature. Cross-references to other entries are indicated by SMALL CAPITALS.

Above: Red Square, Moscow, USSR. Top right: Golden Gate Bridge, San Francisco, USA.

Contents

A-Z gazetteer

All the nations of the world, their major cities, towns and geographical features, described and mapped in nearly 8000 separate entries

Major natural features

Their structure, how they were formed and their importance in shaping our environment

World records

The biggest, longest, widest, highest and deepest in both the natural and man-made worlds

The nations of the world and their dependencies

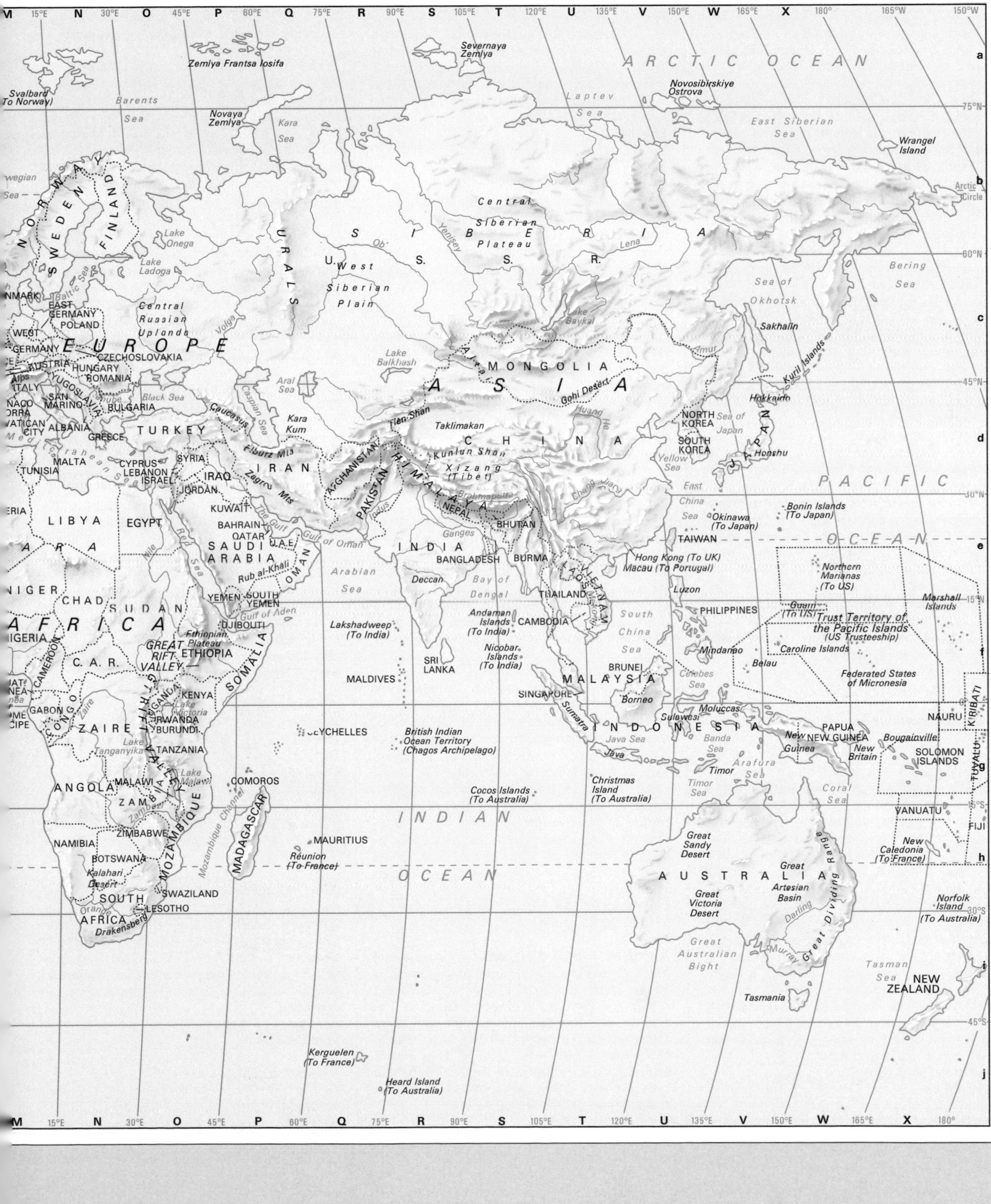

ARCTIC OCEAN
Severnaya Zemlya
Zemlya Frantsa Iosifa
Svalbard (To Norway)
Barents Sea
Novaya Zemlya
Kara Sea
Laptev Sea
Novosibirskiye Ostrova
East Siberian Sea
Wrangel Island
Arctic Circle
NORWAY
SWEDEN
FINLAND
Lake Onega
Lake Ladoga
Baltic Sea
URALS
S I B E R I A
Central Siberian Plateau
U. S. S. R.
West Siberian Plain
Ob'
Yenisey
Lena
Bering Sea
Sea of Okhotsk
Sakhalin
Kuril Islands
Lake Baykal
Amur
DENMARK
EAST GERMANY
POLAND
WEST GERMANY
Central Russian Uplands
Volga
EUROPE
CZECHOSLOVAKIA
AUSTRIA
HUNGARY
ROMANIA
Alps
ITALY
YUGOSLAVIA
SAN MARINO
MONACO
ANDORRA
VATICAN CITY
ALBANIA
BULGARIA
GREECE
Black Sea
Danube
Lake Balkhash
Altai
MONGOLIA
ASIA
Gobi Desert
Aral Sea
Caspian Sea
Caucasus
Kara Kum
Tien Shan
Taklimakan
CHINA
Huang He
NORTH KOREA
SOUTH KOREA
Sea of Japan
JAPAN
Hokkaido
Honshu
Yellow Sea
TURKEY
CYPRUS
SYRIA
LEBANON
ISRAEL
JORDAN
IRAQ
IRAN
Elburz Mts
Zagros Mts
AFGHANISTAN
PAKISTAN
Kunlun Shan
Xizang (Tibet)
HIMALAYA
NEPAL
BHUTAN
Brahmaputra
Chang Jiang
East China Sea
PACIFIC OCEAN
Bonin Islands (To Japan)
Okinawa (To Japan)
TUNISIA
MALTA
Mediterranean Sea
LIBYA
EGYPT
KUWAIT
BAHRAIN
QATAR
U.A.E.
The Gulf
Gulf of Oman
SAUDI ARABIA
Rub al-Khali
OMAN
Nile
Red Sea
Indus
Ganges
INDIA
BANGLADESH
BURMA
TAIWAN
Hong Kong (To UK)
Macau (To Portugal)
Northern Marianas (To US)
NIGER
CHAD
SUDAN
YEMEN
SOUTH YEMEN
Gulf of Aden
Arabian Sea
Deccan
Bay of Bengal
LAOS
VIETNAM
THAILAND
Luzon
PHILIPPINES
Marshall Islands
AFRICA
NIGERIA
DJIBOUTI
Ethiopian Plateau
ETHIOPIA
GREAT RIFT VALLEY
SOMALIA
C. A. R.
CAMEROON
Lakshadweep (To India)
Andaman Islands (To India)
CAMBODIA
Mekong
South China Sea
Guam (To US)
Trust Territory of the Pacific Islands (US Trusteeship)
Caroline Islands
Nicobar Islands (To India)
SRI LANKA
MALDIVES
BRUNEI
MALAYSIA
Mindanao
Belau
Federated States of Micronesia
Celebes Sea
SINGAPORE
Borneo
Sumatra
KENYA
UGANDA
Lake Victoria
RWANDA
BURUNDI
GABON
CONGO
ZAIRE
Zaire
SEYCHELLES
British Indian Ocean Territory (Chagos Archipelago)
Sulawesi
Moluccas
INDONESIA
Java Sea
Java
Banda Sea
PAPUA NEW GUINEA
New Guinea
New Britain
Bougainville
SOLOMON ISLANDS
NAURU
KIRIBATI
TUVALU
Lake Tanganyika
TANZANIA
ANGOLA
MALAWI
Lake Malawi
ZAMBIA
Zambezi
COMOROS
Mozambique Channel
MADAGASCAR
Timor
Arafura Sea
Timor Sea
Coral Sea
Christmas Island (To Australia)
Cocos Islands (To Australia)
VANUATU
FIJI
INDIAN OCEAN
ZIMBABWE
MOZAMBIQUE
NAMIBIA
BOTSWANA
Kalahari Desert
MAURITIUS
Réunion (To France)
Great Sandy Desert
AUSTRALIA
Great Artesian Basin
Great Dividing Range
New Caledonia (To France)
SOUTH AFRICA
SWAZILAND
LESOTHO
Orange
Drakensberg
Great Victoria Desert
Darling
Murray
Norfolk Island (To Australia)
Great Australian Bight
Tasman Sea
NEW ZEALAND
Tasmania
Kerguelen (To France)
Heard Island (To Australia)
M N O P Q R S T U V W X
15°E 30°E 45°E 60°E 75°E 90°E 105°E 120°E 135°E 150°E 165°E 180° 165°W 150°W
a b c d e f g h i j
75°N 60°N 45°N 30°N 15°N 0° 15°S 30°S 45°S

A to Z guide to the nations of the world, places of interest and geographical terms in common use

aa Hawaiian term (pronounced ah-ah) for LAVA flow that has solidified into jagged, block-like masses. In English, it is called 'block lava'.

Aachen (Aix-la-Chapelle) ***West Germany*** City near the junction of the borders of Belgium and the Netherlands, about 70 km (43 miles) west of Bonn. It was the capital of Charlemagne, the Frankish king from AD 768 to 814, who conquered most of Europe and was crowned first Holy Roman Emperor by Pope Leo III in Rome in 800. The emperor's marble throne and a gold bust containing his relics are in the cathedral, built around the chapel which gives the city its French name. Thirty emperors were crowned there, from 1349 to 1531.

Aachen is a university city and spa, and its springs, the hottest in northern Europe, were used by the Romans. Aachen possesses numerous engineering and food-processing industries, and hosts international horse trials. Each year the city presents a Charlemagne award for an outstanding contribution to European unity.

Population 250 000
Map West Germany Bc

Aargau ***Switzerland*** Northern canton in the most level part of Switzerland between Basle and Zürich. Its fertile valleys, cut by tributaries of the River Aare, support dairy and fruit farming, while its mineral springs and small towns attract tourists. The principal town and capital is Aarau (population 16 000), a military and manufacturing centre about 40 km (25 miles) west of Zürich, which makes precision instruments and casts bells. The inhabitants of the canton, which covers 1404 km² (542 sq miles), are mainly German-speaking Protestants.

Population 465 600
Map Switzerland Ba

Aasiaat ***Greenland*** See EGEDESMINDE

Aba ***Nigeria*** Industrial town about 450 km (280 miles) east and slightly south of the capital, Lagos. It is growing rapidly as state and private firms strive to escape the overcrowded industrial centres, particularly Lagos. Aba lies near the oil fields of the Niger delta and Port Harcourt, and its products include cement, textiles, plastics, shoes, foodstuffs and chemicals.

Population 210 700
Map Nigeria Bb

Abadan ***Iran*** Oil-refining port on an island in the Shatt al Arab waterway, about 50 km (30 miles) north of The Gulf. It also has a petrochemical complex, producing plastics and detergents. Abadan has been badly damaged during the war between Iran and Iraq, which started in 1980. Previously it was the world's largest exporter of petroleum products.

Population 296 000
Map Iran Aa

Abéché ***Chad*** Capital of the south-eastern prefecture of Ouaddai, about 650 km (405 miles) east and slightly north of the capital, Ndjamena. It lies in a semidesert region, and is a trading centre for nomad caravans, with markets and mosques. Gum arabic is produced locally, but a Chinese-financed groundnut-growing scheme failed.

Population 54 000
Map Chad Bb

Abeokuta ***Nigeria*** Industrial town about 75 km (45 miles) north of the capital, Lagos. It was founded in 1830 as a stronghold against invasion by Dahomeyans from what is now Benin. Its inhabitants adopted Western-style education after European missionaries arrived in 1847, and it became for a time the largest town in West Africa.

Abeokuta now has many traditional craftsmen, including blacksmiths, carpenters, potters and dyers, and large-scale industries such as quarrying, plastics, food processing and breweries. However, most people still earn their living in the production of cocoa and of palm oil, used in margarine and cooking fats.

Population 301 100
Map Nigeria Ab

Abercorn ***Zambia*** See MBALA

Aberdare Mountains ***Kenya*** Mountain range to the north of Nairobi rising to 3994 m (13 104 ft) at Mount Lesatima, and dropping steeply in the west into the Great Rift Valley. Aberdare National Park occupies 788 km² (304 sq miles) of the lower slopes. Britain's Queen Elizabeth II was visiting the Treetops Hotel in the park when she acceded to the throne in February 1952.

Map Kenya Cb

Aberdeen ***Hong Kong*** Small industrial town on the south-western coast of HONG KONG ISLAND. It is separated from the small island of Ap Lei Chau by a narrow stretch of water. Developed as a fishing village after the arrival of the British in Hong Kong in the mid-19th century, Aberdeen is famous for its concentration of boat people – Chinese who live on junks and sampans in the harbour – and for its floating restaurants. Some of these restaurants are like multistorey palaces, ablaze with lights after dark.

Population 92 554
Map Hong Kong Cb

Aberdeen ***United Kingdom*** City on the east coast of Scotland, in Grampian region. It is known as the Granite City because much of it is built of the locally quarried stone. It has a university, founded before 1500, and three cathedrals, the oldest of which, St Machar's, dates back to the 14th century. It is also a fishing port, and thrived in the late 1970s and early 1980s as the headquarters and transport centre for the North Sea oil industry.

Population 214 000
Map United Kingdom Db

Abertawe ***United Kingdom*** See SWANSEA

Abha ***Saudi Arabia*** Capital of Asir province, 575 km (357 miles) south of Jeddah. Set 2100 m (6890 ft) above sea level in the Asir mountains, Abha has a temperate climate with light summer rains and winter mists. The old Arab town – one of the most picturesque in Saudi Arabia – has acquired a number of modern buildings, including a hospital and government offices.

Population 31 000
Map Saudi Arabia Bc

Abidjan ***Ivory Coast*** Capital city standing amid forests and lagoons on the coast. Its economy and strategic importance were boosted in 1950 when the VRIDI CANAL was cut through a huge sandbar, which had blocked access to the open sea. The canal has turned Abidjan into a substantial, sheltered deep-water port, stimulating the whole economy of the Ivory Coast.

Today it is a cosmopolitan city of many faces: the modern skyscrapers of the commercial centre, the Plateau; the administrative heart of the Presidency with marvellous views of Banco Bay; the exclusively African suburb of Adjamé; the elegant, lagoon-side suburb of Cocody Bay with its hotels and embassies; and Treichville, the crowded market and entertainment centre. To the south is an industrial zone and port at the old suburb of Marcory and the new one of Koumassi. There is a busy food and cloth market at Adjamé and local craft shops at the Plateau.

The government plans to move the capital to YAMOUSSOUKRO, but no date has been set for the formal changeover.

Population 1 850 000
Map Ivory Coast Bb

Abilene ***USA*** City in the Midwest state of Kansas, about 225 km (140 miles) west of Kansas City. It developed in the 19th century as a cattle town and rail centre.

Population 6400
Map United States Gc

ablation Loss of snow and ice from the surface of an ice sheet or glacier, chiefly by melting and evaporation.

Åbo ***Finland*** See TURKU

Abomey ***Benin*** Trading and administrative town, about 110 km (68 miles) north of Cotonou. Founded in 1658 by the Fon tribe as the capital of their kingdom of Dahomey, Abomey is rich in historic buildings, relics and craftsmen.

Population 45 000
Map Benin Ab

Aboukir ***Egypt*** See ABU QIR

abrasion Wearing away of part of the earth's surface by an abrasive material, such as sand, carried along by wind, water or moving ice.

Abruzzi ***Italy*** Highland region covering 10 794 km² (4167 sq miles) facing the Adriatic Sea in central Italy. In the past, it has been off the tourist track because of its remoteness, but new motorway links have now made the area accessible from Rome. The uplands are being developed for skiing and summer tourism, while the coastal resorts such as PESCARA, Francavilla al Mare and Ortona are becoming increasingly popular. Abruzzi embraces the GRAN SASSO and Maiella massifs, the highest points of the Apennine mountains at 2912 m (9554 ft) and 2793 m (9163 ft).
Population 1 246 600
Map Italy Dc

Abruzzo National Park ***Italy*** A 292 km² (113 sq mile) park in the Apennine mountains rising to 2247 m (7372 ft) at Mount Petroso. It was founded in 1923 to safeguard rare animals such as the Marsican bear – of which only about 100 are left – the Abruzzo chamois, the Apennine wolf and the wild cat.
Map Italy Dd

Abu Dhabi ***United Arab Emirates*** Largest and richest of the seven emirates, and the prime mover in the formation of the UAE in 1971. Situated on the south-west coast of The Gulf, Abu Dhabi covers 67 350 km² (26 000 sq miles) Its oil reserves are estimated at 30 000 million barrels and it also has huge gas reserves. The main industry is petroleum, and the emirate controls a large gas-processing plant on Das Island in The Gulf. More traditional industries include fishing and the harvesting and marketing of pearls. Abu Dhabi City is the capital of the emirate and the UAE. It has grown dramatically over the last 25 years and is one of The Gulf's most modern cities – with wide boulevards, tall office blocks and high-rise apartments. Little remains of the old town except for the Old Palace. There are three modern palaces and the modern Great Mosque.
Population (emirate) 535 700; (city) 244 000
Map United Arab Emirates Ba

Abu Qir (Aboukir, Abukir) ***Egypt*** Mediterranean bay 20 km (12 miles) north-east of the port of Alexandria. It is the site of the Battle of the Nile. In August 1798, the British admiral Horatio Nelson sailed into the bay and destroyed the French fleet of Napoleon Bonaparte, effectively ending Napoleon's plans to conquer the Middle East. In 1801, the British general Sir Ralph Abercromby landed a force at the village of Abu Qir in the bay, and drove the last of the French out of Egypt.
Map Egypt Bb

Abu Simbel ***Egypt*** Site of two huge temples hollowed out of a sandstone cliff beside the River Nile, near the border with Sudan. The temples were built for the pharaoh Rameses II, who ruled for 66 years during the 13th century BC. Inside one temple are four standing figures of the pharaoh as Osiris, the god of the afterlife, each 9 m (30 ft) high. At the entrance are four seated figures of the king, each 20 m (66 ft) high. In 1966, the complete temples and statuary were cut from the rock face and raised to a higher site, above the flood level of Lake Nasser created by the ASWAN HIGH DAM.
Map Egypt Bd

Abuja ***Nigeria*** Town some 525 km (325 miles) north-east of the capital, Lagos. It lies in federal territory, and in 1976 was chosen as the country's future capital because of its central position and the congestion in Lagos.
Map Nigeria Bb

Abukir ***Egypt*** See ABU QIR

Abyan ***South Yemen*** District along the coast to the north-east of Aden. It is one of the country's richest agricultural regions, and is being developed as a cotton-growing area. Irrigation and land-reclamation projects have been financed by the Kuwait Fund for Arab Economic Development. The Abyan dam is making further agricultural projects possible. The district's extensive limestone deposits are used to make cement.
Map South Yemen Ab

Abydos ***Egypt*** Ruins of an ancient city and royal burial place on the west bank of the River Nile, about 450 km (280 miles) south of Cairo. Abydos was a cult centre of Osiris, the Egyptian god of the afterlife. Between 1600 and 1200 BC, pharaohs, who presented themselves as the earthly embodiment of the god, had temples and tombs built there. The most splendid temple is that of the pharaoh Setekhy I, father of Rameses II.
Map Egypt Bc

abyssal Term applied to the depths of the oceans between 1000 m and 3000 m (3280 ft and 9842 ft) below the surface. Abyssal also applies to the animal life of those depths.

abyssal plain Large, fairly level area, about 4000 m to 6000 m (13 123 ft to 19 685 ft) below the surface of the oceans, that makes up most of the ocean floor.

Acadia National Park ***USA*** Rugged area of forest and hills on Mount Desert Island, Isle au Haut and the coast of Maine, about 580 km (360 miles) north-east of New York city. It was established in 1919 as Lafayette National Park – the first United States national park east of the Mississippi river – and covers 156 km² (60 sq miles).
Map United States Mb

Acapulco ***Mexico*** Seaport and world famous beach resort on the Pacific coast in the state of Guerrero. Acapulco enjoys year-round sunshine and temperatures of around 27°C (80°F), and a score of sandy beaches stretch for 16 km (10 miles) along its coastline. Luxury hotels are set against a majestic backdrop of mountain slopes and evergreen tropical vegetation, including coconut groves.

From 1565 to 1815 galleons sailed annually from Acapulco to Manila in the Philippines, exporting silver in exchange for porcelain, iron, silks and spices. Today the city exports cotton, fruit, hides and tobacco.

In 1985 an earthquake measuring 8.2 on the RICHTER SCALE struck off the coast north-west of Acapulco, damaging hundreds of buildings in Mexico City, 290 km (180 miles) to the north. Acapulco itself was relatively unscathed.
Population 800 000
Map Mexico Cc

Accra ***Ghana*** The country's capital. It stands on the coast, on a dry plain more suited to livestock than crops, and the site has been occupied by the Ga people since at least the 16th century. Their fishing, trading and farming economy was flourishing when the British, Dutch and Danes arrived in the 17th century to establish forts and trading posts. The increase in trade prompted the transfer of the seat of colonial administration to Accra from Cape Coast – 120 km (75 miles) to the south-west – in 1876. A railway line inland was built soon afterwards.

The city's old core lies around the harbour, with the business and shopping district to the east. Beyond, the elegant former dwellings of colonials, now occupied by the wealthier Ghanaians, are set in spacious grounds – a stark contrast to the cramped homes of the poorer African residential areas.

There are many monuments to independence and unity such as the State House, a vast modern conference centre, and Independence or Black Star Square, where all the main ceremonies are held. The city has a university, museum and zoo. Hotels and restaurants abound. Shopping for luxury goods can be a problem simply because they are in short supply, but the colourful local markets are well stocked with cotton batik, baskets, woodcarvings, and hand-made gold and silver jewellery.
Population (Greater Accra) 1 045 400
Map Ghana Ab

acid rock An igneous rock containing 10 per cent or more free quartz. Silica (silicon dioxide or quartz) was formerly thought of as an acidic oxide, and an acidic rock was defined as one containing more than 66 per cent silica. The name has been retained.

Aconcagua ***Argentina*** Mountain about 1040 km (645 miles) west of Buenos Aires. It is the second highest peak in the Western Hemisphere after Ojos del Salado, rising to 6960 m (22 835 ft). The western slopes lie in Chile, but the summit is wholly in Argentina.
Map Argentina Bb

Aconcagua ***Chile*** Region in the central valley, also known as Valparaíso. It has some of the country's most fertile soil and produces much of Chile's fruit, wine, hemp and tobacco. The principal towns are VALPARAISO, the capital, Quillota, Los Andes and San Felipe.
Population 1 205 000
Map Chile Ac

Açores ***Portugal*** See AZORES

Acre ('Akko) ***Israel*** Port on the Bay of Haifa, 13 km (8 miles) north of the city of Haifa. It has been a trading port since the days of the Phoenicians (9th century BC), and was known as Ptolemais in New Testament times. Acre changed hands frequently during the Crusades and was besieged unsuccessfully by Napoleon during his Syrian campaign of 1799. The remains of an underground 13th-century Crusader town are still visible today, including the huge crypt of the Knights of St John who occupied the town from 1191 to 1291.
Population 37 700
Map Israel Ba

Afghanistan

FOUGHT OVER FOR THOUSANDS OF YEARS, THIS WILD MOUNTAINOUS LAND IS ONCE AGAIN EMBROILED IN BITTER WARFARE WITH AN INVADING NEIGHBOUR – RUSSIA

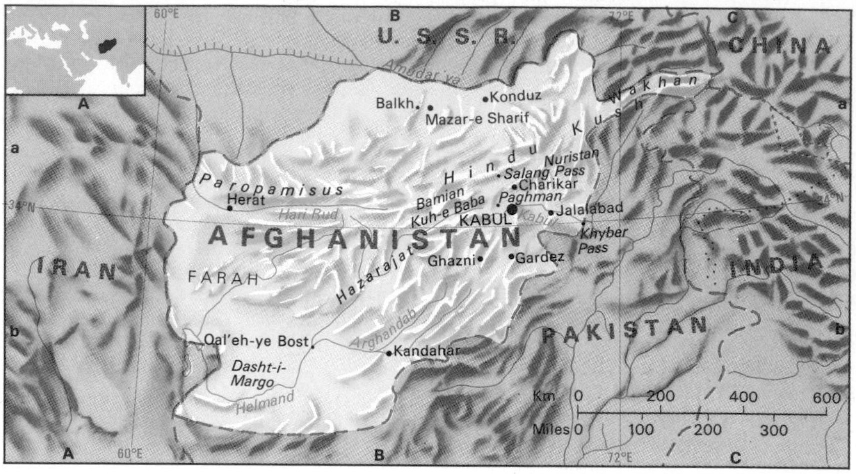

Afghanistan, ribbed and girdled by majestic mountains, is the hinge of Asia. On it pivots the gateway to India through the KHYBER PASS, and the back door to Russia – and so it has been traversed and fought over by invading armies since earliest times. Darius (about 548-486 BC) of Persia, Alexander the Great (356-323 BC), the Arabs, and the Tatars under Genghis Khan (1162-1227) and Tamerlane (1336-1405) all occupied the country. The British fought three times to maintain their hold on the Khyber Pass. And then, on Christmas Day 1979, the Russians came.

They came to shore up a weak Marxist government; they stayed to endure the guerrilla warfare of fierce tribesmen, used to handling guns since they were children, used to blood feuds and tribal battle. Only two things brought the tribes together: they were Muslims, and it was their country. Five years after the Russians landed, the sound of gunfire was still being heard in the capital, KABUL, and despite their modern weapons there were large areas the Russians could not control. And the whisper began that Afghanistan could be Russia's Vietnam.

Afghanistan has an area of some 650 000 km² (250 000 sq miles), about the size of Texas, and a normal population of about 18 million. Following the Russian invasion, some 3 million people crossed the borders, mainly into Pakistan, to escape. Another 2 million have crowded into the towns, where they are safe from Russian bombing if not from guerrilla guns. No one – not even the Russians – ventures far out of town at night, for armed tribesmen lurk not far away.

Afghanistan is a land of wild contours, hidden green valleys and baked mud villages. Little of the country is below 1200 m (about 4000 ft) and much of it remains unexplored. The mountains of the HINDU KUSH, known locally as 'The Theatre of Heaven', form a giant barrier across the centre of the land. The second highest mountain system in the world, its jagged peaks rise to 7690 m (25 229 ft) in the east, forming a moon landscape whose barren heights and windswept plains have always held a fascination for travellers. To the north lies the most productive farming area – along the banks of the AMUDAR'YA which forms a 960 km (595 mile) border with the USSR. Much of the south and south-west is desert.

The climate is one of extremes. Summer temperatures can soar to 49°C (120°F) in the south; winter temperatures may fall to −26°C (−15°F) in the mountains. The weather can vary by as much as 30°C (54°F) in one day, and fierce winds bring terrible dust storms. What rain there is falls between March and May, but for most of the year there are clear brilliant skies.

It is one of the world's poorest countries. Only about 12 per cent of its people are literate and the average expectation of life is 37 years. Only about 12 per cent of the land is suitable for agriculture, yet 70 per cent of the workers are peasant farmers. Two and a half million are nomads, following their flocks of fat-tailed sheep and goats.

The Pathans (Pashtuns) are the largest ethnic group, accounting for half the population. With other groups – Tadzhiks, Hazaras, Turkmen, Uzbeks, Nuristanis and Baluchis – they make a list which sounds like the roll call of invading armies that have passed through the land, each leaving its own settlers. The result is a mixture of many races and tongues.

Tribal rivalries have always made the country difficult to govern. Each group maintains its own individualism, ready to quarrel with its neighbours and unwilling to submit to any central government – at least not without constant protest. Only in recent years have attempts been made to move the country from its state of feudal tribalism. A five-year plan launched in 1956 to improve roads and develop mines and industry was followed by another which opened up parts of the country hitherto inaccessible.

In this development Afghanistan was careful not to favour East or West. A good example is the road, nearly 1600 km (1000 miles) long, that loops south from HERAT to KANDAHAR and then north to Kabul. It forms part of the Trans-Asian Highway and was built partly by the Americans and partly by the Russians, then was maintained by the Chinese.

Modern Afghanistan dates from the middle of the 18th century when Ahmad Shah, a tribal leader, established a united state covering most of present-day Afghanistan. For most of the 19th century the country was subjected to big-power pressures from Tsarist Russia in the north and the British in India to the east. Suspicions that its rulers were favouring the Russians led to the first two Afghan Wars with Britain (1838-42 and 1878-9). British forces were defeated both times, but returned in 1880 to enforce a treaty that gave them control of the Khyber Pass. By playing Britain and Russia off against each other, Afghanistan managed to remain independent, though under British influence.

In 1919 Amanullah Khan (who styled himself king after 1926) invaded India when his demand for complete independence was not met, and after inconclusive fighting (the third Afghan War) this was granted in 1921. Afghanistan was neutral in both World Wars, but relations with Pakistan have been strained since the partition of India and Pakistan's independence; the Afghans wanted the Pathans of the North-West Frontier province to be given the chance to join Afghanistan.

THE RUSSIAN INVASION

In 1973 King Muhammad Zahir was overthrown in a coup, and the monarchy ended after nearly 50 years. Afghanistan was declared a republic. Five years later President Muhammad Daoud was murdered in a second coup, which led to a Marxist government under Muhammad Taraki. In September 1979, Taraki was ousted in another coup and replaced by Hafizullah Amin. But, bitterly opposed by the people of the countryside, the government found it could maintain itself only by calling on Russia for help. The Russians were happy to respond, realising an ambition to control their own 'back door' that goes back to Tsarist days. In December 1979 they invaded the country, deposed Amin and installed Babrak Karmal in his place. He was replaced as leader of the Afghan Communist Party by Major-General Najibullah in May 1986.

The rivalry between clans and their readiness to engage in fights or blood feuds now found a new outlet. The country was split between Marxist supporters, backed by the Soviet army, and resistance fighters providing fierce if uncoordinated opposition. Sometimes they fought among themselves. The 115 000 Russian troops, hampered by mountainous terrain and poor roads, found they were facing a brutal and bitter war against tribesmen renowned for their guerrilla tactics, who took to the hills to harass and strike.

The Russians relied mainly on bombing and ground sweeps, only to find that the guerrillas faded away when large numbers of soldiers appeared. They intensified the bombing to include crop-growing areas as well as villages: fields, stocks and herds were blitzed so that sowing times were missed and in some areas there was serious danger of famine. There were reports of atrocities on both sides. But the war has been costing the Russians the equivalent of US$1500 million a year, and they can be sure only of controlling the towns.

AGRICULTURE AND INDUSTRY

In good times Afghanistan is able to produce food to sustain its people, although the balance of nature is always precarious. Wheat, fruit, cotton, vegetables and sugar, delicious melons, plums, quinces, apricots and figs ripen in the warm autumn days, and the peaches and grapes of Kandahar are famous. Many medicinal herbs, little known in the West, are found. The country has three rare domesticated animals: the yak, the two-humped Bactrian camel and the karakul lamb, whose soft curly fleece is used for the treasured Persian lamb furs. Afghan rugs and sheepskin coats are also much valued.

Manufacturing is little developed, but includes textiles, cement, leather goods, glassware, bicycles and food products. There is a wide range of minerals, including coal, iron ore and copper, but most are so inaccessible that it is uneconomic to exploit them. For instance, it is not worth extracting iron ore found at 3660 m (12 000 ft) when it can be obtained more easily elsewhere. One of the few exceptions is the natural gas found around Sheberghan and Sar-e Pol in the far north. This is extracted and exported to the USSR. Some of the finest lapis lazuli, a blue semiprecious stone, comes from the highlands of Badakhshan and the Panjshir valley in the north-east.

For most Afghans, life is hard and simple. Some live in mountain valleys so high that they are cut off by snow for months. Sometimes whole communities will move from one village to another in search of better land. The herdsman with a rifle slung across his shoulder may look rugged and romantic in his turban and robes as he tends his flocks, but many people live in poverty, and when drought hits the land they suffer desperate hunger.

The people are overwhelmingly Muslim, and education, where it exists, is often in the hands of the mullahs (religious leaders). Families are large and patriarchal, but up to half the children die before they reach the age of five. The first woman was elected to the parliament in 1965, but society remains male-dominated. Except in the towns, most women appear veiled in public; some may not be allowed out by their husbands at all; and in some parts a wife can be bought unofficially for as little as £3.

Rice or nan (a pancake-shaped bread) and mutton, followed by fruit, is the main food, and green tea, served with much sugar, the favourite drink. The teahouse is the place to meet friends and gossip, or entertain strangers. Afghans, even the poorest, are very hospitable, and strangers may be embarrassed to be given a needy family's choicest food. The bazaars and mosques are the real centres of activity; few prices are fixed and in the bazaars bargaining is automatic.

The great attraction for the visitor lies in the wild nature of the land: few countries have more rugged mountain scenery, higher passes or more stunning views. Kabul has a fascination of its own: it has some modern hotels and housing, but the interest lies in the old quarter's narrow, crooked streets and bazaars, and seeing the races who mix there.

Here are Tadzhiks from the region round Herat; Uzbeks and Turkmen from the north-east, including the Wakhan, the narrow strip of Afghanistan jutting north-eastwards between the USSR and KASHMIR; Hazaras, of Mongolian origin, from the central mountains; and Pathans from the east and south.

In Kabul, legacies of former conquerors include the tomb and gardens of Babur, who captured the city in 1504 and made it the capital of the Mongol Empire, and the mausoleum of Nadir Shah of Persia, who took Kabul in 1738. To the east the spectacular Khyber Pass, narrow and steep-sided, runs through the Safed Koh mountains to link PESHAWAR, in Pakistan, with Kabul.

In the west, Herat lies on the old trade route from Persia to India. It is noted for its citadel, markets and decorated *gharries* (horse-drawn cabs). Its Great Mosque was founded in the 12th century, and there are exquisite minarets. North-west of Kabul, the small town of BAMIAN was once a Buddhist stronghold: its deserted Valley of the Gods, with huge stone-carved Buddhist figures, is awe-inspiring. The SALANG TUNNEL, at nearly 3400 m (11 150 ft), leads through the mighty Hindu Kush to the northern plains – and to the USSR, whose leaders must now be pondering the lessons to be learnt from the Afghan Wars.

▲ HOLY WAR A guerrilla leader briefs his men in the mountains west of Kabul. The Muslim 'Mujahidin' are waging a *jihad* ('holy war') against the Afghan and Soviet Communist forces.

AFGHANISTAN AT A GLANCE

Area 652 090 km² (251 772 sq miles)
Population 15 040 000
Capital Kabul
Government One-party republic (under Russian control)
Currency Afghani = 100 puls
Languages Pushtu, Dari (Persian)
Religion Muslim
Climate Continental; average temperature in Kabul ranges from −8 to 2°C (18-36°F) in January to 16-33°C (61-91°F) in July
Main primary products Sheep, goats, fruit, wheat, cotton, nuts, vegetables; natural gas, iron, copper
Major industries Agriculture, fur and leather products, carpets, textiles, cement, glassware, bicycles, food processing
Main exports Fruit, nuts, natural gas, carpets, lambskins, cotton
Annual income per head (US$) 160
Population growth (per thous/yr) 17
Life expectancy (yrs) Male 36 **Female** 39

Ad Dakhla (Villa Cisneros) ***Western Sahara*** Principal port on the Atlantic coast, lying on Rio de Oro Bay, and founded by Spanish colonists in 1884. The bay provides an open anchorage for shipping, and the port has fuel-oil storage facilities for bunkering.
Population 5570
Map Western Sahara Aa

Ad Dawhah ***Qatar*** See DOHA

Adamawa Massif ***Cameroon*** See ADOUMAOUA MASSIF

Adam's Peak (Samanala; Sri Padastanaya) ***Sri Lanka*** Mountain 72 km (45 miles) east of Colombo. It rises to 2243 m (7359 ft) and is sacred to the island's Buddhists, Hindus, Christians and Muslims. A mysterious depression on the summit is shaped like a giant human footprint 1.6 m (5.25 ft) long. Buddhists believe it was made by Prince Siddartha Gautama (about 563-483 BC), the founder of Buddhism, while Hindus attribute it to Shiva. Muslims hold that Adam stood there on one foot for 1000 years as a penance after his expulsion from the Garden of Eden, while Roman Catholics credit it to St Thomas, who preached in the region.
Map Sri Lanka Bb

Adana ***Turkey*** City about 400 km (250 miles) south of the capital, Ankara. It was founded by the Hittites around 1400 BC and conquered by Alexander the Great in 334 BC. The 2nd-century stone bridge across the Seyhan river, which runs through the city, was built by the Roman emperor Hadrian, and other ancient buildings include 14th-century and 16th-century mosques and a 15th-century covered bazaar.

The surrounding lowland province of the same name produces cotton, textiles, cement, agricultural machinery and vegetable oils. It covers an area of 17 253 km^2 (6661 sq miles).
Population (city) 776 000; (province) 1 757 100
Map Turkey Bb

Addis Ababa ***Ethiopia*** National capital, standing at the centre of the country at an altitude of over 2440 m (8000 ft), and within the homeland of the ruling Amhara people. The city developed gradually during the 19th century, and became the capital in 1896. It has an eight-sided Coptic cathedral, built in 1896; Africa Hall, built for the inauguration of the Organisation of African Unity in 1963; and the Mercato market, one of the largest on the continent. Other features of interest include the old imperial palace, the parliament buildings and the university (1961).
Population 1 500 000
Map Ethiopia Ab

Adelaide ***Australia*** State capital of South Australia lying on Gulf St Vincent in the south-east of the state. It follows closely the original 1833 design by Colonel Light, Adelaide's founder, the broad streets of its city centre being completely surrounded by a ring of spacious parks. The Torrens river, dammed to form a lake in one of the parks, separates the city centre from the northern suburbs. Behind the eastern suburbs the Mount Lofty Ranges rise to 725 m (2379 ft) at Mount Lofty.

Historic buildings include the Town Hall (built in 1863-6), Government House (1838-40) and Parliament House (1883-1939). The city is also a port and has a wide range of industries, such as motor-body assembly, railway workshops and the manufacture of electrical goods, household appliances and textiles. Its Festival Complex, overlooking the lake, has housed an arts festival, attracting performers from many countries, every second year since 1960.
Population 969 000
Map Australia Fe

Adelsberg ***Yugoslavia*** See POSTOJNA

Aden ***South Yemen*** The country's capital and main port, Aden lies 160 km (100 miles) east of the Bab al Mandab strait. Aden has been a trade centre since antiquity. From the 7th to 16th centuries it was held by Muslim Arabs, until it fell to the Ottoman Turks in 1538. British control began in 1802 when a treaty was signed between Britain and Turkey to check Aden-based piracy. In 1839 Aden was annexed by Britain. Under British rule it became a coaling station on the sea route to India, and after the opening of the Suez Canal in 1869 its trade expanded rapidly. In 1937 it became a crown colony. During the Second World War it became a fortified British naval base. In 1960 the people put up a fierce fight for independence, until the British finally left in 1968.

It has an oil refinery, which was taken over by the government when the British left, but is operating at only a fraction of its capacity. Aden's other industries include the manufacture of cement blocks, tiles, bricks, salt and food processing. A former Royal Air Force base at Khormaksar – 11 km (7 miles) north of the city – is now an international airport. The present city consists of three sections: Crater, the old quarter, which was built in the crater of an extinct volcano; at-Tawahi, the business section; and Ma'alah, the harbour region.
Population 264 326
Map South Yemen Ab

Aden, Gulf of Major shipping lane linking the RED SEA (via the narrow strait of Bab al Mandab) to the ARABIAN SEA. Geologically, the gulf is still young and is still getting wider as Arabia drifts north-eastwards away from Africa. The gulf's chief ports are ADEN (South Yemen), DJIBOUTI and BERBERA (Somalia).
Length 880 km (547 miles)
Width (at mouth) 480 km (298 miles)
Map South Yemen Ab

Adige ***Italy*** Second longest river in the country, flowing 410 km (255 miles) from its source in the Alps near where the Swiss, Austrian and Italian borders meet, to the Adriatic Sea.
Map Italy Cb

Adirondack Mountains ***USA*** Range in north-east New York state covering some 19 500 km^2 (7500 sq miles) and rising to 1629 m (5344 ft) at Mount Marcy. More than 200 lakes and hundreds of streams make it a popular holiday area.
Map United States Lb

Adoumaoua (Adamawa) Massif ***Cameroon*** Plateau partly covered with volcanic rocks in the central Cameroon. Some 150-300 km (95-185 miles) wide, it divides the country into two, and is the source of most of the country's rivers. The valleys are densely wooded, and the grazing land between is used by about 70 000 families, mainly of the Islamic Fulani tribe, with their herds of zebu cattle. There are reserves of bauxite (aluminium ore) on the massif.
Map Cameroon Bb

Adrar ***Algeria*** Oasis town deep in the Sahara desert about 540 km (335 miles) south of Ain Sefra. It is a major staging post at the beginning of the Route du Tanezrouft, heading south to Gao in Mali, and one of the most barren of all Saharan routes.
Population 7500
Map Algeria Ab

Adriatic Sea Arm of the Mediterranean, lying between the Balkan and Italian peninsulas. The sea is 770 km (478 miles) long with an average width of 160 km (99 miles). The Strait of OTRANTO, 76 km (47 miles) wide, separates Italy and Albania in the south. The Adriatic, especially the scenic Dalmatian coast of Yugoslavia, is a popular tourist area.
Area 160 000 km^2 (41 400 sq miles)
Map Italy Ec

Aegean Sea Arm of the MEDITERRANEAN lying between Greece, Turkey and Crete. The narrow DARDANELLES in the north-east – only 1.2 km (0.7 miles) wide in places – is the gateway to the BLACK SEA. It separates the Aegean from the Sea of Marmara. The Aegean is studded with islands, many of which have associations with Greek legend and history. They consist of three groups: the CYCLADES, Sporades and DODECANESE. These hilly peaks of a sunken plateau cover a total land area of 9122 km^2 (3521 sq miles) and are fringed with whitewashed towns and dotted with the ruins of Byzantine monasteries. Seafaring, tourism, and the cultivation of vines, olives and fruits are the main activities. Earthquakes and volcanoes, notably at SANTORINI, are reminders that this is an unstable part of the earth's surface.
Area 179 000 km^2 (69 110 sq miles)
Map Greece Db

Aegina (Aíyina) ***Greece*** Island resort in the Saronic Gulf 30 km (19 miles) south-west of Piraeus. Its pine-forested interior drops down to fine bathing beaches. The hilltop Temple of Aphaia, dating from the 6th century BC, is a classic example of Doric architecture. The sculptures that once stood in it are now housed in a Munich museum. Aegina is famed for its pistachio nuts.
Population 11 200
Map Greece Cc

aeolian Relating to the wind. Aeolian deposits are fine rock particles that have been carried by the wind (such as the sand of deserts and dunes) and dust known as LOESS.

aeration zone See WATER TABLE

afforestation Conversion of open land, particularly heathland and moorland, into forests, usually by planting fast-maturing coniferous trees. In Britain, in particular, the timber produced constitutes a useful crop on marginal land.

Afghanistan See p. 12

Africa Covering more than 30 million km² (11 million sq miles), Africa is, after Asia, the world's second largest continent. It is nearly bisected by the Equator, and both northern and southern Africa extend to the mid-30s of latitude so that its climatic zones mirror each other from the tropical forests of the equatorial region through semitropical savannahs, dry steppelands and desert to the Mediterranean climates of the extremities. More than 550 million people live in Africa, but it is lightly peopled, the average density being less than 20 per km² (8 per sq mile). The birthrate is high, but is no longer matched by high death rates, and the rate of population increase is rising sharply. For the most part, the people are poor, and disease, famine and drought are all too prevalent.

Most of Africa consists of high plateaus which descend steeply to generally narrow coastal regions. Waterfalls and rapids flowing off the plateaus made exploration difficult, since the only way into the continent for explorers was by sailing upriver, and for long this contributed to the continent's isolation. There are few natural harbours and the coastline is very short in relation to the landmass; 14 out of the 47 mainland African nations are landlocked. Four of the world's greatest rivers are in Africa: the NILE, ZAIRE (Congo), NIGER and ZAMBESI.

There are some majestic mountains, often made more so by their isolation in vast plains. In the north the Atlas range stretches 2250 km (1400 miles) from the Atlantic to Tunisia and reaches up to 4167 m (13 671 ft) at its highest

point. In East Africa, the Ethiopian highlands form a great massif of intersecting ranges, whose loftiest peak is Ras Dashen (4620 m, 15 157 ft). Farther south along the East African Rift Valley are the Ruwenzori mountains, permanently snow-covered, and climbing to 5110 m (16 765 ft), Mount Kenya (5200 m, 17 060 ft), Mount Meru (4565 m, 14 977 ft), and the mighty Kilimanjaro, Africa's highest mountain at 5895 m (19 340 ft).

The East African Rift Valley, which is part of the GREAT RIFT VALLEY system, is the world's most spectacular geological depression. It stretches from Beira in Mozambique in south-east Africa, and divides in two north of Lake Malawi. One branch runs north as far as Lake Albert (Mobutu), while the other, eastern branch becomes the trench of the Red Sea and continues into south-west Asia. Along the Rift Valley are some of the world's largest lakes: Malawi (30 040 km², 11 600 sq miles), the 676 km (420 mile) long Tanganyika and the sea-like Victoria – 70 484 km² (26 828 sq miles).

Much of central Africa is occupied by a great expanse of tropical rain forest, second in size only to that of the Amazon basin. It covers much of the coastlands of West Africa from The Gambia to Cameroon, and covers Gabon, Congo and half Zaire.

Nearly one-third of the continent is desert. The SAHARA, the planet's largest arid area, alone covers 25 per cent of Africa – some 7.7 million km² (about 3 million sq miles) – while to the south are the Namib and Kalahari. Because of climatic changes and overgrazing, the Sahara is advancing some of its borders at an estimated 5 km (3 miles) a year.

Africa is the hottest of the continents. Two-thirds of it is tropical or subtropical. The forests receive heavy rains that leach the soil, while other parts of the continent get less than 250 mm (10 in) of rain a year. Devastating droughts are frequent. In 1983-5 both the north – especially Ethiopia – and also a great belt of land across the south of the continent suffered one of the most savage in history.

Area 30 334 592 km² (11 712 252 sq miles)
Population 568 million

Afsluitdijk ***Netherlands*** Dam, 30 km (18 miles) long, lying 62 km (39 miles) north of Amsterdam. It was completed in 1932 to seal off an arm of the North Sea known as the Zuiderzee, and create IJSSELMEER, a freshwater lake.

Map Netherlands Ba

aftershock Vibrations of the earth's crust after the main waves of an earthquake have ceased. The shocks are caused by minor adjustments in the ruptured rocks, and may go on for weeks.

Agadez ***Niger*** Oasis town in the centre of the mountainous region of Aïr, 380 km (235 miles) east of the border with Mali. Standing some 520 m (1705 ft) above sea level, Agadez has been since the 16th century a regular stopping place for Saharan travellers, including the nomadic Tuareg herdsmen.

A small but growing number of tourists visit the town to see its leather and silver craftsmen, beautiful Sudanic architecture, clean sandy streets and magnificent mud mosque.

Population 21 000
Map Niger Ab

Agadir ***Morocco*** Port on the Atlantic coast 120 km (75 miles) south of Essaouira, founded by the Portuguese in the early 16th century. In 1911 it was the scene of a sabre-rattling international incident when Germany sent the gunboat *Panther* to Agadir during the struggle with France for control of north-west Africa. In 1960 it was virtually destroyed by an earthquake which killed 20 000 people. The new town, built just south of the old, is a popular resort with 8 km (5 miles) of beaches. It is overlooked by the ruins of a 16th-century fortress.

Population 111 000
Map Morocco Ba

Agana ***Guam*** The territory's capital, on the west coast. The main port, Apra Harbor, is 10 km (6 miles) south-west.

Population 900

agate A fine-grained QUARTZ marked with vari-coloured bands or with irregular clouding.

age Division of an epoch in the geological timescale (see p. 246).

agglomerate Mass of volcanic rock consisting mainly of fragments, larger than 2 mm (about 0.08 in) in diameter, which have been ejected from an explosive volcano.

Aggtelek ***Hungary*** Village near the Czech border in the Borsod limestone region, 40 km (25 miles) north of Miskolc. It is noted for its spectacular caves – part of a system which extends under the border to Domica in Czechoslovakia.

Map Hungary Ba

Agno ***Philippines*** River 205 km (128 miles) long in northern Luzon. It rises in the mountains of the Cordillera Central, flows southwards across fertile plains (which it feeds with silt during annual floods) and empties into Lingayen Gulf, an arm of the South China Sea. It is a rich source of freshwater fish, and its upper reaches have been dammed for hydroelectric power at Binga and Ambuklao.

Map Philippines Bb

Agra ***India*** Mogul city about 170 km (105 miles) south and slightly east of Delhi. It contains what is said to be the world's most beautiful building – the Taj Mahal. It is a mausoleum built in 1630-48 in white marble to inter and commemorate Mumtaz-i-Mahal, the wife of Shah Jehan. Originally the mausoleum was inlaid with both precious and semiprecious stones, but the precious stones were stolen during the troubled period of the 18th century. Shah Jehan was deposed by his son Aurangzeb and locked in the Red Fort within sight of the Taj Mahal. The Red Fort is itself a citadel of fine Mogul buildings. Stone inlay work is still a speciality of Agra.

Population 747 000
Map India Cb

Agram ***Yugoslavia*** See ZAGREB

Agrigento ***Italy*** One of the great cities of classical antiquity, founded as Akragas in 581 BC on the south coast of Sicily. It prospered especially under the rule of Theron during the early 5th century BC, when it had a population of 200 000 – four times its present size – and when it was the home of the Greek philosopher Empedocles. He is said to have died by falling into the crater of Mount Etna to prove his immortality. The city was colonised by the Romans, but the majority of its surviving monuments are Greek. Most of these are to be found in the Valley of Temples in a superb setting of olive and almond orchards.

► LASTING MEMORIAL The Mogul emperor Shah Jehan (1592-1666) built the Taj Mahal mausoleum at Agra for his favourite wife. He used craftsmen from many lands, including Italy and Turkey.

The present town has a medieval cathedral, founded in the 11th century, a late 13th-century convent and an archaeological museum.

In contrast to the prosperity of 2500 years ago, when the surrounding area produced an abundance of wine, olive oil, wheat, livestock and sulphur for export, the economy of Agrigento province today is stagnant; it is one of the country's poorest. Farming is in decline as people leave the land, and there is little industry.

Population 53 400

Map Italy Df

Aguascalientes ***Mexico*** State in the north-west, on Mexico's central plateau. It is mountainous in the west and east, and flat in the centre, which is agricultural. Its name means 'hot waters', after the hot mineral springs of its capital, which has the same name. Silver mining prompted the development of the town during colonial times, but it is now a centre for an area producing grapes, other fruits and wine.

Population (state) 556 000; (city) 239 000

Map Mexico Bb

Agulhas, Cape ***South Africa*** Africa's southernmost point, 170 km (105 miles) south-east of Cape Town, where the Atlantic divides from the Indian Ocean.

Map South Africa Bc

Ahaggar ***Algeria*** See HOGGAR

Ahmadabad (Ahmedabad) ***India*** Textile city in Gujarat state, about 440 km (273 miles) north of Bombay. In 1918 it was the scene of a violent anti-British rebellion, and in 1930 the Indian nationalist leader Mahatma Gandhi founded an *ashram* (commune) there – and began his campaign of civil disobedience against British rule in India.

There are numerous fine buildings and monuments, including the 15th-century Great Mosque, the mid-19th-century Hathi Singh Jain temple, and the tomb of Ahmad Shah, the Muslim ruler of Gujarat who founded Ahmadabad in 1411. The city has a university (1949).

Population 2 548 000

Map India Bc

Ahmadi ***Kuwait*** Set in an oasis 35 km (21 miles) south of Kuwait city, Ahmadi was developed in the 1950s following the discovery of the nearby oil field. Until then, Ahmadi was a maze of small roads with single-storey houses, green gardens and trees. Today it is a bustling oil town and commercial centre.
Population 25 000
Map Kuwait Bb

Ahmedabad ***India*** See AHMADABAD

Ahvaz (Ahwaz) ***Iran*** Inland port about 150 km (95 miles) north of The Gulf, at the highest navigable point on the Karun river. The town stands on the site of a city dedicated to Ahura-Mazda (the Zoroastrian god of goodness and light) by the kings of the Sassanid dynasty (3rd-7th century AD).
Population 471 000
Map Iran Aa

Ahvenanmaa (Åland) ***Finland*** Province in the central Baltic at the entrance to the Gulf of Bothnia, consisting of 6554 islands, more than 100 of them being inhabited. It harbours a wealth of remains from the Stone, Bronze and Iron Ages. Except for Maarianhamina, the capital and principal port, each parish has a medieval church and at the village of Bomarsund there is a 13th-century castle, Kastelholm, and an immense fortification system, built by the Russians in the 1830s when Finland was a Russian Grand Duchy, and destroyed by the British and French in 1854 during the Crimean War.

The people are Swedish-speaking, but despite a referendum in 1917, when a majority voted to secede to Sweden, the League of Nations – forerunner of the United Nations – confirmed Finnish sovereignty in 1921. The islands do, however, have a large measure of home rule, including their own flag and their own representatives on the Nordic Council – the annual assembly of parliamentary representatives from Denmark, Finland, Iceland, Norway and Sweden. The main industries are fishing, farming and tourism.
Population 22 800
Map Finland Bc

Aigoual, Massif de l' ***France*** Tree-covered group of mountains in southern France, containing the second highest peak – Mont Aigoual (1567 m, 5141 ft) – in the Cévennes mountains. It is one of the wettest and foggiest places in France, despite being only 70 km (45 miles) from the Mediterranean.
Map France Ed

Ain Salah ***Algeria*** Junction post, 694 km (431 miles) north of Tamanrasset, on the Route du Hoggar – the most frequently used north-south route in the Sahara. The road branches there, with one track leading north of El Golea, the other heading west to Adrar. Ain Salah is a popular stopping point for travellers.
Population 9300
Map Algeria Bb

Aïr ***Niger*** Mountainous area in the Sahara desert in north-central Niger. It extends for about 400 km (250 miles) from north to south, and about 240 km (150 miles) east to west. Its highest peak is Mount Gréboun at 1944 m (6378 ft). Ancient cave paintings in the mountains indicate that the region has had human occupants for at least 5000 years. The pictures portray the area as having had a much milder, wetter climate then than it does today.
Map Niger Ab

air mass Large body of air with only small horizontal variations of temperature, pressure and moisture content, usually covering hundreds of square kilometres, and bounded by FRONTS. Air masses are usually classified according to the regions where they originate, such as polar (cold) or tropical (warm), and according to whether they are maritime (moist) or continental (dry). These terms are combined, for example, in polar maritime (cold and moist) and in tropical continental (warm and dry).

airstream Moving current of air at various altitudes, generally originating in an air mass. Drawn-out wisps of cirrus cloud show the presence of such currents at about 6 km (4 miles) above the earth's surface.

Aisén ***Chile*** Least developed and most sparsely populated of the country's regions, covering 107 153 km² (41 372 sq miles) in the extreme south. It is poorly endowed with natural resources and the economy relies on sheep and cattle. COIHAIQUE is the capital, 67 km (42 miles) from Puerto Aisén, the main port.
Population 69 800
Map Chile Ad

Aitutaki ***Cook Islands*** Second most populous island in the group, covering 18 km² (7 sq miles) and lying 225 km (140 miles) north of Rarotonga. It is hilly, rising to 120 m (400 ft), and has a reef enclosing a large lagoon. It produces most of the islands' banana crop, which is mainly exported to New Zealand.
Population 2400
Map Pacific Ocean Fc

Aix-en-Provence ***France*** First Roman settlement in Gaul and now a thriving university city, 25 km (16 miles) north of Marseilles in Provence. By the 10th century it had become the capital of the Counts of Anjou, who ruled Provence. It remained so until Provence was united with France in 1482. The city retained its own parliament until 1789. An ancient cathedral and many fine 17th and 18th-century mansions survive, some now housing museums. An international music festival is held each year.
Population 129 800
Map France Fe

Aix-la-Chapelle ***West Germany*** See AACHEN

Aix-les-Bains ***France*** See BOURGET, LAC DU

Ajaccio ***France*** Main port and capital of Corse-du-Sud department on the west coast of Corsica. It was founded by Genoese colonists in 1492 and became French in 1768 when Genoa sold the island to France. The French emperor Napoleon was born there in 1769, and many members of his family are buried in the Palais Fensch, which also houses a fine collection of Italian paintings.
Population 55 300
Map France Hf

Ajanta ***India*** Small hill town about 350 km (217 miles) north-east of Bombay. A group of ancient Buddhist cave temples lies some 8 km (5 miles) to the north. The 29 caves, carved from solid rock, date from 200 BC to AD 650 and are up to 20 m² (215 sq ft) in size and 10 m (33 ft) high. Their halls and sleeping quarters are richly decorated with carvings and fresco paintings.
Map India Bc

Ajman ***United Arab Emirates*** The smallest of the seven emirates at only 65 km² (25 sq miles). It is surrounded by the emirate of SHARJAH and has two inland enclaves – Musfat and Manama. The Musfat enclave lies at the foot of the HAJAR MOUNTAINS on the Oman border and has valuable reserves of marble and mineral water. The main port and capital, Ajman, lies 11 km (7 miles) north-east of Sharjah. Its wharves service all the emirates, and a shipyard builds and repairs small vessels of up to 5000 tonnes. Traditional industries include fishing and the building of dhows (single-masted Arab boats).
Population (emirate) 42 000; (town) 27 000
Map United Arab Emirates Ba

Ajmer ***India*** Industrial city about 340 km (211 miles) south-west of Delhi. It is noted for its mosques, forts and artificial lake, the Ana Sagar dating from 1135, which is a birdwatcher's paradise. In the Aravalli range to the west is the sacred Lake Pushkar, beside which, according to legend, the Hindu god Brahma was born. Each October hundreds of thousands of Hindus gather beside the lake to honour Brahma and to attend the ox and camel fairs.

Ajmer is rich in modern temples – as the original temples were destroyed by Muslim rulers in the 17th century. The city's manufactures include salt, cotton cloth, footwear, oils and soaps – and many people are employed in the large railway workshops.
Population 375 600
Map India Bb

Ajuda ***Benin*** See OUIDAH

Akademgorodok ***USSR*** See NOVOSIBIRSK

Akaroa ***New Zealand*** Small settlement on South Island, about 40 km (25 miles) south-east of Christchurch. French settlers bought land there in 1838, but British sovereignty had been established when they arrived in 1840.
Population 700
Map New Zealand De

Akhelóös ***Greece*** Second largest river of Greece after the Aliákmon. It rises in the PINDHOS mountains and flows 220 km (136 miles) to the Ionian Sea. It is harnessed to produce hydroelectric power at the Akhelóös Dam, built in the late 1950s.
Map Greece Bb

Akita ***Japan*** City near the north-west coast of Honshu island, about 440 km (275 miles) north of Tokyo, the national capital. It was founded in AD 734 as a garrison town. It is now a busy centre of industry and commerce, manufacturing food and timber products.
Population 296 400
Map Japan Dc

▲ **CAMEL TRAIN A Tuareg caravan makes its way through the Aïr Mountains, which rise abruptly from the sands of the Sahara in northern Niger.**

Akjoujt ***Mauritania*** See NOUAKCHOTT

'Akko ***Israel*** See ACRE

Akmolinsk ***USSR*** See TSELINOGRAD

Akosombo ***Ghana*** Dam built across the Volta river gorge in 1966 to provide the country's electricity. It is 75 m (246 ft) high, and behind it lies Lake VOLTA. The small town of Akosombo nearby has a yacht club and hotel with a magnificent view of the dam and the lake.
Map Ghana Bb

Akron ***USA*** City in Ohio about 45 km (28 miles) south of Cleveland, on the canal joining the Ohio river to Lake Erie. It is a world centre of rubber manufacturing. About 14 km (9 miles) to the north-east is Kent State University, where four students were shot dead by National Guardsmen during campus riots in May 1970 in protest against American military involvement in Cambodia.
Population (city) 226 900; (metropolitan area) 650 100
Map United States Jb

Akrotiri ***Cyprus*** Bay at the southern extremity of the island; its cliffs are the nesting place of the rare Eleonora falcon and the home of the monastery of St Nicholas of the Cats, founded in AD 325. There is a British military base on the shore of a salt lake to the west of the bay.
Map Cyprus Ab

Aksai Chin ***India*** Cold, high desert plains, covering some 37 550 km^2 (14 500 sq miles) north of the Himalayas. Although India claims them as part of the state of JAMMU AND KASHMIR, they have been occupied since the early 1950s by the Chinese, who have built a road across them. Subsequent protests helped spark off the Indo-Chinese War of 1962, in which India was defeated.
Map India Ca

Aksum ***Ethiopia*** See AXUM

Akure ***Nigeria*** Capital of Ondo state and a cocoa-marketing centre about 210 km (130 miles) north-east of the capital, Lagos. It is also a tourist centre for the seven Olumirin waterfalls in the Erinoke-Ijesha gorge of the Osse river nearby to the east.
Population 114 400
Map Nigeria Bb

Akureyri ***Iceland*** Port at the head of the 60 km (37 mile) long Eyjafjördur inlet. It is the third largest town in Iceland after Reykjavík and Kópavogur. Its industries include fish processing and farm products, ship repairing and textiles. The town is also a winter sport centre and the point of departure for tourist excursions to MYVATN lake.
Population 13 750
Map Iceland Ca

Akyab (Sittwe) ***Burma*** Port in Arakan state. It grew in importance during the British colonial period between 1824 and 1948, and is now a major rice market. It is also the state capital.
Population 143 200
Map Burma Bb

Al Asnam (Ech-Cheliff, Orleansville) ***Algeria*** An important administrative and communications centre in the west, halfway along the Cheliff valley between Algiers and Oran. It was founded in 1843 as a French military base, and is the centre of a large agricultural region. It was hit by earthquakes in 1954 and 1980.
Population 125 000
Map Algeria Ba

Al Ayn ***United Arab Emirates*** Town and oasis 160 km (100 miles) west of the city of Abu Dhabi. The town – noted for its greenery – is growing and has a large cement plant capable of producing up to 750 000 tonnes a year. Al Ayn contains the palace of the crown prince of the United Arab Emirates (UAE), as well as three historic forts: Jahili Fort, built in 1898 and now the headquarters of the UAE Defence Force; Al Ayn Fort, built in 1910; and Muwaiji Fort.

A museum in the town contains finds from nearby archaeological sites, including household utensils and spearheads, some of which are 5000 years old. The Bronze Age site at nearby Hili contains reconstructed burial mounds and huge stone slabs arranged in a circle – similar to Stonehenge – though only one stone is standing. Al Ayn Zoo houses rare species of animals from The Gulf region which have been threatened with extinction. They include oryxes (a type of antelope), Persian fallow deer and Asiatic wild asses.
Population 103 000
Map United Arab Emirates Ba

Al Basrah ***Iraq*** See BASRA

Al Fawr ***Iraq*** See FAO

Al Fahayhil ***Kuwait*** See FAHAHEEL

Al Fujayrah ***United Arab Emirates*** Second smallest of the seven emirates, covering 117 km² (45 sq miles). It is the only emirate on the coast of the Gulf of Oman (the other six are on The Gulf). In 1976 a road was built through the HAJAR MOUNTAINS to connect Al Fujayrah with the rest of the UAE. The town of Al Fujayrah has a small airport, and is overlooked by a 200-year-old fort, which is now a museum.
Population (emirate) 38 000; (town) 760
Map United Arab Emirates Ca

Al Furat ***Middle East*** See EUPHRATES

Al Hasa ***Saudi Arabia*** Subdivision of Eastern Province, known as the emirate of Al Hasa, on the east coast and containing some of the world's largest oil fields. It is also the site of the world's biggest oasis, the Al Hasa oasis, where thousands of litres of water flow from wells every minute, and feed over 12 000 hectares (nearly 30 000 acres) of surrounding land. Major irrigation and drainage schemes are being employed to grow rice, wheat, sesame, citrus fruits, peaches, figs and vegetables. The main oasis town is Hofuf.
Population 250 000
Map Saudi Arabia Bb

Al Hillah ***Iraq*** Regional capital on the Euphrates river, about 100 km (60 miles) south of the national capital, Baghdad. It manufactures textiles and leather goods, trades in cereals and dates, and has a huge casino on the riverfront. About 32 km (20 miles) north is the Hindya Barrage, which diverts most of the Euphrates' waters to a channel farther west for irrigation.
Population 129 000
Map Iraq Cb

Al Jazirah ***Sudan*** See EL GEZIRA

Al Kazimiyah ***Iraq*** See KADHIMAIN

Al Khums ***Libya*** See HOMS

Al Manamah ***Bahrain*** Capital and port at the north-east tip of Bahrain Island. It is the administrative area for the oil industry, an important banking centre for The Gulf region and a centre for boat-building and fishing. A city has existed there since the 15th century, but the oldest buildings are 18th century. Today Al Manamah is a lively mixture of East and West, with a bustling Arab bazaar and some fine modern buildings.
Population 115 054
Map Bahrain Ba

Al Mawsil ***Iraq*** See MOSUL

Al Qurnah ***Iraq*** City near the legendary site of the Garden of Eden about 400 km (250 miles) south-east of the capital, Baghdad.

The marshland area to the north of the city is the home of the Marsh Arabs. The region, which is rich in wildlife, particularly birds, contains more than 1000 artificial islands constructed of mud and papyrus reeds by the Marsh Arabs, and each supporting huts made of reeds and reed matting. Transport is by reed boats.
Map Iraq Cc

Alabama ***USA*** Southern state which became part of the USA in 1819 after periods of French, English and Spanish rule. It was a Confederate stronghold in the Civil War (1861-5) and was for long a centre of resistance to black civil rights. It was a cotton-growing state, but cotton is no longer a main crop and today the chief farm products are chickens, cattle, soya beans, peanuts and maize. Coal is mined; it fuels the iron and steel industry in the city of Birmingham. Oil and natural gas are also produced. MONTGOMERY is capital of the state, which covers 133 667 km² (51 609 sq miles).
Population 4 021 000
Map United States Id

Alamo, The ***USA*** See SAN ANTONIO

Alamut ***Iran*** See QAZVIN

Åland ***Finland*** See AHVENANMAA

Alaska ***USA*** State covering the north-west tip of North America and separated from the rest of mainland USA by Canada. Alaska is the largest state of the Union, with an area of 1 518 800 km² (586 400 sq miles). It includes the Aleutian Islands, the most westerly of which – Attu – is just 90 km (55 miles) across the Bering Strait from the USSR. Formerly known as Russian America, Alaska was bought from Tsarist Russia in 1867 for US$7.2 million, and became America's 49th state in 1959. The discovery of gold in the late 19th century and oil in the 20th both prompted population booms as tens of thousands of prospectors headed for the desolate wastes of the area. Today, the main products also include timber, fish and quartz.

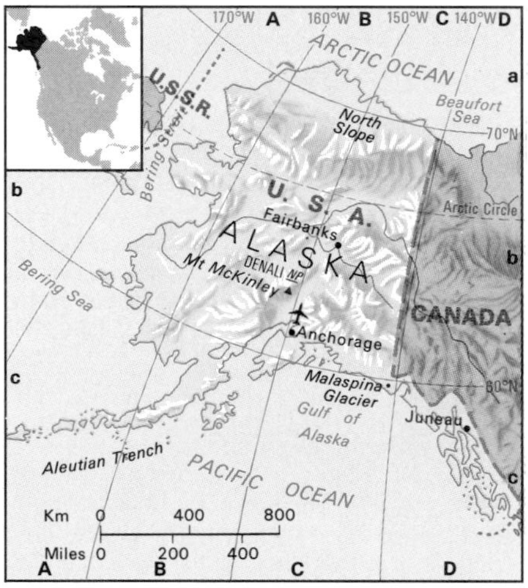

The southern mountains of the Alaska Range – a growing tourist attraction – contain North America's highest peak, Mount McKinley (6194 m, 20 320 ft), in DENALI NATIONAL PARK. The state capital is JUNEAU.
Population 521 000

Alaska, Gulf of Lying in the north-eastern PACIFIC OCEAN, it extends between the Alexander Archipelago and the Alaska Peninsula. Beneath it the Pacific plate is slipping under the North American plate, causing volcanic eruptions and earthquakes. A severe earthquake in 1964 caused a tsunami, which battered the ports of ANCHORAGE, Seward and Valdez, all in southern Alaska. The earthquake and the giant wave killed 178 people.
Area Over 5200 km² (2000 sq miles)
Map Alaska Cc

Alassio ***Italy*** Elegant Riviera resort about 85 km (52 miles) south-west of Genoa. It has a good beach and a mild climate. The old part of the town contains shady alleys packed with shops, lively bars and many churches.
Population 12 700
Map Italy Bb

▼ RIVER OF ICE A glacier wends its way through the St Elias Mountains to the Pacific Ocean in Alaska's Glacier Bay National Park.

Alava ***Spain*** One of the three provinces of northern Spain making up the Spanish part of the BASQUE REGION, which is called Vascongadas in the Basque language.

Population 260 600

Map Spain Da

Alba Iulia ***Romania*** Town and capital of Alba county, on the River Mures in south-west TRANSYLVANIA. Alba Iulia was the capital of the Roman colony of Dacia, and of Transylvania in the 16th and 17th centuries. In 1918 the unification of Transylvania and Romania was proclaimed in the hall of the casino, now the Museum of the Union. Four years later, the town's Orthodox cathedral was the scene of the coronation of Ferdinand I, who was crowned King of All Romania after the country had doubled in size with the acquisition, after the First World War, of Transylvania, the Banat (an agricultural region, formerly in southern Hungary) and Bessarabia. There is an early 18th-century citadel built by the Holy Roman Emperor Charles VI (1685-1740).

Population 50 900

Map Romania Aa

Albacete ***Spain*** Market town about 225 km (140 miles) south-east of Madrid, manufacturing chemicals, soap, knives and scissors. Its museum contains 3rd-century Roman dolls with jointed limbs, found locally. The surrounding province of the same name is sheep country, and also produces, cereals, vines and olives.

Population (town) 177 100; (province) 334 500

Map Spain Ee

Albania See p. 22

Albanian Alps (Bjeshkët e Némuna, Prokletije) ***Albania/Yugoslavia*** Steep-sided mountains lying between the River Drin and Yugoslavia, and dropping precipitously to the Shkodër lowlands. The forested slopes, deeply cut by short rivers, reach their highest point at Jezercë (2694 m, 8838 ft).

Map Albania Ba

Albano ***Italy*** Lake 6 km² (2.3 sq miles) in an area 20 km (12 miles) south of Rome. It was formed by the fusion of two ancient volcanic craters and is fed by underground springs. Its artificial outlet was made by the Romans in the late 4th century. The lake is popular for watersports, including international events. It is overlooked by the town of Albano Laziale and the Pope's summer retreat, Castel Gandolfo.

Map Italy Dd

Albania

A COMMUNIST STATE WHOSE ANXIETY TO PRESERVE THE PURITY OF ITS IDEOLOGY HAS MADE IT A POCKET OF THE PAST

The people of Albania have always been physically isolated in their mountainous country, but since the Second World War they have been virtually cut off from all contact with the outside world. Foreign news and books are restricted, foreign travel is forbidden, and although there are tourists, the Albanians are not encouraged to talk to them. Since 1944 the country has been a Stalinist-Communist state run until 1985 by Enver Hoxha (pronounced Hodja) and now by his successor as head of state, Ramiz Alia.

The land rises to 2694 m (8838 ft) in the north ALBANIAN ALPS, with broad grassy valleys running southwards into Greece and marshy plains on the coast of the Adriatic Sea. Its population is 97 per cent Albanian, with small groups of Greeks, Serbs and Gypsies. The Albanian language is of Indo-European origin and divides into two dialects – Gheg in the north and Tosk in the south.

After 400 years of Turkish rule, Albania became independent in 1912. Ahmed Zogu, who was prime minister from 1922, declared himself King Zog, and was supported by a fellow dictator, Mussolini of Italy. Germany and Italy invaded during the Second World War; the country was retaken by native Communist guerrillas led by Hoxha and helped by Britain and the USA. This led to the establishment of the new state.

All relations with Western nations were broken off after the war, and although Albania has diplomatic connections with 100 other countries, Britain and the USA are not among them. Relations with neighbouring Yugoslavia have also deteriorated, and there is a continuing dispute over the Yugoslav province of KOSOVO, which has a large Albanian population.

THE CHINESE TAKE OVER

Russia was Albania's chief ally, market and source of investment until the Russian leader Josef Stalin died in 1953 and the new leader, Nikita Khrushchev, denounced Stalinism in 1961. Hoxha, for whom Stalin was a hero, thought Russia's new Communism was ideologically impure and switched his allegiance to China. Almost overnight, Chinese technicians replaced Russians on engineering and industrial projects in Albania.

The Chinese style of Communism, with decentralised administration encouraging rural development, suited Albania better than centrally controlled Soviet-style industrialisation. However, in the 1970s disillusionment set in with China, too.

Although factories were built and railways developed during the Stalinist era, it is agriculture which provides Albania's main success story, with results achieved mainly by hard work. The country manages self-sufficiency in food, despite having very backward agricultural techniques. There is hardly any modern machinery. Mules and oxen pull carts and ploughs but most farming is done by hand, with hoes and spades. Two-thirds of the people live in villages, and most of them work on collective farms. Hours are long, and shirkers may be sent to jail. All children of school age have to spend a month of every year working in the fields.

There are no private cars in Albania. People get about by horse or bicycle, with the exception of party officials, who may have the use of an Italian Fiat or, if they rank highly enough, a West German Mercedes limousine. Communist ideology does not extend to the choice of motor cars.

FREE EDUCATION, CHEAP HOUSING

The towns, even the capital city of TIRANE, seem silent and empty with so little traffic. There are buses, sometimes with pictures of their destinations on the front for illiterate members of the population. One of the achievements of the new regime is that illiteracy is falling; education is free, as is health care. Housing is cheap, costing about 5 per cent of most people's income, and Albania has no inflation; prices have hardly changed since the 1950s.

Other notable absences in Albania are unemployment, foreign debts, drug addiction and alcoholism, corruption – and religion. The state was officially declared atheist in 1967. It had been predominantly Muslim, and mosques and churches became museums and schools, warehouses and stables. Women were liberated from the veil, and given equal pay, jobs in the government – and the possibility of being conscripted into the army.

Social attitudes have not necessarily kept pace with new practices; it is mostly men who spend their evenings in the bars (where singing is forbidden), while the women still have domestic responsibilities after work hours. And they still produce more children than in any other European country, creating an exceptionally young population.

There are signs that the country is opening up more to the West, possibly because of disillusionment over new trends in Chinese Communism. Albania needs markets abroad for its economic development. Its railway system is at last being connected to that of neighbouring Yugoslavia and a new highway between southern Albania and northern Greece was opened in 1985, so that there are now at least land routes in and out of the country – even if everyone is not free to use them.

ALBANIA AT A GLANCE

Area 28 748 km² (11 100 sq miles)

Population 3 033 300

Capital Tiranë

Government One-party Communist republic

Currency Lek = 100 quintars

Language Albanian

Religion Officially atheist, but minority still Muslim

Climate Mediterranean; average temperature in Tiranë ranges from 2-12°C (36-54°F) in January to 17-31°C (63-88°F) in August

Main primary products Cereals, potatoes, grapes, olives, tobacco, cotton, timber; petroleum and natural gas, chrome, copper, asphalt

Major industries Agriculture, petroleum refining, mining, food processing, cement, fertilisers, textiles, tobacco processing

Main exports Nonferrous metal ores, petroleum and petroleum products, food, clothing, tobacco and products

Annual income per head (US$) 1300

Population growth (per thous/year) 22

Life expectancy (yrs) Male 68 **Female** 72

Albany ***Australia*** Port and oldest town in Western Australia, 390 km (242 miles) south-east of Perth. Originally called Frederickstown, it was founded as a British penal colony in 1826 in order to forestall a possible French settlement on the west coast. It became a whaling port and a coaling station for ships trading between Australia and Europe. Its chief industries today are fish and meat canning.

Population 15 200

Map Australia Be

Albany ***USA*** Capital of New York state and the oldest city in the country still operating under its original charter, granted in 1686. It stands on the Hudson river, about 210 km (130 miles) north of New York city.

Population (city) 99 500; (metropolitan area) 842 900

Map United States Lb

Alberobello ***Italy*** Town in Apulia, 55 km (35 miles) south-east of Bari. It has been classified a national monument because of its *trulli*, ancient conical stone houses of uncertain origin. They have whitewashed walls and grey-tiled roofs topped with crosses and other symbols.

Population 9800

Map Italy Fd

Albert Canal ***Belgium*** Canal linking the Meuse river at Liège to Antwerp about 130 km (80 miles) away.

Map Belgium Ba

Albert Edward Nyanza ***Uganda/Zaire*** See EDWARD, LAKE

Albert, Lake (Albert Nyanza, Lake Mobuto Sese Seko) ***Uganda/Zaire*** Lake covering some 5180 km^2 (2000 sq miles) in the western arm of the East African section of the Great Rift Valley. It is fed by the Semliki river, which drains Lake Edward, and by the Victoria Nile, which flows from Lake Kyoga, Uganda; it is drained by the Albert Nile which, in Sudan, is called the Bahr el Jebel.

A part of the Nile system, the lake was discovered in 1864 by the British explorers Sir Samuel Baker and his wife, who named it Albert Nyanza after the Prince Consort, husband of Queen Victoria. It is also known as Lake Mobuto Sese Seko, after Zaire's president who took office in 1965. The lake contains crocodiles and hippopotamuses, while around its shores are many other species, including elephants and rhinoceroses.

Map Zaire Ca

Albert National Park ***Zaire*** See VIRUNGA NATIONAL PARK

Albert Nyanza ***Uganda/Zaire*** See ALBERT, LAKE

Alberta ***Canada*** Most westerly of the prairie provinces, covering 661 190 km^2 (255 285 sq miles) and with a landscape ranging from prairies to lofty mountain peaks. Agriculture includes grain crops and beef and dairy cattle. There is extensive forestry and large-scale output of oil, natural gas and coal, and there are unexploited oil reserves in deposits in the north. Western Alberta is a tourist area, with large areas of wooded country fringing the Rocky Mountains. The main cities are EDMONTON, the capital, and CALGARY.

Population 2 238 000
Map Canada Dc

Albertville ***Zaire*** See KALEMIE

Albi ***France*** Capital of the Tarn department, on the Tarn river 70 km (44 miles) north-east of Toulouse. It is an attractive town, with a 13th to 15th-century red-brick cathedral whose interior is adorned with designs by various Italian artists and a huge painting of the Last Judgment. The artist Henri de Toulouse-Lautrec (1864-1901) was born in Albi and some of his paintings are displayed in the bishop's palace.

Population 62 600
Map France Ec

albite A widely distributed feldspar (a compound of sodium, aluminium and silica) found mainly in rocks such as granite.

Ålborg ***Denmark*** Fourth largest of Denmark's cities and third largest port, sitting astride the narrows of LIMFJORDEN, in northern JUTLAND, the fiord which runs from the North Sea to the Kattegat. It is linked to Nørresundby, on the north shore. Historic features by the waterfront include the Gothic-style Cathedral of St Budolfi (Botolph) and Jens Bang's house, one of the finest Renaissance homes in Denmark. Other places of interest include Lindholm Høje, a Viking burial site dating from about 1100; North Jutland Museum of Art; Tivoli Korolinelund; Helligåndsklosteret, a 15th-century cloister containing early 16th-century frescoes; and the Ålborg Zoo. There is an airport, a university, several technical schools and the world-famous art gallery, Nordjyllands Kunstmuseum. To the east of the city beside Limfjorden are Denmark's largest cement plants. It has also acquired an international reputation for its akvavit (Danish schnapps).

Population 160 000
Map Denmark Ba

Albufeira ***Portugal*** Small fishing town on the south coast of the Algarve, about 30 km (20 miles) west of Faro. It is now a tourist resort, with a bustling daily market and plenty of nightlife. A tunnel through the cliffs leads to a sandy beach below a Moorish castle dating from the 8th to the 13th centuries.

Population 12 000
Map Portugal Bd

Albuquerque ***USA*** Resort city on the Rio Grande river in New Mexico, about 350 km (220 miles) north of the Mexican border. It was originally a Spanish settlement and is now the seat of the University of New Mexico.

Population (city) 350 600; (metropolitan area) 449 400
Map United States Ec

Albury-Wodonga ***Australia*** Industrial twin cities straddling the Murray river and the state border, Albury being in New South Wales and Wodonga in Victoria. In the 1970s Albury-Wodonga was designated an inland growth centre. In addition to providing services for the surrounding farming country, the twin cities contain numerous industries, including textiles, furniture making, brickworks and engineering. The cities lie on the major transport routes between Sydney and Melbourne.

Population 50 000
Map Australia Hf

Alcalá de Henares ***Spain*** Picturesque town, much devastated in Spain's Civil War, 29 km (18 miles) north-east of Madrid. It is the birthplace of writer Miguel de Cervantes (1547-1616) author of *Don Quixote*.

Population 142 900
Map Spain Db

Aldabra ***Seychelles*** Oval coral atoll in the Indian Ocean, north of Madagascar, covering 155 km^2 (60 sq miles). Channels divide the ring into four low islands that stand about 30 m (100 ft) above the sea. They have remained isolated from any landmass throughout their existence, and as a result are populated by some unique animals. The atoll, a World Heritage site managed by the Seychelles Island Foundation, is the home of some 150 000 giant land tortoises and the world's rarest bird – the Aldabran brush warbler, of which there are no more than about a dozen specimens. Another rare bird, the Aldabran rail, is the last remaining flightless bird living on the atoll.

Map Indian Ocean Bc

Alderney ***English Channel*** See CHANNEL ISLANDS

Aleksandropol ***USSR*** See LENINAKAN

Aleksandrorsk ***USSR*** See ZAPOROZH'YE

Alençon ***France*** Capital town of the Orne department in the fertile countryside of the Sarthe valley, 48 km (30 miles) north of Le Mans. It was once an important lace-making centre, producing *point d'Alençon* work. A castle and the town's Church of Notre Dame both date from the 14th and 15th centuries.

Population 32 500
Map France Db

Alentejo ***Portugal*** Southern province to the east of the capital, Lisbon, and immediately north of the Algarve, occupying nearly one-third of the country (24 683 km^2, 9142 sq miles). The main rivers are the Tagus, Sado and Guadiana. It was the scene of numerous battles in the Middle Ages, first against the Moors, then against Spanish invaders. Its mostly low, rolling landscape is dotted with cork oaks and wheat fields. To the east, the hills of the Serra de São Mamede rise to 1025 m (3362 ft). Following Portugal's political revolution in 1974, the province's large estates were turned into cooperatives, but many have now returned to private ownership.

Population 511 600
Map Portugal Bd

Aleppo (Halab) ***Syria*** Industrial city and the capital of Aleppo province, 120 km (75 miles) north east of Latakia. Aleppo dates from at least 2000 BC, and together with Damascus claims to be the oldest continuously inhabited city in the world. At the centre lies its citadel ringed with ramparts. Nearby, the modern bus station opens onto a large market, where the covered main souk (bazaar) is more than 800 m (2625 ft) long.

Population 905 944
Map Syria Ba

Aletsch Glacier ***Switzerland*** Longest glacier in the Alps, about 24 km (15 miles) long and covering 129 km^2 (50 sq miles). It skirts the JUNGFRAU, about 60 km (38 miles) south of Lucerne and terminates near the Rhône Valley at Brig. The peak of Aletschhorn nearby is 4195 m (13 763 ft) high.

Map Switzerland Ba

Aleutian Islands See ALASKA

Aleutian Trench Lying in the North PACIFIC OCEAN, the Pacific plate is slipping beneath the North American plate along the trench. The descending plate is melted, producing molten magma which fuels the volcanoes in the Aleutian island chain lying north of the trench.

Greatest depth 7443 m (24 419 ft)
Map Alaska Ac

Alexandretta ***Turkey*** See ISKENDERUN

Alexandria ***Egypt*** Largest port and second largest city of Egypt after the capital, Cairo. It lies at the western edge of the Nile delta about 185 km (115 miles) north-west of Cairo. It is a modern industrial city with a deep-water dock handling more than 75 per cent of Egypt's import and export trade, with many factories,

▲ **PEACEFUL SHORE Atlantic waves break in a cove between tall cliffs near the ancient port of Lagos in the Algarve, Portugal's busiest tourist area.**

and a flourishing tourist trade beside its long Mediterranean beaches.

The city was founded by the Greek conqueror Alexander the Great in 332 BC as the capital of his empire, and is named after him. For more than 600 years it was one of the major cities of ancient times. The first lighthouse, built in about 270 BC of white marble, stood on the island of Pharos in the bay and was one of the Seven Wonders of the Ancient World. The city's library of Greek manuscripts, founded in 300 BC, made it the intellectual heart of the Graeco-Roman civilisation. After the death of the Egyptian queen Cleopatra in 30 BC, Alexandria became the seat of government of Rome's Middle Eastern empire. Decline set in after AD 300, when Constantinople (Istanbul) became the capital of the Byzantine Roman empire; and later silting up of the Nile delta led to Alexandria's collapse as a port.

The city's regeneration began only in the 19th century, after Muhammad Ali, the governor who gained Egypt's independence from the Turkish Ottoman Empire, had built the Mahmudiya Canal, opened in 1847. The canal linked the city to the river and the sea again. In 1882 the city was bombarded by the British fleet and was then occupied to quell an outbreak of riots.

Alexandria now has tanneries, shoe-making factories, vehicle-assembly plants, an oil refinery and chemical plants, and handles the bulk of Egypt's cotton trade.

Population 2 320 000

Map Egypt Bb

Alexandropol ***USSR*** See LENINAKAN

Alföld, Great (Nagyalföld) ***Hungary*** The monotonously flat 'Great Plain' east of the Danube river, covering 51 800 km² (19 995 sq miles) – more than half the country. The Tisza river meanders across the plain from north to south, cutting it roughly in half. The soils are mostly fertile black earths, and the prairie-like *puszta* (meaning 'grasslands') that once covered most of the plain were the traditional home of shepherds and of the colourful *csikos* ('horse-riders').

However, over the past 100 years, there have been dramatic changes. Most of the puszta, apart from the area known as HORTOBAGY, have been ploughed and divided into vast farms or market gardens growing wheat, corn, potatoes and fruit. It produces sugar beet, flax, hemp and tobacco, which are Hungary's chief industrial crops. Villages, often surrounded by extensive orchards, tend to be large, and many of the ancient towns, such as Békéscsaba, Cégled and Debrecen are 14th-century market towns. Szeged has been inhabited since the 8th century. Roads and railways radiate across the Great Alföld from the capital, Budapest, but travel in other directions is often slow.

Map Hungary A/Bb

Alföld, Little (Kisalföld) ***Hungary*** Triangular plain covering the north-west of the country and bounded in the south-east by the Bakony hills. It covers an area of 7000 km² (4320 sq miles). It is similar to the Great Alföld, and is a major producer of maize, sugar beet and wheat. GYOR is the regional capital.

Map Hungary Ab

Algarve ***Portugal*** Province, 4960 km² (1915 sq miles) in area, stretching along the south Atlantic coast. It was the last part of Portugal to be wrested from the Moors (1253). The Algarve is now the country's most popular holiday area. In the west there are beaches backed by cliffs. In the east, sandbanks protect the shore. There are colourful fishing villages on the coast, while orchards of almonds, figs and citrus trees grow inland.

Population 323 500

Map Portugal Bd

Algeciras ***Spain*** Port and winter resort on the country's southern tip, with a large export trade in cork from nearby forests. It was originally settled by Spanish refugees from Gibraltar after the rock was captured by the English in 1704. In 1906 an International Conference to decide the future of Morocco was held at Algeciras.

Population 86 100

Map Spain Cd

Algeria See p. 26

Alghero ***Italy*** Tourist resort in north-west Sardinia, traditionally popular with English visitors. Alghero's culture still reflects its colonisation by Spain in the 14th century. The people speak Catalan and there are many 14th-century Gothic buildings, including the cathedral and the bastions that partially encircle the old town.
Population 36 400
Map Italy Bd

Algiers (El Djazair, Alger) ***Algeria*** The country's largest city, chief port and capital, lying at the centre of the Mediterranean coast. Founded on four islands (now joined to the mainland) by the Phoenicians about 2000 BC, it became a thriving port in Roman times, only to vanish with the fall of the Roman Empire. It was refounded by Muslim Arabs in the 10th century. From the 16th to 18th centuries, the Muslim Barbary pirates, who preyed on Christian shipping in the Mediterranean, made it the centre of their activities. In 1830 France invaded Algeria, which was declared French territory in 1848. Under the French, Algiers became a modern commercial port, increasing its population from a few thousand in 1830 to over 1 million by 1970. Wine, citrus fruits and iron ore are important exports, and the city has chemical, light engineering and consumer goods industries.

The Kasbah (Arab city), with its fort and old houses that have remained in the same families for generations, is a maze of twisting, picturesque streets. The Museum of Popular Arts stands at the foot of the Kasbah. The modern town centres round the Admiralty building. To the east of the city is the fishing port and market. Above the city and overlooking the bay is the former Summer Palace of the Governor; the Bardo Museum of Ethnography, which is housed in an 18th-century Turkish villa; and the Stephane Gsell Museum which contains Roman antiquities, Islamic art and archaeological remains. There are several other museums and fine mosques.
Population 1 800 000
Map Algeria Ba

Algonquin ***Canada*** Provincial park, covering 7537 km² (2910 sq miles) and containing 2500 lakes, in a beautiful wilderness in south-east ONTARIO. Most of the park is accessible only by canoe or on foot.
Map Canada Hd

Aliákmon ***Greece*** Longest river of Greece: 297 km (184 miles). From a source on the PINDHOS mountains, it carves narrow gorges through the highlands of MACEDONIA, and is dammed to generate hydroelectric energy at the Aliákmon Barrage (built in 1973) before emptying into the Aegean Sea.
Map Greece Ba

Alicante ***Spain*** Mediterranean port about 125 km (78 miles) south of the port of Valencia, exporting wine, citrus fruits, olive oil and almonds. Its industries include metalworking, textiles, chemicals, tobacco products, oil refining and tourism. The adjacent province of the same name is mainly barren, but irrigated areas produce wine and citrus fruits.
Population (town) 251 400; (province) 1 148 600
Map Spain Ec

Alice Springs ***Australia*** Town in the Northern Territory, standing in the Macdonnell Ranges in the great central desert, almost in the middle of the continent. It was founded as a staging point for the overland telegraph line in the 1870s and is now a centre for cattle, minerals and tourism. It is situated on the Stuart Highway, which crosses the country from Darwin in the north to Port Augusta in the south. Alice Springs plays host each year to the Henley-on-Todd boating regatta, which is modelled on the annual event at Henley-on-Thames in England. The difference is that the Alice races take place when the Todd river is dry; competitors run along the dusty riverbed holding bottomless boats around them. The town is also the headquarters of the Flying Doctor Service and the School of the Air – radio-linked classes for children on remote outback properties without access to any schools.
Population 18 400
Map Australia Ec

Aligarh ***India*** University town about 115 km (70 miles) south-east of Delhi. It contains the Muslim University, founded in 1920 from the Anglo-oriental college founded in 1875 by the Muslim reformer and educationalist Saiyad Ahmad Khan, who modelled the buildings on the colleges of Oxford and Cambridge. It has a metal works and factories making carpets and cotton.
Population 321 000
Map India Cb

Ali-Sabieh ***Djibouti*** Settlement on the rail link between the port of Djibouti and the Ethiopian capital, Addis Ababa. It has been overwhelmed in the 1980s by tens of thousands of refugees fleeing drought and civil war in Ethiopia.
Map Ethiopia Ba

Aliwal South ***South Africa*** See MOSSELBAAI

Al-Kharijah ***Egypt*** See EL KHARGA

Al-Kharj ***Saudi Arabia*** Resort region 90 km (56 miles) south of the capital, Riyadh, through which pass the main road and railway to Dammam. It is a playground for the capital, with palm groves, gardens and other summer picnic spots. It also has a number of farms, including the Saudi Agriculture and Dairy Company which is said to be the largest dairy project in the Middle East. Al-Kharj is also one of the country's major wheat-producing areas, relying heavily on extensive overhead sprinkler irrigation. The main towns are As Salamiyah, Al Yamamah and Sulaymaniyah.
Map Saudi Arabia Bb

Alkmaar ***Netherlands*** Town 33 km (20 miles) north and slightly west of Amsterdam. It is a tourist centre, with many of its 14th to 17th-century buildings and canals intact. The weekly cheese market is still held from the end of April to September, when its dealers, dressed in 17th-century costume, move the round, yellow cheeses on sledges. The town's manufactures include paper, furniture and clothing.
Population 67 000
Map Netherlands Ba

Allada ***Benin*** Market town about 50 km (30 miles) north-west of Cotonou. It was the capital of an old kingdom decimated by slave traders in the 18th and 19th centuries.
Map Benin Bb

Allahabad ***India*** City at the confluence of two holy rivers, the GANGES and the YAMUNA, midway between Delhi and Calcutta. Each year during the Hindu month of Magh hundreds of thousands of Hindus visit the city to bathe in the holy waters and so purify their souls. In 1954 there was a stampede of pilgrims to enter the waters and some 350 were crushed to death.

The Aryans, a prehistoric people from central Asia who invaded India around 1500 BC, founded a city on the site called Prayag, 'the place of sacrifice', which is still revered. The Buddhist emperor Asoka built a stone edict pillar there in 242 BC which still stands today, and the Mogul emperor Akbar built the fort there in 1575. In the 19th century, Allahabad became a British provincial capital, and was the scene of fierce fighting in the 1857 Indian Mutiny. The Indian National Congress (later to become the ruling Congress Party of India) was founded there in 1885.

There is a Great Mosque and a university, founded in 1887. The main industries are flour milling and textiles, and there is a flourishing trade in sugar.
Population 650 000
Map India Cc

Allegheny Mountains ***USA*** Part of the Appalachian system stretching about 800 km (500 miles) from south-west Virginia to northern Pennsylvania, and forming the most north-westerly main range of the system. They vary in height from about 600 m (2000 ft) to 1481 m (4860 ft) at Spruce Knob in West Virginia. They contain rich timber and coal resources.
Map United States Jc

Allgäu (Algäu) ***West Germany*** Alpine region of south-west Bavaria between Lake Constance, on the Swiss border, and the town of Füssen, about 75 km (45 miles) to the east. It is a skiing area, and round the town of Kempten lies Germany's main cheese-producing area.
Map West Germany De

Allier ***France*** Navigable river, 410 km (255 miles) long. It rises in the southern highlands of the Massif Central and flows north through a series of basins and the fertile Limagne plain, to join the Loire near Nevers.
Map France Ee

alluvial cones and fans Fan-shaped accumulations of ALLUVIUM deposited by mountain streams at the mouths of ravines onto adjacent plains. Fans have gentle slopes, up to 6 degrees, while cones are steeper, up to 15 degrees. They are often found in arid and semiarid areas.

alluvial plain See FLOOD PLAIN

alluvial terrace See RIVER TERRACE

alluvium Eroded particles of rock, usually sand and silt, transported by a river and deposited on its bed, its flood plain or in its delta or estuary. Much of the world's most fertile land consists of alluvium.

Algeria

FOLLOWING A SAVAGE WAR OF INDEPENDENCE WITH FRANCE, ALGERIA HAS NOW BECOME A MAJOR POWER ON THE AFRICAN CONTINENT

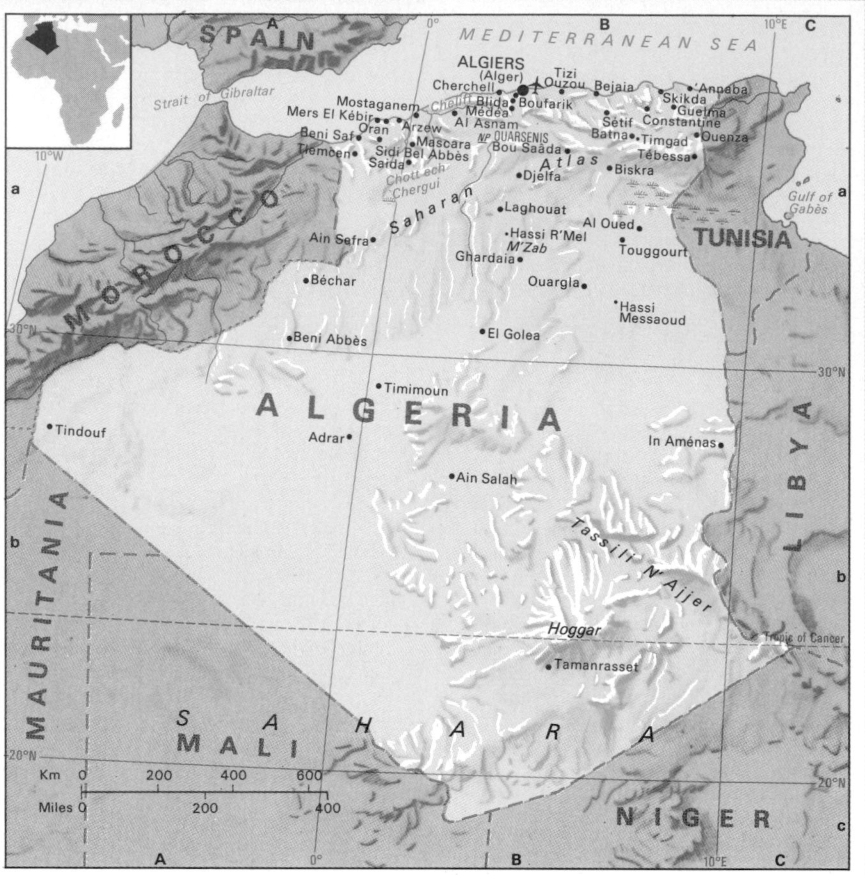

Providing a gateway to the SAHARA, the world's largest desert, Algeria is remembered as the home in times past of the elite French Foreign Legion. Immortalised in P.C. Wren's novel *Beau Geste*, the Legion fought savage Berber tribesmen under a blazing sun in the arid desert wastes. But today the desert has a new image.

Over four-fifths of Algeria is covered by the Sahara, and it was beneath its unproductive surface that enormous oil and gas fields were discovered in the 1950s. Today, Algeria is the third largest oil producer in Africa – only Nigeria and Libya produce more. Reserves of crude oil are estimated at 9000 million barrels, and Algeria also possesses one of the largest reserves of natural gas in the world – a staggering 3090 thousand million m^3 (109 122 thousand million cu ft).

THE FACE OF THE LAND

Algeria is a huge country – the second largest in Africa (only Sudan is bigger) and the tenth largest in the world. It has two main geographical divisions – the ATLAS mountains in the north and the Sahara to the south. The Atlas mountains – several parallel ranges running east-west with valleys between them – consist of three main areas: the Tell Atlas near the coast, the high plateaus, and the Saharan Atlas.

The Tell Atlas form a belt of narrow ranges, plateaus and massifs, between which are fertile lowlands, including the valley of the Cheliff, Algeria's longest river, the alluvial Mitidja plain around the capital ALGIERS, and the ANNABA plain, all of which form Algeria's chief farming regions. The Tell Atlas contain most of Algeria's people and all its larger towns – ORAN, CONSTANTINE, AL ASNAM (Ech Cheliff, Orleansville), BLIDA, Annaba (Bone), as well as Algiers.

To the south are the high semiarid plateaus, containing the Chott ech CHERGUI and the Chott al Hodna, large desert depressions with salt marshes.

Farther south lie the Saharan Atlas mountains, which extend east of the Chott al Hodna as the Aures Massif, rising to 2326 m (7631 ft). A line of oases, among which BISKRA is the most heavily populated, lies along the southern flanks of the Saharan Atlas.

The Sahara itself stretches for nearly 5000 km (over 3000 miles) from the Atlantic to the Red Sea, and dominates 11 African countries. In the Algerian Sahara great dunes of sand, called *ergs*, rise like the waves of a rolling sea in the north and east; the rest is covered with gravel or rocks. In the southern desert, the HOGGAR Massif reaches 2918 m (9573 ft) at Mount Tahat.

The climate on the coast is similar to that of the European countries that fringe the Mediterranean to the north – hot, dry summers and cool, moist winters. Summer temperatures can climb to 32°C (90°F) when the hot, dry, dusty wind known as the sirocco blows from the Sahara. Most of the rain falls between October and March. In the mountains the weather is cooler and affected by altitude; some of the highest peaks are snow-covered in winter. In the Sahara there is little or no rain; temperatures may reach 49°C (120°F) during the day but fall to 10°C (50°F) at night.

ALGERIA'S MANY INVADERS

The earliest known inhabitants were the Berbers, whose kingdom flourished from about 2400 BC. About the 3rd century BC, Numidia, as the Berber Kingdom was called, came under attack from the Carthaginians and later the Romans. Roman rule ended when Vandals from Spain swept across the country in AD 431. In the following century the area was partly conquered by the Byzantines. Successive waves of Arabs swept over the country in the 7th century and introduced the Islamic religion and the Arabic language. Islam soon became dominant but the Berber language survived mainly in the mountains in the east of the country. From 1519, the Ottoman Turks held the coastlands, but their control was, at best, weak.

In 1830, on the pretext that the coastline was a haven for pirates – it was – French troops invaded the country, and in 1831 the French Foreign Legion was formed. Decades of fighting against the Arabs and Berber tribesmen followed. Most of the country was subdued by the early 1870s, though it was not until 1902 that France achieved control of the whole territory. In the period that followed, the French treated northern Algeria as an extension of France rather than a colony and settlers (*colons*) poured into the country.

After the Second World War, a nationalist movement, which had emerged during the 1920s and 1930s, began to grow in strength. One major group, the FLN (the National Liberation Front), became dominant and in 1954 launched a bitter struggle for independence. By 1956 it controlled most of the countryside and was making terrorist attacks on the towns. The French fought back ruthlessly with 500 000 troops to protect the million settlers. In 1958, the French army in Algeria, dissatisfied with the conduct of the war, rebelled and helped General de Gaulle to return to power. Both the army and the colons believed that de Gaulle would keep Algeria French. However, despite last-ditch stands, including counter-terrorism by the OAS (a secret organisation of army officers and colons), de Gaulle realised that Algeria could not be controlled by force indefinitely, and in 1962 independence was granted. The antagonism felt for the French by the Algerians resulted in a million French settlers and 150 000 Jews fleeing to France.

▲ DESERT HEIGHTS Glowing pinkish-brown in the sunlight, the Hoggar Massif, in the south of the Algerian Sahara, reaches a peak of over 2900 m (more than 9500 ft) – the highest point in the country.

The first president of the new republic was Ahmed Ben Bella, an erratic left-wing radical. In 1965 he was deposed by his defence minister, the austere Houari Boumedienne. Foreign petroleum companies were taken over and in 1976 Algeria, already a one-party socialist state, adopted a National Charter which reaffirmed the goal of building a socialist society. A National Assembly of 261 members was elected the following year. Since Boumedienne's death in 1978, President Chadli has relaxed Algeria's rigid socialist policies, allowing some private enterprise to develop.

OIL-BASED ECONOMY

Industry – especially oil and gas production – dominates the economy. Algeria survived the oil recession and the consequent fall in crude oil prices better than some oil-producing countries because she concentrated on exporting natural gas and, more recently, refined products rather than crude oil. Indeed, her crude oil refining capacity – 464 700 barrels a day – is the largest in Africa. Oil and gas account for 98 per cent of all exports and provide over 65 per cent of revenue.

Oil revenues and loans from abroad have financed major industrial developments, which include iron and steel plants as well as a substantial petrochemical industry. Iron ore is mined near BENI SAF, Zaccar Timezrit and near the eastern border at Ouenza and Boa Khadra. There is a steel complex at Annaba, gas liquefaction plants at ARZEW and SKIKDA, and several oil refineries and petrochemical plants. Algeria is one of the few African countries to have its own car-manufacturing industry and the country makes one-third of its commercial vehicles. Food processing, building materials, chemicals and textiles are also important industries. Most of the larger industries are state-controlled but lighter industries remain in private hands.

A large proportion of Algerians (30 per cent) are farmers, although arable land accounts for only 3 per cent of the country's area. The rich coastland produces wheat, vines and olives, as well as early fruit and vegetables for the European market. Dates are also an important agricultural export – Algeria is the third largest producer in the world. In the mountains, where about 1 million people, mostly Berbers, live, more than 13 million sheep, as well as cattle and goats, graze on the plateau grasslands.

Most of the arable land (60 per cent) is privately owned. After independence in 1962 emphasis was put on cooperative farming, and there are even cooperative villages for the nomads in the oases. The rapid growth of industry has led to large numbers of people migrating from the countryside to urban areas. By the beginning of the 1980s over 60 per cent of the population were living in towns. Now they are faced with increasing housing shortages, and many unemployed workers have migrated to France to find jobs.

ALGERIA AT A GLANCE

Area 2 381 741 km² (919 590 sq miles)
Population 22 710 000
Capital Algiers
Government One-party socialist republic
Currency Dinar = 100 centimes
Languages Arabic; Berber and French also used
Religions Muslim (99%), Christian (1%)
Climate Mediterranean on the coast, hot and dry in the south. Average temperature in Algiers ranges from 9-15°C (48-59°F) in January to 22-29°C (72-84°F) in August
Main primary products Cereals, grapes, olives, citrus fruits, dates, vegetables, livestock, timber, fish; oil and natural gas, iron ore, zinc, phosphates, copper, lead
Major industries Oil and natural gas production and refining, petrochemicals, mining, cement, iron and steel, fertilisers, transport equipment, machinery, agriculture, wine production, food processing
Main exports Crude oil, petroleum products, natural gas, wine, fruit, vegetables, dates
Annual income per head (US$) 1920
Population growth (per thous/yr) 31
Life expectancy (yrs) Male 55 **Female** 58

Alma-Ata ***USSR*** Capital of Kazakhstan, 75 km (46 miles) north of Lake Issyk-Kul' near the Chinese border; formerly Vernyy. A commercial and industrial centre in an agricultural and fruit-growing region, it trades in wheat, sugar beet, apples and grapes. Its industries include machinery, railway equipment, tanning, sawmilling, spinning, tobacco and food processing.

Alma-Ata was founded by the Russians as a fort in 1854 and grew rapidly after 1926 and the building of the Turkestan-Siberian railway. In 1887 and 1911 it was virtually destroyed by earthquakes, but each time it has been rebuilt.

Population 1 046 100
Map USSR Id

Almadies Point ***Senegal*** See VERDE, CAPE

Almería ***Spain*** Ancient port about 275 km (170 miles) east and slightly north of Gibraltar. It was a base in the Middle Ages for Moorish pirates and, after the Moors were expelled from Spain in 1492, the 16th-century cathedral was built to double as a fortress against Moorish raids. Modern Almería processes salt and exports grapes, iron and lead from the mountainous province of the same name.

Population (town) 140 900; (province) 405 500
Map Spain Dd

Almourol ***Portugal*** Romantically located castle on an island in the Tagus river, about 105 km (65 miles) north-east of the capital, Lisbon. It was built in 1171 by the Knights Templar, the Crusaders who helped to drive the Moors from Portugal in the 13th century. The island was given to the knights as a reward by King Alfonso 1 (ruled 1139-85).

Map Portugal Bc

Al-Oued (Al-Wad) ***Algeria*** Oasis town in the north-east corner of the country near the border with Tunisia. It is a major source of dates.

Population 52 000
Map Algeria Ba

alp A shoulder high on a mountain side, especially the gentle, grassy slope above a U-shaped glaciated valley, often used as summer pasture. The term originally applied to such high slopes in the ALPS, but is now also applied elsewhere.

Alps ***Southern Europe*** Magnificent complex of mountains, lakes and glaciers, the Alps stretch for over 800 km (500 miles) in a great crescent 160 km (100 miles) wide across southern Europe. They start near the Ligurian coast behind Monaco and end in northern Yugoslavia. In between they take in south-eastern France, northern Italy, Switzerland, south-west Germany and Austria.

The scenic attractions, winter sports and mountaineering activities they provide have lured generations of tourists and sportsmen from all parts of the world. Chamonix, Interlaken, Grindelwald, St Moritz, Zermatt and a score or more other resorts are internationally famous. Geneva, Como, Lucerne, Garda and half-a-dozen more stunning lakes are strewn like jewels among the dazzling peaks. Highest peak of all is Mont Blanc (France), 4807 m (15 771 ft) near Geneva; more daunting challenges can be found on the Eiger (Switzerland), 3970 m (13 025 ft) in the Bernese Oberland, and the mighty Matterhorn (Switzerland-Italy), towering 4480 m (14 690 ft) near Zermatt.

Nine major passes, six long tunnels (one under Mont Blanc itself) and numerous gaps ensure that the Alps do not form a total barrier between north and south. Cattle and goats graze the summer pastures below the snowline; wine is grown on the warmer, lower slopes. Many of the region's rivers and torrents have been harnessed to generate hydroelectric power for towns and villages.

The complex folds, ridges and pinnacles of the Alps were formed by earth movements about 25-30 million years ago, and further shaped by glaciation around 2 million years ago. Rocky debris scoured from the land by glacier movement created moraines, or dams, to form long, deep lakes. Glaciers – the largest of which is the 26 km (16 mile) Aletsch, in the Bernese Oberland – are the sources of many streams and rivers.

Map Europe Dd

Alsace ***France*** North-eastern region and former province, bordered by the Rhine and Vosges Mountains. With LORRAINE it was occupied by Germany from 1871 to 1919 and from 1939 to 1944. Its architecture and traditions reflect both French and German influences.

The fertile plain of Alsace is part of the Rhine rift valley. Famous white wines are produced from Riesling, Gewürtztraminer and Sylvaner grapes in villages on the eastern slopes of the Vosges. The Grand Canal d'Alsace, running parallel to the Rhine, is navigable and there are 10 power stations on the Rhine which generate hydroelectric power for the region, which is a major industrial area. The main cities are STRASBOURG, MULHOUSE and COLMAR. Alsatian dogs – also known as German Shepherds – got their name from Alsace, where they were first bred and where they are used as sheepdogs.

Population 1 566 000
Map France Gb

Altai (Altay) ***Central Asia*** Group of mountain ranges where the borders of China, the USSR and western Mongolia meet. It stretches more than 1600 km (1000 miles) south-eastwards from southern Siberia into the Gobi Desert. There are many remote summits over 3000 m (9840 ft), including Mount Belukha (4506 m, 14 783 ft) in the USSR and Taban Bogdo Ula (Youyi Feng, Mount Kuytun), which reaches 4356 m (14 290 ft) where the three borders meet.

Map Asia Gd

Altamira ***Spain*** See SANTANDER

Altay Kray ***USSR*** Administrative territory (*kray*) of the RSFSR, in the basin of the upper Ob' river bordering on China and Mongolia. Deposits of gold, silver, lead, zinc and copper are mined in the south, in the ALTAI mountains. The north is cultivated for wheat, maize, sugar beet and oilseeds. BARNAUL is the main city.

Population 2 728 000
Map USSR Jc

Altiplano ***Bolivia*** Windswept, treeless plateau with an average altitude of 3670 m (12 040 ft). It lies between Bolivia's two main mountain ranges: the Western and Eastern Cordilleras of the Andes. It covers about 9 per cent (102 300 km², 39 000 sq miles) of the country and contains half its people. It is mainly arid except around Lake TITICACA in the north. The typical animal found there is the llama, which is a source of meat, wool and leather, and is also used as a high-altitude pack animal. The potato, which is native to the region, is the staple diet of most of the Indian population.

Map Bolivia Bb

altocumulus A middle altitude cloud formation of bands of dense, fleecy balls or rolls occurring 3-6 km (nearly 2-4 miles) above the earth. It usually indicates changeable weather.

altostratus A middle altitude cloud formation of thick, grey, or bluish sheets or layers, occurring 3-6 km (nearly 2-4 miles) above the earth. It usually indicates the approach of a warm front, and the arrival of rain within a few hours.

Altötting ***West Germany*** Bavarian town of religious pilgrimage near the Austrian border about 80 km (50 miles) east of Munich. Each year about 300 000 people visit its 9th-century church to pray at a black statue of the Virgin Mary, dating from the 14th century.

Population 11 000
Map West Germany Ed

Altun Shan ***China*** Mountain range with peaks of more than 4000 m (13 000 ft) forming the northern edge of the Tibetan Plateau.

Map China Cd

Aluvihara ***Sri Lanka*** See MATALE

Amacuro Delta ***Venezuela*** See ORINOCO

Amager ***Denmark*** Flat and fertile island in the Öresund (Danish Sound), immediately to the south of COPENHAGEN. Much of it is occupied by commercial and urban development which is an extension of the capital. Kastrup, Denmark's principal airport, built in 1925, lies here, only 10 km (6 miles) from the centre of Copenhagen. The Dutch were invited in 1521 to settle and reclaim low-lying Amager from the sea. The island has thriving horticulture and market-gardening businesses as well as specialised handicrafts such as embroidery, weaving and ceramics. Dragør, the ferry port to Sweden, retains in its architecture something of the island's earlier picturesque qualities.

Population 178 000
Map Denmark Cb

Amalfi ***Italy*** Historic town, about 60 km (38 miles) south-east of Naples, reached by a coast road that is considered to be the most beautiful in Italy. It was one of the earliest Italian maritime republics, flourishing as early as the 9th century. Today it is an attractive tourist resort hemmed in by rocky hills and cliffs, which have prevented modern growth.

Amalfi's jewel is its cathedral, founded in the 9th century, remodelled in a striking Arab-Norman design in 1206 and modified again in the 18th century. The crypt holds the tomb of St Andrew, brother of St Peter.

Population 6100
Map Italy Ed

Amarillo ***USA*** City in north-west Texas about 110 km (70 miles) east of the New Mexico border. It is a centre for the surrounding grain and livestock producing area.
Population (city) 162 900; (metropolitan area) 190 300
Map United States Fc

Amasya ***Turkey*** Market town about 300 km (185 miles) east of the capital, Ankara, and the capital of a province of the same name. The town was the capital of the kingdom of Pontus, which flourished from the 4th to the 1st century BC, and the tombs of the Pontic kings, including that of Mithradates the Great, can be seen hewn in rocky hills overlooking the town. The surrounding province produces apples, wheat, onions, tobacco, opium, wool, hemp, sugar beet and small amounts of lead, gold and silver.
Population (town) 53 200; (province) 358 900
Map Turkey Ba

Amazon ***South America*** Second longest river in the world after the Nile. It is about 6440 km (4000 miles) long from its source in the Peruvian Andes to its delta in the northernmost corner of Brazil, and drains an area almost the size of Australia. For nearly half its length it flows through the world's largest rain forest – a jungle covering some 6.5 million km² (2.5 million sq miles), an area larger than the whole of Western Europe. In recent years vast swathes of the forest have been destroyed as developers have moved in after timber and minerals, and to clear the land for agriculture. As the forest died so did the Indians who lived there. Once they numbered more than a million; now only tens of thousands survive. However, attempts are being made to limit further destruction of the forest and protect the remaining Indians.

Numerous huge rivers swell the Amazon's volume: the Negro, Branco, and Japurá from the north and Juruá, Purus, Madeira, Tapajós, Tocantins and Xingu from the south. From the Peruvian border the main river is known as the Solimões until it joins the Negro – itself 18 km (11 miles) wide – near the city of Manaus. Because of their speed the two great rivers flow side by side for about 6 km (4 miles) – the acid Negro stained black like tea by the rotting leaves in the swamps of its early course through Colombia, contrasted with the yellow brown Solimões – before the waters finally merge.

The huge river is navigable by ships of up to 6000 tonnes throughout its length in Brazil and beyond the border to the Peruvian jungle port of Iquitos, 3200 km (2000 miles) from the sea. For much of this course, from the ships sailing down the middle of the river, it is difficult to make out the banks because they are so far apart. The Amazon delta is a series of channels with Marajó island, the size of Switzerland, separating the farthest channels. The mouth of the main channel alone is about 50 km (30 miles) wide, and the volume of silt in the water is so great that it stains the sea for 300 km (186 miles) offshore.
Map Brazil Cb

Ambato ***Ecuador*** City 140 km (85 miles) south of Quito, at the foot of the inactive 6310 m (20 700 ft) volcano, Chimborazo. It has been largely rebuilt since an earthquake in 1949 destroyed much of the city and killed about 6000 people. The city is an important manufacturing centre for rugs and textiles.
Population 221 000
Map Ecuador Bb

amber Translucent, yellowish fossilised resin from coniferous trees, found chiefly along the shores of the Baltic Sea. It is used for making ornaments and jewellery.

Ambon ***Indonesia*** Island of 813 km² (314 sq miles) in the Maluku group, south of the Philippines. Since the 16th century, it has been famous as the centre of the European spice trade, and cloves and nutmeg are still two of its main products. The Ambonese are one of Indonesia's few Christian communities. They resisted incorporation into the state of Indonesia in 1950 and many fled to the Netherlands, where they proclaimed a Republic of the South Moluccas and committed periodic acts of terrorism.
Population 73 000
Map Indonesia Gc

▼ IMPENETRABLE FOREST A tributary snakes its way to the mighty Amazon – the easiest route for travellers in the vast rain forest of northern Brazil.

Amboseli ***Kenya*** Lake on the Tanzanian border, south of Nairobi. It is a dry bed of soda most of the year. The name has been given to a 380 km² (147 sq mile) national park, east of the lake, and to a 3000 km² (1160 sq mile) game reserve, which is the home of Masai pastoralists and a safari centre for tourists.

Map Kenya Cb

Ambre Mountains National Park ***Madagascar*** Humid forest park covering 18 200 hectares (about 44 970 acres) in the north which is noted for lemurs and orchids. The volcanic hills rise to 1475 m (4839 ft).

Map Madagascar Aa

Amer ***India*** A series of hilltop palaces of the Maharajah of Jaipur, dating from about 1500 onwards, north of the city of Jaipur. It has ingenious means to store rainwater, in a hollowed-out mountain top which has been roofed over. Amer has its own cannon foundry. A massive cannon said to be capable of hurling a ball 40 km (25 miles) stands, untested.

Map India Bb

America, Central Nearly a thousand years before Europeans came to what was to them the New World, the Mayan Indians had evolved a sophisticated culture in Central America, with advanced art, architecture, engineering, mathematics and astronomy. Their still-impressive buildings survive in the peninsula of Yucatán, and other parts of Southern Mexico, and in much of Guatemala and western Honduras. In Mexico they built their splendid city states of Chichén Itza and Uxmal, which were supported by a well-developed economy based on agriculture, commerce and crafts. But by 1519, when the Spanish conquistadors arrived, the Mayans were already in decline.

Central America is sometimes defined as the land once occupied by the Mayas, but is more conveniently defined as the area between the northern border of Mexico and the border between Panama and Colombia in the south. It therefore includes the nations of Mexico, Guatemala, Belize, Honduras, El Salvador, Nicaragua, Costa Rica and Panama, a largely uneasy grouping of buffer states along the isthmus between North and South America.

A chain of volcanic mountains along the western side of Central America connects the Coast Range of California to the South American Andes. Many of the volcanoes are still active, among them Tajumulco in Guatemala (4210 m, 13 810 ft), the highest point in Central America. The whole area is subject to earthquakes. One particularly calamitous shock hit Mexico City in September 1985, killing about 10 000 people and destroying hundreds of buildings. Earthquakes caused serious damage to Nicaragua's capital, Managua, and to Guatemala City, in 1917, 1918, 1931 and 1972.

The climate of the area is tropical, although the mountains and plateaus are cooled by the altitude. North-east trade winds bring rain from the Gulf of Mexico, which falls on northern slopes. Around July the rainfall is particularly heavy, San Juan del Norte (Nicaragua) receiving a summer total of 3632 mm (143 in). Much of Central America's vegetation is dense rain forest. Cotton, coffee, sugar, beef, hardwoods and fishing have been mainstays since colonial times, and the gold and silver mines that enticed Spanish settlers are still worked.

Area 2 500 680 km² (965 240 sq miles)
Population 108 million

America, North Some old-style textbooks still have it that America is a single continent. Most modern geographers divide the landmass into two, North and South America, with CENTRAL AMERICA often separately identified. North America alone is a grand enough concept, stretching as it does more than 6500 km (over 4000 miles) north to south, from the Arctic Ocean to the Gulf of Mexico, and more than 8000 km (5000 miles) from the western extremity of Alaska to Newfoundland.

Despite intensive development, and overcrowding in some places, North America is relatively empty, with virgin wildernesses of tundra, mountain and desert.

The Atlantic coast in the east has a narrow coastal plain, narrowest and most densely populated in the north, where it is also most indented. Behind this, from Newfoundland to Alabama, lie chains of mountains: in the south, the Blue Ridge, the Alleghenies and Cumberlands form parts of the Appalachian system, while farther north lie the Catskills and the New England ranges. Beneath the western edge of the Appalachians lie rich coal deposits.

North of the east coast mountains, and

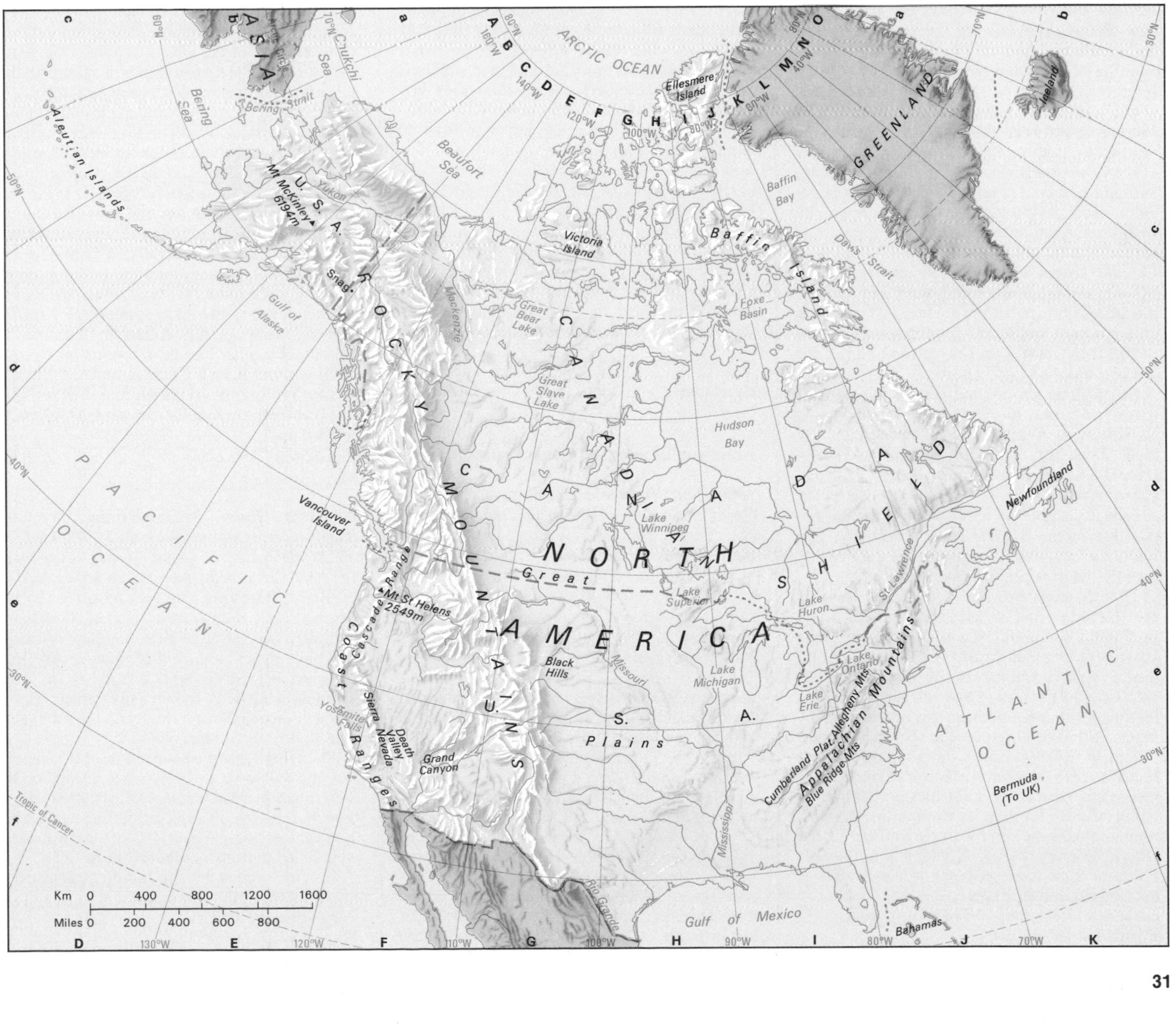

occupying a vast area of eastern Canada and the north-eastern USA, is the Canadian Shield, a rugged 4.8 million km^2 (1.9 million sq mile) plateau of low, worn hills. The southern part of the Shield is clothed in immense coniferous forests; to the north is the tundra, where the subsoil is permanently frozen, permitting only a growth of lichens and mosses.

West of the eastern mountains lie the great central lowlands of North America, extending from the Mackenzie river in Arctic Canada to the subtropical Mississippi Basin in the southern United States. The eastern part of the lowlands divides into a dairy-farming belt in the north, a corn (maize) belt in the centre, and a cotton-growing area in the south with fruit, nut and rice specialities. Farther west, and straddling the Canada-US border, is a great area of spring and winter wheat, grown on what was once tall grass prairie. Farther west again, stretching from Alberta in the north to Texas in the south, is semi-arid cattle country, which it is unwise to plough or plant except where irrigation water is available.

Far to the south, the 6019 km (3741 mile) long Mississippi-Missouri system empties into the Gulf of Mexico, a beach-lined bay of the Atlantic sheltered by the long, low arm of Florida. Below the waters of the Gulf of Mexico lies a rich reservoir of oil, one of many such deposits which make North America a leading producer of crude oil. The south-western USA is arid, and driest in the deserts of Nevada and southern California. Across the region runs the monstrous slash or scar of the Grand Canyon of the Colorado River, 347 km (216 miles) long, and up to 29 km (18 miles) wide. Farther north lie the Rocky Mountains; west of those are the desert basins and then, guarding the approaches to the Pacific coast, the Sierra Nevada and Cascade ranges. The mountainous areas of British Columbia, Washington state and Oregon are famous for timber production, while central and southern California grow huge quantities of soft fruits, citrus, grapes, cotton and rice for US consumption and export.

North America has more than its share of nature's dramas. In 1980, the volcanic Mount St Helens in Washington state spectacularly blew off its top. The Gulf of Mexico usually presents the eastern seaboard with at least one savage hurricane a year, and in San Francisco pessimists expect to be swallowed up at any moment in the San Andreas Fault – the line where two continental plates meet and, consequently, an area prone to earthquakes.

The climate, too, has a tendency towards the dramatic, and is extremely varied. Snag, in Canada's Yukon, boasts North America's record low temperature of −62°C (−80°F), and Death Valley, California, its record high of 56.6°C (134°F). The High Plains have dust storms in summer and blizzards in winter. New York, like Washington, can be steamy and humid in August, but it has a January wind honed to razor sharpness. Montreal and Buffalo record huge snowfalls, and Chicago's midwinter temperatures plunge in its mid-continental position and with the wind whipping off the frozen lake. Ellesmere Island, reaching out towards northern Greenland, is polar, while Florida is subtropical and swampy.

Area 19 137 368 km^2 (7 388 945 sq miles)
Population 265 million

America, South The continent – or arguably, subcontinent – is usually considered to include all the countries below the southern border of Panama. An alternative term, Latin America, is generally held to include not only South America, but also all the nations of CENTRAL AMERICA – an area where most of the countries have ties of Spanish or Portuguese language and culture.

Unquestionably, South America's most striking feature is the great mountain chain of the Andes which runs from Venezuela in the north along the Pacific coast to Tierra del Fuego in the south, a distance of more than 8000 km (5000 miles). This cloud and snow-capped range spreads out to 600 km (373 miles) wide in Bolivia. Several of South America's greatest rivers, the Orinoco and the Amazon among them, rise in the Andes.

In the north and east of the continent are the Guiana and Brazilian highlands, a mass of block-like mountains and plateaus, some of which end in abrupt scarps. In Brazil lies the vast Amazon basin, laced by the mighty river and its many tributaries.

Rainfall varies greatly throughout the continent. The Amazon's rainfall is the heaviest (1500-2500 mm, 60-100 in a year), deposited by trade winds from the Atlantic. By the time they reach Peru and Chile, however, these winds have dropped their moisture, and there are parts of the Atacama desert in Chile that have had no rain at all within living memory. Even the Peruvian capital, Lima, has no more than 40 mm (1.5 in) of rain a year.

Savannah – tropical grassland – lies to the north and south of the Amazon rain forest. Farther south still, the rainfall is heavier, the grassland more lush, and the soil more fertile. This area includes the *campo limpo* of the extreme south of Brazil, the purple grassland of Uruguay, and the *pampa* of Argentina. A large part of it is farmed for cereals, and it incorporates some of the finest grazing land anywhere in the world.

Area 17 798 500 km^2 (6 872 000 sq miles)
Population 276 million
Map See p. 30

American Samoa See SAMOA, AMERICAN

Amersfoort *Netherlands* Industrial city 42 km (26 miles) south-east of Amsterdam. The medieval city centre was once moated and walled, but now only the gates remain. It is dominated by the 95 m (312 ft) Tower of Our Lady, built in the 15th century. The city's manufactures include metal goods, chemicals and foodstuffs.

Population 87 500
Map Netherlands Ba

amethyst Purple or violet variety of transparent quartz, used as a gemstone. This results from manganese in the quartz.

Amiens *France* Capital of the Somme department and an industrial and administrative city on the Somme river, 115 km (72 miles) north of Paris. It was the scene of the signing in 1802, during the Napoleonic Wars, of a treaty between Britain, France and their allies. However, the result was barely a year's peace in the war, which ended only with Napoleon's defeat at the Battle of Waterloo in 1815. During the Second World War, the capture of Amiens by the Germans in 1940 led to the German blockade of the Allies in northern France and the evacuation of British, French and Belgian troops from Dunkirk. The Gothic cathedral of Notre Dame is the largest in northern France. The city's industries include textiles, clothing, tyres and chemicals. A university was opened there in 1964.

Population 159 600
Map France Eb

Amindivi Islands *India* See LAKSHADWEEP

Amman *Jordan* Capital and administrative and business centre of the country, situated 38 km (23 miles) north-east of the Dead Sea. Unlike the ancient capitals of other Arab countries, Amman is a new town. Before 1875, what is now Amman consisted solely of the ruins of the once prosperous Roman city of Philadelphia, itself built on the site of the Biblical city of Rabbat 'Ammon. Popular lore has it that Circassian settlers arrived in the area in 1871; certainly they had built a village there by the 1880s. It developed a reputation as a commercial centre that was much enhanced by the completion of the northern section of the HEJAZ Jordanian Railway which passed only 5 km (3 miles) east of the village in 1905. The status of Amman was settled in 1921 when Emir Abdullah, after trekking round the new emirate, established himself in Amman and made it his capital.

Amman, which is built on seven hills, is expanding rapidly as it caters for an increasing flow of business, banking and diplomatic visitors, as well as many Palestinian refugees. It is the hub of the country's road and rail communications and has a new international airport (Queen Aliya). Despite its modernity, it is easy to get lost in the city's many narrow, twisting streets and alleys. To cater for its growing cosmopolitan community, there are a number of special schools, such as the American Community School, as well as a range of foreign institutes – including the British Council, and the French, American, Spanish, Turkish and Russian cultural centres. Amman also houses the University of Jordan.

Population 1 232 600
Map Jordan Ab

Amritsar *India* Manufacturing city and religious centre of the Sikhs, about 380 km (236 miles) north-east of Delhi. It was founded in 1577 by the fourth guru (spiritual leader) of the Sikhs, Ram Das – and contains the Golden Temple, the Sikh's most sacred Gurdwara (temple). The shrine, a large complex of buildings built around a sacred rectangular lake, was occupied by armed Sikh extremists in 1983. The following year Indian government troops stormed the temple and drove out the occupants in a bloody encounter.

An earlier violent episode in the city's history was the massacre in 1919, when troops under the British general Reginald Dyer fired into a crowd of unarmed people protesting against the antisedition laws. About 400 of the protestors were killed and many others injured.

The city is renowned for its crafts, particularly fine woollen cloths and carpets.

Population 595 000
Map India Bb

Amsterdam ***Netherlands*** The country's capital, commercial centre, second port and second largest city after Rotterdam. It is built on piles, in sand and mud, where the Amstel river flows into IJsselmeer, once a branch of the sea, and is named after a protective dam built in about 1240. Access for oceangoing ships is by the Noordzeekanaal, opened in 1876, and from the Rhine and Maas (Meuse) rivers by canals.

The old city, with its multitude of little bridges over encircling canals, dates mostly from 1650 to 1720, when it prospered as the capital of a worldwide commercial empire. Although its citizens suffered greatly during the German occupation from 1940 to 1945, the city itself escaped virtually unscathed, and attracts many tourists. Its diamond-cutting trade serves a worldwide market, and its other industries include shipbuilding, engineering, textiles and publishing.

Among the places of interest are the Rijksmuseum (with the world's greatest collection of Dutch paintings); Rembrandt House, where the painter Rembrandt van Rijn (1606-69) once lived; the house of Anne Frank (whose diary tells how the Jewish girl and her family hid there during the German occupation); the 17th-century royal palace; and the 17th-century University of Amsterdam.

Population (city) 687 400; (Greater Amsterdam) 936 400

Map Netherlands Ba

Amsterdam Island ***Indian Ocean*** About 30 people man a hospital, research stations and administrative offices on this near-barren volcanic island in the French Southern and Antarctic Territories. Only 54 km² (21 sq miles) in area, and with a peak rising 911 m (2988 ft) near the centre, Amsterdam Island supports colonies of sea lions and herds of wild cattle. With its neighbouring island of St Paul, 100 km (62 miles) south, it lies far in the southern Indian Ocean, about 3700 km (2300 miles) south-east of Madagascar.

Portuguese sailors are thought to have discovered it in 1522, but Anton Van Diemen, a Dutchman, was first to visit the island, in 1633, naming it New Amsterdam. France claimed it in 1843.

Map Indian Ocean Cd

Amudar'ya ***USSR/Afghanistan*** River in central Asia, 2620 km (1630 miles) long. It rises in the Pamir mountains and flows generally west, forming much of the border between the USSR and Afghanistan. It then turns north-west and drains into the Aral Sea by way of a 160 km (100 miles) long delta. Its middle course provides water for the Kara Kum canal and irrigates parts of the Turkmen and Uzbek deserts. The river, known to the Greeks, Romans and Persians as the Oxus, is navigable for 1450 km (about 900 miles) from its mouth.

Map Afghanistan Ba; USSR Hd

Amundsen Sea Situated between Cape Dart and Thurston Island in ANTARCTICA, the sea is part of the South PACIFIC OCEAN. The Norwegian explorer Nils Larsen, who explored the area in 1929, named it after his compatriot Roald Amundsen (1872-1928), the first man to reach the South Pole.

Map Antarctica Fc

Amundsen-Scott Station ***Antarctica*** American scientific base and landing strip at the geographic SOUTH POLE. It was founded in 1957.

Map Antarctica d

Amur (Heilong Jiang) ***USSR/China*** River, 4510 km (2800 miles) long, formed on the Soviet-Chinese border by the union of the Shilka and Argun rivers. The Amur flows south-east along the border then north-east through Soviet territory, entering the Sea of Okhotsk near the port of Nikolayevsk.

Map China Ja; USSR Mc, Oc

amygdule Cavity in lava, caused by an escape of gas or steam, and filled by a mineral, such as quartz, agate or calcite. A lava cavity which is not filled is called a scoria.

An Bhlarna ***Ireland*** See BLARNEY

An Bhóinn ***Ireland*** See BOYNE, RIVER

An Cabhán ***Ireland*** See CAVAN

An Clár ***Ireland*** See CLARE

An Clochán ***Ireland*** See CLIFDEN

An Iarmhí ***Ireland*** See WESTMEATH

An Life ***Ireland*** See LIFFEY

An Mhi ***Ireland*** See MEATH

An Najaf ***Iraq*** Regional capital about 160 km (100 miles) south of the national capital, Baghdad. It is a place of pilgrimage for Shiite Muslims. Inside its golden-domed mosque is the shrine of Ali Ibn Abi Talib, son-in-law of Muhammad, the founder of Islam.

Population 180 000

Map Iraq Cc

An Tsionainn ***Ireland*** See SHANNON, RIVER

anabatic wind Local wind blowing up a valley during the day to replace the mountain air that rises in convection currents as the sun heats the slopes. Also called up-valley wind.

Anadyr' ***USSR*** Town, formerly Novomariinsk, in the Soviet Far East on the estuary of the Anadyr' river. Its industries include coal mining and fish canning. The Anadyr' river, 1116 km (694 miles) long, rises in the mountains south of the Chukot Range and flows north-east to the Gulf of Anadyr' in the Bering Sea.

Map USSR Sb

Anatolia ***Turkey*** Name used for the whole of Asiatic Turkey and also for the central wheat-growing plateau which makes up about 25 per cent of Turkey and rises to between 800 and 2000 metres (2625 and 6562 ft). The plateau is crisscrossed by ravines and dotted with volcanic peaks. It was a battleground for numerous kingdoms of Asia Minor from the time of the Hittites (2000 to 1180 BC) until the 11th century AD, when it was conquered by the Seljuk Turks. The area embraces the site of CATAL HUYUK, where some of the world's earliest farmers built a city 8500 years ago.

Map Turkey Bb

Ancash ***Peru*** Western department lying between La Libertad and Lima. The dominant feature is the Santa river which cascades from its source 4000 m (13 125 ft) up in the mountains at the lagoon of Conocha to the sea at the port of Chimbote. The famous archaeological site of CHAVIN DE HUANTAR lies 64 km (40 miles) from the provincial capital, Huaráz. There are glacial lakes at Llaca and Llanganuco and, in the lower Santa valley, the remains of the 'Great Wall of Peru' which dates from the time of the Chimú Indians who flourished during the 13th to 15th centuries.

Population 818 300

Map Peru Bb

Anchorage ***USA*** Largest city in Alaska. It lies at the head of Cook Inlet on the south coast, about 465 km (290 miles) from the Canadian border. Anchorage is also a port handling gold, oil and other minerals produced in the region, and is a growing tourist centre.

Population 226 700

Map Alaska Cb

Ancona ***Italy*** Seaport 180 km (110 miles) east of Florence on the 'elbow bend' of the Adriatic coast. Its name derives from the Greek word for elbow, *ankon*. It was founded by Greek colonists and became an important Roman and medieval port. Rail and road links in the 19th century led to new industrial expansion – and fishing is major occupation.

The finely preserved Arch of Trajan, built in AD 115, stands on the quayside among railway sidings. Ancona's 11th-13th-century cathedral church is an interesting fusion of Byzantine, Romanesque and Gothic styles built on the plan of a Greek cross. The medieval quarter, which still shows some scars from an earthquake in 1972, has many churches and palaces, including Santa Maria della Piazza – 13th century but of 5th-6th century origin – and the 15th-century Loggia dei Mercanti. The Mole Vanvitelliana is an 18th-century fort built in the shape of a pentagon in the harbour.

Population 105 200

Map Italy Dc

Andalucía ***Spain*** Mountainous region in the south of the country, covering 87 268 km² (33 694 sq miles). Settled by the Phoenicians about 1000 BC, it was later ruled by the Carthaginians, the Romans and finally (until 1492) the Moors. It is rich in minerals – including lead, copper and cinnabar – and grows olives, grapes, oranges and lemons. Its cities include SEVILLE, the regional capital, MALAGA, GRANADA and CORDOBA.

Population 6 441 800

Map Spain Cd

Andaman and Nicobar Islands ***India*** Two groups of islands in the eastern Bay of Bengal, covering 8249 km² (3185 sq miles) opposite the coast of Burma. They were given a unified administration by the British in 1872 and have been used as a penal colony from 1857 to the present day. Much of the area consists of evergreen forest, and the economy is based largely upon timber and copra. The capital of both groups is Port Blair on South Andaman Island.

Population 188 700

Map India Ee

Andorra

SMALL SKI RESORTS AND DUTY-FREE SHOPS BRING IN REVENUE TO THIS FEUDAL STATE LYING IN A POCKET OF THE PYRENEES

High in the eastern Pyrenees, between France and Spain, lies the tiny state of Andorra, governed according to a system that dates from feudal times. According to tradition, the state was granted independence in exchange for help against the invading Moors by Charlemagne, the first Holy Roman Emperor. It became a co-principality in 1278, shared by the French Comte de Foix and the Spanish Bishop of (Seo de) Urgel. The legacy of this arrangement is government by French and Spanish delegations (the co-princes are the President of France and, still, the Bishop of Urgel), and the obligation to pay feudal dues to France of 960 francs and to the bishop of 460 pesetas every second year. A local general council, which can propose legislation, is elected by those of 21 or over who are second and third-generation Andorrans, and all first-generation Andorrans of foreign parentage aged 28 and over – in effect, control is in the hands of the heads of established families.

The land consists of high valleys carved out by glaciers and drained by tributaries of the Balira, which itself flows into the Segre river in Spain. Its mountain peaks reach heights between about 1800 and 3000 m (about 5900 and 9800 ft), and the scenery, which looks rather like the Swiss Alps, is spectacular. Cold winters and mild sunny summers add to the principality's attractions. About 6 million tourists come here every year to ski in the winter, walk in the mountains in the summer – and take advantage of the duty-free goods on sale. Indeed, tourism and the duty-free trade are Andorra's principal source of income – along with revenues from the sale of postage stamps and from advertising on its radio station, Radio Andorra, which broadcasts throughout Europe.

Because of substantial immigration in the 1960s and 1970s only about one-third of the population of 49 500 are now native-born Andorrans. The official and most used language is Catalan – it resembles Provençal and is also spoken in north-eastern Spain and the adjoining area of France – but French and Spanish are also common.

PEOPLE OF THE VALLEYS

The majority of the people are Roman Catholic. They live in six valleys, including the capital, the market town of ANDORRA LA VELLA. Some who do not cater to tourists for a living raise sheep and cattle on the high pastures – which are common land – and spend their summers in temporary homes on the mountainsides. Some grow cereals, vegetables, potatoes or tobacco in small, walled or terraced fields on the steep slopes. The tobacco is sold in the duty-free shops. Smuggling over the borders into France and Spain is also a thriving, though illegal, industry.

Many Andorrans feel that the powerful traditional family groupings and feudal system of government have held them back from the advances made by the rest of Europe. For example, women were not given the right to vote here until 1970, and political parties are not legally recognised. Now there are moves for a more democratic system of government to be introduced.

ANDORRA AT A GLANCE

Map France De
Area 464 km² (179 sq miles)
Population 49 500
Capital Andorra la Vella
Government Co-principality governed jointly by delegations from the French government and Spanish Bishop of Urgel
Currency Spanish peseta, French franc
Languages Catalan (official), French, Spanish
Religion Christian (99% Roman Catholic)
Climate Temperate, but modified by altitude. Average temperature in Les Escaldes ranges from −1 to 6°C (30-43°F) in January to 12-26°C (54-79°F) in July
Main primary products Sheep, cattle, tobacco, potatoes, cereals (barley, rye), vegetables
Major industries Tourism, duty-free trading
Main exports Postage stamps
Annual income per head (US$) 9000
Population growth (per thous/yr) 54
Life expectancy (yrs) Male 70 **Female** 76

Andaman Sea An arm of the INDIAN OCEAN which is bordered in the west by the Andaman and Nicobar islands, and in the east by Burma and Thailand. The Strait of MALACCA in the south – a key shipping lane – leads to the South CHINA SEA. The sea's chief port is RANGOON.
Area 777 000 km² (297 572 sq miles)
Map Indian Ocean Db

Andes ***South America*** Mountain chain that extends like a spine for some 7250 km (4500 miles) along the entire west coast of South America, from Colombia to Tierra del Fuego. Created by the collision of the Nazca Plate of the eastern Pacific with the South American continent (see PLATE TECTONICS), it contains numerous volcanoes and is subject to violent earthquakes. The highest peaks are both in Argentina: OJOS DEL SALADA (7084 m, 23 241 ft) and ACONCAGUA (6960 m, 22 835 ft). Many of the continent's major rivers, including the AMAZON and ORINOCO, rise there.
Map South America Bc, Bd

andesite Fine-grained volcanic rock named after its presence in the volcanoes of the Andes in South America. As a LAVA it is less fluid than BASALT and creates spectacular volcanoes.

Andhra Pradesh ***India*** State in the south-east covering 275 088 km² (106 184 sq miles) and bordering on the Bay of Bengal. In 1956 it took over much of the princely state of HYDERABAD, whose main city is now the state's capital. Its main products are cotton, wheat and rice. The low-lying coastal districts are liable to cyclone damage. The principal language is Telugu.
Population 53 549 700
Map India Cd

Andorra la Vella ***Andorra*** The capital, situated in the country's principal valley, though this lies 1059 m (3475 ft) above sea level. Surrounded by stunning mountain scenery, it is also a shoppers' paradise of duty-free goods, and a constant stream of cars makes parking difficult in the narrow streets. It is practically joined to the larger town of Les Escaldes, a spa with hot sulphur springs.
Population 14 000
Map France De

Andropov ***USSR*** City on the Volga river at the south end of the Rybinsk reservoir. Formerly called Shcherbakov, then Rybinsk, it was renamed in 1984 after President Yuri Andropov (1914-84). It has engineering, boat-building, timber-machinery, food and leather industries.
Population 249 000
Map USSR Ec

Andros ***Bahamas*** Largest of the islands, covering 4144 km² (1600 sq miles) and lying about 56 km (35 miles) south-west of Nassau. Much of the interior is densely forested, and the western shore is fringed by mangrove swamps. The main occupations are farming and tourism.

Scuba diving amid spectacular underwater scenery is popular. A special feature of Andros is the 118 Blue Holes – deep, circular holes, possibly collapsed caves, which rise up through the coral inland and on the coast. They are filled with seawater and are often more than 60 m (200 ft) deep. There is a joint British-American naval base near Andros Town.
Population 8900
Map Bahamas Bb

Angara ***USSR*** Major tributary – 1825 km (1135 miles) long – of the YENISEY river. It flows from the south-west corner of Lake Baikal, past Irkutsk to join the Yenisey at Strelka. It is navigable for most of its course, and has several large hydroelectric power stations.
Map USSR Lc

Angel Fall ***Venezuela*** World's highest waterfall, 979 m (3212 ft) high. Named after the American airman James Angel, who discovered them in 1937, the falls plunge from the lip of the Auyán-Tepuí plateau in the Guiana Highlands – bouncing only once, off a ledge.
Map Venezuela Bb

Angers ***France*** Capital of Maine-et-Loire department, in the former province of Anjou, on the Maine river, 80 km (50 miles) north-east of Nantes. It is a commercial and industrial centre whose industries include food processing, wine, textiles, machinery and electronics. There are a number of medieval buildings, including a vast 13th-century chateau built for Louis IX which now houses a museum of tapestries.
Population 141 100
Map France Cc

Angkor ***Cambodia*** Ruins of the rich Khmer civilisation that dominated north-western Cambodia and eastern Thailand for about 500 years until the early 15th century. Its extensive network of temples and man-made lakes and irrigation canals was discovered, hidden by centuries of jungle undergrowth, in 1860. More than 600 Hindu temples, some as big as cathedrals, all with elaborately carved and decorated towers, were eventually uncovered and restored.

The two largest complexes were at Angkor Wat – meaning 'temple city' in the Khmer language – and at Angkor Thom. At the heart of each complex is a pyramidal temple-mountain with five ornate, soaring towers. The rectangular lakes – reservoirs also known as *barais* – and the network of canals functioned mainly to re-create on earth an image of the Hindu-Buddhist universe. Some images of the Buddha are thought to be portraits of Khmer kings.

The Angkor complex appears to have been abandoned by the Khmers after the Thais attacked Cambodia. Since 1975, when Pol Pot led a Communist takeover of the country, the temples of Angkor, believed to be the largest single group of religious buildings in the world, have been inaccessible to travellers. The jungle has taken over again, and archaeologists fear that fast-growing foliage and water erosion may have damaged the monuments.

Map Cambodia Ab

Anglesey ***United Kingdom*** Island of 715 km² (276 sq miles) off the coast of north-west Wales, separated from the mainland by the Menai Strait – which narrows to only 200 m (650 ft) and is spanned by two bridges – and forming part of the county of Gwynedd. It was the sacred stronghold of the Celtic priests, the Druids, when their order was destroyed by the Romans in AD 61. It remains a focus for Celtic and Welsh traditions and culture. It has splendid holiday beaches but is often regarded by outsiders simply as a stepping stone to the ferry port of Holyhead.

Population 69 000

Map United Kingdom Cd

Anglo-Normandes, Iles ***United Kingdom*** See CHANNEL ISLANDS

Angola See p. 36

Angoulême ***France*** A historical fortified city and capital of Charente department, on a limestone hill overlooking the Charente river 105 km (65 miles) north-east of Bordeaux. It was an episcopal see in AD 379, and has many ancient houses and a 12th-century cathedral; this was restored in the 19th century.

Population 109 400

Map France Dd

Anguilla ***Leeward Islands*** Coral island, 91 km² (35 sq miles) in area, some 250 km (155 miles) east of Puerto Rico. It was named by the Italian explorer Christopher Columbus in 1493 from the Spanish word *anguila*, meaning 'eel' – probably because of its long, narrow shape.

First colonised by the British in 1690, the island has, since 1980, been a self-governing dependency of the UK. British troops intervened in 1969 to restore legal government after Anguilla broke away from domination by the St Kitts' administration. The people are mostly of African descent. Farming is poor, due to thin soil and scant rainfall. There is some fishing, and salt is produced by evaporating seawater, but a large part of the island's income is money sent home by some 20 000 Anguillans living in Britain and the USA. Tourism is limited by poor air communications, despite fine beaches and clear seas. The capital is The Valley.

Population 6500

Map Caribbean Cb

Angus ***United Kingdom*** See TAYSIDE

Anhui ***China*** Province of 130 000 km² (50 000 sq miles) in east central China, astride the Chang Jiang and Huai rivers. The poorly drained northern part of the province in the North China Plain is prone to floods. Wheat and rice are grown in the Chang Jiang valley. The southern part of the province is renowned for its mountain scenery, particularly the HUANG SHAN range. Although Anhui is primarily an agricultural province, it has reserves of coal, iron ore and copper. Cities include Hefei, the provincial capital, and Bengbu, a centre for the engineering industry.

Population 48 030 000

Map China He

Anjou ***France*** Former western province of France that straddles the lower Loire valley. In the 12th century it was part of the Plantagenet domains incorporating England, Normandy and Aquitaine. It was claimed by Louis XI of France in 1481. ANGERS is the region's main market for local fruit, vegetables and wine – including the renowned Anjou Rosé.

Map France Cc

Anjouan ***Comoros*** See NZWAMI

Ankara ***Turkey*** The country's capital, lying at the centre of the Anatolian plateau 350 km (220 miles) east and slightly south of Istanbul. It was founded in the 8th century by the Phrygians, who came from Thrace about 1200 BC and occupied central Turkey. The city is Turkey's second largest after Istanbul, and is dominated by the hilltop mausoleum of Kemal Ataturk (1881-1938), the nationalist leader who founded modern Turkey after the collapse of Ottoman power in the First World War.

Its main industries are cement, textiles, leather goods and mohair – made from the wool of angora goats, from which the city probably gets its former name (Angora).

Population 2 252 000

Map Turkey Bb

Anloga ***Ghana*** Coastal town on Keta Lagoon about 120 km (75 miles) east of the capital, Accra. It is growing in importance at the expense of its neighbour Keta (population 12 000), which lies about 150 km (95 miles) east of Accra, on a narrow sandbar east of the lagoon. The land on which Keta stands is being eroded steadily by the sea, and part of the town is already underwater. Anloga is now taking over Keta's traditional market for cotton cloth, fish, onions and salt – produced by evaporation from the lagoon.

Population 15 000

Map Ghana Bb

Annaba (Bone) ***Algeria*** Town and seaport on the Mediterranean coast, near the border with Tunisia. It is Algeria's most important port after Algiers and Oran. It was built by Arabs in the 7th century on the site of Aphrodisium, the ancient Roman port of Hippo, which was the bishopric of St Augustine in the 5th century. Iron ore, phosphates, wine and cork are the chief exports. The town manufactures chemicals and is the site of Algeria's only iron and steel plant. It is an administrative and transport centre, and has an airport. The Roman site of Hippo Regius, a rich city of Roman Africa until AD 300, lies about 2 km (1 mile) south and many of the archaeological finds are displayed in the museum at Annaba.

Population 245 000

Map Algeria Ba

Annam ***Vietnam*** Name of the central MIEN TRUNG region in French colonial times.

Annamite Chain ***South-east Asia*** See MIEN TRUNG

Annapolis ***USA*** State capital of Maryland, standing on Chesapeake Bay about 48 km (30 miles) east of the national capital, Washington. It was settled about 1648 and became the Maryland colony's capital in 1694. In 1786, representatives of a number of states of the Union met in the city for the Annapolis Convention, a precursor of the Constitutional Convention, held the following year in Philadelphia. A naval school founded in Annapolis in 1845 became the nucleus of the US Naval Academy.

Population 31 900

Map United States Kc

Annapurna ***Nepal*** Mountain in the Great Himalayas, about 175 km (110 miles) north-west of Kathmandu. At 8078 m (26 504 ft), it is the eighth highest mountain in the world. It has claimed one life for every climber who has reached the summit – 31 by late 1984. First up was a French expedition in 1950. The huge, precipitous south face of the main peak, Annapurna I, seems to tower above the town of Pokhara, though it is some 40 km (25 miles) distant. West, across the gorge of the Kali Gandak river, is the even higher peak of Dhaulagiri, 8172 m (26 810 ft).

Map Nepal Aa

Annecy ***France*** Capital of Haute-Savoie department, at the northern end of Lake Annecy, 100 km (60 miles) east of Lyons. It manufactures linen, yarn, paper and precision instruments. A massive chateau, old houses in narrow streets, with mountain scenery nearby, have made the town popular with tourists.

Population 115 700

Map France Gd

Anshan ***China*** Industrial city of Liaoning province about 180 km (110 miles) from the North Korean border. Iron and steel manufacturing was begun there during the Japanese occupation of the north-eastern region of Dongbei (Manchuria) between 1931 and 1932. Today, Anshan produces one-fifth of China's crude steel output of 37 million tonnes.

Population 11 000 000

Map China Ic

Angola

PROSPERITY FROM DIAMONDS, OIL AND OTHER MINERALS IS AROUND THE CORNER – ONCE CIVIL WAR CEASES

Endowed with great mineral wealth and tremendous agricultural potential, the West African country of Angola should be able to provide a prosperous living for little over 8 million people in a territory covering 1 246 700 km^2 (481 350 sq miles). Instead, more than two decades of civil war have caused hardship and poverty.

The conflict started as a war of independence in this vast land of equatorial forests and savannah-covered plateaus. Three separate tribal groups rebelled against the Portuguese colonial government in the 1960s. The rebels combined to form a government after independence in 1975, but they were soon fighting again among themselves.

This battle for power was won by the Popular Movement for the Liberation of Angola (MPLA), who have formed the government since 1976. But their main rivals, the National Union for the Total Independence of Angola (UNITA), continued to fight them in a fierce guerrilla war and to control about one-third of the country in the south and the east.

Inevitably, the war has brought in bigger powers on both sides. The Marxist MPLA has been helped by an estimated 30 000 Cuban soldiers and Russian arms and equipment, plus Russian and East German soldiers stationed in Angola as 'military advisers'. Pro-Western UNITA has had support from the United States and military assistance from South Africa.

South Africa sent troops into southern Angola to support UNITA and to fight one of its own enemies, the South West African People's Organisation (SWAPO); the SWAPO guerrillas use bases in Angola to raid Namibia, South Africa's satellite state. The South African troops were withdrawn in 1985, as part of a complicated scenario in which the Angolan Government was offering to send its Cuban troops home and control SWAPO, while pressing for Namibian independence. Meanwhile, UNITA has fought on under the leadership of Dr Jonas Savimbi.

The war has been disastrous for the people of Angola. The territory held by UNITA has continually waxed and waned, and the military activity has disrupted peasant and commercial farming over a wide area. Some villages have no longer been able even to grow the crops they live on, such as maize, cassava, beans, millet and sorghum. Hundreds of thousands of people have suffered hardship, many of them needing food aid to survive. Many villagers have fled to the towns; the capital, LUANDA, for instance, has grown from 480 000 people at independence to over a million.

COLONIAL HERITAGE

The Portuguese first landed on the coast in the 1480s and began to settle in the territory in the late 1500s. They established the capital city of Luanda on the western coastal plain.

At first the Portuguese had a happy relationship with the African tribes. They converted the Bakongo king, Afonso, to Catholicism, and another tribesman was consecrated as a bishop. But by the late 16th century the Portuguese were raiding the tribes for slaves to ship to South America. The Bakongo were the main victims; their numbers were greatly depleted and their kingdom was destroyed. It is estimated that 3 million slaves were taken from the 16th to the early 19th centuries.

It was not until the 19th century that Portugal began to develop the interior of the country. Roads and railways were built, especially the great BENGUELA railway which runs for 1350 km (840 miles) from LOBITO, on the Atlantic coast, to the border with Zaire in the north-east. Diamond mines were opened in the LUNDA region in the north-east in the 1920s; oil production began near Luanda in the 1950s and off the coast of CABINDA in the 1960s. But by then nationalist movements were growing in this land where 300 000 Portuguese settlers were ruling 2.7 million black Africans.

The complexity of Angola's politics is partly a result of the diversity of its population, with many different ethnic groups having different allegiances. Rebellion began in the north, among the Bakongo. A Bakongo nationalist movement, the National Front for the Liberation of Angola (FNLA), under Holden Roberto, led uprisings in 1961 and was one of the most important independence movements at this stage. In the late 1970s the FNLA threw in its lot with the ruling MPLA party.

The MPLA draws its support from a group of tribes in central Angola, including the Mbundu, Mbaka and Ndongo. These people have had the most contact with Europeans and are the best-educated tribes.

The largest group is the Ovimbundu – 35 to 40 per cent of the people. They mostly live on the Planalto, the central plateau which has the best climate and soils and so has always

▲ MARX IN AFRICA Posters extol Marx, Lenin, Communist heroes and the benefits of peace under Marxist rule in the southern port of Namibe. The year is 1980, when the country was not yet totally won for the revolution.

attracted African and European settlement. Traditionally an agricultural society, the Ovimbundu also controlled trade routes from the Atlantic across the central savannah and they were secure and prosperous before the Europeans arrived. Under colonial rule, they became involved in commercial farming and many of them were employed on the Benguela railway.

At first the Ovimbundu had little to do with the independence movement, but they grew restless as they were forced into villages which the Portuguese felt they could control more effectively. They also lost much land to the Europeans in the 1960s and 1970s. Today UNITA depends largely on their support.

The Lunda and Quioco tribes occupy north-eastern Angola. These people were once hunter-gatherers who also captured people

from other tribes to sell as slaves. They are now mostly farmers.

In the south, a group of related tribes, the Nganguela, have retained more of their culture than other tribes. The pastoralist Herero inhabit the dry south-west, and the Ambo the dry steppe in the south-east, growing sorghum and millet and herding cattle.

The conflict has not only disrupted the lives of these people, but almost put a stop to farming, especially in the once-rich agricultural region of the south and east. Commercial agriculture and industry had already suffered at independence, when almost all the Portuguese who ran the concerns left the country.

The new government nationalised the diamond and oil industries – although the multinational Gulf oil company remains a major partner. Many smaller businesses had to be taken over because the owners had disappeared, but there was a shortage of people to manage them, particularly since education for the Africans had always been limited. The civil war only added to these difficulties.

Angola's economy does have one success story, however, in the oil industry. It is by far the most important aspect of the economy, bringing in hundreds of millions of dollars and 80 per cent of export revenue. Production has been increasing steadily from offshore fields which are mainly north of the ZAIRE (Congo) river mouth, in Cabinda province.

Vast mineral wealth remains unexploited, including reserves of manganese (in the Cassala-Quitungo district between Luanda and N'dalatando), copper (near NAMIBE, Ngunza Kabolo and the Cambambe dam area) and phosphates. Angola has a high capacity for producing hydroelectric power, and the Cambambe dam on the CUANZA river already provides the city of Luanda with its electricity. On the CUNENE river, the Ruacana dam was built jointly by the South Africans and the Portuguese before independence, with plans for irrigation as well as hydroelectric power.

Angola's physical environment provides great prospects for economic development. But with the USSR backing its left-wing forces, and Western support for its right-wing opposition, there is political deadlock. Without peace and stability, the rich potential of Angola can only go to waste.

ANGOLA AT A GLANCE

Area 1 246 700 km² (481 350 sq miles)
Population 8 170 000
Capital Luanda
Government Marxist one-party state
Currency Kwanza = 100 lwei
Languages Portuguese (official) and Bantu languages
Religions Christian (70% Roman Catholic, 20% Protestant), tribal
Climate Tropical, with summer rains. Average temperatures in Luanda range from 18-23°C (64-73°F) in August to 24-30°C (75-86°F) in March
Main primary products Maize, cassava, bananas, palm oil, coffee, cotton, sisal, timber, fish; oil, diamonds, iron ore
Major industries Agriculture, mining, forestry, fishing, chemicals, tobacco, food processing
Main exports Crude oil and petroleum products, coffee, diamonds, sisal, palm oil, fish
Annual income per head (US$) 670
Population growth (per thous/yr) 27
Life expectancy (yrs) Male 41 **Female** 43

Antakya (Antioch) ***Turkey*** City bordering Syria about 500 km (300 miles) south of the capital, Ankara. Founded in 300 BC, it gained a reputation as a city of luxury and pleasure and became known as the 'queen of the east'. In the 1st century AD it was visited by St Peter, and his grotto is just outside the city. The city became part of Syria in 1919, but was restored to Turkey 20 years later. Little of its former splendour remains following a series of earthquakes, but the Hatay museum houses one of the world's finest collections of Roman mosaics.

The surrounding province of Hatay covers an area of 5403 km² (2086 sq miles) and produces cotton and grain; some iron is mined.

Population (city) 109 200; (province) 1 034 000
Map Turkey Bb

Antalya ***Turkey*** Port and holiday resort on the Mediterranean Sea about 400 km (250 miles) south of the capital, Ankara. It was founded in about 150 BC by Attalus II of Bergama, one of the many princes striving to control the country. It has a 2nd-century decorated marble gate built to commemorate a visit by the Roman Emperor Hadrian.

Antalya's main industries are grain, timber, canning, flour milling and tourism. The neighbouring lowland province of the same name covers an area of 20 591 km² (7950 sq miles) and produces grain and timber.

Population (city) 258 100; (province) 892 200
Map Turkey Bb

Antananarivo ***Madagascar*** Capital of the republic, founded in the 17th century and formerly called Tananarive. It stands in the centre of the island on a hill overlooking a fertile rice-growing area. It is the country's industrial heart, producing tobacco, processed food, textiles and leather goods. A university was founded there in 1961.

Many public buildings date from the French occupation after 1895. Tourist attractions include an old royal palace and the colourful Zoma, or market, which on Fridays expands to fill most of the city centre.

Population 663 000
Map Madagascar Aa

Antarctic Circle Latitude 66° 32′ south, along which, at the southern summer solstice (about December 21), the sun does not set, and at the winter solstice (about June 21) the sun does not rise all day.

Antarctic Ocean The 'Antarctic Ocean', or 'Southern Ocean', are names sometimes given to the waters around ANTARCTICA. But most geographers regard these waters as southern extensions of the ATLANTIC, INDIAN and PACIFIC oceans. The region is the only place in the world where a boat can sail in a straight line – east-west – indefinitely without hitting land.

Antarctic Peninsula ***Antarctica*** A largely ice-covered and mountainous 1300 km (808 mile) long extension of Antarctica towards South America. The jagged coast is flanked by ice shelves and numerous offshore islands. It is disputed territory claimed by the United Kingdom, Argentina and Chile. It is sometimes called the Graham Land Peninsula.

Map Antarctica Db

Antarctica Barren, empty and bitterly inhospitable, Antarctica was the scene, some 75 years ago, of one of the most dramatic contests in the history of world exploration – the race to reach the SOUTH POLE. In 1912, two rival expeditions – a British team led by Captain Robert F. Scott and a Norwegian led by Roald Amundsen – reached the Pole within 34 days of each other. With greater experience and superior technique – using dog sledges rather than relying on human muscle – Amundsen won the race. But the tragic deaths of the five British men after reaching the Pole and discovering their defeat won them first place for heroism.

The culminating feat of Antarctic exploration was the first crossing of the continent, via the South Pole, by the multinational Commonwealth Trans-Antarctic Expedition in 1957-8. Today, most Antarctic researchers work for months or years at a time at permanent bases.

They work in remarkable international harmony despite the competing national interests that characterised early Antarctic history. This is due to the successful operation of the Antarctic Treaty, signed in 1959 by 12 nations who had some scientific interest in Antarctica.

This idealistic arrangement silenced the territorial claims of seven nations – Argentina, Australia, Chile, France, New Zealand, Norway and the UK – who each wanted a wedge-shaped piece of the Antarctica pie, in some cases overlapping. The treaty has kept the peace, but it will be reviewed in 1991, and several Third World nations have criticised its restricted membership. The possibility of exploiting Antarctic resources – from coal and iron ore to rich fisheries – is one reason for this criticism.

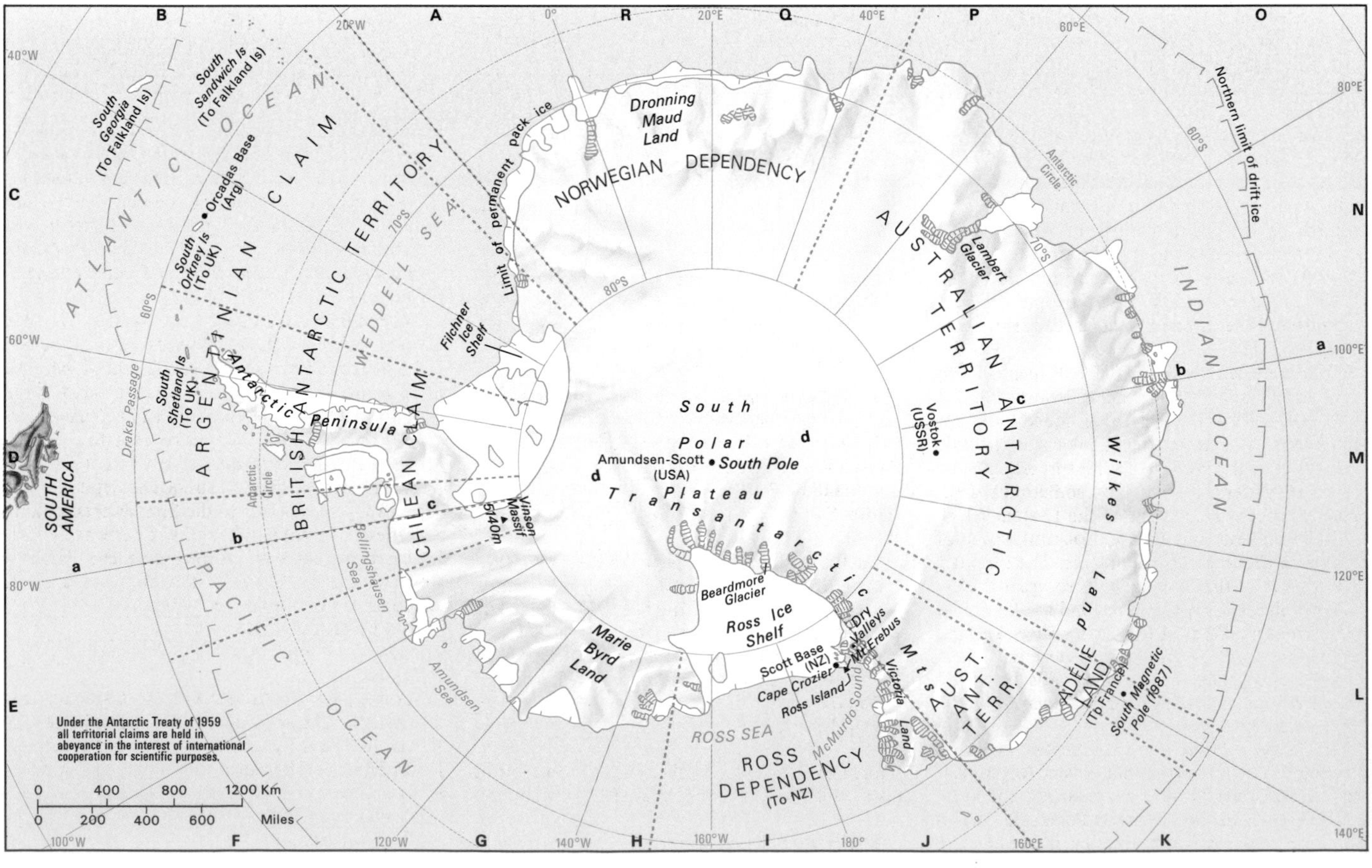

Antigua and Barbuda

WHERE NELSON'S FLEET ONCE RULED THE WAVES, ANTIGUA NOW ATTRACTS TOURISTS TO ITS HARBOURS

Set on the eastern flank of the eastern Caribbean's LEEWARD ISLANDS, Antigua and Barbuda occupies a strategic position that has long been recognised by the major powers. The tiny state comprises three islands – the two in its title plus the uninhabited Redonda.

In the 18th century, Antigua was one of Britain's vital naval bases in the West Indies, and its strategic importance continues in this century. The USA built the island's large Coolidge Airport during the Second World War, to defend the Caribbean and Panama Canal, and today operates two military bases and a transmitter for its international broadcasting service, Voice of America, on Antigua.

Antigua's original inhabitants were Arawak Indians, of whom none survives today; they were killed or taken away as slaves by Carib tribesmen from South America. Columbus discovered and named Antigua in 1493, and in 1632 it was colonised by the British. They brought slaves from Africa whose descendants account for most of the present population of 80 000 people – although there are also tiny minorities of British and Asian stock. In 1981, Antigua and Barbuda became an independent member of the Commonwealth with the British monarch as its head of state.

The staple of Antigua's economy today is tourism. Even for the Caribbean, the island is exceptionally well endowed with beaches, all of them palm-fringed with sparkling white or pink sand. The island started to exploit tourism in the 1960s, and the development has been discreet – only one hotel has more than 100 rooms and many have fewer than 50. There are two gambling casinos which attract wealthy tourists, mostly from North America.

Before tourism, Antigua's economy was almost exclusively based on sugar, which provided three-quarters of its export earnings. In the 1960s, however, faced with low world sugar prices and high wage demands by local labour unions, the group of mostly white landowners who controlled 60 per cent of the arable land threatened to close down the industry. They were bought out by the government in 1967, but four years later another government closed the industry down, contributing to a 25 per cent unemployment rate. Subsequent governments have encouraged farmers to grow food crops, including some cane for local needs and Sea Island cotton for export to Japan. This kind of cotton plant, which produces long, fine, high-quality fibres, is well-adapted to Antigua's relatively dry climate – it receives about 1000 mm (40 in) of rain a year, less than most West Indian islands.

The low, heavily wooded island of Barbuda has a population of 1500 people, most of whom engage in small-scale farming and fishing, especially for lobsters for export. Tourists come for the beaches and the coral reefs surrounding Barbuda, which have a number of spectacular wrecks of sailing ships. On the island itself the wildlife includes wild pigs and English fallow deer. They were introduced by the Codrington family who from 1685 to 1870 leased Barbuda from the British Crown for the rent of 'one fat pig' a year.

Most of Antigua and Barbuda's places of historic and architectural interest are on the larger island, Antigua. ST JOHN'S, the capital, has some fine 18th-century administrative buildings and a splendid Baroque cathedral, built in the 1840s. Sugar mills still dot the island's landscape – many of them now converted into houses – and in the south-east of the island are ENGLISH HARBOUR and Nelson's Dockyard, named after Admiral Horatio Nelson, the British naval hero whose ships used the base. It has many imposing 18th-century naval stores and houses, all of which have been well restored.

ANTIGUA AND BARBUDA AT A GLANCE

Map Caribbean Cb
Area 442 km² (171 sq miles)
Population 80 000
Capital St John's
Government Parliamentary monarchy
Currency East Caribbean dollar = 100 cents
Language English
Religion Christian (mainly Anglican)
Climate Tropical; average temperature at St John's ranges from 21°C (70°F) to 30°C (86°F)
Main primary products Cotton, fruit, vegetables, sugar cane, fish
Major industries Tourism, cotton production, food processing, clothing manufacture, fishing
Main exports Cotton, rum, clothing, shellfish
Annual income per head (US$) 1520
Population growth (per thous/yr) Static
Life expectancy (yrs) Male 71 Female 73

Antarctica has the harshest climate in the world. Temperatures of −89.6°C (−129.3°F) have been recorded in the dark winter, and winds of 145 km/h (90 mph) can blow for 24 hours at a stretch. In a whiteout, when driving snow hides everything, it is possible to be lost only a few metres from a camp hut. Even on calm and sunny days, when people can get sunburned, the temperatures hardly rise above freezing point.

The continent is covered by 90 per cent of the world's snow and ice. Ice forming on the SOUTH POLAR PLATEAU is slowly squeezed under its own accumulating weight outwards towards the edges of the landmass. The surface appears as a featureless ice cap, but sometimes great glaciers are created, such as the LAMBERT (the world's biggest) and the BEARDMORE, route of many journeys to the South Pole. The quantity and power of the moving ice is so great that in several places it is pushed beyond the landmass onto the sea. There it floats as an ice shelf, as in the ROSS and WEDDELL seas, terminating in towering ice cliffs. From these cliffs huge icebergs break off and are carried away into the ANTARCTIC OCEAN.

In some places, ice-free peaks protrude through the white crust that covers Antarctica; the highest peak is the VINSON MASSIF (5140 m, 16 864 ft). And the Antarctic Peninsula, which stretches towards South America, is free of snow in some areas, giving geologists access to rocks which have yielded important fossil and mineral finds. It has been discovered that much of Greater Antarctica – the part of the main landmass east of the Greenwich Meridian – is a high continental shelf, above sea level under the ice. To the west, Lesser Antarctica is a continuation of the South American ANDES, and would become a series of islands if the ice vanished.

Area 14 000 000 km² (5 400 000 sq miles)

Antequera ***Spain*** Textile town about 112 km (70 miles) north-east of Gibraltar. Nearby are the Menga Galleries, the site of mysterious prehistoric standing stones.

Population 35 200
Map Spain Cd

Antibes ***France*** Seaport and luxury resort town on the Côte d'Azur, 16 km (10 miles) south-west of Nice. It trades in dried fruit, olives, tobacco, perfumes and wine. There are Roman remains of an ancient settlement. The museum contains works by Pablo Picasso.

Population 63 250
Map France Ge

anticline A FOLD with strata sloping downwards on both sides from a common crest.

anticyclone Large high-pressure system in the atmosphere with pressure diminishing outwards from the centre. There is little vertical air movement, making for still weather conditions and poor visibility. In Europe, summer anticyclones give long, sunny but hazy days. In winter, anticyclones bring very cold weather. Autumnal fogs are usually associated with anticyclones.

Antigua ***Guatemala*** City 20 km (12 miles) west of the capital, Guatemala City. It became the country's second capital in 1543 after Ciudad Vieja, 6 km (4 miles) away between the Agua (Water) and Fuego (Fire) volcanoes, was destroyed by an earthquake. Antigua was also destroyed by earthquake in 1773, after which Guatemala City was founded as capital.

Antigua was one of the first planned colonial cities in the Americas, and some fine early Spanish buildings survive, including the Palace of the Captains General. The city was damaged again by earthquake in 1976.

Population 33 300
Map Guatemala Ab

Anti-Lebanon (Jebel esh Sharqi) ***Lebanon*** Mountain range rising to 2740 m (about 9000 ft) and running north-south along the border with Syria. The range forms the eastern side of the BEQA'A valley, and ends in the south at Mount HERMON.

Map Lebanon Bb

Antioch ***Turkey*** See ANTAKYA

Antioquia ***Colombia*** The country's most populous state, covering 63 612 km^2 (24 560 sq miles) in the north-west. It is centred on Medellín, Colombia's second largest city. Coal and gold are mined, and coffee and bananas grown.
Population 3 740 000
Map Colombia Bb

Antivari ***Yugoslavia*** See BAR

Antofagasta ***Chile*** Northern region encompassing much of the ATACAMA DESERT. Most of its towns depend on water piped from the foothills of the mountains. Nitrates and copper are the basis of the economy. The open-cast copper mine at Chuquicamata, 220 km (135 miles) north-east of the city of Antofagasta, is the largest in the world. The city of Antofagasta, the region's capital, is an important rail centre and exporter of Chilean copper.
Population (region) 342 000; (city) 170 000
Map Chile Ab

Antrim ***United Kingdom*** Borough and county in Northern Ireland. The town stands near the north-east corner of Lough Neagh. There is a 9th-century round tower north of the town and a fine early 18th-century courthouse.

The county of Antrim (excluding Belfast and Lisburn) covers 2831 km^2 (1093 sq miles). It stretches north to the Atlantic Ocean, where its magnificent cliff scenery includes the GIANT'S CAUSEWAY. The chief town is Ballymena.
Population (borough) 45 500
Map United Kingdom Bc

Antwerp (Antwerpen; Anvers) ***Belgium*** Principal port and capital of the northern province of Antwerp, 45 km (28 miles) north of the capital, Brussels, and the country's second largest city after the capital. It lies on the Schelde river estuary 88 km (55 miles) from the sea. Heavily bombed during the Second World War, Antwerp now has to compete for trade with the Dutch ports of Rotterdam and Amsterdam. Its main industries are diamond cutting, sugar and tobacco manufacture. The narrow streets of the old town contain a 14th-century cathedral, a 16th-century castle (now a museum) and a 16th-century town hall.
Population (city) 488 000; (metropolitan area) 923 000
Map Belgium Ba

Anuradhapura ***Sri Lanka*** Ancient city about 160 km (100 miles) north-east of Colombo. The Sinhalese founded the city in about 500 BC, and made it their royal capital until the Tamil invasions of the 11th century. Its ruins then lay beneath the jungle until the British uncovered them in the 19th century. There is a wealth of fine carving, temples, reservoirs and the dome-shaped Buddhist shrines known as stupas. One of the largest stupas, Jetawanarama Dagoba, is built of bricks and is 76 m (249 ft) high. The city's sacred bo tree, more than 2000 years old, is said to have been grown from a cutting taken from the tree in India under which the founder of Buddhism, Prince Siddartha Gautama, found enlightenment in the 6th century BC.
Population 36 200
Map Sri Lanka Ba

Anvers ***Belgium*** See ANTWERP

anvil cloud A wedge-shaped cloud formed by the flattening of the top of a large convective cloud as it reaches the base of the stratosphere (see ATMOSPHERE), spreading in the direction of high-level winds. Anvil clouds are normally storm clouds, bringing heavy showers.

Anyang ***China*** City of northern Henan province about 470 km (300 miles) south-west of Beijing (Peking), in the WEI valley. From 1300 to 1066 BC, Anyang was the capital of the Shang civilisation of ancient China.
Population 500 000
Map China Hd

Aomori ***Japan*** Capital of Honshu island's northernmost district. It is a ferry port on the island's north coast, and was extensively damaged during the Second World War. The city has been rebuilt and is home to the Nebuta festival in August, when papier-mâché dragons are paraded through the streets.
Population 294 000
Map Japan Db

Aóös Potamós ***Greece/Albania*** See VIJOSE

Aosta ***Italy*** Iron and steel town in the French-speaking Valle d'Aosta 80 km (50 miles) north of Turin. It is a tourist stage on the roads through the Grand St Bernard Pass into Switzerland and through the Little St Bernard Pass and the Mont Blanc tunnel into France.

The modern city faithfully preserves the Roman street pattern. Its Roman monuments include the theatre, the Praetoria Gate, the Arch of Augustus and remains of the Forum. In the centre of the town is the large Piazza Chanoux, fronted by the 19th-century town hall. The Romanesque cathedral is nearby.
Population 37 200
Map Italy Ab

apatite Naturally occurring form of calcium phosphate. It is the main source of phosphates for agricultural fertiliser.

Apatity ***USSR*** Mining town beside Lake Imandra, 120 km (75 miles) north of the Arctic Circle in the Kola peninsula. Nearby is the world's largest deposit of apatite – a phosphate mineral used to manufacture fertilisers.
Population 74 100
Map USSR Eb

Apennines ***Italy*** Mountains forming the backbone of Italy, giving the peninsula its distinctive boot-like shape. They start behind the Genoese riviera and wind 1280 km (800 miles) to the tip of the Italian toe. The highest peak is Monte Corno, 2912 m (9554 ft), north-east of Rome.
Map Italy Cb

Aphrodisias (Geyre) ***Turkey*** Ruined Roman city about 500 km (300 miles) south-west of the capital, Ankara. It was named after Aphrodite, the Greek goddess of love. Aphrodisias was virtually unknown until the 1960s.
Map Turkey Ab

Apia ***Western Samoa*** Capital and main port, on the north coast of Upolu island. The main industries are timber milling, handicrafts and the processing of copra and cacao for export. In March 1889, during a power struggle for the control of Samoa, four German and American warships were wrecked in Apia harbour during a fierce hurricane, two more were beached and damaged, and only one British ship survived; 144 people died. Robert Louis Stevenson, author of *Treasure Island*, was buried on the slopes of Mount Vaea, south of Apia, in 1894.
Population 34 800
Map Pacific Ocean

Apo, Mount ***Philippines*** See DAVAO

Apollonia ***Libya*** Ancient town of Cyrenaica, 30 km (19 miles) east of Al Bayda. It was the port for the Greek city of CYRENE and the birthplace about 276 BC of the astronomer Eratosthenes, who first measured the size of the earth.
Map Libya Ca

Appalachian Mountains (Appalachians) ***North America*** Eastern ranges stretching 2570 km (1600 miles) south-west from the St Lawrence river in Canada to the American state of Alabama. They rise to 2037 m (6684 ft) at Mount Mitchell in North Carolina and are rich in timber and, in places, coal. The attractive ranges are dotted with national parks.
Map United States Jc

Appenzell ***Switzerland*** Canton divided into the 'half-cantons' of Appenzell Inner-Rhoden, which is mainly Catholic, and Appenzell Ausser-Rhoden, which is Protestant. This was the result of the Wars of Religion in the late 16th century. It lies in scenic, mountainous country south of Lake Constance. The industrial town of Herisau (population 15 000) is the capital of Appenzell Ausser-Rhoden. The resort of Appenzell (5000) is the capital of Appenzell Inner-Rhoden. The canton covers 415 km^2 (160 sq miles).
Population (Inner-Rhoden) 13 200; (Ausser-Rhoden) 49 000
Map Switzerland Ba

Appleton layer The F layer. See ATMOSPHERE

Apulia ***Italy*** See PUGLIA

Apurímac ***Peru*** **1.** Southern department in the Andes. It is one of the country's remotest and poorest regions, the economy depending largely on a declining sugar-cane industry. Its capital, Abancay, is an administrative centre. **2.** River, 918 km (570 miles) long, rising in Lake Vilafro in the Andes. It flows north to unite with the Urubamba river at Atalaya and form the Ucayali river, an upper tributary of the Amazon. For short stretches in its lower course it is called the Ené and Tambo. The Ené valley is a major coffee-growing region.
Population (department) 323 300
Map Peru Bb

Apuseni Mountains ***Romania*** Mountain range in western TRANSYLVANIA, rising to 1848 m (6063 ft). Known also as the Bihor Mountains, the range contains deep gorges, caves and wooded valleys, cut into the soft limestone rocks by rain and river.
Map Romania Aa

Aqaba ***Jordan*** The country's only port, situated 100 km (62 miles) south of Ma'an. Set at the head of the Gulf of Aqaba on the Red Sea, it has handled a great deal of transit trade for Iraq since the start of the Gulf War in 1980. It is also the main export base for Jordan's phosphates industry. There are plans to expand Aqaba on three fronts: the port; the industrial zone; and the growing resort area. The tropical climate provides ideal holiday weather for nine months of the year.

In ancient times, Aqaba (then called Elath) was on the Arabian-Egyptian caravan route, and nearby archaeological remains show it was an important settlement in 1000 BC. After spells under the Roman and Byzantine empires, Aqaba was taken in AD 639 by Omar, who made it Muslim. It flourished as a port for pilgrims to Mecca, and for a brief period in the 12th century was occupied by the Crusaders. However, it lost much of its importance with the opening of the Suez Canal in 1869.

Population 35 000
Map Jordan Ab

Aqaba, Gulf of An arm of the RED SEA which is part of the GREAT RIFT VALLEY system, and separates Egypt's Sinai peninsula from north-western Saudi Arabia. In the north are the Israeli port of ELAT and Jordan's only port, AQABA. The gulf was an important waterway in Biblical and Roman times. Egypt's blockade of the gulf in 1967 was a major cause of the Six Day War, when Israel occupied Sinai.

Dimensions 160 km (98 miles) long, 19-27 km (12-16 miles) wide
Map Egypt Cb

aquifer Water-bearing rock or rock formation.

Aquitaine ***France*** Lowland agricultural region – formerly a duchy and kingdom with slightly different borders – in the south-west, roughly triangular in shape and crossed by the Garonne river. It covers 41 407 km² (15 987 sq miles). BORDEAUX and for a period TOULOUSE were its capitals. The marriage in 1152 of Eleanor of Aquitaine to Henri Plantagenet, who became Henry II of England in 1154, heralded 300 years of English rule that ended with the Hundred Years' War (1337-1453). Vine growing flourished to meet English demands, and the Bordeaux wine trade remains important.

Population 2 657 000
Map France Cd

Ar Raqqah ***Syria*** Town on the east bank of the Euphrates river, 120 km (75 miles) north-west of Dayr az Zawr. Since the completion of the Euphrates Dam – 60 km (37 miles) down-river – in 1978, Ar Raqqah has received a new lease of life. Irrigation has made possible farming and cattle-rearing, among other agricultural developments. Ar Raqqah contains the remains of a 9th-century palace built by the Abbassids – an Arabic dynasty that took its name from its founder, Al-Abbas (AD 566-652), uncle of the prophet Muhammad.

Population 218 000
Map Syria Bb

Ar Rusayris ***Sudan*** See ER ROSEIRES

Arabian Gulf See GULF, THE

Arabian Sea Part of the INDIAN OCEAN, which is still dotted with Arab sailing vessels, or dhows. The sea has two arms: the oil-rich GULF and the Gulf of ADEN. Leading ports include ADEN (South Yemen), BOMBAY and COCHIN (both India), and KARACHI (Pakistan).

Area 3 860 000 km² (1 490 000 sq miles)
Map Indian Ocean Cb

Arafura Sea An extension of the south-western PACIFIC OCEAN lying between Australia and New Guinea. In the west it merges into the TIMOR SEA, while the TORRES STRAIT in the east leads to the CORAL SEA.

Dimensions 1290 km (700 miles) long, 560 km (350 miles) wide
Map Indonesia Hd

Aragats ***USSR*** See ARMENIA

Aragón ***Spain*** Thinly populated region in the north-east of the country, covering 47 669 km² (18 405 sq miles). It became an independent kingdom in 1035 and was united with CATALONIA in the 12th century and with CASTILE in the 15th century. It contains the modern provinces of HUESCA, TERUEL and ZARAGOZA.

Population 1 213 100
Map Spain Eb

Arakan ***Burma*** Coastal state of 36 778 km² (14 200 sq miles) on the Bay of Bengal. It includes several offshore islands, of which the largest are Ramree and Cheduba. Arakan is cut off from the rest of the country by the heavily forested Arakan hills, which rise over 1700 m (5600 ft). AKYAB is the state capital.

Population 2 046 000
Map Burma Bb

Aral Sea (Aral'skoye More) ***USSR*** One of the world's largest inland seas – 65 500 km² (25 300 sq miles) – lying between Kazakhstan and Uzbekistan. It is fed by the Amudar'ya and Syrdar'ya rivers and is rich in carp and sturgeon. The main ports on the frequently stormy sea are Aral'sk and Bugun'.

Map USSR Gd

Aran Islands (Oileáin Arann) ***Ireland*** Three limestone islands in Galway Bay, 8 km (5 miles) from the mainland. The islands – Inishmore, Inishmaan and Inisheer – have been occupied since prehistory by fishermen and subsistence farmers. Fishermen still put to sea in canvas and tar currachs. Seaweed and sand have been laid on the bare limestone to make fields on which to grow potatoes, and other crops grow behind high walls. Remains of a powerful pre-Christian fortress, Dun Aengus, stand on cliffs 90 m (295 ft) high, on Inishmore. Tourism flourishes in this Irish-speaking area.

Population 1380
Map Ireland Bb

Ararat (Büjük Agri Dagi) ***Turkey*** Volcanic mountain massif close to the Russian and Iranian borders, some 1050 km (650 miles) east of the capital, Ankara. It is the Biblical landing place of Noah's Ark. The two main peaks are Great Ararat, which at 5165 m (16 945 ft) is Turkey's highest mountain, and Little Ararat at 3907 m (12 818 ft).

Map Turkey Cb

Arauca ***Colombia/Venezuela*** Tributary of the ORINOCO river, rising in Colombia and flowing about 1000 km (620 miles), of which 790 km (490 miles) are navigable. It forms part of the Venezuela-Colombia border.

Map Venezuela Bb

Araucanía ***Chile*** Region in the south, and one of the most picturesque lake districts in the world. Perhaps the best known of the lakes is Villarrica and its resort Pucón, both overshadowed by the active Villarrica volcano (2840 m, 9317 ft). Araucanía is pioneer country – a land of medium-sized farms and a growing timber industry. The capital is TEMUCO.

Population 693 000
Map Chile Ac

Arbe ***Yugoslavia*** See RAB

Archaeozoic era The older Precambrian era, when the first life forms evolved. The name comes from the Greek and means 'primeval life'. The rocks of this period are at least 1750 million years old. See AGE OF THE EARTH, p. 246.

Archangel (Arkhangel'sk) ***USSR*** Major port on the North Dvina delta, 52 km (32 miles) from the White Sea (a branch of the Barents Sea). It is ice-free for six months of the year. Its major export is timber, and its industries include shipbuilding and pulp and paper making. There is a 16th-century monastery named after the Archangel Michael, from whom the town takes its name. A British, French and American expeditionary force was based there in 1918-19 helping the White Russian army against the Soviet forces in the civil war that followed the Russian Revolution.

Population 403 000
Map USSR Fb

archipelago Group of islands clustered closely together.

Arcot ***India*** Agricultural town about 100 km (60 miles) west of the city of Madras. It is the former capital of the Nawabs, native rulers, of Carnatic – an independent region in south-east India bordering on the Bay of Bengal. In the 18th century it was the scene of bitter fighting between the British soldier and statesman Robert Clive and the occupying French.

Population 38 800
Map India Ce

Arctic Only the Inuit (Eskimos) have learned to make their homes in the white wilderness of the high Arctic – the lands and the frozen sea north of the ARCTIC CIRCLE. And only a handful of people from any nation have reached the NORTH POLE at its heart, though thousands fly near it regularly on scheduled airline flights.

Apart from a number of American, Canadian and Russian floating research stations, nobody lives on the ice pack. But people have for centuries made their homes along the Arctic coasts and in the tundra regions inland. Some of these polar peoples, such as the Inuit of North America and GREENLAND and the Aleuts of the Aleutian Islands off Alaska, traditionally live by hunting. Others, including the nomadic Sami (Lapps) of northern Scandinavia and the Chukchis of north-eastern SIBERIA, keep herds of reindeer, from which they get meat, milk, fur, leather and (in the form of dried dung) fuel.

The Polar Inuit, the world's most northerly people, live in a land of permanent ice in THULE, north-west Greenland, less than 1600 km (1000 miles) from the Pole. Farther north still is the world's most northerly land: a tiny island off northern Greenland called Oodaq.

Despite the ferocious climate of the Arctic, some outsiders have moved in. The motive for the early explorers was largely the hunt for a NORTHWEST PASSAGE – a route around North America which would nearly halve the sea distance from Europe to the Far East. In 1903-6 the Norwegian explorer Roald Amundsen made the first voyage through the Passage, but it was not until September 1969 that a giant ice-breaking oil tanker, the SS *Manhattan*, became the first ship to batter its way straight through the pack ice over the route.

Air, road and rail links have so far been more successful in opening up the lands of the Arctic to exploitation. In northern Alaska and Canada, the impetus has come largely from the discovery of Arctic oil fields, particularly on the North Slope of Alaska in 1968. The links have also made it easier for tourists to visit the far north. On the other side of the Arctic, some 3000 Russian and Norwegian miners on SPITSBERGEN already produce between them about 800 000 tonnes of coal a year from the bleak islands' huge reserves. And drilling ships from both nations are hunting for oil and gas in the chilly waters of the surrounding Barents and Norwegian seas.

Arctic Circle An imaginary circle around the earth at latitude 66° 32′ north, with the North Pole at its centre. It represents the southernmost latitude at which, at the northern summer solstice – about June 21 – the sun does not set, and at the winter solstice – about December 21 – the sun does not rise all day.

arctic climates Those climates in which no month has a higher average temperature than 10°C (50°F). They are divided into tundra climates, where in summer temperatures there is some growth of vegetation; and polar climates, where no growth is possible. Similar climates occur at high altitudes as well as high latitudes.

Arctic Ocean The world's smallest ocean, surrounded by northern Europe, Asia and North America. The North Pole is near its centre, and pack ice, 2-3 m (7-10 ft) thick, covers much of the ocean throughout the year.

Many rivers flow into the Arctic and it has a lower salt content than the other oceans. The strongest currents flowing out of the Arctic are the East Greenland and Labrador currents, which carry icebergs from GREENLAND into the North ATLANTIC OCEAN. Balancing these outflowing cold currents is the warm water of the North Atlantic Drift which flows into the Arctic between Iceland and Scotland. Some water also flows through the BERING STRAIT, between Alaska and the USSR. The straits are key summer trade routes for Russian merchant ships.

Area 14 056 000 km² (54 270 000 sq miles)
Greatest depth 5570 m (18 274 ft)

Ardabil (Ardebil) ***Iran*** Carpet-making market town about 56 km (35 miles) west of the Caspian Sea. Founded in the 5th century, it was destroyed by the Mongols in 1220, then became a place of pilgrimage in the late 13th century devoted to the Sunni mystic, Safi al-Din, who was born there in 1253. In 1828 the town was occupied by the Russians, who carried off his splendid library, once the greatest in Persia. The mausoleum of Shah Ishmail (1486-1524), the founder of Shiite Islam, is in the town.

Population 222 000
Map Iran Aa

Ardèche ***France*** River 120 km (75 miles) long. Rising in the Cévennes mountains, it runs for 30 km (18 miles) of its course through deep limestone gorges as it descends to join the Rhône river. The most spectacular feature of these gorges is the 34 m (112 ft) high Pont d'Arc, a natural limestone bridge which arches some 59 m (194 ft) across the river.

Map France Fd

Ardennes (Ardenne) ***Western Europe*** Forested plateau in south-east Belgium, Luxembourg and north-east France. The plateau rises in eastern Belgium to form the Hautes Fagnes (Hohes Venn) mountain range, which is largely a nature reserve, with Botrange the highest peak at 694 m (2277 ft). The sparsely populated Ardennes, with their severe winters, support forestry, limestone quarrying and tourism. During the Second World War the Battle of the Bulge, a desperate but unsuccessful attempt by German troops to break through Allied lines, took place in the Ardennes.

Population 350 000
Map Belgium Ba

Ardennes ***France*** See CHAMPAGNE-ARDENNES

Arecibo ***Puerto Rico*** Seaport on the north coast, about 70 km (43 miles) west of SAN JUAN. The main industry is rum distilling.

Population 87 000
Map Caribbean Bb

Arenal ***Costa Rica*** Volcano 170 km (106 miles) north-west of the capital, San José. About 100 people died in 1968 when the 1633 m (5358 ft) high volcano erupted for the first time in 500 years.

Map Costa Rica Ba

Arendal ***Norway*** Old-established port, chartered in 1723, on the Skagerrak coast of south Norway, 65 km (40 miles) north-east of Kristiansand. It specialises in the timber trade and shipbuilding, and has a ferry to Denmark.

Population 25 100
Map Norway Cd

Arequipa ***Peru*** Southern department, lying in a fertile plain fringed by volcanic cones. The provincial capital, also Arequipa, lying at the foot of the dormant volcano El Misti (5822 m, 19 100 ft) was an Inca town, refounded by the Spanish in 1540. It is sometimes known as the 'White City' because most of the buildings are of white volcanic stone. A fine example is the Convent of Santa Catalina, built in 1580.

Population (department) 706 600; (city) 448 000
Map Peru Bb

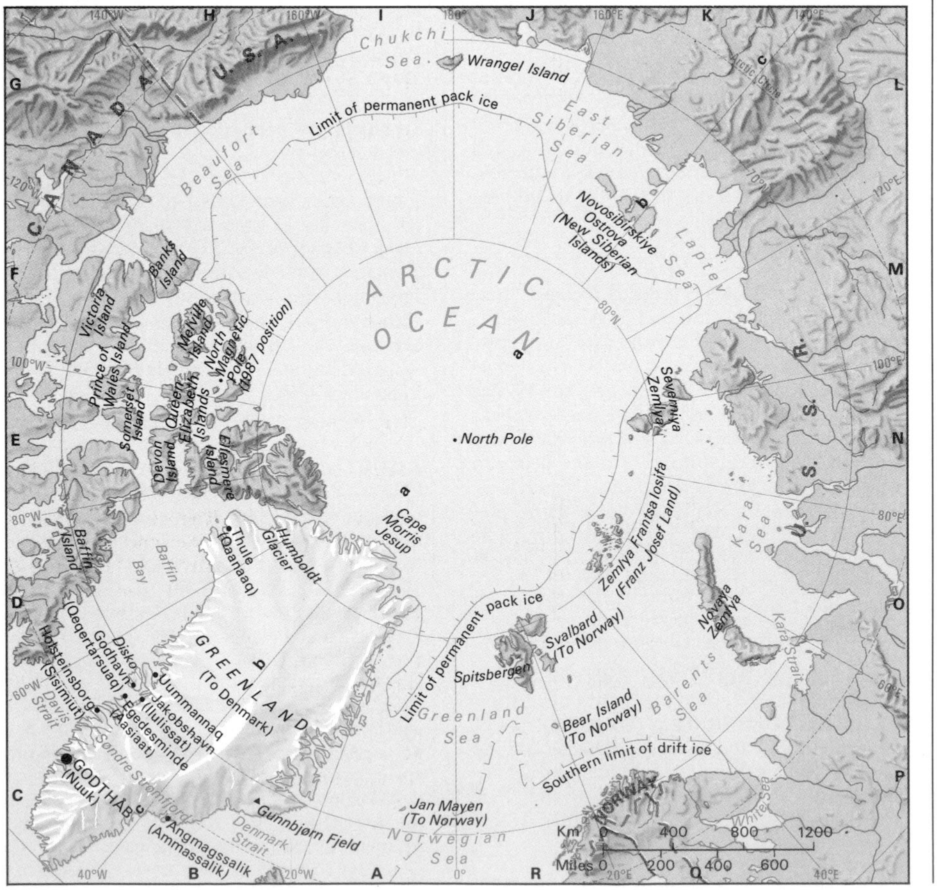

arête Sharp, narrow mountain ridge or spur, specifically a ridge between two adjacent CIRQUES, such as Striding and Swirral Edges on Helvellyn in the English Lake District.

Arezzo *Italy* City and province in Tuscany, 60 km (35 miles) south-east of Florence. The centre of the medieval city contains the church of Santa Maria, a rare example of unmodified Tuscan Romanesque, whilst the Gothic church of San Francesco has frescoes by Piero della Francesca (1420-92). The house of the poet Petrarch, born in Arezzo in 1304, stands near the cathedral. The Archaeological Museum is built on the remains of a Roman amphitheatre. Shoes, clothing and furniture are the main industries. The province is agricultural, peppered with hilltop villages and small market towns.
Population (city) 91 700; (province) 313 600
Map Italy Cc

Argentina See p. 44

Argos *Greece* Town built on the ruins of a 7th-6th century BC city on a plain 10 km (6 miles) north-west of the port of Nauplia in the north-east Peloponnese. The marketplace and theatre of the Mycenaean settlement, dating from the 13th to 14th century BC, have been excavated, but two citadels are partly buried by the modern town. Legend says Jason set sail from the port with the Argonauts to find the Golden Fleece.
Population 20 700
Map Greece Cc

Argyllshire *United Kingdom* See HIGHLAND; STRATHCLYDE

Argyrokastron *Albania* See GJIROKASTER

Århus *Denmark* Once a Viking trading post, now the country's second largest city and port after Copenhagen, and the capital of Jutland. It has an interesting Old Town (Den gamle By) with a collection of 17th and 18th-century buildings re-erected in an open-air museum in the botanical garden. It also has a fine cathedral, Denmark's first modern university, founded in 1934, a music academy, concert hall, and a prehistoric museum which houses the 2000-year-old 'Grauballe Man'; his remains were found preserved in the bog of Grauballe Fen, which lies 40 km (25 miles) west of Århus, in the 1950s. Industries include clothing factories, foodstuff and paper manufacturing.
Population 246 700
Map Denmark Ba

Arialkhan *Bangladesh* Major branch of the Padma river (see JAMUNA).
Map Bangladesh Cc

Arica *Chile* Coastal town near the Peruvian border. It is connected by rail to the Bolivian capital La Paz, and handles half Bolivia's foreign trade. Its mild climate, almost constant sunshine and golden beaches have made the town a popular resort. The Cathedral of St Marcos in the town's Plaza de Armas is built on an iron frame designed by the French engineer Gustave Eiffel (1832-1923).
Population 123 200
Map Chile Aa

A DAY IN THE LIFE OF A NAVAHO INDIAN

The Grand Canyon is striped pink and orange in the late afternoon sun, as Rose makes her last sale of the day. Her stall, perched on the edge of the canyon, offers a reasonable living out of selling Indian crafts to the tourists.

Spread over the table in front of Rose are moccasins, chokers, pendants, purses and earrings. The designs are Navaho, but are increasingly influenced by the fashions of modern America.

These half-Indian, half-commercialised artefacts are fitting symbols of her life: a compromise between Indian traditions and the Coca-Cola culture. Rose is dressed in an acrylic sweater and slacks; her little boy Chuck, in T-shirt and jeans, fiddles with his personal stereo. Rose grew up on a reservation, among a people who had not yet adjusted to the new life imposed on them by outsiders. The reservation land to which their open plateau had been reduced could not support the seminomadic existence the Navaho once knew; herding sheep and goats. And yet most of those who tried to find a job outside the reservation faced discrimination and failure.

Life is easier now. Indian rights to own their land have been recognised in Congress, and Rose's husband Jack has a job as a driver on a large cattle ranch. They also have a small income from their share in the lease of mineral rights for a power station and a coal mine on a piece of Navaho land. Industrial developers can only lease this land; they will never be allowed to buy. Rose and Jack built a neat, wood-lapped bungalow on the Navaho reservation where they grew up. They campaign to protect this arid land from overexploitation and overgrazing, motivated by the Navaho belief that the earth is like a living being; if you take something out of the ground, it is like digging into your own skin. Their house, though, has no piped water or mains sanitation. They make do with an earth closet, and with taking a truck to the trading post to fill a tank of water.

At home Rose spends the evening assembling more necklaces, and tells her son in Navaho, the Athapascan language they speak at home, stories about the famous Navaho chiefs, and about the spirits of the mountains. Sometimes Chuck hears them at school too; he goes to the reservation school, with Indian teachers and classmates. But mostly his lessons are the same as white children's. Rose hopes he will go to college one day. But he won't lose his Navaho identity. He will marry an Indian girl.

With Chuck in bed, Rose and Jack discuss the threat of another oil well on their land. They have already won many battles, for the preservation of the rights and territory of the Navaho, and are confident that they can win another.

Arizona *USA* A state since 1912 and one of the last two states to become a part of the continental USA. (The other was New Mexico.) It lies in the south-west, and has some of the most famous canyon scenery in the world, including the GRAND CANYON. The south is drained by the Salt and Gila rivers which provide hydroelectricity and irrigation. Rainfall is low and there are extensive areas of desert and scrub. The chief farm products are cattle, dairy goods, cotton and vegetables. Arizona is the nation's principal producer of copper. PHOENIX is the capital of the state, which covers 295 024 km² (113 909 sq miles). Early in the 16th century, the Apache people drifted down into what is now Arizona and New Mexico from their original home in western Canada.
Population 3 137 000
Map United States Dd

Arkansas *USA* **1.** Central southern state lying immediately west of the Mississippi river. The broad plain of the Mississippi covers the east, from which the land rises to the forested Ozark Plateau and Ouachita Mountains. The state – whose name is usually pronounced 'Arkansaw' – has an area of 137 539 km² (53 104 sq miles) and produces oil, gas and bauxite (aluminium ore). The main farm products are broiler chickens, soya beans, rice and cotton. The capital is LITTLE ROCK.

2. River rising in the Rocky Mountains of Colorado and flowing about 2335 km (1450 miles) east then south-east through Kansas and Oklahoma into Arkansas, where it joins the Mississippi.
Population (state) 2 359 000
Map (state) United States Hc; (river) United States Fc, Hc

Arkhangel'sk *USSR* See ARCHANGEL

Arlanda *Sweden* See STOCKHOLM

Arlberg Pass *Austria* Mountain pass, rising to 1793 m (5882 ft), in western Austria, which links Vorarlberg with the Tyrol. The Arlberg massif forms part of the watershed between the Rhine and the Danube basins. A 10 km (6 mile) long railway tunnel, opened in 1884, runs beneath the pass between Stuben in Vorarlberg and St Anton in the Tyrol.
Map Austria Bb

Arles *France* Chief town of the Roman province of southern Gaul, situated in south-east France on the east bank of the Grand Rhône. It is now a centre for agricultural marketing and tourism. A vast Roman arena survives, and there are museums depicting Roman and Provençal life in the past. The Dutch painter Vincent Van Gogh (1853-90) spent the last and most productive years of his life in the town – completing, in one frenzied 15 month period there, more than 200 pictures.
Population 50 800
Map France Fe

Arlit-Akokane *Niger* Remote region about 275 km (170 miles) north-west of the oasis of Agadez. It is named after two towns: Arlit and Akokane. Uranium has been mined in the region since 1971.
Map Niger Ab

Argentina

A SPECTACULAR COUNTRY ON A GRAND SCALE, WHERE AN ELEGANT SOCIETY HAS BEEN LIVING BEYOND ITS MEANS

Most Argentines think of themselves as different from other South Americans. They consider themselves more sophisticated and better educated. Indeed, European influence is very strong, and the capital city of BUENOS AIRES, where one-third of the population lives, is like a European city. Most of the people are descended from European immigrants, and there is a substantial British community in Buenos Aires – a legacy of the days when the British arrived to build the railways.

Once Argentina was more prosperous than the other South American countries, but it has failed to fulfil its promise – and the rest of the continent has been catching up.

In 1914, Argentina was the source of much of the world's meat and cereals, the focus of a major migration of people from Southern Europe, and seemed destined to join the ranks of the developed nations. Argentina was never to achieve this status, but it was left with a sense of its own importance which has not lessened now that it is only a middle-ranking nation in terms of power and wealth.

Argentina is the eighth largest country in the world (five times the size of France), has a population of almost 30 million, and a capital which is one of the world's great cities. The first Latin-American Nobel Prize went to an Argentinian, Carlos Saavedra (a Peace Prize in 1936).

Argentina rose to prosperity from unprepossessing beginnings. The Spanish conquerors found no gold or silver, and little to interest them in the area that is now Argentina. Until 1713, Argentina was administered from LIMA, the capital of PERU, under the government of Spain. Independence was declared in 1816. The new republic of Argentina included parts of PARAGUAY, BOLIVIA and URUGUAY, but these three soon became separate nations.

WEALTH FROM THE PLAINS

The main source of wealth for Argentina was the vast fertile plains of the PAMPAS, which occupies the central part of the country. The major exports were at first wool and mutton from these grasslands, and then beef and hides. Argentina also became one of the world's greatest sources of wheat and maize.

The great period of economic growth occurred between 1850 and 1930. Britain poured money into the country, the railways were built and a tide of immigrants arrived, mostly Spaniards and Italians fleeing from poverty in Europe. The newcomers greatly outnumbered the native Indians and the *mestizos* – mixed-blood descendants of Indians and early Spanish settlers.

The towns grew, especially Buenos Aires as the seat of national government and the port for the increasing exports and imports. The boom was marred by the First World War in Europe, and was virtually ended by the worldwide depression of the 1930s.

From 1914 onwards, Argentina started to establish manufacturing industries: iron and steel, cars, clothes and electrical goods. By 1950 the agricultural society had turned into an industrial one, and most Argentines earned their living in the cities. But the manufacturing sector was not able to compensate for a decline in agriculture as Argentina's share of world food exports shrank dramatically.

Since the 1930s Argentina's economy has been the source of grave social and political problems: inflation has recently reached 800 per cent a year, and today Argentina has a huge national debt amounting to US$1500 for every man, woman and child in the country. Some blame this decline on the populist leader Juan Perón, whose economic policy after the Second World War favoured urban workers, to the detriment of agriculture. Still others blame British and United States corporations for draining profits from the country. Others blame conservative policies and successive military regimes.

POLITICAL TURBULENCE

Whatever its causes, economic trouble has always accentuated the political turbulence of Argentina. Governments were short-lived until the era of Colonel Perón and his wife Eva, from 1946 to 1955. Perón was a nationalistic, authoritarian president, who endeared himself to industrial workers by raising their

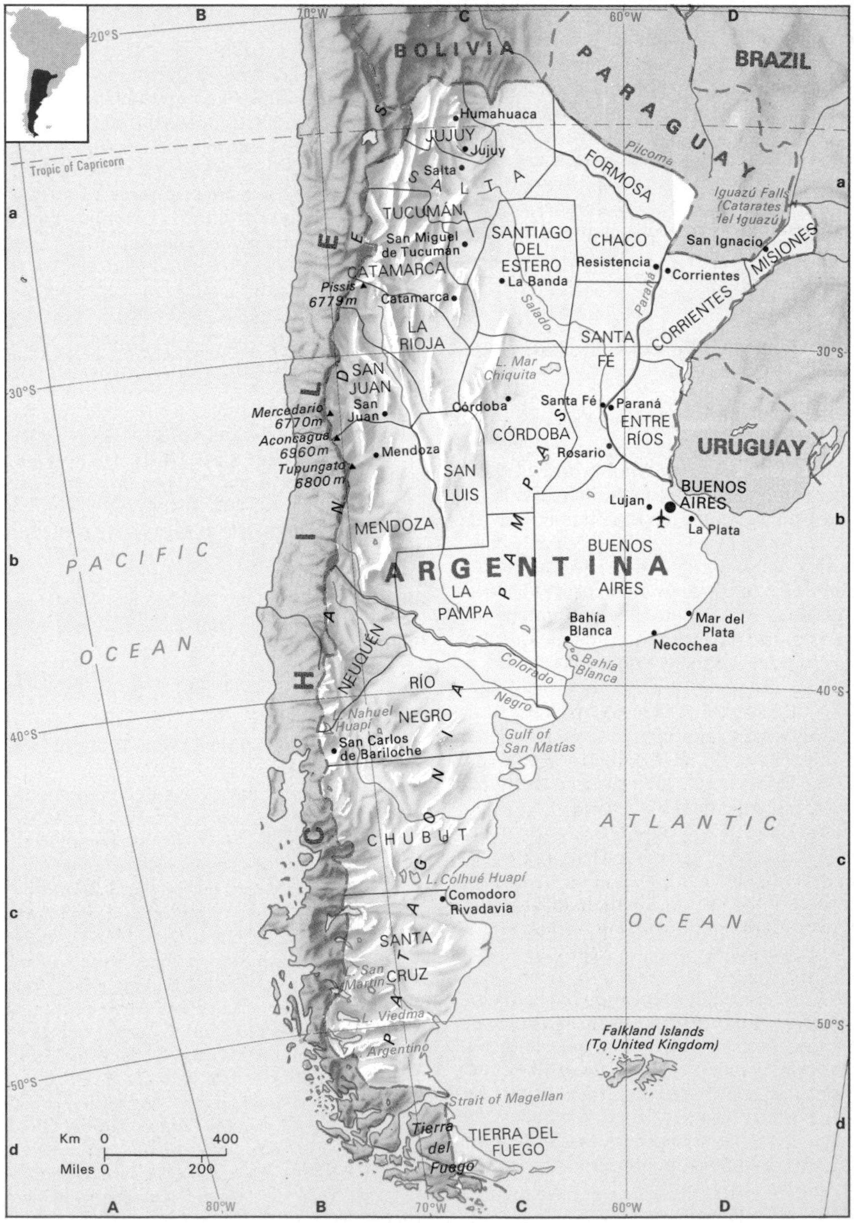

wages repeatedly even though productivity was falling. His neglect of agriculture, however, helped towards an economic decline. Perón's position was weakened by the death of his wife – a popular figure with the working masses – and a quarrel with the Roman Catholic Church, and in 1955 he was deposed by a military and civilian revolution.

A series of military regimes followed, until in 1973 Perón made a short-lived and controversial comeback. He was succeeded after his death in 1974 by his second wife, Maria, who was herself ousted in 1976. The followers of Perón are still one of Argentina's main political parties, although Maria resigned from leadership of the party in 1984.

Another succession of military governments was ended by the disastrous Falklands venture in 1982. An army invaded the British Falkland Islands, which Argentina calls the Malvinas and claims as her territory. British forces reinvaded during a ten-week war and established a garrison on the islands.

The conflict cost Argentina 750 lives, half of them when the cruiser *General Belgrano* was sunk by a British submarine. The defeat led to the ousting of a military junta headed by General Leopoldo Galtieri and a return to civilian rule after a democratic election.

The new president, Raúl Alfonsin, ordered an inquiry into atrocities committed by previous military regimes. The inquiry revealed that 9000 people had been killed, a further 9000 imprisoned and tortured, and 2 million had fled the country. Most of these were young people, and it is said that an entire generation disappeared.

LAND OF THE GAUCHOS

The geography of Argentina is on a grand scale. To the west, the massive chain of the ANDES forms the border with Chile and Bolivia. It is the longest chain of mountains in the world, containing South America's highest mountain, OJOS DEL SALADO. To the south stretches the huge windswept expanse of PATAGONIA, and the breathtaking mountains and lakes around BARILOCHE. It is the geography and culture of the central plains, the pampas, that are best known abroad – a vast area of flat land that accounts for most of the population outside the cities.

The pampas are the home of Argentina's cowboy heroes, the *gauchos*, who ride the plains on horses that never need to jump a fence. Here there are hardly any trees, only miles of grass punctuated by pylons and windpumps for underground water. A few trees may shelter the villages of the pampas, and eucalyptuses often line the approaches to the cattle ranches where the gauchos live and work. The ranches are known for their great barbecues, where whole cows are roasted over open fires. The gauchos, mostly mestizos, wear dress which is romantic to foreigners: baggy knee-length trousers called *bombachas*, with a striped *chiripá* scarf wrapped around the waist and between the legs.

Away from the pampas are several other Argentinas with different lifestyles. The north-west resembles the high Bolivian plateau or *altiplano*. The population here is still mostly Indian and mestizo, and the economy agricultural. The land is lush with subtropical vegetation, and for this reason the area is known as the Garden of Argentina. The colonial towns with their Spanish churches and cathedrals recall the days when this region thrived before the rise of Buenos Aires and the shift of activity southwards.

The north-east is cattle and cotton country: the GRAN CHACO plains. Farther east, between the River PARANA and the River URUGUAY, is the country which grows bitter maté tea – the gauchos' favourite drink. And farther east still, at the point where Argentina, Brazil and Paraguay meet, lies one of South America's greatest natural wonders. In the middle of the jungle, where the orange-red earth seems rusted by spray, the IGUACU FALLS thunder 64 m (210 ft) over a giant horseshoe 4 km (2.5 miles) wide – a wall of water, higher than Niagara and several times as wide.

Lying in the foothills of the Andes in the west is the wine-growing region around MENDOZA. This fertile area, watered by the Mendoza river, is known as the Garden of the Andes. Not only are fine wines produced here, but there are extensive market gardens, as well as some 5 million olive trees.

Three days' train journey south-west of Buenos Aires, the region around Bariloche rivals Switzerland with the grandeur of its blue lakes and snowcapped Andes, its waterfalls, glaciers and forests.

South from this alpine region the plateaus of Patagonia stretch their desolate way to TIERRA DEL FUEGO, through areas where colonies of elephant seals bask on the pebble beaches. In 1865 settlers from Wales came to farm sheep in Patagonia. Today, in remote parts of the southern Andes, at Puerto Madryn and the town of Esquel, two languages are spoken: Spanish and Welsh.

For all its beauty and wealthy, naturally productive land, Argentina has consistently failed to produce what it needs to maintain reasonable living standards. The country faces an austere future as it struggles to resolve its economic problems and shake off the vestiges of authoritarian government.

▲ MAN AND HIS DOGS Sheep rearing has been the main occupation on the barren tablelands of Patagonia since Welsh settlers arrived there in the early 19th century.

ARGENTINA AT A GLANCE
Area 2 766 889 km² (1 068 296 sq miles)
Population 31 700 000
Capital Buenos Aires
Government Federal republic
Currency Peso = 100 centavos
Language Spanish
Religions Christian (90% Roman Catholic, 2% Protestant), Jewish (2%)
Climate Subtropical in the north to subarctic in the south. Average temperature in Buenos Aires ranges from −5 to 14°C (23-57°F) in June to 17-29°C (63-84°F) in January
Main primary products Wheat, maize, potatoes, sorghum, sugar, grapes, apples, citrus fruit, soya beans, sunflower seed, cotton, cattle, sheep, fish, timber; coal, oil and natural gas, uranium, iron ore, lead, zinc
Major industries Steel, food processing, textiles, chemicals, vehicles, machinery, petroleum refining, mining, fishing, forestry, wood pulp and paper, wine
Main exports Wheat, maize, meat, soya beans, animal and vegetable oils, animal foodstuffs, chemicals, fruit and vegetables, leather, machinery, sugar, wool
Annual income per head (US$) 2500
Population growth (per thous/yr) 16
Life expectancy (yrs) Male 69 **Female** 73

Armagh ***United Kingdom*** City and county in Northern Ireland. The city lies 16 km (10 miles) from the Irish border. It has two cathedrals and is the ecclesiastical capital of the whole of Ireland. There is evidence of settlement from as early as 600 BC and the Irish patron saint, Patrick, is said to have founded the see here in the 5th century. The city has the province's oldest observatory and the only planetarium, as well as two museums. Navan Fort, 3 km (2 miles) to the west, was probably a stronghold of the northern Celts until the 5th century AD.

The county of Armagh, with an area of 1254 km² (484 sq miles), is Northern Ireland's smallest. There is rich agricultural land and fruit growing is important. There are textile and other industries in Lurgan and Portadown, both of which are expanding urban areas.

Population (city) 12 700
Map United Kingdom Bc

Armagnac ***France*** Plateau about 6000 km² (2300 sq miles) in area in Gers department, sloping from the Pyrenees towards the Garonne river, crossed by valleys with steep eastern sides. The region's main products are cereals, fodder, red and white wines – and a renowned brandy, named after the plateau, which is distilled around the town of Condom.

Map France De

Armenia (Armeniya) ***USSR*** Smallest of the Soviet Socialist Republics – 29 800 km² (11 500 sq miles) – bordering Turkey and Iran. It is a rugged mountainous region with many peaks above 3000 m (10 000 ft), the highest being Mount Aragats at 4090 m (13 418 ft). The chief products are cotton, fruits, rice and tobacco. There are copper, zinc and lead deposits, and mining is important. The main cities are the capital, YEREVAN, LENINAKAN and Kirovakan.

Armenia – established about the 8th century BC as an independent kingdom – originally included a much larger area centred on Mount ARARAT, the traditional resting place of Noah's Ark. In the 4th century AD it became the first country in the world to make Christianity the state religion; the Armenian Church remains a distinct sect of Orthodox Christianity. But the territory was partitioned in AD 387 between Persia and the Roman Empire, and there followed centuries of wars, occupations, massacres and deportations. It is now divided between Turkey, Iran and Soviet Armenia.

Population 3 267 000
Map USSR Fd

Armero ***Colombia*** Town in the west-central department of Tolima, 80 km (50 miles) north-west of Bogotá. It lies to the east of the volcanic peak of Nevado del Ruiz. In November 1985 the volcano erupted, causing a huge mud avalanche that buried Armero, killing about half of its population of 21 000 and destroying most of its houses and buildings.

Map Colombia Ba

Armidale ***Australia*** University town and agricultural centre in the NEW ENGLAND region of New South Wales, about 370 km (230 miles) north of Sydney. It trades in high-quality wool, fruit and timber.

Population 19 700
Map Australia Ie

Armorican Relating to Armorica, the part of north-west France now known as BRITTANY, or its people or language. In geology, it refers to the time about 200 to 250 million years ago when mountains were formed here and elsewhere in the world. It is also called HERCYNIAN.

Arnhem ***Netherlands*** Provincial capital of Gelderland, 82 km (51 miles) south-east of Amsterdam and situated on a northern branch of the Rhine, the Lek. It was the scene of a battle in September 1944, when British and Polish paratroops were dropped to capture its bridge and secure a path for an Allied advance in the northern Netherlands. The airborne troops were faced by unexpected German armoured forces, and were heavily defeated after a courageous stand.

Now rebuilt, the town is a tourist centre, with the Dutch national open-air museum. It produces rayon, electrical goods and ships.

Population (city) 128 600; (Greater Arnhem) 290 700
Map Netherlands Ba

Arnhem Land ***Australia*** Area in the north of the Northern Territory, bordering the Gulf of Carpentaria. It is named after the Dutch ship *Arnhem* which carried the expedition that discovered the area in 1623. It is an Aboriginal reserve covering an area of 96 089 km² (37 100 sq miles), and is the largest such reserve in Australia. Following the discovery of minerals, such as manganese, bauxite and uranium, mining agreements have been made with the Aboriginal people. Kakadu National Park, in the west, contains some fine examples of ancient Aboriginal art. It includes broken mountains and vast plains inhabited by crocodiles, water buffaloes, wallabies and numerous birds.

Map Australia Ea

Arno ***Italy*** Principal river of Tuscany, flowing 245 km (152 miles). Its upper reaches run through Apennine trenches parallel to the sea, but from Pontassieve, 20 km (12 miles) upstream of Florence, it flows more directly to the coast through Florence, Empoli, Pontedera, Cascina and Pisa. It has an uneven flow, but has a tendency to flood from time to time. It did so in the 12th, 14th and 16th centuries, when bridges in Florence were swept away, but the worst flood occurred in November 1966 when water in the city's streets rose to 6 m (20 ft) and many art treasures were destroyed or severely damaged.

Map Italy Cc

Arras ***France*** Capital of Pas-de-Calais department, on the navigable Scarpe river 40 km (25 miles) south-west of Lille. Its industries include engineering, brewing, sugar refining, vegetable oils and agricultural equipment. The old town was devastated during the First World War, but its town hall and other monuments have been restored to evoke its Flemish past – it was part of Flanders under the Spanish Hapsburgs in the 16th and 17th centuries. The word 'arras' for a wall-hanging or screen made from tapestry comes from the phrase *drap d'Arras* (cloth of Arras), because the town was a famous tapestry centre during the Middle Ages.

Population 84 400
Map France Ea

artesian structure Structural formation in sedimentary rocks, usually folded, that holds water in an AQUIFER below a layer of impermeable rock. Due to the curvature of the aquifer, the water at the edge of the structure is higher than that at the centre. The water at the centre, therefore, is under a head of pressure from the height of the water table at the edge of the aquifer. If a well is sunk into the aquifer, the pressure forces the water to the surface.

Arthur's Pass ***New Zealand*** Mountain pass through the Southern Alps of South Island, linking the west coast by road and rail with Christchurch. It is approached from the west through the Otira Gorge.

Map New Zealand Ce

Artois ***France*** Territory in the north acquired by France from Spain in 1659, occupying a chalk plateau between Flanders and Picardy. It was the scene of three First World War battles between September 1914 and October 1915. Many military cemeteries and monuments may be visited there. Isolated farms and villages are set in rolling farmland producing cereals, sugar beet and fodder for dairy cattle.

Map France Ea

Aruba ***West Indies*** Island of 193 km² (75 sq miles) in the Caribbean, west of Curaçao and 30 km (19 miles) off the north coast of Venezuela. Industries include oil refining, petrochemicals, fertilisers and tourism. The capital city is ORANJESTAD. Until January 1, 1986, it was part of the NETHERLANDS ANTILLES, when it adopted a separate status. It has its own flag, uses the Aruba guilder instead of the Netherlands Antilles guilder as currency, and aims to achieve total independence by 1996.

Population 67 000
Map Caribbean Dc

Arunachal Pradesh ***India*** Union territory in the north-east, covering 90 650 km² (34 991 sq miles) and bordering on Tibet. It consists mainly of mountains and dense jungle, but despite this China used it as an invasion route in the 1962 Indo-Chinese War.

The tribal inhabitants live off the soil and have little contact with the outside world. The capital is the town of Itanagar.

Population 631 800
Map India Eb

Arunta Desert ***Australia*** See SIMPSON DESERT

Arusha ***Tanzania*** Regional capital in the north-east, about 100 km (60 miles) south of the Kenyan border. Arusha is a tourist centre at the foot of Mount Meru (4565 m, 14 977 ft), and lies in a flourishing coffee-growing region.

Population 55 300
Map Tanzania Ba

Arzew ***Algeria*** Seaport 35 km (22 miles) east of Oran and one of the landing places of the American army in November 1942. It is a terminus for oil and gas pipelines from the Saharan fields, and is of key industrial importance with natural gas liquefaction and petrochemical processing plants.

Population 21 000
Map Algeria Aa

Ascension Island ***South Atlantic*** Strategically vital volcanic speck in the ocean – one of the British St Helena Dependencies – that became an important air-sea staging post in the British recapture of the FALKLAND ISLANDS in 1982. Only 88 km² (34 sq miles) in area, it lies 1126 km (700 miles) north-west of St Helena. The United States maintains a satellite tracking station there. Vegetation is sparse, but islanders grow fruit and vegetables, and raise livestock. Wildlife includes goats, rabbits, partridges, sea turtles and sooty terns.

Population (civilian) 1625

Map Atlantic Ocean Ee

Aschaffenburg ***West Germany*** Port on the Main river about 35 km (22 miles) south-east of Frankfurt. The castle palace of Johannisburg, built in 1605-15, was for more than two centuries the seat of the local rulers, the Electors of Mainz. The town's industries today include paper making, textiles and machine tools.

Population 60 000

Map West Germany Cd

Ascoli Piceno ***Italy*** Town and tourist resort about 160 km (100 miles) north-east of Rome. It is rich in Roman remains such as the Solesta bridge, the Gemina gate, the theatre and, above all, in the grid street layout. The cathedral, with an 11th-century crypt and a Renaissance façade, the town hall and the baptistry stand in the Piazza Arringo, the Roman forum.

Population 53 500

Map Italy Dc

Aseb (Assab) ***Ethiopia*** Red Sea port in the province of Eritrea, about 60 km (35 miles) north of the border with Djibouti. It has the country's only oil refinery.

Population 25 000

Map Ethiopia Ba

Ashdod ***Israel*** Port and industrial centre on the coast about 30 km (19 miles) south of Tel Aviv. Industries include a synthetic fibre plant. Ashdod is the site of an ancient Philistine harbour, fort and temple to the sea god Dagon. It was also the site of the world's longest siege. According to the Greek historian Herodotus, it held out for 29 years in the 7th century BC before surrendering to Egyptian invaders.

Population 68 900

Map Israel Ab

Ashkhabad ***USSR*** Capital of Turkmenistan and the southernmost Soviet city, lying in a fertile oasis east of the Caspian Sea and 40 km (25 miles) from the Iranian border. Its industries include food processing, silk, textiles, carpets, brewing, metalware and glass.

Population 347 000

Map USSR Ge

Ashqelon ***Israel*** Port and resort about 50 km (30 miles) south of Tel Aviv. Originally a Philistine city-kingdom, it was captured by the Romans and then, in the 12th century, by the Crusaders. Today it is the Mediterranean terminal of an oil pipeline from the Red Sea. There are archaeological remains from many eras and an ancient synagogue.

Population 54 700

Map Israel Ab

Asia This, in human terms, is the ancient continent; the fabled continent, too, since for so much of European history it seemed to lie just at the edge of belief, between reality and myth. Many wonders and tall tales came out of it. Marco Polo, the 14th-century traveller in Asia and a fairly accurate reporter, said on his deathbed: 'I did not tell the half of what I saw' – for fear, apparently, of raising howls of derision.

Even in these prosaic days, the facts are breathtaking enough. The largest of the continents, Asia incorporates about one-third of the world's landmass and something like 60 per cent of its people. It embraces the permafrost of Siberia, the Malaysian jungles, the Arabian deserts and the symmetrical volcanoes of Japan. There is Everest, soaring to 8848 m (29 028 ft) and earth's lowest surface point, the Dead Sea, 396 m (1229 ft) below sea level. It takes in two-thirds of the USSR, the world's largest country, and all of China, the most populous nation. Babylon, Assyria, Persia, Arabia, the Levant, India, Mesopotamia, China and the Indus Valley were the nurseries of civilisation, and out of them emerged the great religions – Buddhism, Hinduism, Islam and Christianity.

If the USSR is excluded, Asia can be divided into four regions, each with its own definite characteristics. There is EAST ASIA (the Far East), which includes China and Japan; SOUTH ASIA, whose chief feature is the Indian subcontinent; SOUTH-EAST ASIA of the Pacific archipelagoes and Indo China; and the MIDDLE EAST (West Asia) that borders on the Mediterranean and the Red Sea.

Area 43 549 924 km² (16 814 625 sq miles)

Population 2960 million

Map See p. 48

Asia, East (Far East) Despite modern communications, the term 'Far East', with its hints of utter remoteness, remains rooted in European consciousness. Until the late 19th century, the region's comprehension of the West was sketchy. Most of the representatives of the West it met seemed bewildering barbarians.

East Asia embraces China, Mongolia, Korea, Taiwan, Japan, Macau and Hong Kong. The eastern boundary is probably the thousands of tiny islands that lie off Taiwan and Japan; the western is the Pamir knot, the gigantic crinkle of mountains that holds up the heavens over central Asia. The plateaus of Tibet are encircled by two great mountain chains – the Himalayas and the Kunlun, which extends onward into China.

The climate of the Far East is as extreme as its landscape. The chief feature is the monsoon, which brings the region most of its rain. From June to September, for example, it brings 1220 mm (48 in) of rain to Tokyo. Parts of western China and Mongolia, on the other hand, receive less than 20 mm (0.75 in) in the whole year. Such conditions have created the Gobi and the Taklamakan deserts, which can support only wandering herdsmen. Some narrow coastal strips, plains and valleys in Taiwan, Korea and China, however, well watered by the monsoon, grow prodigious quantities of rice, while colder northern China grows wheat and other cereals. In the highlands there is little cultivation and in some areas, especially in eastern China, formerly densely wooded foothills have been eroded by over-intensive agriculture.

Eastern Asia contains some of the world's most thinly populated areas as well as some of the most crowded. Mongolia has 1 inhabitant per km² (about 3 people per sq mile) while Macau's density is 18 750 people per km² (50 000 per sq mile). Some of the world's largest cities are in the Far East, most obviously in China and particularly Japan, 76 per cent of whose people are now urban dwellers.

Area 11 790 302 km² (4 552 240 sq miles)

Population 1263 million

Map See p. 48

Asia, South The region is a great kite-shaped wedge jutting into the Indian Ocean. It is some 3000 km (1800 miles) long, and about the same again at its broadest. It reaches from Pakistan in the west, through India and the Himalayan states of Nepal and Bhutan to Bangladesh in the east and the island-nation of Sri Lanka to the south. It is the most densely populated region in the world – a teeming diversity of peoples with 18 principal languages, innumerable dialects and many religions.

Geographically, the area is no less diverse, but is generally divided into three major parts: the northern mountains, the Indus-Ganges plain, and the southern peninsula, to which is added Sri Lanka. The Himalayas are young mountains, still being shoved and lifted by continental drift, and consequently display more than 30 peaks that rise to over 8600 m (28 200 ft). The high tops are a wilderness of snow, ice and naked rock, but lower down there are alpine pastures, coniferous woods, bamboo and rhododendrons, and tropical forests.

South of the foothills are the great plains, stretching from Assam to the Punjab, then south to Pakistan and the near-deserts of Rajasthan. The northern plain is the land of the rivers, the Ganges, the Brahmaputra, the Yamuna and their many tributaries that all converge to run into the Bay of Bengal. The north and west are mainly fed by irrigation schemes, but the centre is watered by the summer monsoons. Here is the strong heart of India – the most fruitful agricultural lands, the most productive industries and the vibrant cities.

The third part, the Deccan, is a country of rolling hills and wide river valleys, bordered to the west by the spectacular peaks of the Western Ghats and to the east by the lower, interrupted range of the Eastern Ghats. The black volcanic soil and the red loam of the Deccan produce – with a boost from irrigation canals – beans, millet, cotton and other crops. To the south of the Deccan lie the scenic splendours of Sri Lanka. Geologically, it is a splinter of India, with plains around its central mountains, and sits on India's continental shelf.

The average population density of the subcontinent is more than 200 per km² (530 per sq mile). Some areas are relatively empty – the northern foothills, for example, where there are fewer than 10 people per km² (26 per sq mile). Throughout the subcontinent, the majority of people live in rural areas – in Bangladesh over 90 per cent of them. But population growth also sends many of them to the cities, most of which are desperately overcrowded.

Area 4 488 638 km² (1 733 180 sq miles)

Population 1008 million

Map See p. 48

Asia, South-east The term South-east Asia was coined about 1900 as a convenient means of defining a trading area, though long before that the Chinese and the Japanese both used a similar name, meaning 'the southern seas', to describe the region.

The area is extremely complex and diverse, reaching eastwards from the borders of India and embracing Burma, Thailand, Laos, Cambodia, Malaysia, Singapore, Brunei, Indonesia, Vietnam and the Philippines. At their heart is a tangle of seas and straits – the Strait of Malacca between Malaysia and Sumatra, the Java and South China seas at the centre, the Makassar Strait (also called the Ujung Pandang strait) between Borneo and Sulawesi and the Banda, Sulu and Celebes seas to the east. Each of the mainland nations has a major river valley, the Irrawaddy of Burma, Thailand's Chao Phraya, the Mekong that runs through Laos, Cambodia and south Vietnam and the Red in north Vietnam. They pursue their courses between forested mountains in the upper courses and end in great plains, subject to frequent flooding.

Suitably for such a dramatic collection of peninsulas and islands, South-east Asia is bordered by oceanic trenches of staggering depth, the Mindanao Trench which plunges down to about 10 670 m (35 000 ft) and the rather shallower chasm of the Java Trench. The wide mountains that run through Sumatra, Java and the Lesser Sunda Islands are volcanic, 70 of which have erupted in the past 200 years.

South-east Asia has two distinct types of climate. The Equator neatly bisects Borneo and Sumatra, and there and on nearby islands the climate is hot and humid all the year round. To the north and south, however, it is influenced by the monsoons that bring heavy rain to the mainland nations above the Equator from May to October, while the islands to the south get their rain from November to April. The climatic patterns are reflected in the vegetation. The equatorial area, up to a height of 1000 m (about 3300 ft), is covered by the richest rain forest in the world; above are temperate forests and their more open dwarf trees and shrubs. Below, by the coasts, lie mangrove swamps. In the monsoon areas, with their wet and dry seasons, there are forests of teak and other deciduous trees, pine forests higher up, and wooded savannah.

The people of South-east Asia are as diverse as their world. It is hardly surprising. The islands of Indonesia alone stretch for more than 5000 km (3000 miles) over the ocean, and harbour 300 ethnic groups who speak 250 languages. There is, too, an astonishing array of faiths. Buddhism is all-pervading in Burma and Thailand, and used to be in Vietnam, Laos and Cambodia before the Communist ascendancy. Islam is the main religion in Malaysia, Brunei and Indonesia, while the Philippines are largely Roman Catholic, Bali is Hindu, and tribal gods are worshipped by the people of the hills.

Farming, for subsistence or cash, is the background to the region's way of life, except in the

city state of Singapore, one of the world's great trading centres. Some 12 million families in South-east Asia live by subsistence and shifting cultivation, while the main cash crops are palm oil, rice, sugar, coffee and rubber. Maize, tea, cotton, copra and spices are also profitable exports; so are hardwoods, though there is some danger of over exploitation. Malaysia and Indonesia supply about half the world's tin, and also possess tungsten and nickel. Vietnam has coal, the Philippines copper in plenty and Burma and Thailand produce the major share of the world's sapphires and rubies. Indonesian oil and natural gas are beginning to benefit the nation's development plans, while in Brunei they are the mainstay of the country's wealth.

Area 3 939 498 km² (1 521 042 sq miles)
Population 417 million

Asir ***Saudi Arabia*** Mountainous coastal region, now part of Southern Province, in the south-east of the country, covering 104 000 km² (40 130 sq miles). The Asir mountain range runs for some 370 km (230 miles) along the coast, with an average height of 1830-2134 m (6000-7000 ft). Though the passes are rugged and difficult to negotiate, this once almost inaccessible region is now linked by road with Mecca and Medina.

The region has Saudi Arabia's highest rainfall – 370 mm (15 in) a year – and there is much fertile agricultural land. Most of the country's wheat, maize and sorghum (a species of grass grown for grain or forage) come from there. The farming potential has been increased by the Jizza dam in the mountains, which have rich copper and nickel deposits.

Map Saudi Arabia Bc

Asmara (Asmera) ***Ethiopia*** Chief city of Eritrea province, about 700 km (435 miles) north of the capital, Addis Ababa, and 100 km (60 miles) from the Red Sea coast. It was the Italian colonial headquarters from 1889 to 1941. The United States built Africa's largest military communications centre there in the 1950s. Today it is a garrison town at the forefront of the war against guerrillas seeking Eritrean independence. It has been severely hit in the 1970s and 1980s by drought and famine.

Population 430 000
Map Ethiopia Aa

Aso ***Japan*** National park in Kyushu island, about 60 km (37 miles) south-west of the resort of Beppu. It covers 731 km² (282 sq miles) and contains the world's largest volcanic crater, on Mount Aso. The crater, which contains five peaks, measures a total of 24 km (15 miles) north to south and 18 km (11 miles) east to west.

Map Japan Bd

Assab ***Ethiopia*** See ASEB

Assad, Lake ***Syria*** Lake, south-east of Aleppo, created by the building of the Euphrates Dam (1968-78). It is 80 km (50 miles) long and will eventually irrigate about 640 000 hectares (1 581 400 acres). The dam supplies 70 per cent of the country's hydroelectric power.

Map Syria Ba

Assam ***India*** North-eastern state covering 99 680 km² (38 476 sq miles) bounded by Bhutan and Arunachal Pradesh to the north, Nagaland and Manipur to the east, and Meghalaya to the south. Constituted in 1950, its people are fiercely independent and resent the Bengali immigrants who later moved into the area. As recently as 1984, the Assamese have massacred the immigrants with traditional bows and arrows. Assam lies in the Brahmaputra valley, which is prone to violent earthquakes. Its main products are oil, tea, rice, silk, cotton, timber and jute. The chief town is Gauhati.

Population 19 900 000
Map India Eb

Assen ***Netherlands*** Market town about 130 km (80 miles) north-east of Amsterdam. It was founded in the 13th century around a nunnery, which now serves as public buildings. Nearby are 4000-year-old megaliths (huge standing stones like those at Stonehenge in England), the 'Giants' Caves' mentioned by the Roman historian Tacitus (about AD 55-117), and a grand prix motorcycle circuit.

Population 45 900
Map Netherlands Ca

Assisi ***Italy*** Hillside town about 130 km (80 miles) north of Rome, immortalised by St Francis, patron saint of animals and founder of the Roman Catholic Church's largest monastic order, the Franciscans. He was born there in 1182 and his life is recorded in a series of frescoes painted by Giotto in 1296 in the upper part of the massive Basilica of St Francis.

The 12th-century convent of St Damian stands 2 km (1.25 miles) to the south. It was built on the site where St Francis experienced the vision of the Crucifixion. In an oak forest 4 km (2.5 miles) east is the Eremo delle Carceri, the saint's retreat, marked by a simple church surrounded by a 15th-century convent and hermits' grottoes. The ornate church of Santa Maria degli Angeli is built over the saint's death chapel, the Porziuncola, in a valley 5 km (3 miles) below the town.

Population 24 400
Map Italy Dc

Assyria ***Middle East*** Empire of the ancient world whose capital was NINEVEH. It was founded some time before 3000 BC and controlled much of western Asia. Despite its armed might it was later conquered by the Babylonians and then the Romans. It stretched along the eastern bank of the middle TIGRIS river, in what is now part of modern Iraq.

asthenosphere The part of the mantle in the earth's interior that lies below the lithosphere. It is of low rigidity, thought to be caused by heating and partial melting. See CONVECTION CURRENTS

Asti ***Italy*** City about 40 km (25 miles) south-east of Turin. Its buildings are mainly 18th and 19th century, but there are several interesting medieval monuments, including the large 14th-century cathedral, the 13th-15th century church of San Secondo and the 12th-century Baptistry of San Pietro. The Piedmont province of Asti is noted for its vines, and produces copious amounts of sparkling *Asti Spumante* wine.

Population (city) 75 900; (province) 214 200
Map Italy Bb

Astrakhan ***USSR*** Port on the Volga river delta near the Caspian Sea and 375 km (235 miles) south-east of Volgograd. It trades in timber, fruit, cotton, grain, cereals and rice. Its industries include fishing (especially for sturgeon, prized for its caviar, or roe), cotton, shipbuilding and the processing of karakul lamb's-wool – used to make the curly fur named after the town. Astrakhan was founded by the Tatars in the 13th century. It has a 16th-century fortress built by Ivan the Terrible, who captured the city from the Tatars in 1556.

Population 487 000
Map USSR Fd

Asturias ***Spain*** Region in the north of the country, on the Bay of Biscay. It covers 10 565 km² (4080 sq miles) of a mostly mountainous land. Largely agricultural, it has also deposits of lead, iron and copper. Its capital is OVIEDO.

Population 1 227 000
Map Spain Ba

Asunción ***Paraguay*** National capital and only large city in the country. It was founded as a fort in 1537 by Spanish colonists at the junction of the Paraguay and Pilcomayo rivers. The city was named because work on the fort started on the Feast of the Assumption (Asunción in Spanish), August 15.

Little of the present city is older than the 19th century, although it is still laid out on a colonial grid pattern with tree-lined avenues. The government palace and buildings around the Plaza de los Héroes are modelled on Parisian public buildings. There are a number of open-air restaurants where the country's distinctive harp music can be heard. The main industries are cotton textile manufacturing and processing agricultural products.

Population 456 000
Map Paraguay Bb

Aswan ***Egypt*** City on the east bank of the River Nile about 700 km (435 miles) south of the capital, Cairo. It lies near the quarries from which the ancient Egyptians took granite for obelisks and statues at Luxor and Karnak, about 200 km (125 miles) to the north. Aswan is now a summer and winter resort. It has steel and textile industries, and some granite is still quarried.

It stands by one of the most verdant stretches of the Nile, where cataracts tumble among rocky islands. One, Plants Island, is a forest of rare and introduced plants. On Elphantine Island is the Nilometer, the gauge by which, early in June each year, the ancient Egyptians measured the rate of rise as the river began its annual flood.

Population 200 000
Map Egypt Cc

Aswan High Dam ***Egypt*** One of the world's largest dams, 13 km (8 miles) south of the city of Aswan. It holds back the flood waters of the Nile, controlling absolutely for the first time the flooding of hundreds of kilometres of the Nile Valley to the north. The dam was completed in 1971. Its span is 3600 m (11 812 ft), it is 114 m (375 ft) high, and it holds back Lake Nasser, which covers about 5180 km² (2000 sq miles) of Egypt and the Sudan.

Map Egypt Cd

Atacama ***Chile*** Northern region including the southern part of the Atacama Desert. Its prosperity was originally founded on the silver mines at Chañarcillo but today relies on both minerals and farming. Copper produced at Potrerillos and El Salvador is exported from the ports of Barquito and Chañaral. Iron ore from El Algarrobo, in the wine-producing Huasco valley, is exported from Huasco and Los Lozes. The other major port, Caldera, serves the capital, COPIAPO, 80 km (50 miles) away. Copiapó and Caldera were linked by the Southern Hemisphere's first railway in 1852.

Population 183 000

Map Chile Ab

Atacama Desert ***Chile*** Wasteland covering some 132 000 km² (51 000 sq miles) in the north of the country. It contains some of the driest spots on earth – parts of the desert went without rain for 400 years from 1570 to 1971. But it is rich in nitrates, which are exported for fertilisers, and in iodine, used in medical drugs.

Map Chile Bb

Atakora (Atacora) Highlands ***West Africa*** Uplands stretching across central Togo and continuing into north-west Benin and south-east Burkina. Their peaks form the highest points in Togo (850 m, 2789 ft) and Benin (835 m, 2739 ft). The highlands are laced with spectacular gorges, waterfalls, lakes and rivers.

Map Benin Aa

Atbara ***Sudan*** Industrial town about 350 km (220 miles) north and slightly east of the capital, Khartoum. It lies on the junction of two major road and railway lines to Khartoum (from Wadi Halfa and Port Sudan). Its main industries include railway engineering and cement works and the government railways headquarters are situated in the town.

The town stands at the confluence of the Nile and the Atbara river – also called the Black Nile – which flows 1120 km (695 miles) north-west from Ethiopia.

An Anglo-Egyptian army defeated followers of the Sudanese religious leader, the Mahdi, there in 1898.

Population 73 100

Map Sudan Bb

Athabasca ***Canada*** River which rises in the Rocky Mountains and flows north 1231 km (765 miles) to Lake Athabasca, which drains into the Slave river. Important oil-sand deposits flank the river in ALBERTA.

Map Canada Dc

Athens ***USA*** City 100 km (60 miles) east and slightly north of Atlanta in north-eastern Georgia. It is the seat of the University of Georgia, the oldest state university in the country, whose charter dates back to 1785.

Population (city) 42 500; (metropolitan area) 136 600

Map United States Jd

Athens (Athínai) ***Greece*** Capital and largest city of Greece on the Attic plain beside the Saronic Gulf. It is a sprawling, bustling, noisy city of concrete high-rise blocks and car-choked streets – around the tranquil, classical majesty of the Acropolis. Ancient Athens, built on the rock of the Acropolis 156 m (512 ft) high, was a powerful Hellenic city-state in the 8th century BC, and for almost 1000 years was the classical centre of Western civilisation. Until the Roman general Sulla sacked the city in 86 BC, it was supreme in the arts, philosophy, science, literature and drama, unlike its main rival among the Greek states, militaristic Sparta.

Most of the magnificent architectural remains on the Acropolis were built during the Golden Age of the 5th century BC, under the political leadership of Pericles. These include the Parthenon (447-438 BC), Propylaia (437-432 BC), Temple of Athena Nike (427-424 BC), Erechtheion (395 BC), Theatre of Dionysus (6th-5th century BC), and the Temples of Theseion (449 BC) and Olympeion (550-510 BC). Many philosophers made their home there, Socrates, Plato and Aristotle being the best known.

Athens went into further decline after it was sacked by Germanic warrior tribes in AD 267,

▼ RUINS OF PASSION Plays were staged in the theatre dedicated to Dionysus, the god of wine and ecstasy, for the great spring festival in ancient Athens.

and it became a provincial Byzantine town. The Turks, who occupied Athens in 1458, were not driven out until 1833, and the following year Greece's new king, Otto, made it the capital of united Greece.

Most of the present city is modern. Athens is now the main banking, shopping and communications centre of Greece, and a major industrial centre with its port at PIRAEUS.

Population (metropolitan area) 3 027 300; (city) 885 700

Map Greece Cc

Atherton Tableland ***Australia*** Plateau region in northern Queensland to the west of Cairns, covering about 32 000 km² (12 350 sq miles) at the northern end of the Great Dividing Range. It is mostly about 600-900 m (2000-3000 ft) high and produces beef, peanuts, tobacco, maize and dairy goods. The area's rain forests are a tourist attraction.

Map Australia Gb

Athlone (Baile Atha Luain) ***Ireland*** Market town in Westmeath at an ancient crossing of the River Shannon, 113 km (70 miles) west of Dublin, and almost the exact centre of Ireland. It is a fishing and cruising centre, with a 13th-century castle and ruined Franciscan abbey. It is named after Luain, who ran the riverside inn – Átha Luain, means 'Luain's Ford'.

Population 9410

Map Ireland Cb

Athos (Ayion Oros) ***Greece*** Self-ruling community of more than 20 monasteries on and around Mount Athos, 2033 m (6670 ft) high, on the most eastern promontory of the CHALCIDICE peninsula. Áthos, founded in AD 963, has an all-male population of 1472 living in Byzantine monasteries and observing the Julian calendar, which is 13 days behind the Western European calendar. No women or 'beardless' boys are allowed, and the number of foreign visitors is restricted to ten a day. Its archives contain Byzantine art and manuscripts.

Map Greece Da

Atitlán ***Guatemala*** Mountain lake 80 km (50 miles) west of the capital, Guatemala City, and 1562 m (5125 ft) above sea level. The lake is surrounded by volcanic cones soaring to 4000 m (13 000 ft). Atitlán, described by the English writer Aldous Huxley (1894-1963) as the world's most beautiful lake, is 18 km (11 miles) long, 10 km (6 miles) at its widest and 300 m (990 ft) deep. Hot springs around the lake, and underwater thermal currents, constantly change the shades of blue and green in the water. The biggest of the 12 settlements is Panajachel which has a Sunday market. A boat ride across the lake is the village of Santiago. The village of Sololá, at an altitude of 2113 m (6933 ft), has a spectacular view of the lake.

Map Guatemala Ab

Atlanta ***USA*** Capital and largest city of Georgia, in the north-west of the state, about 90 km (55 miles) east of the Alabama border. It was founded as a small rail town in 1837 and grew rapidly as a transport and cotton-manufacturing centre. It was a chief arsenal of the Confederacy during the Civil War (1861-5), and was burnt to the ground by General Sherman's Union troops in 1864. Today the city is the industrial and financial centre of the south-eastern states. It has one of the country's largest and most modern airports. The writer Margaret Mitchell (1900-49) lived there and set her epic novel *Gone with the Wind* in the area.

Population (city) 426 100; (metropolitan area) 2 380 000

Map United States Jd

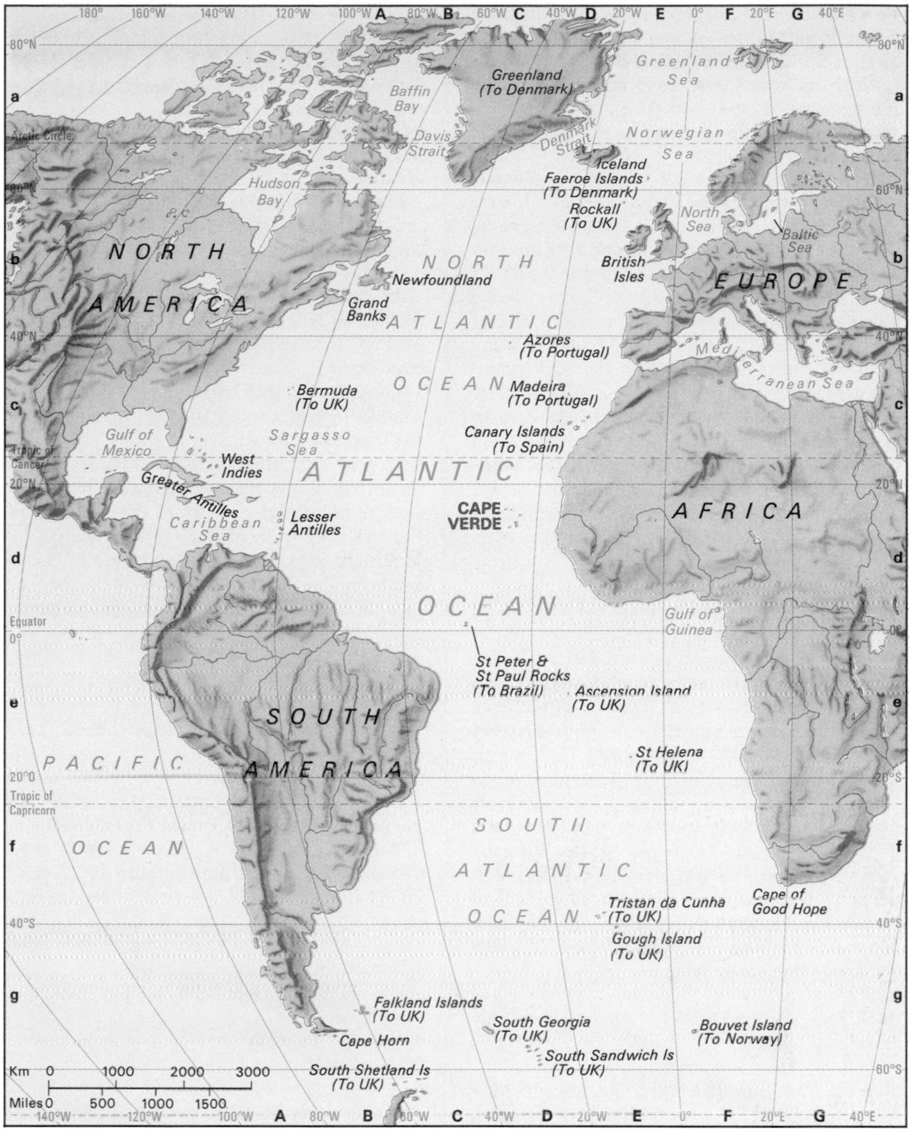

Atlantic Ocean The world's second largest ocean, after the Pacific. It is named after the Atlas Mountains in north-west Africa, which marked the western boundary between the known and the unknown world for the Ancient Greeks. It includes the Gulf of MEXICO and the CARIBBEAN, MEDITERRANEAN, NORTH and BALTIC seas.

The Atlantic was formed within the past 175 million years as continents drifted apart. It is still getting wider, by 10-20 mm (0.4-0.8 in) a year, and spreading outwards from the S-shaped ridge that runs through the centre of the ocean. A deep rift valley, 24-48 km (15-30 miles) wide, runs north-south through the centre of the ridge. As the continental plates of the sea floor move apart, new rocks are formed in this valley. Parts of the huge central Atlantic ridge reach the surface as volcanic islands, including ICELAND, the AZORES, ASCENSION, ST HELENA and TRISTAN DA CUNHA.

The Atlantic also contains abyssal plains and shallow continental shelves. Unflooded parts of the continental shelves form islands such as Newfoundland and the British Isles.

The average depth of the Atlantic is 3330 m (10 925 ft). It has, on average, the warmest and saltiest waters of any ocean. In the South Atlantic the waters circulate in an anticlockwise direction (the South Equatorial Current, the Brazil Current and the Benguela Current). In the North Atlantic, the circulation is clockwise (the North Equatorial Current, the Gulf Stream and the Canary Current). Within these circulatory systems (or gyres) are areas of comparatively calm water, such as the SARGASSO SEA. The North Atlantic Drift, an extension of the Gulf Stream, conveys warm water into the ARCTIC OCEAN, while the East Greenland and Labrador currents bring cold Arctic water into the Atlantic.

The Atlantic provides about one-third of the

world's fish and shellfish catch. The leading fishing grounds are over the North Atlantic's continental shelves, including the Grand Banks off Newfoundland and the Dogger Bank in the North Sea. The continental shelves contain oil reserves, notably in the Gulf of Mexico and the North Sea.

Area 82 217 000 km² (317 340 000 sq miles)
Greatest depth 9200 m (30 200 ft)

Atlas Mountains ***North Africa*** Mountain system extending from the Atlantic coast of Morocco north-eastwards some 2250 km (1400 miles) to northern Tunisia. It consists of several roughly parallel ranges. In Morocco these include (from south to north) the Anti-Atlas, High Atlas and Middle Atlas; and in Algeria the Saharan Atlas (which includes the Amour mountains and extends to the Aurès Massif in the east) and, farther north, the Tell, or Maritime, Atlas. The highest peak in North Africa – Djebel Toubkal (4167 m, 13 671 ft) – stands in the High Atlas of Morocco.

Map Africa Ca

atmosphere See panel opposite

atmospheric pressure Pressure exerted by the weight of the atmosphere on the earth, measured by a barometer and expressed in millibars (1000 millibars is equivalent to approximately 750.1 mm, or 29.53 in, of mercury). The average pressure at sea level is 1013.25 millibars, equivalent to the weight of a column of 760 mm (29.92 in) of mercury, or a pressure of about 1033 grams per cm² (14.66 lb per sq in). Pressure varies on the earth's surface with latitude, temperature and altitude, and these differences in pressure are responsible for winds blowing from areas of high to areas of low pressure, and thus for major climatic differences.

atoll Ringlike coral reef that almost or entirely encloses a lagoon.

Attock ***Pakistan*** Town and fort about 80 km (50 miles) west of Islamabad, at a strategic crossing point of the Indus river where the flow is constrained by a series of gorges. The clifftop fort was built by the Mogul emperor Akbar the Great who ruled from 1556 to 1605. The British built a rail and road bridge across the river in the 19th century. The bridge has survived floods which have raised the water level by as much as 30 m (100 ft).

There are several oil wells in the area and the town has a refinery.

Population 40 000
Map Pakistan Da

Auckland ***New Zealand*** The country's largest and liveliest city, chief port and former capital. Lying on a narrow isthmus between Manukau and Waitemata harbours in the north of North Island, it has spread over many extinct volcanoes. The city was founded in 1840 by British settlers on a Maori site and has the largest Polynesian population – some 57 000, including many Pacific islanders – of any city in the world. The Auckland War Memorial Museum houses one of the world's finest collections of Polynesian, especially Maori, artefacts.

Maoris sold the site to the British in 1840 for clothes, food, tobacco and about £30 sterling. Today the port exports dairy produce, wool, hides and timber, and imports machinery, oil and fertilisers. Its industries include textiles, chemicals, food processing, engineering, metal-working and vehicle assembly.

Population 769 600
Map New Zealand Eb

Audenarde ***Belgium*** See OUDENAARDE

augite Dark green to black PYROXENE mineral.

Augsburg ***West Germany*** Bavarian city about 56 km (35 miles) north-west of Munich, where in 1555 the right of Europe's Protestants to freedom of worship was recognised at the religious treaty, the Peace of Augsburg. Romans founded the city in 15 BC and named it after the Emperor Augustus.

It grew rich trading between Italy and the north, and in the 16th century was the banking capital of northern Europe. One family, the Welsers, was granted virtual sovereignty over Venezuela after the Emperor Charles V could not repay his debts to them; another, the Fuggers, financed the Holy Roman Emperors for centuries.

The painters Albrecht Dürer (1471-1528) and Hans Holbein the Younger (about 1497-1543) were born in the city, and works by them hang in the Schaezler Palace. Rudolf Diesel (1858-1913), who invented the engine named after him, was also born there, as were the aircraft designer Willy Messerschmitt (1898-1978) and the playwright Bertolt Brecht (1898-1956).

Augsburg's cathedral, which dates from the 9th to the 14th centuries, claims to have the oldest stained-glass windows in Europe, fitted in the 11th century. The city has been a main textile-producing centre for almost 500 years, but now engineering industries have become relatively more important to the wealth of Augsburg.

Population 245 000
Map West Germany Dd

Augusta ***Italy*** Fortified peninsula town on the east coast of Sicily. It was founded by the Holy Roman Emperor Frederick II of Swabia in 1232. It is now one of the Mediterranean's biggest oil refining and petrochemicals centres.

Population 38 900
Map Italy Ef

Augusta ***USA*** City in the state of Georgia on the Savannah river, which forms the border with South Carolina. It was founded as a military base in 1736. Today it is a market and textile manufacturing centre and river port.

Population (city) 46 000; (metropolitan area) 368 300
Map United States Jd

Augustow (Avgustov) ***Poland*** Boating and sailing centre amid the forested Suwalki lake district in the north-east, about 35 km (22 miles) west of the border with the Soviet Union. The Augustow Canal crosses the marsh and forest wilderness on the border. It was built in 1824-39, during the Russian occupation, to link the Vistula and Neman river systems, bypassing east Prussia, which was then held by Germany.

Population 25 600
Map Poland Eb

Aurangabad ***India*** Textile town about 240 km (150 miles) east and slightly north of Bombay. In the 17th century it was the capital of the Mogul emperor Aurangzeb, and many of its buildings date from then. It is also noted for its Buddhist cave temples.

Population 284 600
Map India Bd

aureole Zone around an igneous rock intrusion which has been altered by the heat and chemicals generated during the intrusion of the hot molten rock (magma). It can be traced around the margins of a BATHOLITH, such as Dartmoor.

Aurès Massif ***Algeria*** See ATLAS MOUNTAINS

aurora Brilliant sheets of coloured light appearing in the skies at high latitudes. They are caused by ultraviolet radiation and electrically charged particles from the sun interacting with the earth's atmosphere. The radiation and particles are drawn by the earth's magnetic field towards the North and South poles, where they produce brilliant green, blue, white or red flashes by ionising gases in the earth's atmosphere, mainly at altitudes of some 95-145 km (60-90 miles). In the Northern Hemisphere the display is called the Aurora Borealis, or Northern Lights; in the Southern Hemisphere it is called the Aurora Australis.

Auschwitz (Oswiecim) ***Poland*** Chemical-producing town 54 km (33 miles) west of the city of Cracow. It was the site of the largest Nazi concentration camp, which operated from June 14, 1940, to January 27, 1945. Auschwitz was actually a group of three main camps – Oswiecim, Brzezinka (Birkenau) and Monowice (Dory) – with 39 smaller camps nearby.

More than 4 million people of 39 nationalities were shot, gassed, starved or tortured to death in Auschwitz, and its crematoria burnt up to 12 000 bodies daily. Today it is preserved as the National Museum of Martyrology, which, with the world's largest burial ground, at Brzezinka, is a place of pilgrimage.

Population 35 600
Map Poland Cc

Austerlitz ***Czechoslovakia*** See SLAVKOV

Austin ***USA*** Capital of Texas, in the central south of the state about 370 km (230 miles) west and slightly north of Houston. It is the commercial centre of a large farming and ranching region, and has the main campus of the University of Texas.

Population (city) 397 000; (metropolitan area) 645 400
Map United States Gd

Austral Islands ***French Polynesia*** See TUBUAI ISLANDS

Australasia Term used for Australia, New Zealand and sometimes Papua New Guinea, all of which are also part of OCEANIA.

Australia See p. 56

Australian Alps ***Australia*** See SNOWY MOUNTAINS

THE DIFFERENT LAYERS OF THE ATMOSPHERE

Earth's atmosphere is a multi-layered mixture of gases bound to the planet by gravity. The protection it affords against harmful rays from the sun, debris from outer space and cosmic rays (the source of which is unknown) makes life on earth possible. Solar radiation and cosmic rays are absorbed or scattered harmlessly; meteors and meteorites mostly burn up in the frictional heat generated by their entry into the atmosphere. Without this protection, the planet would also be subjected to extremes of temperature, for the atmosphere dissipates the sun's heat by day and insulates at night.

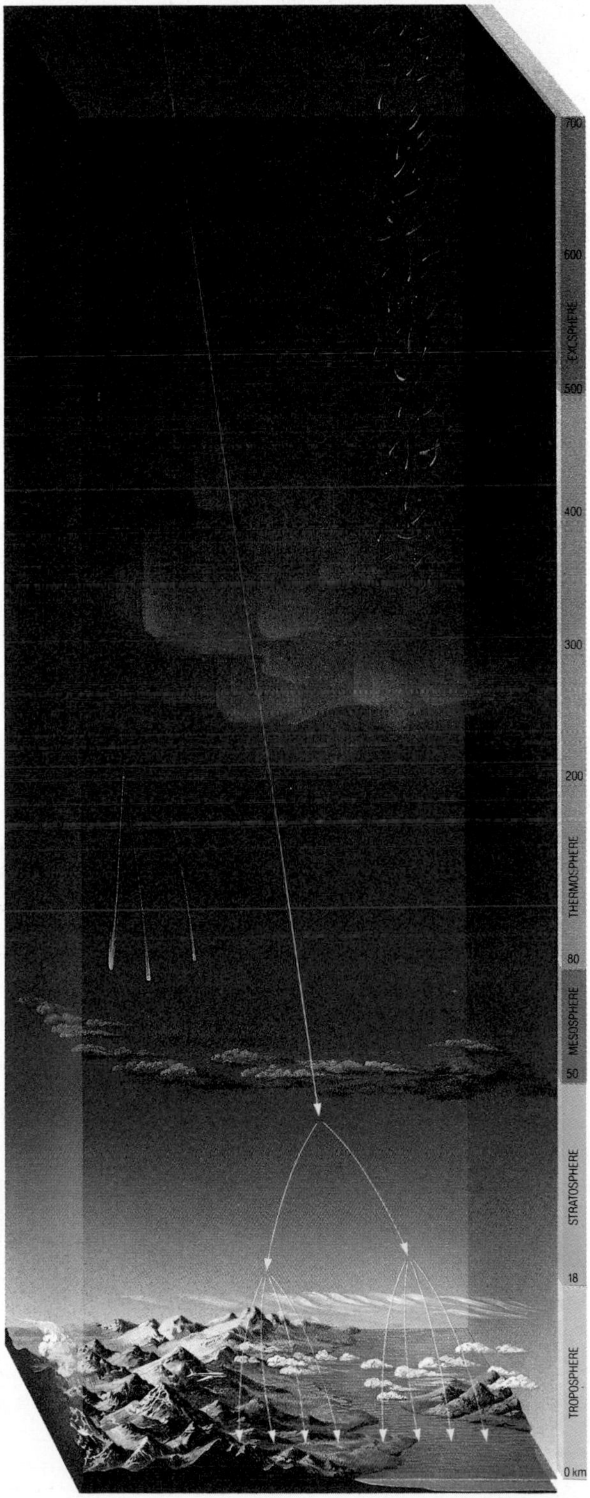

EXOSPHERE Starts about 500 km (approx. 310 miles) above earth and extends via the magneto sphere into space at around 2000 km (1250 miles). In the very rarified atmosphere above about 2100 km (1500 miles) hydrogen particles predominate. Down to about 965 km (600 miles) helium and hydrogen are equal, while below this level there is some oxygen. Heavy solar radiation and cosmic rays penetrate to the ionosphere below.

THERMOSPHERE Extends from 80 km (50 miles) up to the exosphere. Gas molecules (mainly oxygen and nitrogen) within it are broken down by intense solar radiation, producing charged particles called ions. Glowing lights – aurorae – in earth's high latitudes result from these electrical disturbances due to the impact on the oxygen and nitrogen molecules of electrons and protons from the sun. Layers of ions at 100 – 300 km (about 60 – 190 miles) reflect radio waves.

MESOSPHERE A 30 km (20 mile) deep layer between the ionosphere and stratosphere. Temperature here falls sharply from 10°C (50°F) at 50 km (30 miles) to −80°C (−112°F) at 80 km (50 miles).

STRATOSPHERE Contains a band of ozone – oxygen with three, rather than two, atoms – which is vital to life on earth. It absorbs and filters out most of the solar ultraviolet radiation which is deadly to living organisms. Here, above the tropopause, temperature rises from −55°C (−67°F) at around 15 km (10 miles) to 10°C (50°F) at 50 km (30 miles). Meteors burn up in the frictional heat generated by their passage through a denser atmosphere. Meteoric dust forms noctilucent clouds. Here, too, cosmic rays are absorbed and scattered.

TROPOSPHERE The bottom layer of the atmosphere, and the most dense. It rises to an average of 15 km (10 miles) above earth, and all normal clouds and weather patterns are formed within it. Temperature falls about 2°C per 305 m (4°F per 1000 ft), stabilising at −55°C (−67°F) at the top, or tropopause, where high winds flow horizontally – the jet streams used to speed airliners.

There are five major layers in the atmosphere. In ascending order they are troposphere, stratosphere, mesosphere, thermosphere and exosphere. Certain zones of nuclear, chemical and electrical activity occur within them, and some of the boundaries between layers in the atmosphere vary with latitude and with season. Other layers within the five major layers include the ionosphere and the magnetosphere.

Gases in the atmosphere include nitrogen (78 per cent), oxygen (21 per cent), argon (0.9 per cent) and carbon dioxide (0.03 per cent) with minute traces of helium, hydrogen, krypton, methane, neon, ozone and xenon. Generally they are mixed in these proportions up to about 60-80 km (40-50 miles) high, but there is a concentration of ozone in the stratosphere at 25-27 km (15-17 miles) in low latitudes, falling in altitude in high latitudes. The lower atmosphere contains varying amounts of water vapour, which is virtually absent above about 8-18 km (5-11 miles).

Above the mesosphere, in the lower thermosphere, the atmosphere is mainly oxygen and nitrogen, but above 200 km (124 miles) oxygen predominates. In the exosphere, oxygen, hydrogen and helium constitute the very tenuous atmosphere. Ionised particles become more numerous and above 2000 km (1250 miles) only electrons are present. This is the magnetosphere, with an increased incidence of electrons at 3000 and 15 000 km (1850 and 9300 miles) – the Van Allen radiation belts. It is this part of the atmosphere that is influenced by the earth's magnetic field.

The whole atmosphere envelope weighs about 5000 million tonnes, but three-quarters of its mass is concentrated below the height of the world's loftiest mountain peak – Everest (8848 m, 29 028 ft). This weight of air exerts a constant pressure of about 1 tonne on the human body at sea level.

However, air pressure varies; it increases as cold air sinks downwards and becomes more dense, and falls as warm air expands and rises. These changes in pressure affect the earth's weather systems (see p.704). The density of the atmosphere also decreases rapidly with height – at 16 km (10 miles) above sea level, density is only one-tenth of that at sea level; at 32 km (20 miles) it is only one-tenth of that at 16 km.

Within the ionosphere there are several distinct layers of electrically charged particles called ions. Radio waves, which travel in straight lines, can be transmitted around the earth's curvature by bouncing them off ionised layers. The highest – and the most highly ionised – is the F2 layer, just above 300 km (186 miles). Below is the F1 layer, at about 150 km (92 miles). This is the lowest level at which satellites can orbit, because the atmosphere below is too dense for free flight. At 90-160 km (56-100 miles) is the E (Heavyside-Kennelly) layer, quite strongly ionised in daylight, but dissipating at sunset. In the mesosphere, at around 50-90 km (30-56 miles) is the D layer, where ionisation is weak and radio waves are poorly reflected.

Australia

A VAST AND SPARSELY PEOPLED ISLAND CONTINENT WHERE A YOUNG, COMPETITIVE SOCIETY FINDS AFFLUENCE AND FREEDOM

In the 1960s, Australia was dubbed 'the Lucky Country' – though not without some irony, the author implying that its prosperity owed more to luck than good judgment. That luck has not yet run out, although Australia today may be facing a more difficult period, with relatively high unemployment, a drop in overseas demand for its export commodities and a weakened currency. But despite these problems it is still a fortunate country by almost any criterion, with most of its 16 million people enjoying a life of affluence and freedom matched in few other parts of the world.

The good fortune stems from abundant mineral resources and an agricultural sector that, in most years, produces a large surplus of food for export. The wealth created by these primary industries is spread among a relatively small population and supports an easy-going lifestyle that contains many echoes of European and North American (particularly Californian) ways but has also developed a distinctive flavour of its own. However, the affluence and freedom have largely bypassed one group of Australians: the Aborigines who inhabited the country for some 50 000 years before the first Europeans settled there in 1788.

THE GREAT SOUTH LAND

Whether you consider Australia – the 'South Land' – to be the world's largest island or its smallest continent, it is undoubtedly a land of superlatives. Despite its great size and reliance on rural production, it is one of the most highly urbanised countries in the world, with two-thirds of the population living in the eight biggest cities, all but one of these – the federal capital, CANBERRA – being situated on the coast.

It is difficult for most Europeans or even North Americans to appreciate the size and emptiness of Australia. PERTH, the capital of WESTERN AUSTRALIA, probably qualifies as the most isolated city in the world. It is over 2700 km (nearly 1700 miles) from its closest neighbour, ADELAIDE, capital of SOUTH AUSTRALIA – farther than from London to Moscow or New York to Denver. Moreover, there is only one town approaching a population of 20 000 in between. On the route north from Adelaide to DARWIN at the 'top end' of the NORTHERN TERRITORY – a distance of some 3500 km (2200 miles) – the only town is ALICE SPRINGS, with 18 000 people.

Large parts of Australia's huge 'red centre' – particularly in the western half – are an inhospitable wasteland of desert and mallee (scrub) with little or no permanent population. Even in the semiarid but more productive rough grazing country farther east and north, neighbours may live 100 km (60 miles) or more apart and measure their stock density in terms of hectares or acres per head of cattle rather than vice versa. In these isolated settlements, the Flying Doctor brings medical services, the School of the Air teaches by radio and stockmen muster cattle by helicopter.

Superlatives apply to the climate, too. In the centre – and even in parts of the temperate south-east inland from the GREAT DIVIDING RANGE – droughts are commonplace, and years can pass without a drop of rain. Then huge downpours can cause extensive (though not usually very destructive) flooding. In the north and north-east, such flooding is an annual occurrence during the summer 'wet' any time between October and May; then roads may be impassable and tropical cyclones (hurricanes) may batter coastal settlements – like Cyclone Tracy, which devastated Darwin on Christmas Day, 1974. Only in relatively restricted areas, notably along the south-east coast, is rainfall at all reliable.

Australia's image as a sunburnt land contains a lot of truth, and as one consequence QUEENSLAND ('the Sunshine State') has the unenviable record of the world's highest incidence of skin cancer. Cloncurry in north-central Queensland holds Australia's temperature record: 53.1°C (127.6°F). Yet there are also ski resorts in the SNOWY MOUNTAINS (part of the Great Dividing Range), within 400 km (250 miles) of both SYDNEY and MELBOURNE, the two biggest cities.

It is in this south-eastern part of the country, generally well-watered and (except in the mountains) temperate, that most Australians live – in a broad arc from BRISBANE (the Queensland capital) in the north, through Sydney (the capital of NEW SOUTH WALES), Canberra and Melbourne (the capital of VICTORIA), to Adelaide. The only other major communities are Perth, HOBART in the southern island state of TASMANIA, and on parts of the Queensland coast from Brisbane through ROCKHAMPTON and TOWNSVILLE to CAIRNS.

Geologically, Australia is an ancient, stable land long worn down by wind and water, so that today a mere 13 per cent of its area is above 500 m (1640 ft). The main mountain range – the Great Dividing Range which runs down the entire east coast – peaks at Mount KOSCIUSKO (2230 m, 7316 ft) near the border of New South Wales and Victoria. There are numerous smaller ranges, such as the HAMERSLEY RANGE in the PILBARA of Western Australia, the MACDONNELL RANGES of the Northern Territory and the Musgrave and FLINDERS ranges of South Australia, but they reach only 1000-1500 m (3300-5000 ft). On a smaller scale, there are spectacular gorges in most of these ranges whose stark beauty is often highlighted by the red and yellow rocks.

The Australian coastline – estimated to extend more than 36 000 km (22 500 miles), much of it superb sandy beaches – has its own spectacular features. Greatest of all is the GREAT BARRIER REEF, a vast complex of coral reefs extending 2000 km (1250 miles) up the Queensland coast almost to Papua New Guinea. The biggest of its kind in the world, in some places it is 300 km (190 miles) from the shore and over 500 m (1650 ft) thick.

Although most of the continent consists of extensive plains and low undulating downland – a somewhat monotonous landscape – there are also spectacular forests, especially along the east coast, in Tasmania and in the far south-west. Areas with heavy rainfall – both temperate and tropical – have dense rain forests with their closed canopy of foliage and climbing plants that scramble up the tall tree trunks to the light. In drier areas there are more open sclerophyll forests, whose trees have leathery leaves to conserve moisture.

Among the best known of these are the acacias (wattles) and the numerous species of eucalypts – over 600 have been identified. Many eucalypt leaves (which vary from blue-grey to dark green in colour) give off droplets of aromatic oil, causing a bluish haze in such areas as the BLUE MOUNTAINS west of Sydney. Many species also have thick, hard bark able to withstand the bush fires that are a perennial hazard in most areas.

Some native Australian plants positively benefit from fire; the seed heads of some banksia trees, for example, open and scatter the seeds only after a fire. Others merely adapt for survival – as with the baobab, whose swollen trunk stores water, and the numerous ephemeral plants of the dry heart, whose seeds may remain dormant for years until rare rains enable them to germinate, grow and carpet the ground with flowers, all within a few days. There are even fish whose eggs are believed to remain similarly dormant until it rains.

UNIQUE ANIMAL LIFE

Australia's long isolation from any other landmass has influenced evolution and given it unique zoological marvels. Some 200 million years ago, Australia was part of the great southern continent of GONDWANALAND, with Antarctica, South America and Africa. Small, primitive, warm-blooded animals lived there, and continued to do so and to evolve when Gondwanaland broke up, isolating Australia. They were the ancestors of today's egg-laying monotremes and pouched marsupials; almost everywhere else the more advanced placental mammals, which bear well-developed live young, ousted them.

Marsupial species are as diverse as the tree-climbing possums and cuddly looking koala, the ground-burrowing wombat and the heavy-haunched kangaroos and smaller wallabies – as well as marsupial 'mice' and 'cats', which superficially resemble their (unrelated) placental counterparts.

Among the monotremes, which hatch from leathery eggs, are the echidna or spiny anteater – well described by this latter name – and the duck-billed platypus. When British naturalists first saw this, with its flat bill, small clawed and webbed feet and furry body, they thought it was a taxidermist's joke.

Australia's bird life is also distinctive, if more closely related to birds elsewhere. There are two species of large flightless birds – counterparts of the African ostrich – the emu and the cassowary. The lovely lyrebirds are

▲ **SAILS IN THE SUN A lone surfboarder skims past the sail-like roofs of the Sydney Opera House, a stone's throw from where the first European settlers landed – the outdoor life beside the symbol of Australia's new cultural awareness.**

renowned mimics whose males display their lyre-shaped tail plumage during mating displays. But most characteristic of all are the numerous species of Australian parrots, ranging from budgerigars to large cockatoos.

Although Australia lacks large land predators like the lion, bear and wolf – marsupial carnivores are small and inefficient hunters – dangerous creatures abound. There are large crocodiles in the north, and venomous spiders and snakes (including some of the most dangerous of both) in all areas. Sharks appear off the entire coast, but most popular beaches are protected by nets and beach patrols, and fatal shark attacks are rare. More common is the painful and often fatal sting of the stonefish.

Feral animals have had a strong impact on the Australian environment and native fauna. Feral pigs, donkeys, cats and dogs all thrive in the wild with few natural enemies. Rabbits, introduced by immigrants in 1858, were a severe blight – competing with stock for grass – until myxomatosis and feral cats and foxes controlled their numbers. Dingoes, Australia's 'native' dogs, probably introduced thousands of years ago, are regarded as pests by sheep farmers since one dingo can kill 100 sheep in a night, but they are probably less harmful than domestic dogs gone wild. The depredations of dingoes and hybrid dogs (for they interbreed) led to the building of the world's longest fence (8500 km, 5300 miles) in South Australia and Queensland.

Australia has the world's only wild camels, the descendants of animals that played an important part in the exploration and development of the inland areas in the 19th century. Some are today exported to the Middle East for breeding. A more unwelcome import is the large and poisonous cane toad, introduced to control a sugar cane beetle pest in the 1930s and now teeming over large areas of north-eastern Australia.

CONVICTS AND FREE SETTLERS

The concentration of people in south-eastern Australia began after a Pacific voyage by the British navigator Captain James Cook. He was not the first European to 'discover' Australia, however. The 17th-century Dutch seamen Willem Janszoon and Abel Tasman explored the coasts of the GULF OF CARPENTARIA and Van Diemen's Land (now Tasmania) respectively. The Englishman William Dampier surveyed the west Australian coast in 1699, but all these explorers wrote off New Holland (as Australia was then called) as of little value for trade or colonisation.

Cook had the good fortune to hit upon a much more hospitable part of the coast. In 1770, returning to England from a scientific expedition to Tahiti and after charting New Zealand, he made landfall at an inlet on the east coast that he named BOTANY BAY because of its profuse and unusual plant life. He went on to chart almost the entire east coast (and was nearly wrecked on the Great Barrier Reef), claiming it for Britain and naming it New South Wales. Soon after, Britain lost its American colonies and, faced with overcrowded prisons, decided to start a penal colony at Botany Bay. The First Fleet of 11 convict ships arrived in January 1788, found the land around Botany Bay too marshy, and landed instead a little farther north at a spot they named Sydney Cove, on the superb natural harbour of Port Jackson – which Cook had failed to explore.

Life for the first colonists was extremely tough, and the settlement barely survived. But it did, and over the next 50 years other penal settlements were founded on NORFOLK ISLAND and near present-day Hobart and Brisbane; free settlers began to arrive (and by the early 1830s outnumbered the convicts); and the future cities of Perth, Adelaide and Melbourne were founded. These communities looked back to Britain for trade and aid, developing port facilities that became the gateways into Australia and the main means of communication between themselves. They also became the focus of overland communications – by road and, from later in the 19th century, rail – that linked them with the small agricultural

and mining settlements that spread over the country. Inevitably, manufacturing industry largely concentrated in these future state capitals and many immigrants went no farther.

This is why Sydney, Melbourne, Brisbane, Adelaide, Perth and Hobart are so much bigger than the second cities of their states and why Australia has few large provincial centres. There are exceptions, notably Canberra (created as the national capital in the 20th century), NEWCASTLE and WOLLONGONG in New South Wales (important industrial centres built on coalfields), and GEELONG in Victoria. A large part of Australia's wealth is created beyond the urban frontier and outside the notice of many Australians. The worst drought of this century, which ended in 1983 after temporarily crippling much of the farming sector, occurred without affecting the daily lives of 90 per cent of the population.

A high proportion of agricultural output is exported – for example, half of the annual beef total of 1.3 million tonnes and 95 per cent of the annual wool clip of 730 000 tonnes. Coal has overtaken wool as the main export, but Australia still has 140 million sheep and produces a quarter of the world's wool, accounting for nearly 9 per cent of the country's export earnings.

The huge sheep and cattle stations (farms) extend over large parts of the interior, though in many areas the grazing is not rich enough to support a high density of stock. But since many sheep stations exceed 500 000 hectares (1.2 million acres) and the largest cattle station is over 3 million hectares (7.4 million acres, almost as big as Belgium), enormous numbers can be raised even at only one sheep per 3 hectares (7.4 acres) or one head of cattle per 50 hectares (125 acres).

These are figures for the drier areas. Where the rainfall is higher and the grass more abundant, stock raising is much more intensive – especially in the south-east and the extreme south-west. Generally speaking, cattle are better able than sheep to adapt to hot conditions, and nearly half the country's 22 million beef cattle are reared in Queensland and the Northern Territory. The national herd used to be much bigger, but numbers declined by over 10 million during the 1979-83 drought. Some 2.8 million dairy cattle graze mainly in eastern New South Wales and Victoria.

Cereal growing in Australia is dominated by wheat, most of it produced to the west of the Great Dividing Range, in a belt from southern Queensland to Victoria, and in Western Australia. Annual output is about 15-20 million tonnes, most of which is exported, making Australia one of the world's leading suppliers of wheat. Other cereal crops include barley, sorghum, oats and about 750 000 tonnes of rice a year, mostly in the Riverina area of New South Wales but also in northern Queensland and the new ORD river irrigation project in the north of Western Australia.

Because of the great range of climates, crops are equally diverse. The warmer areas of New South Wales and Queensland produce cotton, sugar cane, pineapples, bananas, peanuts, avocados and Australia's native macadamia nuts. Fruits flourish – especially peaches, apricots, grapes and citrus fruits along the MURRAY and Murrumbidgee rivers, and apples in Tasmania. All states produce wine – both ordinary and finer vintages – the older-established wine areas including South Australia's BAROSSA VALLEY (settled largely by German immigrants in the last century) and the HUNTER VALLEY near Newcastle. In recent years the industry has boomed, with many new vineyards. Timber is another important product, especially in Tasmania.

HUGE MINERAL WEALTH

In 1946, agriculture accounted for more than 75 per cent of Australia's export earnings. Today that proportion has been halved, although the volume is not very different: farm output has been overtaken by greatly increased exports of minerals. Massive, easily worked coal deposits provide the biggest single export, sold mainly to Japan. Hydroelectric power (mainly from the Snowy Mountains and Tasmania) and limited reserves of oil and natural gas combine with coal to make Australia self-sufficient in energy. The country is also a leading producer of iron ore, copper, lead, zinc, manganese, nickel, tungsten, uranium, tin and bauxite (aluminium ore). As icing on the mineral cake, there is silver, platinum and gold, and such strategically important minerals as zircon, titanium and cobalt.

A number of substantial towns – notably BROKEN HILL and MOUNT ISA – owe their existence to mining. Along 160 km (100 miles) of coast at WEIPA on Cape YORK Peninsula the world's largest bauxite deposit is worked. Iron-ore mining has been the basis of enormous developments since the 1960s, especially in the MIDDLEBACK RANGE of South Australia and the KIMBERLEY and Pilbara regions of Western Australia. Although the worldwide recession in the steel industry has hit production, Australia remains the major supplier to Japan and South Korea and is supplying millions of tonnes to China. Uranium – of which Australia has one-third of the world's proven reserves – is also facing a difficult market, political and environmental pressures at home combining with reduced European and North American nuclear power programmes.

The mineral that put Australia on the map was gold. The first major strike was in New South Wales in 1851, but this was soon followed by bigger finds at BALLARAT and BENDIGO in Victoria. People flocked to the goldfields from around the world, Victoria's population quadrupling between 1851 and 1855 and the capital, Melbourne, growing from a small town to a thriving city. Forty years later gold was discovered at KALGOORLIE in Western Australia, where mining still goes on. In fact, the increase in gold prices in recent years has resulted in some old mines being revived, but Australia today produces only 1.5 per cent of the world's gold.

THE FORGOTTEN AUSTRALIANS

Much of Australia's recently found mineral wealth lies beneath land that is far from centres of European settlement – indeed, in places that few white people have visited and most Australians know only from their television screens. This is the true outback, a term that does not refer to a specific place but to any area isolated from civilisation. It is the ancestral home of many of Australia's largely forgotten and least privileged people, the Aborigines. The conflict between their claims to their own land and the mining companies' desire to exploit it has resulted in one of the most controversial issues in contemporary Australian politics: land rights.

Although Captain Cook declared the country empty, there were in 1770 at least 300 000 Aborigines and, in far northern Cape York, TORRES STRAIT Islanders, a Melanesian group. The Aborigines had been in continuous occupation for some 50 000 years, as shown by burial sites and other archaeological evidence. Indeed, rock paintings at Obiri Rock, east of Darwin, are 10 000 years older than the Egyptian pyramids and even predate the LASCAUX cave paintings in France. When Britain set up its Australian colonies, all land became Crown 'property' without compensation for the Aborigines.

The European settlers also brought diseases such as measles, whooping cough and smallpox that devastated the native population. A greater tragedy was the colonists' persecution of the Aborigines. Many were murdered, under the excuse that they posed a threat to the colony and were in any case less than human, and they were exploited as forced labour. Confiscation of land was justified on the basis that the Aborigines did not 'use' the land efficiently. Perhaps 250 000 died between 1788 and the 1920s as a direct result of European colonisation; the victims included the entire full-blood population of Tasmanian Aborigines, a distinct ethnic group.

By 1930, only about 70 000 Aborigines remained, but their population is now over 160 000 and increasing rapidly. They still suffer discrimination and poverty, however, with an average family income half that of Australians as a whole, an infant mortality rate more than double, an unemployment rate four times higher, and life expectancy 20 years shorter.

Until the 1940s about half the Aborigines of the tropical and desert regions followed their old lifestyle of hunting and gathering, activities feasible in these barren lands only at very low population densities. Except in New South Wales, Tasmania and Victoria, many of them lived on reserve lands segregated from white Australians. In the past 20 years economic and population pressures have led many of the reserve Aborigines to migrate to towns and cities, and try to adjust to urban life. However, such 'detribalised' people lack a cohesive social structure, and many succumb to alcoholism and petty crime. Only about 40 per cent still live in rural areas.

Whether the land-rights campaign will reverse this trend remains to be seen. Over one-third of the Northern Territory (including AYERS ROCK) has been transferred under 'inalienable title' to Aborigines and nearly one-fifth of South Australia. However, many non-Aborigines resent this, and a political dispute still rages as to whether Aboriginal landowners should be able to veto mining developments on their territory. Pro-land-rights campaigners point to the Aborigines' deep – almost mystical – relationship to the land that has supported them for millenia.

In the mainstream of Australian life there have been great changes in the population mix since the end of the Second World War. In 1945, out of the 7.5 million population (less than half the present figure) over 90 per cent were of British or Irish origin. Since then a large influx of Italian, Greek, Yugoslav, Dutch, Polish, Turkish, German, Maltese and many other immigrants has begun to turn Australia into a multicultural society. This has been particularly true since the mid-1960s, when the so-called 'White Australia' immigration policy was abandoned. Now there are significant communities of Lebanese, Vietnamese and Cambodian settlers – many of them former refugees – as well as Cypriots, Egyptians, Pacific islanders, New Zealanders and many others. (There has been a sizable number of Chinese since Gold Rush days.)

The ties with Britain, once so strong, are now much weaker, although British and Irish stock still predominates. Once the emphasis was on integrating the 'new Australians' into the Australian way of life. In the last 20 years, 'old' Australians have increasingly appreciated how the non-English-speaking immigrants can enrich Australian life. Where once steak and chips was standard restaurant fare, there is now cooking as diverse as Japanese, Cambodian, Fijian, Mexican, Caribbean, African and Lebanese, as well as every type of European. In literature and the arts, from folk festivals to newspapers and broadcasting, multiculturalism is official policy. Altogether, some 100 languages are spoken in today's Australia, and many of them can be read in daily or weekly newspapers or heard on radio stations or the television channel of the Special Broadcasting Service.

But as the population mix has grown, so has some racial tension – largely due, many observers believe, to competition for jobs. Immigration was reduced as unemployment rose in the 1980s, and many recent immigrants have fared less well than the previous generation. Naturally more prominent in a still predominantly white society, Asian immigrants have been the objects of prejudice once shown to Italians and Greeks.

STATE AND FEDERAL GOVERNMENT

Australia as a nation is less than a century old. The six states had been self-governing since the 1850s (1890 in the case of Western Australia) when they formed the Commonwealth (federation) of Australia in 1901. The

◀ **RED CENTRE Ayers Rock – sacred Uluru to Aborigines – glows in the sunset, adding to its mystic quality. Like the nearby Olgas, it is the tip of a sandstone hill half-submerged in a semiarid plain.**

constitution divides power between the federal parliament (consisting of the Senate, with equal representation for each state, and the House of Representatives, with membership in proportion to the states' population) and the states, each one having its own parliament. Elizabeth II is head of state and is represented by the governor-general and in each state by a governor. The AUSTRALIAN CAPITAL TERRITORY (in which Canberra stands) was formed in 1911 and the Northern Territory was also placed under direct federal government control; the latter has had limited self-government since 1978.

Power in Canberra has been held for varying periods by the Labor Party and by a coalition of the Liberal Party (conservative) and National (formerly Country) Party, which is strongest in rural communities. There was a constitutional crisis in 1975 when the governor-general, Sir John Kerr, dismissed Gough Whitlam's Labor government after the Senate threatened to block budget legislation. Many Australians felt that Kerr exceeded his powers, and the incident increased support for those – still a minority – who want Australia to become a republic.

After seven years in opposition, Labor returned to power in 1983 under Bob Hawke, a former trade union leader on the right or moderate wing of his party. His economic policies, in particular, have been more centrist than those of most previous Labor administrations. The rise in support for feminism, nuclear disarmament and environmental issues has complicated the political scene, with the rise of minor parties.

The federal and state governments face growing social problems, especially drug abuse, an increase in organised crime (with its involvement in drugs, illegal gambling and other rackets), and corruption among the professions, police and in politics. However, good relations between the federal government and trade unions, leading to an agreement on wages and prices, have reduced strikes and inflation and allowed the economy to flourish despite high interest rates.

State and federal governments have often been at loggerheads, especially where different parties are in control. Recent areas of conflict have included welfare and education, which is partly funded from Canberra but run by the states. Schooling is compulsory to the age of 15 or 16, depending on the state. Nearly one-quarter of all schools are non-government, the majority Roman Catholic.

THE OUTDOOR LIFE

Australia is culturally and materially part of the Western world. The average family regards a car as a necessity – there are more cars per head than in any major country except the USA – and owns a range of modern gadgetry and electronic goods (increasingly imported from Japan). Over two-thirds of households own or are buying their own home, mostly in huge sprawling suburbs.

Australians love the outdoor life, and millions spend sunny Sundays at the beach or having a barbecue in a suburban park. The fine climate provides opportunities for most sports – from surfing to skiing, football to fishing – giving an outlet for Australians' highly developed competitive urges, whether as participants or spectators. There is enthusiastic support for cricket (especially against the old enemy, England), rugby football (League and Union), Australian Rules football (using an oval pitch and ball), soccer, tennis and horse-racing. Australians are keen gamblers, wagering an average of A$400 (about US$300) per person each year on sporting events, lotteries and other forms of betting.

Nor is the artistic world ignored, despite Australia's old image overseas as a cultural wilderness. Most of the major cities have multifunction cultural centres, with theatres, art galleries and concert halls. Home-grown orchestras, opera, theatre and ballet companies and a thriving film industry, offer Australian performers increasing opportunities, but remnants of the 'cultural cringe' – a kind of national inferiority complex regarding the arts, once strong – still linger on despite a more aggressive promotion of Australia's national identity.

AUSTRALIA AT A GLANCE

Area 7 682 300 km² (2 966 140 sq miles)

Population 15 850 000

Capital Canberra

Government Federal parliamentary monarchy

Currency Australian dollar = 100 cents

Languages English, Aboriginal languages; numerous others among immigrant groups

Religion Christian (52% Protestant, 25% Roman Catholic, 3% Greek Orthodox)

Climate Ranges from tropical monsoon to cool temperate; large areas subtropical or warm temperate; much of centre desert or semiarid. Average temperature in Sydney ranges from 8-16°C (46-61°F) in July to 18-26°C (64-79°F) in January.

Main primary products Sheep, cattle, pigs, wheat, barley, oats, sorghum, rice, sugar cane, fruit, vegetables, tobacco, timber, fish; coal, iron ore, bauxite, copper, lead, zinc, nickel, uranium, tungsten, tin, manganese, gold, diamonds, opals, oil and gas

Major industries Agriculture, mining, iron and steel, aluminium refining and smelting, vehicles, machinery, processing of food, wool and hides, forestry, fishing, oil and gas production and refining, chemicals, cement, light engineering

Main exports Coal, wool, wheat, iron ore, alumina and aluminium, meat (mainly beef), other minerals, petroleum products, manufactured goods

Annual income per head (US$) 9518

Population growth (per thous/yr) 13

Life expectancy (yrs) Male 72 **Female** 77

Australian Capital Territory ***Australia*** Region around the capital, CANBERRA, bought from the state of New South Wales in 1911 by the Commonwealth of Australia government following the federation of the colonies in 1901. It is located about 250 km (150 miles) south-west of Sydney and covers 2432 km² (939 sq miles), including a 73 km² (28 sq mile) naval base at Jervis Bay on the New South Wales coast. The main rural activity is sheep farming.
Population 240 000
Map Australia Hf

Austria See p. 60

Auvergne ***France*** Largely rural region of some 26 000 km² (10 000 sq miles) in the Massif Central, where winters are harsh. It consists of glacially eroded uplands, volcanic peaks and rocky lowlands. The highest peak is the Puy de Sancy (1885 m, 6185 ft). Farming and tourism are the main occupations in the thinly populated mountains, with rich farming and modern industries such as rubber and electrical goods in the low-lying Limagne plain. The region contains many spas, including La Bourboule, Le Mont-Dore, and VICHY. CLERMONT-FERRAND is the main city.
Population 1 333 000
Map France Ed

Auyuittuq ***Canada*** National park covering 21 470 km² (8290 sq miles) that was established in 1972 on the Cumberland Peninsula of eastern Baffin Island. It contains spectacular fiords as well as barren tundra and glaciers.
Map Canada Ib

Avarua ***Cook Islands*** See RAROTONGA

Avebury ***United Kingdom*** Prehistoric monument in the English county of Wiltshire, about 120 km (75 miles) west of London. It is the largest stone circle in Europe, 410 m (1350 ft) across, with some 154 standing boulders; another 100 pairs lead off in an avenue to the south to a smaller circle, now destroyed. The circle was built in about 2500 BC, probably as a religious centre, and now encloses the village of Avebury. There are enormous earthworks, including burial mounds, nearby.
Map United Kingdom Ee

Aveiro ***Portugal*** Quiet town on a lagoon, about 60 km (35 miles) south of the city of Oporto. It was a prosperous cod-fishing port until a violent storm silted up the harbour in 1575. The fishing industry revived in the 19th century, when a canal was built to the sea. Saltpans and rice cultivation are other mainstays of the town's economy, and there is a steelworks at Ovar, about 30 km (20 miles) to the north.
Population 29 200
Map Portugal Bb

Avgustov ***Poland*** See AUGUSTOW

Avignon ***France*** Capital of Vaucluse department, near the confluence of the Rhône and Durance rivers, 85 km (53 miles) north-west of Marseilles. It was the seat of seven popes during the 14th century and of the Avignon popes during the Western Schism (1378-1417).

Today, Avignon is one of the liveliest cities in France, to which it was annexed in 1790. It flourishes on commerce and tourism, and holds a drama festival each July. The Palace of the Popes still stands. Also of interest are the city's well-preserved ramparts, built by Popes Innocent VI and Urban V between 1355 and 1365. Part of the famous bridge, immortalised in the folk song, also survives. The English philosopher John Stuart Mill (1806-73) is buried in the Avenue Cemetery.
Population 177 500
Map France Fe

Ávila ***Spain*** Town about 95 km (60 miles) west of Madrid. It was the birthplace and home of St Teresa (1515-82). After an ecstatic vision she founded the Discalced (barefoot) Order of Carmelites and many of the town's religious buildings are associated with her. Ávila, founded in 1090, still retains its 2.5 km (1.5 miles) of walls with 88 round towers. Its surrounding province, of the same name, is the highest in Spain, mostly lying over 1000 m (3280 ft) above sea level.
Population (town) 41 800; (province) 179 000
Map Spain Cb

Avon ***United Kingdom*** County covering 1338 km² (517 sq miles) of western England named after the river which flows through it. It was formed in 1974 from parts of Gloucestershire and Somerset to give administrative unity to the towns of BRISTOL and BATH, formerly split between the two counties. It includes the Mendip Hills and the southern Cotswolds. In the west, it stretches to the shore of the Bristol Channel, where there is a string of seaside resorts and commuter towns for Bristol, such as Weston-super-Mare.
Population 936 000
Map United Kingdom De

▲ STARK SENTINELS Part of the huge prehistoric stone circle at Avebury. Stones came from the nearby Marlborough Downs and weigh up to 40 tonnes each.

Awash ***Ethiopia*** River rising 80 km (50 miles) west of the capital, Addis Ababa, and flowing about 600 km (375 miles) into Lake Abbe on the border with Djibouti. It provides hydro-electricity for the capital and vital irrigation on the arid eastern plains.
Map Ethiopia Bb

Axios (Vardar) ***Yugoslavia/Greece*** Major Balkan river that rises in Yugoslavia and flows 388 km (241 miles) through Macedonia to the Thermaic Gulf near Saloniki. Its valley has long been a trade and invasion route, and its waters were diverted in the 1950s and 1960s to irrigate the large plain of Macedonia.
Map Yugoslavia Fd; Greece Ca

Axum (Aksum) ***Ethiopia*** One of Africa's most important historic towns, in Tigray province about 500 km (310 miles) north-east of the capital, Addis Ababa. It is thought to have been the royal city of the legendary Queen of Sheba, who visited the King of Israel, Solomon, in the 10th century BC, and to have been the city where the Jewish Ark of the Covenant was once kept. The ark was believed to have been taken from the temple in Jerusalem by Emperor Menelik I, the son of Solomon and the Queen of Sheba. It was certainly a royal capital from about 500 BC, and a centre of Christianity in AD 300. The town has many granite obelisks up to 20 m (65 ft) high. The obelisks are carved with pictures of multistorey houses of a type still found across the Red Sea in Yemen, from where the early kings came.
Population 20 000
Map Ethiopia Aa

Austria

ONCE THE HUB OF AN EMPIRE THAT DOMINATED CENTRAL EUROPE, THIS ALPINE REPUBLIC STANDS AS A POLITICAL BUFFER BETWEEN EAST AND WEST

The wall of mountains that runs across the centre of Austria dominates the nation's economy as well as its scenery. The land of the edelweiss, of dirndl dresses and leather breeches (lederhosen), draws around 14 million visitors a year, making tourism a major industry. They come in the summer to wander the forests, pastures and mountains around scores of glittering lakes and hundreds of tumbling rivers. And they come in the winter to swoop above more than 50 ski resorts such as St Anton, KITZBUHEL and INNSBRUCK, one of the venues for the Winter Olympics.

For the Austrians themselves – a conservative and largely Roman Catholic people – the mountains are an irreplaceable national asset, treasured and protected from garish modernity. Farmers, for instance, are subsidised by the government to encourage them to maintain the scenic appeal of their land. And many villages lay down stringent bylaws to make sure that new buildings fit in with the local style of the traditional wood-faced, broad-eaved chalet. In addition, some 3000 km² (nearly 1200 sq miles) of the ALPS have been set aside as protected areas.

Even for the people of Austria's larger towns and cities, the mountains are an irresistible lure. There are some 300 000 second homes in Austria – one for about every 25 people – and most are hillside chalets used for weekends and holidays. The mountains are also a source of energy for factories and towns: around two-thirds of Austria's power is generated from hydroelectric dams fed by streams from the Alpine snows.

SMALL IS BEAUTIFUL

Industrially, Austria practises the 'Small is beautiful' philosophy of the 20th-century German-born economist Ernst Schumacher. In the whole country, only 77 firms employ more than 1000 people; the average size of a factory workforce is just 20. The farms, too, tend to be small. Despite a spate of farm mergers in recent years, there are still 124 000 farms – most run by a single family.

The formula seems to work. Dairy products, beef and lamb from the hill farms supply the cities and contribute to exports as well. Truckloads of wheat and other crops flow in from the eastern lowlands. Wine, predominantly white, pours from the vineyards, which are mostly in LOWER AUSTRIA (Niederösterreich) and BURGENLAND.

One of the big difficulties for a small country in a world of big businesses is how to compete effectively when you have no really large home market. Austria is, after all, only twice the size of its neighbour, Switzerland. The Austrian solution has been to increase specialisation and the quality of products.

The iron and steel industry, for instance, could hardly compete with the big producers of West Germany, and so the answer of the state-run steel company has been to develop more efficient ways of making steel. As a result, the steel towns of LINZ and Donawitz have given their name to one of the most up-to-date methods of steel-making in the world: the Linz-Donawitz process.

Nearly half of Austria's annual production of 3.1 million tonnes of pig iron and 4.6 million tonnes of crude steel is exported. Other successful export industries are optical instruments, which are assembled near SALZBURG, and paper, which is manufactured near GRAZ from Austria's abundant supplies of timber. Forests cover more than 37 per cent of Austria – a greater proportion than in any other European country except Sweden and Finland.

ATTRACTIONS FOR INDUSTRY

Large multinational firms have also been attracted to the country in recent years by the low strike record of the labour force. There were only two strikes in 1982, the second lowest total in Europe after Switzerland. The large firms, such as General Motors and Philips, frequently supply parts to the West German economy rather than build complete assemblies in Austria.

Unemployment is very low compared with other Western European countries and, standing at 3.7 per cent, it is bettered only by Switzerland, Iceland, Norway and Sweden. This has been achieved partly because the working week is slightly shorter than elsewhere: an average of 36-37 hours.

Despite the modernity of Austria's industries, the attachment to local customs is still strong. In the rural areas, men commonly wear lederhosen and jackets or capes made of a woollen material (loden). Women's regional costumes – including the full-skirted, tight-bodiced dress known as the dirndl – can still be seen on feastdays and holidays.

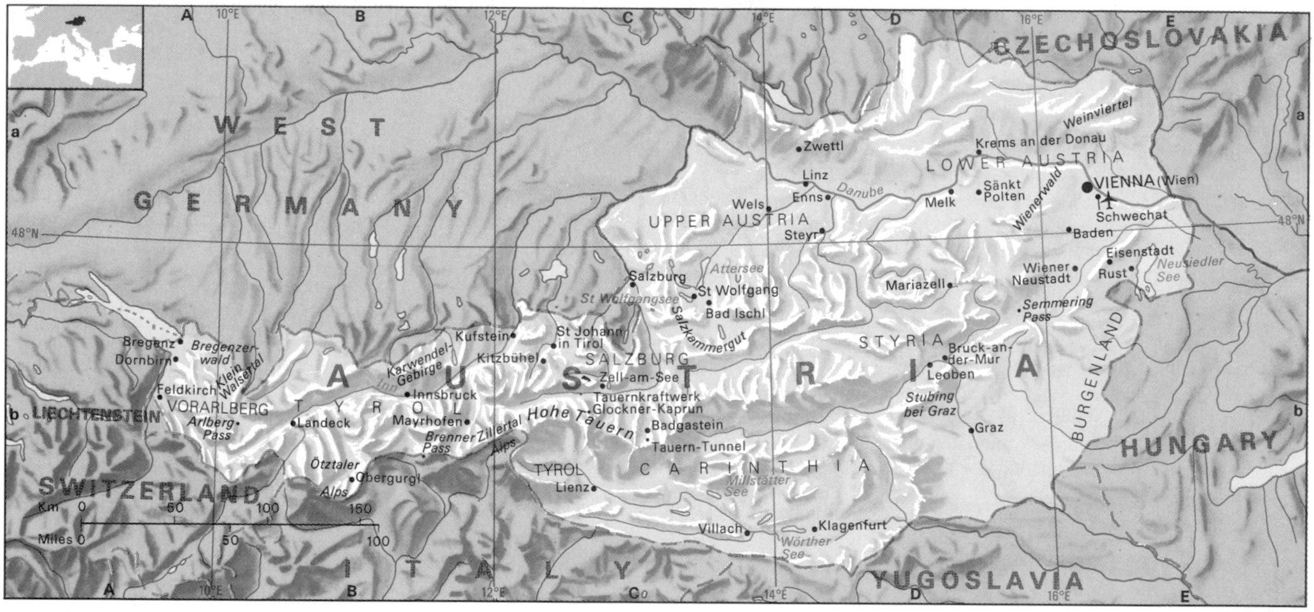

INFLUENCE OF THE CHURCH

With 88 per cent of the population Roman Catholic, the Church is a powerful influence. Crucifixes and statues of the Virgin are conspicuous outdoor features, especially in the mountain villages. The rural areas also contain the best examples of the traditional farmhouses, which vary from region to region. Examples of each type can be seen together in the Österreichisches Freilichtmuseum (Austrian open air museum) near Graz.

The country's emphasis on regionalism – typical of a mountain culture where each valley developed largely in isolation from its neighbours – is reflected in Austria's political structure. The country is a federation of nine *Länder*, or states, each with its own government. Each *Land* is represented in the Federal Council, which is responsible for national and international affairs. National government has been in the hands of the Austrian Socialist Party since 1970. It is heavily involved in providing public housing and in the administration of 82 profitable nationalised industries ranging from steel and chemicals to transport and banking.

Internationally, Austria is firmly neutral – a political buffer between the NATO countries of Western Europe and the Warsaw Pact countries of the Soviet bloc. It is not a member of NATO or the Common Market, nor of their eastern counterparts. As a result, the capital, VIENNA, has become one of the world's great meeting places. The International Centre, in Vienna's Donaupark (Danube Park), is the third major UN complex in the world after New York and Geneva. International agencies, such as the Industrial Development Organisation, the International Atomic Energy Agency and the Organisation of Petroleum Exporting Countries, all have their headquarters there. As a neutral state, Austria also attracts refugees – particularly from Communist Czechoslovakia. Not surprisingly, Vienna is host to a number of refugee agencies, among them the Intergovernmental Committee for European Migration and the United Nations Relief Agency for Palestine Refugees.

Neutrality, like the sparkling elegance of Vienna, is a product of Austria's turbulent history. Settled originally by Celts in about 500 BC, the land that is now Austria was fought over by Romans, Vandals, Visigoths, Huns, Hungarian Magyars and Germanic tribes. Then, in 1246, a remarkable Swiss-Alsatian family came to power – the Hapsburgs. They made Vienna their capital and built around it an empire that by 1530 included Hungary, the western part of Czechoslovakia, northern Italy, the Netherlands and the whole of Spain. The head of the family in each generation was usually elected Holy Roman Emperor, and Hapsburg power endured for more than 600 years.

SHOT THAT STARTED A WAR

Conflict and bitterness between the various nationalities of the empire came to a head on June 28, 1914, when a pistol shot echoed round the world from SARAJEVO, now in Yugoslavia. The shot, fired by a Serbian nationalist, killed the heir to the Hapsburg throne, Archduke Franz Ferdinand. Austro-Hungarian forces invaded Serbia; other European nations took sides in the conflict, and started the First World War.

Hapsburg power did not survive the upheaval. By the treaty of Saint Germain-en-Laye in 1919, seven-eighths of the Hapsburgs' Austro-Hungarian Empire was parcelled up among its neighbours. One-eighth, the German-speaking rump, became the republic of Austria. The republic was a disaster. Economically, it teetered on the brink of bankruptcy, and chronic unemployment led to political unrest. Faction-fighting between socialists and conservatives led to outright civil war in 1934 and a brief right-wing dictatorship under Engelbert Dollfuss. Nazi sympathisers assassinated him after only a few months, however, and in 1938 Adolf Hitler – himself an Austrian by birth – sent in German troops to occupy and annex the country.

▲ SKI PARADISE Colourful 16th-century houses line the north bank of the Inn at Innsbruck below the Nordkette, one of Tyrol's many skiing areas. The city, popular with both winter and summer visitors, hosted the Winter Olympics in 1964 and 1976.

Liberated by the Allies in 1945, Austria was allowed to set up its own government again in 1955, after guaranteeing that the country would be strictly neutral – an undertaking which remains in force today.

MAGNIFICENT CAPITAL CITY

Vienna, straddling the great waterway of the River DANUBE, is full of echoes of the past. Largely and magnificently rebuilt by the Hapsburg emperors in the late 18th and early 19th centuries, the city is laid out on an imperial scale quite disproportionate to Austria's present size and wealth. More than 4.5 million people visit the city each year – more than twice as many as live there. Many of the tourists are from Eastern Europe, particularly Hungary, Austria's former imperial partner. The attractions are the Hofburg and the Schönbrunn palaces, the cavernous Stephansdom, the great church of Karlskirche and the lavish Opera, home of the Viennese waltz made famous by two composers, a father and a son both named Johann Strauss. This city of music inspired other composers: Mozart, Beethoven, Brahms, Schubert, Haydn and Mahler all lived there for long periods.

Vienna has less highbrow attractions, too: comfortable pavement cafés and coffee houses, and jovial *Heurigen* – small winebars which are often in farmhouses near the city and which specialise in selling new wines.

AUSTRIA AT A GLANCE

Area 83 853 km² (32 375 sq miles)
Population 7 540 000
Capital Vienna
Government Parliamentary republic
Currency Schilling = 100 groschen
Languages German (99%), plus 1% Magyar and Slovene
Religion Christian (88% Roman Catholic)
Climate Temperate continental; average temperature in Vienna ranges from −4 to 1°C (25-34°F) in January to 15-25°C (59-77°F) in July
Main primary products Cattle, sheep, wheat, maize, potatoes, hay and fodder, barley, sugar beet, vines, temperate fruits, timber; oil and natural gas, lignite, iron ore, magnesite, graphite
Major industries Agriculture, iron and steel, machinery, forestry and wood products, chemicals, textiles, oil and gas production and refining, wine, beer, food processing
Main exports Iron and steel, machinery, timber and wood products, chemicals, textiles, meat and dairy produce
Annual income per head (US$) 7737
Population growth (per thous/yr) Declining
Life expectancy (yrs) Male 69 **Female** 77

Ayacucho ***Peru*** Southern department in the Andes mountains which was the scene on December 9, 1824 of the Battle of Quinua, in which Antonio José Sucre defeated the Spaniards and won independence for Peru from Spain. Its capital, also Ayacucho, founded in 1539 by the Spaniards, produces textiles, wine and pottery. The site of Huari near Quinua has impressive pre-Columbian circular buildings, probably dating from the 8th century.

Population (department) 503 400; (city) 69 500
Map Peru Bb

Ayers Rock ***Australia*** One of the largest monoliths in the world, lying in the Northern Territory 310 km (200 miles) south-west of Alice Springs. It rises 348 m (1142 ft) above the surrounding plain and is more than 8 km (5 miles) around the base.

The rock, the summit of a vast buried sandstone hill, is an Aboriginal sacred site, and caves in the rock contain Aboriginal paintings. It is named after Henry Ayers, a 19th-century premier of South Australia. In 1985, ownership of the rock was restored to the local Aborigines, whose name for it is Uluru.

More than 50 000 people visit the red rock each year. It is at its most spectacular at sunset, when it glows like a burning coal on the flat desert. Uluru National Park, covering an area of 1325 km² (512 sq miles) encompasses Ayers Rock and the nearby OLGAS.

Map Australia Ed

Ayion Oros ***Greece*** See ATHOS

Áyios Nikólaos ***Greece*** Town on the Mirabello Gulf on the northern coast of Crete. Fishing boats fill its little harbour and tourists swim and windsurf from the nearby sandy beaches at Elounda to the north and Lerapetra to the south. At Kritsá, 11 km (7 miles) to the south, is a 13th-century frescoed church and the 7th-century BC city of Lato.

Population 8100
Map Greece Dd

Ayr ***United Kingdom*** Port, holiday resort and former county of south-west Scotland, in Strathclyde region 50 km (30 miles) south-west of Glasgow. The poet Robert Burns was born in the Ayr suburb of Alloway in 1759; his cottage can be visited.

Population 49 000
Map United Kingdom Cc

Ayutthaya ***Thailand*** Town 80 km (50 miles) north of Bangkok, and the royal capital from 1350 to 1767, when it was sacked by the Burmese. The town lies at the confluence of the Chao Phraya, Lopburi and Pa Sak rivers, which are joined to form a moat around the town.

As a royal capital, Ayutthaya became a busy international port, and several European trading posts were set up outside its walls in the 17th century. They traded in teak, sandalwood, sugar, leather, ivory, hides, silks and local handicraft goods, and also in goods brought in from Japan and China. A late 17th-century Dutch visitor recorded that the gilding on the roofs of the temples and pagodas reflected the light so strongly that it hurt the eyes even from several kilometres away.

Most of this magnificence was destroyed, together with manuscripts recording Thailand's early history, in the Burmese invasion. The pretext for the war between the two hostile kingdoms was the refusal of the Thai king to let his rival have one of his white elephants, which were treasured by royalty for their rarity and as symbols of good luck.

Some treasures, including some of the temples with their gold and silver ornamental contents, did survive the invasion. With the old city walls, they make Ayutthaya a tourist centre today. The town is also a market for rice grown in the surrounding province of the same name.

Population (town) 113 300; (province) 644 100
Map Thailand Bc

Azad Kashmir ***Pakistan*** Disputed area in the former princely state of JAMMU AND KASHMIR which has been governed by Pakistan since a United Nations ceasefire between India and Pakistan in 1949. Azad means 'free'. The Karakoram Highway, built with Chinese help, passes through the region to connect Pakistan with the Xinjiang region of China. The area is dominated by the towering peaks of the Himalayas, including K2 – at 8611 m (28 250 ft), the world's second highest mountain after Everest.

Map Pakistan Da

Azarbaijan ***Iran*** Mountainous region in north-west Iran, separated from Soviet Azerbaidzhan by the Araks river. Its highest point is the volcano of Sabalan, east of the city of Tabriz, at 4811 m (15 783 ft).

The region is divided into two provinces: Azarbaijan Gharbi (capital Orumiyeh), bordering Turkey and Iraq; and Azarbaijan Sharqi (capital Tabriz), bounded by the Caspian Sea. The region's main farm products are sheep, wheat, cotton and tobacco.

The principal language of the region, Azeri, derived from Turkish, is also spoken across the border in the USSR. The population includes Armenians in the north and Kurds in the south.

Population 4 613 000
Map Iran Aa

Azerbaidzhan (Azerbaydzhan) ***USSR*** Constituent republic of the USSR in the east Caucasus, bordering Iran and the Caspian Sea; area 87 000 km² (33 600 sq miles). It is semi-arid and produces cotton, wheat, maize, potatoes, tobacco, tea and citrus fruits with the aid of irrigation. The chief towns are the capital, BAKU, Kirovabad and Sumgait. The republic's Apsheron oil field was, until 1950, the richest in the USSR, when it was overtaken by the Volga-Urals field. Oil was being collected at Baku before AD 800, and burning natural gas seeping from the ground gave Azerbaidzhan its name – 'Land of Flames'. Iron ore deposits are worked at Dashkesan. The people speak their own language and are predominantly Muslim. They have fought invaders – Persians, Arabs and Tatars – through much of their history. The territory was divided between Russia and Persia (Iran) in the 19th century.

Population 6 506 000
Map USSR Fe

Azores (Açores) ***Portugal*** Block of volcanic islands in the North Atlantic, 1290 km (800 miles) off the west coast of Portugal. It consists of three groups: Flores (148 km², 57 sq miles) and Corvo (18 km², 7 sq miles) in the west; Faial (165 km², 64 sq miles), Pico (430 km², 166 sq miles), São Jorge (104 km², 40 sq miles), Graciosa (46 km², 18 sq miles) and Terceira (545 km², 210 sq miles) in the centre; and São Miguel (770 km², 297 sq miles) and Santa Maria (74 km², 29 sq miles) in the east.

They were colonised by the Portuguese in the mid-15th century. The economy is based on agriculture, fishing and in the past whaling, and the islands are becoming increasingly popular as a holiday resort.

Portugal's highest mountain, Pico Alto (2351 m, 7712 ft), is on the island of Pico. The capital of the Azores is Ponta Delgada on São Miguel.

Population 336 100
Map Atlantic Ocean Dc

Azov ***USSR*** Small town and fishing port near the mouth of the Don river where it drains into the shallow Sea of Azov. The town, near the site of the ancient Greek colony of Tanais, was fortified by the Genoese in the 13th century and made a trading port for Oriental goods. The Mongol ruler Tamerlane (about 1336-1405) sacked it in 1395, and in 1739 it became part of Russia. Its industries today include canning, textiles and farm machinery. The town's boats fish the Sea of Azov for herring, grey mullet, anchovy and sturgeon – the source of caviar.

Population 76 000
Map USSR Ed

Azov, Sea of Shallow sea connected to the BLACK SEA by the Strait of Kerch'. Its leading ports include KERCH', ROSTOV-NA-DONU, TAGANROG and ZHDANOV in the USSR. Fishing is important in the south, but the evil-smelling marshes and lagoons at the western end have earned the sea the name Sivash (Putrid Lake).

Area 36 260 km² (14 000 sq miles)
Greatest depth 15 m (49 ft)
Map USSR Ed

Azuay ***Ecuador*** Province of the southern Sierra containing the Cuenca basin, one of the country's most fertile areas. The provincial capital is CUENCA.

Population 441 000
Map Ecuador Bb

azurite Azure-blue, glassy mineral composed of copper carbonate. It is a source of copper and is used as a gemstone.

Baalbek ***Lebanon*** Situated in the Beqa'a valley, 64 km (40 miles) north-east of Beirut, Baalbek was a city of some size and importance in ancient times. It was a centre of worship of the Semitic sun-god Baal, which led the later Greek colonisers to call the city Heliopolis – 'City of the Sun'.

Baalbek has some of the most impressive Roman ruins in the world. The Great Temple dedicated to the god Jupiter (the supreme Roman deity) has six huge columns – 27 m (88 ft) high and 7 m (22 ft) round – still standing. Originally, it was surrounded by 54 such columns. There is also an almost intact temple to Bacchus, the god of wine, and a temple to Venus, the goddess of love. The site contains the world's largest cut masonry stone, weighing 1500 tonnes and 18.2 m (60 ft) long.

Map Lebanon Ba

Babia Góra (Babia Hora) ***Czechoslovakia/ Poland*** See BESKIDY

Babylon ***Iraq*** Ruined city dating from before the 18th century BC, about 90 km (55 miles) south of the capital, Baghdad. It is the site of the Hanging Gardens, one of the Seven Wonders of the Ancient World, and was rebuilt in the 6th century BC by Nebuchadnezzar II. The remains of the king's palaces are also visible, as are the foundations of a 90 m (300 ft) high ziggurat, or stepped pyramid, built by Nebuchadnezzar and thought by some scholars to be the inspiration for the Biblical story of the Tower of Babel. The Macedonian conqueror Alexander the Great died in the city in 323 BC.
Map Iraq Cb

backing Anticlockwise change of direction of a wind, for example, from south-west through south to south-east. A change in the opposite direction is called veering.

Bacolod ***Philippines*** See NEGROS

Bad Hofgastein ***Austria*** See BADGASTEIN

Bad Homburg ***West Germany*** Spa town 14 km (9 miles) north of Frankfurt am Main. In the 1890s, the Prince of Wales (later King Edward VII of England) borrowed the headgear of a local militiaman and popularised the soft-felt Homburg hat.
Population 50 700
Map West Germany Cc

Bad Ischl ***Austria*** Spa town on the Traun river, 45 km (28 miles) south-east of Salzburg. It was the summer residence of the Emperor Franz Josef (1830-1916), whose court was attended by such eminent operetta composers as Johann Strauss and Franz Lehár. Lehár's house is open to the public, and each year the town holds an operetta week.
Population 13 100
Map Austria Cb

Badacsony ***Hungary*** Extinct volcano rising to 464 m (1522 ft) on the north shore of Lake Balaton in western Hungary. Striking columns of black basalt rock, up to 61 m (200 ft) high, and fantastic lava formations adorn its flat top. The rich volcanic soils of the area produce fine dessert grapes and Badacsonyi white wines. The small, lakeside resort village of Szigliget nearby is tucked beside a sandy beach below a Baroque 18th-century castle on a hilltop 183 m (600 ft) above the lake.
Map Hungary Ab

Badajoz ***Spain*** Town about 325 km (200 miles) south-west of Madrid on the Portuguese border. Its industries include brewing, distilling and food processing. Badajoz has many Roman remains, including two aqueducts. It was the capital of a Moorish kingdom in the 11th century, and was bloodily stormed and pillaged by the British under Sir Arthur Wellesley, later Duke of Wellington, in 1812 when they seized it from the French in the Peninsular War.

Badajoz province is Spain's largest at 21 657 km² (8361 sq miles).
Population (town) 114 400; (province) 635 400
Map Spain Bc

Baden ***Austria*** Spa town in Lower Austria, 26 km (16 miles) south-west of Vienna. Its hot sulphur springs have been used for medicinal purposes since Roman times. The German composer Ludwig van Beethoven lived in the town in the early 19th century and there is a museum devoted to his life and works. After the Second World War, Baden was the headquarters of the Soviet Zone of occupation from 1945 to 1955.
Population 23 300
Map Austria Ea

Baden-Baden ***West Germany*** Spa town on the Rhine valley slopes of the Black Forest, 70 km (43 miles) west of Stuttgart. The Romans built the first baths there above its hot springs. The town's modern bathhouse dates from 1821.

In the 19th century, Baden-Baden was known as the summer capital of Europe, because of the number of royalty visiting it. At its notorious casino, closed in 1872, the Russian novelist Feodor Dostoevsky (1821-81) gambled wildly while writing *The Idiot*.
Population 50 000
Map West Germany Cd

Baden-Württemberg ***West Germany*** Industrial and farming state in the south-west corner of the country, flanked by France and Switzerland. Its capital is STUTTGART. It contains the BLACK FOREST and Swabian JURA.
Population 9 241 000
Map West Germany Cd

Badgastein ***Austria*** Small spa town and ski resort on the northern edge of the Hohe Tauern mountains 76 km (47 miles) south of Salzburg. About 8 km (5 miles) to the north is the hot spring spa of Bad Hofgastein.
Population 5600
Map Austria Cb

badlands Area of barren land characterised by roughly eroded ridges, peaks and plateaus. The name is derived from the Badlands of the western United States – in Nebraska, South Dakota and Wyoming – where the large stretches of heavy clay have been dramatically eroded.

Badlands National Park ***USA*** Area of severely eroded land covering 985 km² (380 sq miles) in south-western South Dakota. Its steep gullies between flat-topped hills are etched out of brightly coloured, layered sandstones. The park contains many prehistoric animal remains.
Map United States Fb

Baegdu Son ***China/North Korea*** See PAEKTU SAN

Bafatá ***Guinea-Bissau*** Commercial centre for the interior, at the highest navigable point on the Gêba river, about 110 km (70 miles) from Bissau. It is in the midst of an area producing groundnuts, cattle and, recently, cotton and tobacco. It was one of the last Portuguese strongholds before independence.
Population (region) 116 000
Map Guinea Ba

Baffin Bay Part of the ATLANTIC OCEAN, separating Baffin Island from Greenland, and named after the English navigator William Baffin (1584-1622) who explored it in 1616. The cold Labrador Current, carrying dangerous icebergs, flows through the bay, which is only ice-free for a short time in summer.
Dimensions 1130 km (700 miles) long, 110-640 km (68-400 miles) wide
Map Canada Ia

Baffin Island ***Canada*** Fifth largest island in the world at 507 451 km² (195 927 sq miles), in the Canadian arctic archipelago. It is inhabited mainly by Inuit (Eskimos), who live mostly in settlements scattered along the coast. Trapping and fishing are the chief occupations.
Map Canada Ha

Bafing ***West Africa*** See SENEGAL (river)

Bafoussam ***Cameroon*** Industrial town and provincial capital 220 km (135 miles) north-west of the capital, Yaoundé. The town stands in an area of subsistence farming, with some coffee also grown as a cash crop. It is processed in Bafoussam for export. The Bamendjing dam with its restful lake is a tourist attraction.
Population 75 000
Map Cameroon Bb

Bagamoyo ***Tanzania*** Port about 60 km (37 miles) north-west of Dar es Salaam, and opposite the island of Zanzibar. In the 19th century it was a centre of the Arab slave and ivory trades, and a base for European explorers, but is now overshadowed by Dar es Salaam.
Population 25 000
Map Tanzania Ba

Baghdad ***Iraq*** National capital, straddling the banks of the Tigris river. Founded in about AD 762, it became the city of the legendary tales of the *Thousand and One Nights*, which includes the stories of Aladdin, Sinbad the Sailor and Ali Baba and the 40 Thieves.

Although it was known as The City of Peace, Baghdad has a violent history. It was sacked in 1258 by the Mongols, conquered by the Tatar warlord Tamerlane in 1401, captured by the Ottoman sultan Suleiman the Magnificent in 1534, taken again by the Turks in 1638, and captured by the British in the First World War. In 1921 it became the capital of independent Iraq. Its industries today include distilling spirits for tourists (Muslims do not drink), tanning, tobacco and food processing, and the manufacture of clothing and cement.

The city's places of interest include colourful street bazaars, the 13th-century Mustansiriya School (Muslim University), remains of the city's ancient walls and gates, and the 14th-century Khan Murjan, the only completely roofed khan (inn) in Iraq.
Population 3 300 000
Map Iraq Cb

Baguio ***Philippines*** Summer capital and popular hill resort on Luzon, in the southern Cordillera Central about 210 km (130 miles) north of MANILA. It is set in pine forests and has many official buildings, hotels, restaurants and parks. Thanks to the cool climate, temperate fruit and vegetables grow well. About 75 km (47 miles) north-east, at Banaue, are extensive rice terraces which are some 2000 years old.
Population 119 009
Map Philippines Bb

Bahamas

PLAYGROUND OF THE NORTH AMERICANS, THIS CARIBBEAN ARCHIPELAGO HAS DEVELOPED TOURISM ON A VAST SCALE

Nearly 500 years ago, the Italian navigator, Christopher Columbus, made his first landfall in the New World on San Salvador (also called Watling Island) in the Bahamas. Precisely where on the island he landed is not known, however; San Salvador has no fewer than four monuments, each claiming to mark the exact spot where the great discoverer stepped ashore.

San Salvador is just one of the 700 islands and 2000 coral cays – or islets – that compose the Bahamas. Stretching in a 1200 km (about 750 miles) arc from the eastern edge of Cuba to within 80 km (50 miles) of south-eastern Florida, they range in size from ANDROS island which measures 4144 km² (about 1600 sq miles) to tiny uninhabited dots in the ocean. The two most populated islands are NEW PROVIDENCE, where the capital NASSAU lies, and GRAND BAHAMA. New Providence, the more densely populated of the two with over 150 000 people, is slightly smaller than Grand Bahama, with just over 27 000 people. Together they account for nearly 75 per cent of the Bahamas' population.

DESCENDANTS OF SLAVES

More than 80 per cent of the Bahamas total population of 237 000 people are the descendants of slaves brought from Africa, largely in the 18th century. The original inhabitants of the islands were Arawak Indians, but they were all taken away by the Spanish to work as slaves in HISPANIOLA. The Spanish themselves never settled in the Bahamas; the first European settlers were a curious combination of Puritans, escaping religious persecution in 17th-century England, and pirates, taking advantage of the thousands of hidden, well-protected anchorages. They were later joined by Loyalists escaping from America after the American Revolution (1775-83) and later by Southerners escaping from the American Civil War (1861-5). The Bahamas were a British colony until 1973 when they became an independent member of the Commonwealth.

The islands have few natural resources and for long most Bahamians had to look to the sea for at least part of their livelihood. They were fishermen as well as small-scale farmers, and the only industry of importance was sponge fishing in the coral reefs that surround the islands. But since the 1950s there has been a dramatic change in the country's economy. Taking advantage of closeness to North America, a hot, sunny, though humid climate, white sand beaches and waters kept warm by the Gulf Stream, the Bahamians and businessmen from outside the country have developed tourism on a vast scale. Now, over 2 million

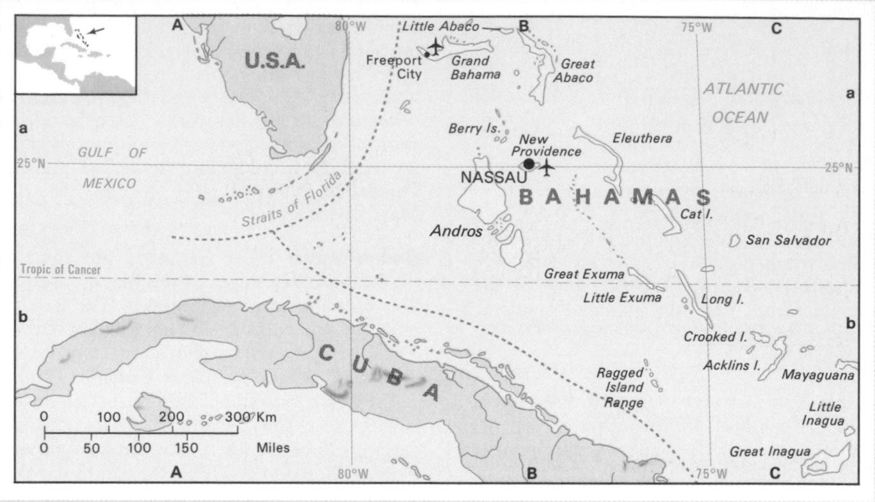

tourists, most of them from North America, visit the Bahamas each year and the tourist industry is the mainstay of the country's economy, employing two-thirds of the workforce.

One of the most remarkable projects of these years has been the FREEPORT CITY/ Lucaya development near the western end of Grand Bahama. Until the 1950s, this was a barren, unpromising area mostly covered with pine forest. But then an American financier named Wallace Groves came to an agreement with the government to develop the area for both tourism and industry. Freeport City/ Lucaya today is the largest single resort complex in the West Indies, with luxury hotels, casinos, villas, fine beaches and even a fanciful International Bazaar, which comprises 4 hectares (10 acres) of shops and restaurants reproducing settings from all over the world – an English pub, a Parisian café and a setting from Thailand, for example. There is also a deep-water harbour with a number of industrial enterprises set round it, including a cement works, a pharmaceutical plant, a major oil refinery and oil transshipment terminals.

On INAGUA, the southernmost inhabited islands, there is one of the world's largest complexes for producing salt by solar evaporation. A plant on New Providence processes frozen lobster tails for export to the USA, and the government is encouraging agriculture on a number of islands, especially Andros, to make the country more self-sufficient in food. At the same time, Bahamian laws, which exempt individuals and corporations from income and inheritance taxes, have attracted more than 360 banks and trust companies to the country. Nassau is a Eurodollar trading centre that compares with London.

The main centres in the Bahamas are New Providence and Grand Bahama, which have been developed for tourism and business, but life on many of the other islands, known as the Family or Out Islands, has been little affected by the changes of recent years.

Off ELEUTHERA, for example, to the east of New Providence, is the little island of Spanish Wells. Its white inhabitants are the descendants of the first Puritan settlers and have managed to retain their separate identity for nearly 300 years. Craftsmen on the Abaco group of islands, between Eleuthera and Grand Bahama, are still building boats as their ancestors have done for generations; and George Town, in the EXUMA group and almost exactly on the Tropic of Cancer, still keeps the character of a tiny island capital.

Other islands include Long Cay, with a population in 1980 of 33 people, and the tiny Bimini group which lie closest to Florida. Here the American writer Ernest Hemingway lived for a while and the black actor Sidney Poitier spent much of his childhood. CAT ISLAND, on the eastern edge of the Bahamas, has their highest point, Mount Alvernia – only 63 m (207 ft) above sea level – and Andros, west of New Providence, has a number of blue holes. Blue holes – found throughout the Bahamas but particularly common on Andros – are large, steep-sided inland tunnels where the ocean rises through the coral foundations of the island. Even more impressive, though, is the Andros Barrier Reef off the island's eastern coast. It is 200 km (about 125 miles) long and is second only in size to Australia's Great Barrier Reef.

BAHAMAS AT A GLANCE

Area 13 935 km² (5380 sq miles)
Population 237 000
Capital Nassau
Government Parliamentary monarchy
Currency Bahamian dollar = 100 cents
Language English
Religion Christian (47% Nonconformist, 25% Roman Catholic, 21% Anglican)
Climate Mild and subtropical; average temperature at Nassau ranges from 18-25°C (64-77°F) in February to 24-32°C (75-90°F) in August
Main primary products Timber, fish and shellfish
Major industries Tourism, oil refining, fishing, forestry
Main exports Crude oil (re-export), petroleum products, chemicals, seafood, rum, salt, aragonite (limestone)
Annual income per head (US$) 4200
Population growth (per thous/yr) 20
Life expectancy (yrs) Male 62 **Female** 65

Bahariya Oasis (Behariya) ***Egypt*** One of the great oases of the Western Desert, around the little town of Bawiti, about 280 km (175 miles) south-west of Cairo. The land is watered from bountiful artesian wells and good crops of dates are produced. Manganese and iron deposits have been found near by, but have not yet been developed.
Map Egypt Bb

Bahawalpur ***Pakistan*** Town and former princely state on the Sutlej river, about 345 km (215 miles) south-west of Lahore. Most of the area is in the Thar desert, which extends into India, but cotton is grown by the river.
Population 178 000
Map Pakistan Cb

Bahia ***Brazil*** See SALVADOR

Bahia Blanca ***Argentina*** Port and inlet about 550 km (340 miles) south-west Buenos Aires. The port's main trade is in grain and fruit being shipped from the southern pampas.
Population 223 800
Map Argentina Cb

Bahr el Ghazal ***Sudan*** Province covering some 134 700 km² (about 52 000 sq miles) in the south-west, beside the border with the Central African Republic. The provincial capital is Wau. The province is mainly a cattle-raising area and millet, oil seeds and bananas are grown. Rice is also produced at a government-sponsored project at Uwaye. It also contains the Southern National Park, in which is found a rich variety of species from elephants to rare butterflies. The vegetation is woodland or savannah with mahogany, shea, khaya and other types of trees.

The Bahr el Ghazal river (240 km, 150 miles long) – which runs through the province – mostly disperses into the Sudd swampland. Only in years of heavy rainfall does it flow on to join the White Nile at Lake No, about 700 km (435 miles) south and slightly west of the capital, Khartoum.
Population 1 493 000
Map Sudan Ac

Bahrain Island ***Bahrain*** Largest of the 33 islands which make up the independent state of Bahrain. It lies in The Gulf, 25 km (16 miles) off the Al Hasa coast of Saudi Arabia, to which it is linked by a 26 km (16 mile) long causeway. Oil was found there in 1931. Although the island is mainly an arid limestone plateau, it is irrigated by wells, making agriculture possible. Its highest point is Jabal ad Dukhan, which rises 137 m (449 ft) above sea level.
Population 216 815
Map Bahrain Ba

Baia Mare ***Romania*** City and capital of the north-west region of MARAMURES, about 40 km (25 miles) south of the Russian border. It is set near a lake in a wooded section of the Carpathians. It has a 15th-century belfry, an 18th-century cathedral, and an artists' quarter with a school of painting which has been renowned in Romania since the 19th century. There are also zinc and lead smelting plants.
Population 117 800
Map Romania Aa

Bahrain

LOOKING TO THE FUTURE, BAHRAIN IS SEEKING TO ATTRACT NEW INDUSTRIES TO OFFSET ITS DEPENDENCE ON OIL REVENUES

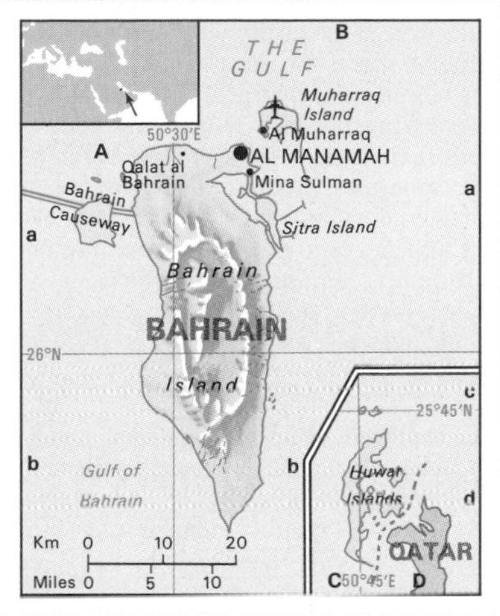

Known historically by the people of The Gulf as 'the island of a million palms' because of its abundant freshwater springs, Bahrain was the first Arab country to strike oil – in 1931. Today it has ambitions to make itself the 'Singapore of The Gulf' and diversify its economy before the oil runs out. Already it has an economically attractive free port and industrial zones Mina Sulman and North SITRA – where no customs duties are charged on transshipped goods or on goods manufactured within the free area. There is also an aluminium smelting and processing plant, the largest industrial complex in The Gulf not connected with the oil industry (apart from being fuelled by local natural gas); a petrochemical industry; a major shipbuilding and repair yard handling supertankers; and a satellite communications centre. Emphasis is also being placed on establishing Bahrain as a regional centre for offshore banking, and several prominent international banks now operate there.

The emirate comprises 33 low-lying islands, of which Bahrain Island is the largest (562 km², 216 sq miles). It is connected to its neighbouring island of Muharraq by a 2.4 km (1.5 mile) causeway, while on its east coast another new causeway links Bahrain to Sitra Island. Apart from a fertile strip in the north, Bahrain Island consists of limestone rock covered by varying depths of sand which is too poor and saline to support anything more than tough desert plants. Drainage schemes have been started to reduce the salinity. A fertile green belt has been created in the northern region with the addition of fertile soil from other parts of the island, manures and chemical fertilisers. Although Bahrain has underground springs (mostly in the north), much of its water is obtained from desalination plants.

The islands are oppressively hot during the period between May and September, with an average temperature of over 35°C (95°F) during the day accompanied by high humidity. But the weather cools down quickly from October, after which December to March is the coolest period (21°C, 70°F).

Most of the population lives on the islands of Bahrain and Muharraq; about one-third are immigrants, among them Omanis, Indians, Pakistanis and Iranians, attracted by tax-free incomes. The Bahrainis themselves are Arab Muslims. All are ruled by the Amir, who governs with advice from his ministers.

Oil revenues account for some three-quarters of the country's total revenue. Much of it is spent on building schools, colleges, hospitals and other public services. Education and health care are free, and much housing is subsidised. Pensions and sickness and unemployment benefits are provided.

Apart from the new industries, traditional occupations still survive – though on a much smaller scale than in years gone by. These include pearl fishing – Bahrain was once famed for the quality of its pearls – dhow building, basket and cloth weaving and pottery making. With the advent of new irrigation schemes, farming has expanded – not only are dates grown around the oases, but figs, pomegranates, bananas and a whole spectrum of vegetables are cultivated on reclaimed land.

Like the other small states of The Gulf, Bahrain maintains close relations with Saudi Arabia, which it sees as the best guarantor of stability and protection. Indeed, a physical link has now been established between the two countries with the opening of a US$900 million, 25 km (16 mile) causeway. But the oil recession in The Gulf is hitting the country, and unemployment is emerging as a significant problem.

BAHRAIN AT A GLANCE

Area 661 km² (255 sq miles)
Population 443 200
Capital Al Manamah
Government Monarchy
Currency Bahrain dinar = 1000 fils
Languages Arabic, English widely spoken
Religions Muslim (85%), Christian (7%)
Climate Very dry; very hot summers with high humidity. Average temperature ranges from 14-20°C (57-68°F) in January to 29-38°C (84-100°F) in August
Main primary products Dates, vegetables, livestock, fish; crude oil
Major industries Crude oil production and refining, aluminium smelting, shipbuilding and repairs, petrochemicals, banking
Main exports Crude oil and petroleum products, aluminium
Annual income per head (US$) 6900
Population growth (per thous/yr) 38
Life expectancy (yrs) Male 65 **Female** 69

Baikal (Baykal) ***USSR*** Freshwater lake covering 31 500 km² (12 150 sq miles) in south-east Siberia, near the Mongolian border. Fed by 336 rivers and drained by only one, the Angara, it is in volume the world's largest body of fresh water. It holds 23 000 km³ (5520 cu miles) – as much as all five of North America's Great Lakes combined. It is also the world's deepest lake and is as much as 1940 m (6365 ft) deep in one place.

Map USSR Lc

Baile Átha Cliath ***Ireland*** See DUBLIN

Baile Átha Luain ***Ireland*** See ATHLONE

Baile Átha Troim ***Ireland*** See TRIM

Bairiki ***Kiribati*** See TARAWA

Baja (Lower) California ***Mexico*** Rugged, mountainous peninsula of desert and semidesert extending southwards into the Pacific Ocean from the US border, mostly separated from the rest of Mexico by the Gulf of California. It is about 1300 km (nearly 800 miles) long and on average 80 km (50 miles) wide, and rises to a height of 3096 m (10 157 ft) in the Sierra San Pedro Martir. Baja California is divided into the states of Baja California Norte ('North') and Baja California Sur ('South').

Agriculture is confined to an irrigated area in the north and the valleys, where cotton, maize and wheat are grown. Since 1973 a new highway running the length of Baja California from Tijuana in the north to Cabo San Lucas in the south has encouraged the development of a thriving tourist industry along miles of sandy coastline.

Baja California Sur contains Vizcaino Desert and the salt flats of GUERRERO NEGRO. Its capital, La Paz, was discovered in 1535 by Hernando Cortez, the Spanish conqueror of Mexico.

Population (Norte) 1 200 000; (Sur) 200 000

Map Mexico Ab

Bakarganj ***Bangladesh*** See BARISAL

Bakhtaran ***Iran*** Trading city – whose name was recently changed from Kermanshah – about 400 km (250 miles) south-west of the capital, Tehran, on ancient caravan routes between Tehran and the Iraqi capital of Baghdad. It is the centre of a grain-producing area which is home to many Kurdish people, and there is an oil refinery.

Population 531 400

Map Iran Aa

Bakony (Bakony Forest) ***Hungary*** Range of forested hills in the north-west of the country between Lake Balaton and the Little Alföld, rising to 704 m (2310 ft) at Kőrishegy peak. Its pretty valleys and ruined castles, its lakeside resorts and its extinct volcanoes, such as the Badacsony, make it a popular holiday area during the summer months.

Map Hungary Ab

Baku ***USSR*** Port and capital of AZERBAIDZHAN, on the west coast of the Caspian Sea. It is the centre of a major oil field, and its industries include oil refining, oil-field equipment, chemicals, shipbuilding, textiles and cement. A pipeline runs to the port of Batumi on the Black Sea, carrying oil for export. Baku was founded in the 11th century, and has palaces and mosques from the 14th to 15th centuries.

Population 1 661 000

Map USSR Gd

Bakwanga ***Zaire*** See MBUJI-MAYI

Balaklava ***USSR*** See SEVASTOPOL'

Balaton, Lake ***Hungary*** Central Europe's largest lake and a major holiday area covering 601 km² (232 sq miles). It is 110 km (68 miles) south-west of the capital, Budapest. The beautiful north shore, 80 km (50 miles) long, is lined with volcanic, forested and vine-clad hills, sandy coves and picturesque villages. By contrast the south shore is flat with a shallow beach packed with camp sites, modern hotels, private holiday homes and villas – now nationalised as workers' holiday centres. Water sports are popular in summer, although motorboats are prohibited. Winter facilities include ice-skating, ice-sailing and ice-fishing.

Map Hungary Ab

Balatonfüred (Füred) ***Hungary*** Oldest and most elegant of the health resorts on the north shore of Lake Balaton, 110 km (68 miles) south-west of the capital, Budapest. Its rambling streets climb vine-clad hillsides, and there are 11 medicinal springs used for treating heart and nervous disorders. One of the springs bubbles through a fountain built in 1800 in Spa Square (Gyógy Tér), the town's centre. In summer, colourful equestrian pageants and jousts are held in the grounds of 15th-century Zichy Castle in the old town of Nagyvazsony, 16 km (10 miles) to the west.

Population 12 000

Map Hungary Ab

Balboa ***Panama*** See CANAL ZONE

Bâle ***Switzerland*** See BASLE

Balearic Islands ***Spain*** Island group off the east coast and one of Europe's prime holiday areas, with sweeping beaches, dramatic mountain scenery, and a mild climate. The group consists of Majorca (Mallorca), Minorca (Menorca), Ibiza (Iviza), Formentera, Cabrera and many islets, making up the province of Baleares, whose capital is Palma, in Majorca. Majorca also has the highest peak of the archipelago – Puig Mayor, at 1445 m (4731 ft). Besides tourism, there are fishing, pottery and metalware industries, and vines, cereals, olives and almonds are grown.

The islands have had a host of rulers, including Carthaginians in the 5th century BC, Romans, the German Vandals, Greeks and Arabs. The group became Spanish after it was seized from the Moors by the King of Aragon in the 13th century. In ancient times the islands' warriors were renowned as 'stone-slingers' – *balearii* in Latin, hence the name Balearic.

Population 685 000

Map Spain Fc

Bali ***Indonesia*** Volcanic island of 5591 km² (2159 sq miles) off the east coast of Java. It is one of the world's great tourist haunts because of the preservation of the Hindu civilisation of ancient Indonesia with its rich traditions in art, architecture and music. Kuta Beach is the main tourist resort and DENPASAR the capital.

Population 2 470 000

Map Indonesia Ed

Balikpapan ***Indonesia*** Port and city of East Kalimantan on Borneo. It is the supply centre for local oil fields.

Population 280 700

Map Indonesia Ec

Balkan Mountains (Stara Planina) ***Bulgaria*** Range of mountains extending for more than 320 km (200 miles) across central Bulgaria from the Yugoslav border to the Black Sea. They are crossed by 20 passes, including the SHIPKA PASS and the ISKUR Gorge. The highest point is Botev Peak, 2375 m (7792 ft).

Map Bulgaria Ab

Balkans ***Europe*** Mountainous peninsula in the south-east between the Adriatic and Ionian seas in the west and the Black and Aegean seas in the east. It encompasses ALBANIA, BULGARIA, GREECE, ROMANIA, YUGOSLAVIA and European TURKEY.

Balkh ***Afghanistan*** Ancient ruined city 20 km (12 miles) west of Mazar-e Sharif in the north of the country. In pre-Christian times it was an important staging post on the caravan route between China and Rome. In the 7th century AD it became a cultural centre of Islam, but was destroyed by the Mongol ruler, Genghis Khan, in the early 13th century. It was rebuilt by his successor Tamerlane (about 1335-1405) and his son, but by the 18th century it was a ghost town – although still with a large bazaar where caravans stopped.

Map Afghanistan Ba

Balkhash ***USSR*** Freshwater lake near the Chinese border, covering 17 000-22 000 km² (6500-8500 sq miles), depending on the season. It is frozen for five months of the year, but provides a quarter of all the trout caught in the USSR. The town of the same name on the lake's northern shore is a centre for processing copper ore mined nearby.

Population 81 000

Map USSR Id

Ballarat ***Australia*** City in Victoria, 112 km (70 miles) west of Melbourne. It was founded in 1851 at the start of the Victorian gold rush, and this era of the city's life has been recreated at Sovereign Hill Memorial Park, which is a major tourist attraction. A monument at the site of the Eureka Stockade marks the place where local gold miners temporarily defied government tax collectors in 1854. This rebellion – and the flag the miners raised – became symbols of resistance to British rule in Australia. Ballarat is a major market and commercial centre, and is the third largest urban area in Victoria.

Population 62 600

Map Australia Gf

Ballymena ***United Kingdom*** Borough in County Antrim in Northern Ireland. It lies 16 km (10 miles) north of Lough Neagh in a fertile farming region and hosts a large annual agricultural show. It has linen and wool fibre and engineering industries. Ballymena has a population largely of Scottish origin from 17th-century settlement, and is still largely Presbyterian.

Population 55 400
Map United Kingdom Bc

Balsas ***Mexico*** Major river, 700 km (435 miles) long, rising as the Atoyac in the SIERRA MADRE ORIENTAL in the state of Tlaxcala. It flows south and then west, forming the boundary between the states of Guerrero and Michoacán, to Petacalco Bay on the Pacific. In 1966 a dam, El Infiernillo, was built near its mouth, since when it has been extensively used for hydroelectric power and irrigation.

Map Mexico Cc

Baltic Sea Lying almost landlocked in northern Europe, it is linked to the NORTH SEA by narrow straits between Denmark and the Scandinavian peninsula. The Baltic was once a freshwater lake, formed when glaciers melted at the end of the last Ice Age, about 10 000 years ago. Because so many rivers flow into it, it still has a low salt content, especially in the east.

There are hardly any tides, which makes navigation easy though storms can be a danger. On the other hand, much of it freezes in winter and icebreakers are needed to keep the shipping lanes open. Viking longships once looted the Baltic coastlands (between the 7th and 10th centuries), but in the late Middle Ages, the Hanseatic League (a confederation of German merchant cities) dominated Baltic trade.

Branches of the Baltic include the Gulf of BOTHNIA, the Gulf of FINLAND, the Gulf of Riga and the Gulf of GDANSK.

Area 422 000 km^2 (162 930 sq miles)
Greatest depth 439 m (1440 ft)
Map Europe Ec

Baltimore ***USA*** Largest city of Maryland, standing on Chesapeake Bay, about 60 km (37 miles) north-east of the national capital, Washington. It was founded by the Provincial Assembly in 1729, and has developed as a port and shipbuilding centre. In 1830 it had the country's first railway passenger station. Baltimore's industries include steel and chemicals. The city contains the country's oldest Roman Catholic cathedral, is the seat of Johns Hopkins University, founded in 1876, and has many fine houses and public buildings which escaped a devastating fire that destroyed more than 1000 of the city's buildings in 1904.

Population (city) 763 600; (metropolitan area) 2 244 700
Map United States Kc

Baluchistan ***Pakistan*** Province of the south-west, covering 347 190 km^2 (134 050 sq miles). The tribes who live there have spread into both Iran and Afghanistan, paying little attention to the international borders drawn up under British rule. The province – whose capital is QUETTA – is mountainous, rocky, hot and arid. Goats and sheep are the mainstay of the economy, but vegetables are grown in irrigated valleys. Oil and gas have recently been discovered. The fiercely independent tribesmen have often taken up arms against the government.

Population 4 332 000
Map Pakistan Bb

Bamako ***Mali*** National capital, standing on the Niger River. It became Mali's first city in 1908 after a railway link was built from the town of Kayes near the Senegal border in 1904. The line was extended to the Senegalese capital, Dakar, in 1923. Bamako city centre is attractive and spacious. The roads are wide, tree-lined and usually made of sand. Elegant buildings from the French colonial era, now mainly used as offices, stand along them. There are also many fine traditional mud brick buildings, such as the old market. The Great Mosque, to which thousands throng each Friday, is modern, built with Arab money, but it re-creates the traditional style.

The administrative quarter stands on top of the escarpment of the Manding Mountains north of the city and has commanding views. Housing for the majority of Malians is cramped, mainly of mud brick and with few facilities, in marked contrast to European buildings. The capital is the main market for cattle and for kola nuts, which are used in making soft drinks. Light industries are growing between the Sotuba and Koulikoro roads.

Population 405 000
Map Mali Bb

Bamberg ***West Germany*** Port on the Regnitz river about 50 km (30 miles) north of Nuremberg. It is connected to a waterway network 3500 km (2200 miles) long, which when completed will link the North Sea with the Black Sea through canals joining the Rhine, the Main and the Danube rivers. The four spires of the Romanesque cathedral, started in the 11th century, soar over the town. Pope Clement II is buried in the choir. The Bible of Alcuin (735-804), the Northumbrian theologian who became an adviser to Emperor Charlemagne, is in the city library.

Population 70 000
Map West Germany Dd

Bambouk Mountains ***Mali*** Mountain range in the far west. The mountains were once a major source of gold and were largely responsible for West Africa being known as 'The Land of Gold' by Arab historians in the Middle Ages.

Map Mali Ab

Bamenda ***Cameroon*** Provincial capital in the west, about 270 km (165 miles) north-west of the capital, Yaoundé. It stands on the high, cool Bamenda grasslands and is a market point for tea, coffee and bananas. It also has craft industries producing traditional masks, figurines and musical instruments of wood, pottery and stone carvings. The nearby Bamenda Falls are a source of hydroelectric power.

Population 50 000
Map Cameroon Bb

Bamian ***Afghanistan*** Valley in the Hindu Kush mountains, known as 'The Valley of the Gods'. Situated 112 km (69 miles) north-west of Kabul, and 2583 m (8474 ft) above sea level, it is the site of the small town of Bamian with a hotel, teashops and a bazaar. In the cliffs on the north side of the valley are carved two giant Buddhas: one is 53 m (175 ft) high, the other 35 m (115 ft). There are many cliffside caves where Buddhist monks once lived.

Map Afghanistan Ba

Bamingui Bangoran National Park ***Central African Republic*** Savannah-covered park of 40 000 km^2 (15 440 sq miles) in the north. Its wildlife includes antelopes, buffaloes, lions, leopards and rhinoceroses.

Map Central African Republic Ba

▼ BALI DANCERS The Barong (Balinese lion-dog) fights the Rangda (widow-witch) in this dance symbolising the eternal conflict between good and evil.

Ban Chiang ***Thailand*** Silk-weaving village 480 km (300 miles) north and slightly east of Bangkok. It caused an archaeological sensation in 1974, when Thai and American scholars there found bronze artefacts believed to date from 3600 BC. Until then, archaeologists had almost universally believed that metalworking began in Mesopotamia in the Middle East several hundred years later.

Banaba (Ocean Island) ***Kiribati*** Small (5 km², 2 sq mile) raised atoll about 450 km (280 miles) south-west of Tarawa. Between 1900 and 1979 its rich deposits of phosphate rock were mined by a British company, leaving little productive land. The Banaban people were resettled on Rabi island, Fiji, after the Second World War, but have been claiming substantial compensation from the British government.

Map Pacific Ocean Dc

Banana ***Zaire*** Seaport and site of deep-water harbour on the Atlantic coast, near the outlet of the Zaire river. It is linked by rail to Boma and Kinshasa.

Map Zaire Ab

Bananal ***Brazil*** River island 480 km (300 miles) north-west of the capital, Brasília. At 20 000 km² (7720 sq miles) – slightly smaller than Wales – it is the world's largest river island and splits the Araguaia river for about 500 km (310 miles) of its course. It gets its name from the giant wild banana trees which have developed in the hot and humid climate. The island is so large that it has rivers of its own, some of them more than 300 km (186 miles) long.

Map Brazil Cc

Banco National Park ***Ivory Coast*** Wildlife region just north-west of Abidjan. It was set up in 1926 during the French colonial period and became established as a national park in 1953. It covers 300 km² (116 sq miles) and includes areas of virgin tropical forest as well as an arboretum. Elephants, hippopotamuses, monkeys and birds roam the park, though the area is of more interest for its trees.

Banda Sea Part of the PACIFIC OCEAN in south-eastern Indonesia. The movements of continental plates in the area cause frequent undersea earthquakes. The Banda Islands in the sea are all of volcanic origin.

Area 738 150 km² (285 020 sq miles)

Greatest depth Over 6400 m (20 990 ft)

Map Indonesia Gd

Bandama ***Ivory Coast*** One of the country's three major rivers with the SASSANDRA and the KOMOE. With its two main tributaries, the Bandama Blanc and the Bandama Rouge, it drains some 50 per cent of the country. It rises as the Bandama Blanc in the highlands near Korhogo and flows south for 1050 km (643 miles to enter the Atlantic Ocean at Grand Lahou. The Bandama Rouge rises in the north-west highlands and meets the Bandama Blanc about 33 km (20 miles) west of Bouaké. A hydroelectric plant at KOSSOU, just north of the confluence, provides power to Bouaké and the central part of the country. The river irrigates large areas of the rural centre.

Map Ivory Coast Bb

Bandar Abbas ***Iran*** Port on the Strait of Hormuz, opposite Oman. It was established in 1623 by Shah Abbas I to replace the island port of Hormoz, captured by the Portuguese in about 1614. During the 17th century Bandar was Persia's main port, trading with India. Today it is the focal point of the trade routes of south Iran, and its industries include cotton milling, fishing and fish canning.

Population 89 200

Map Iran Bb

Bandar Seri Begawan ***Brunei*** National capital of the Sultanate, set on the Brunei river in the midst of an alluvial coastal plain. Once known as Brunei town, it is dominated by the vast Omar Ali Saiffudin mosque, the largest in South-east Asia, while the enormous new royal palace is nearby. This and the modern commercial district of the town contrast markedly with the traditional river-bank homes on stilts of Kampong Ayer, a traditional village now part of the capital.

Population 50 000

Map Brunei Ab

Bandiagara Scarp ***Mali*** Rockface south of the Niger River with an almost sheer drop of 244 m (800 ft).

Map Mali Bb

Bandipur ***India*** Wildlife sanctuary 80 km (50 miles) south of Mysore. It is noted for its tigers, elephants and leopards. It was formerly a game reserve for the Maharajah of Mysore in the 19th and early 20th centuries.

Map India Be

Bandundu ***Zaire*** Capital of Bandundu region in the south-west and a busy commercial centre. Formerly called Banningville, it lies about 145 km (90 miles) from the Congo border on the Kwilu river just above its junction with the Kasai river. Bandundu region covers 295 658 km² (114 124 sq miles). In the north it is low-lying and densely forested and includes swamps and Lake Mai-Ndombe. The land rises to the savannah-covered south. The mostly Bantu-speaking peoples include the Teke in the north, Kwango in the centre and Lunda in the south.

Population (town) 120 000; (region) 4 120 000

Map Zaire Ab

Bandung ***Indonesia*** Provincial capital of West Java. Its main industry is textiles. In 1955 the Indian Prime Minister Pandit Nehru, Premier Chou-Enlai of China and Indonesia's President Sukarno founded the non-aligned movement at a conference in the city.

Population 1 462 700

Map Indonesia Cd

Banff ***Canada*** The country's oldest national park, established in 1885 in south-west Alberta. It covers 6640 km² (2564 sq miles) and contains spectacular ROCKY MOUNTAINS scenery. Lake Louise is a major attraction. The town of Banff is the main tourist centre.

Map Canada Dc

Banffshire ***United Kingdom*** See GRAMPIAN

Banfora ***Burkina*** South-western market town on the Comoé river at the foot of the Banfora Escarpment. It lies on the railway running from Burkina's capital, Ouagadougou, to Abidjan, the capital of Ivory Coast. The town is a collecting point for groundnuts and has a new sugar mill. Wild honey is abundant and the area is rich in iron ore, although it has yet to be mined because of prohibitive transport costs.

Map Burkina Aa

Banfora Escarpment ***Burkina*** White sandstone escarpment 170 m (560 ft) high, which forms the southern edge of the Sikasso Plateau in south-west Burkina. It boasts spectacular waterfalls and stunning views from the top.

Map Burkina Aa

Bangalore ***India*** Rapidly expanding industrial city in south-central India, about 260 km (160 miles) west of the port of Madras. It was founded in the 16th century and today has booming aircraft and electronics industries. It also manufactures machine tools, farming implements, textiles, footwear and paper.

In 1964 two flourishing educational centres were opened there – Bangalore University and the University of Agricultural Sciences. Bangalore is also the site of the All-India Institutes of Science and of Management.

Population 2 921 800

Map India Be

Bangkok (Krung Thep) ***Thailand*** The kingdom's capital, a wild and noisy city whose streets seem to be clogged with traffic almost around the clock and some of whose centre is garish with neon-lit strip clubs. The city sprawls beside the Chao Phraya river, about 30 km (19 miles) from the Gulf of Thailand. The old city, founded as a royal capital in 1782, is a place of *klongs*, or canals, and *wats* – temples or monasteries – of which there are some 300. This 'Venice of the East', as it was once known, is surrounded by a chaotic and mushrooming metropolis where about 6 million people live. Despite growing industries, the city has been unable to provide adequate productive employment for the many migrants who have arrived over the past 25 years. Many of the poorer people thus find work in marginal service industries, some of which have given Bangkok notoriety as the so-called 'sex capital of the world'.

The city is also rich in Thai culture. The Grand Palace, begun in 1782 as the home of the Thai royal family, houses the Wat Phra Keo temple containing the 460 mm (18 in) high Emerald Buddha. The statue is fashioned from translucent jasper, though when or where it was made is a mystery. The temple of Wat Po contains the Reclining Buddha, 50 m (164 ft) long, and Wat Trimitr has a statue known as the Golden Buddha; the figure is said to contain more than 5 tonnes of gold.

The whole range of Thai culture is exhibited in the nearby Rose Garden. There is traditional dancing and Thai boxing, which allows kicking as well as punching.

Bangkok is not, strictly, the capital's correct name, although it is universally used by foreigners. Bangkok means 'village of the wild plum'; it is the name of part of the Thon Buri side of the river. Thais call the city Krung Thep – meaning 'the city of angels'.

Population (metropolitan Bangkok) 5 900 000

Map Thailand Bc

Bangladesh See p. 70

Bangor *United Kingdom* Northern Ireland's largest seaside resort, about 20 km (12 miles) north-east of the capital, Belfast. Bangor's abbey is built on the site of a 6th-century monastery, which during the Dark Ages sent missionaries to help keep Christianity alive in Europe. There are some engineering factories, but Bangor is largely a residential centre for Belfast. Its harbour is an important yachting centre, and the town and district have abundant sporting facilities, mostly golf courses.
Population 46 500
Map United Kingdom Cc

Bangui *Central African Republic* National capital and port on the Ubangi river, from which the city gets its name. It handles much of the country's cotton and timber for export. The modern city centre has a colourful market, and a university which was founded in 1970.
Population 387 000
Map Central African Republic Ab

Bangweulu *Zambia* Lake covering 9800 km^2 (3784 sq miles) about 500 km (310 miles) north and slightly east of the capital, Lusaka. In 1873 the Scottish missionary David Livingstone travelled through the swampland south-east of the lake on his last journey of exploration before his death later that year.
Map Zambia Bb

Bani Suwayf *Egypt* See BENI SUEF

Baniyas *Israeli-occupied Syria* Village 65 km (40 miles) south-west of Damascus in a region occupied by Israel since 1967. It is the site of the ancient Greek city of Paneas, which was renamed Caesarea Philippi by the Romans. There is a temple built by Herod the Great, and a shrine to the Greek God Pan. Near the Kibbutz Dan, to the south-west of the village, are the thermal springs which feed the headstreams of the River Jordan.
Map Syria Ab

Baniyas *Syria* Port and terminal for the Iraq pipeline from Kirkuk, situated 144 km (89 miles) south-west of Aleppo. Spread around a small bay, Baniyas is overshadowed by a refinery and huge oil storage tanks. On the surrounding hills are the white domes of tombs, and on the highest hill is the huge Fortress of Marqab. Built of black basalt, with 14 towers, it was an important military post for the Crusaders. In 1140 it was re-fortified by the Knights Hospitallers, or Knights of St John (an order founded in the 11th century to care for poor or sick pilgrims in Jerusalem). It could house 1000 people as well as the defenders, and held enough provisions to withstand a five-year siege. Since ancient times Baniyas has been noted for its gardens, and tall cypresses protect the town and its surrounding orchards from the sea winds. In the Middle Ages the port was used to export wool, but today the harbour is silted up and used mainly by fishermen.
Map Syria Ab

Banja Luka *Yugoslavia* City in north-west Bosnia, 148 km (92 miles) north-west of Sarajevo. A road and rail junction on the Zagreb-Sarajevo route, it lies where the Vrbas gorge through the Dinaric Alps opens onto the Sava plain. Banja Luka, a spa since Roman times, has Roman Catholic and Orthodox cathedrals, and many mosques built by the Turks who fortified the town in the late 16th century.
Population 183 600
Map Yugoslavia Cb

Banjarmasin *Indonesia* Provincial capital and fishing port of South Kalimantan on Borneo. It is the centre of a farming region which produces coconuts, rubber and rice.
Population 381 300
Map Indonesia Dc

Banjul *The Gambia* Capital and commercial centre of The Gambia, standing on the low sandbanks at the mouth of the Gambia River. British traders built a fort there in 1816 to control the slave trade. Merchants and missionaries followed, and the town that grew was named Bathurst after Earl Bathurst (1762-1834), then Secretary of State for the Colonies. It was renamed Banjul, meaning 'bamboo', at Independence in 1965.

The town has been improved to a certain extent, by comparison with the 1930s when flooding and overcrowding resulted in dysentery, yellow fever and malaria. There are wide streets in the centre, which has been redeveloped to alleviate some of the overcrowding. However, in general, housing is poor – much of it is built from corrugated iron. Water comes from standpipes and electricity supplies are minimal.

Despite the poverty Banjul is being developed as a tourist resort, and is becoming popular with Europeans wanting a winter tan or a chance to explore a largely undeveloped part of Africa. A number of good hotels have been built alongside the miles of sandy beach and warm seas near the city. The rains arrive from June to October in Banjul, with August being the wettest month.
Population 42 000
Map The Gambia Ab

banner cloud Type of cloud that forms in clear skies on the lee side of a mountain as air rising to pass over the peak cools.

Banningville *Zaire* See BANDUNDU

Banska Bystrica *Czechoslovakia* Regional capital 165 km (103 miles) north-east of Bratislava. It was a prosperous copper and silver-mining town in the Middle Ages, and has fine Gothic churches and town hall, and metal, textile and woodworking industries. The town was the centre of the Slovak uprising against the Nazis (August 29 to October 27, 1944).
Population 73 300
Map Czechoslovakia Db

Bantry (Beanntrai) *Ireland* Market town, tourist centre and fishing port on Bantry Bay, 72 km (45 miles) south-west of the city of Cork. The French sailed into Bantry Bay in two unsuccessful attempts at landing – in 1689 to support James II, and with Wolfe Tone, the exiled Irish nationalist, during the 1798 rebellion. Offshore, Whiddy Island was developed as an oil terminal. Bantry House is home of the Earls of Bantry.
Population 2960
Map Ireland Bc

Baoshan *China* Industrial town 8 km (5 miles) north of Shanghai. It is the site of China's most modern iron and steel works.
Population 8000
Map China Ie

Baqubah *Iraq* Market town in the oil and fruit producing province of Diyala. It lies about 45 km (25 miles) north-east of the capital, Baghdad.
Population 40 000
Map Iraq Cb

bar Ridge of mud, sand or shingle extending across a river mouth, harbour or bay, and lying roughly parallel to the coast.

Bar (Antivari) *Yugoslavia* Montenegrin fishing port 48 km (30 miles) from the Albanian border. Largely rebuilt after an earthquake in 1979, it is now a ferry port for Bari (Italy) and Corfu, and terminus of the spectacular Belgrade-Bar railway, which threads through tunnels, along ravines and over gorges across the wildest, most rugged part of the Dinaric Alps. Ulcinj, an old town only 18 km (11 miles) from Albania, has 15 km (9 miles) of palm-fringed sandy beaches.
Population 3600
Map Yugoslavia Dc

Barania Gora *Czechoslovakia/Poland* See BESKIDY

Barbados See p. 73

Barbuda *Antigua and Barbuda* Island covering 160 km^2 (62 sq miles), lying about 43 km (27 miles) north of Antigua. It has wide beaches and a coral reef. The capital and the island's only settlement, Codrington, takes its name from an English family who owned the island between the middle of the 17th century and 1860, when it became a British colony. The islanders grow crops – mainly maize and peas – and rear livestock for local consumption.
Population 1500
Map Caribbean Cb

Barcelona *Spain* Seaport and industrial city on the northern Mediterranean coast. It is Spain's second largest city after Madrid and is noted for its fascinating central avenue, the Ramblas, with its bookstalls, flowerstalls and caged birds for sale. Its industries include textiles, machinery, oil refining, chemicals, plastics, leather goods and tourism. It exports textiles, machinery, wine, olive oil and cork.

In the old city are Roman remains and 14th-century Gothic churches and palaces. The newer districts are dominated by the flamboyant buildings of the Spanish sculptor-architect, Antonio Gaudi (1852-1926), who echoes the shapes of natural phenomena, such as icicles, in his designs. His most outstanding building in Barcelona is the Sagrada Familia (Sacred Family) cathedral, begun in 1883.

The Barcelona province is the most densely populated and highly industrialised in Spain. It was a Republican bastion in the Civil War (1936-9).
Population (city) 1 754 900; (province) 4 618 700
Map Spain Gb

Bangladesh

THE GOLDEN LAND OF HISTORY HAS BEEN ROBBED OF ITS WEALTH AND RAVAGED BY NATURAL DISASTERS

Once a land of fabulous wealth, Bangladesh was for centuries known as Golden Bengal (Sonar Bangla). But rapacious rulers have squeezed it too hard, and it is now one of the poorest countries in the world.

Bangladesh – from 1947 to 1971 East Pakistan – is almost unbelievably flat. Apart from hills covered in bamboo forests in the south-east, the country is virtually one huge delta formed by the GANGES, BRAHMAPUTRA and Meghna rivers. It is so flat that from the air it is sometimes difficult to see where the land ends and the sea begins. The country is subject to devastating floods, and cyclones (hurricanes) that sweep in from the Bay of Bengal regularly exact huge death tolls. In 1970 a combined cyclone and tidal wave killed an estimated 300 000 people.

Large and small rivers snake over the landscape, often unstable and shifting in their courses. Whole villages may be washed away overnight as the Brahmaputra shifts its banks. Villages have to be ready for rivers suddenly to rise 6 m (20 ft) or more. Most villages are built on mud platforms to keep them above water, and the individual houses are raised still higher on plinths.

The rainfall, sun and silt make the land productive: it is often possible to grow two or three crops a year. Irrigation could help to grow more crops in some areas during the long dry season from November to March, but there is a shortage of power to drive irrigation pumps.

WORLD-FAMOUS COTTON

Bangladesh became world famous for cotton in the 18th century; the special cotton cloth called muslin was named after the Muslims of DHAKA who wore it. The country is also renowned for jute, grown by small farmers who also grow rice as a subsistence crop to feed their families. Rice crops were heavily taxed by the Mogul emperors (1526-1857) and by the British when they ruled the territory as part of India (1857-1947).

Bangladesh is mainly an Islamic country, and its Muslim women are seldom seen away from the home. The rice fields, for instance, are strictly the province of the men and boys. Around the homesteads there are many trees and bushes, shading small plots. Here the women tend fruits and gourds, and grow spices such as turmeric, garlic and chilli.

The homestead may be picturesque, with cows quietly chewing in the shade of a shelter made of bamboo overgrown with pumpkin vines, but everything is there for a purpose. Trees provide fodder from their leaves, timber for building and for fuel, and fruits such as mangoes. They provide juices to preserve fish nets from rotting, or sap from which palm sugar is made. The bamboo plant is used in building for posts and rafters, for plough tackle and mallet handles, and split and woven to make walls, fish nets and water scoops.

Cow dung is very important: it is used for manure and for burning, but even more importantly for the preparation of the threshing floor. A normal floor of mud would crack all over and steal the grains, so the top is skimmed with a slurry of mud and dung, which dries smooth and hard without a crack.

The men do back-breaking work clearing weeds in the fields (although in the paddy fields the standing water inhibits unwanted plants), but even the weeds are used – as cattle fodder. Rice once provided most of the fodder, but modern rice varieties have shorter straws and do not yield so much (modern rice strains also need fertilisers, which kill fish that swim in the paddy fields).

Many farmers grow jute as a source of cash. But they also use the woody stems for fuel and for fencing, and the young green leaves as a vegetable. They sell the fibre from the bark – stripped by a process called retting (prolonged soaking in water). The fibre goes to make rope and sacks, and backing for carpets.

A COUNTRY DIVIDED

Linguistically, culturally and geographically, Bangladesh and the Indian state of West Bengal are a single region. In former times they were either a single kingdom or a single vice-royalty of one of the Indian empires. But under the Islamic empires of India, the two Bengals became divided by religion, the majority in the east becoming Muslims, while the majority in the west remained Hindu.

The British ruled Bengal as a united province, except for a short period at the beginning of the 20th century. But it was divided on independence, and the eastern, mainly Moslem, half became the eastern wing of Pakistan.

The marriage of West and East Pakistan was short-lived and ended in violence. At independence in 1947 East Pakistan's jute was the major export, but the East received very little of the revenue it raised. The Pakistani Government was dominated by the West, particularly by Punjabis, during years of military dictatorships. One of the greatest provocations to the East was a campaign to impose Urdu as the national language, thus excluding Bengali from use for official purposes.

A movement for self-rule in East Pakistan was led by Sheik Mujibur Rahman. In the 1970 elections, which introduced civilian government to Pakistan, his Awami League won the greatest number of seats, all of them in the East. The second largest party was led by Zulfikar Ali Bhutto, whose Pakistan People's Party dominated the West.

The two failed to negotiate a constitution, and there was a rebellion in East Pakistan. The West Pakistani army embarked on a ruthless campaign to put down the insurrection, and more than a million people were killed. Millions more fled over the border to India to

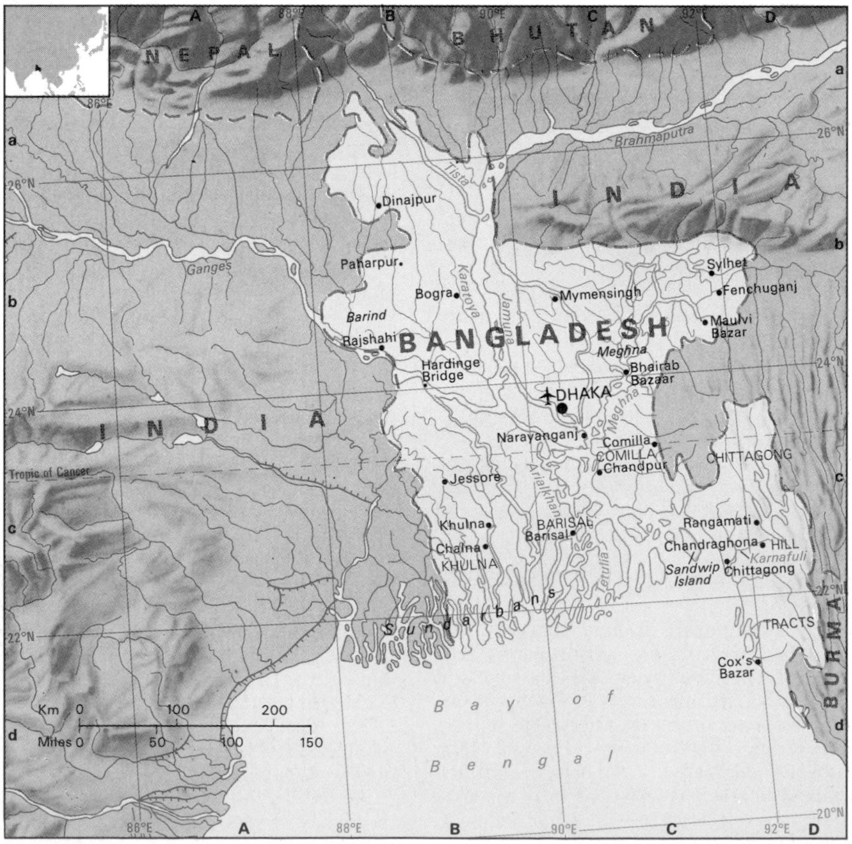

▲ RIVER OF LIFE Slow-moving boats carry grain and other vital foods on the Ganges river near the capital, Dhaka. The boats are built to a design that has not changed for hundreds of years.

escape the slaughter. India entered the conflict and defeated the West Pakistani army of occupation. In 1971 Sheik Mujibur Rahman became the leader of the new independent state of Bangladesh. He was assassinated in 1975, and most governments since then have been military dictatorships.

Many of Bangladesh's present problems were forecast by Muhammad Ali Jinnah, founder and first president of the original two-winged Pakistan. He said that the country should take in the whole of Bengal, including the metropolis and port of Calcutta in West Bengal, and not just the eastern part. He believed that without Calcutta, where the wealth of Bengal was concentrated, the territory would become a poverty-stricken rural slum. Today, many people think this is just what Bangladesh has become.

It is still an overwhelmingly rural country – fewer than 10 per cent of the population live in towns. It has a very low level of literacy, and a high population growth (more than 3 per cent per year). Very few villages are served by roads, and more than 60 per cent of the country's internal trade is carried by boat. Nevertheless, there is some modern industry – a steel rolling mill and oil refinery at CHITTAGONG and textiles at Dhaka. A massive find of natural gas beneath the delta provides energy and raw material for the manufacture of fertiliser. Some cement is also produced.

The health standards of the rural areas are low. The principal scourges are malnutrition and stomach diseases, the most feared of which is cholera.

There is poverty in the towns as well as in the hinterland. The capital, Dhaka (formerly Dacca), which has grown to a city of 2 million since independence in 1971, is thronged with poor people trying to make something to live on. Many of them ply for hire as rickshaw pullers. Others are street hawkers, selling wares that would be rubbish in most cities: empty bottles, stripped-down parts from discarded cigarette lighters, and piles of old reading glasses. Those who have nothing to sell or are crippled or blind will beg.

For all this, Bangladesh is renowned for its culture, and for producing men of great intellect. Unfortunately, too many of the well-educated Bangladeshis live abroad.

An urban educated elite holds power in Bangladesh, even though rural villages send representatives to district development boards. The townspeople run a bureaucracy which is supposed to be helpful, but is often too elaborate to be useful. Development programmes have given some help, including wells for irrigation and trials of new varieties of rice and wheat.

Bangladesh's future depends greatly on relationships with its neighbours, particularly India and Nepal. The three countries need to agree on a policy for the Brahmaputra-Ganges basins, to control floods, provide electricity and improve water supplies. The future also depends upon agricultural improvements for the millions of sharecroppers and small farmers, and a reduction in the birthrate.

BANGLADESH AT A GLANCE

Area 144 000 km² (55 598 sq miles)
Population 104 250 000
Capital Dhaka
Government Muslim republic under martial law
Currency Taka
Language Bengali
Religions Muslim (83%), Hindu (16%), Buddhist, Christian
Climate Tropical; monsoon from June to September; dry season between January and March. Average temperature in Chittagong ranges from 13-26°C (55-79°F) in January to 25-31°C (77-88°F) in June
Main primary products Rice, jute, sugar cane, tea, wheat, tobacco, cattle, timber, fish; salt, white clay, glass sand, natural gas
Major industries Jute spinning, textiles, sugar refining, tea, fertiliser, paper and leather processing, fishing
Main exports Jute, leather, tea
Annual income per head (US$) 117
Population growth (per thous/yr) 28
Life expectancy (yrs) Male 47 **Female** 48

Barcelos ***Portugal*** Market town on the Cavado river, about 40 km (25 miles) north of the city of Oporto. The remains of a 15th-century palace overlook the river. The town produces painted pottery, especially the decorated cockerels which are a national emblem.
Population 10 800
Map Portugal Bb

barchan Crescent-shaped sand dune, concave on the side sheltered from the prevailing wind, which can exceed 30 m (about 100 ft) in height. It forms when the wind blows predominantly from one direction. The dune advances as the sand is blown from the windward side of the dune to fall down the dune face.

Barclay, Mount ***Liberia*** The country's first rubber plantation, established on a hill of the same name near Monrovia, the capital, by a British company in 1907. It is now part of the Firestone plantation centred on HARBEL.
Map Liberia Aa

Barcoo ***Australia*** See COOPER CREEK

Barents Sea Part of the Arctic Ocean covering the continental shelf north of Norway and European USSR. The North Atlantic Drift keeps its southern waters relatively ice-free, but dense fogs are common. Named after the Dutch explorer Willem Barents (1550-97), the sea has oil reserves and abundant fish.
Dimensions 1290 km (800 miles) long, 1050 km (650 miles) wide
Greatest depth About 180 m (590 ft)
Map USSR Ea

Bari ***Italy*** Adriatic seaport about 240 km (150 miles) east of Naples. Its industries include oil refining and engineering products. Bari first grew to importance as the Byzantine capital of southern Italy. Later it was conquered by the Normans, who built a great square castle, the cathedral and a church which is the prototype of all the Norman churches in the region.

The Fascist dictator Benito Mussolini (1883-1945) made Bari his headquarters for imperial ventures in the eastern Mediterranean, putting up many buildings in the grand style.
Population 370 800
Map Italy Fd

Bari Doab ***Pakistan*** Fertile region between the Sutlej and Ravi rivers in the Punjab. It was once semiarid wasteland, but the British built the Upper Bari Doab Canal in 1857 to irrigate the area, and its resulting prosperity stimulated the growth of the city of Lahore.
Map Pakistan Db

Bariloche (San Carlos de Bariloche) ***Argentina*** Holiday resort at the centre of Argentina's lake district, about 1350 km (840 miles) south-west of Buenos Aires. Set beside the lake of Nahuel Huapí (area 531 km^2, 205 sq miles) and studded with Alpine-style chalets, it is a ski resort in winter. In summer, holidaymakers go there to explore the surrounding national park – 7850 km^2 (3030 sq miles) of forests and mountain meadows rich in fish, red deer, swans, geese and ducks.
Population 60 000
Map Argentina Bc

Barind ***Bangladesh*** Region in the Ganges delta between the Ganges and Jamuna rivers. It is one of two parts of the delta (the other is Madhupur, north of Dhaka), which are about 50m (164 ft) higher than the rest. Heavy clay soils make agriculture more difficult than elsewhere in the delta, but the extra height means that the areas are not afflicted with floods. Part of Madhupur is still covered by forest.
Map Bangladesh Bb

Barisal (Bakarganj) ***Bangladesh*** **1.** Densely settled district covering 7299 km^2 (2818 sq miles) close to the Bay of Bengal. One-third of the area's agricultural workers are landless labourers living at risk of starvation or arrested development. **2.** River port and chief town of Barisal district, near the Tetulia mouth of the Ganges river. It exports fish, and produces rice. The town is noted for a natural phenomenon known as the 'Barisal guns' which sounds like cannon-fire and is thought to be caused by earth movements deep underground.
Population (district) 4 667 000; (town) 142 100
Map Bangladesh Cc

Barletta ***Italy*** Adriatic coast town about 50 km (30 miles) north-west of Bari. It was an important port in Byzantine and Norman times. The Holy Roman Emperor Frederick II (1194-1250) built the imposing castle, which was added to by the Spanish in the 16th century. Outside the church of San Sepolcro is a 5 m (16 ft) bronze statue of a 4th-century Roman emperor known as Il Colosso.
Population 83 700
Map Italy Fd

Barnaul ***USSR*** Capital of the Altay territory, on the Ob' River 175 km (110 miles) south of Novosibirsk. Its industries include engineering, diesel motors, textiles and timber.
Population 568 000
Map USSR Jc

Barnsley ***United Kingdom*** Town in South Yorkshire 18 km (11 miles) north of Sheffield. Originally a market town, it was for a century a leading centre of England's coal industry. Now, however, all the mines close to the town have shut – as has most of the maze of railway lines that used to serve them.
Population 74 000
Map United Kingdom Ed

Baroda ***India*** See VADODARA

Baroghil Pass ***Afghanistan/Pakistan*** Strategic pass in the Hindu Kush, connecting the northern tip of Pakistan and Azad Kashmir to the Wakhan region of Afghanistan. The pass was closed after the USSR invaded Afghanistan in December 1979.
Map Pakistan Da

Barossa Valley ***Australia*** Area in South Australia about 50 km (30 miles) north of Adelaide. Established by German immigrants in the 19th century, it is the principal non-irrigated wine-producing area of Australia, producing 10 per cent of the country's wine.
Map Australia Fe

Barotseland ***Zambia*** See WESTERN PROVINCE

Barquisimeto ***Venezuela*** Old city that has rapidly expanded into a modern industrial centre, about 270 km (170 miles) west and slightly south of the capital, Caracas. Its main industries are meat-packing, cement manufacture and sugar refining.
Population 600 000
Map Venezuela Ba

barrage Artificial obstruction in a watercourse, used especially to store water for irrigation and to prevent flooding. Similar structures are usually called dams if hydroelectricity is also generated.

Barranquilla ***Colombia*** Port on the Caribbean coast near the mouth of the Magdalena river. It was founded in 1629 and remained a small port until the 20th century. Rapid growth in manufacturing industries such as textiles, chemicals and food processing have helped to make it one of Colombia's largest cities and one of its major ports. Barranquilla has five universities, and each year in the week before Lent it holds a wild carnival, when it calls itself *la ciudad loca* – 'the crazy city'.
Population 1 067 000
Map Colombia Ba

barrier beach Long, narrow bar of sand built up parallel to a coastline by wave action, and separated from the coastline by a lagoon. Also called barrier islands, barrier beaches are common along the eastern coast of the USA from New Jersey to Florida.

barrier reef Long, narrow CORAL REEF close to a coastline, but separated from it by a lagoon too deep or not clear enough for coral to grow.

Barrow Island ***Australia*** Island, with an area of approximately 250 km^2 (97 sq miles) off the north-west coast of Western Australia, about 1200 km (750 miles) north of Perth. It contains one of Australia's main oil fields.
Map Australia Bc

Barumini ***Italy*** Village in the rolling wheatlands of central Sardinia, 60 km (37 miles) north of Cagliari. It is the site of Nuraghe Su Nuraxi, the most important Sardinian nuraghe (megalithic structure of Bronze Age round towers and dwellings). The central tower, made of uncemented volcanic blocks between the 13th and 9th centuries BC with later modifications, is surrounded by 50 dwellings built about the 7th century BC.
Population 1530
Map Italy Be

basalt Fine-grained, dense, dark volcanic rock of BASIC composition and often having a glossy appearance. In its molten state it is highly fluid, and usually forms lava flows erupting from volcanic fissures. When it cools rapidly it may form hexagonal columns as at the GIANT'S CAUSEWAY.

basaltic layer Crustal layer which forms the floors of the ocean basins. It is also the name sometimes given to the denser and lower of the two layers of the continental crust. The upper, less dense, layer is called the sial or granitic layer.

Barbados

'BIMSHIRE' – CARIBBEAN ISLAND OF CRICKET AND RUM – REMAINS A TRADITIONAL OUTPOST OF BRITAIN IN THE TROPICS

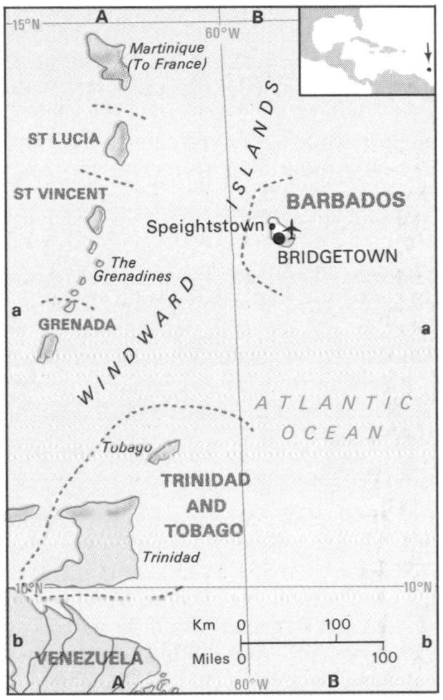

The Barbadian capital of BRIDGETOWN has a Trafalgar Square, just like London; and rising in the centre of the square, behind iron railings, is a statue of Nelson placed there in 1813.

Barbados has an unmistakably English air and is sometimes called Little England, or Bimshire, as if it were an English county (Bim is slang for Barbados). Churches that could have come straight from the English shires scatter its countryside and, like the English, Barbadians (Bajans, they call themselves) have an instinctive respect for tradition. Bridgetown's Harbour Police, for example, still wear sailor suits and straw hats.

Barbados, the easternmost island of the West Indies, lying well outside the arc of the Lesser Antilles, has had an untroubled history. In 1627 the first English settlers arrived. They founded Holetown on the island's western coast and, 12 years later, Barbados's House of Assembly met for the first time. The island then remained a British colony until 1966, when it became an independent nation within the Commonwealth with the British sovereign as its head of state.

Most of Barbados, which is 34 km (21 miles) long and 23 km (14 miles) wide, is relatively flat. Its highest point, Mount Hillaby, near the centre, reaches only 340 m (1116 ft). The cooling North-east Trade Winds are felt in most places and temperatures rarely go above 30°C (86°F) or below 22°C (72°F). There are only two seasons: the dry (from December to May) and the wet (June to November).

Barbados is densely populated with, on average, about 600 people per km² (1550 per sq mile) – compared with 230 people per km² (596 per sq mile) in the United Kingdom. As a result of this dense population, large numbers of Barbadians – most of them descended from slaves brought from Africa in the 17th and 18th centuries to work on sugar plantations – have long had to emigrate.

For those who remain in Barbados, however, there is a stable and relatively high standard of living. Although much of the economy is still controlled by a tiny white majority, including the descendants of the British planters who for long directed the island's affairs, political power is now in the hands of the black majority. There is an efficient two-party system; and the social services are well run. The standard of education is high.

The Barbadian economy was built almost exclusively on the production of sugar, and its by-products rum and molasses. The industry has declined considerably in importance in recent years, but sugar is still the principal export and in the early 1980s some 60 per cent of the island was still devoted to the cultivation of sugar cane. To reduce imports, the government is encouraging farmers to grow food for local consumption – including such staples as yams and sweet potatoes.

The mainstay of the economy today is tourism. Indeed, Barbados has attracted visitors for more than 200 years, and many invalids came to the island for its pleasant climate. In 1968 tourism overtook sugar as the main source of foreign currency and by 1980 about 370 000 tourists a year – most of them from Canada and the USA – were arriving.

The island is almost entirely surrounded by spectacular pink and white sand beaches and coral reefs. The principal tourist area is the Platinum Coast – a sheltered strip on the western shore north of Bridgetown.

At the same time that tourism began to develop, the government started to promote industry. There are now over 100 factories, largely concentrated in nine industrial parks. Some of them produce electrical goods for export. Unemployment, however, is high.

▲ NATIONAL PASSION Youths take a break during sugar-cane cutting. In Barbados, as in the rest of the English-speaking Caribbean, devotees of cricket abound. Sir Garfield (Garry) Sobers, captain of the West Indies team from 1965 to 1974, was born on the island in 1936.

BARBADOS AT A GLANCE
Area 430 km² (166 sq miles)
Population 253 000
Capital Bridgetown
Government Parliamentary monarchy
Currency Barbados dollar = 100 cents
Language English
Religion Christian (70% Anglican)
Climate Tropical maritime; average temperature in Bridgetown ranges from 21-28°C (70-82°F) in February to 23-30°C (73-86°F) in June to September
Main primary products Sugar cane, livestock (sheep, pigs, goats, poultry), fish and shellfish
Major industries Tourism, agriculture, sugar refining, molasses, rum, processed foods, textiles, electrical parts
Main exports Sugar, molasses, rum, clothing, electrical parts
Annual income per head (US$) 3780
Population growth (per thous/yr) 3
Life expectancy (yrs) Male 69 **Female** 73

Basel ***Switzerland*** See BASLE

Bashkiria (Bashkir Republic) ***USSR*** Oil-rich autonomous republic of the RSFSR covering 143 500 km² (55 400 sq miles) in the western foothills of the Ural mountains. Its main industries are oil, coal, steel, chemicals, timber, paper and electrical engineering. Sugar beet, potatoes and cereals are grown. The republic's capital is UFA. The people are mainly Russians, Tatars and Bashkirs (formerly nomadic Muslims).

Population 3 860 000

Map USSR Gc

basic rock Dark-coloured igneous rock containing no free quartz. It usually has a low silica content, up to 55 per cent, as opposed to an ACIDIC ROCK.

Basilan ***Philippines*** Island, covering about 1280 km² (495 sq miles), and city (formerly called Isabel) off the southern tip of Zamboanga peninsula of Mindanao. Geographically part of the Sulu Archipelago, it is administered from Mindanao; its people are mainly Muslim. Basilan produces timber, rubber, palm oil, coconuts and other crops; many people are fishermen.

Population (city) 171 000

Map Philippines Bd

Basilicata ***Italy*** Remote region of mountains and desolate hill country at the bottom of the Italian 'boot'. The region covers 9992 km² (3850 sq miles): the provinces of POTENZA, with the regional capital Potenza, and MATERA. It is agriculturally poor except for the irrigated coastal plain.

Population 618 200

Map Italy Ed

basin 1. Large depression on the earth's surface, which may be structural or caused by erosion. **2.** Area drained by a river and its tributaries.

Baska ***Yugoslavia*** See KRK

Basle (Bâle; Basel) ***Switzerland*** Mainly German-speaking city on the northern border. It is a river port at the highest point of navigation on the Rhine, and is surrounded by cherry and vine-growing country. Founded in Roman times, Basle is now second in size and manufacturing importance only to Zürich of all Swiss cities. It is a world centre of the pharmaceutical business, produces silk, and has a publishing industry. Exceptionally rich in museums and art galleries, Basle houses the national fairgrounds, where the annual Swiss Trade Fair is held, a Gothic minster, and Switzerland's oldest university, founded in 1459. The canton of Basle covers 465 km² (165 sq miles).

Population (city) 200 000; (canton) 223 400

Map Switzerland Aa

Basque Region ***France/Spain*** Area of south-western France and northern Spain, centred on the western end of the Pyrenees, whose native people have a unique language – unrelated to any other European tongue – and a strong tradition of independence. It corresponds roughly to the French department of Pyrénées-Atlantiques and the Spanish provinces of Alava, Guipúzcoa and Vizcaya. The main cities are BAYONNE in France and BILBAO in Spain. The Basques keep their traditions alive in folk dancing, costume and the game of *pelota*. An underground Basque Separatist movement – whose aim is a Basque state independent of both France and Spain – has carried out a number of shootings and bombings in recent years.

Map France Ce; Spain Da

Basra (Al Basrah) ***Iraq*** The country's chief port, exporting dates, cereals, wool and oil. It lies on the Shatt al Arab waterway, close to The Gulf, about 550 km (340 miles) south-east of the capital, Baghdad. It has an oil refinery built in 1974. Founded in AD 638 by Caliph Omar I, it was the legendary starting point for the voyages of Sinbad the Sailor.

Population 1 200 000

Map Iraq Cc

Bass Strait Situated between mainland Australia and Tasmania, the stormy strait contains offshore oil deposits. It was named after the British naval surgeon George Bass, who explored it in 1798-9.

Dimensions 290 km (180 miles) long, 130-240 km (80-150 miles) wide

Map Australia Hf

Bassar (Bassari) ***Togo*** Main town of the Bassari people, about 350 km (220 miles) north of the capital, Lomé. It is a market where cattle, millet and sorghum from the north are exchanged for palm products, rice, root crops, kola nuts and coconuts from the south.

Nearby are the forest reserves of Fazao and Malfakassa, which contain warthogs, monkeys, antelopes, buffaloes, cheetahs and leopards. There are also deposits of iron ore in the area, though they are not yet exploited.

Population 17 800

Map Togo Bb

Basse Casamance ***Senegal*** National park covering some 5000 hectares (12 360 acres), about 60 km (37 miles) west of the river port of Ziguinchor in the south-west of the country. It extends to within 12 km (7 miles) of the sea. It specialises in preserving threatened species, and contains warthogs, monkeys, baboons and a wide range of birds – including guinea fowl, pheasants, snipe, ducks and bustards.

Map Senegal Ab

Basse Terre ***Guadeloupe*** Seaport and capital of Guadeloupe, situated on the south-west coast of Basse Terre island. It stands at the foot of the Soufrière volcano, which erupted in 1797 and 1836 and was last mildly active in 1956. From the Soufrière, a chain of mountain peaks extends some 30 km (20 miles) up the spectacular western side of the island.

Basse Terre is the administrative centre of the French-ruled department of Guadeloupe, and its old town, dating from the 1640s, contains many French colonial buildings.

Population (island) 141 000; (town) 14 000

Map Caribbean Cb

Bassein ***Burma*** City in the Irrawaddy delta, about 140 km (85 miles) west of the capital Rangoon. It is the centre of the country's rice trade and has rice mills and storage silos.

Population 355 600

Map Burma Bc

Basseterre ***St Kitts and Nevis*** Seaport and capital, lying on the south-west coast of St Kitts. The chief industry is sugar refining, and sugar is the main export. Founded in 1627, the town was rebuilt after a fire in 1867. About 13 km (8 miles) to the west is Brimstone Hill, a fortress built on top of a 215 m (700 ft) rock. One of the major historical monuments in the West Indies, it took a century to build, and was abandoned in 1852.

Population 16 000

Map Caribbean Cb

Bastia ***France*** Principal commercial and industrial city of Corsica, 105 km (65 miles) north-east of Ajaccio. The old town is a maze of narrow streets. Fishing boats and yachts fill the old port, while ferries and commercial ships use the new harbour. Exports include wine and fish.

Population 51 700

Map France He

Bastogne ***Belgium*** Town in the Ardennes region 69 km (43 miles) south of the city of Liège. It was the focus of the Second World War Battle of the Bulge in 1944 when the refusal of American troops to surrender it thwarted a German attempt to break through Allied lines.

Population 11 000

Map Belgium Ba

Bas-Zaire ***Zaire*** Smallest and westernmost of Zaire's eight administrative regions, covering 53 920 km² (20 813 sq miles). It includes the towns of Matadi (its capital), Boma, Mbanza-Ngungu and Banana. Behind the narrow coastlands, the land rises to the forested African plateau. The region contains the massive Inga hydroelectric scheme on the Zaire river.

Population 1 922 000

Map Zaire Ab

Bata ***Equatorial Guinea*** Port and chief town of mainland Equatorial Guinea. It has no harbour, but there is offshore anchorage.

Population 30 700

Map Equatorial Guinea Ab

Bataan ***Philippines*** Peninsula and province of Luzon about 50 km (30 miles) long and 25 km (15 miles) wide, forming the western boundary of Manila Bay; off the southern tip is CORREGIDOR. Partly mountainous and densely wooded, Bataan was the site of fierce battles as American and Filipino troops resisted Japanese forces in 1942 during the Second World War. After its fall on April 9, 1942, prisoners of war were subjected to the infamous Death March to a prison camp near Cabanatuan, about 130 km (80 miles) north-east. Today Bataan is being developed as an industrial and free trade zone.

Population 263 300

Map Philippines Bc

Batangas ***Philippines*** Province and city in south-western Luzon, about 96 km (60 miles) south of Manila, facing the island of Mindoro. It is mountainous, with the active Taal volcano on an island in a huge crater lake (Lake Taal) in its north. Once important for growing coffee, the province contains a major steel works and oil refinery.

Population (province) 1 032 000; (city) 143 700

Map Philippines Bc

▲ **CITY AT WAR Iraq's charismatic president Saddam towers over shoppers in Basra on the Shatt al Arab, the waterway that the Iraqis and Iranians have fought over since September 1980.**

Batanghari ***Indonesia*** Largest river of Sumatra. Most of its 980 km (600 mile) length is navigable.
Map Indonesia Bc

Batéké Plateau ***Congo*** A vast plateau in the south-east of the country stretching northwards some 200 km (125 miles) from Brazzaville. The region, deeply dissected by river gorges and valleys, is generally unproductive. Covered largely by savannah, with strips of forest along its rivers, the plateau contains impressive amphitheatres, formed by huge rockfalls eroded from steep hillsides. The Téké people, a Bantu-speaking group, occupy the area.
Map Congo Ab

Bath ***United Kingdom*** Beautiful city in south-west England on the River Avon, 20 km (12 miles) south-east of Bristol. The suburbs cling to steep hills overlooking the centre. It owes its fame to the Romans, who exploited its hot mineral springs and called it Aquae Sulis, and to 18th-century developers who made it a fashionable watering place by building a stone city of squares and crescents as fine as anything in Britain. The dandy Richard 'Beau' Nash (1674-1762) brought the elite of society to Bath, where he set new standards of dress.

Despite fierce opposition from conservationists, some of this Georgian city has been demolished to make way for new houses and offices because its narrow valley restricts expansion. But much remains: the Roman baths have been excavated; the Pump Room, renovated in 1796, stands across a square from the 15th to 16th-century abbey; and there is a costume museum in the Assembly Rooms (built 1769-71). There is a modern university outside the city.
Population 85 000
Map United Kingdom De

batholiths Large irregularly shaped masses of intrusive igneous rock, usually granite. They extend to unknown depths and only their uppermost parts are exposed on the earth's surface by erosion. A batholith of relatively small size is sometimes called a stock.

Bathurst ***Australia*** City on the Macquarie river in New South Wales 160 km (100 miles) west of Sydney. It is the commercial centre for a sheep, wheat, beef and fruit producing area, and a tourist centre with good fishing. It was founded in 1815, only two years after the first European crossing of the GREAT DIVIDING RANGE, and is the country's oldest inland town.
Population 19 600
Map Australia He

Batinah ***Oman*** Narrow coastal plain facing the Gulf of Oman, about 20 km (12 miles) wide and 350 km (217 miles) long. Protected by the HAJAR MOUNTAINS, it is the most fertile part of the country, with a good supply of underground fresh water from oasis springs. Date palms, limes, oranges and breadfruit are among the most common crops. Because of its fertility and also the good fishing, the coast has attracted many invaders. Traces of ancient civilisations, some going back to the 3rd millenium BC, can be found throughout the area.
Map Oman Aa

Baton Rouge ***USA*** State capital of Louisiana, on the Mississippi river about 240 km (150 miles) from its mouth, but easily reached by ocean-going vessels. It is the seat of Louisiana State and Southern universities and a busy trade centre.
Population (city) 238 900; (metropolitan area) 538 000
Map United States Hd

Battambang ***Cambodia*** Western province, bordering Thailand. It is Cambodia's main rice-growing region, and lies beside the shores of the lake of Tonle Sap. Historically it is a disputed area between Thailand and Cambodia. The chief town, also called Battambang, is on the railway line between the Thai capital, Bangkok, and the Cambodian capital, Phnom Penh.
Population (province) 551 860; (city) 40 000
Map Cambodia Ab

Batu Caves ***Malaysia*** Series of spectacular caves in the foothills of the Main Range, 11 km (7 miles) north of the capital, Kuala Lumpur. The caves, Malaysia's most popular tourist attraction, are reached by a climb of 272 steps cut into the hillside. They open out, through brightly lit caverns of colourful stalagmites, into an underground temple dedicated to the Hindu god Subrahmanya. Beyond a cavern called the Dark Cave is the Art Gallery, a chamber with Hindu carvings and sculpture.

In January or February each year, the caves are the site of the festival of Thaipuram, when Hindu mystics demonstrate their faith by piercing their flesh with metal skewers, apparently without feeling pain.
Map Malaysia Bb

Batumi ***USSR*** Seaport, tourist resort and naval base in Georgia, on the east coast of the Black Sea 20 km (12 miles) from the Turkish border. It is connected by rail and oil pipeline with Tbilisi and Baku on the Caspian Sea. Its main industries are oil refining, chemicals, engineering, furniture and fruit and vegetable canning. It exports oil, manganese and citrus fruits.
Population 130 000
Map USSR Fd

Bauchi ***Nigeria*** Tobacco and tin-producing state covering 18 750 km² (7170 sq miles) in the north-east. Its capital, the town of Bauchi, lies some 820 km (510 miles) north-east of the national capital, Lagos. It grew rapidly after the rail link to Port Harcourt was completed in 1914 and a massive demand for tin arose in the First World War.
Population (city) 60 800; (province) 2 432 000
Map Nigeria Ba

Bautzen ***East Germany*** Historic town midway between Dresden and the Polish border. It is the centre of the Slavonic people called Sorbs, whose traditional colourful costumes, elaborate lace ruffs and ornate hats are still worn on special occasions. Industries include steel, railway vehicles and electrical appliances.
Population 51 210
Map East Germany Dc

bauxite Principal ore of aluminium. It is composed mainly of aluminium hydroxide. It forms as a result of leaching of the soil in tropical conditions where the clay minerals are decomposed and the silica is removed in solution, leaving an insoluble, alumina-rich soil. The name derives from Baux-en-Provence, France, which had a tropical climate 60 million years ago when its deposits were formed.

Bavaria (Bayern) ***West Germany*** The country's largest state, covering 70 553 km² (27 241 sq miles) in the south-eastern corner bordering Austria, Czechoslovakia and East Germany. It is a region of mountains, lakes, forests, farms, medieval towns and Baroque churches.

Industry clings to the two major cities: NUREMBERG in the north; and MUNICH, Bavaria's historic capital, in the south. Toys, textiles, cars, paper, chemicals, porcelain, glass and musical instruments are made.

Bavaria was ruled independently by the Wittelsbach family from 1180 until 1918 – first as a duchy, then, from 1806, as a kingdom. It is a staunchly Catholic province, and has Catholic universities at Munich and WURZBURG.
Population 10 958 000
Map West Germany Dd

Bay For physical features whose names begin 'Bay', see main part of name.

Bay, Laguna de ***Philippines*** Shallow crescent-shaped lake in central LUZON just south-west of Manila. It is the largest lake in the Philippines, covering 890 km² (344 sq miles), and is surrounded by fertile low-lying plains which are often flooded in the wet season. The plains on the southern side of the lake form the province of Laguna, whose fertile soil produces coconuts, rice, sugar cane and maize. The International Rice Research Institute at Los Baños, on the southern shore, has developed high-yielding, fast-maturing 'miracle' rice varieties.
Map Philippines Bc

bay bar Ridge of mud, sand or shingle, extending across a bay and usually linking two headlands.

Bayamon ***Puerto Rico*** Second city after San Juan and, by tradition, the first European settlement on the island; Spaniards settled there in 1508. It stands on the fertile plain, 10 km (6 miles) west of San Juan, in an area producing pineapples, citrus fruits, tobacco, sugar and coffee.
Population 202 500
Map Caribbean Bb

Bayern ***West Germany*** See BAVARIA

Bayeux ***France*** Market town on the Aure river, 27 km (17 miles) west of Caen in Lower Normandy. It was the first town to be liberated during the Second World War, in June 1944. The 800-year-old Bayeux Tapestry, a strip of embroidered linen 70 m (230 ft) long and 500 mm (20 in) deep, recording the Norman Conquest of England, is on display in the town's cultural centre.
Population 15 300
Map France Cb

Baykal ***USSR*** See BAIKAL

Baykonyr (Baikonur) ***USSR*** See KAZAKHSTAN

Bayonne ***France*** Capital of the French Basque region, on the Adour river just inland from the Gulf of Gascony. Its ancient cathedral and low arcaded streets contrast with the busy port and modern industries, including food processing and aircraft. Bayonets take their name from the town, where they were first made in the 17th century.
Population 129 730
Map France Ce

bayou Marshy creek or backwater along a river or coast, chiefly in southern USA.

Bayreuth ***West Germany*** Town in northern Bavaria, about 65 km (40 miles) north-east of Nuremberg, where the composer Richard Wagner (1813-83) built the Festspielhaus, or festival theatre, in 1876 for his operatic works, such as *The Ring*, to be performed in. A summer season of his work is held there each year. Wagner and his wife Cosima, daughter of the composer Franz Liszt (1811-86), are buried in the garden of their villa in the town.

Textiles are made in Bayreuth and it produces almost half of Germany's cigarettes.
Population 71 800
Map West Germany Dd

Bcharre (Basharri, Besharre, Bisharri) ***Lebanon*** Small town 32 km (20 miles) south-east of Tripoli. It stands in the Lebanon Mountains, 1828 m (5997 ft) above sea level. Bcharre is the birthplace of the poet and mystic Kahil Gibran (1883-1931), author of the international best-seller *The Prophet* – a study of the power of love.

The town is close to the Grove of Giant Cedars – one of the country's few remaining stands of cedar trees, though it is feared that these have been attacked by a disease and may not survive. It was of Lebanon cedar that Solomon (about 1015-977 BC), King of Israel, built the Great Temple in Jerusalem.
Map Lebanon Ba

Beanntrai ***Ireland*** See BANTRY

Bear Island (Bjørnøya) ***Norway*** Uninhabited island covering 178 km² (69 sq miles) in the Arctic Ocean, 386 km (240 miles) north of Norway. It has been recognised internationally as Norwegian territory since 1925, after a long history of disputes over territorial rights.
Map Arctic Rb

Beardmore Glacier ***Antarctica*** One of the world's largest known valley glaciers, about 200 km (125 miles) long and an average of 15 km (9 miles) wide. It descends at a rate of about 1 m (3 ft) per day from the South Polar Plateau to the Ross Ice Shelf, between the Queen Maud and Queen Alexandra ranges in the vast Transantarctic Mountains.
Map Antarctica d

Beaufort Scale Scale ranging from 0 to 12, devised in 1805 by the British Admiral Sir Francis Beaufort (and revised in 1926), to register the various velocities of wind, according to their effect at sea. Thus at scale 5, fresh breeze, 'white horses' appeared on the surface of the water. Today, the scale relates to exact wind speeds – 0 represents calm with little or no wind; 1 is 'light air', with winds of 1-5 km/h (1-3 mph); while 12 represents a hurricane with winds above 119 km/h (73 mph). See chart opposite.

Beaufort Sea A mostly ice-covered part of the ARCTIC OCEAN lying between northern Alaska and Banks Island (Canada). It was named after Admiral Sir Francis Beaufort (1774-1857), originator of the Beaufort scale of wind speeds. The Beaufort Sea has no islands and is partly unexplored, but large oil deposits have been found off the coasts of Alaska and north-western Canada. However, severe weather conditions make extracting and transporting the oil difficult. Depths of more than 3660 m (12 000 ft) have been recorded in the north.
Map Canada Aa

Beaujolais ***France*** Region on the north-east edge of the Massif Central, on the Saône river between Mâcon and Lyons. It is noted for its wines. The Beaujolais mountains, making up much of the region, rise to 1012 m (3320 ft).
Map France Fc

Beaune ***France*** Town in the Côte d'Or department, 35 km (22 miles) south of Dijon. It is the centre of the Burgundy wine-producing district, and manufactures casks and agricultural equipment. The town has 15th-century ramparts and the charity hospital – the Hôtel-Dieu or Hospice de Beaune – has been open since 1443; it is supported by a wine auction which is held every November.
Population 21 100
Map France Fc

Beauvais ***France*** Capital of Oise department, on the Thérain river, 65 km (40 miles) north-west of Paris. It manufactures blankets, carpets, chemicals and tractors; other industries include food processing and mechanical engineering. There is a 13th-century cathedral, with massive buttresses, that survived German bombing in 1940 which destroyed most of the old town.
Population 57 600
Map France Eb

THE BEAUFORT WIND SCALE

Scale No.	Description	Speed (km/h)	(mph)	Observed effects
0	Calm	0	0	Smoke rises vertically
1	Light air	1–5	1–3	Wind direction shown by smoke drift, but not by vane
2	Light breeze	6–11	4–7	Wind felt on face; leaves rustle; vane moves
3	Gentle breeze	12–19	8–12	Leaves and small twigs in motion; flags are extended
4	Moderate breeze	20–29	13–18	Raises dust; small branches move
5	Fresh breeze	30–39	19–24	Small trees sway; small crests on waves on lakes
6	Strong breeze	40–50	25–31	Large branches in motion; wind whistles in telephone wires
7	Near gale	51–61	32–38	Whole trees in motion
8	Gale	62–74	39–46	Breaks twigs off trees
9	Strong gale	75–87	47–54	Slight structural damage to houses
10	Storm/full gale	88–101	55–63	Trees uprooted; considerable structural damage
11	Violent storm	102–119	64–73	Widespread damage
12	Hurricane	Above 119	Above 73	Devastation

Béchar ***Algeria*** Oasis town with a picturesque fort, in the north-west Algerian Sahara, 467 km (290 miles) south-west of Oran. Béchar is an administrative centre at the beginning of the Route du Tanezrouft, one of the main trans-Saharan routes to the south, and also the start of the western trans-Saharan route to TINDOUF and Mauritania. The town is served by air and by rail; the railway ends just beyond Béchar at Kenadsa, where coal is mined.

Population 65 100
Map Algeria Aa

bed Layer or stratum of rock, usually sedimentary rock, divided from the layers above and below by bedding planes.

Bedford ***United Kingdom*** Town and county of south-central England. The county town stands on the Great Ouse river, 70 km (45 miles) north of London. The religious writer John Bunyan (1628-88) wrote *Pilgrim's Progress* in Bedford in 1675 during a six-month spell in jail for his nonconformist teachings.

The county covers 1235 km² (477 sq miles) and is largely agricultural, but there are extensive brickfields and engineering industries. The town of LUTON produces motor vehicles. The 18th-century stately home of Woburn Abbey, which now has a safari park, is one of Britain's major tourist attractions.

Population (county) 517 000; (town) 75 000
Map United Kingdom Ed

bedrock Solid unweathered rock that lies beneath all the layers of broken and weathered rock (regolith) and soil on the earth's surface.

Beechworth ***Australia*** Town in Victoria, 240 km (150 miles) north-east of Melbourne, in the foothills of the Victorian Alps. It was the centre of the Ovens gold-mining district in the 1850s and has several fine buildings dating from then, including a powder magazine (used to store explosives). Wangaratta, 32 km (20 miles) farther west, is now the main centre of the area, which produces wool, wheat, tobacco and grapes. The outlaw Ned Kelly (1855-80) and his gang operated in the district in the 1870s.

Population 3100
Map Australia Hf

Beersheba (Be'er Sheva) ***Israel*** Modern capital of the Negev (the southern part of Israel), about 72 km (45 miles) south-west of Jerusalem. According to tradition, it is the spot where Abraham pitched his tent 3800 years ago. The Ben Gurion University is at Beersheba. There are also archaeological remains and a Bedouin market. Its industries include glass and pottery.

Population 114 300
Map Israel Ab

Behariya ***Egypt*** See BAHARIYA OASIS

Behistun ***Iran*** See BISOTUN

Beijing (Peking, Peiping) ***China*** The country's capital is, in part, a paradox: an imperial city carefully preserved and tended by an anti-imperial Communist government. Lying on the extreme north-western edge of the North China Plain, it is the ancient seat of government and a modern industrial and trading city.

Beijing did not become a seat of government until the 10th century AD when the Khitans, a semi-nomadic Mongol people, made it their headquarters. In the 13th century, the first Yuan dynasty emperor, Kublai Khan (1215-94), grandson of the great Mongol warlord Genghis Khan, established his capital on a site to the north of the walls of the present-day city.

In 1421, Beijing became the imperial capital of the Ming dynasty (1368-1644). The third Ming emperor, Yong Le (ruled 1402-28), laid out a spacious walled city which, like many ancient Chinese cities, had its walls and streets orientated towards the cardinal points of the compass. Within the walls was the Imperial City and within that a further enclosure known as the Forbidden City – the emperor's court from which ordinary people were excluded on pain of death. It is now open to tourists. Beijing has remained the capital of China since then, except for the period 1928-49, when the Nationalist government was transferred to NANJING.

Today, Beijing spreads across 16 807 km² (6490 sq miles) in northern China, some 140 km (87 miles) from the shores of the Bo Hai (Gulf of Chihli). Its broad straight streets are thronged by bicycles, buses and pedestrians (private car ownership is practically unknown in China). Most government buildings stand in Zhongnanhai Park. There are more than 50 institutes of higher education, including Beijing University and Qinghua University. Qinghua University is a centre for scientific research. Industry, based in the outer suburbs, ranges from engineering and iron and steel to refining.

Above all, Beijing is a city of great historical and cultural importance. In the middle is Tiananmen Square, which covers 40 hectares (100 acres) and is one of the largest public squares in the world. It is used as a vast arena for political rallies and military parades. On the square's western side is the Great Hall of the People, built in 1959, where the National People's Congress meets. To the south stands the massive Memorial Hall of Chairman Mao, built in 1977 to house the remains of Communist China's founding father.

Beyond Meridian Gate, north of the square, lie the palaces of the Forbidden City, the largest group of classical Chinese buildings in existence. They include the Hall of Supreme Harmony, formerly the main throne room of the Chinese emperors, and the Palace of Heavenly Purity, where emperors presided over affairs of state.

To the north is Jing Shan, sometimes called Prospect Hill, the site of the Pavilion of a Thousand Springs, which commands a magnificent view – air pollution permitting – of the Forbidden City's golden roofs.

Beijing has many parks, some of them laid out as private gardens by emperors. The largest is the Temple of Heaven Park. Among its 15th-century temples stands the Qiniandian, or Hall of Prayer for Good Harvest, a 41 m (134.5 ft) high, cone-shaped building with triple eaves and a beautiful deep blue roof.

Beijing is in an area rich in history. The GREAT WALL OF CHINA snakes along the mountains about 80 km (50 miles) to the north. The road to the wall passes the Valley of the Ming Tombs, which contains the imposing mausoleums of 13 Ming emperors. The Summer Palace of the Dowager Empress Ci Xi, an autocrat who governed China as regent from 1861 to her death in 1908, stands in the city's north-west suburbs. It is a late 19th-century complex of ornate halls, pavilions and pagodas set among artificial lakes.

Population 9 231 000
Map China Hc

Beira ***Mozambique*** Deep-water port in the south of the country, about 190 km (120 miles) south-west of the Zambezi river mouth. Founded in 1891 on the site of an ancient Arab settlement, it developed as an outlet for the landlocked countries in the interior, and is now linked by rail and oil pipeline to Zimbabwe and by rail to Malawi. Beira exports minerals and cash crops, including cotton and sugar.

Population 113 800
Map Mozambique Ab

Beira ***Portugal*** Region in central Portugal divided into three provinces. The province of Beira Alta (10 525 km², 4063 sq miles) lies south-east of the city of Oporto and produces Dão wine. To the south, separated from it by the mountains of Serra da Estrela, is the province of Beira Baixa (6675 km², 2576 sq miles), which includes Covilhão, a town producing woollen textiles. Castles, guarding the frontier with Spain, are dotted along the eastern border of both provinces.

To the west of Beira Baixa is the coastal province of Beira Litoral (6755 km², 2607 sq miles), which lies due south of the city of Oporto. Rice is grown in the low-lying areas near the coast, and vines, olives, wheat and maize are grown inland.

Population (Beira Alta) 629 200; (Beira Baixa) 234 200; (Beira Litoral) 1 059 300

Map Portugal Bb

Beirut ***Lebanon*** Capital and principal port lying astride the small Beirut river in the shelter of the Lebanon Mountains. The city controls the main road and railway routes along the coast, as well as the links heading east through the Baider Pass in the mountains to Syria and Damascus.

Beirut was founded by the Phoenicians around the 14th century BC, and eventually it became an important Greek and Roman trading centre. It was also noted as a place of learning – the ruins of a Roman law school can still be seen. The tradition of learning continues in its four universities: the Lebanese University, the Arab University, the Université Saint-Joseph and the American University of Beirut. The city fell to the Muslim leader Saladin in the Crusades, then to the Ottoman Turks in the 16th century. In 1918, it was captured by the French and became the capital of the new state of Lebanon in 1920.

Beirut prospered as the chief financial and trade centre of the Middle East until the start of the present factional strife. Since 1975, civil war and the Israeli invasion of Lebanon have taken their toll – much of the city is badly damaged and continued sporadic outbursts of fighting have driven away business and tourism.

However, Beirut is still a free port and has a free market in gold, silver and currency. Though much of its business life has been halted – many of its leading banks, offices and hotels have been destroyed – the smaller businesses and workshops, particularly those engaged in textile and clothing manufacture, engineering, food processing and furniture, continue to operate. Whatever else may have changed, the city's climate has not. It is still possible, between December and March, to ski in the snow-clad mountains nearby in the morning and swim in the warm Mediterranean in the afternoon.

Population 938 000

Map Lebanon Ab

Beit Eddine (Bayt Ad Din) ***Lebanon*** Settlement and former capital when Lebanon was under the rule of the Ottoman Turks. It lies in the Lebanon Mountains 48 km (30 miles) south-east of Beirut and 792 m (2600 ft) above sea level. Beit Eddine contains the well-preserved palace of the Turkish Emir Beshir Chehab, Prince of Lebanon from 1804 until his death in 1840. One wing of the palace is now a folklore museum, and it is used as a summer residence for Lebanon's president.

Map Lebanon Ab

A DAY IN THE LIFE OF A BEIRUT SHOPKEEPER

Mitri Farah is an Orthodox Christian living in the western, Muslim-controlled sector of Beirut. The city is divided by the 'Green Line', the front line between warring Christian and Muslim militia groups. There has been serious conflict since 1975. Some of Mitri's relatives think him mad to live in the west, but at least it is still possible for Christians and Muslims to live together here. Besides, it was on this side of the city that Mitri's father established the tailor's shop that Mitri took over and in which he now sells fashionable imported clothes for the young.

Mitri and his wife Najla rent a three-bedroomed apartment in a modern seven-storey block in the fashionable district of Hamra.

This morning, as every morning, Mitri breakfasts on strong coffee and a large piece of pitta bread from the local bakery, spread with *za'tar*, a mixture of thyme and olive oil, made by his mother to her own special recipe. He kisses goodbye to Najla and his two children, Ghassan and Leila, and takes a shared taxi to his shop.

The children go by school bus to the Orthodox Christian school in the heart of the Hamra district. Mitri is increasingly anxious for their safety. He remembers the time when he received a visit at the shop from militiamen who asked him for a 'voluntary' contribution to the 'war effort'. Mitri told them he did not believe in war. That night they blew in his shop front with dynamite. So he paid his 'voluntary' contribution, repaired his shop and restocked.

Business is hard, too. Mitri has to pay dues to the militia controlling the port to clear his consignments of clothes, and then has to get them across the Green Line. Every day there is the danger of damage to his shop in another gun battle. Like everyone else here, Mitri wonders how long it must all go on.

Mitri goes home for lunch at 2 pm. The children are safely back from school (their day runs from 7.30 am to 1 pm), and Najla has cooked a family favourite, *Mahshi baazinjan*, stuffed aubergines. After a rest Mitri goes back to open up the shop again until 6 pm. Then he heads straight home before dark. He and Najla used to visit friends in the evenings, but these days it is too dangerous to go out after dark, and they only see their Christian friends at church on Sundays now. Their social circle consists of other people in the apartment block. After sitting together in candle-lit cellars during shelling, the people in the building have come to seem like family, even though most of the others are Muslim. The only evening entertainment for the Farahs alone is the TV or the video; mostly they watch foreign films.

Beit She'arim ***Israel*** Archaeological site about 15 km (10 miles) west of Nazareth. The site's remains include a 2nd-century synagogue and 3rd-century catacombs. In the 2nd century AD, when the area was under Roman rule, it was the seat of the Sanhedrin (Israel's High Court of Judges) following the destruction of the Temple in Jerusalem in AD 70.

Map Israel Ba

Beitbridge ***Zimbabwe*** Town on the border with South Africa. The Limpopo river marking the border is crossed by a road and rail bridge built in 1929 and named after Alfred Beit, who paid for it. It is the main route for freight between Zimbabwe and South Africa.

Population 1900

Map Zimbabwe Bb

Bejaia (Bougie) ***Algeria*** Minor seaport in Carthaginian and Roman times, lying on the Mediterranean coast, 152 km (94 miles) north-west of Constantine, in the middle of 30 km (19 miles) of fine beaches. In the 5th century, it became the fortified capital of Genseric the Vandal for a brief period before falling into decline. Refounded by the Berbers in the 11th century, it became a major port and cultural centre in north-west Africa. Today, Bejaia is a pipeline terminal, handling oil from the great HASSI MESSAOUD field. The port also exports fruits, cereals, olive oil, iron ore and phosphates.

Population 126 000

Map Algeria Ba

Békéscsaba ***Hungary*** Chief town of the south-eastern county of Békés, 178 km (111 miles) south-east of Budapest. It is a typical town of the Great ALFOLD (Plain), where single-storey, farm-like houses predominate. It is also a centre of Slav culture, where descendants of 18th-century Slovak settlers paint their houses pastel blue for distinction. Machinery and tool-making industries have been added to traditional textile making and food processing.

Population 70 100

Map Hungary Bb

Belait ***Brunei*** River about 160 km (100 miles) long in western Brunei which gives its name to the country's largest administrative district (area 2727 km², 1053 sq miles), through which it flows. The coastal area of Belait district, in the south-west of the country, is the traditional oil-producing area, but the interior, composed mainly of low forested hills, is hardly developed.

Population (district) 53 000

Map Brunei Ab

Belém ***Brazil*** River port on the north coast, near the mouth of the Amazon. The cathedral is the site of the Cirio, a religious festival held each October. The festival honours a statue of the Virgin Mary which is said to have several times been removed from its original site – only to return miraculously on each occasion. The city's principal products are rubber, nuts, tropical hardwoods, fish and vegetable oils.

Population 934 000

Map Brazil Db

Belau

TOURISTS DIVE AMONG HAUNTED WRECKS AND CORAL WONDERS

Ghost-grey warships haunt the coral canyons of Belau's reefs and lagoons; bombers and fighters bask beneath weed-fronded fathoms. Terrible battles raged in this Pacific paradise in the Second World War. Over 15 000 men died in ten bloody weeks, as Americans stormed the beaches, jagged ridges and cave-riddled Japanese redoubts of Peleliu (Beleliu) and Angaur, the two southernmost islands of the group. The Japanese fought almost to the last man: 13 600 died, and the Americans lost nearly 2000 – all for control of two dots of land that guard a strategic passage between the Pacific and Indian oceans.

In 1947 the islands, also known as Palau, became part of the UN Trust Territory of the Pacific Islands under US administration. In 1981 they became internally self-governing. A compact of 'free association' with the USA for 50 years was approved by Belauan voters in 1986. This provides for much needed US aid to Belau in return for control of the republic's defence, and land for military use if desired.

The capital, KOROR, with one of the Pacific's few natural deep-water harbours, is on an island of the same name – though at 10.5 km² (3.6 sq miles) it is not the largest; this is Babelthuap, 396 km² (153 sq miles), where coastal plains surround a jungle interior. Peleliu and Angaur are the other main islands. The people are largely Micronesian.

Copra is a major export, but tourism is increasingly important, and there are several world-class hotels. The old battlefields are one attraction, but many more tourists are drawn by the superb scuba diving, dazzling white beaches and the relaxed island lifestyle.

BELAU AT A GLANCE
Map Pacific Ocean Bb
Area 494 km² (191 sq miles)
Population 14 000
Capital Koror
Government Republic
Currency US dollar = 100 cents
Languages Local dialects, English
Religion Christian
Climate Tropical
Main primary products Fish, coconuts, cassava, fruit, vegetables, poultry, hogs
Major industries Fishing, agriculture, tourism
Main exports Copra, coconut oil, fish
Annual income per head (US$) 1100
Population growth (per thous/yr) 27
Life expectancy (yrs) Male 63 **Female** 66

Belfast ***United Kingdom*** Northern Ireland's capital, parts of which are divided into bitterly opposed sectarian groups. The city stands at the mouth of the River Lagan. Its name – *Béal Feirsde* in Irish, meaning 'mouth of the sand-pit' – first appears in the 15th century, but it was not until 1613 that the town received its first charter. By the beginning of the 19th century the population had grown to some 20 000, due largely to the cotton industry, which was later replaced by linen manufacture. Shipbuilding developed during the 19th century when the River Lagan was improved for navigation and land reclaimed for vast shipyards, notably those of Harland and Wolff. Engineering remains a major industry, especially aircraft manufacture.

The sectarian troubles that have dogged the city date back to the coming of Scottish and English settlers during what is called the Ulster Plantation in the 17th century. They introduced a new religious outlook into Catholic Ireland, and the rise of Belfast as an industrial centre in the 19th century drew in both Protestants and Catholics, sparking explosive rivalry.

The city hall, an ornate building whose interior has much Greek and Italian marble and a fine banqueting hall, stands at the heart of Belfast. Other gems include: Queen's University, built in 1849 as a Tudor-style college; the Ulster Museum and Art Gallery in the Botanic Gardens; Custom House on Donegal Quay, built of golden-coloured stone in Italian style; the Presbyterian General Assembly's headquarters; and the government buildings built in 1928 at Stormont to the east. The Grand Opera House, built in the 19th century, has a splendid rococo interior. The fine modern Romanesque Church of Ireland cathedral is now completed, after taking 80 years to build.

Population 360 000
Map United Kingdom Cc

Belfort ***France*** Town and region on the Savoureuse river, 40 km (25 miles) south-west of Mulhouse. The region remained French after 1870, when neighbouring Alsace was occupied by Germany. The town lies in a gap between the Vosges and Jura mountains, on an important trade route which links France to West Germany and Switzerland. Its industries include electrical and mechanical engineering and textiles.

Population (town) 76 200; (region) 132 000
Map France Gc

Belgium See p. 80

Belgrade (Beograd) ***Yugoslavia*** Capital of Yugoslavia and of Serbia, founded by Celts on a strategic site where bluffs rise 60-70 m (about 200-230 ft) above the junction of the Sava river with the Danube. The city, whose name means 'white citadel', was developed, in turn, by Romans, Slavs, Byzantines, Turks, Austrians and Serbs to guard the Morava-Vardar route to the Aegean against invaders from the north. The huge Kalemegdan fortress, now enclosing a park with pavilions, a military museum and keep, stands on the original site above the rivers.

A new city centre of government buildings was developed to the south between the World Wars, and after 1945 Belgrade expanded as a transport, international trade and industrial centre. An oil refinery was built, along with a new airport and factories producing machinery, electrical goods, farm equipment, consumer durables and foods.

Little remains of the old city. In the Second World War Belgrade was devastated by the Germans who captured it in April 1941. Yugoslav and Soviet forces liberated the city in October 1944. The 19th-century neo-Baroque cathedral has fine icons and tombs. The Church of Sv Marko (St Mark) stands beside Tasmajdan Park. Belgrade's university dates from the 1860s and the national museum is one of Europe's finest.

Population 1 407 100
Map Yugoslavia Eb

Belize See p. 82

Belize City ***Belize*** Former national capital beside the Caribbean Sea, and the country's commercial centre, with about one-third of its population. It is the main port for Belize and parts of Mexico, exporting sugar, timber, fruit and maize. The town consists largely of wooden buildings raised on stilts above the mangrove swamp on both banks of Haulover Creek.

After serious hurricane damage in 1961, the government was moved to Belmopan in 1970. Belize is at the mouth of the Belize river. Inland along the river are the remains of Maya settlements dating from before the 10th century AD. Altun Ha, 50 km (30 miles) north of Belize city, is the largest of the Maya sites.

Population 55 000
Map Belize Bb

Belle-Ile ***France*** Island, 16 km (10 miles) long, 12 km (7.5 miles) south of the Quiberon peninsula in Brittany. Its mild climate and varied landscapes of cliffs, coves, sandy bays and tiny fishing harbours make it a popular destination for artists and tourists.

Population 4200
Map France Bc

Bellingshausen Sea Lying off Antarctica in the South PACIFIC OCEAN between Thurston and Alexander islands. It was explored by a Russian expedition led by Captain von Bellingshausen in 1821.

Width 1130 km (700 miles)
Map Antarctica Ec

Belluno ***Italy*** Province and town about 80 km (50 miles) north of Venice. The town stands on a bluff overlooking the Piave river, which forms the central corridor of this mountainous province. The province stretches north into the Alps and the Dolomites, with many peaks exceeding 3000 m (10 000 ft) and ski resorts such as CORTINA D'AMPEZZO. The city's Venetian-style architecture includes the porticoed Palazzo dei Rettori, begun in 1491.

Population (town) 36 500; (province) 217 800
Map Italy Da

Belmopan ***Belize*** Newly developed capital, about 75 km (47 miles) south-west of Belize City. The colony's government moved there in 1970 to be outside the hurricane zone. The town, which contains a number of modern public offices built in the style of Maya architecture, is inhabited mainly by civil servants.

Population 5000
Map Belize Bb

Belgium

EUROPE'S BATTLEGROUND HAS SURVIVED AGAINST THE ODDS, AND PROSPERS AS THE CENTRE OF EUROPEAN UNITY

Among the countries of Europe, Belgium is a surprise. Not on account of its scenery, which is rarely spectacular; nor for its people, who are pleasantly average West Europeans with an above-average taste for beer and fried potatoes; not even for its small size since, at about 30 500 km^2 (11 600 sq miles), it is ten times as big as Luxembourg to the south.

The surprise stems simply from its existence – that there should be a Kingdom of Belgium at all. It was born of a revolution within the larger Kingdom of the Netherlands. It had to import a king from Germany (Leopold I, 1831-65), since it had no royalty of its own. It brought together two quite dissimilar groups, with different languages, and has lived through decades of tension and rivalry between them. Divided permanently now into two regions, with a special status for the capital, BRUSSELS, Belgium has surprisingly survived for 155 years.

UNHAPPY HISTORY OF CONFLICT

The Low Countries, comprising Belgium, the Netherlands and Luxembourg, have a more chequered history than almost any other European area of similar size. They have been held by Burgundians, Spaniards, Austrians and Frenchmen. They have been fought over by Protestants and Catholics, revolutionaries and imperialists: Belgium, in particular, has an unhappy history as the ground where other nations stage their battles.

The last three conflicts in this 'Cockpit of Europe' were the Waterloo campaign, when British and Prussians fought the French; the muddy trench warfare of the First World War, when the British and French fought with the Belgians against the Germans, and the Battle of the Bulge in the Second World War, when Americans and British fought the Germans.

The territories and boundaries of Belgium took a long time to congeal into their modern form. The present borders date only from 1919, and pay small regard to people. Leaving aside 40 000 German-speakers brought in by boundary changes round EUPEN and MALMEDY in 1919, and discounting for the moment several thousand people who speak a few ancient dialects, Belgium's 9.8 million population divides into two: Flemish-speakers and French-speakers (who are, confusingly, known as Walloons, although hardly any of them speak that language). The Flemish outnumber the Walloons by about six to four, and no other fact about modern Belgium is so important.

The long-standing geographical divide between Flemish and French runs from east to west roughly through Brussels. In the north Flemish is spoken; the language is virtually identical with Dutch, and the area shares the cultural and religious traits of the southern Netherlands. South of the line, French is spoken and the cultural ties are with France.

HOLDING THE NATION TOGETHER

Nothing would appear simpler than to divide the country along the language line, and apportion the halves to the Netherlands and France. But for 155 years the Belgians have been opposing that division, and this despite the fact that, since the First World War, holding the nation together has been an uphill struggle.

In 1830, when Belgium became independent, French was the language of government, society and scholarship, and Flemish considered fit only for peasants. The economic powerhouse of Belgium was in the south – on the coalfields that ran from MONS, through CHARLEROI to LIEGE, and this was one of the first industrialised areas in Europe.

Time brought changes – the decline of the southern coalfield; the search by newer industry for more space to develop; the impact of different birthrates. The birthrate in French Belgium fell; that of the Flemings was, and is, one of the highest in Europe. Furthermore, a Flemish cultural movement grew up to preserve and spread the language, and its success was marked by a law of 1932 making the Belgian government officially bilingual.

This achieved, the Flemish majority was set to consolidate its position. After years of turbulence, a law of 1970 effectively divided the country into four. In former German areas on the eastern frontier, German is the official language. Brussels, with its 1 million inhabitants, remains officially bilingual, with every notice and street sign printed in both Flemish and French. The nine provinces are divided – four Flemish in the north, four French in the south – and in each group only one language is official. The ninth province is divided in two.

To consolidate their progress, Flemish Belgians had to counter the former bias towards French in science, literature and higher education; otherwise, Flemish-speakers would still be debarred from reaching the highest levels in academic and industrial life. So it was further enacted that every order given in a plant or office in a Flemish province must be in Flemish: everything from invoices to notices about not smoking. At the same time the universities, except that of Brussels, became single-language. Belgium's oldest university, LOUVAIN – now Leuven and Flemish – was split in two, library and all, and the second half was established south of the language line, to operate in French.

Today, Belgium has, in effect, three governments: a national government with separate ministries of education and separate cultural budgets, a Flemish regional government and a Walloon regional government.

NATURAL DIVIDING LINE

There is a natural dividing line across Belgium as well as a language line, but the two do not coincide. The natural divide is the trenchlike valley of the MEUSE river. South of the river lie the uplands – among them the ARDENNES – rising by gentle slopes to almost 700 m (2300 ft): a region of sparse settlement, harsh winters, forests and tourism, where many old farm buildings have been turned into second homes and holiday centres. North of the river, where the rim of the valley rises to about 200 m (660 ft), there begins a slope which continues unbroken for 150 km (95 miles) to the sea. The slope is covered alter-

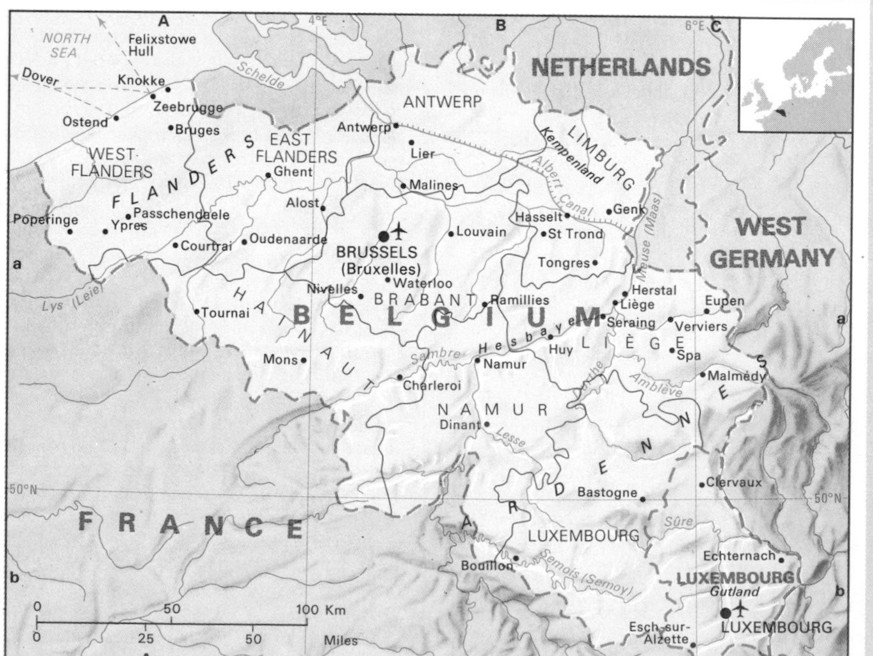

▲ GOTHIC GEM The Grand Place at the heart of Brussels, with its elegant pavement cafés, comes to life after dark. The richly carved Town Hall (left) was built between 1402 and 1454, and is topped by a statue of the archangel Michael, the patron saint of the city.

nately by fertile loams and barren sands or gravels; the coastlands are flat and grassy. This is the Belgium of dense population and ancient cities; of tiny farms alongside modern factories; of bicycles and canals.

In between the two lies the Meuse Valley, the industrial heart of Belgium from the time when its Iron Age metalworkers first began making swords until the closure in the 1960s and 1970s of its hopelessly uneconomic coal mines. Steel output at Liège and Charleroi reached nearly 13 million tonnes in 1970 (10.1 million in 1983). The valley floor became cluttered with industry and its paraphernalia. Today, the mines have gone and Belgium's coal now comes from the KEMPENLAND, farther north. Some of the industry has gone, too, but the Meuse Valley remains a place apart from the lands to north or south.

The cities of northern Belgium were among Europe's earliest boom towns, drawing their wealth from trade and textiles. In spite of repeated devastation by war, they possess a legacy of marvellous old streets, squares and churches. Some of them, like ANTWERP and GHENT, have survived to become the metropolitan centres of today. Others, of which the largest is BRUGES, have central areas little changed since medieval times.

Belgian farming is focused on livestock production, and most of the field crops are for feeding animals. There are areas that specialise in orchards, hops or sugar beet, and the glasshouses around Brussels are famous for their plants and vegetables. But many farmers have left the land for urban occupations and many others are part-timers; less than 3 per cent of the workforce is in agriculture, and farm labourers are hard to find.

Belgian industry has long and proud traditions, particularly in view of its small home markets. It survived the Second World War almost intact and then set itself, in the words of one Belgian, to produce 'calves with five feet' – manufactured goods in which ingenuity counted for much more than mass production. In the old wool cities of the north there are new textile industries producing a range of goods from carpets to clothing. In the relatively empty spaces of the Kempenland there are steel, chemical and car industries. In the Meuse Valley the traditional skills in engineering and the manufacture of arms live on. The great port of Antwerp processes imports from the tropical world, builds and repairs ships, and cuts diamonds in competition with its rival, AMSTERDAM.

Brussels is not only a city of a million people and many fine buildings, it is also, in a sense, the capital of Western Europe. The unity which Brussels is unable to impose upon the Belgians is found, paradoxically, in the larger setting of the European Economic Community (EEC). This owed its origins to a kind of pilot project – the agreement of exiled wartime governments of Belgium, the Netherlands and Luxembourg to form an economic union, Benelux. With the concept expanded to embrace France, Italy and West Germany in the 1950s, Britain, Denmark and Ireland in 1973 and, more recently, Greece, Spain and Portugal, the city of Brussels has flourished as the administrative headquarters of Western Europe's supranational government.

Today, Brussels is no longer a smaller, less glamorous version of Paris, or merely a place to change trains on the way to more exotic places, but a centre of business, information and policy making in its own right.

BELGIUM AT A GLANCE
Area 30 513 km² (11 781 sq miles)
Population 9 860 000
Capital Brussels
Government Parliamentary monarchy
Currency Belgian franc = 100 centimes
Languages Flemish, French, German
Religion Christian (90% Roman Catholic)
Climate Temperate; average temperature in Brussels ranges from −1 to 4°C (30-39°F) in January to 12-23°C (54-73°F) in July
Main primary products Cereals, sugar beet, potatoes, livestock, timber; coal
Major industries Iron and steel, light and heavy engineering, textiles, petroleum refining, chemicals, cement, glass, food processing, diamond cutting, coal mining
Main exports Chemicals, motor vehicles, foodstuffs, iron and steel, petroleum products, machinery, cut diamonds
Annual income per head (US$) 7870
Population growth Static
Life expectancy (yrs) Male 71 **Female** 76

Belize

BRITISH TROOPS GUARD ITS BORDERS, AND ALSO WATCH OUT FOR DRUG SMUGGLERS AND GUN RUNNERS

British troops have been stationed in Belize ever since it gained independence from Britain in 1981. Self-governing since 1964, the country was previously British Honduras, and had been a British colony since 1862. Today, the military presence is intended to protect Belize from the territorial claims of neighbouring Guatemala, which date back to 1821.

However, border patrols also limit the smuggling of marijuana, which is one of Belize's main sources of income (although it is illegal and bypasses the national coffers). And the patrols provide a barrier to arms smuggling through Belize from Cuba to El Salvador; like drug running, this brings money and employment to a country that is short of both.

A small but increasingly important legal source of income and work is now provided by tourists, drawn to Belize by the splendid Caribbean beaches and the ruined temples of the Mayan civilisation buried deep in the tropical forest. The Mayas flourished between about AD 300 and 900.

The coast of Belize is approached through 550 km² (210 sq miles) of coral cays and reefs. The coastal belt and the northern half of the country are low-lying and swampy. Behind the coast the land rises through a series of ridges, some scrub-covered, others densely forested. Southern Belize consists mainly of a broad plateau backed by the Maya Mountains, a succession of serrated ranges with peaks of up to 1100 m (3600 ft).

The new capital, BELMOPAN, in the mountainous and jungle-covered interior, was established in 1970 after the former capital, BELIZE city, was badly damaged by a hurricane in 1961.

The climate is warm and humid, with cooling sea breezes. Rainfall is heavy, especially towards the south, but there is a relatively dry period from February to May. Hurricanes can occur in late summer. Dense tropical forest covers the wetter plains and highlands, with scrub-type savannah in the drier places. The forest provides valuable hardwoods such as mahogany, logwood, lignum-vitae (guaiacum) and cedars.

The country was settled in the 17th century by pirates and by British woodcutters from Jamaica. Black slaves were imported later to work on sugar plantations. Today, 81 per cent of the people are black, Creole (a mixture of black and white races) or *mestizo* (a mixture of white and native Indian). There are minorities of Indians (descendants of the Mayas) in the interior and of Europeans, mostly in the towns.

Just under one-third of the workforce is employed on the land, growing the main export crops of sugar cane, citrus fruits (particularly grapefruit), bananas and coconuts. Others work in forestry or fishing.

However, the people are poor and making a living is difficult. Industries are underdeveloped and many people have emigrated, especially to the United States, to find work.

BELIZE AT A GLANCE
Area 22 963 km² (8866 sq miles)
Population 165 000
Capital Belmopan
Government Constitutional monarchy
Currency Belize dollar = 100 cents
Languages English, Creole, Maya, Spanish
Religions Christian (62% Roman Catholic, 22% Protestant)
Climate Subtropical; average temperature in Belize City ranges from 19-27°C (66-81°F) in January to 24-31°C (75-88°F) in August
Main primary products Sugar cane, citrus fruits, maize, rice, bananas, timber, coconuts, fish
Major industries Agriculture, sugar refining, rum, forestry, fishing, clothing
Main exports Sugar, citrus fruits, timber, bananas, fish products, molasses, clothing
Annual income per head (US$) 870
Population growth (per thous/yr) 23
Life expectancy (yrs) Male 65 **Female** 69

Belo Horizonte ***Brazil*** The country's third largest city after São Paulo and Rio de Janeiro. It lies 720 km (450 miles) south-east of the national capital, Brasília. It is the capital of the state of Minas Gerais – a region rich in gold and gemstones. The city's manufactures include cars, steel, textiles, cement and electric trains.
Population 1 777 000
Map Brazil Dc

Belorussia (Byelorussia; White Russia) ***USSR*** Largely agricultural constituent republic of the USSR bordering Poland. It occupies a low-lying forested plain. The main industry is the production of peat, which is the republic's major fuel; others are oil refining and engineering. Agricultural produce includes cereals, oilseeds, potatoes and livestock. The chief cities are the capital, MINSK, Bobruysk, GOMEL, VITEBSK and MOGILEV.

During the Second World War, as the gateway to Moscow, Belorussia suffered the full force of Nazi Germany's invasion; about one in every four of its people perished. After the war it was greatly increased in size by the addition of a large area of north-eastern Poland. The republic now covers 207 600 km² (80 150 sq miles). It is one of only two Soviet republics which have separate votes in the United Nations in addition to that of the Soviet Union as a whole; the other is the Ukraine.
Population 9 878 000
Map USSR Dc

Belostock ***Poland*** See BIALYSTOK

Belvoir, Vale of ***United Kingdom*** Lowland area some 20 by 10 km (12 by 6 miles) in central England, straddling the borders of Leicestershire and Nottinghamshire. It is overlooked by medieval Belvoir (pronounced 'Beaver') Castle, which was extensively restored in the 19th century. The vale is famous for Stilton cheese and is the site of Britain's newest coalfield, still under development.
Map United Kingdom Ed

ben Scottish term for peak, used as a prefix to many Scottish mountain names; for example, Ben Nevis, Ben More.

Ben Nevis ***United Kingdom*** See GRAMPIAN

Benares ***India*** See VARANASI

Bendigo ***Australia*** City, in Victoria, 130 km (80 miles) north-west of Melbourne. It was founded in 1851 during the Victorian gold rush and became one of the richest towns in the country. The Central Deborah Mine, which closed in 1954, is now restored as a museum. Its other industries include iron foundries and potteries. Originally called Sandhurst, the town was officially named Bendigo in 1891, after a local shepherd who took his nickname from the English pugilist Abednego William Thompson.
Population 52 700
Map Australia Gf

Benevento ***Italy*** Ancient city about 50 km (30 miles) north-east of Naples. It has a splendid Roman theatre, which could seat 20 000, the 1st century AD Arch of Trajan, and the Basilica of Santa Sofia, founded in 700 AD.
Population 76 600
Map Italy Ed

Bengal See BANGLADESH; WEST BENGAL

Bengal, Bay of Arm of the INDIAN OCEAN bounded by Sri Lanka, India, Bangladesh, Burma and the Andaman and Nicobar islands. It is known for its violent tropical cyclones, which have capsized large vessels and caused terrible floods in the delta of the GANGES and BRAHMAPUTRA rivers. A cyclone in 1970 killed an estimated 1 million people in Bangladesh, while 15 000 perished in floods caused by another cyclone in 1985.
Width 1600 km (990 miles)
Map India Dd

Benghazi ***Libya*** Major port and chief town of CYRENAICA at the eastern end of the Gulf of Sirte, situated on a tongue of land separated from the mainland by a salt lagoon and stretch of marsh. It has one of the country's two international airports and is a road and rail centre. Benghazi served as one of Libya's two capitals during the monarchy but lost importance after the 1969 revolution, when Tripoli was declared the sole capital.

Greek colonists established a settlement, Euhesperides, on the outskirts of the town, but this was abandoned in the 3rd century BC. Modern Benghazi stands on the site of the Graeco-Roman town of Berenice. A great deal of building was carried out by the Italians during their colonial occupation from 1912 to 1943. The city changed hands several times during the North African campaigns of the Second World War and was badly damaged. Many of the Italian and earlier Turkish buildings were destroyed, though a few remain in the old town.

Industries include tanneries, a shoe factory, leatherworks and textiles, carpets, olive oil, pasta and cement. The University of Garyounis is located here.

Population 266 196
Map Libya Ca

Benguela ***Angola*** Port about 410 km (255 miles) south of the capital, Luanda, and the capital of a province of the same name. Founded in 1617, it now has sugar-milling and fish-drying plants, and a rail link to the port of Lobito, 29 km (18 miles) away. Lobito is at one end of a 1415 km (880 mile) long railway network linking the inland republics of Zaire and Zambia to the coast. The railway is vital to Angola for the taxes it can charge to both countries on cargoes carried along it, and to Zaire and Zambia because the line is one of their main export trade routes. Since 1975 the railway has been a regular target for sabotage by rebel UNITA guerrillas.

Population (city) 41 000; (province) 475 000
Map Angola Ab

Benguela Current ***South Atlantic*** Cold ocean current flowing northwards along the west coast of southern Africa, and named after the Angolan port of Benguela. See OCEAN CURRENTS

Beni Abbès ***Algeria*** Oasis town about 160 km (100 miles) south of Béchar, set among date-palm groves at the beginning of the Route du Tanezrouft, the westerly trans-Saharan route to the south. Huge sand dunes surround the town, which has become a major centre for desert studies, including dune control and reclamation.

Population 4300
Map Algeria Aa

Beni Suef (Bani Suwayf) ***Egypt*** Trading town beside the Nile about 100 km (60 miles) south of the capital, Cairo. It processes cotton and other products from nearby El Faiyum.

Population 117 910
Map Egypt Bb

Benidorm ***Spain*** Holiday resort which sprang up after the Second World War around a Mediterranean fishing hamlet about 100 km (62 miles) south of the port of Valencia. During summer months its population swells to almost 1 000 000.

Population 25 600
Map Spain Ec

Benin

SLAVING AND VOODOO SACRIFICES ARE PART OF A SINISTER HISTORY, BUT BEAUTIFUL BEACHES AND A WEALTH OF WILDLIFE ARE ATTRACTING TOURISTS

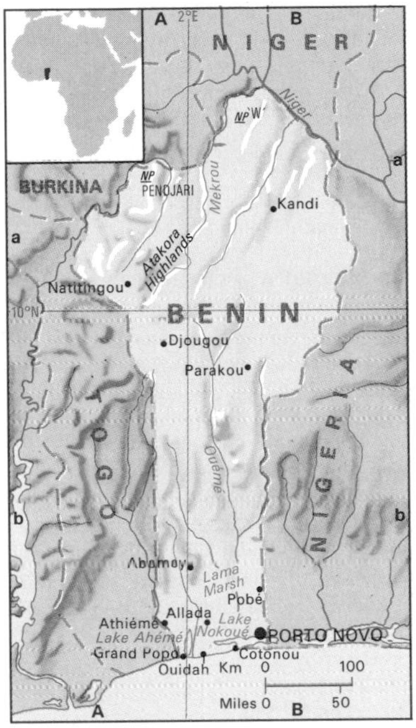

Shaped like an ice-cream cone standing on the southern coast of West Africa, Benin is one of the region's smallest but most diverse nations. Its short southern coast has idyllic beaches of white sand, backed by coconut groves and pounded by surf from the Bight of Benin.

Behind the coast lie tranquil lagoons such as Nokoue and PORTO NOVO, bordered by picturesque villages where square houses are built high on stilts beside the water. Rows of elegant dugout canoes rest by the water's edge – for here people live mostly by fishing, ferrying and, recently, tourism.

North of the lagoons are the fertile Terre de Barre lands where much of the tropical rain forest has been cleared for cultivation. These in turn give way to a deeply cut plateau covered with plantations of oil palm, coffee, oranges and tobacco.

In the north-west, from NATITINGOU to Pendjari Park, the ATACORA mountains are grassy, savannah-covered plateaus up to 800 m (2600 ft) high, separated by steep forested valleys.

Most of the people live in the equatorial south where the Fon and the Yoruba tribes predominate. Many are subsistence farmers, growing yams, cassava, sweet potatoes, maize, rice, groundnuts and vegetables for their own use.

With at least nine rainy months a year, giving an annual rainfall of 1500 mm (59 in) in the west and south, there is little fear of crop failure. In the grassy savannahs north of the Atacora, the rains last 'only' five to six months. Here, farmers in scattered villages plant millet, sweet potatoes and sometimes other vegetables, with a few cash crops such as cotton and groundnuts.

In north and central Benin the principal tribes are the Bariba – great horsemen – and the Somba, who build square, fortress-like houses. The nomadic Fulani range through the northern territory.

Benin, which was called Dahomey until 1975, was once a source of slaves for the slave traders. It was also notorious for the human sacrifices demanded by its traditional animist religions – particularly Voodoo.

Rival African kingdoms flourished in the area in the 18th and 19th centuries. It was occupied by the French in 1892, and became a territory of French West Africa in 1904.

Independence from France in 1960 led to lengthy regional power struggles, but after several coups political stability came in 1972, with the takeover by Mattheu Kerekou, who established a one-party Communist state. However, the country remains poor, over-dependent on cotton, palm oil and oilseed exports. Most of the country's trade passes through the port of COTONOU. Efforts to diversify the economy are suffering from a lack of foreign investment, though foreign aid is supporting the construction of a large hydroelectric plant at Nangbeto on the border with Togo.

Tourism is one major growth area, and although hotels and other facilities are few except in coastal towns, Benin has two of West Africa's most beautiful wildlife parks – the Pendjari, and the 'W' shared with Niger and Burkina. Both are home to hippopotamuses, crocodiles, elephants, lions, cheetahs, buffaloes and antelopes.

BENIN AT A GLANCE

Area 112 622 km² (43 484 sq miles)
Population 4 140 000
Capital Porto Novo (official), Cotonou (economic)
Government One-party republic
Currency CFA franc = 100 centimes
Languages French; also Fon, Yoruba and other African languages
Religions Animist (70%), Christian (15%), Muslim (13%)
Climate Tropical; average temperature in Cotonou ranges from 23°C (73°F) in August to 28°C (82°F) in March
Main primary products Cassava, yams, maize, sorghum, coffee, cotton, palm products, groundnuts
Major industries Agriculture, palm oil and palm-kernel oil processing, textiles, cotton ginning, beverages
Main exports Cotton, palm oil, palm kernels, cocoa (grown in Nigeria)
Annual income per head (US$) 260
Population growth (per thous/yr) 31
Life expectancy (yrs) Male 48 **Female** 52

Benin, Bight of Bay in the Gulf of GUINEA once known as the Slave Coast. It extends along the south coast of West Africa between Ghana and the NIGER RIVER delta.

Map Nigeria Ab

Benin City ***Nigeria*** Capital of Bendel state and centre of the Bini (or Edo) people, about 240 km (150 miles) east of the national capital, Lagos. It was founded in the 10th century and was the capital of the Kingdom of Benin, whose territory in the 15th and 16th centuries stretched westwards to include what is now Benin, and eastwards to the Niger river. The city was visited by the Portuguese in 1485 and was soon exporting slaves, leopard skins, elephant tusks, pepper and coral beads in return for such goods as guns and gunpowder, used by the Beninese for slave raiding.

Benin city is now a market for palm oil (used in margarine, cooking fats and soaps) and rubber. The old city, strongly influenced by the neighbouring Yoruba people, produced exquisite cast bronzes and carved ivories, and today its hundreds of craftsmen include tailors, woodcarvers and furniture makers.

Population 161 700

Map Nigeria Bb

Benioff zone Zone of earthquake activity that marks the path of a descending oceanic plate as it moves deeper into the earth's interior.

Benmore ***New Zealand*** See WAITAKI

Benue (Bénoué) ***Cameroon/Nigeria*** River some 1390 km (865 miles) long. It rises on the Adoumaoua Massif of central Cameroon and flows north, then west to the Niger river in Nigeria. A project based on the Lagdo (or Lagdom) Dam near the town of Garoua in Cameroon will irrigate 80 000 hectares (200 000 acres) for the cultivation of cereals, including rice, and sugar cane, and provide hydroelectric power. The Nigerian part of the river is navigable for about ten weeks each year, and also provides irrigation water and power.

Map Cameroon Bb

Benxi ***China*** Industrial city of Liaoning province, about 175 km (110 miles) from the North Korean border. It is a major producer of iron and steel and lies close to a coal-mining area.

Population 1 200 000

Map China Ic

Beograd ***Yugoslavia*** See BELGRADE

Beppu ***Japan*** Resort and spa on the north-east coast of Kyushu island. Beppu attracts millions of visitors each year to its 3000 hot springs and mud pools, which have names such as 'lake of blood' and 'waterspout of hell'.

Population 134 800

Map Japan Bd

Beqa'a (Biqa) ***Lebanon*** Fertile valley running down the centre of the country between the Lebanon Mountains and the Anti-Lebanon Mountains. It is about 16 km (10 miles) wide and 160 km (100 miles) long. Lebanon's two most important rivers rise in the Beqa'a: in the north, the ORONTES flows into Syria; and in the south, the LITANI flows south and then west. The waters of the rivers allow cotton and other crops, including citrus fruits, to be grown. In addition, the valley receives some 380 mm (15 in) of rain a year and is the country's main cereal-growing area.

In Roman times, Beqa'a was one of the granary regions supplying the empire, and now it is also noted for its wine. Today it is under the control of Syrian forces, and during the Israeli occupation of Lebanon the valley was the scene of much fighting.

Map Lebanon Ab

Berbera ***Somalia*** The country's second largest port after the capital, Mogadishu, standing on the north coast on the Gulf of Aden. It ships hides, gum and animals, mainly to Arab countries. The population of the town, which was founded before the 13th century, was swollen in the 1980s by refugees from the drought-stricken interior.

Population 40 000-70 000

Map Somalia Aa

Berchtesgaden ***West Germany*** See MUNICH

berg Term used in Germany for a single hill or mountain, such as Feldberg, and in South Africa for a range of mountains, for example Drakensberg.

Bergama (Pergamum) ***Turkey*** Town near the Aegean coast about 250 km (150 miles) south-west of Istanbul. It became part of the Roman Empire in 133 BC at the height of its prosperity. Among its many classical remains are shrines dedicated to the Greek deities Asclepius, the god of medicine, and Athena, the goddess of wisdom and the arts, and a terraced gymnasium.

Population 56 610

Map Turkey Ab

Bergamo ***Italy*** City about 50 km (30 miles) north-east of Milan. Bergamo is divided into the lower town, devoted to commerce and industry, and a smaller upper town, Bergamo Alta, which is enclosed within Venetian walls and holds its historic monuments. These include the Communal and Gombino towers, the medieval Ragione Palace, Santa Maria Maggiore church, and the startling Renaissance Colleoni chapel, built by a powerful local family in red and white marble. Works by painters of the Bergamese school are on display at the Carrara Academy, just below the old town. There is a theatre named after the opera composer Gaetano Donizetti (1797-1848), who began and ended his life in Bergamo.

Population 121 800

Map Italy Bb

Bergen ***Belgium*** See MONS

Bergen ***Norway*** Second largest city in Norway after Oslo and the largest until the 19th century. Bergen, on the coast of south-west Norway, was founded by King Olaf III in about 1070. It is a busy port and tourist centre. Bergen was a centre of the Hanseatic League – a mercantile association of mostly north German ports – until 1560. The old town focuses on the medieval harbour, near the former homes of Hanseatic traders and the restored Hall of King Haakon IV (1204-63). There is a 13th-century cathedral and some fine medieval churches, museums and galleries.

The Norwegian composer Edvard Grieg (1843-1907) was born at Troldhaugen, near Bergen, and the town has an annual music festival. Bergen's theatre, built in 1850, gained widespread recognition under such directors as the Norwegian playwright Henrik Ibsen (1828-1906), author of *Hedda Gabler*.

The city's main industries are shipbuilding, production of steel, electrical equipment, office machinery and textiles, oil refining and fish processing, and it is a centre for the North Sea oil industry. The Bergensbanen (Bergen railway) is a scenic rail route over the mountains to Oslo.

Population 181 300

Map Norway Bc

Bergen op Zoom ***Netherlands*** Town 108 km (67 miles) south and slightly west of Amsterdam, and close to the Belgian border. Formerly a fortified seaport, Bergen has been barred from the sea by a dyke which is part of the Delta Plan for land reclamation, but has canal links to the Rhine, Maas and Schelde rivers. Much of the medieval centre survives, and industries include sugar refining and iron making.

Population 45 200

Map Netherlands Bb

bergschrund Crevasse at the head of a glacier separating the moving ice from the ice adhering to the valley walls.

bergwind Hot, dry wind that blows from the plateau in South Africa down to the coast, warming as it descends.

Bering Sea Situated in the North PACIFIC OCEAN and bounded in the south by the ALEUTIAN and Komandorskiye islands. It is named after a Dane, Vitus Bering (1681-1741), who, sponsored by the Russian Tsar Peter the Great (1672-1725), explored the area. He entered the Bering Strait in 1728 and proved that Asia and North America were separate continents.

Area 2 269 000 km² (956 700 sq miles)

Greatest depth 5121 m (16 801 ft)

Map Alaska Ac

Bering Strait Separating the USSR from Alaska, the strait links the BERING and CHUKCHI seas. During the Pleistocene Ice Age (which began about 1.75 million years ago and ended about 10 000 years ago) the sea level fell, creating a land bridge between the continents. Crossing this bridge, perhaps 25 000 years ago, the ancestors of the American Indians began their colonisation of North America. In the middle of the strait, only 4 km (2.5 miles) apart, are the US Little Diomede and the Russian Big Diomede islands.

Width 88 km (55 miles)

Greatest depth 52 m (170 ft)

Map Alaska Ab

Berkshire ***United Kingdom*** County of southern England covering 1256 km² (485 sq miles), mostly to the south of the River Thames. The west has areas of fine open downland and chalk hills, while the east is lightly industrialised around the county town, READING, and is acquiring a reputation as England's Silicon Valley.

▲ TALL SHIPS' HAVEN The medieval town of Bergen, with its 12th-century Mariakirke and houses of Hanseatic merchants, nestles beside the harbour. The modern port lies across the water.

The county is often known as Royal Berkshire because it includes WINDSOR, with its royal castle.

Population 708 000
Map United Kingdom Ee

Berlin, East ***East Germany*** National capital since the division of Germany into East and West in 1949. It is a heavily industrialised city, producing machinery, electrical goods, textiles, furniture and food, but is overshadowed completely by its infamous wall built in 1961 to prevent its citizens from fleeing to West Berlin. The adjacent mined belt, cleared of buildings and overseen by watchtowers, has brought concentration-camp landscapes to the capital. But many of the historic buildings on the Unter den Linden – the beautiful linden trees were cut down under Adolf Hitler's regime – have been restored, including the opera house, national gallery, the Pergamon Museum (containing treasures from classical Greece, such as the altar of Zeus) and Humboldt University.

Population 1 196 870
Map East Germany Cb

Berlin, West ***West Germany*** Western part of the former German capital city, now an isolated enclave in the heart of Communist East Germany and 160 km (100 miles) from the border of West Germany. It remained a part of the West under the YALTA agreement of 1945.

Despite the difficulties of land communications, West Berlin is a lively, brightly lit outpost of the Western economic system. It is administered by a mayor and city parliament, but is still divided into the three sectors created at the end of the Second World War in 1945 – the French sector in the north, the British in the centre and the American in the south. Each country keeps a military presence in its sector.

From June 1948 until October 1949, Communist officials shut all roads and rail links between West Germany and West Berlin, but the blockade was thwarted by a massive Allied airlift.

In 1961, the East Germans built a wall dividing the West from the Russian sector of the city, in a bid to stop a steady drain of East German people to the more affluent West. Now all legal road crossings by holders of non-German passports must be made through Checkpoint Charlie, a heavily guarded crossing point near the Brandenburg Gate. West Berlin, once Germany's greatest industrial centre, had to be almost entirely rebuilt after the massive damage of 1945. Today its industries include chemicals, electronics and publishing.

Berlin was founded more than 1000 years ago when Slavs and, later, Teutons settled at what was then a marshy crossing of the river Spree. From the early 15th century, Hohenzollern princes ruled the Brandenburg province from Berlin, until it became the capital of united Germany in 1871. The Tiergarten, a 255 hectare (630 acre) former royal park, lies against the Berlin Wall in the centre of the city alongside the Brandenburg Gate, a partly restored triumphal arch erected in the 18th century, and the old Reichstag, the former parliament, which was burnt down in 1933, just after the Nazi leader Adolf Hitler became Chancellor. The Kurfürstendamm, once described by the American novelist Thomas Wolfe (1900-38) as the longest coffee shop in Europe, is the shopping and entertainment centre of the city. The world's first motorway, the Avus Autobahn opened in 1921, is still in use.

Population 1 900 000
Map West Germany Eb

Bermuda ***Atlantic Ocean*** Group of 150 small islands, only about 20 of which are inhabited, in the western Atlantic. Bermuda is situated about 920 km (575 miles) east of Cape Hatteras in North Carolina, the nearest point on the North American mainland. The largest island, Great Bermuda, is 21 km (13 miles) long, and is linked to the other main islands by bridges and causeways; their total land area is 53 km² (21 sq miles).

The islands were discovered by a Spaniard, Juan Bermúdez, in 1503, but were uninhabited until 1609 when a party of British colonists led by Sir George Somers was shipwrecked there. In 1684, the British Crown took over the settlement, and today Bermuda is a self-governing colony with one of the oldest parliaments in the Commonwealth, dating from 1620. It adopted a new constitution in 1968. More than half the inhabitants are descendants of African slaves, introduced to work on the land in the years before the abolition of slavery in 1834.

The hilly, limestone islands are the caps of ancient volcanoes rising from the seabed. Bermuda's mild Gulf Stream climate, together with its proximity to the USA, is the basis of a flourishing tourist industry. And a substantial income comes from the US naval and air bases on the islands, leased to the USA for 99 years in 1941. Many foreign banks and financial organisations operate to take advantage of the lenient tax laws for individuals. The capital, HAMILTON, stands beside a deep, land-locked harbour on the main island of Great Bermuda.

Population 55 000
Map Atlantic Ocean Bc

Berne (Bern) ***Switzerland*** Capital city on a narrow promontory in the gorge of the River Aare. Berne is the seat of national government, with national museums and libraries, but Zürich is the business and industrial centre of Switzerland. The Old Town of Berne was founded in the 12th century, but most of its buildings date from the 18th century. It is almost unspoilt, with its 15th-century minster, clock tower, arcades and fountains, and the bear pit near the river where the bears cavort for the tourists. This animal is Berne's heraldic symbol, and the city gets its name from the German *Bär*, meaning 'bear'.

The surrounding canton of Berne covers an area of 6050 km² (2336 sq miles). The canton's main crops are corn, potatoes and sugar beet, along with dairy produce. Its main industries are metalworking, watchmaking, and tourism, especially around lakes Brienz, Thun and Biel.

Population (city) 150 000; (canton) 926 900
Map Switzerland Aa

Bernese Oberland ***Switzerland*** Mountain area south of the national capital, Berne. It embraces the central Alps from Lake Geneva in the west to Lake Lucerne, and stretches southwards to the Rhône Valley. Its heart is the massif dominated by the Jungfrau (4158 m, 13 642 ft) and Finsteraarhorn (4274 m, 14 022 ft) mountains.

The area is one of the world's great tourist regions, with lakeside resorts such as Interlaken and mountain resorts such as Adelboden, Grindelwald and Gstaad.

Map Switzerland Aa

Berre, Étang de ***France*** Shallow lake, 20 km (12 miles) long and 5-13 km (3-8 miles) wide, linked to the Gulf of Fos midway between Marseilles and Arles. Deep navigation channels reach its northern shores from the Mediterranean. Industries along the lake shore include oil refining, petrochemicals, engineering and steelmaking.

Map France Fe

Berry ***France*** Ancient province dating from the 8th century and occupying some 7000 km² (2700 sq miles) on a low plateau just north of the Massif Central. It is drained by the Cher, Creuse and Indre rivers. It contains numerous large farms which produce wheat and fodder. Bourges is the main town.

Population 320 000
Map France Ec

Berwickshire ***United Kingdom*** See BORDERS

Berwick-upon-Tweed ***United Kingdom*** Northernmost English town, situated at the mouth of the River Tweed in Northumberland, about 5 km (3 miles) south of the Scottish border. It changed hands 13 times in the turbulent border history of the two countries. It is one of Britain's few remaining walled towns, the largely intact walls being last rebuilt in the 1760s. It is best known for the Tweed's salmon and for the Royal Border Bridge, a 28 arch structure built 38 m (125 ft) above the river by the railway engineer Robert Stephenson (1803-59) in 1850.

Population 12 000
Map United Kingdom Ec

beryl Mineral, essentially aluminium beryllium silicate, usually bluish-green or green and occurring in hexagonal prisms. It is used as a gemstone and is also the chief source of the metal beryllium.

Besançon ***France*** Fortress town originally settled in the Iron Age, on the flanks of the Jura mountains 75 km (47 miles) east of Dijon. Its massive citadel was built by Sébastien de Vauban, a 17th-century military engineer, to defend the eastern flanks of France from the Hapsburg Empire. There are also Roman ruins, including forts, an amphitheatre, aqueduct and a triumphal arch. Industries include textiles and watchmaking.

Population 120 800
Map France Gc

Beskidy (Beskids; Beskydy) ***Czechoslovakia/Poland*** Series of forested ranges along the border of Czechoslovakia and Poland which form the outer arc of the Carpathian Mountains. The highest peak is Babia Góra (Babia Hora), which rises to 1725 m (5659 ft) on the border between the two countries. The Pieniny, a forested chain of limestone crags, lie on the border farther east and rise to 982 m (3222 ft) at Trzy Korony (the Three Crowns). They are cut by the DUNAJEC gorge, and include a national park, which spans the border. Tourism has developed since the 1960s with the building of a 137 km (85 mile) circular road, the 'Beskidy Loop', from the town of Lesko in Poland.

Map Czechoslovakia Db; Poland Dd

Bessarabia ***USSR*** See MOLDAVIA

Bet Guvrin ***Israel*** Ancient city in the foothills of Judaea, about 40 km (25 miles) south-west of Jerusalem. It has Roman, Byzantine and Crusader remains. At the Biblical town of Maresha, nearby, there are 2nd-century burial caves.

Map Israel Ab

Bethlehem ***Israeli-occupied Jordan*** Town in the WEST BANK area, situated 8 km (5 miles) south-west of Jerusalem. It lies on a hill amid green, fertile country, and looks out to the Dead Sea and beyond. Bethlehem is regarded by Christians as the birthplace of Jesus Christ, and as a result the town has become an important pilgrimage and tourist centre.

From this small town in Judaea, the Hebrew king David started his conquest of the northern kingdom of Israel, making Jerusalem the capital and religious centre of the new state. Bethlehem fell into obscurity, ignored by all but those who looked to the town for the coming of the Messiah. The traditional place of the Nativity was desecrated by the Roman emperor Hadrian, who established a grove sacred to the god Adonis. The Byzantine emperor Constantine, a Christian, replaced the heathen grove in AD 315 and built the Church of the Nativity in its place. The church, rebuilt in the 6th century, is now shared by monks of Latin, Greek and Armenian orders. The site of the manger in which Jesus was born is claimed to be the grotto beneath the church.

Population 30 000
Map Jordan Ab

Béthune ***France*** Market and industrial town 27 km (17 miles) north-west of Arras. The old town was destroyed during the First World War but has been rebuilt. Its industries include chemicals, plastics and tyres.

Population 259 700
Map France Ea

Beuthen ***Poland*** See BYTOM

Beyrouth ***Lebanon*** See BEIRUT

Béziers ***France*** Leading centre of the Languedoc wine trade, on a hill 60 km (37 miles) south-west of Montpellier. It was once surrounded by defensive walls, which were eventually destroyed in 1632. The town is dominated by the massive Gothic church of Saint Nazaire. Most occupations are linked to the wine trade.

Population 83 200
Map France Ee

Bhadgaon (Bhaktapur) ***Nepal*** Town 16 km (10 miles) east of Kathmandu. Many of the brick and timber buildings are medieval, ornately decorated with gods and animals. There are Buddhist temples and monasteries, and a fine museum of Nepalese paintings. The town is also noted for pottery.

Population 40 100
Map Nepal Ba

Bhadra ***India*** See TUNGABHADRA

Bhairab Bazaar ***Bangladesh*** Town built at a crossing point of the Meghna river. The King George VI Bridge, which carries the Dhaka-Chittagong railway, is one of only two major river bridges in Bangladesh (despite the country having many rivers), the other being the Hardinge.

Population 63 700
Map Bangladesh Cb

Bhakra Dam ***India*** Dam on the Sutlej river in the Himalayan foothills, about 205 km (127 miles) east of the city of Amritsar. It was completed in 1963 and is 226 m (742 ft) high and 518 m (1700 ft) long. It is part of a project to provide water for the RAJASTHAN Canal, the world's longest irrigation canal, which flows into the THAR desert. The main canal is 692 km (426 miles) long, but its total length with all its distributaries is 6993 km (4346 miles). Below the dam is the Nangal Barrage, completed in 1952, which is used for hydroelectric power.

Map India Bb

Bhaktapur ***Nepal*** See BHADGAON

Bhamo ***Burma*** Town on the Irrawaddy river in north-eastern Burma, about 30 km (20 miles) from the Chinese border. It is the up-river terminus for steamers on the Irrawaddy and an important staging post for trade with China along the Burma Road, built between the two countries in the 1930s.
Population 65 000
Map Burma Cb

Bhilai ***India*** Steel town in central India, about 210 km (130 miles) east of the city of Nagpur. Its vast steel works, the country's largest, was built with Russian help in the 1950s.
Population 174 600
Map India Cc

Bhlarna ***Ireland*** See BLARNEY

Bhopal ***India*** Industrial city in central India, about 300 km (186 miles) north-west of the city of Nagpur. It produces heavy electrical equipment. Until 1956 it was the capital of the princely state of Bhopal, and the Nawab's lakeside palace is a tourist attraction. In December 1984 Bhopal was the scene of the world's worst chemical disaster, at the Union Carbide plant, when some 3000 people were killed and many others disabled by a gas leakage.
Population 671 000
Map India Bc

Bhubaneshwar ***India*** Pilgrimage city about 320 km (198 miles) south-west of Calcutta. It has some 500 Hindu temples, including the Great Temple, dating from about AD 650. This contains a granite statue of the god Shiva, the Lord of the Universe, which is bathed daily with water, milk and *bhang* (a marijuana derivative). Bhubaneshwar is the capital of Orissa state.
Population 219 200
Map India Dc

Bhumiphol Dam ***Thailand*** The country's first dam, built in 1964 across the Ping river. The dam provides hydroelectricity and water for irrigation; its output is 550 megawatts.
Map Thailand Ab

Biafra ***Nigeria*** See IBOLAND

Biala Podlaska ***Poland*** Town situated 145 km (90 miles) east of the capital, Warsaw, and about 40 km (25 miles) from the Soviet border. It has fine Gothic and Baroque buildings dating from the 16th and 17th centuries, when it thrived on the wealth of the powerful Radziwill family, and on the production of carpets, cloth and glass.
Population 44 400
Map Poland Eb

Bhutan

THE DRAGON KING RULES A REMOTE MOUNTAIN LAND WHICH FEW OUTSIDERS ARE ALLOWED TO SEE

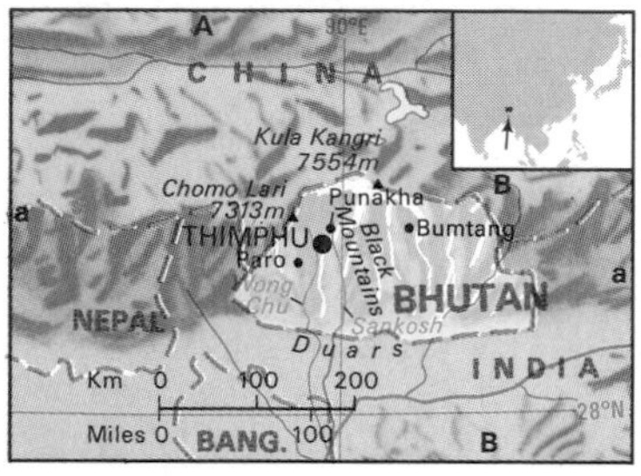

By the standard measures of our so-called advanced civilisations, the people of Bhutan are among the poorest on earth. But there is no starvation in their remote Himalayan country; there is no unemployment, no begging and virtually no crime. They are wary of wealth, because of the damage it would cause to their cultural traditions.

Although there is money in Bhutan – silver coins, until paper was recently introduced – most trade is conducted by barter.

Bhutan, known as the Land of the Dragon, rises from the jungle-covered foothills, which overlook the BRAHMAPUTRA river, to the white southern slopes of the HIMALAYAS. Heights range from 2000-7500 m (6500-24 500 ft). There was a state here as long ago as the 10th century, but apart from conflicts in defence of its territory, its contacts with the outside world have been few and cautious. Bhutan has in many ways stayed locked in the Middle Ages – a real-life version of the fictional Shangri-la.

In a nation of about 1.4 million people, there are 1300 fortress monasteries, or *dzongs*, which dominate the countryside as castles once did in Europe. Most of the people are Buddhist – although more primitive pagan beliefs influence their religion. The brightest sons of families are sent, sometimes when they are only three years old, to live in the dzongs and become monks. At colourful religious festivals, masked dancers enact legends in which Buddhist heroes vanquish the forces of evil.

Monks ruled the country until 1907, when they were replaced by an absolute monarch, the Dragon King. The present ruler, the youthful Jigme Singye Wangchuk, follows the regal customs established by his great-grandfather, dispensing justice and even marital advice to all supplicants.

The king is slowly opening Bhutan's doors to allow the outside world a glimpse through the crack. A strictly limited 1500 tourists are allowed in each year, and mountaineers are given access to a new Himalayan climb every two years, as long as they carry their own equipment and do not use the Bhutanese as porters. The few tourists and the sale of postage stamps supply most of the foreign money that comes in – an amount so small that in 1983 the purchase of a second 17 seater aircraft for the new national airline exhausted the country's meagre foreign exchange reserves.

About 95 per cent of the Bhutanese workforce are farmers. Yaks on the high pastures provide milk, cheese, meat and hair (used to make ropes from which many local bridges are made). Lower down potatoes, wheat and barley are grown, and lower still, rice is produced. But only 9 per cent of the country is cultivated, and 70 per cent remains forested. The demand for new farmland is low, partly because many sons become monks. Little forest has been cleared, and the Bhutanese hills have not been exposed to soil erosion as elsewhere on the southern Himalayan slopes.

BHUTAN AT A GLANCE
Area 47 000 km² (18 000 sq miles)
Population 1 447 000
Capital Thimphu
Government Constitutional monarchy
Currency Ngultrum (paper), tikehung (silver); Indian rupee
Languages Dzongkha, Nepali, English (all official)
Religions Buddhist (70%), Hindu (20%), Muslim (5%)
Climate Temperate; average temperature in Thimphu ranges from 4°C (39°F) in January to 17°C (63°F) in July
Main primary products Rice, millet, wheat, barley, maize, potatoes, oranges, apples, cardamom, timber, yaks; dolomite (a form of limestone), coal
Major industries Agriculture, forestry, handicrafts, mining
Main exports Postage stamps, rice, dolomite (limestone used in cement), handicrafts, fruits, coal
Annual income per head (US$) 85
Population growth (per thous/yr) 21
Life expectancy (yrs) Male 46 **Female** 44

Bialowieza ***Poland/USSR*** Forest astride the Polish-Soviet border east of Poland's capital, Warsaw. It was made a national park in 1921, 580 km² (220 sq miles) of it in Poland, and 700 km² (260 sq miles) in the USSR. European bison, elks, wild tarpan ponies, deer, wild boar and lynx wander between oak, fir, pine, spruce and magnificent hornbeam trees up to 40 m (130 ft) tall.
Map Poland Eb

Bialystok (Belostock) ***Poland*** Textile city in north-east Poland about 55 km (35 miles) from the Soviet border. It lay in the old Polish province of Podlasie, and is the site of the 'Podlasian Versailles' – an elegant Renaissance palace of 1728-58, which was built by the Branicki family and resembles the palace of the French kings near Paris.
Population 240 300
Map Poland Eb

Bianco, Monte ***France/Italy*** See BLANC, MONT

Biarritz ***France*** Fishing port and coastal resort 8 km (5 miles) south-west of Bayonne. It became fashionable after Napoleon III, nephew of Napoleon Bonaparte, visited it in the 1850s.
Population 26 700
Map France Ce

Bicol ***Philippines*** Winding peninsula forming

the south-eastern part of LUZON. It is mountainous, with several active volcanoes. Its isolation and poor internal communications resulted in slow economic development until recent years, but it has long produced copra and Manila hemp and has important copper, lead and zinc ore deposits. The main centres are Legaspi, capital of Albay province in the south-east, and Naga, capital of Camarines Sur province in the north-east; Naga was one of the earliest Spanish settlements in the Philippines. Larap, in Camarines Norte province, has the Philippines' most important iron-ore workings. Tiwi, 45 km (28 miles) north of Legaspi, has thermal springs and a geothermal power plant. The southern part of the peninsula is drained by the 120 km (75 mile) long Bicol river.

Map Philippines Bc

Bidar (Vidarba) ***India*** Fortified town about 500 km (310 miles) east and slightly south of Bombay. It was the capital of the Bidar kingdom in the 16th century. Bidar is noted for its traditional craft of inlaying iron with gold and silver to make decorative ware.

Population 177 400

Map India Bd

Bié ***Angola*** See KUITO

Bielawa ***Poland*** See DZIERZONIOW

Bielefeld ***West Germany*** Industrial city about 90 km (55 miles) south-west of Hanover. It produces machinery and glass.

Population 310 000

Map West Germany Cb

Bielsko-Biala (Bielitz) ***Poland*** Industrial city in southern Poland, at the foot of the Carpathian mountains, 42 km (26 miles) south of the city of Katowice. The twin towns of Bielsko and Biala, united since 1950, are separated by the Biala river. The city, which specialises in making high-quality woollen fabrics, now also produces textile machinery, electrical goods and engines for Polski Fiat cars.

Population 172 000

Map Poland Cd

Big Bend National Park ***USA*** Area covering 2866 km² (1106 sq miles) of Texas in a large bend of the Rio Grande river on the Mexican border. It is a place of desert, canyons and mountains.

Map United States Fe

bight Indentation in the sea coast, similar to a bay but with a wider, gentler curve.

Bihac ***Yugoslavia*** See GRMEC PLANINA

Bihar ***India*** Densely populated north-eastern state covering 173 877 km² (67 117 sq miles). Although rich in minerals and rice – and with many coal mines and steel plants – it is the country's poorest state and most of its inhabitants are illiterate or backward. Wheat is grown, but due to a chronic food shortage, grain has to be imported into the state.

Some 80 per cent of the people are employed on the poorly irrigated land, and most of them live in crowded rural settlements on the Ganges plain. The main towns are Bihar Sharif, a place of pilgrimage for Muslims with its holy graves and mosques, Patna and Gaya.

Population 69 914 700

Map India Dc

Bijapur (Vijayapura) ***India*** Muslim city about 386 km (240 miles) south-east of Bombay. It has many fine mosques, forts and mausoleums – one of which, the Gol Gumbaz, is noted for its domes. The central dome itself, at 38 m (125 ft) diameter, is larger than St Paul's, London (33 m, 108 ft), but smaller than St Peter's, Rome (42 m, 138 ft).

Population 147 300

Map India Bd

Bikini ***Marshall Islands*** Atoll 750 km (470 miles) north-west of Majuro where the USA tested 23 nuclear weapons (including the first hydrogen bomb) in 1946-62. The people were resettled on another island, but sued the US government for compensation. In 1983 they were promised more than US$150 million over 15 years; restoration of the atoll could cost US$42 million.

Map Pacific Ocean Db

Bilbao ***Spain*** North-eastern port and industrial city 11 km (7 miles) south of the Bay of Biscay, to which it is linked by the canalised River Nervion. It exports iron ore, lead and wine and is capital of the Basque province of Vizcaya.

Population 433 000

Map Spain Da

bill Long, narrow promontory or small peninsula.

billabong Australian term for a stagnant pool or backwater extending from the main stream of a river.

Billings ***USA*** City on the Yellowstone river in Montana, 85 km (53 miles) north of the Wyoming border. It is a market, servicing and processing centre for the surrounding farms.

Population (city) 69 800; (metropolitan area) 118 700

Map United States Ea

Billund ***Denmark*** See VEJLE

Binga Mountain ***Mozambique*** See CHIMOIO

Bingerville ***Ivory Coast*** Colonial-style town overlooking the Ebrié Lagoon, 16 km (10 miles) from the capital, Abidjan. It was the capital of Ivory Coast from 1900 to 1934. The town, named after the country's first French governor-general, Louis Binger, has botanical gardens and an agricultural research centre.

Population 20 000

Map Ivory Coast Bb

Bintan ***Indonesia*** Island of the Riau archipelago immediately south of Singapore. It covers 1075 km² (415 sq miles) and has rich deposits of bauxite (aluminium ore).

Map Indonesia Bb

Bintulu ***Malaysia*** Coastal town in Sarawak about 350 km (220 miles) north-east of Kuching. Its petrochemicals industry processes natural gas from the offshore Luconia fields for export to Japan, and to process into artificial fertiliser.

Population 59 000

Map Malaysia Db

Bío-Bío ***Chile*** **1.** Largest river in the country, 380 km (235 miles) long. It rises in the Andes of central Chile, near the Argentine border, and flows north-west to enter the Pacific near Concepción. Its upper tributaries are harnessed to provide hydroelectric power. The river forms a natural divide between central and southern Chile. **2.** Central region of Chile that is heavily industrialised around Arauco and CONCEPCION, with coal mines at Lota, Lebu and Coronel and a steel plant at Huachipato. Near Los Angeles, in the foothills of the Andes, is an area of outstanding natural beauty typified by the lake of Laja, its river and falls, and the towering volcanoes of Antuco, Tolhuaca and Copahué, which rises to 2969 m (9741 ft).

Population 1 517 000

Map Chile Ac

Bioko ***Equatorial Guinea*** Tropical volcanic island of 2017 km² (780 sq miles) in the Gulf of Guinea. Formerly called Fernando Póo, it was occupied by Portugal, Spain and Britain before independence in 1968. The national capital, Malabo, stands on the island, which produces cocoa, bananas, coffee, palm products, sugar and timber.

Population 57 000

Map Equatorial Guinea Aa

Biorra ***Ireland*** See BIRR

biotite Dark green to black mica found in igneous and metamorphic rock.

Biratnagar (Morang) ***Nepal*** Industrial city in the far south-east of Nepal, near the border with Bihar in India, about 120 km (75 miles) south-west of Darjiling. Like other border towns, Biratnagar trades heavily – and mostly illegally – with India, untaxed goods disappearing over forest and country trails. A hot and humid place, not on the tourist itinerary, it has jute and chemical factories, and a sugar mill.

Population 90 000

Map Nepal Ba

Birim Valley ***Ghana*** The country's largest diamond field. It lies in the Birim river valley, about 80 km (50 miles) west of the southern part of Lake Volta. The field, discovered in 1919, yields 90 per cent of Ghana's total production. The diamonds are of industrial grade.

Map Ghana Ab

Birkenau ***Poland*** See AUSCHWITZ

Birkenhead ***United Kingdom*** See WIRRAL

Birmingham ***United Kingdom*** Second largest city of the UK after London, lying in west-central England. It is often called 'the heart of England', and is the main city of the West Midlands conurbation that extends north-west to Wolverhampton. It was little more than a market town before it was overwhelmed by a wave of industrialisation in the 19th century. With coalfields to the east and west, and a maze of canals linking mines to factories and factories to seaports, the area became known as the Black Country. One of the earliest railways linked Birmingham with London, 160 km (100 miles) away to the south-east.

The engineer James Watt produced the world's first rotative steam engines – the power source of the Industrial Revolution – in Birmingham in the 1780s. Today the city has an almost endless list of industries, including jewellery, guns, brassware, buttons, wire, toys and motor vehicles. Chocolate manufacture created a whole suburb, Bournville.

Its industrial importance led to the building of grandiose public offices in the 19th century, such as the town hall, museum and art gallery. The Anglican cathedral is 18th century and its Roman Catholic counterpart dates from 1841. Some of Britain's earliest urban planning and legislation originated in Birmingham.

Despite the decline of many of its older industries, it remains one of Britain's leading commercial and industrial centres. It is home for the National Exhibition Centre, next to its airport, and it has two universities and a polytechnic.

Population 920 000
Map United Kingdom Ed

Birmingham ***USA*** Largest city of Alabama. It lies in the centre of the state in a valley of the Appalachian foothills. The city grew rapidly during the Civil War (1861-5) when local iron ore and coal were first utilised on any scale. Steel is still produced and has been joined by a range of other industries, including textiles and chemicals. The city was the scene of black civil rights riots in the 1960s.

Population (city) 279 800; (metropolitan area) 895 200
Map United States Id

Birr (Biorra) ***Ireland*** Georgian market town, 113 km (70 miles) west and slightly south of Dublin. It was formerly known as Parsonstown, after the Parsons family who became Earls of Rosse. Several were noted astronomers, and the 17th-century castle, on the western edge of the town, contains the tube of a telescope which when built in 1845 was the world's largest. The castle has 200-year-old box hedges which, at more than 9 m (30 ft), are said to be the highest in the world.

Population 3680
Map Ireland Cb

Biscay, Bay of Inlet of the ATLANTIC, called the Gulf of Gascony by the French, and noted for its high tides, sudden, severe storms caused by westerly and south-westerly gales, and strong currents. Lord Byron wrote of the 'rude' winds in 'Biscay's sleepless bay'. The bay extends between the Isle d'Ouessant in BRITTANY, in north-western France, and Cape Ortegal in north-western Spain.

Greatest width 480 km (300 miles)
Greatest depth 366 m (1200 ft)
Map Europe Cd

Biskra ***Algeria*** Oasis town and communications and administrative centre, 190 km (118 miles) south-west of Constantine, just south of the Aurès mountains. Biskra lies at the start of the main eastern Algerian route to the far south. Since Roman times, when it was the military base of Vescera, Biskra has been a fashionable resort because of its good climate and hot springs. At the entrance to the town is the spa Hamma Salahine, famous for its sodium sulphate waters, which are the hottest in the world after those of Iceland. The remains of the Roman town lie nearby, as well as a number of ancient dolmens (prehistoric stone tombs). A main attraction is the huge palm groves following the course of the oasis.

Population 90 500
Map Algeria Ba

Biskupin ***Poland*** One of Europe's best-preserved prehistoric swamp towns. It was discovered in 1933 on a peninsula in Lake Biskupin, 84 km (52 miles) north-east of the city of Poznan. The oval settlement covering 40 hectares (100 acres) dates from about 550 BC, and was built by the Lusatians, the forerunners of Slav civilisation. An earth-and-timber rampart, incorporating more than 35 000 stakes, encloses 12 timber-paved streets lined by 105 wooden huts and workshops. It has a museum.

Bismarck ***USA*** State capital of North Dakota, standing on the Missouri river 240 km (150 miles) south of the Canadian border. It is an agricultural processing and servicing centre.

Population (city) 47 600; (metropolitan area) 86 100
Map United States Fa

Bismarck Archipelago ***Papua New Guinea*** Group of mainly volcanic islands off the north-east coast of New Guinea. It includes NEW BRITAIN, NEW IRELAND and the Admiralty Islands (whose main island is Manus). The economy relies largely on agriculture and fishing; there are extensive coconut and cocoa plantations. The population is primarily Melanesian, but there is a substantial Chinese and European element. The main town is RABAUL in East New Britain.

Population 315 000
Map Papua New Guinea Ba

Bismarck Range ***Papua New Guinea*** Mountain range in the north-east of New Guinea island. Its highest point, and the highest point in the country, is Mount Wilhelm at 4509 m (14 793 ft).

Map Papua New Guinea Ba

Bismarck Sea Arm of the PACIFIC which was the scene of a victory by US aircraft over a Japanese naval force in 1943. It is bounded by the BISMARCK ARCHIPELAGO and New Guinea.

Width 800 km (500 miles)
Map Papua New Guinea Ba

Bisotun (Behistun) ***Iran*** Sugar-refining town, about 400 km (250 miles) south-west of the capital, Tehran. Outside the town, carved into a cliff face at a height of 60 m (200 ft), there is a bas-relief more than 2000 years old depicting the triumph of the Persian king, Darius I (ruled 522-486 BC), over his enemies. The inscription, repeated in three languages (Persian, Akkadian and Elamite), enabled 19th-century scholars to decipher Babylonian cuneiform script. At the foot of the cliff there are two Parthian bas-reliefs, dating from the 2nd century BC and the 1st century AD.

Map Iran Aa

Bissau ***Guinea-Bissau*** The country's capital, main commercial city and port, on the right bank of the Gêba river estuary. Founded by the Portuguese as a trading post and fort in 1692, it became the administrative capital of Portuguese Guinea in 1941. With good port facilities, it grew rapidly after 1945. Bissau retained this dominance, although MADINA DO BOE was named as the country's new capital in 1980; it still has most of the country's few manufacturing industries – the processing of farm products and timber – and handles 85 per cent of its overseas trade.

An attractive town, Bissau retains some of its colonial splendour, with wide tree-lined streets and a seaside promenade. In spite of its poverty, peeling paintwork and crumbling masonry, it has a charming atmosphere and several points of interest, including Fort Sao José (built in 1693), the museum (with a good collection of early African carvings and artefacts) and the massive Roman Catholic cathedral. There are magnificent beaches nearby.

Population 109 000
Map Guinea Aa

Bitola (Bitolj; Monastir) ***Yugoslavia*** The country's southernmost city lying 110 km (68 miles) south of Skopje in the fertile Pelagonia Valley in the republic of Macedonia. A major Turkish headquarters from 1820, where Ataturk (1881-1938), founder of modern Turkey, trained at the military academy, it became Macedonia's largest town. The Balkan, World and Greek Civil wars brought decline, but subsequent drainage and farm improvements in Pelagonia produced wheat, maize, sugar beet, cotton, rice, tobacco and poppies for the city's processing plants. Other new factories produce carpets, clothing, leather, beer, metal goods and household appliances. Bitola is also a tourist centre for Lake Prespa, Lake Ohrid and the national park around Mount Perister, which rises to 2600 m (8530 ft) west of the city. The park is noted for magnificent beech and black pine forests, trout-stocked rivers and streams, bears, boar and deer. Just south of the city lie the ruins of Heraclea (Lyncestis), a Roman town with outstanding mosaics.

Population 137 800
Map Yugoslavia Ed

Bitter Lakes ***Egypt*** Two lakes – the Great and Little Bitter Lakes – lying between Isma'iliya and Suez and forming part of the Suez Canal. Before the canal was cut in 1859-69, the lakes were marshy depressions 8-12 m (26-36 ft) below sea level.

Map Egypt Cb

bitumen Group of solid or liquid hydrocarbons which occur naturally as asphalt (pitch), tar and crude oil, as in Trinidad's Pitch (or Asphalt) Lake.

Biu Plateau ***Nigeria*** Granite plateau rising to about 700 m (2300 ft) in the north-east. It is dotted with extinct volcanoes, and its farmers produce millet, vegetables and cattle to feed themselves.

Map Nigeria Ca

Biwa ***Japan*** The country's largest lake. It lies on Honshu island, just north-east of Kyoto,

and covers 675 km² (260 sq miles). The reflection of Hikone Castle in Biwa's shimmering waters is a famous attraction.
Map Japan Cc

Bizerte (Bizert, Binzert) ***Tunisia*** Port and heavily fortified naval base situated on Tunisia's northern coast. Lying on a channel leading from Lake Bizerta, it is the site of one of the earliest Phoenician settlements in 800-700 BC. It became a Roman colony (Hippo Zaritus), was taken by the Arabs in the 7th century, by Spain in 1535 and then, in 1881, by the French who turned it into a major naval base. Bizerte was fought over in the Second World War and captured from the Germans by the Americans in 1943. Among the places of special interest are the old port area, a huge 16th-century fort and the 17th-century Great Mosque.

Claimed to be the safest harbour on the coast, Bizerte handles iron ore exports. Its chief industry is oil refining; cement and steel are also produced. Major industries are also located at nearby Inenzel Bourguiba.
Population 94 600
Map Tunisia Ad

Bjeshkët e Némuna ***Albania/Yugoslavia*** See ALBANIAN ALPS

Björneborg ***Finland*** See PORI

Black Country ***United Kingdom*** English Midlands industrial area centred on the city of BIRMINGHAM. The area got its name because of its concentration of coal mines and factories. It stretches approximately from Castle Bromwich in the east to WOLVERHAMPTON in the north-west, and encompasses most of the West Midlands. Today, clean air legislation and the decline of mining and traditional small metal-working industries have lightened the landscape and left a decreasing legacy of dereliction.
Map United Kingdom Ed

Black Forest (Schwarzwald) ***West Germany*** Pine-clad mountainous region in the south-west corner of the country, extending between the Swiss border and the city of Karlsruhe, 160 km (100 miles) to the north. Its highest point is Feldberg, at 1493 m (4898 ft).

The forest contains several spa towns. It is crisscrossed by paths, and there are many campsites, youth hostels and holiday homes. There is scattered farming in the valleys around distinctive Black Forest buildings, with their eaves stretching almost to the ground. The main towns on its fringes are Freiburg im Breisgau, Pforzheim, and Baden-Baden.
Map West Germany Be

Black Hills ***USA*** Range of hills on the border between the states of South Dakota and Wyoming, rising to 2207 m (7242 ft) at Harney Peak. The hills were sacred to the Indian tribes, but were settled by white people in the 1870s when gold was discovered; it is still mined.
Map United States Fb

black ice, glaze ice A thin, glassy coating of ice on the ground or on exposed objects, caused by freezing weather after a partial thaw, or by the freezing of rain or fog droplets on impact with a cold or freezing surface. Also called glazed frost, silver frost and silver thaw.

▼ LIGHT AND SHADE Sunny pastures lie along the Wiese valley which winds through the Black Forest. Dark conifers clothe the heights – hence the name.

Black Mountains ***Bhutan*** Mountain range, rising to almost 5000 m (16 400 ft), and forming a spur running south from the main Himalayas and effectively cutting Bhutan in two. The eastern half is largely inhabited by Indian peoples, mostly from Assam. The western half has a more Tibetan culture. Until recently there was no road over the range, but now one crosses the Pere-La Pass between Wangdu-Phodrang and Tongsa. At the northern extremity of the spur – and within the main Himalayan range – in Kula Kangri, the highest peak in Bhutan (7554 m, 24 783 ft).
Map Bhutan Ba

Black Mountains ***United Kingdom*** Range on the border of England and southern Wales, to the north-east of the River Usk valley. It is good pony-trekking country, rising to 811 m (2660 ft). The Black Mountain is an upland of similar height, 40 km (25 miles) farther west, beyond the BRECON BEACONS.
Map United Kingdom De

Black River (Song Da) ***China/Vietnam*** River rising in the Chinese province of Yunnan and flowing generally south-east for some 800 km (500 miles) to join the Red River, about 50 km (30 miles) west of Hanoi.
Map Vietnam Aa

Black Sea Lying between Europe and Asia, the tideless Black Sea, a landlocked arm of the MEDITERRANEAN, is the base for a Russian fleet. Because large rivers flow into it, the water in the top 90 m (295 ft) has a low salt content. Underlying this layer is salty water brought in by an underwater current from the Mediterranean. The salty water lacks oxygen and contains little life, while the top layer is inhabited by many kinds of fish. The Romans called the sea Pontus Euxinus (the friendly sea). But severe storms and fogs make it dangerous in winter.
Area 461 000 km² (175 960 sq miles)
Greatest depth 2243 m (7359 ft)
Map Europe Gd

Black Volta ***Burkina*** The main headstream of the VOLTA river in western Burkina. It flows north-east, then east and south, forming part of the Burkina-Ghana border, then part of the Ivory Coast-Ghana border. It then swings eastwards to Lake Volta, some 645 km (400 miles) from its source. Its northern tributary, the Sourou (or Bagué), is the site of one of the Burkina government's most successful self-help projects: a barrage has been built across the river by local labour to store water for irrigation.
Map Burkina Aa

Blackburn ***United Kingdom*** Town in the English county of Lancashire 32 km (20 miles) north-west of Manchester. It dates back to Saxon times, and had a church in 596, but grew as a cotton town in the 18th and 19th centuries. There is a textile museum, but many of the mills have closed.
Population 88 000
Map United Kingdom Dd

Blackpool ***United Kingdom*** Largest holiday resort in Britain, on the coast of Lancashire 65 km (40 miles) north-west of Manchester. Its beaches are backed by the greatest concentration of entertainment facilities in Britain, which include the 157 m (515 ft) high tower, a half-size copy of the Eiffel Tower in Paris.

Population 147 000
Map United Kingdom Dd

Blagoevgrad ***Bulgaria*** Capital of Blagoevgrad province, on the Struma river 70 km (44 miles) south of Sofia. Industries include ceramics, textiles, wood and tobacco processing. The centre of the tobacco-growing Struma Valley, it gives easy access to the Rila mountains. Nearby, too, are the small town of Melnik – once a feudal fortress and now an artists' retreat – and the ruins of the Roman spa of Sandanski, said to be the birthplace of Spartacus, the rebel gladiator. Blagoevgrad, formerly Gorna Dzhumaya, was renamed in 1950 after Dimitar Blagoev (1856-1924), founder of the Bulgarian Marxist movement.

Population 68 000
Map Bulgaria Ab

Blagoveshchensk ***USSR*** City and capital of the far eastern region of Amur, on the Amur river near the Chinese border. Its industries include food processing, flour milling, sawmilling, machinery, footwear and furniture.

Population 192 000
Map USSR Nc

Blanc, Cape ***Mauritania/Western Sahara*** Narrow peninsula about 50 km (30 miles) long. The frontier between Mauritania and Western Sahara runs down the centre.

Map Mauritania Ba

Blanc, Mont (Monte Bianco) ***France/Italy*** Highest mountain in Western Europe at 4807 m (15 770 ft). It stands on the French-Italian border at the head of the Valle d'Aosta, about 105 km (65 miles) north-west of Turin. Its lower slopes are a popular playground for skiers, with resorts such as COURMAYEUR and Entrèves in Italy and CHAMONIX in France. A road tunnel runs 12 km (7.5 miles) underneath the mountain between Courmayeur and Chamonix.

Map Italy Ab

Blantyre ***Malawi*** Capital of Southern province and the country's largest city, standing 1100 m (3600 ft) up in the Shire Highlands. It was formerly an African village abandoned because of internal strife. Scottish missionaries used the deserted huts to found a mission in 1876 and named it after the Scottish birthplace of the explorer and missionary David Livingstone (1813-73).

The town was joined in 1956 with Limbe, a town 8 km (5 miles) away on the Beira-Lake Malawi railway. Blantyre serves a major farming region and has food processing, textile and cement manufacturing industries.

Population 333 800
Map Malawi Ac

▲ PERPETUAL SNOW AND ICE The Mont Blanc Massif has the highest peaks in the Alps, and its Bossons Glacier (right) slowly descends towards the French resort of Chamonix.

Blarney (An Bhlarna) ***Ireland*** Town 8 km (5 miles) north-west of the city of Cork. Visitors to Blarney Castle who kiss the Blarney Stone, built into battlements on the castle's 15th-century keep, are said to acquire eloquence. The legend arose from a 16th-century Lord of Blarney, Cormac McCarthy, who used his verbosity to frustrate English demands for land reforms. Elizabeth I is said to have remarked: 'It's all Blarney, what he says, he never means.' The eloquence is not guaranteed, but the view is splendid.

Population 1130
Map Ireland Bc

Blato ***Yugoslavia*** See KORCULA

Bled, Lake (Bledsko Jezero) ***Yugoslavia*** Glacial lake in the Julian Alps in the republic of Slovenia, 48 km (30 miles) north-west of Ljubljana. The skiing centre of Bled, lying at 475 m (1558 ft) on the north shore, has a crimson-spired church on an islet and a Baroque castle clinging to a lakeside crag.

Map Yugoslavia Ba

Bledow Desert (Pustynia Bledowska) ***Poland*** The only true sand desert in Europe. It lies in southern Poland about 40 km (25 miles) east of the city of Katowice. During the Second World War, the German general Erwin Rommel (1891-1944) trained his Afrika Korps on the 32 km² (12 sq miles) of shifting dunes, where mirages are common.

Blenheim (Blindheim) ***West Germany*** Village on the Danube river about 90 km (55 miles) north-west of Munich, where, in 1704, a British army led by John Churchill, 1st Duke of Marlborough, routed French and Bavarian troops during the war of the Spanish Succession.
Population 1600
Map West Germany Dd

Blida (El Boulaida) ***Algeria*** Administrative and trading centre on the Algiers-Oran railway. It lies 48 km (30 miles) south-west of Algiers on the Mitidja Plain, in the middle of a citrus fruit growing area, and trades mainly in oranges and wine. It also has some light industry and manufactures flour, olive oil and soap.
Population 195 000
Map Algeria Ba

Blindheim ***West Germany*** See BLENHEIM

blizzard Violent windstorm accompanied by intense cold and driving, powdery snow or ice crystals.

block faulting Fracturing of part of the earth's crust into separate blocks by faulting.

block lava See AA

block mountain A mountain mass standing up prominently either because it has been elevated by earth movements between FAULTS or because the surrounding area has sunk between faults. The Harz Mountains and Black Forest in Germany, and the Vosges in France are all block mountains. It is also called a HORST.

blocking high An anticyclone that remains relatively stationary and so blocks the passage of approaching DEPRESSIONS, or cyclones.

Bloemfontein ***South Africa*** Capital of Orange Free State and judicial capital of South Africa. The city was founded around a fort in 1846. Its historic buildings include the old Raadsaal (1849), where the original Orange Free State Volksraad, or council, met. Bloemfontein is a commercial centre, and makes furniture, glassware and textiles.
Population 256 000
Map South Africa Cb

Blois ***France*** Capital of Loir-et-Cher department, on the Loire river 55 km (34 miles) south-west of Orléans. It manufactures aircraft, precision instruments and footwear. In the 16th and 17th centuries Blois was a favourite haunt of French kings, who stayed in the great 13th to 15th-century chateau – one of the finest in the Loire valley.
Population 61 100
Map France Dc

Blood River ***South Africa*** Short tributary of the Buffalo river in central Natal, where on December 16, 1838, Zulus under Dingaan were defeated by Voortrekker settlers. About 3000 Zulus were killed and 4 Boers were wounded. The battle is commemorated in a South African public holiday known as the Day of the Covenant, or Dingaan's Day.
Map South Africa Db

blowhole Almost vertical vent that reaches from the roof of a coastal cave to the cliff top. Incoming tides compress the air in the cave, blowing it out through the vent. Water may also be blown out.

Blue Mountains ***Australia*** Part of the GREAT DIVIDING RANGE in New South Wales, about 65 km (40 miles) inland from Sydney. Rising to 1100 m (3609 ft), the region is a plateau intersected by sheer sided deep valleys. It takes its name from the blue haze given off by the eucalyptus trees, and which can be seen from a distance. The rugged terrain, numerous waterfalls and striking views make the mountains a popular tourist area.

The range was first crossed by European colonists in 1813; until then, it represented a barrier

▼ DREAMY BLUES Australia's Blue Mountains take their name from the blue haze of oil droplets given off by their eucalyptus trees. It clothes the landscape and can be seen for miles.

to westward expansion of the colony. The Blue Mountains National Park, covering 1113 km² (430 sq miles), was established in 1959.
Map Australia Ie

Blue Mountains ***Jamaica*** Thickly wooded mountain range, about 50 km (30 miles) long, in the eastern end of the country. Several peaks rise from a 1500 m (4900 ft) ridge, the highest being Blue Mountain Peak (2256 m, 7402 ft). High quality coffee is grown on the lower slopes.
Map Jamaica Ba

blue mud Ocean-floor mud found on the continental slopes. It is washed down from land, and gets its dark slate-blue colour from tiny particles of iron sulphide.

Bluefields ***Nicaragua*** Caribbean port about 320 km (199 miles) east of the capital, Managua. It exports tropical hardwoods such as mahogany, rosewood and black walnut. The mainly Creole inhabitants, descended from Africans and European settlers from Jamaica, speak English, the town having grown around a number of 18th-century British settlements.
Population 20 000
Map Nicaragua Ba

bluff Steep headland, promontory, river bank or cliff often formed by a river cutting into the valley side on the outside of a MEANDER.

Bluff ***New Zealand*** Industrial town on the southern tip of South Island. It has the country's only aluminium smelter, which uses bauxite imported from Australia and is run by hydroelectricity generated in the Fiordland mountains.
Population 2720
Map New Zealand Bg

Bo ***Sierra Leone*** Southern regional capital, about 180 km (115 miles) south-east of the national capital, Freetown. Bo, at the edge of the Sewa river swamps, is a market town for local crops, including ginger and oil-palm kernels. It produces soap and palm oil.
Population 150 000
Map Sierra Leone Ab

Bo Hai (Po Hai) Arm of the YELLOW SEA off the east coast of China. It was formerly called the Gulf of Chihli. It contains oil fields.
Dimensions 480 km (300 miles) long, 290 km (180 miles) wide
Map China Hc

Bobo Dioulasso ***Burkina*** The country's second largest town after the capital, Ouagadougou. It lies some 310 km (193 miles) south-west of the capital, and is a trade and industrial centre with groundnut-crushing mills, soap works, cotton ginneries and textile factories. Bobo was the terminus of the railway from the Ivory Coast capital, Abidjan, and the capital of Upper Volta from 1934 to 1954. It has a mosque, botanical gardens, lively markets and abundant restaurants and bars.
Population 148 000
Map Burkina Aa

bocage Type of farmland divided into small fields by hedges and trees, or dry-stone walls, applied especially to north-west France.

Bochum ***West Germany*** Industrial city of the Ruhr, 20 km (12 miles) west of Dortmund. It produces coal, steel, cars and household goods.
Population 410 000
Map West Germany Bc

Bodensee ***West Germany*** See CONSTANCE

Bodh Gaya ***India*** The most important shrine of Buddhism – though the place is not visited to the same extent as Muslim Mecca or Hindu Varanasi. Here, about 500 BC, the Lord Buddha achieved enlightenment while sitting in contemplation under the banyan, or fig tree.

The various world sects of Buddhism maintain temples here, so that it is possible to hear and see the chanting and services of Tibetan, Chinese, Japanese, Thai and Sri Lankan monks in close proximity.

The town is small and no industries have sprung up to destroy the peace of the temples.
Population 15 700
Map India Dc

Bodø ***Norway*** Seaport, administrative centre and 'midnight sun' tourist resort some 80 km (50 miles) north of the Arctic Circle, and about 180 km (110 miles) south-west of the port of Narvik. The city is the capital of the county of Nordland. The port's main cargoes are copper ore and marble, exported from mines at Sulitjelma and quarries at Fauskeidet, Salten.
Population 27 600
Map Norway Db

Bodrum ***Turkey*** Market town about 580 km (360 miles) south-west of the capital, Ankara. Formerly known as Halicarnassus, it is the site of the Mausoleum of Halicarnassus, built in the 4th century BC and one of the Seven Wonders of the Ancient World. The tomb was put up by the widow of Mausolus – a local king whose name is the origin of mausoleum, the word for a grandiose tomb – and was destroyed by an earthquake some time before the 15th century. The town was the birthplace of the Greek historian Herodotus (about 484-425 BC).
Population 13 090
Map Turkey Ab

bog Area of permanently waterlogged ground with a surface layer of decaying vegetation, particularly sphagnum moss which forms highly acid peat.

Bog of Allen (Móin Aluine) ***Ireland*** A large peat bog covering about 950 km² (370 sq miles) in the eastern counties of Kildare, Offaly, Laois and Westmeath. Its deep, easily worked deposits are used in peat power stations including one at Allenwood west of Dublin.
Map Ireland Cb

Boğazkale (Bogazkoy) ***Turkey*** Mountain village about 145 km (90 miles) east of the capital, Ankara. It is on the site of the ancient city of Hattusas, capital of the Hittite Empire from 1400 to 1200 BC. Among the Hittite remains are huge double-walled fortifications and an underground tunnel, Yen Kapi, which reveals signs of an earlier Iron Age period of occupation. Many of the objects found at the site are now in a museum in the village.
Map Turkey Ba

Bogor ***Indonesia*** City of West Java, about 80 km (50 miles) south of the national capital, Jakarta. It was a Dutch colonial hill station and has a botanical garden which is rich in tropical plants, including the parasitic *Rafflesia arnoldi* (the monster flower). The plant has the world's largest flower – the remaining vegetative organs are so reduced that the plant exists only as a fungus-like growth – which can be as much as 300-900 mm (1-3 ft) across and weighing up to 7 kg (15 lb).
Population 250 000
Map Indonesia Cd

Bogotá ***Colombia*** The country's capital and largest city, and capital of Cundinamarca department on a fertile plateau 2800 m (9200 ft) up in the eastern Andes. In 1538, the Spanish conquistador Gonzalo Jiménez de Quesada, in search of Eldorado, the legendary city of gold, marched onto the plateau which was the home of the Chibcha Indians. After conquering the Indians, he founded the city, calling it Sante Fé de Bogotá – Sante Fé after his birthplace in Spain, and Bogotá from Bacatá, which was the Indian name for the region.

From a population of 100 000 in 1905, Bogotá has grown into an expansive modern city, whose industries include textiles, vehicles, engineering and chemicals. It also has a reputation for lawlessness.

The city centre is a mixture of Spanish colonial architecture and soaring skyscrapers laid out in a grid pattern. The house of the Venezuelan-born Simón Bolívar (1783-1830), who liberated Colombia from Spanish rule, stands just outside the city centre. The gold museum contains more than 20 000 gold artefacts created by the Chibcha Indians. The museum also contains one of the world's largest emeralds, which was mined in Boyacá to the north.

The spectacular Tequendama Falls are 30 km (19 miles) south-west of Bogotá. The river Bogotá, a tributary of the Magdalena, rushes through a gorge 18 m (60 ft) wide and plunges 145 m (475 ft) off the edge of the plateau.
Population 5 789 000
Map Colombia Bb

Bogra ***Bangladesh*** Chief town of Bogra district, which covers 3888 km² (1501 sq miles) in the centre of the country. Parts of the district alongside the Brahmaputra river are prone to both flood and erosion, and whole villages have lost their lands. The Brahmaputra Right Bank Project, begun in 1963, built 217 km (135 miles) of embankment, but has met with only partial success.
Population (district) 2 728 000; (town) 68 200
Map Bangladesh Bb

Bohemia (Cechy) ***Czechoslovakia*** Mountain-girt westernmost region of the country. It is the heartland of Czech culture, and includes the capital, PRAGUE. Bohemia's forests and fertile Vltava and Elbe river basins produce timber, flax and hops, and there are deposits of coal, graphite, iron and uranium. Industrial cities include HRADEC KRALOVE, PLZEN, USTI NAD LABEM, LIBEREC and KLADNO, and it has many fine castles, old towns and spas, including KARLOVY VARY (Carlsbad) and MARIANSKE LAZNE (Marienbad).
Map Czechoslovakia Ab

Bohemian Paradise ***Czechoslovakia*** See JICIN

Bohemian Switzerland ***Czechoslovakia*** See USTI NAD LABEM

Bohinj, Lake ***Yugoslavia*** See JULIAN ALPS

Bohol ***Philippines*** One of the VISAYAN ISLANDS covering 3862 km² (1491 sq miles) to the east of Cebu; also the name of a province including some nearby minor islands and of the sea between Bohol and Mindanao. The island is hilly, the so-called Chocolate Hills (which are brown in summer) being a tourist attraction. The main crops are coconuts and rice.

Population (province) 759 370

Map Philippines Bd

Boirinn ***Ireland*** See BURREN, THE

Boise ***USA*** State capital of Idaho. It lies on the Boise river in the south-west of the state, about 60 km (37 miles) east of the Oregon border. The city is a centre of the mining industry, and is also famous for its hot springs.

Population (city) 107 200; (metropolitan area) 189 300

Map United States Cb

Bois-le-Duc ***Netherlands*** See 'S-HERTOGENBOSCH

Bokaro ***India*** Steel city about 260 km (161 miles) north-west of Calcutta. Its Russian-built steel plant is destined to be the largest and most important in south Asia.

Population 264 500

Map India Dc

Boksburg ***South Africa*** Mining town 21 km (13 miles) east of Johannesburg, which has the world's largest gold mine, covering 4900 hectares (12 108 acres). Coal is also mined there.

Population 156 800

Bol ***Chad*** Polder (area of reclaimed land) on eastern shore of Lake Chad. It is a fertile agricultural area where millet, sorghum, vegetables, wheat and cotton are grown. Much produce is smuggled across the lake to Nigeria, where higher prices can be obtained.

Map Chad Ab

Bol ***Yugoslavia*** See BRAC

Bolama ***Guinea-Bissau*** A former colonial capital, on the beautiful island of the same name south of the mouth of the Gêba river. It still serves as a port for local produce, including palm oil (used in margarine and cooking fats), coconuts and fruit, but is rather tumbledown.

Map Guinea Aa

Bolgatanga ***Ghana*** Capital of the Upper Region, about 30 km (20 miles) from the border with Burkina, north and slightly east of the national capital, Accra. It began as a trading centre and grew rapidly after 1937 when the roads were improved to take motor transport. Today it markets millet, sorghum, groundnuts, vegetables, cattle and animal products.

Population 20 000

Map Ghana Aa

Bolívar ***Venezuela*** Cattle-ranching state in the south-east. It covers 239 250 km² (92 374 sq miles) – an area the size of Britain. It has rich deposits of iron ore and bauxite (an ore of aluminium) which are exploited, and reserves of oil which are not. The provincial capital and largest city is Ciudad Bolívar.

Population 750 000

Map Venezuela Bb

Bologna ***Italy*** City about 80 km (50 miles) north of Florence. In pre-Roman times, Bologna was the site of the Etruscan town of Felsiena. The Roman colony founded there in the 2nd century BC was called Bononia, from which the city's name derives. By the 11th century, Bologna had emerged as one of northern Italy's first independent city-states; today it is the capital of the EMILIA ROMAGNA region.

Central Bologna preserves its Roman-style grid layout. It is surrounded by boulevards that mark the line of the medieval walls. Its chief landmarks are the twin 'leaning towers', survivors of more than 200 fortresses that once stood there. The centre also contains the church of San Petronio, one of northern Italy's largest Gothic buildings. The nearby church of San Domenico contains the tomb of St Dominic, founder of the Dominican monastic order, who died in Bologna in 1221.

Bologna's university, founded in 1088, is reputedly the oldest in the world. The poets Dante (1265-1321) and Petrarch (1304-74) were among its early scholars. Bologna is an agricultural, engineering and electrical equipment centre – and is noted for its pasta, which accounts for its nickname of *La Grassa*, the 'Fat City'. An international children's book fair is held there each spring.

Population 455 900

Map Italy Cb

Bolovens Plateau ***Laos*** Fertile plateau in the south, rising from the Mekong river plains east of Pakse to a height of 1570 m (5150 ft). It has the highest yearly rainfall in Laos (4064 mm, 160 in), and grows coffee, cotton and tobacco.

Map Laos Bc

Bolsena ***Italy*** Volcanic crater lake 80 km (50 miles) north of Rome. Its waters are rich in fish, especially eels. The lake covers 114 km² (44 sq miles) and reaches a depth of 151 m (495 ft).

Map Italy Cc

bolson Basin of inland drainage in an arid or semi-arid region, particularly among the high plateaus of the south-west United States, often containing a salt lake and ALLUVIAL FANS.

Bolton ***United Kingdom*** Cotton town 16 km (10 miles) north-west of Manchester, formerly in the English county of Lancashire but now part of Greater Manchester. It was a birthplace of the Industrial Revolution for it was the home of Samuel Crompton (1753-1827), who in 1779 invented the spinning mule – one of the machines that revolutionised the cotton industry. Crompton's home, Hall-i'th'-Wood (Hall in the Wood), is a 15th-century half-timbered building that now houses a folk museum. The town also has a fine art gallery.

Population 147 000

Map United Kingdom Dd

Bolu ***Turkey*** Leather-making and timber town about 135 km (85 miles) north-west of the capital, Ankara.

The surrounding province of the same name, which covers 11 481 km² (4433 sq miles), includes the forested Bolu mountains. Its main products are cereals, flax, tobacco and opium – used to make the painkilling drug morphine.

Population (town) 52 055; (province) 510 240

Map Turkey Ba

Bolzano (Bozen) ***Italy*** Alpine city about 55 km (34 miles) south of the Austrian border and the BRENNER PASS. The arcaded Via dei Portici is the market centre of the town, which produces steel, aluminium and vehicle parts. The area is a meeting place between German and Italian culture, language, food and architecture. This is seen in its museums, libraries and cultural institutions, in its music and folklore, in its Gothic cathedral and many churches, and in its farms and villages.

Population 104 600

Map Italy Ca

Boma ***Zaire*** Port on the north bank of the Zaire river estuary about 80 km (50 miles) from the Atlantic coast. It was a bustling slave centre from the 16th to the 18th centuries. It now exports agricultural produce, such as coffee, cotton and rubber as well as timber. Between 1885 and 1908 it was the capital of the Congo Free State (controlled by King Léopold II of Belgium) and then was the capital of the Belgian Congo colony until the 1920s.

Population 32 000

Map Zaire Ab

Bombay ***India*** City on the Arabian Sea, the capital of Maharashtra state and former capital city of the now defunct Bombay State. It is India's most important commercial and industrial city, having usurped Calcutta. In the early 16th century the Portuguese had a small trading colony there on Bombay Island, which was part of the dowry brought by Catherine of Braganza when she married Charles II of England in 1661. Although possessing a superb natural harbour, the island (now a peninsula) was hot, humid, low and swampy, and the hinterland hemmed in by the majestic cliffs of the Western Ghats, towering 1524 m (5000 ft) high, just inland.

Bombay grew in prominence when first roads and then railways were forced up the western Ghats, and trade in cotton became possible; but it has always lacked local power supplies. Its importance grew enormously after the Suez Canal was opened in 1869, when it became the Gateway of India. There is a triumphal arch by the waterfront known as India Gate to commemorate the arrival of King (Emperor) George V and Queen (Empress) Mary in 1911.

The recent development of an off shore oil industry has further boosted Bombay's significance. The range of industries now includes virtually any consumer goods – cars, textiles, a flourishing film industry and much else.

Although in the port at Elephanta Island there are Hindu cave temples well worth visiting, the city is noted more for its Victorian buildings than either Hindu or Muslim. Courthouses, schools, railway stations and museums are all built in Victorian versions of European and Oriental styles. It is (*continued on p. 96*)

Bolivia

PARADOXICALLY, EXTENSIVE MINERAL WEALTH HAS BROUGHT ONLY GRIEF AND POVERTY TO BOLIVIA'S MASSES

A world record inflation rate measured in thousands of percentage points is Bolivia's recent dubious claim to fame. Poverty exists in many South American countries but nowhere is it worse than in this landlocked Andean nation. Despite extensive mineral resources, the country and its people remain among the poorest in Latin America. The cause has been two wars and political instability since independence in 1825 which has resulted in 190 governments in 150 years.

The Indian peasants, who form 50 per cent of the population, live in thatched mud huts and raise just enough food – mainly maize and potatoes – to feed their families. Few have any education, and most cannot read or write. Their main source of income was from working in the tin mines or the gas fields. Now, however, world prices for tin and natural gas have plunged and the economy is in grave trouble. Mines are being shut and workers laid off.

In the 16th century the Spanish conquistadores conquered the country, robbed the Indians of their land, and forced them to work as slaves in the silver mines. Independence from Spain was achieved in 1825 but disputes among the liberators brought little improvement for the people or the country. Bolivia lost more than half its territory to more powerful neighbours – including its valuable nitrate deposits and Pacific port of ANTOFAGASTA to Chile in the War of the Pacific (1879-83). An even larger area – most of the eastern lowland Chaco region – was taken by Paraguay in the Chaco War of 1932-5.

The only political change that brought any real benefit to the Indians was the revolution of 1952. It led to land reforms, some measure of emancipation for the Indians and the nationalisation of the tin industry. Until then, monopoly control of tin had been exercised by three large companies, two of them foreign owned. However, nationalisation has failed to transform either the industry or the economy.

Most of the poverty is concentrated in the west of the country, where half the population lives on the *altiplano*, the 3700 m (12 000 ft) high plateau between two chains of the Andes. Here the Aymaras scratch a living. Though the altiplano is the most densely populated part of Bolivia, it is still a vast emptiness, largely untouched by man. Vegetation is sparse – just coarse grasses and a few shrubs. Condors sweep through the heights, and vicuñas, llamas and alpacas graze the land.

LA PAZ, the highest capital city in the world, lies on the altiplano to the south-east of Lake TITICACA – the world's highest navigable lake.

To the east lie the lowlands that comprise 70 per cent of Bolivia's land but contain only 20 per cent of the population. Beyond the foothills of the Andes, the northern tropical forests yield timber and brazil nuts. The rest of the lowlands – which stretch south to the Chaco – are wooded savannah and vast natural pasture lands, inhabited by great herds of semi-wild cattle.

Between the altiplano and the lowlands, fertile valleys – the Yungas – cut through the eastern chain of the Andes (the Cordillera Real) in the north, while in the south is the puna highland area, suitable only for grazing. The high Yunga valleys are subtropical, and farmers produce citrus fruits, bananas, pineapples, avocados and many other fruits. The timber includes cedar, mahogany, walnut, cinchona (source of quinine) and dyewoods. Sugar cane, tobacco, cocoa and coffee grow readily in the lower Yungas, and it is here also that coca leaves – from which cocaine is derived – have become an important but illegal crop. Though tin once comprised 33 per cent of Bolivia's exports, a vast illicit economy has now evolved centred on coca growing. Cocaine may account for two-thirds of the country's exports, and the narcotics trade has become a major influence on Bolivia's economic and political life.

▼ ON THE ALTIPLANO Aymara Indians bring their sheep to market on the high plateau. Barren soils, cold climate and irregular rainfall make their lives hard.

BOLIVIA AT A GLANCE

Area 1 098 581 km² (424 163 sq miles)
Population 6 360 000
Capital La Paz
Government Parliamentary republic
Currency Peso = 100 centavos
Languages Spanish, Quechua, Aymara
Religion Christian (95% Roman Catholic)
Climate Tropical; cooler at altitude. Average temperature in La Paz ranges from 1-17°C (34-63°F) in July to 6-19°C (43-66°F) in November
Main primary products Potatoes, maize, sugar cane, rice, cassava, coffee, llamas, alpacas; tin, oil and natural gas, copper, lead, zinc, antimony, bismuth, tungsten, silver, gold, sulphur, iron
Major industries Mining and smelting, oil and gas production, textiles, handicrafts, food processing, cement
Main exports Tin, antimony, tungsten, zinc, silver, lead, natural gas
Annual income per head (US$) 600
Population growth (per thous/yr) 26
Life expectancy (yrs) Male 49 Female 53

▲ **OFF TO WORK People hurrying to offices and shops in central Bombay crowd a narrow street, where a bewildering array of shop signs festoons the balconies of the tenements.**

also the only city in India with significant daily commuting to the office district by train.

Bombay has become the centre of the Parsi community, the Zoroastrians from Persia, who are immensely influential in large Indian companies.

Population 8 243 400
Map India Bd

Bomu (Mbomou; Mbomu) ***Equatorial Africa*** River about 805 km (500 miles) long, forming part of the frontier between Zaire and the Central African Republic. It rises near the Sudan border and joins the Uele at Yakoma to form the Ubangi. The Bomu was discovered in 1877 by a Greek explorer named Potagos.

Map Zaire Ba

Bon, Cape (Ras el Tib) ***Tunisia*** Extreme point of the 80 km (50 mile) long Maouin Peninsula that protrudes from the north-east corner of Tunisia into the Mediterranean and shelters the harbour of Tunis. The eastern end of the ATLAS chain of mountains finally fades away in a series of low hills along the peninsula and ends in the rocks of El Haouria.

The region is dominated by its links with the Phoenicians, who founded nearby CARTHAGE; most of the small towns on the peninsula have the ruins of ancient Phoenician counterparts alongside them. In the 14th century the area attracted Arab refugees from ANDALUCIA, fleeing from Spanish attempts to Christianise them. The peninsula's mild climate has made it a vast fruit and flower garden dotted with vineyards, citrus fruit groves and market gardens, developed during the period of Italian colonisation. It also has excellent beach resorts.

Map Tunisia Ba

Bonaire ***Netherlands Antilles*** Arid island of 288 km² (111 sq miles) in the Caribbean, east of Curaçao and off the north coast of Venezuela. The capital is Kralendijk on the south-west coast (population 2500). Its chief industries are textiles and clothing, salt – which is produced by evaporating seawater – and tourism.

Population 9700
Map Caribbean Bc

Bondi Beach ***Australia*** World-famous ocean beach in the eastern suburbs of Sydney, New South Wales, about 6 km (4 miles) from the city centre. It is a classic surfing beach and attracts thousands of city dwellers. Australia's first official surf lifesaving club was formed at Bondi in 1906.

Map Australia Ie

Bondoukou ***Ivory Coast*** Market town on the eastern border with Ghana, 340 km (211 miles) north-east of Abidjan. One of the country's oldest towns, it was founded in AD 1466 by traders on the caravan route to the Niger River in the north. The French introduced cocoa as a crop in 1914.

Population 22 000
Map Ivory Coast Bb

bone bed Layer of sedimentary rock containing fossil bones, teeth and scales of vertebrates, especially fishes. It may be the result of a rapid catastrophe, such as an underwater earthquake, that killed all life simultaneously.

Bong Mountains ***Liberia*** Range of mountains in western Liberia which rise to 645 m (2146 ft) some 80 km (50 miles) north-east of the capital, Monrovia. The range contains substantial reserves of iron ore. The mining settlement of Bong Town (population 12 000) has grown up since an iron ore mine was opened in the hills in 1965.

Map Liberia Ba

Bonin Islands (Ogasawara-shoto) ***Japan*** Volcanic island chain stretching 800 km (500 miles) due south of Tokyo.

Population 2300
Map Pacific Ocean Ca

Bonn ***West Germany*** National capital on the bank of the Rhine, about 30 km (18 miles) south-east of Cologne. Before 1945, Bonn was a dignified university town and site of the Baroque palace of the Archbishop-Electors of Cologne. The last of them, keen on music,

A DAY IN THE LIFE OF A BOMBAY DABBAWALLAH

Most office workers in Bombay never eat out or go home for lunch. Instead, their lunch arrives from home. Alok Chaudhry is one of the 2300 members of the Union of Tiffinbox Suppliers who form a network of relay teams that collect 100 000 lunch-boxes in mid-morning from Bombay's suburbs and delivers them to offices in the city centre by lunchtime. The home-cooked lunches of curry, rice, vegetables and chapatti arrive still warm in aluminium containers stored in tin boxes, or dabbas.

It is for a teammate in the lunch-box relay race that Alok is waiting, by the clock at Malad railway station, 32 km (20 miles) from Bombay, at 11 am. Fifty other white-hatted dabbawallahs or lunch-carriers wait with him.

Reflecting on his role as a link between wives at home and working husbands, Alok thinks wistfully of his own wife. Menakshi, in their village near the town of Poona, four hours away from his rented tenement room in Malad. He sees her and their three children only three or four times a year, but at least he has a job and can send his family 15 rupees a month out of the 90 he earns.

Alok's reverie does not last long. Punctually, at 11.02, a bicycle hung with dabbas wheels into the station, ridden by Alok's friend, Rajnish. The dabbawallahs' rendezvous are always punctual. Suburban homes expect the first in the chain of lunch-carriers at 10 am every day. These dabbawallahs arrive on foot and on time. Rajnish, second in the chain, collects his lunch-boxes from them at the first meeting point at 10.20.

Rajnish is already unhooking dabbas from his bicycle as Alok greets him. All the dabbas are marked with signs – dots, crosses, strokes and circles – the address that guides them through the stages of their journey. Written addresses would be useless; many of the dabbawallahs, like 64 per cent of Indians, cannot read. Alok is among these, despite a few childhood years at his village school. He selects the 39 dabbas marked with a red dot and packs them onto a wooden rack, then he boards the crowded train.

At each stop other dabbawallahs get on and off the train and Alok checks his tins for symbols which mean they must be handed over to continue their journey. At midday the train draws into Churchgate station. The last of Alok's dabbas are collected and he can stop to eat his own lunch – chapatti and vegetables wrapped in newspaper. But his work is not over. The whole process of collection and delivery is reversed with the empty tins, so that the right dabba is returned to the right home by 4 pm.

had an organist whose son was the composer Ludwig van Beethoven (1770-1827). His birthplace in Bonngasse is now a museum.

The destruction of Berlin and the division of Germany led to the choice of Bonn as the new capital in 1949. The West German Parliament settled into a new Bundeshaus, formerly a teachers' training college. And diplomats poured into the Rhine valley, overflowing into nearby Bad Godesberg, an ancient spa town.

Population 300 000
Map West Germany Bc

Bonny ***Nigeria*** Seaport village on the east side of the Niger river delta with a sandy beach bordered by mangrove swamps. An oil terminal was opened in 1961 and enlarged in 1966.
Map Nigeria Bc

Bonsa Valley ***Ghana*** The country's second largest diamond field after the Birim Valley. It lies some 220 km (135 miles) west of the capital, Accra. Mining is frequently carried out by individuals or small family concerns. The diamond-bearing gravels are dug from shallow pits by hand, washed by the women and the industrial diamonds picked out by hand.
Map Ghana Ab

Bophuthatswana ***South Africa*** Former Bantu homeland in seven separate areas in the western Transvaal, northern Cape and Orange Free State and covering a total of 44 109 km² (17 030 sq miles). It was declared an independent republic in 1977, but only South Africa recognises it. The people are mostly Tswanas. Cattle are reared, and chrome and platinum mined. The capital is Mmabatho.
Population 1 935 000
Map South Africa Bb

bora Cold, violent, north or north-east wind blowing along the Dalmatian coast of Yugoslavia and in northern Italy. It occurs mainly in winter when atmospheric pressure is high over central Europe and the Balkans and low over the Mediterranean.

Borås ***Sweden*** Manufacturing town 69 km (43 miles) east of Gothenburg. The centre of the country's textile industry, it was founded in 1622.
Population 99 963
Map Sweden Bd

Bordeaux ***France*** Seaport and capital of the Gironde department, on the Gironde estuary about 95 km (60 miles) from its mouth. The surrounding area – the Bordelais – is noted for its wines, known as Bordeaux or claret, and an important wine-exporting trade was established when the city was in English hands between 1154 and 1453. There are several medieval churches, an 11th to 15th-century cathedral, and a large university which was founded in 1441. Apart from wine, the city's main industries are shipbuilding, marine engineering, oil refining, chemicals and food processing.
Population 650 125
Map France Cd

Bordelais ***France*** See BORDEAUX

Borders ***United Kingdom*** Region of southern Scotland covering 4662 km² (1800 sq miles) on the boundary with England. It was for centuries the scene of bloody clashes between both sides until the kingdoms were united in 1707. It covers the old counties of Berwick, Roxburgh, Selkirk and Peebles, plus part of Midlothian.

This hill country is famous for sheep, the ruined abbeys at Melrose, Jedburgh and Dryburgh, and its associations with the poet and novelist Sir Walter Scott (1771-1832), who set much of his work in the region. He is buried in Dryburgh Abbey.
Population 101 000
Map United Kingdom Dc

Bordighera ***Italy*** Riviera resort 10 km (6 miles) from the French border. There are panoramic corniche roads and luxury villas with tumbling gardens of subtropical vegetation. The English nonsense poet Edward Lear (1812-88) lived there in his last years.
Population 18 000
Map Italy Ac

bore High wave, like a wall of water, travelling upstream in the tidal reaches of certain rivers, caused by the surge of a flood tide upstream in a narrowing estuary or by colliding tidal currents. The bore, sometimes called an eagre, gradually diminishes in height and eventually dies out. There is a bore on the River Severn in England where, during spring tides, it is often 1 m (over 3 ft) high and may be more than 2.5 m (nearly 9 ft). On some occasions it reaches beyond Gloucester, 33 km (21 miles) upstream.

boreal forest Collective name given to the northern forests of Canada and Eurasia which consist mainly of conifers, but which also include hardy deciduous species, such as birch.

Borecka Forest See MASURIA

Borgå ***Finland*** See PORVOO

Borkou-Ennedi-Tibesti (BET) ***Chad*** By far the country's largest prefecture, covering 600 350 km² (231 175 sq miles) in the north – almost half the country's total area. Its capital is Faya-Largeau. Although part of the French colony of Chad, its intensely Muslim people always resisted French rule. It was ceded to the Ndjamena government but remained a centre of rebellion, and in the civil war beginning in 1980 was controlled by Libyan-backed anti-government forces.

Most of its people are nomadic herdsmen who keep camels and goats, but semisedentary groups grow millet, wheat, vegetables and dates around oases that have not dried up.
Population 88 000
Map Chad Ab

Borkum ***West Germany*** See FRIESIAN ISLANDS

Borneo ***South China Sea*** One of the world's largest islands, bordering the South China Sea. With a total area of 751 900 km² (290 320 sq miles) Borneo is divided between a northern section, formerly under British control, and a much larger southern part, formerly Dutch and now ruled by INDONESIA. The northern coastline is now divided between the Malaysian states of SARAWAK and SABAH and the independent sultanate of BRUNEI. The two parts are separated by a mountainous area which stretches north-east to the massif of Mount Kinabalu. Much of the coastal part of Borneo is low-lying, swampy land which has only recently been developed.

Map Indonesia Db

Bornholm ***Denmark*** Island (587 km², 227 sq miles) in the south Baltic Sea, consisting mostly of rugged granite rocks containing kaolin, from which pottery is made. It has fine examples of round, fortified medieval churches and the ruins of the impressive late medieval castle of Hammershus. The little island of Christiansø, off the north-east coast, was also fortified principally because of recurrent conflict with Sweden. Bornholm's principal town and ferry port is RONNE and the fishing port is Neksø. Agriculture and tourism are the chief sources of income. The island was invaded by Germany during the Second World War and liberated by the Russians, who occupied it until April 1946.

Population 47 300
Map Denmark Cb

Borno ***Nigeria*** The country's largest state covering 116 400 km² (44 942 sq miles) in the extreme north-east. It is sparsely populated, and most of its people are farmers who produce millet, sorghum, groundnuts, vegetables and cattle. Maiduguri is the capital. The vast Bornu plain beside Lake Chad is dominated by the Kanuri people. It was the centre of the powerful Bornu Empire, which flourished in the 10th and 11th centuries, and joined the Kingdom of KANEM in the 13th century to form the mighty Kanem-Borno Empire.

Population 2 998 000
Map Nigeria Ca

Bornu ***Nigeria*** See BORNO

Borobudur ***Indonesia*** Magnificent Buddhist temple 40 km (25 miles) north-west of the city of Yogyakarta in Central Java. It was built between AD 750 and 850. The shrine, which consists of a series of terraces with carvings depicting Buddhist legends, was restored with United Nations help in the 1960s and 1970s.

Map Indonesia Dd

Borovets ***Bulgaria*** Winter sports resort, on the northern slopes of the Rila mountains. It has good pine-forested walking country, and is a departure point for Musala (2925 m, 9596 ft), the highest peak in the RHODOPE MOUNTAINS. Bulgaria's oldest mountain resort, it has several palaces and villas (now holiday homes) reflecting its former glory as the haunt of royalty and aristocracy before the Second World War.

Map Bulgaria Ab

Borsod ***Hungary*** Attractive limestone upland, largely covered by deciduous forest, lying along the Czech border north of the city of Miskolc. The spectacular cave system at AGGTELEK lies near the border. Sarospatak, the region's picturesque capital, 70 km (43 miles) north-east of Miskolc, has a population of 15 300 and a Calvinist college, now a state school. Iron ore mined at the town of Rudabanya, in the region, provides the raw material for the iron and steel works in Miskolc and the town of Ozd.

Map Hungary Ba

Börzsöny ***Hungary*** See MATRA

Bosanski Brod (Brod) ***Yugoslavia*** Major road and rail junction and port on the Sava river in Croatia, 190 km (about 120 miles) north-west of Belgrade. It has large railway engineering works.

Population 106 400
Map Hungary Db

Bosnia Herzegovina (Bosna Hercegovina) ***Yugoslavia*** Most central of the country's six republics, covering an area of 51 129 km² (19 736 sq miles). It is mainly mountainous and deeply cut by rivers flowing north to join the Sava river or south to join the Neretva. In Bosnia, the densely forested Dinaric Alps give way northwards to rolling farmlands famous for plums and plum brandy (*slivovica* or *rakija*). Herzegovina is limestone plateau country, pitted by large depressions, which are fertile after winter flooding.

Turkish invasions from 1386 left ruined castles or turrets on almost every hill or rocky outcrop in the republic. Bosnia finally fell to the Turks in 1463, Herzegovina followed in 1482 and the Turks remained in control until 1878. The Turkish influence was enormous. Today 40 per cent of the people are Muslims and there is a wealth of Islamic art and architecture, including mosques, covered bazaars, baths, fountains and bridges in towns such as Sarajevo, the republic's capital, Banja Luka, Mostar, Travnik and Visegrad.

Industries are based on the republic's coal, iron ore, bauxite and timber, especially around Zenica and Tuzla. The population, 32 per cent of which is Serb and 18 per cent Croat, has grown rapidly – by 10 per cent between 1971 and 1981 – outpacing the provision of jobs.

Population 4 124 000
Map Yugoslavia Cb

Bosporus ***Turkey*** Strait 29 km (18 miles) long and up to 4 km (2.5 miles) wide linking the Black Sea and the Sea of Marmara; formerly known as the Bosphorus. It is strategically important as the only outlet for the USSR's Black Sea fleet to the Mediterranean. ISTANBUL is at the strait's southern end.

The name means 'ox ford', after the legend of Io, the maiden loved by the presiding Greek god Zeus. In mythology, she swam across the strait after he turned her into a heifer. Today cars can drive across the Bosporus on a suspension bridge which was opened in 1973.

Map Turkey Aa

boss Small BATHOLITH with a roughly circular cross-section of about 100 km² (roughly 40 sq miles) such as Shap in Cumbria, England, and the northern part of Arran, Scotland.

Boston ***USA*** Port and state capital of Massachusetts, about 300 km (185 miles) north-east of New York city, and the largest city in New England. English colonists first landed there in 1630 – ten years after the Pilgrim Fathers established the nearby PLYMOUTH colony – and in 1632 it became capital of Massachusetts Bay Colony. The city was the scene of the so-called Boston Tea party in 1773, when a party of angry citizens, disguised as Indians, boarded tea ships and dumped their cargo overboard in a gesture of defiance against economic dominance from Britain.

Boston today is the centre of a conurbation that spreads north into southern New Hampshire. Its long-established clothing, footwear and textile industries have declined and been replaced by a wide range of newer industries, including electronics. The city is an important financial, business and educational centre, and is the seat of Harvard University (the oldest in the USA, founded in 1636) and the Massachusetts Institute of Technology. The American writer and poet Edgar Allan Poe was born there in 1809.

Population (city) 570 700; (metropolitan area) 2 820 700; (conurbation) 4 026 600
Map United States Lb

Bosumtwi, Lake ***Ghana*** Mysterious circular lake 21 km (13 miles) south-east of Kumasi. It is about 10 km (6 miles) across and more than 72 m (236 ft) deep. No one knows how it was formed, but a meteorite or volcanic explosion followed by subsidence are two common explanations. Much folklore and superstition are associated with the lake. Fishermen, forbidden to use boats on it, sit astride planks instead.

Map Ghana Ab

Botany Bay ***Australia*** Coastal inlet lying in the southern suburbs of Sydney, discovered by the English explorer Captain James Cook in April 1770. It was so named by Cook because of the rich variety of plant life growing there. The bay was planned to be a British penal colony in 1788, but conditions were unsatisfactory so the site was transferred to Port Jackson (now Sydney Harbour).

Map Australia Ie

Bothnia, Gulf of Northern arm of the BALTIC SEA covered by ice for up to five months every year. Its shoreline is slowly rising, ever since the end of the last Ice Age 10 000 years ago – a consequence of the removal of the tremendous weight of the ice sheets.

Dimensions 640 km (400 miles) long, 80-140 km (50-90 miles) wide
Map Sweden Cc

Botosani ***Romania*** Capital of the province of the same name in northern MOLDAVIA, 40 km (25 miles) west of the Russian border. Founded in the 14th century, Botosani has a medieval church. Now the trading centre of a fertile agricultural area, the town manufactures textiles, clothing and processed food.

Population 79 000
Map Romania Ba

Bottrop ***West Germany*** City of the industrial Ruhr near the border with the Netherlands. It produces coal, textiles and steel.

Population 117 000
Map West Germany Bc

Bou Saâda ***Algeria*** Oasis town in the Ouled Nail mountain region, 250 km (155 miles) south-east of Algiers on the road to Biskra. With its old streets, citadel, the Ouled Attik and El Nekla mosques, and its many crafts shops, it has become a major tourist centre.

Population 50 000
Map Algeria Ba

Bouaké ***Ivory Coast*** Second largest city in the country, lying about 300 km (nearly 185 miles) north-west of Abidjan. It was established by the French as a military post in 1899 on the site of a former slave market. By the 1970s it had become the focus of commerce and transport in the interior. The city is a busy market for local produce, and cotton, sisal, rice and tobacco are processed there. It is a collecting point for cocoa and coffee crops which are sent to Abidjan for export. The city also contains the country's oldest textile mill (1922). There are many mosques as well as a Benedictine monastery – the city has long been a focal point for Roman Catholic and Protestant missions.

Population 640 000

Map Ivory Coast Bb

Bouba Ndjida National Park ***Cameroon*** Wildlife reserve on the border with Chad covering about 2200 km² (850 sq miles). There is a camp on the banks of the Mayo Lidi river where guides are available, and the park is rich in such wildlife as giraffes, buffaloes, lions, leopards, cheetahs, rhinoceroses, elephants and many varieties of antelope, including the large and magnificent Derby eland.

Map Cameroon Bb

Bougainville ***Papua New Guinea*** Easternmost major island of the country, lying some 1000 km (620 miles) north-east of Port Moresby. Geographically it is part of the Solomon Islands but does not belong to the country of that name.

It is a densely wooded, mountainous island, the highest point being the active volcano Mount Balbi (2743 m, 9000 ft). There is a huge Australian-owned open-cast copper mine at Panguna, inland from the capital Arawa. Copra and cocoa are also produced.

Population (with outlying islands) 130 000

Map Papua New Guinea Ca

Bouillon ***Belgium*** Town on the border with France about 110 km (70 miles) south of the city of Liège. Set in the gorge of the Semois river, the town is dominated by a magnificent castle once owned by the Crusader Godfrey of Bouillon (1061-1100).

Population 5500

Map Belgium Bb

Botswana

THE LAND ALONG LIVINGSTONE'S MISSIONARY ROAD, WHERE THE LAST CHIEF MARRIED AN ENGLISH GIRL

The explorer David Livingstone was among several gospel-spreading Britons who trekked into the South African interior in the 19th century. The track they trod became known as the Missionary Road, and it ran through the tribal lands of the people known as the Batswana. The friendly relations established led to the Batswana chief seeking British protection when the Boer republics to the south and west began taking a predatory interest in his territories. Thus encouraged, in 1885 the British set up a protectorate over the whole of a region that was then called Bechuanaland and is now Botswana.

Despite South African takeover bids, the country maintained its ties with Britain until it became independent in 1966 under the presidency of a well-loved chief, Sir Seretse Khama, who died in 1980. As a young man studying in Britain in 1948, Khama had married an English girl, Ruth Williams, who still lives in Botswana.

The early British administrators found themselves responsible for a large, land-locked country, sparsely populated and seemingly with little economic potential. Much of the western and south-western areas formed part of the KALAHARI desert, a vast sea of shifting red sands, with some sparse covering of grass and bush.

The eastern border zone against South Africa is the most populated area, because the land is fertile and has enough rain to raise livestock and grow crops.

To the north the landscape changes dramatically into the huge OKAVANGO DELTA of the Kavango river, fed by waters that rise in neighbouring Angola, and by heavier rainfall. This marshy area is a haven for wildlife, the home of elephants, lions, leopards, buffaloes, waterbucks, zebras and crocodiles. The delta forms an intricate pattern of channels, with palm-covered islands. In good rainy seasons it drains into the vast Makgadikgadi Pans, and when these fill with water there are spectacular parades of flamingos and pelicans.

The people of Botswana are mainly farmers. By tradition their social status and wealth is judged by the size of their cattle herds. The population has always been small – it is now just over 1 000 000 – and although some white settlers established huge commercial ranches, colonial history largely followed a path of benign neglect.

After independence in 1966, Botswana's fortunes were changed dramatically by the exploitation of mineral resources. Diamonds became the major foreign currency earner and copper-nickel was also exported. The traditional meat exports were maintained and coal mining added to a temporary economic boom, aided

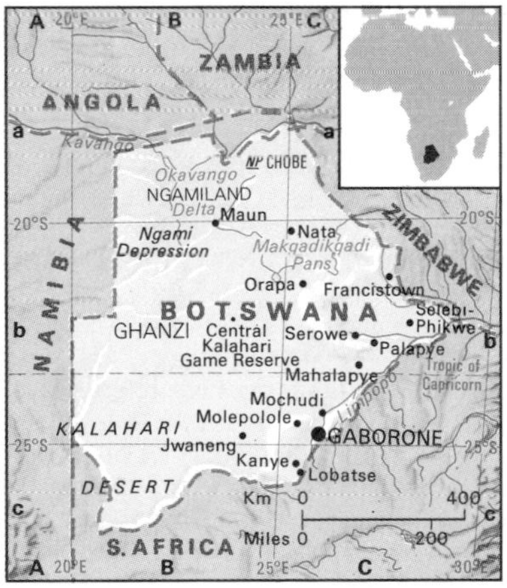

by foreign investment. Recent years have brought a decline, as world markets became depressed, and several dry years in succession have severely reduced grain crops.

There are eight major Batswana tribes; the largest is the Ngwato and the others are the Kwena, Ngwaketse, Tawana, Kgatla, Lete, Rolong and Tlokwa. Water and good pasture are crucial factors in their daily lives and settlement patterns, although the inhospitable Kalahari is still inhabited by a few thousand hunter-gatherer Bushmen.

An unusual feature of Batswana settlement patterns is the vast villages such as SEROWE and MOCHUDI, each with populations around 20 000. In the past it has been customary for households to have three different homes: one in the village, one on the family's agricultural land up to 45 km (28 miles) distant, and a third at far-off cattle posts, where young men spent most of the year tending herds.

The fickle rainfall makes it necessary for families to earn money from activities other than farming – for example, brewing, trading or building.

The village centre is the *kgotla*, the meeting place where the chief and headmen used to order the times of sowing and harvesting. Poor households could borrow cattle from the chief, which they tended on his behalf in return for the first calf. Nowadays the chief's power is waning because people are beginning to leave the villages to live permanently on their land – partly to escape the pressure on resources of a growing population; perhaps also to avoid the village headmen's traditional right to call on them for labour.

The main towns are GABORONE, expanded as a capital city after independence, the old colonial FRANCISTOWN, Lobatse which has a large abattoir, and the mining towns of Selebi-Phikwe and ORAPA.

The present government, under President Quett Masire, is a multi-party democracy. It has made notable protests against apartheid despite its dependence on South Africa for export routes and most imports, as well as employment for about 40 000 workers.

BOTSWANA AT A GLANCE

Area 600 372 km² (231 805 sq miles)

Population 1 100 000

Capital Gaborone

Government Parliamentary republic

Currency Pula

Languages English (official) and Setswana

Religions Christian (15%), remainder tribal

Climate Subtropical and dry; average temperature in Francistown ranges from 5-23°C (41-73°F) in June to 18-31°C (64-88°F) in December/January

Main primary products Cattle, sheep, maize, sorghum; diamonds, nickel, copper

Major industries Cattle rearing, mining, meat processing, tourism

Main exports Diamonds, beef, nickel, copper

Annual income per head (US$) 930

Population growth (per thous/yr) 33

Life expectancy (yrs) Male 56 **Female** 62

Boukra (Bu Craa) ***Western Sahara*** Area containing one of the richest deposits of phosphates in the world, situated south of LAAYOUNE in the north of the territory. Reserves are estimated at 600 million tonnes, but production, begun by the Spanish, was temporarily halted by the war between Morocco and the Polisario guerrillas of the Western Saharan independence movement. Morocco, which now controls the area, reopened production in 1982.
Map Western Sahara Aa

boulder A large individual fragment of rock, greater than 250 mm (about 10 in) in diameter. It is larger than a COBBLE.

boulder clay Mass of unsorted material consisting of a mixture of clay loaded with sand, gravel and boulders, carried by a glacier and deposited when the ice melts. It is also known as till.

Boulogne-sur-Mer ***France*** Largest French fishing port, in the north of the country 32 km (20 miles) south of Calais. Ferries and hovercraft provide good links to Britain.
Population 98 900
Map France Da

Bourbonnais ***France*** Historic province of central France, 7340 km² (2833 sq miles) in area, and crossed by the Allier river. It has been in existence since the 10th century and was the original seat of the royal family of Bourbon. Woodlands, small fields and hedged pastures where Charolais beef cattle are raised, make up its countryside. The chief towns are Moulins, Vichy and Montluçon.
Population 370 000
Map France Ec

Bourg-en-Bresse ***France*** Chief town of the Ain department, about 60 km (37 miles) north-east of Lyons. It has a busy livestock market and a 16th-century Gothic church, now a museum. Its industries include furniture and ironworks.
Population 43 700
Map France Fc

Bourges ***France*** Chief town of Cher department, at the confluence of the Yèvre and Auron rivers 65 km (40 miles) west of Nevers. It has a vast cathedral, ancient town houses and museums. Nearby there are modern industries, including metalworking, aircraft, missiles, textiles and tyres.
Population 92 200
Map France Ec

Bourget, Lac du ***France*** Largest lake in France, covering 45 km² (17 sq miles) and set in a steep-sided valley in the northern Alps. The spa town of Aix-les-Bains is on its eastern shore.
Map France Fd

Bourke ***Australia*** Town in New South Wales, on the Darling River, 625 km (390 miles) north-west of Sydney. It is known as the 'gateway to the Outback' and was first established as a fortified storage depot in 1835, to guard against attacks by Aborigines. It is the centre of one of the world's richest wool areas.
Population 3300
Map Australia He

bourne An intermittent or seasonal stream, especially in the chalk regions of southern England, which may flow in winter when the water table rises above the valley floor level.

Bournemouth ***United Kingdom*** Seaside town in the county of Dorset 150 km (93 miles) south-west of London. In 1850 it had fewer than 1000 inhabitants but has grown to be one of Britain's largest and best-equipped resorts. Its conference centre is a favourite for political and trade union conventions. Just to the west is the port of Poole on a large, shallow harbour; the port dates from the 13th century and today is mainly an anchorage for leisure craft.
Population 145 000
Map United Kingdom Ee

Boyacá ***Colombia*** Highland state covering 23 189 km² (8951 sq miles) in the eastern range of the Andes. The capital is TUNJA. There are wheatfields and olive groves in the warmer valleys towards the Magdalena river in the west, and at Muzo, some 80 km (50 miles) west of Tunja, is the Western world's largest emerald mine. In the north-east, at Paz de Río, is Colombia's major iron and steel works.
Population 1 200 000
Map Colombia Bb

Boyne, River (An Bhóinn) ***Ireland*** River, which rises in the Bog of Allen about 95 km (60 miles) west of Dublin and flows about 115 km (70 miles) north-east to the Irish Sea at Drogheda. It is noted for its salmon. In the river valley 6 km (4 miles) west of Drogheda on July 1, 1690, the Protestant William of Orange defeated the Catholic James II at the Battle of the Boyne, ensuring the Protestant succession to the British throne. West of the battlefield is the prehistoric passage grave of Newgrange.
Map Ireland Cb

Boyoma Falls ***Zaire*** Seven cataracts above the river port of Kisangani, and the point where the Lualaba river becomes the Zaire. There is a fall of 61 m (180 ft) over a distance of 90 km (56 miles). Formerly called the Stanley Falls, they were explored in 1877 by the British adventurer Sir Henry Morton Stanley, who successfully established that the Lualaba river was the headstream of the Zaire river. On reaching the cataracts, his men dismantled his boat and carried it and all their supplies through thick forest and over difficult terrain, fighting off attacks by cannibals. Today a railway connects Kisangani with Ubundu above the cataracts.
Map Zaire Ba

Bozen ***Italy*** See BOLZANO

Brabant ***Belgium*** Central province covering 3358 km² (1297 sq miles) around the capital, Brussels. A fertile lowland supporting cattle farming and market gardening, it is the country's most densely populated area. The historic duchy of Brabant arose in 1190 and spread into the South Netherlands. The Hapsburgs took it in the 15th century. Its prosperity in the Middle Ages was founded on wool and other textiles. A revolt by South Brabant against Dutch rule in 1830 led to Belgian independence.
Population 2 200 000
Map Belgium Ba

Brac (Brazza) ***Yugoslavia*** The largest Dalmatian island, covering 394 km² (152 sq miles), lying off the resort of Split. Its best-known export is milk-white marble, used in Diocletian's Palace (at Split), the White House in Washington, and the high altar of Liverpool's Catholic cathedral. The island's terraced slopes produce olives, figs, almonds and grapes, and fishing remains important, but labour is scarce as young people migrate to mainland cities. With excellent beaches, as at Bol, tourism is developing, but fresh water has to be shipped from the mainland. Supetar is the chief town.
Population 14 200
Map Yugoslavia Cc

Bracciano ***Italy*** Volcanic crater lake about 30 km (20 miles) north of Rome. It covers 57 km² (22 sq miles) and is fed by both streams and underground springs. It reaches a maximum depth of 165 m (541 ft).
Map Italy Dc

brackish Term used to describe water that is slightly salty, but with a salt content less than that of seawater.

Bradford ***United Kingdom*** City in the county of West Yorkshire, England, 50 km (30 miles) north-east of Manchester. It has been the headquarters of the English wool trade since the 19th century, and its prosperous past is reflected in the grandiose Italian-Gothic town hall. One of its many now-defunct mills, Moorside, has been turned into an industrial museum. It has a university, and the 15th-century Church of St Peter was designated a cathedral in 1919.
Population 281 000
Map United Kingdom Ed

Bragança ***Portugal*** Medieval town 170 km (106 miles) east of the city of Oporto. The old walls and castle overlook the modern town. The House of Braganza ruled Portugal from 1640 to 1910, the title Duke of Braganza being given traditionally to the heir to the Portuguese throne.
Population 13 900
Map Portugal Cb

Brahestad ***Finland*** See RAAHE

Brahmaputra (Yarlung Zangbo) ***China/India/Bangladesh*** River, 2900 km (1802 miles) long, rising on the north (Chinese) side of the Tibetan Himalayas (where it is called the Yarlung Zangbo) and flowing east to the Dihang Gorges, 5500 m (1800 ft) deep, where it cuts south across the mountains, then west to join the Ganges in Bangladesh, where its main channel is known as the Jamuna. Its discharge is the greatest of any Indian river, but extremely variable between the dry and wet seasons. At Gauhati in Assam, where it is already 1.5 km (1 mile) wide, it will rise 5 m (16 ft) during the monsoons. The course is constantly changing, as the river moves huge quantities of silt towards the Bangladesh delta, often released by landslips caused by earthquakes. In Tibet there is a navigable section of 640 km (397 miles) at an altitude of 3650 m (12 000 ft). Development of the lower basin will depend upon cooperation between India and Bangladesh.
Map India Eb

Braila ***Romania*** Inland port and capital of the province of the same name on the banks of the River Danube, 140 km (87 miles) west of the Black Sea port of Sulina, Braila has been an important port since the late 14th century. It was under Turkish occupation for 300 years. Its manufactures include food, textiles, rolling stock and metalwork. The centre of the city is dominated by the spire of a 19th-century church.

Population 214 900

Map Romania Ba

Bran Castle ***Romania*** Grim and imposing castle in TRANSYLVANIA 120 km (74 miles) north of Bucharest. Built in 1377, on crags overlooking a narrow valley, Bran is associated with the bloodthirsty, 15th-century ruler Vlad Tepes – known as Vlad the Impaler because of his practice of impaling unwelcome visitors on stakes, and upon whom the fictional vampire Count Dracula is said to have been based. The castle was once owned by Tepes' grandfather, and in the 1460s Tepes himself was said to have been imprisoned there by King Matthias I Corvinus of Hungary – to which Bran then belonged. The castle is now a museum.

Map Romania Ba

Brandenburg ***East Germany*** Port and industrial town on the Havel river about 60 km (35 miles) west of Berlin. It produces tractors, steel, machines, clothes, leather and river boats.

Population 95 100

Map East Germany Cb

Brandon ***Canada*** Manitoba's second largest city, on the Assiniboine river 200 km (124 miles) west of Winnipeg.

Founded by the Canadian Pacific Railway in 1881, it has a university and a residential school for Indians. It is the centre of a rich farming area and industries include oil refining, chemicals and farm implements.

Population 36 240

Map Canada Fd

Brasília ***Brazil*** National capital, 970 km (603 miles) north-west of Rio de Janeiro. Founded in 1960, this strikingly modern city was built on the largely unpopulated open plateau known as the Cerrados. The basic shape is that of an aeroplane, the wings formed by high-rise residential blocks and the fuselage by the Avenue of Ministries, culminating, at the nose, in the most impressive buildings around the Square of the Three Powers: the presidential palace (the Planalto), supreme court, and Congress. The square's name is a reference to the three arms of government: executive, judiciary, and legislature. The main buildings were designed by the Brazilian architect Oscar Niemeyer (1907-).

Only light industry is permitted and, although the city's airport has frequent connections to all parts of the country, Brazil's principal international airport remains in Rio de Janeiro.

Population 412 000

Map Brazil Dc

Brasov ***Romania*** Major industrial city and capital of the province of the same name, 140 km (87 miles) north-west of the capital Bucharest. Brasov has large Hungarian and German minorities. Its manufactures include tractors, lorries, machinery, tools, cement, glass, paper and chemicals. The country's largest city after Bucharest, Brasov was founded in 1211 by the order of Teutonic knights. It is set in a steep valley in TRANSYLVANIA and is overlooked by Mount Tämpa, 957m (3140 ft) high. Brasov's Orthodox church dates from 1392.

Population 304 700

Map Romania Ba

Bratislava (Pressburg) ***Czechoslovakia*** The capital of Slovakia and the country's second largest city after the national capital, Prague. It lies on the Danube river near the Hungarian and Austrian borders and was part of Hungary from 906 to 1918, when Czechoslovakia became independent. It was then given its present name after Bretislav, its 5th-century Slav founder. Following the Turkish invasions Bratislava was the capital of Hungary from 1541 to 1784 and the seat of its Diet (parliament) until 1848. Many Hungarian kings were crowned in the city's Gothic cathedral of St Martin. The Komensky university dates from 1467.

Outside the city centre lie post-1950 suburbs, with chemicals, oil refining, rubber, electrical

▼ CROWN OF THORNS The roof of the Cathedral of Brasília was designed to represent Christ's crown; the vast, circular nave is below ground level.

goods and engineering industries. About 16 km (10 miles) north is the village of Stupara, with a Roman camp nearby.

Population 402 000

Map Czechoslovakia Cb

Bratsk ***USSR*** Industrial town on the Angara river about 460 km (285 miles) north-west of Irkutsk. Its industries, based on the USSR's largest hydroelectric power station (capacity 4500 MW), include cement works, sawmilling, paper and aluminium refining.

Population 236 000

Map USSR Lc

Braunschweig ***West Germany*** See BRUNSWICK

Brazil See p. 104

Brazza ***Yugoslavia*** See BRAC

Brazzaville ***Congo*** National capital, set in luxuriant tropical forest on the Malebo (Stanley) Pool, a broad section of the ZAIRE (Congo) river, opposite the Zairean capital, Kinshasa. Founded in 1880 by the French explorer and coloniser Pierre Savorgnan de Brazza, Brazzaville became capital of French Equatorial Africa in 1910 and expanded rapidly after the Second World War when it was the headquarters of the Free French Forces in Africa. In 1944, the city was the site for a major conference, set up on General de Gaulle's initiative, at which the postwar future of France's overseas colonies was planned.

Brazzaville is a major river port at the lowest navigable point on the Zaire-UBANGI river systems, and is linked to the port of POINTE-NOIRE by the Congo-Océan railway. It has become a transshipment point and nearly half of the goods handled are for other Central African countries. Industries include railway repair works, shipyards and factories producing consumer goods. Brazzaville has an international airport at Maya-Maya. The University of Marien-Ngouabi was founded in 1972 as the National University. Near the city are scenic rapids on the lower part of Zaire river, the BATEKE PLATEAU and the village of Linzolo, site of Congo's first Christian church.

Population 500 000

Map Congo Bb

breccia Rock consisting of angular fragments of different rocks cemented together. See CONGLOMERATE

Brecknock ***United Kingdom*** See POWYS

Brecon Beacons ***United Kingdom*** Twin sandstone peaks in the south of the county of Powys, forming the highest massif of South Wales; Pen y Fan is 886 m (2907 ft) high. The name 'Beacons' comes from their ancient use for displaying signal flares. They lie in the 1344 km² (519 sq mile) Brecon Beacons National Park, a paradise for hikers, pony trekkers and naturalists. Many of the hillsides are grazed by sheep but are also rich in wild plants and animals – particularly birds.

Map United Kingdom De

Breconshire ***United Kingdom*** See POWYS

Breda ***Netherlands*** City 88 km (55 miles) south of Amsterdam. It is a route centre, and in former times was heavily fortified, withstanding several sieges in the 16th to 18th centuries. The medieval castle now houses the Dutch military academy. The city's manufactures include metal goods, chocolate and artificial silk. In 1660, the exiled Charles II of England issued his Declaration of Breda, offering amnesty to all those who had opposed him and his father, Charles I, excluding those specifically excepted by parliament.

Population (city) 118 800; (Greater Breda) 153 500

Map Netherlands Bb

breeze Wind between Force 2 (light breeze; 6-11 km/h, 4-7 mph) and Force 6 (strong breeze; 40-50 km/h, 25-31 mph) on the Beaufort Scale.

Bregenz ***Austria*** Capital of Vorarlberg state and a port and yacht harbour, at the eastern end of Lake Constance, 125 km (78 miles) west of Innsbruck. Bregenz began as a Celtic settlement and was later an important Roman camp. It has a museum of Celtic and Roman remains. In the Middle Ages it was ruled by the Counts of Bregenz. The town's industries include food, electrical goods and textiles.

Nearby is the dense and beautiful Bregenzerwald (Bregenz Forest), which lies between Lake Constance and the 1679 m (5508 ft) high Hochtannberg Pass. The highest peaks in the area are Hoher Ifen (2232 m, 7322 ft) and Kanisfluh (2047 m, 6715 ft).

Population 24 800

Map Austria Ab

Breiđamerkurjökull ***Iceland*** Glacier on the south-east coast, covering an area of 250 km² (97 sq miles). It is part of Europe's largest glacier, the huge VATNAJOKULL.

Map Iceland Cb

Brela ***Yugoslavia*** See MAKARSKA

Bremen ***West Germany*** Port and moated city beside the river Weser, 95 km (59 miles) south-west of Hamburg. It is the capital of the state of Bremen, established in 1947. The city's wealth grew from 1358, when it became one of the leading members of the Hanseatic League, a powerful group of Baltic trading cities. Bremen is today the country's second biggest port after Hamburg, and its industries include shipbuilding, food processing, oil, textiles, electronics and cars.

Medieval ramparts, now turned to flower gardens, encircle the cathedral, town hall and marketplace of the old city. A 9 m (30 ft) high statue, dating from 1400, of Roland, nephew of the Holy Roman Emperor Charlemagne, stands beside the town hall.

Population (city) 550 000; (state) 682 000

Map West Germany Cb

Bremerhaven ***West Germany*** North Sea port at the mouth of the River Weser, 55 km (34 miles) north of Bremen, from which it takes its name. The harbour, which can handle ships of any size, is also Europe's biggest fishing port, with huge weekday auctions of the catch.

Population 135 000

Map West Germany Cb

Brenner Pass ***Austria/Italy*** Lowest of the main transalpine routes – 1380 m (4528 ft) high – which links Innsbruck (Austria) with Bolzano (Italy). A railway runs through the pass and a motorway passes over it. The pass became the border between Austria and Italy after the First World War. In the early years of the Second World War the village of Brennero, at the Italian end, was often used as a meeting-place by Hitler and Mussolini.

Map Austria Bb; Italy Ca

Brescia ***Italy*** City 40 km (25 miles) east of Milan. It has two fine cathedrals, one dating from the 9th century and the other from the 17th century – and a 12th-century palace. Its manufactures include silks and linens, firearms and metal goods.

Population 206 500

Map Italy Cb

Breslau ***Poland*** See WROCLAW

Bresse ***France*** Fertile agricultural area east of the Saône valley, composed of pastures, lakes and woods. Dairy and poultry farming and cheesemaking are the main occupations.

Map France Fc

Brest ***France*** Largest city of Finistère department, near the country's westernmost tip. Shipyards, naval schools and an arsenal were founded in the 17th century. The city was rebuilt after repeated bombing raids during the Second World War. Its main activities include the naval base, shipyards, a university and modern engineering and chemical factories. Its 15th and 16th-century fortress, containing the Naval Museum, and some 17th-century fortifications remain.

Population 205 600

Map France Ab

Brest (Brzesc) ***USSR*** Port on the Bug river where it forms the border with Poland. Brest trades in timber, grain and cattle. Its industries include sawmilling, cotton spinning, food processing and engineering.

Founded by Poland in the 11th century as Brest-Litovsk, it was annexed by Russia in 1795. It was the scene of the signing of the Russo-German peace treaty in March 1918, under which the Russians sacrificed much territory. From 1919 to 1939 it reverted to Poland, but after being occupied by the Germans for most of the Second World War, it once again became part of the USSR.

Population 214 000

Map USSR Dc

Brezhnev ***USSR*** Industrial town on the Kama river about 900 km (560 miles) east of Moscow. Formerly called Naberezhnyye Chelny, it was renamed in 1984 after President Leonid Brezhnev (1906-82), to commemorate its rapid growth during his leadership into one of the world's largest truck manufacturing centres.

Population 414 000

Map USSR Gc

brickfielder Hot, dry, dusty wind blowing in south-eastern Australia, especially in summer. It blows from the north, ahead of a depression or trough of low pressure, with temperatures soaring to more than 38°C (100°F) for days.

A DAY IN THE LIFE OF A BARBADOS BOAT AGENT

John Selby, one of the island's small white minority, drives ten-year-old Peter and nine-year-old Sally to the primary school in Christchurch parish at 8.30 am. He turns his Datsun onto the coast road, and heads north through the resort of Hastings to Bridgetown, the capital of Barbados. He thinks about his children's prospects in the secondary school common entrance exams. His wife Shirley says he pushes the kids too hard, but he wants them to go on to be professionals.

Selby has two 9 m (30 ft) motorboats, paid for with money inherited from his father, a Londoner who came to the island when it was a British colony. Some 400 000 tourists visit Barbados every year, swamping the 270 000 locals but providing jobs and spending millions of dollars. Selby competes with a dozen major boat agents and scores of small-timers to earn 2000 Barbadian dollars (about US$1000) – a month's profit from tourists' fishing trips, snorkelling expeditions and rides to the smaller islands.

As he drives past the sugar-cane fields, palm trees and wooden houses, he has an eye on the fish in the street markets and an ear to the cricket news on the radio. His small office is tucked away amid colonial buildings in Broad Street, just off bustling Trafalgar Square, the hub of activity of the town's 100 000 people. Penny, his part-time helper, has sorted messages, bookings and bills for him to attend to.

He walks over to the harbour where Johnny and Sammy, his black fishermen, are settling tourists onto the boats and topping up the refrigerators with canned drinks. He gets on well with these two, and cares about their families in a way that erases any lingering ill feelings about the history of slavery in their distant past – some 80 per cent of Barbadians descend from Africans.

He sees the boats out. They will be back by late afternoon with a cargo of blue marlin, barracuda, red snapper, yellowtail or amberjack, and of tourists with beaming faces who think it good fishing for the 350 dollars (US$175) each group of six has paid. Selby drives home for lunch with Shirley. She has laid out chicken, cheese, salad and tropical fruits on the terrace beside the bearded fig tree and the orchid plants.

Meeting the boats at 4 pm, he finds some tackle has been lost, and some damaged, and goes with Johnny to one of the fishing shops off the square for replacements. So he gets home late at 6 – and finds his children have not done their homework; as a punishment, he cancels their weekend visit to the cinema.

BRIDGING THE GAP BY ROAD AND RAIL

The world's first bridge with a main span of more than 1.6 km (1 mile) is soon to be opened in Japan. The Akashi-Kaikyo suspension bridge, over the Akashi Strait, will be a double-decker – with a railway on one level and a motorway on the other. Work on the bridge began in 1978 and is due to end in 1988. The longest multi-span bridge is the Second Lake Pontchartrain Causeway in Louisiana, USA; it is 38.4 km (23.9 miles) long.

SOME OF THE WORLD'S LONGEST BRIDGE SPANS

Name	Location	Span (m/ft)	Open
Akashi-Kaikyo (suspension)	Akashi Strait, Japan	1780/5840	1988
Humber Estuary (suspension)	Hull, UK	1410/4626	1981
Verrazano Narrows (suspension)	New York, USA	1298/4260	1964
Golden Gate (suspension)	San Francisco, USA	1280/4200	1937
Quebec Railway (cantilever)	Quebec, Canada	549/1800	1917
Forth Rail (cantilever)	Firth of Forth, UK	518/1700	1889
New River Gorge (steel arch)	West Virginia, USA	518/1700	1977
Bayonne (steel arch)	New York, USA	504/1652	1931
Sydney Harbour (steel arch)	Sydney, Australia	503/1650	1932

Bridgeport ***USA*** City on the coast of the state of Connecticut, opposite Long Island. Its manufactures range from electronics and electrical goods to helicopters, hot-air balloons and stage sets for Broadway. It is also a popular residential area for commuters to New York.

Population (city) 142 100; (metropolitan area) 441 500

Map United States Lb

Bridgetown ***Barbados*** Capital, commercial centre and chief port, with a deepwater harbour, founded by the British in 1628 on the south-west coast of the island. A port of call for cruise ships, it also exports the island's sugar crop, rum and molasses, as well as serving as a trans-shipment point for other places in the region.

Population 97 500

Map Barbados Ba

Brie ***France*** Agricultural area east of Paris, between the Marne and Seine rivers. It produces the cheese named after the area.

Map France Eb

brigalow scrub Type of scrub, consisting mainly of a low-growing species of acacia known as brigalow, which covers much of central Queensland in Australia.

Brighton ***United Kingdom*** English coastal city 75 km (45 miles) south of London, in the county of East Sussex. It was originally a small fishing port called Brighthelmston which grew in the 18th century with the then novel habit of sea-bathing. The Prince Regent, later King George IV (1762-1830), patronised the town and commissioned the magnificent Indian Mogul fantasy, the Royal Pavilion, finally completed in 1815. It immediately became a fashionable resort with fine terraces of elegant houses. The modern University of Sussex is nearby.

Population 149 000

Map United Kingdom Ee

Brijuni (Brioni) ***Yugoslavia*** See PULA

Brindisi ***Italy*** Adriatic port on the heel of Italy, with sea links to Greece. The old town, on a knob-like peninsula between two creeks, has a castle built by the Holy Roman Emperor Frederick II in 1227, and an 18th-century Baroque cathedral rich in 12th-century mosaics. Modern suburbs have spread since the city became a petrochemicals centre in 1960.

Population 92 000

Map Italy Fd

Brisbane ***Australia*** State capital of Queensland, standing on the Brisbane river, about 20 km (12 miles) from its mouth. It was founded in 1824 as a penal colony and became the capital on Queensland's foundation in 1859. It was largely destroyed by fire in 1864. The Old Observatory, a massive tower built as a windmill by convicts in 1829, is among the few surviving original buildings.

Industries include textiles, agricultural machinery, sugar and oil refining, and shipbuilding. The University of Queensland, founded in 1909, stands in a park beside the river.

Population 942 400

Map Australia Id

Bristol ***United Kingdom*** City in the county of Avon and once a principal port in the west of England, 170 km (105 miles) west of London. However, the River Avon, on which it stands, is too narrow to handle large ocean-going ships, and Avonmouth – downriver on the Bristol Channel – has become its main outlet.

Although Bristol was severely damaged in the Second World War, it is particularly rich in old churches. Its cathedral dates back to the 12th century and the Church of St Mary Redcliffe was built between the 13th and 15th centuries. The suburb of Clifton has a suspension bridge 80 m (260 ft) high over the river gorge; it was started by the engineer Isambard Kingdom Brunel (1806-59) and completed after his death. The gorge's 18th and 19th-century houses give it, from the river, a Mediterranean look. The University of Bristol was founded in 1909. The suburb of Filton has an aircraft factory.

Population 399 000

Map United Kingdom De

Brazil

COFFEE BUILT THIS ENORMOUS COUNTRY; NOW IT IS TRYING TO RUN ON SUGAR. IT IS THE WORLD'S BIGGEST DEBTOR – BUT CONTAINS FABULOUS RICHES

There is fabulous RIO DE JANEIRO, its carnival, its Copacabana and Ipanema beaches, and its Sugarloaf Mountain. There are the limitless, primordial rain forests of the AMAZON basin; a new futuristic capital city in the middle of nowhere; empty deserts, lush savannahs, wild highlands. There is an awful lot of coffee, and almost everything else, in Brazil – including over 140 million people, half of them under 25 years old.

Riches it possesses, beyond imagination – but little money. Brazil is the world's biggest debtor, owing over US$100 000 million to foreign banks at the last count. It managed to pay off the US$12 000 million interest in 1984, but inflation ran at 220 per cent.

Surviving, struggling and somehow managing to thrive again, despite all, is a fun-loving, football-crazy, car-mad population of staggering ethnic complexity and every known complexion. The soccer stadiums hold 4 million fans – all at once – and their idol Pelé (now retired) is a world figure. Brazil has won the soccer World Cup three times and has bred two motor-racing world champions, Emerson Fittipaldi and Nelson Piquet. Brazilians may well be the world's greatest consumers of alcohol – when oil became scarce and expensive in the early 1970s they began building cars that run on industrial alcohol made from their vast sugar-cane resources.

Distances do stretch in this enormous land – 4320 km (2700 miles) from north to south and a little more from east to west. It has 7400 km (4600 miles) of coastline and over 43 000 km (27 000 miles) of navigable waterways. It is, indeed, the fifth largest country in the world, covering nearly half of South America, with an area of 8 511 965 km^2 (3 286 488 sq miles). Brazil owes its great size to the fact that the Portuguese 'discovered' it rather than the Spanish. After the subcontinent was divided between Spain and Portugal in 1494, Spain split its share into provinces, which later became independent countries. Portugal, more interested in its eastern discoveries, imposed direct colonial rule when it colonised Brazil in the 16th century, so that despite internal rivalries, it emerged as a single, Portuguese-speaking nation.

The Portuguese colonists mingled with native Indians they found there, not to mention African slaves they imported when the Indians proved reluctant workers. Around 4 million slaves were shipped in between the 16th century and the abolition of slavery in 1888. Massive immigration from Europe, the Middle East and the Far East added to the racial mix between 1820 and 1939. Whole German colonies were founded in the south, and the Japanese grew pepper around the Amazon estuary and started market gardening outside the great city of SAO PAULO. The aftermath of the Second World War brought more immigrants. Today around 55 per cent of the people are white or near-white, 38 per cent of mixed origin, 15 per cent black and the remainder Asians or native Indians. There is an almost total absence of racial prejudice.

A yawning social gulf separates the few rich from the many, many millions of very poor. But rich or poor, they are seen at their extrovert, ebullient and talented best when they celebrate at carnival time. The pre-Lenten festival is celebrated in many Roman Catholic countries, but no place does it quite like Rio.

RIO DE JANEIRO, a city of over 9 million people, lies in a magnificent setting, stretching 20 km (12 miles) along a narrow strip of land between green mountains and blue sea. Sailing into the beautiful harbour, it is easy to believe the local assertion that 'God made the world in six days; the seventh he devoted to Rio'. As though in affirmation a gigantic statue of Christ looks down on the city.

It is the former capital and the major city of the east coast, though not the biggest overall. It lies about midway between SALVADOR in the north and PORTO ALEGRE in the south. Along this entire seaboard the land rises sharply in a great escarpment, leaving a coastal strip averaging only 100 km (60 miles) wide – occupying a mere 7.7 per cent of Brazil's total area, yet supporting 30 per cent of its population.

Again, history as well as terrain is responsible for this: the early Portuguese settlers landed here and immediately began cultivating the rich alluvial soil. Originally much of the region was tropical forest, but most of that has been cleared over the centuries and the area is now Brazil's political and economic heartland. It forms one of the country's five great natural regions, and within it are the states of Minas Gerais, Espirito Santo, São Paulo and Rio de Janeiro, the two last containing the great cities of the same names. The coastal escarpment averages about 790 m (2600 ft) above sea level, while the uplands of the interior rise to about 2130 m (7000 ft).

Largest of Brazil's regions is the 5 300 000 km^2 (2 050 000 sq mile) Amazon river basin, which covers about one-third of the country, in the north and west. This vast lowland area is covered by rain forest and drained by the 6570 km (4080 mile) river and its tributaries such as the Xingu and Negro. Not for nothing is the forest known as the *inferno verde* (green hell), with a year-round temperature of about 27°C (81°F) accompanied by exhaustingly high humidity and an annual rainfall of 3000 mm (118 in). However, the Trans-Amazonian Highway and other new roads and railways have opened it up for development.

► GOLD RUSH Miners swarm over a human ant hill in the Serra Pelada range in Para state. Many came from Brazil's drought-stricken north-east when the huge gold ore deposit was discovered in the 1970s.

Huge areas of the forest have been cleared for crop growing and ranching, as Brazil seeks to improve its economy by encouraging farmers from the desperately poor north-eastern region to move into the cleared lands. However, these schemes are proving disastrous. The luxuriant rain forest appears to be supported by fertile soils, but this is not so. The forest lives on its own dead remains, and removal of the forest yields farmland which becomes barren within two to three years. And big business concerns have displaced small farmers, and incorporated their land into vast ranches. The immigrants from the north-east have exchanged one poverty trap for another.

Displaced Indian tribes, some still living a Stone Age existence, are being absorbed into the mainstream of society – a sometimes painful process which has caused disquiet among those opposed to the disturbance of the Indians' traditional cultures and ways of life, and to changing the ecology of the forest itself. Indeed, there have been times, from colonial days until the present, when the process has seemed more like elimination than assimilation. There were more than a million Indians in Brazil when the Portuguese arrived. Many were killed or made slaves; many more succumbed to imported European diseases, such as influenza, measles and smallpox, to which they had no natural resistance. Even into this century they were still being enslaved to gather wild rubber, and killed when no longer useful. The Brazilian Indian Service has done much to mitigate their lot, but there is again strong evidence of maltreatment, and even murder, in the Amazon clearances, and it is feared that only about 100 000 Indians remain in the whole country.

The impoverished north-eastern region includes large stretches of semidesert, covered by a thorny scrub called *caatinga*; here years can pass with barely a drop of rain, but enough falls on the coastal area to encourage both evergreen and deciduous trees.

The fourth region is the southern plateau, incorporating the states of PARANA, Santa Catarina and Rio Grande Do Sul, on the border with URUGUAY. Coniferous and broad-leaved trees cloak the high country to the north; prairie grasslands similar to the *pampas* of ARGENTINA cover the south. This is *gaucho* country; a fertile temperate land with cattle ranches in the west and huge coffee plantations in the north and north-east.

Finally there is the west-central region, where savannah grasslands scattered with trees roll across another vast plateau. It contains three stock-raising states: MATO GROSSO, Mato Grosso Do Sul and Goiás. And it was in the unpopulated wilderness of Goiás that President Juscelino Kubitschek chose to build his dream city, BRASILIA, which succeeded Rio as the new capital in 1960.

Despite its surrounding shanties, Brasilia remains a modest-sized city of some 1 million – far behind São Paulo, the largest city in Brazil. This is the fastest-growing city in the world, and one of the most powerful financial centres. Its population of over 15 million is increasing at an estimated rate of 150 000 a year.

A BETTER LIFE

But the tough, cosmopolitan, go-getting *Paulistanos*, as the citizens are called, love it all – the bustle, the opportunities for carving out a better life, the cool, temperate climate and easy access to splendid seaside resorts and the port of SANTOS. This, for South America, is where the action is – far removed from the small missionary settlement founded there by Jesuits in 1554, though a replica of their original church has been built. The region all around proved an agricultural gold mine, its rich, red soil perfect – as it turned out – for large-scale coffee growing. A century ago São Paulo really started to expand, as wealth created by coffee was ploughed into new industries. Vast numbers of immigrants arrived from Europe, bringing their own skills and enterprise. Now this city alone accounts for 40 per cent of Brazil's industrial production. The 247 898 km² (about 96 500 sq mile) state of São Paulo, with a population of around 25 million, produces 65 per cent of the nation's industrial output, 62 per cent of its sugar, 50 per cent of its cotton, 33 per cent of its coffee and 90 per cent of its motor vehicles.

It is perhaps fitting that São Paulo was the city where Brazilians first claimed independence from Portugal, in 1822. Their whole history, before and since, has been extraordinary. A sea captain named Pedro Alvares Cabral 'found' the Brazilian coast in 1500 and claimed the country for PORTUGAL, but it was 49 years before the Portuguese got around to appointing a Governor-General. In 1572 the colony was divided into two, with Salvador – capital of what is now the state of Bahia – as the northern capital and Rio the southern. African slaves worked huge sugar plantation-estates for colonists along the coastal strip. Other settlers pressed inland to the São Paulo and Minas Gerais regions, looking for gold, precious stones and more slaves. Dutch attacks on Salvador and RECIFE were beaten off in the mid-17th century, but in general – apart from an unsuccessful revolution in Minas Gerais in 1789 – colonial life changed little until it was changed for ever in 1808, by Napoleon.

In that year the Portuguese royal family fled to Brazil ahead of Napoleon's invading armies – on Britain's advice and escorted by British ships. They landed in Rio, which became the seat of the Portuguese Empire. King João VI remained in Brazil until 1821 (when Napoleon died on ST HELENA island), then returned to Lisbon, leaving his son Pedro as regent. The next year Pedro proclaimed Brazil independent, with himself as Emperor Dom Pedro I.

His reign did not, alas, continue as well as it began. He lost what is now URUGUAY and abdicated in 1831, returning to Portugal to subdue his brother Dom Miguel who had usurped the throne. His five-year-old son, another Pedro, became ruler under a regent. At the tender age of 15, Pedro took over as Dom Pedro II, and immediately embarked upon liberal reform. Communications and schools were improved, corruption was attacked. Immigrants poured in by the thousand. He won a war against a Paraguayan dictator, and was moved to declare that he would rather lose his crown than let slavery continue.

This declaration did, indeed, lose him his crown. His daughter, Princess Isabel, acting as regent during his temporary absence, freed the slaves in 1888. Plantation owners turned against him; so did the military. He was, like his father, forced to abdicate, and a republic was declared in 1889.

AFTER BRAZIL'S EMPEROR

However, Brazil continued to prosper after Pedro's departure – until 1929-30, when the coffee market went the way of most others in the Depression. Dr Getulio Vargas seized power by organising the working classes, and ruled as dictator until he was deposed in 1945. He was legally re-elected president in 1950, but committed suicide four years later, after members of his bodyguard were implicated in a bid to kill a political opponent. Kubitschek, founder of Brasilia, took over in 1955. Inflation beat him and his successors and the military seized power in 1964. General João Figueiredo, last of a line of military presidents, took office in 1979, committed to a slow, step-by-step return to democracy. New parties were formed and finally, in January 1985, the 686-member congressional electoral college voted in 74-year-old Tancredo Neves, the Opposition candidate. It was a landslide victory by 300 votes over his military-backed rival, Paulo Salim Maluf.

In his victory speech, Neves said that despite national frustration over the military-backed administration's refusal to allow public elections, 'we have made the electoral college the very instrument of the government's defeat!' He continued, amid wild applause from Congress delegates: 'I have come to propose economic, social and political change . . .'

Neves never lived to be sworn in. Soon after his triumph it became known that he was seriously ill with an abdominal complaint.

When finally Neves died in April 1985, his doctors, who had seen him through a total of seven operations, revealed that he had suffered from a benign tumour, but also had a long-standing and widespread abdominal infection, and had been taking strong antibiotics 'for months' before going into hospital. To his admirers, this could only mean that, knowing he was ill, Neves put off proper treatment until after the election: a martyr for democracy.

Meanwhile, Vice-President José Sarney had taken over and democracy was marching on. In May, Congress voted unanimously in favour of choosing the next president by direct elections. It also voted to allow new parties to be formed without restrictions – meaning that hitherto-banned Communist parties would be permitted. A US$2500 million aid programme for the poor was announced, including subsidised food for the 13 million worst-off, school meals for 20 million children, 250 000 new homes and 500 000 homes to be connected to mains services.

Hopes are high once more in the nation that had the courage to join the Allies against Germany in both the First and Second World Wars (Brazilian troops fought in Italy in 1944-5). For despite massive debts, the crushing poverty of millions and the glaring social inequalities, the country seems to be on the move again. It has much to offer. Although Brazil has already yielded up fabulous riches, they are but a small fraction of its almost limitless resources.

Its first export was a red wood called *pau-brasil*, used for dyeing in the 16th century – and from which the country takes its name. Next the Portuguese introduced sugar cane from Madeira; and in the following century gold, silver and precious gems of many kinds were found in Minas Gerais. However, it was not until late in the 19th century that Brazil found its greatest asset, coffee, although the plant had been introduced as early as 1727. Brazil is still the world's biggest coffee producer. Rubber grew wild in the Amazon basin, and between 1880 and 1920 the burgeoning tyre industries created an enormous demand until Malaya broke Brazil's monopoly with rubber grown from seeds smuggled out of Brazil by an Englishman.

MASSIVE SCHEMES

Efforts are currently concentrated on creating a basis for more industrial and agricultural expansion. Great new electrification schemes are coming to fruition to provide power. In the north-east a US$4600 million dam at Tucuruí came on stream in January 1985; it was eight years building and by 1989 will be churning out 8000 megawatts – the fourth largest hydroelectric complex ever built. Among areas to benefit from it will be Carajas, site of Brazil's biggest concentration of mineral resources – including the world's largest iron ore reserves and large deposits of bauxite, gold, nickel, copper and manganese. A new railway passes from Carajas through land suitable for livestock and agriculture to the Atlantic port of São Luís. However, Tucuruí almost pales into insignificance alongside the ITAIPU scheme on the Parana river, at the border with Paraguay. When complete in 1988 it will be the world's largest hydroelectric plant, with a capacity of 12 600 megawatts. And ten more hydroelectric schemes are underway.

Money has been poured into mechanising agriculture – particularly in the central and southern areas, where beef production is expanding. The vast undertakings in the Amazon basin are bringing new life to the extraordinary city of MANAUS. At the turn of the century, during the height of the rubber boom, the city was known as 'The Paris of the Tropics'. Here glittering audiences in the huge, ornate opera house would hurl diamonds onto the stage to show their appreciation when great stars of their time came to perform. For even getting there was no mean feat – deep in the jungles of what is now the state of Amazonas, Manaus could be reached only up the Negro river, by way of the Amazon itself – a 1450 km (900 mile) journey from the Atlantic coast. Until recently the only alternative way was by air, but now a new highway connects it into Brazil's main road system.

Ocean-going ships can navigate the Amazon and Negro to dock in this free port, which handles the products of the whole upper Amazon region – rubber (still), rosewood and other timbers, beef and hides, jute, even oil, brought by barge from Peru. Tourism is expanding, and visitors can stroll through fine botanic and zoological gardens and a natural jungle park. The population numbers around 635 000 and temperatures average 27°C (81°F). The city overlooks the Negro river and is laced with numerous creeks spanned by little bridges. But perhaps the most significant development for modern Manaus is the discovery of oil in the area.

Whatever else Brazil may have, it does not have enough oil – only about one-fifth of its needs – hence the development of alcohol as an alternative fuel. The country is also short of coal and natural gas, and currently generates only one-third of its total energy needs. But oil is the major problem: most of Brazil's supplies come from the Middle East, and price increases since the early 1970s have slowed the development of heavy industries, emphasis being switched to agriculture and light industry. The great hydroelectricity schemes are designed to increase self-sufficiency, and the first of several nuclear power stations came on stream in 1983, near Rio (Brazil has large uranium and thorium reserves). The search for new oil and coal sources has been intensified.

However, despite a large-scale population drift from countryside to towns and cities, about 30 per cent of the labour force still work in farming, forestry or fishing. The last available surveys reveal that a quarter of all Brazilians are illiterate, and 2 million children aged 7 to 14 do not even have a school to attend. Even when they have one, only 60 per cent stay long enough to become literate.

Brazil is, according to the Vatican, the largest Catholic nation in the world. However, great numbers of Brazilians wear both a cross and a voodoo charm, and worship African gods as well as the Christian god.

▼ CARNIVAL IN RIO Five frantic days of parades, balls, street dancing and general revelry precede Lent in Rio de Janeiro. The throbbing beat of samba music specially composed for the festival fills the air.

BRAZIL AT A GLANCE
Area 8 511 965 km² (3 286 488 sq miles)
Population 140 650 000
Capital Brasília
Government Federal republic
Currency Cruzado = 1000 cruzeiros
Language Portuguese
Religions Christian (89% Roman Catholic, 12% Protestant), voodoo
Climate Mainly tropical and subtropical, though cooler on the southern coast and on higher lands. Average temperature in Rio de Janeiro ranges from 17-24°C (63-75°F) in July to 23-29°C (73-84°F) in February
Main primary products Cereals, cassava, soya beans, sugar, oranges, cocoa, coffee, rice, cotton, tobacco, bananas, rubber, timber, fish; iron ore, bauxite, manganese, crude oil and natural gas, coal, chromium, nickel, tin, zinc, gold, silver, diamonds, phosphates, salt, quartz crystal, beryllium, graphite, titanium, tungsten, asbestos
Major industries Agriculture, mining, iron and steel, motor vehicles, oil and mineral refining, chemicals, wood pulp and paper, machinery, food processing, consumer goods, textiles, rubber processing, fertilisers
Main exports Machinery, animal feedstuffs, coffee, cocoa, iron ore, motor vehicles, sugar, soya beans and oil, oranges, iron and steel, chemicals, nonferrous metals
Annual income per head (US$) 2100
Population growth (per thous/yr) 23
Life expectancy (yrs) Male 62 **Female** 66

Bristol Channel ***United Kingdom*** Inlet of the ATLANTIC OCEAN, between South Wales and south-western England, where tides often rise by 12 m (40 ft). Periodically a bore (high wave) sweeps up the River SEVERN reversing the flow.
Dimensions 137 km (85 miles) long, 8-69 km (5-42 miles) wide
Map United Kingdom De

British Columbia ***Canada*** Mountainous province on the Pacific coast covering 929 730 km² (358 968 sq miles). Most people live in the valleys, especially in the lower FRASER river region around VANCOUVER. Furs, fish and gold brought the first settlers. Modern mines produce copper, gold, silver and coal. Limited agriculture is concentrated on fruit, vegetables and dairy produce, and there is a salmon-fishing industry. Hydroelectric power has led to the development of electrometallurgical industries. Tourists are attracted by scenic beauty and the mild climate of the south-west. Vancouver is the largest city and VICTORIA the capital.
Population 2 744 000
Map Canada Cc

British Indian Ocean Territory ***Indian Ocean*** Territory comprising the five mid-ocean coral atolls forming the Chagos Archipelago. They total only 52 km² (20 sq miles) in area, and lie 1899 km (1180 miles) north-west of Mauritius. In the 1970s the US Navy built a large base on the biggest island, Diego Garcia, under a British-American joint defence agreement. There is no permanent population.
Map Indian Ocean Cc

British Virgin Islands ***Caribbean*** See VIRGIN ISLANDS

Brittany ***France*** Historic peninsula region of north-western France, with 3500 km (2175 miles) of jagged coastline, formerly known as Armorica. In the 5th and 6th centuries it was occupied by Bretons, a Celtic people who fled the Anglo-Saxon invasion of Britain, and from whom it gets its name. The peninsula became part of France in 1491.

Breton is still spoken in western areas, and many picturesque folk traditions survive – with local festivals of music and folk dancing, and the wearing of headdresses. The main occupations are farming and fishing. Industries include car manufacturing and electronics. RENNES and BREST are the main cities.
Population 2 708 000
Map France Bb

Brno (Brünn) ***Czechoslovakia*** Central industrial city making textiles, machinery, chemicals and armaments – it was the home of the Bren gun. It hosts an international trade fair every autumn at the time of the wine vintage. The city, which is about 190 km (120 miles) south-east of Prague, was the capital of Moravia. Its old town includes Spilberk Castle, built on a hill in the 13th century, a university, and an opera house named after the Czech composer Leos Janácek (1854-1928), who wrote his *Sinfonietta* for the city.
Population 381 000
Map Czechoslovakia Cb

Brod ***Yugoslavia*** See BOSANSKI BROD

Broken Hill ***Australia*** Mining city in New South Wales, 920 km (571 miles) west of Sydney but only 420 km (260 miles) from Adelaide, capital of South Australia. It is known as 'Silver City' and has some of the richest silver, lead and zinc deposits in the world.
Population 26 900
Map Australia Ge

Broken Hill ***Zambia*** See KABWE

Bromberg ***Poland*** See BYDGOSZCZ

Broome ***Australia*** Port on the north-west coast of Western Australia, about 1650 km (1030 miles) north and slightly east of Perth. It was founded in 1883 as a port for the Kimberleys. At about that time pearl grounds were discovered offshore, and once supplied 80 per cent of the world's mother-of-pearl. Broome's main industry today is still the production of mother-of-pearl, and cultured pearls. At the foot of the surrounding sandstone cliffs, the fossilised tracks of dinosaurs are still visible.
Population 3600
Map Australia Cb

brown coal Brown fibrous deposit, intermediate between peat and black coal. See LIGNITE

Bruce, Mount ***New Zealand*** Bird reserve about 100 km (60 miles) north-east of the capital, Wellington, on North Island. Some of the country's rarest birds, such as the flightless takahe and the kokako, are bred there.
Map New Zealand Ed

Bruges (Brugge) ***Belgium*** Historic lace-making city about 90 km (55 miles) north-west of the capital, Brussels. The capital of West Flanders province, Bruges prospered in the 13th and 14th centuries as the centre of the wool trade, but fell into disuse when silt blocked its waterway outlets to the North Sea. A 13 km (8 mile) long deep-water canal – opened in 1903 and linking the city to the port of Zeebrugge – led to Bruges' economic recovery. As well as lace, its main industries today are chemicals and electronics.

Tourists are attracted to its medieval buildings and the quiet canals of the old town.
Population 120 000
Map Belgium Aa

Brugh Na Boinne ***Ireland*** Collection of prehistoric tumuli in the Boyne Valley, 11 km (7 miles) south-west of Drogheda, the chief of which is Newgrange. A passage grave – 85 m (280 ft) in diameter and 13 m (43 ft) high – it dates from about 2500 BC and is made of 180 000 tonnes of stone. The corbelled central chamber, with stones inscribed with spiral patterns, is reached through a passage 20 m (65 ft) long. Other passage graves have been discovered at nearby Knowth and Dowth.
Map Ireland Cb

Brunei See p. 110

Brünn ***Czechoslovakia*** See BRNO

Brunswick (Braunschweig) ***West Germany*** Ancient capital of the Dukes of Saxony, founded in 861 about 55 km (35 miles) east of Hanover. The castle, the cathedral and the cast bronze lion in the main square were built by Henry the Lion (died 1195), most powerful of the Saxon dukes, who was married to Matilda, daughter of the English king, Henry II.
Population 255 000
Map West Germany Db

Brussels (Brussel; Bruxelles) ***Belgium*** National capital in the centre of the country and the administrative headquarters of the European Economic Community since 1958. The city, developed after AD 580 on a marshy island in the River Senne, gets its name from the Flemish words *brock*, meaning 'marsh', and *sali*, meaning 'building'. Most of its surviving buildings, however, date back only to the 15th century. In the walled Lower City, with its narrow streets, the superb central square is still intact. The city hall, built between 1402 and 1480, still dominates the square with its 96 m (315 ft) high spire.

Modern Brussels, where every sign – by law – is in both Flemish and French, is the headquarters of the Common Market's officials and diplomats. It also has the Palais des Beaux Arts (art gallery), the royal library and palace, and the Palais de la Nation (parliament buildings), all mainly constructed during the reign of King Leopold II (1865-1909).

The city's main industries are textiles, chemicals, electrical equipment, machinery, rubber goods and lace. Brussels sprouts – first developed from cabbages in the region in the 13th century – get their name from the city.
Population 1 000 000
Map Belgium Ba

Bryansk ***USSR*** City on the Desna river 340 km (210 miles) south-west of Moscow. It has a locomotive plant, ironworks, flour mills and sawmills, distilleries and glassworks.
Population 424 000
Map USSR Ec

Bryce Canyon National Park ***USA*** Beautiful area of 144 km² (56 sq miles) on the edge of the Paunsaugnut Plateau in south-west Utah. The park has strikingly colourful pillars, stacks and horseshoe-shaped amphitheatres created by erosion.
Map United States Dc

Bu Craa ***Morocco*** See BOUKRA

Bubiyan Island ***Kuwait*** Large island at the head of The Gulf. Bubiyan is a source of conflict with neighbouring Iran, which has long made claims to it. In 1977 a joint committee of the two countries was set up to deal with border claims, none of which have been resolved.
Map Kuwait Bb

Buçaco ***Portugal*** Ancient forest, 30 km (20 miles) north of the university city of Coimbra. There was a Carmelite monastery there from the 6th century until 1834, when the monks were expelled following the abolition of all Portuguese religious orders. It is also the site of the British victory against the French in 1810 during the Napoleonic Wars.

Today the forest's 105 hectares (260 acres) contain some 700 species of plants.
Map Portugal Bb

Brunei

THE 'OIL SHEIKDOM' OF THE FAR EAST, WHERE THE WEALTHY STILL CHOOSE TO LIVE IN PRIMITIVE HOUSES ON STILTS

Oil made Brunei so rich that it has been dubbed the 'Shellfare State' after the oil company that has dominated its economy. Before the 1985-6 oil slump, this Muslim country on the north-west coast of Borneo had an annual income of about US$21 000 per head – one of the highest in the world. There is no income tax. Education and health services are free, and housing loans are available at 0.5 per cent interest.

Brunei is ruled by Sir Hassanal Bolkiah, an autocratic sultan with two wives. There was an election in 1962 when the Brunei People's Party won all the seats, but no parliament was called and a rebellion was put down by British Gurkha troops. The sultan and his brothers continue to rule by hereditary rights stretching back centuries.

The country is divided in two by a tongue of land – part of the Malaysian state of SARAWAK. Broad, tidal swamplands cover the coastal plains of both parts. Inland, much of Brunei is hilly and covered with tropical forest: rainfall is heavy and varies from 2500 mm (98 in) to 5000 mm (197 in).

In the 16th century the sultans of Brunei controlled all of northern Borneo, but their influence declined and in 1847 they signed a treaty with Britain to gain help in fighting pirates. Brunei became self-governing in 1959 and independent in 1984. Prince Charles attended independence celebrations, for which 5000 foreign workers built the sultan a new US$300 million palace.

Oil production began in the 1920s, and oil and natural gas now account for almost all exports. Until oil income slumped, Brunei spent less than half its annual revenue – reserves invested abroad total US$13 billion. Before Brunei struck oil, the economy depended on rubber, but the rubber plantations are now neglected.

There is a drive to make the country self-supporting in food; meanwhile, the sultan has bought a cattle ranch in Australia, larger than Brunei itself, to supply meat.

About 65 per cent of the people are Malay and 20 per cent Chinese. The Chinese, usually denied citizenship, run nearly all the shops. In the forest, Iban tribesmen live in long houses, accommodating 100, and sometimes have human skulls hanging from the roofs from head-hunting days. BANDAR SERI BEGAWAN, the capital, has 58 000 inhabitants and a huge golden-domed mosque.

Many Malays who run expensive cars and motor launches still choose to live in *kampongs* (water villages) such as Kampong Ayer, in primitive houses on stilts. On the busy waterways among the houses, boats ferry children to school, housewives to market and men to their work.

In this strict Muslim state, alcohol is served only in hotels for Europeans, mixed bathing is forbidden on public beaches and there is little night life, even in the capital.

BRUNEI AT A GLANCE

Area 5765 km² (2226 sq miles)
Population 230 000
Capital Bandar Seri Begawan
Government Hereditary sultanate
Currency Brunei dollar = 100 cents
Languages Malay, Chinese, English
Religions Muslim (62%), Buddhist (13%), Christian (8%)
Climate Tropical; very humid and wet. Average temperature ranges from 24°C (75°F) to 30°C (86°F)
Main primary products Rice, cassava, bananas, cattle, fish, timber; oil and natural gas
Major industries Production and refining of oil and natural gas
Main exports Crude oil, natural gas, refined petroleum products
Annual income per head (US$) 21 000
Population growth (per thous/yr) 33
Life expectancy (yrs) Male 70 **Female** 73

Bucaramanga ***Colombia*** Capital of Santander state, about 310 km (190 miles) north-east of Bogotá. It is known as the city of parks because of its green and spacious centre. The city manufactures cigars, cigarettes and straw hats.
Population 516 000
Map Colombia Bb

Bucegi Mountains ***Romania*** Mountain range at the south-eastern end of the Carpathians. The mountains, which rise to 2505 m (8218 ft), are renowned for the fantastic shapes of their eroded summits. The slopes are used for skiing.
Map Romania Aa

Bucharest (Bucuresti) ***Romania*** Known as 'The Paris of Eastern Europe', Bucharest has been the capital of the Romanian state since its formation in 1862. It is a green and pleasant city of spacious parks, broad, tree-lined boulevards, terrace cafés and restaurants, and handsome 19th and 20th-century public buildings. The main thoroughfare of the 'Parisian' area is the elegant Calea Victoriei (Victory Road), which was originally built in 1702.

It is bounded by the Operetta Theatre; the Stavropoleos church with its beautifully decorated walls and carved doors; the History Museum of Romania with its magnificent displays of gold models, jewellery, precious stones, plates, goblets and weapons – some of them dating from the 4th century BC; the early 18th-century Creţulescu church and the former Royal Palace, now the National Art Gallery. Nearby is the Park of Culture and Rest, spread over 210 hectares (470 acres), and containing a superb folk museum housing some 70 farmsteads, watermills, windmills and rural homes from all over the country.

Bucharest, which lies on a tributary of the River Arges, is said to have been founded in the 15th century by a shepherd named Bucur. In 1659 Bucharest became the permanent capital of the Ottoman province of WALACHIA, taking over from the previous capital – the nearby city of TIRGOVISTE.

Bucharest's old town contains the ruins of the 15th-century Princely Palace and the fine 16th-century Curtea Veche (old church).

Since the end of the Second World War (1945) Bucharest has expanded rapidly; its modern outer suburbs bristle with high-rise apartment blocks. Under a controversial project, buildings in a large area of the town centre, including several 18th-century historic churches, have been demolished to make room for a new administrative centre.

Bucharest is the country's most important economic centre. Its industries include power stations, engineering and chemical plants, food processing, textiles and furniture.
Population 1 861 000
Map Romania Bb

Buchenwald ***East Germany*** Town about 80 km (50 miles) south-west of Leipzig. Between 1933 and 1945 it was the site of a Nazi concentration camp in which some 560 000 people died – many the victims of inhuman scientific experiments.
Population 6000
Map East Germany Bc

Buckingham ***United Kingdom*** Town and county of south-central England, to the east of Oxfordshire. The town was devastated by fire in 1725 and never fully recovered its importance, and today the county town is Aylesbury, 60 km (38 miles) north-west of London. The town of Milton Keynes is home of the Open University, which gives courses by radio, television and correspondence, while Buckingham itself houses Britain's only private university. The Chiltern Hills and attractive villages have made the county popular with London commuters.
Population (county) 609 000; (town) 6700
Map United Kingdom Ed

Budapest ***Hungary*** The country's capital situated on both sides of the Danube river in the north of the country, some 40 km (25 miles) from the Czech border. The great city, with one-fifth of the country's people, dominates Hungarian cultural, educational, political and economic life, producing 40 per cent of the country's manufactures.

It has two distinct parts – Buda on the west side of the river, and Pest to the east. They became one city in 1872.

The Romans built a military camp at Buda. King Matthias Corvinus (1458-90) fortified the later medieval town and made it Hungary's capital in the 15th century.

Castle Hill, at Buda's heart, has a mixture of Romanesque, Gothic and Baroque buildings, painstakingly restored after the near-total destruction of 1944-5, when the occupying Germans held out against Soviet forces for seven weeks. Among its narrow streets and steep alleys are: the former royal palace, now a superb complex of museums, galleries and a library; the colourful Matthias (or Coronation) Church; the Fishermen's Bastion, built on Castle Hill over a former fishmarket and fishermen's village in 1903; medieval churches and fine town houses. Gellert Hill to the south is topped by the Liberation Monument 32 m (105 ft) high, which commemorates battles of the Second World War, and has wide views of the city. Buda's Main Street was the scene of the demonstration that led to the Hungarian Uprising of 1956, which was crushed by Russian troops.

Pest, on the plain to the east, a small fortified Roman settlement in the 1st century AD and later a busy trade centre, expanded rapidly after it became part of the capital. It has concentric tree-lined boulevards crossed by avenues such as Rakoczi Ut, the main shopping street, radiating from the Inner City of Pest, which has remnants of its old walls. Within it stands a 12th-century parish church, with a Moslem prayer niche, used as a mosque during the Turkish occupation (1541-1686); the Vigado concert hall, where the composers Liszt, Brahms, Bartok and Kodaly performed; and modern riverside hotels. The enormous, domed parliament building dominates the river front to the north, with ministries and law courts nearby. Pest's underground railway, begun in 1895, is the world's second oldest after London's.

Several elegant bridges – completely restored following wartime destruction – link Buda with Pest. The oldest, the Chain Bridge, was designed and built by the British engineers William and Adam Clarke (1839-49). Margaret Bridge links both sides with the Margaret Island resort.

Extensive suburbs, whose industries include engineering, electrical engineering, food processing, chemicals and oil refining, pharmaceuticals, paper and textiles, ring outer Pest. They contrast with the Buda side of the city – where villas, second homes and resorts for summer walking and winter skiing sprawl in the Buda Hills – and with the charming southern suburb of Budafok, where wine cellars have been dug into the slopes beside the river.

Population 2 064 400

Map Hungary Ab

Budva ***Yugoslavia*** Fishing village and resort on the Montenegrin riviera, 40 km (25 miles) south-west of Titograd. The walled old town has a slender campanile (bell tower) and red-roofed buildings, many of which date from the 15th century. The new town, built after an earthquake in 1979, has hotels spaced among pines and cypresses behind superb sandy beaches and the Adriatic's warmest waters. Lovcen mountain to the north towers 1749 m (5735 ft) over all. About 5 km (3 miles) south-east across the bay is Sveti Stefan (St Stephen). Once a fortified fishing village, Sveti Stefan now has about 50 houses and a chapel huddled on a rock which is joined to the mainland by a causeway. The village is used as a hotel complex.

Population 9000

Map Yugoslavia Dc

Budweiss ***Czechoslovakia*** See CESKE BUDEJOVICE

Buea ***Cameroon*** Town and tourist resort near the coast, on the southern slopes of Mount Cameroon. It was the capital of the German protectorate of Cameroon (Kamerun), and from 1919, of British Cameroon. The beautiful town has marvellous views over the Tiko-Misselelé plains, and much German architecture.

Population 16 000

Map Cameroon Ab

Buenaventura ***Colombia*** The country's largest Pacific coast port, about 330 km (205 miles) west and slightly south of Bogotá. It is the main port for exporting coffee, which is grown in the Andes. It also handles imports for much of the country.

Population 190 000

Map Colombia Bb

Buenos Aires ***Argentina*** Capital city and port on the Rio de la Plata (River Plate). Often called the Paris of Latin America, its traditional architecture is very like that of late 19th-century France. It was founded by Spanish settlers in 1536, and today has a large population of Italians as well. The city centre is the square known as the Plaza de Mayo, around the president's Casa Rosada – 'Pink House'.

The main industries are car making, mechanical engineering, chemicals, textiles, food processing, oil products and paper.

The city's name is a shortened form of its original name: Ciudad de la Santísima Trinidad y Puerto de Nuestra Señora la Virgen María de los buenos aires (City of the Most Holy Trinity, and Port of Our Lady the Virgin Mary of good winds). The name was chosen because the city was founded on Trinity Sunday and because Mary was a patron saint of sailors, who needed favourable winds.

Population (city) 3 325 000; (metropolitan district) 9 948 000

Map Argentina Db

▼ ORTHODOX WORSHIPPERS Sunday church-goers flock to Bucharest's Church of St John on Bulevardul 1848, named for Europe's 'Year of Revolutions'.

Buffalo ***USA*** Port and industrial city at the eastern end of Lake Erie in New York state. The Erie Canal, opened in 1825, links it to the Hudson river and New York city, some 465 km (290 miles) to the south-east. Its industries include grain milling, feed manufacture, chemicals and steel. The Niagara Falls are 25 km (16 miles) to the north.
Population 339 000; (metropolitan area) 1 205 000
Map United States Kb

Bug ***Poland/USSR*** East European river, 813 km (480 miles) long. It rises near the Ukrainian city of L'vov, and flows north to form 200 km (125 miles) of the Polish-Soviet border. It then turns west to the Zegrzynek recreational lake just north of Warsaw, where it joins the Narew and Vistula rivers. It is sometimes known as the Western Bug to distinguish it from the Southern Bug, farther south in the Ukraine.
Map Poland Eb

Bugac ***Hungary*** See KISKUNSAG

Buganda ***Uganda*** Former kingdom of the Baganda (Ganda people) on the northern shore of Lake Victoria. It formed the nucleus of what became by 1914 the British Protectorate of Uganda, and was given a considerable degree of autonomy in the constitution under which Uganda became independent in 1962.

Indeed the *kabaka*, or 'king', of Buganda was the country's first president. But when the *kabaka* was deposed in 1966, the monarchy was abolished, and the kingdom divided into a cluster of districts.
Map Uganda Bb

Büjük Agri Dagi ***Turkey*** See ARARAT

Bujumbura ***Burundi*** National capital and port near the northern tip of Lake Tanganyika. Founded by the Germans in 1899, it now exports coffee, cotton and tea.
Population 180 000
Map Tanzania Aa

Bukavu ***Zaire*** Chief town of Kivu region, formerly called Costermansville. It stands at the southern end of Lake Kivu, beside the Rwanda border. It is a commercial centre, with an international airport, and a resort, recalling the atmosphere of a European spa town.

About 30 km (20 miles) to the north is the Kahuzi-Biega National Park, a sanctuary of the rare mountain gorilla.
Population 209 000
Map Zaire Bb

Bukhara ***USSR*** Central Asian city in Uzbekistan, about 440 km (275 miles) from the Afghan border and about 225 km (140 miles) west of Samarkand. One of the oldest cities in central Asia, Bukhara was an Islamic centre from the 8th to the early 13th century, when it was destroyed by the Mongol chieftain Genghis Khan (about 1162-1227). But it was rebuilt and remains a city of mosques and minarets. It is also a commercial centre with silk, woollen and cotton industries; its carpets are renowned throughout the world.
Population 204 000
Map USSR He

Bukidnon ***Philippines*** Province and plateau area in northern MINDANAO. Long isolated by lack of access and dense forests, its population has grown by more than 25 per cent since 1975. It has good grazing land and produces much maize, sugar cane and pineapples.
Population 532 818
Map Philippines Cd

Bukit Timah ***Singapore*** Central district of Singapore Island, named after the island's highest point, 176 m (577 ft) above sea level. Bukit Timah Road, which passes through the district, links Singapore city with the north of the island and the causeway to Malaysia.
Population 124 120
Map Singapore Ab

Bükk ***Hungary*** Group of wooded limestone hills rising to 959 m (3146 ft) west of the north-eastern city of Miskolc. The rounded hills, deeply cut by trout rivers, descend steeply to the city of Eger and the Great Alföld (Plain) to the south. Szilvasvarad, a small resort 28 km (17 miles) north of Eger, is noted for its waterfalls and for its stud and riding school of Lipizzaner horses. Lipizzaners are the breed used for equestrian displays in the Spanish Riding School in Vienna, Austria.
Map Hungary Ba

Bukoba ***Tanzania*** Regional capital on the western shore of Lake Victoria, about 40 km (25 miles) south of the Ugandan border. It is the centre of a coffee-growing region.
Population 30 000
Map Tanzania Ba

Bulacan ***Philippines*** Province of central LUZON, immediately north of MANILA. Mainly a rice-growing area, it also produces cement. The provincial capital, Malolos, was chosen as the capital of the short-lived Philippine Republic during the revolution of 1896-9; the Republican constitution was drawn up there, in the Barasoain Church, in 1898.
Population 1 050 000
Map Philippines Bc

Bulawayo ***Zimbabwe*** The country's second largest city after the capital, Harare, and chief town of the province of Matabeleland. It is also the largest industrial town in the country. Founded on its present site in 1894, it now produces textiles, tyres, building materials, processed foods, agricultural machinery, furniture and radios. Tourists have been drawn to the city, and to the Khami Ruins, the remains of a 13th-15th century civilisation, 21 km (13 miles) to the west.
Population 414 000
Map Zimbabwe Bb

Bulembu ***Swaziland*** Town (formerly Havelock) very near the border with the South African province of Transvaal, 40 km (25 miles) north of Mbabane. It is the site of one of the world's largest asbestos mines, which was opened in 1939. The asbestos is exported by means of a spectacular 20 km (12 mile) aerial ropeway over the mountains to Barberton, a railhead in South Africa.
Population 4900
Map Swaziland Aa

Bumtang ***Bhutan*** The largest and oldest monastic settlement of central Bhutan. It is 3000 m (9850 ft) up in the mountains, and has two monasteries and a school. It is also the home of some members of the Bhutan royal family.
Population 1000
Map Bhutan Ba

Bundala ***Sri Lanka*** See WIRAWILA TISSA

Bundi ***India*** Small town about 390 km (242 miles) south-west of Delhi. The Indian-born English author Rudyard Kipling (1865-1936) wrote about its magnificent Rajput Palace.
Population 48 000
Map India Bc

Bungoma ***Kenya*** Main town of a densely populated district in western Kenya, bordering Uganda. It lies about 90 km (55 miles) north-east of Lake Victoria, and is a trading centre on a transport route across the border.
Population 35 000
Map Kenya Ba

Bur Safaga (Safajah) ***Egypt*** Red Sea port about 120 km (75 miles) south of the entrance to the Gulf of Suez.
Map Egypt Cc

Buraydah ***Saudi Arabia*** Oasis town situated 450 km (280 miles) north-west of Riyadh. Buraydah is noted for its grain and dates, which are sold in its fine covered market. There are also some attractive old Arab buildings. The inhabitants are more puritanical and reserved than the rest of the country, and visitors are advised to dress modestly and not to take photographs.
Population 75 000
Map Saudi Arabia Bb

Burgan ***Kuwait*** The country's largest oil field, 19 km (11 miles) north-west of Mina al-Ahmadi – to which it is connected by pipeline. Oil was found there in 1938. As the oil rises under its own pressure, production costs are among the lowest in the world. Reserves are estimated at between 40 000 million and 72 000 million barrels.
Map Kuwait Ab

Burgas ***Bulgaria*** Major seaport and capital of Burgas province, on the Black Sea coast 60 km (37 miles) north of the Turkish border. Founded in the 18th century as a fishing village on the site of the medieval village of Pyrgos, it developed as a port and industrial town after Bulgarian liberation from the Turks in 1878, and expanded greatly after Communists took over the country and abolished the tsarist monarchy in 1946. It now has food, machinery and oil-refining industries, and international trade funnels to and from the Bulgarian interior.
Population 183 500
Map Bulgaria Cb

Burgenland ***Austria*** Eastern state noted for its agriculture and wine growing. Until 1921 it was part of Hungary; then it was transferred to Austria following a referendum. However, Burgenland has retained much of its distinctive Hungarian character. Its capital is Eisenstadt.
Population 272 270
Map Austria Eb

Bulgaria

THE GOOD LIFE – AND WIDESPREAD OBESITY – HAVE COME TO A FERTILE LAND THAT NOT LONG AGO WAS A MUSEUM PIECE FROM THE MIDDLE AGES

In less than 50 years, Bulgaria has been transformed from a country of poor peasants into one of the most advanced and prosperous states of Eastern Europe. The town has replaced the village as the centre of Bulgarian life. The seven or eight-hour working day usually starts early and ends in mid-afternoon, leaving time for a long dinner and a siesta before the towns spring to life for the evening.

Centring on the Maritsa river valley, Bulgaria straddles the north-west approaches of the strategic land bridge linking Europe and Asia Minor. It has a coast on the Black Sea which helps to attract 6 million tourists a year, and a climate ranging from Mediterranean in the south-facing valleys to extreme continental in the mountains and the north. It is a fertile land with substantial mineral and water resources, natural beauty and a charm that has survived the dramatic changes of the last 50 years.

The Bulgars are a Turkic-Tartar race who conquered a Slav civilisation between AD 500 and 700. The Bulgars became a ruling aristocracy, but absorbed Slav language and culture. The territory was overrun by the Byzantine Empire in 1018, but the Bulgars returned to power in 1186.

In 1396 the country fell to the Turks, who imposed serfdom and severe taxation. A Bulgarian uprising in 1876 was brutally crushed. International outrage, and Russian defeat of the Turks in 1878, laid the foundation of an independent Bulgarian state and formed the roots of Bulgarian-Russian friendship. The last formal links with Turkey were severed after the First Balkan War of 1912, when combined Bulgar, Greek, Serb and Montenegrin forces ousted the Turks.

Bulgaria was an uneasy ally of Germany in the Second World War. When Russian forces invaded in 1944 the Communists seized power and the tsarist monarchy was abolished. A substantial Turkish minority – about 8·5 per cent of the population – remains to this day.

The relatively good life today has been created by a revolution in farming. Food output has quadrupled since the 1950s when 13 million parcels of peasant land were made into 3453 collective farms. These, in turn, were fused into 792 large units in 1959, averaging 7000 hectares (17 300 acres) each. Workers earn wages for their work on a farm or in a factory, and many families feed themselves from small plots exempted from the collective farming programme, growing fruit and vegetables, and raising poultry and pigs.

The collective farms have increased output by using more machinery, fertilisers and irrigation, specialising by area: irrigated rice along the DANUBE in the north; wheat, maize, sugar beet, vines and fruit on the north-sloping Danube plain; sunflowers around Razgrad in the north-east and roses for perfume in the KAZANLUK basin; cereals, tobacco, cotton and fruit in the broad Maritsa valley. Mulberry trees are grown to feed silkworms around Svilengrad near the frontier with Greece and Turkey, and tobacco crops flourish in the southern valleys. Livestock is reared in the BALKAN and RHODOPE mountains.

While smallholdings have been forged into collective farms, some 160 larger collectives have been developed since the 1960s into agro-industrial complexes. These combine modern, crop-growing estates with food-processing plants covering, for example, the extraction of sugar from sugar beet, cotton ginning (separating the seeds and seed hulls from the cotton fibres), and canning, freezing and packing plants. Employing these advanced agricultural methods has yielded Bulgaria large earnings from exports.

Farm mechanisation has released people to fill jobs in the growing number of mines and factories. One-third of industrial jobs lie within 80 km (50 miles) of the capital, Sofia, but the rest are widely dispersed. A wide range of minerals is processed – coal, natural gas, iron ore, copper, lead, zinc and chrome – and there are abundant water supplies and hydroelectricity for textile, clothing and engineering factories. The large tourist trade on the Black Sea coast, however, has confined industry there to the ports of VARNA (shipbuilding) and BURGAS (petrochemicals and metals).

The urban population has increased from 1 in 5 back in 1946 to more than 3 in 5 today. Sofia, with a population of 1.1 million, is six times larger than 40 years ago. New towns like DIMITROVGRAD and Pernik house the workers mining and processing coal and metal ores.

Today, Bulgaria ranks as the third most urbanised Communist country in Europe, after East Germany and Czechoslovakia, having overtaken Hungary, Romania and the Soviet Union in the 1970s. One reason for the Bulgars' quick progress has been relative freedom from Communist dogma imposed from outside. They adopt a pro-Soviet stance, which makes them more reliable in Soviet eyes and less subject to interference.

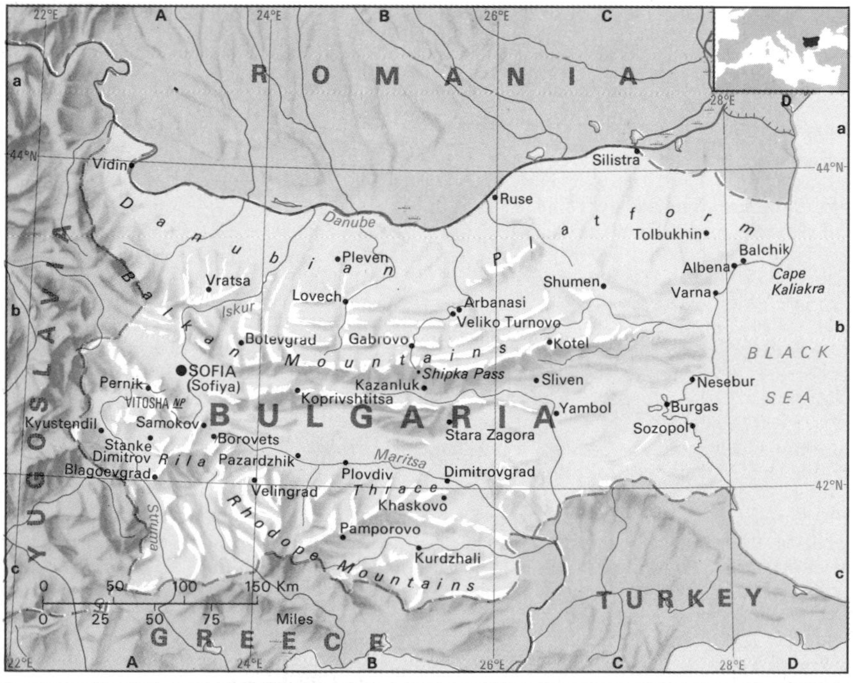

BULGARIA AT A GLANCE

Area 110 912 km² (42 823 sq miles)
Population 9 000 000
Capital Sofia
Government Communist republic
Currency Lev = 100 stotinki
Languages Bulgarian, Turkish
Religions Mainly atheist or non-religious; Christians include 25% Eastern Orthodox, 1% Roman Catholic; also 10% Muslim
Climate Continental in the mountains and north; Mediterranean in the south-facing valleys. Average temperature in Sofia ranges from −4 to 2°C (25-36°F) in January to 16-27°C (61-81°F) in July
Main primary products Wheat, barley, maize, grapes, sunflower seeds, apples, tobacco, cattle, sheep, timber; coal, lignite, oil and natural gas, manganese, uranium
Major industries Agriculture, tobacco processing, cement, iron and steel, coke, machinery, textiles, chemicals, fertilisers, leather goods, forestry, brewing and distilling, wine, oil and gas refining
Main exports Machinery, transport equipment, chemicals, cigarettes, meat, fruit, vegetables, tobacco, wines, spirits
Annual income per head (US$) 2900
Population growth (per thous/yr) 2
Life expectancy (yrs) Male 71 **Female** 76

Burkina

OVERPOPULATION AND OVERCULTIVATION HAVE TURNED THIS ARID FRINGE OF THE SAHARA, FORMERLY UPPER VOLTA, INTO AN INFERTILE DUST BOWL

Nature seems to have loaded the dice against Burkina. The good intentions of the leaders of this landlocked state in West Africa have not overcome the problems of its terrain and climate. Troubles have ranged from unemployment to famine, much of which can be blamed ultimately on the country's harsh environment. A history of human folly has not helped, and the revolutionary government, which in 1984 changed the country's name from Upper Volta to Burkina Faso (translating approximately as 'Republic of Upright Men'), has difficulties to tackle which uprightness alone cannot resolve.

Burkina's problems are partly a legacy of French colonial rule, which lasted from the 1890s until independence in 1960. During this period the French introduced a system of forced labour on the plantations in the neighbouring Ivory Coast, which robbed Upper Volta villages of their menfolk. Shortage of work in the country today means that many of the young men still go across the borders to Ghana and the Ivory Coast to find jobs.

Since Burkina's independence, military and civilian governments have alternated. The coup in August 1983 resulted in a mixed military and civilian government led by Captain Thomas Sankara; this began its work full of optimism for the future, embodied in the change of name and new symbols of national identity – a new flag and new national anthem. The government promised to double the country's workforce by emancipating its women. But the poverty of the land remains a problem, and social conditions – particularly health, housing and water supplies – desperately need improvement. Moreover, 95 per cent of the people are illiterate.

Much of Burkina – the north and west – lies in the SAHEL, the arid fringe of the Sahara. Here there are dusty, grey plains of infertile soil which has been further impoverished by overgrazing and overcultivation.

THE VANISHING TREES

The deterioration is happening quickly. Elders in the capital city of OUAGADOUGOU can remember a time when the central plateau of MOSSI was tree-covered; now there are very few trees in a landscape which has become increasingly arid. In the extreme north the rains have failed year after year (six years running before 1973), and here food shortages have been so bad as to result in famine.

The rivers of the central plateau are the Black, Red and White Voltas – headwaters of the main VOLTA river. They flow sluggishly (and sometimes not at all in droughts) through land which is swampy where it is not dry, towards the border with Ghana. Downstream, they flow more reliably through the wetter south where the landscape, though still flat, is greener with wooded savannah, against which the red soils look deceptively fertile. But the river valleys, which one might expect to be the most populated areas, are deserted: these are the breeding grounds of onchocerciasis (river blindness), bilharzia and malaria.

The country has considerable potential for tourism. The wildlife has had an easier time in Burkina than the 7 million people; the southern savannah is inhabited by elephants, hippopotamuses and many other creatures. There is also some spectacular scenery in the south-west, where plateaus rise to a height of 150 m (about 500 ft) above the plains, and where waterfalls on the BANFORA ESCARPMENT more than justify the long climb.

Some 90 per cent of the people live by farming. Food crops include sorghum, beans and maize, and over 70 per cent of the country's exports are provided by cotton, livestock and oil seeds. Investors in Burkina's agriculture have often insisted on the growing of such cash crops, and this has aggravated the shortage of food for the local people and damaged the soil as well.

Unemployment and poverty have led to lawlessness; theft and muggings are frequent on the trains of Burkina's main railway line, which leads from Ouagadougou through the Ivory Coast to the sea. Besides this danger, the railway is uncomfortable, crowded and smelly. Bush taxis, which drive from town to town, are a far preferable way of travelling long distances.

Most of Burkina's towns are concentrated along the railway line. Ouagadougou is not only the capital of the country but also the centre of the Mossi country – the homeland of the people who make up half of Burkina's population and who had kingdoms there as long ago as the 15th century. But all the people of Burkina are to be seen in Ouagadougou's bustling markets: Tuareg traders from the north in flowing robes and elaborate headdresses; Mossi, Bobo and Dioula farmers in bright cotton boubous or Western clothes; tall, fine-featured Fulani or Peul; and non-Africans, especially French, Americans, and the Lebanese who own many of the plentiful restaurants and bars.

About 90 per cent of the Burkinabé (people of Burkina) live in rural areas. In the villages, the conical huts of each family group are enclosed by a fence that sets them apart from other compounds. For most of these people, even in good years survival through the dry season from November to April is precarious, and is almost wholly dependent on the success of the previous harvest or, in recent years, on foreign aid.

BURKINA AT A GLANCE

Area	274 200 km² (105 869 sq miles)
Population	7 080 000
Capital	Ouagadougou
Government	One-party republic
Currency	CFA franc = 100 centimes
Languages	French (official), African languages
Religions	Tribal religions (65%), Muslim (25%) Christian (10%)
Climate	Tropical; rains between May and November. Average temperature in Ouagadougou ranges from 16-33°C (61-91°F) in January to 26-39°C (79-102°F) in April
Main primary products	Sorghum, millet, maize, rice, livestock, groundnuts, shea nuts, sesame, cotton
Major industries	Agriculture, processed foods, textiles, mining, tyre manufacture
Main exports	Cotton, livestock, sesame, groundnuts, shea nut products, hides and skins, rubber tyres
Annual income per head (US$)	190
Population growth (per thous/yr)	25
Life expectancy (yrs)	**Male** 43 **Female** 45

Burgos ***Spain*** Industrial town about 210 km (130 miles) north of Madrid, whose products range from flour to tyres. Burgos was the birthplace of El Cid (Rodrigo Diaz, Sire of Vivar, about 1060-99), a Spanish national hero who led the Christian fight against the Moors. His remains are in the town's twin-spired 13th-century cathedral, one of Europe's finest Gothic buildings. The surrounding Burgos province is largely forested.

Population (town) 156 500; (province) 363 500

Map Spain Da

Burgundy (Bourgogne) ***France*** Region of hills and valleys in east-central France covering 31 582 km² (12 194 sq miles) on both sides of the Saône river. It has more than 5000 vineyards producing the wines named after the region. For most of the 15th century it formed part of a larger area ruled by the Dukes of Burgundy; much of the area was annexed to France in 1482, other parts being held by Spain. DIJON is the main city.

Population 1 596 000

Map France Ec

Burma See p. 116

Burma Road ***Burma/China*** A 1130 km (700 mile) mountain highway between Lashio, in Burma, and Kunming, in the Yunnan province of China, the Burma Road was built by the Chinese. It was an amazing feat of engineering. Begun in 1937, after the Japanese invaded China, it was completed a year later and was designed to carry in military supplies from Burma. It became increasingly important during the Second World War, when another strategic highway (the Ledo Road) was built in 1942-4 from Ledo, in northern India, to join the Burma Road through Myitkyina. No longer important, both roads are now in a poor state.

burn A Scottish term for a small brook or stream.

Burren, The (Boirinn) ***Ireland*** Barren, windswept limestone region of about 500 km² (195 sq miles) on the west coast, on the south side of Galway Bay. It rises in terraces to Slieve Elva, 345 m (1132 ft) high. Drainage is by underground streams, with fine cave systems. In sheltered areas there is a rich flora and good pastures. There was clearly a large prehistoric population, for there are about 700 stone forts and dolmens – Stone Age burial chambers – including the impressive Cahermancaghten fort near Lisdoonvarna. The Burren is to become a national park with a centre at Kilfenora.
Population 3090
Map Ireland Bb

Bursa ***Turkey*** Winter ski resort and capital of the province of the same name, about 100 km (60 miles) south of the city of Istanbul. Bursa's main products are tobacco, textiles, carpets, silks, satin, tapestry and wine.

Farmers in the surrounding province, which covers an area of 11 053 km² (4268 sq miles), mostly grow fruit, grain, grapes, cotton and tobacco.
Population (city) 614 100; (province) 1 327 800
Map Turkey Aa

Buru ***Indonesia*** Island covering 8800 km² (3400 sq miles) of the Maluku group south of the Philippines. It is used as a penal colony for political prisoners.
Map Indonesia Gc

Burutu ***Nigeria*** Port in the mangrove swamps of the Niger river delta. It was created by the Niger River Transport Company in the 19th century, and has two ocean docks where goods are transferred to and from river boats. Burutu exports groundnuts and cotton brought down the Benue river, from Cameroon and Chad, and is a base for offshore oil drilling.
Map Nigeria Bb

Bury St Edmunds ***United Kingdom*** Cathedral town, also known as St Edmundsbury, in the eastern English county of Suffolk 100 km (62 miles) north-east of London. It is named after the Anglo-Saxon king Edmund, who was killed by the Danes and reburied there in about AD 903.

An abbey was founded in 1020, but only its massive gatehouses remain substantially unaltered. A 15th-century church within the abbey precinct was raised to cathedral status in 1914. The bishopric is shared with Ipswich 40 km (25 miles) to the south-east. The town has many fine old buildings, including the town hall, designed in the 1770s as a theatre by the architect Robert Adam.
Population 29 000
Map United Kingdom Fd

Buryat Republic ***USSR*** Largely agricultural and forested autonomous republic of the RSFSR in the Soviet Far East, lying between Mongolia and Lake Baikal. It covers 351 300 km² (135 600 sq miles). Cattle and sheep farming are the main occupations of the people, many of whom are Buddhists or Shamanists. The republic's capital is ULAN-UDE.
Population 985 000
Map USSR Lc

bush Wild, uncultivated land, particularly a region covered with scrub or woodland, such as the vast uninhabited areas of Australia, New Zealand, South Africa and the United States.

Bushire (Bushehr) ***Iran*** Port on The Gulf, almost due south of the capital, Tehran. It exports wool, rugs and cotton. Giant gas fields lie within about 110 km (68 miles) to the south-east – Kangan (onshore) and Pars (offshore). As yet, they are undeveloped.
Population 57 700
Map Iran Bb

Bushmanland ***Namibia*** So-called homeland set up by South Africa in the wooded savannah region of the north-east for the San (Bushmen) people. Tsumkwe is the only settlement of any size in the region, which covers 18 468 km² (7130 sq miles).
Population 500
Map Namibia Aa

Burundi

A MOUNTAINOUS BUT CROWDED LAND WHERE TWO TRIBES ARE STILL DEEPLY DIVIDED AFTER A BLOODY CIVIL WAR

Tribal warfare resulting from centuries-old enmities killed perhaps 120 000 people – 3 per cent of Burundi's population – in 1972, in a tragic climax to ten troubled years of independence. Today there is still a brooding bitterness in this high, mountainous country in the heart of Africa. It is rooted in the traditional supremacy of the minority Tutsi tribe – a tall people originating from Ethiopia and Uganda who are outnumbered six to one by the shorter Hutu. (The pygmy Twa, the original inhabitants, are now less than 1 per cent of the population.)

The farming Hutu displaced the Twa nearly 1000 years ago. The Tutsi arrived in the 16th century and established a feudal system under their *mwami* (king); the Hutu were reduced to serfdom, tending cattle on behalf of the Tutsi. In 1890 Burundi became part of German East Africa and then, from 1919, part of the Belgian-administered territory of Ruanda-Urundi. Belgium's policy of ruling through the Tutsi aristocracy consolidated the Tutsi elite.

Independent Burundi had seven governments in four years while Tutsi and Hutu vied for power. A failed Hutu coup in 1965 led to Tutsi retaliation and the execution of most Hutu politicians and many others. The next year, after the king's son ousted his father, a military coup established a republic under Michel Micombero, another Tutsi. In 1972 he took dictatorial powers and there was an immediate Hutu uprising in which about 1000 Tutsi were killed. Tutsi retaliation was brutal and systematic.

Some efforts have since been made to heal the wounds left by the war; Tutsi landlords have ceded some land to Hutu tenants. But the Tutsi still run the government and army; when another military coup toppled Micombero in 1976 his successor was Jean-Baptiste Bagaza – yet another Tutsi.

Burundi is a small, densely populated country that is intensely cultivated despite its mountainous terrain. West Burundi extends into the GREAT RIFT VALLEY, through which the Ruzizi river flows into Lake TANGANYIKA. Beyond a ridge to the east lie high grasslands. Much of the country is above 1500 m (4900 ft) and most slopes steeply. Soils are not rich, but there is enough rain to grow crops in most areas and the people feed themselves from tiny plots.

The main subsistence crops are bananas, sweet potatoes, peas, lentils and beans in the mountains, and cassava near the lake and river shores. Some land holdings are big enough to grow coffee as a cash crop, and this is the source of 90 per cent of Burundi's very small export earnings. There are also small crops of tea and cotton. Cattle, goats and sheep are grazed on the grasslands.

One of the sources of tribal tensions may be population density and growth. The estimated average population density of about 180 per km² (460 per sq mile) is extremely high for a country in which 90 per cent of the people live off the land, and the population is growing fast. The capital, BUJUMBURA, is the only large town. Add the problems of isolation – 2000 km (1240 miles) to the Atlantic, 1400 km (870 miles) to the Indian Ocean – and difficult trade routes, and Burundi seems bound to remain dependent on foreign aid for many years.

BURUNDI AT A GLANCE
Map Tanzania Aa
Area 27 834 km² (10 746 sq miles)
Population 4 910 000
Capital Bujumbura
Government One-party republic
Currency Burundi franc = 100 centimes
Languages Kirundi, French, Swahili
Religions Christian (74%), tribal (20%), Muslim (6%)
Climate Equatorial; average temperature in Bujumbura 24°C (75°F)
Main primary products Cassava, sweet potatoes, maize, bananas, beans, coffee, tea, coconuts, cotton; nickel and uranium (both unexploited), tin
Major industries Agriculture, forestry, fishing, beverages
Main exports Coffee, tea, cotton
Annual income per head (US$) 255
Population growth (per thous/yr) 26
Life expectancy (yrs) Male 43 **Female** 47

Burma

*ISOLATIONISM AND THE
GOVERNMENT'S OWN RIGID BLEND OF
SOCIALISM REDUCED THIS BUDDHIST
COUNTRY TO POVERTY, BUT
IMPROVEMENTS MAY BE COMING*

The author Rudyard Kipling once wrote a ballad about the road to MAN-DALAY. The British 14th Army, nick-named 'the Forgotten Army', fought the Japanese to reach the road in the Second World War. It is a highway that runs through the mainstream of Burmese village life. It was a colonial road once, but Burma's rulers today have chosen a different path: the Burmese Way to Socialism, free in their view of both capitalist and communist taint.

Annexed by Britain in 1885 after three border wars, Burma became a province of India. It became a crown colony and was given an elected assembly in 1937. During the war, Burmese forces under the nationalist leader General Aung San fought first for the Japanese, who had promised them independence, then for the British. Independence came in 1948, but Aung San was assassinated in 1947 before he could take office as Burma's first prime minister. Out went Burma from the Commonwealth; a military dictatorship followed in 1962, a one-party state in 1974.

The one party is the Burma Socialist Programme Party, headed until 1981 by Ne Win. The party's blend of socialism, Buddhism and isolationism made Burma one of the ten poorest countries in the world. This was despite great natural resources in teak forests, coal, oil, gas, iron, tin, tungsten, lead, zinc, silver, copper, nickel, rubies, sapphires and emeralds. Rigid economic controls caused great shortages of all goods and there was a thriving black market. Since Ne Win stepped down, to be replaced by San Yu, more liberal policies have gradually appeared and foreign aid and investment encouraged. As a result, the economy has begun to grow much more rapidly, with gross domestic product expanding at 6-7 per cent a year. But the shortages and black market remain. Drug trafficking in opium is a major problem, and the government is also fighting rebel groups who control one-third of the country.

HORSESHOE OF MOUNTAINS

Burma, with an area of 676 552 km² (261 217 sq miles), is the second largest country in South-east Asia after Indonesia. The heartland of the country is the IRRAWADDY valley, home of the Burmans who comprise two-thirds of the country's 38 million population. Tribal peoples live in the foothills of a giant horseshoe of mountains which forms a barrier between Burma and its neighbours.

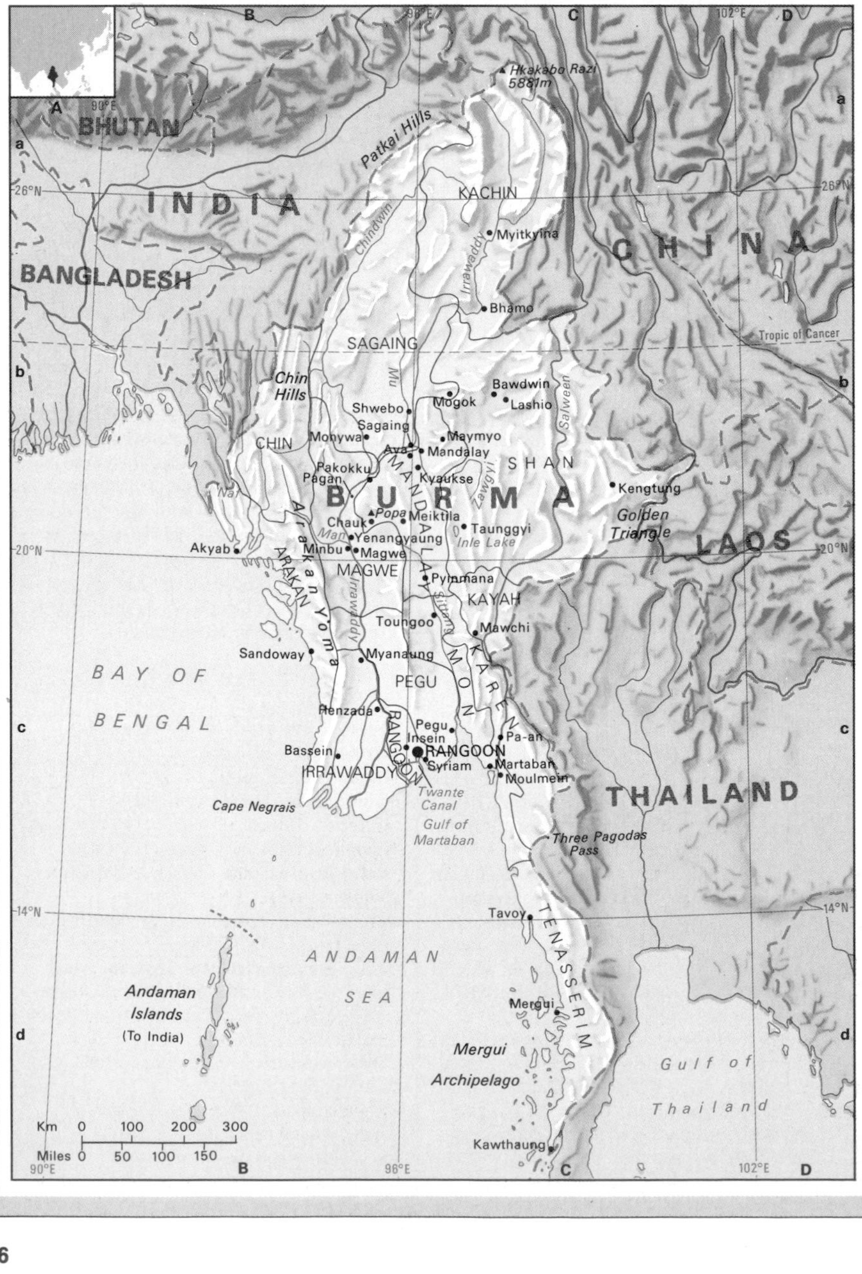

The Irrawaddy, 2010 km (1250 miles) long, flows south through gorges strewn with rapids to Mandalay and RANGOON, the capital. Its delta, still extending into the ANDAMAN SEA, provides ideal swampy land for rice cultivation. Until 1964 Burma was the world's largest rice exporter, and rice still accounts for half the country's export earnings. Rice and salted fish are the staple foods.

Farther upstream, the valley is sheltered from the monsoon by the surrounding hills, and in this 'dry zone' cotton, sesame and beans are interspersed with irrigated ricelands, particularly around Mandalay. At KYAUKSE an irrigation system ordered by King Anawratha, who established Burmese supremacy over the delta 900 years ago, is still in use.

King Anawratha built his ceremonial capital, PAGAN, in the 'dry zone' in the 11th century. It was the Pagan dynasty which introduced Buddhism to Burma. By 1550 the Burmese controlled the present territory and by 1800 they had extended their influence into Assam and Thailand. The last of Burma's kings, Thibaw, was displaced by the British in 1885.

The 'dry zone' was also the home of Burmah Oil, containing the small oil fields which once made Burma the second largest producer in the British Empire. It holds enough oil and natural gas for most of Burma's needs. Near Mandalay the CHINDWIN river joins the Irrawaddy; to the north are found most of Burma's precious stones.

About two-thirds of Burma is covered by trees. In addition to teak forests there are rubber plantations, some oil palms, and other woods. In the north, mountains rise to 5881 m (19 294 ft) at Hkakabo Razi, and great rivers such as the Chindwin and Irrawaddy rise to form torrents, flowing through deep gorges. In the west, the Chin and NAGA hills (3800 m, 12 450 ft) extend towards the isolated Arakan coast, bordering the Bay of Bengal.

In the east, the hills between Burma and

▲ 'CITY OF A THOUSAND TEMPLES' Buddhist temples surround the old Burmese capital of Pagan, founded in AD 849. The city, once a leading centre of Buddhist learning, never recovered from sacking by Mongols in 1287.

Thailand rise to more than 3000 m (10 000 ft), broadening northwards into the SHAN plateau. There is still no way of crossing these hills other than by foot, mule, horse or elephant; the infamous 'Death Railway', built across the THREE PAGODAS PASS by prisoners of war of the Japanese, has been abandoned.

Rangoon has 4 million people. Coming in by air, the traveller's first sight of the capital may be the sun glinting on the golden pinnacle of the Shwe Dagon pagoda, some 100 m (330 ft) high. There are countless monasteries and shrines: Rangoon claims to have the world's largest reclining Buddha, 78 m (256 ft) long and 17 m (56 ft) high.

Away from Rangoon, the road to Mandalay remains in part a narrow, rutted track. Mandalay contains the Arakan pagoda and the ruins of King Thibaw's finely carved and gilded palace, destroyed by shellfire in 1945.

There are about 70 different ethnic groups in the country. The largest after the Burmans (65 per cent) are the hill-dwelling Karens (11 per cent), Shans (8.5 per cent) and Kachins (3.5 per cent). After independence in 1948 the hill tribes objected to domination by the Burmans and some tribesmen took to arms.

There are three major conflicts. In the south, the Karen National Liberation Army has waged a guerrilla war for independence for 30 years; its 10 000 men control a 640 km (400 mile) stretch of land in the mountain zones along the Thai border. In the north and east, the Burmese Communist Party has an army of 12 000 in KACHIN and Shan states, and controls 15 000 km² (5800 sq miles) of land adjoining China. In the same area, feudal warlords battle with the authorities to protect their profitable and deadly trade in opium, grown in the infamous GOLDEN TRIANGLE.

The Burmese Way to Socialism was launched as a do-it-yourself economic system which rejected foreign aid despite the ravages of the Second World War. Agriculture, vital to Burma's prosperity, was government-planned and the price of rice was controlled. A foreign policy of neutrality was adopted. But in 1958 splits in the government saw the first prime minister, U Nu, give way to General Ne Win, who in 1962 staged a military coup. Widespread nationalisation followed, covering mining, forestry, manufacturing and retailing.

The strategy was unsuccessful. Farmers were reluctant to invest in crops for inadequate returns and held back their produce from the government. Industries were unable to get enough raw materials from agriculture. Rebel movements denied the government the timber and minerals of the upland areas.

Rice exports fell from 1.7 million tonnes a year to 170 000 tonnes, leaving Burma with insufficient capital to restore much of its war damage. In 1975, when food shortages brought riots, Burma was forced to seek aid. Foreign aid has brought improvements. Prices for crops have been increased and high-yielding varieties of rice raised production by 60 per cent in eight years. The mining and timber industries are getting more money for development. New oil and gas finds may make the country self-sufficient in these commodities. Education has been reorganised and adult literacy is put at 66 per cent. But the people of Burma still have few freedoms, and fighting the rebels still drains vital resources.

BURMA AT A GLANCE
Area 676 552 km² (261 217 sq miles)
Population 37 660 000
Capital Rangoon
Government One-party socialist republic
Currency Kyat = 100 pyas
Languages Burmese and tribal
Religion Buddhist
Climate Tropical; monsoon from May to September. Average temperature in Rangoon ranges from 18-32°C (64-90°F) in January to 24-36°C (75-97°F) in April
Main primary products Rice, sugar cane, groundnuts, jute, cotton, rubber, timber; oil and natural gas, lead, zinc, tungsten, nickel
Major industries Forestry, oil and gas production, mining, fishing, food processing, textiles
Main exports Rice, teak, jute, rubber
Annual income per head (US$) 160
Population growth (per thous/yr) 20
Life expectancy (yrs) Male 53 **Female** 56

bushveld, bushvelt Area of open, uncultivated country in tropical and subtropical southern Africa whose flora consists mainly of tall grasses, low trees and thorn scrub.

Busia ***Kenya*** Town on the Ugandan border about 45 km (30 miles) north of Lake Victoria. There is some agriculture around the town, but its main business is cross-border trade.
Population 35 000
Map Kenya Ba

Busto Arsizio ***Italy*** Industrial town 30 km (20 miles) north-west of Milan. It makes cotton and rayon, and also textile machinery.
Population 79 800
Map Italy Bb

Butare ***Rwanda*** The country's second largest town after the capital, Kigali, about 20 km (12 miles) from the border with Burundi. It is the seat of Rwanda's only university.
Population 30 000
Map Rwanda Ab

Bute ***United Kingdom*** See STRATHCLYDE

Butha Buthe ***Lesotho*** Administrative and market town and the country's northernmost district. It covers 1767 km² (682 sq miles) of highlands where King Moshoeshoe I founded the Basotho nation in the early 19th century.
Population (district) 92 200; (town) 2500
Map South Africa Cb

butte Small, prominent, flat-topped hill or isolated column of rock found in arid regions that rises abruptly from the surrounding area. It is similar to, but smaller than, a MESA.

Butte ***USA*** Mining city in the Rocky Mountains in south-west Montana, about 110 km (68 miles) from the Idaho border. It is one of the country's leading producers of copper, zinc, lead, silver and gold.
Population 34 500
Map United States Da

Butterworth ***Malaysia*** Port on the north-west coast of Peninsular Malaysia facing Penang Island. Ferries and the Penang Bridge – opened in 1985 and, with a total length of 13.5 km (8.4 miles), the third longest in the world – link the port to George Town, Penang's chief city. Butterworth has an oil refinery, tin smelters and an industrial estate at Mak Mandin. The Royal Australian Air Force shares a base nearby with Malaysian airmen.
Population 78 000
Map Malaysia Ba

Butterworth (Gcuwa) ***South Africa*** Town set in a livestock-rearing and dairy region in south Transkei, 80 km (50 miles) north and slightly east of the port of East London.
Population 29 000
Map South Africa Cc

Buxar ***India*** Small town about 200 km (124 miles) east of the city of Allahabad. Set on the southern bank of the Ganges river, it is of great importance to Hindus because of its connections with the early life of the Hindu god Rama. In 1764 it was the scene of a fierce battle in which the British defeated the Nawab Wazir of Oudh, a major semi-independent province of the Mogul Empire, to secure their position in India.
Population 43 000
Map India Dc

Buzau ***Romania*** Industrial city on the Buzau river, about 100 km (60 miles) north-east of Bucharest. Set in a region of rich oil and gas fields, its industries produce chemicals, textiles, alcohol and processed food.
Population 112 800
Map Romania Ba

Bydgoszcz (Bromberg) ***Poland*** Central city, about 235 km (145 miles) north-west of the capital, Warsaw. It straddles the Brda river. 'Bydgoszcz Venice' – a medieval complex of riverside granaries, mills and a royal mint – survives, along with Gothic churches and parts of the old town fortifications.

The first Nazi mass execution of Polish civilians – some 20 000 men, women and children – began in Bydgoszcz market square on 'Bloody Sunday' in October 1939. Today the city's industries produce furniture, paper, shoes, bicycles, refrigerators, machinery and books.
Population 357 700
Map Poland Bb

Byelorussia ***USSR*** See BELORUSSIA

Byron, Cape ***Australia*** Most easterly point in the country, lying in New South Wales about 50 km (30 miles) south of the border with Queensland.
Map Australia Id

Bystrá, Mount ***Czechoslovakia*** See TATRA MOUNTAINS

Bytom (Beuthen) ***Poland*** Industrial city in the south, 13 km (8 miles) north-west of the city of Katowice. It was founded in 1055 and despite rapid industrialisation, with coal mining, steel making and metal working in the 19th century, the medieval street plan and a restored 13th-century church survive.

The town mines calamine – a zinc ore used in soothing skin lotions and creams.
Population 238 100
Map Poland Cc

Byzantium ***Turkey*** See ISTANBUL

Caacupé ***Paraguay*** Capital of La Cordillera department, about 50 km (30 miles) east of the national capital, Asunción. It was founded in 1770 by the Spanish and is a popular tourist and pilgrimage centre, particularly in December on the Feast of the Blue Virgin of the Miracles. A modern basilica is dedicated to her. The town produces oranges, tobacco, sugar and tiles.
Population 10 000
Map Paraguay Bb

caatinga Thorn forest growing in the semiarid region of north-east Brazil. It forms a virtually impenetrable deciduous jungle from which rise occasional taller drought-resistant trees such as acacias, wax-palms and giant cacti. For three-quarters of the year it is a grey-looking leafless tangle, but after the short and unreliable rains which come in December, it bursts into brilliant colour for two or three months, and then reverts to its dormant state.

Cabañas ***El Salvador*** Mountainous, coffee-producing department north-east of the capital, San Salvador, on the border with Honduras. Its capital is Sensuntepeque (population 10 000).
Population 130 000

Cabinda ***Angola*** Coastal province cut off from the rest of the country by a corridor 30 km (19 miles) wide belonging to Zaire. The provincial capital – also called Cabinda – lies about 390 km (242 miles) north of Luanda. The province produces 75 per cent of Angola's oil following the discovery of offshore fields in 1966. The other main products are natural gas, hardwoods, cocoa, coffee and palm oil.
Population (town) 13 500; (province) 81 000
Map Angola Aa

Cabo Delgado ***Mozambique*** See PEMBA

Cabora Bassa Dam ***Mozambique*** Hydroelectric dam on the Zambezi river about 210 km (130 miles) upstream from the town of Tete. Begun in 1969, it was commercially operational by 1978 – and by the mid-1980s, South Africa took 98 per cent of the output. The Portuguese government is financially responsible for the dam until the end of the century, when the capital investment will have been paid off and Mozambique will take it over. At the moment, Mozambique receives a share of the revenues from the sale of electricity to South Africa.

Cabora Bassa lake, which is some 240 km (150 miles) long, extends from the dam almost to the Zambian border. It covers some 1600 km² (620 sq miles).
Map Mozambique Ab

Cacahuamilpa ***Mexico*** Largest natural caves in central Mexico, hidden in a valley near Taxco, 150 km (90 miles) south-west of Mexico City. They are carved by running water, and are part of a network of underground river channels that crisscross a limestone formation. Among the maze of passageways and chambers is a partly explored gallery 1380 m (4528 ft) long, 100 m (325 ft) wide, and 70 m (230 ft) high, formed by slowly dripping water.

Cáceres ***Spain*** Market town, about 240 km (150 miles) south-west of Madrid, manufacturing cork, leather goods, textiles and fertilisers. Its province, of the same name, produces wool, cereals and *embutidos* – red sausages – for which it is famous.
Population (town) 71 900; (province) 414 700
Map Spain Bc

Cacheu ***Guinea-Bissau*** River of northern Guinea-Bissau, about 200 km (125 miles) long. Cacheu is also the name of a region, and of a town on the left bank of the Cacheu estuary. Founded by the Portuguese in 1630, the town was the capital of Portuguese Guinea from the 17th century until 1879. It declined after this, and declined further during the guerrilla war of independence (1964-74), but remains a regional capital, market town and fishing port.
Population (region) 130 200
Map Guinea Aa

Cádiz ***Spain*** Seaport on the south Atlantic coast, just north of the Strait of Gibraltar. It exports sherry, salt, olive oil and cork produced in its adjacent province, also called Cádiz.

Cádiz is one of Europe's oldest towns, founded in 1100 BC as a trading post by the Phoenicians from the Middle East. It flourished under the Romans, who called it Gades, but declined after the fall of Rome in AD 476. It revived again after the discovery of the New World and became the centre of Spanish trade with the Americas – despite English attacks. In 1587 Sir Francis Drake burned a Spanish fleet there, and in 1597 the Earl of Essex partly destroyed the town.

Population (port) 157 800; (province) 1 001 700
Map Spain Bd

Caen ***France*** Capital of Calvados department in Normandy, 110 km (68 miles) west of Rouen. Parts of the castle built by William of Normandy – the ruler who led the Norman Conquest of England in the 11th century – are still standing. The city was rebuilt after the Second World War following its devastation during the Allied invasion in 1944. The city contains a new university and steelmaking and electrical industries.

Population 187 600
Map France Cb

Caerdydd ***United Kingdom*** See CARDIFF

Caernarfon ***United Kingdom*** See GWYNEDD

Caesarea (Qesaria) ***Israel*** Seaport and resort about 35 km (22 miles) south of Haifa, founded by Herod the Great in the 1st century BC. St Paul baptised the Roman centurion Cornelius in Caesarea and was imprisoned in the city in AD 57-59. Its Roman remains include the harbour and aqueduct and a reconstructed theatre. The walls and moat of a Crusader town on the site still survive, though the town was largely destroyed by Muslims in 1265.

Population 37 000
Map Israel Aa

Cagayan ***Philippines*** Largest river in LUZON, 350 km (220 miles) long, rising in the southern Sierra Madre and flowing northwards through a broad valley (bordered to the west by the Cordillera Central) to the Babuyan Channel near Aparri. It is also called the Rio Grande de Cagayan to distinguish it from the Cagayan river of Mindanao. In all, with its tributaries the Chico and Magat, the river drains some 10 000 km² (3900 sq miles). Rice, tobacco and other crops grow in the Cagayan valley, but there is little irrigation as yet. Cagayan province comprises the lower part of the valley.

Map Philippines Bb

Cagayan de Oro ***Philippines*** City on the north coast of MINDANAO, at the mouth of the Cagayan river. It is the capital of the region and of Misamis Oriental province. It is mainly an administrative, transport and trading centre, and has a fine harbour.

Population 163 000
Map Philippines Bd

Cagliari ***Italy*** City on the south coast of Sardinia, and the capital of the island. It stands on a series of treeless low hills and is a place of stone and of wind. It was founded by the Phoenicians and has a Roman amphitheatre, hewn out of solid rock. To the west of the city, a new canal port serves a large industrial estate where petrochemicals are dominant. The city's salt pans are the biggest in Italy.

Population 232 800
Map Italy Be

Caguas ***Puerto Rico*** Commercial and industrial city in a fertile agricultural region, 25 km (15 miles) south of San Juan. It produces tobacco and sugar, and manufactures leather goods, cigars and cigarettes.

Population 121 100
Map Caribbean Bb

Cainozoic era See CENOZOIC ERA

Cairngorm Mountains ***United Kingdom*** Range within the Grampian Mountains on the border of the Scottish Highland and Grampian regions. The highest peak is Ben Macdui (1309 m, 4296 ft). The area has been developed for winter sports centred at the village of Aviemore. A form of yellow-brown quartz known as cairngorm is found in the mountains, and is used in jewellery.

Map United Kingdom Db

Cairns ***Australia*** Tourist resort and port on the north-east coast of Queensland, about 1400 km (875 miles) north-west of Brisbane. It is a commercial centre for the surrounding sugar-cane country, with a bulk storage terminal for sugar exports. An access point for the northern section of the Great Barrier Reef Marine Park, the city is a well known big-game fishing base.

Population 48 500
Map Australia Hb

Cairo (El Qahira) ***Egypt*** National capital and Africa's largest city, at the head of the Nile delta about 160 km (100 miles) south of the Mediterranean Sea. It is a noisy, traffic-jammed, cosmopolitan city, a jostling crossroad of East and West. A forest of minarets proclaim its Islamic affinity, but beneath them swarms a mixed population of Arabs, Turks, Coptic Christians, Jews, black Africans and Europeans.

It is a city of two hearts. Old Cairo in the south is a walled enclave of old stone houses crowding together over narrow, crooked streets, some still unpaved, of packed bazaars as old as the city, and of the buzz of prayers chanted daily in its 400 mosques. Above it, on Moqattam Hill, dividing the old city from the new, stands the Citadel, a fortress built in AD 1177 by the sultan Saladin. To the north and west is new Cairo, a working city of broad avenues and modern blocks – offices, banks, hotels, government ministries and an opera house.

Cairo was originally a riverside military camp set up in AD 641. In 969 the Fatimid Muslim rulers of Libya conquered Egypt and began building the walled city, which they called al-Kahira, meaning 'The Triumphant One'. Saladin extended the city in the 12th century and it became the capital of the Mameluke sultans until the Ottoman Turks took over Egypt in 1517. Egypt, and Cairo, regained independence under Muhammad (or Mehmet) Ali (1769-1849). Ali, made governor by the Turks in 1806, ruled for 43 years, winning the country autonomy and creating modern Egypt. At EL GIZA,

▼ MUSLIM METROPOLIS The minarets of Sultan Hasan Mosque (left) pierce modern Cairo's skyline at sunset. The mosque was built about 1361 when the city already had nearly half a million people.

A DAY IN THE LIFE OF A RURAL MIGRANT IN CAIRO

Ahmed sits on his bunk bed and chews the end of his pencil. 'I bought a radio today,' he writes, 'and I'm making friends with the three men who share my room.' He is not used to writing letters, and today's letter is especially difficult.

This is the first time he has slept away from his village in the Nile delta in all his 21 years. This morning he stepped off the train in Cairo's central railway station and into a new life. Lured from his village by the hope of work in the city, he ignored the anxious discouragement of his village friends. Now, as he writes, he does not dare admit his growing fear: that a job may not be as easy to find as he had hoped.

It was the images of city life on the village's new communal television that had filled Ahmed's head with dreams. The tall buildings of Cairo held out the promise that there was money there for the making. City-dwellers in their smart Western clothes seemed to proclaim their prosperity by their very appearance. There must be jobs with good wages in the city. In the village the piece of land he shared with his brothers was too small to offer them all a good living. So he took his modest savings, put his spare ankle-length *galabieh* (outer garment) in a bag, made his farewells, and clutching the address of a contact who could provide lodgings, he took the train.

The scale and the bustle of the city surprised him, but he found his way from the station to the contact's address – a small, dark dwelling in a narrow back street. He knocked on the door, asked for a place to stay and was shown to his room, where three other people were talking. He discovered that they all came from different villages in the country, and was impressed to see that all three had smart radios; they must have found a way to make money.

But further enquiry brought disappointing answers. None had a full-time job. Casual work sometimes came their way, or they would get goods from a wholesaler to sell on the street. But they had no trading licences, and there was always the risk of being picked up by the police. Ahmed wondered if he might find a job in a repair shop. But he was a farmer – and what skills could a farmer offer a mechanic?

But there was time yet to sort out these worries. And after all, a few small jobs might be enough. He might be able to save up the deposit on an airline ticket to Saudi Arabia. And everyone knows that a construction worker in Saudi Arabia can earn in a year what it would take a peasant farmer 20 years to earn in Egypt . . .

10 km (6 miles) south of the city on the west bank of the Nile, are the pharaohs' burial pyramids, including the great pyramid of Cheops, and the Sphinx. Cairo's Al-Azhar University, which was founded in 972, is one of the oldest in the world.

Population 8 540 000
Map Egypt Bb

Caiseal ***Ireland*** See CASHEL

Caithness ***United Kingdom*** See HIGHLAND

Cajamarca ***Peru*** Northern department centred on the wide valley of a tributary of the Río Marañón. The surrounding hills support a thriving dairy industry. The capital, Cajamarca, is where the Spanish conqueror Francisco Pizarro captured and killed the ruling Inca, Atahualpa, in 1532. The only Inca buildings to survive are the Ransom Chamber and the Inca Bath. The cathedral and the churches of the towns of Belén and San Francisco are notable examples of colonial architecture.

Population (department) 1 045 600; (city) 75 000
Map Peru Ba

Calabria ***Italy*** Region making up the toe of the Italian boot. It covers 15 080 km^2 (5820 sq miles) and takes in the provinces of COSENZA, Catanzaro and REGGIO DI CALABRIA. This is the poorest of Italy's 20 regions, and there has been much emigration to the more prosperous north. Modern industry hardly exists in Calabria, and agriculture, chiefly olive growing, is hampered by the mountainous terrain. Tourism is an increasingly important spur to its economy.

Population 2 121 700
Map Italy Fe

Calais ***France*** Channel port on the extreme north coast, handling 4 million passengers and 1 million cars travelling to and from Britain each year. Submerged in Roman times, it was established in the 7th century when a channel was cut from a lagoon through the dunes to the sea. In 1347 it fell to Edward III of England after a year-long siege, and was regained by France only in 1558. It was largely rebuilt in the 1940s after its destruction in the Second World War. Leading industries now are clothing, lace-making and food processing.

The French sculptor Auguste Rodin's monument *The Burghers of Calais* commemorates six citizens who offered their lives to spare the town following its capture by Edward III. They, and the town, were saved by Edward's Queen, Philippa, who interceded on their behalf. The monument stands in front of the town hall.

Population 101 500
Map France Da

Calbayog ***Philippines*** See SAMAR

calcite Crystalline form of natural calcium carbonate, the basic constituent of limestone, marble and chalk.

Calcutta ***India*** Port, capital city of West Bengal and India's largest city. It stands in the north-east about 145 km (90 miles) from the mouth of the Hugli river and the Bay of Bengal. It was founded in 1690 by the East India Company as a trading post on the river bank. The settlement prospered until it was attacked by the Nawab of Bengal in 1756 when 126 Europeans, out of a total of 146, suffocated while held captive in a confined room, which became known as the infamous Black Hole of Calcutta. The British soldier Robert Clive came from Madras to retake the town, and in 1757 the British established control of Bengal. Fort William was laid out in parkland in 1758, to establish an impregnable stronghold. A field of fire was kept clear around the fort, so that the Maidan, where the racecourse and cricket fields are sited, was saved from being built over.

From 1774 Calcutta became the capital of British India, and the imperial capital after the proclamation of Queen Victoria as Empress of India in 1877. Its wealth expanded rapidly as a result of trade in cottons, silks, indigo, opium (to China), and later jute and tea. With the coming of jute processing, the town became an industrial centre. In 1911 the imperial capital was moved to Delhi, but this had little effect on Calcutta's prosperity at that time.

When India became independent in 1947 two factors adversely affected the city: first, the arrival of millions of refugees from, and the loss of the natural trading hinterland of, East Bengal (now Bangladesh); and second, the Hugli began to silt up due to the discharge of the Ganges shifting east over a number of years, and could neither provide enough water for the city nor adequate navigation. To counter this problem India has built the Farakka Barrage.

Calcutta still produces and exports jute as well as rice, tea, textiles, chemicals and paper.

Population 9 194 000
Map India Dc

Caldas ***Colombia*** Coffee-growing state covering 7888 km^2 (3045 sq miles) in the centre-west producing 30 per cent of Colombia's coffee, the chief export. The capital is MANIZALES.

Population 760 000
Map Colombia Bb

caldera Large basin-shaped crater formed by the collapse of a volcanic cone, or by a volcanic explosion which removes the top of a cone. The largest known caldera is ASO in Japan.

Caledon (Mohokare) ***Lesothó/South Africa*** River rising in the Drakensberg of northern Lesotho and flowing 480 km (300 miles) south-west to the Orange river in South Africa. It forms much of Lesotho's western border, and its fertile valley, now a maize-growing region, was once inhabited by Bushmen, many of whose cave paintings survive.

Map South Africa Cc

Caledonian Term used for the mountain-building episode during the later Silurian and Devonian periods (see GEOLOGICAL TIMESCALE).

Calgary ***Canada*** Alberta's second largest city, at the junction of the Bow and Elbow rivers. The ROCKY MOUNTAINS can be seen 80 km (50 miles) away and provide a majestic background to the city. Calgary was established as a police post in 1875, and is now the hub of Alberta's oil business and a centre for agriculture, industry and commerce. It is surrounded by rich arable and grazing land, and sources of oil and natural gas. Industries include oil refining, flour milling, timber processing and the manufacture of cement and bricks. The Calgary Stampede is a renowned annual rodeo.

Population 593 000
Map Canada Dc

Cali ***Colombia*** Capital of the rich agricultural department of Valle del Cauca, straddling the Cali river about 300 km (185 miles) south-west of Bogotá. Founded by the Spanish in 1536, Cali became the sedate, colonial capital of a cattle-rearing and cotton and sugar-growing region – until the railway arrived early this century. Since the 1950s it has mushroomed into an industrial centre, and the country's third largest city, making paper, textiles, pharmaceuticals and processed food.

Population 1 755 000
Map Colombia Bb

Calicut (Kozhikode) ***India*** Port in the south-west, about 400 km (248 miles) from the tip of India. It gave its name to calico, the cotton cloth that has been made there since the 17th century. It also manufactures ropes and mats, and exports tea, coffee, coconuts and spices.

Population 546 100
Map India Bc

California ***USA*** Most populous of all the states. It lies on the Pacific coast, and in 1850 became the first in the west to reach statehood, only two years after it was ceded to the USA by Mexico following the Mexican-American War (1846-8) and the almost simultaneous discovery of gold in the SIERRA NEVADA. This prompted a huge rush to the area by money-hungry immigrants – the '49ers.

California's eastern mountains rise to 4418 m (14 495 ft) at MOUNT WHITNEY in the Sierra Nevada. Immediately west is the Central Valley, a 600 km (375 mile) strip of fertile farmland east of the COAST RANGES. Much of the state is prone to earthquakes, especially along the San Andreas Fault on which SAN FRANCISCO lies.

The climate of southern California is Mediterranean-like, with long, hot, dry summers and mild, moist winters, and irrigation is frequently essential for crops. It produces one-tenth of the country's farm output by value, including oranges, grapes, lettuces, cotton, rice and dairy goods. California is a leading oil-producing state and is known for its electronics and computer industries based in the area south of San Francisco known as Silicon Valley.

The state covers 411 015 km^2 (158 693 sq miles) and includes LOS ANGELES, the film capital of the world. SACRAMENTO is the state capital.

Population 26 365 000
Map United States Bc

California, Gulf of Arm of the PACIFIC OCEAN in Mexico which was first named 'The Sea of Cortes', after the Spanish conquistador Hernando Cortés (1485-1547). It was later called 'The Vermilion Sea'. The gulf has oyster beds, pearl and sponge fisheries and big game fish such as sharks and rays. Storms develop quickly and sailing can be hazardous.

Area 161 870 km^2 (62 500 sq miles)
Greatest depth 810 m (2660 ft)
Map Mexico Ab

Callao ***Peru*** Main port of the capital, Lima, which handles most of the country's exports and has a large oil refinery. In the war of independence, the Real Felipe fortress saw the last stand of the Royalists under General Rodil before they surrendered in 1826.

Population 440 500
Map Peru Bb

Callejón de Huaylas ***Peru*** Fertile valley of the Río Santa in Ancash department, lying between the snowless Black Cordillera and the snow-covered range of the Andes, known locally as the White Cordillera, whose highest peak is Huascarán at 6768 m (22 205 ft). Devastating avalanches, most recently as a result of an earthquake in 1970, have taken their toll of the valley's picturesque towns and villages.

Map Peru Ba

calm Force 0 on the Beaufort Scale, when there is virtually no horizontal movement of the air.

Caloocan ***Philippines*** See MANILA

Calvados ***France*** Largely agricultural department in Normandy. The D-Day landings by Allied troops took place on its beaches in 1944. The region is noted for its dairy farming, cider and an apple brandy named after it. The chief town is Caen.

Population 590 000
Map France Cb

Cam Ranh Bay ***Vietnam*** Natural harbour 20 km (12 miles) long and 16 km (10 miles) wide, on the central coast about 290 km (180 miles) north-east of Ho Chi Minh City. It was developed as a base by the United States during the Vietnam War (1965-75), and is now used by the Russian navy and air force.

Map Vietnam Bc

Camargue ***France*** An area of marshes, lagoons and farmland within the Rhône delta. Rice, wheat and animal fodder are grown in the north. To the south are the Etang de Vaccarès – a small man-made lake – and a regional park famed for its wild black bulls, white horses and flamingos.

Map France Fe

Camarines ***Philippines*** See BICOL

Cambodia See p. 122

Cambrian First of the six periods of the Palaeozoic era. See GEOLOGICAL TIMESCALE

Cambrian Mountains ***United Kingdom*** Range forming the backbone of Wales, extending about 140 km (85 miles) between SNOWDONIA in the north and the BRECON BEACONS in the south. The rugged plateau has a number of deep lakes and is cut by many river valleys, but in ancient times was a formidable barrier to invasion from England. The highest peak is Aran Fawddwy (905 m, 2970 ft).

Map United Kingdom Dd

Cambridge ***United Kingdom*** City and county of eastern England with a world famous university dating back to the founding of the first college, Peterhouse, in 1281. The city – 80 km (50 miles) north and slightly east of London – is also an administrative centre, a market town, a processor of farm produce and, most recently, a centre of computer and other high technology industries. Many of these draw on the skills of the university, which has long excelled in science and mathematics.

The earliest students came from Oxford in the 13th century to escape riots there. Although Oxford's colleges are scattered and sometimes hard to find, the glory of Cambridge is focused on a single sweep of architectural wonders along the east bank of the River Cam.

The county – which is mostly agricultural and covers 3409 km^2 (1316 sq miles) – includes the former county of Huntingdon, much of the southern FENS, the cathedral city of ELY and the industrial centre of PETERBOROUGH.

Population (county) 611 000; (city) 96 000
Map United Kingdom Fd

Cameron Highlands ***Malaysia*** Hill station in west Peninsular Malaysia about 50 km (30 miles) south-east of the city of Ipoh. Its cool climate makes it popular with European expatriates. Vegetables and tea are grown.

Population 21 500
Map Malaysia Bb

Cameroon See p. 123

Cameroon, Mount ***Cameroon*** The country's highest volcano, rising to 4095 m (13 435 ft) near the northern end of the coast. It is still active, and its eruptions have made the surrounding plains extremely fertile. Crops include palm oil, rubber, tea, bananas and pepper.

Map Cameroon Ab

Campania ***Italy*** Region covering 13 595 km^2 (5250 sq miles) centred on NAPLES. Campania has two distinct faces. The first is the densely settled coast with its fertile plains around the towns of CASERTA and SALERNO. It includes the thickly populated lower slopes of the active volcano VESUVIUS and the beautiful AMALFI and SORRENTO coasts. The other face is the remote and poor mountainous interior. This area was severely damaged by an earthquake in 1980. The economy of the interior is rural, whereas the coastal towns, especially Naples, contain much industry – chiefly steel, shipbuilding, cars, electrical appliances and food processing.

Population 5 623 400
Map Italy Ed

Campeche ***Mexico*** Picturesque port on the western coast of the YUCATAN peninsula, where, in 1517, the Spaniards first set foot on Mexican soil. The export of logwood – a rare source of dye – found growing in nearby forests made Campeche an exceptionally prosperous city. By the end of the 17th century the thriving port had to be fortified against constant raids by French, Dutch and English pirates; its two main gates and seven of the original eight fortresses still remain. Today Campeche, the tranquil capital of its state, gives the impression of having seen better times. The local economy relies heavily on fishing, textiles, wood and crafts.

Population (port) 99 000; (state) 408 000
Map Mexico Cc

Campi Flegrei ***Italy*** See PHLEGREAN FIELDS

Cambodia

DRAGGED INTO A NEIGHBOUR'S WAR, THIS GREAT EMPIRE OF ANCIENT TIMES IS NOW PROBABLY THE POOREST COUNTRY IN THE WORLD

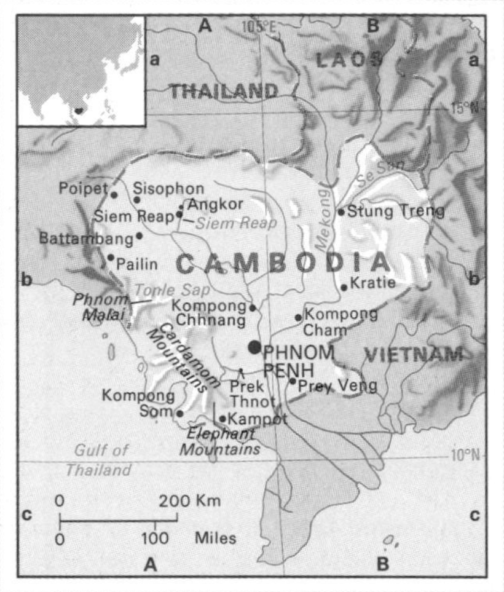

Darkness fell on Cambodia on April 17, 1975, when Pol Pot led his army of Khmer Rouge Communists into the capital PHNOM PENH and started a reign of terror. He claimed to be a political saviour, creating a unique economy based on an agricultural society cut off from world capitalism. But his social experiment brought death and famine to a land which was once the rice bowl of Indo-China.

Under the Pol Pot regime, towns were evacuated and their populations, young and old, healthy and sick, ordered to work on the land. The drive for a self-reliant peasant society received Chinese support, but was carried to brutal extremes. Perhaps a million or more people were executed and half a million died from disease and starvation. Some 600 000 left the country, exiled to Vietnam or to refugee camps in Thailand.

THE ALL-IMPORTANT MEKONG

Cambodia is dominated by the MEKONG river. It floods in the rainy season from May to October, providing fine rice-growing land. In its centre is TONLE SAP (Great Lake), which quadruples in size in the rains and is one of the world's great inland fisheries. Most of the people live in the central lowlands, which are surrounded by forested hills. Rice and maize are the main crops; there are rubber plantations north of Phnom Penh. Temperatures average 27-32°C (81-90°F), and rainfall varies from 1300 mm (51 in) a year in the plains to 5000 mm (196 in) in parts of the south-west.

Cambodia, known to its own people as Kampuchea, was overrun in the 6th century AD by the Khmer tribe from the north, whose empire flourished until about 1450 and created the great Buddhist temple of ANGKOR WAT. The country was then dominated alternately by its neighbours Annam (now Vietnam) and Thailand, but Khmers still make up 90 per cent of the population. In 1863 Cambodia became a French protectorate, and during the Second World War the Japanese occupied most of it. The king, Norodom Sihanouk, proclaimed independence from France in 1944 but the French returned and Cambodia was not independent until 1954.

Sihanouk abdicated to become prime minister and was also head of state from 1960. He tried to keep Cambodia neutral during the Vietnam War, but the North Vietnamese used Cambodian territory to supply their forces in South Vietnam. Cambodia became embroiled in the conflict, and Sihanouk was deposed in 1970 by General Lon Nol, supported by the USA. This move provoked the Khmer Rouge uprising – led by Pol Pot – and the eventual fall of Phnom Penh.

Pol Pot's reign ended in December 1978, when Vietnamese troops invaded. They set up a puppet government which has begun to revitalise the economy. But there is a long way to go. The national income per person in 1982 was estimated at only US$60.

Crops are always uncertain in Cambodia, depending on rainfall and the floods of the Mekong and Tonle Sap. But production was disrupted during the civil war against the Khmer Rouge, and then by the Pol Pot regime. Rice and maize yields remain low and little rubber is produced. The cities, cut off from the countryside by potholed and cratered roads, depend largely on overseas aid.

Industry has been crippled by shortages of power and raw materials. Few shops remain, and most trading is on the streets, with goods smuggled in through Thailand and sold for hoarded gold. In Phnom Penh, once an exquisite colonial city, there are wrecked cars, piles of uncollected rubbish – and Vietnamese troops.

The Vietnamese – supporters of Heng Samrin's puppet government – and the Cambodian army face three resistance groups: the Khmer Rouge, the pro-Western National Liberation Front led by former prime minister Sol Sann, and a small force led by Prince Sihanouk. These form an uneasy coalition, which claims to represent the legitimate government of Cambodia, but they faced an uncertain future when the Vietnamese overran their border camps in 1985.

CAMBODIA AT A GLANCE

Area	181 035 km² (69 898 sq miles)
Population	6 380 000
Capital	Phnom Penh
Government	One-party Communist republic
Currency	Riel = 100 sen
Languages	Khmer (official), French
Religions	Buddhist (90%), Muslim (2%), Christian (1%)
Climate	Tropical; monsoon from April to October. Average annual temperature in Phnom Penh is 27°C (81°F)
Main primary products	Rice, maize, bananas, rubber, livestock, tobacco, jute, timber, fish
Major industries	Agriculture, fishing, forestry
Main exports	Rubber, vegetables
Annual income per head (US$)	60
Population growth (per thous/yr)	21
Life expectancy (yrs)	**Male** 30 **Female** 32

▼ CHILD OF WAR A 12-year-old boy guards a bridge near the Tonle Sap lake in central Cambodia. Many of the orphans created by the fighting that has devastated the country since 1970 have joined guerrilla groups.

Cameroon

DIVERSITY OF LANDSCAPES, A MULTIPLICITY OF TRIBES AND LANGUAGES – BUT ONLY ONE POLITICAL PARTY

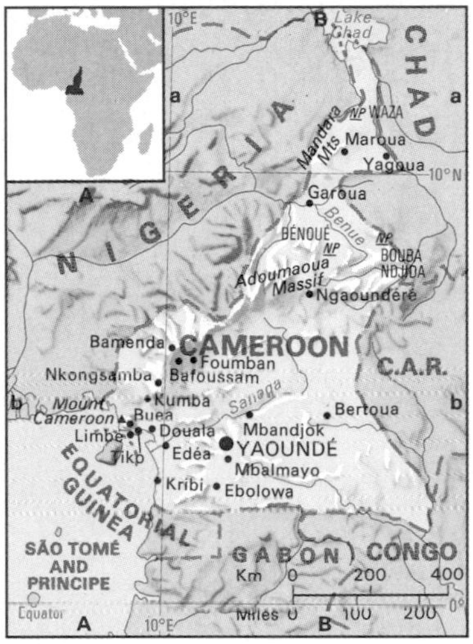

Primitive tribal customs and modern consumer society attitudes meet face to face in the African state of Cameroon. In rural areas, there are memorials and offerings of food to dead ancestors at almost every turn, and fetish figures of wood, clay or metal. While in the towns, such as the capital YAOUNDE, choice foodstuffs are imported (using up large amounts of valuable foreign exchange) to satisfy sophisticated urban palates – even though Cameroon could be self-sufficient in food.

Diversity is perhaps the country's most outstanding quality. Starting in the north, it ranges from low-lying lands on the southern shores of Lake CHAD, through the semidesert SAHEL to dramatic mountain peaks (the Kapsiki mountains north of Garoua). Then come grassy savannahs, rolling uplands (the ADOUMAOUA MASSIF), steaming tropical forests, and neat coffee, cocoa, rubber, banana and hardwood plantations. Finally there are rain-drenched volcanoes – the highest active cone is Mount CAMEROON (4095 m, 13 435 ft) – and palm-fringed beaches at Kribi and LIMBE. Coastal Cameroon is hot, with rain all the year, especially from April to November; Mount Cameroon receives over 10 m (33 ft) of rain annually. The interior is drier.

The people of Cameroon are also a diverse mixture – the country has some 130 ethnic groups, including Fulani herdsmen in the north, the Kirdi peoples whose villages teeter on the cliff tops of the Kapsiki mountains, Bamileke tribes in the south and pygmies in the remotest jungle. The largest language group is formed by about 150 Bantu-language tribes in the forested south. Some of these tribes have fewer than 100 members, but each group speaks its own language.

The Portuguese were the first European arrivals, in the 15th century. The Dutch, Spanish and British came here to snatch slaves until the 19th century, and the country became a German protectorate in 1884. After the defeat of the Germans in the First World War, Cameroon was handed over to the French and British; League of Nations mandates gave France four-fifths of the country in the east and Britain the western strip. The two parts were run separately as French Cameroon and British Cameroon until 1960. The eastern state then became independent as the Cameroon Republic, and was joined in 1961 by the southern section of the British territory to form the Federal Republic of Cameroon. The north opted to join Nigeria, which accounts for the country's strange triangular shape. Full union of the two separately administered parts of the federal republic came about in 1972.

Most Cameroonians live in villages. In the south they grow maize and vegetables to live on, as well as selling the products of the tropical trees – rubber and hardwood timber. In the dry northern savannah and Sahel, drought and hunger are not unknown. There, a single harvest of cereals depends on the short five or six-month rainy season, as do the cash crops of groundnuts and cotton.

Mineral resources – oil, gas and aluminium – are increasing the country's prosperity, although coffee and cocoa still account for nearly half of the exports.

In a diverse country, one aspect of life has no variety – politics. There is a one-party government, and multiplicity of political opinion is regarded as dangerous. Strikes are illegal and the Press is restrained.

CAMEROON AT A GLANCE
Area 475 439 km² (183 570 sq miles)
Population 10 035 000
Capital Yaoundé
Government One-party republic
Currency CFA franc = 100 centimes
Languages French and English (official), African languages
Religions Christian (33%), tribal (25%), Muslim (17%)
Climate Tropical; average temperature in Yaoundé ranges from 18°C (64°F) to 29°C (84°F)
Main primary products Maize, millet, yams, sorghum, cassava, plantains, cocoa, coffee, sweet potatoes, groundnuts, cotton, palm oil, livestock, timber, rubber; petroleum and natural gas, bauxite
Major industries Agriculture, mining, aluminium smelting, food processing, forestry, textiles
Main exports Petroleum, coffee, cocoa, timber, cotton, aluminium and rubber
Annual income per head (US$) 640
Population growth (per thous/year) 27
Life expectancy (yrs) Male 49 **Female** 52

Campina Grande ***Brazil*** Market town 160 km (100 miles) inland from Recife. It is the gateway to the large inland semiarid zone of north-east Brazil known as the *sertão*. The main products are leather, textiles and vegetable oils; there are also light engineering works.
Population 248 000
Map Brazil Eb

Campinas ***Brazil*** Fast-growing modern town about 75 km (47 miles) north-west of the city of São Paulo. There are 25 multinational companies based there, many specialising in electronics. Its other manufactures include paper, textiles, machinery, cotton, maize and sugar.
Population 665 000
Map Brazil Dd

Campine ***Belgium*** See KEMPENLAND

campos Savannah country in central Brazil covering an area of some 2400 by 2750 km (about 1500 by 1700 miles) south of the equatorial forests of the Amazon basin. The vegetation consists of low-growing trees and grass.

Campos ***Brazil*** Industrial town 230 km (144 miles) north-east along the coast from Rio de Janeiro. It services Brazil's principal oil fields in the offshore Campos basin. Apart from oil, its main products are marble, cement and sugar alcohol. Areas of rain forest near the town are the home of the rare golden lion monkey.
Population 349 000
Map Brazil Dd

Can Tho ***Vietnam*** River port on the Mekong delta, about 140 km (85 miles) south-west of Ho Chi Minh City. A waterway network links it to the delta's rice-growing districts.
Population 209 600
Map Vietnam Bc

Cana (Kafr Kanna) ***Israel*** Village 4 km (2.5 miles) north-east of Nazareth, where Christ performed his first miracle, turning water into wine. It was also the home of Nathaniel, traditionally identified with St Bartholomew the Apostle. The village has a 7th-century Franciscan and an 8th-century Greek church.
Map Israel Ba

Canada See p. 124

Canadian Shield ***Canada*** Rocky plateau stretching from the Great Lakes to the Arctic Ocean. It covers 5 million km² (1.9 million sq miles), more than half of Canada, and includes virtually all of Labrador, Quebec and Ontario, part of Manitoba and most of the Northwest Territories. It is rich in minerals and its rivers, lakes and waterfalls have been harnessed for hydroelectric power. The soil is generally poor, and forestry is the major land use.
Map Canada Eb

Canal Zone ***Panama*** The strip of land 8 km (5 miles) either side of the 82 km (51 mile) long PANAMA CANAL in Central America. It covers an area of 1432 km² (553 sq miles) and includes the cities of Cristóbal and Balboa, from where the zone was governed while it was US territory from 1903 to 1979.
Map Panama Ba

Canada

BEARS AND WOLVES ROAM THE WILDERNESS AREAS OF THE WORLD'S SECOND-LARGEST COUNTRY – A LAND OF FROZEN WASTES, FERTILE PRAIRIES AND ABUNDANT WATERS

A land of climatic and geographical extremes and a great diversity of peoples, Canada is the world's second-largest country after the USSR. The Trans-Canada Highway spans 8000 km (5000 miles) from NEWFOUNDLAND to BRITISH COLUMBIA, and it takes four days and five nights to cross the country by train.

Winter temperatures can dip to −62°C (−80°F) in the ARCTIC north, yet summer temperatures of 27°C (81°F) are commonplace in towns and cities to the south. In fact the southernmost part of Canada is at the same latitude as Rome and northern California. Most of the country's 25 million people live within a narrow strip along the American border. Almost two out of every three live in ONTARIO or QUEBEC.

With nearly four-fifths of this enormous land uninhabited, nature lovers can readily explore vast stretches of unspoilt countryside or pitch camp where the only neighbours may be black or grizzly bears. Moose, caribou, elk and wolves also roam the wilderness, and the inland waters abound with salmon, trout, bass and pike. Canada has more lakes than any other country.

The world's most famous police force is the Royal Canadian Mounted Police – the Mounties – who stand as a symbol of Canada in their red ceremonial tunics. Now they are an 8000-strong federal force, but there were only 300 of them when they were formed to check frontier lawlessness in 1873.

The Mounties were involved in one of the most dramatic episodes in Canadian history, the KLONDIKE Gold Rush. Gold was discovered in Bonanza Creek, a tributary of the Klondike River in the YUKON TERRITORY in 1896. When word of the strike reached the outside world it sent 100 000 men, and a few women, stampeding to the wild north-west of Canada. Over six years the Klondike yielded more than US$100 million in gold – worth many times that today. Many people became rich, some went mad with hardship and gold fever, while others died or were killed in a reckless society of gold-robbers, gambling saloons and rowdy dance halls, policed by a few Mounties.

A DIVERSE LAND

Canada is a sovereign state within the Commonwealth. Most of the people are of British (about 10 million) or French (7 million) origin; there are 368 000 native Indians, 25 000 Inuit (Eskimos) and a wide range of immigrant nationalities. Most of the French-speaking people live in the province of Quebec. Strong cultural and commercial influences come across the 4200 km (2600 mile) border with the United States.

Most of the Canadian people and industry are in the GREAT LAKES region and the ST LAWRENCE lowlands. This heartland starts as a narrow strip near Quebec city, broadens out into southern Quebec and extends westwards across Ontario. The St Lawrence river flows through this corridor from the Great Lakes to the Atlantic Ocean.

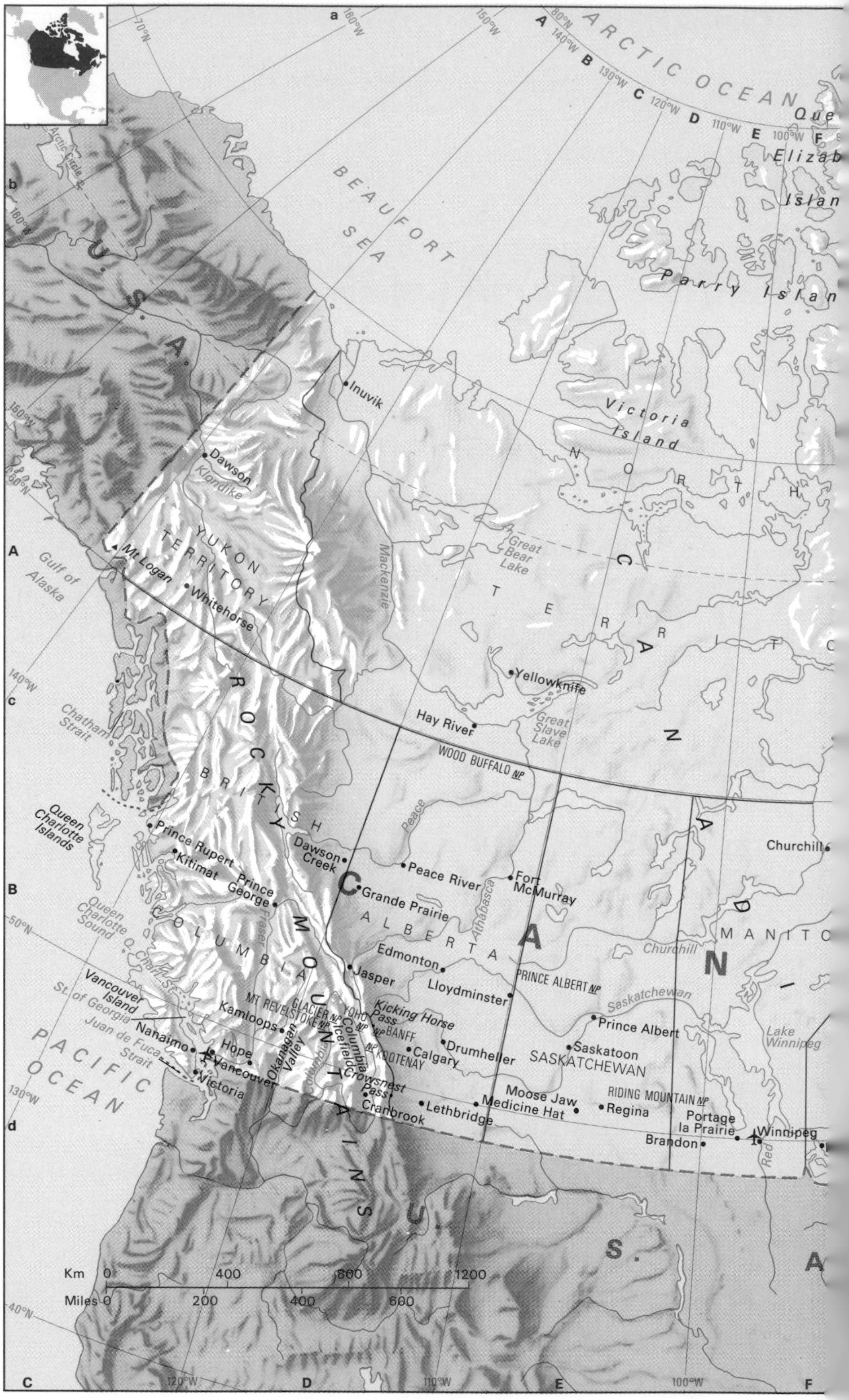

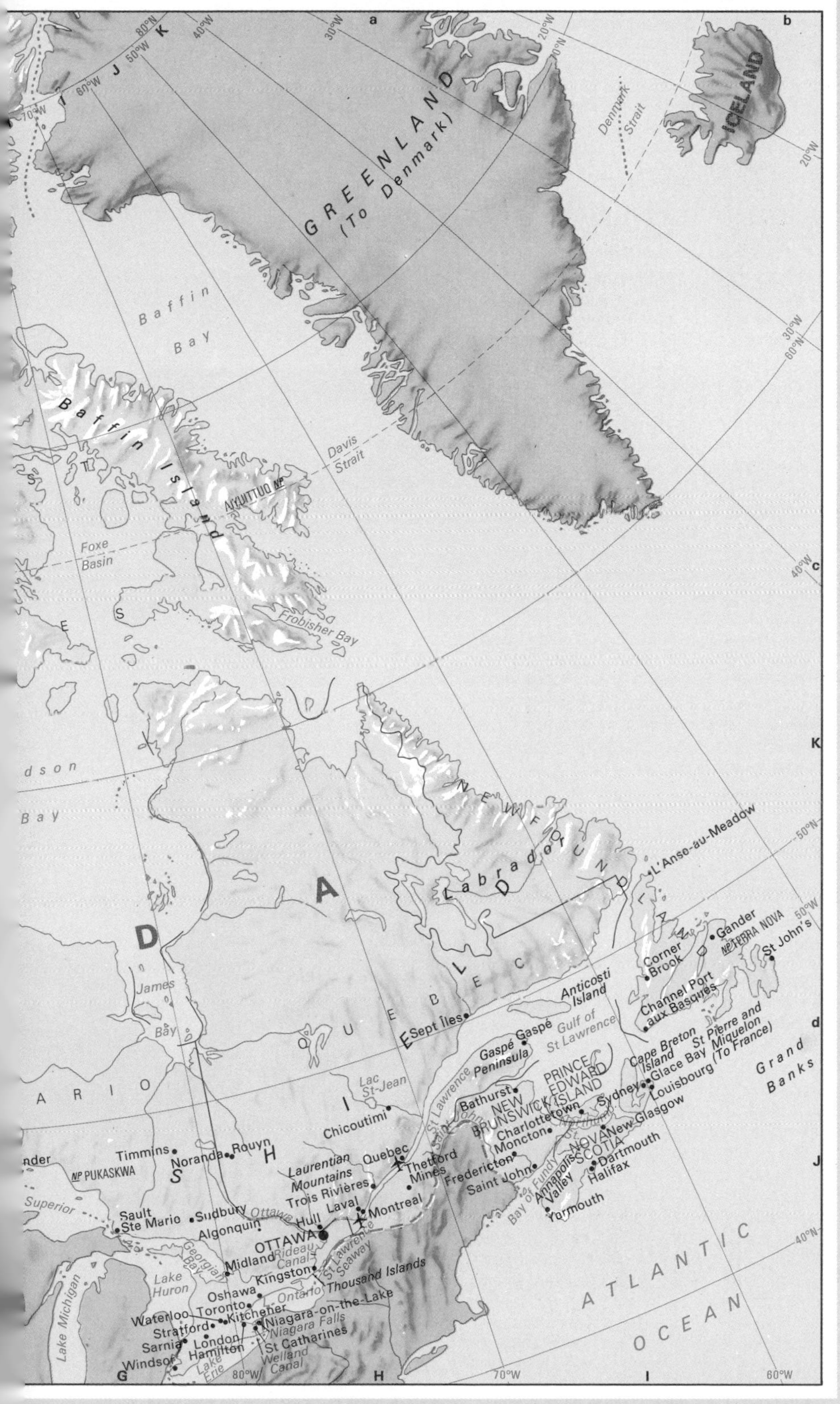

In the east, the APPALACHIAN highlands extend into Quebec's GASPE PENINSULA, NEW BRUNSWICK, NOVA SCOTIA, PRINCE EDWARD ISLAND and Newfoundland. They are part of an old mountain range that extends northwards from the United States, but seldom rises to more than 1000 m (3300 ft). Prince Edward Island in the Gulf of St Lawrence is an exception to the rugged terrain and is the best eastern province for agriculture.

The CANADIAN SHIELD, consisting of old hard rocks extending over more than 4.65 million km² (1.8 million sq miles), forms the bedrock of half of Canada. It is an area of forests, lakes and rocks which sweeps westwards from LABRADOR through most of Quebec and Ontario, embraces the northern parts of the prairie provinces of MANITOBA and SASKATCHEWAN, and extends into the NORTHWEST TERRITORIES. The Shield has valuable mineral and forest resources and potential for hydroelectric power.

The provinces of ALBERTA, Manitoba and Saskatchewan, bordered by the USA to the south and the ROCKY MOUNTAINS to the west, contain Canada's largest farming area. Forests cover the northern parts of them.

Much of the west is more than 2000 m (6600 ft) high; the country's highest peak, MOUNT LOGAN in the Yukon, reaches 5951 m (19 524 ft). A series of mountain ranges called the Western Cordillera run down the west side of the country, covering most of British Columbia, the Yukon and parts of Alberta and the Northwest Territories. The Rocky Mountains form the eastern edge, towering to more than 3000 m (10 000 ft), while the Coast Mountains rise to 4042 m (13 260 ft) at Mount Waddington in British Columbia.

The west coast provides spectacular scenery with fiords, mountains and islands including VANCOUVER ISLAND and the QUEEN CHARLOTTE ISLANDS. The city of VANCOUVER stands in the main lowland area of the west coast, the lower FRASER valley.

The far north of Canada is inhabited mainly by small communities of Inuit. It is wild and inhospitable, with low rainfall, little tree growth and large areas where the land is permanently frozen. But it holds great mineral wealth, including oil and gas. The 25 000 Inuit live in the MACKENZIE delta, on the Arctic islands, and on the mainland coast of Labrador. The 368 000 Indians live mostly in 2250 reserves covering 2.6 million hectares (6.4 million acres). They are widely spread across the country and speak 60 different languages.

The average daytime temperature over most of Canada exceeds 24°C (75°F) in July. But in winter, nearly all the country is snow-covered and has temperatures below freezing point. The St Lawrence river is closed to traffic for three or four months, and most other rivers are icebound.

TRAVELLING THE LAKES AND RIVERS

The St Lawrence has long been an important trade route, especially since the St Lawrence Seaway was built in 1954-9, enabling ocean-going vessels up to about 28 000 tonnes to sail into the Great Lakes. The major rivers in the west are the Fraser and Thompson, which run into the Pacific but are of limited navigational value. Most other rivers flow to virtually uninhabited coasts of the HUDSON BAY and the Arctic Ocean, and are frozen for much of the year.

Nevertheless, the huge network of rivers and lakes played an important role in opening up the country to the first explorers, trappers and fur traders. They were able, with only short hauls overland, to paddle their canoes from the St Lawrence river westward to central Alberta, north to the Mackenzie delta, and back again.

The first European settlers were the French. In 1605 they founded a community that is now Annapolis Royal on the coast of Nova Scotia, and in 1608 they sailed up the St Lawrence to the site that is now the city of Quebec. British settlement started two years later on Newfoundland. By the 1670s French explorers, missionaries, traders and trappers had penetrated west to the MISSISSIPPI headwaters and Manitoba. The British set up the Hudson's Bay Company to trade in pelts and furs, and the scene was set for rivalry between the two powers.

BATTLES OF THE EARLY SETTLERS

The French and British destroyed each other's settlements, each side recruiting Indian tribes as allies. The British navy raided French coastal forts, and French troops attacked British garrisons and trading posts. The issue was finally settled in the Seven Years' War of 1756-63. In 1759, a British army led by General James Wolfe routed the French in the Battle of the Plains of Abraham and took Quebec. The French surrendered MONTREAL in 1760, and the French colonies were ceded in 1763.

Despite Britain's military victory the country continued to support separate British and French cultures, languages and institutions. French settlers far outnumbered the British in Quebec and the surrounding area, so it was decided to recognise the Roman Catholic religion and retain French civil law and language (the Quebec Act, 1774). This separate system continues today.

After the American War of Independence (1775-81) more than 30 000 loyalist refugees moved north from the new republic to New Brunswick, Nova Scotia, Quebec and Ontario, increasing the population fourfold in many areas. From 1815 the British government encouraged settlers, and in the next 40 years more than a million immigrants landed in HALIFAX, SAINT JOHN and Quebec.

MAKING MODERN CANADA

After the American Civil War (1861-5) there were fears that the strengthened United States might try to annex some Canadian territory. As a defence, the provinces of New Brunswick, Nova Scotia, Ontario and Quebec signed the British North America Act, creating the Dominion of Canada, in 1867. At the time, much of the west was in the hands of the Hudson's Bay Company, but in 1869 its territory was bought by the government, and overnight Canada tripled in size.

Two new provinces, Manitoba (1870) and British Columbia (1871), joined the expanding nation, the latter partly because it was promised a rail link to Ontario. The first line from the heartland to the Pacific (the Canadian Pacific Railway) was opened in 1885; the rail route eastwards from Montreal to Halifax had been opened in 1876.

Prince Edward Island joined Canada in 1873, and Saskatchewan and Alberta in 1905. The development of railways and the purchase of the Hudson's Bay Company lands opened the PRAIRIES to settlement and agriculture, and brought more European immigration.

The population doubled between 1900 and 1930. A lot of the newcomers came from central and southern Europe. Many of them stayed in the expanding cities and helped to give Canada its rich ethnic diversity.

In the First World War, Canada sent more than 600 000 men out of a total population of less than 8 million to fight in Europe. Canadians again hurried to the Allied side in September 1939. Canadians also fought in the Korean War and provided peace-keeping forces for Cyprus, Suez and elsewhere.

Many political institutions have British origins. Canada is a constitutional monarchy under Queen Elizabeth II. The Crown's powers are administered by a governor-general, now always a Canadian. There is a federal Parliament in OTTAWA which has two houses. The House of Commons plays the dominant part. It has 282 members who are elected every five years. The upper house, the Senate, has 104 members who are appointed by the governor-general on the advice of the prime minister, and who serve until the age of 75. The two houses have similar powers, but all major legislation is introduced and first debated in the House of Commons. The Senate can veto legislation, but rarely does so.

Each of the ten provinces has its own elected parliament with a single house, but the Yukon and Northwest Territories have only limited self-government. In any case, provincial laws are binding only within the province concerned, whereas federal legislation is nationwide. At times there are conflicts between federal and provincial governments. A recent dispute involved the right to develop oil and other mineral resources off the coast of Newfoundland. The provincial governments of Alberta and Saskatchewan have also been at odds with the federal government over the price of oil produced from their territories.

AGRICULTURE AND NATURAL RESOURCES

More than 80 per cent of Canada's farmland is in the prairies that stretch from Alberta to Manitoba, and two-thirds of farm income is earned by prairie farmers. Despite the importance of agriculture, only about one Canadian in 80 is a farmer. The average farm size is about 225 hectares (555 acres).

Wheat and other grain crops cover three-quarters of the arable land. Canada exports about 80 per cent of its wheat production, which accounts for 20 per cent of the world's wheat exports. In the prairie provinces, tall narrow grain elevators, where the grain harvest is stored, are distinctive landmarks. Most elevators stand beside railways, and most grain is sent to market by rail.

In addition to grain, large prairie areas are devoted to raising cattle. In some other places, high quality farmland is being used for industrial and urban developments, and this is causing concern, particularly in areas like the Niagara peninsula in southern Ontario, where much fruit is grown.

Canada's major natural renewable resource is the forests that extend across the country in a belt 500 to 2000 km (300 to 1250 miles) wide, and cover more than half the total land area. Most woodland is coniferous: fir, pine and spruce. Canada is the world's largest exporter of wood pulp and paper – there are massive mills turning the forests into newsprint for newspapers around the world – and one job in every ten depends on wood. Nearly all the forests are owned by the provincial or federal governments.

Fish helped to attract the first Europeans to Canada, and fishing remains important. About 56 000 commercial fishermen operate

from the east coast in the Atlantic fishing grounds. About 20 000 work from the west coast where salmon is the chief catch. Much fish is exported, mainly to the USA.

Minerals range from asbestos to zinc; the most valuable are oil, natural gas, coal and iron ore. Alberta produces 90 per cent of the country's oil output and nearly half the coal total of 44 million tonnes a year. This mineral-rich province also has natural gas and large potential resources of oil from bituminous tar sands on the ATHABASCA river. The second most important coal province is British Columbia, with nearly one-third of national production. Canada is a net exporter of coal and natural gas, but imports some oil.

Iron ore is found on the north-western side of LAKE SUPERIOR and on the Quebec-Labrador border. The world's chief source of nickel is around SUDBURY, Ontario, and 220 km (135 miles) to the north the town of Timmins is Canada's main gold-mining centre. Asbestos is produced in Quebec, and copper in Ontario and British Columbia. Canada is a leading world producer of zinc and titanium, and also mines uranium, silver and molybdenum.

Electricity has usually been generated from water power, and 70 per cent still comes from hydroelectric stations; one of the world's largest hydroelectric power plants is at the

▲ WINTER'S HOARY GRIP A herd of caribou crosses the frozen tundra near Repulse Bay in the Northwest Territories. Sparse reindeer moss – a type of lichen – is their only winter food.

▲ MAJESTIC GRANDEUR A valley cuts through the Rocky Mountains of Alberta in Banff National Park, Canada's first, established in 1885. Noted for hot springs, ice fields and glacial lakes, it is a game sanctuary and popular summer and winter sports area.

Churchill Falls in Labrador. But potential sites for new dams tend to be in northern regions, far away from the users, so Canada has developed nuclear technology and has four major nuclear power stations.

INDUSTRY AND TOURISM

Nearly one-fifth of Canadian workers are in manufacturing. Industry is highly mechanised and includes petroleum products, car manufacture, food and metal processing. Ontario contains most of the car factories, is the main steel producer and has important aircraft, electronics and electrical machinery sectors. Quebec has textiles, paper and wood products, clothing, chemicals and engineering.

Inevitably, most international trade is with the USA, and nearly half of Canada's manufacturing industry is owned by American firms. US President John F. Kennedy told the Canadian Parliament in 1961: 'Geography has made us neighbours, history has made us friends, economics has made us partners and necessity has made us allies.'

The Americans are largely responsible for a recent surge in tourism, which has taken the industry into Canada's top eight export earners, employing about one person in ten. Vast open spaces draw the fisherman, hunter, canoeist and camper, mostly from the USA, but many from Britain, continental Europe and Japan. The need to safeguard areas of beauty has long been recognised, and the first national park was set up at BANFF in 1885.

NIAGARA FALLS and the Rocky Mountains are world-famous, and tourists are also attracted to the historic city of Quebec, the old part of Montreal, modern TORONTO with its CN Tower – the world's tallest free-standing building – and the fishing ports of the eastern Maritime Provinces.

Travel is mostly by car or aircraft. Railways carry most freight, but passenger trains are few outside the Montreal-Toronto-WINDSOR area. Exceptions are the popular routes through the Rockies, especially the Banff-Vancouver line.

Three-quarters of the 25 million Canadians live in urban areas, and over half in metropolitan areas of more than 100 000 population. Only three metropolitan areas have more than a million people: Toronto, with 3 million, Montreal (2.8 million) and Vancouver (1.25 million). Like most North American urban areas, major Canadian cities have centres dominated by skyscrapers.

CANADA AT A GLANCE

Area 9 976 139 km² (3 851 793 sq miles)

Population 25 650 000

Capital Ottawa

Government Federal parliamentary monarchy

Currency Canadian dollar = 100 cents

Languages English, French

Religion Christian (47% Roman Catholic, 41% Protestant)

Climate Continental; arctic in north; maritime near coast (especially in British Columbia). Average temperature in Ottawa ranges from −15 to −6°C (5-21°F) in January to 15-26°C (59-79°F) in July

Main primary products Cereals, fruit and vegetables, livestock, rapeseed, tobacco, linseed, timber, fish; oil and natural gas, coal, copper, zinc, titanium, iron, lead, asbestos, nickel, salt, uranium, potash

Major industries Agriculture, forestry, paper and other timber products, food processing, iron and steel, engineering, mining, transport equipment, chemicals, fertilisers, oil and gas refining, cement

Main exports Motor vehicles, machinery, cereals and other foodstuffs, natural gas, chemicals, paper, crude oil and products, coal, metal ores, timber, wood pulp

Annual income per head (US$) 10 275

Population growth (per thous/yr) 10

Life expectancy (yrs) Male 73 **Female** 77

Cañar ***Ecuador*** West-central province in the Andes, just north of Azuay. Much of it is cold, dry and windswept, but around the provincial capital, Azogues, in the south-east, the land is intensively farmed. Azogues (population 13 500) is a centre of the straw hat industry. About 25 km (15 miles) north is the small town of Cañar, to the north of which is the Inca fortress of Ingapirca.

Population 181 000

Map Ecuador Bb

Canary Islands ***Spain*** Seven large islands with islets about 95 km (60 miles) off the north-west coast of Africa. Many of the islands are popular tourist resorts, particularly in winter. They are mountainous with a generally mild climate, but are subject to severe drought and tornadoes. Irrigated farms grow grain, bananas, oranges, lemons, vegetables and vines. The islands' industries include fishing and canning.

Known to the ancients as the Fortunate Islands, they were rediscovered and claimed by the Portuguese in 1341, but ceded to the Spaniards in 1479.

The islands of Lanzarote, Gran Canaria and Fuerteventura form the province of Las Palmas. The islands of Tenerife, La Palma, Gomera and Hierro form the province of Santa Cruz de Tenerife. Between them the islands cover an area of 7273 km² (2808 sq miles).

The main towns are Las Palmas (on Gran Canaria) and Santa Cruz (on Tenerife), both of which are duty-free ports. The islands' highest point, and also the highest point in Spanish territory, is Pico de Teide, a volcanic cone on Tenerife, which rises to 3718 m (12 198 ft).

Population 1 444 600

Map Morocco Ab

Canaveral, Cape ***USA*** Sandspit on the east coast of Florida, about 300 km (185 miles) north and slightly west of Miami. It is the site of the John F. Kennedy Space Center, and was for a period called Cape Kennedy. All America's manned space missions have been launched from there.

Map United States Je

Canberra ***Australia*** The national capital, standing in the Australian Capital Territory, about 250 km (155 miles) south-west of Sydney. The city was planned in 1911-13 by Walter Burley Griffin, an American architect, and in 1927 the Australian Parliament moved there from Melbourne. The city's population has increased eightfold since the 1950s, when foreign diplomats and federal government departments moved there from Sydney and Melbourne. The origin of the city's name is uncertain; it may come from an Aboriginal word *canberry*, meaning 'meeting place'.

Population 255 900

Map Australia Hf

Cancún ***Mexico*** Tiny 22 km (14 mile) long island in the Caribbean, linked to mainland Mexico by a narrow causeway. Cancún is hot and sunny, and has fine beaches and clear seas – all of which have helped to make it a successful tourist resort. The splendid Mayan ruins of Tulum and Xel-Ha are nearby.

Population 70 000

Map Mexico Db

Cango Caves ***South Africa*** Limestone caves 26 km (16 miles) north of the town of Oudtshoorn, about 355 km (220 miles) east and slightly north of Cape Town. The caves were once occupied by Stone Age men and were rediscovered in 1780. They are now a National Monument, with chambers known by such evocative names as the Crystal Palace, the Devil's Workshop and the Rainbow Room. Concerts are held in the outermost cave. The caves' full extent is still unknown, but the passages so far discovered have a total length of 3 km (2 miles).

Map South Africa Bc

Çankiri ***Turkey*** Town about 100 km (60 miles) north-east of the capital, Ankara. It contains a huge 16th-century mosque built by Sinan, regarded as one of the greatest Islamic architects. The surrounding province, also called Çankiri, covers 8666 km² (3346 sq miles). Its chief crops are fruits and grains, and asbestos is one of its main minerals.

Population (town) 40 900; (province) 267 300

Map Turkey Ba

Cannes ***France*** Seaport and fashionable resort town on the Côte d'Azur, 28 km (18 miles) south-west of Nice. It manufactures soap, oils and perfumes, and exports fruit and anchovies. There are two casinos, a yachting marina, and an annual film festival each May.

Population 72 800

Map France Ge

Cantabria ***Spain*** Province on the Bay of Biscay around the port of Santander. Its main industries are mining iron, zinc and lead, fishing, rearing cattle and sheep.

Population 510 800

Map Spain Ca

Cantabrian Mountains (Cordillera Cantabrica) ***Spain*** Northern range extending about 480 km (300 miles) west from the Pyrenees. It is rich in minerals, especially coal and iron. The highest peak is Peña Cerredo (2648 m, 8685 ft).

Map Spain Ba

Cantal, Massif du ***France*** Volcanic upland in the Massif Central, about 80 km (50 miles) south and slightly west of CLERMONT-FERRAND. The chief industries are livestock, cheese-making and tourism. The Plomb du Cantal, 1855 m (6086 ft), is the area's highest peak.

Map France Ed

Canterbury ***United Kingdom*** English cathedral city in the county of Kent, 90 km (55 miles) south-east of London. It was the birthplace of the established Church in England, with its first archbishop appointed in AD 601, and is today the seat of the Primate of All England and the world headquarters of Anglicanism. Pilgrims journeyed there for centuries to the shrine of Archbishop Thomas Becket (1118-70), who was murdered in his own cathedral. These pilgrimages were the inspiration for Geoffrey Chaucer's 14th-century poetic masterpiece *The Canterbury Tales*.

The city has retained much of its medieval character, including narrow streets, houses and a large section of the city wall. The cathedral was begun in about 1070, and completed only in 1400. It stands in a wide close, surrounded by old monastic buildings and the King's School. In the countryside outside the city is the modern University of Kent.

Population 36 000

Map United Kingdom Fe

Canterbury Bight Inlet of the South PACIFIC OCEAN on the east coast of SOUTH ISLAND, New Zealand. The chief port is TIMARU.

Width 185 km (115 miles)

Map New Zealand Cf

Canterbury Plains ***New Zealand*** Grasslands on South Island stretching more than 150 km (90 miles) along the east coast, from the Kaikoura Ranges in the north to the Waitaki river. Sheep and cattle have grazed them since the 1840s. The chief town is CHRISTCHURCH.

Map New Zealand Cf

Canton ***China*** See GUANGZHOU

canyon Steep-sided narrow gorge, usually with a river at the bottom, mainly found in arid or semiarid areas, the source of the river being elsewhere in a well-watered area. The most famous is the GRAND CANYON.

Canyon de Chelly National Monument ***USA*** Home of the Navaho Indians, covering 335 km² (129 sq miles) in north-eastern Arizona. The monument contains many Indian remains dating back to AD 350, including cave dwellings in the base of sheer red canyon cliffs.

Map United States Ec

Canyonlands National Park ***USA*** Spectacular area astride the Colorado river in south-eastern Utah. It is dotted with high flat-topped mountains called mesas, deep canyons and natural arches. It covers 1365 km² (527 sq miles).

Map United States Ec

cape Promontory or headland projecting into the sea. For individual capes, see the main part of the name.

Cape Breton Island ***Canada*** Part of Nova Scotia, covering 10 349 km² (3970 sq miles) and linked to the mainland by a causeway over the Strait of Canso. Much of Cape Breton is ringed by the spectacular Cabot Trail road. In the north is the Cape Breton Highlands National Park, 950 km² (367 sq miles), which has forests and impressive cliffs.

Map Canada Id

Cape Coast ***Ghana*** Town on the coast 120 km (75 miles) west of the capital, Accra. It was the British colonial capital until 1876, and is now the capital of Central Region. The castle built there by the British is now used as government offices, while another fort is now a lighthouse. Citrus fruit production and fishing form the basis of the economy.

Population 72 100

Map Ghana Ab

Cape Doctor In South Africa, a humorous name for the strong, south-east wind that blows chiefly in summer in Cape Town, so called because it prevents stagnation of the mountain air and produces stimulating fresh conditions.

Cape Maclear ***Malawi*** Tourist resort on the southern shore of Lake MALAWI. Offshore is Thumbi Island, a nature reserve.
Map Malawi Ab

Cape Peninsula ***South Africa*** Peninsula immediately south of Cape Town. It has fishing villages, resorts, a naval base at Simon's Town, and the Cape of Good Hope Nature Reserve, which covers 77 km^2 (30 sq miles).
Map South Africa Ac

Cape Province ***South Africa*** Largest of the country's provinces, covering 645 767 km^2 (249 331 sq miles). It was first settled in 1652, and early settlers included Dutch farmers and French Huguenots. Britain occupied it in 1795-1803 and again in 1806, creating Cape Colony, which became a province of South Africa in 1910. Cereals, fruits, tobacco and grapes are grown in the fertile coastlands; livestock are reared on inland plateaus, and diamonds, copper and some coal are mined. There are manufacturing centres at Cape Town, East London and Port Elizabeth.
Population 5 373 800
Map South Africa Bc

Cape Town ***South Africa*** City founded in 1652 as a supply depot for Dutch ships and lying at the foot of the majestic Table Mountain. Its castle, dating from 1666, is South Africa's oldest building. Groote Schuur on the south-eastern outskirts was the estate and home of the statesman Cecil Rhodes. It now contains a museum, the University of Cape Town, the Rhodes Memorial and the Groote Schuur hospital. On the eastern slopes of Table Mountain are the Kirstenbosch National Botanic Gardens, founded in 1913, which contain a magnificent collection of South-African flora.

Cape Town is an industrial city and the chief port for the agricultural coastlands. It also houses the nation's Parliament buildings, making it South Africa's legislative capital. Black townships around the city – often scenes of unrest – include Mitchells Plain, Elsiesrivier, Guguletu, Langa and Nyanga.
Population 213 800
Map South Africa Ac

Capernaum (Kefar Nahum) ***Israel*** Ruins of an ancient town on the north shore of the Sea of Galilee (Lake Tiberias) where Jesus spent most of the period of his ministry, and chose his first disciples (Andrew, Peter, James and John). There is a 2nd-century synagogue on a spot where Jesus is said to have preached. Nearby, St Peter's house is being excavated.

Rising to 91 m (368 ft) above Capernaum is the hill known as the Mount of the Beatitudes, where according to tradition Christ preached his Sermon on the Mount. Capping the hill is the 13th-century Franciscan Church of The Beatitudes.
Map Israel Ba

Cap-Haitien (Le Cap) ***Haiti*** Seaport, and second largest city after Port-au-Prince, lying on the north coast. Founded in 1670 by the French, it was razed by French and Haitian troops in 1802. Henri Christophe, the self-proclaimed King of North Haiti, rebuilt much of it, but an earthquake in 1842 caused widespread destruction. Today its modernised harbour handles exports of coffee, sugar cane, bananas, sisal and cocoa.

About 16 km (10 miles) south is the ruined Sans Souci palace, built by Henri Christophe in 1813, and wrecked by the 1842 earthquake. A further 11 km (7 miles) south is another of the king's constructions, the Citadelle, a massive fortress rising from the sheer cliff face, with 200 cannons ranged around its 4 m (13 ft) thick walls. Begun in 1804, it took 13 years to complete and 20 000 slaves died building it.
Population 75 000
Map Caribbean Ab

A DAY IN THE LIFE OF A CAPE TOWN UNIVERSITY LECTURER

Few people can wake up to a view as spectacular as the one Ray Wilson sees from the balcony of his bedroom every morning. His restored Victorian terraced house in the Cape Town suburb of Tamberskloof ('Drummer's Ravine') looks directly up to the majestic flat-topped Table Mountain. Ray joins his wife Harriet and two small sons, Sean and Anton, at the breakfast table. The black domestic, Sarah, serves scrambled eggs to the children, while Ray swallows a cup of coffee.

After breakfast, Ray drives Harriet and the children to a private nursery school on his way to work. Harriet then takes the bus 6.5 km (4 miles) into town, to the Social Services office, headquarters of her job as a social worker. Blacks and whites can now travel on the same buses in Cape Town; segregated buses is one of the aspects of 'petty apartheid' that are disappearing from the South African scene.

Ray, meanwhile, drives to the University of Cape Town (UCT), where he lectures in African History.

UCT was designated a 'white' university by the government, in 1959, but officially is open to students of all races. Most of the students and staff oppose the government and the apartheid system. Students have often been detained for their political activities. Ray himself was held in jail for two months three years ago for refusing to testify against one of his students who was suspected of planting a bomb.

This morning, Ray has a seminar for second-year students on the frontier wars between the British and the Xhosa in the 19th century in the Eastern Cape. There are two coloured students in his class, Alex and Sam, and they are among the most vociferous in the discussion of these events – which is heated but never acrimonious.

Alex and Sam – like the Wilsons' own domestic – could not live where Ray does because Tamberskloof is a whites-only area. They have to commute 32 km (20 miles) every day from the new 'coloured city' of Mitchells Plain. Alex and Sam have been to Ray's house for dinner, but they have problems when they all want to see a film together; they have to go some way to a cinema in the central business district, which is integrated by law, since the local cinemas in the suburbs are segregated.

Ray spends the afternoon in his office, putting the finishing touches to a series of lectures. By 5 pm he has finished, and, before going home for a quiet and relaxing evening, he drives down to Muizenburg Beach for a run and a swim in the warm surf.

Capitol Reef National Park ***USA*** Area of 979 km^2 (378 sq miles) in south-central Utah, astride a tributary of the Colorado river. It contains a 32 km (20 mile) long sandstone cliff that rises abruptly from the desert and contains former Indian dwellings.
Map United States Dc

Capo d'Istria ***Yugoslavia*** See KOPER

Cappadocia ***Turkey*** Volcanic plateau lying between the capital, Ankara, and the town of Kayseri, 250 km (150 miles) to the south-east. Homes have been cut in its soft volcanic rock since at least 400 BC. There are still cave houses in use, including hundreds around the villages of Aucilar and Ürgüp. The Göreme valley contains numerous early Christian chapels and a Byzantine monastic complex, the inside of which is covered with 10th-century frescoes.
Map Turkey Bb

Capri ***Italy*** Beautiful holiday island in the Bay of Naples covering 10.4 km^2 (4 sq miles). It is garlanded with shrubs, vineyards and exotic gardens. The main towns are Capri and Anacapri. The island is renowned for its marine caves, particularly the Blue Grotto. For many years Capri was the home of the English entertainer Gracie Fields (1898-1979), whose villa is now a shrine to her memory. It was also a favourite haunt of the Roman Emperor Tiberius nearly 2000 years ago.
Population 16 500
Map Italy Ed

Caprivi Strip ***Namibia*** Strip of land, about 450 km (280 miles) long and 30-100 km (18-60 miles) wide, between Angola and Zambia to the north and Botswana to the south. It was acquired from Britain by an agreement in 1893 by the German Chancellor Count Leo von Caprivi, to give German South West Africa access to the Zambezi river. It has been the scene of clashes between South African forces and SWAPO (South West Africa People's Organisation) guerrillas since 1972.
Map Namibia Ba

Carabobo ***Venezuela*** State in the north, containing the cities of Valencia – its capital – and Puerto Cabello. In 1821 it was the scene of the battle which assured Venezuela's independence from Spain. There is a large monument to the victory at Campo de Carabobo, south of Valencia. The state's principal products are sugar, citrus fruits and coffee.
Population 1 209 560
Map Venezuela Ba

Cape Verde

A YOUNG ISLAND REPUBLIC OFF THE AFRICAN COAST WHOSE CHANCES OF PROSPERITY ARE CRIPPLED BY DROUGHT

Though superficially possessing all the classic attributes of tropical isles – white, sandy beaches, a backdrop of volcanic mountains and cloudless skies – the 15 islands and islets that make up Cape Verde are far from paradisical. The main problem is the cloudless skies, and the consequent periods of severe drought that since 1900 have brought about the deaths of more than 75 000 of the inhabitants. The last major drought, which lasted from 1968 to 1982, brought agriculture to a virtual standstill, even in the cooler and generally moister uplands, and caused some 40 000 Cape Verdeans to emigrate to Portugal, seeking work. The money they sent back to their families has at least made some contribution to the economy.

Cape Verde is one of the world's smallest nations, and one with a not too promising future due to lack of resources. Lying some 645 km (400 miles) off the West African coast, its principal islands are Santo Antão, Boa Vista, Fogo, São Nicolau, Maio, Sal and São Tiago, on which is situated the capital, Praia; about half the population lives on São Tiago. The coastal plains are arid, the best soils are in the uplands, and the mountains are clothed in thin forest. Some of the mountains reach considerable heights; the highest, Fogo, climbs to 2829 m (9281 ft) and is an active volcano. Though the climate is tropical, it is tempered by north-east winds and is no more than pleasantly warm.

The islands were discovered by the Portuguese explorer Diogo Gomes in 1460, and began to be settled by the Portuguese two years later. For most of their history, the islands' principal use was as a provisioning station for ships sailing to South Africa and Brazil and as a major assembly point for slaves in the Guinea slave trade. They still have some importance as a refuelling base on the sea routes between Europe and South America. But the population, mainly of mixed African and European descent, is still chiefly engaged in farming and fishing. The only minerals produced are salt and a volcanic rock – *pozzolana* – used in the manufacture of cement. In 1975, after five centuries of Portuguese rule, Cape Verde obtained its independence. Its president is the head of the African Party for the Independence of Cape Verde, the only political party. It has a 1100-strong army, a navy of three gunboats and 24 torpedo boats, and a young air force which so far possesses two transport aircraft. There is an active merchant marine of some 20 ships.

Though politically stable, the republic is economically shaky – apart from the chronic drought, the lack of development means that Cape Verde needs to import much of its food, and relies heavily on foreign aid. The best hopes for the future would seem to lie in tourism, fishing and farming, provided an effective means of water conservation and management can be found.

CAPE VERDE AT A GLANCE

Area 4033 km² (1557 sq miles)
Population 321 000
Capital Praia
Government One-party republic
Currency Cape Verde escudo = 100 centavos
Languages Portuguese (official), Crioulo
Religion Christian (96% Roman Catholic, 2% Protestant)
Climate Tropical, but cooled by sea breezes; arid. Average temperature in Praia ranges from 19-25°C (66-77°F) in February/March to 24-29°C (75-84°F) in October
Main primary products Maize, beans, sweet potatoes, yams, livestock, sugar, bananas, coffee; salt, volcanic rock
Major industries Agriculture, fishing, food canning, cement
Main exports Fish, coffee, bananas, salt
Annual income per head (US$) 250
Population growth (per thous/yr) 20
Life expectancy (yrs) Male 60 **Female** 64

Caracas ***Venezuela*** Capital and largest city. Founded in 1567, it was sacked by the English admiral Sir Francis Drake in 1595. It is the birthplace of Simon Bolívar, the 19th-century nationalist leader. Under his leadership, Venezuela was the first colony to successfully rise in revolt against the ruling Spanish. The city lies in a series of valleys about 900 m (3000 ft) above sea level and close to its port, La Guaira, on the Caribbean coast. It is dominated by motorways and skyscraper office blocks, and the extensive shanty settlements of the poor.

There are a number of museums, including the Casa Natal del Libertador, which is a reconstruction of the house in which Bolívar was born. National monuments include the Panteón Nacional, where Bolívar is buried. The cathedral was built by the Spanish in 1614.

The city's main industrial products are foodstuffs, cement, plastics, automobiles, chemicals, textiles and paper.
Population 3 500 000
Map Venezuela Ba

Carboniferous Fifth of the six periods of the Palaeozoic era. See GEOLOGICAL TIMESCALE

Carcassonne ***France*** Ancient walled town, founded in the 5th century BC, lying close to the Aude river, 73 km (45 miles) north-west of Perpignan. Its medieval fortifications – with 54 towers – were restored in the 19th century.
Population 46 000
Map France Ee

Cardamom Mountains ***Cambodia*** Forested range in the south-west, which rises to 1813 m (5948 ft) at Phnom Aural, the highest point in the country. It is Cambodia's wettest area, with more than 3500 mm (140 in) of rain each year. Khmer Rouge guerrillas operate in the hills.
Map Cambodia Ab

Cardiff (Caerdydd) ***United Kingdom*** Capital of Wales, in the county of South Glamorgan. It was founded as a Roman fort on the northern shore of the Bristol Channel, and grew to become the world's greatest coal-shipping port (serving the South Wales coalfields) in the 19th century. Today it has a splendid government complex, a group of landscaped buildings laid out on a formal plan like a small-scale equivalent of the Indian capital, New Delhi. The city is also home for two colleges of the University of Wales, a cathedral in suburban Llandaff, bombed in the Second World War and since rebuilt, a castle and the Welsh National Rugby Stadium, Cardiff Arms Park – arguably the emotional heart of Wales.
Population 281 000
Map United Kingdom De

Cardigan ***United Kingdom*** See DYFED

Cardigan Bay Inlet of the IRISH SEA in west Wales, extending from the tip of the Lleyn Peninsula to St David's Head in Dyfed.
Dimensions 105 km (65 miles) long, 56 km (35 miles) wide
Map United Kingdom Cd

Caribbean Sea Tropical sea, an area of the Atlantic, lying between the north coast of South America, Central America and the Greater and Lesser Antilles of the West Indies. Following its exploration by Christopher Columbus (1451-1506), it became an arena for warring European nations, a haven for pirates and a scene of exploitation which caused the near extinction of the original inhabitants, including the ferocious Caribs, after whom it is named.

Because of the PANAMA CANAL, the Caribbean remains a busy shipping lane and it has become a major tourist area, despite such natural hazards as hurricanes, earthquakes and volcanic eruptions. Sea temperatures near the surface are around 24°C (75°F).
Area 1 943 000 km² (758 400 sq miles)
Greatest depth 7680 m (25 200 ft)
Map See p. 132

Carinthia (Kärnten) ***Austria*** Southern *Land* (state) bordering on Italy and Yugoslavia. Its capital is the industrial city of KLAGENFURT, set amid spectacular mountain scenery. The province produces lead, iron ore and cereals.
Population 536 730
Map Austria Cb

Carletonville ***South Africa*** Town on the west Witwatersrand, formed by the amalgamation of several gold-mining settlements in the 1940s. The Western Deep Levels gold mine there is the world's deepest mine, 3777 m (12 392 ft) deep.
Population 100 200
Map South Africa Cb

Carlisle ***United Kingdom*** English city 16 km (10 miles) south of the Scottish border in the north-western county of Cumbria (of which it is the county town). It stands near the western end of HADRIAN'S WALL, and changed hands

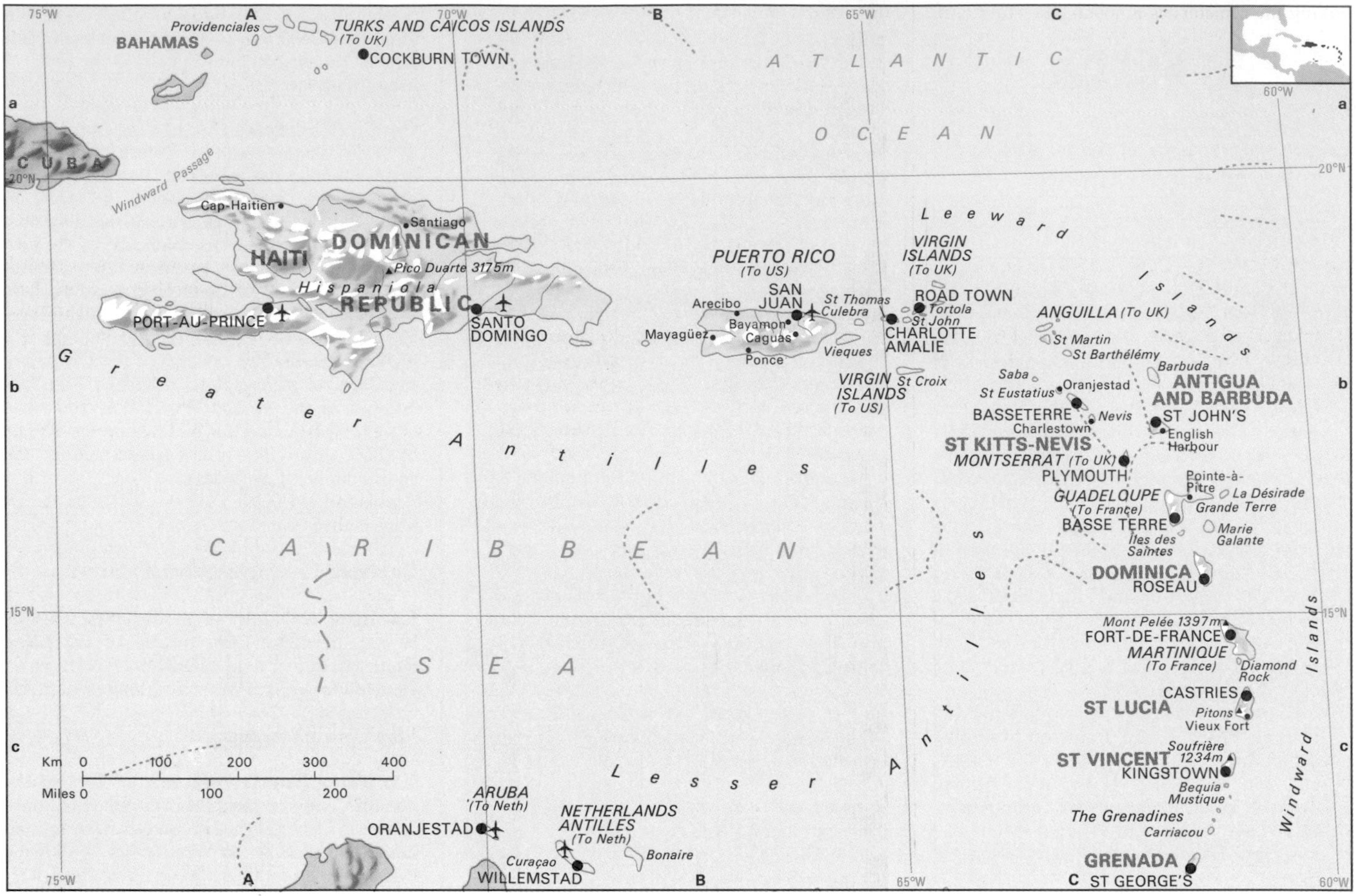

frequently between English and Scots during the Middle Ages. The castle and some parts of the city walls of this frontier fortress remain. There is a 12th to 14th-century cathedral and a 14th-century guildhall. The city flourished in the 19th century as a railway and textile centre.

Population 72 000

Map United Kingdom Dc

Carlow (Ceatharlach) ***Ireland*** Market town and county in the south-east. The county covers 896 km² (346 sq miles) and is mostly a cereal and fodder-producing plain bounded by the Blackstairs Mountains in the south-east. At Browne's Hill is Ireland's largest dolmen – a Stone Age burial chamber – with a 100 tonne capstone. The county town of Carlow lies some 70 km (45 miles) south-west of Dublin.

Population (county) 39 300; (town) 11 700

Map Ireland Cb

Carlsbad ***Czechoslovakia*** See KARLOVY VARY

Carlsbad Caverns National Park ***USA*** Limestone area containing gigantic caves and beautiful rock formations. It lies in south-eastern New Mexico, 5 km (3 miles) from the Texas border. The park covers 189 km² (73 sq miles); the largest cave is more than 800 m (0.5 mile) long and up to 200 m (650 ft) wide.

Map United States Fd

Carlsruhe ***West Germany*** See KARLSRUHE

Carmarthen ***United Kingdom*** See DYFED

Carmel, Mount (Har Karmel) ***Israel*** Mountain ridge running 22 km (14 miles) north-west from the Samarian Hills to Haifa, which stands on its northern slopes. The summit is 528 m (1746 ft) high. To the south-west is the pass of Megiddo. The prophet Elijah confronted the priests of Baal on Mount Carmel, and the Carmelite Order of mendicant friars was founded there in the 12th century.

Map Israel Aa

Carnatic ***India*** Region on the Bay of Bengal. It was seized by the British in 1751 from the Nawab of Carnatic, whose capital was the town of ARCOT. Today the region is part of TAMIL NADU and southern ANDHRA PRADESH.

Map India Ce

Caroline Islands ***Pacific Ocean*** Scattered chain of islands in the western Pacific north of the Equator. Part of the US Trust Territory of the Pacific Islands from 1947, they are now BELAU and the Federated States of MICRONESIA.

Map Pacific Ocean Bb

Carpathian Mountains ***Eastern Europe*** Arc of mountains, more than 980 km (610 miles) long and up to 290 km (180 miles) wide, which forms the Czechoslovak-Polish border and continues through the Ukraine into Romania. It includes the BESKIDY, TATRA MOUNTAINS and the Transylvanian Alps (or South Carpathians), and rises to 2663 m (8737 ft) at Gerlachovsky Stit in Czechoslovakia.

Map Europe Fd

Carpentaria, Gulf of Shallow gulf off northern Australia, forming part of the ARAFURA SEA. It is rich in animal life, including fish, birds, turtles, crocodiles and dugongs.

Dimensions 480 km (300 miles) (west-east), 600 km (375 miles) (north-south)

Map Australia Fa

Carrara ***Italy*** Town 100 km (62 miles) south-east of Genoa, and the most famous centre of marble production in the world. It has been quarried from the white-scarred Apuanian Alps since Roman times. The marble comes from some 300 quarries and is either exported in its crude state or fashioned on the spot by local and visiting craftsmen. Michelangelo (1475-1564) used Carrara marble for his most famous sculptures, including his gigantic *David*, now in the Academia gallery, Florence.

Population 68 500

Map Italy Cb

Carrickfergus ***United Kingdom*** Old seaport in Northern Ireland, standing on the north shore of Belfast Lough. It has an imposing Norman castle and a 16th-century parish church.

Population 19 000

Map United Kingdom Cc

Carrowmore ***Ireland*** Bronze Age cemetery, equivalent to Brittany's Carnac, near the north-west coast, 3 km (2 miles) south-west of Sligo. More than 100 tombs once littered the meadows. Now about 60 remain. The others have been lost to farming and collectors. Overlooking the cemetery is Knocknarea Mountain (328 m, 1076 ft), topped by Maeve's Lump, a huge cairn of stones said to be the burial place of the 1st-century Queen Maeve of Connaught.

Map Ireland Ba

carse Scottish term for an alluvial plain beside a river or an estuary.

Carso ***Yugoslavia*** See KRAS

Carson City ***USA*** Capital of Nevada, standing in the south-west of the state only 20 km (12 miles) from the border with California. The city takes its name from 'Kit' Carson (1809-68), a frontiersman who helped to open up the Wild West, and was a distinguished soldier in the Civil War (1861-5). The city is a popular tourist centre for gambling and the nearby mountains and lakes.

Population 35 900
Map United States Cc

Cartagena ***Colombia*** Historic port and capital of Bolivar department on the Caribbean coast about 95 km (60 miles) south-west of Barranquilla. It was founded by the Spanish in 1533 as their fortress base on the South American continent, and became the largest port for the shipping of gold and silver to Europe.

Despite its four forts ringing the town and harbour – including the massive San Felipe de Barajas on a 42 m (135 ft) clifftop above the sea – it was sacked by the French and English. The old walled city of tiny streets is well preserved. Beyond it is a modern city of docklands and petrochemical plants.

Population 548 000
Map Colombia Ba

Cartagena ***Spain*** Port and naval base on the south-east Mediterranean coast, exporting locally mined lead and iron. Industries include smelting, metalworking, shipbuilding, oil refining, chemicals, glass and bicycle making. Founded in 227 BC by the Carthaginians, Cartagena had a variety of rulers. It was rebuilt after the English sacked it in 1585, and still has an ancient castle and Gothic cathedral.

Population 172 800
Map Spain Ed

Cartago ***Costa Rica*** City and, until 1823, the national capital, about 25 km (15 miles) east of the present capital, San José. The Basilica of Our Lady of the Angels contains La Negrita, an Indian image of the Madonna about 150 mm (6 in) high, which attracts pilgrims from all over Central America. The surrounding province of Cartago was the first area in Costa Rica to be settled by the Spanish, in 1563. It is now a coffee-growing and dairy-farming province, with most farms around the Irazú volcano.

Population (city) 75 500; (province) 270 000
Map Costa Rica Bb

Carthage ***Tunisia*** Ancient port and city, whose ruins stand on a promontory north of Tunis Lagoon. Carthage was founded by colonists from the Phoenician kingdom of Tyre in 814 BC. In the 6th century BC, it invaded and conquered other parts of North Africa as well as Sicily and Sardinia. Later, as its maritime power increased, Carthage came to dominate the western Mediterranean. The city is thought to have contained more than 700 000 people at the height of its wealth and power. However, Carthage came to be involved in a bitter struggle with the expanding Rome (the Punic Wars) and was eventually defeated, after which the city was razed by the Romans in 145 BC. A Roman city was built on the site but this was destroyed too, in 689, by the Arabs. Little remains now apart from the Roman baths of Antoninus, the ruined harbour and an expanse of rubble. Much of the classical site is now a suburb of Tunis.

Map Tunisia Ba

Casablanca (Dar el Beida) ***Morocco*** Largest city and principal port and commercial centre of the country (handling about 75 per cent of Morocco's foreign trade), situated on the Atlantic coast 100 km (60 miles) south of Rabat. It was founded in 1515 by the Portuguese on the site of Anfa, an ancient Berber city they had destroyed in 1468. By 1830 the population was 600 and had reached only 27 000 by 1900. Under the French, who occupied it in 1907, Casablanca grew into a huge city and port.

There is a small medina (old Muslim residential quarter) in the old city with many curio shops, and ramparts surround the harbour. The commercial section is a modern bustling European-style place of wide streets, office blocks and big stores. The city has plenty of green spaces, parks, beaches and tourist facilities. Outside the city are growing numbers of *bidonvilles* (shanty towns), for Casablanca has acted as a magnet for the country's unemployed.

Population 2 140 000
Map Morocco Ba

Cascade Range ***Canada/USA*** Mountain range in north-western North America. It stretches some 1125 km (700 miles) from northern California, through the American states of Oregon and Washington and into southern British Columbia, Canada. Most of the peaks are between 1600 and 2500 m (5250 and 8200 ft) high, but Mount Rainier in Washington state rises to 4392 m (14 410 ft).

Map United States Bb

Cascais ***Portugal*** Fishing port and resort, about 25 km (17 miles) west of the capital, Lisbon. Just to its west is the Boca do Inferno ('Mouth of Hell'), where the sea churns into a deafening cauldron. According to legend, a local fisherman, Alfonso Sanches, reached America in 1482 – ten years before Christopher Columbus. Columbus is said to have planned his voyages west with the help of Sanches's logs.

Population 30 000
Map Portugal Bc

Caserta ***Italy*** City about 30 km (20 miles) north of Naples. Its main claim to fame is the sumptuous Palazzo Reale, built by the architect Vanvitelli for Charles the Bourbon in 1752-74. The palace has 6 storeys, 1200 rooms and 1970 windows, and is surrounded by a Versailles-scale park with grand avenues of trees and huge staircases of water, all profusely adorned with marble statues. Caserta Vecchia, a medieval town with a Romanesque cathedral, stands in the hills above.

Population 66 800
Map Italy Ed

Cashel (Caiseal) ***Ireland*** Market town 79 km (49 miles) north-east of the southern city of Cork, and seat of the kings of Munster for 700 years from about AD 400. On the Rock of Cashel, 61 m (200 ft) high, a 12th-century cross stands near the spot where St Patrick is said to have held aloft a shamrock to illustrate the Trinity, and so gave Ireland its emblem. The cross was erected on the stone on which Brian Boru, the scourge of the Danes, was crowned King of Munster in the early 11th century. Ruined buildings on the Rock include a round tower, cathedral and archbishops' castle. The buildings, burned in the 15th and 17th centuries, were finally abandoned in 1749.

Population 2540
Map Ireland Cb

Caspian Sea Occupying a huge depression in the USSR and Iran, the Caspian Sea is the world's largest inland body of water. It measures about 1200 km (745 miles) from north to south and 352 km (218 miles) at its widest point. Its surface is 28 m (92 ft) below sea level. The URAL and VOLGA rivers flow into it, but water escapes only through evaporation. In recent years, water taken from these rivers upstream has caused a fall in the level of the sea and a reduction in its overall area. The salt content is generally low, except in the shallow eastern gulf, the Kara-Bogaz Gol. Here the rate of evaporation is so high that the salt content is ten times that of normal seawater.

The sea is tideless and its northern parts are frozen for two or three months in winter. Its waters are rich in salmon and sturgeon – the source of caviar.

Area 393 900 km^2 (152 100 sq miles)
Map USSR Fd

Cassai ***Angola/Zaire*** See KASAI

Cassel ***West Germany*** See KASSEL

Cassino ***Italy*** Market town midway between Rome and Naples. It was called Casinum in Roman times. The Benedictine Abbey of Monte Cassino, founded in the 6th century on a peak overlooking the town, was totally destroyed by the Allies when Axis forces defended it for six months in 1944, during the Second World War. It has since been painstakingly rebuilt – as has the town itself.

Population 31 100
Map Italy Dd

cassiterite The chief ore of tin – tin oxide – found in acid, igneous rocks or in alluvial deposits derived from such ore-bearing rocks.

Castel Gandolfo ***Italy*** Town in the hills about 20 km (12 miles) south-east of Rome, overlooking Lake Albano. Its palace is the summer residence of the Pope.

Population 6000
Map Italy Dd

Castellammare di Stabia ***Italy*** Seaport and resort on the Bay of Naples with Greek, Etruscan and Roman origins. The ancient Stabia – the forerunner of the present town – was one of the first sites in the world to be excavated by modern archaeologists.

Population 70 300

Castile (Castilla) ***Spain*** Largely arid region of central and north-central Spain covering some 138 472 km² (53 464 sq miles). Castile was an ancient kingdom which expanded from the 11th century by the conquest of several Moorish kingdoms, and is divided into two parts: Old Castile in the north and New Castile in the south. It formed the nucleus of modern Spain, and Castilian Spanish is regarded as the purest form of the language. The region consists of plains surrounded and divided by mountains, with some fertile areas in the south, and incorporates important cities and provinces, including MADRID, VALLADOLID and TOLEDO.

Map Spain Cb

Castries ***St Lucia*** Seaport and capital city, on the north-west coast. First settled by the British in 1605, it was largely destroyed by a severe hurricane in 1780 and by fires in 1796, 1813, 1927, 1948 and 1951. There are a number of fortifications built by the British in and around Vigie, just to the north, and to the south on the hill called Morne Fortune.

The city's industries include processing limes, sugar, coconut oil and cocoa beans. There is a botanical research establishment.

Population 45 000

Map Caribbean Cc

Çatal Hüyük (Çatalhöyük) ***Turkey*** Site of a Stone Age city about 250 km (150 miles) south of the capital, Ankara. Dating from 6700 BC, it is one of the world's oldest known towns. Its inhabitants lived in mud-brick houses entered through a hole in the roof. The site was discovered by the British archaeologist James Mellaart in the early 1960s.

Map Turkey Bb

Catalonia (Cataluña) ***Spain*** Hilly, industrial autonomous region in the north-east, covering some 31 934 km² (12 330 sq miles) and including the COSTA BRAVA holiday area. Settled by the Phoenicians and the Greeks, it was united with ARAGON in the 12th century and with CASTILE in the 15th. Its historical capital is BARCELONA. It has its own language, Catalan, and was autonomously governed during the Civil War of 1936-9, when it fought against Franco's Nationalist forces. Its manufactures include chemicals, metal goods and textiles, and it produces olive oil, wine, almonds and fruit.

Population 5 958 000

Map Spain Fb

Catamarca ***Argentina*** Hill resort about 900 km (560 miles) north-west of Buenos Aires. Founded in 1683, the town now attracts visitors because of its mountain setting, its hot springs and its local handicrafts, particularly woollen ponchos.

Population 60 000

Map Argentina Ca

Catania ***Italy*** Port and second city of Sicily after the capital, Palermo. It stands on the east coast, where the southern slopes of Mount ETNA meet the sea. The fiery volcano dominates the life of Catania, hanging over it like a rumbling giant. The city has been destroyed twice by earthquakes – in 1169 and 1693 – and once inundated by lava, in 1669. It is a centre for the island's food and light-engineering industries. The city, with its grid plan of about 1700, has many monuments made of black lava.

The Bellini Gardens, named after the opera composer Vincenzo Bellini, who was born in Catania in 1801, provide a retreat from the bustle and traffic. Ursino Castle, built by the Holy Roman Emperor Frederick II in 1232, was reduced to a keep by the 1669 lava flow.

Population 378 500

Map Italy Ef

cataract A series of rapids on a stretch of river as on the Nile. The term was formerly used for any large waterfall.

Catbalogan ***Philippines*** See SAMAR

Catskill Mountains ***USA*** Group of mountains on the west side of the Hudson river valley, about 120 km (75 miles) north-west of New York city. Their beautiful scenery, fish-filled rivers, hotels, picnic areas and winter sports resorts are popular with city dwellers. They rise on average to 900 m (2950 ft), but Slide Mountain is 1281 m (4204 ft) high.

Map United States Lb

Cattaro ***Yugoslavia*** See KOTOR, GULF OF

Cau Mau ***Vietnam*** Swampy, lowland peninsula forming the country's southern tip. River sediments swept by coastal currents make an ever-changing shoreline which, in parts, is estimated to be advancing by up to 75 m (about 250 ft) each year. Boatmen fish the streams and the muddy flatlands, which are fringed by mangrove forests and produce prolific rice crops.

Map Vietnam Ad

Caucasus (Kavkaz) ***USSR*** Region between the Black Sea and the Caspian Sea consisting of the North Caucasus, the Great Caucasus, and Transcaucasia. It covers an area of about 399 500 km² (154 250 sq miles) and is divided between three Soviet republics – Armenia, Azerbaidzhan and Georgia – plus part of the RSFSR.

The North Caucasus is mainly plain, devoted to crops of cereals and cotton. The Great Caucasus – like Transcaucasia – is mountainous. The Caucasus range is generally regarded as part of the natural boundary between Europe and Asia. It contains Europe's highest point: Mount Elbrus (5642 m, 18 510 ft), which lies to the north – the European side – of the main ridge. The mountains also contain deposits of oil, manganese and other minerals.

Map USSR Fd

causse French term for limestone plateau country, specifically the Grands Causses in the south-west of the MASSIF CENTRAL of France. It is very much like KARST in form, but is generally wetter.

Causses ***France*** Limestone plateaus in the southern Massif Central, noted for their gorges. Summers are hot, but winters harsh. Much land in this sparsely populated area is grazed by sheep whose milk is used for Roquefort cheese. The main towns are Millau and Rodez.

Map France Ed

Cauvery ***India*** See KAVERI

Cavan (An Cabhán) ***Ireland*** Rural county and town in the province of Ulster in the north-east. The county covers 1890 km² (730 sq miles) and, dotted with lakes and rivers, is a centre for coarse fishing. Cavan, the county town, lies 103 km (64 miles) north-west of Dublin.

The old town was destroyed in 1690 by Irish followers of William of Orange after the Battle of the BOYNE. A 3rd-century defensive wall, Black Pigs Dyke, runs from the town westwards to Lough Allen.

Population (county) 53 900; (town) 3260

Map Ireland Cb

Caveri ***India*** See KAVERI

caves See opposite and p. 136

Cavite ***Philippines*** Naval base first established by the Spanish on the southern shore of Manila Bay. A walled town with many historic buildings, it was badly damaged during the US invasion of the Philippines in 1944 during the Second World War. On May 1, 1898, American forces destroyed the Spanish fleet off Cavite in the Battle of Manila Bay, helping to ensure American victory in the Spanish-American War. The province of Cavite (capital Trece Martires) extends to Tagaytay City, overlooking Lake Taal in the south.

Population (city) 75 800; (province) 628 400

Map Philippines Bc

Cawnpore ***India*** See KANPUR

cay Small, low islet composed largely of coral or sand. Cays are also known as keys.

Cayenne ***Guyane*** Capital of the country, standing on an island in the estuary of the Cayenne river. It is Guyane's principal port, handling timber, rum, pineapples and shrimps. There are many handsome villas along the beautiful coast. Fort Diamant, built in 1652, and the 18th-century residence of the Prefect (administrator) are among the city's architectural landmarks.

Population 38 000

Map Guyane Bb

Cayman Islands ***Caribbean*** Three islands forming a British Crown Colony north-west of Jamaica. The islands – Grand Cayman, Little Cayman and Cayman Brac – cover a total land area of 260 km² (100 sq miles).

An economic revolution in the 1960s made the islands, formerly occupied with fishing and farming, into an international financial and tourist centre. By 1983, favourable company and financial laws had attracted over 17 500 registered companies and 450 banks. At the same time, tourism – which now provides some 70 per cent of the gross national product – has made beach front land prices on Grand Cayman among the highest in the world.

Immigrants, mostly from Jamaica and North America, make up about one in five of the population. Another one-fifth of the island's people is of European descent, one-fifth is African, and the rest are of mixed race.

Population (Grand Cayman) 17 000; (Little Cayman) 100; (Cayman Brac) 2000

Map Cuba Ab

Ceanannus Mór ***Ireland*** See KELLS

BENEATH THE SURFACE OF THE EARTH

Much of the world's land surface is honeycombed with caves – natural hollows beneath the ground. They mostly consist of air-filled chambers joined by long, low crawlways – or of long, narrow galleries linked by pits. Deeper caves often contain streams, waterfalls and lakes – beneath some of which are complex cave systems. The cave system lying below Mammoth Cave National Park, Kentucky, USA, is the world's most extensive – some 484 km (301 miles) of passages have been measured so far.

THE WORLD'S DEEPEST NATURAL CAVES

Cave	Location	Depth (m/ft)
Gouffre Jean Bernard (Réseau de Foillis)	France	1535/5036
Atea Kananda	Papua New Guinea	1500/4920
Snezhnaya	USSR	1470/4823
Puerta de Illamina	Spain	1338/4390
Sistema Huautla	Mexico	1246/4088
Schwersystem	Austria	1219/3999

Ceatharlach *Ireland* See CARLOW

Cebu *Philippines* One of the central VISAYAN ISLANDS covering 5088 km² (1964 sq miles) to the east of Negros. Long and narrow – about 225 by 40 km (140 by 25 miles) – it has a central mountain chain rising to 1013 m (3324 ft). Once forested, the highlands are now largely cleared and the soil severely eroded in parts. This is partly because Cebu was the first Philippine island to be settled by the Spanish, in 1565, although it was well populated and prosperous long before that.

Its capital, Cebu city on the west coast, was the first capital of the Philippines and even today is second only to Manila as a commercial centre. It is the country's main copra port and also handles trade in maize, rice, fish and other produce, acting as the regional port for the Visayas and northern Mindanao.

Offshore, protecting Cebu harbour, is the small Mactan island, site of the island's airport but better known as the place where the Portuguese explorer Ferdinand Magellan was killed on April 27, 1521 by local natives.

Population (island) 2 092 000; (city) 490 231
Map Philippines Bc

Cechy *Czechoslovakia* See BOHEMIA

Celaque, Sierra de *Honduras* Mountain range 160 km (100 miles) west and slightly north of the capital, Tegucigalpa. It contains the country's highest peak, Cerro de las Minas (2849 m, 9115 ft).

Map Honduras Ab

Celebes *Indonesia* See SULAWESI

Celebes Sea An arm of the PACIFIC OCEAN bordered by Borneo, the SULU ARCHIPELAGO and MINDANAO in the Philippines, the Sangi Islands in the east and Celebes (SULAWESI) in the south. Plate movements along the MINDANAO trench to the east cause earthquakes and volcanic activity in the area.

Dimensions 800 km (500 miles) (east-west), 640 km (400 miles) (north-south)
Map Indonesia Fb

Celje *Yugoslavia* See KARAWANKE ALPS

Celle (Zelle) *West Germany* Town about 35 km (22 miles) north-east of Hanover. It is a horse-breeding centre and has a parade of stallions in the streets each October. The town was largely untouched during the Second World War and contains a castle in late Gothic style with 17th-century Baroque additions, the picturesque old town of half-timbered houses, the 14th and 17th-century town church and a late Renaissance town hall.

Population 75 000
Map West Germany Db

Celtic Sea Oil prospectors, anxious to avoid giving political offence, have named the waters covering the continental shelf between southern Ireland, southern Wales and south-western England the Celtic Sea. Until prospecting began, the area had no specific name.

Map United Kingdom Be

Cenozoic era The most recent geological era in which mammals and flowering plants first flourished. See GEOLOGICAL TIMESCALE

Central African Republic See p. 137

Central Kalahari Game Reserve *Botswana* Part of the KALAHARI desert, a wilderness of sandy savannah with scattered thorn trees and inhabited by a few Bushmen hunting groups. The 51 800 km² (20 000 sq mile) park has no roads and can be entered only with special permission. Wildlife includes gemsbok, hartebeest and springbok. Joined to the park in the south is the Kutse Game Reserve, covering about 2590 km² (1000 sq miles). Visitors to either must have four-wheel-drive vehicles and carry their own supplies, including water.

Map Botswana Bb

Central Region *United Kingdom* Local government area of Scotland, created in 1975 from the former county of Clackmannan and parts of Stirlingshire and Perthshire. It covers 2590 km² (1000 sq miles). The principal towns are STIRLING, Falkirk, and the industrial centre of Grangemouth on the River Forth, with its oil refining and chemical plants. Both shores of the Forth have been extensively mined for coal, and some of Scotland's earliest ironworks were there, but the field has been largely worked out. The region reaches into the Scottish Highlands, culminating at the peaks of Ben More (1174 m, 3851 ft) and Ben Lui (1130 m, 3707 ft).

Population 273 000
Map United Kingdom Cb

▼ AWAY FROM IT ALL Some 300 000 tourists visit the Cayman Islands each year, yet, with spacious beaches such as Grand Cayman's Seven Mile Beach, they remain a Caribbean paradise.

HOW CAVES ARE FORMED

Caves have fascinated men from the days when they used to live in them. The famous painted caverns of Lascaux, in the French Dordogne, were occupied at least 20 000 years ago. But their discovery and opening in 1940 began a slow deterioration of the paintings that has led to them being closed.

Prime cause of the damage was simply contact with the outside atmosphere – its moisture content, the chemicals it contained and its temperature. For these are limestone caverns, owing their very existence to the slow, steady action of water – and the chemicals dissolved in it.

Limestone lies close to the earth's surface in many parts of the world. As rain falls and percolates through soil it gathers carbon dioxide, forming a weak acid which dissolves limestone. Streams penetrate cracks and crevices in the rock, gradually widening them into sinkholes, opening to the surface. Galleries form along horizontal breaks between layers of rock. Sometimes galleries are enlarged to chambers by rock falls from their roofs.

Over thousands of years the rock may become honeycombed into labyrinths of interconnected chimneys and chambers, pothole shafts and galleries. Streams, disappearing down sinkholes, escape through the lower galleries, often emerging some distance away. Seeping water embellishes caverns with elaborate formations as it erodes the limestone: dissolving minerals on the way, it drips from ceilings and spreads across walls, depositing minerals on their surfaces as it evaporates. The deposits take on an array of colours, depending on the minerals that were dissolved. Ceiling drips form stalactites on the roof; stalagmites grow up from the floor; sometimes they meet to form grotesque columns.

On coastlines, relentless tides and pounding seas do the job on many types of rock. Sea salt hastens erosion; molluscs and other sea creatures bore into the rocks; algae and weeds secrete destructive chemicals. But breaking waves do most of the work, driving into fissures and hollows, battering weak points, sucking away debris and boulders as they retreat, then hurling them back to carve sea caves and dramatic overhangs.

Caves may also be formed by volcanic action. Pahoehoes, for example, occur where surface lava cools, and the inner hot, molten lava flows away to leave a cavity. They may also result from molten lava solidifying above a mass of ice, which then melts, leaving a cave below. Or they may be blasted through solidified walls of old lava by new material.

The largest cave is the Big Room in Carlsbad Caverns, New Mexico, which is 1300 m (4265 ft) long, 200 m (656 ft) wide, 100 m (328 ft) high and 400 m (1312 ft) below the ground. Britain's deepest cave is 308 m (1010 ft) Ogof Ffynnon Ddu, in Wales, where a rock column rises to the roof from the middle of an underground lake.

CAVES IN LIMESTONE COUNTRY

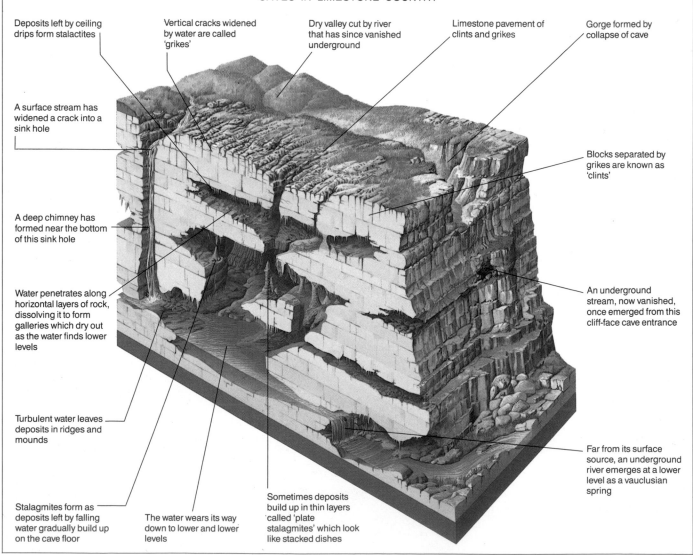

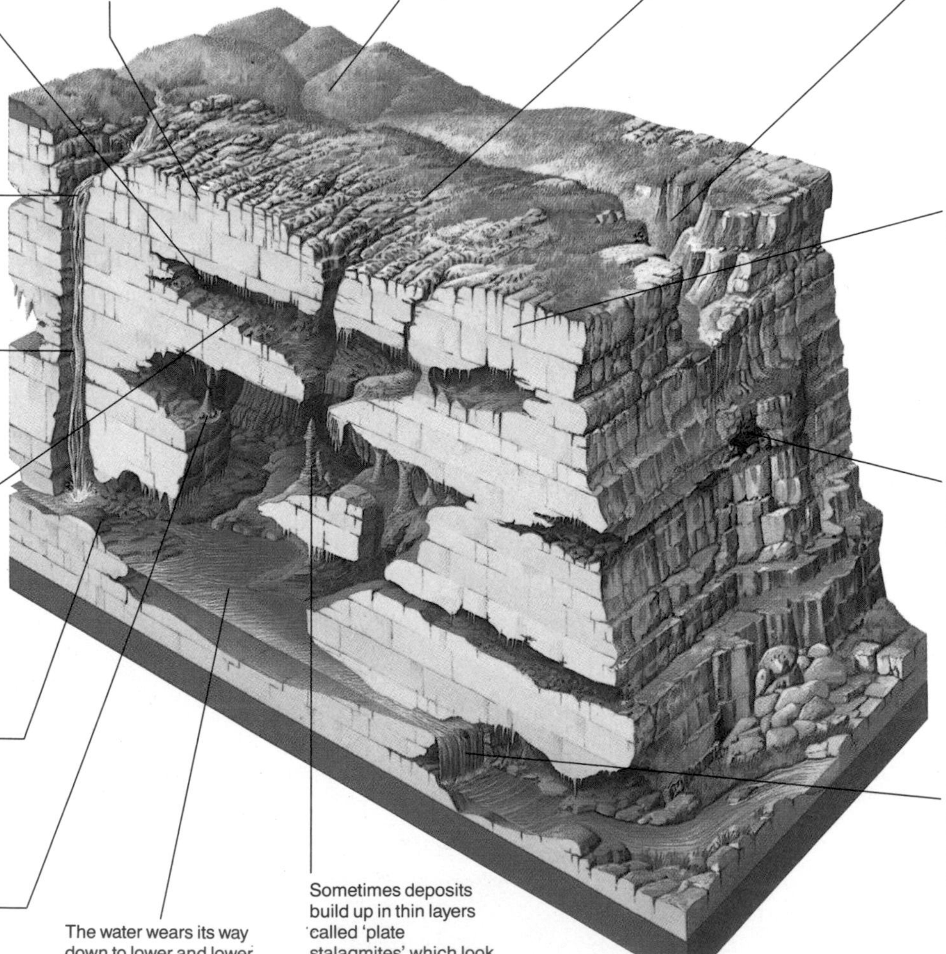

Deposits left by ceiling drips form stalactites
Vertical cracks widened by water are called 'grikes'
Dry valley cut by river that has since vanished underground
Limestone pavement of clints and grikes
Gorge formed by collapse of cave
A surface stream has widened a crack into a sink hole
Blocks separated by grikes are known as 'clints'
A deep chimney has formed near the bottom of this sink hole
Water penetrates along horizontal layers of rock, dissolving it to form galleries which dry out as the water finds lower levels
An underground stream, now vanished, once emerged from this cliff-face cave entrance
Turbulent water leaves deposits in ridges and mounds
Far from its surface source, an underground river emerges at a lower level as a vauclusian spring
Sometimes deposits build up in thin layers 'called 'plate stalagmites' which look like stacked dishes
Stalagmites form as deposits left by falling water gradually build up on the cave floor
The water wears its way down to lower and lower levels

Central African Republic

RECOVERING FROM THE RULE OF A MEGALOMANIAC 'EMPEROR', THE REPUBLIC REMAINS UNDEVELOPED AND POVERTY-STRICKEN

Until 1977, the remote, landlocked Central African Republic in the very middle of Africa was rarely mentioned in the world news. But in that year it sprang into the headlines with the coronation on December 4 of its vainglorious leader, Jean-Bedel Bokassa, who had proclaimed himself emperor (and the country an empire) a year earlier. The bizarre ceremony was estimated to have cost US$22-30 million at a time when most of his people lived on an average income of around US$5 a week.

Formerly known as Ubangi Shari, a territory of French Equatorial Africa, the country became independent as the Central African Republic in 1960. Although it has rich mineral resources – diamonds in particular account for 25 per cent of exports – about 90 per cent of the people lead frugal lives as farmers cultivating small plots of cassava, maize, millet, sorghum, sweet potatoes and yams. Only 5 per cent of the land is cultivated. Livestock rearing is small-scale because of the prevalence of the disease-carrying tsetse fly.

Much of the country is rolling grassland and bush on a high plateau 600-900 m (about 2000-3000 ft) above sea level, with mountainous areas on the western border with Cameroon and the eastern border with Sudan. In the drier north the plateau slopes down to the CHARI (or Shari) river across the northern border with Chad. In the south it slopes towards the UBANGI (Oubangui) river which forms much of the southern border with Zaire and flows into the ZAIRE (Congo). Most of the people live in the west and the hot, humid south and south-west – many of them in the dense south-western rain forest, where it rains nearly every day from March to November. The forest is the source of another major export – hardwood timber. The largely uninhabited eastern half of the country contains two of the three national parks as well as various game reserves.

There are no railways in the republic. Much of the transport is by water – there are some 7000 km (2300 miles) of inland rivers – particularly the Ubangi, which is navigable for most of the year except for a few months at the height of the November-March dry season. BANGUI, the capital and chief port, is on the Ubangi, and from it goods are shipped some 1200 km (740 miles) to Brazzaville in the Congo Republic, then sent 500 km (310 miles) by rail to the Congo port of POINTE NOIRE on the Atlantic coast.

Apart from the Babinga (pygmies) of the rain forest, there are eight main tribal groups, the two largest being the Banda and Baya who make up half the population. Seven of the groups speak Sudanese languages, but the widely spoken national language is Sango, the language of the Ubangians. These, although they make up only 5 per cent of the population, have dominated the government and civil service since independence.

Jean-Bedel Bokassa is a Ubangian, as is his predecessor David Dacko, who was president at independence. Bokassa overthrew Dacko in a military coup in 1966 and ruled for 14 years, during which he reduced the country to bankruptcy and the elephant population by three-quarters. Indeed, much of his personal fortune was made from ivory and diamonds. Bokassa's extravagance, combined with accusations of brutality, led to his downfall. Riots occurred in 1979 when schoolchildren protested at having to buy uniforms costing about US$20. Some 250 children were detained by the police and about 100 of them murdered; Bokassa was said to have personally taken part. He was deposed in a French-supported coup while visiting Libya in September 1979.

David Dacko became president again but unrest continued. Dacko was deposed again in September 1981 in a coup led by General André Kolingba, who has tried to stamp out corruption and foster economic recovery.

Despite long-standing government attempts to eradicate old customs and tribal affinities, traditional values remain strong outside the city and the Westernised elite. Only about one-third of the people can read and write, and more than half follow the old tribal religions. France remains the republic's chief trading partner and the main source of aid and investment. But while foreign interests influence the economy – as in the exploitation of diamonds and in plans to develop vast uranium deposits discovered near Bakouma, 480 km (300 miles) east of Bangui – little wealth reaches the population at large.

CENTRAL AFRICAN REPUBLIC AT A GLANCE

Area 622 984 km² (240 535 sq miles)
Population 2 740 000
Capital Bangui
Government One-party military republic
Currency CFA franc = 100 centimes
Languages French (official), Sango
Religions Christian (35%), tribal religions (57%), Muslim (8%)
Climate Tropical; average temperature in Bangui ranges from 21-29°C (70-84°F) in July/August to 21-34°C (70-93°F) in February
Main primary products Cassava, groundnuts, bananas, plantains, sweet potatoes, maize, millet, coffee, cotton, timber; diamonds, gold, uranium (unexploited)
Major industries Agriculture, forestry, mining
Main exports Timber, coffee, diamonds, cotton, ivory, gold
Annual income per head (US$) 280
Population growth (per thous/yr) 28
Life expectancy (yrs) Male 42 **Female** 44

Centre ***France*** Farmland region to the south of Paris, astride the Loire river. It comprises Beauce, part of the LOIRE valley and BERRY. The main towns are Orléans and Tours.
Population 2 264 000
Map France Dc

Cephalonia (Kefallinia) ***Greece*** Largest of the IONIAN ISLANDS, extending to 782 km² (302 sq miles). Its hilly countryside lined with fir trees, vine and olive groves, and its ragged coast of little bays, draws tourists. Many visit the island's stalactite cave with its coloured underground lake, Melisani. White wine is produced on the island and perfumes are made from its flowers.
Population 27 600
Map Greece Bb

Ceram ***Indonesia*** See SERAM

Ceram Sea See SERAM SEA

Cernauti ***USSR*** See CHERNOVTSY

Cërrik ***Albania*** Principal oil-refining centre of Albania since 1950. It is linked to DURRES port by a railway started by the occupying Italians during the Second World War and completed after the war with Russian aid.
Population 12 500
Map Albania Bb

Cerro de Pasco ***Peru*** Town 4350 m (14 270 ft) up in the Andes mountains, 160 km (100 miles) north-east of Lima. It was founded in 1771 as a silver town, but now copper, zinc, bismuth and tungsten deposits have made it the country's biggest mining centre. A new town, San Juan de Pampa, was founded in 1965, 2 km (1.25 miles) away, to cope with the growth.
Population 71 500
Map Peru Bb

cerussite Naturally occurring lead carbonate. The mineral is the main source of lead.

Cerveteri ***Italy*** Town about 30 km (20 miles) north-west of Rome. There are traces of Etruscan walls in the medieval castle and an Etruscan necropolis to the north, with many splendid monumental tombs dating from the 7th to 1st centuries BC.
Population 12 100
Map Italy Dd

Cervino ***Italy/Switzerland*** See MATTERHORN

Ceské Budejovice (Budweiss) ***Czechoslovakia*** South Bohemian city on the Vltava river, 125 km (78 miles) south of Prague. Budvar beer has been produced in the city since the Middle Ages, and Budweiser is its American version. Just north of the city lies Hluboka Castle, built in Tudor style in the 19th century and modelled on England's Windsor Castle. The village of Husinec to the north-west was the birthplace of John Huss (Jan Hus, about 1369-1415), the Czech hero and religious reformer, who was burnt as a heretic.
Population 92 800
Map Czechoslovakia Bb

Cesky Tesin ***Poland*** See CIESZYN

Cetinje ***Yugoslavia*** Major tourist centre, and until 1945 the capital of the Yugoslav republic of Montenegro. It lies more than 600 m (2000 ft) up in a basin on the KRAS plateau, 18 km (11 miles) north-west of Lake Shkodër. The fortified monastery, home of the town's prince-bishops from 1515 to 1851, has fine collections of icons and books, including one of the first printed in a Slav language (1493).

The Biljarda, a palace built in 1836 for the poet Bishop Petar Njegos II, is now a museum with ethnographic collections and the Museum of National Liberation. The state museum is housed in a second palace, once the home of Nicholas I, Montenegro's last independent ruler (1860-1918). He used to sell the principality's postage stamps from a palace window.
Population 12 100
Map Yugoslavia Dc

Ceuta ***Spanish enclave in Morocco*** Duty-free seaport on the northernmost peninsula of Morocco (Punta Almina), facing Europe. It is administered as part of Spain's CADIZ province. The Mediterranean lies on two sides of the town, and Jebel Musa (Mount Hacho) provides a spectacular view of the STRAIT OF GIBRALTAR; together with GIBRALTAR itself, this 842 m (2762 ft) mountain was known to the ancient Greeks as one of the Pillars of Hercules guarding the entrance to the Mediterranean.
Population 80 000
Map Morocco Ba

Cévennes ***France*** Rocky mountainous area, cut by deep gorges, along the southern fringe of the Massif Central. The highest peaks are Mont Lozère (1702 m, 5584 ft) and Mont Aigoual (1567 m, 5141 ft). Wide stretches of the hills are given over to sheep grazing and forestry. Sunny slopes are, however, terraced for vines, fruit and chestnuts. A national park covers 2360 km^2 (911 sq miles) of farmland and rough country in the area.
Map France Ed

Ceylon See SRI LANKA

Chad, Lake ***Equatorial Africa*** Shallow lake in north-west central Africa, shared by Cameroon, Chad, Niger and Nigeria. Numerous rivers drain into it, but it has no outlet, and during the wet season from May to October it expands, to cover about 26 000 km^2 (19 040 sq miles), making it the ninth largest freshwater lake in the world. Evaporation reduces it to half that size by April, and sometimes it shrinks to 10 000 km^2 (3860 sq miles). Geological evidence shows that in the Ice Ages the lake was as big as 310 000 km^2 (120 000 sq miles) – almost the size of the Caspian Sea – and even in historical times it extended as far as the Bodélé depression. Today its maximum depth is about 7 m (23 ft), and much of its surface is choked with floating vegetation.

The area round Lake Chad has always been sparsely populated because of the arid soils and because it was once a no-man's-land between rival states. However, the lake is much used by animals, including herds of cattle. Rice is traditionally planted on flooded land, and sorghum as the lake recedes. Groundnuts and maize are also grown. The lake is a valuable source of fish, salt and potash.
Map Chad Ab

Chaillu Mountains ***Congo/Gabon*** Range in south-eastern Gabon and south-western Congo. Named after the French-born American explorer Paul du Chaillu, who explored the area in 1855-65, they include Mount Iboundji, Gabon's highest peak at 1575 m (5167 ft). The mountains form a watershed between the N'GOUNIE and OGOOUE river systems, which unite at LAMBARENE.
Map Congo Ab

Chainat ***Thailand*** Market town and province 175 km (110 miles) north of Bangkok. The province is a fertile rice-growing area.
Population (town) 67 700; (province) 335 400
Map Thailand Bb

Chalcidice (Khalkidhikí) ***Greece*** Wooded peninsula in MACEDONIA, with three finger-like small peninsulas reaching into the Aegean Sea south of SALONIKI. They are named Kassándra, Sithoniá and ATHOS. Long, sandy beaches have made Chalcidice a popular tourist resort and there are good camping sites.

On the Gulf of Kassándra, 8 km (5 miles) west of Gerakini, lies the ancient city of Olynthos, destroyed by Philip II of Macedon in 348 BC and unoccupied ever since.
Population 79 000
Map Greece Ca

chalk Soft, whitish limestone consisting almost entirely of calcium carbonate. Chalk is derived chiefly from the remains of small marine organisms.

Challenger Deep See MARIANAS TRENCH

Châlons-sur-Marne ***France*** Capital of Marne department, 150 km (93 miles) east of Paris on the plain of Champagne. Chief town of the ancient Catalaunic tribe, it was occupied and fortified by the Romans. At the Battle of Châlons on plains to the south in AD 451, the Romans defeated the Barbarian invader Attila the Hun. Major activities today include making beer and champagne, food processing, electrical engineering and textiles.
Population 54 400
Map France Fb

Chambéry ***France*** Capital of Savoie department, 45 km (28 miles) north of Grenoble. It manufactures aluminium and cement. The old town, with its arcaded streets and imposing houses, is dominated by a huge 13th-century chateau, once the home of the Dukes of Savoy and now open to the public.
Population 100 340
Map France Fd

Chambord ***France*** Village 45 km (28 miles) south-west of Orléans. It is famous for its chateau, the largest in the Loire valley. The chateau was begun in 1519 as a royal hunting lodge and – although still unfinished – has 440 rooms and 50 staircases.
Map France Dc

Chamonix ***France*** Winter sports resort in the Arve valley, 57 km (35 miles) east of Annecy. It is the nearest town to the highest Alpine peak, Mont Blanc (4807 m, 15 771 ft). Six major glaciers and over 20 smaller ones flow towards its valley. It was the site in 1924 of the first Winter Olympic Games.
Population 11 000
Map France Gd

Champagne-Ardennes ***France*** Region to the east and north-east of Paris, crossed by the Aisne, Aube, Marne and Meuse rivers. The sparkling white wine which takes its name from the Champagne district was first produced there in the 18th century. Cereals, dairy goods and textiles are also produced. RHEIMS is the largest city. The wooded hills in the north of the region extend across the Belgian and Luxembourg borders to form the forest of the ARDENNES.
Population 1 346 000
Map France Fb

Champassak ***Laos*** Southern province straddling the Mekong river. It produces more rice than it needs, and most of the surplus is smuggled across the border to Thailand.
Population 103 000
Map Laos Bc

Champion ***Brunei*** One of three offshore oil fields in the South China Sea to the north-west of an older onshore production area, SERIA. With the South-west Ampa and Fairley fields, Champion now accounts for almost all the country's oil production. All three fields lie between 13-32 km (8-20 miles) north-west of Seria.

Chan Chan ***Peru*** Capital of the pre-Inca Chimú Empire, about 5 km (3 miles) north of Trujillo. The adobe ruins cover some 20 km^2 (8 sq miles). There are nine great citadels, each surrounded by a 10 m (33 ft) wall and 300 m (985 ft) apart. Each compound contains temples, houses, workshops and store rooms, and most have a large well. Relics, including clothes and pottery, are on show in the museums of the capital, Lima.
Map Peru Ba

Chandannagar (Chandernagore) ***India*** Town on the Hugli river, on the north-west fringe of Calcutta. It was founded as a trading post in 1688 by the French East India company. In 1949, whilst still a French colony, it voted to become part of newly independent India.
Population 101 900
Map India Dc

Chad

NATIONAL SURVIVAL IS THE KEY ISSUE IN CHAD, ONE OF THE WORLD'S POOREST NATIONS, WEAKENED BY CIVIL WAR AND DROUGHT

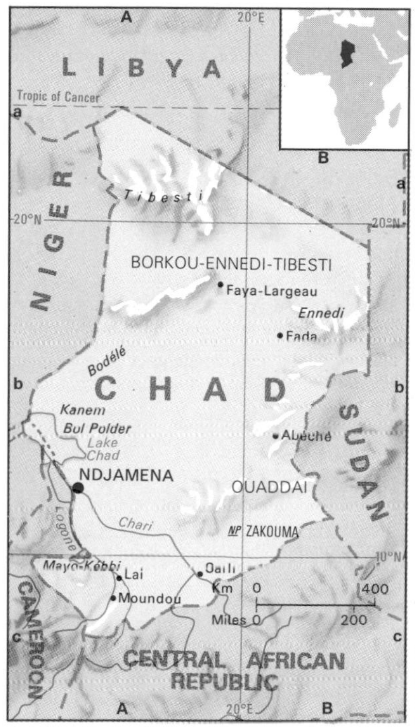

At the root of Chad's tragedy in modern times has been its geographical location. Landlocked in the centre of northern Africa and spanning the semiarid SAHEL, the country extends from the edge of the equatorial forests for some 1800 km (1100 miles) north to the middle of the SAHARA. Chad has suffered only slightly less than Ethiopia in the African drought that began in 1980. And with a journey of more than 1600 km (1000 miles) from the nearest seaports – in some cases through unfriendly neighbouring countries – plus woefully inadequate transport, it has been extremely difficult to ferry in relief aid.

This tragedy has been overlaid, as in Ethiopia, by internal political conflict. Chad stands at the junction of Arabic and black African cultures, and has a multiplicity of ethnic groups which, with entirely different ways of life, religions and allegiances – have long been political rivals. In 1965, the resulting instability boiled over into full-scale civil war.

Chad is more than twice the size of France, its former colonial ruler, and lies mostly in a vast basin draining (when there is rain) into Lake CHAD, on the western border. The most dramatic landforms are the TIBESTI mountains, in the far north, whose twisted peaks rise from the desert sands to more than 3000 m (10 000 ft); the extinct volcano of Emi Koussi, at 3415 m (11 204 ft), is the highest.

The southern part of the country is the least poverty-stricken although it is the most densely populated. Relatively well watered, its wooded savannah has always been Chad's main arable region, but even here the rains failed in 1984. In normal times it is farmed for cotton (the country's main cash crop), millet, sorghum, groundnuts, rice and vegetables, mainly by the predominant Sara, among whom the older people are distinguished by the bold patterns of scars on their faces, a custom that is now dying out. Also in the south and south-west are groups who fish the rivers and Lake Chad for a living. These southerners mainly follow tribal religions or Christianity, while the north is strongly Muslim.

In physical features, the northern peoples resemble the Mediterranean Arabs, though some are very dark-skinned. They are largely nomadic or seminomadic graziers (drought permitting) – of cattle and sheep in the Sahel, mainly of camels in the far north. The drought has driven them farther south or to the outskirts of towns or oases, wherever they can find water.

The Islamic influence is to be seen throughout Chad in the craftsmanship of its superb carpets, handwoven and embroidered cottons and other goods. Much more ancient artworks found on rocks in Tibesti and elsewhere show that, some 3000 years ago, the climate was wet enough for dense vegetation, and for hippos and rhinos to live where there is now desert. Much later, from 1000 to 400 years ago, the Sao people from the Nile Valley settled near Lake Chad and left remains of distinctive pottery. In the 16th and 17th centuries, three major Muslim kingdoms were established in the Sahelian region and became traders in slaves captured in the south. The most durable of these, the Wadai, was conquered by the French only in 1911.

French forces entered Chad in the 1890s and were welcomed as protectors by the southern tribes. Declared a separate French colony in 1920, Chad was granted independence in 1960. The northern prefecture of Borkou-Ennedi-Tibesti – always a centre of resistance to colonial rule – remained under French military control until 1965, when it joined the rest of Chad. Since then there has been continuous instability which developed into all-out civil war.

Libya backed one faction, Egypt and Sudan backed the other, and French peace-keeping troops also became involved. By 1985, when famine rather than politics was the preoccupation of most Chadians, there was an uneasy stalemate, with the Libyan-backed forces controlling the northern third of the country and the capital, NDJAMENA.

CHAD AT A GLANCE

Area 1 284 000 km² (about 496 000 sq miles)
Population 5 380 000
Capital Ndjamena
Government One-party republic
Currency CFA franc = 100 centimes
Languages French, Arabic, African tribal languages
Religions Muslim (52%), Christian (5%), tribal religions (43%)
Climate Tropical; north arid to semiarid, wet season in south, May to October. Average temperature in Ndjamena ranges from 14-33°C (57-91°F) in December to 23-42°C (73-108°F) in April
Main primary products Cotton, cattle, millet, sorghum, rice, cassava, fish; natron (sodium carbonate)
Major industries Agriculture, textiles, food processing, mining, fishing
Main exports Cotton, cattle, meat, processed fish
Annual income per head (US$) 100
Population growth (per thous/yr) 25
Life expectancy (yrs) Male 42 **Female** 44

Chandernagore ***India*** See CHANDANNAGAR

Chandigarh ***India*** City about 230 km (143 miles) north of Delhi. It was the capital of the states of Punjab and Haryana, although it belonged to neither. This came about after the British Punjab was divided between India and Pakistan in 1947, Pakistan retained the capital of Lahore, and India built the new, modernistic city of Chandigarh, designed by the controversial Swiss-born French architect Le Corbusier (1887-1965). However, Indian Punjab was Hindu in the east and Sikh in the west, and it finally split into the western state of Punjab, and what are now the states of Haryana and Himachal Pradesh. Both Punjab and Haryana laid claim to Chandigarh, but it was given to neither and was designated a Union territory, a property of the Federal Government, until 1986, when it was to be ceded to Sikh Punjab. However, Haryana was not satisfied with the compensation offered for its loss and the scheduled transfer was put off.
Population 422 800
Map India Bb

Chang Jiang (Yangtze) ***China*** Third longest river in the world after the Amazon and the Nile. It rises on the Xizang Gaoyhan (Tibetan Plateau) and flows 6380 km (3965 miles) to the East China Sea at Shanghai. The Chang Jiang drains the central part of China and is an important source of irrigation and hydroelectric power. It is navigable for ocean-going ships 1100 km (680 miles) to Wuhan and for smaller craft 2253 km (1400 miles) to Chongqing. The river carries between 60 and 70 per cent of the country's inland waterways trade. Flooding is frequent, but rarely disastrous because of a network of natural regulating lakes on its reaches either side of Wuhan.
Map China Ed, He

Changchun ***China*** Capital city and industrial centre of Jilin province, about 900 km (560 miles) north-east of Beijing (Peking). It produces motor vehicles, locomotives and railway rolling stock. During the 1930s, it was the capital of the Japanese puppet state of Manchukuo.
Population 1 604 000
Map China Ic

Chang-hua ***Taiwan*** Old city near the central west coast amid a wealthy agricultural region that produces rice, asparagus and pineapples. On a hill overlooking the city is a 20 m (66 ft) high statue of Buddha.
Population 1 206 400
Map Taiwan Bb

Changi ***Singapore*** District at the eastern tip of Singapore Island. It is the site of Singapore's new international airport, completed in 1981, partly on land reclaimed from the sea. Changi Jail was a notorious prison for Allied prisoners of the Japanese in the Second World War.
Map Singapore Ab

Changsha ***China*** Capital of Hunan province about 550 km (340 miles) north of Guangzhou (Canton). It is a centre for traditional handicrafts such as ceramics, embroidery and bamboo carving, and for food processing, textiles, machine tools and precision engineering. The late Chinese Communist leader Mao Zedong (1893-1976) was a student and teacher there from 1913 to 1923.
Population 2 638 000
Map China Ge

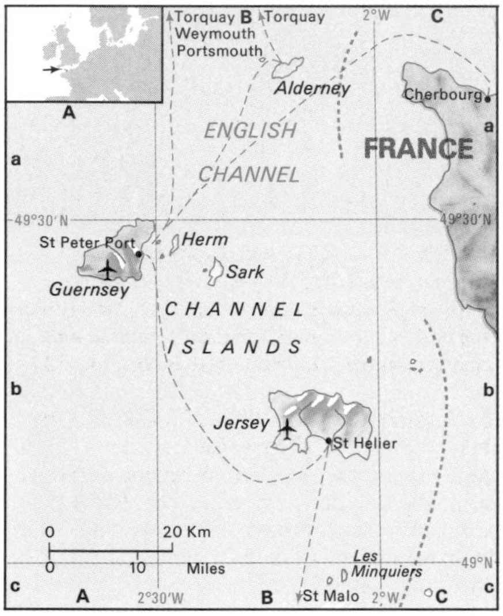

Channel Islands ***English Channel*** Self-governing group of islands under British sovereignty lying just off the north-west coast of France, where they are called Les Iles Anglo-Normandes. The islands were part of the Duchy of Normandy in the 10th century, and were linked with England by the Norman Conquest of 1066.

The largest island, Jersey (117 km², 45 sq miles), is little more than 20 km (12 miles) from the Cotentin peninsula. Most place-names are French, and many of the islanders speak it. Jersey and Guernsey (78 km², 30 sq miles) are large enough for road networks and ports – St Helier on Jersey and St Peter Port on Guernsey. Alderney (8 km², 3 sq miles) has few roads and one village, St Anne, of granite-cobbled streets. There are no cars (only horse and tractor-drawn carriages) on Sark, which covers 5.5 km² (2.1 sq miles), or on tiny Herm.

The mild climate draws many tourists, and the low taxes attract wealthy British businessmen, but immigration is tightly controlled. The islands produce spring flowers and vegetables, especially tomatoes, for the mainland.
Population (Jersey) 77 000; (Guernsey) 55 000; (Alderney) 2000; (Sark) 610

Chanthaburi ***Thailand*** Market town and gem-mining centre 200 km (125 miles) south-east of Bangkok. The surrounding province of the same name produces tropical fruits.
Population (town) 93 250; (province) 381 950
Map Thailand Bc

Chao Phraya ***Thailand*** River, 1000 km (620 miles) long, running from the northern hills southwards through the capital and into the Bight of Bangkok. Its flood plain forms the country's main rice-growing area.
Map Thailand Bc

Chaouèn (Xauen) ***Morocco*** Small, attractive town perched on a rocky prominence in the RIF MOUNTAINS, 60 km (37 miles) south of Tétouan. It has been called the 'town of fountains' because of its many mineral water springs. The centre of the Riff Berbers, Chaouèn dates back to the 15th century, although only the city centre remains from that era. With its many mosques and sanctuaries, it is a Muslim holy city much frequented by pilgrims. It has a rug factory.
Population 24 000
Map Morocco Ba

Chapala, Lake ***Mexico*** The largest lake in Mexico, 2460 km² (950 sq miles), on the high plateau of Jalisco state, near GUADALAJARA. It has tourist facilities and is used for fishing, sailing and waterfowling.
Map Mexico Bb

chaparral Dry scrub vegetation consisting of evergreen scrub oaks, vines and sparse grasses, found especially in the south-western United States and Mexico. It is similar to MAQUIS.

Chari (Shari) ***Equatorial Africa*** River rising in the Central African Republic and flowing some 1000 km (625 miles) north-west to enter Lake Chad. The Chari is Chad's only permanent river, though navigable only in the wet season, and is much valued for irrigation.
Map Central African Republic Aa; Chad Ab

Charleroi ***Belgium*** Industrial city about 50 km (30 miles) south of the capital, Brussels. The main industries are steel, electrical machinery and glassware.
Population 213 000
Map Belgium Ba

Charleston ***USA*** Chief port of South Carolina and one of the country's oldest cities, founded in 1670. The Civil War broke out at nearby Fort Sumter in 1861. A number of old forts and 18th-century buildings remain, and there are many plantation homes in the area. The city gave its name to the popular 1920s dance.
Population (city) 67 100; (metropolitan area) 472 500
Map United States Kd

Charleston ***USA*** State capital of West Virginia. It lies in the west of the state, about 50 km (30 miles) from the Ohio border, and was founded in 1794. The city is the centre of a coal and chemical producing region, and also manufactures glass.
Population (city) 59 400; (metropolitan area) 267 000
Map United States Jc

Charlestown ***St Kitts and Nevis*** Chief town of Nevis, on the south-west coast. It is linked to Basseterre on St Kitts by ferry. St John's Church, in Fig Tree Village to the south of the town, records the marriage of Horatio Nelson, the British admiral, to Frances Nisbet on March 11, 1787. The town's hot springs are the main tourist attraction.
Population 1800
Map Caribbean Cb

Charleville-Mézières ***France*** Capital of Ardennes department on the Meuse river, only about 12 km (7 miles) from the Belgian border. Charleville was founded in the 17th century on the opposite bank of the river from the medieval walled town of Mézières. In 1966 they merged into a single city. The main industry is the manufacture of metal goods.
Population 70 800
Map France Fb

Charlotte Amalie ***US Virgin Islands*** Capital of the territory, standing beside a large and protected harbour on the island of St Thomas. It has some attractive 19th-century buildings.
Population 11 800
Map Caribbean Cb

Charlottetown ***Canada*** Capital of PRINCE EDWARD ISLAND, on its south coast, overlooking Hillsborough Bay. It is a commercial centre built around fishing, tourism and industries producing knitwear, dairy products, canned goods and timber.
Population 45 000
Map Canada Id

Chartres ***France*** Capital of the Eure-et-Loir department, in the plain of Beauce about 80 km (50 miles) south-west of Paris. Its vast cathedral, the sixth on the site, was begun in 1194 and consecrated in 1260. It is famous for the size and beauty of its stained-glass windows – in particular for the luminous colour known as 'Chartres blue' which is used in them. The cathedral's most precious relic is a veil said to have been worn by the Virgin Mary. It was given to the cathedral in AD 876 by Charles the Bald, King of the West Franks and grandson of the first Holy Roman Emperor Charlemagne (ruled 800-814).
Population 80 340
Map France Db

Chatham ***United Kingdom*** See MEDWAY TOWNS

Chatham Islands ***New Zealand*** Island group in the Pacific, about 850 km (530 miles) east of Christchurch. It is made up of Chatham Island, Pitt Island, and several rocky islets. The chief town is Waitangi, on Chatham Island. Sheep farming is the main livelihood.
Population 750
Map Pacific Ocean Ed

▲ **HIGH GOTHIC The cathedral at Chartres – most of it built in less than 30 years – towers over the city, its spires dominating the plain around. Its stained glass is probably the finest in the world.**

Chattanooga ***USA*** Industrial city and railway centre in Tennessee, immortalised in the popular song, *Chattanooga Choo Choo*. It stands on a bend in the Tennessee river on the southern border of the state. The commercial bottling of Coca Cola soft drinks originated there. Near the city is Lookout Mountain (648 m, 2126 ft), which provides spectacular views and was the site of a decisive Union victory in 1863 during the Civil War.

Population (city) 164 400; (metropolitan district) 422 500
Map United States Ic

Chaumont ***France*** Capital of Haute-Marne department, 220 km (137 miles) south-east of Paris. The town manufactures gloves, hosiery and textiles. In 1814 Britain, Austria, Russia and Prussia signed the Treaty of Chaumont in the town binding themselves to pursue the war against the French Emperor Napoleon. French president General Charles de Gaulle (1890-1970) died at Colombey les Deux Églises, 20 km (12 miles) to the north-west. His memorial is a huge Cross of Lorraine on a nearby hillside.

Population 30 600
Map France Fb

Chaux-de-Fonds, La ***Switzerland*** Town about 50 km (30 miles) west of the capital, Berne. It is the centre of the Swiss watchmaking industry, and houses a watch and clock museum and a watchmakers' school. It was the birthplace of the revolutionary modern architect Le Corbusier (1887-1965).

Population 33 000
Map Switzerland Aa

Chaves ***Portugal*** City and spa about 120 km (75 miles) north-east of the city of Oporto and 10 km (6 miles) south of the border with Spain. It was fortified by the Romans and was known as Aquae Flaviae (the Flavian Springs). Roman remains include an 18-arch bridge and two inscribed columns. The city has several fine churches and a castle of the Dukes of Braganza. Chaves is an agricultural and textile centre, and manufactures silks and linens.

Population 11 900
Map Portugal Cb

Chavín de Huantar ***Peru*** Ruined fortress temple 3200 m (10 000 ft) up in the White Cordillera of the Andes, centre of the Chavín culture which flourished from 800 to 200 BC. These South American Indians worshipped a feline god which is sometimes depicted mingled with characteristics of other animals in stone carvings around the temples, pyramids and plazas. The Temple of Lanzón contains a blackish, dagger-shaped stone monolith dating from about 800 BC.

Map Peru Ba

Cheb (Eger) ***Czechoslovakia*** Bohemian bicycle-making town, 5 km (3 miles) from the West German border and due west of Prague. Its medieval town centre survives intact.

The spa of Frantiskovy Lázne (Franzensbad), just to the north, has a mineral water spring with the highest concentration in the world of Glauber's salt – sodium sulphate – which is used as a laxative.

Population 31 300
Map Czechoslovakia Aa

Chechen-Ingush Republic ***USSR*** Autonomous republic of the RSFSR covering 19 300 km^2 (7450 sq miles) on the northern slopes of the Caucasus Mountains, east of the Black Sea. Its industries, based on the Groznyy oil field, include engineering, chemicals and food canning. GROZNYY is the capital.

Population 1 204 000
Map USSR Fd

Cheju Do (Jejudo; Quelpart) ***South Korea*** Semitropical island covering 1828 km^2 (706 sq miles). It lies 90 km (56 miles) off Korea's south coast. Citrus fruits are grown and cattle and Mongolian horses are raised. The island is dominated by Halla san, at 1950 m (6398 ft) the highest mountain in South Korea.

Population 463 000
Map Korea Cf

Chelm (Chelm Lubelski; Kholm) ***Poland*** Rapidly growing industrial city in the south-east, about 25 km (16 miles) from the Soviet border. It was founded in the 13th century, and became a cathedral city and flourishing market for the rich farmlands around. The Polish Committee of National Liberation, the embryo Communist government of postwar Poland, was established in the city in July 1944.

Population 58 100
Map Poland Ec

Chelmno (Kulm; Culm) ***Poland*** Medieval city on a promontory above the Vistula river 220 km (135 miles) north-west of the capital, Warsaw. Its encircling 13th-century ramparts with 17 bastions, its red-brick Gothic churches, and its Renaissance town hall and market square all survive.
Population 20 700
Map Poland Cb

Chelmno-nad-Nerem ***Poland*** See KONIN

Chelmsko Slaski ***Poland*** Small Sudetenland town near the Czech border, about 60 km (37 miles) south-west of the city of Wroclaw. Its market square is surrounded by fine burghers' houses, a Baroque church and arcaded rows of weavers' wooden houses. All date from the 17th and 18th centuries, when the town was a major textile centre. Hand-woven fabrics are still sold.

Chelyabinsk ***USSR*** City on the eastern fringe of the Ural mountains about 200 km (125 miles) south of Sverdlovsk. Its main industries are iron and steel, vehicles, engineering and chemicals.
Population 1 086 000
Map USSR Hc

chemical weathering Processes which cause the break-up of rocks through chemical reactions such as solution, oxidation and hydration.

Chemnitz ***East Germany*** See KARL-MARX-STADT

Chengde ***China*** Town in Hebei province about 270 km (165 miles) north-east of Beijing (Peking). It is the site of the summer home of the Manchu emperors, the dynasty which ruled China from 1644 until 1912, when it was overthrown by republican Nationalists. There are several palaces, pavilions and 18th-century temples, some of them in the Tibetan style.
Map China Hc

Chengdu (Chengtu) ***China*** Capital of Sichuan province, in one of the richest agricultural regions in China. Its industries include silk, electronics and timber processing.
Population 2 470 000
Map China Fe

Chenonceaux ***France*** Village in the Indre-et-Loire department, famous for its elegant Renaissance chateau which stands astride the Cher river. The chateau was begun in 1515 by Thomas Bohler, the French finance minister, and later enlarged by Diane de Poitiers and Catherine de Médicis – respectively the mistress and the wife of the French King Henry II. Every year some 800 000 visitors view its magnificent paintings and tapestries, and walk in the formal gardens.
Map France Dc

Chenstokhov ***Poland*** See CZESTOCHOWA

Cheonam ***South Korea*** See CH'ONAN

Cherbourg ***France*** Channel port and naval dockyard on the north coast of the Cotentin peninsula. It became an important naval base and transatlantic liner port after 1860, but now deals mainly with cargo boats and cross-Channel ferries. The town has been largely rebuilt since 1945.
Population 89 200
Map France Cb

Cherchell ***Algeria*** Coastal town and fishing port, 95 km (59 miles) west of Algiers. It is built on the site of a Carthaginian settlement, which became a Roman town called Caesarea in 25 BC. The town has some fine Roman remains, including a theatre and an arena.
Population 17 000
Map Algeria Ba

Chergui, Chott ech ***Algeria*** Marshy salt lake, 150 km (93 miles) long, situated south-east of Oran in the High Plains of the north-west, between the Tell Atlas and the Saharan Atlas. This large saucer-shaped depression provides rough pastureland for mountain pastoralists.
Map Algeria Ba

Chernigov ***USSR*** Ukrainian town and port on the Desna river, about 130 km (80 miles) north-east of Kiev. It is an agricultural centre and produces textiles, knitwear, footwear, chemicals, musical instruments and electrical goods. It has a 12th-century Byzantine cathedral.
Population 270 000
Map USSR Ec

Chernikovsk ***USSR*** See UFA

Chernobyl' ***USSR*** See KIEV

Chernovtsy (Cernauti) ***USSR*** City in the Ukraine, near the Romanian border and 225 km (140 miles) south-east of L'vov. Its industries include sawmilling, engineering, food processing, textiles and rubber products. Formerly the Austrian city of Czernowitz, it became part of Romania (Cernauti) in 1919 and the USSR (first as Chernovitsy, then Chernovtsy) in 1940.
Population 238 000
Map USSR Dd

Cherrapunji ***India*** Reputedly the wettest inhabited place on earth. The village lies on the south side of the Khasi Hills in the north-eastern state of Meghalaya and receives the full impact of the monsoon rains. The average annual recorded rainfall is 10 874 mm (428 in).
Map India Ec

chert A crystalline form of silica in which the crystals are so small that they can be seen only under a powerful microscope. It usually occurs in sedimentary rocks in bands or as layers of pebbles. FLINT is one variety of chert.

Chesapeake Bay The largest inlet on the east coast of the USA, rich in historical associations. English settlers entered the bay in 1607 and went on to found JAMESTOWN. Captain John Smith explored the bay in 1608, expressing the view that 'heaven and earth never agreed better to frame a place for man's habitation'. Ports include BALTIMORE in the state of Maryland, and NORFOLK and Portsmouth in Virginia. The bay is popular with yachtsmen and is famed for its oysters and other seafood.
Dimensions 314 km (195 miles) long, 5-48 (3-30 miles) wide
Map United States Kc

Cheshire ***United Kingdom*** County covering 2322 km^2 (897 sq miles) of north-west England, stretching from the Welsh border and the Dee estuary across the River Mersey to the industrial centre of Warrington in the north and to the Pennine hills in the east. Its undulating farmland is ideal for dairy herds – hence Cheshire cheese – and there is an unusual wealth of half-timbered houses. CHESTER, the county town, Runcorn, Ellesmere Port, CREWE and Macclesfield are its industrial centres. Rock salt deposits have been worked for centuries at Nantwich and Northwich. Altrincham and STOCKPORT, formerly in Cheshire, were transferred to Greater Manchester in 1974.
Population 933 000
Map United Kingdom Dd

Chester ***United Kingdom*** County town of the north-west English county of Cheshire, 55 km (35 miles) south-west of Manchester. It has been described as the best preserved walled city in Britain, and receives many tourists. It was founded by the Romans as Deva in the 1st century AD. The present walls were built mainly on the foundations of the Roman fortifications and were started in the 10th century.

The city has one of the country's finest Roman amphitheatres, a castle, a red sandstone cathedral dating back to the 12th century, many fine timbered houses and unusual two-storey medieval shopping galleries called The Rows. During the 13th and 14th centuries, the city was a key regional port, but it declined because of silting in the River Dee.
Population 88 000
Map United Kingdom Dd

Cheviot Hills ***United Kingdom*** Range extending about 60 km (37 miles) along the border of Scotland with the English county of Northumberland. They reach 816 m (2676 ft) at The Cheviot. Most of the hills were open moorland until reafforestation began after the First World War, creating among others Kielder Forest – one of the largest man-made forests in Europe – at the south-western end of the range.
Map United Kingdom Dc

Cheyenne ***USA*** State capital of Wyoming, in the south-east of the state, 15 km (9 miles) from the Colorado border. It was founded in 1867 and, as a rail junction, played a key role in opening up the west. Its history is commemorated in an annual Frontier Days festival.
Population 50 900
Map United States Fb

Chiang Mai ***Thailand*** See CHIENG-MAI

Chiang Rai ***Thailand*** Province and market town in the far north, about 60 km (35 miles) from the Burmese and Laotian borders. Farmers in the province mostly grow rice, but much of the town's prosperity comes from cross-border trade and from smuggling links with the heroin-producing GOLDEN TRIANGLE region.
Population (town) 194 850; (province) 980 700
Map Thailand Ab

Chianti ***Italy*** Hilly wine-making district about 40 km (25 miles) south of Florence. It is famous for its full-bodied red wines.
Map Italy Cc

Chiapas ***Mexico*** State near the Guatemalan border in the extreme south of Mexico, on the Gulf of Tehuantepec. Now one of the poorest Mexican states, it was once at the core of the great civilisation of the Mayas which lasted from as early as 1500 BC to the 12th century. Several Indian groups such as the Chamulas, the Zinacantecans and the Huistecos still speak their ancient languages and dress and live according to ancient customs, undisturbed by any outside influences. Although agriculture is backward, Chiapas exports tropical products such as hardwoods, fruit, coffee, cotton, cocoa and rubber. Its capital is Tuxtla Gutiérrez.

Population 2 200 000

Map Mexico Cc

Chiba ***Japan*** Industrial city on Honshu island about 35 km (22 miles) east of Tokyo. Large tracts of land have been reclaimed from Tokyo Bay for this fast-developing city, which specialises in steelmaking and oil refining.

Population 788 900

Map Japan Dc

Chicago ***USA*** The country's third largest city after New York and Los Angeles. It stands in Illinois on the south-west shore of Lake Michigan, and is the country's transport hub. Railways radiate in every direction; it is a major Great Lakes port and is connected to the Mississippi river by a ship canal and the Illinois river; and its airport is the busiest in the world. Its stockyards and grain elevators handle huge quantities of foodstuffs. Industries include iron and steel, and electronics. Chicago is known as 'Windy City', and it became infamous during Prohibition in the 1920s for its alcohol bootleggers, such as the gangsters Al Capone and Bugsy Moran.

Population (city) 2 992 500; (metropolitan area) 8 035 000

Map United States Ib

Chichén Itzá ***Mexico*** Important archaeological site in Yucatán combining building genius of both the Mayas and the Toltecs. Its most remarkable feature is the four-sided El Castillo pyramid, where at the spring and autumn equinoxes, light falling on its carefully constructed terraces casts a shadow image of the feathered Toltec snake god Quetzalcóatl winding down its steps. Among its other notable buildings are the Caracol – an ancient Mayan observatory, and the Toltec Temple of the Warriors – a vast complex of columns in the form of serpents.

Map Mexico Db

Chichester ***United Kingdom*** Cathedral city in southern England and county town of West Sussex, about 90 km (55 miles) south-west of London. It was founded by the Romans as Noviomagus in the 1st century AD and still retains its Roman plan, with an almost complete ring of later city walls. Just to the west are the remains of Fishbourne Palace, a major Roman relic. Chichester's cathedral is mainly 12th and 13th century, with a spire reaching 84 m (275 ft). There are many 18th-century buildings in the area, including Goodwood House and its racecourse. The city's fine, modern Festival Theatre opened in 1962.

Population 25 000

Map United Kingdom Ee

Chichicastenango ***Guatemala*** Market town and religious centre 90 km (56 miles) north-west of the capital, Guatemala City, and 2071 m (6795 ft) up in the Quiché mountains. Each Thursday and Sunday thousands of Indians from the surrounding Maya villages, clad in magnificent costumes, descend on the town. They come for the markets and for the religious ceremonies – part Christian, part pre-Christian – held outside the 16th-century Santo Tomás Church.

Population 56 000

Map Guatemala Ab

▼ PHOENIX ASCENDANT Chicago's skyscrapers, including the 100-storey John Hancock Center, hug the shore of Lake Michigan. The modern city arose from the ashes left by a fire in 1871 which made 90 000 homeless.

Chiclayo ***Peru*** Capital of Lambayeque department, 640 km (400 miles) north of Lima. It was originally an Indian stronghold, then a Franciscan friary and now a bustling modern city which is the centre of a conurbation that includes the towns of Lambayeque and Ferreñafe. Its industries include timber mills and food-processing factories.
Population 280 000
Map Peru Ba

Chiemsee (Chiem) ***West Germany*** Largest lake in Bavaria, 80 km^2 (31 sq miles) in area, about 65 km (40 miles) south-east of Munich. On Herrenchiemsee, one of three islands in the lake, is a palace, built in 1878-85 by Ludwig II of Bavaria and modelled on Versailles, near Paris. Concerts are held in the ballroom on Saturdays in the summer.
Map West Germany Ee

Chieng-Mai ***Thailand*** City in the north-west, about 120 km (75 miles) from the Burmese border. Founded in 1292, it still has some of its medieval walls. The surrounding province grows garlic, tobacco, strawberries and other fruit on irrigated farms. Chieng-Mai is also popular with tourists visiting the tribal areas of the Hmong, Yao and Karen peoples in the surrounding hills.
Population (city) 200 700; (province) 1 261 000
Map Thailand Ab

Chieti ***Italy*** Medieval town about 145 km (90 miles) east and slightly north of Rome. It has Roman remains and a fine cathedral founded in the 11th century with 14th century and modern additions.
Population 55 600
Map Italy Ec

Chihuahua ***Mexico*** **1.** The country's largest state sharing its northern border with New Mexico and Texas in the United States. It is known for a breed of small, hairless dog, which is named after the state. Cattle ranching and mining, especially of zinc, lead and silver, are the main industries. **2.** Capital city of the state of Chihuahua, situated 1524 m (5000 ft) above sea level. It has derived its wealth ever since colonial times from silver and cattle ranching. Chihuahua was the home of Pancho Villa (1877-1923), the revolutionary leader, and Father Miguel Hildago, leader of the Mexican Independence movement, was executed there by the Spaniards in 1811.
Population (state) 2 232 000; (city) 410 000
Map Mexico Bb

Chile See p. 146

Chillán ***Chile*** Town 360 km (225 miles) south of Santiago, in the fertile Central Valley. It is an important agricultural centre. Although it was largely destroyed by earthquakes in 1835 and 1939, it still retains a few of its Spanish colonial buildings, including the birthplace of the revolutionary Bernardo O'Higgins (son of the Irish-born Viceroy of Chile and Peru, Ambrosio O'Higgins), who liberated Chile from Spanish rule in 1817. It has a modern cathedral and a market specialising in local handicrafts.
Population 124 000
Map Chile Ac

Chiltern Hills ***United Kingdom*** Ridge of chalk hills in southern England, arcing around the north and west of London and rising to more than 260 m (850 ft) near Wendover. London's suburbs have now reached into its valleys and have been restricted from further growth only by rural protection laws. The hills run between Luton in Bedfordshire and the River Thames west of Reading.
Map United Kingdom Ee

Chi-lung ***Taiwan*** See KEELUNG

Chimborazo ***Ecuador*** Central province, in the Andes. It includes the impressive snow-capped inactive volcano of the same name, which at 6310 m (20 700 ft) is the highest point in Ecuador and the farthest summit from the centre of the earth. Because of the bulge in the earth at the Equator, Chimborazo's summit is 2150 m (7050 ft) farther from the earth's centre than the summit of Mount Everest, which at 8848 m (29 028 ft) is the world's highest mountain as measured from sea level. The summit of Chimborazo was first reached in 1880 by Edward Whymper, an English wood-engraver and mountaineer. The provincial capital is RIOBAMBA; it lies at a height of 2700 m (8860 ft) on a flat plain fringed by high, extinct volcanic peaks.
Population 330 000
Map Ecuador Bb

Chimbote ***Peru*** The country's largest fishing port, at the mouth of the Río Santa 400 km (250 miles) north of Lima. It is the site of Peru's first iron and steel plant (opened 1958) using ore from southern Peru and which is fuelled by hydroelectric power from Cañon del Pato in the Santa valley.
Population 216 600
Map Peru Ba

chimney Steep vertical cleft in a rock face, of great use to climbers. Sometimes also applied to the vent of a volcano.

Chimoio ***Mozambique*** Market town and capital of the landlocked central province of Manica. Formerly called Vila Pery, the town lies on the railway linking the port of Beira – some 190 km (120 miles) to the south-east – with Harare, neighbouring Zimbabwe's capital. The province covers 61 661 km^2 (23 801 sq miles). It rises to highlands on the Zimbabwe border, where Binga Mountain, Mozambique's highest peak, reaches 2436 m (7992 ft). Subsistence farming and cattle rearing are the chief activities.
Population (town) 5000; (province) 666 800
Map Mozambique Ab

China See p. 148

china clay Greyish-white clay produced by chemical alteration of feldspars in granite. It is also known as kaolin, and is used in porcelain, white earthenware, tiles and as a coating for paper.

China Sea Part of the PACIFIC OCEAN, consisting of two areas separated by the Formosa Strait. The East China Sea (Tung Hai in Chinese) lies between the north coast of China and Japan's Kyushu and Ryukyu islands. The South China Sea (Nan Hai) lies between southern China and south-eastern Asia, Borneo and the Philippines.

Once infested by pirates, the China Sea is a major trading region, with such great ports as SHANGHAI, HONG KONG, MANILA, BANGKOK and SINGAPORE on its shores.
Area (East China Sea) 1 248 000 km^2 (281 850 sq miles); (South China Sea) 2 318 000 km^2 (894 980 sq miles)
Map (East) Asia Kf; (South) Asia Jg

Chindwin ***Burma*** Tributary of the Irrawaddy river. It rises in the far north of the country and flows south, parallel to the Irrawaddy. The Chindwin is navigable by small boats for more than half its 1130 km (700 mile) course. It joins the main Irrawaddy stream near the towns of Myingyan and Pakokku.
Map Burma Bb

chine A fissure in a cliff composed of soft earthy material.

Chingola ***Zambia*** Commercial town about 300 km (185 miles) north and slightly west of the capital, Lusaka. It was founded in 1943 to service the Nchanga copper mine, which has some of Zambia's richest ores.
Population 146 000
Map Zambia Bb

Chinhae (Jinhae) ***South Korea*** City on the south coast, about 310 km (192 miles) south-east of the capital, Seoul. Its natural harbour has been a naval base since 1905.
Population 104 000
Map Korea De

Chinhoyi ***Zimbabwe*** Market town about 100 km (60 miles) north-west of the capital, Harare. The Chinhoyi Caves, 8 km (5 miles) north-west, contain the deep-blue Sleeping Lake about 50 m (165 ft) underground. Sunlight filtering through the entrance illuminates the caves, whose African name, meaning 'pool of the fallen', refers to an incident in about 1830 when Angonni people migrating north attacked local people and hurled their bodies into the caves.
Population 24 000
Map Zimbabwe Ca

Chin-men Tao (Quemoy) ***Taiwan*** Group of two main islands and 12 islets in the Taiwan Strait, just off the south-east coast of the Chinese province of Fujian, close to city of Xiamen (Amoy). Like MA-TSU TAO, they are heavily fortified and defended by Chinese Nationalist forces, and were shelled from the mainland in 1954 and 1958. The non-military inhabitants are farmers and fishermen.
Map Taiwan Ab

Chinnampo ***North Korea*** See NAMP'O

chinook A warm, dry wind, similar to a FOHN, that descends from the eastern slopes of the Rocky Mountains in North America, causing a rapid rise in temperature and rapid melting of snow in winter. Temperatures of 14°C (57°F) have occurred during a chinook at Calgary, Alberta, in January, when the normal average is –10°C (14°F).

Chios ***Greece*** Island in the Aegean Sea 8 km (5 miles) from the Turkish coast, and the traditional home of the poet Homer. Chios flourished in the 5th century BC. It was captured first by Venetians (1204), then Genoese (1261), and finally the Turks (1566). A massacre of the islanders by the Turks in 1822, during the Greek War of Independence, is re-created in a masterpiece by the French artist Eugène Delacroix, *The Massacre of Chios*, which hangs in the Louvre, Paris. Chios became part of modern Greece in 1912. It has an 11th-century convent, Nea Moni, with fine mosaics. Piryí is an attractive medieval fortress-town on the island, with decorated houses and streets spanned by arches. It is a centre for producing mastic, a tree resin used as chewing gum and in varnish. At the town of Mesta, local women still wear the island's traditional costume.
Population 49 900
Map Greece Db

Chipata ***Zambia*** Town near the border with Malawi, about 520 km (325 miles) east and slightly north of the capital, Lusaka. Formerly called Fort Jameson, it was founded in the 19th century as a military post to suppress the slave trade – and has since become the centre of a tobacco-growing region.
Population 15 400
Map Zambia Cb

Chirripó Grande ***Costa Rica*** The country's highest mountain, about 50 km (30 miles) south-east of the capital, San José. It rises to 3820 m (12 533 ft).
Map Costa Rica Bb

Chisinau ***USSR*** See KISHINEV

Chitipa (Fort Hill) ***Malawi*** Town in Northern province, 9 km (6 miles) from the Zambian border. It is the centre of a rural development project launched in 1977 with assistance from the World Bank to increase crop production and improve health and social services.
Population 1400
Map Malawi Aa

Chittagong ***Bangladesh*** Chief city of the Chittagong division and the country's main port, on the Karnafuli river 20 km (12 miles) from its mouth. It exports jute and tea. Industries include oil refining, jute milling, engineering and textile, plywood and tobacco manufacture. Originally part of an ancient Hindu kingdom, Chittagong passed to the Mogul Empire of India in the 16th century. In 1760 it came under British control. The British linked the city to Dhaka, now the capital, with a railway line. Chittagong is the second largest city of Bangladesh. It has a military base, at which the country's former leader, President Ziaur Rahman, was assassinated by a group of army officers during a coup d'etat in 1981.
Population (district) 5 491 000; (city) 1 392 000
Map Bangladesh Cc

Chittagong Hill Tracts ***Bangladesh*** Hilly jungle district east of Chittagong, near the borders with India and Burma. It covers 8679 km² (3351 sq miles) and is mostly inhabited by tribal minorities, including the Mro, Magh, Chakma, Tipra and Tenchungya peoples. By contrast with the mostly Muslim Bengalis, who dominate Bangladesh, the tribal groups are largely Buddhist or followers of tribal animist religions. Most are backward and depend on slash-and-burn cultivation and what they gather in the forest. A significant number were displaced during the late 1950s by the building of the Karnafuli reservoir, completed in 1961.
Population 580 000
Map Bangladesh Cc

▲ RIVER TAXIS Cabbies carry both passengers and goods on the Karnafuli river at Chittagong. Like most of Bangladesh, the port lies in a vast delta washed down from hills far inland.

Chitwan (Royal Chitwan Park) ***Nepal*** The country's first wildlife park, established in 1973 in 932 km² (360 sq miles) of the Tarai jungle, near the Indian border. Among the protected animals there are tigers and rhinoceroses.
Map Nepal Ba

Chobe National Park ***Botswana*** A 1160 km² (500 sq mile) park opened in 1968 in the extreme north-east of the country. There is an enormous variety of animals, including antelopes, elephants, hippopotamuses, lions, leopards and cheetahs. The Chobe river, a tributary of the ZAMBEZI, runs through the park, which is best seen in the dry season – April to November – when animals and birds congregate on the Chobe's banks.
Map Botswana Ba

Cholon ***Vietnam*** Component town of HO CHI MINH CITY, separated from the city proper by a narrow stream. It was set up by Chinese migrants in 1778, and is the city's commercial and industrial quarter.
Map Vietnam Bc

Cholula ***Mexico*** Town built on a vast Toltec Indian archaeological site dating from 500 BC. It lies 8 km (5 miles) west of Puebla in south-east central Mexico. Rising above Cholula is the largest pyramid in the world – 54 m (177 ft) tall – whose base covers 18 hectares (nearly 45 acres). The pyramid was built of earth, clay, brick and stone as a temple to the Toltec god Quetzalcoatl, the feathered serpent. Human sacrifices took place on the summit, which is now crowned with the Chapel of Los Remedios built by the Spanish in colonial times.
Population 160 000
Map Mexico Cc

Ch'onan (Cheonam) ***South Korea*** Railway town about 85 km (53 miles) south of the capital, Seoul. It is on the main line linking the capital with the port of Pusan. The city is a marketplace for locally grown fruit and vegetables.
Population 120 500
Map Korea Cd

Ch'ongju (Jeonju) ***South Korea*** Central city about 100 km (60 miles) south-east of the capital, Seoul. Cigarettes, silk and woollen yarn are manufactured there.
Population 193 000
Map Korea Cd

Chongqing (Chungking) ***China*** Largest city of Sichuan province, and a centre for steel and heavy engineering industries on the Chang Jiang river. It was the headquarters of Chiang Kai-shek's Nationalist government during the Second World War.
Population 2 650 000
Map China Fe

Chonju ***South Korea*** One of the oldest cities in the Korean peninsula, in the fertile south-west, about 200 km (125 miles) south of the capital, Seoul. It produces silk, paper, umbrellas and bamboo goods.
Population 367 100
Map Korea Ce

Chile

BOUNDED BY THE MIGHTY ANDES AND THE RESTLESS PACIFIC OCEAN, CHILE'S DIVERSE SCENERY IS AMONG THE MOST BEAUTIFUL IN SOUTH AMERICA

Like the backbone of a huge animal, Chile runs its long, thin course down the South American continent's broad back – the Pacific coast – for 4200 km (2610 miles), and is never more than 400 km (about 250 miles) and usually less than 200 km (about 125 miles) in width. It has every kind of climate, from deserts to icy wastes. The most hospitable area is in the centre of the country, around the capital, SANTIAGO, especially the coastal lowlands and the great valley which lies between the double chain of the Andes. In the north lies the ATACAMA DESERT, a hostile expanse of sand and pebbles stretching for 1000 km (620 miles). The south, by contrast, is one of the stormiest places, 1000 km (620 miles) of mountains, forest, volcanoes and desolate islands – a rain-drenched wilderness of fire, sea and ice.

Chile's cherished poet and Nobel prize-winner Gabriela Mistral (1889-1957), described her country as a 'synthesis of the whole planet' – a comment on its astonishing variety. It can be regarded as five distinct regions, from north to south – the Norte Grande; the Norte Chico; the Central Valley; the Lake District, and the Far South, an area of mountains, islands, fiords and forests. However, despite this regional variety, the people of Chile all have one thing in common – they are never very far from the sea.

NORTE GRANDE

Despite its barren surface, the Norte Grande contained a treasure worth fighting for, and Chile won the territory from Bolivia in the War of the Pacific (1879-84). The Atacama Desert was a rich source of nitrates – the raw material for fertilisers and explosives. Chile's national coffers were filled with the revenue from nitrate exports until other sources were found and cheaper synthetic nitrates became available in the early 20th century. While the wealth rolled in, boom towns flourished in the desert. Iquique and Pisagua in the far north were grand enough to have their own opera houses. Though the nitrate industry collapsed, mining did not. Copper, which the desert yields in abundance, brought in new riches, and now accounts for 45 per cent of Chile's exports. Today Chuquicamata is the largest open-cast copper mine in the world and supports a town of over 30 000 people. The copper is shipped out through the port of ANTOFAGASTA.

The teeming sea along the coast is rich in tuna and sardines, which are processed by numerous fish-canning factories. The sea also brings to the coastal towns a climate that is more moderate, though still hot, than the extreme conditions inland. The Peruvian Current sweeps cool water northwards along the coast preventing the formation of rain clouds – the cause of the conditions inland. In the foothills of the Andes, the height cools the weather and people grow oranges, lemons and mandarins.

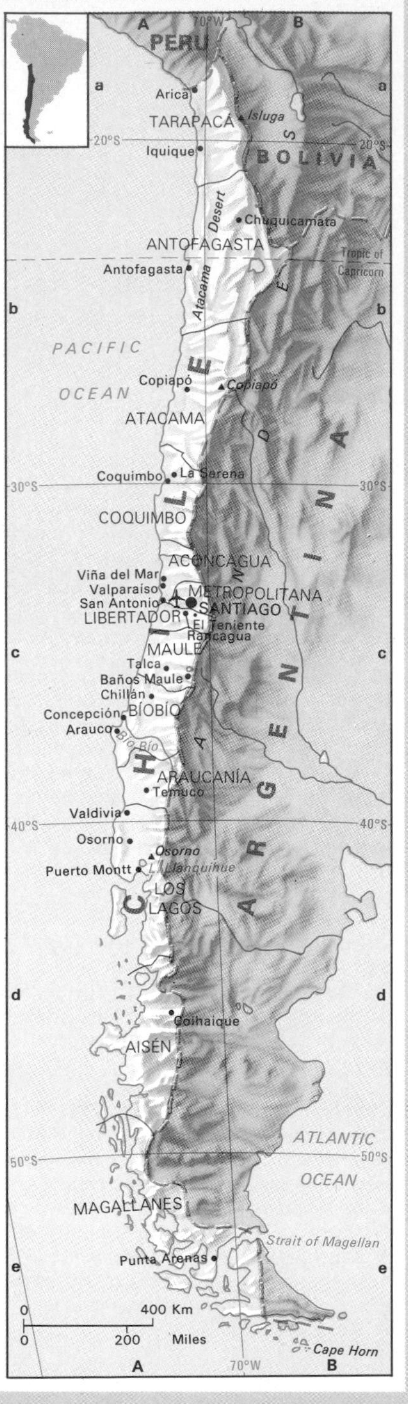

NORTE CHICO

The scrubland of the Norte Chico stretches 470 km (290 miles) south from Copiapo to Illapel, as the desert slowly gives way to the fertile valleys of central Chile. Some copper and iron is mined here, but towards the south the valley dwellers grow fruit, olives, red peppers and barley, as well as alfalfa to fatten cattle.

Four thousand Indians of the Atacama and Calchaquí tribes live in the Norte Chico, in small stone villages where they cultivate maize, beans and tobacco. Some of them are skilled craftsmen who make fine ornaments in silver and copper.

CENTRAL VALLEY

Sixty per cent of Chileans live in the 800 km (500 mile) stretch of land which runs from the valleys of ACONCAGUA province to the River BIOBIO in the province of CONCEPCION. The Central Valley is Chile's heartland, with a mild Mediterranean climate like that of Southern California. The landscape is filled with orchards and rich pastureland, fields of cereals, and Chile's main vineyards.

At Lota and Schwager, just south of Concepción, coal mines extend under the seabed for several kilometres. Inland lies the vast copper mine of EL TENIENTE, one of the largest in the world, where a community of 12 000 miners and their families live on the steep mountain slopes.

The main industrial cities also lie in this area. They include Concepción and the port of VALPARAISO – recently badly damaged by one of Chile's frequent earthquakes. The capital, Santiago, lies 97 km (60 miles) from the sea, in the foothills of the Andes.

Although the central area of Chile is fertile and its climate perfect for growing crops, there is still a great deal of hardship in the country around Santiago. Poverty has brought people flocking from the land into the prosperous industrial cities, causing the present housing problem.

The reason for the penury of farm workers was Chile's traditionally unequal system of land tenure. When the Spanish conquistadores occupied the area in the 16th century, they shared among themselves the land they had seized. Ever since, the farmland has been divided into vast estates owned by the very rich, but farmed by the poor.

During the 1960s, the Christian Democratic president, Eduardo Frei, introduced laws to split up the larger estates and establish a minimum wage structure for farm workers. But the pace of change under Frei did not satisfy the radical element in Chile. He had also made more promises to the poor than he could fulfil, and in 1970 their votes brought in Salvador Allende, to lead the world's first democratically elected Marxist government.

However, only one-third of the electorate voted for Allende, and a sizable majority was opposed to his socialist reforms. The United States, ideologically hostile to Allende's administration, gave financial support to the big economic groups that opposed it. With the state of the economy as an excuse, the military acted. In 1973 they took over the government and killed Allende. A reign of terror began which led to thousands of left-wing supporters losing their lives and perhaps 1 million fleeing the country. A drastic reversal of all socialist policies followed – including the handing back of land to its original owners, and the

▲ **VALLEY OF THE MOON Gaunt and flanked by shadowy mountain peaks, Moon Valley lies in arid isolation in Chile's northerly Atacama Desert – one of the driest regions in the world.**

denationalisation of more industries than Allende had nationalised.

Harsh deflationary measures led to a brief economic revival. However, in the world recession of the late 1970s Chile slumped into inflation, accompanied by high unemployment and worsening living conditions.

The military junta led by President Augusto Pinochet has remained in power since 1973. However, harsh repressive measures taken by the regime, and terrorist activities by its supporters have led to violent opposition.

LAKE DISTRICT

The shockwaves of these political dramas are perhaps less strongly felt in the more sparsely populated areas of the south of Chile. Southwards from the River Biobío to the town of PUERTO MONTT – a distance of 600 km (370 miles) – lies the ravishingly beautiful Lake District. Here is a region of mountains, lakes, rivers, waterfalls and now mostly extinct volcanoes. Its most beautiful lakes are Llanquihue and Todos Los Santos, both overlooked by the snow-capped OSORNO volcano.

Most of the remaining pure-blood Indians live in this region. Driven out of the areas farther north in the late 19th century, they survived to keep a community. The Indians still speak their own language – Araucanian – instead of Spanish.

THE FAR SOUTH

South of Puerto Montt lies the fifth region – a great mountainous, forested wilderness, which is largely uninhabited. The scenery is spectacular, especially when seen by boat from among the fiords and rocky islands.

Most of the population eke out a livelihood from this rainy and desolate land of forests and valleys by rearing cattle and sheep. However, oil and gas were discovered in the southernmost province of Magallanes after the Second World War and now supply half of Chile's oil needs. The capital of the province is PUNTA ARENAS, southernmost city in the world, which rose to importance as a fuelling station on the Magellan Strait before the Panama Canal was opened. The indigenous inhabitants, the Fuegian Indians, who live in or near TIERRA DEL FUEGO, are mostly nomadic. The largest of the region's islands, Chiloé, 50 km (30 miles) south-east of Puerto Montt, is about the size of Corsica. Its community of fishermen and small farmers contend with the fogs and storms, and live mostly on a diet of potatoes, which grow wild in abundance.

Chile's territory does not end at Tierra del Fuego. There are islands in the Pacific and Antarctic that are also possessions of Chile, among which perhaps the most interesting are the JUAN FERNANDEZ archipelago, about 640 km (400 miles) west of Chile, and EASTER ISLAND, 3780 m (2350 miles) out in the Pacific.

CHILE AT A GLANCE
Area 756 945 km² (292 256 sq miles)
Population 12 060 000
Capital Santiago
Government Military-ruled republic
Currency Peso = 1000 escudos
Language Spanish
Religion Christian (90% Roman Catholic)
Climate Varies from dry desert in the north to subarctic in the south. Average temperature in Santiago ranges from 3-14°C (37-57°F) in July to 12-29°C (54-84°F) in January
Main primary products Wheat, maize, rice, sugar beet, potatoes, beans, timber, fruit, grapes, fish; copper, coal, iron, nitrates, molybdenum, zinc, manganese, lead, oil
Major industries Steel, cellulose and wood pulp, cement, food processing, ceramics, glass, forestry, fishing
Main exports Copper, paper and wood pulp, iron ore, nitrates, wine, fruit, fishmeal, processed fish
Annual income per head (US$) 1699
Population growth (per thous/yr) 15
Life expectancy (yrs) Male 65 **Female** 70

China

AFTER CENTURIES OF ISOLATION AND DECADES OF CIVIL STRIFE, THE MOST POPULOUS NATION ON EARTH IS NOW INVITING THE WEST OVER THE WALL TO HELP TO GET ITS ECONOMY ON THE MOVE

Few people believed the 13th-century explorer and merchant Marco Polo when he returned from China to Venice and claimed that he had discovered a land even richer and more sophisticated than his own. Yet he was speaking the plain truth. China's civilisation is one of the oldest in the world – in 1500 BC the Shang dynasty was already ruling a society that worked bronze, knew the art of writing and supported a number of sizable towns. By the 6th century BC, the Chinese regarded themselves as a special people inhabiting the 'Middle Kingdom' at the centre of the universe.

For centuries the GREAT WALL, which writhes 3460 km (2150 miles) from the coast north-east of BEIJING (Peking) to the deserts of Inner Mongolia, kept the surrounding 'barbarians' at bay. Even now, when the wall is a historic monument and tourist attraction, this sense of proud isolation survives, fostering many modern misconceptions.

LAND OF CONTRASTS

For a start, China is not the nation of blue boiler-suited conformity portrayed by the Communist propaganda of the Cultural Revolution. The third largest country in the world after the USSR and Canada, it is a land of contrasting landscapes, peoples and lifestyles. Even its cooking – recognised as among the most sophisticated in the world – is far more varied than most Westerners realise, with at least four major regional styles. Although officially atheist, China has large numbers of Buddhists, Muslims and Christians as well as followers of two home-grown faiths – Confucianism and Taoism.

And its population of 1000 million, which accounts for one-quarter of mankind, embraces 55 officially designated minorities, such as Mongolians, Tibetans, Kazakhs, Uygurs, Zhuangs, Yis and Miaos. Apart from the languages spoken by minorities such as these, Chinese itself has many spoken dialects, which nevertheless share the same written ideograms and thus can be understood by all who read Chinese.

The country can be divided into distinct western and eastern halves, and then further into four contrasting quadrants.

Western China accounts for roughly half of the nation's territory, yet, due to the inhospitable terrain, it houses a mere 5 per cent of the population. In the north, the deserts and steppes of XINJIANG (Sinkiang) stretch into neighbouring Mongolia and roll into the Central Asian republics of the Soviet Union. Nomadic peoples, such as the Uygurs, the Kazakhs and the Tajiks, herd cattle, sheep and horses, or settle beside oases along the fringes of the TAKLIMAKAN desert and, as followers of Islam, worship in mosques – a way of life that has been uninterrupted for centuries.

In the south-west, the ice-capped peaks of XIZANG (Tibet) – many of them over 6000 m (20 000 ft) high – form one of the most remote wildernesses in the world, sparsely inhabited by a people who practise their own version of Buddhism.

Eastern China divides into two very different regions either side of a line running eastwards from the QIN LING mountains (near the city of XIAN) to the sea. To the north much of the land is comparatively inhospitable, with low rainfall and temperatures that rise in summer to 30°C (86°F) and sink in winter to well below freezing. Nowhere are these extremes more uncomfortable than in Beijing, the capital, which swelters during the summer months, is raked by winds from the icy wastes of Mongolia and Siberia in winter, and is carpeted by a yellow dust blown during the spring from the GOBI DESERT.

It is in the plains of northern China, on either side of the HUANG HE (Yellow river), that the country's population is most concentrated. In an area about the same size as France live 250 million people – four times France's population. Throughout the north there is very little rain, and the summer growing season is too short to permit widespread rice cultivation. Staple crops are therefore wheat, millet, maize and gaoliang, a Chinese variety of sorghum.

Whereas the north is generally brown, dusty and crisscrossed by cart tracks, the land to the south of the line is green and well-watered, studded with lakes and traversed by an intricate pattern of rivers, creeks and canals. This is the China of most Westerners' imagination – a land of paddy fields and terraced hillsides, water buffaloes and farmers in broad-brimmed hats. The fertile, mountain-ringed province of SICHUAN (Szechwan) alone supports 100 million people, while the mountains of neighbouring YUNNAN are a botanist's paradise, home of many unique plants.

Rice is the staple crop of the south, but the happy combination of climate and landscape also favours cultivation of a wide range of subsidiary crops, including tea, oilseed plants, a range of vegetables, sugar cane and citrus fruits. In contrast to the north, which produces only one crop a year or three crops every two years, parts of the south produce two, three or even four main crops a year.

The well-watered lowlands in the valley of the CHANG JIANG (Yangtse river) and the warm and humid plains which surround the delta of the ZHU-JIANG (Pearl river) contain two of China's greatest cities, SHANGHAI and GUANGZHOU (Canton) respectively. But rather than the cities, it is the countryside, with its teeming villages and carefully tilled fields, that epitomises this part of China. Even today, farmers and their families make up over 75 per cent of the entire population.

China's history over the past 100 years has been dramatic. After thousands of years of proud isolation under a series of dynastic emperors (including a century of Mongol rule instituted by Kublai Khan), the country was desperately slow to react to the threat posed by the Western imperial powers in the 19th century. In 1838 the emperor attempted to eliminate the lucrative traffic in opium enjoyed by European traders – with unfortunate results. China was humiliated by the British in the ensuing Opium War of 1839-42 and

▲ INSPIRATION The weird hills of Guilin in south-western China look down on the town of Yangshuo, isolated by a moat of flooded paddy fields. Tourists can now cruise the Li Jiang river through this fantastic landscape which has inspired Chinese artists for generations.

forced to cede HONG KONG. This was the beginning of a long, disastrous period in the nation's fortunes. Not only did China fall increasingly under the control of the West and later Japan, but it was also split from within.

THE TURBULENT YEARS

In 1911-12, a failed attempt by the Nationalist Party leader, Sun Yat-sen, to unify the country under a republic left China without an effective central government; and for the next 37 years Chinese Communists, Nationalists and warlords fought a bitter and destructive struggle for power. On the verge of defeat in 1934, the Communist Red Army battled through the encircling Nationalist troops and embarked on a 9500 km (5900 mile) fighting retreat, known as the 'Long March'. Some 100 000 people (85 000 troops and supporting administrative staff) set out, but only 8000 survived the year-long ordeal.

Internal confusion was compounded when Japan, in pursuit of imperial ambitions, invaded in 1937, crippling cities and communications and causing massive loss of life. The defeat and withdrawal of the Japanese at the end of the Second World War left the same two rival groups – Nationalists and Communists – contending for power in China. Chiang Kai-shek and the Chinese Nationalist Party, or Kuomintang, relied on the support of a middle class dominated by landlords, generals, industrialists and bankers. The Chinese Communist Party, led by Mao Ze-dong (Mao Tse-t'ung), appealed to the peasants who then, as now, made up the great majority of the Chinese population.

When civil war between the Communists and the Nationalists broke out in 1947, the outcome was inevitable – although it cost 11 to 12 million lives. Sweeping all before them, the Communist armies swiftly occupied the entire mainland. Chiang and his followers fled to the island of TAIWAN and on October 1, 1949, the People's Republic of China was proclaimed.

From 1949 to 1976 China was dominated by the wayward genius of Mao. Under his charismatic leadership, it turned first to the Soviet Union for advice on building a new socialist state – a recipe that called for the nationalisation of industry and the collectivisation of agriculture. This extension of the Communist Party's rule over all aspects of Chinese life seemed, however, an ominous development to the West; and the Chinese takeover of Tibet in 1950, followed by its military intervention in the war in Korea and later border clashes with India, did nothing to dispel the fear that Mao's regime was a threat to world peace.

The alliance between China and the Soviet Union proved to be short-lived. Mao abandoned the Russian approach to modernisation and in 1958 launched the 'Great Leap Forward' – a once-for-all attempt to mobilise the population and resources of China towards rapid economic growth. Peasants were organised into self-sufficient people's communes and pressured into meeting production targets laid down by the party.

Mao's great leap was a disaster. The havoc it wrought was partly repaired by forsaking some of Mao's original ideals: incentives were reintroduced and the communes which were the essence of his peasant paradise were less rigidly enforced. However, this pragmatism in economic policy did not extend to the realm of politics. Mao was concerned that a self-perpetuating bureaucratic and intellectual elite would emerge in China, as he believed it had in the Soviet Union. For this reason, he launched the Great Proletarian Cultural Revolution in 1966 – a gigantic upheaval that was meant to transform China into a truly classless society.

Teachers and technicians, artists and engineers, and in particular moderate party and state officials were denounced as 'class enemies', sacked from their jobs and sent off to the countryside to work in the fields. Gangs of young thugs were mobilised into the makeshift Red Guards. As they roamed from city to city, they dispensed a summary justice to those whom they believed to be the 'bourgeois elements' of the counter-revolution. One of their aims was to create as much chaos as possible; by 1968, however, Mao had had enough. He ordered the army to restore order; and it was now the turn of millions of Red Guards to be sent off for re-education by hard labour. But the Cultural Revolution continued at a less violent level for another eight years, until Mao's death.

Today the Cultural Revolution is referred to by the government as the 'decade of destruction', and the blame is put on the 'Gang of Four' – a group of extreme radicals led by Mao's wife – rather than on Mao himself.

Surprisingly, the Cultural Revolution had little effect on the conduct of Chinese foreign policy – which, if anything, became more diplomatic in tone. Hostility towards the Soviet Union remained acute and, despite opposition to American policy in South-east Asia, the Chinese leadership began to make friendly overtures towards the United States. In 1972 President Nixon visited Beijing and the door for détente was open.

Relations with Japan and the countries of western Europe steadily improved, and Chinese trade with capitalist countries began to grow. As one modern Chinese saying goes: 'Our minds are on the left, but our pockets are on the right.'

In ancient times earthquakes were seen as heralding the end of a dynasty, and the disastrous TANGSHAN earthquake, which preceded Mao's death in 1976, did just that. Power passed to a group of moderates led by Deng Xiaoping, the Communist Party vice-chairman. The new leadership jettisoned the ideological obsessions that had racked China's domestic policy for over 20 years and, in a dramatic about-turn, embarked on the wholesale modernisation of the country. For the first time since 1949 the West and Japan were invited to supply the expertise, capital and technology needed for economic growth. At the same time, foreign tourists were once again welcomed (not least as a source of foreign exchange) and material incentives were introduced to raise production within the country. As a result, China's economy has been transformed since the late 1970s.

AN ECONOMY ON THE MOVE

Agricultural output has soared; foreign trade has taken off; and attempts have been made to improve industrial organisation. China hopes to achieve advanced nation status by the year 2049, the hundredth anniversary of its revolution.

Education is one of the keys to this advance. Today, nearly every Chinese child attends school and more than a million students attend the country's 700-plus universities and institutes of higher education. Even so, illiteracy is widespread, especially among the old.

In the long term, China has sufficient fuel and mineral resources to support the growth of an industrial economy, but the colossal investment needed holds back the exploitation of these fuel and mineral resources. Many of China's coal mines lack modern machinery; the oil fields of the future will be expensive to develop; and many of the most valuable mineral deposits are to be found in the icy wastes of the Tibetan plateau.

China's greatest single resource is undoubtedly its people, but the rate of population growth has become a great cause for concern. In 1949 the population stood at 542 million, but by 1969 it had risen to 807 million and was growing at over 2 per cent a year – enough to make it multiply by more than seven times over the next 100 years. So in 1979 the government set a quota of one child per couple, enforced with penalties such as reduced food rations and virtually compulsory abortion. As a result, it hopes to keep China's population below 1200 million until the year 2000. One unforeseen effect of this policy, highlighted in 1986, was the emergence of a generation of over-indulged, over-fed only children.

Thanks to a massive health care programme, life expectancy has more than doubled from the 1949 average of 32 years. Health care is in the hands of about a million doctors who practise Western-style medicine and about 300 000 practitioners of traditional healing methods. Acupuncture is widely used by both groups. The treatment involves inserting needles up to 230 mm (9 in) in length into the patient's body. Specific locations are punctured to produce specific effects. Western observers have seen the technique used, apparently effectively, to anaesthetise and to treat many conditions including stomach ulcers, rheumatism and anxiety.

In the short term, living standards are likely

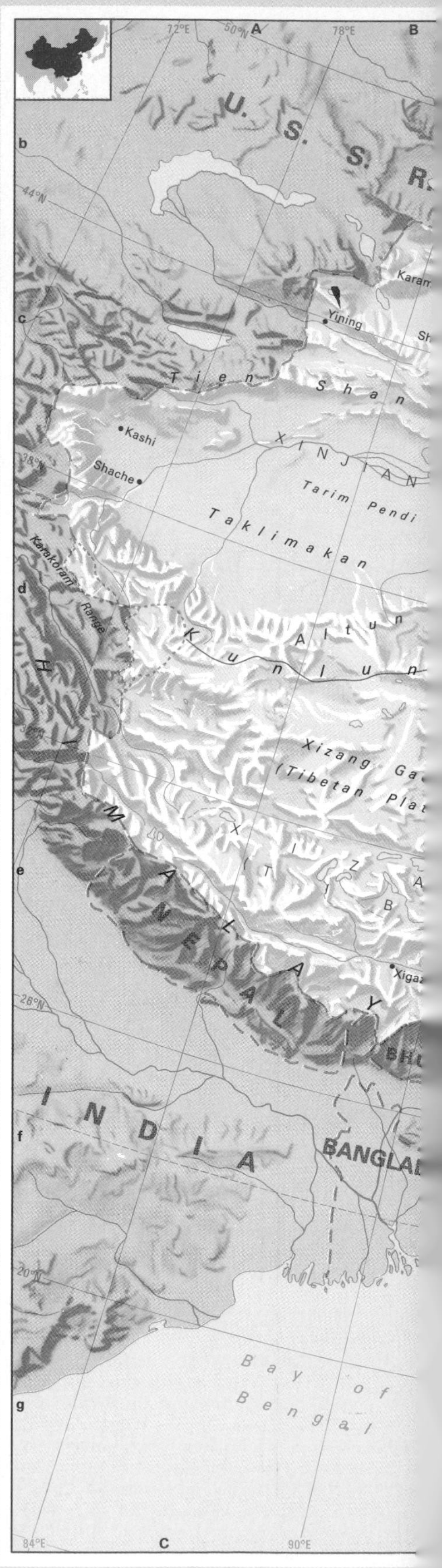

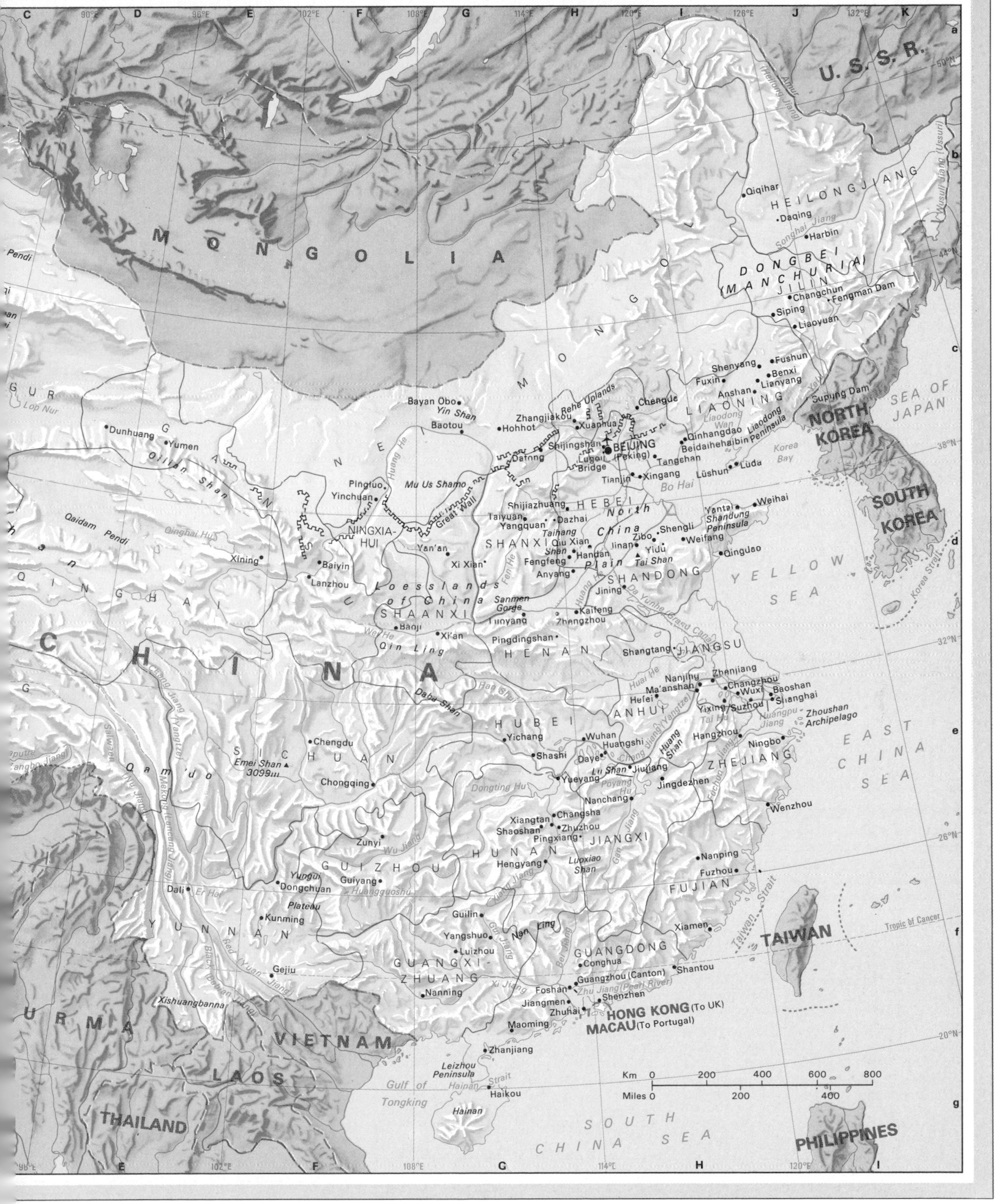
U. S. S. R.
M O N G O L I A
C H I N A
NORTH KOREA
SOUTH KOREA
SEA OF JAPAN
YELLOW SEA
EAST CHINA SEA
SOUTH CHINA SEA
TAIWAN
PHILIPPINES
VIETNAM
LAOS
THAILAND
HONG KONG (To UK)
MACAU (To Portugal)
HEILONGJIANG
DONGBEI (MANCHURIA)
JILIN
LIAONING
HEBEI
SHANXI
SHANDONG
SHAANXI
HENAN
JIANGSU
ANHUI
HUBEI
SICHUAN
HUNAN
JIANGXI
ZHEJIANG
FUJIAN
GUIZHOU
YUNNAN
GUANGXI-ZHUANG
GUANGDONG
NINGXIA-HUI
QINGHAI
Qiqihar
Daqing
Harbin
Changchun
Fengman Dam
Siping
Liaoyuan
Shenyang
Fushun
Benxi
Fuxin
Anshan
Lianyang
Supung Dam
Chengde
Rehe Uplands
Zhangjiakou
Xuanhua
Hohhot
Bayan Obo
Yin Shan
Baotou
Datong
Shijingshan
BEIJING (Peking)
Lugou Bridge
Tangshan
Qinhangdao
Beidaihehaibin
Liaodong Peninsula
Liaodong Wan
Korea Bay
Lüshun
Lüda
Tianjin
Xingang
Bo Hai
Great Wall
Mu Us Shamo
Pingluo
Yinchuan
Shijiazhuang
Taiyuan
Yangquan
Dazhai
Taihang Shan
North China Plain
Yantai
Weihai
Shandong Peninsula
Shengli
Zibo
Weifang
Jinan
Yidu
Tai Shan
Qingdao
Qiu Xian
Handan
Fengfeng
Anyang
Jining
Yan'an
Xi Xian
Loesslands of China
Sanmen Gorge
Luoyang
Kaifeng
Zhengzhou
Baoji
Xi'an
Qin Ling
Wei He
Pingdingshan
Shangtang
Huang He
Da Yunhe (Grand Canal)
Dunhuang
Yumen
Qilian Shan
Qaidam Pendi
Qinghai Hu
Lop Nur
Xining
Baiyin
Lanzhou
Nanjing
Zhenjiang
Changzhou
Ma'anshan
Wuxi
Baoshan
Hefei
Shanghai
Yixing
Suzhou
Tai Hu
Huangpu Jiang
Zhoushan Archipelago
Hangzhou
Ningbo
Huai He
Han Shui
Daba Shan
Chengdu
Emei Shan 3099
Chongqing
Yichang
Shashi
Wuhan
Huangshi
Daye
Huang Shan
Lu Shan
Jiujiang
Jingdezhen
Yueyang
Dongting Hu
Poyang Hu
Nanchang
Chang Jiang (Yangtze)
Fuchun Jiang
Wenzhou
Xiangtan
Changsha
Shaoshan
Zhuzhou
Pingxiang
Zunyi
Wu Jiang
Hengyang
Luoxiao Shan
Gan Jiang
Nanping
Fuzhou
Yungui Plateau
Dongchuan
Guiyang
Dali
Er Hai
Kunming
Gejiu
Qamdo
Mekong (Lancang Jiang)
Salween
Xishuangbanna
Red (Yuan Jiang)
Black (Babian Jiang)
Guilin
Yangshuo
Luizhou
Nan Ling
Gui Jiang
Bei Jiang
Xi Jiang
Nanning
Conghua
Guangzhou (Canton)
Shantou
Xiamen
Taiwan Strait
Tropic of Cancer
Foshan
Zhu Jiang (Pearl River)
Jiangmen
Shenzhen
Zhuhai
Maoming
Zhanjiang
Leizhou Peninsula
Hainan Strait
Haikou
Gulf of Tongking
Hainan
Korea Strait
Ussuri (Wusuli Jiang)
Amur (Heilong Jiang)
Songhai Jiang
Km 0 200 400 600 800
Miles 0 200 400

to rise as much due to Deng Xiaoping's agricultural reforms as to population control. Since 1979 the regimentation of farm production through communes has gradually been replaced by the 'responsibility system', which has transformed China's peasants into tenant farmers. Either on their own or in small working groups, farmers have to meet a production quota but can then sell the surplus on the open market. Rural markets do a roaring trade in pigs, chickens, fruit and vegetables, all of which were in short supply during the days of Chairman Mao. Some of the profits are then used to set up small village factories and workshops, producing umbrellas, fabrics or farm tools.

The responsibility system has also been applied to industry. The result has been an improvement in the quality of manufactures and a wider range of products. China now makes more steel than any country in western Europe and is the world's leading producer of cotton textiles. The consumer goods industries, which languished from neglect under the puritan Mao, are growing too. Over a million colour television sets are made each year, and Chinese-made washing machines, electronic watches and tape recorders are to be seen in big city stores.

One problem that will be solved only by massive investment is the transport system. Most freight is hauled by steam trains, but the railways, even when supplemented by water transport on a network of rivers and canals, are simply unable to shift the growing traffic generated by China's economic expansion.

Cars are rare. Most are used by government officials or as taxis. Until recently few have been made in China and fewer imported. But a joint venture with the German Volkswagen company in Shanghai is said to be currently producing 20 000 vehicles a year. The bicycle is the most widespread form of personal transport – there are estimated to be more than 100 million in the country – though mopeds are coming into wider use.

Despite Deng Xiaoping's liberal reforms, there is no escaping the fact that modern China is a Marxist state with a strong central government that controls the economic, political and even personal activities of over a thousand million people.

▲ WOMEN'S WORK Two to a shovel – one pushing, one pulling – a gang of women spread hard core for a new road near Lhasa, in Tibet. Since the Chinese takeover in 1950, a network of modern roads has been constructed.

A social revolution is in progress as elements of Western culture – from Coca-Cola to Mozart, banned as decadent during the Cultural Revolution – are allowed to infiltrate alongside traditional art forms such as Peking opera.

On the other hand, freedom of expression and freedom of action, although considerably greater than under Mao, remain limited. To change jobs, get married or simply travel to another city, an individual must ask for permission from the local *danwei* or work unit. The minutest details of everyday life are under constant scrutiny from the street committees which keep watch on every block of flats. And should a person be foolish enough to break the rules, he can be punished – usually by ostracism or other social sanctions.

Although the general standard of living has certainly improved since the late 1970s, modernisation has tended to accentuate the regional differences that exist in China. Not everyone has benefited equally.

Southern China, with its fertile soil and kind climate, has benefited most from the agricultural reforms and from the 'open door' policy towards the West. Southerners with relatives in Hong Kong are assured a steady supply of cash and consumer goods. Just across the border from Hong Kong in SHENZHEN, one of several 'special economic zones' established with the help of foreign investment, workers take home wages that are the envy of Chinese labourers in most parts of the country. However, in the northern and western provinces, every day is a struggle against a harsh, unyielding soil.

The changes of the last decade partly explain the recent rise in crime – from violent robbery to petty theft. This trend is particularly noticeable among those who, consigned to the provinces during the Cultural Revolution, have drifted back to the cities where they make up a large and disaffected class of urban unemployed. The government's reaction has been severe, and thousands of criminals have been summarily executed.

On balance, most of China's changes have had happier results. China's economy is on the move and, although income per head is a mere 3 per cent of Japan's, its population is better fed, better housed and probably more content than at any time since the Revolution. Its relations with capitalist countries have steadily improved and are balanced by a thaw in its hitherto icy relations with the Soviet Union. So long as the pragmatic domestic and foreign policies are preserved, there is every reason to believe that China will achieve its goal of advanced nation status by 2049.

CHINA AT A GLANCE

Area 9 596 961 km² (3 705 387 sq miles)

Population 1 050 000 000

Capital Beijing

Government One-party Communist republic

Currency Yuan = 100 fen

Languages Chinese (including Mandarin and other dialects), Tibetan, Uygur, Mongolian, Zhuang, Miao, Yao

Religions Confucian (20%), Buddhist (6%), Taoist (2%), Muslim (2%), Christian (1%)

Climate Continental, with extremes of temperature; temperate or subtropical and humid in the south-east and central south, dry in the north and north-east. Average temperature in Shanghai ranges from 1-8°C (34-46°F) in January to 23-32°C (73-90°F) in July/August

Main primary products Wheat, rice, cotton, sugar cane, sugar beet, soya beans, jute and hemp, livestock, tobacco, tea, timber, fish; oil and natural gas, coal, iron ore and other minerals

Major industries Iron and steel, cement, fertilisers, mining, agriculture, machinery, machine tools, vehicles, textiles

Main exports Crude oil and products, minerals, foodstuffs, textile yarns and fabrics, clothing, chemicals, machinery, coal

Annual income per head (US$) 223

Population growth (per thous/yr) 9

Life expectancy (yrs) Male 66 **Female** 69

Chonnam ***South Korea*** See KWANGJU

Chontales ***Nicaragua*** Cattle-raising department in the central highlands east of the capital, Managua. The departmental capital is Juigalpa, 95 km (59 miles) east of Managua.
Population (department) 99 000; (town) 20 000
Map Nicaragua Aa

Chorzow ***Poland*** Manufacturing and coal-mining city in the south, 8 km (5 miles) north-west of the city of Katowice. The Kosciuszko Steelworks, formerly the Prussian royal foundry established in 1795, dominates the 19th-century city centre. There are chemical works and a tram-assembly plant.
Population 145 000
Map Poland Cc

Chott al Hodna ***Algeria*** See ATLAS MOUNTAINS

Chott el Jerid ***Tunisia*** Huge salt lake, about 4920 km² (1900 sq miles) in area and 16 m (52 ft) below sea level, lying in the west-centre of the country. Legend has it that it is the birthplace of the Greek goddess Athena. Now freshwater irrigation schemes have been applied in the region to help eliminate the salt in the ground and so increase the productive area for date palm cultivation.
Map Tunisia Aa

Christchurch ***New Zealand*** Largest city on South Island, on the edge of the Canterbury Plains. It is the centre of one of the country's most productive wheat, grain and sheep-rearing regions. Its industries include food processing and canning, woollen mills, railway workshops, carpets, fertilisers, footwear and furniture. It was founded in 1851 as a Church of England settlement, and some of its finest buildings date from then, such as the Provincial Government Buildings (1859-65) and the Gothic-style cathedral (1864-1904).

Often called the 'Garden City' of New Zealand, it contains about 400 hectares (1000 acres) of public gardens and parkland, including the Queen Elizabeth II Park created for the 1974 Commonwealth Games.
Population 300 000
Map New Zealand De

Christiania ***Norway*** See OSLO

Christiansand ***Norway*** See KRISTIANSAND

Christmas Island ***Indian Ocean*** Island about 400 km (250 miles) south of the western end of Java. Covering about 142 km² (55 sq miles), it was annexed by Great Britain in 1888 and incorporated in the former Straits Settlement in 1889. During the Second World War it was occupied by Japan (1942-5) and passed into Australian control in 1958. The inhabitants are mostly Malays and Chinese who work the large deposits of phosphates.
Population 3500
Map Indian Ocean Ec

Christmas Island (Kiritimati) ***Kiribati*** Atoll in the Line Islands more than 3200 km (2000 miles) east of Tarawa. Its land area of 432 km² (167 sq miles) is the largest of any atoll in the world. It was discovered by the British navigator Captain James Cook on Christmas Eve, 1777. In 1956-62 the island was the base for numerous British and US tests of nuclear weapons. Contamination has now dispersed and there are copra and shrimp industries; tourism is developing.
Population 1300
Map Pacific Ocean Fb

Chubu Sangaku (Japan Alps) ***Japan*** Mountainous national park in the centre of Honshu island, about 200 km (125 miles) north-west of the capital, Tokyo. It covers 1698 km² (656 sq miles) and is very similar in appearance to the European Alps. It contains two of the highest peaks in Japan, Mount Hotaka (3190 m, 10 466 ft) and Mount Yari (3180 m, 10 434 ft).
Map Japan Cc

Chulung Pass ***India/Pakistan*** Mountain pass over 4000 m (13 000 ft) high in the Karakoram Range, on the 1949 ceasefire line between Indian and Pakistani Kashmir. It is in the shadow of Gasherbrum, 8068 m (24 470 ft).
Map Pakistan Da

Chungking ***China*** See CHONGQING

Chung-yang Shan-mo ***Taiwan*** Principal range of Taiwan's central mountains and an area of spectacular mountain scenery. It has several peaks over 3000 m (10 000 ft) high, Yu Shan being the highest (3997 m, 13 113 ft).
Map Taiwan Bc

Chuquicamata ***Chile*** Mining town in the Andes, on an arid plateau nearly 3000 m (9850 ft) high in Antofagasta region. It contains the world's largest open-cast copper mine. The ore is processed on site before being transported to the port of Antofagasta for export.
Map Chile Bb

Churchill ***Canada*** **1.** Seaport and railway town on Hudson Bay at the mouth of the Churchill river in Manitoba. Churchill has the best harbour on Hudson Bay and is an administrative and tourist centre. **2.** River, 1600 km (993 miles) long, rising near the Saskatchewan-Alberta border and flowing eastwards to Hudson Bay at Churchill.
Population 1186
Map Canada (town) Fc; (river) Ec

Chuvash Republic ***USSR*** Autonomous republic of the RSFSR in the Volga valley, about 600 km (375 miles) east of Moscow; it covers 18 300 km² (7050 sq miles). Its main occupations are farming, forestry and timber processing, engineering, chemicals and textiles. The capital is the university city of Cheboksary.
Population 1 314 000
Map USSR Fc

Ciarraí ***Ireland*** See KERRY

Cicmany ***Czechoslovakia*** Slovak mountain village 137 km (85 miles) north-east of Bratislava. It is preserved as a working folk museum: its people still wear traditional, richly embroidered costumes and decorate the outsides of their wooden houses with carvings and paintings.
Map Czechoslovakia Db

Ciechanow ***Poland*** Old market town in central Poland, 99 km (62 miles) north of the capital, Warsaw. Its massive, rectangular castle with two round towers was begun in the 11th century, and left a ruin by invading Swedes in 1650; it is now being restored.
Population 37 600
Map Poland Db

Cienfuegos ***Cuba*** Seaport and capital of Cienfuegos province, on Cienfuegos Bay 225 km (140 miles) south-east of Havana. Industries include sugar refining, coffee processing, cigar and soap making. The port handles sugar, tobacco, coffee and molasses.
Population (city) 115 000; (province) 332 000
Map Cuba Aa

Cieszyn ***Poland*** Polish part of the city of Teschen (or Tesin) straddling the Olza river, which now forms part of the Polish-Czech border. The city, the old cultural capital of Upper (or Austrian) Silesia, lies 64 km (40 miles) south-west of Katowice. Cieszyn's 'old town', centred on Castle Hill, has a 14th-century tower – the remnant of a prince's castle – and an 11th-century Romanesque chapel, which is one of Poland's oldest stone buildings.

Cieszyn manufactures metal goods, chemicals and electrical goods. The Czech part of the city is called Cesky Tesin.
Population 35 000
Map Poland Cd

Cilipi ***Yugoslavia*** See DUBROVNIK

Cill Airne ***Ireland*** See KILLARNEY

Cill Chainnigh ***Ireland*** See KILKENNY

Cill Dara ***Ireland*** See KILDARE

Cill Mhantáin ***Ireland*** See WICKLOW

Cimpulung Moldovenesc ***Romania*** A timber-processing centre in the north of the country, 50 km (30 miles) south of the border with Russia. It is a furniture-making centre, and the town contains the remarkable Museum of Wooden Spoons, which contains some 5400 individually carved spoons from different regions of Romania and various countries of the world – including France and the USA.
Population 20 000
Map Romania Ba

Cincinnati ***USA*** City on the Ohio river, in the south-west of the state of Ohio, close to the borders with Kentucky and Indiana. It is an industrial centre, and its university, founded in 1819, is one of the oldest west of the Appalachian Mountains.
Population (city) 370 500; (metropolitan area) 1 673 500
Map United States Jc

cinder cone Conical hill of fragmentary material around a volcanic vent.

cinnabar Heavy reddish mineral form of mercuric sulphide, the principal ore of mercury.

Cinque Ports ***United Kingdom*** Group of towns on the English Channel coast of Kent and Sussex that were granted special privileges by King Edward the Confessor in the 11th century in return for defending the coast. There were originally five (hence the name, from the French word *cinq*) – DOVER, Sandwich, Romney, HASTINGS and Hythe – but they were later joined by Winchelsea and Rye.

Circeo National Park *Italy* Park covering 83 km² (32 sq miles) near the coast about 75 km (45 miles) south-east of Rome. It was founded in 1934 to protect the classic Mediterranean vegetation of the Pontine Marshes when the rest were drained and settled. Its forest is a unique area of untouched native vegetation and its lakes contain a great variety of birds and marsh vegetation.
Map Italy Dd

Circum-Pacific Belt Narrow zone around the edge of the Pacific Ocean coinciding with the outer boundary of the adjoining Pacific, Nazca and Cocos plates of the earth's lithosphere. Because of the intense volcanic activity in the area it is also known as the Ring, or Circle, of Fire.

cirque A steep-walled semicircular rock basin occurring above or at the upper end of some mountain valleys. It results from erosion by ice, and in areas still under glaciation contains compacted snow and ice which is often the source of a glacier. Elsewhere the basin may contain a small lake or tarn. It is also called a coire, corrie, cwm or kar.

cirrocumulus A high cloud formation of fleecy, small white balls arranged in ranks 8 km (5 miles) or more above the earth. It usually indicates unsettled weather to come and is also known as 'mackerel sky'.

cirrostratus A high cloud formation like a fine whitish veil 8 km (5 miles) or more above the earth. If it grades into ALTOSTRATUS, it may herald the approach of a warm front, and the approach of rain.

cirrus High clouds at altitudes of about 8 km (5 miles), seen as wisps of white in a blue sky. They often herald a depression.

Ciskei *South Africa* Bantu homeland for the Xhosa people in eastern Cape Province. It was declared an independent republic in 1981, but is recognised only by South Africa. It covers a total of 8300 km² (3205 sq miles). Despite the dry climate, cereals and vegetables are grown, and pineapples and timber exported. Bisho is the capital. The town of Alice has the non-racial University of Fort Hare.
Population 728 400
Map South Africa Cc

Citlaltépetl (Orizaba) *Mexico* Volcanic peak, 5747 m (18 855 ft) high, in the Sierra Madre Oriental mountains in Veracruz state. It is Mexico's highest point and was first scaled in 1873. It is sacred to Mexican Indians, who believe it contains the spirit of the feathered snake-god, Quetzalcoatl.
Map Mexico Cc

Citta Vecchia *Malta* See MDINA

Ciudad Bolívar *Venezuela* City on the south bank of the Orinoco river, about 400 km (250 miles) south-east of the capital, Caracas. Previously called Angostura, it was renamed after the Venezuelan liberator Simon Bolívar, who captured the town in 1817. The cocktail ingredient Angostura bitters originated there.
Population 200 000
Map Venezuela Bb

Ciudad Guayana *Venezuela* Industrial city, founded in 1961, on the Orinoco river some 500 km (310 miles) south-east of the capital, Caracas. It contains the country's principal steel and aluminium plants.
Population 325 000
Map Venezuela Bb

Ciudad Real *Spain* Market town about 160 km (100 miles) south of Madrid, trading in cereals, olive oil and wine. Its industries include flour milling, brandy distilling and textiles. The surrounding province of the same name is mostly a farming region, but mercury and lead are also mined.
Population (town) 51 200; (province) 468 300
Map Spain Dc

Clackmannanshire *United Kingdom* See CENTRAL REGION

Clare (An Clár) *Ireland* County on the west coast, bounded by the Shannon and its fertile valley to the east and south, and includes the barren limestone BURREN in the north. Eamon de Valera (1882-1975), the Irish statesman who fought against Britain during the Easter Rebellion in 1916, became MP for East Clare in 1917 and later became Prime Minister of the Irish Free State and President of Ireland. The county town is Ennis (Innis).
Population 87 500
Map Ireland Bb

clastic rock Rock composed of fragments of other rock that have been broken off, transported and deposited elsewhere. Clastic rocks include sedimentary rocks, agglomerates, breccias, tuff and conglomerates.

clay Fine-grained sedimentary rock with particles less than $^1/_{256}$ mm (0.00015 in) across. The term is also used for any sediment which is plastic when wet.

Clermont-Ferrand *France* Capital of the Puy-de-Dôme department, 135 km (84 miles) west of Lyons. The towns of Clermont and Montferrand were united in 1731. Main occupations include a tyre works, food processing and tourism.
Population 262 175
Map France Ed

Cleveland *United Kingdom* County covering 583 km² (225 sq miles) of north-east England, named after the Cleveland Hills on its southern border. It was carved out of the counties of Durham and Yorkshire in 1974 to unite the urban and industrial areas along the River Tees. The town of STOCKTON-ON-TEES is an old port and market town, which was transformed after 1825 by the world's first steam railway, set up by the locomotive pioneer George Stephenson. The iron and steel town of MIDDLESBROUGH on the south side of the Tees was founded in 1830. There is a vast chemical complex at Billingham, a modern steel works at Redcar on the coast, and chemical works at Hartlepool.
Population 565 000
Map United Kingdom Ec

Cleveland *USA* Port and industrial city in Ohio, on the south shore of Lake Erie. It was founded in 1796, and handles coal and iron ore. Its varied industries include steel and car manufacture, oil refining and chemicals. It has three universities.
Population (city) 546 500; (metropolitan area) 1 867 000
Map United States Jb

Clifden (An Clochán) *Ireland* Main town of Connemara, on the west coast, 66 km (41 miles) north-west of the city of Galway. It is a tourist and fishing centre, and manufactures tweed. The remains of Marconi's first transatlantic wireless station – destroyed during the Irish Civil War of the 1920s – are nearby. At Derrygimlagh Bog, south of the town, there is a memorial to John Alcock and Arthur Brown, British aviators who landed there in 1919 after making the first non-stop flight across the Atlantic.
Population 1380
Map Ireland Ab

climate The average weather conditions of a region over a period of 30 to 35 years, particularly of temperature and rainfall, including their seasonal variation. The type of climate depends on latitude, altitude, continental or maritime location, configuration of the land, soils and vegetation. Climatologists also study long-term trends, including those of the past.

climax vegetation Natural plant community that eventually becomes dominant, stable and mature. It is the 'natural vegetation' before interference by man.

clints Irregularly shaped blocks in a limestone area making up a flat, exposed 'pavement', separated by deep grooves called GRIKES.

Clonmacnoise (Cluain Mhic Nóis) *Ireland* Religious site on the River Shannon, about 120 km (75 miles) west of Dublin. Founded in AD 548 by St Ciaran, it became a monastery and a seat of learning despite Viking, Anglo-Norman and English attacks. The site, abandoned in 1552 after the English finally sacked it, has a cathedral, seven churches, two round towers, three high crosses, and a castle dating from the 11th century.
Map Ireland Cb

Clontarf *Ireland* Northern suburb of Dublin, lying on Dublin Bay. There the Irish defeated the Danes on Good Friday, 1014, and broke Viking power in Ireland. The Irish king Brian Boru, then an elderly man, was slain at the moment of victory. His son and grandson also died in the battle.
Population 19 000
Map Ireland Cb

cloudburst Sudden, concentrated downpour of rain, often associated with a thunderstorm.

clouds Clusters of very small droplets of water, or, at high altitudes, ice crystals suspended in the air. They are named according to their height, thickness and shape.

clough A steep-sided, narrow valley.

Cluain Mhic Nóis ***Ireland*** See CLONMACNOISE

Cluj-Napoca ***Romania*** Industrial city in north-west TRANSYLVANIA 130 km (80 miles) south of the border with Russia. In Roman times, on the same site, there was a town called Napoca. In the 15th century, under Hungarian rule, Cluj became a cultural, educational and economic centre. There is still a strong Hungarian atmosphere in the city, particularly in its architecture. And a Baroque, 18th-century Hungarian palace is now an Art Museum.

Cluj-Napoca has the country's largest Roman Catholic church, St Mikhail's, founded in the 14th century, the gothic building had several parts added subsequently. Most of its sculptures and paintings were destroyed during the Reformation period. Cluj is the country's second university town (after the capital) and has a well-stocked botanic garden. The city's industries include metalworking, engineering, chemical and ceramic manufacturing.

Population 260 000
Map Romania Aa

Cluny ***France*** Town 75 km (47 miles) north and slightly west of Lyons. Its Benedictine abbey, founded in 910 by William I, Duke of Aquitaine, became a major intellectual and artistic centre in medieval Christendom. Much of the vast church was demolished in 1798. Over 100 000 people visit the remnants each year. There is a 13th-century church and some old houses, including the one where the painter Pierre Paul Prud'hon (1758-1823) was born.

Population 4700
Map France Fc

cluse French term for a steep-sided valley cutting across a limestone ridge, especially in the Jura Mountains and the fore-Alps of Savoy.

Clutha ***New Zealand*** South Island river rising in Lake Wanaka and flowing some 320 km (200 miles) into the Pacific near Balclutha, in the island's south-eastern corner. Alluvial gold was found in the river during the 1860s, but today it is important for hydroelectric power.

Map New Zealand Bf

Clwyd ***United Kingdom*** County of north-east Wales, created in the 1970s from the old county of Flint and parts of Denbigh and Merioneth. It covers 2425 km^2 (936 sq miles) and stretches east to the suburbs of the English city of Chester and the banks of the River Dee, and west to include the popular seaside resorts of Rhyl and Colwyn Bay. There is an old coalfield around Wrexham and fertile farmland in the Dee valley, but most of the area is hill country or open moorland given over to sheep farming. The county town is Mold.

Population 395 000
Map United Kingdom Dd

Clyde ***United Kingdom*** River in south-west Scotland, entirely within Strathclyde region. It rises in the Lowther Hills and flows 170 km (105 miles) north and north-west across the Lanarkshire coalfield to Glasgow, where its estuary, the Firth of Clyde, stretches more than 100 km (60 miles) to the sea. It is navigable to Glasgow and was one of the world's great shipbuilding centres.

Map United Kingdom Cc

Clydebank ***United Kingdom*** See GLASGOW

Cnoc ***Ireland*** See KNOCK

coal Natural dark brown to black combustible deposit formed from the remains of swamp vegetation which has been converted to a solid state by the pressure of rocks deposited on top. As the vegetation compacted, water and gases were driven off leaving mainly carbon. Stages in this conversion are represented by peat, lignite, coal and anthracite.

Coanza ***Angola*** See CUANZA

coast Zone of land next to the sea, including the land behind the cliffs or their equivalent as well as the beach and the shore. See EROSION

Coast Ranges ***USA*** Mountains stretching about 1600 km (1000 miles) along the Pacific coast. They extend from Puget Sound on the Canadian border south to the vicinity of Los Angeles, California, where they join the Sierra Nevada. Here they reach 3301 m (10 831 ft) in the San Jacinto Mountains, but most of the ranges are less than 760 m (2500 ft) high.

Map United States Bc

Coatbridge ***United Kingdom*** See STRATHCLYDE

cobble Rounded or semi-rounded rock fragment, larger than a pebble but smaller than a boulder, with a diameter between 64 and 256 mm (about 2.5 and 10 in).

Cóbh ***Ireland*** Town on Great Island, the largest island in Cork Harbour on the south coast, with road and rail connections to the mainland. Victims drowned in the sinking of the *Lusitania*, a Cunard passenger liner torpedoed by a German submarine off Old Head of Kinsale in 1915, are buried at Clonmel 4 km (2 miles) from the town. A monument stands beside the harbour, which was once a regular port of call for transatlantic liners. Offshore is Spike Island, whose fort was a Victorian prison which was recently reopened, but largely destroyed in jail riots in 1985.

Population 6600
Map Ireland Bc

Coblenz ***West Germany*** See KOBLENZ

Coburg (Koburg) ***West Germany*** Bavarian town on the Itz river about 90 km (55 miles) north of Nuremberg. It was the seat of the Dukes of Saxe-Coburg-Gotha, and their ducal palace in the town dates from 1549. Queen Victoria of England married one of the family, Prince Albert, in 1840. Coburg has metal, glass and ceramics industries, and also produces toys.

Population 45 000
Map West Germany Dc

Cochabamba ***Bolivia*** City and department 230 km (144 miles) south-east of the capital, La Paz, on the slopes of the Eastern Cordillera bordering the tropical lowlands. The department's fertile valleys produce grain, fruit and beef for the towns of the Altiplano plateau.

The city was founded in 1574 by the Spanish adventurer Sebastián Barba de Padilla. The central square, with galleries on all four sides, is typically Spanish. The city has a number of museums, including one which specialises in pre-Columbian relics from the cities of Incallacta and TIAHUANACO.

Cochabamba is a manufacturing centre with car assembly, cement, textiles, shoes, furniture, fruit canning and milk processing plants – and its oil refinery supplies much of the petrol and kerosene Bolivia needs.

Population (city) 282 000; (department) 908 700
Map Bolivia Bb

Cochin ***India*** Port on the Arabian Sea about 230 km (143 miles) from the country's southern tip. It was first visited by Portuguese explorers in 1500, who set up the earliest European settlement there. The Portuguese navigator Vasco da Gama, who discovered the sea route to India, died there on Christmas Day 1524, aged about 55. The port now exports copra, rubber, tea and nuts. Its industries include petrochemicals and shipbuilding.

Population 551 600
Map India Be

Cochin China ***Vietnam*** Name of the MEKONG DELTA region in French colonial times.

Cockburn Town ***Turks and Caicos Islands*** Capital of the Caribbean island group, on the island of Grand Turk. It is the main financial and business centre.

Population 3200
Map Caribbean Aa

Cockpit Country ***Jamaica*** Deeply eroded limestone region in the north-west of the island, to the south-east of Montego Bay. Among the limestone features of caverns, underground streams and potholes are steep-sided hollows, or 'cockpits', some of which are 150 m (nearly 500 ft) deep. The thickly wooded, jagged slopes of the area became a hiding place for slaves who had escaped from their Spanish masters. When the English captured the island from the Spanish in 1655, the fugitive slaves carried on a successful guerrilla war against the new conquerors that resulted in a cease-fire on the ex-slaves' terms in 1739. Today, the area is sparsely populated by the descendants of the slaves, called Maroons, who live in the Cockpit free of taxation and other government interference, with their rights guaranteed by treaty.

Population 5000
Map Jamaica Ba

Cocos Islands (Keeling Islands) ***Indian Ocean*** Group of 28 small coral islands about 1100 km (683 miles) south-west of Java. Covering 14 km^2 (5.5 sq miles) of land, they were discovered in 1609 by William Keeling, a mariner with the East India Company. They became a British possession in 1857 and were put under Australian control in 1955. However, from 1886 to 1978 the Clunies-Ross family owned all the

islands. The inhabitants are mostly Malays who work in the coconut plantations. In 1914, during the First World War, the islands were the scene of the sinking of the German cruiser *Emden* by the Australian cruiser *Sydney*.
Population 700
Map Indian Ocean Dc

Cod, Cape ***USA*** Area of sand dunes, lakes and salt marshes on the coast of Massachusetts, forming a narrow curving peninsula that largely encloses Cape Cod Bay, to the south-east of Boston. It was here that the Pilgrim Fathers first landed in 1620 before setting up the PLYMOUTH Colony to escape religious persecution in England. Much of it is preserved as the Cape Cod National Seashore. The area is popular with summer holidaymakers.
Map United States Mb

Cognac ***France*** Historic medieval town on the south bank of the Charente river. It once belonged to the English king, Richard the Lionheart. Its fame derives from its brandy, which has been distilled there for 400 years. Three-quarters of all Cognac is exported.
Population 32 400
Map France Cd

Coihaique ***Chile*** Capital of Aisén region in the south. It was founded in 1931 in a grassy valley on the east flank of the Andes. It is a growing tourist centre.
Population 35 700
Map Chile Ad

Coimbatore ***India*** Cotton textile and engineering city about 300 km (186 miles) from the southern tip of India. It also produces sugar and coffee.
Population 920 400
Map India Be

Coimbra ***Portugal*** University city about 175 km (110 miles) north-east of the capital, Lisbon, on a hill overlooking the Mondego river. The university was founded in 1290, in Lisbon, moved to Coimbra in 1308, and became established there permanently in 1537. There is a fortress-like cathedral, built in the 12th century. The older part of the city is a maze of steep, narrow, cobbled streets.
Population 71 800
Map Portugal Bb

coire See CIRQUE

col Pass across a ridge or range of mountains providing a routeway from one side to the other.

cold climates Those climates with a long cold season – having more than 6 months with temperatures below 6°C (43°F).

cold continental climate Climate in high latitudes, namely at about 60°N as in Siberia, with a long, severe winter, light rainfall but snow in winter. The climate corresponds to the boreal forest region of the Northern Hemisphere.

cold desert Term applied to tundra and polar regions where vegetation is non-existent or restricted by low temperatures.

cold front Surface of separation between a cold air mass and a warm air mass which the colder air is undercutting and uplifting. As the front moves, colder air replaces the warm air at ground level, hence its name. Uplift of the warm air produces cumulus and cumulonimbus CLOUDS along the line of the front, often accompanied by heavy showers of rain or even thunderstorms.

cold wave Sudden burst of cold weather brought by a cold AIR MASS following a DEPRESSION.

Colditz ***East Germany*** Town about 40 km (25 miles) south-east of Leipzig. Its forbidding castle was used by the Nazis as a prison for Allied officers between 1939 and 1945. The inmates several times disproved its 'escape-proof' reputation. The town has a 16th-century town hall, and makes machinery and china.
Population 2500
Map East Germany Cc

Coleraine ***United Kingdom*** Town in Northern Ireland, 76 km (47 miles) north-west of the capital, Belfast. The town is located at the lowest crossing point of the River Bann. It is a vigorous market centre with some engineering, textile and agricultural industries, and a port for goods traffic. In 1968, Coleraine became the site of the New University of Ulster. Mountsandel Fort, 2 km (1 mile) south, was a Celtic and then Norman stronghold. It also has Stone Age remains dating back to about 7000 BC.
Population 16 000
Map United Kingdom Bc

Colima ***Mexico*** **1.** Small state on the Pacific coast in south-west Mexico, which produces citrus fruit, cotton, sugar cane, tobacco and rice, as well as making a living from the sea. Its main port, Manzanillo, is a modern, bustling tourist resort, with fine beaches. **2.** Capital of Colima state, 64 km (40 miles) inland from Manzanillo. It is a sleepy colonial town with graceful arcades, and markets selling cattle and agricultural produce.
Population 368 000
Map Mexico Bc

Colmar ***France*** Capital of Haut-Rhin department, on the Lauch river 38 km (24 miles) north of Mulhouse. The old town has narrow streets with attractive timbered houses. There are textile mills, and a wine festival is held in the late summer.
Population 84 000
Map France Gb

Cologne (Köln) ***West Germany*** City, inland port and road and rail junction on the banks of the Rhine, 25 km (15 miles) north and slightly west of the capital, Bonn. The Romans established a walled colony there in AD 50, and the city takes its name from *colonia*, the Latin word for 'colony'. Remains of the Roman governor's residence have been found beneath the City Hall. On the south side of the cathedral is an excellent Romano-Germanic museum with many relics of Roman times, while the Wallraf-Richartz art gallery contains one of Germany's finest picture collections.

Cologne's cathedral, which took 600 years to

A DAY IN THE LIFE OF A WEST GERMAN SCHOOLTEACHER

The alarm clock wakes schoolteacher Klaus von Harlessem at 6.35 each morning in his comfortable, five-bedroomed house in Marienburg – a prosperous suburb on the outskirts of Cologne, West Germany. A neighbour in a luxury villa around the corner is a former West German president, but Klaus's terraced house – which he is buying for DM130 000 (US$51 000) – is more modest.

Klaus, 39, tiptoes downstairs to make his own breakfast; his wife Sibylle, 34, who runs a translation agency, is still asleep. So, he hopes, is their only child, 18-month-old Nils.

The school day begins early in West Germany. The first lesson is at 7.55 am in the Freirherr von Stein Secondary School in Bonn, about 19 km (12 miles) away down the motorway, where Klaus teaches English, geography and religion. To get there, he has a choice of transport: the family's ageing Volkswagen camper van, or a smart Passat car bought outright this year.

The school – which Klaus reaches in about 17 minutes in the Passat – has some 400 boys and girls aged from 10 to 16 years. His first stop is the staff room. The atmosphere among the 22 teachers is fairly formal. Some shake hands on arrival and address each other by their surnames. 'I like this old-fashioned formality,' says Klaus. 'It reflects a serious approach to teaching.'

On two mornings of the working week, Klaus's first lesson is English – which he teaches to mixed classes of all ages. As a form master, he is also in charge of Class 8B, consisting of 23 pupils mostly aged 14. A few years ago, classes of 40 and more were commonplace; but the declining birthrate in West Germany has reduced their size.

The babble of teenage chatter dies away as Klaus enters and says in English, 'Good morning'. The class replies in unison, 'Good morning, sir'. Discipline is no problem. 'I can only recall one incident when some youths from another school beat up one of our boys,' says Klaus. 'I talked to them and in the end they offered to work for 20 hours in a hospital and pay some compensation to the victims' parents for medical treatment.'

Klaus teaches for 27 hours a week in the mornings only, including alternate Saturdays. His gross salary is DM4486 (US$1760) a month, which is reduced to DM3158 (US$1240) after deductions for tax and the mortgage on his home.

His school and family are the centre of Klaus's interests. 'I'm content with my lot,' he says, sipping lemonade on his back lawn, 'and I certainly would not describe myself as upwardly mobile.'

build, was completed only in 1880 and survived the bombing of the Second World War. It is one of Europe's most splendid Gothic buildings, with twin towers topped by lacy spires rising to 157 m (515 ft). It contains relics claimed to be of the Three Kings of the Nativity. The city's university, founded in 1388, is one of Europe's oldest. Konrad Adenauer (1876-1967), West Germany's first Chancellor, was a former mayor of the city.

Population 932 400

Map West Germany Bc

Colombia See p. 158

Colombo *Sri Lanka* The country's largest city and capital. It lies on the west coast and is the major port. handling 90 per cent of Sri Lanka's overseas trade. The city, founded by Arab traders, was captured by the Portuguese in 1517. The Dutch ousted the Portuguese in 1658, and built a fort in the central district now called Fort. The British took over in 1796, and Colombo expanded rapidly when railways were built to new tea plantations inland. A new harbour was added in the 19th century.

The city has many fine avenues, parks, old colonial buildings and Hindu and Buddhist temples, as well as newer skyscrapers. It has excellent bazaars and gemstone markets. However, there are slums too, and – on the swampy ground around the city – squatter settlements. A new seat of government is being built at the town of Kotte to the east. Colombo has three universities and one of the finest zoological gardens in Asia.

Population 600 000

Map Sri Lanka Ab

Colón *Panama* The country's second largest city after the capital, Panama City. Colón lies at the Caribbean end of the Panama Canal, about 80 km (50 miles) north-west of Panama City. Its name means 'Columbus' and its port and twin town is Cristobal – meaning 'Christopher' – so that the two places take their names from America's discoverer.

Colón's free-trade zone is a showpiece of the Panamanian economy, but away from the business centre, with its imposing civic buildings, there are extensive shanty towns.

Population 117 000

Map Panama Ba

Colón Archipelago *Pacific Ocean* See GALAPAGOS ISLANDS

Colonial National Historical Park *USA* See JAMESTOWN; WILLIAMSBURG; YORKTOWN

Colorado *North America* River rising in the Rocky Mountains north-west of Denver, in the US state of Colorado and flowing about 2330 km (1450 miles) south-west to the Gulf of California in Mexico. It flows through deep gorges, including the GRAND CANYON, and is dammed in several places, including the Hoover Dam.

Map United States Dd, Ec

Colorado *USA* Western state covering 270 000 km² (104 247 sq miles) between Wyoming to the north and New Mexico to the south. The eastern plains average a height of 1500 m (4900 ft), but in the west there are the Rocky Mountains with some 1100 peaks above 3000 m (10 000 ft) and 50 higher than 4265 m (14 000 ft); the highest is Mount Elbert (4399 m, 14 431 ft). Cattle, wheat and maize are the main farm products, and oil, gas and coal the chief minerals, with some silver, lead and molybdenum (used in steel making).

The fine scenery and climate attracts tourists, and also immigrants and high-technology industries, and Colorado has one of the country's fastest-growing populations. Aspen is one of the USA's largest ski resorts. DENVER is the state capital, and at COLORADO SPRINGS is the headquarters of the North American Air Defense Command (NORAD).

Population 3 231 000

Map United States Ec

Colorado Plateau *USA* Area of spectacular scenery in Colorado, Utah, New Mexico and Arizona. The plateau, 610-3660 m (2000-12 000 ft) above sea level, is composed of sedimentary rocks, and is deeply cut by the canyons of the Colorado river and its tributaries, including the Grand Canyon.

Map United States Dc

Colorado Springs *USA* Resort and spa city in Colorado, about 100 km (60 miles) south of the state capital, Denver, at the foot of Pikes Peak (4300 m, 14 109 ft). The US Air Force Academy is nearby.

Population (city) 247 700; (metropolitan area) 349 100

Map United States Fc

Columbia *Canada/USA* River 1950 km (1210 miles) long, which rises in eastern British Columbia. It flows north, then turns south through Washington state in the USA and empties into the Pacific at Portland. It has been harnessed for irrigation and hydroelectricity and is a major tourist attraction.

Map Canada Dd; United States Ba

Columbia *USA* State capital of South Carolina, about 170 km (105 miles) north-west of Charleston. It was burnt down in 1865 during the Civil War. Today it is a manufacturing and market centre and is the seat of the University of South Carolina.

Population (city) 98 600; (metropolitan area) 433 200

Map United States Jd

▲ DOWNTOWN COLOMBO Grand offices built during British rule house government firms and the city's many banking, insurance and brokerage houses.

Columbia, District of *USA* See WASHINGTON DC

Columbia Icefield *Canada* Largest permanent icefield outside the Arctic and Antarctic, covering 300 km² (116 sq miles). It lies in the ROCKY MOUNTAINS and can be reached from the Icefields Parkway, near the boundary of the Banff and Jasper national parks. Tourist excursions are made by snow tractor.

Map Canada Dc

Columbus *USA* State capital of Ohio. It lies in the centre of the state, about 160 km (100 miles) from Lake Erie. The city is an industrial centre, particularly for metalworking and printing, and has three universities.

Population (city) 566 100; (metropolitan area) 1 279 000

Map United States Jb

columnar structure Natural structure of columns, such as the Giant's Causeway in Northern Ireland, formed when a pattern of regular joints developed in an igneous rock as it cooled and contracted.

Comayagua *Honduras* Town about 65 km (40 miles) north-west of Tegucigalpa, which replaced it as the national capital in 1880. Founded in 1537, Comayagua has a wealth of Spanish architecture, including a 17th-century cathedral containing a 12th-century Moorish clock. It is the centre of a cattle-raising area.

Population 28 120

Map Honduras Ab

Colombia

ONCE FAMOUS FOR EMERALDS AND GOLD – NOW NOTORIOUS FOR DRUGS AND VIOLENCE – COLOMBIA IS A COUNTRY WHOSE TROUBLES HAVE TAUGHT IT THE VALUE OF COMPROMISE

Crime and violence go hand in hand with the illegal drugs trade and cast a long shadow over Colombia. Its inhabitants have learned to live with the perils involved and know how to avoid them; it is more likely to be tourists who find themselves in trouble, especially in the cities where pickpocketing, thieving and mugging are all too frequent occurrences. But for those who tread carefully, Colombia is a fascinating and beautiful country, with a rich mixture of cultures. Traditions survive, from the Indians who preceded colonisation, from the Spanish colonists, and from the African slaves brought by the Europeans; the influence of these cultures is apparent in the country's food, music, art, architecture and literature. Recently, modern industry has been added to this mixture, as Colombia's recipe for prosperity.

Thirty-five thousand relics of Colombia's past prosperity are on show at the Gold Museum in the capital city of BOGOTA. Before the Spanish came to Colombia in the 16th century, the area was inhabited by Indian tribes who left an impressive legacy. The Chibchas in particular, who lived in the area around Bogotá, were skilled craftsmen in gold. Their jewellery and ceremonial objects were encrusted with emeralds, for which Colombia is renowned; indeed, the jewels are still mined at Muzo near Chiquinquira, and one of the world's largest unpolished emeralds is in the Gold Museum. Today's Indians still make fine artefacts in gold and silver.

The Chibchas gave rise to the legend of *El Dorado*, meaning 'the gilded one' in Spanish. Chibcha chiefs underwent a ritual in which they were smeared with resin and then rolled in gold dust; clad in nothing but this precious powder they dived into the sacred Lake Guatavita and were ceremonially cleansed. Rumours of this ceremony reached the gold-hungry Spaniards, who found not only gold but also a rich, fertile land and beautiful scenery. Colombia has almost every kind of climate and country, with the snow-covered peaks of the ANDES mountains in the west, grassland in the centre, and tropical jungle to the east and south. There is even desert, between the Andes and the Pacific coast.

The Andes take the form of a three-pronged fork, with three ranges – the Cordilleras Occidental, Central and Oriental – running north and largely disappearing towards the Caribbean. Between the forks are the fertile valleys of the Cauca river, with rich volcanic soils, and the MAGDALENA river, which join before flowing into the Caribbean at BARRANQUILLA, Colombia's chief port. Large estates in this area account for many of Colombia's agricultural exports. Sugar is grown in the Cauca Valley, while cotton and cattle come from the Magdalena Valley and the coastal plain. By contrast coffee, Colombia's main export crop (excluding, possibly, drugs), is grown on small farms in the mountains. Small farms also raise most of the home produce, including rice, maize, barley, fruit and potatoes.

Half the country lies east of the Andes. Much of this land takes the form of hot plains covered with tropical grassland. Farther east, in the Amazon basin, tropical forest takes over. To the west of the mountains, the Pacific coastal plain rises to the Serranía de Baudó.

There are no seasons here; climate depends on altitude, so that much of the hot Pacific coast is desert, the slopes of the Andes offer eternal spring, and the summits are wintry and frozen. In the north, along the Caribbean coast, it is summer all year.

This wide range of climate supports an extraordinary range of plants and wildlife. There are many species of orchids, one of which is Colombia's national flower. It is said, too, that Colombia has more species of birds than any other country in the world – some 1500, including North American migrants.

But it is the illegal drugs trade that has brought undesirable fame to Colombia. Marijuana is widely grown in the northern coastal region, and cocaine is produced from Colombian coca plant leaves. Furthermore, large amounts of coca leaves are imported illegally

▶ HIGH-LEVEL FARMING Indian farmers grow beans, barley, wheat and potatoes on the uneven fields lying on the high, arid plateau surrounding the city of Pasto in south-west Colombia. Pasto is the commercial centre for the agricultural area, which rises to about 2560 m (8400 ft) above sea level.

from Bolivia and Peru for processing and exporting as cocaine. Obviously, the drug mafias do not provide figures for their trade, but it is believed that drugs are a major export. The US Narcotics Bureau has been working with the Colombian Government in an attempt to stamp out the trade by destroying crops of marijuana, dumping hoards of processed cocaine in the Caribbean and investigating the drug syndicates.

SHARING THE POWER

Bogotá is the seat of one of the most unusual governments in the world. The democratically elected president appoints a cabinet of which half is from his own party, and half is from the opposition. All public appointments are similarly divided; half the government employees, from the highest civil servants to the road sweepers, belong to the Conservative Party, and half belong to the Liberals. This system results from a long history of civil conflict. Between 1899 and 1902 the two parties fought the War of a Thousand Days, in which 100 000 people were killed. Then again, between 1948 and 1953, they clashed in a terrible civil war known as 'The Violence' in which 300 000 Colombians died. To end this strife, the two parties agreed to their unique system of power sharing. The agreement to govern in this way ended formally in 1974, but still continues in spirit.

Today a new enemy faces this united government; guerrillas fighting for social reforms, and especially for a fairer distribution of farmland. Large estates hold most of the land – notably the sugar plantations and the cotton and cattle farms. In contrast, many peasant farmers have so little land that they can hardly grow enough to live on. As a consequence, people are drifting away from the countryside and 70 per cent of the people now live in towns.

The main cities of Bogotá, MEDELLIN, CALI and Barranquilla, are busy sophisticated centres with impressive architecture and modern roads. But they all suffer from unemployment and housing shortages.

In part, the social problems are due to a high birthrate, though family planning programmes are slowly beginning to work. They have lowered the average number of children a woman expects to have in her lifetime from seven (in the 1960s) to six (in the 1970s).

The country's future depends on its ability to generate more exports. Coffee is still the most important – every car that enters Colombia is sprayed against diseases which might damage or destroy the coffee crop.

Oil was discovered recently near the border with Venezuela and provides a potential export along with natural gas. From the mid-1980s Colombia will probably become South America's leading coal exporter, from the giant deposits found near the Caribbean town of Riohacha. Nickel from the department of Cordoba is also being launched onto the export market.

Many of Colombia's manufactures, including iron and steel, cars, refrigerators and other domestic appliances, are in the hands of international corporations, though the state is involved in some important industries, such as textiles.

Colombia's wealth is still under-exploited, not least its potential for tourism. Among the many interesting places is the fortified city of CARTAGENA on the Caribbean coast.

Colombia is a land of festivals. Medellín, the orchid capital, has its own flower festival, and Caribbean Baranquilla goes wild with parades and parties at carnival time. The south has a particular reputation for fun, and the first week of the year is one long fiesta. January 5 is known as 'Black Day', when practical jokers dip their hands in grease and smear it on each other's faces. The next day is 'White Day' – when a great deal of flour or talcum powder is thrown at passers-by.

The influence of Spain is clear from another popular pastime – bullfighting. The bullfighting highlight is the second week in January, when the mountain town of MANIZALES has a programme of first class fights.

The plains – *llanos* – east of the Andes, are rich in wildlife, including jaguars, pumas, deer, boars and tapirs. But visitors have to beware of snakes, alligators and piranha fish. Piranhas can strip a deer – or a man – to a skeleton in a few minutes.

COLOMBIA AT A GLANCE
Area 1 138 914 km² (456 535 sq miles)
Population 30 125 000
Capital Bogotá
Government Republic
Currency Peso = 100 centavos
Language Spanish
Religion Christian (97% Roman Catholic, 1% Protestant)
Climate Tropical on coast, temperate on plateaus. Average temperature in Bogotá (altitude 2610 m, 8563 ft) ranges from 10-18°C (50-64°F) in July to 9-20°C (48-68°F) in February
Main primary products Coffee, rice, potatoes, cassava, sugar, bananas, maize, cattle, timber; oil and natural gas, emeralds, platinum, gold, silver, iron ore, coal
Major industries Agriculture, mining, iron and steel, food processing, cement, oil refining, paper, textiles
Main exports Coffee, textiles and clothing, fruit and vegetables, sugar, chemicals, cotton, machinery
Annual income per head (US$) 1100
Population growth (per thous/yr) 21
Life expectancy (yrs) Male 61 **Female** 65

combe See COOMBE

Comilla ***Bangladesh*** **1.** District in the south-east near the Indian border consisting chiefly of a level, alluvial plain cut by rivers. Low, forest-clad hills rise in the east, where some tea and cotton are grown. Rice, jute, oilseeds and vegetables are the main crops on the plain. The district covers 6599 km² (2548 sq miles).

2. Chief town of Comilla district and site of the Rural Development Academy, which was established about 1965 and became something of an international show-piece. Comilla trades in rice, jute, oilseeds, cane and bamboo, hides and skins, and tobacco. Industries include engineering, metal goods, a match factory, a jute mill and soap. The town is noted for the many tanks (reservoirs) nearby. There are over 400 and one, dating from the 15th century, is 1.6 km (1 mile) in circumference.

Population (district) 6 881 000; (town) 126 100
Map Bangladesh Cc

Communism Peak (Pik Kommunizma) ***USSR*** The country's highest peak, rising to 7495 m (24 590 ft) in the Pamir mountains near the Afghan border. Until 1957 it was called Stalin Peak in honour of Soviet leader Joseph Stalin (1879-1953), but was renamed after his regime was posthumously denounced by the Soviet authorities.

Map USSR Ie

Como ***Italy*** Resort town at the south-western tip of Lake COMO, about 35 km (22 miles) north of Milan. The centre of the city preserves the street plan of the Roman fort. The Fascist dictator Benito Mussolini (1883-1945) was executed near Como by Italian partisans just before the end of the Second World War.

Population 95 200
Map Italy Bb

Como, Lake ***Italy*** Alpine lake about 35 km (22 miles) north of Milan, covering 146 km² (56 sq miles). It is shaped like an inverted Y, with the towns of Lecco and COMO on its southern prongs. Bellagio, a beautiful little resort town, stands on a promontory at the fork of the Y. The lake reaches a depth of 410 m (1345 ft).

Map Italy Ba

Comodoro Rivadavia ***Argentina*** See PATAGONIA

Comorin, Cape ***India*** Southernmost tip of India, where the Arabian Sea meets the Bay of Bengal. The cape is named after its temple to the Hindu virgin goddess Kumari.

Map India Be

Compiègne ***France*** Historic town surrounded by woodland on the Oise river 72 km (45 miles) north-east of Paris. Joan of Arc was imprisoned in the 12th-century tower there before her execution at the stake in 1431. In the 18th century Compiègne's ancient chateau was enlarged into a royal palace, which became the favourite residence of Emperor Napoleon III. The armistice ending the First World War, and that between France and Germany in 1940, was signed just outside the town.

Population 43 300
Map France Eb

Comoros

TROPICAL ISLANDS WHOSE PEOPLE CHOSE A PRECARIOUS AND LONELY INDEPENDENCE

The volcanic Comoro Islands are one of the world's least known sovereign states. They lie in the Indian Ocean, between Mozambique and the northern tip of Madagascar, and comprise four main islands. Three of them have recently changed their names following independence: NJAZIDJA (formerly Grande Comore), MWALI (formerly Mohéli) and NZWAMI (formerly Anjouan). The easternmost island of MAYOTTE remains a French dependency.

France had administered, but largely neglected, the Comoros since the middle of the 19th century. When the three islands became independent in 1975 they became the Federal Islamic Republic of the Comoros.

They are mostly forested, but small areas are cultivated. The land mostly belongs to foreign plantation owners and small, traditional power groups.

In their turn, Malay, Malagasy, Persian, Arab and African peoples have come to the islands, resulting in a population of mixed blood. Islamic culture and religion are dominant.

The economy is precarious; it is increasingly dependent on foreign aid and the republic is plunging into debt, with little prospect of escape. The population is growing fast, and many emigrants seek work in Madagascar, East Africa and France.

COMOROS AT A GLANCE

Map Madagascar Aa
Area 1862 km² (719 sq miles)
Population 422 000
Capital Moroni
Government Federal republic
Currency CFA franc = 100 centimes
Languages Swahili, Arabic, French
Religions Muslim (99%), Christian (1%)
Climate Tropical, with heavy monsoon rains November to April; average temperature in Moroni ranges from 19-27°C (66-81°F) in August to 24-31°C (75-88°F) in March
Main primary products Vanilla, cloves, coconuts, rice, maize, sweet potatoes, cassava, timber
Major industries Agriculture, fishing, forestry, food processing, perfume oil extraction
Main exports Vanilla, cloves, perfume oils, copra, coconut fibre
Annual income per head (US$) 260
Population growth (per thous/yr) 29
Life expectancy (yrs) Male 44 **Female** 48

Compostela ***Spain*** See SANTIAGO DE COMPOSTELA

Conakry ***Guinea*** The country's capital, on the Atlantic coast, partly on Tumbo island and partly on the adjacent mainland, the two being linked by causeway. Originally a Sousou settlement, the present city was laid out by the French in 1889. The administrative and commercial centre, on the island, has broad, straight streets and tree-lined avenues, while the main residential areas – where grand colonial villas contrast sharply with the crowded housing of the local people – lie on the mainland. Here also is a large market and the National Assembly.

Conakry has a deep natural harbour sheltered by the offshore Iles des Los. Its trade was limited until the early 1950s, when iron ore was first exported. As the terminus of Guinea's main railway, it now handles most of the country's exports of bananas, citrus fruits, pineapples, kola nuts, groundnuts and coffee.

Population 763 000
Map Guinea Bb

Conamara ***Ireland*** See CONNEMARA

Concepción ***Chile*** Province and city in Bío-Bío region, about 420 km (260 miles) south of Santiago. It is one of the country's most heavily industrialised areas, with coalfields, an oil refinery, a steel plant at Huachipato and excellent port facilities at Talcahuano, which is also a fishing and whaling centre. The city was founded in 1550 on the Bío-Bío river, and despite severe earthquakes – most recently in 1939 and 1960 – it has expanded to become one of Chile's largest cities, with industries including wool and cotton textiles, glass and cement. It has an attractive central square – the Plaza de Armas – the colonial Planchada Fort and a modern university. San Pedro, across the river, has lagoons and forests and is a tourist resort.

Population (province) 762 800; (city) 210 000
Map Chile Ac

Concord ***USA*** State capital of New Hampshire. It lies in the south of the state, about 100 km (60 miles) north and slightly west of the city of Boston. In the early 19th century Concord gave its name to a type of stagecoach. Quarries nearby provided stone for the Library of the Congress building in Washington DC.

Population 30 900
Map United States Lb

condensation Process by which a substance changes from the vapour state to the liquid state. For instance, clouds are formed by the condensation of water vapour into water droplets in the atmosphere. Cooling reduces the capacity of air to hold water vapour and sufficient cooling will cause SATURATION followed by condensation. Occasionally condensation does not take place and the air is then described as supersaturated.

condensation trail, vapour trail, contrail Long cloud-like trail that forms behind an aircraft, particularly a jet, resulting from condensation of water vapour produced by the combustion of its fuel. Trails sometimes form at the wing tips of high-flying aircraft passing through a region of supersaturated air.

A DAY IN THE LIFE OF A CHILEAN WINEGROWER

Pedro Antonio Perez Sanchez is a winegrower just outside the town of Concepción in southern central Chile. From his estate, nestling between the Andes mountains and the Pacific Ocean, come some of Chile's 1000 or so varieties of wine.

The 162 hectare (400 acre) estate gives Pedro more than enough to do. Much of his time is spent touring the estate on horseback checking the vines, testing the grapes, overseeing the maintenance of the land and generally directing the work of the 20 or so Indians and Spanish Chileans he employs. These workers live in small houses on the estate and are paid partly in money and partly in wine for their own consumption. Occasionally, Pedro has to drive in his ageing Ford the 48 km (30 miles) into Concepción, Chile's third largest industrial city, to arrange some business affairs or to see his younger brother, Juan Alberto, a lawyer who takes care of the legal side of the estate for Pedro.

Slowly, the grapes ripen on the vines and the highlight of the year approaches – the harvest at the end of March or the beginning of April. For a while there is no rest for anyone. The estate's regular workers are supplemented by casual labourers. When the harvesting work is complete Pedro throws a big feast which lasts three or four days. He provides wine, and spreads of chicken, fish and beef dishes, stuffed eggplant, tomatoes and peppers, beans cooked in every possible way, cheese and fruit. The guests sing, play guitars and tambourines, and dance in the open air.

But at the end of a normal working day, Pedro aims to return straight home – where his wife, Maria Isabel, will be preparing dinner, a lavish and leisurely meal of Spanish cooking. Home is a large whitewashed building of two storeys made of sun-baked bricks in the Spanish colonial style, with a verandah round the outside. The house, with its nearby outhouses and stables, is not only the hub of activity on the estate but also the centre of a close-knit family life which Pedro and Maria share with their three youngest children – two boys and a girl – and Pedro's parents.

Over dinner the children chatter about their day at the private high school in Concepción, where one of Pedro's employees drives them every morning. After they have gone off to watch television or to do their homework, Pedro and his father sit and discuss estate business or the latest news of Pedro's two eldest sons, Jose Antonio and Felipe Alberto, in the capital, Santiago, where Jose is at university and Felipe at military training college.

conglomerate Rock composed of rounded rock fragments larger than 2 mm ($^1/_{16}$ in) in diameter cemented together by calcium carbonate, silica, iron oxide, clay or some other fine-grained material. It is commonly known as puddingstone.

Congo See p. 163

Congo ***Equatorial Africa*** Old name for the ZAIRE river, still used in the People's Republic of the Congo.
Map Congo Bb

coniferous forest Forest consisting mainly of evergreen cone-bearing trees with needle-shaped leaves and softwood timber. See also TAIGA and BOREAL FOREST

Connaught (Cuige Chonnacht) ***Ireland*** One of Ireland's four ancient provinces, lying in the west of the country. The other three are Leinster, Munster and Ulster. 'To Hell or Connaught', was Oliver Cromwell's cry as he drove off landowners there to make space for his land grants to soldiers and supporters who had helped finance his conquest of all Ireland following the rebellion of 1641-2.
Population 423 900
Map Ireland Bb

Connecticut ***USA*** North-eastern state on the Atlantic coast opposite Long Island. It was first settled by Europeans in 1635-6 and was one of the 13 original states. Connecticut covers 12 973 km² (5009 sq miles) and is an undulating, wooded area divided by the Connecticut river. Agriculture is based on dairy produce, fruit, vegetables and tobacco. Industries include aerospace technology, armaments and machinery. The state capital is HARTFORD. Many people in the south-western part of the state commute into New York city.
Population 3 174 000
Map United States Lb

Connemara (Conamara) ***Ireland*** Wild, sparsely populated region of lakes, mountains and streams north of Galway Bay on the west coast. Connemara is dominated by The Twelve Pins – a group of hills which rise to 730 m (2395 ft) – and by the Maumturk Mountains (668 m, 2193 ft). Tough Connemara ponies are traded at an annual summer show and fair at CLIFDEN.
Map Ireland Bb

Conrad discontinuity The discontinuity within the earth's crust between the granitic and lower layers at which the velocity of earthquake waves changes abruptly.

Constance (Konstanz, Bodensee) ***West Germany/Switzerland*** Lake covering 540 km² (208 sq miles) and straddling the border. The Rhine flows east to west through the lake, which is up to 245 m (800 ft) deep. The tourist resort of Constance (population 68 400) is on the south shore of the lake, a part of Germany surrounded by Swiss territory.

The Council of Constance, held at the town hall in 1414-18, ended the schism in the Roman Catholic Church which, from 1378, had led to the existence of two rival papacies, one in Avignon and one in Rome. In 1417 unity was restored with the election in Constance of Pope Martin V. The council also condemned the reformer John Huss, who attacked the buying of church offices. Huss was burnt at the stake in 1415 and his ashes scattered on the lake. A memorial to airship inventor Count Ferdinand von Zeppelin (1838-1917), born in the town, stands by the lakeside quay.
Map West Germany Ce

Constanţa ***Romania*** Major port, resort and industrial city on the Black Sea, 200 km (124 miles) east of Bucharest. It was founded in the 6th century BC by the Greeks, who called it Tomis. Then, in the 4th century AD it was renamed after the Roman emperor Constantine the Great. In AD 8 the Roman poet Ovid was banished to the city by the Emperor Augustus for some undisclosed indiscretion. Ovid spent the last nine years of his life in exile in the area, writing pathetic poems to Rome – collected as *Tristia*, or 'Sadness' – pleading futilely to be forgiven. Today there is a square named after the poet, containing a statue of him.

Nearby is the National History and Archaeological Museum, which has exhibits going back to the later Stone Age (4000-3000 BC) – including some superb Greek statuettes. There are also the remains of Roman walls, shops, warehouses, baths and a magnificent mosaic measuring 2000 m² (21 528 sq ft).
Population 283 600
Map Romania Bb

Constantine (Qacentina) ***Algeria*** Third city of Algeria, and the capital of the eastern part of the country. It is a centre of communication situated in a spectacular setting on a rocky plateau surrounded on three sides by the Rhummel river whose gorges are a natural wonder. The city was rebuilt in AD 312 by the Roman emperor Constantine on the site of Cirta, the ancient capital of the Princes of Numidia, which was destroyed in AD 311. Following the fall of the Roman Empire, it was occupied by the Vandals and then the Byzantines before it was taken by the Arabs in 710.

When the French moved into Algeria in 1830, Constantine put up bitter resistance but finally fell in 1837. The old Arab quarter has preserved its character and colour. The leatherwork of its saddlers and shoemakers is famous, and the city is also a centre for the grain trade and for woollen textiles. Roman remains include arcades, part of an aqueduct and a bridge.
Population 452 500
Map Algeria Ba

Constantinople ***Turkey*** See ISTANBUL

constructive plate margins Boundary zone between plates which are moving apart. See PLATE TECTONICS, OCEAN RIDGE

Contadora, Isla de ***Panama*** One of the smallest of the Pearl Islands, an archipelago of 227 islands, largely uninhabited, 75 km (47 miles) off Panama City in the Pacific Ocean. Huge pearls were found offshore until the beginning of the 20th century, when the oysters died out because of over fishing. Contadora now attracts deep-sea fishermen, and holds world fishing records for red snapper, marlin and sailfish.
Map Panama Ba

continent One of the seven larger, unbroken landmasses into which the earth's surface is divided. The seven are: AFRICA, NORTH AMERICA, SOUTH AMERICA, ANTARCTICA, ASIA, AUSTRALIA and EUROPE.

continental air mass An AIR MASS originating over a continent.

continental divide An extensive stretch of high ground which separates rivers flowing to opposite sides of the continent. In North America the Great Divide, formed by peaks of the Rocky Mountains, separates rivers flowing west to the Pacific and east and south to the Atlantic.

continental drift See p. 164

continental shelf Region at the edge of a continent extending from the shoreline to the first steepening of the continental slope. Where there is no obvious beginning of the continental slope, the edge of the continental shelf is set at the 200 m (about 660 ft) depth line.

continental slope Steep underwater slope between the continental shelf and the continental rise, sloping at between 3 degrees and 6 degrees to the horizontal.

contour Line drawn on a map to connect places at the same selected height above mean sea level. See ORDNANCE DATUM

convection currents Circulating currents in the air, water or the mantle of the earth caused by local heating. Heated material expands, and rises, allowing cooler material to flow in below to take its place.

convection rain Rainfall resulting when warm moist air from a heated land surface rises, expands and then cools, causing condensation into cumulus clouds, within which some of the droplets grow large enough to fall as rain.

Coober Pedy ***Australia*** Opal-mining town in the Stuart Range in the centre of South Australia, about 750 km (470 miles) north-west of Adelaide. Many of its inhabitants live in underground homes carved out of the hillsides to insulate them from the fierce desert sun. The name comes from Aboriginal words meaning 'hole in the ground'.

Population 2000

Map Australia Ed

Cook, Mount ***New Zealand*** Highest mountain in the country, rising to 3765 m (12 352 ft) in the Southern Alps of South Island, about 200 km (125 miles) west of the city of Christchurch. It is a popular venue for mountaineering and skiing. Maoris call the triple-peaked mountain Aorangi – meaning 'Cloud-Piercer'.

Map New Zealand Ce

Cook Islands See p. 165

Cook Strait Narrow strait between NORTH ISLAND and SOUTH ISLAND, New Zealand, noted for its strong currents. It was charted by the British explorer Captain James Cook in 1769.

Narrowest width 26 km (16 miles)

Map New Zealand Ed

cool temperate climates Mid-latitude (about 50°N and 40°S) with a marked seasonal temperature rhythm and a winter that imposes a resting period on plants. There are two main types. The first, cool temperate maritime with rainfall throughout the year, is found on the western margins of continents – for example, the British Isles. The second, cool temperate continental with mainly summer rainfall and a large range of temperature, is found both in the interiors of continents where rainfall is light, as in the Canadian prairies, and on the eastern sides of continents, as in the maritime provinces of Canada and New England in the USA.

There are, however, *(continued on p. 165)*

THE STRUCTURE OF THE CONTINENTS

The outer layer of the solid crust on which we live is composed of 92 per cent igneous and metamorphic rock (much of which is granite), and 8 per cent sedimentary rocks. This rides on top of a denser layer whose composition is not altogether certain, but from a very few outcrops that occur here and there in the world, it appears to be a metamorphosed form of the crust that underlies the oceans. The foundations of continents, comprising both layers, may be nearly 60 km (40 miles) thick, but those of the oceans, consisting only of a denser layer of basalt, may be no more than 5 km (3 miles).

The continental structure is by far the more ancient and more complex. Broadly, it is made up of three main geological features – shields, continental platforms and mountain belts. Shields, such as the Canadian Shield and much of the Australian interior, are areas of low-lying, ancient and much eroded rock. Platforms, the plains that make up most of the remainder of continental interiors, are areas of shield overlaid by up to several thousand metres depth of sediments. Mountain belts are caused by the collision of two plates and the ensuing folding and volcanic activity. In the case of young mountain belts, like the Andes, volcanic and earthquake activity continues still, while older fold mountains, such as the Western Highlands of Scotland, are now completely stable.

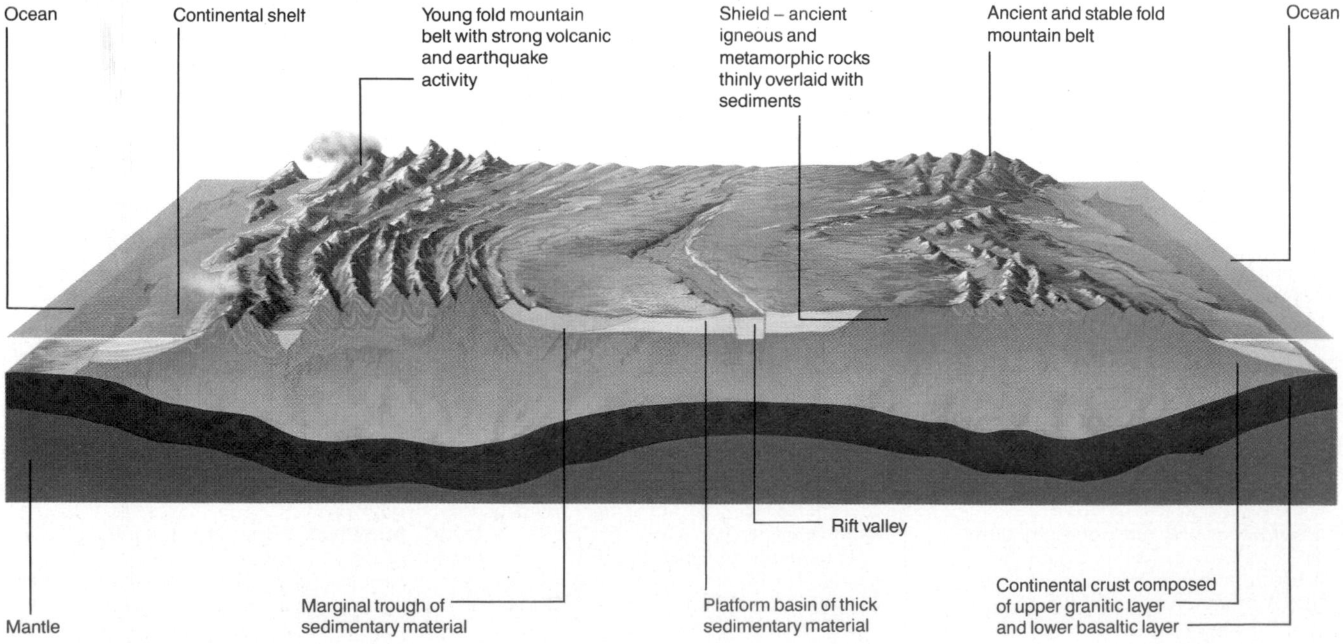

Congo

OIL DISCOVERIES BOLSTERED THE AILING ECONOMY OF AFRICA'S FIRST DECLARED COMMUNIST STATE

Straddling the Equator in west-central Africa, the People's Republic of the Congo has suffered acute political instability since achieving independence from France in 1960. A left-wing group seized power in 1963, while the 1960s and 1970s were marked by coups, countercoups and political assassinations.

However, after 1979, when the Congolese approved their fourth national Constitution

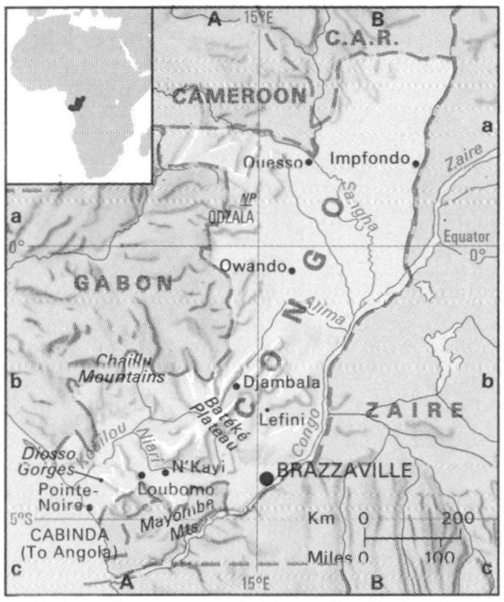

since independence, Congo entered a politically calmer and economically more active phase. The country's oil industry, which had declined in the late 1960s, was brought to life again with the discovery of vast offshore oil deposits in 1969. By 1984 oil accounted for over 90 per cent of export revenues and Congo's economy was on a firmer footing again until the oil price drop of 1985-6. Though the country was the first in Africa to declare itself a Communist state, its pursuit of Marxism-Leninism has not prevented it from turning to the capitalist West, particularly France, in search of finance to help develop local industry. In 1983 Congo's foreign debt totalled US$1370 million.

First occupied by France in 1880, Congo was the poorest country in French Equatorial Africa and one-third of its people still eke out a meagre living by farming. Even so, less than 2 per cent of the land is cultivated, much of the soil being poor because nutrients are washed out by heavy rainfall, which averages 1780 mm (70 in) a year in the north and 1200 mm (47 in) in the south.

Behind the coastal plain are forested ridges, plateaus, mountains and some fertile valleys, notably the Niari basin in the south-west, which is linked to the capital, BRAZZAVILLE, and the country's main port, POINTE-NOIRE, by railway. The north is a vast plain drained by the Sangha, Likouala and other tributaries of the UBANGI and Congo (ZAIRE) rivers. Swamps and flooded forest cover much of this sparsely populated region.

Timber, including the valuable hardwood limba and softwood okoumé, is the second most important product after oil. Diamonds (from Zaire), sugar and coffee are also exported. Cash crops, such as coffee and cocoa, are grown mainly on large plantations. But most Congolese farms are small, worked by women, and devoted mainly to food crops.

▲ THREE MEN IN A BOAT The setting sun lays a golden carpet across the Congo river as three Congolese traders make their way home by canoe after exchanging goods with a supply boat.

Farmland is generally held in common by the village or clan. But traditional society, with its belief in folk medicine and tribal religions, was weakened in the colonial era and is largely rejected by the governing elite. This consists of army officers, officials and politicians, most of whom are young, educated and French-speaking, as also is the small urban middle class. Christianity has its strongest roots in the urban areas.

Apart from its class system, the Congolese are also divided into several ethnic groups. Besides about 12 000 Binga (pygmies), 7500 Europeans and a few northerners who speak Sudanese languages, most people speak Bantu languages. The largest Bantu cluster is the Kongo group, which accounts for about 45 per cent of the population. Most Kongo are farmers, living between Brazzaville and the sea. The Téké dialect cluster includes the peoples of the plateaus east and north of Brazzaville, where they farm, hunt and fish. This second group makes up 20 per cent of the population, as also does the Sangha cluster in the north. The Ubangi (or M'bochi) cluster in central Congo forms another 10 per cent.

Tourism is on a small scale, though Congo has a wide range of wildlife which includes elephants, rhinoceroses, giraffes, cheetahs and, in the forests, gorillas, wild boars and bongos – a type of antelope. Some wildlife is protected in the Lefini and Divenie game reserves and in Odzala National Park.

CONGO AT A GLANCE

Area 342 000 km² (132 000 sq miles)

Population 1 850 000

Capital Brazzaville

Government One-party Marxist republic

Currency CFA franc = 100 centimes

Languages French (official), local languages including Kongo, Lingala, Téké, Sanga

Religions Christian (50%), tribal religions (47%), Muslim (3%)

Climate Tropical; average temperature in Brazzaville ranges from 17-28°C (63-82°F) in July to 22-33°C (72-91°F) in April

Main primary products Cassava, sweet potatoes, groundnuts, pineapples, bananas, plantains, sugar cane, coffee, cocoa, timber; crude oil, lead, copper, zinc, gold

Major industries Crude oil production and refining, mining, forestry, food processing, textiles, cement, chemicals

Main exports Crude oil, timber and timber products, coffee, cocoa, diamonds (re-exports from Zaire)

Annual income per head (US$) 950

Population growth (per thous/yr) 30

Life expectancy (yrs) Male 58 **Female** 62

CONTINENTAL DRIFT EXPLAINS EARTH'S ANCIENT MYSTERIES

In our short lives, the pattern of earth's landmasses and oceans seems the very symbol of permanence. To be the first to consider that it might not be permanent at all calls for an imagination and intellect of formidable proportions. Both qualifications were possessed to an admirable degree by the 17th-century scholar and statesman Francis Bacon, who noticed the jigsaw-like fit between the west coast of Africa and the east coast of South America, and deduced from this that the two must at one time have been joined.

In 1858, a French scientist, Antonio Snider, advanced the theory of an ancient splitting of the landmasses by pointing out that fossils found in European and American coalfields were identical, and must have evolved in a common environment. But it was not until 1915 that these and other observations were united by a German meteorologist, Alfred Wegener, to confirm his own theory of Continental Drift – that long ago, all the continents had been joined together, but over hundreds of millions of years had somehow split and drifted apart.

Certainly, his summing up went far towards explaining some puzzling coincidences – the identical rock formation and structures, and numerous fossils that are found in countries now widely separated by oceans. It also answered the mystery of ancient climatic differences, such as the rocks scarred by long-vanished glaciers in present deserts; the continent in which the desert now lies had drifted closer to the Equator.

Plausible though this seemed, the central problem remained; was it possible for solid continents to wander about the oceans, and if so what was the motive force? Then in the 1950s and 1960s, geophysicists offered a new explanation.

Far from being immovably anchored, earth's crust – continents and ocean beds – is 'afloat' on a lower layer of partially molten rock which they called the asthenosphere. This rock, which would be fully molten were it not for the pressures to which it is subjected, wells up in a semifluid state through the ridges and trenches of the ocean floor, constantly expanding it. The expansion thrusts against the landmasses causing them to move, infinitely slowly, upon the asthenosphere. (See also JOURNEY TO THE CENTRE OF THE EARTH, p. 200.)

Armed with this explanation, and with earth's geological patterns, it is theorised that up to about 200 million years ago there existed only one supercontinent which has been christened Pangaea. This split into halves, called Laurasia and Gondwanaland, which divided again into the present continents. The process is by no means finished, as is apparent from vast crustal splits such as the Great Rift Valley of Africa. But why Pangaea broke up, instead of moving as one mass, remains unknown.

PANGAEA – THE SUPERCONTINENT OF 200 MILLION YEARS AGO

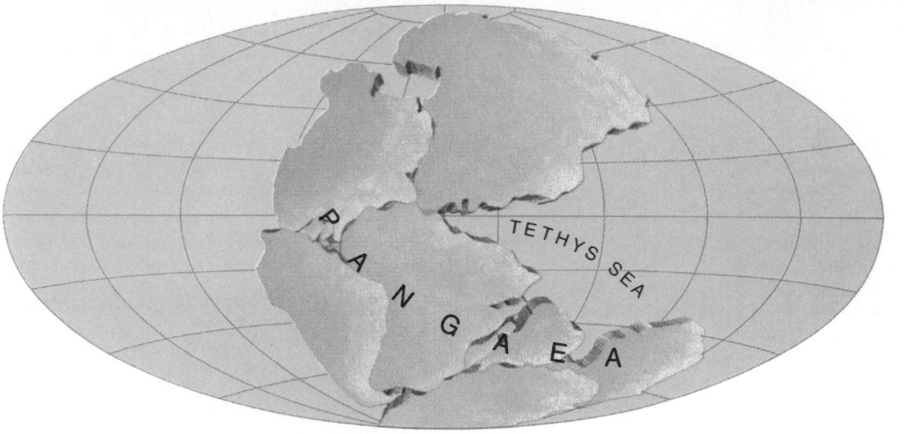

Until about 200 million years ago, scientists believe, earth's landmasses were combined in a single continent now called Pangaea. Some cataclysm split the supercontinent and began shaping the present world, though not all movement began at the same time or rate.

SUBOCEANIC FORCES SEND THE LANDMASSES WANDERING

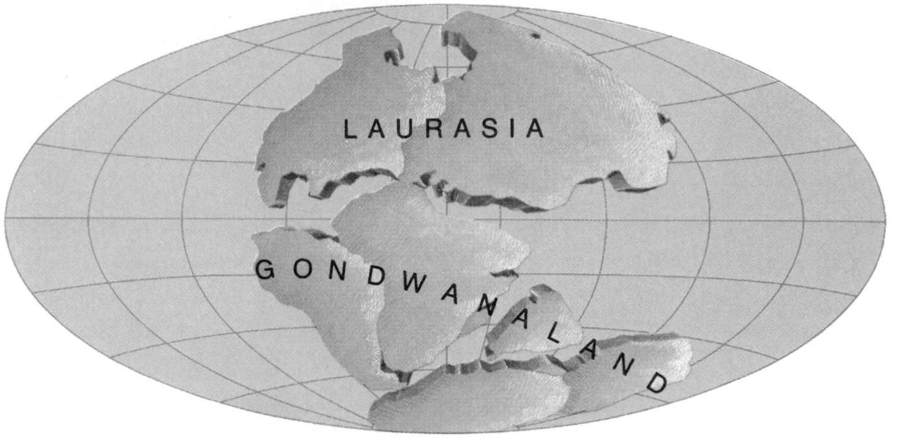

About 180 million years ago, Pangaea divided into Laurasia and Gondwanaland, separated by the Tethys Sea. Laurasia included N. America, Europe and Asia, while Gondwanaland consisted of S. America, Africa, Australia, India and Antarctica.

TOMORROW'S WORLD – 50 MILLION YEARS HENCE

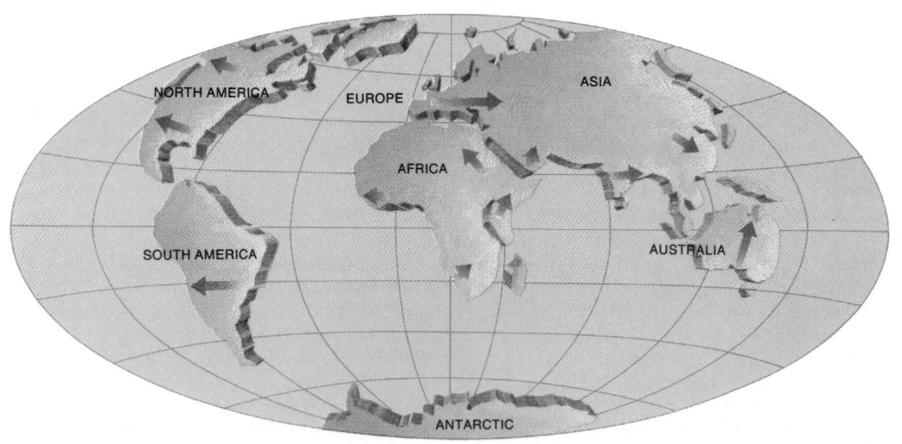

If the present movement continues – actually no more than a few millimetres a year – the world in 50 million years' time will look something like this. The Atlantic will have widened, a great rift will open in eastern Africa, and Australia will move towards Asia.

many sub-types, with significant differences between areas, which fit broadly into these two general categories. In the Southern Hemisphere, the narrowing of the continents southwards and the greater extent of ocean has the effect of reducing the areas of cool temperate climate and of moving them nearer the Equator.

Coolgardie ***Australia*** See KALGOORLIE

coombe In southern England, a short valley or hollow in the side of a hill, commonly found in chalk country, or a short steep valley running down to the sea.

Cooper Creek ***Australia*** River rising in central Queensland and flowing during wet seasons about 1420 km (880 miles) south-west to Lake Eyre in South Australia. Its upper course is known as the Barcoo river, a name formerly applied to the whole course. The explorers Robert O'Hara Burke and William John Wills, who led the first expedition to cross Australia from south to north, starved to death near its banks on their return journey in 1861 – after missing rescuers by a few hours.

Map Australia Gd

Coorg ***India*** See KODAGU

Copacabana ***Bolivia*** Picturesque red-roofed town on Lake Titicaca. It contains a shrine built around the Dark Virgin of the Lake, a statue carved by the Indian artist Tito Yupanqui in 1576.

Population 14 200

Copacabana ***Brazil*** See RIO DE JANEIRO

Copán ***Honduras*** Splendidly preserved Maya city near the Guatemalan border, 225 km (140 miles) north-west of the capital, Tegucigalpa. The city, which reached its peak between AD 450 and 800, was abandoned early in the 9th century. The Great Plaza is dotted with elaborately carved stone figures.

Map Honduras Ab

Copenhagen (København) ***Denmark*** Seaport and capital city on the island of ZEALAND facing Sweden 20 km (12 miles) away across the Öresund, the principal entrance to the Baltic Sea. Founded in the 13th century, it was attacked by Sweden in the 17th century and bombarded by British ships during the Napoleonic Wars. In the mid-19th century the ramparts within which it was still confined were largely removed. It has a beautiful waterfront, the setting for the well-known statue of Hans Andersen's Little Mermaid, and along the coast of the Öresund modern developments complement older villas. Canals and moats add to the attractions of the city.

There is a rich architectural heritage, from the Dutch Renaissance of the Børsen (Stock Exchange) and Rosenborg Castle, through the Baroque style of Our Saviour's Church and Charlottenborg Castle, now the seat of the Art Academy, to the rococo of Amalienborg Palace and the classical style of late 18th-century town houses, offset by imaginative contemporary architecture such as the Carlsberg brewery's bottling plant.

The city centre is distinguished by the university, dating from 1497; by Christiansborg – the seat of government which was rebuilt after a 19th-century fire; by the national theatre (with its old-established ballet); by the massive brick city hall with its famous astronomical world clock; and by the Tivoli Gardens, a summer amusement park. Other places of interest include the Ny Carlsberg Glyptotek (Egyptian, Greek, Etruscan and Roman collections); the Toy Museum; Rosenborg Castle; the Circus Building and Zoo; the Zoological Museum; the Open Air Museum; and the City Museum and National Art Gallery.

Manufactures include ships, furniture, clothing, silverware and porcelain from the 200-year-old Royal Copenhagen factory. There is an airport and a harbour, which includes a 50-year-old free port, which profits from the large Baltic hinterland.

Population 641 900

Map Denmark Cb

Cook Islands

BEARING THE NAME OF THE PACIFIC'S BEST KNOWN EXPLORER, THESE SMALL, SCATTERED ISLANDS RETAIN LINKS WITH NEW ZEALAND

Captain James Cook discovered and charted many of the islands that bear his name on his second voyage of Pacific exploration in 1773. He was lucky not to miss them altogether, for the 15 islands total only 240 km² (93 sq miles) – RAROTONGA, the largest (which Cook in fact missed), is only 10 km (6 miles) across – and they are scattered over some 2.2 million km² (850 000 sq miles) of the Pacific Ocean west of Tonga and Samoa. The first European to see any of the islands was the Spaniard Alvaro de Mendana, who sighted Pukapuka island in August 1595.

They lie in two main groups, over 1100 km (700 miles) apart. The northern group consists of seven low coral islands, six of them atolls with central lagoons. Most of the southern (or 'lower') group are volcanic, but only Rarotonga is really mountainous, reaching 652 m (2139 ft) at Te Manga; these are also the most fertile islands, and have 90 per cent of the land area and 85 per cent of the population.

The Cook Islands became a British protectorate in 1888, and in 1901 became New Zealand territory. But the link with New Zealand is much older and in the reverse direction, for the Cooks were a major stepping-off point for the Polynesian colonisation of New Zealand some 600 years ago, and Cook Islanders are closely related to Maoris. Muri Beach on Rarotonga is said to be the starting place of the Maori's journey to New Zealand. In 1965 history turned full circle and the islands became independent 'in free association' with New Zealand, and Cook Islanders have all the privileges of New Zealand citizenship. New Zealand looks after foreign affairs and defence, and supplies much material and managerial aid, but the Cook Islands parliament can break these ties at any time. Meanwhile, they remain a part of the Commonwealth, with Elizabeth II as head of state.

In fact, more Cook Islanders (about 25 000) live in New Zealand than in their own country, and the money they send back home helps to support the islands' economy. This depends on agriculture, largely for export to New Zealand. Citrus fruits and pineapples (exported canned or as juice), tomatoes and bananas are grown in the fertile southern islands; coconuts are the main crop in the northern group. Tourism, although growing, is still on a fairly small scale and largely restricted to Rarotonga (which has an international airport) and AITUTAKI. This is not for want of natural beauty – Rarotonga, in particular, is one of the most beautiful islands in Polynesia – nor are the people unfriendly (quite the contrary). The happy, charming Cook Islanders simply do not want their small islands overrun by thousands of visitors.

COOK ISLANDS AT A GLANCE

Map Pacific Ocean Fd
Area 240 km² (93 sq miles)
Population 17 700
Capital Avarua
Government Parliamentary monarchy
Currencies Cook Islands and New Zealand dollars = 100 cents
Languages English, Cook Islands Maori
Religion Christian (70% Protestant)
Climate Tropical; average temperature at Avarua ranges from 18-25°C (64-77°F) in July to 23-29°C (73-84°F) in January
Main primary products Coconuts, citrus fruits, pineapples, pawpaws, tomatoes, bananas, yams, taro, cassava, fish
Major industries Agriculture, fruit processing, clothing manufacture
Main exports Copra, fresh and canned tropical fruit, vegetables, clothing
Annual income per head (US$) 1100
Population growth (per thous/yr) Declining due to emigration

Copiapó ***Chile*** Capital of Atacama region, standing south of the Atacama Desert on the Copiapó river. When silver was discovered in 1832 at nearby Chañarcillo, Copiapó became one of the country's most prosperous cities. The first railway in the Southern Hemisphere was completed in 1852 to link its copper mines with the port of Caldera 80 km (50 miles) west.

Population 70 000

Map Chile Ab

Copperbelt ***Zambia*** The country's smallest province, covering an area of 31 328 km² (12 096 sq miles) about 225 km (140 miles) north of Lusaka. It has some of the world's largest known copper reserves – more than 882 million tonnes – around the towns of Ndola, Kitwe-Nkana, Chingola and Chililabombwe. The area also produces cobalt, lead and zinc.

Population 1 249 000

Map Zambia Bb

coppice Small wood, often of oak or hazel, periodically cut back almost to ground level to encourage growth from the stumps to provide wood for fencing and fuel.

Coquimbo ***Chile*** Region in central Chile, between the Atacama Desert and the Chilean heartland around Santiago and Valparaíso. The economy relies on copper and manganese mining, agriculture, sheep and cattle. The chief towns are the capital LA SERENA, its port Coquimbo and Ovalle. The region is noted for La Silla Observatory near La Serena, and for its pilgrimage to the shrine of the Virgen del Rosario de Andacollo, about 16 km (10 miles) south of the town of Coquimbo.

Population 420 000

Map Chile Ac

coral Lime-secreting marine polyp found living in colonies in tropical seas, characterised by calcareous skeletons massed in various shapes and often forming reefs or islands. The term is also used for the hard, rocklike structure which is the accumulation of their skeletons.

coral reef Marine ridge or mound built up along a shoreline by CORAL polyps. Fringing reefs border the land or are separated from it by a shallow lagoon. Barrier reefs (such as the GREAT BARRIER REEF) are separated from the land by a deep lagoon. See also ATOLL

Coral Sea Arm of the PACIFIC OCEAN with rather indeterminate boundaries; but most geographers agree that it washes the shores of north-eastern Australia (the GREAT BARRIER REEF), Papua New Guinea, Vanuatu and New Caledonia. From May 4 to 8, 1942, it was the scene of a naval battle fought by US carrier-borne aircraft, which probably saved PORT MORESBY, New Guinea, and possibly Australia from a Japanese invasion force.

Map Pacific Ocean Cc

Corcaigh ***Ireland*** See CORK

cordillera Series of broadly parallel mountain ranges, especially the main mountain system of a large landmass. In South America the term applies to a small mountain range.

▲ DEEP BLUE SEA The Coral Sea laps Australia's Great Barrier Reef, 1900 km (1200 miles) long and strewn with islets and atolls. Ships reach port via 'passes' through the reef.

Cordillera Cantabrica ***Spain*** See CANTABRIAN MOUNTAINS

Cordillera Central ***Philippines*** Major mountain range of northern Luzon, lying between the Cagayan valley and South China Sea. The range extends about 320 km (200 miles) north to south and is up to 88 km (55 miles) wide. It consists of three parallel ranges: the Malayan, Central and Polis. Mount Pulog, the highest peak (2928 m, 9606 ft), is in the Polis range.

The mountains form a massive barrier, making access to many areas difficult, and only small groups of hill people, known as the Igorots, live there; they are ethnically distinct from the Malay Filipinos. BAGUIO, summer capital of the Philippines, lies in the southern foothills of the mountains.

Map Philippines Bb

Córdoba ***Argentina*** The country's second largest city after Buenos Aires, about 660 km (410 miles) north-west of the capital. The city's factories make cars, textiles and foodstuffs. Farmers in the surrounding province of the same name mostly grow wheat, maize and soya beans, and rear livestock.
Population (city) 969 000; (province) 2 408 000
Map Argentina Cb

Córdoba (Cordova) ***Spain*** City about 115 km (70 miles) north-east of the southern city of Seville. It is remarkable for its Moorish atmosphere. Its cathedral was the main mosque of western Islam until the Christians captured the city in 1236. Modern Córdoba's industries include engineering, brewing, distilling and tourism. Its surrounding province, also called Córdoba, is mainly agricultural, but coal, copper, lead and silver are also mined.
Population (city) 284 700; (province) 717 200
Map Spain Cd

core Central part of the earth's interior lying below its MANTLE.

Corfu (Kérkira) ***Greece*** Most northerly of the IONIAN ISLANDS, and their capital. The island was the principal Venetian arsenal in Greece from AD 1386 to 1797, and Italian is still spoken. Venetian fortifications and two citadels with magnificent views dominate the harbour town of Kérkira. Corfu was the birthplace of Count Capodistria (1776-1831), the first president of Greece from 1827 to 1831. Beaches at Palaiokastritsa on the west coast have good hotels, camping sites and a marina. Corfu produces red wine, fruit and olives.
Population 97 100
Map Greece Ab

Corinth (Kórinthos) ***Greece*** Modern town in the north-east Peloponnese about 3 km (2 miles) from the Corinth Ship Canal, and 5 km (3 miles) east of the ancient city which was destroyed by earthquake in 1858. From its hilltop site, the old town dominated the 6 km (4 mile) wide isthmus and only land route into the Peloponnese. Archaeologists have uncovered the forum, theatre, a temple of Apollo and the Sacred Fountain of Peirene. Corinth is now a port and centre of a thriving trade in currants, made from dried grapes. The resort of Loutraki to the north is famed for its spring water. The Corinth Canal, built 1882-93, cuts through the isthmus, linking the Gulf of Corinth with PIRAEUS. It can take ships of up to 10 000 tonnes.
Population 22 700
Map Greece Cc

Cork (Corcaigh) ***Ireland*** County with an area of 7459 km² (2880 sq miles), and the Republic's second largest city – after the capital, Dublin – on the River Lee about 220 km (140 miles) south-west of Dublin. The old city and heart of Cork is on an island in the Lee, originally a Viking stronghold, invaded by Anglo-Normans in AD 1172. Historically Cork has an unenviable reputation of backing losers: Perkin Warbeck, rebel pretender to the English throne in 1492 – as a result of whose defeat the city's mayor lost his head; the royalists in the English Civil War against Oliver Cromwell; and James II, who was defeated by William of Orange at the Boyne in 1690. Another mayor was a victim of the fight for Irish independence: Terence McSwiney died on hunger strike in Brixton Jail, London, in 1920.

Cork is the county town, and has a Protestant and a Catholic cathedral, a fine customs house and modern city hall. It is a port, and has part of the National University of Ireland. The city produces beer, spirits, chemicals, cars, clothes and hosiery. Cork Harbour, to the south of the city, is a centre for oil exploration off the Irish coast. The county has broad river valleys and a flourishing dairying industry based on cooperative creameries and factories.
Population (county) 402 300; (city) 136 300
Map Ireland Bc

Cornwall ***United Kingdom*** County of 3546 km² (1369 sq miles) occupying the south-west tip of England. It belongs to the Celtic fringe of western Europe and has many physical and cultural similarities to Brittany in France and to Wales. It has its own language, and trueborn Cornishmen regard it as 'going up to England' to cross the River Tamar into Devon.

Tin and copper mined in Cornwall were traded with the Phoenicians of the Mediterranean as long ago as the 6th century BC, and continued to be exploited under the Romans. Today, virtually all the mines are closed, but china clay is quarried around St Austell. The economy relies mainly on tourists attracted by the splendid beaches and spectacular coastal scenery, including LIZARD POINT and LAND'S END. Thanks to its mild climate and subtropical vegetation, the south coast is known as the Cornish Riviera, while the Isles of SCILLY have some of the mildest winters in the British Isles. The county town is Truro.
Population 432 000
Map United Kingdom Ce

Coromandel Coast ***India*** Coast in the south-east, of which the main ports are Madras and Pondicherry. Its low, sandy shoreline is constantly lashed by heavy seas, particularly from October to April – the season of the monsoon. The coast has no good natural harbours.
Map India Ce

Coromandel Peninsula ***New Zealand*** Mountainous area of North Island stretching from the town of Waihi to Cape Colville, about 65 km (40 miles) east of Auckland. European traders first went there for its magnificent forests, especially the huge kauri trees, but in the 1890s for gold as well. Timber, sheep and cattle provide the chief livelihoods today and tourism is growing. The red flowers of the pohutukawa trees along the coast are a magnificent sight around Christmas.
Map New Zealand Eb

corona A series of coloured circles surrounding the sun or moon, ranging from blue (inside) to red (outside), and caused by the diffraction of light by waterdroplets in middle altitude clouds.

Coronet Peak ***New Zealand*** Mountain on South Island rising to 1646 m (5400 ft) above Queenstown, some 160 km (100 miles) north-west of the port of Dunedin. It is one of the country's leading ski resorts.
Map New Zealand Bf

Corpus Christi ***USA*** Industrial port on the coast of Texas, about 200 km (125 miles) north of the Mexican border. It is an oil refining and petrochemicals centre, and processes fish.
Population (city) 258 100; (metropolitan area) 361 300
Map United States Ge

Corregidor ***Philippines*** Island covering 5 km² (2 sq miles) at the entrance to Manila Bay, off the southern tip of the BATAAN peninsula. It was first fortified to guard Manila by the Spanish in the 18th century. American forces built extensive fortifications and tunnels in the rock after 1900. When Bataan fell to the Japanese army in April 1942, during the Second World War, American and Filipino troops on Corregidor continued to fight, under heavy bombardment, until May 6, 1942.
Map Philippines Bc

corrie See CIRQUE

Corsica (Corse) ***France*** The country's largest island, covering 8680 km² (3350 sq miles). It lies in the Mediterranean, 160 km (100 miles) from the south-east coast of mainland France. It was ruled by the Genoese for 400 years until 1768, when it was sold to France.

The interior, rising to 2710 m (8890 ft) at Monte Cinto, is occupied by maquis scrub and woodland. A regional park covers some 200 km² (77 sq miles) on the west coast, and includes the Golfe de Porto, a beautiful coastline of red rocks. The eastern plain has rich farmland, and fishing villages and small resorts are found along the coast. The main towns are AJACCIO, birthplace of the French Emperor Napoleon (1769-1821), and BASTIA.
Population 240 000
Map France He

Cortina d'Ampezzo ***Italy*** Ski resort in the DOLOMITES about 120 km (75 miles) north of Venice. It stands 1210 m (3970 ft) up in the mountains and is surrounded by peaks of more than 3200 m (10 500 ft). It was the site of the 1956 Winter Olympics.
Population 7810
Map Italy Da

Corubal ***Guinea/Guinea-Bissau*** One of the main rivers of Guinea-Bissau, flowing about 400 km (250 miles) on a winding course from the Fouta Djallon of western Guinea across southern Guinea-Bissau to the estuary of the Gêba. The Corubal was formerly called the Rio Grande, and in Guinea it is sometimes called the Koumba. It is a busy transport artery.
Map Guinea Ba

corundum Extremely hard mineral (aluminium oxide), which sometimes contains iron, magnesia or silica, occurring in various gemstone varieties such as ruby and sapphire, and in a common grey, brown or blue form that is used chiefly as an abrasive.

Corunna (La Coruña) ***Spain*** Seaport on the country's north-west corner specialising in sardine fishing and canning, but also manufacturing cigars, cotton goods and glassware. The Spanish Armada called at Corunna during its disastrous voyage to invade England in 1588, and in 1809 it was the scene of a British victory over French troops. The town's Tower of Hercules, 58 m (190 ft) high, is the only Roman lighthouse still in use in the world.

Its farming and fishing province, also called Corunna, has Spain's heaviest rainfall, averaging 800 mm (32 in) a year.

Population (town) 232 400; (province) 1 083 400

Map Spain Aa

Cosenza ***Italy*** Calabrian city at the junction of the Busento and Crati rivers, about 230 km (145 miles) south-east of Naples. It dates back to pre-Roman times, and flourished as a centre of learning until the 15th century. It has a Norman-Swabian castle and a recently restored Gothic cathedral. The Visigoth king Alaric I was buried in AD 410, after sacking Rome, in the bed of the Busento river after his followers temporarily diverted its waters. Cosenza's industries include furniture and textiles.

Population 106 100

Map Italy Fe

Costa Brava ***Spain*** Holiday area of cliffs and bays – its name means 'wild coast' – along the northern Mediterranean shore, north-east of Barcelona. Adjoining it is the Costa Dorada ('Golden Coast'), stretching to south of Tarragona. The Costa del Azahar ('Orange-blossom Coast') is next, in the provinces of Castellón and Valencia, then comes the Costa Blanca ('White Coast') reaching almost to Almería. The Costa del Sol ('Sun Coast') is the south-facing stretch of Mediterranean coast with the resorts of Malaga and Marbella, and the Costa de la Luz ('Coast of Light') is the southern Atlantic coast beyond.

Map Spain Gb

Costa Smeralda ***Sardinia*** The 'Emerald Coast' – named after the colour of the sea – stretches for about 50 km (30 miles) along the coast of north-east Sardinia. It is a centre for yachting, water sports and the luxury tourist trade. The main resorts are Baja Sardinia, Porto Cervo and Cala di Volpe.

Map Italy Bd

Costermansville ***Zaire*** See BUKAVU

Cotabato ***Philippines*** Region in south-western MINDANAO comprising the swampy lowlands of the Mindanao river and its tributaries, and the surrounding forested uplands. Inaccessible and malaria-infested until the late 1930s, development has led to a rapid population increase. Most newcomers have been Christian Filipinos from the VISAYAN ISLANDS, who now outnumber the original Muslim Maguindanao people. In 1971, a previously unknown tribe of about 30 people, the Tasaday, were discovered in an isolated rain-forest area of southern Cotabato.

Population (Cotabato province) 472 302

Map Philippines Bd

Côte d'Azur ***France*** The heart of the French Riviera, one of Europe's prime tourist areas. Sprawled along the sunny Mediterranean, it contains rocky headlands, bays, beaches and yacht marinas. Its main resorts are CANNES and NICE.

Map France Ge

Costa Rica

A HAPPY ODDITY IN CENTRAL AMERICA: A STABLE DEMOCRACY THAT HAS DISBANDED ITS ARMY TO PRESERVE THE PEACE

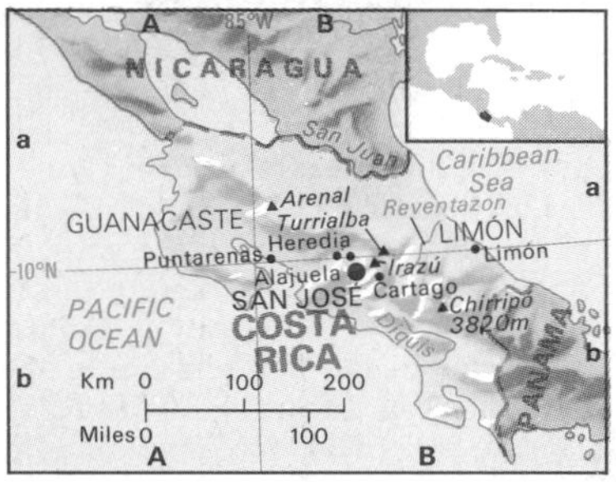

In a region of political instability and military rule, Costa Rica is renowned for its stable democracy. All men and women over 18 years are obliged to vote, and their voting actually determines which party takes power. Civilian government has lasted since 1902. The last civil war occurred in 1948. It lasted only two months, and afterwards the army was disbanded, so military coups are now impossible.

Forming a bridge between the Pacific Ocean and the Caribbean, Costa Rica is also a buffer between less tranquil countries: Panama to the south and war-torn Nicaragua to the north. Costa Rica – the 'Rich Coast' – has a tradition of neutrality which its population enthusiastically supports. The only cloud on its horizon is the risk of neighbouring conflicts spilling over its borders.

Much of the country consists of volcanic mountain chains, running north-west to south-east, that reach their highest point at Chirripó (3820 m, 12 533 ft). Over half of the 2.7 million people live in the Valle Central, a highland basin which was the first area settled by the Spanish in the 16th century. The rich volcanic soils of the upland areas are good for coffee growing and the slopes provide lush pastures for cattle.

Lowland swamps cover most of the two coasts; the Pacific coast is wetter and cooler than the Caribbean. The north-west region is partly savannah and partly lowland forest. Increasingly people are settling along the river and mountain valleys. About four-fifths of the country remain forested.

Costa Rica differs from its neighbours in that 80 per cent of the people are white and most of the rest *mestizo* (of mixed European and native Indian ancestry). When Europeans arrived in the 16th century they introduced diseases, including measles, to which the Indians had no resistance. Thousands died and they have never recovered their numbers.

Costa Rica was the first country in Central America to grow coffee and bananas commercially, which are still its major agricultural exports. An enlightened 19th-century government offered free land to anyone willing to grow coffee for export. This policy made Costa Rica prosperous, and avoided the Central American pattern of a few rich landowners and a large class of poverty-stricken peasants. Growth brought benefits to almost everyone, until the recent recession. Today, the economy suffers from high inflation and a large foreign debt.

There is potential for further prosperity from large deposits of bauxite and iron ore. The mountainous terrain provides attractive scenery for a growing tourist industry and hydroelectric power which has made the country almost self-sufficient in electricity.

COSTA RICA AT A GLANCE

Area 50 700 km² (19 575 sq miles)

Population 2 720 000

Capital San José

Government Parliamentary republic

Currency Colón = 100 céntimos

Language Spanish

Religion Roman Catholic

Climate Tropical; temperate in highlands. Average temperature in San José ranges from 14-24°C (57-75°F) in December to 17-27°C (63-81°F) in May

Main primary products Coffee, bananas, maize, rice, cattle, oranges, sugar cane, cocoa; gold, silver, sulphur, bauxite, iron ore

Major industries Agriculture, aluminium smelting, fishing, chemicals, fertilisers, textiles, forestry, food processing

Main exports Coffee, bananas, beef, sugar, cocoa, textile yarns and fabrics

Annual income per head (US$) 1100

Population growth (per thous/yr) 25

Life expectancy (yrs) Male 71 **Female** 75

Cotonou ***Benin*** Business capital of Benin and deep-water port west of the outlet of Lake NOKOUE. Well served by local and international transport links, its flourishing industries, based in a suburb east of the lake's outlet, produce palm oil, cement, soap, furniture, beer and soft drinks, and assemble cars. The city founded by the French in the early 20th century has a wealth of hotels, restaurants, boutiques and colourful markets. It also has beautiful beaches, but undercurrents can make swimming dangerous.

Population 488 000

Map Benin Bb

Cotopaxi ***Ecuador*** West-central province, lying in the Andes to the south of Pichincha. It is dominated by Mount Cotopaxi which, at 5896 m (19 344 ft) is the world's highest active volcano. It last erupted in 1877, setting off avalanches of mud that reached the provincial capital, Latacunga, 30 km (19 miles) to the south and destroyed large areas of farmland.

Population 280 000

Map Ecuador Bb

Cotswold Hills ***United Kingdom*** Highest section of a range of limestone hills which stretch across England from the county of Avon in the south-west to beyond the Humber estuary in the north-east. The hills rise to 329 m (1080 ft) at Cleeve Cloud near Cheltenham in Gloucestershire. Their eastern slopes feed the headwaters of the River Thames. The open, plateau-like summits, steep, wooded valleys and stone-built villages are a major tourist attraction.

Map United Kingdom De

couloir Deep mountainside gorge or gully, especially in the Alps.

country rock The existing rock into which igneous rocks have intruded.

Courmayeur ***Italy*** Oldest of the country's alpine and ski resorts. It stands in the French-speaking Valle d'Aosta at the foot of Mont BLANC, about 100 km (60 miles) north-west of Turin. It was first frequented by the Piedmontese nobility seeking escape from the summer heat in the early 19th century.

Population 2740
Map Italy Ab

Courtrai (Kortrijk) ***Belgium*** Textile town near the French border, about 80 km (50 miles) west of the capital, Brussels. Originally a medieval cloth-making centre using Flanders wool, it turned to linen made from local flax in the 17th century. Modern synthetics are now manufactured along with lace.

Population 76 000
Map Belgium Aa

Coventry ***United Kingdom*** Car-making city in the West Midlands of England, 30 km (20 miles) south-east of Birmingham. Founded by the Saxons in about the 7th century, it was immortalised by the legend of Lady Godiva, who persuaded her husband, Earl Leofric, to ease the tax burden on the citizens in the 11th century by riding naked through the streets.

The city was devastated by bombing in 1940-1. The architect Sir Basil Spence designed a magnificent new cathedral and the artist Graham Sutherland provided a stunning tapestry of Christ 28 m (75 ft) high for the altar. Tourism, pedestrian shopping precincts, a polytechnic and the University of Warwick (1965) have provided new life for the old city.

During the Civil War of the 1640s, Coventry was a Parliamentary stronghold. Royalists captured in the Midlands were imprisoned in the 14th-century Church of St John in the city's Bablake district. They were 'sent to Coventry' in the words of an account written in 1647 – a phrase that came to mean 'shunned by society'.

Population 315 900
Map United Kingdom Ed

Cox's Bazar ***Bangladesh*** Small fishing town and coastal resort near the Burmese border, renowned for its cigars. The beach stretches for 90 km (56 miles) and is one of the longest in the world; but it is as yet hardly developed. Small islands offshore were badly hit by a cyclone in 1985.

Population 29 800
Map Bangladesh Cd

Cozumel ***Mexico*** Caribbean island just off the north-east coast of Yucatán. In Mayan legend, it was the home of the god of the Sun; now it is a tourist resort. Just offshore is the second largest coral reef in the world – Palancar Reef.

Map Mexico Db

Crac des Chevaliers (Qal'at al Hisn) ***Syria*** Magnificent ruined Crusader castle set 650 m (2132 ft) above sea level 65 km (40 miles) west of Homs, Crac des Chevaliers controlled the gap between the Nusayriyah mountains to the north and the Lebanese mountains which linked inland Syria with the coast. For two centuries it played a major role in the Crusader-Muslim Wars. Much restoration work has taken place since the 1930s, but the vast evocative remains of ramparts, towers, halls and fortifications still grip the imagination.

Map Syria Bb

Cracow (Krakow) ***Poland*** The country's main tourist city, and third largest city (after Warsaw and Lodz). It lies in the south, 251 km (156 miles) south-west of the capital. Cracow grew where the Vistula river narrows between limestone hills, at the crossroads of major trade routes. It was Poland's capital from 1138 to

▼ CASTLE OF THE KNIGHTS The Knights Hospitallers of St John built Crac des Chevaliers (1142-1271) to hold a garrison of 2000. Its two concentric, towered walls withstood 12 sieges.

1596. Its university, the country's oldest, dates from 1364. Nicolaus Copernicus (1473-1543), the Polish astronomer whose principle that the earth orbits the sun forms the basis of modern astronomy, was a student there from 1491 to 1494.

According to legend, Krak, a cobbler, founded Cracow and became its first prince after killing a virgin-devouring dragon living in Wawel Cave. Above, on Wawel Hill commanding the Vistula narrows, stand a royal castle and cathedral erected by King Casimir the Great (1330-70) on the remains of a 10th-century fort and church. The castle, rebuilt in the 16th century around an arcaded courtyard, has 71 restored rooms, and houses the world's only surviving 16th-century Arras tapestries.

The cathedral, former see of Cardinal Wojtyla (who became Pope John Paul II), contains the tombs of Polish kings, and those of St Stanislaus, a former bishop and the country's patron saint, and the nationalist Tadeusz Kosciuszko (1766-1817), who led an abortive rising against occupying Russians in 1793.

The central market square, in the old part of the city north of Wawel Hill, is dominated by the 16th-century Cloth Hall, the Town Hall tower (1383), and the Church of the Virgin Mary. The church has an altarpiece carved of linden wood in the 15th century. Every hour (and at noon on Polish radio), a bugle call is sounded from the church tower. The call is always broken off in mid-blast – a custom commemorating a bugler whose throat was pierced by a Tatar arrow in 1241 while he was trying to alert the city to a Tatar attack.

Cracow's main industries today include steel (at nearby NOWA HUTA), chemicals, printing and ceramics. Each June the International Short Films and Polish Film Festivals are held in the city, along with an international exhibition of graphic arts and a folk arts fair.

Population 520 700
Map Poland Cc

crag and tail Glacial landform, consisting of a steep-faced rocky outcrop on one side (the crag) with a gentle slope on the other where there are deposits of moraine (the tail).

Craigavon ***United Kingdom*** See LURGAN

Craiova ***Romania*** Rapidly expanding industrial city in the south-west of the country, 55 km (34 miles) north of the border with Bulgaria. Craiova produces agricultural machinery, cars, railway equipment, textiles, food and electrical goods. In the city's main square there is a magnificent green-roofed, 19th-century palace which is now a museum of the arts. There are also several old churches and mansions.

Population 228 000
Map Romania Ab

crater Bowl-shaped depression at the mouth of a volcano or caused by collision of a meteorite with the earth.

crater lake Lake which has formed in the crater of an extinct volcano.

Crater Lake National Park ***USA*** Area of 641 km² (247.5 sq miles) on the crest of the Cascade Range in south-west Oregon, about 90 km (55 miles) north of the Californian border. It is centred on a lovely blue lake, 589 m (1932 ft) deep and about 10 km (6 miles) across, in the crater of an extinct volcano. The lake is encircled by walls of multicoloured lavas, 650 m (2130 ft) high.

Map United States Bb

Craters of the Moon National Monument ***USA*** Volcanic area covering 217 km² (84 sq miles) in south-eastern Idaho, about 145 km (90 miles) north of the Utah border. Its lava flows, fissures, cinder cones and craters resemble the surface of the moon.

Map United States Db

craton Large section of the earth's crust that has remained stable and immobile for millions of years. It is also called a shield or kraton.

Cremona ***Italy*** City about 75 km (45 miles) south-east of Milan. It is famous for its violins: the Stradivari, Amati and Guarneri families worked there between the 16th and 18th centuries, and violins are still made. Cremona's civic pride is its 13th-century cathedral, Gothic inside with a Renaissance façade, which has a 111 m (365 ft) high bell tower, the Terrazzo.

Population 80 800
Map Italy Cb

Cretaceous The third period of the Mesozoic era in the earth's time scale. This was the period during which the chalk was formed. See GEOLOGICAL TIMESCALE

Crete (Kriti) ***Greece*** The largest Greek island, lying in the south Aegean Sea. Crete was the heart of the rich Minoan civilisation of 3000-1100 BC. Its huge palace at KNOSSOS has been excavated and restored. The island (8366 km², 3229 sq miles) rises east to west to a limestone backbone peaking at Mount Ida (2456 m, 8058 ft) in the White Mountains. The scenery varies from wild gorges to fertile plains where rare herbs and wild orchids abound.

Crete flourished under Venetian occupation (1210-1669), producing some of Greece's finest medieval painting, poetry and drama. It was impoverished under Turkish rule (1669-1896), and was united with Greece in 1912-13. In May 1941, there was fierce fighting in the north of the island, when in the first massive paratroop operation of the Second World War, German forces seized Maleme airfield and went on to defeat 40 000 British, Commonwealth and Greek troops. Besides tourism, industry today includes weaving and embroidery. The main town and capital is HERAKLION.

Population 502 100
Map Greece Dd

crevasse Deep vertical crack or fissure in a glacier. See GLACIATION

Crewe ***United Kingdom*** English industrial town and railway centre in Cheshire, 45 km (28 miles) south of Manchester. The railway works were opened in 1840, and turned a small market town into a flourishing engineering centre.

Population 48 000
Map United Kingdom Dd

Crimea (Krym) ***USSR*** Peninsula and resort area in the Ukraine, separating the Sea of Azov from the Black Sea. It consists mainly of dry but fertile steppe land. The Crimean Mountains run for 150 km (95 miles) along the southern coast, providing shelter for the resort region known as the Soviet Riviera, which is centred on YALTA. The range's highest peak is Roman Kosh, which reaches 1545 m (5069 ft). The industrial city of SIMFEROPOL' is the capital.

The Crimea, which was settled in the 6th century BC by Greek traders, became part of Russia in 1783. In 1845-6 it was the scene of the Crimean War between Russia and the forces of England, France and Turkey. The ill-fated Charge of the Light Brigade – in which 247 English cavalrymen perished out of a total of 673 – took place near SEVASTOPOL, today one of the area's main towns. Many of the wounded were tended by the English nurse Florence Nightingale (1820-1910), who took 38 nurses to the Crimea and set up hospitals at Balaklava and Scutari (now USKUDAR in Turkey).

During the Second World War the Crimea was occupied by German troops in 1941-4. It was made a region of the Ukraine in 1954.

Population 2 309 000
Map USSR Ed

Cristóbal ***Panama*** See COLON and CANAL ZONE

Crna Gora ***Yugoslavia*** See MONTENEGRO

Crnojevica ***Yugoslavia*** See SHKODER, LAKE

Croagh Patrick (Cruach Phádraig) ***Ireland*** Holy mountain just south of Clew Bay on the west coast. It is the site of an annual pilgrimage believed to be the oldest in the Western world. On the last Sunday in July each year thousands walk, many barefoot, to the summit (765 m, 2510 ft), where Mass is celebrated in a small, modern chapel. Ireland's patron saint, St Patrick, is said to have prayed and fasted on the summit for 40 days and nights during the Lent of AD 441.

Map Ireland Bb

Croatia (Hrvatska) ***Yugoslavia*** Second largest (in area and population) of the country's six republics, after Serbia. It lies in the northern half of the country, between Hungary and the Adriatic Sea, and covers 56 538 km² (21 824 sq miles).

Predominantly rural until 1945, the republic now has industries based on timber, oil and gas from SLAVONIA, hydroelectricity, bauxite, limestone and Bosnian iron ore in towns that include Zagreb, the capital, Osijek, Bosanski Brod, Varazdin and Sisak. The chief ports are Pula, Rijeka, Split, Zadar, Sibenik, and tourism is a major industry in Istria and DALMATIA.

Croatia was politically linked with Hungary from 1102 to 1918. This gave it ties with Western Europe and the Catholic Church – like the Yugoslav republic of Slovenia. In contrast, most of Serbia has historical ties with Eastern Europe and the Orthodox Church, Bosnia Herzegovina shows strongly the influence of Muslim Turkey, while in the rest of the country both East European and Muslim trends are prominent.

Population 4 601 500
Map Yugoslavia Bb

Crocodile ***Southern Africa*** See LIMPOPO

Crocodilopolis ***Egypt*** See EL FAIYUM

Cronstadt ***USSR*** See KRONSHTADT

Crotóne ***Italy*** Coastal town on the sole of the boot of Italy. It was famous in antiquity as the ancient Greek colony of Croton, founded in 710 BC. The 6th-century BC mystic and philosopher Pythagoras lived and taught there. It was the site of one of the ancient world's great schools of medicine, and from 588 BC it was renowned for its successes in the Olympic games. The town's wrestler Milo was Olympic champion six times. Today it is an industrial town relying on its zinc works for prosperity.

Population 26 700
Map Italy Fe

Crowsnest Pass ***Canada*** Pass rising to 1357 m (4450 ft) through the Rocky Mountains in south-east British Columbia. In 1897 the Canadian Pacific Railway received government funds to build a railway through the pass.

Map Canada Dd

Crozet Islands ***Indian Ocean*** There are five islands and 15 islets in this archipelago, set in the Indian Ocean about midway between Antarctica and Madagascar. Main islands are Apostles, Pigs and Penguins in the western part of the group; and Possession and Eastern in the eastern part. Volcanic in origin and rugged, their steep cliffs have seen many a shipwreck. They are in the French Southern and Antarctic Territories, and France maintains a scientific and meteorological station on rocky Possession Island, which rises to 934 m (3064 ft).

The islands were discovered in 1772 by Captain Marion-Dufresne and his ship's mate, Crozier, who annexed them for Louis XV. They cover an area totalling 300 km² (116 sq miles).

Map Indian Ocean Be

Crozier, Cape ***Antarctica*** Eastern tip of Ross Island, abutted by the Ross Ice Shelf. The pressure of ice against land squeezes the ice into vast ridges. The cape is the home of large colonies of emperor and Adélie penguins.

Map Antarctica Jc

Cruách Phádraig ***Ireland*** See CROAGH PATRICK

crust Solid exterior portion of the earth above the Mohorovičić Discontinuity. See JOURNEY TO THE CENTRE OF THE EARTH (p. 200)

Crystal Mountains ***West Africa*** Low range in north-west Gabon and south-east Equatorial Guinea. They form part of the edge of the African plateau, where rivers plunge down through rapids and waterfalls. The highest point is Mount Koudaké (980 m, 2920 ft).

Map Gabon Ba

crystalline rocks Rocks in which the constituent minerals form a mass of crystals which have solidified from a molten state or have been produced by metamorphism. Hence, igneous and metamorphic rocks are crystalline.

Cuando Cubango ***Angola*** Province whose capital, Menongue, lies about 800 km (500 miles) south-east of Luanda. Since Angola became independent in 1975, the province has been controlled by rebel UNITA guerrillas. Subsistence crops are grown, including maize, millet, beans and groundnuts.

Map Angola Bb

Cuango ***Angola/Zaire*** See KWANGO

Cuanza (Coanza; Kuanza; Kwanza) ***Angola***
1. River which flows for about 970 km (600 miles) from the central plateau to the Atlantic 56 km (35 miles) south of the capital, Luanda. The river is navigable for about 190 km (120 miles) upstream and in the 19th century was a main route to the interior for explorers.

2. Name of two central provinces on either bank of the river. The capital of Cuanza Norte, N'dalatando, is about 190 km (120 miles) east and slightly south of Luanda. The province produces beans, coffee, cotton, palm products and tobacco. Cuanza Sul's capital, SUMBE is about 270 km (170 miles) south of Luanda. This province's main products are coffee, cotton and palm oil.

Population (Cuanza Norte) 290 000; (Cuanza Sul) 459 000
Map Angola Aa, Ab

Cuba See p. 172

Cubango ***Southern Africa*** See KAVANGO

Cuchilla de Haedo ***Uruguay*** Range of hills running from the Brazilian border for 200 km (125 miles) across the north and west of the country between the valleys of the Negro and Uruguay rivers. At its highest, it reaches 700 m (2300 ft). A second and lower range of hills, the Cuchilla Grande, lies east of the Negro river.

Map Uruguay Ab

Cúcuta ***Colombia*** Capital of Norte de Santander department, on the Venezuelan border, about 410 km (255 miles) north-east of Bogotá. It was from Cúcuta in 1813 that the Venezuelan-born revolutionary leader Simón Bolívar (1783-1830) launched his troops into the battle for Colombia's independence from Spanish rule. The town was destroyed by earthquake in 1875, but was rebuilt in its former style. Today it is the centre of a cattle-rearing region where tobacco and coffee are also grown.

Population 516 000
Map Colombia Bb

Cuelap ***Peru*** Ruins of a walled fortress city discovered in 1843 near Tingo in the northern department of Amazonas. It is thought to be the last outpost of a light-skinned race, known as the Chachas or Sachupoyans, who were retreating from the invading Incas in about 1460.

Map Peru Ba

Cuenca ***Ecuador*** Capital of Azuay province, founded by the Spanish in 1557 on the banks of the Rio Torrebamba, about 330 km (205 miles) south of Quito. The new and old colonial cathedrals stand in the Parque Calderón, the central square. The city also has many fine colonial churches and Inca ruins. It manufactures panama hats, textiles and leather, and is a major centre for Ecuadorian handicrafts.

Population 272 500
Map Ecuador Bb

Cuenca ***Spain*** Picturesque town 139 km (86 miles) east of Madrid, on the confluence of the rivers Júcar and Huecar. Its cathedral dates from the 12th century. The town is noted for its 'hanging houses', perched on a river cliff. About 30 km (18 miles) north-east is the 'Enchanted City', an area of limestone rocks eroded into weird shapes. The surrounding province of Cuenca is sparsely populated and is largely agricultural, producing wheat and wines.

Population (town) 42 000; (province) 211 000
Map Spain Db

Cuernavaca ***Mexico*** Old and exotic resort town of tropical flowers and luxurious homes, known as the 'City of Eternal Spring' because of its warm, sunny climate. It lies about 80 km (50 miles) south of MEXICO CITY, amid scenic mountains, and is the capital of the state of Morelos. The town has a 16th-century palace, an impressive Baroque cathedral and 18th-century botanical gardens. About 70 km (44 miles) to the south-west are the CACAHUAMILPA caves, the largest system in Mexico.

Population 557 000
Map Mexico Cc

cuesta Spanish term that has been widely adopted for an upland area with a gentle slope on one side, known as the dip slope, and a much steeper one on the other side, known as the scarp face or escarpment. This common relief feature is usually found where there is a tilted outcrop of resistant sedimentary rocks. The dip slope corresponds with the tilt of the rocks, while the scarp face is cut by erosion across them. The North Downs and the Chilterns in England are examples.

Cuige Laighean ***Ireland*** See LEINSTER

Cuige Mumhan ***Ireland*** See MUNSTER

Cuige Ulaidh ***Ireland/United Kingdom*** See ULSTER

Culebra ***Puerto Rico*** Small island, 26 km² (10 sq miles) in area, off the east coast. It is noted for its crescent-shaped beaches and the underwater life on its reefs. It has a deep and sheltered harbour.

Population 1300
Map Caribbean Bb

Culiacán (Culiacán-Rosales) ***Mexico*** Capital of Sinaloa state in north-west Mexico. It is a mining city and an agricultural centre which provides the rest of Mexico with most of its winter vegetables.

Population 358 000
Map Mexico Bb

Culloden ***United Kingdom*** Moor 8 km (5 miles) east of Inverness in the Scottish Highlands. The last great battle on British soil took place there in 1746. After a bloody struggle, the English army under the Duke of Cumberland defeated the forces of Bonnie Prince Charlie, the Young Pretender, forcing him to flee and putting an end to the Stuart family's attempt to regain the British Crown. Some 1200 Highlanders died, but only 50 Englishmen were killed in the battle.

Map United Kingdom Cb

Cuba

A MARXIST STATE IN THE CARIBBEAN THAT BROUGHT THE GREAT POWERS TO THE BRINK OF NUCLEAR WAR

In 1962 Cuba became the fuse that nearly sparked off a third world war. Its Marxist leader, Fidel Castro, had accepted plans to establish Russian missile bases on the island – a mere 145 km (90 miles) from the Florida coast of the USA.

The US navy mounted a blockade to stop Russian ships taking missiles to Cuba, and the American and Russian leaders, John F. Kennedy and Nikita Khrushchev, glowered at each other across their nuclear weapons. As the world watched with bated breath, Khrushchev backed down. Since then, Cuba has continued to loom large in Communist intrigues, exporting revolution and aid to Marxists in Third World countries as far apart as Angola, Ethiopia and Nicaragua.

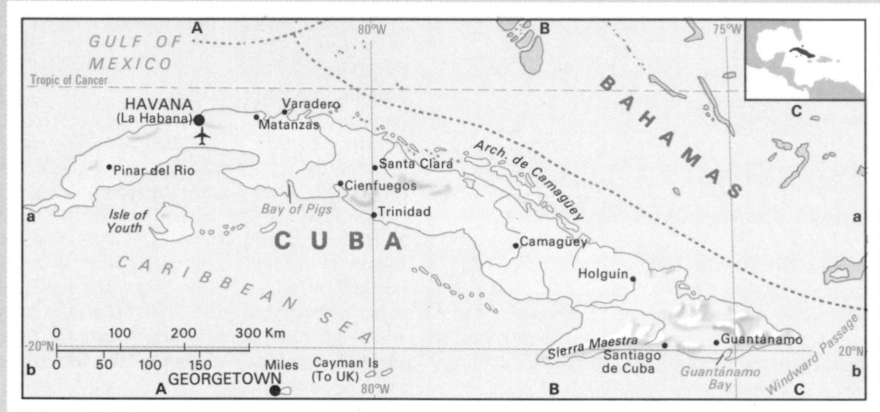

When Columbus landed on Cuba in 1492, he found it inhabited by Indians whose ancestors had come from South America several centuries before. There were about 100 000 of them on the island when the Spanish built their first city at Baracoa in 1512, but by 1570 the Indians had disappeared – the victims of genocide, overwork and disease. Apart from brief British rule in 1762, Spain retained its hold on Cuba until defeated by the USA in the Spanish-American War of 1898.

US troops occupied Cuba until 1902, when an independence treaty was implemented. American interests continued to dominate Cuban economic life – and the island became a popular resort for Americans attracted by its sunny beaches, exotic night clubs and casinos. Meanwhile, the country was ruled by a series of dictators. The last was Fulgencio Batista, whose corrupt and intolerant regime lasted on and off from 1933 until 1959.

That year, a group of revolutionaries led by Fidel Castro, a law student, and his lieutenant, Che Guevara, ousted Batista after a two-year guerrilla campaign. Mass arrests and executions followed, and thousands of Cuban exiles poured into the USA. Cuba turned to Russia for economic aid and Castro proclaimed a Communist state. In 1961, American CIA-trained Cuban exiles landed in the BAY OF PIGS in an unsuccessful attempt to launch a rebellion, while the following year saw the Russian missile crisis. Since then Cuba has tried to re-establish links with the USA.

Cuba is as big as all the other 3700 Caribbean islands put together. Its 10 million people make up one-third of the total population of the West Indies. The island is 1250 km (780 miles) long and 30 to 190 km (20 to 120 miles) wide, and consists mostly of extensive plains.

The climate is warm, although the northwest has winter spells below 10°C (50°F). Most areas have more than 1140 mm (45 in) of rain a year. Many hurricanes pass close by but few cause serious damage. Most of Cuba's soil is fertile, and the only surviving large wooded areas are in the mountains covering a quarter of the island.

Before 1959 foreign interests controlled the economy, owning 75 per cent of arable land, 90 per cent of essential services and 40 per cent of the sugar industry. The Communists tackled corruption and unemployment; wages were regulated and ownership of production was largely taken over by the state. About 70 per cent of Cuba's trade today is with Communist countries. Russia buys sugar and nickel at well over the world price, and supplies cut-price oil and cheap loans.

Sugar provides 80 per cent of Cuba's export earnings and the island is the world's third largest commercial producer. Latin America's first railway was built in 1837 to link HAVANA, the capital, with the canefields. Before the revolution nearly all the cane was grown by small tenant farmers, whose land was owned by the sugar companies. Today, all the factories and most of the land is owned by the state, and new dams provide irrigation water for drought years.

Since 1971 there have been steps to diversify agriculture. Tobacco was once the second farm export, but has been overtaken by citrus fruits. To reduce dependence on imported food, rice growing and the rearing of beef and dairy cattle have been increased. Modern fishing fleets have increased the catch sevenfold since 1959.

Cuba has the fourth largest nickel reserves in the world, and this is the second most valuable export. The most important industry is still processing sugar and other foodstuffs, but factories also make cement, farm machinery, fertilisers, paper, textiles and footwear. Invest-

▲ WORLD FESTIVAL OF YOUTH Young leftists form a portrait of Che Guevara, Fidel Castro's charismatic aide. They were among the 18 500 young people from 140 countries who attended the festival in Havana in 1978.

ment in new hotels has revitalised tourism.

About half of Cuba's population are mulattos – mixed-blood descendants of African slaves and Europeans. Most others are descended from Spanish immigrants, but there is a small Chinese minority. Two-thirds of the people live in towns. They do not share the stark poverty of most Latin American countries. They have free education and health services and a guaranteed basic wage, but they also have an austere lifestyle similar to that of the Communist countries in Eastern Europe.

CUBA AT A GLANCE

Area 114 524 km² (44 218 sq miles)

Population 10 215 000

Capital Havana

Government One-party Communist republic

Currency Peso = 100 centavos

Language Spanish

Religion Christian (predominantly Roman Catholic)

Climate Tropical; average temperature in Havana ranges from 18-26°C (64-79°F) in January to 24-32°C (75-90°F) in July

Main primary products Sugar cane, rice, sweet potatoes, cassava, maize, oranges, tobacco, coffee, livestock, fish; nickel, copper, cobalt, chrome

Major industries Agriculture, mining, food processing, tobacco products, metal refining, chemicals, cement, fertilisers, textiles, fishing

Main exports Sugar, nickel, copper, fish, fruit and vegetables, tobacco, coffee, chemicals

Annual income per head (US$) 1400

Population growth (per thous/yr) 11

Life expectancy (yrs) Male 71 **Female** 75

Culm ***Poland*** See CHELMNO

Cumaná ***Venezuela*** Capital of the state of Sucre, about 320 km (200 miles) east of the national capital, Caracas. Founded by Spaniards in 1520, it is the oldest Hispanic city in South America. Straddling both banks of the Manzanares river, it is a port exporting tobacco, cacao (from which cocoa is made), fish, cotton textiles and coffee.

Population 200 000

Map Venezuela Ba

Cumberland ***United Kingdom*** See CUMBRIA

Cumberland Gap ***USA*** Pass through the Cumberland Plateau of the Appalachians, near the junction of Virginia, Kentucky and Tennessee. It was discovered in 1750 and was used by the pioneer Daniel Boone as he opened up the Wilderness Road in 1775. The pass, 500 m (1640 ft) above sea level, lies in a national historical park covering about 82 km² (32 sq miles).

Map United States Jc

Cumbria ***United Kingdom*** County of north-west England, covering 6809 km² (2629 sq miles) on the Scottish border. It is made up of the former counties of Cumberland, Westmorland and the part of Lancashire known as Furness, and includes Britain's LAKE DISTRICT. The old Cumberland coalfield is on the coast, with the industrial towns of Whitehaven and Workington and the shipbuilding centre of Barrow-in-Furness. The county town is CARLISLE. Hill-farming and Lake District tourism are the main livelihoods inland.

Population 483 000

Map United Kingdom Dc

Cumbrian Mountains ***United Kingdom*** See LAKE DISTRICT

cumulonimbus cloud A cloud that builds up to a great height, rising in towering, turbulent clusters from a flat base, often spreading out at the top into an anvil-shaped head. They grow out of cumulus clouds in unstable air, bringing heavy showers and thunderstorms.

cumulus Vertical cloud formation of fluffy cauliflower-shaped clusters rising from a flat base, often seen during the middle and latter part of the day and usually associated with sunny spells. They are distinguished from cumulonimbus by their rounded tops.

Cundinamarca ***Colombia*** Central department that includes the country's capital, Bogotá. It contains a variety of climatic areas from the tropical to the very cold, consequently a wide range of crops are grown, including coffee, maize, bananas and potatoes.

Population 5 527 000

Map Colombia Bb

Cunene (Kunene) ***Angola/Namibia*** River rising in west-central Angola and flowing some 970 km (603 miles) south-west, then west along the Angola-Namibia border to the Atlantic Ocean. Hydroelectricity is produced at the Matala Dam in Angola, and at the 107 m (351 ft) high Ruacana Falls on the border.

Map Namibia Aa

Curaçao ***Netherlands Antilles*** Largest, most populous and most highly developed of the Netherlands Antilles, lying off the northern coast of Venezuela between Aruba and Bonaire. Its 444 km² (171 sq miles) are rocky, and there is little agriculture. The chief industries are oil refining, electronics and tourism. Phosphates are mined for export. There is a deepwater harbour, the Schottegat, in the south-west beside WILLEMSTAD, the capital of the island and of the Netherlands Antilles. An orange-flavoured liqueur, Curaçao, is named after the island, where it was first made.

Population 170 000

Map Caribbean Bc

Curepipe ***Mauritius*** Town, health resort and sugar production centre set in the middle of the island. It is about 16 km (10 miles) south of Port Louis at a height of about 500 m (1650 ft).

Population 70 000

Curitiba ***Brazil*** Regional capital 330 km (206 miles) south-west of the city of São Paulo. Founded in 1693, it is today one of the country's most prosperous and rapidly developing industrial cities. Products include tea, tobacco, furniture and metals.

Population 1 442 000

Map Brazil Dd

current See OCEAN CURRENTS (p. 483)

Curtea de Arges ***Romania*** Town on the River Arges, 150 km (93 miles) north-west of Bucharest. In the 14th century Curtea de Arges became the capital of the former principality of Walachia. It contains the ruins of the princely palace. The Church of St Nicolae Donnesce is the oldest feudal monument in southern Romania (14th century). The church of the monastery, built by one of the ruling princes in the 16th century and restored in 1875, contains the tombs of the first monarchs of Romania.

Population 25 000

Map Romania Aa

Curzola ***Yugoslavia*** See KORCULA

Cuscatlán ***El Salvador*** Department to the east of San Salvador producing cotton, sugar cane and livestock. Its capital is Cojutepeque (population 35 000).

Population 156 000

Cuttack ***India*** City on the Mahanadi river about 325 km (202 miles) south-west of Calcutta. It is built on a spit of land between two distributaries, and is liable to flood. Cuttack produces exquisite silver and gold filigree jewellery and also trades in oilseeds and rice.

Population 327 400

Map India Dc

Cuzco ***Peru*** Department and city in the southern Andes mountains. The high pastures support sheep and alpacas (long-haired relatives of the llama, kept for their wool), and the fertile valleys are planted with maize, barley and potatoes.

The departmental capital, Cuzco, stands 3500 m (11 480 ft) above sea level and was the religious and administrative centre of the far-flung Inca Empire between the 11th and early 16th centuries. The area is rich in Inca history. The ruins of Sacsahuamán and Salamancu are nearby, and to the north-west the famous 'Inca trail' leads from Ollantaytambo to Machu Picchu – 'the Lost City of the Incas'.

The main square of the city was the centre for many Inca ceremonies. The magnificent Baroque cathedral and Church of the Compãnia, built in 1668, stand on the site of the Temple of Viracocha and an Inca palace. The city's heritage is reflected in its festivals, which range from the Catholic Corpus Christi (2nd Thursday after Whitsun) to Inti Raymi, the Inca Festival of the Sun, held about the same time.

Population (department) 832 500; (city) 184 600

Map Peru Bb

cwm See CIRQUE

Cyclades ***Greece*** Group of 30 islands in the centre of the Aegean Sea, lying in a circle around Delos. They were vital links in the early sea-trading routes to the east, producing a race of sailors from around 2600 BC. The islands were occupied by Venetians in 1204, then Turks in the 16th century, before becoming part of independent Greece in 1832.

MIKONOS is the most popular island with tourists, NAXOS is the largest and Ermoúpolis on the island of Síros is the group's capital. Soil on the islands is thin, so there is little agriculture, but their coastlines and romantic settings make them attractive to tourists.

Population 88 400

Map Greece Dc

cyclone Storm accompanied by strong winds revolving round a centre of atmospheric low pressure. There are two types. In temperate latitudes a cyclone is known as a DEPRESSION; in the tropics, as a TROPICAL CYCLONE – usually very violent.

cyclonic disturbance See DEPRESSION

cyclonic rain, frontal rain Precipitation which marks the fronts and the warm sector of a depression in temperate latitudes, as one air mass overrides or undercuts another.

Cyrenaica ***Libya*** One of Libya's three main historical regions occupying the eastern half of the country (855 000 km², 330 116 sq miles in area). BENGHAZI is its chief town. Most of the province consists of desert, but there is fertile land along the Mediterranean coast, and the northern slopes of the coastal ridges are the only region of Libya with forest vegetation. The country's main oil fields are in the province.

In the 19th century the Cyrenaican Bedouins embraced the teachings of the Sanusi movement, a militant Muslim sect, and remain its most numerous adherents in the region. During the Italian occupation (1912-43), they were in the forefront of the resistance movement.

Today much of the life of the province remains nomadic or semi-nomadic: herding, farming and fishing. The main agricultural products are barley, wheat, grapes, olives, dates and tuna fish. The Jebel Akhdar in the north is an important agricultural area. About 23 per cent of the country's people live in the province.

Map Libya Cb

Cyprus

AN ISLAND THAT LIKES TO BE VISITED, MEDITERRANEAN CYPRUS HAS BEEN HOST TO GODS AND CRUSADERS, SAINTS AND HEROES – AND THOUSANDS OF HOLIDAYMAKERS

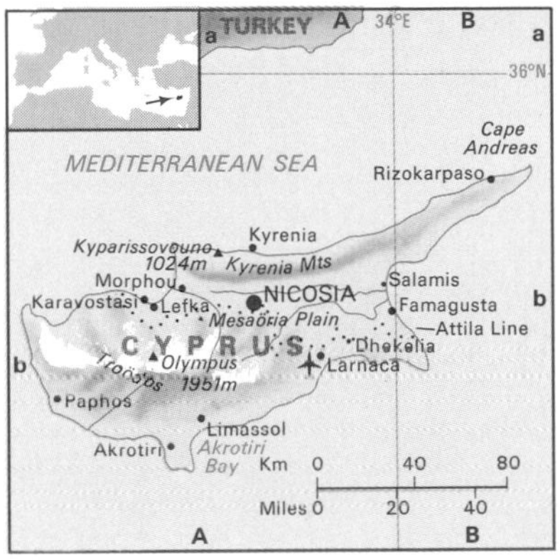

Golden Aphrodite, Greek goddess of love, was born of the seafoam that hisses on the beach of Achni on the south coast of Cyprus. Mortal men, too, have known Cyprus for a very long time. Idols have been unearthed that were carved by Stone Age sculptors 8000 years ago and more, and Cypriot mines – which are still worked – provided copper for the entire eastern Mediterranean area in the 3rd millennium BC. The Greek name of the metal and the island – *Kypros* – are the same.

Then too, lying as it does on a maritime crossroads between Europe, Asia and Africa, Cyprus was always assured of a variety of visitors and conquerors. Mycenaean Greeks established city-kingdoms there in about 1500 BC, and thereafter the island was in turn annexed by the empires of Egypt, Assyria, Persia, Rome and Byzantium. Christianity was brought to the island by St Paul and St Barnabas in the 1st century AD. In 1191, it was occupied by Richard I ('the Lionheart') of England, who married Princess Berengaria of Navarre at LIMASSOL, and handed the island over to the Knights Templar, a military and religious order founded to protect pilgrims to the Holy Land. The French Lusignan dynasty were kings of Cyprus from the 12th century, until they were ousted by the Venetians in the 16th century. They in turn were expelled by the Turks who governed Cyprus until 1878, when they leased it to the British who made the island a Crown Colony in 1925.

The majority of the population is of Greek descent, with a minority (18 per cent) of Turkish descent. From about 1900, there was a strong movement among Greek Cypriots for unification with the homeland of their ancestors. In the 1950s this escalated into the terrorist movement EOKA – *Ethniki Organosis Kypriakou Agonos*, or National Organisation of Cypriot Struggle – which carried out a campaign of assassination against both the British and the Turkish Cypriots. It partially gained its ends in that, in 1960, the island became an independent republic within the Commonwealth, with the Greek Cypriot Archbishop Makarios as president.

Three years later, however, civil war broke out between the Greek and Turkish Cypriots. It was quelled by a United Nations peacekeeping force, but in 1974 the situation was still sufficiently uneasy for Turkey to mount a full-scale invasion of the island under the pretext of protecting Turkish Cypriots from the Greeks. Turks occupied the northern part of the country including the towns of FAMAGUSTA and KYRENIA, and expelled some 200 000 Greek Cypriots from their homes to make way for settlers from Anatolia. Hundreds of Greek Cypriots were killed or injured and more than 1600 are still missing. In 1975, Turkey declared the part of Cyprus occupied by its troops to be separate and independent.

The island has a long, thin panhandle and is divided from west to east by two parallel ranges of mountains, which are separated by a wide central plain open to the sea at either end. The south-westerly range, the Troödos mountains, are covered with pines, cypresses and cedars and culminate in the 1951 m (6401 ft) Mount Olympus. At that height, there is sufficient snow for skiing in January and February, but otherwise, the summers are hot and dry and autumn and winter warm and damp. This delightful climate contributes towards the island's astonishing variety of crops – early potatoes, vegetables, cereals, tobacco, olives, carobs, bananas, table grapes and wine grapes for the rich, strong wines, sherries and brandies for which Cyprus has been famous since classical times. Fishing also has some significance in the island's economy, as do copper, asbestos and chrome, though the mining industry has deteriorated since the Turkish invasion. Additional income is provided by the British bases at DHEKELIA and AKROTIRI.

But, above all, the island depends on its visitors. Unfortunately, at the moment, it is no longer possible to view Cyprus as a whole. Visitors who have landed in the Greek sector are not permitted to pass over into the Turkish portion, and vice versa. However, though some of the most important tourist sites, such as the ruins of SALAMIS near Famagusta, are now in the Turkish zone, it is the Greek-Cypriot sector that is most popular. Indeed, tourism has led a recovery in the economy of the Greek sector since 1974, while that of the Turkish-held area has stagnated.

▲ IMPREGNABLE FORTRESS The 12th-century castle overlooking Kyrenia harbour in northern Cyprus has withstood six sieges.

CYPRUS AT A GLANCE

Area 9251 km² (3572 sq miles)
Population 678 000
Capital Nicosia
Government Parliamentary republic
Currency Cyprus pound = 100 cents
Languages Greek, Turkish, English also used
Religions Christian (77% Greek Orthodox), Muslim (18%)
Climate Mediterranean; average temperature in Nicosia ranges from 5-15°C (41-59°F) in January to 21-37°C (70-99°F) in July
Main primary products Potatoes and other vegetables, grapes, citrus fruits, wheat, carob beans, olives; chromium, copper, asbestos, iron
Major industries Agriculture, mining, textiles and clothing, cement, petroleum refining, wine making, tourism, fishing
Main exports Clothing, potatoes, cement, cigarettes, wine, citrus fruits, footwear
Annual income per head (US$) 3040
Population growth (per thous/yr) 13
Life expectancy (yrs) Male 71 **Female** 75

Czechoslovakia

THIS CENTRAL EUROPEAN STATE, CREATED IN 1918, HAS ESTABLISHED ITSELF AS ONE OF THE MOST SUCCESSFUL ECONOMIES IN THE SOVIET BLOC

A glance at the map of Europe shows Czechoslovakia landlocked at the heart of the continent. It is shaped like a rambling east-west bridge linking the USSR and West Germany. Czechoslovakia is the only country having a border with both these states – Russia, the power base of the Warsaw Pact, and West Germany, the front line of NATO. This explains much about recent experiences of the people, particularly the Russian reaction to their liberalisation attempts in 1968 under the reformist leader Alexander Dubcek.

Yet cultural, political, religious and economic ties and traditions have long linked Czechoslovakia to the West. It is a predominantly Roman Catholic country. Its capital, PRAGUE, became one of the most splendid cities in Europe in the 14th century, and (with Vienna) was a seat of the Holy Roman Emperors. Tomas Masaryk, architect of the first Czechoslovak republic after the First World War ended in 1918, orientated his country firmly to the West.

But outside events soon impinged, especially the worldwide economic crisis and the growing might of Nazi Germany. Czechoslovakia lost its border regions to Germany after the Munich crisis of 1938 and was completely overrun the next year. After the Second World War Russia was the dominating outside influence, and after a fragile coalition period in 1945-8 Czechoslovakia found itself with a Communist government, playing the role of 'a guard on the borders of the Socialist community'. The Iron Curtain had reversed the country's historic orientation.

FRICTION BETWEEN PEOPLES

In addition to outside pressures, there was internal friction in this fragile 20th-century union of two distinct peoples: the Czechs in the west (in BOHEMIA and MORAVIA) and the Slovaks in the east (in SLOVAKIA).

Prague is in Czech territory, and for most of the 1920s and 1930s the government was dominated by Czechs, who in 1918 were educationally and economically far ahead of the Slovaks. There was little choice but to send qualified Czech personnel into Slovakia to run the administration. However, with rapidly improved education and the emergence of a new Slovak intelligentsia, Slovaks began to become restless over Czech influence in their domestic affairs. (During the Second World War, Slovakia was a separate state that cooperated fully with Nazi policies.)

In 1969, under the government of Gustav Husak (installed by the Russians after the 1968 uprising), the Slovaks gained some autonomy when Czechoslovakia became a federation of 'two equal and brotherly nations' – the Czech and Slovak Socialist Republics.

Czechs and Slovaks have contrasting national characteristics, distinct languages and separate histories. Both endured centuries of outside domination, but whereas the Czechs flourished under Austrian (Hapsburg) rule in the 19th century – having full representation in the Vienna parliament – the Slovaks were exposed to 'Magyarisation' under the Hungarians. By 1914 they were experiencing serious national decline – with 47 per cent illiteracy, for example, and no university. Despite rapid development between 1918 and 1938, when the Second World War ended in 1945, Slovakia still lagged far behind the rest of Czechoslovakia.

This great disparity in living conditions had several origins. Slovakia, cut off from Poland on the north by the BESKIDY and TATRA mountains, comprises low alp-like CARPATHIAN country. The Slovaks never enjoyed the political independence, economic growth and cultural renaissance that enabled the Czechs to develop trade, crafts, education, art and culture, especially from the 11th to the 15th century and again in the 19th. This flowering was most prominent during the Golden Age under Charles (Karel) I (1347-78).

In addition, German colonists, invited by Wenceslas (Vaclav) I (1230-53) and his son Ottokar II (1253-78) to settle in Bohemia, brought wealth and new skills, helping to establish urban life in the province. They became known as *Sudetendeutsche*, after the SUDETY area of north-east Bohemia where many lived. In the 1930s, they numbered 3 million out of the country's total population of 14 million, and were ruthlessly used by Hitler, the German Nazi leader, to foment confrontation with the Czechs. The Munich agreement between Germany, France, Italy and Britain in 1938 allowed Hitler to march into the Sudetenland, where the majority of the population were German-speaking. In 1945 the victorious allies agreed that all Germans should be expelled from the country.

The German exodus left serious labour shortages in parts of Bohemia, which had been among the most advanced industrial, agricultural and urban areas of the country. Czech and Slovak peasants and workers were resettled in the deserted towns and villages, but not in sufficient numbers.

Population growth among Czechs was con-

sistently lower than for Slovaks, and offered no solution to the manpower shortage. In the early postwar years the birthrate among Slovaks was almost double the rate for Czechs, and the Slovak proportion of the total population rose from 27 per cent in 1950 to over 32 per cent in 1980. The Czech population has been in decline for many years, and the population of the whole country is now increasing only slowly.

SUCCESS STORY

In order to overcome the labour shortages, the government began taking the work to the workers. Many textile, leather, food and machinery factories were moved from Bohemia eastwards to towns in Slovakia. But the most significant action was the consolidation of 1.3 million peasant holdings into less than 2000 state and collective farms. This and mechanisation reduced the need for labour, and today 800 000 farm workers produce more than 2.5 million did in 1950.

Economic development in Slovakia has been a success story for the postwar government. There have been huge investments in new construction, damming rivers to generate hydroelectricity, sinking mines to extract metals, and building factories. This has brought jobs in steel, engineering, aluminium, chemicals (using Russian oil and gas) and textiles. Dense forests cover 40 per cent of Slovakia and support expanding sawmills, papermills and furniture manufacture.

Improved drainage in the west Carpathian valleys and irrigation in the dry southern plains bordering Hungary have raised the output of grapes, sugar beet, maize, tobacco and many other crops, as well as poultry and pork. In the mountains traditional sheep rearing and tourism are expanding.

The Slovaks' living standards have risen sharply. BRATISLAVA (Slovakia's capital) and KOSICE have mushroomed as thousands of rural migrants have moved into their massive housing estates. But most Slovak towns remain small, drawing commuters from villages in the densely populated valleys.

In Bohemia and Moravia the manpower shortage has led to the introduction of labour-saving machinery in the nationalised industries, including coal, metallurgy, engineering, the famous Jablonec glassworks in northern Bohemia, papermills, the huge (formerly Bata) shoe factory at Gottwaldov, and armaments. New industries have made better use of resources in the interests of self-sufficiency. Hydroelectricity from Bohemian rivers powers textile, clothing, shoe, food and electrical factories. Recently, more investment has gone into electronics factories and research establishments in and near Prague, which can attract highly trained workers, many recruited from the country's 17 universities.

While money and effort have been poured into industry, the towns themselves have sometimes been neglected. This has especially affected towns in Bohemia, some of them living museums of Romanesque, Gothic, Renaissance, Baroque, Art Nouveau and Art Deco architecture. However, a restoration programme has started, and most towns have modern hospitals, schools, sports halls, cinemas and libraries. New sanatoriums and tourist facilities have been built in old-established spas like KARLOVY VARY (Carlsbad), MARIANSKIE LAZNE (Marienbad) and Piestany.

One of the most dramatic periods in Czechoslovakia's recent history followed the election of Alexander Dubcek as leader of the Czechoslovak Communist Party in January 1968. Dubcek, a Slovak, quickly began to liberalise the country and raised hopes of a return to more democratic government called 'socialism with a human face'. But in August Warsaw Pact troops invaded with tanks and took control of the country despite widespread anti-Soviet demonstrations. Dubcek was replaced as party leader, and the new Husak regime reverted to mass political purges, arrests, religious persecution and exiling people – though not on the scale of the 1950s. Dubcek served briefly as ambassador to Turkey before he was expelled from the Communist Party in 1970.

The Czech struggle for social progress has deep historical roots, stretching back to St Wenceslas (the original 'Good King') in the 10th century. At the Charles University of Prague, the first in central Europe, Jan Hus (1370-1415) preached freedom of thought and unleashed the reformation among the Czechs.

Later there followed four centuries of Austrian Hapsburg domination (1526-1918). Czech resistance was ruthlessly crushed, especially during the Thirty Years' War (1618-48). However, over the centuries the Czechs developed passive resistance as part of their reaction to oppression, venting their feelings through literature (in the works of such writers as Franz Kafka, who wrote in German, Karel Capek and the 1984 Nobel prizewinner Jaroslav Seifert) and music (Dvorak, Smetana, Janacek and others). In Czechoslovakia today there is an underground civil rights movement called Charter 77.

▲ GOOD KING WENCESLAS
Czechoslovakia's patron saint looks down on Wenceslas (Vaclav) Square in Prague. According to legend, the murdered 10th-century Christian king will rise and lead his people in their hour of greatest need.

CZECHOSLOVAKIA AT A GLANCE
Area 127 889 km² (49 378 sq miles)
Population 15 550 000
Capital Prague
Government One-party Communist republic
Currency Koruna (crown) = 100 haler
Languages Czech, Slovak, Hungarian
Religion Christian (67% Roman Catholic, 10% Protestant)
Climate Continental, with hot summers and cold winters; average temperature in Prague ranges from −4-1°C (25-34°F) in January to 14-23°C (57-73°F) in July
Main primary products Potatoes, sugar beet, cereals, livestock; coal, lignite, iron ore, uranium, mercury, antimony, magnesium
Major industries Iron and steel, transport equipment, machinery, cement, chemicals, fertilisers, textiles
Main exports Machinery, motor vehicles, iron and steel, chemicals, footwear, textile yarns and fabrics, food, clothing, coal, railway vehicles
Annual income per head (US$) 4500
Population growth (per thous/yr) 3
Life expectancy (yrs) Male 70 **Female** 74

Cyrene ***Libya*** Ancient ruined city, 224 km (139 miles) from Benghazi in the north-east of the country. The original capital of Cyrenaica, founded in 630 BC by the Greeks, it flourished for 1000 years as the centre of a Greek colony before being devastated by an earthquake in AD 364. At its height, Cyrene was a city of 100 000 inhabitants and included an important medical school. Lying a few miles inland, it was served by the port of APOLLONIA.

The city's ruins include the Roman baths, the sanctuary of Apollo, the Greek theatre, the temple of Zeus, the house of the 2nd-century Jew Jason Magnus (one of the authors of the Old Testament book of Maccabees), tombs cut out of solid rock and an early church containing fine mosaics.

Map Libya Ca

Czech Socialist Republic (Ceskoslovensk Socialistická Republika) ***Czechoslovakia*** Westernmost of the country's two constituent republics – the other is Slovakia. It covers 78 864 km² (30 449 sq miles) – an area the size of Scotland – and includes BOHEMIA and MORAVIA. PRAGUE is the capital.

Population 10 291 900

Map Czechoslovakia Ba

Czechoslovakia See p. 176

Czestochowa (Chenstokhov; Czenstochau) ***Poland*** Industrial city and the country's chief pilgrimage centre. It lies in the south, 64 km (40 miles) north of the city of Katowice. Pilgrims flock to the richly adorned Jasna Gora (Hill of Light) Monastery founded in 1382 and now a symbol of Polish nationalism. Its tower, 105 m (344 ft) tall, dominates the city, and its treasures include the Black Madonna, a Byzantine painting of our Lady of Czestochowa, said to have miraculous powers.

The mainly modern industrial area around the monastery produces iron and steel, textiles and paper.

Population 244 100

Map Poland Cc

D layer See ATMOSPHERE

Da Nang (Tourane) ***Vietnam*** Chief port of central Vietnam, about 600 km (375 miles) north-east of Ho Chi Minh City. It was an important US military base during the Vietnam War (1965-75), and the base is now used as an international airport.

Population 365 200

Map Vietnam Bb

Da Yunhe (Grand Canal) ***China*** The country's longest canal, a 1200 km (750 mile) link between the Chang Jiang river and Tianjin (Tientsin). It was built mainly during the Tang (AD 618-906) and Song (960-1270) dynasties.

Map China Hd

Dabrowa Gornicza (Dombrau) ***Poland*** Southern coal and steel town, 13 km (8 miles) north-east of the city of Katowice. Its first mine opened in 1796 to work the world's thickest surface coal seam; the seam is 20 m (66 ft) thick, and is now preserved as a geological spectacle.

Population 141 600

Map Poland Cc

▲ A WHIFF OF AFRICA Dakar's Kermel market near the harbour sells mainly flowers and baskets. Like the Sandaga market, with its dried fish and spices, it is smelled long before it is reached.

Dacca ***Bangladesh*** See DHAKA

Dachau ***West Germany*** Market town in Bavaria about 15 km (10 miles) north and slightly west of Munich. In 1935, the Nazi rulers of Germany opened their first concentration camp there. The camp's site is preserved as a memorial to the estimated 70 000 people who died in its confines.

Population 35 000

Map West Germany Dd

Daejeon ***South Korea*** See TAEJON

Dagestan ***USSR*** See MAKHACHKALA

Dahlak Islands ***Ethiopia*** Red Sea coral islands off the coast of Eritrea province. There are two large and 124 smaller ones. They provide an anchorage for Soviet warships in exchange for Soviet military aid. There is no permanent population.

Map Ethiopia Ba

Dairbhre ***Ireland*** See VALENCIA

Dakar ***Senegal*** The country's capital, on the south side of the Cape VERDE peninsula. The French founded the city in 1857 on the site of a fishing village opposite the island of Goree.

Dakar has one of the finest Atlantic harbours in Africa, and became the capital of colonial French West Africa in 1902 and of Senegal in 1958. In the late 1970s and early 1980s the city had been overtaken by Ivory Coast's Abidjan as the leading city of former French Africa, but Dakar is bigger; it is one of the largest industrial centres in West Africa. It has a naval base and a railway yard, and its manufactures include footwear, textiles and soap. Tourism is a major industry – the town successfully combines French and African cultures and is West Africa's second most sumptuous capital after LAGOS in Nigeria.

Population 1 000 000

Map Senegal Ab

Dal, Lake ***India*** Lake cradled in the folds of the Himalayas in Kashmir about 645 km (400 miles) north and slightly west of Delhi. It became popular as a cool, hill station during the British Raj, and a major tourist centre.

Most visitors stay on richly carved houseboats, and are attended by vendors paddling *shikara*, small vessels similar to Venetian gondolas. There are many floating vegetable beds, constructed from reeds, and cropped to take fresh food to the city of SRINAGAR. The lake shore is lined with Mogul gardens and old and new mosques.

Map India Ba

Dalarna ***Sweden*** Central province covering 29 236 km² (11 288 sq miles) to the north-west of Stockholm. Largely forested, it has old-established metalworking and softwood processing industries. The province's main tourist area is around Lake SILJAN. The people of the province are noted for their traditional and colourful costumes and handicrafts.

Population 284 400

Map Sweden Bc

dale Broad, open valley especially in Yorkshire, Derbyshire and the English Lake District.

Dallas ***USA*** Commercial city in north-eastern Texas, about 110 km (68 miles) south of the Oklahoma border. It shares, with its neighbour FORT WORTH, one of the world's busiest airports. Dallas city covers some 860 km² (332 sq miles), and is a centre of the oil and electronics industries. President John F. Kennedy was assassinated there in November, 1963.

Population (city) 974 200; (metropolitan area) 2 203 700; (conurbation) 3 348 000

Map United States Gd

A DAY IN THE LIFE OF A SENEGAL BUS DRIVER

Sheikh takes his prayer mat out of the locker he has been using for ten years, ever since he started work as a driver with Sotrac, Dakar's public transport company. It is 1.20 pm, and he has just finished his seven-hour shift behind the wheel of a 75 passenger bus. He enters the depot yard, to perform his ritual ablutions, say his prayers, and have his lunch of *mafe* – meat and groundnut stew with rice.

After a rest, he walks to Dakar's main coach station, where his afternoon job begins. Sheikh drives a taxi on a set route from the coach station to the suburb of Pikine, about 20 km (12 miles) out from the centre, and back, stopping whenever a passenger wants to get out, or whenever he is hailed. Each passenger pays 200 CFA francs (about 50 US cents).

Sheikh waits only a few minutes at the coach station before his friend Abdoulaye turns up in his battered Datsun. Sheikh takes his place in the driver's seat.

At about 8 pm, after filling the tank, Sheikh returns the car and gives Abdoulaye the agreed rental of 5000 francs (about US$15). He keeps the rest of the money he has made.

On his way home, Sheikh stops to give some of his earnings to his second wife, Fatou, who still lives in her parents' compound along the road from Sheikh's. She and Sheikh have a baby daughter, Awa. Fatou will continue to live with her father's family until Sheikh can afford a bigger house.

Sheikh has three children by his first wife Aminata, who is waiting for him when he gets home. Sheikh treats both his wives equally, although Aminata deserves more respect, socially, because of her seniority. Sheikh's marriage to Aminata was arranged in his hometown of Kaolock, and she brought him a dowry of clothes and cattle. Fatou, who was Sheikh's personal choice, he met in Dakar, and her dowry was 200 000 francs (about US$600).

The two-bedroomed house that Sheikh lives in is made of cement bricks, and stands in a compound of five houses, shared by five families (12 adults and 15 children altogether). The two showers and two lavatories are communal.

While he sits down to his dinner of fish and rice, Aminata tells him about her day. She did well at the market, where three women came to her to have their hair plaited. The money she made will go towards shoes for their eldest child, Ibrahim, aged six, who starts school in September.

As the evening draws on, visitors and the neighbours who share Sheikh's compound come visiting. A fire is lit in the yard, and tea utensils are brought out. Everyone chats until late into the night.

Dallols, The ***Niger*** Series of wide dry valleys carved by rivers before the Sahara began to turn into a desert some 5000 years ago. The ancient river courses wind across the north-west from the border with Mali to the Algerian border. In places, acacias and palm trees line the former river banks, where water is often close to the surface.

Map Niger Ab

Dalmatia (Dalmacija) ***Yugoslavia*** Adriatic coast and islands of Croatia, stretching 375 km (about 230 miles) south-east from the Kvarner Gulf to the Gulf of Kotor. The Dinaric Alps, through which there are few paths, border the region inland. Below the mountains is a narrow coastal strip on which grapes, olives, walnuts and citrus fruits are grown.

Dalmatia is a leading Yugoslav source of cement, marble, aluminium, wine and olive oil. Its sunny beaches, coves and islands, and its charming old towns and villages – many of them once ruled by Venice – have generated a booming tourist industry. Rab, Trogir, Losinj, Krk, Split and Dubrovnik are the major resorts.

Map Yugoslavia Bc

Damanhur ***Egypt*** Town in the heart of the cotton-growing region of the Nile delta, 61 km (38 miles) south-east of the port of Alexandria. It processes cotton, and is on the Mahmudiya Canal, which links the Nile to Alexandria.

Population 175 000

Map Egypt Bb

apartments, contemporary university buildings and handsome, wide boulevards. There is also a rapidly growing industrial area. Historically, Damascus was first mentioned in 2500 BC on clay tablets discovered in Mari (an ancient Mesopotamian city, now Tel-Hariri, Syria). According to these early records, it was the town of Shem (Noah's son) and had become an important city by 1900 BC. Built on hills some 690 m (2264 ft) above sea level, it is now a major road and rail crossroads, linked with Amman, Baghdad and Beirut. There is an international airport. The city's main industries are silver and metal-working, silk weaving, glassblowing, cement manufacture and sugar refining.

Among its antiquities are the Great Mosque, built in AD 708 with a 130 m (426 ft) long prayer hall; the tomb of Saladin, the Muslim political and military leader (about 1137-93); and the ruins of a Roman temple to Jupiter. The old city is largely ringed by 13th-century ramparts and contains a number of Christian churches. In the heart of the modern city is a garden area watered by the Barada river which flows through Damascus. The museums include the National Museum of Syria and the Museum of Epigraphy (ancient inscriptions), housed in a 15th-century madrassa (religious school).

Population 1 042 000

Map Syria Bb

Damavand ***Iran*** See ELBURZ MOUNTAINS

Damietta ***Egypt*** See DUMYAT

HARNESSING THE EARTH'S RIVERS

One of the world's earliest major dams was built across the Garawi valley in Egypt about 3000 BC. Used to regulate flood waters, it was some 106 m (350 ft) long and was made of packed earth with masonry walls. Egypt is also the home of one of the world's finest modern concrete dams – the Aswan High Dam, 114 m (375 ft) high and 4 km (2.5 miles) long. The world's largest concrete dam and largest concrete structure is the Grand Coulee on the Columbia river in Washington State, USA. It contains some 8 092 000 m³ (10 585 000 cu yds) of concrete.

THE WORLD'S GREATEST DAMS

Dam	Location	Statistics
Grand Coulee	USA	(length) 1272 m/4173 ft (height) 167 m/548 ft
New Cornelia Tailings (most massive)	USA	(volume) 209 501 000 m³/ 274 015 735 cu yds
Yaciretá-Apipé (longest)	Paraguay-Argentina	(length) 72 km/45 miles
Grand Dixence (highest)	Switzerland	(height) 285 m/935 ft

Damaraland ***Namibia*** Homeland for the Damara (or Bergdama) people in the north-west. It ranges from desert in the west to savannah grasslands in the east and covers 46 560 km² (17 977 sq miles). The capital is Khorixas (population 5400).

Population 12 400

Map Namibia Ab

Damascus (Dimashq) ***Syria*** Capital of the country, situated on the fringe of an oasis, the Ghouta, 92 km (57 miles) east of the Mediterranean coast. It is a busy modern city, containing new high-rise hotels, office blocks,

Dammam ***Saudi Arabia*** Oil centre on The Gulf coast, situated 400 km (249 miles) east of Riyadh. Oil was found in the area in 1940, and in the 1950s Dammam became the capital of the Eastern Province. The city centre retains some of the traditional Arab atmosphere with *shisha* (pipe-smoking) houses. Shisha pipes, along with gold jewellery and antiques, are sold in the main bazaar, the Souk al-Hareem. Veiled women sell cloth in the bazaar – but, apart from that, women are seldom seen in public. After JEDDAH, Dammam is the country's second port.

Population 150 000

Map Saudi Arabia Cb

Damodar ***India*** River flowing 519 km (322 miles) from the north-east region of Chota Nagpur down to the plains of Bengal, creating its own delta west of the Ganges delta. Its valley has been made into the industrial heartland of India, and since 1948 has been developed by the Damodar Valley Authority. Large dams control floods, store water for industry and agriculture, and provide power for the states of Bihar and West Bengal.
Map India Dc

Dampier ***Australia*** Iron ore port, built since 1965, on the north-west coast of Western Australia about 1250 km (780 miles) north of Perth. Natural gas from offshore deposits is piped to Dampier.
Population 2400
Map Australia Bc

Danakil ***Ethiopia/Djibouti*** Arid region, about 500 km (over 300 miles) from north to south, and 150 km (nearly 100 miles) from east to west, between the Ethiopian highlands and the Red Sea. Its only inhabitants are nomads tending flocks of sheep and goats.
Map Ethiopia Ba

Danube ***Europe*** Second longest European river after the Volga, flowing some 2850 km (1770 miles) from its source in the Black Forest to its mouth on the Black Sea. It passes through eight countries: West Germany, Austria, Czechoslovakia, Hungary, Yugoslavia, Romania, Bulgaria and the USSR. And three capitals – Vienna (Austria), Budapest (Hungary) and Belgrade (Yugoslavia) – stand on its banks.

It has long been one of Europe's main water transport arteries, being navigable as far upstream as Ulm in West Germany. It has some 300 tributaries and its delta covers an area of about 2600 km² (1000 sq miles).
Map Europe Ed, Fd

Danube Delta ***Romania*** See DOBRUJA

Danubian Platform ***Bulgaria*** Undulating region in the north, between the River Danube and the Balkan Mountains and stretching 450 km (280 miles) west to east from Yugoslavia to the Black Sea. The platform's western section is crossed by rivers flowing northwards to the Danube. To the east of VELIKO TURNOVO province, the region is higher, drier and steppelike. Bulgaria's leading farming region, it produces wheat, maize, sugar beet, sunflowers, fruits and vines on very large state farms. The main towns are Vratsa, Pleven, Lovech, Razgrad, Shumen and Tolbukhin. Less industrialised than towns in other parts of the country, they process the farm produce and service the farms. Vidin, Ruse and Silistra in the north of the platform are river ports on the Danube.
Map Bulgaria Ab

Danzig ***Poland*** See GDANSK

Daphni ***Greece*** Hilltop monastery dating from the 5th and 6th centuries, 11 km (7 miles) west of Athens. Its 11th-century Byzantine church has some of the finest mosaics in Greece. A wine festival takes place annually in September and October.
Map Greece Cb

Daqing ***China*** Oil field of Heilongjiang province. Oil was discovered there in 1956 and the field, the biggest in China, now accounts for over a million barrels a day, almost half the country's output.
Map China Ib

Dar El Beida ***Morocco*** See CASABLANCA

Dar es Salaam ***Tanzania*** Largest town, chief seaport and former capital of Tanzania, 45 km (28 miles) south of the island of Zanzibar. Its name is of Arabic-Swahili origin and means 'haven of peace'. Dar es Salaam was founded in 1862 by the Sultan of Zanzibar (which joined with Tanganyika to become Tanzania in 1964). It was captured by the British in 1916 during the First World War. Its port, which is linked by rail to Zambia, handles most of Tanzania's exports, including sisal and oil. It also handles much of landlocked Zambia's trade. The city produces more than half of Tanzania's manufactures, and its industries include paint, oil refining, textiles and footwear.
Population 757 346
Map Tanzania Ba

Dar'a (Deraa) ***Syria*** Town and major road and rail junction, 96 km (60 miles) south-west of Damascus. Dar'a is the centre of a noted cereal-growing and agricultural area. It lies about 8 km (5 miles) north of the border with Jordan, and the two countries are setting up a duty-free zone to the south of the town.
Population 282 000
Map Syria Bb

Dardanelles (Canakkale Bogazi) ***Turkey*** Narrow strait linking the Aegean Sea with the Sea of Marmara. It was called the Hellespont by the ancient Greeks, and has been of great strategic and commercial importance throughout recorded history. The city of Troy prospered at its western entrance, and the Persian King Xerxes crossed it with a bridge of boats in an unsuccessful attempt to conquer Greece in 480 BC. Today it is the outlet for the USSR's Black Sea fleet.

The GALLIPOLI peninsula on its northern shore was the scene of bitter fighting between Allied forces – including many Australians and New Zealanders – and the Turks in 1915-16 during the First World War.
Dimensions 72 km (45 miles) long, 1.2-8 km (0.7-5 miles) wide
Map Turkey Aa

Darfur ***Sudan*** Western region divided into two provinces – Northern Darfur and Southern Darfur – covering 317 809 km² (122 706 sq miles) along the Chad border. The features include a 600-900 m (2000-3000 ft) plateau in the east, low hills in the west and north, and volcanic uplands – the Marra Mountains rising to 3022 m (9915 ft) – in the centre. Broadly, the vegetation is tropical savannah.

The regional capital is El Fasher. Many of the region's people are nomadic horsemen, who muffle their faces Tuareg-style and raise sheep and camels. Gum arabic – a resin used as a thickener in jellies and glues – is also produced there.
Population (region) 3 094 000
Map Sudan Ab

Darhan ***Mongolia*** The country's second largest city, in the Haraa Gol valley on the Trans-Mongolian Railway, 188 km (117 miles) north of the capital, Ulan Bator. Its prosperity is due to coal which has been mined there since the 1960s. It is a centre for the manufacture of cement, firebricks, textiles, and iron and steel. Its coal-burning power station provides electricity for northern and central Mongolia.
Population 52 000
Map Mongolia Db

Darién ***Panama*** The eastern province of the Panama isthmus, on the border with Colombia. It is a lightly populated, hot, wet forest country and is the only point where there is a break in the Pan-American Highway which runs from Alaska to Chile. A road that will bridge the gap is slowly being driven through its remote jungles, where only a few Chocó Indians live.
Population 36 100
Map Panama Ba

Dariya ***Saudi Arabia*** Ruined town and former capital, situated 30 km (19 miles) north-west of Riyadh. In 1818 Dariya – the home of a fundamentalist Islamic sect called the Wahhabis – was destroyed by Egyptian-Turkish forces led by Ibrahim Pasha (son of Mohammed Ali, the ruler of Egypt). The attackers were opposed to the sect's founder Abd-al Wahhab, who had sought the protection of the sheik of Dariya. Among the town's remains are the old fort and palace. Egyptian cannonballs, fired during the siege, are still lying in the streets.
Map Saudi Arabia Bb

Darjiling (Darjeeling) ***India*** Town and hill station in northern West Bengal, by the Nepalese border. It stands at 2150 m (7053 ft) on a mountain ridge and became the summer capital of the government of Bengal during British rule. It is renowned for the tea estates that developed around it, which provide some of the world's finest brews. The town has unrivalled views of some of the main Himalayan peaks, including KANGCHENJUNGA. A narrow-gauge railway climbs from the plains to the town.
Population 282 200
Map India Db

Darling ***Australia*** An inland river system draining an area of southern Queensland and western New South Wales, comprising 650 000 km² (247 905 sq miles). For the major part of its 3057 km (1900 mile) course the river runs across arid plains and its flow is erratic – low in times of drought, and flooding the surrounding countryside in times of heavy rain. It is navigable only in certain sections and at certain times. The river has several names, being called the Barwon before it becomes the Darling just upstream from Bourke. It joins the Murray near Wentworth, and together the two rivers form the country's longest river system.
Map Australia Ge

Darling Downs ***Australia*** Fertile farming district in Queensland, west of the Great Dividing Range to the west of Brisbane. It is drained by tributaries of the Darling river and produces many grain crops, cattle and sheep. There are also oil and natural gas deposits and the area has vast reserves of coal. The district's main towns are Warwick (population 9200) and TOOWOOMBA.

Map Australia Hd

Darmstadt ***West Germany*** Former capital of the Grand Duchy of Hesse, 27 km (17 miles) south of Frankfurt am Main. It is in the centre of a wine-producing district, and makes machinery and chemicals. A 16th-century town hall survives. After the Second World War, the capital of Hesse was moved to Wiesbaden.

Population 140 000

Map West Germany Cd

Darnah ***Libya*** See DERNA

Dartmoor ***United Kingdom*** Hilly moor in the county of Devon in south-west England. Much of it is taken in by the Dartmoor National Park, which covers 945 km² (365 sq miles). The rolling moorland and bogs are occasionally broken by rocky outcrops, called tors, the highest of which is High Willhays at 621 m (2037 ft). Sheep and wild ponies graze on the moor, but these are under threat from the increasingly heavy tourist traffic. The area was once extensively mined for tin, lead and copper. Dartmoor Prison, built in the early 19th century originally to house French prisoners of the Napoleonic Wars, is at Princetown, the largest town on the moor, 425 m (1400 ft) up.

Map United Kingdom De

Darwin ***Australia*** Capital of the Northern Territory and the main port for the north of the country. Founded in 1869, it was a headquarters for Allied forces during the Second World War and was bombed by the Japanese in 1942. It was almost completely destroyed by a cyclone on Christmas Day 1974, but has since been rebuilt and prospered with exploitation of the Northern Territory's mineral wealth.

Population 50 000

Map Australia Ea

Datong (Tatung) ***China*** Coal-mining and industrial city of Shanxi province about 280 km (175 miles) west of Beijing (Peking). It is an important railway centre where locomotives are assembled.

Population 800 000

Map China Gc

datum level The zero from which altitudes on land and depths of the sea are measured.

Daugavpils (Dvinsk) ***USSR*** Town in eastern Latvia, 220 km (136 miles) south-east of Riga. Built on the West Dvina river, it is a trading centre for grain, flax and timber. Its main industries are railway engineering, textiles and food. Until 1893 it was the German town of Dünaburg.

Population 123 000

Map USSR Dc

Dauphiné ***France*** Historic region and former province in the south-east. It was bought by Philip VI of France in the 14th century from the count whose title was Dauphin Humbert II, on condition that the heir apparent to the French throne should adopt the title attached to the land and govern Dauphiné as a separate province. The main occupations today are farming, forestry, tourism and metal and electronic industries using hydroelectric power. GRENOBLE is the largest city.

Map France Fd

Davao ***Philippines*** City and region (comprising three provinces) in south-central MINDANAO. Davao city has overtaken CEBU city as the Philippines' second-largest urban area, and its overall land area of 1937 km² (748 sq miles) makes it one of the world's most extensive cities. It is a prosperous trade centre and port for the region, which produces much maize, Manila hemp (an industry begun by Japanese early this century), pineapples, bananas, timber and other crops, as well as fish from Davao Gulf. Long ruled by the Sultans of Mindanao, the region became a

▼ ROYAL FOREST Rocks dot Yellowmead Down on Dartmoor, which was once a 'royal forest' – a hunting preserve of Saxon kings. It was the setting for Conan Doyle's *The Hound of the Baskervilles*.

Spanish-controlled province only in 1849. At the end of the Second World War there was severe fighting between American and Japanese forces before Davao city was liberated. About 30 km (18 miles) south-west is Mount Apo (2954 m, 9690 ft), a dormant volcano and the Philippines' highest mountain.

Population (3 provinces) 1 825 000; (city) 540 000

Map Philippines Cd

Davis Strait Broad strait, named after the English navigator John Davis who discovered it while searching for the NORTHWEST PASSAGE in 1587. It lies between BAFFIN ISLAND and Greenland, and is an arm of the ATLANTIC OCEAN. It is navigable in late summer and autumn, although the Labrador Current sweeps ice down its western shore.

Dimensions 640 km (400 miles) long, 290-640 km (180-400 miles) wide

Map Canada Jb

Davos ***Switzerland*** One of the largest Swiss mountain resorts, 1500 m (4920 ft) up, and about 115 km (72 miles) south-east of the city of Zürich. Its dry and sunny climate made it a health resort, especially for sufferers from tuberculosis, and it is famous as a winter sports centre. It has the largest ice-rink in Switzerland, and a pretty lake at the end of its valley (the Davosersee).

Population 12 500

Map Switzerland Ba

Dawson ***Canada*** Settlement in the Yukon territory at the junction of the Yukon and Klondike rivers, which was the focus of the Klondike gold rush in 1896. The population rose from a few hundred to 20 000 as prospectors poured in and Dawson became a boom town. It subsequently declined, but is still a mining and tourist centre.

Population 700

Map Canada Bb

Dayr az Zawr ***Syria*** Town on the EUPHRATES, 272 km (169 miles) south-east of Aleppo. Dayr az Zawr is the most important town in eastern Syria, and a vital road junction between the region and southern Turkey and Iraq. It is set 600 m (1968 ft) above sea level. Since the completion of the Euphrates Dam in 1978, agriculture has brought new prosperity. The town's main industries are flour-milling and tanning.

Population 332 000

Map Syria Cb

Dead Sea ***Israel/Jordan*** The world's saltiest body of water, called Bahret Lut or 'the Sea of Lot' in Arabic, lying on the Israel-Jordan border in the arid northern extension of the GREAT RIFT VALLEY. No fish can live in its waters, which have a salt content of over 25 per cent. Salt, potassium, magnesium and bromine are extracted from it at a nearby Israeli chemical plant. The lake's surface is the lowest point on the earth's surface – 396 m (1299 ft) below the Mediterranean. The Dead Sea covers 1049 km² (395 sq miles) and is up to 400 m (1312 ft) deep in the north, where the river Jordan enters, but only 2 m (6.6 ft) deep in the south. There is no outlet, and water is lost only through evaporation – hence the salinity. The Jewish sect that left the Dead Sea Scrolls lived in caves to the north-west.

Map Israel Bb

Dean, Forest of ***United Kingdom*** Ancient English forest, mainly of oak and beech trees, covering some 110 km² (42 sq miles) between the Severn and Wye rivers south-west of Gloucester. Since ancient times, the local people – known as Foresters – have mined iron ore and coal, and grazed their stock in the forest which is now a forest park.

Map United Kingdom De

Death Valley ***USA*** Depression in eastern California, near the Nevada border. It is the lowest point in the Western Hemisphere, 86 m (282 ft) below sea level, and is a desert area of salt beds and borax formations. The valley lies in a national monument covering 8399 km² (3243 sq miles).

Map United States Cc

▼ SEA OF SAND A tract of sand dunes laps the rocky walls of Death Valley – North America's hottest and driest place, where in some years there is no measurable rainfall and summer daytime temperatures often exceed 49°C (120°F).

Deauville ***France*** Elegant coastal resort of Lower Normandy, on the Seine estuary opposite Le Havre. It has golden beaches, a casino, marina and racetrack. Dives, 15 km (9 miles) to the west, is where William the Conqueror assembled his fleet for the invasion of England in 1066. In the church is a list of the noblemen who accompanied him.

Population 4770

Map France Db

Debrecen ***Hungary*** Economic and cultural focus of eastern Hungary and chief town of the county of Hajdu-Bihar, situated about 190 km (120 miles) east of Budapest. It is the largest city on the Great ALFOLD (Plain), in a rich crop growing and stock-raising area, and produces foodstuffs, chemicals and electrical goods.

Medieval Debrecen, protected by surrounding low-lying marshes and lakes, survived Turkish threats and Catholic domination by its Austrian Hapsburg rulers to become Hungary's Protestant capital – the 'Calvinist Rome'. And it was here in 1848 that the nationalist leader Lajos Kossuth (1802-94) declared Hungary's independence from Austria.

The old city has the air of an overgrown village, with many ancient buildings, including the Old County Hall, decorated with ornaments of majolica – a type of coloured earthenware. It is also noted for its 16th-century Calvinist College – now a university – the Great Church, and the colourful Arany Bika (Golden Bull) Hotel, built in art nouveau style.

Population 207 000

Map Hungary Bb

Deccan ***India*** The peninsula of India, and more especially the rocky plateau that forms most of it. Parts of it receive little rain, and an international research centre on semiarid tropical crops has been set up in the city of Hyderabad. The area has a few small-scale dams and reservoirs which allow a limited amount of rice to be grown. The wetter districts produce grains such as millet, and cotton is also grown, especially in the south.

Map India Bd

deciduous forest Forest consisting mainly of broadleaved trees that lose their leaves at one season of the year. In a monsoon forest the leaves are shed during the dry season to protect the trees against excessive loss of moisture by transpiration. In cool temperate regions, such as north-west Europe, Chile and New Zealand, the leaves fall during the autumn and the trees remain dormant during the winter.

Dedza ***Malawi*** Town in Central province, 80 km (50 miles) south-east of the capital, Lilongwe, near the border with Mozambique. It is the centre of an agricultural region producing tobacco, subsistence food crops, including maize, and timber. It stands at the foot of the beautiful, wooded Mount Dedza, which rises to 2170 m (7120 ft). The mountain has fine views of Lake Malawi and is a popular tourist attraction.

Population 2300

Map Malawi Ab

deepening In meteorology, decreasing atmospheric pressure at the centre of a depression. It is the opposite of filling.

deforestation Process of felling and clearing wooded areas, usually to extend agriculture.

Deganya ***Israel*** Israel's first kibbutz, established in 1909, near the south-west shore of the Sea of Galilee (Lake Tiberias), about 25 km (15 miles) east of Nazareth. The ruined Canaanite city of Bet Yerah is nearby.

Map Israel Ba

degradation See DENUDATION

Dehra Dun ***India*** Town in a valley at the foot of the Himalayas, about 210 km (130 miles) north-east of Delhi. It stands 700 m (2296 ft) above sea level and, according to legend, it was the home of Siva, the Hindu god of destruction and reproduction.

The British established a military academy there in 1934 which became the Sandhurst of India. And the Indian president's Sikh bodyguard is based in the town.

Population 293 000

Map India Cb

Delaware ***USA*** State on the Atlantic coast covering 5328 km² (2057 sq miles) on the east of the peninsula between Delaware and Chesapeake bays; it is the country's second smallest state after Rhode Island. It was first permanently settled by Europeans in 1638, and was one of the original 13 states. Broiler chickens, soya beans, maize and dairy produce are the agricultural mainstays, and chemicals the chief industry. The state capital is DOVER.

Population 622 000

Map United States Kc

Delaware Bay Inlet between NEW JERSEY and DELAWARE on the USA's eastern seaboard, discovered by the English navigator Henry Hudson in 1609. It was easily approachable and so became an important anchorage in the early days of European colonisation.

Dimensions 84 km (52 miles) long, about 19 km (12 miles) wide

Map United States Kc

Delft ***Netherlands*** City 55 km (34 miles) south-west of Amsterdam which has been renowned for its fine glazed pottery, normally decorated in blue, since the end of the 16th century. The artist Jan Vermeer (1632-75) lived there, and much of the old city still resembles scenes in his paintings with its old churches, marketplace and 1618 town hall. William the Silent, Prince of Orange (1533-84), great grandfather of the prince who became William III of England in 1689, lived at the Prinsenhof (now a museum) and was assassinated there. Delft's manufactures include spirits, cables and penicillin.

Population 86 300

Map Netherlands Ba

Delhi ***India*** Capital city of the Federal Republic of India, situated on the banks of the Yamuna river, on the borders between the northern states of Uttar Pradesh and Haryana. The city is in the strategic region linking the INDUS and GANGES valleys, bounded on the south by the Aravalli Range and on the north by the Himalayas. It is just south of the town of Panipat, where many renowned battles have been fought between (*continued on p. 186*)

A DAY IN THE LIFE OF AN INDIAN MONEYLENDER

Anjali stands at the doorstep of her brick house in a village near Delhi and surveys the dusty street. Breezes and passing feet keep blowing dust into her doorway. Her servant will have to brush it out again.

Her servant is a Harijan woman – an Untouchable – one of a group of people traditionally regarded as unclean, who are the lowest members of the Hindu caste system.

As a moneylender, Anjali's livelihood depends on the Harijans, because they are too poor to have any possessions to pledge as security, so banks refuse to lend them money. If a breadwinner in a local Harijan family falls ill, the family are forced to borrow money for food from Anjali. Often, too, they have to borrow for dowries for their daughters. And the Harijan stonebreakers who live in Anjali's village get into debt because of the high prices charged by contractors to transport the stone they quarry. Anjali feels justified in charging 25 per cent interest because of the risk in lending to people who have no security, though it means her creditors are often in debt to her for the whole of their lives.

Three of Anjali's seven children, aged between 12 and 16, still live at home. Her husband died ten years ago, so she has to support them alone. But for her moneylending, she would have to depend on her two eldest sons – one a refrigeration engineer, the other a newspaper reporter. But they, and her two eldest girls, are married, and have children of their own to look after.

Soon Anjali starts to tour the village. She must check that everyone who owes her money is working, and that none is ill and unable to keep up the payments, in which case she will have to raise their interest when payments are resumed.

As she does her rounds today she has one more worry than usual. Recently an aid project has bailed out some of her stonebreaker creditors, giving them the money, in an interest-free loan, to pay off debts in one go. The aid workers have also showed the stonebreakers how to form cooperatives for transporting the stone they quarry. This means that they do not have to pay outside contractors at the rate that forces them to borrow. At the same time there are women from upper castes in Delhi who encourage the Harijans to aspire beyond their traditional place as the outcasts of Indian society.

It makes Anjali nervous, that the whole order of things might be disrupted. She consoles herself with the thought that change always occurs slowly in India. Her livelihood must be safe for some time yet.

Denmark

A HISTORIC HOME OF THE VIKINGS, NOW FAMOUS FOR DESIGNERS, ARCHITECTS, HANS CHRISTIAN ANDERSEN AND LEGO BUILDING BRICKS

Internationally, Denmark enjoys a reputation as a comfortable and contented welfare state, where people live long and happy lives. It has an egalitarian society, but retains the romantic trappings of a monarchy. It is peace-loving, yet expects all men to undertake two years' national service.

This fortunate country, lying between the North Sea and the entrance to the Baltic Sea, is one of Europe's smallest states, slightly bigger than Switzerland. In medieval times (9th-12th century) it was a major European realm, exacting a tribute – called *Danegeld* – from England in return for restraining its seagoing warriors, the Vikings. It once embraced provinces that today form Sweden, Norway (ceded to Sweden in 1814), Schleswig-Holstein (lost to Prussia in 1864) and Iceland, which became independent in 1944. The Faeroe Islands and the vast, largely ice-covered island of Greenland remain Danish, but are largely self-governing.

LAND AND SEA

Denmark itself consists of a western peninsula and an eastern archipelago of 406 islands of which 89 are populated. It is a low-lying land – the highest point is 171 m (561 ft) above sea level. It is the most dissected country in Europe and has an immensely long coast – 7300 km (about 4500 miles).

Denmark has been peopled since the end of the Ice Age, some 10 000 years ago. Many artefacts from the Stone Age are displayed in museums, and in 1950 the body of an Iron Age man was found perfectly preserved after 2000 years in a peat bog at Tollund in JUTLAND. The land itself is neither scenically exciting nor rich in resources. It is a homely country, with good soils lying on chalk rocks and a cool temperate climate which favours a wide variety of plants and crops. Rain and snow fall mostly in the south-west, where the average fall is about 750 mm (30 in) annually.

The country faces the North Sea in broad beaches which are the longest in Europe. Artificial harbours, such as ESBJERG in the south and Hanstholm in the north, had to be built because the coast offers little natural shelter for ships. Behind a coastal rampart of dunes lie flat, sandy heathlands, largely reclaimed from the sea, and peat bog, much of which has been drained. Plantations and belts of spruces and pines act as windbreaks against the strong westerly winds.

East Jutland offers a landscape of steep, low hills, with beech-clad slopes, its coast broken by broad inlets. At the heads of these inlets are old ports and market towns as well as Jutland's biggest cities, ARHUS and ALBORG.

JIGSAW OF ISLANDS

East of Jutland lies a jigsaw of islands. The second largest, FUNEN (Fyn), is linked to Jutland by bridges. Funen is called The Garden of Denmark because of its flowery villages with their pastel-painted farmhouses. To the east is the largest island, Zealand (Sjælland), linked to Funen by a shuttle service of train and car ferries.

The capital, COPENHAGEN, is on the eastern shore of Zealand, as far away as possible from the Jutland mainland, but strategically placed on the Öresund strait between Denmark and Sweden. At the northern approach to the Öresund stands ELSINORE (Helsingør), the setting for Shakespeare's tragedy *Hamlet, Prince of Denmark*. The most isolated Danish island, Bornholm, is several hours' ferry journey across the Baltic from the capital.

SUCCESS IN AGRICULTURE

Denmark is an agricultural country. Three-quarters of the land is cultivated, mostly by the rotation of grass, barley, oats, sugar beet and fodder beet, but animal husbandry has been the mainstay of rural Denmark for the past century. Its best-known products are bacon and butter from 12 million pigs and 2.8 million cows. Farms are small, with the great majority ranging from 15 to 30 hectares (37.5-75 acres), and are mostly owned and worked by single families.

The success of Denmark's agriculture is due to cooperation, mechanisation and knowledgeable farmers. For a century the industry has benefited from cooperative processing and marketing of farm products, purchasing of farm supplies and financing of farm needs. Denmark made mechanisation cheaper by manufacturing its own tractors, harvesters and milking machines appropriate to smallholdings and limited purses; this blossomed into an important export industry in farm machinery.

Farmland is regarded as an important national asset. All young farmers must study to earn a 'green licence', without which they cannot own farmland. The small farmer is protected from urban investors: any property of more than 5 hectares (12 acres) must be held by someone whose principal occupation is farming. Even the development of second homes in rural areas is now strictly controlled.

INDUSTRIAL EXPORTS

Despite Denmark's rural image, only 5 per cent of the population are engaged in farming. The overwhelming majority are town dwellers engaged in manufacturing and service activities. Many factories are small, sited in the countryside, and employ fewer than 100 people, but exports from these small factories exceed those of agriculture.

In spite of its limited range of raw materials, Denmark sells a wide range of manufactured exports, thanks largely to a strong tradition of imaginative design. This is revealed, for example, in furniture and kitchenware, the silverware of Georg Jensen, and Royal Copenhagen porcelain.

Unexpected industries have sprung up in towns on the heaths of Jutland; where cottagers once made stockings for export from the wool of their sheep, their descendants may be working in one of Europe's largest shirt factories at Herning. A carpenter's workshop created the children's constructional toy called Lego, which is sold throughout the world. An important pharmaceutical industry has developed; Denmark is the world's chief exporter of insulin.

Services and ideas are also exported. Danish bridge and harbour construction, system-built factories and housing components have found wide markets, especially in the Middle East, South-east Asia, Africa and South America.

Danish architects are world famous, most notably Jørn Utzon, who designed Sydney Opera House in Australia.

All this economic success is accompanied by political stability. Coalition governments, elected by a system of proportional representation, place a high value on individual liberty. Denmark is also a strong supporter of international agencies and generous in its aid to the developing world. As a member of the Nordic Council, Denmark collaborates closely in social, economic and cultural fields with its neighbours, Sweden, Norway, Iceland and Finland.

MAGNIFICENT CASTLES

The happy picture is reflected in a landscape of well-manicured farmland and woodland, brick-built farmhouses with capacious barns, whitewashed churches, handsome manor houses and magnificent castles. In the towns, public gardens are overflowing with flowers; red and white flags (the national colours) flutter from the main buildings; and bright shop fronts convey an air of calm prosperity. Cyclists stream through the streets; there are nearly 400 000 cycles in use, in addition to a car for every two or three people. Danish food is another happy experience – the open sandwich (*smørrebrød*) piled high with the products of farm and fishery – extravagant pastries, akvavit, cherry brandy and lagers.

In addition to material prosperity, Denmark has a rich intellectual tradition. It is found in the influential philosophy of Søren Kierkegaard (born of the Lutheran Church to which almost all Danes nominally belong); in the atomic physics of Niels Bohr; and in the sombre music of Carl Nielsen (which balances the more light-hearted world of ballet created by August Bournonville).

For all its appeal to the outsider, Denmark poses its leaders with plenty of problems. Foremost is its limited range of natural resources and especially its lack of energy sources. Economic expansion since the 1950s was based largely upon cheap oil imports, and the big price increases of the 1970s have completely changed the picture. However, reserves of oil and gas occur in the North Sea, and the first pipeline ashore was operating in 1985. Coal has to be imported from Germany and Poland to fuel power stations at Copenhagen and Esbjerg. And, with a number of large oil refineries, a shipbuilding industry which is geared to construct giant tankers, and a heavy investment in merchant shipping, Denmark is always anxious about oil prices, and about the general level of world trade.

Nor is agriculture able to escape from economic changes. The surpluses produced by Danish farmers have added to overproduction within the European Economic Community. Like other members, Denmark has been required to reduce its output of dairy products, and farmers' incomes have fallen.

Urban and industrial expansion, and modern methods of farming, have raised environmental problems. Denmark has a Ministry of Pollution, with water resources a matter of prime concern, and nature conservation a close second. National parks and conservation areas are preserving natural habitats, but it is too late for the storks – once the symbol of Denmark. Only a few score of pairs now arrive on their annual migration, because drainage of bogs has reduced their feeding grounds.

Running the Danish economy becomes ever more expensive. The network of state railways may be trimmed, but few economies can be made to costly ferry services. The contributions of the NATO alliance weigh heavily. Generous unemployment benefits produce tensions, even though the number of unemployed is declining. Denmark's welfare structure cannot be maintained without heavy personal taxation – to which there is opposition.

▲ GRUNDTVIG CHURCH P.V. Jensen Klint, Denmark's leading early 20th-century architect, designed this church built in Copenhagen in 1921-6. It is named after Bishop N.F.S. Grundtvig (1783-1872) who brought about a revival of the Danish Lutheran Church.

DENMARK AT A GLANCE
Area 43 000 km² (16 629 sq miles)
Population 5 100 000
Capital Copenhagen
Government Constitutional monarchy
Currency Krone = 100 øre
Language Danish
Religion Christian (Evangelical Lutheran)
Climate Temperate maritime; average temperature in Copenhagen: −3-2°C (27-36°F) in February; 14-22°C (57-72°F) in July
Main primary products Cattle, pigs, poultry, cereals, potatoes and other root crops, fodder, fish; some oil and natural gas
Major industries Agriculture, food processing, fishing, engineering, shipbuilding, petroleum refining, chemicals, furniture
Main exports Meat, dairy produce, animal feedstuffs, fish, machinery, electronic equipment, chemicals, furniture, leather
Annual income per head (US$) 11 436
Population growth (per thous/yr) 1
Life expectancy (yrs) Male 73 **Female** 77

invading powers from the North-west Frontier and the civilisations of the Ganges Plains.

The city has two distinct parts: Old Delhi and New Delhi. Old Delhi is a Muslim city centred around the Red Fort built by Emperor Shah Jehan between 1636 and 1658. The streets of Old Delhi are narrow and bustling: the beauty and serenity lies inside the courts of the main buildings. New Delhi was proclaimed the capital of India by the British in 1911, but the newly built city was not officially inaugurated until 1931. It was designed by the British architect Sir Edwin Lutyens (1869-1944), and is tree-lined and spacious.

It is built around Rajpath Avenue, 5 km (3 miles) long and 0.8 km (0.5 mile) wide, known previously as Kingsway, which leads to the huge Viceroy's Palace, now the residence of the Presidents of India (Rashtrapati Bhawan). To one side is the parliament, the Lok Sabha. Government buildings flank all sides of the Rajpath. Just to the north is Connaught Place, an elegant ring of colonnaded shops and offices, now being replaced by skyscrapers. In the southern part of New Delhi are Palam airport, new universities and the arenas built in 1982 for the Asian Games.

Places of interest include the Red Fort, the Jami Majid mosque, the Rashtrapati Bhawan (outside only) and Rajpath Avenue, the Qutb Minar, at 22 m (73 ft) high thought to be the highest free-standing stone tower in the world, Tughlaqabad (a ruined fortified city), and the astronomical observatory of Jai Singh, the Maharajah of Jaipur.

In 1948 Mahatma Gandhi, who had been trying to achieve a harmonious relationship between Muslims and Hindus, was assassinated in Delhi by a Hindu fanatic. A national shrine now marks the place of his cremation at Ram Shat. In 1984 Mrs Indira Gandhi, Prime Minister of India and the daughter of Pandit Nehru, was assassinated in Delhi by militant Sikhs.

Delhi's industries – many of them located in satellite townships in the outlying areas – include chemicals, textiles, rubber goods and various handicrafts.

Population 5 729 300
Map India Bb

dell Small, secluded, wooded valley.

Delos (Dhílos) ***Greece*** Smallest of the Cyclades islands and legendary birthplace of the god Apollo. Delos has the most extensively excavated archaeological site in Greece. Ancient sanctuaries, temples, colonnades and monuments in an area 1200 m (0.75 mile) long have been uncovered. The island was a Greek political and religious centre as early as the 7th century BC.

Map Greece Dc

Delphi ***Greece*** Impressive ruins of the Temple of Apollo and the seat of his oracle, built into the craggy slopes of Mount Parnassus, 166 km (102 miles) north of Athens. It was founded in the 2nd millenium BC and its oracle became the most important in classical Greece. The English word delphic, for an ambiguous statement, comes from the cryptic utterances of the soothsayers and priests of Delphi. One such example was 'The wooden walls will save Athens from the Persians'. It is thought to have referred to the Greek fleet that defeated the invasion fleet of the Persian King Xerxes at the Battle of Salamis in 480 BC. The temple, theatre, stadium and marketplace cling to the crevasses and precipices of the mountain (2457 m, 8061 ft high), overlooking the olive-forested ravine above the Itea plain. A museum at the site has many findings from the ruins, including a bronze charioteer by Sotades (4th century BC). Rugs and embroidery are made at Arakhova 9 km (5.5 miles) east.

Map Greece Cb

delta An alluvial area, often triangular, at the mouth of a river. It forms where the flowing water is slowed down on contact with the sea or lake water and its capacity to transport material is suddenly reduced. Sediments built up in the mouth cause repeated subdivision of the river channels.

Delta Irrigation Scheme ***Senegal*** See RICHARD TOLL

Demerara ***Guyana*** River 320 km (200 miles) in length which flows into the sea at the capital, Georgetown. Farmers along its banks grow the sugar named after it.

Map Guyana Ba

Demilitarised Zone (DMZ) ***North Korea/ South Korea*** No-man's-land boundary area between North and South Korea. The zone straddles the cease-fire line agreed at the end of the Korean War (1950-3). It is 4 km (2.5 miles) wide and winds across the Korean peninsula from coast to coast for 243 km (150 miles).

Map South Korea Cc

Den Bosch ***Netherlands*** See 'S-HERTOGENBOSCH

Den Haag ***Netherlands*** See HAGUE, THE

Denali National Park ***USA*** Area of 24 413 km² (9426 sq miles) in southern Alaska. It centres on Mount McKinley, which the local Indians call Denali, and was formerly called the Mount McKinley National Park. The mountain, which rises to 6194 m (20 320 ft), is North America's highest peak. The park has many lakes and glaciers, spectacular mountain scenery and tundra landscape, and wildlife including bears, mountain sheep, moose and caribou.

Map Alaska Bb

Denbigh ***United Kingdom*** See CLWYD; GWYNEDD

dene **1**. Wooded valley with steep sides. **2**. Group of sand hills near the sea.

Denmark See p. 184

Denmark Strait Sea passage between Greenland and Iceland containing the cold East Greenland Current. Here, on May 24, 1941, the 'unsinkable' German battleship *Bismarck* sank the British battle-cruiser *Hood*. British warships and aircraft sank the *Bismarck* off the French coast on May 26.

Dimensions 480 km (300 miles) long, 290 km (80 miles) wide
Map Atlantic Ocean Bc

Denpasar ***Indonesia*** Seaport and capital of the island of BALI. It is a market and light industrial centre. The museum contains many art treasures, and just outside the town is the Bali Cultural Centre with an art gallery and large, open-air theatre. Nearby are the tourist resorts of Kuta and Sanur.

Population 82 140
Map Indonesia Ed

denudation The wearing down of the land by the combined processes of weathering, mass movement, transport of loose material and erosion by wind, ice and water. Armed with the loose material they are transporting, they erode the surface of the ground over which they move. Also called degradation.

Denver ***USA*** State capital of Colorado, standing at 1610 m (5280 ft) just east of the Rocky Mountain foothills, about 140 km (85 miles) south of the Wyoming border. It was founded during a gold rush in 1858. The city's manufactures include aircraft, chemicals and electronics, and it has the largest centre of federal government operations outside the national capital, Washington; they include the Mint.

Population (city) 504 600; (metropolitan area) 1 582 500
Map United States Ec

depression, cyclonic disturbance Near-circular cell of low pressure in the atmosphere in which pressure increases outwards and into which winds spiral (anticlockwise in the Northern Hemisphere) as the depression fills. Depressions move more rapidly than anticyclones, bringing unsettled weather, stronger winds, fronts and PRECIPITATION. The winds circulate in the opposite direction to those in an anticyclone.

Dera Ismail Khan ***Pakistan*** Town on the west bank of the Indus river, at a strategic crossing point by a pontoon bridge, about 330 km (205 miles) west of Lahore. It was founded in 1469 by Ismail Khan, a Baluchi ruler, but was destroyed by floods and refounded in 1823. It produces fine woodwork, ivory and glassware, and has a lively bazaar. The town is also a trading centre for grain and oilseeds.

Population 116 000
Map Pakistan Cb

Derby ***Australia*** See KIMBERLEY PLATEAU

Derby ***United Kingdom*** City and county of north-central England. The city, 55 km (35 miles) north-east of Birmingham, has been renowned since the 18th century for its silk and Crown Derby porcelain, and is also the headquarters of Rolls-Royce cars and aircraft engines. The Church of All Saints was raised to cathedral status in 1927.

The county includes the coalfields around Chesterfield and the hills of the southern Pennines – a major tourist area. They include the PEAK DISTRICT National Park, rising to 636 m (2086 ft) at Kinder Scout, and the spa towns of Buxton and Matlock Bath. The county stretches north to the outskirts of Manchester and Sheffield, and covers 2631 km² (1016 sq miles).

Population (county) 911 000; (city) 215 000
Map United Kingdom Ed

Derna (Darnah) ***Libya*** Second most important port of CYRENAICA after Benghazi, situated in the north-east 100 km (62 miles) east of Al Bayda. In recent years the port has been reconstructed and modernised. The town was the scene of much fighting during the Second World War and changed hands several times. The nearby oasis produces dates and bananas. Industries include cement, food processing, and Libya's first ready-made clothes factory.
Population 37 716
Map Libya Ca

Derry ***United Kingdom*** See LONDONDERRY

Des Moines ***USA*** State capital of Iowa. It stands in the middle of the state on the Des Moines river, about 190 km (120 miles) east of the Missouri river. The city is a manufacturing, trade, financial and cultural centre in an area producing maize and dairy goods.
Population (city) 190 800; (metropolitan area) 377 100
Map United States Hb

desert Area where rainfall is so low that vegetation is scant or absent. See FEATURES OF THE DRY DESERTS (p. 188).

The SAHARA is a trade-wind desert – a region of high atmospheric pressure, where air which has risen and shed its moisture in the Tropics descends to earth, and where trade winds warm and dry as they blow towards the Equator. Rainfall is less than 250 mm (10 in) a year.

Deserts such as the NAMIB and ATACAMA occur on the western margins of the continents where cold currents flow. Local on-shore winds are warmed on crossing the coast and retain their moisture. Deserts such as the GOBI lie in continental interiors remote from the sea.

In the hot deserts the high temperatures cause high evaporation. The cloudless skies give rise to rapid heat loss from the earth by radiation after sunset. Temperatures may soar to over 37°C (100°F) during the day, then fall below freezing at night. Rain occurs in isolated showers, sometimes years apart. Cold deserts occur in polar regions such as Antarctica and Greenland. These are high pressure areas of descending air and outflowing winds all the year.

desert climate (hot) A climate characterised by high temperatures, scanty, irregular rainfall and evaporation so high that the rainfall is insufficient to support vegetation typical of other regions in the same latitude.

desert pavement Desert surface of closely packed smooth pebbles where winds have removed fine material between them and windborne sand has ground and polished their tops.

desert vegetation Sparse cover of drought-resistant plants that survive in areas of very low and irregular rainfall. Drought-resistant plants are adapted in various ways to withstand prolonged dry spells. Their characteristics include small, waxy leaves and thick, fibrous bark to reduce transpiration, bulbous trunks to store water, and extensive root systems to locate GROUND WATER.

Dessau ***East Germany*** Industrial town about 100 km (60 miles) south-west of the capital, East Berlin. It was devastated by Allied bombing during the Second World War because of armaments industries there. Today it produces machinery, precision instruments and vehicles.
Population 103 800
Map East Germany Cc

destructive plate margin Another name for a BENIOFF ZONE. See also PLATE TECTONICS

Detmold ***West Germany*** Town on the northern flanks of the TEUTOBURGER WALD, about 80 km (50 miles) south-west of Hanover. Its 16th-century castle was the home of the Princes of Lippe, the local rulers. On a hill 6 km (4 miles) south of the town is a huge monument to Arminius, or Hermann, a local hero, who annihilated three Roman legions in the district in AD 9.
Population 66 600
Map West Germany Cc

Detroit ***USA*** City in Michigan that is sometimes called the car capital of the world. It stands on the Detroit river, one of the world's busiest inland waterways, between Lake Erie and Lake St Clair, across the river from the Canadian city of Windsor. Detroit was founded in 1701 but its strategic importance soon led to its rapid expansion. Today it is the headquarters of the General Motors, Ford and Chrysler car companies, and a busy industrial port. Much redevelopment has taken place, especially along the waterfront, where the hotel, office and shopping complex of the Renaissance Centre stands. The city has one of the Western world's worst crime rates, with more than 500 murders a year.
Population (city) 1 089 000; (metropolitan area) 4 315 800
Map United States Jb

THE ARID FACE OF THE EARTH

About one-third of the world's land surface is made up of deserts – both hot and cold – in which the land is partially or totally barren, as in the Sahara and the polar regions. One of the world's largest and most arid hot deserts – the Rub al-Khali in Arabia – has droughts which sometimes last for years. At the other extreme – the cold deserts – are the Dry Valleys east of McMurdo Sound in Victoria Land, Antarctica, where no rain is believed to have fallen for 2 million years.

THE WORLD'S LARGEST DESERTS

Desert	Location	Area (km²/sq miles)
Sahara	Africa	9 100 000/3 500 000
Great Australian	Australia	3 830 000/1 480 000
Gobi	Asia	1 295 000/500 000
Rub al-Khali	Arabian Peninsula	650 000/251 000
Kalahari	Southern Africa	520 000/200 000
Kara Kum	USSR	340 000/130 000
Taklimakan	China	327 000/126 000

Deva ***Romania*** Metalworking town and important road and rail junction on the River Mures. It lies in the west of the country, 145 km (90 miles) north of the border with Yugoslavia. Deva's 13th-century fortress was held by the national hero Michael the Brave, who briefly united the provinces of Moldavia, Transylvania and Walachia before his defeat by the Turks at Deva in 600. A museum of archaeology housed in a 17th-century castle contains Dacian and Roman remains.
Population 71 000
Map Romania Aa

Deventer ***Netherlands*** Town founded in the 8th century on the IJssel river, 87 km (54 miles) east and slightly south of Amsterdam. It was a medieval centre of learning, and has several old churches and a library with manuscripts dating from the 10th century. It produces tapestries, carpets and *Deventerkoek* (honey gingerbread) and has modern textile and brick plants.
Population 64 500
Map Netherlands Ca

Devon ***United Kingdom*** County of south-west England with an area of 6715 km² (2593 sq miles). The north coast, on the Bristol Channel, has steep cliffs while the south, on the English Channel, is deeply indented with many natural harbours. They include the naval base of PLYMOUTH. It has a long tradition of seamanship; the English admiral and navigator Sir Francis Drake (about 1540-96), who sailed around the world and later defeated the Spanish Armada, was a Devon man. It also has rich farmland and is famous for its cattle and clotted cream. Tourists flock to TORBAY and to the uplands of DARTMOOR and EXMOOR, both of which are national parks. The county town is the cathedral city of EXETER.
Population 980 000
Map United Kingdom De

Devonian Fourth period in the Palaeozoic era of the GEOLOGICAL TIMESCALE named after the English county of Devon where marine rocks of the period were first studied.

dew Water droplets on the ground or on plants which have condensed from water vapour in the air, usually during a clear night when the surfaces of the ground or plants have cooled below the dew point of the air around. See CONDENSATION

dew point The temperature to which air has to be cooled to cause saturation – the state when it is holding as much water vapour as is possible at that temperature. Any further cooling will lead normally to condensation.

Dhahran ***Saudi Arabia*** Commercial and business centre situated 5 km (3 miles) south of Dammam. Dhahran is less a town than a giant complex of oil company offices, schools, hospitals and recreational facilities. It is also the site of the country's Petroleum and Minerals University, an eye-catching piece of contemporary architecture. In addition, there are the provincial offices of the Ministry of Petroleum and Petromin (the domestic oil marketing organisation), and an Oil Exhibit Centre depicting the history of Saudi Arabia's oil industry.

Population 25 000

Map Saudi Arabia Cb

Dhaka (Dacca) ***Bangladesh*** The country's capital, which was also the capital of Bengal – under the Mogul emperors of India – for a little over 100 years from 1575 until it was superseded by Calcutta. Dhaka's most interesting buildings date from this period, including the Lal Bagh Fort and several mosques. Other striking buildings, including the supreme court and older parts of the university, dated from 1905 to 1911 when, under the British, Dhaka was the capital of East Bengal. During this period, beautifully shaded avenues were planted, though some of the shade cover is now being felled to meet modern traffic conditions. The city grew again as capital of East Pakistan after India was partitioned in 1947. But its fastest growth has come since 1971 as capital of Bangladesh. Its new parliament buildings and vast railway station are ultra-modern in style.

The city occupies a low-lying site on the Buriganga, one of the medium-sized rivers of the Ganges delta. Much of the city and the surrounding district – which is also called Dhaka and covers 7470 km² (2884 sq miles) – is prone to flooding. Even so, many flood-prone areas have been settled by squatter colonies. The city relies on bicycle rickshaws more than any other form of transport. Its industries

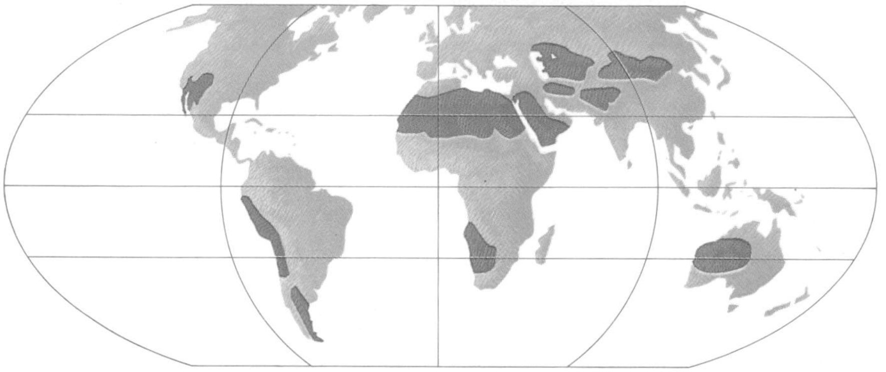

FEATURES OF THE DRY DESERTS

Two belts of arid desert girdle the earth on either side of the Equator. The northern belt extends through Mexico and America's south-west, across North Africa, through the Middle East and on to the Gobi Desert of central Asia. The southern belt stretches from South America's Pacific coast, across south-west Africa to central Australia. Scouring winds work on the face of the deserts, sandblasting rocky outcrops into bizarre shapes, and blowing sand dunes along in ever-changing patterns.

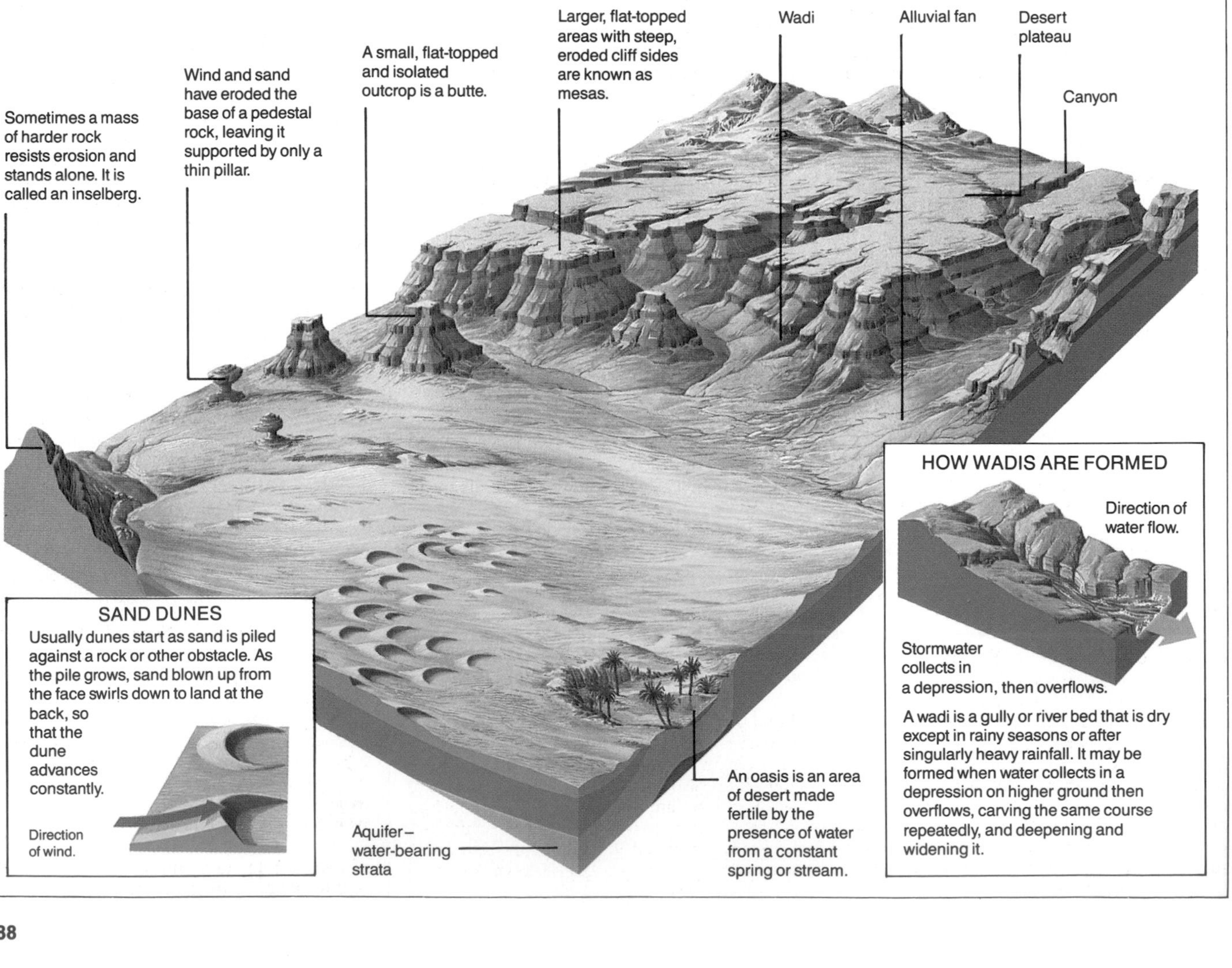

Sometimes a mass of harder rock resists erosion and stands alone. It is called an inselberg.
Wind and sand have eroded the base of a pedestal rock, leaving it supported by only a thin pillar.
A small, flat-topped and isolated outcrop is a butte.
Larger, flat-topped areas with steep, eroded cliff sides are known as mesas.
Wadi
Alluvial fan
Desert plateau
Canyon
SAND DUNES
Usually dunes start as sand is piled against a rock or other obstacle. As the pile grows, sand blown up from the face swirls down to land at the back, so that the dune advances constantly.
Direction of wind.
Aquifer – water-bearing strata
An oasis is an area of desert made fertile by the presence of water from a constant spring or stream.
HOW WADIS ARE FORMED
Direction of water flow.
Stormwater collects in a depression, then overflows.
A wadi is a gully or river bed that is dry except in rainy seasons or after singularly heavy rainfall. It may be formed when water collects in a depression on higher ground then overflows, carving the same course repeatedly, and deepening and widening it.

include textiles and leather goods. Apart from the University of Dacca, it has a University of Science and Technology (1961).
Population (district) 10 014 000; (city) 3 458 600
Map Bangladesh Cc

Dhanbad *India* Coal-mining city in the industrial Damodar Valley about 230 km (143 miles) north-west of Calcutta. A college of coal mining was founded there in 1926.
Population 433 100
Map India Dc

Dhaulagiri *Nepal* See ANNAPURNA

Dhekelia *Cyprus* South coast British military base on Larnaca Bay. It was built in the 1950s and is now one of the UK Sovereign Base Areas on the island.
Map Cyprus Ab

Dhílos *Greece* See DELOS

Dhodhekánisos *Greece* See DODECANESE

Dhofar *Oman* Southern province that is the only place in the Arabian peninsula to have reliable monsoon rains; they last from late June to September. The Dhofar Mountains, rising from a fertile plain with wooded hills, run parallel to the coast, and reach a height of 2000 m (6561 ft). The region is a source of frankincense (an aromatic resin from trees of the genus *Boswellia*) which is used as incense. Dhofar traded in incense with the people of North Yemen as far back as the 10th century BC, and today much of the incense goes to India. More than half of the province's 70 000 inhabitants live in its capital, Salalah. From 1964 to 1975, Yemeni-backed rebels fought for Dhofar's independence from the rest of Oman.
Map Oman Ab

Diamantina *Brazil* Diamond-mining town 190 km (119 miles) north of the city of Belo Horizonte and named after the diamond industry which has flourished there since 1729. Juscelino Kubitschek (1902-76), president of Brazil in the late 1950s and founder of the purpose-built national capital, Brasília, was born in the town.
Population 37 000
Map Brazil Dc

diamond Crystalline form of carbon found in some igneous rocks and alluvial deposits. It is used as a gemstone, for cutting and as an abrasive. It is the hardest known mineral and is generally colourless. However, it may be tinted yellow, orange, blue, brown or black by impurities.

diapir Body of rock formed by material piercing the earth's crust from below and forcing it upwards into a dome. The material may be hot molten rock, producing a diapir with an igneous core, or a sediment which flows under high pressure. A common type of diapir has a core of salt and is known as a salt dome.

diastrophism Process of deformation of the earth's crust which produces major features such as continents, mountains and ocean basins. See PLATE TECTONICS

diatom ooze Silica sediment consisting of the cases of minute single-celled aquatic organisms known as diatoms.

diatomaceous earth Fine, powdered siliceous earth, composed of the remains of single-celled organisms known as diatoms. It is used for polishing, as an absorbent in the manufacture of high explosives and as an insulating material. It is also known as diatomite or kieselguhr.

Dien Bien Phu *Vietnam* Small market town near the Laotian border and some 275 km (170 miles) west and slightly north of the capital, Hanoi. The defeat of French troops at the town in May 1954 effectively ended the eight-year Indo-China war and brought to an end French colonial rule in Vietnam. The battlefield is now preserved as a historical site.
Map Vietnam Aa

Dieng *Indonesia* Volcanic plateau in Central Java. It has many megalithic monuments.

Dieppe *France* Channel fishing and ferry port and seaside resort in Upper Normandy, at the mouth of the Arques river. The town was rebuilt after the Second World War, in which it was severely damaged. It now handles more than 500 000 ferry passengers a year.
Population 42 300
Map France Db

Digne *France* Spa town 115 km (72 miles) north-east of Marseilles. Fruit and lavender are grown in the surrounding countryside.
Population 16 400
Map France Gd

Dijon *France* Capital of Côte d'Or department and of the former Duchy of Burgundy, about 250 km (155 miles) south-east of Paris. It has many attractive old houses, some dating from the 15th century, and is noted for its wine trade, university and food fair held each November. Celebrated local specialities include mustard, gingerbread and cassis (a blackcurrant liqueur).
Population 221 900
Map France Fc

dike See DYKE

Dilmun *Bahrain* See QALAT AL BAHRAIN

Dimitrovgrad *Bulgaria* Industrial town on the Maritsa river, 16 km (10 miles) north of Khaskovo. The chief industry is chemicals, especially fertilisers. It manufactures cement and pottery, and there is some coal mining and food processing. Founded in 1947, the town was named after Georgi Dimitrov (1882-1949), the country's first Communist leader.
Population 51 000
Map Bulgaria Bb

Dinajpur *Bangladesh* Chief town of Dinajpur district near the Tista river in an isolated zone at the foot of the Himalayas. The district covers 6566 km² (2535 sq miles), much of it a dry, sandy alluvial fan. Agriculture in the district has advanced considerably with the development of underground irrigation since independence.
Population (district) 3 200 000; (town) 96 300
Map Bangladesh Bb

Dinant *Belgium* Tourist town about 80 km (50 miles) south of the capital, Brussels, in the gorge of the River Meuse. It stands at the foot of sheer limestone cliffs, 100 m (328 ft) high, crowned by a citadel dating from 1821, built by the Dutch on what has been for centuries a defended site. In 1914, Dinant suffered great damage by the advancing German armies and again in 1940.
Population 12 000
Map Belgium Ba

Dinaric Alps (Dinara Planina) *Albania/Yugoslavia* Series of fold mountain ranges and plateaus along the Adriatic coast, including the KRAS plateau, the mountain ranges of VELEBIT and Prokletije, and the mountains of Biokova, Lovcen and DURMITOR.
Map Yugoslavia Bb

Dinder National Park *Sudan* Nature reserve covering some 6475 km² (2500 sq miles) on the Ethiopian border, about 520 km (320 miles) south-east of the capital, Khartoum.
Map Sudan Bb

Dingaan's Kraal *South Africa* Site of the capital of the 19th-century Zulu king Dingaan in eastern Natal, about 145 km (90 miles) north of Durban. On February 5, 1838, the Voortrekker leader Piet Retief and 70 of his men were slaughtered on Dingaan's orders in the royal kraal. The original huts were burnt down in reprisal ten months later, but many have been rebuilt as a film location.
Map South Africa Db

Dinosaur National Monument *USA* Area covering 827 km² (319 sq miles) across the northern end of the state border between Utah and Colorado. It is a semiarid wilderness where, preserved in the alluvial deposits of a prehistoric river, many dinosaur remains were found.
Map United States Eb

diorite Coarse-grained, intrusive igneous rock, between acidic and basic, containing largely plagioclase feldspar and iron-magnesium silicates. See ROCK

Diosso Gorges *Congo* Formation about 23 km (14 miles) north of POINTE-NOIRE, consisting of a series of spectacular and deeply dissected cliffs facing the Atlantic Ocean. About 50 m (165 ft) high, the cliffs contain rocks coloured red, white and yellow by iron and other oxides. The Diosso Amphitheatre, the most impressive section, is an area which has been eroded into a huge natural bowl.
Map Congo Ab

Diourbel (Djourbel) *Senegal* Market town for a groundnut area, 140 km (87 miles) east of the national capital, Dakar. It was linked by rail to Dakar and Mali in 1905, and has food processing and craft industries. The town is also an Islamic religious centre. There is an agricultural research station at the nearby town of Bambey.
Population 50 600
Map Senegal Ab

dip Angle between the inclination of rock strata and the plane of the horizontal.

Dir ***Pakistan*** Mountainous princely state in the extreme north-west, covering 5286 km² (2041 sq miles) on the border with Afghanistan. It became part of Pakistan in 1947, but retained considerable autonomy. In 1948, fighting broke out with the neighbouring principality of Chitral over a royal betrothal, and the Pakistani army had to intervene. The Pakistani government deposed the ruler in 1960 in favour of one of his sons, on the pretext of preventing an Afghan incursion. The area is next to the Konar valley in Afghanistan, scene of bitter resistance to the Russian invasion since 1979, and remains semi-independent.
Population 500 000
Map Pakistan Ca

Dire Dawa ***Ethiopia*** Modern city about 350 km (220 miles) east of the capital, Addis Ababa, on the railway line to Djibouti. It has large textile and cement factories.
Population 85 000
Map Ethiopia Bb

Dirschau ***Poland*** See TCZEW

discontinuity Boundary across which the internal character of the earth changes abruptly. The two main discontinuities are the Mohorovičić, or Moho, between the earth's crust and mantle, and the Gutenburg between the mantle and core. See EARTH

Discovery Bay Bay on the south coast of Australia, at a point where the states of SOUTH AUSTRALIA and VICTORIA meet. It was visited in 1802 by both the French navigator Nicolas Baudin and the British explorer Matthew Flinders.
Dimensions 72 km (45 miles) long, 13 km (8 miles) wide
Map Australia Gf

Disko (Qeqertarsuaq) ***Greenland*** Largest of Greenland's islands, covering an area of 8580 km² (3311 sq miles) off the west coast. Its hills rise to a maximum height of 1919 m (6296 ft).
Map Greenland Cc

Disney World ***USA*** Development in central Florida, about 25 km (16 miles) south-west of Orlando. The grandiose scheme involves some 113 km² (44 sq miles) of the state's lake country, and includes residential and industrial estates. Tourists flock to the first section to be opened – a 'vacation kingdom', that includes an amusement park, hotels and artificial lakes and lagoons. Epcot (Experimental Community of Tomorrow) is a futuristic computer-controlled fantasy world.
Map United States Je

Disneyland ***USA*** Popular amusement park in southern California opened in 1955. It is sited in Anaheim, part of Los Angeles, and was designed by the film animator Walt Disney (1901-66) as a series of 'themelands', including Adventureland, Fantasyland, Main Street USA, and Frontierland.
Map United States Cd

divide Alternative name for a WATERSHED.

Diyarbakir ***Turkey*** City, and capital of the province of the same name, about 750 km (465 miles) south-east of the capital, Ankara. It markets grain, wool and mohair produced on nearby farms, and makes textiles and leather goods. It is noted for the ancient black basalt walls, almost 5 km (3 miles) long, around the old town. The largely agricultural province covers 15 355 km² (5928 miles) and produces cereals, fruits, cotton and tobacco.
Population (city) 305 300; (province) 941 500
Map Turkey Cb

Djakovo ***Yugoslavia*** See SLAVONIA

Djanet ***Algeria*** See TASSILI N'AJJER

Djelfa (El-Djelfa) ***Algeria*** Town in the Sahara Atlas Mountains at the end of the railway line from Algiers, on the middle route southwards to TAMANRASSET. It is an administrative centre and a market for the Sahara nomads. It has dolmens (prehistoric stone burial chambers) and salt rock formations.
Population 55 000
Map Algeria Ba

Djenne ***Mali*** Town 380 km (240 miles) north-east of the capital, Bamako. It is thought to have been founded by the Songhai in the 8th century and has always been an important centre for Islamic teaching and for tribal craftsmen producing leather goods, cloth and blankets. It is also the market where north and south meet. The magnificent mosque was rebuilt by the French in 1907 in the original traditional style. Djenne is the site of considerable archaeological finds, including pottery, jewellery and leatherwork which had been preserved in sand.
Population 7000
Map Mali Bb

Djibouti ***Djibouti*** National capital and port on the Gulf of Aden. Economically, it depends largely on its role as a transit port handling trade with Ethiopia.
Population 200 000
Map Ethiopia Ba

Djourbel ***Senegal*** See DIOURBEL

Dnepr ***USSR*** See DNIEPER

Dnepropetrovsk ***USSR*** City on the Dnieper river 195 km (120 miles) south-west of Khar'kov; formerly Ekaterinoslav. It is the centre of the Ukrainian wheat trade and manufactures iron, steel and machinery.
Population 1 140 000
Map USSR Ed

Dnestr ***USSR*** See DNIESTER

Dnieper (Dnepr) ***USSR*** Third longest European river (after the Volga and Danube), at 2285 km (1420 miles). Rising in the Valdai Hills west of Moscow, it flows generally south to its broad estuary on the Black Sea, passing the cities of Smolensk, Kiev, Dnepropetrovsk and Zaporozh'ye. It is navigable in its upper course for eight months of the year and the lower course for nine months. It is dammed for hydroelectric power, and connected by canals with the West Dvina and Bug rivers.
Map USSR Ed

Djibouti

THIS POOR DESERT COUNTRY OF NOMADIC TRIBES IS A MAGNET FOR STARVING REFUGEES

Nomadic tribes roam the hot, inhospitable desert country of Djibouti. It is a land of 23 200 km² (9000 sq miles) with an important port, also called DJIBOUTI, on the Gulf of Aden. The land is mainly basalt plains, with mountains rising over 1500 m (4900 ft). Djibouti became an independent republic in 1977 after 96 years as a French colony.

The climate is among the world's hottest, temperatures averaging 30°C (86°F) and often exceeding 40°C (104°F). Inland, rain is rare and even on the coast the average annual rainfall is only 130 mm (5 in). Only 10 per cent of the land can be farmed, even for seasonal grazing, and the country has great difficulty in producing food for its own modest population of 305 000, let alone a further 100 000 who arrived in the early 1980s as refugees from Ethiopia.

The native population, most of which is Muslim, is composed of two main groups: the Afars in the north and west, and the Issas (Somalis), in the south. Both are nomadic and, with their sheep and goats, often cross into Ethiopia or Somalia in search of water and grazing. Crops are grown and cattle are grazed around a few oases, but most food for the urban population must be imported.

Cattle, hides and skins are the chief exports, but most of the country's foreign exchange is earned by the port of Djibouti, which is also the capital and the only city. The port's strategic position between the RED SEA and the INDIAN OCEAN makes it an important staging post on a major shipping route.

DJIBOUTI AT A GLANCE
Map Ethiopia Ba
Area 23 200 km² (9000 sq miles)
Population 305 000
Capital Djibouti
Government One-party republic
Currency Djibouti franc = 100 centimes
Languages Arabic and French (official), Somali, Afar
Religion Muslim
Climate Very hot and dry; average temperature in Djibouti city ranges from 23-29°C (73-84°F) in January to 31-41°C (88-106°F) in July
Main primary products Sheep, goats, cattle, dates, fish; salt
Major industries Stock rearing, processing hides and skins, fishing
Main exports Livestock, hides and skins
Annual income per head (US$) 1010
Population growth (per thous/yr) 34

Dniester (Dnestr) ***USSR*** River rising in the Carpathians of the western Ukraine. It winds south-eastwards for 1411 km (877 miles) through the Ukraine and Moldavia, and finally enters the Black Sea 30 km (18 miles) south-west of Odessa. Between 1918 and 1940 it formed the border between Romania and the USSR.

Map USSR Dd

Doab ***India*** Territory between the Yamuna river and the Ganges, stretching from the Himalayan fringe at the town of HARIDWAR to the confluence at ALLAHABAD. The Moguls built canals there, and so did the British from 1848 onwards. The railways followed, and together the two technologies transformed agriculture and brought prosperity to the area.

Map India Cb

Dobreta-Turnu-Severin ***Romania*** Industrial city built on terraces above the Danube river, where it forms the border with north-east Yugoslavia. The spectacular Iron Gate – a gorge whose surging rapids feed a vast hydroelectric power station – lies 18 km (11 miles) north-east.

Nearby are the ruins of the Roman town of Dobreta, complete with port and baths, and the remains of a huge bridge across the Danube built by the Roman emperor Trajan in the 2nd century AD. The city's industries include food processing, shipbuilding and woodworking.

Population 83 000

Map Romania Ab

Dobrna ***Yugoslavia*** See KARAWANKE ALPS

Dobruja (Dobrogea) ***Romania*** Fertile region lying between the Danube river and the Black Sea, of which CONSTANTA is the main resort. Covering 23 258 km² (8980 sq miles), Dobruja contains huge fields of cereals, the Murfatlar vineyards and the Danube Delta – a major wildlife sanctuary and game preserve. The delta consists of lakes, channels, marshes, reeds and forest. Birdwatchers can see more than 300 bird species, including herons, pelicans and waders. Fishermen can catch perch, carp and sturgeon, and hunters pursue boar, deer and wild cats.

Map Romania Bb

doctor, the Any of several local winds, such as the HARMATTAN of West Africa, that mitigate extremely unhealthy conditions.

Dodecanese (Dhodhekánisos) ***Greece*** Twelve islands in the south-east Aegean Sea, stretching from SAMOS in the north to the largest of them, RHODES, in the south. They are popular boating and holiday centres. The islands were occupied by Italy from 1912 until the Second World War. They formally passed to Greece in 1947. The name of the group comes from *dodeka*, the Greek word for 'twelve'.

Population 145 000

Map Greece Ec

Dodoma ***Tanzania*** Town about 400 km (250 miles) west of Dar es Salaam, which it superseded as the capital in 1975, largely because of its central location. A master plan for a city of some 400 000 people has been drawn up.

Population 45 703

Map Tanzania Ba

Dogger Bank Large submerged sandbank in the centre of the NORTH SEA about 180 km (110 miles) off the north-east coast of England. The bank is a famous fishing ground, especially for cod. (The word *dogger* means 'cod' in Dutch.) A major naval battle between a British fleet under Admiral Beatty and a German fleet under Admiral Hipper was fought there in 1915.

Depth Mostly 18-36 m (60-118 ft)

Map United Kingdom Fc

Doha (Ad Dawhah) ***Qatar*** Capital, main port and commercial centre of Qatar. A growing, modern city, it has areas named after the country's old tribal families who moved in when it became the capital in 1971. The old fishing port is still used by dhows (single-masted Arab sailing vessels), while the new port handles exports and imports. An elegant corniche, or coast road, has been built – and there is an international airport. Doha has two museums.

Population 180 000

Map Qatar Ab

Doi Inthanon ***Thailand*** The country's highest mountain 40 km (25 miles) west of the northern city of Chieng-Mai. The 2590 m (8288 ft) peak and its surrounding waterfalls have become popular with tourists since a road was built almost to the summit in the mid-1970s.

Map Thailand Ab

doldrums A zone of sultry calms and light breezes within a few degrees of the Equator, where the north-east and south-east trade winds converge. Intense heat allied with low pressure often causes strong updraughts of air, resulting in sudden squalls and heavy tropical rain. The zone was avoided by sailing ships because they could become becalmed for days.

dolerite Basic, medium-grained, intrusive igneous rock, composed mainly of feldspars, pyroxene and sometimes olivine.

dolomite Yellowish or brownish mineral which is a double carbonate of calcium and magnesium. Also a rock, often called dolomitic limestone or magnesian limestone, containing more than 15 per cent magnesium carbonate.

Dolomites (Alpi Dolomitiche) ***Italy*** Range of dolomitic limestone mountains between the upper Adige and Piave valleys of north Italy. The limestone erodes into fantastically rugged shapes, including pinnacles, tooth-edged ridges, deep gorges and precipitous cliffs. On its southern side the highest peak, Marmolada (3342 m, 10 964 ft), drops sheer by some 610 m (2000 ft). The range is a climber's paradise, and the gentler, lower slopes offer good skiing. Cortina d'Ampezzo is the main resort.

Map Italy Ca

Dombrau ***Poland*** See DABROWA GORNICZA

Dominica See p. 192

Dominican Republic See p. 193

Domrémy-la-Pucelle ***France*** Village 45 km (28 miles) south-west of Nancy, where Joan of Arc was born in 1412.

Map France Fb

Don ***USSR*** River, 1870 km (1165 miles) long, rising in hills about 200 km (125 miles) south of Moscow. It flows south to the city of Voronezh, then south-east to a canal link with the Volga near Volgograd, then south-west to the Sea of Azov. It is navigable as far upstream as Voronezh, but is closed by ice for three to four months each winter.

Map USSR Fd

Donbass ***USSR*** See DONETS BASIN

Doncaster ***United Kingdom*** Centre of one of England's most productive coalfields, about 28 km (17 miles) north-east of Sheffield, South Yorkshire. It is also an important railway junction with workshops for building and repairing rolling stock, and since 1776 has been the venue of the St Leger horse-race.

Population 82 000

Map United Kingdom Ed

Dondra Head ***Sri Lanka*** The island's southernmost point, sacred to both Hindus and Buddhists. There are remains of the 7th-century Hindu shrine of Maha Vishnu Devala whose gilded roof was once a landmark for passing ships. The shrine was destroyed by the Portuguese in the 16th century.

Donegal (Dún na Ngall) ***Ireland*** Most northerly county in Ireland, in the province of Ulster, with a wild and jagged coastline facing the North Atlantic. It covers 4830 km² (1865 sq miles), and its mountains and lakes attract tourists, especially anglers. It has good dairy farmland known as the Lagan in the east, along the border with Northern Ireland. On the poorer west coast the Gaelic language and culture are still entrenched. The county is renowned for its textiles, especially Donegal tweed. The small market town of Donegal lies at the head of Donegal Bay. Lifford is the county town.

Population (county) 125 100; (town) 2870

Map Ireland Ca

Donets Basin (Donbass) ***USSR*** Largest Soviet coalfield and one of the country's leading industrial areas, straddling the eastern part of the Ukraine and the Rostov region, south of the Donets river (a tributary of the Don). Coal, discovered in 1721 and mined from 1800, became a vital fuel with the building of the railways in the 1870s. There is a vast metal industry using iron ore from the adjoining KRIVOY ROG fields. The basin's output today is some 200 million tonnes of coal and over 50 million tonnes of steel a year. The region's chief city is DONETSK; others include MAKEYEVKA and VOROSHILOVGRAD.

Map USSR Ed

Donetsk ***USSR*** City in the Ukraine, about 255 km (160 miles) south-east of Khar'kov; formerly Stalino. It was founded in 1872 by John Hughes, a British industrialist, to make iron rails for the Russian railways. It is now the chief industrial centre for the DONETS BASIN, with coal, iron, steel, machinery, chemicals, cement and clothing industries.

Population 1 064 000

Map USSR Ed

Donga Ridge ***Nigeria*** See UDI PLATEAU

Dominica

FORESTED, RUGGED AND MOUNTAINOUS – THE CARIBBEAN ISLAND THAT COLUMBUS DISCOVERED ON A SUNDAY

The most northerly of the WINDWARD ISLANDS in the West Indies, Dominica is the least changed since the Caribbean was first settled by Europeans. Christopher Columbus gave the island its name because he discovered it on a Sunday (in Spanish, *Domingo*) in 1493. Fierce disputes between the British and French over the island's ownership were finally settled in 1805, when the French left the island after extracting a ransom of £12 000 (then US$53 000) from the British.

Dominica became an independent member of the Commonwealth in 1978, with an elected president as head of state. Many of the island's place names remain French and some of the inhabitants, who are mostly descendants of African slaves, still speak a French patois.

On a reserve in the north-east of the island there are several hundred Caribs, the native Indians who originally inhabited the islands and after whom the Caribbean is named. But they have lost most of their original culture and language, and live in much the same way as their black neighbours.

THREE YOUNG VOLCANOES

Dominica is the most rugged island in the eastern Caribbean. Only 3 per cent of the land is flat. The rest consists of three relatively young but inactive volcanoes, the highest of which is Morne Diablotin (1447 m, 4747 ft). There are several hot springs in the Valley of Desolation in the south. Most of the island has more than 2500 mm (98 in) of rain a year and even on the sheltered leeward coast it rains two days out of three.

The steep mountain slopes are difficult to farm and three-quarters of Dominica is still forested. The island did not attract settlers as early as others in the Caribbean and it still has a lower population density and poorer standard of living than most West Indian countries. Overseas aid is important, and was running at US$80 million a year in the early 1980s.

Agriculture provides almost all the exports. Dominica is the largest Windward Islands exporter of bananas to the UK. Copra, citrus fruits (especially limes), cocoa, bay leaves and vanilla are also revenue earners. Manufacturing is mostly confined to small-scale processing of agricultural products.

Dominica is not a typical tourist island, but its rugged beauty and dense forests attract adventurous travellers. Until 1982 ROSEAU, the small capital, was a difficult two-hour drive from the only airport, but the new Canefield airport is only five minutes' drive from the town. The old fort which once defended Roseau has been made into a hotel. Ships calling at Roseau have to anchor offshore. Portsmouth, in the north-west, has a better harbour but surrounding swampland has hindered the town's growth.

DOMINICA AT A GLANCE

Map Caribbean Cb
Area 751 km² (290 sq miles)
Population 74 000
Capital Roseau
Government Parliamentary republic
Currency East Caribbean dollar = 100 cents
Languages English, French patois
Religion Christian (80% Roman Catholic)
Climate Tropical; average temperature ranges from 20-29°C (68-84°F) in January to 23-32°C (73-90°F) in June
Main primary products Bananas, coconuts, citrus fruits, cocoa, beef cattle, pigs; pumice
Major industries Agriculture, food processing, soap, essential oils
Main exports Bananas, copra, citrus fruits, essential oils, cocoa, spices, soap
Annual income per head (US$) 850
Population growth (per thous/yr) Declining
Life expectancy (yrs) Male 57 **Female** 59

Dongbei (Manchuria) ***China*** Region in the north-east, covering 780 000 km² (310 000 sq miles). It is made up of the provinces of HEILONGJIANG, JILIN and LIAONING. From 1931 to 1945, Dongbei was the Japanese puppet state of Manchukuo. It is a land of forests and prairies and has important reserves of oil, coal and iron ore which have played a major part in the development of China's heavy industry.
Population 87 962 000
Map China Ib

Doornik ***Belgium*** See TOURNAI

Dorchester ***United Kingdom*** County town of Dorset, about 38 km (24 miles) west of Bournemouth. It was founded by the Romans as Durnovaria in AD 70, but the massive earthworks of Maiden Castle, 3 km (2 miles) southwest, date back to the late Stone Age around 2000 BC. This was later fortified to house a township of some 5000 people that was overrun by the Roman 2nd Legion in AD 43.

The Tolpuddle Martyrs – six pioneer trade unionists – were imprisoned at Dorchester in 1834 but released two years later after prolonged local protests.
Population 15 000
Map United Kingdom De

Dordogne ***France*** River of western France 475 km (295 miles) long. It rises in the Massif Central, cuts through the limestone plateau of Périgord and continues west to join the Garonne north of Bordeaux, emptying into the Bay of Biscay. It crosses the department of the same name, whose capital is Perigueux and which contains the renowned LASCAUX caves.
Population (department) 375 000
Map France Dd

Dordrecht (Dordt; Dort) ***Netherlands*** Industrial city and medieval port in the delta of the Rhine and Maas rivers, 19 km (12 miles) south-east of Rotterdam. Its products include tobacco, timber, metal goods, chemicals, linen and glass.
Population (city) 107 600; (Greater Dordrecht) 198 600
Map Netherlands Bb

Dorset ***United Kingdom*** County of southern England immortalised in the works of the poet and novelist Thomas Hardy (1840-1928). It covers 2654 km² (1025 sq miles) along the English Channel coast. Inland, the county is a deeply rural area of chalk downs and wooded valleys, while the coast is remarkable for the variety of its features. It stretches westwards from the resort of BOURNEMOUTH past Poole Harbour, with its narrow entrance and many islands; along the rocky and beautiful shore of the so-called Isle of Purbeck (on which lies one of Britain's oil fields); past Lulworth Cove to the port of Weymouth and the island and naval base of Portland; then along 15 km (10 miles) of Chesil Beach to Lyme Regis, with its complex of fallen cliffs. Dorset is a popular tourist area. The county town is DORCHESTER.
Population 618 000
Map United Kingdom De

Dort ***Netherlands*** See DORDRECHT

Dortmund ***West Germany*** Inland port and second largest city of the industrial Ruhr, 32 km (20 miles) east of the main Ruhr city, Essen. It is Europe's largest producer of beer and has one of its biggest steel plants. The city has been almost entirely rebuilt since the Second World War. Its Westfalenhalle is one of the largest sports halls in Europe, seating 23 000 people.
Population 620 000
Map West Germany Bc

Dortmund-Ems Canal ***West Germany*** Canal opened in 1899 to link the expanding industries of the Ruhr with the North Sea at the port of Emden, near the Dutch border. The canal joins the River Ems at Meppen, 115 km (71 miles) north of Dortmund.
Map West Germany Bb

Douai ***France*** Town on the Scarpe river, 30 km (19 miles) south of Lille. The Old Testament of the Douai Bible, still used by Roman Catholics, was published there in 1609-10. The New Testament for the same Bible was published at Reims in 1582.
Population 204 700
Map France Ea

Douala ***Cameroon*** The country's chief port and trade centre. It is a busy, densely peopled town of great contrasts. Almost anything can be bought there, from local charms and potions to Parisian haute couture, for there are several colourful markets and also expensive shops. The tall, modern buildings of the port contrast with a host of older, poorer buildings clustered along narrow streets. Visitors can escape the sticky heat and the flies on the cool slopes of Mount Cameroon or on the beaches of black volcanic sand at Limbe (Victoria).
Population 800 000
Map Cameroon Ab

Dougga (Thugga) ***Tunisia*** Ruined Roman city 110 km (about 70 miles) south-west of Tunis. It lies on a much older site founded by native Berbers and later occupied by Phoenician colonists. Built on the top of a plateau, the town was at its height under the Romans in the 2nd century AD, although it continued to flourish well into the Arab-Islamic period. Considered to be Tunisia's best preserved Roman site, Dougga's theatre is almost intact and the capitol is one of the finest in Africa. Other monuments include a forum, temples to Jupiter, Juno and Minerva as well as to other Roman gods, public baths, triumphal arches, private houses and streets. There is also a 2nd-century BC Carthaginian mausoleum. The surrounding plain is dotted with more Roman remains.

Map Tunisia Aa

Douro (Duero) ***Spain/Portugal*** River, 895 km (555 miles) long, which rises in north-central Spain and flows westwards to cross northern Portugal and enter the Atlantic downstream from OPORTO. It forms part of the border with Spain. The Portuguese section is dammed in several places to provide hydroelectric power, and its steep schist slopes between Pêso da Régua and the Spanish border form the area where grapes for port wine are grown.

Map Portugal Cb

Dominican Republic

AFTER YEARS OF REVOLUTION AND DICTATORSHIP, DEMOCRACY HAS ARRIVED, THOUGH POVERTY IS STILL WIDESPREAD THROUGHOUT THIS CARIBBEAN NATION

Poverty, revolution and dictatorships are the ingredients of this state which shares the West Indian island of Hispaniola with HAITI. Unlike the Negroes of French-speaking Haiti, the mixed population of the Dominican Republic shares the language, religion and lifestyles of Latin America.

Sailors left behind by Christopher Columbus following the wreck of one of the ships on his first voyage in 1492 founded La Navidad, the first European settlement in the Americas. Most of the early settlers left to seek their fortunes in Mexico, Panama and Peru, and for three centuries the country was an unimportant Spanish colony. Within 50 years of Spanish settlement, the 300 000 Tainos Indian population either died of European diseases, to which they had no inherited immunity, or were slaughtered.

After a troubled and violent history, including spells of French, Haitian and again Spanish rule, the country finally gained its independence in 1865. Over the next 50 years there were 28 revolutions and 35 governments, and the United States occupied the republic from 1916 to 1924 to bring stability. Dictators ruled the country until it became a democracy in 1966 following a second US military landing.

The western part of the country is dominated by four almost parallel mountain ranges. In the Cordillera Central, Pico DUARTE (3175 m, 10 417 ft) is the highest point in the West Indies. The wide fertile Cibao valley lies between the two most northerly mountain ranges. The south-east is occupied by fertile plains.

Though much of the land is fertile, only about one-third of the cultivable area is farmed. Sugar is the most valuable crop and a mainstay of the republic's economy. It is grown mainly on plantations in the southern plains. Coffee, the second export crop, is cultivated on smallholdings. Both crops are grown with the help of migrant labour from Haiti. Cocoa and tobacco are also produced.

Gold – the country has the largest operating gold mine in the Caribbean – silver, platinum, nickel and aluminium ores are mined and, in some cases, refined on the island. Apart from mining, the main industries include food processing and making consumer goods for the home market. Some new industries, mostly foreign owned, have been built in duty-free industrial zones, where they also benefit from the plentiful cheap labour. They are concerned mainly with processing imported, semi-finished goods for re-export. In spite of these developments, the people remain poor.

The country has fine beaches and the government is developing tourism fast. New resorts are being built at La Romana, Punta Cana and Puerto Plata. East of La Romana, a village in 16th-century Spanish style has been built as a tourist attraction. SANTO DOMINGO is the main port and capital. For many years it was called Ciudad Trujillo after the long ruling dictator Rafael Trujillo who retained power, even when not holding the presidency, from 1930 until his assassination in 1961. The colonial part of the city has been extensively rebuilt and restored. The cathedral, built between 1512 and 1540, is one of the oldest European-style buildings in the Americas.

DOMINICAN REP. AT A GLANCE

Map Caribbean Ab

Area 48 442 km² (18 703 sq miles)

Population 6 766 000

Capital Santo Domingo

Government Parliamentary republic

Currency Peso = 100 centavos

Language Spanish

Religion Christian (95% Roman Catholic)

Climate Tropical; cooler in mountains. Average temperature in Santo Domingo ranges from 19-29°C (66-84°F) in January to 23-31°C (73-88°F) in August

Main primary products Sugar, coffee, cocoa, tobacco, rice, maize; bauxite, gold, silver, platinum, nickel, salt

Major industries Agriculture, tourism, textiles, cement, food processing (sugar, rum, molasses), tobacco products, mining and metal refining, petroleum products

Main exports Sugar, gold, silver, nickel, coffee, cocoa, bauxite, tobacco

Annual income per head (US$) 1190

Population growth (per thous/yr) 27

Life expectancy (yrs) Male 60 **Female** 64

Dover ***United Kingdom*** Town in the English county of Kent, 105 km (65 miles) south-east of London, which has been the principal Channel port since Roman times, being the nearest to the French port of Calais, 34 km (21 miles) away. A Roman lighthouse, Saxon church and Norman castle stand on its famous white cliffs. There are monuments to Captain Matthew Webb who, in 1875, became the first man to swim the English Channel, and to the French aviator Louis Blériot, who in 1909 became the first man to fly across it.

Population 33 000

Map United Kingdom Fe

Dover ***USA*** State capital of Delaware, about 135 km (85 miles) east and slightly north of the national capital, Washington. It was established in 1683 by the Quaker settler William Penn, who founded the state of Pennsylvania, and lies in the heart of a prosperous fruit and vegetable producing region.

Population 22 500

Map United States Kc

Dover, Strait of Busy waterway separating England from France and connecting the ENGLISH CHANNEL to the NORTH SEA. It was called Fretum Gallicum by the ancient Romans and is known as the Pas de Calais to the French. It was first swum by Captain Matthew Webb in 1875. In 1940, it was the scene of the British evacuation from DUNKIRK.

Narrowest width 34 km (21 miles)

Map United Kingdom Fe

Down ***United Kingdom*** County covering 2448 km² (945 sq miles) in Northern Ireland's south-east corner beside the Irish Sea. It includes the Mourne Mountains, which rise to 852 m (2795 ft). There is evidence of early prehistoric settlement along the coast, and monasteries at Bangor, Nendrum and Movilla date from the time of Saint Patrick, in the 5th century.

Population 362 100

Map United Kingdom Cc

Downs, The ***United Kingdom*** Two roughly parallel ranges of low chalk hills – less than 300 m (1000 ft) high – in southern England. The North Downs extend about 130 km (80 miles) in a curve from the county of Surrey across north Kent to the English Channel coast at Dover's famous white cliffs. The South Downs are about 110 km (70 miles) long and stretch from south-east Hampshire across southern Sussex to the cliffs of Beachy Head, east of Brighton. Between them is the rich farming country – once forested – of The Weald. The name 'downs' is also used in other parts of England for rolling treeless grasslands.

Map United Kingdom Ee

downthrow The vertical distance between the level of a rock on one side of a FAULT plane and that of its lower continuation on the other – the 'downthrow side'.

Drake Passage Waterway linking the PACIFIC and ATLANTIC oceans between CAPE HORN and the SOUTH SHETLAND ISLANDS. It was discovered by Sir Francis Drake on his voyage round the world in 1577-80.
Width 640 km (400 miles)
Map Antarctica Da

Drakensberg ***South Africa/Lesotho*** Mountain range stretching 1125 km (700 miles) from east Transvaal, through Orange Free State, Natal and Lesotho, to east Cape Province. It is part of the GREAT ESCARPMENT, and gets its name from African legends which describe it as the home of dragons. The highest peaks are on the Lesotho-Natal border, where Thabana Ntlenya (on the Lesotho side) rises to 3482 m (11 424 ft) and Champagne Castle (on the Natal side) 3374 m (11 069 ft).
Map South Africa Cc

Drava (Drave; Drau) ***Austria/Hungary/Yugoslavia*** Major central European river, 718 km (447 miles) long. It rises in East Tyrol and flows east past Villach in Austria to enter Yugoslavia west of Dravograd. It is dammed for hydroelectric power above Maribor, forms part of the Hungarian-Yugoslav border farther along its course, and joins the Danube 23 km (14 miles) east of the Yugoslav city of Osijek.
Map Yugoslavia Cb

Drenthe ***Netherlands*** Farming province covering 2681 km² (1035 sq miles) in the north-east. It produces cereals, potatoes and raises cattle, and Assen is its capital.
Population 424 700
Map Netherlands Ca

Dresden ***East Germany*** Historic city – the former capital of Saxony – in the south-east, 40 km (25 miles) from the border with Czechoslovakia. It is a nuclear research and medical centre and produces precision and machine tools, aircraft, instruments, electronics, business machines, textiles and porcelain.

The restored 19th-century Semper opera house was reopened on February 13, 1985, to commemorate the 40th anniversary of the city's destruction by Allied bombs in the Second World War, and to mark the completion of the painstaking reconstruction of the old Baroque city once known as 'Florence on the Elbe'. Many of its churches, the magnificent Zwinger Palace and the Semper Gallery – containing collections of china, porcelain and old masters' paintings – have also been restored.
Population 522 500
Map East Germany Cc

drift 1. Bank or pile of sand or snow – for example, heaped up by currents of air or water. 2. Rock debris eroded and transported by ice, then deposited either by the ice or by meltwater. In a wider sense, the term includes ALLUVIUM. 3. In South Africa, a ford in a river, where it is also called a drif.

Drin ***Albania/Yugoslavia*** River, 280 km (174 miles) long, which empties into the Adriatic Sea from the marshy SHKODER lowlands in north-west Albania. Its northern mouth, the Buenë river, forms part of the Albanian-Yugoslav border. Higher up its course, it is a torrential river running through deep gorges and with many tributaries. At several points the Drin has been harnessed to produce hydroelectricity.
Map Albania Ca

Drina ***Yugoslavia*** River, 346 km (215 miles) long, forming much of the boundary between Serbia and Bosnia Herzegovina. Formed near Sutjeska from the Piva and Tara rivers, it flows swiftly north-east through a spectacular limestone gorge with rapids – popular for rafting and kayaking – past Foca to VISEGRAD, through '*slivovica* (plum-brandy) country' to the Sava. Reservoirs behind hydroelectric dams just below Visegrad and at Zvornik are used for water sports.
Map Yugoslavia Db

drizzle Fine, continuous rainfall in which raindrops are less than 0.5 mm (0.02 in) in diameter, usually associated with a warm FRONT.

Drogheda (Droichead Átha) ***Ireland*** Port on the estuary of the River Boyne about 40 km (25 miles) north of Dublin. It was the meeting place of Ireland's medieval parliaments. The town was sacked in 1649 after a siege by Oliver Cromwell, Lord Protector of England, and most of the 2000 defenders were massacred. The handful of survivors were shipped to the West Indies. Drogheda surrendered to the British again in 1690 after William of Orange defeated James II at the Battle of the Boyne. In St Peter's Church is preserved the head of St Oliver Plunkett, Archbishop of Armagh, and Primate of All Ireland from 1669. He was executed at Tyburn in London in 1681, after being implicated in the fictitious popish plot to murder Charles II. Drogheda's manufactures include textiles, beer, footwear, chemicals, foodstuffs and cement.
Population 22 200
Map Ireland Cb

drought Any prolonged period of dry weather, specifically defined as *absolute drought*, when the daily rainfall is less than 0.25 mm (0.01 in) for a period of at least 15 consecutive days; *partial drought*, a period of at least 29 consecutive days during which the mean daily rainfall does not exceed 0.25 mm (0.01 in); and *dry spell*, a period of at least 15 consecutive days during which the daily rainfall is less than 1.0 mm (0.04 in). In the USA, a *dry spell* is a period of 14 days when there is no measurable rainfall.

drowned valley A valley that has been filled with water from the sea or a lake either because of land subsidence or the rising of the sea.

drumlin Small, streamlined hill composed of glacial DRIFT, which has been shaped by ice moving over it. Drumlins are generally egg-shaped – the blunt end facing the direction from which the ice advanced. Most are 0.4-0.8 km (0.25-0.5 mile) long and about 30m (100 ft) high. See GLACIATION

Drvar ***Yugoslavia*** Small timber-processing town in west Bosnia, lying on the Una river amid forested mountains. Its full name is Titov Drvar to commemorate Yugoslavia's former Communist leader Marshal Tito (1892-1980) whose general headquarters was based in the town during the Second World War. At one stage of the war, Tito and his men sought refuge in a nearby cave during a protracted battle with occupying Nazi forces. The cave, Titova Pecina (Tito's Cave), is now a place of pilgrimage.
Population 6417
Map Yugoslavia Cb

dry spell See DROUGHT

dry valley Valley that was formerly cut by running water but which now has no river running through it. Such valleys are usually found in limestone regions and were cut when the climate was wetter and the water table was higher, or when the subsurface was frozen and impermeable during a colder climate. The permeable limestone has since allowed the water to sink underground, leaving the valley dry.

Dry Valleys (McMurdo Oasis) ***Antarctica*** Name commonly given to an ice-free area in Antarctica covering about 3000 km² (1160 sq miles). It comprises a series of extremely arid valleys near McMurdo Sound where no rain has fallen for at least 2 million years. There are several lakes such as Lake VANDA – hence the alternative name of McMurdo Oasis.
Map Antarctica Jc

Duars ***Bhutan*** Plains region on the border with Assam, India. Much of it was annexed by the British in 1854 following a series of short wars, and its savannah grasslands and jungle were cleared for tea plantations. The annexed section is now part of India. The remaining Bhutanese part is less developed, yet sustains nearly a quarter of Bhutan's population of 1.25 million. The region gets its name from a Bhutanese word meaning 'doorway', because the torrential rivers which flow onto the plain have cut gaping ravines into the wall of mountains alongside.
Population 300 000
Map Bhutan Aa

Duarte, Pico ***Dominican Republic*** Mountain peak (formerly Monte Trujillo) in the Cordillera Central 3175 m (10 417 ft) high. It is the highest point in the West Indies.
Map Caribbean Ab

Dubai ***United Arab Emirates*** Second largest of the seven emirates, Dubai has grown wealthy on oil. Fishing is an important industry, and Dubai city is a busy seaport and modern capital. It also has an international airport. In the 1960s the city consisted of one and two-storey houses constructed from coral limestone, some with wind towers – an architectural feature unique to this part of the world, and one of the earliest forms of air conditioning in which wind is diverted off the tower wall into the house to cool it. There was no electricity until 1961 and no mains water until 1968. During the 1970s there was urban development including the building of many high-rise office blocks. The tallest building is the 39-storey International Trade and Exhibition Centre. Despite its modernity, Dubai retains its links with the past. The port is still used by dhows (single-masted Arab coastal sailing vessels) and served by abras (water taxis). There is an early 19th-century fort and some old Arab houses.
Population (emirate) 296 000; (city) 265 700
Map United Arab Emirates Ba

Dubbo ***Australia*** Country town in New South Wales, 285 km (180 miles) north-west of Sydney. It is the business centre for the surrounding sheep and cattle country and its stock markets are the largest in the state. It is the home of the Western Plains Zoo, the country's largest open-range zoo.

Population 24 000

Map Australia He

Dublin (Baile Atha Cliath) ***Ireland*** Capital of the Republic of Ireland, astride the River Liffey on the east coast. Vikings settled at the river mouth in the 9th century and remained until Ireland's warrior king, Brian Boru, drove them out after the Battle of Clontarf (1014). The Normans made Dublin their centre for the conquest of Ireland in 1170. By the end of the Middle Ages, however, the area under English control had been reduced to a small area around Dublin – the Pale. Those in the wild lands beyond were deemed to be 'beyond the pale'.

Dublin is largely a Georgian city. The Mansion House has been the Lord Mayor's residence since 1715 and it was from there that Ireland's Declaration of Independence was made in 1919. Leinster House, where the Dail, or Parliament, meets was started in 1744. The Custom House (1791) and Four Courts (1786) date from the same period. O'Connell Street, one of the widest and most gracious thoroughfares in Europe, runs north from the Liffey past the Post Office, headquarters of the insurgents during the Easter Rising of 1916. Dublin Castle, partly Norman, was the seat of British administration in Ireland for centuries. It has been restored as government offices. South of the Liffey is St Patrick's Cathedral, which has a 14th-century tower, topped with an 18th-century spire.

Dublin has rich links with literature. Jonathan Swift (1667-1745), author of *Gulliver's Travels*, was born in Hoe's Court, and was Dean of the cathedral for 32 years from 1713. The playwright George Bernard Shaw (1856-1950) was born in Synge Street and writer Oliver Goldsmith (about 1728-74) was a student at Trinity College. Nearby is Merrion Square, where another playwright, Oscar Wilde (1854-1900), was born. James Joyce (1882-1941), author of *Dubliners* and *Ulysses*, was born in the city too.

In the library at Trinity College is the *Book of Kells*, an 8th-century illustrated book of the Gospels, and 11th-century Irish king Brian Boru's harp – used as the trademark for Dublin's most famous product, Guinness stout, which is brewed near Hueston Station, one of the main rail terminals. Over Hueston Bridge is the road to Phoenix Park – at 7.1 km² (2.75 sq miles) the largest urban park in Europe.

Dublin's strategic position facing Britain, its harbour, airport and road network, make it the focal point of Irish trade and industry. Electrical goods, metals, food and printed materials are produced, and it is the major ferry terminal for travellers in and out of the country and the starting point for most tourists.

The surrounding county of Dublin has an area of 922 km² (356 sq miles).

Population (county) 1 002 000; (city) 525 400

Map Ireland Cb

Dubrovnik (Ragusa) ***Yugoslavia*** Medieval port and tourist resort on a rocky headland in Dalmatia, 145 km (90 miles) south of Sarajevo. The old city, defended by the 16th-century Revelin fortress to the east, is for pedestrians only. Its double ramparts with 20 towers and bastions enclose a Baroque cathedral and exquisite churches, monasteries, palaces, fountains and red or yellow-roofed houses.

Dubrovnik became fabulously wealthy from trade during the Middle Ages, and was virtually an independent city-state from 1205 to 1808 when it was conquered by Napoleon. The Congress of Vienna assigned it to Austria in 1815. Known as the 'South Slav Athens', it was renowned for its art and literature from the 15th to 17th centuries. It has several museums, a

▼ HONOURED PAST A new mosque built in the traditional local limestone to a conventional Arab design stands in a modern suburb of Dubai – a city made magnificent on large oil revenues.

cable car up Mount Srdj behind the city, an international airport at Cilipi to the south-east, a rail link with Sarajevo and ferries to Venice, Rijeka, Split and Piraeus.
Population 31 200
Map Yugoslavia Dc

Duchcov ***Czechoslovakia*** See TEPLICE

Duero ***Spain*** See DOURO

Duisburg ***West Germany*** The largest inland port in Europe, at the junction of the Rhine and Ruhr rivers 18 km (11 miles) west and slightly south of Essen. The canals of the Ruhr valley carry iron ore and coal to the huge steel plants, chemical works and shipbuilding yards situated alongside the port. The Flemish map-maker Gerhard Mercator (1512-94), who invented the map projection named after him, is buried in the town.
Population 541 800
Map West Germany Bc

Dukhan ***Qatar*** Oil town situated 80 km (50 miles) west of the capital, Doha, in the centre of the Dukhan oil field. The field produces about half of the country's oil output of some 85 million barrels a year. The first well was drilled in 1940, but because of the Second World War exports did not get underway until 1949. The field measures about 55 km (34 miles) by 9 km (6 miles) and has estimated reserves of 2400 million barrels. The Khuff natural gas field started production in 1978, and a liquefaction plant processes the gas for export.
Map Qatar Ab

Duluth ***USA*** Industrial port in Minnesota, at the western end of Lake Superior. It handles grain, iron ore, oil and coal, and makes steel.
Population (city) 85 600; (metropolitan area) 253 800
Map United States Ha

Dumaguete ***Philippines*** See NEGROS

Dumbier ***Czechoslovakia*** See TATRA MOUNTAINS

Dumfries and Galloway ***United Kingdom*** Region of south-west Scotland, covering 6475 km² (2500 sq miles) and taking in the former counties of Dumfries, Kirkcudbright and Wigtown. The south coast, along the Solway Firth, is corrugated with bays, many of which shelter small resorts. The port of Stranraer on Loch Ryan is the principal crossing point to Northern Ireland. Inland, the hill country – chequered with forestry plantations – rises to 843 m (2766 ft). Dumfries (population 29 000), on the River Nith, is the county town.
Population 146 000
Map United Kingdom Cc

Dumyat (Damietta) ***Egypt*** Town on the Dumyat river, the main eastern branch of the Nile delta, 13 km (8 miles) from its mouth. It makes textiles from the cotton grown in the delta region. A new port complex is under construction with 27 berths planned, some of which are already in use.
Population 110 000
Map Egypt Bb

Dún Dealgan ***Ireland*** See DUNDALK

Dún Laoghaire ***Ireland*** Port in the south-east of the Dublin conurbation, and the car-ferry terminal for the Holyhead (Wales) service. The harbour designed by John Rennie was started in 1817 as a famine relief work and completed by convicts. *Dún Laoghaire*, which is Gaelic for Dun Leary, fort of Leary, was known as Kingstown after George IV's visit in 1821. It resumed its original name with Gaelic spelling in 1920.
Population 54 500
Map Ireland Cb

Dún na Ngall ***Ireland*** See DONEGAL

Dünaburg ***USSR*** See DAUGAVPILS

Dunajec ***Poland*** River of southern Poland, 247 km (153 miles) long. It rises in the Tatra Mountains and flows north to join the Vistula river about 60 km (37 miles) north-east of the city of Cracow. The Dunajec cuts a spectacular gorge, up to 300 m (985 ft) deep, through the Pieniny hills.
Map Poland Dd

Dunaújváros ***Hungary*** The country's first and largest postwar new town, on the west bank of the Danube river 58 km (36 miles) south of Budapest. It was formerly the small, sleepy village of Dunapentele. However, from 1950 it was transformed into the major iron, steel and cement town of Sztalinvaros, named after the Soviet leader, Joseph Stalin (1879-1953). The name was changed after 1956, when Stalin was disgraced. A museum traces the town's planning and growth.
Population 65 000
Map Hungary Ab

Dunbartonshire ***United Kingdom*** See STRATHCLYDE

Dundalk (Dún Dealgan) ***Ireland*** Market town and port 77 km (48 miles) north of Dublin near the border with Northern Ireland.
Population 25 600
Map Ireland Ca

Dundas Strait Sea passage between Melville Island and the Cobourg Peninsula in north-central Australia linking the TIMOR SEA to the Van Diemen Gulf. It was named after a short-lived convict settlement, Fort Dundas, on Melville Island, which was itself named after Admiral Henry Dundas, 1st Viscount Melville (1742-1811).
Dimensions 64 km (40 miles) long, 29 km (18 miles) wide
Map Australia Ea

Dundee ***United Kingdom*** Port on the east coast of Scotland standing on the Firth of Tay, 60 km (37 miles) north of Edinburgh. The site has been occupied since before the birth of Christ, but the growth of the city came with shipbuilding, engineering and jute mills in the 19th century. Today the mills are largely silent, but its other industries are flourishing in the dock area and the suburbs. The rail bridge over the Tay collapsed in a gale in 1879, plunging a train into the river with the loss of 75 lives. The replacement was joined, in 1966, by a road bridge. Dundee is famous for its rich cakes, and (since the 18th century) for orange marmalade.
Population 180 000
Map United Kingdom Db

dune Hill or ridge of wind-blown sand, especially one with no vegetation.

Dunedin ***New Zealand*** South Island port, 306 km (190 miles) south-west of the city of Christchurch. Its industries include ship-repairing, brewing, chemicals and furniture. It exports frozen meat, wool and dairy products. The town was founded in 1848 by Scottish Presbyterians and became the country's leading settlement when gold was discovered in the area in 1861.

Its wealth was used to put up many fine public buildings, including First Church (1868-73), the university (1878) – modelled on Glasgow's in Scotland – and the elegant Gothic railway station (1902). There are also several museums depicting Maori culture and life in the days of the early European settlers.
Population 110 000
Map New Zealand Cf

Dunfermline ***United Kingdom*** See FIFE

Dungannon ***United Kingdom*** Market and textile town in County Tyrone, Northern Ireland, about 55 km (35 miles) west and slightly south of the capital, Belfast. Dungannon was the seat of the O'Neills, kings of Ulster for some 500 years until the end of the 16th century. The town's Royal School was founded by James I in 1608.
Population 8500
Map United Kingdom Bc

Dunkirk (Dunkerque) ***France*** Channel ferry port 40 km (25 miles) north-east of Calais. In May and June 1940, during the Second World War, some 225 000 soldiers of the retreating British Expeditionary Force and 112 000 French and Belgian soldiers escaped from its beaches in an armada of small boats in the face of the German army's advance. The town was almost completely destroyed by bombing in 1940 and has been rebuilt as one of France's most carefully planned industrial and residential towns. Major industries include steel and aluminium manufacturing, and oil refining.
Population 196 600
Map France Ea

Dunkwa ***Ghana*** One of the largest towns in the thinly populated south-west, about 170 km (105 miles) west and slightly north of the capital, Accra. Dunkwa is the chief town of the Denkyira people and a market for firewood, plantains, rice, cassava, vegetables and palm wine. It prospered after the Dunkwa-Awaso railway opened in 1944 to carry bauxite (aluminium ore) from near Awaso, timber, and gold from mines at Bibiani.
Population 15 000
Map Ghana Ab

Durance ***France*** River of Provence 380 km (236 miles) long. It flows through the southern part of the French Alps to join the Rhône near Avignon.
Map France Fe

Durango ***Mexico*** State in northern Mexico crossed by the Sierra Madre mountains in the west. It has huge reserves of gold, silver and iron. One of the largest iron-ore deposits in the world lies at Cerro el Mercado, a hill just north of the state capital, Victoria de Durango. During Spanish colonial rule, it was a wealthy religious and political centre. It is now a centre for the surrounding mining and farming district.

Population (state) 1 200 000; (city) 209 000
Map Mexico Bb

Durban ***South Africa*** Major port and largest city of the province of Natal, on South Africa's Indian Ocean coast. It was first called Port Natal, but was renamed in 1835 after Sir Benjamin D'Urban, then the governor of Cape Colony. Durban is a tourist resort as well as an industrial city, with fine beaches and wildlife reserves. Its black townships, hit by political unrest in the mid-1980s, include KwaMashu, Inanda, Phoenix and Umlazi.

Population 960 800
Map South Africa Db

Durham ***United Kingdom*** City and county of north-east England. The city, 22 km (14 miles) south of Newcastle, is one of the most spectacular in Britain, set in a narrow loop of the River Wear, whose deep wooded gorge separates the old and new towns. It is dominated by its massive Norman cathedral, burial place of the English historian and monk the Venerable Bede (673-735), which towers over the castle and the narrow streets of the medieval town. The castle is now part of the city's university.

The county covers an area of 2436 km² (941 sq miles). Its economy in the 18th and 19th centuries was dominated by coal, but mining has declined and the steel industry at the town of Consett has been closed down. Parts of the county were cut off in 1974 to form the new counties of CLEVELAND and TYNE AND WEAR.

Population (county) 607 000; (city) 26 000
Map United Kingdom Ec

Durmitor ***Yugoslavia*** National park covering 320 km² (124 sq miles) between the torrential Piva and Tara rivers in the KARST of central Montenegro. Mount Durmitor itself, scene of the Montenegrin uprising in 1941, is a table-like plateau rising to 2522 m (8274 ft). It is speckled with lakes, forests and, in parts, snow for year-round skiing. To the east lies Zabljak, at 1450 m (4757 ft) Yugoslavia's highest town, with traditional log cabins and modern hotels. To the west is the Mratinje hydroelectric dam.

Map Yugoslavia Dc

Durrës ***Albania*** Port founded in the 7th century BC by the Greeks on a sheltered Adriatic bay. It was taken successively by the Illyrians (312 BC), Romans (230 BC), Ostrogoths (AD 481), Bulgars (10th century), Byzantines (12th century), Venetians (1394) and Turks (1501) who held it until 1912. It is now Albania's leading port, but retains its medieval walls and towers. Durrës serves the country's main industrial region: railways run to Tiranë and Elbasan, and the port is the focus for grain mills, and rubber, fertiliser, machinery and shipbuilding industries.

Population 75 900
Map Albania Bb

Dushanbe ***USSR*** Capital of Tadzhikistan, a modern industrial city 150 km (95 miles) north of the border with Afghanistan. From 1929 to 1961 it was called Stalinabad (in honour of the Russian leader Joseph Stalin, 1879-1953), and grew from a village of 6000 inhabitants in 1926 to the well-laid-out city of today. Its main industries are cotton and silk textiles, engineering and food processing.

Population 539 000
Map USSR He

Dusky Sound ***New Zealand*** Inlet on the Fiordland coast of South Island about 310 km (193 miles) west of Dunedin. It was one of the anchorages used by the English explorer Captain James Cook in 1773. Australian-based sealers arrived there about 1792 and built the first European-style houses in the country.

Map New Zealand Af

Düsseldorf ***West Germany*** City on the river Rhine about 34 km (21 miles) north and slightly west of Cologne. It is the commercial and cultural centre for the industrial Ruhr, with a progressive film industry and an avant-garde art movement; it was the birthplace of the poet Heinrich Heine (1797-1856). Its main street, the Königsallee, has a lake in the middle, part of the old town moat. Northern Germany's main stock exchange is situated in Düsseldorf. Its industries include engineering and chemicals.

Population 579 800
Map West Germany Bc

dust Small particles of matter which are fine enough to be carried by the wind. There are various sources of dust, such as volcanic activity, forest fires, and smoke from industry.

dust bowl Barren area in semiarid regions produced by excessive wind erosion of the soil, especially after the removal of vegetation by overgrazing or badly managed cultivation. The dust bowl that developed in the United States in the 1930s stretched from Kansas through Oklahoma, Texas, Colorado and into New Mexico.

dust devil Small, short-lived whirlwind that swirls dust, debris and sand up into the air. It is caused by intense local heating of the earth's surface resulting in convection.

Dux ***Czechoslovakia*** See TEPLICE

Dvina ***USSR*** Name of two unconnected river systems in the USSR. The North (Severnaya) Dvina drains a large part of northern European Russia. It is 1320 km (820 miles) long, and is navigable between May and November from Archangel on the White Sea to the industrial town of Kotlas, 470 km (290 miles) upstream. Once used by trappers, it is now used by forestry companies to transport timber by raft.

The West (Zapadnaya) Dvina flows 1020 km (635 miles) from the Valdai hills to the Baltic Sea at Riga. It was part of the 'water road' used by Viking traders travelling south-east from Scandinavia to Turkey in the Middle Ages, but is now little used. Both rivers flood widely when the ice melts in April.

Map (North Dvina) USSR Fb; (West Dvina) USSR Dc

Dvinsk ***USSR*** See DAUGAVPILS

Dwarka ***India*** One of Hinduism's seven holy cities and legendary capital of the god Krishna, set on the west coast about 510 km (317 miles) north-west of Bombay. It has a temple of Krishna, who is said to have sought shelter there after fleeing from Mattra. The temple is closed to non-Hindus.

Population 21 400
Map India Ac

Dyfed ***United Kingdom*** County of south-west Wales covering 5765 km² (2226 sq miles). It was created in the 1970s from the counties of Cardigan, Carmarthen and Pembroke and named after an old Celtic kingdom. Hill farms and open moorland of the Cambrian Mountains dominate the area, but it is the coastline for which Dyfed is famous. MILFORD HAVEN is one of Britain's finest and deepest natural harbours, and the cliffs to the west are some of the most spectacular in Britain. Aberystwyth and Tenby are the main seaside resorts. There are two university colleges, at Aberystwyth and Lampeter, and the former town houses the National Library of Wales. Britain's smallest city, St David's, with its cathedral dedicated to the patron saint of Wales, stands at the most westerly point. According to Welsh legend, the great Celtic wizard Merlin was born at Carmarthen, now the county town.

Population 377 000
Map United Kingdom Cd

dyke 1. Earth and rock embankment built to protect low-lying land from flooding. Examples include those protecting the POLDERS of the Netherlands, totalling 1300 km (over 800 miles). **2**. A LEVEE. **3**. A drainage ditch or watercourse. **4**. Man-made earthwork, such as Offa's Dyke on the border of Wales, originally a boundary. **5**. A sheet-like mass of igneous rock that cuts through the horizontal structure of the rock into which it was intruded.

Dzerzhinsk ***USSR*** Industrial city – a centre of the chemical industry – dating from the 1940s, 30 km (18 miles) west of Gor'kiy. It was built as an 'overflow' for Gor'kiy, and its population consists largely of workers from the factories which line the nearby Oka river and the Moscow-Gor'kiy railway.

Population 272 000
Map USSR Fc

Dzhambul ***USSR*** Industrial town in southern Kazakhstan, 250 km (155 miles) north-east of Tashkent. Its main products are metals, fertilisers, processed food, refined sugar and textiles.

Population 298 000
Map USSR Id

Dzierzoniow (Reichenbach) ***Poland*** Textile town in south-west Poland, 48 km (30 miles) south-west of the city of Wroclaw. It was a flourishing maker of linen by 1500, and with adjoining Bielawa became Lower Silesia's main textile centre from 1700.

Population 37 600
Map Poland Bc

Dzungaria ***China*** See XINJIANG UYGUR AUTONOMOUS REGION

E layer See ATMOSPHERE

eagre See BORE

earth See HOW THE EARTH WAS FORMED (opposite) and JOURNEY TO THE CENTRE OF THE EARTH (p. 200)

earth pillar Column of soft, earthy material such as clay, capped by a boulder which has protected it from erosion. It can be up to 60 m (about 200 ft) tall.

earthquake Series of shock waves, which travel through the earth and along its surface, caused by a sudden release of energy in the CRUST or the upper MANTLE. Some earthquakes are so slight that they can be detected only by the most sensitive instruments; others, such as the Assam earthquake of 1897, which devastated 388 000 km^2 (150 000 sq miles) of north-east India and was felt over an area of almost 11 million km^2 (4.25 million sq miles), can unleash the energy of several nuclear bombs.

Many shallow earthquakes are caused by movements of FAULTS in the earth's crust, while more severe ones occur at much greater depths related to SUBDUCTION ZONES below the earth's surface. How these deeper earthquakes are produced is not yet known. Earthquake zones or belts coincide most often with PLATE MARGINS, which suggests that the zones are closely related to PLATE TECTONICS. One of the belts, the so-called Ring of Fire, encircles the Pacific. The second belt stretches west through China along the Himalayas and across Iran and then to the north and south of the Mediterranean.

There are two ways of measuring an earthquake – by its intensity, or size, and by its magnitude, or strength. Intensity is determined by assessing the degree to which shaking is felt by people in the area, the amount of damage to buildings and other man-made structures, and the extent of visible deformation of the earth. The scale most commonly used to measure intensity is the MODIFIED MERCALLI SCALE.

Earthquake magnitude is determined by the size of the seismic waves, that is the vibrations emitted by the release of energy. The waves are measured on a RICHTER SCALE.

The centre of an earthquake is called the focus and it is from the focus that seismic waves are emitted. All foci so far located have been in the upper 720 km (440 miles) of the earth.

The earthquake's epicentre is the point on the earth's surface directly above the focus. Intensity decreases outwards from the epicentre, but since buildings constructed on soft soil suffer more damage than buildings on hard rock, two separate points at the same distance from the epicentre can have a different intensity. The measurement of intensity therefore varies from place to place. Magnitude, on the other hand, is a single measurement for each earthquake.

MEASURING AN EARTHQUAKE'S MAGNITUDE

About a million earthquakes occur throughout the world each year, and their magnitude is measured on the Richter scale devised in 1935 by the Californian seismologist, Charles Richter. The scale is logarithmic, so that an increase of one point means that the force of the earthquake is ten times greater than the preceding number. An earthquake with a magnitude of 8, therefore, is ten times more powerful than one of magnitude 7, and 100 times more powerful than one of magnitude 6.

MAJOR EARTHQUAKES OF THE 20TH CENTURY

Place	Magnitude	Deaths	Date
China	8.6	180 000	1920
Japan	8.6	48	1968
Alaska	8.5	178	1964
Chile	8.5	4000-5000	1962
China	8.3	200 000	1927
Japan	8.3	99 000	1923
USA (San Francisco)	8.3	700	1906
Mexico	8.1	12 000	1985
China	8.0	655 000	1976
Iran	8.0	189	1977
Ecuador	7.9	600	1979
New Zealand	7.9	255	1931
China	7.8	240 000	1976
Algeria	7.7	3500	1980
Iran	7.7	15 000	1978
Peru	7.7	50 000-70 000	1970
USA (California)	7.7	11	1952
Italy	7.5	83 000	1908
Pakistan	7.5	20 000-60 000	1935
Chile	7.4	177	1985
Iran	7.4	12 000	1968
Italy	7.2	3000	1980
Yugoslavia	6.0	1000	1963

East Anglia ***United Kingdom*** Area of eastern England comprising the counties of NORFOLK and SUFFOLK, and parts of CAMBRIDGESHIRE and ESSEX. It was an Anglo-Saxon kingdom and is now one of Britain's leading producers of wheat and other cereals. It is relatively flat and thinly populated.

Map United Kingdom Fd

East London ***South Africa*** Seaport, founded in 1836 at the mouth of the Buffalo river. It serves the farming region of south-eastern Cape Province, Ciskei and Transkei. The coastline near East London contains superb beaches whose waters are warmed by the Mozambique current.

Population 164 200

Map South Africa Cc

East Rand ***South Africa*** Part of the Witwatersrand east of Johannesburg. It became a major gold-mining area after 1923, when the mines in the Central Rand became exhausted – and is now a thriving industrial area.

East Siberian Sea Part of the ARCTIC OCEAN lying between the coast of north-eastern SIBERIA, Wrangel Island and NOVOSIBIRSKIYE OSTROVA (New Siberian Islands). Its northern limit is the edge of the Arctic continental shelf. The East Siberian Sea is only navigable in August and September, when it is ice-free.

Map USSR Pa

Easter Island (Isla de Pascua; Rapa Nui) ***Pacific Ocean*** Tiny, remote volcanic island of 120 km^2 (46 sq miles) in the Pacific, 3780 km (2350 miles) from the South American coast. It was discovered in 1722 by the Dutch admiral Jacob Roggeveen, who named it after the day he landed, and was annexed by Chile in 1888. The island has some of the most fantastic monuments in the world – over 600 strange stone faces, some 20 m (65 ft) tall, which were carved by the islanders' Polynesian predecessors over 1000 years ago; together with other carved stonework, they are all that remain of their ancient civilisation. An airport was opened in 1967, and flights link the island with Tahiti and with Santiago, Chile.

Population 1300

Map Pacific Ocean Hd

ebb tide Receding tide between high water and the succeeding low water.

Ebolowa ***Cameroon*** Market town about 110 km (70 miles) south of the capital, Yaoundé, on the main road to Gabon. It is a collection centre for timber and cocoa for export, and its future depends largely on improving the quality of cocoa from smallholdings where trees, frequently aged and diseased, need to be replaced. The town is also a tourist centre for the surrounding forests, rich in animal and bird life, and the spectacular mountain scenery of Ako Akas and Mbilibékon abyss.

Population 30 000

Map Cameroon Bb

HOW THE EARTH WAS FORMED

About 4600 million years ago, part of a vast nebula of dust and gas surrounding the sun began to condense into a ball of matter – planet earth.

There are several theories about exactly how this happened. But it is generally assumed that particles collided, or were attracted to each other by electrostatic action, to form a nucleus which became progressively more dense. As this gaseous mass grew, it began to exert a gravitational pull, which in turn attracted more material. As gravity increased, it drew heavier matter to the centre, while lighter materials collected around. The whole mass condensed into liquid form, then began to solidify.

Whether this mass was hot or cold is also the subject of debate. Some hold that it began cold, but heated up to a molten state as materials at the centre became more and more compressed, and radioactive decay set in. Others believe that the primitive mass coalesced into a molten ball, which has been cooling ever since.

However, there is no doubt that earth was once extremely hot, and then cooled. As it did so minerals started to crystallise and globules of iron and nickel sank to the centre, where a core of great density was forming. Outside this core, the next layer of material was building up – silicates and sulphides which would later be compressed into the dense mantle of rock which now surrounds the core. Finally the lightest materials – more silicates – rose to the surface of the ball and, as they cooled, hardened into earth's crust.

By this time the lighter gases had floated off into space again and others had been absorbed, leaving almost no atmosphere. So, with a minimum of atmospheric friction to burn them up, enormous meteorites plunged unhindered through the newly forming crust to add even more materials to the mixture. But all the time the crust was deepening and becoming more solid. Gases, including carbon monoxide and carbon dioxide, escaped from the interior furnaces to create an atmosphere of sorts. The crust in turn developed in two layers – a top one of lighter rocks, with a denser layer underneath.

It is thought that at this stage a vital new ingredient was exuded into the atmosphere – water vapour. As earth continued to cool, and the atmosphere with it, this water vapour condensed and fell as rain.

Meanwhile, surface rock was cracked and melted again and again, as earth's seething interior erupted in volcanic spasms. But gradually, over millions of years, large, stable rafts of granitic material accumulated, floating around on the molten matter beneath them. Some collided and joined up – the primitive continents. Rains falling on these landmasses formed lakes and streams, and flowed down onto the low-lying areas between the continents to form the first oceans.

Materials carried and dissolved into the oceans mixed in a salty soup of – among many elements – boron, bromine, chlorine, iodine and sulphur. In the seas and in the thicker soups of warm, volcanic pools, primitive living cells developed. And about 2000 million years ago, plant cells capable of photosynthesis evolved. They began adding oxygen to the atmosphere, in which an ozone layer built up to defend earth from radiation, so allowing a rapid expansion of animal life.

Dust and gas in the solar nebula condense into a molten ball. Gases are absorbed or lost into space, so there is no atmosphere.

As earth cools a crust forms, but intense volcanic activity in the super-heated regions below expels gases in large quantities through the crust, to create a primitive type of atmosphere – but one which still lacks oxygen.

Further cooling frees water vapour into the atmosphere, and finally temperatures drop sufficiently (to 100°C or 212°F) for the thick clouds to condense into rain. Great lightning storms charge the atmosphere with electricity and the rains gather new chemicals as they fall to earth. Organic compounds are formed by lightning and radiation.

The storms subside and as water evaporates organic compounds are concentrated. Self-sustaining cells evolve. Plant cells capable of photosynthesis follow, and add growing quantities of oxygen into the air. High in the atmosphere a belt of ozone forms, to shield earth from damaging solar radiation – and allow its new life forms to multiply.

Ebro ***Spain*** River in the north-east, 909 km (565 miles) long, flowing from the Cantabrian Mountains to the Mediterranean south of Tarragona. The Battle of the Ebro in 1938, in which 75 000 Republicans died, was a decisive victory for the Nationalists in the Civil War which ended a year later.
Map Spain Eb

Echmiadzin ***USSR*** Ancient town and Christian religious centre in Armenia, 15 km (9 miles) north of the Turkish border; formerly Vagarshapal. It was founded in the 6th century BC, and its magnificent cathedral was originally built in AD 301. Christ is said to have descended from heaven and struck the site with a hammer of gold, and the cathedral – rebuilt in the 6th to 7th centuries – attracts thousands of pilgrims each year.
Population 37 000
Map USSR Fd

Écrins, Parc des ***France*** Largest of the French national parks, covering 918 km² (354 sq miles), about 60 km (37 miles) south-east of Grenoble. It contains outstanding alpine glaciers, lakes and gorges, with four peaks rising close to 4000 m (13 120 ft): Barre des Écrins (4102 m, 13 458 ft); La Meije (3983 m, 13 067 ft); L'Ailefroide (3953 m, 12 969 ft); and Pelvoux (3946 m, 12 946 ft).
Map France Gd

Eden, Mount ***New Zealand*** Highest of the volcanic cones around Auckland on North Island, rising to 196 m (643 ft) and giving panoramic views across the city.
Map New Zealand Eb

Eder ***West Germany*** River rising in the Rothaer hills north-west of Frankfurt am Main and flowing east then north for about 175 km (110 miles) to join the Fulda river near the town of Kassel. A dam built on the river and another on the Möhne, both providing hydroelectric power, were destroyed in 1943 by Britain's Royal Air Force in the 'Dambusters' raid, using a bouncing bomb invented by the British engineer Barnes Wallis (1887-1979).
Map West Germany Cc

Edinburgh ***United Kingdom*** Capital of Scotland, near the southern shore of the Firth of Forth. Writers and visitors long ago exhausted the adjectives to describe the city: the majesty of its castle, the quality of its cultural life – which led to its nickname, the Athens of the North – even the bitterness of the wind when it blows off the North Sea.

Although it has been without a resident monarch since 1603 or a Scots parliament since 1707, it is not short of splendid buildings. Parts of the castle date back to about 1100. Holyroodhouse was the seat of the last *(continued on p. 202)*

JOURNEY TO THE CENTRE OF THE EARTH

Earth's outer skin, called the crust, is a complex structure forming part of the lithosphere. The continental crust is 92 per cent igneous or metamorphic rock, much of it granitic, and 8 per cent sedimentary rock. Beneath extensive flat areas of land such as the North American prairies, the crust is about 30-35 km (19-22 miles) thick. But under mountain ranges it increases to as much as 80 km (50 miles). The crust under oceans is largely basaltic with a thin layer of sediment, and is only about 5 km (3 miles) thick.

Separating the crust from the next layer, the mantle, is a zone known as the Mohorovičić discontinuity in which primary shock waves from EARTHQUAKES speed up dramatically.

The mantle extends to a depth of 2900 km (1800 miles). The uppermost mantle is solid, but between depths of 75-250 km (about 50-155 miles), there is a semifluid transition zone (the asthenosphere). The mantle below is solid again and is called the mesosphere.

Beneath the mantle lies the core. The outer layer – about 2100 km (1305 miles) thick – is molten and probably consists of nickel-iron. It is in constant motion, and electrical currents generated and circulating within it account for earth's magnetic field. The inner core of the earth, about 2740 km (1700 miles) in diameter, is believed to be a solid ball of nickel-iron squeezed to unimaginable densities by a pressure of 3800 tonnes per cm² at a temperature of around 4000°C (7200°F).

THE LAYERS OF THE EARTH

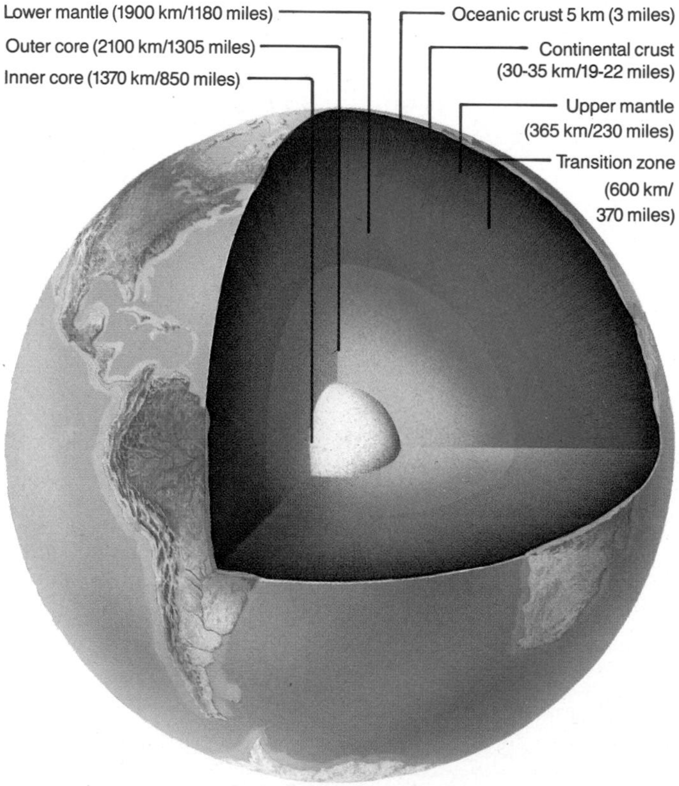

SOLID AND SEMI-FLUID LAYERS

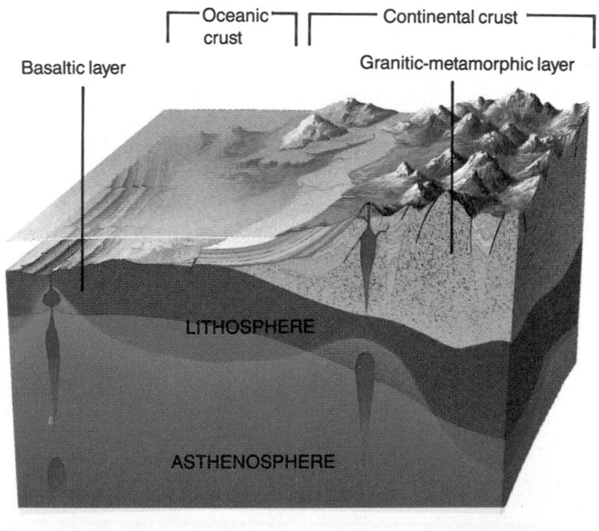

An alternative division of earth, emphasising physical rather than chemical properties, shows the solid lithosphere (which includes the crust and part of the upper mantle) surrounding a semi-fluid asthenosphere. This in turn surrounds a solid mesosphere, which goes down to the core. The asthenosphere lies between depths of 75 km (47 miles) and about 250 km (155 miles). It is also known as the low velocity layer, because the speed at which the seismic waves of earth tremors travel – generally increasing with depth – is slowed by a few per cent in the asthenosphere. However, seismic waves speed up as they cross the Mohorovičić discontinuity, between the crust and mantle, and as they penetrate the transitional zone.

Ecuador

WHERE POLITICS FLARE UP AS FREQUENTLY AS THE VIOLENT ERUPTIONS FROM THE VOLCANOES THAT MARK THE LANDSCAPE

The country's political scene is almost as volatile as the landscape, which contains over 30 active volcanoes. Like the volcanic eruptions in the mountains, which sometimes shower nearby villages with hot ashes, there are regular explosions as the military powers usurp a civilian government, hold power for a period, and then are replaced by another civilian government.

The main source of political strife is related to geography. In effect, Ecuador is two countries, west and east, with different climates, different terrain, and different customs and attitudes. The west of the country is a hot coastal plain, and is the seat of industry and commerce as well as an agricultural region producing the major exports of bananas, coffee and cocoa. In this region lies the port of GUAYAQUIL, Ecuador's largest city, now much modernised by an oil boom, resulting from a discovery in the early 1970s. The coastal people are go-ahead, willing for change and interested in economic progress. But their neighbours to the east feel differently.

Running down the middle of Ecuador, from north to south, are two ranges of the ANDES, divided by a central plateau known as the Sierra. The plateau is broken into small, isolated river basins where most of the people live.

Life in the mountains is slower, and less obviously affected by recent change and modernisation. The capital city of QUITO retains the air of a Spanish colonial town, and the Indian villages of the surrounding country seem to belong to an even earlier age. The inhabitants of the Sierra are conservative, by comparison to the coastal people, and the two groups consistently support different political groups. The problem for every government is to reconcile these opponents.

The eastern oil region is yet another Ecuador, with only 2 per cent of the population and so underdeveloped that it does not take much part in the political dramas of the coast and the Sierra. The area, known as the ORIENTE, consists of tropical savannah and dense jungle around the headwaters of the AMAZON. East of Baños lies some of the last really wild, unexplored territory in the world. The region is valuable because, apart from oil, it offers major opportunities for exploitation, both of the forest (Ecuador already exploits balsa wood and rubber) and its minerals.

Ecuador's greatest problem is poverty. It has always been one of South America's poorest countries, but when oil was discovered in the 1970s this seemed to promise salvation. It did bring about rapid growth, especially in the cities, but its benefits did not reach all the population. As elsewhere in South America, recession in the 1980s brought expansion to a sudden halt, and today Ecuador faces the common Latin American diet of debt (over US$7600 million), financial advice from the International Monetary Fund, and drastic devaluations of the currency – by 14 per cent in early 1986.

Furthermore, those who did not profit from the oil boom while it lasted, still suffer. Impoverished rural Indian families leave the countryside in the hope of making money and build makeshift homes on the fringes of cities such as Guayaquil which is surrounded by shanty towns. While 1 per cent of Ecuador's landowners have large estates comprising 40 per cent of the land, three-quarters of all land holdings are smaller than 5 hectares (12 acres) – too small for the average Indian farmer to make a living. Those with plots on the steep hillsides of the Andes also have to cope with destructive soil erosion. In parts of the highlands, life is particularly hard; infant mortality amongst Indians is high (9 per cent) and life expectancy low (under 50 years).

About 25 per cent of Ecuadoreans are pure Indian, but they have very little share of power or wealth. Once the Indians had a great empire: in the early 16th century the Incas invaded from Peru, overwhelming the local Indian tribes. But the Inca rule was short-lived; the Spanish arrived in 1534 and made the Indians slaves for three centuries. When independence from Spain was achieved, it merely meant the transfer of power from the Spanish to the *mestizos* – the mixed-blood descendants of the conquerors. Today, the mestizos make up 55 per cent of Ecuador's population of 9.1 million, and they are still the privileged group.

▲ WOMEN AT WORK Indian women often work as shepherds in the Andes of Ecuador, and children are carried to work on their mothers' backs. Life is hard, and many Indians chew coca leaves – a source of cocaine – to appease their hunger.

ECUADOR AT A GLANCE

Area 283 561 km² (109 483 sq miles)

Population 9 120 000

Capital Quito

Government Parliamentary republic

Currency Sucre = 100 centavos

Languages Spanish, Indian languages

Religion Christian (96% Roman Catholic, 2% Protestant)

Climate Tropical; cooler at altitude. Average temperature in Quito (altitude 2850 m/9350 ft) ranges from 8°C (46°F) to 21°C (70°F)

Main primary products Rice, maize, cassava, potatoes, bananas, oranges, coffee, cocoa, sugar cane, fish; crude oil and natural gas

Major industries Agriculture, crude oil production and refining, cement, petrochemicals, food processing

Main exports Crude oil, cocoa and products, bananas, coffee, processed fish

Annual income per head (US$) 1292

Population growth (per thous/yr) 27

Life expectancy (yrs) Male 60 **Female** 64

Scottish monarchs (including Mary, Queen of Scots, who was executed in England in 1587), and is now owned by the British Crown. The city is also rich in artistic attractions, with its National Gallery, Royal Scottish Academy, Royal Scottish Museum and Museum of Modern Art. There are two universities, three cathedrals and a number of theatres, but no opera house – Edinburgh people must travel to Glasgow for that. Each year in August and September, the Edinburgh International Festival fills not only these theatres but every church hall, basement and, indeed, many pavements with performers and audiences.

The Old Town's spine is the 'Royal Mile, – Lawnmarket, the High Street and Canongate – which runs down from the castle, past St Giles Cathedral, built in the 14th and 15th centuries, to Holyrood. The New Town lies to the north. It dates from the 18th and early 19th centuries and is laid out in formal streets (notably Princes Street), squares and crescents. North again is the port of Leith. Arthur's Seat, a 251 m (823 ft) hill of volcanic rock, overlooks the city.

Population 439 000

Map United Kingdom Dc

Edirne ***Turkey*** Town on the Greek border about 250 km (155 miles) west of Istanbul. Until the 1930s it was called Adrianople after the Roman Emperor Hadrian (AD 76-138). It is dominated by the Selimiye mosque and its four large minarets. The town's main industries are textiles, carpets, leather goods, soap and perfumes. The surrounding province of Edirne, which covers 6266 km² (2419 sq miles), produces wheat, rye, fruits and rice.

Population (city) 86 700; (province) 388 900

Map Turkey Aa

Edmonton ***Canada*** Capital of ALBERTA and one of Canada's largest cities. It was founded in 1795 as a trading post and fort of the Hudson's Bay Company, and is now the retailing outlet for a rich agricultural area. Edmonton is a major cultural, educational and financial centre. Oil was discovered nearby in 1947.

Population 657 000

Map Canada Dc

Edo ***Japan*** See TOKYO

Edward, Lake (Lake Rutanzige) ***Uganda/Zaire*** Lake covering about 2135 km² (820 sq miles) in the Great Rift Valley south of the Ruwenzori mountains, at a height of about 915 m (3000 ft). The Rwindi and Rutshuru rivers flow into it and the Semliki river drains it, flowing north into Lake Albert; hence Lake Edward is one of the sources of the Nile.

It was explored in 1889 by the British adventurer Sir Henry Morton Stanley, who named it after Albert Edward, the then British Prince of Wales. Uganda's President Amin renamed it Lake Idi Amin Dada, but the former name was restored when Amin was deposed in 1979. There is no navigation on the lake.

Map Zaire Bb

Éfaté ***Vanuatu*** See PORT-VILA

Efes ***Turkey*** See EPHESUS

Eger ***Czechoslovakia*** See CHEB

Eger ***Hungary*** Beautiful Baroque city on the southern slopes of the Bükk hills, 40 km (25 miles) south-west of the north-eastern city of Miskolc. It is the chief town of Heves county, and the centre of a wine-making district which produces Egri Bikaver (Bull's Blood) wine.

The Turks occupied Eger from 1596 to 1687, and it has Europe's northernmost surviving minaret. It also has a 16th-century castle standing at the end of a street of exquisite 17th and 18th-century palaces and houses. The city's cathedral, rebuilt early this century, is one of Hungary's largest churches. The city is also a medicinal spa for rheumatic sufferers.

Population 63 600

Map Hungary Bb

Egmont, Mount ***New Zealand*** Dormant volcano rising to 2518 m (8261 ft) on North Island, about 220 km (140 miles) north and slightly west of the capital, Wellington. It is noted for its symmetry and beauty. According to Maori legend, the cone rises in splendid isolation from other volcanoes on the island because of a lovers' quarrel between Taranaki (Mount Egmont) and Mount Tongariro.

Map New Zealand Ec

Egypt See p. 204

Eifel ***West Germany*** Volcanic highland between the lower Moselle river and the borders of Luxembourg and Belgium. It is a sparsely inhabited region of moors and forests rising to more than 700 m (2300 ft), though damming parts of the Roer river has created lakes which attract visitors.

Map West Germany Bc

Eiger ***Switzerland*** Mountain 3970 m (13 025 ft) high, about 60 km (35 miles) south-east of the capital, Berne. Its rugged north face – the Nordwand – is notorious for the number of climbers killed while trying to scale it – more than 50 between the first ascent in 1938 and 1986 – and has been nicknamed *mordwand* – 'murder face'.

Map Switzerland Ba

Eilat ***Israel*** See ELAT

Eindhoven ***Netherlands*** Industrial city near the Belgian border, about 110 km (70 miles) south and slightly east of Amsterdam. It is the headquarters of Philips, the electrical and electronics firm, founded there in 1891, and has the Philips Museum and a technical university.

Population (city) 194 600; (Greater Eindhoven) 373 900

Map Netherlands Bb

Einsiedeln ***Switzerland*** Small town about 30 km (19 miles) south-east of Zürich. It is Switzerland's chief place of pilgrimage for Catholics, with the Black Madonna, a 15th-century woodcarving which has survived many fires and attempts at pillage. It is kept in a Benedictine monastery church, a splendid 18th-century Baroque complex. The Protestant Reformer Ulrich Zwingli (1484-1531), parish priest in 1516-18, was angered by what he saw as the superstition surrounding the Madonna.

Population 9000

Map Switzerland Ba

Eisenach ***East Germany*** Resort town in the south-west, 12 km (8 miles) from the West German border. The religious reformer Martin Luther (1483-1546), one of the founders of Protestantism, lived there during his schooldays. It was also the birthplace of the composer Johann Sebastian Bach (1685-1750). Its manufactures today include cars, machinery, metal, timber, textiles, electronics and chemicals.

Population 51 000

Map East Germany Bc

Eisenstadt ***Austria*** Capital of Burgenland state, 40 km (25 miles) south of Vienna. Until 1921, Burgenland was part of Hungary. Eisenstadt became its capital in 1925. Until then it had been dominated by the wealthy and influential Esterházy family. In 1761 the Esterházys engaged the composer Joseph Haydn as their musical director, and he worked in the ornate Esterházy Castle for almost 30 years. The castle's Haydn room is open to the public, and the composer himself is buried in the nearby Bergkirche, or town church. Haydn's house is now a museum. The town is the centre of a fertile wine and fruit region.

Population 10 200

Map Austria Eb

Eisleben ***East Germany*** Town about 60 km (35 miles) west and slightly north of Leipzig. The religious reformer Martin Luther (1483-1546) was born there and eventually returned to die there; his homes are now preserved as museums. The town's industries include textiles, furniture, cigars, clothing and copper smelting.

Population 27 400

Map East Germany Bc

Ekaterinburg ***USSR*** See SVERDLOVSK

El Alamein ***Egypt*** Desert village on the Mediterranean coast 104 km (64 miles) south-west of the port of Alexandria. It gave its name to a decisive Second World War battle. There, in October-November 1942, the British general Bernard Montgomery turned back the advance of the German general Erwin Rommel's Afrika Corps troops on Cairo, the Allies' Middle East headquarters. Near the village are English, German and Italian war cemeteries.

Map Egypt Bb

El Borma ***Tunisia*** The country's principal oil field, situated in the far south on the Algerian border. It was discovered in 1966 and produced 3.5 million tonnes a year by 1969 – enough to make Tunisia self-sufficient in petroleum. Crude oil from the field is piped to Sakhira, where some is exported and some transported by sea to BIZERTE for refining and export.

Map Tunisia Aa

El Faiyum (Fayum) ***Egypt*** Nearest oasis to the Nile Valley on a plateau about 100 km (60 miles) south-west of the capital, Cairo. It lies in a fertile, cotton and fruit-growing stretch of the Western Desert. Nearby is Crocodilopolis, the ancient centre of Egyptian worship of the crocodile god, Sebek. El Faiyum was a seat of the pharaohs for a period of the Middle Kingdom, about 1990-1785 BC.

Population 167 080

Map Egypt Bb

El Gezira (Al Jazirah) ***Sudan*** Cotton-growing clay plain covering about 6300 km² (2430 sq miles) between the Blue and White Nile rivers immediately south of Khartoum. The Gezira Scheme, begun in 1925, is Sudan's major irrigation scheme. It involves numerous canals, which are supplied from the Sennar and ER ROSEIRES dams on the Blue Nile. The scheme is based on cotton crops supplemented by groundnuts, wheat and rice (alongside the old-established crops of millet and beans).

Map Sudan Bb

El Ghor Canal ***Jordan*** Important irrigation canal running for 69 km (43 miles) along the east bank of the Jordan river and completed in 1964. It is fed throughout the year with water from the YARMUK river, and has enabled fruit and vegetables to grow extensively in the Jordan Valley.

Map Jordan Aa

El Giza ***Egypt*** Cairo suburb and site of the Egyptian pyramids, built more than 4000 years ago. The pyramids are the oldest of the Seven Wonders of the ancient world and the only one still surviving. The three main pyramids tower over the desert 15 km (9 miles) south-west of the capital.

The largest is the tomb of Cheops, or Khufu, who reigned in Egypt in about 2650 BC. It stands 137 m (450 ft) high and covers an area of 5 hectares (13 acres) at its base. Its sides face directly north, south, east and west. Near it stands the sphinx, carved about 3500 years ago from a single outcrop of sandstone. It is 73 m (240 ft) long and 20 m (66 ft) high, with the body of a lion and the face of Cheops' son, the pharaoh Chephren (Khafre), whose tomb is in another of the pyramids. The third pyramid, less than half the size of the other two, is that of Mycernus, Chephren's son. The three huge monuments are surrounded by numerous other smaller pyramids, temples and tombs of Egyptian nobles and court officials.

The suburb itself contains a traditional Muslim quarter alongside government offices, embassies, the University of Cairo and modern villas. Industry includes textiles, footwear and brewing; much of the country's film industry is located there.

Population 1 230 500

Map Egypt Bb

El Jadida ***Morocco*** Atlantic resort 100 km (60 miles) south-west of Casablanca on the site of an important 16th-century Portuguese trading post; formerly Mazagan. The old town has imposing ramparts from this period that were once considered impregnable. However, it fell to the Moroccans in 1769 after long, hard fighting. Much of the old town remains, including the Portuguese underground reservoir.

Population 82 000

Map Morocco Ba

El Jem ***Tunisia*** Small town halfway between Sousse and Sfax, rising abruptly from the surrounding land. It is famous for its spectacular Roman amphitheatre, the third largest in the world with space for 30 000 spectators, which is better preserved than the Colosseum in Rome.

Population 12 800

Map Tunisia Ba

El Kharga (Al-Kharijah) ***Egypt*** Oasis town about 500 km (310 miles) south-west of the capital, Cairo. It lies in a green basin about 320 km (200 miles) long and 50 km (30 miles) wide on an ancient caravan route from Libya to the Nile Valley and makes cotton textiles. Around the oasis large-scale desert reclamation using deep underground water for irrigation produces dates, cotton, wheat, rice, barley, bananas and vines.

Population 17 000

Map Egypt Bc

El Mansura ***Egypt*** City on the east bank of the Damietta, the main eastern branch of the Nile delta, about 140 km (85 miles) north of the capital, Cairo. It was the site of a battle in AD 1250, at which Crusaders, led by Louis IX of France, were overcome by the Arab army of Turan Shah, effectively ending the Crusades. The city's name comes from an Arabic word meaning 'triumph'. The main industries today are cotton, linen and the making of sailcloth.

Population 323 000

Map Egypt Bb

El Misti ***Peru*** See AREQUIPA

El Niño Name given to the abnormal warming of the surface waters of the eastern tropical Pacific Ocean which has catastrophic effects on the weather. Normally the highest temperatures and lowest pressure are over Indonesia and northern Australia, and the easterly trade winds pile up warm surface water in the western Pacific. When the low-pressure area changes location and the trade winds weaken or reverse, warm water wells eastwards, thickening the warm surface layers off Ecuador and Peru. The phenomenon occurred six times between 1902 and 1985, and because it usually happened at Christmas, Peruvian fishermen called it El Niño ('The Infant' in Spanish).

The warm water displaces the cool waters of the Peruvian Current, with their wealth of microscopic plants and animals on which shoals of anchovies feed, and was the major factor in the collapse of the Peruvian fishing industry in 1972. Another El Niño began in early 1982 and by March 1983 the trade winds had reversed to bring heavy rains and floods that devastated coastal Peru and Ecuador.

El Obeid (Al Ubayyid) ***Sudan*** Capital of Northern Kordofan province, about 370 km (230 miles) south-west of the capital, Khartoum. The town is encircled by a forest reserve to give it some protection from dust storms. It has the world's largest market for gum arabic, which is collected from trees in the region for use as a thickener. It also trades in oil seeds – used to make cooking fats and lubricants – groundnuts, millet and livestock. An Anglo-Egyptian army was wiped out nearby in 1883 by followers of the Sudanese religious leader, the Mahdi.

Population 139 500

Map Sudan Bb

El Oro ***Ecuador*** Province south of the Gulf of Guayaquil on the Peruvian border. Bananas, its main crop, are exported from Puerto Bolívar. The provincial capital is MACHALA.

Population 336 000

Map Ecuador Ab

El Paso ***USA*** City in the extreme west of Texas, on the Rio Grande, which forms the Mexican border. A mission was established nearby in 1682. The city is one of the main crossing points into Mexico.

Population (city) 463 800; (metropolitan area) 526 500

Map United States Ed

El Qahira ***Egypt*** See CAIRO

El Salvador See p. 207

El Sumidero ***Mexico*** Spectacular canyon 42 km (26 miles) long and 1800 m (6000 ft) deep, gouged out by the Río Grijalva in CHIAPAS state. In the 16th century, El Sumidero witnessed the tragic suicides of over 1000 Indians who jumped to their deaths rather than become slaves of the Spaniards. Their act shamed the conquistadores into retreating from the territory to let the Indian people live in peace.

Map Mexico Cc

El Tajín ***Mexico*** Ruined capital of the Totonac civilisation (6th to 10th centuries) in the forests of the humid coastal state of VERACRUZ. It is a vast city, at the centre of which lies the Pyramid of the Nichos. The number of niches in its seven storeys equals the number of days in the calendar year. Every Sunday, circus 'Flyers' attached to wound-up ropes leap from a platform at the top of a 30 m (98 ft) mast beside the pyramid. They whirl to a stop, sometimes head-down, just short of the ground.

Map Mexico Cb

El Teniente ***Chile*** The world's largest underground copper mine, lying 2750 m (9020 ft) up in the Andes in the Libertador region, about 80 km (50 miles) south-east of Santiago. Ores have been mined there since 1912. They are refined locally and also provide molybdenum, which is used in producing alloy steels. The refined products are exported from the port of San Antonio, on the coast about 130 km (80 miles) to the north-west.

Map Chile Ac

Elat (Eilat) ***Israel*** Southernmost town and port, at the head of the Gulf of Aqaba, the north-eastern arm of the Red Sea. Elat is Israel's only southern outlet. It is also a seaside resort, with an airport, marina and underwater observatory. About 25 km (15 miles) to the north is the Timna Valley, which includes King Solomon's Mines, the remains of an Egyptian temple, modern copper and manganese mines, and curiously shaped rock formations known as King Solomon's Pillars.

Population 18 800

Map Israel Ac

Elaziğ ***Turkey*** Market town and capital of a province of the same name, about 600 km (370 miles) east of the capital, Ankara. It trades mostly in fruit and cotton grown in the surrounding mountains of the province, which covers 9205 km² (3554 sq miles).

On the Firat (EUPHRATES) river near the town is the Keban dam, the country's largest hydroelectric scheme, which was opened in 1973.

Population (town) 181 500; (province) 494 000

Map Turkey Bb

Egypt

DEPENDENT ON THE NILE AS IT WAS IN THE DAYS OF THE PHARAOHS, EGYPT CONTAINS ASTONISHING CONTRASTS BETWEEN RICH AND POOR, ANCIENT AND MODERN

Ancient kingdom of the pharaohs, Egypt is the site of one of the world's oldest civilisations, with a recorded history of 5000 years. In contrast, there are also modern wonders, such as the High Dam at ASWAN on the NILE. Egypt is the doorway between Africa and Asia, with access to the MEDITERRANEAN and the RED SEA, and as such has been a tempting prize for invaders for centuries. In recent times, the strategic importance of the SUEZ CANAL placed it in the forefront of international affairs.

It is a land of astonishing contrasts between rich and poor, the lush Nile valley and the surrounding desert, wretched hovels and huge monuments to the past. CAIRO is an enormous, overcrowded, sprawling city of 10 million people, one of the most cosmopolitan anywhere in the world. ALEXANDRIA, on the Mediterranean coast, has a population of 4 million.

The Greek historian Herodotus, writing 2500 years ago, called Egypt 'the gift of the Nile', for its existence depends on the waters of that great river, at 6695 km (4160 miles) the longest in the world. The rich soils deposited by flood waters along the banks of the Nile have supported large populations since history began; the Nile Delta is one of the world's most fertile agricultural regions. Ninety-six per cent of the people live in the Delta and the 20 km (12 mile) wide strip along the river from Cairo to Sudan, making a population density of over 1300 per km² (3370 per sq mile).

THE LAND AND ITS PEOPLE

Only a fraction of Egypt's 1 002 000 km² (386 900 sq miles) is settled or cultivated – the 36 000 km² (about 14 000 sq miles) along the Nile and around oases. The country consists of two deserts divided by the Nile valley. The Western Desert – the north-eastern part of the SAHARA – is, apart from the mountainous south-west, generally low and undulating country. There are several extensive oases – BAHARYA, Dakhla, Farafra, El Kharga and SIWA – and large, wind-scoured depressions such as the QATTARA DEPRESSION, which drops to 133 m (436 ft) below sea level.

The smaller Arabian, or Eastern, Desert rises steeply from the Nile valley to a bare, broken plateau. This slopes upwards to a high mountain range which borders the RED SEA. At the northern tip of the Red Sea, the SINAI peninsula – a wedge-shaped corner of land – links Egypt and Africa with Israel and Asia. On it lies Egypt's highest peak, Gebel Katherina, 2637 m (8651 ft) high.

It is a hot, dry land, with little rain except on the Mediterranean coast. The average maximum summer temperature in Cairo is 36°C (97°F); in winter it is 18°C (64°F). The figures for Aswan in the south are 40.6°C (105°F) and 24°C (75°F). Two winds are a curious feature of the climate. A north wind blows all year and since earliest times has enabled ships to sail up the Nile against its strong currents. But between April and June a hot south wind, the Khamsin, may blow, carrying sand and producing a yellow fog that may obscure the sun for days, while temperatures may rise by up to 17°C (30°F) in two hours.

In recent years Egypt has made big strides in industrial development and today is the second largest industrial nation on the African continent after South Africa. But the population increases even faster than the economy – each year, Egypt gains another 1.25 million people. If this rate is maintained, the population could well reach 70 million by the end of the century. The country is always running – or so it seems – to catch up with its ever-growing population.

The original Egyptians were a distinctive race: a blend of African and Asian peoples of ancient lineage among whom the Arabs came as conquerors and settlers. But over the centuries the people have become a mixture. In the desert live the Bedouin Arabs and the Berbers (a fair-skinned aboriginal people), and in the far south are small numbers of Nubians – a people of mixed Arabic, ancient Egyptian and Negro blood.

Despite recent advances, Egypt remains one of the world's poorest countries. There are great extremes of wealth: the *fellahin*, the peasants of the rural areas, earn only one-tenth as much as a craftsman. At the top of the scale are the businessmen, industrialists and wealthy government officials in the cities.

The fellahin, who live along the Nile and in the Delta, make up two-thirds of the population. It is their labour across the centuries that has provided the wealth on which Egypt's achievements rest. Many have never left the villages where they were born. Their farming methods and outlook remain much the same as centuries ago. Their hard lives, their poverty, their diseases (such as bilharzia, caused by intestinal flukes carried by water snails) have changed little, even though they may now have radio or television. They eat simply – their staple diet is rice, bread and beans. There is still much illiteracy despite great strides in education in the last 25 years.

Over 90 per cent of the people are Muslims but Egypt is not an extreme Islamic society, despite some rise of Islamic fundamentalism in the 1970s. As one might expect in the land of Cleopatra, women are more emancipated than anywhere else in the Arab world. They dress modestly but do not use the veil and can compete for a wide range of jobs – from waitresses to the highest positions in state enterprises. Thirty places are reserved for women in Egypt's parliament, the People's Assembly, in which half of the 458 seats must be held by workers or farmers. Polygamy, though still acceptable as part of Islam, is dying out for economic and social reasons.

Apart from Muslims there are an estimated 3 to 5 million Coptic Christians who claim to be descendants of the original inhabitants before the coming of the Arabs. Although a minority, they are important politically and have been the focus of sectarian conflict.

LONG ROAD TO FREEDOM

Egyptians are long used to outsiders. Foreigners come now as tourists, as businessmen or as technicians working for the United Nations or other international agencies, but for centuries they came as conquerors. From 525 BC, when the Persians occupied the country, Egypt was dominated by outside powers. In turn came the Greeks under Alexander the Great, the Romans and the Arabs. In 1517 Egypt became part of the Turkish Ottoman Empire. Napoleon's invasion in 1798 brought Egypt firmly to the attention of Europe.

In 1854 the French engineer Ferdinand de

Lesseps was granted a concession to build the Suez Canal, over 160 km (100 miles) long, shortening the long sea voyage from the Far East. It was not a new idea; a canal as far as Lake Timsah was built by the pharaohs around 2100 BC and extended by the Persian conqueror Xerxes I, but it fell into disuse in the 8th century AD. In 1882 Britain occupied Egypt, beginning an uneasy 75 year semi-colonial relationship. British troops remained to defend Egypt and the canal even after Egypt officially became independent again in 1922. One of the greatest artillery battles of the Second World War was fought in 1942 at EL ALAMEIN in the Western Desert.

In the 1950s the nationalist cry of 'Egypt for the Egyptians' was finally translated into effective action. In 1952 King Farouk was forced to abdicate. Four years later, after British forces had left, President Nasser nationalised the Suez Canal. A triple attack by Britain, France and Israel to regain control of the vital waterway was ended by world pressure – and Egypt was its own master.

Upon the crest of the nationalist wave, Nasser had launched a huge programme of modernisation, symbolised by the Aswan High Dam. Opened in 1970, the dam created one of the largest man-made lakes in the world, Lake Nasser. By regulating the waters of the Nile, the dam has allowed vast tracts of land to be reclaimed from the sandy desert, and also generates hydroelectric power for industry. Those people who live along the Nile no longer have to cope with annual flooding, and in some areas they can grow two crops a year by controlled irrigation.

But not all the hopes raised by the dam have been fulfilled. Large portions of reclaimed land have returned to the desert through inefficient farming and lack of technical expertise – the wrong crops were grown

▲ TOMBS OF KINGS The sun rises behind the pyramids at El Giza. That built about 2570 BC for Chephren (right) was originally 143.5 m (471 ft) high and is made of limestone blocks weighing up to 15 tonnes. It stands by the Great Pyramid of his father, Khufu.

or badly drained areas became saline, for example. There have been other problems – for instance, the lack of flood-borne nutrients has made it necessary to use artificial fertilisers on the land beside the river, and clay is no longer available for brick making.

Two other major desert reclamation schemes are the New Valley project and the Tahrir (Liberation) Province project. The New Valley consists of a line of oases in the Western Desert paralleling the Nile valley. Supplies of subterranean water for irrigation are obtained by drilling to a depth of about

900 m (3000 ft). The Tahrir scheme is located on the western edge of the Delta, midway between Cairo and Alexandria, and relies on both Nile water and wells.

The land by the Nile produces rice, maize, wheat, vegetables and fruit. Here, in contrast to the desert, vegetation is abundant: date palms, tamarisk, mimosa, eucalyptus, cypress and jasmine. Cotton forms the main cash crop and foreign exchange earner.

Cattle, buffaloes, sheep, goats and camels are the main domesticated animals. Camels, used as beasts of burden, can survive without water for several days, carry loads of 230-270 kg (about 500-600 lb) and cover 50 km (30 miles) a day. Most wildlife is relatively small: wild cats, hares, hyenas, jackals and foxes. Boars are found in the Delta and there is a great variety of bird life throughout the Nile area – geese, ducks, quail, cuckoos, nightingales and sparrows.

Egypt has a growing and sophisticated industrial sector producing iron and steel, aluminium, chemicals, tools, furniture, glass and textiles. The economy has been boosted by the discovery of oil – not large amounts, but enough to supply Egypt's needs and leave a surplus for export. The Suez Canal provides finance from shipping tolls. More than 60 ships use the canal daily, paying over US$900 million per year in dues.

One result of Egypt's strategic position has been its ability to attract large-scale international assistance. Nasser was determined that Egypt should go her own way – to be non-aligned. But the growth of population and four wars with Israel since that country became independent in 1948 (the first took place before Nasser came to power) slowed economic progress. His successor, President Sadat, took two immensely brave steps: first, talking with Israel and, second, signing the Camp David Agreement in 1979 under which Israeli troops withdrew from Sinai.

Others had different ideas: Sadat was assassinated by Muslim extremists and in 1981 was succeeded by the vice-president, Hosmi Mubarak. The new president continued Sadat's policies of maintaining contact with Israel and seeking a peaceful solution to the Arab-Israeli problem.

There is an undeniable Western façade to the cities: educated Egyptians are likely to speak English or French in addition to Arabic. There is a widely varied Press, often lively but periodically subject to censorship. Cairo is the second largest publishing centre in the Middle East after Tel Aviv, and Egypt has extensive radio and television services with programmes in many languages. There is also a substantial film industry, though the films are mainly unsophisticated and geared for popular consumption. The products of all these media are distributed throughout the Arab world and are very influential.

FOR THE VISITOR

Tourists flock to see Egypt's pyramids and other historical treasures. The pyramids of ancient Egypt were royal tombs, built on the west side of the Nile, from the 3rd dynasty of pharaohs (about 2680 BC) to the 18th dynasty (about 1570 BC). Many of the pharaohs built their own pyramids, in which their mummified bodies might be preserved for eternity.

EL GIZEH is the centre from which to visit the pyramids and the sphinx. The three pyramids of Gizeh are the largest and finest of their kind: the Great Pyramid of Khufu or Cheops (about 2590 BC) is one of the Seven Wonders of the Ancient World. The largest pyramid ever built, its mass of limestone blocks covers an area the size of 13 standard football pitches. It was originally 147 m (482 ft) high and 230 m (755 ft) along each side of its base. The sphinx, a mythical beast with a man's head and a lion's body, symbolised the pharaoh Khafre, Khufu's son, as an incarnation of the sun god Ra.

From LUXOR and KARNAK you can see numerous temples and burial grounds, including the Valley of the Tombs of the Kings. But Egypt is also of great interest as a modern place to visit; the sheer movement and vitality of its great cities present excitement and colour. Bargaining is the natural way to do business, whether for taxi rides or shopping at the bazaars. Tourists who look 'fair game' may find themselves asked to pay steep prices or pestered for '*baksheesh*' (alms). Cairo is rich in mosques and museums, and Alexandria has ancient Christian catacombs (underground burial chambers). A camel ride, a Cairo shopping expedition and a trip down the Nile should be part of every visitor's itinerary, while many visit the El Alamein battlefield 110 km (70 miles) from Alexandria.

EGYPT AT A GLANCE
Area 1 002 000 km² (386 900 sq miles)
Population 49 560 000
Capital Cairo
Government Parliamentary republic
Currency Egyptian pound = 100 piastres
Languages Arabic (official), Berber, Nubian, Beja
Religions Muslim (90%), Christian
Climate Hot and dry; average temperature in Cairo ranges from 8-18°C (46-64°F) in January to 21-36°C (70-97°F) in July
Main primary products Rice, cotton, wheat, maize, sugar cane, barley, beans, lentils, millet, onions, Egyptian clover; cattle, buffaloes, sheep, goats; oil and natural gas, iron, phosphates, sea salt
Major industries Oil refining, cement, textiles, iron and steel, fertilisers, processed foods
Main exports Crude oil, cotton, textiles, rice, fruit, vegetables
Annual income per head (US$) 690
Population growth (per thous/yr) 26
Life expectancy (yrs) Male 55 **Female** 59

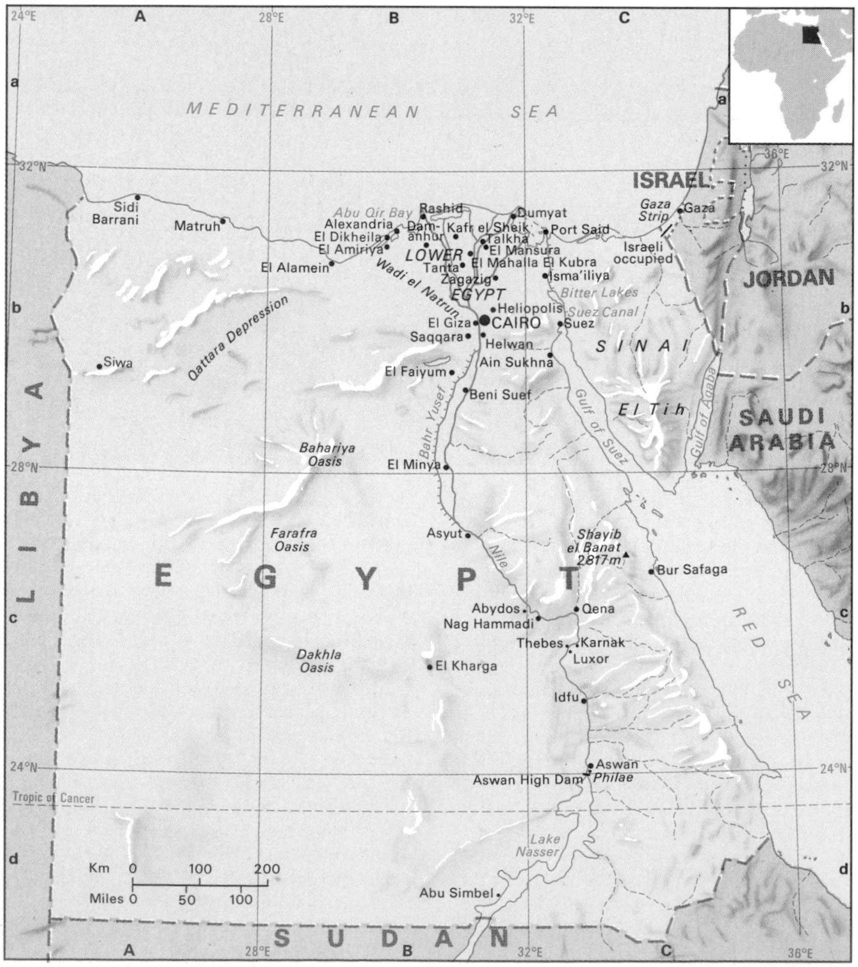

El Salvador

WAR-TORN AND VERGING ON ECONOMIC RUIN, EL SALVADOR IS PROPPED UP BY THE USA TO PREVENT COMMUNISM SPREADING IN CENTRAL AMERICA

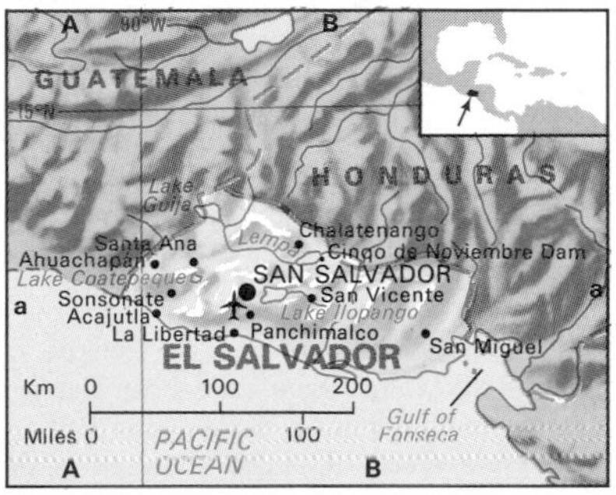

The struggle for power in the small Central American nation of El Salvador has resulted in the death of thousands of its people every year since 1979, and has driven 500 000 people from their homes and into refugee camps. Successive governments during the early 1980s were accused of abusing human rights by torture, 'disappearance' of political activists and executions without trial.

Then, as many Salvadoreans died through assassinations and massacres as were killed by direct military action. Among the many victims was the leading spokesman for human rights, the Roman Catholic Archbishop Oscar Romero, gunned down by right-wing terrorists while conducting mass in the cathedral in the capital, SAN SALVADOR, in March 1980.

The government has been threatened by left-wing guerrillas, the FMLN (Farabundo Marti National Liberation Front), backed by the left-wing regime in Nicaragua. The USA, fearing the spread of Communism northwards to its own borders, has supported the Salvadorean Government with substantial economic and military aid. But between 1982 and 1984 the ultrarightist Roberto d'Aubuisson held power, and his brutality caused an outcry among Americans who objected to their government's support for his regime.

In 1984 El Salvador had its first democratic presidential election, and the more moderate Christian Democrat leader José Napoleon Duarte took office. He hoped to end the deadlock with the guerrillas by inviting them to negotiate and stand for election, thereby entering the legitimate democratic process. But the guerrillas saw this as a tactic to disarm them, and continued to fight for power without an election. Meanwhile, d'Aubuisson's military-dominated party, Arena, lurked in the wings, accusing Duarte of being a Communist himself. Even under Duarte's presidency, Arena has often thwarted economic and judiciary reforms in the Legislative Assembly.

A HISTORY OF CONFLICT

Conflict and repression were not new to Salvadoreans when the latest civil war began. After it became fully independent in 1838, internal strife and military dictatorship became the pattern for the country. There were also external conflicts – in 1969, a Salvadorean army invaded neighbouring Honduras after a World Cup soccer match between the two nations. It was a two-leg fixture, played in the two capitals, and on each occasion the visiting supporters were subjected to ill-treatment and violence. This sparked the war, but the underlying cause was the illegal migration of some 300 000 Salvadoreans into Honduran territory over the previous 20 years. The Salvadorean invasion lasted for only five days, but the resulting uneasy peace was broken in 1976, when troops from both countries clashed on the border.

The Honduras Government suspects that El Salvador is seeking more living space – and Honduras is the obvious choice. El Salvador is the most densely populated country in Central America with 248 people per km² (642 per sq mile), and one of its main exports is people. There is too little work and food to support all 5.2 million; civil war has damaged agriculture, industry and services. Apart from US and West German aid, money is no longer invested from abroad.

The land is owned by a tiny elite of white people, although 87 per cent of Salvadoreans are *mestizo* – of mixed Spanish and Indian descent – and the rest are pure Indian. There are many social problems; many of the people are illiterate and undernourished. Farms and industries damaged in the fighting have had no interlude of peace in which to get production restarted, so unemployment has increased and inflation is high. In recent years the incomes of the people have gradually declined in real terms.

The major towns support a few industries – mainly food processing, textiles, cement and chemicals manufacture. Mineral output is negligible. The roads between towns are poor and mainly unmetalled. Like the railways, they are often disrupted by the fighting.

About half the people work in agriculture, which brings in a quarter of the national income. Coffee accounts for about 40 per cent of the country's exports, followed by cotton (declining), sugar (increasing) and textiles, while the people grow rice to feed themselves and rear livestock (beef is also exported). Timber is a minor export, and El Salvador is the world's leading producer of balsam gum (used in patent medicines and perfumes). Soil fertility is a problem since the slopes of the country's volcanic mountains have been cleared of natural forest – mostly for fuel – causing severe erosion.

Two volcanic ranges run from east to west across El Salvador, divided by the LEMPA river valley. The river turns southwards halfway along the southern range, cutting through the mountains and dividing the country into western and eastern parts. Beyond the southern range, the Lempa's large, sandy delta opens into the Pacific. In the wet season, between May and October, it floods. The ranges nearest the sea form clusters of volcanic cones, several of which are active. Most of the country's settlements are in the cooler, temperate highlands rather than on the hot, humid coast.

Before the Spanish came in the 16th century, El Salvador had a thriving civilisation of Pipil Indians, who were related to the Aztecs. They called the territory *Cuscatlan* (Land of Jewels). Monuments of this civilisation remain, notably the El Tazumal ruins at Chalchuapa. Today, the idea of El Salvador as a land of riches seems strange and remote.

EL SALVADOR AT A GLANCE

Area 21 041 km² (8124 sq miles)
Population 5 220 000
Capital city San Salvador
Government Parliamentary republic
Currency Colon = 100 centavos
Languages Spanish, Indian dialects
Religion Christian (97% Roman Catholic)
Climate Tropical; cool in the highlands. Average temperature in San Salvador is 22-24°C (72-75°F) all year
Main primary products Coffee, cotton, maize, rice, sorghum, sugar, beans, cattle, timber, shrimps
Major industries Agriculture, food processing, textiles and clothing, chemicals, forestry, fishing, electricity
Main exports Coffee, cotton, sugar, textile yarns and fabrics, chemicals, clothing
Annual income per head (US$) 686
Population growth (per thous/year) 28
Life expectancy (yrs) Male 62 **Female** 65

Elba ***Italy*** The third largest Italian island after Sicily and Sardinia, covering 223 km² (86 sq miles) and lying off the Tuscany coast. Its capital is Portoferráio, where the French emperor Napoleon I (1769-1821) was exiled in 1814-15, after his disastrous Russian campaign. The island, now a tourist spot, was an iron-working centre for both the ancient Greeks and the Etruscans, and the Romans built many villas.
Population 28 400
Map Italy Cc

Elbe ***Central Europe*** River rising in north-east Czechoslovakia and flowing 1160 km (720 miles) across East and West Germany into the North Sea just north of the West German city of Hamburg. It is navigable for about 940 km (585 miles) from its mouth.
Map East Germany Bb, Cc

Elblag (Elbing) ***Poland*** Baltic seaport and industrial city, 48 km (30 miles) south-east of Gdansk. Teutonic Knights founded the city in the 12th century, but since the devastation of the Second World War, only key monuments have been restored, including the Church of St Nicholas and its 95 m (312 ft) tower. These remnants contrast strikingly with postwar housing and shops, and factories making turbines and equipment for the Gdansk shipyards. The Elblag Canal, 92 km (57 miles) long, links the city with the Masurian lakeland to the east.
Population 115 900
Map Poland Ca

Elbrus, Mount ***USSR*** Highest peak in Europe, standing in the Caucasus on a northern spur from the main Caucasus range, which forms part of the boundary between Europe and Asia Minor. It consists of two volcanic peaks; the western summit is 5642 m (18 510 ft) high, the eastern a little lower.

Map USSR Fd

Elburz Mountains ***Iran*** Range of mountains running between the capital, Tehran, and the Caspian Sea. The highest peak – and the highest point in Iran – is Damavand (5670 m, 18 600 ft), a snow-capped volcano, now extinct.

Map Iran Ba

Eldoret ***Kenya*** One of the country's largest towns. It lies on the main road and railway to Uganda, about 260 km (160 miles) north-west of Nairobi. It was a European settlers' town which declined after independence in 1963. Several industries, including textiles and clothing, were introduced in the 1970s and Kenya's second university after Nairobi was set up in the town.

Population 70 000

Map Kenya Ca

Elephanta Island ***India*** Island in Bombay harbour, about 10 km (6 miles) from the city and reached by excursion boats. It was named during the British Raj after a rock carved into an elephant shape, but which has since collapsed. The island is known in India as Lenen, and has a complex of Hindu cave temples dating from AD 450-75, with sculptures portraying Hindu gods and their stories.

Eleuthera ***Bahamas*** Island in the north-central group of the archipelago, and one of the first to be colonised by the British in the 17th century. About 160 km (100 miles) long and mostly only 3 km (2 miles) wide, Eleuthera has a narrow strip of fertile land surrounded by dunes and beaches of pinkish sand. It is one of the major resort islands of the Bahamas. Commercial farming benefits from the island's proximity to the capital, Nassau.

Population 9500

Map Bahamas Ba

Elgon, Mount ***Kenya/Uganda*** Extinct volcano sometimes known as Wagagai which rises to 4321 m (14 176 ft). The fertile lower slopes are intensively cultivated with coffee and bananas.

Map Uganda Ba

Elisabethville ***Zaire*** See LUBUMBASHI

Ellesmere Island ***Canada*** Mountainous snow and ice-covered island in the Canadian arctic archipelago, separated from Greenland by a narrow strait. Only a few Inuit (Eskimos) live there, apart from the staff of Canadian-US research bases at Eureka and Alert. Cape Columbia, the northernmost point of Canada, is just 756 km (470 miles) from the North Pole. It was from there that Robert Peary set off to conquer the Pole in 1909. Ellesmere Island covers an area of 197 000 km² (75 780 sq miles).

Map Canada Ga

Ellice Islands See TUVALU

Elsinore (Helsingør) ***Denmark*** Town founded in the Middle Ages on the island of ZEALAND overlooking the mouth of the Öresund, the main shipping channel leading to the Baltic Sea. Sweden lies on the far side of the 4 km (2 mile) wide channel. Kronborg Castle – the setting for William Shakespeare's play *Hamlet* – dominates the town. The castle was built in the 16th century to guard the channel, and to extract tolls from ships heading for the Baltic. Today the town is Denmark's main ferry port for Sweden.

Population 65 200

Map Denmark Ca

Elvas ***Portugal*** Fortified hilltop town on the Spanish border, about 180 km (112 miles) east of the capital, Lisbon. The Portuguese wrested it from the Moors in 1226 and it was captured by Spain in 1580. The fortifications date mostly from the 17th century and there is a 16th-century aqueduct. Its main manufacture is jewellery, and the area is noted for its plums.

Population 12 700

Map Portugal Cc

El-Wanza ***Algeria*** See OUENZA

Ely ***United Kingdom*** English cathedral city in Cambridgeshire, 22 km (14 miles) north and slightly east of Cambridge. It used to stand on an island – the Isle of Ely – in the heart of The Fens, and could be reached only by boat or by three narrow causeways. For this reason it was used as a refuge for five years by the English warrior Hereward the Wake after the Norman invasion of 1066. The island disappeared when The Fens were drained. The cathedral, built of stone delivered by boat, is one of Britain's most beautiful. It was built in several distinct styles between the 11th and 14th centuries, and has a unique octagonal tower. The King's School next to the cathedral was founded in AD 970.

Population 10 000

Map United Kingdom Fd

Emden ***West Germany*** Port in the north-west corner of the country facing Dutch territory across the estuary of the Ems river. Silting in the 16th century led to the river changing course, and Emden is now linked to the North Sea and to the Ruhr industrial region by canal.

Population 50 000

Map West Germany Bb

emerald Brilliant, transparent green form of beryl, used as a gemstone.

Emi Koussi ***Chad*** See TIBESTI

Emilia Romagna ***Italy*** Region in north-central Italy centred on its capital, BOLOGNA. It covers 22 123 km² (8542 sq miles) of the Po valley and Apennines, and is a leading producer of apples, pears, peaches, sugar beet, pigs and cattle. The region was once a Fascist stronghold – the Italian dictator Benito Mussolini (1883-1945) was born there in the village of Predappio – but today 'Red' Emilia is the most fervently Communist region in Italy.

Population 3 943 000

Map Italy Cb

Emmental ***Switzerland*** Valley of the River Emme, especially famous for its cattle and its cheese. The Emme, about 80 km (50 miles) long, rises east of the capital, Berne, and flows into the River Aare at Gerlafingen, near Solothurn.

Map Switzerland Aa

Ems ***West Germany*** River rising east of the north-western city of Münster and flowing for about 370 km (230 miles) into the North Sea near the Dutch border. It is linked to the industrial Ruhr through the Dortmund-Ems Canal.

Map West Germany Bb

Enewetak ***Marshall Islands*** Atoll in the northern Marshalls used, like Bikini 320 km (200 miles) to the east, for US nuclear weapons tests in the 1940s and 1950s. The people were evacuated but, after a cleanup operation, were allowed to return in 1980 despite fears that radioactivity remained. In 1983 they decided to make Ujelang atoll their permanent home.

Map Pacific Ocean Db

Engadin ***Switzerland*** Swiss part of the valley of the River Inn, which rises near the Italian border and flows north-east into Austria. The resort-filled Ober, 'Upper', Engadin is dominated by St Moritz; the Unter, 'Lower', Engadin, also a recreation area, contains the Swiss National Park. This extends over 168 km² (65 sq miles) and consists of mountains reaching to about 3000 m (10 000 ft) and forests.

Map Switzerland Ca

Engel's ***USSR*** Industrial town on the Volga river, 320 km (200 miles) north of Volgograd. It was founded in 1747 as Pokrovskaya Sloboda, and was renamed Engel's in 1932 after the German political philosopher Friedrich Engels (1820-95), co-author with Karl Marx (1818-83) of *The Communist Manifesto*. The town's manufactures include leather goods, chemicals, railway wagons and machinery.

Population 175 000

Map USSR Fc

England ***United Kingdom*** Largest constituent country of the UNITED KINGDOM, covering some 130 357 km² (50 331 sq miles). It was settled in pre-Christian times by the Celts and was later conquered by, among others, the Romans, Danes and finally the Normans. Its capital is LONDON and its main industrial cities include BIRMINGHAM, LEEDS, LIVERPOOL, MANCHESTER and SHEFFIELD.

Population 46 795 000

Map United Kingdom Ed

English Channel Waterway between England and France, which is called La Manche (the sleeve) by the French and is part of the ATLANTIC OCEAN. Its western entrance, between LAND'S END in south-western England and the Ile d'Ouessant (Ushant), off Brittany, is 180 km (110 miles) across. To the north-east, the Channel narrows to 34 km (21 miles) at the Strait of DOVER, which was formed about 7500 years ago when ice sheets melted at the end of the Ice Age and the sea level rose. The world's busiest sea passage (some 350 ships pass through each day), the English Channel is lined with ports and resorts on both French and English coasts.

Area 77 700 km² (30 000 sq miles)

Greatest depth 172 m (565 ft)

Map France Ba

English Harbour ***Antigua and Barbuda*** - Former British naval dockyard in the south of Antigua where the English admiral Horatio Nelson and his fleet were based for a time in the late 18th century. It is now a tourist attraction and yachting centre. The dockyard has been restored to its 18th-century appearance, and Nelson's house is a museum.
Map Caribbean Cb

Enna ***Italy*** City and resort known as the 'navel of Sicily', about 100 km (60 miles) south-east of Palermo. Enna stands on a plateau 1000 m (3300 ft) high, and contains one of the island's most impressive medieval fortresses – the Castello di Lombardia, which rises out of sheer rock. Enna trades in rock salt and sulphur.
Population 28 900
Map Italy Ef

Ennis ***Ireland*** See CLARE

Enns ***Austria*** Small town at the confluence of the Enns and Danube rivers, 18 km (11 miles) south-east of Linz. One of the country's oldest towns, Enns received its charter in 1212. It has a Gothic parish church, a 16th-century fortress, and parts of its medieval walls are still standing.
Population 9700
Map Austria Da

Enschede ***Netherlands*** Industrial town, east and slightly south of Amsterdam and only 5 km (3 miles) from the West German border. It has grown rapidly since 1900 as the 'Manchester of the Netherlands' – the centre of the country's cotton industry.
Population (town) 144 900; (Greater Enschede) 247 800
Map Netherlands Ca

Entebbe ***Uganda*** Town on a peninsula jutting into Lake Victoria, 33 km (20 miles) south-west of the capital, Kampala. It was Uganda's capital from 1894 until independence in 1962, and has the country's main airport where, in July 1976, Israeli commandos freed 100 Jewish hostages being held by Arab gunmen.
Population 30 000
Map Uganda Ba

Entre Deux Mers ***France*** Triangular area in Aquitaine, just east of the junction of the Dordogne and Garonne rivers. It is famed for the dry white wines named after it.
Map France Cd

Enugu ***Nigeria*** Capital of Anambra state, on the Udi Plateau 440 km (275 miles) east of the national capital, Lagos. It lies on the country's only large coalfield, with estimated reserves of 72 million tonnes. Most of the coal goes to the railways or power stations, but demand has fallen with increasing use of diesel locomotives and the opening of the Kainji Dam hydroelectric station.
Population 222 600
Map Nigeria Bb

Eocene Second epoch in the Tertiary period of the Cenozoic era of the earth's time scale. See GEOLOGICAL TIMESCALE

Eochaill ***Ireland*** See YOUGHAL

Eolian Islands (Lipari Islands) ***Italy*** Group of volcanic islands off the north coast of Sicily. They are named after Aeolus, the Greek god of the four winds. Since 1970 there has been a boost to the islands' fortunes, due largely to tourism. The main islands are LIPARI and Salina. VULCANO attracts tourists for its volcanic mudbaths, Panarea for its scenery and STROMBOLI for its active volcano.
Population 12 500
Map Italy Ee

Ephesus (Efes) ***Turkey*** Ruined city overlooking the Aegean Sea about 400 km (250 miles) south of Istanbul. It was the site of the Temple of Artemis (or Diana), one of the Seven Wonders of the Ancient World. The marble temple – dedicated to the Greek goddess of the hunt – was built in the 6th century BC by the fabulously wealthy Lydian king Croesus, who ruled the city. The temple was destroyed by invading Goths in the 3rd century AD.
Map Turkey Ab

epicentre Point on the earth's surface directly above the focus, or origin, of an earthquake.

Epídhavros (Epidaurus, Hieron Epidaurou) ***Greece*** Ancient site devoted to Asclepius, the Greek god of medicine. It lies 37 km (23 miles) south of Corinth. Its theatre, dating from the 4th century BC, is the best preserved in Greece, with a circular stage and remarkable acoustics. An international festival of drama, opera and music takes place there in summer.
Map Greece Cc

Epirus (Ipiros) ***Greece*** Region of north-west Greece covering 9203 km² (3552 sq miles), bordering Albania and the Ionian Sea, but cut off from the rest of the country by the Pindhos mountain range. It is a hilly region with the highest rainfall in Greece. Agriculture and pasture farming thrive in the valleys. In 278 BC, its King Pyrrhus (319-272 BC) won a battle against the Romans so costly in officers and men that it gave rise to the description Pyrrhic victory, a victory that might just as well have been a defeat. Turkey occupied Epirus from AD 1449 until 1912. Its chief town is Yannina.
Population 324 500
Map Greece Bb

epoch Third category of the subdivisions of geological time – era, period, epoch, age. See GEOLOGICAL TIMESCALE

Eptanisos ***Greece*** See IONIAN ISLANDS

Equateurville ***Zaire*** See MBANDAKA

equatorial climate Type of climate occurring in lowland areas close to the Equator. There is virtually no seasonal change. There are 12 hours of daylight all the year, and monthly mean temperatures average 26°C (79°F). Rainfall occurs each afternoon, and averages more than 2000 mm (79 in) a year.

equatorial forest Part of the tropical rain forest found within a few degrees of the Equator where there is no seasonal change. It is noted for its enormous diversity of species.

▼ NO MEAN CITY Majestic ruins conjure up the wealthy city of Ephesus, once visited by St Paul.

Er Rachidia (Ksar es Souk) ***Morocco*** Berber oasis town on the southern flanks of the High Atlas mountains 240 km (150 miles) south-east of Meknès, in the Ziz valley. It is a transport centre and market for wool, dates, olive oil and esparto grass. The town lies in a huge palm grove and has a kasbah (old walled quarter) and fortress.

Population 45 710

Map Morocco Ba

Er Roseires (Ar Rusayris) ***Sudan*** Blue Nile port near the Ethiopian border, handling cotton, sesame seeds and cereal exports. To the south is the Er Roseires dam, which provides irrigation for part of the Managil extension of the Gezira Scheme, and for the new Rahad Scheme producing cotton, millet, groundnuts and clover.

Population 16 400

Map Sudan Bb

era The major division of geological time, which is subdivided into periods. See GEOLOGICAL TIMESCALE

Erawan ***Thailand*** The most productive offshore gas field in the Gulf of Thailand, 500 km (310 miles) south of Bangkok. Its daily output is 17-20 million m³ (600-700 million cu ft).

Ercolano ***Italy*** See HERCULANEUM

Erdenet ***Mongolia*** Mining town about 220 km (135 miles) north-west of the capital, Ulan Bator, between the Selenge and Orhon rivers. The mine, Erdenetiyn-Ovoo, produces large quantities of copper and molybdenum (a metallic element used to toughen alloy steels), and is said to contain half the known copper deposits in Asia. It was opened in 1976 with substantial Russian aid, and accounts for 30 per cent of Mongolia's total export earnings.

Population 36 000

Map Mongolia Db

Erebus, Mount ***Antarctica*** Active volcano on Ross Island, 3794 m (12 447 ft) high. In November 1979, a New Zealand airliner carrying tourists on a scenic trip crashed into its side and all 257 people aboard died.

Map Antarctica Jc

Erfoud ***Morocco*** Gateway to the Sahara, a town situated at the end of the Ziz valley, 280 km (175 miles) south-east of Meknes. An old fortress dominates the town. Nearby are the ruins of the 15th-century town of Sijilmassa, once the northern end of the caravan route along which gold was brought from the empire of Ghana. At Rissani, to the south, the founder of the present Moroccan royal dynasty was born in the 17th century.

The Ziz river cuts a deep, cliff-lined valley in the surrounding near-desert, and supports palms and other brilliant green vegetation along its banks. It flows for about 280 km (175 miles) through southern Morocco.

Map Morocco Ba

Erfurt ***East Germany*** Tourist resort in the south-west, about 60 km (37 miles) from the West German border. It produces heavy machinery, typewriters, electrical goods, shoes and clothing, but it is as a historical treasure chest that the city is best known. The religious reformer Martin Luther (1483-1546) was a monk at the Augustinian monastery there for five years. There is a Scottish church built in 1200 and a 13th-century Dominican monastery, whose gardens founded the city's flower, vegetable and seed industry. The 14th-century Kramer Bridge is covered by half-timbered houses and shops.

Population 215 000

Map East Germany Bc

erg Originally an area of sand dunes in the Sahara, but now used for any sand desert.

Erie ***USA*** The only Great Lakes port in Pennsylvania, standing on the lake of the same name 185 km (115 miles) north of Pittsburgh. It handles coal, grain, timber, oil and chemicals, and makes machinery, boilers, electric locomotives and paper products.

Population (city) 117 500; (metropolitan area) 282 100

Map United States Jb

Equatorial Guinea

THOUGH ITS ECONOMY IS IN RUINS, FREEDOM IS SLOWLY RETURNING TO A LAND THAT SUFFERED YEARS OF TYRANNICAL RULE

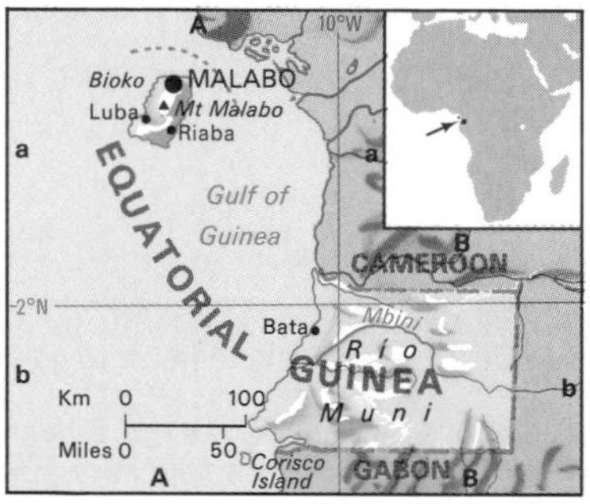

Lying about 200 km (125 miles) north of the Equator, on the hot, humid coast of West Africa, the Republic of Equatorial Guinea is one of the poorest countries in a poor continent. Yet before independence from Spain in 1968, this territory, which is nearly the size of Belgium, had a thriving economy based on cocoa, coffee and timber. But in the hands of the first president, Francisco Macias Nguema (1968-79), the economy collapsed.

Torture, forced labour, systematic murders of intellectuals, the persecution of foreigners and the Churches, together with increasing isolationism from the outside world were features of Nguema's 11 year rule. In 1979, when he was deposed and executed after a coup led by a nephew, a United Nations report described the nation as 'decomposed'. Since 1979 the government has released political prisoners, restored religious freedom, and sought to restore the shattered economy with overseas aid from East and West, but especially Spain.

Equatorial Guinea consists of two contrasting regions. The largest, on the African mainland, is called RIO MUNI and comprises 93 per cent of the country. It has extensive tropical forests and savannah-covered plateaus. There are few roads, and foreign visitors are almost unknown. Included in this region are three offshore islets: Corisco, Great Elobey and Small Elobey.

The second region consists of the islands of BIOKO (called Fernando Póo in Spanish times) north of the mainland off the coast of Cameroon, and Annabon, 640 km (400 miles) to the south-west.

RICH POTENTIAL

The fertile volcanic island of Bioko contains the country's capital, MALABO, which stands beside a volcanic crater flooded by the sea. The island is the centre of the country's cocoa production, and has even higher temperatures and higher rainfall than the hot, humid mainland.

The original inhabitants of Bioko, the Bubi, speak a Bantu language, like most Equatorial Guineans. Bioko also has a few thousand Fernandinos, Creole descendants of slaves freed by Britain in the 19th century. Before 1975-6, 30 000 Nigerians lived on the island, working on the cocoa plantations. But they were evacuated after maltreatment and violence from the island's Fang rulers. The Fang are the dominant Bantu-speaking group on the mainland, who, after 1968, extended their authority to Bioko.

Equatorial Guinea has considerable tourist potential, though it is difficult to reach. The natural beauty of Bioko, praised among others by the Victorian traveller Mary Kingsley, and the abundant wildlife on the mainland are potential attractions.

EQUATORIAL GUINEA AT A GLANCE	
Area	28 051 km² (10 828 sq miles)
Population	289 000
Capital	Malabo
Government	One-party republic
Currency	CFA franc = 100 centimes
Languages	Spanish (official), African languages including Fang
Religions	Christian (80% Roman Catholic), tribal
Climate	Tropical; average temperature in Malabo ranges from 21°C (70°F) to 32°C (90°F)
Main primary products	Sweet potatoes, cassava, bananas, coffee, cocoa, coconuts, timber
Major industries	Agriculture, forestry, fishing, food processing
Main exports	Cocoa, timber, coffee
Annual income per head (US$)	140
Population growth (per thous/yr)	25
Life expectancy (yrs)	**Male** 43 **Female** 45

Erie, Lake ***Canada/USA*** Second smallest of the five Great Lakes after Lake Ontario, covering about 25 670 km^2 (9910 sq miles). It drains north through the Niagara river (with Niagara Falls) into Lake Ontario and is linked to that lake by the Welland Canal. Lake Erie is fed by the Detroit river, Lake St Clair and St Clair river from Lake Huron. Much of the southern shore is heavily industrialised, with the cities of Toledo, Cleveland and Buffalo.

Map United States Jb

Erie Canal ***USA*** Waterway linking the port of Buffalo on Lake Erie with the Hudson river at Albany, a distance of about 550 km (340 miles). The original canal – some 30 km (20 miles) longer – was opened in 1825 and was a key route in the expansion westwards. It was improved in the early 20th century and became the main part of the New York Barge Canal system, which has links with Lake Ontario.

Map United States Kb

Eritrea ***Ethiopia*** Northern province (117 400 km^2, 45 316 sq miles) along the Red Sea, regarded by many Eritreans as a nation under Ethiopian colonial rule. The people are ethnically different from Ethiopians, though they do not form a single distinct ethnic group. It was controlled by Italy from 1890 to 1941 and administered by Britain for 11 years, before the United Nations decided that it should be federated with Ethiopia. There has been civil war since 1962 when Ethiopia took complete control. The province has been devastated by drought and famine as well as war in the 1970s and 1980s, and many inhabitants have fled to Sudan.

Population 3 000 000

Map Ethiopia Aa

Erlangen ***West Germany*** University town in Bavaria 20 km (12 miles) north of Nuremberg. The town, which has an extensive electrical industry, was the birthplace of George Simon Ohm (1787-1854), the physicist who gave his name to the electrical unit of resistance.

Population 100 000

Map West Germany Dd

erosion Wearing away of the earth's surface by rivers, sea, ice and wind. See ROCK CYCLE; THE CHANGING PATTERN OF THE COASTS (p. 212)

erratic Large piece of rock that has been transported some distance from its source, particularly one carried by a glacier and left standing when the ice has melted. It is different in composition from the rocks in its present location – a geological alien.

Erzgebirge (Ore Mountains) ***Czechoslovakia/East Germany*** Forested hills stretching 130 km (80 miles) along the border. They rise to 1244 m (4081 ft) and have deposits of iron ore, silver, lead, copper, pitchblende and uranium which are still worked. It is a popular tourist area.

Map East Germany Cc

▲ ICE AND FIRE The name of volcanic Mount Erebus comes from Greek mythology, where it denoted the place of darkness between this world and Hades – the abode of departed spirits.

Erzurum ***Turkey*** Market town about 845 km (525 miles) east of the capital, Ankara. It is a trading centre for cattle, grains, furs and vegetables produced on farms in the surrounding province of the same name. Erzurum province, which covers 25 066 km^2 (9678 sq miles) also produces sugar beet, wheat and barley. The town's main industries are iron, copper, sugar refining and tanning.

There are well-preserved Byzantine walls around Erzurum and various Seljuk remains, such as the Great Mosque (which dates from 1179) and the mausoleum of the 12th-century emir, Sultan Turbesi.

Population (town) 252 700; (province) 875 000

Map Turkey Cb

Esbjerg ***Denmark*** Port on the west coast of Jutland in the lee of the holiday island of Fanø. Its original harbour was built in 1868, primarily as a gateway for exports to Britain (to which a ferry still operates). Today, it is the world's largest export harbour for butter and bacon, and Denmark's biggest fishing port. It also has an airport, an art centre, a maritime museum and a university.

Population 80 300

Map Denmark Bb

THE CHANGING PATTERN OF THE COASTS

The coast is where both the destructive and constructive energies of the seas come together, and the land forever dies and is perpetually born again. Some coastal landscapes feature deep indentations (rias) which are actually ancient river valleys drowned by the sea when the glaciers of the Ice Age melted; at others, where the land is slowly springing back from its Ice Age depression, the sea has retreated almost out of sight. But the swiftest and most dramatic alterations are wrought by the sea, which on one hand may carve away a cliff in a night, and on the other, lay down a new sandy beach.

Oceanic waves are the most powerful machines on earth. Driven by the wind, the wave motion that strikes western Europe, for example, begins off the tip of South America, gathering strength on its 10 000 km (6200 mile) journey. What happens at its journey's end depends very much on the terrain. On a gently shelving coastline, the wave motion will be transformed to a current as the waves break, with the main force directed towards the headlands, while the lesser flow deposits sand and sediment, picked up from the seabed, into bays.

Rock types have a great effect upon coastal scenery. Hard rock – granite and certain limestones, for example – yields slowly, even to a storm's fury, but nevertheless, the sea undercuts it, carving caves, arches, stacks and cliffs, which may reach up to more than 600 m (2000 ft). The effect upon softer rock is quite different. There, the sea constantly wears away rock and debris, causes landslides and hollows out bays. Cliffs are low and tumbled, and the detritus is washed along the coast to create beaches, gravel banks and pebble ridges. Plants may take precarious root in these, but unless there is a swift build-up of sediments and protective banks, they are liable to be washed away by the next storm.

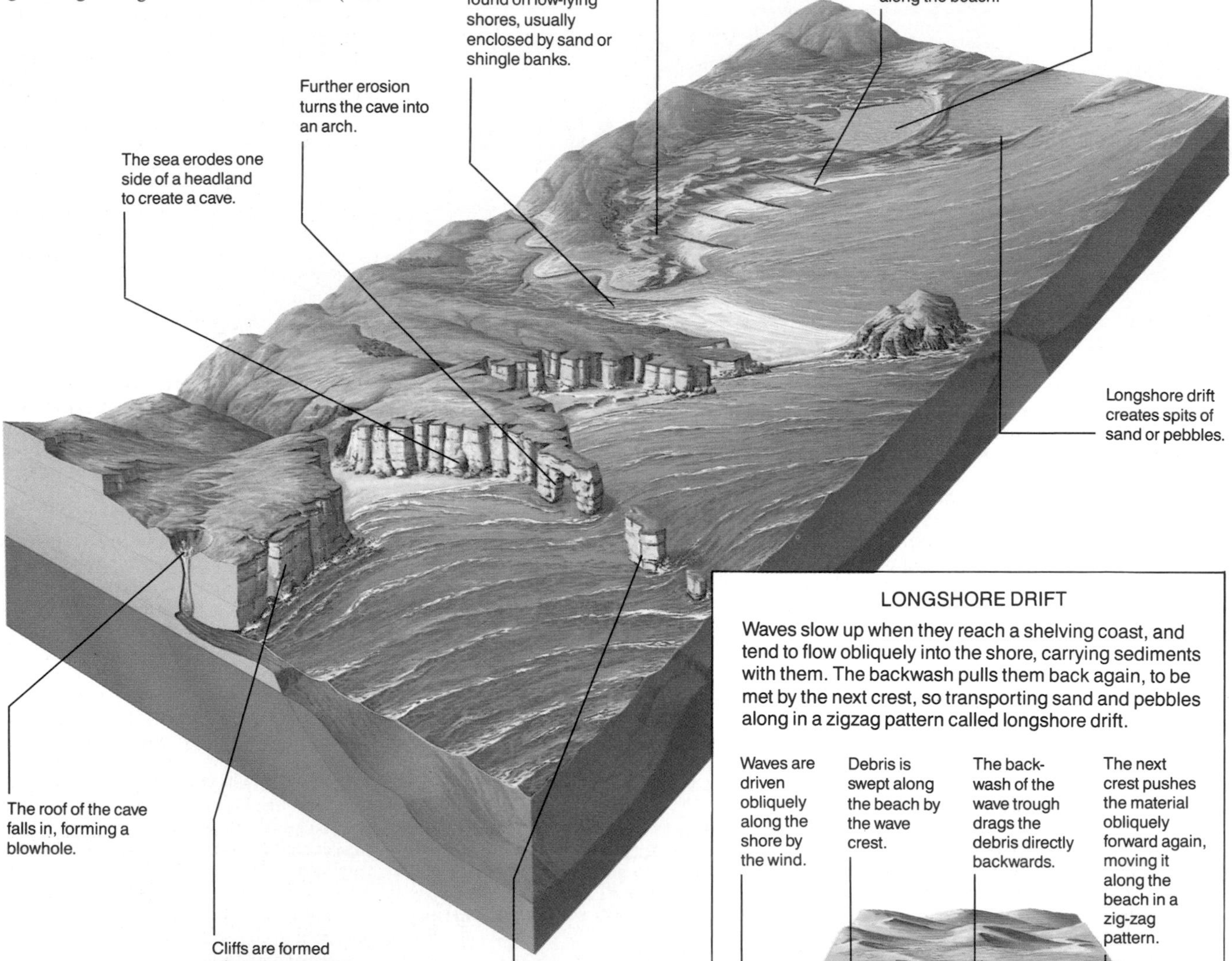

Salt marshes are found on low-lying shores, usually enclosed by sand or shingle banks.
Sand dunes are formed when beach sand is driven inland by the wind.
Groynes have been constructed to counteract the drift of sand and shingle along the beach.
Sand and shingle bars washed up by the sea form a lagoon.
Further erosion turns the cave into an arch.
The sea erodes one side of a headland to create a cave.
Longshore drift creates spits of sand or pebbles.
The roof of the cave falls in, forming a blowhole.
Cliffs are formed when storm waves attack the headland above normal high tide level, so causing landslides.
A stack is formed by the total collapse of an arch.

LONGSHORE DRIFT

Waves slow up when they reach a shelving coast, and tend to flow obliquely into the shore, carrying sediments with them. The backwash pulls them back again, to be met by the next crest, so transporting sand and pebbles along in a zigzag pattern called longshore drift.

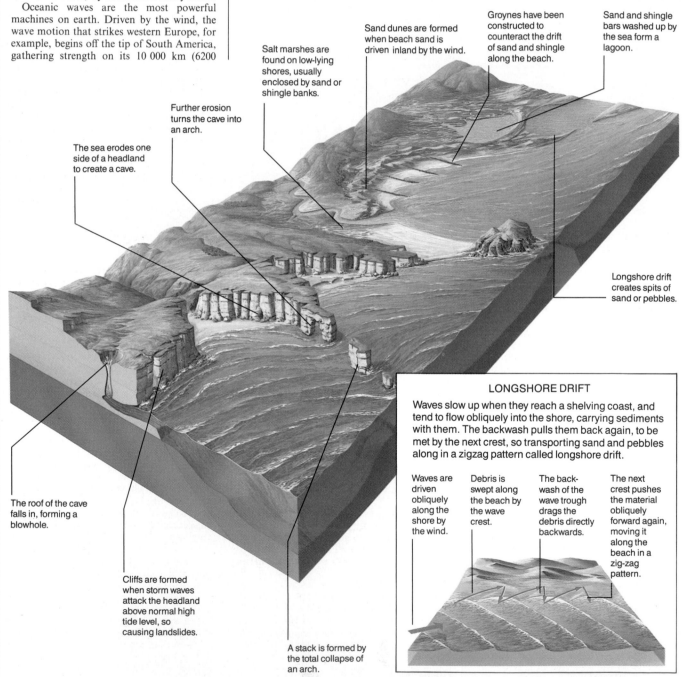

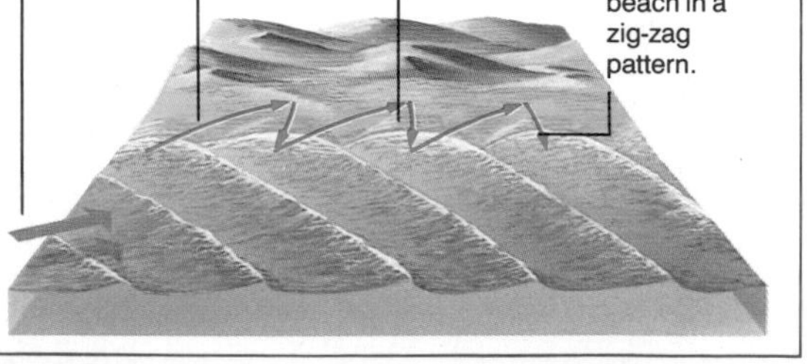

Waves are driven obliquely along the shore by the wind.
Debris is swept along the beach by the wave crest.
The back-wash of the wave trough drags the debris directly backwards.
The next crest pushes the material obliquely forward again, moving it along the beach in a zig-zag pattern.

Esbo ***Finland*** See ESPOO

escarpment Steep slope or cliff resulting from erosion or faulting. The steep slope of a cuesta is also called an escarpment.

Escaut ***Belgium*** See SCHELDE

Esch-sur-Alzette (Esch; Esch sur l'Alzette) ***Luxembourg*** Coal mining and steel town near the French border, 16 km (10 miles) south-west of the capital, Luxembourg. It is the Grand Duchy's second largest town after the capital.
Population (canton) 115 000
Map Belgium Bb

Escorial, El (San Lorenzo del Escorial) ***Spain*** Town, 40 km (25 miles) north-west of Madrid, containing a magnificent palace, monastery and mausoleum, built between 1563 and 1584 for the Spanish king, Philip II. A library and art treasures were added later.
Population 9500
Map Spain Cb

Escuintla ***Guatemala*** Market town in tropical lowlands 40 km (25 miles) south-west of the capital, Guatemala City. It is the departmental capital of the country's most prosperous region, also called Escuintla, which produces 75 per cent of the country's livestock, 20 per cent of its coffee, 85 per cent of its cotton and 80 per cent of its sugar.
Population (town) 92 500; (department) 412 800
Map Guatemala Ab

Esfahan (Isfahan) ***Iran*** City about 350 km (220 miles) south and slightly east of the capital, Tehran. The riverside city, 1402 m (4600 ft) above sea level, has several graceful bridges, including the 16th-century Bridge of Thirty-three Arches and the 17th-century Khaju Bridge, which has two tiers of arcades. The city's products include carpets and rugs, textiles, hand-printed cotton fabrics and metalwork, especially silver filigree.

In 1388 the Tatar warlord Tamerlane pillaged the city. He massacred 70 000 of its inhabitants and is said to have built a tower with their skulls. Among the city's places of interest today are the Great Bazaar, more than 200 mosques and mausoleums, 28 *madrasehs*, or schools, and numerous palaces.
Population 926 700
Map Iran Ba

esker, eskar Twisting narrow ridge of sand and gravel deposited by a stream flowing beneath a glacier or in the ice of a glacier.

Eskilstuna ***Sweden*** Industrial city beside Lake Mälaren, 115 km (71 miles) west of Stockholm. It has engineering plants and steelworks, and specialises in the making of cutlery.
Population 88 528
Map Sweden Cd

Eskisehir ***Turkey*** Spa town on the Porsuk river about 210 km (130 miles) west of the capital, Ankara. It has hot sulphur springs, and meerschaum, a clay used in making smokers' pipes, is quarried nearby. The town's main industries are sugar refining and the manufacture of agricultural machinery, cement and textiles.

The surrounding province of the same name, which covers 13 652 km² (5271 sq miles), produces sugar beet, oats, wheat and barley.
Population (town) 367 300; (province) 597 200
Map Turkey Bb

Esmeraldas ***Ecuador*** North-western province lying on the Pacific coastal plain. It has tropical rain forest which extends to the foothills of the Andes. It is a mainly agricultural area, growing tobacco, coconuts, cocoa, rice and bananas. The forests yield balsa wood, hardwoods, and tagua nuts (which are used as a substitute for ivory). There are gold mines nearby, and an oil terminal for a pipeline from the Oriente region. The provincial capital, also called Esmeraldas, at the mouth of the Río Esmeraldas, is the chief trading centre and port for the region.
Population (province) 248 000; (city) 141 000
Map Ecuador Ba

Espírito Santo ***Brazil*** See VITORIA

Espiritu Santo ***Vanuatu*** Largest island of Vanuatu lying about 240 km (150 miles) north-west of the capital, Port-Vila. Copra is produced for export on mainly French-owned plantations. The island – known locally as Santo – was the scene in 1980 of a rebellion aimed at securing separate independence from Vanuatu. After the Franco-British colonial administration failed to intervene, the rebels, led by native politician Jimmy Stevens and supported by about 50 armed Europeans, were beaten with the aid of troops from Papua New Guinea.
Population 23 000
Map Pacific Ocean Dc

Espoo (Esbo) ***Finland*** District and town about 20 km (12 miles) west of Helsinki. The district includes Tapiola, a postwar garden city and the home of Helsinki technical university.
Population 152 900
Map Finland Cc

Essaouira ***Morocco*** Port and resort on the Atlantic coast between Safi and Agadir; formerly Mogador. Phoenicians and Carthaginians occupied the site from the 5th century BC until the Romans arrived about 25 BC. The town with its 18th-century fortress stands on a low peninsula, and the harbour is sheltered by offshore islets and a rocky headland.

Sardines are abundant offshore and there is a busy fish market. Industries include sugar refining, fish canning and tanning. The town is noted for its fine crafts, especially inlaid cabinet-work.
Population 42 000
Map Morocco Ba

Esseg ***Yugoslavia*** See OSIJEK

Essen ***West Germany*** Largest city of the RUHR industrial region, about 60 km (37 miles) north of Cologne. It is a coal-mining centre and has huge iron and steel works; it is also a centre of commerce and finance. It was heavily bombed in the Second World War.

Modern Essen was built on the armaments empire of the Krupp family. The Krupps built suburbs such as Margaretenhöne for their workers and the Villa Hügel for themselves, overlooking Lake Balderney. The mansion grounds are now a city park and there is a museum of industry and Krupp technology.

The city's cathedral was consecrated in 873, and at the nearby hamlet of Werden is Germany's oldest parish church, St Lucius, dedicated in 995.
Population 635 200
Map West Germany Bc

▼ AUSTERITY AND MAJESTY The spartan quarters of Spain's pious king, Philip II, are preserved in the sumptuous monastery-palace at El Escorial where he died in 1598. It has some 300 rooms.

Ethiopia

TELEVISION BROUGHT THE WORLD TO THE RESCUE OF A COUNTRY WHERE MILLIONS FACED STARVATION WHILE THE GOVERNMENT CONCENTRATED ON FIGHTING A WAR

The disastrous famine of the 1980s brought the East African state of Ethiopia to the forefront of the world stage. Famine is not unusual in the area, but this was the worst of recent times and it was brought to the attention of people all over the world by television.

By mid-1985 more than 250 000 people had died from starvation. Many of them were young children, and poignant pictures of their plight had appeared night after night on television screens. Consciences were pricked and a massive international relief operation gathered momentum.

There are three main reasons for famines in Ethiopia. Firstly, while some areas are always well-watered others are regularly afflicted by severe drought, and extensive mountain terrain hinders the transport of food to stricken areas. Secondly, most Ethiopians are so poor that they cannot afford to buy food when they lose the crops they grow to feed themselves. Thirdly, there is long-standing civil war and guerrilla conflicts in several areas, and the government has devoted most of its energies and financial resources to these battles.

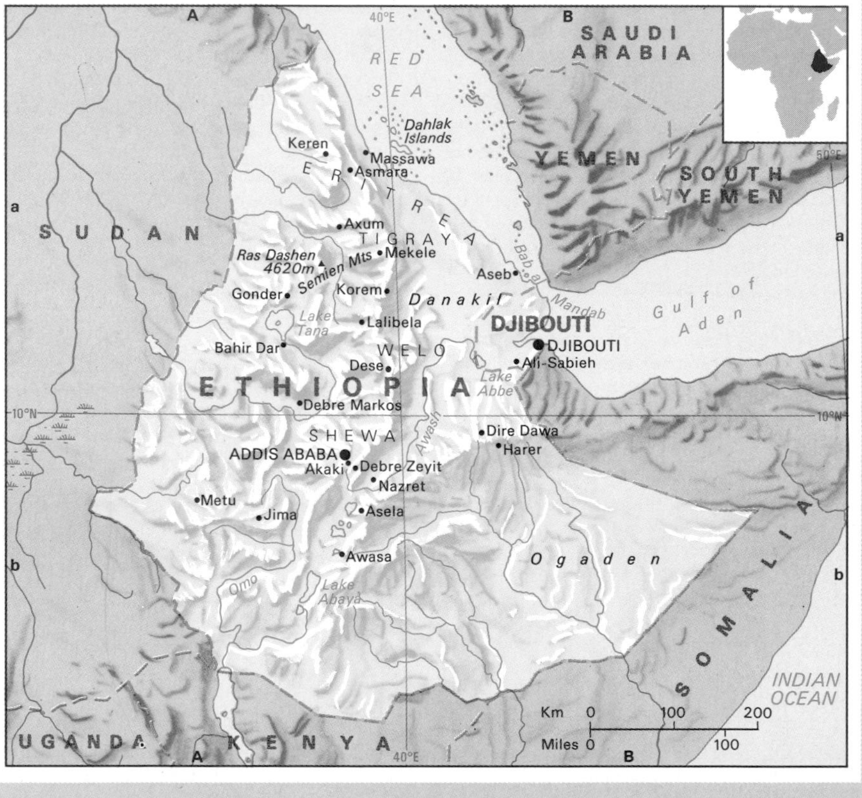

By the mid-1980s the drought had become the longest-running of recent times. It had hit different regions in different years, but had continued intermittently since 1972. Many local economies had been destroyed as crops failed and livestock perished. In addition, the war had disrupted food production by taking men away from their farms, and was hindering relief efforts.

Politically and historically, Ethiopia (formerly known as Abyssinia) is highly distinctive among African countries, for it has existed as a nation in some form for over 2000 years and was strong enough in the late 19th century to resist Italian attempts at colonial occupation. The Italians ruled ERITREA in the north from 1890, and eventually took over Ethiopia in 1936, but they were forced out of both five years later by Ethiopian and Allied forces.

Emperor Haile Selassie ('The Lion of Judah') was restored to the throne he had lost to the Italians in 1936, and this re-established a line of monarchs descended from King Solomon (972-932 BC). Britain administered Eritrea until 1952, when it was united with Ethiopia, though with considerable autonomy until 1962.

In 1974 Haile Selassie was overthrown by a military regime headed by a group known as the *Dergue* and led by Colonel Mengistu Haile Mariam. They have swept away the former feudal aristocracy and have adopted a Marxist-Leninist ideology, with land-reform policies and a political structure based partly on peasant associations.

Religion is important to most Ethiopians. Christianity was established as early as the year AD 330 around AXUM in the north, while Islam followed a few centuries later. The population is divided between the two faiths. Ethiopian Orthodox, or Coptic, Christianity was formerly the official religion, but the military regime does not support any.

Ethiopia is also one of Africa's largest countries, with an area of 1.22 million km^2 (over 470 000 sq miles) and a population of 42 million. Most of the country is highlands, rising to 4000 m (13 000 ft). It drops sharply towards the Sudan border in the west and towards the DANAKIL lowlands and the Red Sea in the north-east. The land falls more gently towards Kenya in the south and the OGADEN desert and Somalia in the east.

The highlands are divided by the northern end of the East African GREAT RIFT VALLEY, which forms a series of lakes. The main rivers are the Abbai, which flows into Sudan and is also known as the Blue Nile, and the Awash, which drains the northern part of the Rift Valley and is used for irrigation before it disappears into the Danakil sands. Rain falls mainly in the west, but is unreliable. Drought sometimes stops crops growing and heavy rain, often in thunderstorms, at other times erodes soil from slopes and reduces the area where crops can be planted.

Most Ethiopians live in the highlands, in family clusters of 10 to 20 huts, without any amenities such as electricity and piped water, often miles from the nearest shop or health centre. They must provide almost all their own basic needs.

Their farms generally occupy less than 1 hectare (2.5 acres), often in scattered fragments. An occasional farmer may have a pair of oxen and a plough, but many – especially at lower altitudes – have only hoes for working their land. The main crops are *tef* (a local grain), barley and wheat in the cool highlands, maize and sorghum on the lower land, and *inset* (a relative of the banana) in well-watered parts of the south. Yields are usually low because the soil is poor, and fertiliser is too expensive.

The lower dry areas such as the Ogaden are more sparsely populated by nomadic tribes who move across country with their camels, cattle, sheep and goats in search of water and grass. Some of them have permanent homes where they live for part of the year, planting sorghum in the hope of rain.

The chief agricultural export is coffee, grown mainly on small farms especially in Kefa, Sidamo, Wollega and Illubabor regions in the west and south. A few large plantations, some financed by foreign firms, produce commercial crops like sugar and cotton.

There are few countries of Ethiopia's size with so few known mineral resources, but there is a vast potential for hydroelectricity if there is ever enough industrial development to need it.

There may be undiscovered mineral wealth, including oil, but for the present, employment outside agriculture is confined to a small manufacturing sector covering building materials, footwear, tyres, food processing and textiles,

▲ **FLIGHT FROM FAMINE Thousands of starving refugees have taken shelter at the famine relief camp at Korem in east-central Ethiopia. Driven from their rural homes by drought, they have increased the population of the village threefold to 30 000.**

and to service industries such as hotels and shops in the towns. The towns are scattered across the country, and ADDIS ABABA, with a population of 1.5 million, is by far the largest. It is the capital city for a highly centralised national administration, and easily the most important commercial and industrial centre. A number of factories are sited between the capital and Nazret, 100 km (60 miles) away, and there are a few at ASMARA and at DIRE DAWA.

Asmara, in Eritrea, is the second biggest city, with 430 000 inhabitants; most of the other large provincial towns have 50 000 to 80 000 people. About 15 per cent of Ethiopians live in the towns and enjoy a higher income, more food and better housing than the rural people.

Ethiopia is among the world's ten poorest countries, and by some yardsticks is at the bottom of the list. For instance, there is only one doctor for every 72 000 people, which is the worst figure in the world. Medical needs are greatly increased by famine.

At the same time the state has maintained an army of at least 300 000, representing a vast drain of manpower from the land, and large areas have been affected by armed conflict between the army and a number of regional liberation movements: in Eritrea and TIGRAY in the north (where separatists want an independent Eritrean state), in the Ogaden (long claimed by neighbouring Somalia, which aids the local rebels), and in parts of the south. Despite Soviet military aid for the Addis Ababa government, rebels control large areas, including most of Eritrea and Tigray, and the government has been accused of withholding famine relief from these areas.

The environmental, economic and political problems of the country are interrelated, as is illustrated by the official resettlement programme. There is simple logic in encouraging movement from overcrowded and degraded land in the north to unoccupied, potentially productive areas in the south-west, but this is increasingly seen as a move to crush rebellion in the north. People arriving at northern food-distribution centres have sometimes been forcibly moved long distances to different regions, even when this meant dividing families. It is alleged that most of the people moved in this way are men of military age who are seen as potential recruits for the rebels.

Half a million Ethiopians have trekked into Sudan as refugees. More than 250 000 Ethiopians have died. But the Ethiopian military government and the regional rebels seem far more concerned with fighting each other than with fighting famine.

By the mid-1980s the government was beginning to show more concern about the famine and international aid was pouring into the country from governments and international agencies, from long-established charities and impromptu groups drawn from Western show business. The immediate problem was to supply food and transport it to the starving millions, but the more important longer-term aim must be to try to rebuild Ethiopia's shattered agriculture.

ETHIOPIA AT A GLANCE
Area 1 221 900 km² (471 800 sq miles)
Population 42 580 000
Capital Addis Ababa
Government Marxist military republic
Currency Birr = 100 cents
Languages Amharic, Galla, Sidamo, Arabic
Religions Muslim (45%), Christian (40%), tribal (15%)
Climate Temperate on plateau, hot in lowlands. Average temperature in Addis Ababa ranges from 5°C (41°F) to 25°C (77°F)
Main primary products Wheat, barley, maize, sorghum, millet, coffee, cotton, sugar cane, beans and peas, cattle, timber; salt
Major industries Agriculture, food processing, textiles, cement
Main exports Coffee, hides and skins, beans, cotton, sesame seeds
Annual income per head (US$) 126
Population growth (per thous/yr) 7
Life expectancy (yrs) Male 39 **Female** 43

Essequibo ***Guyana*** River, at around 1000 km (625 miles) the country's longest. It rises in the Guiana Highlands on the Brazilian border and at its mouth, west of the capital, Georgetown, it is nearly 50 km (30 miles) wide.
Map Guyana Bb

Essex ***United Kingdom*** English county, 3674 km² (1419 sq miles) in area, immediately east of London on the north bank of the River Thames. Much of its original landscape of low hills and woodland, such as Epping Forest, has been largely swallowed up by the growth of the capital, but the north-east of the county retains its rural character. The industrial centres are the New Towns of Harlow and Basildon, and the Thames-side port of Tilbury. Farther down the river estuary is Southend-on-Sea, one of Britain's biggest seaside resorts, while the ferry port of Harwich is at the far north-east tip of the county. Colchester was the chief town from pre-Roman times and is now the seat of the University of Essex, but today the county town is Chelmsford.
Population 1 492 000
Map United Kingdom Fe

Esslingen am Neckar ***West Germany*** Town 10 km (6 miles) south-east of Stuttgart. Industrial and residential expansion has made it little more than a suburb of that city.
Population 87 300
Map West Germany Cd

Estonia ***USSR*** Smallest of the Soviet Baltic republics, lying north of Latvia. Estonia consists mainly of boulder-strewn lowland, but also includes over 800 islands in the Baltic. It covers 45 100 km² (17 400 sq miles), and until 1945 was noted mainly for its cattle and dairy produce. Its industries now include food processing, electrical engineering and shipbuilding, and its people have the highest standard of living in the USSR. The chief towns are the capital, TALLINN, the university town of TARTU and the port of Pärnu on the north-east coast of the Gulf of Riga.

Estonians form a distinctive group with their own language related to Finnish and Hungarian, and have kept their own strong culture despite 700 years of foreign domination; their only brief period of independence was between 1920 and 1940. Estonia was in turn ruled by or divided between Russia, Poland, Sweden and Germany; it was Russian territory from 1721 to 1918. The rise of a nationalist movement in the 19th century culminated in a war of independence against occupying Germans in 1918-20. But the secret Russo-German pact of 1939 promised Estonia to the USSR, and the Russians occupied it and made it one of their republics the next year. Nazi Germany again occupied Estonia in 1941-4, killing 125 000 Estonians and deporting thousands to Germany. It was liberated and forcibly reincorporated in the USSR during the Second World War, but this annexation is not recognised as legal by some Western governments.
Population 1 518 000
Map USSR Dc

Estoril ***Portugal*** Seaside resort about 24 km (15 miles) west of the capital, Lisbon. It has a casino, parks, fine beaches and a golf course. It was originally developed because of the reputed healing qualities of its waters.
Population 24 300
Map Portugal Bc

Estremadura ***Portugal*** Coastal province immediately west of the capital, Lisbon, stretching northwards. It has numerous small fishing villages and seaside resorts, and it contains much of the country's heavy industry, including shipbuilding, iron and steel, cement and chemicals. Farmers in the region mostly grow wheat, maize, olives and vines. The chief cities of the province, which covers 5345 km² (2064 sq miles), are Lisbon and Setúbal.
Population 2 656 180
Map Portugal Bc

Estremoz ***Portugal*** Fortified town about 130 km (80 miles) east of the capital, Lisbon. There is a castle, with a 13th-century keep and 17th-century ramparts. Estremoz is noted for its pottery. Marble is quarried near the town.
Population 7300
Map Portugal Cc

estuary Tidal mouth of a river where sea water mixes with fresh water.

Esztergom ***Hungary*** Attractive Baroque city and the country's ecclesiastical capital. It lies at the foot of the Pilis volcanic hills which look across the Danube river into Czechoslovakia, 43 km (27 miles) north-west of Budapest. The Romans first founded and fortified the city's site, and here in the year 1000 Hungary's first king, St Stephen, was crowned. The present magnificent, domed cathedral, incorporating the fine 16th-century Bakocz chapel, dates from the 19th century and is Hungary's largest church. The Cardinal-Primate's Palace houses a superb Museum of Christian Art, with splendid European paintings and French tapestries.
Population 31 400
Map Hungary Ab

étang French term for a shallow pool or lake lying among sand dunes or beach gravels, especially those of the Landes coast of France.

etesian winds Strong north or north-west winds blowing from the high-pressure areas in the eastern Mediterranean at intervals during the summer. They strengthen in the afternoons and often die away towards evening.

Ethiopia See p. 214

Etna ***Italy*** One of the world's most active volcanoes and the largest in Europe, rising in east Sicily to about 3323 m (10 902 ft). Its base is 200 km (125 miles) in circumference and there are more than 200 cones scattered about its slopes. The biggest recorded eruptions occurred in 1329, 1669, 1910-11, 1928, 1950-1 and 1971.

The lower slopes on the south and east sides are densely settled, with rich plantations of vines, citrus trees and temperate fruits. The wooded upper slopes have been developed for skiing and as a summer retreat from the coastal heat.
Map Italy Ef

Etosha Pan ***Namibia*** Shallow lake, about 160 km (100 miles) south of the Angolan border, which is dry for most of the year. Its many waterholes attract vast numbers of animals, including elephants, giraffes, kudus, leopards, lions, oryxes, springboks and zebras. The 8565 km² (3307 sq mile) Etosha National Park, founded in 1958, includes the lake.
Map Namibia Aa

Euboea (Évvoia) ***Greece*** Island in the west Aegean Sea, separated from the mainland by a narrow strait, bridged at Evripos with a 60 m (197 ft) span. The capital, Khalkís, is an industrial centre with a yachting harbour. Inland there are mountain ranges with pine-forested slopes, dotted with the ruins of 13th to 15th-century Venetian towers, Frankish castles from Crusader days, and Byzantine monasteries. Tourists visit its charming villages such as Prokópion and Limni, and the beaches at Káristos, Kími and Stíra. Ancient remains, including a theatre and gymnasium, have been uncovered at Eretria, 14 km (9 miles) south of Khalkís.

Red wine, figs, honey and livestock are produced – and the island gets its name from the ancient reputation of its cows: Euboea comes from Greek words meaning 'good cattle'. Cement is also made on the island and fine marble is quarried.
Population 188 400
Map Greece Cb

Eupen ***Belgium*** Manufacturing town 45 km (28 miles) east of the city of Liège. The town lay inside Germany until 1919, when it was ceded to Belgium by the Treaty of Versailles. Many of its inhabitants are still German-speaking. Manufactures include tanning, brewing, needle and wire making, domestic appliances and food products.
Population 15 000
Map Belgium Ca

Euphrates (Al Furat; Firat) ***Middle East*** Longest river in western Asia (2720 km, 1690 miles), rising in the Armenian highlands of Turkey between the Black Sea and Lake Van. It has two branches: the Kara, with its source near Erzurum, and the Murat near Mount Ararat. They flow south-west and join near Keban to form the Euphrates proper (known in Turkey as the Firat).

The river flows for 1096 km (681 miles) in Turkey before crossing the Syrian desert from north-west to south-east. Its Arabic name is Al Furat. For much of the way through Syria the Baghdad-Aleppo highway runs beside it.

After the Syrian border town of Abu Kamal, the Euphrates enters Iraq. It finally joins the TIGRIS river above Al Qurnah to form the SHATT AL ARAB, 64 km (40 miles) north-west of Basra. The river – which can be navigated only by shallow-draught vessels – is used for irrigation in Syria and Iraq. Together with the Tigris, it was the cradle of some of the earliest civilisations, and the ruins along its course include those of BABYLON, SUMER and UR.
Map Middle East Ba

Europe If continents have specialities, then that of Europe would appear to be political and economic systems. The notion of Democracy was born in Europe, and so were the notions of Capitalism and Communism. The physical line between these two 'isms', the Iron Curtain,

is one of the most rigid boundaries in the world.

There are eight countries within Communist Eastern Europe – Albania, Bulgaria, Czechoslovakia, East Germany (the German Democratic Republic), Hungary, Poland, Romania, Yugoslavia – and a large part of a ninth nation, the USSR. All, to a greater or lesser degree, are under Soviet influence – although some, and especially Yugoslavia, have shown a considerable degree of independence.

Eastern Europe is an enormous area, stretching from the Adriatic to the Ural Mountains, and from the Arctic Ocean to the Caspian Sea. The most characteristic landscape is rolling plain – coniferous forest and parkland, arable farmland and endless vistas of grass. There are great and famous rivers too – the Elbe and the Danube, the Don, the Dnieper and the Volga – and formidable mountain ranges from the Transylvanian Alps and the Caucasus to the Urals. The climate is transitional between Atlantic and extreme Continental, with winters getting steadily colder towards the east. The Black Sea, however, has delightfully warm summers, while the Adriatic, as Western holidaymakers in Yugoslavia know, has a Mediterranean climate.

These differences are reflected in a wide range of farming activities. About 45 per cent of East Europe is cultivated, and some 20 per cent – chiefly in Hungary and Romania – is grazing land. Rye and oats along the Baltic give way to wheat and maize to the south, and potatoes and sugar beet are grown in the cooler, moister areas – particularly East Germany, the USSR and Poland. More specialised crops include Czechoslovakian hops; wine grapes from Hungary, Bulgaria, Romania and Yugoslavia; citrus fruits and olives from the Adriatic; and Balkan cotton and tobacco.

In the post 1945 era, heavy industry in the region was initially developed at the expense of living standards. A mass of new mines and steel and chemical plants appeared, though Poland and Czechoslovakia both maintained their prewar positions as Eastern Europe's leading producers of iron and steel outside the USSR. East Germany's war-ravaged industry has long been rebuilt, and the country is now highly productive, particularly in the chemical field. Hungary specialises in engineering and electronics while Yugoslavia has looked to the West to find markets for its clothes, cars, furniture, household appliances and food exports. Industrial growth in Albania, Romania and Bulgaria has been remarkable since 1945, but still lags behind that of the older developed nations. Many of the Eastern European Communist nations are members of the Council for Mutual Economic Assistance (COMECON). This international organisation coordinates economic policy and trade amongst its members. Despite all this activity, the living standards in most of the region still remain below those of most of Western Europe.

The peoples of Eastern Europe suffered terribly during the Second World War. Civilian casualties ran to more than 30 million, and whole populations were transported as slave labourers; nevertheless, the ancient ethnic patterns of the nations remain virtually unaltered. About 60 per cent of Eastern Europeans are Slavs, who probably originated somewhere to the north of the Carpathians. The East Slavs became the Russian peoples (who also possess some Viking blood), the West Slavs include the Poles, Czechs and Slovaks, while the Bulgarians and peoples of Yugoslavia had their beginnings in the South Slavs. The remaining 40 per cent of the population are mostly Germans, Hungarian Magyars, Romanians and Albanians, plus sizable minority groups of gypsies, Jews, Turks and Vlachs. They are all divided by an extraordinary mixture of tongues. Albanian is not related to any other living language, Romanian is a wandered-off Latin tongue and Hungarian is distantly related to Finnish. The Slav languages have much in common, but are not mutually intelligible. The Bulgarians, like the Russians, use the Cyrillic alphabet, while the principal language of Yugoslavia is called Serbian when written in Cyrillic and Croatian when the Roman alphabet is used. German is the most widely used common language, resisting all efforts of Russian to supplant it.

While the Eastern Bloc contains a sizable

▲ ANGRY MOUNTAIN The snow-capped, fire-belching summit of Mount Etna reaches into the clouds. More than 140 recorded eruptions — many of them highly destructive — have not driven farmers from its fertile volcanic soils.

amount of the largest nation in the world and a great deal of open space, Western Europe, by contrast, embraces such tiny nations as Monaco, Andorra, Liechtenstein and San Marino, and is densely populated.

Western Europe consists of 24 independent nations that occupy a number of peninsulas and islands jutting out from the main Eurasian land mass. Much of its jagged shape, at least to the north, is due to the grinding and tearing of the Ice Ages, while the other physical features were created by the long-ago upthrust of mountain chains. Some of these are very ancient indeed, like the Precambrian rocks in the north of Scotland – 580 million years old. Others are more youthful, like the Alps, which are no more than 35 million years old. In between, there are plains like those of northern France and Belgium, plateaus such as that of central Spain and basins like those in which Paris and London lie. All this produces an enormous, and often beautiful, variety of scenery. The climate is generally kindly, since it is influenced by the North Atlantic Drift, a branch of the Gulf Stream.

Exceptions are northern Sweden and Finland, whose climate is Arctic, and the North German Plain, where it is more Continental, with hot summers and cold winters.

Conditions suit farming, and outside Scandinavia, where little agricultural land is available, about 50-75 per cent of each country's land area is devoted to it. Livestock, root crops, orchards, cereals, feedstuffs and such industrial crops as oil seed rape are all important, while to the sunny south, the vineyards of France, Italy, Spain and so on produce about 60 per cent of the world's wine. Olive oil and citrus fruits are also major products of the Mediterranean area.

An abundance of energy sources, raw materials, manpower and ingenuity made Western Europe a great industrial region at a relatively early stage. Its Industrial Revolution, which began in Britain in the 18th century, spread outwards and came eventually to affect the lives of everyone worldwide. The first, and still important, energy source was coal. It was supplemented later by hydroelectric power obtained from the region's great rivers, by imported oil and nuclear power, and latterly by North Sea oil and natural gas. Iron ore is the most important non-fuel mineral mined, and steel production, traditionally at least, the most important basic manufacture. Other major products include textiles, machinery, cars, aircraft, ships, foodstuffs, consumer goods and electronics.

Down the years, Western Europe gave much to, and took much from, the world at large. But since the Second World War, its nations have relinquished their empires, and surrendered their status as world powers to the Americans and the collective comfort of the North Atlantic Treaty Organisation, the mutual defence against the Eastern Bloc. Among themselves, they have instituted the European Economic Community which is still attracting new members.

Area 10 498 000 km^2 (4 053 300 sq miles)

Population 682 million

Europoort ***Netherlands*** Port and industrial complex created since 1958 as the 'gateway to Europe'. It lies on the coast, on the Nieuw Waterweg (New Waterway), 28 km (17 miles) west of central Rotterdam. Europoort can handle the world's largest bulk carriers of oil, ore and grain, and has refineries for oil and nonferrous metals, and shipbuilding, chemicals, engineering and food processing plants.

Map Netherlands Bb

eustatic change Worldwide rise or fall in sea level. Glacial-eustatic change is caused by the melting or growing of the polar ice caps; otherwise changes in the configuration of the ocean basins are responsible.

evaporite Sedimentary rock made up of minerals such as rock salt or gypsum that has been formed by evaporation of salt water.

evapotranspiration Combined loss of water by evaporation from the ground and by transpiration from plants.

Everest, Mount ***Nepal/China*** The world's highest mountain, rising 8848 m (29 028 ft) at the eastern end of the Great Himalayas, on Nepal's unmarked border with Chinese Tibet. Its height was first measured in 1852, and it was named after Sir George Everest, Surveyor-General of India at the time. The Tibetans call the mountain Chomolungma – Goddess Mother of the World – or in Chinese Qomolangma.

Doubts arose later about that first measurement. A mass the size of Everest can exert its own gravitational pull, so the British survey team were not sure that their spirit levels were true – meaning that their theodolites, too, might be giving an inaccurate reading. Because of this they took six measurements and worked out the average: exactly 29 000 ft. More recent measurements by Chinese and Indian surveyors have put the height at 8849 m (29 032 ft) – but the official height remains a metre less than this. In fact, the changing thickness of ice and snow at the summit continually alters the height.

Everest was first climbed by the New Zealand mountaineer Sir Edmund Hillary and Tenzing Norgay, a Sherpa guide, in 1953. Since then there have been more than 130 successful ascents, including four by women and five without oxygen. However, about 50 have also died attempting the climb. Despite that, there is a queue – currently seven years long – of mountaineers eager to try.

Only two expeditions are allowed on the mountain at any one time, and they must take different routes; and the fierce weather permits only three 'climbing seasons' – April-May, before the monsoons; October, after the rains; and December-January, in winter.

Map Nepal Ba

Everglades ***USA*** Largest area of subtropical wilderness in the country – including swamps and forest – lying in southern Florida, south of Lake Okeechobee. Its survival has been threatened at times by water extraction for agricultural and urban use. The area is rich in wildlife and there are extensive areas of saw grass and mangrove. Some 5668 km^2 (2188 sq miles) in the far south form a national park.

Map United States Je

Évian-les-Bains ***France*** Health resort with mineral springs on the south shore of Lake Geneva. It produces bottled mineral water, named after the resort.

Population 6200

Map France Gc

Évora ***Portugal*** Capital of the province of Alentejo, about 110 km (70 miles) east of the national capital, Lisbon. It was captured by the Romans in 80 BC and has a Roman aqueduct and temple of Diana, the goddess of hunting. There is also a 13th-century Gothic cathedral. From the 14th to the 16th centuries Évora was the residence of the Kings of Portugal. Today it is mainly an agricultural market, although there is some cork, leather and woollen-carpet production.

Population 34 000

Map Portugal Cc

Evpatoriya ***USSR*** See YEVPATORIYA

Évreux ***France*** Capital of Eure department in Upper Normandy, on the Iton river 90 km (56 miles) west of Paris. It was founded in pre-Roman times and is one of the oldest towns in France. The Romans occupied the town for nearly five centuries until it was seized by Clovis, King of the Franks, in the late 5th century. There are Roman ruins, an ancient Norman church and a Gothic cathedral. New factories house engineering and electrical firms which have moved from Paris since 1955.

Population 57 300

Map France Db

Evros (Maritsa; Meriç) ***South-east Europe*** River, about 480 km (300 miles) long. It rises in Bulgaria's Rila Mountains and flows east then south-east past Pazardzhik, Plovdiv and Svilengrad to the border with Greece and Turkey near Edirne, Turkey. It then forms the Greek-Turkish border and flows south and south-west to enter the Aegean Sea in a delta some 11 km (7 miles) wide about 5 km (3 miles) west of Enez. It is dammed to provide hydroelectric power and used to irrigate fields of cotton, cereals, tobacco, fruits and rice on the plain of THRACE. Its main tributaries are the Bacha, Asenovitsa, Arda, Topolnitsa, Stryama, Tundzha and Ergene rivers.

Map Bulgaria Bb; Greece Ea

Evry ***France*** Capital of Essonne department and one of France's new towns, 30 km (19 miles) south-east of Paris. It was founded in 1970.

Population 29 600

Map France Eb

Évvoia ***Greece*** See EUBOEA

Exeter ***United Kingdom*** City and port on the River Exe in south-west England, 250 km (155 miles) from London, and the county town of Devon. The city walls date back to Roman times. It held out for two years after the Norman Conquest in 1066 and only fell after an 18 day siege. The cathedral, which survived the bombing that flattened parts of the city during the Second World War, is largely 13th century, and the guildhall is pre-15th century. The city's quay has some fine 19th-century buildings and a splendid maritime museum. The University of Exeter occupies a beautiful site on the north side of the city.

Population 102 000

Map United Kingdom De

Exmoor ***United Kingdom*** Area of moorland pasture and heathland in the north of the English counties of Somerset and Devon, near the Bristol Channel coast. The highest point is Dunkery Beacon (520 m, 1705 ft), and along most of the coast there are steep cliffs. The heathland is the home of red deer, ponies, sheep and grouse, and was the setting for R.D. Blackmore's novel *Lorna Doone*. Exmoor National Park covers 685 km² (265 sq miles).

Map United Kingdom De

exosphere See ATMOSPHERE

extrusion Movement of magma through volcanic craters and fissures in the earth's crust out onto the earth's surface, forming extrusive igneous rock. This cools very quickly on the surface, thus forming small crystals which give it a fine-grained or glassy texture.

extrusive rock See EXTRUSION

Exuma ***Bahamas*** Island group in the west-centre of the archipelago, consisting of Great Exuma, Little Exuma and adjacent cays (low reefs and banks). They total about 335 km² (130 sq miles). The main occupations are fishing, farming and tourism.

Population 4000

Map Bahamas Bb

eye The circular area of relative calm at the centre of a CYCLONE, where the atmospheric pressure may be very low.

eyot Small island, especially in a river.

Eyre, Lake ***Australia*** Salt lake in South Australia, covering about 8900 km² (3400 sq miles). It lies about 700 km (430 miles) north and slightly west of Adelaide. It is named after the explorer Edward Eyre, who discovered it in 1840. The lake – Australia's largest lake and, at 16 m (52 ft) below sea level, the continent's lowest spot – is mostly barren salt flats. It has filled only three times this century, in 1950, 1974 and 1984. In 1964 Donald Campbell set up a new world land speed record of 701.030 km/h (429.311 mph) on Lake Eyre, in his turbine powered *Bluebird*.

Map Australia Fd

Ezulwini Valley ***Swaziland*** Scenic valley between Mbabane and Manzini, whose name means 'place of heaven'. It is Swaziland's leading tourist area, especially for South African visitors, and has a string of hotels, a casino, night club and cinema. Outdoor attractions include the majestic Mantenga Falls on the Little Usutu river and the Mlilwane Wildlife Sanctuary, which protects animals such as white rhinoceroses, giraffes, zebras, elands, buffaloes, crocodiles and antelopes within its 44 km² (17 sq miles).

Map Swaziland Aa

F layer See ATMOSPHERE

Fada ***Chad*** Oasis settlement on the Ennedi plateau, some 900 km (560 miles) north-east of the national capital, Ndjamena. It is a stronghold of the Toubou nomads and became involved in the 1980s civil war.

Map Chad Bb

Faeroe Islands (Føroyar) ***Denmark*** Group of 18 islands in the North Atlantic between the Shetlands and Iceland. Their total area is 1399 km² (540 sq miles). The largest island is Streymoy (Strømø), on which the capital, Tórshavn (with 12 800 inhabitants), is located, and the highest point is Slaettaratindur (882 m, 2894 ft) on Eysturoy (Østerø). The islands are largely composed of volcanic rock, heavily gouged and eroded by glaciers, and the rock makes a dramatic, often forbidding, backdrop to the scattered fishing and farming settlements. Around the fringes of the islands the high cliffs are the haunt of millions of sea birds.

The principal occupation is farming, with open-range sheep rearing dominant, and fishing, with herring, cod and haddock as the main catch. The islanders also carry out an annual 'round-up' of small pilot whales. Most of the exports consist of frozen, salted and processed fish, fish meal and fish oil.

The islanders, who have had a measure of home rule since 1948, have their own language, Faeroese, springing from Old Norse, in which there is a growing literature, and their own flag, postage stamps and currency notes.

Population 44 500

Map Europe Cb

Fagatogo ***American Samoa*** See PAGO PAGO

Fahaheel (Al Fuhayhil) ***Kuwait*** Rapidly expanding port 30 km (18 miles) south of Kuwait city, of which it is virtually a suburb. Until the 1960s it was no more than a Bedouin fishing village. Since then the development of a vast gas liquefaction project has brought prosperity and foreign workers – mainly Koreans and Filipinos – to the town.

Population 42 200

Map Kuwait Bb

Failakka Island (Faylakah) ***Kuwait*** Island 19 km (11 miles) from the mainland, east of Kuwait Bay. Archaeological remains show that between 3000 and 1200 BC there was a thriving community on the island, which had close trading links with the cities of the TIGRIS and EUPHRATES and with Dilmun (see QALAT AL BAHRAIN). Today the island is being developed as Kuwait's main leisure centre and tourist resort with sailing clubs, museums, a hypermarket, and a regular ferry service to the mainland.

Population 4900

Map Kuwait Bb

Fair Isle ***United Kingdom*** Isolated island, lying between the Orkneys and Shetlands, 37 km (23 miles) from the nearest land. The islanders are mainly farmers and fishermen, and the women still produce traditional Fair Isle knitwear with its intricate coloured patterns, possibly of Norse origin.

Population 75

Map United Kingdom Gh

Fairbanks ***USA*** Largest inland town in Alaska, standing in the centre of the state, 190 km (120 miles) south of the Arctic Circle on the Trans-Alaska (oil) Pipeline. It is a trade, tourist and university centre.

Population 26 600

Map Alaska Cb

Faisalabad (Lyallpur) ***Pakistan*** City about 120 km (75 miles) west of Lahore. Its prosperity dates from the opening of the Lower Chenab Canal in 1892. It stands in a wheat and cotton area, but the area's fertility is now threatened by a rising water table and increasing salinity of the soil.

Population 1 092 000

Map Pakistan Db

Falkirk ***United Kingdom*** See CENTRAL REGION

Falkland Islands (Islas Malvinas) ***South Atlantic*** Two large and about 200 small islands, covering in all 12 173 km² (4700 sq miles), the Falklands are a British Crown Colony lying about 770 km (480 miles) north-east of Cape Horn. The two main islands are East and West Falkland. The islands are believed to have been discovered in 1592 by the English navigator John Davis, but the first settlers were the French. They established themselves at Port Louis on East Falkland in 1764, which they sold to the Spanish in 1767. A British garrison, which had been on West Falkland since 1765, clashed with the Spaniards and was eventually withdrawn in 1774. In 1828, the newly independent Argentina established a settlement on East Falkland. This was destroyed by the US warship *Lexington* on the grounds that the settlers were

attacking American seal hunters in the region.

The British resumed occupation of both islands in 1833 and have maintained them as a British colony ever since. Argentina has continued to claim the islands, which it calls Islas Malvinas. The prospect of rich oil deposits offshore aggravated the long-standing dispute between Britain and Argentina, and in April 1982 the Argentines made a surprise attack and occupied the islands. In June, the Argentines surrendered to a British military expeditionary force after a brief, bloody war in which 225 British servicemen were killed and Argentina lost 725 men. Britain continues to maintain a military garrison of about 4000 men.

The capital, Port Stanley, lies on the largest island, East Falkland, where Britain has built a major airfield to accommodate large jet airliners. South Georgia, a whaling settlement 1288 km (800 miles) to the south-east, and the South Sandwich Islands farther south – a group of small volcanic islands – are Falklands Dependencies.

The main islands are bleak, rocky moorlands and bogs, swept by high winds and heavy rainfall. Sheep farming is the sole industry. The population is mostly of British stock and nearly all were born there. The Falklanders are anxious to remain under the British Crown; in the meantime Argentina continues to pursue its case through the United Nations.

Population 1800

Map Argentina Dd

fall line The point at which a river crosses from hard rock to a softer rock is often marked by a waterfall or rapids. A line or zone linking such falls on nearly parallel rivers is known as a fall line. The rivers of the south-east USA drop from the Appalachians along such a line.

False Bay (Vals Bay) Inlet of the South ATLANTIC OCEAN, east of the Cape of GOOD HOPE, South Africa. It was so called because navigators sometimes mistook it for nearby TABLE BAY. False Bay was explored in 1488 by the Portuguese navigator Bartolomeu Dias, who was seeking a sea route to India. The water in the bay comes from the INDIAN OCEAN and is about 5°C (9°F) warmer than the Atlantic water in Table Bay, which partly explains why tourists flock there in summer.

Dimensions 32 km (20 miles) long, 29 km (18 miles) wide

Map South Africa Ac

Falster *Denmark* Island linked to the main Danish island of ZEALAND to its north by Farøbroen, Europe's second longest bridge (after Öland Bridge); it spans Storstrømmen and is 3.2 km (about 2 miles) long. The island covers 514 km² (198 sq miles). Nykøbing is the main town, and Denmark's southernmost ferry port, Gedser, links the island with Warnemünde (East Germany) and Travemünde (West Germany).

Population 45 900

Map Denmark Cb

▲ YOMPING GROUND Along this road on East Falkland, British troops did their 'yomping' – fast footslogging with full kit – on the way to capture Mount Harriet above Port Stanley in 1982.

Famagusta *Cyprus* East coast resort and port exporting fruit and vegetables, and manufacturing clothing. Since 1974, the town has been under Turkish occupation. The 16th-century walls surrounding the old city were built by the Venetians. Its last Venetian governor, Bragadin, was skinned alive by the Turks after they captured the town in 1535. The Turkish admiral in charge of the operation is said to have used the skin as a pennant.

In the old city there is a 16th-century palace, now a prison, a Turkish museum, a citadel and a lively market.

Population 39 500

Map Cyprus Ab

fan See ALLUVIAL FAN

Fanling *Hong Kong* Town in the NEW TERRITORIES 35 km (22 miles) north of KOWLOON. Fanling is one of several new towns built to relieve the pressure of population in the main built-up areas. It incorporates the old villages of Fanling and Sheung Shui, and has a target population of 226 000.

Population 87 900

Map Hong Kong Ca

Fao (Al Faw) ***Iraq*** Port at the mouth of the Shatt al Arab waterway on The Gulf. It is the site of a deep-water tanker terminal, serving the country's southern oil fields. But it also has a pipeline to Kirkuk in the northern fields, through which oil can flow in either direction. The pipeline was so designed to provide an alternative export route via Syria for the southern oil if The Gulf were ever blocked. However, the port and its installations were damaged in 1986 during the war with Iran.
Map Iraq Dc

Faradofay ***Madagascar*** See TAOLANARO

Farafangana ***Madagascar*** Indian Ocean port about 430 km (265 miles) south of the capital, Antananarivo. It handles coffee, hides, rice, vanilla, cloves, pepper and waxes.
Population 136 000
Map Madagascar Ab

Farakka Barrage ***India*** Low barrage across the River Ganges, 12 km (7.5 miles) west of the Bangladesh border which also carries the Assam railway line from Calcutta. The barrage diverts water into the Hugli river, so keeping Calcutta open to navigation. It was opened in 1971 and became fully operational in 1975.
Map India Dc

Faro ***Portugal*** Capital of the Algarve, Portugal's southernmost province, about 90 km (56 miles) east of Cape St Vincent, the country's south-west tip. Faro airport serves the many tourist resorts of the region. The town has several industries, including salt, fishing, cork and food processing. The large sandy beach nearby is popular with holidaymakers.
Population 28 200
Map Portugal Cd

Faroe Islands ***Denmark*** See FAEROE ISLANDS

Fatehpur Sikri ***India*** Former royal capital of the Mogul emperor Akbar, about 150 km (93 miles) south of Delhi and a popular tourist attraction. Akbar built the town about 1580, and it was abandoned after his death in 1605. A perfectly preserved wall some 10 km (6 miles) in circumference shelters the present-day city on three sides: the fourth side was open to a lake 32 km (20 miles) in circumference, which is now dry.
Population 20 000
Map India Bb

Fátima ***Portugal*** Village about 105 km (65 miles) north of the capital, Lisbon. A shrine in the village has been a place of pilgrimage since 1917, when the Virgin Mary is said to have appeared to three children on six occasions with messages calling for world peace.
Population 6500
Map Portugal Bc

fault Fracture of rock structures as a result of stresses or strains within the earth's crust, along the plane of which the rocks have been displaced in relation to each other. See also THE RESTLESS EARTH (opposite)

Faya-Largeau (Faya-Abouchar) ***Chad*** Oasis settlement about 780 km (485 miles) north and slightly east of Ndjamena. It is the capital of Borkou-Ennedi-Tibesti prefecture, and the main centre for the northern half of the country. Faya-Largeau is the focus of several marked and unmarked desert tracks, and a trading centre where nomadic camel herders exchange dates for cereals and other goods from the south.

In the civil war that raged in Chad in the 1980s Faya-Largeau became a main stronghold of Libyan-backed antigovernment forces.
Map Chad Ab

Faylakah ***Kuwait*** See FAILAKKA ISLAND

Fayum ***Egypt*** See EL FAIYUM

Fdérik ***Mauritania*** Iron-mining town about 625 km (390 miles) north-east of the capital, Nouakchott, near the border with Western Sahara. The iron ore has been mined since 1963, and is exported via the 670 km (420 mile) railway to the port of Nouadhibou.
Map Mauritania Ba

Fehmarn ***West Germany*** Island in Kiel Bay, 1 km (less than 1 mile) off the coast of Holstein. Fehmarn has now been linked by bridge to the mainland, which has not only brought an increase of visitors to its fine beaches, but has also created a new direct road-and-ferry route (European Highway E4) between Hamburg and Copenhagen. The island is West Germany's largest, covering 185 km^2 (71 sq miles).
Population 15 000
Map West Germany Da

feldspar, felspar A group of widely distributed rock-forming minerals consisting of silicates of aluminium combined with potassium, sodium, calcium and rarely barium, and probably comprising half of the earth's crust. Feldspars divide into two groups, the potassium feldspars – silicates of aluminium and potassium – the most common of which are orthoclase and microcline; and the plagioclase group – silicates of aluminium and sodium or calcium, the most common being albite and anorthite.

fell Upland stretch of open country, or moor, especially in northern England.

felspar See FELDSPAR

fen Tract of low-lying swampy land in which peat is accumulating beneath the surface, but in which the ground water is alkaline or neutral. The name is applied especially to the region around The Wash in eastern England.

Fengfeng ***China*** Important coal-mining centre of Hebei province, about 250 km (155 miles) south of Beijing (Peking). It supplies high-quality coking coal for the domestic steel industry and for export.
Map China Hd

Fens, The ***United Kingdom*** Low-lying area of eastern England to the south and south-west of The Wash, mainly in Cambridgeshire, Lincolnshire and west Norfolk. It has alternated between sea and land for centuries as the sea level fluctuated. The present area, parts of which are below sea level, is the product of extensive drainage, dyking and reclamation, especially since the 17th century. The land is made up of peat and silt. It is enormously fertile when drained and is devoted to arable crops.
Map United Kingdom Fd

Fergana ***USSR*** **1.** Irrigated agricultural region in south-eastern Uzbekistan, near the Chinese border. It consists mainly of a broad, fertile valley set 300-1000 m (1000-3300 ft) above sea level and surrounded by two arms of the Tien Shan range with peaks over 5500 m (18 000 ft) high. It produces cotton, rice, silk and a variety of fruits – including grapes, apricots, peaches and plums. There are also metal and chemicals industries. Its main cities are Kokand, the centre of the cotton industry, the railroad centre of Andizhan, and NAMANGAN, which manufactures cotton and food products. **2.** Industrial city in the region, 235 km (145 miles) east of Tashkent.
Population (region) 1 944 000; (city) 180 000
Map USSR Id

Fermanagh ***United Kingdom*** County area of Northern Ireland in the south-west corner of the province. It has many forests and lakes, including Upper and Lower Lough Erne. Marble Arch is a fine natural bridge formed from limestone in a mountainous area with numerous caves, 15 km (9 miles) south-west of Enniskillen. The main town is Enniskillen (population 11 500) and the chief industries are agriculture, forestry and tourism.
Population 51 400
Map United Kingdom Bc

Fernando Póo ***Equatorial Guinea*** See BIOKO

Ferrara ***Italy*** City about 90 km (55 miles) south-west of Venice. Most of Ferrara still lies within its 11 km (7 mile) long medieval walls, and the massive, moated Castello Estense, begun in 1385 and completed nearly 200 years later, stands in the centre. It is one of many monuments in the city associated with the Dukes of Este, who ruled Ferrara from the 14th to the 16th centuries, making it a famous centre for music, printing and art.

Close by the Este Castle are two fine medieval buildings: the Town Hall, built in 1243, and the Romanesque-Gothic cathedral. There are many fine Renaissance palaces, including the exquisite Palazzo Schifanoia, with 15th-century frescoes depicting the months, and the Palazzo dei Diamanti, named after its 12 600 diamond-shaped facing stones. Today, Ferrara's economy is based upon its food and petrochemical industries.
Population 150 300
Map Italy Cb

Fertile Crescent ***Middle East*** Historical agricultural area of the Middle East, shaped like a narrow crescent, running from the NILE valley northwards through Palestine, passing through the west and north of Syria, then turning south-eastwards through Iraq.

The Fertile Crescent receives only 203-381 mm (8-15 in) of rain a year; but since ancient times its agriculture has been greatly helped by irrigation – especially along the Nile, TIGRIS and EUPHRATES valleys.
Map Middle East Ba

THE RESTLESS EARTH

The pattern of relief, the shape and distribution of hills and valleys, mountains and lakes, is an amalgam of both the constructive and destructive forces working within the earth.

Hills, mountains and valleys are associated with faults and folds in the earth's crust. The folding and faulting help to construct the uplands, which streams immediately begin to destroy.

Folds are waves in the structure, and faults are cracks; both result from the enormous pressures exerted by the movement of the crustal plates (see PLATE TECTONICS). The greatest pressures are exerted at plate margins where two plates meet and push the land up into mountain ranges. Conversely, where plates separate, they pull the land apart, causing long depressions like the rift valleys of East Africa. Simple folds usually occur in young rocks, and complex folds in older strata which have been longer exposed to earth's movements and may have been refolded many times. As distance increases from the plate margins, the folds gradually die out, as in the foothills of the Himalayas.

When rocks can take no more bending, they break, so forming a fault. If this is simply stretched, a normal fault results, but if breaks are later compressed, then the resulting features are known as reverse or thrust faults. Faults and folds frequently occur together in the same region. Such areas too are often prone to earthquakes, especially in those where new faulting takes place along the line of an older, buried fault.

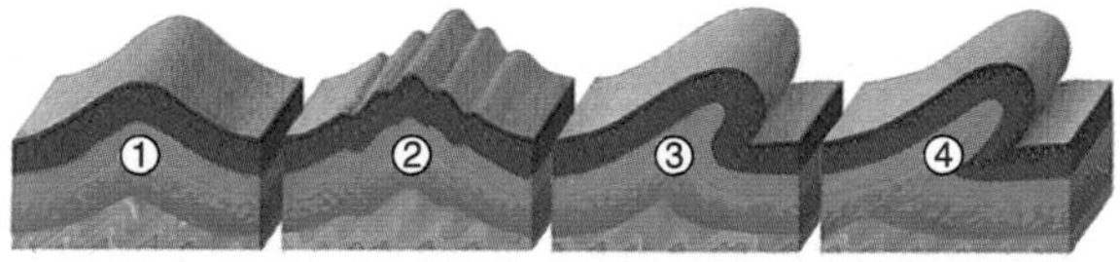

A single, simple upfold of strata is known as an anticline (1). An anticlorum (2) is a large anticline incorporating a number of smaller folds. Overfolds (3) are anticlines which have been pushed over. Sometimes overfolded strata are sheared – the overthrust layer is called a nappe (4). Strata under tension

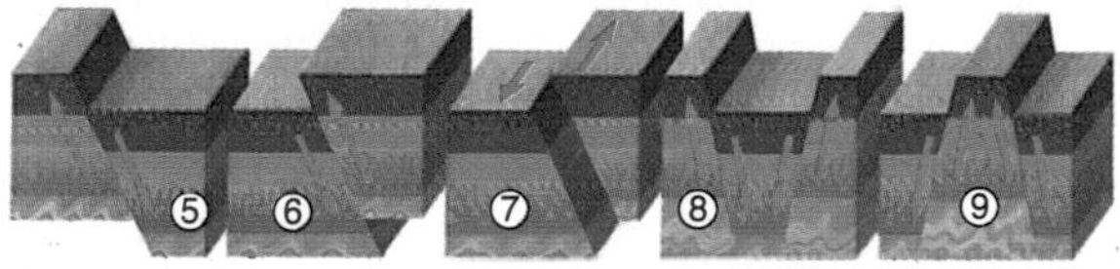

(5) may result in a normal fault. Compressed rocks produce a reverse fault (6), while horizontal movement may cause a strike-slip fault (7). A rift valley (8) occurs when land slips downwards between two parallel faults. When land is uplifted between two parallel faults (9), the formation is known as a horst or block mountain.

FOLDS AND FAULTS

Faulting and folding are produced by vertical and horizontal stresses in the earth's crust. Fold mountains are caused by movements in the crustal plates; examples are the Jura of France and Switzerland, the Alps and the Urals in the USSR. Fault-block mountains occur when a block between faults is uplifted and tilted, although the block may well be the remnant of an earlier fold mountain; the Sierra Nevada of California was formed in this way.

STRIKE-SLIP FAULTING

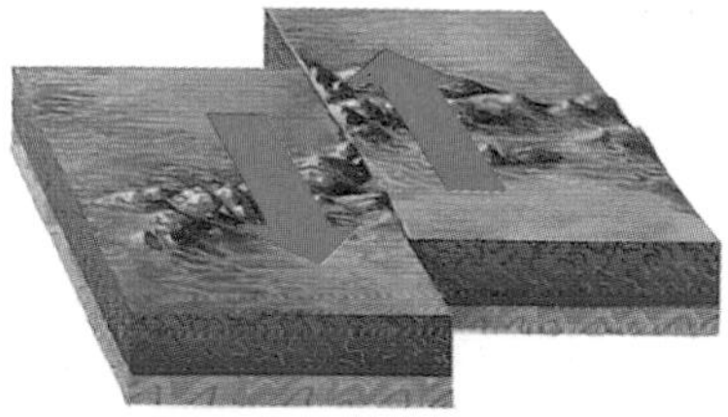

Also known, more graphically, as tear faults, strike-slip faults are dislocations of the earth's crust in which there is horizontal movement along the opposing STRIKE surfaces of the break. The faults are at least as deep as the foundations of the crust, and it is thought that some of them may form such ocean trenches as the Puerto Rico Trench in the Western Atlantic.

THE FAULT THAT THREATENS CITIES

The great majority of earthquakes that afflict the world occur round the fringes of the Pacific, where several crustal plates meet. One of these junctions, between the North American and Pacific plates, is the San Andreas Fault which cuts through the San Francisco peninsula. This makes the area prone to earthquakes, the most notorious of which was the San Francisco earthquake of 1906, which killed 700 people. Many seismologists believe that further earthquakes are imminent. In the San Andreas Fault, whose line can be clearly seen from the air, the Pacific Plate to the west moves at a rate of about 10mm a year relative to the North American Plate. In the last few million years their relative lateral movement has been several tens of kilometres, and eventually the peninsula may be entirely sheared from the continent and moved bodily north.

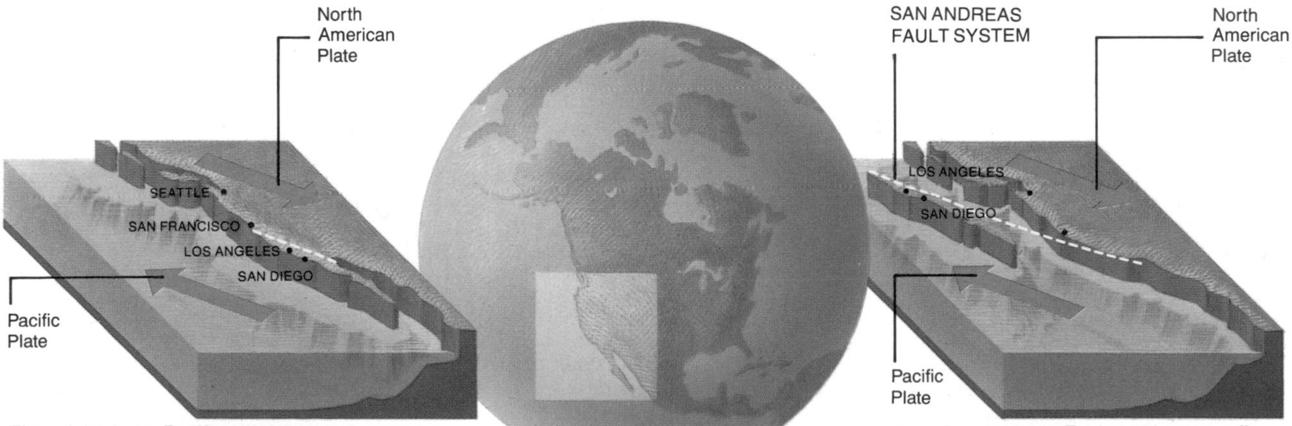

Though both the Pacific and North American plates are moving north-west, the Pacific Plate moves faster. They therefore seem to be moving in opposite directions.

If the San Andreas Fault continues to slip at its present rate, in about 50 million years, Los Angeles will be on an island off the Canadian west coast.

Fertöd ***Hungary*** Superbly restored Baroque palace built by the aristocratic Esterházy family 27 km (17 miles) south-east of the city of Sopron near the Austrian border. Dating from the mid-18th century, the horseshoe-shaped palace, with 126 rooms, stands amid colourful formal gardens. The Austrian composer Franz Joseph Haydn was court conductor for the Esterházys from 1761 to 1790, and is commemorated in a museum and summer music festival. Nearby are a fine Baroque inn and Haydn's house.
Map Hungary Ab

Fès (Fez) ***Morocco*** One of Morocco's four imperial cities (old royal capitals), today a commercial city and major communications centre lying 200 km (125 miles) south-east of Tangier in a valley north of the Middle Atlas Mountains. It remains Morocco's cultural and spiritual capital and has one of the largest mosques in the world – Qaraouyine or Qarawiyin – which is also one of the world's oldest universities, founded about AD 850.

Fès is one of the sacred cities of Islam, and was founded in 808 by Moulay Idriss II, an Arab prince. At its height in the 14th century, when Fez Jedid (new Fès) was built, its fame spread far beyond Africa as a place of learning, attracting students from all Muslim countries and parts of Europe. Today it is a bustling city with many mosques, shrines, colleges, palaces, markets and gateways. Excellent brocades and silks are available and Fès is also a centre for leather goods, carpets and musical instruments. The royal palace, Dar el Makhzen, dates from the 13th century. The medina (ancient Muslim residential quarter), with its labyrinthine streets reached through gateways in its walls, is the largest in Morocco.
Population 448 823
Map Morocco Ba

Fezzan ***Libya*** Historical region in the south-west of the country, 725 km² (279 sq miles) in area. Its population – only 2 per cent of the country's whole – live mainly in oases, where date palms grow. It is very dry, and consists mostly of desert rock and gravel wastes and sand dunes. The chief town is Sebha.
Map Libya Bb

Fife ***United Kingdom*** Ancient kingdom and county of eastern Scotland that remains a modern administrative region. It stands on a peninsula between the firths, or estuaries, of the Forth and Tay – an area of 1308 km² (505 sq miles). It contains the town of Dunfermline, capital of Scotland in the 11th century and the birthplace of Charles I in 1600, and many ruined castles. It also contains the town of St Andrews, famous for its university – the oldest in Scotland, founded in 1411 – its 12th-century cathedral ruins, and the Royal and Ancient Golf Club, one of the ruling bodies of world golf, whose Old Course was the world's first.

The south and west of the region have a long history of coal mining, but only a handful of pits now survive. The principal town is Kirkcaldy, where the pioneer economist Adam Smith was born in 1723. It is now the headquarters of a number of computer firms. The administrative centre is Glenrothes.
Population 344 000
Map United Kingdom Db

Figuig ***Morocco*** Oasis on the edge of a great stretch of barren desert in the extreme east of Morocco, on the border with Algeria and linked to major desert caravan routes.
Population 14 480
Map Morocco Ba

Filadelfia ***Paraguay*** See GRAN CHACO

Filchner Ice Shelf ***Antarctica*** Large body of floating ice lying at the head of the Weddell Sea on the Atlantic coastline of Antarctica. It was discovered by the German explorer Wilhelm Filchner in 1912.
Map Antarctica Cc

filling In meteorology, a term used to express increasing atmospheric pressure at the centre of a depression. It is the opposite of deepening.

finger lake Long, narrow lake occupying a U-shaped valley carved by a glacier, as in the English Lake District and the Finger Lake District of New York state in the United States. The valleys may either have been over-deepened to form the lake basin, or the lakes may be dammed behind by a terminal moraine.

Finistère ***France*** Windswept department in the far west. It has rugged cliffs, wide bays and many fishing harbours. BREST, on the western coast, is the largest town.
Population 828 000
Map France Ab

Finisterre, Cape ***Spain*** Rocky headland in the north-west corner, the westernmost part of the Spanish mainland.
Map Spain Aa

Finland See p. 226

Finland, Gulf of Shallow eastern arm of the BALTIC SEA, between Finland and the USSR, which freezes over in winter. Shoals and rocks also make navigation difficult. The chief ports are HELSINKI (Finland) and LENINGRAD, TALLINN and VYBORG (USSR).
Dimensions 400 km (250 miles) long, 20-130 km (12.5-80 miles) wide
Map Finland Cc

Finnmark (Finmarken) ***Norway*** Northernmost county of Norway and strategically important to the North Atlantic Treaty Organisation, which has military installations here, because of its border with Russia. Its inhabitants include Lapps and Finns, and the main occupations are fishing, raising reindeer, farming and iron mining. Vadsø is the county capital.
Population 77 400
Map Norway Ga

fiord, fjord Long, narrow coastal inlet, bordered by steep hills, found especially in Norway and Alaska. Fiords are the result of a rise in sea level causing penetration into the U-shaped valleys eroded by former glaciers.

Fiordland National Park ***New Zealand*** Remote area along the south-west coast of South Island covering some 12 000 km² (4635 sq miles). It consists of rugged mountains, precipitous valleys, roaring torrents and waterfalls,

A DAY IN THE LIFE OF A MOROCCAN OASIS FARMER

With eager eyes, Mustapha scans the desert horizon as the sun rises behind him. Some time today, or perhaps tomorrow, a camel train will come into view, led all the way from Mauritania and Mali by his friend Farouk. It is one of the last caravans that still journey from the coastal ports to the southern Sahara; lorries now carry supplies across the desert more economically.

Mustapha's wife brings him a bowl of water and he washes his feet, hands, neck and mouth according to Muslim ritual. Prostrating himself on a small mat in the direction of Mecca, he says his prayers. Breakfast is a substantial meal, before the heat of the day, consisting of maize, dates and nuts. Today, some of his children are breaking down the low mud walls of one of the shallow irrigation canals that run out from the lake at the heart of the oasis. Water cascades from the canal into the fields of maize chosen for today's irrigation. At the same time the other children are building up walls that were broken down yesterday to flood other fields.

Mustapha's wife goes out to weed the crops, and as he watches her, he notices a dark line on the horizon and can just make out the train of camels and men. He sets out to bring back his old friend and the merchants, whom he entertains in his main room, each of them sitting crosslegged upon hand-embroidered cushions. Mustapha's wife, back from weeding, brings thick, sweet coffee in small porcelain cups with brass holders. She retreats to the kitchen when she has served it, leaving the men to talk.

The camel men would once have been Mustapha's only source of news from distant oases and from the outside world. Today he hears it all on his transistor radio. Nevertheless, local gossip, the state of the animals and distant grazing lands never reach the air waves, and talk of them occupies Mustapha and his friends for several hours.

Then Farouk, dressed in traditional loose-fitting blue-dyed clothes, begins to put up his tent of dark brown, camel-hair cloth.

Meanwhile, Mustapha's eldest son, Muhmoud, comes over to bid him farewell. Muhmoud wears an open-necked shirt and the latest in casual Western trousers. He will never be a farmer, nor indeed the leader of a camel train. He aspires to be a lawyer. He tells his father that he is ready to leave for the start of another term's study in the city.

Having said his goodbyes, Muhmoud strides to his motorbike. Mustapha and Farouk watch the trail of dust he raises disappear into the distance.

Fiji

THE CRY OF 'MBULA' – 'WELCOME' – GREETS TOURISTS FOLLOWING IN THE FOOTSTEPS OF CAPTAIN BLIGH TO THIS FORMER HOME OF FIERCE CANNIBALS

The most economically advanced and racially mixed nation in its region – and one of the largest – Fiji is strategically placed at the crossroads of the western Pacific. Today, Fiji allows Western navies to dock in its ports, and the most striking impression on visitors – there are 200 000 tourists a year – is of the broad smiles and friendliness of the Fijians. But this is a far cry from only 200 years ago, when they were feared and savage cannibals.

With native Fijians – mainly frizzy-haired Melanesians, some with Polynesian blood – outnumbered by descendants of indentured sugar plantation labourers from India, the scene could be set for serious racial strife. But – thanks to great efforts by Britain in colonial days and by politicians of all races since – this has been avoided. Indeed, Prime Minister Ratu Sir Kamisese Mara has made racial harmony his main aim.

The electoral system allows no race to dominate, while giving the native Fijians a special place in their own country. They own all the land – their name for themselves, *Taukei*, means 'the Owners' – while the Indians, many of them first-class businessmen, dominate commerce and the professions as well as working on the land. The Indians have also provided a political awareness that has made Fiji a regional leader.

There are some 320 islands and atolls in the group, plus another 480 islets, but only 150 are inhabited. The two main islands, VITI LEVU and VANUA LEVU, account for 87 per cent of the land area. Extinct volcanoes give these dramatic skylines, the tallest being TOMANIIVI (Mount Victoria) rising to 1324 m (4344 ft) on Viti Levu, where SUVA, the capital lies.

Coral reefs fringe most of the islands, which have tropical rain forests on the side facing the south-east trade winds; elsewhere there are grassy plains and clumps of palmlike pandanus trees. The hilly landscapes of the main islands were once heavily forested, but many trees have been felled or burnt to create grazing land on which soil erosion is now a growing problem.

Aided by year-round temperatures which rarely fall below 16°C (60°F), tropical sunshine tempered by ocean breezes, and splendid coral strands, tourism is now the major industry. More traditional industries have centred around the cultivation of sugar cane.

Dutch explorer Abel Tasman was the first European visitor, in 1643; Britain's Captain Cook called in 1774; but the islands were not fully explored until 1792, when Captain William Bligh returned to the islands he had seen from an open boat after the mutiny on the *Bounty*. Whaling ships and traders seeking sandalwood and bêche-de-mer (sea cucumbers) arrived – bringing minor European ailments that proved disastrous for the islanders.

Guns were sold to these aggressive and volatile people, exacerbating inter-island wars to such a degree that in 1874 Britain stepped in. Partly at the urging of European settlers and partly to ward off French and American interest in the islands, it accepted the offer of cession made by the chief warlord, King Cakobau, who needed money for warfare. Fiji became a colony and Britain gradually imposed order. The islands became an independent member of the Commonwealth, with Elizabeth II as head of state, in 1970.

FIJI AT A GLANCE

Map Pacific Ocean Dc
Area 18 376 km² (7095 sq miles)
Population 715 000
Capital Suva
Government Parliamentary monarchy
Currency Fiji dollar = 100 cents
Languages English, Fijian, Hindustani, Urdu, Tamil, Chinese
Religions Christian (51%), Hindu (40%), Muslim (8%)
Climate Tropical; average temperature in Suva ranges from 20-26°C (68-79°F) in August to 23-30°C (73-86°F) in February
Main primary products Sugar, cassava, coconuts, rice, sweet potatoes, ginger, cocoa, tobacco, livestock, timber, fish, gold
Major industries Agriculture, tourism, mining, sugar refining, food processing, forestry, fishing
Main exports Sugar, gold, molasses, canned fish, coconut oil and copra, ginger
Annual income per head (US$) 1800
Population growth (per thous/yr) 21
Life expectancy (yrs) Male 70 **Female** 74

dense forests, tranquil lakes and imposing cliffs. The region, which was named after its resemblance to the fiords of Norway, is the home of the beautifully coloured takahe (*Notornis mantelli*), one of the world's rarest birds. The bird is about the size of a small turkey – and there are only about 100 left.
Map New Zealand Af

Firat ***Middle East*** See EUPHRATES

Firenze ***Italy*** See FLORENCE

firn, névé Compacted snow accumulated over the years in successive layers in hollows high in the mountains, firn forms masses of granular, whitish ice, the source of glacier ice.

firth Scottish word for a long narrow arm of the sea, such as the Firth of Forth.

fissure eruption Volcanic eruption in which lava is emitted through a long crack in the earth's crust. See VOLCANO

Fiume ***Yugoslavia*** See RIJEKA

fjord See FIORD

flagstone Hard, fine-grained sedimentary rock, usually sandstone or sandy limestone, that splits or cleaves easily along the bedding planes and is hard enough to be used for building or paving.

Flanders (Flandre, Vlaanderen) ***Belgium*** Coastal region in the north-west, formerly a medieval county and now two provinces covering a total of 6115 km² (2361 sq miles). It includes the cities of BRUGES and GHENT, which developed in the 13th and 14th centuries as centres of the European wool trade.

Now the region's prosperity is based mostly on farming the coastal *polders* – land reclaimed from the sea – together with an old-established textile industry and numerous modern plants manufacturing electrical goods and machinery. The region was the scene of almost continuous fighting during the First World War.
Population 2 400 000
Map Belgium Aa

flash flood Sudden, violent flood in a normally dry valley after heavy rain.

Flensburg ***West Germany*** Baltic port in Schleswig-Holstein and the most northerly city in the country, barely 2 km (1 mile) from the Danish border. In 1920, after the First World War, Flensburg voted to be part of Germany; the farming people to the north chose Denmark. In April 1945 it was the last retreat of the Nazi government. Admiral Dönitz, Führer for ten days, stayed in the town's 16th-century Glucksburg Castle, while negotiating the German surrender with Allied leaders.
Population 87 900
Map West Germany Ca

Flinders Range ***Australia*** Mountains in South Australia stretching more than 350 km (220 miles) north from Spencer Gulf. The highest point is St Mary Peak (1188 m, 3898 ft).
Map Australia Fe

flint Hard form of CHERT usually found in nodules along the bedding planes of chalk. It fractures along shell-shaped planes and forms sharp edges. This, together with its hardness, made it of great importance to early man for tools.

Flint ***United Kingdom*** See CLWYD

flood plain Flat land on the borders of a river, especially the lower reaches, over which alluvium is deposited when the river floods.

flood tide The rising tide between low and high water, or the inflowing tidal stream in an inlet.

Florence (Firenze) ***Italy*** Capital of Tuscany and one of the great artistic centres of the world, standing on the Arno river about 230 km (145 miles) north-west of Rome. The site was occupied in Roman times, but it was only after the 12th century that the city grew in wealth and power. Wool was the medieval city's staple industry. Florence's cultural splendour reached its peak during the 15th century after it had eclipsed rival city-states such as Pisa and Siena. After earlier turmoil, it (*continued on p. 230*)

Finland

ONCE PART OF SWEDEN, ONCE PART OF RUSSIA, THIS LAND OF LAKES AND FORESTS CLINGS TENACIOUSLY TO ITS INDEPENDENCE BETWEEN EAST AND WEST

Lying at the eastern limit of Western Europe, Finland is a frontier country standing close to the Communist heartland. Here the technological 20th century meets a wilderness of forest and Arctic wasteland. Indeed, two-thirds of Finland is covered by lakes and forests, and the snows that fall in November do not melt until April or May. But it is this austere beauty which inspired the music of Jean Sibelius (1865-1957) and the clean lines of the architecture and furniture of Alvar Aalto (1898-1976) and Eero Saarinen (1910-61).

Most of Finland is low-lying although it rises to over 1000 m (3300 ft) in the north in Lapland, and there are extensive peat bogs and swamps. The extensive forests, from which one-fifth of Finland's main export – paper – is produced, consist mainly of spruce where the soil is clay and fragrant pine on the drier sands and gravels; the slender birch thrives almost everywhere.

At the end of the last great Ice Age when temperatures rose, the vast ice sheet retreated and left immense quantities of glacial debris on the ancient granite bedrock. Thousands of lakes were formed in the process. As the ice melted, the sea level rose. But, relieved of the weight of ice, the land is today rising at a rate of as much as 800 mm (30 in) each century along parts of the west coast on the Gulf of BOTHNIA, an arm of the BALTIC SEA. As a result, land is being gained from the sea at the rate of 1 km^2 every ten years. Lapping the shores of this land of lakes, even the Baltic resembles a lake: it has virtually no tides and is not very salty.

Across the Gulf of Bothnia, and forming Finland's western boundary, lies Sweden; the eastern boundary of the country is shared – for 1269 km (about 790 miles) – with the USSR. The Soviet city of LENINGRAD, on the Gulf of Finland, lies only 300 km (190 miles) from the Finnish capital, HELSINKI. Norway, reaching across the top of Scandinavia, forms the northern boundary of Finland. One-third of Finland lies inside the ARCTIC CIRCLE.

The Finns experience great extremes of climate between winter and summer. The citizens of Helsinki in the south may sometimes have to endure winter cold as low as −30°C (−22°F) in January, and yet find themselves basking in temperatures as high as 30°C (86°F) in July. Summer days are long and light, but in the winter the days are short and gloomy. Inside the Arctic Circle the sun can shine at midnight for two months in the summer but some winter days see only an hour or two of daylight.

Autumn is short and spring comes late; frosts in early summer are a widespread hazard. It is not surprising that Finland is sometimes known as the 'land of three winters' – autumn winter, high winter and spring winter. During the long, harsh winter everything freezes, even the Baltic; March is a brilliant month, with dazzling sun on ice and snow.

THE PEOPLE OF FINLAND

The people of Finland probably settled their country about 2000 years ago, coming from beyond the URAL mountains in Russia. Their language is closely related to Estonian but bears no resemblance to any other European tongue except Hungarian.

Coastal areas were colonised in the 12th century by Swedish speakers from the west, and today Finland is officially a bilingual country although only 6 per cent of the inhabitants speak Swedish. This figure includes all the people of the province of AHVENANMAA (Åland) in the south-west archipelagoes between Sweden and Finland, who inhabit what amounts to a piece of Sweden in what is nominally Finland. In 1920 the province voted to belong to Sweden, but for strategic reasons the League of Nations insisted that Ahvenanmaa should stay Finnish. It was, however, given considerable provincial autonomy – for example, the province has its own flag and issues its own stamps. Its people are exempt from Finnish military service and Finnish speakers are not allowed to buy land in Ahvenanmaa unless they have lived there permanently for at least five years.

In the north of Finland there are some 2500 Lapps, a minority of whom are reindeer herders, though the traditional image has changed. Some reindeer are herded on skidoos (motorised sledges), and some Lapps, who live largely from fishing, hire seaplanes to take them to otherwise inaccessible lakes. Many of the Lapps still speak their native language, which is taught in school.

Finland is not a country to which other people have emigrated to any great extent over the centuries (though it was for long periods ruled by others). As a result, the Finns share a physically similar appearance, and something of a 'family likeness' can be detected among people throughout the whole of the country.

There is not a great deal of religious diversity. The major Church is evangelical Lutheran, though there is a Greek Orthodox

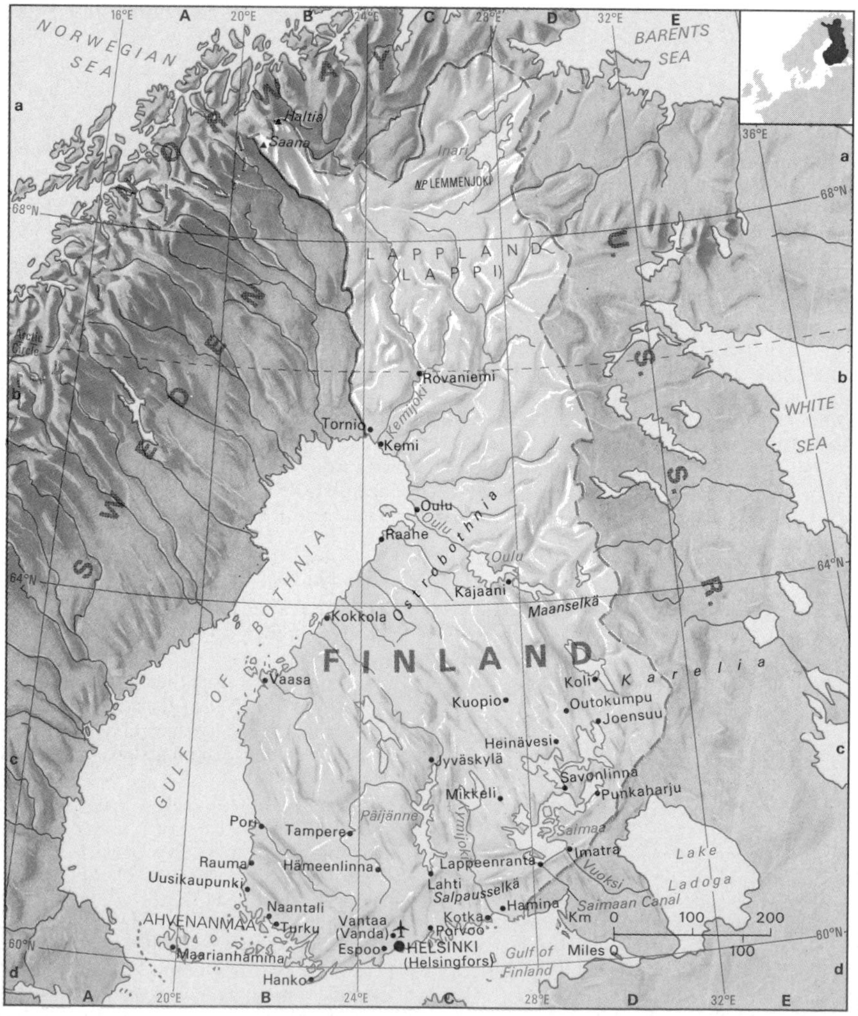

minority of some 58 000 owing allegiance to the patriarch in Constantinople (Istanbul).

Christianity arrived in the 12th century, when the Swedes conquered Finland and started to convert the people. The Swedish occupation affected mainly the south-western coastal areas of the Gulf of Bothnia and the islands of Ahvenanmaa. Finland became part of Sweden in 1540 when Finns were given equal rights with the Swedes. In 1809 it changed hands, and was ceded to Russia to become a grand duchy. Over a century of Russian rule ended on December 6, 1917, when Finland declared independence.

NATION REBORN

In November 1939, after the Finns had rejected Russian territorial demands, the USSR attacked Finland and forced it to give up large areas; over one-tenth of its land in all, including most of the province of KARELIA, half of the north-eastern parishes of Salla and Kuusamo, and a corridor of land leading to the Arctic coast.

Finland tried to regain the lost territory after Hitler invaded Russia in 1941, but was forced to agree an armistice in 1944 and to push the German troops out of the country. Some 400 000 displaced people had to be rehabilitated, and Finland was also obliged to pay huge reparation in the form of ships and machinery to Russia. The pressure this caused on Finland's economy led to a complete restructuring from agriculture to engineering, which has ultimately turned out to the country's great advantage.

Since the Treaty of Paris in 1947, Finland has a pact of friendship and cooperation with the USSR, which has become its principal trading partner. At the same time, Finland maintains strict neutrality in international affairs. It is a member of the Nordic Council of Scandinavian countries, and a member of the European Free Trade Association along with Norway, Sweden, Iceland, Switzerland and Austria, as well as having a special relationship with the European Economic Community.

Since the Second World War, Finland's transformation from a rural to an industrial and urban country has been staggering. Its economic growth rate, estimated at 3 per cent between 1983 and 1987, is even surpassing that of Japan.

Most people live in the south-western third of the country, where the three biggest cities have grown up: Helsinki, TAMPERE and TURKU. But enough people are still farmers for Finland to be largely self-sufficient in foodstuffs, and to have such a surplus of dairy produce that the government has paid farmers to reduce the dairy herds. Most of the country people work on their own small farms, often combining agricultural work with forestry. To the north, where there is so much space – only two inhabitants to the km² – a higher proportion of the land is forested; it is here too that the reindeer are herded. Most of the crops are grown in the south-west, where the climate is mildest and where many fine estates were established in the 19th century.

Since 1900, Finland has developed a widespread network of cooperatives for marketing farm produce, for bulk-buying of farm equipment, fertilisers and seed, and for organising finance and insurance. Agricultural schools and colleges have also been established.

▲ DYING WAY OF LIFE A Lapp family build a tent of reindeer skins in the tundra of the far north. Few Lapps now live the 2000-year-old traditional nomadic life in which reindeer provided food, clothing, building material, herding equipment and fat for lamps. Many of those who do depend on summer tourism.

Around the coasts, fishing and farming are combined. However, the big commercial fleets account for most of the country's catch. Some of the fishermen-farmers also breed animals such as mink and silver fox for fur. The business is so big now that the country has become the largest exporter of farmed furs in the world.

NATURAL BOUNTY

In the autumn and summer there is a harvest of wild berries and mushrooms, simply waiting to be picked. Finland's natural bounty also includes a wealth of game birds and wildfowl, and the lakes and rivers are rich in pike, perch, powan, trout and salmon, although pollution – caused by waste from wood processing, fertilisers and acid rain – is becoming a problem. Almost every Finnish family has its hunting and fishing enthusiasts, and some 70 000 licences to shoot elk are issued every year to keep down their numbers, because of the damage they cause to crops and young trees. Bears, which were once hunted, are now protected. Wolves come out of the forests, and often make local headlines; their victims are usually the reindeer.

Finland still lives to a large extent by its forests, although the picturesque days of the seasonal employment of hordes of lumberjacks, and large-scale floating of logs down the rivers, are over. Now skilled workers are employed in the forests all year, although felling still takes place mostly in winter; huge rafts of logs are sometimes floated across the lakes, but much of the timber is moved by lorry.

Most of the forests are privately owned and held in small lots, and widespread education in forestry has improved yields enormously. Continuous surveys of forest areas, both by air and on land, ensure that the timber is not cut down faster than it is replaced. Advances in forestry are matched by those in the factory. New processes and diversification have led to new types of paper, wallboards, laminates and prefabricated timber components, as well as chemical products, all of which swell the export market.

Not only has the timber-processing industry advanced recently. The complete revision of industry which followed the last war led to investment in metallurgy, electrical manufactures and engineering, which now employ most of the country's industrial workers. For example, there is a new state-owned iron and steel complex in the north at Rautaruukki.

Shipbuilding has boomed, prompted in the 1950s by Finland's need for an effective fleet of icebreakers. Icebreakers are now built for other countries too, including an entire fleet for the Russians to use in the Arctic. Shipyards in Helsinki and Turku also build ferries and cruise liners, and recently mustered the expertise of several nations to build the P. & O. cruise liner *Royal Princess*.

Entire industrial plants have been exported from Finland, including power stations and factories for processing softwoods. The biggest of these undertakings has been the construction of an iron and steel complex in Soviet Karelia, with a Finnish labour force who camped on site and travelled home for weekends. Finnish skills are also exported: Finnish banking representatives work in all the world's major banking centres, and Finnish design of clothing, glass, porcelain, sports equipment, kitchenware and electronic goods has not only made for a good export trade but influenced design elsewhere.

Finland's trading links with the USSR are strengthened by its need for Russian energy supplies. Only one-third of its energy needs can be supplied by domestic sources. It has already developed almost all possible hydroelectric power, and readily available peat has a limited use. Finnish nuclear reactors are producing power, and the Finns seem to have no qualms about establishing more of them. But meanwhile the energy gap is closed by imports of crude oil and natural gas from the USSR, and of coal from Poland and West Germany.

In the areas where seasonal unemployment is worst (the construction industries suffer especially from the long winters), new industries and public works have been set up to ease the situation. Financial support is provided on a graduated scale, so that most goes to the needy north-east and the least to the wealthier south-west. Developments are complemented by an elaborate conservation programme, with 22 national parks and over a thousand protected areas, as well as controls on industrial pollution. More Finns are building summer homes, and the people have not lost their appreciation of their country's exceptional landscape.

A SPORTING LIFE

Life for Finns follows the rhythm of the seasons. They are great sportsmen, with sports halls in many towns and saunas to relax in after physical exertion. One-third of them can claim certificates of proficiency in skiing. There are ski jumps on most village skylines, and schools have a skiing holiday, which coincides with the national winter sports festival in Lahti in February. The long summer days are to be made the most of, and the working hours are fewer then, while holidays in educational institutions run from the end of May to mid-August. A great summer exodus from the towns takes place, as boats and cars leave for summer cottages and family farms. Comfortable town apartments are exchanged for frequently primitive rural retreats, in response to the call of the wild.

A high point of the year is the festival of midsummer, with its bonfires. In August, the Finns bid the summer farewell at crayfish parties, feasting under the last of the lingering sunshine. The Swedish-speaking people's festival of Santa Lucia, in December, illuminates the dark winter with a candle-lit celebration. The long winter ends with Walpurgis Night on April 30, a festival observed especially by students, who sing in the summer on the eve of the May Day public holiday.

NATIONAL PRIDE

For each day as well as for festivities, Finns have their own version of the Scandinavian cold table, or *smörgåsbord – voileipäpöytä* in Finnish – and their own vodka, which is sold, along with all alcohol, at shops owned by a state monopoly. Beer is popular, coming in three different strengths. But the Finns also drink curdled milk, prepared in different regional ways by half a dozen different bacterial cultures, and they consume so much coffee that Finland even has an ambassador in Colombia, from which the coffee comes.

On all festive and formal occasions, especially on Independence Day, December 6, the blue and white Finnish flag is flown, and the music of Sibelius is not far away. Finland is proud of itself. And its pride is justified, since a century ago the country was on the famine fringe of Europe, poverty stricken and ruled as a dependent duchy of imperial Russia. Today it has found a national identity, achieved a high standard of living and managed to create a flourishing economy in a harsh environment with limited resources.

FINLAND AT A GLANCE
Area 338 142 km² (130 557 sq miles)
Population 4 910 000
Capital Helsinki
Government Parliamentary republic
Currency Markka = 100 penni
Languages Finnish, Swedish
Religion Christian (92% Lutheran, 1% Greek Orthodox)
Climate Temperate, with cold winters. Average temperature in Helsinki ranges from −9 to −4°C (16-25°F) in February to 12-22°C (54-72°F) in July
Main primary products Timber, cereals, potatoes, livestock, fish; copper, zinc, iron, chromium, lead
Major industries Forestry, timber products including wood pulp and paper, machinery, shipbuilding, clothing, chemicals, fertilisers
Main exports Paper, machinery, ships, timber, chemicals, clothing, wood pulp, petroleum products
Annual income per head (US$) 8685
Population growth (per thous/yr) 4
Life expectancy (yrs) Male 72 **Female** 77

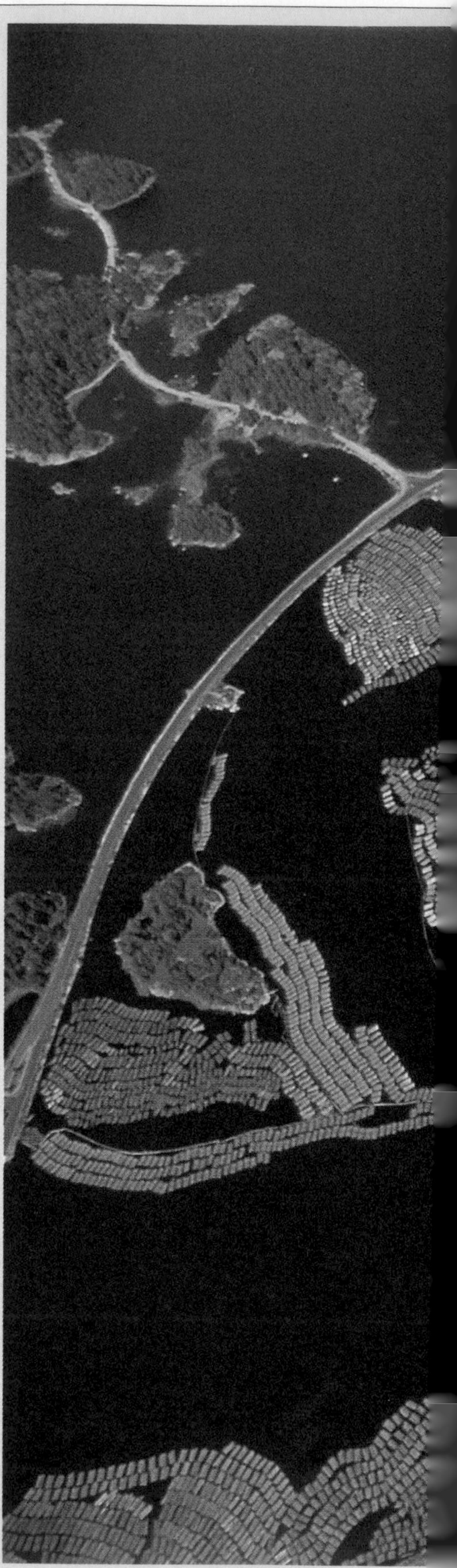

► BLUE AND GREEN JIGSAW Finland's 55 000 lakes cover nearly one-tenth of the country, and forests, dominated by conifers, over a half. Cut logs are floated to the timber mills, and wood products – paper, paper board, timber, wood pulp, veneers and plywood – account for one-third of the country's exports.

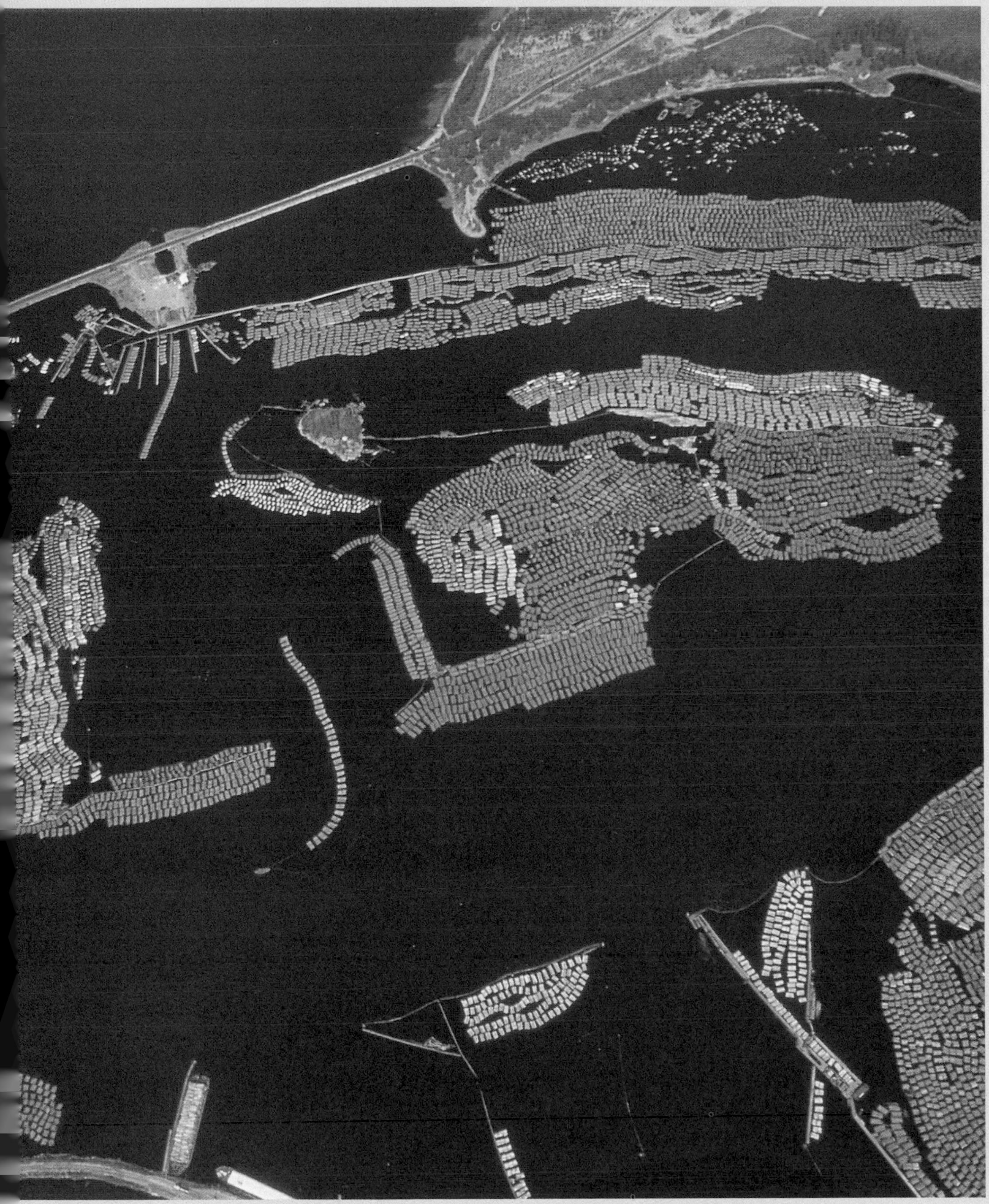

enjoyed the stable rule of the Medici family and later was briefly capital of Italy (1865-70), succeeding Turin and being succeeded by Rome.

Florence gave Italy its language, its art and its philosophy. The city's cultural refinement – still evident – developed under the patronage of the Medicis in the 15th and 16th centuries, when such artists as Giotto, Brunelleschi, Donatello, Botticelli and Michelangelo flourished there. Even today works of art adorn its public squares, streets and buildings. The Piazza della Signoria, for example, is a veritable open-air museum of sculpture, while priceless frescoes and oil paintings are found in the many churches. The Uffizi and Pitti palaces have perhaps the finest collections of Renaissance paintings in the world. Michelangelo's *David* and his sculptures of slaves are in the Accademia gallery.

Many of the city's buildings are themselves works of art, and often seem to be designed less for business or worship than for the display of art treasures. Michelangelo described the Baptistry's bronze doors (by Ghiberti and Pisano) as the 'gateway to Paradise'. Nearby are the Campanile (Bell Tower) by Giotto and the Duomo – the cathedral of Santa Maria del Fiore – with its huge dome, designed by Brunelleschi.

Other historic buildings include the church of San Lorenzo (also by Brunelleschi) with its tombs of the Medicis by Michelangelo; the Palazzo Vecchio (1299-1314), one of the greatest Gothic palaces in Italy and the seat of the Florentine Republic until the 16th century; the Ponte Vecchio (1345), the oldest bridge over the Arno; and the church of Santa Croce (1294), which has frescoes by Giotto and the tombs of Michelangelo, the political theorist Machiavelli, the scientist Galileo, and the operatic composer Rossini.

Florence's present-day prosperity rests on its range of economic interests: banking, publishing, the university, tourism, food and drink, textiles, fashion clothing, leather goods, jewellery, furniture, metallurgy and light engineering.

Population 453 300

Map Italy Cc

Flores ***Indonesia*** Mountainous volcanic island of 14 250 km² (5500 sq miles) in the Lesser Sundas group, between Java and Timor. Maize, groundnuts (peanuts) and coconuts are the main crops. About half the population has been converted to Christianity.

Population 803 000

Map Indonesia Fd

Flores Sea Indonesian sea lying between SULAWESI (Celebes), FLORES and SUMBAWA islands. Underneath Flores and Sumbawa, the northward-drifting Indo-Australian continental plate is sliding beneath the south-eastern part of the Eurasian plate, causing earthquakes and volcanic activity.

Width 240 km (150 miles)

Map Indonesia Ed

Florianópolis ***Brazil*** Port and regional capital 480 km (300 miles) south of the city of São Paulo. It stands on Santa Catarina island and is linked to the mainland by a bridge. The natural beauty of the island makes it popular with tourists. The surrounding region produces coal, textiles, wheat, apples and vegetables.

Population 189 000

Map Brazil Dd

Florida ***USA*** The 'sunshine state', in the extreme south-east of the country. The area was ceded by Spain in 1819 and became a state in 1845. Florida covers 151 670 km² (58 560 sq miles), and is mostly a low-lying peninsula which separates the Atlantic from the Gulf of Mexico. The state is dotted with lakes, the largest of which is Lake Okeechobee. The Florida Keys, a string of coral islands, stretch 225 km (140 miles) south-west from MIAMI BEACH, to KEY WEST.

The state's subtropical climate is ideal for citrus, vegetable and salad crops. Manufacturing is mainly food processing, electronics, and the making of beach-wear. There are extensive deposits of phosphates, used in fertilisers. Tourism throughout the year is a boom industry and Florida is an increasingly popular place for retirement. Its tourist attractions include the EVERGLADES National Park, DISNEY WORLD and John F. Kennedy Space Center at Cape CANAVERAL. The state capital is TALLAHASSEE.

Population 11 366 000

Map United States Je

▼ DRAMA IN TRANQUILLITY White ibises grace a quiet pool in Florida's Everglades, but the peaceful scene belies their constant life and death struggle with predatory alligators and panthers.

Florida, Straits of Straits between the FLORIDA Keys and Cuba, and between south-eastern Florida and the Bahamas containing the warm, fast-flowing Florida Current. This current flows from the Gulf of MEXICO, but becomes the Gulf Stream when it emerges into the ATLANTIC. Buccaneers once lurked around the western end of the straits, waiting to plunder treasure-laden galleons on their way from Mexico to Spain.
Width Up to 145 km (90 miles)
Map Bahamas Ab

fluorite, fluorspar White or colourless mineral, calcium fluoride, often tinged with impurities of many colours, found in veins or as deposits from hot gases in the earth. Blue John is a blue variety found in Derbyshire, England.

Flushing (Vlissingen) ***Netherlands*** Port on the island of Walcheren about 75 km (45 miles) south-west of Rotterdam, with ferry services to Sheerness, England. It has been a Dutch naval base since the 16th century, and is also a resort, and fishing and shipbuilding centre.
Population 46 400
Map Netherlands Ab

Fly ***Papua New Guinea*** River flowing some 1200 km (750 miles) south and east from the Victor Emanuel Range in the west central part of the country through swampy lowlands to the Coral Sea at the Gulf of Papua. Its mouth is more than 50 km (30 miles) wide. The river is navigable by medium-sized craft for more than 1000 km (620 miles).
Map Papua New Guinea Ba

flysch Group of sediments or sedimentary rock consisting of shales, muds and marls, interspersed with layers of conglomerate and coarse sandstone. They are produced by the rapid erosion of fold mountains, while the mountains are being uplifted. The sediments are deposited in adjacent troughs and are then themselves deformed by the continuing folding.

focus Centre of an earthquake and the point from which seismic waves are emitted.

fog Cloud-like masses of water droplets and smoke or dust particles close to the ground, limiting visibility to less than 1 km (0.6 mile). Fog is formed when air is cooled below its DEW POINT, and the three main circumstances in which it occurs are: (i) When warm moist air passes over a cooler sea or land surface – called advection fog. (ii) In calm conditions when air cools because the land on which it rests is cooled by radiation – called a radiation fog. (iii) When warm, moist air meets cold air along a FRONT – frontal fog.

Foggia ***Italy*** City about 125 km (78 miles) north-east of Naples. It is set in an area rich in wheat, olives and vines; its economy depends upon food processing.
Population 158 400
Map Italy Ed

föhn, foehn Descending warm, dry wind coming off the leeward side of a mountain range, especially off the northern slopes of the Alps. It causes rapid melting of snow in winter, with avalanches. The air cools and drops its moisture on the windward side of the range, and being drier, it then warms faster as it descends.

fold Bend in rock strata resulting from compression by forces in the earth's crust. See THE RESTLESS EARTH (p. 223)

fold mountains Mountains formed by folding and uplifting of the earth's crust. They are usually associated with DESTRUCTIVE PLATE MARGINS. The Alps, Himalayas and Andes are fold mountains.

Fontainebleau ***France*** Town near the west bank of the Seine, 55 km (35 miles) south-east of Paris. It is famous for its 16th-century chateau, once a home of French kings and now the official summer residence of the French presidents. Set in formal gardens, it is visited by more than 270 000 people each year. The Forest of Fontainebleau, covering 170 km^2 (65 sq miles), is one of the most beautiful in France, and has inspired many famous landscape painters of the Barbizon school – named after a local village – such as Corot and Millet.
Population 39 400
Map France Eb

foreland 1. Low promontory projecting into the sea, such as North Foreland in Kent. 2. An ancient continental mass forming one side of a geosyncline onto which mountain-building movements have pushed up folded ranges. 3. An area flanking a mountain range onto which glaciers moved down during the Ice Age, such as the Bavarian and Swiss Forelands.

foreshock Minor earthquake heralding the approach of a much larger shock.

foreshore Area stretching from the lowest low-water line to the average high-water line.

Forli ***Italy*** City about 80 km (50 miles) north-east of Florence. It has much 1930s grandiose Fascist-style architecture, and the vast 17th-century Palazzo Paolucci was restored by the dictator Benito Mussolini (1883-1945) – who was born in the village of Predappio, 10 km (6 miles) south of the city.
Population 110 800
Map Italy Db

Formosa See TAIWAN

Føroyar ***Denmark*** See FAEROE ISLANDS

Fort Dauphin ***Madagascar*** See TAOLANARO

Fort Hill ***Malawi*** See CHITIPA

Fort Knox ***USA*** See LOUISVILLE

Fort Lamy ***Chad*** See NDJAMENA

Fort Lauderdale ***USA*** Atlantic resort city in Florida, about 40 km (25 miles) north of Miami. It has one of the world's largest yacht marinas, and a network of over 435 km (270 miles) of natural and artificial waterways crisscrosses the city. It was founded as a military base in 1838.
Population (city) 149 900; (metropolitan area) 1 093 300
Map United States Je

Fort McMurray ***Canada*** Town which stands amid vast oil-sand deposits, at the junction of the Clearwater and Athabasca rivers in north-east Alberta. It is the terminus of a water transport system through to the Arctic.
Population 31 000
Map Canada Dc

Fort Worth ***USA*** City in north-east Texas. It lies in the west of the Southwest Metroplex – the Dallas-Fort Worth conurbation with more than 3 million people. Fort Worth is an agricultural centre, with oil refining, aircraft and car-making industries.
Population (city) 414 600; (metropolitan area) 1 144 400
Map United States Gd

Fortaleza ***Brazil*** Port and regional capital 2312 km (1435 miles) north-east of the national capital, of Brasília. A population explosion in the years 1979-84, following a massive influx of migrants from the drought-stricken interior to its *favelas* (shanty towns), now makes it the largest city in the northern half of Brazil. It has also developed a thriving tourist industry with fine sea food still brought in by traditional *jangada* sailing rafts.

On either side of the city, massive sand dunes stretch along the coast, and inland there are orchards growing cashew nuts and tropical fruits.
Population 1 309 000
Map Brazil Eb

Fort-de-France ***Martinique*** Capital and port, standing on the island's west coast beside one of the best harbours in the West Indies. It was originally only an administrative and military centre, but following the destruction of ST PIERRE by a volcanic eruption in 1902, it also became the chief commercial and cultural town of the island.
Population 100 000
Map Caribbean Cc

Foshan ***China*** City with varied light industries, including textiles, in Guangdong province. It stands on one of the most fertile plains of southern China.
Population 500 000
Map China Gf

Fouta (The Futa; Fouta Toro) ***Senegal*** flood plain beside the Senegal river along the Mauritanian border. Toucouleur and Fulani tribes people grow groundnuts, millet and rice on the highly prized land, which is fertilised by silt each year when the river floods. However, an irrigation scheme upstream due to be completed by the end of the 1980s will control the river's flow, end the flooding and reduce traditional cropping patterns. Bakel, Matam, and Dagana are the region's largest towns.
Map Senegal Ba

Fouta Djallon (Futa Jallon) ***Guinea*** Deeply dissected plateau in the west, rising in a steep escarpment on the west and dropping more gently to the savannah plains of Upper Guinea to the east. Over 900 m (3000 ft) high in places, it is extensively used for livestock rearing, particularly by the Fulani people, while the deep and fertile valleys support crops, including bananas and fonio (a small millet). The region contains immense deposits of bauxite (aluminium ore) including those at Boké, Dabola, Fria, Kindia and Tougué.

Map Guinea Ba

Fouta Toro ***Senegal*** See FOUTA

Foveaux Strait Strait separating SOUTH ISLAND, New Zealand, from STEWART ISLAND. A ferry runs from Bluff, South Island, to Halfmoon Bay, Stewart Island. Fishing and oyster cultivation are important activities.

Width 24-34 km (15-21 miles)

Map New Zealand Ag

Fox Glacier ***New Zealand*** Mountain glacier near Mount Cook in the Southern Alps of South Island. Like the Franz Josef Glacier, 40 km (25 miles) to the north-east, it has thinned and retreated in recent years. The glaciers are unusual in that they descend into lush forests. They move 0.5-4.5 m (1.5-15 ft) a day.

Map New Zealand Ce

Foxe Basin Arm of the ATLANTIC OCEAN between northern Canada's Melville Peninsula and BAFFIN ISLAND, named after Luke Foxe who explored it in 1631. In the south, the 320 km (200 mile) long Foxe Channel leads into HUDSON BAY.

Dimensions 480 km (300 miles) long, 180-230 km (110-140 miles) wide

Map Canada Hb

France See p. 234

Franceville ***Gabon*** Capital of the south-eastern province of Haut-Ogooué, lying on a tributary of the Ogooué river. It was founded in 1880 by the Italian-born explorer Pierre Savorgnan de Brazza and is now the main town and trading centre in a region rich in manganese and uranium. In 1979, the Supernational Centre for Medical Research of Franceville was inaugurated; it is unique in Africa.

Population 35 000

Map Gabon Bb

Franche Comté ***France*** Region of eastern France, comprising the agricultural plateaus of the Upper Saône, the Belfort Gap, and the forests and high pastures of the Jura mountains. It became French territory in 1674, when Spain ceded it to France after the Thirty Years War. Main activities are dairy farming, cheesemaking, forestry, watchmaking and winter tourism. BESANCON is the chief city.

Population 1 084 000

Map France Fc

Francistown ***Botswana*** Centre of trade, industry and tourism in the north-east, 360 km (220 miles) from Gaborone near the border with Zimbabwe. It was founded as the centre of the Tate Concession, a mining and farming area of more than 5200 km² (2000 sq miles) ceded to British interests in 1887 and named after the Tate river. The town was named after a prospector, Daniel Francis. It reached its peak as a mining town in the 1890s, but declined as the gold mines were exhausted.

Population 36 000

Map Botswana Cb

Frankfort ***USA*** State capital of Kentucky, on the Kentucky river in the north of the state, about 60 km (37 miles) south of the Indiana border. Its main manufactures are clothing, shoes and whiskey.

Population 26 800

Map United States Jc

Frankfurt (Frankfurt am Main) ***West Germany*** City in the state of Hessen, beside the river Main about 155 km (95 miles) south-east of Cologne. Frankfurt – the name means 'ford of the Franks' – has been a major trading city since the Romans threw a bridge across the Main 2000 years ago.

In 1152 the German princes known as Electors met in St Bartholomew's Cathedral and chose Friedrich Barbarossa as Holy Roman Emperor. For the next 400 years, emperors were elected at Frankfurt – but until 1562 they were always crowned at Charlemagne's old capital, Aachen.

In today's West Germany, it serves as the commercial and financial capital which BONN is not. It has the principal stock exchange, the headquarters of the national bank and the largest number of national and international fairs, including the autumn Book Fair, the largest in the world. It is also the headquarters of the US forces in Europe. Frankfurt operates the nation's major international airport and lies at the centre of West Germany's autobahn and rail network.

The walled medieval town, destroyed by bombing in 1944, has been replaced by tower blocks and skyscrapers of concrete and glass. Only the medieval cathedral, the 15th-century houses around cobbled Romerberg Square, and the home of the writer Johann Wolfgang Goethe (1749-1832), creator of *Faust*, have been restored. Flower gardens now trace the path of the old ramparts. The opera house has been reconstructed as a conference centre. The city also has one of the finest zoos in the world – with a huge walk-through aviary where visitors mix with the free-flying birds.

Population 614 700

Map West Germany Cc

Frankfurt an der Oder ***East Germany*** Town on the Polish border, east of the capital, Berlin. It stands on the River Oder and produces metal and electrical goods and processed food.

Population 84 800

Map East Germany Db

Franklin ***Australia*** River in south-western Tasmania flowing 118 km (73 miles) into the Gordon river. A combination of high rainfall, luxuriant forests and rugged scenery combine to make it one of the most beautiful in the country. Plans to dam it and the lower Gordon for hydroelectric power were dropped after worldwide protests from conservationists.

Frantiskovy Lazne ***Czechoslovakia*** See CHEB

Franz Josef Glacier ***New Zealand*** See FOX GLACIER

Franz Josef Land ***USSR*** See ZEMLYA FRANTSA-IOSIFA

Franzensbad ***Czechoslovakia*** See CHEB

Frascati ***Italy*** Resort town in the Alban Hills 20 km (12 miles) south-east of Rome. It is surrounded by vineyards and sumptuous 16th-century villas, and has been a retreat for the citizens of Rome for 2000 years.

Population 18 700

Map Italy Dd

Fraser ***Canada*** River which rises in the Rocky Mountains near the Yellowhead Pass and winds its way to the Strait of Georgia south of Vancouver. It is 1370 km (850 miles) long, but only the lowest 145 km (90 miles) are navigable. The river is famous for salmon.

Map Canada Cc

Fraser Island (Great Sandy Island) ***Australia*** Largest sand island in the world, covering 1840 km² (710 sq miles). It lies off the coast of Queensland some 190 km (120 miles) north of Brisbane. The island's sand dunes rise, in places, to 290 m (950 ft), and inland they are covered by rain forest, eucalyptus trees and heath. The sands of the island, much of which is a national park, also contain minerals such as titanium and zircon; plans to mine the sand in the mid-1920s were successfully opposed by conservationists.

Map Australia Id

Frauenburg ***Poland*** See FROMBORK

Fredensborg Palace ***Denmark*** See HILLERØD

Fredericton ***Canada*** Capital of New Brunswick. It has two universities, an art gallery, and a theatre. Chief industries are timber processing and shoe manufacture. Nearby, at Oromocto, is a Canadian forces base.

Population 43 723

Map Canada Id

Frederiksborg Castle ***Denmark*** See HILLERØD

Fredrikstad ***Norway*** Historic seaport in south-east Norway on the east shore of Oslofjord, covering the approaches to the Swedish border. Founded in 1567, it contains the remains of a fortress which was an important defence post between the 14th and early 19th centuries, when Norway was united with Denmark. The port's major exports are timber, softwood products and chemicals.

Population 27 600

Map Norway Cd

Freeport City ***Bahamas*** Main port, industrial town and tourist centre of GRAND BAHAMA island in the north-west of the archipelago. With the adjacent Lucaya area it is the biggest tourist resort complex in the West Indies. It grew rapidly in the 1960s with the construction of a deepwater harbour. There is also a cement works and an oil refinery.

Population 17 000

Map Bahamas Ba

freestone Fine-grained, even-textured rock, usually sandstone or limestone, that can be cut in any direction without shattering or splitting.

Freetown ***Sierra Leone*** The country's capital, chief port and largest city. It stands on the north

▲ **ORIENTAL EXTRAVAGANZA El Casino, built in Moorish style and now renamed Princess Casino, is a major attraction of Freeport City on Grand Bahama. Red London double-decker buses take visitors on tours of the island.**

side of the SIERRA LEONE PENINSULA. Like many other West African cities, it has distinctive quarters: Kru Town, where the Kru, a native ethnic group, live; Portuguese Town; Maroon Town, where maroons – fugitive slaves from Jamaica – settled in the 19th century; and modern residential areas.

Freetown was founded by the British in 1787 as a settlement for freed slaves, and was named for that reason. In the 19th century it was used as a base for suppression of the slave trade, and in the 20th century as a wartime naval base. There is little industry apart from rice and groundnut (peanut) mills, and a small industrial estate at nearby Wellington. Fourah Bay College, founded in 1827, stands on Mount Aureol to the south.

Population 316 300
Map Sierra Leone Ab

Freiberg ***East Germany*** Town in the south-east, midway between Dresden and Karl-Marx-Stadt. Silver has been mined there since 1168, and it has the world's oldest mining academy, founded in 1765. Lead, zinc and iron pyrites were also mined until 1968.

Population 51 300
Map East Germany Cc

Freiburg (Freiburg im Breisgau) ***West Germany*** Largest city of the Black Forest on the western slopes of the Rhine rift valley, 110 km (68 miles) south-west of Stuttgart. Its 13th-century cathedral is the seat of an archbishopric and its university dates from 1457.

Population 175 000
Map West Germany Be

Fréjus ***France*** Small town in Provence region, 25 km (16 miles) south-west of Cannes. It was founded in the 1st century BC by the Roman general Julius Caesar. It now produces wine, olive oil and cork.

Population 51 500
Map France Ge

Fremantle ***Australia*** See PERTH

French Polynesia ***Pacific Ocean*** French overseas territory consisting of some 130 islands in the south-central Pacific, about halfway between Australia and South America. They have a total land area of 3941 km² (1522 sq miles) but are spread over 4 million km² (1.5 million sq miles) of ocean. There are five main island groups. The Windward Islands (including TAHITI, the main island, which contains the capital PAPEETE) and Leeward Islands together make up the SOCIETY group. These and the MARQUESAS are all volcanic, with steep mountains, as are most of the TUBUAIS and the small GAMBIER ISLANDS. The TUAMOTUS, however, are low-lying atolls.

They all form part of POLYNESIA, and 78 per cent of the people are Polynesians, most of them Christians (over half Protestant) who speak Tahitian, although French is the official language. They are renowned for their beauty and for the hip-swaying *tamure* dance. Another 10 per cent are French and some 7 per cent are descendants of Chinese workers imported in 1865-6 by a British-owned plantation.

Spanish conquistadores from Peru discovered the Marquesas in 1595, but kept their discovery a secret to keep out other European powers. Tahiti was discovered by the English navigator Samuel Wallis in 1767 and independently by Louis de Bougainville (who claimed it for France) in 1768. They returned with stories of lush landscapes and friendly – not to say amorous – people that created the image of a South Sea island paradise perpetuated by artists and writers ever since.

After years of persuasion, the tribal chiefs of Tahiti accepted a French protectorate in 1842. The last Tahitian king, Pomare V, ceded his kingdom entirely to France in 1880, and by 1900 the French controlled all the islands. They acquired the status of an overseas territory in 1958, but gained increased autonomy with a Territorial Assembly in 1984. There are some calls for complete independence, but the Tahitians know that their standard of living depends on French government expenditure, especially on military facilities to support its nuclear weapons tests at MURUROA.

Phosphate mining stopped in 1966, but the islands export copra and coconut oil (coconut palms grow on the lowland fringes of most islands and on the atolls), cultured pearls and pearl-shell (a newly established industry), vanilla and citrus fruits. However, tourism brings in three-quarters of all non-military foreign earnings; the main tourist destinations are Tahiti itself, nearby Moorea, and Bora-Bora in the Leeward Islands.

Population 169 000
Map Pacific Ocean Fc

French Southern and Antarctic Territories ***Indian Ocean*** The territories comprise Adélie Land, in Antarctica, and several islands and island groups in the southern Indian Ocean south of latitude 37°S. The islands include the Kerguelen and Crozet groups, with Amsterdam and St Paul Islands. Administration is from Paris, and the population consists of staff manning scientific research and meteorological stations scattered through the territory, and a hospital on Amsterdam Island.

fresh breeze Force 5 on the Beaufort Scale, wind speed 30-39 km/h (19-24 mph).

Fresno ***USA*** City in the San Joaquin Valley in California, about 265 km (165 miles) south-east of San Francisco. It is sometimes called the 'grape capital of the world' because it is the heart of the country's largest vine-growing area.

Population (city) 267 400; (metropolitan area) 564 900
Map United States Cc

fret Mist which forms at sea and suddenly extends a short distance inland.

Fribourg (Freiburg) ***Switzerland*** Capital of the canton of the same name 27 km (17 miles) south-west of Berne. The town stands dramatically in and above a bend in the gorge of the Saane (Sarine) river and has in its medieval centre, among many old churches and towers, a fine 13th-century cathedral.

The canton of Fribourg straddles the country's language frontier: about three-quarters of the inhabitants speak French, the rest German. It is famous for cheeses, particularly Gruyère – inexactly named after the village of Gruyères. The canton's area is 1670 km² (648 sq miles).

Population (town) 38 500; (canton) 190 000
Map Switzerland Aa

Friedrichshafen ***West Germany*** Port on the north shore of Lake Constance. Friedrich I of Württemberg created the city named after him in 1811 by combining the ancient city of Buchhorn and the monastery-village of Hofen. It is the birthplace of the Zeppelin airships, and their history is outlined in the Zeppelin Museum. The main industries are tourism and the manufacture of precision instruments.

Population 51 400
Map West Germany Ce

France

A COUNTRYSIDE RICH IN COLOUR AND VARIETY, PUNCTUATED BY PICTURESQUE VILLAGES, ANCIENT TOWNS AND BUSTLING CITIES – THIS IS THE KALEIDOSCOPE OF FRANCE

Picture a Parisian terrace café in the busy Champs-Elysées. Turn your head one way towards the Louvre, one of the world's greatest museums of art and culture; then look in the other direction and you glimpse the Arc de Triomphe, built for the Emperor Napoleon – a symbol of centuries of French history and power.

For many visitors this is France. But with all its attractions, PARIS is only the heart of a country whose language and culture flourish all over the world.

Beyond the capital another France soon appears – vast, and largely rural. Covering 551 000 km² (212 740 sq miles), including CORSICA but not 'overseas departments' such as Martinique, it is the largest country in Europe outside the USSR, but with only 55.3 million inhabitants it is much more sparsely peopled than its neighbours. Wide expanses of farmland produce a rich variety of vegetables, fruit, meat, cheese and wine – all vital ingredients in the renowned French cuisine.

FROM COUNTRY TO TOWN

As recently as 1940, almost half the nation lived in the countryside and one-third of French people worked on the farms. Things have changed rapidly since then. New jobs in factories and offices have attracted millions away from the land so that now only one-fifth of French people are country dwellers and many of these drive to work in town. One worker in three has a job in a factory and one in every two works in an office or a shop.

Only one Frenchman in ten is a farmer, and he works much more land than his father or grandfather did. Almost everywhere farms have been enlarged, mechanised and made more productive. By 1980 about half a million of France's 1.2 million holdings were 20 hectares (50 acres) or more, and many marketing cooperatives had been formed. In the Mediterranean south, a large number of irrigated orchards and vineyards have been planted; and in the heat of summer, vast fields of shimmering wheat stretch to the far horizon in the PARIS BASIN.

The controversial common agricultural policy brought in by the European Economic Community, which France helped to found in 1957, aided this transformation and enriched many French farmers. But it has also contributed to 'mountains' of surplus butter and 'lakes' of milk and wine.

France's various regions are arranged for the most part like a giant amphitheatre facing west. Its lower levels are the great basins of northern and south-western France and the peninsula of BRITTANY. It rises to the MASSIF CENTRAL and its upper tiers contain the mountains of the ALPS, JURA and PYRENEES – all part of the frontiers of France.

Farming is possible in all parts of France despite variations in rainfall and temperature. The Atlantic Ocean warms the country's western shores in winter and produces moist temperate conditions ideal for rearing livestock. Eastern France has cold winters and hot summers that are typical of central Europe.

In the south, high summer temperatures and drought make irrigation necessary for good harvests of cereals and fruit. But rosemary, thyme and other aromatic herbs flourish on dry Provençal hillsides. In the Massif Central and the higher mountains the winters are much colder and agriculture more restricted than their southern location might suggest.

The Paris Basin, which contains large expanses of fertile ploughland, is crossed by major rivers including the SEINE and MARNE. Paris grew up near their junction and has been a stimulus for agricultural improvement over the centuries. The most modern and productive farms in France are found here.

AQUITAINE, to the south-west, is the second large lowland in France, crossed by the broad valleys of the GARONNE and DORDOGNE. Here farming is less modern, but the vineyards around BORDEAUX produce some of the world's finest wine.

Windswept Brittany, with its old hard rocks, was once a backward farming area, but is now known for early vegetables and intensive factory farming of pigs and poultry.

The ancient Massif Central is built of comparable rocks, but has been dubbed 'the dead heart of France'. High plateaus, steep slopes, long-dead volcanoes, deep valleys and a harsh climate combine to form a difficult environment for farming. Now mixed husbandry is giving way to livestock farming, and uncultivated land is being planted with trees or left to revert to scrub.

▲ OPEN FOR BUSINESS Small family shops still flourish in France, and the patronne of this chemist's-cum-ironmonger's shop in Saint-Rémy, Provence, is ready to sell anything from household detergents and paint to face-creams and mineral water.

Conditions are harsher still on the high ridges and in the deep valleys of the Jura, the Pyrenees and the French Alps which contain Western Europe's highest challenge to climbers, Mont BLANC (4807 m, 15 770 ft).

The valleys of the RHONE and SAONE provide a break in the ascending steps of the giant amphitheatre. For thousands of years traders and travellers have used these natural routes which now contain motorways, railways and gas pipelines. The Rhône valley houses the nation's second largest city, LYONS, at the point where its turbulent waters meet the Saône. It flows on past terraced vineyards, powers hydroelectric works and provides irrigation before reaching its delta and the Mediterranean Sea. Here is the CAMARGUE with its ricefields and marshes, home of flamingos and wild horses.

To the east is MARSEILLES, France's third largest city after Paris and Lyons, and the centrepiece of the Midi. This southernmost fringe of the country sweeps from the millionaires' paradise of the COTE D'AZUR, through rocky coves and inland valleys to PROVENCE, abundant with flowers, fruit and vegetables. It continues to the plain of LANGUEDOC, with its vines, orchards and new holiday centres.

THE DELIGHTS OF FOOD

French cuisine offers a delightful way of experiencing the richness and variety of the nation's farming. Despite the appearance of fast-food chains in large cities, the preparation of good wholesome food is still regarded as an art. Bistros and restaurants offer excellent and often inexpensive menus which provide a good introduction to regional tastes. Many dishes are famous throughout the world, for example: *bouillabaisse* (a rich fish and seafood stew from the Midi), *boeuf à la bourguignonne* (tender beef, mushrooms and BURGUNDY wine) and *quiche lorraine*, a delicate blend of eggs, cream, cheese and gammon in pastry from the north-east.

France produces a quarter of the world's wine and every restaurant has a cellar of wines from simple *vins de table* to superior vintages. Special meals are not complete without CHAMPAGNE wine or COGNAC spirit. But drink is also a national problem: nearly 20 000 French people a year die from conditions brought on by alcohol, and 40 per cent of hospital patients show some alcoholic symptoms.

Shopping for the freshest of ingredients and skilfully preparing them are essential features of French family life. One-fifth of the average household budget goes on food. Most towns have magnificent markets crammed with colourful stalls, laden with local produce and goods brought from other regions. There are mouth-watering displays of cooked meats, pastries and locally made chocolates with temptation to buy at every turn.

The urban French do not forget their rural roots. Regular rush hours take place on Friday nights as Parisian families drive to enjoy *le weekend* out of town. Some have their own country cottages; and restoring an old house or barn as a second home is a serious hobby for many families. France now has more than 2 million second homes, old and new, favourite places for summer holidays – usually taken in August – as well as weekends. Growing numbers of city people retire to the villages where they were born.

THE IMPRINT OF HISTORY

France's countryside and provincial towns still bear the imprint of centuries past, long before they became part of a unified nation. In the days of Charlemagne (*c* 742-814), who founded the first great French Empire, feudal lords built castles and used them to defend their territory. During the Renaissance of the 16th century, Italian craftsmen were employed to build gracious chateaux, often set in elegant gardens and parks.

These were the homes of royalty and nobility until the Revolution of 1789 which led to the execution of King Louis XVI. The

Loire valley contains particularly handsome examples such as Amboise, ANGERS, CHAMBORD and CHENONCEAUX.

Thousands of parish churches and scores of beautiful cathedrals were built because of the wealth and power of the Roman Catholic Church and the piety of the French people. Round arches and domes are typical of the solid Romanesque churches erected during the 11th and 12th centuries and are found across much of southern France, especially along pilgrim routes leading to the shrine of the apostle St James at SANTIAGO DE COMPOSTELA in northern Spain. Nowadays the main place of pilgrimage is LOURDES, where 3 million people go each year.

Tall pointed arches and large stained-glass windows are seen in the soaring Gothic cathedrals of the 12th and 13th centuries in many towns of northern France. The magnificent cathedral at CHARTRES incorporates such austere features along with later, more flamboyant, wavy-lined decoration. But the Revolution stripped away much of the Church's power. Today 86 per cent of the people are Catholic, but only 14 per cent regularly attend mass. Although the Church outlaws contraception and abortion, the state legalised the first in 1967 and the second in 1974. One in four marriages ends in divorce, and there are 22 abortions for every 100 births.

After decades of official disapproval, there is now a revival of interest in regional languages, and Breton and Corsican broadcasts are heard for a few hours each week. Commercial recordings of folk music and publications in regional languages demonstrate that *le folklore* has become fashionable. Some groups have sought to use their cultural identity as grounds for obtaining a much greater say in regional affairs.

THE RISE OF INDUSTRY

Until the 20th century many industries were rural in character and most towns were small market centres closely linked to farming. France did not experience the great wave of factory building and urban growth that transformed Victorian England. But a handful of mining districts, textile towns and ironmaking centres in northern and eastern France did mushroom at that time. Their pit machinery, grim factories and workers' housing can still be seen in northern LORRAINE and the NORD-PAS-DE-CALAIS.

Paris was the exception. The city's powers were greatly enhanced after 1789 when provincial parliaments were abolished and the running of the nation's economic and cultural life was centralised in the capital. It later became the focal point of a star-like railway network which increased the city's influence. In recent years France has developed one of the most efficient state railway systems in Europe, and the TGV (high-speed train) competes with road and air travel in maintaining links with the provinces. High-speed trains between Paris and Lyons can cruise at 280 km/h (170 mph).

In 1801 Paris housed more than half a million people. During the next hundred years it attracted the poor, who came in search of work, as well as scholars, artists and rich investors. New ideas were sparked off, stimulating innovations in manufacturing, fashion, culture and knowledge. For centuries French had been the language of diplomacy and educated people throughout the world looked to Paris for inspiration. The music of Berlioz, Saint-Saëns and Bizet; the paintings of Degas, Cézanne, Monet, Renoir and Toulouse-Lautrec; the literary work of Balzac, Hugo, Zola, Flaubert and Maupassant represented the peak of creativity in the 19th century.

By 1850 the population of Paris had trebled and two decades of frenetic expansion began. This was masterminded by the Emperor Napoleon III and the city planner Baron Haussmann, who wanted the capital to be both efficient and elegant, with impressive public buildings. New markets, churches, schools and hospitals were built and parks laid out, and inner Paris was transformed into the beautiful city that visitors enjoy today.

The First World War with Germany (1914-18) cost France 1 385 000 men killed. The Second World War (1939-45) cost a further 536 000 lives – men, women and children. But, despite frequent changes of government, a new France grew out of the disasters of the two wars. The birthrate, low in the thirties, rose after 1945 in an impressive 'baby boom', although there are still more French women than men. Birthrates tumbled from 21.1 per 1000 at the peak of the baby boom in 1948 to 13.6 in 1976, though they have risen since. Almost one-sixth of the population is over 65, and France is on the way to becoming a country of the elderly.

POST-WAR ECONOMIC DEVELOPMENT

National planning was introduced in 1947 to steer economic recovery and stimulate further growth. Heavy investment, including aid from the USA, enabled France to catch up on lost time and become an industrial nation almost overnight. Regional planning followed in the 1950s to deal with three big problems – too many new jobs in the congested capital; too few jobs in the rural provinces where farms were shedding labour fast; and not enough jobs in old industrial areas where modern equipment was reducing the need for manpower in mines, textile mills and steel works.

In 1954 France lost Indo-China to the Vietnamese Communists; in 1956 MOROCCO and TUNISIA were given independence, but rebellion in ALGERIA in 1958 brought General de Gaulle, hero of the Free French resistance to the Germans, to power.

De Gaulle was president for ten years. In 1962 he gave independence to Algeria and about 1 million people of French extraction flocked to mainland France.

From 1957 free movement of workers between the EEC countries enabled France to import labour from Italy to help build new factories, homes and roads and to work in manufacturing and service activities. Later the French widened their policy of recruitment to attract migrants from Spain, Portugal and Algeria, and by 1975 there were 3.5 million foreigners in France.

This increase in working population helped the industrial economy to flourish. Old industries were brought up to date and new ones began, making cars, aircraft, electrical goods, electronics and other products.

In the early 1970s, 40 per cent of all French workers had jobs in factories. Many of the workers were housed in large blocks of flats in the towns. The first flats were cramped and poorly equipped, and the estates lacked adequate shops and schools. Conditions began to improve in the mid-1960s and more comfortable but expensive flats were built for sale or rent to the new rising class: office workers, managers, civil servants and teachers.

Massive new office complexes, such as La Défense in western Paris and La Part-Dieu at Lyons, were built. Single-family houses went up in the outer suburbs of French towns for families who wanted a house and garden of their own rather than an apartment in a large concrete block. Today the French are a nation of townspeople and 10 million live within 50 km (30 miles) of the centre of Paris.

The entry of workers from non-EEC countries was halted in the mid-1970s and some foreigners accepted 'golden handshake' payments to return home. But foreign wives and children have continued to join their menfolk in the poorer quarters of French towns.

The government is faced with the task of creating a society catering for several nationalities, and this is particularly difficult with unemployment rising fast. In the last few years deep-rooted xenophobia has surfaced, and politicians urging the expulsion of foreigners have been returned in elections.

As early as 1968, when strikes and demonstrations disrupted everyday life throughout the country, many French people were expressing their discontent with the capitalist system and concern over social and environmental issues. Some of their fears came true after the oil crisis in the 1970s when the Gaullist President Giscard d'Estaing and Prime Minister Barre introduced tough measures to trim surplus manufacturing capacity, especially in the iron and steel industries. Unemployment started to rise and most French electors felt it was time for a change.

THE ADVANCE OF SOCIALISM

In May 1981 a Socialist, François Mitterrand, was elected president. He promised devolution of power, nationalisation and a generous range of social welfare schemes. In the next two years new laws turned some of the promises into reality. Some power was handed down to elected councillors in the nation's 95 *départements* and Corsica was chosen to establish the first, experimental, regional assembly. Nationalisation put one-fifth of manufacturing capacity and most financial institutions under state control.

The Socialists increased family allowances, pensions and minimum wages, reduced the working week and introduced five weeks' holiday with pay and earlier retirement. Women's rights were advanced, the death penalty abolished and controversial projects for new military training grounds and extra nuclear power stations were scrapped. But money for these changes had to come from the taxpayer.

▲ **MIRROR IMAGE The handsome and Italianate Chateau de la Roche – in Poitou, in western France – is reflected in its ornamental lake. It is one of hundreds of chateaux which survived the savage pillage and destruction of the French Revolution.**

Early in 1983, in financial trouble, the president had to introduce an unpopular austerity programme to cut state expenditure and social benefits. Taxes and health charges were raised. Workers felt betrayed and trade unions organised strikes. The president issued an ultimatum to nationalised industries to improve their efficiency. Many workers in steel, shipbuilding, chemicals, car manufacturing and mining faced redundancy.

This reversal of policy angered the Communist Party, the second largest in Western Europe, and Communist members of Mitterrand's administration resigned in July 1984. Unemployment rose from 1.7 million (8 per cent) when the president came to power to 2.55 million (11.3 per cent) in early 1985. More than a million people under 25 were jobless. Young people who had never worked, and those who had interrupted periods of employment, suffered because of the way the benefit system worked. Nearly a million people without work received 'little or nothing' – and the French began to talk about the 'new poor'.

Nearly all the Socialist government's policies changed in practice. Devolution ran into trouble when the Corsican experiment failed and elections for other regional assemblies were put off until 1986. Mitterrand's proposal to control Church schools, which educate one-sixth of French children, met strong opposition and the whole idea was dropped. In the light of these and many other problems Mitterrand had to accept that his Socialist plans could not be applied to France in the 1980s. The French people did not want wholesale changes. In the 1986 general election the Gaullists and their right-wing allies won power, and Prime Minister Jacques Chirac began undoing the socialist policies – despite the fact that Mitterrand was still president.

On the whole, the French family is now healthier and better fed, housed and educated than ever before. It has more free time for holidays, a second home, hobbies, sports and other leisure activities. French men are avid supporters of Association and Rugby football and are fascinated by the annual round-France cycling race (the Tour de France). Older men spend much of their leisure time playing cards in cafés and *boules* (an informal version of bowls) in parks and open spaces throughout the land. French women enjoy a reputation for elegance and good dress sense.

Men and women are traditionally pictured sitting with friends under the bright umbrellas of street cafés, leisurely sipping their drinks, watching the world go by and setting it to rights. But the picture of prosperity may be changing under the pressure of international economics. The big questions in everyone's mind focus on unemployment, racism and what will happen in national politics in the final years of the 20th century.

FRANCE AT A GLANCE

Area 551 000 km² (212 740 sq miles)

Population 55 300 000

Capital Paris

Government Presidential-parliamentary republic

Currency Franc = 100 centimes

Languages French; some Basque, Breton, Catalan, Corsican, Provençal

Religion Christian (mainly Roman Catholic)

Climate Temperate, with hot, dry summers on the Mediterranean coast. Average temperature in Paris ranges from 1-6°C (34-43°F) in January to 14-25°C (57-77°F) in July

Main primary products Cattle, sheep, pigs, poultry, sugar beet, wheat, barley, maize, oats, potatoes, grapes and wine, fruit, vegetables; timber; fish; iron ore, bauxite, coal, uranium, salt, oil and natural gas, potash

Major industries Iron and steel, engineering, chemicals, textiles, electrical goods, cars, aircraft, cement, aluminium, agriculture, perfume, forestry, fishing, food processing, oil and gas refining

Main exports Cars, chemicals, iron and steel, textiles, leather goods, electrical equipment, wine, cereals, processed foods, petroleum products, clothing

Annual income per head (US$) 8850

Population growth (per thous/yr) 4

Life expectancy (yrs) Male 73 **Female** 78

▲ **MAGNIFICENT OBSESSION The Japanese love affair with Mount Fuji goes back to earliest times. On a clear day its perfect cone can be seen from the Tokyo-Nagoya railway, and every summer thousands climb to the top.**

Friesian Islands ***North Sea*** Fringe of islands along the North Sea coast. All are sandbars running parallel to the coast and separated from it by 5-30 km (3-20 miles) of shallows. Many of the islands are tourist resorts and some of them are used by nudists. They are divided into three groups. The West Friesian Islands are Dutch, the largest islands being Texel and Terschelling. The East Friesian Islands belong to West Germany, and include Borkum and Norderney. The North Friesian Islands beyond the estuary of the Elbe are split between West Germany and Denmark, and Sylte and Heligoland (Helgoland) are German. Numerous ferries link the islands to the mainland.

Population (Dutch Friesians) 20 000; (German Friesians) 30 000; (Danish Friesians) 3750

Map Netherlands Ba

Friesland (Vriesland) ***Netherlands*** Northern province covering 3808 km² (1470 sq miles), including most of the West Friesian islands off its coast. The province owes its name to the Germanic Frisian tribe which settled there in the 1st century AD. The West Frisian language, distinct from Dutch, is in common usage. The region has many small lakes, which make it popular with tourists. Agriculture, with the breeding of Frisian dairy cattle, is the principal industry. Leeuwarden, the provincial capital, is the only large town.

Population 595 200

Map Netherlands Ba

Friuli-Venezia Giulia ***Italy*** Region covering 7846 km² (3029 sq miles) in the north-east corner of the country, bordering Yugoslavia and Austria. The region has three languages: Ladin, a tongue similar to Latin, in the north, Italian in the south and Slovenian along the border with Yugoslavia. Its capital is the port and shipbuilding centre of TRIESTE. It is noted for its textiles and dairy products.

Population 1 229 900

Map Italy Da

Frobisher Bay ***Canada*** Inlet of the Atlantic Ocean, penetrating 230 km (143 miles) into the south-east coast of Baffin Island.

Map Canada Ib

Frombork (Frauenburg) ***Poland*** Town 64 km (40 miles) east of Gdansk dominated by a majestic 14th-century cathedral. Polish astronomer Nicolaus Copernicus (1473-1543) was a canon at the cathedral when he put forward his belief that the earth was not the fixed centre of the universe but revolved around the sun. A tower in the cathedral courtyard now houses a museum devoted to him.

Map Poland Ca

front Surface of separation between different air masses, either marking a major zone of atmospheric activity, as at the polar front, or, on a smaller scale, occurring within a depression as a cold, warm or occluded front.

frontal rain See CYCLONIC RAIN

frost 1. Deposit of minute ice crystals on the ground, plants and buildings, formed from frozen water vapour. **2.** The atmospheric condition when the temperature is at or below the freezing point of water (0°C, 32°F).

frost hollow Low-lying area which may experience frost while surrounding higher areas are frost free, or which may have more severe frost than surrounding areas. The frost is caused by heavy cold air sinking into the hollow and being unable to escape. Also called frost pocket.

frost shattering Weathering process which causes rocks to disintegrate through water within the joints and pores freezing and expanding, thus splitting it into fragments.

Frunze ***USSR*** Capital of Kirghizia, an industrial city between Lake Balkhash and the Chinese border. It stands at the foot of the snowcapped Kirghiz range, over 4400 m (14 450 ft) high. It was originally called Pishpek and was renamed Frunze in 1926 in honour of its most distinguished native son, General Mikhail Frunze (1885-1925), one of the fathers of the Red Army. The modern city has tree-lined boulevards and spacious parks. Its industries include food processing, textiles, agricultural machinery, and metal and electrical products.

Population 577 000

Map USSR Id

Fuchun Jiang (Tsien Tang Kiang) ***China*** River of Zhejiang province. It is the main transport waterway in the province, flowing more than 400 km (250 miles) into the East China Sea about 50 km (30 miles) south of Shanghai.

Map China He

Fuengirola ***Spain*** See MALAGA

Fuji, Mount (Fuji-yama) ***Japan*** Volcanic mountain about 100 km (62 miles) south-west of the capital, Tokyo, and the country's best known natural feature. Fuji's majestic cone – dormant since 1707-8 – is Japan's highest peak at 3776 m (12 389 ft) and is sacred to the Japanese. It has been celebrated for centuries in Japanese paintings and verse. The peak lies within the Fuji-Hakone-Izu National Park.

Map Japan Cc

Fujian ***China*** Province in the south-east of the country, covering 120 000 km² (46 350 sq miles). It lies beside the East China Sea opposite Taiwan. It is a predominantly mountainous area, rich in timber. The subtropical climate allows rice, sweet potatoes, tea and citrus fruit to grow in the valleys. Fishing has always been the chief occupation but the subsistence lifestyle forced many people to emigrate to South-east Asia and, in the 19th century, to Hawaii and California. The capital is Fuzhou.

Population 24 800 000

Map China Hf

Fukui ***Japan*** Market town near the north coast of central Honshu, about 320 km (200 miles) west of the capital, Tokyo. It was badly damaged in the Second World War and by an earthquake in 1948. It is noted for its traditional figured silk products, to which a modern rayon-weaving industry has been added.

Population 250 300

Map Japan Cc

Fukuoka ***Japan*** Largest city of Kyushu island, standing on the island's north-west coast. A commercial and industrial centre, it is also the western terminus for the *shinkansen*, or 'bullet train', from Tokyo, which reaches a maximum speed of 210 km/h (130 mph) on the 1176.5 km (735 mile) journey taking 6.5 hours.

Population 1 160 400

Map Japan Bd

Fulda ***West Germany*** **1.** Town about 80 km (50 miles) north-east of Frankfurt am Main. Its Baroque cathedral has the staff and relics of St Boniface, the English-born missionary whose followers founded a Benedictine monastery on the site in the 9th century. **2.** River rising 20 km (12 miles) to the south-east of the town and flowing about 220 km (137 miles) north to join the Weser near the town of Kassel.

Population 55 400

Map West Germany Cc

fumarole Hole or vent in the earth's surface from which issue hot liquids and gases such as steam, hydrochloric acid, sulphur dioxide and ammonium chloride. Fumaroles are usually associated with volcanic activity, as in Alaska's Valley of Ten Thousand Smokes.

Funafuti ***Tuvalu*** Main atoll and capital of Tuvalu, lying about 1000 km (620 miles) north of Fiji. The atoll measures 20 by 18 km (12 by 11 miles), but the land area is only 2.4 km² (0.9 sq miles).

Population 2600

Map Pacific Ocean Dc

Funchal ***Portugal*** Capital of the island of Madeira in the Atlantic, about 140 km (85 miles) off the west coast of Morocco. The town lies in a large natural amphitheatre above a bay. A tourist resort – especially in the winter – it has a casino and is a port of call for cruise ships.

Population 45 600

Map Morocco Aa

Fundy, Bay of ***Canada*** Bay between New Brunswick and Nova Scotia. It has the highest tidal range in the world, the water rising up to 15 m (50 ft) from low to high tide. Fundy National Park covers 206 km² (79.5 sq miles) on the northern coast of the bay.

Dimensions 160 km (100 miles) long, 100 km (60 miles) wide at entrance

Map Canada Id

Fünen (Fyn) ***Denmark*** The nation's second largest island (2976 km², 1048 sq miles). It dominates the Fünen group of 5 large and 24 small islands, which are clustered between the LITTLE BELT (Lille Bælt) and the GREAT BELT (Store Bælt) channels at the entrance to the Baltic Sea. Eighty per cent of the land is cultivated, with a large proportion under horticultural crops such as flowers. For this reason the island is often called the 'garden of Denmark' – and its hills, less than 130 m (430 ft) high, are humorously called the Fünen 'alps'. Almost all of its principal settlements have medieval foundations and many have well-preserved half-timbered houses. In addition there are Stone Age burial chambers and Viking remains, including the Ladby-skibet, the burial ship of a Viking chief (about AD 950).

On the west of the island is the medieval ferry port of Middelfart; in the east the fortified town and ferry port of Nyborg. To the south there are pleasant resorts and towns such as Svendborg and Fåborg. ODENSE, in the centre, is the island's main administrative, educational and industrial city.

Population 433 800

Map Denmark Bb

funnel cloud Whirling, dark grey cloud at the centre of a waterspout or tornado, sometimes extending downwards as far as the earth's surface.

Füred ***Hungary*** See BALATONFURED

Furka Pass ***Switzerland*** Mountain pass about 85 km (52 miles) south of Zürich, connecting the Rhône and Rhine valleys. The pass is 2436 m (7992 ft) high.

Map Switzerland Ba

Fushun ***China*** Mining city in Liaoning province. Coal and oil shale are mined, providing raw materials for local engineering and petrochemicals industries.

Population 1 800 000

Map China Ic

Futa, The ***Senegal*** See FOUTA

Futa Jallon ***Guinea*** See FOUTA DJALLON

Futuna ***Pacific Ocean*** See WALLIS AND FUTUNA

Fuzhou ***China*** Capital of Fujian province. It is a thriving port at the mouth of the Min river on the strait separating China and Taiwan. Its main industries are food processing and the manufacture of paper and chemicals.

Population 1 050 000

Map China He

Fyn ***Denmark*** See FUNEN

gabbro Coarse-grained, basic intrusive rock composed largely of plagioclase FELDSPARS and PYROXENES, sometimes with other minerals. Its coarse grain is due to slow cooling deep in the earth rather than on the surface.

Gabès ***Tunisia*** Seaport and oasis town on the Gulf of Gabès, 320 km (200 miles) south of Tunis. It forms the entry point to the south and the desert. Apart from exporting locally grown fruit, and hides and wool, the port also exports phosphates. There is also a fishing industry. Local craft goods include carpets, wicker work, wood, amber, leather and silver. The nearby Chenini oasis produces dates, oranges, peaches, apricots and pomegranates.

Population 64 500

Map Tunisia Ba

Gabès, Gulf of Inlet of the MEDITERRANEAN SEA off southern Tunisia, called Syrtis Minor by the ancient Romans, and distinguished from Syrtis Major, or the larger Gulf of SIRTE to the east. Jerba island in the Gulf of Gabès has recently become a popular tourist centre. GABÈS and SFAX are the chief ports on the coast.

Dimensions 100 km (60 miles) long, 100 km (60 miles) wide

Map Tunisia Ba

Gabon See p. 240

Gaborone ***Botswana*** Capital of Botswana since 1965, since when it has grown rapidly. It is now a well planned, attractive city, containing the University of Botswana and the National Museum and Art Gallery. There are several hotels and a casino. Because it is only about 20 km (12 miles) from the border with South Africa, it has a large exile community. South African commandos raided the town in June 1985, claiming that guerrillas were operating from Gaborone.

Population 79 000

Map Botswana Cb

Gabrovo ***Bulgaria*** Industrial town and capital of Gabrovo province, on the Yantra river 96 km (60 miles) north-east of Plovdiv. Industries include textiles, footwear, furniture, sawmilling

Gabon

A POOR, FORESTED COUNTRY THAT STRUCK IT RICH WITH OIL – BUT NEVER QUITE MADE GOOD

Dense tropical forest covers three-quarters of the hot, humid Gabonese Republic, straddling the Equator in west-central Africa. Timber was virtually its only resource until the 1960s, and at independence from the French in 1960 Gabon seemed to be yet another thinly populated African country with a stagnant, near-bankrupt economy and an uncertain future. Its people suffered from acute malnutrition and endemic diseases. Then Gabon struck oil.

By the mid-1980s, Gabon was Africa's sixth largest oil producer, and other minerals were being exploited. It had become the fourth most important manganese producer in the world. Uranium is mined around MOUNANA, and the manganese is mined at MOANDA, both towns on the BATEKE PLATEAU in the south-east. There are huge deposits of iron ore in the north-east. With a small population and big export earnings from oil and metal ores, Gabon seemed on the brink of a miraculous transformation.

But prosperity for the nation did not bring prosperity to all its inhabitants. There were good intentions: in 1976 an ambitious four-year plan was announced with a budget of US$32 000 million, to bring Gabon's inadequate roads and railways up to date, and to encourage local industry. Instead, most of the money was squandered on such projects as the building of the US$120 million presidential palace, and few people benefited from the country's new wealth. The mining industries continued to employ a small number of people, and the lives of most Gabonese have been largely unaffected by the mineral windfalls.

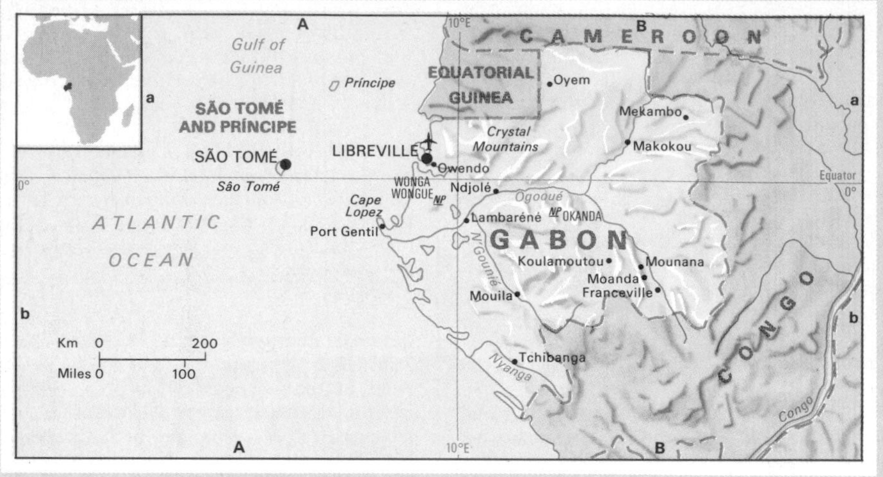

Because of the dense, hardwood forests and the many rivers which have cut deep valleys in the interior plateaus and highlands (including the CRYSTAL MOUNTAINS and the CHAILLU MOUNTAINS in south-central Gabon), transport links are poor. As a result, the country's tourist potential has not been developed. Wildlife flourishes in the national parks and game reserves, but elephant poaching is a problem.

Gabon's history has in many ways been a sad one, but it has two notable monuments to altruistic and humanitarian actions. One is Albert Schweitzer's hospital at LAMBARENE, built in 1913 as part of the fight against poverty and disease which earned him the Nobel Peace Prize in 1952. The other is the capital city of LIBREVILLE, which was founded by the French in 1849 as a home for slaves rescued from illegal slaving ships.

As a one-party state, Gabon has not been so noted for its freedom since independence, but it has at least been comparatively stable.

GABON AT A GLANCE

Area 267 667 km² (103 346 sq miles)
Population 1 020 000
Capital Libreville
Government One-party republic
Currency CFA franc = 100 centimes
Languages French (official), Bantu languages
Religions Christianity (50%), rest mainly tribal, with 2000 Muslims
Climate Equatorial; average temperature in Libreville ranges from 24°C (75°F) to 27°C (81°F).
Main primary products Cocoa, coffee, palm oil, bananas, sugar cane, cassava, plantains, maize, rice, livestock, timber, petroleum, manganese, uranium, iron ore
Major industries Mining, petroleum production and refining, agriculture, forestry, fishing, food processing
Main exports Petroleum, manganese, uranium, timber
Annual income per head (US$) 3040
Population growth (per thous/yr) 31
Life expectancy (yrs) Male 47 **Female** 51

and leather goods. The first Bulgarian national school was opened here in 1835. Every odd-numbered year, in May, an international festival of comedy and satire is held in the House of Humour. In the mountains around Gabrovo are picturesque villages such as Bozhentsi and Shipka, the Dryanovo Monastery, and the open-air museum of Etur depicting 19th-century life, arts and crafts.
Population 83 000
Map Bulgaria Bb

Gabu ***Guinea-Bissau*** Region and town in the wooded savannah of the north-east, some 160 km (100 miles) from Bissau. Formerly called Nova Lamego, the town was one of the final Portuguese strongholds before independence, the local Muslim aristocracy of the Fula tribe having cooperated with the Portuguese in an attempt to preserve their social privileges. Now it is the centre of an agricultural region producing groundnuts, millet, guinea corn, vegetables and, in recent years, cotton and tobacco. These are transported by road to Bissau via Bafatá.
Population (region) 104 200
Map Guinea Ba

Gaeseong ***North Korea*** See KAESONG

Gafsa ***Tunisia*** Oasis town 185 km (115 miles) west of Sfax. It is a major road and rail junction on the Sfax-Tozeur railway, and is also the centre of a phosphate mining region which accounts for 85 per cent of the country's output. Founded by the Romans, Gafsa later became a Berber stronghold. Indeed, it is one of the few places in Tunisia where there are still traces of the Berber language and its traditions. Today its protective walls of rose-pink stone enclose a town of solid brick buildings decorated in bold geometric patterns. The local craftsmen specialise in carpets, rugs and blankets, and the oasis produces dates, olives and citrus fruit.
Population 61 000
Map Tunisia Aa

Gaillimh ***Ireland*** See GALWAY

Galápagos Islands (Colón Archipelago) ***Pacific Ocean*** Group of volcanic islands on the Equator, 1100 km (680 miles) west of Ecuador – of which they are a province. They were discovered by the Spanish in 1535 and annexed by Ecuador in 1832. The capital is San Cristóbal Island, also called Chatham Island (population 2000). Four other islands have permanent human populations.

The islands were visited by the English naturalist Charles Darwin in 1835, during his historic voyage on the *Beagle*. The animal life he found there provided part of the inspiration for his theory of evolution, expounded in *The Origin of Species* (1859). Because of their unique species of plant and animal life – including the almost-extinct giant tortoise – 90 per cent of the land area is now a national park.
Population 6200
Map South America Ac

Galati ***Romania*** Major inland port on the lower Danube, 180 km (112 miles) north-east of Bucharest. With its steel works, shipyard and major rail junction, Galati is an important industrial centre. Its manufactures also include machinery, oil products, synthetic rubber, textiles and food. Galati has a number of 17th-century churches and two fine museums.
Population 261 000
Map Romania Ba

gale Very strong wind, varying between Force 8 and 10 on the BEAUFORT SCALE. A gale (Force 8) has winds of 62-74 km/h (39-46 mph). Force 10 is a full gale or storm with winds of 88-101 km/h (55-63 mph). A near gale (Force 7) has winds of 51-61 km/h (32-38 mph).

galena Soft and heavy grey mineral, essentially lead sulphide, found usually in carboniferous limestone. It is the chief ore of lead.

Galicia ***Spain*** Region and former kingdom in the north-west, covering 29 434 km² (11 365 sq miles). It was colonised by the Goths in the 6th century and was united with CASTILE in the 11th. Bounded by the Bay of Biscay on the north and the Atlantic Ocean on the west, it has a thriving fishing industry. Its chief river is the Miño, which forms part of the Portuguese border, and its main towns CORUNNA, SANTIAGO DE COMPOSTELA and VIGO.

Population 2 754 000

Map Spain Aa

Galicnik ***Yugoslavia*** See MAVROVO

Galilee ***Israel*** Northern region bounded by Lebanon to the north and to the east by Syria and Jordan. The majority of Israel's Arab population is concentrated in this area. The northern part of Galilee is hilly, but farther south there is fertile agricultural land. The main towns are TIBERIAS and NAZARETH. The region is closely associated with Christ's teachings and ministry and includes the Sea of Galilee (Lake Tiberias).

Map Israel Ba

Galle ***Sri Lanka*** Old colonial city on a rocky headland 105 km (65 miles) south and slightly east of Colombo. Formerly known as Point de Galle, it was founded as a walled seaport by the Portuguese around 1600. In 1643, it was taken by the Dutch, and the fort they built survives. Galle has many fine colonial mansions, and its craftsmen specialise in lacemaking, ebony carving and gem cutting. There are good beaches and diving nearby.

Population 77 200

Map Sri Lanka Bb

gallery forest Dense tangle of trees lining the banks of a river in open savannah country. Along the smaller streams, the trees meet overhead, giving the appearance of a gallery.

Gallipoli (Gelibolu) ***Turkey*** Peninsula on the northern side of the DARDANELLES strait some 200-250 km (125-155 miles) west and slightly south of Istanbul. It extends for about 80 km (50 miles) between the Dardanelles and the Gulf of Saros. Its beaches were the scene of heavy fighting in 1915 and 1916, during the First World War, when Allied troops – notably Australians and New Zealanders – tried unsuccessfully to take control of the strait from the Turks (who were fighting on the side of Germany). Gallipoli is also the name of a fishing port (population 16 900) on the Dardanelles coast; it was the first place in Europe to be conquered by the Turks, about 1356.

Map Turkey Aa

Galloway ***United Kingdom*** See DUMFRIES AND GALLOWAY

A DAY IN THE LIFE OF A KIBBUTZNIK

When the alarm clock wakes Amos, its luminous hands in the dark read 4 am. Beside him, his wife Ilana stirs, turns over, and goes back to sleep. Amos dresses quickly and slips out to the cowsheds, through the kibbutz of Kfar Hanossi which has been his home for five years.

The community of 550 people lies 5 km (3 miles) from the town of Rosh Pinah near the Sea of Galilee in northern Israel, on land which was once swamp. The kibbutz (Hebrew for 'gathering') was founded in 1948 and is one of 250 such cooperative settlements in Israel – communities born of necessity in the days when teamwork was imperative to build a nation out of a wilderness.

By the time the sun rises at 6 am, Amos and his four fellow dairy workers have milked 50 cows. Israel has some of the most modern milking methods in the world, and the yield is high. They go on to clean out the byres, and then wash before going into the dairy to share out the rest of the day's chores over a cup of coffee. At 7.30 they join friends and families in the communal dining room for a buffet breakfast of eggs, cereals, salads, fruit, rolls and goat's yoghurt.

Ilana soon slips away to feed her two-year-old daughter, Rinat, in the baby house. Amos joins her there to see their child, before leaving her in the capable hands of trained childminders.

While Ilana goes to work in the kitchens, Amos takes a tractor to check the cowsheds, and then drives out to the hayfields. He and his friend Gideon load heavy bales of fodder onto the tractor trailer all morning. For both of them it is hot work as the sun rises higher, and they are glad when it is lunchtime and the day's work is over.

After lunch, which is again in the communal dining room, anyone who can, takes a siesta. Overhead fans whisper in the silent children's houses. At 4 pm the children wake and parents come to collect them. In minutes, the centre of the kibbutz is full of noise and shouting. Amos and Ilana take Rinat home to play on their lawn. Later, as the sun begins to sink over the wheatfields, they take Rinat for a walk around the kibbutz. Work is progressing on new flats – the kibbutz always seems to be growing. Amos jingles as he walks, with tokens for the kibbutz shop in his pocket. He has no salary, merely an allowance, but the kibbutz provides for all his needs, so money is not a worry, nor a matter of great interest. The kibbutz belongs to everyone who lives in it.

Back at the baby house, Ilana puts Rinat to bed. Watching her, Amos knows that they made the right decision when they chose to live on a kibbutz.

Galveston ***USA*** Port in Texas on a coastal island in the Gulf of Mexico, about 80 km (50 miles) south-east of Houston. It is linked to the mainland by a causeway, and handles cotton and grain, although it has declined since the opening of the Houston Ship Canal in 1914.

Population (city) 62 400; (metropolitan area) 215 400

Map United States He

Galway (Gaillimh) ***Ireland*** County on the north side of Galway Bay covering 5940 km² (2293 sq miles). Its county town is the city of Galway, 196 km (122 miles) west of Dublin. In the 13th century the city – then walled – became an isolated Anglo-Norman colony led by families known as the 14 tribes. Despite violent opposition from local people in the countryside, they dominated the area for four centuries, and even held out for the Royalists against Oliver Cromwell until 1652. According to local legend, one Anglo-Norman Mayor, James Lynch Fitz-Stephen, condemned and executed his own son for murder in 1493 – giving his name to the term 'Lynch mob'. A stone near St Nicholas's Church marks the site of the gallows.

Galway is a port; it has a modern Catholic cathedral and is the site of part of the National University of Ireland. There are a wide variety of industries and a strong tourist trade.

Population (county) 171 800; (city) 37 700

Map Ireland Bb

Gambia ***West Africa*** One of Africa's major rivers, some 1130 km (700 miles) long. It rises in the Fouta Djallon in Guinea, and is fundamental to the existence of The Gambia, Africa's smallest mainland country, which occupies a narrow strip of land on either side of the river. In the dry season from December to April the Gambia is salty as far as the town of Kuntaur, about 200 km (125 miles) upstream, hampering irrigation projects.

The river grows from a width of about 30 m (100 ft) at Koina in the east, to an estuary which is more than 1.6 km (1 mile) wide at Banjul, the capital of The Gambia. There, the Gambia is up to 9 m (30 ft) deep, but a sandbar restricts access to ships of less than 25 000 tonnes.

Map The Gambia Ab

Gambia, The See p. 243

Gambier Islands ***French Polynesia*** Small group some 1600 km (1000 miles) south-east of Tahiti. The main island is Mangaréva, once a centre of missionary activity in eastern Polynesia. It and the other small islands are enclosed by a reef forming a lagoon some 26 km (16 miles) wide.

Population 600

Map Pacific Ocean Gd

Gan ***Maldives*** Small coral island in the southern Maldives atoll of Addu, 644 km (400 miles) south-west of Sri Lanka. The island was an important Royal Air Force staging post during the later years of Britain's protectorate over the Maldives which ended in 1965. The present government hopes to expand air transport to encourage tourism in the area.

Gand ***Belgium*** See GHENT

Gandak ***Nepal/India*** River 690 km (428 miles) long, formed by the union of several other rivers – including the Kali Gandak and Buri Gandak – in central Nepal. It drains the whole west-central area of the country, and flows into the Bihari plains of India – bringing considerable flooding on occasion – before entering the Ganges at Patna. On its journey through Nepal, the Kali Gandak passes between the mountains of Annapurna and Dhaulagiri, forming the world's deepest valley, 5.5 km (3.5 miles) below the 8172 m (26 810 ft) high peak of Dhaulagiri.

Map Nepal Ba

Gander ***Canada*** Town in eastern Newfoundland with a major airport which used to be a busy refuelling stop for transatlantic airliners. The airport has declined in importance because modern aircraft can usually cross the Atlantic without refuelling, but it is still the centre for North Atlantic air traffic control.

Population 10 405
Map Canada Jd

Gandhara ***Pakistan*** Ancient kingdom and province in the area of Peshawar and Islamabad, straddling the Punjab and North-West Frontier Province. The Macedonian conquerer Alexander the Great overran it in 327-326 BC, and it has also been conquered by Persians and Parthians. Alexander's influence was strong and a Graeco-Buddhist culture emerged, marked by the remains of a 2nd-century BC university at the capital, TAXILA.

Map Pakistan Da

Gandhinagar ***India*** New town planned and developed since 1961, about 460 km (285 miles) north of Bombay. It is the capital of Gujarat, the home state of the political and religious leader Mahatma Gandhi (1869-1948).

Population 62 400
Map India Bc

Ganges (Ganga) ***South Asia*** The most holy river of Hinduism, 2525 km (1568 miles) long. It rises in the Indian Himalayas near GANGOTRI mountain and flows down through the Ganges plain, where it is harnessed for irrigation and power. The new low barrage at the town of Haridwar has a narrow slit in the middle, so that some of the water passes in its natural state to the sea. On its right bank it is joined by the Yamuna; on its left bank it is joined by other great rivers flowing from the Himalayas, such as the Kosi. It then flows into Bangladesh just downstream from the FARAKKA BARRAGE. In Bangladesh the Ganges is known as the Padma until it joins the Jamuna (Brahmaputra), after which the combined river is called the Meghna. The delta of the Ganges is the world's biggest, and is a highly active zone with new islands emerging in the Bay of Bengal.

Along the length of the Ganges are holy shrines, most notably at VARANASI and ALLAHABAD. Hindus believe the water is so holy that it will keep clean forever. They also wish their cremation ashes to be scattered on the river in the belief that their souls will go straight to Heaven. Although the river contains all kinds of pollutants – industrial and human – from the many major cities on its banks, it has a remarkable ability for cleaning itself.

Map India Dc

Ganges-Kobadak ***Bangladesh*** Irrigation scheme in western Bangladesh. Water is pumped from the Ganges to flow south through the Kushtia canal into parts of the delta that have become silted up over the centuries, as water flow has moved to rivers farther east. A secondary aim of the scheme is to keep salt water from intruding upriver from the Bay of Bengal during the dry season. The scheme should irrigate some 90 000 hectares (220 000 acres).

Gangneung ***South Korea*** See KANGNUNG

Gangotri ***India*** Mountain in the north about 40 km (25 miles) south of the Chinese border. It is 6614 m (21 695 ft) high, and its glaciers are the source of the Ganges. It has a Hindu temple, visited by pilgrims, at 3145 m (10 320 ft).

Map India Cb

Gangtok ***India*** Capital of the former Himalayan kingdom of Sikkim, a state of north-east India since 1975. A road climbs 1768 m (5800 ft) to the town from the plains. But from Gangtok itself only tracks go higher into the mountains, including one that leads into Tibet. Visitors to Gangtok need special passes.

Population 37 800
Map India Db

Gansu ***China*** Province in the north-west, shaped like a long narrow corridor and lying between Nei Mongol and the Qilian Shan mountains. It is a semiarid region covering 450 000 km² (170 000 sq miles), taking in parts of the LOESSLANDS OF CHINA and the Gobi Desert. It is crossed by the Silk Road, a centuries-old trade route to central Asia by which the Venetian explorer Marco Polo travelled to China in the 13th century. Wheat, millet and gaoliang (a kind of sorghum) are grown in the loesslands. The province is rich in oil, iron ore, coal and copper. Its capital is Lanzhou.

Population 19 600 000
Map China Ec

Ganvié ***Benin*** Picturesque lakeside village on Lake Nokoue. The houses are built on stilts, an arrangement that formerly exempted their owners from taxation. It was argued that because they did not actually live on the ground, they could not be taxed as local residents. The village can be visited by dugout canoe from the port of Cotonou.

Gao ***Mali*** Town about 960 km (600 miles) down the Niger River from the capital, Bamako. It was once the capital of the Songhai Empire, but declined after the Moors occupied it in 1591. Gao is a major market for local produce and an important river port, the terminus of river steamers from Bamako. The population has soared recently because of the vast number of nomads who have lost their herds and their traditional living in the drought that has devastated the north since the early 1970s.

Its historic buildings include the Mosque of Kankan Moussa and the tomb of the Askia Dynasty – rulers of the former Mali Empire.

Population 50 000
Map Mali Cb

Gap ***France*** Capital of the department of Hautes Alpes, 75 km (about 50 miles) south-east of Grenoble. It manufactures gloves and textiles. It guards the Bayard Pass, the route taken by Napoleon on his return to Paris in 1815 from his exile on Elba.

Population 32 100
Map France Gd

Garda, Lake ***Italy*** Alpine resort lake about halfway between Milan and Venice. It covers an area of 370 km² (143 sq miles), making it Italy's largest lake. Its beauty and mild climate make its shores a miniature riviera, with luxuriant vegetation and plantations of olive and citrus trees.

Map Italy Cb

Gargano ***Italy*** Mountainous promontory, about 50 km (30 miles) long, extending into the Adriatic Sea about 300 km (190 miles) north-west of the heel of Italy. It rises to 1056 m (3465 ft) and is flanked by sandy beaches. Among its villages is the medieval pilgrimage centre of MONTE SANT'ANGELO.

Map Italy Ed

Garmisch-Partenkirchen ***West Germany*** Twin towns at the foot of the Bavarian Alps, about 75 km (45 miles) south-west of Munich. The towns form one of West Germany's leading winter sports resorts. A railway leads to the Zugspitze, at 2963 m (9721 ft) West Germany's highest mountain.

Population 28 000
Map West Germany De

garnet One of several widely distributed silicate minerals found in igneous and metamorphic rock, coloured red, brown, black, green, yellow or white and used as gemstones and abrasives.

Garonne ***France/Spain*** River, 575 km (355 miles) long, in the south-west. It rises in the Spanish Pyrenees, and flows north-east to Toulouse. It then turns north-west and flows on past Bordeaux to join the Dordogne at the Gironde estuary, which empties into the Bay of Biscay. The middle and lower sections of the Garonne valley contain rich farmland producing wheat, maize, fruit, vegetables and vines from which fine Bordeaux wines are made.

Map France Cd

garrigue Impoverished scrub, similar to maquis, found in the drier limestone areas of the Mediterranean region. It consists of stunted trees and low, drought-resisting plants, many of which are aromatic, with much bare earth. See MEDITERRANEAN WOODLAND

Gascony (Gascogne) ***France*** Territory in the extreme south-west, held by the English from 1152 to 1453. Its name comes from the Vascones, a Spanish tribe who conquered it in the 6th century AD and set up the Duchy of Vasconia (Gascony).

Map France Ce

Gasherbrum ***China/Pakistan*** Mountain on the disputed border between Pakistani Azad Kashmir and Chinese-occupied territory in the Karakoram Range. At 8068 m (26 470 ft), it is Pakistan's third highest peak after neighbouring K2 and Nanga Parbat.

Map Pakistan Da

The Gambia

A TINY, POVERTY-STRICKEN COUNTRY WITH FEW RESOURCES, DEFENDING ITSELF AGAINST ENGULFING SENEGAL

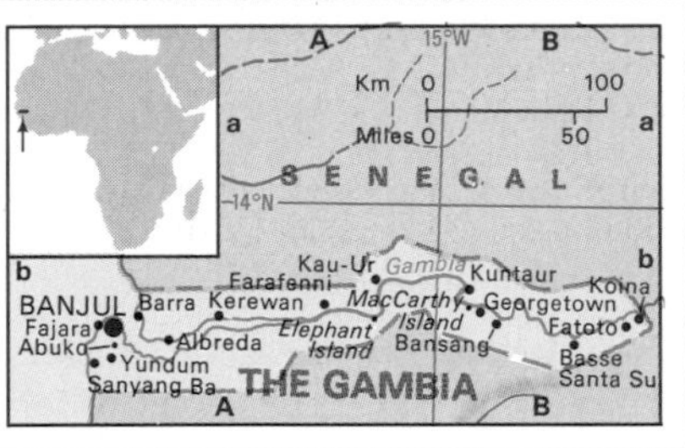

The smallest country in Africa, The Gambia pokes into the west coast of the continent like a crooked finger. The finger is 48 km (30 miles) wide at the coast, narrowing to only 24 km (15 miles) as it protrudes 320 km (200 miles) inland. The country is divided along its entire length by the GAMBIA river, and the division is emphasised by the lack of a single bridge over the river; there are only two main ferry crossings.

The Gambia is a little English-speaking sliver left behind by British colonists, and in danger of being swallowed up by the French culture of its giant neighbour, Senegal, which surrounds it on all sides except for the sea.

This incongruous remnant of British rule in Africa is a legacy of the rivalry between Britain and France for control of West African trade. The British arrived in 1664 in the wake of Portuguese traders and displaced a group of Baltic German settlers from a fort on James Island in the Gambia estuary, 30 km (19 miles) from the sea. Ivory, ebony, gold and slaves were the lure for the Europeans in those days, and these were exchanged for salt, iron, firearms and gunpowder. Fort James was the collection point for slaves shipped by the British to the Americas in the 18th century, until the slave trade was abolished in 1807.

TERRITORY UNDER SIEGE

For centuries The Gambia was accustomed to being under siege – either by pirates, or by the French (who made six attempts to take the country), or by local tribal kings who cut off supplies from time to time. But Britain managed to maintain its hold on this important trade outlet, which was finally established as a colony in 1888, while the land upriver became a protectorate in 1894. In 1889 a treaty between Britain and France defined the border between The Gambia and French-controlled Senegal.

In 1963 The Gambia was given self-government and in 1965 became fully independent. Several attempts have been made at a merger between The Gambia and Senegal but the Gambians continue to hang on to their separateness, because their way of life and language are so different. Indeed, one Senegalese government official remarked: 'The trouble with the Gambians is that they are so British.'

However, a loose confederation – Senegambia – has been established. It came about in 1981 when an attempted coup against the government led the president, Sir Dawda Jawara, to accept the help of Senegalese troops. The troops are still present in The Gambia as 'confederal' troops, of the Senegambia Confederation, a defence alliance established in 1982. Negotiations for an economic agreement are also under way, although the 777 000 Gambians are anxious that its terms should not treat them as the 'poor relations'.

The Gambia is, however, by far the poorer of the two countries. In the capital, BANJUL (formerly Bathurst), which lies on a sandy island at the mouth of the Gambia river, the majority of Africans live in closely packed, corrugated iron dwellings. For most of them, the only source of water is a communal standpipe in the street. Outside the city centre, the suburbs of Cape St Mary and Fajara are a stark contrast. Here, wide, tree-lined roads where the ruling British once lived now contain the beautifully kept offices of government departments, embassies and the houses of diplomats and foreign visitors.

RELICS OF BRITISH INFLUENCE

Evidence of British influence is apparent everywhere, from Banjul's once elegant cricket ground to the ornamental cannons that are dotted about the city centre. Even the Gambian army is still organised and drilled on British military principles. But these are only superficial relics of more prosperous days.

Today, the most luxurious existence in The Gambia is in the increasingly popular hotels on the seafront, where tourists can step straight out of their rooms onto white, sandy beaches. It is always hot on the coast – temperatures averaging around 24°C (75°F) in January and 29°C (84°F) in June. The dry season stretches from November to May with the main rains between June and October – the annual average rainfall is about 760-1140 mm (30-45 in). The local markets are geared to the tourist trade, selling carved wooden figures, tribal masks, and brightly printed batik cloth imported from the Ivory Coast alongside the vegetables and fish that provide the daily food of the Gambians.

THE GREAT RIVER

Boats are important in a country dominated by a river, and there is a thriving local fishing industry using dugout canoes called *pirogues*. Most of the catch comes from the sea, though some 800 tonnes of freshwater fish are landed annually.

The best way to see The Gambia is by boat too. A river boat, the *Lady Chilel Jawara*, sails weekly from Banjul. It leaves the sandy coast and the mangrove swamps of the estuary, and travels through the *banto faros* (meaning 'beyond the swamp' in the Mandingo language), which is flooded by the river for part of the year, towards the sandstone uplands on which Basse Santa Su lies. At each river port – such as Kerewan and Georgetown – people, goats, chickens and luggage are noisily loaded and unloaded. All along the banks and on the midstream islands are flocks of exotic birds.

It is also possible to drive the length of The Gambia on a tarmac road, skirting the edges of villages and passing alternately through cultivated land and scrub country, where cattle, sheep and goats browse. The red tracks that lead off this road to the villages are iron hard in the dry season, but turn to quagmires when the rains begin.

Most Gambians live precarious lives in the villages, tending a few animals and trying to grow enough millet and sorghum to feed themselves. If they are lucky, they vary their diet with maize, rice, vegetables and fruit. But drought and overgrazing have meant that the last few years have disappointed farmers' hopes. Life expectancy here is low – only about 36 years.

Groundnuts are The Gambia's main export, and are often grown by these same village farmers for cash. No one can keep track of the quantity produced, because when prices are higher in The Gambia than in neighbouring Senegal, groundnuts are smuggled in for sale, and The Gambia's production figures are inflated. There is a great deal of smuggling between the two countries whenever there is a profit to be made; border patrols have a difficult job because the boundaries of The Gambia divide ethnic groups living on both sides of the border: the Mandingo (or Malinke), Fulani (or Peul), Wolof, Dioula (or Jola) and Serer tribes among others.

In some ways this division between The Gambia and Senegal makes little sense. The Gambia's difficult times mean that large numbers of the fast-growing population are having to emigrate to find jobs – often in Britain or France. Economically, The Gambia is struggling, and depends heavily on foreign aid, mainly from West Germany, China and Britain. Even its tourism brings in less than might be hoped: most visitors come on package tours and pay for everything before they arrive. But most of the diverse people of this country are still reluctant to forfeit their national status and their common identity: The Gambia is determined to go it alone.

THE GAMBIA AT A GLANCE
Area 11 295 km² (4361 sq miles)
Population 777 000
Capital Banjul (formerly called Bathurst)
Government Parliamentary republic
Currency Dalasi = 100 bututs
Languages English (official), African languages
Religions Muslim (85%), tribal, Christian
Climate Tropical; temperature at Banjul ranges from 15-31°C (59-88°F) in January, and 23-32°C (73-90°F) in June
Main primary products Groundnuts, cassava, millet, rice, cattle, fish, timber, palm kernels
Major industries Agriculture, food processing, fishing, forestry
Main exports Groundnuts and derived products, fish, hides and skins
Annual income per head (US$) 350
Population growth (per thous/yr) 35
Life expectancy (yrs) Male 35 **Female** 37

Gaspé ***Canada*** **1.** Peninsula projecting eastwards into the Gulf of St Lawrence in eastern Quebec. The explorer Jacques Cartier landed here on his first voyage in 1534 to claim the area for France. The peninsula is mainly forested and is a summer resort. **2.** Town at the eastern end of the Gaspé peninsula. It is a railway terminus and its industries include timber and tourism.
Population (town) 17 260
Map Canada Id

Gateshead ***United Kingdom*** See NEWCASTLE UPON TYNE

Gatooma ***Zimbabwe*** See KADOMA

Gauhati ***India*** North-eastern town in Assam on the Brahmaputra river, about 70 km (43 miles) south of the border with Bhutan. It is the centre of a flourishing tea industry established in 1838. It also trades in cotton, jute and rice – and there are flour mills and oil refineries.
Population 123 000
Map India Eb

Gävle ***Sweden*** Seaport and industrial city on the Baltic coast 180 km (112 miles) north of Stockholm. It exports iron ore and softwoods, and manufactures textiles, chemicals and beer.
Population 87 700
Map Sweden Cc

Gaza ***Gaza Strip*** Mediterranean seaport about 65 km (40 miles) south-west of Tel Aviv. It was a Philistine city-kingdom in the 1st millennium BC and it was there that Samson pulled down the Temple of Dagon, crushing to death his Philistine captors and himself.
Population 118 300
Map Egypt Cb

Gaza ***Mozambique*** See XAI-XAI

Gaza Strip ***Middle East*** Territory occupied by Israel since the Six-Day War of 1967. It is an area 42 km (26 miles) long and 6-8 km (4-5 miles) wide running along the Mediterranean coast up to the Egyptian frontier. The area had been part of Palestine, and when the State of Israel was created in 1948 it was left under Arab control and its administration was handed to Egypt. After 1948 large numbers of Palestinian refugees moved in, and it is still dominated by eight large Palestinian refugee camps. More than 500 000 people live there and about 45 per cent of the adult population work in Israel, travelling there daily by bus and car. Some Israeli settlements and an Islamic university have been established in the Strip.
Population 510 000
Map Egypt Cb

Gazankulu ***South Africa*** Former Bantu homeland in north-east Transvaal which became a self-governing, but not independent, state in 1973. It covers 7410 km² (2860 sq miles) and is populated by the Shangaan and related Tsonga peoples. Giyani is the capital.
Population 582 500
Map South Africa Da

Gaziantep ***Turkey*** Town on the border with Syria about 500 km (310 miles) south-east of the capital, Ankara. Its main manufactures are processed agricultural produce, textiles and goatskin rugs. The neighbouring province of the same name, which covers 7642 km² (2951 sq miles), is the centre of Turkey's pistachio nut farming belt.
Population (town) 466 300; (province) 953 900
Map Turkey Bb

Gcuwa ***South Africa*** See BUTTERWORTH

Gdansk (Danzig) ***Poland*** The country's largest seaport and shipbuilding centre, 354 km (220 miles) north-west of the capital, Warsaw. It is the focus of Trojmiasto (Tri-City) – the combined urban area of Gdansk, Sopot and Gdynia, which stretches for 40 km (25 miles) along the Gulf of Gdansk, and has a million inhabitants.

Gdansk was already a large Polish stronghold, fishing port, and crafts and amber-trading town 1000 years ago. At different times since then, it has been ruled by Poles, Teutonic Knights and Germans. For 28 years (1807-15 and 1919-39) it was the Free City of Danzig. Hitler's claim to the city in 1939 led to the German invasion of Poland and the start of the Second World War. A member of the Hanseatic League of mostly north German towns from 1343, Gdansk prospered on trade. Shipbuilding began only in 1850, and most of its other main industries – engineering, chemicals and food processing – began only after 1950.

Gothic churches, a Renaissance town hall and the 14th-century Great Mill are the pride of the oldest parts of the city. The medieval layout survives, with nine parallel streets ending at the river with watergates, where goods were once loaded and unloaded. Each August, the city holds the Dominican Fair – a good time to buy souvenirs and antiques.

Three crosses stand outside the Lenin Shipyard, one of the city's largest. In 1980, the yard was the birthplace of Solidarity, Eastern Europe's only independent trade union, banned in 1981, but still in existence. The crosses were set up to commemorate workers killed in civil demonstrations following massive increases in food prices in 1970.
Population 464 500
Map Poland Ca

Gdansk, Gulf of (Gulf of Danzig) An inlet of the BALTIC SEA in north-eastern Poland. It contains GDANSK and GDYNIA, two of Poland's most important ports.
Dimensions 64 km (40 miles) north-south, 60 km (37 miles) east-west
Greatest depth 113 m (370 ft)
Map Poland Ca

Gdynia (Gdingen) ***Poland*** Baltic seaport 16 km (10 miles) north-west of Gdansk. Founded as a fishing village in 1224, the port now handles about 18 million tonnes of cargo a year.
Population 240 200
Map Poland Ca

Gêba ***Guinea/Senegal/Guinea-Bissau*** West African river, about 190 km (120 miles) long. It rises in the Fouta Djallon of western Guinea, and flows through southern Senegal, then through Guinea-Bissau to the Atlantic Ocean via a wide estuary. Gêba is also the name of a former capital of Portuguese Guinea (now Guinea-Bissau), situated near the upper limit of navigation on the river, near present-day Bafatá. The town was abandoned as the capital because it was in a swampy, malaria-infested area, and is today virtually in ruins.

The Gêba river is Guinea-Bissau's main transport artery. The country's leading commercial centre and port, Bissau, lies on the estuary.
Map Guinea Aa

Geelong ***Australia*** Port and largest provincial city in Victoria, about 60 km (38 miles) south-west of Melbourne, and once the port of entry to the Victorian gold fields. It has one of the world's largest and most modern bulk wheat terminals, and is a major wool-selling centre. Its manufactures include woollen textiles, automobiles, cement and agricultural equipment.
Population 142 000
Map Australia Gf

Geirangerfjord ***Norway*** A 15 km (9 mile) long branch of Storfjord, south-east of Ålesund in west Norway. The fiord, carved by Ice Age glaciers, is claimed by many Norwegians to be the nation's most beautiful stretch of water, and it attracts thousands of tourists each summer. All along its length, it is edged by sheer rock walls laced with ribbons of waterfalls – among them the gauzy Bridal Veil, and the multiple cascades of the Seven Sisters. The cliffs above the fiord's mirror-smooth surface are some 400 m (1310 ft) high, and they extend for as much as 300 m (980 ft) below the water.
Map Norway Bc

Gela ***Italy*** Town on the south coast of Sicily, about 140 km (87 miles) south-east of Palermo. Gela has grown rapidly since a petrochemical plant was built there in the 1960s. It was originally a Greek city called Gelon, and there are extensive archaeological sites to the east and west of the town.
Population 74 800
Map Italy Ef

Gelderland (Guelderland) ***Netherlands*** The country's largest province, covering 5129 km² (1980 sq miles). It lies in the south-east, along the West German border, between the Maas and IJssel rivers. The region is mostly agricultural, with areas of forested hills and heathland, popular with tourists. However, numerous industries have developed along the Waal and Lek branches of the Rhine, especially around Nijmegen and the provincial capital Arnhem.
Population 1 727 500
Map Netherlands Ba

Gelibolu ***Turkey*** See GALLIPOLI

Gelsenkirchen ***West Germany*** Coal-mining town in the industrial Ruhr, 24 km (15 miles) west of Dortmund.
Population 290 000
Map West Germany Bc

Genck ***Belgium*** See GENK

General Santos ***Philippines*** Port and capital of Cotabato del Sur province, southern MINDANAO, formerly called Dadiangas. It serves the rehousing projects of the Koronadal valley.
Population 149 396
Map Philippines Cd

Geneva (Genève; Genf) ***Switzerland*** City at the western end of Lake Geneva (also known as Lac Léman), where the River Rhône flows out of the lake. Founded by the Romans, Geneva became a powerful, independent city under its prince bishops in the Middle Ages and under its merchant rulers in the 15th and 16th centuries. It was seized by revolutionary France in 1798, but joined the Swiss confederation in 1815 after the fall of Napoleon. The city is still surrounded on three sides by French territory.

Geneva was one of the principal centres of the Reformation, putting the austere religious theories of John Calvin (1509-64) into political practice in the 16th century. In 1865 the International Red Cross was founded in Geneva by Henri Dunant (1828-1910), and since then the city has become a centre for international organisations including UN agencies and negotiations concerned with politics, trade, health, science and peace. As a conference centre it provides five-star comfort in a scenic setting – with a 145 m (476 ft) fountain shooting from the lake.

The canton of the same name around the city grows grapes, fruit and vegetables. The district covers 282 km^2 (109 sq miles).

Population (city) 165 000; (canton) 364 800
Map Switzerland Aa

Genk (Genck) ***Belgium*** Industrial town about 30 km (19 miles) north of the city of Liège. Developed in the 1920s as a coal-mining village, Genk now also produces steel and cars.

Population 62 000
Map Belgium Ba

Genoa (Genova) ***Italy*** Seaport in the north-west about 120 km (75 miles) south-east of Turin. It has been a trading centre and naval base since the early Middle Ages, and remains Italy's leading port. The explorer Christopher Columbus was born there in about 1451. Giuseppe Mazzini (1805-72), the republican leader of the Risorgimento ('Resurgence') movement, which paved the way for Italian unification, made the city his headquarters.

Today, Genoa's economy relies on heavy industries, principally steel and shipbuilding, which are declining. The lack of level land prevents the expansion of new industries.

Population 760 300
Map Italy Bb

Genoa, Gulf of Part of the LIGURIAN SEA, washing the shores of north-western Italy. The port of GENOA, which was for long a great maritime power, occupies a central position on the coast of the gulf. The Riviera di Levante, east of Genoa, and the Riviera di Ponente, extending to the French border to the west, are dotted with resorts.

Map Italy Bb

Genova ***Italy*** See GENOA

Gent ***Belgium*** See GHENT

geo, gio Long, narrow, steep-sided inlet running inland from the edge of a cliff, probably caused by the collapse of the roof of a cave worn by the waves along a line of weakness such as a major JOINT or minor FAULT.

geode A small, usually spheroidal hollow in a rock, lined with crystals.

geographic poles The North and South Poles, the points at which the axis of rotation (the imaginary line about which the earth rotates) reaches the earth's surface. They are not identical with the GEOMAGNETIC POLES.

geological cycle See ROCK CYCLE

geological timescale The division of geological time into chronological units. See AGE OF THE EARTH (p. 246)

geomagnetic poles The points at which the axis of the earth's magnetic field reaches the earth's surface. The axis lies at an angle of 11 degrees to the axis of rotation, therefore the geomagnetic poles do not coincide with the geographic poles. At present the north geomagnetic pole lies at 79°N, 70°W and the south geomagnetic pole lies at 79°S, 110°E, but both are moving very slowly. Over the past 150 years, for example, the latitude of the poles has remained fairly constant, but they have moved westward by about 6 degrees of longitude. As well as varying in location, the axis of magnetism has also undergone reverses of polarity. The effects of these reversals on the alignment of magnetic particles in rocks is important evidence supporting the theory of plate tectonics.

geomagnetism The magnetism of the earth.

George ***South Africa*** Resort beneath the Outeniqua mountains about 375 km (230 miles) east of Cape Town, and 8 km (5 miles) from the Indian Ocean. It was founded in 1812 and named after Britain's King George III. The town, noted for its wildflowers, is on the Garden Route – a scenic highway linking the seaport of Mosselbaai to Port Elizabeth.

Population 62 300
Map South Africa Bc

George Town ***Malaysia*** Port and capital of Penang Island, on a headland facing mainland Malaysia 300 km (186 miles) north of the capital, Kuala Lumpur. The city grew beside Fort Cornwallis, founded by Captain Francis Light of the British East India Company in 1786 to guard the island's harbour. Today most of the harbour trade has moved to the mainland port of Butterworth, 16 km (10 miles) away.

George Town, with its noisy mixture of East and West, Malay and Chinese, is now primarily a tourist centre. It has streets of Chinese shops, Buddhist temples and easy access to the hills and beaches of the island.

Population 250 600
Map Malaysia Ba

Georgetown ***Guyana*** National capital and the country's largest city. It lies on the Atlantic coast and is largely below sea level, protected by dykes. The city was founded by the French in 1781, developed by the Dutch, who drained its marshy site with a canal system, and seized by the British in 1812. Today it is an administrative and commercial centre with some small-scale industry producing beer, soft drinks and timber products.

Population 200 000
Map Guyana Ba

Georgetown ***USA*** Part of Washington, DC, the national capital. It lies on the north bank of the Potomac river, and was already a thriving town when the national capital was created in 1790-1800. Georgetown still retains many of its fine terraced houses and cobbled streets and much of its old, small-town elegance. Georgetown University was founded in 1789.

Map United States Kc

▼ GOTHIC STRENGTH The massive Church of St Mary dominates Gdansk's old city, painstakingly restored since 1945. The tower of the 105 m (344 ft) long brick building is 78 m (256 ft) high.

AGE OF THE EARTH

Earth, it seems, was not created in 4004 BC – the date postulated by the 17th-century Archbishop James Ussher – but a great deal earlier. Most geologists and physicists, with the aid of radioactive dating, are reasonably agreed that our planet evolved about 4600 million years ago. As to how it evolved, or assumed its present form with crust, mantle and core, there is considerably less unanimity.

That some rocks are older than others was discovered in the late 18th century when a few people of inquiring mind began to study rock strata and the fossils embedded in them. These gave some idea of the relative ages of rocks, and names were assigned to distinguish them; but their ages in years was incalculable. With the discovery of radioactivity in 1896, however, and the subsequent realisation of its presence in rocks, a 'time ruler' was established. Each radioactive element in a rock decays and transforms itself into a new element at a constant, measurable rate. By measuring the amount of original element left in a rock, and the amount of new element, it is possible to calculate the date at which the rock was formed. Thus it has been discovered that the oldest rocks, part of the original crust, are about 3900 million years old.

GEOLOGICAL TIMESCALE

Era	Period – and epoch	Began (millions of years ago)	Activity
Archaeozoic	Precambrian	4600	Origin of the earth. Formation of crust, continents and oceans.
		3900	Oldest known rocks.
		3300	Origin of life. Formation of present-day atmosphere.
Palaeozoic	Cambrian	570	
	Ordovician	500	
	Silurian	430	Life comes ashore from the seas.
	Devonian	395	Caledonian mountains raised.
	Carboniferous	345	
	Permian	280	Appalachians and Central European mountains raised.
Mesozoic	Triassic	225	Urals raised. Pangaea starts to break up (200-180 million years) and North Atlantic starts to open.
	Jurassic	190	South Atlantic starts to open (140-130 million years).
	Cretaceous	135	India splits from Antarctica (105-100 million years).
Cenozoic	Tertiary – Palaeocene	65	Formation of the Rocky Mountains.
	– Eocene	53	Australia splits away from Antarctica (45 million years).
	– Oligocene	37	India collides with Asia (30 million years). Formation of the Himalayas and the Alps.
	– Miocene	26	
	– Pliocene	12	
	Quaternary – Pleistocene	2	Earliest forms of hominid appear.
	– Holocene		Started 15 000 years ago.

Georgia ***USA*** South-eastern state covering 152 490 km² (58 876 sq miles) immediately north of Florida. It was founded as an English colony, named after King George II, in 1733, and became one of the original 13 states. The land rises from the Atlantic in the south-east to the Blue Ridge in the Appalachian Mountains in the north.

With mild winters and warm (but hurricane-prone) summers, the state formerly produced tobacco and cotton on large plantations. Today the main products are broiler chickens, peanuts, soya beans, textiles, transport equipment, food products, and wood products from the extensive forests, as well as tobacco. Tourism is expanding. ATLANTA is the state capital, and SAVANNAH the chief port.

Population 5 837 000

Map United States Jd

Georgia (Gruziya) ***USSR*** Largest, richest and most productive of the Caucasian republics, lying south of the Caucasus range at the eastern end of the Black Sea. It covers an area of 69 700 km² (26 900 sq miles), and is rich in citrus fruits, tea, wine, tobacco, wheat, barley, and flowers and herbs from which perfume is made. There are also large deposits of manganese ore. Among the republic's other industries are coal, timber, machinery, chemicals, silk, food processing and furniture. Its main cities are the capital, TBILISI, the Black Sea port of BATUMI, and the industrial city of KUTAISI.

Georgians have a strong native culture, with their own language and alphabet, and a long literary tradition; most are Orthodox Christians. They are famed for their longevity, with more than 2000 people (one in every 2500) over 100 years old.

Georgia was the land of the Golden Fleece in Greek mythology. It was at times under Roman rule, and was invaded by the Persians in the 5th to 6th centuries AD, followed by the Arabs. From the late 10th century it was again independent, but then suffered invasions of Tatars, Persians and Turks. It came under firm Russian control after 1800 and remained so despite several uprisings.

Its most famous native son was the dictator Joseph Stalin (1879-1953), architect of the Soviet system and leader of the USSR through the Second World War. He was born at Gori, 65 km (40 miles) north-west of Tbilisi.

Population 5 976 000

Map USSR Fd

Georgia, Strait of Channel between south-western BRITISH COLUMBIA and VANCOUVER ISLAND, Canada, forming part of the scenic inland water route (the Inside Passage) which extends from SEATTLE, in the American state of Washington, to ALASKA.

Dimensions 240 km (150 miles) long, 30-64 km (19-40 miles) wide

Map Canada Cd

geosyncline Major elongated depression in the earth's crust, stretching up to hundreds of kilometres, which sinks as it is filled with sediment worn away from the landmasses on either side of it. Geosynclines are often uplifted to produce FOLD MOUNTAINS.

geotherm A surface that links all the points in the earth, or in part of the earth, at which the temperature is the same.

Gerlach Peak (Gerlachovsky Stít) ***Czechoslovakia*** See TATRA MOUNTAINS

German Bight Part of the NORTH SEA, off the north-western coast of West Germany. The Germans call it the Deutsche Bucht, but it is also known as the Heligoland Bight, after the island of HELIGOLAND which was a fortified German naval base in both World Wars. The fortifications were destroyed in 1947 and the island is now a tourist area.

Map West Germany Ca

Germany, East (German Democratic Republic) See p. 248

Germany, West (Federal German Republic) See p. 250

Germiston ***South Africa*** Gold-mining city on the East Rand, 13 km (8 miles) south-east of Johannesburg. Its ore refinery, the world's largest, handles 70 per cent of the Western world's production of gold bullion.

Population 166 400

Map South Africa Cb

Gerona ***Spain*** Industrial town 88 km (55 miles) north-east of the Mediterranean port of Barcelona. It manufactures textiles, chemicals, electronic equipment and soap, and has a 14th-century cathedral.

Its province, of the same name, produces cereals, fruit and wine.

Population (town) 87 600; (province) 467 900

Map Spain Gb

Gettysburg National Military Park ***USA*** Site of a decisive battle in the Civil War where, on July 1-3, 1863, Union forces, led by General Meade, defeated the Confederates under General Lee and stopped Lee's advance into the North. It lies in southern Pennsylvania, about 105 km (65 miles) north and slightly west of the national capital, Washington. More than 30 000 men were killed or badly wounded in the clash. In the cemetery four months later, President Abraham Lincoln delivered his Gettysburg Address, pledging that 'government of the people, by the people, for the people, shall not perish from the earth'.

Map United States Kc

Geyre ***Turkey*** See APHRODISIAS

geyser Natural hot spring that intermittently ejects a column of water and steam into the air. Geysers occur in active or recently active volcanic areas.

Geysir ***Iceland*** One of the largest geysers in the world, in the centre of Iceland's major group of hot springs, 85 km (53 miles) north-east of Reykjavik. The geyser spouts every 80-90 minutes up to a height of 100 m (330 ft), and has been the object of scientific investigation for more than two centuries. The English word geyser comes from its name.

Map Iceland Bb

Gezhouba ***China*** Site of what is to be the country's largest dam. Building across the Chang Jiang river began in 1970. When it is completed, the dam will supply 14 000 million kilowatt-hours of electricity a year.

Ghadames ***Libya*** Route centre in the west of the country, 760 km (472 miles) south-west of Tripoli. It is an ancient pre-Roman focus for caravan routes across the desert. Known as the 'pearl of the desert', Ghadames is one of the oldest Berber oases in the Sahara and was captured by the Romans in 20 BC. It became a citadel of Islam, and no European entered it before the present century. The oasis, which is famed for its hot springs, produces apricots, dates and figs, and has an airfield.

Population 6300

Map Libya Aa

Ghana See p. 256

Gharb ***Morocco*** See RHARB

Ghardaia ***Algeria*** The capital and most impressive of the five towns of M'ZAB. It is a road junction and airport 500 km (310 miles) south of Algiers on the Route du HOGGAR across the SAHARA. The town, built in the 11th century, possesses a remarkable mosque with an imposing minaret; the houses and mosques are white. It has several hotels, and an active, mixed market dealing in produce brought by caravan from all over the Sahara and the deep south, making it an attractive stopping place for tourists. Nearby is the holy town of Beni Isguen where prayers are said from midday to 3.30 pm, when its gates are closed to strangers, who are forbidden to stay overnight.

Population 65 000

Map Algeria Ba

▲ VIOLENT BUILDER Castle Geyser is slowly raising a 'castle' of deposits around its rim as it shoots up to 30 m (100 ft) into the air in Yellowstone National Park, USA.

ghat 1. In India, a mountain pass or mountain chain. **2**. In the West Indies, a ravine.

Ghats ***India*** Two mountain ranges in the south of the country: the Eastern Ghats and the Western Ghats. Between the two ranges lies the peninsula of DECCAN.

The Eastern Ghats have an average height of 457-609 m (1500-2000 ft), and run for about 805 km (500 miles) along the south-east and east coast from south of Cuttack to Mount Anai Mudi. Set among them is the Holy Shrine of TIRUPATI.

The Western Ghats, with an average height of 914-1524 (3000-5000 ft), extend for some 1287 km (800 miles) along the south-west and west coast from the River Tapti to Mount Anai Mudi.

Map India (Eastern Ghats) Cd; (Western Ghats) Bd

Ghawar ***Saudi Arabia*** World's largest oil field lying inland along The Gulf, to the west of Qatar. It stretches for 241 km (150 miles) from north to south, and is up to 35 km (22 miles) wide. Ghawar can produce up to 5 million barrels of oil a day.

Map Saudi Arabia Bb

Ghazni ***Afghanistan*** Town and trading centre for wool and fruit, 120 km (74 miles) south-west of Kabul. It became Muslim in the 9th century AD, and in 977 was made the capital of the Ghaznavid dynasty. It is renowned for its bazaar and the handicrafts sold there – and also for the delicious Alubekhara plums which grow in the district.

Population 32 000

Map Afghanistan Bb

Ghent (Gent; Gand) ***Belgium*** Medieval city about 60 km (35 miles) north-west of the capital, Brussels. It stands on a number of islands connected by bridges over the Lys and Schelde rivers. The capital of East Flanders province, Ghent flourished on the wool trade of the 13th and 14th centuries, and still has a remarkable collection of old buildings including the moated castle of 's Gravensteen – dating from the 9th century and rebuilt in its present form in 1180 – the seat of the Counts of Flanders.

Ghent is today the centre of the flower seed and bulb market, and a port linked by the Ghent-Terneuzen canal, 29 km (18 miles) long, across Dutch territory to the North Sea. John of Gaunt (Ghent), son of Edward III of England, was born here in 1340.

Population (city) 235 000; (metropolitan area) 490 000

Map Belgium Ba

Giant's Causeway ***United Kingdom*** Northern Ireland's most famous tourist attraction, on the north coast about 80 km (50 miles) north-west of the capital, Belfast. It is a spectacular complex of hexagonal basalt columns, stretching some 275 m (900 ft) along the coast and up to 150 m (500 ft) into the sea. The tallest is the Giant's Organ, 12 m (39 ft) high.

The most valuable treasure recovered from the invading Spanish Armada of 1588 was found there on the wreck of the galleon *Girona* in 1967. The gold and silver hoard, including coins and personal jewels, is now in the Ulster Museum in Belfast.

Map United Kingdom Bc

gibber Australian term for a stone or rock, especially one polished by the wind.

gibber plain Type of gravel desert in arid parts of Australia.

gibli, ghibli Local name for the hot southerly sirocco blowing across Libya and Tunisia.

East Germany

THE COMMUNIST HALF OF DIVIDED GERMANY HAS BEEN RESTORED TO ECONOMIC POWER BY ITS OWN EFFORTS FROM THE ASHES OF THE SECOND WORLD WAR

The country's name, 'German Democratic Republic' (GDR), is something of a misnomer. Divided from Western Europe by electrified fences and minefields, East Germany has seen little in the way of democracy. Total Russian domination since the end of the Second World War has helped to frustrate any attempt at reunification of the two German peoples – indeed, one of the USSR's major fears is the military revival of a united Germany. But East Germany has succeeded in engineering its own revival, and by the efforts of its people has become one of the most highly developed countries in the Eastern Bloc.

The East Germans are renowned for their athletic prowess. They win ten times as many Olympic medals per head of population as the Soviet Union. The GDR has fostered outstanding opportunities for sports, recreation and culture. There are three main reasons.

First, the GDR sought to demonstrate the superiority of Communism in caring for all aspects of human welfare, compared with the capitalism of West Germany where money was seen to rule people's lives.

Second, it seemed a practical way of offering a more attractive and healthy life to East Germans, both to divert attention from their walled-in predicament and to counteract the lure of West Germany. Sport helps to erase memories of war, and of the exodus of 3 million people to West Germany from 1949 until 1961, when the East Germans stopped the last gap with the BERLIN Wall.

Third, the development of a sporting super-race was a peaceable way of asserting German superiority over other Communist states.

The landscapes of the GDR have much in common with those of northern West Germany. The BALTIC coast is flanked by long, soft white sandy beaches and spits, but RUGEN is an island wonderland of rocks, tall cliffs and bays. Inland, in MECKLENBURG, a string of lakes – relics of the last Ice Age – stretches among woods and rolling farmland between Schwerin and Eberswalde. This separates fertile wheat, sugar beet and maize-growing eastern areas from the poorer west, which yields rye and potatoes.

Southwards, between HELMSTEDT-Marienborn and FRANKFURT AN DER ODER, lie the featureless farmlands surrounding MAGDEBURG, and around BRANDENBURG, POTSDAM and Berlin, forbidding, thick pine forests. These are crossed by the Havel and Spree rivers whose lakes provide highly prized recreation areas for the city-dwellers.

To the south lie the lowlands, hills and valleys of Saxony, cut by the ELBE river from DRESDEN downstream to DESSAU. Outside East Berlin, this is the economic heart of the GDR. Densely populated, fertile agricultural areas, once with many small farms, were made into collective farms in the 1950s; production recovered only slowly from the change. These areas are punctuated by the heavily industrialised Dessau–HALLE–LEIPZIG triangle; the Cottbus mining area; and a string of major industrial, commercial and cultural cities along the southern autobahn between GORLITZ on the Polish frontier and EISENACH near West Germany: Dresden, KARL-MARX-STADT, ZWICKAU, Gera, JENA, WEIMAR, ERFURT and Gotha.

Encircling these lowlands is a crescent of imposing, forested uplands, cut by steep river valleys and cradling reservoirs behind tall dams. The areas of natural beauty include the HARZ mountains, THURINGER WALD, ERZGEBIRGE (Ore Mountains) and the peaks of Saxon Switzerland along the Czech border.

Of all Germans, the people of the GDR have paid the fullest price for Hitler's war in Europe and the Nazi atrocities. In 1945, defeated Germany lost territory in the east to Poland and the USSR. The rest was divided into four occupation zones – British, French, American and Russian. The capital, Berlin, was controlled by the four powers but was an island inside the Soviet zone.

The four powers could not agree on an overall policy for Germany, and the 'cold war' began between the USSR and the Western allies. The Russians blockaded West Berlin for ten months in 1948-9, but were foiled by a massive airlift of supplies.

In May 1949 East Germany adopted its own constitution and began to complete its political and physical isolation from the West. In 1952 a zone 4.8 km (3 miles) wide – with electrified fences, minefields, watchtowers and military patrols – was established along the entire border with West Germany to stop refugees leaving the GDR. The last loophole, from East to West Berlin, was closed in 1961 by the Berlin Wall.

In contrast to West Germany, which received massive aid from the United States, the GDR had little help after the Second World War. From 1945 to 1954 the Russians exacted huge reparations, dismantling factories and power stations, reducing all double-tracked railways to single tracks, and taking scarce machinery from East German factories. This punitive policy was softened in the 1950s after East German workers rioted over shortages of food, clothing, heating and housing. Other causes of unrest were persecution of the Churches by the Russians, and the enforced collectivisation of agriculture.

Although the riots brought some relief from oppression, there was still no help for the East Germans. They were left to rebuild their economy by their own efforts. They faced great handicaps. They had lost many people, in war and in the postwar exodus, so were left with an ageing and slowly declining population. Labour and machinery shortages slowed the reconstruction of devastated cities such as Berlin, Dresden, Dessau and Magdeburg, and curtailed the expansion of farm and factory output.

INDUSTRIAL RECONSTRUCTION

Before 1945 the eastern half of Germany was largely an agricultural area. It had few natural resources apart from lignite (brown coal) and potash, and was ill-fitted for the postwar plans laid down by the USSR for the development of heavy industry in the 1950s and early 1960s.

An iron and steel industry was created at Eisenhuttenstadt (formerly Stalinstadt) on the ODER river in the east, and at Calbe in the west, to make the GDR self-sufficient in pig iron and steel. But it led to expensive imports of iron ore, coal and coke, carried from Poland or 1600 km (1000 miles) by rail from the Soviet Ukraine. In the 1960s and 1970s, imports of oil and gas by pipeline from the Volga-Ural regions made the GDR further dependent on the USSR.

The carving up of Germany had cut off the GDR from its former main ports of HAMBURG (now in West Germany) and Stettin (now SZCZECIN in Poland). The ports at ROSTOCK and STRALSUND had to be developed, and rail and road connections improved.

A new surge of economic progress started when Russia changed the rules in 1959. Nikita Khrushchev, the new Soviet leader after the death of Stalin, cancelled policies which said that Soviet Bloc countries should be self-sufficient. In future the GDR had to concentrate on specialised production for Communist markets; the East Germans were to use their manufacturing skills to pay for imports from the USSR.

Rapid growth in brown coal output from huge open-cast mines supported the expansion of electricity supplies. Chemical plants grew up making fertilisers, films, dyes, electrochemicals, synthetic rubber and glass fibre. A big potash industry was developed around STASSFURT.

The hardworking Germans quickly revived old skills and adopted new ones. State industries produced machinery, tools, instruments, optical goods, electrical and electronics products. Factories and research laboratories for these goods are based in cities with excellent sports and cultural facilities such as Berlin, Leipzig, Dresden, Jena, Dessau, Eisenach and Erfurt. More recently, manufacturing spread to centres of the rural north, like Schwerin.

A common language has given the GDR access to technological and scientific work published in West Germany. Manufacturing, agriculture, transport and other industries have benefited from West German progress. Closer ties between West and East Germany were initiated by their respective leaders, Willy Brandt and Erich Honecker, in the 1970s. As a result West Germany (and the European Common Market) were opened to GDR exports.

By 1984 the East Germans had gained a technological lead over other countries of the Council for Mutual Economic Assistance (COMECON) – the Soviet Bloc equivalent of the Common Market.

The GDR always stresses its Communist nature. The state runs all production, trade, services, welfare, sports, and most apartment housing in towns. Rebuilt cities contrast starkly with those in West Germany; they have plainer, repetitive architecture, wider open spaces, and more prominent cultural centres.

Recently, painstaking care has been lavished on restoring or renovating major legacies of German culture. Among these are the historic sites at EISLEBEN, WITTENBERG and Eisenach associated with Martin Luther, initiator of the Reformation – the religious revolution that took place in Western Europe in the 16th century.

To the tourist from the West, East Germany can present a dour picture. An 'iron curtain' of high concrete walls, electrified fences and watchtowers with armed guards is the visitor's first sight of the border. Access to East Germany is only at specified points, such as the famous Checkpoint Charlie in Berlin.

After the neon lights and traffic jams of West Berlin, East Berlin looks shabby and underpopulated. No traffic jams here – there is a waiting list of up to ten years for a car. Family businesses and inns are missing; the quality of some goods in state shops is poor.

Against this, the prices of food, transport and cultural visits to the theatre and cinema are low. Almost all households have a television set (often tuned to the more exciting West German programmes), a refrigerator and a washing machine. Travel abroad is not generally permitted, but state holidays within the Eastern Bloc are arranged for workers at low prices.

▲ OUT OF THE ASHES The Elbe flows under the Charles (Carol) Bridge in the heart of rebuilt Dresden. Five Allied bombing raids in spring, 1945, destroyed 60 per cent of the city.

EAST GERMANY AT A GLANCE

Area	108 333 km² (41 827 sq miles)
Population	16 700 000
Capital	East Berlin
Government	Communist republic
Currency	GDR mark = 100 pfennig
Language	German
Religion	Christian (47% Protestant, 7% Roman Catholic)
Climate	Continental; average temperature in Berlin ranges from −0.5°C (31°F) in January to 19°C (66°F) in July
Main primary products	Wheat, rye, barley, oats, sugar beet, potatoes, timber; lignite, potash, cobalt, bismuth, arsenic, antimony
Major industries	Machinery, tools, precision instruments, optical equipment, chemicals, textiles, railway equipment, iron and steel
Main exports	Machinery, precision instruments, railway equipment, chemicals, potash, lignite, textiles
Annual income per head (US$)	6000
Population growth (per thous/yr)	Static
Life expectancy (yrs)	**Male** 71 **Female** 75

West Germany

THE FEDERAL REPUBLIC OF GERMANY HAS EMERGED FROM THE RUBBLE OF THE SECOND WORLD WAR TO CREATE EUROPE'S MOST POWERFUL INDUSTRIAL ECONOMY

The history of Europe has been, by turns, glorious and tormented, but nowhere have the glory and the torment been more interwoven than in Germany. The marvellous achievements of musicians, scholars and builders have been matched by horrors – the ravages of marching armies, century after century; the peacetime collapse of the currency following the First World War; cruel dictatorship and the destruction of a nation's cities in the Second World War; and, in the aftermath, the separation of parents from children and brothers from sisters by the world's most fearsome frontier, the Iron Curtain. The territory that is Germany has moved back and forth, eastwards and westwards, as the Germans and the neighbouring Slavs and French have redrawn their frontiers over the centuries.

Even Germany's emergence as a nation has been a dramatic story of conquest, division, reunion and change. First there were Germanic tribes, such as the Teutons and Cimbri who fought the Romans, and the Alemanni, Lombards, Goths and Franks who fought the Romans and each other. Then there was a kind of political crazy quilt called the Holy Roman Empire, founded by the Frankish King Charlemagne in AD 800. However, challenges from the papacy and powerful local rulers brought total disunity by the 15th century. The religious wars of the 16th and 17th centuries, in which Catholic fought Protestant and which culminated in the Thirty Years' War (1618-48), resulted in a collection of petty German states. Prussia and Bavaria emerged as the strongest of these following the 1848 'year of revolutions'. In 1871 they and 23 other German states formed a new empire headed by Wilhelm I of PRUSSIA, whose capital was BERLIN. By 1914, at the outbreak of the First World War, this stretched from METZ in LORRAINE to Memel (present-day KLAIPEDA in the USSR) on the Baltic. It had an area of 540 858 km² (208 825 sq miles) and 41 million people.

▲ **GATEWAY TO THE WORLD A forest of wharf cranes used for loading and unloading ships stands beside the Elbe river in the port of Hamburg. It is the country's busiest port, handling 17 000 oceangoing ships a year. The city's massive television tower rises to 271 m (890 ft) above the ground.**

From that point onwards, Germany has resembled a concertina playing a central European tune. After the First World War, in 1919, it was squeezed down under the Treaty of Versailles to 468 620 km² (180 935 sq miles) by loss of territory in the east (including West Prussia) and west (ALSACE, Lorraine and the SAARLAND). (It also lost its African and Pacific colonies.) Germany expanded again in 1938-43 to huge proportions, incorporating Austria, Czechoslovakia and Poland, and overrunning most of Europe from the Atlantic coast to MOSCOW and from Norway to Greece. It was squeezed again in 1945, at the end of the Second World War, so that only 357 000 km² (137 800 sq miles) or 76 per cent of its 1914 territory remained. And that has been divided into two – East Germany and West Germany – by wall and wire and tracker dogs. Few states in the world have gained so much, and lost so much, in the short span of their emergence.

THE POSTWAR PATTERN

Today's West Germany is therefore merely a remnant of a remnant of everything which has at one time been German soil. In 1945, the victorious Allies, remembering how they had failed to enforce the conditions of their peace settlement on Germany after 1918, decided in 1945 to occupy the country. It was divided into four zones, one each for the USA, the USSR, Great Britain and France. The capital Berlin, although within the Russian zone, was divided into four sectors to match.

Almost at once disagreements between the USSR and the Western powers began to emerge. The Soviet zone in the east began to follow the path towards Communism, while the three western zones were drawn together in mutual opposition to the East. The boundary of the Soviet zone, drawn originally to demarcate military areas, became a divide between two opposing political and economic systems, dubbed by British politician Winston Churchill as the Iron Curtain.

The three main western zones of occupation – the American, British and French – plus the Saarland, which was incorporated in 1957, became the 248 687 km² (96 018 sq miles) of territory that make up West Germany. Since 1949 it has been known as the Federal Republic (*Bundesrepublik*) of Germany, and since 1955 it has enjoyed full sovereignty under that name. This area includes the 480 km² (185 sq miles) of West Berlin – again, the American, British and French sectors – which have a special status; West Berlin is not legally a part of the Federal Republic but it does send (non-voting) delegates to the West German parliament. The long axis of the old Germany ran from east to west, but that of the Federal Republic runs from south to north – from the Bavarian ALPS to the Danish border, 850 km (530 miles) in all. Most of the old east-west routes have been barred by East Germany and Czechoslovakia; their frontiers can be crossed only at designated points.

As with German territory, so with German people. The population of the Federal Republic is just over 61 million. Less than four-fifths of it, however, is native to West Germany. Most of the remainder consists of immigrants from the East or their descendants. When the Allies reduced Germany's territory at the end of the Second World War, they agreed also to squeeze in the German population of all those lands – East PRUSSIA, SILESIA, part of POMERANIA and Brandenburg, and the SUDETENLAND – that were being handed over, or handed back, to Poland and Czechoslovakia. Germans had settled throughout eastern Europe; now they were to be returned to base – 12 million of them (2 million others died or disappeared along the way). West Germany took 9 million and they have been joined over the years, since Germany's division, by roughly 3 million East Germans.

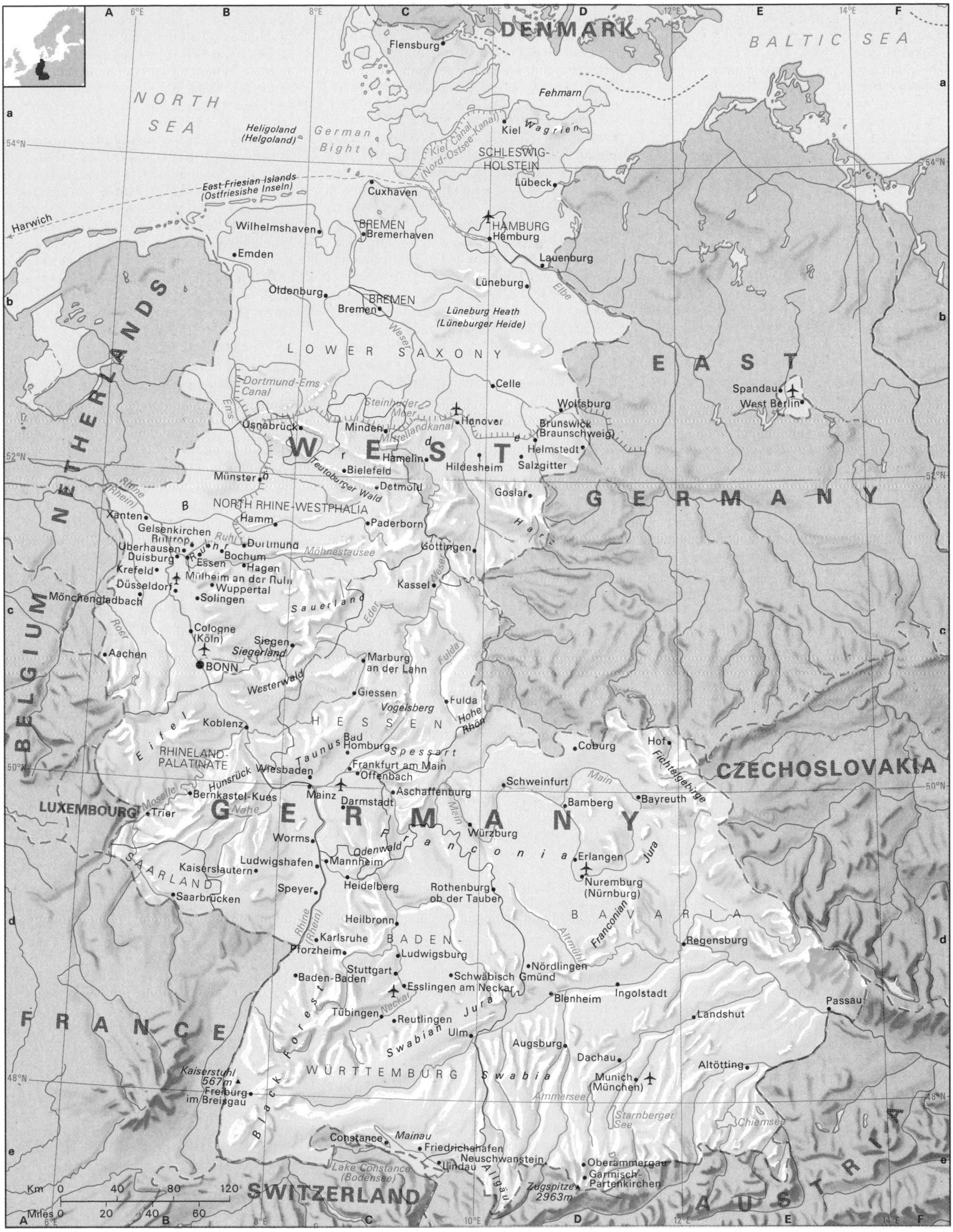
DENMARK
BALTIC SEA
NORTH SEA
Flensburg
Fehmarn
Kiel
Wagrien
Heligoland (Helgoland)
German Bight
Kiel Canal (Nord-Ostsee-Kanal)
SCHLESWIG-HOLSTEIN
Lübeck
East Friesian Islands (Ostfriesishe Inseln)
Cuxhaven
Harwich
Wilhelmshaven
BREMEN
Bremerhaven
HAMBURG
Hamburg
Emden
Lauenburg
Lüneburg
Oldenburg
BREMEN
Bremen
Elbe
Lüneburg Heath (Lüneburger Heide)
Weser
LOWER SAXONY
NETHERLANDS
EAST
Spandau
West Berlin
Dortmund-Ems Canal
Celle
Ems
Steinhuder Meer
Wolfsburg
Osnabrück
Minden
Mittellandkanal
Hanover
Brunswick (Braunschweig)
WEST
Helmstedt
Hameln
Hildesheim
Salzgitter
Bielefeld
Teutoburger Wald
Münster
Detmold
Goslar
Rhine (Rhein)
GERMANY
NORTH RHINE-WESTPHALIA
Harz
Xanten
Hamm
Paderborn
Gelsenkirchen
Ruhr
Dortmund
Oberhausen
Duisburg
Essen
Bochum
Möhnestausee
Göttingen
Krefeld
Hagen
Düsseldorf
Wuppertal
Kassel
Mönchengladbach
Solingen
Sauerland
Roer
Eder
Cologne (Köln)
BELGIUM
Siegen
Siegerland
Aachen
Marburg an der Lahn
Fulda
BONN
Westerwald
Giessen
Fulda
Vogelsberg
Hohe Rhön
HESSEN
Koblenz
Eifel
Bad Homburg
Spessart
Coburg
Hof
RHINELAND-PALATINATE
Taunus
Fichtelgebirge
Hunsrück
Wiesbaden
Frankfurt am Main
Offenbach
CZECHOSLOVAKIA
Main
Schweinfurt
Moselle
Bernkastel-Kues
Mainz
Aschaffenburg
LUXEMBOURG
Trier
Nahe
Darmstadt
Bamberg
Bayreuth
GERMANY
Würzburg
Worms
Odenwald
Franconia
Erlangen
Jura
SAARLAND
Kaiserslautern
Ludwigshafen
Mannheim
Nuremburg (Nürnberg)
Speyer
Heidelberg
Rothenburg ob der Tauber
Saarbrücken
BAVARIA
Franconian
Heilbronn
Altmühl
Karlsruhe
BADEN-
Regensburg
Pforzheim
Ludwigsburg
Stuttgart
Nördlingen
Baden-Baden
Schwäbisch Gmünd
Esslingen am Neckar
Blenheim
Ingolstadt
Black Forest
Neckar
Passau
FRANCE
Tübingen
Reutlingen
Swabian Jura
Landshut
Ulm
Augsburg
Dachau
WÜRTTEMBURG
Swabia
Altötting
Kaiserstuhl 557m
Munich (München)
Freiburg im Breisgau
Ammersee
Starnberger See
Chiemsee
Constance
Mainau
AUSTRIA
Friedrichshafen
Neuschwanstein
Oberammergau
Lindau
Lake Constance (Bodensee)
Allgäu
Garmisch-Partenkirchen
Zugspitze 2963m
SWITZERLAND
Km 0 40 80 120
Miles 0 20 40 60
A B C D E F
a b c d e
6°E 8°E 10°E 12°E 14°E
54°N 52°N 50°N 48°N

West Germany has therefore had to find homes in a war-shattered land for an additional 12 million inhabitants.

There is also a third element in the population: the foreign workers (*Gastarbeiter*). In the postwar years, with over 5 million Germans dead and reconstruction under way, West Germany brought in labourers from Turkey and Yugoslavia, Italy, Greece and Spain. In 1983, these workers and their families numbered 4.5 million, or 7.4 per cent of the population, one-quarter of them under 16; one-third of the total were Turks. Recruitment ceased with rising unemployment in the early 1970s (there have been 2 million jobless since 1981), and various incentives were offered to get the workers to return to their own countries, but not before massive social problems – particularly in housing, education and employment of school-leavers – had arisen in the foreign community.

The Federal Republic has eight states (or provinces) and two city-states, forming ten *Länder*, each with a provincial government. The federal capital is BONN, a former small university town which has had to expand to house government and parliament. The Länder vary considerably in size: BAVARIA covers over 70 551 km² (27 240 sq miles) and contains nearly 11 million inhabitants; NORTH RHINE-WESTPHALIA is smaller but has nearly 17 million people, whereas the Saarland covers a mere 25 713 km² (993 sq miles) and the city-state of BREMEN counts only 682 000 inhabitants. Disparities of this kind were even more marked in the old German Empire, and the federal system of government goes some way towards levelling differences in regional wealth or economic strength.

The regions and landscapes of West Germany fall into three main divisions: the Alps and their Foreland, the Middle German Uplands and the North German Lowlands.

SOUTHERN PEAKS

Only the northernmost fringe of the Alps lies within Germany – the limestone mountains which culminate in the ZUGSPITZE (2963 m, 9721 ft), on the Austrian border. This is, nevertheless, enough to give West Germany a share of alpine scenery and ski resorts (of which the best-known is GARMISCH-PARTENKIRCHEN). The foothills and valleys contain the Bavarian lakes and the fantastic castles of Bavaria's Ludwig II (1845-86), patron of the composer Wagner. The foothills give way northwards to a broad slope, surfaced with materials swept down by ice and water from the mountains. This is the Alpine Foreland, stretching north to the River DANUBE, in places barren and stony, in others covered by bogs called *moos*.

In the centre of the Foreland lies MUNICH (München), third largest city after West Berlin and HAMBURG, and capital of the former kings of Bavaria. Like its neighbour AUGSBURG and the Danube ports of REGENSBURG and PASSAU, Munich was old and wealthy when Berlin was still a village.

North of the Danube lie the hills and basin lands of central West Germany. The region is like a badly laid pavement, on which a series of generally smooth paving stones are tilted in different directions, with wide cracks and hollows between them. The 'paving stones' represent uplifted sections of an old surface, raised by mountain building long before the Alps appeared. Between and around them are draped younger formations.

The principal uplands include the BLACK FOREST and the Schwäbische Alb (Swabian Alps or Jura) in the south-west, the Bohemian Forest along the Czech border and nearby Bavarian Forest, the HARZ mountains which straddle the border between East and West Germany, and the uplands of the middle RHINE basin – EIFEL, Hunsrück, TAUNUS, Saarland and WESTERWALD – between which the Rhine cuts its famous gorge. The highest points are 1493 m (4898 ft) in the Black Forest, 1457 m (4780 ft) in the Bavarian Forest and 1142 m (3747 ft) in the Harz mountains.

Some of the highest points and isolated hill masses of the region are the product of volcanic activity. The upper levels are largely forested, and valleys are deeply cut into these old blocks. Sunny, sheltered and fertile slopes, however, support the vineyards that produce the famous wines of the Rhine, MOSELLE and NECKAR. Between the uplands, the fertile basins form the cores of the old German states of SWABIA, Franconia and Thuringia.

Middle Germany is, in fact, a very complex region in both its physical and its political structure. The political fragmentation of past centuries produced numerous duchies, grand duchies and margraviates, each fiercely status-conscious and each centred on one or more towns. Some of the towns grew into cities, notably those such as NUREMBERG and FRANKFURT AM MAIN which became important trade centres, or those such as WURZBURG and MAINZ where the Church was the dominant force. But many others served no special purpose and, with the passing of the small political units, they have little function today except to attract tourists. Most of today's major cities lie on, or close to, the water routes: Ludwigshafen, MANNHEIM, Mainz and KARLSRUHE in the Rhine valley, STUTTGART on the Neckar and Frankfurt on the MAIN (to which Nuremburg is linked by canal).

Timber and mineral ores in the uplands formed an early basis for industry in central West Germany, but the materials for heavy industry have had to be imported into the region from outside, except for coal, in the case of the Saarland. Nevertheless, industry grew with local initiative: Germany's first railway was built in 1835 from Nuremberg to Fürth. Stuttgart was the birthplace of the motor car in the 1880s and southern Germany was making Mercedes and BMWs long before Volkswagen went into production in LOWER SAXONY.

THE GREAT NORTHERN PLAIN

The line along which the Middle German Uplands give way to the North German Lowlands is, on the whole, abrupt. Movement is easy on the plain and difficult within the upland, so that this edge or dividing line is followed by important routeways. It is also marked by a line of major coalfields, where industry has developed and cities have grown up. Among the coalfields are those of the AACHEN area and the greatest field in Europe – the RUHR, which is just east of the Rhine and is West Germany's industrial heart. Along this same line, also, are found huge deposits of brown coal, or lignite, west of COLOGNE and within East Germany. The cities on the line include Roman foundations such as Aachen and Cologne, and products of 19th-century industrial growth such as ESSEN, DUSSELDORF and DORTMUND. The densest population is found along this junction between upland and plain; farther north, outside the port-cities, the region is deeply rural.

Northern Germany is composed of materials either deposited by ice from the north during the last Ice Age or by melting ice water and winds during and after the last Ice Ages. Some of this material is very fertile, and (with Bavaria and BADEN-WURTTEMBERG) supports the major part of West German agriculture. Elsewhere it is sand or gravel, and in low-lying areas there is extensive peat bog. Towards the coast, reclamation from the sea has been going on for centuries, as in the neighbouring Netherlands, by a painstaking process of dyking and draining. Beyond the present shore lie the East and North FRIESIAN ISLANDS, forming a coastal barrier and providing holiday beaches.

Reclamation has also taken place inland. Peat bogs have been drained for agriculture, and crops or forest have been planted on the sandy heathlands. During the late 1940s and 1950s, this process was pushed ahead to help accommodate the German refugees from the East, many of whom were farmers and knew no other life. A programme of 'inner colonisation' on the marginal lands of the North German Plain was carried through. The landscape transformation has been so far-reaching that it has been necessary to preserve, as a nature reserve, the largest stretch of the original heathland, the LUNEBURG Heath south of Hamburg.

The emptiness of much of the North Sea coast stands in marked contrast to the growth of the port-cities: Hamburg (which is by far the largest) on the ELBE estuary, Bremen, with its outport BREMERHAVEN, and WILHELMSHAVEN, the former German naval base. The cities of Hamburg and Bremen are, in fact, city-states and have the status of Länder of the Federal Republic. This status they jealously guard together with their title of *Hansestadt* or Hanseatic city, to commemorate their membership of the great medieval trading association, the Hanseatic League.

North and east of Hamburg stretches the Land of SCHLESWIG-HOLSTEIN, bounded to the east and west by the sea and northwards by the Danish frontier. Here, glacial deposition has produced a hilly landscape with numerous lakes, and one part of the east coast has the nickname Holstein Switzerland. It is mainly a farming area, with KIEL as capital of the Land and Kiel and LUBECK, both Hanseatic cities, as its ports on the Baltic side. The Kiel Canal (in German, *Nord-Ostsee Kanal*) links the Baltic with the North Sea.

Fishing is still important for ports on both

▲ MAD KING'S DREAM
Neuschwanstein Castle perches on a lofty crag above the Lech valley in the foothills of the Bavarian Alps. It was built for King Ludwig II of Bavaria, and completed shortly before he drowned himself in the nearby lake in a fit of depression in 1886. The marble fantasy was erected at a cost of more than 6 million gold Marks purely for pleasure – it was never defended.

the North Sea and Baltic. However, German fish catches have fallen since the late 1960s, largely because of smaller catches from deep-sea fisheries (those outside these seas). Bremerhaven, followed by Cuxhaven, are the main centres of the fish-processing industry, which now also handles imported fish.

A STORMY HISTORY

At no time since the modern German nation was founded in 1871 under the political leadership of Chancellor Otto von Bismarck has it been possible for the rest of Europe to ignore Germany. Bismarck secured Germany against its great rival France by a series of alliances, but its surging commercial growth and colonial expansion in the late 19th century, during the reign of Kaiser (Emperor) Wilhelm II, alarmed its neighbours and led ultimately to the First World War. There then followed defeat, the end of the German Empire, reparations, partial military occupation, a collapse of the currency and the wobbly democracy of the WEIMAR republic (1919-33). Finally there came the worldwide depression after 1929.

It was out of this succession of disasters that Adolf Hitler, with his National Socialist (Nazi) party, undertook to lead Germany. After narrowly winning an election in 1932, Hitler's government was voted dictatorial powers. He dealt with unemployment by public works (including the world's first motorways), national service and rearmament. He adopted the doctrine of Aryan racial supremacy – idealising the fair-haired blue-eyed Nordic type – and made the Jews the scapegoats for all Germany's problems. Political opponents, trade unionists, cultural minorities and Jews were collected together in concentration camps, treated as slave labour and ultimately killed in huge numbers; some 6 million European Jews died in the so-called 'Holocaust'. His expansionist foreign policy – first to recoup Germany's losses under the Treaty of Versailles (1919) and then to reunite all German-speakers in a greater German empire – led to annexations and invasions of neighbouring territories to gain *Lebensraum* (living space). It plunged Europe and much of the rest of the world into war again in 1939.

With the destruction of the Nazi regime in 1945 went destruction of cities, industries and transport. Germany in 1945 was a chaotic wasteland. While every potholed road carried columns of ragged marchers, some fleeing their homes and some returning to look for them, the occupying armies strove to impose some order and restart essential services. Cigarettes and other barter goods were the only currency of value; the black market usually the only source of supply. Until the currency reform of the virtually worthless German mark and its replacement by the new *Deutsche Mark* in 1948, the prospect of recovery in the nation's economic life looked slender indeed.

AN ECONOMIC MIRACLE

The renewal of West Germany after 1948 was, however, rapid. Through a combination of American aid, hard work and shrewd economic policies, recovery was achieved sooner than anyone could have expected. Agriculture was intensified to feed a much denser population. Industry was rebuilt and, before long, West Germany became Europe's greatest industrial power. Production of the goods which had made Germany's prewar reputation – machinery, chemicals, electrical equipment, cars, whole steelworks even – was higher than ever; all except armaments. The network of motorways was greatly extended and is still growing. Rivers and canals were restored and widened and today carry almost as much of Germany's freight as goes by rail. Railways have been largely electrified, and a network of trans-Europe and inter-city services set up. The West German Deutsche Mark is one of Europe's strongest currencies.

The people have changed also. Although they are still among the world's greatest beer drinkers and meat eaters, they have lost the image of the shaven-headed army officer and the dumpy *Hausfrau*. Many of today's West Germans are as athletic, figure-conscious and stylish as anyone in Europe; as keen on diet, health farms, sport and environmental concerns; people who – and this has not changed – reverence nature, spend time hiking in field and forest, and agonise over acid rain and its destruction of their woodlands.

With growing prosperity and the spread of car ownership, the West Germans have become obsessive travellers and tourists. They are encountered everywhere, but the majority head south – to the Alps in winter and the Mediterranean in summer.

A RICH CULTURAL HERITAGE

Within West Germany itself, while the war years of 1939-45 resulted in tragic losses of historic buildings, there has been so much careful restoration since that the list of cathedrals, ducal palaces and walled towns for the tourist to visit is enormous – including around 370 castles, palaces and stately country houses.

Earlier aristocratic patronage helped to spread theatre and music far and wide through Germany. Small towns have civic theatres, even opera houses, and large ones their orchestras. In West Germany, BAYREUTH has its Wagner festival each summer and OBERAMMERGAU its Passion Play every ten years. These events, however, are eclipsed in size and participation by the great open-air and public occasions: by the trade fairs of Frankfurt and HANOVER, and by the huge carnival celebrations that precede Lent in Munich and Cologne.

German education, especially on the technical side, has always been highly regarded. Alongside old universities such as those of HEIDELBERG (founded 1386) and Cologne (1388), which have long attracted scholars from all over the world, there are newer foundations – for example three modern universities (at BOCHUM, Dortmund and Essen) in the Ruhr region – a total today of 64 universities and over 100 higher technical institutes. There is a wide range of research establishments, notably over 50 institutes bearing the name of the atomic physicist Max Planck.

During its short, postwar history, the government of the Federal Republic has been in the hands of a chancellor (the equivalent of a prime minister) and his cabinet, drawn from a parliamentary majority provided by one or more of the three main parties – Social Democrat, Christian Democrat (known in Bavaria as the Christian Social Union) and Free Democrat. The whole serve under an elected president, who represents the republic in international relations. West Germany's postwar politicians had little experience of a working democracy, but have managed to achieve a large measure of stability in the past 30 years. The trade unions also played a part in the country's recovery; they have not formed religious or political pressure groups, as in France or Italy, and are not splintered into dozens of rival unions.

Outside parliament itself pressures on the government in recent years have come from two main sources. One is the pacifist Green (or Ecological) Party – antinuclear, antipollution, anti the consumer society – which is now represented in parliament. The other is urban terrorism, whose activities reached a peak in the 1970s with the assassinations and bombings carried out by the notorious Baader-Meinhoff group.

A NEW ROLE IN THE WORLD

Apart from the immense tasks of recovery and reconstruction, the West German government has had to identify its role in Europe. There have been two choices. One option was to throw in its lot with West European nations (which would automatically rule out the possibility of reunification with East Germany); the other was to opt for a strict neutrality and so be able to achieve reunion (but only on terms acceptable to the USSR). Under successive governments, the West Germans have followed the first alternative.

Having made that decision, West Germany had to choose whether to join with other West European nations in the military alliance of NATO and subsequently join the economic – and eventually political – alliance of the EEC, or whether to stand apart from its former Western enemies. In other words, was Germany to be German first and European second, or the reverse?

Under Konrad Adenauer, who was chancellor from 1949-63, an effective decision was taken to choose Europe, and to start by healing the centuries-long hostility towards France. Then, as a full member of the Western alliance (from 1955) and EEC (from 1958), the Federal Republic returned to its original problem – that of regularising relations with the Eastern Bloc.

Under the leadership of Willy Brandt, who was chancellor from 1969-73, the Federal Republic pursued its so-called *Ostpolitik* (eastern policy). In 1970-2, West Germany signed treaties with East Germany, the Soviet Union and Poland, agreeing the basis of relations between itself and those countries, recognising the inviolability of postwar borders. That between the two Germanies enabled more West Germans to visit East Germany, and in 1973 they were both admitted to the United Nations. In 1980 an agreement provided funds to improve road, rail and waterway links between the two Germanies.

Agreements with other Eastern Bloc countries followed. A series of canals between the Rhine and Danube river basins in West Germany were begun in the 1970s. They are part of a plan to create an inland waterway linking the North Sea to the Black Sea.

Trade between the two Germanies now totals US$4600-5800 million each year. However, the possibility of reunification is politically remote and is of little concern to most younger Germans who never knew the united empire and whose loyalties are much more local. But radio and television know no Iron Curtain, and the two Germanies can and do meet and mingle on the air waves and in the sports arena.

Meanwhile, the new West Germany, with its rather loose federal structure, has to deal with internal pressures and regional interests. Just as the long axis of the country now runs north-south, so regional rivalries in the country set the north against the south – Bavaria and Baden-Württemberg against North Rhine-Westphalia, Munich and Stuttgart against Cologne and the Ruhr cities with their great industrial strength.

Perhaps this is why so many functions of government and business – the central stock exchange, the supreme court, West Germany's main international airport – have come to be located close to the 'border' between north and south, in the Frankfurt-DARMSTADT-Karlsruhe area. Bavaria has been rich and powerful for long enough to resent any hint of patronising or dictation from the north. The north and east of West Germany are mainly Protestant; the south and west largely Catholic. The two groups balance each other overall, but are unevenly spread. Internal unity within the nation cannot be taken for granted.

WEST GERMANY AT A GLANCE

Area 248 687 km² (96 018 sq miles) including West Berlin (480 km², 185 sq miles)

Population 61 009 700

Capital Bonn

Government Federal republic

Currency Deutsche Mark = 100 Pfennig

Language German

Religions Christian (49% Protestant, 45% Roman Catholic), Muslim minority

Climate Temperate; average temperatures in Frankfurt am Main range from −1 to 20°C (30 to 68°F)

Main primary products Cereals, potatoes, sugar beet, fruit and vegetables, livestock, milk and milk products, grapes, wine, timber, fish; coal, lignite, oil and natural gas, iron ore, potash, zinc, lead, salt

Major industries Iron, steel and nonferrous metals, mining, petroleum products, chemicals, motor vehicles, aircraft, machine tools, electrical and electronic equipment, computers, textiles, ceramics, glass, cement, agriculture, food processing, forestry, fishing

Main exports Machinery, motor vehicles, chemicals, iron and steel, food, small metal manufactures, textile yarns and fabrics, instruments

Annual income per head (US$) 9341

Population growth (per thous/yr) Declining

Life expectancy (yrs) Male 71 **Female** 76

Gibraltar

THE BARBARY APES LOOK ON WHILE THE DEBATE CONTINUES OVER BRITISH AND SPANISH RIGHTS TO THE ANIMALS' ROCKY HOME

On the south coast of Spain a sheer limestone rock rises to 425 m (1394 ft) from a promontory overlooking the entrance to the Mediterranean Sea. The promontory is now the British Crown Colony of Gibraltar, valued as a strategic naval and air base and communications centre for monitoring naval traffic.

Since the ancient Greeks settled there, Gibraltar has seen many invasions. In 711 the Moors arrived from North Africa, and it was they who gave the rock the name Jebel Tariq – the Rock of Tariq – from which 'Gibraltar' is derived. The peninsula remained under the Moors almost continuously until 1462, when the Spanish drove them out. It was captured by the British and Dutch in 1704, during the War of the Spanish Succession, and was recognised as a British colony in 1713 by the Treaty of Utrecht. Over the years it has been besieged by the Spanish and the French. Today, Spain still lays claim to it.

In a United Nations referendum in 1967 all but 44 of over 12 000 voters stated that they would rather be ruled by Britain than by Spain. Britain's respect for the wishes of the Gibraltarians about their nationality is written into the country's constitution.

In 1969 General Franco, the Spanish head of state, closed the border completely between Spain and Gibraltar, so that visitors from one country to the other had to travel via Tangier in Morocco, and planes to Gibraltar had to make a tricky and steep landing to avoid Spanish airspace. In 1982 Spanish and Gibraltarian pedestrians were allowed to walk through the border. Then, in 1985, as Spain prepared to join the European Economic Community, the road was opened completely. A wave of tourists passed through both ways, but especially into Gibraltar where shoppers could take advantage of duty-free goods.

But some Gibraltarians are worried about the effects of the new agreement between Britain and Spain, which will allow Spanish workers, previously excluded from Gibraltar, to find jobs there. Unemployment in Gibraltar is only 5.3 per cent; in Andalucia, across the border, it is 30 per cent.

Gibraltarians are a blend of Moorish, British, Maltese, Asian, Genoese and Spanish descent, and they speak both English and Spanish. The majority are Roman Catholic. Since Gibraltar has no agriculture or mineral resources, most of the inhabitants depend for their living on the port, dockyards and NATO bases. Although the British naval presence in Gibraltar is much reduced from its peak before the Second World War, the Strait of Gibraltar is one of the world's busiest waterways, with a ship passing every six minutes.

The Rock is well known for its colony of wild barbary apes – which has grown from only five in 1923 to 53 in 1985. Some see the opening of the border as the first stage in a gradual handover of Gibraltar to the Spanish – but folklore relates that the British will stay in Gibraltar as long as the apes do.

GIBRALTAR AT A GLANCE

Map Spain Cd
Area 6.5 km² (2.5 sq miles)
Population 31 200
Capital Gibraltar Town
Government Self-governing British Crown Colony
Currency Gibraltar pound = 100 pence; UK and Spanish currency also used
Languages English, Spanish
Religion Christian (74% Roman Catholic)
Climate Warm temperate; westerly winter winds bring rain. Average temperature ranges from 8-16°C (46-61°F) in February to 20-29°C (68-84°F) in August
Main primary products None
Major industries Ship repairs, ship bunkering, beverages and food processing, tourism and service industries
Main exports Re-exports of petroleum products, tobacco and other manufactures, wines and spirits
Annual income per head (US$) 4400
Population growth (per thous/yr) 9
Life expectancy (yrs) Male 71 Female 76

Gibraltar, Strait of Strategically important strait linking the MEDITERRANEAN SEA with the ATLANTIC OCEAN. A strong surface current, about 80 m (some 260 ft) deep, flows through it from the Atlantic into the Mediterranean, replacing water which is lost by evaporation. At greater depths, a counter-current of dense, salty water flows outwards from the Mediterranean. In the Second World War, this undercurrent was used by German U-boats to carry them noiselessly past the Allied blockade into the Atlantic.
Dimensions 58 km (36 miles) long, 13-43 (8-27 miles) wide
Average depth 366 m (1200 ft)
Map Spain Ce

Gibson Desert ***Australia*** Mainly arid area in central Western Australia, extending about 400 km (250 miles) north-south and 800 km (500 miles) east-west. It now constitutes the Gibson Desert Nature Reserve. The terrain consists largely of parallel gravelly ridges up to 15 m (59 ft) high. There is sparse scrubby vegetation and virtually no population. The desert was crossed by the explorer Ernest Giles in 1874, and named by him after a member of the expedition who died on the crossing.
Map Australia Cc

Giessen ***West Germany*** University town on the Lahn river about 50 km (30 miles) north of Frankfurt am Main. It produces precision instruments.
Population 75 000
Map West Germany Cc

Gifu ***Japan*** Castle town and regional capital in central Honshu island, about 270 km (170 miles) east of the capital, Tokyo. It is noted for its manufacture of traditional paper parasols and lanterns, and is popular for the nighttime spectacle of fishing with cormorants on the Nagara river.
Population 411 700
Map Japan Cc

Gijón ***Spain*** Seaport and industrial town on the Bay of Biscay, producing machinery and hardware, chemicals, cement, glass and pottery. It exports coal and iron ore and has a fishing fleet.
Population 256 000
Map Spain Ca

Gilbert Islands See KIRIBATI

Gilgit ***Pakistan*** District, river and town in the far north of the country, at the centre of some of the highest and most beautiful mountain scenery on earth. The Silk Road, a trade route used by the Venetian adventurer Marco Polo on his way to China in the 13th century, ran through the area. The Karakoram Highway into western China passes there today, winding among peaks of more than 8000 m (26 250 ft).
Map Pakistan Da

gill, ghyll 1. Swift-flowing mountain stream. **2**. A ravine.

Gillingham ***United Kingdom*** See MEDWAY TOWNS

Gimchaeg ***North Korea*** See KIMCH'AEK

Gippsland ***Australia*** Region of south-eastern Victoria, stretching from Melbourne to the border with New South Wales. Brown coal is open-cast mined in the Latrobe Valley and used for power generation in the region. There are major oil fields in Bass Strait, to the south. Along the south-east coast is the Ninety Mile Beach with, behind it, an extensive lake system popular with holidaymakers.
Map Australia Hf

Gir Forest ***India*** Game reserve about 25 km (40 miles) north of the western seaport Diu on the Gulf of Khambhat. It is the main home of the Indian lion, the only Asian species of lion.
Map India Ac

Girna ***India*** Sacred mountain about 385 km (240 miles) north-west of Bombay. It is 1150 m (3772 ft) high and contains temples, cave dwellings and a fort dating from early Muslim and Hindu times. There are rock-carved decrees dating from 272 BC, the time of Asoka, a Buddhist Emperor of India. Visitors can reach the top by way of steep and precipitous paths.
Map India Ac

Gironde ***France*** Estuary, some 80 km (50 miles) long, draining the Dordogne and Garonne rivers. It extends from the junction of the two rivers, about 20 km (12 miles) north of Bordeaux, to the Bay of Biscay. It is 10 km (6 miles) wide near its mouth.
Map France Cd

Ghana

GREAT EXPECTATIONS HAVE FADED IN THE FORMER GOLD COAST

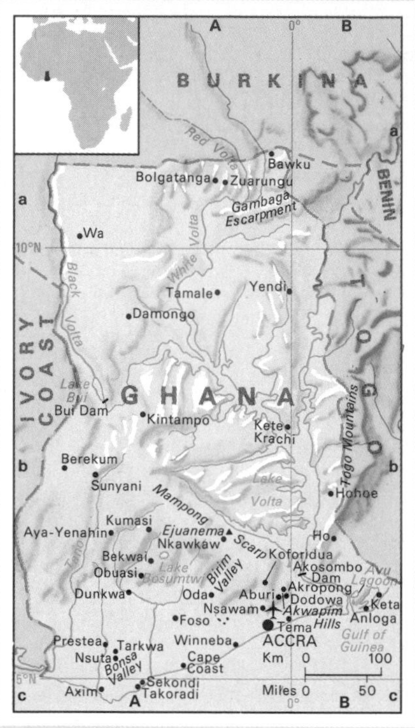

When Ghana became the first black African state to achieve independence from European colonial rule, in 1957, waves of excitement rippled through the entire continent. The atmosphere in the newly independent state was electric. Colourful celebrations in the towns and the countryside, with singing, drumming and dancing, expressed the ecstasy of a people who had lived under British colonial rule for decades.

As an additional boost to morale, the country's name was changed. It was formerly the Gold Coast, and the European appellation recalled a history often painful to the African. Trade in gold with the Portuguese in the 15th century had developed by the 17th century into the slave trade, involving the Dutch, the British, the Danish and the French.

The new name, Ghana, was borrowed from the great African empire which flourished in Mali and Senegal between 700 and 1100, and was intended to inspire a comparable greatness in this new state on the southern coast of West Africa.

But the euphoria and the expectation that Ghana would soon become a prosperous, Western-style, industrialised nation turned to disappointment. The new government's control was undermined by opposition from traditional chiefs and big farmers, while repressive laws to contain the opposition tarnished the idealistic image of the ruling Convention People's Party. Its leader, Kwame Nkrumah, kept out of the fray, and continued to inspire other nationalist movements abroad with his denunciations of imperialism. But money was spent on prestige projects which ignored the country's real needs, debts to the West went unpaid, and Nkrumah was deposed by a military coup in 1966. He was overthrown partly because of his dictatorial and self-indulgent style of government; in 1966 he expanded his personal guard into a regiment.

A series of coups followed which swung Ghana from one political extreme to another, and through them all corruption and mismanagement damaged the economy. The unfavourable economic climate abroad made things worse, and it is now many years since Ghana has been able to balance its budget. As a result, there has been rampant inflation and a chronic shortage of consumer goods. Even food is so scarce that restaurants in the capital city of ACCRA have usually run out of supplies by early afternoon. There is high unemployment (11 per cent of the workforce are jobless), particularly in the cities and towns which now house some 37 per cent of Ghana's people.

WHITE BEACHES AND BLUE LAGOONS

Visitors to the country will find many attractions including palm-fringed beaches of white sand along the Gulf of GUINEA in the south. There are peaceful blue lagoons where the great VOLTA river and the smaller rivers to the west of it meet the sea.

In their long, elaborately decorated dugout canoes – many of which now have outboard motors – the local people fish the lagoons and the sea, and operate waterborne taxi and freight services along the coast. Much of the coastal region is farmed, and around Keta lagoon, in the east, there is a unique area where shallots are intensively cultivated. On the flat coast farther west, near Elmina and around Cape Coast and Apam, people build shallow pools with low mud walls, which trap the seawater at high tide. When the water evaporates, it leaves a crust of salt, for which there is a ready market.

The coast is humid, its stickiness relieved slightly by the dry and dusty Harmattan wind blowing from the north in December and January. The beaches reach northwards to steaming tropical evergreen forests of hardwoods and silk-cotton trees, so called because their fruit, the kapok, yields a silky fibre used for stuffing cushions and upholstered furniture. This exotic forest, with its brilliant birds and flowers and winding creepers, gives way to deciduous woodland farther north, where the climate is hotter and drier.

In this region, parts of the original forest were chopped down by itinerant farmers who cultivated a piece of land until its fertility was exhausted, then moved on to clear another piece of the forest. These methods are being replaced by garden areas in which cocoa and palm oil are produced. In the west, original rain forest on the Ashanti plateau has been cleared to grow cocoa. In the east more areas of the old forests have been destroyed by fires in periods of drought.

Still farther north, woodland gives way to the tropical savannah of the upper Volta basin. This is the home of antelopes and buffaloes, elephants, lions and wild hogs. At the northern end of Lake VOLTA, the landscape becomes harsh and barren, and near the northern frontier with Burkina (once Upper Volta), it becomes an arid plain relieved by shea trees, whose nuts yield an edible yellow fat, and acacias and huge baobabs with their swollen, water-storing trunks. The nuts, gums and resins of these trees are collected and sold by the local people. Billowing clouds of dust are stirred up by herds of cattle, sheep and goats on the move in search of pasture. The land is overgrazed and soil erosion is also a growing problem.

In the north-east, years of drought have added to agricultural problems. The crops that people live on, as well as those they might sell – including millet, sorghum and groundnuts – have given poor harvests. Food shortages throughout the country have been made worse in urban areas because Nigeria has repatriated 300 000 Ghanaians – many of whom had gone there to work.

TRIBAL LIFE IN THE VILLAGE

About two-thirds of Ghanaians live in villages. In the north their homes tend to be round huts with flat or conical roofs made of mud, sticks or any other available local materials. The huts of an extended family are grouped around a yard and surrounded by a fence or wall. The yard is used for keeping poultry, storing grain and tethering animals overnight. Ancestor worship is still practised in these villages, involving ceremonies for the departed spirits of the dead, in which weeping alternates with dancing.

Farther south are the tribal lands of the Akan and the Fante – who once traded with the Europeans in slaves captured from other tribes. Here, too, are the Ashanti, whose kingdom dominated the area in the 17th century. The dwellings in this region are more sophisticated than in the north, often rectangular with gable roofs of corrugated iron or, near the coast, of split bamboo and palm leaves. As in the north, family homes are grouped together round a courtyard.

The ethnic groups of the north, which include the Gonja, Wala, Dagomba and Mamprussi, adhere most closely to their traditional way of life. They were the least exposed to European influence, as the colonists virtually ignored the area. As a result, the level of economic development in southern Ghana is far higher than in the north. The British promoted cocoa, rubber, palm oil and coffee production in the southern forests. These crops are still grown, side by side with food for the local people, so almost anywhere in the forest zone, women (more frequently than men) can still be seen planting, hoeing or weeding plots of yams, cassava, maize and vegetables under cocoa, rubber or palm trees.

Twenty years ago Ghana's cocoa crop was the biggest in the world. But drought and bushfires hit cocoa production and at the same time prices in the world market fell. In Ghana today there are further discouragements for cocoa growers. The roads badly need repair, and the harvested crop may not reach a sales depot for many months – and even when it does payment is extremely slow. For all these reasons producers have been reluctant to replant cocoa, preferring to grow food crops such as maize and cassava which can be sold locally. Ghana now produces about one-eighth of the world's cocoa, and the government is now trying to redevelop the country's ailing industry to boost its export earnings.

Ghana has another important source of wealth in minerals, especially manganese and bauxite as well as, to a lesser extent, gold and diamonds, which were mined even before the colonial era. Although large companies now work the mines, in the BIRIM VALLEY for instance, the traditional methods are still used as well by some small family businesses. In the BONSA VALLEY the local people, mainly women, pan for diamonds in the gravels deposited by the river.

Bauxite (aluminium ore) mining is now being developed at Kibi, to supply the aluminium smelter at TEMA which has up to now been fed with imported ore.

▲ ON WITH THE NEW Awnings and brightly coloured umbrellas have replaced the traditional thatched roofs of the stalls in Kumasi market. Most men now wear 'Western' dress, and many women have discarded the 'kente', the wax-painted cloth worn over the shoulder.

Ghana's most ambitious project since independence was the Volta lake, made by damming the Volta river. It is one of the largest man-made lakes in the world. Over 8480 km² (3251 sq miles) of land were flooded, and 76 000 people had to be rehoused. Now rice is grown around the lake on irrigated land.

The lake's dam, the AKOSOMBO, lies in the south, and could produce enough hydroelectric power for Ghana and some of its neighbours. In recent years, however, the water level in the lake has been low and sometimes insufficient to turn the turbines.

Roads and railways are very poor in Ghana, but ferries run to and fro across the Volta lake. Crowds of people, their luggage tied in brightly coloured bundles, their goats, chickens and in some cases their furniture too, gather noisily to board the ferries. And in the crush on board, as the boat passes the canoes of silent fishermen, the people converse noisily – invariably about politics.

Most of Ghana's towns lie in the south, the majority of them on the coast. They combine wealth and poverty. Relics of the colonial era remain – two-storey buildings with broad verandahs at the front and back, supported by tall wooden or stone pillars. But in cities such as the capital, Accra, dramatically rapid growth has turned the outskirts into an unplanned sprawl. Farther inland, cities such as KUMASI, once the capital of the Ashanti kingdom, retain more of their African character.

LOVE POTIONS AND FOLK MEDICINES

All the Ghanaian towns and cities have their market days – cheerful and noisy occasions when business and pleasure meet. In the larger towns the markets sell anything from vegetables to love potions and elaborate medicines based on the folklore of primitive tribal religions. Dominating the markets are the famous and formidable 'market mamas' – women whose business acumen is rarely equalled, and whose political clout at local, regional, and, in some cases, national level is considerable.

None of the various governments of Ghana's chequered political career since independence has managed to come to grips with Ghana's social and economic problems. Under the current regime of Flight Lieutenant Jerry Rawlings, concerted efforts are being made to restore the responsibility for local development to the local people. Some progress is being made, but with cocoa production in trouble, other parts of the economy need to make much greater headway to justify the optimism with which Ghana celebrated independence.

GHANA AT A GLANCE
Area 238 537 km² (92 099 sq miles)
Population 13 590 000
Capital Accra
Government Republic military-civilian junta
Currency Cedi = 100 pesewas
Languages English (official), Akan, Ga, Ewe
Religions Christian (42%), tribal (38%), Muslim (12%)
Climate Tropical; average temperature in Accra ranges from 22°C (72°F) to 31°C (88°F)
Main primary products Cassava, taro, maize, yams, bananas, cocoa, sorghum, timber; gold, diamonds, manganese, bauxite
Major industries Agriculture, bauxite refining, steel, oil refining, food processing, cement, vehicle assembly, forestry, mining
Main exports Cocoa, gold, timber, diamonds, manganese ore, bauxite and alumina
Annual income per head (US$) 370
Population growth (per thous/yr) 30
Life expectancy (yrs) Male 53 **Female** 56

Gisborne ***New Zealand*** Seaport and resort on Poverty Bay, about 320 km (200 miles) south-east of Auckland. It exports meat, dairy produce and wool from local farms. The town is a centre for tourists exploring the surrounding East Cape area, whose forests and mountains rise to a peak of 1754 m (5754 ft) at Mount Hikurangi.
Population 32 000
Map New Zealand Gc

Giuba ***Ethiopia/Somalia*** See JUBBA

Giulie, Alpi ***Yugoslavia*** See JULIAN ALPS

Gizycko ***Poland*** See MASURIA

Gjirokastër (Argyrokastron) ***Albania*** Picturesque town 350 m (1148 ft) up on the fringes of Albania's central uplands and south-west highlands. The townsfolk are a mixture of Albanians and ethnic Greeks, living in houses which spread over slopes beneath a fortress built by the Ottoman Turks under Ali Pasha in 1811. The town was the scene of a Greek victory over the Italians in 1940-1, during the Second World War. Recently, light industries have been introduced, partly stimulated by the new road link with Yannina, in Greece.
Population 24 400
Map Albania Cb

Glace Bay ***Canada*** Coal-mining town on the Atlantic coast of Nova Scotia. The Italian inventor Marconi set up his first transatlantic wireless service there on October 17, 1907.
Population 21 470
Map Canada Jd

glacial lake Any lake formed in a depression cut by ice or held up by deposits of MORAINE left by ice. Also those confined by the presence of ice itself, as in a tributary valley blocked by a glacier. See GLACIATION

glaciation 1. The formation, movement and retreat of glaciers and ice sheets. **2**. The overall effects on a landscape, both erosion and deposition of debris, brought about by glacial action. See HOW GLACIERS CHANGE THE LANDSCAPE (opposite)

glacier 1. Huge mass of ice, originating from compacted snow, forming FIRN, moving slowly in a continuous stream down a valley under its own weight. **2**. Ice sheet that has spread out from a central mass and covers a large part of a continent – for example, Antarctica (see opposite).

Glacier ***Canada*** National Park with magnificent glacial and alpine scenery, covering 1350 km² (521 sq miles) in the Selkirk Mountains of southern British Columbia.
Map Canada Dc

Glacier National Park ***USA*** Area of jagged peaks, icefields, glaciers and waterfalls in the Rocky Mountains in northern Montana, on the Canadian border. Rivers rising there drain west to the Pacific Ocean, north to Hudson Bay in Canada, and south to the Gulf of Mexico. The park covers 4101 km² (1583 sq miles).
Map United States Da

Gladstone ***Australia*** Port in Queensland, 440 km (275 miles) north-west of Brisbane. Coal mined in the hinterland is the main export, and bauxite (aluminium ore) is brought in to be processed into alumina in the world's largest plant. The port is also a base for tourists visiting the Great Barrier Reef resort of Heron Island.
Population 22 000
Map Australia Ic

Glåma (Glomma) ***Norway*** The country's longest river, rising near the border with Sweden in the county of Sør-Trøndelag. It then flows towards the south for 587 km (364 miles), through Østerdalen in the south-east, and then into Oslofjord at Fredrikstad. In its lower reach, below Lake Øyeren, it is known as the Glomma. Timber is floated down the river and its falls are harnessed for hydroelectricity near Oslofjord.
Map Norway Cc

Glamorgan ***United Kingdom*** Former county of south Wales and the most densely settled and industrialised part of the principality. It was split into three administrative counties in the 1970s. Mid Glamorgan is the biggest, covering 1019 km² (393 sq miles). It takes in what are known as The Valleys, which extend, like the fingers of a hand, north from the coast into the coal-mining flanks of the hills. Names like the Rhondda and Merthyr Tydfil are synonymous with Welsh mining the world over. South Glamorgan (416 km², 161 sq miles) centres on the city of CARDIFF and the port of Barry. West Glamorgan (815 km², 315 sq miles) includes the industrial complex of SWANSEA, Port Talbot and Neath, and the beautiful Gower peninsula with its beaches and holiday homes.
Population (Mid Glamorgan) 536 000; (South Glamorgan) 392 000; (West Glamorgan) 367 000
Map United Kingdom De

Glarus ***Switzerland*** Capital of the canton of the same name on the Linth river about 50 km (30 miles) south-east of Zürich. The canton is an alpine region, rising to 3600 m (11 800 ft), with forests and high pastures where cattle and sheep are reared. Protestant reformer Ulrich Zwingli (1484-1531) was priest at Glarus in 1506-15. The town was largely destroyed in 1861 by a fire fanned by a violent FOHN wind.
Population (canton) 36 200; (town) 6000
Map Switzerland Ba

Glasgow ***United Kingdom*** Scotland's largest city, lying on the River Clyde, and the centre of the administrative region of Strathclyde. Glasgow's starting point was a 12th-century cathedral, in what is now the east end of the city, built on the remains of a 6th-century church. From there, the city grew steadily westwards, following the Clyde toward the sea. Eventually, Glasgow grew downstream past Govan and Clydebank, with their shipyards, overran the separate burgh of Paisley with its abbey, cathedral and textile mills, and created a virtually continuous built-up area for 25 km (16 miles) along the river.

Glasgow's great prosperity began with 18th-century overseas trade, but shipbuilding soon followed. 'Clyde-built' became a mark of quality for steamships the world over from about 1820. So, too, in the field of rail transport, did the name of Springburn, in the northern suburbs, from which locomotives were exported equally widely. The stately ocean liners known as the Cunard 'Queens' were built at Clydebank. But since the Second World War these major industries have fallen into a steep decline.

Central Glasgow today is mainly 19th century in layout and appearance, with George Square and the City Chambers at its heart, its streets laid out in a formal grid pattern. West of the centre are the higher-priced residential areas of the past century and the cultural amenities that went with them: the University of Glasgow (founded in 1451, but on its present site since 1870), the botanical gardens, the Kelvin Hall exhibition centre, and the art gallery with a fine collection of British and European paintings. A second university, Strathclyde, was founded in 1964.
Population 751 000
Map United Kingdom Cc

glass Hard, shiny volcanic rock which has cooled too quickly to have a crystalline structure.

Glastonbury ***United Kingdom*** Town in the county of Somerset in western England, 185 km (115 miles) from London. It stands at the foot of a steep, isolated hill. A small chapel built at Glastonbury in early times was replaced by a monastic church which burned down in 1184. This was replaced in turn by a huge monastery which was largely destroyed in 1539. Its chief remains are a fine Lady Chapel and a superb late 14th-century kitchen.

After the late 12th century, two legends grew up about Glastonbury: that St Joseph of Arimathea, who buried Christ, built a wattle church there and housed in it the Holy Grail – the chalice used by Christ at the Last Supper; and that King Arthur of the Round Table was buried in the monastery in the 6th century.
Population 7000
Map United Kingdom De

Glatz ***Poland*** See KLODZKO

Gleiwitz ***Poland*** See GLIWICE

Glendalough (Gleann dá Loch) ***Ireland*** Valley in the Wicklow Mountains 39 km (24 miles) south of Dublin, where the hermit St Kevin set up a monastery and seat of learning in the 6th century. Remains include a round tower, a 10th-century cathedral and St Kevin's Kitchen, an early Irish oratory, now a museum.
Map Ireland Cb

Glittertind ***Norway*** The nation's highest mountain, at 2470 m (8104 ft). It is in the Jotunheimen region in south Norway, 112 km (70 miles) north-west of Lillehammer.
Map Norway Cc

Gliwice (Gleiwitz) ***Poland*** Southern manufacturing city, 23 km (14 miles) west of Katowice. Despite its industrialisation, the old town retains much of its beauty, with medieval walls and many buildings dating from the 16th to the 18th centuries. Gliwice produces beer, textiles, iron, steel and chemicals.
Population 211 200
Map Poland Cc

HOW GLACIERS CHANGE THE LANDSCAPE

A glacier is like a river in slow motion – it flows downhill, it erodes the ground it passes over and carries the eroded material with it, then finally deposits the debris when it melts and its force is spent.

A glacier is a major tool in sculpting the surface of the land. At its head, as it moves downhill from a mountain top, it scoops out a steep-sided amphitheatre-like hollow, or CIRQUE. On some mountains, several glaciers may have carved a group of cirques around the summit, separated from each other by sharp ridges called ARETES.

As the glacier moves, it picks up rock fragments and carries them along with it. The fragments act like grit in glasspaper and erode the underlying rocks, gouging them with parallel scratches, or striations. These marks show the course of a glacier long after it has vanished.

The scouring action of a glacier transforms V-shaped river valleys into U-shaped ones by carving away the sides and levelling the floors. The ends of spurs between tributary valleys are removed, leaving HANGING VALLEYS with their outlets high above the main valley floor. Hillocks and outcrops of rock over which the glacier has passed are often left with gently sloping sides facing the glacier's advance, and rugged slopes on the lee side where the glacier has plucked away material. They are called roches moutonnées – meaning 'sheep-like rocks' in French – because of their resemblance to recumbent sheep.

Unlike a river, a glacier does not deposit the debris it carries until it melts or recedes. The eroded material, known as TILL, ranges from boulders to fine dust. It may be transported many hundreds of kilometres by the glacier. If a glacier has melted gradually and steadily, the till will be spread over a wide area and is called ground MORAINE. But if a glacier has remained stationary for a long time before retreating, the till builds up along its leading edge. When the ice melts, the till is left as a ridge called a terminal moraine.

Glaciers also produce lateral moraines. These are composed of materials that fall or are scraped from the valley sides and build up along the outer edges of a glacier. When two glaciers meet, their two inside lateral moraines merge to form a medial moraine down the centre of the combined glacier.

KAMES – long, low hills or hummocks of fine sediment – are left where streams issued from the edge of an ice sheet. KETTLES are ice-filled depressions covered by till that remain after an ice sheet has retreated. The ice eventually melts and leaves a bowl-shaped hollow lined with debris. DRUMLINS are small hillocks, which may rise to 100 m (about 330 ft), composed of fine till. They are shaped like half an egg with the thick, steep end facing the flow of ice. Long, winding ridges of sand and gravel, known as ESKERS, are deposited by streams flowing beneath an ice sheet.

A valley glacier is not a smooth sheet of ice. Its surface is covered by debris that falls from the sides of the valley down which it moves. Cracks, or crevasses, appear in the surface when the glacier is 'stretched' as it moves over a hummock in the ground or turns a sharp corner. A *bergschrund* is a crevasse formed around the upper rim of the ice within a cirque – the huge weight of the glacier tears the ice away from the steep rock face. The scouring action of the glacier as it moves down the valley leaves many depressions in the ground. After the glacier has melted, some of these may fill with water and form cirque lakes, or tarns, in the mountains, and ribbon lakes where the valley is dammed by terminal moraines.

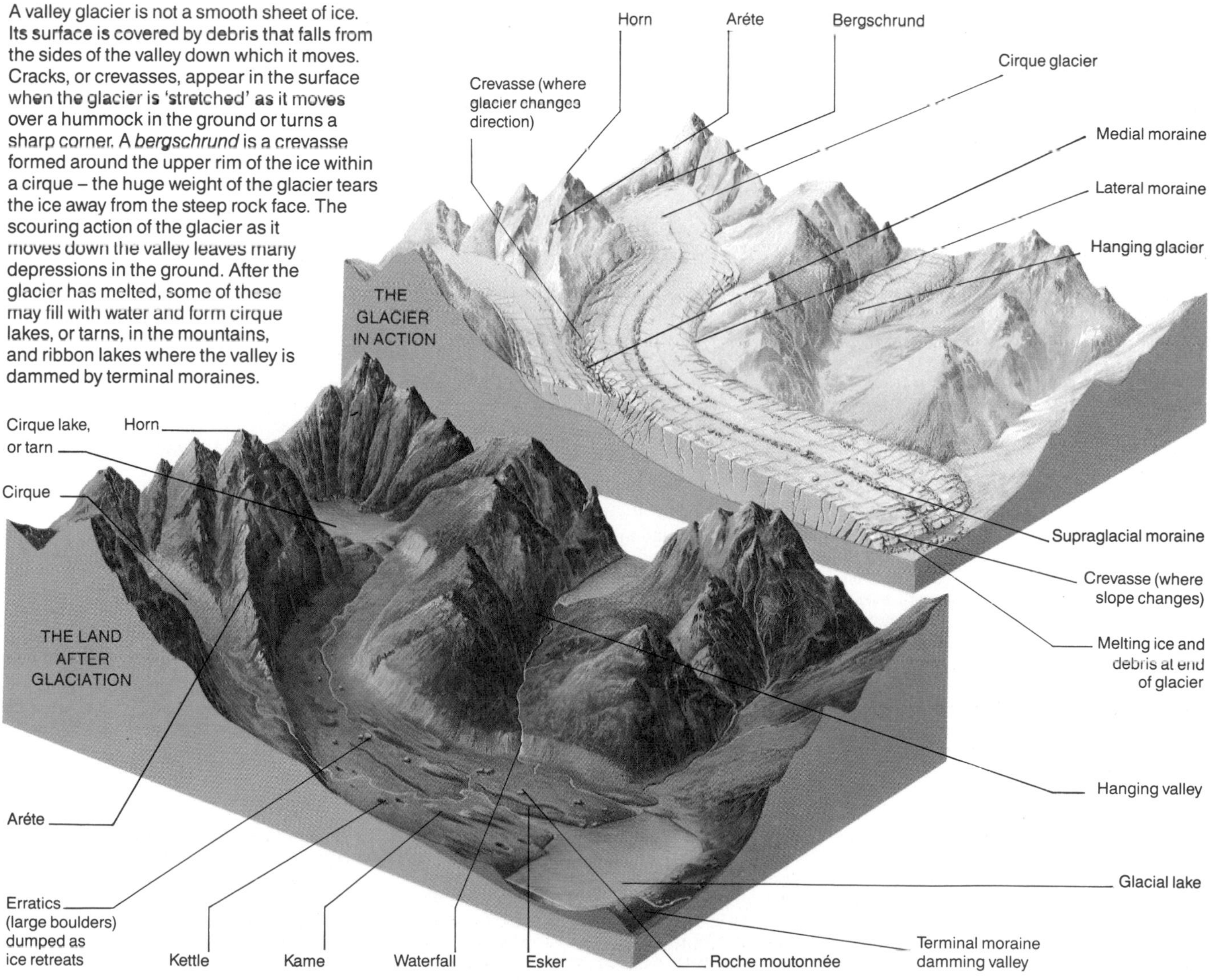

Glomma ***Norway*** See GLAMA

Gloucester ***United Kingdom*** City and county of western England. The city was the Roman town of Glevum and stands on the River Severn, 150 km (93 miles) west and slightly north of London. The cathedral is one of the most beautiful buildings in Britain. It was founded in the 11th century and completed in the 16th. The cloisters are particularly fine. It is one of the venues, with Hereford and Worcester, of the annual Three Choirs Festival. The city is a centre for aircraft manufacture.

The county has an area of 2638 km^2 (1019 sq miles). It includes the Forest of DEAN on the Welsh border, the beautiful 18th-century spa town of Cheltenham – birthplace of the composer Gustav Holst in 1874 – the COTSWOLD HILLS, and Cirencester, the second largest town in Roman Britain.

Population (county) 508 000; (city) 95 000
Map United Kingdom De

Gnesen ***Poland*** See GNIEZNO

Gniew ***Poland*** Medieval northern town, 62 km (39 miles) south of the port of Gdansk. Its 13th-century red-brick castle, which still towers over the town, was one of the strongholds of the Order of Teutonic Knights in 1282. Attractive arcaded houses surround the market square.

Map Poland Cb

Gniezno (Gnesen) ***Poland*** Ancient town in central Poland, 45 km (28 miles) north-east of the city of Poznan. It was the country's first capital, in the 11th century, and remains of an 8th-century fort have been found on the town's Gora Lecha (Hill of Lech). According to legend, the fort was founded by Lech, the leader of the Polan tribe, which founded the Polish state and gave the country its name. Gniezno gets its name from the word *gniazdo*, meaning 'the nest' – a reference to the white eagle, symbol of the tribe and of Poland. The massive cathedral, standing where a church has stood since 973, contains the remains of St Adalbert (Wojciech), Poland's first patron saint.

Population 66 100
Map Poland Bb

Goa ***India*** Territory on the west coast covering 3702 km^2 (1429 sq miles) and lying about 400 km (250 miles) south of Bombay. Together with Daman – on the Gurarat coast 160 km (100 miles) north of Bombay – and the small island of Diu off the Kathiawar peninsula to the north-west, it forms a Union Territory of India. All three were formerly Portuguese possessions; Goa was seized by the naval officer Alfonso de Albuquerque in 1510, the other two territories in the 1530s.

After four centuries of Portuguese rule, Goa's population is racially mixed and predominantly Catholic. St Francis Xavier, who died in 1558, was buried in Old Goa, the abandoned former capital noted for its Baroque cathedrals. The capital is now the port of PANAJI. Goa, Daman and Diu were annexed by India in 1961 and Goa has since developed into a major tourist attraction – partly because of its superb beaches.

Population 1 007 800
Map India Bd

Gobi Desert ***Mongolia*** Wasteland covering 1 295 000 km^2 (500 000 sq miles), about half the country. It extends south into the Nei Mongol Autonomous Region of China. The desert is between 1000 and 2000 m (3300-6600 ft) above sea level, and ranges from sand in the west to stone in the east. Its climate is one of extraordinary extremes – temperatures can rise to 45°C (113°F) in summer and can fall to −40°C (−40°F) in winter.

The main route across the desert, followed by a road and by the Trans-Mongolian Railway, runs south-east from the capital, Ulan Bator. The recent discovery of coal at Tavan-Tolgoyt, in the middle of the Gobi, has encouraged hopes that other minerals may be present.

Map Mongolia Db

Godavari ***India*** River rising in the Western GHATS near Bombay, and flowing 1465 km (910 miles) to form a constantly growing delta in the Bay of Bengal. The delta is laced with irrigation canals, and the region is becoming prosperous by improved rice farming – although the coastal areas are at risk from cyclones and tidal waves. It is one of the seven sacred rivers of the Hindus.

Map India Cd

Godthåb (Nuuk) ***Greenland*** Largest town and capital of Greenland, on the south-west coast. Founded by the Danes in 1721, it is the oldest Danish settlement in the country. Godthåb is the seat of the national council and the supreme court. It has fish canning and freezing plants, and a heliport.

Population 10 500
Map Arctic Cc

Godwin Austen ***China/Pakistan*** See K2

Golan Heights ***Syria/Israel*** High, bare hills rising to 2225 m (7300 ft) in south-west Syria and overlooking the Sea of Galilee and Israel. Because of their dominance, they are of great military and strategic importance. The western slopes and highest crest were captured by the Israelis during the 1967 Arab-Israeli War, and were annexed by Israel in 1981.

Map Syria Ab

Golconda ***India*** Ruined town and fortress about 4 km (2.5 miles) north-west of the city of Hyderabad. It stands on a rocky mountain 152 m (500 ft) above the plain and was the capital of the kingdom of Kutb Shahi (1507-1687). It protected the nearby diamond mines, from which, some claim, came the magnificent Koh-i-noor diamond, though its history is controversial before 1849, when it was acquired by the British and placed with Queen Victoria's crown jewels. In 1937 it was incorporated into the queen's crown for the Coronation of Queen Elizabeth, consort of George VI. The fortified remains are well preserved and visitors can also see the tombs of the Kutb Shahi kings.

Map India Cd

Gold Coast ***Australia*** A series of beach resorts stretching along some 32 km (20 miles) of the southern Queensland coast from the New South Wales border to within about 60 km (38 miles) of Brisbane. Its main centres include Coolangatta and Surfers Paradise. Developed from a series of small settlements since the late 1940s, it is now Australia's ninth biggest 'city'.

Population 135 400
Map Australia Id

Golden Triangle ***South-east Asia*** Name recently given to the largely mountainous border area where Thailand, Laos and Burma meet. Inhabited mainly by hill peoples who grow the opium poppy, it is one of the world's main sources of opium and of refined morphine, although the trade is mainly controlled by Chinese and Shan warlords who run private armies to protect their interests.

Map Burma Cb

Gomel' ***USSR*** Industrial city in Belorussia, 210 km (130 miles) north of Kiev. It was a Lithuanian possession until 1772, when Russia took control. Gomel' was virtually destroyed during the Second World War, and was rebuilt to become a centre for timber products, machinery, textiles and footwear.

Population 452 000
Map USSR Ec

Gomera ***Spain*** See CANARY ISLANDS

Gonder ***Ethiopia*** City about 400 km (250 miles) north and slightly west of the capital, Addis Ababa. It was the royal town of the Abyssinian Empire from about 1700 to 1855, and several rocky outcrops within the city are topped by the castles of successive emperors. It is a predominantly Christian community and has seven impressive churches.

Population 80 000
Map Ethiopia Aa

Gondwanaland Southernmost of the two ancient continents into which Pangaea, an earlier supercontinent, divided about 180 million years ago. The northernmost was Laurasia. See CONTINENTAL DRIFT

Gongju ***South Korea*** See KONGJU

Good Hope, Cape of ***South Africa*** Storm-blown tip of Cape Peninsula, about 50 km (30 miles) south of Cape Town. The Portuguese navigator Bartolomeu Dias, who became in 1488 the first European to sight it, called it the Cape of Storms. But King John II of Portugal renamed it Good Hope, 'for the promise it gave of finding India'.

Map South Africa Ac

Goodwin Sands ***English Channel*** Stretch of shoals and sandbanks off the coast of Kent in south-eastern England. Numerous ships have been wrecked on the sands, which are partly exposed at low water and are marked by lightships and buoys.

Length 16 km (10 miles)
Map United Kingdom Fe

Gora Kalwaria ***Poland*** Pilgrimage town in eastern Poland, set amid orchards 34 km (21 miles) south-east of the capital, Warsaw. Its name means 'Mount Calvary'. It was begun in 1672 by a bishop of Poznan in western Poland, and laid out in the shape of a cross, its founder's idea of the plan of ancient Jerusalem.

Map Poland Dc

▲ **SILK ROAD Nomads make their way along this ancient trade route below the Altun Shan at the edge of the Taklimakan Desert – an extension of the Gobi in Xinjiang, China.**

Gorakhpur ***India*** City in the north about 65 km (40 miles) south of the Nepalese border. Traditionally, it is the place of Buddha's death and cremation in about 483 BC. It has numerous old Buddhist temples and ruins. Today Gorakhpur manufactures textiles and dyes and has railway workshops.

Population 307 500
Map India Cb

Gordon ***Australia*** River rising in the highlands of south-west Tasmania and flowing 200 km (124 miles) north-west through ancient forests of Huon pine – now protected – to Macquarie Harbour. It has been dammed to form Lakes Gordon and Pedder, in order to generate hydro-electric power, but plans for further dams downstream were dropped in 1983 after worldwide protests by conservationists.

Map Australia Hg

Gorée ***Senegal*** Volcanic islet and town off the coast beside Dakar, the capital. It was captured in turn by the Portuguese (15th century), Dutch (16th century), and by the French in 1677. The town was a flourishing base for the slave trade and a provisioning port on the route to India. It is now a quiet fishing port and tourist resort. Visitors can still see the dungeons where slaves were kept before shipment across the Atlantic. Three museums in the town trace the island's life and history.

Map Senegal Ab

Gorgan (Gurgan) ***Iran*** Town in the former province of Gorgan, about 50 km (30 miles) east of the south-eastern corner of the Caspian Sea. It trades in cotton, wheat and salt. About 30 km (19 miles) to the north of the town lie the remains of Alexander's Wall, a line of earthworks, once 70 km (43 miles) long, built during the Sassanid dynasty (3rd-7th century AD). They were mistakenly named by later historians, who believed that they had been constructed by Alexander the Great. The Gorgan river, 240 km (150 miles) long, which drains into the Caspian Sea just north of the town of Bandar -e Torkeman, is one of the few Iranian rivers that never dries up.

Population 88 400
Map Iran Ba

gorge Deep, steep-sided rocky river valley, formed where river erosion cuts down more rapidly than WEATHERING and MASS MOVEMENT can wear back the sides.

Gori ***USSR*** See TBILISI

Gor'kiy (Gorky) ***USSR*** Major industrial city and port on the Volga river, 380 km (235 miles) east of Moscow. It was founded as a fort in 1221 and grew into a riverside market centre. In the 19th century it was the site of Russia's largest trade fair. Originally called Nizhni Novgorod, it was renamed in 1932 in honour of the Russian author Maxim Gorky (1868-1936). He was born and bred in the city's slums and described the plight of the poor in works such as his autobiographical *Childhood*, published in 1913-14. Today Gor'kiy is a closed city that foreigners are not allowed to enter, because it is a place of internal political exile. The Soviet dissident and scientist Andrei Sakharov was banished there in January 1980.

Gor'kiy makes Volga cars, as well as riverboats and hydrofoils. Its other industries include oil refining and the manufacture of aircraft, diesel motors, machine tools, paper and agricultural equipment.

Population 1 392 000
Map USSR Fc

Görlitz ***East Germany*** Industrial city on the Neisse river (which forms the Polish border) east of Dresden. Its products include railway vehicles, machinery, precision instruments, textiles and furniture. In the part of the city east of the Neisse – which has belonged to Poland since 1945 and is called Zgorzelec – are a ruined 12th-century castle and many 16th-century houses.

Population (East German) 80 200
Map East Germany Dc

Goroka ***Papua New Guinea*** Capital of Eastern Highlands province, about 400 km (250 miles) north-west of Port Moresby. The fertile valleys of the region, discovered by outsiders only in the 1930s, have extensive coffee plantations. At the biennial Goroka Show villagers from the surrounding area celebrate at a 'sing-sing', wearing traditional dress.

Population 19 000

Map Papua New Guinea Ba

Gorongosa National Park ***Mozambique*** See SOFALA

Goslar ***West Germany*** Former imperial capital at the foot of the Harz mountains, about 70 km (43 miles) south-east of Hanover, and 15 km (9 miles) from the East German border. The Holy Roman Emperor Heinrich II built a palace there early in the 11th century, lured by the wealth of the lead and silver mines in the Harz mountains. The cathedral, rebuilt in 1820, still has Heinrich's bronze throne, the only imperial throne remaining in West Germany apart from Charlemagne's at Aachen.

Population 54 000

Map West Germany Dc

Göta Kanal ***Sweden*** Waterway, using rivers, lakes and canal, which links the city of Gothenburg on Sweden's south-west coast with the Baltic. The waterway follows the Göta river to Lake Vänern, then via canals and stretches of river to Lake Vättern. It emerges on the east coast near Söderköping. In total, the waterway is some 611 km (380 miles) long. The 19th-century Scottish engineer Thomas Telford was in charge of the project, which was opened in 1832. The Göta river section, 90 km (56 miles) long, is the only part navigable by ocean-going ships.

Map Sweden Bd

Gothenburg (Göteborg) ***Sweden*** The country's second largest city after Stockholm and its main seaport. It lies on the Kattegat channel at the mouth of the Göta river. It was founded in 1624 near the site of a medieval settlement, and the city plan – with its defensive canal system – was based on the Dutch model. Its harbour, ice-free throughout the year, is a base for ferries serving Denmark and across the North Sea to Britain. The main industries today are car and ballbearing production, foodstuffs, textiles and clothing.

Gothenburg is an important seat of learning: Gothenburg University was founded in 1891, and the country's oldest technical university – Chalmers' University of Technology – was founded in 1829 by a donation of an English engineer named William Chalmers. The city also has an oceanographic institute, a medical research centre, a botanical garden, museums and theatres.

Population 425 500

Map Sweden Ad

Gotland ***Sweden*** The largest island in the Baltic, covering an area of about 3140 km² (1210 sq miles), 150 km (93 miles) south of Stockholm. It has been part of Sweden since 1645. It was an important trade centre in Viking times and reached its peak of prosperity in the Middle Ages. It was once a haunt of pirates.

Its ports include the walled city of VISBY, Slite and a number of fishing harbours. The island has 90 medieval churches and is rich in archaeological remains, including some 200 runic stones – stones inscribed with ancient Scandinavian symbols. The island's limestone, the basis of a modern cement industry, has been exported throughout the Baltic. Fishing, sheep and sugar beet contribute also to the economy, but Gotland's main industry is tourism.

Each summer the island is the setting for the Stånga Games. Among the events are an early form of tennis, a throwing contest in which the competitors hurl a heavy stone disc, and a contest resembling the Scottish Highland event of tossing the caber.

Population 56 100

Map Sweden Cd

Göttingen ***West Germany*** University town about 90 km (55 miles) south of Hanover. England's George II, as Elector of Hanover, founded the university in 1737. The physicist Max Planck (1858-1947), who devised the Quantum Theory, one of the foundations of modern atomic physics, was a student and is also buried at Göttingen.

Population 138 000

Map West Germany Cc

Gouda ***Netherlands*** Cheese-making town about 42 km (26 miles) south and slightly west of Amsterdam. Much of the medieval town centre with its encircling canals (originally defensive moats) and marketplace survive.

Population 59 200

Map Netherlands Ba

Governador Valadares ***Brazil*** Mining town 240 km (150 miles) north-east of the city of Belo Horizonte. Its mines produce gemstones, including emeralds, amethyst, topaz, agate, aquamarine and tourmaline.

Population 197 000

Map Brazil Dc

Gozo ***Malta*** Second largest island of the Maltese archipelago, separated from the main island of Malta by a channel 5 km (3 miles) wide which also contains the 3 km² (1.1 sq mile) islet of Comino. Gozo is still substantially free of tourist trappings and presents the restful, unpretentious nature of Mediterranean island life at its best. The people are modest, friendly and dignified. Most live from farming and fishing, and almost all of the women make lace. The island covers 67 km² (26 sq miles) and is noted for its oranges, potatoes, onions, carrots and broad beans. Its rough wines have a legendary potency, as many visitors have found to their cost. Ferries to Malta run from the main port, Mgarr, 8 km (5 miles) by road from the capital, Victoria.

Most of Gozo's dozen or so villages are strung out on rocky ridges overlooking broad, carefully cultivated valleys. Huge Baroque churches dominate the landscape. The one at Xewkija will hold 12 000 people – even though the entire village population is only 2600. Several modest hotels and guesthouses cater for visitors. There is a large sandy beach at Ramla Bay on the north coast, and smaller beaches at Marsalforn, Xlendi and Dwerja. Dwerja is the scene of the phenomenon of Qawra, or Inland Sea, where water rushes through a natural archway into a bay encircled by cliffs. Other places of interest include the impressive prehistoric Ggantija temples and the pilgrimage church of Ta Pina, built between 1920 and 1936 on the site of a 16th-century chapel.

Population 23 500

Map Malta Ca

Graaff Reinet ***South Africa*** Beautifully preserved market town, founded in 1786, on the Great Karoo plateau about 190 km (120 miles) north-west of Port Elizabeth. In 1795, a group of Afrikaner burghers occupied the town and proclaimed a short-lived republic.

Population 29 800

Map South Africa Bc

Gracanica ***Yugoslavia*** See PRISTINA

Gradec (Slovenj Gradec) ***Yugoslavia*** See KARAWANKE ALPS

Grahamstown ***South Africa*** Cathedral and university town in eastern Cape Province, about 135 km (85 miles) south-west of the port of East London. It is nicknamed the City of Saints because of its many churches and church schools. Grahamstown was founded as a military post in 1812, and was attacked often by Xhosa warriors during the Kaffir War between Africans and the European settlers.

Population 25 100

Map South Africa Cc

Grampian ***United Kingdom*** Administrative region in east central Scotland covering 8550 km² (3301 sq miles) and taking in the former counties of Aberdeen, Kincardine and Banff, plus part of Moray. It is named after the Grampian Mountains, which extend into neighbouring regions and rise to 1344 m (4409 ft) at Ben Nevis, Britain's highest peak, in Highland region; they also include the ski resorts of the Cairngorm massif. In the east they descend gradually to a coastal fringe of fertile land, the home of the Aberdeen Angus breed of cattle. The region's best-known product is whisky and its best-known inhabitants are the royal family, whose castle at Balmoral is about 70 km (43 miles) west of the North Sea oil port of ABERDEEN, the main administrative centre.

Population 497 000

Map United Kingdom Db

Gran Canaria ***Spain*** See CANARY ISLANDS

Gran Chaco ***Paraguay*** Wilderness covering some 240 000 km² (92 660 sq miles) – an area about the size of Britain – on the borders of Argentina and Bolivia. Territorial disputes with Bolivia led to the Chaco War (1932-5), which Paraguay won – but at the cost of 36 000 men.

The north and west of Chaco is largely arid thorn scrub. Towards the east and south there is cattle pasture and forests famous for the quebracho, or 'axe-breaker' tree, a source of tannin.

The population is sparse, with some Indian tribes and isolated settlements of Mennonites, a Christian fundamentalist sect originally from Germany. The main town is Filadelfia.

Population 100 000

Map Paraguay Ab

Gran Pajatén ***Peru*** Magnificent ruined city of a pre-Inca civilisation, discovered by the American explorer Gene Savoy in 1964 in the forest of the Urubamba river valley. There are many circular stone buildings, temples and blocks of carved stone depicting human figures and condors.
Map Peru Ba

Gran Paradiso National Park ***Italy*** Oldest of the country's national parks, founded in 1922, about 50 km (30 miles) north-west of Turin. It covers 620 km² (240 sq miles) of the Graian Alps, rising to the Gran Paradiso peak at 4061 m (13 323 ft). The park protects rare animals and birds such as the ibex, chamois and royal eagle.
Map Italy Ab

Gran Sasso ***Italy*** Part of the APENNINE mountains about 120 km (75 miles) north-east of Rome whose Monte Corno (2912 m, 9554 ft) is the highest point in the range.
Map Italy Dc

Granada ***Nicaragua*** The country's oldest Spanish settlement, founded in 1523 at the northern end of Lake Nicaragua, about 50 km (30 miles) south-east of the capital, Managua. Many of its colonial buildings survive, including the church of La Merced (Mercy), completed in 1781. Its industries include the manufacture of furniture, clothing, cotton, rum and soap.
Population 60 500
Map Nicaragua Aa

Granada ***Spain*** Tourist resort and city about 195 km (120 miles) north-east of Gibraltar. It is famous for its Moorish architecture, notably the Alhambra, a magnificent palace built between 1248 and 1354. The city also has a 16th-century cathedral which contains a royal chapel, with the tombs of the Catholic monarchs Ferdinand and Isabella.

Modern Granada manufactures textiles, paper and liqueurs as well as catering for tourists. One of the many tourist attractions is the flamenco dancing of the gypsy community living in cave dwellings in the nearby mountain of Sacromonte. Granada province produces cereals, sugar cane and tobacco.
Population (city) 262 200; (province) 761 700
Map Spain Dd

Grand Bahama ***Bahamas*** One of the largest islands in the Bahamas, covering 1370 km² (530 sq miles). It lies in the north-west of the group, only 100 km (62 miles) off the coast of Florida. The interior is heavily forested, and the island was largely undeveloped, with a population of about 8000, until the early 1960s. Then a group of American, Bahamian and European businessmen began to develop it as a tax-exempt tourist, industrial and commercial area, mainly in and around the main town of FREEPORT CITY and its suburb Lucaya. Now it has the second largest population in the Bahamas.
Population 27 000
Map Bahamas Ba

Grand Banks Part of the continental shelf off south-eastern NEWFOUNDLAND where the warm GULF STREAM and cold Labrador Current meet. The mixing of these waters causes fogs, particularly in the north-east. Icebergs are another hazard, especially in spring. This is one of the world's leading fishing grounds, although overfishing led the Canadian government to introduce restrictions in 1977. Oil was discovered in the late 1970s.
Dimensions 680 km (422 miles) east-west, 560 km (348 miles) north-south
Depth Mainly 90-180 m (295-590 ft)
Map Canada Jd

▲ MIGHTY WORK The Colorado river carved its vast Grand Canyon, 6-29 km (4-18 miles) wide, as the land slowly rose during earth movements, and still erodes huge amounts of debris daily.

Grand Canal ***China*** See DA YUNHE

Grand Canyon ***USA*** Spectacular gorge of the Colorado river in north-western Arizona. It is about 350 km (220 miles) long and up to 1870 m (6135 ft) deep, and is cut in multicoloured rock strata laid down over more than 1000 million years. The most beautiful stretch of the canyon walls lies within a national park covering 4931 km² (1904 sq miles). Almost 3 million people visit the canyon each year.
Map United States Dc

Grand Popo ***Benin*** Port, market and tourist resort about 22 km (14 miles) from the border with Togo. It was founded in 1727 by refugees fleeing the slave traders in Ouidah. It is one of two places along the coast where there is a channel linking the coastal lagoons and the sea.
Map Benin Ab

Grand Rapids ***USA*** City in western Michigan, about 40 km (25 miles) east of Lake Michigan. It derives its name from falls on the Grand river. The city was founded as a fur trading post in 1826 but soon developed as a centre of furniture manufacture, and held its own furniture fair annually from 1887 until the early 1960s. Its main museum contains streets re-creating the period 1870-1900.
Population (city) 183 000; (metropolitan area) 626 500
Map United States Ib

Grand St Bernard Pass ***Switzerland*** Alpine pass between Switzerland and Italy, about 25 km (15 miles) east of Mont Blanc. The area is famous for its St Bernard dogs, which formerly rescued many travellers from snowdrifts. The successors of the monks who bred the dogs have a hospice near the 2469 m (8100 ft) high summit of the pass. Today's travellers use a 5.8 km (3.6 mile) tunnel under the pass.
Map Switzerland Ab

Grand Terre ***Guadeloupe*** See GUADELOUPE

Grand Teton National Park ***USA*** Mountainous area covering 1241 km² (479 sq miles) in western Wyoming, immediately south of Yellowstone National Park. It is the most spectacular part of the Teton Range, which rises to 4196 m (13 766 ft) at the Grand Teton, and is dotted with lakes, streams, extensive forests and areas of sage brush.
Map United States Db

Grande Comore ***Comoros*** See NJAZIDJA

Grande Prairie ***Canada*** City which is the business centre for the PEACE RIVER agricultural district of north-west Alberta. Farming started in the area around 1900.
Population 24 260
Map Canada Dc

granite A coarse-grained igneous rock consisting largely of quartz, feldspars and micas.

granitic layer Upper, less dense of the two layers into which the continental portions of the earth's CRUST are divided.

Grasse ***France*** Manufacturing town and resort on the Côte d'Azur, 12 km (7.5 miles) north of Cannes. It is set among orange groves and fields of flowers, from which essences are extracted to make perfumes. The French artist Jean Fragonard (1732-1806) was born in the town, and there is a museum there commemorating his life and work.

Population 38 400

Map France Ge

grassland Region where the vegetation consists mainly of grass, the rainfall being too light for forest growth. The main types of grassland are tropical grassland, or savannah, mid-latitude (temperate) grassland which includes the STEPPES, PRAIRIES and PAMPAS, and mountain grasslands, a zone above the tree line on mountains. These grasslands are mostly natural, though they have probably been modified and extended by man over thousands of years – particularly by the use of fire.

Graubünden (Grisons) ***Switzerland*** Largest of the 26 cantons, covering a mountainous area of 7106 km² (2744 sq miles) in the south-east. The main occupations in the canton, which is home to most of Switzerland's Romansch-speakers, are forestry, tourism and dairying.

Population 172 200

Map Switzerland Ba

gravel Mass of rock fragments ranging in diameter from 2 to 4 mm (0.08 to 0.16 in).

Graz ***Austria*** The country's second largest city and provincial capital of Styria, 140 km (87 miles) south-west of Vienna. Graz grew up as a settlement on the trade route from Italy to Hungary, and became a fortress town in the 12th century, guarding the northern route through the Alpine foothills. It has an 11th-century castle, a 15th-century Gothic cathedral and a 17th-century arsenal. Its manufactures include paper, shoes, clothing and chemicals. There are also steel foundries, ironworks and railway workshops. The German astronomer Johannes Kepler, whose work confirmed that the planets revolve around the sun, lived in Graz from 1594 to 1600.

Population 243 000

Map Austria Db

Great Artesian Basin ***Australia*** The world's largest underground natural reservoir, lying beneath an area of about 1.7 million km² (656 000 sq miles). It stretches from the Gulf of Carpentaria into South Australia and New South Wales and gathers its water from the porous rocks west of the Great Dividing Range. About 5000 flowing boreholes have been constructed to tap the basin's water. Only around half of these are still flowing, with a total discharge rate of 1500 million litres (330 million gallons) a day for irrigation, livestock and domestic use. Without them, farming Australia's dry interior would be impossible.

Map Australia Fc

Great Australian Bight Arm of the INDIAN OCEAN lying off southern Australia. It is partly bordered by the inhospitable NULLARBOR PLAIN. It was discovered by a Dutch navigator, Captain Thyssen, in 1627, but he turned back before reaching the fertile lands of SOUTH AUSTRALIA. The coast was charted by the English explorer Captain Matthew Flinders in 1802.

Dimensions 1160 km (720 miles) east-west, 350 km (218 miles) north-south

Map Australia De

Great Australian Desert ***Australia*** Vast arid heart of the island continent, covering some 3 830 000 km² (1 480 000 sq miles) and comprising several distinct areas, including the GIBSON DESERT, GREAT SANDY DESERT, GREAT VICTORIA DESERT, NULLARBOR PLAIN and SIMPSON DESERT.

Great Barrier Island ***New Zealand*** Main island of the Hauraki Gulf, 100 km (60 miles) north-east of Auckland on North Island. It covers an area of 285 km² (110 sq miles) and is a mixture of pasture, scrub and forest. Tryphena is the island's main town.

Population 300

Map New Zealand Eb

Great Barrier Reef ***Australia*** Largest coral reef system in the world, stretching about 2000 km (1250 miles) along the coast of Queensland in the Coral Sea. The reef, which is up to 500 m (1650 ft) thick, extends up to 300 km (190 miles) from land in the south, but much less in the north. Its waters are dotted with more than 600 islands, some of which have been developed as tourist resorts. There are some 350 species of corals, and the reef is rich in colourful fish.

Map Australia Ga

Great Bear Lake ***Canada*** Fourth largest lake in North America (31 153 km², 12 028 sq miles). It lies in the Mackenzie district of the NORTH-WEST TERRITORIES and drains into the Mackenzie river. The lake, which is rich in fish, has a maximum depth of 413 m (1356 ft) and is frozen for about eight months of the year.

Map Canada Db

Great Belt (Store Bælt) ***Denmark*** Central channel through the Danish archipelago, with the islands of FUNEN and LANGELAND to the west and ZEALAND and LOLLAND to the east. It is 115 km (71 miles) long, 11-35 km (7-22 miles) wide and 20-58 m (66-190 ft) deep.

Map Denmark Bb

Great Bitter Lake ***Egypt*** See BITTER LAKES

Great Britain ***United Kingdom*** The island comprising ENGLAND, SCOTLAND and WALES.

Great Dividing Range ***Australia*** Mountain chain running almost the full length of eastern Australia from Queensland to Victoria. The range includes the SNOWY MOUNTAINS, BLUE MOUNTAINS and ATHERTON TABLELAND. The range's, and the continent's, highest peak is Mount Kosciusko, 2230 m (7316 ft).

Map Australia Gb, Hf

Great Dyke ***Zimbabwe*** A Precambrian intrusion of igneous rock forming a ridge stretching more than 500 km (310 miles) across the high plains of central and southern Zimbabwe. The ridge contains many important minerals including chrome, asbestos and gold.

Map Zimbabwe Ba

Great Escarpment ***South Africa*** The uptilted rim of the broad plateau forming most of southern Africa, with a steep drop reaching 1500 m (4921 ft) in the east. Its highest region is the DRAKENSBERG mountains.

Map South Africa Cc

Great Fish ***South Africa*** River rising in the east Cape Province and flowing south-east 640 km (397 miles) to the Indian Ocean. Its lower course forms the western border of the republic of Ciskei. Historically, the river was the frontier between European and Xhosa territory.

Map South Africa Cc

Great Glen ***United Kingdom*** See NESS, LOCH

Great Lakes ***North America*** Series of five lakes drained by the St Lawrence river to the Atlantic Ocean. Four – HURON, SUPERIOR, ERIE and ONTARIO – straddle the Canadian-US border. Lake MICHIGAN lies within the USA.

Map United States Ia

Great Plains ***North America*** Vast area of undulating grassland stretching south-east from the Mackenzie Mountains in north-west Canada to the US state of Texas; it includes the PRAIRIES. It is bordered by the foothills of the Rocky Mountains, and stretches eastwards almost to the Great Slave Lake and Lake Winnipegosis in Canada, and to the Missouri river in the USA. The Great Plains are up to 1000 km (625 miles) wide in the USA. They were the domain of buffalo and the Indian tribes until the middle of the 19th century when wagon routes, followed by railways, led to the colonisation of the area by the white man. Livestock, grain, oil and coal are the chief products today.

Map United States Fa

Great Rift Valley ***Middle East/East Africa*** Great depression running about 6400 km (4000 miles) from the Jordan valley through the Red Sea to Mozambique in south-east Africa (where it is sometimes known as the East-African Rift Valley). It consists of a series of geological faults caused by huge movements of the earth. For much of its length, its traces have been lost by erosion; in some parts, however, such as southern Kenya, cliffs rise thousands of metres. The Great Rift Valley floor reaches its highest levels in parts of central Kenya, where it rises to more than 1830 m (6000 ft); its lowest point is the DEAD SEA, whose surface lies 396 m (1299 ft) below sea level.

Map Middle East Bb; Africa Fe, Gd

Great Salt Lake ***USA*** Salt lake in central Utah, west of the Wasatch Range and north-west of Salt Lake City. Its area fluctuates according to the flow of the rivers emptying into it, reaching a maximum of about 5200 km² (2000 sq miles).

Map United States Db

Great Sandy Desert ***Australia*** Arid region in the north of Western Australia, stretching inland from the coast towards the state boundary with the Northern Territory.

Map Australia Cc

Great Sandy Island ***Australia*** See FRASER ISLAND

▲ **EMPEROR'S DUAL PURPOSE Under China's first emperor, the Great Wall kept out barbarians and was also a communications link; messages were passed by smoke signals in the day and by fire at night.**

Great Slave Lake ***Canada*** The deepest lake in North America (614 m, 2015 ft), covering 28 570 km^2 (11 030 sq miles) in the southern Northwest Territories. It is drained by the MACKENZIE river. Gold was discovered on the northern shore in the 1930s and the mining town of YELLOWKNIFE grew up.

Map Canada Db

Great Smoky Mountains ***USA*** Range in the Appalachian Mountains, straddling the border of North Carolina and Tennessee. They rise to 2025 m (6643 ft) at Clingmans Dome. The range is enclosed in a 2092 km^2 (807 sq mile) national park, which contains a wide variety of plants and wildlife.

Map United States Jc

Great Victoria Desert ***Australia*** Huge area of parallel sand dunes straddling the state borders of Western Australia and South Australia.

Map Australia Dd

Great Wall ***China*** The longest fortification in the world, winding a total of 3460 km (2150 miles) from Shanhaiguan, close to the gulf of Bo Hai, to Yumen in the Gobi Desert, with branches adding a further 2865 km (1780 miles). Many died constructing it. Most of the wall was completed in earth and stone by an army of conscript labour during the Qin dynasty (221-206 BC), but it was strengthened by later emperors. It was intended to protect China against nomads from the north, though it did not entirely prevent later invasions. Most of the present wall was constructed during the Ming dynasty (1368-1644). The best-preserved section is at Nankou Pass, 80 km (50 miles) north of Beijing (Peking). Parts of it are 9 m (30 ft) high and wide enough for a column of soldiers. Despite a popular belief that it is visible from the moon, it is not in fact visible from even a fraction of that distance. US astronaut Alan Bean exploded the myth on a lunar mission in 1969.

Map China Gd

Great Yarmouth ***United Kingdom*** Port and seaside resort at the mouth of the River Yare, 175 km (110 miles) north-east of London in the English county of Norfolk. It thrived as a port for 100 years before receiving a new boost as a centre for developing the southern North Sea natural gas fields in the 1960s.

Population 48 000

Map United Kingdom Fd

Great Zimbabwe ***Zimbabwe*** Historic site about 300 km (185 miles) south of the capital, Harare. It is the largest of about 150 dry-stone village sites in the country, which were occupied by the Shona-Karanga civilisation that flourished between about AD 1200 and 1450. The site is 28 km (17 miles) south-east of the town of Masvingo (formerly known as Fort Victoria).

Great Zimbabwe's buildings include a hilltop fort, and a circular temple with a conical tower 11 m (36 ft) high. The country took its name from the site.

Map Zimbabwe Cb

Greece See p. 266

greenhouse effect Term used to describe the heating of the earth's surface by the retention of infra-red radiation. When the ground gives off heat as infra-red radiation, some of it is absorbed by water vapour and carbon dioxide in the air and is radiated back to the ground, warming it further. Some scientists think that this could lead to the melting of the polar ice caps and the inundation of the low-lying – and most heavily populated – parts of the world.

Greece

LAND OF LEGENDARY BEAUTY AND OLD GLORIES, STRUGGLING TO MAKE ITS WAY IN THE MODERN WORLD

A 17th-century travel writer said of Greece: 'The mind is fascinated, the body faints, the eyes become moist and are delighted.' It is a legendary country which attracts lyrical prose. It has great natural beauty, a rich architectural heritage in monuments which are famous throughout the world, and priceless art treasures.

Some 2500 years ago, Greece was the centre of the Western world, where European civilisation and the concept of democratic government were born. Today, Greece is suffering economic hardship, and its famous ruins symbolise the 20th-century poverty of a once-great nation.

In between, Greece has had a tumultuous history. It flourished even under the Roman Empire, which ruled Greece from 146 BC, and even more in the Byzantine period from AD 330. But in 1460, seven years after the fall of Constantinople to the Turks, Greece became a Turkish province and remained under Turkish rule for nearly 400 years – a period that has coloured Greek politics and foreign affairs to this day. After an eight-year war (1821-9), Greece became an independent kingdom, but its present frontiers were not stabilised until after the Second World War.

AFTERMATH OF WAR

Greece sided with the Allies in the First World War, and won further territory from the Turks, who had sided with Germany. Italian and German troops occupied Greece during the Second World War, but were driven out by the British and Communist-led Greek resistance fighters in 1944. For years after the war, Communist guerrillas fought in the mountains against nationalist troops in a bloody civil war that cost 120 000 lives. American aid helped to ensure defeat of the guerrillas under General Markos, as did President Tito of Yugoslavia's quarrel with Moscow, which closed the Yugoslav border to support for the guerrillas.

Political instability after the civil war led eventually to a military coup in 1967 and the establishment of a republic in 1973. Meanwhile, relations with Turkey worsened over rights to oil discoveries under the AEGEAN SEA and the problem of CYPRUS, with its majority Greek and minority Turkish population. Turkey invaded northern Cyprus in 1974 and the military regime in Greece fell in a bloodless counter-coup. Democracy had returned to the country that invented the word. It is, however, a fragile democracy; elections are held every four years, but the Press is not entirely free.

Greece rejoined NATO (after a six-year gap) in 1980, and became a member of the European Economic Community in 1981. But the Socialist government in power since 1981 has maintained friendly relations with the Eastern Bloc and has threatened from time to time to withdraw from NATO and to close US military bases in Greece. At the root of the dispute is American friendship with (and NATO membership of) the old rival, Turkey.

Modern Greece has been described as 'a poor nation with all its ribs showing'. It is an evocative description, covering economics and geography. The ribs are the mountain ranges that dominate the country. The PINDHOS mountains divide Greece from the Albanian border in the north to the Gulf of CORINTH in the south, peaking at Smolikas (2637 m, 8651 ft). Another rib runs southwards from Yugoslavia, soaring to Mount OLYMPUS (2917 m, 9570 ft) and extending to the Aegean Sea.

About 70 per cent of the land is hilly, with harsh mountain climates and poor soils, but agriculture is the chief activity. Until 1950 most of the people were involved in subsistence farming, producing enough food for themselves but with no surplus for sale. There has since been a steady migration to the towns, but 30 per cent of the population still lives in the countryside.

Forests cover large areas in the highlands of MACEDONIA, THRACE and Pindhos. Woodland in lower areas has been reduced by forest fires during long, dry summers and by felling to provide fuel and clear land for pasture.

Dry Mediterranean vegetation covers southern and central Greece, with bushes and open woods. The island of CRETE has an astonishing variety of wild plants and flowers; there are more than 100 species with medicinal properties. Back on the mainland, Attiki is renowned for herbs and wild flowers which provide the nectar for excellent honey. Firs are common in the mountains; oaks, beeches and chestnuts lower down.

In coastal waters, sea sponges are an important crop in the DODECANESE islands, and sea anemones and urchins are a tourist attraction among the submerged rocks. Dolphins sometimes follow ships along the coasts.

FARMERS AND SAILORS

Large-scale agriculture is concentrated on the fertile plains on the eastern coasts. Crops include tomatoes, cotton and cereals in THESSALY, fruit trees in the river valleys of AXIOS and ALIAKMON in Macedonia, tobacco in the Strymon valley in Thrace, and vineyards, orange and olive groves in the PELOPONNESE and Crete.

Communications between most regions are poor because of distance and mountains. The sea unifies the country, and more than 89 per cent of imports and exports are carried by water. On land, the coastal roads are the main means of communication between the principal towns and the countryside. Greece was, in 1869, one of the last European countries to start building railways.

The IONIAN SEA in the west and the Aegean in the east contain 2000 Greek islands. Fewer than 200 are inhabited, but they contain 11.3 per cent of the population. The main tourist islands, such as CORFU, Crete, MIKONOS and RHODES, have international airports, while many others handle domestic flights only. The rest of the islands use PIRAEUS, the port of ATHENS, as their link with the outside world.

As the southernmost part of the BALKANS, the Greek peninsula is the most south-easterly extension of Europe – cut off by its Communist neighbours from West European countries, 1300 km (800 miles) away by road. Again, sea and air provide important links; Italy is less than 200 km (125 miles) away by sea.

Greek poverty sank to a modern nadir in about 1950, when most families were still farming small patches of poor-quality land. Industry consisted of small processing plants for tobacco, leather, cotton and foodstuffs. There was little money to develop natural resources. Over the next 25 years a development programme was pushed through, financed largely from Western Europe and the USA. Large irrigation schemes helped agriculture, and new power stations laid the foundations for industries such as shipbuilding, aluminium, steel, plastics, chemicals and electrical engineering.

But oil price rises and the world recession of the mid-1970s slowed development. Far too much manufacturing work is still carried out manually, and technology lags behind many other countries. So Greece remains relatively poor, with an income per head about half that of most of its partners in the EEC. Membership of the NATO military alliance increases the economic burdens of a poor country; the purchase of sophisticated armaments takes 8 per cent of the gross national product.

While tourism brings a steady income, the economy needs foreign exchange from other sources. Communist countries have been some of the biggest customers for Greece's agricultural products and the ship repair yards in SIROS, in exchange for electricity, telecommunications equipment and oil.

The merchant shipping business has enjoyed mixed fortunes in the last 40 years, but with a seafaring tradition stretching back to the mythical Odysseus, Greece today has the second largest merchant fleet after Liberia, with one-ninth of the world's tonnage. There are some very rich Greek shipowners, but many ships are owned by foreigners, using Greece as a tax haven and taking their profits out of the country.

PEOPLE AND POLITICS

Greek political views are intense, and electoral campaigns divide the nation. Patronage is rife, and victorious politicians are often expected to give their supporters rewards such as appointments in the civil service.

Domestic politics are concerned mainly with urban problems. Industrial development since 1950 has caused a rapid move to the towns and an explosion of housebuilding. The capital, Athens, has increased its population sixfold since 1945, and now has more than 30 per cent of the total population, 60 per cent of manufacturing capacity, and practically all government and business administration. More than half of the 10 million population lives in the six largest cities: Athens, SALONIKI, PATRAS, HERAKLION, VOLOS and Larisa.

People have moved from the rural areas

▲ CROWNING GLORY The Parthenon stands on the Acropolis or 'high city' in Athens. It was built as a temple to the goddess Athena in 438 BC, and was destroyed when Venetians besieged the city in 1687.

and from the islands, except those where tourism has become a source of income, as in Mikonos, Kos, Rhodes, Corfu and Crete. Many of the city immigrants are youngsters escaping from a life of hard labour on poor farms. There is not enough work for them all in the towns, but the hardships of unemployment are cushioned because Greeks live in extended families of two or three generations, all helping each other. Towns are becoming overcrowded, with many suburbs lacking sewerage and piped water. Health care and education are inadequate; there are not enough hospitals and schools.

Many more young people have emigrated, mostly to the USA, West Germany, Canada, Australia and Africa. Some return to Greece, having saved enough to buy a house, a car, perhaps to start a small business.

Greece attracts more than 5 million tourists a year, and three out of every four come from Western Europe. Most visitors go to Athens or the major islands. In some historic towns, handsome mansions and houses have been replaced by featureless tourist apartments, but most coasts have not been disfigured by high-rise flats. Tourists have recently discovered unspoilt islands in the Ionian Sea, such as LEVKAS, Paxos and ZAKINTHOS, and in the Aegean (Amorgos, Kea, PATMOS, Skopelos).

ISLANDS FOR TOURISTS

The Greek islands have a variety of rocks which give them distinctive characters: limestone in the Ionian islands, marble in Skiros, PAROS, Ivios, Anafi and TINOS, granite in Mikonos, pumice and lava in SANTORINI. Some islands are waterless and bare like Patmos, others have picturesque wooded hills like AEGINA or sheltered and watered valleys like CEPHALONIA.

The Ionian islands of Corfu and Levkas are known for their green landscape, Zakinthos for its fragrant gardens and ITHAKA for its miniature harbours. Most of the CYCLADES are barren, with white houses and windmills. CHIOS, LESBOS, SAMOS and Ikaria are fertile with abundant springs and thick woods. Amorgos and Santorini have dramatic and colourful cliffs. Set in transparent seas, Rhodes and Crete are sprinkled with beautiful historic towns surrounded by orchards, vineyards, fields of wheat and tree-clad mountains.

Rural settlements are dotted over much of the mainland and the islands. Most houses are built of stone and timber, with tiled roofs, wooden pillars and balconies hanging over narrow, sloping streets. Inland villages develop around the *plateia*, the lively central square; the centre of most coastal towns is the harbour.

The mountain ranges of central and southern Greece form a mass of jagged, rocky prongs, and descend through craggy ridges to sea-girt peninsulas and promontories, as in the southern Peloponnese and CHALCIDICE. Deep gulfs and inlets penetrate the land, so that no place in Greece is more than 70 km (45 miles) from the sea. The narrow coastal plain meets the sea in a fretwork of bays, some of which are inaccessible from land. Elsewhere, gulfs are bounded by mountains cradling tiny harbours and fishing villages, or bordered by pine-clad slopes.

Each season lends its own charm to the landscape. Spring comes early in April and myriads of Mediterranean plants burst into a blaze of colour. The hot season starts in May and builds to an average summer temperature of 27°C (81°F), with heatwaves reaching 40°C (104°F). Cool breezes may temper the heat on the coasts and the islands, but the clear air and bright sunshine bring out all the beauty of the landscape. Winters are warm and wet in the lowlands, but there is snow in the mountains.

While the Greek summer is a great tourist attraction, the country's rich history brings visitors all year. The grandeur of ancient civilisations is recalled by spectacular ruins. Most are in the central and southern parts of Greece and the islands. The Bronze Age (2000-1100 BC) produced the splendid Minoan palaces of KNOSSOS, Faistos and Mallia in Crete, the citadels of MYCENAE and Tirins, and the city of ancient Thira on Santorini. Athens contains the most important group of architectural landmarks of the Classical period (500-30 BC) at the Acropolis.

The ancient sites of DELPHI and OLYMPIA and the ruins of the sacred city of DELOS are great tourist attractions. The theatre of EPIDHAVROS, which overlooks the Argolic plain, and the Roman Odeon of Herodes Attikos at the foot of the Acropolis are used for international festivals of drama and dance.

Byzantine monuments include elegant churches and monasteries decorated with mosaics. Saloniki, which was the second city of Byzantium, has a rich heritage of these stylish churches, and the monasteries of DAPHNI and Osios Loukas are famous for their art treasures.

Venetian and Frankish occupation early in the 13th century enriched Greece's capes and islands with picturesque castles and walled cities such as MONEMVASIA, NAXOS, Koroni and, most beautiful of all, Rhodes. Museums contain magnificent collections.

COUNTRY AND TOWN LIFE

Greeks are traditionally associated with folk dancing and singing, often in picturesque national dress. Their biggest folk festivals are linked with the Greek Orthodox religion, which is followed by 98 per cent of the people. Easter Sunday, especially, is a day of great rejoicing when families gather in town and village squares to sing and dance, and roast lambs on spits. (It is often a different day from that celebrated by the Western churches.) On saints' days, fairs are held outside monasteries and churches. There are two national holidays in honour of the Virgin Mary – on March 25 (which coincides with Independence Day) and August 15, when there is a national pilgrimage to the monastery on the island of Tinos.

The people of Greece are united by religion and by the modern Greek language of *dhemotike*, derived from the ancient language and spoken by almost everyone. An artificial language, *katharevousa*, was introduced for use in government documents and by officials, but is now dying out. Dialects survive in many rural areas, but Ancient Greek is incomprehensible to the layman, and is taught only in high schools and universities.

Rural life is simple, and recreation is mostly confined to Sundays, feast days and weddings. Older people pass the time sitting and talking in the village square. Traditional costume is still worn by older people on festive occasions.

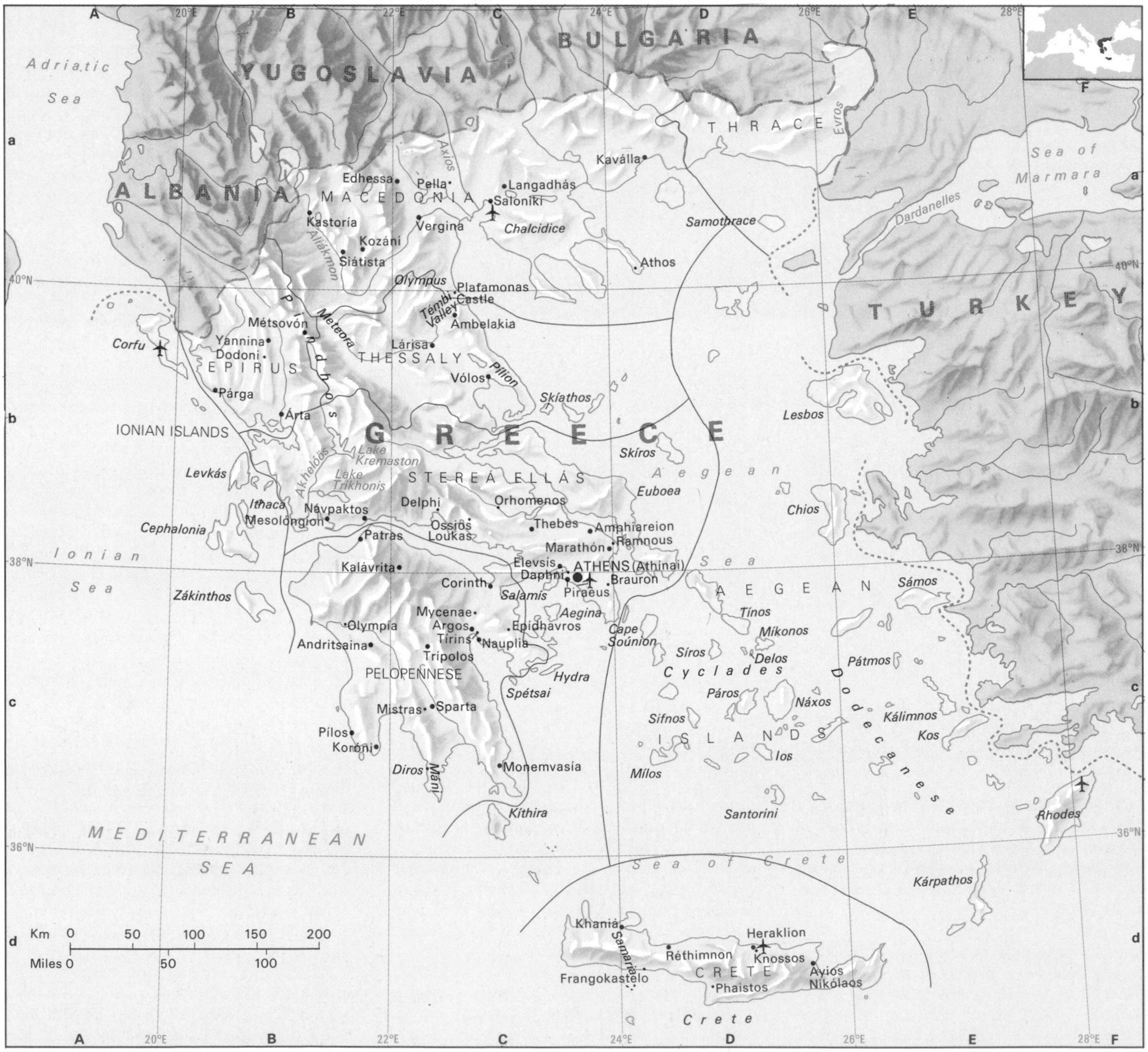

▲ VIOLENT BIRTH A chapel overlooks the tranquil harbour of the island of Santorini (Thira) in the Aegean Sea. The bay formed when the volcanic island blew apart about 1400 BC. The resulting shock wave, sea wave and hail of ash devastated much of Crete, 120 km (75 miles) to the south.

There are great regional variations in embroidery, colours and patterns; old handwoven costumes, coloured with plant dyes, are valued by collectors. Handwoven carpets and blankets are still made in Thrace, Crete and central Greece, assisted by state subsidy.

In the cities and towns, social life for older people is centred on the cafés; eating out at *tavernas* is still a widespread custom for those who can afford it. In the few traditional tavernas that survive, one can still see men dancing by themselves – the *zeimbekiko*. Poorer city families eat at home, but the meal is an important family gathering.

When people from rural areas move to the towns, their values and family behaviour change little. A village girl may use her parents' savings to buy a flat in Athens as a step towards marriage. But she will probably marry a country boy who is using his family's resources to study in Athens and seek a job in the civil service. Many families deprive themselves to send their children to live in Athens. In return, children tend to work hard at school, and are disciplined and obedient at home. Young people usually need their parents' permission to go out in the evenings, take a job or get married. The father is the head of the family and makes decisions for all its members. This is a rural lifestyle, but it is only slowly dying out in urban environments.

Women suffer discrimination in many ways. A bride is expected to provide a dowry, perhaps a flat or the equivalent cash. After marriage, a wife must obey her in-laws as well as her husband. In many rural areas, single girls are not expected to appear in public unless accompanied by a male relative. Married women are not supposed to go to work, have guests at home, or meet friends outside, without their husbands' permission. Many family businesses use their womenfolk as unpaid labour, leaving them with most of the work while the male head of the family idly sits on the pavement or a balcony.

It is part of the Greek character to 'watch the world go by'. The country faces a huge task in trying to create a modern industrial economy while tradition is dying so hard.

GREECE AT A GLANCE
Area 131 986 km² (50 960 sq miles)
Population 10 000 000
Capital Athens
Government Parliamentary republic
Currency Drachma = 100 lepta
Language Greek
Religions Christian (98% Greek Orthodox), Muslim (1%)
Climate Mediterranean; average temperature in Athens ranges from 6-13°C (43-55°F) in February to 23-33°C (73-91°F) in July
Main primary products Wheat, tobacco, cotton, sugar beet, olives, grapes, fruit, vegetables; iron, bauxite, nickel, zinc, lead, lignite, chrome, marble, magnesite, barytes, salt, crude oil
Major industries Agriculture, textiles, cement, fertilisers, chemicals, steel, tobacco products, food processing, aluminium smelting, consumer products, wine, shipbuilding, tourism
Main exports Fruit and vegetables, petroleum products, textiles, clothing, iron and steel, tobacco, cement, aluminium, chemicals, metal ores
Annual income per head (US$) 3600
Population growth (per thous/yr) 6
Life expectancy (yrs) Male 71 **Female** 75

Greenland

THE WORLD'S CHILLIEST NATION, KNOWN TO ITS INHABITANTS AS KALAALLIT NUNAAT, IS GOING OUT INTO THE POLITICAL COLD TO CARVE ITSELF A NEW POLAR ROLE

On February 1, 1985, the European Common Market lost half of its territory when Greenland became the first nation to leave the EEC since the community's foundation in 1958. The gesture, by the people of the world's largest island, summed up the determination of Greenlanders to go it alone economically as well as politically.

The concertina-playing Lutheran priest who steered the island out of the EEC – Prime Minister Jonathan Motzfeldt – also wants Greenland to cut its remaining political ties with Denmark, ending an association that dates back to the 14th century, when Norway and its possessions, including Greenland, came under Danish rule.

The island, which its inhabitants call Kalaallit Nunaat (Land of the Greenlanders), has had home rule since 1979, but is still nominally part of Denmark. The people are a mixture of Eskimo and Danish extraction. 'We just don't belong to Europe', says Motzfeldt. 'We are a polar people. Our ties are with the Eskimos of Canada, Alaska and Siberia. We want to make our own future.'

The task will not be easy. The rocky and mountainous island is nearly 50 times the size of Denmark – with only one-hundredth of its population. All but one-seventh of its vast area is permanently covered with an ice cap whose average depth is 1500 m (nearly 5000 ft). Crop-growing on any scale is impossible, and the only livestock farming of any commercial consequence is sheep-farming. All but the southern end is inside the Arctic Circle. And even near the southern tip, at the capital GODTHAB (or Nuuk, as Greenlanders call it), the sun does not rise until 10 am in winter, and it sets again at 2 pm.

Godthåb, with a population of 10 500 – about the size of a small British town – is home to nearly one Greenlander in five. Like the other major towns on the island, it is on the west coast, warmed by southerly currents from the Atlantic. By contrast, a northerly current from the Arctic Ocean chills the east coast so that pack ice makes it inaccessible for most of the year.

PROBLEMS OF MODERN SOCIETY

The capital's inhabitants, like other urban Greenlanders, live uneasily between the modern and primitive. Secure jobs in fish factories – Greenland's only major manufacturing industry – have drawn many Eskimos to the towns. Families that would once have wandered the ice as nomadic hunters with harpoons, kayaks, huskies and rifles now get their food from supermarkets kept stocked by freezer ships from Denmark. But these well-meant developments have carried a price. Greenland's suicide rate is the highest in the world, approximately one per thousand, and alcoholism is frighteningly common. Moreover, a high degree of sexual freedom – partly a legacy of Eskimo tradition – has given the capital a high rate of venereal disease.

▲ SILENT BEAUTY Huge, desolate ice-fields sweep down to seas around Greenland. Sparkling icebergs like this one were once the 'snout', or seaward end of a glacier, which has broken off in a process called 'calving'.

The island's leaders are confident that Greenland can survive on its own. It has fishing grounds producing abundant catches of cod and shrimp. It exports lead and zinc from a rich mine near Uummannaq on the north-western coast; it has untapped reserves of uranium in the south; and some oil companies believe there may be oil reserves beneath the eastern coast. In addition, it benefits economically from two American bases at SONDRE STROMFJORD and THULE, part of the USA's nuclear early warning system. There is also a growing tourist trade. Holidaymakers are flown by helicopter to villages along the west coast (there are few roads in Greenland outside the towns) to sample the sea-fishing, or to explore on foot and sledge the fringes of the island's empty and scantily mapped interior.

There is some precedent for the islanders' confidence. The first European to visit the island was a Norseman named Eirikr Raudi (Eric the Red), who sailed from Iceland in AD 982. He was sufficiently confident about its future to make his home there; his farmstead, at Brattahlid, is now a tourist attraction. Eirikr also named the island Greenland – as a means, he admitted later, of making it sound more attractive to potential settlers. His optimistic strategy worked. In 986, 35 small shiploads of pioneers followed him to Greenland, founding a Norse colony which lasted for 500 years.

GREENLAND AT A GLANCE

Map Arctic Cb
Area 2 175 600 km² (840 000 sq miles), of which 341 700 km² (131 930 sq miles) are ice free
Population 54 600
Capital Godthåb (also known as Nuuk)
Government Self-governing part of Denmark
Currency Danish Krone = 100 øre
Languages Greenlandic (Eskimo) and Danish
Religion Christian (Evangelical Lutheran)
Climate Polar; average temperature in the north (Thule) ranges from −23°C (−9°F) in February to 5°C (41°F) in July, and in the south from −8°C (18°F) to 10°C (50°F)
Main primary products Fish, sheep; zinc, lead
Major industries Fishing, mining, fish and food processing, hides and skins
Main exports Fish and fish products, cryolite (used in making aluminium), zinc, lead
Annual income per head (US$) 9600
Population growth (per thous/yr) 12
Life expectancy (yrs) Male 60 **Female** 67

Greenland Sea Southern arm of the ARCTIC OCEAN lying north of Iceland and north-east of Greenland. Pack ice covers large areas and the East Greenland Current carries ice southwards.
Map Arctic Ocean Ab

Greenwich ***United Kingdom*** See LONDON

Greenwich meridian Line of LONGITUDE through Greenwich, England, taken as the prime or zero meridian – 0° east and west – for world-wide navigation and timekeeping.

gregale Strong blustery north-east wind, sometimes accompanied by showers, blowing in the central Mediterranean during the winter, when a high atmospheric system develops over the Balkans with low pressure over the Sahara.

Greifswald ***East Germany*** Baltic port north of the capital, East Berlin. It was founded by Dutch merchants in 1240 and today is the country's smallest university town. To the east is Peenemünde, the site where V-rockets were developed during the Second World War.
Population 63 900
Map East Germany Ca

Grenadines ***West Indies*** Group of about 600 small islands in the WINDWARD group between ST VINCENT and GRENADA. The largest is CARRIACOU. The northern part of the group belongs to St Vincent, the southern part to Grenada. The Grenadines are a yachtsman's paradise, and tourism has brought some prosperity.
Population 14 000
Map Caribbean Cc

Grenoble ***France*** Commercial and manufacturing city about 100 km (60 miles) south-east of Lyons. The writer Stendhal (1783-1842), author of the novel *The Red and the Black* (1830), was born there. It manufactures textiles, chemicals and electrical goods, and has a large university. In the 19th century Grenoble industrialists pioneered the generation of hydroelectricity.
Population 396 800
Map France Fd

greywacke Dark, very hard sandstone, consisting of angular rock particles, up to 2 mm (0.08 in) in diameter, cemented together.

Grijalva ***Mexico*** Major river 480 km (300 miles) long, flowing from the highlands of CHIAPAS state northwards across the humid lowlands of TABASCO state into the Gulf of Campeche. It is also known as the Tabasco. There are several hydroelectric power projects along its course – including the Nezahualcóyotl Dam in Chiapas, the largest in Mexico. The vast river basin is a rich agricultural area.
Map Mexico Cc

grikes, grykes Deep fissures caused by the widening of joints by solution. They develop on horizontal limestone surfaces between CLINTS and are commonly found in the Pennines in northern England, the eastern Alps, the Yugoslavian karst and many other limestone areas.

Grimsby ***United Kingdom*** Fishing port on the south shore of the Humber estuary, 24 km (15 miles) south-east of Kingston upon Hull in the English county of Humberside. Flemish fishermen began using the port in the 15th century, but new harbour works and railway links to the rest of the country greatly boosted it in the 19th. It became one of the world's great fishing ports, with a fleet of some 250 vessels landing up to 200 000 tonnes of fish each year. However, it was badly hit by the 'cod war' fisheries dispute with Iceland in the 1970s and today's fleet is much smaller.
Population 92 000
Map United Kingdom Ed

Grenada

A BEAUTIFUL ISLAND OF SPICES WHERE MARXISTS FELL OUT AND TROOPS WENT IN

The spice island of Grenada in the Caribbean made world headlines in 1983, when United States marines and troops from other Caribbean countries invaded the country to topple a Marxist government. Grenada had moved to the left after a bloodless coup in 1979. The new prime minister, Maurice Bishop, established links with Cuba, Libya and Algeria. The invasion came after he and three Cabinet ministers were murdered by rival Marxists who then seized power. Since then, much foreign aid has been directed to the island in an effort to ensure both a stable elected government and a sound economy. The first elections for eight years were held in 1984.

The most southerly of the Windward Island chain, Grenada is one of the most beautiful West Indian islands, with an attractive wooded landscape. The island consists of the remains of extinct volcanoes, the youngest and highest of which is Mount St Catherine (840 m, 2756 ft). Two of the old volcanic craters have become the Grand Etang and Antoine lakes. The only flat land is along the lower courses of the rivers and along part of the coast.

Grenada was sighted by Christopher Columbus in 1498. The first serious attempt at settlement in 1609 was prevented by the Carib Indians who lived there. The French established a settlement in 1650, wiping out the Caribs, but Grenada was captured and ceded to Britain in 1783. With the small neighbouring islands of CARRIACOU and Petit Martinique, it became an independent member of the Commonwealth with the Queen as head of state in 1974.

Agriculture is the island's main industry, with cocoa, nutmegs, bananas and mace the chief exports. Most manufacturing is concerned with the processing of crops, but there are factories which brew beer and make clothes and furniture. About 30 per cent of the work force have no full-time job.

Tourism is an important source of foreign income. The wooded hillsides, the long coral beach of Grand Anse and the beautiful setting of the capital, ST GEORGE'S, are the main attractions. Much of the town was rebuilt after a hurricane destroyed it in 1955. Its harbour has berths for ocean-going ships.

Cruise ships call at St George's, but the number of visitors used to be restricted by lack of an international airport. The Cubans built a long runway for the new Point Salines airport, causing fears that it would be used by Communist military aircraft. This was one reason for the invasion in 1983. The airport, which was financed by the USA, was opened in 1984 and can handle long-haul flights, increasing tourist potential.

Carriacou is the largest of the small GRENADINE islands north of Grenada, most of which belong to ST VINCENT. Tourism and boatbuilding are the main activities. Petit Martinique is uninhabited.

GRENADA AT A GLANCE
Map Caribbean Cc
Area 344 km² (133 sq miles)
Population 87 600
Capital St George's
Government Parliamentary monarchy
Currency East Caribbean dollar = 100 cents
Languages English, French patois
Religion Christian (64% Roman Catholic)
Climate Subtropical; average temperature in St George's is 26-28°C (79-82°F) all year
Main primary products Cocoa, nutmegs, mace, bananas
Major industries Agriculture, food processing, tourism
Main exports Cocoa, bananas, nutmegs, mace
Annual income per head (US$) 890
Population growth (per thous/yr) Declining
Life expectancy (yrs) Male 60 **Female** 65

Grindelwald ***Switzerland*** Resort town at 1393 m (4570 ft) altitude and some 55 km (35 miles) south-east of the capital, Berne. Its closeness to the Bernese Oberland – the EIGER, Mönch and JUNGFRAU tower over it – has made it popular with tourists for more than a century.
Population 3600
Map Switzerland Ba

Grisons ***Switzerland*** See GRAUBUNDEN

grit, gritstone Coarse sedimentary rock, usually sandstone, composed of angular grains with a diameter of 0.06-2 mm (0.0025-0.08 in). It occurs widely in the Pennines in northern England and elsewhere.

Grmec Planina ***Yugoslavia*** Wild, forested limestone massif in western Bosnia between the Sana and Una rivers, rising to 1604 m (5262 ft). A winding road links the ancient fortified towns of Bihac at the north-west end and Kljuc to the south-east, through an area noted for bullfights and rich in monuments to the partisans who fought there during the Second World War.
Map Yugoslavia Cb

Grodno ***USSR*** Industrial city in the lowlands of Belorussia beside the Polish border. It dates from the 12th century and was the capital of

Lithuania in the 13th century. Its main products are electrical equipment, textiles, leather, tobacco and food.
Population 239 000
Map USSR Dc

Groningen ***Netherlands*** Regional centre for the north-east, 145 km (90 miles) north-east of Amsterdam. Its medieval moats and fortifications have been converted into canals and parks, and there are several old squares, museums, and a university dating from 1614. The city's manufactures include printed books and sugar. It is the capital and only town of any size of the farming province of Groningen. This has a land area of 2593 km² (1001 sq miles), including five of the Friesian Islands, and produces sugar beet, cereals, potatoes, dairy goods and horses.
Population (city) 166 900; (Greater Groningen) 205 700; (province) 560 700
Map Netherlands Ca

Grootfontein ***Namibia*** Cattle-rearing district around a town of the same name. The town was founded in the 1890s around a German garrison. It stands some 340 km (210 miles) north-east of the capital, Windhoek. The world's largest known meteorite, weighing an estimated 59 tonnes, was found in 1920 on Hoba West farm, about 20 km (12 miles) west of the town. The meteorite, a mass of iron and nickel, was 2.75 m (9 ft) long and 2.43 m (8 ft) wide. The district covers 26 520 km² (10 239 sq miles).
Population (district) 21 900; (town) 7600
Map Namibia Aa

Grossglockner ***Austria*** See HOHE TAUERN

grotto A natural or man-made cave.

ground frost Term used to indicate a ground level temperature of 0°C (32°F) or below, while the air temperature remains above freezing point.

ground water 1. Any water beneath the earth's surface. 2. Region of subsurface water beneath the WATER TABLE, including underground streams. It forms the saturation zone in which all pore spaces and fissures are filled with water.

groyne Narrow jetty or wall built out into the sea from the shoreline to arrest the movement of waterborne sand or gravel along the shore and thereby to protect beaches from erosion.

Groznyy ***USSR*** Oil city in the north Caucasus foothills, 145 km (90 miles) west of the Caspian Sea, and capital of the Chechen-Ingush Republic. It was founded in 1818 as a fort and developed after the discovery of oil in 1893. It has oil pipelines to the Caspian port of Makhachkala and to Tuapse on the Black Sea. Its other industries include petrochemicals, textiles and processed food.
Population 389 000
Map USSR Fd

Grudziadz ***Poland*** Northern industrial town about 100 km (60 miles) south of the port of Gdansk. It was developed as a fortress by the Teutonic Knights in 1291. Today its factories make metals, beer, footwear, rubber and farm machinery, and centuries-old riverside granaries are still in use.
Population 92 800
Map Poland Cb

Grünberg ***Poland*** See ZIELONA GORA

Grunwald (Tannenberg) ***Poland*** Battlefield near the village of Stebark in northern Poland, about 140 km (85 miles) south-east of the port of Gdansk. On July 15, 1410, a Polish-Lithuanian-Tatar force led by King Wladyslaw Jagiello defeated the Teutonic Knights there. And in August 1914, during the First World War, German troops defeated Russian forces. A granite obelisk and a museum mark the site.
Map Poland Db

Gruziya ***USSR*** See GEORGIA

Gstaad ***Switzerland*** Fashionable skiing resort 1050 m (3444 ft) up in a valley about 55 km (35 miles) south and slightly west of the capital, Berne. It is also a popular summer resort for tourists visiting the Bernese Oberland.
Population 2500
Map Switzerland Aa

Guadalajara ***Mexico*** Capital of the state of Jalisco. It lies on a plateau 1580 m (nearly 5190 ft) above sea level, and is the main industrial, commercial and agricultural centre of the western highlands. In colonial times it was reputed to be 'more Spanish than Spain itself'. Much of its colonial heritage survives; including parks, fountains, squares, churches, a majestic 16th-century cathedral and the very ornate government palace. There is also an 18th-century university. The city has a gracious, provincial atmosphere, despite its skyscrapers and its factories which make, for instance, shoes and textiles. It is also noted for its crafts: leatherwork, embroidery, glass, weaving and ceramics.
Population 2 300 000
Map Mexico Bb

Guadalajara ***Spain*** Market town about 50 km (30 miles) north-east of Madrid, trading in cereals, wine and olives produced on farms in the surrounding province of the same name.
Population (town) 57 000; (province) 143 100
Map Spain Db

Guadalcanal ***Solomon Islands*** Island in the south of the group on which stands the Solomon Islands' capital, HONIARA. It was the scene of bitter fighting between Japanese and US forces from August 1942 to February 1943. This battle was one of the turning points of the Second World War. The island has extensive rice paddies and plantations of oil palms.
Population 63 000
Map Pacific Ocean Cc

Guadalquivir ***Spain*** River, 580 km (360 miles) long, flowing south-west into the Gulf of Cádiz on the southern Atlantic coast.
Map Spain Cd

Guadalupe Mountains ***United States*** Range on the southern border of New Mexico with Texas, reaching 2667 m (8751 ft) at Guadalupe Peak. Guadalupe Mountains National Park contains important Permian limestone fossil beds.

Guadeloupe ***West Indies*** Group of islands in the Leeward chain in the eastern Caribbean. They were discovered by Christopher Columbus in 1493, but the native Carib Indians (who called the main island Karukera) resisted Spanish would-be colonists. Eventually, in 1635, French settlers occupied the islands and apart from short periods of British occupation (and one of Swedish control), Guadeloupe has remained under French rule. In 1946 it became an overseas department of France, with full representation in the National Assembly.

The main island of Guadeloupe is made up of two sections separated by a narrow strait, Rivière Salée (Salt River), about 6 km (3.5 miles) long. The western section, Basse Terre, is one of a chain of volcanic islands in the eastern Caribbean. The eastern section, Grande Terre, is part of an arc of limestone islands. Basse Terre contains Soufrière, an active volcano 1467 m (4813 ft) high that is the highest point in the eastern Caribbean. The other islands in the group are nearby MARIE GALANTE, La Désirade and the tiny Iles des Saintes, plus ST BARTHELEMY and ST MARTIN – both more than 200 km (125 miles) north-west. In all, the islands total 1702 km² (657 sq miles).

Agriculture is important and the main crops are bananas, sugar cane and aubergines. The chief industries are sugar refining and rum distilling. The year-round sunshine, good sea bathing and beautiful mountain scenery attract many tourists – especially from France and the United States. Most of the islands' inhabitants are black or mulatto, and are descended from African slaves. However, the populations of St Barthélémy and of Terre de Haut in the Iles Des Saintes are descended from 17th-century settlers from northern France. French is the official language throughout the islands, but many people speak a Creole dialect.

The seaport of BASSE TERRE is the capital of Guadeloupe, but the largest town is the commercial centre and port of POINTE-A-PITRE on Grande Terre.
Population 328 400
Map Caribbean Cb

Guadiana ***Spain/Portugal*** River, some 820 km (510 miles) long, which rises south-east of the Spanish capital, Madrid. It forms parts of the southern Spanish-Portuguese border and flows into the Atlantic at the Gulf of Cádiz.
Map Spain Cc

Guaíra Falls ***Paraguay*** See ITAIPU

Guajira ***Colombia*** Department with a thinly populated, largely desert peninsula jutting into the Caribbean Sea in the extreme north of the country. The capital is Ríohacha, a treeless city and port about 220 km (140 miles) north-east of Barranquilla. Marijuana is widely, though illegally, grown in the area and smuggled out for sale in North America. Deposits of coal and natural gas are being developed.
Population 284 000
Map Colombia Ba

Guam ***Mariana Islands*** Largest and southernmost of the Mariana Islands, covering 549 km² (212 sq miles) and lying about 2250 km (1400 miles) south-east of Okinawa, Japan. Inhabited by Micronesian people for at least

3000 years, it was discovered by Ferdinand Magellan in 1521 and claimed by Spain in 1565. Half today's population is Chamorro – of mixed, mainly Filipino, Indonesian and Spanish descent. Guam became American territory after the Spanish-American War in 1898 and, apart from Japanese occupation in 1941-4, has remained so. It has internal self-government, but a 1982 referendum favoured Commonwealth status like that of the neighbouring NORTHERN MARIANAS.

Over one-third of the land is a US military base, and this dominates the local economy, ensuring a far higher standard of living than elsewhere in Micronesia. Nevertheless, in 1979 voters overwhelmingly demanded a return of much of this land to civilian use. The second major source of income is tourism. Guam is a transport hub of the western Pacific and is second only to Hawaii as a Pacific tourist destination; two-thirds of the visitors come from Japan.

Although parts of the island are highly developed, American-style, there are also dense jungles (in one of which a Second World War Japanese soldier was found still hiding in 1972), beautiful beaches (some with crashing surf) and sleepy South Sea island villages – as well as architectural relics of Spanish rule. Some copra is produced, but most food is imported.

Population 112 000

Map Pacific Ocean Cb

Guanajuato ***Mexico*** Capital of Guanajuato state, and one of the most striking colonial towns of central Mexico. Its picturesque houses and winding, cobbled streets nestle in the bottom of a narrow gorge in the mountains of the Sierra Madre. Founded in 1554, Guanajuato soon became a boom town when silver was discovered in the surrounding countryside. By the end of the 17th century it was the richest city in Mexico. The legacy of its prestigious past now includes the gleaming white university, the ornate Juarez Theatre, and the Alhondiga de Granaditas – an old grain warehouse converted into a museum dedicated to the heroes of the Independence movement. Guanajuato provides a charming backdrop to the annual Cervantes Festival – a three-week international event of the performing arts.

Population (state) 3 006 000; (town) 45 000

Map Mexico Bb

Guangdong ***China*** Mountainous province in the south covering 210 000 km^2 (81 000 sq miles). Part of it lies within the tropics, and the growing season in the valleys of the Xi, Bei and Dong rivers and in the delta of the Zhu Jiang lasts all year. Farmers can grow three crops of rice annually, as well as sugar cane, sweet potatoes and tropical fruits. The population is concentrated in the cities, of which the capital, Guangzhou (Canton), is the largest. Guangdong is an increasingly prosperous province. Its economy has benefited from recent changes in Chinese agricultural policy allowing farmers to sell their surplus at a profit instead of growing it only for the state, and from increased trade with nearby Hong Kong and Macau. It will get a further fillip if oil exploration offshore is successful.

Population 56 810 000

Map China Gf

Guangxi-Zhuang ***China*** Autonomous region on the north of the Gulf of Tongking bordering Vietnam. The interior is hilly and agriculture is confined to narrow river valleys and the coast. The region covers 230 000 km^2 (890 000 sq miles) and straddles the Tropic of Cancer. Two crops of rice are grown each year, as well as sugar cane, tobacco and tropical fruits. Several ethnic minorities closely linked to the peoples of South-east Asia live there, including 12 million Zhuang, who are related to the Thais. The regional capital is Nanning.

Population 34 700 000

Map China Ff

Guangzhou (Canton) ***China*** Capital of Guangdong province and the country's sixth largest city. It lies in the fertile delta of the Zhu Jiang river. It rose to prominence as a maritime trade centre during the Han dynasty (206 BC-AD 220). Portuguese traders arrived in the city in 1514 and were followed by British merchants in the 17th century. Later, the city became a centre for the opium trade, encouraged by Britain but outlawed by the Chinese government, whose attempt to suppress the traffic in the city in 1839 led to the Opium War with Britain. After three years the Chinese were defeated and opium traders flourished. Today, Guangzhou is one of China's most cosmopolitan cities. Its industries include food processing, paper making, chemicals and textiles.

Population 5 350 000

Map China Gf

guano The droppings of seabirds or bats, accumulated along dry coastal areas or in caves. It is a rich source of phosphate of lime, used in fertiliser.

Guantánamo ***Cuba*** Town in the extreme south-east, 65 km (40 miles) east of Santiago de Cuba. It is the commercial centre and capital of an agricultural province, with sugar-milling, coffee-roasting, tanning and confectionery industries. Guantánamo Bay, on which the town stands, is the site of a United States naval base.

Population (city) 205 000; (province) 469 000

Map Cuba Ba

Guatavita la Nueva ***Colombia*** New town about 75 km (45 miles) north-east of Bogotá. The town was built in colonial style to replace an older town submerged by a hydroelectric scheme in 1967. Near the town – in the crater of an old volcano – is Lake Guatavita, source of the legend of El Dorado – 'The Gilded One'. Chibcha Indians lived there before the Spaniards arrived in the 16th century, and once a year the ruler was coated in resin and gold dust and then leapt into the lake from a raft, as an offering to tribal gods. He would emerge cleansed of the gold.

Gold objects were also thrown into the lake as offerings, and several unsuccessful attempts have been made to drain it to find the treasure. It was this story that lured the conquistadores into the heart of South America. A gold model of a raft, found in the lake, is in the museum at Bogotá.

Population 15 000

Map Colombia Bb

Guatemala See p. 275

Guatemala City ***Guatemala*** National capital and the largest city between Mexico and Colombia. It sprawls across a plateau 1500 m (4920 ft) high, about 80 km (50 miles) inland from the Pacific Ocean. It is ten times larger than any other town in Guatemala, and is responsible for half the country's industrial output.

The city was founded as Guatemala's third capital in 1776 after earthquakes had destroyed the earlier capitals of Antigua and Ciudad Vieja. It was itself virtually destroyed in 1917-18, by earthquakes which lasted on and off for six weeks, and was again seriously damaged by an earthquake in 1976.

Today Guatemala City is a modern, commercial city with good shops and a colourful market. The National Archaeological Museum has an impressive collection of finds from Maya sites. The city's university of San Carlos is one of the oldest in the Americas, founded in 1676. Its manufactures include textiles, foodstuffs and silverware.

Population 1 329 600

Map Guatemala Ab

Guayaquil ***Ecuador*** Capital of Guayas province, on the Guayas river near its mouth on the Gulf of Guayaquil. It is the country's largest city and chief port, whose exports include bananas, cocoa, sugar, coffee and cattle. Its manufactures include textiles, leather goods, cement and iron products.

The original European settlement, which was destroyed by Indians, was founded just east of the present site by the Spanish conquistador Sebastián de Belalcázar in 1535. In 1537, the Spanish explorer Francisco de Orellana established the present town. It was destroyed by earthquake in 1942, since when much of the town has been rebuilt.

In 1822 Guayaquil was the scene of the fateful meeting between Simón Bolívar and José de San Martín, the principal leaders of the struggle for South American independence from Spain. The two leaders disagreed on policy and San Martín retired from active service in the wars of liberation, leaving Bolívar to complete the struggle alone.

Population 1 223 500

Map Ecuador Bb

Guayaquil, Gulf of Large bay on the Pacific coast of Ecuador which was visited in 1526 by the Spanish conquistador Francisco Pizzaro. In the American Indian town of TUMBES on the south shore he found a temple covered with gold. This confirmed his belief that South America was rich with treasure, and led to the Spanish conquest of the region. The two main ports on the bay are Guayaquil and Puerto Bolívar.

Width 180 km (112 miles) at the mouth

Map Ecuador Ab

Guayas ***Ecuador*** Coastal province which produces more than half the country's bananas, as well as rice, sugar cane, cotton and, farther inland, coffee and cocoa. Most of the produce is taken to the provincial capital, GUAYAQUIL, by road or rail, but some is carried by small boats on the Guayas river. The Guayas River Agency, covering the whole river basin – and therefore parts of nine provinces – is developing an irrigation project at Daule which will reduce the danger of seasonal flooding in the valley. There are substantial oil and natural gas deposits in the river basin.

Population 2 117 000
Map Ecuador Ab

Gubbio ***Italy*** Well-preserved medieval town about 160 km (100 miles) north of Rome. It has a church and cathedral, both of the 13th century, and an elegant 15th-century ducal palace. Its chief product is ceramics.

Population 32 000
Map Italy Dc

Guelderland ***Netherlands*** See GELDERLAND

Guernica ***Spain*** Town 19 km (12 miles) east of the northern port of Bilbao and known as the 'holy city' of the Basque people. Their parliament met there – under an oak tree – from the Middle Ages until the 19th century.

The town was destroyed in 1937 by German bombers supporting the Nationalists in the Civil War, and became a symbol of the war's horrors when it inspired Pablo Picasso's painting, *Guernica*, now at the Prado Museum in Madrid. Today the town makes metal goods and furniture.

Population 17 836
Map Spain Da

Guernsey ***English Channel*** See CHANNEL ISLANDS

Guerrero ***Mexico*** **1.** Mountainous state with a long Pacific coastline in south-west Mexico. It contains the famous beach resorts of ACAPULCO and ZIHUATANEJO, and the beautiful 18th-century silver-mining town of TAXCO near its northern border. Guerrero's capital, Chilpancingo, is principally an agricultural centre, although it has a prestigious university and industry is developing. **2.** Town on the Conchos river in Chihuahua state. Horses are raised there and cereals and fruit are grown in the surrounding districts.

Population (state) 2 360 000; (town) 35 630
Map Mexico Bc

Guerrero Negro ***Mexico*** Marshes on the BAJA CALIFORNIA peninsula which are the largest salt flats in the world. Oil and natural gas have recently been discovered there.

Map Mexico Ab

Gui Jiang ***China*** River of the Guangxi-Zhuang Autonomous Region, flowing 350 km (220 miles) to join the Xi Jiang at Wuzhou.

Map China Gf

Guiana Highlands ***South America*** Vast, thinly populated area south of the Orinoco river, extending eastwards from eastern Venezuela across northern Brazil, Guyana, Surinam and Guyane. Its forested hills contain many minerals, including bauxite (an ore of aluminium), iron ore and manganese. Their highest point is Roraima, 2810 m (8565 ft), a flat-topped mountain which lies at the intersection of the boundaries of Brazil, Venezuela and Guyana. The highlands are the source of many rivers including those of the Amazon system to the south and of the Orinoco system to the north.

Map Venezuela Bb

Guildford ***United Kingdom*** See SURREY

Guilin ***China*** City of the Guangxi-Zhuang Autonomous Region, about 400 km (250 miles) north-west of Guangzhou (Canton) and close to some of the most spectacular mountain scenery in China.

On either side of the Gui Jiang river to the south of the city stretches an extraordinary landscape of natural domes and towers, some rising almost vertically 100 m (330 ft) from the shimmering green paddy fields. The hills are riddled with caves and garlanded with trees, orchids and vines. Their haunting beauty has been immortalised in Chinese landscape paintings, and attracts many tourists.

Population 300 000
Map China Gf

Guimarães ***Portugal*** Town about 40 km (25 miles) north of the city of Oporto. It was the ancient capital of Portucale, the heartland from which Portugal was reconquered from the Moors between the 11th and 13th centuries. The old castle, with a 10th-century tower, and the 15th-century palace of the Dukes of Braganza still survive. The city has textile and craft industries.

Population 22 000
Map Portugal Bb

Guinea See p. 276

Guinea ***West Africa*** Name derived from an ancient African kingdom, and at one time applied to the whole coastal region of West Africa from present-day Senegal to Angola. Much of this 6250 km (3900 mile) coast is hot, humid jungle. The region today includes the states of EQUATORIAL GUINEA, GUINEA and GUINEA-BISSAU.

Guinea, Equatorial See EQUATORIAL GUINEA

Guinea, Gulf of Arm of the ATLANTIC OCEAN off the south coast of West Africa extending from Cape Palmas, Liberia, to Cape Lopez, Gabon. It includes the bights of BENIN and Biafra.

Length about 1900 km (1180 miles)
Map Africa Cd

Guinea Highlands ***West Africa*** Mountainous region in Guinea extending into Sierra Leone, Liberia and Ivory Coast. The mountains are generally rounded and – thanks partly to their inaccessibility – are heavily forested and contain much wildlife. The peaks include some of West Africa's highest points, including Guinea's Mount Nuon, 1825 m (5987 ft), and Mount Nimba, 1752 m (5748 ft). There are coffee, tea and kola plantations in some areas, and rice is the staple crop of the local Kissi people. There are iron-ore deposits in the Nimba Mountains.

Map Guinea Cb

Guinea-Bissau See p. 277

Guipúzcoa ***Spain*** One of the three BASQUE REGION provinces, on the Bay of Biscay. At 1997 km² (771 sq miles) it is Spain's smallest province, but densely populated.

Population 693 000
Map Spain Da

Guiyang ***China*** Capital of Guizhou province, about 400 km (250 miles) north of the border with Vietnam. It is an important railway junction, with rolling-stock factories and an iron and steel works.

Population 1 260 000
Map China Fe

Guizhou ***China*** Mountainous province in the south, covering 170 000 km² (65 600 sq miles). It is part of a great highland plateau stretching into neighbouring Yunnan province. The climate is mild and wet. Rice and wheat are grown in the fertile, narrow valleys and livestock is reared in the mountains. Many of Guizhou's inhabitants are not ethnic Chinese: the Buyi and Zhuang, for example, who live in the mountains, are related to the Thais of South-east Asia. Guizhou is one of the most remote, backward and sparsely inhabited provinces of China. Its capital is Guiyang.

Population 27 310 000
Map China Fe

Gujarat ***India*** Western state covering 196 024 km² (75 665 sq miles) on the borders with Pakistani Sind. It was formed in 1960 when Bombay state was divided between the Guajarati-speaking north and the Mahrathi-speaking south. It is overwhelmingly Hindu, though there are significant Muslim representations in, for example, the city of AHMADABAD.

Population 34 085 800
Map India Ac

Gujranwala ***Pakistan*** Town in the Punjab about 65 km (40 miles) north of Lahore. The greatest of the Sikh kings, Ranjit Singh, was born there in 1780. A memorial to him commemorates his cremation in Lahore in 1839, at which his concubines threw themselves onto their husband's funeral pyre.

Population 597 000
Map Pakistan Da

Gulag Archipelago ***USSR*** See NORIL'SK

gulch Deep rocky ravine or valley, especially one cut by a torrent, in the western United States.

Gulf For physical features whose name begins 'Gulf', see main part of name.

Gulf, The Variously called the Arab, Arabian or Persian Gulf, it is linked to the ARABIAN SEA by the Strait of HORMUZ and the Gulf of OMAN. The Gulf has become an important shipping lane for tankers and other cargo vessels because of the tremendous wealth of the oil-producing nations that surround it. In many places the water is so shallow that ships of more than 5000 tonnes cannot get within 8 km (5 miles) of the shore, and the shoals and reefs on the Arabian coast, where famous pearl oyster beds lie, are hazards to shipping. Several of The Gulf states have constructed artificial 'islands' offshore that are linked to the mainland by submarine pipelines. This enables tankers up to 500 000 tonnes deadweight to load crude oil from the inland oil fields.

Area 233 100 km² (90 000 sq miles)
Greatest depth 91 m (300 ft)
Map Saudi Arabia Cb

Guatemala

A COUNTRY OF BEAUTIFUL MOUNTAINS AND LAKES, WHOSE TRANQUILLITY IS DISTURBED BY A BITTER CIVIL WAR

The mountainous land of Guatemala is regularly shaken by earthquakes and by political unrest. A ridge of volcanoes parallel to its Pacific coast marks a line of physical instability which has devastated successive earlier capitals – first ancient Ciudad Vieja in 1542, then colonial Antigua in 1773, and the present capital, GUATEMALA CITY, in 1917, 1918 and 1976.

Politics are also unstable: governments have mostly been established by armed coups, and their regimes have been unjust and cruel. Rebellious left-wing guerrillas are now fighting the right-wing government, and the non-combatant population have become victims.

Forty-five per cent of Guatemalans are pure-blood Indians, descendants of the Mayans whose ruined cities, built 2000 years ago, still stand in remote lowlands in the north. Most of the rest are *mestizos* (of mixed Indian and Spanish blood). The population is less varied than elsewhere in Central America because few Spaniards settled in the country and no slaves were brought in.

The indigenous population is divided culturally into *ladinos*, who are Westernised, and *indigenas*, who have kept their ancient customs and languages. The culture of the

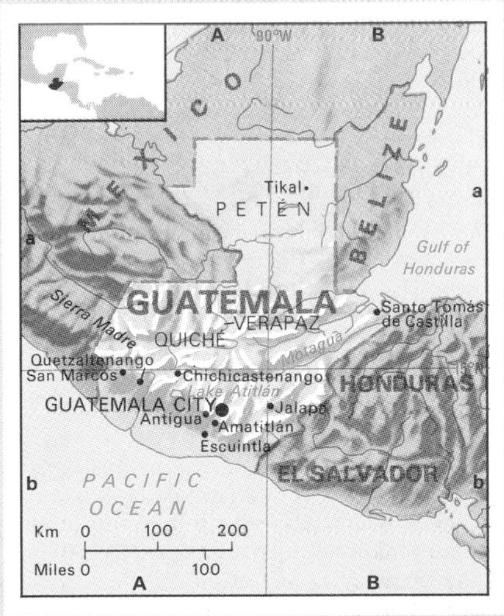

indigenas is threatened by the civil conflict. Villages have been destroyed and villagers killed by government and rebel forces.

Guatemala is one of the most militarised countries in Central America, and government forces have little regard for human rights. In an attempt to isolate possible supporters of the anti-government guerrillas, rural Indians are being resettled into new villages and refugee camps, breaking up their communities.

The Indians are extremely poor. Roman Catholic missionaries who tried to improve their lot were regarded as subversives by the government, and had to flee the country. In their place came North American evangelists who support the government – and a quarter of Guatemalans are now born-again Christians.

Although there is widespread poverty, Guatemala is one of the richest countries in Central America. Coffee made the country – and the coffee producers and traders – prosperous and still dominates the economy. It is grown on the lower slopes of the volcanic highlands; the heights, though cool and inhabited by 65 per cent of the people, are almost uncultivated. A 50 km (30 mile) strip of the hot Pacific coast produces sugar, cotton, cattle and bananas.

▲ LIFE AMONG THE RUINS A market is held in the remains of a Jesuit monastery in Antigua, the old capital wrecked by an earthquake in 1773. The Indians sell vegetables, and bags and women's tunics woven in bright colours.

The banana crop was once of major importance, and allowed the United Fruit Company to influence Guatemalan politics. In 1954 the company helped to oust President Jacobo Arbenz in an armed coup backed by the United States. The Arbenz administration was thought to be Communist, principally because it seized land from the great estate owners to give to the landless Indians.

Repressive military regimes have been in power ever since; originally they returned land to the landlords, but gradually some is being given to the Indians, to help rectify unequal distribution.

Civil conflict has almost destroyed tourism, and has dragged the economy down. The land could generate more wealth; there is oil, zinc, lead and nickel, and hydroelectric power potential. The volcanic subsoil is fertile, and although one-third of the country is uninhabited and very little cultivated, crops could be grown almost everywhere.

Guatemala City is growing, with a drift of people from the countryside over the past 30 years swelling the population to 1.2 million – ten times the size of any other town in the country. It is the seat of industry – textiles, paper and pharmaceuticals – and the core of economic life.

To improve access to the Caribbean, Guatemala wants a longer coastline, and regularly threatens to invade neighbouring Belize to obtain it. While defending Belize, Britain is arbitrating between the two countries.

GUATEMALA AT A GLANCE

Area 108 889 km² (42 042 sq miles)
Population 8 600 000
Capital Guatemala City
Government Military-dominated republic
Currency Quetzal = 100 centavos
Languages Spanish, local Indian languages
Religions Christian (predominantly Roman Catholic), tribal religions
Climate Tropical, temperate in highlands. Average temperature in Guatemala City ranges from 12-23°C (54-73°F) in January to 16-29°C (61-84°F) in May
Main primary products Coffee, bananas, cotton, sugar, maize, cattle, tobacco, timber; crude oil, nickel, lead, zinc
Major industries Crude oil refining, pharmaceuticals, food processing
Main exports Coffee, bananas, cotton, sugar, meat, chemicals
Annual income per head (US$) 1050
Population growth (per thous/yr) 31
Life expectancy (yrs) Male 57 **Female** 60

Guinea

LONG ISOLATED THROUGH FRENCH RETRIBUTION AND ITS LEADER'S PARANOIA, THE FIRST INDEPENDENT STATE OF FRENCH-SPEAKING AFRICA STRUGGLES FOR SELF-SUFFICIENCY

A green, lush and beautiful country with perhaps the greatest agricultural potential of all the former French territories in West Africa, Guinea was for 20 years after independence hidden from the outside world by its own isolationist policies – a reaction to the abrupt withdrawal of French aid. At the same time, its people suffered extreme poverty and political persecution meted out by its revolutionary socialist leader Ahmed Sékou Touré. From 1978 the dire need for aid forced a more 'open door' policy, and after Touré died in 1984 the new military government began reforms that could at last speed development.

▲ POVERTY WITHOUT LIBERTY
A poster of Guinea's first president, Sékou Touré, adorns a wall in Conakry. He led Guinea to independence in 1958 saying: 'We prefer poverty in liberty to wealth in slavery.' He brought them poverty and oppression.

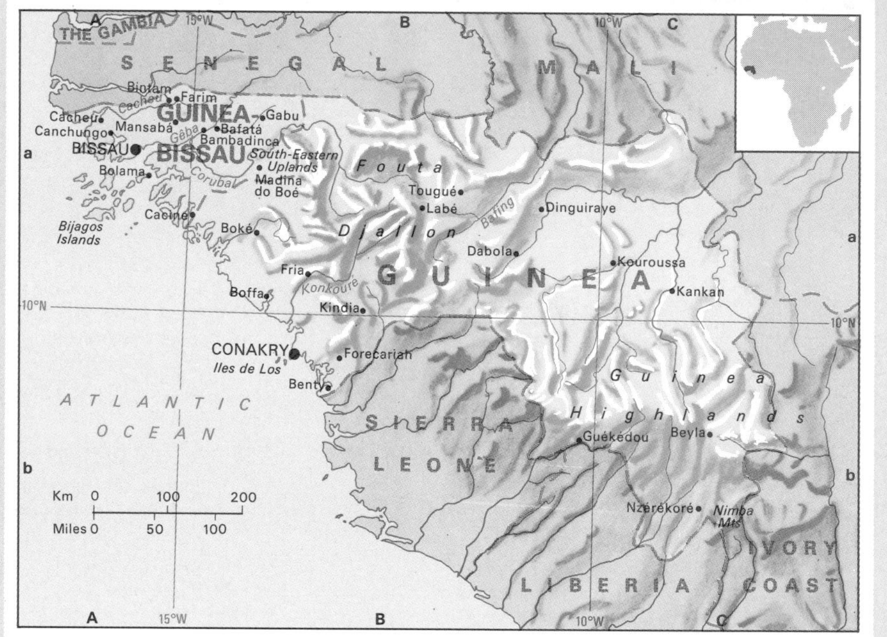

Located on the bulge of West Africa, Guinea is about the size of the United Kingdom. Dense mangroves edge the deeply indented 280 km (175 mile) Atlantic coastline, but parts of the swamps and the forested coastal plains have been cleared to grow rice, cassava, yams, maize and vegetables. Sheer sandstone escarpments rising from the plains mark the edge of the FOUTA DJALLON, where level plateaus – often rising to over 900 m (3000 ft) – are cut by fertile valleys. Despite the often bare red dusty surface, the plateaus provide grazing for the unique dwarf N'Dama cattle, while crops such as fonio (a small millet), cassava, vegetables and fruit (especially pineapples and bananas) grow in the valleys.

East of the Fouta are the savannah plains of Upper Guinea, where livestock and cereals are the main products. To the south-east rise the rounded, forested hills of the GUINEA HIGHLANDS; here coffee and kola nuts are important cash crops and rice is also grown.

Guinea has a number of ethnic groups. The Susu predominate near the coast, the Kissi in the Guinea Highlands and the Tenda in the east. The Fulani are pastoralists of the Fouta Djallon and Upper Guinea, while the Malinke (Mandinka) are traders who live throughout the country. As in much of West Africa, early European involvement was restricted to coastal trading and slave raiding. Trading posts were set up, and in the 19th century France gradually assumed control.

In a referendum in 1958, Guineans chose total independence against membership of a French-African community. France's reaction was to break off diplomatic relations, stop all aid and assistance, withdraw investment and close its market to Guinean produce. Instantly crippled, Guinea took aid from wherever it could – mainly the USSR.

Guinea's main source of income – and its main hope for the future, apart from revitalising agriculture – is its mining industry. The country has some of the world's biggest reserves of high-grade bauxite (aluminium ore), is rich in iron ore and still produces diamonds in the Guinea Highlands. But development is hampered by poor transport.

GUINEA AT A GLANCE

Area 245 857 km² (94 926 sq miles)
Population 5 890 000
Capital Conakry
Government Military-dominated republic
Currency Guinean franc
Languages French (official), eight official local languages
Religions Muslim (69%), tribal religions (30%), Christian (1%)
Climate Tropical; wet season from May to November. Average temperature in Conakry ranges from 22°C (72°F) to 32°C (90°F)
Main primary products Rice, maize, cassava, bananas; bauxite, iron ore, diamonds
Major industries Agriculture, mining, bauxite refining, fishing
Main exports Bauxite, alumina, diamonds
Annual income per head (US$) 290
Population growth (per thous/yr) 27
Life expectancy (yrs) Male 42 **Female** 44

Guinea-Bissau

EARNING LESS IN A YEAR THAN MOST WESTERNERS DO IN A WEEK, THE PEOPLE OF THIS SMALL COUNTRY STRUGGLE TO FEED THEMSELVES

Guinea-Bissau (formerly Portuguese Guinea) is one of the smallest and poorest nations of West Africa. Yet it is a peaceful country with stunning scenery. It rises gently from the deeply indented and island-fringed coastline (with a dense cover of forest) to wooded savannah on the low inland plateau. Three main rivers form natural highways into the interior. The 15 ethnic groups include the Balante, Fulani, Mandjako, Malinké and Pepel.

In 1963, the Partido Africano da Independencia da Guiné e Cabo Verde (PAIGC) began a war of independence for Guinea-Bissau and the nearby Portuguese colony of Cape Verde Islands (aiming to combine them as one independent state). The Portuguese finally recognised Guinea-Bissau's independence in 1974 and that of the Cape Verde Islands a year later.

Eleven years of war on top of a century of Portuguese rule left the country impoverished. The PAIGC government has concentrated on agricultural development, aiming for self-sufficiency in food, but progress was set back by a drought in 1977. A military coup in 1980 changed little (except to scotch plans to merge Guinea-Bissau and Cape Verde).

GUINEA-BISSAU AT A GLANCE
Map Guinea Aa
Area 36 125 km² (13 948 sq miles)
Population 874 000
Capital Bissau
Government Military-dominated republic
Currency Peso = 100 centavos
Languages Portuguese, Criolo (Creole), African languages
Religions Tribal (65%), Muslim (30%), Christian (5%)
Climate Tropical, heavy rains June to November; average temperature ranges from 24 to 27°C (75 to 81°F)
Main primary products Cereals, coconuts, groundnuts, rice, palm kernels, timber, fish; bauxite
Major industries Agriculture, fishing, forestry, beverages
Main exports Fish, groundnuts, palm kernels, timber
Annual income per head (US$) 180
Population growth (per thous/yr) 19
Life expectancy (yrs) Male 39 **Female** 43

Gulf Stream ***Atlantic Ocean*** Warm ocean current of the North Atlantic, flowing from the Gulf of Mexico north-eastwards up the east coast of the USA. After passing Newfoundland, it turns more to the east and is properly known as the North Atlantic Drift. This tempers the climate of north-western Europe and also helps to ensure copious rainfall.

Gullfoss ***Iceland*** Waterfall, 50 m (164 ft) high, on the Hvitá river about 100 km (60 miles) east of REYKJAVIK, the capital. *Foss* is the Icelandic word for 'waterfall'.

Map Iceland Bb

gully Small waterworn channel on a hillside, cut by a rapidly flowing stream. Gullies also form on poorly farmed agricultural land during 'gully-erosion'.

Gulu ***Uganda*** Town about 270 km (168 miles) north of the capital, Kampala, and the regional centre for the Acholi people. It stands in a mainly subsistence farming and cattle-herding district, but some cotton is produced commercially. There is a large military base nearby.

Population 15 000

Map Uganda Ba

Gunnbjørn Fjeld ***Greenland*** Highest mountain in Greenland, at 3702 m (12 146 ft). It rises on the east side of the island in King Christian IX's Land.

Map Arctic Bc

Gunsan ***South Korea*** See KUNSAN

Gunung Agung ***Indonesia*** Sacred Hindu mountain of Bali, rising to 3142 m (10 308 ft). An eruption in 1963 killed over 1500 people during an important religious festival.

Map Indonesia Ed

Gunung Kerinci (Kerintji) ***Indonesia*** Active volcano and the highest mountain on Sumatra. It rises to 3805 m (12 484 ft) in the centre of the island.

Map Indonesia Bc

Gunung Merapi ***Indonesia*** Volcanic peak of Central Java, one of the most active on the island. It rises to 2911 m (9550 ft) some 30 km (18 miles) north of Yogyakarta.

Gurgan ***Iran*** See GORGAN

Guri Dam ***Venezuela*** Hydroelectric scheme on the Caroní river south of Ciudad Guayana. It is Venezuela's largest producer of hydroelectric power.

Map Venezuela Bb

Guru Sikhar ***India*** Outlying mountain of the Aravalli Range, on the Rajasthan border with Gujarat. It stands at 1722 m (5648 ft) and is noted for its 11th-13th century Dilwara Jain temples, which contain some of the finest carvings in India. (The Jains were a breakaway group from Hinduism.)

The mountain is one of the haunts of the sacred monkeys called Hanumń langurs. The monkeys, which are revered by Hindus, are found in forests on the mountain and allowed to roam through the temples and outlying villages, stealing food from general stores and raiding crops at will. They live in groups of about 20 to 30, and are noted for their slender bodies.

Map India Bc

▼ HOLY MOUNTAIN The cone of Gunung (Mount) Agung – called 'the navel of the world' – towers over Bali's rice fields. Besakih, the island's most important Hindu temple, stands halfway up.

A DAY IN THE LIFE OF A GUYANAN CATTLE RANCHER

From the window of the ranch manager's office, which is part of his home, Duane Mendes surveys the 100 000 acres of gently undulating savannah where his cattle roam. In the distance he can just make out the dark line of the edge of the jungle, inhabited by American Indians, many of whom have worked for Duane's team of several dozen employees, making excellent *vaqueros* or cowhands.

But Duane has business to attend to. Beside his office stands a new four-wheel-drive Jeep. But as this is the wet season it could get bogged down in the mud in minutes. So Duane walks to the corral, to get his horse. As he mounts, he is joined by vaqueros from the southern part of the ranch, who have come to report that a few of the cattle are suffering from worms. Duane issues instructions for the medicines they are to be given, and then gallops down to the airstrip to meet the plane from the capital, Georgetown, which will carry away a cargo of Duane's beef.

There is no suitable road for taking the beef to Georgetown's market, and Duane has persuaded the owners of the company he works for that the most efficient way to transport beef is by air. The beef cattle are killed in the ranch's abattoir, and this way the meat can still be fresh when it gets to market.

Duane's grandparents were Portuguese, but he feels fully Guyanan himself – speaking English as do the rest of this former British colony. He went to college in Brazil and in Canada, where he learned the modern methods of ranch management. Now he has sent his son, Andrew, aged 20, to college, to ensure that he has the same advantages.

Late in the afternoon Duane returns to his office, having seen off the last planeload of beef. He wants to check if a letter has arrived from his son. There is a glow of red light in his office, cast by the fire burning over part of the ranch's grassland – one of Duane's modern methods designed to encourage the regrowth of the grass. It seems to symbolise the new hope that has characterised Guyana since the country achieved independence in 1960 – the hope that has spurred ranchers like Duane to make the most of their land.

But there is no letter from Andrew. Duane hopes it is because he is working so hard; he knows himself how much there is to learn in this adventurous but demanding career. Thinking of his boy, he goes into dinner with his wife and three teenaged daughters. Perhaps the girls will marry men who can help on the ranch too. The future he contemplates, as he beams at his family, is a rosy one.

Gusau ***Nigeria*** Northern industrial town and market about 730 km (450 miles) north-east of the capital, Lagos. It deals in hides and skins, cotton, groundnuts, tobacco and kola nuts (which are chewed as a mild stimulant), and makes textiles.
Population 111 400
Map Nigeria Ba

Güstrow ***East Germany*** Farming town 160 km (100 miles) north-west of the capital, East Berlin. It has a fine range of 16th-century buildings and is a tourist centre for both the Baltic coast and the lakes of MECKLENBURG.
Population 39 100
Map East Germany Cb

gut Narrow channel opening into the sea or a large estuary, especially in the eastern United States.

Gutland ***Luxembourg*** The southern, and more fertile, two-thirds of the Grand Duchy. The northern third – Ösling – is part of the Ardennes plateau.
Map Belgium Cb

Guyane ***South America*** Still living down the horrors of Devil's Island, Guyane is a remnant of European colonisation in South America. The country, which is largely covered with equatorial forest, is still an overseas department of France, with elected representatives in Paris and no plans for independence. Its economy relies on subsidies from France, and it exports very little of its own besides shrimps, an industry developed by American companies. Most of its food and necessities are imported.

Devil's Island, off the coast, was until 1945 a notorious penal settlement for some of France's most desperate criminals. More recently France has established a rocket-launching station at Kourou, expanding the town from a population of 600 to 6000.

Guyane has a total area of some 89 941 km² (34 726 sq miles) and 73 000 inhabitants, more than half of them concentrated in the handsome French-style capital of CAYENNE (from which the pepper gets its name). Few slaves were brought to Guyane because the sugar industry was never greatly developed, but two-thirds of the population are descendants of African slaves from neighbouring Guyana and Surinam. The rest are of Amerindian, Chinese and French descent, together with a small number of Laotians, refugees from war-torn Indo-China, who were settled on farms in Guyane in the late 1970s.

Five thousand civil servants have recently been sent from France to Guyane in a drive to boost its population. It is hoped to develop tourism and bauxite mining, and to exploit the hardwoods of the extensive and well-watered jungle interior.
Population 73 000

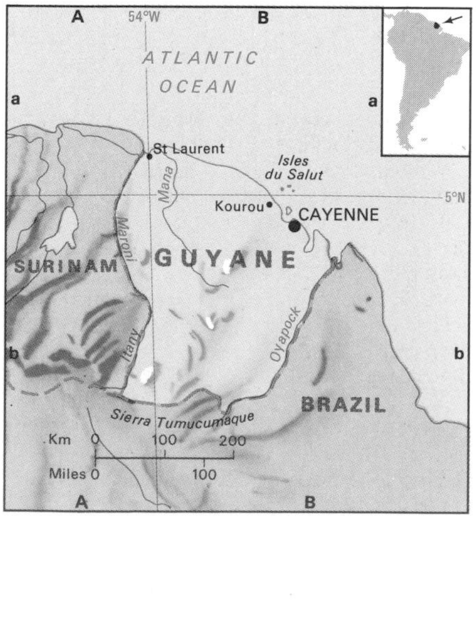

Guyenne ***France*** South-western former province covering some 41 000 km² (15 800 sq miles) between the Gulf of Gascony and the south-western part of the Massif Central. BORDEAUX is the chief city.
Population 2 660 000
Map France Cd

guyot Flat-topped submarine mountain, possibly a volcano truncated by waves, rising from the floor mainly of the Pacific Ocean.

Gwalior ***India*** City about 280 km (174 miles) south and slightly east of Delhi. On a sandstone plateau overlooking Gwalior is a medieval Hindu fort, near which are some impressive rock sculptures of Jain (a religion between Buddhism and Hinduism) saints. The statues, which were carved around AD 1450, are some 10 m (33 ft) high.

The city's main industries are textiles, footwear, leather goods and pottery. There are also flour and oilseed mills.
Population 555 900
Map India Cb

Gwangju ***South Korea*** See KWANGJU

Gwent ***United Kingdom*** County in south-east Wales, formerly known as Monmouthshire, which was renamed in the 1970s after an ancient Welsh principality. It covers 1376 km² (531 sq miles) and centres on the industrial city of Newport at the mouth of the River Usk. The spectacular Wye Valley forms much of its border with England.
Population 440 000
Map United Kingdom De

Gwynedd ***United Kingdom*** County of north-west Wales, created out of the former county of Caernarfon, plus parts of Denbigh and Merioneth, and named after the ancient Celtic kingdom that dominated the country after the Romans left in the 5th century. It has an area of 3868 km² (1493 sq miles) and includes the island of ANGLESEY, the Lleyn peninsula and SNOWDONIA National Park. There are superb medieval castles at Conwy, Caernarfon and Harlech, and many seaside resorts, of which the largest is Llandudno. Tourists flock to the narrow-gauge mountain railway at Ffestiniog and the rack railway to the summit of Snowdon, at 1085 m (3560 ft) Wales's highest mountain. Holyhead on Holy Island is the principal port for crossing to the Irish Republic.
Population 232 000
Map United Kingdom Cd

Gyeongju ***South Korea*** See KYONGJU

Guyana

A PIECE OF THE BRITISH WEST INDIES ON THE CONTINENT OF SOUTH AMERICA, BEDEVILLED BY RACIAL PROBLEMS AND ECONOMIC DECLINE

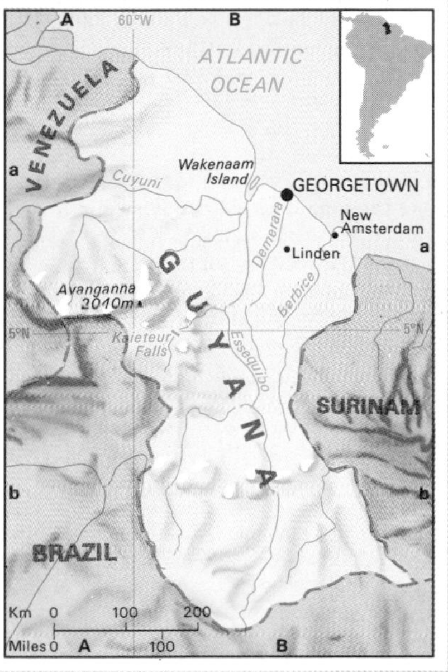

The Spanish adventurers turned up their noses at Guyana, as a land with muddy harbours and no gold. The Dutch, French and English all held territory in the area, which continued to change hands until Britain gained control in 1814 of the area that later (1831) became British Guiana. Independent since 1966, Guyana is physically part of South America, but is culturally linked with the West Indies. It is the only English-speaking country in South America.

Gold was eventually found in small quantities, but the real wealth of Guyana is its bauxite (aluminium ore), on which the economy depends, and the sugar and rice it grows. The land could be much more productive, but large areas of the country are unexploited. The south-west is high savannah, from which the jungle slopes down towards the coast. The land is intersected by the numerous rivers which gave Guyana its name – it means 'land of many waters'.

The jungle and rivers could produce minerals, hardwood and hydroelectric power, but 90 per cent of the people live and work on the narrow coastal plain. Here, often on land reclaimed from tidal marshes and mangrove swamps, rice is grown and vast plantations produce sugar, as they have done since the Dutch brought slaves from Africa in the 17th century.

LAND OF SIX PEOPLES

The descendants of these slaves are one of the six racial groups that make up Guyana's diverse population. When slavery was abolished in 1833, the British colonists were deprived of a labour force. Foreign labourers were offered five-year contracts to work on the plantations, with the opportunity to stay in Guyana afterwards. This brought an influx of Portuguese and Chinese workers, who still form small communities in Guyana.

A steady flow of poor Asian Indians arrived during the second half of the 19th century, and they are now the largest racial group in the country, still providing the labour for the sugar farms and rice paddies. People of mixed race make up a fifth group.

A sixth group in this 'land of six peoples', as Guyana calls itself, are the original inhabitants, now swamped by the tide of immigrants; only 60 000 Carib Indians are left, some living as nomads in the rain forests.

The capital city of GEORGETOWN is the best-preserved town of wooden architecture in the Caribbean. Its civic buildings are all made of white-painted wood, and its cathedral, rising to 39.6 m (132 ft), is one of the tallest wooden buildings in the world.

The country is deeply divided, racially and politically. The two main opposing parties are the black-dominated People's National Congress (nominally socialist), which has held power since 1964, and the (Asian) Indian-dominated People's Progressive Party (which propounds a Marxist doctrine). Their rivalry has caused economic and social problems. Antagonism between the Asians, who control a large proportion of the nation's commerce, and blacks led to clashes and bloodshed in the 1960s.

Bauxite companies were nationalised in 1971, and the state took over the sugar industry in 1976. But political competition diminished the benefits this might have brought. The People's National Congress used these industries to increase its political support by employing many more people without improving productivity. Very little has been done to promote economic growth, and today the economy is in a crisis.

Unemployment is at 40 per cent, and the three main exports of sugar, bauxite and rice are being produced at well below their full potential. Foreign investment has been curtailed, and there is a desperate shortage of foreign exchange.

There are also severe restrictions on imports, so that spare parts for machinery have to be smuggled in from Suriname and Brazil. Imports such as wheat flour are classified as luxuries and banned; there is no shortage of food, but there is limited variety. Morale in the country is low. The Asians complain of discrimination and the state grows more repressive. Health care and education, which were once good by South American standards, are in decline. Many people are emigrating, mostly to Britain and the United States, and new uncertainties followed the death in 1985 of President Forbes Burnham, who led the country for 21 years.

GUYANA AT A GLANCE

Area 215 000 km² (83 000 sq miles)
Population 801 000
Capital Georgetown
Government Parliamentary republic
Currency Guyana dollar = 100 cents
Languages English (official), Creole
Religions Christian (57%), Hindu (33%), Muslim (9%)
Climate Tropical; average temperature in Georgetown ranges from 23°C (73°F) to 31°C (88°F)
Main primary products Sugar, rice, timber, oranges; bauxite, gold, diamonds, semiprecious gemstones
Major industries Agriculture, mining, forestry, sugar refining, timber milling
Main exports Sugar, bauxite, alumina, rice
Annual income per head (US$) 428
Population growth (per thous/yr) 4
Life expectancy (yrs) Male 68 **Female** 73

Györ ***Hungary*** Regional capital of the Little Alföld and administrative centre of Györ-Sopron county, 108 km (67 miles) west of Budapest. The city is a busy river port at the confluence of the Rába and Danube rivers. It is also a major industrial centre, whose many products include textiles, spirits, foodstuffs, and Ikarus buses made in the former Rába railway works.

The old town at the heart of the city clusters around Kaptalan Hill, site of the Roman fortress of Arrabona, which guarded a route along the Danube Valley. The old town's elegant Baroque buildings include: the cathedral, built originally in the 12th century, which contains the elaborate golden reliquary of St Ladislas; the 15th-century bishop's palace; an 18th-century Carmelite church; and Napoleon's House museum, where the emperor lived in 1809.

Pannonhalma Abbey, a large Benedictine foundation 21 km (13 miles) south-east of Györ, is also largely Baroque in style, but retains its 10th-century church. The abbey still maintains a grammar school and an extensive library which contains some unique early medieval documents.

Population 127 000
Map Hungary Ab

gypsum A white EVAPORITE mineral, calcium sulphate, found in clays and limestones, and used in the manufacture of cements and plasters, especially plaster of Paris.

gyre Circular flow of water in each of the great ocean basins of the world, produced by the combined effects of prevailing winds and the earth's rotation. In the Northern Hemisphere, each gyre flows clockwise, while in the Southern Hemisphere the flow is anticlockwise. The Gulf Stream is part of the North Atlantic gyre. See OCEAN currents.

Gyula ***Hungary*** Town on the Great Alföld (Plain) near the Romanian border about 90 km (55 miles) north-east of the city of Szeged. It has a fine medieval castle, and since 1960 has become a leading spa.

Population 35 200
Map Hungary Bb

Ha Khotso ***Lesotho*** Village about 40 km (25 miles) east of the capital, Maseru. Nearby is a rock shelter called Ha Baraona, meaning 'home of the little Bushmen', which contains a superb collection of rock paintings.
Map South Africa Cb

Ha'apai ***Tonga*** Large group of low-lying coral islands about 160 km (100 miles) north of Nuku'-alofa. The main town is Pangai on Lifuka island. The notorious mutiny on HMS *Bounty* took place on April 28, 1789, at Tofua in the west of the group.
Population 11 700
Map Pacific Ocean Ec

haar Local name for a cold mist or fog off the east coast of England or Scotland, usually in spring or early summer.

Haarlem ***Netherlands*** Provincial capital, about 18 km (11 miles) west of the national capital, Amsterdam, and the centre of Holland's tulip-growing region. Its 15th-century Church of St Bavo is one of the largest in the country, and its museums include one devoted to the painter Frans Hals (about 1580-1666) who lived and died in the city.
Population (city) 154 300; (Greater Haarlem) 219 400
Map Netherlands Ba

Haast Pass ***New Zealand*** Mountain pass, 560 m (1840 ft) high, through the Southern Alps of South Island. A road was built through the pass in 1965. It provides a low-level link between Queenstown in the central Otago region and the west coast. As both areas are of little economic significance, the route is primarily an all-weather tourist one.
Map New Zealand Bf

Habbaniyah ***Iraq*** Lake south of the Euphrates river, 100 km (60 miles) west of the capital, Baghdad. It is 20 km (12 miles) long and 5-10 km (3-6 miles) wide. A British airfield built on its north shore in the Second World War has been converted into a tourist village.
Map Iraq Bb

Hachinohe ***Japan*** Fishing port in the north of Honshu island, about 70 km (43 miles) south-east of Aomori. It is also an industrial centre specialising in fish products, steel, paper and pulp. Local craftsmen carve *yawata-uma* – traditional brightly painted wooden horses.
Population 241 400
Map Japan Db

hade Angle of inclination from the vertical of a VEIN, FAULT or LODE.

Hadhramaut ***South Yemen*** Long, wide valley running parallel with the coast of the Gulf of Aden and between 160 km (100 miles) and 320 km (200 miles) inland. It is some 560 km (348 miles) long. Grain, dates and sesame are grown in the upper and middle parts of the valley – which have torrential, seasonal rain. But the barren eastern section is largely uninhabited. Many of the inhabitants have emigrated, to form communities in other Middle East countries and in Indonesia.
Map South Yemen Aa

Hadrian's Wall ***United Kingdom*** Roman fortification built on the orders of Emperor Hadrian (ruled AD 117-38), who visited Britain in 122. It stretches 117 km (73 miles) from the Solway Firth near Carlisle in the west to Wallsend near the east coast, and formed for centuries the north-west frontier of the Roman Empire and the border between England and Scotland. It was about 4.5 m (15 ft) high and had milecastles housing troops every Roman mile (1480 m, 4860 ft), with watchtowers in between. There were also 17 forts along the wall. It fell into disrepair after the Romans left Britain in the 5th century, but many stretches still remain, particularly in the county of Northumberland.
Map United Kingdom Dc

Hagen ***West Germany*** Ruhr steel town, about 35 km (22 miles) south-east of Essen.
Population 210 000
Map West Germany Bc

Hague, The ('s-Gravenhage; Den Haag; La Haie) ***Netherlands*** The Dutch seat of government since 1580, and also capital of Zuid-Holland province. It lies about 55 km (35 miles) south-west of Amsterdam and includes the seaside resort of Scheveningen. The city's name 's-Gravenhage ('the counts' private domain') derives from the former hunting lodge of the Counts of Holland in a wooded area known as the Haghe ('hedge'). In 1248 Count William II built a castle there, and around it grew a palace, the Binnenhof. An artificial lake, the Hofvijer, was dug beside the palace in 1350. These now lie at the heart of The Hague, with government buildings and embassies around. Beyond is a spacious city, with canals, parks and woods separating one district from another.

The Hague is an international city, with the International Courts of Justice, established there in 1922. It is also a fashionable place for conferences, with the Netherlands Congress Centre, and has the offices of multinational business firms, including Royal Dutch Shell.

Among many places of interest are the Groote Kerk (Great Church) dating from 1399; the town hall (1565); the Palace of Peace (housing the International Courts of Justice); the Mauritshuis Royal Art Gallery; the Royal residence, 'The House in the Wood' (1647) and its pleasure gardens; and the Madurodam model village.
Population (city) 449 300; (Greater Hague) 674 500
Map Netherlands Ba

Hahotoé ***Togo*** Phosphate-mining region in the south, just north of Lake Togo. Its reserves are estimated to be 100 million tonnes.
Map Togo Bb

Haifa ***Israel*** Capital of Haifa district, and the country's chief port, on the Mediterranean coast about 85 km (55 miles) north of Tel Aviv. It is set at the foot of Mount Carmel, the summit of which can be reached by a cable car. Among the many places of interest are the university, zoological gardens, maritime and art museums, and a temple of the Bahai religious sect. Haifa's industrial products include textiles, chemicals, glass, soap and electrical equipment.
Population 224 700
Map Israel Aa

hail, hailstones Precipitation in the form of pellets of ice. Droplets of water that condense in the lower part of a thundercloud may be carried by ascending air currents to a level where they freeze into ice pellets. The strong convection currents prevent the hailstones from falling until they have grown much larger. In a turbulent cloud current, small hailstones can be tossed up and down between the layers of moisture and ice many times, growing larger all the time. Hailstones more than 700 grams (1.5 lb) in weight and 175 mm (7 in) in diameter have been recorded.

Hail ***Saudi Arabia*** Town and oasis situated 250 km (155 miles) north-west of Buraydah. It lies on the old caravan and pilgrim route from Iraq to MEDINA and MECCA. It was the former capital of the Ibn Rashids, who were the main opponents of the Saud family, and the Wahhabis (a fundamentalist Muslim sect that now predominates in Saudi Arabia). In 1921 King Abdul-Aziz besieged and took Hail and incorporated the region into the new kingdom of Saudi Arabia. Hail has two impressive forts built by the Ibn Rashids and a market specialising in Bedouin tents, rugs and pots.
Population 45 000
Map Saudi Arabia Bb

Hainan ***China*** Island in the South China Sea off the coast of Guangdong province. Hainan is a tropical island covering 33 670 km^2 (13 000 sq miles) with more than 1000 km (620 miles) of largely unspoilt coastline. Coffee and rubber plantations are being developed in the interior, and oil exploration is underway offshore. Its chief city is Haikou.
Population 5 400 000
Map China Gg

Hainaut (Henegouwen) ***Belgium*** Industrial province covering 3787 km^2 (1462 sq miles) along the south-west border with France. The coalfield which crosses the area from east to west once fuelled (1850-1950) metalworking, textile and chemical industries around the main towns of Charleroi, Mons and Tournai, but the economy has become depressed since the worked out coal mines closed in the 1970s.
Population 313 000
Map Belgium Aa

Haiphong ***Vietnam*** Port and second largest city of the country after Ho Chi Minh City. It lies about 90 km (55 miles) east of the capital, Hanoi. Heavily damaged by bombing during the Vietnam War (1965-75), it is now being rebuilt.
Population 1 379 000
Map Vietnam Ba

Hajar Mountains ***Oman/United Arab Emirates*** Range running parallel with the Gulf of Oman and forming the spine of the MUSANDAM peninsula. In its mid-point, the mountains rise to Jebel Akhdar (the Green Mountain), crowned by the 3018 m (9000 ft) high peak Jebel Shams. Below Jebel Akhdar is the long oasis of Rustaq, containing numerous villages and five watercourses. The Sumail Gap separates the Western Hajar from the Eastern Hajar Mountains.
Map Oman Aa

Haiti

VOODOO AND BOGEY MEN SUPPORTED THE DICTATORS OF THE WORLD'S FIRST BLACK REPUBLIC

Dr François 'Papa Doc' Duvalier ruled the West Indian republic of Haiti from 1957 to 1971, using voodoo (a religious cult from Africa based on sorcery and fetishism) and his special police, the *Tontons Macoutes* (bogey men), to subjugate the population. When Papa Doc died, he was succeeded as president-for-life by his 19-year-old son, Jean-Claude ('Baby Doc'). His regime was little less oppressive than his father's; it collapsed in January 1986, and Baby Doc and his family fled the country. He was succeeded by army leader General Henri Namphy, but unrest and instability continued as some Duvalier aides clung to their positions of power.

Haiti, the poorest country in the Americas, is part of the island of HISPANIOLA (the remainder is the Dominican Republic). The island was discovered by Christopher Columbus in 1492 and settled by the Spanish. But in 1697 French pirates in the western part of the island forced Spain to cede this area to France. The French sugar planters imported slaves from Africa, and made Saint Domingue, as it was then called, a prosperous colony.

In 1791 the 500 000 slaves revolted, and in 1804 they declared their independence, forming the world's first black republic. Haiti also ruled Santo Domingo, now the Dominican Republic, from 1822 to 1844. Haiti had 22 dictators between 1843 and 1915, when the United States moved in to bring about stability. The US occupation ended in 1934.

More than 90 per cent of Haitians are of African descent and the rest are mostly of mixed race. French is the official language, but 90 per cent of the people speak Creole, a French patois. Catholicism is the official religion, but voodoo is widely practised. About 80 per cent of the people live in rural areas and, although education is free, less than half the children go to school. Yet the country has a strong artistic and literary tradition.

Hispaniola is the second largest West Indian island after Cuba, and the most mountainous. The mountain ranges are separated by deep valleys and plains; the highest point in Haiti is La Selle (2680 m, 8793 ft).

Agriculture is the chief occupation but there is much rural unemployment. The average holding is only 1.5 hectares (3.7 acres) and many farmers grow only enough to feed their own families. Large estates grow coffee, sugar and sisal for export. Most of the forests have been cleared, causing severe soil erosion and declining crop yields.

The supply of cheap labour has attracted US investment in assembly plants which produce goods for export, including textiles, sports and electrical goods, but foreign aid is vital for the country's economy.

Before recent upheavals, tourism was the largest source of foreign income after farm exports. American visitors were attracted by the scenery, the culture and almost complete absence of street crime. Most stayed in the capital, PORT-AU-PRINCE, and its hillside suburb of Pietonville. The Sans Souci palace and the spectacular Citadelle near CAP-HAITIEN in the north were the chief attractions. Both were built by a former waiter who became King Henri Cristophe of northern Haiti (1806-20).

HAITI AT A GLANCE

Map Caribbean Ab
Area 27 750 km² (10 714 sq miles)
Population 5 900 000
Capital Port-au-Prince
Government Military-dominated republic
Currency Gourde = 100 centimes
Languages French, Creole
Religions Christian (84% Roman Catholic, 14% Protestant); most also follow voodoo
Climate Tropical; average temperature in Port-au-Prince ranges from 25°C (77°F) in January to 29°C (84°F) in July
Main primary products Bananas, cassava, maize, rice, sorghum, coffee, sugar cane, sisal, timber; bauxite, copper (unexploited)
Major industries Agriculture, mining, textiles, food processing, cement, assembly of imported parts, forestry, tourism
Main exports Coffee, light industrial products, sugar, sisal, bauxite, essential oils
Annual income per head (US$) 290
Population growth (per thous/yr) 19
Life expectancy (yrs) Male 52 **Female** 55

Hakodate ***Japan*** Port and gateway to Hokkaido on the island's southernmost coast. Open to foreign trade since the 19th century, it has the country's only western-style castle. Hakodate is linked to the island of Honshu by the Seikan rail tunnel, at 54 km (33.5 miles) the world's longest underwater tunnel. In parts, the tunnel runs 100 m (330 ft) below the seabed.
Population 319 200
Map Japan Db

Hakone ***Japan*** Mountainous region of central Honshu island lying south-west of the capital, Tokyo, between Mount Fuji and the Izu peninsula. It is a popular summer holiday destination, scattered with volcanoes and hot springs in the central section of the Fuji-Hakone-Izu National Park. The region's main town is Odawara on Sagami Bay (population 185 900).
Map Japan Cc

Hakusan ***Japan*** Mountainous national park, 474 km² (83 sq miles) in area, on Honshu island, facing the Sea of Japan about 235 km (145 miles) west of the capital, Tokyo. One of the park's most popular attractions is Shirakawa village, which has enormous wooden houses topped by thickly thatched roofs.
Map Japan Cc

Halas ***Hungary*** See KISKUNSAG

Halemaumau ***Hawaii*** Lava pit (or fire pit) of the Kilauea Crater of Mauna Loa volcano on Hawaii island. It has been the main source of lava erupted from the volcano, and during eruptions contains a lake of molten lava up to 75 hectares (185 acres) in area.
Map Hawaii Bb

Halicarnassus ***Turkey*** See BODRUM

Halifax ***Canada*** Capital and largest city of Nova Scotia, and Canada's main ice-free Atlantic port. Its strategic position made it an important wartime naval base. James Wolfe, the British general, sailed from here to capture Quebec in 1759, and in both World Wars Halifax was a gathering place for Atlantic convoys. The first transatlantic steamship service started between Halifax and Great Britain in 1840. Today the city is important for bulk cargoes and container traffic, and is an educational, cultural and industrial centre, with iron foundries, oil refineries and factories making soap and shoes.
Population (city) 114 595; (metropolitan area) 278 000
Map Canada Id

Halifax ***United Kingdom*** Town in West Yorkshire, between Bradford and Huddersfield, that has been an important English wool-trade centre since the 15th century. Piece Hall – rebuilt in 1770 and now converted for craft shops and galleries – was a manufacturers' trade hall where weavers sold their 'pieces' (lengths) of cloth.
Population 88 000
Map United Kingdom Ed

halite, rock salt A white evaporite mineral, sodium chloride (common salt), found in sedimentary rocks, such as sandstone and shale. It is a source of salt and chlorine.

Halland ***Sweden*** South-western farming county covering 4785 km² (1847 sq miles). It was ceded by Denmark to Sweden in the 17th century. The chief city is HALMSTAD.
Population 244 078
Map Sweden Bd

Halle ***East Germany*** Town and inland port on the Saale river, about 30 km (20 miles) north-west of Leipzig. It was founded as a fortress in AD 806 and became a salt-mining centre. The composer George Frederick Handel (1685-1759) was born there. Modern industries include printing, machinery, rubber, cement, paper and petroleum products.
Population 236 500
Map East Germany Bc

Halmstad ***Sweden*** Seaport and industrial city on the Kattegat coast, 150 km (93 miles) south-east of Gothenburg. It has a ruined medieval castle. Its industries include shipbuilding, steel, paper, textiles and brewing.
Population 76 600
Map Sweden Bd

halo Circular band of light, sometimes coloured, seen around the sun or moon, caused by the refraction of light by ice particles in cirrostratus clouds. Halos herald bad weather.

Hälsingborg ***Sweden*** See HELSINGBORG

hamada Rock desert, particularly in the Sahara, which has been swept clear of sand and dust by the wind.

Hamadan (Hamedan) ***Iran*** City about 300 km (185 miles) south-west of the capital, Tehran. It produces carpets, rugs and leather goods, and also trades in wool, hides and skins. Hamadan was the capital of the Medes in the 6th century BC and of the Achaemenid Persian Empire (559-330 BC). The city has a large, colourful bazaar and numerous monuments, including the tomb of the Persian physician and philosopher Avicenna (AD 980-1037), and the mausoleum of the Biblical characters Esther and her cousin and guardian, Mordecai.

Population 234 500

Map Iran Aa

Hamah ***Syria*** Commercial and industrial city on the Orontes river, 125 km (78 miles) south of Aleppo. The centre of Syria's growing steel industry, Hamah also has tanning and weaving industries and trades in cereals and fruit. It lies 308 m (1010 ft) above sea level and is surrounded by orchards and market gardens.

Hamah is noted for the huge wooden water-powered wheels which lift water from the Orontes for irrigation. The use of such wheels dates back to ancient times, but the present ones date mostly from the Middle Ages. The smallest of the wheels is 10 m (33 ft) in diameter, and the sound of their creaking can be heard throughout the town. Among the town's places of interest are the Great Mosque with its 14th-century water wheel.

Population 514 750

Map Syria Bb

Hamamatsu ***Japan*** City near the southern coast of Honshu island about 215 km (135 miles) south-west of the national capital, Tokyo. Its main products are pianos and motorcycles. In May a 'battle of the giant kites' festival is held, in which rival teams try to sever the strings of one another's kites.

Population 514 100

Map Japan Cd

Hamar ***Norway*** Market town on the west shore of Lake Mjøsa in south central Norway. The town was founded in 1152 by the only Englishman to become pope – Adrian IV (1154-9) – while he was a cardinal bishop in Scandinavia. The town's industries include the processing of wood and of farm produce.

Population 27 000

Map Norway Cc

Hamburg ***West Germany*** Second largest city in the country after West Berlin and its biggest port, on the Elbe river about 110 km (68 miles) from its mouth. Hamburg, founded by the Holy Roman Emperor Charlemagne in 810, was a leading member of the Hanseatic League, the medieval trading association of north European towns. Its modern docks cover an area of over 100 km² (39 sq miles) on both sides of the Elbe, and it has petrochemical, electronics and food processing industries.

For centuries, Hamburg was an independent city and it keeps some autonomy as a *Land* (state) of the Federal Republic. It has a long musical tradition. The composers Johannes Brahms (1833-97) and Felix Mendelssohn (1809-47) were born there and George Frederick Handel (1685-1759) was a violinist in the opera orchestra before he moved to England. The Second World War song *Lili Marlene* was written by a Hamburg man, Hans Leip. In St Pauli district, west of the city centre, is the Reeperbahn, or Rope Walk, an infamous street of nightclubs and brothels.

Population 1 617 800

Map West Germany Db

Hamedan ***Iran*** See HAMADAN

Hämeenlinna (Tavastehus) ***Finland*** Capital of the province of Häme, 98 km (60 miles) north and slightly west of the capital, Helsinki. It was the birthplace of Jean Sibelius (1865-1957), composer of *Finlandia*, the symphonic tone poem he wrote in 1900 which became Finland's unofficial national anthem. The town's 13th-century castle was built by Swedes when Finland was part of Sweden. The town now manufactures textiles and plywood. Nearby is Aulanko, a large pleasure park.

Population 42 400

Map Finland Bc

Hamelin (Hameln) ***West Germany*** Carpet-making town on the Weser river, 49 km (30 miles) south-west of Hanover, site of the legend of the Pied Piper. In AD 1284, a ratcatcher is said to have lured with his pipes a plague of rats from the town into the Weser to drown. When he was not paid he piped away the town's 128 children – leaving one crippled boy – and they were never seen again.

The basis of the legend is long forgotten; the so-called ratcatcher's house in the town dates only from 1600.

Population 56 300

Map West Germany Cb

Hamersley Range ***Australia*** Mountains in the PILBARA region of Western Australia, rising to 1235 m (4052 ft) at Mount Bruce. The range contains rich deposits of iron ore. Two of the peaks – Mount Whaleback and Mount Tom Price – are composed almost entirely of high-grade ore and are being systematically demolished. Most of the central section of the range constitutes the Hamersley Range National Park, established in 1969.

Map Australia Bc

Hamhung (Hamheung) ***North Korea*** Industrial port and provincial capital on the east coast: 180 km (112 miles) north-east of the national capital, Pyongyang. Heavily bombed in the Korean War (1950-3) and since rebuilt, it now manufactures machinery, chemicals and fertiliser.

Population 420 000

Map Korea Cc

Hamilton ***Bermuda*** Capital and main port of Bermuda. Large cruise ships dock alongside The Front, the town's main shopping and commercial street. Founded in 1790, Hamilton was made the capital in 1815 and became a free port in 1956.

Population 3000

Hamilton ***Canada*** Large industrial city on the shore of Lake Ontario. Steel, automobiles and chemicals are its main products. It was first settled in 1778 and it became a city in 1846. It is the home of the Royal Botanical Gardens, historic Dundurn Castle and McMaster University. The city extends from the lake shore onto the Niagara escarpment 100 m (328 ft) high.

Population (city) 306 430; (metropolitan area) 542 090

Map Canada Hd

Hamilton ***New Zealand*** Market and industrial town on North Island, about 120 km (75 miles) south of Auckland. Its industries include food processing, the manufacture of agricultural equipment, meat freezing and electronics. The University of Waikato was founded there in 1964.

Population 97 900

Map New Zealand Eb

Hamilton ***United Kingdom*** See STRATHCLYDE

Hammamet ***Tunisia*** Resort town and port on the southern coast of the peninsula ending in Cape Bon. It has long been renowned as a beauty spot patronised by the famous, and has splendid beaches and many hotels which cater for European visitors. It is also the venue for an annual international cultural festival held in July. There is a small medina (the old Muslim quarter), a souk (market) for brassware and carpets, a 15th-century fort, the Great Mosque, and a museum containing a collection of traditional Tunisian costumes.

Population 30 500

Map Tunisia Ba

Hamma Salahine ***Algeria*** See BISKRA

Hammerfest ***Norway*** Fishing port and one of the most northerly towns in the world, on Kvåløy Island, some 200 km (125 miles) south-west of North Cape. It lies 460 km (285 miles) north of the Arctic Circle and has 'midnight sun' from mid-May to the end of July. The town was rebuilt after the Second World War, when the Germans destroyed most of the province of FINNMARK. Hammerfest manufactures cod-liver oil and exports fish and furs.

Population 7400

Map Norway Fa

Hampshire ***United Kingdom*** County covering 3772 km² (1456 sq miles) of southern England on the English Channel coast. It used to include the Isle of Wight, now a separate county. The main centres are the ports of SOUTHAMPTON and PORTSMOUTH. Aldershot, in the north, is the home base of the British army. The middle of the county is a rural haven of chalk downland and quiet trout streams, focused on the ancient city of WINCHESTER. The old royal hunting grounds of the NEW FOREST, now a popular holiday region, are in the south-west.

Population 1 500 000

Map United Kingdom Ee

Han Shui ***China*** River of central China, flowing about 1130 km (700 miles) south through Hubei province to join the Chang Jiang river at the city of Wuhan.

Map China Gd

Hangchow ***China*** See HANGZHOU

hanging valley Tributary valley whose floor is much higher than that of the main valley, so that its river enters the main valley by a waterfall or rapids. It is usually caused by deeper glacial erosion of the main valley.

Hangzhou (Hangchow) ***China*** Capital of Zhejiang province 160 km (100 miles) south-west of Shanghai. It was the capital of southern China from 1127 until the Mongol conquest in 1279. It is now an administrative and industrial centre.
Population 1 105 000
Map China Ie

Hankow ***China*** See WUHAN

Hannover ***West Germany*** See HANOVER

Hanoi ***Vietnam*** Capital city in the north of the country on an arm of the Red River. It was chosen as the capital in 1010, when it was called Thang Long, 'City of the Soaring Dragon'. Its present name was first recorded in 1831 and means 'City Inside the River'. The heart of the modern city was laid out by the French during the colonial period, but it was heavily bombed during the Vietnam War (1965-75) and is still being restored. Its main manufactures are cars, machine tools, cement, textiles and chemicals.
Population 2 570 900
Map Vietnam Ba

Hanover (Hannover) ***West Germany*** Capital of Lower Saxony, about 100 km (62 miles) south-east of Bremen. It was the city of the Electors of Hanover – one of whom, George I, a grandson of James I, became England's king in 1714. The British astronomer Sir William Herschel (1738-1822), who discovered the planet Uranus, was born in Hanover. Today the city is a cultural and educational centre and also manufactures chemicals, textiles, rubber and steel, and holds a trade fair each spring.
Population 514 000
Map West Germany Cb

Hanshin ***Japan*** Industrial zone on the north-east shore of the Inland Sea, around the cities of Osaka and Kobe. It accounts for about one-sixth of the country's manufactures. Main products include iron and steel, chemicals and machinery.
Map Japan Cd

Haora (Howrah) ***India*** Large industrial town on the west bank of the river Hugli opposite Calcutta. It has railway workshops, jute and cotton mills, paper factories and iron and steel rolling mills.
Population 744 400
Map India Dc

Har Horin ***Mongolia*** See KARAKORUM

Har Karmel ***Israel*** See CARMEL, MOUNT

Harappa ***Pakistan*** Site of an ancient city, about 80 km (50 miles) south of Lahore. Harappa was probably occupied from 2500 to 1500 BC. It covered an area 5 km (3 miles) around and had a heavily fortified castle. The city was built of brick, but much of it had been ransacked for railway ballast and local buildings before proper archaeological excavations began in the 1920s. The name Harappan has been given to the Indus valley civilisation, whose greatest city was MOHENJO DARO.
Map Pakistan Db

Harare ***Zimbabwe*** National capital, formerly called Salisbury after the 19th-century British Prime Minister, Lord Salisbury. Founded in 1890 on a site 1470 m (4822 ft) above sea level, it is now a modern city, many of whose streets are lined with jacaranda trees that flower in spring, summer and early autumn. Harare has the world's largest tobacco market and produces a wide range of products, including bacon, beer, building materials, flour, furniture, sugar and tobacco. It is the home of the National Parliament and the University of Zimbabwe.
Population 656 000
Map Zimbabwe Ca

Harbel ***Liberia*** Headquarters and industrial centre of the American Firestone Plantation Company, 45 km (28 miles) east of the capital, Monrovia. It lies in one of the world's largest and most modern rubber plantations, which covers 290 km² (110 sq miles). Harbel itself has one of the world's biggest rubber-processing factories, and most of the services of a city – including a hydroelectric power station, a radio station, schools, churches, hospitals, a brick-works and a soft-drink bottling plant.
Map Liberia Aa

Harbin ***China*** Capital of Heilongjiang province about 500 km (300 miles) north of the North Korean border. It is a major railway junction and a centre for food processing and heavy engineering industries.
Population 2 100 000
Map China Jb

Hardangerfjord ***Norway*** One of Norway's longest fiords – 145 km (90 miles) – running east and south-east of Bergen. It is the site of several hydroelectricity plants and some of Norway's best-known tourist resorts. To the east is Hardangervidda, an extensive plateau.
Map Norway Bc

hardness Relative resistance of a mineral to scratching, as measured by Mohs Scale.

Hardwar ***India*** See HARIDWAR

hardwood Timber from broadleaved trees, as opposed to conifers. The temperate varieties include oak, beech, cherry, maple and walnut, while from tropical forests come ebony, ironwood, mahogany, rosewood and teak. Most temperate hardwood trees are deciduous, shedding all their leaves at one time. In equatorial forests, there is no seasonal leaf fall – leaves persist for a period and are shed irregularly.

Hargeysa ***Somalia*** Northern market town, about 130 km (80 miles) south-west of the port of Berbera. Nomadic herdsmen trade livestock there. It was the capital of British Somaliland from 1941 to 1960.
Population 50 000-80 000
Map Somalia Ab

Harghita Mountains ***Romania*** Ridge of volcanic mountains in eastern TRANSYLVANIA, 168 km (104 miles) south of the border with Russia. Stretching for 50 km (30 miles) from north to south, the mountains rise to 1800 m (5905 ft).
Map Romania Ba

Hari Rud ***Afghanistan/USSR*** River, 1125 km (700 miles) long, which rises in the Kuh-e Baba mountains, about 150 km (93 miles) west of Kabul, and flows west through the fertile Herat Valley. It curves north to form part of the border with Iran, and ends at the remote oasis of Tedzhen in Soviet Turkmenistan.
Map Afghanistan Ba

Haridwar (Hardwar) ***India*** Holy town on the Ganges about 165 km (102 miles) north-east of Delhi. Each year more than 2 million Hindu pilgrims come here to bathe, and to spread the ashes of cremated relatives. The bathing ghat (area) contains a footprint said to be that of the Hindu god Vishnu.
Population 145 900
Map India Cb

Harkany ***Hungary*** See MOHACS

harmattan Dry, dusty north-east wind blowing from the high atmospheric pressure areas of the Sahara towards humid low pressure areas over the West African coast.

Harper (Maryland) ***Liberia*** Port in the south-east of the country, 25 km (16 miles) from the Ivory Coast border. Founded in 1834 as an independent colony, it became part of Liberia in 1857 and was named after Robert C. Harper, a member of the American Colonisation Society which founded Liberia in the early 19th century as a home for freed US slaves. Harper's modern development began after the establishment of a US-owned rubber plantation near Gedetarbo on the Cavally river, 25 km (16 miles) to the north. The port's main exports now are rubber, cocoa and timber.
Population 14 000
Map Liberia Bb

Harrisburg ***USA*** State capital of Pennsylvania, about 150 km (95 miles) west and slightly north of the city of Philadelphia. It was founded in the early 18th century as a trading post on the Susquehanna river. Nearby coal mines were the basis of its industries, which now produce machinery, aircraft parts and wood products.

In 1979 it was the scene of a near-disaster when a fault in the nearby Three Mile Island nuclear power plant caused a partial meltdown of a reactor's core. Some radioactivity escaped but the accident reportedly caused no deaths.
Population (city) 52 100; (metropolitan area) 570 200
Map United States Kb

Harrogate ***United Kingdom*** Town in North Yorkshire, 30 km (18 miles) west of York, that became one of England's most renowned spas in the 19th century. As demand for 'the cure' – drinking and bathing in its sulphur, iron and magnesia spring waters – declined, the town developed conference and exhibition facilities.
Population 66 000
Map United Kingdom Ed

Hartford ***USA*** State capital of Connecticut, on the Connecticut river about 150 km (95 miles) north-east of New York city. It is a centre for firearms, aircraft engines, office equipment and domestic appliances manufacturing, and of the insurance industry.
Population (city) 135 700; (metropolitan area) 1 030 400
Map United States Lb

Haryana ***India*** Agricultural state to the north-west of the capital, Delhi, covering 44 212 km² (17 066 sq miles). It was formed in 1966 and its wheat farming depends upon new 'green revolution' technology – the use of high-yielding seeds, fertilisers and new cultivating techniques – and irrigation from canals built since 1947. In 1986 its capital, CHANDIGARH, was to be transferred to the adjacent state of Punjab, but the plans were delayed after protests. Haryana has been promised a new capital.
Population 12 922 600
Map India Bb

Harz ***East Germany/West Germany*** Range of forested mountains on both sides of the border between West and East Germany. The mountains, which have mines and many summer tourist resorts, were immortalised in the 19th century in the German writer Johann Goethe's poetic drama *Faust*. Their highest point is the Brocken, in East Germany, which rises to 1142 m (3747 ft).
Map East Germany Bc

Hasselt ***Belgium*** Industrial town 66 km (42 miles) east of Brussels and capital of the province of Limburg. The medieval heart of the town is surrounded by modern suburbs and an industrial belt along the Albert Canal, which runs past the town on the northern side. Main industries are brewing, distilling and agricultural machinery. Some 6 km (4 miles) east of Hasselt is the estate of Bokrijk, with a large open-air museum of Flemish rural life.
Population 64 000
Map Belgium Ba

Hassi Messaoud ***Algeria*** Centre of the Algerian oil industry, situated in the middle of the Sahara some 320 km (about 200 miles) south of Biskra. Pipelines begin at Haoud El Hamra and take the oil to the northern ports of ARZEW, BEJAIA and SKIKDA. The field is also linked by pipeline to the HASSI R'MEL gas field. Since oil was discovered in the 1950 s, water has also been found and the area now has swimming pools and tree plantations.
Map Algeria Ba

Hassi R'Mel ***Algeria*** One of the world's largest natural gas fields with reserves estimated at about 1.2 million m³ (42.38 million cu ft). The field lies in the Sahara about 100 km (62 miles) north of Ghardaia, and pipelines take the gas to the ports of ARZEW, 416 km (260 miles) to the north-west and SKIKDA, about 590 km (370 miles) to the north-east.
Map Algeria Ba

Hastings ***New Zealand*** Town on North Island, about 250 km (155 miles) north-east of the capital, Wellington. It stands inland from Hawke Bay and is the centre of a rich fruit-growing and pastoral region. Settled in 1864 on land leased from local Maoris, it was named after Warren Hastings (1732-1818), Britain's first Governor-General of India. Like NAPIER, about 20 km (12 miles) to the north-east, it was rebuilt after an earthquake in 1931.
Population 52 600
Map New Zealand Fc

Hastings ***United Kingdom*** Sussex coast town, 85 km (53 miles) south-east of London, best known for the battle in 1066 when the invading Normans under William the Conqueror defeated the English under King Harold. This actually took place near Battle, 10 km (6 miles) inland. The old part of Hastings, with narrow winding streets, half-timbered houses and, on the beach, tall wooden 16th-century huts for drying nets, preserves the atmosphere of a medieval fishing village. It was one of the CINQUE PORTS and is overlooked by the remains of the castle built by William I in 1067.
Population 77 000
Map United Kingdom Fe

▼ SLEEPING GIANT Heleakala, a dormant volcanic crater of Kolekole (3055 m, 10 023 ft), lies in a national park on the island of Maui in Hawaii. The crater covers 49 km² (19 sq miles).

Hat Yai ***Thailand*** Town in the far south, about 40 km (25 miles) from the Malaysian border. It is a rubber processing town which is also popular with Malaysian tourists.
Population 261 400
Map Thailand Bd

Hatteras, Cape ***USA*** Easternmost tip of Hatteras Island, one of a 160 km (100 mile) string of islands along the stormy Atlantic coast of North Carolina. It was notorious for shipwrecks in the days of sail but is now guarded by a lighthouse. Much of Hatteras island is preserved as a National Seashore.
Map United States Kc

Hattusas ***Turkey*** See BOGAZKALE

Hauraki Gulf Bay on the north-eastern coast of NORTH ISLAND, New Zealand. The port of AUCKLAND stands on an isthmus between the MANUKAU and Waitemata harbours, two inlets of the Hauraki Gulf.
Dimensions 55 km (34 miles) (east-west), 42 km (26 miles) (north-south)
Map New Zealand Eb

Hausaland ***Nigeria*** Northern region inhabited by the Hausa peoples, and centred on the Sokoto river basin in the north-west and the great plateau to the east. It is mostly dry and flat, and rises from 600 m (1970 ft) in the west to 1050 m (3445 ft) near the adjoining Jos Plateau to the south-east. Hausaland is one of Nigeria's main cattle producing areas, and crops include millet, sorghum, groundnuts, cotton, vegetables and, in the river valleys, rice. The main towns are Kano, Sokoto and Kaduna.
Map Nigeria Ba

Hautes Fagnes ***Belgium*** See ARDENNES

Haut-Zaire ***Zaire*** The country's largest administrative region, covering 503 239 km² (194 300 sq miles) in the north-east. It lies mostly in the humid Zaire river basin, with highlands overlooking the Great Rift Valley system in the east. Forests blanket the centre and south, with savannah in the north and mountain vegetation in the east. The economy depends mainly on plantation crops, including cotton. There are scattered bands of pygmies, but most people belong to various Bantu or Sudanese farming groups. The capital is Kisangani.
Population 4 542 000
Map Zaire Ba

Havana (La Habana) ***Cuba*** **1.** Seaport and capital of Cuba, and the largest city in the West Indies, situated on the north-west coast. The whole metropolitan area forms the province of Cuidad de la Habana. The harbour, one of the best in the western hemisphere, was fortified by the Spanish in the 16th century. An entrance canal, guarded by the late 16th-century Castillo del Morro, leads into the broad interior haven. To the west of the harbour, the Río Almendares gently meanders through the city and down to the sea. The old city, with its 16th and 17th-century buildings, which reflect its Spanish past, is the commercial centre of the country. The new city is mainly residential, and reflects American influence rather than Spanish. In 1898 the harbour was the scene of the blowing up of the US battleship *Maine*, which precipitated the Spanish-American War, and resulted in US military occupation of Cuba until 1902, when it became independent.

Havana exports sugar, tobacco products, coffee and tropical fruit, and handles most of the country's imports. Industries include oil refining, cigar making (the world-famous Havanas), textile mills, clothing and packing plants.

2. Province of Cuba lying near the west end of the island. Most of its economy is based on agriculture – mainly tobacco, sugar cane, citrus fruits, cattle, rice and coffee – while copper and iron are found in the highlands. The capital is Güira de Melena (population 42 000).

Population (metropolitan area) 1 925 000; (Havana province) 586 000

Map Cuba Aa

Hawaii ***Pacific Ocean*** State of the USA made up of 122 islands in the North Pacific. They were settled by Polynesian people (originally from South-east Asia) 1000-1500 years ago. The English navigator Captain James Cook discovered them in 1778 and named them the Sandwich Islands. They remained independent throughout the 19th century but, at the request of the Hawaiian people, were annexed to the USA in 1898. They became the 50th state in 1959.

The coral and volcanic islands, of which 14 are inhabited, lie over 3700 km (2300 miles) from the city of San Francisco on the US mainland. There are many volcanoes, including MAUNA KEA (4206 m, 13 796 ft) and MAUNA LOA (4169 m, 13 667 ft), both on the largest island, also called Hawaii. Agriculture is subtropical, and sugar cane, pineapples and coffee are the chief products. There are several defence bases, including PEARL HARBOR. The state, which has a total land area of 16 705 km² (6450 sq miles), is popular with tourists. HONOLULU on the island of OAHU is the state capital.

Population 1 054 000

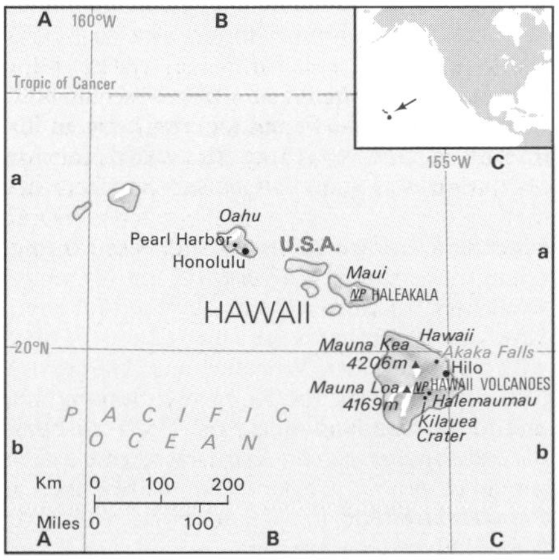

Hawaii Volcanoes National Park ***Hawaii*** Park covering 990 km² (382 sq miles) on the island of Hawaii. It includes the volcano MAUNA LOA (4169 m, 13 677 ft), with its active craters of Kilauea and Mokuaweoweo. The park has much tropical vegetation and birdlife, and fantastic lava formations.

Map Hawaii Bb

Hawke Bay Bay on the east coast of NORTH ISLAND, New Zealand, named in 1769 by Captain James Cook after the British First Lord of the Admiralty, Sir Edward Hawke (1705-81). Its southern tip, Cape Kidnappers, was so named because it was where Maoris seized a boy from Cook's ship. The boy escaped.

Dimensions 80 km (50 miles) long, 56 km (35 miles) wide

Map New Zealand Fc

Hawke's Bay ***New Zealand*** District of North Island centred on Hawke Bay 320 km (200 miles) south-east of Auckland. It is a farming area covering 11 289 km² (4358 sq miles) and producing sheep, fruit, vegetables and wine.

Population 148 400

Map New Zealand Fc

Hazarajat ***Afghanistan*** Bleak and forbidding mountain region in the centre of the country, to the west of Ghazni. Its valleys are fertile and are inhabited by the Hazaras, a Mongol people believed to have settled there after the invasion of Genghis Khan in the early 13th century.

Map Afghanistan Bb

haze Mass of atmospheric water droplets with dust, smoke and salt which restricts visibility to 1-2 km (0.6-1.2 miles).

Heard Island ***South Indian Ocean*** Uninhabited island about 4000 km (2500 miles) south-west of the Western Australian capital of Perth. The island, which is 40 km long and 17 km wide (25 miles by 11), contains an active volcano: Mawson Peak (2745 m, 9006 ft), named after the Australian scientist and Antarctic explorer Douglas Mawson (1882-1958). With the Macdonald Islands, about 40 km (25 miles) to the east, it is administered as an Australian external territory.

Map Indian Ocean Ce

heat equator Zone of maximum heating on the surface of the earth which migrates seasonally, following the position of the overhead sun. It moves farther north and south over areas of land than over the oceans, since the land heats up more quickly than water.

heath, heathland Extensive tract of open, uncultivated land, often on sandy soil and covered with shrubby plants of the heather family.

heave Horizontal displacement of rock strata by an inclined FAULT.

Heaviside-Kennelly layer See ATMOSPHERE

Hebei ***China*** Province in the north of China, covering 180 000 km² (70 000 sq miles) of the northern half of the North China Plain. It surrounds Beijing (Peking) and includes part of the mountainous region between China and the Mongolian plateau. The climate is extreme. The winters are bitterly cold; the summers are uncomfortably hot and sometimes bring disastrous droughts. Despite its harsh environment, Hebei is densely populated, particularly in the lowland areas, where wheat and cotton are grown. There are also abundant coal reserves which are used to fuel iron and steel works and other heavy industries.

Population 51 046 400

Map China Hc

Hebrides ***United Kingdom*** See WESTERN ISLES

Hebron ***Israeli-occupied Jordan*** Town in the WEST BANK area, situated 35 km (21 miles) south-west of Jerusalem. The centre of a rich agricultural area, its main industry is glass-making. Hebron, said to be one of the world's oldest cities, is the traditional burial place of Abraham – the father of the Hebrew people – in the nearby Cave of Machpelah. Other Biblical figures said to be buried there are Isaac, Jacob, Leah, Rebecca and Sarah.

Population 38 000

Map Jordan Ab

Heerlen ***Netherlands*** Former coal and steel-producing town in Limburg, in the extreme south-east just 5 km (3 miles) from the West German border. Since 1975, when the Dutch closed all their coal mines and switched to mainly home-produced natural gas, it has been hard pressed to find replacement jobs.

Population 92 200; (Greater Heerlen) 265 000

Map Netherlands Bb

Hefei ***China*** Capital of Anhui province about 450 km (280 miles) west of Shanghai. Its industries include steel and aluminium production, and engineering.

Population 1 484 000

Map China He

Heidelberg ***West Germany*** Town on the Neckar river some 18 km (11 miles) south-east of Mannheim. It has the country's oldest university, founded in 1386, and the ruins of a 13th-century castle overlook the town. From the 13th to the 18th centuries Heidelberg was the residence of the Electors of the Palatinate. The chemist Robert Bunsen (1811-99), who invented the Bunsen burner, was a professor at the university.

Population 130 000

Map West Germany Cd

Heilbronn ***West Germany*** Wine-making town in the valley of the Neckar river, about 40 km (25 miles) north of Stuttgart. It also has leather, paper, chemical and ironware industries.

Population 110 000

Map West Germany Cd

Heilong Jiang ***China/USSR*** See AMUR

Heilongjiang ***China*** The country's most northerly province, covering 460 000 km² (178 000 sq miles) and bordering on the Soviet Union. Its climate and soil are similar to those of Alberta, Canada, and, like the Canadian prairie province, the Chinese province is a major wheat-growing region. Because of this, farms are larger

and more mechanised than in the rest of the country. Other crops include soya beans and sugar beet, and sunflowers grown for oilseed. The province has much timber, but its most important natural resource is oil. Its capital is Harbin.
Population 31 690 000
Map China Ib

Hejaz ***Saudi Arabia*** Red Sea coastal region and former province (now renamed Western Province), stretching from The Gulf of Aqaba in the north to Asir in the south. Once an independent kingdom, it was conquered and became part of the dual kingdom of Nejd and Hejaz, formed in 1926 (which became Saudi Arabia in 1932). The Hejaz mountains run parallel to the coast, rising to a peak of 2174 m (7000 ft). The area is mainly arid, but in the south there are two particularly fertile areas, yielding wheat, millet and dates. The disused Hejaz railway also runs parallel to the coast to Medina. The railway featured prominently in the Arab campaign of Colonel T.E. Lawrence (Lawrence of Arabia) during the First World War.
Map Saudi Arabia Ab

Hekla ***Iceland*** Active volcano in the south of the country about 112 km (75 miles) east of REYKJAVIK. Its crater is some 5 km (3 miles) across and the mountain is nearly 1491 m (4892 ft) high. There have been 24 recorded eruptions, the last in 1970, and they have left huge lava fields covering 420 km² (162 sq miles).
Map Iceland Cb

Helder, Den ***Netherlands*** Port and the country's main naval base. It lies on the north-western tip of the mainland, and its ferries connect with the Friesian Island of Texel.
Population 63 700
Map Netherlands Ba

Helena ***USA*** State capital of Montana, in the foothills of the Rocky Mountains, about 175 km (110 miles) east of the Idaho border. Gold was discovered in what is now the city centre in 1864. Today Helena is a centre of metal smelting, agriculture and tourism.
Population 24 600
Map United States Da

Heligoland (Helgoland) ***West Germany*** Island 1 km² (0.5 sq miles) in the North Sea some 70 km (44 miles) from the mouth of the River Elbe. It was held by the British until 1890, when it was handed over to Germany. It was used as a naval base during the two World Wars. Heligoland is now a duty-free zone.
Population 2000
Map West Germany Ba

Heliopolis ***Egypt*** North-eastern suburb of Cairo, on the site of an ancient holy city (referred to in the Bible as Ob). It was dedicated to the sun-god Ra, and its temple became an important library. Two of its stone obelisks – called Cleopatra's needles – now stand in London and New York city.
Map Egypt Bb

Helmand ***Afghanistan*** The country's longest river, 1100 km (688 miles), rising in the Hindu Kush west of Kabul. It flows south-west across the country, through the Hazarajat region and on to the Iranian border – where it turns north and drains into the swamps of Lake Harmum.
Map Afghanistan Bb

Helmstedt ***West Germany*** Town at the principal crossing point between West and East Germany, east and slightly south of Hanover.
Population 26 000
Map West Germany Db

Helsingborg ***Sweden*** Main ferry port to Denmark, at the narrowest part of ORESUND, 4.5 km (3 miles) wide, opposite the Danish town of Elsinore. Its manufactures include rubber, cotton, sugar, pottery and beer.
Population 105 500
Map Sweden Bd

Helsingfors ***Finland*** See HELSINKI

Helsingør ***Denmark*** Sea ELSINORE

Helsinki (Helsingfors) ***Finland*** Capital of Finland, founded in 1550 on a peninsula in the Gulf of Finland. It was made capital in place of Turku after 1809, when Russia annexed Finland from Sweden, of which Finland had been a part since the 12th century.

Helsinki, planned upon a gridiron pattern of streets, rapidly outstripped other Finnish towns in size to become the largest port and manufacturing centre. The city's industries include shipbuilding, printing, sugar refining and the manufacture of textiles and porcelain. One major shipyard dock is completely enclosed to protect workers from the bitter winter weather, and the Arabia porcelain factory is the largest ceramic plant in Europe.

In addition to the neo-classical buildings of the city centre, there are a number of other attractive public buildings, among them the neo-classical cathedral, the railway station (1916), the grey granite parliament building, the Olympic stadium and the marble-faced Finlandia Concert and Congress Hall. There is also an international airport and an expanding network of underground trains. Many of the city's inhabitants live in a semi-circle of suburban developments set in the forested hinterland and along the coves and headlands of the waterfront. Helsinki has Finland's oldest and biggest university founded as a technical university in 1849 – and many museums and galleries.
Population 482 900
Map Finland Cc

hematite Red, grey or black mineral, ferric oxide, found mostly in igneous rock and veins. It is an important source of iron.

Henan ***China*** Province covering 160 000 km² (62 000 sq miles) in central China. Eastern Henan is low-lying and is drained by tributaries of the Huang He and Huai He rivers. The western half is hilly and includes part of the LOESSLANDS OF CHINA as well as the foothills of the Daba Shan. Henan is a major wheat area and also produces tobacco, sesame, rapeseed and peanuts. The floods that used to plague the province have been ended by river control schemes, particularly along the Huang He.
Population 71 890 000
Map China Gd

Henegouwen ***Belgium*** See HAINAUT

Henzada ***Burma*** Town on the Irrawaddy delta in the south of the country, 125 km (78 miles) north-west of Rangoon. An important rice trading centre, it is linked by ferry to the country's main rail network.
Population 283 600
Map Burma Bc

Heraclea Lyncestis ***Yugoslavia*** See BITOLA

Heraklion (Candia; Irklion) ***Greece*** Main port and capital of the island of CRETE. In the 9th century AD it was used as a base by Arab pirates and slave traders. The Venetians took over in 1204. Heraklion fell to the Turks in 1669 after 22 years of siege. It became part of Greece in 1913. There are Venetian fortifications dating from the 14th to the 17th centuries, Venetian churches and a Venetian castle commanding the city's twin harbours. Minoan vases, statues and frescoes, many from the excavated palace of KNOSSOS, 5 km (3 miles) to the south, are in the city museum.

Heraklion was the birthplace of the painter Domenikos Theotokopoulos, better known as El Greco (1541-1614), and of the novelist Nikos Kazantzakis (1885-1957), the author of *Zorba the Greek*.
Population 111 000
Map Greece Dd

Herat ***Afghanistan*** City on the Hari Rud river 220 km (137 miles) north of Farah. Known as 'The Gateway to Afghanistan', Herat is 914 m (2998 ft) above sea level, and is enclosed by 15th-century walls. The city is dominated by its spectacular citadel, and is noted for its mosques, palaces and ancient tombs. Its main products are textiles and carpets.
Population 150 500
Map Afghanistan Ba

Hercegovina ***Yugoslavia*** See BOSNIA HERZEGOVINA

Herculaneum (Ercolano) ***Italy*** Graeco-Roman archaeological site on the Bay of Naples at the foot of Mount VESUVIUS. It was smothered by a volcanic eruption in AD 79 which buried it in a sea of mud and ash up to 25 m (82 ft) thick. The excavations, entered from the modern town of Ercolano, lie well below ground level. There are several fine black and white mosaic floors and many well preserved public buildings, shops, houses and statues.

Hercynian Belonging to a phase of mountain building in the late Palaeozoic era, between 200 and 250 million years ago. Also called Variscan and ARMORICIAN. See GEOLOGICAL TIMESCALE

Hereford ***United Kingdom*** City (and former county) of western England, 75 km (47 miles) south-west of Birmingham. Its beautiful sandstone cathedral shares the Three Choirs Festival with its neighbours, Worcester and Gloucester. The city stands on the River Wye, among hills and fertile farmland where the Hereford breed of cattle originated and where the region's cider is made.
Population 48 000
Map United Kingdom Dd

Hereford and Worcester ***United Kingdom*** County of the west of England on the Welsh border, combining the two former counties of its name. It covers 3927 km² (1516 sq miles), stretching north to the edge of the Black Country near the city of Birmingham and south-east into the Vale of Evesham, one of Britain's leading fruit-growing and market gardening areas. Tourism is centred on the valley of the River Severn and on the Malvern Hills, which rise to 425 m (1394 ft) and were the home of the composer Edward Elgar (1857-1934).
Population 648 000
Map United Kingdom Dd

Hereroland ***Namibia*** Two homelands – East and West Hereroland – covering a total of 68 449 km² (26 428 sq miles) in the north-east. They were set up in the 1970s for the Herero people and consist of wooded and partly shrub-covered savannah lands. Cattle and sheep farming is the main occupation.
Population 19 700
Map Namibia Ab

Hermon, Mount ***Lebanon*** Peak 44 km (27 miles) south-east of the port of Sidon, standing 2814 m (9232 ft) above sea level close to the meeting point of the borders of Lebanon, Syria and Israel. Known to the Arabs as Jebel esh Sheikh, it is also called the Old Man Mountain and Snow Mountain. Its summit is snow-covered for most of the year, and its three peaks tower over the ruins of the Biblical town of Dan. Mount Hermon's lower slopes are well-wooded, and it is surrounded by the ruins of ancient temples – mainly to Baal, the chief god of the Phoenicians. The mountain is the source of the River Jordan.
Map Lebanon Ab

Hertford ***United Kingdom*** Town and county of southern England. The town, 32 km (20 miles) north of London, has the ruins of a castle founded in the 10th century by the son of the Saxon King Alfred the Great. Queen Elizabeth I (1533-1603) spent her childhood at nearby Hatfield House. The 'garden cities' of Letchworth and Welwyn were built in the 1920s as an experiment in planned living. In the centre of the county, whose area is 1634 km² (631 sq miles), lies the Roman city of Verulamium, today the cathedral city of ST ALBANS.
Population (county) 980 000; (town) 22 000
Map United Kingdom Ee

Herzegovina ***Yugoslavia*** See BOSNIA HERZEGOVINA

Hessen ***West Germany*** Hilly state covering 21 112 km² (8151 sq miles) of central West Germany between the Rhine valley and the East German border. Its capital is the spa town of WIESBADEN.
Population 5 500 000
Map West Germany Cc

Hexrivier Valley ***South Africa*** Ancient route of Bushmen hunters, along which they left many cave paintings. The route, in south-west Cape Province, was followed in the 19th century by white settlers heading into the interior. The area grows most of South Africa's dessert grapes.
Map South Africa Ac

Hidalgo ***Mexico*** Small agricultural and silver-mining state north of Mexico City, with impressive Toltec remains, dating from about the 12th century, at Tula.
Population 1 622 000
Map Mexico Cb

Hieron Epidaurou ***Greece*** See EPIDHAVROS

high An area of high atmospheric pressure. See ANTICYCLONE

high water Highest point reached by the sea in any one tidal cycle.

Highland ***United Kingdom*** Administrative region covering 26 136 km² (10 091 sq miles) of northern Scotland – the largest in the country. It takes in the former counties of Inverness, Nairn, Ross and Cromarty, Sutherland and Caithness, plus parts of Argyll and Moray. The sparsely populated Highlands have suffered through a turbulent history of depopulation and clearance, but have been regenerated today – by forestry, tourism and industries based on the North Sea oil fields. Although Ben Nevis, the highest mountain in Britain at 1344 m (4409 ft), is in the Grampian Mountains, it falls within the Highland region. The main town is INVERNESS.
Population 196 000
Map United Kingdom Cb

Highlands, The ***United Kingdom*** The part of Scotland north and west of an approximate line joining the Firth of Clyde and the town of Stonehaven on the east coast at the foot of the Grampian Mountains. It includes the most rugged of Scotland's mountains but also some low-lying country, especially near the coast. Its symbols are heather and tartan, kilt and clan. The area to the south of the line – known as the Lowlands, although it contains some substantial hills – has long been culturally distinct.
Map United Kingdom Cb

highveld Term used in southern Africa for open grassland found between about 1500 and 1850 m (4900 and 6100 ft) above sea level. Trees are few because of the altitude, the large range of temperatures each day (especially in winter, when frosts occur), and the high frequency of droughts. Corresponding to the pampas, prairies and steppes in other continents, the highveld is used to rear cattle and for arable farming.

Hildesheim ***West Germany*** Cathedral town 30 km (19 miles) south and slightly east of Hanover. The cathedral was started in the 9th century but was mostly rebuilt after damage in the Second World War. The bronze doors and a chandelier remain from the 11th century.

In the cloisters is a rose which folklore claims to have been growing there since 815, when a chapel was first built on the site.
Population 101 000
Map West Germany Cb

hill fog Low cloud enveloping high ground.

Hillerød ***Denmark*** Small tourist town that has grown up around the magnificent Frederiksborg Castle, on a lakeside 36 km (22 miles) north-west of Copenhagen. The Renaissance castle, where the kings of Denmark used to be crowned, was built by Christian IV in the 17th century. It was gutted by fire in 1859, but was restored by private subscription in 1860-75 and is now Denmark's National History Museum. About 9 km (6 miles) north-east of Hillerød is the 18th-century Fredensborg Palace, the spring and autumn home of the Danish royal family, which is open to the public in July. The park surrounding it is open all year round.
Population 23 500
Map Denmark Cb

Hilo ***Hawaii*** Largest settlement on the island of Hawaii. The city is also a port, and was devastated by a tsunami (a large wave created by an under-sea earthquake) in 1946.
Population 35 300
Map Hawaii Bb

Hilversum ***Netherlands*** Town about 25 km (15 miles) south-east of the capital, Amsterdam. It has a 'garden city' appearance, and has grown in the 20th century as a commuter town for the capital and as the headquarters of the Dutch broadcasting system.
Population (town) 89 500; (Greater Hilversum) 106 800
Map Netherlands Ba

Himachal Pradesh ***India*** Mountainous state bordering on China and covering 55 673 km² (21 490 sq miles). It was created as a territory from 30 former hill states in 1948. In 1966, parts of Punjab were added to it, and it was made a state in 1970. Tea, rice, maize and wheat are grown there. Its capital is the hill town and resort of SIMLA.
Population 4 280 800
Map India Ba

Himalayas ***Southern Asia*** The world's mightiest mountain range, curving across southern Asia for some 2400 km (1500 miles). It stretches between the Indian states of Jammu and Kashmir in the west and Assam in the east, and along the way sweeps right across northern Nepal.

The range is divided into three sections: the Great Himalayas in the north, which include Mount EVEREST; the Lesser Himalayas in the centre; and the Outer Himalayas in the south, which include the Siwalik Range. The Himalayas have an average height of about 6100 m (20 000 ft), and of the 109 peaks in the world rising to more than 7315 m (24 000 ft), above sea level, 96 are in this range.

The name 'Himalayas' comes from the Nepalese words *him*, meaning 'snows' and *alya*, meaning 'home of'. Some of the Western World's favourite flowering shrubs were introduced from the Himalayas, including some fine species of rhododendrons in the 18th and 19th centuries.
Map Asia Fe

Himeji ***Japan*** Steelmaking and oil-refining city and port on the Inland Sea about 75 km (47 miles) west and slightly north of the city of Osaka. It is overlooked by the beautiful White Heron Castle, dating from the 14th century, with its five-storey main tower, 38 separate buildings and 21 gates.
Population 452 900
Map Japan Bd

Hims ***Syria*** See HOMS

Hindu Kush ***Central Asia*** The western extension of the Himalaya Mountains, stretching 600 km (373 miles) from the PAMIRS in the north-east to the KUH-E BABA mountains in the west. The name Hindu Kush means 'Destroyer'. Many of the passes are more than 2500 m (8200 ft) high. The highest peak in the range is Tirich Mir (7690 m, 25 229 ft), lying in the east in Pakistan. To the west, the mountains form a great almond-shaped range. There the average height is 3200 m (10 500 ft). Since ancient times the lower passes have been used by armies heading south to invade India.

Map Afghanistan Ba

Hirmand ***Afghanistan/Iran*** Huge arid basin, 240 km (150 miles) long, to the east of the Dasht e Lut desert, on the countries' border. The basin contains Lake Hamun, situated partly in Iran and partly in Afghanistan. Several rivers flow into it, but none flow out.

Map Iran Ca

Hirosaki ***Japan*** University town of northern Honshu, about 35 km (20 miles) south-west of the port of Aomori. It is dominated by the 17th-century Hirosaki Castle, which is famous for the cherry blossom in its gardens – among the best in Japan.

Population 176 100

Map Japan Db

Hiroshima ***Japan*** Regional capital in western Honshu straddling the Ota river delta at the western end of the Inland Sea. It is now a car manufacturing and shipbuilding centre. The Peace Memorial Park and the ruined dome of the Municipal Industrial Promotions Hall are reminders of the world's first atomic bomb that was dropped on the city on August 6, 1945, killing 78 000 people and destroying 80 per cent of the city's buildings. August 6 is commemorated each year with religious ceremonies and rallies.

Population 899 400

Map Japan Bd

Hispaniola ***West Indies*** Second largest island in the Caribbean after Cuba, covering about 76 200 km² (29 400 sq miles). It lies between Cuba and Puerto Rico, and comprises HAITI on the west and the DOMINICAN REPUBLIC on the east. On it stands Pico Duarte, formerly Monte Trujillo (3175 m, 10 417 ft), the highest point in the West Indies. The island was discovered by Christopher Columbus on his first voyage to the Americas in 1492 and settled by the Spanish a year later.

Population 12 600 000

Map Caribbean Ab

Hispur Glacier ***Pakistan*** Longest glacier in the Himalayas and, at 50 km (30 miles), one of the longest in the world outside the polar regions. It lies in the Karakoram Range in Azad Kashmir.

Map Pakistan Da

Histria ***Romania*** Extensive Greek, Roman and Byzantine archaeological site overlooking the Black Sea, 50 km (30 miles) north of Constanta. Histria, which dates from the 7th century BC, contains ruins of a temple to Aphrodite, the Greek goddess of love, Greek baths, and Roman and Byzantine homes and shops.

Map Romania Bb

Hitachi ***Japan*** City on the east coast of Honshu island about 125 km (78 miles) north-east of the capital, Tokyo. Originally a copper-mining area, the city now produces heavy electrical machinery.

Population 206 100

Map Japan Dc

Hluhluwe Game Reserve ***South Africa*** Reserve in KwaZulu (Zululand), about 280 km (175 miles) north-east of Durban and noted for its white and its black rhinoceroses. Other species in the reserve, which covers 231 km² (89 sq miles), include antelopes, baboons, buffaloes, cheetahs, crocodiles, giraffes, leopards, wart hogs and wildebeests. Hluhluwe and the nearby St Lucia and Umfolozi reserves are the oldest wildlife sanctuaries in Africa. Hluhluwe (pronounced *shloo-shloo-wee* with the accent on the second syllable) was founded in 1897.

Map South Africa Db

Ho Chi Minh City (Saigon) ***Vietnam*** The country's largest city, lying in a rich plain, some 50 km (30 miles) from the southern coast. The city was the French colonial capital and the capital of South Vietnam. It was laid out as a French-style city, with broad tree-lined boulevards and pavement cafés, and became the shopping and trading centre for the people of wealthy rubber and rice plantations of the plains and the Mekong river delta.

It kept the name Saigon until the collapse of the US-backed South Vietnam regime in 1975, when the two Vietnams were unified under the Hanoi government. Then it was renamed Ho Chi Minh City after the Communist leader Ho Chi Minh (1890-1969), who was president of North Vietnam from 1954 until his death.

In the late 1960s, a huge influx of people from the war-torn countryside flooded the city, and many of its former open spaces have now been filled with commercial and residential development. Industry still prospers in the Chinese town of Cholon beside the city, and the deep-water port (still called Saigon) on Khanh Ho island in the delta handles almost all the trade of the southern part of the country.

Population 3 500 000

Map Vietnam Bc

Ho Chi Minh Trail ***South-east Asia*** Complex of roads, trails and paths, cutting through the hills of the Truong Son and leading from the northern part of Vietnam through Laos to southern Vietnam.

Although some of the routeways date from French colonial times, the 'trail' became important in the Vietnam War as the supply route for the insurgent movement in the south. Heavily bombed by the US airforce from bases in Thailand, the 'trail' nevertheless was maintained and today has been upgraded into a drivable road along large sections of its length.

hoar frost Frozen dew or water vapour that forms a white surface on objects near the ground. It forms instead of dew when the dew point is below freezing point (0°C, 32°F).

Hobart ***Australia*** Capital of Tasmania, lying at the mouth of the Derwent river, and one of the country's oldest cities. It was founded in 1804 as a port, and whaling centre, and many early 19th-century buildings remain. The Tasmanian Museum and Art Gallery in the city depicts the culture of the Tasmanian Aboriginal people, who became extinct in 1876 when the last known full-blood Aborigine, Truganini died. Today the port handles fruit, grain, wool, minerals and timber. Industries include flour mills, sawmills and iron foundries.

Population 173 700

Map Australia Hg

Hodeida (Al Hudaydah) ***Yemen*** The country's main industrial and commercial port, situated 144 km (87 miles) south-west of SAN'A. Its principal exports are coffee, dates, senna, myrrh and sesame. It has an oil refinery, an oil-blending plant, a textile mill and two cotton-cleaning plants. Other industries include cigarettes, soft drinks, an oxygen plant and a small aluminium products factory; there are also dairy and fruit-juice factories. Hodeida's new harbour was completed with Russian aid in 1962, and has been expanded since then. There is also an international airport.

Population 128 000

Map Yemen Ab

Hofuf ***Saudi Arabia*** Capital of Al Hasa, the world's largest oasis, Hofuf is situated 112 km (70 miles) south-west of Dhahran. It is an important trading centre and deals in locally grown cereals, fruit and dates. The city is a charming patchwork of alleyways, mud-plastered buildings, and Turkish-style forts and walls. There is a weekly camel market at which women at wooden spindles sell newly spun camel wool and sheep wool. Local tailors are renowned for making braid-trimmed overcoats called 'bishts'.

Population 102 000

Map Saudi Arabia Bb

hogback, hog's back Sharp ridge with steeply sloping sides, produced by the erosion of the broken edges of highly tilted strata on one side and the dip of the rocks on the other. It is a steep form of CUESTA.

Hoggar (Ahaggar) ***Algeria*** Mountain massif of great beauty with moonlike landscapes, in the far south-east of the country. The highest point is Mount Tahat (2918 m, 9573 ft) near Tamanrasset. The region is the home of the nomadic Tuareg people, noted for their excellent basket work.

Map Algeria Bb

Hohe Tauern ***Austria*** Range of mountains in the eastern Alps, between Carinthia and Tyrol. The highest point is Grossglockner (3797 m, 12 460 ft), Austria's highest mountain. Farther east is the Niedere Tauern range, which reaches 2863 m (9393 ft).

The Hohe Tauern National Park, opened in 1971 and the country's largest, covers 2000 km² (770 sq miles). It contains the Pasterze Glacier, 10 km (6 miles) long and covering 24 km² (9 sq miles). The Grossglockner Hochalpenstrasse toll road, which begins at Zell-am-See, runs through some of the finest scenery in Austria.

▲ **SOLDIERS IN WAITING Youths like these in Ho Chi Minh City join the Militia, a paramilitary defence force, at 15. At 17 they will be conscripted for three years' national service.**

It reaches a height of 2505 m (8218 ft) at the Edelweiss Spitze before descending to Lienz. It is usually blocked by snow from November to May.

The 8.5 km (5.3 mile) long Tauern railway tunnel, opened in 1909, was for long the major route through the eastern Alps, but the Katschberg tunnel, 35 km (22 miles) farther east, now carries motorway traffic. There is a major hydroelectric power station, with three dams, in the Kaprun valley south of Zell-am-See.

Map Austria Cb

Hohes Venn ***Belgium*** See ARDENNES

Hohhot ***China*** Capital of the Nei Mongol Autonomous Region (Inner Mongolia). It lies 400 km (250 miles) west of Beijing (Peking). Its industries include food processing and the manufacture of chemical fertilisers.

Population 1 130 000

Map China Gc

Hokitika ***New Zealand*** Port on the west coast of South Island 160 km (100 miles) north-west of Christchurch. It grew up during the gold rush in the 1860s. There is a greenstone (nephrite) factory that produces trinkets for tourists. Kaniere, 5 km (3 miles) to the east, was the site of the country's last surviving gold dredge.

Population 3400

Map New Zealand Ce

Hokkaido ***Japan*** The nation's second largest island, after Honshu. It is rugged, and covers 78 509 km² (30 312 sq miles) in the far north, where winters are bitter. Hokkaido is sparsely populated, and its main products are paper, pulp, butter, cheese, sugar and fish. It is home to the distinctive Ainu people. Sapporo is the chief town, and the highest peak is Mount Asahi (2290 m, 7513 ft).

Population 5 679 400

Map Japan Db

Holguín ***Cuba*** Capital of Holguín province, in the fertile plateau region, 105 km (65 miles) north-west of Santiago de Cuba. It manufactures furniture and tiles, and exports tobacco, coffee, sugar, cereals, beef and hides through the port of Gibara, 30 km (19 miles) to the north.

Population (city) 186 000; (province) 911 000

Map Cuba Ba

Holland ***Netherlands*** Name commonly given to the Kingdom of the Netherlands, but more accurately applied to two north-western provinces: Noord-Holland, including the Friesian Island of Texel (2963 km², 1134 sq miles), and Zuid-Holland (3344 km², 1291 sq miles). Most of the area lies below sea level, protected by dykes and coastal dunes. It is the heart of the country, with 38 per cent of its people, and Randstad – the ring of cities including Amsterdam, Rotterdam, The Hague (the largest Dutch cities), Haarlem and Utrecht. Industries in the growing towns compete for space with market gardeners and bulb growers.

Population (Noord-Holland) 2 308 000; (Zuid-Holland) 3 129 900

Map Netherlands Ba

Hollokö ***Hungary*** See MATRA

Hollywood ***USA*** Northern suburb of Los Angeles, California. It has long been the centre of the country's film industry, and more recently of its television industry. Not only are the studios concentrated in the area, but many stars live in the luxurious suburb of Beverly Hills just to the south-west.

Map United States Cd

holm 1. Small island in a river. **2.** Low land near a stream.

Holocene Second epoch of the Quaternary period of the Cenozoic era of the earth's time scale. See GEOLOGICAL TIMESCALE

holt Small wood, or wooded hill.

Holy Cross Mountains (Gory Swietokrzyskie) ***Poland*** Range of hills in the south, about halfway between Warsaw and Cracow. Thick larch and fir forests clothe its ridges. Many summits are topped by churches or monasteries, including the Benedictine Abbey of the Holy Cross on Lysa Gora, at 612 m (2008 ft) the highest hill. The main town is KIELCE.

Map Poland Dc

Holy Island ***United Kingdom*** See GWYNEDD; NORTHUMBERLAND

Homburg ***West Germany*** See BAD HOMBURG

Homonhon ***Philippines*** See VISAYAN ISLANDS

Homs (Al Khums) ***Libya*** Seaport and market town 104 km (65 miles) east of Tripoli, at the foot of the Jebal Nafusah escarpment. Founded by the Turks, it became important after 1870 for its exports of esparto grass (used for rope, shoes and paper). Today, the town is a tourist base for visiting the ruins of LEPTIS MAGNA, a Roman walled city lying about 3 km (2 miles) to the east. Its principal activities are olive-oil refining, cement manufacture, tuna-fish processing and esparto processing.

Population 89 000

Map Libya Ba

Homs (Hims) ***Syria*** Industrial city on the Orontes river, situated 140 km (87 miles) north of Damascus and an important road and rail centre. The city's industries include oil refining, sugar-beet processing, textiles, jewellery and metallurgical works.

In ancient times Homs (then called Emesa) was noted for the worship of the sun god Baal. It was the birthplace in AD 204 of the Roman Emperor Heliogabalus, and the Church of St Elian commemorates a Roman governor of Emesa who became a Christian martyr. Today there are a number of churches for the Christian community, and in 1957 several catacombs (early Christian underground burial chambers) were discovered. The city's most interesting monument is the Tomb of Khaled Ibn Al-Walid, the warrior who brought the Islamic religion to Syria in AD 636.

Near Homs is the even more ancient site of Kadesh, scene of a battle in 1288 BC between the Hittites and the Egyptians.

Population 414 401

Map Syria Bb

Honduras See p. 29

Honduras, Gulf of (Bay of Honduras) Inlet of the CARIBBEAN SEA, washing the shores of Honduras, Guatemala and Belize. It was explored by the Spanish in the 1520s.

Map Honduras Aa

Hong Kong See p. 292

Hong Kong Island ***Hong Kong*** One of the three territories which make up the Crown Colony of Hong Kong, the other two being the peninsula of KOWLOON and the NEW TERRITORIES. Britain acquired the island under the Treaty of Nanking in 1842 as a base from which to provide protection for British traders operating in Chinese waters. Separated from the mainland by Victoria Harbour, the island is very hilly, the highest point being Victoria Peak (551 m, 1808 ft). On the northern shore stands the city of VICTORIA, capital of the colony. The southern part of the island with its deep bays and inlets contains scattered towns and settlements such as ABERDEEN, Repulse Bay and STANLEY. The island covers an area of 80 km² (31 sq miles).

Population 1 184 000

Map Hong Kong Cb

Honiara ***Solomon Islands*** National capital, on the north coast of Guadalcanal island. It was developed after the Second World War near the site of the wartime air base of Henderson Field, now the country's main airport.

Population 23 500

Honolulu ***Hawaii*** State capital and chief port of Hawaii. It stands on the south coast of the island of Oahu, just east of the US Navy base of Pearl Harbor – the bombing of which by the Japanese in December 1941 brought America into the Second World War. The renowned Waikiki Beach, with its numerous tourist hotels, is in the eastern part of the city, which is backed by mountains of the Koolau Range.

Population (city) 373 000; (metropolitan area) 805 300

Map Hawaii Ba

Honshu ***Japan*** Largest of the Japanese islands, covering 230 988 km² (89 185 sq miles) in the centre of the country. The country's political, economic and cultural life revolves around its great cities of Tokyo, the national capital, Kyoto, Yokohama, Nagoya and Osaka – all on or near the south coast. A mountainous backbone divides the island, and the Pacific Ocean side enjoys hot summers, while the west on the Sea of Japan gets heavy winter snow.

Population 96 685 000

Map Japan Cc

Hooghly ***India*** See HUGLI

Hook of Holland (Hoek van Holland) ***Netherlands*** Rotterdam's coastal port, at the mouth of the Nieuw Waterweg (New Waterway), about 25 km (15 miles) west of the city centre. It is the terminus for North Sea ferries.

Map Netherlands Ba

Hoorn ***Netherlands*** Former seaport, now a town on the west side of IJsselmeer, about 30 km (20 miles) north-east of Amsterdam. It is popular with tourists, for much of the prosperous port of the 17th century, when the Netherlands built a vast empire in the East Indies, survives intact. Hoorn was the birthplace of the explorers Willem Schouten, who named Cape Horn (the southern tip of South America) after the town in 1616, and Abel Tasman, who discovered Tasmania (named after him) in 1642.

Population 26 000

Map Netherlands Ba

Honduras

THIS POOR REPUBLIC, WITH AN ECONOMY BASED ON BANANAS, IS USED BY THE UNITED STATES AS A TRAINING GROUND FOR GUERRILLAS

From the 1920s until the late 1940s, Honduras was the world's leading banana exporter – the original banana republic. Foreign-owned companies – particularly the United Fruit Company of the United States – ran the industry and the government, and had the power to determine national policy. Dependent on a single crop of limited value whose profits went abroad, Honduras did not prosper. Produce is now more diverse, though bananas are still the principal export, but the country remains the poorest in mainland Central America.

Four-fifths of Honduras is mountain country indented with river valleys running down to the short Pacific coast in the south and the longer banana-growing Caribbean coast in the north. The capital, TEGUCIGALPA, is surrounded by mountains. Its name means 'silver hills' in a language of the Indians (Mayas) who lived there before the Spanish arrived in the 16th century. The country became a fully independent republic in 1838.

Ninety per cent of the 4.5 million Hondurans are *mestizos* – of mixed Spanish and Indian blood. The rest are of Indian or African origin. Most are peasants producing maize, rice and beans for their own use; only one-third of the people live in towns. Although education is free, half the people cannot read.

The country is sparsely populated. It is five times as large as neighbouring El Salvador, but has a smaller population. Some areas of Honduras can only be reached by air. The forested and swampy Mosquitia in the far north-east has only one road. In this remote region the USA 'secretly' trains a guerrilla army of counter-revolutionaries (*Contras*) to fight the Communist *Sandinista* government in neighbouring Nicaragua.

Honduras has so far escaped the serious political unrest in Nicaragua, El Salvador and Guatemala. A brief war with El Salvador in 1969 was started by a World Cup football match. It was a two-leg fixture, played in the two capitals. On each occasion the visiting supporters complained of ill-treatment and violence, including rape. The allegations sparked off military action by both countries, who were already at daggers drawn over the migration of Salvadoreans into rural areas of Honduras.

This century's major domestic disturbance was a strike in 1954 of 40 000 workers of the United Fruit Company, which secured improved conditions and reduced the company's political power.

Only 22 per cent of the land is cultivated. Mountainous terrain and tropical forest make agricultural development difficult, but the late 1970s saw great progress in cattle raising and forestry.

HONDURAS AT A GLANCE

Area 112 088 km² (43 266 sq miles)

Population 4 543 000

Capital Tegucigalpa

Government Parliamentary republic

Currency Lempira (peso) = 100 centavos

Languages Spanish, Indian dialects

Religion Christian (96% Roman Catholic)

Climate Tropical on coast; temperate inland. Wet season May to November. Average temperature in Tegucigalpa ranges from 4-27°C (39-81°F) in February to 12-33°C (54-91°F) in May

Main primary products Bananas, coffee, maize, beans, rice, sugar cane, tobacco, timber, shellfish, cattle, fruits; lead, zinc, silver, gold, tin

Major industries Agriculture, mining, forestry, cement, fishing

Main exports Bananas, coffee, meat, sugar, natural ores, chemicals, hardwoods

Annual income per head (US$) 612

Population growth (per thous/yr) 34

Life expectancy (yrs) Male 56 **Female** 60

Hopetown ***South Africa*** Town in Cape Province, 120 km (75 miles) south-west of Kimberley. The first diamonds in South Africa were discovered in Hopetown in 1867. The town is now a livestock-rearing and wool-producing centre.

Population 11 600

Map South Africa Bb

Hormuz, Strait of (Strait of Ormuz) Strait separating Oman and the United Arab Emirates from Iran. It was a haunt of pirates from the 7th century BC until the 19th century. Today the Strait's strategic significance lies in its position at the southern outlet to the oil-rich GULF.

Width 60-100 km (37-62 miles)

Map United Arab Emirates Ca

horn Pyramidal peak formed when several CIRQUES develop back to back, leaving a central mass with prominent faces and ridges, such as the Matterhorn in Switzerland.

Horn, Cape (Cabo de Hornos) ***Chile*** Most southerly point of South America. It is a rocky headland 424 m (1391 ft) high on Horn Island in the Magallanes region of Tierra del Fuego. It was sighted by the English adventurer Sir Francis Drake in 1578 and was named by the Dutch explorer Willem Schouten in 1616, after his home town of Hoorn. As an alternative route to the Strait of Magellan on the northern side of Tierra del Fuego – despite its heavy seas and gale-force winds – it has played a major role in the world's seafaring history until the opening of the Panama Canal in 1914.
Map Chile Be

horse latitudes Zones of high air pressure around latitudes 30 degrees North and 30 degrees South, which move north and south with the overhead sun. The trade winds and the westerlies blow outwards from the horse latitudes. On land these zones are associated with deserts, while at sea they are areas of calms. In the days of sail, horses – if part of the cargo – were often thrown overboard from becalmed ships in order to save drinking water – and this practice may be the origin of the name.

horst Block of the earth's crust upthrust between two parallel FAULTS. It is the opposite of a graben. See RIFT VALLEY

Hortobágy ***Hungary*** Europe's only surviving area of steppe or prairie – vast, flat grassland – outside the USSR. It covers some 10 000 km² (about 3900 sq miles) west of the city of Debrecen in eastern Hungary. The drainage of its salt marshes and shallow lakes has provided farmland for crops, including rice and cotton. Even so, the Hortobágy remains a traditional *puszta* (Hungarian for 'grassland') of the Great ALFOLD (Plain). And it has something of the flavour of the American West and of central Asia with its seemingly endless spaces roamed by shepherds and their flocks, splendid horses and haughty *csikos* (horsemen). The Nagycsarda (Great Inn), dating from 1699, stands at the centre of the lonely region by the nine-arched bridge built in the 1830s over the Hortobágy canal. The inn is noted for its cuisine of local specialities.
Map Hungary Bb

hot spring, thermal spring Natural spring continuously discharging water that is hotter than 37°C (99°F). Hot springs are usually associated with volcanic action.

Houston ***USA*** Largest city in Texas. It is linked to the Gulf of Mexico by the 80 km (50 mile) long Houston Ship Canal, and is a port and oil refining and petrochemical centre. Houston is also the headquarters of NASA (the National Aeronautics and Space Administration), and has a World Trade Centre.
Population (city) 1 705 700; (metropolitan area) 3 164 400
Map United States Ge

Hovenweep National Monument ***USA*** Area of 204 hectares (505 acres) in the extreme south-east of Utah. It contains the remains of six groups of cliff dwellings, towers and pueblos, some dating back to prehistoric times.
Map United States Ec

Howrah ***India*** See HAORA

Hradec Králové (Königsgrätz) ***Czechoslovakia*** Regional capital of eastern Bohemia, about 105 km (65 miles) east of Prague. The city has a fine castle and a 14th-century Gothic cathedral of red brick. It manufactures musical instruments, machinery, chemicals and timber products.
Population 98 100
Map Czechoslovakia Ba

Hrvatska ***Yugoslavia*** See CROATIA

Hsin-kao Shan ***Taiwan*** See YU SHAN

Huai He (Hwai-Ho) ***China*** River flowing for about 1100 km (690 miles) over a vast alluvial plain south across Jiangsu province, joining the Chiang Jiang east of ZHENJIANG.
Map China Hd

Hua-lien ***Taiwan*** One of the few cities on Taiwan's east coast, about 120 km (75 miles) south of T'ai-pei. It is the largest port on the Pacific coast, and its industries include cement, fertilisers, sugar refining and pineapple canning; it is famous for its marble and jade products.

T'ai-lu-ko Gorge, one of the most scenic spots in Taiwan, is 20 km (12 miles) to the north. A section of the East-West Highway runs through the gorge.
Population 359 000
Map Taiwan Bc

Huambo ***Angola*** Railway town and provincial capital founded in 1912, about 520 km (320 miles) south-east of the national capital, Luanda. It was formerly known as Nova Lisboa.

The savannah-covered province, also called Huambo, has Angola's highest peak, Mount Moco, which rises to 2620 m (8595 ft). The province's main products are coffee, grain and cattle.
Population (town) 61 900; (province) 838 000
Map Angola Ab

Huancavélica ***Peru*** South central department in the Andes. Mercury was discovered here in the 16th century and was used in the production of silver amalgam at Potosí. The department capital, also Huancavélica, 240 km (150 miles) south-east of Lima, was founded in 1570 by Francisco de Toledo, Spanish Viceroy of Peru, under the name Villa Rica de Oropezza. It retains much of its Spanish colonial heritage in its cathedral and its 16th and 17th-century churches.
Population (department) 346 800; (city) 21 100
Map Peru Bb

Huang He (Hwang Ho; Yellow River) ***China*** The country's second longest river after the CHANG JIANG. It rises in the mountains of the west and flows 4670 km (2900 miles) into the gulf of Bo Hai about 320 km (200 miles) from Beijing (Peking). Its name, which means Yellow River, comes from the huge amount of silt it carries – so much, in fact, that the river has raised the channel it flows in well above the surface of the surrounding North China Plain. Much of the silt ends up in the gulf of Bo Hai, but the delta is not particularly fertile owing to salt accumulation in the soil. The river was formerly nicknamed China's Sorrow because of its tendency to flood, but the construction of artificial dykes since 1949 has greatly reduced this danger.
Map China Fc, Hd

Huang Shan ***China*** Beautiful range in Anhui province about 30 km (19 miles) south of Jiuhua Shan. The peaks form great towers, with gnarled pine trees growing out of rock crevices, and are frequently swathed in mists and cloud. The Huang Shan peaks, up to 1841 m (6039 ft) high, are a favourite subject of Chinese landscape painters.
Map China He

Huánuco ***Peru*** Central department producing sugar cane and tea. The department capital, also called Huánuco, is an attractive Spanish colonial town standing 1812 m (5945 ft) above sea level. The archaeological sites of the Inca city of Huánuco Viejo and the Temples of Kotosh built during the almost 3000-year-old Chavín culture are nearby.
Population (department) 484 800; (city) 61 800
Map Peru Ba

Huaráz ***Peru*** Capital of Ancash department on the Río Santa at an altitude of 3060 m (10 040 ft). It was severely damaged by an earthquake in 1970 which left 20 000 people dead. It has now been completely rebuilt. The archaeological museum has a fine collection of antiquities from the Chavín culture, which flourished between 700 and 200 BC.
Population 44 900
Map Peru Ba

Huascaran ***Peru*** The country's highest mountain, 6768 m (22 205 ft), overlooking the Río Santa valley, and one of 20 peaks exceeding 6000 m (19 685 ft) in the White Cordillera range. The Huascaran National Park, established in 1975, covers the whole of the White Cordillera mountains above 3900 m (12 795 ft).
Map Peru Ba

Hubei ***China*** Province of central China straddling the Chang Jiang river and covering 180 000 km² (69 500 sq miles). The Chang Jiang has cut deep gorges through the western mountains, part of a series of spectacular canyons beginning upstream in Sichuan province. The climate is moist and warm and the soil fertile, making Hubei an important producer of rice, soya beans, wheat, maize, tea and cotton. The capital is Wuhan.
Population 46 320 000
Map China Ge

Huddersfield ***United Kingdom*** West Yorkshire town, about 24 km (15 miles) south-west of Leeds, that prospered from the 19th century on wool. Hand-woven cloth had been produced in cottages and farms for centuries, but mechanisation brought the huge steam-powered mills. Many of these survive still, but some have been converted to other uses and their chimneys smoke no longer. The market hall, town hall and grand Victorian railway station reflect the former prosperity of the textile industry.
Population 124 000
Map United Kingdom Ed

Hong Kong

ONE OF THE WORLD'S MOST SUCCESSFUL ECONOMIES, THIS TINY BRITISH COLONY – A SHOWCASE OF CAPITALISM – WILL BE HANDED OVER TO ITS GIANT COMMUNIST NEIGHBOUR IN 1997

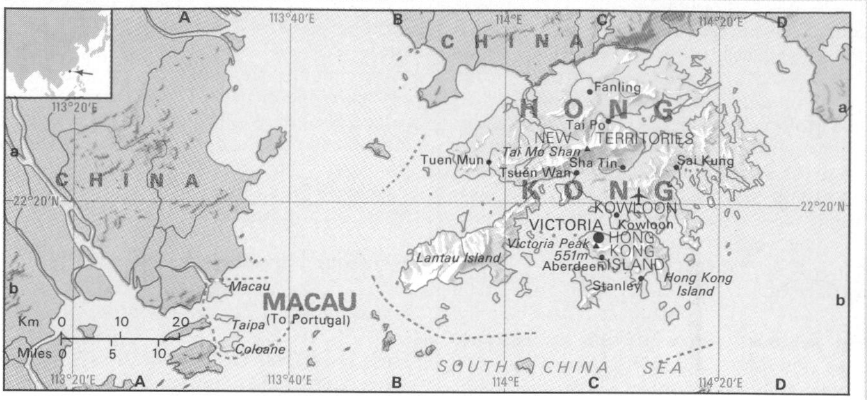

The time: a summer midnight. The place: a street corner on Nathan Road in downtown KOWLOON, heart of Hong Kong. The air is sticky and warm on the skin. Lights blaze from every window. Huge neon advertising signs compete with dazzling shopfronts against a background of high-rise blocks hung with washing. Construction workers swarm across floodlit towers of scaffolding made of bamboo canes as thick as a man's leg. Buses bulging with passengers snarl past, bumper to bumper with hooting trucks, delivery vans and limousines. Motorcycles, bicycles and handcarts – but only rarely a rickshaw these days – clatter beside and between them.

In a hole-in-the-wall café, tucked beside a six-storey emporium, mah-jong players slap down their ivory tiles, the staccato clacks punctuating the syncopated rhythm of flip-flop sandals on the pavements outside. On a side street, an elderly Chinese man in vest and shorts lies asleep beneath an awning, oblivious to the pedestrians eddying around his wooden camp bed. Overhead, a few stars twinkle palely through the damp, exhaust-laden air – vanishing from time to time as another jet thunders over the rooftops on its climb out of Kai Tak airport.

WORKING ROUND THE CLOCK

This bewildering, electrifying bustle goes on round the clock in Hong Kong. For the city never sleeps. Thousands of flats have small workshops in the back, so that moonlighting workers can go on making money long into the night. Order a made-to-measure shirt or suit in Hong Kong one afternoon, and it is ready for a final fitting the next morning. Much of its clothing, however, is made specially for export to Europe, North America and Australia, and for home consumption many cheap but serviceable garments are imported from China. At the other end of the scale, another popular line in imports is expensive French wines and perfumes for sale to tourists and rich businessmen.

Many international business empires have offices in this flourishing business centre. A glass-fronted skyscraper completed in 1985 for the Hong Kong and Shanghai Banking Corporation is reported to have cost US$615 million, making it the most expensive building in the world.

The colony's remarkable history began in 1842 when – after a defeat in the Opium Wars – China ceded to Britain a largely barren rock at the mouth of the ZHU JIANG (Pearl River) south of GUANGZHOU (Canton). In 1860, after more fighting, a mainland peninsula opposite the rock was ceded to Britain as well. In 1898, a much larger area backing onto the peninsula, together with numerous outlying islands, were leased by Britain from China for 99 years.

The rock is now HONG KONG ISLAND, a hill fringed by golden beaches and studded with luxurious blocks of flats. The peninsula is now Kowloon, which, with more than 72 000 people per square kilometre (720 per hectare, 290 per acre), is one of the most densely populated places on earth. And the larger area making up more than 90 per cent of the colony's land is known as the NEW TERRITORIES, the steep-sided valleys of the mainland part dotted with factories and new towns.

Beyond the New Territories is the vastness of Communist China. Between the two is a river not much wider than a country lane, and a high barbed-wire fence erected by the British to keep refugees from Communism out – an ironic reversal of the European Iron Curtain built by Communist governments to keep would-be refugees in.

WINNING COMBINATION

China is vital to Hong Kong's prosperity. All but 2 per cent of its people are Chinese, and much of the raw material that Hong Kong transforms into a fountain of products for the world comes from China. Hong Kong itself has no natural resources – no minerals and no fuels. Even its water comes largely from reservoirs across the Chinese border.

Hong Kong has only three assets: its position, close to the main trading routes of the Pacific; its magnificent natural harbour; and its people. This combination has proved to be a winner. Hong Kong exports more radios than any other country in the world – including even the USA and Japan. It handles more international trade than far larger countries, outranking Brazil, India, South Africa and China. And its economy has been growing, after allowing for inflation, at nearly 10 per cent a year – mainly financed by massive reinvestment of profits. This rate of growth exceeds that of the world's industrial giants. As a result of this success, Hong Kong is almost as important to China's prosperity as China is to Hong Kong's.

Hong Kong's economic achievements may be the best guarantee of its political future. In 1997, Britain's 99 year lease on the New Territories runs out – and with it Britain's sovereignty over the entire colony. On July 1, 1997, under an agreement signed in Beijing (Peking) in 1984, China will take over. But it is not a simple handover. Under the agreement, China has guaranteed to preserve Hong Kong's freewheeling capitalist economy for 50 years: a political recognition of Hong Kong's economic strength. Although Britain would be powerless to intervene if a new generation of Chinese leaders decided to change the rules, China's own need for cash for development may well be enough to restrain Communist politicians until well into the 21st century from disturbing a goose that lays so many golden eggs.

HONG KONG AT A GLANCE

Area 1067 km² (412 sq miles)
Population 5 580 000
Capital Victoria
Government British Crown Colony
Currency Hong Kong dollar = 100 cents
Languages Chinese (Cantonese), English
Religions Buddhist, Christian, Taoist, Confucian
Climate Subtropical; monsoon from May to September. Average temperature ranges from 14°C (57°F) in February to 29°C (84°F) in July/August
Main primary products Pigs, poultry, fruit, vegetables, fish
Major industries Textiles, clothing, electronic goods, clocks and watches, toys, plastic products, ship repair, aircraft engineering, iron and steel rolling, fishing
Main exports Textiles, clothing, electronic goods, clocks and watches, toys, plastic products, light metal products
Annual income per head (US$) 5965
Population growth (per thous/yr) 16
Life expectancy (yrs) Male 73 **Female** 77

▶ HUB OF TRADE Junks and flat-bottomed sampans add to the colour and bustle of Hong Kong harbour, one of the best and busiest in the world. Some are home to families of Hong Kong Chinese; many act as lighters to load and unload ships at anchor. The largest junks – many from China – are trading vessels in their own right.

Hudson ***USA*** River rising in the Adirondack Mountains of New York state and flowing about 500 km (310 miles) south to the Atlantic at New York city. With the ERIE CANAL it links the Atlantic with the Great Lakes.
Map United States Lb

Hudson Bay A vast bay in east-central Canada which has been part of Canada since 1869. The bay, and the 724 km (450 mile) long Hudson Strait, which links it with the ATLANTIC OCEAN, were discovered by the English navigator Henry Hudson in 1610. CHURCHILL, Manitoba, is the bay's main port. Here ships take on cargoes of livestock products and cereals brought by rail from central Canada.
Area 1 233 000 km² (476 000 sq miles)
Greatest depth 259 m (850 ft)
Map Canada Gb

Hue ***Vietnam*** Ancient imperial capital, about 80 km (50 miles) north-west of the port of Da Nang. It was the seat of conquering Chinese warlords in 200 BC, and 400 years later became the capital city of invading Cham tribes from the region of present-day Cambodia.

The heart of the present city, built in the early 19th century, is a moated enclave beside the River of Perfumes. It covers an area of 7 km² (3 sq miles) and includes palaces, tombs, temples, and the Forbidden City, once the seat of Vietnamese emperors.

Hue declined, however, after the French colonists moved their capital to Saigon, now Ho Chi Minh City, in 1883. And its grandeur has been allowed to deteriorate since the Communist takeover in 1975.
Population 190 100
Map Vietnam Bb

Huelva ***Spain*** Port about 95 km (60 miles) north-west of Cádiz on the south Atlantic coast. It exports copper, iron and manganese mined in its adjacent province of the same name. It also has a tuna and sardine fishing fleet. Christopher Columbus planned his 1492 voyage of discovery to the West while staying at the nearby La Rabida monastery.
Population (town) 127 800; (province) 414 500
Map Spain Bd

Huesca ***Spain*** Mountain town about 225 km (140 miles) north-west of the Mediterranean port of Barcelona, marketing fruit, wine and cereals and making agricultural machinery, cement, pottery, cloth and leather.

Its surrounding province, also called Huesca, contains Pico d'Aneto, on the French border, the highest point in the Pyrenees at 3404 m (11 168 ft).
Population (town) 44 400; (province) 219 800
Map Spain Ea

Hugli (Hooghly) ***India*** One of the arms of the GANGES river in the Ganges delta in the north-east. It splits off as the Bhagirathi from the Ganges just above the Bangladesh border and flows for about 193 km (120 miles) into the Bay of Bengal. The river banks north of Calcutta – known as Hooghlyside – are heavily built up with jute and other industrial factories. There is a barrage at FARAKKA which aids the river's flow.
Map India Dc

Huíla ***Angola*** Hilly province whose capital, Lubango, is about 680 km (420 miles) south of the national capital, Luanda. There are large iron deposits at Cassinga, about 280 km (175 miles) east and slightly south of Lubango.
Population 492 000
Map Angola Ab

Huila ***Colombia*** Rice-growing and cattle-rearing department in the south-west. The capital is Neiva, which lies about 240 km (150 miles) south-west of Bogotá. Huila is named after the volcanic peak of Nevado de Huila, which reaches 5750 m (18 865 ft).
Population 566 000
Map Colombia Bb

Hull ***Canada*** City across the Ottawa river from Ottawa, to which it is connected by five bridges. It manufactures pulp, paper and newsprint, and many people are employed in federal government offices. More than 88 per cent of the people are French Canadians.
Population 56 225
Map Canada Hd

Hull ***United Kingdom*** See KINGSTON UPON HULL

Humber ***United Kingdom*** Estuary of the Ouse and Trent rivers on the east coast of England, some 225 km (140 miles) north of London. It is 60 km (35 miles) long and is crossed by a suspension bridge 1410 m (4626 ft) long, opened in 1981, to the west of Kingston upon Hull.
Map United Kingdom Ed

Humberside ***United Kingdom*** County of 3512 km² (1356 sq miles) in eastern England, taking in parts of Yorkshire and Lincolnshire around the Humber estuary. It centres on the port of KINGSTON UPON HULL and includes the steel town of Scunthorpe and the ports of GRIMSBY and Immingham.
Population 854 000
Map United Kingdom Ed

Humboldt Glacier ***Greenland*** Largest glacier in Greenland, 114 km (71 miles) long and 95 km (59 miles) wide. The glacier, in the north-west of the country, is named after the German explorer and naturalist Alexander von Humboldt (1769-1859), a renowned writer on America, who is sometimes called the founder of modern physical geography.
Map Arctic Db

humidity Water vapour content of a mass of air, given either as absolute humidity, expressed in grams of water per cubic metre of air, or as relative humidity, in which case the water vapour content is expressed as a percentage of the total amount of water vapour required to saturate the air at a stated temperature.

Hunan ***China*** Province covering 210 000 km² (81 000 sq miles) in central China. It lies south of the great Dongting lake on the rich alluvial plains of the Xiang Jiang river with hills in the south-west. It was settled by the Chinese in the early 15th century, and because of its warm, humid climate and fertile soils it developed rapidly into a major rice area. Today, it supplies one-sixth of the country's rice, mainly from the densely settled lowlands around Dongting lake and the Xiang floodplain. The province also produces tungsten, used for lamp filaments and steel production, and other metal ores. The capital is Changsha.
Population 52 230 000
Map China Ge

Hunedoara ***Romania*** Iron and steel town in the west of the country, 120 km (75 miles) north of the border with Yugoslavia. It has a massive medieval fortress built largely by the Romanian nobleman and ruler of Transylvania Iancu Hunedoara, who in 1456 defeated the Turks at Belgrade. The castle was badly damaged by fire in the early 1850s but has since been restored.
Population 85 800
Map Romania Aa

Hungary See p. 296

Hungnam ***North Korea*** East-coast fishing port about 200 km (125 miles) north-east of the capital, Pyongyang. It has become an industrial centre making machinery, chemicals and textiles since the 1920s, through the development of local coal deposits and hydroelectric power from the nearby Bujon river.
Population 143 600
Map Korea Cc

Hunter Valley ***Australia*** Wine, sheep and coal-producing area in New South Wales. The valley, about 100 km (60 miles) north of Sydney, covers about 30 000 km² (11 600 sq miles). The Hunter river, 462 km (287 miles) long, rises in the Great Dividing Range and flows east to the Pacific at Newcastle.
Map Australia Ie

Huntingdon ***United Kingdom*** See CAMBRIDGE

Huntly ***New Zealand*** Mining town on North Island, about 80 km (50 miles) south of Auckland. It is the centre of the country's most important coalfield.
Population 6500
Map New Zealand Eb

Huntsville ***USA*** City in northern Alabama, about 30 km (20 miles) from the Tennessee border. It is an old agricultural town with textile industries, but its main employer is the Redstone Arsenal and its associated rocket and space equipment plants.
Population (city) 149 500; (metropolitan area) 210 000
Map United States Id

Huron, Lake ***Canada/USA*** One of the Great Lakes, lying between Ontario, Canada, and the US state of Michigan. It is supplied by both Lake Superior and Lake Michigan, and drains via the St Clair river and Lake St Clair into Lake Erie. About 60 per cent of its approximately 59 570 km² (23 000 sq mile) area is in Canada.
Map United States Jb

hurricane Name used for a CYCLONE in the North Atlantic (especially in the Gulf of Mexico and West Indies). Also the term for the highest wind speed on the BEAUFORT SCALE, in excess of 119 km/h (73 mph).

Hutt Valley ***New Zealand*** Industrial centre and commuter area for the capital, Wellington. Settlers first went there in 1839, but moved to Wellington because of regular flooding by the Hutt river. It was an agricultural area until residential overspill from the capital made it effectively a part of the city. Industries today include cars, heavy engineering and textiles.

Population 94 700

Map New Zealand Ed

Hvar (Lesina) ***Yugoslavia*** Rocky, forested Dalmatian island lying off the tourist resorts of the Makarska riviera. Covering 287 km² (111 sq miles), it is Yugoslavia's Madeira, producing wine, honey, grapes, olives, figs, dates and marble. Hvar town has a Baroque cathedral, a 16th-century fortress and a Franciscan monastery. Starigrad is the island's main ferry port.

Population 20 000

Map Yugoslavia Cc

Hwang Ho ***China*** See HUANG HE

Hwange ***Zimbabwe*** Coal-mining town, formerly called Wankie, near the country's western tip, over 500 km (310 miles) west and slightly south of the capital, Harare. Hwange National Park, formerly the Wankie National Park, covers 14 651 km² (5657 sq miles). It lies south-east of the Victoria Falls and has 107 recorded animal species, including buffalo, eland, elephant, giraffe, kudu, wildebeest and zebra. It also has some 400 species of birds.

Population 39 000

Map Zimbabwe Ba

Hyderabad ***India*** Walled city and capital of the state of Andhra Pradesh, about 500 km (310 miles) north-west of Madras. The former capital of the Nizam of Hyderabad's princely state, it was forcibly taken over by India in 1948.

The city has many fine Muslim buildings including the Charminar tower, a 16th-century civic monument, standing 46 m (150 ft) high. There is also a splendid artificial lake, mosques, tombs and museums. It is a thriving centre of craft industries including metal inlay work, carpets and silks. There are many major colleges and research institutions, and an open-air zoo.

Population 2 093 500

Map India Cd

Hyderabad ***Pakistan*** City about 160 km (100 miles) north-east of Karachi, at the head of the Indus delta. In its modern form the city dates from 1782, when it was planned by the Afghan prince Sarfaraz Khan. Its fort is now a ruin, but there are several surviving royal tombs. The University of Sind is in the city. A few kilometres to the north-west is the site of a battle in 1843 when the British under Sir Charles Napier defeated Mir Sher Mahomed Khan to annex Sind province. The battle site, by the village of Miani, is marked by a monument.

Population 795 000

Map Pakistan Cc

Hydra (Ídhra) ***Greece*** Island resort in the Saronic gulf, 71 km (41 miles) south-west of Piraeus. Hydra is barren but unspoilt, and has no motor traffic. Its harbour town has 18th-century mansions, small hotels and a marina.

Population 2700

Map Greece Cc

hydrological cycle, water cycle Endless series of changes which water undergoes between the sea, air and land. The water in the sea evaporates and becomes water vapour, which condenses into clouds and falls back to earth in the form of rain, running off into rivers and so back into the sea.

hydrosphere All the waters of the earth, including oceans, rivers, lakes, streams, snow, ice, water vapour and underground water.

Hyères, Iles d' ***France*** A group of three small islands off the coast of Provence. Ile de Porquerolles has some tourist facilities, Ile de Port Cros is a 16 km² (6 sq mile) national park, and Ile du Levant is a naval base and a holiday resort for naturists.

Map France Ge

▼ ASTRONAUT'S EYE VIEW The hurricane named 'Gladys' appeared as a vast spiral of clouds over the Gulf of Mexico to the crew of the Apollo 7 spacecraft in October 1968.

Hungary

TANKS QUELLED THE 1956 RISING, BUT TODAY'S GENERATION HAS FOUND THE PATH TO PROSPERITY

In the autumn of 1956, Russian troop reinforcements were sent into Hungary to quell a national uprising that threatened to take the country out of the Eastern Bloc. The streets of the capital, BUDAPEST, echoed to the crackle of gunfire as unarmed Hungarians (later joined by some Hungarian troops) faced Soviet armoured divisions. The rebels were demanding the withdrawal of all Soviet troops from the country, the withdrawal of Hungary from the Warsaw Pact and free elections.

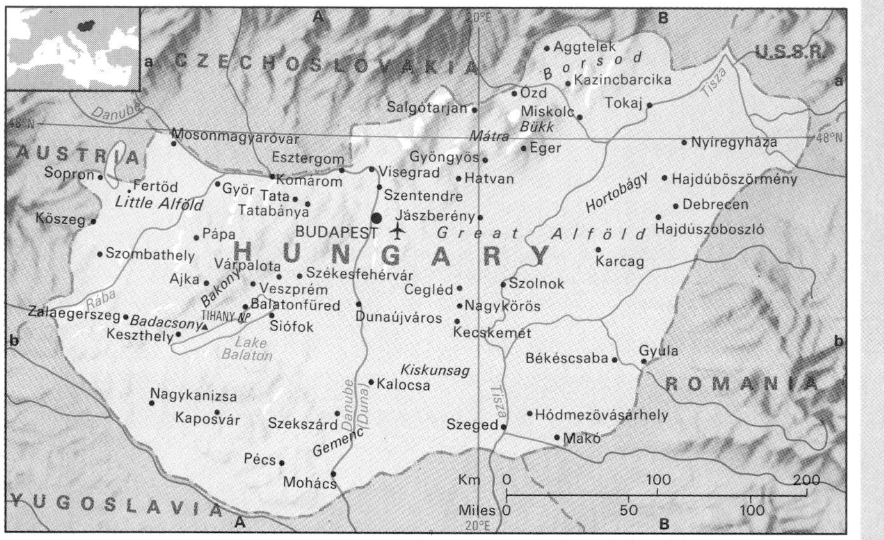

The Hungarian uprising was crushed so brutally because the Soviet leader, Nikita Khrushchev, was determined to maintain control over a country occupying a key strategic position. The Hungarian government of the day, led by Imre Nagy – a nationalist who had taken over from pro-Russian Matyas Rakosi in 1953-5 and came to power again in 1956 – was regarded by the Russians as an unreliable ally.

In retrospect, it is tragic that the events of 1956 cost so much bloodshed – about 3000 Hungarian dead and over 19 000 injured – and the emigration of 200 000 Hungarians who fled to the West. For the new leader who was installed with Soviet support, Janos Kadar, has gradually led Hungary to a degree of liberalisation linked with economic progress.

Landlocked in the heartland of Europe, Hungary is dominated by the Great ALFOLD (Nagyalföld, or Great Plain) to the east of the DANUBE, which runs north to south across the country. Most of the plain is fertile agricultural land. Two more lowlands lie west of the Danube. The Little ALFOLD (Kisalföld, or Little Plain) in the north-west is separated from the Transdanubian plains by the Central Highlands. In the south, the Mecsek hills rise to 681 m (2234 ft). In the shadow of the BAKONY hills, which dominate the Central Highlands, lies Lake BALATON – the largest lake in Central Europe, covering 601 km² (232 sq miles). The highlands continue in the north of Hungary, sweeping round to the east beyond the 300 m (1000 ft) deep Danube gorge. Here they form an extension of the CARPATHIAN mountains, and include the MATRA range, with Hungary's highest point.

Hungarian winters are severe as cold air flows down from the surrounding mountains. However, in summer the mountains shield the plains and temperatures average 22°C (72°F) in July, though May and June are often cold and wet. Western Hungary is wetter than the east, where summer droughts occur.

The country's history goes back over a thousand years to the foundation of the state. In 896, the Magyars (pronounced 'modjars'), nomadic horsemen from the Russian steppes, swept into the Danube valley and, under their semi-legendary leader Arpad, occupied territory which is now south-western Hungary. The first Hungarian kingdom was established under the Magyar King (later Saint) Stephen (1000-38). The kingdom expanded but there were frequent wars with the Turks, who finally conquered the central plain and ruled for 150 years during the 16th and 17th centuries.

The Turks were challenged by the Austrian Hapsburgs, who took over the country in 1699. A war of independence against Austria was crushed in 1848 with Russian help, but it was in the interests of the Magyars to make common cause with the Austrians, and in 1867 a dual monarchy was established. Hungary became an autonomous partner in the Austro-Hungarian Empire, which also included Czechs, Slovaks and Serbs.

The empire finally collapsed after the First World War, and Hungary was left with one-third of its former territory – the rest going to Czechoslovakia, Romania and Yugoslavia. In the 1930s Hungary came under German influence, and in the Second World War the Hungarians fought with Germany against the Russians. After the war, the country became a Soviet satellite.

ECONOMIC BOOM

Hungary has enjoyed more of an economic boom and less foreign indebtedness in the 1970s and 1980s than many of its Communist neighbours (especially Poland, Romania and Yugoslavia). The secret of success lies in the Hungarian government's own blend of policies since the mid-1960s. It combines progressive incentive schemes for management and workers with the kind of improvements made by the Bulgarians in agriculture and by the East Germans in industry, and allows some elements of private production.

When Hungary became Communist in 1948 after the Second World War, land ownership was more feudal than elsewhere in Eastern Europe. One per cent of those who gained a living from farming owned 57 per cent of the land. The largest landowners were the Roman Catholic Church and the leading Magyar noble family – the Esterhazys. The estates either became state farms (now 11 per cent of all land) or were divided among 640 000 landless labourers and smallholders. In 1950, a start was made in collectivising all peasant land into agricultural cooperatives. Today, only 6 per cent remains privately owned.

After initial neglect, money has been poured into agriculture since the mid-1960s. Enormous strides were made in mechanising farm work. Use of fertilisers soared from 13 kg per hectare (about 12 lb per acre) in 1938 to 300 kg per hectare (about 270 lb per acre) in the 1980s, supported by large-scale expansion of the chemical industry. Irrigation, extended from 14 000 hectares (about 35 000 acres) in 1945 to 450 000 hectares (over 1.1 million acres), brought new land into cultivation along the Danube, TISZA and Körös rivers and the most arid areas of the Great Plain – the HORTOBAGY and Hajdusag.

As a result, yields of cereals used for making bread rose in many areas. However, the country still continues to import cereals so that land can be employed for more intensive farming. The policy has been to expand output and improve the quality of specialised produce in which Hungary is competitive in both Western and Eastern European markets. The cultivation of grain for animal fodder has often displaced bread cereals to ensure big increases in the output of ham, pork, beef, salami, milk, skins and wool. Rice production has also soared.

Large areas, especially in the KISKUNSAG between the Danube and Tisza rivers, are now devoted to a variety of crops whose Hungarian harvest time falls between those of Mediterranean and northern Europe, or for which Hungary has environmental advantages. They include such vegetables as cabbages, lettuces, onions and red and green peppers; grapes for wine; and fruits like melons, apricots, peaches, plums, cherries and strawberries.

In recent years, Hungary's industrial path

has been radically altered. Under the first five-year plan (1951-5), Rakosi tried to turn Hungary into what he boldly termed 'a country of iron and steel'. There was, however, not enough iron ore and coal in the northern mountains to support a large-scale steel industry, and these had to be imported from the Ukraine. The plan was an expensive failure.

Later five-year plans have been more realistic; industries have been developed where adequate natural resources exist – for example, food production from central, eastern and southern areas, and aluminium production in the Bakony mountains. Greater use has been made of traditional skills in textiles, leather and ceramics. Newer industries have also been promoted, such as the manufacture of electrical and electronic equipment. In addition, recent reforms have given managers of individual concerns more responsibility for making decisions on product lines, investment and finding markets at home and abroad.

SOCIAL UPHEAVAL

Yet Hungary is not without its problems, most of which are historically based. Orthodox Communists in some other Eastern European countries oppose its liberal management, banking and foreign trade policies. With poor domestic energy resources, Hungary depends on these neighbours to supply oil, gas, coal and electricity in exchange for food and manufactured goods.

Another problem is the huge cost of providing essential services in the Great Plain. Here, widely scattered towns – often no more than large villages such as Hodmezovasarhely, developed for defence during or after the Turkish occupation – and remote farmsteads require modern amenities, cultural facilities and social services.

Lack of essential services has caused many to move, leaving farming to the old. To stem the tide and retain a productive work force on the land, Hungary in the 1960s provided pensions and other welfare rights for farm workers in cooperatives equal to those enjoyed by state employees – the first East European country to do so.

Most rural migrants flocked to Budapest, which houses 2.2 million people (one-fifth of Hungary's population) and provides nearly 30 per cent of the country's jobs; it has a serious housing shortage. Most people there live in high-density areas of flats ranging from 19th-century inner city art nouveau to 1960s suburban socialist utilitarian.

Today, car ownership has soared, giving Budapest problems of parking, congestion, noise and pollution. Many managerial, administrative, professional, scientific and industrial workers own second homes. This has been made possible because of rising real incomes, cheap housing and public utilities and free medical care and education; and because of the increasing number of families in which husband and wife both go out to work.

Hungarians are hearty eaters and many drink heavily. Paprika is present in most national dishes. Tokaj and 'Bull's Blood' are among the country's excellent wines. Cafés and restaurants everywhere are lively.

The Hungarians have managed to retain their culture through a turbulent history, including subservience to and revolt against Turks and Austrians. Their Magyar language – a unique tongue related only to Finnish – exotic food, ancient folk music and unusual rhythms immortalised in the works of Liszt, Bartok and Kodaly sharply distinguish them from their German, Slav and Romanian neighbours. These were shaped 15 centuries ago when the Magyars still roamed the Ural-Altai steppes of southern Russia. Hungary's Great Plain, truly flat, seemingly endless, is remarkably similar to those steppes.

Yet Hungary is not just plains, though its varied landscapes are all small in scale. Picturesque rolling country in Transdanubia in the west is dotted with old towns, spas, orchards and vineyards, and merges eastwards into the Bakony hills. These thrust their spurs and remarkable hill shapes, including extinct volcanoes, towards Lake Balaton. This, the largest lake and international tourist playground in Central Europe, has splendid beaches, modern holiday resorts, old towns with cool wine cellars, and the beautiful countryside of the Tihany peninsula. Northern Hungary is different again – a chain of thickly forested hills and mountains which cradle spas, attractive villages and the vineyards of EGER and TOKAJ. In the south, towns such as PECS contain 16th-century mosques – reminders of the Turkish occupation of the country.

▲ THE BLUE-GREY DANUBE The vast 19th-century neo-Gothic Parliament Building stands beside the river in Budapest. In 1944, the Nazis blew up the nearby Margaret Bridge at the height of a rush hour, with much loss of life, and it has been rebuilt.

HUNGARY AT A GLANCE

Area 93 032 km² (35 920 sq miles)
Population 10 620 000
Capital Budapest
Government One-party Communist republic
Currency Forint = 100 filler
Language Magyar (Hungarian)
Religions Christian (55% Roman Catholic, 22% Protestant), Jewish (1%)
Climate Continental; average temperature in Budapest ranges from −4 to 1°C (25-34°F) in January to 16-28°C (61-82°F) in July
Main primary products Cereals, potatoes, sugar beet, fruit and vegetables, grapes, livestock, timber; bauxite, coal, lignite, oil and natural gas
Major industries Agriculture, iron, steel, textiles, chemicals, machinery, transport equipment, forestry, timber products, mining
Main exports Food, machinery, chemicals, motor vehicles, clothing, iron and steel
Annual income per head (US$) 1900
Population growth (per thous/yr) Declining
Life expectancy (yrs) Male 67 **Female** 75

Iasi ***Romania*** City and capital of Moldavia until 1862, 11 km (7 miles) west of the Russian border. Despite being ravaged by Tatars (1513), Turks (1538) and Russians (1686), Iasi retains many ancient buildings, including the magnificent Trei Ierarhi church (the Three Hierarchs), built in 1639, and the 17th-century Golia Monastery. There is also the neo-Gothic Palace of Culture built at the turn of the century. In front of it is the statue of Stephen the Great (1457-1504), the Moldavian prince most successful in resisting the Turkish advance.

All this contrasts markedly with the modern high-rise districts created with the growth of Iasi's electrical, chemical, textile, tobacco and engineering industries.

Population 271 400
Map Romania Ba

Ibadan ***Nigeria*** Capital of Oyo state and the country's second city after the national capital, Lagos, some 120 km (75 miles) to the south-west. Ibadan is Nigeria's intellectual heart, with a world-renowned university. It is also a market for cocoa, palm oil (used in margarine, cooking fats and soaps) and kola nuts.

Population 1 009 000
Map Nigeria Ab

Ibagué ***Colombia*** Market town and capital of Tolima department, about 140 km (85 miles) west of Bogotá, to which it is linked by road and rail. It is the centre of a cattle-rearing and coffee-growing district and its industries include flour milling, brewing and making leather goods. Renowned as the musical capital of Colombia, it has fine orchestras and choirs and holds a folk festival each June.

Population 361 000
Map Colombia Bb

Ibar ***Yugoslavia*** Serbian river, 276 km (172 miles) long. It rises north of the market town of Pec and flows east, then north through a lovely gorge to the Western (Zapadna) Morava river at the city of Kraljevo. It is rich in carp and pike and ideal for canoeing.

Map Yugoslavia Ec

Ibarra ***Ecuador*** Capital of Imbabura province on the Pan American Highway 90 km (55 miles) north-east of Quito. It lies in a deep fertile basin in the Andes and is the market centre for the surrounding agricultural region, trading in coffee, sugar cane and cotton. It manufactures textiles and furniture. It is also a tourist centre, noted for its woodcarvings.

Population 59 000
Map Ecuador Ba

Ibb ***Yemen*** Small town 60 km (37 miles) north of TAIZ. Ibb is the capital of the Green Valley, an area noted for its superb scenery of contrasting mountains and valleys.

Population 19 700
Map Yemen Ab

Ibiza (Iviza) ***Spain*** One of the BALEARIC ISLANDS, with a port of the same name which exports figs, raisins, pine timber and salt. The town and the island are also fashionable tourist resorts.

Population (town) 25 500; (island) 45 100
Map Spain Fc

Iboland ***Nigeria*** Tribal homeland in the south-east. It is one of the country's most densely populated areas, with up to 400 people per square kilometre (1036 per sq mile), and despite careful farming techniques, intense cultivation has resulted in soil erosion. Palm oil (used in margarine and cooking fats) is the main product, but cocoa, rubber, maize, cassava, rice and vegetables are also grown. Enugu is the main town. In 1967 the Ibo people seceded from Nigeria as the independent republic of Biafra, and the resulting civil war, which lasted until 1970, and famine killed a million people.

Population 10 000 000
Map Nigeria Bb

Içá ***Brazil*** See PUTUMAYO

Ica ***Peru*** South-western department, producing cotton, and grapes for wine and brandy making. The area is rich in pre-Columbian history, particularly at Paracas, Nazca and Tambo Colorado. The Ica valleys are irrigated by water diverted from Lakes Choclocha and Orchococha. The department capital, also Ica, stands on the Ica river 273 km (170 miles) south-east of Lima. It has a university, and a fine archaeological museum.

Population (department) 433 900; (city) 117 800
Map Peru Bb

ice ages See GRIP OF THE ICE AGES (opposite)

ice fall Broken, tumbled mass of ice where a glacier steepens and crumbles.

ice floe Level expanse of floating ice. It is also called simply a floe.

ice sheet, ice cap Vast, continuous expanse of land ice, such as that covering the Antarctic continent.

ice shelf Thick, floating ice sheet attached to a coastline.

iceberg Massive floating body of ice that has broken away from a glacier or ice sheet. Icebergs move under the influence of currents and winds, and can be hazardous to shipping. The largest ever seen, 335 km (208 miles) long and 97 km (60 miles) wide, was sighted in the south Pacific Ocean in 1959. The tallest, 167 m (550 ft), was recorded off western Greenland in 1959.

Iceland See p. 300

icing Formation of ice on exposed objects, especially ice formed from moisture in the atmosphere on an aircraft or ship.

Idaho ***USA*** North-western state covering 216 413 km² (83 557 sq miles). It is mountainous, with the rugged Bitterroot Range of the Rockies in the east and the high plateau of the Snake river in the south and west. The area was not permanently settled until the 1860s, when minerals were found. Today the chief products are silver, lead and phosphates (used in fertilisers) and potatoes, cattle and wheat. A third of the state's income comes from tourism. BOISE is the state capital.

Population 1 005 000
Map United States Ca

Idfu ***Egypt*** Town on the Nile, 123 km (76 miles) north of Aswan. It is the site of a temple to the falcon-headed god Horus, built in about 250 BC and one of the best preserved of ancient Egypt's monuments. The people of Idfu trade in cotton, cereals and dates, and make pottery.

Population 27 000
Map Egypt Cc

Idhra ***Greece*** See HYDRA

Idi Amin Dada, Lake ***Uganda/Zaire*** See EDWARD, LAKE

Idlib ***Syria*** Commercial town 60 km (37 miles) south-west of Aleppo. It is set in a fertile agricultural region producing fruit, olives, cereals and tobacco.

Population 428 000
Map Syria Bb

Ieper ***Belgium*** See YPRES

Ife ***Nigeria*** Town 170 km (105 miles) north-east of the capital, Lagos. It is probably the oldest town of the Yoruba people. They trace their ancestry back to Oduduwa, a tribal chief who is said to have founded the town between the 7th and 10th centuries. By the 13th century it was producing exceptionally fine bronzes. Today Ife is a cultural and religious centre with a university, museum, and, it is said, more than 400 gods. It also markets cocoa and timber.

Population 209 100
Map Nigeria Ab

Ifni ***Morocco*** Fishing is the main industry of this arid, 1502 km² (580 sq mile) former Spanish province on Morocco's Atlantic coast, just opposite the Canary Islands. Spain established a fortified trading, slaving and fishing settlement there in 1476, abandoned it half a century later, then reclaimed it from Morocco in 1860. It was finally returned to Morocco in 1969. The mainly Berber population of about 46 000 also raises camels, sheep and goats. About 16 000 of them live in Sidi Ifni, the former provincial capital, a small port engaged in fishing and trading with the Canaries. The town is also a handicrafts centre for the region.

Population (town) 16 300
Map Morocco Ab

Igaraçu ***Brazil*** Town 30 km (18 miles) west of the city of Recife, near the country's north-eastern tip. Much of the town is protected as a national monument, and the church of São Cosme and São Damião – dating from 1535 – is the oldest in Brazil. The town's main products are sugar, coconuts, yams, cassava and paper.

Population 74 000
Map Brazil Eb

igneous rock Crystalline rock that has been formed by the solidification of magma, molten matter formed within the earth's crust and upper mantle. Major igneous rocks include BASALT, DIORITE, GRANITE and SERPENTINE.

Iguaçu Falls (Iguazú Falls) ***Brazil/Argentina*** Waterfall on the border of Brazil and north-eastern Argentina, and 19 km (12 miles) from the Paraguayan border. The Iguaçu river

THE GRIP OF THE ICE AGES

During the last 2 million years, earth has been deep-frozen nine times, and through the study of glacial scarring on ancient rocks, it is apparent that similar Ice Ages have occurred at intervals over something like 930 million years. Tentative theories have been advanced to explain these phenomena – periodic wanings of the sun's heat, volcanic dust blocking the sun's rays, a decrease in the atmosphere's carbon dioxide content accelerating the heat loss to outer space, and wobbles of the earth as it orbits the sun. Earth's most recent icy blanket covered Antarctica, parts of New Zealand, Patagonia and the southern Andes, the Caucasus and Himalayas, and large parts of northern Eurasia and North America. The ice sheet covering the North Sea and Britain down to the line of London's northern suburbs retreated some 10 000 years ago. Today the considerable remains of this last Ice Age are the ice caps of Antarctica and Greenland. The results of this glaciation may be seen in ice-carved valleys, and in uplifted coastlines as the land slowly rose after being relieved of the weight of the ice.

The vast ice sheets locked up enormous quantities of water, so much that sea levels stood at least 135 m (442 ft) below present tidal marks. What are now continental shelves were then dry land, and many offshore islands, including Britain, were part of adjacent landmasses, and Russia and North America were joined at what is now the Bering Strait. At these times, lower sea levels allowed the movement of people and animals between the landmasses. Whether earth has seen its last Ice Age is not known; most likely we live in an inter-glacial period, one of many in earth's history, and in ten, or twenty, or perhaps a hundred thousand years, the ice will return. But a warmer planet would be equally disastrous. The polar ice caps would melt, raising the sea level by some 60 m (200 ft) and submerging many major cities including New York, Rio de Janeiro and London.

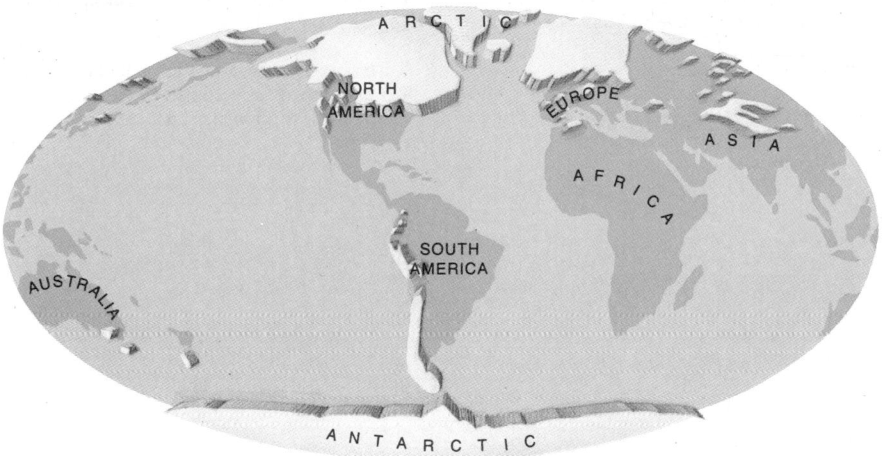

During the most recent Ice Age, which ended some 10 000 years ago, much of the Northern Hemisphere, and part of the Southern, was covered by glaciers and ice sheets. Their courses can be traced from the debris and scarrings they left on the land.

is fed by 30 tributaries on its way from its source near the Brazilian city of Curitiba, and is 4 km (2.5 miles) wide just before it plunges 80 m (about 260 ft) down a crescent-shaped cliff which is half as wide again as Niagara.

At times of peak flow, enough water tumbles over the falls to fill more than six Olympic-sized swimming pools every second. But in some years rainfall is so slight that the river can dry up altogether, as it did for a month in 1978. The largest of the 275 cascades in the falls, the Devil's Throat, is on the border line in mid-river. It can be reached from the Argentinian side by a long catwalk over the water and by boat from the Brazilian side.

The Iguaçu river goes on to flow into the Paraná at the Paraguayan border.

Map Brazil Cd

Ijebu Ode ***Nigeria*** Market town in Yorubaland, 75 km (47 miles) north-east of the capital, Lagos. It handles cocoa, palm oil (used in margarine and cooking fats), rubber, kola nuts (chewed as a mild stimulant) and timber, and makes wood products and bicycles. In addition there are some 1000 one-man craft workshops.

Population 110 300
Map Nigeria Ab

IJmuiden ***Netherlands*** See NOORDZEEKANAAL

IJssel (Issel; Yssel) ***Netherlands*** River some 115 km (70 miles) long that acts as a mouth of the Rhine. It flows out of the Lower Rhine near the city of Arnhem and runs north into IJsselmeer.

Map Netherlands Ca

IJsselmeer ***Netherlands*** Former arm of the North Sea, 95 km (60 miles) long and up to 60 km (37 miles) wide. Known as the Zuiderzee, it was formed by floods in the 13th century. Land reclamation began in 1920, and the arm was finally sealed off by the Afsluitdijk (dam) in 1932 to form IJsselmeer, a freshwater lake fed by the IJssel river.

Four polders (dyked areas of reclaimed land) covering a total of 1815 km² (700 sq miles) are now farmland. Work on a fifth polder, the Markerwaard (or West Polder) covering 400 km² (155 sq miles), is under way, and 1400 km² (540 sq miles) behind the Afsluit-dijk will remain as the IJsselmeer.

Map Netherlands Ba

Ikopa ***Madagascar*** River rising south of the capital, Antananarivo, and flowing 400 km (250 miles) north-west into the Betsiboka. The river's source is the Mantasoa reservoir, completed in 1939, which supplies water to the capital and to the surrounding rice fields.

Map Madagascar Aa

I-lan ***Taiwan*** City in north-eastern Taiwan, 40 km (25 miles) south-east of T'ai-pei. The coastal plain on which it lies is one of the main rice-growing areas of eastern Taiwan, and I-lan acts as its commercial and distribution centre.

Population 100 000
Map Taiwan Bb

Ile, Iles For physical features whose names begin 'Ile' or 'Iles', see main part of name.

Ilebo ***Zaire*** Town, formerly called Port Francqui, about 580 km (360 miles) east of Kinshasa, the capital. It lies on the Kasai river just above its confluence with the Sankuru. It is the terminus of a railway from the mineral-rich Shaba region, and transports copper and other products to Kinshasa by the Ilebo-Kinshasa river route, Zaire's business waterway.

Population 40 000
Map Zaire Bb

Ile-de-France ***France*** Region of 12 012 km² (4638 sq miles) around Paris. It contains a rich variety of farmland and forests, and many chateaux, including those at FONTAINEBLEAU and VERSAILLES.

Population 10 073 000
Map France Eb

Ilesha ***Nigeria*** Cocoa-marketing town in Yorubaland, about 200 km (125 miles) north-east of the capital, Lagos. Missionary influence there was strong and, unlike other Yoruba towns, Christians far outnumber Muslims – by about six to one.

Population 266 700
Map Nigeria Ab

Iligan ***Philippines*** Industrial centre on the north coast of MINDANAO, 50 km (31 miles) south-west of Cagayan de Oro. It is the capital of Lanao del Norte province, and has chemical, steel and other industrial plants fed with hydro-electric power from the 97 m (318 ft) high Maria Cristina Falls on the Agus river, which are within the city boundaries.

Population 167 358
Map Philippines Bd

Iceland

THE LAND IS STILL YOUNG AND THE MOUNTAINS ARE STILL MOVING, BUT THE GOVERNMENT IS A MODEL OF STABILITY

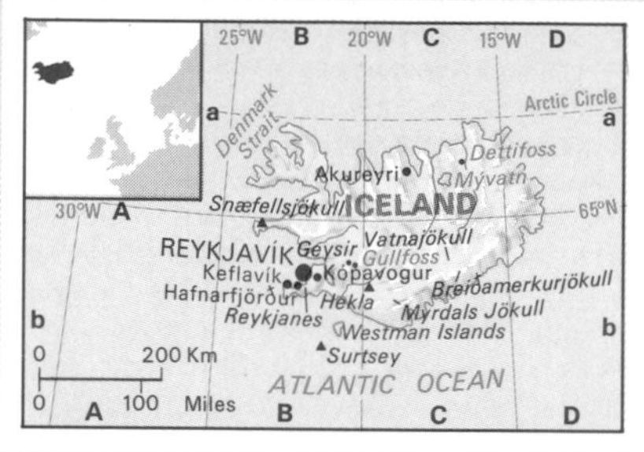

When American astronauts wanted to simulate walking on the moon, they went to the lunar-like landscapes of Iceland. It is mostly a desolate country of black rock, lava fields and glaciers, described as 'a desert of the ocean' by early churchmen who visited the country. The tourist industry highlights the drama of volcanoes, geysers and pools of boiling mud, and calls it 'The Land of Ice and Fire'.

In human terms this strange island just south of the Arctic Circle has a long history, but it is geologically young. If the aeons of time in which the earth was formed were compressed into one day, Iceland would be born at five minutes to midnight.

It is an unstable part of the earth's crust, and the activity which formed the country continues. Iceland has over 100 volcanoes, at least one of which can be guaranteed to erupt every five years; new lava flows cover the old lava, and change the face of the island. In 1963 a new volcano, SURTSEY, emerged from the sea off the south coast; in 1973 lava from Helgafell on the island of Heimaey, one of the WESTMAN ISLANDS, virtually destroyed one of Iceland's largest towns, and the 5300 inhabitants had to be evacuated overnight.

DESOLATE LANDSCAPE

One-ninth of the country is still covered in ice and snowfields. The glaciers include VATNAJOKULL, the largest in the Northern Hemisphere outside Greenland. The ice cover waxes and wanes, and its retreat over the last century has exposed the relics of farms which were covered in the Middle Ages.

There are more than 700 hot springs, which provide three-quarters of Iceland's population with central heating. At Hankadalur the springs can be heard grumbling underground, and every two minutes the Stokkur geyser erupts, belching clouds of steam up to 22 m (70 ft) high.

The base-rock of Iceland is basalt, which has weathered to a bluish-black colour and produced a central plateau with forbidding cliffs. It is a high country – over one-third lies above 600 m (1970 ft) – with limited coastal lowlands. Lava flows have been eroded into fantastic shapes by water and weather. The land is treeless except for some birch scrub and a few coniferous plantations. Most of the coastline is eroded into deep fiords.

The sea makes Iceland habitable. The chief lowland, which is the most densely populated area, lies in the south-west corner where the North Atlantic Drift warms the shores and keeps them free from ice. In winter, average temperatures in this area hover around freezing point; in July and August, they rise to 12°C (54°F). The summer days have short nights, and the winter days are short and gloomy. Average precipitation is high at 800 mm (30 in) a year, most of it falling as snow in the uplands.

The economy of Iceland is based on its sea fishing industry, which accounts for 70 per cent of exports and employs 14 per cent of the workers. Over the past 50 years, fishing has grown from a small-scale family activity, often run in conjunction with farming, to a large-scale commercial enterprise with trawlers and processing plants owned by big companies or cooperatives.

Only 20 per cent of the country is agricultural land, and only 1 per cent is cultivated, mainly for fodder and root crops. Most is used for sheep and cattle. The country is self-sufficient in meat and dairy products, and wool sweaters and sheepskin coats help to diversify exports.

A quarter of the labour force works in energy production and manufacturing. They process food and fish, make cement and refine aluminium from imported ore. The people are voracious readers – perhaps because of those long winter nights – and their needs are met by a lively graphics and printing industry.

Iceland was settled by Vikings from Norway in 874, but recent archaeological discoveries suggest there were earlier settlements. In 930, the settlers created the world's first parliament (the Althing). Members met in the open air, in the lakeside arena at Thingvellir, overshadowed by lava crags.

Iceland united with Norway in 1262, and in 1380 both countries came under the Danish crown. During the Second World War, Iceland was occupied by Allied troops and in 1944 the country became independent.

Icelanders enjoy remarkably high standards of living, education, social security and health care. Life expectancy is high, despite Europe's highest consumption per head of animal fats and sugar. The people are proud of their egalitarian society and their cultural heritage. They have ambivalent views about the American base on their island at KEFLAVIK, the more so because, although Iceland is a member of the NATO military alliance, it has no military forces of its own.

ICELAND AT A GLANCE

Area 103 000 km² (39 768 sq miles)
Population 243 000
Capital Reykjavík
Government Parliamentary republic
Currency Krona = 100 aurar
Language Icelandic
Religion Christian (Evangelical Lutheran)
Climate Cold temperate; warmed by the North Atlantic Drift. Average temperature in Reykjavik ranges from −2 to 2°C (28-36°F) in January to 9-14°C (48-57°F) in July
Main primary products Fish, potatoes, sheep, dairy cattle, poultry
Major industries Fishing, fish and food processing, agriculture, cement, aluminium smelting
Main exports Fish, fish products, aluminium, wool and sheepskin products
Annual income per head (US$) 8900
Population growth (per thous/yr) 10
Life expectancy (yrs) Male 75 **Female** 80

▼ HOT AND COLD Icy glacial meltwaters snake among hot springs at Landmannalaugar in southern Iceland.

Illampu ***Bolivia*** Spectacular mountain with a permanent snowcap, rising to 6485 m (21 276 ft) above sea level over the town and valley of Sorata about 80 km (50 miles) north-west of the Bolivian capital, La Paz.

Map Bolivia Bb

Illimani ***Bolivia*** Triple-peaked mountain which rises to 6402 m (21 004 ft) and towers over the capital, La Paz. It has the largest glacier in tropical Latin America.

Map Bolivia Bb

Illinois ***USA*** Midwestern state bordered by the Mississippi and Ohio rivers, and in the north-east by Lake Michigan. It covers 146 076 km² (56 400 sq miles) and its gently undulating, fertile lowlands produce maize, soya beans, pigs and cattle. Coal is mined in the south and there is also some oil production. Manufacturing, whose main product is machinery, employs more than a million people. CHICAGO is the chief city and SPRINGFIELD the state capital.

Population 11 535 000

Map United States Hb

ilmenite Lustrous brownish-black mineral, titanium iron oxide, found in basic IGNEOUS ROCK, in VEINS and in SEDIMENTS derived from rocks bearing these ores.

Ilo ***Peru*** Major fishing port at the mouth of the Moquegua river in the south-west, with a fishmeal processing industry. It exports avocados, wine and copper mined at Toquepala and Cuajone in the river valley. A smelter 8 km (5 miles) north of Ilo produces copper ingots.

Population 31 700

Map Peru Bb

Ilocos ***Philippines*** Narrow coastal region of north-western LUZON between the Cordillera Central and South China Sea. It produces mainly tobacco and rice, but is densely populated and has long provided settlers for other parts of the country.

The coast is rugged and scenic, with fine beaches. Bauang, south of the port of San Fernando, is the Philippines' most important beach resort area. Ilocos has many fine old stone churches, especially in Vigan. Laoag, in the north, is a lively commercial centre.

Population 4 230 000

Map Philippines Bb

Iloilo ***Philippines*** Capital of Iloilo province on the south-west coast of Panay. It is a regional centre and major port of the western VISAYAN ISLANDS. The area produces mainly rice and sugar, and has many fine Spanish colonial churches and other buildings.

Population (province) 1 300 000; (city) 244 800

Map Philippines Bc

Ilorin ***Nigeria*** Capital of Kwara state, some 260 km (160 miles) north and slightly east of the national capital, Lagos. It is a fascinating mixture of Hausa-Fulani, Yoruba and European colonial cultures. Traditional industries include pottery, weaving, dyeing, shoemaking and carpentry, and modern factories refine sugar and make cigarettes and matches.

Population 335 400

Map Nigeria Ab

Imatong Mountains ***Sudan/Uganda*** Range, 150 km (93 miles) long. Its highest point, Kinyeti, is Sudan's highest mountain at 3187 m (10 456 ft).

Map Sudan Bc

Imatra ***Finland*** Town in south-east Finland about 35 km (20 miles) north-east of the town of Lappeenranta and near the border with the USSR. It lies near some of the country's largest sawmills, and manufactures cellulose and chemicals, copper and iron. The tumbling waters of the nearby Imatra cataract on the Vuoksi river are a tourist attraction, but the rapids have been tamed somewhat by a hydroelectric dam.

Population 35 300

Map Finland Dc

Imbabura ***Ecuador*** North-central province lying mainly in the Andes with a narrow westward extension into the coastal plain. It is crossed by the Pan-American Highway and dotted with volcanic crater lakes. The population, comprising mainly native Indians and descendants of Black African slaves, is concentrated in the Ibarra basin, where there is some light industry. The province produces maize, wheat, barley, potatoes, sugar cane and cotton. Dairy cattle, sheep and llamas are pastured. The provincial capital is IBARRA.

Population 248 500

Map Ecuador Ba

Imjin ***North Korea/South Korea*** River about 160 km (100 miles) long rising in North Korea's southern mountains. It crosses the DEMILITARISED ZONE, and flows into the Yellow Sea west of the South Korean town of Munsan via an estuary shared with the Han. It was the scene of a heroic stand by British troops – the 1st Glosters – in 1951 during the Korean War.

Map Korea Cc

Imperia ***Italy*** Seaport and resort on the Gulf of Genoa about 40 km (25 miles) east of the French border. It is noted for its olive oil and flowers.

Population 41 800

Map Italy Bc

impermeable rock See PERMEABLE ROCK

impervious rock See PERVIOUS ROCK

Impfondo ***Congo*** Capital of Likouala region in the far north of the country. Formerly Desbordesville, it is a landing point for steamboat passengers on the UBANGI river, and serves as a trading and transport centre for the forested region surrounding it.

Population 30 000

Map Congo Ba

Imphal ***India*** City and capital of the north-east state of Manipur, about 60 km (37 miles) west of the Burmese border. It is set in a beautiful valley tucked in a fold of mountains. The Japanese attacked and laid siege to it in 1944, during the Second World War. The siege was raised after three months, marking the end of the planned Japanese invasion of India. Foreigners need special permission to visit Imphal.

Population 156 600

Map India Ec

In Aménas (In Amnas) ***Algeria*** Important staging post and road junction, near the Libyan border, on the route southwards to the TASSILI N'AJJER mountain region. It has an airport and is the centre of an oil and gas region, with an oil pipeline running to Sakhira in Tunisia.

Population 7500

Map Algeria Bb

Inagua ***Bahamas*** Two islands, Great Inagua and Little Inagua, in the extreme south of the archipelago. On Great Inagua, the third largest island of the Bahamas, salt is produced by evaporating seawater. The chief town is Matthew Town in the extreme south-west. Little Inagua is uninhabited.

Population 1100

Map Bahamas Cb

Inari (Enare) ***Finland*** The largest of the lakes, in Finnish Lappland, covering 1085 km² (419 sq miles). It is fed by the Ivalo river from the south, and drains by way of the Paats river to the Arctic Ocean.

Map Finland Ca

Inch'on (Incheon) ***South Korea*** Second largest port in the country after Pusan. It lies beside the Yellow Sea, 39 km (24 miles) west of the capital, Seoul. Inch'on was opened to international commerce in 1876. Near the port American troops made a successful amphibious landing during the Korean War (1950-3). The city makes iron and steel, glass, plywood, cars and chemicals.

Population 1 083 900

Map Korea Cd

incised meander River MEANDER which has been cut deeply into its original valley floor. The river's power to erode is increased, and it maintains the same pattern of the meander at progressively lower levels. Examples include the Wye near Chepstow and the Dee near Llangollen in Wales, and the Moselle near Trier in West Germany.

India See p. 302

Indian Desert ***India/Pakistan*** See THAR

Indian Ocean Third largest ocean, lying between Asia, Australia, Antarctica and Africa. Surface temperatures range from 27-29°C (81-84°F) in the shallow waters in the north-west to −2°C (28°F) in the far south. The ocean includes the RED SEA, the oil-rich GULF, the ARABIAN SEA, the Bay of BENGAL and the ANDAMAN SEA.

The Indian Ocean was formed in the last 170 million years as the ancient continent of GONDWANALAND broke up. The Indian subcontinent, which was formerly joined to Africa, Australia and Antarctica, drifted north and collided with the Eurasian continental plate.

CONTINENTAL DRIFT is still changing the shape of the ocean. OCEAN FLOOR SPREADING is occurring along the central ocean ridges on the sea floor, which extend from the Red Sea into the Arabian Sea. The Mid-Indian Ridge divides into two parts around latitude 20 degrees south. One part bends eastwards, bypassing the south of Australia, and links with the Pacific-Antarctic Ridge. The (*continued on p. 307*)

India

DEMOCRACY HAS TRIUMPHED IN THIS VAST LAND OF RURAL POVERTY AND NOISY, OVERCROWDED CITIES

More than 700 million people live in India, and there will be 1000 million by the end of the century if present trends continue. It is a country with huge problems, centred mainly on poverty and population growth. But it is an astonishingly stable democracy, a model in many ways for other developing nations.

The British, who governed this huge land from 1757 until independence in 1947, were the overlords of the whole subcontinent that now embraces not only India but also PAKISTAN and BANGLADESH.

The HIMALAYAN mountains lie in the north, forming a barrier between India and China (including Tibet). The Himalayas are the world's youngest and highest mountains. They run into two other mountain chains to the west: the KARAKORAM, which lie in KASHMIR, and the HINDU KUSH, in Afghanistan. Kashmir is claimed by both India and Pakistan, and has been an area of conflict between the two countries since 1947.

The lower slopes of the mountains are covered in lush vegetation, and the area around DARJILING is one of the world's best tea-growing areas. The climate is almost tropical, with lots of sunshine and an abundance of water, from the monsoons and from numerous rivers flowing from the snows of the Himalayas. The monsoon rains last from June to September, bringing high humidity and slightly cooler days.

At the foot of the Himalayas the land changes abruptly into a huge plain, drained by two great rivers – the INDUS and the GANGES. This is the most densely populated part of India. It fills one-third of the country and extends 3500 km (2200 miles) from the PUNJAB in the west to BENGAL in the east. Although it receives enormous amounts of water from the monsoon-fed rivers and the occasional flooding caused by melting Himalayan snows, irrigation is still needed. However, it is one of the most fertile areas in the world, producing rice and, in the western areas of UTTAR PRADESH, HARYANA and the Punjab, sugar cane and wheat.

The plains run south to the Vindhya mountains, which border the DECCAN plateau – the oldest part of India with rocks more than 600 million years old. The Deccan extends to the southern tip of the country.

▲ CROWNING GLORY The Maharajah's Palace, rebuilt in 1897, stands on Chamundi Hill in the centre of Mysore city. It contains a fabulous gold and ivory throne – a gift of the Mogul emperor Aurangzeb (1658-1707).

The human inhabitants of India have varied origins. Before 2000 BC the people were mostly dark-skinned tribes, including the Dravidians. The course of history was changed when fair-haired and blue-eyed Aryan tribes moved down from the north-west. The beliefs of these invaders – recorded in religious texts called the Vedas – formed the basis of the great Hindu religion. Four thousand years later in modern India the hymns of the Vedas are still recited by Hindu priests (Brahmins).

The Aryan peoples had a three-tier social system of warriors, priests and common people. These were the origins of the caste system in Hinduism. Caste dictated what work you did, whom you married and even whom you could eat with, and the same restrictions applied to your descendants. Outside the main castes were the people whose occupations made them the 'Untouchables'. They were responsible, for instance, for the removal of dead bodies, or of human excrement.

In the modern republic of India, caste has little legal significance, although it is still a paramount influence on society. To maintain the distinctiveness of each caste, nearly all marriages are still arranged.

MUSLIM WEALTH AND ARTISTRY

The rival Muslim religion was established in India in 1192 by a Muslim prince from Afghanistan, Muhammad Ghuri. He conquered the territory when he defeated the Hindu ruler, Prithviraja, in a historic battle near DELHI. The ruling class of India remained Muslim until the British took control.

Under the Mogul emperors (1526-1761), Muslim art flourished and gave the world priceless paintings and buildings, including arguably the most beautiful building on earth, the TAJ MAHAL. Europeans were fascinated by the wealth and products of this empire, which attracted traders from Britain, the Netherlands, Portugal and France. The British East India Company was established in 1600, and over the next 100 years held considerable power over the country.

The Mogul Empire collapsed from within, riven by wars of succession. The British took sides in the conflict and became a governing power after Robert Clive defeated the Nawab of Bengal at the Battle of PLASSEY in 1757. The British presence was challenged by the Indian Mutiny in 1857, after which the British government took complete control and remained in power for another 90 years, developing railways and irrigation canals and expanding the port cities of CALCUTTA, MADRAS and BOMBAY. This period brought little change for the rural population. In states where Britain ruled indirectly through the Indian princes, it was as if time stood still.

BIRTH PANGS OF FREEDOM

The final struggle that brought independence was led by Mohandas Karamchand Gandhi, who had risen to power and earned the title of Mahatma ('Great Soul') through his 50 year struggle to help the poor of India, especially the Untouchables. He insisted that the independence campaign be non-violent and, though there were sporadic outbreaks of violence, he largely succeeded in this aim until the last terrible birth pangs of freedom.

Independence struck the fetters from two Indian nations: the Hindus who had clamoured for self rule for 750 years and the Muslims, dispossessed by the British. To meet their different ambitions, the subcontinent was partitioned.

The Muslims, led by Muhammad A. Jinnah, were given a new state called Pakistan. It was a country with two 'wings' – one in the Indus Valley, and the other in Bengal – separated by more than 1600 km (1000 miles) of Indian territory.

As these two parts were torn from India in August 1947, tensions between Hindus and Muslims erupted into wholesale massacres. One of the princely states, Kashmir, wavered between joining India or Pakistan. The two countries immediately went to war over the issue, which is still unresolved.

Mahatma Gandhi was assassinated by a Hindu fanatic in 1948. The first Prime Minister of India after independence was Pandit Nehru, who was committed to modernising his country, and modifying the caste system. He held power from 1947 to 1964, and was succeeded by Lal Bahadur Shastri, whose period of office saw war with Pakistan in 1965.

Nehru's daughter, Mrs Indira Gandhi (no relation to the Mahatma), became Prime Minister in 1966 and maintained stable government for 18 years, despite controversial measures such as material incentives for sterilisation. In 1971 her government fought another war with Pakistan which led to the foundation of Bangladesh from the eastern wing of Pakistan. In 1975 Mrs Gandhi declared a state of emergency – said by critics to have been characterised by brutal suppression of opposition.

The next violent crisis involved the Sikhs, a religious minority living mainly in the wealthy Punjab state, who wanted their own separate

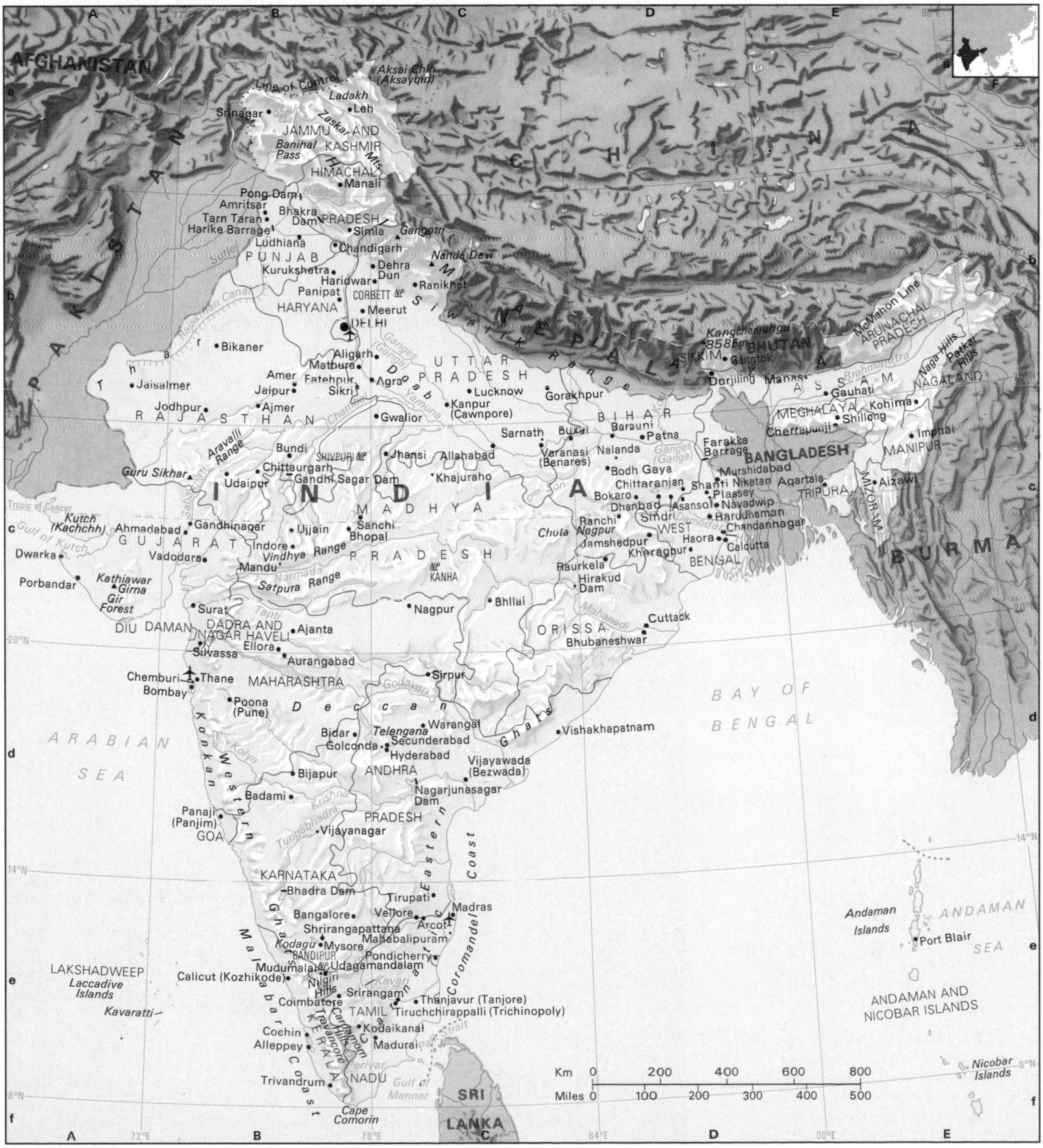

nation. Sikh terrorists used Sikh temples as their strongholds, and on June 5, 1984 the Indian Army stormed 37 temples, including the Sikhs' most sacred place, the Golden Temple at AMRITSAR.

The government said that 500 terrorists and civilians died in the Golden Temple – including the Sikh leader, Jarnail Singh Bhindranwale – and 92 soldiers were killed. Unofficial sources said that about 1200 people died. Four months later Mrs Gandhi was assassinated, and two Sikh officers in her personal security guard were accused of her murder.

Mrs Gandhi's son Rajiv was appointed as her successor and his position was confirmed by a sweeping victory at a general election soon afterwards.

POOR PEASANT FARMERS

The drive for modernisation started by Nehru, and continued in six successive five-year plans, has made India the tenth most important industrial power in the world. Today, India supplies 80 per cent of her own energy needs, mainly from coal, and 73 per cent of her oil needs from her own offshore oil wells. The government controls all heavy industry, but there is growing relaxation of central control, and most consumer goods are produced by private enterprise.

But the India that Mahatma Gaṇdhi loved was based on village life, and this is still true despite the great industrial revolution. About 75 per cent of India's people live in the villages, and 70 per cent of employment is still in agriculture. Many of the farmers have smallholdings of less than 2 hectares (5 acres). Others own no land at all. They till the soil, but rent the land, paying the landlord either in cash or with part of their harvest (sharecropping). This sharecropping system is blamed for the low levels of investment and productivity in much of Indian agriculture, and is one reason for the widespread rural poverty. Nevertheless, there has been a 'Green Revolution' in Indian agriculture in recent years. This has given the government a reserve stock of grain, held against the inevitable bad harvest, which comes when the monsoon rains are inadequate – as happens from time to time.

India has 200 million people living below the country's own official poverty line, and nearly all of these are in rural areas. Besides the often exploited sharecropper, there are landless peasants, employed day by day. Some will work for food and a place to sleep; others for cash when the farmers need help, as at transplanting time.

Many of the rural poor do not even have access to village land on which to put a small mud hut. The government has recently conducted a search for homestead sites for them.

Most families cook on dried dung or wood fires, and there is no longer enough of either

► SILVER STREET Merchants sell cloth and silver in Old Delhi's Chandni Chowk (Silver Street). It leads to the Red Fort, the palace of Shah Jehan begun in 1638, and its balconied houses were built when the street was widened to take his great processions.

KHUSH-DIL HOTEL
KUMAR & CO.
CHAINA RAM

fuel. Women and children forage for hours for a few sticks or in the hopes of finding an unclaimed cowpat. The alternative is to own an animal, but it is only the farmer with some hectares of land who has enough rice-straw and other fodder to keep a cow. Fodder is so scarce that stray cows are shooed out of the countryside and find their way into the towns, where they are allowed to wander virtually at will and graze on rubbish tips.

WHY THE COW IS SACRED

The cow is the sacred animal of Hinduism and it is easy to see why. Cows and bullocks are an essential part of the rural economy. They pull the ploughs, help to puddle rice fields and pull carts to market. In some places they even grind corn as they did in Biblical times, pushing a millstone round and round. Cows also provide a little milk, but their yield is very small by Western standards.

India has more cattle per head of the population, and per square kilometre than any other country. But there is barely enough food for them and their quality is correspondingly low. Cows are never slaughtered – the taking of animal life is forbidden by the Hindu religion. This tenet is so strictly adhered to that even in times of famine people starve to death while cattle survive.

Tractors could take over much of the cows' work, but are economical only on large farms. Tractors are still quite rare in India, and are concentrated in prosperous areas which have good water supplies and lower population densities – such as Punjab and Haryana.

To improve agricultural productivity the government is developing irrigation. Only half of the irrigation potential has been used so far: the government wants to develop the other half by the early 21st century. Some of the irrigation schemes are massive: the RAJASTHAN project, for instance, includes a canal hundreds of kilometres long, and thousands of kilometres of distribution channels which take water to the THAR desert. New roads and new towns have to be constructed and millions of people will depend on the scheme.

Rice is the best known of India's food grains, but in the north and west wheat is the staple, the basis for chapattis and other unleavened bread that is an important food. In much of the Deccan plateau, coarse grains such as sorghum or millet are grown. Most farmers grow vegetables, herbs and spices.

MARKET DAY IN THE VILLAGE

Nearly all rural people go to weekly markets, although they may have little surplus to sell, and will probably have to carry their wares on their heads for many kilometres. On market day a township of stalls will sprout around a small village or town. It is a day for gossip, perhaps for sounding out the marriage market for one's children. It is a time to seek medicines, many of them of a traditional herbal kind.

Some women will go with their husbands to a jeweller and spend thousands of rupees on gold and silver bangles. Even families who are nearly destitute will buy jewellery – as an investment to be worn until the terrible day when it has to be sold in the face of starvation.

Towns as well as villages face great pressure on resources. There is not enough public transport and rickshaws are too expensive for the poor, so everything has to be within walking distance. Each quarter of the town will have its own bazaar, where vegetables can be bought daily. Many small traders and artisans live over their shops, in one part of town the metal workers, in another the grain traders, and in another the truck operators.

Buildings are low, the streets narrow, with open sewers and flooded gutters. But the density of human settlement is higher than in Western skyscraper cities.

The houses of the well-to-do have often turned their backs on the outside world: all the rooms will open onto an internal courtyard. It will be the home of an extended family, including grandparents and uncles, brothers, sisters and cousins.

Poverty is worst in the countryside, but it is much more visible in towns, where poor people are concentrated. Towns become more and more crowded as people stream in from the countryside, seeking any opportunity to make a little money. Overcrowding is increased by the number of animals that join the crush in the streets – monkeys, goats, chickens, and the sacred cows.

The urban poor build shacks out of mud, tin and scraps of wood, usually on land which they occupy illegally. Some will pay rent to racketeers just to sleep on the pavement.

Sometimes a group of poor people organise a swoop on a building site, then defy eviction. Such squatters are adept at stealing power from overhead cables: more than 30 per cent of Calcutta's electricity is stolen.

But squatters cannot steal sewers that should have been laid before they raised their makeshift shacks, so sanitation is the worst problem. The shanty people cook their evening meals on fires fuelled by coal dust, dung, oily rags and rubbish, and the smog that results is the worst in the world.

THE ROAR OF THE TRAFFIC

Indian cities have also developed some of the world's worst noise pollution. From every street corner loudspeakers blare music at maximum volume. Hundreds of shouting sellers of lottery tickets add to the din. Every car and every truck repeatedly blasts its horn.

The buses make their contribution to both noise and smog, leaning at crazy angles with people hanging outside from the window bars and with the engine belching black diesel clouds in the faces of the rickshaw pullers. Some of the better-off middle classes dash through the traffic on motor scooters.

The educated classes mostly seek salaried jobs with big companies or with government. Pay is low, perhaps 2000 rupees ($160) a month, and city living costs are high.

The uneducated poor who live in shanty towns may work in factories, but more often they have their own workshops, turning waste materials into useful items. Old tin cans are beaten into cooking pots, and old car tyres used to sole shoes.

Long-distance travel between cities is still mostly by train, although coaches compete on some routes.

Hundreds of different languages are used in India, and the government has tried to promote Hindi as the official language. Southerners who do not use Hindi as a mother tongue have resisted attempts to downgrade English, so it is still accepted as an equal with Hindi.

Primary education is free and theoretically compulsory, but many children do not attend school, particularly in rural areas where they may be needed on the farms. Equipment standards are very low, and the government does not provide books. About 36 per cent of the population are literate.

ENDURING DEMOCRACY

Despite poverty, and low levels of literacy and education, India has an impressive record as a democratic state. It has never suffered an army coup, nor a transfer of power between parties except by general election.

There are two major tiers of government, at federal and state level. Some states are bigger than any nation in Europe – Uttar Pradesh has 130 million people.

Life is easier for the central government if its own party holds power at state level. The centre can also hold power through the administration: the senior civil servants of each state belong to a nationwide Indian Administrative Service, and they are always chosen from a different state from that in which they serve. In the army, the local commanders will also be out-of-state people. This makes it easier to exercise impartial authority.

Despite all her problems, modern India must be acclaimed as a major triumph – a synthesis of Eastern spiritualism and tradition and Western political liberalism.

INDIA AT A GLANCE

Area 3 166 829 km² (1 261 816 sq miles) excluding parts of Jammu and Kashmir occupied by Pakistan and China

Population 778 520 000

Capital New Delhi

Government Federal republic

Currency Rupee = 100 paisa

Languages Hindi and English official, but at least 13 other languages used including Urdu

Religions Hindu (83%), Muslim (10%), Christian (3%), Sikh (2%), Buddhist (1%), Jain (1%)

Climate Tropical; monsoon from June to September. Average temperature in New Delhi ranges from 7-21°C (45-70°F) in January to 26-41°C (79-106°F) in May

Main primary products Rice, wheat, sugar cane, barley, sorghum, millet, potatoes, tea, groundnuts, cotton, jute, pulses, vegetables, fruit; coal, iron ore, oil and gas, bauxite, chromite, copper, manganese, gemstones

Major industries Textiles, iron and steel, transport equipment, chemicals, fertilisers, machinery, oil refining, agriculture, cement, coke, food processing, beverages

Main exports Textiles, food (including fish, tea), machinery, gemstones, iron ore, leather

Annual income per head (US$) 217

Population growth (per thous/yr) 21

Life expectancy (yrs) Male 52 **Female** 53

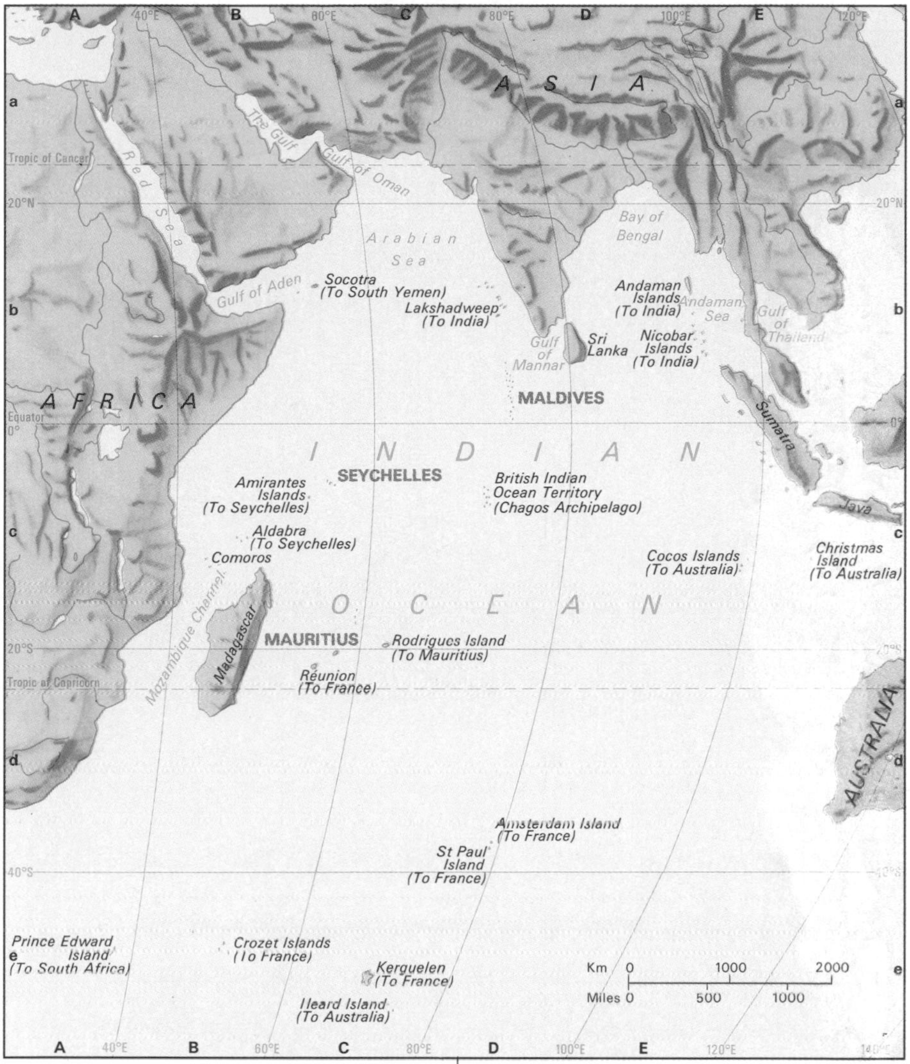

South - west Indian Ridge curves around the south of Africa and links up with the Atlantic-Indian Ridge. The Indian Ocean also contains the deep JAVA TRENCH, where the Indo-Australian continental plate is sliding under the Eurasian plate.

South of the Equator, the surface water circulates in an anticlockwise direction, through the West Australian, South Equatorial and Mozambique currents. North of the Equator, the monsoon causes seasonal changes in the directions of surface currents.

Fishing is less developed than in the ATLANTIC or PACIFIC oceans. The main fishing grounds are in the Arabian Sea and the Bay of Bengal. Oil is extracted in The Gulf and manganese nodules are scattered over large areas of the ocean floor, especially in the Red Sea.

Arab, Indian and Chinese trading ships have sailed across the ocean since early times. European trade increased after the Portuguese explorer Vasco da Gama discovered the sea route around Africa to India in 1498. But this shipping route was partly superseded by the SUEZ CANAL, opened in 1869. The development of oil resources in The Gulf region greatly increased the ocean's strategic importance.

Area 73 481 km² (28 370 sq miles)
Greatest depth 8047 m (26 400 ft)

Indian summer Spell of fine, sunny and dry weather that sometimes occurs in late autumn in Britain and North America.

Indiana ***USA*** Midwestern state covering 93 994 km² (36 291 sq miles) east of Illinois, between Lake Michigan and the Ohio river. It has much good farmland producing maize, soya beans, pigs and cattle. Coal is mined in the south-west and there is also oil production and limestone quarrying. Electrical and transport equipment and steel are the main manufactures. INDIANAPOLIS is the state capital.

Population 5 499 000
Map United States Ic

Indianapolis ***USA*** State capital of Indiana. It lies in the centre of the state, about 220 km (135 miles) south-east of Lake Michigan. The city is an agricultural and industrial centre, and venue of the annual Indianapolis 500 car race.

Population (city) 710 300; (metropolitan area) 1 194 600
Map United States Ic

Indonesia See p. 308

Indore ***India*** Cotton textile city about 650 km (403 miles) south and slightly west of Delhi. The former capital of the princely state of Indore, it is now one of the major cities of the state of Madhya Pradesh. It has many buildings from the colonial period of the British Raj, when it was the focus of British political activity in central India. Daly College there was established in the 19th century to educate the sons of Indian princes.

Population 829 300
Map India Bc

Indus ***South Asia*** One of the great rivers of Asia, rising north of the Himalayas at 5200 m (17 000 ft) on Mount Kailas in Tibet (where it is called the Shiquan He) and flowing 2740 km (1700 miles) between mountains, through gorges up to 3000 m (10 000 ft) deep and across blistering deserts to the Arabian Sea south of Karachi. The river and its tributaries, such as the Jhelum, Chenab and Sutlej, are vital sources of irrigation water for both Pakistan and India, who signed an agreement in 1960 on sharing their flow. The Indus is dammed at TARBELA for hydroelectric power, and there are numerous irrigation barrages as it crosses the Punjab and Sind. The river's flow is diminished once it reaches the plains because of evaporation and irrigation, and because of the sponge-like effect of the river's alluvial deposits, which are up to 1.6 km (1 mile) deep.

A great civilisation, centred on MOHENJO DARO and dating back 4000 years, once flourished along the river's banks. The Indus valley is an important transport route, especially between Sind and the Punjab.

Map Pakistan Cb, Da

Infanta ***Philippines*** See QUEZON CITY

Inga Dam ***Zaire*** One of the world's largest hydroelectric projects, on the Zaire river about 230 km (145 miles) south-west of Kinshasa. Begun in 1968, it has a potential of 30 million kilowatts, and has already made possible a great expansion of industry in Bas-Zaire region.

Map Zaire Ab

Ingolstadt ***West Germany*** Industrial town on the Danube river about 72 km (45 miles) north of Munich. It is the pipeline terminus for oil supplies from the Mediterranean Sea, and has oil refineries, a car industry, and electronics and textile factories.

Population 90 800
Map West Germany Dd

Inhambane ***Mozambique*** Seaport on the site of an ancient Arab settlement at the tip of a peninsula, some 450 km (280 miles) north-east of the South African border. The port is the capital of Inhambane province, which covers 68 615 km² (26 485 sq miles). Most people in the province belong to the Tsonga and the Chopi tribes. The chief products are cashew nuts, cotton, rice and sugar, and, in the drier interior, cattle.

Population (town) 27 000; (province) 1 037 500
Map Mozambique Ac

inland sea Extensive isolated expanse of water having no link with the open sea, and often associated with distinctive species of animals that have bred there in isolation. Lake Baikal in the USSR is an example.

Indonesia

A SPRAWLING MASS OF VOLCANIC ISLANDS THAT IS OCCUPIED BY ONE OF THE MOST DIVERSE AND EXOTIC NATIONS ON EARTH

Scattered across the Indian and Pacific oceans in a vast emerald crescent, Indonesia is made up of 13 677 islands, making it by far the world's biggest island chain. Other statistics are nearly as impressive. With over 176 million people, it is the world's fifth largest country. And about 90 per cent of its population follow the Islamic faith, making it the world's largest Muslim country. From the tip of SUMATRA in the north-west to IRIAN JAYA in the south-east the islands of Indonesia curve through more than 5000 km (about 3200 miles) of ocean – farther than the distance across the continental United States. It has almost as many tribal and ethnic groups as there are days in the year (360), and between them they speak more than 250 languages and dialects.

Indonesia has more than 100 active volcanoes, and the turbulence of nature has been echoed by the turbulence of the country's political life. A war against an independence movement in East TIMOR is estimated to have cost at least 100 000 lives in the past ten years. Maintaining the unity of this sprawling, many-peopled nation of dense jungle, smouldering volcanoes and stunning scenery is one of the most vital tasks facing its leaders. Small wonder that the nation's official motto is *Bhinneka tunggal ika* ('Unity in diversity'). Since independence was won from the Dutch, in 1949, two strong men, President Sukarno and President Suharto, have in turn striven to unite this vast republic.

WHERE THE LAND TREMBLES

Indonesia's land area is more than 2 million km² (about 780 000 sq miles), but its islands are scattered over 8 million km² (more than 3 million sq miles). The biggest landmass is KALIMANTAN, Indonesia's share of the island of Borneo, followed by Sumatra, Irian Jaya (formerly West New Guinea) and the irregularly shaped SULAWESI (Celebes). JAVA, though fifth in size, is the dominant island and by far the most heavily populated. It contains more than 90 million people.

This emerald chain of mountainous islands is a series of high points on undersea extensions of mainland Asia and Australia. The high points, formed at about the same time as the Himalayas, are the islands. Kalimantan, Sumatra, Java, LOMBOK and BALI are all peaks on the SUNDA SHELF, a largely submerged extension of Asia. Irian Jaya and its attendant islands are on the SAHUL SHELF, which links New Guinea and Australia. Sulawesi and the MOLUCCAS are on a third extension – from Japan and the Philippines.

All the larger islands have a volcanic mountainous area flanked by coastal plains. Steep-sided mountain ranges face the deep seas, and sloping lowlands fall away gently to the shallow seas covering the continental shelves. The highest mountains are PUNCAK JAYA (5039 m, 16 531 ft) and Gebel Daam (4922 m, 16 148 ft) on Irian Jaya. There are peaks above 3000 m (over 10 000 ft) on Java, Sumatra, Sulawesi, Lombok and Bali.

Some of Indonesia's 100 or more active volcanoes have been exceedingly destructive. In 1883 KRAKATAU, an island volcano in the SUNDA STRAIT between Java and Sumatra, exploded after two centuries of inactivity, killing 36 000 people, forming new islands with its lava and scattering debris as far as Madagascar. The sound of the explosion, the most violent ever recorded, was heard nearly 5000 km (about 3000 miles) away.

In the ten years to 1983, there were 21 eruptions. Galunggung in west Java, for instance, blew a dense cloud of ash into the upper atmosphere and nearly brought down a passing airliner. Earthquakes are frequent in the southern islands – in 1976 and 1977, 6302 people were reported killed or missing, and 4158 were injured by earthquake motion.

These massive volcanic eruptions have produced a jagged and fragmented landscape – a feature which is accentuated by the geological action known as fracturing in some mountain areas. There, huge movements within the earth's crust have created plateau-like uplands and rift valleys – such as the one that slices through the northern Barisan Mountains on Sumatra. Because the volcanic soil is easily eroded, the heavy rains wash masses of material off the highland areas, resulting in a rapidly changing landscape in terms of geological time. For example, in southern Kalimantan, northern Sumatra and southern Irian Jaya, coastlines swollen by debris washed from the hills push steadily seawards, creating ever-increasing swamplands.

Rainfall, often falling as torrential downpours, is more than 3000 mm (118 in) per year, while temperatures average 27°C (81°F). Because it is so hot and humid throughout its huge area, Indonesia contains more equatorial rain forest than any other country. However, in many areas, sections of the dense forest have been extensively cut for valuable hardwoods, leaving exposed fragile soils that are easily washed away.

Only on Java and the LESSER SUNDA ISLANDS – which lie in the path of dry air moving north from Australia, creating a relatively dry season from April to October – is there any variation in the dominant rain forest. The lower humidity allows mixed deciduous forest – with evergreen teak a major component – to grow. But even here, more and more timber has been cleared to provide land to resettle and feed the population.

The islands have been inhabited since long before historic times. In 1890, near Trinil village in east Java, Dr Eugene Dubois found fossils of Java Man (*Homo erectus*) dating back 500 000 years. Stone Age people spread from mainland Asia in several waves between 2500 and 1000 BC and in the 1st century AD Indian traders and Hindu and Buddhist priests from the Indian subcontinent arrived and introduced their culture. The remains of their many temples can be seen today.

Muslim traders, possibly from Gujurat in India, introduced Islam to the country in the 14th century; the new religion spread rapidly but Bali remained an outpost of Hinduism. Eventually, in 1509, the first Europeans – the Portuguese – established a foothold in Indonesia and its spice trade. Spain, Holland and England also competed for the lucrative spice business but the Dutch East India Company triumphed over its rivals, and gained control of JAKARTA, which it called Batavia, in 1619. Dutch government rule followed, apart from a spell when Britain's Sir Stamford Raffles took over the islands in the Napoleonic Wars.

FORGING NATIONAL UNITY

Indonesia was occupied by the Japanese in the Second World War, but the Dutch returned in 1945 to reclaim the colony. Open fighting broke out with the independence movement in 1947 and, after a two-year war, the Dutch conceded and left the islands. Ahmed Sukarno became president.

Sukarno gave Indonesia a national identity but his ambitious plans brought economic failure. He sent paratroops into Dutch New Guinea, which eventually became Irian Jaya; and he openly waged war with Malaysia, the formation of which in 1963 he strongly opposed.

Under Sukarno's leadership, the country was tolerant of Communism, but in 1965 an attempted Communist coup, which involved Indonesian Communists and in which six generals were assassinated, was quelled by the army. The people rose up and in three months massacred an estimated 500 000 Communists and their supporters; whole villages were wiped out. A new strong man emerged in the turmoil, Lieutenant-General Suharto, who replaced Sukarno as president in 1967.

Suharto ended the confrontation with Malaysia, restored links with the West and dealt with inflation of 600 per cent. He launched a series of five-year plans to increase food production and tackle unemployment. He also extended Indonesia's territory in 1975 by invading and annexing East (Portuguese) Timor where civil war was raging. However, the East Timor independence movement, Fretilin, continued to fight a guerrilla war against Indonesian troops into the 1980s.

President Suharto has been faced by political unrest too. In September 1984 at least 30 people died in the capital, Jakarta, when troops fired on rioters. Opposition has come from fundamentalist Muslims.

One of the greatest problems facing the government is overpopulation, which is being dealt with by migration schemes and family planning campaigns. Already 3 million people

◄ BENEATH THE VOLCANO Smoke funnels from Merapi – one of 35 active volcanoes on the island of Java – as workers toil in the outlying rice fields. The mountain rises to some 2911 m (9550 ft) and its last major eruption was in 1867.

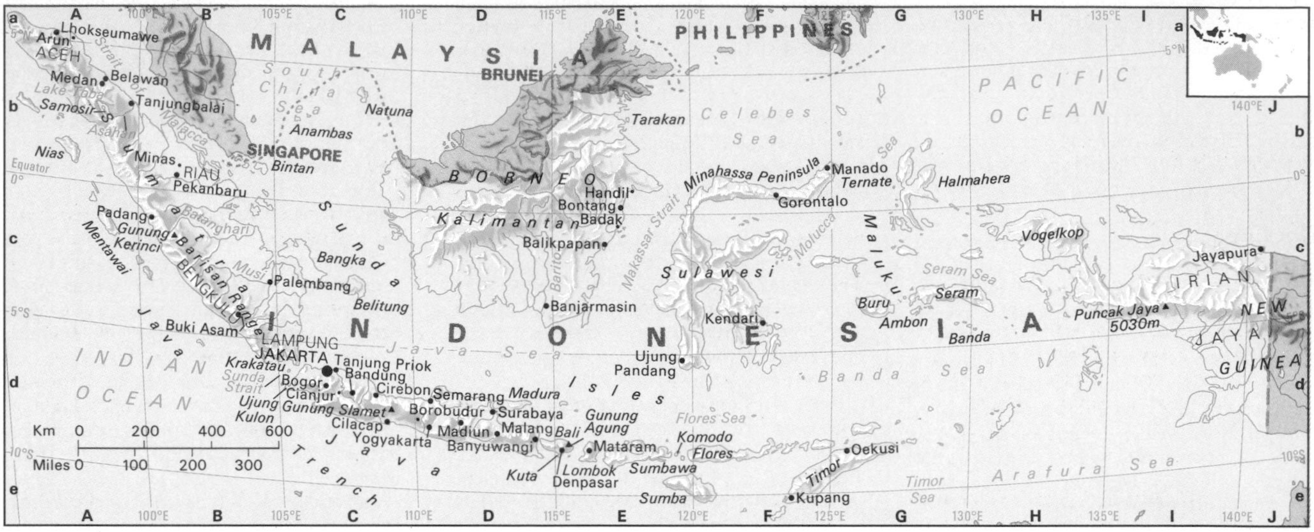

have been moved to less densely populated areas, particularly Sumatra and Irian Jaya.

Only 6000 of the 13 677 islands are inhabited. Overcrowded Java, with only 7 per cent of the land, contains 60 per cent of the population. The island has 690 people per km² (about 1790 per sq mile) – in irrigated areas the figure rises to 2000 (5180). Thanks mainly to the Dutch, who kept down local tribal warfare and improved productivity, the population has been growing for almost two centuries. Improvements in health care have accelerated the increase, putting pressure on the land available for cultivation. More than one-third of rural families are landless, and the average holding is only half a hectare (just over an acre).

Despite increased irrigation and the development of high-yielding varieties of rice, Java's fertile rust-coloured soil, enriched by volcanic residues, is in danger of becoming exhausted. In 1980, 47 per cent of rural Javanese were below the poverty line, and many have left for the cities. The capital, Jakarta, is the largest city in South-east Asia, with 7 million people, but only has housing and proper sewerage and other services for less than 1.5 million. The rest are crammed into ramshackle shanty towns, especially along the railway tracks. Regional centres such as SURABAYA and BANDUNG have also grown rapidly and have populations of over 2 million and 1.5 million respectively.

In contrast, parts of Kalimantan and Irian Jaya remain wildernesses with fewer than 10 people per km² (26 per sq mile). Here the forests are difficult to clear and unhealthy for settlement, with malaria a problem. Apart from scattered forest clearance for logging and shifting cultivation, settlement of the outer islands has been traditionally restricted to the coasts. The people make a living from fishing and trading in local forest produce.

The main problem for Indonesia's leaders since independence has been to draw together the different lands and people which the country inherited from the Dutch. The population falls into two main groups, Malayan and Papuan, with a transitional area in central Indonesia. The main ethnic groups include the Javanese and Sundanese in Java; the Acehneses, Bataks and Minangkabaus in Sumatra; the Madurese in Madura; the Sasaks in Lombok; the Ambonese in the Moluccas; the Timorese on Timor; the Menadonese and Buginese in Sulawesi; the Dyaks in Kalimantan; and the Irianese in Irian Jaya. There are also about 3 million Chinese.

Attempts to bring unity to these many different groups have included the adoption of one language – Bahasa Indonesia, similar to Malay – as the country's official tongue. Agreement has been reached with Malaysia on a common Roman script for both countries. Modern technology also has helped to unify the nation. Indonesia was the fourth country in the world to operate a domestic communications satellite system, which carries TV and radio to all Indonesia's 27 provinces.

Oil was discovered in Indonesia in 1885, but only in the 1960s did the country become a major producer: output is about 1.5 million barrels a day. Large deposits of natural gas have been found in Aceh and eastern Kalimantan, and Indonesia is now the world's biggest exporter of liquefied natural gas.

Tin has long been mined in the RIAU archipelago, and Indonesia's reserves remain the largest in the world. There is nickel in Sulawesi and the Moluccas, copper in Sumatra and Irian Jaya, bauxite in the Riau islands and coal in Kalimantan.

Indonesia is one of the world's leading rubber producers and timber is also an important export. There are plantations of sugar cane, coffee, tea, tobacco, palms and spices. Fields of rice provide the main food crop, and maize, cassava, sweet potatoes and soya beans are also grown for the home market. There is an abundance of tropical fruits, including bananas, pineapples, papayas and mangoes. Fishing has great potential from the well-stocked seas around the islands, but the methods used are often primitive and not geared to a modern export market.

These resources, as yet not fully developed, make Indonesia a storehouse of energy and valuable raw materials, and offer the country a solid base for economic development. Earnings from oil and gas exports have already been invested in irrigation in Java and in expanding the plantations of rubber, tea and oil palms in the outer islands. The country has begun to build up its heavy industry, using its huge mineral resources, and powered by its natural gas and the hydroelectric power generated in Sumatra. Technical knowledge provided by the other members of the Association of South-east Asian Nations (Malaysia, the Philippines, Singapore and Thailand) should help Indonesia to develop more advanced industries such as electronics and machine-tool manufacture.

INDONESIA AT A GLANCE
Area 2 027 087 km² (782 660 sq miles)
Population 176 740 000
Capital Jakarta
Government Military-ruled republic
Currency Rupiah = 100 sen
Languages Bahasa Indonesia (official), also many other languages and dialects
Religions Muslim (88%), Christian (9%), Hindu (2%), Buddhist
Climate Tropical/monsoonal; average year-round temperature in Jakarta ranges from 23°C (73°F) to 33°C (91°F)
Main primary products Rice, maize, cassava, sweet potatoes, soya beans, sugar, bananas, palm oil, copra, rubber, coffee, tobacco, tea, groundnuts, fish, timber; oil and natural gas, tin, nickel, copper, bauxite, coal
Major industries Agriculture, oil and gas production and refining, mining, forestry, fishing, textiles, transport equipment assembly, food processing, paper, cement, matches, tyres, glass, chemicals, fertilisers
Main exports Crude oil, natural gas, refined petroleum products, timber, rubber, tin and other metal ores, coffee, fish products, tea, tobacco
Annual income per head (US$) 520
Population growth (per thous/yr) 21
Life expectancy (yrs) Male 52 **Female** 55

Inland Sea (Seto Naikai) ***Japan*** Almost enclosed area of sea between the islands of Honshu, Shikoku and Kyushu. It stretches nearly 500 km (310 miles) between the Honshu cities of Osaka in the east and Shimonoseki in the west, and forms the Seto Naikai National Park. The sea, although heavily polluted in places by coastal industries, has many areas of outstanding beauty, including thousands of pine-clad islands, with small fishing villages under steep cliffs.

Map Japan Bd

Inle ***Burma*** Lake in Shan state, about 200 km (125 miles) south of Mandalay. A major tourist attraction, the lake lies in a valley close to the state capital, Taunggyi. Buddhist temples line its banks and 'leg-rowers' – boatmen who row standing up, with a paddle strapped to one leg – fish the waters.

Map Burma Cb

Inn ***Switzerland/Austria/West Germany*** River rising at the Lake of Sils in south-east Switzerland. It flows for some 510 km (317 miles) into south-west Austria (passing through Innsbruck) and on through the Bavarian Alps into West Germany – where it joins the Danube at the city of Passau. There are more than 20 hydroelectric plants along its course.

Map Austria Bb

Inner Mongolia ***China*** See NEI MONGOL AUTONOMOUS REGION

Innsbruck ***Austria*** Tourist resort, manufacturing city and the capital of Tyrol, on the Inn river 140 km (87 miles) south-west of Salzburg. In 1809 it was the headquarters of Andreas Höfer, the Austrian hero who led an unsuccessful revolt of Tyrolese peasants against Napoleon's Confederation of the Rhine, which Austria had joined after capitulating to the French. A statue of Höfer in the city marks the event.

Badly damaged in the Second World War, Innsbruck was extensively rebuilt. Its major buildings include the 18th-century imperial palace – the Hofburg, and a16th-century Franciscan church (the Hofkirche), in which stands the magnificent tomb of the Holy Roman Emperor, Maximilian I (1459-1519). However, the emperor's body is buried at Weiner Neustadt. The city's industries include metal products, textiles, bookbinding and bellmaking.

Population 116 110

Map Austria Bb

inselberg Isolated domed hill left by erosion in a desert or semiarid region.

insolation weathering The weathering process that occurs when rocks or rock minerals expand and contract at different rates as a result of heating by the sun during the day and cooling at night.

Interlaken ***Switzerland*** Leading Alpine resort in the Bernese Oberland 42 km (26 miles) south of the capital, Berne. It lies between the lakes of Thun and Brienz, and is famous for its view of the Jungfrau massif. The town manufactures textiles, clocks and watches.

Population 13 000

Map Switzerland Aa

intertropical convergence zone Zone of ascending, moist air and low pressure where the trade winds converge. This zone migrates north and south of the Equator with the heat equator. It is characterised by thick clouds and thunderstorms, and the summer monsoon rains of India and South-east Asia are associated with it. It was formerly known as the intertropical front.

intrusion A body of INTRUSIVE ROCK.

intrusive rock An IGNEOUS ROCK formed when magma flows or forces its way into a crack, cavity or other space below the earth's surface and solidifies there. The body of rock so formed is called an intrusion; BATHOLITHS, DYKES and SILLS are common examples.

Inuvik ***Canada*** Administrative, communications and fur-trading centre on the eastern side of the Mackenzie delta in the NORTHWEST TERRITORIES. It was laid out as a model Inuit (Eskimo) community between 1954 and 1962.

Population 3145

Map Canada Bb

Invercargill ***New Zealand*** Town on an inlet of the Foveaux Strait in the extreme south of South Island. It is the centre of a sheep-farming and dairy-farming area, and the bulk of its trade goes through the deep-water port of BLUFF, about 25 km (15 miles) to the south. Its industry is based on fertiliser and meat freezing works.

Population 54 000

Map New Zealand Bg

Inverness ***United Kingdom*** Northernmost major town of Scotland, on the River Ness some 130 km (80 miles) west and slightly north of Aberdeen. As a result it is the administrative, transport and supply centre for the whole of the Highland region; it is also a major tourist centre, just to the north-east of Loch NESS. Nothing remains of the old castle of Inverness, where Duncan, the Scottish king was murdered in 1039 by Macbeth – a deed immortalised in the Shakespeare play. About 8 km (5 miles) east of the town is CULLODEN, where an English army defeated Bonnie Prince Charlie's forces in 1746 in the last battle on British soil.

Population 40 000

Map United Kingdom Cb

inversion layer A layer of the atmosphere in which temperature increases with height, contrary to the usual pattern. The condition is often associated with major high pressure systems and prevents dust escaping from the lower layers of air and so causes haze. A global inversion layer exists in the stratosphere. The atmosphere is not normally warmed directly by the sun, but indirectly by the radiation of solar energy from the earth's surface. An exception occurs in the stratosphere where ozone filters out ultraviolet radiation, holding the heat, and producing an inversion layer with a temperature of 80°C (176°F). The air below this layer has a temperature of −5°C (23°F).

inversion temperature Increase of temperature with height above the earth's surface, contrary to the normal conditions of temperature falling with height. The increase may occur just above the earth's surface or at altitudes far above the surface. A surface inversion may be the result of rapid heat loss from the ground by radiation at night, especially when the air is calm and the sky clear, or by warm air being cooled over a cold surface. A high altitude inversion may occur when a cool air mass undercuts a warm one at a cold front, or a warm air mass overrides a cold one ahead of a warm front in a depression; this is called an OCCLUSION.

Iona ***United Kingdom*** Scottish island of 8 km² (3 sq miles) off the west coast of the island of Mull. The Irish evangelist St Columba founded a monastery there in AD 563, and his missionaries spread Christianity throughout pagan Scotland. An abbey was built in the Middle Ages and 60 Scottish, Norse and Irish kings are said to be buried there. A religious brotherhood founded in 1938 has restored the abbey and seeks to re-create the island's spirit of piety.

Map United Kingdom Bb

Ionian Islands (Eptanisos) ***Greece*** The seven main islands of the Ionian Sea – CORFU, CEPHALONIA, LEVKAS, ITHACA, ZAKINTHOS, KITHIRA and Paxoi. They have a mild climate, lush vegetation, sandy beaches, fishing ports and timeless villages. The islands have been settled since the Homeric age (11th-12th century BC). They were ruled in turn by the Romans, Venetians (1203-1797), French (1807-15) and British (1815-64), after which they were returned to Greece. Tourism, olive and fig production and fishing are the main activities. The Greek name for the islands, Eptanisos, comes from words meaning 'seven islands'.

Population 182 700

Map Greece Ab

Ionian Sea Lying between southern Italy and Greece, the sea includes some of the deepest parts of the MEDITERRANEAN. It is rich in classical and mythological associations. The sea was named after Io, mistress of the god Zeus, who changed her into a white cow to conceal her from Hera, his jealous wife. Tormented by a gadfly sent by Hera, Io wandered around the shores of the sea until Zeus finally turned her back into human form.

Greatest width 676 km (420 miles)

Map Europe Ee

ionosphere See ATMOSPHERE

Ios (Nios) ***Greece*** One of the CYCLADES islands in the Aegean, and the place where the poet Homer is said to have died in the 8th century BC. Its only town stands on a hill, surrounded by oak forests, windmills and 400 chapels. Red wine production, dairy farming and tourism are the main industries.

Population 1450

Map Greece Dc

Iowa ***USA*** Midwestern state covering 145 791 km² (56 290 sq miles) between the Mississippi and Missouri rivers. More than 90 per cent of it is farmland, producing pigs, cattle, maize and soya beans. Iowa is second only to California by value as a producer of agricultural equipment and as a processor of food. DES MOINES is its capital.

Population 2 884 000

Map United States Hb

Ipek ***Yugoslavia*** See PEC

Ipiros ***Greece*** See EPIRUS

Ipoh ***Malaysia*** Tin-mining and commercial city about 195 km (122 miles) north-west of the capital, Kuala Lumpur. It is the capital of Perak state and is noted for its temples and caves.

Population 293 850
Map Malaysia Bb

Ipswich ***United Kingdom*** County town of Suffolk in eastern England, 110 km (68 miles) north-east of London, and a port on the River Orwell. Like the rest of East Anglia, the area was densely settled by the Anglo-Saxons. The treasure-filled buried ship of one of their chieftains was found at Sutton Hoo, less than 15 km (9 miles) away, in 1939. The city flourished in the Middle Ages because of Suffolk's wool industry. Britain's entry into the Common Market in 1973 renewed east coast trade links with mainland Europe and increased the port traffic of Ipswich and its neighbours, Felixstowe and Harwich.

Population 120 000
Map United Kingdom Fd

Iquique ***Chile*** Capital of Tarapacá, the country's most northerly region. Its protected harbour was at one time a major exporter of minerals but is now the country's most important fishing port. A clock tower and the Casino Español – built in the Moorish style of Spain's Alhambra Palace in Granada – stand on the central plaza.

Population 113 000
Map Chile Ab

Iquitos ***Peru*** Port on the Amazon river and capital of Loreto department, 1024 km (640 miles) north-east of Lima. It was founded in 1864 on the site of an old Jesuit mission and became the chief port for the region during the rubber boom of the late 19th century. It declined after 1912 when rubber production dropped, but regained importance in the 1950s with the economic development of eastern Peru, when oil and mineral reserves were exploited in the surrounding jungle. Iquitos is now the cultural, religious and tourist centre of eastern Peru. It is connected to Lima by road and air.

Population 178 700
Map Peru Ba

Iran See p. 314

Irapuato ***Mexico*** Developing city in the central Mexican state of GUANAJUATO. It is a market town for the surrounding agricultural area, and is renowned for its strawberries and sweetmeats. Heavy industries are now being developed under the federal government's plans to move industry into the provinces.

Population 300 000

Iraq See p. 316

Irazú ***Costa Rica*** Volcano, 3432 m (11 260 ft) high, about 30 km (19 miles) east and slightly north of the capital, San José. Between March 1963 and February 1965, the volcano was in almost constant eruption. The eruptions were not fatal, but residents of the capital took to wearing bandanas, goggles and even gas masks in the streets to protect themselves from the volcano's acrid fumes and dust. Today a paved road leads to the crater's rim, and on a clear day tourists at the summit can see both the Caribbean Sea and the Pacific Ocean.

Map Costa Rica Bb

Irbid ***Jordan*** Town and grain-trading centre, situated 70 km (43 miles) north of AMMAN. It is set 529 m (1735 ft) above sea level. In 1976 it became the home of the new Yarmuk University. Its places of interest include the Great Mosque and some Roman tombs and statues. Just outside the town is a huge Bronze Age archaeological site more than 3000 years old.

Population 150 000
Map Jordan Aa

Ireland See p. 318

Irian Jaya ***Indonesia*** Largest of the nation's provinces, covering the western half of the island of New Guinea, an area of 410 660 km^2 (158 556 sq miles). The coast is swampy and the inland is mountainous, rising to more than 4500 m (14 750 ft). Copper, nickel and oil are produced. Headhunting was formerly practised in many areas. JAYAPURA is the main town.

Population 2 584 000
Map Indonesia Ic

Iringa ***Tanzania*** Regional capital about 420 km (260 miles) south-west of Dar es Salaam. It is the centre of a maize-growing area.

Population 57 182
Map Tanzania Ba

Irish Sea Separating Ireland from Great Britain, the sea was named Oceanus Hibernicus by the Romans. It is connected to the ATLANTIC OCEAN by the North Channel to the north and St George's Channel to the south. Passenger and cargo services operate between Holyhead (Wales) and DUN LAOGHAIRE (Ireland) and between LIVERPOOL and DUBLIN.

Area 103 600 km^2 (40 000 sq miles)
Average depth 61 m (200 ft)
Map United Kingdom Cd

Irkutsk ***USSR*** Industrial and cultural centre of eastern Siberia, on the Angara river near Lake Baikal. It was founded in 1652 as a garrison town and has grown rapidly since the arrival of the Trans-Siberian Railway in 1898. It has a university and is a key Siberian research and educational centre.

There are richly carved and colourfully painted traditional wooden houses, a fine Catholic church built in the 18th century by Poles who were exiled there by the Tsarist government, and a cluster of monastic buildings at the riverside. Irkutsk's industries – powered by a large hydroelectric station on the Angara – include machine tools, trucks, chemicals, metals, aircraft, oil refining, timber and electrical equipment. It is also a market for furs and skins trapped in the region's vast forests.

Population 590 000
Map USSR Lc

Irrawaddy ***Burma*** River running the length of Burma. One of the world's great rivers, the Irrawaddy is about 2000 km (1250 miles) long and drains the 415 000 km^2 (160 000 sq miles) of the Burmese heartland. It rises in two major headstreams, the Mali Hka and the Nmai Hka, in the northern hills, and flows southwards through a series of narrow gorges before spreading into the lowlands near Mandalay.

The delta is Burma's main rice-growing zone, and in places the river is so wide that it contains large seasonal islands. The islands emerge as the river level drops during the dry season, and they are used to grow crops – primarily rice – before the floods of the wet season return. Irrawaddy division, covering 35 139 km^2 (13 567 sq miles), administers the delta area.

Population (division) 4 991 000
Map Burma Cb, Bc

Irtysh ***China/USSR*** River rising in the western Altai mountains in north-west China. It flows west across the Chinese-Russian border into Kazakhstan, through Lake Zaysan and then north-west into the Ob' river at Khanty-Mansiysk in western Siberia. It is navigable for most of its 4440 km (2760 miles), and links several ports – including Semipalatinsk, Pavlodar, Omsk and Tobol'sk.

Map USSR Ic

Irún ***Spain*** Basque town on the French border adjoining the Bay of Biscay. It has iron foundries, tanneries, paper mills and medicinal springs.

Population 53 500
Map Spain Ea

Isabel ***Philippines*** See BASILAN

Isalo National Park ***Madagascar*** Savannah and woodland area covering 81 540 hectares (201 485 acres) about 400 km (250 miles) south-west of the capital, Antananarivo. The park is rich in rare animals such as lemurs. Its many caves include the Portuguese Grottoes, used as a refuge by shipwrecked Portuguese sailors in the 16th century.

Map Madagascar Ab

Isandhlwana ***South Africa*** Village in KwaZulu (Zululand), about 175 km (110 miles) north and slightly west of Durban, where Zulu forces of about 17 000 men under their chief Cetewayo defeated the British on January 22, 1879. Both sides lost about 1000 men each. There is a museum at the site.

Map South Africa Db

Ischia ***Italy*** Scenic volcanic island of 46 km^2 (18 sq miles) in the Bay of Naples. The highest point is Mount Epomeo, 788 m (2585 ft). Its main resorts are Ischia Porto, a spa, and Ischia Ponte, a fishing village guarded by an offshore castle.

Population 43 900
Map Italy Dd

Iseo ***Italy*** Alpine lake about 65 km (40 miles) north-east of Milan. Vines, olives and fruit trees grow in the mild climate around the 62 km^2 (24 sq miles) lake, and there are many small resorts.

Map Italy Cb

Isère ***France*** River, 290 km (180 miles) long, in the south-east. It rises in the Alps near the Italian border, then meanders west and south-west to join the Rhône 6 km (4 miles) north-west of Valence. There are many hydroelectric power stations along the upper Isère, in the tributary valleys of the Arc and Arly.

Map France Fd

Ise-shima ***Japan*** National park, 555 km² (214 sq miles) in area, in southern Honshu island, on the Shima peninsula on the south-west side of Ise Bay. It is the site of the sacred Shinto shrines of Ise, the most revered in Japan. The unpainted cypress-wood shrines, said to have first been erected in the 3rd century AD, are pulled down and replaced with identical buildings every 21 years – most recently in 1973.

The holiest of all Shinto objects, also at Ise, is a mirror said to have been looked into by the sun goddess Amaterasu soon after the world was created. Until 1945, the Japanese imperial family claimed to be directly descended from the goddess – and Japan is known as the Land of the Rising Sun.

Toba, just east of Ise, is the base for the firm of Mikimoto, a leading producer of cultured pearls. Oysters are also raised for pearls at Ago Bay to the south.

Map Japan Cd

Iseyin ***Nigeria*** Textile town on the north-western edge of Yorubaland, about 175 km (110 miles) north of the capital, Lagos. Its industry is based on imported yarns, and there is a popular nightly market.

Population 153 100

Map Nigeria Ab

Isfahan ***Iran*** See ESFAHAN

Ishikari ***Japan*** Longest river on Hokkaido. It rises in the west of the island, and flows about 262 km (164 miles) south then west through a densely populated plain to the Sea of Japan.

Map Japan Db

Ishinomaki ***Japan*** Fishing port on the east coast of Honshu island, about 40 km (25 miles) north-east of the city of Sendai. The port is a base for Japan's deep-sea fishing fleet.

Population 122 700

Map Japan Dc

Iskenderun ***Turkey*** Mediterranean port, formerly known as Alexandretta, about 450 km (280 miles) south-east of the capital, Ankara. Founded in the 4th century BC by Alexander the Great, and named after him, it is now a naval base and a steel town. There is a Crusader fortress at Toprakkale, about 50 km (30 miles) north, and several 16th-century Ottoman buildings – including a mosque and covered bazaar – at Yakacik, just to the east of the port.

Population 173 600

Map Turkey Bb

Iskur ***Bulgaria*** River, 402 km (250 miles) long, rising in the Rila mountains. It flows north through Sofia, then cuts a deep gorge 70 km (43 miles) long, before meandering across the DANUBIAN PLATFORM to join the Danube on Bulgaria's northern border with Romania.

Map Bulgaria Ab

Isla For physical features whose names begin 'Isla', see main part of name.

Islam Barrage ***Pakistan*** Barrage (formerly known as the Pallah) on the Sutlej river, 250 km (155 miles) south-west of Lahore. It was built in 1927 to irrigate fields along the lower river. In 1929 the dam's central section was destroyed in a flood and had to be rebuilt.

Map Pakistan Db

Islamabad ***Pakistan*** National capital since 1967 when it displaced Karachi, which is the capital of Sind province. It is a new city, built in the early 1960s just to the north of Rawalpindi, in a newly created Federal Capital Territory covering 907 km² (350 sq miles). Its name means 'city of Islam'.

Population (city) 201 000; (territory) 340 000

Map Pakistan Da

ISLANDS IN THE SUN AND ICE

An island is a mass of land – smaller than a continent – which is entirely surrounded by water. Therefore Australia, which is generally regarded as a continent, does not qualify. But Great Britain – consisting of England, Scotland and Wales – does. Indeed, with an area of 229 870 km² (88 730 sq miles), it is the largest island in Europe.

THE WORLD'S LARGEST ISLANDS

Island	Location	Area (km²/sq miles)
Greenland	North America	2 175 600/839 700
New Guinea	Australasia/Asia	808 510/312 085
Borneo	Asia	757 050/292 220
Madagascar	Africa	594 180/229 355
Baffin Island	Arctic Ocean	507 451/195 927

island arc Curved string of elongated islands found round some landmasses, such as the Aleutians and the Japanese islands. Island arcs are associated with destructive plate margins in oceanic areas. See PLATE TECTONICS

Islands, Bay of ***New Zealand*** Inlet on the north-east coast of North Island, where the first major contacts between Maoris and Europeans took place. The Reverend Samuel Marsden, a British missionary, established the first organised European settlement in New Zealand at Rangihoua in 1814. It was a centre for whalers and traders during the 1830s and was where the Treaty of Waitangi, between the Maoris and the British, was signed in 1840.

Under the treaty, the leaders of the northern Maori tribes ceded New Zealand to Britain. However, open warfare between the two cultures broke out in the Bay of Islands War a short while later, when the Maoris revolted over loss of trade to the colonial capital at Auckland, 200 km (125 miles) to the south. After a number of inconclusive battles between British troops and Maori warriors, an uneasy peace was established in 1847.

The bay, which covers 400 km² (154 sq miles) and contains more than 20 islands, is now a tourist area for sailing and big-game fishing.

Map New Zealand Ea

Isle For physical features whose names begin 'Isle', see main part of name.

Isma'iliya ***Egypt*** Town set up as a market and construction camp beside the Suez Canal in 1863. It is on the north-west shore of Lake Timsah, 72 km (45 miles) south of Port Said. It was badly damaged during the Arab-Israeli wars of 1967 and 1973-4.

Population 146 000

Map Egypt Cb

isobar Line on a weather map connecting points of equal atmospheric pressure.

isohyet Line on a climatic map joining places with the same rainfall.

isoseismal line Line joining points on the earth's surface at which the intensity of an earthquake is the same, usually a circle round the earthquake's EPICENTRE.

isostasy State of equilibrium or balance in the earth's CRUST between highlands and lowlands. Mountain masses of less dense rock of the granitic layer have a compensatory 'root' of similar rock which penetrates deeply into denser layers of the earth's crust. They 'float' on this heavier basaltic layer. The ocean basins on the other hand are floored almost directly by the denser basaltic layer of the crust. See THE STRUCTURE OF THE CONTINENTS (p. 162).

isostatic readjustment Vertical movements in the earth's CRUST to restore balance (isostasy) in response to changes in the load which have been caused by erosion, deposition and melting of ice sheets.

isotherm Line on a weather map joining places that have the same temperature.

Israel See p. 322

Issel ***Netherlands*** See IJSSEL

Isselmeer ***Netherlands*** See IJSSELMEER

Issyk-Kul' ***USSR*** Large mountain lake in the central Asian republic of Kirghizia, about 105 km (65 miles) from the Chinese border. It is 702 m (2303 ft) deep and covers an area of 6280 km² (2424 sq miles). Its surface is 1609 m (5279 ft) above sea level. The hot springs rising from its depths make it a fisherman's paradise – rich in perch, pike and carp. There are also medicinal springs used to treat rheumatism.

Map USSR Id

Iran

RELIGIOUS FANATICISM AND NATIONALIST FERVOUR IN A FUNDAMENTALIST ISLAMIC SOCIETY HAVE GIVEN BIRTH TO A DANGEROUS STORM CENTRE IN THE MIDDLE EAST

Early in 1979 in Paris, a tall, forbidding, bearded figure stepped onto an Air France jet bound for Tehran. The Ayatollah Khomeini, nearly 80, a Muslim leader banned from Iran by the Shah (King), was on his way back from exile to head a religious revolution.

The Peacock Throne was empty. Muhammad Reza Shah, the country's exiled ruler, had used Iran's vast oil wealth to modernise and industrialise. He had wanted to 'drag Iran into the 20th century', but had gone too far. There had been savage riots inspired by religious leaders and political opponents. This was the time the Ayatollah chose to turn the clock back and found a 'true' Islamic republic.

The word Iran comes from the Persian name for the country, 'Arian', and was adopted in 1935; before then it was known as Persia. Cyrus the Great founded the Persian Empire in the 6th century BC, with a domain stretching from India to the Mediterranean. The empire's 2500th anniversary in 1971 was celebrated by the last Shah and other royalty in pomp and at vast expense at the site of PERSEPOLIS, the ancient capital of Persia, in the desert near SHIRAZ. Among the treasures displayed was the splendid Peacock Throne, captured in India (together with the fabulous Koh-i-noor diamond) during a campaign by Nadir Shah in 1738. The throne, used by subsequent rulers, disappeared during the Ayatollah's revolution.

The country facing these traumas is a harsh land, seven times the size of Britain, stretching from the CASPIAN SEA (which Iran shares with the USSR) to The GULF. Its 1 648 000 km^2 (636 293 sq miles) are dominated by mountain ranges in the north and west, while the centre is a vast expanse of desert. The desert is so hot that almost no life can survive; The Gulf is little cooler. When winter comes, terrible dust storms sweep the plains. In the north there is a danger of earthquakes; in 1978, 20 000 died in a quake in the province of KHORASAN.

Most of the population of 46.5 million live in the mountainous north and west where the capital, TEHRAN, with an estimated population of 6 million, is situated. The ELBURZ MOUNTAINS, with forests ranged on their slopes, rise to 5771 m (18 933 ft). Only the Caspian plains, which receive adequate rainfall all year round, are good agricultural land.

Wheat is the main crop in the north and north-west, and rice in the Caspian region. Barley, maize, cotton, sugar beet, tea and tobacco are all grown; other products are fruit, nuts and dates. Salmon, carp, trout and pike are found in the rivers running into the Caspian Sea, and sturgeon (from which caviar is obtained) in the Caspian itself. In Iran there are almost as many sheep as people; nomads, who wander with their sheep and goats in the mountains and highlands, form 5 per cent of the population.

Oil was first discovered and exploited by the Anglo-Persian Oil Company (later to become BP) in the south-western province of KHUZESTAN in 1908. But Iran has other valuable minerals: coal, iron ore, copper, lead and salt. Its natural gas reserves are second only to those of the USSR. Precious stones – turquoise, emeralds, topaz and sapphires – are found, mainly in the north-east.

As with neighbouring Afghanistan, Iran's strategic position ensured it the attentions of outside powers, especially Britain and Russia, and their forces occupied parts of the country, on and off, until after the Second World War. Meanwhile, its own political history has been a chequered one. In 1921 Reza Khan, an Army officer, staged a coup against the government and, after serving under the then Shah, Ahmad, supplanted him in 1925. Reza Khan then made himself Shah.

In the Second World War Reza Shah was forced to abdicate in favour of his son, Muhammad Reza Shah. The latter had to flee briefly in 1953 when the elderly Premier Mossadegh, leader of the extremist National Front, tried to nationalise Iran's oil industry, but Mossadegh's ideas were ahead of his time and a boycott of Iranian oil by Britain and other countries led to his overthrow. The Shah returned and regained absolute power.

Meanwhile, the Shah had launched his own revolution. He was determined to turn the old-fashioned peasant and Islamic society into a modern Western-style state in his lifetime. He aimed to expand industry, end feudalism, redistribute land to the peasants, and use the country's oil (10 per cent of the world's oil resources) to finance his dream.

Modernisation forged ahead during the 1960s and 1970s. However, the Shah's land reforms generally benefited the wealthier peasants at the expense of the poorer, landless labourers. This poorer class left tilling the fields to work in the new factories. A huge new middle and working class grew up, and the country changed from a predominantly rural society to one in which 40 per cent of the people lived in towns. But the Shah's reforms were accompanied by growing repression.

He did not understand the forces of opposition and hatred he had unleashed – particularly among orthodox Shiite Muslims, the majority religion. By 1977 the rich – the so-called 'thousand families' who had owned more than 80 per cent of the finest land – were sensing the end and sending money abroad at an estimated US$1000 million a month. Riots broke out and in January 1979 the Shah left the country for good. Within a fortnight the Ayatollah Khomeini, exiled for 14 years, had returned in triumph.

Two-thirds of Iran's people are descended from the ancient Persians, but there are sizable minorities. A quarter are of Turkish stock, and there are Kurds in the north-west, Arabs on The Gulf and Baluchis in the east.

The mosque, the bazaar and the home are the three centres of Iranian life. The vast majority of the people are Shiites, the second of the two great branches of the Islamic religion. They differ from the other main group, the Sunni Muslims, in believing that Ali, son-in-law of the Prophet Muhammad, ranks as his first successor; the Sunnis place

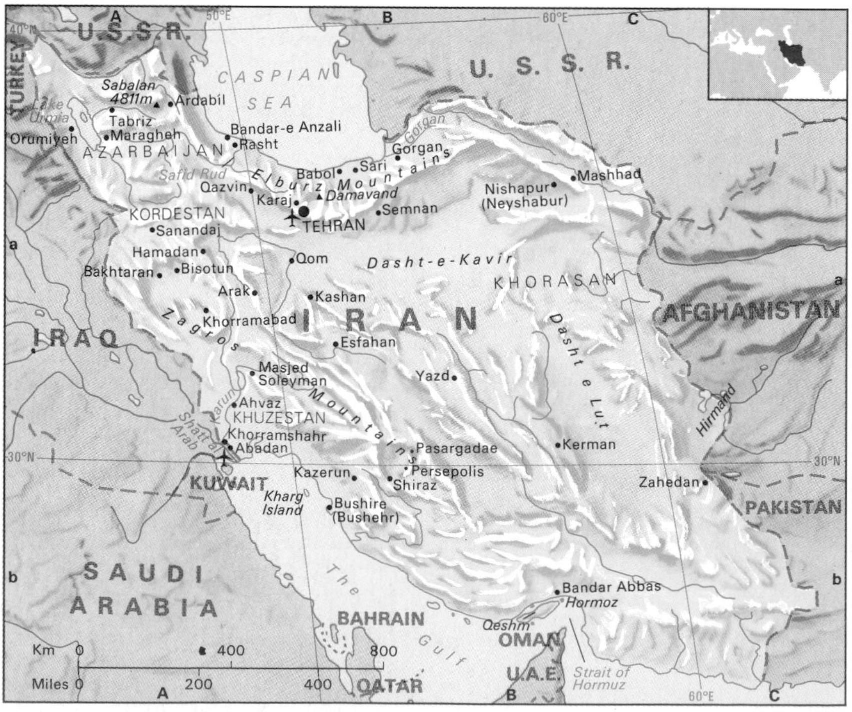

▲ TO DIE FOR ALLAH Thousands of young Iranians congregate in Tehran in support of Ayatollah Khomeini and his war with Iraq. In one speech he told them: 'The purest joy is to kill and be killed for Allah.'

Ali fourth. The Kurds and Arabs are Sunnis. Other religious groups include the Zoroastrians, who believe that life is a constant struggle between the powers of good and evil and have bases in YAZD and KERMAN.

Against this background, the Shiite Ayatollah Khomeini set to work to make Iran a fully Islamic state, purged of Western influences. Power was vested in the clergy who followed Khomeini, under the slogan *Vilayat-i faqih* ('guardianship of the just'). In 1979 a referendum was held and, by an overwhelming majority, the people voted in favour of Iran becoming an Islamic republic.

The ancient penal code of Islam, providing for punishments ranging from whippings and amputations for theft to executions for adultery, was resurrected. Security men, the Komiteh, revolutionary guardians of Islamic morality, raided homes to seek out offenders. At least 5500 people were executed by firing squad including supporters of the Shah, political opponents and offenders against the moral code. In addition, thousands of political prisoners were held in jail. Alcohol was banned. Women were ordered to cover their hair and wear concealing dress. Drug addicts and homosexuals were executed.

In 1980 the Shah died in exile in Egypt. Western Europe, dependent on The Gulf states for 30 per cent of its oil, and Japan (60 per cent dependent) listened nervously as Khomeini outlined his economic plans. In 1977, he argued, Iran had produced 6 million barrels of oil, but only enough food for a month. It had built assembly industries for the West instead of independent industry. The cry went up 'Is this not putting all our requirements at the mercy of the foreigners?' Many of the Shah's schemes were dropped and the emphasis switched to food production.

In September 1980 Iraq invaded Iran. The Shah had forced Saddam Hussein, the Iraqi leader, to relinquish sovereignty over the eastern half of the SHATT AL ARAB river dividing the two countries; now, with Iran in turmoil, seemed a good time to get it back. The Iraqis had early successes, crippling Iran's oil production, but the Iranians, swayed by the fiery rhetoric of their religious leaders, fought back. Against superior Iraqi firepower they staged 'human wave' assaults: teenage boys carrying plastic keys to heaven attached explosives to themselves and dived under tanks.

The Iranians turned back the tide, recaptured lost territory and the conflict seesawed along the border. After five years of war Iranian casualties were variously estimated at between 150 000 and 630 000, and refugees totalled 2 million. Tehran's cemetery, Behestee-Zahra, the 'paradise of flowers', grew into a city in its own right. Surrounding the grim centrepiece of a fountain flowing simulated blood, lay the graves of those slain in battle, with framed portraits of themselves – or their mutilated bodies.

Violence was not only external. The left-wing Mujahaddim opposition group carried out terrorist attacks. In 1980 two bomb blasts in Tehran killed the president, Muhammad Ali Rajaie, Prime Minister Bahonar, the chief justice, Ayatollah Behesti, and 74 others. Cracks appeared in the revolution: there were signs that some, at least, of the mullahs wanted to come to terms with the discredited middle classes.

IRAN AT A GLANCE
Area 1 648 000 km² (636 293 sq miles)
Population 46 500 000
Capital Tehran
Government Islamic republic
Currency Rial = 100 dinar
Languages Farsi (official), Kurdish, Baluchi and Turkic languages
Religions Muslim (99%); Christian, Jewish, Zoroastrian minorities
Climate Continental; average temperature in Tehran ranges from −3 to 7°C (27-45°F) in January to 22-37°C (72-99°F) in July
Main primary products Sheep, cattle, goats, wheat, sugar beet, barley, rice, cotton, tea, tobacco, timber, fish; crude oil and natural gas, coal, salt, chrome, lead, copper, iron
Major industries Oil and gas refining, steel, electrical equipment, cement, textiles, sugar, flour, furniture, building materials, fishing
Main exports Crude oil and refined products, carpets and rugs, textiles, raw cotton, fruit, leather goods
Annual income per head (US$) 2500
Population growth (per thous/yr) 32
Life expectancy (yrs) Male 57 **Female** 59

Iraq

FABLED SITE OF THE GARDEN OF EDEN AND ONCE THE GRANARY OF THE ANCIENT WORLD, IRAQ HAS SQUANDERED ITS OIL WEALTH ON A BLOODY WAR WITH IRAN

Land of the ancient civilisations of Babylon and Assyria and of the fabled *Thousand and One Nights* of Haroun al-Raschid, Iraq is twice blessed with oil, and with water from two major rivers. Its oil reserves, discovered in 1926, are among the largest in the world and supply almost half its revenue. The rivers, the TIGRIS and EUPHRATES, support a band of fertile country which cuts through the centre of a desert land. The Garden of Eden, the first home of man according to the Bible, is said to have been sited near BASRA, where the rivers meet.

Oil is the basis of Iraq's wealth, and the waters are its sustenance. Yet the rapid progress made in improving living standards since oil prices boomed in the 1970s was put in jeopardy in the 1980s because of a war with IRAN, and a dramatic fall in oil prices in 1986. The war was costing US$1000 million a month, and over 100 000 Iraqis have been killed or wounded.

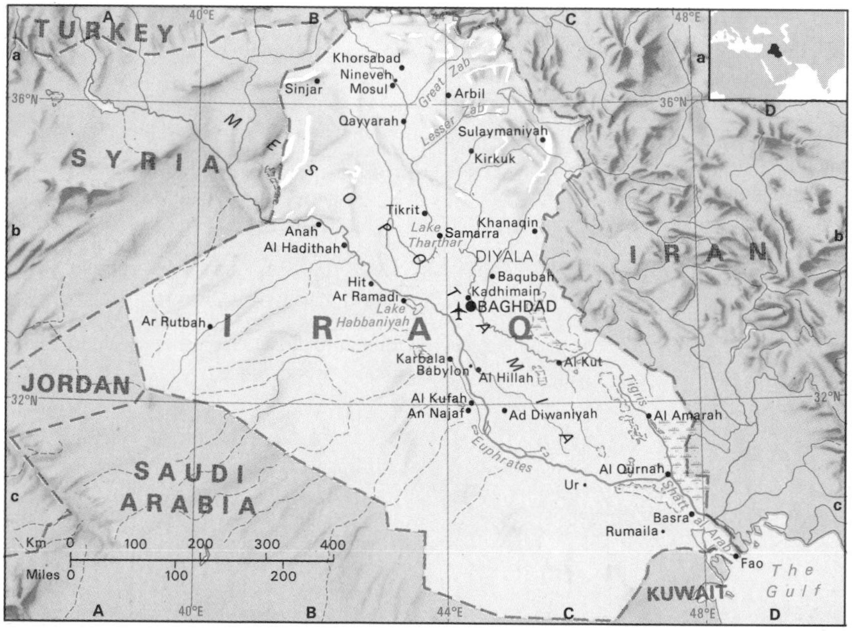

LAND AND HISTORY

Iraq forms a wedge of land, 434 925 km² (167 925 sq miles) in area, between Syria and THE GULF. It is landlocked except for its outlet to The Gulf at the SHATT AL ARAB, the channel formed by the junction of the Tigris (1900 km, 1160 miles long) and the Euphrates (2700 km, 1650 miles). In the north, the mountains of KURDISTAN, up to 4168 m (13 674 ft) high, stretch into Turkey and Iran. They are snow-covered for up to six months of the year, but blossom with millions of flowers in the spring. In the west is the great Syrian Desert, home only of nomadic shepherds.

For centuries, Iraq was known as Mesopotamia, the 'land between the rivers'. Only 6 per cent of the land is fertile, and in April and May the annual flooding of first the Tigris, then the Euphrates takes place, and the year's round of irrigation and growth begins. The country produces wheat, barley, rice, tobacco, cotton, vegetables and fruit, and Iraq is one of the world's largest producers of dates. The date palm, whether clustered in oases or growing along the banks of the Shatt al Arab, is the most distinctive of Iraq's sparse vegetation.

Everywhere, apart from the northern highlands, is murderously hot – up to 50°C (122°F) in summer, and cold in winter. From May to October the heat is so great that no one travels unnecessarily, and since ancient times the people have retired to underground rooms in the middle of the day.

Historians dispute whether it was Egypt's NILE or Iraq's rivers that cradled the world's first civilisation. Iraq was the home of the Sumerians, who developed a writing system before 3000 BC. They were followed by the Babylonians and the Assyrians, and later the Arabs who conquered the country in the 7th century AD. The BAGHDAD of Haroun al-Raschid was a famous centre for learning and the arts. The *Arabian Nights* stories told by Scheherazade to her husband, Schariar – including those of Aladdin, Ali Baba and Sinbad the Sailor – have become world classics.

In the 16th century Iraq fell to the Turkish Ottoman Empire, and the Turks were ousted only when the British invaded the country, aided by Arab troops led by T.E. Lawrence (Lawrence of Arabia) during the First World War. After the war Iraq became a British mandate and a kingdom under Faisal I. The mandate ended when the country became fully independent in 1932. The monarchy ended in a bloodbath during a revolution in 1958 when Arab nationalism was sweeping the Middle East. In July of that year, King Faisal II and his entire household were murdered by revolutionaries. Ten troubled years of coups and counter-coups followed before the socialist Ba'athist party emerged supreme.

In the mid-1970s, like other oil-producing countries, Iraq profited from the great oil boom. The KIRKUK and MOSUL oil fields of the north and that of Basra in the south were augmented by a new field found near Baghdad in 1975. By 1979 the country was producing 3 million barrels of oil a day. Prosperity descended on the land: cars, consumer goods, major development programmes providing new schools, hospitals, roads and factories. A free health service was established and unemployment virtually abolished by creating unproductive jobs.

That year Saddam Hussein, long regarded as the strong arm of his government, became president. Under him is a revolutionary council of nine members. A national assembly of 250 members is elected every four years, though the choice of candidates is limited. The only legal party is the National Progressive Front, composed of Ba'ath and Kurdish parties.

In 1980 Hussein launched a war against Iran which, having got rid of the Shah, was in the throes of Ayatollah Khomeini's revolutionary changes. It was the culmination of years of feuding over their shared river border along the Shatt al Arab. Khomeini's young armies resisted fanatically, and the war dragged on. Development programmes slowed down, and oil production fell to only 1 million barrels a day.

THE PEOPLE OF IRAQ

Iraq reflects its history in the wide variety of ethnic types in its population, and this has resulted in sectarian stresses that, according to some people, make the country almost ungovernable. Although three-quarters of the people are Arab, these are split between the Sunni (the minority, though they have always held political power) and Shiite sects. The other various minorities include the blood of races from the distant past. The Kurds of the north, a fierce warrior people, descend from the ancient Medes who once ruled Persia, and their armed struggle for independence has long drained Iraq's resources. The broad flat face seen in Sumerian carvings is still found among the Marsh Arabs who live in raft houses among the reed beds and lagoons of southern Iraq, much as they did thousands of years ago.

The Sabians are a sect whose hero is John the Baptist, forerunner of Christ; their chief rite is frequent baptism. The Yazidis, whose religion contains pagan, Islamic and Christian elements, take their name from the Zoroastrian stronghold of YAZD in Iran.

Arabic is the main language; Kurdish and Turkish are spoken by sizable minorities. Many people can speak English, partly the result of British influence and also because it is the international language of oil. The professional classes, such as civil servants, enjoy club life (also a British tradition); an invitation to a stranger to accompany his host to his club is a courteous sign of acceptance. Iraqi women have more freedom than in many Arab countries: in the towns they walk about in Western clothes. More are going to university and (especially since the war) obtaining responsible jobs.

Iraq is a country of young people – half the population of 16 million is under 15 years of age. It can be a very colourful country. When spring arrives, the women of the north don beautiful costumes, every colour of the rainbow, with bracelets and bangles to enhance the effect.

FOOD AND DRINK

The most common food dishes are kebabs or *tikka* (small pieces of grilled lamb or chicken) with rice. Yoghurt is often a side dish with meals, vegetables and fruit are plentiful and there is excellent bread. The Iraqis are great tea and coffee drinkers, usually taking it very sweet; the teahouse is a popular meeting place.

Most Iraqis, as Muslims, do not take liquor, but the country makes its own brands of beer and adopts the attitude that those who wish to drink should not be forbidden. There are two kinds of *arak* (rice wine), and local wines. In most towns there are certain hotels and restaurants which serve drinks openly, but in the two holy cities of KARBALA and AN NAJAF drink is banned. This is in deference to the shrines of Ali, son-in-law of Muhammad the Prophet, in An Najaf, and of the Shiite Muslim leader Husein in Karbala.

PROBLEMS OF MODERN IRAQ

One of the disputes that divides the Arab world is about the importance of Ali. The Sunni, who comprise about 85 per cent of all Muslims, regard Ali as the fourth successor to Muhammad; the Shiites say he should rank first. Khomeini's Iran is Shiite, and fanatical; it condemns Iraq's President Hussein, a Sunni, as an unbeliever – and, to compound his sins, someone who expelled Khomeini when he was in exile in Iraq in 1978.

Iran constantly urges Iraq's milder Shiite majority to rise up against Hussein; so far they have shown little inclination to do so. Paradoxically, Iran has frequently shelled BASRA, the main Shiite city.

Apart from the war, Iraq faces the problem of rapid change from a traditional, rural society to a modern, sophisticated, town-dwelling one. The urban population has quadrupled in the last 25 years and now accounts for more than 60 per cent of the population. Most of them are in the capital, Baghdad, which has a population of more than 3 million and most of the country's industry. The manufactures include carpets, leather goods, textiles, cement, tobacco products and *arak*. Basra, with a population of under half a million, is Iraq's only port. It has many oil refineries and normally exports petroleum products, wool, grain and dates, but because of the war oil is being exported by pipeline through Turkey.

▲ WAR OF ATTRITION An Iraqi soldier at the front near Basra. When Iraq's Saddam Hussein invaded Iran in 1980 to secure the Shatt al Arab waterway, he expected an easy victory, but the Iranian troops, believing that death in the war would open the doors of an Islamic heaven, resisted fanatically.

FOR THE VISITOR

Iraq possesses many major tourist attractions from its long past, and tourism has become an important cash earner. Baghdad contains mosques, museums and many old buildings and monuments. The bazaars, especially the great covered bazaar, are vibrant with life, with shops and stalls selling craftware – copper and silver goods, and handwoven rugs – spices, fruit, vegetables and other local produce. For the more athletic, there is skiing in the hills on the north-east border and water skiing and sailing at the tourist centre at HABBANIYAH.

The ancient city of NINEVEH, on the Tigris opposite Mosul, was the capital of the Assyrian Empire. Excavations there have revealed palaces with magnificent sculptures. BABYLON, on the Euphrates 90 km (56 miles) south of Baghdad, became legendary for its colour and luxury in the days of King Nebuchadnezzar, six centuries before Christ. Its walls, palace and processional way were decorated with coloured brick; its Hanging Gardens were one of the Seven Wonders of the Ancient World. The Sumerian city of UR, 310 km (190 miles) south-east of Baghdad, was the home of Abraham, the founder of Judaism.

Iraq, a socialist country with a powerful military ruling caste, is one of the most isolated states in the Arab world, partly because it occupies the north-eastern boundary of that world. The war with Iran may yet bring greater damage or (as has happened in some Muslim countries) a move towards a more fundamentalist view of Islam, which will threaten the present regime.

Despite the recent economic progress, there remain huge differences between the town dwellers and the peasants and nomads. But the long-term outlook for Iraq, with its oil and other minerals such as sulphur and copper, must be bright. It has a growing industrial base, sufficient to alarm the Israelis into bombing an unfinished nuclear power reactor in 1980. It could produce enough food for its people and have a surplus for export. And when the oil runs dry, Iraq will still have the fertile land fed by the rivers on which its history has been built.

IRAQ AT A GLANCE
Area 434 925 km² (167 925 sq miles)
Population 16 000 000
Capital Baghdad
Government One-party socialist republic
Currency Dinar = 20 dirhams = 1000 fils
Languages Arabic; also Kurdish, Turkish and Assyrian
Religions Muslim (90%), Christian (10%)
Climate Hot summers, cool winters; average temperature in Baghdad ranges from 4-16°C (39-61°F) in January to 24-50°C (75-122°F) in July/August
Main primary products Sheep, dates, cereals, tomatoes, watermelons, cotton, cattle, goats, camels; crude oil and natural gas
Major industries Oil and gas production and refining, cement, chemicals, agriculture, food processing, brick making, textiles, leather
Main exports Crude oil and refined products, cotton, dates, cement, hides, wool
Annual income per head (US$) 2000
Population growth (per thous/yr) 33
Life expectancy (yrs) Male 55 **Female** 58

Ireland

FIFTEEN CENTURIES OF FAITH AND POETRY OVERLAID BY EIGHT CENTURIES OF STRIFE HAVE LEFT THEIR MARK ON THE LAND AND ON THE PEOPLE

Not for nothing is Eire – to use Ireland's ancient, Celtic name – called the Emerald Isle. Its stunning, shamrock green is the product of the mild, humid and ever-changing climate created by Atlantic weather systems washing over one of Europe's most westerly islands. Warm fronts sweep in from the ocean at any time of the year, drenching the countryside beneath sheets of low cloud. At such times a well-soaked stranger seeking solace in a snug bar may be surprised to be told that 'it's a fine, soft day, praise be to God'. Then a cold front will arrive, bringing bright skies and drifting showers that endow the landscape with a sparkling brilliance of colour.

The Republic of Ireland occupies four-fifths of an island that bears roughly the shape of a gigantic saucer, with a broken chain of mountains forming an irregular rim around the coast. The central plain is largely limestone, covered mainly with boulder clay, providing good farmland and pasture.

In the mountainous west, full face to the Atlantic, the rainfall varies from 1016 mm (40 in) on the coast to twice that or more in the mountains. And there, in the far south-west of KERRY, are Ireland's highest peaks. Carrauntoohil, tallest of all, rears to 1041 m (3414 ft) and Brandon Mountain near Dingle falls sharply to coastal cliffs from its 953 m (3127 ft) summit. Winter temperatures average 4-7°C (39-45°F) with the highest figures in the west, which has little frost. Summer temperatures range from 14°C (57°F) in the north to 16°C (61°F) in the south.

The country's history is a good deal less temperate. The stormy story of its fight for independence ended just over 60 years ago, and the Republic ranks among Europe's youngest and free nations. Yet its contribution to Western culture is immeasurable. From Swift and Sheridan to Yeats and Shaw, from Oscar Wilde and James Joyce to Samuel Beckett, Ireland's literary giants have enthralled readers and theatregoers in many countries.

Today the island is experiencing difficult times. Eight bitter and often bloody centuries have left the island divided into two countries – the Republic, which is predominantly Roman Catholic, and Northern Ireland, which is two-thirds Protestant and still under British rule. Although DUBLIN politicians advocate a united Ireland, apart from a small minority, neither they nor their supporters condone terrorism, and the Irish Republican Army (IRA) is an illegal organisation in both the Republic and Northern Ireland.

Ireland's tragedy has been that she has had more than her share of history. The land became a centre of Christian faith and learning, a realm of kings and poets, after it was converted by St Patrick in AD 432. But three centuries later its prosperity lured Vikings, who came to raid and stayed to settle. One of Ireland's first national heroes, Brian Boru, defeated the Vikings at CLONTARF in 1014, but this taste of freedom did not last long. In 1168 Ireland's Norman Conquest began when England's Henry II sent his knights across the Irish Sea.

This was the beginning of a turbulent relationship with Ireland's powerful neighbour, marked by misunderstanding, religious intolerance, rebellion, repression, evictions, atrocities, retaliations – and occasional outbreaks of goodwill.

The list of Irish grievances and misfortunes is long. In the 16th and 17th centuries, Irish lands were confiscated, and 'plantations' of Protestants from England and Scotland were settled in ULSTER, in the north-east – a policy that produced Northern Ireland's Protestant majority. Oliver Cromwell left a name that is still detested for the ruthlessness with which he crushed a Catholic and Royalist rebellion in 1649. Catholics were forbidden to exercise their religion after defeat by William III's forces at the Battle of the BOYNE in 1690.

An abortive uprising by Wolfe Tone and his United Irishmen in 1798 was the curtain raiser to another troubled century. In 1801 an Act of Union came into force under which Ireland's Parliament was abolished and Irish MPs were elected to the British House of Commons at Westminster. The temper of the times in England was anti-Catholic, and Catholics were not allowed to hold public office. But in 1828 the Catholic Daniel O'Connell was elected MP for County CLARE, an event which forced the British Government to accept Catholic emancipation the following year.

The harsh facts of economics took over from politics in the middle of the century when Ireland's potato crop fell victim to blight – a fungal disease that rotted the tubers in the ground or in store. Before the Great Famine, which began in 1845, the land supported a population of more than 8 million, half of them living on home-grown potatoes. By 1848 Ireland had lost over a million people by starvation, typhoid or emigration.

In 1847 alone, 250 000 died and 3 million were being supported by public funds. Many landlords sacrificed fortunes trying to save their tenants; others simply evicted them. The famine years launched a decline in population, at first rapid, then more gradual from the 1890s. But over 2 million have emigrated since 1900. Their departure has etched clear marks on the Republic's countryside. Throughout the province of CONNAUGHT and the counties

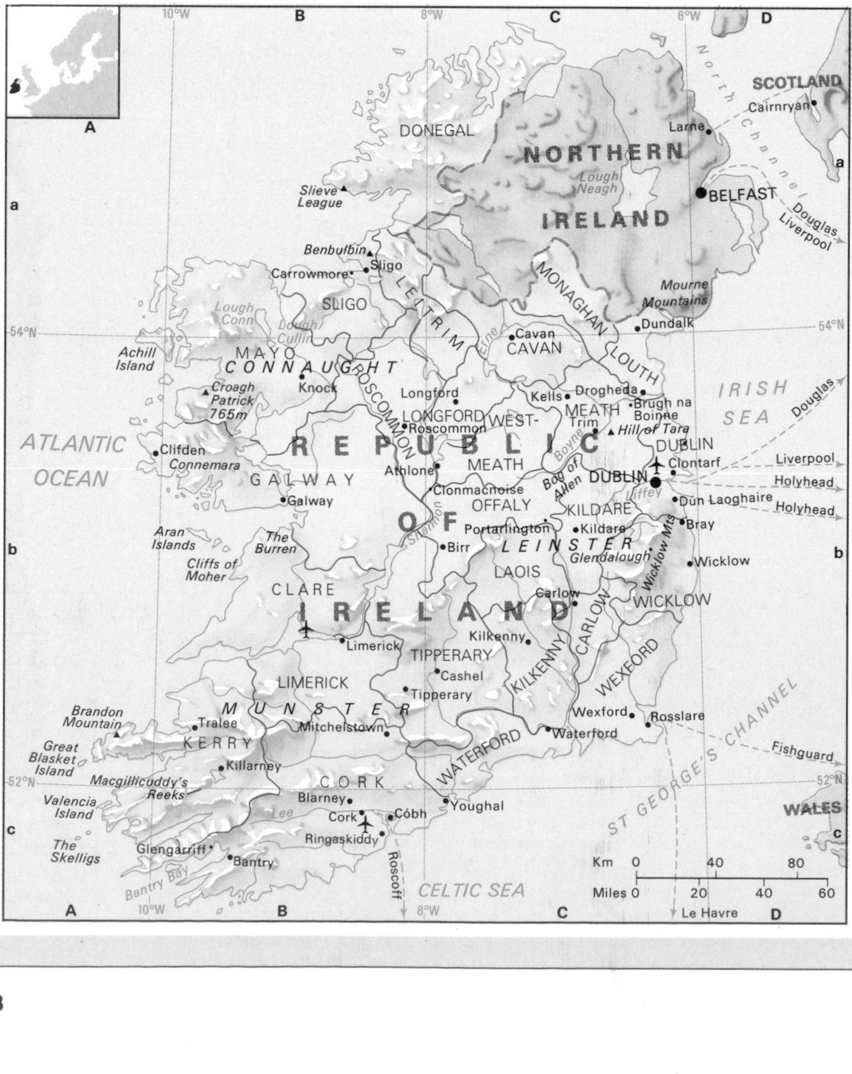

of Kerry and DONEGAL, ruined and deserted hamlets, abandoned fields and forgotten peat diggings tell the tale, though in recent years farms have been united to give their holders a more adequate living.

On the eve of the First World War, the British Government seemed at last ready to grant Home Rule – but there were always Protestants in Ulster who were prepared to take up arms rather than submit to Catholic rule. 'Ulster will fight, and Ulster will be right', was the slogan of their leader, the Protestant southern Irishman Sir Edward Carson. And fight they did – but not against other Irishmen. Many of the southern Irish laid aside their grievances on the outbreak of the World War, and 250 000 of them volunteered to fight alongside Ulstermen in the British army.

Even so, there was enough discontent for Germany to think it worth while supporting an insurrection against British rule – the Easter Rising of April 1916. It was crushed, and 90 rebels were condemned to death, of whom 15 were executed. Eamon De Valera, who was born American, was one of those reprieved.

By now, a nation was struggling to be born. In 1918 the political movement called *Sinn Fein* ('Ourselves Alone') proclaimed an Irish Parliament, the *Dáil Eireann*, with De Valera as president. The Irish Republican Army declared war on the British, and Britain responded by drafting ex-servicemen into the Royal Irish Constabulary. These auxiliaries were the notorious Black and Tans (named after the colour of their uniforms) who fought terror with terror.

In 1921, the IRA leader Michael Collins put his signature to a treaty with Britain which gave Ireland independence – with the exception of six of the nine counties of Ulster. 'Early this morning I signed my death warrant', he wrote to a friend. And he was right, for before a year had passed, he had been assassinated in a bitter civil war between those Sinn Feiners who accepted the treaty and those who did not.

The Irish Free State came into formal existence in 1922, at first as a dominion within the British Commonwealth. De Valera, who became president in 1932, set about demolishing the country's final links with Britain. Finally, in 1949, the Free State was proclaimed a republic, no longer within the Commonwealth. In 1955 came acceptance into the United Nations; and in 1973, membership of the European Economic Community (Common Market).

The demand for unification of all Ireland is written into the country's constitution, and inexorably, the Republic has been drawn in to share the agony of Northern Ireland. The latest attempt to solve Northern Ireland's problems peacefully was a pact in late 1985 between the British and Irish governments, giving the Republic a consultative role in running Northern Ireland, with a view to securing the rights of both the Catholic and the Protestant communities there, and improving co-operation in economic development and such matters as law and policing. Only time will tell whether this bid for tolerance will succeed.

▲ CITY OF MEMORIES The sun sets on Dublin and the shimmering River Liffey. On the right stands the dome of the Four Courts, headquarters of the judiciary. Time's imprint is everywhere in the city, whose mixture of elegance and shabbiness casts a potent spell, as many writers have testified.

Despite the fertility of its land, the Republic is one of the poorer countries in Western Europe, with a per capita income less than half that of France. Peat, created by a moist climate acting on ill-drained tracts of glacial clays, covers one-seventh of the Republic's 70 283 km^2 (27 136 sq miles), and is the country's chief indigenous fuel source. Five million tonnes a year are cut by a government-sponsored corporation; 15 major peat works and briquette factories process it, and peat-fired power stations generate electricity. However, present reserves may last only 20 years.

Government policies have been directed towards industry since 1932, partly to reduce emigration. This was running at 40 000 a year in the 1950s but has since declined. Hardly any left in the 1980s, as unemployment soared

in Britain. Dublin and its suburbs have a population of 915 000 – about a quarter of the national total. Its industries include brewing, food production, light engineering, clothing and glass making. The capital is also the main publishing and printing centre, and focus of radio, television and other communications industries. Despite efforts to distribute industry more widely, Dublin still has half the nation's manufacturing output. In fact the Republic still ranks among the least industrialised countries in Europe.

The country's lack of energy resources (apart from peat) and remoteness from major population centres and markets slowed industrial progress in the developing Europe of the 19th century. Coal – essential ingredient of the times – had to be imported, and most of it came from Britain. An electrification programme started in the 1920s with a hydroelectric scheme on the River SHANNON. Later schemes harnessed other rivers, so that today 75 per cent of the country's water-power potential is used. Peat firing of power stations began in the 1940s, but the demand for electricity has increased so much that more than half is produced with coal and oil – mainly imported.

Some of Ireland's most ambitious enterprises are state operated; the Aer Lingus airline,

◄ **LAKELAND MAGIC Lush green countryside and high mountains encircle Killarney's Upper Lake. Tourists who throng the little town of Killarney can travel to the lake in pony-drawn jaunting cars.**

Irish Shipping and the CIE (*Coras Iompair Eireann*) road-rail system are major companies. *Bord Grain* markets surplus cereals and *Bord Bainne* dairy products; sugar and fertiliser production are in state hands. An Industrial Development Board was set up in 1950, using tax concessions to attract foreign firms. Perhaps the most striking venture was the creation of a whole new industrial town beside Shannon international airport, on the west coast – a customs and tax-free zone which encouraged industries producing a wide range of goods from radios to machine tools and textiles. Unfortunately the development of long-range jets means that transatlantic flights no longer need to refuel at Shannon, which is now more of a flight-control centre.

However, the country still manages to profit handsomely from a thriving import trade in tourists. An estimated 17 million people of Irish descent live abroad, and many dream of returning to their roots. Millions do so every year. They have included two American presidents – John F. Kennedy and Ronald Reagan – in recent times. Others go just for the wonderful, unspoilt beauty of the land and the lakes; the friendly, hospitable people, the relaxed atmosphere and change of pace. There were 1.1 million visitors from Britain alone in 1984, by far the largest proportion (46 per cent) of the total from all countries. Another half million (24 per cent) were from Northern Ireland and 14 per cent from America and Canada.

Visitors find a land where the Catholic faith is enshrined in the constitution, and whose citizens voted overwhelmingly against the introduction of divorce in 1986. They find, too, a land divided into provinces which were once its ancient kingdoms. To the north and west is wild, beautiful Connaught, including the counties of SLIGO, MAYO and GALWAY and the blue lakes and matching blue mountains of CONNEMARA; south, in MUNSTER, the long Atlantic-probing peninsulas of CORK and Kerry are backed by towering peaks of the Caha and Slieve Mish mountains, and MACGILLICUDDY'S REEKS. The heart of LEINSTER – MEATH, LAOIS (pronounced 'Leesh'), CARLOW and WICKLOW – has rolling hills and prosperous farmlands full of small towns and hamlets. And so to the mild east coast between Dublin and WEXFORD, where small harbours are interspersed with sandy beaches.

At an anglers' paradise called Pontoon, between loughs Conn and Cullin, in Mayo, a licence is required to fish for salmon, but only after you have caught one. And that, like a fine, soft day, is very Irish.

IRELAND AT A GLANCE
Area 70 283 km² (27 136 sq miles)
Population 3 630 000
Capital Dublin
Government Parliamentary republic
Currency Punt = 100 pighne
Languages English, Irish
Religion Christian (94% Roman Catholic, 3% Church of Ireland, 1% Presbyterian)
Climate Temperate, warmed by the North Atlantic Drift. Average temperature ranges from 4-7°C (39-45°F) in January/February to 14-16°C (57-61°F) in July/August
Main primary products Cereals, potatoes, sugar beet, vegetables, livestock, fish; peat, natural gas, lead, zinc, barytes, gypsum
Major industries Agriculture, food processing, machinery, chemicals and fertilisers, textiles, clothing, tourism
Main exports Machinery, chemicals, meat, dairy produce, textile yarns and fabrics, instruments, livestock
Annual income per head (US$) 4355
Population growth (per thous/yr) 10
Life expectancy (yrs) Male 71 **Female** 75

Israel

THE PROMISED LAND OF THE JEWS AND BIRTHPLACE OF JUDAISM AND CHRISTIANITY HAS BEEN THE FOCUS OF CONFLICT IN THE MIDDLE EAST FOR NEARLY FOUR DECADES

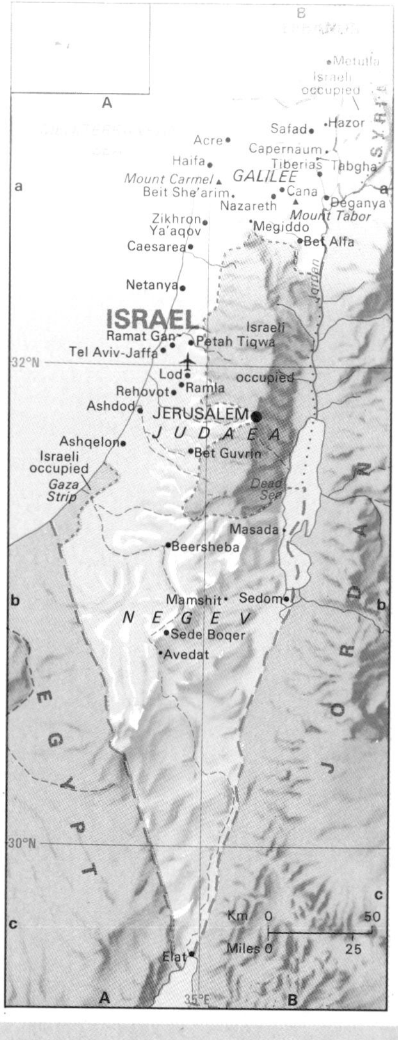

Born in strife, the modern state of Israel has been almost constantly at war with its powerful Arab neighbours, many times its size, ever since its creation in 1948. The triumphant return of Jews to their historic homeland ended 18 centuries of exile, but the land they returned to had long been the home of the Palestinian Arabs, and they made determined attempts to strangle the tiny state at birth. Five wars and countless acts of terrorism followed as Arab countries and Palestinians rose up against it and Israel tried to secure its frontiers.

Israel occupies a long narrow stretch of land at the south-eastern corner of the Mediterranean. This hilly country has as its natural eastern boundary the GREAT RIFT VALLEY, the great fault which split the earth's surface from Syria to Africa millions of years ago. The valley contains the River JORDAN, which flows first through subtropical vegetation, then arid wasteland to the DEAD SEA, 394 m (1293 ft) below sea level. The triangular wedge of the NEGEV Desert ends at the Gulf of AQABA, at the head of which is Israel's southern outlet, the small port of ELAT.

Israel's northern half is temperate and fertile, while the south is arid and barren. Most of the country's 4.1 million citizens live on the coastal plain bordering the Mediterranean, where TEL AVIV is the leading commercial city. The greenest part is GALILEE, though here, as in all Israel, no rain falls in summer. The Sea of GALILEE, lying among the northern hills, is Israel's main reservoir of fresh water. This is piped to the Negev – the desert and semidesert home of wandering Bedouin Arabs – and it is the Israelis' proud boast that they have 'made the desert bloom' with sunflowers, sugar beet and other crops.

The Jews are a clever, restless, vibrant people. In the 40 years since modern Israel was founded, large numbers of them have left Europe, North America, the Middle East and Africa to make new lives in their historic homeland. The Jewish population was 650 000 in 1948; today it is more than 3.5 million – one-fifth of the world's Jews. In addition there are 650 000 non-Jews – Arabs, Druse and others.

It is a young country – about one-third of the population is under 14. The people are highly educated; many of those who flocked to Israel brought with them a wealth of skills, and it is not uncommon for, say, a truck driver to have a degree. It is also one of the most polyglot societies on earth; there are, for instance, Iraqi, Ethiopian, Moroccan and Russian Jews, and many others.

FIGHT FOR ECONOMIC SURVIVAL

Israel's economic and military survival has largely been made possible by the massive help it has received from the USA, both at government level and from the powerful American Jewish community. Israel is virtually self-supporting in foodstuffs, and a major exporter of agricultural produce. Its Jaffa oranges are famous throughout Europe, and it exports many other fruits and vegetables.

Much of its agriculture is based on methods employed at the *kibbutz* (collective settlement) or *moshav* (cooperative village). The *kibbutzim*, with their organised communal life, are designed as agricultural collectives, but are also geared, especially in border areas, for rapid defence in times of war.

There are 270 kibbutzim, with a total population of more than 110 000, scattered around the country. They produce 40 per cent of the country's agricultural output, but now there is hardly one without a factory as well – making electronic or irrigation equipment, processed foods, farm machinery or furniture.

A 1951 law gave Israeli women equal status with men, in theory at least – and there are women in uniform. Men and women serve side by side in the Israeli defence forces. The men do three years' national service, single women two. Married women are exempt.

There are 140 000 Israelis in the armed forces. Nearly a quarter of the national budget is spent on defence, and this puts a heavy strain on the economy. There are constant, huge balance of payments deficits, which are made good by aid from the USA, and Israel's inflation rate reached a staggering 1000 per cent a year in 1984.

THE ARAB-ISRAELI CONFLICT

The reason for Arab hostility is rooted in history. For over 2600 years there was no independent Jewish state. From the conquest of ancient Israel by the Assyrians in 722 BC until the emergence of modern Israel in 1948, the land was part of some other empire. Jewish revolts against the Romans in AD 70-73 and 132-5 led to the devastation of the land and the dispersal (Diaspora) of the Jews. The result was 1800 years without a homeland, so that Jews in many lands were to take on the flavour of the countries of their adoption, though never losing their identity. Often they were subjected to persecution.

Israel was renamed PALESTINE by the Romans; it was taken by the Muslims; the Crusaders fought to free the Christian holy places from Muslim control. Then it became part of the Turkish Ottoman Empire until occupied by the British during the First World War.

In 1917 the British foreign secretary, Arthur Balfour, promised British support for a Jewish national home in Palestine. But Britain had also agreed to back the creation of independent Arab states, which the Arabs claimed included Palestine. The League of Nations authorised a British mandate in 1922. During and after the Second World War, Jewish survivors of Nazi persecution in Europe flocked to Palestine, and Zionists (Jewish nationalists) adopted terrorist tactics against the British, who were trying to restrict immigration. Following a United Nations vote to establish Arab and Jewish states in Palestine, the British left in 1948.

Egypt, Jordan, Lebanon, Syria and Iraq attacked at once; the Israelis lost the old part of Jerusalem but gained territory elsewhere. In 1956, provoked by invasion threats, Israel struck at Egypt and occupied the Gaza strip and the Sinai peninsula, but eventually withdrew. In the Six Day War of 1967 it again took Sinai and also the GOLAN HEIGHTS from Syria and the WEST BANK area and old Jerusalem from Jordan. In 1973 Egypt and Syria hit back on the Jewish atonement day of Yom Kippur. Egyptian and Israeli forces both crossed the Suez Canal, making gains; the Syrians were driven back. Some withdrawals followed, but Israel continued to hold Sinai until peace moves by President Sadat of Egypt resulted in the Camp David Agreement with Israel in 1979, when Sinai was handed back to Egypt.

The problem of 2 million Arab Palestinian refugees in the areas surrounding Israel remains. The Palestine Liberation Organisation (PLO), recognised by Arab countries,

▶ HOLY CITY Behind the onion domes of St Mary Magdalene, a Russian church in Jerusalem, rears the Dome of the Rock, built on what Muslims revere as the site from which Muhammad rose to heaven. In a city of holy places, the main focus of Jewish devotion is the Wailing Wall.

has made many terrorist attacks on Israel, and Israel has retaliated by bombing Palestinian refugee camps. PLO forces, led by Yasser Arafat, were driven out of LEBANON in 1982 by an Israeli invasion aided by right-wing Lebanese Christian militiamen, but remain active elsewhere. The operation, involving a lengthy stay in Lebanon and the massacre by right-wing Lebanese of more than 700 Palestinian refugees in Israeli-occupied BEIRUT, brought worldwide criticism and political controversy in Israel; so did the settlement of Jews on the West Bank. In 1985, the Israelis withdrew from Lebanon.

LIFE IN ISRAELI CITIES

About 90 per cent of Israelis live in urban areas, such as Jerusalem, TEL AVIV-JAFFA and HAIFA. Until about 1875 Jerusalem – special to people of differing beliefs – consisted only of a walled city with four distinctive quarters: Jewish, Christian, Muslim and Armenian. Then the growing Jewish population began to establish new neighbourhoods outside the walls.

After Israel's war of independence in 1948, Jerusalem was divided for nearly 20 years. The Old City was in the Jordanian sector and Israelis were barred from it. Since the 1967 war the city has been reunited under Israeli rule and the walled sector extensively renovated. Here is the seat of government, with the Knesset (parliament) of 120 members elected by proportional representation; to date no party has had a clear majority.

City life in Israel is generally informal. Parks and playgrounds lie outside the city limits and planning aims at separating residential areas from commercial and industrial zones. Diamond polishing and finishing is the most valuable industry, but there is growing emphasis on research and development in high technology fields such as electronics, medical equipment, computers and alternative energy.

The country is geared to tourism – in part a deliberate propaganda exercise to show Israel to best advantage despite its hostile environment. The Israelis, backed by thousands of years of faith, represent one of the oldest cultures in the world: yet there is also a younger, almost brash nationalism, a pride in the achievements of this small state surviving against great odds. This pride leads to charges that service is often grudging. Prices are high, but few small countries have so much to offer the visitor.

What of the country's future? Modern Israel seems here to stay – yet so far only Egypt of the Arab states has even talked to the Israelis. The Palestinian refugee problem has not been solved, and until this happens Israel will continue to be on a war footing, ready to fight for its life.

ISRAEL AT A GLANCE

Area 20 700 km² (7992 sq miles) as defined in the 1949 armistice agreements

Population 4 150 000

Capital Jerusalem

Government Parliamentary republic

Currency Shekel

Languages Hebrew, Arabic

Religions Jewish (83%), Muslim (11%), Christian/others (6%)

Climate Subtropical; average temperature in Jerusalem ranges from 5-13°C (41-55°F) in January to 18-31°C (64-88°F) in August

Main primary products Livestock, citrus fruits, grapes, wheat, sugar beet, cotton, olives, figs, vegetables; potash, crude oil, phosphates, bromine, natural gas

Major industries Agriculture, mining, food processing, textiles, clothing, leather goods, transport equipment, aircraft, chemicals, metal products, machinery, diamond cutting, fertilisers, cement

Main exports Finished diamonds, textiles, fruit, vegetables, chemicals, machinery, fertilisers

Annual income per head (US$) 4900

Population growth (per thous/yr) 16

Life expectancy (yrs) Male 72 **Female** 76

Italy

FROM SOUTH TO NORTH, OLIVES TO OLIVETTIS, THE LANDSCAPES AND LIFESTYLES PRESENT A HARMONY OF CONTRASTS – ALL UNMISTAKABLY ITALIAN

In Italy, history is all around you. Its cities are so venerable that they seem to have existed since time began. In fact its network of towns and roads was formed mostly in the times of Imperial Rome. The glories of Rome, its styles and ideas, were revived in the early years of the Italian Renaissance, an unprecedented flowering of the arts and sciences that began in the 14th century and continued for 250 years. Between them, Rome and the Renaissance made an incalculable contribution to Western civilisation – a contribution that continues to be provided by today's Italy, especially in the areas of style and design.

Yet, despite its antiquity, Italy as a political unit is little more than a century old. After the fall of the Roman Empire in AD 476, the peninsula fragmented politically. The Lombards overran the north, established their capital at PAVIA, and gained footholds in the south at SPOLETO and BENEVENTO. Much of the rest of Italy stayed under the Eastern Roman Empire. Around the 8th century new powers emerged: the Papal States along the ROME-RAVENNA axis; the trading city of VENICE; the Frankish Empire which absorbed much of the north; and the Arabs in the south. From the 12th century on, 200 or more city-states flourished between Rome and the Alps: this was the great communal age which led into the Renaissance. Meanwhile the south, after a brief spell of splendour under the Normans, sank into colonial obscurity under Spanish rule. The 16th and 17th centuries were periods of economic decline throughout the whole of the peninsula, with Austria emerging as the key imperial power in the north. Thus, virtually all parts of Italy changed hands several times between the fall of the Roman Empire and unification.

After 15 centuries of political chaos, along came Napoleon, whose brief period of supremacy struck a spark of nationalism which gave rise to the *Risorgimento* – the unification movement. Unification took 50 years, spearheaded by three of Italy's most famous figures, Mazzini the theorist, Garibaldi the buccaneer and Cavour the politician. The first all Italian parliament met in TURIN in 1861, in 1865 the capital moved to FLORENCE and in 1870 finally to Rome – the 'Eternal City' being the last major piece of Italian territory to be absorbed into the new kingdom.

A century or more of unity has seen Italy make steady and occasionally spectacular economic progress. The more shameful aspects of this include Italy's attempts at creating an empire in Africa, and the inter-war

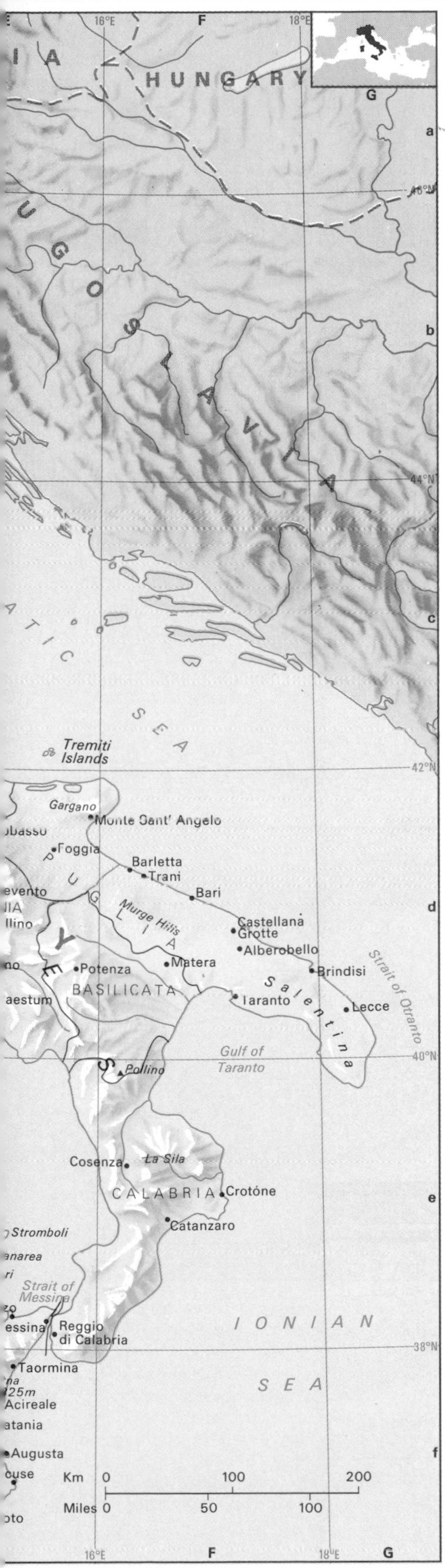

Fascist period, when Benito Mussolini ruled for more than 20 years. In 1946 Italy became a republic, and the 1948 elections ushered in the Christian Democrats who have dominated Italian politics ever since.

A GLITTERING MOSAIC

The cities of the north, such as Turin and MILAN, have an industrial and commercial mentality that rivals any in Europe. Living standards are high; the rates of car and second home ownership, for example, are higher than in Britain. Excellent railways – with notably cheap fares – and first-rate motorways provide swift, convenient transport links throughout the north.

The south – the region they call the *Mezzogiorno*, 'land of the midday sun' – is a complete contrast. Here medieval concepts of honour and behaviour dictate a fierce loyalty to family and a lingering suspicion of outsiders. In SICILY, NAPLES and CALABRIA secret societies such as the Mafia and Camorra are social plagues. Peasant farmers wrest a living from eroded hillsides. Here there is resignation and a degree of hopelessness; waiting for someone else to do something is an all-too-prevalent attitude in this sunbleached land.

But despite differences, north or south, it is the legacies of history, the culture, the works of civilised and sophisticated men that are the abiding glories. In the south, at PAESTUM, Selinunte and AGRIGENTO, are Greek temples as fine as those of Greece, as well as irrigation systems introduced by Arab colonists a thousand years ago. In central Italy stands Florence, cradle of the Renaissance, and Rome itself. Around them stretch landscapes redolent of Leonardo da Vinci, Dante and St Francis, where hilltop towns and hillside terraces bear the fruits of centuries of labour.

With the exception of GENOA, most of the major northern towns are on the margins of the great plain drained by Italy's largest river, the PO. One line of towns is strung like beads along the Via Emilia, the arrow-straight Roman road that stretches from Milan to the ADRIATIC coast: PIACENZA, PARMA, REGGIO NELL'EMILIA, MODENA, BOLOGNA, Imola, Faenza and FORLI. The road meets the coast at the popular resort of RIMINI. Bologna, chief city of this group, disputes with Capua in Campania the distinction of being the oldest town in Italy: people have lived there continuously for nearly 3000 years.

On the northern side of the plain is another row of towns at the mouths of Alpine valleys. They, too, are ancient in foundation, but many have developed important industries over the last century. Turin (Fiat cars), IVREA (Olivetti typewriters and electronics), Biella (wool), VARESE (engineering), COMO (silk) and the engineering towns of Lecco, BERGAMO and BRESCIA are examples. A third row, between the 'high' and 'low' plain, includes Vercelli, Novara, Milan, PADUA and TREVISO.

Then there are the cities on or near the Po river itself: PAVIA, CREMONA, MANTUA and FERRARA. Finally, sited on the fringes of the delta, VENICE and RAVENNA are unique: Venice, the Queen of the Seas, because of its imperial history, artistic heritage and island site, Ravenna because of its role as a Byzantine capital.

Apart from a group of famous valley towns in northern TUSCANY (Florence, PISTOIA, PISA, LUCCA), hill settlements are characteristic of central Italy. Most of them are in southern Tuscany, UMBRIA and MARCHE. Star-shaped URBINO, spreadeagled along its ridges, or triangular SIENA are distinctive forms. Others, such as PERUGIA and ASSISI, cling picturesquely to their slopes.

The south, too, has its hill towns, the most dramatic being ENNA and Erice in Sicily. But most of its major cities are coastal and have Graeco-Roman origins – among them are Naples, BRINDISI, TARANTO, REGGIO DI CALABRIA, MESSINA, CATANIA, SYRACUSE and Agrigento. Byzantine rule continued the classical traditions of many southern cities but, later, feudalism caused widespread stagnation of urban and rural life.

In the Middle Ages Norman rulers built many castles and churches in southern Italy, especially in PUGLIA and in Sicily. But earthquakes have caused much destruction and subsequent rebuilding, notably in eastern Sicily in the late 17th century, at Messina and Reggio di Calabria in the early 20th century, and in Naples since 1980. The character of many southern cities has also been changed as a result of post-1960 policies of industrial development, as in Naples (Alfa Romeo cars), Taranto (steel) and Syracuse (petrochemicals).

VILLAGE LIFE, ITALIAN-STYLE

Italy's urban, and urbane, way of life – at its best in cities such as Bologna and Florence which have not become too industrialised – extends also into rural areas where about 30 per cent of the population still live, and about 10 per cent of the national labour force works on the land. Most farmers do not live in isolated farmsteads or hamlets, but cluster in bustling 'town-villages' which may, especially in the south, contain as many as 60 000 people.

Life in these villages tends to follow a pattern, as do the villages themselves. At the village centre is a large, open, paved space, the *piazza*. Facing onto this at one end will be the façade of a church; nearby will be the *municipio*, the town hall. Other public buildings include the *carabinieri* (police) station. The carabinieri are the traditional police force, present in even the smallest villages. The more recently instituted *polizia* are found mainly in provincial capitals and larger towns. The bank and a few bars and shops are also likely to be set around the piazza, along with the large houses of the local landowning and professional families.

The changing scene in the piazza has a precise daily rhythm which presents the observer with a pageant of everyday life Italian style. At dawn the farmers appear, snatching a cup of coffee on the way to their fields; some still ride donkeys or horses, especially in the south. Next, the children pass through – school in Italy starts between 8 and 8.30 am. The younger children, smart in smocks and ribbons, are on their way to the local elementary school; the older ones may take a bus to a

secondary school or college in a neighbouring town. Shops, too, are open early – before 8 am – by which time the noise of craftsmen and mechanics is added to the cacophony. Once a week a market fills the square.

Later on the professional workers – doctors, lawyers and so on – take the stage, along with office workers.

Few women are seen – especially in the south, where there is still prejudice against women being out too much on their own. For teenagers and young unmarried adults, contacts between the sexes still depend, to a certain extent, on the rituals of the village carnival, churchgoing and the *passeggiata* (evening stroll). The women's domain lies in the streets leading off the piazza, where the food shops are found.

Around 1 pm the piazza fills up again with children returning from school, office workers going home and shopkeepers leaving their counters. But they do not linger long in the heat of the sun; besides, lunch awaits. During the next few hours the piazza is deserted, save perhaps for a few children. For after lunch comes the *siesta*.

Some offices and most shops open again from late afternoon to mid-evening. Then, around 7 to 8 pm (earlier in winter) the piazza and the main streets come alive again. This is the time for the passeggiata, the traditional strolling up and down, arm-in-arm with family and friends; a time for relaxed chatting and greeting, for showing off one's best clothes, and for teenagers and young people to exchange playful banter. Only wives and mothers miss the passeggiata – they are at home preparing the evening meal.

Later, after the meal, the piazza may return

◀ **A GATHERING OF GONDOLAS Their prows curved like the necks of proud black swans, some of Venice's 300 or so gondolas line up in the San Marco Canal – with St Mark's Cathedral in the background. They have been painted black since 1562, when a city law was passed prohibiting too much pomp.**

to life – especially on summer evenings, as men gravitate towards the bars and young people and engaged couples stroll about.

Italy is rather a rigidly stratified society, with serious conflict between classes and little opportunity for climbing the social ladder. Centuries of foreign rule, the hereditary power of Italy's aristocratic families and chronic unemployment have been significant factors in perpetuating the deep chasms separating social groups. There is a long history of class conflict, from the brigandage of the past to the modern urban terrorism of the Red Brigades, who in 1978 kidnapped and murdered the former prime minister, Aldo Moro.

In the past Italy's educational system did much to preserve social inequalities, with its emphasis on classical languages and history, and with learning by heart taking precedence over imaginative project work. Teachers tended to come from middle-class backgrounds, and to impart a culture which was alien to children from the working classes, from the south, or from areas of relatively closed and traditional culture.

But since 1968 there have been dramatic changes – best demonstrated by the fact that by 1985 half the students at university were from working-class or peasant backgrounds. And Italy has one of the highest proportions of university students among Common Market countries.

BOOM TIME IN THE CITIES

Many social problems result also from the great expansion of cities after the Second World War. Rapid urban growth has taken place in most parts of the country, but has been more dramatic in the north because of the migration of job-seekers from the south. The proportion of Italians living in towns of at least 100 000 people increased dramatically from 35 per cent in 1951 to 50 per cent by 1981. The outer suburbs of large cities seem particularly soulless, consisting of large, tightly packed blocks of flats with little public open space and few communal services. Municipal housing schemes are few, and poorer people live in dilapidated tenements, seen at their worst in central Naples.

This city growth has resulted from the remarkable industrialisation of Italy since 1950. Before then Italy had a mainly rural economy. Earlier industrial phases, such as the medieval textile industries of Venice and Florence or the late 19th-century textile and engineering industries of LOMBARDY, had made little difference on a national scale.

In the immediate postwar period, the economic prospect again looked dreary. Armies had tramped over the countryside and bombardment had destroyed towns and industries; 2 million unemployed – among them many former soldiers and landless peasants – threatened revolution. Living standards were abysmally low compared with those of northern European countries. Yet, astonishingly, Italy underwent a modern renaissance, rising like a phoenix from the ashes of war. There were two main developments.

First came the industrial boom of the 1950s and 1960s, with a rate of expansion second only to that of Japan. Italy's gross national product doubled from 1950-62, and seemed well on the way to repeating the performance in subsequent years. Then recession struck in the early 1970s. World oil prices quadrupled, administering a massive shock to an economic system founded largely on imported fuels. Italy produces only tiny quantities of coal and oil, although it generates considerable hydroelectric power in the Alps, and extracts natural gas in the lower Po Valley.

The second development was Italy's entry into the Common Market (of which it was a founder-member). This enlarged the markets open to Italy's rapidly expanding industries – especially for cars, electrical goods, clothing and shoes. Also, large numbers of Italian workers could cross frontiers freely in search of jobs.

Two great population movements began: some workers emigrated to other Common Market countries – especially to France and West Germany – during the 1950s and 1960s, and at the same time more than 2 million people moved from rural areas – especially the south – to Rome and the 'Industrial Triangle' of Milan, Turin and Genoa.

Not until the 1980s did these migrations come to an end. Now emigrants are returning from abroad to their home villages. Cities such as Milan, Turin, Genoa, Venice and TRIESTE have stopped growing as people choose to remain in the countryside or settle in smaller towns.

UNSTABLE GOVERNMENT

It is difficult to reconcile the dynamism of Italy's economy for most of the postwar period with the stagnation of its political system. Here is a classic case of multiparty political stalemate.

There are eight main national parties: Communist, Socialist, Social Democrat, Republican, Christian Democrat, Liberal, Monarchist and Neo-Fascist. No party, except the Christian Democrats in 1948, has won a majority of votes in any postwar election.

The main opposing forces are the Christian Democrats – strongly allied to the Catholic Church – and the Communists. They have tended to poll about 40 and 30 per cent of the votes respectively. Government has been based on a pattern of continuously shifting coalitions, with Christian Democrats always at the hub and Communists excluded. The instability of government is notorious; each cabinet has lasted on average only ten months. Yet the same politicians seem always to be involved – they just change roles.

An important postwar political trend has been the rise of regional governments. This has taken place in two phases. First, four regions – Sicily, SARDINIA, VALLE D'AOSTA and TRENTINO ALTO ADIGE – received special status between 1946 and 1949. They were joined by FRIULI-VENEZIA GIULIA in 1963. This act was a recognition of their cultural distinctiveness as islands or frontier regions.

The second stage came in the early 1970s when individual governments were created in the other 15 regions. The hope was that regional devolution would lessen the inertia inherent in a centralised political system, and allow more local participation in dealing with issues closer to the hearts of the people.

With one or two exceptions, this hope has

▲ **HEART OF A CITY A pavement artist draws religious pictures – including 'The Heart of Mary' – outside Milan's vast cathedral, the Duomo. Made of pink marble, it can hold 20 000 people.**

not been realised; regional administrations have struggled unsuccessfully with central government to define their respective roles, and have still to gain true independence of action.

A journey through Italy is a scenic idyll. The northern borderlands embrace spectacular Alpine scenery culminating in the Mont BLANC massif on the French frontier and the MATTERHORN on the Swiss. Small glaciers survive in places, and splendid ski slopes have popularised winter resorts such as COURMAYEUR, in the AOSTA valley, and CORTINA D'AMPEZZO in the DOLOMITES.

Flat-bottomed valleys seam the Alpine massif. Glacial action has deepened and then dammed many of these valleys, to produce a series of lakes – MAGGIORE, COMO, GARDA.

Some widely differing Italians live there, too – 100 000 French-speakers in the Valle d'Aosta, 300 000 German-speaking people in BOLZANO province and a few thousand Slovenes and Croatians around Trieste and Gorizia near the Yugoslav border.

The Alps merge abruptly into the great northern plain, a 45 000 km² (17 000 sq mile) triangle of lowland which gradually sinks and widens towards the Adriatic. This is the nation's economic heartland, containing most of its larger cities, industries and productive farmland.

Somewhere behind Savona, a port on the LIGURIAN SEA, the Alps give way to the APENNINES, which then wind southwards down the entire length of the country.

And as you go south, so the climate seems to grow perceptibly warmer, the sky bluer and the people more outgoing and gregarious. The trees change from the pines of the Alps and poplars of the plain to more Mediterranean species – evergreen oaks and cypresses then, continuing southwards, olives. Farther south still, almonds, oranges and lemons reflect the influence of the southern summer drought and mild winter.

The Apennines are mainly sandstones and limestones – young mountains where the earth's crust is still active, as the frequent tremors and earth movements testify. Italy's most recent major earthquake, in November 1980, struck in the heart of the southern Apennines east of Naples.

The western flank of the Apennines is also volcanic, from the extinct volcanoes and volcanic plateaus of southern Tuscany and LAZIO to the cone of VESUVIUS (at present quiescent but terrifyingly active in the past – as in AD 79, when it destroyed POMPEII) and then, via the volcanic LIPARI ISLANDS, to Sicily's ETNA, bonfire of Europe.

On their eastern, Adriatic flank the Apennines are less dramatic, with smooth, rounded outlines. Landslides and soil erosion tear great gashes in their clay soils and the rural economy in regions like MOLISE and BASILICATA is among the poorest in the country.

Finally there are the two island regions: Sicily an exaggerated microcosm of everything Italian; Sardinia a law unto itself – isolated and boulder-strewn, where shepherds carry on their fiercely traditional way of life.

ITALY AT A GLANCE

Area 301 225 km² (116 304 sq miles)

Population 57 320 000

Capital Rome

Government Parliamentary republic

Currency Lira = 100 centesimi

Languages Italian, some French, German and Slovene

Religion Christian (83% Roman Catholic)

Climate Mediterranean; average temperature in Rome ranges from 4-11°C (39-52°F) in January to 20-30°C (68-86°F) in July

Main primary products Cereals, potatoes, sugar, vegetables, soft fruits, citrus fruits, grapes, olives, livestock, fish; oil and natural gas, asbestos, potash, iron ore, zinc, mercury, sulphur, marble

Major industries Iron and steel, motor vehicles, chemicals, oil and gas refining, machinery, textiles, clothing, agriculture, food processing, wine, fishing, tourism

Main exports Machinery, chemicals, petroleum products, motor vehicles, clothing, food, textile yarns and fabrics, iron and steel, footwear

Annual income per head (US$) 5512

Population growth (per thous/yr) 3

Life expectancy (yrs) Male 72 **Female** 76

Ivory Coast

A FORMER FRENCH COLONY THAT HAS ENJOYED PROSPERITY AND STABILITY UNDER ONE RULER SINCE INDEPENDENCE IN 1960

From the 15th to the 19th centuries the few Europeans who landed on the Ivory Coast's tropical shore came only to trade in elephants' tusks – which gave the land its name – and in slaves. Meanwhile, in the jungle and savannah interior, those people who escaped the slavers continued to farm or wage war on each other, far removed from Western influences. But the French arrived and colonised the country between 1893 and 1960. Their influence was strong, and today the independent state is one of the richest, most Westernised and most modern in West Africa.

The first European visitors were discouraged by the rocky cliffs of the south-west coast, along the Gulf of GUINEA; but farther east are coastal plains which today are the country's most prosperous region. The former capital city of ABIDJAN lies there, on the edge of vast lagoons, its administrative and financial heart in the French quarter (known as 'the Plateau'), a forest of modern skyscrapers.

A tourist centre has grown up by the lagoons where there are vast sandy beaches lined with palm trees – an area now known as the African Riviera. The interior of the Ivory Coast is less affluent, but has its oases of prosperity, such as BOUAKE, on the central plateau of the country, and Ferkessedougou in its more sparsely forested north. The small town of YAMOUSSOUKRO has recently become the new seat of government, and boomed into a prestigious, though still half-built, capital.

The reason for the general prosperity – and particularly for the boom of Yamoussoukro – has been the government in power since independence in 1960. It has been dominated by one man, President Félix Houphouet-Boigny. His outlook is modern, moderate and pro-Western, and he has pursued a policy of liberal capitalism, encouraging Western investors to put their money into the Ivory Coast by placing few restrictions on the import or export of capital. The response to this has been staggering, and French business in particular has poured into the country.

The wealth from foreign investment has, however, benefited the few much more than the many. With Abidjan now one of the most expensive cities in the world, life is even harder for those who profited least from the country's 'economic miracle'.

Some improvements, though, are for the good of all. Roads are well-kept and the Ivory Coast has the greatest length of metalled road in West Africa. There is an efficient rail service too, from Abidjan up to the northern border with Burkina (once Upper Volta).

JUNGLE AND SAVANNAH

Riding the railway northwards, the visitor passes through the humid tropical jungle of the south, where teak and mahogany grow, along with iroko trees whose wood was once used as a substitute for teak – felling is now prohibited. Oil palms and raffia palms provide useful products.

The Ivory Coast is an agricultural land that has nudged back the natural forest to grow cocoa, coffee, rubber, bananas and pineapples. The country is now the world's largest producer of cocoa and the fourth largest producer of coffee – and together these two crops bring in half the country's export revenue. But beyond the plantations nestle villages which have seen little of this revenue, where the people still lead a largely self-sufficient, traditional existence. They grow their own food, usually cassava – the sturdy root from which tapioca comes – vegetables and rice. They sell the produce of a few cocoa, rubber and banana trees, but village life is still hard, and the environment harsh. Most people at some time in their lives contract at least one of many debilitating diseases such as malaria, yellow fever and bilharzia.

North of the humid lowland plain the land rises gradually to a savannah plateau, and becomes increasingly dry. The far north is grassy and treeless. Here the rainy season is shorter than in the south, so there is only one harvest a year – of millet, vegetables and the syrup-producing sorghum grass. But there is also the cotton crop and groundnuts to sell. Towards the northern border with Mali, around KORHOGO, the dry months bring intense heat. The parched brown plains are often covered with a haze of dust which settles on everything and hangs in the air

In the west lies quite a different land – the forested MAN Mountains, the eastern flank of a group of mountains centred in neighbouring Guinea. Metalled roads are opening up this once isolated area, in particular bringing in tourists.

The 60 diverse ethnic groups of the Ivory Coast have no common language but French, and their loyalties to their groups are still stronger than their loyalty to the new nation. They have a very rich and varied culture: the Baoulé in the centre are skilled woodcarvers and goldsmiths, living in homes with intricately carved doors. Even more elaborate doors guard the homes of the Senoufo.

Despite its metalled roads, dams and hydroelectric power stations (which will soon be able to meet the country's total energy needs), its oil refinery and deep-water ports at Abidjan and San Pedro, the Ivory Coast is still underdeveloped. Drought in the 1970s and early 1980s reduced harvests and depressed the economy. But in 1983 the rains returned and this – combined with rises in the prices of coffee and cocoa – has caused an upturn in the economy.

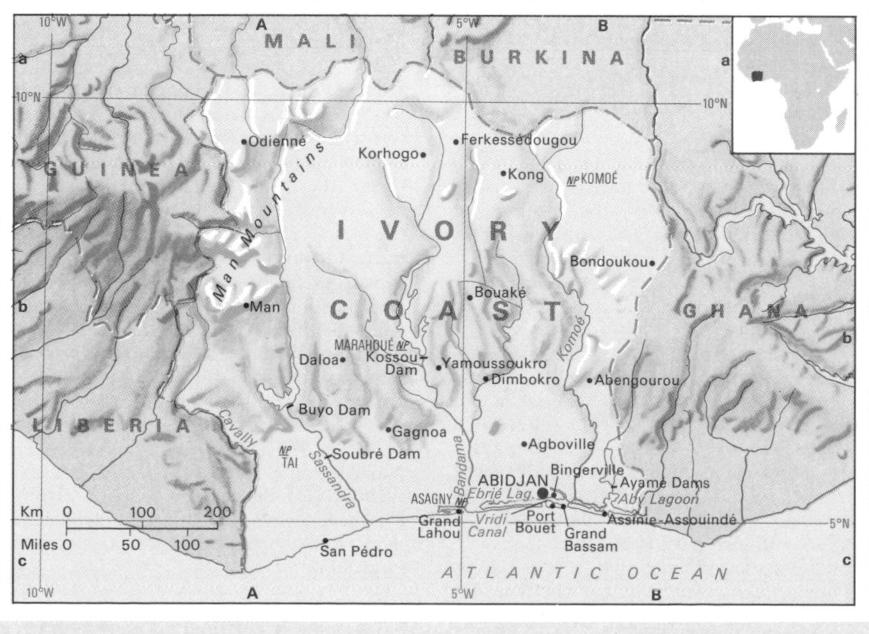

IVORY COAST AT A GLANCE
Area 322 463 km² (124 503 sq miles)
Population 10 460 000
Capital Yamoussoukro
Government One-party republic
Currency CFA franc = 100 centimes
Languages French and tribal languages
Religions Animist (63%), Christian (12%), Muslim (23%)
Climate Tropical; average temperature in Abidjan ranges from 22°C (72°F) to 32°C (90°F)
Main primary products Cocoa, coffee, cotton, bananas, pineapples, rubber, sugar, rice, cassava, yams, timber; crude oil, diamonds, cobalt, uranium
Major industries Agriculture, food processing, textiles, leather goods, forestry, mining, petroleum refining
Main exports Cocoa, coffee, timber, petroleum products, cotton
Annual income per head (US$) 1030
Population growth (per thous/yr) 40
Life expectancy (yrs) Male 46 **Female** 48

Istanbul *Turkey* [illegible] Black Sea [illegible] networks.

Founded by the Greeks in about 600 BC, it is the only city [illegible] to be built on two continents, Europe and Asia. Formerly called Byzantium and then Constantinople, it was the capital of the Byzantine Empire for 1000 years, and of the Ottoman Empire for 500 years. Finally, in 1923, Kemal Ataturk, founder of modern Turkey, moved the capital 350 km (220 miles) east to ANKARA.

Among the city's historic sites are the walled old town, built on the arm of the Bosporus called the Golden Horn, St Sofia (in turn a church, mosque, now a museum), the Blue Mosque, the Mosque of Suleiman the Magnificent, the Topkapi Palace, the Aqueduct of Valens, and the hospital at Scutari (present-day USKUDAR) where the British nurse Florence Nightingale tended the wounded during the Crimean War (1853-6).

The city's main industries are shipbuilding, textiles, pottery, tobacco, cement, glass and leather goods.

Population 5 858 600
Map Turkey Aa

Istra (Istria) ***Italy/Yugoslavia*** Peninsula on the north Adriatic coast, stretching 97 km (60 miles) south from Trieste, Italy, on its western neck, and about 80 km (50 miles) from Rijeka, Yugoslavia, on the eastern side. PULA is the chief town. Most of the peninsula was ceded by Italy to Yugoslavia in 1946.

Map Yugoslavia Ab

Itaipu ***Brazil/Paraguay*** Dam on the Paraná river, straddling the Brazil-Paraguay border about 720 km (450 miles) west of the city of São Paulo. The dam, completed in 1982, is due to be fitted with 18 giant turbines by 1988. When they are in place, the dam will be able to generate 12 600 million watts of electricity, making it the world's largest power station.

The lake behind the dam has, however, submerged the homes of 200 000 people and drowned the Guairá Falls 190 km (118 miles) upstream. The falls – now marked only by an underwater cliff – were once the world's largest body of falling water. The flow over them was more than double that over the Niagara Falls, and 12 times greater than the Victoria Falls.

Map Brazil Cd

Italy See p. 324

Ithaca (Itháki) ***Greece*** One of the IONIAN ISLANDS, kingdom of Homer's legendary hero Odysseus. The main harbour town of Vathí lies on a sheltered inlet. The island's rocky coast has many small bays.

Population 3650
Map Greece Bb

Itsukushima ***Japan*** See MIYAJIMA

Ituri Forest *Zaire* Named after its chief river, Ituri Forest covers an area of between 30 000 and 54 000 km² (11 580 and 20 850 sq miles) in north-east Zaire. It is the largest primeval forest outside the Amazon basin, and its dense canopy of [illegible] foliage blocks out sunlight from the dank forest floor.

In 188[illegible] the British adventurer Sir Henry Morton Stanley became the first European to cross this twilight world from east to west. The journey took five months, and the expedition, which suffered 180 deaths in its final stages, was frequently attacked by pygmies using poisoned arrows. Pygmies, locally called Mbuti, are still the region's main inhabitants. They have preserved their traditional hunting and gathering lifestyle, resisting intrusions by supplying their Bantu-speaking neighbours with all the forest products they need.

Much of the Ituri Forest remains unexplored, but tourism has begun. Visitors, who approach the region from Kisangani or Goma, can sometimes catch a glimpse of such rare animals as an okapi or a mountain gorilla, as well as many more common species.

Map Zaire Ba

Ivanovo ***USSR*** A major textile and clothing city in the centre of a rich flax-producing area, 240 km (150 miles) north-east of Moscow. Ivanovo was formed as Ivanovo-Voznesensk in 1871 by joining two villages, and by 1905 it had more than 20 textile mills. That year 80 000 mill workers went on strike for more money and better conditions, and formed Russia's first workers' councils, or 'soviets'. Other products include machinery and chemicals.

Population 476 000
Map USSR Fc

Iviza ***Spain*** See IBIZA

Ivory Coast See p. 329

Ivrea ***Italy*** Town at the foot of the Alps about 50 km (30 miles) north of Turin. It is the headquarters of the Olivetti office machinery empire. The higher, old town has a 10th-century Romanesque cathedral and a cylindrical towered 14th-century castle.

Population 27 700
Map Italy Ab

Iwaki ***Japan*** City on the east coast of Honshu island, about 175 km (110 miles) north-east of the capital, Tokyo. Created in 1966 out of the former mining towns of the Joban coalfield, it is now a centre of the chemicals industry.

Population 350 600
Map Japan Dc

Iwo Jima ***Japan*** Largest of the Volcano Islands, covering 21 km² (8 sq miles) in the western Pacific about 1250 km (776 miles) south and slightly east of Tokyo. In 1945, towards the end of the Second World War, it was the scene of fierce and bloody fighting between US and Japanese forces. Some 22 000 Japanese troops were killed or captured, and the 21 000 US casualties included more than 4500 killed.

The US forces occupied Iwo Jima and, after the war, America administered the Volcano Islands. They were returned to Japan in 1968.

Map Pacific Ocean Ca

Ixtaccihuatl *Mexico* Spectacular snow-capped volcano 5286 m (17 342 ft) high which stands beside its twin, Popocatepetl, in the Anahuac valley south of MEXICO CITY. It is sometimes called 'the Sleeping Lady' because its outline resembles a reclining woman.

Izhevsk *USSR* See USTINOV

Izmir (Smyrna) *Turkey* Port on the Aegean coast about 335 km (210 miles) south-west of Istanbul. Founded by the Greeks in about 1000 BC, it was captured by the Ottoman Turks in 1424. Invading Greeks occupied the city in 1919. It was retaken by the Turks in September 1922, and a few days later was almost completely destroyed by fire. Most of the Greek population fled.

Izmir produces soap, dyes, tobacco, textiles, leather goods, olive oil, silk, cotton, carpets, wool, copper, pharmaceuticals, figs, sponges and raisins. The neighbouring province of the same name, which covers 12 825 km² (4952 sq miles), is rich in minerals including silver, mercury, iron, manganese and chromium.

Population (city) 1 489 800; (province) 2 315 800
Map Turkey Ab

Izmit (Kocaeli) ***Turkey*** Naval base on the Sea of Marmara about 90 km (55 miles) south-east of Istanbul. It is the capital of Kocaeli province, which covers 3986 km² (1539 sq miles) and grows tobacco and olives. Izmit's main industries are chemicals, cement and paper.

Population (town) 236 100; (province) 753 400
Map Turkey Aa

Izu ***Japan*** Peninsula on the south coast of Honshu island between Sagami and Suruga bays, in the south of the Fuji-Hakone-Izu National Park. Its rugged coastline and the hot springs at the town of Ito attract day-trippers from the capital, Tokyo, about 100 km (60 miles) to the north-east. Ito, on the west coast of the peninsula, was the home of Will Adams, an English seaman who was shipwrecked there in the early 17th century. He was the inspiration for the hero in James Clavell's novel *Shogun*.

Map Japan Cd

Izumo ***Japan*** Town near the north coast of western Honshu island about 115 km (70 miles) north and slightly east of the city of Hiroshima. It is the site of the country's oldest Shinto shrine, the Izumo Taisha, dating from the 1st century AD. In legend the Shinto gods gather there each October to discuss marriage arrangements for single people.

Population 80 700
Map Japan Bc

Jabal ad Duruz (Jebel al-Arab) ***Syria*** Mountainous district in the extreme south of the country near the border with Jordan. Its highest point is the extinct volcano of Jabal ad Duruz, 1735 m (5691 ft) high, the lava from which has formed a series of sheets, cones and caves. The west side of the district is partly fertile and produces cereals, but the east side is a barren wilderness, the refuge of bandits and outcasts.

Map Syria Bb

Jabal al Nusayriyah ***Syria*** Mountain range running parallel to the coast for some 113 km

(70 miles). The mountains rise from about 1200 m (3937 ft) to about 1500 m (4921 ft), with a high point of 1562 m (5125 ft). Dotted along the range there are a number of ruined Crusader forts.
Map Syria Bb

Jablonec ***Czechoslovakia*** See LIBEREC

Jachymov ***Czechoslovakia*** Mining town on the East German border west and slightly north of Prague. Silver mining began there in 1516. From 1519, the town – then called Sankt Joachimsthal – minted silver *Joachimsthalers*. These coins were known as *thalers* for short – the origin of the word 'dollar'.

The pioneering physicists Marie and Pierre Curie used pitchblende (uranium ore) from Jachymov for their investigations into radioactivity, and Marie (1867-1934) went on to isolate the elements radium and polonium from it after her husband's death in 1906. Production of the ore for radium began in 1908, and since 1949 the town has provided much of the Eastern Bloc's uranium. Jachymov is also a spa with radioactive water.
Map Czechoslovakia Aa

Jackson ***USA*** State capital of Mississippi, about 260 km (160 miles) north of New Orleans. It was founded as a French trading post, Le Fleur's Bluff, in the late 18th century, and is now an industrial, educational and administrative centre. Nearby is an 89 hectare (220 acre) hydraulic model of the Mississippi river basin used to study flood control.
Population (city) 208 800; (metropolitan area) 382 400
Map United States Hd

Jacksonville ***USA*** Port and major transport centre in north-east Florida, about 40 km (25 miles) south of the Georgia border. It stands on the St Johns river, 45 km (28 miles) from the Atlantic Ocean, and is an exporter of forest and agricultural products, and also has naval installations. Its manufactures include paper, chemicals and cigars.
Population (city) 578 000; (metropolitan area) 795 300
Map United States Jd

jade Either of two distinct minerals, JADEITE and NEPHRITE, used as gems or in carved ornaments. Jadeite is the more highly prized.

jadeite Green or white mineral, sodium aluminium silicate, found in IGNEOUS and METAMORPHIC rocks. It is a form of JADE.

Jadotville ***Zaire*** See LIKASI

Jaén ***Spain*** Town about 290 km (180 miles) south of Madrid, manufacturing chemicals, brandy and leather goods. Its province of the same name contains lead mines.
Population (town) 96 400; (province) 627 600
Map Spain Dd

Jæren ***Norway*** Peninsula covering 1700 km² (656 sq miles) in the extreme south-west of Norway, and often called 'a little bit of Denmark attached to Norway' because of its absence of hills. It has been farmed intensively since prehistoric times and now specialises in dairy products and vegetable cultivation.
Map Norway Bd

Jaffna ***Sri Lanka*** Port in the north-west, about 300 km (185 miles) north of Colombo. It lies on a peninsula less than 70 km (45 miles) from the Indian coast. Tamils from India captured Jaffna several times over 2000 years ago, and established a flourishing kingdom in the 13th century; and it has been the headquarters of the island's Tamil population ever since. The Portuguese took the port in 1617, but were ousted about 40 years later by the Dutch, one of whose forts is preserved as the city's police station. Jaffna's exports include fruits, cotton, tobacco and timber.

In normal times, Jaffna is also a popular tourist centre, with fine sandy beaches. But in the 1980s the peninsula became a base for Tamil guerrillas seeking an independent Tamil state. Provoked by terrorist acts, government forces bombed Jaffna's city-centre in May 1986.
Population 118 200
Map Sri Lanka Ba

Jaipur ***India*** City and capital of the north-west state of Rajasthan, about 240 km (150 miles) south-west of Delhi. It was founded in 1728 by Maharaja Jai Singh II, who ordered it to be painted in pink, the colour which his astrologers declared auspicious for him. It is one of the most beautiful cities in India.

One frontage of the royal palace is renowned for its many stone filigree windows from which it is known as the Palace of the Winds. It is the symbol of the modern state. Inside the city walls are numerous other magnificent buildings, including the finest of Maharajah Jai Singh's open-air observatories. There the heavens were charted on stone geometres (gnomons), some of which are up to 30 metres (98 ft) tall. Outside the walls are the spacious government headquarters and modern suburbs. Jaipur is noted for its carpets, jewellery and paintings, including some exquisite miniature work.
Population 1 015 200
Map India Bb

Jajce ***Yugoslavia*** Former Bosnian capital amid forested hills, 53 km (33 miles) south of the city of Banja Luka. Its chalets, Turkish-style houses and mosques, and campanile (bell tower) cluster around a hilltop fort, near where the Pliva river tumbles into the Vrbas by a spectacular waterfall. Yugoslavia's former Communist leader Marshal Tito (1892-1980) proclaimed Yugoslavia a federal, socialist republic at Jajce on November 29, 1943.
Population 9100
Map Yugoslavia Cb

Jak ***Hungary*** See SZOMBATHELY

Jakarta ***Indonesia*** National capital, in western Java, and the most heavily populated city in South-east Asia. It was the centre of the Dutch East India Company's trading empire from 1621, when it was known as Batavia. The old centre of the city is marked by typical Dutch architecture – bridges, pleasant squares and elegant public buildings. However, the city has grown beyond this central area into a sprawling conurbation surrounded by shanty towns. Tea and rubber are exported; there are textile mills, iron foundries and sawmills.
Population 6 503 000
Map Indonesia Cd

▼ IN THE PINK The Palace of the Winds (Hawa Mahal) in the pink-painted city of Jaipur acquires a rosy magic at sunset each day.

Jamaica

VIOLENCE AND POVERTY LIE BEHIND THIS CARIBBEAN ISLAND'S IMAGE AS A LUSH AND PEACEFUL PLAYGROUND FOR TOURISTS

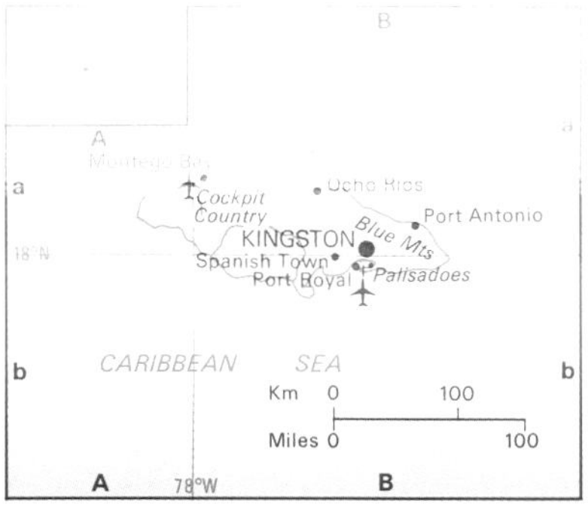

Nearly a million tourists a year visit Jamaica. They sip rum punches in the blazing Caribbean sun, ride bamboo rafts down lazy rivers, and dance through balmy nights to the throbbing beat of reggae discos. Along the way, they spend some US$400 million a year.

These prosperous visitors stay in luxurious hotel complexes near palm-fringed beaches and find it easy to imagine that Jamaica is an island of peace and lotus-eating plenty. But they see only one face of this turbulent island.

Behind the façade of the beaches is Jamaica's less happy face: one lined by hard work, and lean from the struggle to survive. Among the hills that rise in the centre of the island from the mostly narrow coastal flatlands, farmers' wives rise before dawn to walk – sometimes for hours – to reach the nearest road where they can get a bus to market. Maids in the plush hotels of MONTEGO BAY, OCHO RIOS, PORT ANTONIO and Negril earn in a week about the same as the price of a glass of whisky from

[illegible] too expensive because Jamaica has no cheap supplies of energy to power the smelters. Production of bauxite and alumina dropped by nearly 40 per cent in the early 1980s, but has been rising since 1984 when the Jamaican Government negotiated agreements with Colombia, the USA, the USSR and Yugoslavia.

Under Prime Minister Edward Seaga, leader of the conservative Jamaica Labour Party which came to power in 1980, the government has encouraged foreign investment and agricultural diversification. The aim is to raise the Jamaicans' standard of living and diminish the country's massive foreign debts. In 1984, interest payments alone on these debts amounted to 40 per cent of Jamaica's total export revenues. In 1985 the country was one of the largest recipients of aid from the USA.

New crops, such as winter vegetables, flowers, fruit and honey, are being developed for export alongside the more traditional products: sugar and bananas, molasses and rum, peppers and ginger, and cocoa and coffee. The bulk of the coffee harvest is the type known as Blue Mountain coffee, rated by some gourmets as the world's best, which is grown in the hills of the same name at the eastern end of the island.

There are less orthodox crops, too. Cannabis – known locally as ganja – is widely, though illegally, grown in the inland hills and hawked more or less openly in bars and on beaches. In addition, an estimated 1900 tonnes of it is smuggled into the USA each year,

[illegible] inhabitants, the Arawak Indians, who called it Xaymaca, meaning 'land of wood and water'. Christopher Columbus, who in 1494 claimed the island for Spain (it remained Spanish until the British captured it in 1655, and remained British until independence in 1962), described it simply as 'the fairest isle that eyes have beheld'.

JAMAICA AT A GLANCE

Area 10 991 km² (4244 sq miles)

Population 2 470 000

Capital Kingston

Government Parliamentary monarchy

Currency Jamaican dollar = 100 cents

Language English

Religions Christian (18% Baptist, 17% Church of God, 15% Anglican, 8% Roman Catholic, 6 % Seventh Day Adventist, 6% Methodist, 5% Presbyterian), Rastafarian

Climate Tropical at sea level. Average year-round temperature in Kingston ranges from 19°C (66°F) to 32°C (90°F)

Main primary products Sugar, bananas, coffee; bauxite

Major industries Mining, agriculture, tourism, sugar refining, molasses, rum, tobacco products, cement, petroleum refining

Main exports Bauxite, alumina, sugar, chemicals, petroleum products, bananas, other fruit and vegetables

Annual income per head (US$) 1160

Population growth (per thous/yr) 16

Life expectancy (yrs) Male 69 **Female** 73

Jalalabad (Djelalabad) ***Afghanistan*** Town on the Kabul river, 120 km (74 miles) east of Kabul, commanding the head of the KHYBER PASS. It is the capital of the country's most densely populated province, Nangarhar. Founded in 1560, the town trades in fruit, sugar and timber. Bitter fighting took place there in the 19th-century Afghan Wars.

Population 57 800

Map Afghanistan Ba

Jalapa (Jalapa Enríquez) ***Mexico*** Capital of VERACRUZ state, in the Sierra Madre Oriental mountains, 130 km (80 miles) from the Gulf of Mexico. It is a typical colonial town, with cobbled streets, squares and gardens, and has earned a reputation for its interest in ballet and the performing arts. There is a university (opened in 1944). The main occupations are sugar refining, flour milling and coffee roasting. Jalapa gave its name to a purgative produced there from the roots of a climbing plant.

Population 212 800

Map Mexico Cc

Jalisco ***Mexico*** Principal state in western Mexico, whose capital is GUADALAJARA. It is the heartland of Spanish culture in Mexico and is renowned for its conservatism and traditional ways. It is commercially busy – with silver and copper mining in the mountains inland, and cereals, coffee, tobacco and cotton grown in the valleys and on the coast. Cattle are reared, and Jalisco is also the land of the blue maguey cactus – the source of tequila and mescal, Mexico's best known alcoholic drinks. The state has a number of beaches, such as that at Puerto Vallarta.

Population 4 600 000

Map Mexico Bb

James Bay ***Canada*** Shallow southern arm of HUDSON BAY which was discovered by Henry Hudson in 1610. It was here in 1611 that Hudson's crew mutinied and cast him adrift along with his son John and seven loyal men. They had no food, and were never heard of again. The bay was named after Thomas James, another searcher for the NORTHWEST PASSAGE, who explored the area in 1631 and 1633.

Dimensions 440 km (273 miles) long, 217 km (135 miles) wide

Map Canada Gc

Jamestown ***USA*** Ruins of the first permanent English settlement in the USA, beside the James river in Virginia, about 185 km (115 miles) south of the national capital, Washington. The village, founded on May 13, 1607, lies in the Colonial National Historical Park.

Map United States Kc

Jammu and Kashmir ***India*** North-western, former princely, state bordering on Chinese Tibet. The majority of the population is Muslim, but in 1947 the ruling Hindu maharajah refused to accede the state to either India or Pakistan. War between the countries resulted, followed in 1949 by a ceasefire agreement dividing the area between Pakistani AZAD KASHMIR and Indian Kashmir. There was further fighting in 1965 and 1971, and from the mid-1950s China occupied a portion on its borders.

India claims all 222 236 km² (85 783 sq miles) of the state but occupies only about half of this. SRINAGAR, popular with tourists, is its summer capital and Jammu (population 164 200) the winter capital. The main products are wool, fruit, maize, wheat and rice.

Population (Indian Kashmir) 5 987 400
Map India Ba

Jamshedpur ***India*** Steel city in the north-east, about 210 km (130 miles) west of Calcutta. It developed after its first steel plant was founded in 1907-11 by Jamshedi Tata, a Parsi from Bombay who was the first private businessman to build such a plant in India. At that time the site – rich in iron ore – was just a remote jungle valley. Today Jamshedpur – the centre of a backward tribal area – is a hive of blast furnaces, coke ovens, machine factories, metal works, and steel and iron concerns.

Population 669 600
Map India Dc

Jamuna ***Bangladesh*** River forming the main stream of the BRAHMAPUTRA from the Tista to the Ganges. The Ganges and Jamuna together form the Padma; this flows into the Meghna river.

Map Bangladesh Bb

Jan Mayen ***Norway*** Island in the Arctic Ocean, 540 km (335 miles) north-east of Iceland, annexed by Norway in 1929. The island, which covers 380 km² (146 sq miles), is dominated by Beerenberg, an extinct volcano 2277 m (7470 ft) high. It is the site of a weather station, whose operators are the only inhabitants.

Map Arctic Ab

Janakpur ***Nepal*** Lowland town and religious centre in the south-east of the country, 10 km (6 miles) north of the Indian border. Janakpur is centred around (and named after) the Hindu Janaki temple and contains many tanks used for ritual bathing. The town is visited by Hindu pilgrims from all over India as well as Nepal.

Population 143 000
Map Nepal Ba

Japan See p. 334

Japan Alps ***Japan*** See CHUBU SANGAKU

Japan, Sea of Arm of the PACIFIC OCEAN, between Japan, Korea and the USSR, which contains important fishing grounds, where anchovy, herring and mackerel are caught. The Tsushima branch of the warm Japanese Current flows northwards through the sea, while a cold countercurrent brings ice to the mainland shore.

Area 1 008 000 km² (389 200 sq miles)
Greatest depth 4000 m (13 123 ft)
Map Japan Bb

Jarash (Jerash) ***Jordan*** Town containing well preserved remains of a provincial Roman settlement, situated 38 km (24 miles) north of Amman. Jarash has paved Roman streets lined with hundreds of carved columns. There are ruts in the streets made by Roman chariots, and under the town the 2000-year-old drainage system still works. There are also theatres, temples, plazas, baths and public buildings and a triumphal arch commemorating a visit by the Emperor Hadrian in AD 129. In the Hippodrome there are stone polo goal posts erected after the Persian invasion of 614. The city – which was founded in 322 BC by soldiers of Alexander the Great – was abandoned in AD 747 after a series of earthquakes.

The modern town dates from the late 19th and early 20th centuries.

Population 29 000
Map Jordan Aa

Jaroslaw (Jaroslau; Yaroslav) ***Poland*** Agricultural town in the south-east, about 30 km (18 miles) from the USSR border. In the Middle Ages, it was a flourishing marketplace at the crossroads of the Rhineland-Russia and Baltic-Black Sea trade routes. The town still has Renaissance and 18th-century Baroque buildings, including some fine merchants' houses.

Population 38 600
Map Poland Ec

Jarrow ***United Kingdom*** See NEWCASTLE UPON TYNE

Jars, Plain of ***Laos*** Central part of the TRAN NINH plateau, characterised by narrow river valleys and limestone and sandstone hills rising to about 2000 m (6560 ft) above sea level. The region derives its name from the hundreds of prehistoric stone funerary jars found there in the 19th century by the French. It was the scene of fierce fighting between government and Communist Pathet Lao troops during the civil war of the early 1960s. The plain is of little economic or agricultural value.

Java (Jawa) ***Indonesia*** Island covering 130 987 km² (50 574 sq miles) and containing some 60 per cent of the republic's people. Rice is grown on its fertile volcanic soils. It has 112 volcanoes. Many of them are active, including the Semeru, the highest at 3675 m (12 057 ft). The island is home to the national capital, Jakarta. Apart from agriculture, there are textile mills, car factories and smaller manufactures.

Population 91 269 600
Map Indonesia Cd

Java Sea Area of the PACIFIC OCEAN lying between Java and Borneo, with Sumatra in the west and the FLORES SEA in the east. In the Battle of the Java Sea on February 27, 1942, the Japanese defeated an Allied fleet and opened the way for their invasion of Java.

Area 310 800 km² (120 000 sq miles)
Map Indonesia Cc

Java Trench Deep trench in the INDIAN OCEAN off southern Indonesia. Here the Indo-Australian continental plate is sliding under the Eurasian plate (see PLATE TECTONICS). As it descends, earthquakes occur and the lower plate melts, partly due to friction, creating molten magma to fuel Indonesia's many volcanoes. The 1883 volcanic explosion at KRAKATAU (between Java and Sumatra) was the most momentous volcanic catastrophe in modern times.

Greatest depth 7450 m (24 440 ft)
Map Indonesia Bd

Jaworzno ***Poland*** Industrial town about 25 km (15 miles) east of the city of Katowice. It is one of the country's largest coal mining and electricity generating centres, and also produces lead and zinc.

Population 93 900
Map Poland Cc

Jayapura ***Indonesia*** Town in Irian Jaya province situated on the north coast of New Guinea. It is the administrative centre of the province and was formerly known as Sukarnapura, Kota Baru and Hollandia.

Population 45 800
Map Indonesia Jc

Jbail ***Lebanon*** Historic seaport lying 29 km (18 miles) north along the coast from the capital, Beirut. Site of the Biblical Gebal, the place is claimed to be the oldest continuously inhabited town in the world. The ancient Greeks called it Byblos, and papyrus (used to make a kind of paper) was given its early Greek name *byblos* because it was exported from Egypt through the port to the Aegean. As a result, the term *byblos* became synonymous with book and the word 'Bible' was derived from it.

The massive walls of the ancient city date from 2800 BC and nearby traces of civilisation go back 9000 years – including late Stone Age and Bronze Age remains. A necropolis dating from about 3500 BC contains some 1450 burial jars, and excavations have uncovered the remains of a Greek temple and citadel. The modern town of Jbail has suffered considerable damage during the recent civil war.

Population 1200
Map Lebanon Aa

Jeddah (Jidda) ***Saudi Arabia*** Major port on the Red Sea, situated 70 km (44 miles) west of Mecca. It is the main arrival place for pilgrims to Mecca, and its deep-water port has more than 40 berths. In addition, its international airport with its impressive, ultra-modern terminal buildings, can handle and accommodate thousands of pilgrims a day. Jeddah's shore is lined with high-rise office blocks, and a new town and suburbs are being built. Among the most notable constructions is the spectacular Corniche, lined with sculptures by internationally acclaimed artists.

A walled city some 300 years old, Jeddah has some of its old town still intact – including coral-fronted houses with wooden balconies, some well-preserved merchants' houses with elaborate carvings, fretwork and moulding (for example, Beit Nassif), and the House of the Tree, built in the 1850s, which took its name from a huge tree alongside that was a local landmark. It was then the city's tallest building and King Abdul-Aziz, the first King of Saudi Arabia (1932-53), used to sit on the roof and enjoy the panoramic views of the city. It is now a folk museum.

Population 750 000
Map Saudi Arabia Ab

Japan

THE SUN OF ECONOMIC PROSPERITY HAS RISEN WITH A NEW BRILLIANCE FOR ONE OF THE MOST ADAPTABLE NATIONS ON EARTH

Strung across the eastern margin of Asia, the island nation of Japan has performed an economic miracle. In little over a century, an isolated feudal backwater with few natural resources has been transformed into a dynamic industrial superstate with a growth rate of around 5 per cent a year. Japan dominates the North Pacific area, the world's most vigorously expanding economic region. Its influence is felt worldwide, its modern and sophisticated products being prized in markets everywhere.

A plain recital of the statistics indicates the dynamism of modern Japan. With less than 3 per cent of the world's population, Japan enjoys nearly 10 per cent of the world's income as measured by gross domestic product. It leads the world in both shipbuilding and car making, and runs a close second to the USSR in the production of crude steel. Every third television set in the world is made in Japan. The Japanese dominate world markets in audio, video and photographic equipment and many uses of microelectronics, are at the forefront as the world moves into the Information Age, and lead the field in the development of industrial robots.

Small wonder that in 1984 the nations of the European Economic Community were US$10 000 million and the USA a staggering US$33 000 million in the red in their trade with Japan.

In 1945 Japan became the first nation to experience a nuclear holocaust when atomic bombs were dropped on HIROSHIMA and NAGASAKI. Emperor Hirohito told his people that the war had developed 'not necessarily to our advantage'. Japan's economy lay in ruins, her people malnourished, demoralised and suffering for the first time in their history the humiliation of occupation by a foreign power. The Emperor, once regarded as a god, renounced divinity and assumed the role of a constitutional monarch.

Yet today the Japanese can be proud of an economy that is the second largest in the free world, and is still expanding at rates few other industrial nations dream of achieving. The rate of unemployment, at less than 3 per cent, is almost unrivalled and income per head (US$9684) is higher – and more equitably distributed – than in many European countries.

Dismissed by many as a mere imitator, Japan now leads the world in many areas of new technology. Furthermore, the price paid for economic prosperity has been small. The people are healthy; no major nation can boast a lower infant mortality rate or a greater life expectancy. They are united; few social inequalities can divide a nation in which more than 90 per cent of people regard themselves as middle class and where opportunities for advancement are open to nearly everyone, thanks to a uniform (but fiercely competitive) education system that has given the Japanese a literacy rate of 99 per cent.

The price of individual failure can be heavy – suicide is not infrequent, especially among young people. Yet strikes are rare (and largely ritual affairs), cities have few slums although homes are cramped.

▲ MOUNTAIN-GIRT SEA The fine gateway to a Shinto shrine stands on the shore of Miyajima (Itsukushima), an island in the Inland Sea near the city of Hiroshima. The island is popular with tourists and has many ancient temples.

Most remarkable of all, however, the Japanese have sacrificed little of their cultural identity, which they cherish above all things. Deeply patriotic by nature, they derive great satisfaction from their economic achievements. Nevertheless, [illegible] their future [illegible] with Europe and America reveal how vulnerable Japan can be to external forces. This vulnerability is compounded by severe environmental pressure at home arising from Japan's physical geography.

LIFE ON THE PLATE EDGE

The Japanese archipelago owes its existence to the cataclysmic forces that are unleashed wherever the giant plates making up the earth's crust move against each other. Six chains of steep, serrated mountains form the broken skeleton of the country. Studded with volcanoes (more than 60 of them active), bathed in mists and steam, draped in delicate forest greens, the mountains endow Japan with a rare scenic beauty, but little else.

There are few workable deposits of minerals or fuels, and the silts washed down by tumbling mountain streams have rarely created more than a narrow and fragmented coastal plain. Less than a quarter of the land is flat enough for agriculture or human settlement, and in the south spectacular flights of mountain terraces are still cultivated.

Much of the land upon which the modern cities and factories stand has been painstakingly reclaimed from the sea, at the risk of damage from storm and *tsunami* – giant waves resulting from earthquake action. Earthquakes are frequent and widespread, and though most are mild the threat of catastrophe is ever present. TOKYO, the capital and largest city, was severely damaged by an earthquake in 1923, and a repeat is forecast before the end of this century. Heavy rains and hot summers allow rice to be grown on most of the lowlands, although chill breezes can damage crops in the north. The north-eastern 'snow country' is heavily blanketed in winter, and typhoons (hurricanes) in summer and early autumn can devastate town and countryside anywhere.

Within this naturally hazardous environment live more than 120 million people; and although overall population density is not quite the highest in the world, the ratio of people to habitable land certainly is. Before the modernisation of industry most of them worked on the land, and as recently as the mid-1950s many still lived in rural areas. However, the rapid growth of industry and commerce after the Second World War generated an enormous drift away from the country, and now 75 per cent of the people live in cities.

Nearly a quarter of the population has now settled within 50 km (30 miles) of central Tokyo, and even the more distant provincial towns have grown rapidly. City centres have been transformed by modern department stores, hotels, shops and offices, but the suburbs beyond can be chaotic. Few streets bear names, and a custom of numbering houses according to when they were built rather than location adds to a visitor's confusion.

Japan's modern history could be said to

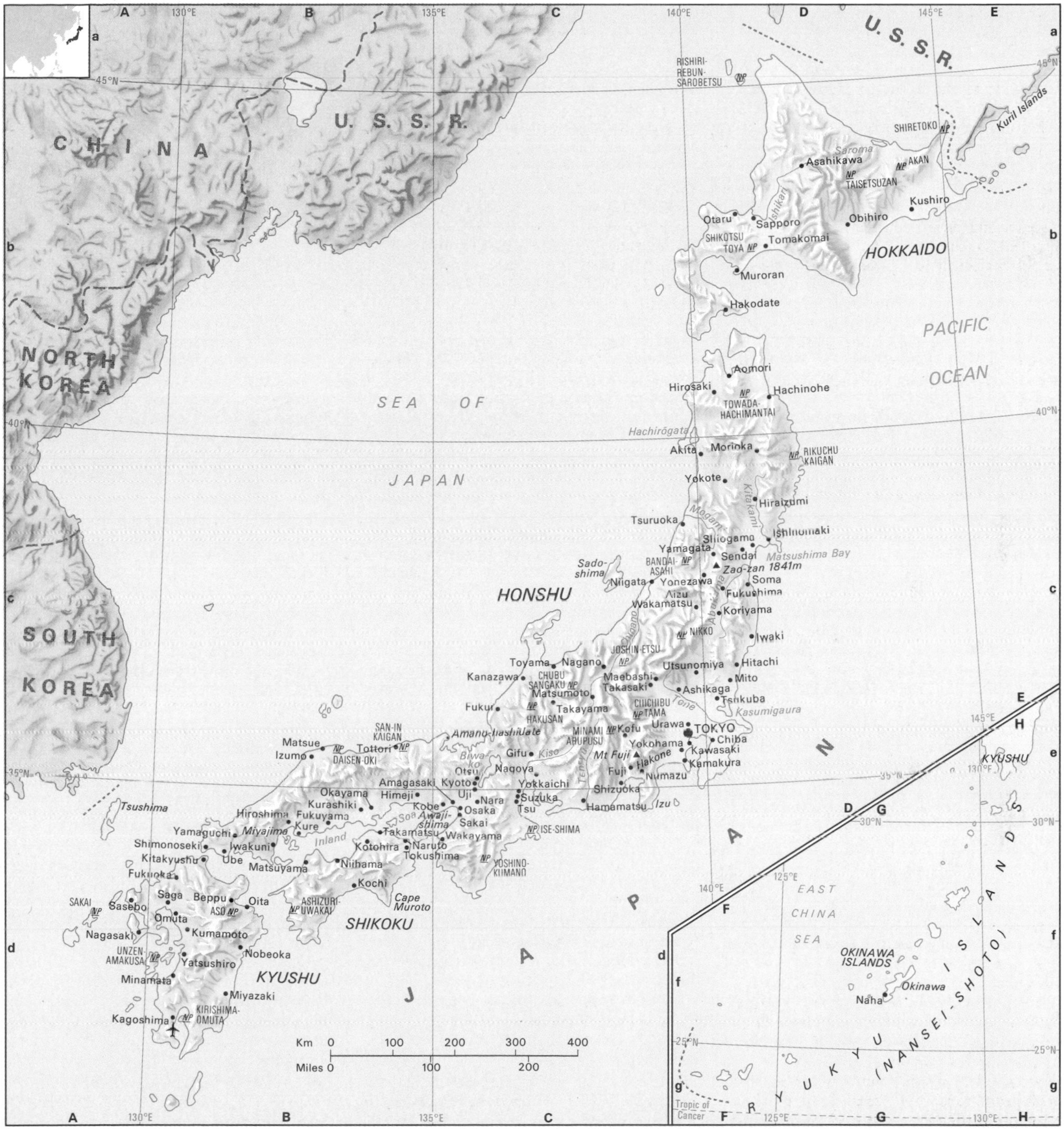

have started in the late 16th century, when a long civil war exposed the people to disturbing foreign influences. The victorious Tokugawa shoguns (military dictators) set out to check the spread of alien ideas, and of Christianity in particular. Almost any form of contact with the outside world was forbidden, Christianity was suppressed and the country remained in virtual seclusion for over two centuries.

Thus isolated, the Japanese had time to perfect many elements of their unique culture. By the middle of the 19th century, however, when the arrival of an American fleet under Commodore Matthew Perry finally awoke Japan to the renewed threat from without, it was technologically backward and quite unable to defend itself. Only a concerted effort to emulate the foreigners offered any hope of escape from eventual domination by them.

At first Japan was compelled to sign treaties with the foreign powers on disadvantageous terms, but this bought time in which to modernise the country. In 1868 the last shogun was overthrown, an imperial government was established and a crash programme of modernisation began.

Foreign experts advised on all aspects of government, economy and defence, and bright young men were sent overseas to learn all they could from the West (as they still do). Educational reforms led to a rapid spread of literacy and fostered a deepened sense of nationalism based upon reverence for the emperor and respect for ancient Confucian ideals. Despite having few natural resources, Japan was soon powerful enough to defeat both China (1894–5) and Russia (1904–5) and began to build an overseas empire.

By the 1920s Japan had gained considerable international prestige and influence, but the great depression of the 1930s soon brought devastation to the economy. Foreign trade slumped with the collapse of vital overseas markets, and eventually the government fell under the influence of military right-wing extremists. A Japanese invasion started a war with China in 1937, and in December 1941 attacks upon Pearl Harbor, in Hawaii, and European colonial outposts in the Far East signalled Japan's entry into the Second World War on the side of Hitler's Germany.

Brilliant initial successes led to dreams of a 'Great East Asian Coprosperity Sphere' – dreams rudely shattered when Japan lost control of the sea lanes vital for the import of raw materials. Despite fierce (and often suicidal) resistance, defeat was inevitable, and was only hastened by the atomic bomb. By August 1945 the goal of 'Rich country, strong army' had brought mass destruction to the homeland.

Japan remained occupied until 1952 – long enough to establish another foreign import: democracy. Then, militarism having been discredited, it emerged with a new goal – 'Rich country, no army'. By the mid-1950s, Japan had not only recovered, but was achieving phenomenal rates of economic growth, averaging more than 10 per cent a year during the 1960s, and around 5 per cent since then. With a unique provision in its constitution forever renouncing war as an instrument of policy, Japan has kept defence spending to a minimum, preferring instead to remain on friendly terms with other nations. Its small 'Self-Defence Forces' are backed up by substantial American military bases.

COMPANY LOYALTY

Japan's remarkable economic performance can be attributed to many factors, first among which is the quality of its workforce – more reliable, conscientious and better educated than those of many competitors. Although education is compulsory only from 6 to 15 years of age, more than 90 per cent of children finish high school to sixth-form standard. A third of them go to university, where many obtain degrees in subjects immediately relevant to industry's needs. Training continues throughout a worker's career, and university graduates destined to manage in industry must first perform even the most basic jobs.

Workers are given responsibility for the quality of their work, and take pride in what they do. In larger firms, a guarantee of lifetime employment offers job security in exchange for total commitment to the company. When changing circumstances call for workers to move, they are transferred from one place or job to another with little friction. Company union representatives are told about management priorities, and cooperate to avoid wasteful disputes.

Each factory or office is built up from small 'sections', which act both as production units and as a focus of social activity – a vital function in a culture where 'belonging' is important. Loyalty within sections extends outwards to the company as a whole, and finds expression in collective holidays and excursions – even in singing the company anthem.

Investment in new industrial technology has been helped by high levels of personal savings, low company dividends and a willingness by all to consider long-term gains at the expense of short-term performance. The influential Ministry of International Trade and Industry guides industry on the best technologies to adopt, while encouraging less successful firms to reduce capacity or move up market.

Until recently Japanese manufacturers looking for a better product relied heavily upon improving imported technology. Now they give fundamental research a higher priority.

Bereft of most industrial raw materials, Japan has to import oil, coking coal and metallic ores, and so benefited greatly from low commodity prices during the 1960s. Economic growth halted briefly in 1974, following the rise in world oil prices, but Japan adapted well to an era of expensive energy. Energy efficiency was improved and sources of supply diversified to include nuclear power. Emphasis shifted from heavy industries such as steel and shipbuilding to new high-technology sectors, in which the quality of its workforce puts Japan at an advantage. The switch also brought improved processes to older industries, leading to a cut in environmental pollution and a productivity boost that has helped to protect these industries from Third World competitors.

Manufacturers have also benefited from the size and sophistication of the home market. Japanese demand quality, and this in turn helps to ensure high standards in exports. Although exports represent only about one-tenth of the Gross National Product – lower than for comparable European countries – they are concentrated in rapid-growth sectors. A vital ingredient of Japan's success therefore has been the openness of world markets.

Equally, Japan stands to lose much from protectionism, especially in Europe and America. To forestall this, Japanese companies have been investing heavily in overseas factories. This also lessened the impact of the yen's steep rise in the mid-1960s (thus increasing Japanese export prices), which was only partly offset by falling oil costs.

Concern over trade has enhanced the importance of diplomacy, and Japan is fortunate that its civil service standards are very high. Equally, it benefits from a stable and democratic political regime, the ruling Liberal Democratic Party having remained in office for over three decades.

COUNTRYSIDE AND TOWN

Politically the government's most solid support lies in rural areas; despite depopulation they exert great influence. Fewer than 5 million households remain on the land, and two-thirds of these earn more from part-time work in factories and offices than from the fields. The average farm of 1.4 hectares (3.5 acres) may be uneconomic, but subsidies that raise rice prices to several times the world level (a reward in part for political loyalty) help to ensure rural prosperity, besides causing an annual rice surplus. (Japanese rice yields per hectare are among the highest in the world.)

Actually the Japanese eat less rice every year, as incomes rise and preferences shift to alternative foods. The government keeps urging farmers to rear animals and grow fruit and vegetables, but nearly half of all farmland is still devoted to rice. The result is that two-thirds of the nation's maize, wheat and other grain foods has to be imported. Meat is also a major import.

Japan's agriculture is highly advanced. Extensive use is made of fertilisers and miniature farm machinery designed for small fields. On higher ground vegetables and fruits are grown in wonderful variety and abundance, including apples, oranges, cabbages, potatoes, persimmons, loquats, giant radishes and burdock roots.

Fishing communities along the coasts have always given Japan most of its protein foods. While larger ports send boats thousands of kilometres across open sea, others depend upon local catches using traditional methods. There are fish farms wherever shallow and sheltered waters permit, producing oysters and edible seaweeds as well as fish. Japan remained a whaling nation until 1985.

In the city, a day is as likely to start with toast and bacon as with a bowl of rice and fish. White-collar workers – known as 'salarymen' – don business suits and cram into crowded suburban trains for the long and uncomfortable journey into town. Offices are large and open-plan. New technology has speeded the basic office processes somewhat – in particular, electronic word processors and facsimile machines have eased the problems of handling Japanese writing, with its thousands of characters.

For the factory hand, conditions relate closely to the size of the company. Workers in large, automated assembly plants often spend the day tending robots that do all the hard labour. Those not so fortunate, working for tiny subcontractors, turn out components in cramped workshops, earning less and enjoying much less job security. Like other industrial nations, Japan has a rapidly expanding service sector.

During the hectic 1960s, an ideal husband spent little time at home, his devotion to work being complete; after long hours at the office he would be expected to spend the evening drinking with his boss and other colleagues. Changing attitudes have made it more acceptable to spend time at home with wife and children. At the same time, despite social customs that discourage working wives, many married women now have part-time jobs.

Neverthelesss, the lack of career opportunities leads many mothers to devote themselves instead to their children's education. These 'education mamas' can be merciless in their efforts to ensure that vital examinations are passed, for doing well remains virtually the only passport to a successful future. Many children also go to evening 'cramming schools', leaving little time for relaxation and play. Failure to pass an entrance examination may bring disgrace upon the child's family.

A typical home is small by European or

American standards, but as incomes rise so larger houses and apartments become the norm. Thatched roofs are still found in remote areas, but in the cities most new homes are in tall apartment buildings known confusingly as 'mansions' or in comfortable estates of two-storey housing. Land is expensive, so gardens are tiny if well manicured.

Nevertheless, at least one room is decorated in traditional style, with reed *tatami* mats on the floor and a small alcove where a favourite painting or calligraphy scroll can be hung. This room keeps the household in daily contact with its cultural roots. The typical Japanese is both a Shintoist and a Buddhist; an artist or intellectual may be a Christian too. Few Japanese are ardently religious, however, though there are exceptions in aggressive new sects such as the Buddhist *Sokkagakkai* that have gained strength since the war. But joining in semireligious festivals is popular and widespread. Colourful shrines and temples are kept in perfect order, and are open to visitors, the beautiful and mysterious gardens of the austere *Zen* sect being a particular attraction.

AFTER A DAY'S WORK

What spare time is available is spent on a rich variety of leisure pursuits. An office worker returning home enjoys a quick, scalding bath, dines, then perhaps spends the evening playing with some new consumer gadget – likely these days to be an electronic keyboard or a microcomputer. Television may be showing baseball or one of the six annual *sumo* wrestling tournaments; or perhaps one of many programmes designed to improve the viewer's golf. The Japanese are great golfers, but only the rich and powerful ever see much of a real golf course – a place where much of the nation's business is said to be transacted. For most, a multistorey driving range must suffice. Traditional arts, such as calligraphy or the tranquil tea ceremony, are also popular.

Instead of returning home, a 'salaryman' may spend an entire evening in town with colleagues, exploring the 'pleasure quarter', a noisy and garish maze of narrow side streets lined with bars, burlesques and tiny restaurants. Here it is possible to let off a little steam, secure in the knowledge that whatever happens all will be forgiven (perhaps even forgotten) in the morning. But a visitor must take care, for prices are tailored to the expense account, not the product.

In the main shopping arcades, nestled between stylish boutiques and expensive department stores, coffee bars with exotic Western names offer relaxation for a tired office girl; a chance for the student to learn the fate of his favourite comic-book hero, or a rendezvous for young lovers – important now that fewer marriages are arranged.

There are some, of course, who reject this typical lifestyle. There has been a worrying increase in juvenile crime in recent years. Others suffer from general discrimination, such as the 1.2 million descendants of feudal outcasts, who face many social and economic barriers – especially in marriage. Similar problems confront a 700 000-strong Korean minority, many of them forcibly removed to Japan during the war. Nevertheless, for most people life is comfortable and fulfilling.

▲ SHORT BREAK City-dwellers relax beside a bathing pool in Tokyo. The average Japanese worker takes only 101 days off a year, including weekends, compared with a European's 140.

JAPAN AT A GLANCE

Area 372 313 km² (143 750 sq miles)

Population 121 400 000

Capital Tokyo

Government Parliamentary monarchy

Currency Yen

Language Japanese

Religions Shinto (79%), Buddhist (81%), Christian (1%) (a majority is both Shinto and Buddhist)

Climate Temperate, monsoonal; average temperature in Tokyo ranges from −2 to 8°C (28-46°F) in January to 22-30°C (72-86°F) in August

Main primary products Rice, potatoes, sweet potatoes, tea, fruit, vegetables, edible seaweeds, timber, poultry, fish and shellfish; coal, sulphur

Major industries Iron and steel, nonferrous metals, ships, cars, motorcycles, electrical and electronic equipment, textiles, oil refining, petrochemicals, fertilisers, cement, optical goods, watches, fishing, food processing, forestry

Main exports Motor vehicles, machinery, electrical and electronic goods, iron and steel, chemicals, instruments, textiles, ships, metal goods

Annual income per head (US$) 9684

Population growth (per thous/yr) 6

Life expectancy (yrs) Male 74 Female 80

Jefferson City ***USA*** State capital of Missouri, on the Missouri river about 170 km (105 miles) west of St Louis. It is a tourist centre for the Lake of the Ozarks to the south-west.
Population 35 000
Map United States Hc

Jejudo ***South Korea*** See CHEJU DO

Jena ***East Germany*** Town about 80 km (50 miles) south-west of Leipzig. The French emperor Napoleon I decisively defeated Prussia there in 1806. The university, founded in 1557, had on its staff the philosopher Georg Hegel (1770-1831) and the poet and dramatist Johann Schiller (1759-1805), and among its students was the 19th-century political theorist Karl Marx. In 1846, the optician Carl Zeiss opened a workshop in the town, and started what became Europe's biggest optical instruments industry. Precision instruments, optical products, cameras, chemicals and pharmaceuticals continue the tradition today.
Population 107 700
Map East Germany Bc

Jenolan Caves ***Australia*** Extensive limestone cave system in the Great Dividing Range 115 km (72 miles) west of Sydney. The caves, which are open to visitors, have been known to Europeans since the 1840s when a settler found them while hunting a bushranger who used them as a hideout.
Map Australia Ie

Jeonju ***South Korea*** See CH'ONGJU

Jerba ***Tunisia*** Island in the Gulf of Gabès, 67 km (42 miles) south of Sfax and connected to the mainland by a causeway. It is said to be the mythical land of the Lotus Eaters visited by Ulysses. It has an airport and is a resort area with fine beaches, good hotels, colourful and varied souks (markets) and Roman remains, including two forts and a triumphal arch. Its soil is fertile, and dates, olives and oranges are grown. Apart from tourism and farming, sponge fishing, textile manufacture and pottery are the other main occupations.
Population 92 300
Map Tunisia Ba

Jerez de la Frontera ***Spain*** Town just north of Cádiz, on the south-west Atlantic coast, from which the name sherry derives. It produces and stores the wine and brandies in numerous *bodegas* (cellars), and makes bottles and casks.
Population 176 200
Map Spain Bd

Jericho ***Israeli-occupied Jordan*** Oasis town in the WEST BANK area, situated 22 km (14 miles) north-east of Jerusalem. Set 251 m (825 ft) below sea level, it is the site of one of the world's oldest cities, dating back to about 7000 BC.
Population 15 000
Map Jordan Ab

Jersey ***English Channel*** See CHANNEL ISLANDS

Jerusalem (Yerushalayim) ***Israel/Israeli-occupied Jordan*** Capital of Jerusalem district, and capital of Israel since 1950 (though not accepted in international law), lying about 30 km (20 miles) west of the northern end of the Dead Sea. A holy city for Jews, Christians and Muslims, it has been entirely controlled by Israel since the Six-Day War of 1967 (before which Jordan held the Old City). Its Arabic name is El Quds Esh Sherif.

Conquered from the Philistines by David in 1000 BC, Jerusalem is the site of Solomon's First Temple, destroyed by the Babylonian King Nebuchadnezzar in about 586 BC, and of the Second Temple, destroyed by the Roman Emperor Titus in AD 70. The city was captured by the Arabs, who built the Dome of the Rock (Mosque of Omar) on the site of the temple in AD 632. Jerusalem was taken by the Crusaders in 1099, then recaptured by the Muslim leader Saladin in 1187. The existing wall of the Old City was built by the Ottoman Sultan Suleiman the Magnificent in the 16th century.

Inside the Old City is the Western (Wailing) Wall, the only remnant of the Second Temple; the El-Aqsa Mosque; the Via Dolorosa (Sorrowful Way, or Way of the Cross); and the Franciscan Church of the Holy Sepulchre, built on the spot where Christ was buried. The Old City includes Jewish, Christian, Muslim and Armenian quarters, and a colourful market.

Just outside the Old City wall, to the south-west, is Mount Zion, with the traditional tomb of David and the Room of the Last Supper. To the north-west is Mea Shearim (the ultra-orthodox Jewish quarter), and the tombs of the Kings of Judah, where King David is now believed to be buried. The Jewish quarter has been almost entirely rebuilt since 1967 and large new housing complexes surround the city.

Outside the Old City wall to the east is the Mount of Olives, with its ancient Jewish cemetery, and the site of the Garden of Gethsemane, where Judas betrayed Jesus.

At En Karem, on the outskirts of Jerusalem, are John the Baptist's birthplace and the new Hadassah Medical Centre, remarkable for the stained-glass windows of its synagogue. The windows were designed by the Russian-born Jewish painter Marc Chagall (1887-1985).

Among the city's many museums are the Israel Museum, the National Museum with the Shrine of the Book, which houses the Dead Sea Scrolls, and Yad Vashem, the memorial to Jewish victims of the Nazi Holocaust. Public buildings in the modern part of the city include the Knesset (Parliament), the President's residence, the Hebrew University and the Jewish Agency (World Zionist Organisation) complex.
Population 446 500
Map Israel Bb

Jessore ***Bangladesh*** Chief town of Jessore district, which covers 6573 km² (2538 sq miles) in the west. It trades in jute, rice, tamarind, tobacco and sugar cane. Industries include rice and linseed mills. As one of the principal garrison bases of the Pakistani army and lying only 20 km (12 miles) from the Indian border, it was one of the first targets of the Indian army in the 1970-1 war, when it was besieged. It used to be headquarters of the indigo dye industry which migrated to India at the time of Partition (1947).
Population (district) 4 020 000; (town) 149 400
Map Bangladesh Bc

jet Dense black lignite, capable of taking a brilliant polish, used for jewellery.

jet stream Strong westerly wind blowing at high altitude, usually about the boundary level (called the tropopause) between the troposphere and the stratosphere in the earth's atmosphere. Jet streams, with winds of more than 160 km/h (100 mph), are used by easterly flying aircraft to conserve fuel.

Jezzine ***Lebanon*** Once a popular resort, the small stone-built town perches almost 1000 m (3280 ft) up in the southern Lebanon Mountains, about 16 km (10 miles) inland from the port of Sidon. Magnificent walks amid splendid mountain scenery once drew tourists, and the town also had a reputation for making fine silver-plate cutlery. However, it is now the centre of a Christian enclave in the continuing civil war, its modest population swollen by thousands of refugees and militiamen.
Map Lebanon Bb

Jhansi ***India*** City with large railway works about 380 km (236 miles) south-east of Delhi. Its early 17th-century Mogul fort, still standing though modernised in the 19th century, was the scene of a bloody massacre in 1857, when the Sepoys (Indian soldiers under British command) rebelled. The British regained the fort after much fierce fighting.
Population 284 100
Map India Cc

Jhelum ***India/Pakistan*** One of the five main Punjabi rivers, flowing some 720 km (450 miles) through the Vale of Kashmir and the northern Punjab to join the Chenab about 200 km (125 miles) west of Lahore. It is dammed for irrigation at Mangla.
Map Pakistan Da

Jiangsu ***China*** One of the country's richest provinces, covering 100 000 km² (38 600 sq miles) on the northern side of the fertile Chang Jiang delta. This low-lying area is studded with lakes and crisscrossed by canals and waterways. Jiangsu is only about half the size of Britain, yet its population exceeds that of the whole of the UK. It is a major rice and cotton area, but the most striking feature of its recent economic development has been the growth of its small rural industries, mainly textiles, agricultural machinery and food processing. These along with its established urban industries such as textiles and engineering, have made it China's second largest concentration of industry after neighbouring Shanghai. The capital is Nanjing.
Population 60 521 000
Map China Hd

Jiangxi ***China*** Province in the south, covering 160 000 km² (39 000 sq miles) and centred on the basin of the Gan Jiang river. To the north, Jiangxi opens onto the valley of the Chang Jiang (Yangtze river) and Poyang Hu, China's largest lake. The climate is warm and moist and its fertile soil is used to grow rice, tea and cotton. Coal is mined at Pingxiang and clay at Gaoling mountain near JINGDEZHEN. The capital is Nanchang.
Population 32 290 000
Map China He

Jicin ***Czechoslovakia*** Bohemian market town founded in 1302, 76 km (47 miles) north-east of

Prague. Jicin and Turnov, to the north-west, are the main tourist centres for the 'Bohemian Paradise' (Cesky Raj) – an area of picturesque wooded hills with outcrops of volcanic basalt and sandstone eroded into fantastic shapes.

Population 17 000

Map Czechoslovakia Ba

Jih-yüeh T'an ***Taiwan*** See SUN MOON LAKE

Jilin (Kirin) ***China*** Province covering 180 000 km² (69 500 sq miles) of central Dongbei (Manchuria). Its climate is extreme: winter temperatures fall far below freezing while summers are sunny and warm. Unlike the neighbouring provinces of Heilongjiang and Liaoning, mineral resources are scant in Jilin. Agriculture is the main occupation; it is China's leading producer of soya beans and also grows wheat, millet and sugar beet. The capital is Changchun.

Population 22 502 000

Map China Ib

Jima ***Ethiopia*** Market town about 230 km (145 miles) south-west of the capital, Addis Ababa. It is the centre for a coffee-growing area.

Population 70 000

Map Ethiopia Ab

Jinan ***China*** Capital of SHANDONG province about 360 km (225 miles) south of Beijing (Peking). It has long been renowned for its silk industry. Food products are also made.

Population 3 202 000

Map China Hd

Jingdezhen ***China*** City famous for ceramics in Jiangxi province. It is the country's leading centre for the manufacture of porcelain. China clay comes from nearby Gaoling mountain, from which the English word 'kaolin' is derived.

Map China He

Jinhae ***South Korea*** See CHINHAE

Jinja ***Uganda*** Town on the Nile, beside the northern shore of Lake Victoria, about 80 km (50 miles) east of the capital, Kampala. It was a cotton centre in the early 20th century, and became the country's second largest town after the capital in the 1950s with the building of the Owen Falls Dam just to the north-west, on the Nile. The dam was East Africa's first major source of hydroelectricity. Today Jinja's factories, set up to take advantage of the dam's power, make textiles and light engineering products.

Population 100 000

Map Uganda Ba

Jiujiang ***China*** Ancient city founded by the Han dynasty (206 BC-AD 220) on the south bank of the Chang Jiang river. It contains the beautiful Yanshui Pavilion, built at that time on an island in the middle of Lake Gantang.

Population 150 000

Map China He

Joal-Fadiout ***Senegal*** Tourist resort formed from two settlements some 115 km (70 miles) south-east of Dakar.

The ancient port of Joal, first visited by Europeans in the 15th century, was the birthplace of Léopold Sédar Senghor, the country's first president (1960-80). Nearby are large banks of shells built up by the sea, and on one of these banks stands the village of Fadiout.

Map Senegal Ab

João Belo ***Mozambique*** See XAI-XAI

▼ TWINKLING CITY A television tower dominates downtown Johannesburg, whose streets were laid out soon after the discovery of gold-bearing quartz rock on the Witwatersrand in 1886.

João Pessoa ***Brazil*** State capital and seaside resort 100 km (62 miles) north of the city of Recife. It has a number of beautiful beaches lined with coconut palms. The beach of Tambaú sweeps round to the cliffs of Cabo Branco – the easternmost point of South America.

The city lies close to the eastern end of the Trans-Amazonian Highway, and has a number of churches dating from the 16th century. The city is named after a governor of the state who was assassinated in 1930. Its main industries are textiles, plastics and paper.

Population 331 000

Map Brazil Eb

Jodhpur ***India*** Walled city on the edge of the THAR desert, about 470 km (292 miles) south-west of Delhi. Formerly the capital of the princely state of Jodhpur, it was founded by Rao Jodha, its first maharajah, in 1459.

The fort, built on rocky cliffs, is reached by a zigzag ascent through numerous gateways. At the last gate are the handprints of the widows of various Hindu maharajahs who performed *suttee* – burning themselves to death on the funeral pyres of their dead husbands. The last time this happened was in 1845.

The city is a centre for equestrian sports, and the close-fitting riding breeches called jodhpurs take their name from there.

Population 506 300

Map India Bb

Jogjakarta ***Indonesia*** See YOGYAKARTA

Johannesburg ***South Africa*** Largest city in the country, built near the spot where gold was discovered in 1886. In 100 years, Johannesburg has developed from a mining camp into a modern city, the financial and commercial hub of the prosperous WITWATERSRAND. Gold is still mined there and white heaps of spoil from the mines rise like hills alongside the city centre. Johannesburg also produces chemicals, foodstuffs, furniture, machinery, metal goods, pharmaceuticals and textiles.

Non-whites who work in the city live in segregated areas, including the Asian suburb of Lenasia and the vast black township of SOWETO on the extreme south-west of the city limits – which gets its name from an abbreviated form of the words South-West Townships.

Population 1 536 500

Map South Africa Cb

John o'Groats ***United Kingdom*** Small Scottish settlement in the north-east, traditionally regarded as the northernmost point of the British mainland, although it is actually about 4 km (2.5 miles) farther south than Dunnet Head, 20 km (12 miles) to the west. It is named after a Dutch family, De Groot, who lived there.

Map United Kingdom Da

Johnston Island ***Pacific Ocean*** Atoll about 1250 km (800 miles) south-west of Hawaii. Formerly part of Hawaii, it became a US naval base in 1934. It has been used as a base for nuclear weapons tests and currently for storing nerve gas; plans for building a disposal plant for the deadly poisonous gas have been announced by the US government.

Population (mainly military) 400

Map Pacific Ocean Eb

Johor (Johore) ***Malaysia*** Most southerly state of Peninsular Malaysia, extending over 18 986 km² (7330 sq miles). It was an independent sultanate until European colonisation in the early 19th century. Much of its forest land has been cleared in recent years and palm oil is now the chief product, along with rubber and pineapples. Some bauxite is mined.
Population 1 602 000
Map Malaysia Bb

Johor Bahru ***Malaysia*** Capital of Johor State situated on Johor Strait. It is rapidly developing as a port, exporting palm oil from new plantations to the north. The older part of the town has a palace of the sultans of Johor. The ruins of the ancient capital of the sultanate lie 30 km (19 miles) north at Johor Lama.
Population 246 400
Map Malaysia Bb

Johor Strait (Tebrau Strait) Narrow arm of the sea between the southern tip of the Malay peninsula and Singapore Island. It is crossed by a road and rail causeway.
Dimensions 48 km (30 miles) long, 1-5 km (0.6-3 miles) wide
Map Singapore Aa

joint Fracture or crack in a rock mass along which no movement has occurred, caused by rocks contracting on cooling or drying-out, or by folding.

Jokkmokk ***Sweden*** Small community on the Arctic Circle, 825 km (515 miles) north of Stockholm. A centre of Lapp culture, Jokkmokk contains the Lapp People's College and the Lappland Museum.
Map Sweden Cb

Jolo ***Philippines*** Town and administrative centre for the Sulu Archipelago, on the north coast of the agriculturally rich Jolo island. The island, whose fertile volcanic soil produces manioc and a wide range of fruits, covers 894 km² (345 sq miles), and contains many extinct volcanoes, the highest of which is Mount Tumatangas, 812 m (2664 ft).

The town, a walled Islamic stronghold, is a port of entry and a busy fishing and pearling centre, although much of this activity has been disrupted in recent years as Jolo has become a focus of Muslim nationalism.
Population 52 500
Map Philippines Bd

Jonglei Canal ***Sudan*** A cutting, about 360 km (223 miles) long, across a loop of the White Nile about 800 km (500 miles) south of the capital, Khartoum. It is being built to reduce evaporation and conserve the waters of the White Nile from loss in the SUDD swamps.
Map Sudan Bc

Jönköping ***Sweden*** Industrial city, dating from the 13th century, at the south end of Lake Vättern, about 150 km (95 miles) east of Gothenburg. The mechanised production of safety matches started there in 1845, and it has a match museum. It also produces machinery, chemicals and footwear.
Population 107 362
Map Sweden Bd

Jordan ***Middle East*** River which rises on the slopes of Mount Hermon, near the Lebanese border with the Israeli-occupied GOLAN HEIGHTS and flows southwards through the Sea of Galilee (Lake Tiberias) into the Dead Sea. It is 256 km (159 miles) long and is used for hydroelectric power and for irrigation. The part between the Sea of Galilee and the Dead Sea forms the border of Jordan with Israel and the Israeli-occupied WEST BANK.
Map Israel Ba

Jordan Valley ***Jordan/Israel*** Fertile valley in which wheat and barley have been grown since 8000 BC. Watered by the Jordan river and the EL GHOR CANAL, the valley is 100 km (62 miles) long and contains more than 30 farming villages on the Jordanian side of the river. In the western part of the valley, in the Israeli-occupied WEST BANK area, both agriculture and industry have stagnated. Agricultural development on the Jordanian side of the river has been greatly aided by the building of the King Talal Dam on the ZARQA river.
Map Jordan Aa

Jos ***Nigeria*** Capital of Plateau state, on the Jos Plateau, about 710 km (440 miles) north-east of the national capital, Lagos. It was founded by European colonists at the end of the 19th century, and developed as the centre of the country's tin-mining industry. Jos was also a popular hill resort, and still is. Its museum contains fine sculptured heads and terracotta figures from the civilisation that flourished at nearby Nok from 500 BC to AD 200.
Population 145 400
Map Nigeria Ba

Jos Plateau ***Nigeria*** Uplands covering some 7780 km² (3000 sq miles) in the centre of the country, with an average height of about 1260 m (4135 ft). It is littered with huge granite outcrops, extinct volcanic cones and waterfalls such as the Assob and Kuru falls on the steep plateau edge, about 50 km (30 miles) south of the town of Jos. The striking scenery makes the plateau a leading resort area. It is also a source of tin, with mines at Bukuru near Jos, and of columbite (a mineral used in alloy steels).
Map Nigeria Ba

Joshin-etsu Kogen ***Japan*** National park, 1889 km² (729 sq miles) in area, just north-west of the town of Nagano in central Honshu. Its alpine landscape rises to 2446 m (8025 ft) at Mount Myoko. The park is a popular summer resort for the inhabitants of Tokyo, about 155 km (95 miles) to the south-west.
Map Japan Cc

Joshua Tree National Monument ***USA*** Area of 2266 km² (875 sq miles) in California, about 160 km (100 miles) east of Los Angeles. It contains Joshua trees (tree-like species of yucca), cacti and other desert plants, and strikingly eroded granite outcrops.
Map United States Cd

Jostedalsbreen ***Norway*** Largest glacier on mainland Europe, covering 816 km² (315 sq miles). It lies just north-east of Sognefjord. The glacier, which is 97 km (60 miles) long and 24 km (15 miles) wide, consists of a central ice cap more than 320 m (1050 ft) thick and about 25 tongues which radiate down the surrounding valleys. At its summit the ice is 2040 m (6700 ft) above sea level.
Map Norway Bc

Jotunheimen ***Norway*** Mountain region with an impressive cluster of peaks in south central Norway. The name means 'the home of the giants'. Glittertind and Galdhopiggen (2469 m, 8100 ft) are the two highest peaks in the region, and nearby Lake Gjende is a tourist attraction.
Map Norway Cc

Juan de Fuca Strait US-Canadian sea passage, between VANCOUVER ISLAND and the WASHINGTON mainland, giving SEATTLE, VICTORIA and VANCOUVER access to the PACIFIC OCEAN. It was reputedly named after a Greek navigator who used the name Juan de Fuca while serving in the Spanish navy in 1592.
Dimensions 160 km (100 miles) long, 18-27 km (11-17 miles) wide
Map Canada Cd

Juan Fernández Islands ***Pacific Ocean*** Archipelago belonging to Chile, named after the Spaniard Juan Fernández, who landed there in 1572. The islands – Más Afuera, or Alejandro Selkirk, Más á Tierra, or Robinson Crusoe, and Santa Clara, or Goat Island – lie in the Pacific 650 km (400 miles) west of the port of Valparaiso. The English writer Daniel Defoe based his novel *Robinson Crusoe* on the real-life adventures of a Scottish seaman, Alexander Selkirk, who lived on Más á Tierra in 1704-9. The main occupation is lobster fishing.
Population 550
Map Pacific Ocean Id

Juba ***Sudan*** Market town about 1200 km (720 miles) south of the capital, Khartoum. It is the capital of Sudan's Eastern Equatoria province, which stretches along the national borders of Kenya, Uganda and Zaire. The townspeople are mainly black and non-Muslim, unlike the Arab people of the north. Most are pagan though there are some Christians.
Population 83 800
Map Sudan Bc

Jubail ***Saudi Arabia*** Port and growing industrial complex on The Gulf, situated 85 km (53 miles) north of Dammam. A former fishing village, Jubail began expanding in the late 1970s, when the government started work on the complex, planned to have refineries, petrochemical industries, a steel mill, fertiliser plant, aluminium smelter and other factories.
Population 50 000
Map Saudi Arabia Bb

Jubba (Giuba) ***Ethiopia/Somalia*** River rising in the Ethiopian highlands and flowing southwards for 1200 km (750 miles). It reaches the Indian Ocean near the southern Somali port of Kismaayo. Its year-round water is used to irrigate farms along its valley.
Map Somalia Ab

Júcar ***Spain*** River, 505 km (314 miles) long, entering the Mediterranean at Cullera just south of the port of Valencia.
Map Spain Ec

Jordan

AN OASIS OF STABILITY WITHIN THE TROUBLED MIDDLE EAST, JORDAN HAS TAKEN ON A MEDIATOR'S ROLE

The development of modern Jordan has been dominated by one man for over 30 years – its ruler King Hussein. His courage has seen the country through constant political crises. Jordan remains a frontline state in relation to its neighbour Israel, and among the Arab countries, no other has shown a greater interest in negotiating a lasting settlement to the Arab-Israeli conflict.

The area was long ruled by the Ottoman Turks. The emirate of Transjordan, created in 1923, remained under British protection until 1946 when it achieved full independence as the Kingdom of Jordan. Almost immediately, the new state became embroiled in the 1948 Arab war against Israel. In the armistice that followed, the West Bank (of the River Jordan), which had been occupied by Palestinian Arabs for centuries, was annexed by Jordan. At the same time, Jordan received Palestinian refugees fleeing from Israel. Later, in the Six-Day War of 1967, Jordan lost the West Bank to Israel and gained even more refugees.

King Hussein allowed refugee militants from the Palestine Liberation Organisation (PLO) to raid Israel from bases in Jordan. However, the guerrillas defied the government's authority, and Hussein's army acted in 1971 and expelled them from the country. The bitter conflict that ensued ended with the expulsion or imprisonment of the guerrillas until Hussein declared a general amnesty for all political prisoners and exiles in 1973. The following year, Jordan acknowledged the PLO as the sole legitimate representative of the Palestinians.

Today, over half of Jordan's population is Palestinian. Many live in the ten UN refugee camps. All have been granted Jordanian citizenship and all take a full part in the political and economic life of the kingdom.

The rich Arab states, such as Kuwait and Saudi Arabia, give substantial economic aid to strengthen Jordan as an Arab country bordering Israel. The recent civil war in Lebanon has allowed Jordan's capital AMMAN to replace the Lebanese capital, BEIRUT, as the financial centre of the region. Jordanians have a reasonably high standard of living and 70 per cent of adults are literate. Although they have no oil they refine it, and have an efficient system of intensive agriculture.

Only one-fifth of Jordan is fertile enough to be farmed; the rest is desert. The fertile areas include the disputed West Bank and the valley of the Jordan. To the east lie the hills and mountains of Biblical history which are also fertile. All these areas have cool winters with moderate rainfall.

There is a great difference between the lives of the town dwellers (including Palestinian refugees), the settled rural population of the villages, and the traditional nomadic Bedouin (or *Bedu* in Arabic) tribes of the desert. But there is one common bond – they are, except for a small Christian community, all Muslim.

▲ ROCK CITY The city of Petra, carved in the red sandstone walls of a desert gorge, prospered on a caravan route for more than 1000 years from 300 BC.

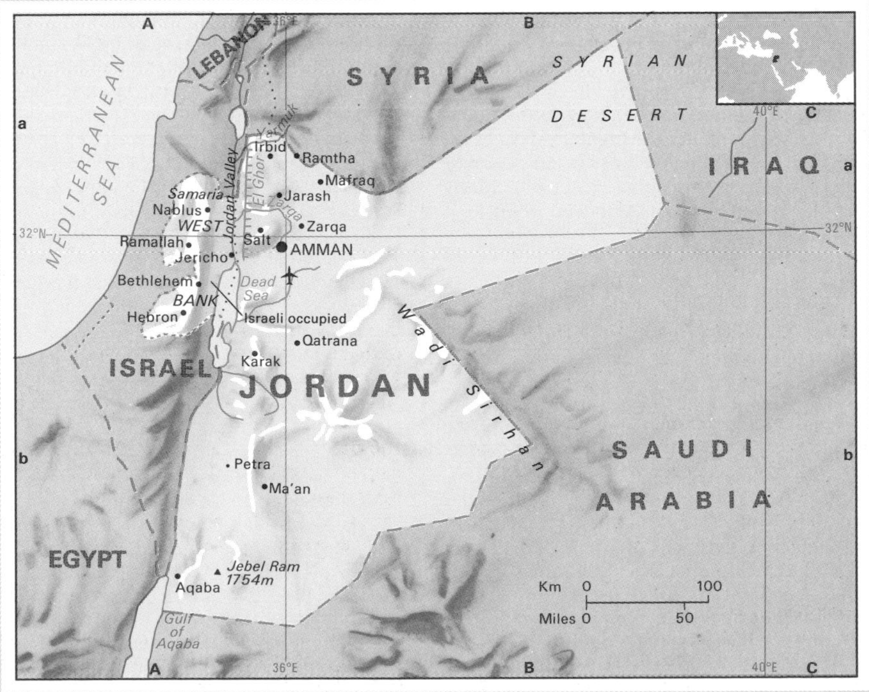

JORDAN AT A GLANCE

Area 97 740 km² (37 730 sq miles), including about 5880 km² (2270 sq miles) on West Bank

Population (East Bank) 2 900 000; (West Bank) 960 000

Capital Amman

Government Parliamentary monarchy

Currency Dinar = 1000 fils

Language Arabic; English widely understood

Religions (East Bank) Sunni Muslim (90%), Christian (8%)

Climate Hot and dry; cool in winter. Average temperatures in Amman: 4-12°C (39-54°F) in January; 18-32°C (64-90°F) in August

Main primary products Vegetables, fruits, olives, wheat; phosphates, potash

Major industries Agriculture, mining, fertilisers, chemicals, cement

Main exports Phosphates, chemicals, fruit and vegetables, potash

Annual income per head (US$) 1937

Population growth (per thous/yr) 36

Life expectancy (yrs) Male 60 **Female** 64

Judaea ***Middle East*** The southern part of ancient Palestine, under the rule of the Persians, Greeks and Romans, stretching from the Mediterranean to the Jordan river and Dead Sea. It was bordered on the south by Sinai and on the north by Samaria, and formed an area divided today between Israel and the Israeli-occupied WEST BANK area. The highest point is Mount Hebron (1013 m, 3323 ft), about 30 km (19 miles) south-west of Jerusalem. Much of the region consists of densely populated farmland.
Map Israel Ab

Jujuy ***Argentina*** See SAN SALVADOR DE JUJUY

Julian Alps (Alpi Giulie; Julijske Alpe) ***Yugoslavia*** Mountain range of north-west Slovenia. It has 250 peaks exceeding 2000 m (about 6560 ft), including Triglav, Yugoslavia's highest mountain, and is crossed by the forested valleys of two main rivers: the upper Soca; and the Trenta, which cradles Lake BLED and the wilder Lake Bohinj to the west. A major skiing area, its resorts include Bled and Bohinjska Bistrica to the west, and Mount Vogel above Lake Bohinj. Triple-peaked Triglav (meaning 'the three-headed') rises to 2863 m (9393 ft) above the Triglav National Park, Yugoslavia's largest. Established in 1924, this protects 780 km² (301 sq miles) of peaks, forests, pastures noted for wild flowers, and seven small, blue lakes; it is best reached from Lake Bohinj.
Map Yugoslavia Aa

Jumla ***Nepal*** Town in the west of the country about 350 km (215 miles) north-west of Kathmandu. Until the 18th century, Jumla was the capital of Nepal – then an independent mountain kingdom, newly opened to foreigners. The town leads to some of the best mountain forest trekking in Nepal and to some of the finest Himalayan scenery.
Population 10 000
Map Nepal Aa

Jumna ***India*** See YAMUNA

Juneau ***USA*** Capital of Alaska, lying in the south-eastern 'panhandle' of the state, about 400 km (250 miles) north-west of the state's southern boundary. It was a gold rush town in the 1880s, but today it is a port and fishing centre. It is linked to the rest of the state only by sea and air, and a site for a new capital at Willow, north of Anchorage, was chosen in 1976, but no action has yet been taken.
Population 23 800
Map Alaska Dc

Jungfrau ***Switzerland*** Magnificent mountain in the Bernese Oberland about 60 km (37 miles) south-east of the capital, Berne. It is 4158 m (13 642 ft) high. First climbed in 1811 by two Swiss brothers, Rudolf and Heironymus Meyer, it was tamed by a railway, passing mostly through tunnels, which took 16 years to build, opening in 1912. The railway is the highest line in Europe. It climbs to the Jungfraujoch, a mountain saddle where there is a station at 3450 m (11 320 ft). From there a cable car rises a further 120 m (394 ft) towards the summit. The 9 km (5.5 mile) long railway starts in the nearby village of Kleine Scheidegg.
Map Switzerland Aa

Junggar Pendi (Dzungarian Basin) ***China*** Vast area of grassland and desert in the Xinjiang Uygur Autonomous Region of about 880 000 km² (340 000 sq miles). It is remote and inhospitable but has rich oil deposits, which are being exploited at the city of Karamay.
Map China Cb

jungle Dense tropical forest.

Junín ***Peru*** Central department which produces half the country's wheat. Its capital, Huancayo (population 164 950), stands on the Mantaro river, about 195 km (120 miles) east of Lima, to which it is linked by the highest railway in the world. The railway reaches 4817 m (15 806 ft) above sea level at La Cima, on the Morococha branch, and 4781 m (15 688 ft) in the Galera tunnel, on the main line. The town of Jauja, 40 km (25 miles) to the north of Huancayo, was the Spanish conquistador Francisco Pizarro's capital in 1534 before he founded Lima the following year.
Population 852 200
Map Peru Bb

Jura ***France/Switzerland/West Germany*** Crescent-shaped mass of limestone plateaus, valleys and mountain ridges, extending 240 km (150 miles) either side of the Franco/Swiss border. The crescent's southern end is east of the city of Lyons. The highest peak is the Crêt de la Neige (1723 m, 5653 ft), just north-west of Geneva. Dairy farming – especially for cheese – forestry, winter sports, and watch and clock-making are the main occupations of the region's inhabitants. The French department of Jura occupies the western slopes.

The Jura continue north-eastwards in West Germany on the other side of the Black Forest. The Swabian Jura (Schwäbische Alb) and its continuation, the Franconian Jura (Fränkische Alb), run more than 320 km (200 miles), ending east of Nuremberg, and reach a height of about 900 m (2950 ft).
Population (department) 243 000
Map France Gc

Jura ***Switzerland*** Watchmaking canton at the eastern end of the Jura Mountains. It covers 837 km² (323 sq miles) around the cantonal capital, Delémont (population 12 500), which is some 30 km (19 miles) south-west of Basle.
Population 65 000
Map Switzerland Aa

Jurassic Second period of the Mesozoic era in the earth's time scale. See GEOLOGICAL TIMESCALE

Jurong ***Singapore*** District in the south-west of Singapore Island containing the country's main industrial area. There are more than 1400 factories in an area of 20 km² (8 sq miles), employing 112 000 people – 39 per cent of the country's industrial workforce. Products include ships, steel rods and pipes, chemicals, pharmaceuticals, plastics, cement, bricks, cables, textiles and tyres. The area is a free trade zone, and has a major new port.

A massive new town development surrounds the industrial area and includes parks, gardens and a bird sanctuary.
Map Singapore Ab

A DAY IN THE LIFE OF A SINGAPOREAN FINANCIER

Robin Lee will make a fortune before he is 50, some 14 years from now. Like many of his fellow 2.5 million Singaporeans he tells himself this daily as he drives his Toyota to meetings, or takes a train to Malaysia, or jets to Bangkok and Hong Kong.

Singapore's energetic and ambitious people, three-quarters of whom are Chinese like Robin, have transformed a marsh and jungle island no bigger than a big Western city into one of the world's most dynamic states. Its people are young – almost half were not born when it became independent in 1965. The growing finance firm where Robin is a rising star funds the construction of oil rigs and factories – and Robin handles a large number of clients.

Like 77 per cent of Singaporeans, he lives in a government-built flat, with his wife Rose and two children, James and Susan, aged six and seven. The modern three-bedroomed flat in Jurong cost Robin 80 000 Singapore dollars (US$37 585) six years ago when he bought it from Singapore's public housing authority, the Housing and Development Board. He paid for it with part of his savings in the Central Provident Fund – the government's compulsory savings scheme. A quarter of a Singaporean's income goes into the fund, but he gets it back with interest on retirement.

Robin studied finance at the National University and at work speaks English, the language of administration. He also speaks three of Singapore's six Chinese dialects plus Malay, the national language, and has some experience of engineering, thanks to his two years' military service. He looks forward to a fat year-end bonus on top of his 2500 Singapore dollars (US$1175) a month, and to the day when he will start his own consultancy. Rose earns 1000 Singapore dollars (US$470) as a well-paid secretary while the children are at school. The couple earn enough to go gambling occasionally. Robin likes to put money on the horses at the Singapore Turf Club. He also gambles a little with friends at mahjong – a game with four players which resembles cards, but with the motifs on small, hard chips.

At mid-morning Robin drives back to Jurong estate to see a factory site. After his tour, he takes his client out to a Chinese lunch. He spends the afternoon on figures and paperwork, and on the telephone, forgetting the time until seven, when Rose calls and reminds him they – and the children – have a dinner appointment at eight with friends in a Western restaurant. There is plenty of choice – cosmopolitan Singapore is the culinary capital of South-east Asia.

Jutland ***North-west Europe*** Peninsula over 400 km (250 miles) long between the North Sea and the Baltic. Most of it comprises the mainland of Denmark, where it is called Jylland; the southern section forms part of the West German state of SCHLESWIG-HOLSTEIN. The border, which crosses the peninsula between Flensburg and the North Sea coast, was settled with Germany after a referendum in 1920.

The peninsula tapers northwards to The Skaw (Grenen). The North Sea side of the peninsula lacks natural harbours. It contrasts with the deeply indented and sheltered Kattegat coast in the east. In the south-west, the remains of a much older coastline, breached largely by storms in medieval times, are marked by the North FRIESIAN ISLANDS. Large areas of salt marsh between the islands and the modern coastline have been drained and reclaimed.

With the ceding of the duchies of Schleswig and Holstein to Prussia after the war of 1864, Denmark lost one-third of its 19th-century territory. By way of compensation, the Danes began a vigorous campaign to reclaim barren heathlands in central Jutland, and have turned some 10 000 km^2 (3860 sq miles) into productive farmland.

Jutland gave its name to the only major naval battle between the British and German fleets in the First World War. It was fought off the peninsula's west coast in the summer of 1916 and resulted in Britain retaining control of the North Sea.

Map Denmark Ba; West Germany Ca

Jwaneng ***Botswana*** Southern town about 110 km (68 miles) west of Gaborone. Its diamond mine is owned jointly by Botswana and De Beers, the South African mining company whose former chairman, Harry Oppenheimer, said it was 'probably the most important kimberlite pipe [diamond deposit] discovered anywhere in the world since the original discoveries in KIMBERLEY, South Africa'. The mine, which went into full production in 1982, is a source of high-quality gemstones accounting for almost half of Botswana's exports.

Population 5600

Map Botswana Bb

K2 (Godwin Austen; Qogir Feng) ***China/ Pakistan*** Second highest mountain in the world after Everest, rising to 8611 m (28 250 ft) on an extremely inaccessible part of the disputed border between Azad Kashmir and China. It was first climbed by an Italian team in 1954.

Map Pakistan Da

Kabah ***Mexico*** Ruined city of the Mayan civilisation (which flourished in the 7th to 10th centuries) in southern YUCATAN, 19 km (12 miles) from Uxmal. The principal building is the Palace of the Masks, which is faced with hundreds of stone reliefs of the rain-god Chac.

Map Mexico Db

Kabalega Falls ***Uganda*** Waterfall on the Nile north-east of Lake Albert. The waterfall, formerly known as the Murchison Falls, and the wildlife of the surrounding national park were Uganda's prime tourist attractions before the eight-year dictatorship of President Idi Amin put foreigners off visiting the country.

Map Uganda Ba

Kabul ***Afghanistan*** The country's largest city and capital, and capital of Kabul province on the Kabul river, 190 km (118 miles) west of the Khyber Pass. At a height of almost 1830 m (6000 ft), it commands the mountain passes to the north and west – and has been on the route of several major invasions, including those of Alexander the Great in the 4th century BC and the Indian emperor Babur, who in 1504 made Kabul the capital of the Mogul Empire. His tomb lies in the western outskirts of the city. In 1842, during the first Afghan War, Kabul was occupied and partially destroyed by the British. It was rebuilt and modernised in the late 19th century by the emir, Abder-Rahman Khan. Its most colourful street is the Jodi Maiwand in which is the bustling main bazaar. Kabul is the country's cultural and commercial centre. Its main manufactures are textiles, footwear, vehicle components and timber products.

Population 1 036 400

Map Afghanistan Ba

Kabul ***Afghanistan/Pakistan*** River rising in the Hindu Kush west of the Afghan capital, Kabul, and flowing about 480 km (300 miles) generally eastwards to join the Indus between Peshawar and Pakistan's capital, Islamabad. It is dammed at Warsak, north-east of the Khyber Pass, to provide hydroelectric power and some irrigation water. Although it is not navigable, its lower reaches are used for logging.

Map Afghanistan Ba; Pakistan Ca

Kabwe (Broken Hill) ***Zambia*** The country's oldest mining town, about 100 km (60 miles) north of the capital, Lusaka. The mine was opened in 1902 and produces high-quality lead, vanadium (a metallic element used in rust-resistant high-precision tools) and zinc.

Population 143 600

Map Zambia Bb

Kachchh ***India*** See KUTCH

Kachin ***Burma*** State covering 89 042 km^2 (34 379 sq miles) in the far north of the country between India and China. Burma's highest peak, Hkakabo Razi, which rises to 5881 m (19 294 ft), is there. The thickly forested hills are largely inhabited by the Kachin people, some of whom are still at war with the central government over their demands for independence. The lowlands have been settled by Shans and Burmese, who have developed a jade-mining industry near the town of Mogaung, and sugar-cane growing around the state capital MYITKYINA. It is also an opium-growing area.

Population 904 000

Map Burma Ca

Kadhimain (Al Kazimiyah) ***Iraq*** Holy city of the Shiite Muslims, now a suburb just north of Baghdad. It contains an elaborate mosque, a shrine of former Shiite imams built in 1515, whose domes and minarets are coated with gold.

Map Iraq Cb

Kadoma ***Zimbabwe*** Formerly a gold-mining town called Gatooma, about 140 km (87 miles) south-west of the capital, Harare, it is now the centre of the cotton textile industry. It produces goods from locally grown and imported cotton.

Population 44 600

Map Zimbabwe Ba

Kaduna ***Nigeria*** Capital of the central province of the same name, on the Kaduna river some 625 m (390 miles) north of the national capital, Lagos. It was founded by the British in 1917, and became the capital of the former vast Northern Region. The town lies at the heart of Nigeria's main cotton-growing area, and produces cotton textiles, beer, soft drinks, bicycles, ammunition and shoes.

Population (town) 202 000; (province) 4 500 000

Map Nigeria Ba

▼ RIVER OF LIFE AND DEATH The Kabul river valley is a precious green oasis in a dusty, dry land, but it has seen much fighting between Afghan guerrillas and government forces.

▲ **SHIPS OF THE DESERT Camels cross the shifting sands of the Kalahari in Botswana. Dr Livingstone's party abandoned their crossing in 1849 because the wagons sank into the sand.**

Kaesong (Gaeseong) ***North Korea*** City about 150 km (93 miles) south-east of the capital, Pyongyang. It was the capital of Korea during the Koryo dynasty (AD 918-1392), but its prosperity waned after Seoul, now in South Korea, was made the capital in 1394. Today, textiles and porcelain are Kaesong's chief industries.

Population 140 000

Map Korea Cd

Kafr Kanna ***Israel*** See CANA

Kafue ***Zambia*** River, 970 km (600 miles) long, which rises close to the border with Zaire, about 420 km (260 miles) north and slightly west of the capital, Lusaka. It flows into the Zambezi south-east of the capital. Lake Kafue, just south of Lusaka, was created when the river's gorge was dammed at the confluence with the Zambezi; a hydroelectric station was opened there in the 1970s. The river supports some fishing, but navigation is restricted by rapids.

A national park named after the river and covering 22 400 km² (8649 sq miles) was set up in 1951. Its wildlife includes elephants, lions, rhinoceroses, crocodiles and the fish eagle, Zambia's national emblem.

Map Zambia Bb

Kagan Valley ***Pakistan*** Beautiful mountain area north of Islamabad which is renowned for its lakes and fishing. The 4200 m (13 780 ft) high Babusar Pass links it to the Indus valley.

Map Pakistan Da

Kagera ***Rwanda/Tanzania/Uganda*** River some 530 km (330 miles) long. It rises in south-east Rwanda and for more than 150 km (95 miles) forms the border with Tanzania. It then turns eastwards to form part of the Uganda-Tanzania border, and then flows through Tanzania to Lake Victoria at the border with Uganda. It is the longest headwater of the Nile. The Kagera National Park covers an area of 2500 km² (965 sq miles) on the west bank of the river, in north-east Rwanda. It abounds in big game, crocodiles and water birds.

Map Tanzania Ba

Kagoshima ***Japan*** City and port on Kagoshima Bay in southern Kyushu island. It lies in the shadow of an active volcano, Sakurajima (1118 m, 3668 ft). The city's inhabitants carry umbrellas even in fine weather to protect themselves from showers of fine ash thrown out by the volcano. The mountain was a small island in Kagoshima Bay until 1914, when a tremendous eruption hurled out enough lava and ash to link it with the mainland.

Population 530 500

Map Japan Bd

Kahuzi-Biega National Park ***Zaire*** See BUKAVU

Kaieteur Falls ***Guyana*** Spectacular waterfall on the Potaro river about 230 km (145 miles) south-west of the capital, Georgetown. The river, which is more than 90 m (295 ft) wide at the brink of the falls, plunges down a cliff as a solid wall of water – with an uninterrupted drop of 226 m (741 ft), more than four times the height of North America's Niagara. The waterfall, whose existence became known to the outside world after a British geologist, C. Barrington Brown, visited it in 1870, is now the centrepiece of Guyana's only national park.

Map Guyana Bb

Kaifeng ***China*** Ancient city in Henan province, about 650 km (400 miles) south of Beijing (Peking). It is noted for its handicrafts and for Xiangguosi monastery, a Buddhist sanctuary founded in the 6th century AD.

Population 500 000

Map China Hd

Kaikoura Ranges ***New Zealand*** Two mountain chains (Kaikoura Range and Seaward Kaikoura Range) on South Island, about 175 km (108 miles) north of Christchurch. The highest point is Mount Tapuaenuku at 2885 m (9465 ft). Much of the area is given over to cattle and sheep.

Map New Zealand De

Kaingaroa Forest ***New Zealand*** One of the largest man-made forests in the world, about 200 km (120 miles) south-east of Auckland on North Island's VOLCANIC PLATEAU. It covers some 200 000 hectares (495 000 acres). The area was planted – mostly with radiata pines – during the Depression years of the 1930s as part of a government relief work scheme.

Map New Zealand Fc

Kainji Reservoir ***Nigeria*** One of Africa's largest man-made lakes. It is on the Niger river in western Nigeria and covers 1295 km² (500 sq miles), stretching 137 km (85 miles) downstream from the town of Yelwa to the Kainji hydroelectric dam. After this was completed in 1968, 150 villages were flooded and 50 000 people had to be resettled. The project also supplies irrigation water and aids flood control, and the reservoir provides fish.

Map Nigeria Aa

Kairouan ***Tunisia*** The holiest Muslim city in Africa and the fourth holiest in the world after MECCA, MEDINA and JERUSALEM, situated 57 km (35 miles) west of Sousse. For the faithful, seven trips to Kairouan are equivalent to one visit to Mecca. It was founded in AD 671 by Okba Ibn Nafaa, an Arab warrior who had known Muhammad. Beginning as a strategic military outpost it blossomed into the greatest Islamic cultural centre in the MAGHREB, producing Tunisia's finest romantic poets and experts in religious law.

Kairouan's buildings reflect the clean, unfussy classic lines of the early Islamic style of architecture. There are imposing city walls and ramparts dating from 1052 and the Medina – the old Muslim quarter – which has many Islamic shrines, mosques and minarets, is a busy place of cafés and rug-weaving shops. There is a variety of fascinating souks (markets) offering rugs, carpets and leatherware, for which the city is famous. Among the places of interest are the beautiful 9th-century Great Mosque (the oldest in Africa); the 9th-century Thleta Bibane Mosque; the Sidi Sahbi Mausoleum – according to legend Sidi Sahbi was a holy man who preserved three hairs from the prophet Muhammad's beard, and so was called 'The Barber' – and the splendid 13th-century *zawia* (courtyard) of the house of Abid El Ghariani.

Population 72 300

Map Tunisia Ba

Kaiserslautern ***West Germany*** Industrial and university city about 100 km (60 miles) south-west of Frankfurt am Main. It produces machinery, cars, textiles, tobacco and beer.

Population 103 000

Map West Germany Bd

Kaiserstuhl ***West Germany*** Group of volcanic hills in the Rhine rift valley near the Black Forest town of Freiburg. The name of this prominent, abrupt feature means 'Emperor's Chair'. Its highest point is 567 m (1860 ft). The slopes are terraced for vine growing.

Map West Germany Bd

Kajaani ***Finland*** City founded in 1651 beside rapids just south-east of Oulu lake in the centre of the country. It is a trade and transport centre, and its industries include sawmills, wood-pulping, papermills and cellulose. It has the remains of a late-medieval castle.

Population 35 900

Map Finland Cb

Kalahari ***Southern Africa*** Arid, semi-desert region of sand and dried-up salt pans, covering 520 000 km² (200 800 sq miles). Temperatures are highest in November-January when, as the 19th-century Scottish explorer and missionary David Livingstone noted: 'The very flies sought the shade and the enormous centipedes coming out by mistake from their holes were roasted to death on burning sand.' It is peopled mostly by the Khoi-San – better known as Bushmen and Hottentots. Wildlife reserves include the Central Kalahari, the Makgadikgadi Pan, and the Kalahari Gemsbok National Park in north Cape Province and south-west Botswana.

Map South Africa Ba

Kalamazoo ***USA*** City in southern Michigan, about 210 km (130 miles) west of Detroit. It serves the surrounding agricultural region and makes paper, pharmaceuticals and vehicles. West Michigan University is in the city.

Population (city) 77 200; (metropolitan area) 215 200

Map United States Ib

Kalávrita ***Greece*** Rail terminal and modern city in the northern Peloponnese, 75 km (47 miles) from Corinth. Its cathedral's clock stands permanently at 2.35, the time on December 13, 1943, when the building was set on fire by German soldiers who then massacred the town's 1436 males, as retaliation for guerrilla activities in the region. The monastery of Ayia Lavra, founded in AD 961, lies 7 km (4.5 miles) to the south. It was the starting point in 1821 for the revolution that eventually won independence for Greece from Turkish occupation. Kalávrita is the end of a narrow-gauge railway line, built in the late 19th century, that runs from Dhiakoptón, on the Gulf of Corinth, through gorges, mountain passes, bridges and tunnels.

Map Greece Cb

Kalawewa Tank ***Sri Lanka*** Reservoir in central Sri Lanka, about 140 km (85 miles) north-east of Colombo. It originally covered 18 km² (7 sq miles), and was formed when King Dhatu Sena dammed the Kala river in AD 460. The tank is now used as part of a new irrigation scheme designed to encourage settlers to move into the region.

Map Sri Lanka Ba

Kalémié ***Zaire*** Town founded in 1891 as Albertville. It is a transport centre on the west shore of Lake Tanganyika, linked by rail to Kabalo, 250 km (155 miles) to the west, on the Lualaba river. Lake steamers operate to Kigoma in Tanzania, which is connected by rail to Dar es Salaam. Kalémié is thus a centre of foreign trade. It also has fishing and various manufacturing industries, including textiles and cement.

Population 61 000

Map Zaire Bb

Kalgoorlie ***Australia*** Gold and nickel mining town in Western Australia, 540 km (340 miles) east of Perth. Gold was found there in 1893, and Kalgoorlie and its twin town of Boulder have since produced more than 1000 tonnes of gold. Mining still continues, and the historic Hainault mine is preserved as a tourist attraction. Coolgardie, 40 km (25 miles) to the south-west, was the site of the first gold discovery in the area, in 1892. It is now virtually a ghost town. Kambalda, to the south, is a new nickel-mining centre.

Population 19 800

Map Australia Ce

Kalimantan ***Indonesia*** The Indonesian part of Borneo, covering 538 718 km² (208 000 sq miles) of the mainly mountainous island. Forestry, oil and natural gas are the economic mainstays.

Population 6 724 000

Map Indonesia Dc

Kálimnos (Calino) ***Greece*** Island in the DODECANESE group in the east Aegean Sea. The menfolk of the island sail off each May to North Africa to dive for natural sponges, which are brought back to be processed in the chief town, also Kálimnos. Kálimnos lies south of the island of Léros, where the Roman general Julius Caesar was caught by pirates in 75 BC. Caesar raised his ransom, raised a naval force, captured his captors and had them crucified.

Population 26 900

Map Greece Ec

Kalinin ***USSR*** Industrial city and river port straddling the Volga, 160 km (100 miles) north-west of Moscow. It was founded in 1181 as a fort, and was rebuilt after a devastating fire in 1763. As centre of the Tver principality until 1485, it was originally called Tver, but was renamed Kalinin in 1932 in honour of its most renowned native son, Mikhail Kalinin (1875-1946), Soviet head of state 1937-46. The city's industries include textiles, electrical and mechanical engineering, rolling stock and rubber.

Population 437 000

Map USSR Ec

Kaliningrad (Königsberg) ***USSR*** Commercial port and naval base near the Baltic Sea, 30 km (18 miles) north of the Polish border. It was founded as Königsberg in 1255 by Teutonic Knights, and in the Middle Ages was the capital of the German state of East Prussia. During the Second World War it fell to the Russian army after a long and devastating siege. It was ceded to Russia at the 1945 Potsdam Conference between Britain, the USA and USSR. The following year the city was renamed Kaliningrad after President Mikhail Kalinin (1875-1946).

Built on the Pregel (Pregolya) river, Kaliningrad is linked to the Baltic port of Baltiysk by a canal 42 km (26 miles) long. Most of the postwar city lies to the north of the original town, of which little remains. It is the USSR's third largest fishing port and also makes ships, paper and machinery. The German philosopher Immanuel Kant (1724-1804) was born in Königsberg and taught at the former university there. He was buried in the 13th-century cathedral, which was destroyed during the Russian siege.

Population 380 000

Map USSR Dc

Kalisz (Kalisch) ***Poland*** Industrial city about 105 km (65 miles) south-east of the city of Poznan. It has textile, metal, engineering and piano-making industries.

Population 102 900

Map Poland Cc

Kalmar ***Sweden*** Seaport on the south-east coast, 390 km (242 miles) south of Stockholm. It is opposite the isle of Öland, to which it is linked by a 6 km (4 mile) long bridge. Its medieval castle – rebuilt in the 16th and 17th centuries – was where the Union of Kalmar, uniting Sweden, Denmark and Norway, was signed in 1397. As well as shipyards and railway workshops, Kalmar has match factories and paper mills.
Population 54 165
Map Sweden Cd

Kalmyk Republic ***USSR*** Autonomous republic of the RSFSR to the north-west of the Caspian Sea, covering an area of 75 900 km² (29 300 sq miles). It consists mainly of arid lowland, and is the home of the nomadic Kalmyk tribesmen, who drive their flocks of merino sheep, pigs, cattle and camels between winter and summer pastures. The republic is partly irrigated by the Volga and yields fodder and melons. The capital is Elista (population 78 000), which produces leather, woollen textiles and food from the region's livestock.
Population 315 000
Map USSR Fd

Kalocsa ***Hungary*** City in the centre of the country, about 110 km (70 miles) south of Budapest. It lies 6 km (4 miles) east of the Danube in a particularly rich part of the Great ALFOLD (Plain) which specialises in growing the fiery red spice, paprika. Kalocsa's wealth of fine Baroque buildings includes an 18th-century cathedral and archbishop's palace, with a library containing many rare manuscripts. The city is also noted for its vividly coloured embroideries and brightly painted houses.
Population 20 000
Map Hungary Ab

Kaluga ***USSR*** Industrial city and railway junction on the Oka river, 160 km (100 miles) south-west of Moscow. Its industries include iron and steel, railway equipment, matches, glass, perfume, footwear, clothing and food.
Population 291 000
Map USSR Ec

Kamakura ***Japan*** Town on Sagami Bay, about 48 km (30 miles) south-west of the capital, Tokyo. The seat of Japan's shogun chiefs in the 12th and 13th centuries, it is now a centre of Zen Buddhism. Since 1900 it has become a prosperous dormitory town for Tokyo.
Population 175 500
Map Japan Cc

Kambui Hills ***Sierra Leone*** Low plateaus and hills rising to 320 m (1050 ft) in the extreme south-east astride the Moa river. The hills are rich in minerals, especially diamonds, which are found in the river gravels.
Map Sierra Leone Ab

Kamchatka ***USSR*** Peninsula in the far north-east of the USSR, separating the Sea of Okhotsk from the Bering Sea and the Pacific. It is about 1200 km (750 miles) long, up to 560 km (350 miles) wide and covers an area of about 350 000 km² (135 000 sq miles) – about the size of Japan. It contains 22 active volcanoes, the highest of which is Klyuchevskaya Sopka, at 4750 m (15 580 ft). The inhabitants' main occupation is fishing – for crabs, salmon, herring and cod. The peninsula's largest port is PETROPAVLOVSK-KAMCHATSKIY on the south-east coast.
Population 422 000
Map USSR [illegible]

kame Sand and gravel deposited as a cone where a stream issues from an ice sheet, and left either as a hummock or a steep-sided ridge as the ice retreats.

Kamina ***Zaire*** Transport centre in Shaba region, about 420 km (260 miles) north-west of the city of Lubumbashi, on the railway to Ilebo. It is also the centre of a farming region producing cotton, tobacco and vegetables, and some of these are processed in Kamina.
Population 57 000
Map Zaire Bb

Kamloops ***Canada*** City at the junction of the north and south Thompson rivers, at the heart of an area which has gold and copper mines and is the centre of British Columbia's cattle industry. It was founded as a trading post in 1812.
Population 65 000
Map Canada Cc

Kamnik ***Yugoslavia*** See KARAWANKE ALPS

Kampala ***Uganda*** National capital since independence in 1962. It grew up next to the site, known as Mengo, where the *kabaka* (king) of Buganda had his palace. The British built a fort there in 1890, and the town that grew around it became the country's commercial centre because of its position near Lake Victoria, in the country's most prosperous region.

The city is built on hills, two of which are occupied by the cathedral, hospital and schools of the Protestant and Roman Catholic churches. Another has the main mosque, and a fourth has the university. Most of the city's factories – which produce goods for local consumption – are built on drained swampland between the hills.

Many of the poorer inhabitants live in shanty towns around and between the main buildings.
Population 500 000
Map Uganda Ba

Kamphaeng Phet ***Thailand*** Market town 320 km (200 miles) north-west of the capital, Bangkok. It was founded in the 13th century and parts of the original settlement are still visible. Farmers in the surrounding province of the same name mostly grow rice, sugar cane and mung beans (used in Chinese cooking as bean sprouts).
Population (town) 19 600; (province) 634 900
Map Thailand Ab

Kampong Ayer ***Brunei*** See BANDAR SERI BEGAWAN

Kananga ***Zaire*** City founded as Luluabourg in 1894, about 750 km (465 miles) west of Lake Tanganyika. It is a centre of the Luba people and capital of Kasai Occidental region. It is also a commercial and transport centre on the Lulua river and on the Lubumbashi-Ilebo railway. The city, which has grown rapidly since 1960, serves the diamond-producing region around Tshikapa to the south-west. The main products processed include coffee and cotton.
Population 704 000
Map Zaire B[illegible]

Kanazawa ***Japan*** Port [illegible] coast of Honshu island, about 29[illegible] km ([illegible] miles) north-west of the capital, Tokyo. It is a beautiful castle town largely untouched by modernisation. Traditional Japanese gardens of lakes, waterfalls and azalea groves fill the castle grounds. The nearby Myoryuji temple, built in 1643, was the house of the Ninja, a secret sect widely employed as spies and assassins because of their skills of concealment.
Population 430 500
Map Japan Cc

Kanchanaburi ***Thailand*** Town about 120 km (75 miles) west of the capital, Bangkok. The surrounding province, which lies along the Burmese border, is mostly devoted to growing sugar cane. Tourists visit the area to see the railway that occupying Japanese built in the Second World War over the River KWAI and the two nearby war cemeteries.
Population (town) 139 100; (province) 612 750
Map Thailand Ac

Kanchenjunga ***India/Nepal*** See KANGCHENJUNGA

Kandahar ***Afghanistan*** Capital of Kandahar province and the country's second city, 470 km (292 miles) south-west of Kabul. It has long been an important crossroads town, and is the only stopover on the road between Kabul and Herat, 440 km (273 miles) to the north-west. The city is thought to have been founded by Alexander the Great in the 4th century BC. It changed hands many times, and in the 16th century it was conquered by Babur (about 1482-1530), founder of the Mogul empire of India. In 1748 it became the first capital of Afghanistan. The old city is dominated by the mosque and tomb of Ahmed Shah, who created the first Afghan federation in 1747. Hewn out of a rock, at the top of a flight of 42 steps, is the throne of the emperor, Babur. Kandahar's main industries are textiles and fruit canning.
Population 191 400
Map Afghanistan Bb

Kandy ***Sri Lanka*** Former royal capital of the Sinhalese kings. It stands beside an artificial lake 465 m (1525 ft) up in the central mountains, about 95 km (60 miles) north-east of Colombo. The Sinhalese kings were established in Kandy by the 14th century, and from the 16th century they held off Portuguese, Dutch and British invaders until 1815, when the last king ceded the city to Britain.

Today Kandy is one of the world's most sacred Buddhist sites and a symbol of Sri Lankan nationalism. A tooth believed to be from Siddartha Gautama, who founded Buddhism in the 6th century BC, is kept in the 16th-century Dalada Maligawa ('Temple of the Tooth') and is carried in procession through the city at an annual festival in August. The tooth is kept concealed in seven jewel-encrusted caskets, each nested inside a larger casket like Russian dolls. The country's national museum is near the temple in what were once the quarters of the royal concubines, and the remains of the Sinhalese royal palace now hold an archaeological museum. Sri Lanka's largest university and the Royal Botanical Gardens are at Peradeniya, 6 km (4 miles) west of Kandy.

Population 101 300
Map Sri Lanka Bb

Kanem ***Chad*** Region (and present-day prefecture) in western Chad, just north-east of Lake Chad. It was the site of a powerful kingdom which was founded in the 9th century, and converted to Islam in the 11th century. The kingdom later joined with the Nigerian kingdom of Bornu to form Kanem-Bornu, an empire which lasted until conquered by Europeans in the 19th century. Kanem was ruled as a separate French protectorate and became part of Chad only in 1958. Mao is the prefecture capital.

Population 200 000
Map Chad Ab

Kanem-Bornu ***Nigeria*** See BORNO

Kangaroo Island ***Australia*** Mineral-rich island 145 km (90 miles) long and 50 km (30 miles) wide, off the South Australian coast south-west of Adelaide. It produces salt and gypsum. Barley and sheep are the main agricultural products. At the western end of the island is Flinders Chase, a wildlife reserve.

Population 3300
Map Australia Ff

Kangchenjunga ***India/Nepal*** Mountain on the border between Nepal and Sikkim, at 8585 m (28 165 ft) the third highest peak in the world. It was first climbed in May 1955 by a British expedition led by Charles Evans. The headwaters of the river TISTA and one of the major tributaries of the river KOSI flow from the mountain's glaciers.

Map India Db

Kangerlussuaq ***Greenland*** See SØNDRE STRØMFJORD

Kangnung (Gangneung) ***South Korea*** Rail terminus on the north-east coast, almost due east of the capital, Seoul. Lake Kyongpo, north of the city, is a bathing resort in summer and used by skaters in winter.

Population 74 000
Map Korea Dd

KaNgwane ***South Africa*** Small black state in east Transvaal, formerly a Bantu homeland and originally called Swazi after its people. It covers 3910 km^2 (1510 sq miles), and is nominally self-governing. In the early 1980s, an offer by South Africa to cede it to neighbouring Swaziland was dropped after opposition from KaNgwane and KwaZulu leaders.

Population 458 000
Map South Africa Db

Kankan ***Guinea*** Main town of Upper Guinea, on the Milo river, a tributary of the Niger about 485 km (300 miles) east and slightly north of the capital, Conakry. It is the eastern terminus of the railway from Conakry and the focus of the main roads linking Conakry with Mali, Liberia and the Ivory Coast. Kankan is a trading centre for local produce, especially rice, and has a lively walled market. However, the town itself is rather like an enormous village, with huts and crops grown by the roadside.

Population 265 000
Map Guinea Ca

Kano ***Nigeria*** Capital of the state of the same name on the Niger border. It is Nigeria's third city after the national capital, Lagos, 830 km (515 miles) to the south-west, and Ibadan. Kano is the largest of the medieval Hausa cities, and by the 12th century was a thriving trading centre at the end of a Saharan caravan route. The old town, with its bustling market selling traditional leather goods, baskets, cloth and jewellery, survives. Modern factories produce peanut oil, corned beef, textiles, shoes and soft drinks. Kano is a Muslim religious and educational centre, and also a tourist city with a modern night life.

Population (city) 475 000; (province) 5 775 000
Map Nigeria Ba

Kanpur (Cawnpore) ***India*** Industrial city in the central-north about 70 km (43 miles) south-west of Lucknow. Its products include wool, cotton, leather and sugar. In July 1857, 1000 British soldiers and their families were massacred there during the Indian Mutiny after being promised safe conduct on surrender.

Population 1 639 100
Map India Cb

Kansas ***USA*** One of the Great Plains states, covering 213 064 km^2 (82 264 sq miles) of undulating prairie. It has hot dry summers and cold winters, and is the nation's chief state for producing wheat. Other farm products include sorghum, maize and cattle for both meat and dairy products. The state has gas fields and some oil, and its industries include aircraft and vehicle making and food processing. TOPEKA is the state capital.

Population 2 450 000
Map United States Gc

Kansas City ***USA*** Industrial city straddling the Missouri river and divided between the states of Kansas and Missouri. Its position on a river crossing made it a gateway to the West, and it became the principal market for wheat and livestock in the southern Great Plains. Grain milling, meat packing, agricultural equipment, vehicle assembly, chemicals and engineering are the main industries.

Population (city) 603 600; (metropolitan area) 1 476 700
Map United States Hc

Kansk ***USSR*** Industrial city in central Siberia, about 800 km (500 miles) east and slightly north of Novosibirsk. It was founded as a fort in 1628 and became a station on the Trans-Siberian Railway in the 1890s. The present city grew from the development of nearby coal mining after 1950. Its other industries include machinery, chemicals, food processing and textiles.

Population 103 000
Map USSR Kc

Kao-hsiung ***Taiwan*** Major industrial city and leading seaport, on the south-western coast. Taiwan's second largest city, it has a large iron and steel works, an oil refinery, an aluminium processing plant, one of Asia's biggest shipyards and one of the largest shipbreaking yards in the world, and an international airport. Kao-hsiung imports crude oil, iron ore and timber, and exports manufactured goods and agricultural products. The city's manufactures range from electronics and precision instruments to plastic goods and toys.

Population 1 269 000
Map Taiwan Bc

Kaolack (Kaolak) ***Senegal*** The country's second largest city after Dakar, which lies some 150 km (95 miles) to the north-west. Kaolack is the commercial and cultural centre of Senegal's main groundnut-producing region.

Population 107 000
Map Senegal Ab

kaolin See CHINA CLAY

kaolinite White or grey mineral, hydrated aluminium silicate, formed by the decomposition of granite. It is the principal constituent of kaolin or CHINA CLAY.

Kaposvár ***Hungary*** City and administrative centre of the south-western county of Somogy, 50 km (about 30 miles) south of Lake Balaton. It is a market centre for the picturesque fruit-growing Transdanubian hills to the east and for livestock farms on the plains to the west. Postwar light industries producing textiles, shoes, foodstuffs and electrical goods have brought considerable expansion over the past 25 years.

Population 74 000
Map Hungary Ab

Kaptai Dam ***Bangladesh*** See KARNAFULI

Kapuni ***New Zealand*** Gas field 200 km (120 miles) north and slightly west of the capital, Wellington. Natural gas was discovered there in 1959. It is linked to Auckland and Wellington by pipeline. Many New Zealand cars are now converted to run on compressed natural gas.

Map New Zealand Ec

kar See CIRQUE

Kara Kum (Karakumy) ***USSR*** The country's largest desert, covering an area of some 340 000 km^2 (130 000 sq miles) along the Iranian and Afghan borders east of the Caspian Sea. It consists mostly of sand and includes much of Turkmenistan. The Kara Kum ship canal (still under construction) will run for 1400 km (870 miles) from the Amudar'ya river, east of the desert, to the port of Krasnovodsk on the Caspian Sea.

Map USSR Gd

Kara Sea Shallow sea forming part of the ARCTIC OCEAN between the NOVAYA ZEMLYA and SEVERNAYA ZEMLYA islands. The OB', YENISEY and other Siberian rivers flow into it, but ice covers its surface all year except for a few weeks in late summer. It was reached in 1597 by a Dutch expedition led by Willem Barents, who died there.

Depth Mostly less than 200 m (656 ft)
Map USSR Ha

▲ **HEIGHT OF DESOLATION The bleak Skigar valley lies beyond the Pass of Strangdokmola in the Karakoram; perpetually snow-capped Koser Gunge rises to 6400 m (21 000 ft).**

karaburan Strong, hot, dust-laden north-east wind blowing in the Tarim basin of central Asia during spring and summer.

Karachi ***Pakistan*** Capital of Sind province and Pakistan's largest city, a port and industrial city on the Arabian Sea. In 1800 it was a mere fishing hamlet on a desert shore. It grew in the 19th century after the British built a railway up the Indus river valley, and became the country's leading port because it was far enough from the Indus delta to avoid silting up.

Karachi was the national capital from the creation of Pakistan in 1947 until 1967, when the new city of Islamabad became the capital. The port exports cotton and wheat, and has metal and engineering industries.

Population 5 103 000
Map Pakistan Cc

Karaganda ***USSR*** Industrial and mining city in Kazakhstan, 560 km (350 miles) south of Omsk. It started life in 1857 as a mining village with 150 inhabitants and grew along with the local coal industry. Since the Second World War Karaganda has become Kazakhstan's second largest city. Its other main industries are iron and steel.

Population 608 000
Map USSR Id

Karak ***Jordan*** Town lying 22 km (13 miles) from the south-east edge of the Dead Sea, and 90 km (60 miles) south of Amman. It is a road junction and busy commercial centre and market. Karak was originally an Italian Crusader town, and has a castle overlooking the Dead Sea valley, high town walls and winding, medieval streets. There is an Italian hospital run by nuns.

Population 12 000
Map Jordan Ab

Karakoram ***China/India/Pakistan*** Mountain range in northern Kashmir, stretching some 400 km (250 miles) across disputed Indian and Pakistani territory along the border – also disputed in parts with Chinese Tibet. It is linked with the Himalayas to the south and south-east and – with the Pamirs to the north-west, the Hindu Kush to the west and the Kunlun Shan range to the east – the range forms a mountain knot known as 'The Roof of the World'. There are 33 peaks higher than 7300 m (23 950 ft), including the world's second highest peak, K2, and several passes above 5500 m (18 050 ft). The Karakoram also has one of the longest glaciers outside the polar regions, HISPUR.

Map China Ad; Pakistan Da

Karakorum (Har Horin) ***Mongolia*** Historic site 322 km (200 miles) west of the capital, Ulan Bator, in the valley of the Orhon river. In 1235, Ogodai Khan, the son of the conqueror Genghis Khan, built his headquarters there and Karakorum became capital of the Great Khans. The settlement was abandoned in the 16th century as Mongol power collapsed and only a few scattered ruins remain.

Map Mongolia Db

Karamoja ***Uganda*** Arid region in the north-east, on the Kenyan border. It is sparsely inhabited, largely by Karamojong and Jie peoples, who grow some sorghum but are primarily cattle-herders. Severe famine hit the area in 1979, and in the wake of Uganda's recent coups, banditry was rife for a time.

Population 300 000
Map Uganda Ba

Karawanke Alps (Karawanken; Caravanche) ***Austria/Yugoslavia*** East-west mountain range rising to 2558 m (8392 ft) at Grintavec in Yugoslavia, where Kranjska Gora and Planica to the west are the main ski resorts. The range extends about 370 km (230 miles) eastwards through the Yugoslav republic of Slovenia as the Pohorje spur in the north (with the ski resort of Gradec and spa of Rogaska Slatina), and the Savinjske spur in the south (with the skiing centre of Kamnik and spas including Lasko and Dobrna). The fertile Savinja basin between the spurs centres on the prosperous town of Celje (population 30 900). A spa since Roman times, it makes beer, textiles, zinc and machinery.

Map Yugoslavia B[illegible]

Karbala ***Iraq*** Regional capital about 90 km (55 miles) south-west of the national capital, Baghdad. It houses the tombs of Hussein bin Ali and his brother Abbas – grandsons of Muhammad, the founder of Islam – who were killed there in the Battle of Tuff in AD 680. It trades in dates, hides and wool, and provides facilities for pilgrims visiting the tombs.

Population 107 500
Map Iraq Cb

Karelia ***Finland/USSR*** Historically, a buffer province between Finland and Russia, which has, at different times, been part of both countries. When the frontier was last adjusted, in 1944, Finland had to cede most of Karelia and was left only with what is now the province of North Karelia (Pohjois-Karjala). Its capital is Joensuu. The rest forms the Karelian Autonomous Soviet Socialist Republic of the USSR.

Population (North Karelia) 177 200
Map Finland Dc

Karelia (Karjala) ***USSR*** Autonomous republic of the RSFSR lying just south of the Arctic Circle between Finland, the White Sea (an inlet of the Barents Sea) and Lake Ladoga. It covers an area of 172 400 km² (66 500 sq miles) and consists largely of some 50 000 lakes, dense forest, marshes and rivers. The eastern part of the region has belonged to Russia since the 12th century. The western part – which belonged to Finland from 1920 to 1939 – was occupied by the Russians after Finland's defeat in the Russo-Finnish War of 1939-40, and was ceded to them in 1944.

The republic has flourishing timber, fishing and hunting industries, and copper and iron are mined. Transport is aided by a canal from the White Sea to Lake ONEGA, linking with the Volga-Baltic waterway. The main towns are the capital and ship-repairing centre of PETROZAVODSK, and the ports of Kem' on the White Sea and Sortalava on Lake Ladoga.

Population 769 000
Map USSR Eb

Karen ***Burma*** Hilly state formerly called Kawthulei. It lies in the south-east of the country, bordering Thailand. The fierce Karen people of this isolated region have been fighting for separation from Burma since Burma gained independence in 1948. Pa-an, capital of the state – which covers 30 383 km² (11 731 sq miles) – is linked to the outside world by the Salween river. Most of the Karens are subsistence farmers, largely growing rice.

Population 1 057 500
Map Burma Cc

Kariba ***Zambia/Zimbabwe*** Hydroelectric dam on the Zambezi river about 150 km (93 miles) south of the Zambian capital, Lusaka. It has created a lake covering 5200 km² (2000 sq miles) that is the largest man-made expanse of water in Africa after Lake Volta, Ghana.

The lake, whose southern shoreline is largely

in Zimbabwe, reached its present size in 1961 after Operation Noah rescued thousands of wild animals trapped by the rising waters. The lake is being developed as a tourist attraction, but crocodiles make bathing inadvisable.

Map Zambia Bb

Karjala ***USSR*** See KARELIA

Karl-Marx-Stadt ***East Germany*** City in the south-east, formerly called Chemnitz after the river that runs through it. Textiles have been produced there since the 14th century, and in the 19th century the machine-construction industry was established. The first German machine tools and the first German steam locomotive were made there. Though heavily damaged in the Second World War, the city has been largely restored. Textile and engineering industries have now been augmented by chemicals and electronics manufacturing. Some of its buildings date back to the 12th century and rub shoulders uneasily with modern monolithic architecture. The medieval town hall, with its vaulted Ratsherrestube hall of 1557, is for instance flanked by the massive new town hall which was completed in 1911.

Population 319 000

Map East Germany Cc

Karlovy Vary (Carlsbad; Karlsbad) ***Czechoslovakia*** Health spa about 115 km (70 miles) west of the capital, Prague. It was founded to cure digestive disorders by King Karel (Charles) I of Bohemia (1347-78) – crowned Holy Roman Emperor as Charles IV in 1355 – and is named after him.

The town's present luxurious hotels, sanatoriums, baths and colonnades date mostly from Karlsbad's 19th-century heyday when it was patronised by royalty, aristocrats and the famous from all over Europe, including the German composer Johannes Brahms (1833-97), the Polish composer Frédéric Chopin (1810-49) and the Russian poet Alexandr Pushkin (1799-1837). Karlovy Vary hosts an international film festival each year, and manufactures fine porcelain and crystal objects.

Population 59 200

Map Czechoslovakia Aa

Karlskrona ***Sweden*** The country's main naval base, on the Baltic coast about 220 km (136 miles) east of Helsingborg. The layout of its streets and dockyard – built on the mainland and five neighbouring islands – dates from the late 17th century. Its dry docks are hewn out of solid granite.

Population 59 393

Map Sweden Bd

Karlsruhe (Carlsruhe) ***West Germany*** University and industrial city in the Rhine valley about 65 km (40 miles) north-west of Stuttgart. It was laid out in the shape of a wheel in 1715, as a new capital city for Karl Wilhelm, ruler of Baden. His palace stood at the hub, and the avenues and streets radiated outwards like spokes. The palace is now a museum. Karlsruhe has become an inland port and industrial centre with oil refineries, nuclear research laboratories and bicycle and motorcycle industries.

Population 275 000

Map West Germany Cd

Karlstad ***Sweden*** Provincial capital of Varmland, on the north shore of Lake Vänern, 305 km (190 miles) west of Stockholm. The treaty ending the union of Sweden and Norway was signed there in 1905. It manufactures softwood products, chemicals, textiles and clothing.

Population 74 439

Map Sweden Bd

Karnafuli ***Bangladesh*** River draining the Chittagong Hill Tracts and reaching the sea at Chittagong. The river estuary provides the principal harbour area. To provide power and increase the water depth available at Chittagong in the dry season, the river has been dammed upstream of the city at Kaptai. Boats travelling up and down the river are lifted over the dam by a crane.

Map Bangladesh Cc

Karnak ***Egypt*** Temple site on the east bank of the Nile about 500 km (310 miles) south of Cairo. The ruins at Karnak, the most extensive group of surviving ancient buildings in the world, represent the creative peak of the brilliant New Kingdom, whose pharaohs ruled Egypt from Thebes for more than 500 years. Most of the buildings date from about 1560-1090 BC, and are temples to Amon, the supreme god of the 18th dynasty. The colonnaded hall, largely built for the warrior king Rameses II, who reigned for 66 years in the 13th century BC, is probably the largest single room in a religious building, and could hold Paris's Notre Dame Cathedral with space to spare. It has 134 pillars, many bearing inscriptions praising Rameses. Beyond, in an avenue of pillars stretching more than 400 m (1300 ft), stands an inscribed obelisk, shaped from a single block of granite quarried at Aswan. It was erected by Queen Hatshepsut, the dynasty's only queen, who ruled around 1500-1480 BC.

Map Egypt Cc

Karnataka ***India*** State in the south-west covering 191 791 km² (74 031 sq miles). It was formed in 1956-60 from parts of the older states of Mysore, Hyderabad and Bombay. It includes the distinctive coastal Konkan area and parts of the Western GHATS and DECCAN plateau. Its capital is the city of BANGALORE.

Population 37 135 700

Map India Bd

Kärnten ***Austria*** See CARINTHIA

Karoo (Karroo) ***South Africa*** Two semi-desert regions of Cape Province named from Karoo, a Hottentot word meaning 'thirstland'. The Great Karoo lies between the Great Escarpment and the Swartberg mountains, the Little Karoo between the Swartberg and Langeberg ranges. These arid regions – which form plateaus broken by hills and KOPJES – are used as grazing lands for sheep and, when irrigated, make fertile farming land.

Map South Africa Bc

Karpacz ***Poland*** Ski resort in the Karkonosze hills. It lies at about 600 m (1970 ft) on wooded slopes of the Jelenia Gora valley in south-west Poland. There is a bobsleigh run, and a ski lift up nearby Mount Sniezka (1602 m, 5257 ft) on the Czech border.

The town contains a 13th-century Norwegian church. The church originally stood beside a lake in southern Norway. It so impressed Frederick William IV, King of Prussia (1840-61), that he asked for it to be dismantled and re-erected on its present site – then under Prussian control – in 1841.

Population 6000

Map Poland Ac

Karroo ***South Africa*** See KAROO

Kars ***Turkey*** Fortified town near the north-eastern border with the USSR, about 1000 km (620 miles) east of the capital, Ankara. It has been occupied by Russia three times and was ceded to the Russians with the surrounding province of the same name in 1878, but returned to Turkey in 1921. The town produces dairy goods, textiles and carpets, and the mountainous province, which covers 17 379 km² (6710 sq miles), has large salt deposits.

Population (town) 70 400; (province) 728 700

Map Turkey Ca

karst Barren limestone or dolomite region in which erosion has produced rock PAVEMENTS, GRIKES, SINKHOLES, underground streams and caverns.

Karst ***Yugoslavia*** See KRAS

Karun ***Iran*** River rising in the Zagros Mountains south-west of the capital, Tehran, and flowing 848 km (527 miles) to the Shatt al Arab waterway near the tip of The Gulf. It is navigable for 150 km (95 miles), from the Shatt al Arab to the inland port of AHVAZ.

Map Iran Aa

Kasai (Cassai) ***Angola/Zaire*** River 2150 km (1350 miles) long. It rises in the eastern plateaus of Angola and flows east and then north, forming part of the Angola-Zaire border, then north and west to its confluence with the Zaire. The upper course is punctuated by waterfalls and rapids, but the 772 km (479 miles) below the river port of Ilebo form a navigable waterway, providing an outlet for the mineral wealth of the Shaba region.

Map Zaire Ab

Kasai Occidental (Western Kasai) ***Zaire*** Administrative region covering 156 967 km² (60 589 sq miles) in south-central Zaire, with its capital at Kananga. Other towns include Tshikapa and Ilebo. The region produces gem diamonds and gold, but the main activity is farming, producing coffee, cotton, palm products and livestock. Most people belong to the Lulua group.

Population 2 935 000

Map Zaire Bb

Kasai Oriental (Eastern Kasai) ***Zaire*** Administrative region covering 168 216 km² (64 931 sq miles) in south-central Zaire, with its capital at Mbuji-Mayi. Inhabited mainly by people of the Luba group, the region produces most of the world's industrial diamonds. Coffee, cotton, palm products, rubber and livestock are the main agricultural products.

Population 2 337 000

Map Zaire Bb

Kasama *Zambia* Market town about [illegible] (435 miles) north-east of the capital, [illegible]. It is the regional capital of Northern Province and has a coffee [illegible].
Population [illegible]
Map Zambia [illegible]

Kashan *Iran* City on the edge of the [illegible] Kavir desert, about [illegible] south of the capital, Tehran. Its chief products are carpets, cotton, silk and copperware. The city's skyline is dominated by a 13th-century minaret 45 m (148 ft) high.

Just to the west of Kashan, Shah Abbas I (ruled 1587-1629), who is buried in the city, laid out his magnificent *Bagh-i Shah* (Garden of the King), where cypress trees over 300 years old, fountains and tiled canals remain.
Population 84 600
Map Iran Ba

Kashi (Kashgar) *China* City in the Xinjiang Uygur Autonomous Region, about 200 km (125 miles) from the border with the USSR. It was formerly an important oasis on the Silk Road trade route to central Asia. Now it is the economic centre of southern Xinjiang.
Map China Ac

Kashmir *India/Pakistan* See AZAD KASHMIR; JAMMU AND KASHMIR

Kassala *Sudan* Market town about 400 km (250 miles) east of the capital, Khartoum. It is the capital of a cotton-growing province of the same name, which has become home to thousands of refugees fleeing the Ethiopian civil war in the 1980s. It is linked by road, rail and air to the capital, Khartoum.
Population (town) 143 000
Map Sudan Bb

Kassel (Cassel) *West Germany* Commercial and industrial city on the Fulda river about 155 km (95 miles) north-east of Frankfurt am Main. It has machinery, precision instruments, electronics and textile industries. The brothers Jakob and Wilhelm Grimm, who worked in the city library, lived in Kassel from 1798 to 1828, during which time they published their fairy tales.
Population 190 400
Map West Germany Cc

Kasserine (Al-Qasmayn) *Tunisia* District capital and major road and rail junction, 216 km (134 miles) south-west of Tunis. It lies at the foot of Jebel Chambi and close to the El Douleb oil field. The rolling land to the south and north of Kasserine and the narrow valleys hedged in by steep, rocky cliffs, which form the Kasserine pass, became the scene of furious attack and counter-attack in 1943 between Rommel's Afrika Korps and the Allies. The Germans were eventually defeated and Kasserine became one of the turning points in the Allied victory in North Africa. Today, the town is the centre of an irrigated agricultural area. It manufactures wood pulp and cellulose.
Population 47 700
Map Tunisia Aa

Kastoría *Greece* Town on a peninsula in Lake Kastoría, Macedonia, 221 km (137 miles) west of Salonika. Since 200 BC Kastoría has been occupied successively by Romans, Bulgars, Serbs and Turks (1385-1913), who made it a centre for the Middle Eastern trade in fur. [illegible] Byzantine churches with medieval frescoes, the oldest – Anargyron and Taxiarchis – dating from the 10th century. The nearby village of Argos Orestikon is famed for its woollen carpets. Kastoría holds a Fur Fair in March.
Population 20 700
Map Greece Ba

Kastrup *Denmark* See AMAGER

Kasvin *Iran* See QAZVIN

katabatic wind Valley wind blowing downhill during the night, caused by the flow downward of dense air chilled by radiation on the upper slopes. See also ANABATIC WIND

Katanga *Zaire* See SHABA

Katherine Gorge *Australia* Magnificent gorge of the Katherine river, rich in wildlife, in the Northern Territory about 275 km (170 miles) south-east of Darwin. The gorge is about 12 km (7.5 miles) long and up to 100 m (330 ft) deep; the water level may rise 18 m (60 ft) in the wet season. An oasis in an arid region, the gorge teems with amphibians, reptiles (including snakes and crocodiles), birds, kangaroos and wallabies.
Map Australia Ea

Kathmandu (Katmandu) *Nepal* The country's capital and also the name of the fertile valley around the city, in central Nepal 90 km (55 miles) from the Indian frontier. The city centre – around Durbar Square – is a maze of narrow streets, wood-balconied houses, markets and both Hindu and Buddhist temples. Farther out, there are modern European-style buildings, including the Royal Palace – with some 1700 rooms – built around 1900. On a hill overlooking the city is the Buddhist temple and monastery of Swayambhu, whose terraces are the haunt of monkeys. The temple is topped by a four-sided stupa, or shrine, each face of which is painted with an eye to symbolise the all-seeing wisdom of the Buddha.

Kathmandu was founded in AD 723 and became the capital in the late 18th century. In 1934 it was badly damaged by an earthquake. In the 1960s it became a favourite destination for the world's hippies, who travelled there to sample the cheap and readily available drugs. However, tough new drug laws have put the trade into decline.
Population (city) 195 260; (district) 400 000
Map Nepal Ba

Katowice (Kattowitz) *Poland* Southern industrial city in Upper Silesia, 72 km (45 miles) north-west of the city of Cracow. Its main industries are coal, zinc, steel and engineering. In the city centre there is a striking three-winged monument to the Silesian Uprisings of 1919-21, when the city's Polish majority rebelled successfully against German rule.
Population 361 300
Map Poland Cc

Katsina *Nigeria* Northernmost town in Hausaland, only 20 km (12 miles) from the border with Niger. Founded in the 11th century at the end of a Saharan caravan route, it flourished after the fall of the Songhai Empire to the west in 1591 with an influx of Songhai scholars and craftsmen, especially leather workers. Caravans brought salt, slaves, leather, kola nuts and European goods. However, Katsina was sacked during a religious war in 1807, and today remains a rural town (marketing cattle, hides, leather goods, cotton and groundnuts) and a Muslim seat of learning.
Population 145 500
Map Nigeria Ba

Kattegat (Cattegat) Strait between south-western Sweden and the east coast of Denmark's JUTLAND peninsula, forming part of the sea passage between the NORTH and BALTIC seas. Its name means 'cat's throat' in both Swedish and Danish. Ferries run from Frederikshavn and Grenå in Denmark to Varberg and Gothenburg in Sweden. Sailing is popular in summer.
Dimensions 220 km (nearly 140 miles) long, 34-160 km (21-100 miles) wide
Map Denmark Ba

Kattowitz *Poland* See KATOWICE

Katyn *USSR* Forest and village about 20 km (12 miles) west of Smolensk. It was the site of the massacre in 1940, during the Second World War, of some 4500 Polish army officers. They and about 11 000 others (whose whereabouts are unknown) were taken prisoner when the Russian army occupied eastern Poland in 1939. Their graves were discovered in 1941, after the Germans took Smolensk, but did not become generally known until two years later; they had all been shot in the back of the head. The Polish government-in-exile accused the Russians of the atrocity; the Russians accused the Nazi Germans, but refused to allow an independent investigation by the International Red Cross.
Map USSR Ec

Kaunas (Kovno) *USSR* Industrial city in the Baltic republic of Lithuania, about 210 km (130 miles) east of Kaliningrad. From 1920 to 1940 it was the capital of independent Lithuania, before that republic was annexed by Russia. Under Soviet rule Kaunas has flourished as a university town, river port, railway junction and industrial centre. Its main products are metal goods, electrical machinery, textiles and paper.
Population 400 000
Map USSR Dc

Kavála *Greece* City and port of MACEDONIA, 163 km (101 miles) east of Saloniki. The Roman fleet of Brutus and Cassius landed at the town after the assassination of Julius Caesar and before the Battle of Philippi, which took place 15 km (9 miles) north-west in 42 BC. Kavála has a well-preserved Byzantine citadel with an aqueduct, a mosque and old Turkish houses, baths and shops. It exports tobacco, clothing, wood, furniture and plastics, and there are offshore oil wells at Prinos. At Philippi, St Paul preached for the first time in Europe and wrote an epistle to its inhabitants (the Philippians).
Population 56 400
Map Greece Da

Kavango ***Namibia*** Homeland, formerly called Okavango, set up in 1970 for the Kavango people on the Angolan border. This subsistence farming area, covering 50 955 km² (19 674 sq miles), is centred on the capital, Rundu (population 15 000).

Population 54 000
Map Namibia Aa

Kavango (Cubango) ***Southern Africa*** River, formerly known as the Okavango, which rises in central Angola, where it is called the Cubango. It flows generally south-east about 1600 km (1000 miles) along part of the border of Angola and Namibia, and into Botswana, where it drains into the OKAVANGO DELTA.

Map Angola Ab; Namibia Aa

Kaveri (Caveri; Cauvery) ***India*** River in the south, rising in the DECCAN plateau and flowing south-east then east for 800 km (497 miles) to the Bay of Bengal. It passes through the country's first gorges to be dammed for hydroelectric power (1902). The large Mettur Dam built in 1925 is 2150 m (7053 ft) long and 94 m (308 ft) high. It was the largest in the world when built and was the first combined power and irrigation dam in India.

Map India Cc

Kavkaz ***USSR*** See CAUCASUS

Kawasaki ***Japan*** Industrial city in the Tokyo-Yokohama conurbation. It has oil refineries, chemical plants and steelworks built on land reclaimed from Tokyo Bay.

Population 1 088 600
Map Japan Cc

Kawthulei ***Burma*** See KAREN

Kayah ***Burma*** Tiny eastern border state of 11 733 km² (4530 sq miles) which straddles the Salween river. A hydroelectric power station at Lawpita Falls on the Baluchaung river in the state supplies electricity for most of southern Burma. The inhabitants, mostly Karens, are largely subsistence farmers, growing rice in the narrow upland valleys. Loikaw is the state capital.

Population 168 400
Map Burma Cc

Kayes ***Mali*** Town about 100 km (60 miles) from the border with Senegal. It was the terminus of the rail link between the Niger and Senegal rivers until the railway was extended to the Senegalese capital, Dakar, in 1923. It is also a market for cattle, sheep, animal skins, groundnuts and gum.

Population 45 000
Map Mali Ab

Kayseri ***Turkey*** Textile town about 230 km (140 miles) south-east of the capital, Ankara. It is the capital of an agricultural province of the same name. The town lies at the foot of an extinct volcano, Erciyas Dagi (3916 m, 12 848 ft). Its main industry is the weaving of rugs and carpets. The province, which covers 16 917 km² (6532 sq miles), produces fruit, raisins, millet and rye.

Population (town) 378 500; (province) 870 900
Map Turkey Bb

Kazakhstan ***USSR*** Constituent republic of the USSR stretching from the Caspian Sea to north-western China. It covers 2 717 000 km² (1 050 000 sq miles) – almost the area of India – and consists mainly of dry plains with a central, mineral-rich plateau. Kazakhstan's inhabitants include nomadic herdsmen who roam the undeveloped land, workers on the state-owned farmland surrounding the city of TSELINOGRAD and workers in industrial cities such as KARAGANDA and the capital ALMA-ATA.

It is the USSR's major stock-raising area, and produces wheat, cotton, fruit and other crops on irrigated land. Mines produce coal, iron, lead and other minerals, and industries include food, chemical and metal processing. The republic also contains the district of Baykonyr (Baikonur), some 240 km (150 miles) north-east of the Aral Sea. Baykonyr is the main Russian missile and rocket-testing site, and the launching pad for most of the USSR's space probes and satellites.

Kazakhstan contains over 100 nationalities in all; over one-third of the people are native Kazakhs, who are Muslims and skilled horsemen. The region was invaded by Tatars in the 13th century and became Russian territory during the 18th and 19th.

Population 15 654 000
Map USSR Gd

Kazan' ***USSR*** Capital of the Tatar Republic, an industrial city 700 km (434 miles) east of Moscow. It was founded in the 13th century as Bolgar by the Golden Horde, the Mongol army that swept over eastern Europe. In 1552 it was captured and destroyed by Tsar Ivan the Terrible (1530-84). He built the city's kremlin (fortress), which is still standing. Its university was founded in 1804; students included the author of *War and Peace*, Leo Tolstoy (1828-1910), and the revolutionary leader Vladimir Lenin (1870-1924).

Set on a tributary 5 km (3 miles) from the Volga, Kazan' is the industrial centre of the booming Volga-Urals region. Its traditional industries of leather, silk and fur goods have been joined by chemicals, electrical equipment, shipbuilding, machinery, tools, paper and oil refining.

Population 1 039 000
Map USSR Fc

Kazanluk ***Bulgaria*** Town, 47 km (29 miles) south of Gabrovo on the Sofia-Burgas and Ruse-Kurdzhali rail and road routes. It is the capital of the Valley of Roses (between the Balkan Mountains and Sredna Gora Mountains) and the centre of the attar of roses industry. Attar of roses is an important ingredient for many perfumes. Rose petals are harvested in May and June, each day's crop being gathered before the sun robs the petals of their oil. On the Tundzha river, 16 km (10 miles) to the south-west, is the reservoir of the Georgi Dimitrov hydroelectric power station, beneath which lies the ancient Thracian capital of Sevthopolis.

Population 61 000
Map Bulgaria Bb

Kebnekaise ***Sweden*** Peak in the Kjölen Mountains in north-west LAPPLAND, and – at 2111 m (6945 ft) – the highest point in Sweden.

Map Sweden Cb

Kecskemét ***Hungary*** City in the centre of the country on the Great Alföld (Plain), 84 km (52 miles) south-east of Budapest. It is the market centre of a rich agricultural area producing fruit and livestock, a potent apricot brandy known as *barackpalinka* and leather goods. The city is also the regional capital of KISKUNSAG – the area between the Danube and Tisza rivers. Kecskemét was the birthplace of the composer Zoltan Kodaly (1882-1967), and there is a music institute named after him. Its centre is popular with artists, and contains several buildings in the distinctive Hungarian art nouveau style developed by the architect Ödön Lechner in the 1890s. The city also has two fine Baroque churches and a restored synagogue.

Population 101 300
Map Hungary Ab

Kedah ***Malaysia*** State on peninsular Malaysia's north-west coast, covering 9425 km² (3639 sq miles). It has the country's largest area of rice lowland, in the irrigated plain of the Muda river, near the state capital Alor Setar. Rubber is grown on hill slopes north of Kedah Peak.

Population 1 117 600
Map Malaysia Ba

Kediri ***Indonesia*** Town of East Java. It is the centre of a rice-growing region and is famous for its handicrafts such as basketwork, silverwork, rattan and batik.

Population 200 000

Keeling Islands ***Indian Ocean*** See COCOS ISLANDS

Keelung (Chi-lung) ***Taiwan*** Major seaport and industrial centre on the north coast about 30 km (20 miles) north-east of T'ai-pei. One of the oldest settlements in Taiwan, it was occupied by Spanish and Dutch colonists in the 17th century and became a flourishing port during the period of Japanese colonial rule over Taiwan (1895-1945). In parallel with the Taiwanese economy, it has grown rapidly since the mid-1960s and now handles more than 30 million tonnes of cargo a year. The coastline to the north-west is famed for its sculptured rock formations.

Population 351 200
Map Taiwan Bb

Kefallinia ***Greece*** See CEPHALONIA

Kefar Nahum ***Israel*** See CAPERNAUM

Keflavík ***Iceland*** Town and site of the country's main civil airport, about 35 km (22 miles) south-west of REYKJAVIK. The airport, part of which is a US military base, was built in 1941 as a wartime staging post for the Allies. There is a fishing harbour nearby.

Population 6700
Map Iceland Bb

Keihin ***Japan*** The country's prime industrial zone. It lies in south-east Honshu island and includes the cities of Tokyo and Yokohama. It grew rapidly in the 1930s as the country expanded its armaments industry ahead of the Second World War. Since 1945 it has switched to car manufacturing, printing and publishing, shipbuilding, oil refining and steelmaking.

Map Japan Cc

Kékes ***Hungary*** See MÁTRA

Kelang ***Malaysia*** River rising in central Peninsular Malaysia and flowing through the national capital, Kuala Lumpur, into the Strait of Malacca at Port Kelang (formerly Port Swettenham). This modern container port has been expanded to reduce Malaysia's dependence on Singapore. The town of Kelang, 19 km (12 miles) upstream from the port, is a dormitory area for the capital. The river valley is a developing industrial area.
Population (town) 192 100
Map Malaysia Bb

Kelantan ***Malaysia*** Densely populated rice-growing state in north-east Malaysia on the Thai border, covering 14 943 km² (5769 sq miles). Development of the region – one of Malaysia's poorest – has been slow, in part because of its isolation and also because Kelantan, heartland of Islamic fundamentalism in Malaysia, did not vote for the ruling government coalition in the 1960s. There are, however, some land-development projects in the jungle-covered interior and a new highway now links it directly with the west coast.
Population 895 400
Map Malaysia Bb

Kelibia ***Tunisia*** Charming fishing port and tourist centre on the south-eastern shore of the Maoun Peninsula, which ends at Cape Bon. It is dominated by a great Roman castle that overlooks the port. Tomatoes, oranges and grapes are grown in the surrounding country and muscatel wine is made there. Farther along the coast stands the Cape Bon lighthouse.
Population 24 600
Map Tunisia Ba

Kells (Ceanannus Mór) ***Ireland*** Ancient town 58 km (36 miles) north-west of Dublin. St Columba founded a 6th-century monastery there which became a refuge for Iona monks. It was dissolved in 1551, but a round tower, high crosses and other buildings survive. The 8th-century *Book of Kells*, an illuminated edition of the Gospels made by the monks, and one of the most precious books in the world, is kept in the library at Trinity College, Dublin.
Population 2600
Map Ireland Cb

Kelsty ***Poland*** See KIELCE

Kemerovo ***USSR*** Industrial city in southern Siberia, 195 km (120 miles) east and slightly north of Novosibirsk. It developed mainly after 1928 with the opening of the nearby Novokuznetsk coalfield. As well as coal and coke, its industries include chemicals, pharmaceuticals, plastics, paints and farm machinery.
Population 502 000
Map USSR Jc

Kemi ***Finland*** Seaport founded in 1869 on the north coast of the Gulf of Bothnia, about 25 km (15 miles) from the Swedish border. It lies at the mouth of the Kemi river, Finland's richest source of hydroelectric power. The river, which is 512 km (318 miles) long, rises in Finnish Lappland. The town is one of the main centres of Finland's timber industry.
Population 26 500
Map Finland Cb

Kempenland (Campine) ***Belgium*** Heath and forest-covered plateau about 45 km (28 miles) north of the city of Liège, containing Belgium's only working coalfield. The region, which extends just across the Dutch border, covers about 6000 km² (2300 sq miles). The largest town is Genk.
Population 200 000
Map Belgium Ba

Kempton Park ***South Africa*** Industrial town 19 km (12 miles) north-east of Johannesburg. One of the world's largest dynamite factories is at nearby Modderfontein just to the west.
Population 289 300
Map South Africa Cb

Kenana ***Sudan*** Vast, state-run sugar plantation, covering some 336 km² (130 sq miles), lying about 290 km (180 miles) south of the capital, Khartoum, in southern El Gezira, just east of the White Nile. It was officially opened in 1951 and is one of the world's largest sugar complexes with finance from the Sudanese,

▼ FANTASTIC FLORA Giant plants such as 9 m (30 ft) tall groundsels grow on the upper slopes of Mount Kenya and other high peaks in equatorial Africa.

Kuwait and Saudi Arabian governments, and other international groups. Kenana draws its irrigation water from the White Nile and in 1984-5 produced over 300 000 tonnes of sugar.

Kenema ***Sierra Leone*** Eastern regional capital in the Kambui Hills, some 225 km (140 miles) south-east of the national capital, Freetown. Kenema is a market for locally produced cocoa, palm oil (used in margarine and cooking fats), coffee, ginger and timber, and for diamonds which are sifted from local river gravels. The town has a sawmill and furniture factory. There are also government and missionary secondary schools, a Catholic teacher training college for women, a technical institute, a government library and a hospital.

Population 14 000
Map Sierra Leone Ab

Kengtung ***Burma*** Town in Shan state, one of the areas making up the lawless opium poppy growing area known as the GOLDEN TRIANGLE. The town is reputed to be the centre of the illegal opium trade carried out in the surrounding hills east of the Salween river.

Population 174 000
Map Burma Cb

Kénitra (Mina Hassan Tani) ***Morocco*** Port 40 km (25 miles) north of Rabat on the Sebou river; formerly Port Lyautey. Founded by the French in 1913, it is the main outlet for the RHARB plain and the agricultural areas of Fès and MEKNES. Its industries include textiles, fish processing, fertilisers and tobacco manufacture. Nearby is Mehdiya, a former Carthaginian trading post and the site of the 1st-century Roman port of Thamasada; it is now a beach resort.

Population 189 000
Map Morocco Ba

Kennedy, Cape ***USA*** See CANAVERAL, CAPE

Kent ***United Kingdom*** County in the south-east corner of England, covering 3732 km² (1441 sq miles) and crossed by the chalk hills of the North DOWNS. The centre of the county, called The Weald, was formerly densely wooded. Kent is known as the 'Garden of England' because of its orchards, hops and spring flowers. The pastures of Romney Marsh are noted for sheep. Industries such as paper making and cement line the Thames and Medway rivers. The MEDWAY TOWNS also used to have a naval dockyard. Margate and Ramsgate are popular tourist resorts, while the ports of DOVER and Folkestone handle most of Britain's cross-Channel traffic. Inland, tourists flock to the cathedral city of CANTERBURY. The county town is Maidstone.

Population 1 494 000
Map United Kingdom Fe

Kentucky ***USA*** South-central state covering 104 623 km² (40 395 sq miles) of mostly rolling Blue Grass country, between the Cumberland Mountains in the south-east and the Ohio river to the north. Tobacco, whiskey, horses, machinery, foodstuffs and cigarettes are the main products. The state is also the country's largest coal producer. FRANKFORT is the state capital.

Population 3 726 000
Map United States Ic

Kenya See p. 354

Kenya, Mount ***Kenya*** Long-extinct volcano rising to 5200 m (17 058 ft), high enough to have glaciers between its jagged peaks even though it lies almost on the Equator. The slopes between 4600 and 3200 m (15 100 and 10 500 ft) consist of moorland dotted with giant heather, groundsel and lobelia. Below that there is forest, including a zone of bamboo. Below about 1700 m (5600 ft), the southern and eastern slopes are farmed by Kikuyu, Embu and Meru people. The main crops are maize and beans, with coffee and tea as cash crops. The drier western slopes have grain and livestock farms. The first recorded ascent of the mountain was by the British geographer Sir Halford Mackinder in 1899.

Map Kenya Cb

Kerala ***India*** South-western state bordering on the Arabian sea, covering an area of 38 863 km² (15 005 sq miles). It has a high rainfall, in many places above 4000 mm (157 in) annually and much tropical rain forest. The coastal area has many lagoons and channels, and most people travel by small boats.

Its capital is the port of TRIVANDRUM, and the inhabitants make their living mainly from fishing, rice, timber, rubber and tea.

Population 25 453 700
Map India Be

Kerch' ***USSR*** Crimean port between the Sea of Azov and the Black Sea. One of Russia's oldest cities, it was founded by the Greeks in the 6th century BC under the name Pantikapaion. It was later ruled by Rome, Genoa, Turkey and (from 1771) Russia. The area is rich in archaeological finds, including the remains of a Greek acropolis and Roman burial mounds and catacombs; articles from these are kept in the local history museum. Kerch' is one of the Crimea's chief fishing bases, and exports iron ore. Industries include iron, chemicals and shipbuilding.

Population 166 000
Map USSR Ed

Kerguelen ***Indian Ocean*** Largest of a group of about 300 islands, islets and rocks, Kerguelen is part of the French Southern and Antarctic Territories, and lies about 5310 km (3300 miles) south-east of Cape Town, in the southern Indian Ocean. It was aptly named Desolation Island by the French navigator Kerguélen-Trémarec, who discovered it in 1772. Much of the western part of its 3414 km² (1318 sq miles) is a huge glacier; snowfields carpet the centre, giving way to peat marshes and glacial lakes. The island is mountainous, rising to 1850 m (6070 ft) at Mount Ross in the south.

Kerguelen has belonged to France since 1893, but was not occupied effectively until 1949. A permanent base and scientific centre, Port-aux-Français, was built the following year and now has a staff of around 75.

Map Indian Ocean Ce

Kericho ***Kenya*** Highland town about 195 km (120 miles) north-west of Nairobi. It is the centre of Kenya's leading tea-growing district.

Population 40 000
Map Kenya Cb

Kérkira ***Greece*** See CORFU

A DAY IN THE LIFE OF A KENYAN FARMER'S WIFE

Under the pure blue Kenyan skies, another long day is beginning for Wamboi Muthoka, a 25-year-old farmer's wife from Kenya's largest tribe, the Kikuyu. She and her husband, Kitau, 27, own a coffee farm on the slopes to the south of Mount Kenya, and have six children, all aged under eight. The young woman labours from dawn to dusk on the farm. As well as planting, weeding or harvesting, Wamboi has to prepare food for her family and take care of them.

About half of the 1.6 hectare (4 acre) farm is given over to coffee, and the rest is planted with maize, bananas, beans and other vegetables, which make up the family's staple diet. The coffee crop brings in hard cash, equivalent to around US$100 a year.

The first task of Wamboi's day is to feed and milk the family's two cows. Mostly they are fed on maize stalks, supplemented by any available vegetation. Then Wamboi packs off her two eldest children, aged six and seven, to the nearby primary school, a timber hut that all the local people helped build a few years ago. Although primary school is free in Kenya, the cost of uniforms and books is a constant strain on family finances. (Fortunately shoes are not obligatory.) Despite the cost, few people question the need for an education – which can earn their children a prized government job in Nairobi – even if they have had no education themselves. Wamboi had none, and Kitau only a few years of primary schooling.

Wamboi's four other children are not yet old enough for primary school. The latest addition to the family, a six-month-old girl, goes everywhere strapped tightly to her mother's back, while the others spend the day doing odd jobs or running around the farm. Wamboi's mother and her husband's parents also live on the farm. The whole family is housed in two grass-roofed huts made of wood and mud. Kitau has just one wife – some of the older Kikuyu have several.

When her day's work in the fields is done, Wamboi goes down to the stream, 20 minutes away, to bring back water in a huge jar which she balances on her head. She also has to go out in search of firewood – an increasingly difficult task, as nearby trees are felled for fuel, and collecting wood these days can take as much as an hour and a half. Wamboi carries the firewood in a large bundle on her back supported by a leather strap around her forehead.

She and her family will then spend the evening either chatting with their neighbours, or listening to the 'Voice of Kenya' on their large old radio.

Kenya

THE CRADLE OF MANKIND TWO MILLION YEARS AGO THAT IS NOW COPING WITH THE FORMATIVE YEARS OF NATIONHOOD

For a country whose history as a single entity extends back less than 100 years and whose life as a nation is a bare quarter-century old, it may seem paradoxical that Kenya is often described as the 'cradle of mankind'. It was here – on the shores of Lake TURKANA (formerly Lake Rudolf) in the country's remote north-west – that the skull and leg-bones of '1470 Man' (the number is a museum catalogue reference), who probably lived some 2 million years ago, were found. Anthropologists believe that Modern Man descended from such a creature.

Before the 19th century, the only part of Kenya known to the outside world was its coast – more than 500 km (310 miles) of white beaches, fringed by palm trees, fanned by the trade winds and protected from the Indian Ocean by a continuous flank of coral reef. From medieval times onwards, traders from Europe, Asia and Arabia came here in search of slaves and ivory. Some of them settled, and today's coastal population, the Swahili, are a racial blend of Arabs and black Africans.

It was not until the late 19th century that British explorers ventured into the interior, and even then their interest was not so much in Kenya itself as in charting a route from the coast to the British protectorate in Uganda. To underline this point, the railway that was subsequently built between MOMBASA on the coast and KISUMU on Lake VICTORIA was called the Uganda Railway; NAIROBI, now Kenya's capital and one of the largest cities in East Africa, was the site chosen for the railway workshops and main depot.

The British were, however, quick to realise that some of the land they had come upon, almost by accident, was in fact very fertile. By 1914 several thousand British farmers had settled in the south-western region. They appropriated large tracts of well-watered and volcanic soil from the Kikuyu and Masai peoples and, in what became known as the 'White Highlands', employed cheap African labour to produce crops such as coffee. The resentment caused by this colonisation erupted in 1952 when the Mau Mau, a terrorist organisation which bound its Kikuyu members to secrecy with ritual oaths, began its campaign of violence.

The south-west is, to this day, Kenya's most

▶ GLORIOUS DAWN The flamingos of Lake Nukuru awaken at first light. The lake – protected as a national park – lies in the Great Rift Valley, a vast geological fault which slashes 6400 km (4000 miles) across Africa, cutting through Kenya's highlands.

densely inhabited region, accounting for the bulk of the country's population and almost all its economic production. In contrast the rest of Kenya is very arid. In the east is a wilderness of dry scrubland and desert, dissected by the country's longest river, the TANA, and inhabited by a small nomadic population. The north-west is similarly dry, although it does include Lake Turkana. The lake which was once described as a 'sinister jade sea set in black, purple and blood-red lava' lies in the GREAT RIFT VALLEY.

As a single entity Kenya is entirely a colonial creation, first under the control of the British East Africa Company then, from 1895, as a British protectorate. It was proclaimed a colony by the British in 1920 and granted its independence in 1963. Its first president, Jomo Kenyatta, who had himself been released from prison only two years before as leader of the Mau Mau, succeeded in keeping the tribal tensions which have always threatened to divide Kenya well beneath the surface. Despite an attempted coup by junior air force officers in 1982, his successor, President Daniel arap Moi, has since 1978 continued to steer the country smoothly into the 1980s.

Nevertheless, Kenya is still a collection of different peoples closely identified with specific territories rather than an integrated nation. About 1000 years ago, Bantu-speaking migrants – the ancestors of the present-day Kikuyu, Kamba and Luhya tribes – entered Kenya from the south up the coastal belt. Then, 500 years later, other peoples, including the nomadic Masai, entered Kenya from the north. Later, the Luo, who are now seen as the main political rivals of the Kikuyu, entered from the west. Finally, there are still some 78 000 Asians in the towns – the descendants of those who were originally brought from India to help build the railway but stayed on to play a vital role in shaping the trading economy in the early years of the century.

The current dominance of the Kikuyu in Kenyan national affairs reflects not only their numbers – although only 20 per cent of the population, they are by a good margin the biggest ethnic group – but also their relatively high levels of education and income and their occupation of the national heartland which includes the capital.

PROBLEMS OF POPULATION GROWTH

Although it affects directly only a very small proportion of the population, tourism is Kenya's third most important source of foreign exchange after coffee and tea. But the tourists who visit Kenya are appreciating only selected aspects of the country. They may not be aware that recent droughts – though not typical and not on the Ethiopian scale – have brought widespread hardship, and that in 1984, in the worst drought for over 50 years, thousands of cattle died.

One of Kenya's most pressing problems is its population growth. Currently it is increasing at over 4 per cent a year, one of the highest rates in the world. If this trend continues, the population will double every 20 years. In recent years public health programmes have brought a sharp reduction in the death rate but, despite a more active family planning programme than in most parts of Africa, there has been no compensating drop in the birthrate. As a result, half of Kenya's population is under 15 years old and, unless enough jobs can be created, the people of Kenya will soon outstrip the country's economic capacities.

Not only is the population growing too fast, it is also unevenly distributed. A bare 20 per cent of the land is arable. Over one-third of the population is concentrated in a very restricted part of the central highlands, and a further third is concentrated in the extreme west, near Lake Victoria. Both these regions are now experiencing a severe shortage of land and, as people attempt to cultivate more intensively or spread onto land that is only marginally suited to cultivation – or encroach onto national parks – there are bound to be environmental problems such as soil erosion.

In the countryside there are wide disparities of income, despite the great structural changes to the rural economy that have been made since independence. During the colonial period, cultivated land consisted of large European-run farms, some producing coffee and tea for export, and small plots where African farmers scratched enough to feed the family. In the 1950s there developed smaller African-run coffee and tea farms. Then, after 1963, most of the settler farms changed hands: some were bought outright by the new African elite; others were used in resettlement schemes for the landless; and many began to be farmed by consortia and cooperatives. However, these changes have not eliminated the disparities in income.

It is for this reason that many Kenyans migrate to the towns, which are growing rapidly although they still account for only 15 per cent of the population. The cities are still essentially centres of administration and commerce, but industry is growing, with a special emphasis on agricultural processing and producing substitutes for imported consumer goods. The country's largest single factory is an oil refinery in Mombasa which meets all the petroleum fuel requirements of Kenya, Uganda, Rwanda and Mauritius with capacity to spare.

Kenya is poorly endowed with minerals, however, and the country has only very limited hydroelectric potential; for most people the only available fuel is wood.

The country's greatest tourist attraction and a major natural resource is its wildlife. It was therefore a far-sighted move – and one applauded by conservationists all over the world – when, in 1977, the government banned hunting. Today, over 250 000 tourists a year visit AMBOSELI, TSAVO and a dozen other national parks and game reserves.

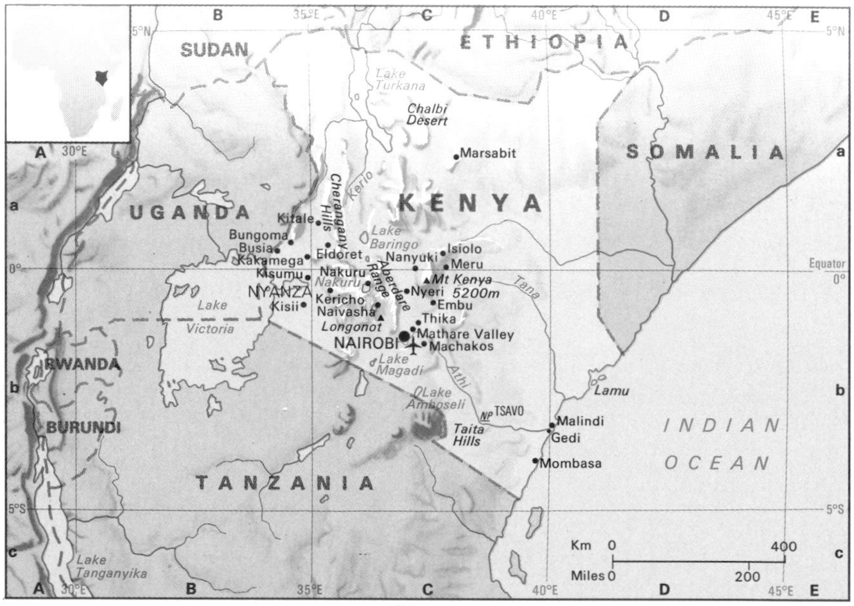

KENYA AT A GLANCE
Area 582 646 km² (224 960 sq miles)
Population 21 040 000
Capital Nairobi
Government One-party republic
Currency Kenya shilling = 100 cents
Languages Swahili, English, tribal languages
Religions Christian (66%), tribal religions (26%), Muslim (6%)
Climate Tropical; hot and humid on coast, temperate inland, dry to the north. Average temperatures in Nairobi range from 11-21°C (52-70°F) in July to 13-26°C (55-79°F) in February
Main primary products Maize, millet, beans, cassava, potatoes, sweet potatoes, sugar, coffee, tea, cotton, sisal, cattle, pyrethrum (flower used in insecticides)
Major industries Agriculture, oil refining, cement, food processing, tourism
Main exports Coffee, tea, petroleum products, fruit and vegetables, sugar, chemicals, cement, hides and skins
Annual income per head (US$) 315
Population growth (per thous/yr) 42
Life expectancy (yrs) Male 54 **Female** 58

Kerkouane ***Tunisia*** Impressive ruins of a Carthaginian town at the eastern end of Cape Bon which fell to the Romans in the Punic Wars (264-146 BC). Nothing was built on the old site, which remains intact. Excavations have revealed luxurious houses, shops and workshops for the production of the purple dye known as Tyrian purple. The town was a retreat for Carthaginian aristocrats, and appears to have been hastily abandoned at the coming of the Romans.

Kermadec-Tonga Trench Ocean trench in the floor of the south-western PACIFIC, extending from just north of New Zealand to south of Western Samoa. It is flanked to the west by the volcanic Kermadec and (farther north) Tonga islands. Along the trench, the westward-drifting Pacific plate is descending beneath the Indo-Australian plate (see PLATE TECTONICS), causing earthquakes and the volcanoes which make this area part of the Pacific 'ring of fire'. The trench was discovered in 1895.
Greatest depth Kermadec Trench 10 047 m (32 963 ft); Tonga Trench 10 822 m (35 499 ft)
Map Pacific Ocean Ed

Kerman ***Iran*** Capital of Kerman province, about 820 km (510 miles) south-east of the national capital, Tehran. Its products include shawls, carpets and brassware. Founded in the 3rd century, the town has numerous ancient buildings, including the Friday Mosque (built in 1348) and the Pamenar Mosque, which has a portal decorated with 14th century mosaics. The surrounding province covers an area of 220 150 km² (85 000 sq miles).
Population (town) 238 800; (province) 1 085 000
Map Iran Ba

Kermanshah ***Iran*** See BAKHTARAN

Kerry (Ciarraí) ***Ireland*** County covering 4701 km² (1815 sq miles) on the south-west coast with mountainous peninsulas jutting out into the Atlantic Ocean. The county town Tralee stands at the neck of Dingle Peninsula, which rises to 953 m (3127 ft) at Brandon Mountain. Sandy beaches, cliffs, lakes and lush unspoilt countryside make Kerry popular with tourists.
Population 122 800
Map Ireland Bb

Keszthely ***Hungary*** Medieval town, cultural centre, and modern resort with fine sandy beaches on the west shore of Lake Balaton in western Hungary. Its attractions include: the great 18th-century Baroque Festetics Palace, richly furnished and now a venue for musical performances; collections of folk arts in the Balaton Museum; and numerous facilities for water sports. The town's prestigious Agricultural University, originally founded as the Georgikon College in 1797, is the oldest agricultural university in Europe.
Population 22 600
Map Hungary Ab

Ketrzyn (Rastenburg) ***Poland*** Market town in the north-east, west of Lake Mamry and about 30 km (20 miles) south of the Russian border. Near it is Poland's largest state farm-industrial complex, which produces wheat, sugar beet and cattle. Ketrzyn lies in the former German province of East Prussia, and about 8 km (5 miles) from the town are the remains of the 'Wolf's Lair' – the massive concrete bunker headquarters of Adolf Hitler's general staff, where senior officers tried and failed to assassinate the Nazi leader on July 20, 1944.
Population 27 200
Map Poland Da

▲ LONELY HEADLAND A constant procession of Atlantic rollers breaks below lofty Slea Head at the tip of the Dingle Peninsula in Kerry.

kettle, kettle hole Depression left in a mass of glacial DRIFT, apparently formed by the melting of an isolated block of ice.

Key West ***USA*** Southernmost settlement of the continental USA. The city lies in the state of Florida at the end of the chain of coral islands known as the Florida Keys, about 210 km (130 miles) south-west of Miami. It is a fishing port and tourist centre connected to the mainland through the Keys. The home of writer Ernest Hemingway (1899-1961) is a museum.
Population 24 900
Map United States Jf

Khabarovsk ***USSR*** Second largest city in the Soviet Far East after Vladivostok, standing 35 km (22 miles) from the north-eastern tip of China on the Amur river. It has an embankment walk with views across a treeless plain into China. It is a centre for the local fur trade, and its industries include oil refining, machinery, shipbuilding, furniture and timber.
Population 569 000
Map USSR Od

Khabur ***Turkey/Syria*** River rising from tributaries in south-east Turkey and flowing 320 km (200 miles) south-east into Syria. It irrigates the desert area in the north-east and ends in the Euphrates just below Dayr az Zawr. The most important town on the river is Al Hasakah. The Abbasid Bridge, built across the Khabur in the 8th century, is considered a perfect example of early Arab architecture.
Map Syria Cb

Khajuraho ***India*** Greatly revered temple site about 150 km (93 miles) south of the city of Kanpur in north-central India. There were originally 50 Hindu and Jain (a religion between Hinduism and Buddhism) temples dating from the 10th century. They were scattered over 20 km² (8 sq miles), and only 22 of them have survived. The sculpture on many of the temples is highly erotic.
Map India Cc

Khamis Mushait ***Saudi Arabia*** Rapidly growing military and air force town in the Asir mountains, situated 30 km (19 miles) east of Abha. It is the home of the King Faisal Military City (a major army base and training centre) and the headquarters of the Saudi Arabian air force. There is a strong European atmosphere due to the many expatriates who work there.
Population 54 000
Map Saudi Arabia Bc

Khammouan ***Laos*** Province of central Laos, stretching from the Mekong river to the Vietnamese border. It is the main routeway through to Vietnam, with two passes through the Truong Son chain of mountains that separate the two countries.
Population 259 000
Map Laos Bb

Khaniá (Chania) ***Greece*** Port on the north-west coast of CRETE. Known in ancient times as Cydonia, the town was occupied by the Venetians (1252-1645), who called it Canea, and then the Turks (1645-1898). Khaniá still retains its old fortified quarter with a 14th-century rampart and there are Venetian churches and a Turkish lighthouse. Kastélli, a 16th-century Venetian town with a fortress, stands on a sandy beach 42 km (26 miles) to the west.
Population 62 000
Map Greece Dd

▲ **NORTH-WEST FRONTIER A motor road snakes up the Khyber Pass in Pakistan. British infantry once toiled over the rugged terrain to subdue fierce Pathan tribesmen and secure the frontier.**

Kharg Island (Khark) ***Iran*** Small island about 3 km (2 miles) long in The Gulf, off the Shatt al Arab waterway, about 30 km (20 miles) from the coast. Iran's largest oil terminal on the island has been repeatedly bombed during the war between Iran and Iraq which began in 1980.

Map Iran Bb

Khar'kov ***USSR*** Second largest city in the Ukraine after Kiev, about 640 km (400 miles) south of Moscow. It expanded rapidly after 1880 as an industrial centre and market city serving the Ukraine's fertile farming region and the Donets Basin coalfield. Today its factories produce tractors, combine harvesters, mining machinery, tools, railway wagons, electrical equipment and foodstuffs.

Population 1 536 000

Map USSR Ed

Khartoum ***Sudan*** National capital, at the confluence of the Blue and White Niles, and one of the world's hottest capital cities, with average temperatures reaching 41°C (106°F) in June and 32°C (90°F) in January.

It was founded by the Egyptian ruler Muhammad Ali in 1823, destroyed by the religious leader, the Mahdi, in 1885 after a siege in which the former governor of Sudan, General Gordon, was killed; and was rebuilt by the British after the defeat of the Mahdi's followers by Lord Kitchener in 1898. Kitchener laid out the new city centre in the form of a Union Jack, and there are still many fine examples of colonial architecture, besides mosques, bazaars, hotels and a national museum.

The city centre lies between the two rivers on a curving strip of land shaped rather like an elephant's trunk – and its name comes from the Arabic *Ras-al-hartum*, meaning 'end of the elephant's trunk'. Khartoum North, across the Blue Nile, is a modern city of factories and offices and the country's industrial capital. The city's industries include food processing, pharmaceuticals, leather and textiles. OMDURMAN across the White Nile is the old Arab town. Hence the conurbation – Khartoum, Khartoum North and Omdurman – is known as 'The Three Towns'.

Population (Khartoum) 561 000; (Khartoum North) 341 200; (Omdurman) 526 300

Map Sudan Bb

Khaskovo ***Bulgaria*** Industrial town and capital of Khaskovo province, 56 km (35 miles) south of Stara Zagora in fertile THRACE. Industries include silk, textiles, tobacco products, food processing, wine and metal goods. After the two World Wars, many refugees from Macedonia and Grecian Thrace settled there.

Population 88 700

Map Bulgaria Bc

Khawr Fakkan ***United Arab Emirates*** Small port on the east coast of the Gulf of Oman, 25 km (15 miles) north of Al Fujayrah. It is part of the enclave belonging to the emirate of SHARJAH on The Gulf. The natural harbour is used by dhows (single-masted Arab coastal vessels), and there are plans to turn it into a container port for imports to the UAE. It has a backdrop of mountains.

Population 2000

Map United Arab Emirates Ca

Khayelitsha ***South Africa*** New black township on the False Bay coast, 40 km (25 miles) south-east of Cape Town.

Kherson ***USSR*** Port and shipbuilding city in the Ukraine on the Dnieper river 25 km (16 miles) from the Black Sea. It was founded in 1778 as a fort and naval base by the Russian soldier and statesman Prince Gregory Potemkin (1739-91) for the empress Catherine the Great, and much of his fort – including its walls, gates and arsenal – survives. Its industries include textiles, machinery, oil refining and food processing.

Population 340 000

Map USSR Ed

Khiva ***USSR*** Town on the banks of the Amuda-r'ya river in Uzbekistan, about 250 km (160 miles) south of the Aral Sea. It was once a major centre of the slave trade, and remains a largely intact medieval city, with mosques, minarets, bazaars, gates and shaded courtyards. Many of its wooden houses are exquisitely carved, while other buildings are painted or faced with ceramic tiles in vivid floral or geometrical designs. It produces silk, cotton and carpets.

Population 24 200
Map USSR Hd

Khodzhent ***USSR*** See LENINABAD

Khojak Pass ***Pakistan*** Pass 2200 m (7220 ft) up in the mountains of Baluchistan, on the road between Quetta and Kandahar in Afghanistan. Underneath, there is a railway tunnel.

Map Pakistan Cb

Kholm ***Poland*** See CHELM

Khon Kaen ***Thailand*** Regional capital and university city about 400 km (250 miles) north-east of Bangkok, in one of the country's poorest regions. Farmers in the surrounding province of the same name mostly eke out a living by growing rice, cassava and sugar cane.

Population (city) 215 800; (province) 1 521 400
Map Thailand Bb

Khone Falls ***Laos*** Complex group of waterfalls on the Mekong river in the extreme south of the country. The falls are a formidable barrier to navigation on the river. Before the Laotian civil wars of the 1960s and 1970s, the falls were bypassed by a railway portage system, but this has been abandoned, leaving Laos with no outlet to the sea along the river.

Map Laos Bc

Khor (Al Khawr) ***Qatar*** The state's second city, situated 40 km (25 miles) north of the capital, Doha, on the shore of a narrow inlet. Khor used to be a major fishing harbour, and dhows (single-masted Arab coastal vessels) still bring their catches there. The seashore is being turned into a leisure area, and the city has numerous modern buildings – including schools, a hospital and a sports complex. Khor is a major archaeological site; discoveries include burial chambers and bones dating back to 7000 BC. The Museum of Archaeology contains information on the sites.

Map Qatar Ab

Khorasan ***Iran*** Mostly arid north-eastern province, 313 337 km^2 (120 980 sq miles) in area, bordered by the USSR and Afghanistan. Its capital is the holy city of MASHHAD. Cereals, cotton, tobacco, sugar beet, fruit and nuts are grown around oases such as Birjand and Qayen. The main industries are rug making, food processing and tanning.

Population 3 267 000
Map Iran Ba

Khorramabad (Khurramabad) ***Iran*** Wool and fruit market town about 400 km (250 miles) south-west of the capital, Tehran. Its massive citadel was the stronghold of the Lurs, a fiercely independent tribe who robbed and raided the surrounding area for several centuries until Shah Abbas I captured the citadel in the 17th century.

Population 105 000
Map Iran Aa

Khorramshahr ***Iran*** Inland port on the Shatt al Arab waterway, near the border with Iraq. It is the terminus of the Trans-Iranian railway and normally handles almost half of Iran's international trade, though this has been severely disrupted by the war with Iraq. Exports include cotton, dates, hides and skins.

Population 146 700
Map Iran Aa

Khouribga ***Morocco*** A centre of phosphate mining and associated industries, 105 km (65 miles) south-east of Casablanca.

Population 128 000
Map Morocco Ba

Khulna ***Bangladesh*** Coastal district in the south-west bordering India. Although most of its people were Hindu, Khulna was awarded to predominantly Muslim Pakistan after Partition in 1947. Many of the Hindus fled to India during the riots which followed Partition. In 1970 the district was struck by one of the worst cyclones of this century when at least 250 000 people died, mostly drowned by huge tidal waves. The port of Khulna is the chief town. It exports timber and builds ships.

Population (district) 4 329 000; (town) 646 400
Map Bangladesh Bc

Khumbu ***Nepal*** **1.** Mountain region in the Himalayas that includes Mount EVEREST. **2.** A glacier which flows from a shoulder of Everest over the dangerous Khumbu icefall.

Map Nepal Ba

Khunjerab Pass ***China/Pakistan*** See MINTAKA PASS

Khurramabad ***Iran*** See KHORRAMABAD

Khuzestan (Khuzistan) ***Iran*** South-western province, covering 64 651 km^2 (24 962 sq miles) between the Zagros Mountains and the Iraqi border and The Gulf. It is the main source of Iran's oil and produces dates, melons, vegetables, rice and cotton. Most of Iran's Arab minority, who are Sunni Muslims, live in the province. AHVAZ is the capital. SUSA, the capital of the Elamite kingdom some 4000 years ago, lies between the mountains and The Gulf.

Population 2 177 000
Map Iran Aa

Khyber Pass ***Afghanistan/Pakistan*** Most important pass between the two countries, carrying a modern road and railway line and an ancient caravan route across the Safed Koh mountains between Peshawar in Pakistan and the Afghan capital, Kabul. It is about 50 km (30 miles) long and reaches a height of about 1070 m (3520 ft). It has been a major trade and invasion route since ancient times; it may have been used by the Macedonian conqueror Alexander the Great in 327 BC and the Mongol Tamerlane in 1398; it was certainly of great importance to the British during the Afghan Wars in the 19th century.

The pass was closed briefly after the Russian invasion of Afghanistan in 1979 but is open now. By 1984, Pakistan had received over 3 million Afghan refugees – 20 per cent of the Afghan population – by this route.

Map Afghanistan Ba; Pakistan Ca

Kicking Horse Pass ***Canada*** Route through the Rocky Mountains between Alberta and British Columbia. It rises to 1627 m (5338 ft), and carries the highest stretch of the Canadian Pacific Railway.

Map Canada Dc

Kiel ***West Germany*** Port on an estuary opening into Kiel Bay in the extreme north, and capital of the northernmost state of Schleswig-Holstein. It has the country's largest shipyards and is an international yachting centre. The Kiel Canal, 45 m (148 ft) wide and 14 m (46 ft) deep, was opened by Kaiser Wilhelm II in 1895 and runs for 96 km (60 miles) from Kiel to the mouth of the Elbe, north of Hamburg. It provides a passage for ocean-going ships between the Baltic and the North Sea.

Population 248 400
Map West Germany Da

Kielce (Kelsty) ***Poland*** Industrial city about 145 km (90 miles) south of the capital, Warsaw. The city has copper mines and marble quarries, and its factories produce electrical goods, machinery and foodstuffs.

Population 197 000
Map Poland Dc

Kiev (Kiyev, Kyiv) ***USSR*** The country's third largest city (after Moscow and Leningrad), on the banks of the Dnieper river, about 750 km (470 miles) south-west of Moscow. The Russians call it 'the Mother of Cities', for it is one of the oldest cities in the land, dating from the 6th century. Kiev became the capital of the powerful Kiev-Rus principality (the infant Russian state) in 882, and its Grand Prince, having been converted, made Christianity the state religion in 988. The Mongols sacked the city in 1240, and it was part of Lithuania and then Poland before it was restored to Russia in 1654.

The city grew rapidly once the rail links with Moscow (1863) and Odessa (1870) were forged. It was the scene of bitter fighting in the aftermath of the Russian Revolution of 1917, and was severely damaged during the Second World War, when 200 000 people died and the city centre was razed. It was rebuilt with wide, tree lined boulevards and parks which enhance sur viving relics such as the 11th-century St Sophia's Cathedral, now a museum.

Kiev is now the capital of the Ukraine, with engineering, electrical, chemicals, food-processing and precision tools industries. Its university was founded in 1834 and is now an academy of science.

In April 1986 the world's worst recorded civil nuclear accident occurred at a power station at Chernobyl', about 90 km (55 miles) north of Kiev. One of the plant's four reactors overheated and exploded, releasing tonnes of radioactive debris into the atmosphere; the cloud of fallout spread over much of the Ukraine, parts of Poland and Scandinavia, and as far away as Britain. The immediate danger was not contained until the following month, by which time thousands of people had been evacuated from the area. A number died of radiation exposure – notably firemen who fought the blaze that followed the explosion – and a large area of farmland was made unusable.

Population 2 411 000
Map USSR Ec

Kigali ***Rwanda*** National capital standing in the centre of the country. Before independence in 1962, the Belgians administered the country from the present capital of Burundi, Bujumbura. Since becoming Rwanda's capital Kigali's population has soared from the colonial figure of some 5000.
Population 170 000
Map Rwanda Ba

Kigoma ***Tanzania*** Capital of Kigoma region, on the east shore of Lake Tanganyika about 50 km (30 miles) south of the border with Burundi. It is the terminus of the railway from Dar es Salaam and the main transit port for Burundi. It ships timber, cotton and tobacco. About 8 km (5 miles) to the south is the old fishing settlement of Ujiji, where the British explorer and journalist Henry Morton Stanley found the missing Scottish missionary and explorer Dr David Livingstone on October 28, 1871.
Population 50 000
Map Tanzania Aa

Kildare (Cill Dara) ***Ireland*** County covering 1694 km² (654 sq miles) in the south-east, in the flattest part of Ireland. The Grand Canal, built in the 18th and early 19th centuries to carry trade between Dublin and Shannon, passes through the county, and now carries holiday boats. Naas is the county town.

The market town of Kildare, 48 km (30 miles) south-west of Dublin, is the heart of Ireland's horse-racing industry, and home of the National Stud and Horse Museum. The Irish Derby is held at The Curragh just to the east each June.
Population (county) 104 100; (town) 3150
Map Ireland Cb

Kilimane ***Mozambique*** See QUELIMANE

Kilimanjaro ***Tanzania*** Africa's highest mountain in the north-east of Tanzania near the Kenyan border. Kilimanjaro has two peaks – Kibo, which rises to 5895 m (19 340 ft), and Mawenzi, rising to 5149 m (16 893 ft). Both were once active volcanoes, and although only 3 degrees south of the Equator, the summit of Kibo is permanently covered in snow. Lower down, on the fertile slopes of the mountain below 2000 m (6560 ft), bananas, coffee and maize are grown.

Kilimanjaro was first conquered in 1889 by two Germans – Ludwig Purscheller and Hans Mayer – who scaled Kibo. Mawenzi was first climbed in 1912 by another German, Fritz Klute. Today Kibo is climbed by several thousand people each year.
Map Tanzania Ba

Kilkenny (Cill Chainnigh) ***Ireland*** City 102 km (63 miles) south-west of Dublin amid a hilly county of the same name which covers 2062 km² (796 sq miles). The city was an Anglo-Norman stronghold, and its medieval castle, round tower, and two cathedrals still brood over close-packed streets. Parliaments met there, and in 1366 the Statute of Kilkenny made it illegal for the English to marry the locals.

Black marble was once mined in the surrounding hills. Now the city processes food and makes textiles and shoes.
Population (county) 70 800; (city) 10 100
Map Ireland Cb

Killarney (Cill Airne) ***Ireland*** Tourist centre and market town in County Kerry, 77 km ([illegible] miles) west and slightly north of the southern city of Cork. It is surrounded by mountains and lakes, and jaunting cars – open, two-wheeled carts with back-to-back seats, and drawn by ponies – carry visitors on trips through the fine valleys to the south of the town. Ireland's highest range, Macgillicuddy's Reeks, lies to the south-west, topped by the country's loftiest peak, Carrauntoohil (1041 m, 3414 ft).
Population 7700
Map Ireland [illegible]

Kilmain ***Mozambique*** See QUELIMANE

Kilmarnock ***United Kingdom*** See STRATHCLYDE

Kilwa Kisiwani ***Tanzania*** See KILWA KIVINJE

Kilwa Kivinje ***Tanzania*** Small market town on the coast, about 210 km (130 miles) south of Dar es Salaam. It was established in the 1830s by residents of Kilwa Kisiwani, who abandoned their island settlement about 30 km (20 miles) to the south in favour of a mainland site. Kilwa Kisiwani was founded in the 11th century by people from Shiraz in Persia and became the centre of a sea-trading area stretching south into present-day Mozambique. In 1505 it was largely destroyed by the Portuguese. It then became an Arab slave-trading centre until it was abandoned by its inhabitants. Its impressive remains include the Great Mosque, a palace and a Portuguese fort, dating from the 11th to 14th centuries.

Just north of Kilwa Kisiwani is Kilwa Masoko, a new administrative centre with a small airstrip.
Population 10 000
Map Tanzania Ba

Kimberley ***South Africa*** Town in north Cape Province, just west of the Orange Free State border and known as the diamond capital of the world. The Kimberley Mine, called the Big Hole, is the world's deepest open-cast mine. During the 44 years the mine was open, between 1871 and 1915, more than 20 million tonnes of earth were removed from it – mostly by men with picks and shovels. The hole – which is 1.6 km (1 mile) around and a total of about 1200 m (4000 ft) deep – also produced about 3 tonnes of diamonds. It is now part of the Kimberley Mine Museum. Kimberley has four other diamond mines and also produces asbestos, gypsum and manganese.
Population 153 900
Map South Africa Bb

Kimberley Plateau ***Australia*** Region – also known simply as the Kimberleys – of low, rugged hill blocks cut by gorges, covering about 420 000 km² (162 000 sq miles) in the far north of Western Australia. Monsoon rains fall from November to April (when the Fitzroy and ORD rivers become torrents) and are followed by dry, hot winter months. Several thousand Aborigines live in the Kimberleys, their last Western Australian stronghold.

The highest peak is Mount Ord (936 m, 3071 ft) in the King Leopold Ranges. Derby (population 3000) is the main town. Beef cattle are raised in the region and diamonds are mined at Argyle. There is cultured pearl farming at Cygnet and Kuri Bays.
Map Australia [illegible]

kimberlite The greenish-blue intrusive igneous rock in which diamonds are found, as at Kimberley in South Africa.

Kimch'aek (Gimchaeg) ***North Korea*** East-coast fishing port, 350 km (218 miles) from the capital, Pyongyang. It has fish oil and fish-processing factories, and also smelts iron.
Population 265 000
Map Korea [illegible]

Kincardineshire ***United Kingdom*** See GRAMPIAN

Kindia ***Guinea*** Hill town about 120 km (75 miles) north-east of the capital, Conakry, at the edge of the Fouta Djallon, about 400 m (1300 ft) above sea level. It was developed as a health resort for Europeans after the opening of the railway to Conakry in 1904, and it still has many colonial-style buildings, including offices, prisons, schools and housing. There are numerous streams and waterfalls in the region. Kindia markets locally grown fruits including bananas, and has a fruit research centre and a branch of the Pasteur Institute (researching into apes).
Population 79 900
Map Guinea Ba

Kindu ***Zaire*** River port on the Lualaba about 370 km (230 miles) west of the Burundi border. It is the terminus of a railway that extends southwards to the mineral-rich Shaba region with a branch line east to Kalémié on Lake Tanganyika. Below Kindu, the river is navigable as far as Ubundu (Ponthierville), which is linked by rail to the river port of Kisangani.
Population 50 000
Map Zaire Bb

Kings Canyon National Park ***USA*** Area of 1863 km² (719 sq miles) in the Sierra Nevada in California, 300 km (190 miles) north of Los Angeles. It has two canyons – each about 1220-1525 m (4000-5000 ft) deep – on the Kings river, and many giant sequoia trees.
Map United States Cc

Kingston ***Canada*** City on Lake Ontario at the mouth of the Cataraqui river. Founded in 1783 by loyalist refugees from the American Revolution, Kingston was the capital of Canada from 1841 to 1844. It is an important grain processing and shipping port, and has industries making locomotives, ships, diesel engines and aluminium products. It has Queen's University and the Royal Military College of Canada.
Population 52 600
Map Canada Hd

Kingston ***Jamaica*** Capital and chief port, lying 65 km (40 miles) from the eastern end of the south coast, with a deep, landlocked harbour. The harbour is enclosed by the Palisadoes, a 13 km (8 mile) long sandspit extending from the coast east of the city. The island's main international airport lies in the middle of the spit, and PORT ROYAL stands at the western end.

Kingston was founded by the British after

Port Royal was destroyed in an earthquake in 1692, and became the commercial capital in 1703 and the island's political capital in 1872. An earthquake and fire in 1907 destroyed many buildings in the town. A few architectural relics remain among the modern buildings, most notably Headquarters House, formerly the seat of government; Devon House, a restored mansion; and Gordon House, which houses Jamaica's legislature. On the eastern outskirts stands Rockfort, a 17th-century moated fortress.

Today, the city is a busy commercial centre whose industries include oil refining, food processing, tobacco manufacture, clothing and shoes. It is also the centre of the Jamaican coffee trade. A government-owned railway connects the city with most of the island's 14 parishes.

Population 700 000
Map Jamaica Ba

Kingston upon Hull (Hull) ***United Kingdom*** Port on the north shore of the Humber estuary in the county of Humberside. Its docks were founded by Edward I in 1293 and the town has been associated with ships ever since. Whaling and coastal fishing have come and gone, and most of today's business is connected with traffic across the North Sea. The city was devastated by bombing in the Second World War and has been largely rebuilt, with many new industries, such as engineering and chemicals, and a university. One of the buildings to survive was the home of William Wilberforce, champion of the campaign to abolish slavery, who was born in Hull in 1759. The Humber suspension bridge links the city with the river's south bank. It has a span of 1410 m (4626 ft) and was opened in 1981.

Population 270 000
Map United Kingdom Ed

Kingstown ***Ireland*** See DUN LAOGHAIRE

Kingstown ***St Vincent*** Seaport and capital, standing on the south-west coast at the head of Kingstown Bay. Its botanic gardens, established in 1765, are the oldest in the Western hemisphere, and it was to obtain breadfruit plants to propagate at the gardens and provide food for plantation slaves that Captain Bligh made his ill-fated voyage on the *Bounty* in 1787. Exports include arrowroot, bananas and coconuts. The chief industries are processing arrowroot, copra, fruit and sugar.

Population 22 800
Map Caribbean Cc

Kinross ***United Kingdom*** See TAYSIDE

Kinshasa ***Zaire*** Capital of Zaire and the largest city in Central Africa. It stands on the south bank of Malebo (formerly Stanley) Pool, a broad stretch of the Zaire river, some 400 km (250 miles) from the Atlantic.

Before the arrival of the British adventurer Sir Henry Morton Stanley in 1877, the site of the city contained two riverside villages: Kintambo and Kinshasa. In 1881 Stanley acquired the rights to Kintambo and named it Léopoldville after his patron Léopold II of the Belgians. After 1898, when the population was barely 5000, a railway was completed from Léopoldville to the seaport of Matadi, by-passing the Livingstone Falls and enabling Léopoldville to develop quickly as a commercial centre. It became the national capital in the 1920s.

After Zaire became independent in 1960, the city grew rapidly from a population of 400 000 to more than 2 million in 1981. There are, however, few European residents. Renamed Kinshasa in 1966, it has become the focus of communications that extend far inland along the river and rail network. There is also an international airport. The city is a major industrial centre and its food processing, woodworking, textile, chemical and other industries have increased rapidly since the opening of the nearby Inga Dam hydroelectric project.

The city centre has attractive modern buildings, spacious parks and broad boulevards, contrasting with the extensive African township around it. Tourist attractions include the colourful central market, the National Museum and the Beaux Arts Academy. Kinshasa contains the Université National du Zaire, one of the country's three universities.

Population 2 444 000
Map Zaire Ab

Kirghizia (Kirghiz Republic) ***USSR*** Mountainous republic in the south-central USSR, covering an area of 198 500 km² (76 600 sq miles) on the border with north-western China. It consists mainly of pastureland for sheep, goats and cattle, but rises to spectacular peaks such as Pobedy (Tomur; 7439 m, 24 406 ft) in the Tien Shan mountains on the Chinese border.

About half of the republic's inhabitants are Mongoloid Kirghiz people, who have their own language and are traditionally pastoralists. Since 1926 cultivation on irrigated land in the FERGANA region has produced fruit, cereals, sugar and cotton. Its other main products are tobacco, silk, coal, agricultural machinery and textiles. The capital is the industrial city of FRUNZE.

Population 3 886 000
Map USSR Id

Kirin ***China*** See JILIN

Kiritimati ***Kiribati*** See CHRISTMAS ISLAND

Kirkcaldy ***United Kingdom*** See FIFE

Kirkcudbright ***United Kingdom*** See DUMFRIES AND GALLOWAY

Kirkenes ***Norway*** See VARANGERFJORD

Kirkuk ***Iraq*** Regional capital and oil town, some 240 km (150 miles) north of the capital, Baghdad. It is also an agricultural and market centre, and trades in salt, sheep, grain and fruit.

Population 650 000
Map Iraq Cb

Kirov ***USSR*** Industrial city about 790 km (490 miles) north-east of Moscow. It was founded in 1174 on the Vyatka river by traders. In 1934 the city – formerly Khlynov, then Vyatka – was renamed in memory of Sergei Kirov (1886-1934), a locally born Communist leader who was assassinated that year. Its industries include agricultural equipment, chemicals, metalworking and heavy machinery.

Population 407 000
Map USSR Fc

Kiribati

THESE FAR-FLUNG PACIFIC ISLANDS STRADDLE BOTH THE INTERNATIONAL DATE LINE AND THE EQUATOR

The flag of Kiribati shows a frigate bird flying over a sun rising from a blue sea. It is a fitting symbol for these sunny coral islands in the middle of the Pacific Ocean. They were known as the Gilbert Islands until their independence in 1979. Their new name is pronounced 'Kirribass' – the closest the Gilbertese language can get to 'Gilberts'.

They consist of 33 islands and atolls, including BANABA (formerly Ocean Island), the Phoenix Islands and some of the Line Islands. Together with the Ellice Islands (now Tuvalu), most of these were a British protectorate from 1892 and a colony from 1916. The Ellice Islands broke away in 1975. Altogether, Kiribati spreads over some 5 million km² (2 million sq miles) of the Pacific. The Japanese occupied the islands during the Second World War, but they were recaptured by US troops after the brutal battle of TARAWA in November 1943.

The indigenous people are Micronesian (unlike those of Tuvalu, who are Polynesian) and Christian following the influence of European missionaries. Fish is their staple food. The soil is poor and the most common vegetable is the coarse *babai* (taro), which is grown in deep pits. Coconut palms grow prolifically on most of the islands. From 1900 to 1980 Banaba yielded vast amounts of phosphates for export, but these have now been exhausted. Kiribati has negotiated fees for the fleets of other countries, including the USSR, to fish in its waters. It is heavily dependent on overseas aid.

KIRIBATI AT A GLANCE

Map Pacific Ocean Db
Area 717 km² (277 sq miles)
Population 63 000
Capital Tarawa
Government Parliamentary republic
Currency Australian dollar = 100 cents
Languages Gilbertese, English
Religion Christian (50% Roman Catholic, 45% Protestant)
Climate Tropical; average temperature in Tarawa is 26-32°C (79-90°F) all year
Main primary products Breadfruit, pandanus, coconuts, vegetables, fish
Major industries Copra processing, handicrafts, fishing, fish farming
Main exports Copra, handicrafts, fish, postage stamps
Annual income per head (US$) 480
Population growth (per thous/yr) 16
Life expectancy (yrs) Male 50 **Female** 54

Kirovabad *USSR* [illegible] food processing, agricultural machinery, cotton-seed oil and soap. Nearby is the mausoleum of the 12th-century Persian poet Nizami, who was born in the town.
Population [illegible]
Map USSR [illegible]

Kirovograd *USSR* City in the southern Ukraine, some 250 km (155 miles) north-east of the Odessa, formerly called Elizavetgrad. It is the centre for a rich farming region, and produces agricultural equipment.
Population 253 000
Map USSR Ed

Kirşehir *Turkey* Carpet-making and prayer rug-making town about 135 km (85 miles) south-east of the capital, Ankara. It is the capital of a heavily forested province of the same name. The province, which covers 6570 km² (2537 sq miles), produces grain, linseed oil and mohair.
Population (town) 65 900; (province) 261 500
Map Turkey Bb

Kisalföld *Hungary* See ALFÖLD, LITTLE

Kisangani *Zaire* Capital of Haut-Zaire region. It lies on the Zaire river just below the BOYOMA FALLS, about 530 km (330 miles) west of the Ruwenzori mountains on the Uganda border.

Formerly called Stanleyville, it was founded in 1898 on the site of a trading post set up by the British adventurer Sir Henry Morton Stanley in 1883. It is now a busy commercial, fishing, manufacturing and tourist centre, and a trans-shipment point for goods arriving by rail from the south or by river boat from KINSHASA.
Population 339 000
Map Zaire Ba

Kishinev (Chisinau) *USSR* Capital of Moldavia, 160 km (100 miles) north-west of the Black Sea port of Odessa. It was founded in 1466 and was part of Romania until 1812 and from 1918 to 1940. The city, devastated in 1944, has now been rebuilt. A fine 19th-century Orthodox cathedral lies at the heart of the city, which has food processing, wine, tobacco and textile industries.
Population 605 000
Map USSR Dd

Kiskörös *Hungary* See KISKUNSAG

Kiskunhalas *Hungary* See KISKUNSAG

Kiskunsag *Hungary* The section of the Great ALFÖLD (Plain) which lies between the Danube and Tisza rivers in central Hungary. A flat, sandy region, it produces table wines; apricots, peaches and other tree fruits; tobacco; and poultry and livestock. KECSKEMET is the regional capital.

The town of Kiskunhalas (or Halas) lies 55 [illegible] place of the Hungarian poet and patriot Sandor Petofi, one of the leaders of the 1848 Hungarian rebellion against Austrian rule.
Map Hungary [illegible]

Kismaayo (Kismayu) *Somalia* [illegible] between the capital, Mogadishu, and the Kenyan border. It exports bananas and has meat canning factories.
Population 40 000-70 000
Map Somalia Ac

Kistna *India* See KRISHNA

Kisumu *Kenya* One of the nation's largest towns, on the north-east shore of Winam (formerly Kavirondo) Gulf off the east side of Lake Victoria, 260 km (160 miles) north-west of Nairobi. Kisumu was the original terminus of the Uganda Railway, and the busiest port on the lake, until the break-up of the East African Community in the 1970s reduced lake traffic with Tanzania and Uganda. It is now a minor manufacturing centre.
Population 200 000
Map Kenya Bb

Kitakyushu *Japan* Industrial city at the northern tip of Kyushu island. Formed in 1963 by the merger of five towns, it is now Japan's leading iron and steel producer, and also manufactures chemicals and machinery.
Population 1 056 400
Map Japan Bd

Kitchener-Waterloo *Canada* Twin cities in southern Ontario about 100 km (62 miles) west of Toronto. Their industries include furniture, cars, meat packing, distilling, clothing and wood products. Kitchener was settled in 1805 by Mennonites from Pennsylvania, then by Germans. It was known as Berlin till 1916 and still celebrates an Oktoberfest annually. Waterloo has two universities.
Population 288 000
Map Canada Gd

Kíthira (Cerigo) *Greece* Outlying IONIAN ISLAND 14 km (9 miles) south-east of the Peloponnese, and according to classical mythology, the birthplace of the goddess Venus. Its entire population was sold as slaves by the Turkish pirate Barbarossa in 1537. The main town has a Venetian castle and there are large caverns nearby. The chief businesses are tourism and agriculture. The island's honey is an important export.
Population 3350
Map Greece Cc

Kitwe-Nkana *Zambia* Copper-mining town founded in the 1930s about 290 km (180 miles) north of the capital, Lusaka.
Population 314 800
Map Zambia Bb

Kitzbühel *Austria* [illegible]
Population [illegible]
Map [illegible]

Kivu *Zaire* Administrative region covering [illegible] in eastern Zaire. Its western boundaries lie within the Zaire river rain forest zone, but to the east mountains border the Great Rift Valley, which encloses Lakes Edward, Kivu and Tanganyika. Eastern Kivu is popular with tourists and contains the Kahuzi-Biega, Virunga and part of the Maiko national parks. The chief towns are Bukavu, the capital, Goma and Kindu.
Population 4 714 000
Map Zaire Bb

Kivu, Lake *Rwanda/Zaire* Lake covering 2850 km² (1100 sq miles) in the Great Rift Valley. It is Africa's highest lake with its surface 1460 m (4790 ft) above sea level. The only outlet is the Ruzizi river, which drains into Lake Tanganyika. Unlike other East African lakes, Lake Kivu has no crocodiles or hippopotamuses, and contains few fish.
Map Zaire Bb

Kiyev *USSR* See KIEV

Kizil Irmak *Turkey* The country's longest river. It flows for 1130 km (700 miles) from its source near Sivas, some 340 km (210 miles) east of Ankara. It enters the Black Sea at Cape Bafra about 350 km (215 miles) north-east of the capital.
Map Turkey Ba

Kladno *Czechoslovakia* Bohemian industrial town about 25 km (15 miles) north-west of Prague. Coal has been mined there since 1720, and the town also produces coke and steel.

The new village of Lidice stands 8 km (5 miles) to the west. Beside it, roses cover the remains of the old village, which was destroyed on June 9-10, 1942, in reprisal for the assassination in Prague of Reinhard Heydrich, the merciless German SS Deputy Protector of Bohemia and Moravia. The Nazis shot all the men of Lidice, shipped the women and children to death camps and razed the village to the ground.
Population 72 700
Map Czechoslovakia Ba

Klagenfurt *Austria* Capital of Carinthia *Land* (state), 100 km (62 miles) south-west of Graz. Much of Klagenfurt was destroyed by fire in 1514, and its outstanding buildings – including a cathedral – date mostly from the mid to late 16th century. Its products include wood, iron, leather and matches.
Population 86 000
Map Austria Db

Klaipeda (Memel) *USSR* Chief Lithuanian port and a naval base on the Baltic coast about 180 km (110 miles) south-west of Riga. Icebreakers are used between December and March to keep the port open. A canal links it to the mouth of the Neman (Memel) river nearby.

The city, founded as a fort in 1252, later became a member of the Hanseatic League – a trading federation of mostly north German and other Baltic Sea towns. It was German territory (as Memel) for much of its history, but was independent from 1924 to 1939 and was occupied by Russian troops in 1945. It is now an industrial city producing plywood, cellulose, paper, textiles, foodstuffs and canned fish. It exports oil carried by pipeline from the Volga-Ural oil field. Klaipeda craftsmen also carve jewellery from amber – a fossil resin found along the Baltic coast.

Population 191 000
Map USSR Dc

Klis ***Yugoslavia*** See SPLIT

Kljuc ***Yugoslavia*** See GRMEC PLANINA

Klodzko (Glatz) ***Poland*** Manufacturing city and tourist centre 84 km (52 miles) south and slightly west of Wroclaw. It was founded in 981 to defend the Cracow-Prague route over the Sudety mountains, and lies in the 'Valley of Health', a fertile basin of rolling farmlands and forests. The valley has many spas – known in Polish as *zdroj*. Klodzko produces metals, paper, textiles, timber and foodstuffs. The Skull Chapel (1776-8) at Czermna, about 2 km (1 mile) to the north, is lined with the skulls and bones of 3000 people, mainly victims of the Thirty Years' War (1618-48).

Population 29 300
Map Poland Bc

Klondike ***Canada*** River, 160 km (100 miles) long, in Yukon Territory, flowing west to join the Yukon river at Dawson. It gave its name to the gold-producing region on Bonanza Creek (a tributary) where gold was discovered in 1896. In the next two years nearly 30 000 prospectors joined the gold rush, but by 1910 only 1000 were left as gold became more difficult to extract. Some mining went on until 1966.

Map Canada Bb

kloof Gorge, ravine or pass in South Africa.

Knock (Cnoc) ***Ireland*** Village 60 km (37 miles) north and slightly east of the city of Galway on the west coast, where a vision of the Virgin Mary was seen on a church wall in 1879. A huge modern basilica has been built near the site, and 2 million pilgrims visit it each year. Pope John Paul II prayed there in 1979. An international airport was built in the 1980s.

Population 1400
Map Ireland Bb

Knokke ***Belgium*** Coastal resort 18 km (11 miles) north of the city of Bruges. It is one of a string of former fishing villages and draws visitors for its horse-riding facilities, golf courses and casino.

Population 30 000
Map Belgium Aa

knoll Small, rounded hill or mound.

Knossos ***Greece*** The largest excavated palace of the Minoan civilisation, 5 km (3 miles) south-east of the present-day capital of Crete, Heraklion. It was first built in 1950 BC, rebuilt in 1700 BC and destroyed by fire in 1400-1380 BC. Knossos was the cultural and economic centre of one of the first great Mediterranean civilisations. The ruins comprise a stone-built rectangular complex, laid out on four levels around a square covering 1325 m² (1584 sq yds). The British archaeologist Sir Arthur Evans began excavations in 1899, spending large sums of his own money on the project. Now an area of 20 000 m² (24 000 sq yds) has been uncovered, revealing many chambers, workshops, and a throne room with the alabaster throne of King Minos – master, according to Greek mythology, of the bull-headed Minotaur which wandered the underground labyrinth below the palace. Many of the artefacts found at Knossos are now displayed at Heraklion.

Map Greece Dd

Knoxville ***USA*** Industrial city in eastern Tennessee, about 70 km (45 miles) south of the Kentucky border. It is the centre for tobacco growing in the eastern Tennessee river valley, and also produces textiles, furniture and large amounts of marble. The headquarters of the Tennessee Valley Authority are there.

Population (city) 174 000; (metropolitan area) 589 400
Map United States Jc

Koba National Park ***Senegal*** See NIOKOLO-KOBA NATIONAL PARK

Kobe ***Japan*** Port on Honshu island, on Osaka Bay at the eastern end of the Inland Sea. One of the world's largest container terminals, the city also makes ships, steel – and some of the country's best *sake*, or rice wine.

Population 1 410 800
Map Japan Cd

København ***Denmark*** See COPENHAGEN

Koblenz (Coblenz) ***West Germany*** Wine-making city at the junction of the Rhine and Moselle rivers, 53 km (33 miles) south-east of the capital, Bonn. It has been a fortress town since Roman times. The German publisher Karl Baedeker (1801-59), now famous for his series of guidebooks, had his offices in Koblenz.

Population 113 000
Map West Germany Bc

Koburg ***West Germany*** See COBURG

Kochi ***Japan*** Port and market city on the south coast of Shikoku island. It is noted for the snow-white 17th-century castle, for its mastiff-sized Tosa fighting dogs, and for breeding ornate roosters with tails up to 8 m (26 ft) long.

Population 306 000
Map Japan Bd

Kodagu (Coorg) ***India*** Mountainous region on the south-west coast about 800 km (497 miles) south of Bombay. Its capital is the town of Madikeri and it contains the Nagarhole Wildlife Sanctuary, known for its wild elephants, panthers and deer.

Map India Be

Koforidua ***Ghana*** Capital of Eastern Region, and market town at the foot of the Kwahu uplands, about 55 km (35 miles) north of the capital, Accra. It is a centre for cocoa, palm oil – used in making margarine – and kola nuts, which are chewed as a mild stimulant.

Population 60 300
Map Ghana Ab

Kofu ***Japan*** City in the mountains of Honshu island, about 105 km (65 miles) west of the capital, Tokyo. It produces silk and wines.

Population 202 400
Map Japan Cc

Kohima ***India*** Hill town and capital of the state of Nagaland, about 110 km (68 miles) west of the Burmese border. It stands at 1500 m (4921 ft) and was the furthest that the Japanese drove into India in the Second World War. The Japanese captured it after a fierce battle in June 1944, and it was recaptured by Anglo-Indian forces later that month.

Population 67 200
Map India Ec

Kokkola (Gamlakarleby) ***Finland*** Seaport founded in 1620 on the Gulf of Bothnia, 112 km (70 miles) north-east of the port of Vaasa. It exports timber and manufactures chemicals, metal goods, clothing and sports goods.

Population 34 500
Map Finland Bc

Kokoda Trail ***Papua New Guinea*** Jungle track across the Owen Stanley Range north-east of Port Moresby. It was the site of bitter fighting during the Second World War, after Japanese troops following the trail from their landing bases on the north coast of New Guinea threatened to capture Port Moresby. Several thousand Japanese troops died, and about 4700 Australians and 2800 Americans were killed or wounded in the fighting. There is a war memorial at the village of Kokoda at the northern end of the trail.

Kola (Kol'skiy) Peninsula ***USSR*** Arctic peninsula near the Norwegian border in the far north-west of European Russia, between the Barents and White Seas. Lapps herd reindeer in the tundra of the north or work in the great coniferous forests of the south. Apatite (calcium phosphate) is mined at APATITY and Kirovsk for the production of fertilisers. The ice-free port of MURMANSK is the main town and capital of the region incorporating the peninsula, which covers some 129 500 km² (50 000 sq miles). It is an important strategic area with major military bases.

Map USSR Eb

Kolarovgrad ***Bulgaria*** See SHUMEN

Kolberg ***Poland*** See KOLOBRZEG

Kolding ***Denmark*** Town in east JUTLAND lying at the head of a 10 km (6 miles) long inlet of Little Belt. A market town since the Middle Ages, it exports fish, grain and cattle. There is a large castle ruin, Koldinghus, founded in 1208 and destroyed by fire in 1808. The town's Geographical Garden contains about 2000 plants from all over the world, among which are more than 200 different rose varieties.

Population 56 300
Map Denmark Bb

Shaba region in the south-east, about 240 km (150 miles) north-west of the city of Lubumbashi. It is also a commercial and tourist centre.
Population 80 000
Map Zaire Bc

Kolyma ***USSR*** River of north-east Siberia about 2600 km (1600 miles) long. It rises in the gold-bearing Kolyma mountains and flows north-west and north-east into the East Siberian Sea. From October to June, Arctic ice holds back its water, causing widespread flooding in summer and vast ice jams in winter.
Map USSR Qb

Komarno ***Czechoslovakia*** The country's southernmost town. It lies in Slovakia at the junction of the Danube and Vah rivers. A busy river port, it builds river boats and small seagoing vessels, and is linked by a road bridge with the Hungarian town of Komárom.
Population 35 400
Map Czechoslovakia Dc

▼ MUSLIM RELIC The former mosque built by a local warrior – a Muslim convert who fought at Constantinople in 1453 – still dominates old Korcë.

tries of Leningrad, the Volga valley and the Ural mountains. Timber rafted down the Sysola and Vychegda rivers (both tributaries of the North Dvina) is processed at Syktyvkar, the republic's capital, and the towns of Ukhta and Sosnogorsk are expanding producers of oil and of natural gas which is piped to Western Europe. Coal is mined at Vorkuta.
Population 1 197 000
Map USSR Gb

Kommunizma, Pik ***USSR*** See COMMUNISM PEAK

Komodo ***Indonesia*** Island of 520 km² (200 sq miles) in the Lesser Sundas group west of Flores. It is the home of a giant monitor, the 3 m (10 ft) long Komodo dragon, which is the world's largest lizard.
Map Indonesia Ed

Komoé National Park ***Ivory Coast*** The country's largest park (1 150 000 km², 444 000 sq miles) established in 1926 in the north-east around the headwaters of the KOMOE river. Elephants, buffalo, deer and wild boar roam the wooded savannah. The park has hills reaching 640 m (2100 ft).
Map Ivory Coast Bb

Population 820 000
Map Cambodia Bb

Kompong Chhnang ***Cambodia*** Central province between the Tonle Sap river and the capital, Phnom Penh. It is mainly a rice-growing area, with some fishing along the river.
Population 273 000
Map Cambodia Ab

Kompong Som ***Cambodia*** The country's main seaport, on the Gulf of Thailand. It has grown rapidly since road and rail links over the Elephant Mountains with the country's heartland were opened in the 1960s. Cement works and an oil refinery have been built nearby.
Population 53 000
Map Cambodia Ab

Komsomol'sk-na-Amure ***USSR*** One of the largest cities of the Soviet Far East, a port on the Amur river some 880 km (550 miles) north-east of Vladivostok. Komsomol (Young Communist League) pioneers founded the city amid remote forests and swamps in 1932, and it grew rapidly with factories making heavy machinery and equipment, foodstuffs, clothing, footwear and paper. Shipbuilding, oil refining and fishing are also important. It is on a branch of the Trans-Siberian Railway from Khabarovsk.
Population 291 000
Map USSR Oc

Kongju (Gongju) ***South Korea*** Ancient town, and capital of the Paekche kingdom from AD 475 to 538. It lies 120 km (75 miles) south of the capital, Seoul.
Population 17 000
Map South Korea Cd

Kongsberg ***Norway*** Town in south Norway, 70 km (44 miles) south-west of Oslo. It was founded in 1624 as a silver mining town, but the silver ran out in 1957; the mines are now a museum and tourist attraction. The town has a magnificent 18th-century church. Kongsberg's manufactures include small arms, electronics and – as the site of Norway's mint – money.
Population 14 200
Map Norway Cd

Kongsvinger ***Norway*** Town in south-east Norway, in the valley of the Glåmma river, about 75 km (47 miles) north-east of Oslo. It has a ruined fortress, dating from 1683, and a special commercial relationship with the

Swedish town of Arvika. The Norwegian and Swedish governments have promoted the twinning of the towns, with financial assistance for trade and special arrangements for the exchange of goods across the border.

Population 10 300

Map Norway Dc

Königsberg ***USSR*** See KALININGRAD

Königsgrätz ***Czechoslovakia*** See HRADEC KRALOVE

Konin ***Poland*** Coal mining and aluminium-smelting city, 89 km (55 miles) east and slightly south of the city of Poznan. The town of Chelmno-nad-Nerem, about 25 km (16 miles) east of Konin, is where the Nazis first used mobile gas-chamber lorries to murder Jews; 360 000 of them died there between 1941 and 1944.

Population 73 000

Map Poland Cb

Konkouré ***Guinea*** The main west-flowing river of Guinea. It rises in the Fouta Djallon and flows through a narrow, deep valley for about 250 km (160 miles) to the Atlantic Ocean just north of the capital, Conakry. Power stations at Souapiti and Amaria on the lower reaches may eventually generate almost half Guinea's hydroelectricity.

Map Guinea Ba

Konstanz ***West Germany*** See CONSTANCE

Konya ***Turkey*** Carpet-making town about 235 km (145 miles) south of the capital, Ankara. It is the capital of a province of the same name. The town was the cultural centre of the Anatolia region during the 13th century, when the sect of Whirling Dervishes was founded there by the Sufi Muslim mystic Jalal ad-Din Rumi. The sect holds its main festival in the town each December, when the devotees literally whirl themselves into a mystical trance.

The province covers 47 420 km² (18 309 sq miles) and its farmers raise sheep, goats, horses and camels, and grow grain.

Population (town) 438 900; (province) 1 791 600

Map Turkey Bb

Kootenay ***Canada*** National park covering 1380 km² (533 sq miles) on the western slopes of the Rocky Mountains in south-east BRITISH COLUMBIA. It contains snow-capped peaks, tree-covered valleys, hot springs and geysers.

Map Canada Dc

Kopar ***Yugoslavia*** See KOPER

Kópavogur ***Iceland*** Town just south of the capital, REYKJAVIK. Since the Second World War it has developed into the country's second largest town after the capital. This is mainly because it is an outlier of the capital in a favoured part of the country.

Population 14 400

Map Iceland Bb

Koper (Kopar; Capo d'Istria) ***Yugoslavia*** Adriatic seaport and resort on the Istra peninsula, 15 km (9 miles) south-west of Trieste, Italy. It was ruled by Venice for more than 500 years and by Austria from 1815 to 1910, when the Italians occupied it. Made part of the Free Territory of Trieste in 1945, it became Yugoslavian in 1954, giving Slovenia a port. Since then a new harbour and car, truck and motorcycle plants have been built.

The Venetian old town, once an island, has massive fortifications, a Gothic-Renaissance cathedral and palace, and several campanili (bell towers). The little salt-producing port of Piran, 9 km (6 miles) to the west, with a web of arched alleyways and piazzas, contains a copy of St Mark's campanile in Venice.

Population 16 700

Map Yugoslavia Ab

kopje Small, but prominent, isolated rocky hill in southern Africa.

Korcë ***Albania*** The country's largest mountain town, standing at 855 m (2805 ft). It is the centre of an extensive wheat and sugar-beet growing basin in the south-east highlands and Macedonian lakes region. The population is mixed, and includes Romanian-speaking Vlachs. Albania's only pre-war hydroelectric power station is here, and since 1950 Korcë has grown into a modern town with sugar refining, brewing, knitwear and glassmaking factories. Copper and lignite are mined nearby.

Population 63 600

Map Albania Cb

Korcula (Curzola) ***Yugoslavia*** Lush Dalmatian island covering an area of 273 km² (105 sq miles) that commands the Dubrovnik-Split sea route. The island was known in ancient times as Corcyra Nigra, and was held by Venice until 1815. It was at the 13th-century sea battle of Curzola between Venice and Genoa that the Venetian explorer Marco Polo was captured by the Genoese. Polo's account of his travels in Asia was written while he was being held in a Genoese jail.

Long a community of sailors, shipbuilders and fishermen, Korcula also produces vines, olives, and fine white marble from the ancient quarries at Vrnik. Blato, the largest settlement, is famous for its *Kumpanjinja*, a traditional battle dance performed each April. A similar ceremony is staged each week in summer at the old walled port of Korcula. Tourism is developing fast on the island. Fresh water, once scarce, is now piped from the mainland.

Population 30 000

Map Yugoslavia Cc

Korea, North See p. 366

Korea, South See p. 367

Korea Strait Sea passage between the south coast of South Korea and Japan, linking the CHINA SEA to the Sea of JAPAN. The island of TSUSHIMA divides the strait into two parts, the eastern section of which is sometimes called the Tsushima Strait. In 1905, in the Battle of Tsushima Strait during the Russo-Japanese War, the Japanese nearly annihilated the Russian Baltic Fleet which had sailed halfway round the world to confront the Japanese navy.

Width 97 km (60 miles) (western channel), 64 km (40 miles) (eastern channel)

Map Korea De

Korem ***Ethiopia*** Mountain town about 400 km (250 miles) north of the capital, Addis Ababa. Its population has tripled since 1982 as refugees from the drought-hit countryside have arrived in their thousands. It was television reports from Korem which sparked off Irish singer Bob Geldof's pop music appeals, Band Aid and Live Aid, in 1984-5.

Population 30 000

Map Ethiopia Aa

Korhogo ***Ivory Coast*** Capital of the northern savannah region, and main town of the Muslim Senoufo people, 190 km (118 miles) north-west of Bouaké. Millet, sorghum, rice and yams are the staple crops, and cotton is the main cash crop. Korhogo is losing its importance to the nearby rail centre of Ferkessedougou.

Population 65 000

Map Ivory Coast Ab

Kórinthos ***Greece*** See CORINTH

Koror ***Belau*** Capital and main port, scenically situated on the lagoon facing the Floating Garden Islands, eroded formations of coral limestone. Industries include copra and fish processing. From 1914 to 1945, Koror was the capital of all Japanese-administered Micronesia and had a Japanese population in 1935 of over 25 000.

Population 8000

Kortrijk ***Belgium*** See COURTRAI

Kos (Cos) ***Greece*** One of the larger DODECANESE islands in the eastern Aegean Sea. It was the home of Hippocrates (about 460-380 BC), the father of medicine, who founded a medical school on the island. The harbour, Mandraki, is guarded by a 14th-century Frankish castle. The town also has a mosque dating from 1786 and, in the same square, a gigantic plane tree with a trunk 13.5 m (44 ft) in girth and some branches held up by props. A fountain plays beneath it. A Doric temple to Asclepius, the god of medicine, is 6 km (4 miles) south-west of the town.

The volcanic island of Nísiros, with natural sulphur springs, lies south of Kos. In mythology, the island is said to have been a piece of Kos which the sea god Poseidon flung at Polybotes, the Titan.

Population 20 300

Map Greece Ec

Kosciusko, Mount ***Australia*** The country's highest mountain. It rises to 2230 m (7316 ft) in the Snowy Mountains near the state border between New South Wales and Victoria.

Map Australia Hf

Kosi ***Nepal/India*** River flowing for some 480 km (300 miles) through eastern Nepal and northern India. It is formed by the confluence of two minor rivers: the Sun Kosi and the Arun. The Sun Kosi, which drains much of eastern Nepal, plunges through rugged gorges before emerging as the Kosi on the plains of northern India to join the Ganges river. It is dammed for hydroelectric power and flood control on the Nepali side of the border, but massive flooding still occurs in India in heavy monsoons.

Map India Db; Nepal Ba

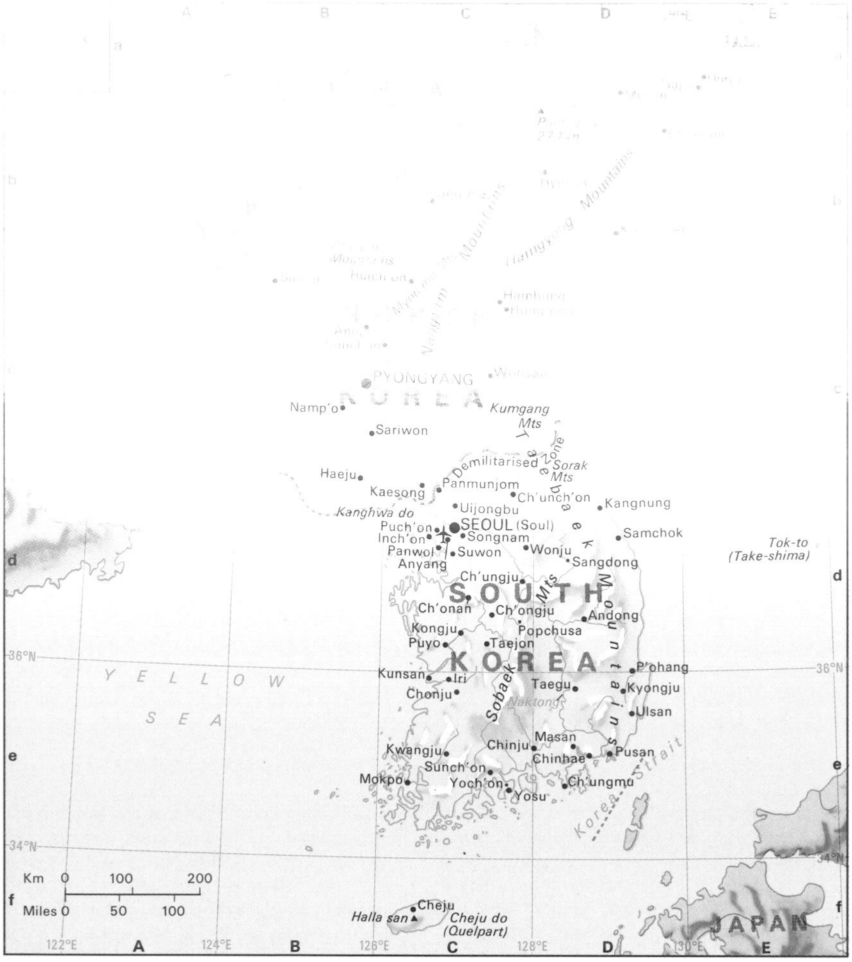

North Korea

AN ISOLATED STATE DEDICATED TO THE GLORY OF COMMUNIST PARTY BOSS KIM IL SUNG, NORTH KOREA LAGS ECONOMICALLY FAR BEHIND ITS VITAL SOUTHERN NEIGHBOUR

Remote, closed and hostile to Western contact, North Korea is a virtual dictatorship under the control of its leader Kim Il Sung. It is a Communist state of the Stalinist school, seemingly dedicated to two main aims: reunification (under Communist control) with its neighbour SOUTH KOREA, and the glorification and perpetuation of the Kim dynasty. Since the armistice that ended the Korean War on July 27, 1953, little has been achieved towards the first aim, abusive propaganda and the occasional 'incident' in the DEMILITARISED ZONE along the ceasefire line being punctuated by more conciliatory spells. But on the second, signs indicate that Kim's son, Kim Jong Il, is being groomed as the next head of state.

With North Korea, however, it is difficult to be sure of anything. Few Western visitors are allowed entry, and contacts even with fellow Communist states are kept to a minimum.

Whatever its shortcomings, North Korea undoubtedly has provided its population with adequate housing, basic education and health care, but the economy is stagnant. Planners press for the exploitation of minerals and the production of munitions, steel and chemicals at the expense of consumer goods, and spending on the 750 000-strong armed forces takes up nearly 15 per cent of all state expenditure, reducing what is available for productive investment. The collective farms, into which the Communist government reorganised agriculture, have proved unwieldy and inefficient. In most respects, life in this rigid, authoritarian and tightly disciplined society is drab and unexciting.

North Korea has some 55 per cent of the Korean peninsula's land area, stretching from the YALU and Tumen rivers, which form its border with CHINA and, for an 18 km (11 mile) stretch in the east, with the USSR, south to the Demilitarised Zone mostly just north of the 38th Parallel. It is a mountainous country and is especially rugged in the far north, where one peak exceeds 2500 m (8200 ft). Three-quarters of the land consists of highlands or scrubland and only 18 per cent is cultivated. As in South Korea, the main lowlands lie to the west.

The climate is also similar to that of the South but slightly cooler, much less humid in summer and even colder in winter. In the interior, freezing air from central Asia can make January temperatures drop below −18°C (0°F), and rivers and lakes freeze over. There is generally less rainfall than in the South, but enough to grow rice, the main grain crop.

By contrast with its southern neighbour, however, North Korea is quite well endowed with fuel and mineral resources. Although it has no oil, coal mines can supply 70 per cent of its energy needs. Coal is supplemented by hydroelectric power, generated mainly along the Yalu river. There are substantial iron ore deposits around Musan and the capital, PYONGYANG, and other mineral resources include copper, lead, zinc, tungsten and graphite. Textile weaving is the most important light industry.

NORTH KOREAN HISTORY

Much North Korean industry was developed by the Japanese, who occupied the peninsula from 1910 to 1945. (For earlier history, see SOUTH KOREA.) With the defeat of Japan in 1945, Russian occupying troops opened the way for a Communist state to be formed in 1948 under Kim Il Sung, a former partisan. The Korean War began with an attack by the North on the South in June 1950. North Korea's cities were devastated, and a million of its people were killed or fled to the South.

After the war, reconstruction centred on heavy industry and collective farms. The country's prosperity peaked in the mid-1960s, but then declined to well below that of the South.

Meanwhile, Kim Il Sung built up a personality cult that made those of Stalin and Mao Ze-dong seem modest by comparison, with vast monuments and garish portraits of the 'Great Leader' in city streets.

There have been signs of a less isolationist attitude. North Korea sorely needs contacts to get the modern technology it needs.

NORTH KOREA AT A GLANCE

Area 122 098 km² (47 142 sq miles)
Population 20 540 000
Capital Pyongyang
Government Communist republic
Currency North Korean won = 100 chon
Language Korean
Religions Shamanist, Chundo Kyo, Buddhist
Climate Continental; average temperature in Pyongyang ranges from −8°C (18°F) in January to 24°C (75°F) in August
Main primary products Rice; coal, iron ore
Major industries Agriculture, iron and steel, engineering, cement, chemicals
Main exports Metals and metal ores (mainly iron), chemicals
Annual income per head (US$) 620
Population growth (per thous/yr) 23
Life expectancy (yrs) Male 61 **Female** 65

South Korea

INSPIRED BY ONE NEIGHBOUR, THREATENED BY ANOTHER, SOUTH KOREA HAS A VIGOROUS, BOOMING ECONOMY WHICH IS OVERSHADOWED BY POLITICAL DISSENT

Korea used to be called 'Land of the Morning Calm', but there is precious little tranquillity today in this dynamic industrial country of east Asia. From the capital SEOUL, in the northwest, to the main port of PUSAN in the southeast, factories belch smoke into the sky, while bulldozers and excavators grind across the landscape, clearing the way for further industrial development and preparing sites for new motorways and housing schemes. The beauties of rural Korea can still be found – especially in the mountainous east – and there are historical sites, notably at KYONGJU, dating back more than 1000 years. But signs of dramatic change are apparent everywhere, for South Korea's economy is expanding at a rate that makes even Japan's growth seem sluggish by comparison.

Like that of Japan, South Korea's economic miracle has been achieved on the basis of very slender natural resources. A country slightly larger than Portugal (though far more densely populated), it occupies the southern half of the Korean peninsula, stretching some 400 km (250 miles) from the KOREA STRAIT to the DEMILITARISED ZONE bordering NORTH KOREA. It is predominantly mountainous, resembling a great block tilted towards the west, so that the highest mountain ranges run north and south along the east coast. Only 23 per cent of its surface area is cultivated, the lowlands being mainly in the south and west. Here also are the main areas of settlement, where population densities are extremely high – higher even than in Japan.

Domestic coal and hydroelectric power help to meet the burgeoning demand for energy, but South Korea has no oil and relies on imported fuels for three-quarters of its energy needs. Apart from large tungsten deposits, it is poor in commercially viable minerals, and its rapidly growing industries depend heavily on imported raw materials.

Farmers have to cope with a monsoonal climate that veers between the extremes of icy winters – when winds from central Asia whip southwards across the peninsula – and hot, humid summers. (Average August temperatures in Seoul, for example, rise to an uncomfortable 31°C, 88°F.) Nevertheless, the growing season is long and wet enough for good yields of rice, the staple crop.

BETWEEN TWO POWERFUL NEIGHBOURS

As a political entity, South Korea dates from the end of the Second World War. From 1910 to 1945 it was simply the southern – primarily agricultural – portion of a Japanese colony comprising the whole Korean peninsula. Its geographical location as the bridge between China and Japan has always ensured its central role in east Asian affairs, and Japanese domination was merely one stage in a historical progression. For most of its history China has been the dominant neighbour, but Korean culture itself has proved remarkably resilient through the centuries.

Korea was first united under the Silla emperors, based at Kyongju, in the 7th century, when the kingdoms of Kogyuro (in the north) and Paekche (in the south) were conquered with Chinese help. Culture, art and the Buddhist religion flourished under the Sillas. The Koryo kings, whose capital was at PYONGYANG (now in North Korea), ousted the Sillas in 935 and ruled until overrun by Mongol tribes in the 13th century.

In 1392, the Yi (or Choson) dynasty, allied to the Ming rulers of China, re-established self-rule; their capital was Seoul. Again there was a cultural blossoming, including the introduction of Confucianism and the invention of a phonetic alphabet (unlike the picture writing of China and Japan), movable type for printing and iron-clad warships. But in the 17th and 18th centuries internal Chinese strife and then Japanese expansionism again brought foreign domination.

After the Sino-Japanese war of 1894-5 and Russo-Japanese war of 1904-5, Japan annexed Korea in 1910 and ruled totally until American troops landed in the south in August 1945. Russia occupied the north and Korea was divided by the 38°N line of latitude – the 38th Parallel – into two zones of occupation. This split was meant to be temporary, but the deepening of the cold war led to two rival states – one Communist, the other a liberal democracy – being established, fixing the division. In June 1950 the armies of North Korea launched a massive surprise attack on the South, initiating the Korean War (1950-3). (See also NORTH KOREA.)

The North Koreans were repelled by United Nations (mainly American) forces but were reinforced by China. The eventual outcome was an armistice, and a ceasefire line near the starting point.

BUILDING AN INDUSTRIAL NATION

Its economy largely agricultural and its cities and communications devastated by the war, South Korea in 1953 was exhausted and impoverished. American aid enabled the country to survive, but it was only in the 1960s that its fortunes began to improve. An export promotion policy directed energies towards developing light industries such as textiles and clothing, and cheap electrical and electronic products. From the early 1970s, heavy industry grew rapidly, making South Korea a significant producer of steel, ships (with the world's largest shipyard), motor vehicles and chemicals.

Japanese investment and know-how aided industrial development, and in many ways the South Korean industrial miracle reflects and has been inspired by postwar Japanese expansion (although at a political and social level there have been tensions between the two countries dating from wartime experiences). As in Japan, the population is well educated, hardworking and loyal to its employers. Strikes are rare and there is a crucial labour-cost advantage over many competitors. In shipbuilding, for example, wages are roughly one-third of those in Japan.

THE TWO KOREAS

Government planning has helped to steer the economy towards success, but perhaps the most important factor of all is the people's attitude towards their North Korean rivals. Not only are South Koreans determined to outperform the North on all fronts, but the ever-present threat of invasion has instilled an unusually strong sense of discipline and seriousness of purpose.

To counter Northern ambitions, South Korea maintains a well-equipped army of 540 000, backed by 40 000 American ground troops. But all is not well on the home front. Although theoretically a democracy, South Korea has been ruled by a succession of powerful and authoritarian military leaders since 1961. Political opponents have been harassed and persecuted, demonstrations against the regime vigorously and sometimes brutally suppressed. Demands for a more liberal form of democracy grow, however, and long-term political stability is by no means assured. Moreover, large overseas debts and the threat of protectionist measures against South Korean goods in foreign markets could undermine its apparently robust economy.

For the present, however, the country offers a spectacular example of rapid transformation from rags to riches. Its people enjoy a level of prosperity undreamt of a generation ago. Their hard work and determination have brought South Korea a long way towards joining the ranks of the world's advanced nations – a status it hopes to advance further by staging the 1988 Olympic Games.

Relations between North and South Korea have been slightly more relaxed recently. In 1985 the two sides agreed on limited reunion visits for families separated by the border.

SOUTH KOREA AT A GLANCE
Area 98 992 km² (38 221 sq miles)
Population 43 280 000
Capital Seoul
Government Republic
Currency South Korean won = 100 chon
Language Korean
Religions Buddhist, Christian, Confucian, Chundo Kyo
Climate Far south warm, temperate; rest continental. Average temperature in Seoul ranges from −9-0°C (16-32°F) in January to 22-31°C (72-88°F) in August
Main primary products Rice, fish; tungsten
Major industries Agriculture, textiles and clothing, iron and steel, shipbuilding, chemicals, electrical and electronic goods
Main exports Clothing and textiles, electronics equipment, iron and steel, ships, food products
Annual income per head (US$) 1528
Population growth (per thous/yr) 15
Life expectancy (yrs) Male 64 **Female** 69

▲ **CHINA'S BACKDOOR The skyscrapers of Hong Kong's Kowloon city stand on a tip of the Chinese mainland across the fine harbour from Victoria (foreground), the colony's business capital.**

Kosice ***Czechoslovakia*** Regional capital of eastern Slovakia, in the south-east about 25 km (15 miles) from the Hungarian border. It lies in a wine-producing area. Since 1950, Kosice's population has quadrupled with the building of steel, metal, engineering, chemicals and food industries, and a university and other cultural facilities.

The city has a 14th-century Gothic cathedral, and the macabre Miklus's medieval prison, complete with instruments of torture.

Population 214 300
Map Czechoslovakia Eb

Kosovo (Kosmet) ***Yugoslavia*** Autonomous province in south-west Serbia covering an area of 10 887 km² (4202 sq miles). It lies on the Albanian border, and three in four of its people are ethnic Albanians. After the Battle of Kosovo in 1389, the Turks held the area until 1912, and it remains an economic backwater despite its well-watered, fertile lowlands and rich deposits of silver, chrome, lead and zinc in its mountains. However, since the 1950s determined efforts have made some improvements to the province's agriculture, and industries, but its rapid population increase – 28 per cent between 1971 and 1981 – has led to considerable unemployment.

Tourists are now being encouraged to visit the region for skiing or to enjoy the mountain scenery; to tour its fine Orthodox monasteries such as Decani, Peć, Prizren and Gracanica; and to see some of Yugoslavia's oldest oriental monuments, including a 14th-century sultan's tomb at Kosovo Polje, and two medieval mosques at PRISTINA, the capital of the province.

Population 1 584 000
Map Yugoslavia Ec

Kosovo Polje ***Yugoslavia*** See PRISTINA

Kosrae ***Federated States of Micronesia*** Most easterly state of the federation, consisting of the single atoll of Kosrae (pronounced 'Korsh-eye') lying some 500 km (310 miles) south-east of Pohnpei. It was a whaling port in the last century, but most people today are subsistence farmers and fishermen.

Population 5600
Map Pacific Ocean Db

Kossou ***Ivory Coast*** Dam 1.6 km (1 mile) long and 60 m (196 ft) high on the BANDAMA river near the town of YAMOUSSOUKRO. It was designed to produce enough hydroelectric power for the whole of the country's needs, but production has been hit by droughts which have reduced water levels. The government rehoused 75 000 people in order to build the dam. The man-made lake behind the dam is being turned into a tourist centre.

Map Ivory Coast Ab

Kostroma ***USSR*** One of the oldest cities in Russia, founded in 1152 about 315 km (195 miles) north-east of Moscow. It began as a fortress built to guard trade routes along the Kostroma and Volga rivers. The kremlin (fortress) and cathedral still survive. It has been a centre of the linen industry since the 16th century, and most of the city's workers are still employed in its linen textile mills, which use locally grown flax.

Population 267 000
Map USSR Fc

Kota Baharu ***Malaysia*** Port and capital of Kelantan state on the north-east coast of peninsular Malaysia, just south of the Thai border.

Population 170 600
Map Malaysia Ba

Kota Kinabalu ***Malaysia*** Capital of Sabah state on the north-west coast of Borneo. The town, formerly called Jesselton, was badly damaged during the Second World War, and has been rebuilt. Mount Kinabalu, about 55 km (34 miles) east, is 4094 m (13 432 ft) high.

Population 108 800
Map Malaysia Ea

Kotka ***Finland*** Seaport founded in 1879 on the Gulf of Finland, in the south-east about 50 km (30 miles) west of the Russian border. It is Finland's busiest port after the capital, Helsinki, exporting timber and importing fuel and machines. Industries include paper, pulp, sugar, machine workshops and farm processing. In 1789, during a naval battle between Russia and Sweden in the vicinity, more than 100 ships foundered, and the sunken wrecks have yielded underwater archaeologists a rich harvest. Nearby is Langinkoski, once a summer residence of the Russian royal family and now a nature reserve.

Population 59 400
Map Finland Cc

Kotor, Gulf of (Boka Kotorska) ***Yugoslavia*** Bay enclosed by high mountains on the coast of Montenegro, making one of the world's finest natural harbours. The small medieval seaport of Kotor (Cattaro), on the south side of the gulf, was ruled by Venice from 1420 to 1797. It has a fine 12th-century cathedral, 14th-century mint and a seaman's guild founded in the early 9th century, but it was badly damaged by earthquakes in 1979 and 1981. The road from the port to the former Montenegrin capital of Cetinje inland twists through 32 hairpin bends up the side of Lovcén mountain, offering breathtaking views over the gulf.

Map Yugoslavia Dc

Kotte ***Sri Lanka*** Town being developed as a new seat of national government, 11 km (7 miles) east of central Colombo, the present capital. When completed, it is due to be renamed Sri Jayawardhanapura after the 15th-century Sinhalese name of the old capital at Kotte. Meanwhile some government departments have moved to Kotte from Colombo, which is due to remain the country's commercial capital.

Population 135 600

Map Sri Lanka Ab

Koudougou ***Burkina*** Market town for export crops on the edge of the Mossi Highlands, about 75 km (45 miles) west of the capital, OUAGADOUGOU. It lies on the country's only railway and has some industry, including a shea butter mill (for extracting edible fat from the fruits of the shea tree), a cotton ginnery and a textile complex.

Population 44 000

Map Burkina Aa

Kouilou (Kwilu) ***Congo*** River, 320 km (200 miles) long, rising on the Batéké Plateau in the south-west, near the Gabon border. Its upper section (called the Niari) flows south, then west, then north west, before turning south-west at Makabana and flowing (as the Kouilou) to the Atlantic. Gorges along its path provide potential sites for hydroelectric power stations. Large plantations in the Niari Valley (Congo's most fertile region) produce export crops including cocoa, coffee, palm products, sugar and tobacco, which are shipped to Pointe-Noire by the Congo-Océan railway.

Map Congo Ab

koum, kum Sandy desert with dunes and sand-seas in Turkestan, equivalent to the ERG of the Sahara.

Koumba ***Guinea*** See CORUBAL

Kourou ***Guyane*** Coastal town 56 km (35 miles) west of the capital, Cayenne. Since 1968 it has been the European Space Agency's rocket-launching site. The space centre covers some 120 km² (46 sq miles) beside the coast. The site was chosen because it is less than five degrees from the Equator, where the earth is moving fastest as it spins. The extra speed acts like a natural slingshot, enabling rockets to get into orbit with less fuel.

Population 7000

Map Guyane Bb

Kovno ***USSR*** See KAUNAS

Kowloon ***Hong Kong*** City and one of the three constituent territories of the British Crown Colony of Hong Kong, the others being HONG KONG ISLAND and the NEW TERRITORIES. The acquisition in 1860 of the small peninsula (which is part of the Chinese mainland) gave the British control over Victoria Harbour, the vast sheltered anchorage which separates Kowloon from Hong Kong island. Kowloon, which occupies a large part of the peninsula, developed in the 19th century as a harbour settlement.

Today it is an industrial region and an important commercial centre. It is also one of the most crowded areas on earth – in places, there are 87 000 people per square kilometre (225 000 per sq mile). Manufactures include toys, cotton goods, footwear, cement and rope. Kowloon has an ocean terminal with the world's third largest container port, the colony's Kai Tak international airport and a railway link to Guangzhou (Canton) in China.

Population 800 000

Map Hong Kong Cb

Kpalimé (Palimé) ***Togo*** Market town in the Togo Highlands, about 105 km (65 miles) north-west of the capital, Lomé. It is the focus of Togo's main cocoa-producing area, and also handles coffee and palm oil (used in margarine and cooking fats). The town, which is dominated by Mont Agou, is being developed as a tourist resort. Local attractions include waterfalls at Kpimé and Akrowa.

Population 27 500

Map Togo Bb

Kra ***Burma / Thailand*** Isthmus linking peninsular Malaysia with mainland South-east Asia. It is also the name of a river, about 60 km (37 miles) long, flowing into the Andaman Sea. The isthmus, some 200 km (125 miles) long, is an average of only about 50 km (30 miles) wide.

Map Thailand Ad

Kragujevac ***Yugoslavia*** Commercial town 97 km (60 miles) south-east of Belgrade. It was the capital of Serbia (1818-39), and later became the site of Serbia's first schools, theatre, newspaper and printing press. A monument and park commemorate 7000 men and boys aged 14 to 70 who were shot by the Nazis on October 21, 1941 as a reprisal for partisan resistance. The town produces Zastava cars.

Population 164 800

Map Yugoslavia Eb

Krak des Chevaliers ***Syria*** See CRAC DES CHEVALIERS

Krakatau (Krakatoa) ***Indonesia*** Volcanic island between Java and Sumatra. It is the remains of a much larger volcano which collapsed in what is regarded as the world's greatest natural explosion on August 27, 1883. The shock wave of the explosion was felt as far away as California, and giant waves, or tsunamis, created by the blast killed some 36 000 people.

Map Indonesia Cd

Krakow ***Poland*** See CRACOW

Kraljevo ***Yugoslavia*** Large industrial and market town in Serbia at the junction of the Ibar and Western (Zapadna) Morava rivers 124 km (77 miles) south of Belgrade. Its factories produce machinery, railway vehicles and metal goods. Serbia's medieval kings were crowned at Kraljevo – the name means 'kings' place' – and from 1945 to 1962 the town was called Rankovicevo after Aleksandar Rankovic, a leading Serbian Communist disgraced in 1962. The stately ruins of Maglic, a 15th-century archbishop's castle, crown a hilltop overlooking the Ibar 16 km (10 miles) to the south-west. Kalenic monastery, famous for its fine 15th-century frescoes, lies 30 km (19 miles) north-east.

Population 121 600

Map Yugoslavia Ec

Kranjska Gora ***Yugoslavia*** See KARAWANKE ALPS

krans In South Africa, an overhanging, sheer wall of rock or precipice.

Kras (Karst, Carso) ***Yugoslavia*** Bare, dry limestone plateau in the Dinaric Alps near the Italian border, north of Trieste and east of Gorizia on the Soca (Isonzo) river. Rising to 1495 m (4908 ft), it is noted for its caves, such as those around POSTOJNA, its waterfalls and underground rivers. Karst, the German name for the plateau, is also applied to the much larger limestone area stretching the length of western Yugoslavia. The word KARST has come to be used by geographers to describe any deeply eroded limestone area.

Map Yugoslavia Ab

Krasnodar ***USSR*** City in the Caucasus, about 80 km (50 miles) north-east of the Black Sea. Founded in 1793 as the Cossack fort of Ekaterinodar, it is now the hub of the rich Kuban' river plains, which grow melons, tomatoes, tobacco, sunflowers, sugar beet and wheat, and of the west Caucasus wine-producing and cattle-rearing area. It also produces machinery, textiles and chemicals.

Population 604 000

Map USSR Ed

Krasnoyarsk ***USSR*** One of the largest cities in Siberia and capital of the administrative territory of the same name. Cossacks founded it in 1628 as a fort beside a crossing of the Yenisey river, about 620 km (385 miles) east of Novosibirsk. The city is on the Trans-Siberian Railway, and since the 1930s has become the focus of a huge territory which is rich in gold, timber, coal, copper, iron ore and uranium. To the south is the Krasnoyarsk Dam (capacity 6000 MW), one of the world's largest hydroelectric plants.

Population (city) 860 000

Map USSR Kc

Kratie ***Cambodia*** Northern rubber-growing province. The provincial capital, also called Kratie, lies on the Mekong river, at its highest navigable point.

Population (province) 136 000; (town) 12 000

Map Cambodia Bb

Krefeld ***West Germany*** Industrial town about 30 km (19 miles) west and slightly south of Essen, and the centre of the country's silk industry. It produces hosiery and textiles.

Population 224 000

Map West Germany Bc

Krishna (Kistna) *India* [illegible]
Map [illegible]

Kristiania *Norway* [illegible]

Kristiansand (Christiansand) *Norway* [illegible]
Population [illegible]
Map Norway Bd

Krivoy Rog *USSR* City in the Donets Basin of the southern Ukraine, some 250 km (155 miles) north-east of the Black Sea. Cossacks founded it in the 17th century, and it remained a small village until high-grade iron ore was discovered nearby in the 1880s. Since then Krivoy Rog has become one of the world's biggest iron-mining centres, producing over 125 million tonnes of ore per year. Its other products include steel, coke, mining machinery, diamond drills, foodstuffs and cement.
Population 680 000
Map USSR Ed

Krk (Veglia) *Yugoslavia* Fertile Croatian island covering 408 km² (158 sq miles), lying off Rijeka in the northern Adriatic. It is the largest Yugoslav island. Vines, olives and figs grow on the sheltered western side and the island is renowned for its red wines. The island's main town, also called Krk, is a medieval, walled seaport with a Romanesque cathedral, a magnificent bishop's palace and a rapidly developing tourist industry. The island was settled by Croats in 1059, and the Bascanska Ploca (Baska Tablet) of stone, found in the town of Baska, is inscribed with the oldest surviving example of the Croatian language. It dates from the 11th century. The original stone is now in Zagreb, but there is a copy in the town church.
Population 1500
Map Yugoslavia Bb

Kronshtadt *USSR* Naval base on Kotlin Island in the Gulf of Finland, 25 km (16 miles) from Leningrad. The Russian tsar Peter the Great (1682-1725) captured the island from the Swedes (who called it Kronslott) in 1703 and began the construction of its fortress seven years later. Its defences played a key role in the Siege of Leningrad during the Second World War, when the city held out against German forces from September 1941 to January 1944.
Map USSR Dc

Krosno *Poland* Town about 150 km (95 miles) east and slightly south of the city of Cracow. It has fine 16th to 18th-century pastel-painted houses set around a central square. Krosno is at the centre of Poland's Carpathian oil field. There is a museum of kerosene lamps – which were invented in Poland in 1853.
Population 42 900
Map Poland Dd

Kruger National Park *South Africa* [illegible]
Map [illegible]

Krugersdorp *South Africa* [illegible]
Population [illegible]
Map South Africa [illegible]

Krung Thep *Thailand* [illegible]

Krusevac *Yugoslavia* Industrial and market town on the Western (Zapadna) Morava river in Serbia, 153 km (95 miles) south-east of Belgrade. It produces cars, and its restored castle dates from the 14th century.
Population 164 800
Map Yugoslavia Ec

Kruszwica *Poland* See KUJAWY

Krym *USSR* See CRIMEA

Ksar es Souk *Morocco* See ER RACHIDIA

Kuala Lumpur *Malaysia* National capital, at the junction of the Kelang and Gombak rivers, 30 km (19 miles) upstream from the Strait of Malacca. It was founded in 1857 as a tin-mining camp, and its name comes from Malay words meaning 'Muddy River Mouth' – a reference to the river bottom from which tin ore was scooped by giant dredges. The mining camp grew rapidly to become the economic centre of Selangor state, Malaysia's most developed region. In 1895, it was chosen as the capital of the Federated Malay States.

Since independence in 1963, the city has grown. Its modern government complex, surrounded by lake gardens, mixes uneasily with old temples and mosques, Oriental shopping streets, a magnificent, white-domed Victorian railway station and new housing blocks. KL, as it is called, also has its share of squatter colonies.
Population 937 900
Map Malaysia Bb

Kuala Terengganu *Malaysia* Capital of Terengganu state on Malaysia's east coast, about 140 km (85 miles) south of the Thai border. It has no rail links, but the east coast highway has been improved, opening communications. Hydrocarbons have been discovered offshore.
Population 186 700
Map Malaysia Ba

Kuantan *Malaysia* Capital of Pahang state on the east coast, about 200 km (125 miles) east of Kuala Lumpur, the national capital. A new trans-Peninsular road now links Kuantan with the industrialised west. A port has been built to export palm oil, used in margarine and cooking fats, produced on plantations inland.
Population 136 700
Map Malaysia Bb

Kuanza *Angola* See CUANZA

Kuching *Malaysia* [illegible] developed 11 km (7 miles) downstream from Kuching at Pending.
Population 120 000
Map Malaysia Db

Kufra (Kufrah) *Libya* Group of five oases about 48 km (30 miles) long and 20 km (12 miles) wide in the Cyrenaican desert, 920 km (572 miles) south-east of Benghazi. They have long been an important caravan centre on the trans-Saharan route from Benghazi to Chad. In 1895, the area became the headquarters of the Sanusi, a Muslim religious fraternity. They strongly resisted the Italian colonial invasion in 1912, and it was not until 1931 that the Italians occupied the oases after prolonged bombing.

Kufra produces barley and the country's best dates, as well as traditional hand-woven textiles. It is an area selected for major agricultural improvements using pumped water (there are large underground water reservoirs). Some 10 000 hectares (about 25 000 acres) of irrigated land produces fodder for sheep rearing, and another 5000 hectares has been allotted to 900 small farms. Iron ore and low grade manganese deposits have been found nearby.
Map Libya Cb

Kuh-e Baba *Afghanistan* Massive mountain range separated from the Hindu Kush by the Bamian valley. Its spectacular passes include the Unai, 3100 m (10 170 ft), and the Hajigak, 3250 m (10 660 ft). The highest peak is Shah Fuladi, 5600 m (18 370 ft). The range is the source of the Hari Rud and Kabul rivers.
Map Afghanistan Ba

Kuito (Bié Silva Porto) *Angola* Market town about 550 km (342 miles) south-east of the capital, Luanda. It is the capital of Bié province. Lying at the heart of one of the country's most fertile regions, the town handles coffee, meat, rice, beeswax and cereals.
Population (town) 25 000; (province) 650 000
Map Angola Ab

Kujawy *Poland* Ancient province straddling the Notec river in central Poland. Kruszwica, the regional capital, 105 km (65 miles) north-east of the city of Poznan, was a cradle of the Polish state. Its Romanesque church – built in 1120-40 – stands on the site of a 10th-century wooden cathedral built by Mieszko I, Poland's first ruler, who accepted Christianity for himself and his people.
Map Poland Cb

Kukës ***Albania*** Main town of north-east Albania, straddling the confluence of the Black and White Drin rivers. It lies in a basin beneath the high, dazzling white gypsum and marble Korab mountains. The town is in an area of chrome and copper mines, and has ore-processing plants built by Italians before the Second World War. The plants were expanded after 1948 with Czech and Soviet aid in the wake of the Communist takeover of Albania in 1946.

Population 12 500
Map Albania Ca

Kula Kangri ***Bhutan*** Highest mountain in the country, 7554 m (24 783 ft) high. It is in the Himalayan range near the border with Chinese Tibet.

Map Bhutan Ba

Kulm ***Poland*** See CHELMNO

Kumamoto ***Japan*** City near the west coast of Kyushu island, about 75 km (47 miles) east of the port of Nagasaki. It manufactures small electrical components, supplying parts for other factories on Japan's 'silicon island'. Its spectacular 17th-century castle has been rebuilt and is now a military museum.

Population 555 700
Map Japan Bd

Kumasi ***Ghana*** Capital of Ashanti region. It has been capital of the Ashanti people since the 17th century and was one of the first towns to develop in Ghana's southern forest. The Ashanti took control of the surrounding area so they could channel as much trade as possible through their capital. Even today Kumasi is known as the 'City of the Golden Stool' – a symbol of the Ashanti royal line – and the tribal king still has his court at Manhiya Palace. The town is said to have Ghana's largest market. It also has a museum, open-air theatre, crafts workshop, model farm and a zoo. It is linked by road and rail to the capital, Accra, about 200 km (125 miles) to the south-east.

Population 415 300
Map Ghana Ab

Kumrovec ***Yugoslavia*** Croatian village, 40 km (25 miles) north-west of Zagreb, where President Tito was born in a red-tiled farmhouse, now a museum.

Map Yugoslavia Ba

Kumana ***Sri Lanka*** See RUHUNA

Kundelungu National Park ***Zaire*** Founded in 1970 in Shaba region, south-west of Lake Mweru on the Zambian border, the park includes Lofoi Falls, some forest country, and savannah on the scenic Kundelungu plateau. Wildlife includes antelopes, cheetahs, elands, elephants, hartebeests and zebras.

Map Zaire Bc

Kunene ***Angola/Namibia*** See CUNENE

Kunlun Shan ***China*** Mountain range with peaks of more than 7000 m (23 000 ft). It stretches eastwards from the Karakoram Range, forming the northern edge of the Xizang Gaoyuan (Tibetan Plateau).

Map China Bd

Kunming ***China*** Capital of Yunnan province. It is an important railway junction and trading centre. Its industries include iron and steel, engineering and truck assembly.

Population 1 930 000
Map China Ff

Kunsan (Gunsan) ***South Korea*** Port on the west coast, on the south bank of the Kuan Estuary. Paper, plywood, rubber, chemicals and textiles are made there.

Population 154 000
Map Korea Ce

Kuopio ***Finland*** City and tourist resort, founded in 1782 on the west coast of Kallavesi lake, about 340 km (210 miles) north-east of the capital, Helsinki. It is the seat of the Orthodox archbishopric and has a museum devoted to the Orthodox church in Finland. There is also a Lutheran cathedral, and a university. The city's main manufactures are timber, softwood and clothing.

Population 77 300
Map Finland Cc

Kupang ***Indonesia*** Capital and main port of East Nusa Tenggara province at the western tip of the island of Timor.

Population 123 400
Map Indonesia Fe

Kurashiki ***Japan*** City on Honshu island about 160 km (100 miles) west of Osaka. Old Kurashiki, an important port during feudal times, is carefully preserved. Its quiet canal, spanned by 'willow pattern' bridges, is shaded by weeping willows and flanked by solid white warehouses with heavy black-tiled roofs. New Kurashiki, which sprawls across land reclaimed from the Inland Sea, is an important industrial centre specialising in oil refining, petrochemicals, iron and steel and rayon.

Population 411 400
Map Japan Bd

Kurdistan ***Middle East*** Area of central Asia inhabited by the highly independent Kurds – a nomadic, stock-raising Muslim people. Kurdistan comprises south-east Turkey, north-east Iraq and north-west Iran. There are immense variations in the estimates of the numbers of Kurds, ranging from 6.9 million to 16.4 million: 3.2-8 million in south-east Turkey, 1.8-5 million in north-west Iran, and 1.55-2.5 million in northern Iraq. An independent Kurdistan was proposed in the 1920 Treaty of Sèvres, between the First World War Allies and Turkey, but the idea came to nothing when Turkey became a republic in 1923. In 1944-5, Russia helped to establish an unofficial Kurdish 'republic' at Mahabad in north-west Iran. Since then, Turkey, Iraq and Iran have repressed the Kurds, giving rise to periodic disturbances and rebellions. In the early 1970s there was a major uprising of Kurds in Iraq. This led to the creation of the Kurdish Autonomous Region, which recognises the special status of the Kurds.

Map Middle East Ba

Kurdzhali ***Bulgaria*** Main regional town of the eastern Rhodope Mountains, 55 km (34 miles) north of the Greek border. Since 1950, when it was a quiet and largely Turkish-style town of 9000 inhabitants, mostly involved in tobacco farming and processing, it has grown into a major lead and zinc mining and processing centre. Power for the new industries comes from two hydroelectric dams close to the town.

Population 59 000
Map Bulgaria Bc

Kure ***Japan*** Town on the Inland Sea, about 20 km (12 miles) south-east of Hiroshima. A former naval base with an excellent natural harbour, it now has modern steel mills and builds some of the world's largest ships.

Population 226 500
Map Japan Bd

Kuril (Kurile) Islands ***Japan/USSR*** Chain of 56 volcanic islands – many uninhabited – stretching from the Japanese island of Hokkaido north-east to the Kamchatka peninsula. They have a total land area of 15 600 km² (6020 sq miles). The southern islands (Kunashir, Etorofu, Shikotan and the Habomai) were seized from Japan by the USSR in 1945, but the Japanese still regard them as theirs.

Map Japan Eb

Kuril Trench An extension of the Japan Trench in the Pacific Ocean floor adjacent to the KURIL ISLANDS and the USSR's KAMCHATKA peninsula. These form part of the area known as the Pacific 'ring of fire', where earthquakes and volcanoes are caused by the Pacific plate meeting and descending beneath the Eurasian plate (see PLATE TECTONICS).

Greatest depth 10 498 m (34 442 ft)
Map USSR Pd

Kursk ***USSR*** Industrial city about 450 km (280 miles) south of Moscow. It was established on the Seym river in the 9th-11th centuries, but was destroyed by Tatars in 1240 and not rebuilt until 1586. The biggest tank battle of the Second World War – involving a total of 5700 tanks – took place around Kursk in 1943. It ended in Russian victory, but almost the whole city except for the centre was destroyed. The parts that survived included the 18th-century St Serge Cathedral, which has a finely carved 17 m (46 ft) high screen. Near Kursk is one of the world's richest iron-ore deposits, containing 160 000 million tonnes.

Population 423 000
Map USSR Ec

Kurukshetra ***India*** Holy Hindu town about 140 km (87 miles) north of Delhi. There Lord Krishna (God) is said to have revealed himself to the prince Arjuna before a mighty battle (about 1000 BC) in which many died. The battle, and the moral discourse in duty and obligation, right and wrong, that resulted from it, are related in the *Bhagvad-Gita* (the Song of the Lord), Hinduism's most sacred and philosophical book.

Population 186 100
Map India Bb

Kushiro ***Japan*** Fishing port on the south-east coast of Hokkaido island. A pulp and paper industry has developed on the basis of local timber and coal resources.

Population 214 500
Map Japan Db

Kuta Beach ***Indonesia*** See BALI

Kutaisi ***USSR*** Georgia's second largest city after Tbilisi. It stands on the Rioni river in the south-west of the republic, 85 km (53 miles) east of the Black Sea and some 70 km (40 miles) from the Turkish border. Its factories make vehicles, mining machinery, chemicals, silk, clothing, footwear, furniture and canned goods. It contains the remains of a fortress built on the site in the 1st century AD and of an 11th-century cathedral.

Population 210 000

Map USSR Fd

Kutch (Kachchh) ***India*** Coastal district some 550 km (340 miles) north-west of Bombay, covering an area of 44 185 km² (17 060 sq miles). It borders the Gulf of Kutch and consists largely of a vast, swampy depression, the Rann, and it floods from both sea and rivers in the monsoon period. At other times it is dry, salty and inhospitable. Part of it is claimed by Pakistan, which waged an unsuccessful tank campaign there against India in 1965.

Population 696 500

Map India Ac

Kutna Hora ***Czechoslovakia*** Former silver-mining town in Bohemia about 65 km (40 miles) south-east of Prague. Mining began there in the 13th century and the Royal Mint was established in the 14th-century Vlassky Dvur ('Italian Court'), where craftsmen brought in from Florence designed and struck silver coins. The process is now re-created in a coin museum. The sumptuous Gothic cathedral dedicated to St Barbara – the patron saint of gunners and miners – has a chapel with 15th-century wall paintings of the craftsmen at their trade. The mines were gradually exhausted, and the mint closed in the 18th century. The town now has engineering, metal, textile and food industries.

Population 21 400

Map Czechoslovakia Bb

Kuwait ***Kuwait*** Situated at the southern entrance to Kuwait Bay, the country's capital stands on the site of a 16th-century Portuguese fort. In the early 1960s it consisted of one main street and a few thousand people living on the edge of the desert. During the next 20 years it was transformed into a wealthy, expanding city of high-rise buildings, large office blocks, boulevards, parks and gardens. The modern city has been laid out with roads radiating outwards from its centre. Its principal buildings include the Seif Palace (the Amir's headquarters), built in 1896, and the Kuwait Towers (1970s), which contain restaurants, an indoor garden, and an observatory from which there are panoramic views of the city. Its museums include the Kuwait Natural History Museum and the Kuwait Museum (history). Building materials and chemicals are manufactured in the city, and there is a large fishing industry.

Population (Kuwait and suburbs) 360 000

Map Kuwait Bb

Kuybyshev ***USSR*** Industrial city and Volga river port 860 km (535 miles) south-east of Moscow. It was the wartime Soviet capital between October 1941 and August 1943, during the German invasion of Russia. Founded as Samara in 1586, it was renamed in 1935 after the Revolutionary leader Valerian Kuybyshev (1888-1933). It now produces precision machinery, transport equipment, petrochemicals, timber, foodstuffs and consumer goods. The Volga river was dammed near the city in the 1960s to form the Kuybyshev reservoir – at 6450 km² (2490 sq miles) the largest in area in the USSR and second largest in the world after Lake Volta in Ghana.

Population 1 251 000

Map USSR Gc

Kuznetsk ***USSR*** See NOVOKUZNETSK

Kwai (Khwae) ***Thailand*** Two tributaries of the Mae Khlong river, the Kwai Yai (Big Kwai) and the Kwai Noi (Little Kwai) both about 150 km (93 miles) long. The Bridge over the River Kwai, famous from the film of the same name, was built between 1941 and 1944, during the Second World War, over the Kwai Yai, 100 km (60 miles) west of Bangkok. The railway it carried – which was planned by the Japanese to continue into Burma – was never completed. But the bridge, and the cemeteries filled with those who died during its construction, remain to attract visitors. An estimated 110 000 Allied prisoners of war and slave labourers died during the construction of the bridge and the 415 km (257 miles) of line.

Map Thailand Ac

Kwajalein ***Marshall Islands*** The world's largest atoll, in the Ralik Chain about 450 km (280 miles) north-west of Majuro. Its 280 km (175 mile) long reef encloses a lagoon covering 2800 km² (1100 sq miles). It is leased by the USA under a 30 year agreement from 1982 for use as a test target for missiles fired some 8000 km (5000 miles) from California.

Population 8000

Map Pacific Ocean Db

KwaNdebele ***South Africa*** Semi-autonomous black state in the Transvaal north-east of Pretoria, set up in 1977 for the South Ndebele people. It covers 3410 km² (1317 sq miles). The capital is Siyabuswa.

Population 380 000

Map South Africa Cb

Kwangju (Gwangju; Chonnam) ***South Korea*** Provincial capital about 275 km (170 miles) south of the national capital, Seoul. It lies on fertile lowlands in the once remote province of Cholla. Modern communications have ended Kwangju's isolation, but its tradition of independence broke out in violent student riots against the Seoul government in 1980. The city's industries include motor vehicles and textiles.

Population 727 600

Map Korea Ce

Kwango (Cuango) ***Angola/Zaire*** River about 1100 km (680 miles) long. It rises in the central plateaus of Angola, and flows northwards to form part of the Angola-Zaire border before reaching its confluence with the Kasai river near Bandundu. Most of its course in Zaire is navigable.

Map Zaire Ab

Kwanza ***Angola*** See CUANZA

A DAY IN THE LIFE OF A PALESTINIAN MIGRANT IN KUWAIT

Like most of the 300 000 Palestinians in Kuwait, Faisal came here to find work. Fifteen years ago he left the refugee camp where he grew up, at Kalandia in the Israeli-occupied West Bank, and settled in a Palestinian quarter of Kuwait city: Nouqra Hawalli. He works for an air-conditioning maintenance firm, his three children go to the local school, which is funded by the Palestinian Liberation Organisation, and he and his wife Zahra now feel well established.

Faisal spent his entire childhood in a refugee camp. His parents were forced to leave their home when the land was captured by the Israelis in 1948, and have lived in the camp ever since. Faisal went to secondary school and on to a United Nations vocational training centre for refugees nearby. But when he qualified, he knew there would be no jobs in the West Bank. Since military occupation in 1967, economic growth has been stifled. If Faisal had gone into Israel to find work, he might have found manual labour – but pay for Palestinians is notoriously low.

The journey to work from his rented apartment in Hawalli is easy. He gets a lift from a friend, one of the many Palestinians who drive the 'One-Eights' – pick-up trucks used as taxis and named after some long-forgotten model of truck first used for the purpose. The journey takes 20 minutes or so, and Faisal is at work in the industrial quarter of Shuwaikh by 8 am.

Today, Thursday, work finishes at 1 pm; Friday is the day off for everyone in Muslim Kuwait. Then Faisal has an errand to run for his wife: vegetables are cheaper in the downtown souk, or bazaar, than in the stores around Hawalli. He has his pay in his pocket – not a bad wage, although nearly half of it goes on rent.

Faisal is worried that he will have to find a new place to live. His apartment block is being demolished under the city's redevelopment plan. It looks as though he and his family will end up in one of the districts near the airport which are mostly occupied by Palestinians – and where there is likely to be accommodation they can afford. It will give the children a difficult journey to school, and he will have farther to travel to Shuwaikh.

Determinedly, he forces his thoughts in another direction. Tomorrow he has promised Zahra and the children that they will have a picnic in the desert. But oh, he thinks, how much less enjoyable the oppressively hot desert is than the breezy olive groves of home . . .

Kuwait

WHERE OIL HAS TURNED TO GOLD AND AN ENLIGHTENED GOVERNMENT HAS ESTABLISHED A SUBSIDISED OASIS OF PROSPERITY

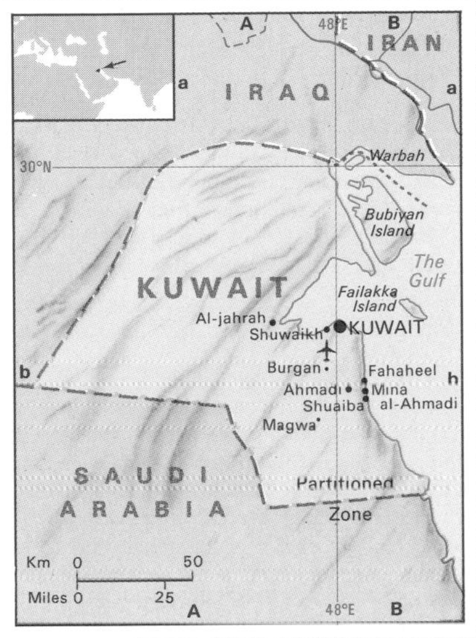

The tiny Arab state of Kuwait, on The GULF, has one of the most comprehensive welfare systems in the world. Rich with oil, an enlightened government has set out to distribute its wealth to the local population and the average income of equivalent to US$15 770 a year (1982) is one of the world's highest. Education and health care are free to all nationals; food, water, electricity and transport are subsidised.

And yet it is not paradise, for other prices can be high. A middle-class suburban house costs the equivalent of more than US$1 million, and flats are expensive to rent, but the government is providing homes for young couples at nominal rents. Meanwhile, oil production has been cut back to one-third of the 1979 figure and revenue has fallen, as world demand and prices have decreased.

The country, of only 17 818 km² (6879 sq miles), consists of the city of KUWAIT at the southern entrance of Kuwait Bay and a small wedge of undulating desert between Iraq and Saudi Arabia. It was founded in the 18th century by Anaizah tribesmen, driven from the Arabian desert by drought. The ruling house of Al Sabah, which came to power in 1756, still provides the country's emir, who is advised by a partly elected national assembly.

At the end of the 19th century Kuwait came under British protection, which lasted until 1961. Neighbouring Saudi Arabia and Iraq have at times claimed the territory and even invaded, the latter as recently as 1973, but Kuwait remains stable and independent, defended by small but well equipped armed forces.

RANGE OF INDUSTRIES

Kuwait is a commercial crossroads, alive with development: housing, factories, power stations, schools, hospitals, public buildings and hotels – all paid for by oil. Pearls and camel hides were once the main exports. Oil was found in 1938 and Kuwait's reserves of 90 000 million barrels are the third largest in the world.

Its growing range of industries, mostly oil-related, are highly dependent on foreign technicians and imported skills. Of the almost 2 million people, nearly 60 per cent are immigrants, who provide 75 per cent of the work force but do not share in the state's distribution of oil wealth. They include Palestinians, Egyptians, Iranians, Yemenis, Indians, Pakistanis, Europeans and Americans.

It is an arid hot land in summer, with monthly mean temperatures of up to 45°C (113°F); winters are cool and sometimes there is frost. Little rain falls and Kuwait has the world's largest saltwater distillation plants; they produced 545 million litres (120 million gallons) a day in 1984. Surprisingly there is a variety of flowers and animals. A total of 373 desert flower species, appearing in the short spring and summer, have been listed. As houses and gardens have sprung up, more migrant birds have been recorded.

Only 8.4 per cent of the land can be farmed. Cultivated land is found around oases, but there is some pasture for sheep and goats. Most food is imported, but the government is financing wheat growing and other schemes. Hunting has reduced the desert wildlife, but there are gazelles, the ratel (a badger-like animal), the fennec fox and some wolves, jackals, hares and sand rabbits.

Kuwaiti men traditionally meet in the evening in *diwaniyas* (large tents) to sit and talk and drink bitter coffee flavoured with cardamom seeds. Alcohol is banned, even for embassies. Soccer is popular and the national team has been Gulf champions five times. Women wear Western dress, take jobs and drive cars, but are not allowed to vote or hold parliamentary posts.

Parliament has 66 members, 50 of whom are elected; there are free elections but no political parties. Kuwait has the most independent Press in the Arab world, and is the only Gulf state to have full diplomatic relations with Russia.

Not much of the old Kuwait is left, although the city has its ancient gates, museums with Islamic art treasures, and the royal palace. The shopper is more likely to find the latest Western goods in the *souk* or bazaar than anything traditional. FAILAKKA ISLAND, in Kuwait Bay, has magnificent beaches; and excavations there have revealed 3000-year-old Bronze Age dwellings.

Kuwait, keeping output low to maintain its reserves, has a falling oil revenue, a population composed mainly of immigrants, and a strong minority of Shiite Muslims, the sect which rules in Iran. The divide between liberalism and traditionalism, the desert tribal culture and today's urban life, may well present problems for the future.

KUWAIT AT A GLANCE

Area 17 818 km² (6879 sq miles)
Population 1 986 000
Capital Kuwait (city)
Government Constitutional monarchy
Currency Dinar = 1000 fils
Languages Arabic (official), English
Religion Muslim
Climate Hot and dry; average temperature in Kuwait City ranges from 8-18°C (46-64°F) in winter to 29-45°C (84-113°F) in summer
Main primary products Sheep, goats, dates and other fruit, vegetables, fish; oil and natural gas
Major industries Oil and gas production and refining, boat building, fishing, processed foods, chemicals, ammonia fertilisers, cement, seawater desalination
Main exports Crude oil and oil products, natural gas
Annual income per head (US$) 15 770
Population growth (per thous/yr) 62
Life expectancy (yrs) Male 68 **Female** 73

KwaZulu ***South Africa*** Fragmented black state covering 32 390 km² (12 503 sq miles) in northern Natal. Formerly called Zululand after its inhabitants, it is self-governing but not independent. Its capital is Ulundi. The region's chief cash crops are cotton and sugar.
Population 4 186 000
Map South Africa Db

Kwilu ***Congo*** See KOUILOU

Kwinana ***Australia*** Port and industrial town in Western Australia, on the natural harbour of Cockburn Sound about 20 km (12 miles) south-west of Perth. An important industrial centre, Kwinana handles bauxite (aluminium ore) and nickel, produces steel, and refines oil.
Population 12 500
Map Australia Be

Kyaukse ***Burma*** Town in central Burma about 50 km (30 miles) south of Mandalay. It is the centre of an intensively cultivated area producing rice, cotton, chillies and groundnuts (peanuts). Kyaukse had well developed irrigation systems 900 years ago that are still in use.
Population 115 700
Map Burma Cb

Kyiv ***USSR*** See KIEV

Kyoga, Lake ***Uganda*** Area of shallow water and swamps fed by the Nile river, about 110 km (69 miles) north of Lake Victoria. It stretches more than 200 km (125 miles) from east to west, forming a barrier to communications.
Map Uganda Ba

Kyongju (Gyeongju) ***South Korea*** South-eastern city 80 km (50 miles) north of the port of Pusan. It was the capital of the Silla kingdom from 57 BC to AD 935 and still contains many buildings from that period. The Pulguk-sa temple, south-east of the city, was built in 751.
Population 108 000
Map Korea De

Kyoto ***Japan*** City 45 km (28 miles) north-east of Osaka near the southern end of Biwa-ko lake in central Honshu. The cultural treasure house of Japan, it was the imperial capital for more than 1000 years until 1868. Kyoto was the only major Japanese city to escape damage in the Second World War. It still retains the atmosphere of traditional feudal Japan. It abounds with temples, shrines and museums. The Zen Buddhist temple of Ryoanji, on the city outskirts, contains one of the world's oldest stone gardens. Laid out in 1499, the garden consists of 15 large stones set in a 'sea' of grey-white gravel. There are no plants, other than a few mosses, but the gravel is raked every day. It exists as an aid to meditation.

Population 1 479 100
Map Japan Cc

Kyrenia ***Cyprus*** North coast port and market town for the neighbouring farming district. Since 1974 it has been under Turkish occupation. Founded by the ancient Greeks, the town has a 12th-century castle and Byzantine chapel, and the surrounding countryside is dotted with the remains of medieval abbeys and fortresses.

The Kyrenia Mountains stretch for 160 km (100 miles) along the coast just inland from the town. Their highest point is Kyparissovouno at 1024 m (3360 ft).

Population 23 000
Map Cyprus Ab

Kyushu ***Japan*** The country's third largest island, after Honshu and Hokkaido. Kyushu covers 43 065 km² (16 627 sq miles) and is dotted with many active volcanoes. It enjoys a mild climate, but agriculture is declining as more of the islanders move to the major cities, the largest of which are Fukuoka, Kirakyushu and Kumamoto. Instead, Kyushu has become a centre of Japan's electronics industry, earning the nickname of 'silicon island'.

Population 13 276 000
Map Japan Bd

Kyustendil ***Bulgaria*** Spa town and capital of Kyustendil province, 65 km (40 miles) south-west of Sofia. It trades in wine, fruit and tobacco, and is the market and processing centre for Bulgaria's major fruit-growing area. The spa's hot sulphur springs were used by the Roman Emperor Trajan in the 1st century AD, and have drawn visitors ever since. To the north of the town, through the Struma Gorge, is Zemen Monastery, which has 14th-century frescoes.

Population 55 500
Map Bulgaria Ab

La Ceiba ***Honduras*** North-western port exporting mostly bananas and pineapples from farms inland.

Population 68 900
Map Honduras Aa

La Condamine ***Monaco*** The 'town centre' of the principality – the main business area, where many Monegasques also live – on the north-western side of the port.

Population 12 500

La Coruña ***Spain*** See CORUNNA

La Goulette (Halq al-Wadi) ***Tunisia*** Seaport town and beach resort, 11 km (7 miles) north-east of the capital, Tunis, to which it is connected by a deepwater channel. It was the original harbour for the capital before the channel was dug in 1893 and Tunis was able to take ocean-going vessels. A new port to serve the capital was built at La Goulette in 1968. Nevertheless, many port services remain at the old port of Tunis, close to the heart of the city. La Goulette has a massive kasbah (Muslim residential quarter) and the main street contains many cheap seafood restaurants.

Population 61 700
Map Tunisia Ba

La Guaira ***Venezuela*** Main port, about 13 km (8 miles) to the north of the capital, Caracas, but some 37 km (23 miles) by a winding mountain railway line. However, a new road completed in the 1950s is more direct. The port builds ships and has dry dock facilities as well as a good harbour.

Population 30 000
Map Venezuela Ba

La Haie ***Netherlands*** See HAGUE, THE

La Libertad ***Peru*** North-western department, producing most of the country's sugar cane. The crop is exported mainly through the port of Salaverry, 500 km (310 miles) north of Lima. Salaverry also handles copper from mines 120 km (75 miles) inland. The ruins of the pre-Inca Chimú city of CHAN CHAN lie 5 km (3 miles) north of the department capital, Trujillo.

Population 962 900
Map Peru Ba

La Mancha ***Spain*** Arid treeless plateau – 680 m (2230 ft) high and some 4800 km² (18 530 sq miles) in area – about 160 km (100 miles) south of Madrid. It was the setting for the mythical exploits of the knight Don Quixote, in the 17th-century novel by Miguel de Cervantes.

Map Spain Dc

La Oroya ***Peru*** Major lead, tin and silver mining and smelting centre, lying in a deep narrow valley at the fork of the Mantaro and Yauli rivers, 160 km (100 miles) east of Lima.

Population 36 000
Map Peru Bb

La Pampa ***Argentina*** See PAMPAS

La Paz ***Bolivia*** National capital and department and the highest capital city in the world, 3627 m (11 900 ft) above sea level.

The city – which lies in a valley at the foot of ILLIMANI, sheltered from the winds of the surrounding ALTIPLANO plateau – was founded in 1548 by the Spanish adventurer Alonso de Mendoza. It flourished as a colonial town on the route from the silver mines in the south-east to the Peruvian capital, Lima. It was one of the centres of resistance during the War of Independence between 1809 and 1825. After victory in the war it was named La Paz de Ayacucho in honour of the decisive battle at Ayacucho in Peru. It has been the seat of government since 1898.

Steep streets climb from the business and wealthier residential areas at the bottom of the canyon to the terraced heights where most of the Indian population live. Woollen goods made from alpaca wool, embroidered shawls and leather goods are on sale in the city's markets, which are crowded with bowler-hatted Indian women. The church and monastery of San Francisco, consecrated in 1778, is one of the few good remaining examples in the country of colonial architecture.

The city's main industries include textiles, electrical appliances, chemicals, furniture and glass. It also handles lead and silver for export.

Population (department) 1 913 200; (city) 900 000
Map Bolivia Bb

La Plata ***Argentina*** Port on the Rio de la Plata (River Plate) 56 km (35 miles) south-east of Buenos Aires. Its main cargoes are oil and frozen meat. The huge estuary, which covers 35 000 km² (13 500 sq miles) was the scene of the Second World War Battle of the River Plate.

Population 455 000
Map Argentina Db

La Rochelle ***France*** Seaport, tourist centre and capital of Charente-Maritime department, 120 km (75 miles) south of Nantes. A leading port in the 14th-16th centuries, it became a stronghold of the Protestant Huguenots during the 16th-century Wars of Religion, but was captured by the Catholics in 1628 after a 14 month siege. It now manufactures fertilisers, plastics and cement, and other industries include shipbuilding, fish canning and sawmilling. The new port of La Pallice is 5 km (3 miles) to the west.

Population 104 700
Map France Cc

La Serena ***Chile*** Capital of Coquimbo region, 400 km (250 miles) north of Santiago. It was founded by the Spaniards in 1543 and named after the Spanish birthplace of the Conquistador Pedro de Valdivia. During the 16th and 17th centuries it was a target of Indian and pirate attacks.

La Serena stands at the mouth of the Elqui river, an area renowned for the quality of its peaches and oranges. It was rebuilt in colonial style in the 1950s.

Population 91 800
Map Chile Ab

La Skhirra ***Tunisia*** See SAKHIRA

La Spezia ***Italy*** Major naval base and port about 75 km (47 miles) south-east of Genoa. It has the country's most important naval arsenal and naval museum.

Population 115 200
Map Italy Bb

La Unión ***El Salvador*** Department on the south-east coast, growing cotton and sugar cane. Its capital, also called La Unión (population 32 000), is a deep-water harbour on the Gulf of Fonseca near the Honduran and Nicaraguan borders.

Population 224 100

laagte, laegte In South Africa, broad hollows, less well defined than valleys, between wide rises in a comparatively flat landscape.

Laatokka ***USSR*** See LADOGA, LAKE

Laâyoune ***Western Sahara*** Capital town of the territory situated in the north-west 20 km (12 miles) from the Atlantic coast. Its name means 'the springs' in Arabic. It is served by a small port, and there is also an airport. Because of the war with Polisario guerrillas fighting for independence, the Moroccans have built a 34 km (21 mile) system of perimeter defences round the town. They have also built new homes in the town to induce nomadic tribesmen to settle.
Population 38 000
Map Western Sahara Aa

Labasa ***Fiji*** See VANUA LEVU

Labé ***Guinea*** Town in the Fouta Djallon, about 260 km (160 miles) north-east of the capital, Conakry, on the road north from Mamou towards Senegal. Its cool climate and mountain scenery made it a popular resort among French colonists. It markets livestock and other produce, and there are plantations of bitter oranges, jasmine and coffee nearby.
Population 79 700
Map Guinea Ba

Labrador ***Canada*** The mainland section of the province of Newfoundland, covering an area of 295 800 km² (112 826 sq miles) and separated from the island part by the Strait of Belle Isle. It is a plateau area of the CANADIAN SHIELD and varies in height from 200 to 1000 m (656 to 3280 ft). The north is tundra, the south is covered by coniferous forests, and mineral deposits include iron ore. Indians and Inuit (Eskimos) engaged in fishing and trapping live in small communities on the coast. Goose Bay airport and the mining towns of Labrador City and Wabush are inland.
Map Canada Ic

Labuan ***Malaysia*** Island 75 km² (35 sq miles) in area, lying 55 km (35 miles) off the coast of Brunei. It has a fine port, Victoria Harbour, which was once a key port in the China trade. It is now a free port.
Population 26 400
Map Malaysia Ea

Laccadive Islands ***India*** See LAKSHADWEEP

laccolith, laccolite Mushroom-shaped body of IGNEOUS ROCK intruded between layers of sedimentary rock which it has pushed up.

Lacq ***France*** Village in the extreme south-west where oil and natural gas were discovered in 1951. Sulphur, used for chemical manufacture, is also extracted from the gas. The nearby town of Mourenx (population 9000) was built in 1957-61 to house the oil-field's workers.
Map France Ce

Ladakh ***India*** District of JAMMU AND KASHMIR in the western Himalayas. Its AKSAI CHIN area has been occupied by the Chinese since the 1950s.
Map India Ba

Ladoga, Lake (Ladozhskoye Ozero; Laatokka) ***USSR*** The largest lake in Europe, covering 18 390 km² (7100 sq miles) and lying just north-east of Leningrad. Its north and west shores belonged to Finland from 1919 to 1940. It was a major trade route even in Viking times, and a canal system was begun in the 18th century along its southern shore to protect ships from the lake's frequent storms. This is now part of the Volga-Baltic waterway, which connects with the Baltic-White Sea Canal. From October to April each year the lake is frozen, and vital winter supplies for Leningrad were sledged across it when the besieged city held out against German forces from September 1941 to January 1944.
Map USSR Eb

▲ FASHIONS NEW AND OLD A mini-skirted Aymara woman of La Paz carries her baby on her back; another sports the traditional full skirt and bowler hat. Half the city's people are Indians.

Ladysmith ***South Africa*** Town in Natal, about 190 km (120 miles) north-west of Durban. Founded in 1850, it is named after the wife of Sir Harry Smith, then governor of Cape Colony. In the Anglo-Boer War (1899-1902), British forces were besieged there from November 2, 1899 to February 28, 1900.
Population 21 900
Map South Africa Cb

Lae ***Papua New Guinea*** The country's second largest and fastest-growing town, on Huon Gulf about 320 km (200 miles) north of the capital, Port Moresby. Its port handles gold, timber, plywood, cattle, cocoa and coffee. The town was devastated by bombing in the Second World War, and was completely rebuilt. It is the country's major industrial centre, with many light industries, and contains Papua New Guinea's University of Technology. Its tropical heat and heavy rainfall – 4570 mm (180 in) a year – mean that flowers grow everywhere.
Population 62 000
Map Papua New Guinea Ba

Laghouat ***Algeria*** Administrative centre, and staging post on the Route du Hoggar 400 km (about 250 miles) south of Algiers. It is the first oasis in the desert after the Saharan Atlas mountains. The town was founded in the 11th century by the Hilians – a Berber tribe that had fled from the north – and is the main market and commercial centre for the Ouled Nail Mountains Confederation of tribes.
Population 45 000
Map Algeria Ba

lagoon Body of salt water separated from the sea by sand or shingle BARS OR CORAL REEFS.

Lagos ***Nigeria*** National capital and port on an island in the south-west. It was founded by the Yoruba people in the 13th century as a refuge from warring tribes. Today it is the most congested city in West Africa. Overcrowded slums with open sewers contrast with modern houses, and squalid winding streets with the new main roads. The streets are notorious for muggings and theft. The administrative and business heart of the city is on Lagos Island where there are large shops – not always well stocked – and high-rise office blocks. Two road bridges link the island to the mainland and to the expensive and elegant residential areas of Ikoyi Island and Victoria Road. But because most people live on the mainland, there is appalling traffic congestion on the bridges. All moves to ease the congestion have failed and there are plans to move the administrative capital to ABUJA. Industry is located mainly in modern estates, and products include foodstuffs, metal goods, chemicals, textiles, furniture and glass.

Tourist attractions in the city include the 18th-century Oba's (King's) Palace on Lagos Island, good beaches, polo, golf, fishing and sailing. Local craftsmen produce dyed cotton, handwoven cloth, leather goods and jewellery.
Population 1 477 000
Map Nigeria Ab

LAKES AND INLAND SEAS

There are two main types of natural lake: salt and freshwater. Inland seas – such as the Caspian – have no outlet; water is lost only by evaporation so that salinity gradually builds up. Freshwater lakes have outflowing streams, and the amount of water entering them usually exceeds that lost by evaporation, so they do not become salty.

LARGEST INLAND WATERS OF THE WORLD

Lake	Location	Area (km²/sq miles)
Caspian Sea (salt)	USSR/Iran	393 900/152 100
Superior (freshwater)	USA/Canada	82 400/31 800
Victoria (freshwater)	Kenya/Uganda/Tanzania	69 500/26 800
Aral Sea (salt)	USSR	65 500/25 300
Huron (freshwater)	USA/Canada	59 570/23 000
Michigan (freshwater)	USA	57 750/22 300
Tanganyika (freshwater)	Zaire/Tanzania/Zambia/Burundi	32 900/12 700

Lahej ***South Yemen*** Town and district of the same name 10 km (6 miles) north-west of Aden. The district is one of the country's richest agricultural and cotton-growing areas.
Population (town) 35 000
Map South Yemen Ab

Lahore ***Pakistan*** City once known as 'the Queen of British India', when it was capital of the imperial Punjab. Since Pakistan's independence in 1947, the city has continued to grow despite losing much of its surrounding land to India in border changes. The Mogul emperors built many fine palaces, gardens and mosques there in the 16th and 17th centuries, and it prospered both under the Sikh kings of the Punjab and after the British takeover in the 1840s. It retains many of the tree-lined malls of fine public buildings from imperial days. Outside the museum stands the great cannon by which the English writer Rudyard Kipling's fictional hero Kim played. Muhammad Ali Jinnah, leader of the Muslim League, made a speech there in 1940 when he proclaimed the foundation of Pakistan as the League's goal.

Lahore, which is about 270 km (170 miles) south-east of the national capital, Islamabad, remains the capital of the Pakistani province of Punjab. It has metalworking, engineering, chemicals, textiles, leather and other industries.
Population 2 922 000
Map Pakistan Db

Lahti ***Finland*** Modern industrial town, about 100 km (60 miles) north and slightly east of Helsinki. Most of its factories are involved in the processing of timber and the manufacture of wood products. The town is also the setting for Finland's annual winter sporting competitions. The ski jumps, located on the great ridge called SALPAUSSELKA, command magnificent views of the country's central lake district.
Population 94 100
Map Finland Cc

Laibach ***Yugoslavia*** See LJUBLJANA

lake Inland body of fresh or salt water, varying in size from a few tens of square metres to thousands of square kilometres, and in depth from a few metres to more than 1000 m (about 3300 ft). Lakes occupy impermeable basins formed by earth movements, erosion, deposition or volcanic activity. The lake with the largest area in the world is the Caspian Sea (393 900 km², 152 100 sq miles). The deepest lake, more than 1770 m (5807 ft) deep, is Lake Baikal in Siberia.

Lake District ***United Kingdom*** Scenic area of lakes and mountains in the north-west English county of Cumbria; it stretches about 40 km (25 miles) in each direction. The Cumbrian Mountains, which form its core, have many peaks including England's highest, Scafell Pike (978 m, 3208 ft). The major lakes radiate out from this core: WINDERMERE and Coniston to the south, ULLSWATER to the north-east, Derwent Water and Bassenthwaite Lake to the north. The district has rich literary associations, notably with the 19th-century Lakeland Poets, of whom William Wordsworth (1770-1850) is the best known. For over a century the Lake District has been one of Britain's major tourist areas, despite being one of the wettest. It has been made a national park, but remains threatened by the hordes of visitors and by exploitation of its water resources.
Map United Kingdom Dc

Lake of the Woods ***Canada/USA*** Extremely irregular lake dotted with some 17 000 islands and covering about 4390 km² (1695 sq miles). It lies on the US-Canadian border, mostly in the Canadian province of Ontario, with a small part in Manitoba and some in the US state of Minnesota. It drains to the north-west into Lake Winnipeg. It is a popular tourist area.
Map Canada Fd

Lakshadweep ***India*** Territory in the south Arabian Sea made up of the Laccadive, Minicoy and Amindivi Islands. The islands – with a total land area of 32 km² (12 sq miles) – consist mainly of low coral atolls enclosing lagoons. The inhabitants live by fishing and coconut farming – although about one-third of the coconuts are eaten from the trees by the islands' population of rats.

The majority of the people are Muslim, but in the Minicoy there is a Buddhist culture in which women are the dominant sex.
Population 40 300
Map India Be

Lalibela ***Ethiopia*** Mountain town about 320 km (200 miles) north of the capital, Addis Ababa. It is renowned for its 11 Christian churches carved out of solid rock in the 12th and 13th centuries.
Map Ethiopia Aa

Lalita Patan ***Nepal*** See PATAN

Lalitpur ***Nepal*** See PATAN

Lambaréné ***Gabon*** Capital of Moyen-Ogooué province, situated on a large island in the Ogooué river, 160 km (100 miles) east of PORT GENTIL. It gained fame as the site of a hospital founded in 1913 by the German-born philosopher, musician, missionary doctor and 1952 Nobel Peace Prize winner, Albert Schweitzer (1875-1965). The hospital still stands, though Lambaréné now has another modern hospital. The town is a trading centre with sawmills that process timber for shipping downriver by barge.
Population 28 000
Map Gabon Bb

Lambayeque ***Peru*** Department in the north-west. It is the country's largest rice producer and a major sugar-cane producer, particularly in the irrigated sandy lowlands near Chiclayo, the department capital. Irrigation is turning the Sechura Desert on the western coastal plain into cultivable land. The Chancay and Leche valleys have several pre-Inca Mochica and Chimú sites. The Bruning museum in Lambayeque town contains historical ceramics and artefacts.
Population 674 500
Map Peru Aa

Lambert Glacier ***Antarctica*** The world's largest known glacier, flowing through the Prince Charles Mountains to the Amery Ice Shelf in east Antarctica. It is at least 400 km (250 miles) long and is fed by several smaller glaciers. Where it reaches the Amery Ice Shelf it is 200 km (125 miles) wide and flows at about 750 m (2460 ft) per year. It was discovered by an Australian aircraft crew in 1956.
Map Antarctica Oc

Lampang ***Thailand*** Market and timber town 535 km (332 miles) north of the capital, Bangkok. The town's timber yards store teak cut from nearby forestland before it is transported south by truck. Elephants, used in the logging camps, are trained at a school near the town. The main crop in the surrounding province of the same name is rice.
Population (town) 215 800; (province) 734 100
Map Thailand Ab

Lamphun ***Thailand*** Northern province and town 585 km (365 miles) north and slightly west of Bangkok. The town was the capital of a kingdom ruled by the Mon people who lived in the area more than 1000 years ago. One of their town moats is still visible. Lamphun is now the centre of the longan fruit-growing industry.
Population (town) 154 300; (province) 395 200
Map Thailand Ab

Lampung ***Indonesia*** Province of southern Sumatra, covering 35 367 km² (13 655 sq miles). Coffee, pepper and rubber are produced for export.
Population 4 624 800
Map Indonesia Bc

Lanao ***Philippines*** Volcanic plateau comprising the provinces of Lanao del Norte and Lanao del Sur, in north-central MINDANAO. It has several volcanic peaks, Mount Ragang being 2815 m (9235 ft) high, and the Philippines' second-largest lake, Lake Lanao. Possibly formed by the collapse of another volcano, the lake is 340 km² (131 sq miles) in area. Originally forested, much of the area is now farmed for maize and other crops, largely by Christian settlers from the VISAYAN ISLANDS. The original inhabitants, the Maranos, are strict Muslims.
Population (Lanao del Norte) 461 000; (Lanao del Sur) 405 000
Map Philippines Bd

Lanark ***United Kingdom*** See STRATHCLYDE

Lancashire ***United Kingdom*** County of north-west England, formerly including Furness (now part of CUMBRIA) and the areas around LIVERPOOL and MANCHESTER but now covering 3043 km² (1175 sq miles) to the north of these cities. It was a cradle of the Industrial Revolution in the 18th and 19th centuries, and became the world's major cotton centre before the First World War in mill towns such as PRESTON (now the county town), ROCHDALE (now in Greater Manchester), BLACKBURN and Burnley. Most of the great mills are now closed or converted to other businesses. The Lancashire coalfield has declined, and towns such as WIGAN (also now in Greater Manchester), which once supplied fuel to the cotton mills, have been hit by unemployment. The county's seaside resorts, such as BLACKPOOL, and the PENNINE hills are tourist attractions.
Population 1 378 000
Map United Kingdom Dd

Lancut ***Poland*** Small south-eastern town, about 160 km (100 miles) east of the city of Cracow. It contains Poland's most magnificent Baroque palace, built between 1629 and 1641 for the powerful and aristocratic Lubomirski family. The ivy-clad building, which is topped by onion-shaped domes, is now a museum, with fine collections of paintings, sculpture, clocks, furniture, tapestries, carriages and coaches.
Population 15 600
Map Poland Ec

land breeze A wind blowing from the land towards the sea or a lake at night or in the very early part of the day, following greater cooling of the land by radiation during the night. The opposite of a sea breeze.

Landes ***France*** Coastal department of some 9237 km² (3566 sq miles) in the south-west. The name also applies to the wider area of sandy coastal plain. It was good only for rearing sheep until 19th-century landowners planted pine trees and drained coastal marshes. In the 20th century, some pines were cleared for farming, and the main activities are now forestry, farming and tourism; the fine beaches at Arcachon are particularly popular.
Population (department) 297 400
Map France Cd

Land's End ***United Kingdom*** Most westerly point of mainland England, in the county of Cornwall. Its treacherous granite cliffs are visited by thousands of tourists each year.

The Wolf Rock lighthouse, 13 km (8 miles) offshore, and the Longships lighthouse, 2.4 km (1.5 miles) away, were built in the 19th century to warn ships away from the dangerous waters around the headland. These waters have claimed scores of ships and hundreds of lives over the centuries.
Map United Kingdom Ce

landslide, landslip Sudden fall of masses of rock and soil down a cliff or hillside due to heavy rain acting as a lubricant or to the vibrations of an earthquake.

Langeland ***Denmark*** Island in the FUNEN group, covering 283 km² (109 sq miles) and linked by a bridge to Fünen. The island's beaches are popular holiday spots. The chief town is Rudkøbing.
Population 17 700
Map Denmark Bb

Langkawi ***Malaysia*** Island in the Straits of Malacca beside the Thai border. It is a tourist resort and has a number of fishing villages.
Population 28 350
Map Malaysia Aa

Languedoc-Roussillon ***France*** Wine-producing region – noted particularly for its red table wines – stretching behind the Mediterranean coast west of the Rhône river to the Pyrenees. The historical province of Languedoc occupied a much larger area than the present region. MONTPELLIER is the main city. Besides wine, its industries are farming and tourism.
Population 1 927 000
Map France Ee

L'Anse-au-Meadow ***Canada*** Historic site in northern Newfoundland where Vikings from Greenland landed in the late 10th century, forming possibly the first European colony in North America. They arrived in *knorrs* (heavy longships) and built seven turf houses. The ruins of their settlement were discovered in the 1960s and are open to view.
Map Canada Jc

Lansing ***USA*** State capital of Michigan. It lies 130 km (80 miles) west and slightly north of Detroit, and is a car-making centre where, in 1887, one of the first American cars was built. Michigan State University is at East Lansing.
Population (city) 128 000; (metropolitan area) 416 200
Map United States Jb

Lantau ***Hong Kong*** Largest (150 km², 58 sq miles) of the many small islands of the NEW TERRITORIES which belong to the British Crown Colony of Hong Kong. It lies to the west of Victoria Harbour. It is a mountainous but tranquil island, fringed by beautiful beaches, and contains the Po Lin Buddhist monastery.
Population 17 100 (including associated islands)
Map Hong Kong Bb

Lanzarote ***Spain*** See CANARY ISLANDS

Lanzhou ***China*** Capital of Gansu province, 1200 km (750 miles) south-west of Beijing (Peking) on the Huang He river. It is an important communications centre for western China and contains oil refining, chemicals and woollen textiles plants.
Population 2 260 000
Map China Fd

Laoag ***Philippines*** See ILOCOS

Laois (Leix) ***Ireland*** Farming county of 1719 km² (664 sq miles) about 65 km (40 miles) south-west of Dublin. It was formerly known as Queen's County after Mary I, for during her reign there was an unsuccessful attempt to start an English community there. The county town is Port Laoise, once known as Maryborough.
Population 51 200
Map Ireland Cb

Laon ***France*** Capital of the Aisne department, overlooking the plain of Champagne 120 km (75 miles) north-east of Paris. Old 17th-century houses and narrow streets cluster around a 12th-century cathedral.
Population 29 800
Map France Eb

Laos See p. 378

Lappeenranta (Villmanstrand) ***Finland*** Lakeside port, founded in 1649, on the south shore of Saimaa lake in south-east Finland, about 25 km (15 miles) from the Russian border. It has one of Finland's largest concentrations of softwood processing plants. Other manufactures include chemicals, cellulose and machinery.
Population 54 000
Map Finland Dc

Lappland ***Scandinavia*** Region largely inside the Arctic Circle, with no formal boundaries, traditionally inhabited by the Lapp people. It stretches across the north of Norway, Sweden and Finland. The territory has mining and forestry activities as well as fishing and reindeer herding. It includes several large national parks.
Map Scandinavia Ab

Laptev Sea Shallow part of the ARCTIC OCEAN (formerly called the Nordenskjold Sea) named after the 18th-century Russian navigator Dmitri Laptev. It lies between the SEVERNAYA ZEMLYA and NOVOSIBIRSKIYE OSTROVA islands, with eastern Siberia to the south and the edge of the continental shelf to the north. The Laptev, or Dmitri Laptev, Strait links the sea to the EAST SIBERIAN SEA. The Laptev Sea is icebound and unnavigable for ten months every year.
Depth Up to 100 m (328 ft)
Map USSR Na

Laos

A LANDLOCKED, UNDERDEVELOPED, WAR-RAVAGED COUNTRY TRIES A MODIFIED BRAND OF COMMUNISM

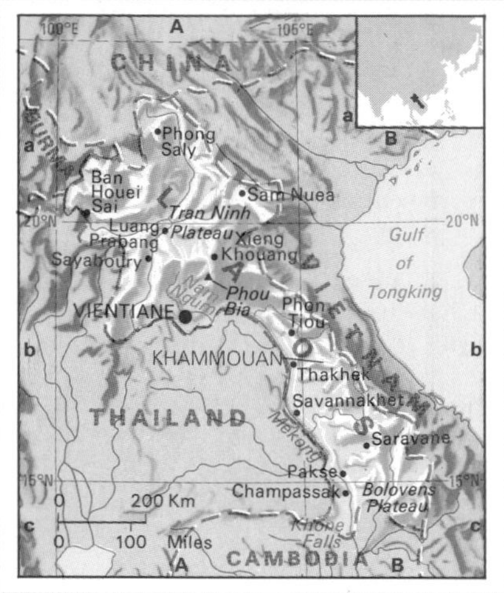

This small country has long suffered the twin fates of isolation from economic development and interference by outside forces. Progress was also hindered by centuries of internal political division. The French colonial era brought peace, but little development.

Laos, independent after 1954, faced crippling problems. Ruggedly mountainous and with primitive communications, no industry and old-fashioned agricultural techniques, it needed peace and national unity. Instead, its scattered ethnic groups became the focus of political rivalries. Civil war was escalated by foreign aid to rival factions.

A Communist regime won control of the whole country in 1975. But the ravages of war, the withdrawal of American aid, the new regime's doctrinaire socialist policies, and a crippling drought in 1977, all hindered development. Over 300 000 Laotians fled the country.

Since 1980, more liberal government policies and massive Soviet and Vietnamese aid have produced an upturn. But Laos remains dependent most of all on the weather. Good monsoon seasons give self-sufficiency in rice, the staple diet. Drought or flood can mean disaster, and diversion of development funds to pay for rice imports.

The country's mountainous terrain broadens in the north to a wide plateau, some 2000 m (6560 ft) above sea level, which contains the Plain of JARS. In the south, the ANNAM mountains reach 2500 m (8202 ft) on the Vietnam border. On the western side are the country's only lowlands, the alluvial plains of the MEKONG. This is Laos's principal river and its main communications artery since there are no railways and few roads. On its north bank, on the border with Thailand, lies VIENTIANE, the capital and largest city. It is the country's main trade outlet, via Thailand.

Laos exports timber, coffee, tin and electricity. The black market also exports opium and gold. All manufactured goods must be imported, and Laos has a constant foreign trade deficit. The country's principal crop is rice, mostly still grown on small peasant plots. Little goes to market, for many farms are too small to produce a surplus, and unrealistically low prices are enforced by the government.

The main ethnic group are the Lao, who make up about 45 per cent of the population. They are Buddhists, live mostly along the Mekong valley, and their language is the official language. The rest of the population is made up of hill-dwelling groups, such as the Tai, Khmu, and the Meo-Yeo peoples.

Tai tribes from the north settled Laos in the 10th and 11th centuries. By the 14th century a powerful kingdom, Lan Xang, had developed, but feudal rivalries and intervention by Burma, Siam (Thailand) and Vietnam weakened it. By the 18th century it had split into two rival states, both of which Siam dominated in the 19th century.

Following French colonial expansion in Vietnam and Cambodia, Laos became a French protectorate in 1893, and remained under French rule until March 1945, when the Japanese overthrew the French. After the defeat of Japan in August that year, a provisional government of Pathet Lao (meaning 'State of Laos') was formed under Prince Souphanouvong, but in 1946 the returning French rejected its demands for independence and established the Kingdom of Laos under French control. In 1950 the kingdom became a sovereign state, with Souvanna Phouma as premier, but French influence remained strong.

Meanwhile the exiled Souphanouvong formed an alliance with Vietnam's Ho Chi Minh. The forces of the former (the Pathet Lao) and of Ho (the Viet Minh) united in Laos to set up a Pathet Lao government, controlling much of the east by 1954, when the French abandoned the country.

After 1955 the royal government received massive American support. In 1975 the Pathet Lao forced the abdication of the king, and Souphanouvong became president.

The new government made some progress, but attempts to impose a socialist system on the country's agriculture and trade proved disastrous. Since 1980 private trade has come back to life: less than one-fifth of peasants are in cooperatives. But Laos continues to suffer from its natural handicaps and its war-torn history.

LAOS AT A GLANCE

Area 236 800 km² (91 400 sq miles)
Population 3 888 700
Capital Vientiane
Government Communist republic
Currency Kip = 100 at(t)
Languages Lao, also tribal languages
Religions Buddhist (58%), tribal religions, Christian
Climate Tropical. Average temperature in Vientiane ranges from 14 to 34°C (57 to 93°F)
Main primary products Rice, maize, vegetables, tobacco, coffee, cotton; tin
Major industries Agriculture, mining and ore processing, forestry and timber milling
Main exports Timber, coffee, electricity
Annual income per head (US$) 80
Population growth (per thous/yr) 22
Life expectancy (yrs) Male 42 **Female** 45

Larache (El Araïche) ***Morocco*** Spanish-style fishing port on the Atlantic coast, 70 km (45 miles) south-west of Tangier. It exports wool, hides, cork, wax, fruit and vegetables. In the town lies the 17th-century Stork Castle, so called because of the birds that nest on its ramparts. On the hillside to the north is Lixus, an ancient settlement containing megalithic, Phoenician and Roman remains.
Population 64 000
Map Morocco Ba

Laramie ***USA*** Market town in a timber, mining and stock-raising area of south-east Wyoming, about 35 km (22 miles) north of the Colorado border. It was founded in 1868 as a depot by the Union Pacific Railroad. It is the seat of the state's university and a tourist centre.
Population 25 300
Map United States Eb

Larap ***Philippines*** See BICOL

Laredo ***USA*** Main frontier city on the Texas-Mexico border, on the Rio Grande about 475 km (295 miles) south-west of Houston. It was settled by Spaniards in 1755 and occupied by Texas Rangers in 1846; the Mexican War (1846-8) established it as part of Texas, but it still retains a Mexican flavour. It thrives on export-import trade and tourism; the surrounding area produces cattle, oil and gas.
Population 118 200
Map USA Ge

Larnaca ***Cyprus*** South-east coast port trading in local farm produce such as cereals, potatoes and livestock and making brandy, tobacco and soap. The town contains a Muslim shrine to Hala Sultan, a female relative of the Prophet Muhammad. According to Christian legend, Lazarus, who was raised from the dead by Jesus, lived in Larnaca after his miraculous revival. Since the closure of Nicosia airport (1974), the Greek-Cypriots have developed Larnaca as the main international tourist airport.
Population 48 400
Map Cyprus Ab

Larne ***United Kingdom*** Manufacturing town and port in Northern Ireland, 34 km (21 miles) north-east of the capital, Belfast. Car and passenger ferries link the town to Stranraer in Scotland, 60 km (37 miles) away. Its manufactures include radio and television sets.
Population 18 200
Map United Kingdom Cc

Las Palmas de Gran Canaria ***Spain*** Largest city and chief port of the CANARY ISLANDS, exporting sugar, tomatoes, almonds and

bananas. It is a popular year-round resort, and has fishing, fish-processing, craft and boat-building industries.
Population 366 500
Map Morocco Ab

Las Vegas ***USA*** Gamblers' paradise in the desert in the extreme south-east of Nevada. It began as a Mormon settlement in 1855-7 and was re-established in 1905 as a small agricultural town. But because of the state's liberal gaming laws it soon became world famous for its casinos, nightclubs and entertainment.
Population (city) 183 200; (metropolitan area) 536 500
Map United States Cc

Lascaux, Grottes de ***France*** Caves in the Vézère valley near the Dordogne river in the south-west. The caves – discovered by four boys out walking with their dog on September 12, 1940 – contain what is now regarded as the world's finest collection of prehistoric art. The paintings, on the walls of the caves, depict bulls, bison, deer, horses and hunters. Their Stone Age creators lived between 20 000 and 15 000 years ago. The caves were closed to the public in 1963 to preserve them from damage. A facsimile, Lascaux II, was opened nearby in 1983; it was created over ten years by artists using similar techniques and materials to those employed by the Stone Age painters.
Map France Dd

Lassen Volcanic National Park ***USA*** Area of 433 km² (167 sq miles) in the Sierra Nevada mountains of California, about 300 km (185 miles) north-east of San Francisco. It contains Lassen Peak (3187 m, 10 457 ft), the largest plug dome volcano in the world, and a fine range of lava flows, hot springs, steam vents and lakes.
Map United States Bb

Latakia (Al Ladhiqiyah) ***Syria*** Seaport and capital of the district of the same name, 100 km (62 miles) north-west of Hamah. Most of the country's exports and imports pass through Latakia, which has a duty-free zone. Its main industry is the processing of tobacco. The bustling modern city, now the centre of a rich agricultural area, dates from Roman times, and there are remains from the reign of the Emperor Septimius Severus (AD 193-211).
Population (city) 204 000
Map Syria Ab

laterite Hard, reddish-brown earth formed in tropical areas with marked wet and dry seasons by the LEACHING of weathered rock material high in iron content.

latitude Angular distance of a point north or south of the equator. It is the angle at the centre of the earth between a line to the point and a line to a point on the equator due north or south of it. See also PARALLEL.

Latium ***Italy*** See LAZIO

Latvia (Latviya; Lettland) ***USSR*** One of three Soviet republics on the Baltic Sea, sandwiched between Estonia and Lithuania and covering 63 700 km² (24 600 sq miles). For almost the whole of its recorded history, Latvia has been ruled by foreigners – Vikings, Danes, Lithuanians, Germans, Poles, Swedes or Russians – and often divided in several parts. It had only a brief period as an independent republic, from 1920 to 1940. It was occupied by Russians in 1940 and by Germans in 1941-4; it was finally annexed by the USSR in 1944 – a move regarded as illegal by some Western governments – and many people fled to Sweden or West Germany.

Latvia's people have lived mainly by traditional forestry, fishing and livestock rearing. Industrialisation began only after the Second World War, but it is now one of the most industrialised republics of the USSR. Cities such as the capital RIGA, DAUGAVPILS, Ventspils and LIEPAJA now produce high-quality textiles, machinery, electrical appliances, paper, chemicals, furniture and foodstuffs.

Well over half the population are native Latvians (Letts), with their own language and strong culture. They are mainly Lutherans.
Population 2 589 000
Map USSR Dc

Lauenburg ***Poland*** See LEBORK

Lauis ***Switzerland*** See LUGANO

Launceston ***Australia*** North Tasmanian port at the head of the Tamar river estuary, about 160 km (100 miles) north of Hobart, and the state's second largest city. It was founded in 1805 by Colonel William Patterson and was originally named Pattersonia, but was later named Launceston after Governor King's birthplace in Cornwall, England. It is often called the 'Garden City' because of its beautiful parks and gardens. It exports agricultural produce and timber. Its industries include hydroelectricity, textiles, and flour milling.
Population 65 000
Map Australia Hg

Laurasia The northernmost of the two ancient continents into which Pangaea, an earlier supercontinent, divided about 200 million years ago. The southernmost was Gondwanaland. See CONTINENTAL DRIFT

Laurentian Mountains ***Canada*** Range in southern Quebec province which rises to 1166 m (3825 ft) at Mount Raoul-Blanchard. It is an important area for winter sports, and there are a number of major resorts in the mountains.
Map Canada Hd

Lausanne ***Switzerland*** French-speaking city on the northern shore of Lake Geneva. It is a centre for international fairs and conferences, with a university and numerous professional schools, theatres and museums. Medieval streets surround the town hall and the cathedral, which dates from the 13th to the 16th centuries. Ouchy, once a hamlet, is now a suburb and Lake Geneva's busiest port. The city is the capital of the French-speaking Protestant canton of Vaud, formerly called Leman.
Population (city) 140 000; (canton) 530 000
Map Switzerland Aa

Lautoka ***Fiji*** The country's second largest city and port, lying on the north-west coast of Viti Levu island. In fact its exports – mainly sugar – exceed those of the capital, Suva. It is the centre of an extensive sugar cane growing region, and has one of the largest sugar mills in the Southern Hemisphere. Offshore from Lautoka are a number of tourist resort coral islands, while 20 km (12 miles) south is Nadi (pronounced 'Nandi'), Fiji's main international airport and one of the busiest in the South Pacific.
Population 29 000

lava Molten rock that has issued from a VOLCANO or fissure in the earth's surface.

▼ SUNSEEKERS' CITY Tourists and small boats crowd the beach at Las Palmas. Founded in 1478, it developed as a modern port from 1883 and as a luxury resort after 1950.

lava tube Tube formed by the cooling and solidification of the outer surface of a lava stream while the molten lava inside has flowed away.

Laval ***Canada*** City in Quebec province, separated from neighbouring Montreal by the Rivière des Prairies. The city was created in 1965 from 14 communities on Ile Jésus.
Population 268 300
Map Canada Hd

Lazio (Latium) ***Italy*** Region covering 17 203 km² (6642 sq miles) in central Italy. Lazio's richest agriculture is found in the reclaimed coastal lowland – such as the Pontine Marshes – about 50 km (30 miles) south of Rome.
Population 3 076 000
Map Italy Dc

Le Havre ***France*** The nation's second largest port after Marseilles, sited on the north bank of the Seine estuary on the English Channel coast. Founded in 1517 by Francis I, it rose to prominence in the 19th century, handling transatlantic and tropical cargoes. It was largely destroyed in 1944, during the Second World War. Reconstruction began in 1946, giving the city wide boulevards, modern buildings and a greatly enlarged port. Its main industries now are shipping, banking, oil refining, chemicals, engineering and car assembly.
Population 255 900
Map France Db

Le Mans ***France*** City about 185 km (115 miles) south-west of Paris. A 24 hour race for sports cars is held there each year. The city, which was the birthplace of Henry II – the first Plantagenet King of England – contains a fine 11th to 15th-century cathedral and an 11th to 12th-century church.
Population 194 000
Map France Db

Le Puy ***France*** Manufacturing city among eroded volcanic hills about 110 km (70 miles) south-west of Lyons. The 12th-century cathedral, Notre Dame, contains a statue of the Black Madonna, which is the object of an annual pilgrimage on the Feast of Assumption (August 15). The church of St Michel-d'Aiguilhe stands on an 80 m (260 ft) rock pinnacle.
Population 26 000
Map France Ed

Le Touquet-Paris-Plage ***France*** Fashionable seaside resort on the English Channel coast, 50 km (30 miles) south of Calais. It has sandy beaches, casinos, a racecourse and yacht harbour.
Population 8000
Map France Da

leaching Removal of soluble mineral salts from soil or rock by percolating water. Soils become progressively less fertile with leaching.

Lebanon See p. 382

Lebanon Mountains (Jebel Liban) ***Lebanon*** Range of mountains running for about 160 km (100 miles) parallel to the Mediterranean coast – from Tripoli in the north to Tyre in the south. The highest point is Qornet es Saouda 3087 m (10 125 ft) at the northern end of the range. There are large springs some 1200 m (3930 ft) to 1500 m (4900 ft) up, which are the source of a number of small rivers.

The western slopes facing the Mediterranean are particularly well watered, and farming is possible as high as 1525 m (5000 ft). Olives and soft fruits are grown there.
Map Lebanon Ab

Lebombo (Lobombo, Lubombo) Mountains ***Southern Africa*** Low range about 660 m (2000 ft) high, running north-south along the boundary of South Africa and Mozambique. They continue southwards through eastern Swaziland, where their flat-topped, grassy surface is used mainly for cattle rearing and arable farming. The climate is cool and pleasant, contrasting with that of the warm, arid LOWVELD to the west. In Swaziland three rivers – the Usutu, Mbuluzi and Nkalashane – have cut scenic gorges through the mountains. One of the gorges, on the Mbuluzi river in the north-east, is the route used by the Swaziland-Maputo railway which links the country's interior with the Mozambique coast.
Map Swaziland Ba

Lebork (Lauenburg) ***Poland*** Market town in Pomerania, about 65 km (40 miles) north-west of the Baltic port of Gdansk. It has a 15th-century castle.
Population 31 000
Map Poland Ba

Lebowa ***South Africa*** Black state in north Transvaal, created in 1969 as a homeland for the North Sotho people. It consists of two large, unconnected areas, and other small ones and covers a total of 22 476 km² (8678 sq miles). The capital is Seshego, but a new capital, Lebowakagomo, is planned. The main activities are subsistence farming and rearing livestock.
Population 2 246 000
Map South Africa Ca

Lecce ***Italy*** Beautiful Baroque city in the heel of Italy about 12 km (8 miles) from the Adriatic. Lecce was important in Roman and Norman times, but reached its zenith under Spanish influence in the 16th-18th centuries. Then its best buildings – from the cathedral and church of Santa Croce to palaces and ordinary homes – were built from honey-coloured limestone. The town also has a Roman amphitheatre, a tall Roman column brought from the Appian Way, and a 16th-century castle. It trades in olive oil, wine and tobacco.
Population 97 200
Map Italy Gd

Ledenika Cave ***Bulgaria*** See VRATSA

Leeds ***United Kingdom*** City of West Yorkshire, England, 60 km (37 miles) north-east of Manchester. It rose to prominence in the 19th century as one of the country's leading producers of ready-made clothing, textile machinery and railway equipment. But it also developed badly overcrowded slums. These have, however, been improved or cleared since the Second World War, and the city centre is now a modern business area mostly reserved for pedestrians. The massive town hall, built in 1858, reflects its 19th-century wealth in the same way that its civic centre, television studios and university mirror its 20th-century growth.
Population 450 000
Map United Kingdom Ed

Leeuwarden ***Netherlands*** Provincial capital of Friesland about 110 km (70 miles) north-east of Amsterdam. Its canal-ringed centre has several old churches, and the leaning tower of one left standing when the church was demolished in 1595. The Friesian Museum is there, and the town makes dairy products, textiles and paper.
Population 85 100
Map Netherlands Ba

Leeward and Windward Islands ***Caribbean*** The Leeward Islands are the northern group of the Lesser Antilles – so called because they are more sheltered from the prevailing north-easterly winds than the more southerly group, the Windward Islands. The Leewards run south-east from the Virgin Islands to Guadaloupe; the Windwards in a crescent from Dominica to Grenada, near the north-east coast of Venezuela.
Map Caribbean Cb

Leeward Islands ***French Polynesia*** See SOCIETY ISLANDS

Legaspi ***Philippines*** See BICOL

Leghorn (Livorno) ***Italy*** Port and industrial city about 150 km (93 miles) south-east of Genoa. Its industries include shipbuilding, engineering, electrical equipment, chemicals, marble, copper and glass.
Population 175 300
Map Italy Cc

Legnica (Liegnitz) ***Poland*** Industrial city 63 km (39 miles) west of the city of Wroclaw. In 1241, at Legnickie Pole (Field of Legnica), 4 km (2.5 miles) to the east, advancing Tatar hordes fought the Poles, whose leader Prince Henryk the Pious was killed. Though the Polish army was defeated, the westward advance of the Tatars was stopped. A Gothic chapel is now the battlefield museum. Legnica is a copper smelting and sulphuric-acid producing centre, and has a large Soviet army base.
Population 96 500
Map Poland Bc

Leicester ***United Kingdom*** City and county of the eastern Midlands of England, which now includes Rutland – formerly Britain's smallest county – and covers 2553 km² (986 sq miles). It stretches from the hills of Charnwood in the west, with their ancient forests, to the uplands of Rutland in the east, the site of Britain's largest man-made lowland lake, Rutland Water, with a shoreline of 38 km (24 miles).

The city – 140 km (87 miles) north and slightly west of London – was the Roman town of Ratae and has the remains of its Roman baths. Simon de Montfort, who forced his brother-in-law, Richard III, to hold the nation's first Parliament in 1275, was Earl of Leicester. Today Leicester is a centre for hosiery, textiles and engineering. The Church of St Martin was raised to cathedral status in 1926, and a local college to university status in 1957.

Loughborough, 16 km (10 miles) north of Leicester, has a leading technological university. Melton Mowbray, to the east, is renowned for its pork pies

Population (county) 864 000; (city) 280 000
Map United Kingdom Ed

Leiden (Leyden) ***Netherlands*** University city about 35 km (20 miles) south-west of Amsterdam. Its university dates from 1575. The city was the birthplace of the painter Rembrandt (1606-69), and the Pilgrim Fathers, seeking to escape religious persecution in England, spent 11 years there before sailing to America in 1620.

Population (city) 103 800; (Greater Leiden) 175 500
Map Netherlands Ba

Leinster (Cuige Laighean) ***Ireland*** The most fertile of Ireland's four historic provinces (the others are Connaught, Munster and Ulster), occupying the south-east of the country.

Population 1 786 500
Map Ireland Cb

Leipzig ***East Germany*** Southern city which the 18th-century German writer Johann Goethe called 'Paris in miniature'. It is the country's second largest city after the capital, East Berlin, 180 km (112 miles) to the north-east.

It is the cultural centre of the country. The religious reformer Martin Luther (1483-1546) was active there; Goethe (1749-1832) and his fellow poet Johann Schiller (1759-1805) were both students at Leipzig University. Music thrived through the genius of Johann Sebastian Bach, who lived and worked there for 27 years until his death in 1750. There is a memorial to him in the Thomaskirche. The operatic composer Richard Wagner was born in the city in 1813. Today, the Leipzig Gewandhaus Orchestra is one of the finest in the world.

Leipzig is also a publishing, printing and industrial centre, producing agricultural machinery, chemicals and paper. Its international trade fair was first held in 1165.

Population 559 000
Map East Germany Cc

Leitrim (Liatroim) ***Ireland*** Thinly populated county in the north-west, with a short Atlantic coastline. An area of bogs and mountains covering 1525 km² (589 sq miles), it is split almost into two by Lough Allen. The county town is Carrick on Shannon.

Population 27 600
Map Ireland Ba

Leix ***Ireland*** See LAOIS

Lek ***Netherlands*** See RHINE

Lelystad ***Netherlands*** New town about 40 km (25 miles) north and slightly east of Amsterdam, on the East Flevoland polder, an area of low-lying land reclaimed from the IJSSELMEER in 1957. The town, named after Cornelius Lely, the statesman-engineer who conceived the reclamation project begun in 1920, is the administrative centre of the IJsselmeer polders.

Population 52 300
Map Netherlands Ba

Léman, Lac ***Switzerland*** See GENEVA

Lemberg ***USSR*** See L'VOV

Lempa ***El Salvador*** The country's longest river, at 260 km (160 miles).

Map El Salvador Ba

Lena ***USSR*** East Siberian river, about 4270 km (2650 miles) long. It rises at 1810 m (5940 ft) in the mountains along the west shore of Lake Baikal, and tumbles northwards over rapids to the small town of Kachug, where it becomes navigable. It then flows north-east and north past the city of Yakutsk to enter the Laptev Sea via a large delta. The river is frozen from October or November until May or June.

Map USSR Nb, Lc

Lenin Peak (Pik Lenina) ***USSR*** The Soviet Union's third highest peak after Communism Peak and Pik Pobedy. It soars to 7134 m (23 405 ft) in the Pamir mountains near the western tip of China. It is named after the Russian revolutionary leader Vladimir Lenin (1870-1924).

Map USSR Ie

Leninabad ***USSR*** One of central Asia's oldest towns, formerly known as Khodzhent. It lies in Tadzhikistan, about 145 km (90 miles) south of Tashkent. The town grew around a fortress founded by the Macedonian conqueror Alexander the Great (356-323 BC). Now renamed after the Communist leader Vladimir Lenin (1870-1924), it produces traditional ceramics, woodcarvings, embroideries and silks.

Population 147 000
Map USSR Hd

Leninakan ***USSR*** Armenian city, formerly called Aleksandropol or Alexandropol. It lies on the flanks of Mount Aragats (4090 m, 13 418 ft), 15 km (9 miles) from the Turkish border. It was founded in 1837 by Armenian refugees fleeing Turkish persecution and now manufactures textiles, carpets, copper products and machinery.

Population 220 000
Map USSR Fd

Leningrad ***USSR*** Former Russian capital and the country's second largest city after Moscow. It was the capital from 1712 to 1918, when it was known until 1914 as St Petersburg, and then until 1924 as Petrograd. Its final name honours Vladimir Lenin (1870-1924), one of the leaders of the Russian Revolution.

The tsar Peter the Great (1682-1725) founded the city at the head of the Gulf of Finland, on about 100 marshy islands in the delta of the Neva river. Today the flood-prone city – sometimes called the 'Venice of the North' – is linked by nearly 700 bridges. The tsar's architects, engineers and artisans – mostly from Western Europe – gave central Leningrad an air of grandeur, with well-planned squares, parks and wide avenues between neo-classical Baroque buildings, often pastel-painted with gilt spires and domes.

The Peter and Paul Fortress, the old heart of the city beside the Neva, encloses a cathedral and was once used as a political prison. The author Maxim Gorky (*continued on p. 384*)

A DAY IN THE LIFE OF A LENINGRAD SHIPYARD WORKER

At 6.15 on a Friday morning Alexei Ivanovich Kirsanov is standing on the platform of the marble-columned metro station in the Leningrad suburb where he lives. He does not wait long for a train; as well as costing only 5 kopecks (about 5 US cents) on a flat-rate system, the service is frequent and efficient; but the journey takes 40 minutes as the housing block in which he and his wife, Tanya, live is quite a distance from the Lenin shipyard, where he works as a nautical engineer.

The job came automatically when he graduated from the city institute five years ago. It is one of the blessings of living in Russia, Alexei thinks as he opens his daily *Pravda* and reads about massive unemployment and strikes in capitalist countries. He may grumble about shortages in the shops, but at least he and Tanya have jobs.

He earns an average of 200 roubles (US$210) a month, while she, as a teacher, gets 170 (US$175). Their combined salary easily covers the rent (25 roubles – US$30 – a month), transport (heavily subsidised) and food. And like a lot of his friends, he supplements his income by doing some moonlighting, and the extra money has bought him luxury items such as a stereo.

As usual, he eats a cheap lunch in the shipyard's canteen, washing down the potatoes, gristly meat and carrots with a glass of watery fruit juice because under the new laws alcohol is banned from factories and other workplaces.

Alexei and Tanya have a three-year-old daughter, Katya. Their flat has only one bedroom and Katya sleeps in the sitting room. With Tanya working as well, they have to take their daughter to a day centre.

As he leaves work around half past three, Alexei is thinking about the wooden *dacha* (cottage) where he and his family will spend their weekend. How much easier it will be when he gets the car for which he has been waiting for over a year. When Katya grows up, he hopes life will be a little easier for her than it is for her mother.

As he is walking towards their block, Alexei spots a queue and instinctively joins it. 'What are they selling?' he asks the lady in front of him as he takes out the string bag, called an *avoska* ('maybe bag'), which every Russian carries just in case there is something worth buying somewhere. 'Grapes' comes the reply, and sure enough behind a makeshift stall stands a swarthy Georgian with a suitcase full of grapes from the warm south, the profits of which will more than cover his air fare. Alexei smiles to himself; yes, what a nice surprise for Tanya.

Lebanon

TRAGEDY OF A COUNTRY IN CHAOS AS DOZENS OF PRIVATE ARMIES FIGHT OUT COMMUNAL AND RELIGIOUS RIVALRIES

Once the jewel of the eastern Mediterranean, the playground of the rich with a distinctly French flavour, this sad little country has been under violent assault for a decade. The Palestine Liberation Organisation (PLO), a Syrian army, an Israeli occupation force, Western and United Nations peace-keeping forces have all been involved; Lebanese have fought Lebanese, and any stranger entering a town must, for his own safety, find out which local militia is in control.

Lebanon is a spectacularly lovely country whose cosmopolitan people were until recently among the most prosperous in the Arab world. The Lebanese are shrewd businessmen and so adaptable that for many years the country was a kind of middleman in a sea of Middle Eastern trouble. Lebanon's main ports, the capital BEIRUT and TRIPOLI, acted as transit centres for the countries of the interior – Syria, Jordan and Iraq. Beirut was the banking centre of the Middle East, and the home of entrepreneurs who could arrange every kind of business deal.

All this has changed. The country is heavily in debt, inflation is rising steeply and at present the people's main aim is just to survive in a land where sniping, bombing, kidnapping and assassination are the daily norm. They have carried on as traders and businessmen in circumstances in which most others would give up in despair. Perhaps they inherited this fortitude from their highly successful trading predecessors, the Phoenicians.

Lebanon has an area of 10 400 km² (4015 sq miles), with a 240 km (150 mile) coastline on the Mediterranean. Inland from a narrow plain, spectacular mountains, snow-covered in winter, rise to a maximum of 3086 m (10 125 ft). In the east of the country these shelter the fertile BEQA'A valley – now, all too often, the scene of bitter fighting.

The coast has a warm Mediterranean climate, with dry summers. In the mountains, the snow lies from December to May so that in winter it is possible to swim in a warm sea and then go 10 miles inland to ski. Evergreen forests grow in the mountains. The evergreen oak is a familiar tree in much of the country, and Lebanon is famous for its giant cedars, although now only a small number are left in groves over 900 m (2950 ft) above sea level which are carefully protected.

Lebanon has about 2.6 million people, and is truly a country of minorities. Although some 90 per cent of the population are Arabs (including a sizable Palestinian minority), there are also Armenians, Assyrians, Kurds, Greeks, Turks and Jews. More significant, however, is the religious breakdown, since this governs political affiliations. Christians, once just over half the population, now make up 44 per cent but comprise four main sects. Most powerful are the Maronites, who are in union with the Roman Catholic Church but have their own Eastern rites and Patriarch; there are also Catholic, Orthodox and Armenian Christians.

There are three Muslim sects: the majority Shiites, now becoming increasingly militant, the Sunnis, who formerly outnumbered the Shiites, and the Druse. The Druse believe in reincarnation, do not fast during the Muslim holy month of Ramadan or go on pilgrimage to MECCA, and show equal hostility to Christians and other Muslims.

This mixture of races and beliefs led inexorably to civil strife. In addition, the country's position adjoining Syria and Israel, the two most bitter opponents in the Middle East, meant that Lebanon was bound to become involved, whatever it did to avoid trouble.

About one-fifth of the employed people work in agriculture, near the coast and in the Beqa'a valley. The country produces a variety of fruits and vegetables such as olives, grapes, figs, apricots, apples, onions and potatoes, as well as cotton, tobacco and sugar beet. The coast is fertile and the steep valleys are suitable for growing olives and vines. Industry is on a small scale. Manufactures include cement, fertilisers, jewellery, sugar and tobacco, and there are oil refineries at Tripoli and SIDON.

Two-fifths of the population are under 15 years old. The Lebanese are better educated than most Arabs: free primary education was introduced in 1960 and literacy is about 75 per cent. Many people speak French or (increasingly) English in addition to Arabic.

A HISTORY OF STRIFE

In past centuries, most major armies in the area passed through Lebanon, while those who fled before them often sought refuge in the mountains. The ancient Phoenicians reached their peak of power in the 12th to 9th centuries BC, leaving the towns of TYRE and Sidon as their monuments. Assyrians, Babylonians, Egyptians, Persians, Greeks and Romans fought in turn for the rich land. Later came the Arabs, who introduced Islam in the 7th century AD, followed by the Crusaders and the Turks.

In the 1860s the Maronite Christians appealed for aid after a series of massacres by the Druse, and Britain and France prevailed on the Turks to set up a Christian government. When the Ottoman Empire broke up after the First World War, Lebanon came under French control, and the French left their influence in language, food and customs.

Independence was declared by the Free French in 1941, and foreign troops left in 1946. Under an unwritten agreement of 1943, government posts were shared out among the various religious groups. Thus the president was always a Maronite, the premier a Sunni and the speaker of the Chamber of Deputies a Shiite; other government posts and seats in the Chamber reflected the sectarian make-up of Lebanese society. However, these proportions did not remain static and by the mid-1960s the now-predominant Muslims – and particularly the Shiites – were demanding more power, as befitted their numbers; the Maronites resisted change. Another factor was the arrival of Palestinian refugees and, more seriously, the PLO after it was expelled from Jordan in 1970.

PRIVATE ARMIES

Lebanon took little part in the Arab-Israeli wars of 1948 and 1956, and none in the wars of 1967 and 1973. But clashes between private armies of Muslims, Druse, Christians and militant Palestinians led to civil war in 1958 and again in 1975. Attacks by the PLO across the Israeli border in 1969 and 1973 brought reprisals from the Israelis and Lebanese forces led by Christians of the right-wing Falangist Party.

From 1975 the fragile political set-up in the country began to collapse in the face of Christian-Muslim rivalries and mistrust, and Lebanon began to slip into a more or less permanent state of civil war. In 1977, 40 000 Syrian troops entered the country to support the Muslims; the next year Israel invaded south Lebanon to establish a buffer zone between itself and PLO groups who frequently raided over the Israeli border. The United Nations sent a temporary peace-keeping force and the Israelis withdrew.

By 1980 there were 33 private armies or militias. In 1982 the Israelis, supported by Falangists in the south, mounted a full-scale invasion of southern Lebanon to eject the PLO forces under Yasser Arafat, first from Beirut to Tripoli and then from the country altogether. The same year the Falangist president, Bachir Gemayal, was murdered after three weeks in office and succeeded by his brother, Amin.

A Western peace-keeping force from the USA, Britain, France and Italy was withdrawn after suicide attacks by Muslim extremists on French and American troops. By 1984 the Israelis and Syrians each controlled 40 per cent of the country, leaving 20 per cent in the hands of various factions and militias and the legal Lebanese government virtually powerless. Beirut was a divided city: the Christians held the east, the Muslims the west of the capital. Sunni Muslims were strong in Tripoli, Shiites in the Beqa'a and the south, and Druse in the Chouf mountains. The government was an uneasy coalition, excluding Walid Jumblatt, the Druse leader, but including Nabih Berri, the Shiite militia chief, as Minister of Justice.

The civil wars were fought with savagery. Falangists massacred Palestinians in refugee camps; Shiites murdered Christians in their village strongholds. The country became an astonishing patchwork of no-go areas. Towns were divided into sections where particular Christian or Muslim or Druse militants held sway; once-prosperous places became ghost towns. People became used to life made up of sudden violence, then business as usual. Businessmen from abroad made sure that trusted contacts made their arrangements and met them on arrival.

In 1985 the Israelis withdrew their forces, except from a 'security zone' in the south. The Syrians remained. The international community, though it offers political initiatives and economic aid for reconstruction, has so far failed to halt the fighting or produce any workable solution. One estimate puts the damage to the country at US$33 000 million. Meanwhile, the people of Lebanon suffer privation and hardship as the destruction continues and the casualty lists grow.

FOR THE VISITOR

When tourists feel safe to go to Lebanon again there is much worth seeing. The climate makes it a delightful place for a holiday and one-fifth of the country's income used to come from tourism. Visitors were offered a pleasing mixture of Arab and French cuisine, beach resorts, old Crusader castles at Sidon and Byblos (now JBAIL), and ancient Phoenician (Byblos) and Roman sites (the chariot circus at Tyre), superb scenery (especially the view across the Beqa'a valley from the Dah el Beidar pass in the Lebanon range) and winter sports in the mountains.

Among the places of historic interest are the two Phoenician cities of Tyre and Sidon, BAALBEK with its Roman temple to Jupiter, and the cedars at BCHARRE and QORNET ES SAOUDA. Beirut had a reputation for arranging most kinds of pleasure – at a price. But today half the hotels and banks have been damaged or destroyed in repeated battles, and the airline and airport periodically close down.

LEBANON AT A GLANCE
Area 10 400 km² (4015 sq miles)
Population 2 600 000
Capital Beirut
Government Parliamentary republic
Currency Lebanese pound = 100 piastres
Languages Arabic (official), French, English
Religions Muslim (55%, including 6% Druse), Christian (44%)
Climate Mediterranean; cool in highlands. Average temperature in Beirut ranges from 11-17°C (52-63°F) in January to 23-32°C (73-90°F) in August
Main primary products Citrus fruits, apples, grapes, potatoes, sugar beet, wheat, olives, cotton, goats, sheep; some iron ore
Major industries Trade and banking, cotton yarn, textiles, agriculture, food processing, cement, tobacco manufacturing, fertilisers
Main exports Fruit and vegetables, chemicals, machinery, metals, textiles and clothing, cement, tobacco
Annual income per head (US$) 1100
Population growth (per thous/yr) 7
Life expectancy (yrs) Male 65 **Female** 68

▼ PROLONGED AGONY A car bomb of 1981 explodes in Beirut, torn by strife since 1976. The city is split by the so-called 'green line', with two main warring factions: Christians to the east in the once prosperous business sector, and Muslims to the west.

(1868-1936) was imprisoned there in 1905. The old city also includes the sumptuous Winter Palace of the tsars – now the Hermitage Museum, with its superb collection of paintings – and the Kirov Theatre, home of the Kirov Ballet. The main shopping street, Nevsky Prospekt, is named after a national hero, Alexander Nevsky (1220-63), who defeated invading Swedes on the Neva in 1240.

The city was the scene of the Decembrist uprising against the tsar in 1825, and the Russian Revolution of 1917 began there. In the Second World War, German and Finnish forces besieged it for almost 900 days, from September 1941 to January 1944, and more than a million inhabitants died – 640 000 of them from starvation. More than 500 000 were buried in the world's largest cemetery, in the city.

Leningrad is now, with Moscow, the USSR's most important cultural centre and a popular destination for tourists, as well as being one of the country's leading manufacturing cities and its main export port. Its major industries include engineering, shipbuilding, metal refining, chemicals, paper, textiles, clothing and food processing.

Population 4 832 000

Map USSR Ec

▼ MURDERED TSAR Leningrad's traditionally styled Church of the Resurrection was built on the spot where Alexander II was killed in 1881.

Lens ***France*** Industrial city in northern France, surrounded by pit villages, slag heaps and deserted coal mines, closed since the 1970s. Modern industrial zones around the city produce chemicals, car parts, electrical and metal goods, and clothing.

Population 327 400

Map France Ea

lenticular cloud Lens-shaped cloud, often associated with squally winds over hills and mountains.

León ***Mexico*** Commercial city in GUANAJUATO province on the fertile plain of the Gómez river in the central highlands of Mexico. It suffered a terrible flood at the end of the 19th century, but now thrives as Mexico's shoe-making capital. Tanneries and factories making leather goods stand beside textile and flour mills. The industries grew out of the old crafts of handworking leather and traditional embroidery, both of which still flourish.

Population 675 000

León ***Nicaragua*** The country's second largest city, after the capital, Managua. Lying near the northern Pacific coast, León has Nicaragua's oldest university – founded in 1804 – and Central America's largest cathedral, built between 1746 and 1846. It was the national capital for 300 years until 1858, when the seat of government was moved to Managua. Its industries include cotton, timber, food processing, and the production of fertilisers and insecticides.

The surrounding department of the same name produces cotton and sugar cane.

Population (city) 158 600; (department) 248 700

Map Nicaragua Aa

León ***Spain*** Ancient city about 290 km (180 miles) north-west of Madrid, founded by the Romans and capital of the Spanish kingdom of ASTURIAS – later named León – which flourished in the 10th to 12th centuries. Today it is the centre of an agricultural province of the same name, and manufactures textiles, pottery and leather goods.

Population (town) 131 100; (province) 518 000

Map Spain Ca

Léopoldville ***Zaire*** See KINSHASA

Lepanto ***Greece*** See NAVPAKTOS

Leptis Magna ***Libya*** Ruined ancient port near Homs, 123 km (76 miles) east of Tripoli. Leptis, which was founded by Phoenicians about 600 BC, was a leading Carthaginian city by the 2nd century BC. Unlike CARTHAGE, it was not destroyed by the Romans during their conquest of the Carthaginians in the Third Punic War (149-146 BC), and Leptis Magna became a major outpost of Roman Africa. Annexed by the Romans in 46 BC, it became important as a major trade centre and port for goods coming across the Sahara. The city reached the height of its influence in the 2nd century AD and was the birthplace in AD 146 of the Roman Emperor Septimius Severus.

The well-preserved remains include forums, baths, temples, walls, arches and an impressive theatre.

Map Libya Ba

Lérida ***Spain*** Town 134 km (83 miles) west of the Mediterranean port of Barcelona. It has a tanning industry and manufactures textiles and paper. The surrounding province of the same name produces wine and olive oil. It was the scene of a nine-month battle in 1937-8, during the Spanish Civil War.

Population (town) 109 600; (province) 355 500
Map Spain Fb

Lerma (Río Grande de Santiago) ***Mexico*** Mexico's longest river flowing west from the Sierra Madre range to Lake Chapala, where it becomes the Río Grande de Santiago, and onwards to the Pacific in northern JALISCO state. The Lerma irrigates the central highlands for agriculture, and since 1951 its headwaters have supplied Mexico City's huge population with water.

Map Mexico Bb

Lesbos (Mylíni) ***Greece*** Island in the Aegean Sea 10 km (6 miles) from the coast of Turkey. It is the birthplace of Sappho (about 612-580 BC), classical Greece's greatest poetess. Her passionate poetry – which seems to have been directed at her female admirers – has given us the word lesbian, for a female homosexual, after the name of the island.

Lesbos is a fertile land of hot springs, olive groves and sandy beaches. The chief town, Mitilíni, on the eastern side of the island, and its harbour are overlooked by the huge castle of Gattelusi, which dates from 1373. The castle was built by a Genoese family that ruled the island from AD 1355 to 1462, when the Turks occupied it. South of Sigri on the west coast of Lesbos is a petrified forest formed 800 000 years ago when conifers and sequoias were buried in volcanic ash. A rare species of salamander and blind mice live there.

The island has several Byzantine churches, monasteries and an archaeological site at Eressós, where Sappho was born.

Population 41 900
Map Greece Eb

Lesina ***Yugoslavia*** See HVAR

Leskovac ***Yugoslavia*** See MORAVA

Lesse ***Belgium*** Ardennes river (80 km, 50 miles long), which flows underground for part of its course. At the village of Han, about 60 km (35 miles) south and slightly west of the city of Liège, it has created spectacular limestone caves: the Grottes de Han. Visitors enter the caves through one of the river's older and now dry channels, and leave, farther along its course, by boat. The largest of the caverns is the Salle du Dome – a vaulted chamber some 150 m (490 ft) across and 130 m (425 ft) high. A museum in the village contains relics retrieved from the caves. Among them are Bronze Age artefacts dating from about 500 BC.

Map Belgium Ba

Lesser Sunda Islands ***Indonesia*** See NUSA TENGGARA

leste Hot, dry, southerly to easterly wind blowing from the Sahara and experienced in Madeira. It is often dust-laden and precedes a depression.

Lesotho

WITH FEW NATURAL RESOURCES, LESOTHO IS AN ECONOMIC AND POLITICAL HOSTAGE TO ITS GIANT NEIGHBOUR, SOUTH AFRICA

Snow-capped mountains and treeless uplands cut by spectacular gorges and isolated valleys cover two-thirds of the small kingdom of Lesotho (formerly Basutoland), perched fortresslike on southern Africa's HIGHVELD. Landlocked Lesotho – about the size of Belgium – is entirely surrounded by the Republic of South Africa, and is in the dilemma of being economically dependent on its giant neighbour yet strongly opposed to the republic's policy of *apartheid.*

Characteristic of the people of Lesotho – the Basotho (or Basuto) – are the coloured blankets they wear. They are a shield against the cold winters and the frequent summer hailstorms and thunderstorms.

The Basotho were forged as a nation in the early 19th century. Groups of tribesmen fleeing from the expanding Zulu nation gathered under the outstanding leadership of King Moshoeshoe I in the sheltering mountains, where they made the flat-topped, impregnable THABA BOSIU (Mountain of Night) their citadel. However, they were unable to resist the incursions of the Boers, who took much of their arable land. As a result Moshoeshoe was forced to ask for British protection. The country became a British protectorate in 1868 but remained under its indigenous chiefs. Full independence finally came in 1966. Today, King Moshoeshoe II is Head of State, but power lies essentially with a military council which early in 1986 deposed Chief Leabua Jonathan, prime minister since independence, who had maintained power by suspending the constitution during 1970-3.

As a result of the influence of 19th-century French Protestant missionaries, Lesotho has one of the highest literacy rates in Africa – 68 per cent for women and 44 per cent for men (more girls attend school because boys often herd livestock). Yet it is one of the world's poorest countries. There are few industries, and the mountainous terrain allows only about one-eighth of the land to be cultivated, the chief crop being maize. Yields are low, partly because of soil erosion caused by steep slopes and over-grazing by large herds of sheep, goats and cattle. Countermeasures are now being taken, including reducing the number of animals and introducing mixed farming. Wool, mohair and diamonds are exported, but most of the country's foreign exchange comes from remittances sent home by Lesotho workers in South Africa. Much of the Lesotho government revenue comes from duties raised through the Southern African Customs Union, which is controlled by South Africa. Overseas aid is provided by Britain.

About 200 000 of the population of some 1.5 million migrate to South Africa to earn a living, chiefly in the gold and coal mines of the TRANSVAAL, although they are being hit by South Africa's foreign labour cut backs. Providing alternative employment for its growing labour force and alternative sources of foreign exchange are now some of Lesotho's major problems.

Lesotho is under constant political and economic pressure from South Africa to sign a pact ending any form of support for the African National Congress (a banned South African political party which wages a violent campaign against apartheid).

One area in which Lesotho's fortunes are improving is tourism. Many of the tourists are South Africans attracted by the casinos in the capital, MASERU (gambling is prohibited in South Africa). Another potential source of income is the ORANGE river, which rises on the Lesotho side of the DRAKENSBERG range. Damming its headwaters could provide power needed by South Africa's industrial Transvaal. The two countries have agreed to the study of such a scheme.

LESOTHO AT A GLANCE

Map South Africa Cb
Area 30 355 km² (11 720 sq miles)
Population 1 550 000
Capital Maseru
Government Parliamentary monarchy
Currency Loti = 100 lisente
Languages Sesotho, English
Religions Christian (80%), tribal religions (20%)
Climate Continental; temperature in Maseru ranges from −3 to 17°C (27-63°F) in July to 15-33°C (59-91°F) in January
Main primary products Wheat, maize, sorghum, pulses, livestock; diamonds
Major industries Agriculture, tourism
Main exports Wool, mohair, diamonds
Annual income per head (US$) 370
Population growth (per thous/yr) 25
Life expectancy (yrs) Male 51 **Female** 54

Leszno (Lissa) ***Poland*** Beer-brewing and china-making town 66 km (41 miles) south of the city of Poznan. It was settled by Protestants from the Czech province of Moravia in the 16th century, and became the centre of Polish Calvinism. The Czech theologian and educational reformer Comenius (Jan Komensky, 1592-1671) made his home there. The town is also a market for cereals and cattle.

Population 52 900
Map Poland Bc

Lethbridge ***Canada*** City in southern Alberta on the Oldman river, 177 km (110 miles) southeast of Calgary. Established in 1870 as a coal-mining settlement, it is now the centre of a large ranching and irrigated agricultural area, whose products it processes. The city's university opened in 1967.

Population 54 070
Map Canada Dd

Lettland ***USSR*** See LATVIA

Leuven ***Belgium*** See LOUVAIN

levanter, levante Strong easterly or east-north-easterly wind blowing across south-east Spain and the Balearic Islands, and in the Strait of Gibraltar.

leveche Hot, dry, southerly wind blowing from Morocco across to the coast of southern Spain, and similar to the SIROCCO. It precedes an advancing depression and often carries a great deal of dust. Under certain conditions the dust reaches northern Europe and has been known to cause pink 'blood' rain.

levee Bank of a river built up above the flood plain by the deposition of sediment close to the channel during flooding. Levees are formed naturally, but may be strengthened by man as protection against flooding. Wholly artificial levees are also built for the same purpose.

Levkás ***Greece*** Island in the Ionian Sea separated from the mainland only by a canal first cut in ancient times and reopened in 1905. In the south of the island, at Cape Doukáton, is the Leucadian Rock from which the poet Sappho is said to have leapt to her death in about 580 BC in despair at being rejected in love.

Population 20 000

Map Greece Bb

Lexington ***USA*** City in central Kentucky, about 120 km (75 miles) south of Cincinnati, Ohio. It is a tobacco market and horse-breeding centre.

Population (city) 210 200; (metropolitan area) 327 200

Map United States Jc

Leyden ***Netherlands*** See LEIDEN

Leyte ***Philippines*** Large irregular-shaped island covering 7213 km² (2785 sq miles) in eastern VISAYAN ISLANDS, south-west of Samar (to which it is linked by a bridge). Rugged and rather undeveloped, it receives heavy rains and many typhoons from the Pacific. In the coastal lowlands, rice, maize, sugar and other crops grow. It is best known as the landing place (just south of the provincial capital, Tacloban) of American invasion forces on October 20, 1944, during the Second World War. Over the next week, the Americans won a decisive air-sea battle against the Japanese fleet in Leyte Gulf, to the east.

Population 1 480 000

Map Philippines Bc

Lhasa ***China*** Capital of the Xizang Autonomous Region (formerly Tibet) and one of the most remote cities in the world. It lies in a broad valley on a tributary of the Yarlung Zangbo (Brahmaputra) river, 3800 m (12 450 ft) up between the Himalayas and the Nyainqêntanglha Range, and is separated from eastern China by the rugged wilderness of the Xizang Gaoyuan (Tibetan Plateau). Lhasa contains many Buddhist temples and the Bodala (Potala) palace, the former residence of the Dalai Lama, who fled to India in 1959 after the defeat of a Tibetan rebellion against Chinese rule.

Population 120 000

Map China De

Lhotse ***Nepal/China*** Twin-peaked mountain in the Himalayas on the Nepali frontier with Tibet, just south-west of Mount Everest. Lhotse I rises to 8501 m (27 890 ft) – 118 m (387 ft) higher than Lhotse II.

Map Nepal Ba

Liaoning ***China*** Province in the north-east covering 140 000 km² (540 000 sq miles). It is the southernmost of the three provinces of Dongbei (Manchuria) and borders on North Korea. The fertile Liao river basin forms the centre of the province with low hills on either side. Tobacco, apples and pears are grown in the province, which is unusually well endowed with minerals. The great coalfields of Fushun, Fuxin and Benxi are among the largest and richest in China. There is iron ore at Anshan, Benxi and Liaoyang, and there are also valuable reserves of manganese and molybdenum, used in steel production. As a result, Liaoning is China's leading producer of iron and steel, aluminium and heavy machinery. The capital is Shenyang.

Population 34 426 000

Map China Ic

Liatroim ***Ireland*** See LEITRIM

Libau ***USSR*** See LIEPAJA

Liberec (Reichenberg) ***Czechoslovakia*** Industrial city about 20 km (12 miles) from both the East German and Polish borders. It produces 'Liaz' lorries and diesel engines, textiles, textile machinery, clothing, foodstuffs and furniture. To the south and west lies Bohemia's main glass-making area, which exports glass and glass jewellery around the world. The area's main towns are Jablonec, Novy Bor and Zelezny Brod.

Population 99 600

Map Czechoslovakia Ba

Libertador (O'Higgins) ***Chile*** Region just south of Santiago taking in Cachapoal and Colchagua provinces. The economy is based on copper mining at EL TENIENTE and pastoral and stock farming around RANCAGUA, the regional capital, and San Fernando. Rodeo championships are held in the towns. There is a hydroelectric plant on the Rapel river. The region's full name, El Libertador General Bernardo O'Higgins, commemorates Chile's liberator from Spanish rule.

Population 585 000

Map Chile Ac

Libreville ***Gabon*** Capital of Gabon and of Estuaire province, founded on the north shore of the Gabon river estuary in 1849 as a settlement for freed slaves. It has expanded rapidly since the exploitation of Gabon's mineral resources in the 1960s. It is now a major port, exporting hardwoods, rubber and palm products, an industrial centre, and the educational hub of the country, with the Université Omar Bongo (founded in 1970 and named after Gabon's second president). The city retains a French character, though many modern structures now tower above French colonial buildings. There is an international airport.

Population 308 000

Map Gabon Aa

Libya See p. 388

Libyan Desert ***Libya*** See SAHARA

Lidice ***Czechoslovakia*** See KLADNO

Lidköping ***Sweden*** Small manufacturing town on the south-east shore of Lake Värnen, about 130 km (80 miles) north-east of Gothenburg. It is set in fertile farmland amid prehistoric remains, medieval churches and old villages. Rörstrand porcelain is made in the town.

Population 35 212

Map Sweden Bd

Liechtenstein See p. 390

Liège (Luik; Lüttich) ***Belgium*** City on the Meuse river, 95 km (60 miles) east of Brussels. It gives its name to a Belgian province 3862 km² (1491 sq miles) in extent, and is also claimed by French-speaking Belgians as the capital of French-language culture in Belgium, since Brussels, the national capital, is a bilingual city.

Owing to frequent attacks on Liège through its history, there are few medieval buildings surviving; the great cathedral of St Lambert was destroyed in the spreading fires of the French Revolution in 1792, but the palace of the bishops (built in the 1530s, re-faced in the 1730s) survived and today serves as the law courts.

Both upstream and downstream from the city centre are industrial zones containing much of Belgium's iron and steel industry. The downstream zone also comprises the port of Liège, the entrance to the ALBERT CANAL and the arms-manufacturing town of Herstal. Coal mines and slag heaps were formerly features of the landscape around the city, but mining has now ceased along the Meuse. The space formerly taken by the coalfield has been occupied by new industrial estates and the motorway junctions that make Liège a focus of trans-European routes.

Population (city) 203 000; (metropolitan area) 609 000; (province) 1 000 000

Map Belgium Ba

Liegnitz ***Poland*** See LEGNICA

Lienz ***Austria*** Main town of the east Tyrol at the junction of the Isel and Drava valleys, 80 km (50 miles) west of Villach. The area was cut off from the rest of Austria after the First World War, when South Tyrol was ceded to Italy. In 1967, Felber-Tauern Tunnel connected the east Tyrol with the rest of the Austrian Tyrol. Despite the growth of winter sports, its remoteness has resulted in people leaving the region.

Population 11 700

Map Austria Cb

Liepaja (Lepaya; Libau) ***USSR*** Baltic port, naval base and industrial city in Latvia, 195 km (120 miles) west of Riga. Liepaja's harbour is ice-free all year, and in winter it handles Riga's traffic, for the Latvian capital is on the ice-bound Gulf of Riga. Liepaja's products include iron and steel, heavy machinery, shipbuilding, leather and spirits.

Population 111 000

Map USSR Dc

Lietuva ***USSR*** See LITHUANIA

Liberia

IN A LAND OF FREED SLAVES, AN ELITIST SOCIETY INFLICTED FORCED LABOUR ON THE NATIVE TRIBES

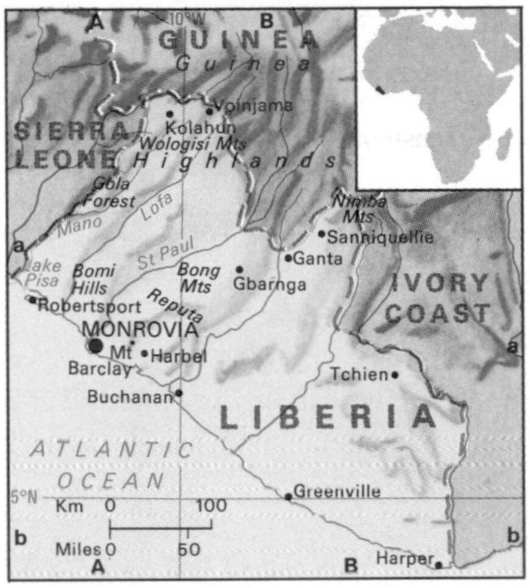

Africa's oldest independent republic and the only African country never ruled by a foreign power, Liberia was created in 1847 by freed slaves returning from America. With the help of philanthropic societies, including the American Colonisation Society, more than 11 000 Africans were repatriated to Liberia by 1860.

From the start, life was difficult for them. The beautiful but treacherous coast, with rocky cliffs and lagoons enclosed by sand bars, offered few landing places. Even when ships found one, the new settlers had much to endure. Torrential rain, flies and stinging insects, mysterious and deadly diseases took their toll. The bright red soils proved not so fertile as they looked.

However, those who stuck it out and settled formed the core of an elite who have long dominated the 30 or so other ethnic groups – made up mostly of the Kpelle, Bassa and Gio peoples. Unfortunately the Afro-Americans' treatment of the indigenous people has left much to be desired. Their forced-labour policies have been strongly criticised in the West. So has their large flag-of-convenience merchant navy, for the lack of control exercised over its 2600 or so foreign-owned ships.

But the country is beautiful. The narrow coastal strip is a man-made mosaic of tropical rain forest and savannah where the Kru and Vai tribes who live there have cleared areas of woodland to grow crops. Ample rainfall from April to October ensures that it is green all year. Inland, the land rises to a densely forested plateau, dissected by deep, narrow valleys.

Farther inland the scenery becomes more spectacular: there are magnificent waterfalls such as St John's; the NIMBA MOUNTAINS, rising to 1752 m (5748 ft) and rich in iron ore; and the 1380 m (4525 ft) Wologisi range. Towards the border with Guinea, where the temperature range is greater and the humidity lower than on the coast, the forest is broken by large areas of savannah.

Since 1926, shortly after the Firestone Rubber Company of America was given a concession to plant rubber in return for revenue, Liberia's governments have pursued an 'open door' policy, inviting foreign investment. Rubber plantations developed rapidly in the forest zone, and as Firestone's interests have grown, other multinational companies have followed them in.

Firestone is no longer Liberia's largest employer and rubber producer, but the company's 290 km² (112 sq miles) plantation at HARBEL – named after Harvey and Isabelle Firestone, the founders of the company, is still one of the largest in the world. The company has set good examples as an employer – Liberian workers have been well trained and educated, and good health facilities have been provided for them.

Roots nurtured in America are clearly evident in Liberia: buildings in early American styles; churches of many denominations; and mostly American cars. In towns American food is popular, and shops, boutiques and supermarkets are well stocked.

But this image of prosperity is deceptive as most Liberians remain poor. The government is in debt and exports such as rubber, iron ore, timber and diamonds are sold mainly in their raw, unprocessed state, which limits their value. Around 70 per cent of the 2.3 million population eke out a living from agriculture, mostly on poor land left over from foreign concessions and large state farms. Rice is sown on the flood plains of the many rivers, and efforts are being made to increase production. Cassava, yams, maize, rice and vegetables are grown under small stands of rubber, cocoa and banana trees, and this diet is supplemented with fish, meat and fruit.

Money is earned by womenfolk selling surplus fruit and vegetables in markets and along the roadside. Liberia could, indeed, be self-sufficient in basic crops; but vast amounts of scarce foreign exchange go on imports of food and consumer goods to satisfy the tastes of townspeople, many of whom are foreign workers.

Tourist facilities are moderate. Hotels are plentiful – though rarely cheap. A variety of restaurants range from good European and American cuisine to 'cookshops' serving local African dishes. Apart from a few casinos and nightclubs, entertainment is very much do-it-yourself. The beaches are beautiful and swimming is also excellent in the lagoons, such as Lake PISA. But in rivers and lakes inland there is a risk of bilharziasis – a disease carried by a tropical, parasitic worm.

Long-distance travel can be difficult, as public transport is poor, but cars can be hired in the capital, MONROVIA. Roads are good, but not always in convenient places. For example, there is no good road parallel with the coast.

Forest and animal reserves are magnificent. A rich variety of wildlife includes the rare dwarf hippopotamus, several species of antelopes and anteaters, elephants, crocodiles, monkeys and chimpanzees.

Until 1980, strong governments dominated by the descendants of American-born Liberians ensured political stability within a democratic framework. But in 1980, a military coup was led by Master Sergeant Samuel Doe. President W.R. Tolbert was shot dead during the coup and 13 other members of his government were executed later. Doe was promoted to general and became head of state. Foreign investment fell away after the coup, but as tensions eased the investment climate has improved.

LIBERIA AT A GLANCE

Area 111 370 km² (43 000 sq miles)
Population 2 300 000
Capital Monrovia
Government Military-ruled republic
Currency Liberian dollar = 100 cents, US dollar also used
Languages English, local languages
Religions Tribal (75%), Christian (10%), Muslim (15%)
Climate Tropical; average temperature in Monrovia ranges from 22°C (72°F) to 31°C (88°F)
Main primary products Rice, cassava, bananas, rubber, palm kernels, maize, coffee, cocoa, timber; iron ore, gold, diamonds
Major industries Mining, forestry, agriculture
Main exports Iron ore, rubber, timber, diamonds, coffee, gold, cocoa
Annual income per head (US$) 360
Population growth (per thous/yr) 33
Life expectancy (yrs) Male 52 **Female** 55

Liffey (An Life) ***Ireland*** River which rises in the Wicklow Mountains and runs 80 km (49 miles) in a loop northwards to Dublin, passing under the city's ten bridges and into the Irish Sea. It feeds the Lacken reservoir and hydroelectric scheme in the mountains.

Map Ireland Cb

lightning Large-scale natural electrical discharge in the atmosphere in the form of a visible flash of light. The discharge, which results from a build-up of opposing electrical charges in two regions, may be between two clouds, between cloud and air, within a cloud, or from cloud to ground. The commonest forms of lightning are fork lightning and streak lightning with fewer branches. Sheet lightning is a discharge which gives a cloud a white sheet appearance. Ball lightning is a rare cloud-to-ground discharge in which a luminous ball appears near the point of impact on the ground, moves about for a few seconds and then may explode.

Lightning Ridge ***Australia*** Opal field in New South Wales about 570 km (360 miles) north-west of Sydney. It produces some of the world's finest stones, including the rare 'black' opal.

Population 1100
Map Australia Hd

lignite Brown coal, intermediate between peat and the more familiar black coal. It is formed in the process by which vegetable matter is converted into coal.

Libya

AN ANCIENT DESERT COUNTRY WHICH WAS CATAPULTED INTO THE 20TH CENTURY BY OIL MONEY AND A CONTROVERSIAL COLONEL

The Socialist People's Libyan Arab Jamahiriyah and its leader, Colonel Moammar Gaddafi, have played a provocative role on the world stage over the past 15 years. They have usually been cast as villains by the nations of the West.

Libya was one of the militant members of the Organisation of Petroleum Exporting Countries (OPEC) who pushed through big price rises that shocked Western economies in the 1970s. Libyan money has been used to support guerrilla movements around the world. The country is criticised as a training ground for guerrilla warfare and terrorism, and Libya has been blamed for a number of outrages. American anger against Gaddafi – called a 'mad dog' by President Reagan – reached a climax when US bombers based in Britain attacked 'terrorist-related' targets in Libya on April 15, 1986. Many countries expressed strong disapproval of the attack.

At home, Gaddafi is seen by many of his people as a folk hero, and by some as a dangerous tyrant. Because of assassination attempts he made his home in a fortified barracks outside TRIPOLI, the capital.

Gaddafi's power at home and abroad rests on oil. Libya was one of the world's poorest countries until oil was discovered in 1958. Crude and refined oil now accounts for 98 per cent of exports, giving Libyans the highest income per head on the African continent. Gaddafi, who led an army coup to end the monarchy in 1969, has used oil revenues to improve life for Libyans in his 'Socialist Cultural Revolution', and also to finance political, military and terrorist forays worldwide.

NATION OF YOUNG PEOPLE

The revolution has produced a decentralised form of democracy with many powers devolved to people's congresses in each community. This gives representation to everyone in a diverse society of townspeople, farmers and nomads. There is increasing political awareness – especially among young people. The population is very young; over 55 per cent of the people are under 18, thanks to a high birthrate and improved medical care that has reduced the number of infant deaths. The population is increasing at a rate of 6.5 per cent annually, which is one of the highest rates in the world.

This young land of young people has a very ancient history. There were ancient Egyptian and Greek settlements on the north-east coast; the Phoenicians and then the Carthaginians settled in the north-west. The Romans came, and adapted some of the Greek buildings. The finest Roman ruins in Libya – and perhaps in Africa – are on the coast at LEPTIS MAGNA, with walls, baths, arches, temples and forums. The Romans were followed by the Vandals and the Byzantines before Arab invasions and migrations started in the 7th century. The Arabs brought Islam and converted the native Berbers. In the mid-16th century the Turks brought the area under Ottoman rule. The final colonisation, by the Italians, began in 1911, and it took 30 years to suppress fierce opposition by the desert tribes.

During the Second World War, the British Eighth Army fought the Italians and Germans over the coastal region – TOBRUK was a major battlefield. After the war Britain and France occupied the country until the United Kingdom of Libya was established under King Idris in 1951.

The Sahara covers much of Libya. Apart from desert oases, the only green areas are the Mediterranean scrublands of the north-west and the forested hills of the north-east, near BENGHAZI. Between these two, where two-thirds of the people live, the desert reaches the shore of the Mediterranean along the Gulf of SIRTE. It extends southwards, over plateaus and depressions, for 800 km (500 miles).

The interior of Libya has had some of the highest recorded temperatures of anywhere in the world, ranging up to 60°C (140°F) in the shade, although the desert nights can be bitterly cold. The north-east and the north-west enjoy a Mediterranean climate of hot, dry summers and cool, moist winters. Highlands in these areas receive much of the rain and provide the country's most fertile soils.

Most of the Libyan people are Arabs, though there still remain distinct North African native Berber groups and Berber-speaking villages. Among the minority groups are the Tuaregs, a nomadic Berber people herding their animals in the desert.

Virtually all Libyans are Muslims. Gaddafi believes that Islam is the best religion for the world, but he is not a strict fundamentalist. Some Muslim rules are rigidly applied – alcohol is strictly forbidden. In other ways Gaddafi's policies are contrary to Muslim traditions. He believes in the liberation of women, and his army contains highly trained female soldiers. Among his personal bodyguards are skilled, gun-carrying girls in uniform, known as the 'Green Nuns of the Revolution'.

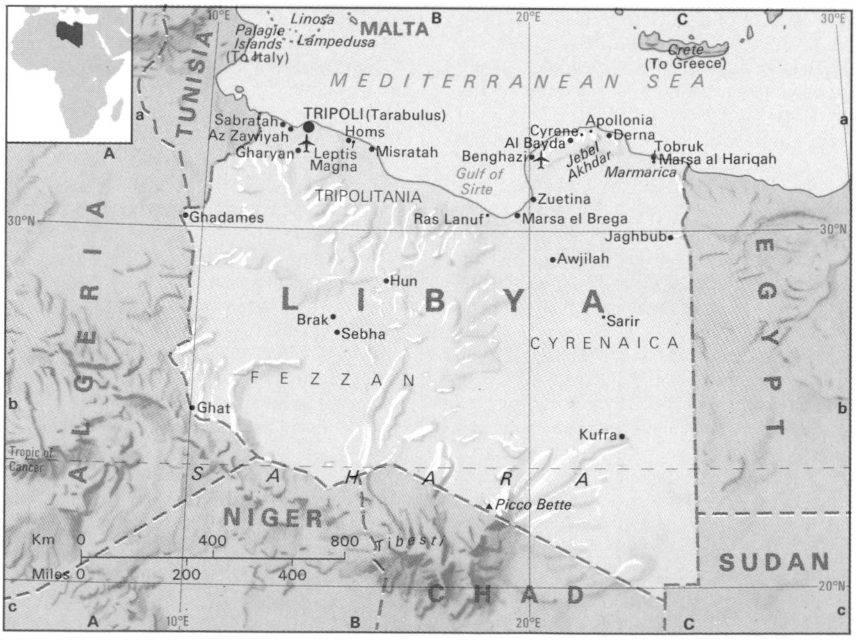

About 14 per cent of the people work on the land, although agriculture contributes no more than 2 per cent to the gross domestic product (GDP). The main agricultural region lies in the triangle formed by the towns of Tripoli, HOMS and Gharyan, which lie in the north-west. The country's main crops are barley, dates, fruit, olives, groundnuts and wheat. There are about 6 million sheep in Libya, as well as goats and some cattle; there is an export trade in hides, skins and hairs. Vineyards which once produced Libyan wine have been affected by the banning of alcoholic drinks, and production has declined.

GADDAFI'S DREAM

Gaddafi has pursued a dream of ending poverty in Libya. Huge irrigation projects are creating fertile land from the desert, with wells tapping underground reservoirs. There is also a massive scheme to build the 'great man-made river', a system of pipelines to transport water from these reservoirs to the towns and farms of the coast. The country has rapidly been industrialised, with new factories producing petrochemicals, steel and cement.

Inevitably the rapid changes have brought problems. The growth of the oil and other industries has prompted a drift to the towns by people seeking work, and this has caused social problems, particularly a shortage of housing. But social service programmes were launched, including free education, free medical services and old age care.

The housing programme is very important to Gaddafi. Born of a nomadic Arab family, he grew up in tents in the desert. When he proclaimed himself Leader of the Revolution, he vowed that he would not move his mother out of her tent and into a house until every Libyan had a home. No one quarrels with housing for the poor, but parts of Libya's revolution have been more controversial. Gaddafi abolished savings accounts, and most of the legal system. He also nationalised urban property, foreign trade and the retail trade.

EVIDENCE OF UNREST

Having provided free education, Gaddafi decided to close all primary schools and make parents responsible for their children's early education. The idea failed in the face of furious opposition from parents. A proposal to ban luxury cars was also abandoned, but not before youths full of revolutionary fervour had destroyed any they could find.

Most of Gaddafi's random austerity measures were accepted by a people who had fared very well under the revolution. But declining oil revenue has hit the economy and there have been serious shortages of foods. There is evidence of unrest: Gaddafi's jails may hold up to 2000 political prisoners; opponents have been publicly hanged; and after the American air raid it was reported that a military junta had reduced Gaddafi's powers.

Arab nationalism, anti-imperialism and anti-Israeli feeling have been the dominant themes in Gaddafi's foreign policy. Anything vaguely connected with Israel is confiscated from travellers arriving in Libya, as is anything anti-Libyan or offensive to Islamic beliefs, such as alcohol or pornographic material. He has ambitions to unite the Arab countries, although his claims to some of their territories and his support for rebel groups have prompted divisions rather than unity.

In recent years, Libya has backed one of her territorial claims by invading Chad, to the south. Gaddafi still claims – and holds – the Aozou strip in northern Chad. And, though France and Libya negotiated a joint withdrawal of their troops from Chad in 1984, there is evidence that Libyan soldiers are still present. Until 1984, Gaddafi was supporting Polisario guerrillas against the Moroccan government, but he suddenly changed sides after an approach from King Hassan of Morocco, and the two former enemies are now on good terms. Libya and Sudan have been bitter enemies for years, each supporting anti-government forces in the other country. Libyan troops helped to maintain the tyrannical regime of Idi Amin in Uganda until it was overthrown in 1979. Libya has sent arms and money to help Muslims and Leftist militias in Lebanon, to Eritreans fighting the Ethiopian government, to anti-American forces in Nicaragua and to the Irish Republican Army fighting the British in Northern Ireland.

LIBYA AT A GLANCE
Area 1 759 540 km² (679 358 sq miles)
Population 4 260 000
Capital Tripoli
Government Military-ruled republic
Currency Dinar = 1000 dirhams
Language Arabic
Religions Muslim (99%), Christian (1%)
Climate Hot and dry in the south; Mediterranean on the coast. Average temperatures in Tripoli: 8-16°C (46-61°F) in January; 22-30°C (72-86°F) in August
Main primary products Wheat, barley, olives, dates, citrus fruits, groundnuts, tobacco; oil and natural gas
Major industries Oil and gas production and refining, agriculture, food processing, textiles, handicrafts, cement, fishing
Main exports Oil, natural gas and petroleum products
Income per head (US$) 8600
Population growth (per thous/year) 65
Life expectancy (yrs) Male 55 **Female** 59

▼ CLOSED FOR BUSINESS Shuttered shops fill Tripoli's market. With chronic shortages, their keepers have nothing to sell, and under Gaddafi's revolutionary policy, can hire no staff.

Liechtenstein

THE PRINCIPALITY THAT COUNTED ITS TROOPS HOME FROM THE WAR AND FOUND IT HAD GAINED AN EXTRA MAN

Among the micro-states of Europe, Liechtenstein is most people's favourite. Andorra is too remote, and Monaco is too expensive. Liechtenstein is just right – a Swiss motorway runs within a stone's throw of the border, and the main railway from Zurich to Austria crosses it.

The principality of Liechtenstein lies on the east bank of the Upper RHINE river. It uses Swiss currency and belongs to the Swiss customs union, but is otherwise independent. The population speaks German and a local dialect. The country's ruling family is of Austrian origin. There is a one-chamber parliament in VADUZ, the capital.

Some companies have their headquarters in Liechtenstein so that they can profit by the country's lenient tax laws. This is good for the state's revenue, as is the trade in the state's decorative postage stamps. More than 75 000 visitors a year also means big business. Beautiful scenery and fine skiing are attractions.

Its territory is roughly 25 km (16 miles) from north to south, with a width of 6 km (3.7 miles). Agriculture is carried on, but over the past 30 years Liechtenstein has been moving into industry, producing quality metal goods, precision instruments, pharmaceuticals and ceramics.

Liechtenstein made an odd population gain in 1866 when it sent men to fight for Austria against Italy in the Tyrol. The contingent is said to have returned with one man more than when it set out.

LIECHTENSTEIN AT A GLANCE

Map Switzerland Ba
Area 160 km² (61 sq miles)
Population 28 500
Capital Vaduz
Government Parliamentary monarchy
Currency Swiss franc = 100 centimes
Languages German, Alemannic dialect
Religion Christian (86% Roman Catholic, 9% Protestant)
Climate Temperate
Main primary products Dairy cattle
Major industries Textiles, ceramics, metalworking, precision instruments, processed foods, pharmaceuticals, light machinery, business services, tourism
Main exports Machinery and transport equipment, metal manufactures, chemicals, postage stamps
Annual income per head (US$) 15 800
Population growth (per thous/yr) 18
Life expectancy (yrs) Male 65 **Female** 74

Liguria ***Italy*** Region in the north-west covering 5415 km² (2091 sq miles) around the Gulf of Genoa. The coast is sheltered by mountains, and because of the mild climate and beautiful scenery many fashionable resorts have opened up since the 19th century – among them Ventimiglia, BORDIGHERA, SAN REMO, ALASSIO, Albenga, RAPALLO and Santa Margherita. Industry is centred on the ports of GENOA, Savona and LA SPEZIA and includes steel, shipbuilding and oil refining.
Population 1 772 000
Map Italy Bb

Ligurian Sea Branch of the MEDITERRANEAN named after the Italian province of LIGURIA, which borders it in the north. TUSCANY is to the east and CORSICA to the south. The Ligurian Sea includes the Gulf of GENOA. Parts of the sea have been badly polluted by industrial effluents and untreated sewage.
Depth More than 2830 m (9285ft)
Map Italy Ac

Likasi ***Zaire*** Mining, industrial and commercial city in Shaba region in the south-east, about 115 km (70 miles) north-west of the city of Lubumbashi. Formerly called Jadotville, it lies on the Lubumbashi-Ilebo railway, and its industries include copper and cobalt refining, railway engineering, chemicals and metalwork.
Population 190 000
Map Zaire Bc

Likoma ***Malawi*** Island of 18 km² (7 sq miles) in Lake MALAWI just 10 km (6 miles) from the Mozambique shore. It was the site of a Church of England mission founded in 1885. The village of Likoma – population 7900 – is a fishing port. Its impressive cathedral, built in 1911, contains a crucifix carved from the wood of the tree in Zambia beneath which the heart of the missionary and explorer David Livingstone (1813-73) is buried. Its twin isle, Chisumulu, is 10 km (6 miles) west.
Population 9160
Map Malawi Ab

Lille Bælt ***Denmark*** See LITTLE BELT

Lille-Roubaix-Tourcoing ***France*** Conurbation near the Belgian border, 210 km (130 miles) north of Paris. Lille produces textiles, chemicals, metal goods and food products. Roubaix and Tourcoing, north-west of Lille itself, produce woollen yarn, carpets and textiles. Many of the towns' workers commute daily across the international frontier.
Population (conurbation) 945 600
Map France Ea

Lilongwe ***Malawi*** National capital since 1975 and the country's second largest city after BLANTYRE. It was chosen to replace ZOMBA as capital because of its position in the centre of Malawi. It is an attractive, well-planned city which is growing quickly. The railway was extended 111 km (69 miles) west from Salima to Lilongwe in 1979, and extended another 120 km (75 miles) to Mchinji on the Zambian border in 1981.

Lilongwe is a trading centre, serving a fertile area producing maize, peanuts and tobacco. It is also the centre of a 445 000 hectare (1 100 000 acre) rural development project launched in 1977 to improve crop production and provide social services.
Population 172 000
Map Malawi Ab

Lima ***Peru*** National capital and capital of Lima department, on the Rímac river about 13 km (8 miles) east of its port, Callao. It was founded by the Spanish conquistador Francisco Pizarro in 1535, and is now the economic and cultural centre of Peru and the fifth largest city in South America. Its University of San Marcos, founded in 1551, is the continent's oldest. Most of the Spanish colonial buildings, including the cathedral, the Torre Tagle Palace and the House of the Inquisition, lie close to the Plaza de Armas, now dominated by modern skyscrapers. Pizarro's home was built here, on a site now occupied by the Palacio de Gobierno (Government Palace). Lima has spread to meet its port, Callao, and together they form a vast conurbation housing 70 per cent of the country's manufacturing industry including textiles, clothing, pharmaceuticals and food processing.
Population 5 500 000
Map Peru Bb

Limassol ***Cyprus*** The island's main port and its second largest town after the capital, Nicosia. It is a major tourist centre and produces brandy, cigarettes and perfume. Since the Turkish occupation of northern Cyprus in 1974, the town's economy has expanded considerably. Richard the Lionheart, King of England, used it as his base for the Third Crusade in 1191. In 1815 an earthquake killed all but 150 of its inhabitants, and the town did not fully recover until the British took it over in the 1870s.
Population 107 200
Map Cyprus Ab

Limbang ***Borneo*** River about 150 km (95 miles) long in northern Borneo, the lower course and estuary of which form the enclave of Malaysian territory that divides the two wings of Brunei. The estuary is extremely swampy.
Map Brunei Ab

Limburg (Limbourg) ***Belgium*** Flemish-speaking province lying between the city of Liège and the Dutch border, and covering 2422 km² (935 sq miles). The capital is Hasselt. Its main industries – chemicals, nonferrous metals, car assembly, electrical goods – are based on the province's Kempenland coalfield, but there is some forestry among the heathlands.

Limburger cheese was originally made in the town of Limbourg, which now lies in the province of Liège about 110 km (70 miles) east and slightly south of Brussels. The town was the seat of the former Duchy of Limbourg; the area divided in 1839 to form the Belgian and Dutch provinces of the same name.
Population (province) 727 000
Map Belgium Ba

Limburg ***Netherlands*** Hilly province which covers 2209 km² (853 sq miles) in the extreme south-east. It contained the country's only coalfield, until it was closed in 1975. The capital is MAASTRICHT, and the province today produces ceramics, cement and chemicals.
Population 1 080 500
Map Netherlands Bb

Limerick (Luimneach) ***Ireland*** City and port on the south bank of the River Shannon, 87 km (54 miles) north of the southern city of Cork. Viking invaders founded the city in 922, but it was later sacked by the medieval Irish king Brian Boru. Still later, the Anglo-Normans occupied the city and King John (1199-1216) built the castle there which still carries his name. Modern Limerick is mostly Georgian, and built on a grid system. It has a wide scope of agricultural industries, and is noted for ham and bacon. Limerick is also a commercial centre, and the county town of the surrounding county of the same name.

The county covers 2686 km^2 (1037 sq miles), and includes the west of the Golden Vale, noted for its livestock and dairy products.

Population (county) 161 700; (city) 60 700
Map Ireland Bb

limestone SEDIMENTARY ROCK, chiefly calcium carbonate, the main mineral of which is CALCITE although DOLOMITE may also be present. It is formed from the skeletons and shells of small marine organisms, or by deposition from solution. It is used as building stone, for making lime and cement, and in smelting iron.

Limfjorden ***Denmark*** A broad channel across north JUTLAND, extending from the Kattegat in the east nearly to the North Sea in the west. The 1 km (0.6 mile) long Thyborøn canal links its western end with the North Sea. Denmark's largest cement plants are located on the channel's shores, and its waters are celebrated for their oysters.

Map Denmark Ba

Limoges ***France*** Town on the Vienne river 180 km (110 miles) north-east of Bordeaux. It is renowned for both its enamel work and its porcelain. Limoges enamel was well established in the 13th century, and reached its peak in the work of the 16th-century painter Léonard Limousin; however, it declined after the late-16th-century Wars of Religion, when Limoges was devastated. The porcelain industry, using local china clay, began in the 18th century and is still an important employer. Other industries include tanning, shoemaking and vehicle manufacture; uranium is mined nearby.

The town has a 13th to 16th-century cathedral, a museum of porcelain, and an art gallery containing many works by Auguste Renoir (1841-1919), who was born in Limoges.

Population 144 100
Map France Dd

limon Fine-grained deposit in France and Belgium from which brown, loamy soils have developed. It is of wind-blown origin, deposited under humid conditions, and possibly re-deposited by running water. See also LOESS

Limón ***Costa Rica*** Agricultural province extending along the country's entire Caribbean coast. It grows bananas, cacao (used to make cocoa), abaca (a Manila hemp) and sugar cane. Its capital is the town of Limón, the country's main Caribbean port, built on the site of an Indian village which was visited by the explorer Christopher Columbus in 1502.

Population (province) 187 000; (town) 48 200
Map Costa Rica Ba

▲ DUAL PURPOSE An enterprising citizen of Lima combines his shoe-shine booth with a newsstand so that his customers may polish up their knowledge of sporting and current affairs.

Limousin ***France*** Region of rocky plateaus in the north-west of the Massif Central. It rises to 978 m (3208 ft) at the Millevaches plateau. Limousin bulls – used by stock-breeders all over the world – are reared in the region. LIMOGES is the main city.

Population 737 000
Map France Dd

Limpopo ***Southern Africa*** River, 1610 km (1000 miles) long, and also known as the Crocodile in its upper reaches. 'The great grey-green, greasy Limpopo', as the English writer Rudyard Kipling (1865-1936) described it, rises on the Witwatersrand near Johannesburg in South Africa and flows north-west, then north-east to form the country's borders with Botswana and Zimbabwe. In its lower course, it flows sluggishly through Mozambique to the Indian Ocean 85 km (53 miles) north-east of Maputo, Mozambique's capital. The river is navigable for about 100 km (60 miles) up to a dam near the town of Guija.

Map Mozambique Ac

limpo Type of SAVANNAH in Brazil consisting mainly of tall, coarse grasses with relatively few bushes or trees.

Lincoln ***United Kingdom*** City and county of eastern England. The city has a fine 11th-century castle and stands on a ridge above the River Witham, 190 km (120 miles) north of London. Its cathedral, one of the largest in Britain, has been rebuilt several times since the 11th century because of fire and earthquake damage, and it lost its main spire in a storm in the 16th century. The modern city is an industrial centre specialising in farm machinery and agricultural supplies. The first military tank was built there during the First World War.

The county – 5885 km^2 (2272 sq miles) in area and largely flat apart from The WOLDS – has rich farmland and a host of seaside resorts, including Skegness. Part of north Lincolnshire was incorporated in the new county of HUMBERSIDE in 1974.

Population (county) 558 000; (city) 77 000
Map United Kingdom Ed

Lincoln ***USA*** Capital of Nebraska, in the south-east of the state, about 75 km (47 miles) south-west of the Omaha. It is an agricultural marketing and servicing city, and a financial centre. It has two universities.

Population (city) 180 400; (metropolitan area) 203 000
Map United States Gb

Lindau ***West Germany*** Tourist town on the north shore of Lake Constance near the Austrian border. It has a casino, and also textile, machinery and electronics industries.

Population 23 500
Map West Germany Ce

Lindisfarne ***United Kingdom*** See NORTHUMBERLAND

line squall Band of extremely stormy weather bringing with it sudden, strong, gusty winds along a line sometimes as much as 500 km (about 310 miles) in length. Line squalls are associated with COLD FRONTS and are characterised by dark rolls of cloud and downpours of rain and hail.

Line-of-Rail ***Zambia*** Densely populated strip of land along the railway which links the capital, Lusaka, with Maramba in the south and the northern copper-mining towns, and runs for about 700 km (435 miles). The railway, built at the beginning of the 20th century, has drawn people to it – rather like the Nile river in Egypt – leaving the remainder of the country comparatively sparsely populated.

Map Zambia Bb

Lingayen ***Philippines*** Trading port and capital of Pangasinan province, on Lingayen Gulf at the mouth of the Agno river in LUZON. The centre of a rich farming area, it was first visited by Chinese, Japanese and other barter traders during the 14th century. Spanish and later American colonists developed the town, but it was badly damaged in the Second World War. It was the first landing point on Luzon by Japanese forces in 1941, and also by American forces in 1945.

Population 56 300

Map Philippines Bb

Linköping ***Sweden*** Industrial city, 198 km (123 miles) south-west of Stockholm. It was an important religious centre in the Middle Ages and has a magnificent 12th to 15th-century cathedral. It manufactures cars, aircraft and railway rolling stock.

Population 116 834

Map Sweden Bd

Linz ***Austria*** The country's third largest city and capital of Upper Austria, 160 km (100 miles) west of Vienna. Straddling the Danube, Linz has been an important river port and trading centre since the Middle Ages. In 1939 Herman Goering, the head of Nazi Germany's Luftwaffe (air force), established a major steelworks in the city – which resulted in its being heavily bombed in the Second World War. As well as iron and steel, Linz now manufactures chemicals, textiles, tobacco and fertilisers. Adolf Hitler, the German dictator, spent his early schooldays at Linz.

Population 201 500

Map Austria Da

Lion, Golfe du (Gulf of Lions) Section of the MEDITERRANEAN SEA between the Franco-Spanish border and TOULON, France. The MISTRAL, the cold, northerly wind, is funnelled through the RHONE valley and reaches a great speed over the gulf, causing problems for boats, especially small ones. MARSEILLES (France) is the gulf's chief port.

Map France Ee

Lipari ***Italy*** Principal island of the EOLIAN group off the north-east coast of Sicily. It covers 38 km² (15 sq miles) and has 12 extinct volcanoes. Half the island's inhabitants live in Lipari town. There are four other villages and many scattered farms. The main occupations are fishing, agriculture (chiefly vines) and tourism. Obsidian (black volcanic glass) was mined in the Stone Age.

Population 10 700

Map Italy Ee

Lisbon (Lisboa) ***Portugal*** Capital, port and the country's largest city, located on the west coast. It stands on the north bank of the Tagus river, about 15 km (9 miles) from the sea. It was from Lisbon that the Portuguese navigator Vasco da Gama sailed for India in 1497.

According to legend, the city was founded by the Homeric hero Ulysses (also known as Odysseus). Most modern authorities attribute its foundation to the Phoenicians around 1200 BC. It was conquered by a variety of peoples, including the Greeks, Carthaginians, Romans and Visigoths, before the Moors overran the region in AD 714.

In 1147 Alfonso I, the first King of Portugal, captured the city for the Portuguese and in 1255 Alfonso III made it his capital in place of the city of Coimbra. In the 14th and 15th centuries Lisbon was one of the most magnificent cities in Europe, but on All Saints' Day 1755 much of the city was destroyed by an earthquake and associated wave, followed by fire. About 60 000 people died in the disaster. The city was rebuilt by the Marquis of Pombal, and the 18th-century town plan is still evident today.

One of the best views of Lisbon is from the re-named Ponte 25 de Abril, the city's first bridge across the Tagus; 2278 m (7474 ft) long, the bridge was completed in 1966. The modern centre of the city is Pracça Dom Pedro IV, known as Black Horse Square because of the equestrian statue standing in it. To the east lies the Alfama, or medieval town, with its 12th-century cathedral and narrow streets. Above these is the Castelo de São Jorge, fortified by the Visigoths in the 5th century and modified first by the Moors, then by Alfonso I. Among other places of interest are the Jeronimos Monastery; the Tower of Belem, with carved balconies and domed turrets; the Museum of Ancient Art; and the Monument to the Discoveries, commemorating the great Portuguese explorers.

The city's main industries are textiles, soap, flour, steel, sugar and oil refining, and shipbuilding.

Population 817 600

Map Portugal Bc

Lisburn ***United Kingdom*** Market town in Northern Ireland, 10 km (6 miles) south-west of the capital, Belfast. It is a centre of the linen industry. Christ Church Cathedral, in the town's market square, dates from the early 17th century and was rebuilt in 1708.

Population 40 000

Map United Kingdom Bc

Lissa ***Poland*** See LESZNO

Litani ***Lebanon*** River in southern Lebanon which rises near Baalbek and flows south between the Lebanon Mountains and the Anti-Lebanon Mountains. The Litani turns sharply west near the town of Marjayoun, plunges through a deep gorge in the Lebanon Mountains and enters the Mediterranean 10 km (6 miles) north of the port of Tyre.

Map Lebanon Ab

lithification Process by which loose sediment is consolidated into hard rock.

lithosphere Solid outer layer of the earth lying above the semi-fluid asthenosphere. It includes the earth's crust and upper mantle. See JOURNEY TO THE CENTRE OF THE EARTH (p. 200)

Lithuania (Lietuva; Litva) ***USSR*** Largest and southernmost of the three Soviet republics bordering the Baltic Sea, covering 65 200 km² (25 200 sq miles) to the south of Latvia. Most of it is lowland covered by forest and swamp. However, the eastern part is an area of forested sandy ridges, with many lakes. Rye, barley, sugar beet, flax, meat, milk and potatoes are the main farm products, and the republic's industries produce ships, building materials, heavy machinery, linen and other textiles, timber and paper. Lithuanian craftsmen make jewellery from amber, a fossilised resin found along the Baltic coast. The main towns are the capital VILNIUS, KAUNAS, the port of KLAIPEDA and Siauliai

The Lithuanians settled the area before 1100, and in the 13th-14th centuries the area, which included much of the Ukraine, became first the Lithuanian Principality and eventually part of a Roman Catholic Polish-Lithuanian empire. However, the empire dwindled with German, Swedish and Russian invasions, and by 1795 all of Lithuania was under Russian rule. Resistance led to peasant revolts and large-scale emigration to North America in the late 19th century, and the birth of a nationalist movement that was active into the 1950s. Lithuania was independent between 1920 and 1940, but was occupied by Russians in 1940, then by Germans in 1941-4 and finally annexed by the USSR in 1944 – a move not recognised by some Western governments. Lithuanians remain staunchly Roman Catholic, with their own language and strong culture.

Population 3 539 000

Map USSR Dc

Little Belt (Lille Bælt) ***Denmark*** The channel between the Danish peninsula of JUTLAND and the FUNEN island group. It is some 130 km (80 miles) long, and varies in width from 30 km (19 miles) to 600 m (nearly 2000 ft). It was bridged for the first time in 1935.

Map Denmark Bb

Little Rock ***USA*** Capital of Arkansas, on the Arkansas river in the centre of the state. It was founded in 1814 and is now an industrial centre producing textiles, electronics and foodstuffs. The city achieved worldwide notoriety in 1957 when President Dwight D. Eisenhower sent in federal troops to enforce a Supreme Court ruling outlawing racial segregation in schools.

Population (city) 170 100; (metropolitan area) 492 700

Map United States Hd

littoral current See LONGSHORE CURRENT

Litva ***USSR*** See LITHUANIA

Liverpool ***United Kingdom*** Port city of north-west England on the north side of the Mersey estuary, formerly in Lancashire but now the main centre of the county of Merseyside. It flourished in the late 17th and 18th centuries on trade with the New World and Africa, first in slaves and sugar, then in cotton. In the 19th century, it became the main port for emigrants to the United States not only from Britain but also from Scandinavia and Germany. Its docks and quays stretched along the Mersey for 12 km (7 miles) while, on the south shore, the town of Birkenhead grew to provide further docks and, after 1824, shipbuilding yards. Liverpool and Birkenhead were linked by a 4.8 km (3 mile) long road tunnel in 1934 and by a second opened in 1971.

Liverpool has declined seriously since the Second World War. There are two large cathedrals – the Roman Catholic one a masterpiece of modern design – many fine civic buildings and a university, but the transatlantic passenger traffic now goes by air, cotton is no longer king, unemployment is high, and the docks are all but empty and have been partly reclaimed for a park. For many, the city's greatest claim to fame is the Beatles pop group – John Lennon, Paul McCartney, George Harrison and Ringo Starr, who were all born there and established their careers at its Cavern Club.

Population 497 000
Map United Kingdom Dd

Livingstone ***Zambia*** See MARAMBA

Livingstone Falls ***Zaire*** A series of 32 cataracts between Kinshasa, the capital, and Matadi on the lower Zaire river. The series descends 260 m (853 ft) in 354 km (220 miles) through the Crystal Mountains.

Map Zaire Ab

Livingstonia ***Malawi*** Town about 400 km (250 miles) north of the capital, Lilongwe, and named after the explorer David Livingstone. There are splendid views of Lake Malawi, which Livingstone discovered in 1859. The town was founded as a mission in 1875 and stands about 1370 m (4500 ft) above sea level. Coal has been found nearby.

Map Malawi Ab

Livorno ***Italy*** See LEGHORN

Liw ***Poland*** One of eastern Poland's oldest settlements, 83 km (52 miles) east of the capital, Warsaw. Its massive, 14th-century brick fortifications, which rise from poppy fields, are the remnants of the third stronghold to be built on the site. The building is now a museum.

Map Poland Eb

Lizard Point ***United Kingdom*** Most southerly point on the British mainland, jutting into the English Channel from the county of Cornwall. The cliffs soar to 55 m (180 ft) on its rugged coast, which is corrugated with coves and harbours. A lighthouse has overlooked the treacherous waters around the headland since 1619. Goonhilly Downs inland is a space satellite communications station.

Map United Kingdom Cf

Ljubljana (Lyublyana; Laibach) ***Yugoslavia*** The capital of Slovenia lying on the Sava river, 121 km (75 miles) west of Zagreb. It is also a road and rail junction, a tourist centre and an industrial city producing machinery, textiles, porcelain, paper, furniture, shoes, chemicals and leather goods. Ljubljana grew around the 12th-century hilltop Grad fortress, which commands splendid views of the surrounding basin and the Karawanke and Julian Alps to the north and west respectively.

Population 305 200
Map Yugoslavia Ba

Llangollen ***United Kingdom*** Small town on the River Dee in the Welsh county of Clwyd, 50 km (30 miles) south of Liverpool. It has become renowned since 1947 for its annual summer eisteddfod – an international festival of folk music, dancing and poetry that attracts thousands of performers from many lands.

Population 3200
Map United Kingdom Dd

llanos Term originally meaning 'plains', now applied chiefly to the tropical grassland or savannah of the Orinoco basin and the Guiana Highlands in South America. In particular, it is applied to the Llanos of central Venezuela, lying between the Cordillera de Mérida range and the Orinoco river. This is a rich grazing area, watered by the Orinoco and its tributaries.

Map Venezuela Ab

Llanquihue ***Chile*** Largest lake in the country, covering 878 km² (339 sq miles). It lies in the southern region of Los Lagos in a spectacular setting of snow-capped mountains, notably the 2660 m (8727 ft) high Osorno volcano. It is a popular fishing and water sports resort. On its western shore, near Puerto Varas, is one of South America's oldest archaeological sites.

Map Chile Ad

loam Soil containing sand, silt and clay in roughly equal proportions.

Loanda ***Angola*** See LUANDA

Lobamba ***Swaziland*** Town lying 16 km (10 miles) south-east of Mbabane. It is the site of the Swazi Parliament and the seat of the Queen Mother, called the *Ndlovukazi* (Swazi for 'Great She-Elephant'). Lobamba is famed for two ceremonies, both of which take place over several days. The Reed Dance, which is held in August-September, honours the Queen Mother, while the Incwala Kingship ceremony in December-January honours the King. The ceremonies include spectacular dancing in traditional costumes, singing and feasting. Visitors are welcome for most of the time, but they must leave during taboo parts of the ceremony. On the sixth day of the Kingship ceremony, warriors build a fire with wood collected from the hills. Dancing and singing round the fire, they call on the spirits of ancestors to send rain to douse the fire as a sign of favour for the coming year.

Population 5800
Map Swaziland Aa

Lobito ***Angola*** Port about 390 km (240 miles) south of the capital, Luanda. Founded in 1834, it grew rapidly after the completion of the Benguela Railway in the 1920s. It now handles much of the country's agricultural exports, and minerals from Zaire and Zambia. It also manufactures building materials, boats and textiles.

Population 59 000
Map Angola Ab

Lobombo Mountains ***Southern Africa*** See LEBOMBO MOUNTAINS

Locarno ***Switzerland*** Italian-speaking winter and health resort – known as 'the Nice of Switzerland' – at the northern end of Lake Maggiore. There are boat connections to Italy along the lake. The pilgrimage church of the Madonna del Sasso, founded in 1480 and enlarged in 1616, stands above the town.

Population 15 000
Map Switzerland Ba

loch In Scotland, a lake or a long narrow arm of the sea.

Lod (Lydda) ***Israel*** Town with an international airport, about 15 km (10 miles) south-east of Tel Aviv. Its factories produce aircraft parts and electronic equipment. Lod was visited by St Peter, who told Aeneas to get up and walk as Jesus had cured him of his paralysis. In the 12th century the city was destroyed by the Muslim leader Saladin, then rebuilt by the English king Richard I (Richard the Lionheart), during the Third Crusade.

Population 41 400
Map Israel Ab

lode **1.** Mineral deposit contained in hard rock, usually in the form of a group of VEINS. **2.** In East Anglia, England, an artificial watercourse.

lodestone, loadstone Strongly magnetised, pure form of MAGNETITE.

Lodz ***Poland*** The country's second largest city after the capital, Warsaw. It lies in central Poland, 121 km (75 miles) south-west of Warsaw. Founded in 1423, the town was turned by occupying Russians in the 19th century into a polluted agglomeration of spinning, weaving, dyeing and clothing industries, with unsanitary housing, especially in the Jewish ghetto, which was razed by the Nazis in 1944. Today Lodz's modernised factories turn out textiles, clothing and chemicals, and electrical, engineering and photographic goods.

Population 848 500
Map Poland Cc

Loei ***Thailand*** Province and town near the Laotian border, 450 km (280 miles) north of the capital, Bangkok. Cotton is the province's main crop.

Population (town) 102 100; (province) 507 700
Map Thailand Bb

loess, löss Deposit of fine-grained, friable, porous silt-like dust, yellowish to grey in colour, generally deposited by the wind after being winnowed from deserts or the debris left by the retreat of the last ice sheets. Limon in France and Belgium and brick-earth in Britain are substantially the same material, possibly redeposited to some extent by water.

Loesslands of China Geographical region of northern China, forming a great plateau stretching through the provinces of Shanxi, Shaanxi, Ningxia-Hui and eastern Gansu. Its yellowish LOESS soil – derived from the wind-blown sands of the Gobi Desert – is fertile but erodes easily. Most of the eroded soil is carried by rain into the HUANG HE – resulting in its name, which means 'Yellow River'. Drought is a perennial threat to crop production throughout the loesslands. In parts of Shanxi and Shaanxi, farmers used to live in caves dug into the soft loess earth beneath their fields.

Map China Fd

Lofoi Falls ***Zaire*** Waterfall in the south-east, on the Lofoi river, about 90 km (55 miles) north of the city of Lubumbashi. It has a sheer drop of 340 m (1116 ft), and is now protected in the Kundelungu National Park.

Map Zaire Bc

Lofoten ***Norway*** Arctic island group off the north-west coast of Norway. The islands, which cover a total area of 1667 km² (644 sq miles), stretch 190 km (118 miles) along the north side of the Vestfjorden channel. Their cod fisheries are among the richest in the world. Hundreds of fishing boats converge on the islands at the height of the cod season in February and March. The main ports for fish processing are Svolvaer and Henningsvær.

Just off the most southerly major island in the group – Moskenes Island – a powerful tidal current sweeps through the strait between the southern tip of Moskenes and a nearby islet. Imaginative descriptions of the current's effects by such writers as the American Edgar Allan Poe (1809-49) and the Frenchman Jules Verne (1828-1905) have made the current's name – the Moskenesstraumen (Moskenes Current) or, more familiarly, the Maelstrom – synonymous with a malevolent whirlpool capable of swallowing entire ships. In reality, even small boats can often cross it safely. However, some combinations of wind and tide can make the Maelstrom a terrifying hazard to navigate.

Population 26 200

Map Norway Da

Logan, Mount ***Canada*** The country's highest peak – and the second highest, after Mount McKinley, in all North America – at 5951 m (19 524 ft). It stands in the Saint Elias Mountains in south-west Yukon.

Map Canada Ab

Logroño ***Spain*** Walled town about 250 km (149 miles) north of Madrid, on the River Ebro. It was taken from the Moors by Alfonso VI in 1095. During the Middle Ages it was an important station on the pilgrims' route to Santiago. The town's main products are textiles, fruit preserves and equipment for winemaking. It is the capital of La RIOJA province.

Population 111 000

Map Spain Da

Loire ***France*** Longest river in France (1020 km, 635 miles), rising in the southern Massif Central. It flows north through gorges to Orléans and then sweeps westward through a wide valley to the Bay of Biscay west of Nantes. The middle valley – known as the 'garden of France' is lined by many elegant chateaux, including those at Amboise, Blois and Chambord. Several nuclear power stations are sited along the middle stretch of the river. Vines, fruit and vegetables are grown, and mostly white wines such as Muscadet, Saumur, Pouilly and Vouvray are produced.

Map France Cc, Fd

Loire, Pays de la ***France*** Region of western France, covering 32 083 km² (12 387 sq miles). It is centred on the city of NANTES.

Population 2 931 000

Map France Cc

Loja ***Ecuador*** Capital of the most southerly province of the same name. It stands on the Pan-American Highway and the Zamora river about 200 km (125 miles) south-east of Guayaquil, and is the commercial centre for the surrounding agricultural and dairy farming region. It is noted for its law school founded in 1897 (now a university).

Population (province) 358 600; (city) 66 000

Map Ecuador Bb

Lolland ***Denmark*** Third largest Danish island covering 1234 km² (476 sq miles). Its broken, indented coastline contains a land of forests in the north and east, while the marshy south is protected from flooding by dykes and sand dunes. The main crops from its fertile soil are sugar beet and tobacco. Places of interest

▼ FAIRYTALE CASTLE The medieval Château de la Roche towers over the Loire river at Roanne. Most later chateaux of the region were designed more for beauty than defence.

include the 12th-century royal palace, Alholm Castle in Nysted; it now houses the largest collection of veteran cars in Europe. The island contains many fine medieval buildings including both manor houses and churches. The principal towns are Nakskov, Maribo and Sakskøbing.

Population 81 800

Map Denmark Bb

Lomami *Zaire* River of central Zaire, about 1450 km (900 miles) long. It rises about 450 km (280 miles) north-west of the southern city of Lubumbashi, and flows north to the Zaire near Isangi. Its lower 400 km (250 miles) are navigable.

Map Zaire Bb

Lombardy (Lombardia) *Italy* The country's most heavily populated region, covering 23 854 km² (9210 sq miles) of the Italian Alps and the North Italian Plain. Lombardy is Italy's industrial powerhouse, prominent in all main sectors of manufacturing. Agriculture and dairy farming are also highly productive, due to careful control of water in the low plains of the great Po river.

Major cities include MILAN, the regional capital, BERGAMO, BRESCIA, COMO, CREMONA, MANTUA, MONZA, PAVIA and VARESE. The town of Magenta, 25 km (16 miles) west of Milan, was the scene of a victory of the French and Italian armies over Austrian troops in 1859, during the campaign to unify Italy. Some 9000 men died in the battle, which was so bloody that Magenta gave its name to a shade of red.

Population 8 898 700

Map Italy Bb

Lombok *Indonesia* Island of 5435 km² (2098 sq miles) lying across a 35 km (22 mile) wide strait from Bali. Through the strait runs the imaginary Wallace's Line, suggested by the British biologist Alfred Russel Wallace (1823-1913) as dividing the Asian biogeographical realm from the Australasian. Thus Lombok has marsupial creatures related to kangaroos, but Bali to the west does not. Lombok is mountainous, reaching 3726 m (12 224 ft). Rice and coffee are the main crops.

Population 1 300 200

Map Indonesia Ed

Lomé *Togo* The country's capital and by far its largest town. Lomé, on the coast beside the Ghana border, became the colonial capital of German Togoland in 1897. It now has an international airport, and a modern deep-water port which exports phosphates, cocoa, coffee, cotton and palm oil (used in margarine and cooking fats). The port handles much of the transit trade from the land-locked states of Mali, Burkina and Niger. It also has an attractive palm-lined seafront, and good hotels with a variety of entertainment and sports facilities.

The local two-storey market stocks cloth from all over West Africa, gold and silver jewellery, and local craft work. The city's university was founded in 1965.

Conferences at Lomé in 1975 and 1979 drew up conventions or treaties on aid and trade between the European Economic Community and many Third World countries.

Population 283 000

Map Togo Bb

Lomond, Loch *United Kingdom* Lake in Scotland, 28 km (17 miles) north-west of Glasgow. It covers 70 km² (27 sq miles) and stretches 35 km (22 miles) from Ardlui in the north to Balloch. The loch lies in the shadow of 973 m (3192 ft) high Ben Lomond and is dotted with islands. It is known all over the world through the words of the folk song *The Bonnie Banks o' Loch Lomond*, and is a popular tourist attraction.

Map United Kingdom Cb

London *Canada* Industrial and financial centre of south-western Ontario. London was planned as a regional capital in 1793, but was not settled until 1826. A university was established in 1878. Industries make food and paper products, locomotives, textiles and refrigerators.

Population 284 000

Map Canada Gd

London *United Kingdom* National capital and major metropolitan area, with 32 boroughs plus the City of London itself spreading over 1580 km² (610 sq miles) on both banks of the River Thames. The Romans founded Londinium in AD 43, in what is now the 'square mile' of the City, the business heart of the capital. They also built the first London Bridge, on a site where, until 1749, stood London's sole bridge across the Thames. In the 11th century a royal palace and then a minster (abbey church) were built some 3 km (2 miles) to the west at what became known as Westminster. Thus were created twin cities that grew in parallel and did not really merge until the 17th century: the City itself the commercial capital, Westminster the centre of royal and later political power. The movement westwards was greatly accelerated by the Great Fire of 1666 which devastated most of the City.

The western part of central London now contains the Houses of Parliament, government departments, Buckingham Palace and other royal homes, the main national museums, concert halls and art galleries, and – in the 'West End' – the major theatres, restaurants and shops. In the west too are the principal parks, of which Hyde Park with Kensington Gardens is the largest. The City, developed over the centuries as an international centre of banking and commerce, had to its east the roots of its wealth: the docks from which ships traded all over the world. The active docks have now moved downriver, to be replaced by modern industrial, leisure and tourist developments, but the City retains its role, if not its appearance; redevelopment after the bombing of the Second World War has all but submerged its greatest glory – Sir Christopher Wren's St Paul's Cathedral, built after the Great Fire between 1675 and 1710 – in a sea of office blocks.

The Thames snakes through London in graceful, sweeping curves and is spanned by six road bridges between the Tower of London – built next to the City by William the Conqueror after 1066 – and the Houses of Parliament alone. It was the capital's main transport artery, for local as well as international trade, before the railways arrived. It has long posed a threat to London of surge-tide flooding. This risk was eliminated only in 1984 with the completion of a movable barrier downstream at Woolwich. Today travellers in London have the choice of moving on often congested streets, many of them rather narrow compared with those of other major cities, or of using one the world's most comprehensive networks of surface and underground railways; the river is hardly used at all for personal transport.

The rail network was built to serve an ever-growing capital from the mid-19th century, and in turn encouraged urban growth as workers moved out to commuter suburbs. As it grew, London swallowed communities that were once outlying country towns and villages. Many, such as Greenwich (with the old Royal Observatory buildings, the Royal Naval College, National Maritime Museum and *Cutty Sark* clipper ship), are full of historic interest. Many have become dormitory suburbs, distinguishable only by their shopping centres. But other former villages – many with literary or artistic associations, such as Chelsea, Hampstead and Highgate – have retained their local identity.

Few major capitals so dominate the cultural and commercial life of their nation as does London, and this helps to explain the vast influx of visitors – tourists and business people alike – who arrive each year. In 1985, with 9 million overseas visitors, London topped the international tourist league table, and another 14 million Britons visited the capital.

Visitors discover a population that is cosmopolitan and lively to a degree – who are pleased to help strangers but not afraid to tell the world their opinions from atop a soapbox at Speakers' Corner in Hyde Park. They find traditions that have changed little in centuries – from the pageantry of the Changing of the Guard at Buckingham Palace to the humble street markets – alongside a vitality that keeps London a trendsetter in fashion and the arts. They observe a cultural life that has few equals, including one of the world's great Opera Houses at Covent Garden, over 40 major theatres and many smaller ones, and concerts by five full-time symphony orchestras and countless other groups performing music from madrigals to Pop.

Above all, however, they view London's physical heritage: the great buildings, art collections and museums. They include great religious buildings – St Paul's Cathedral, Westminster Abbey and the Roman Catholic Westminster Cathedral – Westminster Hall and the Houses of Parliament, and those associated with the royal family: the Tower of London (where, among many other things, the crown jewels can be seen), St James's Palace (1532), Kensington Palace (1605) and the 19th-century Buckingham Palace (where the Queen's Gallery and royal stables are opened to the public).

Major museums include the British Museum (with its superb collection of antiquities and associated British Library), Victoria and Albert Museum (fine and applied arts of all kinds), Bethnal Green Museum of Childhood and the specialised museums of Science, Natural History and Geology. There are numerous art galleries, of which the most famous are the National Gallery, Tate Gallery (emphasising modern art), National Portrait Gallery, Royal Academy, Wallace Collection and Courtauld Institute (with many Impressionist and Post-Impressionist works). And there are countless lesser buildings of note: churches, houses, even street furniture.

Population 6 755 000

Map United Kingdom Ee

Londonderry ***United Kingdom*** Northern Ireland's second largest city after the capital, Belfast. The city stands on the River Foyle 100 km (60 miles) north-west of the capital. The city's original name, Derry – from the Irish *Doire*, 'a place of oaks' – is derived from the wood where a monastery was founded by St Columba in AD 546. It is now no more than a meagre ruin.

Derry was granted its first charter in 1604 and in 1613 was given to the City of London as a plantation – an area to be colonised. Funds for the plantation were raised by the London Livery companies – as a result the name was changed to Londonderry. The massive walls, which date from the period of the plantation, were also funded by the London companies. The Church of Ireland cathedral was completed in 1633. The Roman Catholic cathedral stands on a hilly site opposite it.

The city was besieged in 1641 and 1649 by Royalist forces during the English Civil War, and again in 1688-9, when it held out for 105 days against an army supporting the deposed Catholic king, James II. At one stage Governor Lundy was prepared to let the king's troops into the city, but the gates were closed by 13 apprentices and Londonderry stood firm.

The county of Londonderry covers 2076 km² (801 sq miles) in north-west Northern Ireland. Much of it is given over to farming, especially dairying and cereal crops. Traces of what may be the oldest dwellings in the British Isles have been found in the lower valley of the Bann river in the east of the county.

Population (city) 62 000; (county) 84 000
Map United Kingdom Bc

Londrina ***Brazil*** Modern city about 400 km (250 miles) west of São Paulo. Its prosperity is based on light industries whose products include soft drinks, coffee and vegetable oils. A British company was responsible for much of its early development – hence the name.

Population 302 000
Map Brazil Cd

Long Island ***USA*** Island forming part of New York state covering 3685 km² (1423 sq miles) and extending 190 km (118 miles) east and slightly north from New York city. The city boroughs of Queens and Brooklyn lie at the western end. The rest is given over largely to commuter communities, farms and summer homes, although industry has grown rapidly since the Second World War. The southern coast has long sandy beaches, and there is good sport and commercial fishing.

Map United States Lb

Long Island Sound Busy inlet of the ATLANTIC OCEAN between CONNECTICUT and LONG ISLAND, New York. It is linked to New York Bay by the East river.

The Dutch merchant Adriaen Block explored the north coast in 1614. He sailed up the Connecticut river which flows into the sound and was impressed by the valley's beauty and potential. His enthusiastic report encouraged the establishment of Dutch settlements in the region. Today the sound is lined with residential areas, resorts and yachting centres.

Dimensions 145 km (90 miles) long, 5-32 km (3-20 miles) wide

Longford (Longfort) ***Ireland*** Market town 109 km (68 miles) north-west of Dublin. It is the county town of a cattle-raising county of the same name covering 1044 km² (403 sq miles). The writer Oliver Goldsmith (about 1728-74) was born in the county, at Pallas, about 20 km (12 miles) south of Longford, and his poem *The Deserted Village* is set in the area.

Population (county) 31 100; (city) 4330
Map Ireland Cb

longitude Angular distance of a point east or west of the GREENWICH MERIDIAN. It is the angle at the centre of the earth, as measured at the Equator, between the Greenwich meridian and the meridian through the point.

longshore current, littoral current Current that flows close to and parallel with the shore, produced by the waves moving into the coast at an angle.

longshore drift Movement of material along a beach by the action of waves breaking at an angle to the shoreline. The rush of water up the beach (the swash) pushes sand and gravel obliquely up the beach, then the backwash pulls it straight down, producing a sideways drift of the beach deposits.

Lop Buri ***Thailand*** Town about 130 km (80 miles) north of the capital, Bangkok, whose history dates back to the 8th century when it was ruled by the Khmer people of Cambodia. Today the town has a large Thai military base. The chief crops of the surrounding province – also called Lop Buri – are maize and cotton.

Population (town) 36 600; (province) 680 000
Map Thailand Bc

Lop Nur ***China*** Remote salt lake covering 2000 km² (770 sq miles) of the TARIM PENDI. It lies in the empty desert country of central Xinjiang Uygur Autonomous Region and has been used by China as a testing ground for nuclear weapons.

Map China Dc

lopolith Saucer-shaped body of intrusive igneous rock lying between rock strata.

Lord Howe Island ***Australia*** Forested island, covering 16 km² (6 sq miles) in the Pacific Ocean, about 800 km (500 miles) north-east of Sydney. It is administered as part of New South Wales and is a popular tourist resort.

Population 300
Map Pacific Ocean Cd

Lorelei ***West Germany*** See RHINE

Loreto ***Italy*** The country's most celebrated pilgrimage town, about 20 km (12 miles) south-east of Ancona. Each year thousands of people visit a house there in which Jesus is said to have grown up. It was supposedly brought from Nazareth in Israel by angels in 1294. The house lies inside the Santuario della Santa Casa – the Sanctuary of the Holy House.

Population 10 600
Map Italy Dc

Loreto ***Peru*** Northernmost department bordered by Ecuador, Colombia and Brazil. It is a jungle region, crossed by many tributaries of the Amazon and separated from coastal departments by the Andes. It produces rubber, timber and palm oil, used for making soap. There are also deposits of gold, iron and oil. Ocean-going vessels travel up the Amazon from the Atlantic 3680 km (2285 miles) away to the department commercial centre and capital, Iquitos.

Population 445 400
Map Peru Ba

Lorraine ***France*** Region in the north-east of the country, alongside the West German border. With ALSACE, most of Lorraine was annexed by Germany in the Franco-Prussian War (1870-1) but restored to France in 1919, after the end of the First World War. The region, which covers 23 547 km² (9092 sq miles), includes several First World War battlefields, among them VERDUN. NANCY and METZ are the main cities. The double-barred Cross of Lorraine was an important symbol of the Free French forces during the Second World War.

Population 2 320 000
Map France Fb

Los Angeles ***USA*** City in southern California, about 200 km (125 miles) from the Mexican border. It is the centre of one of the largest urban areas in the world, which sprawls over much of southern California and contains more than 12 million people. Even the city and metropolitan area of Los Angeles proper have population figures second in the USA only to those of New York. However, it is almost entirely a city of suburbs, with no real centre.

The original settlement, founded in 1781, was near the present downtown business district by the Los Angeles river, some 24 km (15 miles) from the Pacific Ocean. As the city has prospered, it has grown beyond the surrounding hills, although it is still contained by the San Gabriel, San Bernardino and Santa Ana mountains, which help to trap the bane of the city – smog. Rich oil fields, film making centred on the suburb of Hollywood, the aircraft industry and tourism are the main livelihoods. It is a major port and has several universities.

Robert Kennedy, brother of murdered President John F. Kennedy, was assassinated in the city in 1968 while campaigning for the presidency.

Population (city) 3 096 700; (metropolitan area) 7 901 200; (conurbation) 12 372 600
Map United States Cd

Los Baños ***Philippines*** See BAY, LAGUNA DE

Los Lagos ***Chile*** Region constituting the southern limits of the unbroken settled area of central Chile, which stretches from Santiago. Some of the most beautiful of the country's lakes lie inland in OSORNO province, and there are ski resorts on the slopes of the Andes. The economy is based on tourism. The regional capital, PUERTO MONTT, is a fishing port.

Population 844 000
Map Chile Ad

Lot ***France*** River, 481 km (299 miles) long, of the south-west. It rises in the southern Massif Central and flows westward to join the Garonne 30 km (19 miles) west of Agen.

Map France Dd

Lothian ***United Kingdom*** Scottish region based on the capital, EDINBURGH, and covering 1756 km^2 (678 sq miles). The name is taken from an ancient kingdom also known in the Middle Ages as Lyonesse. In west Lothian, there are a number of small industrial towns struggling for new employment following the decline of coal mining, including Livingston and the vehicle-making town of Bathgate.

Population 745 000

Map United Kingdom Dc

Lötschberg ***Switzerland*** Mountain massif about 60 km (37 miles) south and slightly east of the capital, Berne. It was virtually impassable until 1913 when a railway between the towns of Thun and Brig was built through the Lötschberg Tunnel, 14.6 km (9 miles) long.

Map Switzerland Aa

lough In Ireland, a lake or a long narrow arm of the sea; equivalent to the Scottish loch.

Louisbourg ***Canada*** Small port on the east coast of CAPE BRETON ISLAND with the remains of an impressive fortress built by the French in 1720-45 and destroyed by the British in 1758. The site is a national historic park.

Population 1600

Map Canada Jd

Louisiana ***USA*** Fertile southern state covering 125 675 km^2 (48 523 sq miles) mainly between the Mississippi river, the Gulf of Mexico and the state of Texas. Most of the state is lowland, and the Mississippi, flanked by levees (banks) to reduce flood risks, is above the level of the land on either side. The climate is mostly subtropical, and rice, sugar cane, cotton and soya beans are grown. Other products include sulphur, oil and natural gas, chemicals and foodstuffs.

The name Louisiana – after the French King Louis XIV – originally applied to the entire Mississippi river basin, which was first settled by the French in 1699. After losing the territory to England and Spain, then regaining the Spanish (western) part in 1800, the French sold the remainder to the USA in 1803 for US$15 million (the Louisiana Purchase). In 1812 the southern part of this territory became the state of Louisiana, but many French and Spanish influences can be seen to this day. NEW ORLEANS is the main city and BATON ROUGE the state capital.

Population 4 481 000

Map United States Hd

Louisville ***USA*** City in Kentucky, on the Ohio river where it forms the northern border of the state. It is an agricultural marketing centre and manufactures tobacco, whiskey and domestic appliances. It is also the home of the Kentucky Derby horse race. About 40 km (25 miles) south-west of the city is Fort Knox, the country's gold-bullion vault.

Population (city) 289 800; (metropolitan area) 962 600

Map United States Ic

Lourdes ***France*** Centre of pilgrimage on the fringes of the Pyrenees about 130 km (80 miles) south-west of Toulouse. The shrine was founded after a peasant girl named Bernadette Soubirous said she had seen the Virgin Mary there 18 times between February and July 1858. More than 4 million pilgrims, some of them seriously ill or handicapped, now visit the shrine each year. Since 1858, more than 300 apparently miraculous cures have been reported. There is a vast underground basilica and several hospitals.

Population 17 600

Map France Ce

Lourenço Marques ***Mozambique*** See MAPUTO

Louth (Lú) ***Ireland*** Smallest county in Ireland, covering 823 km^2 (318 sq miles), about 40 km (25 miles) north of Dublin. Carlingford Lough lies to the north beyond the Cooley Mountains, which rise to 590 m (1963 ft). The county produces cereals, textiles and footwear. The county town is Dundalk.

Population 88 500

Map Ireland Cb

Louvain (Leuven; Löwen) ***Belgium*** City on the Dijle river, 24 km (15 miles) east of Brussels, and the seat of the country's best-known university (founded in 1425). Despite serious damage to the city in the wars of 1914 (when the university library was burnt down) and 1940, Louvain retains it medieval, perfectly circular form, with its walls replaced by boulevards, its marketplaces and churches. Because, however, it lies north of the language line (see BELGIUM), Louvain has become Leuven and the university wholly Flemish-speaking. A new, French-speaking university, Louvain-la-Neuve, has been set up 25 km (16 miles) south, near Wavre. Louvain's main industries are brewing, flour milling and light engineering.

Population 113 000

Map Belgium Ba

low See DEPRESSION

Löwen ***Belgium*** See LOUVAIN

Lower Austria (Nieder-Österreich) ***Austria*** The country's largest province, known as 'The Cradle of the Nation'. It produces timber, livestock, grain and wine. Lower Austria covers 19 172 km^2 (7402 sq miles) in the hilly north-east. It contains Vienna, which although administered as a separate province is both the national capital of Austria and capital of Lower Austria province.

Population 439 500

Map Austria Da

Lower Egypt ***Egypt*** That part of Egypt north of Cairo including the Nile delta and the Mediterranean coast. Upper Egypt is the southern part, covering the whole region south of the Nile delta as far as the Sudanese border. Union of the two, in about 3100 BC, was followed by the 30 dynasties of pharaohs that ruled Egypt until Alexander the Great's conquest in 332 BC.

Map Egypt Bb

Lower Saxony (Niedersachsen) ***West Germany*** State stretching from the North Sea coast southwards to the borders of Hessen and North Rhine-Westphalia. Except for the south-east, it is part of the North European Plain, a mosaic of fertile farmland, sandy heaths and marshy fenland and peat bog. Some of West Germany's best agricultural land – producing wheat, rye, potatoes and fodder crops – lies between the cities of BRUNSWICK and MINDEN. HANOVER is the state capital.

Population 7 216 000

Map West Germany Cb

Lowicz ***Poland*** Market town 76 km (47 miles) south-west of the capital, Warsaw. On Sundays and for special festivals, many townswomen wear traditional regional costumes of gaily striped aprons and tall headdresses. Nieborow, a magnificent Baroque palace (1695-7), stands in fine formal gardens 10 km (6 miles) to the south-east. Owned by the powerful and aristocratic Radziwill family from 1774 to 1945, it is now part of the National Museum.

Population 26 900

Map Poland Cb

Lowlands, The ***United Kingdom*** See HIGHLANDS, THE

lowveld Also known as bushveld, lowveld is found in parts of subtropical southern Africa between about 300 and 900 m (990 and 2950 ft) above sea level. It includes regions of dry wooded savannah and scrub, as found in South Africa's Kruger National Park.

Loyalty Islands ***New Caledonia*** Major outlying islands of the territory, consisting of three main islands about 100 km (60 miles) east of New Caledonia itself. The people are mostly Melanesians, with some Polynesians on Ouvéa. The principal crop is coconuts.

Population 15 500

Map Pacific Ocean Dd

Lú ***Ireland*** See LOUTH

Lu Shan ***China*** Mountain range south of the Chang Jiang river in Jiangxi province; the highest peak reaches 1474 m (4836 ft). It is a popular summer holiday destination. The Lu Shan botanical garden, on the Peak of the Nine Strange Things, is one of the finest in China.

Map China He

Lualaba ***Zaire*** River rising near the Zambian border, about 120 km (75 miles) west of the Lubumbashi, and flowing 1800 km (1120 miles) northwards. It is navigable between Bukama and Kongolo, Kasongo and Kibombo, and Kindu and Ubundu. Below Ubundu are the Boyoma Falls, where the Lualaba becomes the Zaire. The Scottish missionary and explorer Dr David Livingstone (1813-73) thought that the Lualaba was a source of the Nile. But in 1876-7 the British adventurer, Sir Henry Morton Stanley, followed its course and proved that it was a headstream of the Zaire.

Map Zaire Bb

Luanda (Loanda) ***Angola*** National capital and port founded by Portuguese settlers in 1576. The castle of St Michael (built in 1638) is still a landmark. Today, Luanda is a busy port with food processing, paper, woodworking and textile plants, and an oil refinery. The neighbouring province, also called Luanda, produces coffee, cotton, palm products and sisal.

Population (city) 700 000; (province) 561 000

Map Angola Aa

▲ FORTRESS CITY The fortifications of the old city of Luxembourg still stand above the steep walls of the Alzette gorge, and are now topped by a spectacular 'corniche' road.

Luang Prabang ***Laos*** Province and town on the Mekong river in northern Laos, about 220 km (140 miles) north of the capital, Vientiane. The town was the nation's royal capital during the French colonial period and until 1975, and the heart of Laotian Buddhism. Prior to that it had also been the seat of the principality of Luang Prabang. But its importance has declined since the Communist takeover in 1975 because the monarchy has been abolished and Buddhism discouraged.

Population (province) 450 000; (town) 45 000

Map Laos Ab

Luapula ***Zambia*** North-eastern province bordering Zaire and covering 50 567 km² (19 524 sq miles). Its capital is the town of Mansa, about 450 km (280 miles) north and slightly east of the national capital, Lusaka. The province is named after the Luapula river, which marks the Zaire border. The main occupations of the thinly populated area are arable and livestock farming and fishing.

Population 413 000

Map Zambia Bb

Lubango ***Angola*** Market town about 680 km (420 miles) south of the national capital, Luanda. The surrounding province of Huíla produces coffee and cattle – but farming has been disrupted in the 1980s by raiding parties of rebel UNITA guerrillas and by South African troops pursuing SWAPO guerrillas across the Namibian border.

Population 31 700

Map Angola Ab

Lübeck ***West Germany*** Port on the Trave river 20 km (12 miles) from its mouth on the Baltic Sea, and 60 km (38 miles) north-east of Hamburg. Lübeck was founded in 1157 by Henry the Lion, Duke of Saxony, and rapidly became a thriving Baltic port, trading in timber. It was one of the leaders of the Hanseatic trading league in the 13th century. Today its shipping trade is mostly to and from Scandinavia, and its industries include shipbuilding, engineering, textiles, chemicals and food processing.

Willy Brandt, the West German Chancellor between 1969 and 1974, was born in Lübeck; so was the writer Thomas Mann (1875-1955), author of *Death in Venice* (1913).

Population 80 000

Map West Germany Db

Lublin (Lyublin) ***Poland*** Eastern Poland's largest city, 153 km (95 miles) south-east of the capital, Warsaw. It is the marketplace for a fertile farming region producing meat, wheat, flour, sugar, beer, tobacco products and vegetable oils. The city's factories also make agricultural equipment and other machinery.

Lublin was the seat of two independent Polish governments; the first was set up on November 7, 1918, as a temporary socialist government; and the second, which developed into the country's present-day government, was established on July 22, 1944, as the German army retreated. It has five academic institutions, including the Catholic University – the only one of its kind in Poland – and the Marie Curie-Sklodowska University, named after the Polish-born scientist (1867-1934) who discovered radium.

The Museum of Martyrology and a monument in the south-eastern suburb of Majdanek are grim reminders of Nazi terror. They stand on the site of an extermination camp where 370 000 Poles, Russians, Jews and people of 17 other nationalities were murdered between 1941 and 1944.

Population 320 000

Map Poland Ec

Lubombo Mountains ***Southern Africa*** See LEBOMBO MOUNTAINS

Lubumbashi ***Zaire*** Capital of Shaba region in the south-east, Zaire's chief mining region. Founded in 1910 as Elisabethville, it lies only 30 km (20 miles) from the Zambian border, and is an industrial city, with copper and cobalt refining, and processing and consumer industries. Four rail lines connect Lubumbashi to the Atlantic and Indian oceans.

Population 600 000

Map Zaire Bc

Lucca ***Italy*** Beautiful walled Tuscan city and market centre about 60 km (38 miles) west of Florence. Its early wealth came from silk and banking, and is expressed in the city's many 12th-century Romanesque churches. The narrow streets are rich in medieval palaces, towers and attractive tenements. Surrounding the old town are 16th-century fortifications, with avenues of trees facing a grassy moat. The operatic composer Giacomo Puccini (1858-1924) is Lucca's most renowned native son. The surrounding region is noted for its olive oil.

Population 89 100

Map Italy Cc

Lucerne, Lake (Vierwaldstätter See; Lac des Quatre-Cantons) ***Switzerland*** Alpine lake in central Switzerland. The German and French versions of the name mean 'The Lake of the Four (Forest) Cantons'. It was on the shores of the lake, in 1291, that officials of the four original Swiss cantons – Nidwalden, Obwalden, Schwyz and Uri – signed a pact of unity against the country's Austrian rulers. The pact set up the 'Everlasting League', which was the embryo of the modern state of Switzerland.

Surrounded by mountains and fringed with woodland, the lake is one of the country's most popular and beautiful tourist spots. It is often known simply as Lake Lucerne, after the city of LUCERNE on its north-west tip. It covers an area of 113 km² (44 sq miles).

Map Switzerland Ba

Lucerne (Luzern) ***Switzerland*** Lakeside city about 65 km (40 miles) east and slightly north of the capital, Berne. Large sections of the town's medieval walls and watch towers are still standing, as are the 14th-century covered wooden footbridges over the Reuss river, which flows through the town from the lake.

Lucerne also has a Renaissance town hall, a 17th-century cathedral and the giant Lion of Lucerne (1820-1). The dying lion, 9 m long by 20 m high (30 ft by 65 ft), is carved on a wall of rock as a monument to the Swiss Guards of the French King Louis XVI, more than 700 of whom were massacred defending the Tuileries Palace in Paris in 1792 during the French Revolution.

The wheat-growing canton of Lucerne stretches across 1492 km² (576 sq miles) to the north, south and west of the city.

Population (city) 67 500; (canton) 303 900

Map Switzerland Ba

Lucknow ***India*** Capital of the northern state of Uttar Pradesh, in the Ganges river valley about 435 km (270 miles) east and slightly south of Delhi. Its old buildings include the Pearl

Palace of the rulers of the former Mogul province of Oudh, and the Residency, where in 1857 (during the Indian Mutiny) a British force held out from July until finally relieved in mid-November. The city is a centre for Indian handicrafts, where *bidri* – the ancient art of silver inlay on gunmetal – is being revived.
Population 1 007 600
Map India Cb

Lüda (Dalian) ***China*** City and port in Liaoning province. It stands on the Liaodong peninsula and is a centre for heavy engineering.
Population 4 000 000
Map China Ic

Lüderitz ***Namibia*** Lobster-fishing port on the arid southern coast about 230 km (145 miles) from the South African border. It was named after Adolf Lüderitz, a Hamburg merchant who bought the site and persuaded Germany to take it under Imperial protection in 1884. This was Germany's first step in colonising South-West Africa. The port also serves the nearby diamond-mining zone, to which access is heavily restricted.
Population 16 700
Map Namibia Ab

Ludhiana ***India*** Agricultural trading town in the north-western state of Punjab, about 290 km (180 miles) north and slightly west of Delhi. Its agricultural university is recognised throughout the world for its research into wheat cultivation.
Population 607 000
Map India Bb

Ludlow ***United Kingdom*** English town in the Midland county of Shropshire, 3 km (2 miles) from the Welsh border. It is rich in Tudor and Georgian buildings and has a fine 11th to 14th-century castle. Many of its houses were built between the 16th and 18th centuries by wealthy landowners and merchants.
Population 7500
Map United Kingdom Dd

Ludwigshafen ***West Germany*** Town and port on the Rhine opposite Mannheim, about 80 km (50 miles) south of Frankfurt am Main. It grew around a chemical works and also produces glass and steel.
Population 163 000
Map West Germany Cd

Lugano (Lauis) ***Switzerland*** Lakeside tourist resort near the Italian border in the extreme south of the country. It is overlooked by the hills of Monte San Salvatore and Monte Brè, both more than 900 m (nearly 3000 ft) high.
Population 20 000
Map Switzerland Ba

Lugansk ***USSR*** See VOROSHILOVGRAD

Lugo ***Spain*** Town in the north-west – about 80 km (50 miles) south-east of Corunna – with Roman walls and towers. It trades in cattle and has tanning, flour-milling and textile industries. The surrounding province of Lugo supplies fish and timber.
Population (town) 74 000; (province) 399 200
Map Spain Ba

Lugou Bridge (Marco Polo Bridge) ***China*** Bridge 15 km (9 miles) south-west of Beijing (Peking). It was the site of the outbreak of war between China and Japan in 1938, and is said to be the modern version of a bridge crossed by Marco Polo during his travels in China in the 13th century.
Map China Hc

Luik ***Belgium*** See LIEGE

Luimneach ***Ireland*** See LIMERICK

Luleå ***Sweden*** Seaport at the head of the Gulf of Bothnia, some 946 km (588 miles) north-east of Stockholm. Its harbour is icebound for most of the winter, but when it is open it exports about one-third of Lappland's output of iron ore – as well as wood pulp and timber. It has a large iron and steel mill, and there is a notable collection of Lapp handicrafts and costumes in the Norrbotten Museum.
Population 66 600
Map Sweden Db

Luluabourg ***Zaire*** See KANANGA

Lumbini ***Nepal*** Grove in the modern village of Rummindei, formerly Kapilavastu, which was the birthplace of Prince Siddhartha Gautama (about 563-488 BC), the founder of Buddhism, who came to be known as the Buddha – a Sanskrit title meaning the 'Enlightened One'. His birthplace – near the Indian border about 200 km (125 miles) west and slightly south of Kathmandu – is marked by a stone pillar erected by the Buddhist Indian emperor Ashoka in about 250 BC and is now a shrine.
Map Nepal Aa

Lund ***Sweden*** City in the extreme south of Sweden, 16 km (10 miles) north-east of Malmo. It was founded in the early 11th century by the Danish King Canute, who also ruled England between 1016 and 1035. It has a fine Romanesque cathedral, and its university, established in 1668, is the oldest in the country after Uppsala.
Population 82 000
Map Sweden Be

Lüneburg ***West Germany*** Canal port 40 km (25 miles) south-east of the city of Hamburg.

On Lüneburg Heath, a sandy, partly forested area south of the town, the British Field-Marshal Bernard Montgomery received the surrender of German troops in May 1945, ending the Second World War in Europe. Most of the heath is a nature reserve; part is used as a NATO military training area.
Population 60 600
Map West Germany Db

Luoxiao Shan ***China*** Range of mountains in Jianxi province. It was used as a refuge by Communist rebels under the late Mao Zedong from 1927 until they broke out from the Nationalist general Chang Kai-Shek's encirclement in 1934. That was the first step of the Long March, when the rebels fought their way 9500 km (6000 miles) to safety in Yan'an.
Map China Ge

Luoyang ***China*** Ancient city in Henan province dating from 2100 BC. Luoyang, with Anyang and Zhengzhou, was one of the main centres of Shang civilisation, the earliest in China. From 770 BC to 221 BC the city was capital of the Eastern Zhou kingdom – one of the successor states to the Shang. Many important archaeological sites have been opened up in the area. It is now a centre for the manufacture of bearings, tractors and mining machinery.
Population 500 000
Map China Gd

Lurgan ***United Kingdom*** Market town in Northern Ireland, about 30 km (20 miles) south-west of the capital, Belfast. It has been a centre of the linen industry since damask weaving began in 1691. The small lanes leading from the main street contained weaving shops in the 18th century. Lurgan merged with Portadown in 1973 to form the new city of Craigavon.

The small community of Waringstown, about 5 km (3 miles) south-east of Lurgan, is named after Samuel Waring – who, using methods similar to those he had seen in Holland and Belgium, introduced the weaving of damask (reversible material with a pattern made by different weaves) to the district.
Population 21 000
Map United Kingdom Bc

Lusaka ***Zambia*** The country's capital and centre of the north-south railway and east-west road network. Founded in 1905 to serve the lead mine at Kabwe, it became the capital 30 years later. The city's main buildings include the National Assembly, the president's State House and an Anglican cathedral. Lusaka is at the centre of a fertile agricultural region rich in cotton, maize, tobacco and dairy products. It also has car and textile plants.
Population 538 500
Map Zambia Bb

Lüshun (Port Arthur) ***China*** Naval base at the tip of the Liaodong Peninsula. It was Chinese until 1898 and then passed to Russia on a 25-year lease cut short by Japan's victory in the Russo-Japanese War of 1904-5. The lease was transferred to Japan, which held on to the base after expiry of the lease. It was occupied by the Soviet Union from 1945 to 1955, when it was restored to China.
Map China Lc

Luton ***United Kingdom*** Industrial town in the English county of Bedfordshire, 50 km (30 miles) north of London. It is the site of one of London's airports. The principal industry is the manufacture of cars and trucks.
Population 165 000
Map United Kingdom Ee

Luxembourg See p. 400

Luxembourg ***Belgium*** South-eastern province adjoining the independent Grand Duchy of the same name. It is mostly covered by the forested upland of the Ardennes.
Population 222 000
Map Belgium Cb

Luxembourg ***Luxembourg*** Capital of the Grand Duchy, consisting of the old city, on a cliff above the Alzette river, and the newer city

Luxembourg

EUROPE'S LAST INDEPENDENT DUCHY IS A SMALL STATE THAT PLAYS A BIG ROLE IN INTERNATIONAL AFFAIRS

To be a delegate from the Grand Duchy of Luxembourg must be one of life's loneliest experiences. So small is this state, yet so active has it been in West European affairs, that it is always represented on committees, but seldom rates more than a single delegate.

It became a Grand Duchy in 1354, and is the only one of hundreds of independent duchies that once made up west-central Europe that has survived to become a member of today's United Nations. With its hereditary grand duke and its parliamentary government, it was a founder member of the North Atlantic Treaty Organisation (1949) and of the European Economic Community (1957). It is the seat of the European Court of Justice, the Secretariat of the European Parliament, and is a major international banking centre, with 114 different banks. In 1944, together with Belgium and the Netherlands, it formed the Benelux economic union – which, in 1960, became the world's first totally free international goods and labour market.

The northern part of Luxembourg is hilly, picturesque country. It includes the southern edge of the ARDENNES and EIFEL uplands, and is drained by the Sûre (Sauer) river. The southern part is physically part of French LORRAINE: it is lower and more fertile, and beds of iron ore form the basis of a substantial steel industry. On the east, Luxembourg is bordered by the MOSELLE river and the lower Sûre in whose valleys wines are produced.

Luxembourgers speak both French and German but, for preference, they use their own Germanic dialect. Their heavy industry and international institutions have attracted many foreign workers, who make up almost a quarter of the population.

In May 1940, during the Second World War, Nazi Germany overran Luxembourg and made it part of the Third Reich. About 10 000 of its inhabitants were forced to fight for Hitler, while others joined the Resistance against the occupying Germans. In 1986 West Germany agreed to pay Luxembourg DM12 million (about US$5.2 million) by way of reparation.

Europe's first commercial radio station, Radio Luxembourg, went on the air in 1934 and is still broadcasting today. Its studios stand in some scenic gardens near the Old Town in the capital, Luxembourg.

LUXEMBOURG AT A GLANCE

Map Belgium Bb
Area 2586 km² (998 sq miles)
Population 367 400
Capital Luxembourg
Government Constitutional monarchy
Currency Luxembourg and Belgian francs
Languages Letzeburghish; French, German and English widely spoken
Religions Christian (93% Roman Catholic, 1% Protestant)
Climate Temperate; average temperature in the capital ranges from −1 to 2°C (30-36°F) in January to 13-23°C (55-73°F) in July
Main primary products Cereals, potatoes, grapes, dairy cattle; iron ore
Major industries Iron and steel, chemicals, banking, food processing, metal products and engineering, agriculture, wine making
Main exports Steel and other metals, chemicals, international financial services
Annual income per head (US$) 11 800
Population growth (per thous/yr) 1
Life expectancy (yrs) Male 71 **Female** 76

to the east and south. The city's two halves are linked by viaducts. The old city, dating from the 10th century, centred round the remains of the old fortress whose foundations are Roman and Frankish, contains the ducal palace (1572), the town hall where the parliament sits, and the gothic cathedral of Notre Dame (1613-23).

The new city, founded in the 19th century, and its north-eastern suburb of Kirchberg house several European Community institutions such as the Court of Justice, the Investment Bank and the Coal and Steel Community. There are also numerous conference and exhibition centres and Europe's oldest commercial radio station, Radio Luxembourg, which broadcasts to Britain, France, the Netherlands and West Germany.
Population 78 000
Map Belgium Cb

Luxor ***Egypt*** Site of temple monuments and cenotaphs to the pharaohs of the New Kingdom, on the south side of their ancient capital, Thebes. The site is on the east bank of the Nile, about 500 km (310 miles) south of the present capital, Cairo. The temples, dating mostly from about 1400 BC, are linked to the temples at Karnak, 3 km (2 miles) to the north, by an avenue of sphinxes, each with a ram's head. The name Luxor is derived from the Arabic word for palaces. There is a museum at the site, and hotels for tourists.
Map Egypt Cc

Luzern ***Switzerland*** See LUCERNE

Luzon ***Philippines*** The largest island of the Philippines, some 800 km (500 miles) long and covering 104 688 km² (40 420 sq miles). It accounts for about one-third of the country's land area and more than half of its people. It is the Philippines' most northerly major island, and is very irregular, with several natural harbours. It extends into a narrow peninsula, Bicol, in the south-east. It is geographically diverse, with high mountain ranges (notably the Cordillera Central, the Sierra Madre and the Zambales) and volcanic peaks, rich rice-growing lowlands and barren uplands where people can barely scrape a living. Luzon also contains the bulk of the country's industry, centred on the vast metropolitan area of the capital, MANILA, and has important mineral and timber resources. The official language of the Philippines, Pilipino, is based on Tagalog, spoken by people of the Manila region, but several other languages are spoken in more isolated areas.
Population 29 400 000
Map Philippines Bb

L'vov (Lwow; Lwiw; Lemberg) ***USSR*** Industrial city in the western Ukraine about 470 km (290 miles) west of Kiev. A Ukrainian prince founded L'vov in 1256, but it was taken by the Poles in 1340. The city was part of Austria from 1772 until 1918, then again returned to Poland until 1939. L'vov reflects all this today, with a mixture of Ukrainian, Polish and Austrian building styles side by side with modern architecture. Among its oldest buildings are the remains of a 13th-century fort, a Roman Catholic cathedral (dating from the 14th to the 17th centuries), an Armenian cathedral (1370-1493) and some old Russian Orthodox churches. Its main activities are food processing, manufacturing chemicals and oil refining.
Population 688 000
Map USSR Dd

Lyallpur ***Pakistan*** See FAISALABAD

Lydda ***Israel*** See LOD

Lyons (Lyon) ***France*** The country's second largest city after Paris, at the junction of the Rhône and Saône rivers. It is a major trade centre with the oldest bourse (stock exchange) in France, dating from 1506; it is also regarded as the gastronomic centre of France. Its main industries are metallurgy, vehicle manufacture, chemicals and textiles. The city's historic buildings include the cathedral, built between the 12th and the 15th centuries, the Basilica of Notre Dame de Fourvière, opened in 1894, and the Palais Saint Pierre (1659-85).
Population 1 236 100
Map France Fd

Lyublin ***Poland*** See LUBLIN

Lyublyana ***Yugoslavia*** See LJUBLJANA

Ma'an ***Jordan*** Market town and road and rail junction, situated 115 km (71 miles) south of Karak. It is the administrative centre for southern Jordan and a commercial centre trading mainly in grain. Its main industry is tobacco. Formerly a staging post for pilgrims to Mecca, most of its old town is built of clay; an irrigation system allows gardens to be cultivated. There are the remains of an 18th-century Turkish fort and a rest-house. It also has an airport.
Population 12 000
Map Jordan Ab

Maas ***Western Europe*** See MEUSE

Maastricht ***Netherlands*** Provincial capital of Limburg, on the Maas (Meuse) river about 170 km (105 miles) south and slightly east of the national capital, Amsterdam. Lying in a narrow strip of the Netherlands between Belgium and West Germany, it is a multilingual town. On the main square is the country's oldest church,

St Servatius, founded in the 6th century. The city's products include ceramics, cement, paper and glassware.
Population (city) 112 600; (Greater Maastricht) 156 500
Map Netherlands Bb

Macau See p. 402

Macdonnell Ranges ***Australia*** Central Australia's major mountain system. They form parallel ridges in the Northern Territory, running east and west from Alice Springs. The highest point is Mount Ziel (1510 m, 4954 ft). Gold has been found at Arltunga to the east.
Map Australia Ec

Macedonia ***Greece*** Largest region of Greece, bounded by Albania, Yugoslavia and Bulgaria in the north, and covering an area of 34 177 km² (13 192 sq miles). It is a geographically varied region of forested mountains including PINDHOS, OLYMPUS and Vérmion; many lakes; broad fertile plains; and three long rivers, the Aliákmon, Axios and Strimón.

The main cities are SALONIKI, KAVALLA, Sérrai and Drama. Wildlife is abundant and includes brown bears, boars, jackals, wildcats, wild pigs, roe deer, wild goats, wolves, storks, vultures, hawks, golden eagles, herons, pelicans and flamingos. Macedonia is also rich in minerals, producing copper, gold, iron, lead, magnesite, silver and chrome, and is the country's major power-producing and oil-refining region. Agriculture in Macedonia is the most varied and productive in Greece. Maize, beans, vegetables, rice, fruit, tobacco, timber and livestock are raised.

Until the 4th century BC, Macedonia was on the fringe of Hellenic culture, but under Philip II it became the dominant and unifying power of Greece, and then the springboard from which his son, Alexander the Great (336-323 BC), built his world empire. It was later held by Romans, Goths, Huns, Slavs and Crusaders, and was under Turkish rule from 1371 until the Balkan Wars (1912-13). It was then split between Greece and Serbia (now in Yugoslavia), a small area going to Bulgaria. The Greek region is now a southern bastion of NATO.
Population 2 122 000
Map Greece Ba

Macedonia (Makedonija) ***Yugoslavia*** Southernmost Yugoslav republic bordering Greece and covering an area of 25 713 km² (9928 sq miles). It consists mainly of plateaus at 650 to 950 m (about 2130 to 3120 ft); it is bordered and broken by high mountains, including the SAR PLANINA and Galicica, and by lake basins. SKOPJE is the capital.

Macedonia was a backward region of Yugoslavia between the World Wars, and its development dates only from 1945. Since then irrigation and farm mechanisation have increased the republic's output of fruit, vegetables, sugar beet, grapes, cotton, rice and tobacco. Industries, greatly diversified since the opening in the 1950s of the Mavrovo hydroelectric dam, include textiles and the processing of chrome, lead and iron ores.
Population 1 912 200
Map Yugoslavia Ed

Maceió ***Brazil*** Port and capital of Alagoas state, 190 km (119 miles) south of the coastal city of Recife. It is a popular area for tourists, with white sand beaches backed by coconut palms and the lagoons from which the state gets its name (*lagoa* is the Portuguese word for 'lagoon'). Apart from sugar, the area's dominant industry, the port's main products are salt and machinery. The state – one of Brazil's smallest – covers 27 731 km² (10 704 sq miles), making it almost the size of Belgium.
Population 401 000
Map Brazil Eb

Macerata ***Italy*** City in the Marche region about 175 km (110 miles) north-east of Rome. Macerata was founded in the 11th century; its old town is enclosed within 14th-century walls and has several late medieval and Baroque buildings. The university dates from 1290.
Population 43 600
Map Italy Dc

Macgillicuddy's Reeks (Na Cruacha Dubha) ***Ireland*** Highest mountain range in Ireland. It lies in the south-west, about 105 km (65 miles) west of the city of Cork, and rises to 1040 m (3414 ft) at Carrauntoohil.
Map Ireland Bb

Machu Picchu ***Peru*** Ruined Inca city in Cuzco department. It stands at 2280 m (7480 ft) above the Urubamba river on the saddle of a mountain. The complex of palaces, houses, plazas and terraces, linked by stairways, was one of the last strongholds of the Incas before they were destroyed by the Spaniards in the 16th century. The abandoned city lay undiscovered until 1911 when an American archaeologist, Hiram Bingham, found it.
Map Peru Bb

Mackay ***Australia*** Port on the east coast of Queensland, about 800 km (500 miles) north-west of Brisbane. It is at the centre of a sugar-cane area and processes and exports one-third of the country's total crop. The port also handles some of the coal from open-cast mines 200 km (124 miles) inland, shipping it to Japan, and is a base for exploring the Great Barrier Reef. About 100 km (60 miles) north, off Proserpine (population 3000), are the Whitsunday Islands, many of them tourist resorts.
Population 31 600
Map Australia Hc

Mackenzie ***Canada*** River which rises in the Great Slave Lake in the Northwest Territories and flows 1800 km (1120 miles) to enter the Arctic Ocean in a 16-pronged delta. It was first traced to the Arctic by Sir Alexander Mackenzie, the Canadian explorer. The river is navigable to the Great Slave Lake from June to October. Much of the Mackenzie valley is heavily forested. There is some fur trading, but the economy of the area is dominated by large oil and gas fields.
Map Canada Cb

▼ CITY OF 3000 STEPS The ruins of Machu Picchu – the Inca city that the plundering Spanish conquerors never found – cascade down 13 km² (5 sq miles) of terraces built high in the Andes.

Macau

ONCE A SEEDY OUTPOST OF EMPIRE, PORTUGAL'S FORMER CASINO COLONY IS BOOMING

The tiny Portuguese enclave of Macau on the China coast offers some bizarre contradictions. Although Chinese make up 98 per cent of the population, Portuguese cultural influence shows everywhere. Sometimes Chinese and Portuguese traditions blend, with curious results.

Macau consists of a peninsula bordering the Chinese mainland province of GUANGDONG, together with two small islands – Taipa and Coloane – which are joined by a causeway and bridge. There is no airport, and most visitors arrive by jetfoil from HONG KONG, which lies 64 km (40 miles) to the east across the ZHU JIANG (Pearl River) estuary.

The Portuguese first used Macau as a trading post in 1521. They occupied it in 1557, and used it as a base for trading and missionary activities in China. Portuguese sovereignty over the territory, formally recognised by China in 1885, lasted until 1974. Macau was then redefined as a Chinese territory under Portuguese administration. In 1985 negotiations began for the return of Macau to Chinese rule in 1997.

Macau was never able to compete commercially with the dynamism of Hong Kong, and for many years it remained a seedy and neglected backwater, its economy dominated by gambling (still popular with visitors). Since the mid-1970s, Macau has emerged as an industrial and commercial centre. Manufactures include clothes, toys and artificial flowers – all produced cheaply because wages are poor for thousands of refugees from mainland China.

MACAU AT A GLANCE

Map Hong Kong Ab
Area 15.5 km² (6 sq miles)
Population 406 000
Capital Macau City
Government Chinese territory administered by Portugal
Currency Pataca = 100 avos
Languages Cantonese, Portuguese
Religions Buddhist (77%), Christian (5%), Confucian
Climate Subtropical
Main primary products Fruit, vegetables, livestock, fish
Major industries Textiles, light manufacturing, tourism
Main exports Clothing, textile yarns and fabrics, artificial flowers, toys
Annual income per head (US$) 2100
Population growth (per thous/yr) 34
Life expectancy (yrs) Male 68 **Female** 73

Mackenzie Basin ***New Zealand*** Sheep-farming region in the centre of South Island, about 200 km (124 miles) south-west of the city of Christchurch. Overlooked by Mount Cook, it covers an area of 2000 km² (772 sq miles).
Map New Zealand Be

mackerel sky Sky covered with cirrocumulus or altocumulus cloud, resembling the markings on a mackerel. It is usually seen in summer during unsettled weather.

McKinley, Mount ***USA*** See DENALI NATIONAL PARK

McMahon Line ***China/India*** Part of the boundary between Tibet in China and what is now the Indian territory of Arunachal Pradesh, from the Bhutan border north-eastwards to the Brahmaputra river. It was agreed by Tibet (then an independent country) and Great Britain at Simla in 1914, and named after Sir Henry McMahon, the chief British negotiator. China disputed the line then, and after the Communist takeover in 1949, and it was one of the causes of the 1962 war between the two countries.
Map India Eb

McMurdo Oasis ***Antarctica*** See DRY VALLEYS

McMurdo Sound A branch of the ROSS SEA forming a channel between ROSS ISLAND and Victoria Land, Antarctica. It was discovered in 1841 by the British explorer Sir James Clark Ross, who named it after one of his officers. The British explorers Ernest Shackleton (1874-1922) and Robert Falcon Scott (1868-1912) and the Commonwealth Trans-Antarctic Expedition of 1957-8 established their bases alongside the sound. The American McMurdo Base, on Ross Island, is the largest Antarctic base; it is staffed by about 750 people in summer.
Dimensions 148 km (92 miles) long, 37-74 km (23-45 miles) wide
Map Antarctica Jc

Mâcon ***France*** Market centre of the Mâconnais wine area 60 km (37 miles) north of Lyons. It has restored medieval wine cellars, which are open to visitors.
Population 39 900
Map France Fc

Mactan ***Philippines*** See CEBU

Madain Saleh ***Saudi Arabia*** Important archaeological site situated 800 km (497 miles) north of Jeddah. It consists of about 100 tombs – many with ornate façades – hewn out of the face of some sandstone cliffs. The tombs date from about 100 BC to AD 100, when the town of Al-Hejr stood on the site. It was the main stopping-place on the caravan route between the now-ruined city of PETRA, in south-west Jordan, and the Arabian peninsula.
Map Saudi Arabia Ab

Madang ***Papua New Guinea*** Port on the north coast of New Guinea, about 500 km (310 miles) north-west of Port Moresby. The main exports are copra, coconuts, coffee and cocoa. Industries include timber milling, tobacco processing and light engineering. A lighthouse on the coast is a memorial to the 'Coastwatchers' – volunteers who served the Allies during the Second World War by reporting Japanese troop and ship movements from behind enemy lines.
Population 34 250
Map Papua New Guinea Ba

Madara ***Bulgaria*** See SHUMEN

Madeira ***Brazil*** River, one of the Amazon's principal tributaries. The Madeira, fed by tributaries from the Andes, flows 2013 km (1251 miles) to join the Amazon east of the city of Manaus. Near the Bolivian border, prospectors still pan gold from the river.
Map Brazil Bb

Madeira ***Atlantic Ocean*** Group of volcanic islands belonging to Portugal, 900 km (560 miles) south-west of its capital, Lisbon. The islands, probably known to the Genoese from the early 14th century, consist of Madeira (740 km², 286 sq miles), Porto Santo (42 km², 16 sq miles) and the small uninhabited and barren island groups of the Desertas (Dezerte) and the Selvagens. A popular winter resort, the island of Madeira has a mild climate, subtropical vegetation and picturesque scenery. The capital is Funchal and the island rises to 1861 m (6106 ft) at Pico Ruivo. The main crops are grapes for Madeira wine, bananas and sugar cane.
Population (Madeira) 248 500; (Porto Santo) 4400
Map Morocco Aa

Madhya Pradesh ***India*** The largest Indian state, covering 443 446 km² (171 170 sq miles) in the centre of the country. It includes the hills south of the Ganges plains and much of the Narmada river basin. About one-third of the area is still forested. Farming is the mainstay of the economy, producing wheat, rice, maize, sugar cane, oilseeds and cotton. The state has a diverse population, including Bhil people who retain their tribal way of life, but most people speak Hindi. Bhopal is the capital.
Population 52 178 800
Map India Cc

Madina do Boé ***Guinea-Bissau*** Small town about 150 km (95 miles) east of Bissau. In 1980 it was designated to become the country's administrative capital.
Map Guinea Ba

Madison ***USA*** State capital of Wisconsin, standing on the isthmus between Lake Mendota and Lake Monona, about 190 km (120 miles) north-west of Chicago. The city was founded in 1836 and is at the heart of a rich dairy-farming area. It is also the seat of the University of Wisconsin, established in 1836.
Population (city) 170 700; (metropolitan area) 333 000
Map United States Ib

Madras ***India*** Capital of Tamil Nadu state and the country's main east-coast port. Its harbour is entirely artificial, protected by a breakwater begun in the late 19th century, for it lies on the sandy Coromandel Coast.

The city was founded as Fort St George by the British East India Company in 1639, and prospered on the cotton (*continued on p. 404*)

Madagascar

AN AFRICAN ISLAND, ENDOWED WITH ABUNDANT NATURAL RESOURCES, THAT HAS STILL TO DEVELOP ITS FULL POTENTIAL BEFORE POVERTY IS ERADICATED

Although it lies only about 400 km (250 miles) off the East African coast, the island of Madagascar is more closely related in its people, culture and language to the islands of Indonesia some 5000 km (about 3000 miles) eastwards across the Indian Ocean than to Africa. The islanders probably owe the Asian element in their ancestry to migrants who began settling there some 2000 years ago. Madagascar is unusual in that all 18 different tribal groups speak the same language – Malagasy. There are local dialects but they are mutually intelligible.

Madagascar is the world's fourth largest island (after Greenland, New Guinea and Borneo, but excluding Australia). Portuguese explorers discovered it in 1500, and it came under French rule from 1896 until 1960.

The island is distinctive not only in its people but also in its wildlife. Its animals (and plants) evolved in long isolation after the island broke away from the African continent some 150 million years ago.

Despite its abundant mineral resources, Madagascar is one of the world's poorest countries. Most of its 10.2 million people work on the land and more than 80 per cent grow only enough to feed themselves. The staple food is rice. A decline in the rice crop in recent years, caused partly by persistent bad weather and soil erosion, and exacerbated by an expanding population, led to rice rationing in 1980 and the economy's collapse.

A series of high savannah-covered plateaus occupy the centre of the island, and to the east the forested mountains fall steeply to the hot, humid coast. To the west and south-west the land falls more gradually through drier grassland and scrub.

▲ HIGH CAPITAL Antananarivo lies on a ridge 1249 m (4094 ft) above sea level, and much of the city dates from the French occupation of the island from 1896 to 1960.

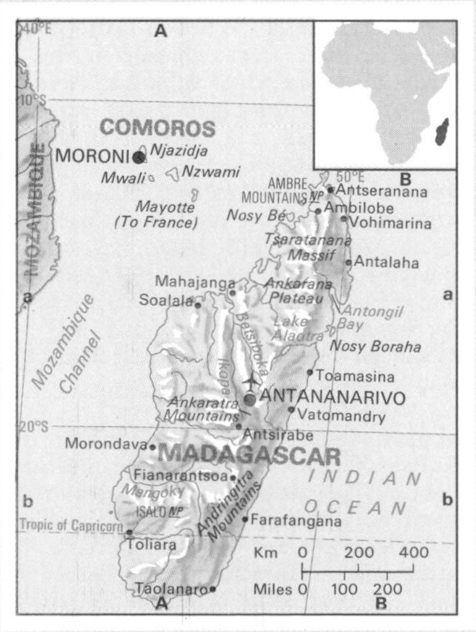

Cassava, a starchy root, is almost as widely grown as rice, but only about 5 per cent of the total land area is cultivated. Some 58 per cent is pasture, and cattle outnumber the human population. Most cattle are low-quality animals, and the government is trying to improve breeds and encourage beef farming. The belief in large numbers of cattle as a sign of wealth and status is an African element in the Malagasy culture.

Coffee is the main export earner, along with cloves, vanilla and sugar. Industry, apart from mining, includes an oil refinery at TOAMASINA.

Half the people follow local religions. However, Protestantism has taken root in the uplands and Roman Catholicism is the religion of many *côtiers*, or coast dwellers.

Most of the people live on the central plateau and east coast. The largest tribal group, the light-skinned Merina (or Hova) of Indonesian origin, form nearly a quarter of the population. They inhabit the central plateau. The Betsimisaraka are côtiers. They have controlled the government since 1960, and there is antagonism between them and the plateau people. The coastal groups are of mixed Indonesian, negro and Arab ancestry.

Upon independence in 1960, Madagascar became known as the Malagasy Republic, but the country's constitution was changed by referendum in 1975, shortly after the military-socialist government of Lieutenant Commander Didier Ratsiraka took power, and it is now the Democratic Republic of Madagascar. President Ratsiraka was re-elected for a second seven-year term in 1982.

MADAGASCAR AT A GLANCE

Area 587 041 km² (226 657 sq miles)

Population 10 200 000

Capital Antananarivo

Government Marxist republic

Currency Madagascar franc = 100 centimes

Languages Malagasy (official), French

Religions Tribal religions (50%), Christian (40%), Muslim (7%)

Climate Tropical; cooler at altitude. Temperatures in Antananarivo: 9-20°C (48-68°F) in July; 16-27°C (61-81°F) in December

Main primary products Rice, cassava, mangoes, vanilla, bananas, potatoes, sugar cane, maize, coffee, pepper, cattle, timber; graphite, chromium, zircon, beryl, garnet

Major industries Agriculture, oil refining, forestry, food processing, mining

Main exports Coffee, vanilla, cloves, petroleum products, meat, fish, sugar

Annual income per head (US$) 300

Population growth (per thous/yr) 28

Life expectancy (yrs) Male 46 **Female** 49

trade and textile making. It expanded as the most important town of British India, and declined when Calcutta grew. However, Madras has grown rapidly since Indian independence in 1947, with new engineering and car plants. It is also the centre of the Tamil film industry.

Madras is India's fourth largest city (after Calcutta, Bombay and Delhi). It has the country's oldest town charter, granted in 1688. It also has India's oldest English church, St Mary's, founded in 1678, where Robert Clive (Clive of India, 1725-74), the British soldier and statesman, was married in 1753.

Population 4 289 300
Map India Ce

Madre de Dios ***Peru*** South-eastern department, crossed by several rivers. Its capital, Puerto Maldonado (population 12 700), stands at the confluence of the Tambopata and Madre de Dios rivers. Gold is mined in the region. Its forests, with timber and brazil nut resources, are giving way to pasture for beef cattle.

The Madre de Dios river, 960 km (600 miles) long, rises in the south-east and flows north then east into Bolivia to join the Beni.

Population 33 000
Map Peru Bb

Madrid ***Spain*** Capital and largest city, in the centre of the country. Its name may be derived from the Arabic name *Medshrid*, which comes from *materia*, meaning 'timber' – a reference to the forests that grew there when the Moors built a fortress on the site in the 9th century.

Today it is Spain's second largest industrial centre after Barcelona, manufacturing aircraft, electrical equipment, agricultural machinery and leather goods. It is the home of the Spanish royal family, and among its many museums, the Prado art gallery is outstanding.

Madrid was captured from the Moors by Castilians in 1083 and made capital of Spain in place of Valladolid by Philip II in 1561. Because of its remoteness, however, it became Spain's largest city only after the coming of the railways in the 19th century. Madrid suffered shelling and bombing during the Civil War (1936-9). The surrounding province named after the city is a dry plain whose farmers grow wheat, vines and olive groves.

Population (city) 3 188 300; (province) 4 727 000
Map Spain Db

Madura ***Indonesia*** Farming and tourist island of 5290 km² (2042 sq miles) off the north-east coast of Java. Cattle, teak, copra and coconut oil are important products.

Population 1 860 000
Map Indonesia Dd

Madurai ***India*** Southern cotton-manufacturing city, about 215 km (135 miles) north and slightly east of the country's southern tip. It is also an ancient Tamil city and seat of learning. For more than 14 centuries from the 5th century BC it was the capital of the Pandya dynasty, which conquered northern Ceylon (Sri Lanka). The city has several palaces, and the Great Temple, a complex covering 57 hectares (141 acres), with the Hall of 1000 Pillars.

Population 907 700
Map India Ce

▲ TEMPLE ON THE SHORE The host of intricately carved Hindu shrines at Mahabalipuram were built for the powerful Pallava princes. They include the wave-washed Shore Temple.

Mae Mo ***Thailand*** Town 550 km (340 miles) north of the capital, Bangkok. It is in the centre of a large deposit of lignite (brown coal), which was being developed in the mid-1980s to reduce Thailand's dependence on imported oil. The Mae Mo river, which runs through the town, is 60 km (37 miles) long, and drains into the Wang river south of Lampang.

Population 28 700
Map Thailand Ab

Mae Nam Khong ***Thailand*** See SALWEEN

Mae Suai ***Thailand*** Remote northern town 700 km (435 miles) north of the capital, Bangkok. The town's economy is largely based on black market trade with nearby Burma in consumer goods into Burma and cattle, jade and precious stones into Thailand. The surrounding district of the same name is agricultural, growing mostly rice.

Population 65 500
Map Thailand Ab

Maebashi ***Japan*** City in central Honshu about 100 km (60 miles) north-west of the capital, Tokyo. Once an important silk town, it now depends largely on local commerce.

Population 277 300
Map Japan Cc

Maelstrom ***Norway*** See LOFOTEN

maestrale Cold northerly to north-westerly wind blowing in northern Italy, corresponding to the MISTRAL of France.

Mafikeng (Mafeking) ***South Africa*** Town formerly known as Mafeking and renamed when it was transferred from the Republic of South Africa to the black state of Bophuthatswana in 1980. It lies about 260 km (162 miles) west of Pretoria. The town survived the longest siege of the Anglo-Boer War (1899-1902) when its British garrison, under Colonel (later Lord) Robert Baden-Powell – subsequently the founder of the Boy Scouts – held out from October 14, 1899 to May 17, 1900.

Until 1965 Mafikeng was the seat of the British administration of Bechuanaland, now Botswana. It was unique in lying outside the country it administered. It is now the centre of a cattle-rearing and dairy-farming district.

Population 29 400
Map South Africa Cb

Mafraq ***Jordan*** Town and road junction of routes from Iraq and The Gulf, situated 50 km (31 miles) north-east of Amman. It is also on the railway from Amman to Syria and has a small airport. There is a Turkish castle dating from the 16th century.

Population 26 000
Map Jordan Ba

Magadan ***USSR*** Port and industrial city on the north shore of the Sea of Okhotsk in the Soviet Far East. It was founded in 1933 on a fine harbour ringed by saw-toothed mountains, and has large fish-canning and engineering plants. Magadan is a busy naval base, and also the administrative centre of a huge region covering 1 200 000 km² (463 000 sq miles) – more than five times the size of Britain. The region is rugged, with bitter winters, and is inhabited by Chukchi, Evenky and Yakut peoples, who live mostly by reindeer herding. The region also produces gold and tin.

Population (region) 510 000; (city) 138 000
Map USSR Qc

Magadi, Lake ***Kenya*** Lake in the Great Rift Valley, about 85 km (53 miles) south-west of Nairobi. The lake is covered by a white layer of trona, a sodium-based mineral, which is pumped ashore and exported through Mombasa as soda ash, which is used in glass manufacture. The small town of Magadi is a railhead on the east side of the lake.

Map Kenya Cb

Magallanes ***Chile*** Southernmost region of the country, stretching down to Cape HORN. The major industries are sheep farming and oil. All the country's oil fields are in Tierra del Fuego and north of the Strait of Magellan. They produce one-third of Chile's needs. The discovery of coal at Pecket and the development of tourism based at the capital, PUNTA ARENAS, have added to the region's economic importance.
Population 133 000
Map Chile Ae

Magat ***Philippines*** A tributary of the Cagayan river, in northern LUZON, that has been developed for hydroelectric power.
Map Philippines Bb

Magdalena ***Colombia*** The country's longest river. It rises in the Andes south of Popayán and flows north for some 1550 km (965 miles) between the eastern and central ranges of the Andes. It drains into the Caribbean Sea near Barranquilla. Until the country's network of roads spread during the 20th century, the Magdalena was the main artery of communication between the coast and the interior. Goods were carried to and from Barranquilla by boat.
Map Colombia Bb

Magdeburg ***East Germany*** City 120 km (75 miles) south-west of the capital, East Berlin. It was founded in the 9th century on the River Elbe and has become the country's busiest river port. The city was badly damaged during the Second World War, but its cathedral, dating from the 13th century, has been lovingly restored. Its industrial products include machinery, chemicals and precision instruments.

In 1657, the German physicist Otto Von Guericke carried out the Magdeburg hemispheres experiment there. He pitted horses against a vacuum created by pumping air from between two fitted copper hemispheres. It took 16 horses to separate them.
Population 289 000
Map East Germany Bb

Magellan, Strait of Strait separating mainland South America from TIERRA DEL FUEGO, and linking the ATLANTIC and PACIFIC oceans. It was discovered in 1520 by the Portuguese navigator Ferdinand Magellan, who wept with joy when his three ships emerged safely into the Pacific. Overlooked by snow-capped peaks and strewn with islands, the strait has great scenic beauty, but navigation for sailing ships was made difficult by the high tidal range (resulting in fast currents) and frequent fogs. It became more important as a shipping route when steamships came into use, but completion of the PANAMA CANAL in 1914 provided a short cut between the Atlantic and Pacific and greatly reduced the importance of the strait. It is now an international waterway, although it lies within Chile's territorial waters.
Dimensions 560 km (350 miles) long, 3-32 km (2-20 miles) wide
Map Chile Be

Maggiore, Lake ***Italy/Switzerland*** Alpine lake covering 212 km² (82 sq miles) between northern Italy and southern Switzerland. Its 372 m (1220 ft) deep basin was gouged by a glacier. The surrounding land, although mountainous, is agricultural and has many lakeside villas and hotels. Resorts such as Verbania, Stresa and Pallanza stand on the western shore. LOCARNO is the main town on the Swiss part of the lake.
Map Italy Ba

Maghreb (Maghrib) ***North Africa*** Region of north-west Africa comprising the states of ALGERIA, TUNISIA and MOROCCO. The name comes from the Arabic *Djezira el Maghreb*, meaning 'Island of the West'. This island of land between the Sahara and the Mediterranean is the western extremity of the Arab world. It has a cultural as well as geographical unity, since all three countries were ruled by France.

magma Molten matter formed within the earth's crust or upper mantle, which may rise towards the earth's surface and consolidate on cooling to form IGNEOUS rock.

magma chamber Reservoir in the earth's crust where magma that has risen from greater depths is stored. It is believed that volcanoes are connected to magma chambers, and that these are the source of the material that the volcanoes ultimately eject onto the surface.

magnesian limestone See DOLOMITE

magnetic poles See GEOMAGNETIC POLES

magnetite Black mineral, iron oxide, often occurring with titanium or magnesium, and an important ore of iron. A magnetically polarised piece of the mineral is called a LODESTONE.

magnetosphere Outermost part of the earth's atmosphere, which blends into interplanetary space, and in which charged particles are trapped and dominated by the earth's magnetic field. See ATMOSPHERE

Magnitogorsk ***USSR*** Industrial city on the south-eastern slopes of the Ural mountains, about 385 km (240 miles) south of Sverdlovsk. It was established in 1930 as a new town for workers brought in to the local iron mines and steelworks. Apart from iron and steel, the city's products include military vehicles and weapons, mining and transport equipment and glass.
Population 421 000
Map USSR Gc

Magwe ***Burma*** Market town on the Irrawaddy river, some 230 km (140 miles) south-west of Mandalay, and capital of the division of the same name covering 44 820 km² (17 305 sq miles). The town lies in one of the country's driest areas, near Burma's most important gas and oil fields at Man and Yenangyaung.
Population (division) 3 241 000
Map Burma Bb

Mahabalipuram (Seven Pagodas) ***India*** Ancient port and modern tourist centre on the east coast, about 40 km (25 miles) south of Madras. Its Hindu religious complex dating from the 7th century includes caves, temples, a rock relief depicting the 'Penance of Arjuna' (a legendary prince who fought at KURUKSHETRA), and the remains of seven pagoda-like temples made of single, huge blocks of stone.
Map India Ce

Mahajanga ***Madagascar*** Port about 400 km (250 miles) north-west of the capital, Antananarivo. It was formerly called Majunga and was founded in about 1700 by Arab traders. Its manufactures today include processed foods, textiles and tobacco.
Population 66 000
Map Madagascar Aa

Maharashtra ***India*** West-coast state in which Marathi is the dominant language. Its 307 690 km² (118 768 sq miles) lie mainly on the dry, western Deccan plateau, where farming is poor, concentrating on grains such as millet, except where irrigation makes sugar cane and cotton growing profitable. The wetter coastal strip below the Western Ghats produces rice. Manufacturing, mainly textiles, machinery and food processing, is concentrated in the cities of Thane, Poona and Bombay, the capital.
Population 62 784 200
Map India Bd

Mahaweli Ganga ***Sri Lanka*** The island's longest river. It rises in the wet central mountains and flows 330 km (205 miles) generally northwards across a dry plain to Koddiyar Bay, near Trincomalee. The river has more water than can be used by farmers along its lower valley, so a new irrigation scheme transfers water to neighbouring rivers such as the Kala Oya and Maduru Oya. The scheme also feeds water to ancient reservoirs, known as tanks, and generates electricity. The Victoria Dam is the largest dam of the scheme.
Map Sri Lanka Bb

Mahé ***Seychelles*** Chief island of the group, situated in the western Indian Ocean. Covering 153 km² (59 sq miles), it makes up half the total land area of the Seychelles. Its main town is the seaport of VICTORIA. Mahé is mountainous, with a highest point of some 905 m (2969 ft).
Population 55 000

Mahón ***Spain*** Seaport and capital of Minorca in the BALEARIC ISLANDS. It exports wine, cheese, brandy and agricultural produce, and manufactures shoes and jewellery. It was a British naval base in the 18th century.
Population 23 000
Map Spain Hc

Mahore ***Indian Ocean*** See MAYOTTE

Mähren ***Czechoslovakia*** See MORAVIA

Maiduguri ***Nigeria*** Capital of Borno state, about 100 km (60 miles) south-west of Lake Chad. It is an attractive town, with wide tree-lined avenues and a bustling daily market, but it stagnated until the railway arrived in 1964. Now it is the meeting place of roads from Cameroon and Chad and an industrial centre with government mills producing palm oil, a footwear factory and cotton ginneries.
Population 225 100
Map Nigeria Ca

Main ***West Germany*** River rising near Bayreuth in north-eastern Bavaria, and flowing for 524 km (325 miles) through Würzburg and Frankfurt to join the Rhine at Mainz.
Map West Germany Cd

Maine ***USA*** Most north-easterly state, covering 86 027 km² (33 215 sq miles). Most of it is mountainous, and it rises to 1605 m (5267 ft) at Mount Katahdin in the east. It has many lakes. Much of the state is given over to forestry. Farming (dairy produce, eggs and potatoes), fishing and tourism are also sources of income. AUGUSTA is the state capital and PORTLAND the largest city.
Population 1 164 000
Map United States Ma

Mainz (Mayence) ***West Germany*** Capital of Rhineland-Palatinate state and port at the meeting point of the Rhine and the Main rivers, 30 km (19 miles) west and slightly south of Frankfurt am Main. Mainz has been a market town since Roman times, and in August-September each year it holds Germany's biggest wine market. The city has a university founded in 1477 and makes machinery, wine and paper.

Johann Gutenberg (1398-1468) printed the first Bible using movable type at his works in Mainz in 1452. There is a museum of printing with a Gutenberg Bible of 1452-5 near the cathedral and a reproduction of Gutenberg's printing house.
Population 185 000
Map West Germany Cd

Majorca (Mallorca) ***Spain*** Largest of the BALEARIC ISLANDS (3639 km², 1405 sq miles).
Population 460 000
Map Spain Gc

Majunga ***Madagascar*** See MAHAJANGA

Majuro ***Marshall Islands*** Atoll in the Ratak Chain in the east of the territory. Three linked islands at the eastern end – Dalap, Uliga and Darrit – form the Marshall Islands capital; they were badly damaged by a huge wave in 1979.
Population 8700
Map Pacific Ocean Db

Makarska ***Yugoslavia*** Ancient Dalmatian port, now a popular tourist resort, 56 km (35 miles) south-east of Split. Its palm-lined streets and seafront lie beneath the magnificent crags of Biokova mountain which rises to 1962 m (6435 ft). The town has retained its old-world charm by keeping modern hotel developments at a distance – at Tucepi, Podgora, Zivogosce, Baska Voda, and at Brela, a former fishing hamlet just to the north.
Population 7000
Map Yugoslavia Cc

Makassar ***Indonesia*** See UJUNG PANDANG

Makassar Strait (Macassur, Makasar Strait) Broad strait running between Borneo and SULAWESI (Celebes) in Indonesia. It links the CELEBES SEA to the JAVA and FLORES seas. A five-day battle between a Japanese convoy and American and Dutch air and naval forces was fought in the strait in January 1942, but the Allies failed to stop the Japanese invasion of Borneo.
Dimensions 960 km (595 miles) long, 130-370 km (81-230 miles) wide
Map Indonesia Ec

Makedonija ***Yugoslavia*** See MACEDONIA

A DAY IN THE LIFE OF A YUGOSLAV HOTEL OWNER

As Ivo Ivic sits, at 9 am, behind the counter of his 'Hotel Adriatica', one German couple checks out and another arrives. The Germans invaded Yugoslavia before Ivo was born on the sun-scorched Makarska Riviera of Dalmatia, 40 odd years ago. But today the Germans have made him rich. Ivo did four years' welding in Munich, saved 60 000 Deutsche marks and built his guest home almost from scratch five years ago. He speaks German almost as well as Serbo-Croat, and many of his customers are German.

He checks off his shopping list; he has already been to the market this morning and met the fishing boats to stock up, while his wife Anka (whom he met in Munich) saw to the guests and supervised the dining room. His little hotel is in a quiet bay south of Makarska, screened from the towering, grey wall of the Biokova mountain by pine trees.

The sun has lifted the early haze, and from the palm-lined waterfront Ivo glimpses the isles of Brac and Hvar. His wife checks the German couple into one of 12 double rooms (75 marks or US$29, with full board) and tells them about the sauna and the bar with disco music. The bill is rarely paid in dinars. Ivo will take 90 000 marks (nearly US$35 000) this summer, but in the winter he lives from the dinars earned by opening as a café. The state is Communist but encourages small businesses like this, which cater for Yugoslavia's 7 million tourists a year. Ivo hopes to retire in style, in another 15 or 20 years. His sons Slobodan and Mate will have been seen through Belgrade University and one of them will take over the hotel. His daughter Iva, now seven, will find a good match.

Business mixes with pleasure when he slips away at 11 am to meet friends in a café in Makarska – the local health inspector and a policeman. He treats them to slivovitz brandy, lunch and Kutjevo wine, and invites them to watch soccer on his video on Saturday.

Meanwhile, Anka serves lunch for the hotel guests. Fish stew and grilled mackerel are the main courses, with local cheese and olives or pancakes to follow. Ivo comes home at 2 pm to read the newspaper, and take a nap.

At 5 pm he does a couple of chores – repairs to a door handle and a washbasin. He plays bowls on the terrace with his cousins, who work in the hotel, before helping his wife take the dinner orders. The Ivic family have their own dinner later, at 10 pm, under the bougainvillea. As usual, they play cards and drink with Ivo's visitors. The family sleep on the roof these summer nights because the hotel rooms are all in use.

Makeyevka ***USSR*** Industrial city in the southern Ukraine, near the city of Donetsk in the Donets coal-mining basin. It was founded as a steel town in 1899, and also produces machinery and chemicals.
Population 448 000
Map USSR Ed

Makhachkala ***USSR*** Port and industrial city on the west coast of the Caspian Sea, north of the Caucasus range. It has been rebuilt following a devastating earthquake in 1970, and has textile, footwear and cement industries.

Makhachkala is also the capital of Dagestan, an autonomous republic of the RSFSR with an area of 50 300 km² (19 400 sq miles). This is a mountainous region which is home to some 30 nationalities. It was annexed from Persia in 1723. The area was the setting of the novel *Hadji Murad*, by Leo Tolstoy (1828-1910).
Population (republic) 709 000; (city) 269 000
Map USSR Fd

Ma-kung ***Taiwan*** See PESCADORES

Makurdi ***Nigeria*** Capital of Benue state, about 580 km (360 miles) east and slightly north of the national capital, Lagos. It is the flood-season headquarters of Benue river steamers and stands on the railway linking Kaduna and Port Harcourt, and on the road between Jos and Enugu. The town is a market centre and its manufactures include cigarettes and soft drinks.
Population 86 800
Map Nigeria Bb

Malabar Coast ***India*** The coastal strip of what is now the south-western state of Kerala. It was known to Greek, Roman and earlier Middle Eastern voyagers, who could sail there and back because of the seasonally changing wind patterns. With more than 3000 mm (120 in) of rain a year, its natural cover is tropical rain forest.
Map India Be

Malabo ***Equatorial Guinea*** National capital and chief port of Bioko island. It was formerly called Santa Isabel, and stands on the north of the island, on the rim of a volcanic crater which has been breached by the sea to create a natural harbour. Malabo has a cathedral built in 1916, and exports include cocoa and hardwood.
Population 37 200
Map Equatorial Guinea Aa

Malacca ***Malaysia*** See MELAKA

Malacca, Strait of Sea passage between the west coast of Peninsular Malaysia and Sumatra linking the ANDAMAN and South CHINA SEAS. With Singapore at its southern end, it is one of the world's busiest shipping lanes, where liners and cargo ships sail alongside traditionally rigged Arab dhows, Chinese junks and Malay praus. On average, 150 major ships a day pass through the strait.
Dimensions 800 km (500 miles) long, 50-180 km (31-112 miles) wide
Map Malaysia Ab

malachite Green mineral, copper carbonate, used as a source of copper and for ornamental stoneware.

Málaga ***Spain*** Port and tourist resort on the south Mediterranean coast. It produces building materials, foodstuffs, fertilisers, textiles, beer and wine. Fruit and nuts are exported from its adjacent province, an area of lush subtropical vegetation which includes the resorts of Fuengirola, Torremolinos and Marbella.
Population (town) 503 300; (province) 1 036 300
Map Spain Cd

Malanje (Malange) ***Angola*** Market town and railhead linked to the national capital, Luanda, 350 km (215 miles) to the west. Malanje is the capital of a province of the same name. The province's farmers produce coffee and cotton.
Population (town) 31 600; (province) 551 000
Map Angola Aa

Mälaren ***Sweden*** Lake stretching about 110 km (70 miles) westwards from Stockholm, and covering 1140 km² (440 sq miles). Its shores and over 1000 islands – which contain historic ruins, castles, palaces and residential villas –are a popular weekend destination for Stockholm families.A canal links it to the Baltic Sea.
Map Sweden Cd

Malaspina Glacier ***USA*** One of the world's largest ice sheets, lying in the St Elias Mountains just west of Yakutat Bay on the south coast of Alaska. It is more than 300 m (1000 ft) thick and covers some 3900 km² (1500 sq miles).
Map Alaska Cc

Malawi (Nyasa, Niassa), Lake ***East Africa*** Third largest of East Africa's lakes after VICTORIA and TANGANYIKA, covering 23 300 km² (9000 sq miles) and about 80 km (50 miles) across at its widest point. Most of the lake lies within Malawi, part is in Mozambique and part borders Tanzania.

When the Scottish missionary and explorer David Livingstone first visited the lake in 1859, it was busy with Arab dhows carrying cargoes of slaves. But today fishing vessels and tourist boats dominate the lake. Livingstone named it Nyasa because of a misunderstanding. He was told the word when he asked local people what the lake was called. Nyasa was not, however, the proper name, simply a word meaning 'lake'.

For some only partly understood reason, the level of the lake varies considerably. In 1915 it reached a low of 469 m (1540 ft) above sea level, compared with a high of 475 m (1560 ft) in 1980. These rises, plus occasional storms, have caused serious erosion at some lakeside resorts, including Karonga and Salima.
Map Malawi Ab

Malaysia See p. 408

Malbork (Marienburg) ***Poland*** Amber collecting and processing town 58 km (36 miles) south-east of the Baltic port of Gdansk. Its restored castle, built by the Teutonic Knights between 1274 and 1400, is one of Europe's largest medieval fortresses. The fortress consists of three turreted strongholds, one of which houses the Amber Museum. The museum contains a display of jewellery and other items made in the town from the fossil resin, which is found on the nearby Baltic coast.
Population 36 600
Map Poland Ca

Malawi

WHERE LAKES, RIVERS, WATERFALLS, WOODS AND MOUNTAINS PROVIDE A PARADISE FOR THE ADVENTUROUS TOURIST

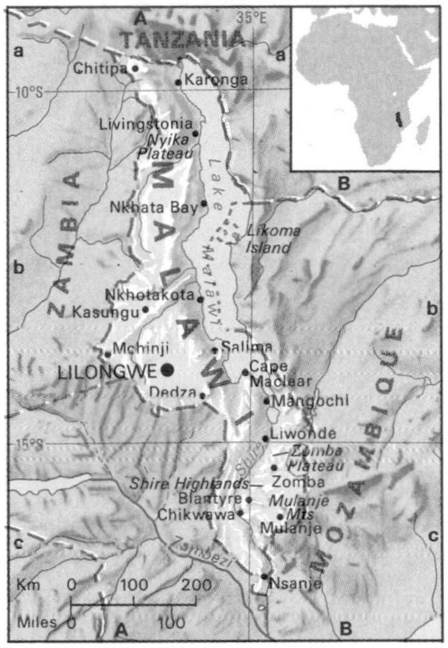

The name Malawi means 'the lake where sun-haze is reflected in the water like fire', and it is the vast and splendid Lake MALAWI – the third largest lake in Africa – from which the country takes its name. Nyasa, the name originally given to the lake by the explorer David Livingstone after he discovered it in 1859, is actually a local word meaning 'lake', or 'broad water'. As Nyasaland, the country was under British rule from 1891 to 1964.

Malawi's great natural beauty lies not only in its shimmering lake but in its park-like grasslands and woodland scenery. In the south, the SHIRE river with its spectacular waterfalls flows through a valley overlooked by wooded, towering mountains.

For all its scenic diversity, Malawi lies in a narrow strip never more than 160 km (100 miles) wide, and the country's share of Lake Malawi covers one-fifth of the area. The lake lies in a deep valley, and is renowned for its golden beaches and teeming fish and bird life. Crocodiles have been removed from resort areas such as Salima, making them safe for water sports and swimming.

Although Malawi is a tourist's paradise, tourism is hampered by strict laws – women are not allowed to wear trousers, or skirts above the knee, and men are forbidden to wear their hair longer than collar length. These restrictions – relaxed slightly in some resorts and the three main national wildlife parks of Nyika, Kasungu and Lengwe – reflect the attitudes of Malawi's Presbyterian president, Dr Hastings Kamuzu Banda.

Dr Banda has been president since the country became independent in 1964, and had himself declared president for life by his parliament in 1971. This was not difficult since he is also head of the Malawi Congress Party, the only party represented in the National Assembly. Malawi, one of the world's poorest nations, contains some unexpected features for a black-ruled nation of southern Africa. For example, there are cordial diplomatic links with South Africa, which is Malawi's major trading partner, supplying 40 per cent of imports as well as financial aid and numerous key personnel.

Agriculture is the predominant occupation, and many Malawians live on their own crops, especially maize. The government encourages wealthy Malawians to invest in plantation farming to increase agricultural exports. This has resulted in more and more land being held by an elite group, and has caused an acute land shortage for small-scale peasant farmers.

Apart from maize, crops also include groundnuts, coffee, beans, cassava, millet, and rice on the lake shores, as well as tobacco and tea – the main exports – grown on large privately owned estates. Most tea estates are on lush, terraced hillsides in the south, and tobacco is grown mainly on the fertile central plateau. Exporters are very much reliant on the two railway lines across Mozambique to the coast, but these are often disrupted by Mozambiquan guerrillas.

Malawi has deposits of bauxite and coal but little is mined because they lie in inaccessible areas. Manufacturing, using hydro-electricity, has been developed rapidly in recent years, but imports of manufactured goods are still large.

More than two-thirds of Malawi's inhabitants live in the central and southern regions – most in rural areas. Nearly all (98 per cent) belong to one of nine tribes – Chewa, Nyanja, Lomwe, Yao, Tumbuko, Sena, Tonga, Ngoni and Ngonde. The most numerous are the Chewa, who live in the central region.

MALAWI AT A GLANCE
Area 118 484 km² (45 747 sq miles)
Population 7 290 000
Capital Lilongwe
Government One-party republic
Currency Kwacha = 100 tambala
Languages Chichewa, English
Religions Christian (60%), tribal religions (20%), Muslim (16%)
Climate Tropical; cooler in the highlands. Average temperature ranges from 14-24°C (57-75°F) between November and April to 19-32°C (66-90°F) May to October
Main primary products Maize, rice, cassava, sorghum, millet, beans, groundnuts, sugar, tea, tobacco, cotton
Major industries Agriculture, food processing, brewing, cement, tourism
Main exports Tobacco, sugar, tea, groundnuts, cotton, textiles, cereals
Annual income per head (US$) 190
Population growth (per thous/yr) 33
Life expectancy (yrs) Male 43 **Female** 45

Malaysia

AN UNLIKELY UNION OF LANDS AND PEOPLES HAS PRODUCED A STATE THAT IS ONE OF THE WORLD'S ECONOMIC SUCCESSES

Jungle-clad mountains, dense and humid forests, vast plantations of rubber trees and oil palms and miles of golden beaches make up Malaysia. In a combination of heat and high humidity a thousand different kinds of orchid bloom. Lush and tropical, Malaysia is one of the most successful of the developing countries: the world's largest producer of rubber, palm oil and tin. It is a major source of hardwoods and has recently begun to export oil and gas.

All this has resulted from the unlikely union of West (now Peninsular) Malaysia and East Malaysia (SABAH and SARAWAK on the island of Borneo). They are separated by 700 km (about 440 miles) of the South China Sea; they had no ties with each other, apart from the accident of British colonial rule; and the west was far more developed than the east.

Malaysia's success has been achieved by 16 million people who are a potentially explosive cocktail of races: Malays (48 per cent), Chinese (34 per cent), Indians (9 per cent) and a variety of other indigenous peoples. The Malays are politically dominant while the Chinese are economically powerful.

Much of Malaysia is mountainous and forested: the mountains rise to 2189 m (7182 ft) at Gunung Tahan on the Malay peninsula, and to 4094 m (13 432 ft) at Gunung Kinabalu in Sabah. Rainfall is high everywhere, averaging 2540 mm (100 in) a year.

The ancestors of the Malays came from Yunnan in Southern China, and moved into the peninsula in about 2000 BC. Arab traders from The Gulf converted the Malays to Islam in the 15th century. Soon after, the Europeans arrived. The Portuguese (1509) and the Dutch (1641) in turn occupied MELAKA (Malacca) on the south-west coast, before the British took over in 1824. Earlier, the British had established trading settlements on PENANG Island in 1786 and SINGAPORE in 1819. In 1826 the three were combined as the Straits Settlements. In 1874 Britain established control over PERAK and SELANGOR, and by 1930 had absorbed the whole Malay peninsula.

The two territories of East Malaysia, however, did not come under formal British rule until after the Second World War. An Englishman, James Brooke, put down a revolt against the Sultan of Brunei in 1841; he was rewarded by being made the Rajah of Sarawak. His successors ruled there until the Japanese invasion in the Second World War, at the end of which the rajah gave the country to the British Crown. In the case of Sabah, it was a British company – the British North Borneo (Chartered) Company – that administered the country from 1882 to 1942.

Anticolonial movements had grown in the Malay peninsula during the 1920s and 1930s, influenced by the pressures for independence in India. The Japanese occupation of Malaya (1942-5) taught the Chinese Communist guerrillas that Westerners were not invincible, so giving a new impetus to the independence movement. After the war, groups of guerrillas took to the jungle again, waging a terrorist campaign against the British. The Emergency, as it was called, lasted from 1948 until 1960, when the rebels were finally suppressed.

Britain realised, however, that the situation could not be resolved by military means alone. In 1957 the Malay peninsula became the self-governing Malayan Federation. This was expanded in 1963 to include Singapore, Sarawak and Sabah in the Federation of Malaysia. But tension soon led to Singapore's departure and the loss of the federation's principal port and industrial centre.

Nine of Malaysia's 13 states have a sultan or hereditary head of state, the other four have governors appointed by the king. Every five years the nine hereditary rulers elect a king: in 1984 they elected King Iskander, Sultan of JOHOR. There is a House of Representatives with 154 members and a Senate of 58 members, and each state has its own constitution and state government. Malaysia is a member of the Association of South-east Asian Nations (ASEAN).

Malaysia's economic success since independence – an industrial growth rate of 7.4 per cent a year between 1970 and 1982, and one of the world's strongest currencies – has been achieved by developing its wealth of natural resources. This is particularly true of Sabah and Sarawak across the South China Sea which, at the time of union, were much less developed than West Malaysia.

Apart from coastal oil and small timber handling ports, the Borneo territories were almost entirely underdeveloped when independence came. Tropical rain forest was everywhere, including large tracts of swamp forest along the coast of Sarawak; much of the interior was inaccessible and occupied only by a sparse population of farmers who lived by slash and burn agriculture, and dwelt communally in so-called longhouses. Along the coast lay the scattered villages of the Sea Dayaks, or Iban, who lived off the forest and by fishing but whose ancestors were pirates.

Today, Sabah and Sarawak have grown rich on logs and sawn hardwoods, sold to Japan and the American west coast. Felling and the rapid expansion of the road network have cut great swathes through the forest.

Peninsular Malaysia, on the other hand, already had a thriving commercial economy based on rubber growing and tin dredging at independence. Rubber was introduced into the Malay peninsula in 1877 after Sir Henry Wickham, a British explorer, had smuggled seeds from Brazil (which monopolised the rubber trade) to England. About 20 seedlings were sent to Malaya and from them vast plantations developed. The British, taking advantage of good communications along the railway lines already used for tin exports, rapidly expanded rubber cultivation to cover the western foothills of the peninsula. Here much of the tropical rain forest – the natural vegetation of a country within five degrees latitude of the Equator – was largely replaced by an artificial forest of rubber trees and, later, of oil palms.

The local population followed the lead of the British planters and investors. The typical Malay *kampong* (village), especially in the centre of the peninsula, lines the roadway between a narrow belt of riceland in the valley and rubber smallholdings on the hillsides.

The eastern side of the peninsula remained largely untouched by the British planters. But since independence, the government has opened up vast tracts, mainly for oil-palm growing. Blocks of smallholdings have been organised on estate lines by the Federal Land Development Authority (FELDA), which provides service vehicles and processing plants.

Tranquillity remains in the small kampongs along the river valleys, where rice is grown and cattle are reared, and in the fishing villages of the TERENGGANU coast, where *perahu* – outrigger canoes with triangular sails – line the shore. But irrigation projects have begun to modernise rice cultivation and local fishermen face competition from larger fishing boats. Tourists in their chalets, too, have begun to disturb the giant turtles whose nocturnal egg laying is one of the area's attractions.

RACIAL HOTCHPOTCH

In the eastern villages, the Muslim Malays dominate: the largest building is invariably the mosque. But in the older estate areas of the west coast, rubber tapping is carried out mainly by Indians, descendants of workers originally brought to the country by the British. In the ports, there are Hindu temples and Tamil shop signs.

In the states of Penang, Perak, Selangor and Melaka, the Chinese make up the majority. The earliest Chinese came to the peninsula to trade in the 17th century, settling particularly in Melaka, then controlled by the Dutch. But most arrived two centuries later when Chinese workers were brought in to dredge for tin. The industry also attracted business interests from China and Europe.

Although some of the larger enterprises around the capital KUALA LUMPUR and in the Kinta valley depend on foreign capital, in general the Chinese control the industry. This control has increased since the richest deposits, which made Malaysia undisputed leader of the world tin industry, have begun to give out.

With their tin profits, the Chinese moved on to control other parts of the Malaysian economy. In South-east Asia most 'Chinatowns' are an enclave within the indigenous cities, but in Malaysia many towns *are* Chinese. The people live in shophouses – three or four-storey buildings in terraces which are used as home, factory and shop by the hard-working Chinese residents. In their midst are Buddhist and Taoist temples.

Only in recent years has the economic power of the local Chinese been challenged by the Malays. In 1969 racial tensions led to riots, and the Malay parties' vote dropped at elections. The then prime minister, Tunku Abdul Rahman (a Malay), concluded that the economic gap between the *bumiputras* (indigenous Malays) and the richer communities had to be reduced.

In 1970 the Malays owned 2 per cent of the shares in limited companies. The government's aim is to increase the Malays' share to 30 per cent by 1990. The Chinese would own and manage 40 per cent, and others (including foreigners) the remaining 30 per cent. By 1985 the Malays' share had grown to 21 per cent.

This 'new economic policy' brought many younger Malays into the towns to seek jobs in Malaysia's growing industries, especially on the island of Penang, as well as in the KELANG valley leading from Kuala Lumpur to the country's main port, Port Kelang. Kuala Lumpur, the capital, is now a modern maze of skyscrapers and highways surrounded by industrial estates and leafy suburbs. Towards Kelang, new towns house workers in electronics and other industries. But the migration to the towns has been frowned on by Muslim traditionalists, anxious for the younger generation to maintain strict Muslim values.

▲ VANISHING TRANQUILLITY
Motorised as well as traditional fishing craft ply from this village in Terengganu in eastern Peninsular Malaysia, and increasing numbers of tourists are visiting the area's fine beaches.

MALAYSIA AT A GLANCE

Area 330 434 km² (127 581 sq miles)
Population 16 010 000
Capital Kuala Lumpur
Government Parliamentary monarchy
Currency Ringgit (Malaysian dollar) = 100 sen
Languages Malay, Chinese, English, Tamil and indigenous languages
Religions Muslim (50%), Buddhist (26%), Hindu (9%), Christian (4%)
Climate Tropical; monsoon from October to February in the east and from May to September in the west. Average daily temperature in Kuala Lumpur is 22–32°C (72–90°F) all year; the highlands are cooler
Main primary products Rice, palm oil, rubber, timber, cocoa, pepper, pineapples, fish; crude oil, natural gas, tin, bauxite, iron ore, copper
Major industries Agriculture, crude oil production, mining, food processing, forestry, fishing, rubber, tyre manufacture, cement, electronics, chemicals, textiles
Main exports Crude oil, petroleum, liquefied natural gas, timber, electronic components, palm oil, rubber, tin, pepper, cocoa, pineapples, bauxite, iron ore, copper
Annual income per head (US$) 1610
Population growth (per thous/yr) 22
Life expectancy (yrs) Male 63 **Female** 67

Maldives

A DELECTABLE SCATTERING OF JEWEL-LIKE ISLANDS ACROSS THE INDIAN OCEAN, THE REPUBLIC OF MALDIVES IS AN UNSPOILT PARADISE OF PALMS AND BEACHES

Travellers who visit the Maldives find them an unspoilt earthly paradise of waving palms, glittering white coral sand and translucent blue-green seas. The republic has no direct taxation, no political parties, has almost no crime and shuts down its television at 9.30 in the evening. All these attractions may, paradoxically, prove a mixed blessing, for in the age of modern communications travellers are soon followed by tourists. Isolation, however – the 2000-odd atolls that comprise the republic lie some 640 km (400 miles) south-west of Sri Lanka – may postpone the day when the Maldives become just another group of holiday islands.

Only about 220 of the atolls are inhabited, and nowhere does the land stand more than 1.5 m (5 ft) above the sea. The people, mostly of far-off Indian or Sri Lankan descent, were ruled by sultans of the Didi family from about 1100 until 1968, when a republic was proclaimed. Once Buddhists, the islanders were converted to Islam by a Moroccan saint in 1153. From 1887 to 1965, the Maldives were a British protectorate.

The main occupations are fishing and farming. The Maldivians were always great seafarers, and used to carry cowrie and tortoise shells, dried fish and ambergris to India. Now the chief export is canned or frozen tuna, most of which goes to Japan. Rice, the islanders' staple food, has to be imported. The capital, MALE, was founded by Portuguese traders in the 16th century.

MALDIVES AT A GLANCE
Map Indian Ocean Cb
Area 298 km^2 (115 sq miles)
Population 183 000
Capital Malé
Government Republic
Currency Rufiyaa = 100 laari
Languages Divehi (a form of Sinhalese), English
Religion Muslim
Climate Tropical; average temperature ranges from 25°C (77°F) to 29°C (84°F)
Main primary products Fish, fruit, vegetables, coconuts
Major industries Fishing, agriculture
Main exports Canned, processed fish
Annual income per head (US$) 180
Population growth (per thous/yr) 30
Life expectancy (yrs) Male 46 **Female** 48

Malé ***Maldives*** Principal atoll of the Maldives, little more than 2.6 km^2 (1 sq mile) in land area, the country's capital for centuries and the only urban area in the islands. It is a trade as well as administrative centre, and fish (bonito and tuna), copra and other coconut products are shipped out. Tourism is becoming increasingly important, and Malé's airport, Hulule, has been extended to take jumbo-sized jets. However, although there are some hotels and guesthouses in the capital, most visitors continue to one of the many resort islands.
Population 20 000

Malines (Mechlin; Mechelen) ***Belgium*** Cathedral city 32 km (20 miles) south of the city of Antwerp, and Belgium's religious capital since the 15th century. The cathedral of St Rombout dates from the 13th century; it has a tower nearly 100 m (330 ft) high, and a painting of the Crucifixion by the Flemish artist Van Dyck (1599-1641). Once famous for its lace, the city now makes tapestries, furniture (particularly chairs) and beer.
Population 65 000
Map Belgium Ba

mallee scrub Dense, scrubby thicket of low-growing eucalyptus bushes, found in parts of the semiarid, subtropical regions of south-east and south-west Australia where the rainfall is only 200-350 mm (8-14 in) a year.

Malmédy ***Belgium*** Tourist town in the Ardennes hills, about 40 km (25 miles) south-east of the city of Liège. Formerly in Germany, it was ceded to Belgium in 1919 by the Treaty of Versailles. It was the scene in 1944 of the Malmédy Massacre, in which about 100 US prisoners of war were shot by the Germans. The motor-racing Grand Prix circuit at Francorchamps is near the town.
Population 6000
Map Belgium Ca

Malmö ***Sweden*** Port on the coast of the Öresund channel which separates Sweden from Denmark. The country's third largest city after Stockholm and Gothenburg, Malmö stands almost opposite the Danish capital, Copenhagen, with which it is linked by ferry. The city was ruled by the Danes until 1658, when Charles Gustav X seized it for Sweden. It has a 14th-century church and a 16th-century castle, in which the Earl of Bothwell – the third husband of Mary, Queen of Scots – was held in protective custody from 1567 to 1573. He had fled to Malmö in 1567 after being implicated in the assassination of Mary's second husband, Lord Darnley.

Malmö's industries include food processing, shipbuilding, cement, chemicals and textiles.
Population 229 900
Map Sweden Be

Malolos ***Philippines*** See BULACAN

Malopolska ***Poland*** See POLAND, LITTLE

Malta See p. 412

Maluku (Moluccas) ***Indonesia*** Archipelago of 74 504 km^2 (28 766 sq miles) south of the Philippines. Between the 16th and 19th centuries, the islands were known to Europeans as the Spice Islands; they include Halmahera, Ambon, Seram, Buru and the Aru group. The main exports are timber, spices (including cloves, mace and nutmegs), fish and copra.
Population 1 411 000
Map Indonesia Gc

Mammoth Cave National Park ***USA*** Park covering 205 km^2 (79 sq miles) on the Green river in western Kentucky, about 125 km (78 miles) south of Louisville. It contains a spectacular system of huge underground caverns and passages, with a total length of more than 484 km (301 miles).
Map United States Ic

Mamry, Lake ***Poland*** See MASURIA

Man ***Ivory Coast*** Administrative and commercial capital of the country's western region, 460 km (286 miles) from Abidjan. The town is beautifully set in a lush valley of the Man Mountains near the borders of Guinea and Liberia. It is best reached by air, although road links are rapidly improving.

Among nearby tourist attractions are a jagged rock known as the Dent de Man (Tooth of Man); a bridge made out of liana vines at Danané; and a luxury hotel designed to resemble an African village at Gouessesso, 50 km (31 miles) to the north.
Population 80 000
Map Ivory Coast Ab

Man, Isle of ***Irish Sea*** Island, midway between England and Ireland, that is a dependency of the British Crown but is not part of the United Kingdom. It covers 585 km^2 (226 sq miles) and rises to 621 m (2036 ft) at Snaefell. The principal town and ferry port is Douglas. The island has low taxes and its own parliament, the Court of Tynwald, which dates from the 9th century and is one of the oldest in the world; it includes the House of Keys, a representative assembly.

The island is popular with tourists, especially during the TT (Tourist Trophy) motorcycle races, held each summer on a 60 km (37 mile) circuit. The Manx language, a Celtic tongue, is almost extinct, but tailless Manx cats, which originated on the island, are prized by cat fanciers all over the world.
Population 66 000
Map United Kingdom Cc

Manabí ***Ecuador*** Coastal province to the west of Cotopaxi. The region, which is subject to periodic droughts, produces the toquilla palm from whose leaves panama hats are made. The seaport of Manta, about 32 km (20 miles) west of Portoviejo (the provincial capital), is a processing centre for vegetable oils and cotton. It is also an important tuna and game-fishing port and handles coffee and bananas for export.
Population 885 000
Map Ecuador Ab

Manado ***Indonesia*** Provincial capital of North Sulawesi, standing at the north-east tip of the island. It is a fishing port and a processing centre for coconuts, cloves and nutmeg.
Population 129 910
Map Indonesia Fb

Mali

THE SAHARA IS ADVANCING IN THIS DROUGHT-STRICKEN COUNTRY, WHICH ONCE BOASTED SOME OF THE RICHEST CIVILISATIONS IN AFRICA

Great empires once flourished on the plains of Mali, where now there are scattered villages in a parched, dusty wilderness. The mighty kingdom of Ghana ruled these lands as long ago as the 4th century, and the Mali empire which succeeded it reached its peak of wealth and culture in the 13th and 14th centuries. Mali was renowned for its gold, and was a great trading nation. The name of the legendary Tuareg city of TIMBUKTU (Tombouctou), founded in the 11th century, retains its mystique to this day. But now there is neither gold nor glory. The SAHARA of northern Mali is encroaching southwards, and the modern state is one of the poorest countries in the world, suffering famine, disease and abject poverty.

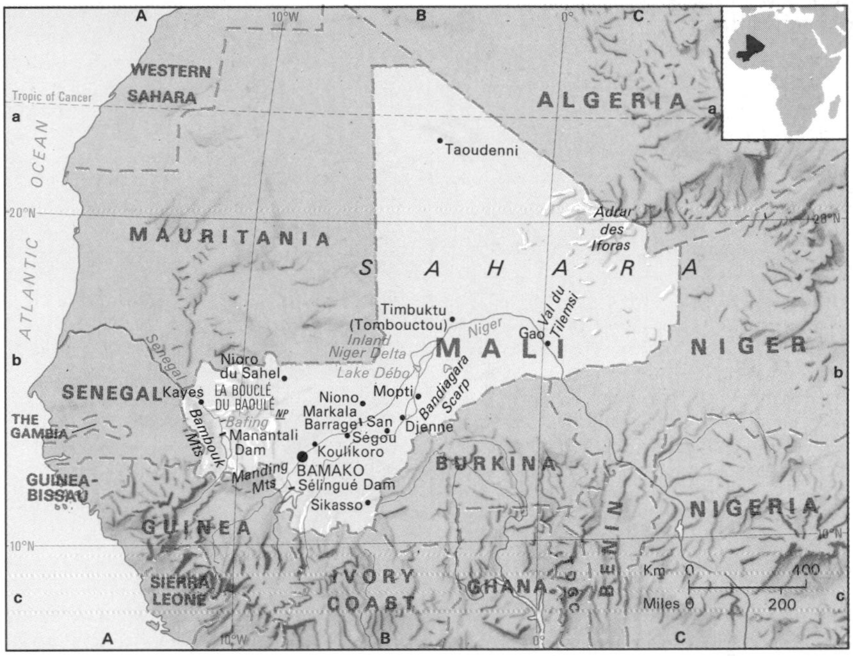

▲ **POT MARKET Pottery traders' canoes, hollowed out tree trunks known as pirogues, line the bank of the Niger at the town of Mopti.**

After Moroccan occupation in the 16th century, the French conquered the country in 1893. They called it Soudan; others knew it as French Sudan; the name Mali came with independence in 1960.

The land is mostly vast and monotonous plains and plateaus. The few high spots in this unvaried landscape are far apart.

In the south there is some rain and the plains are covered by grassy savannah with scattered trees. This peters away to the north where the barren and virtually rainless Sahara covers half the country.

Mali's people include many ethnic groups, most of whom are farmers or herders of animals. The Bambara, Songhai, Marka and Malinké are mostly farmers and river fishermen. Life for most of them revolves around the vast inland delta of the NIGER river where flooding waters the land enough to grow crops without irrigation. The river and its tributaries also provide the best transport route.

In addition, almost every family keeps cattle, goats or sheep. The animals provide milk and meat, as well as wool, hides and dung for fuel and fertiliser. Camels and oxen are used for transport or as pack animals. The nomadic Tuareg and Fulani tribesmen herd their animals across the northern plains.

A prolonged drought that first swept across West Africa in the early 1970s has brought grief to nomads and farmers alike in Mali. It has extended the desert south into southern Mali, turning land that used to provide grazing into useless dust. Thousands of cattle died, crops failed, and thousands of people died of famine and disease.

Gold is now only mined in small quantities. Iron ore and bauxite have been discovered but are not yet mined. Economic problems led to the present military rule, but since President Moussa Traoré assumed power in 1968, politics have remained fairly stable.

MALI AT A GLANCE

Area 1 240 000 km² (478 767 sq miles)

Population 7 910 000

Capital Bamako

Government One-party republic under military rule

Currency CFA franc = 100 centimes

Languages French (official), Bambara, Fulani, Mandingo

Religions Muslim (90%), Animist (9%), Christian (1%)

Climate Hot and dry; temperatures at Bamako range from 16-33°C (61-91°F) in January to 24-39°C (75-102°F) in April

Main primary products Millet, maize, vegetables, rice, groundnuts, cotton, sorghum, sugar cane, livestock, fish

Major industries Agriculture, processed foods, tanning, fishing

Main exports Cotton, livestock (cattle), groundnuts, rice

Annual income per head (US$) 170

Population growth (per thous/yr) 23

Life expectancy (yrs) Male 43 **Female** 46

Malta

AN ISLAND FORTRESS FACED A NEW CHALLENGE WHEN THE BRITISH SAILED AWAY – NOW THE BRITONS ARE FLYING BACK, SEEKING SUN AND TAX RELIEF

▲ VENETIAN AIR Boats resembling Venetian gondolas line a quiet quayside of Grand Harbour. The 16th-century houses show both European and Middle Eastern influences.

In the time of Classical Greece, nearly 3000 years ago, Malta was known as 'the Navel of the Inland Sea'. Its position, standing sentinel across the narrows between Sicily and Tunisia which divide the Mediterranean, has always governed its destiny.

British naval and strategic interests largely moulded the island's character, developing the Grand Harbour area beside the capital, VALLETTA, far beyond the capacity of the island's meagre resources. The Valletta conurbation, crowded onto finger-like peninsulas and spreading inland, has over 200 000 people – two-thirds of the total population. There are 1342 people per km² (3475 per sq mile) in Malta – one of the world's highest population densities. A second island, GOZO, has only 355 people per km² (920 per sq mile).

Overcrowding has not obliterated fascinating vestiges of Malta's far-distant past. The oldest – such as the cave dwellings at Ghar Dalam in the south – are 5800 years old.

Inevitably, as a small, strategically placed island, Malta has a history of foreign domination, successively under the Phoenicians, Carthaginians, Romans, Arabs, Normans and Spaniards. Then followed a period of more formal colonial history, first under the Knights of St John (1530-1798), then fleetingly under the French (1798-1800), and finally under the British (1800-1964).

Throughout these four centuries Malta was used as an island fortress, first as the southern bastion of Christian Europe, later as a guardian over the short sea route to India after the SUEZ CANAL was opened in 1869. During the Second World War it was an Allied base controlling the central Mediterranean and threatening enemy supply routes to North Africa. The island's two great moments in history were in 1565, when it resisted the Turks during the Great Siege, and in the Second World War when it suffered many months of fierce German and Italian bombardment. The islanders's bravery in withstanding this latter onslaught earned the whole island the George Cross.

Until the late 1950s the British military base was the mainstay of the Maltese economy, accounting for over half of Malta's foreign exchange earnings and a quarter of its civil employment. Then Britain began running down the base, and by 1979 had withdrawn altogether. Those 20 years saw great changes in Malta's economy. The naval dockyard was converted for commercial shipbuilding and repairs, light industries were encouraged to develop and tourism boomed – from 20 000 visitors in 1960 to 730 000 in 1980; the island also became a haven for retired Britons seeking sunshine and low taxes. Nevertheless, nearly half of the working population is still engaged in agriculture.

At the same time there were political changes. After Britain rebuffed a suggestion that Malta should be made an integral part of the United Kingdom, full independence came in 1964; it became a republic in 1974. Under the mercurial Dom Mintoff, prime minister from 1971 until his retirement in 1984, Malta adopted a fiercely independent stance, which included economic and trade agreements with China and Libya. Mintoff's Labour Party and the politically powerful Catholic Church (aligned with the Nationalist Party) are the two main protagonists in the islands' post-war political scene.

The Maltese are passionate Catholics, and see their religion as a fundamental link with European culture. But their way of life is highly individualistic, linked to the special character that comes from being islanders. They have their own language, Maltese, universally spoken but not written until the present century.

The landscape is flattish, rocky, bare save for a sprinkling of evergreen carob and olive trees, and rather untidy. Shortage of water and porous limestone bedrock account for the sparseness of the vegetation, and pose problems for farming, industry and the hotel trade. Gozo is greener and more fertile.

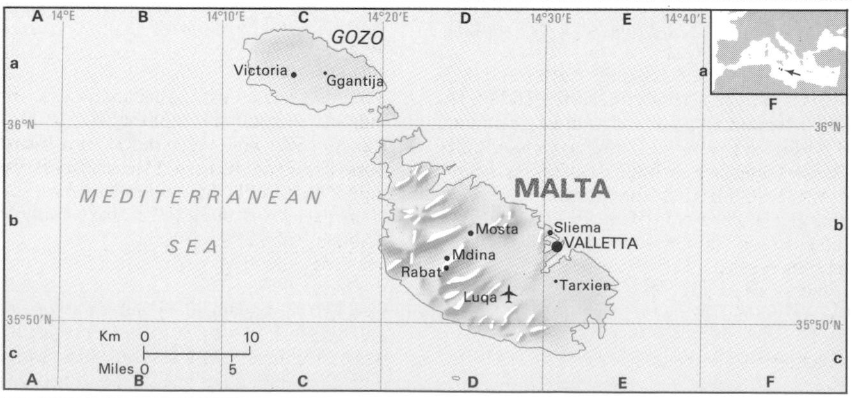

MALTA AT A GLANCE
Area 316 km² (122 sq miles)
Population 355 000
Capital Valletta
Government Parliamentary republic
Currency Maltese lira = 100 cents = 1000 mils
Languages Maltese, English
Religion Christian (97% Roman Catholic)
Climate Mediterranean; average temperature in Valletta ranges from 10-14°C (50-57°F) in January to 23-29°C (73-84°F) in August
Main primary products Vegetables, fruit, pigs, poultry, wheat
Major industries Agriculture, shipbuilding and repairs, textiles, clothing, electronic equipment, foodstuffs, tourism
Main exports Textiles, clothing, machinery, ships, instruments
Annual income per head (US$) 3368
Population growth (per thous/yr) 5
Life expectancy (yrs) Male 70 **Female** 74

Managua ***Nicaragua*** National capital city, on the southern shore of Lake Managua, only 55 m (175 ft) above sea level. Despite its humid climate and its record of earthquake disasters – 8000 people died in the last one in December 1972 – Managua contains one-fifth of the country's population.

It was established as capital in 1858 as a compromise choice between conservative Granada, to the south, and liberal León, to the north. It is now a commercial and industrial city, producing some 60 per cent of the nation's goods by value. These include cotton goods, soft drinks and processed foods. The surrounding fertile volcanic lowlands produce maize, beans, rice, sugar, bananas and sesame seeds.

Lake Managua covers 1049 km² (405 sq miles) and has on its shoreline 1000-year-old footprints preserved in lava and probably made by Indians fleeing an eruption of the nearby Masaya volcano.

Population 630 000
Map Nicaragua Aa

Manapouri ***New Zealand*** Lake covering some 142 km² (55 sq miles) in the Fiordland National Park in the south-west of South Island. At the western end of the lake, a tunnel almost 10 km (6 miles) long draws water from the lake to generate hydroelectricity. In Maori legend, the hills around the lake contain the grave of a legendary princess who died trying to rescue her sister. The lake's Maori name means 'Lake of the Sorrowing Heart'.

Map New Zealand Af

Manaus ***Brazil*** Capital of Amazonas state, on the Amazon river, about 1440 km (900 miles) north-west of the national capital, Brasília. It is now a free port, with duty-free shopping. Much of the city is modern, but there are traditional houses on stilts along the waterside and a 19th-century floating dock – a reminder that Manaus is still a major port despite being 1600 km (1000 miles) from the sea.

There is also the Teatro Amazonas opera house, where the French actress Sarah Bernhardt and the Italian tenor Enrico Caruso both performed. The magnificent building was financed by rubber merchants who had made fortunes after 1839, when the invention of vulcanisation made rubber – then a Brazilian monopoly – indispensable to wheeled transport all over the world. The opera house was completed in 1896. But soon afterwards the Amazon rubber market collapsed with the development in Malaysia of rubber plantations grown from seeds smuggled out of Brazil in 1876 by an English botanist, Sir Henry Wickham.

Amazonas is Brazil's largest state, covering 1 564 445 km² (603 875 sq miles) of mostly equatorial jungle. The main products of the state – which is about three times the size of France – are timber, rubber, jute, nuts and vegetable oils.

Population (city) 635 000; (state) 1 406 000
Map Brazil Bb

Manawatu ***New Zealand*** River on North Island, rising about 130 km (80 miles) north-east of the capital, Wellington. It flows 182 km (113 miles) through the Manawatu gorge to the Tasman Sea.

Map New Zealand Ed

▲ BOOMING CITY Traditional houses on stilts line a quiet creek in Manaus. The city is enjoying a 20th-century boom as the key centre and port in the opening up of the vast Amazon basin, whose jungle is never far away.

Manche, La See ENGLISH CHANNEL

Manchester ***United Kingdom*** City of north-west England, formerly in Lancashire but now headquarters of the county of Greater Manchester, which extends over 1286 km² (497 sq miles). It was a Roman fort called Mancunium and has a 15th-century cathedral, but its rise as one of the major cities of Great Britain dates from the growth of the mechanised cotton industry in the 18th century. (Flemish weavers had settled there as early as the 14th century.) This transformed a rural area into one dotted with mill towns, such as BOLTON, Bury, ROCHDALE, OLDHAM and BLACKBURN. Manchester dominated them all, overrunning Salford, on the opposite bank of the River Irwell, and merging with STOCKPORT, 10 km (6 miles) to the south-east. The Manchester Ship Canal linked the city with the sea, 55 km (35 miles) away, in 1894.

Most of Manchester's public buildings date from the 19th century; they include the magnificent town hall and the Free Trade Hall. The Royal Exchange (the city's stock exchange) was opened in 1921 – just in time, ironically, to preside over the decline of the cotton industry that had sparked the city's growth. Today, the city is a banking centre second in England only to London. It has two universities (plus another at Salford) and two of the country's finest libraries, the Rylands and the Central Reference Library. It is also home of one of Britain's most famous soccer teams, Manchester United, and of one of its most famous orchestras, the Hallé.

Population 448 000
Map United Kingdom Dd

Manchuria ***China*** See DONGBEI

Mandalay ***Burma*** Major city and traditionally the cultural heart of the country; also capital of the division of the same name covering 37 024 km² (14 295 sq miles) in the centre of Burma. Mandalay's royal palace, built between 1853 and 1878, was destroyed during the Second World War, and much of the old city around it was burnt down in 1982. But the commercial hub around Zegyo market is still a centre for the gem trade – rubies in particular.

The most important of Mandalay's many temples are the Arakan pagoda – named after the principal Buddha image which was brought from Arakan state in western Burma – and Kuthodaw, with its mass of small pagodas carved with the texts of Buddhist scriptures. Mandalay Hill, on which stand numerous Buddhist temples, can be climbed by a shrine-lined stone staircase for impressive views of the city.

Population (city) 417 300; (division) 4 581 000
Map Burma Cb

Mandara Mountains ***Cameroon/Nigeria*** Ranges of rounded granite hills and extinct volcanic peaks rising to 1500 m (about 4920 ft) on the northern part of the border between Cameroon and Nigeria. Erosion on the Kapsiki Plateau in Cameroon has left a beautiful landscape of rock pinnacles, where Kirdi tribes people, known for their unique dancing, live in cliff-hanging villages. These and the area's forests, streams, waterfalls and wildlife are a growing tourist attraction.

Map Cameroon Ba

Mandu ***India*** Ruined fortified city in central India, on a rocky promontory of the Vindhya Range, 250 km (150 miles) north-east of Bombay. From the early 15th century it was the capital of the Muslim state of Malwa, and its main mosque is one of the finest examples of Afghan architecture in India.

Map India Bc

manganese nodule Irregularly shaped, crumbly fragment of rock containing on average 20 per cent manganese, 6 per cent iron and 1 per cent of nickel and copper, found on the ocean floors and likely to become an important commercial resource in the future.

mangrove swamp Tidal mud flat covered chiefly by mangroves – tropical evergreen trees and shrubs with stilt-like roots – occurring mainly near the Amazon and Niger deltas and along the coasts of Sumatra and Borneo.

Manica ***Mozambique*** See CHIMOIO

Manila ***Philippines*** National capital, the Philippines' chief commercial, financial and industrial centre and one of the major cities of Asia, lying on the eastern shore of Manila Bay, west-central LUZON, at the mouth of the Pasig river. The bay (area 2000 km², 770 sq miles) is a superb natural harbour – a major reason why the Spanish colonists chose it as the site for their capital of the islands in 1574. The old walled district of Intramuros, beside the Pasig, dates back to this era, but most of the old Spanish buildings (and much of the rest of Manila) were destroyed in the Second World War. During this century, Manila has expanded widely north and south of the river. Today the urban area of so-called Metro Manila covers 635 km² (245 sq miles) and incorporates several fringing cities (such as Pasay, Makati and Mandaluyong in the south, the former capital QUEZON CITY and San Juan in the east, and Caloocan in the north) as well as the 14 municipal districts of Manila proper.

Each district has its own character, reflecting the Philippines' diversity and extremes of wealth and poverty. The inner northern district of Tondo was a squalid, densely populated slum until recent improvements aided by the World Bank. Nearby Binondo is Manila's Chinatown, while the Ermita and Malate areas, behind the sweep of Roxas Boulevard along the bay front south of Intramuros, form the main tourist zone of hotels, restaurants and nightclubs. Sampaloc is chiefly a centre of higher education, with several private universities, while at nearby Santa Mesa is the Malcañang, the presidential palace, originally a Spanish aristocrat's mansion. Makati is the city's financial district and also has fashionable homes of the well-to-do.

Nearby, but outside Metro Manila itself, are the Second World War battlefields of BATAAN and CORREGIDOR, the CAVITE naval base, the large lake of Laguna de BAY and the Taal volcano in BATANGAS province.

Population (city) 6 000 000; (Metro Manila) 8 000 000
Map Philippines Bc

▲ TAXI! American army surplus jeeps – gaudily painted and bedecked with signs, stickers and accessories – ply as taxis (or jeepneys) in Manila's streets.

Manipur ***India*** State covering 22 327 km² (8618 sq miles) of hill country on the Burmese border. Most of its people still lead a tribal way of life. Imphal is its capital.

Population 1 421 000
Map India Ec

Manisa ***Turkey*** Market town about 300 km (185 miles) south and slightly west of Istanbul. Farmers in the surrounding province of the same name mostly produce grain, olives, raisins and tobacco. The province has an area of 13 810 km² (5332 sq miles).

Population (town) 126 320; (province) 1 050 000
Map Turkey Ab

Manitoba ***Canada*** Most easterly of the prairie provinces, covering an area of 650 087 km² (250 998 sq miles) with WINNIPEG as its capital. The first agricultural colony was established there by Scots in the province's RED RIVER Valley in 1812. Two-thirds of Manitoba lies in the CANADIAN SHIELD area, and one-sixth of its area is covered by lakes. The prairies produce grain crops, beef and dairy cattle. Minerals include oil, nickel, copper and zinc. Industries produce processed food, agricultural equipment, clothing and furniture. Many East Europeans have settled in the province. The largest group is from the UKRAINE, and a Ukrainian festival is held every year at Dauphin.

Population 1 026 000
Map Canada Fc

Manizales ***Colombia*** City about 175 km (110 miles) north-west of Bogotá. It is perched on both sides of an Andean ridge at a height of 2126 m (6975 ft). It is the business heart of Caldas, Colombia's biggest coffee-growing region. A coffee festival is held each January.

Population 259 000
Map Colombia Bb

Mannar, Gulf of (Gulf of Manaar, Manar) Inlet of the INDIAN OCEAN between Sri Lanka and the Indian state of TAMIL NADU. The northern boundary is Adam's Bridge, which according to legend was built by the mythological Hindu hero Rama. It is, in fact, a chain of sandbanks between Pamban Island, in India, and Mannar Island, Sri Lanka.

Dimensions 160 km (100 miles) long, 130-270 km (80-168 miles) wide
Map Sri Lanka Aa

Mannheim ***West Germany*** Port and industrial city at the junction of the Rhine and Neckar rivers about 70 km (43 miles) south of Frankfurt am Main. The city centre, built by the Electors of the Palatinate province in the 18th century, is laid out in a fan-shaped pattern radiating from the Baroque palace, now a university. At Mannheim in 1885, the engineer Karl Benz made a petrol-driven, three-wheeled carriage – the first true motor car.

Population 300 000
Map West Germany Cd

mantle Middle layer of the earth, between the crust and the core. See JOURNEY TO THE CENTRE OF THE EARTH (p. 200)

Mantua (Mantova) ***Italy*** City built on the marshy land in the middle of the North Italian Plain, about 120 km (75 miles) east and slightly south of Milan. During its rule by the Gonzaga dynasty in the 15th and 16th centuries, it became a renowned centre of art and learning. It has many fine buildings, including the cathedral, replanned by the architect Giulio Romano (1492-1546) after a fire in 1545. The Roman poet Virgil (70-19 BC) was born nearby and lived in the city for most of his life. Mantua's industries include tourism, printing, tanning and processing farm products.

Population 60 400
Map Italy Cb

Manukau Harbour Largest of the two harbours on which New Zealand's port of AUCKLAND stands. Its name is Maori for 'place for wading birds', and it is largely surrounded by mud flats and sandbanks. A sandbar at its entrance restricts shipping. A narrow isthmus separates Manukau Harbour from the economically more important Waitemata Harbour.

Area Manukau Harbour 394 km² (152 sq miles); Waitemata Harbour 200 km² (77 sq miles)
Map New Zealand Eb

Manzini ***Swaziland*** Town (called Bremersdorp until 1960) founded as a trading post in 1885 and lying 41 km (25 miles) south-east of Mbabane. With the nearby industrial township of Matsapa, it is now a leading agricultural, commercial, industrial and trading centre. Its industries include brewing, cotton ginning and processing, meat and dairy products.

Population 32 700

Map Swaziland Aa

Maoming ***China*** City of Guangdong province, 320 km (200 miles) south-west of Guangzhou. It is close to oil-shale deposits and is emerging as an important mining centre.

Population 200 000

Map China Gf

Maputo ***Mozambique*** Capital of Mozambique. A major seaport formerly known as Lourenço Marques, it became the national capital in 1907, despite not being in a central position – it lies in the extreme south, only some 100 km (60 miles) from the South African border. Maputo stands on a large natural harbour, and its modern development, based on transit trade, began in 1895 when it was linked by rail with Johannesburg in South Africa. Today, Maputo is also Mozambique's leading manufacturing centre, with food processing, textile and iron industries. But its tourist industry collapsed when most of its 250 000 Europeans left in the wake of independence in 1975.

Maputo is also the capital of the country's smallest and southernmost province, also called Maputo. It covers 25 756 km² (9941 sq miles), and produces cotton, maize, sugar and rice.

Population (city) 785 500; (province) 511 000

Map Mozambique Ac

maquis (macchia) Drought-resistant scrub of small myrtle and wild olive trees and aromatic shrubs, usually found on acid soils in parts of the Mediterranean region. Maquis, replaced by garrigue on limestone soil, corresponds to the chaparral of the south-western United States and Mexico. See MEDITERRANEAN WOODLAND

Mar del Plata ***Argentina*** Beach resort about 400 km (250 miles) south-east of Buenos Aires. Famous in South America for its hectic nightlife and one of the world's largest casinos, the city attracts more than 3 million visitors during its summer season from December to March.

Population 424 000

Map Argentina Db

Maracaibo ***Venezuela*** Oil city on the north-west shore of Lake Maracaibo, about 500 km (310 miles) west of the capital, Caracas. Maracaibo has become the country's second largest city after the capital since the discovery of the lake's oil fields in 1914.

Population 1 100 000

Map Venezuela Aa

Maracaibo, Lake ***Venezuela*** Shallow inland sea opening into the Caribbean, on the extreme north-west tip of Venezuela. It covers 13 280 km² (5127 sq miles), but is nowhere more than 35 m (115 ft) deep. The lake contains the country's largest oil fields, providing 70 per cent of Venezuela's oil.

Map Venezuela Ab

Maracay ***Venezuela*** Capital of Aragua state, about 50 km (30 miles) west of the country's capital, Caracas. It was the favourite city of General Juan Vicente Gómez, the dictator who ruled Venezuela from 1908 until 1935. Several of his grandiose architectural follies still stand, including an unfinished opera house, a bull ring modelled on that in Seville, and the huge triumphal arch of the Gomez mausoleum. The city's main industries are cigarette production, meat packing, textiles, paper and chemicals.

Population 387 700

Map Venezuela Ba

Marajó ***Brazil*** Island in the mouth of the Amazon covering 47 964 km² (18 514 sq miles). Switzerland would fit into it with room to spare. Much of the island's flat plains are now cattle and buffalo ranches.

Population 50 000

Map Brazil Db

Maralinga ***Australia*** Area in South Australia in the Great Victoria Desert, where British atomic bombs were tested in the 1950s.

Map Australia Ee

Maramba (Livingstone) ***Zambia*** Tourist town on the border with Zimbabwe, about 380 km (236 miles) south-west of the capital, Lusaka. The capital of Northern Rhodesia from 1911 to 1935, it has a museum devoted to the life and travels of the Scottish medical missionary Dr David Livingstone (1813-73), after whom the town was named. Today it is the capital of Zambia's Southern Province, producing cotton, maize, tobacco and sugar cane. The VICTORIA FALLS are 11 km (7 miles) south-west of the town.

Population 72 000

Map Zambia Bb

Maramures ***Romania*** North-west province on the borders of Russia and Hungary, noted for its rich and varied peasant culture. It consists mainly of well-wooded mountains, fertile plains and unspoilt villages – Maramures represents the best of 'old' Romania. Richly decorated peasant costumes are a daily sight, and the village houses have ornately carved wooden porches and gateways.

Population 517 000

Map Romania Aa

Maranhão ***Brazil*** North-eastern state covering 328 660 km² (126 860 sq miles). Despite a booming aluminium industry, the state remains one of the poorest areas in the country.

The state capital is the port of São Luís, about 480 km (300 miles) east of Belém. The port was founded by the French in the 17th century as St Louis, and many of its tile-faced colonial buildings survive.

Population 4 000 000

Map Brazil Db

Marañón ***Peru*** A headstream of the Amazon, about 1650 km (1025 miles) long. It rises in the Andes at Lake Huaiuash and flows north then north-east across the lowlands of Loreto department to join the Ucayali and become the Amazon. It is navigable as far as the Pongo de Manseriche gorge in north-west Peru.

Map Peru Ba

Marathón ***Greece*** Small town 5 km (3 miles) north of the battlefield where the Athenians routed the Persians in 490 BC. After the battle a courier named Pheidippides is said to have run the 42 km (26 miles) to Athens without stopping. On arrival, he gasped out his message: 'Rejoice, we conquer!' – and dropped dead, possibly from heat stroke. His feat gave the name marathon to the long-distance race which is a main feature of the Olympic Games.

Map Greece Cb

Marbella ***Spain*** See MALAGA

marble METAMORPHIC ROCK formed by recrystallisation of limestone under great heat and pressure, used in sculpture and for decorative purposes.

Marble Bar ***Australia*** See PILBARA, THE

Marburg ***Yugoslavia*** See MARIBOR

Marburg an der Lahn ***West Germany*** University town on the Lahn river, about 80 km (50 miles) north of Frankfurt am Main. Timbered houses climb the side of a hill beneath its medieval castle. Each hour during daytime, the figure of a herald on the town hall clock blows a Trumpet of Justice. Field-marshal Paul von Hindenburg (1847-1934), President of the Weimar Republic (1925-34), is buried in Marburg's 13th-century Gothic cathedral of St Elizabeth.

Population 76 300

Map West Germany Cc

Marche (Marches) ***Italy*** Rural region covering 9694 km² (3743 sq miles) on the Adriatic coast, extending about 160 km (100 miles) south-east from San Marino. It is covered with well-tended farms, a patchwork of vines, olives, cereals and livestock. There is also a host of craftsmen making shoes, paper and musical instruments. The capital is the port of ANCONA.

Population 1 424 000

Map Italy Dc

Marco Polo Bridge ***China*** See LUGOU BRIDGE

Mardan ***Pakistan*** Agricultural town with one of Asia's largest sugar factories, about 120 km (75 miles) north-west of Islamabad. Nearby are Buddhist monuments, and one of several carved stone pillars erected by the Buddhist Indian emperor Ashoka in about 250 BC.

Population 148 000

Map Pakistan Da

Mareb ***Yemen*** See MARIB

mare's tails Drawn-out, wispy cirrus cloud, a sign of strong winds in the upper atmosphere. They often herald the approach of bad weather.

Margarita ***Venezuela*** Largest island in the Nueva Esparta group which occupies an area of 920 km² (355 sq miles) in the Caribbean about 290 km (180 miles) east of the capital, Caracas. In a duty-free zone, with many superb beaches and enviable climate, the island has become a popular tourist resort. Its capital is La Asunción (population 14 100), but the largest town is Porlamar (population 22 500).

Map Venezuela Ba

Marshall Islands

ONCE THE SITE OF NUCLEAR TESTS, THE ISLANDS NOW DEPEND ECONOMICALLY ON AN AMERICAN MILITARY BASE

There are some 1250 islands and reefs in the Marshalls, part of MICRONESIA in the western Pacific Ocean, yet they total just 181 km² (70 sq miles) and have a population of 37 000. The islands lie 3200 km (2000 miles) south-west of Hawaii, and form two chains: Ralik and Ratak. The largest atoll is KWAJALEIN, 120 km (75 miles) long, but the most famous are BIKINI and ENEWETAK, where 64 nuclear weapons were exploded in tests by the USA between 1946 and 1958.

An internally self-governing republic since 1979, the islands' economy is dependent on aid from the USA totalling $750 million, in exchange for the 30 year use of a military base on Kwajalein. The base is used to recover missiles launched over the Pacific in test flights from California, 6700 km (4200 miles) away. Most of the Marshall Islanders – mainly Micronesians plus some Polynesians – live in villages, farming and fishing for their food, as they have always done, and growing coconuts for export.

The islands were named after John Marshall, a British sea captain who visited them in 1788, although they had been discovered by a Spanish expedition in the 1520s. They became a German protectorate in 1899, but were occupied by the Japanese during the First World War. The Marshalls were Japanese by a League of Nations mandate until US forces captured them during the Second World War. In 1947 they became part of the US Trust Territory of the Pacific.

MARSHALL ISLANDS AT A GLANCE

Map Pacific Ocean Db
Area 181 km² (70 sq miles)
Population 37 000
Capital Majuro
Government Parliamentary republic
Currency US dollar = 100 cents
Languages Marshallese, English
Religion Christian
Climate Tropical; average temperature 27°C (81°F)
Main primary products Coconuts, cassava, sweet potatoes, fish
Major industries Agriculture, fishing, coconut processing
Main exports Copra, coconut oil
Annual income per head (US$) 1100
Population growth (per thous/yr) 32

Maria Cristina Falls ***Philippines*** See ILIGAN

Mariana Islands ***Pacific Ocean*** Chain of mountainous islands some 2400 km (1500 miles) east of the northern Philippines. They consist of the US territory of GUAM plus the Commonwealth of the NORTHERN MARIANAS.
Map Pacific Ocean Cb

Marianas Trench Curving trench on the seabed near Guam in the north-western PACIFIC OCEAN. It is a PLATE MARGIN and contains Challenger Deep, the deepest known spot in the oceans, which was discovered by HM Survey Ship *Challenger* in 1951. In January 1960, the US Navy bathyscape *Trieste*, manned by the Swiss Dr Jacques Piccard and the American Lt Donald Walsh, reached a point in the deep 10 915 m (35 810 ft) down. Echo-soundings by a Russian survey ship during the same year established the current record depth.
Greatest depth Challenger Deep 11 033 m (36 198 ft)
Map Pacific Ocean Cb

Marianske Lazne (Marienbad) ***Czechoslovakia*** Spa town in Bohemia 12 km (7 miles) from the West German border. It was patronised by England's Edward VII (1901-10), the German composer Richard Wagner (1813-83) and the German writer Johann Wolfgang von Goethe (1749-1832). Today, visitors come to the spa for treatment for respiratory, nervous and skin disorders, or to play golf, fish, walk in the surrounding hills, or attend conferences or the annual international film festival.
Population 18 400
Map Czechoslovakia Ab

Maria-Teresiopel ***Yugoslavia*** See SUBOTICA

Mariazell ***Austria*** Winter and summer resort in the state of Styria and the country's main pilgrimage centre, 95 km (59 miles) south-west of Vienna. It was founded by Benedictine monks in 1157 and is famous for its torchlight processions. In the Gnadenkirche (Pilgrimage church) is a 12th-century wooden statue of the Virgin Mary (Our Lady of Mariazell), the object of veneration probably since the 16th century, when a Hungarian victory over the invading Turks was attributed to the intervention of the Virgin Mary.
Map Austria Db

Marib (Mareb) ***Yemen*** Ruined city situated 140 km (87 miles) east of SAN'A. Marib was the capital of the kingdom of Sheba around 1000 BC. Among its remains are the Temple of Bilquis (the Muslim name for the Queen of Sheba) and parts of the 5th-century BC walls. A temple near the walls has been turned into a mosque. To the south of the town are the ruins of a 550 m (1800 ft) stone and masonry dam, also dating from the 5th century BC.
Map Yemen Ba

Maribor (Marburg) ***Yugoslavia*** Major industrial city on the Drava river in Slovenia, close to the Austrian border. Founded in the 10th century around a castle and citadel, it prospered on Austrian trade with Venice. Its modern products include automobiles, railway equipment, steel and nonferrous metal goods, leather goods, tobacco, textiles and foodstuffs. A pretty town at the foot of a spur of the Karawanke Alps, Maribor is a tourist centre for Alpine ski resorts and nearby spas. Ptuj, 24 km (15 miles) to the south-east, has a fine medieval castle, and there are large wine cellars at the nearby towns of Ljutomer and Ormoz.
Population 185 700
Map Yugoslavia Ba

Marie Galante ***Guadeloupe*** Island situated 25 km (15.5 miles) south of Grande Terre island. It is the largest of Guadeloupe's dependent islands, and its capital is Grand Bourg on the south-west coast. The chief crop is sugar cane, though its production has much declined since the 19th century.
Population 13 700
Map Caribbean Cb

Marienbad ***Czechoslovakia*** See MARIANSKE LAZNE

Marienburg ***Poland*** See MALBORK

Marigot ***Guadeloupe*** See ST MARTIN

marin Warm, moist south-east wind that blows across the coast of southern France, especially in spring and autumn, as a result of a DEPRESSION in the Gulf of Lions.

Marinduque ***Philippines*** Island covering 958 km² (370 sq miles) in the Sibuyan Sea between MINDORO and southern LUZON. It has important iron-ore mines. Its capital, Boac, and the towns of Gasan and Mogpog are the scene each year on Good Friday of a re-enactment of the crucifixion of Jesus.
Population 163 000
Map Philippines Bc

Maritsa ***South-east Europe*** See EVROS

marl A clay or mudstone with a lime content. It is placed on sandy soils to improve texture.

Marlborough Sounds ***New Zealand*** Rugged coastline at the north-eastern tip of South Island. The English explorer Captain James Cook anchored there on his visits to New Zealand in the 1770s. Its beaches, woods and pastures make it a popular tourist attraction, especially by boat. The longest inlet, Pelorus Sound, stretches for more than 55 km (34 miles).
Map New Zealand Dd

Marmara, Sea of Known to the Turks as Marmara Denizi, the sea is part of the strategically important waterway that links the BLACK SEA (via the Bosporus) with the MEDITERRANEAN (via the Dardanelles). Its coasts are lined by fishing villages, romantic ruined castles and gleaming Turkish tombs. Istanbul commands the entrance to the Bosporus.
Area About 11 140 km² (4300 sq miles)
Greatest depth About 1370 m (4495 ft)
Map Turkey Aa

Marmarica ***Libya*** North-east coastal district of CYRENAICA, consisting of a desert plateau of ridges no more than 200 m (656 ft) high, that extends into Egypt parallel to the Mediterranean coast. Through the centuries, many armies have fought in the region – Romans, Egyptians, Libyans, Arabs and modern Europeans. In the Second World War, much of the fighting during 1942-3 took place there. TOBRUK is the chief town.

Map Libya Ca

Marne ***France*** River, 525 km (326 miles) long, of north-central France. It rises in the Plateau de Langres 70 km (43 miles) north of Dijon and flows north and then west to join the Seine in eastern Paris. It was the scene of two major German defeats in the First World War – in September 1914 and July-August 1918.

Map France Eb

Maroua ***Cameroon*** Chief town of the far north. It is a market for the region's cattle, cotton, groundnuts, millet and sorghum, with a cattle-breeding research station. Maroua is also the country's main Islamic centre, dominated by the Fulbé people. Its buildings, mostly of mud, are in traditional style, and its peaceful, relaxed atmosphere contrasts with the modern cities farther south.

Population 70 000
Map Cameroon Ba

Marquesas Islands ***French Polynesia*** Group of about ten rugged islands with deep fertile valleys lying some 1400 km (875 miles) north-east of Tahiti. Tai-o-haé on Nuku Hiva island is the capital. The islands were discovered by the Spanish in 1595. They had a population of 50 000 in 1800 but this dropped to only 2000 in the 1920s, the result of 'blackbirding' (kidnapping by Europeans for forced labour) and disease. The French painter Paul Gauguin was buried on Hiva Oa island in 1903.

Population 6500
Map Pacific Ocean Gc

Marrakech ***Morocco*** A former capital, from which Morocco takes its name, founded in the 11th century by the Berber Almoravid dynasty. Surrounded by a vast oasis of palm groves, the town lies on a major crossroads at the foot of the High ATLAS MOUNTAINS 220 km (140 miles) south of Casablanca. Berber, Arab and African cultures meet at this point in a dazzle of colour, sound and activity throughout the city.

The Djemaa el Fna square is full of noise and movement: snake charmers, acrobats and storytellers. In the souks (bazaars) craftsmen produce leatherwork, silver, brass, pottery and jewellery. Weavers abound – the city is especially renowned for its carpets and rugs.

Other places of interest include the 12th-century city walls with their gateways, the 70 m (230 ft) minaret of the 12th-century Koutoubia mosque, the Saadian (Sovereign's) Mausoleum and the Ben Youssef *medersa* (religious college).

Population 440 000
Map Morocco Ba

Marsa el Brega (Mersa el Brega) ***Libya*** Oil port on the Gulf of Sirte. The first Libyan oil terminal to be built, it was opened in 1961, and is the foremost petrochemical centre in the country. It has a refinery, gas liquefaction plant, and ammonia and urea plants. It is also the site of Libya's first Technical University.

Population 7100
Map Libya Ba

▲ PARIS IN THE CARIBBEAN Balata Church, a replica of Sacré Coeur in Paris, overlooks Martinique's capital, Fort-de-France, and Les Trois Ilets, where Napoleon's Joséphine was born.

Marsa Matruh ***Egypt*** See MATRUH

Marsala ***Italy*** Wine-making and fishing town at the western tip of Sicily. Marsala wine was first produced by English families who settled there in the 18th century. The Italian patriot Giuseppe Garibaldi (1807-82) and his thousand red-shirted volunteers landed there in 1860 to begin his campaign to unite Italy.

Population 79 900
Map Italy Df

Marseilles (Marseille) ***France*** The country's premier port and third largest city after Paris and Lyons, on the Mediterranean coast east of the delta of the Rhône river. Commerce has flourished there since Greek traders arrived in 600 BC. Today the city is a cosmopolitan mix of Mediterranean peoples including many Arabs, and sprawls up the slopes above its small natural harbour, to the south of the modern port.

In the centre, Marseilles is a jumble of narrow streets and alleyways. But farther out there are broad streets and modern buildings; they include the Unité d'Habitation, an apartment block designed by the Swiss-born architect Le Corbusier (1887-1965) and completed in 1952, that is regarded as a landmark of modern architecture. The Château d'If – a castle built in the early 16th century on an island in the harbour – was the real prison in which the fictional Count of Monte Cristo was locked away in the novel by Alexandre Dumas (1802-70).

The French national anthem, *La Marseillaise*, was named after a troop of volunteers from the city who marched to Paris in 1792 to join the revolution against France's old monarchical and aristocratic regime. The men from Marseilles sang the song as they entered the capital, and the stirring words and music caught on.

Population 1 110 500
Map France Fe

marsh Area of temporarily flooded land beside a river or lake, characterised by water-loving vegetation such as reeds, and often silty. The term is used loosely to describe any area of low-lying, wet land such as a fen, swamp or bog.

Martaban ***Burma*** Town on the Gulf of Martaban, an inlet of the ANDAMAN SEA to which the town has given its name. The town lies about 160 km (100 miles) east of Rangoon. Vast deposits of natural gas have recently been discovered in the gulf. Martaban is the terminus of the railway from Rangoon. A ferry across the estuary of the Salween river takes passengers on to the city of Moulmein and the railway southwards.

Population 15 000
Map Burma Cc

Martinique ***West Indies*** One of the Windward group of islands in the eastern Caribbean, made up of three groups of volcanoes and covering an area of 1079 km² (417 sq miles). The highest peak, Mont PELEE, reaches 1397 m (4583 ft).

The tropical and humid island was home of the fierce Carib Indians, who called it Madinina. It was discovered by Christopher Columbus in 1493, but Indian resistance discouraged colonisation until the French settled there in 1635. Apart from short periods of British occupation it has remained under French rule. In 1946, it became an overseas department of France with local self-government and full representation in the National Assembly in Paris.

Bananas, sugar cane – originally produced with slave labour – and pineapples are the main export crops. Sugar refining, the distilling of rum, cement manufacture and petroleum products are the main industries. Tourists, mainly from France and the United States, bring in some additional revenue, but the island depends heavily on financial support from the French government. FORT-DE-FRANCE is the capital and main port.

Population 330 000
Map Caribbean Cc

Maryland ***Liberia*** See HARPER

Maryland ***USA*** East coast state straddling Chesapeake Bay and covering 27 394 km² (10 577 sq miles) north-westwards to the Appalachian Mountains; its southern boundary follows the Potomac river. It was first settled in 1634 and was one of the original 13 states. Chickens, dairy goods, soya beans, maize, coal, metals, electronics, processed foods and steel are the chief products. With some of the suburbs of the national capital, WASHINGTON, in the state, tourism is also a source of revenue. The state capital is ANNAPOLIS, and BALTIMORE is the largest city.
Population 4 392 000
Map United States Kc

Masada (Mezada) ***Israel*** Hilltop fortress about 55 km (35 miles) east of Beersheba, overlooking the Dead Sea. In AD 73 it was the scene of the last stand of the Zealots, an extreme Jewish sect, in the Jewish revolt against the Romans. After a siege lasting two years, the 960 remaining defenders committed suicide rather than surrender. Many of the buildings have been restored, including the palace of Herod the Great (who ruled from 37 to 4 BC), the synagogue and storehouses. There is a cable car to the 49 m (160 ft) high summit.
Map Israel Bb

Masan ***South Korea*** City and port about 60 km (35 miles) west of the southern port of Pusan. Developed during Japanese colonial rule (1905-45), it is now the centre of a free-trade zone set up by the government to promote exports. Its industries include flour milling, textiles and electrical machinery.
Population 386 700
Map Korea De

Masbate ***Philippines*** Island of northern VISAYAN group, covering 3269 km² (1262 sq miles) between Panay and the southern tip of Luzon. Mainly rolling hills, it is best known for cattle-raising, but maize is also grown.
Population 585 000
Map Philippines Bc

Maseru ***Lesotho*** National capital, about 1500 m (4921 ft) above sea level. It has light industries whose products include candles, carpets, sheepskin goods and pottery, and is popular with tourists from South Africa to which it is linked by rail. The city lies in a district of the same name, which covers 4279 km² (1652 sq miles), and whose main products are maize, sorghum, wheat, beans and peas.
Population (district) 277 800; (city) 45 000
Map South Africa Cb

Masherbrum ***Pakistan*** Mountain rising to 7821 m (25 660 ft) in the Karakoram Range in Azad Kashmir, just south of K2.
Map Pakistan Da

Mashhad (Meshed) ***Iran*** Holy city and capital of Khorasan province, about 750 km (470 miles) east of the national capital, Tehran. It trades in carpets and cotton goods, and produces leather, flour, sugar and rugs. It is a place of pilgrimage for Shiite Muslims, because the Shiite religious leader Imam Ali Reza was buried there in 817.
Population 1 120 000
Map Iran Ba

Mashonaland (Shonaland) ***Zimbabwe*** Northern region consisting of three provinces covering a total area of 112 089 km² (43 278 sq miles) around the national capital, Harare. It is the home of the Shona people, the country's largest language group.
Population 2 837 400
Map Zimbabwe Ca

Masjed Soleyman (Masjid Suleiman) ***Iran*** Town about 250 km (156 miles) south and slightly west of the capital, Tehran. It is at the centre of Iran's oldest oil field, which began production in 1908. The surrounding area has numerous archaeological remains, including a huge stone terrace, built by the Parthians (200-100 BC). It is approached by a flight of steps 25 m (82 ft) wide.
Population 77 200
Map Iran Aa

Mason-Dixon Line ***USA*** The southern boundary of Pennsylvania, surveyed in the 1760s by two English astronomers (Charles Mason and Jeremiah Dixon) in order to settle a dispute between the proprietors of the Pennsylvania and Maryland colonies. Before and during the Civil War (1861-5) it represented the boundary between slave-owning and non-slave states, and subsequently in a general way that between the North and South of the USA.
Map United States

mass movement, mass wasting Downward movement of weathered material on a slope under the influence of gravity, usually lubricated by rain or melting snow.

Massa ***Italy*** Chemical-producing town 40 km (25 miles) north-west of Pisa. Marble is also quarried there.
Population 66 600
Map Italy Cb

Massachusetts ***USA*** State covering 21 386 km² (8257 sq miles) in the north-east, on the Atlantic coast. In 1620 the Pilgrim Fathers landed in Cape Cod Bay and established their colony of PLYMOUTH, and the city of BOSTON was founded ten years later. Massachusetts is mostly lowland, rising to 1064 m (3491 ft) in the Berkshire Hills in the west.

Like most New England states, it has few industrial minerals. Former industries producing textiles, clothing and footwear have declined and have largely been superseded by those making electrical equipment, electronic instruments and machinery. Agriculture concentrates on dairy produce, market gardening and cranberries, and the coast and the Berkshires are tourist areas. Boston is the state capital.
Population 5 822 000
Map United States Lb

Massachusetts Bay Inlet of the ATLANTIC OCEAN between Cape Ann and Cape COD in the north-eastern USA where the Pilgrim Fathers landed in 1620 and founded PLYMOUTH. Soon afterwards other Puritans founded SALEM and BOSTON on the bay. It has two arms: Boston Bay and Cape Cod Bay.
Length 105 km (65 miles) (north-west to south-east)
Map United States Lb

Massawa ***Ethiopia*** Main Red Sea port of Eritrea province, standing on an island linked by a causeway to the mainland. Today, it handles mainly imports of a wide assortment of goods. Its prosperity has been severely affected by the civil war.
Population 35 000
Map Ethiopia Aa

massif **1.** Large plateau-like mass of uplands usually with clearly defined margins, often formed by FAULTS. **2.** Compact group of connected mountains forming a distinct portion of a mountain range.

Massif Central ***France*** Upland, largely pastoral area in south-central France. It covers 85 000 km² (32 800 sq miles) – about one-sixth of the country. The highest peak is the Puy de Sancy (1885 m, 6184 ft).
Map France Ed

Masuria (Mazuria; Masuren; Pojezierze Mazurskie) ***Poland*** Northern lakeland, east of the Vistula river. Its ridges, hills of gravel and long, winding lakes were left by glaciers retreating at the end of the last Ice Age some 10 000 years ago. The lakes include Lake Sniardwy (Spirding), Poland's largest, which covers 106 km² (41 sq miles) and is a nesting place for grey herons.

With the Suwalki lake district to the north-east, the region has more than 1000 lakes; and a labyrinth of connecting rivers and canals makes it a superb holiday area. There are some 90 nature reserves. Wild European bison are protected in the Borecka Forest on the Soviet border, north-east of Lake Mamry, and Europe's largest herd of elk can be seen in the Red Swamp (Czerwone Bagno) reserve near Augustow. The resort of Gizycko (Lötzen; population 26 300) lies on an isthmus near Lake Mamry, Poland's second largest lake. The lake, which covers 105 km² (about 40 sq miles), is alive with cormorants by day, and at night tourists hunt for crayfish by torchlight.
Map Poland Db

Matabeleland (Ndebeleland) ***Zimbabwe*** South-western region consisting of two provinces covering a total of 132 087 km² (51 000 sq miles) around the city of Bulawayo. It was a centre of guerrilla activity against the former white regime in the 1970s. The provinces are the home of the Matabele (or Ndebele) people.
Population 1 379 400
Map Zimbabwe Ba

Matadi ***Zaire*** Capital of Bas-Zaire region, and a seaport on the Zaire river estuary, at the upper limit of navigation for ocean-going ships. Built on a hillside at the base of the Crystal Mountains, it was founded by the English adventurer Sir Henry Morton Stanley as a trading station in 1879. It is now a busy commercial city with varied industries, including pharmaceuticals and printing. The Matadi-Kinshasa railway skirts the Livingstone Falls, which begin just above Matadi.
Population 162 000
Map Zaire Ab

Matale ***Sri Lanka*** Town founded by the British in the central mountains, 20 km (12 miles) north of the royal Sinhalese city of Kandy. Some 3 km (2 miles) to the north lies Aluvihara, an ancient Buddhist rock temple. There monks are still copying onto palm leaves the story of the Buddha's many incarnations – a project that has so far taken more than 135 years. The project is designed to replace a set of leaves written in 43 BC and destroyed accidentally in 1848 during a skirmish between rebel Sinhalese tribesmen and British troops.

Population 29 700
Map Sri Lanka Bb

Matanzas ***Cuba*** Seaport and capital of Matanzas province, situated in the fertile sugar cane region, 90 km (55 miles) east of Havana. It exports sugar and the coarse, red fibre of the henequen plant, used in making rope and twine, and manufactures rayon, shoes and fertilisers. Farther east, 35 km (22 miles), is the beach resort of VARADERO.

Population (city) 112 000; (province) 568 000
Map Cuba Aa

Matera ***Italy*** Town near the 'arch' of the Italian boot, some 200 km (125 miles) east of Naples. Its most outstanding feature is its numerous cave dwellings cut into the side of a limestone gorge and known as *Sassi*.

Population 52 200
Map Italy Fd

Mathura (Muttra) ***India*** Northern city on the Yamuna river, about 130 km (80 miles) south and slightly east of Delhi. It is revered by Hindus as the birthplace of the legendary hero and god, Krishna, and has relics from local Hindu, Buddhist and Muslim kingdoms. Its Buddhist artists produced some of India's finest sculptures from the 3rd century BC to the 6th century AD.

Population 159 500
Map India Bb

Mato Grosso ***Brazil*** Plateau region bordering Bolivia and Paraguay, and covering a total of 1 232 000 km² (475 700 sq miles) – an area larger than France and Spain combined. The region's main towns are Cuiabá and Campo Grande. Rice, soya beans, wheat and livestock are the chief sources of income. The region gets its name from its characteristic landscape: *mato grosso* means 'thick scrub'.

Population 2 500 000
Map Brazil Cc

Matopo Hills ***Zimbabwe*** Granite range south of the city of Bulawayo. The hills, which rise to 1554 m (5098 ft), have been settled for about the last 40 000 years. The tomb of Cecil Rhodes (1853-1902), South African statesman and developer of Rhodesia, stands at World's View in Matopos National Park.

Map Zimbabwe Bb

Mátra ***Hungary*** Group of volcanic mountains near the Czechoslovakian border north-east of Budapest. They reach 1015 m (3330 ft) at Kékes, Hungary's highest peak, and drop steeply to the Great Alföld (Plain) to the south. Vineyards clothe the lower slopes, with extensive beech and oak forests above. A popular holiday area for winter skiing and summer walking, the Mátra have several fine resorts, including Mátrafüred, lying at 400 m (about 1300 ft), to the north of Gyongyos, the regional centre.

The Börzsöny hills, which rise to 936 m (3071 ft), are an extension of the Mátra westwards to the Danube. There are numerous ruined castles, popular summer resorts above the great Danube bend, and, in the Nograd Valley, the pretty villages of the Palocs – one of Hungary's most distinctive ethnic groups – who retain their colourful folk customs and costume. Hollokö is a delightful Paloc village.

Map Hungary Aa

Matruh (Marsa Matruh; Mersa Matruh) ***Egypt*** Harbour town on the Mediterranean coast about 300 km (185 miles) west of the port of Alexandria. In a cave near the town, General Erwin Rommel, known as the Desert Fox, is said to have had his headquarters while vainly trying to stem the British breakout after the battle of El Alamein in October-November, 1942. The cave is now a war museum.

Population 27 900
Map Egypt Ab

Ma-tsu Tao (Matsu Islands) ***Taiwan*** Group of Islands lying 19 km (12 miles) off the coast of the Chinese province of Fujian, near Fuzhou. Like Chin-men Tao the islands are heavily fortified by the Chinese Nationalist Government.

Map Taiwan Ba

Matsue ***Japan*** City near the north-west coast of Honshu island, some 130 km (80 miles) north of Hiroshima. Known as the 'city of water', it stands on the Ohashi river which links Shinji lake to the Naka lagoon. In July, thousands of candle-lit paper lanterns, commemorating the souls of the dead, are set adrift on the waters of Shinji during the Toro Nagashi festival.

Population 140 000
Map Japan Bc

Matsushima ***Japan*** Bay on Honshu's east coast just north-east of the city of Sendai. The bay's name means 'islands of pines' – a reference to more than 260 forested rocky islets that characterise the area. The waters are farmed for seaweed and oysters.

Map Japan Dc

Matsuyama ***Japan*** Port and city on the north-west coast of Shikoku island overlooking the Inland Sea. Among its industries are oil and petrochemical plants built on reclaimed land. Nearby is Dogo Spa, a hot-spring resort with a popular public bathhouse.

Population 426 600
Map Japan Bd

Matterhorn (Cervino) ***Italy/Switzerland*** One of the world's most spectacular and distinctive mountain peaks, on the frontier, about 5 km (3 miles) south of the Swiss town of Zermatt. It was eroded by ice to the shape of a steep, bent pyramid – with a particularly sheer east face. The mountain, 4477 m (14 688 ft) high, was first conquered in 1865 by a party organised by the English mountaineer Edward Whymper (1840-1911). On the way down four of the party slipped and fell to their deaths. The rope broke, saving the lives of Whymper and two of his guides.

Map Switzerland Ab

Maui ***Hawaii*** Second largest island in the US Pacific Ocean state. It includes the 3055 m (10 023 ft) peak of Kolekole, and has a flourishing tourist industry.

Map Hawaii Ba

Maule ***Chile*** Region of wine, wheat and livestock in the basin of the Maule river. It also embraces some of the country's lake district and most spectacular mountain scenery. The summer resort of Constitución, surrounded by pine-covered hills at the mouth of the Maule river, lies 90 km (55 miles) west of the regional capital, TALCA, where Chilean independence was declared in 1818.

Population 724 000
Map Chile Ac

Mauna Kea ***Hawaii*** Extinct volcano rising to 4205 m (13 796 ft) on the island of Hawaii. It is the highest point in the Hawaiian islands.

Map Hawaii Bb

Mauna Loa ***Hawaii*** Volcano which rises to 4169 m (13 677 ft) on the island of Hawaii. One of its craters, Kilauea, is active.

Map Hawaii Bb

Mauritania See p. 420

Mauritius See p. 421

Mavrovo ***Yugoslavia*** National park covering 686 km² (265 sq miles) in north-west Macedonia on the Albanian border. It includes the craggy SAR PLANINA, and Korab mountain, rising to 2764 m (9068 ft), and contains beech and fir forests harbouring bear, lynx, fox and deer. In the south of the park is the spectacular Radika river gorge. There are several villages rich in folklore, including Galicnik, which is noted for its traditional costumes and a multiple wedding ceremony held each year in July.

Map Yugoslavia Ed

Mayagüez ***Puerto Rico*** Seaport, industrial city and cultural centre on the west coast, 115 km (72 miles) from San Juan. It processes and exports sugar, coffee, pineapples and citrus fruit. It is the centre of the island's knitwear industry, which supplies the US market. The city has a university college of agriculture and mechanics founded in 1763.

Population 101 000
Map Caribbean Bb

Mayence ***West Germany*** See MAINZ

Maymyo ***Burma*** Hill town about 50 km (30 miles) east of Mandalay. Its cool climate attracted the British, who developed it after 1886 as Burma's summer capital.

Population 100 800
Map Burma Cb

Mayo (Meigh Eo) ***Ireland*** Rugged county covering 4831 km² (1865 sq miles) on the Atlantic. Its west coast is fringed by islands, and inland it is dotted with lakes. The main activity is agriculture, but the land is poor. Some textiles are produced. The county town is Castlebar (population 6410).

Population 114 700
Map Ireland Bb

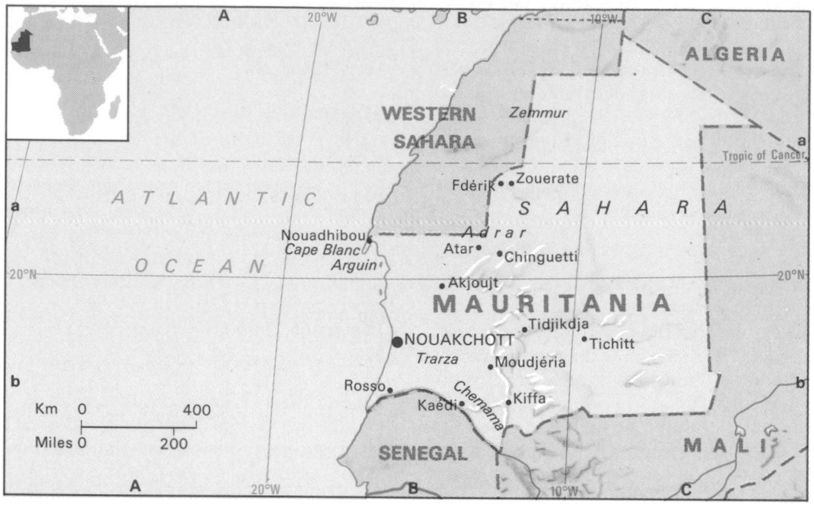

Mauritania

A MUSLIM LAND WHERE MOORS AND BLACK AFRICANS MEET AND MIX – AND BOTH SUFFER THE DEPRIVATIONS OF DROUGHT

Prolonged droughts have devastated the way of life of Mauritania's people. Today its traditional economy is in ruins. It can produce little of the food it needs, and relies heavily on foreign, particularly Arab, aid.

Mauritania is nearly twice the size of France, yet it has less than 2 million people. It links the Arab MAGHREB with black West Africa. The SAHARA covers the north, and here the only settlements are those around oases such as Atar, where millet, dates and some vegetables can be grown. Temperatures rise well over 30°C (86°F) during the day, but can fall to freezing point at night. In the south is the main agricultural region, the fertile valley of the SENEGAL river, and between the two lie the drought-stricken SAHEL grasslands.

About two-thirds of the people are of Moorish (Arab/Berber) or mixed Moorish and black descent. The Bidan (white Moors) have pale skins, and legend has it that the traditional pale blue robe is worn to enhance their colouring. The Harattin (black Moors) have the same features, but much darker skins, and were slaves of the Bidan in times past. The remaining Mauritanians are black Africans, who live in the south and include the Peul and Wolof. However, Islam is a unifying force, for the faith is almost universal and Islamic law is applied. Despite anti-slavery laws, an estimated 100 000 people remained in bondage in 1981.

The majority of the people – the Moors and Peul – are traditionally nomadic or partly nomadic herders of goats, sheep, cattle or camels. However, severe drought since the 1970s has killed 50 to 70 per cent of the nation's animals. In 1963, 83 per cent of Mauritanians were to some extent nomadic, but by 1980 the figure had fallen to 25 per cent. The rest had settled along the Senegal, near remaining oases, or in urban areas where unemployment is high. Vast shanty towns have sprung up around all the towns, particularly NOUAKCHOTT, the capital.

Iron ore found near Zouerate and copper deposits near Akjouit have provided the main exports, and despite the fact that two-thirds of the country is virtually desert, Mauritania is one of the Sahel's richer countries. Development of its mineral resources and fishing industry is the best hope for its future.

Originally a Berber kingdom, Mauritania became a French protectorate in 1903, and a colony in 1920. It emerged as an independent republic in 1960, but a military coup occurred in 1979. Links with France are gradually being reduced as the country strives for recognition as a true Arab nation. In 1979, Mauritania withdrew its claim in Western Sahara.

MAURITANIA AT A GLANCE

Area 1 030 700 km² (397 953 sq miles)
Population 1 690 000
Capital Nouakchott
Government Military
Currency Ouguiya = 5 khoums
Languages Arabic, French (official), African languages
Religion Muslim
Climate Hot and dry, summer rains near coast. Average temperature in Nouakchott ranges from 13-28°C (55-82°F) in December to 24-34°C (75-93°F) in September
Main primary products Livestock, millet, sorghum, dates, fish; iron ore, copper, phosphates, salt
Major industries Agriculture, mining, fishing
Main exports Iron ore, processed fish
Annual income per head (US$) 400
Population growth (per thous/yr) 20
Life expectancy (yrs) Male 43 **Female** 46

Mayon ***Philippines*** Active volcano 2462 m (8077 ft) high on the Bicol peninsula of southern LUZON, just north of Legaspi. It is a symmetrical cone, the name coming from a local dialect word for 'beautiful'. It has erupted nearly 40 times since 1800.
Map Philippines Bc

Mayotte (Mahore) ***Indian Ocean*** French dependency in the COMOROS island group covering 373 km² (144 sq miles). A French possession since 1843 and one of the group's four main islands, Mayotte voted to remain French when the other Comoros voted for independence in 1974. It is valued by France as a naval base. The main town is Mamoutsou.
Population 60 000
Map Madagascar Aa

Mazar-e Sharif ***Afghanistan*** Manufacturing town 304 km (190 miles) north-west of Kabul. Its industries include flour milling, brick-making and textiles, and it trades in carpets and skins. It contains the tomb of Ali, the son-in-law of Muhammad, and is a place of pilgrimage for Shiite Muslims. The town is known for the game of bazkashi, which is played on horse-back – with the riders trying to gain possession of the body of a dead goat or lamb.
Population 110 400
Map Afghanistan Ba

Mazatlán ***Mexico*** Fast-growing beach resort at the mouth of the Gulf of California, in the state in Sinaloa. It was once an important port for trade with the Philippines, and is now a haven for thousands of tourists. They are drawn by Mazatlán's swimming, surfing, deep-sea fishing contests and exuberant carnivals.
Population 202 000
Map Mexico Bb

Mazovia (Mazowsze) ***Poland*** Ancient, semi-independent principality in eastern Poland, which joined the Polish kingdom in 1529. It is a flat lowland, with the capital, Warsaw, at its centre. There are vast pine forests punctuated by sand dunes, and peat bogs rich in bilberries and cranberries. There are also areas of loamy cropland and orchards.

Traditional Mazovian thatched cottages, often crowned by storks' nests, with gardens full of huge sunflowers and hollyhocks, can still be found in places. A lively dance developed in the region gets its name from the Polish word *mazurka*, meaning 'a Mazovian woman'.
Map Poland Db

Mazuria ***Poland*** See MASURIA

Mbabane ***Swaziland*** Capital of the country, situated in west-central Swaziland. It was founded as a trading station, and became capital of the British Protectorate of Swaziland in 1902 when British administrators set up their headquarters there. Although Parliament meets in nearby Lobamba, Mbabane has remained the national capital and the country's chief commercial centre. Occupying an attractive position overlooking the Ezulwini Valley about 1150 m (3770 ft) above sea level, it lies on Swaziland's main east-west road.
Population 36 000
Map Swaziland Aa

Mauritius

NEW INDUSTRIES ARE GROWING ON A TROPICAL ISLAND WHERE SUGAR NO LONGER GUARANTEES THE SWEET SMELL OF SUCCESS

An overpowering odour of molasses surrounds the island of Mauritius. It used to be the sweet smell of success when the cane sugar industry and its by-products of molasses and rum kept the island in relative prosperity. But changing world markets have affected the economy and the country is now deeply in debt.

Sugar still dominates the island. The cane plantations cover 45 per cent of the land and use one-third of the labour force. But new industries have been developed and sugar accounts for only 60 per cent of the island's exports, compared with 98 per cent in former days.

Mauritius now has a manufacturing sector which employs 30 000 people, mainly making textile goods and fertilisers, but 14 per cent of workers are unemployed.

The development costs of an alternative national livelihood are the main reason for the country's foreign debt. The situation is getting worse because imported food and fuel cost far more than export earnings, though a newly developed tourist industry is beginning to make a significant contribution – it is already the third largest source of foreign exchange after sugar and clothing exports.

The population has increased rapidly in recent years, largely because the eradication of malaria has reduced the death rate. Population density exceeds 547 per km² (1417 per sq mile), and the capital city of PORT LOUIS is growing to form a conurbation with the towns of Beau Bassin and CUREPIPE.

With all its problems Mauritius is a beautiful island with superb tropical beaches; a pear-shaped paradise in the Indian Ocean with great potential for tourism. It is a volcanic island, with many craters surrounded by lava flows. The undulating northern plain rises gradually to the central plateau, at one point 823 m (2700 ft) high, before dropping sharply to the south and west coasts. The climate is varied, but mainly hot and humid, and the prevailing south-east winds bring an annual rainfall of 5000 mm (200 in) to the windward slopes. Fast-flowing rivers have been harnessed to produce hydroelectric power. The well-watered, fertile soil was once covered with luxuriant natural vegetation, but much of this has given way to the sugar plantations.

SUCCESSION OF SETTLERS

Mauritius was known to 10th-century Arabs, but the first recorded settlement was by the Dutch in 1598, who called the island after their ruler, Prince Maurice of Nassau. The Dutch abandoned the settlement in 1710, the French arrived in 1715, and the British seized the island in 1810. Mauritius, together with the small island of RODRIGUES, became an independent Commonwealth state in 1968.

Despite its European colonial history, two-thirds of the population are of Indian extraction – descendants of contract workers who were brought in to work the plantations after the abolition of slavery in 1835. The rest are mostly Creole descendants of African slaves and French settlers. As economic problems grow, Mauritius faces increasing political and social tensions between these two groups.

MAURITIUS AT A GLANCE

Map Indian Ocean Bd
Area 1865 km² (720 sq miles) excluding Rodrigues and dependencies
Population 1 020 000
Capital Port Louis
Government Parliamentary monarchy
Currency Mauritius rupee = 100 cents
Languages English, French, Creole, Hindi
Religions Roman Catholic, Hindu, Muslim
Climate Subtropical; average temperature ranges from 17-24°C (63-75°F) in August to 23-30°C (73-86°F) in January
Main primary products Sugar cane, tea, tobacco, potatoes
Major industries Agriculture, sugar processing, molasses, rum distilling, textiles, clothing, fertilisers, electronic equipment, tourism
Main exports Sugar, molasses, rum, clothing, textile yarns and fabrics
Annual income per head (US$) 1050
Population growth (per thous/yr) 9
Life expectancy (yrs) Male 64 **Female** 68

Mbala ***Zambia*** North-eastern market town near the Tanzanian border. Formerly known as Abercorn, it serves a region producing coffee, fruit, maize and tobacco. The 221 m (725 ft) high Kalambo Falls plummet over a cliff at the southern end of Lake Tanganyika, about 30 km (19 miles) west of the town.
Population 5200
Map Zambia Ca

Mbale ***Uganda*** One of Uganda's largest towns, about 190 km (120 miles) north-east of the capital, Kampala. Founded in 1902, it is the regional centre for the east of the country, and handles coffee grown on the slopes of Mount Elgon to the east.
Population 40 000
Map Uganda Ba

Mbandaka ***Zaire*** Capital of Equateur region, about 580 km (360 miles) north-east of Kinshasa. Formerly called Equateurville and later Coquilhatville, it is a commercial and industrial city, serving an extensive region producing coffee, rice, rubber and timber. It has food processing and printing industries.
Population 160 000
Map Zaire Aa

M'Banza-Congo ***Angola*** Historic town about 290 km (180 miles) north and slightly east of the capital, Luanda. Formerly the capital of the powerful Kongo kingdom, it was flourishing in the 15th century when early Portuguese settlers arrived. They named it São Salvador.
Population 12 700
Map Angola Aa

Mbanza-Ngungu ***Zaire*** Commercial and industrial town, formerly called Thysville. It lies about 120 km (75 miles) south-west of the capital, Kinshasa, and has railway workshops.
Population 84 000
Map Zaire Ab

Mbeya ***Tanzania*** Regional capital in the south-west, about 90 km (55 miles) north-west of Lake Malawi (Nyasa). It markets tobacco, tea, coffee, cattle and goats.
Population 100 000
Map Tanzania Ba

Mbomou ***Equatorial Africa*** See BOMU

Mbuji-Mayi ***Zaire*** Capital of Kasai Oriental region, about 615 km (380 miles) west of Lake Tanganyika. Formerly called Bakwanga, it is the centre of a major diamond-mining region, which produces, by weight, about 75 per cent of the world's industrial diamonds.
Population 383 000
Map Zaire Bb

Mdina (Notabile) ***Malta*** Walled town in the west of the island, which as Citta Vecchia was the capital until it was superseded by Valletta in 1570. It is less than 500 m (550 yds) square, but has many fine palaces and religious buildings which date mainly from the 15th century and are still occupied by aristocratic families.
Population 925
Map Malta Db

Mead, Lake ***USA*** Artificial lake covering some 600 km² (230 sq miles) on the Arizona-Nevada border. It was formed by damming the Colorado river by the Hoover Dam.
Map United States Dc

mean sea level Average level of the sea, calculated from a long series of continuous records of tidal oscillations, and used as a basis from which to measure the height of land masses.

meander One of a series of sweeping curves in the course of a river across its FLOOD PLAIN.

Meath (An Mhi) ***Ireland*** Fertile grassland county covering 2336 km² (902 sq miles) north-west and north of Dublin. It is known as Royal Meath because it includes Tara, ancient seat of the Irish kings. There are textile and engineering industries and Ireland's largest metal-mining complex at Navan, the county town.
Population 95 400
Map Ireland Cb

Meaux ***France*** Chief town of the Brie cheese-making district on the Marne river, 40 km (25 miles) east of Paris. The town is also noted for the production of mustard.
Population 56 750
Map France Eb

▲ **RICE BOWL A vast patchwork of glittering green rice paddy-fields covers the valley of the Mekong and Tonle Sap rivers in Cambodia.**

Mecca (Makkah) ***Saudi Arabia*** Capital of the Western Province (formerly HEJAZ), situated 64 km (40 miles) east of Jeddah. It lies in a narrow valley overlooked by hills crowned with small forts. Mecca was the birthplace of Muhammad, the founder of Islam, and is the religion's most holy city. All Muslims try to make the *haj* (pilgrimage) to Mecca at least once in their lives, and between 800 000 and 1.5 million pilgrims visit the city each year.

Muhammad was born there about the year 570 and it was his home until 622, when he fled to Medina – 350 km (217 miles) to the south – to avoid religious persecution. Mecca contains the Kaaba, a shrine which houses the sacred Black Stone, towards which followers of Muhammad face, wherever they are in the world, when they pray. The Black Stone is said to have been given to the prophet Abraham by the archangel Gabriel. The city is also the site of the University of Umm Al Qura. Like Medina, the city is protected on behalf of the Muslim world by the Saudi royal family. Non-Muslims are forbidden to enter Mecca; any caught within its walls face heavy penalties and deportation. The city's commercial life revolves almost entirely around the pilgrims.

Population 375 000

Map Saudi Arabia Ab

Mechelen ***Belgium*** See MALINES

Mechlin ***Belgium*** See MALINES

Mecklenburg ***East Germany*** Lakeland region covering 26 694 km² (10 307 sq miles) between the capital, East Berlin, and the Baltic Sea. It is a popular tourist area.

Population 2 106 900

Map East Germany Bb

Medan ***Indonesia*** Provincial capital of North Sumatra. Local produce includes coffee, tea, palm oil, latex, cinnamon and tobacco.

Population 2 378 000

Map Indonesia Ab

Médéa (Lemdiyya) ***Algeria*** District administrative centre in the middle of a vine and fruit-growing area in the Tell Atlas mountains, 24 km (15 miles) south-west of Blida. It is situated 914 m (3000 ft) above sea level at the foot of Djebel Nador. It makes small plastic goods.

Population 90 000

Map Algeria Ba

Medellín ***Colombia*** Second largest city of the republic, about 240 km (150 miles) north-west of the capital, Bogotá. It is known as the city of eternal spring, because the combination of its altitude (1486 m, 4875 ft) and tropical location give it a warm climate all year round.

Medellín, founded in 1616 in the Aburrá valley, was the country's first city to develop a modern industrial base. It began producing textiles in the late 19th century, and still makes 80 per cent of Colombian production. Medellín produces plastics and processes chemicals and food, and employs one in four of the country's manufacturing workers.

It is also a city of flowers and is the self-proclaimed orchid capital of the world.

Population 1 998 000

Map Colombia Bb

Medias ***Romania*** Industrial town in central TRANSYLVANIA about 190 km (120 miles) south of the Russian border. Medias, which grew from an 11th-century craftsmen's village, retains many of its old streets, a fortified church built in the 14th and 15th centuries, and the remains of other medieval buildings. Its manufactures include machinery, shoes, glass, china, porcelain and wine.

Population 68 400

Map Romania Aa

Medicine Hat ***Canada*** Main industrial and commercial city of south-eastern Alberta, standing on the South Saskatchewan river, 265 km (165 miles) south-east of Calgary. It exploits rich gas reserves and has flour and timber mills, grain elevators and iron foundries.

According to legend, the town stands on the spot where a Cree medicine man lost his hat in the river while bolting from Blackfoot braves who had slaughtered his tribe.

Population 40 385

Map Canada Dc

Medina (Al Madinah) ***Saudi Arabia*** Oasis city situated 350 km (217 miles) north of Mecca. After Mecca, it is the country's second holiest city. In AD 622 Muhammad, the founder of Islam, fled here from Mecca, where he was being persecuted for his religious beliefs. His flight (known as the Hegira) took place on September 22 – later marked as the beginning of the Islamic calendar. Muhammad's tomb lies within the city's main mosque. Medina also contains the Islamic University, established in 1962. The city trades in cereals, dates and fruit.

Population 210 000

Map Saudi Arabia Ab

Mediterranean climate Warm temperate climate on the western margins of continents, and around the Mediterranean Sea, characterised by hot, dry, sunny summers and moist, warm winters.

Mediterranean Sea Lying between the continents of Europe and Africa, the sea is strictly an arm of the ATLANTIC OCEAN to which it is linked by the Strait of GIBRALTAR. In the north-east it is connected to the BLACK SEA. A submerged ridge between Sicily and Tunisia divides the sea into two parts. In the west are the TYRRHENIAN and LIGURIAN seas and the BALEARIC ISLANDS Basin. In the east are the ADRIATIC, IONIAN and AEGEAN seas. The average depth is 1370 m (4495 ft) and the deepest spot is 4846 m (15 900 ft) down in the Hellenic Trough off south-western Greece.

Many sea-floor features are the result of compression caused by the northward drift of the African plate which has pushed against the Eurasian plate (see PLATE TECTONICS). These plate movements have also thrown up the Alps, the Apennines and other fold mountains, and still cause earthquakes and volcanic activity.

Evaporation and an undercurrent of dense, salty water to the Atlantic removes water from the Mediterranean three times as fast as it is replaced by rain or by rivers. This makes the Mediterranean saltier than the Atlantic and also, to maintain the level, results in a constant inflow of surface water from the Atlantic.

The Mediterranean is almost tideless and so toxic materials tend to accumulate in coastal waters, where they have been discharged. This has caused severe pollution in places, affecting both tourism and the fishing industry. As a result, the Mediterranean nations in 1982 agreed a plan to clean up their sea.

The Mediterranean has been a busy shipping lane for at least 5000 years, and the opening of the SUEZ CANAL in 1869 placed it on the main sea route from northern Europe to Asia.

Area 2 505 000 km² (967 230 sq miles)

Map Africa Da

Mediterranean woodland Wooded areas of drought-resisting species of evergreen trees such as the wild olive, cork oak and conifers together with flowering, often aromatic, shrubs. This type of vegetation is found around the Mediterranean and on the western margins of continents in subtropical latitudes where similar climates prevail. Grazing and deforestation by fire has caused degeneration into areas of scrub such as MAQUIS and GARRIGUE.

Médoc ***France*** One of the greatest wine-producing areas in the world, on the west bank of the Gironde river some 30 km (20 miles) north-west of Bordeaux. It is particularly famous for its red wines, named after the region. Among the renowned vineyards are those of the Châteaux Lafite, Latour, Margaux and Mouton-Rothschild.

Map France Cd

Medway Towns ***United Kingdom*** Riverside towns of Rochester, Chatham and Gillingham in the south-east English county of Kent, about 50 km (30 miles) from London. Rochester was a Roman walled city at the River Medway's lowest bridge point; some parts of the walls remain. There is a splendid 12th-century Norman castle and a cathedral founded in 604 and rebuilt between the 12th and 15th centuries. It is Britain's second oldest bishopric after Canterbury, 40 km (25 miles) to the east. Chatham and Gillingham shared the naval dockyard opened in the 16th-century reign of Elizabeth I and closed in the 1980s, during that of Elizabeth II. Both towns depended heavily on the dockyard for employment and income, and its closure created an urgent need for new jobs. The Medway is 112 km (70 miles) long and flows north and north-east into the Thames estuary.

Population 210 000
Map United Kingdom Fe

Meerut ***India*** Northern industrial city, about 60 km (40 miles) north-east of Delhi. From 1806 it was a British military base, where the Indian Mutiny broke out in 1857. The city's manufactures include textiles, sugar, chemicals and leather goods.

Population 536 600
Map India Bb

Meghalaya ***India*** North-eastern state comprising the Garo, Khasi and Jaintia Hills, which are inhabited mainly by Garo, Khasi and Jaintia tribes people.The south side of the hills is one of the world's wettest places, where the city of Cherrapunji has more than 10 150 mm (400 in) of rain a year. Most of the people are farmers, but the state's main products also include coal, corundum (a very hard mineral used in abrasives) and forest products. Meghalaya covers 22 429 km^2 (8658 sq miles), and its capital is Shillong.

Population 1 335 800
Map India Ec

Megiddo ***Israel*** Archaeological site in the Jezreel Valley, about 15 km (10 miles) south-west of Nazareth. Remains have been found of a fortress built by King Solomon in the 10th century BC and of stables built by King Ahab in the 9th century BC. According to Christian tradition, it is at Megiddo that the battle of Armageddon will be fought on the Day of Judgment.

Map Israel Ba

Meigh Eo ***Ireland*** See MAYO

Meiktila ***Burma*** Textile town in central Burma, about 130 km (80 miles) south of Mandalay. A new textile mill built in 1958 and a series of irrigation reservoirs excavated in the 1960s have boosted the development of the town as the centre of a cotton-growing region.

Population 229 600
Map Burma Bb

Meissen ***East Germany*** Town on the River Elbe 20 km (12 miles) north-west of Dresden. It is world renowned for its porcelain, commonly known as Dresden China. The town is dominated by its fine Gothic cathedral and by Albrechtsburg Castle (1471-85) where, between 1710 and 1863, the delicate blue and white tableware was made. It is now a museum devoted to the product which made it famous.

Population 38 200
Map East Germany Cc

Mekambo ***Gabon*** Small town in the north east, 177 km (129 miles) north-east of Makokou. It is a minor trading centre for rubber gathered in the surrounding forests. The region around Mekambo contains some of the world's richest iron-ore deposits, discovered in 1971. These will be exploited when the projected Trans-Gabon railway opens in the 1990s.

Population 2000
Map Gabon Ba

Mekele ***Ethiopia*** Chief town of war-torn Tigray province, about 500 km (310 miles) north of the capital, Addis Ababa. It is the site of the castle of Yohannes IV, the Emperor of Ethiopia in the 1870s and the 1880s.

Population 50 000
Map Ethiopia Aa

Meknès ***Morocco*** Historic city, 58 km (36 miles) south-west of Fès, founded by the Berbers in the 11th century. Many of the buildings still standing date from the 13th and 14th centuries. In the 17th century Sultan Moulay Ismail made Meknès his capital and built his magnificent Dar El Kebira Palace, a sort of Versailles of the Arab world, which was damaged by an earthquake in 1755. Meknès lies within a huge triple wall – 40 km (25 miles) of battlements and towers – containing the medina (old Muslim residential quarter) and palace. In the palace complex are the ruins of stables that once housed 12 000 horses. Other places of interest include the Bab Mansour gateway, the Kouba El Khiyatine (reception place for foreign ambassadors) and the mausoleum of Sultan Moulay Ismail. Today, the town is a rich market centre; its main industries are traditional – pottery, leatherwork and carpets.

Population 320 000
Map Morocco Ba

Mekong ***South-east Asia*** South-east Asia's largest river with a total length of 4184 km (2562 miles) and a drainage basin covering 310 000 km^2 (nearly 12 000 sq miles). For almost half its length the Mekong flows in a series of deep gorges from its origin in Chinese Tibet as far as the mountains of northern Laos. South of Luang Prabang the valley is much more open, although interrupted by rocky structures and major waterfalls, but it is only below Kratie in Cambodia that it forms an extensive alluvial plain. Following its confluence with the Tonle Sap river near Phnom Penh, the Mekong enters its extensive delta which forms the whole of the extreme south of Vietnam, flowing out to sea in two main branches.

Fed by melting snows from the Tibetan mountains, the Mekong is a river with wide fluctuations in flow and its floods have been destructive throughout history, even making colonisation of the swampy delta difficult. Since the mid-1950s there have been plans to control the river and put its resources to work for man's benefit, but the Mekong is an international river, forming the border of Thailand and Laos for much of its middle course, before flowing through Cambodia and Vietnam. Political divisions have prevented any development of the main stream although the United Nations Mekong Committee, set up to coordinate its development, has done useful work on tributaries, especially in Thailand and Laos.

Map Asia If

Mekong Delta (Mien Nam) ***Vietnam*** The southern plain of Vietnam from the south coast to the foothills of the Truong Son (Annamite) mountain chain. It was colonised by the French (who called it Cochin China) in 1862, and large, prosperous rice and rubber plantations were set up. Its buoyant economy – sharply different from the small-scale farming of the north – was one of the factors that led to Vietnam's partition in the mid-1950s and to the subsequent Vietnam War (1965-75).

Map Vietnam Bc

Melaka (Malacca) ***Malaysia*** Tourist resort on the Strait of MALACCA, about 120 km (75 miles) south-east of the capital, Kuala Lumpur. Founded in about 1400, the town became one of the Far East's chief trading centres. It was colonised in turn by the Portuguese, Dutch and, in 1824, by the British – and buildings put up by each of the colonial powers still survive. The town lost its main source of income when trade shifted to Singapore, farther south, and when silting blocked the entrance to the harbour.

The state of Melaka, of which the town is the capital, extends to 1657 km^2 (640 sq miles) and is largely agricultural, growing rice and rubber. Malacca canes – walking sticks made from the stems of the Asian rattan palm – get their name because many of them were exported through the port.

Population (state) 465 350; (town) 87 500
Map Malaysia Bb

Melanesia ***Pacific Ocean*** One of the three main regions of the Pacific islands (with Micronesia and Polynesia) which are based on ethnic differences. Melanesians are more or less dark-skinned people with frizzy hair who arrived in the Pacific from South-east Asia in waves of migration beginning some 10 000 years ago. They spread from New Guinea (now IRIAN JAYA and PAPUA NEW GUINEA, where even earlier peoples, the Papuans, were already settled) as far east as FIJI and south to NEW CALEDONIA.

Melanesia (which means 'black islands') also includes the SOLOMON ISLANDS and VANUATU.

Traditional Melanesian society is based on village units, each often with its own language, and generally without hereditary kings or chiefs. Elaborate tribal religions involve initiation ceremonies, sacred places and objects, and ancestor worship. The sight of Western material wealth – especially during the Second World War – led to the growth in many parts of Melanesia of 'cargo cults', religious movements based on a belief in the return of the ships and aircraft with fabulous cargo.

Population 5 265 000
Map Pacific Ocean Cc

Melbourne ***Australia*** Australia's second biggest city and state capital of Victoria, standing astride the Yarra river at the head of Port Phillip Bay. It was founded in 1835 by settlers from Tasmania and grew as a result of the Victoria gold rush during the 1850s, when Victoria also became a separate colony. Many buildings date from this period, such as the State Parliament House and the Treasury (1853) and Melbourne University (1855). There are also two modern universities, Monash (1958) and La Trobe (1964). The city centre is laid out on a grid pattern, with spacious avenues. The Botanical Gardens, founded in 1845, cover 43 hectares (106 acres) and contain more than 12 000 plant species from around the world.

Melbourne is one of Australia's main financial, commercial and industrial centres. It exports wheat, wool, meat, fruit and dairy products. Its industries include electronics, engineering, car manufacture, metallurgy, chemicals, textiles and food processing.

Population 2 700 000
Map Australia Gf

Meleda ***Yugoslavia*** See MLJET

Melilla ***Spanish enclave in Morocco*** Seaport 220 km (137 miles) east of Ceuta on the Mediterranean coast. The old walled town stands on a massive rock rising from the sea; the modern quarter lies south and west on the mainland. The entire enclave covers 12 km^2 (about 4.5 sq miles) and exports iron ore from the nearby Beni bu Ifrur mines. The principal industry is fishing.

Population 70 000
Map Morocco Ba

Melnik ***Bulgaria*** See BLAGOEVGRAD

Melos ***Greece*** See MILOS

Memel ***USSR*** See KLAIPEDA

Memphis ***USA*** City on the Mississippi river in the extreme south-west of Tennessee; the metropolitan area extends across the river into Arkansas. It rose to prominence during the steamboat days as a cotton market. Much remains of the plantation era, but today the city is an industrial, commercial, university, agricultural and medical as well as market centre. The Nobel Peace Prize winner and civil rights leader Dr Martin Luther King was assassinated there in 1968.

Rock star Elvis Presley (1935-77) was born in the town of Tupelo in Mississippi, about 150 km (93 miles) south-east of Memphis, but spent most of his life in the city. His Memphis mansion, 'Graceland', is now a museum devoted to his life and his music.

Population (city) 648 000; (metropolitan area) 934 600
Map United States Hc

Menai Strait Strait separating Anglesey from mainland Wales, crossed by a graceful road-carrying suspension bridge built by the Scottish engineer Thomas Telford in 1826. The present rail bridge was built on the surviving stone and brickwork of Robert Stephenson's Britannia Tubular Bridge (1850), burnt down in 1970.

Dimensions 22.5 km (14 miles) long, 180-1100 m (590-3600 ft) wide
Map United Kingdom Cd

Mendip Hills ***United Kingdom*** See SOMERSET

Mendoza ***Argentina*** City about 1000 km (620 miles) west of Buenos Aires, at the foot of the Andes. Founded in 1561, it was destroyed by an earthquake in 1861, and has since been largely rebuilt. It is the centre of Argentina's principal wine-producing region.

Population 600 000
Map Argentina Cb

Mergui ***Burma*** Port in the Tenasserim region, in the extreme south of the country. Fishermen from the port work the pearling grounds among the Mergui archipelago offshore, while cargo boats use the docks to collect rubber and coconuts from nearby plantations.

Population 199 800
Map Burma Cd

Meriç ***South-east Europe*** See EVROS

Mérida ***Mexico*** Gracious colonial capital of the state of YUCATAN, built on the site of a Mayan city by the Spanish conquerors in 1542, after one of the most violent encounters between the Spanish and the Mayas. Its colonial buildings include a splendid cathedral, the richly ornamented Casa de Montejo, and the chapel of Cristo de las Ampollas. Mérida is the gateway to the magnificent Mayan ruins of Yucatán.

Population 424 500
Map Mexico Db

Mérida ***Spain*** Market town about 265 km (160 miles) south-west of Madrid. The town is the site of two irrigation dams built by the Romans in the 2nd century AD and still in use. The only major maintenance work the dams have needed in 1800 years has been the renewal of their stone facings, carried out in the 1930s.

Population 41 800
Map Spain Bc

Mérida ***Venezuela*** Capital of Mérida state, high in the Andes, about 500 km (310 miles) south-west of the national capital, Caracas. Founded in 1558, it is a mixture of colonial and modern architecture around 21 parks. They include the Parque de las Cinco Repúblicas, which contains soils from the five countries liberated by the Venezuelan nationalist Simon Bolívar (1783-1830).

Population 110 000
Map Venezuela Ab

meridian A line of LONGITUDE, an imaginary north-south line drawn between the poles and crossing the Equator at right angles.

Merioneth ***United Kingdom*** See CLWYD; GWYNEDD

Meroe ***Sudan*** Ruined city of ancient Nubia about 280 km (175 miles) north of the capital, Khartoum, on the east bank of the Nile. It was the capital of an Ethiopian kingdom between about 700 and 300 BC, and of the kingdom of Cush between 300 BC and AD 350. Excavations have revealed streets and buildings that indicate a heavily populated city which was well developed industrially and artistically. The remains include a riverside quay, palaces, a temple, and a copy of a Roman public bath.

Map Sudan Bb

Mers El Kébir (El-Marsa El-Kebir) ***Algeria*** Former fortified seaport and French naval base to the north-west of Oran, overlooked by the 16th-century Santa Cruz fortress perched on the hills inland. It was under Spanish control until it was taken by the French in 1830. After the fall of France to Germany in 1940 the harbour was the scene of Operation Catapult, when the British destroyed most of the French warships to prevent them from falling into enemy hands. Today it is an Algerian naval base.

Population 7800
Map Algeria Aa

Mersa el Brega ***Libya*** See MARSA EL BREGA

Mersa Matruh ***Egypt*** See MATRUH

Mersey ***United Kingdom*** River about 110 km (70 miles) long of north-west England, formed by the confluence of the Etherow and Goyt at Stockport in Greater Manchester. It flows into the Irish Sea downstream of Liverpool. Its estuary is of great commercial importance, allowing ocean-going ships to reach the ports of Liverpool, Birkenhead and (via the Manchester Ship Canal) Manchester.

Map United Kingdom Dd

Merseyside ***United Kingdom*** County of north-west England, formed in 1974 out of parts of Lancashire and Cheshire, and covering 652 km^2 (252 sq miles). It is centred on the ports of LIVERPOOL and Birkenhead on either side of the River Mersey. It also takes in the town of St Helens, with its glass industry dating from the 18th century.

Population 1 501 000
Map United Kingdom Dd

Mersin (Içel) ***Turkey*** The country's largest port on the Mediterranean Sea, about 400 km (250 miles) south and slightly east of the capital, Ankara. The port handles cotton, wool, chrome, fruit, cereals and timber, and has an oil refinery. During winter nomadic Turkman tribesmen pitch their goatskin tents around the city to take advantage of the mild climate.

Population 314 100
Map Turkey Bb

mesa Flat-topped elevation with one or more clifflike sides, common in the south-western United States.

Mesa Central ***Mexico*** Vast mountain plateau in the centre of Mexico, fringed by the Sierra Madre Oriental mountains in the east, and the Sierra Madre Occidental in the west. It is about 1130 km (690 miles) long, and has an average height of between 1220-2440 m (4000-8000 ft). The highest point is the volcano Citlaltépetl, 5747 m (18 855 ft), Mexico's tallest peak. Mining was once the mainstay of the region's economy, but irrigated agriculture in the fertile basins is now more important.

Meseta ***Spain*** Central plateau covering about three-quarters of the country (about 350 000 km², 135 000 sq miles).
Map Spain Ca

Meshed ***Iran*** See MASHHAD

Mesolóngion (Missolonghi) ***Greece*** Town on a large lagoon west of Stereá Ellás on the north side of the Gulf of Patras. The town withstood three sieges by the Turks (1822-6) during the Greek struggle for independence. Many of the heroic defenders broke out in 1826. Those who remained lit the arsenal and blew themselves up. The English poet Lord Byron landed at the town in January 1824 to help the Greeks in their fight for independence, but died of fever in April. His heart is buried in a mound beside the Heroon, the monument to the fallen.
Population 10 200
Map Greece Bb

Mesopotamia ***Middle East*** Oil-rich region between the Euphrates and Tigris rivers, stretching from the Armenian Mountains in the north to The Gulf. Most of it lies in modern Iraq. The region's ancient civilisations are credited with the development of writing, mathematics and astronomy, and with the first use of the wheel.
Map Iraq Bb

mesosphere **1.** See ATMOSPHERE. **2.** Solid part of the earth's mantle, between the 'soft' middle mantle – the asthenosphere – and the core. See JOURNEY TO THE CENTRE OF THE EARTH (p. 200)

Mesozoic era Third great era in the evolution of life, when the dinosaurs and other reptiles were dominant. See GEOLOGICAL TIMESCALE

Messina ***Italy*** Seaport near the north-east tip of Sicily, facing the toe of Italy, 6 km (4 miles) away across the Strait of Messina. It handles most of the traffic crossing the strait, which was the location in Greek mythology of the sea monster Scylla and the whirlpool Charybdis.

The city was founded in the 8th century BC, but its period of greatest prosperity as a trading centre began with the Crusades in the 12th century and lasted until the 14th. Cholera, earthquakes and war killed many thousands of its citizens over the centuries, and only a handful of ancient buildings have survived the earthquakes common on Sicily. The oldest is the 12th-century church of Annunziata dei Catalani. The cathedral is a 20th-century reconstruction of the original Norman building.
Population 266 300
Map Italy Ee

Messina, Strait of (Gulf of Messina) Lying between the Italian mainland and Sicily, linking the IONIAN and TYRRHENIAN seas. Its strong currents and numerous whirlpools gave rise to legends about monsters, including Charybdis, whom Zeus turned into a whirlpool, and Scylla. Scylla had six heads, each with three rows of teeth, which were ideal for plucking rowers from passing ships.
Dimensions 56 km (36 miles) long, 6-48 km (4-30 miles) wide
Map Italy Ee

▲ BRIGHT LIGHTS Luna Park still comes to life after dark. The fun fair and the vast Palais Theatre made St Kilda Melbourne's most popular seaside playground at the turn of the century.

Mestre ***Italy*** See VENICE

Meta ***Colombia/Venezuela*** River, 1000 km (620 miles) long, rising in the eastern Andes, near Bogotá. It flows north-east, then east, forming part of the Colombia-Venezuela border, to Puerto Carreño, where it joins the Orinoco.
Map Colombia Bb

metamorphic rock Igneous or sedimentary rock that has been altered in chemical composition or texture, or both, by heat, pressure or the chemical action of migrating fluids – or by a combination of any of these. Examples include granulite (which resembles granite but is a metamorphosed form of sandstone), MARBLE (metamorphosed limestone) and SLATE (which was once shale).

metamorphic zone Region where metamorphism has taken place.

metamorphism The process by which rocks within the earth's crust are altered, by heat, pressure and chemical action.

Meteora ***Greece*** Eroded valley in THESSALY, on the eastern side of the Pindhos mountains, that is remarkable for its monasteries. From the valley floor, huge iron-grey towers of rock rise almost perpendicularly for up to 300 m (950 ft). In the 14th century Byzantine monks built 24 monasteries on these precarious but impregnable towers. Today, five remain, the largest being Great Meteoron.

The monasteries were once accessible only by windlasses, ropes and cut steps, but now narrow paths wind up the towers past caves where hermits once lived. The monasteries contain Byzantine artefacts, including rare books; some are open to the public.
Map Greece Bb

meteorite Piece of solid material from space which passes through the atmosphere without being completely burnt up and reaches the surface of the earth.

Metropolitana (Santiago) ***Chile*** Central region taking in the capital, SANTIAGO, and the provinces of Melipilla and Maipo. The town of Maipú, just south-west of Santiago, was the site of the liberator José de San Martín's victory over the Spaniards in 1818. It has monuments to commemorate the battle. Wheat, fruit and livestock are the main products of the countryside, supporting the industries of Santiago and its port Valparaíso. The ski resort of Farellones lies in the Andes nearby.
Population 4 572 000
Map Chile Ac

Metz ***France*** City straddling the Moselle river near the German border, about 280 km (175 miles) east of Paris. A military strongpoint since Roman times, it was fortified in the 17th century and again when it was under German rule from 1871 to 1918; parts of the walls and fortifications are still standing. The city lies in a fertile area producing Moselle wines, and at the heart of the iron and steel region of Lorraine.
Population 194 800
Map France Gb

Mexico

VIOLENCE BUILT MEXICO, WHERE EVEN THE GROUND TREMBLES. ITS PAST WAS BIZARRE AND TRAGIC; ITS PRESENT SEES HOPES OF PROSPERITY FADING

More than 17 million people live in MEXICO CITY, capital of Mexico. It is the largest and fastest-growing city on earth. So rapidly is Mexico's population expanding that its people are flooding into the United States – either legally or illegally. Over 10 million Mexicans are estimated to be legally resident in the USA. No one knows how many more have crossed the border off the record.

Mexico can be a wonderful place to visit – as booming tourist figures prove. It can also be a wonderful place to live. Sadly for many, it is also a great country to get out of. For despite – and perhaps because of – social and political reforms won by centuries of blood and tears, the population is outstripping the country's ability to support its people. Every year more than 2 million newborn Mexicans must be provided for. Mexicans joke that in crossing the US border they are just winning back California – which used to be part of their country, along with Texas, Arizona and New Mexico.

America responds with strong border patrols – and counter-invasions by millions of *gringo* tourists bound for ACAPULCO and other superb resorts on the Pacific coast. Tourist attractions include more than 10 000 awesome monuments to ancient civilisations that pepper the country, and the Latin attitude that makes for maximum enjoyment of the many feast days and fiestas throughout the calendar.

Nowhere are these festivals celebrated with more enthusiasm than in Mexico City itself. Appalling slums it may have; air made barely breathable by millions of smoking exhausts; whole areas destroyed in the 1985 earthquake or subsiding gently into the spongy ground. But style, exuberance, treasures of art and architecture and a talent for living it also has.

The city's 2357 km^2 (nearly 910 sq miles) embraces the last great temple of Montezuma's Aztec capital, Tenochtitlan, the palace of his Spanish conqueror Cortés, Olympic sports stadiums and the world's biggest bullring, seating 60 000. Modern buildings decorated by Diego Rivera and other Mexican masters of the monumental mural live with Spanish-colonial churches and the quiet, elegant plazas of secluded neighbourhoods that could be in another time, another place.

Mexico was literally born of violence, for it lies on a particularly sensitive part of the earth's crust and remains prone to earthquakes. It is a land of volcanic mountain ranges and high plateaus, with lowlands only along the Pacific coastline, the Gulf of Mexico and the YUCATAN peninsula. The spectacular 5452 m (17 887 ft) volcano POPOCATEPETL, dormant since 1802, looms south of the capital; beyond it rises 5699 m (18 697 ft) CITLALTEPETL (Pico de Orizaba), the highest peak in Mexico and perpetually snow-capped. About 70 per cent of the country is over 500 m (1640 ft) above sea level. In the north-west the SIERRA MADRE OCCIDENTAL mountains lie between the coast and the great central plateau, a region of high plains and great deserts, merging into broad valleys, lakes and dormant volcanoes towards the south. The SIERRA MADRE ORIENTAL range rises between the eastern edge of the plateau and the coast. South of Mexico City the SIERRA MADRE DEL SUR extends to the Guatemalan border, broken by the narrow isthmus of Tehuantepec.

ANCIENT LEGACIES

The first recorded civilisation in the territory was that of the Olmecs, who lived near the eastern coast from 1300 BC and left many carvings and sculptures. Next came those great empire-builders the Mayas of Yucatán, who flourished from about AD 300 and spread through the central uplands and south into Guatemala, Honduras and El Salvador. In the early 13th century, the Aztecs swept down from the north to dominate much of the area. Their astonishing cities, fabulous wealth and horrific religious ceremonies amazed and appalled even the invading Spaniards of the 16th century.

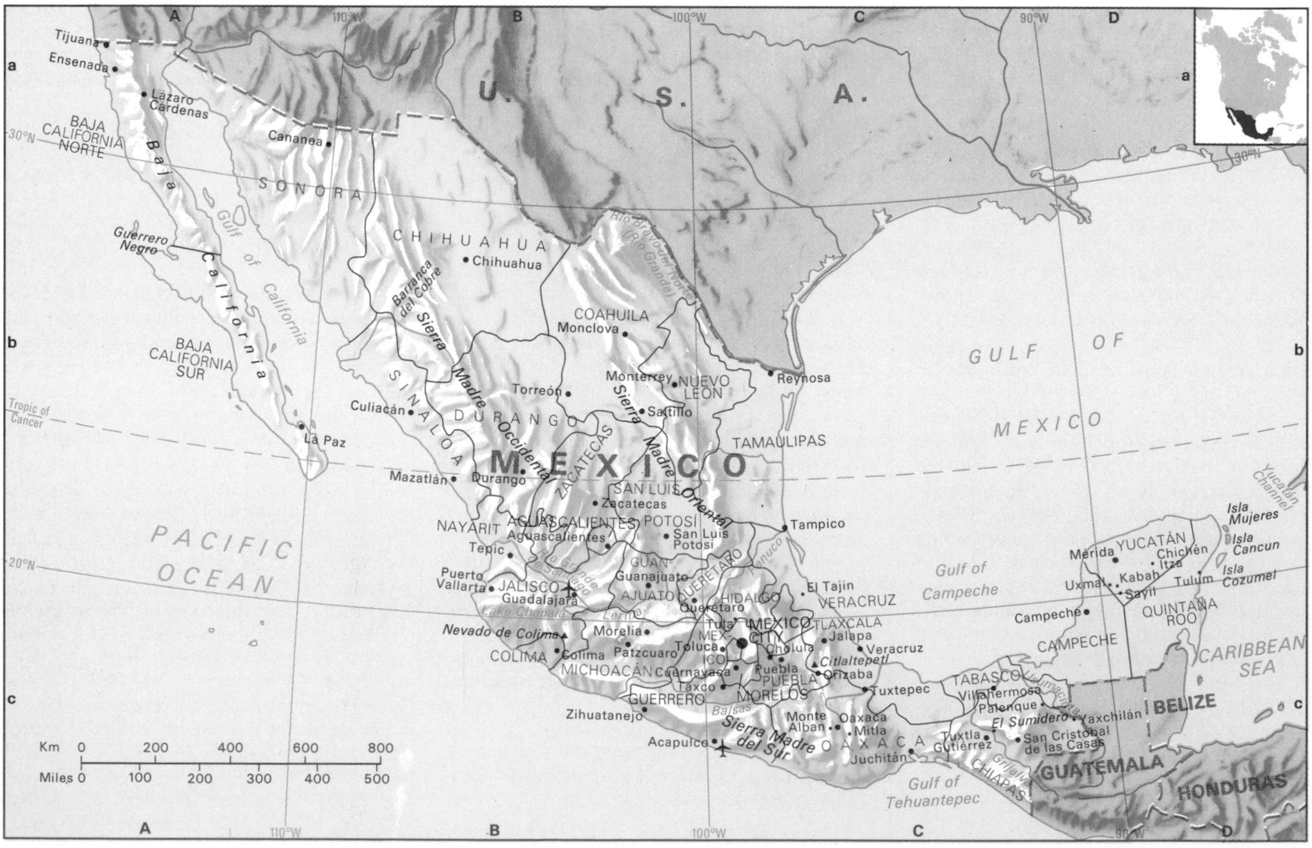

Blood ran down the steep steps of the Aztecs' pyramid temples as grotesquely masked priests wielding obsidian knives slashed open the chests of their victims and tore out their still-beating hearts. Countless thousands died to propitiate hundreds of gods and goddesses. In 1487, when the Aztecs dedicated their Great Temple of Tenochtitlan, they are said to have sacrificed people four at a time from dawn to sunset for four days.

But one of their most potent gods, the plumed serpent Quetzalcóatl, helped in their downfall. The god embodied the spirit of a once all-too-human ruler, a fair-bearded dedicated drunk who fled his throne in disgrace, but was expected back some day, from the east. When a bearded white-skinned man in a shiny metal suit turned up in 1519, the Aztec emperor, Montezuma II, thought Quetzalcóatl had come home. The blond stranger turned out to be the Spaniard Hernán Cortés, who with his soldiers and his priests would kill, torture, enslave and 'convert' the Aztecs and other peoples of Mexico with a zeal ferocious enough to match their own.

It took Mexico three centuries to cast off the iron grip of Spain, and almost another one to free the people of other assorted adventurers and oppressors. Even now, foreign interests loom large in the economy and the government has imposed austerity measures in the face of runaway inflation and huge debts.

Little remains of Montezuma's magnificent island capital on Lake Texcoco – or even of the lake itself – which has been submerged, built over and lost in the growing sprawl of today's capital. The site of the Great Temple is being excavated, but the most important building in the vicinity today is a Spanish-built cathedral, which was started in 1573.

Not only the works of the Aztecs survive, so does their language – Náhuatl – as well as the monuments and languages of the Maya, Zapotec, Otomí and Mixtec peoples of pre-Columbian Mexico. From these tongues are derived the 59 different dialects spoken among over 3 million of the country's 25 million or so 'Indians', who form about 30 per cent of the population; another 55 per cent are *mestizo* – of mixed European and Indian descent – and 16 per cent are *criollos*, of European stock.

OLD COUNTRY OF YOUNG PEOPLE

Mexico has been called an old country of young people: 48 per cent under 15 years old and more than 50 per cent under 20. Throughout the present century the birthrate has been high; population is currently growing at 2.5 per cent a year, compared with 0.9 per cent in the United States and 0.1 in the United Kingdom. Rapid economic growth in a half century of relative calm is a major cause of this growth.

Extensive oil discoveries and increasing industrialisation have caused a massive shift of people from countryside to towns and cities – about 65 per cent now live in urban areas, one-fifth of them in or around the capital. The population of GUADALAJARA city has topped 2 million; MONTERREY and PUEBLA have both passed the million mark. Many millions have simply exchanged rural poverty for urban deprivation, but despite low incomes and poor housing they are in many ways better off: most houses have electricity, most children go to school, and social services and health care are more efficient.

▲ NATURE'S HARSHNESS Barren granite mountains stretch inland from the Pacific coast on the Baja (Lower) California peninsula. Sparse sage scrub is the only natural vegetation that can survive on the poor desert soils, which in some years receive no rain at all.

The great oil finds were on the east coast around the city of TAMPICO and in the state of TABASCO and offshore in the Gulf itself. Mexico became one of the world's major oil producers, with reserves second only to those of Saudi Arabia. However, great industrial and public service projects financed by investments and foreign loans geared to the oil boom turned sour in the late 1970s and early 1980s as oil prices began falling and interest rates soared. In 1982, inflation zoomed from a previously 'normal' 10 to 20 per cent to 99 per cent, and foreign debts equalled a staggering 82 per cent of the gross national product. In a bid to stem the tide of cash pouring out of the country, the government nationalised Mexico's 54 private banks. Further economic measures reduced inflation to 56 per cent by 1984. Tightened controls on spending and some international financial diplomacy have won a breathing space for interest repayments.

However, Mexico is among many nations to suffer, and is better equipped than most to win through today's economic turbulence. Industrial output in 1981 included over 7 million tonnes of steel, about 20 million tonnes of cement and 600 000 motor vehicles: the country still turns out most of its own 'consumer durables'. Oil production in 1984 was 2.76 million barrels a day and Mexico remains, as it has been for centuries, one of the leading producers of silver – nearly 2000 tonnes a year from 70 mines. Copper and sulphur are also mined in large quantities. And despite industrialisation, agriculture continues to be a major factor in the economy, though now employing only about one-third of the workforce.

LANDLESS FARMERS

Mexico still grows most of its own food and is, indeed, supplying the United States on an increasing scale. Maize and beans are the staples and there is large-scale cattle ranching; fruit and vegetables are grown in the north-west and are an important export. That the land supports such a minor part of the workforce reflects both the natural difficulties that beset farmers on much of the cultivable land, and the disappointing outcome of numerous attempts to redistribute the land since the end of Spanish rule. Apart from the western highlands and the jungles of the south-east, the country gets little rainfall – and almost none at all across the whole of the great central plateau. Most of the cultivable land needs irrigation, and vast government projects have almost quadrupled irrigated areas to 3.5 million hectares (8.5 million acres) since the 1950s.

Despite independence, revolution and the stirring exhortations and promises of a succession of charismatic leaders, most of the Mexican people never did win their promised lands. And now there are simply too many of them to make practicable any scheme to share what is – or is ever likely to be – available. Right up to the revolution of 1910, just 1 per cent of the population owned 97 per cent of the land, 96 per cent of the people owned 1 per cent of the land, and 88 per cent of the peasant farm workers had no land.

COSTLY WAR OF INDEPENDENCE

Napoleon's invasion of Spain in 1808 started unrest within the colony of Mexico, which scented independence and won it in 1821 after a war which lasted 11 years and cost 600 000 lives. In 1836 the cattle and cotton barons of Texas, angered by the abolition of slavery, rebelled and declared their own independence. The resulting conflict developed into the Mexican War against the United States (1846-8). Mexico was heavily defeated and had to cede Texas, Arizona, California and New Mexico to the US for little more than US$28 million.

Conflict continued as Benito Juárez introduced reforms during two spells as president. His presidency was interrupted by a regime headed by Archduke Maximilian of Austria, who was installed as emperor by a French invasion force supported by conservative Mexicans.

A paternalistic but autocratic dictator, Porfirio Diáz, ruled from 1876 to 1910 and saw huge foreign investments pour in, 24 000 km (nearly 15 000 miles) of railway laid, industries and mineral resources developed and a civil service created. Unfortunately the resultant prosperity did not extend to the peasant masses; their land was systematically stolen and many of them were reduced to semi-slavery in mines and on plantations. A law passed by Diáz in 1883 resulted in 55 000 km^2 (214 285 sq miles) of 'national' land being sold off at bargain prices to rich criollos, foreigners and his own personal friends. One wealthy family ended up with 50 ranches covering 30 000 km^2 (11 583 sq miles).

There was increasing unrest, and finally revolution. When it came in 1910 it was confused, horrific and long-drawn-out, as peasants, smallholders, bandits, Indians and even middle-class ranchers rose in fury. Diáz fled to exile in Paris and a succession of revolutionary leaders came to the fore. They included the teenage bandit Pancho Villa, who built a superb guerrilla army with guns and artillery captured in daring raids on the military trains of the government troops.

At appalling cost, the revolution triggered real social upheavals. A constitution of 1917 heralded major land reforms, advanced social welfare schemes, compulsory free education, minimum wages and the right to strike. Many of these aims would not be achieved until President Lázaro Cárdenas held office, between 1934 and 1940. Only then were the great estates parcelled out into *ejidos*, or common lands, and education made widely available.

Oil fields and railways were nationalised; power stations and the major airlines followed. Land reform remains a continuing process and agricultural development in some areas – such as the north-west – has been spectacularly successful. However, this has benefited mainly those companies or middle-class families which have sufficient capital to invest in irrigation schemes, fertilisers and mechanisation. Outside the growth areas, rural life continues to be something of a struggle.

PRESERVING THE INDIAN CULTURE

Perhaps due to its exceptionally violent history, Mexico has not attracted the mass immigration from Europe that has done so much to shape the rest of the continent. And given the large native population present when Cortés landed – estimated at 7 or 8 million – the country has retained a distinctively Indian identity. Indeed, ever since the Zapotec-born President Juárez's time, there has been a pronounced national bias in favour of the Indian culture and race in official pronouncements and in public monuments. How deeply this feeling really permeates is hard to assess, since Mexico retains a highly stratified, class-based society in which pale faces seem more often than not to occupy the positions of real power. It may also be more than a coincidence that the majority of people in the poorest regions of the south are Indian.

Indeed, the socialist ideology proclaimed by successive revolutionary and elected governments is embarrassingly at odds with the glaring contrasts in income and wealth that are the actualities of life in Mexico. That political stability has been maintained is due largely to the skill of the majority party in co-opting opposition groups while retaining the support of its traditional allies. The Institutional Revolutionary Party (PRI) has controlled Mexican politics since it was formed in 1929. Its power base is a triple alliance with labour, the peasantry and state employees. The PRI is now so closely intertwined with the government apparatus that it is difficult to untangle them: thus the party supports the state and vice versa.

Since opposition parties were banned until a few years ago and are still very weak – as recent elections showed – policies are determined by the results of power struggles within the PRI. The most vital of these conflicts is about choosing a party candidate for the presidency. Since the PRI candidate always wins, and the president has almost absolute power, the choice is critical both to the country and the party, for the successful candidate holds office for six years. However, he is then barred from ever seeking re-election – one of the few rules in Mexican politics.

Culturally, the country is very rich. It has produced not only Rivera but two other great masters of the mural – José Clemente Orozco and David Alfaro Siqueiros; internationally famous writers such as Octavio Paz and Carlos Fuentes; modern architects such as Luis Barragán and Pedro Ramírez Vázquez. It possesses a wealth of Indian cultures which contribute to contemporary song and dance, as performed by Mexico's renowned folklore ballet company. And, as in the beginning, it has its numberless treasures of other, unknown masters – the sculptors, artists and architects of the early civilisations.

MYSTERY OF THE MONUMENTS

To many visitors the ruined cities, spectacular pyramids and huge, mysterious monuments of ancient peoples are the abiding glories of Mexico. They survive in great numbers, but perhaps the most striking are the huge pyramids at TEOTIHUACAN, about 55 km (34 miles) north-east of Mexico City. This was the cultural centre of a trading empire that flourished between 300 BC and AD 750. Archaeologists believe that as many as 200 000 people lived in its grid network of streets covering 21 km^2 (8 sq miles) – making the city larger than Imperial Rome. The Pyramid of the Sun once soared in four massive terraces to a height of 65 m (213 ft); its neighbouring Pyramid of the Moon, though smaller, is still 60 m (197 ft) high. The city was destroyed by a mysterious fire and had been deserted for a thousand years when the Spaniards first saw it.

Maya ruins include the magnificent city-complexes of CHICHEN ITZA and Uxmal, in the jungle of the Yucatán peninsula; PALENQUE, in CHIAPAS, is another. Maya beginnings go back to 2000 BC, but their golden age was AD 250-900. They were sophisticated mathematicians and astronomers, with an accurate calendar calculated back more than 3000 years, which took leap years into account. An observatory survives in Chichén Itzá; in a carved crypt at Palenque was found the skeleton of a 7th-century ruler covered with jade

▲ **NATURE'S FURY Rescue workers and survivors gather beside a ruined hotel in Mexico City after the major earthquake of 1985. Over 7000 lives were lost, but more than a week later a number of victims, including several newborn babies, were still being rescued from the debris.**

jewellery – necklaces, bracelets and rings. The Mayas' hieroglyphic writing is, unfortunately, still only partly understood.

MONTE ALBAN, in OAXACA, was the capital of the Zapotec culture. Its huge main square, measuring 600 by 400 m (1970 by 1310 ft) was formed by levelling an entire hilltop. Pyramids, terraces, sculptures, tombs and staircases are dramatic evidence of how the place looked at the height of Zapotec power, in the 9th century AD.

Mexico's ancient cities once blazed with colour – vast murals, vividly decorated statues, the feathered capes and plumed headdresses of rulers, priests and warriors. Some of that colour survives in peeling frescoes, on pottery and on the fierce images of the gods they worshipped – faded now but bright enough to show why the modern Mexican still loves a bold stroke of the paintbrush, and where Rivera and his fellow artists found much of their brilliant inspiration.

MEXICO AT A GLANCE
Area 1 972 547 km² (761 600 sq miles)
Population 81 650 000
Capital Mexico City
Government Federal republic
Currency Peso = 100 centavos
Languages Spanish, Indian languages
Religion Christian (Roman Catholic)
Climate Tropical and temperate according to altitude. Average temperature in Mexico City ranges from 6-19°C (43-66°F) in January to 12-26°C (54-79°F) in May
Main primary products Maize, sorghum, wheat, barley, rice, cotton, sugar, coffee, beans, fruits, cattle; uranium, copper, iron, coal, lead, zinc, silver, gold, oil and natural gas, aluminium, phosphates
Major industries Oil and natural gas production and refining, agriculture, mining, iron and steel, aluminium refining, vehicles, cement, machinery, textiles, pottery
Main exports Oil and gas, nonferrous ores, machinery and industrial goods, coffee, chemicals, cotton, fruit and vegetables, shrimps
Annual income per head (US$) 2090
Population growth (per thous/yr) 25
Life expectancy (yrs) Male 64 **Female** 68

Meuse (Maas) ***France/Belgium/Netherlands*** River 935 km (580 miles) long. It rises in Lorraine, eastern France, and flows north, through the edge of the Ardennes upland, into Belgium and on to the Netherlands, where it is known as the Maas. It shares its delta with the Rhine, with which it is linked by canals. It is navigable (with lateral canals in places) from the sea up into France, where it is linked by canals eastwards with the Rhine and westwards with the rivers of the Paris Basin. It carries heavy barge traffic (coal, iron ore, stone, sand and oil), especially through Liège, where it meets the Albert Canal from Antwerp. Its valley is lined with towns – Verdun, Sedan, Namur, Liège, Maastricht – whose names recur in Europe's wars, fortress towns in what has long been a political buffer zone.

Map Belgium Ba

Mexico See p. 426

Mexico ***Mexico*** State west of MEXICO CITY's Federal District, in the mountainous centre of the Republic of Mexico. A quarter of Mexico City's inhabitants and many of its industries have spilled over into Mexico state, but the state capital, Toluca, 56 km (35 miles) south-west of Mexico City, is a busy commercial centre in its own right.

Population 8 500 000

Map Mexico Cc

Mexico, Gulf of World's largest gulf, linked to the ATLANTIC OCEAN by the Straits of FLORIDA and to the CARIBBEAN SEA by the YUCATAN CHANNEL. Its warm waters were alive with pirates and adventurers in the late 18th and early 19th centuries. Its broad continental shelves contain important oil deposits. Warm waters from the Gulf of Mexico contribute to the GULF STREAM via the Florida Current.

Area 1 544 000 km² (596 140 sq miles)

Greatest depth 4377 m (14 360 ft)

Map South America Ab

Mexico City ***Mexico*** Capital of the Republic of Mexico and of Mexico City Federal District. It lies 2200 m (7350 ft) above sea level on the southern end of Mexico's high central plateau, where the air is exceptionally thin.

The homes of Mexico City, many of them single-storey houses or shanty dwellings, sprawl over a vast area across the plateau, beyond the Federal District and into the state of Mexico. The city has approximately 17 million inhabitants and is constantly growing. Over half the Mexicans employed in industry work there, in metallurgical, chemical, food-processing, electronics, tyre-making and textile factories, in cement plants and in paper mills. But in spite of this activity, large numbers of the people live in abject poverty. Many try to make a living as street-sellers, rubbish pickers, scavengers or prostitutes. There are countless beggars in the city, and numerous people live in squalid housing that they built themselves.

The excessive concentration of industry has added to the air pollution and the city is often veiled in a haze of pink smog. In September 1985 an earthquake, measuring 8.1 on the RICHTER SCALE, destroyed hundreds of buildings in the city. Over 7000 people were killed.

But in spite of this depressing picture Mexico City contains pockets of great beauty and historical interest. The city was built on the ancient Aztec site of Tenochtitlan, the ruins of which are being excavated near the central plaza. The plaza itself is surrounded by magnificent colonial buildings, and another fine square, the Plaza de Las Tres Culturas (Square of the Three Cultures), contains examples of Aztec, Baroque and modern architecture – illustrating Mexico's three main cultural influences. The imposing modern buildings of the university campus are covered with colourful murals depicting Mexico's struggle for independence and the subsequent revolution. Evidence of the nation's rich early history is preserved in the city's famous Museum of Anthropology, which houses over 60 000 exhibits of ancient treasures.

Some of Mexico City's residential areas are notable, too, for their grandeur or picturesqueness. The houses around Chapultepec Park are among the most luxurious in the country, and lie close to the elegant shopping area of the *Zona Rosa* (Pink Zone). The neighbourhood of San Angel in the south of the city has a different kind of charm, with a village atmosphere, especially on Saturdays when its craft bazaar is held.

Population 17 000 000

Map Mexico Cc

Mezada ***Israel*** See MASADA

Miami ***USA*** City in south-east Florida, separated from the Atlantic Ocean by Biscayne Bay and a fringe of spits and islands. It has developed in the 20th century into an industrial, financial and business centre, and its international airport and port are always busy. It also has one of the highest crime rates in the country. It is the southernmost large city of the continental USA, and its proximity to the Caribbean has resulted in a large community of Cuban exiles from the Castro regime.

Population (city) 372 600; (metropolitan area) 1 706 000

Map United States Je

Federated States of Micronesia

ONE OF THE WORLD'S YOUNGEST NATIONS IS A SCATTERING OF HUNDREDS OF ISLANDS OVER THOUSANDS OF KILOMETRES

Spread over a vast area of the western Pacific to the north of Papua New Guinea are the 600-odd islands and atolls of one of the world's youngest independent nations, the Federated States of Micronesia, or FSM. After a succession of colonists – from Spain, Germany and Japan – followed by US rule as part of the Trust Territory of the Pacific Islands and 16 years of negotiations, the US Congress in 1985 approved a compact of free association for the FSM (as it did for the neighbouring Marshall Islands). The agreement gives self-government to the islands while safeguarding American defence interests. The final hurdle is United Nations approval.

The new nation stretches more than 3200 km (2000 miles) from end to end, yet the total land area – 701 km² (271 sq miles) – is only a quarter that of Rhode Island, America's smallest state. One island – POHNPEI, on which the capital, Kolonia, stands – accounts for almost half the land. The country consists of four states – from east to west, KOSRAE, Pohnpei, TRUK and YAP. Farther west still is the Republic of Belau, which decided in 1978 to go its own way to independence. The FSM and Belau together make up the Caroline Islands, a name given by the Spanish.

They first claimed the islands in the late 17th century, but the forebears of the present native Micronesian population arrived there from Asia at least 2500 years earlier. And some 600 years ago an unknown civilisation on Pohnpei built a community – Nan Madol – of stone buildings on artificial islands separated by a system of canals. Today the normal accommodation for many islanders is a wooden hut.

The FSM has received (and will continue to receive) substantial American economic aid. In the bigger towns many people follow an Americanised way of life, but elsewhere the lifestyle has been touched remarkably little by centuries of colonial rule. Here, the people are subsistence farmers and fishermen. Copra is the main cash crop, and there are rich tuna fishing grounds offshore, exploited mainly by American, Japanese and Korean fleets. However, the FSM has established a 200 nautical mile (370 km) economic zone, bringing in revenue from the fishing. The nation's main economic prospects lie in the growing development of fishing and tourism.

FEDERATED STATES OF MICRONESIA AT A GLANCE

Map Pacific Ocean Cb

Area 701 km² (271 sq miles)

Population 89 000

Capital Kolonia

Government Federal republic

Currency US dollar = 100 cents

Languages English, Micronesian dialects

Religion Christian (mainly Roman Catholic)

Climate Tropical; average temperature at Kolonia is 26-27°C (79-81°F) all year

Main primary products Coconuts, yams, cassava, tropical fruits, fish

Major industries Copra, agriculture, fishing, food processing

Main exports Copra, fish, handicrafts

Annual income per head (US$) 424

Population growth (per thous/yr) 22

Life expectancy (yrs) 66

Miami Beach *USA* Booming resort in southern Florida, on an island 5 km (3 miles) off the Atlantic coast. It is connected by four causeways to the mainland city of Miami, and is a concentration of hotels, apartment blocks and holiday homes. A string of resorts stretches almost unbroken for 140 km (85 miles) north from Miami Beach to Palm Beach and beyond.

Population 95 800

Map United States Je

mica Any of a group of complex silicate minerals occurring as thin flaky sheets in igneous and metamorphic rock. The two main members are muscovite, which contains aluminium and potassium, and biotite, containing aluminium, potassium, magnesium and iron.

Michigan *USA* State made up of two peninsulas on the Great Lakes: Lower Michigan, flanked by Lakes Huron and Michigan; and Upper Michigan, between Lakes Superior, Michigan and Huron. The latter is more hilly, forested and less populated. The lower peninsula has some forest but is more fertile, and dairy produce, maize, beef, fruit and vegetables are the mainstays of farming. Iron ore, oil and gas are produced, and manufactures include cars and accessories – especially in and around DETROIT – machinery and processed foods. The state covers 150 780 km² (58 216 sq miles) and the state capital is LANSING.

Population 9 088 000

Map United States Ia

Michigan, Lake *USA* The only one of the Great Lakes wholly in the USA. It covers about 57 750 km² (22 300 sq miles), and CHICAGO and MILWAUKEE are its chief ports.

Map United States Ib

Michoacán *Mexico* State in the south-west, on the Pacific coast. It is a place of lush woodland, volcanoes, hot springs, lakes and rivers – and the home of the Tarascan Indians who still practise traditional customs and crafts. Lakeside towns such as Pátzcuaro; the old Indian capital of Tzintzuntzan and the picturesque colonial capital, Morelia, are popular with tourists. Farming and iron mining are important, and Michoacán's pottery and lacquerware are famous.

Population 3 200 000

Map Mexico Bc

Micronesia *Pacific Ocean* Geographically one of the three main regions of the Pacific islands (with Melanesia and Polynesia) which are based on ethnic differences. The name means 'small islands', and it consists of thousands of tiny islands and atolls, mainly in the North Pacific. It stretches from BELAU in the Caroline Islands to the MARSHALL ISLANDS and KIRIBATI (which straddles the Equator); also included are GUAM, the Federated States of MICRONESIA, NAURU and the NORTHERN MARIANAS.

The Micronesians first settled these islands from Asia about 3000 years ago. They are copper-coloured with straight black hair, and have probably interbred with the Polynesians. They had hereditary chiefs but in most cases no priests or sacred sites, and were great seafarers.

Population 315 000

Map Pacific Ocean Cb

microseism Faint, recurrent tremor of the earth's crust, due to natural causes such as wind and waves, or to traffic and machinery.

Middelburg *Netherlands* Provincial capital of Zeeland and resort on the island of Walcheren, some 130 km (80 miles) south-west of Amsterdam. Most of the old core, ringed by canals, has been successfully rebuilt after destruction in the Second World War.

Population 38 700

Map Netherlands Ab

Middle East As a geographical definition, 'the Middle East' has undergone a change of meaning. Before the First World War, Western Europe distinguished the Near East from the Middle East. The Near East meant the old Ottoman Empire: Turkey and the countries bordering the eastern Mediterranean; while the Middle East defined Afghanistan and the lands about The Gulf. When the Ottoman Empire expired in the 1920s, the Near East ceased to be a definite unit and the countries that once comprised it are now generally included in the Middle East. Today this collective label covers Turkey, Syria, Jordan, Israel, Lebanon, Afghanistan, Iran, Iraq, Saudi Arabia, South Yemen, Yemen, Oman, the United Arab Emirates, Qatar, Bahrain and Kuwait. As a political concept, the term also includes Egypt and Cyprus.

The northern and southern extremities of the Middle East are two great peninsulas: the Turkish peninsula between the Black Sea and the Mediterranean, and the Arabian peninsula bordered by the Red Sea, the Arabian Sea and The Gulf. The Caspian Sea forms part of Iran's northern border and its shore therefore marks part of the northern limit of the Middle East.

The region includes some of the world's highest and most spectacular mountains. In Turkey two great ranges run from west to east: the Pontus Mountains in the north and the Taurus Mountains in the south.They meet in the mountainous tangle of the Armenian Knot, whose greatest peak is Ararat (5165 m, 16 945 ft). From the Knot, the Elburz Mountains stretch across northern Iran to merge with the western and lower extremity of the high ranges of Afghanistan. The most spectacular of these, the Hindu Kush, rise to over 7500 m (24 600 ft) and are the western extremity of the Himalayan and Pamir system of central Asia. Some of the passes through these Afghan ranges soar to 5300 m (17 000 ft). Finally, the Zagros range, rising to 4548 m (14 921 ft) at Zardeh Kuh, runs south-east from the Armenian Knot through Iran to form a barrier that faces Iraq and The Gulf.

The names of the region's two great rivers, the Tigris and the Euphrates, ring through the history of mankind, for it was in the land between them, about 6000 years ago, that Western civilisation may well have begun. The Euphrates, 2720 km (1690 miles) long, rises in Turkey, then flows through Syria and Iraq, irrigating the farmlands of three countries. The Tigris, 1899 km (1180 miles) long, also rises in Turkey, but for much of its course flows through Iraq. There are a number of rivers in Afghanistan and Iran, the Helmand for example, which have no outlet to the sea; a number of others dry out for part of the year. The vast Arabian peninsula, however, possesses virtually no rivers at all.

The feature that gave rise to the early Middle Eastern civilisations is the FERTILE CRESCENT. Extensive irrigation schemes based on the Tigris and the Euphrates, and reinforced by a moderate rainfall, have made this one of the most productive regions in the world. Long ago, it provided perfect conditions for the agricultural communities that grew into the city-kingdoms of Sumer, Babylon and Assyria.

But the climate of much of the Middle East is harsh. The region presents some of the hottest places on earth: the Rub al-Khali (Empty Quarter) of southern Arabia, for example, the An Nafud desert in the north of the peninsula, or the Dasht-e-Kavir, the central desert region of Iran, where temperatures as high as 55°C

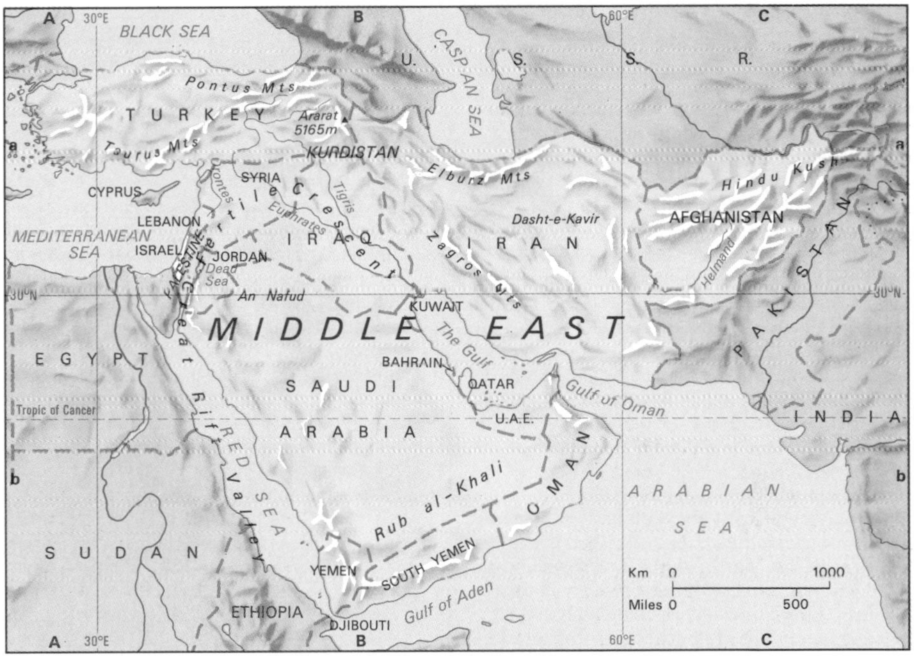

▲ **ARTISTS' RETREAT A gleaming white windmill stands over the resort of Míkonos. Like the houses, it contrasts with the deep blue sea, providing stunning scenes for painters to capture.**

(130°F) in the shade have been recorded. The climate of central Turkey and that of much of Iran and Afghanistan is continental with searing heat in summer and temperatures as low as −37°C (−35°F) in winter. Vegetation conforms to the climate: Turkey, northern Iran and Syria have conifer and pine forests, the Mediterranean coastal countries are typified by scrubland. In The Gulf and Arabia, most of the little that grows is found in the oases. The rest of the land is desert.

Until very recently the economy of these arid regions was not unlike that described in the Old Testament – a world of peasant farmers and nomadic herdsmen. But the discovery of oil and later of other minerals has completely transformed the area. The Gulf contains the greatest oil reserves in the world, which have brought vast wealth and modern development.

Area 7 916 058 km² (3 056 405 sq miles)
Population 176 million

Middle West ***USA*** See MIDWEST

Middleback Ranges ***Australia*** Hills rich with iron ore about 260 km (160 miles) north-west of Adelaide in South Australia. Ore from mines such as Iron Knob and Iron Monarch is sent some 80 km (50 miles) by rail to the port of Whyalla on Spencer Gulf for smelting.
Map Australia Fe

Middlesbrough ***United Kingdom*** Iron and steel-making town and port on the River Tees, 70 km (45 miles) north of York; formerly in North Yorkshire, it is now in the county of Cleveland. It was a tiny village in 1830, when the Stockton and Darlington railway reached it. Thanks to the shipping of coal from the south Durham coalfields, the working of iron ore in the Cleveland Hills to the south of the town and the establishment of its iron and steel industry (now severely depressed), it had grown by 1870 to a population of 40 000.

The 19th-century town, laid out in a grid pattern, connects with the older market town of STOCKTON-ON-TEES over the Tees to the west via the Transporter Bridge – unique in Britain – with its travelling suspended roadway.
Population 149 800
Map United Kingdom Ec

Middleveld In southern Africa, a region of undulating grassland and scrub-covered rands (low, hilly ridges) and kopjes (small, isolated hills) which occur between about 900 and 1500 m (2950 and 4900 ft) above sea level. It is mainly cattle-rearing country.

Midi-Pyrénées ***France*** The nation's largest region, covering 45 348 km² (17 647 sq miles) and centred on the southern city of TOULOUSE. It stretches from the Massif Central to the Pyrenees.
Population 2 326 000
Map France Dd

Midlands, The ***United Kingdom*** The central region of England, generally considered to take in the counties of DERBY, LEICESTER, NORTHAMPTON, NOTTINGHAM, STAFFORD, WARWICK and WEST MIDLANDS. It is Britain's industrial heartland, including the cities of BIRMINGHAM, COVENTRY, Derby, Leicester, Nottingham and STOKE-ON-TRENT.
Map United Kingdom Ed

Midlothian ***United Kingdom*** See BORDERS; LOTHIAN

mid-ocean ridge Line of submarine mountains rising from the ocean floor, notably in the Atlantic and Indian oceans, and probably due to volcanic action. Mid-ocean ridges are associated with constructive plate margins. See PLATE TECTONICS

Midway ***Pacific Ocean*** Atoll containing two islands about 1850 km (1150 miles) north-west of Honolulu, Hawaii. An American possession since 1867, it is a US naval and air base. The Battle of Midway in June 1942, when a Japanese invasion fleet was defeated, was the first decisive American victory in the Second World War.
Population (mainly military) 2200
Map Pacific Ocean Ea

Midwest (Middle West) ***USA*** The country's fertile heartland around the upper Mississippi and Ohio rivers. It includes the states bordering the Great Lakes (Ohio, Indiana, Michigan, Illinois, Wisconsin and Minnesota), and Iowa and Missouri to the west. (Kentucky, Kansas, Nebraska and North and South Dakota are sometimes included as well.) In the east farming concentrates on the growing of maize (corn), on which cattle and pigs are fattened, and the area is sometimes called the Corn or Feed Belt. Manufacturing, originally based on farm products, thrives on local coal, oil and gas production.
Map United States Hb

Mien Bac (Bac Bo) ***Vietnam*** Northern region of Vietnam, including Hanoi and the whole of the Red River valley plain, stretching from the northern slopes of the Truong Son chain to the Chinese border. In French colonial times it was called Tonkin, from Dong Kinh ('Eastern Capital'). It was the heart of Communist-led North Vietnam during the Vietnam War.
Map Vietnam Aa

Mien Trung ***Vietnam*** Hilly central region, called Annam in French colonial days, between the plain of the Mekong river delta and the valley of the Red River, south of the capital, Hanoi. It consists almost entirely of the Truong Son (Annamite Chain), about 1100 km (700 miles) of jungle-covered mountains stretching along the Laos border. The chain's highest point is Ngoc Linh (2598 m, 8524 ft), west of the coastal town of Quang Ngai. The hills are inhabited by people sometimes called *montagnards*, who have mixed Malayo-Polynesian and Mon-Khmer origins. They grow rice and tea.
Map Vietnam Bb

migmatite Composite rock body formed when an existing rock, usually a METAMORPHIC ROCK, is invaded by another rock material, such as magma. The reaction between them changes the structure and texture of the existing rock.

Mihintale ***Sri Lanka*** Sacred hill some 310 m (1020 ft) high and 12 km (7 miles) east of the ancient Sinhalese royal city of Anuradhapura. In 247 BC, Mahinda, son of India's Buddhist emperor Ashoka, converted the Sinhalese king Devanamplyatissa to Buddhism on the hill. Since then, Mihintale has been a home for monks and has attracted pilgrims to its many rock shrines; these are linked by a stone staircase of 1840 steps. Since the 10th century, Sinhalese laws have preserved the hill's forest and wildlife, making it probably the world's longest surviving wildlife park.
Map Sri Lanka Ba

Mikinai ***Greece*** See MYCENAE

Míkonos ***Greece*** Island in the CYCLADES group, north of Náxos. The island resort of the

same name, which can be reached by boat from Athens, is a town of winding lanes and white-washed houses. It has a double-storey, domed church. Apart from tourism, the economy of Míkonos relies on wines, figs and manganese.
Population 5500
Map Greece Dc

Milan (Milano) ***Italy*** The country's leading industrial and commercial city, standing on the north-western part of the North Italian Plain. It was capital of the Western Roman Empire in the 4th-5th centuries, then flourished as a city-state in the 11th-13th. During the Renaissance period (14th-16th centuries) its reputation for art, culture and politics spread throughout Europe. Finally, it developed through the industrial revolution in the 19th century to become Italy's most prosperous city in the 20th.

With three universities and a polytechnic, newspapers and publishing houses, theatres, museums and art galleries, offices, stock exchange, trade fair and industries – including motor cars, machinery, chemicals, textiles, clothing and printing – Milan is the only Italian city to be fully integrated into the network of European cultural, industrial and financial capitals. Rome is Italy's administrative capital, but Milan creates its wealth.

Milan has a medieval centre surrounded by successively more recent suburbs. The Piazza Duomo marks the middle of the city where the cathedral explodes skywards in a riot of pinnacles and spires. This is Italy's largest Gothic building and the third largest cathedral in the world; it was started under Gian Galeazzo Visconti in 1386 and not completed until the 19th century. The Galleria, a huge iron and glass arcade built in 1865, leads to the renowned La Scala opera house; the Brera, the city's main art gallery, is a little to the north. The Sforza Castle, second of Milan's great public monuments and the major legacy of the Renaissance period, fronts a large park to the west. The convent of the Renaissance church of Santa Maria delle Grazie, built between 1466 and 1492, is the home of Leonardo da Vinci's fresco the *Last Supper*.
Population 1 530 300
Map Italy Bb

Milford Haven ***United Kingdom*** Town and fine natural harbour at the south-west corner of Wales, in the county of Dyfed. Since 1960 it has been developed as a major oil port, with several refineries and petrochemical factories.
Population 14 000
Map United Kingdom Ce

Milford Sound ***New Zealand*** Glacier-sculpted inlet of the Tasman Sea on South Island, about 260 km (160 miles) north-west of Dunedin. The spectacular cone of Mitre Peak, rising 1692 m (5551 ft) from the sea, dominates the entrance. The area's high rainfall – around 2650 mm (246 in) a year – feeds hundreds of tumbling streams and waterfalls, among them the 158 m (518 ft) high Bowen Falls. The sound is a major tourist attraction, as is the 55 km (34 mile) long Milford Track – described as the most beautiful walk in the world.
Map New Zealand Af

Mílos (Melos) ***Greece*** One of the CYCLADES – an arid, volcanic island with hot springs. It was in the sea near the ruined 700 BC city of Melos that the ancient Greek sculpture known as the Venus de Milo was found in 1820. The statue is now in the Louvre museum, Paris. The island's main port is Adhámas, 4 km (2 miles) from the biggest town, Plaka, which has a 13th-century church and a Venetian castle. Near old Mílos is a small, hillside Roman theatre.
Population 4550
Map Greece Dc

Milwaukee ***USA*** Chief port and industrial centre of Wisconsin. It stands on the west shore of Lake Michigan. Many German immigrants settled there and made the city a beer-brewing centre. Other industries include steel, electrical equipment, and engineering and transport equipment.
Population (city) 620 800; (metropolitan area) 1 393 800
Map United States Ib

Mina al-Ahmadi ***Kuwait*** The country's main oil terminal, situated 40 km (25 miles) south of KUWAIT city. It handles some of the world's largest tankers. The Kuwait Oil Company's refinery there is the largest in the country, and can process up to 300 000 barrels a day.
Map Kuwait Bb

Mina Hassan Tani ***Morocco*** See KENITRA

Minamata ***Japan*** Fishing port and industrial town on Yatsushiro Bay on the west coast of Kyushu island. Minamata disease, for which the town became notorious, has claimed 200 lives since the early 1950s. The disease was caused by methyl mercury wastes discharged into the bay. The mercury poisoned the fish, and many of the people who ate them.
Population 36 500
Map Japan Bd

Minami Arupusu ***Japan*** National park in central Honshu island, about 130 km (80 miles) west of the capital, Tokyo. It contains Mount Shirane (3192 m, 10 472 ft) the second highest peak in Japan after Mount Fuji. It is also known as the Southern Alps National Park.
Map Japan Cc

Minas Gerais ***Brazil*** See BELO HORIZONTE

Minch, The Channel separating mainland Scotland from the Outer Hebrides, also called the North Minch. This is to distinguish it from the Little Minch between North Uist and Benbecula and Skye to the south. Both channels have many associations with the Jacobite Rebellion of 1745 and the subsequent wanderings of Bonnie Prince Charlie.
Width North Minch 56 km (35 miles); Little Minch 24-40 km (15-25 miles)
Map United Kingdom Bb

Mindanao ***Philippines*** Second-largest island covering 94 631 km² (36 537 sq miles) of the Philippine archipelago and its most southerly major island, lying south of the VISAYAN ISLANDS. It is also the name of the island's longest river (about 320 km, 200 miles long).

The island is an irregular triangle in shape, extending into the curving peninsula of Zamboanga in the west. Largely mountainous (Mount Apo near DAVAO being the Philippines' tallest peak), it also has swampy lowlands in the Agusan valley and the COTABATO region (where the Mindanao river flows). It receives more rain than most of the Philippines, and year-round cultivation is possible. As a result it is the most important island for growing pineapples, maize and copra, and also produces much coffee, cacao and Manila hemp. In addition, it has the Philippines' most important mineral deposits – including iron, nickel, copper, silver and gold.

As a result of these rich resources, many migrants from the other islands have settled in Mindanao since the 1930s. This has led to tension between the newcomers (mainly Christians) and the predominantly Muslim local people (known to other Filipinos as Moros), and even occasional violence. Development continues, however, and as a result Davao is now the Philippines' second-largest city.
Population 11 100 000
Map Philippines Cd

Minden ***West Germany*** Port on the Weser river about 60 km (37 miles) west and slightly south of Hanover. It has a 13th-century cathedral. The town manufactures machinery, textiles, glass, ceramics and furniture. In the Seven Years' War (1756-63), British and Hanoverian armies defeated the French at Minden in August 1759.
Population 75 900
Map West Germany Cb

Mindoro ***Philippines*** Island covering 9736 km² (3759 sq miles) south-west of LUZON. It comprises two provinces, Mindoro Occidental and Mindoro Oriental. One of the least developed of the Philippines' islands until recent years, it remains heavily forested and malarial. The centre is mountainous, and most people live on the eastern and western coastal plains, where rice is the main crop. Sugar cane grows near San Jose in the south. The forests yield valuable timber, but parts have been protected as nature reserves.
Population (Occidental) 145 700; (Oriental) 329 000
Map Philippines Bc

Mindoro, Sea of See SULU SEA

mineral Inorganic chemical element or compound that occurs naturally in the earth. Minerals are distinguished from one another by their chemical formula, crystal structure, colour, lustre, hardness and degree of transparency. Rocks are aggregates of minerals.

mineral spring Spring of water that contains an appreciable amount of mineral salts, such as iron compounds, magnesium chloride and sodium chloride. Such springs are often used for medicinal purposes.

Minho ***Portugal*** Coastal province 4928 km (1902 sq miles) in area, to the north of the city of Oporto. Its capital is the town of Braga. The main produce is maize, beans, grapes and Vinho Verde wines.
Population 965 700
Map Portugal Bb

Minho (Miño) ***Spain/Portugal*** River which rises in north-west Spain and flows south-westwards to the Atlantic. Some 275 km (170 miles) long, it marks part of the northern border of Portugal.
Map Portugal Ba; Spain Ba

Minicoy Islands ***India*** See LAKSHADWEEP

Minneapolis ***USA*** City on the west bank of the Mississippi river, and the business, cultural, financial and industrial centre of Minnesota. The Falls of St Anthony – a series of rapids with a total fall of about 25 m (82 ft) – determined the site of the city because they provided power for the original timber and flour mills. Minneapolis adjoins its twin city of ST PAUL.
Population (city) 358 300; (metropolitan area) 2 230 900
Map United States Hb

Minneriya Giritale ***Sri Lanka*** See POLONNARUWA

Minnesota ***USA*** State on the Canadian border west of Lake Superior. Although it was explored in the 17th century, it did not become a state until 1858. The north is heavily wooded and dotted with lakes – hence Minnesota's nickname, 'state of ten thousand lakes'. Iron ore and manganese are both mined in the east. The state covers 217 736 km² (84 068 sq miles) and produces dairy goods, soya beans, maize, cattle and wheat. Industries include machinery and food processing. The state capital is ST PAUL.
Population 4 193 000
Map United States Ga

Miño ***Spain/Portugal*** See MINHO

Minorca (Menorca) ***Spain*** Second largest of the BALEARIC ISLANDS after Majorca, 702 km² (271 sq miles) in area.
Population 50 200
Map Spain Gc

Minquiers, Les ***Channel Islands*** Group of uninhabited islets, reefs and shoals in the Gulf of St Malo.
Map Channel Islands Bc

Minsk ***USSR*** Capital of Belorussia and a major industrial centre, some 260 km (160 miles) from the Polish border. It was founded in 1067 as Menesk, became part of Lithuania in 1326, was later ceded to Poland, then to Russia in 1793. By 1939, 40 per cent of its people were Jewish, but virtually all of them were killed after invading Germans transferred them to concentration camps during the Second World War. Since 1945 new printing and timber plants have been set up, along with factories making trucks, tractors, machinery, radios and television sets, clothing and textiles.
Population 1 442 000
Map USSR Dc

Mintaka Pass ***China/Pakistan*** Pass 4709 m (15 500 ft) up in the mountains on the border of Azad Kashmir and western China. It lies on a spur of the Karakoram Highway. The main Karakoram Highway runs over the Khunjerab Pass (4602 m, 15 100 ft) to the east.
Map Pakistan Da

Miocene Fourth epoch of the Tertiary period in the Cenozoic era of the earth's time scale. See GEOLOGICAL TIMESCALE

mirage Optical phenomenon, particularly in hot deserts, in which an image, often inverted, is produced as a result of refraction of light by layers of air with differing densities. The commonest form involves an image of the sky, producing the illusion of water.

Miranda ***Venezuela*** Northern state which includes part of the metropolitan region of the national capital, Caracas. It covers an area of 7950 km² (3069 sq miles). The capital is Los Teques.
Population 1 618 000
Map Venezuela Ba

Misiones ***Argentina*** North-eastern province tucked between Paraguay and Brazil. The provincial capital Posadas (population 140 000), on the Paraguayan border, is about 840 km (520 miles) north of Buenos Aires. The province's name is derived from the Jesuit missions set up in the 17th and 18th centuries to convert the Guaraní Indians to Christianity. The area is subtropical and heavily forested. The port of Posadas is on the Paraná river. The province produces yerba maté, tung oil, tobacco, citrus fruit, timber and fish.
Population 590 000
Map Argentina Da

Miskolc ***Hungary*** The country's second largest city after Budapest, lying in the north-east, 137 km (85 miles) north-east of the capital. Miskolc is the administrative capital of Borsod-Abauj-Zemplen county, and is a major industrial city with iron and steel, engineering, food processing and textile plants. The city spreads across the Sajó river valley and up into the Borsod upland to the north. The southern suburb of Tapolca is a popular spa.
Population 210 000
Map Hungary Ba

Misratah ***Libya*** Oasis town and seaport 140 km (87 miles) east of Tripoli, dealing in dates and cereals. It is the capital of the Misratah division in the north-west. Other industries are iron and steel, carpet making and traditional hand-woven textiles. An oil refinery is under construction, and there are plans to make Misratah a major industrial city with a population of 180 000.
Population 103 400
Map Libya Ba

Mississippi ***USA*** **1.** Southern state covering 123 585 km² (47 716 sq miles) between the Mississippi river, the state of Alabama and the Gulf of Mexico. The land is everywhere below 250 m (820 ft). Cotton, rice, soya beans and broiler chickens are the farming mainstays. Gas and oil are produced and there are refining and engineering industries. JACKSON is the state capital.

2. River rising in the northern state of Minnesota and flowing 3779 km (2348 miles) south to the Gulf of Mexico. It is the second largest US river after the MISSOURI, which is its main tributary. Together, the Mississippi-Missouri form the world's fourth longest river system, flowing a total of 6019 km (3740 miles) and draining some 3.25 million km² (1.25 million sq miles). Other major tributaries of the Mississippi include the ARKANSAS and OHIO rivers. It has long been a major transport artery, and has been navigated by steamboats since about 1820; it is still used to carry bulky cargoes. It is prone to serious flooding, but has been stabilised recently by massive control schemes.
Population (state) 2 613 000
Map (state) United States Hd; (river) United States Hb, Hd

Missolonghi ***Greece*** See MESOLONGION

Missouri ***USA*** **1.** Midwest state bounded by the Mississippi river in the east and crossed by the Missouri river, which joins the Mississippi near ST LOUIS. The state covers 180 487 km² (69 686 sq miles) and stretches south into the Ozark Plateau, which rises to 540 m (1772 ft) in the south-west and forms a tourist area. Cattle, pigs, dairy produce and soya beans are the chief sources of farm income. Lead and zinc are mined, and manufactures include transport equipment, foodstuffs and chemicals. The state capital is JEFFERSON CITY, but the largest cities are St Louis and KANSAS CITY. About 150 km (95 miles) north-west of St Louis, on the Mississippi, is the town of Hannibal, the boyhood home of Samuel Clemens (1835-1910), better known as the writer Mark Twain.

2. Longest river in the country, rising in Montana and flowing east then south-east 3969 km (2466 miles) into the MISSISSIPPI just above St Louis. It has been dammed in places for flood and navigation control.
Population (state) 5 029 000
Map (state) United States Hc; (river) United States Ea, Hc

mist Mass of water droplets in the lower layers of the atmosphere caused by the condensation of water vapour in the air, reducing visibility to 1-2 km (0.6-1.2 miles). If visibility is less than 1 km the mist becomes a FOG.

mistral Strong, cold, dry northerly to north-westerly wind blowing from the central plateau of France towards a low-pressure system over the Mediterranean Sea, experienced especially in the Rhône Valley and the Gulf of Lions. It corresponds to the MAESTRALE of Italy.

Mistras ***Greece*** Ruined Byzantine town on a hilltop beside Mount Taíyetos, 7 km (4 miles) east of Sparta. It was ruled by Byzantine princes from the 13th to the 15th centuries and became an intellectual centre and meeting place of the Western and Byzantine cultures. A palace, churches, monasteries, houses and Turkish baths of the period survive, with fine stone and brickwork, frescoes, paintings and icons. The Byzantine philosopher Plethon Gemistos (about 1355-1452), who advocated a return to the Ancient Greek religion and influenced many writers in Renaissance Europe, lived there.
Map Greece Cc

Mitchell, Mount ***USA*** Highest peak in the country east of the Mississippi river. It rises to 2037 m (6684 ft) in the Appalachian Mountains in North Carolina.
Map United States Jc

Mitilíni ***Greece*** See LESBOS

Mitla ***Mexico*** Ancient ruined city of the Zapotec Indians, in the south-eastern state of Oaxaca. Mitla, which means 'Place of the Dead', was built more than 2000 years ago over catacombs consecrated by the Zapotecs. Stone fretwork and geometrical mosaics adorn its long, low palaces, and its Hall of Columns has six enormous pillars, each one a single stone.
Map Mexico Cc

Mito ***Japan*** City near the east coast of Honshu island, about 100 km (60 miles) north-east of the capital, Tokyo. Its fine landscape garden, the Kairakuen, is noted for plum blossoms and a poetry-reading pavilion.
Population 229 000
Map Japan Dc

Miyajima (Itsukushima) ***Japan*** Small island in the Inland Sea, about 19 km (12 miles) south-west of the city of Hiroshima. It is dotted with shrines and is the scene of many colourful festivals, including Tamatori Ennen in August, in which young men fight through water to retrieve a sacred ball. See JAPAN (picture)
Map Japan Bd

Miyazaki ***Japan*** Market town on the east coast of Kyushu island. The town is known as the 'Hawaii of Japan' for its sunny climate. A Shinto shrine at Miyazaki commemorates Jimmu Tenno, the legendary first emperor of Japan, who reigned in the 7th century BC.
Population 268 300
Map Japan Bd

Mizoram ***India*** Union territory covering 21 081 km^2 (8137 sq miles) of wet, jungle-clad hills on the Burmese border. It is inhabited by Mizo and Lushai tribespeople, who live by shifting agriculture or rice growing. Some are fighting a guerrilla war for independence.
Population 493 800
Map India Ec

Mjøsa ***Norway*** The nation's largest natural lake (366 km^2, 141 sq miles), lying 55 km (34 miles) north of Oslo. It is surrounded by an extensive cultivated area, growing grain and potatoes, and small manufacturing towns.
Map Norway Cc

Mljet (Meleda) ***Yugoslavia*** Southernmost Dalmatian island, lying north-west of Dubrovnik. Covering 98 km^2 (38 sq miles), it is now a pine-forested national park. There are three seawater lagoons, and in Veliko Jezero ('Great Lagoon'), 46 m (151 ft) deep, is an islet on which stands a 12th-century monastery, now a hotel. Mongooses have been introduced on the island to control the large number of snakes. Mljet may have been the Biblical site of Melita where St Paul was shipwrecked.
Map Yugoslavia Cc

Mo i Rana ***Norway*** Town some 30 km (20 miles) south of the Arctic Circle, where a major state-owned iron and steelworks was built in the 1950s to combat local unemployment. The Dunderland valley, which the town lies near, has been mined intermittently for a century, but most of the ore for the works is brought from the Sydvaranger mine in Finnmark. Energy comes from the nearby Røssaga hydroelectric power station and from imported coal. By-products of the works include zinc, copper and sulphur. The town also has port facilities.
Population 20 700
Map Norway Db

▲ **MAJESTIC OLD LADY A century ago, swashbuckling 'Mississippi gamblers' rode on the *Creole Queen*, a stern-wheeler steamboat. Now it carries tourists exploring near New Orleans.**

Moanda ***Gabon*** Modern town with an airport, situated in the south-east 50 km (31 miles) north-west of Franceville. It lies in an area containing some of the world's largest manganese deposits, which were first mined in 1951. Production in 1981 amounted to 1.5 million tonnes. The manganese is refined at Moanda and taken to Brazzaville in neighbouring Congo by cable railway. From there it is carried by rail to the port of Pointe-Noire.
Population 6500
Map Gabon Bb

Mobile ***USA*** Port on the Gulf of Mexico in Alabama. Shipbuilding, chemicals and bauxite (aluminium ore) processing are the basis of its industries.
Population (city) 204 900; (metropolitan area) 465 700
Map United States Id

mobile belt Any long, narrow region of the earth's crust in which structural or mountain-building activities are currently taking place; for example, the zone along which two lithospheric plates are colliding, or a GEOSYNCLINE which is still subsiding and receiving more sediment.

Mobuto Sese Seko, Lake ***Uganda/Zaire*** See ALBERT, LAKE

Moçambique ***Mozambique*** Ancient Arab port some 580 km (360 miles) south of the Tanzanian border. It lies on a small coral island and was visited by the Portuguese explorer Vasco da Gama in 1498. The Portuguese established a stronghold in 1507 and Moçambique became a provisioning port for Portuguese fleets. It served as national capital until 1907.
Population 12 200
Map Mozambique Bb

Moçamedes ***Angola*** See NAMIBE

Moche Pyramids ***Peru*** Largest pre-Columbian structures in South America. They are made of adobe bricks and stand in a valley 5 km (3 miles) south of Trujillo. The biggest, Huaca del Sol, is 48 m (158 ft) high, 228 m (748 ft) long and 135 m (443 ft) wide. They are relics of the Mochica culture, which flourished from AD 100 to 600.
Map Peru Ba

Modena ***Italy*** Industrial city 38 km (24 miles) north-west of Bologna. It has a handsome Romanesque cathedral, built between 1099 and 1148, and a colossal 17th-century palace, now a military academy. Its manufactures include high-performance sports cars, agricultural machinery, machine tools, other metal goods and leather.
Population 178 300
Map Italy Cb

Modified Mercalli scale A 12-point scale designed in 1931 to measure the intensity of an earthquake, replacing the earlier 10-point Rossi-Forel scale. It is related to the amount of damage at a particular point, and decreases with distance from the earthquake's focus, unlike the absolute RICHTER SCALE of magnitude.

Mogadishu (Mogadiscio; Muqdisho) ***Somalia*** National capital and chief port, on the south-east coast. It exports livestock and sugar. The town was seized by the Sultan of Zanzibar in 1871, sold to the Italians in 1905, and taken by the British in 1941 during the Second World War. It was a ready-made capital on independence in 1960. The old quarter, known as Hammar Wein, is the remnant of an Arab settlement dating back more than a thousand years.
Population 400 000
Map Somalia Ab

Mogami ***Japan*** River 230 km (143 miles) long in the north-west of Honshu island. It flows north-west to the Sea of Japan at the town of Sakata.
Map Japan Dc

Mogilev ***USSR*** Industrial city on the Dnieper river in Belorussia, some 515 km (320 miles) south-west of Moscow. The city was badly damaged in the Second World War, and nine in ten of its people died. It has since been rebuilt as a key industrial centre making metal goods, building machinery, chemicals, artificial fibres, leather goods, beer and wine.
Population 334 000
Map USSR Ec

Mogok ***Burma*** Mining town on the edge of the eastern Burmese hills, about 125 km (75 miles) north-east of Mandalay. The town is the centre of Burma's ruby-mining industry, which – despite primitive methods – makes an important contribution to the country's exports.
Population 79 500
Map Burma Cb

▲ CITADEL BY THE SEA Phoenician traders colonised Monaco-Ville's rocky base. Genoese settlers fortified it in 1215 and the ruling house of Grimaldi gained control in 1297.

Mohács ***Hungary*** City and port on the west bank of the Danube river in the south of the country, 12 km (7 miles) from the Yugoslav border. Here on August 29-30, 1526, King Louis of Hungary and his 20 000-strong army were heavily defeated by forces of the Turkish Sultan, Suleiman I (the Magnificent). The Ottoman victory resulted in the Turkish domination of Hungary for more than a century and a half, and paved the way for the Turkish sieges of Vienna in Austria. The event is commemorated by an annual carnival in the city. The tables were turned 162 years later when the retreating Turks were defeated in 1687 at the Second Battle of Mohács fought at Harkany to the south-west.
Population 21 500
Map Hungary Ab

Mohammedia ***Morocco*** Port and seaside resort 25 km (16 miles) north of Casablanca. It was formerly settled by the Portuguese, and some of their buildings can still be seen in the Fedala Kasbah (old town). In 1942 it was one of the landing places for American troops in the North African campaign. The port is famed for its fish and has some fine fish restaurants. It is also the site of a major oil refinery.
Population 105 100
Map Morocco Ba

Mohave Desert ***USA*** See MOJAVE DESERT

Mohenjo Daro ***Pakistan*** One of the world's greatest archaeological sites, about 200 km (125 miles) north of Hyderabad. A city of the Indus civilisation, it lasted 1000 years from 2500 BC, and some of the remains are 4000 years old. It had wide streets, good sewers and large houses. A huge public bath – 12 by 7 m (39 by 23 ft) – is faced with gypsum mortar and backed with bitumen to make it watertight. Most buildings were built with fired bricks of uniform size and quality. A public granary was built with ventilation passages below the floors to keep the grain in good condition.

It seems that the city traded with China and the Middle East. But it was abandoned, apparently suddenly, in about 1500 BC for reasons that are still uncertain. Military conquest by invading Aryans, economic decline and an earthquake have all been suggested. The collapse may have been accelerated by a river flood which washed water and mud over the city at about that time and may have driven people away. The site now faces the risk of a second destruction because of the rise in the water table and soil moisture brought about by irrigation canals. The United Nations agency UNESCO has launched a rescue plan to counteract this.
Map Pakistan Cb

Möhnestausee ***West Germany*** Man-made lake created by damming the Möhne river in the hills 30 km (18 miles) east of the Ruhr industrial complex. The Möhne dam, and another on the Eder river to the south-east, was destroyed by Barnes Wallis's bouncing bomb in the British 'Dambusters' raid of 1943. The dam was repaired by the end of the war.
Map West Germany Cc

Moho Abbreviation for the Mohorovičić discontinuity, the boundary which separates the earth's crust from the mantle. See JOURNEY TO THE CENTRE OF THE EARTH (p. 200)

Mohokare ***Southern Africa*** See CALEDON

Mohole Hole which scientists planned to drill through the earth's crust and into the mantle to obtain samples of rocks. The project, which was conceived in the 1950s, was not completed, and the idea was abandoned in 1966 because of rising costs and disagreements among the scientists concerned. It did help, however, to stimulate more limited projects to drill into the basaltic crust of the ocean floors, such as the Deep Sea Drilling Project.

Mohs scale of hardness A 10-grade scale used to measure and specify the hardness of a mineral, devised in the 19th century by the German mineralogist Friedrich Mohs. Each mineral in the scale can be scratched by those having higher numbers, and can scratch those having lower numbers. The hardest mineral, with a 10 rating, is diamond; the softest, with a 1 rating, is talc.

Móin Aluine ***Ireland*** See BOG OF ALLEN

Mojave Desert (Mohave Desert) ***USA*** Wilderness covering about 38 850 km² (15 000 sq miles) in southern California, between Death Valley and the city of Los Angeles. It is the site of Edwards Air Force Base, main landing site for US space shuttle missions.
Map United States Cc

Mokanji Hills ***Sierra Leone*** Upland region west of the Jong river, some 176 km (110 miles) south-east of the capital, Freetown. Most of the country's exports of bauxite (aluminium ore) is mined in the hills.
Map Sierra Leone Ab

molasse Soft predominantly sandstone sediment deposited in the Alpine Foreland by streams bringing eroded material down from the rising Alps. The term is also used for similarly formed deposits of much greater age such as the Newark Sandstone of the eastern United States, derived from the Appalachians, and the Old Red Sandstone of Britain, derived from the Caledonian Mountains.

Moldau ***Czechoslovakia*** See VLTAVA

Moldau ***USSR*** See MOLDAVIA

Moldavia ***Romania*** Fertile north-west region containing the rugged eastern CARPATHIANS. Moldavia is noted for its beautifully decorated monasteries and vividly painted houses. Established as a principality in the 14th century, largely under Hungarian control, it was taken over by the Turks in the 16th century and, with WALACHIA, became the core of Romania in 1862. The region consists mainly of rich farmland and covers 38 046 km² (14 690 sq miles).
Map Romania Ba

Moldavia (Moldova; Moldau) ***USSR*** Second smallest – after Armenia – and most south-westerly of the USSR's 15 republics. It covers 33 700 km² (13 000 sq miles), and is also the country's most rural and most densely populated republic.

Moldavia was settled by the Romans in the 1st century AD and by Slavic peoples in the 9th-10th centuries. It was founded as a principality in the 14th century, and parts of it have since been ruled by Romania and Turkey. Bessarabia, the part of Moldavia between the Prut river (which forms the present Romanian border) and the Dniester river, rises to 429 m (1407 ft) and is cut by steep-sided ravines. But the Dniester plains of Moldavia have a warm climate and fertile soil; they produce grapes, tree fruits, strawberries, walnuts and honey. The farm crops include wheat, maize, sunflowers, sugar beet and tobacco. Food processing, distilling, wine making and engineering industries are concentrated in KISHINEV, the Moldavian capital, and in the towns of Bel'tsy, Tiraspol and Bendery.
Population 4 083 000
Map USSR Dd

Molise ***Italy*** Rural region covering 4438 km² (1714 sq miles) immediately south of ABRUZZI, between the Adriatic Sea and the Apennines. Its mainly peasant population makes its living from mixed farming – vines, wheat, vegetables, sheep and goats.
Population 332 900
Map Italy Ed

Molotov ***USSR*** See PERM'

Molucca Sea Indonesian sea, an arm of the PACIFIC OCEAN between SULAWESI (Celebes) and the MALUKU islands. It is linked in the south to the BANDA and SERAM seas. The first expedition to circumnavigate the globe sailed into the sea in 1521, soon after the death of its commander, Ferdinand Magellan.
Map Indonesia Fc

Moluccas ***Indonesia*** See AMBON; MALUKU

Mombasa ***Kenya*** Second largest city of Kenya after the capital, Nairobi, and the main port for Uganda and Rwanda, as well as Kenya. Mombasa stands on an island with a natural harbour in two sheltered creeks. Kilindini Harbour, to the south, has deep-water berths for 15 ocean-going vessels. The town was occupied by medieval Arab traders, then the Portuguese, before becoming part of a British protectorate in 1895. In the 1890s a railway was built, linking Mombasa with inland Kenya and Uganda. There is an oil refinery on the mainland beside the harbour, and tourist hotels line the coast north and south of the city.
Population 500 000
Map Kenya Cb

Mon ***Burma*** State in the south of the country covering 12 297 km² (4748 sq miles). The capital is MOULMEIN.
Population 1 682 000
Map Burma Cc

Møn ***Denmark*** Island in south-eastern Denmark covering 216 km² (83 sq miles). The chief town is Stege. At the eastern end of the mostly low-lying island (the name Møn is pronounced to rhyme with 'earn') the land rises above the Baltic to form a line of chalk cliffs – a famed beauty spot – some 7 km (4 miles) long and in places more than 120 m (390 ft) high. One of these yellow and white cliffs, down which runs water from a spring, has been named the Weeper. Another is known as the Queen's Chair: a clump of beech trees at the top forms the queen's crown, and a waterfall flowing down the side gives the impression of a shimmering gown.
Population 12 300
Map Denmark Cb

Monaco-Ville ***Monaco*** Ancient capital of the tiny principality, occupying a rocky headland projecting into the Mediterranean south of the harbour. It contains the 16th-century royal palace of the ruling Grimaldi family, the 19th-century cathedral and Monaco's noted oceanographic museum. The commune itself has narrow streets of old houses.
Population 1250
Map France Ge

Monaghan (Muineachán) ***Ireland*** Rural county covering 1291 km² (498 sq miles) on the border with Northern Ireland. It is part of the province of Ulster and, with nearly 200 lakes, is a popular area for coarse fishing. The county town of the same name lies 6 km (4 miles) from the border.
Population (county) 51 200; (town) 6170
Map Ireland Ca

Monastir ***Tunisia*** Small coast town and tourist centre south of Sousse. It was the birthplace in 1903 of President Habib Bourguiba, who was voted life president in 1975. Among the places of interest are the presidential palace and Bourguiba's mausoleum which has been prepared already for his eventual death. Other attractions include an 8th-century fortress in which son et lumière performances are held in summer; an Islamic Museum; the Great Mosque dating from the 9th to the 11th centuries; and the Habib Bourguiba Mosque, an excellent example of modern Islamic architecture. The town has undergone dramatic redevelopment since independence, and is a showpiece for urban development. The small port is now dominated by a tourist complex and marina.
Population 39 000
Map Tunisia Ba

Monastir ***Yugoslavia*** See BITOLA

Monaco

THE WHEEL OF FORTUNE KEEPS ON TURNING FOR A FAIRYTALE PRINCIPALITY WHICH THRIVES ON FUN AND GAMES

Sunlight sparkling on the blue Mediterranean, luxurious yachts bobbing in the marinas, and bronzed people basking on the beaches – these are the main ingredients of the heady cocktail that is Monaco. To which must be added the risks and romance of the casino, the thrills and spills of the Monte Carlo Rally and Monaco Grand Prix and the fairytale marriage, in 1956, of Prince Rainier and Grace Kelly, the beautiful American film star (there is tragedy too: Princess Grace died in a car crash in 1982).

There are only 4500 Monégasque citizens but another 23 500, mainly French and Italian, live in this tiny country in the south-east corner of France, with which it has a customs union.

Members of the Grimaldi dynasty – originally a Genoese family – have ruled Monaco since medieval times. But their country attracted little interest until the second half of the 19th century, when the casino and first hotels were built.

There are four distinct districts. The old town of MONACO-VILLE, on a rocky promontory, houses the royal palace and the cathedral. MONTE CARLO retains its world-famous casino, but has changed dramatically as stately houses have given way to high-rise apartments and hotels. Between the two is LA CONDAMINE, with thriving businesses, shops, banks and attractive residential areas. Fontvieille has new marinas and light industries.

The Monégasques pay no taxes in this rich little country, but neither do they share all the fun – they are not allowed to gamble in the casino.

MONACO AT A GLANCE
Map France Ge
Area 1.9 km² (0.73 sq miles)
Population 28 000
Capital Monaco-Ville
Government Constitutional principality
Currency Monégasque and French francs = 100 centimes
Languages French, Monégasque
Religion Christian (90% Roman Catholic, 5% Protestant)
Climate Mediterranean; average temperature ranges from 8-12°C (46-54°F) in January to 22-26°C (72-79°F) in August
Major industries Tourism, gambling, banking, perfumery, plastics
Annual income per head (US$) 11 000
Population growth (per thous/yr) 12
Life expectancy (yrs) Male 73 **Female** 78

Mönchengladbach ***West Germany*** Industrial city 25 km (16 miles) west of Düsseldorf, to the south-west of the Ruhr industrial region. The town grew as a textile centre around a 10th-century Benedictine abbey and expanded rapidly in the 19th century with the growth of Germany's chemical and engineering industries.

Population 257 000

Map West Germany Bc

Monclova ***Mexico*** Important industrial town in the northern state of Coahuila. Together with Monterrey (in neighbouring Nuevo Leon) and Saltillo it marks out a triangle that constitutes one of the most productive mining and industrial regions in Mexico. The area yields zinc, copper, silver and lead, and Monclova has a huge steel mill, iron foundries, railway workshops, and a range of manufacturing industries.

Population 174 000

Map Mexico Bb

Moncton ***Canada*** City on the Petitcodiac river in south-east New Brunswick. It is the transport hub of the Maritime Provinces (New Brunswick, Nova Scotia and Prince Edward Island). Moncton has meat-packing plants and manufactures steel, chemicals and woollens. The town has Canada's only French-language university outside QUEBEC.

Population 54 750

Map Canada Id

Mondego ***Portugal*** Longest river (220 km, 137 miles) entirely within Portugal. It rises in the Serra da Estrela, in the province of Beira Alta, and reaches the Atlantic at the town of Figueira da Foz, about 150 km (95 miles) north of the capital, Lisbon.

Map Portugal Bb

Monemvasía (Napoli di Malvasia) ***Greece*** Venetian walled town built on a high crag rising from the sea, with a single entrance on a causeway linking it with the mainland. It lies on the south-east coast of the Peloponnese. Founded in the 8th century by the Byzantines, it became a flourishing wine centre, exporting Malmsey wines (after the town's medieval name, Malmesia) all over Europe.

Map Greece Cc

Mongolia

A CENTURIES-OLD NOMADIC LIFESTYLE IS VANISHING AS CITY LIFE CALLS THE HERDSMEN IN FROM THE HILLS

The traditional circular domed tents (*yurts*) of Mongolia's nomadic herdsmen can still be seen, but they are likely to be equipped with television sets and transistor radios. There are fewer yurts because a growing number of Mongolians live in cities: the capital, ULAN BATOR, alone, holds more than 25 per cent of this huge country's sparse population.

Most of Mongolia is over 1500 m (4900 ft) above sea level. The north-western area consists of remote mountain ranges, including the Hangayn Mountains and the ALTAI, which includes the country's highest peak, Taban Bogdo Ula (4356 m, 14 290 ft). To the south-east, mountains give way to grass-covered steppes and the desert wastes of the GOBI. Most of Mongolia's territory is pasture – cultivated land accounts for less than 1 per cent of the surface.

The environment is harsh and demanding – some land is permanently frozen and some is sandy desert – and the climate is extreme. Rainfall is low and unreliable.

The country was once part of China, and divided into Inner Mongolia and Outer Mongolia. Inner Mongolia remains Chinese; Outer Mongolia's emergence as an independent country dates from the Russian-backed revolution in 1921 which established the Communist Mongolian People's Republic in 1924. Russia dominates the economy of Mongolia and stations troops there to watch the northern frontiers of China.

For many centuries, the herding of horses, cattle, sheep, yaks, and camels has been the mainstay of the economy. But since the Second World War there have been significant changes.

The mining industry, developed with lavish Russian aid at centres such as ERDENET and DARHAN, rivals animal husbandry as a major source of national income. The nomadic life of the herdsmen has given way to raising livestock on cooperative ranches.

MONGOLIA AT A GLANCE

Area 1 565 000 km^2 (604 247 sq miles)

Population 1 960 000

Capital Ulan Bator

Government Communist republic

Currency Tugrik or togrog = 100 mongo

Languages Mongolian (official), Kazakh

Religions Shamanist (31%), Buddhist (2%), Muslim (1%)

Climate Dry and cold; average temperature in Ulan Bator ranges from −32 to −19°C (−26 to −2°F) in January to 11-22°C (52-72°F) in July

Main primary products Livestock, cereals; coal, lignite, petroleum, copper, molybdenum, gold, fluorspar

Major industries Agriculture, food processing, textiles, mining, processing hides and skins

Main exports Minerals, livestock, wool, hides, meat

Annual income per head (US$) 990

Population growth (per thous/yr) 27

Life expectancy (yrs) Male 65 **Female** 65

Mongu ***Zambia*** Market town on the edge of the Zambezi river floodplain, about 550 km (340 miles) west of the capital, Lusaka. Local farmers make their living from maize, millet, goats and sheep.

Lealui, a village 18 km (11 miles) west of Mongu, is the home of the chief of the Lozi people, who, to escape river floods during the January-July rainy season, moves his family by royal barge to higher ground at Limungula in a ceremony called Ku-omboka.

Population 24 900

Map Zambia Bb

Monmouth ***United Kingdom*** See GWENT

Monrovia ***Liberia*** The country's capital, largest city and major seaport. It lies in the west, about 75 km (45 miles) from the Sierra Leone border at the mouth of the Mesurado river, which almost encircles the built-up area. The city was founded by the American Colonisation Society in 1822 as a home for freed US slaves. It gets its name from James Monroe (1758-1831), the US President of the day.

The American influence in Monrovia is still strong. There are broad avenues, thronged with large American cars and lined by modern buildings; and there are the stone houses of the earliest settlers, built in the elegant styles of the American South. But there are also shanty towns, mud huts and slums.

The city has modern cement, pharmaceuticals, paint and fish-processing industries, and a mill that produces palm oil – used in margarine and cooking fats. The port handles exports of iron ore, rubber, logs and timber, and its main imports are machinery and transport equipment. Monrovia is a free port, and is developing a considerable import-export trade.

On Providence Island, the settlers' first landing place, there is a museum of memorabilia and native crafts, and an amphitheatre where African music and dancing are staged. Just outside the city, which has several good hotels, there are fine sandy beaches with swimming and boating, and tours can be made of the huge rubber plantation at HARBEL.

Population 425 000

Map Liberia Aa

Mons (Bergen) ***Belgium*** Medieval town 55 km (34 miles) south-west of the capital, Brussels. It has a Gothic cathedral, and no less than six important museum collections. Located on a coalfield, it developed industries, and in the 19th century was an important textile centre. But today the mines are closed and industry – textiles, sugar refining, engineering – is of minor significance.

On August 23, 1914, Mons was the scene of the first encounter in the First World War between a British army and the advancing Germans. With the formations on either side of them falling back, however, the British were obliged the following day to begin the long 'retreat from Mons' that forced them back almost to Paris before they could counter-attack. The battle gave rise to the legend that heavenly forces – the 'Angels of Mons' – had been seen coming to the aid of the British and standing between them and the Germans.

Population 100 000
Map Belgium Aa

monsoon Regular seasonal reversal of the wind system bringing summer rainfall to India, other parts of Asia, northern Australia and Africa. The reversal occurs because of atmospheric pressure differences over large landmasses and over the oceans. In winter the landmass is dominated by anticyclonic systems and winds blow from the continent, but by early summer the heat from the sun warms the land, thins the air above and produces low pressure over the landmass. Winds blowing from the sea into the low-pressure area bring moisture with them that falls as the monsoon rains.

monsoon forest Form of tropical forest found in areas that have a marked dry season, such as the monsoon regions of Burma, Indonesia, Thailand, India and northern Australia. The rainfall is sufficient during the wet season to support the growth of trees, but they shed their leaves during the dry season. New foliage appears when the monsoon rains begin.

montaña Forested slopes of the tropical eastern Andes, particularly in Peru.

Montana *USA* Sparsely populated north-western state along the Canadian border. Almost half its 381 087 km² (147 138 sq miles) lie within the Rocky Mountains and the rest is in the high plains to the east. Wheat, cattle, barley and sugar beet are the main farm produce, while minerals extracted include coal, oil, natural gas and some copper. HELENA is the state capital.

Population 826 000
Map United States Da

Monte Albán *Mexico* Site of an ancient Zapotec Indian religious centre, known as the 'White Acropolis', on a steep hill overlooking Oaxaca city. Occupied originally in 300 BC and rebuilt in AD 600, it was the ruling city of south-central Mexico until the 10th century. Palaces, pyramids, observatories, sanctuaries, dwellings and terraces cover an area of 39 km² (15 sq miles). In the centre lies a great plaza 300 m (1000 ft) long and 200 m (660 ft) wide. Tombs here have yielded priceless treasures of funeral urns, goblets and jewels.

Map Mexico Cc

Monte Bello Islands *Australia* Remote uninhabited archipelago in the Indian Ocean off the coast of Western Australia. The first British atomic bomb was tested there in 1952.

Map Australia Bc

▲ KEEPING DRY Workers near Bombay take a break from transplanting rice during the monsoon. It is back-breaking work, and their straw protectors shield them when bending to the task.

Monte Carlo *Monaco* Home of the famous casino and main tourist area of the tiny principality, with numerous luxury hotels. The casino, facing the Mediterranean on one side and beautiful gardens on the other, was opened in 1856; it has gaming rooms and an ornate theatre. Nearby are conference and exhibition centres.

Population 13 200
Map France Ge

Monte Sant'Angelo *Italy* Town and pilgrimage centre standing at nearly 800 m (2025 ft) on the Gargano peninsula – the 'spur' of Italy – about 170 km (105 miles) north-east of Naples. It has grown up around the sanctuary of the Archangel Michael, founded in the 5th century.

Population 16 500
Map Italy Ed

Montecatini Terme *Italy* Health resort about 40 km (25 miles) west and slightly north of Florence. It is noted for its baths and thermal springs.

Population 21 300
Map Italy Cc

Montego Bay *Jamaica* Second largest town and seaport (after Kingston), standing on the north-west coast 130 km (80 miles) from Kingston. It was originally the site of a large Arawak Indian village, visited by the explorer Christopher Columbus in 1494. The port exports bananas. It has exceptionally fine bathing beaches, and the town's prosperity is largely derived from the luxury tourist trade.

Population 70 000
Map Jamaica Ba

Montélimar *France* Town about 130 km (80 miles) south of Lyons. It is best known for the manufacture and sale of nougat made with locally grown almonds.

Population 30 200
Map France Fd

Montenegro (Crna Gora) *Yugoslavia* Smallest of the country's republics, covering 13 812 km² (5331 sq miles). It is composed mainly of deeply eroded limestone mountains, where a kingdom established in the 14th century and ruled by prince-bishops never totally submitted to the Turkish invaders. Along the Adriatic coast, the wilds around the Gulf of Kotor in the north merge with the tourist riviera at Budva, and farther south lie olive and citrus groves behind fine sandy beaches backed by palms and walnut trees. Inland, Lake Shkodër and the capital TITOGRAD, formerly Podgoritsa, are framed by high, rocky mountains, including Lovćen. Farther inland rise the republic's highest, most densely forested mountains, cut by the deep gorges of the Moraca, Tara, Piva and upper Lim rivers. These shelter architectural gems such as the town of Pljevja, 48 km (30 miles) south of Visegrad, both with Orthodox monasteries and Turkish mosques and houses. Montenegro produces livestock, maize, wheat, tobacco, copper, bauxite and lead.

Population 584 300
Map Yugoslavia Dc

Monterey *USA* Pacific Ocean resort in California, about 135 km (85 miles) south-east of San Francisco. It is popular with writers and artists, and holds a jazz festival each year.

Population 28 700
Map United States Bc

Monterrey *Mexico* The country's third largest city, after Mexico City and Guadalajara. It is capital of Nuevo León state and lies 690 km (430

miles) north and slightly west of the national capital. A major industrial centre, it produces 75 per cent of Mexico's iron and steel as well as lead, chemicals and a wide range of manufactures. It was captured by American forces in the Mexican War (1846-8).
Population 1 916 500
Map Mexico Bb

Montevideo ***Uruguay*** Capital city and port on the northern bank of the River Plate, about 100 km (60 miles) across the estuary from Argentina. The city was founded in 1726 by the Spanish, and expanded with the country's growing meat industry. Later, the city grew as a financial centre. Successive waves of Spanish and Italian immigrants have made Montevideo's population – which represents about half the national total – almost exclusively European.

The city has a Mediterranean climate and its shady squares, parks and gardens – notably the El Prado with 850 varieties of roses – and stately 19th-century buildings, its tea rooms and well-kept old cars give it an atmosphere of pre-Second World War Europe. About 90 per cent of Uruguay's trade passes through the port, where the ship's bell of HMS *Ajax* commemorates the Battle of the River Plate, when the German pocket battleship *Graf Spee* was scuttled offshore in 1939.

A late 18th-century fort, now containing a military museum, on a hill 118 m (387 ft) high, west of central Montevideo, has a commanding view over the city and river.
Population 1 500 000
Map Uruguay Ab

Montgomery ***United Kingdom*** See POWYS

Montgomery ***USA*** Capital of Alabama, in the centre of the state about 220 km (135 miles) from the Gulf of Mexico. The city is a business and market centre which once handled large amounts of cotton. New industries such as fertiliser manufacture have been attracted because of reliable sources of hydroelectric power.
Population (city) 185 000; (metropolitan area) 284 800
Map United States Id

Montpelier ***USA*** Capital of the north-eastern state of Vermont. It lies about 80 km (50 miles) from the Canadian border, and is a centre of granite quarrying, insurance, skiing and fishing.
Population 8200
Map United States Lb

Montpellier ***France*** Busy commercial centre 125 km (78 miles) west and slightly north of Marseilles. It trades in wine, and manufactures electronic equipment and computers. Its botanical garden, dating from 1593, is the oldest in France.
Population 225 300
Map France Ee

Montreal ***Canada*** Canada's second largest city after TORONTO, and one of the world's largest French-speaking cities. It was settled by the French in 1642 and two-thirds of the people are Roman Catholics of French origin.

Montreal gets its name from Mount Royal, 200 m (656 ft) high, which dominates the city. Montreal stands at the junction of the St Lawrence and Ottawa rivers and is a major port, although it is 1600 km (1000 miles) from the ocean. The heart of the city is an island, but development has spread northwards to LAVAL and the North Shore and southwards across the St Lawrence. Montreal has four universities, a major financial centre, and industries producing electrical goods, aircraft, railway equipment, clothing and chemicals. Vieux Montreal, the old part of the city near the harbour, has been declared a historic area. Other attractions for the visitor include large, temperature-controlled underground shopping areas, Mary Queen of the World Cathedral, Notre Dame church, and a variety of museums and parks.
Population 2 828 350
Map Canada Hd

▼ GLITTERING METROPOLIS The 48-storey Royal Bank Building (centre) stands at the heart of downtown Montreal. An underground complex of shops, theatres and restaurants lies beneath it.

Mont-Saint-Michel ***France*** Vast granite rock, 78 m (256 ft) high and 850 m (2790 ft) around its base, set amid sandbanks just off the Normandy coast 35 km (22 miles) east of Saint-Malo. It is topped by a 10th to 13th-century Benedictine abbey on a site where St Michael the Archangel is said to have appeared in AD 708. Around the base are medieval walls and towers below a village of close-packed houses. A causeway leads to the rock through partially reclaimed marshland. The abbey, whose spire rises 150 m (492 ft) above sea level, is visited by more than 700 000 tourists each year.
Population 80
Map France Cb

Montserrat ***Spain*** Mountain 1236 m (4055 ft) high and about 50 km (30 miles) north-west of the Mediterranean port of Barcelona. On its foothill is a Benedictine monastery, where pilgrims come to worship at a statue known as the Black Madonna.
Map Spain Fb

Montserrat ***West Indies*** British Crown Colony in the Leeward Islands, south-west of Antigua. It was discovered in 1493 by Christopher Columbus, who named it after the Santa Maria de Montserrat monastery in Spain. British settlers under Sir Thomas Warner colonised the island in 1632; 20 years later a large contingent of Irish arrived from neighbouring St Kitts. Partly because of the Irish connection and partly because of the vivid colour of its tropical landscape, it is sometimes called the Emerald Isle. It became internally self-governing in 1960.

Montserrat is a small rugged island – area 102 km² (39 sq miles) – made up of the remains of three groups of volcanoes, the highest of which, Chance Peak, rises to 914 m (2999 ft). About a quarter of the land is cultivated, and vegetables and fruit are sold to neighbouring islands, particularly Antigua. More than nine in ten of the islanders are descended from African slaves brought to the island to work on the sugar plantations, abandoned in the 19th century. Some Sea Island cotton is grown and woven locally into a high-quality cloth which is either sold to tourists or exported. Cotton was once the island's chief industry.

More recently, some light industry has moved to the island, attracted by tax concessions, and tourists come in increasing numbers, drawn by the lush scenery and the relaxed atmosphere. The small town of PLYMOUTH is the capital.
Population 13 000
Map Caribbean Cb

Monza ***Italy*** Town about 12 km (8 miles) north-east of Milan. Each year the Italian Grand Prix motor race is held on its track. Monza's cathedral houses a gold and gem-encrusted ring reputedly containing a nail from the cross of Christ. The town's manufactures include textiles and carpets.
Population 122 500
Map Italy Bb

moor, moorland Broad tract of open land, often elevated but poorly drained, having patches of heather, coarse grass, bracken or similar vegetation or acid peat bogs.

Moose Jaw ***Canada*** Town in south Saskatchewan, 70 km (43 miles) west of Regina, which serves a large farming area. It has meat packing and dairy plants, flour and timber mills, grain storage and stockyard facilities.
Population 33 940
Map Canada Ec

Mopti ***Mali*** Town about 460 km (285 miles) north-east of the capital, Bamako, on the Bani river near its meeting place with the Niger. It is built on three islands joined by dykes and is sometimes known as the 'Venice of Mali'. Its market deals in cattle and fish, and is a centre for the river trade with the Ivory Coast and Ghana.
Population 54 000
Map Mali Bb

Moquegua ***Peru*** Southern department stretching inland from the Pacific Ocean. Copper is mined at Cuajone and avocados, olives and wine are exported. The department capital, also Moquegua, stands on the Rio Moquegua 75 km (47 miles) north-east of its port, Ilo. It was founded in 1541 and has many splendid Spanish colonial houses along its winding, cobblestoned streets, despite several earthquakes.
Population (department) 101 600; (capital) 22 200
Map Peru Bb

moraine Accumulation of rocks, stones or other debris, carried and deposited by a glacier or ice sheet. See GLACIATION

Morang ***Nepal*** See BIRATNAGAR

Morat ***Switzerland*** See MURTEN

Morava ***Czechoslovakia*** See MORAVIA

Morava ***Yugoslavia*** Main Serbian river system with three major arms. The Western (Zapadna) Morava, 218 km (135 miles) long, rises in the west Serbian mountains. It flows eastwards to join the 318 km (198 mile) long Southern (Juzna) Morava at Stalac, forming the Great (Velika) Morava. The Great Morava flows 221 km (137 miles) north to join the Danube near the city of Smederevo.

With the Axios (Vardar) river, the Morava valleys have always provided easy access between the Aegean coast and the Danube plains, and major road and rail routes follow their paths. Orchards and tobacco fields surround Vranje in the upper Southern Morava valley. Lower down the river's course, below the Grdelica gorge, lie fertile farmlands growing wheat and maize and the industrial towns of Leskovac and Nis. The Western Morava has carved the spectacular Ovcarsko-Kablarka gorge.
Map Yugoslavia Eb

Moravia (Morava; Mähren) ***Czechoslovakia*** Region in the centre of the country between Bohemia and Slovakia, now part of the Czech Socialist Republic, and crossed by the Morava (March) river. The produce of its fertile lands includes cereals, sugar beet, fruit, vegetables, wine and *slivovice* (plum brandy) – Czechoslovakia's national drink. The region has deposits of coal and small reserves of oil and gas; its industrial towns include BRNO, Gottwaldov and OSTRAVA.
Map Czechoslovakia Cb

Moray ***United Kingdom*** See GRAMPIAN; HIGHLAND

Moray Firth Scottish inlet of the NORTH SEA with INVERNESS at its head. It is sometimes defined as the large inlet west of the line drawn between Duncansby Head in the north and Kinnairds Head in the south. In a narrower sense, it is the inlet bounded by the line drawn from Lossiemouth to Tarbat Ness.
Map United Kingdom Db

Mordovia ***USSR*** See SARANSK

Morelia ***Mexico*** Market town, and capital of the south-western state of MICHOACAN, set in scenic mountains. It was founded in 1541 by Mexico's first viceroy, Antonio de Mendoza, who called it Valladolid after his home town in Spain. It was renamed in 1828 after the leader of Mexican independence, José María Morelos. Morelia is notable for its 18th-century aqueduct, Baroque cathedral, courtyards and rose-tinted, carved façades.
Population 230 000
Map Mexico Bc

Morelos ***Mexico*** Small state lying directly south of the Federal District around MEXICO CITY. It was originally the territory of the Tlahuica Indians. At Oaxtepec, near the state capital of CUERNAVACA, the Aztec Emperor Montezuma (1466-1520) had his botanical gardens, which have now been converted into a public recreation centre.

Many wealthy Mexicans have weekend homes in Morelos. The state produces cereals, sugar, rice, coffee, fruit and vegetables.
Population 1 200 000
Map Mexico Cc

Morena, Sierra ***Spain*** Mountain range in the south-west, dividing the Meseta Central from the Guadalquivir valley. The highest peak is Cerro Angulosa, at 1301 m (4267 ft).
Map Spain Cc

Morioka ***Japan*** City in north-east Honshu island, producing traditional cast-iron tea kettles and dyed cotton fabrics. It is the northern terminus for the *shinkansen*, the 210 km/h (130 mph) 'bullet train' from Tokyo, some 460 km (285 miles) to the south.
Population 235 500
Map Japan Dc

Morocco See p. 442

Morogoro ***Tanzania*** Regional capital about 175 km (110 miles) west of Dar es Salaam. It is the centre of Tanzania's sisal industry. Its other products include tobacco and sugar.
Population 90 000
Map Tanzania Ba

Morondava ***Madagascar*** Port on the Mozambique Channel, about 375 km (235 miles) south-west of the capital, Antananarivo. It exports farm produce, including beans, maize, rice and raffia, and has food processing and woodworking industries.
Map Madagascar Ab

Moroni ***Comoros*** Capital of the island group, set on the west coast of NJAZIDJA. It is a fascinating mixture of modern government buildings and old Arab houses and mosques, of broad squares and winding alleys.
Population 20 000
Map Madagascar Aa

Morphou ***Cyprus*** North-western town near the bay of the same name. It is at the centre of an area growing citrus fruits, strawberries and tulips. The Church of St Mamas is dedicated to the well-loved Cypriot saint who in Byzantine times refused to pay taxes, and further showed his contempt for authority by taming a lion and riding on it into the presence of the governor. The governor was so impressed by this that he exempted Mamas from taxation for life.
Map Cyprus Ab

Morris Jesup, Cape ***Greenland*** Most northerly point of the mainland of Greenland. The world's northernmost land is a 30 m (100 ft) wide speck north of the cape: the islet of Oodaaq, off Peary Land. It is 706 km (439 miles) from the North Pole and the sea around it is iced up for most of the year.
Map Arctic a

mortlake An OXBOW lake.

Moscow (Moskva) ***USSR*** Capital of the Soviet Union and of the RSFSR, and the city from which all power flows in one of the world's most centralised nations. It is the home of the glittering and elegant Bolshoi Ballet, and of the feared Lubianka Prison; of special department stores open only to foreigners and high Communist Party officials, and of thinly stocked state shops where housewives queue; of forest suburbs dotted with luxury weekend cottages, or *dachas*, and of bleak workers' suburbs lined with high-rise blocks of flats.

The city straddles the Moskva river, after which it is named, in European Russia. Since the time of Ivan the Great (1462-1505), it has been the cultural and educational capital of the Russian people, even when, in 1712, Peter the Great (1672-1725) moved his court and the country's government to his new capital of St Petersburg (now Leningrad) on the Baltic. In 1918 the leaders of the Russian Revolution restored Moscow as the capital.

The city is first mentioned in Russian chronicles in 1147. In 1156, the Russian prince Yuri Dolgoruky built a wooden *kremlin* there – the word means 'town (*continued on p. 444*)

Morocco

THIS GATEWAY TO NORTH AFRICA STANDS HEIR TO THE MAGNIFICENT MOORISH KINGDOM

Strategically placed at the western entrance to the Mediterranean Sea, Morocco is but a step away from Europe and yet is unmistakably Eastern in flavour. It has long attracted Europeans, from traders to tourists.

But the traffic has been far from one-way. Across the Strait of GIBRALTAR – a mere 13 km (8 miles) wide at one point – the Moors of North Africa poured into Spain in the 8th century, bringing their religion, their architecture and their culture – proud monuments that still stand today. From 1085 the rulers of the newly established Berber Kingdom of Morocco played an important part in the Muslim domination of Spain.

It was a domination that did not last, of course. In time, Spain and particularly France came to rule Morocco – though never so completely as some other North African countries.

Today Morocco is one of only three kingdoms remaining in Africa, a conservative state that tries to maintain close links with other Arab countries yet is firmly pro-Western. However, its claim to the sparsely populated neighbouring territory of WESTERN SAHARA has embroiled Morocco in years of costly warfare.

With an area of some 458 000 km^2 (177 000 sq miles), Morocco has a 1400 km (875 mile) coastline on the Atlantic but only 400 km (250 miles) on the Mediterranean. It is a land of vivid contrasts: high rugged mountains which are snow-capped in winter, the fierce arid SAHARA, and the green and cultivated Atlantic and Mediterranean coasts. The country is split from south-west to north-east by the ATLAS MOUNTAINS, which continue across Algeria. To the north of the mountains are fertile coastal plains along the Atlantic; to the south, the forbidding desert with scattered oases.

The RIF MOUNTAINS in the north – green, rounded and well watered – rise to 2456 m (8058 ft). They are separated from the Middle Atlas by the TAZA gap, an area partly covered by pine, holm oak and cedar trees. Farther south, the High Atlas reach 4167 m (13 671 ft) at Toubkal. In the south-west, the Anti Atlas form an elevated rim at the edge of the Sahara plateau.

The north has a pleasant Mediterranean climate of hot dry summers with clear blue skies and mild moist winters. Average temperatures at TANGIER range from 11°C (52°F) in winter to 29°C (84°F) in summer with 810 mm (32 in) of rain annually; in MARRAKECH the winters are warmer and the summers even hotter. The cool Canaries Current tempers the Atlantic coast climate; at ESSAOUIRA, for example, the average temperature range is 14°C (57°F) to 20°C (68°F), with 330 mm (13 in) of rain. Snow often falls in winter on the High Atlas mountains.

The natural vegetation varies from Mediterranean scrub in the north to mountain forests in the Atlas ranges; about 12 per cent of the country is wooded. In spring, the mountain slopes become carpeted with beautiful alpine wild flowers. On the drier southern flanks of the Atlas, sparse vegetation merges into desert. Among the wild animals are macaque monkeys – the so-called Barbary apes.

Morocco contains about 25 million people. Most are something of a mixture; Arabs, Berbers and, later, negroes, French and Spanish have intermingled. The Berbers are descendants of the original inhabitants who were in this area thousands of years ago and at one time controlled all the land between Egypt and Morocco. Arabic is the official language, spoken by most people except the Berbers, whose ancient tongue still predominates in the central mountains. French is the business language of the cities, but Spanish is also spoken. There are more than 100 000 Europeans (mainly French) in the country.

Nearly all Moroccans are Muslims, but they are more tolerant than in many Arab states. There is no general ban on alcohol, for instance, and a Jewish minority lives peaceably in the country. But Muslim rules about designs on craftwork are strictly enforced: no human or animal figures may be represented.

Morocco is still mainly a farming country – over half of the people work on the land. Wheat, barley and maize are the main food crops, and Morocco is one of the world's chief exporters of citrus fruits. Olives, grapes and many varieties of vegetables are grown. Large flocks of sheep and goats are reared in the mountains. However, the standard of living varies greatly and there is much poverty. When most colonial settlers left, they were replaced by Moroccan landowners, and the big estates were not broken up. Most farmers have only small plots from which they feed themselves and their families, and production has not kept pace with a rising population, so that much food has to be imported.

Morocco's main wealth is derived from phosphates, used for fertiliser manufacture. It is the third-largest producer in the world, after the USA and USSR, and its reserves are the world's largest. Other minerals exist as well – iron, zinc, manganese, lead, cobalt and the only sizable coal stocks in North Africa – but these are comparatively undeveloped.

The country's mixed economy is one of the most complex and advanced in Africa. Morocco is almost self-sufficient in textiles, and cans most of its fruit. It has car assembly plants and soap and cement factories, while its sea-fishing industry is the fifth-largest in Africa. Hydroelectric power provides about 70 per cent of its electricity. Tourism is a major source of revenue.

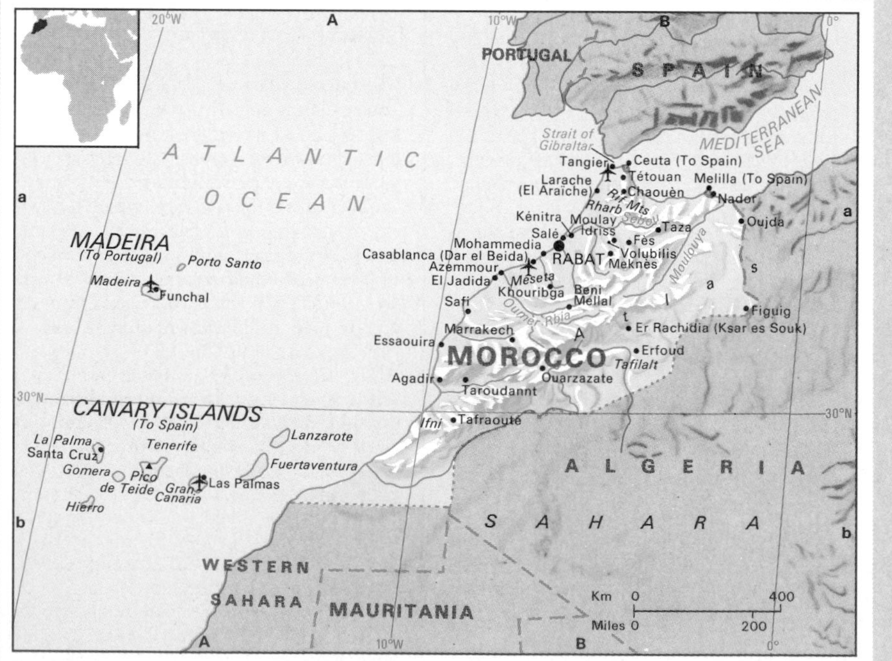

AN ANCIENT KINGDOM

Like other countries of North Africa, Morocco came in turn under Carthage, Rome and then the Vandals. Around AD 685 Arabs from the east swept across the country, bringing the Muslim religion. From 711, many Moroccans joined the Muslim, or Moorish, invasion of Spain, led by Tariq ibn Zaid, which stamped the Arab influence on Andalucia and even briefly reached south-western France. The name Moor itself is derived from Mauri, a North African Berber people. The Almoravid dynasty, which ruled from Marrakech in the 11th century, established a Berber kingdom stretching from central Spain to Senegal.

But conflict arose between the Arabs and Berbers. The Christian kingdoms of northern Spain and Portugal took advantage of the divisions in Moorish power and began to obtain footholds in northern Morocco. In 1492 the Christian armies drove the Moors

▲ DYER'S MARKET Skeins of wool hang out to dry after dyeing in a market in Marrakech. Morocco's *souks* (markets or bazaars) enchant nearly 2 million tourists who visit the country each year.

from their last Spanish stronghold, GRANADA, and then conquered several cities on the Moroccan coast. The Moors had regained most of them by 1578, but CEUTA and MELILLA remain Spanish to this day.

For more than 300 years Morocco remained independent. Then, in the late 19th century, the country became the centre of European imperialist rivalries. France and Spain wanted to divide it, and Germany also tried to intervene. These rivalries were resolved eventually and France occupied most of the country in 1912, leaving Spain in possession of a small protectorate in the north; Tangier became an international zone.

By and large, the next 45 years were successful and prosperous ones for Morocco; the French administration in particular, led until 1925 by General Louis Lyautey, did a lot to modernise the country. The sultan remained, but with little power, and the only serious opposition came from Berber tribes in the hills. Moroccan troops fought alongside French in the First World War and with General Franco in the Spanish Civil War (which began with a revolt at Spanish army bases in Morocco).

In the Second World War, Allied forces landed in Morocco and Algeria in 1942 to drive the Germans out of North Africa, and Allied leaders – Churchill, Roosevelt and de Gaulle – met at CASABLANCA in January 1943. After the war, Sultan Sidi Mohammed ben Youssef became the focus of a popular and unified independence movement. He was able to negotiate Moroccan independence in 1956 without the bitter guerrilla war that convulsed Algeria. Tangier was returned to Morocco, but Spain retained two small enclaves at Ceuta and Melilla; IFNI was ceded in 1969.

The sultan became king in 1957, and was succeeded in 1961 by his son, Hassan, an astute though conservative political leader. There are a dozen political parties and a Parliament with 306 seats, but King Hassan is in effect an absolute monarch and at times has run the country without Parliament.

In 1975 he organised the peaceful 'green march' in which 350 000 unarmed Moroccans entered the mineral-rich Spanish Sahara (now Western Sahara). Spain agreed that the country should be divided between Morocco and Mauritania, but the indigenous Sahrawis formed a movement for self-government, the Polisario Front, and fought a guerrilla war against both countries. Mauritania gave up its claim in 1979; Morocco fortified the few small towns and claimed to have overcome the guerrillas.

The war, though nationally popular, led to bad relations with Algeria, and in 1984 Morocco left the Organisation of African Unity when the OAU recognised the Sahrawi delegation. The same year the conservative King Hassan negotiated a 'treaty of unity' with radical Libya. He caused controversy in 1986 by meeting Israeli premier Shimon Peres.

Although possessing many advantages by comparison with other African states, Morocco also faces numerous social problems. Not least is the rapid growth in population – nearly half the people are under 15 years old. Education is compulsory between the ages of 7 and 13, but only a quarter of children go to secondary school. The adult literacy rate is under 30 per cent and special newspaper and radio campaigns have been started to combat illiteracy. Medical care is poor and there are relatively few doctors.

MOROCCO AT A GLANCE

Area 458 730 km² (177 116 sq miles)
Population 24 950 000
Capital Rabat
Government Constitutional monarchy
Currency Dirham = 100 centimes
Languages Arabic (official), Berber, French, Spanish
Religions Muslim (99%), Christian (1%)
Climate Warm on coast, hot inland. Average temperature in Rabat ranges from 8-17°C (46-63°F) in January to 18-28°C (64-82°F) in August
Main primary products Cereals, pulses, citrus fruits, vegetables, olives, grapes, sheep, goats, poultry, almonds, dates, timber, fish; phosphates, iron, lead, manganese
Major industries Agriculture, food processing, textiles, leather goods, cement, wine, fertilisers, forestry, fishing, mining
Main exports Phosphates, fruit and vegetables, phosphoric acid, metal ores, fish, clothing, petroleum products
Annual income per head (US$) 650
Population growth (per thous/yr) 29
Life expectancy (yrs) Male 55 **Female** 59

▲ **GOOD VINTAGE Sun-drenched vineyards clothe both banks of the Moselle between Trier and Koblenz in West Germany. They produce the famous light, white Moselle wines.**

fortress' – and the city that grew around it prospered at the hub of the river routes across Russia. By 1400, it was the capital of a thriving principality, often known as Muscovy, founded in 1263 by Alexander Nevsky (1220-63). This survived Tatar raids and gradually absorbed neighbouring principalities, such as Kiev and Novgorod. Grand Duke Ivan III (Ivan the Great) consolidated the state in 1480, finally defeated the Tatars and made himself Tsar of all the Russias. In 1571 Tatars captured Moscow and in 1610 it was partly destroyed by Poles, but it survived its two greatest sieges damaged but unconquered – by Napoleon's forces in 1812 and by Adolf Hitler's in 1941.

The city grew rapidly as the country's centralised administrative centre after 1918, and in the 1930s saw large-scale industrial growth, with steel, machinery and vehicle plants. Postwar rebuilding since the 1940s, and new suburbs of workers' apartments with industries such as chemicals, paper, textiles, food processing, furniture and electronics, have pushed the city's limits out to a ring motorway which is 110 km (68 miles) long.

The heart of Moscow is still the Kremlin – the symbol of Russian power and authority, with former royal palaces and the glittering onion-shaped domes of its cathedrals rising above red-brick walls beside Red Square, site of the Lenin Mausoleum and the May Day displays of Russian military muscle. The Kremlin's walls (1485-95), overlooking the Moskva river, are 2.3 km (nearly 1.5 miles) long, with 19 towers. They enclose Kremlin Square, a magnificent assemblage of 15th to 18th-century Russian architecture. Its four cathedrals and church are reminders that Moscow has been the headquarters of the Russian Orthodox Church since 1326. The 19th-century Grand Palace, now the headquarters of the Supreme Soviet of the USSR, and the 18th-century Council of Ministers building – where the revolutionary leader Vladimir Lenin (1870-1924) lived and worked from 1918 to 1924 – also stand on the square.

Away from the Kremlin, mostly grey buildings line avenues radiating from the centre and crossed by ring roads. Among the greyness, however, are over 120 museums and galleries, 35 theatres and concert halls, including the Bolshoi Opera and Ballet Theatre and Moscow Arts Theatre, and 75 cinemas. Farther out, in the nearby Lenin Hills, is the Olympic Village built for the 1980 Games.

Moscow's population has exploded in the past 100 years. In 1871, the city had only 602 000 people, but more than double that by 1912 (1 618 000). And, despite the losses of the Second World War, the 1939 total of 4 542 000 had almost doubled by 1983, when unofficial estimates put the total population of Greater Moscow – the city and its surrounding areas – at 12 000 000, making it Europe's largest city.

Population 8 600 000

Map USSR Ec

Moselle (Mosel) ***Western Europe*** River, 550 km (340 miles) long, that rises in the Vosges mountains of eastern France. It flows north through Lorraine, passing Nancy and Metz, to the West German border. From there it forms part of the frontier between Luxembourg and West Germany, and then descends through a narrow, winding valley to join the Rhine at Koblenz. Locks and other works between Nancy and Koblenz allow 1500 tonne barges to use the river. The sides of its valley, particularly in West Germany, are planted with vines that produce the fruity Moselle white wine. Its chief tributaries are the Meurthe and Saar.

Map France Gb

Moskva ***USSR*** See MOSCOW

moss Acid bog, especially in northern England and Scotland, such as Chat Moss in Lancashire.

Mossamedes ***Angola*** See NAMIBE

Mosselbaai (Mossel Bay) ***South Africa*** Fishing port and resort in the south coast of Cape Province, formerly called Aliwal South. It was visited in 1488 by the Portuguese navigator

Bartolomeu Dias. Twelve years later, a Portuguese sailor left a message in a boot hanging from a milkwood tree, and started a regular postal system. Letters are still posted in the same place, but in a boot-shaped letterbox.

Population 33 200

Map South Africa Bc

Most ***Czechoslovakia*** Coal-mining town in north-west Bohemia, about 15 km (10 miles) from the East German border. In 1975, its late Gothic cathedral was moved, complete, to a nearby site to make way for new mining operations.

Population 62 100

Map Czechoslovakia Aa

Mosta ***Malta*** Small yet bustling town in the middle of the island. It is dominated by the huge dome of its 19th-century neoclassical church, built largely with donations from local emigrants. On display near the altar is the now defused bomb which crashed through the roof during a crowded service in 1942 and miraculously failed to explode.

Population 9030

Map Malta Db

Mostaganem (Mestghanem) ***Algeria*** Trading and fishing port, and administrative capital of the Mostaganem department – a rich agricultural area which is also designated for industrial development. The town lies 80 km (50 miles) north-east of Oran, to which it is linked by rail. Founded in the 11th century, it reached the height of its prosperity under the Turks in the 16th century. Today, it exports citrus fruits, wine, wool, grain and vegetables. The older part is dominated by an 11th-century citadel.

Population 101 600

Map Algeria Ba

Mostar ***Yugoslavia*** Former capital city of Herzegovina, part of the Republic of Bosnia Herzegovina, 80 km (50 miles) south-west of Sarajevo. Surrounded by barren, pitted mountains, Mostar is a mix of grey turrets, white minarets and dark cypresses rising above red-roofed houses. The city – its name means 'old bridge' in Serbian – spreads around a fine 16th-century bridge, which spans the blue Neretva by a single arch, 29 m (95 ft) wide and rising 20 m (66 ft) above the river. The town – Herzegovina's largest – with an ancient bazaar and handicraft quarter, is the centre of a small but rich basin producing high-quality tobaccos once smoked in Turkish and European courts, fine *blatina* and *zilavka* wines, fruit, walnuts and vegetables. Its aluminium plant, which uses local bauxite, and its cotton mills are powered by the nearby Jablanica hydroelectric dam, opened in 1957. Mostar is now also a flourishing tourist centre popular with artists.

Population 110 400

Map Yugoslavia Cc

Mosul (Al Mawsil) ***Iraq*** City on the Tigris river about 350 km (220 miles) north and slightly west of the capital, Baghdad. Mosul has an oil refinery and is a marketplace for the surrounding oil-producing and farming region, trading in grain, fruit, livestock and wool. Its other main industries are flour milling and tanning hides. A mosque in the city is dedicated to the Biblical prophet Jonah, who is said to be buried there. The remains of the ancient Assyrian capital, NINEVEH, lie on the opposite bank of the Tigris.

Population 1 500 000

Map Iraq Ba

Motherwell ***United Kingdom*** See STRATHCLYDE

Moulay Idriss ***Morocco*** Town 50 km (30 miles) west of Fès, lying on the edge of rocky hills among olive groves and cacti. It contains the tomb of Moulay Idriss I, the founder of the Islamic kingdom of Morocco, who died in AD 791. A *moussem* (religious festival) takes place each September in his memory. The remains of VOLUBILIS are nearby.

Map Morocco Ba

moulin Circular sinkhole in the surface of a glacier, or in the bedrock beneath, worn by swirling meltwater falling down a crevasse.

Moulins ***France*** Town and former capital of the Duchy of Bourbon, about 145 km (90 miles) north-west of Lyons. It produces leather, hosiery and furniture, and is an important road and railway junction.

Population 25 550

Map France Ec

Moulmein ***Burma*** City, port and capital of Mon state, about 120 km (75 miles) south-east of the national capital, Rangoon. It is a marketplace for the sugar and rubber grown along the coastal strip at the mouth of the Salween river. It is also a timber centre, processing logs – particularly teak – which are floated down the Salween river from jungle logging sites, where elephants are used to move the giant trees.

Population 203 000

Map Burma Cc

Mount Gambier ***Australia*** City in the extreme south-east of South Australia, about 380 km (240 miles) south-east of Adelaide. It lies at the foot of an extinct volcano which has three crater lakes. The largest, Blue Lake, is 500 m (1640 ft) across and 80 m (260 ft) deep. The almost circular lake is a brilliant blue in summer but fades to a dull grey in winter.

Population 19 900

Map Australia Gf

Mount Hagen ***Papua New Guinea*** Capital of Western Highlands province, standing at over 1700 m (5600 ft) about 1000 km (620 miles) north-west of Port Moresby. It takes its name from a mountain peak 3777 m (12 392 ft) high, 24 km (15 miles) to the north-west. It was first visited by Europeans only in 1933, and was established as a small patrol post around an airstrip, but had long been a trading centre for local tribesmen. It now has factories for processing locally grown coffee, tea and pyrethrum (used as an insecticide). It is the site of a biennial show and 'sing-sing', in alternate years with GOROKA in the Eastern Highlands.

Population 28 000

Map Papua New Guinea Ba

Mount Isa ***Australia*** Mining town in harsh, remote country in Queensland, about 160 km (100 miles) east of the state border with the Northern Territory. It is a major mining centre, producing 70 per cent of Australia's copper in the 1970s. Lead, zinc and silver are also mined in large quantities.

Population 23 600

Map Australia Fc

▼ HIGHLAND FLING Papua New Guinea's highlanders don their finery and gather at Mount Hagen every two years for a sing-sing – two days of parades, gift exchanges, ritual pig kills, feasting and dancing.

Mount Rainier National Park ***USA*** Area covering 976 km² (377 sq miles) in the Cascade Range in western Washington state, about 110 km (70 miles) north of the Oregon border. The park centres on Mount Rainier, a 4392 m (14 410 ft) volcanic peak from which a spectacular series of glaciers radiate. There are also dense forests and wide expanses of alpine meadow with hosts of wild flowers.
Map United States Ba

mountain Mass of land with steep slopes projecting well above its immediate surroundings. A mountain differs from a PLATEAU in that the area of its summit is much less than that of its base, and it generally has steep sides with a large proportion of bare rock. There is no precise definition of the height which distinguishes a hill from a mountain. (For the world's highest mountains, see panel below.)

mountain grasslands Areas above the tree-line in most mountainous regions of the world where a short growing season and moisture from melting snow or summer rain support the growth of grasses. See PAMIR, SAETER and ALP

Mourne Mountains ***United Kingdom*** Range in County Down, Northern Ireland, rising to 852 m (2795 ft) at Slieve Donard. The mountains have several reservoirs and are a popular tourist attraction.
Map United Kingdom Bc

Moxico ***Angola*** Farming province whose capital, LUENA, is about 810 km (503 miles) south-east of the national capital, Luanda. Copper is mined in the province, but the economy has been disrupted by civil war since 1975. Subsistence crops are grown, including maize, millet, beans and groundnuts (peanuts).
Population 213 000
Map Angola Bb

Mozambique See p. 448

Mozambique Channel Strait between Madagascar and the African mainland which is an important route for East African shipping. It contains the warm Mozambique Current.
Dimensions More than 1600 km (1000 miles) long, 400-970 km (250-600 miles) wide
Map Mozambique Ab

Mtskheta ***USSR*** See TBILISI

Mtwara ***Tanzania*** Regional capital and port about 40 km (25 miles) north of the Mozambique border. It handles local export crops such as cashew nuts, and imports such as petroleum products for its hinterland.
Population 60 000
Map Tanzania Cb

Mu ***Burma*** Central river draining the area between the Irrawaddy and Chindwin valleys.
Map Burma Bb

ON THE ROOF OF THE WORLD

The Himalayas, 2400 km (1500 miles) long and varying between 200 and 400 km (125 and 250 miles) wide, contain many of the world's highest peaks – 96 of them are over 7315 m (24 000 ft). The massive range started to form 38 million years ago when the Indian Plate of the earth's crust collided with and moved under the Eurasian Plate and forced the land upwards. This movement continues still.

THE WORLD'S HIGHEST MOUNTAINS

Mountain	Location	Height (m/ft)
Everest	Nepal/China	8848/29 028
K2 (Godwin Austen)	Pakistan/China	8611/28 250
Kangchenjunga	Nepal/India	8585/28 165
Lhotse I	Nepal/China	8501/27 890
Makalu	Nepal/China	8481/27 824
Dhaulagiri	Nepal	8172/26 810

HIGHEST POINTS OF THE CONTINENTS

Continent	Mountain	Height (m/ft)
Africa	Kilimanjaro (Tanzania)	5895/19 340
Antarctica	Vinson Massif	5140/16 863
Asia	Everest (Nepal/China)	8848/29 028
Australia	Kosciusko (New South Wales)	2230/7316
Europe (excluding USSR)	Mont Blanc (France)	4807/15 770
Europe (including USSR west of the Urals)	Elbrus (USSR)	5642/18 510
North America	McKinley (Alaska)	6194/20 320
South America	Ojos del Salado (Argentina/Chile)	7084/23 241

Muara ***Brunei*** Deep-water port on the promontory north of the Limbang estuary, about 16 km (10 miles) north-east of Bandar Seri Begawan. It handles most of Brunei's imports, including foodstuffs, consumer goods and machinery.
Map Brunei Ba

mud Apart from describing any wet, sticky, soft earth, geologists use the term to define a fine-grained sediment of clay and silt, usually with a high percentage of water present, found for instance round the margins of continents and where lakes and rivers have been drained.

mud flat Land covered at high tide and exposed at low tide.

mud pot Pool of boiling mud, usually sulphurous and often brightly coloured, found in areas of volcanic activity, such as in the Yellowstone National Park in Wyoming in the USA.

Muda ***Malaysia*** River rising in the hills of north-west Malaysia on the border with Thailand. It flows about 320 km (200 miles) south-west to enter the Andaman Sea near Sungei Petani. It has been dammed near the town of Alor Setar to irrigate the Kedah plain, where farmers can now grow two rice crops a year.
Map Malaysia Ba

▲ MOUNTAIN PANORAMA The view from 7900 m (about 25 900 ft) up on the south-west face of Mount Everest shows the West Ridge of the mountain (right foreground) and Pumon lying below (centre).

mudstone Fine-grained, compact rock similar to SHALE, but without its capacity to cleave.

Mudumalai ***India*** Game reserve in the Nilgiri Hills of the south, about 430 km (270 miles) south-west of the east-coast city of Madras.
Map India Be

Mufulira ***Zambia*** Copper-mining town about 300 km (185 miles) north of the capital, Lusaka. In 1970 a flood at the mine killed 89 men.
Population 149 800
Map Zambia Bb

Muharraq Island ***Bahrain*** Second largest of the 33 islands which make up Bahrain State, lying 3 km (2 miles) north-east of Bahrain Island. It is connected to Al Manamah on the main island by a causeway. In the old quarter of the main town, Al Muharraq, many of the houses are built around enclosed courtyards. They date from a pearl-fishing boom in the 18th and 19th centuries. Al Muharraq is now a centre for building dhows, single-masted coastal sailing vessels. Its main buildings include the Bahrain Museum and a fine wooden mosque.
Population 61 900
Map Bahrain Ba

Muineachán ***Ireland*** See MONAGHAN

Mukallah ***South Yemen*** Port and the country's second city, lying 480 km (298 miles) east of Aden. It has a fish-canning factory and a fish-freezing plant – and is the centre of South Yemen's fishing industry.
Population 100 000
Map South Yemen Ab

mulga scrub Type of scrub consisting mainly of a species of acacia known as mulga, which covers parts of central and western Australia.

Mülheim an der Ruhr ***West Germany*** Port on the Ruhr river, 10 km (6 miles) west of Essen in the Ruhr industrial area. It has steel and engineering plants.
Population 174 000
Map West Germany Bc

Mulhouse ***France*** Industrial city near the border with West Germany and Switzerland. It was a cotton town in the 19th century. Today its industries include textiles, engineering, car assembly and chemicals.
Population 222 700
Map France Gc

Mull ***United Kingdom*** Mountainous island off the west coast of Scotland, covering 925 km² (357 sq miles) and separated from the mainland by the Sound of Mull. Its highest point is Ben More at 966 m (3169 ft). Sheep, fish and tourism are the basis of the island's economy. The chief settlement is Tobermory. Off the western shore lie the islands of Iona and Staffa, which has a series of extraordinary basalt columns and caverns, including Fingal's Cave, which inspired the Mendelssohn overture *The Hebrides*.
Population 2700
Map United Kingdom Cb

Multan ***Pakistan*** Industrial town about 300 km (185 miles) south-west of Lahore and described by some of its inhabitants as 'the Manchester of Pakistan' because of its cotton industry. It was known to the 4th-century BC Macedonian conqueror Alexander the Great, and was a major Hindu shrine before Muslims destroyed its temples in AD 950; they were restored in 1138. The Sikhs took the city but lost it in turn to the British in the 19th century.
Population 730 000
Map Pakistan Cb

Mozambique

AFRICANS ARE FIGHTING AGAINST MISFORTUNE IN A LAND WHICH EUROPEANS LEFT IN A HURRY

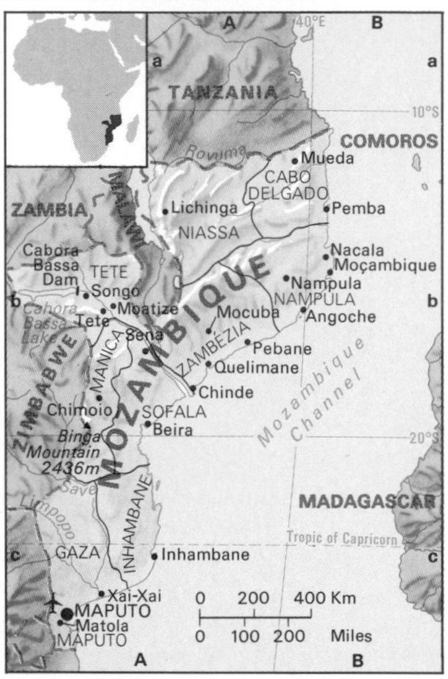

When independence came to Mozambique in 1975, its Portuguese former masters walked out leaving a black African nation with hardly anyone trained to run the farms and the factories.

The Africans had received virtually no education – 99 per cent of them were illiterate. Those who could read and write were mainly youngsters, and during desperate moves to avert economic chaos, young people were appointed to key positions. In 1979 a 19-year-old boy became head of a cotton research centre, and an 18-year-old boy was helping to run one of the country's largest secondary schools.

Mozambique was the last European colony in south-east Africa. It is largely a tropical country with mainly savannah vegetation ranging from open grassland to thickly wooded areas. A coastal plain covers most of the southern and central territory, giving way to the western highlands and north to a plateau including the Nyasa Highlands.

The Portuguese explorer Vasco da Gama visited the territory in 1498, and the first Portuguese settlers arrived early in the 16th century. Very little was done to improve the lot of the Africans. As late as 1970, when most colonial powers had freed their foreign territories, the Portuguese were still dominating the country. About 250 000 Europeans held virtually every professional, managerial and technical post in a country with 8 million Africans.

African discontent had already flared into guerrilla warfare waged by the Mozambique Liberation Front (Frelimo), and this brought some moves towards economic development. But it was the revolution of 1974 in Portugal that gave the final push to independence and the creation of a one-party Marxist state. There was a great exodus of Portuguese, who feared that the Africans would exact revenge for past oppression.

The new government was expecting Communist countries to help Mozambique, but countries like the Soviet Union and Bulgaria showed more interest in the possibility of exploiting the country's natural resources than in aiding development. Mozambique's mineral resources are vast, there are enormous reserves of coal and huge offshore deposits of gas in addition to copper, bauxite (aluminium ore) and other metals.

STARTING FROM SCRATCH

So the new Mozambique had to start from scratch, training managers and technical staff while somehow keeping agriculture and industry going in the meantime. The people, having struggled hard for independence, were enthusiastic and determined. Massive training programmes for literacy and basic health care were started, and Mozambicans with any ability were soon helping to run the new society. The lack of expertise has caused many problems. Some of the worst are in agriculture, which is mainly concerned with cotton, cashew nuts and grain subsistence crops.

Under colonial rule, Portugal had frequently used forced labour to produce key crops. Whole villages would be made to grow cotton, and those who did not fill their quota were often savagely punished.

In the chaos after independence, the government decided that huge state farms were the answer to a desperate shortage of food. These farms failed to produce efficiently, partly because they were short of tractors and pesticides, and partly because inexperienced managers planted crops in the wrong areas and at the wrong times.

The new farm managers eventually began to learn the necessary skills, and the small peasant farms began to make a greater contribution in a country where about 85 per cent of the people are farmers.

But as agriculture began to emerge from crisis, nature took a hand, and the country was struck by the drought that affected much of Africa in the early 1980s. Drought was interrupted by the other extreme of weather in 1984 when the southern Maputo province suffered its worst floods in 30 years. More than 350 000 farmers lost their crops and this disaster, added to years of drought, brought famine which led to more than 100 000 deaths.

Normally conditions are reasonably good for agriculture. Rivers water the plains where most of the cultivation and population are concentrated, and rain falls – mostly as thundery showers – from December to April. It is heaviest in the highlands, where altitude gives cooler temperatures, the north-west highlands receiving more than 1500 mm (59 in) a year. Temperatures on the central coast average 23-27°C (73-81°F) all the year.

Industry and trade brought ideological problems for the Marxist government. It was forced to intervene in hundreds of manufacturing firms abandoned by the Portuguese, in efforts to keep them producing. But although it was opposed to capitalism it was not equipped to take over large sectors of the economy. Over years of crisis it has learnt to take a lenient view of private capital in some areas.

Many of the country's rural shops were previously run by the white settlers, and this retail network collapsed on independence. The government tried to substitute a series of 'People's Shops', but these have not been successful because of lack of expertise. Retailing is also affected by shortage of transport to service the shops, and a lack of foreign exchange to buy imports.

Rural shops are often empty, so the peasant with money in his pocket can find nothing to buy. So he has less incentive to produce surplus crops to sell for cash. The crucial economic links between selling and buying, town and country, and agriculture and industry are weakening. Food rationing has been introduced to combat the shortages, and this has led to the inevitable black market which now accounts for a sizable part of the economy.

THE ENEMY WITHIN

While coping with natural and man-made disasters, the Mozambican Government has also been fighting right-wing guerrillas, the Mozambique National Resistance, known as Renamo or MNR. The MNR has tried to sabotage the new state. Thousands of shops have been destroyed, railways have been blown up, and power lines running from the hydroelectric CABORA BASSA DAM on the LIMPOPO river have been sabotaged. Large areas are unsafe because of MNR activity, and it is even dangerous to travel between major cities. Support for the MNR came from neighbouring South Africa, angered by Mozambican backing for the Communist African National Congress (ANC) in South Africa. In 1983, South Africa bombed Mozambique's capital, Mabuto, allegedly aiming at ANC bases there, but hitting a jam factory. A non-aggression pact between South Africa and Mozambique was signed in 1984.

Another neighbour, Rhodesia (now Zimbabwe), also supported the MNR during the period of illegally declared Rhodesian independence from Britain. The Portuguese had supported the illegal state, but after Mozambique became independent in 1975 it helped to impose international trade sanctions against Rhodesia. One of Mozambique's major sources of revenue is charges on freight carried on the Beira railway for landlocked states of southern Africa; the sanctions against Rhodesia cost Mozambique more than $300 million. Mozambique and Zimbabwe are now on good terms, and troops from Zimbabwe are deployed in Mozambique to protect the Beira railway and the Cabora Bassa Dam.

New problems came out of South Africa when its great mining complexes began to cut back on foreign labour in the mid-1970s. Many Africans from southern Mozambique had worked in the mines, while their wives

▲ SCHOOLING IN THE SUN Outdoor 'classrooms' are not uncommon in Mozambique, and these young students squat informally in the shade of a tree to learn their numerals – written in Arabic and Roman.

and families waited in their home villages for an occasional remittance and the return of their menfolk. Often, there was no money and no man; he might have been killed in a mining accident, or have formed new attachments away from home. But many Mozambican villages depended on this lifestyle, and the labour cutbacks ruined their economy. The result was a huge movement of population, with tens, possibly hundreds, of thousands of people moving to the cities in search of a living. The cities could not cope and the government dealt with the problem in 1983 by sending recent city immigrants to work on the harvest in northern provinces. This drastic but effective solution was called Operation Production.

While battling with a wide range of problems, the Mozambican Government has made great progress in some areas. Huge vaccination programmes have led the way in improving health standards. A massive educational programme is replacing the skills taken out of the country by the retreating Portuguese. And the new society has greatly improved the standing of its women.

Traditionally, women held low status, despite their crucial roles as agricultural workers, water carriers, mothers and cooks. Although they might bear up to ten children (partly because so many died), they had no rights to keep them if the marriage broke down. If a husband abandoned his wife he was not obliged to help to support his offspring.

THE STRUGGLE CONTINUES

New laws have stopped these injustices. Women's participation in political and non-agricultural economic roles has also increased, although strange double standards still operate. During Operation Production, for instance, some women were expelled from cities because they were single women and it was assumed that they must therefore be prostitutes. An important female political official lost her job for adultery, although she had already separated from her husband; much worse behaviour from male officials rarely causes any public action. For women, the Mozambique's slogan of 'A Luta Continua' (The Struggle Continues) is doubly significant.

The struggle does continue: against the MNR; to replace lost skills; and to rehabilitate the economy. Progress is being made. There has been an influx of foreign expertise to help in industry. Mozambique has turned to the West for help, and Britain and the USA are now on good terms with the government of President Machel. And an increasing number of tourists are coming, mainly from South Africa and Zimbabwe, to beaches warmed by the Mozambique Current and national parks rich in wildlife.

MOZAMBIQUE AT A GLANCE
Area 799 380 km² (308 642 sq miles)
Population 14 160 000
Capital Maputo
Government One-party (Marxist) republic
Currency Metical = 100 centavos
Languages Portuguese (official), Bantu languages
Religions Animist (60%), Christian (18%), Muslim (16%)
Climate Humid, tropical; dry season from June to September. Temperatures in Maputo range from 13-24°C (55-75°F) in July to 22-31°C (72-88°F) in February
Main primary products Cotton, cashew nuts, tea, sugar, cassava, cereals, bananas, sisal, groundnuts, coconuts; coal
Major industries Agriculture, textiles, chemicals, food processing, petroleum products, cement, mining
Main exports Cashew nuts, textiles, tea, cotton
Annual income per head (US$) 159
Population growth (per thous/yr) 28
Life expectancy (yrs) Male 47 **Female** 50

Munich (München) ***West Germany*** Capital of Bavaria about 60 km (37 miles) from the Austrian border. Munich has expanded in the past century to become one of Germany's most industrialised cities, making cars, chemicals, cigarettes and beer. At its heart it keeps some of its character as the city of the Wittelsbachs, the princely family that ruled Bavaria as dukes, then kings, from 1180 to 1918. Their palace, built in Renaissance style during the 17th century and rebuilt after it was heavily damaged in the Second World War, has its own theatre, where the 18th-century Austrian composer Wolfgang Amadeus Mozart first produced his opera *Idomeneo*, which he wrote in Munich.

The Baroque Nymphenburg Palace, north-east of the town – summer home of the Wittelsbach family – was built in the 17th century and has a French-style park. Beyond it stands a tower 295 m (960 ft) high, marking the site of the 1972 Olympic Games.

Munich was the birthplace of Adolf Hitler's Nazi movement in the 1920s. About 120 km (75 miles) to the south-east is the alpine resort town of Berchtesgaden; Hitler had a mountain retreat – the 'Eagle's Nest' – above the town, and it was there, in September 1938, that he met British Prime Minister Neville Chamberlain to discuss peace in Europe prior to signing the Munich agreement.

Munich, which calls itself Beer Capital of the World, holds international beer festivals in October (the *Oktoberfest*) and January to February (*Fasching* – meaning 'carnival' in German) each year. Its oldest and largest beer cellar is the Hofbräuhaus, founded by Duke Wilhelm V in 1589 and now the state brewery. The city is also noted for its 15th-century cathedral, university, art galleries and theatres – including the Bavarian State Opera.

Population 1 300 000
Map West Germany Dd

Münster ***West Germany*** Port on the Dortmund-Ems Canal about 70 km (45 miles) north-east of Essen. It is the cultural and economic centre of Westphalia, with varied industries including steel, chemicals and musical instruments. Its university occupies the Baroque palace from which prince-bishops ruled Westphalia until 1803. The treaty ending the Thirty Years' War was signed in Münster's town hall in 1648.

Population 260 000
Map West Germany Bc

Munster (Cuige Mumhan) ***Ireland*** One of Ireland's four historic provinces, in the south-west of the country. The other three are Connaught, Leinster and Ulster.

Population 998 300
Map Ireland Bb

Muqdisho ***Somalia*** See MOGADISHU

Murchison Cataracts ***Malawi*** See SHIRE

Murchison Falls ***Uganda*** See KABALEGA FALLS

Murcia ***Spain*** City about 180 km (110 miles) south of the Mediterranean port of Valencia. It trades in cereals, almonds and citrus fruits, and manufactures silk, textiles, flour, aluminium and leather goods. A former Moorish capital, it is now a mining centre for iron and zinc and a fruit-growing province of the same name.

Population (city) 288 600; (province) 958 000
Map Spain Ed

Murmansk ***USSR*** The world's largest city north of the Arctic Circle. It was founded in 1916 on an estuary of the Kola river about 50 km (30 miles) from the Barents Sea in the extreme north-west of the USSR. It is a major fishing and cargo port – ice-free all year round thanks to the warm ocean current known as the North Atlantic Drift – and an industrial centre producing ships, beer, canned fish and timber. Polyarnyy, nearby, is a naval base.

Population 412 000
Map USSR Eb

Murray ***Australia*** One of the country's longest rivers, at 2570 km (1600 miles). With its extensive system of tributaries (including the Lachlan and Murrumbidgee), the Murray and its main tributary the DARLING (which in turn has many tributaries) drain an area of some 910 000 km² (351 000 sq miles) – one-eighth of the entire country, an area nearly twice the size of France.

The Murray rises in the Snowy Mountains and forms the state boundary between New South Wales and Victoria before entering South Australia and flowing into the sea south of Adelaide. The Murray's and Murrumbidgee's waters are widely used for irrigation, and in places restored paddle steamers re-create for tourists the Murray's transport heyday.

Map Australia Ge

Murree Hills ***Pakistan*** Foothills of the Himalayas to the north-east of Islamabad. They rise to 2150 m (7050 ft), and their forests and mild climate are a major tourist attraction.

Map Pakistan Da

Murshidabad ***India*** Eastern town, about 180 km (110 miles) north of Calcutta. It was founded in 1704 as the capital of the Muslim kingdom of Murshidabad (Bengal), and when the British took over Bengal soon after the Battle of PLASSEY nearby in 1767, it became the administrative centre of British India, with the seat of governor-general from 1773 to 1834. The magnificent palace of the nawabs (Muslim rulers) survives.

Population 356 000
Map India Dc

Murten (Morat) ***Switzerland*** Fortified town on Lake Murten, about 25 km (15 miles) west of the capital, Berne. It still has its 12th to 15th-century town walls and 13th-century castle.

Population 5000
Map Switzerland Aa

Mururoa ***French Polynesia*** Atoll in the Tuamotu archipelago about 1000 km (620 miles) south-east of Tahiti. It has been used as a nuclear weapons testing ground by the French since the 1960s. An underground test in 1985 sparked worldwide controversy when French secret service agents sank the Greenpeace conservation organisation's ship *Rainbow Warrior* in New Zealand as it prepared to sail to the atoll to lead a protest.

Map Pacific Ocean Gd

Musandam ***Oman*** Spectacular mountainous peninsula jutting into the Strait of Hormuz and separating The Gulf from the Gulf of Oman and the Arabian Sea. Though part of Oman, it is divided from the rest of the sultanate by a 70 km (43 mile) wide stretch of the United Arab Emirates (UAE). The isolated settlements on the peninsula are mainly fishing villages. The main town is Khasab, with a shallow harbour and 17th-century Portuguese fort. Until recently there were few proper roads, and most local travelling was done by sea. In 1982 a road was built linking Khasab to Musandam's two other towns, Bakha and Bayah, and on into the UAE.

Map Oman Aa

Musay'id ***Qatar*** See UMM SAID

Muscat ***Oman*** Capital, main port and seat of government of the sultanate; the neighbouring town of MUTTRAH is the commercial centre. Muscat, with its fine natural harbour, grew in influence and power during the Middle Ages and was a thriving city when the Portuguese captured it in 1507. They stayed until the middle of the 17th century, and their city walls and gates are still mostly intact. The city was held by the Persians from 1650 to 1741, after which it became the capital of an independent sultanate. In 1798 the sultan signed a treaty establishing economic and political links with Britain, which lasted until the late 1960s.

Muscat's places of interest include several fine old Arab baits (houses); Bait Nadir (now a museum); Bait Graiza (state apartments); Bait Fransa (French embassy); Bait Zawawi (US embassy); a 16th-century fort and the now royal palace, built on the seafront in the 1970s.

Population (Muscat and Muttrah) 80 000
Map Oman Aa

muscovite Most common form of MICA, found in acid IGNEOUS and METAMORPHIC ROCKS. It ranges from colourless or pale brown to green, has a vitreous lustre and is used as an insulator. It is also called isinglass.

Muscovy ***USSR*** See MOSCOW

muskeg Swamp largely filled with sphagnum moss in the subarctic zone and coniferous forest of northern Canada.

Mustique ***St Vincent*** Privately owned island in the Grenadines forming part of the state of St Vincent. It has been developed as an exclusive tourist resort and provides goods and services for yachtsmen in the area.

Population 200
Map Caribbean Cc

Mutare ***Zimbabwe*** Town, formerly called Umtali, about 220 km (137 miles) south-east of the capital, Harare, near the Mozambique border. It is a tourist resort surrounded by mountains rising, in places, to 2500 m (8200 ft).

Population 69 600
Map Zimbabwe Ca

Muttra ***India*** See MATHURA

Muttrah (Matrah) ***Oman*** New port of greater MUSCAT, situated immediately to the north of the capital. For centuries Muttrah was a separate commercial town, and until 1929 was connected to Muscat by only donkey tracks. The terminus of the caravan routes from the interior, it traded in pearls, dates and fruit. Today the town's numerous *souks* (bazaars) deal in silver, gold, cloth and grain. The seafront is dominated by a 16th-century fort and there are some fine mosques and old Arab merchant houses.

Just to the north-west, in Muttrah bay, is Mina Qaboos, the sultanate's main modern port. Completed in 1974, it has eight deep-water berths where large container ships can be handled. Nearby, the old fishing harbour and fish market still thrives.

Population (Muttrah and Muscat) 80 000
Map Oman Aa

Mwanza ***Tanzania*** Regional capital and the country's second largest town after Dar es Salaam. Mwanza stands on the south shore of Lake Victoria; it makes and exports textiles.

Population 170 000
Map Tanzania Ba

Mweru, Lake ***Zaire/Zambia*** Lake lying 930 m (3051 ft) above sea level, on the border between south-east Zaire and Zambia. It covers about 4920 km² (1900 sq miles). The lake is fed by the Luapula river from the south and drained by the Luvua river, a tributary of the Lualaba. Swamps border the south and east shores. In 1798 a Portuguese official, Francisco Lacerda, became the first European to see the lake, but he died nearby. It was explored by the Scottish missionary Dr David Livingstone in 1867.

Map Zaire Bb

My Lai (Song My) ***Vietnam*** Village in former South Vietnam that was the scene of a massacre in March 1968, during the Vietnam War. American troops killed more than 100 villagers, including women and children, during a 'search and destroy' operation against Viet Cong guerrillas. An American officer was later found guilty of murder, and the case helped to turn public opinion against the war.

Myanaung ***Burma*** Small town on the Irrawaddy river in southern Burma, about 180 km (110 miles) north-west of Rangoon. It grew rapidly during the 1980s as a producer of chemicals synthesised from natural gas. This was found locally in the late 1960s, and is also used to fuel power stations.

Population 183 200
Map Burma Bc

Mycenae (Mikinai) ***Greece*** Excavated, fortified city of Homer's King Agamemnon on a hill 30 km (18 miles) south-west of Corinth in the Peloponnese. Mycenae was occupied from 1580 to 1100 BC, when it was destroyed by the invading Dorians, and was the heart of a rich trading kingdom known as the Mycenaean civilisation after the site. It was excavated from 1874 by the German archaeologist Heinrich Schliemann, who uncovered a palace, temples, a citadel and the remarkable royal chamber tombs containing gold death masks. The largest is named the Tomb of Agamemnon, though it was almost certainly of an earlier date than 1200 BC, when he is believed to have ruled. Many of the finds are now in the Athens museum.

Map Greece Cc

Myitkyina ***Burma*** Town and capital of the state of Kachin in Burma's far north – and the end of the railway line which runs up the centre of the country from Rangoon. It lies in a largely agricultural area of rice and sugar-cane fields, and has a sugar refinery.

Population 103 300
Map Burma Cb

Mymensingh ***Bangladesh*** Chief town of Mymensingh district (area 9668 km², 3733 sq miles) on the old course of the Brahmaputra river. Jute is grown in the area.

Population (district) 6 568 000; (town) 107 900
Map Bangladesh Cb

Mysore ***India*** Southern industrial city, about 400 km (250 miles) west and slightly south of Madras. It was a capital of the wealthy, progressive Muslim state of Mysore from the late 16th century. The palace of its ruling maharajas, a spectacular pavilion of granite, incorporates cast-iron pillars specially made for it in Glasgow (Scotland). A summer palace, now a hotel, stands on a cliff overlooking the city. Mysore is an educational and commercial centre, and produces textiles, fine silk saris, leather goods, chemicals and cigarettes.

Population 479 000
Map India Be

Mytilene ***Greece*** See LESBOS

Mývatn ***Iceland*** Lake covering 37 km² (14 sq miles) in the north of the island. It is popular with scientists and tourists because of hot springs and volcanic craters around its shores, and its bird life. Europe's principal deposits of the mineral diatomite are found in the area.

Map Iceland Ca

M'Zab ***Algeria*** Stony, barren region on a plateau to the north-east of the Great Western Erg in the north Sahara. The area was settled at the beginning of the 11th century by the Mozabites, an austere Muslim sect of Berber origin, who dug wells, planted date palms and built five towns united in a confederation. These towns, of which the best known is Ghardaia, are built on small hills and look like coloured pyramids. The towns have bustling markets and are centres of traditional arts and crafts.

Map Algeria Ba

Na Cruacha Dubha ***Ireland*** See MACGILLICUDDY'S REEKS

Naarden ***Netherlands*** Former seaport on the south side of the IJSSELMEER, 19 km (12 miles) east and slightly south of Amsterdam. The town lies within fortifications built in a perfect star shape by the French military engineer Sébastien de Vauban in the late 17th century. Modern development, including chemical plants, is outside the moat in neighbouring Bussum.

Population 16 400
Map Netherlands Ba

Nabadwip ***India*** See NAVADWIP

Naberezhnyye Chelny ***USSR*** See BREZHNEV

Nabeul ***Tunisia*** Administrative capital of Cape Bon, situated 61 km (38 miles) south-east of Tunis on the Gulf of Hammamet. It is a tourist centre with hotels, restaurants, night clubs, a theatre and good beaches. The manufacture of ceramics, using traditional designs, has been the main industry since Roman times, and there are hundreds of potteries where visitors are welcome. Other crafts include embroidery, needlework and lace-making. Perfume distilleries working from ancient formulas use, among other local natural ingredients, orange-tree blossom and the flowers of geraniums and jasmine.

Population 39 600
Map Tunisia Ba

Nablus ***Israeli-occupied Jordan*** Largest WEST BANK town, situated 48 km (30 miles) north of Jerusalem. The centre of an important agricultural region that has fallen into decline under occupation, it manufactures oil and soap. The university of Al-Abjah is located there.

Population 44 000
Map Jordan Aa

Nadi ***Fiji*** See LAUTOKA

Nadia ***India*** See NAVADWIP

Nador ***Morocco*** Port on the Mediterranean coast, 13 km (8 miles) south of Melilla, handling iron ore from the Beni bu Ifrur mines. The town trades in livestock (especially sheep), cereals and fruit, and is the site of an integrated iron and steel complex.

Population 63 000
Map Morocco Ba

Næstved ***Denmark*** Town in the south of ZEALAND, linked to the sea by a canal. It originated around a Benedictine monastery founded in 1135, near the site of which now stands Herlufsholm, a private residential school. The town developed as a market centre for southern Zealand. Its industries now include paper, glass, pottery, textiles and fishing. There are the remains of a medieval hospital (the Helligåndshuset), now a museum; a Gothic church (St Morten's), and St Peder's (St Peter's) Church, which is the only surviving monastic building.

Population 45 300
Map Denmark Bb

Nag Hammadi ***Egypt*** Industrial town on the Nile about 480 km (300 miles) south of the capital, Cairo. It is on Egypt's main north-south rail line. Nag Hammadi handles produce from the oases to the west, and has textile, sugar and aluminium industries.

Population 19 800
Map Egypt Cc

Naga ***Philippines*** See BICOL

Naga Hills ***Burma/India*** Ranges rising to 3826 m (12 553 ft) on the border between India and Burma. Much of their tropical forest cover has been replaced by dense stands of bamboo as a result of centuries of shifting cultivation by the hill peoples.

Map India Eb

Nagaland ***India*** Eastern state where the Naga peoples are predominant, and both Naga and English are the official languages. It covers 16 579 km² (6399 sq miles) in the southern

NAGA HILLS on the Burmese border. Missionary activity has been intense and as a result about 45 per cent of the inhabitants are Christians, and some have advanced Western educations. The people have little affinity with predominantly Hindu India, and some are waging a separatist war – in ideal terrain.

Population 774 900
Map India Eb

Nagano ***Japan*** City in central Honshu island, 175 km (110 miles) north-west of the capital, Tokyo, and deep in the Japan Alps. Its industries include electrical equipment, foodstuffs and printing. A revered Buddha statue in the city's temple of Zenkoji is put on public display once every seven years – 1987 is a display year.

Population 337 000
Map Japan Cc

Nagasaki ***Japan*** Port and industrial city in the west of Kyushu island, specialising in shipbuilding and heavy engineering and dominated by the Mitsubishi company. It was the target of the second atomic bomb, which was dropped on August 9, 1945 – three days after the Hiroshima bomb. The blast killed 40 000 people and destroyed 40 per cent of the city, though many old buildings survived.

Nagasaki has been a centre of Japanese Christianity since the 16th century, and its Oura Cathedral is dedicated to 26 Christians who were crucified in 1597. For centuries it was virtually the only point of contact between Japan and the outside world. From 1637, all foreigners were excluded from Japan except the Dutch, who were allowed to trade through the small island of Dejima in Nagasaki Bay. The port was not reopened until 1859.

Population 446 300
Map Japan Ad

Nagoya ***Japan*** Manufacturing and commercial city and port on Honshu island, about 270 km (170 miles) west and slightly south of the capital, Tokyo. Its products range from cars to china. It was extensively rebuilt after wartime bombing. The Atsuta shrine in the city houses an ancient sword which is part of the sacred imperial regalia.

Population 2 065 800
Map Japan Cc

Nagpur ***India*** Textile producing, administrative and commercial city on the northern Deccan plateau of central India, about 700 km (435 miles) east and slightly north of the west-coast city of Bombay. The Gonds, a tribal group which are largely outside the mainstream of Indian (Hindu) life, live in the district, which is noted for its sweet oranges.

Population 1 302 100
Map India Cc

Nagyalföld ***Hungary*** See ALFOLD, GREAT

Nagybecskerek ***Yugoslavia*** See ZRENJANIN

Nagycenk ***Hungary*** See SOPRON

Nagyvazsony ***Hungary*** See BALATONFURED

Naha ***Japan*** Capital and main port of OKINAWA. It was once the seat of the Ryukyu dynasty who ruled the islands in the area semi-independently from 1429 to 1879. It was the scene of a heavy battle between US and Japanese forces in May and June 1945.

Population 304 300
Map Japan Gf

Nahuel Huapí ***Argentina*** See BARILOCHE

Nairn ***United Kingdom*** See HIGHLAND

Nairobi ***Kenya*** The national capital, once a leisurely colonial town, now a bustling metropolis whose services are being strained by a rapidly growing population. It lies about 440 km (275 miles) north-west of Mombasa, on the edge of the highlands at an altitude of 1700 m (about 5580 ft). It began as a construction camp on the Uganda Railway in the 1890s, in a no-man's-land between Kikuyu tribal country and grazing land used by the nomadic Masai tribe. Nairobi grew quickly, replacing Mombasa as colonial capital in 1907.

Most of the adult population today are migrants – country people seeking work in light industries and commerce in the south of the city. It is a largely transient population. The people keep close ties with their rural homes and most eventually return to them. Tourism has grown in recent years to become one of Kenya's chief foreign currency earners. Nairobi is the headquarters of two UN agencies concerned with the environment and with human settlements.

Population 1 250 000
Map Kenya Cb

A DAY IN THE LIFE OF A NAIROBI WATCHMAN

In a long, shed-like room on the outskirts of Nairobi 18-year-old Timbau, who rents the room with four other young Masai, warms a paste of maize meal and chopped cabbage over a wood fire on the earth floor. It is late afternoon, and Timbau works at night in the city as a guard and watchman (or *askari*) at the suburban home of a middle-class family.

He always eats before he goes to work. But the pattern of the day and even the food he eats still seem strange and new to him. When he lived with the Masai, who herd cattle over the grassy plains of southern Kenya, his diet was mostly milk and a little cow's blood. His home then was a hut in one of the settlements along the route the Masai travel every year in search of water and grazing. The huts, or *bomas*, are made of mud plastered over a frame of poles, and a group of *bomas* are fenced around with acacia thorn branches, to make a *manyatta*.

Timbau abhors the thought of changes in the way of life of his people, who are under pressure to become settled cultivators. But grazing land is becoming more scarce; some is being fenced off for ranching or farming, and traditional grazing grounds have been made into a national park. Recent droughts have killed cattle so food is short, and the Masai have to supplement their diet with maize meal. If they do not grow crops, they have to buy maize in the markets. Besides, their lives as warrior-nomads have been affected by Western culture – and there is now pressure on them to have an education and own Western goods. All these things have to be paid for, and it was the need to make money that sent Timbau away from his people and into the city.

When Timbau arrived in Nairobi he could speak only Masai, but within a week he had learned enough Swahili to buy things in the local market. Still he feels an outsider, though. Other Africans treat him as a primitive because he wears his red cloak, or *shuka* – and especially because he has gaping holes in his ear lobes, made by the traditional decorative plates used by his people. He has now laid aside the jewellery and beads worn by the Masai, but he still resists Western clothes. His employers seem content to have a red-robed guard armed with a spear and club in the sentry box at the gate of their home.

As he walks the several kilometres from his shanty-district home to work, Timbau thinks that what he most needs to buy out of his first wage packet is a charcoal stove for his room. After that he can send any spare money home. In a few years perhaps he will earn enough to return to what remains of his old life.

Najran ***Saudi Arabia*** Oasis town situated 220 km (about 140 miles) south-east of Abha. An unusual feature is a number of five to eight-storey mud houses, some of which have unique stained-glass windows. The palace of the former emirs, in the Najran valley, is also built of mud. A few kilometres south of Najran are the remains of the town of Ukhdud, which flourished in the early centuries of Christianity. Remains of the 4 m (13 ft) stone walls still stand, and among the carvings is one of two intertwined snakes; the significance is not known.

Population 53 000
Map Saudi Arabia Bc

Nakhodka ***USSR*** Port in the Soviet Far East, on the Sea of Japan. It was built in the 1950s to relieve the overcrowded facilities at Vladivostok, some 90 km (55 miles) to the west, and is now the Trans-Siberian Railway terminal for passengers and freight containers going to and from Japan or Hong Kong. Its industries include fishing and shipbuilding.

Population 148 000
Map USSR Od

Nakhon Pathom ***Thailand*** Province and town about 60 km (37 miles) west of the capital, Bangkok. The province's farmers grow rice, sugar cane, fruit and vegetables for the capital. The town is the site of the country's tallest Buddhist pagoda at 127 m (417 ft).

Population (town) 211 000; (province) 603 700
Map Thailand Bc

Nakhon Phanom ***Thailand*** Agricultural town on the Laotian border, about 630 km (390 miles)

north-east of Bangkok. The town lies in a rice-growing province of the same name. There was a US airbase near the town in the late 1960s and early 1970s during the Vietnam War.
Population (town) 124 600; (province) 582 700
Map Thailand Bb

Nakhon Ratchasima ***Thailand*** Engineering and agricultural market city about 270 km (170 miles) north-east of Bangkok. It is the gateway to the Khorat plateau of the north-east. The surrounding province grows maize, cotton, fruit and rice. It contains some of the most important Khmer monuments in Thailand, notably the 9th-century temple city of Phimai.
Population (city) 366 200; (province) 2 136 900
Map Thailand Bc

Nakhon Sawan ***Thailand*** Province and riverside city, about 240 km (150 miles) north and slightly west of the capital, Bangkok. Bung Boraphet, the country's largest lake at up to 120 km² (46 sq miles), is situated near the city; it contains many crocodiles. The province's farmers mostly grow maize, cotton, soya beans and sorghum.
Population (city) 223 300; (province) 1 037 350
Map Thailand Bb

Nakhon Si Thammarat ***Thailand*** Province and city on the coast of the Gulf of Thailand, about 600 km (375 miles) south of Bangkok. Rubber, tin, tungsten, rice and fruit are the chief local products. The city has several ancient Buddhist temples, notably the Wat Phra Mahathat, and local craftsmen make traditional Thai puppets out of leather and patterned-silver nielloware – silver inlaid with a black alloy.
Population (city) 277 400; (province) 1 332 000
Map Thailand Ad

Nakuru ***Kenya*** One of the country's largest towns, about 140 km (85 miles) north-west of Nairobi. It was the business centre of the Kenya Highlands, which were developed as farmland by European settlers, and is still a regional market town, with food-processing industries. It lies on the highest part of the Great Rift Valley floor, just north of Lake Nakuru, a soda lake occupied by vast flocks of flamingos.
Population 120 000
Map Kenya Cb

Nalanda ***India*** Remains of a massive Buddhist monastic centre or 'university' on the Ganges river plain near the town of Bihar Sharif, about 470 km (290 miles) north-west of Calcutta. It flourished from the 5th to 11th centuries, and included ten monasteries, three libraries and numerous stupas (shrines). Hsuan-tsang, a 7th-century Chinese pilgrim, reported that it had 10 000 student monks from all Asia, and 1500 teachers. Nalanda was sacked by Muslims in about 1200. A site museum houses finds from the ruins, including stone and bronze sculptures.
Population (district) 223 200
Map India Dc

Nal'chik ***USSR*** City on the northern flank of the Caucasus Mountains. It is a tourist resort, spa and university town, and is capital of the Kabardin-Balkar Republic, an autonomous republic of the RSFSR covering 12 500 km² (4800 sq miles). Nearby, the Baksan river valley leads to the twin summits of volcanic Mount ELBRUS (5642 m, 18 510 ft), Europe's highest peak.
Population 222 000
Map USSR Fd

Namangan ***USSR*** Town in Uzbekistan, about 200 km (125 miles) east of Tashkent. It is one of the main towns in the fertile, irrigated and densely populated Fergana valley, and produces cotton textiles, silk, processed foods, wine and beer.
Population 265 000
Map USSR Id

Namaqualand ***Namibia/South Africa*** Region along the Atlantic coast, astride the Cape Province-Namibia border. It is the home of the Namaqua, or Nama, a Hottentot people. The region is arid, but rich in copper and diamonds. Following the brief rainy season in winter, the deserts of Namaqualand are brilliantly carpeted with flowers in springtime – September and October.
Population (in South Africa) 66 400
Map South Africa Ab

Namen ***Belgium*** See NAMUR

Namib Desert ***Namibia*** One of the world's bleakest deserts, with towering sand dunes soaring to 275 m (900 ft). It stretches the length of the Namibian coast – about 1900 km (1180 miles) – and is some 50 to 140 km (30 to 87 miles) wide. It relies almost exclusively on sea mist for moisture, and many living things have developed remarkable survival methods. One species of beetle, for instance, stands on its head in the mists so that condensation will drip down its shell to its mouth.

Plants include the fantastically shaped *Welwitschia mirabilis*, whose frayed leaves, looking like the tentacles of some strange land-dwelling creature, are used to collect moisture from the air. The plant, whose two leaves are constantly growing, can live for more than 100 years.
Map Namibia Ab

Namibe (Moçamedes; Mossamedes) ***Angola*** Port about 700 km (435 miles) south of the national capital, Luanda. It is the capital of a province of the same name. Founded in the 1840s by the Portuguese, the town developed as a fishing port, and as a cargo port after the construction early in the 20th century of a railway to the town of Menongue, about 620 km (385 miles) to the east.

The province has an arid coastal strip below the Serra da Chela mountains, which rise to more than 2100 m (6900 ft). Its main crops are sisal, maize, millet and beans.
Population (town) 233 145; (province) 52 200
Map Angola Ab

Namibia See p. 455

Namp'o (Chinnampo) ***North Korea*** The country's largest port, on the Taedong river 37 km (23 miles) from its mouth on the west coast. The harbour, opened to international trade in 1897, is now the port of the capital, Pyongyang.
Population 130 000
Map Korea Bc

Nampula ***Mozambique*** Trading centre about 150 km (90 miles) inland on the railway linking the seaports of Moçambique and Nacala to the interior. It is the capital of the northern agricultural province of the same name. The province covers 81 606 km² (31 500 sq miles).
Population (city) 126 100; (province) 2 498 400
Map Mozambique Ab

▼ RICHES FROM DESOLATION The barren, shifting sands of the southern Namib Desert, which in some years has no rainfall, yield gem-quality diamonds.

Namur (Namen) ***Belgium*** City about 60 km (35 miles) south-east of the national capital, Brussels, and itself the capital of a province of the same name. Strategically positioned at the confluence of the Meuse and Sambre rivers, it has been the scene of battles and sieges from the 17th century up to 1914. Today it is popular with tourists, and has a number of industries, chiefly food processing and machinery.

The province, which covers 3665 km² (1413 sq miles) is largely occupied by the hills and forests of the Ardennes, with livestock farming and tourism as the chief sources of income.

Population (city) 35 000; (province) 402 000

Map Belgium Ba

Nan ***Thailand*** Mountainous northern province on the Laotian border, named after the Nan river which flows through it. A town of the same name lies in the province about 585 km (365 miles) north of the capital, Bangkok. Both the town and province have a large population of hill tribesmen – such as the Yao, Hmung, Khamu, Htin and Phitong Luang – who cultivate rice and maize.

Population (town) 104 650; (province) 414 000

Map Thailand Bb

Nan Ling ***China*** Mountain ranges between the Chang Jiang river basin and the tropical lowlands of the far south. They stretch for almost 1000 km (625 miles) from the eastern flanks of the Yangui plateau to northern Zhejiang province, reaching 2100 m (6900 ft).

Map China Gf

Nanchang ***China*** Capital of Jiangxi province. It is a centre for the processing of agricultural products and for the manufacture of aircraft and diesel trucks.

Population 2 390 000

Map China He

Nancy ***France*** Capital of Meurthe-et-Moselle department and formerly capital of the Duchy of Lorraine. It is an elegant city, enlarged and beautified in the 18th century by Stanislas Leszczynski, the deposed king of Poland and father-in-law of the French king, Louis XV. Stanislas made Nancy his capital when Louis gave him part of Lorraine. The city's magnificent central square, Place Stanislas, is named after him.

Population 314 200

Map France Gb

Nanda Devi ***India*** The country's second highest peak (after Kangchenjunga on the Sikkim-Nepal border). It rises to 7817 m (25 645 ft) in the Himalayas, about 330 km (205 miles) north-east of Delhi.

Map India Cb

Nanga Parbat ***Pakistan*** One of the world's highest peaks, rising to 8126 m (26 660 ft) and marking the western end of the Great Himalaya Range. It was first climbed by a German-Austrian expedition in 1953.

Map Pakistan Da

Nanjing (Nanking) ***China*** Capital of Jiangsu province in east-central China. It stands on the banks of the Chang Jiang 250 km (155 miles) from Shanghai. Nanjing was the capital of Chiang Kai-Shek's Nationalist government between 1928 and 1937. Its capture by the Japanese in December 1937 was accompanied by savage killing and brutality. Today, the city is an important industrial and trading centre.

Population 3 551 000

Map China Hd

A DAY IN THE LIFE OF A THAI RICE FARMER

Byapha looks out of the doorway of his thatched bamboo hut and surveys Nong Tong, 20 huts in a clearing in the forest in Thailand's Nan province, the homes of his friends and his clansmen who have chosen him as their headman. He must speak for their interests to representatives of a government that has until recently paid little heed to the wishes of the hill tribes of northern Thailand.

But now the government has begun to take notice, as Byapha and others like him have learned the national language and have ventured among the townspeople, who no longer regard them as mere curiosities, the 'embroidery people' in their brilliant costumes. The pink and blue embroidered waistcoat and green pantaloons that Byapha wears were woven on a foot loom and hand-sewn by his wife before their marriage.

Byapha has been willing to adapt to the expectations of outsiders, but he still has a duty to preserve the traditions of his ancestors whose spirits are present around him and whose ashes, brought from the village his clan recently left, lie in the village's new spirit house. His own father's ashes are kept on a shelf in his hut.

It is more than loyalty to tradition that leads Byapha to defend the way of life of his clansmen. He knows that it is the one best adapted to survival. Byapha and his fellow farmers clear land from the forest in order to grow the dry rice on which they live. For two or three years the land is fertile enough to produce a good harvest. Then the villagers must move on; it takes a dozen years of lying fallow for the worked land to regain its former fertility. But if the government forbids his people to clear more land elsewhere, they are obliged to farm their existing slopes until the soil's goodness is used up and the yields are low. Then there is not enough food for everyone.

It is this that has caused many of his clansmen to devote more land to white opium poppies, which survive on quite infertile soil for ten years or more, and provide a crop which fetches consistently high prices. With the money, they can buy the rice they are short of – as well as luxuries of Western culture: radios, and plates and plastic washing bowls and tinned meat.

It is not only the granting of new forest land that Byapha must argue for when he visits the government offices in town. He must resist the influx of impoverished lowland Thais into these hills; their methods of farming speed up the erosion of the soil which is already a problem.

He will return in the evening and report to his fellow villagers – and to his dutifully waiting wife.

Nanning ***China*** Capital of Guangxi-Zhuang Autonomous Region 160 km (100 miles) from the North Vietnamese border. It is a centre for sugar refining.

Population 607 000

Map China Gf

Nansei-shoto ***Japan*** See RYUKYU ISLANDS

Nantes ***France*** Port and capital of Loire-Atlantique department, about 50 km (30 miles) from the mouth of the Loire river. In the 18th century Nantes was France's most important port for trade with the Caribbean and Africa, but a deep-water port was later built at Saint-Nazaire, at the river's mouth. It was at Nantes that Henry IV of France signed the Edict of Nantes in 1598, securing a large measure of religious and civil liberty for the Protestant Huguenots. The edict was revoked in 1685, depriving the Huguenots of these rights.

Population 474 100

Map France Cc

Nan-t'ou ***Taiwan*** City in the western foothills of the island's central mountains, in a region producing rice, sugar, bananas and pineapples. It is a good base for exploring the mountains to the east, and Sun Moon Lake.

Population 100 000

Map Taiwan Bc

Napier ***New Zealand*** Port on North Island, about 270 km (170 miles) north-east of the capital, Wellington. It stands on Hawke Bay and exports fruit, wool, frozen meats and dairy and timber products. Like HASTINGS, about 20 km (12 miles) to the south-west, it was almost totally destroyed in 1931 by an earthquake, when more than 250 people were killed. The town was rebuilt in Art Deco style.

Population 48 800

Map New Zealand Fc

Naples (Napoli) ***Italy*** The country's third largest city and capital of the south as well as of the Campania region, a seaport and tourist centre about 185 km (115 miles) south-west of Rome. The glittering Bay of Naples and the smouldering volcano VESUVIUS provide a dramatic backdrop to the city – which has some of the best museums in the world, as well as some of the worst slums in Europe and an appalling crime rate.

Naples was founded as a Greek colony in the 7th-6th centuries BC and flourished during the Roman Empire and subsequently under Byzantine, Norman, Swabian and Spanish rule. The street plan of the old city reflects the Roman, medieval and Spanish periods. The Roman part, between the church of Santa Maria of Constantinople and the Capuano Castle, contains the cathedral. The medieval parts lead down to the docks and the Castel Nuovo. They are a rabbit warren of densely populated narrow alleys and small squares, but were badly damaged by an earthquake in 1980. (*continued on p. 456*)

Namibia

A DESERT AND HANDFULS OF DIAMONDS CHANGED A LIFESTYLE OF 30 000 YEARS AGO

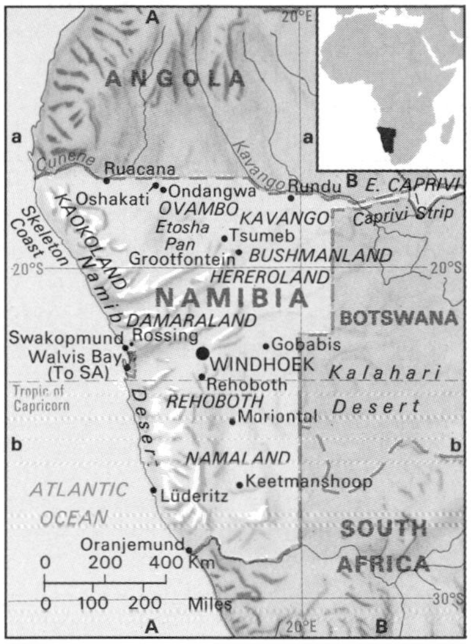

Shovelling away drifting sand from a railway line, a labourer picked up what he called 'a pretty stone' and showed it to his foreman. The foreman rubbed it across the glass of his watch and made a deep scratch. The stone was a diamond. Returning later to prospect the area, the foreman told his men to look for more pretty stones. One of them sank to his knees on the spot and filled both hands with diamonds. They lay 'thick as plums under a plum tree' all around, as though the sky had rained them overnight.

This was in 1908 near Pomona, on the coast of German South West Africa south of LUDERITZ. The territory is now Namibia, a country of staggering mineral riches. In the following six years – up to the start of the First World War – 5 million carats of diamonds were lifted from its deserts; many more have been gathered since. This bizarre, inhospitable land of arid plains and jagged mountain ranges also has some of the world's largest uranium deposits, and vast supplies of copper, tin, lead, zinc and vanadium.

Fewer than 1 200 000 live in Namibia. They include about 75 000 white people, and many African tribes including the Damara, who are believed to have lived in the country for 30 000 years and who probably created the magnificent prehistoric rock paintings found recently in the once-fertile NAMIB DESERT. For all the tribes, life has changed as dramatically since white men came as it did when the desert overtook the land.

If few people live in Namibia, many have coveted the country. As European powers carved up Africa in the late 19th century, Germany declared the territory a protectorate. South African troops ousted them in 1915, during the First World War, and after the war the League of Nations gave South Africa a mandate to govern the country. The mandate stressed the paramountcy of black rights, and the League retained primary responsibility for the country.

When the League was replaced by the United Nations after the Second World War, South Africa refused to recognise the UN trusteeship that succeeded the mandate, and began applying its policy of apartheid – separate development of races – in Namibia. Black nationalists, led by the South-West African People's Organisation (SWAPO), demanded independence and began guerrilla warfare. In 1971 the International Court of Justice ruled South Africa's occupation illegal – but to no avail. Elections held under South African supervision in 1978 were declared null and void by the UN. A multiracial government, comprising a cabinet and 62-member National Assembly, was installed in 1985.

South Africa still rules Namibia and has 100 000 troops there. And the war goes on: SWAPO fighters attack South African troops; South African troops raid SWAPO camps in neighbouring Angola, where Russian and Cuban troops support the Communist government against its own UNITA rebels, which are in turn backed by South Africa.

DESERTS AND MOUNTAINS

Apart from the 77 000 or so Damara, the black population of Namibia includes about 517 000 Ovambo, 98 000 Kavango, 78 000 Herero, 50 000 Nama, 40 000 Caprivians, 7000 Tswana and 29 000 Bushmen of the KALAHARI DESERT. The country, one of the driest on earth, has three main regions. The Namib Desert is a 50-140 km (30-85 mile) wide strip running down the entire Atlantic coastline. The Central Plateau, east of the Namib, has mountains reaching 2000 m (6562 ft) above sea level, rugged outcrops, sandy valleys and plains of poor scrub and grasslands. To the east again, and north, is the Kalahari – sand, patchy scrub, coarse grass and dried-up salt flats.

The Namib has only 50 mm (2 in) of precipitation a year, and most of it comes unusually in the form of fog which rolls in from the sea at night and condenses to form dew as the desert sands cool. Rainfall at the capital, WINDHOEK, in the highlands is higher but still meagre at 200-250 mm (8-10 in).

Cattle, sorghum and maize are raised by the Ovambo in the northern highlands; sheep are farmed in the southern highlands, where 'Persian lamb' skins from the karakul breed are produced for export. Other farmers in this area are the unique Rehoboth Basters, of mixed Nama and European descent, who speak Afrikaans. They number more than 25 000 among about 67 300 people of mixed descent in Namibia, many of the others being fishermen in WALVIS BAY (which is administered as part of Namibia although it is South African territory).

The white population is mostly Afrikaner, with about a quarter being of German descent. The whites live mainly in towns, although some remain on the land, often running large ranches. They enjoy a privileged way of life, in contrast to the blacks, many of whose lives have been shattered by war, causing them to flee south to overcrowded towns. Many blacks have forsaken tribal lifestyles to work on ranches or in mines. Offshore, the cold Benguela Current, which flows north from Antarctica, feeds one of Africa's richest fishing grounds. Mackerel, tuna and pilchards are among the catches brought back to a string of small ports, including Ludernitz and Walvis Bay, both of which have canneries.

Despite all the troubles, tourists keep arriving. They go south to the amazing Fish River Canyon, 900 m (3000 ft) deep and 60 km (37 miles) long, with hot springs and an elaborate spa called Ais-Ais (Very Hot) at the most rugged section. Astonishingly, troops of baboons and herds of zebras and kudu antelopes find a home there, and date palms planted by Germans who hid there during two World Wars still flourish.

In the north is the ETOSHA PAN (Place of Mirages), a huge 4000 km² (1737 sq mile) depression that floods briefly after rains, attracting vast herds of game – elephants, giraffes, wildebeests, zebras and even some rhinoceroses. Lions, leopards and cheetahs arrive to prey on them, and great flocks of flamingos gather in the muddy shallows.

A variety of strange plants include the welwitschia with its huge turnip like root up to 1.2 m (4 ft) across, its two leaves, each 3 m (10 ft) long, and its lifespan of 100 years. The strange plants lure not only botanists: the thorny wild cucumber is both tasty and reputedly aphrodisiac; the marula tree bears fruit that is both aromatic and alcoholic – even bull elephants have been reported tipsy on it.

But exploiting that incalculable mineral wealth is Namibia's real business. Mining and exporting of resources are both banned under international law while South Africa maintains its occupation – but exploitation and trade carry on regardless.

NAMIBIA AT A GLANCE

Area 823 172 km² (317 827 sq miles), excluding Walvis Bay

Population 1 140 000

Capital Windhoek

Government Administered by South Africa

Currency South African rand = 100 cents

Languages English and Afrikaans (official), German and several Bantu

Religions Christian (90%), remainder tribal

Climate Temperate and subtropical, very dry; temperatures in Windhoek range from 6-20°C (43-68°F) in July to 17-29°C (63-84°F) in January

Main primary products Cattle, sheep, maize, millet, sorghum, fish; diamonds, copper, lead, zinc, tin, uranium, vanadium

Major industries Mining, stock rearing, food processing, textiles, ore smelting, fishing

Main exports Diamonds, copper, lead, zinc, uranium, fish, cattle, karakul fur pelts

Annual income per head (US$) 1800

Population growth (per thous/yr) 30

Life expectancy (yrs) Male 50 **Female** 53

The Spanish city stands to the west of the Via Toledo.

The historic buildings of Naples include Castel Nuovo, an imposing castle of 1279-82 which was remodelled in the 15th century; the Floridiana, a splendid early 19th-century villa; the San Carlo Theatre, built in 1737, one of the world's largest opera houses; the Museo Nazionale, with its Graeco-Roman remains, including many statues, mosaics and murals from POMPEII; and the vast Albergo dei Poveri, a former Bourbon poorhouse.

Industries – spread through the urban sprawl around Naples – centre on steel, chemicals, clothing, leather and food processing. Apart from Naples itself, many tourists visit nearby centres such as CAPRI, ISCHIA and SORRENTO.

Population 1 203 900
Map Italy Ed

Napoli di Malvasia ***Greece*** See MONEMVASIA

nappe Almost horizontal over-fold, normally of mountainous proportions, that has also been faulted and forced some distance from its site of origin by forces of compression that caused the folding.

Naqsh-i-Rustam ***Iran*** See PERSEPOLIS

Nara ***Japan*** City in south Honshu island, about 34 km (21 miles) south of Kyoto. Nara was the national capital from 710 to 784, and is second only to Kyoto as a storehouse of national treasures. One of its many temples, Todai-ji, founded in AD 752, contains the world's largest bronze Buddha. The statue is 16.2 m (53 ft) high and weighs more than 450 tonnes; it is housed in the world's largest wooden building, 49 m (161 ft) high. Nara is noted for its calligraphy brushes and ink.

Population 327 700
Map Japan Cd

Narbonne ***France*** Market city near the Mediterranean coast sited about 80 km (50 miles) from the Spanish border. Founded in 118 BC, it is said to have been the first Roman colony beyond the Alps. Its fortified cathedral was started in the 13th century, but remains unfinished. The town also contains many elegant public buildings and squares.

Population 42 700
Map France Ee

Narenta ***Yugoslavia*** See NERETVA

Narew (Narev) ***Poland/USSR*** River, 484 km (301 miles) long. It rises just inside the USSR, about 85 km (55 miles) north-east of the Russian border city of Brest, and joins the Bug river north of the Polish capital, Warsaw.

Map Poland Db

Naruto ***Japan*** Town on north-east Shikoku island at one end of the Onawato Bridge. The bridge, completed in 1986, links Shikoku, via the island of Awaji, to Honshu island across the Naruto Strait. Huge whirlpools, 25 m (82 ft) across, form in the strait with each turn of the tide as water passes between the Pacific and the Inland Sea.

Population 64 300
Map Japan Bd

Narvik ***Norway*** Port in northern Norway on Ofotfjord, east of the Lofoten and Vesterålen islands. Narvik, which has an ice-free harbour, exports iron ore from the Swedish mines at Kirunavaara and Luossavara. A rail link from the mines was built before the First World War and there is also a road to Sweden. In 1940, the port, which was used by the Germans as a naval refuelling base, was the scene of a Second World War battle. A British fleet, headed by the battleship *Warspite*, sailed into the fiord and sank nine German destroyers for the loss of two British ships.

Population 15 000
Map Norway Ea

Nashville ***USA*** State capital of Tennessee and the home of Country and Western music. It was founded in 1779 on the Cumberland river, about 240 km (150 miles) east of the Mississippi river. Apart from recording studios, its industries include footwear, glass, rubber, aircraft components, printing and publishing, and it is also a university centre.

Population (city) 462 500; (metropolitan area) 890 300
Map United States Ic

Nassau ***Bahamas*** Seaport, largest town and capital of the Bahamas, situated in the north of NEW PROVIDENCE Island. Founded by the British in the 17th century, it has many old colonial-style buildings. Its trade increased during the American Civil War (1861-5) when it was a base for ships defying the North's blockade to supply the South. From 1920 to 1933 it was a base for bootleggers during the Prohibition period in the United States. The main industries are tourism and banking; Nassau has become a tax haven for many international companies who have registered offices there, and more than 1.25 million tourists visit it each year – more than anywhere else in the West Indies.

Population 120 000
Map Bahamas Ba

Nasser, Lake ***Egypt*** See ASWAN HIGH DAM

Natal ***Brazil*** City and state capital of Rio Grande do Norte on the country's north-east tip. Its name means 'Christmas', and it was founded on that day in 1597. It still has a fine 16th-century fort. Brazil's experimental rocket base is nearby at Barreira do Inferno.

The people of the state, covering 53 015 km^2 (20 469 sq miles), are called *Potiguares* – meaning 'shrimp eaters' – after the Indian name for the most powerful tribe of Tupi Indians, who occupied the area before European settlers moved in. The state's main products are salt, cement, furniture and cotton.

Population (city) 417 000; (state) 2 085 000
Map Brazil Eb

Natal ***South Africa*** Province in the east covering 86 976 km^2 (33 573 sq miles), including KwaZulu. Its name comes from *Terra Natalis* ('New-born Land') – the name given to it (in reference to the Nativity) by the Portuguese explorer Vasco da Gama, who landed there on Christmas Day, 1497.

In 1800 Natal was inhabited mainly by Zulus. Boer (Afrikaner) farmers settled there from 1837, but the British Cape Colony annexed Natal in 1843. Indentured Indian workers were introduced around 1860, and Indians now make up almost a quarter of the province's population. Mohandas 'Mahatma' Gandhi (1869-1948), who led India's struggle for independence, lived in Natal and practised law there between 1893 and the start of the First World War.

Sugar cane is now the province's chief crop. Natural resources include coal and timber. The main cities are Durban and the provincial capital, Pietermaritzburg.

Population (excluding KwaZulu) 2 841 700
Map South Africa Db

Natitingou ***Benin*** Chief town of the ATAKORA HIGHLANDS about 450 km (280 miles) north of the Atlantic coast and only 27 km (17 miles) from the Togo border. It is the centre of a subsistence farming region producing guinea corn, sorghum, yams and vegetables and maize.

Population 15 500
Map Benin Aa

natural vegetation Wild plant communities that have survived in, or have colonised, an area and have not been planted by man. Few places have remained unaffected by man's activities and the term can rarely be applied in the strictest sense, except in the remotest regions. See CLIMAX VEGETATION

Nauplia (Návplion) ***Greece*** Port on the Gulf of Argos 45 km (18 miles) from Corinth in the north-eastern Peloponnese. It was founded in legend by Nauplios, son of the sea god Poseidon; in fact it was most likely colonised by sea traders. It was a major stronghold of the Venetians during their domination of the Mediterranean (AD 1387-1715), and a vital centre of the Greek revolt against Turkish rule from 1821. From 1828 it was the capital of Greece until 1834, when the newly elected King Otto moved to liberated ATHENS. In 1831, Count Capodistrias, the first president of modern Greece, was shot dead in the town by rival politicians. Nauplia has a museum, the Venetian rock fortress of Palamidi, Byzantine fortifications, and two mosques.

Population 10 600
Map Greece Cc

Nausori ***Fiji*** See SUVA

Navadwip (Nabadwip) ***India*** Eastern town on the Bhagirathi river about 95 km (60 miles) north of Calcutta, and formerly called Nadia. It is the 'Benares of Bengal', a Hindu place of pilgrimage, and a seat of Sanskrit learning.

Population 129 800
Map India Dc

Navajo National Monument ***USA*** Three of the largest and most elaborate Indian cliff dwellings found in the country. They stand in northern Arizona, about 30 km (20 miles) from the Utah state border, and date back more than six centuries. The monument covers 146 hectares (360 acres).

Map United States Dc

Navan ***Ireland*** See MEATH

Navarino ***Greece*** See PILOS

Nauru

THE PROSPERITY OF THIS TINY ISLAND REPUBLIC COMES FROM ITS HUGE PHOSPHATE DEPOSITS

The world's smallest republic is the 21 km² (8 sq mile) Pacific island of Nauru, halfway between Australia and Hawaii, just 40 km (25 miles) south of the Equator. Its inhabitants pay no taxes or import duties, no school fees or health charges, and yet the tiny nation – and particularly the families that own the land – is rich.

This prosperity comes from a treasure chest in the shape of a plateau that rises 60 m (200 ft) above sea level and is bounded by coral cliffs. The plateau contains rich deposits of high-quality phosphate rock, which is sold for fertiliser to Australia, New Zealand, Japan and South Korea. The deposits are expected to run out in the 1990s, but the government is investing overseas.

The island has known foreign occupation several times since it was discovered by the British in 1798. The Germans annexed it in 1888, the Australians took it in 1914, and the Japanese occupied it in 1942. The island was retaken by the Australians in 1945, and became a UN trust territory. It has been independent since 1968, and in 1970 the government took control of the phosphate industry from the British Phosphate Commissioners, a joint Australian-British-New Zealand body.

A narrow band of fertile land between the coast and the uninhabited plateau is cultivated as food gardens by the small Chinese community; the climate is hot and wet, although the rains have been known to fail. The government subsidises imports, so food is cheap. The 5000 native Nauruans are of mixed Micronesian and Polynesian origin; there are also over 2000 migrant phosphate workers from other Pacific islands.

NAURU AT A GLANCE

Map Pacific Ocean Dc
Area 21 km² (8 sq miles)
Population 8100
Government centre Yaren
Government Parliamentary republic
Currency Australian dollar = 100 cents
Languages Nauruan, English
Religion Christian (66% Protestant)
Climate Tropical; average temperature range is 23-32°C (73-90°F) all year
Main primary product Phosphates
Major industry Phosphate mining
Main export Phosphates
Annual income per head (US$) 11 000
Population growth (per thous/yr) 13
Life expectancy (yrs) Male 49 **Female** 62

Navarra (Navarre) ***Spain*** Mountainous north-eastern province on the French border, growing sugar beet, cereals and grapes. The capital is PAMPLONA.
Population 507 400
Map Spain Da

Návpaktos (Lepanto) ***Greece*** Harbour town 170 km (106 miles) west of Athens on the north side of the Gulf of Corinth overlooking the straits where in 1571 a combined Spanish and Venetian fleet under John of Austria routed the Turkish navy in the last sea battle fought by oared galleys. Miguel Cervantes, the Spanish author of *Don Quixote*, fought at this Battle of Lepanto. The town is now a tourist centre with a sports and arts festival each July.
Map Greece Bb

Návplion ***Greece*** See NAUPLIA

Náxos ***Greece*** Largest of the CYCLADES islands, in the south Aegean Sea. Náxos is still largely underdeveloped, with fertile soil, a hilly interior and sandy beaches. Greek mythology says the hero Theseus, slayer of the Minotaur, abandoned Ariadne, a Cretan Princess, there, indicating that it may have been an outpost of the Minoan Empire of Crete. From AD 1207 to 1344 it was the seat of a Venetian duchy. It has a 13th-century cathedral, a French convent and a museum. Red and white wine are produced, as are honey, olive oil, fruits, figs and pomegranates.
Population 14 000
Map Greece Dc

Nayarit ***Mexico*** One of Mexico's smallest states, on the Pacific coast north of JALISCO. Its mineral-rich mountains are inhabited by the Huichol and Cora Indians. Tepic, the capital, lies in a pocket of rain-forest vegetation in Nayarit's humid south-west corner.
Population 781 000
Map Mexico Bb

Nazareth ***Israel*** Capital of the Northern district, about 30 km (19 miles) south-east of Haifa. It was the home of Jesus, where he spent his childhood with his parents Joseph and Mary. Among its many shrines and churches are the Basilica of the Annunciation, built in 1965 on the site where archaeologists believe the Archangel Gabriel appeared to the Virgin Mary; the Church of St Joseph, on the traditional site of Joseph's carpentry shop; and a Greek Orthodox Church of the Assumption, over the well where Mary is said to have drawn water.

The town changed hands many times during the Crusades, was captured by the Turks in 1517 and then taken by the British in September 1918 during the First World War. It produces cigarettes, leather goods and textiles.
Population 46 300
Map Israel Ba

Nazca ***Peru*** Small colonial town in Ica province, on the Pan-American Highway. It is a centre for visiting the Nazca lines, huge geometric and animal patterns, some more than 2 km (1.25 miles) long, etched into the surrounding stony desert. It is not known for sure who drew the lines, but explorers believe they may have been an astronomical calendar made by the people of the Paracas, Nazca and Ayacucho cultures some 5000 years ago.
Map Peru Bb

naze See NESS

Nazwa ***Oman*** Provincial capital of the interior, Nazwa is situated 140 km (87 miles) south-west of the capital, Muscat. Lying behind the centre of the Hajar Mountains, it is linked to the coast by an asphalt road built through the Sumail gap in 1976. The old part of Nazwa dates from the 6th and 7th centuries. The modern town that has sprung up in recent years contains offices, a police academy, hospital and an agricultural college. There is a weekly market with a goat auction, and alfalfa, leather, silverware, spices and animals are traded.
Map Oman Aa

Ndebeleland ***Zimbabwe*** See MATABELELAND

Ndjamena (N'Djamena) ***Chad*** National capital, and river port. It lies on the Chari river near its junction with the Logone, about 80 km (50 miles) south of Lake Chad. Founded as Fort Lamy by the French in 1900, it combines elegant colonial architecture with mud-brick houses and modern buildings. However, much of the city was badly damaged in the civil war that began in 1980. The museum contains relics of ancient civilisations of the region, and there is a university.

Ndjamena is the country's communications centre, with overland routes to Nigeria and Cameroon as well as outlying parts of Chad, and during the recent drought it became the centre for the distribution of relief aid. Drought permitting, groundnuts are grown extensively in the surrounding area and are sent to Ndjamena for processing. The city also has a meat-chilling plant (mainly for export to Zaire), and cattle on the hoof are exported to Nigeria.
Population 303 000
Map Chad Ab

Ndola ***Zambia*** Copper-mining town about 270 km (170 miles) north of the national capital, Lusaka. It is the regional capital of the Copperbelt Province. Founded in 1904, it now has copper and cobalt refineries and steelworks.
Population 282 400
Map Zambia Bb

Neagh, Lough ***United Kingdom*** Largest freshwater lake in the British Isles, covering 381 km² (147 sq miles) in Northern Ireland, 20 km (12 miles) west of the capital, Belfast. Eight rivers flow into the lake but only one, the Bann, leaves it. The lake is Europe's largest source of freshwater eels.
Map United Kingdom Bc

neap tide Tide of lowest range, occurring twice a month, when the sun's gravitational pull is at right angles to that of the moon. See TIDE

Neblina, Pico da ***Brazil*** The country's highest mountain, near the Venezuelan border. It rises to 3014 m (9888 ft). The mountain, whose name means Mist Peak, was not discovered until 1962, and until 1965 it was not clear whether the summit was in Brazil or Venezuela.
Map Brazil Ba

Nebraska ***USA*** Midwest state covering 200 018 km² (77 227 sq miles) between the Missouri river and the high plains of Colorado and Wyoming. It is an agricultural area producing cattle, maize, pigs and soya beans, with industries based mainly on food processing and agricultural servicing. LINCOLN is the capital and OMAHA the largest city.
Population 1 606 000
Map United States Fb

neck Solidified lava that fills the vent of an extinct VOLCANO.

Neckar ***West Germany*** River rising in the Black Forest in the south-west and flowing north for some 365 km (227 miles) to join the Rhine at Mannheim. Its valley produces wine.
Map West Germany Cd

Needles, The ***United Kingdom*** Line of chalk stacks in the English Channel off the western tip of the Isle of Wight.
Map United Kingdom Ee

Nefud ***Saudi Arabia*** Desert of red sand in the north of the country. It covers about 56 000 km² (21 621 sq miles) and extends into southern Jordan. There are a few water wells and a very slight rainfall, which make it possible for nomadic herdsmen to eke out a living.
Map Saudi Arabia Bb

Negeri Sembilan ***Malaysia*** State extending to 6643 km² (2565 sq miles) and lying south-east of the national capital, Kuala Lumpur, in Peninsular Malaysia. Its declining tin-mining industry has been replaced by rice growing in the valleys and rubber trees on the hill slopes.
Population 574 300
Map Malaysia Bb

Negev ***Israel*** Southern part of Israel, covering 12 173 km² (4700 sq miles) bordered by Jordan and Egypt. The region was assigned to Israel at the partition of Palestine in 1948. The main towns are Beersheba in the north and the Red Sea port of Elat, at the country's southernmost tip. Although much of the Negev is desert, irrigation has made agriculture possible and the region now produces cereals, sugar beet, fruit and vegetables.
Map Israel Ab

Negombo ***Sri Lanka*** Fishing port and holiday resort 30 km (19 miles) north of Colombo. It lies between the sea and a lagoon, with a fine sandy beach, old colonial buildings and modern hotels. It is noted for seafood, especially shellfish. Colombo's international airport is at nearby Katunayaka.
Population 61 400
Map Sri Lanka Ab

Negro ***Brazil/Uruguay*** River, 800 km (500 miles) long, rising in Brazil and flowing across Uruguay from the north-east to the town of Soriano, where it joins the Uruguay river. The Negro is dammed near the town of Paso de los Toros, 175 km (109 miles) north-east of the resort of Mercedes, to form one of South America's largest man-made lakes, Embalse del Río Negro (10 360 km², 4000 sq miles).
Map Uruguay Ab

A DAY IN THE LIFE OF A NEBRASKAN FARMER

On an arrow-straight road that stretches over rolling plains from one horizon to the other, Stephen Oman's station wagon heads from his isolated Nebraskan farmstead to the small town of North Platte, an hour away, for the Sunday morning Baptist Church service. Beside Stephen his wife Sue sits stiffly, respectable in her Sunday best, and in the back the children, Glen (12) and Peggy (10), are scrubbed and smart. Even the towering aluminium grain silos they pass beside the road seem to be polished for the occasion.

The silos are monuments to the success of Stephen and his fellow farmers, on land that was once known as the 'Great American Desert'. Stephen's grandfather, an immigrant from Germany, had to gamble his whole future on a good season's rain here. Now, new strains of wheat and new farming techniques have made harvests, and farmers' incomes, more secure. The landscape the Omans cross is chequered black and gold, with alternate strips of waving wheat and bare earth, each strip planted only once in two years in order to conserve moisture in the soil. And as a safeguard against soil erosion, the surface of the earth is disturbed as little as possible, and is not ploughed, but only harrowed and then rolled before planting.

North Platte's New Hope Baptist Church, a neat bungalow, is surrounded by the sedans, station wagons and trucks of Stephen's neighbours. The congregation is large and enthusiastic. Many of the congregation, including Stephen, give 10 per cent of their income to the church.

After the hymns and the sermon, a baptismal service is conducted. The minister immerses a new believer into a sunken pool in the centre of the church. Stephen and Sue both underwent the same ceremony at around the age of 16. Only then can a Baptist be taken fully into the church; children are thought to lack the necessary understanding of the ceremony.

Stephen regards himself as a good Christian. He has strong moral values, is patriotic, Republican, and a great traditionalist. He values discipline.

For the rest of the day, after church, he takes the children to picnic on a bluff near Laramie, overlooking the North Platte river. They say grace before they tuck in to Sue's homemade pies, and sit with their legs dangling in a deep trench in the limestone rock. Stephen explains that it was carved by the iron rims of countless wagon wheels, as pioneers ground their way westward along the Oregon trail.

Stephen still farms this land in the spirit of a pioneer, steadily conquering it with new agricultural techniques.

Negros ***Philippines*** Fourth-largest island of the archipelago, lying in the VISAYAN ISLANDS between Panay and Cebu. It covers 12 704 km² (4905 sq miles), and has a volcanic mountain core that separates the two sides of the island, which are administered as separate provinces – Negros Occidental and Negros Oriental. To the east, the island is mainly mountainous and little developed except in the south around Dumaguete. This part has strong cultural links with Cebu. To the west there are broad plains of rich volcanic soil covered with sugar plantations. These brought the area – and its capital Bacolod, 'the sugar capital of the Philippines' – considerable prosperity until world sugar prices slumped in the 1970s and 1980s. Today, Negros Occidental is one of the Philippines' most depressed areas.
Population (Occidental) 1 930 300; (Oriental) 819 400
Map Philippines Bd

Nei Mongol Autonomous Region (Inner Mongolia) ***China*** Region stretching along the borders of China and the Mongolian People's Republic. It covers 1 200 000 km² (460 000 sq miles) and includes part of the Gobi Desert. It is thinly populated land of vast arid wastes and empty grasslands. Most of the people are ethnic Chinese, with Mongolians, many of them herdsmen of horses, sheep and oxen, making up one-tenth of the population. There are small minorities of Manchus and others. The region is rich in coal, iron ore, asbestos, talc, mica and rare earth oxides. Mining is being developed at centres such as Bayan Obo.
Population 18 510 000
Map China Fc

Neisse ***Czechoslovakia/East Germany/Poland*** East European river, also known as the Lausitzer Neisse or Nysa Luzycka. It rises above the Czech town of Liberec in northern Bohemia, and flows generally north for 256 km (159 miles) to join the Oder river about 155 km (95 miles) south of the Polish port of Szczecin.

From the Czech border it forms part of the 'Oder-Neisse Line', the frontier between East Germany and Poland.
Map Poland Ac

Nejd ***Saudi Arabia*** Region and former province (now renamed Central Province), covering central and eastern Saudi Arabia. Before the First World War it was nominally under Turkish rule, and from 1905 was ruled by ibn-Saud. It was eventually united with HEJAZ as a dual kingdom in 1926, which became the single kingdom of Saudi Arabia in 1932. The Nejd plateau consists of ridges and shallow vales, much of them covered with sand desert. Its chief cities are RIYADH and BURAYDAH, and its main settlements are found in oases.
Map Saudi Arabia Bb

Nelson ***New Zealand*** Port on South Island, situated at the head of Tasman Bay about 120 km (75 miles) west of the capital, Wellington. Settled in 1842, it was named after the British naval hero Lord Nelson (1758-1805). Its industries include timber and food processing, and it is a popular tourist resort.

The Nelson Lakes National Park, about 80 km (50 miles) south-west of the town, covers

some 57 000 hectares (140 850 acres) and contains hundreds of red deer and chamois goats introduced from Europe by settlers.
Population 33 300
Map New Zealand Dd

Nepal See p. 460

nephrite Mottled, often pitted variety of JADE with an oily lustre.

Neretva (Narenta) ***Yugoslavia*** Major river of Herzegovina, 218 km (135 miles) long. It rises 60 km (37 miles) south of Sarajevo and flows north-west, then through the 145 km² (56 sq mile) lake behind the Jablanica hydroelectric dam, and sweeps south past the medieval Turkish towns of Pocitelj and Mostar to the Adriatic. It is noted for trout in its upper reaches, for eels and mullet farther down, and for its spectacular gorges through the deeply eroded limestone mountains of the KRAS region.
Map Yugoslavia Cc

Nesebur ***Bulgaria*** Black Sea resort, 30 km (19 miles) north-east of Burgas, which was developed 2500 years ago as the Greek trade port Mesembria. Crowding a small rocky peninsula linked by a causeway to the mainland, it is now an artists' retreat with attractive old houses and workshops overhanging cobbled streets. Churches, in a variety of Orthodox styles, date from the 5th to 17th centuries.

Pomorie (known as Anhialo to the ancient Greeks), 15 km (9 miles) to the south, is the centre of a wine-growing area. It also has extensive salt pans, and a medicinal mud lake. On the northern side of Nesebur is Sunny Beach (Slunchev Bryag), a popular tourist resort.
Population (Nesebur) 12 000; (Pomorie) 20 000
Map Bulgaria Cb

Ness, Loch ***United Kingdom*** Lake in the Great Glen, a valley that bisects northern Scotland between Inverness and Fort William. It is 37 km (23 miles) long and forms part of the Caledonian Canal, built between 1803 and 1847.

The lake is famous because of the repeated claims that a monster lives in its depths – a dinosaur-like creature numerous visitors claim to have seen. The theory that it could be a survivor from prehistory has been supported by some scientists, but all efforts to find it have been in vain.
Map United Kingdom Cb

ness, naze Headland jutting into the sea.

Netanya ***Israel*** Mediterranean resort, about 30 km (20 miles) north of Tel Aviv. It is the centre of the country's diamond cutting and polishing industry.
Population 107 200
Map Israel Aa

Netherlands See p. 462

Netherlands Antilles ***West Indies*** Dutch territory in the Caribbean consisting of two groups of islands over 800 km (500 miles) apart. CURAÇAO and BONAIRE lie off the north coast of Venezuela, while ST MARTIN (only partly Dutch), ST EUSTATIUS and Saba are in the Leeward Islands. In all, they total 800 km² (308 sq miles). Until January 1, 1986, when it was given separate status, the island of ARUBA formed part of the political group.

They were settled by Dutch colonists in the 1630s, but colonial status was abolished by the Netherlands in 1954. Since then, the islands have been internally self-governing and constitutionally equal with the Netherlands and Surinam. The Dutch Queen Beatrix is head of state and is represented by a governor. The capital is WILLEMSTAD on Curaçao.

The people are mostly of African and Dutch descent. Dutch is the official language, but many people speak Papiamento, a mixture of Dutch, Spanish, Portuguese and other languages. The economy is chiefly based on the refining of oil imported to Curaçao (and Aruba) from Venezuela, and on tourism.
Population 193 000
Map Caribbean Bc, Cb

▼ BIRTH OF A LEGEND Nessie – a 12-15 m (40-50 ft) monster said to live in Loch Ness – first made the news in December 1933.

Nepal

CLOSED TO THE OUTSIDE WORLD UNTIL THE EARLY 1950s, THIS MOUNTAIN KINGDOM IS STILL ADAPTING TO THE 20TH CENTURY

Nepal is a land tilted on its side. It is a long, narrow rectangle on the flanks of the eastern HIMALAYAS, with its northern border running along the mountain tops. The border area is known as the 'Roof of the World' and includes the world's highest mountain, EVEREST (8848 m, 29 028 ft).

▼ TRADE AND RELIGION Bicycle rickshaws bring buyers to a colourful vegetable market in Kathmandu. The pagoda-studded city is the home of Kumari, the young girl whom Nepalis worship as a living goddess.

The rectangle is only 160 km (100 miles) wide; yet in that short distance it drops from the dizzy heights of its northern frontier with Chinese Tibet to an altitude of only 100 m (330 ft) on its southern border with India.

In the TERAI, the flatlands of the south, tigers, elephants and the rare Indian rhinoceros prowl the thick jungle, and rice, sugar cane, jute and oranges grow. Farther north, the foothills blaze in spring with azaleas and rhododendrons, both of which are native to Nepal. North again, yaks browse the high pastures below the eternal snows of the great peaks of Everest, KANCHENJUNGA, Dhaulagiri and ANNAPURNA.

Nepal is one of the world's poorest and least developed nations. Nine in ten of the population eke out a living from the land as peasant farmers and labourers. Only about one person in five can read or write, and the average income per head is about US$170 a year – less than the average weekly income per head in Britain.

The poverty owes as much to the geography of the country as to its politics. Landlocked and largely cut off by the mountains from the trade routes of the Indian plains, Nepal has had little opportunity to develop a manufacturing economy. In addition, it has no significant mineral deposits with which to pay for imports from either of its two giant neighbours – India and China.

Politically, the country developed as a cluster of feudal valley kingdoms, separated from each other by the hills. It still consists of a number of ethnic groups and is still divided between Hindu and Buddhist faiths. Although the Gurkhas conquered the kingdoms and established modern Nepal in 1769, the leaders of each subject kingdom showed more loyalty to their own clans than to the nation. As a result, the political history of Nepal is a web of plots and counter-plots, assassinations and palace intrigues, with noble families conspiring against each other to gain influence at the expense of the central throne. From 1846, these families, chiefly the Ranas, reduced the kings to mere figureheads – until another palace revolution in 1951 restored the power of the monarchy, and began the process of bringing Nepal into the 20th century. However, successive kings have become more autocratic, and political parties are banned.

With Chinese and Indian aid, roads have

been built from the northern and southern borders to KATHMANDU, and along the central valley to link the major towns. Hydroelectric schemes are under way.

Until 1951 the country was closed to foreigners but it is now known as a tourist paradise, with more than 175 000 visitors a year.

It is also famous as the home of the Gurkha warrior. About 80 000 of these fearless fighters – who still carry their traditional curved knife, the kukri – serve in the British army. They have fought alongside the British since the early 19th century, when the British in India made Nepal a protectorate. Another 50 000 serve in the modern Indian army.

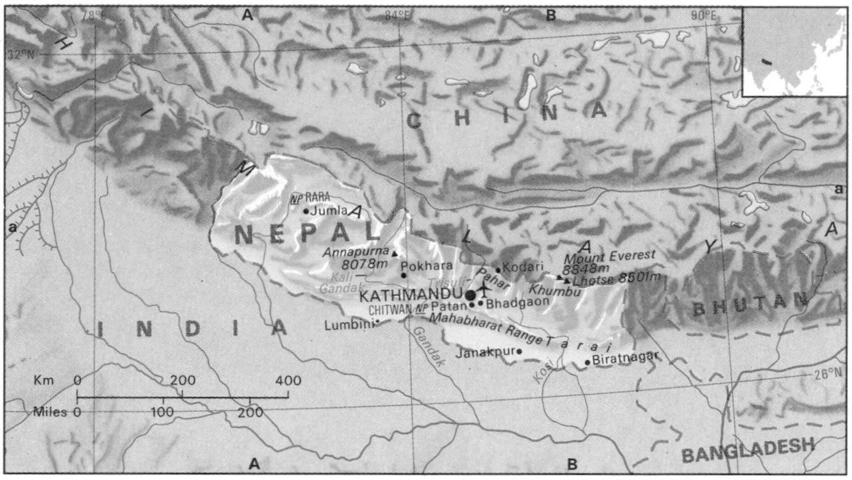

NEPAL AT A GLANCE

Area 145 391 km² (56 136 sq miles)
Population 17 420 000
Capital Kathmandu
Government Constitutional monarchy
Currency Rupee = 100 paisa
Languages Nepali and other local languages
Religions Hindu (90%), Buddhist, Muslim
Climate Temperate; average temperature in Kathmandu ranges from 2-23°C (36-73°F) in January to 20-29°C (68-84°F) in July

Main primary products Rice, wheat, barley, sugar cane, cattle, medicinal herbs, jute, pepper, tobacco, fruit, timber
Major industries Agriculture, jute spinning, sugar milling, textiles, forestry, tourism

Main exports Rice and other farm products, timber, leather goods, jute, sugar, carpets
Annual income per head (US$) 166
Population growth (per thous/yr) 25
Life expectancy (yrs) Male 45 **Female** 44

Netherlands

A WEALTHY NATION WITH THE WORLD'S LARGEST PORT AND A PICTURESQUE PEOPLE WHO SOMETIMES STILL WEAR CLOGS AND EVERYWHERE STILL RIDE BICYCLES

The Kingdom of the Netherlands exists by the whim of the sea and the will of the Dutch who live there. Almost two-fifths of the country is below sea level, including all the main centres of population.

Over the centuries, the sea has sometimes taken from the land, and sometimes given back. In 1421 the sea claimed a huge area of land at the mouth of the RHINE river known ever since as the Hollands Diep. In the 17th century, drainage and dyking added 1120 km^2 (432 sq miles) to the land area, but the 18th century saw only 500 km^2 (193 sq miles) of new land. Another 1170 km^2 (452 sq miles) were gained in the 19th century. But the sea has had its gains too – notably in 1953 – and only in the past 50 years has the balance between gain and loss tipped decisively towards man, with modern machinery aiding great Dutch reclamation works which so far this century have added 2500 km^2 (965 sq miles) to the nation.

This is what comes of establishing the core area of a nation in the delta of not one but two great rivers – the Rhine (with two main branches, the Lek and Waal) and the Maas (MEUSE). As in all deltas at all times, whatever land there is exists through the uneasy equilibrium of three forces – the work of the river, the work of the sea, and the intervention of man in restraining the other two.

The Dutch have had plenty of practice at turning back the sea; enough to tackle some huge reclamation schemes in recent years. One is the IJSSELMEER (Zuiderzee) where, following the building of an enclosing dam in 1932, they have drained four large areas, with the possibility of a fifth, to give themselves an extra 1650 km^2 (637 sq miles) of former seabed to cultivate, not to mention an overspill town for AMSTERDAM called Almere.

Another major project was provoked by the disastrous storm floods of 1953. In the Rhine-Maas delta, there were a number of islands with coasts open to attack by both river and sea. The Delta Plan has sealed these islands from the sea by building barrages across the delta channels, leaving open to wind and tide only the shipping lanes up to ROTTERDAM in the north and the Belgian port of ANTWERP in the south. This has not only made the islands secure, but has also brought them out of isolation by linking them to the mainland and attracting tourists to the area's main man-made lakes.

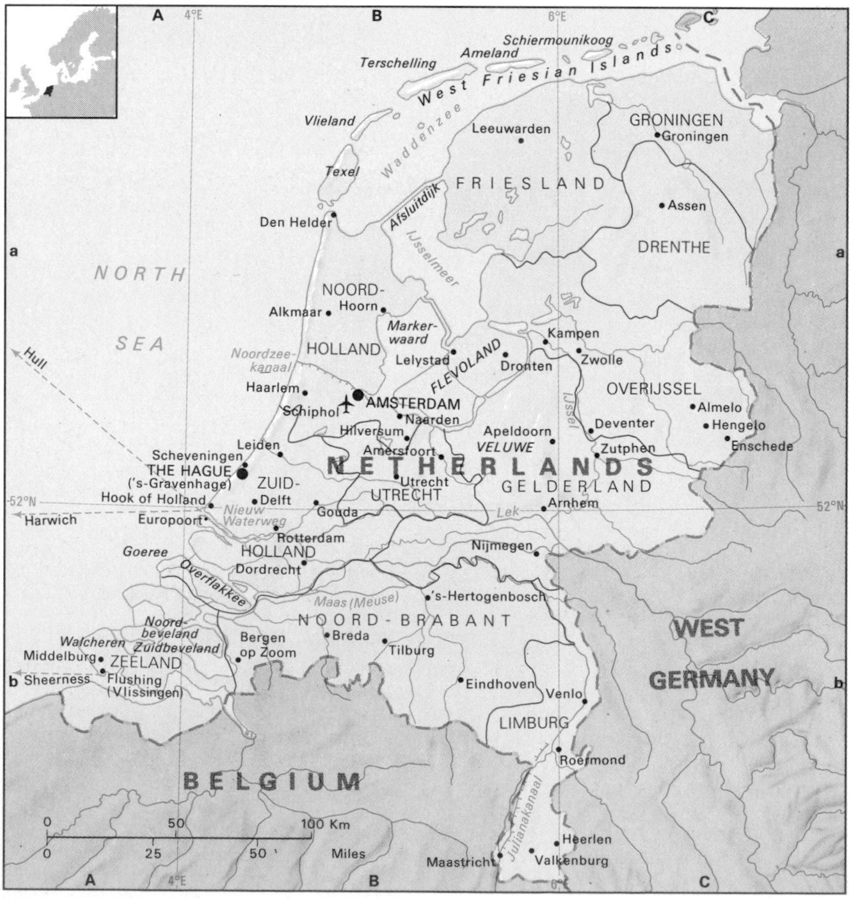

There is another Netherlands – the 'high' country. It lies east and south of the delta, and much of it is low hills of sands and gravels. In the east these materials were brought in by ice from Scandinavia in the Ice Age; in the south they are river terraces of the Rhine and Maas. In the far south, in LIMBURG, an area of chalk rises to 321 m (1053 ft) and is underlain by coal measures.

THE FAMOUS DUTCH BULB FIELDS

That landscape depends very much on the material underlying it. The sands and gravels are infertile: they are largely covered by heath and forest, much of it now protected as nature reserves. The soils of the river valleys are cold and heavy clays, and they lie under water meadows or orchards. The peat of the delta lands is almost all under grass, with a high water table to prevent shrinkage; in fact, 57 per cent of Dutch farm land is under grass. The sea clays of the coast are areas of tillage; they are lighter than the river clays, and are generally cropped. The most famous Dutch landscapes of all are the bulb fields and the areas of horticulture on the line of old sand dunes which lie just behind the present coastal dune barrier from the HOOK OF HOLLAND north to ALKMAAR. Today the most notable feature of this landscape is the sea of glass under which salad vegetables, fruit and flowers are raised.

The name Netherlands (or Low Countries) originally applied to the area now occupied by the modern state, together with Belgium and Luxembourg. In the 19th century first Belgium and then Luxembourg gained independence, leaving a Netherlands kingdom with a Protestant majority and a Protestant monarchy, but a strongly Catholic south.

The Netherlands today are wealthy, even by European standards. But in their first century of independence, between 1600 and 1700, they led the Continent. The young republic attracted refugee artists and craftsmen from all over Protestant Europe. It had military and, especially, naval power, and it went in search of the world's wealth, and founded a great empire overseas. The Dutch East India Company made the money, the great cartographers – Mercator, Ortelius, Blaeu – had provided the maps, and Rembrandt, Hals and Vermeer painted the scene and the players.

Things were never quite so good again; the 18th century brought French armies, and the loss of Belgium in 1830 was a blow. Life became desperate in 1940, when Germany invaded the Low Countries and wiped out the centre of Rotterdam in the first few hours. Four years of German occupation were ended for the southern Netherlands in autumn 1944, but now the Rhine and Maas, which had once protected the Dutch republic, performed the same role for the Germans in the north. German occupation continued through a further terrible 'hunger winter' and, by 1945, few parts of Europe were more denuded of every kind of removable resource than the Dutch heartland. It was from this nadir that the country's post-war recovery has had to proceed.

▲ CITY OF CANALS One of Amsterdam's finest attractions is its network of canals, with their trim bridges and banks of tree-lined streets. Some of the anchored barges serve as floating homes – and visitors can travel on special canal-boats by day and night.

The Netherlands, with 428 inhabitants per km^2 (1108 per sq mile), are the most densely populated country in Europe, but this is only an average figure: the east and south of the country are quite sparsely populated, while the density in South Holland province is 1080 per km^2 (2798 per sq mile). The Dutch population is overwhelmingly concentrated in the cities of the so-called Randstad Holland, a horseshoe of urban areas which starts with DORDRECHT in the south and includes Rotterdam, DELFT, The Hague, LEIDEN, HAARLEM, Amsterdam and HILVERSUM, before it reaches the other tip of the horseshoe at UTRECHT. The diameter of the horseshoe is about 50 km (31 miles).

The whole of the Randstad is below sea level. The drainage must be planned and controlled before anything can be built, let alone the 112 m (367 ft) cathedral tower at Utrecht or the underground railway at Rotterdam.

The cities attract immigrants from the rest of the country and the Randstad is growing. Yet it stands in the midst of some of the world's most productive farmland and market gardens. It is a continuing struggle to preserve agricultural land and recreational areas within and around the horseshoe.

It is not that other parts of the country are empty of opportunity. Boom cities exist in the south, especially EINDHOVEN with its great electrical industry. There is Limburg, with its industries formerly based on local coal, and now on petroleum products. There are the former textile towns of OVERIJSSEL province and there is the natural gas field of the far north-east to attract development. But the Randstad dominates the Dutch economy. Its coastal industries provide refined oil, steel (4-5 million tonnes per year at Ijmuiden), ships and processed tropical produce. It is the administrative and commercial heartland.

Dense settlement demands a dense network of transport links. The Dutch have created the most concentrated motorway network in the world. Their electric railways operate frequent, regular services, though they carry little freight; that goes by water. The canals and, above all, the Rhine are the great arteries of freight traffic – traffic that converges, over oceans, by river and canal, on Rotterdam, the world's largest port. With the mouth of the Rhine under their control, the Dutch are the gatekeepers of Western Europe.

NETHERLANDS AT A GLANCE

Area 41 548 km^2 (16 042 sq miles); land area 33 930 km^2 (13 100 sq miles)

Population 14 520 000

Capital Amsterdam (seat of government, The Hague)

Government Parliamentary monarchy

Currency Guilder or florin = 100 cents

Languages Dutch, Frisian (minority), but English and German widely spoken

Religion Christian (40% Roman Catholic, 24% Dutch Reformed, 9% other Reformed Churches)

Climate Temperate, maritime; average temperature ranges from −1 to 4°C (30-39°F) in January to 13-22°C (55-72°F) in July

Main primary products Cereals, potatoes, sugar beet, fruit, vegetables, livestock, fish; oil and gas, salt

Major industries Agriculture, food processing, iron and steel, chemicals, textiles, clothing, printing, shipbuilding, tobacco processing, diamond cutting, fertilisers, oil and gas production and refining, fishing

Main exports Food (dairy produce, meat, fruit and vegetables), petroleum products and natural gas, chemicals, machinery, textiles, iron, steel, flower bulbs

Annual income per head (US$) 8599

Population growth (per thous/yr) 4

Life expectancy (yrs) Male 74 **Female** 78

Neuchâtel (Neuenburg) ***Switzerland*** City on the western shore of Lake Neuchâtel, about 40 km (25 miles) west of the capital, Berne. It is the capital of the canton of the same name, which extends to 797 km² (308 sq miles). The city produces electrical appliances, watches and jewellery. The town hall and Gothic church date from the 12th century.

Lake Neuchâtel is the largest lake entirely within Switzerland, covering an area of 218 km² (84 sq miles).

Population (city) 38 000; (canton) 155 900
Map Switzerland Aa

Neusander ***Poland*** See NOWY SACZ

Neusatz ***Yugoslavia*** See NOVI SAD

Neuschwanstein ***West Germany*** The dream castle of Ludwig II, king of Bavaria from 1864 to 1886. It was built for him on a craggy outcrop at the foot of the Alps, 90 km (55 miles) south-west of Munich. It looks like the model for all the fairytale castles in children's story books, with tall and slender battlemented towers and pinnacles. It is filled with Bohemian glass, wood-carvings, gilded chandeliers and stone pillars.

It was finished in 1886. But Ludwig, patron of the composer Richard Wagner, lived there for only three months before going mad and drowning himself in a nearby lake.

Map West Germany De

Neva ***USSR*** River which flows just 74 km (45 miles), from Lake Ladoga in north-west Russia through the city of Leningrad to the Gulf of Finland. The Neva is frozen from about December to April, and after the thaw spring floods can raise its height by up to 5 m (16 ft). It is navigable by ocean-going ships, and is a key section in waterways linking the Baltic Sea to the White Sea and Volga river.

Map USSR Ec

Nevada ***USA*** State covering 286 298 km² (110 540 sq miles) largely of desert and mountains, to the east of California, most of it more than 1000 m (3280 ft) above sea level. Ranching, dairying and sheep raising are the main types of farming, which is largely confined to the Still Water Reservoir area and the Colorado and Humboldt river valleys. Copper, gold, mercury, silver, manganese and iron ore are the main minerals. Tourism is another source of income, particularly in LAS VEGAS and RENO. CARSON CITY is the state capital. Some 100 km (60 miles) north-west of Las Vegas is the USA's nuclear weapons test site.

Population 936 000
Map United States Cc

Nevado de Colima ***Mexico*** Smoking, active volcano, 4328 m (14 200 ft) high, in Jalisco state. It erupted in 1941 causing many deaths.

Map Mexico Bc

Nevado de Huila ***Colombia*** See HUILA

Nevado de Toluca ***Mexico*** Snow-capped extinct volcano rising to 4565 m (14 977 ft) above Toluca, capital of MEXICO state. Two lakes, known as the Lake of the Sun and the Lake of the Moon, were formed by melting snow in two craters.

Nevado del Ruiz ***Colombia*** See ARMERO

névé See FIRN

Nevers ***France*** Capital city of Nièvre department and of the former Duchy of Nivernais, on the Loire river about 140 km (90 miles) south-east of Orléans. It has a huge 13th-century cathedral and a 15th-century ducal palace. Its industries include making fine china, earthenware and aircraft parts.

Population 61 250
Map France Ec

Nevis ***St Kitts and Nevis*** Southernmost of the two islands that form the state of St Kitts and Nevis in the Caribbean Leeward Islands. It produces sugar cane. The capital is CHARLESTOWN.

Population 9300
Map Caribbean Cb

New Amsterdam ***USA*** See NEW YORK (city)

New Britain ***Papua New Guinea*** Largest island of the Bismarck Archipelago, covering about 36 500 km² (11 125 sq miles), and lying off the north-east coast of New Guinea. It is mountainous, with active volcanoes, hot springs and peaks reaching 2438 m (8000 ft). The soil, particularly in the north and north-east, is very fertile, and produces coconuts, cocoa, timber and palm oil. The main town is RABAUL.

Population 237 000
Map Papua New Guinea Ba

New Brunswick ***Canada*** Small maritime province covering an area of 73 436 km² (28 354 sq miles) in eastern Canada, between the state of Maine in the USA and Quebec to the west and north-west and Nova Scotia to the south. The French first settled there in 1604, and the territory was in turn part of the French province of Acadia and the British province of Nova Scotia until it was made a colony in 1784. The province has a long coastline and consists mainly of rolling, forested hills. The population is largely rural and there are only three towns with more than 25 000 people. The economy is mainly based on forest products and food processing supported by coal and oil resources. Tourism is becoming increasingly important. FREDERICTON is the capital and SAINT JOHN the largest urban centre.

Population 696 000
Map Canada Id

New Caledonia ***Pacific Ocean*** Group of islands about 1400 km (875 miles) north-east of Brisbane, Australia, which in the mid-1980s were in the sometimes bloody throes of moving towards independence from France, which annexed them in 1853. There were riots in 1981 after the assassination of pro-independence leader Pierre Declercq, and again in 1985 after another liberation leader was shot dead by police. Leading the opposition to independence were the large proportion of the population – over one-third – of French origin. The native population – but less than half the total – are Melanesians known locally as Kanaks.

New Caledonia covers 19 103 km² (7376 sq miles) and is made up of one large hilly, forested island (also called New Caledonia, or La Grande Terre), which covers 16 627 km² (6420 sq miles), and several smaller ones. The latter include the Isle of PINES and the LOYALTY ISLANDS. Almost half the people live in the capital, NOUMEA. The main island has some of the world's largest known nickel deposits; most of the ore is processed before being exported. There are also large deposits of iron ore, chrome and other minerals. Less than 10 per cent of the land can be cultivated, but maize and coffee are important crops; large areas are grazed by cattle, and forestry is important. So was tourism until the unrest of the 1980s.

Population 155 000
Map Pacific Ocean Dd

New England ***Australia*** Plateau region in north-eastern New South Wales, stretching south about 300 km (190 miles) from the Queensland border. Forming part of the Great Dividing Range, the plateau is mostly 1000-1500 m (3300-5000 ft) high, rising to 1608 m (5275 ft) at Round Mountain. Much of it is rich sheep and cattle-grazing country, but areas of dense eucalyptus forest remain in some areas. There are several spectacular waterfalls, including Wollomombi Falls, the highest in Australia at 357 m (1171 ft). The main towns are Tamworth (population 29 600) and ARMIDALE.

Map Australia Ie

New England ***USA*** Region made up of the north-eastern states of Maine, New Hampshire, Vermont, Connecticut, Rhode Island and Massachusetts. The latter three are part of the heavily populated industrial belt stretching along the Atlantic coast from the national capital, Washington, through New York city to Boston. The other three are more rural and less industrialised. The area was settled by Europeans in the 17th century. Their Nonconformist religion, system of government and attention to education have remained strong.

Map United States Lb

New Forest ***United Kingdom*** Ancient English royal hunting forest in south-west Hampshire between Southampton and Bournemouth. It covers 364 km² (141 sq miles) and was placed under forest laws over 1000 years ago. However, many of its great oak trees were felled from the 16th to 18th centuries to build ships, and large areas of the forest are now open heath or grassland. It is a popular tourist area.

Map United Kingdom Ee

New Guinea See IRIAN JAYA; PAPUA NEW GUINEA

New Hampshire ***USA*** One of the original 13 states, first settled in 1623. It covers 24 097 km² (9304 sq miles) in the north-east, between Massachusetts and Maine; its western boundary with Vermont is the Connecticut river. It is mostly woodland. The White Mountains rise to 1917 m (6288 ft) in the north, and there are several lakes, of which the largest is Winnipesaukee. Agriculture concentrates on dairying, greenhouse products and eggs, and machinery and textiles are the main manufactures. CONCORD is the state capital; the only port, Portsmouth, is also a naval base.

Population 998 000
Map United States Lb

New Haven ***USA*** Port city in Connecticut. It was laid out in 1638 by Puritan settlers around nine squares; one of them, The Green, still remains an open space. The city is the seat of Yale University, one of the oldest and most prestigious in America, dating from 1701.

Population (city) 124 200; (metropolitan area) 506 000

Map United States Lb

New Hebrides See VANUATU

New Ireland ***Papua New Guinea*** Mountainous island 320 by 11 km (200 by 7 miles) in the Bismarck Archipelago to the north-east of New Britain. It produces copra and cocoa, and there is a tuna fishing industry. The main town is the port of Kavieng.

Population 70 000

Map Papua New Guinea Ca

New Jersey ***USA*** State covering 20 295 km² (7836 sq miles) in the north-east, on the Atlantic coast between the Hudson and Delaware rivers. It is part of the commuter belt for New York city, and has the highest population density of all the states, with more than 95 per cent of its people living in towns or cities. Nevertheless, the state's farms produce considerable quantities of fruit, vegetables and dairy products for nearby urban areas. Industry is diverse, with shipyards, refineries and chemical plants along the coast and electrical equipment and electronics factories inland. TRENTON is the state capital and NEWARK the largest city.

New Jersey's first European settlers were Dutch, but the area was ceded to England in 1664. It was one of the original 13 states that formed the USA.

Population 7 562 000

Map United States Lc

New Mexico ***USA*** South-western state, on the Mexican border. It was settled by the Spanish in the 16th century and was not ceded to the USA until 1848. It became a state in 1912. New Mexico covers 315 115 km² (121 666 sq miles) and rises in the north and west to more than 3000 m (10 000 ft) in the Rocky Mountains and Colorado Plateau. The state has one of the country's largest Indian populations. Farming is based on cattle, hay and wheat, while gas, oil and coal are the principal minerals. The state capital is SANTA FE and the largest city is ALBUQUERQUE.

Population 1 450 000

Map United States Ec

New Orleans ***USA*** Ocean and river port city in the Mississippi river delta in the southern state of Louisiana. It was founded in 1718 and was under French or Spanish control until 1803. Its chief exports are cotton, rice, oil and petrochemicals, and its manufactures include chemicals and ships. New Orleans is also a busy tourist centre, best known as the birthplace of jazz, and for the Mardi Gras (Shrove Tuesday) parade through the French Quarter, with its narrow streets and houses with iron-trellis balconies. Its renowned Creole cooking blends African, French and Spanish styles.

Population (city) 559 100; (metropolitan area) 1 318 800

Map United States Hd

New Plymouth ***New Zealand*** Market town on North Island, about 260 km (160 miles) north and slightly west of the capital, Wellington. It was one of the earliest European settlements in the country, being founded in 1841. It stands in a rich dairy-farming area and is the supply base for the offshore Maui gas fields.

Population 36 100

Map New Zealand Ec

New Providence ***Bahamas*** Island in the north of the central group of the archipelago. The chief town is NASSAU, capital of the Bahamas. There is some farming, fishing and light industry, but the main occupations are the service industries based in Nassau. The hub of the nation, it has an international airport and sea port. Some 60 per cent of Bahamians live on the island.

Population 150 000

Map Bahamas Ba

New Siberian Islands ***USSR*** See NOVOSIBIRSKIYE OSTROVA

New South Wales ***Australia*** The country's most populous state covering 801 430 km² (309 433 sq miles) bordering the Pacific coast between Victoria and Queensland. It was originally the name of the British colony (founded in 1788) covering the entire eastern half of mainland Australia, but various sections were cut off to form other colonies (later states) during the 19th century.

The state capital, SYDNEY, stands in the fertile coastal plain which lies to the east of the Great Dividing Range. The plains of the far west are semiarid grazing land, but there is richer wheat and intensive grazing country in the central region and along the coast. Dairy products, wheat and other grains, fruit, wine, sheep and cattle are the main agricultural products, and the state produces more than 40 per cent of the nation's wool.

There are large coal deposits in the Hunter Valley near the port of Newcastle and around Wollongong, and at Broken Hill in the far west there are rich reserves of silver, lead and zinc.

Population 5 379 000

Map Australia Ge

New Territories ***Hong Kong*** Most extensive of the three parts of the Crown Colony of Hong Kong, the other two being HONG KONG ISLAND and KOWLOON. Acquired by Britain from China in 1898, on a 99 year lease, the New Territories extend from the peninsula of Kowloon northwards to the Shum Chun river, which marks the border between Hong Kong and China, plus LANTAU and over 200 other outlying islands. Originally inhabited by fishermen and farmers, the once rural landscape of the mainland section is rapidly giving way to wholesale urban and industrial growth. The southern fringe is often referred to as New Kowloon, in recognition of the steady expansion into the New Territories of the built-up area centred in the Kowloon peninsula. Elsewhere in the New Territories are eight new towns – each of them an existing or planned city of considerable size. Many of Hong Kong's largest factories are located in the New Territories.

Population 1 812 000

Map Hong Kong Ca

New Valley Scheme ***Egypt*** Development scheme for the revival of a discontinuous line of oases – Kharga, Dakhla, Farafra, Bahariya and Siwa – in Egypt's Western Desert by tapping deep artesian water in the Nubian sandstone, which lies some 900 m (3000 ft) below ground. Apart from increasing food production, the main objective of the scheme (which has been continuing since the 1960s) is to ease population pressure on the Nile valley and delta regions. Sealed roads have been constructed between newly sited villages, and health and social services have been provided for the villagers. The total area of the oasis depressions is about 30 000 km² (over 11 500 sq miles), and the aim is to reclaim up to 20 per cent of this.

New York ***USA*** **1.** Second most populous state (after California), lying on the Atlantic coast and covering 128 402 km² (49 576 sq miles). It includes LONG ISLAND and the valley of the HUDSON river, stretching north-west to Lakes Erie and Ontario and the Canadian border. The Adirondack Mountains rise to more than 1600 m (5250 ft) in the north-east; in the south lie the Catskill Mountains and part of the Appalachians. The state's chief products are dairy goods, fruit, vegetables, salt, cement, stone, instruments, books and printed material and electrical equipment. NIAGARA FALLS and New York city are the main tourist attractions. ALBANY is the state capital.

It was the Dutch colony of New Netherlands until 1664, when it was annexed by the British and renamed after the Duke of York, brother of King Charles II. New York was one of the 13 original states of the Union.

2. The country's most populous city, known affectionately as the Big Apple, and the centre of a conurbation spreading into neighbouring parts of Connecticut and New Jersey that contains 17.8 million people – one of the very biggest, if not the biggest, in the world.

It lies on the east coast, and began in 1609 as a trading post on Manhattan island at the mouth of the Hudson river. It later grew as the Dutch settlement of New Amsterdam, and its fine natural harbour attracted a flourishing trade even before it was captured and renamed by the English in 1664. It greatest period of growth was, however, in the 19th and 20th centuries, when it was the gateway to the New World for millions of European immigrants. The city has now spread north to Bronx and Yonkers, eastwards across the East river to Queens and Brooklyn on Long Island, and west across the Hudson river to Jersey City and down to Staten Island.

New York is the financial and trade centre of the country, its leading port and an important manufacturing centre. Its heart is still Manhattan, where the financial sector is based in and around Wall Street – the site of a boundary stockade, or wall, built by Dutch settlers in 1653 to protect themselves from raiding English colonists and Indians. Many other businesses have their headquarters in 'midtown' Manhattan, near the southern end of Central Park, while heavier industries stand on reclaimed river marshes. However, some lighter industries, such as clothing, remain near the city centre, whose famous skyline is dominated by such skyscrapers as the 381 m (1250 ft) high Empire State Building and, near Wall Street, the twin

▲ RADIO CITY Two New York policemen patrol 52nd Street in the heart of midtown Manhattan, near the headquarters of several radio and television networks, including CBS.

411 m (1350 ft) World Trade Center towers.

The city is rich in educational and cultural centres, with several universities, numerous museums (including the Metropolitan Museum of Art and the Museum of Modern Art), the Lincoln Center for the Performing Arts (incorporating the Metropolitan Opera House) and a host of theatres in or near the street called Broadway. The United Nations headquarters stands beside the East River.

Population (state) 17 783 000; (city) 7 164 700; (metropolitan area) 8 376 900; (conurbation) 17 807 100

Map (state) United States Kb; (city) United States Lb

New York State Barge Canal ***USA*** See ERIE CANAL

New Zealand See p. 468

Newark ***USA*** City in New Jersey about 16 km (10 miles) west of New York city and forming part of the New York conurbation. It is a port and industrial centre specialising in chemicals. Some of it stands on reclaimed marshland beside Newark Bay. New York city's third airport is there.

Population (city) 314 400; (metropolitan area) 1 875 300

Map United States Lb

Newcastle ***Australia*** Industrial city and port in New South Wales, at the mouth of the Hunter river, 120 km (75 miles) north of Sydney. It exports coal, wool, wheat and dairy produce, its industries include iron and steel, chemicals, shipbuilding, metalworking and textiles. The city was founded as a penal settlement in 1804.

Population 259 000

Map Australia Ie

Newcastle ***South Africa*** Coal-mining and steel-manufacturing town at the foot of the Drakensberg in north-west Natal. Boer troops occupied it in 1899-1900 and briefly renamed it Viljoensdorp.

Population 55 700

Map South Africa Cb

Newcastle upon Tyne ***United Kingdom*** City of north-east England in the county of Tyne and Wear some 400 km (250 miles) from London.

There was a bridge over the River Tyne in the 2nd century AD when the Romans built Hadrian's Wall, whose eastern end was at Wallsend, now a Newcastle suburb. A castle – from which the city got its name – was built in Norman times, and the port was flourishing in the 16th century. The city's growth, however, came with the increased mining and export of coal in the 17th century and, in the 19th century, with the rise of engineering industries, especially shipbuilding. The Tyne was lined with yards from Newcastle, past Jarrow – abbey home of the Venerable Bede (AD 673-735), the father of English historical writing – to Tynemouth with its 11th-century priory. When the great depression of the 1930s hit Britain, no area of the country suffered more deeply than the Tyneside ship towns, especially Jarrow and Gateshead.

Life since then has been a struggle for the area, but few cities have a better record of urban renewal. Newcastle has a spacious 19th-century city centre with grand civic buildings and a medieval church-turned-cathedral. There is a university, industrial estates on both banks of the river, and some surviving shipbuilding. Perhaps the most impressive of the city's buildings

A DAY IN THE LIFE OF A NEW YORK COP

Roll call at the Todt Hill police station, Service Area No 1A, Staten Island, takes place at 7.45 am; Officer Brian Patrick Byrnes, 24, steps out of his car outside the station at 7.40. Inside, a sergeant issues instructions to a dozen officers for their eight-hour shifts. Byrnes is allotted a routine patrol of a multistorey tenement in Stapleton – a mainly black, slum area.

Byrnes rides a three-wheel motor scooter to Stapleton. There, he checks into the room in the tenement which is assigned to the police. In a slum block like this, a careless cop can soon be a dead cop. Byrnes, every nerve alert, takes the lift to the top floor, and, hand on revolver, climbs the stairs to the roof. Nothing suspicious up there. On the way down, by the stairway, he looks and listens at each floor for signs of trouble.

Back on the street, he starts his foot patrol, keeping an eye on upper windows for bottles or other missiles that might be thrown at him. But Byrnes has a quiet day. Passers-by ask him for directions. He takes the number of a speeding car, and radios ahead to colleagues in a squad car – he cannot give chase on his 50 km/h (30 mph) scooter.

Even his quiet days are tense. He knows that when there is trouble it happens suddenly. He has to think fast, keeping in mind the strict rule book that dictates his response to every emergency. Happily, he has never yet had to fire the gun he always carries on duty.

In three years, Byrnes has risen from a starting salary of $18 000 a year to $32 000 – comparable to what many junior executives earn. He gets 29 days' vacation a year, unlimited sick leave, free medical and dental care; and in 20 years, at the age of 44, he can retire on a pension of half his finishing salary and, if he chooses, take another job.

Byrnes graduated in sociology from a Florida college, and went on to the New York College of Criminal Justice and the New York Police Academy. Now he lives in a Staten Island bungalow with his father, a retired fireman, and his mother, a retired schoolteacher. He is the youngest of three children, but the only one still at home. His sister, a nurse, is married; his brother is a chemical engineer. Byrnes thinks he will probably stay at home until he marries.

His girlfriend, Patty, is from New Jersey. He likes to spend his days off with her, and perhaps drive his Plymouth sports car to a mountain lake just over 2 hours away in Pennsylvania, where he keeps a 5 m (18 ft) motorboat. They can spend whole afternoons adrift on the lake, sitting in the inflated tubes of car tyres, sipping beer, listening to the radio, and being spun round dizzily in the wake of other boats.

is the railway station (1850-65), a reminder that the area produced George Stephenson (1781-1848) who built the first steam locomotive for a public railway.
Population 280 000
Map United Kingdom Ec

Newcastle-under-Lyme ***United Kingdom*** See POTTERIES, THE

Newfoundland ***Canada*** The country's most easterly province, comprising a triangular island at the mouth of the Gulf of St Lawrence and Labrador on the mainland. Its area totals 372 000 km² (143 634 sq miles). Newfoundland was discovered in 1497 by the English expedition led by the Italian-born explorer John Cabot, and the first settlers arrived from south-west England in the early 17th century. Much of Newfoundland island is a plateau, rising to 842 m (2762 ft) in the Long Range Mountains, and the coastline is heavily indented. Offshore the Newfoundland Banks yield heavy catches of cod, and the province's economy depends on fishing, forestry and mineral resources including copper and gypsum. Many Newfoundlanders live in small fishing villages. The development of mining and pulp and paper production has created large centres at Corner Brook and Grand Falls. The capital and main city is ST JOHN'S on the east side of the island.
Population 568 000
Map Canada Ic

Newgrange ***Ireland*** See BRUGH NA BOINNE

Newport ***USA*** Port and naval base on Rhode Island, at the mouth of Narragansett Bay, in the state of Rhode Island. It was settled by Europeans in 1639 and was a prosperous trade centre until it was largely destroyed during the American War of Independence (1775-83). It became a fashionable resort in the 19th century, and many millionaires' mansions are now tourist attractions. It is a major yachting centre and was the scene of numerous America's Cup races.
Population 29 900
Map United States

Newport News ***USA*** Major Atlantic port at the mouth of the James river in Virginia. It exports coal and imports ore, and has extensive shipbuilding and repair yards.
Population (city) 154 600; (metropolitan area, with Norfolk) 1 261 200
Map United States Kc

Newry ***United Kingdom*** Port and market town in Northern Ireland about 55 km (35 miles) south-west of the capital, Belfast. It grew up around a Cistercian abbey founded in the mid 12th-century. The Newry Canal, built between 1730 and 1741 to link the Irish Sea and Lough Neagh via the Bann river, was one of the earliest ship canals in the British Isles. Among the town's industries are food processing, linen and clothing.
Population 11 500
Map United Kingdom Bc

Neyshabur ***Iran*** See NISHAPUR

Ngaliema, Mount ***Uganda/Zaire*** See RUWENZORI

Ngami Depression ***Botswana*** Marsh rich in birdlife, which was once a huge lake that the British explorer David Livingstone described as an inland sea. It was fed by the OKAVANGO DELTA and probably covered 52 000 km² (20 000 sq miles). Since the 1890s, papyrus reeds have blocked the main inflow and the marsh has gradually shrunk. A drought in 1965-6 dried it up completely for a time.

The Ngamiland district, which encompasses the Depression and the Okavango Delta, has an area of 109 300 km² (42 200 sq miles). The chief town is Maun.
Map Botswana Bb

Ngaoundéré ***Cameroon*** Town on the ADOUMA-OUA MASSIF, about 450 km (280 miles) north-east of the capital, Yaoundé, to which it is linked by rail. Ngaoundéré is an Islamic centre dominated by the pastoral Peul people, and the smell of its abattoirs hangs over the surrounding countryside.
Population 40 000
Map Cameroon Bb

Ngauruhoe, Mount ***New Zealand*** Active volcano rising to 2291 m (7516 ft) in the Tongariro National Park on North Island, about 250 km (155 miles) north-east of the capital, Wellington. For most of the time it emits only smoke, but occasional violent outbursts send lava flowing down the mountainside. The last big eruption was in 1978.
Map New Zealand Ec

Ngorongoro Crater ***Tanzania*** Large volcanic crater in the north of the country, about 230 km (145 miles) south-east of Lake Victoria. The crater is 18 km (11 miles) across and communities of nomadic Masai tribesmen – together with 100 000 head of their cattle – live on its floor. The crater floor also contains more than 30 000 wild animals, including wildebeests, lions, leopards, elephants, rhinoceroses and hippopotamuses. Ngorongoro has been extinct for several million years. The crater is part of the Ngorongoro Conservation Area, which includes OLDUVAI GORGE.
Map Tanzania Ba

N'Gounié ***Gabon*** River, 440 km (273 miles) long, rising in the Chaillu Mountains and flowing past Mouila, Fougamou and Sindara to its confluence with the OGOOUE river, 19 km (12 miles) above Lambaréné. Its valley is Gabon's leading food-producing region. The river is navigable for much of the year between Sindara and Lambaréné and between Mouila and Fougamou.
Map Gabon Bb

Nha Trang ***Vietnam*** Port and beach resort, about 300 km (190 miles) north-east of Ho Chi Minh City. It is also a market town for farmers growing rice, sugar cane, rubber and coffee in the hills inland.
Population 197 900
Map Vietnam Bc

Niagara Falls ***Canada/USA*** Spectacular falls providing a great tourist attraction on the Niagara river on the Canada-US border. The river, which flows from Lake Erie to Lake Ontario, divides at Goat Island into two falls. The Horseshoe (Canadian) Falls are 54 m (176 ft) high and the crest of the falls is 675 m (2215 ft) wide. The American Falls are 56 m (184 ft) high and 323 m (1060 ft) wide. The river drops 98 m (320 ft) in 48 km (30 miles). The gorge is spanned by bridges and the waters have been harnessed for hydroelectricity.
Map Canada Hd

Niagara-on-the-Lake ***Canada*** Resort at the northern end of the Niagara river, where it enters Lake Ontario near the Canada-US border. It was the first capital of Upper Canada (1792-6), the present-day province of Ontario. The Shaw Festival of plays is held there in the summer.
Population 12 180
Map Canada Hd

Niamey ***Niger*** National capital, about 100 km (60 miles) east of the border with Burkina (formerly Upper Volta). Until 1926, when it took over from ZINDER as the capital, Niamey was no more than a cluster of villages. Today its tree-lined boulevards contain elegant French colonial buildings cheek by jowl with modern office blocks. There are two city markets: the Small Market, which specialises in food; and the Great Market, which sells colourful cloth made in Niger and neighbouring Mali – as well as leather, iron and copper wares.

The magnificent Great Mosque – a large, white, modern building in traditional style – overlooks the Niger river, which flows past the city and which, like the country, is said to get its name from the Tuareg word *n'eghirren*, meaning 'flowing water'. Nearby is the National Museum, set in a 24 hectare (60 acre) park complete with zoo and botanical gardens.
Population 400 000
Map Niger Ab

Niari ***Congo*** See KOUILOU

Niassa, Lake ***East Africa*** See MALAWI, LAKE

Nicaragua See p. 472

Nicaragua, Lake ***Nicaragua*** Largest fresh-water lake in Central America at 8264 km² (3191 sq miles). Many rare and colourful birds are found in the tropical vegetation surrounding it, and its waters also contain salt-water species such as swordfish and sawfish, which reach it from the Caribbean via the San Juan river.
Map Nicaragua Aa

Nice ***France*** The capital of Alpes-Maritimes department, on the Côte d'Azur 25 km (16 miles) from the Italian border. It was annexed by France from the kingdom of Sardinia in 1860, and still has an Italian flavour.

The old town around the harbour is a network of narrow streets and squares with open-air markets. The new town is spread along the beach of the Baie des Anges, and the famous Promenade des Anglais – with its grand hotels – extends for more than 6 km (4 miles) to the airport. The Italian soldier and nationalist Giuseppe Garibaldi (1807-82), whose redshirt supporters helped to unify Italy, was born in the city.
Population 451 500
Map France Ge

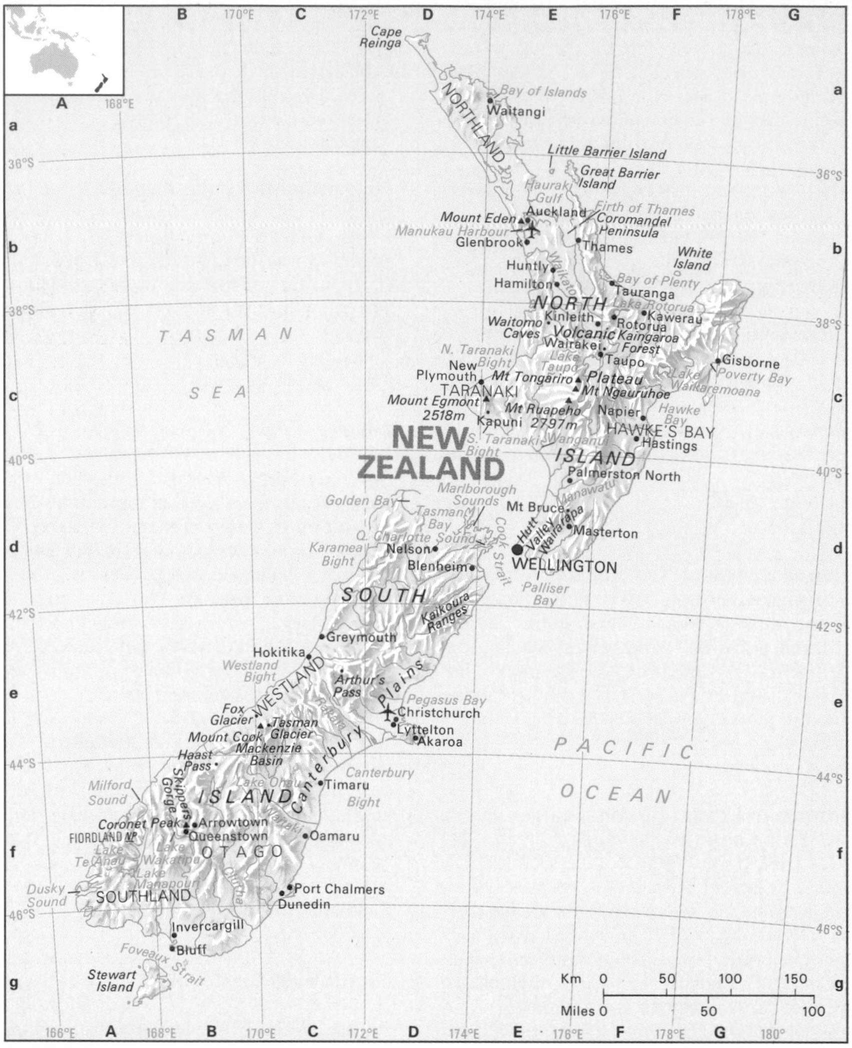

New Zealand

AN URBAN PEOPLE IN A BEAUTIFUL WILD LAND, OUTNUMBERED 24 TO 1 BY THEIR LIVESTOCK, ARE FINDING NEW WAYS TO TRADE AND A NEW NATIONAL IDENTITY

Bonds of trade and kinship tied New Zealand so firmly to Britain that it was long considered as the Britain of the South Pacific. Although it ceased to be a colony in 1907, another 40 years passed before it set up its own foreign ministry and began to develop an independent foreign policy. Moreover, Britain continued to provide a guaranteed market for New Zealand's primary produce.

In fact, not until Britain joined the European Economic Community in 1973 did New Zealand's economic dependence on Britain really begin to decline and the country make a serious effort to diversify its trade. Success in that effort has made New Zealand an affluent modern state with prospering trade with not only Europe but also Asia, the Americas and the Middle East. Its main trading partners today are, in order, Japan, Australia, the USA, Britain, Iran and China.

At the same time, it has become a leader in its own region. In 1971 it hosted the first meeting of the South Pacific Forum, a political grouping of 13 independent countries in the region. New Zealand has been a vocal opponent of French nuclear testing in the Pacific, and in 1985 its government barred nuclear-armed or nuclear-powered ships from its ports – a policy that strained relations with the USA and jeopardised ANZUS, its defence treaty with the USA and Australia. The same year, French agents sank the anti-nuclear protest ship *Rainbow Warrior* at AUCKLAND.

TWO MAJOR ISLANDS

New Zealand consists of two major islands (the NORTH ISLAND and SOUTH ISLAND) and several smaller ones, including STEWART ISLAND and the CHATHAM ISLANDS. Nearly 90 per cent of its people are of European origin, and English is the most widely spoken language. New Zealanders' customs and social structure remain essentially British.

The country lies about the same distance from the Equator as Italy does, but the climate is more temperate, with regular rainfall all year and no extremes of heat or cold. Its terrain also resembles Italy in the nature and variety of its features. In the North Island there is not only fertile grazing land but also active volcanoes, such as the conical Mount EGMONT on the west coast, the majestic snow-capped Mount RUAPEHU – a major skiing area – and Mount NGAURUHOE, which forms an impressive backdrop to Lake TAUPO, a mecca for trout fishermen. Geysers, hot springs and boiling mud attract tourists to ROTORUA.

The mountainous South Island is fringed in the south-east and central east coast by extensive plains where cereals are grown and huge flocks of sheep are grazed. Running the whole length of the island and rising from the plains with astonishing suddenness are the snow-capped peaks of the Southern Alps. Evergreen beech forests clothe the slopes and water abounds: clean, clear lakes such as WAKATIPU and MANAPOURI, lying in ice-scooped hollows, and rushing mountain torrents create a breathtaking landscape.

Twenty of the peaks exceed 3000 m (10 000 ft) – the highest is Mount COOK at 3765 m (12 352 ft). At the higher levels, where the snow is permanent, there are ice fields and glaciers. Some are very large – the Franz Josef and FOX glaciers, dropping to the nearby west coast, and the Tasman, heading east from the main divide, are the most accessible. The area was heavily glaciated 2 million years ago and the south-western coast is deeply cut by fiords. Among the most striking is MILFORD SOUND, where sheer cliffs rise for hundreds of metres.

New Zealand is slightly larger in area than the UK, but has a population of only 3.3 million, three-quarters of whom live in the North Island. Eighty per cent of all the people are town dwellers – nearly half of them in the area around Auckland, the boom city of the North Island. Much of the South Island, which has greater areas of mountains, forests and farmland than the North Island, is very sparsely populated.

Farming and farm-related industries are still the basis of New Zealand's wealth, but they now employ less than 15 per cent of the population. Most of the highly educated and urbanised New Zealanders work in service industries, such as administration, teaching and banking. The small fraction that work in agriculture tend some 70 million sheep and nearly 8 million cattle.

THE MAORI PEOPLE

An important feature of New Zealand's new society is the recent increase in the number of Maoris – the Polynesians who inhabited the islands for at least 700 years before the first Europeans, Dutchman Abel Tasman and his crew, saw them in 1642. In the 19th century it seemed that the Maoris might die out. Their lands had been taken from them and the European way of life had weakened their culture. Tribal warfare with European-introduced firearms and European diseases to which the Maoris had no resistance had reduced their numbers, and their infant mortality rate was high. However, improved health care and a cultural resurgence have been accompanied by an increase in the Maori population.

Now almost one New Zealander in ten is at least half Maori, and many others have some Maori ancestry. There is much intermarriage between Polynesian and European (or 'Pakeha') New Zealanders. Calls have been made for more widespread teaching of the Maori language in schools, and the authorities have recognised Maori land claims – in some cases dating back almost a hundred years.

Early Maori artefacts – tools, weapons and ornaments – are mainly seen in museums now, but the Maori traditions of fine carving and weaving continue. Maoris, like other New Zealanders, take great interest in European art forms; the Maori soprano Dame Kiri Te Kanawa is a world-renowned opera singer. The Maoris, converted to Christianity nearly 150 years ago, have developed their own forms of the religion, incorporating Maori traditions. They are also active in the conventional churches, and a Maori was Anglican primate of New Zealand until 1985, when he was appointed governor-general – the representative in New Zealand of Elizabeth II, who is also Queen of New Zealand.

The political and cultural resurgence of the Maoris has coincided with a wave of immigrants from the Pacific islands, notably Western Samoa, Fiji and the Cook Islands; most of them have settled in Auckland. Relations between the races, while not as harmonious as they were 50 years ago, are still very good by international standards. The urbanisation of the Maori and Pacific islander populations since the Second World War has caused some social problems; nevertheless, New Zealand is basically a caring society and extreme views are exceptional.

INNOVATION AND ADAPTATION

Immigration from Europe, which swelled the population during the 1960s and 1970s, has now slowed; the birthrate has also fallen. As a result, population growth has declined. However, the population is still young – 50 per cent of the people are under 30 – and well educated (10 per cent of school leavers go to university), making for a high-calibre workforce. New Zealanders have always been innovators and are quick to adopt new technologies. Their sheep and cattle rearing benefited from the early use of refrigerated transport, crop spraying and mechanised shearing and milking.

Young and innovative attitudes have also changed national politics over the last decade. The conservative National Party had been in government for all but two three-year terms in the previous 35 years when the Labour party swept them aside in the 1984 election.

This new government faced economic difficulties created partly by the limitations on exports to Britain since 1973 and partly by

▼ ALPINE CREST Fed by heavy snowfalls brought by westerly winds, Fox Glacier stands at the crest of the Southern Alps on New Zealand's South Island. It falls steeply by about 200 m per km (1000 ft per mile).

the increased costs of imported oil. These factors, combined with the baby boom of the 1960s (which put 30 000 new workers on the labour market each year in a country whose labour force numbers less than 1.5 million), resulted in worsening unemployment in the early 1980s. In addition, surplus production of meat and dairy produce in Europe and the USA over several years led to over-supply and price cutting. Despite selling everything it produces, New Zealand has seen its agricultural profits shrink.

To help compensate for these problems, New Zealand has developed several small-scale but efficient secondary industries, such as aluminium and methanol production and the prefabrication of houses for export. It has also diversified its dairy products to include new types of cheese, skimmed milk, casein and baby milk powders, and by aggressive selling has won new markets in the Middle East, Eastern Europe, USSR, China, Japan and Australia. In particular, an agreement on closer economic relations with Australia has in effect enlarged New Zealand's domestic market from 3 million to 18 million. Australia is the most important market for the new manufactured products.

DOMINANT AGRICULTURE, DEVELOPING INDUSTRY

Despite innovations, agriculture still holds the key to the country's future prosperity. New Zealand has some of the finest pastures in the world, thanks to its moist, warm climate. Well-managed farms with sheep, beef cattle and dairy cows produce extremely high yields of wool, meat and milk. New Zealand's tiny population means that demand by the home market is small, and an exceptionally large proportion of the produce is exported. In fact, New Zealand is the world's largest exporter of lamb and dairy products, and is second only to Australia in the export of wool.

Moves have been made to widen the basis for the country's wealth. For example, the timber industry has been developed to produce not only sawn timber (radiata pine, Douglas fir and Corsican pine) but also such products as veneer, chipboard, fibreboard, pulp and paper. Additional land has been planted with vines to support a growing wine industry in the HAWKE'S BAY, MARLBOROUGH SOUNDS and Auckland areas. And exports have been given a filip with increased production of oranges, lemons, grapefruit, kiwi fruit (also known as Chinese gooseberries) and vegetables from the Bay of PLENTY and NORTHLAND, along with apples from the orchards of NELSON and Hawke's Bay.

The seas are also yielding increased riches. New Zealand has declared an exclusive economic zone extending for 370 km (200 nautical miles) beyond its shores, covering much of an extensive submerged continental landmass. This is a rich source of seafood, and New Zealand's seafood exports have multiplied 9 times by volume and 15 times by value since 1975.

Although it has little oil, the country is well endowed with other sources of energy. There are abundant reserves of coal, volcanic areas on North Island produce geothermal energy from the hot springs, and the mountain streams have been harnessed to produce 86 per cent of New Zealand's electrical needs. Offshore natural gas reserves help to cut down oil imports.

Major industrial developments in recent years have included the building of an iron and steel works using iron sands from the west coast of the North Island, and an aluminium smelter, operating on hydroelectric power, that uses imported bauxite. Manufactured goods now account for 30 per cent of New Zealand's exports, compared with only 4 per cent in the mid-1960s.

URBAN CENTRES, PRIMEVAL FORESTS

Much of the manufacturing is done in Auckland, which sprawls along and beyond an isthmus separating the Waitemata and Manukau harbours. The land is dominated by dozens of low volcanic cones, the most striking being Mount EDEN, near the city centre, and Rangitoto, an island in the Hauraki Gulf. Aucklanders, like other New Zealanders, live mainly in detached houses, mostly American or Scandinavian in style. Four out of five New Zealand dwellings are separate houses occupied by one household; half have timber walls and nearly two-thirds are roofed in galvanised steel.

Timber houses are also the rule in windy and earthquake-prone WELLINGTON, which sits astride a fault in the earth's crust (see PLATE TECTONICS) at the southern end of the North Island. The houses perch on steep hillsides overlooking a great harbour and the high-rise offices that form the nation's financial and administrative centre. Although less than half the size of Auckland, Wellington is the capital and seat of New Zealand's one-chamber parliament.

The principal town of the South Island is CHRISTCHURCH, lying on the CANTERBURY PLAINS below the lava flows of a huge extinct volcano; the city's port is in the flooded crater. Christchurch, one of the world's first garden cities, with weeping willows that trail their leaves in rivers meandering through a grid of streets, reminds British visitors of such English towns as Cambridge or Stratford and Americans of their own older East Coast towns. Its air of spacious, languid good living suggests an idealised image of the past.

DUNEDIN, also in the South Island, is still regarded as New Zealand's fourth main centre, although its population (110 000) has been surpassed by new North Island cities – HAMILTON and PALMERSTON NORTH. Dunedin was founded by Scots, but 19th-century gold rushes brought the city a flood of settlers who soon outnumbered the Scots. Gold wealth also gave the city a fine heritage of Victorian buildings.

The isolation of New Zealand for some 70 million years has preserved creatures that exist nowhere else, especially flightless birds such as the long-beaked kiwi, which is the national symbol. The inhabitants have taken its name and refer colloquially to themselves as 'Kiwis'.

In many parts, the most characteristic trees today are introduced – the radiata or Monterey pine and the Monterey cypress, both from California. Vast forests of introduced conifers on the volcanic plateau of the North Island include the world's largest plantation. But in the great national parks the primeval evergreen forests of native conifers and broad-leaved trees are still dense and dark. In places they reach to broad, empty beaches lapped by the sparkling blue of the Pacific Ocean or Tasman Sea.

The splendid scenery of glaciers, mountains, forests and lakes, the empty beaches, clear air and lush pastures continue to justify the title of 'God's own country', bestowed by an early prime minister. The attractions are many, and visitors come in increasing numbers; tourism has become a major earner of foreign exchange for New Zealand.

The local people also relish the space and beauty. Swimming, surfing and sailing are notable pastimes in a sport-mad country where no place is more than 110 km (70 miles) from the sea. There is even greater enthusiasm for horse racing and rugby union; violent passions have been aroused in recent years by the question of whether the national rugby team – called the All Blacks after the colour of their playing gear – should play against South Africa.

In all its interests the country knows no class or colour barriers. Ever since the days of the first settlers, when New Zealanders found that social divisions did not help to farm the land, society has been fair and perhaps uniquely classless. This was the first country to give women the vote – in 1893 – and the Treaty of WAITANGI in 1840 gave Maoris and Europeans equal rights as citizens. This exceptional equity may be the foundation of the new national identity which New Zealanders are building.

NEW ZEALAND AT A GLANCE
Area 268 046 km² (103 493 sq miles)
Population 3 300 000
Capital Wellington
Government Parliamentary monarchy
Currency New Zealand dollar = 100 cents
Languages English, Maori, other Polynesian languages
Religion Christian (81%, mainly Protestant)
Climate Temperate; average temperature in Wellington ranges from 6-12°C (43-54°F) in July to 13-21°C (55-70°F) in January
Main primary products Sheep, cattle, fruit, vegetables, timber, cereals, fish; natural gas, coal, iron sand
Major industries Natural gas processing, paper, iron and steel, aluminium smelting, fertilisers, cement, glass, transport equipment, machinery, agriculture, wool and textiles, meat and dairy processing, wine making, timber milling and wood processing, tanning, fishing, tourism
Main exports Meat, dairy products, wool, timber, wood pulp and paper, timber products, fish, fruit and vegetables, hides and skins
Annual income per head (US$) 6850
Population growth (per thous/yr) 2 (higher for Maori population)
Life expectancy (yrs) Male 72 **Female** 76

Nicosia ***Cyprus*** Capital and largest city. It lies near the centre of the island, and has been divided since the island was partitioned into Greek and Turkish zones in 1974-5, following the Turkish invasion of northern Cyprus. Local industries include footwear, textiles, leather goods, cigarettes and tourism.

The old town centre – which was founded before the 7th century BC – is encircled by 5 km (3 miles) of 16th-century walls and includes, in the northern (Turkish) sector, St Sophia Cathedral, now a mosque, and a 14th-century Armenian church. The city, named after Nike, the Greek goddess of victory, also houses the Cyprus Museum in the southern (Greek) sector, which has a collection of archaeological treasures found on the island and dating back to the Bronze Age.

Population 161 100
Map Cyprus Ab

Nieborow ***Poland*** See LOWICZ

Niedersachsen ***West Germany*** See LOWER SAXONY

Nieuw Amsterdam ***Surinam*** See PARAMARIBO

Nieuw Waterweg (New Waterway) ***Netherlands*** Deep-water channel, about 28 km (17 miles) long, cut between 1866 and 1872 from Rotterdam to the North Sea. It replaced the shallow channels of the Rhine delta, enabling oceangoing vessels to reach Rotterdam. The HOOK OF HOLLAND and EUROPOORT are at its seaward end.

Map Netherlands Bb

Niger ***West Africa*** River some 4170 km (2590 miles) long. Its basin is the largest in West Africa, and the third largest in Africa, after the Nile and Congo (or Zaire). The Niger rises on the Sierra Leone-Guinea border and flows north-east across Guinea into Mali. There, just below the city of Timbuktu (Tombouctou), it makes a great curve south-eastwards, and continues across south-west Niger. It then forms part of the Niger-Benin border and continues into Nigeria, where it sweeps south to the Bight of Benin to empty into the Gulf of Guinea.

The vast Niger delta, an area of mud flats and mangrove swamps covering some 20 000 km² (7720 sq miles), stretches some 300 km (185 miles) north-west from Port Harcourt. Several rivers, besides the Niger, cross the delta, including the Benin, Escravos and Forcados, and they are increasingly blocked by sandbars because of longshore currents. However, new prosperity has come to the delta with the exploitation of its vast oil deposits, which were first discovered in 1956, at the town of Oloibiri.

The Niger is an ancient highway of trade. The traffic in palm kernels (which yield palm oil used in margarine, cooking fats and soaps) from the delta was so great that in the 18th century the British called the delta rivers the 'oil rivers'. The Niger is navigable from the sea to the Nigerian town of Jebba, and for about 1610 km (1000 miles) above the Nigerian town of Yelwa. It is the basis of several hydroelectric and irrigation schemes, including that at KAINJI, and is a valuable source of fish.

Map Africa Dc

A DAY IN THE LIFE OF A GREEK-CYPRIOT SHOPKEEPER

Yiorgos Antoniou's shop in Nicosia sells the fine embroidered linen produced in Cyprus. He also sells hand-loomed fabrics made into dresses, shorts and children's clothes. Many of his customers are local people, but others are European tourists or traders from Arab nations nearby.

This morning, he saunters from his three-bedroom flat in Socrates Street to unlock his small shop, just ten minutes away in Hermes Street. It stands beside a coffee shop. Tables on the street make a meeting place for the neighbouring traders, and Yiorgos sits chatting, with one eye on his wares, drinking coffee and smoking cigarettes, for most of the morning.

The town of Nicosia itself is still divided in two by the frontier between Greek Cyprus and Turkish Cyprus. The border has stood since the troubles began in 1974. In 1983 the northern, Turkish, part declared itself a separate republic, which is not recognised by the Greeks.

Yiorgos's discussions are interrupted whenever a customer appears. Yiorgos is always polite, but haggles shrewdly, making at least a 25 per cent profit on every sale – although his goods are still cheap. After paying overheads, he makes about 7000 Cyprus pounds (US$12 250) a year – enough to live on comfortably. It will also pay college fees for his son Nicos, now 15, in a few years' time. Father and son get on well, and at weekends they swim together on the east coast, south of Famagusta, or escape the excessive heat of summer by travelling to the Troodos Mountains, 60 km (some 40 miles) to the south-west.

At 1.30 pm, with cicadas chirping in the arid and sizzling air, Yiorgos shuts up shop and goes home to his wife Eleni, who has made a lunch of chicken and rice, salad with olives, grapefruit, and a little wine. Yiorgos dozes until four, and then takes a cup of coffee on his terrace, gazing across the rooftops of Nicosia to the hazy purple slopes of the Turkish zone, with its Gothic castles.

He and Eleni go together to the shop, for the evening's business. The alleyways are getting crowded, and foodstalls are setting up everywhere. The two of them get home before nine, just after Nicos and his 13-year-old sister Katerina, who have been at their grandmother's house after school.

Around ten, Eleni and Yiorgos are dressed up and ready to go out for a long, festive meal with friends. Yiorgos, smart in the dark cotton suit he wears to the Orthodox church on Sundays, reflects that the old pleasures of everyday life have not been lost – despite the political upheavals his country has seen.

Niger See p. 473

Nigeria See p. 474

Nijmegen (Nimegue; Nimwegen) ***Netherlands*** City on the Waal (Lower Rhine) south-east of Amsterdam and only 7 km (4 miles) from the West German border. In September 1944 it was the scene of a bitter battle between Allied and German forces for the possession of its bridge over the river. The city suffered extensive damage, but the 16th-century town hall and the 15th-century Church of St Stephen still stand in the marketplace. There are also the preserved foundations of a palace of the Holy Roman Emperor, Charlemagne (742-814). The city has a university, and its manufactured products include machinery, electrical equipment and paper.

Population (city) 147 100; (Greater Nijmegen) 229 400
Map Netherlands Bb

Nikko ***Japan*** Tourist town set in a beautiful mountainous national park, 1407 km² (543 sq miles) in area, in central Honshu island, about 120 km (75 miles) north of the capital, Tokyo. The town contains a large complex of temples and shrines, including the Toshogu, the magnificent mausoleum of the shogun (warlord) Ieyasu. Ieyasu, who ruled Japan from 1603 to 1605, was the founder of the Tokugawa dynasty of shoguns, which kept Japan in feudal seclusion for more than 250 years until 1853. The mausoleum's main building, covered in vermilion lacquer, is decorated with more than 2 million sheets of gold leaf. It is the setting for an annual May pageant in which men parade in the traditional costume of samurai warriors.

Population (town) 23 900
Map Japan Cc

Nikolayev ***USSR*** Ukrainian Black Sea port about 110 km (70 miles) north-east of Odessa. Nikolayev was founded as a naval base at the mouth of the Southern Bug river in 1788, and it grew into a shipbuilding centre as well. Its industries include machinery, chemicals, footwear, clothing and food.

Population 480 000
Map USSR Ed

Nilaveli ***Sri Lanka*** See TRINCOMALEE

Nile (Al-Bahr; Bahr el Nîl) ***Africa*** The world's longest river, flowing northwards about 6695 km (4160 miles) from its farthest head-stream, the Luvironza in Burundi, to the Mediterranean via its delta in north-eastern Egypt.

One of its major sources is Lake VICTORIA, from where it drops (as the Victoria Nile) through Lakes Kyoga and Albert, and then (as the Albert Nile) to the Sudanese border at Nimbule. Winding through the swamps of the SUDD it is called in Arabic the Bahr el Jabel; it then continues to Khartoum as the Bahr el Abiad, but both sections are known in English as the White Nile. At Khartoum it joins the Blue Nile (Abay Wenz; Bahr el Azraq), which rises at Lake TANA in the Ethiopian highlands and flows about 1370 km (850 miles). From Khartoum the river is known simply as the Nile.

The Nile system drains some 2 850 000 km² (1 100 000 sq miles) – about one-tenth of

Nicaragua

AMERICAN MERCENARIES ONCE SEIZED POWER IN NICARAGUA AND THEIR LEADER BECAME PRESIDENT – TODAY, AMERICA IS STILL DEEPLY INVOLVED IN THE LATEST CIVIL WAR

There have been wars of one sort or another in Nicaragua, which became an independent republic in 1838, for over a century. Indeed, the country has been troubled with conflicts and with foreign interference ever since it was conquered by the Spanish in the early 16th century. Today, little has changed, and its left-wing *Sandinista* government is fighting a guerrilla army which survives on aid and armaments sent by the United States.

Nicaragua lies between the Pacific and the Caribbean on the isthmus of Central America, bordered to the north by Honduras and to the south by Costa Rica. The Caribbean coast in the east is forested lowland and the wettest part of Nicaragua. Behind this lies a triangle of volcanic mountains, rising in the north to 2438 m (7998 ft). West of this, a belt of savannah lowland runs parallel to the Pacific coast, south-east from the Gulf of Fonseca. This region, containing the two huge lakes of Managua and Nicaragua, is where nine in ten Nicaraguans live. Between these lowlands and the Pacific coast is another range of volcanic mountains, and the whole area is subject to devastating earthquakes. In both 1931 and 1972 MANAGUA, the capital city, was almost totally destroyed by two disastrous 'quakes.

The east coast was colonised by the British in the 18th century, and Jamaican slaves were brought in to work plantations. Today the east coast population is largely descended from these black slaves and from the original Indian inhabitants of Nicaragua – the Niguiranos, Chorotegas, Chontales and Caribs. The rest of the population (nearly 80 per cent) are *mestizo* – of mixed white and Indian descent. The north-east is the home of the Miskitos, who are descended from a mixture of black slaves and Caribs. These are the people most threatened by the conflict that now racks Nicaragua. It is in the north-eastern hills that government and guerrillas are fighting. Because of this, the Miskitos are being resettled by the government farther south – a move which has attracted international criticism because of the way it has been enforced.

The Sandinista government came to power after a revolution in 1979, which overthrew a repressive military regime that had been dominated by one family – the Somozas – since the 1930s. The Sandinistas are called after General Augusto César Sandino, a patriot who was assassinated by the Somozas in 1934. Before the revolution, the United States backed the Somoza military regime against the left-wing guerrillas. It trained the National Guard – the force of police and soldiers by which the Somozas kept their hold on power. Today the United States supports the counter-revolutionaries, or *Contras* – a number of whom are former members of the National Guard – fearing the spread of Communism through its 'backyard'.

The Contras are trained by US advisers in neighbouring Honduras, and it is with US help that Nicaragua's ports have been mined. The United States, formerly Nicaragua's main trading partner, also introduced a trade embargo in 1985, refusing to accept imports of Nicaraguan sugar.

This is not the first time in Nicaragua's history that the United States has intervened. An American adventurer, William Walker, invaded in 1855, declared himself president, and stayed in power until the US navy ousted him in 1857. In 1912 US marines entered a country torn by conflict between the two major cities of Granada and Léon, and stayed in occupation until 1933. It was the US occupiers that General Sandino fought to remove, and it was they who first appointed a Somoza as head of the National Guard.

The Nicaraguans greeted the fall of the Somozas in 1979 with enthusiasm, but the war that had brought it about cost them dear. Much of the country's agriculture, on which its economy depends, had been destroyed, and continues to be damaged by subsequent fighting. The country's main exports are coffee, cotton, meat and sugar, produced in Nicaragua's lush valleys and fertile plains. Good land is plentiful, but resources are not. Further damage was caused to the economy by the fact that the last General Somoza went into exile with vast sums of Nicaraguan money that he had misappropriated. The Somozas also possessed large tracts of Nicaraguan land, which has now been more evenly redistributed among the people – in small landholdings they work and own themselves, or in collective farms. Health and education reforms were impressive under the Sandinistas, but in parts of Nicaragua health care is still inadequate.

Nicaragua has some deposits of copper, gold and silver, and a few industries such as chemicals and food processing. However, resources are being drained by the continued fighting against the Contras, and with the US embargo curtailing exports, Nicaragua's economy is in crisis. Severe austerity measures have been introduced and there have been huge price rises with wages lagging behind. Nicaraguans are already somewhat cut off from the rest of the world because airline companies will not accept the national currency for air tickets. A black market thrives, and the exchange rate for foreign currency is astronomically expensive for Nicaraguans. The country's deadlock continues as long as the United States sees a chance of overthrowing a government it regards as dangerous.

NICARAGUA AT A GLANCE

Area 130 000 km² (about 50 200 sq miles)

Population 3 140 000

Capital Managua

Government One-party republic

Currency Còrdoba = 100 centavos

Language Spanish; some English and Indian dialects

Religion Christian (95% Roman Catholic, 4% Protestant)

Climate Tropical; temperature throughout the year in Managua ranges from 20°C (68°F) to 34°C (93°F)

Main primary products Coffee, cotton, rice, maize, sugar, bananas, cattle, timber; gold, silver, copper

Major industries Agriculture, chemicals, textiles, cement, mining, food processing, forestry

Main exports Cotton, coffee, meat, sugar, chemicals, fish

Annual income per head (US$) 800

Population growth (per thous/yr) 35

Life expectancy (yrs) Male 55 **Female** 59

Africa. Summer rains and melting snows in Ethiopia cause the annual flooding of the river – now controlled by the ASWAN HIGH DAM – that enabled the flourishing civilisation of ancient Egypt to develop in its valley. The Nile is navigable for most of the way through Sudan and Egypt, except in the low-water season between Khartoum and Lake Nasser (formed by the Aswan High Dam), where there are a series of rapids, the Cataracts.

Map Africa Gb

Nilgiri Hills (Nilgiris) ***India*** Compact group of plateaus rising to 2636 m (8647 ft) at Doda Betta, a precipitous peak about 400 km (250 miles) north-west of the mainland's southern tip. Much is open, rolling country, and before independence in 1947, the British maintained an English-style hunt at UDAGAMANDALAM. About half the area is forested, and there are plantations producing coffee, tea, rubber and medicinal plants.

Map India Be

Nimba Mountains ***West Africa*** Mountain range in south-east Guinea and north-east Liberia, along the Ivory Coast border. It forms part of the Guinea Highlands and rises to 1752 m (5748 ft) at Mount Nimba, where the three frontiers meet. The range contains high-grade iron ore, and diamonds are also exploited. It is rich in wildlife, including hippopotamuses, crocodiles, boars, jackals, various members of the cat family, monkeys and baboons.

Map Guinea Cb; Liberia Ba

Niger

THE SAHARA IS ADVANCING IN A COUNTRY CRIPPLED BY DROUGHT, AND ANCIENT CULTURES ARE IN DANGER

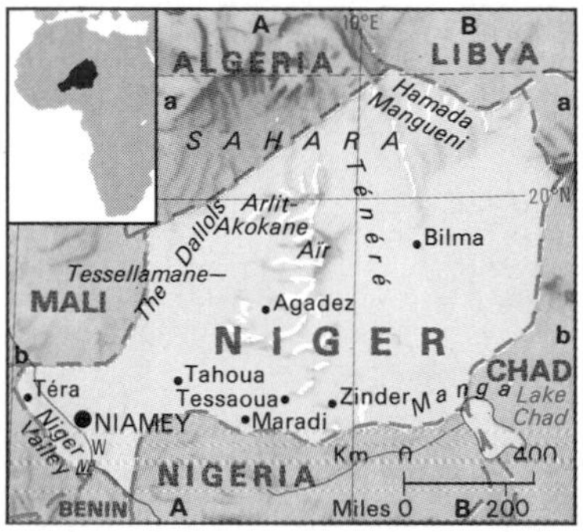

In the 15th century, the landlocked African state of Niger was the site of two rich kingdoms: the Songhai Empire in the south-west and the Hausa states in the south-east. These two corners of the country are still the most densely populated.

In the extreme south-west the NIGER river and its tributaries bring life to the landscape, and the south-east has Lake Chad, but the rest of the country is short of water. Over half of Niger in the north is covered by the encroaching SAHARA, while to the south lies the drought-stricken SAHEL.

The capital NIAMEY stands on the Niger in the south-west. It is an elegant town with wide, tree-lined roads and a modern centre, where busy markets add African colour to the remnants of French colonial rule, pavement cafés and French restaurants. Colonial rule began in the 1890s and ended in 1960.

Near the south-western border, the Niger river passes through the 'W' NATIONAL PARK, so called because the river runs in a 'W' shape. The elephants, buffaloes, antelopes and lions which roam the park's wooded savannah are protected. Before reaching the park, the river runs past villages of square, mud houses with flat roofs and tiny windows. Every settlement has its own mosque, with a tall minaret, elegant despite being made of mud. The people are mostly Muslim, although pagan tribal practices are observed.

Mainly members of the Djerma and Songhai tribes, they fish and farm their own food, growing rice and vegetables on land flooded by the river, and scattering millet and sorghum seed farther afield. In addition, cotton and groundnuts are grown for sale. Harvests away from the river depend on the erratic rains between June and October, and since there has been intermittent drought from 1968, the millet and sorghum crops have failed again and again. Food supplies have been decreasing for years, and even emergency stocks have now come to an end. Hunger drives people to the towns, especially Niamey, in search of work, money and food, and many travel on into neighbouring Nigeria, or even to the Ivory Coast.

The same pattern of drought and migration has affected all the southern region of Niger, eastwards to Lake Chad. Grain stores are low or empty; so are wells and water holes. The people here are Hausa, and they, too, are Muslims with an Islamic society. The movement of population threatens to destroy this culture too, and towns such as ZINDER and MARADI are also overcrowded.

TRADITIONAL GRAZING GROUNDS

Population in the north is concentrated around the AIR mountains, whose jagged peaks rise to 1944 m (6378 ft) from the desert flatness. Valleys are traditional grazing grounds for the camels, cattle, horses and goats of nomadic Tuareg tribesmen. However, since 1973 whole clans have been wiped out by drought. In the mountains there are hot springs, and caves with magnificent rock drawings dating back 5000 years.

Uranium has been mined in the Aïr mountains since the 1970s and is now Niger's main export, superseding groundnuts. The mines have taken metalled roads into the desert where once there were only trails.

A few caravans are still taken into the desert by the Tuaregs: one of their main trails runs from AGADEZ to the oasis of Fachi, where they collect dates, then to Bilma, where they load their camels with salt.

The challenge for the government, a military regime which came to power in 1974 but which, since 1983, has been headed by a civilian, is feeding the people, as drought continues and the desert edges southwards. The immediate aim is to develop agriculture, exploit mineral resources and improve education and health.

NIGER AT A GLANCE
Area 1 267 000 km² (489 191 sq miles)
Population 6 700 000
Capital Niamey
Government Military republic
Currency CFA franc = 100 centimes
Languages French (official), Hausa and other local languages
Religions Muslim (85%), Animist (14%), Christian (0.5%)
Climate Hot and dry; average temperature in Niamey ranges from 14-34°C (57-93°F) in January to 27-41°C (81-106°F) in May
Main primary products Millet, sorghum, groundnuts, cotton, rice, livestock; uranium, tin, phosphates
Major industries Agriculture, textiles, food processing, mining
Main exports Uranium, groundnuts, livestock, hides and skins
Annual income per head (US$) 300
Population growth (per thous/yr) 33
Life expectancy (yrs) Male 43 **Female** 46

nimbostratus Low thick cloud formation forming a uniform dark mass from which rain or snow falls.

nimbus General name for a thick cloud from which rain falls, often used in combination with other cloud names, as in cumulonimbus.

Nimegue ***Netherlands*** See NIJMEGEN

Nîmes ***France*** City overlooking the Rhône river, 100 km (62 miles) north-west of Marseilles. It was an important Roman settlement from 120 BC until the 5th century AD, and many remains survive. They include the Pont du Gard, a three-storey aqueduct across the Gardon Valley, and an amphitheatre that can hold 21 000 people. The manufacture of textiles and clothing is one of the city's chief industries, and denim gets its name from a cloth made there: *serge de Nîmes*.

Population 138 000
Map France Fe

Nimwegen ***Netherlands*** See NIJMEGEN

Nin ***Yugoslavia*** See ZADAR

Nineveh ***Iraq*** Remains of an ancient Assyrian capital near the city of Mosul, some 350 km (220 miles) north and slightly west of the capital, Baghdad. Mentioned often in the Bible, the city was the hub of an empire in the 7th century BC under the Assyrian kings Sennacherib and Assurbanipal. Excavations there have revealed palaces and a library which contained Sumerian and Assyrian dictionaries, giving scholars an invaluable key to these languages.

Map Iraq Ba

Ningbo ***China*** City of Zhejiang province, 60 km (37 miles) east of Hangzhou. It is a major seaport and a growing industrial centre with engineering and light industries.

Population 900 000
Map China Ie

Ningxia-Hui Autonomous Region ***China*** Region of northern China between Nei Mongol and the province of Gansu, stretching south from the Huang He river. The southern part lies in the LOESSLANDS OF CHINA. Ningxia-Hui covers 60 000 km² (23 000 sq miles) and has extensive pastureland which is the home of the argal sheep, kept for its high-quality wool. Irrigation has opened up some of the semiarid land for rice and wheat cultivation, particularly along the Huang He. Ningxia-Hui also has deposits of coal.

Population 3 640 000
Map China Fd

Niokolo-Koba National Park ***Senegal*** Wildlife reserve in the south-east covering about 9000 km² (3475 sq miles). The reserve surrounds the village of Niokolo, about 55 km (35 miles) from the Guinea border, and consists mostly of savannah (tropical grassland) with forest along the Gambia river and its tributaries. It contains 70 species of mammals (including elephants, hippopotamuses and antelopes), 300 species of birds and 60 species of fish and reptiles.

Map Senegal Bb

Nigeria

A COUNTRY WITH A MIXTURE OF DIFFERENT PEOPLES AND THEIR PASTS PAINFULLY REALISING THE REALITIES OF NATIONHOOD

Known as the 'giant of West Africa', Nigeria – the region's largest country and Africa's most populous nation – is less than 30 years old. Changes have been forced at breakneck pace, partly due to political upheaval and partly due to the discovery of oil and natural gas which have come to dominate the nation's economy. LAGOS, a teeming capital of skyscrapers, fly-overs, slums and up to 4 million people, is a testament to Nigeria's new riches, and the city has tended to colour the country's image in the eyes of foreigners. But, despite the blare of klaxons and the roar of cars and taxis, Nigeria is a nation where the brash, bustling present coexists with a very persistent past.

Nigeria is a collection of different peoples and separate cultures thrown together by the British, without consent, and welded into a nation. Although many Nigerians are not keen to admit it, tribal tensions lie beneath the surface and have in the past erupted in terrible conflicts such as the Biafran War of 1967-70.

It is this diversity that has bedevilled all attempts to assess Nigeria's population accurately – since each ethnic group wishes to emphasise its own size. The population is believed to be over 94 million and consists of some 250 ethnic groups. Of these, four – the Hausa and the Fulani in the north, the Yoruba in the south-west and the Ibo in the south-east – account for around 65 per cent of the population.

A VARIED LANDSCAPE

The landscapes that these peoples inhabit are equally varied: from the coastal lands of the south through the tropical forest belts of the interior to the mountains and savannahs of the north. The southern coast is fringed by miles of idyllic sandy beaches, broken only by mangroves at those points where watercourses trickle into the sea. Back from the coast, a maze of creeks and lagoons fans out on either side of the NIGER delta. It was these inland waterways which provided a means of transport into the interior before roads were built. Today, several channels are silted up; but dugout canoes and small boats still carry people and goods between villages and towns such as Badagri and Calabar. Farther north, the forest belt which produces Nigeria's tree crops (cocoa, palm oil, rubber and hardwoods) gradually gives way to grassy parkland. The land rises: past the junction of the two great rivers, the Niger and BENUE, towards the mountains and holiday resorts of the central JOS PLATEAU; and, in the east, towards the steep valleys and wooded slopes of the Cameroon highlands. North of the Jos Plateau, the savannahs become increasingly dry and grassy; and the skeletons of animals are a constant reminder that the drought of the SAHEL can reach this far south.

Inevitably the contrast in landscape between northern and southern Nigeria is reflected in the livelihoods of the country's peoples. The heavy rains that fall in the south encourage the growth of yams, cassava, maize and a wide range of vegetables. Rice is grown on tiny patches of land beside rivers and streams which flood in the wet seasons, and on large stretches of irrigated land.

The north, however, is not so self-sufficient. It is dependent on one rather unreliable rainy season, whereas the south has two. Cereals, such as millet, sorghum and maize, are the staples of the northern economy; with groundnuts and cotton important too. However, the north, unlike the south, is not plagued by tsetse fly because it lacks the swamps and vegetation that are its natural habitat. Livestock can therefore play a vital part in the economy.

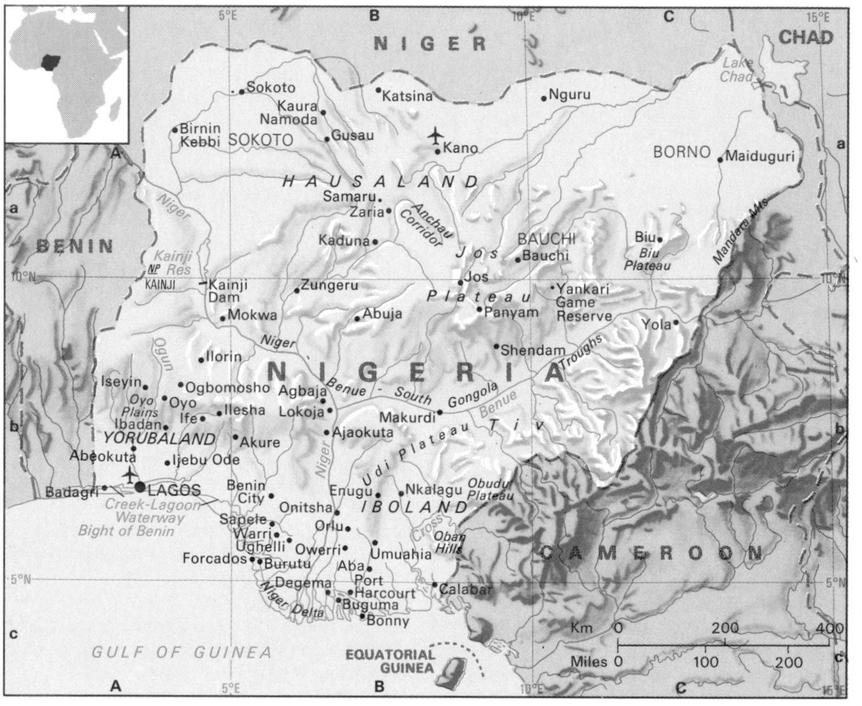

Before the arrival of the European powers in the 15th century, economic activity in West Africa – including what is now Nigeria – focused on the Sahel. City-states, such as SOKOTO, KATSINA and KANO, grew up along the fringes of the SAHARA and thrived on trading northwards across the desert. The Europeans, however, shifted this focus to the coast; and there it has remained ever since. At first they traded in gold and ivory, but it was not long before they switched to slaves.

When the British outlawed the slave trade in 1807, European merchants were forced to turn to other commodities, particularly palm oil. This was used in soap, a product suddenly much in demand by workers in the grimy factories of Britain's Industrial Revolution. Cocoa, rubber, timber, groundnuts and cotton soon followed as leading export products, transforming the economy of Nigeria from one in which land was farmed for subsistence into one in which land was farmed for profit. The British flag followed trade; and in 1914 – purely for the administrative convenience – the northern and southern parts of the country were merged into what is now Nigeria.

The British – who from the start intended to return Nigeria to the Nigerians – adopted a system of indirect rule for its colony. This involved the local chiefs in administration, in gathering raw materials for export and in tax collection – an arrangement which disrupted, but did not totally destroy, the traditional system. Roads and railways were constructed to transport produce to the coast, and today Nigeria has a denser transport network than any other country in West Africa.

INDEPENDENCE ACHIEVED

After years of negotiation, Nigeria was granted independence in 1960. But the nation still consisted of three distinct and hostile regions; north, south-west and south-east. National politics degenerated into a vicious struggle for power between the Hausa-Fulani, Yoruba and Ibo tribes. In June 1967 the eastern region, in which the oil fields were situated, declared its secession from Nigeria as the independent Ibo state of Biafra. A violent civil war broke out and – in one of the most tragic events in the history of modern Africa – over a million people died as the Ibos battled against the federal forces before succumbing to a famine which laid waste their land.

Although Nigeria was bequeathed a

▲ **CLOSE-KNIT COMMUNITY Many Nigerians still follow a tribal way of life in mud-hut villages. They live in large, extended family groups, many of which produce only enough food for their own needs.**

Westminster model of government by the British, military juntas ruled for 15 out of the first 25 years following independence, and accounted for six out of the eight regimes. There was general rejoicing, therefore, when in October 1979 the country returned to civilian rule under President Shehu Shagari. But the jubilation was short-lived. On December 31, 1983, another military takeover put an end to four years of corruption, economic mismanagement and the undermining of the electoral process through violence and fraud. The new government, led by Major General Muhammad Buhari, was pledged to eliminate the sort of ministerial corruption which, it is estimated, has siphoned off US$7000 million from Nigeria into the security of Swiss and London bank accounts. Death sentences by firing squad were mandatory for offences which ranged from armed robbery to tampering with electricity or telephone cables. Jail sentences of 21 years faced those over the age of 18 who cheated in exams. Another coup in August 1985 brought Major General Ibrahim Badamas Babangida to power

Under the British, Nigeria's economy depended on the profitable export of cash crops. As a result, by 1973 Nigeria was, after Ghana, the world's second largest producer of cocoa.

Today, although it is the country's only agricultural export of any significance, cocoa production has declined dramatically. Similarly, until 1972, Nigeria was Africa's leading groundnut producer, but since 1975 no groundnuts have been exported at all. And finally, the country was once a major exporter of cotton and palm oil but is now reduced to importing these very products.

Several factors have conspired to produce this decline. Drought and low prices to the producer fixed by the Nigerian Federal Marketing Board in the early 1970s were an important cause of low output. Initially, however, the drop in revenue from crops was masked by the impact of oil. Between 1973 and 1981 Nigeria was transformed from a rural country with an agricultural economy to one in which oil accounted for 90 per cent of revenues.

Some of Nigeria's new wealth has been channelled back into the country's ailing rural economy – in the form of large, Western-style development projects. But progress has been painfully slow. The Savannah Sugar Company, for example, was taken over by the government in 1972; by 1984 it was on the verge of collapse as sugar imports soared. Rice production has increased through major development schemes but, due to appalling management problems, is still far from meeting domestic demand.

Nigeria's dependence on petroleum inevitably links the country's fortunes with the fluctuations of the world oil market. In 1981, for example, when the oil price slumped, Nigeria was precipitated into a recession from which it had still not recovered by 1986, when a further fall in prices added an even greater burden to the ailing economy.

NIGERIA AT A GLANCE

Area 923 768 km² (356 669 sq miles)
Population 94 300 000
Capital Lagos
Government Federal republic under military rule
Currency Naira = 100 kobo
Languages English (official), Hausa, Ibo, Yoruba
Religions Muslim (48%), Christian (34%), tribal (18%)
Climate Tropical; wettest month June. Average temperature ranges from 23°C (73°F) to 32°C (90°F)
Main primary products Groundnuts, cotton, cocoa, rubber, maize, millet, sorghum, cassava, rice, yams, palm kernels, livestock, timber; oil and natural gas, tin, coal, iron ore
Major industries Oil and natural gas production and refining, agriculture, mining, petrochemicals, fertilisers, motor vehicles, cement, food processing
Main exports Crude oil and refined products, natural gas, cocoa, tin, timber and hides
Annual income per head (US$) 740
Population growth (per thous/yr) 34
Life expectancy (yrs) Male 48 **Female** 51

Nios ***Greece*** See IOS

Nis (Nish) ***Yugoslavia*** Ancient east Serbian city 201 km (125 miles) south-east of Belgrade, situated on the Nisava river near its confluence with the Southern Morava. A strategic Roman town commanding the major trade routes between the central Danube plains and the Aegean and Bosporus, it was the birthplace of Constantine the Great (about AD 280-337), Rome's first Christian emperor. Attacked in turn by Goths, Huns, Bulgars, Hungarians and Byzantines, it nevertheless prospered under the Turks from 1456 to 1877 as a key caravanserai between Constantinople and Hungary. The Turks did have some opposition, though. In 1809 they put up the grim Cele Kula ('Tower of Skulls') near the city. The tower's brickwork is studded with the skulls of 900 Serbian rebels, set there by the Turks as a warning to other dissidents.

Nis is still a hub of central European road and rail routes, and an important manufacturing centre producing machinery, electrical appliances, X-ray equipment, ceramics, tobacco goods and foodstuffs.

Population 230 700
Map Yugoslavia Ec

Nishapur (Neyshabur) ***Iran*** Town about 700 km (435 miles) east of the capital, Tehran. Omar Khayyam, the Persian poet, mathematician and astronomer (about 1050-1123), was born there. His tomb, set in a garden, is about 3 km (2 miles) to the south-east.

The district around the town, which is surrounded by orchards, makes pottery and grows fruit, cereals and cotton. Turquoise, mined in the mountains to the north, is cut and polished in workshops in the town.

Population 59 200
Map Iran Ba

Niterói Bridge ***Brazil*** Bridge linking the city of Niterói with Rio de Janeiro. The bridge is also, and more formally, known as the Ponte Costa é Silva. At 14 km (8.5 miles), it is one of the world's longest bridges and was completed with the aid of British engineers in 1974. At its highest point, the bridge soars 72 m (236 ft) above the waters of Guanabara Bay.

Map Brazil Dd

Nitra ***Czechoslovakia*** Market city about 75 km (50 miles) north-east of Bratislava. It lies among rich farmlands, orchards and vineyards, and has food-processing, brewing, wine-making, engineering and electrical industries. The cathedral incorporates Slovakia's oldest church, consecrated about 11 centuries ago.

Population 82 000
Map Czechoslovakia Db

Niue ***Pacific Ocean*** Raised coral island covering 259 km² (100 sq miles) in the South Pacific about 2100 km (1300 miles) north-east of New Zealand. Its population is Polynesian. It was discovered by Captain James Cook in 1774 and became a British territory in 1900. It was then transferred to New Zealand, but became self-governing in 1974. As with the Cook Islands, New Zealand handles foreign affairs and defence, and contributes aid. The main exports are passionfruit, copra, honey and limes. The population of Niue is dropping as people leave to settle in New Zealand. The capital is Alofi.

Population 3000
Map Pacific Ocean Ec

Nizhni Novgorod ***USSR*** See GOR'KIY

Nizhniy Tagil ***USSR*** Industrial city in the Ural mountains, 125 km (78 miles) north of Sverdlovsk. It was founded in 1725 and is close to gold, copper and iron-ore mines. Its main products are chemicals, machinery, iron and other metals.

Population 415 000
Map USSR Gc

Nizina ***Bulgaria*** See THRACE

Njazidja ***Comoros*** Largest island of the Comoros group, covering 1147 km² (443 sq miles); formerly Grande Comore. The islands' capital, MORONI, is situated on the west coast.

Population 200 000
Map Madagascar Aa

Nkongsamba ***Cameroon*** Picturesque town in a narrow valley about 220 km (135 miles) north-west of the capital, Yaoundé. It is a busy coffee market, and the home of some of the country's richest people. Fog caused by the nearby active volcanoes is common. The area is popular with tourists, who visit the waterfall at Ekom and Mount Manengouba.

Population 80 000
Map Cameroon Bb

Nong Khai ***Thailand*** Mekong river town beside the Laotian border, about 575 km (357 miles) north and slightly east of the capital, Bangkok. The town – which is only about 25 km (15 miles) downstream from the Laotian capital, Vientiane, on the opposite bank of the river – is a railhead for goods being transported to and from Laos. The main crops in the surrounding province of Nong Khai are tobacco and rice.

Population (town) 138 300; (province) 749 800
Map Thailand Bb

Nonthaburi ***Thailand*** Province and town just north of Bangkok that are rapidly becoming engulfed by the growing capital. The province was once renowned for its durians – a fruit said to have the world's most exquisite taste, but which even its devotees agree has a foul smell. But only a few durian orchards have survived the metropolitan sprawl.

Population (town) 31 000; (province) 373 000
Map Thailand Bc

Noord-Brabant ***Netherlands*** Southern province covering 5107 km² (1971 sq miles) on the Belgian border. Much of the south is infertile heath and forest, and sheep and cattle rearing are the main agricultural activities. However, recent industrial development, especially around EINDHOVEN, has brought an increase in population and prosperity. The capital is 's-Hertogenbosch. The province was part of the old duchy of Brabant, the southern part of which now forms Belgium's northern provinces, including Belgian Brabant.

Population 2 094 000
Map Netherlands Bb

Noordzeekanaal ***Netherlands*** Canal opened in 1876 between Amsterdam and the North Sea at IJmuiden. It is about 27 km (17 miles) long, and shortened the sailing route to Amsterdam, formerly reached by the Zuiderzee. Because the land the canal crosses is below sea level, locks lower rather than raise vessels at the entrance.

Map Netherlands Ba

Nordkapp ***Norway*** See NORTH CAPE

Nördlingen ***West Germany*** Medieval town in Bavaria, about 80 km (50 miles) south-west of Nuremberg. Its ancient buildings are still encircled by fortified walls and gates. Each summer it holds a fancy-dress horse race, the oldest festival in Germany, dating from medieval times.

Population 18 300
Map West Germany Dd

Nord-Pas-de-Calais ***France*** Densely populated industrial and farmland region in the far north of the country. Its growth and urbanisation were linked to coal mining, steelmaking, engineering and textiles, but these industries have now declined. New industries include the manufacture of car parts, plastics, chemicals and electronics. It centres on the conurbation of LILLE-ROUBAIX-TOURCOING.

Population 3 933 000
Map France Da

Norfolk ***United Kingdom*** Eastern English county of 5355 km² (2068 sq miles) occupying the northern bulge of East Anglia. Between the coast and the county town, NORWICH, lie the Norfolk Broads, a series of river-linked lakes, partly natural but largely created by medieval peat digging, which are popular for sailing and water sports. The coast is lined with holiday resorts. The largest of these is GREAT YARMOUTH, which is also a fishing port and supply base for North Sea gas rigs.

Population 714 000
Map United Kingdom Fd

Norfolk ***USA*** Port in the Atlantic state of Virginia at the mouth of Chesapeake Bay. Its deep natural harbour has a naval dockyard and docks for coal and grain.

Population (city) 279 700; (metropolitan area, with Newport News) 1 261 200
Map United States Kc

Norfolk Island ***Pacific Ocean*** Island, 35 km² (13 sq miles) in area, lying about 1400 km (875 miles) east of Brisbane, Australia. Discovered by the British explorer Captain James Cook in 1774, it was a penal colony from 1783 to 1803 and was settled by families of mutineers from HMS *Bounty* (who moved there from PITCAIRN) in 1856. It is administered as an Australian external territory with a considerable degree of self-government. Tourism is the main industry. Residents pay no taxes.

Population 2500
Map Pacific Ocean Dd

Noril'sk ***USSR*** Mining town in the Siberian Arctic about 80 km (50 miles) east of the Yenisey river. It was founded in 1935 as part of what became known as the Gulag Archipelago – the chain of prison camps described in the novel of the same name by the exiled Russian author

Alexandr Solzhenitsyn (1918-). Now populated by volunteers rather than exiles and convicts, it lies 380 km (235 miles) north of the Arctic Circle near coal, copper, cobalt, nickel, gold and platinum mines.
Population 183 000
Map USSR Jb

Normandy ***France*** Region in the north-west of the country – between Brittany and Picardy – bordering the English Channel. It was the site in June 1944 of the D-Day invasion of France, which began the liberation of German-occupied western Europe.

The province is noted for its cream, cheeses, cider and calvados, an apple brandy. Its major cities are CAEN, LE HAVRE and ROUEN.
Population 3 006 000
Map France Cb

Norrköping ***Sweden*** Industrial seaport on a Baltic inlet 157 km (98 miles) south-west of Stockholm. Founded in the 14th century, it was burnt by the Russians in 1710 during the Great Northern War of 1700-21. Its manufactures include paper, woollen textiles, chemicals, rubber, and radio and television sets.
Population 118 600
Map Sweden Cd

North Albanian Alps ***Albania/Yugoslavia*** See ALBANIAN ALPS

North Cape (Nordkapp) ***Norway*** One of Europe's most northerly points – 500 km (310 miles) north of the Arctic Circle. It is a promontory rising 300 m (1000 ft) on Magerøya (Mager island), about 95 km (59 miles) north-east of Hammerfest. It is not quite Europe's northernmost cape, being just beaten by the nearby, but less accessible, Knivskjellodden. North Cape is a popular tourist destination for visitors wanting to see the midnight sun in summer.
Map Norway Ga

North Carolina ***USA*** State covering 136 198 km² (52 586 sq miles) beside the Atlantic Ocean south of Virginia. The land rises to the Appalachian Mountains, which reach 2037 m (6684 ft) at Mount Mitchell in the far west. Tobacco, chickens, soya beans, maize and peanuts are the main agricultural products, and phosphates are the chief mineral. North Carolina is the nation's leading producer of textiles and cigarettes, and also has electronics, engineering and chemicals industries. It was one of the original 13 states. RALEIGH is its capital and Charlotte the biggest city.
Population 6 255 000
Map United States Kc

North Dakota ***USA*** Midwestern state on the Canadian border. It covers 183 022 km² (70 665 sq miles) and stretches westwards from the Red River to the high plains on the border of Montana. Agriculture is the main livelihood, producing wheat, beef, oats and barley; coal, oil and lignite are the chief minerals. Originally the home of various Indian tribes, including the Sioux, it became a state in 1889. BISMARCK is the state capital.
Population 685 000
Map United States Fa

North Field ***Qatar*** Natural gas field in The Gulf, lying 70 km (43 miles) north-east of the northern tip of the Qatar peninsula. Discovered in 1972, it has estimated reserves of some 8.5 million million million m³ (300 million million million cu ft) of gas, and could be in production for the next 100 years. The field, one of the largest in the world, covers an area of 2590 km² (1000 sq miles).
Map Qatar Aa

North Island ***New Zealand*** Smaller of the country's two principal islands, covering 114 685 km² (44 281 sq miles). Its population is almost three times that of the larger South Island. The centre of the island is mountainous, but there is also rich pastureland with some 37.5 million sheep – more than half the country's total. The main cities are Auckland, Hamilton and the capital, Wellington.
Population 2 323 000
Map New Zealand Eb

North Land ***USSR*** See SEVERNAYA ZEMLYA

North Pole Point at the northern extremity of the earth's axis of rotation, located in the shifting ice-packs of the ARCTIC. Most authorities agree that the first person to reach the pole was a US naval officer, Robert Peary, on April 6, 1909, but some dispute his claim. Today, thousands of airline passengers fly over or near the Pole on intercontinental flights every day, while in 1958 a US nuclear submarine surfaced through the pack-ice there.
Map Arctic a

North Rhine-Westphalia ***West Germany*** Westernmost state of the republic, formed in 1946 by the union of two ancient provinces on the borders of Belgium and Holland. The union's main effect was to place the huge industrial conurbation of the Ruhr under a single administration.
Population 16 704 000
Map West Germany Bc

North Sea Shallow sea, also once called the German Ocean, lying between Britain and mainland Europe. It is a major shipping lane, a rich fishing ground and a source of oil and natural gas, the main beneficiaries of these fuels being Britain and Norway. The highest tidal range is 4-6 m (13-20 ft) near The WASH. High tides combined with storm winds occasionally raise water levels to dangerous heights. The worst recent floods occurred in the Netherlands and eastern England in January 1953.
Area 575 000 km² (222 000 sq miles)
Greatest depth 661 m (2170 ft)
Map United Kingdom Dc

North Slope ***USA*** Region between the Brooks Range and the Beaufort Sea in northern Alaska with enormous reserves of oil and natural gas. The crude oil is carried by the Trans-Alaska pipeline to the port of Valdez on the Pacific coast. Many caribou live in the region.
Map Alaska Bb

North West Cape ***Australia*** Promontory on the coast of Western Australia, about 1125 km (700 miles) north of Perth. It is the site of a US naval communications base. The greater part of the cape, about 506 km² (195 sq miles), is occupied by the Cape Range National Park.
Map Australia Ac

Northampton ***United Kingdom*** Town and county of central England. The county – 2367 km² (914 sq miles) in area – straddles the south Midlands and is rich in ironstone, which was used to feed the blast furnaces at the steel town of Corby until they were closed. It is a famous shoemaking and fox-hunting area. The town, 105 km (65 miles) north of London, has become a residential area for the capital because of improved motorway and rail links. It has a splendid 18th-century market square, one of the biggest in England, and a round church founded in the 12th century. It was the scene of a battle during the Wars of the Roses in 1460, which resulted in the capture of Henry VI and the Yorkists' proclamation of Edward IV as king.
Population (county) 547 000; (town) 164 000
Map United Kingdom Ed

norther Sudden cold, dry north wind, bringing low temperatures to Texas, the Gulf coast and other parts of southern North America, usually following a depression. It often blows with great violence, reaching up to 100 km/h (over 60 mph), may be accompanied by severe thunderstorms and hail, and temperatures may fall by as much as 20°C (36°F) in 24 hours. When it reaches Central America it is called a *norte*.

Northern Ireland ***United Kingdom*** Constituent province of the UK, covering some 14 121 km² (5452 sq miles). It was established in 1921 when IRELAND divided, and consists of six counties – ANTRIM, ARMAGH, DOWN, FERMANAGH, LONDONDERRY and TYRONE – of the ancient province of Ulster. Since 1972 it has been governed directly from London. Its capital is BELFAST.
Population 1 572 000
Map United Kingdom Bc

Northern Marianas See p. 478

Northern Province ***Zambia*** Province bordering Malawi and Tanzania. The provincial capital, Kasama, is about 660 km (410 miles) north-east of the national capital, Lusaka. The home of the Bemba-speaking people, the province covers 147 826 km² (57 076 sq miles) of largely infertile soil. The most productive farming area is around the town of Mbala, where coffee, fruit, maize and tobacco are grown.
Population 677 900
Map Zambia Cb

Northern Territory ***Australia*** Largely self-governing territory covering 1 346 200 km² (519 770 sq miles) in the central north of the country with a tropical climate. It was originally part of New South Wales and then South Australia before becoming a Commonwealth (federal) government territory in 1911. It is the heartland of Australia's Aboriginal people.

The chief town is DARWIN, where nearly half the population live. The semiarid scrub that covers most of the territory is given over mainly to cattle ranching and mining. One-third is Aboriginal territory. Groote Eylandt, an island in the Gulf of Carpentaria and part of ARNHEM LAND Aboriginal Reserve, contains Australia's

Northern Marianas

A COMMONWEALTH OF THE USA – PERHAPS A FUTURE STATE – NEARER TO ASIA THAN NORTH AMERICA

When the Trust Territory of the Pacific Islands – those far-flung specks in the western Pacific entrusted to American care by the United Nations in 1947 – was negotiating its future in the 1970s, one group of islands opted for close American ties rather than full independence. Thus, in 1978, the Commonwealth of the Northern Marianas was born, with the same self-governing status as Puerto Rico and its people full US citizens. The way is open for full statehood in the future, perhaps in conjunction with GUAM, the southernmost of the Mariana Islands, which is a separately administered US territory.

There are 14 islands and islets in the commonwealth, stretching in a curving chain 725 km (450 miles) long, about 2400 km (1500 miles) east of the northern Philippines. They are volcanic, Agrihan (the highest) reaching 959 m (3146 ft) and Pagan having hot springs and an active volcano that erupted violently in 1981.

The most important islands are SAIPAN, TINIAN and Rota – the first two well known for the fierce Second World War battles when US forces took them from the Japanese in June 1944, and for their subsequent role as military bases. After the war, Saipan became the administrative centre for the whole Trust Territory, and the Marianas people benefited economically as a result.

Most of the people are Chamorros of mixed ancestry, including Spanish and Filipino. The islands were ruled by Spain from 1668 to 1899 and are named after a regent of Spain, Mariana of Austria. There are no pure-bred Micronesians – the indigenous inhabitants. Before the Second World War, Saipan and Tinian grew much sugar cane, but today even subsistence farming is a thing of the past. There are a few full-time farmers and fishermen, and Japanese and Korean boats fish the Marianas' waters.

Most of the workforce is employed by the government or serves the booming tourist industry (in 1984, 131 000 people visited the territory) which provides most of the islands' income. The tourists are mainly Japanese, who come to see wartime relics. It was from Tinian that US B-29 bombers flew to drop atomic bombs on HIROSHIMA and NAGASAKI in August 1945.

The main centre remains Saipan, which contains 86 per cent of the population and the capital, Susupe.

NORTHERN MARIANAS AT A GLANCE

Map Pacific Ocean Cb
Area 476 km^2 (184 sq miles)
Population 20 000
Capital Susupe
Government Self-governing commonwealth of the USA
Currency US dollar = 100 cents
Languages English, Chamorro, Carolinian
Religion Christian (mainly Roman Catholic)
Climate Tropical; average temperature at Susupe ranges from 22-27°C (72-81°F) in January to 24-29°C (75-84°F) in July
Main primary products Coconuts, vegetables, fruit, fish
Major industry Tourism
Annual income per head (US$) 4240
Population growth (per thous/yr) 35
Life expectancy (yrs) Male 69 **Female** 69

largest manganese deposit. Bauxite (an ore of aluminium) is found along the gulf shore, and there are huge uranium reserves in the Alligator river area in Arnhem Land. There have been oil finds offshore in the Timor Sea in an area also claimed by Indonesia.
Population 136 800
Map Australia Eb

Northland ***New Zealand*** Peninsula on North Island, stretching from just north of Auckland to Cape Reinga at the island's northern tip and covering 12 649 km^2 (4883 sq miles). Tourism, agriculture and an oil refinery at Marsden Point are the main industries. The largest city is WHANGAREI.

On the western side is the Waipou State Forest, one of the few remaining kauri pine tree forests in the country. It contains the largest tree in New Zealand, 52 m (170 ft) tall and 14 m (46 ft) in girth. It is known by the Maori name of Tane Mahuta (God of the Forest), and is estimated to be more than 1200 years old.
Population 114 300
Map New Zealand Da

Northumberland ***United Kingdom*** County of north-eastern England, on the border with Scotland, extending over 5033 km^2 (1943 sq miles). Centuries of conflict between the two countries have given the county a frontier spirit and a wealth of medieval castles, such as those at Alnwick and BERWICK-UPON-TWEED. It also has some of the emptiest but loveliest landscapes in England – particularly in the Cheviot Hills, which rise to 815 m (2674 ft), and in the Kielder Forest. A favourite spot with visitors is Lindisfarne, or Holy Island, with its ruined 11th-century Benedictine priory; the area teems with wildfowl and wading birds. The county town is Morpeth.
Population 302 000
Map United Kingdom Ec

North-West Frontier Province ***Pakistan*** Rugged, mountainous province covering 74 521 km^2 (28 773 sq miles) along the northern part of the Afghan border. It was used as a protective buffer zone for the Indus valley by the British and was ruled indirectly through local tribal chiefs. The majority of the inhabitants are Pathans, long renowned for their skills as guerrilla fighters – skills still exercised in their fierce resistance to the Russian occupation of Afghanistan. The Pathan homeland is split by the international border. The most famous of its many passes is the KHYBER.
Population 11 061 000
Map Pakistan Ca

Northwest Passage Northern sea passage joining the Atlantic and Pacific oceans, through Canada's Arctic archipelago and the waters north of Alaska. First sought by 16th-century English explorers, including Sir Martin Frobisher (1576-8), looking for a short sea route to the Orient, it was not discovered until the 19th century. Norwegian Roald Amundsen was first to navigate it, in 1903-6. The first commercial ship to navigate the passage was the US oil tanker, the *Manhattan*, in 1969.

Northwest Territories ***Canada*** Northern area of Canada, including all the country north of 60°N, except for YUKON and the northern tip of Labrador and Quebec. It covers some 3 246 000 km^2 (1 253 400 sq miles) – nearly one-third of Canada. Mountain ridges rise to more than 2700 m (9000 ft) in the west and over 2600 m (8000 ft) on the eastern islands of Baffin and ELLESMERE. Tundra extends over much of the north and east, and agriculture is very limited. About 60 per cent of the population are Inuits (Eskimos), Indians and Métis (mixed French and Indian). In the past, fur trapping and fishing were the main occupations. Today minerals, including lead, gold, zinc, cadmium and silver, are the main source of employment and wealth. YELLOWKNIFE is the capital.
Population 45 740
Map Canada Db

Norway See p. 480

Norwegian Sea Fishing ground for cod and herring in the North ATLANTIC OCEAN, lying off Norway. Despite its northerly latitude, it is usually ice-free, because the relatively warm Norwegian Current (the most northerly extension of the GULF STREAM) flows through it.
Greatest depth 3667 m (12 030 ft)
Map Atlantic Ocean Ea

Norwich ***United Kingdom*** River port in eastern England, 160 km (100 miles) north-east of London. It is the county town of Norfolk, and during the medieval heyday of the East Anglian wool trade was the country's second city after London. It stands on the River Wensum 30 km (20 miles) from the North Sea. Its riches and reputation attracted many skilled immigrants from Europe between the 14th and 16th centuries. A large 12th-century castle – now a museum – overlooks the city and some sections of a 14th-century city wall remain. There are many beautiful churches and streets, such as cobbled Elm Hill. The Anglican cathedral is mainly 15th century, with a spectacular nave roof. Its spire, at 96 m (315 ft), is the country's second tallest after Salisbury. The city is the home of the modern University of East Anglia.
Population 122 000
Map United Kingdom Fd

Notabile ***Malta*** See MDINA

Nottingham ***United Kingdom*** City and county of the northern Midlands of England. The city, 70 km (45 miles) north-east of Birmingham, grew around an 11th-century castle on a crag above the River Trent. Charles I raised his standard there in 1642 at the start of the English Civil War. The city's real growth came in the 19th century, and it now produces lace, hosiery, bicycles and tobacco. It has a university, and was one of the first towns in Britain to ban private motor vehicles from its centre, diverting them to a ring road.

The historic Goose Fair, which dates back to the Middle Ages, is held in Nottingham every October. The city and the remains of Sherwood Forest near it are inextricably linked with the legendary 12th-century outlaw Robin Hood. The county covers an area of 2164 km² (836 sq miles). Its coalfield is today one of the most productive in Britain and is centred on the towns of Mansfield and Worksop.

Population (county) 1 000 000; (city) 277 000
Map United Kingdom Ed

Nouadhibou ***Mauritania*** Port, formerly called Port Etienne, on Cape Blanc peninsula, on the border with Western Sahara. It exports iron ore from the mines at Fdérik, 670 km (420 miles) away by rail, and is also a fishing port, with fish drying, freezing and canning plants.

Population 26 000
Map Mauritania Ba

Nouakchott ***Mauritania*** The country's capital, standing on the Atlantic coast. Built in the 1950s, it is an attractive city of flat-roofed houses, wide sandy streets and beautiful mosques. It was enlarged during the droughts of the 1970s by a massive 'tent city' housing some 300 000 refugees from rural areas.

A small industrial area near the harbour contains a power station and Africa's first water desalination plant, which converts seawater into fresh. The port, which exports ores from the copper-mining centre of Akjoujt, some 260 km (160 miles) to the north-east, was enlarged in the 1970s, and a deep-water harbour is under construction.

Population (excluding refugees) 135 000
Map Mauritania Bb

Nouméa ***New Caledonia*** Capital and main port of the territory, standing near the southern end of New Caledonia island. It is the largest French-speaking city in the Pacific, and has a cosmopolitan atmosphere. There are many tourist hotels on a peninsula to the south of the centre, while on the northern outskirts is the huge smelter that processes most of New Caledonia's nickel ore.

Population 85 000

Nova Lamego ***Guinea-Bissau*** See GABU

Nova Scotia ***Canada*** One of the Maritime Provinces in eastern Canada, comprising a mainland peninsula and CAPE BRETON ISLAND. It gets its name of 'New Scotland' from the charter given to the Scottish poet and statesman, William Alexander, by England's James I (James VI of Scotland) in 1621 for the establishment of a colony in Canada.

The province has wooded hills and valleys, containing numerous rivers and lakes, and a picturesque coastline. Its economic wealth is provided by agriculture, forestry, the mining of coal and gypsum, and industries such as wood processing, chemicals, steel and engineering. Nova Scotia had the first Canadian newspaper (*The Halifax Gazette*, 1752) and the first university (King's College, 1788).

Population 847 000
Map Canada Id

Novaya Zemlya ***USSR*** An almost uninhabited pair of islands (separated by a narrow channel) in the Arctic Ocean between the Barents and Kara seas. They extend from north to south for over 1000 km (625 miles), and cover a total area of 81 400 km² (31 400 sq miles). The tundra-covered land is visited in the summer by fishermen and hunters. Most of the islands' permanent inhabitants work at weather, radar and geological research stations. But there are also Samoyeds, a Mongolian people who herd reindeer and collect the down of eiders (large Arctic sea ducks). The Samoyed dog breed was originally developed by them.

Population 400
Map USSR Ga

Novgorod ***USSR*** City 155 km (95 miles) south and slightly east of Leningrad. Built on the banks of the Volkhov river, Novgorod was founded in AD 859, and is generally regarded as the cradle of Russian civilisation. It became an important trading post on the Viking route from Scandinavia to Constantinople (present-day Istanbul, Turkey), and in the Middle Ages it rivalled Moscow as the country's most influential and prosperous city. But it was overrun by the Moscow Principality in 1478 and largely destroyed by Ivan the Terrible in 1570.

Among Novgorod's historic buildings are the cathedral of St Sophia, built in the mid-11th century, and its kremlin (citadel), built in 1044 and enlarged in 1116. Novgorod today has several light industries, including furniture, fish canning, clothing and chinaware.

Population 215 000
Map USSR Ec

Novi Pazar (Novibazar) ***Yugoslavia*** West Serbian town on the Raska river, 185 km (115 miles) south of Belgrade. It was a cradle of Serbian culture in the Middle Ages, from which period date the 9th-century Church of St Peter, the brilliant 12th-century frescoes in the ruined Djurdevi Stubovi (St Georg's Pillars) monastery, and the superb monastery of Sopoćani, built in 1265. The Turks captured Novi Pazar (the name means 'new bazaar') in the 15th century and made it the capital of one of the Ottoman Empire's administrative regions. The town stagnated after Yugoslav independence, but small-scale timber, food and textile industries are developing, and it is used as a base by tourists visiting the region's monasteries.

Population 28 700
Map Yugoslavia Ec

Novi Sad (Ujvidek; Neusatz) ***Yugoslavia*** Capital of the Vojvodina autonomous province of Serbia, on the banks of the Danube 75 km (47 miles) north-west of Belgrade. The pre-1914 town centre is dominated by the Gothic cathedral's multicoloured roof and spire, while modern flats and factories making farm machinery, textiles, electrical goods, chemicals and foodstuffs stand in village-like suburbs. A university was founded in the city in 1960.

Population 257 700
Map Yugoslavia Db

Novokuznetsk ***USSR*** Industrial city in south-central Siberia, 300 km (185 miles) south-east of Novosibirsk. Novokuznetsk (formerly Kuznetsk, then Stalinsk) was founded in 1617, and had only about 4000 inhabitants until the late 1920s. Then it became the site of the USSR's largest iron and steelworks. Since then more industries have been set up, among them chemicals, alloys, aluminium and heavy machinery. Now it is the largest city of the Kuznetsk Basin industrial and coal-mining region.

Population 572 000
Map USSR Jc

Novosibirsk ***USSR*** One of the largest industrial centres in the USSR, straddling the Ob' river in Siberia, over 2800 km (1750 miles) east of Moscow. It was founded in 1893 and grew rapidly in size and importance – especially during the Second World War, when engineering and electrical factories were moved there in the face of German occupation of the Ukraine. Its products now include machinery, trucks, aircraft, ships, tools, electrical goods, plastics, pharmaceuticals and other chemicals, cloth, knitwear, shoes, timber products, furniture and processed food. It is Siberia's largest scientific research centre, and just to the south is the 'science town' of Akademgorodok.

Population 1 386 000
Map USSR Jc

Novosibirskiye Ostrova (New Siberian Islands) ***USSR*** Group of islands covering 28 000 km² (10 800 sq miles) in the Arctic Ocean between the East Siberian and Laptev seas. They are covered with snow and ice for all but three months of the year and are visited by hunters in pursuit of the Arctic fox. Since 1925 the Soviet Academy of Sciences has run research stations there.

Map USSR Pa

Nowa Huta ***Poland*** Europe's largest new town outside the USSR, now a suburb of the city of Cracow, separated from it by a green belt. Nowa Huta, or 'New Foundry', was begun in 1949 to serve the Lenin Steelworks, built at the same time. The works, the largest in Europe, produce 9 million tonnes of steel a year. The town has a modern church – called after its shape 'The Ark'. The authorities did not want the church built, and it took several battles with the police before it could be completed, in 1977.

Population 219 900
Map Poland Dc

Nowy Sacz (Neusander) ***Poland*** Medieval city 74 km (46 miles) south-east of the city of Cracow. It was founded in the 13th century as a fort on the Dunajec river, then extensively used as a trade route. The old quarter, with its medieval, 17th-century and 18th-century buildings, many with gleaming metal roofs, now attracts artists and tourists. The city is the centre of an apple-growing district.

Population 68 300
Map Poland Dd

Norway

THE NEW HARVEST FROM THE SEA IS OIL AND GAS – AND IT PAYS MUCH BETTER THAN FISHING

The Viking warriors of the land that is now Norway were once among the most feared people in the world. Between the 9th and 11th centuries, places as far afield as North America and Constantinople experienced their impact, while the British Isles suffered their invasions. Today, though Norwegians are still strong and resilient in the face of a harsh environment, they have a far more gentle image abroad.

Norway is a prosperous and picturesque country, welcoming tourists to its attractive villages and spectacular scenery of fiords, cliffs, rugged uplands and forested dales.

It is a narrow, very long country, stretching 1752 km (1100 miles) from Lindesnes, in the south, to the NORTH CAPE. Norwegians like to mention the fact that the northern frontier of Italy is as close to Oslo as the North Cape – and less expensive to reach. In the south, Norway is about 400 km (250 miles) across, but near NARVIK in the north the land narrows to only 6.5 km (4 miles). It has a Siamese twin – Sweden, to which it is joined back to back along a 1619 km (1017 mile) border. In the northern province of FINNMARK the frontier is shared with Finland and the USSR. Even farther north, the islands of SPITSBERGEN, in the SVALBARD archipelago, lie within 1100 km (700 miles) of the North Pole.

The land is rugged and mountainous. Two-thirds stands more than 300 m (1000 ft) above sea level. Massive erosion during the last great Ice Age scoured the fiords of the west coast so that some are the deepest in the world. For instance, the longest fiord, Sognefjord, 184 km (115 miles) long, reaches a depth of 1220 m (4000 ft). Inland, the ice scooped out deep dales which are now ribboned with lakes, one of which, Hornindalsvatn, is the deepest lake in Europe – 500 m (1640 ft).

The ice also sculpted a giant dragon's tail of northern islands, the LOFOTENS, extending 240 km (150 miles) in the Norwegian Sea. Between two of the Lofoten islands the strong and treacherous tidal current known as the Moskenstraumen flows. It is known in literature and legend as the Maelstrom, and is the subject of exaggerated stories of a destructive whirlpool that sucks ships down.

Such dangerous natural phenomena, together with avalanches, landslides, floods and storms at sea, partly explain Norway's rich folklore of the supernatural – with its trolls and giants inhabiting lakes, caves and fiords, and stirring up catastrophes. The Norse mythology from before the 10th century also had its destructive gods, such as Thor, wielding thunderbolts, and Woden, snatching warriors from battle to an afterlife of feasting in Valhalla – the Hall of the Slain.

The Norwegian imagination was perhaps all the more stimulated by sombre conditions: the long dark winters bred storytellers. Although one-third of the country is north of the Arctic Circle, the climate is modified by the North Atlantic Drift off the west coast. This warm current ensures that almost all the coastal waters are ice free in winter, and helps to support a rich variety of fish. Wet west winds bring heavy rain and snow to parts of the country, particularly the west coast; BERGEN in the south-west, for example, receives nearly 2000 mm (79 in) a year. But in some areas only 200 km (125 miles) inland, irrigation is needed to grow crops.

As a compensation for dark winters, northern Norway is one of the best places from which to see the AURORA Borealis (Northern Lights). In summer, Arctic Norway enjoys the beauty of the midnight sun – for three months at TROMSO.

Much of Norway is heavily wooded, but as the climate grows colder farther north, the forests give way to rocky wastes where only mosses and lichens grow, overlooked by snow-clad peaks. Norway has mainland Europe's largest glacier, JOSTEDALSBREEN, with an area of 816 km^2 (315 sq miles).

Most of the 4 million people live in towns, a quarter of them around Oslofjord, TRONDHEIM, the old capital, and BERGEN. The oil boom town of STAVANGER has also drawn people from the countryside. In the last century there was also large-scale migration from the harsh conditions of the countryside.

Only 3 per cent of the land is cultivated and the average holding is fairly small, though many farmers also own forest land, which provides additional income. About a third of all farms are dairy farms and cattle are reared in all parts of the country, even the far north. Sheep farming is widespread – there are 2 million sheep in Norway – and pig and poultry rearing are also important. Some goats are kept in hilly areas and reindeer are herded in the north above the timberline. The most productive agricultural areas are around Oslofjord and Lake MJOSA in the south-east, where the main crops are hay, potatoes, grain, roots for fodder and vegetables. Along the western coasts apples, cherries and soft fruits are grown.

Up to 150 years ago, Norway was a poverty-stricken country, and at the beginning of this century was still one of the poorest countries in Europe. Today, Norway has one of the highest per capita incomes in the world. Much of this new wealth comes from the oil and gas discovered in the North Sea. Mineral fuels – oil and gas – account for 51 per cent of Norway's exports. Norway is also rich in hydroelectric power produced by rivers flowing from the high plateaus. There are important coal mines in Arctic Spitsbergen and a major iron and steel plant at MO I RANA in the north.

The government has also invested in building good roads (which are costly because of the need to tunnel through so many mountains), in providing airfields and airstrips, and in subsidising some 200 car-ferry services and the coastal express steamers.

The growth of Norway's industry started long before the recent energy boom. Aluminium smelting based on hydroelectric power and imported bauxite was first developed in the 1920s. In 1906, Norway's biggest industrial concern, Norsk Hydro, became the first company in the world to make nitrate fertilisers from atmospheric nitrogen, using hydroelectricity, and has expanded into a wider range of products.

SEAGOING TRADITION

Traditionally, Norwegians have looked to the sea for their livelihood. Today, 80 per cent of them live within sight of a coastline which, even disregarding Norway's 50 000 islands, measures some 21 000 km (13 000 miles). So the fishing industry has always been important, and partly to protect fish stocks Norway has steadily extended her territorial waters, which now reach to 200 nautical miles. Lofoten cod fishing in late winter and herring catches off western Norway in the spring have both suffered a decline which is blamed on overfishing. The income of the fishing industry has also been hit by bans on whaling in the North Atlantic and Antarctic.

The seagoing tradition is, however, maintained by Norway's 200 shipping companies and the fifth largest merchant fleet in the world. The ships mainly carry cargo, but some companies have sought to diversify by investing in a fleet of cruise liners which operate principally in Caribbean waters.

The earliest known people of Norway were hunters and gatherers who left rock carvings 5000 years ago. However, the state of Norway is young. From the 14th century the land was part of the Danish realm, and from 1814 until independence in 1905 it was united with Sweden. German forces occupied Norway from April 1940 to the end of the Second World War.

The young nation has a strong sense of national identity, although it is divided over a language problem. In the mid-19th century a new literary language, Landsmål, was fostered to challenge the official language of Riksmål, which had lingered from the days of Danish rule. The two variants burden the schoolroom, complicate signposts and confound mapmakers. There is a third language in northern Norway: that of 25 000 Lapps, about a quarter of whom are involved in traditional reindeer husbandry.

Norway has had some internationally famous heroes such as the Arctic explorers Fridtjof Nansen (1861-1930) and Roald Amundsen (1872-1928), who beat a British party led by Captain Robert Scott in the race to the South Pole in 1911. More recently, the explorer and anthropologist Thor Heyerdahl sailed his raft *Kon-Tiki* from Peru across the Pacific Ocean to the TUAMOTU islands in 1947. The reputation of Norwegian literature was made by the playwright Henrik Ibsen (1828-1906), whose plays portray Norwegian life. The composer Edvard Grieg (1843-1907), whose music is part of the international concert repertoire, was inspired by the folk tunes of his native Norway.

Today, Norwegians have a high standard of living. Few people are very rich (the exceptions are mostly ship-owning families) and few are very poor. The welfare state provides services even to the most isolated of the scattered communities. There is space for everyone and the majority of people own their houses. Most also have leisure homes on the shores of seas and lakes. There is leisure fishing for everyone – its salmon fishing is renowned – and the country has 250 000 pleasure boats. Norway rivals Switzerland for skiing and mountaineering (indeed, Norwegians invented skiing), and woodlands offer elk hunting.

A referendum in Norway voted against membership of the European Economic Community in 1972, but the country is a member of the European Free Trade Association and the Nordic Council (with Denmark, Sweden, Finland and Iceland). Norway was a founder-member of NATO, but is opposed to the stationing of foreign troops and atomic weaponry on its soil.

▲ LAND OF THE VIKINGS Aurlandsfjord branches south from Sognefjord towards the permanent snows of Storeskavlen (1729 m, 5673 ft). The name Viking derives from the word *vik*, an Old Norse term for a creek or inlet.

NORWAY AT A GLANCE

Area 324 219 km² (125 180 sq miles)
Population 4 170 000
Capital Oslo
Government Parliamentary monarchy
Currency Krone = 100 øre
Languages Norwegian (old – Riksmål, new – Landsmål), Lappish and Finnish
Religion Christian (88% Evangelical Lutheran, 1% Pentecostalist)
Climate Temperate; cold in the north. Average temperature in Oslo ranges from −7 to −2°C (19-28°F) in January to 13-22°C (55-72°F) in July
Main primary products Barley, oats, potatoes, livestock, apples, timber, fish; crude oil and natural gas, coal, iron, lead, zinc, copper, nickel, titanium, quartz (silicon)
Major industries Mining, crude oil and natural gas refining, mineral refining, chemicals, shipbuilding, food processing, fishing, forestry, timber products
Main exports Crude oil and natural gas, aluminium and other nonferrous metals, chemicals, ships, machinery, fish, petroleum products, iron and steel, paper, timber products
Annual income per head (US$) 11 273
Population growth (per thous/yr) 4
Life expectancy (yrs) Male 74 **Female** 78

Nu Jiang ***China*** See SALWEEN

nuée ardente Rapidly moving, turbulent, incandescent cloud of gas, ash and rock fragments that flows close to the ground after violent ejection from a VOLCANO, destroying all life in its path – as Mont PELEE on the island of Martinique did in 1902.

Nuevo Leon ***Mexico*** Important northern industrial state with iron, steel, lead and manufacturing industries concentrated in its capital, Monterrey. It is also a citrus-growing region.
Population 2 600 000
Map Mexico Bb

Nuku'alofa ***Tonga*** Chief port and capital, on the north coast of Tongatapu island, between the lagoon and the ocean. Ships tie up at wharves projecting beyond the coral reef. The main industry is copra processing. The Royal Palace, a handsome timber building, was constructed in 1867; nearby are the tombs of former Tongan kings.
Population 21 000

Nullarbor Plain ***Australia*** Arid, treeless, flat and largely uninhabited limestone region along the southern coast of Western and South Australia. It gets its name from Latin words meaning 'no trees'. For much of its length of over 1200 km (750 miles), sheer cliffs drop from the plain into the Great Australian Bight.

The plain is crossed by a sealed road and by the transcontinental railway. One stretch of the line runs dead straight for 500 km (310 miles) – the longest straight track in the world.
Map Australia De

nunatak Isolated mountain peak or hill that projects through the surface of surrounding glacial ice.

Nuremberg (Nürnberg) ***West Germany*** Medieval city in Bavaria, about 150 km (95 miles) north of Munich. It was the home of Tannhäuser, the legendary lyrical poet of the 13th century, and the setting for the 19th-century composer Richard Wagner's opera *The Mastersingers*. Several kilometres of the medieval walls, with rampart walks, still stand, and the city's 11th-century castle and its 14th-century churches and houses have been restored.

The city was used for mass rallies by Adolf Hitler's Nazi party during the 1920s and 1930s, and for the trials of Nazi leaders and war criminals in 1945. Today it is an important industrial centre making electrical equipment, machinery, food products and toys – there is an annual toy fair. Nuremberg also has thriving printing and publishing industries.
Population 468 000
Map West Germany Dd

Nuristan ***Afghanistan*** Well-forested and mountainous region bordering on Pakistan. It produces 80 per cent of Afghanistan's timber, and is noted for intricate wood carvings.
Map Afghanistan Ba

Nürnberg ***West Germany*** See NUREMBERG

Nusa Tenggara (Lesser Sunda Islands) ***Indonesia*** Chain of islands extending east from Java. They include Bali, Lombok, Sumbawa, Sumba, Flores and Timor. The islands are mainly mountainous and volcanic.
Population 8 487 100
Map Indonesia Ed

Nuuk ***Greenland*** See GODTHAB

Nuwara Eliya ***Sri Lanka*** Town at 1884 m (6180 ft) in the central mountains, 100 km (62 miles) east of Colombo, in a region that produces Sri Lanka's best-quality tea. The British established it as a hill station – a refuge from the summer heat of the lowlands – in 1825, but its present buildings are mostly in the later Edwardian English style. The beautifully landscaped town has a lake, park, club, race track, golf course and well-stocked trout streams, and is a good centre for hill walking. Pidurutalagala (2518 m, 8261 ft), Sri Lanka's highest peak, overlooks the town from the north.
Population 21 300
Map Sri Lanka Bb

Nyasa, Lake ***East Africa*** See MALAWI, LAKE

Nyeri ***Kenya*** Prosperous market town and district capital between the Aberdare Mountains and Mount Kenya, about 95 km (60 miles) north of Nairobi. Coffee and tea are grown in the surrounding Nyeri district. Most of its inhabitants are Kikuyu people.
Population 50 000
Map Kenya Cb

Nyíregyháza ***Hungary*** Regional capital of Nyirseg, the country's most important apple-growing region, which lies in the north-east on the Czechoslovak, Soviet and Romanian borders. The city has grown rapidly since 1965 with the establishment of light industries and colleges, and as a tourist centre for nearby Sosto, a spa with hot springs.
Population 115 000
Map Hungary Bb

Nyköping ***Sweden*** Seaport and industrial city 98 km (61 miles) south-west of Stockholm. Founded in the 13th century on the site of an old trading town, it was destroyed by a fire in 1665 and rebuilt on a gridiron plan. It manufactures margarine and tobacco, and has engineering works and electrical and chemical factories.
Population 64 400
Map Sweden Cd

Nysa ***Poland*** River 193 km (120 miles) long. It rises on the Czech border, and flows north-east to join the Oder river about 60 km (35 miles) south-east of the city of Wroclaw.

The city of Nysa straddles the river about 75 km (45 miles) south-east of Wroclaw. Founded as a trading town in the 13th century, Nysa has since become the home of many Jesuit and other Catholic monasteries. It produces bricks, timber, foodstuffs and metal products.
Population 43 500
Map Poland Bc

Nysa Luzycka ***East Europe*** See NEISSE

Nzwami ***Comoros*** Second largest of the Comoros island group, covering 425 km² (164 sq miles); formerly Anjouan. It is noted for the export of ylang-ylang oil, which is used to make perfume. The main town is Mutsamudu.
Population 150 000
Map Madagascar Aa

Oahu ***Hawaii*** Island of 1549 km² (598 sq miles) in the central Pacific US state of Hawaii. It is one of the more westerly and the most highly developed of the Hawaiian islands, and contains the capital, HONOLULU, and the naval base of PEARL HARBOR.
Map Hawaii Ba

Oakland ***USA*** Seaport on San Francisco Bay in the south-western state of California. It prospered as the western end of the transcontinental railway and is an industrial and business centre, whose products include computers, electrical equipment, cars, ships and chemicals.
Population (city) 351 900; (metropolitan area) 1 871 400
Map United States Bc

oasis Area in a desert where sufficient water is available to support plant life, varying in size from a few palm trees round a spring to an area of several hundred square kilometres with a large agricultural population.

Oaxaca ***Mexico*** **1.** Mountainous rural state in southern Mexico. Its highland villages are the home of descendants of the ancient Indian Zapotec and Mixtec tribes, and about one-fifth of the people speak only their Indian language. Although coffee is grown for sale, Oaxaca is on the whole infertile and poverty-stricken.

2. Capital of Oaxaca state on a plateau 1640 m (5000 ft) above sea level, surrounded by the peaks of the southern Sierra Madre. Indians flock into this colonial city on Saturdays to sell their tooled leather, embroidered blouses and woollen ponchos at the market.
Population (state) 2 700 000; (city) 127 000
Map Mexico Cc

Ob' ***USSR*** Siberian river which rises in the Altai mountains near the Mongolian border, flows generally north-west and north, and empties into the Kara Sea, part of the Arctic Ocean. Together with the Irtysh river, which it joins in western Siberia, it forms the longest Russian river system. This extends for a total of 5570 km (3460 miles), making it the fifth longest in the world after the Nile, Amazon, Chang Jiang (Yangtze) and Mississippi-Missouri. Below the city of Tomsk, the Ob' is the main means of transportation in western Siberia – although roads and airstrips are now being built to serve the booming Siberian oil industry. It is also a major source of hydroelectric power.
Map USSR Hb, Jc

Oban ***United Kingdom*** Fishing and ferry port of western Scotland, standing on the Firth of Lorn in Strathclyde region, 100 km (62 miles) north-west of Glasgow. Boats connect it with many of the Western Isles.
Population 8000
Map United Kingdom Cb

Oberammergau ***West Germany*** Alpine town about 70 km (43 miles) south-west of Munich, where, every ten years since 1634, the inhabitants have performed a Passion play in May to celebrate their deliverance from the plague. The performances have become a major international event, attracting thousands of visitors.

Population 4800
Map West Germany De

Obergurgl ***Austria*** Major winter sports and tourist centre in the Tyrol, 50 km (31 miles) south-west of Innsbruck. It lies 1927 m (6322 ft) up in the Ötztaler Alps, and is the highest parish in Austria. The highest hamlet, Rofen, at 2104 m (6903 ft), is nearby.

Population 360
Map Austria Bb

Oberhausen ***West Germany*** Industrial town of the Ruhr between Duisburg and Essen. It produces coal, steel and machinery.

Population 227 000
Map West Germany Bc

obsidian Lustrous, black volcanic glass. It was used in early times for weapons and tools, and is now used for ornaments and jewellery.

Obuasi ***Ghana*** Gold-mining town about 170 km (105 miles) north-west of the capital, Accra. Part of the gold for which the Ashanti kingdom was renowned came from the town. The railway from Sekondi reached Obuasi in 1902 amid speculation that the region would turn out to be as rich as the gold-bearing reefs of the Witwatersrand at Johannesburg in South Africa. The hopes proved to be unfounded, but the country's gold exports today average a healthy US$14.5 million a year – 75 per cent of it coming from Obuasi.

Population 47 400
Map Ghana Ab

occlusion In an atmospheric depression, the overtaking of a warm front by a cold front, which ultimately lifts the warm sector completely off the surface of the earth and forms an occluded front. If the overtaking cold air is colder than the air mass in front, it is called a cold occlusion, if the air is warmer, it is a warm occlusion. If there is no marked difference in temperature, it is a neutral occlusion.

ocean Any of the large bodies of water that cover two-thirds of the earth's surface. Most geographers count four: the ARCTIC, ATLANTIC, INDIAN and PACIFIC; some consider the ANTARCTIC OCEAN to be a fifth. See also HOW OCEAN CURRENTS ARE CAUSED (below)

ocean basin Low-lying part of the earth's crust that is filled, or partially filled, by the salt water of the oceans. See also THE FLOORS BENEATH THE OCEANS (p. 484)

HOW OCEAN CURRENTS ARE CAUSED

There are two main generators of ocean currents. The first is wind. Surface winds whip huge masses of water along with them in drift currents. The second is density difference. Temperature and salinity changes alter the density of the water, making it rise to the surface or sink to the depths, creating vertical circulatory currents. The earth's spin and the shape and position of landmasses affect the direction currents take.

Deflection of moving bodies due to the earth's rotation is known as the Coriolis effect. Currents in the Northern Hemisphere veer right, and those in the south veer left. The effect is more marked as distance from the Equator increases, and results in ocean currents circulating clockwise north of the Equator, and anticlockwise south of it.

The density of seawater increases as it becomes colder or saltier. The hot sun warms equatorial waters, making them less dense, but at the same time evaporation increases their salinity. Ice and cold air chill polar seas. Rain, melting ice and rivers dilute the salinity.

An example of how a landmass can affect a current is the Gulf Stream. Warm Atlantic and Caribbean currents merge off the coast of Florida, are turned north-east by the US coastline and are then launched across the ocean as the North Atlantic Drift, arriving still warm on the western coastlines of Europe.

Where winds are offshore, waters are driven away from the coast. Deep water rises to the surface in compensation, bringing with it many nutrients from the bottom deposits. This explains the rich fisheries, for example, of the Peruvian and Namibian coasts.

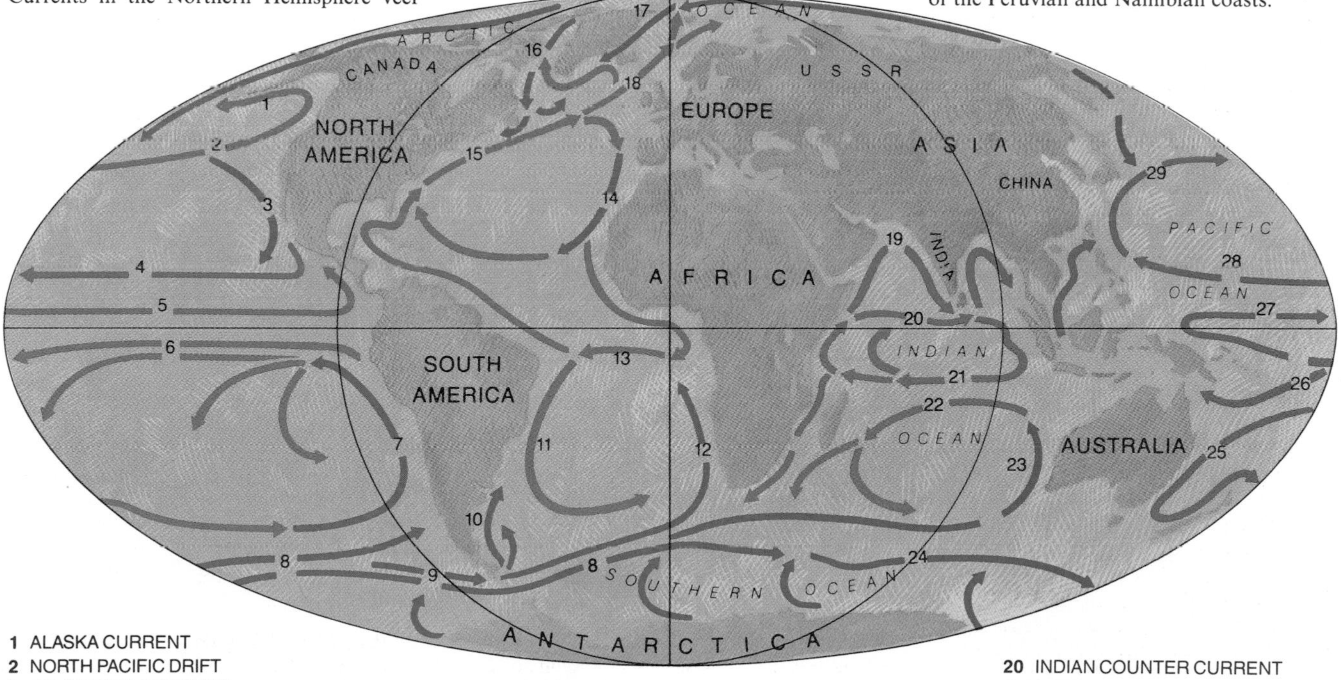

1 ALASKA CURRENT
2 NORTH PACIFIC DRIFT
3 CALIFORNIA CURRENT
4 NORTH EQUATORIAL CURRENT
5 EQUATORIAL COUNTER CURRENT
6 SOUTH EQUATORIAL CURRENT
7 PERU (HUMBOLDT) CURRENT
8 WEST WIND DRIFT
9 CAPE HORN CURRENT
10 FALKLAND CURRENT
11 BRAZIL CURRENT
12 BENGUELA CURRENT
13 GUINEA CURRENT
14 CANARIES CURRENT
15 GULF STREAM
16 LABRADOR CURRENT
17 EAST GREENLAND CURRENT
18 NORTH ATLANTIC DRIFT
19 SOUTH-WEST AND NORTH-EAST MONSOON DRIFT
20 INDIAN COUNTER CURRENT
21 NORTH EQUATORIAL CURRENT
22 SOUTH EQUATORIAL CURRENT
23 WEST AUSTRALIAN CURRENT
24 WEST WIND DRIFT
25 EAST AUSTRALIAN CURRENT
26 SOUTH EQUATORIAL CURRENT
27 EQUATORIAL COUNTER CURRENT
28 NORTH EQUATORIAL CURRENT
29 KURO SHIO

THE FLOORS BENEATH THE OCEANS

The hidden scenery that lies beneath the seas has intrigued mankind since classical times. The insatiably curious Alexander the Great had himself lowered to the seabed in a barrel – though to no great depth – while the legend of the submerged continent of Atlantis continues to fascinate even to the present time.

Deep-diving chambers and instruments and techniques developed in the last decades have given us a reasonable idea of the structure of the deep ocean floors. It is a younger scene than that of the continents, since much of it is being perpetually renewed (see PLATE TECTONICS), and less varied. Nevertheless, it does possess numerous features, such as mountain ranges, chasms, volcanoes and the equivalent of vast, sandy deserts.

The submarine world begins in the shallows at low water mark on the beach and plunges down to such depths as the Pacific's Mariana Trench – 11 033 m (36 197 ft). Continental shelves, part of the continental landmasses, slope gently down from the shore, split here and there by ravines, down which sediments, washed to the coast by rivers, slide into the depths. At a depth of about 200 m (656 ft), the edge of the continental shelf is reached, and the seabed plunges abruptly downwards at an angle of 3 to 6 degrees to the horizontal on the continental slope. The less steep slope of the continental rise – at a gradient of 1 in 100 to 1 in 700 – follows, reaching to an average depth of some 5 km (3 miles) below the surface. There begin the abyssal plains, vast desert-like stretches, inhabited by a few fish, worms and molluscs adapted to these enormous, light-less depths and whose ultimate food source must come from above. In places, nodules of manganese lie scattered over the ocean bed.

From the plains rise seamounts – volcanoes, some of which break the surface as islands. Hawaii Island is the exposed top of a mountain more than 9150 m (30 000 ft) high – more than 200 m (656 ft) higher than Everest. In addition, there are eroded, extinct, flat-topped volcanoes called guyots, together with the ridges that are found in all oceans. These are rugged underwater mountain chains, up to 1000 km (620 miles) across, whose individual peaks sometimes rise several kilometres above the ocean floor. In a few places, like Iceland and Tristan da Cunha, the ridges rise above sea level, and appear as islands. The oceanic ridges are frequently divided and traversed by transform faults.

Immensely long and deep oceanic trenches, often flanked by volcanic islands, border some oceanic regions and cut deep down into the oceanic crust. Through them, the rocks that compose the oceanic plates slide deeper to blend there with sub-crustal magma along what is known as a DESTRUCTIVE PLATE MARGIN, the whole region being a Benioff Zone. A blend of magma and metamorphosed rock re-emerges at the crests of the oceanic ridges and spreads over the ocean floor to maintain the cycle of destruction and renewal.

LANDMARKS OF THE OCEAN FLOOR

About two-thirds of the earth's crust is seabed. All the great oceans show a similar topography, in which continental shelves, slopes and rises drop down from coastal waters to the abyssal plains, from which rise submarine hills, volcanoes and mountain ranges. Until quite recently, less was known about this hidden scenery than the geography of distant planets. The topography of the great depths has been revealed by echo soundings and seismic probes, while some idea of its appearance and composition has been obtained from dredged rock samples, deep-sea drilling and comparison with ancient oceanic rocks now upthrust into the land.

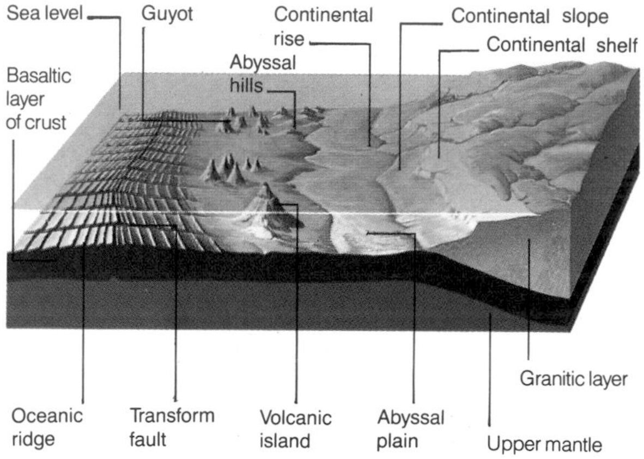

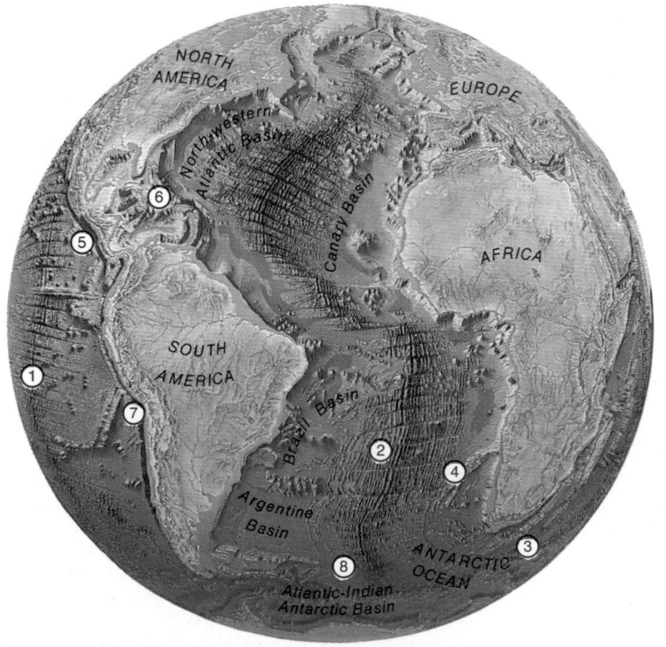

1 East Pacific Ridge
2 Mid-Atlantic Ridge
3 Southwest Indian Ridge
4 Walvis Ridge
5 Middle American Trench
6 Puerto Rican Trench
7 Peru-Chile Trench
8 South Sandwich Trench

9 Mid-Indian Ridge
10 Indian-Antarctic Ridge
11 Kermadec-Tonga Trenches
12 Philippines Trench
13 Marianas Trench
14 Aleutian Trench
15 Kuril-Japan Trenches
16 Sunda-Java Trenches

ocean floor spreading Part of the process in which ocean floors are created and destroyed. Sections of the oceanic crust (plates) move apart and basaltic magma, or molten lava, erupts from the mantle and solidifies, adding new crust to the diverging plates. The zones of separation are known as constructive plate margins and are marked by OCEAN RIDGES. See PLATE TECTONICS

Ocean Island See BANABA

ocean ridge Continuous system of mainly submarine mountains, extending into all the major oceans and running approximately parallel to the ocean margins. The mountains rise to thousands of metres above the ocean floor, some forming islands such as Jan Mayen Island in the Arctic Ocean and Bouvet Island in the South Atlantic. A feature of the mountain ranges is a central rift valley which is a zone of intense earthquake and volcanic activity. Outpourings of basaltic lava create the ridges.

ocean trench Long, narrow, steep-sided depression in the ocean crust, which may be up to 2 km (about 1 mile) deeper than the surrounding ocean floor. Trenches are usually parallel to the coastlines and lie above subduction zones where oceanic plates are descending into the mantle. They are associated with SUBDUCTION ZONES and island arcs. See PLUMBING THE DEPTHS (below)

Oceania ***Pacific Ocean*** Thousands of years ago, voyagers dubbed by some historians 'the Vikings of the Sun' undertook some of the world's greatest sea journeys. No one is sure from where they came, these first colonisers of the Pacific islands of Oceania – probably from South-east Asia but possibly on their primitive rafts from South America. That such journeys were possible was proved when Norwegian explorer Thor Heyerdahl sailed his *Kon-Tiki* raft from Peru to the Tuamotu islands in 1947; and when, some 30 years later, a reconstructed ocean-going canoe sailed 5000 km (3000 miles) from Hawaii to Tahiti without modern navigation aids, as the ancient islanders did.

Oceania encompasses tens of thousands of islands scattered over a vast area of the Pacific Ocean, from Belau 12 500 km (7800 miles) to Easter Island and from Midway 7500 km (4700 miles) to New Zealand. Only about 3000 of the islands are large enough to have names. The term may or may not include Australia, New Zealand and New Guinea – which may be separated and called Australasia.

Unquestionably included are three main island groups: MELANESIA ('black islands', from the inhabitants' skin colour), an arc of islands curving south-east from New Guinea, in which lie the Solomon Islands, Vanuatu and Fiji; MICRONESIA ('small islands'), another arc of small coral islands including the Marianas, Marshalls and Carolines, Kiribati, Nauru and Guam; and, in mid-ocean, the great triangle of POLYNESIA ('many islands'), with New Zealand, Hawaii and Easter Island at its corners and enclosing Samoa, Tonga and Tahiti.

Many of the islands are volcanoes which have risen from the bed of the Pacific near where two plates of the earth's crust collide (see PLATE TECTONICS). This phenomenon is still occurring; in Tonga two volcanoes periodically rise from the sea, to be eroded by the waves. Many islands are subject to eruptions and earthquakes. Most are surrounded by coral reefs – accumulations of the skeletons of tiny creatures.

Explorers came from Europe in the 16th and 17th centuries: Portuguese including Ferdinand Magellan (1480-1521), who was killed by Philippine islanders; Dutch including Abel Tasman (1603-59), the discoverer of New Zealand, New Guinea and Tasmania; and Spanish navigators. There was no established link with Europe until the 18th century, after the voyages of English explorers including Samuel Wallis (1728-95), who discovered Tahiti in 1767, and James Cook, who was killed by Hawaiian natives in 1779. By 1850, many of the major island groups had become colonial outposts of France, Britain, the USA and Netherlands, and missionaries had sown the seeds of the deep-rooted Christianity that is a striking aspect of island life today.

Physical separation meant that island groups developed distinct local cultures, which are now often overlaid with the cultures of European colonists. Tahiti and New Caledonia have a French flavour, and Hawaii – the 50th state – is predictably American. New Zealand and Japan are increasingly prominent in trade in Oceania, but Australia has had the greatest economic influence: most of the goods in the shops are marketed by three Australian companies – Burns Philp, Carpenters and the Steamship Trading Company. Many Australians once worked in the islands for these firms or for Australian banking and insurance companies, but local employment policies of island governments have greatly reduced their numbers.

Copra (dried coconut flesh) and tourism are

▲ FIREWALKERS Fijians on the island of Mbengga in Oceania heat stones for the firewalking ceremony. The walkers will parade around a 4 m (13 ft) circular pit of white-hot stones – amazingly without pain or injury.

PLUMBING THE DEPTHS

The lowest recorded point in the world – the Challenger Deep in the western Pacific – has not been visited by man. Part of the Mariana Trench, it was discovered in 1951 by the British survey ship *Challenger* – and in 1960 a manned American bathyscaphe descended to 10 915 m (35 810 ft). But the record depth of 11 033 m (36 197 ft) was established later that year by a Russian survey ship using echo soundings.

THE DEEPEST POINTS OF THE OCEANS

Ocean	Location	Depth (m/ft)
Pacific	Mariana Trench (Challenger Deep)	11 033/36 197
Atlantic	Puerto Rico Trench	9200/30 200
Indian	Java (Sunda) Trench	8047/26 400
Arctic	Angara Basin	5440/17 850
Antarctic (Southern)	South Sandwich Trench (Meteor Deep)	8428/27 651

the basis of the economies of several small island states. The larger islands have other resources: Fiji sends timber to Australia and New Zealand and sugar to Britain, for instance, New Caledonia has nickel and Nauru has phosphates (used for fertiliser).

The small scale of operations and the distance from major markets are problems for the tiny emerging nations of the area. Tourism provides a partial, but two-edged solution, bringing some destruction of culture and environment with its benefits. A South Pacific Forum has been established to promote economic cooperation between separate islands. Recently, considerable economic change has come to several island states with exploitation of their fishery resources: Fiji has its own fishing industry; the Solomon Islands are in partnership with the Japanese; Kiribati and Tuvalu license fishing fleets from Taiwan, South Korea and Japan; and Kiribati has signed an agreement with the Russians to allow their fishing boats to operate within a 200 nautical mile (370 km) zone around its islands.

Most of the indigenous population still survives on subsistence farming, but there has been a drift towards towns in recent years. Erosion of traditional village life, with its extended families, has resulted in the breakdown of old systems of support for the ill and aged; few island states have the infrastructure or resources to meet this new need. European and Asian immigrants tend to live in the towns and are often a significant section of the population. In Fiji nearly half the people are of Indian origin, French Polynesia has some 25 000 French in a population of about 165 000, while in New Caledonia more than half are non-Melanesians.

During the Second World War, Oceania was a battleground for the Americans and Japanese. Today, the US, Australian and New Zealand umbrellas offer incomplete protection from the fallout of international power struggles. American support for the widely unpopular and unswerving French policy of nuclear testing in the Pacific is seen in Australia, New Zealand and many smaller nations as opening the door to Russian influence. France's dogged and often-reiterated determination to maintain its presence in the area does nothing to smooth relationships within or beyond its Pacific 'external territories'. In French Polynesia and New Caledonia the struggle for independence by radical indigenous groups is bitterly opposed by European settlers. Elsewhere, although ideologies may take second place to personalities, internal politics can be as strident as in other parts of the world, making these palm-fringed coral islets something less than earthly paradises.

Map Pacific Ocean

Ocho Rios ***Jamaica*** Seaport and resort town on the north coast, 90 km (55 miles) east of Montego Bay. Many cruise ships visit the port and it is a tourist centre for visiting the many waterfalls, cascades and limestone scenery in the area. One of the most notable features is the Dunn's River Falls, where a clear mountain stream cascades 180 m (about 600 ft) down a wooded gorge to the sea.

Population 10 500

Map Jamaica Ba

▲ AGE-OLD RITUAL Young Botswanans fish with baskets in the 1600 km (1000 mile) long Kavango river, which flows into the Okavango Delta.

Odawara ***Japan*** See HAKONE

Odense ***Denmark*** Third largest city of Denmark and the administrative centre for the island of FUNEN. An 8 km (5 mile) long ship canal, built in 1797-1804, links the city to Odense Fjord, where there is a major shipyard at Lindø. The city has been a bishop's seat since 1020, and its 13th-century Cathedral of St Knud is one of the finest Gothic buildings in Denmark. Most visitors, however, come to the city because it is the birthplace of the fairytale author Hans Christian Andersen (1805-75). Several museums display mementos of his life. Among the places of interest are Hans Christian Andersen's childhood home and the house in which he later lived; a Railway Museum; and the Fünen Village – an open-air museum of traditional, local country life. The main industries are sugar refining and food processing.

Population 170 200

Map Denmark Bb

Odenwald ***West Germany*** Range of hills about 80 km (50 miles) long, and 32 km (20 miles) wide, east of Mannheim between the Neckar and Main rivers. Its western slopes, along the Bergstrasse ('mountain road'), are sheltered from the east winds and have an exceptionally early spring. They produce tobacco, grapes, peaches, apricots, cherries, almonds and walnuts; the show of early blossom is a great tourist attraction.

Map West Germany Cd

Oder (Odra) ***Czechoslovakia/East Germany/Poland*** River, 912 km (567 miles) long. It rises in the Oder mountains in Czechoslovakia about 60 km (35 miles) south-west of the city of Ostrava. From Raciborz, just inside Poland, it is navigable all the way to the Baltic – some 710 km (440 miles) downstream. The Oder carries about 20 million tonnes of cargo a year, mostly coal, building materials, timber and chemicals. With its tributary, the Neisse, it forms the 'Oder-Neisse Line', the border between Poland and East Germany.

Map Poland Bc

Odessa ***USSR*** Black Sea port and tourist resort in the Ukraine, 160 km (100 miles) north-east of the Romanian border. It has five harbours and is a naval base and the home of Russia's Antarctic whaling fleet. The city stands on a group of hills overlooking the Black Sea, and was built around a 14th-century Tatar fort. It has a rich cultural life, much of which revolves around its handsome 19th-century opera house. It is also an industrial centre, producing steel, ships, machinery, chemicals, foodstuffs and petroleum products.

In 1905 Odessa was the scene of a mutiny aboard the battleship *Potemkin*. The crew mutinied after being served maggot-infested meat, which the ship's doctor pronounced fit to eat. The incident caused a civil uprising in the city – which was bombarded by government forces, killing 6000 of its inhabitants.

In 1925 the Russian film director Sergey Eisenstein (1898-1948) made a feature film, *Potemkin*, about the uprising. One of the film's most celebrated scenes was shot on the Richelieu Steps – named after the French statesman and churchman Cardinal Richelieu (1595-1642) whose statue stands at the top of the steps. The steps – now called the Potemkin Steps – lead down from the upper part of Odessa to the harbour. It was there that the Cossacks of Tsar Nicholas II (1868-1918) slaughtered scores of demonstrators as they marched to support the mutineers aboard the anchored battleship.

Population 1 113 000

Map USSR Ed

Odra ***Eastern Europe*** See ODER

Offaly (Uíbh Fhailí) ***Ireland*** County covering 1998 km² (771 sq miles) in central Ireland, about 50 km (30 miles) west of Dublin. It is mostly grassy plain, with the Slieve Bloom mountains

in the south. Offaly was formerly known as King's County after Philip of Spain, husband of Mary I, queen of England (1553-8). The county town is Tullamore.
Population 58 300
Map Ireland Cb

Offenbach ***West Germany*** Town on the southern outskirts of Frankfurt am Main with a large tanning industry. It has a leather museum and an annual leather fair.
Population 110 000
Map West Germany Cc

Ogaden ***Ethiopia*** Desert region, about 600 km (370 miles) east to west and 400 km (250 miles) north to south, which is claimed by neighbouring Somalia because it is inhabited mainly by Somali nomadic herdsmen. It was the scene of a full-scale war between the two countries in 1977, and the Western Somali Liberation Front is still active there. The Ethiopian government is unwilling to give up the territory, particularly because it may be rich in oil.
Map Ethiopia Bb

Ogbomosho ***Nigeria*** City about 200 km (125 miles) north and slightly east of the capital, Lagos. It was a traditional Yoruba tribal town founded in the 17th century, and is now a busy industrial centre producing foodstuffs, and a market for tobacco.
Population 514 400
Map Nigeria Ab

Ogooué (Ogowe) ***Congo/Gabon*** River, 1200 km (746 miles) long, rising in Congo south of Zanaga, and flowing through the rain forests of Gabon to the Atlantic. It drains some 222 700 km² (85 984 sq miles), and its upper course is marked by rapids and waterfalls, including the Poubara Falls. But it is navigable for part of the year below Ndjolé and navigable throughout the year below Lambaréné. The latter section is Gabon's busiest waterway, especially for timber exports. Near the Atlantic, the river divides into several channels, and has built up a large delta south of Port Gentil.

The Ogooué was explored separately by the Frenchman Paul du Chaillu and the Italian-born Pierre Savorgnan de Brazza (who also found its source), and by the Victorian traveller Mary Kingsley, who braved its rapids in a canoe. She rated the Ogooué as 'the greatest river between the Niger and the Congo'.
Map Gabon Bb

O'Higgins ***Chile*** See LIBERTADOR

Ohio ***USA*** **1.** Northern industrial and farming state covering 106 765 km² (41 222 sq miles) of gently undulating country southwards from Lake Erie to the Ohio river. It was first settled in the 1780s and joined the Union in 1803. Ohio's prosperous farms produce soya beans, maize, dairy goods and beef. Coal, oil and gas are the main mining products and manufactures include machinery, transport equipment and steel. The largest cities are CLEVELAND, CINCINNATI and the state capital COLUMBUS.

2. River used as a key route by settlers travelling to the Midwest. It is formed from the Allegheny and Monongahela rivers, which join at the city of Pittsburgh in Pennsylvania, and flows about 1575 km (980 miles) generally south-west to the Mississippi at Cairo. The Ohio forms the boundary between the states of Ohio, Indiana and Illinois to the north and West Virginia and Kentucky to the south. It is still used for carrying industrial freight.
Population (state) 10 744 000
Map (state) United States Jb; (river) United States Ic, Jc

Ohrid, Lake (Ohridsko Jezero) ***Albania/Yugoslavia*** Yugoslavia's second largest lake (after Lake Shkodër) covering 348 km² (134 sq miles) astride the Albanian border in south-west Macedonia. It is up to 286 m (938 ft) deep, and its abundant trout and eels, fine beaches, mild winters, warm summers, clear water and superb scenery make it Macedonia's leading tourist centre. The ancient town of Ohrid on the east shore has quaint white-painted houses with upper floors jutting out over paved alleys. It also has several fine Byzantine churches. The people still wear ornate folk costumes, and there is a Folk Song and Dance Festival each summer.
Map Yugoslavia Ed

oil and natural gas See THE STORY OF OIL AND NATURAL GAS (p. 488)

oil shale Fine-grained black or dark grey rock rich in kerogen (fossilised organic remains) which, when heated, produces hydrocarbons that can be distilled to produce oil.

Oileáin Arann ***Ireland*** See ARAN ISLANDS

Ojos del Salado ***Argentina/Chile*** Highest mountain in South America at 7084 m (23 241 ft). It stands in the ANDES on the northern border between Argentina and Chile.
Map South America Bc

Ok Tedi ***Papua New Guinea*** Copper and gold mine in the Star Mountains of central New Guinea, near the border with Irian Jaya. It is named after a local river. Its reserves of ore are put at 350 million tonnes.
Map Papua New Guinea Ba

Oka ***USSR*** Volga river tributary which rises in the uplands south of Moscow and flows north and east to join the Volga at Gor'kiy. The Oka, which is 1480 km (920 miles) long, is used for transporting grain and timber.
Map USSR Fc

Okanagan Valley ***Canada*** Fruit-farming area and popular holiday resort in southern British Columbia. Tourist areas include Lake Okanagan and the towns of Penticton and Summerland.
Map Canada Cd

Okavango ***Southern Africa*** See KAVANGO

Okavango Delta ***Botswana*** Swampy delta covering 10 360 km² (4000 sq miles) in the north-west of the country. It is supplied with water by the Kavango river, which rises in Angola. Vast numbers of animals and birds, including hippopotamuses, buffaloes, storks and pelicans, live within its braided channels, and about one-fifth of the delta is protected in the Moremi Game Reserve. Plans for commercial use of the water for irrigation and for livestock ranches have been restricted by the area's remoteness and high evaporation rate. During floods, water drains east through the Botletle river into the Makgadikgadi Pans.
Map Botswana Ba

Okayama ***Japan*** Local capital and commercial centre in south-west Honshu island, about 140 km (90 miles) west of the city of Osaka. Its 'Black Crow Castle' was built in black in 1573 to contrast with the snow-white castle at nearby Himeji. The Korakuen landscape gardens are among Japan's finest, with ancient trees, lawns, ponds and a tiny rice field and tea plantation. Industries range from chemicals to producing traditional *tatami* mats of woven straw.
Population 572 400
Map Japan Bd

Okhotsk, Sea of Arm of the north-western PACIFIC OCEAN, bounded by the KAMCHATKA peninsula and the KURIL ISLANDS. Its broad continental shelves are important fishing grounds. It is also a source of oil for the Soviet Union. In summer it is an outlet for eastern Siberia, but fogs and ice are a hazard in winter. MAGADAN is the chief port.
Area 1 528 000 km² (589 961 sq miles)
Greatest depth 3475 m (11 400 ft)
Map USSR Pc

Okinawa ***Japan*** Main island of the RYUKYU ISLANDS, covering 1176 km² (454 sq miles) in the East China Sea about 500 km (310 miles) south and slightly west of Japan. It was the scene of fierce fighting during the Second World War between US and Japanese troops. Between March and June 1945 it was bombed, invaded and finally overrun by US forces – who established airbases close to the Japanese mainland. After the war, the group was put under US control and the northern section was returned to Japan in 1954. In 1971 the entire group of islands (covering 2388 km², 922 sq miles) was restored to Japanese rule.
Population 1 072 200
Map Japan Gf

Oklahoma ***USA*** South-western Great Plains state covering 173 320 km² (66 919 sq miles) between the Red River and Texas to the south and Kansas to the north. It is gently undulating country, and it was Indian territory until white settlers arrived in 1889. It became a state in 1907. Farming was badly hit in the 1930s when drought created a dustbowl and much soil was blown away by winds; today it relies on cattle, wheat and cotton. Oklahoma is the country's third largest natural gas producing state, and also extracts oil and coal. Industries include oil refining and engineering. OKLAHOMA CITY, the state capital, and TULSA are the biggest cities.
Population 3 301 000
Map United States Gc

Oklahoma City ***USA*** State capital of Oklahoma, lying in the centre of the state. The city is a cattle market and agricultural service centre. Industries include electronics, processing of farm products and petrochemicals.
Population (city) 443 200; (metropolitan area) 962 600
Map United States Gc

THE STORY OF OIL AND NATURAL GAS

Of all the ages of mankind, it seems probable that the Oil Age will be the shortest lived. A few limited uses apart, neither petroleum nor natural gas played much of a part in human affairs until almost within living memory. They now provide more than 70 per cent of world energy, but it is variously prophesied that all reserves will be exhausted by the turn of the century, or at the most, in 40 to 50 years' time.

The first well specifically drilled for petroleum was not sunk until 1859. It was dug by a Colonel E.L. Drake in Pennsylvania to exploit the process recently invented by the Scots chemist James 'Paraffin' Young, who had found a means of distilling paraffin, or kerosene, from crude oil. For both Drake and Young the object of the exercise was the recovery of the by-product paraffin wax, a cheaper material for candle making than beeswax or tallow.

People have been aware of petroleum for a very long time. More than 5000 years ago, the Assyrians, Babylonians and Sumerians used bitumen – a semisolid form of petroleum oil occurring in surface seepages – as mortar, in road making, and as waterproofing in irrigation channels. It was also used to create mosaics, to caulk ships, and in liquefied form as a liniment and a laxative. The ancient Chinese, drilling down to 140 m (450 ft) in search of brine, lit associated natural gas deposits to evaporate the water content from the salt. Oil and gas seepages, probably originally ignited by lightning, at BAKU in what is now the USSR, fuelled the 'eternal fires' of the fire-worshipping early Persians.

Like coal, oil and gas are for all practical purposes non-renewable resources, and like coal too, they are the remains of ancient life forms. But while coal evolved out of the trees and plants of long-ago tropical forests, petroleum has had its beginnings in the seas. It is derived from the remains of simple planktonic life forms, minute plants and animals, although the exact means by which plankton has been converted to petroleum is not completely clear. Plankton were first abundant in the warm seas some 570 million years ago and have continued to be abundant ever since. Over a long period, dead plankton have rained down to form a layer on the seabed which, if near coasts, has become covered by sediments. Over aeons, the sediments and the organic matter have become pressed ever more deeply into the earth's crust by new strata forming above. Millions of years of pressure, combined perhaps with heat from the planet's interior, have forced structural changes upon the organic layer, converting it to liquid and gas. The sediments above have evolved into sedimentary rocks, such as sandstones and limestones, and because of their low density, the liquid oil and gas have moved upwards and sideways to penetrate porous sedimentary rocks. If the deposits have reached the surface in this manner, they were, and are, lost as seepages. But if the upward movement of the oil and gas has been halted by a layer of impervious rock, then subterranean pockets and reservoirs were created, which can be tapped.

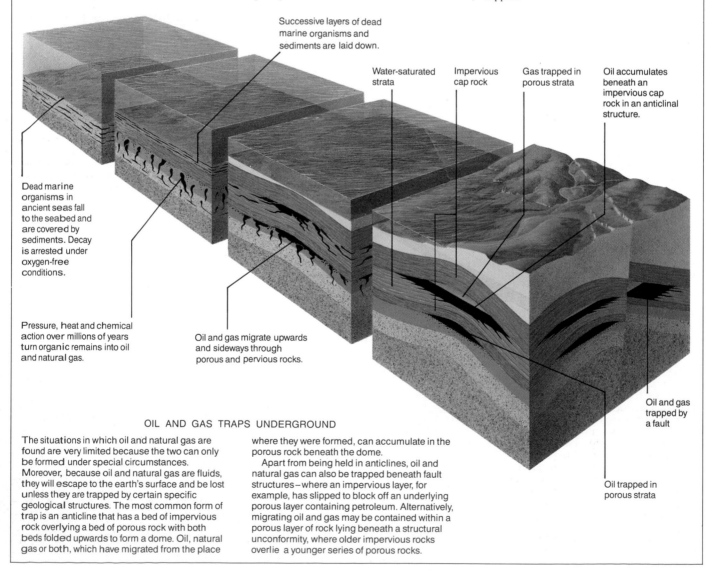

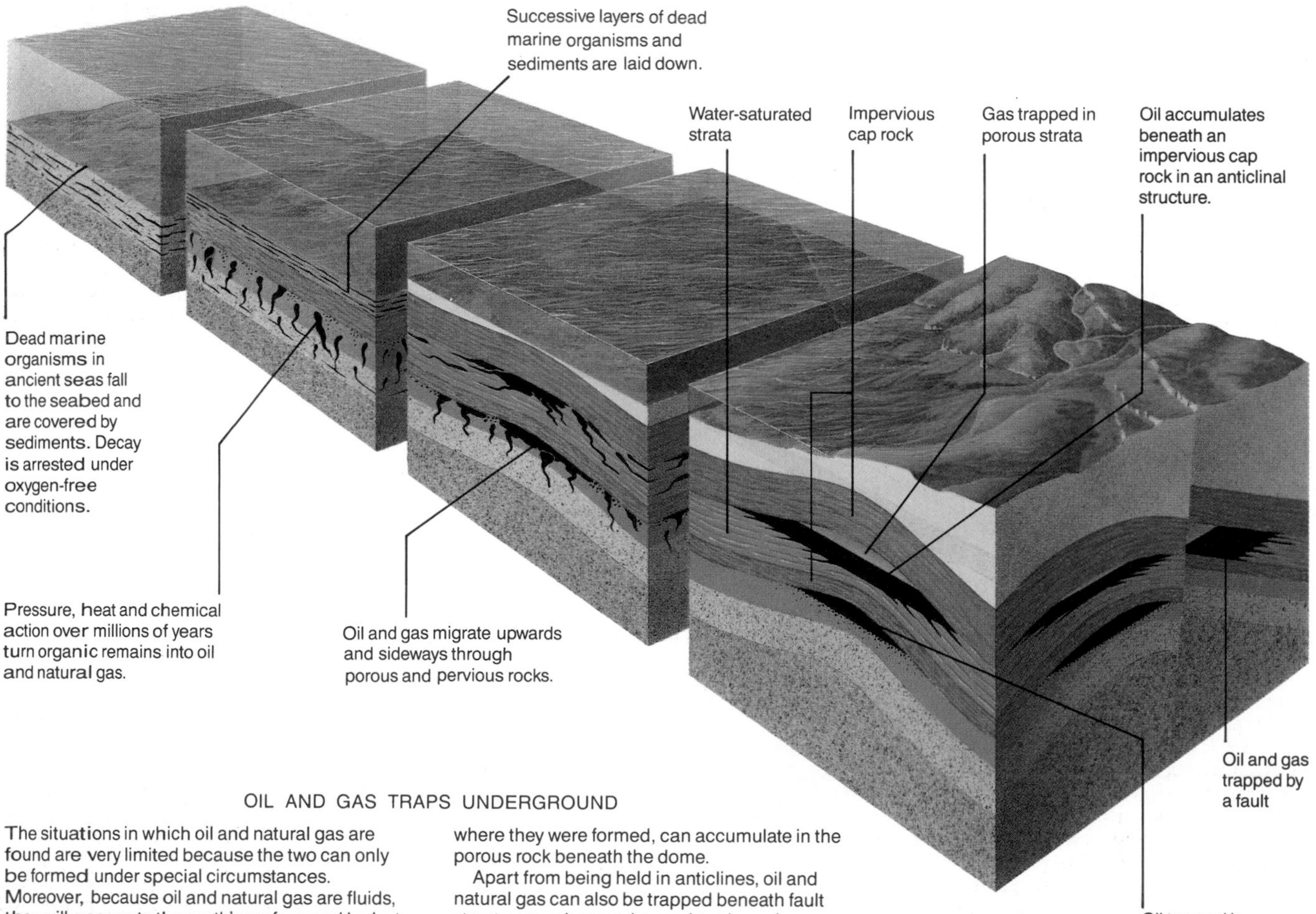
Successive layers of dead marine organisms and sediments are laid down.
Water-saturated strata
Impervious cap rock
Gas trapped in porous strata
Oil accumulates beneath an impervious cap rock in an anticlinal structure.
Dead marine organisms in ancient seas fall to the seabed and are covered by sediments. Decay is arrested under oxygen-free conditions.
Pressure, heat and chemical action over millions of years turn organic remains into oil and natural gas.
Oil and gas migrate upwards and sideways through porous and pervious rocks.
Oil and gas trapped by a fault
Oil trapped in porous strata

OIL AND GAS TRAPS UNDERGROUND

The situations in which oil and natural gas are found are very limited because the two can only be formed under special circumstances. Moreover, because oil and natural gas are fluids, they will escape to the earth's surface and be lost unless they are trapped by certain specific geological structures. The most common form of trap is an anticline that has a bed of impervious rock overlying a bed of porous rock with both beds folded upwards to form a dome. Oil, natural gas or both, which have migrated from the place where they were formed, can accumulate in the porous rock beneath the dome.

Apart from being held in anticlines, oil and natural gas can also be trapped beneath fault structures – where an impervious layer, for example, has slipped to block off an underlying porous layer containing petroleum. Alternatively, migrating oil and gas may be contained within a porous layer of rock lying beneath a structural unconformity, where older impervious rocks overlie a younger series of porous rocks.

Öland ***Sweden*** Long and low-lying island covering 1345 km² (520 sq miles), off the south-east Baltic coast. It is linked by a 6 km (4 mile) long bridge to the mainland port of Kalmar. It has a fishing fleet and there is sheep farming and limestone quarrying. Öland has numerous archaeological features, including a reconstructed Iron Age settlement. It is a popular destination for holidaymakers.

Population 23 800
Map Sweden Cd

Olbia ***Italy*** Port on the north-east coast of Sardinia, about 190 km (120 miles) north of Cagliari. It is the point of entry for most ferry passengers from mainland Italy.

Population 32 600
Map Italy Bd

Oldenburg ***West Germany*** Northern river port and market town on the Hunte river some 40 km (25 miles) west of Bremen. Stone Age funeral mounds of granite boulders dot Ahlhorn Heath, south of the town.

Population 137 000
Map West Germany Cb

Oldham ***United Kingdom*** Mill town on the north-eastern outskirts of Greater Manchester. It had a woollen cloth industry in the 17th and 18th centuries, but this was dwarfed by the cotton industry that developed in the 19th century. Oldham was one of the main cotton centres of Lancashire, but the industry has now declined.

Population 95 000
Map United Kingdom Dd

Olduvai Gorge ***Tanzania*** Prehistoric site in the north of the country about 30 km (20 miles) north-west of the Ngorongoro Crater. Between 1959 and 1961 the British palaeontologist Louis Leakey (1903-72) and his family discovered fossil skulls there thought to be between 1.5 and 2 million years old, and to represent a stage of human evolution prior to *Homo sapiens*. The gorge is up to 150 m (490 ft) deep, and stone tools and the bones of many extinct animals have also been found there.

Map Tanzania Ba

Oléron, Ile d' ***France*** Island 30 km (20 miles) long in the Bay of Biscay to the south of La Rochelle. With an area of 175 km² (68 sq miles) it is France's second largest island after Corsica. It has a mild climate, good fishing, fine beaches and breeding grounds for mussels and oysters.

Population 16 800
Map France Cd

Olgas, The ***Australia*** Cluster of some 28 dome-shaped peaks in central Australia rising to 1069 m (3507 ft) at Mount Olga. They are about 30 km (19 miles) west of Ayers Rock in the Northern Territory. The peaks were first sighted in October 1872 by the explorer Ernest Giles, and it is said that he named them after Queen Olga of Württemberg, at the request of his patron.

Map Australia Ed

Oligocene Third epoch of the Tertiary period of the Cenozoic era of the earth's time scale. See GEOLOGICAL TIMESCALE

Olimbos ***Greece*** See OLYMPUS

Olinda ***Brazil*** Historic city 7 km (4.5 miles) north of the city of Recife. Founded in 1537 by the Portuguese, it is protected by the United Nations Educational, Scientific and Cultural Organisation (UNESCO) as part of the world's cultural heritage. It is so named because early settlers are said to have exclaimed on their arrival at the site 'O linda', meaning 'How beautiful'. One of the first major urban settlements in Brazil, it has beautiful colonial churches – many with distinctive blue and white tiles, notably da Sé (1537), da Misericórdie (1540) and São Francisco (1577).

Olinda has one of the most exciting carnivals in Brazil, held just before Lent, usually in February. Unlike the celebrated carnival in Rio de Janeiro, anyone can join in, between the set-piece groups, and each year about 250 000 people do. They dance the frenetic *Frevo* (beside which the samba seems sedate) through the narrow hilly streets of the old town for at least four days and nights.

Population 282 000
Map Brazil Eb

Ollantaytambo ***Peru*** Site of an Inca tambo (posting station) and fortress, on the Urubamba river, 70 km (45 miles) north of Cuzco. The ruins include flights of terraces, the temple and palace, the Baño de la Ñusta – Bath of the Princess – and a solar observatory. The last Inca leader, Manco Capac, and his soldiers used the fortress in 1537 when they retreated from the Spaniards after the siege of Cuzco.

Map Peru Bb

Oloibiri ***Nigeria*** See NIGER

Olomouc ***Czechoslovakia*** Market town and capital of Moravia in the 11th and 12th centuries, 64 km (40 miles) north-east of Brno. It is at the heart of a fertile grain-growing plain, and has food-processing and engineering industries. The Old Town includes a Gothic cathedral and town hall, and a university founded in 1566.

Population 104 300
Map Czechoslovakia Cb

Olympia ***Greece*** Excavated sanctuary of Zeus, at the foot of Mount Kronos near the Ionian coast of the Peloponnese, and the original site of the Olympic Games. The sanctuary was a tree-lined, sacred grove in which a flame, relighted each spring in a stone enclosure, burned in honour of Zeus. All-male foot races were held in a stadium in the grove from 776 BC and became the Olympic Games. They lasted over 1000 years until AD 393, when the Roman Emperor Theodosius banned them. Excavations by German archaeologists from 1875 revealed the Temple of Zeus, the Altis, or sacred olive grove, the stadium, the hippodrome, where chariot races were held, gymnasium and baths. Excavation of the almost complete village led directly to the revival of the Olympic Games by the French Baron Pierre de Coubertin, at ATHENS in 1896.

Map Greece Bc

Olympia ***USA*** Capital of the north-western state of Washington and port at the head of Puget Sound, more than 250 km (155 miles) from the Pacific Ocean. It is a centre for tourists visiting the state's mountains and forests.

Population (city) 29 200; (metropolitan area) 138 300
Map United States Ba

▼ POW-WOW IN THE DESERT The Aboriginal name for the Olgas is Katatjuta – 'Many Heads' – because of their uncanny resemblance to a group of skulls conferring amid the desert.

Olympic National Park ***USA*** Mountain wilderness of 3628 km^2 (1401 sq miles) in north-western Washington state. It includes Mount Olympus (2428 m, 7965 ft), and has the finest remains of the Pacific north-western rain forest. The park is rich in wildlife and has many glaciers.
Map United States Ba

Olympus (Olimbos) ***Greece*** Highest mountain massif in Greece on the borders of Thessaly and Macedonia, and legendary home of the gods. The massif stretches from the Thessaly coast for 40 km (25 miles) into Macedonia, peaking at Mytikas (2917 m, 9570 ft). Its precipitous, crystalline rocks are broken by thickly wooded ravines. The massif is snow-capped for most of the year, and is now the home of several ski resorts and a military ski school.
Map Greece Ca

Omaha ***USA*** City on the Missouri river on the eastern border of Nebraska. It was the winter quarters of the Mormons in 1846-7 during their migration to Salt Lake City, and became a busy trading post during the gold rush to California two years later. Its industries include railway engineering and food processing. Nearby is the headquarters of the US Air Force Strategic Air Command.
Population (city) 334 000; (metropolitan area) 607 400
Map United States Gb

Oman, Gulf of Arm of the ARABIAN SEA, leading to the Strait of HORMUZ and the oil-rich GULF. It was an important shipping lane in ancient times, when cargoes of ivory, jewels, silks, spices and teak from China and South-east Asia were transferred between ships at Omani ports and then taken on to Babylon. Today huge oil tankers are the most common ships in the gulf.
Dimensions 560 km (350 miles) long, 320 km (200 miles) wide
Map Oman Aa

Omdurman ***Sudan*** City situated across the Nile from the capital, Khartoum, and the seat of the National Assembly, the Sudanese parliament. It is a place of flat-roofed, baked clay houses and narrow streets. It specialises in the production of jewellery, carvings and the delicate working of gold and silver. There are also markets trading in camels, skins and souvenir ostrich eggs.

The city was founded on the site of an insignificant riverside village by the Sudanese religious leader, the Mahdi, in 1884. His tomb lies in the city. The Battle of Omdurman, in which the British overthrew the Mahdi's followers in 1898, was fought near the city.
Population 526 300
Map Sudan Bb

Omo ***Ethiopia*** River rising in the central highlands and flowing southwards for about 800 km (500 miles) into Lake Turkana (Lake Rudolf) on the Kenyan border. Archaeological finds in the lower Omo Valley since 1967 indicate early prehistoric human settlement. One of the oldest discoveries was the jawbone of a man-like ape estimated to be 2.5 million years old.
Map Ethiopia Ab

Oman

IN LESS THAN 20 YEARS AN ENLIGHTENED RULER HAS BROUGHT THIS BACKWARD STATE INTO THE 20TH CENTURY

Ruled by a sultan with absolute power, this country – second largest in the Arabian peninsula – has a population which is 90 per cent Arab. But there are minority groups of Baluchis, from Iran, and Africans descended from former slaves. There are also 266 000 immigrant workers from India, Pakistan, Iran, South Korea and Europe.

Oman is divided in two by the United Arab Emirates. A small mountainous area at the tip of the MUSANDAM peninsula overlooks the strategic Strait of HORMUZ, which controls the entrance to THE GULF. The main part of the country consists of a fertile coastal plain (BATINAH) in the far north, the limestone HAJAR MOUNTAINS rising to 3018 m (9901 ft), and a largely barren plateau with small fishing harbours to the south. The DHOFAR coast in the far south is fertile and has reliable monsoon rain from June to September, and there are good pastures in Dhofar's mountains. Inland the plateau merges into the largely unexplored 'Empty Quarter' in Saudi Arabia.

The old sultanate of Muscat and Oman was one of the most backward Arab States when oil production began in 1967. But in 1970 the present ruler, Sultan Qaboos Bin Said, deposed his father, and since then he has used oil revenues to make huge strides in social development. Today in renamed Oman education is free (but not compulsory), and there are 560 schools – compared with three in the old sultan's time – and 14 hospitals instead of one.

Oil, found mainly in the north, provides 90 per cent of exports, and there are also copper and natural gas. However, 70 per cent of the people live by farming or fishing. With high temperatures and low, generally unreliable rainfall, only 0.1 per cent of the land is cultivated. This area includes the Batinah – a potentially rich garden area, land around mountain villages which have ancient irrigation systems, and sheltered mountain valleys.

The port of MUSCAT was a flourishing trade centre by the 5th century BC. It was a Portuguese base from 1508 to 1648, and the seat of the present dynasty from 1741. In the 19th century the country established a close treaty tie with Britain. It became fully independent in 1971 but has close military links with Britain and the USA.

OMAN AT A GLANCE
Area 300 000 km^2 (116 000 sq miles)
Population 1 276 000
Capital Muscat
Government Absolute monarchy
Currency Omani rial = 1000 baiza
Languages Arabic (official), English
Religion Muslim
Climate Hot summers, mild winters; temperatures in Muscat: 19-38°C (66-100°F)
Main primary products Dates, fruits, vegetables, fish; crude oil and natural gas
Major industries Agriculture, mining, oil refining, copper smelting, fishing, cement
Main exports Crude oil, fish, fruit
Annual income per head (US$) 5200
Population growth (per thous/yr) 39
Life expectancy (yrs) Male 48 **Female** 50

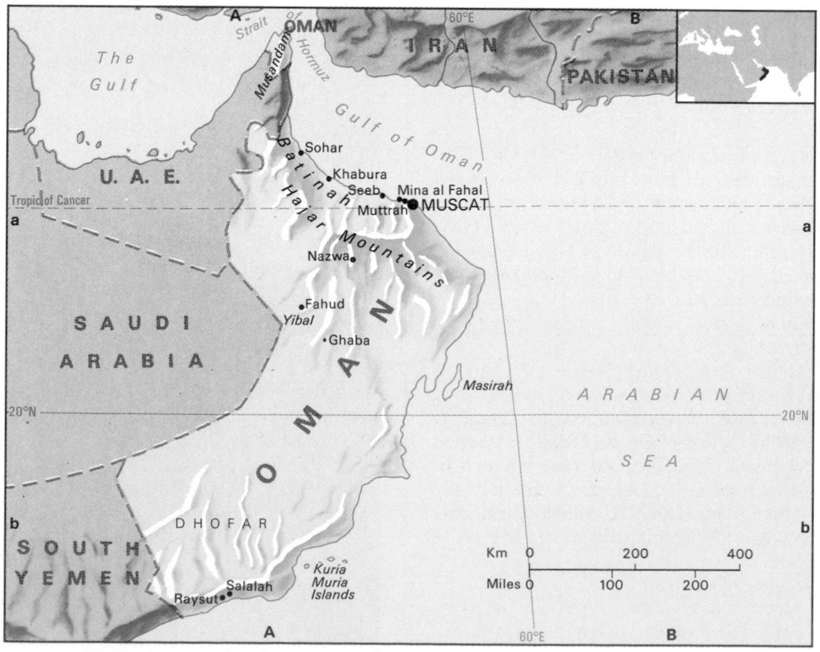

Omsk ***USSR*** Industrial city in western Siberia, about 2200 km (1370 miles) east of Moscow, where the Trans-Siberian Railway crosses the Irtysh river. It was founded in 1716 as a military fort at the confluence of the Irtysh and Om' rivers. During the Second World War Omsk began to make machinery and fertilisers for farms on the plains around it, and its industries now include oil refining, petrochemicals, flour, footwear and textiles.

Population 1 094 000

Map USSR Ic

Onega, Lake (Onezhskoye Ozero) ***USSR*** Second largest lake in Europe after Lake Ladoga, covering 9720 km² (3750 sq miles) in north-western USSR. It is connected by canal with the Baltic Sea, White Sea and Volga river. Its main settlement is the industrial city of Petrozavodsk on the west shore. From there, boats ply to Kizhi island, which has a magnificent open-air museum of wooden churches – including the 22 domed Church of the Transfiguration. The lake, which in places is 120 m (395 ft) deep, is frozen from December to May.

Map USSR Eb

Onitsha ***Nigeria*** Trading town on the east bank of the Niger river, about 360 km (225 miles) east of the capital, Lagos. It lies where the rain forests of the south begin to give way to savannah, and has marketed river-borne goods from both for centuries. Onitsha is now a road focus with one of the few bridges on the lower Niger, and handles palm oil, dried fish, vegetables, rice, cassava and manufactured goods.

Population 262 100

Map Nigeria Bb

Ontario ***Canada*** The country's second largest province, covering 1 068 582 km² (412 580 sq miles) south of Hudson Bay. Northern Ontario is a rocky CANADIAN SHIELD area dotted with lakes, and has abundant forests and minerals. The southern part of the province has gently undulating areas of clays and sands. The southern shoreline stretches 3700 km (2290 miles) along four of the Great Lakes. Agriculture is important in the mild south-west, where crops include wheat, barley, sugar beet, grapes and tobacco. About half of Canada's manufactures come from Ontario, mainly from a southern crescent of territory between Hamilton and Oshawa. Products include steel, cars, electrical goods and chemicals. Ontario is Canada's most populous province. It contains the national capital, OTTAWA, but the largest city is the provincial capital, TORONTO. Tourist attractions include NIAGARA FALLS and 250 000 lakes.

Population 8 625 000

Map Canada Fc

Ontario, Lake ***Canada/USA*** Smallest and most easterly of the Great Lakes, covering about 19 550 km² (7550 sq miles). It is supplied with water from Lake Erie by the Niagara river, with the Niagara Falls, and drains to the Atlantic by the St Lawrence river.

Map United States Kb

Onyx River ***Antarctica*** Longest of the very few rivers in Antarctica. It flows in summer, carrying meltwater from the Lower Wright Glacier about 50 km (30 miles) to Lake Vanda.

▲ **PORT WINE AND ROSES Prince Henry the Navigator (1394-1460), who discovered the Senegal river, was born in Oporto – now renowned for port and fine gardens of roses and camellias.**

oolite Sedimentary rock, usually a limestone, composed of tiny rounded grains embedded in a fine matrix.

Oostende ***Belgium*** See OSTEND

ooze Layer of mudlike sediment covering the floor of oceans and lakes, composed chiefly of remains of microscopic animals. Oozes are named according to the name of the organism whose remains are dominant in them, such as diatom ooze and radiolarian ooze.

opal Variety of silica used as a gemstone. Vivid colours are produced by impurities or by light catching on minute cracks in the stone. Opals range from pearly white, through yellows and reds, to the most valuable black or blue gems.

Oporto (Porto) ***Portugal*** The country's second largest city after Lisbon, lying about 270 km (170 miles) north of the capital on the steep banks of the Douro river. There are three bridges over the Douro: the metal Maria Pia railway bridge completed in 1877 and designed by the French engineer Gustave Eiffel, the designer of the Eiffel Tower in Paris; the Dom Luis I bridge built in 1886, with two layers of roads; and the Arrábida road bridge, completed in 1963.

Before the mouth of the river silted up, the city was a busy port and it still exports wine, including port wine, which takes its name from the city.

Oporto is at the heart of an industrial region producing tyres, chemicals, metals, electrical equipment, car parts, textiles, leather goods, silk, shoes and soap.

Among the city's many places of interest are the 12th-century cathedral; the Clerigos tower – a Baroque church with a granite tower 75 m (246 ft) high; the Gothic and Baroque Church of St Francis; the Soares dos Reis Museum containing pottery, paintings and sculptures; the Ethnographic Museum; and the Palácio da Bolsa – now the stock exchange.

Population 330 200

Map Portugal Bb

Oran (Wahran) ***Algeria*** Port and second city of the country, situated on the north-west coast. The city, overlooked by the 16th-century fort of Santa Cruz on a nearby hill and surrounded by hills planted with orange, lemon and olive groves, was founded in 903 by Arabs from southern Spain. Later it was expanded by the Spaniards who ruled the region from 1509 until it fell to the Turks in 1708. In 1732 it reverted to Spain, and Spanish influence can be seen in a number of buildings and monuments. The area was devastated by an earthquake in 1790, and in 1792 Oran was evacuated and returned to the Turks, who settled a Jewish community there. It began to decline, and in 1831 was occupied by the French, who developed it as a modern port. Industries include cement, chemicals, iron smelting, fruit and fish canning, the manufacture of textiles, glass, footwear and cigarettes. The port exports wheat, vegetables, wine, wool and esparto grass.

Places of interest include the Kasbah (the old Spanish and Arab city); the 18th-century mosque of the Pasha Sidi El Houari; Fort Lamorrue, built by the Spaniards in 1742; the Demaeght Museum with prehistoric exhibits and paintings; and Moorish and Spanish gates.

Population 670 000

Map Algeria Aa

Orange ***France*** Historic town in Vaucluse department, 105 km (65 miles) north-west of Marseilles. Its majestic Roman theatre, seating 10 000, is the best preserved in Europe. The Principality of Orange was annexed by France in the 17th century. It was previously the seat of the Nassau dynasty, members of which were princes of the Netherlands and one of whom, William of Orange, also became King William III of England in 1689.

Population 27 500

Map France Fd

▲ **EXACT COPY It took more than 600 000 men to build Osaka's original castle, which fires in 1615, 1665 and 1867 destroyed. The present building is a reproduction completed in 1931.**

Orange ***Southern Africa*** Longest river of southern Africa, rising in the Drakensberg mountains in north-east Lesotho (where it is called the Senqu), and flowing west to the Atlantic Ocean for 2090 km (1299 miles). Its lower course forms the South Africa-Namibia border. The Orange River Scheme for irrigation and hydroelectric power was launched in 1963, and includes the Hendrik Verwoerd and P.K. le Roux dams.

Map South Africa Ab, Cc

Orange Free State ***South Africa*** Landlocked province of 127 993 km² (49 405 sq miles), excluding Qwa Qwa, on the high plateau between the Orange and Vaal rivers. It became a British possession in 1848 and an independent republic in 1854. In 1900 it was again annexed by Britain, and in 1910 became a province of South Africa, with its capital at Bloemfontein. Gold and coal are the chief resources.

Population 2 080 000
Map South Africa Cb

Oranjestad ***Netherlands Antilles*** **1.** Seaport and capital of the island of ARUBA. The town, which takes its name from the Dutch royal House of Orange, grew around a fort built by the Dutch to protect the island from Spanish and British invasion. The chief industry is oil refining. **2.** Port and capital of ST EUSTATIUS.

Population 1. 10 100 **2.** 1200
Map 1. Caribbean Ac **2.** Caribbean Cb

Orapa ***Botswana*** Site of the world's second largest diamond mine (after Mwadui in Tanzania), discovered in 1967 about 260 km (162 miles) west of Francistown. Mining began in 1971 and is expected to continue for nearly 30 years. Most of the diamonds are used for industrial purposes.

Population 5200
Map Botswana Cb

Orcadas Base ***Antarctica*** The oldest permanently inhabited scientific base in Antarctica, on Laurie Island in the South Orkney group. It was established by a Scottish expedition in 1903, and is now run by Argentina.

Map Antarctica Cb

Ord ***Australia*** River rising in the Kimberley Plateau of Western Australia and flowing 480 km (300 miles) east then north into Cambridge Gulf, 80 km (50 miles) from the Northern Territory border. It has the greatest flow of any Australian river, but this is almost entirely in the summer 'wet'. It is dammed for irrigation to form Lake Argyle, about 475 km (295 miles) south and slightly west of Darwin.

Map Australia Db

Ordesa ***Spain*** Game reserve and national park in the Pyrenees, on the French border about 180 km (110 miles) east and slightly south of the port of San Sebastián. The park lies in a glacier-sculpted valley discovered by a Frenchman – Ramond de Carbonniers – in 1802. The valley, today considered one of Spain's most beautiful spots, is dominated by Monte Perdido (Lost Mountain), 3355 m (11 004 ft).

Map Spain Ea

ordnance datum The mean sea level, calculated from tidal observations at Newlyn, Cornwall between 1915 and 1921, from which all heights on official British maps are derived. The term is abbreviated to OD.

Ordovician Second period in the Palaeozoic era of the earth's time scale, characterised by primitive types of fish. See GEOLOGICAL TIMESCALE

Ordzhonikidze ***USSR*** Industrial city in the Caucasus Mountains, about 160 km (100 miles) north of the Georgian capital Tbilisi. Its products include zinc, lead, glass, chemicals, clothing, footwear and railway wagons. Formerly Vladikavkaz, it was for a long time the main Russian military and political centre in the Caucasus.

Population 300 000
Map USSR Fd

ore Rock containing a mineral or minerals from which metals or other constituents can be economically extracted.

Örebro ***Sweden*** Market town on Lake Hjälmaren, about 196 km (122 miles) west of Stockholm. Founded in the 11th century, Örebro has a 13th-century church and a well-preserved 16th-century castle. It is the centre of Sweden's shoe industry, and it also manufactures paper.

Population 118 000
Map Sweden Bd

Oregon ***USA*** Mountainous state covering 251 180 km² (96 981 sq miles) on the Pacific coast north of California. It rises to 3424 m (11 235 ft) at Mount Hood in the Cascade Range in the north-west. A trading post for furs was set up in 1811, but large-scale immigration by white settlers began in 1842 via the Oregon Trail. This wagon route, some 3200 km (2000 miles) long, began at Independence in western Missouri and followed the Platte and North Platte rivers westwards, crossed the Rocky Mountains at South Pass, and then followed the Snake and Columbia rivers to Fort Vancouver (now the city of Vancouver, in Washington state, across the Columbia from PORTLAND).

Agriculture in Oregon is concentrated in the Willamette and Columbia river valleys, and cattle, wheat, dairy produce, fruit and vegetables are the chief products. Timber, food processing, instrument making and tourism are the main industries. SALEM is the state capital and Portland the major city.

Population 2 687 000
Map United States Bb

Orenburg ***USSR*** River port and industrial city (formerly Chkalov) on the Ural river 360 km (225 miles) south-east of Kuybyshev. It was founded as a fort in 1743, and grew with the arrival of the railway in the early 1870s. It is set in a fertile agricultural district which is also rich in copper, iron ore, nickel, coal and salt; its vast oil and gas resources have been piped to Eastern and Western Europe. Its industries include metal refining, chemicals and food processing.

Population 513 000
Map USSR Gc

Øresund (The Sound) Deepest and most easterly channel connecting the NORTH SEA and the KATTEGAT to the BALTIC SEA. It runs between the Danish island of Zealand and Sweden.

Length About 140 km (90 miles)
Narrowest width 4 km (2.5 miles)
Map Denmark Cb

Oriente ***Ecuador*** Region to the east of the Andes, including the provinces of Morona Santiago, Napo, Pastaza and Zamora Chinchipe. Its northern and eastern borders (with Colombia and Peru respectively) were long disputed until settled at a conference in 1942.
Map Ecuador Bb

Orinoco ***Venezuela*** The country's principal river, about 2200 km (1370 miles) long. It rises in the Sierra Parima in the south, flows north along the Colombian border, then east into the Atlantic through a wide delta. The delta covers some 22 500 km² (8700 sq miles) and consists of innumerable banks and channels.
Map Venezuela Bb

Orissa ***India*** Eastern state covering 155 707 km² (60 103 sq miles) on the Bay of Bengal. Oriya is the official language, but there are numerous tribal groups speaking their own languages, especially in the Eastern Ghats. Rainfall tends to be unreliable, and periodic famines once swept Orissa. It remains underdeveloped, 80 per cent of its people depending on rice growing and 40 per cent of its area still forested. However, more than 50 large industrial projects set up by the government since 1978 include the iron and steel plant at Raurkela. Cuttack is the largest city and Bhubaneshwar the capital.
Population 26 370 300
Map India Cc

Orizaba ***Mexico*** **1.** Town in Veracruz state. It grew around natural springs between the high plateaus of the Sierra Madre Oriental and the humid lowlands of the Gulf Coast. It was severely damaged by an earthquake in 1973, but continues to produce beer, rum, textiles, sugar, coffee and tobacco. **2.** Volcanic peak, better known as CITLALTEPETL.
Population 610 000
Map Mexico Cc

Orkney ***United Kingdom*** Group of about 90 islands and islets, totalling 976 km² (377 sq miles), off the north-east tip of mainland Scotland. A Stone Age settlement has been found at Skara Brae on Mainland island, dating from 2000 BC. The islands were settled by the Vikings in the 9th century and are still rich in Scandinavian remains. During the two World Wars the islands were used as an anchorage by the Royal Navy, and at the end of the First World War the captured German fleet scuttled itself in Scapa Flow – between the islands of Mainland, Hoy and South Ronaldsay – rather than fall into British hands. The islands' capital is Kirkwall on Mainland, which has a 12th-century cathedral. Air services and a two-hour journey by ship links the group with the Scottish mainland. Fishing, cattle rearing and oil industries are the chief sources of income.
Population 19 000
Map United Kingdom Gi

Orlando ***USA*** City in central Florida, about 330 km (205 miles) north and slightly west of Miami. It is the centre of a citrus fruit growing area and is a tourist base for DISNEY WORLD and Cape CANAVERAL.
Population (city) 137 100; (metropolitan area) 824 100
Map United States Je

Orléans ***France*** Capital city of Loiret department, on the Loire 110 km (65 miles) south and slightly west of Paris. In 1429 Joan of Arc, then just 17, forced the English to abandon their siege of the city – a feat for which she became known as the Maid of Orleans. Many old buildings were destroyed in the Second World War but the city, now rebuilt, retains much of its medieval character. Among its modern industries are the manufacture of car parts and chemicals.
Population 225 000
Map France Dc

orogeny, orogenesis Process of mountain building, especially by folding and faulting of the earth's crust.

orographic precipitation Precipitation caused by cooling of moisture-laden air as it rises over a mountain range.

Orontes ***Middle East*** Unnavigable river, 384 km (238 miles) long, rising in the BEQA'A valley in Lebanon and flowing north-east into Syria, where it remains for most of its course. It flows past the cities of Homs and Hamah and is used for irrigation. The river then crosses the Turkish frontier, turns west and enters the Mediterranean north of the Syrian port of Al Ladhiqiyah.
Map Middle East Ba

orthoclase Whitish, red or green potassium aluminium silicate (a feldspar) found in acid IGNEOUS and METAMORPHIC ROCKS, and used in the manufacture of glass and ceramics.

Orumiyeh (Urmia) ***Iran*** City near the Turkish border, about 125 km (80 miles) south-west of the city of Tabriz. According to tradition, it is the birthplace of Zoroaster (about 628-551 BC), the founder of the Zoroastrian religion. The surrounding area supports a population of Armenians, Turks and Kurds.

About 10 km (6 miles) east of the city is Lake Urmia, which, at 4700 km² (1815 sq miles), is Iran's largest lake. The water is too salty for fish, but it is regarded as having curative properties.
Population 164 000
Map Iran Aa

Oruro ***Bolivia*** Department and city 200 km (125 miles) south-east of the capital, La Paz. The department produces about half Bolivia's tin as well as silver, tungsten and antimony from the western hills and slopes of the Cordillera Real range. The city, founded in 1606, is an important mining and rail centre. It is the scene of the country's most famous carnival, La Diablada, which begins on the Saturday before Ash Wednesday each year with a procession of dance groups led by ornately costumed and masked figures representing Satan, Lucifer, St Michael the Archangel, and the Devil's wife. The festival lasts for eight days.
Population (city) 132 200; (department) 385 200
Map Bolivia Bb

Orvieto ***Italy*** Well-preserved medieval city set on a cliff about 95 km (60 miles) north of Rome. Orvieto is known mainly for its white wine and for its Romanesque-Gothic cathedral, which is built in curiously striped stone layers of dark green and white. Tourism is important.
Population 22 800
Map Italy Dc

Osaka ***Japan*** The country's third largest city after Tokyo and Yokohama. It straddles the delta of the Yodo river at the eastern end of the Inland Sea, about 400 km (250 miles) west and slightly south of Tokyo. Osaka has Japan's largest and most sumptuously decorated castle. It also has severe urban congestion, air pollution and land subsidence caused by the extraction of groundwater for its wide variety of manufacturing industries.

The castle, floodlit at night, was built in the late 16th century by the warlord Toyotomi Hideyoshi, thus beginning the transformation of the city into feudal Japan's commercial and financial capital. Its merchants dominated trade and became paymasters to Japan's samurai class. Still a major trade centre, Osaka now has a reputation for hedonism, reflected in its fine food, active nightlife, superb department stores of the Shinsaibashi shopping area and theatres devoted to Japan's traditional but popular *kabuki* dramas and *bunraku* puppet plays.

The city's two main religious sites are the ancient Shinto shrines of Sumiyoshi dating from the 3rd century AD, which are surrounded by hundreds of carved stone lanterns, and the 10th-century Temmangu shrine, where decorated boats parade at festival times.
Population 2 636 300
Map Japan Cd

Oshawa ***Canada*** Industrial city and port on Lake Ontario, 53 km (33 miles) east of Toronto. It is an important car manufacturing centre and also makes glass, textiles, drugs and furniture.
Population 154 200
Map Canada Hd

Osijek ***Yugoslavia*** Croatian city overlooking the Drava river about 30 km (20 miles) from the Hungarian border. It developed around a medieval fort, and has a Gothic cathedral and both Catholic and Orthodox churches, reflecting its mixed Croat, Serb and Hungarian population. As Esseg, the city was part of Hungary from 1699 to 1918, and was the birthplace of Bishop Strossmayer (1815-1905), champion of Croatian independence from the Austro-Hungarian Empire. A major road and rail junction, Osijek is the main market town of Slavonia, processing cereals, fruits, livestock and timber rafted down the Drava – it is the nation's chief producer of matches.
Population 158 800
Map Yugoslavia Db

Ösling ***Luxembourg*** See GUTLAND

Oslo ***Norway*** Capital, largest city, main port and chief industrial centre in the country, lying at the head of Oslofjord on Norway's south-east coast.

The city was founded in 1048 by Harald Hårdråde, king of Norway from 1046 to 1066, and prospered as a trading centre and port, becoming the Norwegian capital in 1299. Norway's principal fortress was built here at Akershus in the Middle Ages. After a disastrous fire in 1624, the city was rebuilt and renamed Christiania (later Kristiania) after Christian IV, then king of both Norway and Denmark, but the name was changed back again in 1925.

The early town's industries developed along the banks of the Aker river, whose successive waterfalls could be used to drive water wheels. Around them, the houses of workers, merchants and government officials spread outwards through the lakes, forests and marshes behind the harbour. As a result of the sprawl, Oslo now covers more than 450 km² (173 sq miles), making it one of the world's largest cities in area. But it retains a rural atmosphere outside the much smaller city centre because of large 'green belt' areas, used today by Oslo residents for hiking and skiing.

The city's spine is the avenue known as Karl Johansgate. It is named after King Karl XIV Johan, a French marshal under Napoleon who was elected Crown Prince of Sweden in 1810 and became King of Norway under the dual monarchy established at the end of the Napoleonic Wars. His palace in central Oslo is still the official residence of Norwegian monarchs. The avenue named after him links the palace to the city's main buildings: the cathedral (built in 1697), in front of which is a statue of King Christian IV; the central market; the original university buildings, founded in 1811; and the Storting, the nation's parliament.

Around this core are theatres, museums and galleries – and nearby one devoted to the work of the Norwegian expressionist painter Edvard Munch (1863-1943). Nearby also is Frogner Park, filled with the works of the Norwegian sculptor Gustav Vigeland (1869-1943), and farther inland the soaring ramps of the ski jumps at Holmenkollen, where an annual jumping championship has been held since 1892.

The island of Bygdøy in the harbour houses a superb maritime exhibition which includes: restored Viking ships; the *Fram*, the ship used by Norwegian explorer Fridtjof Nansen for his voyage through the Arctic ice pack in 1893-6; the *Gjøa*, the ship used by Norway's Roald Amundsen when he became, in 1903, the first man to sail through the North-west Passage; and the *Kon Tiki*, the balsawood raft on which the Norwegian anthropologist Thor Heyerdahl crossed the Pacific from South America to Polynesia in 1947.

Oslo's manufactures include electrical equipment, metal goods, timber, dairy products, machine tools, chemicals, textiles and ships.

Population (city) 448 800; (metropolitan area) 566 500

Map Norway Cd

Osnabrück ***West Germany*** Industrial town about 125 km (78 miles) west of Hanover. It has been a bishopric since the 8th century and has a 13th-century cathedral. Its factories produce steel, cables, textiles, paper, and food products.

Population 160 000

Map West Germany Cb

Osorno ***Chile*** Province in the south-central region of Los Lagos. It has some of the country's most beautiful lakes, including Puyehue and Rupanco, both of which have thermal springs, and Todos Los Santos (All Saints), also known as Esmeralda because of its emerald-green water. The snow-covered slopes of the volcanic Mount Osorno, which rises to 2660 m (8727 ft), are a thriving ski centre. The provincial capital, also called Osorno, was founded in 1553, but was destroyed by Indians half a century later. It was re-established in 1796, and later occupied by German immigrants. It is the site of a 16th-century Spanish fortress.

Population (province) 186 000; (town) 82 000

Map Chile Ad

Ostend (Ostende; Oostende) ***Belgium*** Flemish-speaking port and ferry terminal about 20 km (12 miles) west of Bruges. It is also a naval base and resort. On the seafront, spa baths stand alongside the casino and a villa used by the Belgian royal family.

Population 72 000

Map Belgium Aa

Ostia Antica ***Italy*** Finest surviving Roman city in the country after Herculaneum and Pompeii, remarkable for its public buildings and its beautiful black and white mosaic floors. It stands near the mouth of the Tiber river and was the port of Rome, which lies some 20 km (12 miles) to the north-east. It was destroyed by barbarian invasions and gradually disappeared under mud and wind-blown sand. The nearby village of Ostia, which has an imposing 15th-century castle, was refounded in AD 827.

The excavation of the Roman city started in the early part of this century. The Via delle Tombe leads to the main street of the city, the Decumanus Maximus. The public buildings include the Theatre, the House of Diana (the Roman goddess of hunting) and, facing the Forum (marketplace), the Temple of Vulcan (the god of fire).

Map Italy Dd

Ostrava ***Czechoslovakia*** Coal-mining and manufacturing city in Czech Silesia, the country's main industrial area. The city, the regional capital of northern Moravia, lies about 15 km (10 miles) south of the Polish border, and produces steel, chemicals, metal and heavy engineering goods, building materials and foods.

The Carpathian foothills, south and east of the city, contain the birthplaces of several prominent people: the Austrian psychiatrist Sigmund Freud (1856-1939) at the village of Pribor; the Czech composer Leos Janácek (1854-1928) in Hukvaldy; and Emil Zatopek (born 1922), the Czech long-distance runner who won three gold medals at the 1952 Helsinki Olympic Games, in Koprivnice.

Population 323 700

Map Czechoslovakia Db

Oswiecim ***Poland*** See AUSCHWITZ

Otago ***New Zealand*** District of South Island north-west of the port of Dunedin. It is characterised by flat-topped mountains and glacier-dug lakes. Gold was discovered there during the 1860s, but today sheep and fruit are the economic mainstays. The main town is Alexandra.

Map New Zealand Bf

Otranto, Strait of Sea passage connecting the ADRIATIC and IONIAN seas, and separating the heel of Italy from Albania.

Length 69 km (43 miles)

Map Albania Bb

A DAY IN THE LIFE OF A CANADIAN BUSINESS COUPLE

By 6 am, career wife Monique Webster is wakening twins Daniel and Diane, and her husband, Andrew, 40, is frying bacon and eggs. The Ottawa family is up an hour earlier than usual this frosty October morning; Andrew has to be in Calgary, Alberta – over 2900 km (nearly 1800 miles) away – within six hours. '*Dépêchez-vous*' ('Hurry up'), Monique urges her dawdling three-year-olds. But 'eat slowly', Andrew admonishes later over breakfast. The couple are among some 3 million bilingual Canadians fluent in English and French.

Andrew's four-hour Air Canada flight from the local airport will cross two time zones before landing at 10 am, Calgary time. A middle-level civil servant with Tourism Canada, he is to meet officials of the Calgary Stampede, the world-famous rodeo held each July.

While Andrew is aloft, 38-year-old Monique dresses the twins, stacks plates and coffee mugs in the dishwasher, vacuums the carpeted, two-bedroom, 10th-floor apartment, and does her laundry in the coin-operated machines in the basement. At 12.30 pm she buckles the youngsters in her ageing, Japanese-made station wagon for a 30 minute drive to the *Garderie Paradis des Petits*, a day care centre she owns with Marie-Claire Archambault, a friend from university. The centre yields a modest income, yet lets Monique be with her children and gives her time to do household tasks before the weekend.

Monique's earnings (equivalent to US$7000 a year after taxes) are all saved. Soon she will have the $10 000 down payment for a three-bedroom $70 000 house – modest by Ottawa standards – and $10 000 more for furniture and appliances. The Websters' total income, well above the Canadian average of $30 000 for a family of four, is commonplace in Canada's capital. Taxes take a quarter of Andrew's $40 000 salary, leaving $2500 in monthly take-home pay. After fixed expenses – $500 for rent and $800 for payments on their two cars – there is $1200 for food (which claims about $600 monthly) and for clothes and outings.

On her way home, Monique buys a frozen *tourtière* at the supermarket. She heats the spicy pork pie while the twins watch their favourite television programme, *Polkadot Door*.

Once the children are asleep, Monique relaxes. She is watching the Johnny Carson Show on her bedroom TV set when Andrew telephones. He will be home tomorrow afternoon. Monique tries to share his excitement about the million spectators and 600 cowboys expected at the Calgary Stampede. But she is weary and soon falls asleep.

Ottawa ***Canada*** **1.** Capital of Canada, at the junction of the Ottawa and Rideau rivers in south-east Ontario. The community first developed in 1809 and was initially called Bytown after Lieutenant-Colonel John By, the engineer who built the Rideau Canal. It was renamed in 1854 when it was incorporated as a city, and was chosen by Queen Victoria as Canada's future capital in 1857. Ottawa contains many government offices and the skyline is dominated by Parliament Buildings, rebuilt after a fire in 1916. The capital contains the national museum, national library and public archives. It has many parks and walkways maintained by the National Capital Commission.

2. River that flows through Quebec and Ontario provinces for 1271 km (790 miles). It was an important route for explorers, missionaries and fur traders.

Population (city) 718 000
Map Canada Hd

Ouagadougou ***Burkina*** The country's capital since 1954 when it was linked by rail with ABIDJAN, capital of Ivory Coast. It has been the capital of the Mossi people since the 15th century, and shows a combination of African and French influences. Its buildings of interest include the palace of the Moro Naba, the Mossi emperor; a neo-Romanesque cathedral; and the former French governor's residence. More recently, a huge sports stadium has been built. Tourist attractions are limited, but Ouagadougou is the site of one of the country's few wooded areas, called the Bois de Boulogne after the Parisian park of the same name.

Population 286 500
Map Burkina Aa

Ouargla (Wargla) ***Algeria*** Large oasis town and administrative centre, about 560 km (350 miles) south-east of Algiers, on the cross route that links the central and eastern trans-Saharan routes to the south. It is also a pipeline junction in the middle of the HASSI-MESSAOUD oil region, and is being developed as an industrial area. Water from deep artesian wells has given the town a new lease of life during the past 25 years.

Population 49 000
Map Algeria Ba

Ouarzazate ***Morocco*** Oasis and former colonial French fortress town 130 km (80 miles) south-east of Marrakech on the southern slopes of the High Atlas mountains. The town is noted for its Ouazguita carpets, woven by the women in traditional geometric designs of orange-red on a black background. Nearby to the east stands Taourirt, one of the most impressive kasbahs (old walled quarters) in Morocco.

Population 29 000
Map Morocco Ba

Oudenaarde (Audenarde) ***Belgium*** Cotton textile town 30 km (19 miles) south of the city of Ghent. In 1708 the British under the Duke of Marlborough defeated a French army there during the War of the Spanish Succession.

Population 28 000
Map Belgium Aa

Oudtshoorn ***South Africa*** Town 64 km (40 miles) north of Mosselbaai in southern Cape Province, and the only place in the world where ostriches are bred commercially. It prospered in the 1880s because of a fashion boom in ostrich feathers, which ended with the First World War. It is now largely a tourist centre, although ostriches are still bred on surrounding farms. In recent years the industry has revived somewhat: 90 000 birds on 350 farms produce 2000 kg (4409 lb) of feathers annually. Their meat is also sold, and they are a tourist attraction.

Population 61 500
Map South Africa Bc

Ouenza (El-Wanza) ***Algeria*** Centre of iron ore production, on the railway 120 km (75 miles) south of 'Annaba near the Tunisian border. With neighbouring Bou Khadra it has the oldest iron mines in Algeria, and between them they produce three-quarters of the country's ore, which is exported through 'Annaba.

Population 35 000
Map Algeria Ba

Ouidah (Whydah, Ajuda) ***Benin*** Administrative centre and port about 50 km (30 miles) west of the port of Cotonou. It was originally a Portuguese fort and centre of the Atlantic slave trade during the 18th and 19th centuries. The Portuguese claimed the town until 1961, while the French ruled the rest of Benin until 1960. British and Danish forts nearby are also a grim reminder of the hundreds of thousands of slaves shipped to the New World from Ouidah.

Population 35 000
Map Benin Bb

Oujda ***Morocco*** The eastern gateway to Morocco, and important road and rail junction near the Algerian border, 300 km (190 miles) east of Fès. The town, founded in AD 944, was fought over for centuries by Arabs, Berbers and Turks, earning it the name Medinet el Haira, 'City of Fear'. Little is left of the old town apart from traces of its walls and a fine gateway, the Bab Sidi Abdelwahad. Oujda is now a commercial centre, trading in sheep, wool, cereals, fruit and wine. Nearby is the Sidi Yahya oasis, legendary burial place of John the Baptist and the site of the Battle of Isly where the French defeated the Moors in 1844.

Population 260 000
Map Morocco Ba

Oulu (Uleåborg) ***Finland*** Seaport on the Gulf of Bothnia, about 100 km (60 miles) from the Swedish border, and capital of the province of the same name covering 66 260 km² (23 583 sq miles). Founded in 1605, it is now the largest city in northern Finland, a timber-processing centre, and a gateway for Finnish exports. The city lies at the mouth of the 108 km (67 mile) long Oulu river, which drains the lake of the same name, south-east of the city.

Population 96 400
Map Finland Cb

Ouro Prêto ***Brazil*** Historic town about 70 km (45 miles) south-east of the city of Belo Horizonte. Founded in 1711, and declared part of the world's cultural heritage by the United Nations agency UNESCO, it is an 18th-century time capsule, with cobbled streets and 13 Baroque churches.

Population 30 000
Map Brazil Dd

Ovambo ***Namibia*** Homeland of the country's largest ethnic group, the Ovambos, covering 51 800 km² (20 000 sq miles) in the north of the country. It became Namibia's first homeland in 1967 and was proclaimed a self-governing territory in 1973. Most Ovambo are herdsmen.

Population 545 000
Map Namibia Aa

overburden **1.** Layer of rock that covers a more useful material, such as a coal seam, and which has to be removed in opencast mining. **2.** Any loose material overlying solid rock.

overcast Term used to describe the sky when more than 75 per cent of it is cloud-covered.

Overijssel (Overyssel) ***Netherlands*** Eastern province covering an area of 3925 km² (1575 sq miles) 'over' or east of the IJssel river. It is mostly dairy farmland; but, in the east, are the country's largest textile manufacturing towns: Enschede, Almelo, Hengelo and Oldenzaal. The provincial capital is Zwolle.

Population 1 038 400
Map Netherlands Ca

Oviedo ***Spain*** Northern steel-making city, about 210 km (130 miles) east of Corunna. The 14th-century cathedral contains a cross carried into the Battle of Covadonga in 718 by Don Pelayo, first king of Asturias, at which the Christians began the reconquest of Spain from the Moors. It is the capital of the present-day province of Asturias.

Population (city) 190 100; (province) 1 227 000
Map Spain Ca

Owen Falls ***Uganda*** See VICTORIA, LAKE

Owen Stanley Range ***Papua New Guinea*** Mountains stretching 950 km (600 miles) in the extreme south-east of New Guinea. The highest point is Mount Victoria (4073 m, 13 363 ft).

Map Papua New Guinea Ba

Owendo ***Gabon*** A deep-water port just south of Libreville that has now become a suburb of the capital and its main shipping point.

Map Gabon Aa

Owerri ***Nigeria*** Town just north of the Niger river delta, about 410 km (255 miles) east and slightly south of the national capital, Lagos. It has grown rapidly since it was made the capital of Imo state in 1976, and a twin town is being built across the Nwaori river.

Population 150 000
Map Nigeria Bb

oxbow Horseshoe-shaped lake formed from a U-bend, or meander, in a river, when the river cuts through the neck of the meander and shortens its course.

Oxford ***United Kingdom*** City and county lying in the upper basin of the River Thames. The county stretches as far downstream as the outskirts of Reading and covers 2611 km² (1008 sq miles). It is mainly agricultural with industries concentrated round the city and the towns of Banbury, Didcot and Abingdon. The city of Oxford is 80 km (50 miles) north-west of London. Its university is the oldest in Britain, dating back to at least the 12th century, and has one of the greatest libraries in the world, the Bodleian. Its greatest area of learning has long been the humanities. Its beautiful colleges, mostly of honey-coloured stone, are scattered throughout the inner city, and every narrow street reveals some new architectural treasure. The industrial suburb of Cowley produces cars.

Population (county) 558 000; (city) 117 000

Map United Kingdom Ee

Oxus ***USSR/Afghanistan*** See AMUDAR'YA

Oymyakon ***USSR*** See VERKHOYANSK

Oyo ***Nigeria*** Yoruba tribal town about 170 km (105 miles) north and slightly east of the capital, Lagos. The old town stood 130 km (80 miles) to the north, but was destroyed by Fulani people in a war in the 1830s. Nigeria's first motor road, built in 1905, linked Oyo with the city of Ibadan, some 50 km (30 miles) to the south. Oyo markets tobacco, cotton and indigo dye, and its traditional craft products include beads, embroideries, leather goods and cloth. It lies on the Oyo Plains, which cover most of YORUBALAND, and rise from 120 m (395 ft) in the south (Nigeria's main cocoa-producing area) to 400 m (1310 ft) in the north.

Population 180 700

Map Nigeria Ab

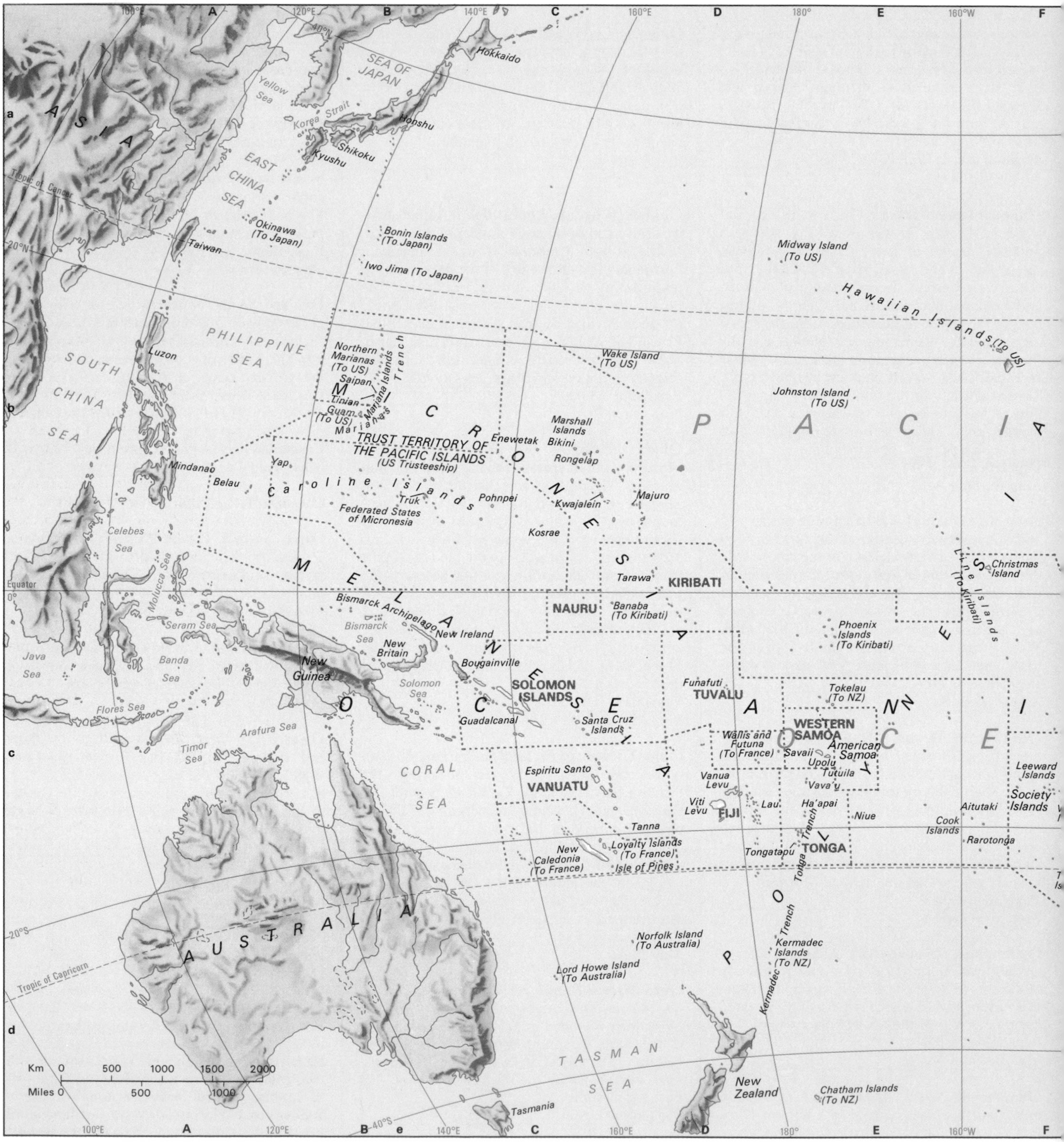

ozone Faintly blue gas with a pungent odour, occurring naturally in the layer of the ATMOSPHERE known as the stratosphere, or ozone layer. It is a form of oxygen in which each molecule contains three atoms of oxygen, unlike the two-atom oxygen in the air we breathe. Ozone is highly poisonous, but it plays a vital part in maintaining conditions suitable for human life on the surface of the earth, as it shields out harmful ultraviolet radiation from the sun, and produces a warm covering, or inversion layer, which limits convectional movements of air to lower atmospheric layers.

Paarl *South Africa* Town at the centre of a major wine-making region 56 km (35 miles) north-east of Cape Town. French Huguenots introduced grapes to the region in 1687. The *Genootskap van Regte Afrikaners* (Institute of True Afrikaners) was founded in the town in 1875 to establish Afrikaans – a variant of Dutch – as a written language.

Population 59 100

Map South Africa Ac

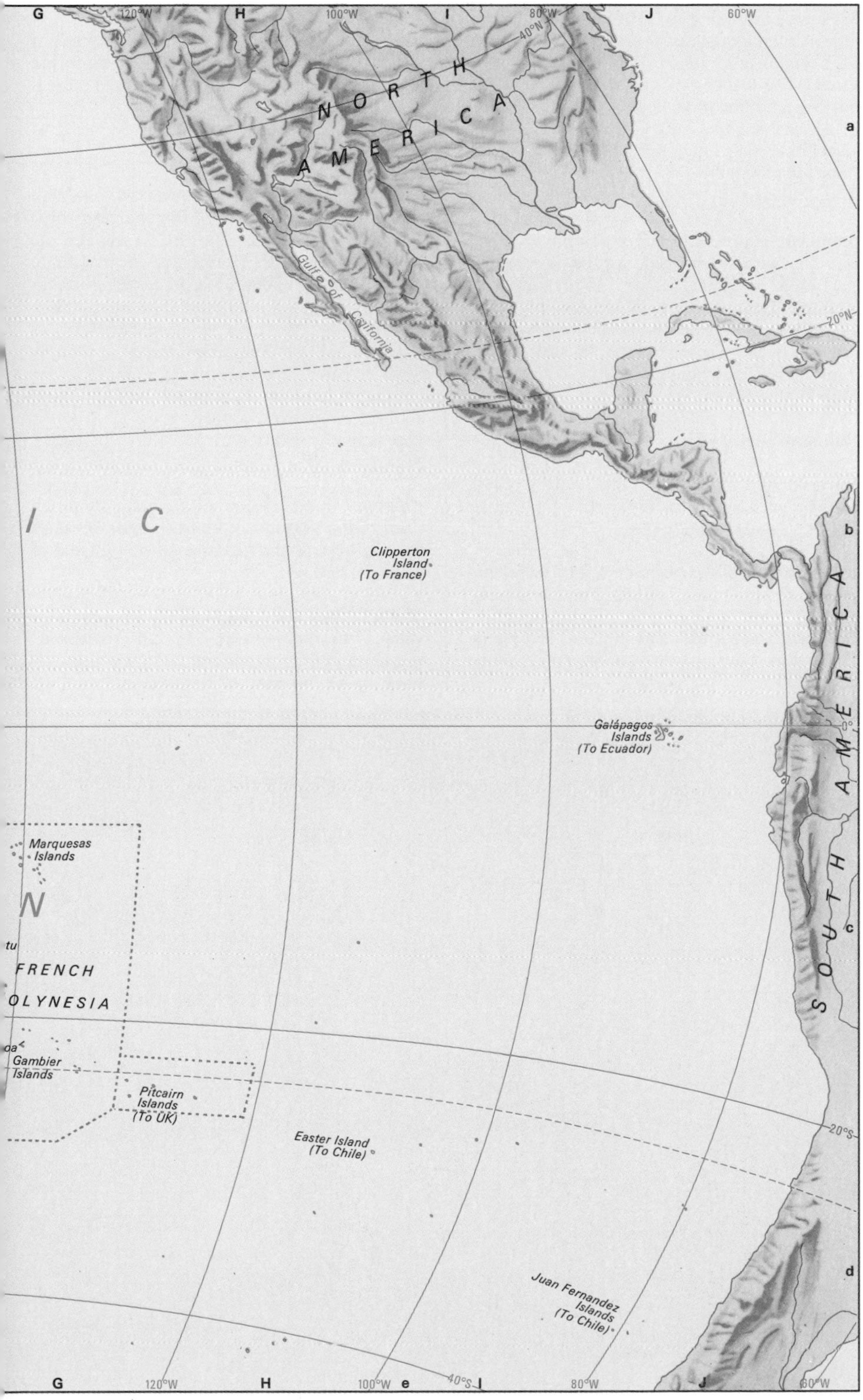

Pacific Ocean Covering a greater area than all the continents put together, the Pacific is the world's largest ocean. It was given its name by the Portuguese navigator Ferdinand Magellan, because of his calm voyage from the tip of South America to the Philippines in 1520-1. The Pacific is by no means always peaceful. The highest wave in an open sea, measuring an estimated 34 m (112 ft) from trough to crest, was recorded there during a hurricane in 1933.

This great ocean includes the BERING SEA and the Sea of OKHOTSK in the north, and the Sea of JAPAN and the CHINA SEA in the west. The Philippine and Indonesian island groups enclose the BANDA, CELEBES, SERAM, FLORES, JAVA and MOLUCCA seas. The ARAFURA, CORAL and TASMAN seas are in the south-west. The islands in the western Pacific are considered part of Asia. But the many volcanic and coral islands of Micronesia, Melanesia and Polynesia are included in OCEANIA.

With an average depth of 4200 m (13 780 ft), the Pacific is the world's deepest ocean. Because of this depth, scientists believe, a smaller proportion of the water evaporates, and thus the ocean remains less salty than the shallower Atlantic.

The main currents in the North Pacific – the North Equatorial Current, the Japan or Kuroshio Current and the California Current – rotate in a clockwise direction, north up the coast of Asia and south down the coast of North America. The circulation in the South Pacific – of the South Equatorial Current, the East Australia Current, the Antarctic Circumpolar Current, and the Peru (Humboldt) Current – is anticlockwise, north up the coast of South America and south past Australia.

The Pacific provides half the world's catch of fish and shellfish. The leading fishing grounds cover the continental shelves in the north-west. The next most abundant waters are off the coast of South America. Oil is extracted near the Asian and North American coasts, while manganese nodules are scattered over large areas of the deep ocean floor – the ABYSS.

Area 165 384 000 km² (63 838 000 sq miles)

Greatest depth Marianas Trench 11 033 m (36 197 ft)

Paczkow *Poland* Town about 75 km (50 miles) south of the city of Wroclaw. It is often known as the 'Polish (or Silesian) Carcassonne', for, like that town in southern France, it is still surrounded by medieval walls, complete with towers and bastions.

Population 10 000

Map Poland Bc

Padang *Indonesia* Provincial capital of West Sumatra. It produces cement and rubber.

Population 480 900

Map Indonesia Bc

Paderborn *West Germany* Town about 100 km (60 miles) south-west of Hanover. It gets its name because the Pader river rises there, bubbling out of more than 100 underground springs. The Frankish ruler Charlemagne met

Pope Leo III there in AD 799 and formed a pact. The next year in Rome Leo crowned Charlemagne the first Holy Roman Emperor.
Population 100 000
Map West Germany Cc

Padma ***Bangladesh*** See JAMUNA

Padua (Padova) ***Italy*** City about 35 km (22 miles) west of Venice. It is one of Italy's most important artistic centres, preserving many medieval palaces, gold-domed churches and priceless works of art.

Roman Padua, Patavium, was one of the most prosperous towns in the empire, and the Roman historian Livy was born there in 59 BC. It was a leading Italian commune during the 11th-13th centuries. The university was founded in 1222, and the astronomer Galilei Galileo (1564-1642) was a lecturer there for 18 years.

The city centre is built around a number of fine piazzas surrounded by medieval palaces. The 13th to 14th-century Palazzo della Regione overlooks the Piazza delle Frutta and the Piazza delle Erbe, the ancient fruit and vegetable markets which are still flourishing. The cathedral dates from the 9th century, but was mostly rebuilt in 1552.

The city's economy is based on food and drink, agricultural machinery, bicycles and motorcycles, electrical goods and textiles.
Population 228 700
Map Italy Cb

Paektu san (Baegdu Son) ***China/North Korea*** Extinct volcano and, at 2744 m (9003 ft), North Korea's highest peak. It stands astride the Chinese border and its crater contains a deep lake.
Map Korea Db

Paestum ***Italy*** Archaeological site by the Gulf of Salerno about 75 km (47 miles) south-east of Naples. The Greeks called it Poseidonia and it was a flourishing Roman colony. It declined because of Saracen attacks and malaria in the surrounding marshes. It has several Doric temples, including the 6th-century BC Basilica and Temple of Ceres, the Roman goddess of agriculture, and the well-preserved 5th-century BC Temple of Neptune, the god of the sea.
Map Italy Ed

Pagan ***Burma*** Ancient capital of the country between 1044 and 1287, when it was sacked by the Mongol horde of the Chinese Emperor Kublai Khan. The Pagan dynasty, which ruled Pagan, introduced Buddhism to the country – and hundreds of large and small stupas (dome-shaped Buddhist monuments) still dot the banks of the Irrawaddy river. Many of the monuments, and some larger temples, were devastated by an earthquake in 1975, but the ruined city – which sprawls across some 16 km² (6 sq miles) about 160 km (100 miles) south-west of Mandalay – remains a major tourist attraction. The largest surviving temple complex is the Ananda monastery, which contains four huge standing statues of the Buddha.
Map Burma Bb

Pago Pago ***American Samoa*** Capital and chief port of the territory, standing at the head of the fine natural harbour of the same name, on Tutuila island. Its name is pronounced 'Pango Pango'. It was a US naval base from 1872 to 1951, and was the setting for Somerset Maugham's famous short story *Rain*; about 5000 mm (197 in) of rain falls each year. Today the main industry is tuna fish canning. Fagatogo, the seat of government, is just to the east of Pago Pago.
Population 3000

Pahang ***Malaysia*** Largest state in Peninsular Malaysia, extending to 35 965 km² (13 886 sq miles). It lies on the east coast between the states of Kelantan and Johor. It is still largely rain forest. The Pahang river, which rises in the north-west of the state, flows 456 km (285 miles) before entering the south China Sea. It is Peninsular Malaysia's longest river.
Population 800 000
Map Malaysia Bb

pahoehoe Hawaiian term for a newly solidified lava flow with a festooned, or rope-like surface.

Painted Desert ***USA*** Wilderness of about 19 400 km² (7500 sq miles) in northern Arizona, just east of the Grand Canyon. Its rock formations are brilliantly coloured.
Map United States Dc

Pakistan See p. 500

Palaeocene First epoch in the Tertiary period of the Cenozoic era of the earth's time scale. See GEOLOGICAL TIMESCALE

Palaeozoic era The earliest era for which there is good fossil evidence of life forms, beginning about 570 million years ago. Fishes, amphibians, primitive reptiles and the first land plants all developed during this time. See GEOLOGICAL TIMESCALE

Palau ***Pacific Ocean*** See BELAU

Palawan ***Philippines*** Narrow sword-shaped island, covering 11 785 km² (4550 sq miles) between the South China Sea and Sulu Sea. Rugged and rather sparsely populated, its main resources are timber and oil in small oil fields recently found off the northern coast. There are few roads or large towns, but abundant wildlife.
Population 300 100
Map Philippines Ad

Palembang ***Indonesia*** Provincial capital of South Sumatra. It was the site of the 7th-century Hindu maritime empire of Sri Vijaya. Today it is an oil and petrochemical centre.
Population 787 200
Map Indonesia Bc

Palencia ***Spain*** Walled town 190 km (118 miles) north of Madrid, on the Carrion river. Founded in the 3rd century BC, it was the site of the first Spanish University, established in 1185. Textiles, chemicals and farming equipment are its main industries. The surrounding province, of the same name, encloses in its southern half the 'Tierra de Campos', probably Spain's major cereal-producing area. Potatoes and sugar beet are also grown, and there are deposits of coal and copper.
Population 75 000 (town); 187 000 (province)
Map Spain Ca

Palenque ***Mexico*** One of the great forgotten cities of the Mayas, deep in the virgin forest on the foothills of the Usumacinta mountains in

▼ PATTERN ON THE PLAINS An aerial view of the harvesting of alfalfa (lucerne grass), used as livestock feed, on the vast, open Pampas of Argentina.

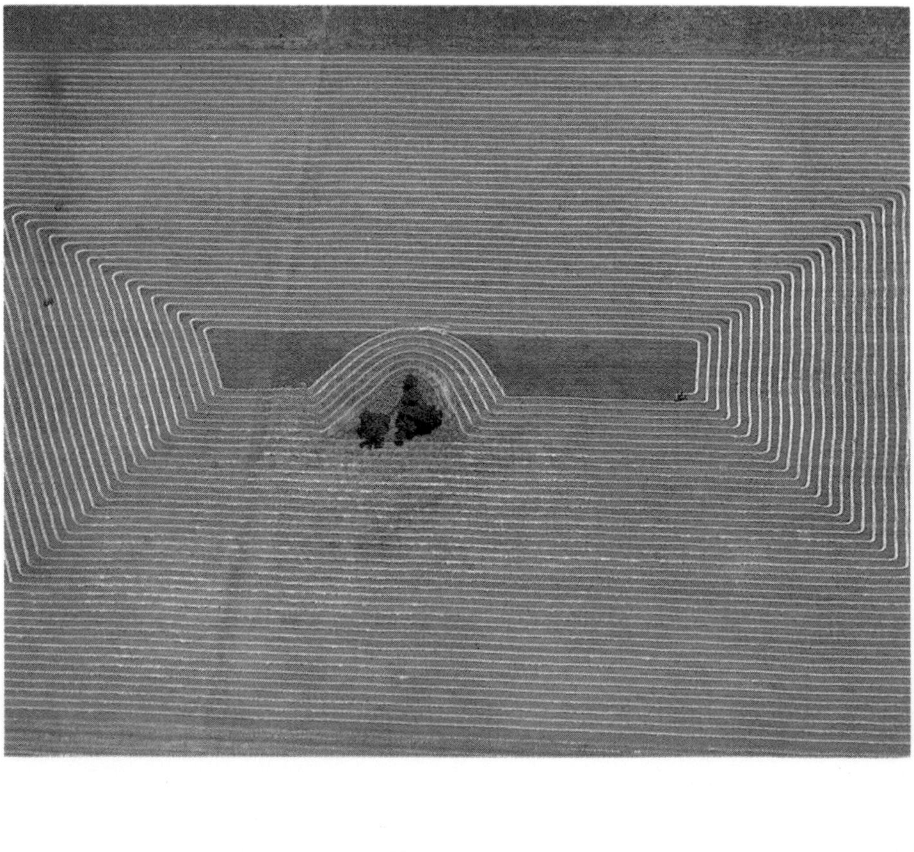

CHIAPAS state. Its temples, palaces and pyramids include the remarkable stuccoed Temple of the Inscriptions, where a royal crypt was discovered, its occupant lying beneath a treasure of jade, with which he had been buried in AD 683.

Map Mexico Cc

Palermo ***Italy*** Main port and capital of Sicily. It stands on the island's north coast surrounded by a fertile plain of orange and lemon groves and sheltered from the north by Mount Pellegrino, 606 m (1988 ft).

The city was founded by the Phoenicians in the 8th century BC, but its great period was under the Arabs between AD 831 and 1072, when many fine buildings were constructed. The Normans preserved much of the Arabs' cultural vitality, but subsequently the city declined. It prospered afresh when it was Spanish viceregal capital in the 17th and 18th centuries, and was transformed into a predominantly Baroque city. The Spaniards founded the university.

Modern Palermo is indelibly associated with the Mafia, which controls many aspects of the city's life and economy.

Population 718 900

Map Italy De

Palestine ***Middle East*** The boundaries of Palestine have varied over the centuries and are disputed today as part of the conflict between Arabs and Jews. In Biblical times it was the Land of Canaan, the Jews' 'Holy Land promised by God', which they invaded under Joshua's leadership in the 15th century BC. The territory included what is now the Jewish state of Israel, and parts of Egypt and Jordan, extending eastwards beyond the Jordan river.

Since Joshua's conquest, Palestine has been ruled by many nations, including Egyptians, Romans, Byzantines, Arabs and Turks. When the Turkish Ottoman Empire broke up after the First World War, a mandate to govern Palestine was granted to Britain, which had invaded the territory in 1917. Later that year the British Foreign Secretary, Arthur Balfour, pledged support for a homeland for the Jews in Palestine, and in 1920 this was reinforced by the Treaty of Sèvres, between the Allies and Turkey.

Jews around the world began the struggle to establish their promised Jewish homeland. The campaign was called Zionism after the ancient Palestinian capital Zion – now modern JERUSALEM. However, Britain was unable to resolve the two conflicting nationalisms, Jewish and Arab, and so placed the problem before the United Nations. The UN agreed that the mandate of Palestine should be divided into two states, one Jewish and one Arab. The state of Israel that emerged in 1948-9 included not only the territory designated by the UN as the Jewish state, but also contained parts of the Arab state. What was left of the UN's proposed Arab state was taken by Jordan (the WEST BANK) and by Egypt (the GAZA STRIP). In 1967 Israel seized the West Bank and the Gaza Strip in the Six-Day War.

As a result of these conflicts, many thousands of Palestinian Arabs became homeless refugees. The Palestine Liberation Organisation (PLO), a former guerrilla movement now officially recognised, is campaigning for a Palestinian state for all Palestinians.

Map Middle East Ba

Palimé ***Togo*** See KPALIME

Palm Beach ***USA*** Atlantic resort in Florida, on an island just offshore, about 100 km (60 miles) north of Miami Beach. West Palm Beach, on the adjacent mainland, is a manufacturing centre for aircraft engines and electronics.

Population (Palm Beach) 10 700; (West Palm Beach) 67 600; (metropolitan area) 692 200

Map United States Je

Palma, La ***Spain*** See CANARY ISLANDS

Palma (Palma de Mallorca) ***Spain*** Port, tourist resort and capital of Majorca and of the Mediterranean BALEARIC ISLANDS. It exports almonds, fruits and wine, and manufactures paper, textiles, shoes, glass and cement.

Population 304 400

Map Morocco Gc

Palmerston North ***New Zealand*** Market town on North Island, about 130 km (80 miles) north and slightly east of the capital, Wellington. Settled in 1866, it was named after the British prime minister, Lord Palmerston (1784-1865). It is now the main town of a rich farming area and has an agricultural research centre, a meat-freezing plant and small-scale agricultural engineering firms. It also has a university.

Population 60 100

Map New Zealand Ed

Palmyra (Tadmur) ***Syria*** Ruined city in an oasis just north of the Syrian Desert. Palmyra, which is 215 km (134 miles) north-east of Damascus, has been greatly restored and has a modern settlement outside its walls.

The city is said to have been built by Solomon in the 1st century BC beside the oasis of Tadmur, where date palms were an important crop. It became the main stopping-place on the caravan route from THE GULF to the Mediterranean. The Romans colonised Tadmur in AD 217 and renamed it Palmyra, 'City of the Palms'. As a desert fortress, it became a powerful buffer between Roman-occupied Syria and Persia (Iran). Later in the 3rd century Queen Zenobia of Palmyra led a revolt against Rome, and in AD 273 the city was pillaged by the Emperor Aurelian. Palmyra never fully recovered, although new fortifications were built and the city was lived in until the 17th century. Known as the 'City of 1000 Columns', it is spread over 6 km² (2 sq miles). Its most impressive remains are the Great Colonnade, which stretched for 1000 m (3280 ft) complete with buildings and monuments; the Great Temple of Bel (Baal), the Babylonian god of heaven and earth; and its largely restored theatre.

Map Syria Bb

pamir Sparse grassland on a high plateau in the mountains of central Asia.

Pamir ***USSR/China/Afghanistan*** Mountainous region of high plateaus, peaks and glaciers in central Asia, forming the north-west limb of the Himalayas. The Pamirs are the USSR's highest mountains, with several peaks rising over 6000 m (20 000 ft). The region is set mainly within the republic of Tadzhikistan, and runs for 225 km (about 140 miles) from north to south and 400 km (245 miles) from east to west. Their highest point and the Soviet Union's highest mountain is COMMUNISM PEAK, at 7495 m (24 590 ft).

Map USSR Ie

Pampas ***South America*** Grassy plains forming the heartland of Argentina, similar to the North American prairies. They cover some 650 000 km² (251 000 sq miles) – an area larger than Spain and Portugal combined – and extend into Uruguay. The region produces well over half of Argentina's livestock and three-quarters of its grain, and is responsible for 90 per cent of the country's agricultural output.

The Argentinian province of La Pampa – 143 440 km² (55 382 sq miles) – takes its name from the plains. Its provincial capital, Santa Rosa (population 52 000), is about 560 km (350 miles) west and slightly south of the national capital, Buenos Aires.

Population (province) 208 000

Map Argentina Cb

pampero Dry, bitterly cold south or south-west wind blowing across the Pampas of Argentina and Uruguay, particularly in summer, corresponding to the southerly buster of Australia.

Pamplona ***Spain*** Town near the western end of the French border. It is famous for the 'Running of the Bulls' each July – when young men run ahead of the animals through the streets. The festival was the setting for *Fiesta* (*The Sun Also Rises*), a novel by the American writer Ernest Hemingway (1898-1961).

Population 183 100

Map Spain Ea

pan Basin or depression in an arid or semiarid region, generally holding water only during the rainy season. Some pans contain water all year but are less permanent and less deep than lakes.

Panaji (Panjim) ***India*** Seaport and capital of the territory of GOA, Daman and Diu, and formerly the capital of all the Portuguese possessions in India. It exports manganese ore, and is the tourist centre for the territory.

Population 77 200

Map India Bd

Panama See p. 502

Panama Canal ***Panama*** Waterway cut across the isthmus of Panama between the cities of Balboa and Colón. It links the Pacific and Atlantic Oceans. The canal was completed in 1914, under the direction of George Washington Goethals, a US army engineer, and is 82 km (51 miles) long. Three groups of locks raise ships of up to 67 000 tonnes to a maximum height of 26 m (85 ft) above sea level. As many as 44 ships a day can pass along the canal, each taking about eight or nine hours to get through. The channel is between 150 and 300 m (490-980 ft) wide and up to 100 m (330 ft) deep.

Gatún Lake, at the highest point, is one of the largest artificial lakes in the world. It supplies the 296 million litres (65 million gallons) of water needed for the locks each time they are opened. Curiously, because of the isthmus's twisting shape, the Pacific end of the canal is east of the Atlantic end.

Map Panama Ba

Pakistan

MUSLIM HOMELAND CONTROLLED BY A POWERFUL ELITE; WHERE STRICT ISLAMIC LAWS ARE ENFORCED ON A RESTIVE MODERN GENERATION

One of history's great caravan routes runs through Pakistan, from the 'crossroads of the world' in Afghanistan, to India and the ancient maritime highways of the Arabian Sea. Armies followed trade, and the territory has been repeatedly invaded and fought over. Thousands of years of turmoil have mixed peoples and cultures to produce a rich diversity.

Pakistan comprises the whole of the INDUS basin, minus a few headwaters in Tibet. But the borders were drawn without regard for ethnic boundaries. They were established by the British, when they ruled all India, and their main concern was military security. As a result, the Pathan tribes of the North-West Frontier straddle the border between Pakistan and Afghanistan, in the north-east the Sikhs' territory is cut in two by the border with India, and in the south-west the Baluchis live on both sides of the border with Iran.

Geographically Pakistan divides into three regions. The largest is the central plain country: the river basin of the Indus and its tributaries. In the north and west are mountains including some of the world's highest peaks; the peak of K2 (Godwin Austen) in Kashmir – a territory also claimed by India – is the second highest in the world at 8611 m (28 251 ft).

Pakistan was born out of Britain's Indian empire when it became independent in 1947. It had been hoped to keep the territory as one state, but it proved impossible to reconcile the religious and cultural differences of the two major factions, the Hindus and the Muslims. So, largely because of the campaigning of the Muslim leader, Muhammed Ali Jinnah, Pakistan was created for the Muslims. The new state was an ungainly creation: two territories separated by a piece of northern India.

The eastern territory, predominantly Bengali, was the more important exporter of Pakistan's early days, supplying most of the world's jute. Little of the resulting wealth was reinvested in the east. The Bengalis became the poor relations of West Pakistan and began to demand their own self-governing state.

In 1971 a guerrilla war broke out in the east, which the Pakistani government tried to repress with force. Millions of refugees poured into India, which intervened and found itself at war with Pakistan. The war resulted in East Pakistan becoming the independent state of Bangladesh in 1971.

From 1971 (West) Pakistan struggled to find its new identity. The four major regional groups are the Punjabis, Baluchis, Sindhis and Pathans, but their societies are overlaid by other groups. Rich urban Muslims, the Muhajirs, migrated from India in 1947 and became a major force, particularly in business.

All these conflicting interests, plus the power struggles between feudal landlords, traditional religious teachers, the new bureaucrats and military officers, were more than democratic institutions could withstand. Since 1958 the military have been in power, except for six years from 1971 to 1977 when Zulfikar Ali Bhutto was at first president, and then prime minister. Bhutto was ousted by a military coup led by General Mohammad Zia-ul-Haq in 1977.

Pakistan began as an exporter of raw

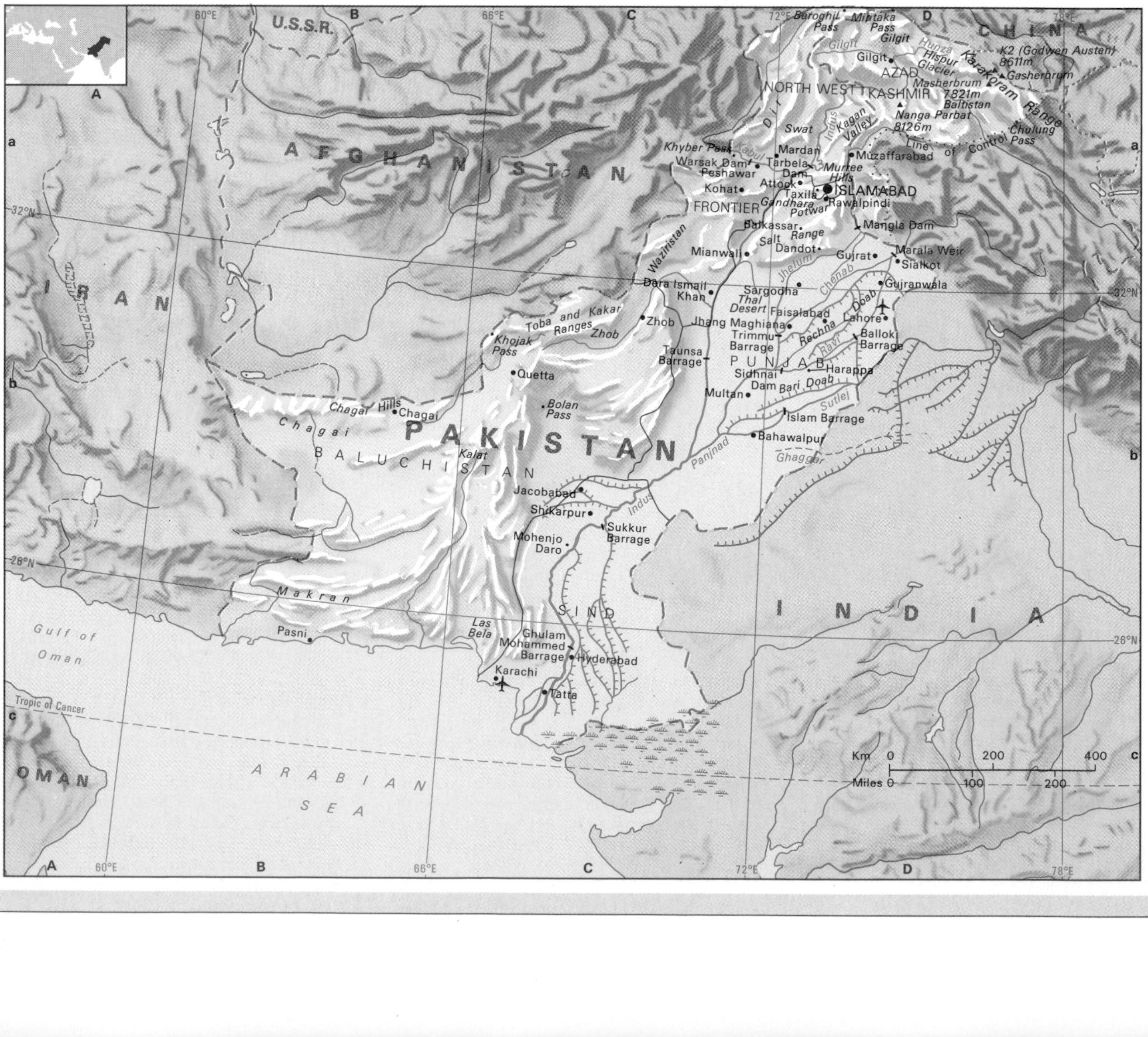

materials and an importer of manufactures. All governments have tried to make the country more self-reliant by industrialisation. Spectacular industrial growth was achieved in the 1960s, thanks largely to massive aid from the USA and from the rich Muslim Arab nations.

Much of the growth was achieved at the cost of agricultural output, although in recent years a better balance has been achieved. Now a considerable amount of Pakistan's exports are manufactures which, like cotton textiles, are often based on her own raw materials. There are traditional craft industries, including carpets and, on the North-West Frontier, the handcrafting of guns. A recent success has been the exporting of sports goods.

The economy is short of power, despite hydroelectric schemes and large reserves of natural gas. About two-thirds of Pakistan's energy is home-produced and one-third imported, mostly in the form of oil. It has proved difficult to maintain growth in manufacturing, which still employs less than 16 per cent of the workforce.

So Pakistan remains overwhelmingly a rural country, and its staple crop is wheat. The central plains are dry, and farmers are dependent on major irrigation schemes. A monsoon season from June to October brings heavy rain to the southern slopes of the Baluchistan and Sulaiman ranges, but only about 150 mm (6 in) of rain per year falls on the plains. During the cold season, from November to February, the winds blow cold and dry; the hot season from March to May is sunny and rainless.

The development and redevelopment of irrigation schemes since independence has mostly been achieved with massive help from the World Bank and the USA. The dams on the Indus at Tarbela and the JHELUM at Mangla are among the largest earth and rock-filled dams in the world. The result has been a steady growth in agriculture and Pakistan is now self-sufficient in food grains.

In a nation of many schisms, the simplest division is between the mountain-dwellers and the peoples of the plains. The hardy, independent mountain people live mostly beyond the control of central government, and conduct their own trade across the frontiers, in guns, modern consumer goods and sometimes in drugs. They grow food to eat, and herd goats and sheep.

The plains people live in mud and brick villages, and worry about the performance of the irrigation system and the state of the cotton or wheat crop.

The people of the mountains and the plains are united by their Islamic religion. But society is split into many castes and tribes. One unit is the *bradri*, or marriage group – a blood group of families which arranges all its marriages within its own membership.

Even though Islam preaches the equality of man, it is common for each different sect in a village to have its own mosque. Each mosque has a resident priest, the *imam*, who calls the faithful to prayer five times a day.

The literacy rate in Pakistan is still low; lower in rural than urban areas, and much lower for women than for men. For many, the imam's teaching may be all they receive, and most of that will be learning the Koran in Arabic by heart.

Women mostly stay at home, out of public view, although sometimes they are seen helping in the fields. The men gather in public, and their favourite evening pastime may be to meet and gossip while smoking a communal pipe called a *hukka*. The men put great emphasis on virility and on male heirs. They may take more than one wife, although this is not as common as it was. Traditional Islam rejects contraception, so the growth rate of population is therefore very high, and the division of land between male heirs over successive generations reduces the sizes of farms.

Most Pakistanis live mainly on *chapattis* (thin unleavened bread), rice and vegetables. The goats and cattle are not very productive, so meat is expensive and dishes like mutton curry or beef kebabs are a rare treat. Alcohol is taboo, and social discourse takes place with a cup of tea.

The growth of industry led to bigger towns and helped to create a new generation of Pakistanis, who are much less traditionalist. For entertainment they look to television and films, but the government tends to monopolise TV for propaganda, and imported foreign films are often second-rate.

The government's protection of new industry and finance in this young country has made some families very rich and powerful. About 20 families, each of them a bradri blood group, control the economy. In conjunction with military leaders they represent a powerful elite, with little inclination to share power. The new urban working class has shown its anger at the system, and civil unrest has toppled governments. But each new regime has been formed from the same old elites.

The present government of President Zia has turned to those same conservative elements of Islam to sustain its authority. Zia has reversed a liberal trend and once again traditional Islamic law prevails.

▲ MOUNTAIN FASTNESS The Chitral river cuts through the Hindu Kush mountains in the far north, where the tribespeople live in fortified villages.

PAKISTAN AT A GLANCE

Area 803 943 km² (310 402 sq miles)

Population 89 831 000

Capital Islamabad

Government Federal Islamic republic

Currency Pakistani rupee

Languages Urdu, English, Punjabi, Sindhi, Pushtu, Baluchi, Bravi

Religions Muslim (97%), Hindu (1.6%), Christian (1.4%)

Climate Subtropical; monsoon from June to October. Average temperature in Karachi ranges from 13-25°C (55-77°F) in January to 28-34°C (82-93°F) in June

Main primary products Cotton, wheat, rice, sugar cane, maize, fruit (dates), tobacco; natural gas, gypsum, iron ore (also bauxite, copper, antimony, magnesite, phosphates – not yet exploited)

Major industries Agriculture, cotton yarn and fabrics, sports goods, petroleum refining, cement, sugar refining, fertilisers, food processing

Main exports Cotton and cotton textiles, rice, hand-crafted carpets and guns, sports goods, leather goods, petroleum products

Annual income per head (US$) 354

Population growth (per thous/yr) 30

Life expectancy (yrs) Male 51 **Female** 50

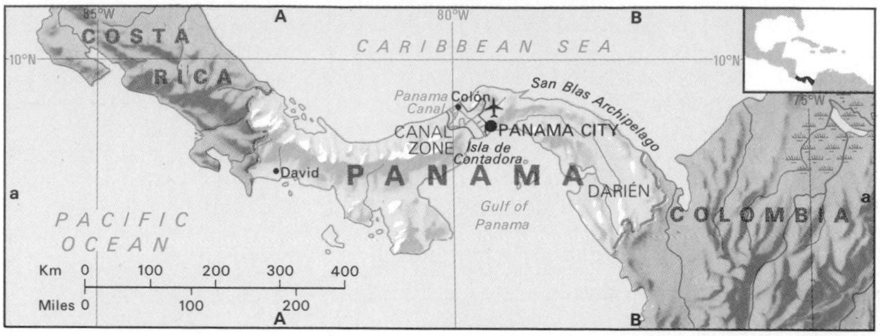

Panama

THOUSANDS OF SHIPS USE PANAMA'S CANAL ANNUALLY, MAKING IT VITAL TO THE COUNTRY'S ECONOMY

A mere 58 km (36 miles) divides the Caribbean and Pacific coasts of Panama at the country's narrowest point. It was from a peak in Darién – as the isthmus was then known – that in 1513 the Spaniard Vasco Núñez de Balboa became the first European to sight the world's greatest ocean. Ever since, Panama has been a routeway from the Caribbean – and Atlantic – to the Pacific.

The country is lucky in having a break in its backbone chain of mountains – a pass only 87 m (285 ft) above sea level, which allowed a relatively easy way across. But dense forest covers much of the country. In addition, Panama was – at least until the beginning of this century – very unhealthy.

At first the crossing was by track, later by railway and finally by canal. Until the early 1800s, silver from Peru was carried across the isthmus to be loaded into Spanish galleons. However, attacks by British 'pirates' such as Henry Morgan and Admiral Vernon helped prevent the route and the colony from prospering – the main towns were looted and burnt on several occasions. By the mid-19th century, however, the railway from COLON on the Caribbean coast to PANAMA CITY on the Pacific was carrying goods and people from America's eastern states to the booming West. Completion of the PANAMA CANAL in 1914 made the country a critical focus of international routes.

Panama achieved freedom from Spain in 1821, at the same time as neighbouring Colombia. Indeed, it remained part of Colombia until 1903, when an American inspired revolution led to a declaration of independence. Panama signed a treaty with the United States giving America control of a 16 km (10 mile) wide 'Panama Canal Zone' for a down-payment of $10 million, plus $250 000 a year.

The Americans set about clearing the country of mosquitos, and completing a canal which the French engineer Ferdinand de Lesseps had started in 1879, but abandoned in 1889. The first ship passed through the 82 km (51 mile) long waterway in 1914, and the project cost $350 million. In 1979 America returned sovereignty of the zone to Panama, which will take full control of the canal on December 31, 1999.

The country's economy is heavily dependent on the canal: it imports much more than it exports, with most of the difference made up by financial services, tourism, and the services and goods it provides for the canal. Nearly a third of the workforce are farmers. Cropland covers only about 5 per cent of the country, but produces the main staple food – rice.

Over the years the government has tried to reduce Panama's dependence on the canal, but it is due mostly to the waterway that this is one of the richer nations in Latin America. A free-trade zone has been set up at Colón. The country is developing as a financial centre, with more than 100 international banks operating there. A huge merchant fleet of about 12 000 ships sails under the Panamanian flag. Most of them do so because registration fees are low and labour laws lenient.

The people are of mixed blood, although there are distinct communities of blacks descended from African slaves, whites and native Indians. The government has for long periods been dominated by the military, but the first elections for 12 years were held in 1980, and the first presidential election for 16 years took place in 1984.

PANAMA AT A GLANCE

Area 78 046 km² (30 134 sq miles)
Population 2 074 700
Capital Panama City
Government Republic
Currency Balboa = 100 centésimos
Languages Spanish (official), English
Religion Christian (93% Roman Catholic, 6% Protestant)
Climate Tropical; average temperature in Balboa Heights ranges from 22°C (72°F) to 32°C (90°F)
Main primary products Bananas, rice, sugar cane, coffee, maize, shrimps, timber
Major industries Agriculture, food processing, petroleum refining, cement, clothing, forestry, financial services
Main exports Petroleum products, bananas, shrimps, sugar
Annual income per head (US$) 1800
Population growth (per thous/yr) 18
Life expectancy (yrs) Male 69 **Female** 73

Panama City ***Panama*** National capital, at the Pacific end of the Panama Canal. It was founded by Spaniards in 1519 on the site of an Indian fishing port. Gold and silver from South America were stockpiled there on their way to Spain. The buccaneer Sir Henry Morgan sacked it in 1671, and although the Spanish rebuilt the city, its fortunes declined. It became the capital in 1903, when Panama gained independence from Colombia, and expanded after the canal was opened in 1914.

Panama City is now a modern metropolis, strongly influenced by United States style. Its main products are oil, plastics, clothing, leather goods, and food and drink.

Population 502 000
Map Panama Ba

Pan-American Highway International highway system linking North and South America. It runs for some 27 350 km (17 000 miles) from north-west Alaska south to Santiago (Chile), east to Buenos Aires (Argentina) and then north to Brasília (Brazil), where it ends. It includes the Inter-American Highway, which extends for about 5470 km (3400 miles) from Nuevo Laredo (Mexico) to Panama City (Panama).

Panay ***Philippines*** Large triangle-shaped island of the western VISAYAN group, covering 11 515 km² (4446 sq miles) north-west of Negros. A combination of rugged mountains (along the west coast), rolling uplands (in the north-east) and coastal lowlands, the island shows extremes of prosperity. It is a major rice producer and grows much sugar. The most prosperous area is around the trading port of ILOILO in the south-east, but Aklan and Antique provinces in the north and west are among the poorest in the Philippines.

Population 2 800 000
Map Philippines Bc

Pangaea Name given to the single supercontinent into which all the world's landmass is thought to have been grouped before it began breaking up into separate continents 200 million years ago. See CONTINENTAL DRIFT

Pangasinan ***Philippines*** See LINGAYEN

Panipat ***India*** Market and textile-making town about 80 km (50 miles) north of Delhi. It was the site of three fateful battles: in 1526, when the Moguls under Baber defeated the Afghan king of Delhi and established their dynasty; in 1556, when Akbar, the great Mogul emperor, defeated a Pathan attack on the new empire; and in 1761, when the Mahrattas, a Hindu race then at the peak of their power in western India, were defeated by invading Afghans and their empire effectively destroyed.

Population 137 900
Map India Bb

Panjim ***India*** See PANAJI

Panjshir ***Afghanistan*** Region and valley about 100 km (60 miles) north-east of Kabul. Since the Soviet invasion of 1979 an estimated 10 000 guerrilla fighters have held out in the 105 km (65 mile) long valley. Most of the civilian residents have fled to Kabul.

Map Afghanistan Ba

Panmunjom ***South Korea*** Village beside the DEMILITARISED ZONE, and next to the meeting place of the Military Armistice Commission formed to supervise the cease-fire at the end of the Korean War (1950-3).
Map South Korea Cd

Pannonhalma ***Hungary*** See GYOR

Pantanal ***Brazil*** Wildlife reserve on the borders with Bolivia and Paraguay. It covers 230 000 km² (88 780 sq miles) – an area about the size of Britain. During the rainy season between October and March, the Pantanal is covered by up to 3 m (10 ft) of water, forming huge lakes interspersed with islands on which cattle and wild animals take refuge.

It is one of the world's richest areas for freshwater fish. There are some 350 species, from the flesh-eating piranha to the pintado, which can weigh up to 80 kg (176 lb). Some 600 species of birds can also be seen, along with alligators, monkeys, boa constrictors, deer, ant-eaters – and the web-footed capybara, the world's largest rodent, which looks like a guinea pig but is the size of a pig.

A railway from the city of São Paulo runs across the region to Corumbá on the Bolivian border.
Map Brazil Cc

Pápa ***Hungary*** A Baroque town with an 18th-century castle, about 130 km (80 miles) west of Budapest. New industries producing electrical equipment and vehicle components have been added to its traditional ones making farm equipment, tobacco products and textiles. In the 16th century, it was a centre of learning and culture, with a Protestant college and a printing house.
Population 34 500
Map Hungary Ab

papagayo Cold, northerly wind which sometimes blows across the Mexican plateau. It is a continuation of the norther and is comparable with the norte of the coastal region.

Papeete ***French Polynesia*** Capital and main port of the territory and of the Society Islands, lying on the north-west coast of Tahiti. It is a busy modern town with a deep-water harbour within its lagoon that is used by cargo and French naval ships as well as cruise ships, traditional inter-island trading schooners and visiting yachts. Its main industries are tourism and copra processing, and the main exports are coconut oil and pearls. It has an international airport 6 km (4 miles) to the south-west, making it a major regional transport centre.
Population 62 700

Paphos ***Cyprus*** South-west coastal town that has become an important tourist centre; it now has an international airport to serve the tourist areas in the west of the island. Paphos is rich in archaeological remains – including a 1st century Roman villa – the House of Dionysus, the Greek god of wine, with a remarkable mosaic floor – a Roman theatre and Roman sea walls, a Byzantine fortress and the so-called Tombs of the Kings, carved in nearby rocks between 1600 and 1050 BC.
Population 20 800
Map Cyprus Ab

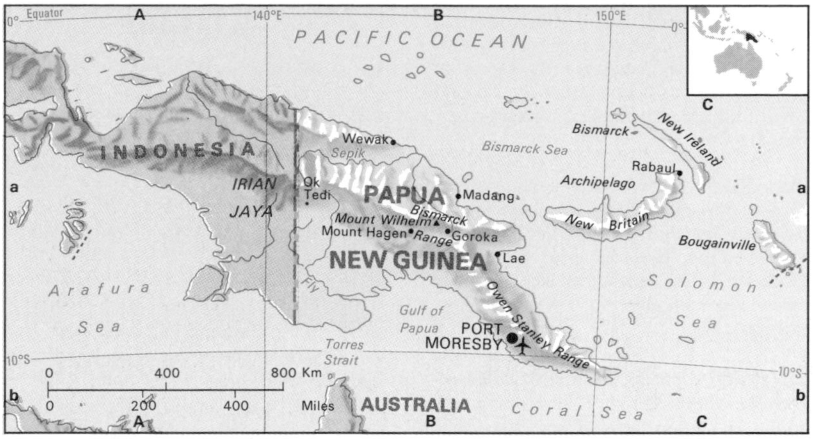

Papua New Guinea

JUNGLE WARRIORS POSE – AND FIGHT – AMONG THE GIANT BUTTERFLIES OF A ONCE CANNIBAL ISLAND

Unique birds of paradise and huge birdwing butterflies with 280 mm (11 in) wingspans live in the dense, tropical rain forests of PNG, as the country is known. So do tribes of fierce, volatile people, some barely touched by the 20th century, partial until quite recent times to such lethal customs as head-hunting and cannibalism. Indeed, tribal warfare still frequently flares up in a few of the more remote highland valleys. But although this is one of the last outposts of Stone Age peoples, the closest most of them come to hunting heads today is when they pose (for a tourist dollar) at the sing-sings at GOROKA or MOUNT HAGEN wearing an elaborate headdress.

Papua New Guinea is a nation in transition from a tribal culture to modern democracy, where rural life was little affected by independence in 1975 but where there is enormous pride in the country and 'Missus Kwin' – Elizabeth II – at its head. As the country finds its feet, however, there are moves towards republican status.

The country embraces the tail half of the bird-shaped island of New Guinea, plus hundreds of other islands, the largest of which are NEW BRITAIN, BOUGAINVILLE (geographically part of the Solomon Islands) and NEW IRELAND. The western half of the main island is IRIAN JAYA, part of Indonesia.

A great spine of jagged, tangled mountains runs down the centre of this island, the world's second largest (third largest if you count Australia), which is part of a volcanic and earthquake belt. Mount Lamington, 1687 m (5535 ft), erupted in 1951; Mount WILHELM, 4509 m (14 793 ft), is the highest peak. Large parts of the highlands are accessible only on foot, by river or by air. Innumerable streams rush down from the mountains to wide, often marshy, lowlands.

The people are largely Melanesians in the coastal areas and islands, and Papuans in the interior. Apart from English, Pidgin (the common language of the whole country) and Motu (spoken by Papuans), there are more than 700 different languages and dialects – the result of groups being isolated from even close neighbours by the difficult terrain. About 125 000 live in and around the capital, PORT MORESBY, but more than 90 per cent of the people are villagers, mostly raising sweet potatoes for their own use.

Some cash crops are grown for sale, such as coffee, cocoa and coconuts; timber is cut for export and a fishing and fish-processing industry is developing. There is some gold and silver mining, but copper is the chief source of mineral wealth, with large-scale workings on Bougainville and in a new gold and copper mine being developed at OK TEDI, on the border with Irian Jaya. Oil exploration is under way, and several natural gas wells are in production.

The country still receives valuable aid – financial and manpower – from Australia, which governed it before independence.

PAPUA NEW GUINEA AT A GLANCE

Area 461 691 km² (178 260 sq miles)
Population 3 396 000
Capital Port Moresby
Government Parliamentary monarchy
Currency Kina = 100 toae
Languages English, Pidgin, Motu and more than 700 native languages and dialects
Religions Christian, tribal
Climate Tropical; average temperature in Port Moresby is 26-28°C (79-82°F) all year
Main primary products Bananas, sweet potatoes, cassava, cocoa, coconuts, coffee, rubber, sago, tea, sugar, vegetables, yams, timber; copper, gold, silver
Major industries Agriculture, forestry, saw milling, mining, food processing
Main exports Gold, copper, coffee, cocoa, timber, palm oil, copra, coconut oil, fish
Annual income per head (US$) 690
Population growth (per thous/yr) 21
Life expectancy (yrs) Male 50 **Female** 52

Paracel Islands ***South China Sea*** Group of low coral islands and reefs about 300 km (185 miles) off the coast of Vietnam. Oil is thought to lie beneath them, and they have large deposits of guano. They were occupied by Japan in 1938 and returned to China after the Second World War, but are also claimed by Vietnam.
Map Vietnam Cb

Paraguay ***South America*** River rising on the Mato Grosso plateau of Brazil. It marks part of the Brazil-Paraguay, and part of the Paraguay-Argentina borders. It flows into the Paraná after 1920 km (1190 miles); its last 1670 km (1035 miles) are navigable.
Map Paraguay Bb

Parakou ***Benin*** Attractive town with wide tree-lined avenues, about 350 km (220 miles) north of Cotonou. Its importance dates from 1936 when it became the inland terminus of the railway from the coast. It trades in groundnuts.
Population 44 000
Map Benin Bb

parallel Line drawn through points on the earth's surface having the same LATITUDE; it is parallel to the Equator.

Paramaribo ***Surinam*** National capital and the country's main port. It was first settled by the British in 1630, but the old town near the port is mostly Dutch. There are many brightly coloured wooden buildings on stilts, and several Dutch-style 18th and 19th-century civic buildings, notably the presidential palace.

The city has Hindu temples, mosques, churches and synagogues reflecting the ethnic and cultural diversity of the people who settled the country under Dutch rule. Across the Surinam river is Nieuw Amsterdam, chief town of the largely Javanese district of Commewijne.
Population 180 000
Map Surinam Aa

Paraná ***Argentina*** Port on the Paraná river about 370 km (230 miles) north-west of Buenos Aires. Its main cargoes are livestock, cereals and fruit from Entre Ríos province.
Population 160 000
Map Argentina Cb

Paraná ***Brazil*** Southern state covering 199 554 km² (77 000 sq miles). It is the country's second largest agricultural producer. Crops include coffee, wheat and soya beans.
Population 7 600 00
Map Brazil Cd

Paraná ***South America*** The continent's second longest river after the Amazon, at about 4200 km (2610 miles). It rises in the Minas Gerais mountains of Brazil and flows generally south into the Rio de LA PLATA (River Plate) above the Argentinian capital, Buenos Aires. It is navigable for most of its length.
Map Brazil Cd

Pardubice ***Czechoslovakia*** Industrial city on the Elbe river, in Bohemia some 95 km (60 miles) east of Prague. It has oil refining, chemicals, machinery and food-processing industries.
Population 93 600
Map Czechoslovakia Ba

Paraguay

HALF WILDERNESS AND HALF GARDEN, PARAGUAY SUPPORTS A PEOPLE WITH A FIGHTING SPIRIT WHICH WAS ALMOST THE DOWNFALL OF THE NATION

Between 1865 and 1870, the young nation of Paraguay, which was becoming an important power in South America, lost some three-quarters of its million-plus population in bloody warfare against the combined forces of Argentina, Brazil and Uruguay. Many of those who died in battle were women. By the war's end, only about 28 000 adult men and 200 000 women remained.

Sixty-two years later, in 1932, Paraguay embarked on another reckless war, this time against Bolivia. The two countries were fighting for the GRAN CHACO territory in the north, which was believed to contain oil. Paraguay won and took half the Gran Chaco, but lost 40 000 men; its hard-won territory has so far yielded no oil.

The wars and a series of repressive governments have left Paraguay among the poorest countries in Latin America, in spite of considerable natural resources. The present government, under its president, General Alfredo Stroessner, has been in office since 1954 and is not noted for respecting human rights.

In view of Paraguay's turbulent history it is not surprising that the economy developed slowly after an initial spurt in the mid-19th century. Its export potential is hampered because it has no coastline, and therefore no deepwater ports, although vessels of up to 1700 tonnes can sail some 1610 km (1000 miles) up the PARAGUAY river to ASUNCION, the capital. Its fertile land is gradually being used more productively, and the Stroessner regime has encouraged investment and growth. In the 1960s cattle ranching became a major export industry.

Paraguay consists of two provinces divided by the Paraguay river. To the west is the Chaco, a vast, river-drained, alluvial expanse of swamp, grassland and scrub forests. The Chaco in turn consists of three regions. Beside the Paraguay is the Low Chaco, open palm forest and marshland where there are huge cattle ranches. The Middle Chaco, mainly scrubland, is the home of the Mennonites, a Christian sect who have developed their highly organised communities since 1900, when they started arriving from Germany, Canada and Russia.

In the extreme north-west, against the Bolivian border, is the Gran Chaco, covered mainly with an impenetrable forest of small thorn trees and designated a national park to preserve the wildlife.

Only 5 per cent of Paraguay's 3.8 million population – the great majority of whom are *mestizos* (of mixed Spanish and native ancestry) – live on the Chaco. The rest live east of the Paraguay river where there are rolling hills, fertile plains, grassland and dense woods. On the river itself lies Asunción, Paraguay's only big city, with a population of 456 000. The fastest-growing town is Ciudad Alfredo Stroessner, which is near to the world's biggest hydroelectric plant at ITAIPU on the PARANA river.

For most of the rural population – about 60 per cent of the total – life is hard. Their land holdings are very small and many have large families to support. The birthrate is high and despite infant mortality of 49 per thousand live births, a tide of youngsters is swelling the population, 70 per cent of which is under 30 years old.

PARAGUAY AT A GLANCE
Area 406 752 km² (157 047 sq miles)
Population 3 820 000
Capital Asunción
Government Republic under dictatorship
Currency Guarani = 100 céntimos
Languages Spanish, Guarani
Religion Christian (96% Roman Catholic, 2% Protestant)
Climate Subtropical; average temperature in Asunción ranges from 12-22°C (54-72°F) in June to 22-35°C (72-95°F) in January
Main primary products Cassava, sugar cane, maize, cotton, cattle, soya beans, rice, tobacco, cotton, fruit, coffee, timber; manganese and iron ore (unexploited)
Major industries Agriculture, food processing, vegetable-oil refining, textiles, cement, forestry, timber products
Main exports Cotton, meat products, tobacco, soya beans, timber, vegetable oils, animal foodstuffs, fruit and vegetables
Annual income per head (US$) 1490
Population growth (per thous/yr) 27
Life expectancy (yrs) Male 63 **Female** 67

Paris ***France*** National capital and one of the world's most elegant cities. From the mannequins of the fashion houses to the can-can dancers of the night-clubs, from the tree-lined 19th-century boulevards to the alleyways of the Latin Quarter, from the glories of the Louvre museum to the vivacity of the pavement cafés, Paris is a city of chic and style.

The city's name derives from the Parisii, a Gallic tribe who settled on an island in the Seine in about the 5th century BC. By the late 13th century, the island – now the Ile de la Cité – contained more than 200 000 people. It also contained the core of the cathedral of Notre-Dame, begun in 1163 though not completed until the 14th century.

As France's medieval kings strengthened their hold over powerful regional aristocrats, Paris went on growing in importance. The French Revolution began in Paris with events that led up to the storming of the Bastille fortress on July 14, 1789. The subsequent turmoil concentrated yet more political power in the capital: France's provincial parliaments were abolished in the same year. When the railways came in the 19th century, they were largely built radiating from the capital – adding to Paris's position as the hub of France.

The present form of the city itself was mostly created in the mid-19th century, and was largely the work of one man: Baron Georges Haussmann (1809-91). Under Napoleon III, the nephew of Napoleon Bonaparte, he redesigned the city, laying out its broad streets, its parks and its avenues like the spokes of intersecting wheels. His new design incorporated some older elements, such as the 17th-century Hôtel des Invalides, the Place de la Concorde and the Champs-Elysées (both dating from the 18th century), and the 50 m (164 ft) high Arc de Triomphe, completed in 1836.

But it also created superb settings for later buildings that have since become Paris's most familiar symbols: the 300 m (984 ft) high Eiffel Tower, named after its designer Gustave Eiffel and built in 1887-9, and the white-domed Basilica of Sacré-Coeur built in the late 19th century atop the artists' hill of Montmartre. Eiffel was also responsible for engineering another city's most familiar symbol: New York's Statue of Liberty, erected in 1886 as a gift from the people of France. The original model from which the giant statue was made still stands on the banks of the Seine in Paris; it was carved by the sculptor Frédéric Bartholdi (1834-1904) and stands 2.7 m (9 ft) high – about one-seventeenth of the height of the New York version.

In the 20th century, the metropolitan area has sprawled across much of the land within a 30 km (20 mile) radius of the Ile de la Cité, embracing old villages, fashionable housing estates and working-class suburbs. Since 1965 a master plan has guided the construction of new motorways, an express railway system, satellite towns, suburban shopping centres and many new housing areas.

But many old neighbourhoods remain, and Parisians say their city is really a collection of 100 villages. Eastern districts still have small workshops, making furniture and clothing. The Latin Quarter, on the left (or south) bank of the Seine, is dominated by the Sorbonne (University of Paris) and by bookshops and publishers. The right bank has the main *haute couture* fashion houses. Elsewhere in the city, there are distinctive Arab, African, Indo-Chinese and Jewish quarters. The presence of immigrant groups adds further diversity to the city, which houses 1 250 000 foreigners, one-third of all those in France. Portuguese, Algerians, Moroccans and Spaniards are the most numerous among the immigrants.

▲ CAPTURED IN OILS An artist on the Left Bank of the Seine in Paris paints the cathedral of Notre-Dame, with its graceful 14th-century spire, just across the Pont de l'Archevêché.

Population (city) 2 188 900; (Greater Paris) 8 761 700
Map France Eb

Paris Basin ***France*** Vast saucer-shaped depression in northern France, containing the capital. It covers 120 000 km² (46 000 sq miles) – almost a quarter of the country – and a much larger proportion of its people. The basin, much of which is richly fertile farmland, is drained by the Seine, Somme, Marne and Loire rivers.
Map France Db

Parma ***Italy*** City about 120 km (75 miles) south-east of Milan. Parma was founded by the Romans in 183 BC and flourished as a trade centre but was destroyed by the Ostrogoth King Theodoric (AD 455-526). It was rebuilt in the Middle Ages. The Parma river bisects the city. The old working-class districts and the partly ruined Palazzo Ducale of the former Dukes of Parma lie to the west. The Romanesque cathedral to the east has a beautiful fresco, *The Assumption of the Virgin*, by Correggio (1494-1534). The city's economy is based on food processing – especially Parmesan cheese and Parma hams.
Population 176 800
Map Italy Cb

Páros ***Greece*** Island of 194 km² (75 sq miles) west of Náxos in the CYCLADES in the south Aegean Sea. Gentle slopes drop down to quiet beaches and the harbour town of Páros, which has a ruined castle dating from AD 1266, and a 10th-century Byzantine cathedral known as Ekatontapyliani, meaning 'hundred gates'.
Population (island) 7900
Map Greece Dc

Parramatta ***Australia*** Originally a farming town in New South Wales on the Parramatta river, now a western suburban satellite of Sydney. Textiles, machinery, cars and flour are produced and it is a major business centre. It was founded in 1788 as the country's second settlement and contains some of Australia's oldest buildings, such as Elizabeth Farm House (1793) and Old Government House (1799-1816).
Population 130 000
Map Australia Ie

Pasadena ***USA*** City in south-west California. It nestles at the foot of the San Gabriel Mountains and forms a north-eastern suburb of Los Angeles. Pasadena has a Tournament of Roses parade each New Year's Day, and the Mount Wilson Observatory stands above the city.
Population 125 000
Map United States Cd

Pasargadae ***Iran*** Ruined city about 100 km (60 miles) north-east of the southern city of Shiraz. It was the capital of Cyrus the Great (ruled 559-530 BC). The remains include Cyrus's palace and his tomb – a huge sarcophagus resting on a pyramid of stepped stone blocks.
Map Iran Ba

Pasay ***Philippines*** See MANILA

Pasco ***Peru*** Central department in the high Andes and one of the country's main mining regions. Iron ore and manganese come from the

valley of the Rió Perené. Copper, zinc, lead, bismuth and silver are mined at the department capital, Cerro de Pasco.
Population 213 200
Map Peru Bb

Pascua, Isla de ***Pacific Ocean*** See EASTER ISLAND

Pas-de-Calais ***France*** Northern department, covering parts of Artois and Picardy. Its chalk cliffs are part of the same geological formation as the white cliffs of Dover, across the Channel in England. The main towns are ARRAS, BOULOGNE-SUR-MER and CALAIS.
Population 1 412 000
Map France Da

Passau ***West Germany*** Town on the Danube river in the south-east corner of the republic on the Austrian border. Its 17th-century Baroque cathedral has the world's largest church organ, completed in 1928. It has more than 17 000 pipes and five keyboards.
Population 52 400
Map West Germany Ed

Passchendaele (Passendale) ***Belgium*** Village about 60 km (35 miles) west of the city of Ghent. It was the scene of heavy fighting in October and November 1917 during the First World War. The battlefields are now marked with monuments to more than half a million Allied and German soldiers who fell there.
Population 4000
Map Belgium Aa

Pasto ***Colombia*** City on a high, dry plateau in the south-western mountains about 90 km (55 miles) from the border with Ecuador. The surrounding countryside is peopled mostly by Indian farmers. The city is flanked by the still active volcano of Galeras (4276 m, 14 028 ft).
Population 258 000
Map Colombia Bb

Patagonia ***Argentina/Chile*** Cold high desert covering the four southernmost provinces of Argentina and part of MAGALLANES region in Chile. It is a vast plateau of some 770 000 km^2 (300 000 sq miles) south of the Rio Negro and Limay rivers. It is virtually barren, except around the river valleys, where grapes and other fruits are grown. Welsh-speaking farmers also raise sheep in the area; they are the descendants of people who emigrated to Argentina in the mid-19th century to preserve the Welsh language and culture in the face of growing Anglicisation of Wales and established settlements such as Puerto Madryn, Trelew and Esquel. Patagonia's chief towns are the Argentinian port of Comodoro Rivadavia (population 120 000) and PUNTA ARENAS in Chile. The region's natural resources include oil, iron, copper and uranium, which are progressively being exploited.
Map Argentina Cc

Patan (Lalita Patan; Lalitpur) ***Nepal*** Oldest town of the Kathmandu valley, just 5 km (3 miles) south of the capital. It is principally a Buddhist temple town, but there are also Hindu and Muslim buildings.
Population 59 000
Map Nepal Ba

Pátmos ***Greece*** Volcanic island among the northernmost of the DODECANESE, in the eastern Aegean Sea. St John the Evangelist was exiled there in AD 81-96, during which time he wrote the Book of Revelation. The battlement-walled monastery of St John, standing on a crag over the town, was founded in AD 1088, and has some richly carved icons and medieval frescoes. In the Convent of the Apocalypse is the cave where St John is reputed to have written the Book of Revelation.
Population 2550
Map Greece Ec

Patna ***India*** Capital of Bihar state, on the south bank of the river Ganges about 460 km (285 miles) north-west of the city of Calcutta. The modern city is just downstream from Patuliputra, the capital of Hindu and Buddhist empires in the 5th to 3rd centuries BC. The city has a huge beehive-shaped granary, 30 m (98 ft) tall, built by an Englishman in 1786 to prevent famine – it was never filled. The many academic institutions include a university and library of rare Persian and Arabic manuscripts, and Hindu and Mogul paintings.
Population 918 900
Map India Dc

Patos, Lagoa dos ***Brazil*** Lagoon in the southern coastal state of Rio Grande do Sul with Porto Alegre at its northern tip. The country's largest salt-water lagoon, it covers 10 144 km^2 (3915 sq miles). Two nearby lagoons to the south, Mirim and Mangueira, extend to the border with Uruguay.
Map Brazil Ce

Patras ***Greece*** Largest city in the PELOPONNESE, on the north-west coast facing the Ionian Sea and the main gateway to Greece from the west. It was founded in 1100 BC. Under Roman, then Venetian, domination it grew into a busy Mediterranean port. Patras is now an industrial centre, processing textiles, leather and olive oil, and exporting wine and currants. It has a university. At the church of Ayios Andreas is a reliquary said to contain the skull of St Andrew, patron saint of both Greece and Scotland.
Population 154 000
Map Greece Bb

Pattani ***Thailand*** Town on the Gulf of Thailand, about 110 km (70 miles) from the country's southern tip. The town, like the surrounding province of the same name, is dominated by Malay-speaking Muslims. The province's main crops are rice, cashew and fruit.
Population (town) 81 000; (province) 478 700
Map Thailand Bd

patterned ground Clusters of stones arranged in stripes or polygons in the surface layer. It is probably caused by intensive freezing and thawing of the ground rearranging the larger rock fragments into these patterns.

Patuliputra ***India*** See PATNA

Pátzcuaro ***Mexico*** Picturesque town on the south shore of Lake Pátzcuaro, deep in the Michoacán mountains of central Mexico. Its Indian name means 'Place of Delights' – possibly because the area has over 250 species of humming birds. The lake, 50 km (30 miles) round, is dotted with seven islands, one of which, Janitzio, has an all-night vigil for All Souls' Day (November 2). Indians illuminate the cemetery with candles and adorn the graves with flowers.
Population 30 000
Map Mexico Bc

Pau ***France*** Pyrenean resort and industrial town about 90 km (56 miles) from the Atlantic coast. It is the capital of Pyrénées-Atlantiques department and was the birthplace of Jean Bernadotte (1763-1844), a lawyer's son who became a Revolutionary general under Napoleon and was adopted in 1810 by the ageing and childless King Charles XIII of Sweden. In 1818 he succeeded to the throne as Charles XIV, founding the present Swedish ruling house.
Population 135 700
Map France Ce

pavement Smooth ground surface of rock formed by wind erosion, weathering or glacial scouring. See DESERT PAVEMENT and CLINT

Pavia ***Italy*** University city about 32 km (20 miles) south of Milan. The ancient city centre preserves the gridiron street plan of the original Roman fort, called Ticinum. The cathedral was started in 1488 and completed, still following the original plans, in 1936. The university was founded in 1361.

Pavia has food industries and manufactures agricultural machinery, textiles, electrical goods and sewing machines.
Population 83 000
Map Italy Bb

Pavlodar ***USSR*** Town on the Irtysh river in Kazakhstan, 385 km (240 miles) south-east of Omsk. It lies in fertile farmland and its main industry is food processing – the growth of which has supported a trebling of the population since 1960.
Population 309 000
Map USSR Ic

Pays Basque ***France*** See BASQUE REGION

Paz de Rio ***Colombia*** See BOYACA

Pazardzhik ***Bulgaria*** Capital of Pazardzhik province, on the Maritsa river, 100 km (62 miles) south-east of Sofia. It serves a fertile agricultural and vine-growing area. Its chief industries are textiles, and rubber and leather goods. There is an underground Church of the Holy Virgin with a magnificent carved icon screen.
Population 79 500
Map Bulgaria Bb

Peace River ***Canada*** **1.** River in western Canada, 1923 km (1195 miles) long, which is a major tributary of the MACKENZIE. Its valley, where cereals are grown, is the most northerly arable farming area in Canada.

2. Town on the Peace River in the centre of an oil, gas and farming area in north-west Alberta. It is the centre of an Anglican bishopric extending to the Arctic Circle.
Population (town) 5900
Map Canada Dc

Peak District ***United Kingdom*** Upland area of moorland, high crags and dales (valleys) at the southern end of the Pennines in Derbyshire. It is popular with hikers and potholers, who explore the numerous underground caves, and its proximity to the industrial areas of Lancashire, South Yorkshire and The Potteries results in many tourists. The highest point, Kinder Scout ('The Peak'), is 636 m (2088 ft) high. Despite most of the area being a national park, fluorite – including the prized coloured form, Blue John – is mined in some places.

Map United Kingdom Ed

Pearl Harbor ***Hawaii*** Deep-water harbour and naval base on Oahu island in the US Pacific Ocean state of Hawaii. It was bombed without warning by the Japanese on December 7, 1941, an event which prompted the USA to enter the Second World War.

Map Hawaii Ba

Pearl Islands ***Panama*** See CONTADORA, ISLA DE

Pearl River ***China*** See ZHU JIANG

peat Partially decomposed, compacted vegetable matter, found in bogs and fens, and used as a soil conditioner and fuel.

pebble Rock fragment 4-64 mm (0.16-2.52 in) in diameter. It is larger than gravel but smaller than a cobble, and usually rounded.

Peć (Pech; Petch; Ipek) ***Yugoslavia*** Ancient market town some 20 km (12 miles) from the north-east tip of Albania. It lies astride the torrential Bistrica river on the western edge of a large and fertile mountain basin. The old Turkish quarters, with narrow streets, bazaars and mosques, contrast with the post-1960 flats and food processing, leather, textile and tobacco plants. Peć was the headquarters of the Serbian Orthodox Church from the 14th to 18th centuries, and the imposing Patriarchate, with three churches, lies at the mouth of the spectacular Rugovo Gorge just west of the town. Decani, medieval Serbia's largest monastery, lies 16 km (10 miles) south of Peć. It was built in the 14th century and has fine frescoes.

Population 111 000

Map Yugoslavia Ec

Pechenga (Petsamo) ***USSR*** Ice-free port on the Barents Sea. It has a large fishing fleet, and refines copper and nickel from local mines. The area was Finnish from 1920, occupied by Germans in 1941, and annexed by the USSR in 1944.

Population 51 000

Map USSR Eb

Pechora ***USSR*** River which rises in the Ural mountains and flows 1809 km (1124 miles) north-west to the Barents Sea. Its basin contains coal, and oil and gas which are piped to Western Europe.

Map USSR Gb

Pécs ***Hungary*** Chief city of south-west Hungary, and administrative centre of Baranya county, 171 km (106 miles) south-west of Budapest. It lies near the old town of Villány on the vine-clad southern flanks of the densely forested Mecsek hills, which reach 681 m (2234 ft). A major manufacturing centre whose products include leather goods, Zsolnay porcelain and Villanyi red and white wines, Pécs has two 'new town' satellites: Uranvaros, where uranium is mined, and Komló, a producer of lignite.

The city – established by the Celts and developed by the Romans – prospered as a trading centre in the Middle Ages, and from those times date Pécs' early Christian catacombs; its magnificent cathedral with four graceful spires; and its university, founded in 1367. The fine Gazi Kassim Pasha and Jakovali Hassan mosques survive from the Turkish occupation (1543-1686), while the imposing Baroque Bishop's Palace dates from the 18th century.

Population 174 500

Map Hungary Ab

pedalfer Soil, rich in iron and aluminium, found in humid regions.

pediment Gently sloping, eroded surface, which may be partially covered with alluvium, found at the base of mountains in dry areas.

Peebles ***United Kingdom*** See BORDERS

Pegasus Bay Bay on the east coast of New Zealand's South Island lying north of Banks Peninsula, which was named after the British botanist Sir Joseph Banks. Banks accompanied Captain James Cook on his circumnavigation of New Zealand in 1769-70. The port of Lyttelton on the south side of the bay serves the city of CHRISTCHURCH.

Dimensions 66 km (41 miles) (north-south), 24 km (15 miles) (east-west)

Map New Zealand De

pegmatite Coarse-grained IGNEOUS ROCK, with crystals at least 100 mm (0.5 in) in diameter.

Pegu ***Burma*** Town and capital of the division of the same name, about 70 km (45 miles) north-east of Rangoon. The town was the capital of the Mon kingdom of Burma at various times between 1250 and 1757 and was Burma's capital from 1531 to 1635, but little remains of its former glory. The Shwethalyaung, a reclining statue of the Buddha over 60 m (200 ft) long and 14 m (46 ft) high, has been extensively renovated. The division, which covers 39 404 km² (15 214 sq miles), is largely agricultural. The main crops are rice and sugar cane.

Population (town) 254 800; (division) 3 800 000

Map Burma Cc

Pekanbaru ***Indonesia*** Provincial capital of Riau in Sumatra. It is an oil port and a processing centre for rubber.

Population 186 300

Map Indonesia Bb

Peking ***China*** See BEIJING

pelagic sediment Sediment formed from dead organisms that inhabit the open ocean, rather than the ocean floor or shoreline areas. These organisms extract minerals such as lime or silica from the ocean water.

Pelagonia ***Yugoslavia*** See BITOLA

Pelée, Mont ***Martinique*** Active volcano in the northern part of the island, 1397 m (4583 ft) high. Its violent eruption on May 8, 1902, destroyed the town of St Pierre on the west coast, 20 km (12 miles) north-west of Fort-de-France. The town's 28 000 inhabitants were killed in just three minutes by an explosive, burning cloud of ash and gas as it rushed down the side of the volcano to the sea. The eruption created a column of lava some 300 m (1000 ft) high, which was destroyed by further volcanic activity in 1929.

Only one person survived in the 1902 eruption – a stevedore who had been jailed for brawling. He was protected by the thick stone walls of the jail in which he had been locked up. Today St Pierre is a small village (population 6000), and the ruins of the old town are a tourist attraction.

Map Caribbean Cc

Pelion ***Greece*** See PILION

Pella ***Greece*** Site of the birthplace and capital of Alexander the Great (356-323 BC). It lies 40 km (25 miles) west of Saloniki, and was rediscovered only in 1957. A number of streets, some with mosaics, have since been uncovered.

Map Greece Ca

Peloponnese (Pelopónnisos) ***Greece*** The southernmost part of mainland Greece connected to northern Greece only by the Isthmus of CORINTH. It gets its name from Pelops, the king who, in mythology, was chopped up and fed to the gods. The Olympic Games were founded to honour his funeral.

The Peloponnese is a land of hills and rugged coastline, bounded by the Aegean Sea to the east and the Ionian Sea to the west. Barren mountains, among them Taíyetos (2404 m, 7885 ft) and Zereia (2377 m, 7798 ft), fall to narrow coastal plains. The Peloponnese was the cradle of Mycenaean civilisation, the forerunner of classical Greece. From AD 267 until the 11th century, the region was ravaged by Goths, Huns, Slavs and pirates; and later it was ruled by Franks (after 1210) and Turks (after 1453). It was united with Greece in 1828. The Peloponnese produces red and white wine, olives and fruit, and makes textiles and paper, and processes food. The main city is PATRAS.

Population 1 012 500

Map Greece Bc

Pelotas ***Brazil*** Agricultural city at the southern end of the lagoon of Lagoa dos Patos, 210 km (130 miles) south of Porto Alegre. It processes locally grown rice, fruit and soya beans.

Population 261 000

Map Brazil Ce

Pemba ***Tanzania*** Island in the Indian Ocean about 60 km (35 miles) north-east of the island of Zanzibar – with which it was a British protectorate from 1890 to 1963. The islands formed the independent country of Zanzibar for four months in 1963-4 before joining mainland Tanganyika to form the republic of Tanzania. Pemba, whose chief export is cloves, covers an area of 985 km² (380 sq miles). Its main town is Wete (population 20 000) on the west coast.

Population 250 000

Map Tanzania Ba

▲ **ROYAL CITY Sunlight highlights the carved gift-bearers lining the stairway of the King's Gateway near the towering Apadana (great audience hall) at ancient Persepolis.**

Pembroke ***United Kingdom*** See DYFED

Penang ***Malaysia*** State on the west coast about 300 km (185 miles) north-west of the capital, Kuala Lumpur. It includes Penang Island and the mainland port of Butterworth, covering in all 1033 km² (399 sq miles). Penang Island is a popular holiday destination, where tourists can choose between the tranquillity of Malay fishing villages and the bustle of Chinese-dominated George Town.

Population 955 600
Map Malaysia Ba

P'eng-hu Lieh-tao (Penghu Islands) ***Taiwan*** See PESCADORES

peninsula Long projection of land into a sea or lake.

Pennines ***United Kingdom*** Range of hills forming the backbone of England. They stretch 225 km (140 miles) north from the River Trent in the Midlands to the Cheviot Hills on the Scottish border. They rise to 636 m (2087 ft) at Kinder Scout in the south, and to 893 m (2930 ft) at Cross Fell in the north. Rough grass, peat bog and forestry plantations cover their slopes, while there is good farmland in the dales (valleys) in the eastern flanks. The range takes in the Northumberland, Yorkshire Dales and PEAK DISTRICT national parks.

Map United Kingdom Dc

Pennsylvania ***USA*** State founded as a colony in 1681 by the English Quaker William Penn, and later one of the original 13 states. It lies in the north-east, and covers 117 412 km² (45 333 sq miles) between Lake Erie and the Delaware river, much of it being in the Appalachian Mountains. Pennsylvania is a centre of coal mining and its industries include the production of steel, machinery, electrical equipment, chemicals, and food processing. Farming is based on dairying, vegetables and fruit.

PHILADELPHIA is the principal city and port, PITTSBURGH the centre of the coal and steel industry, and HARRISBURG the state capital.

Population 11 853 000
Map United States Kb

Pentland Firth Channel between the northern tip of the Scottish mainland and the ORKNEY Islands. It can be difficult to navigate. Many people have drowned in its rough seas and their ghosts are said to call out to sailors passing through the firth on dark nights.

Dimensions 23 km (14 miles) long, 10-13 km (6.2-8 miles) wide
Map United Kingdom Gj

Penzance ***United Kingdom*** Most westerly town in England, on the south coast of the county of Cornwall 12 km (8 miles) from Land's End. It is a rail terminus and sea and air port for services to the Isles of Scilly, and also a tourist resort. St Michael's Mount, a small island just offshore, has a ruined 11th-century monastery and a 15th-century fort.

Population 20 000
Map United Kingdom Ce

Peradeniya ***Sri Lanka*** See KANDY

Perak ***Malaysia*** West-coast state which borders Thailand to the north. It extends over 21 005 km² (8110 sq miles) and embraces the basin of the Perak river. It is the country's major tin-mining area. There are some rubber estates on the Main Range foothills inland.

Population 1 807 400
Map Malaysia Bb

perched block An ERRATIC boulder that has been carried by a glacier or ice sheet and left standing in a precarious, exposed position.

Pereira ***Colombia*** Market town and capital of Risaralda department, about 190 km (120 miles) west of Bogotá. It trades in coffee and cattle, and manufactures textiles, clothing, paper and foodstuffs.

Population 265 000
Map Colombia Bb

Peremyshl ***Poland*** See PRZEMYSL

Pergamum ***Turkey*** See BERGAMA

peridotite Any of a group of coarse-grained ultrabasic rocks composed mainly of olivine, and various pyroxenes and amphiboles, but with less than 10 per cent feldspar. They are called ultrabasic because they have less silica than a BASIC ROCK. They are denser than basic and acid rocks, and are thought to form the outer part of the earth's mantle.

Périgord ***France*** Low limestone plateau in south-west France, about 100 km (60 miles) east of Bordeaux. Caves in the region were inhabited in prehistoric times – as is shown by the LASCAUX cave paintings. The plateau is also noted for its truffles – edible fungi that grow underground and are traditionally collected by using pigs to sniff them out.

Map France Dd

period Unit of geological time, longer than an epoch and shorter than an era. See GEOLOGICAL TIMESCALE

Perister, Mount ***Yugoslavia*** See BITOLA

Perlis ***Malaysia*** The country's smallest state, extending to just 795 km² (307 sq miles) on the north-west coast bordering Thailand. Rice and some sugar cane are grown.

Population 148 500
Map Malaysia Ba

Perm' ***USSR*** Major industrial city in the western Ural mountains, about 280 km (175 miles) north-west of Sverdlovsk. Perm' began its industrial life as a copper-smelting centre in 1723. It grew rapidly after the First World War, when chemical, engineering, vehicle and aircraft industries began to be set up there. It later began to produce steel, paper, timber, petroleum products and textiles. The city lies on three main transport routes: the navigable Kama river, the Great Siberian Highway and the Trans-Siberian Railway across the Urals.

From 1940 to 1957 the city was called Molotov in honour of Vyacheslav Molotov (1890-), the Soviet foreign minister under Joseph Stalin. But with Stalin's death in 1953, Molotov's influence gradually declined and the city reverted in 1957 to its earlier name of Perm'.

Population 1 049 000
Map USSR Gc

permafrost Permanently frozen ground, found only in cold regions at high latitudes, generally within 2000-3000 km (about 1200-1900 miles) of the North Pole. The surface may thaw briefly in summer.

permeable rock Rock that allows water or other liquids to pass freely through it. It may be a PERVIOUS ROCK or a POROUS ROCK.

Permian Sixth period in the Palaeozoic era of the earth's history. See GEOLOGICAL TIMESCALE

Pernambuco ***Brazil*** See RECIFE

Perpignan ***France*** Market town near the Mediterranean coast about 25 km (15 miles) from the Spanish border. It has a cathedral, a chateau and the Palace of the Kings of Majorca, all dating from the 14th century.

Population 140 000

Map France Ee

Persepolis ***Iran*** One of the world's greatest archaeological sites, about 670 km (415 miles) south and slightly east of the capital, Tehran. The Achaemenid emperor Darius I (ruled 522-486 BC) intended Persepolis to be the ceremonial capital of Persia, but before the planned city could be completed it was sacked and burnt by Alexander the Great in 330 BC. Nevertheless, the pillars and other parts of many of the buildings are still standing.

Among the principal remains are the Great Terrace built by Darius, the Gate of Xerxes, the Palace of Darius and Xerxes, Darius's Treasury, the Apadana (great audience hall), the Hall of One Hundred Columns, the tomb of Artaxerxes II and III and that intended for Darius III, who died in the war against Alexander.

The tombs of Darius I and II and those of Xerxes I and Artaxerxes I, hollowed out of a cliff high above the ground, are at Naqsh-i-Rustam, 5 km (3 miles) north of the city.

Map Iran Bb

Persian Gulf See GULF, THE

Perth ***Australia*** State capital of Western Australia. It was founded in 1829 on the Swan river, but grew slowly until it became a convict settlement in 1850. Perth and its port, Fremantle, developed rapidly after gold was discovered 547 km (340 miles) east at Kalgoorlie in 1893. The port handles refined oil, wheat and wool. The University of Western Australia was founded in 1911 and in 1975 Murdoch University was established. Rottnest Island, 20 km (12 miles) off the coast, is a wildlife sanctuary where the quokka, a dwarf rat-like wallaby, lives.

Population 969 000

Map Australia Be

Perth ***United Kingdom*** Scottish city 55 km (35 miles) north of Edinburgh, standing on the salmon-rich River Tay, and former county now divided between CENTRAL REGION and TAYSIDE. It was a strategic fort for the English kings of the Middle Ages in their attempts to conquer Scotland, and was seized by the Scottish leader Robert Bruce in 1311 in his battle against them. The city was the Scottish capital until the murder of James I of Scotland in 1437.

The Scottish Reformation had its birth in Perth in 1559 when the Calvinist John Knox preached his sermon against Catholic idolatry in the Kirk of St John, which still stands in the city centre. The sermon also led to the destruction of the city's four friaries. There is a small port for coastal shipping, but much of the traffic to the oil fields of the North Sea passes through Perth by road to ports farther north.

Population 42 000

Map United Kingdom Db

A DAY IN THE LIFE OF A FRENCH AGRICULTURAL SALESMAN

Pierre Dufour gets up at 7.30 in the morning. He has a busy day ahead and only has time to glance at the newspaper as he sips a strong, black coffee. Pierre, 38, is an agricultural salesman in the south of France. He sells fertilisers, pesticides and machinery to farmers in Languedoc-Roussillon who grow wine grapes, fruit trees, lettuces and tomatoes. Pierre likes his job and is good at it. Once a week he calls his superior in Bordeaux to place his orders. He hopes one day for promotion – to extend his region over a larger area.

This morning Pierre has two clients to see, one in the town of Carcassonne and one in the village of Elne, close to home. The Dufours live in a small but comfortable house in the village of Toreilles, only a few miles from Perpignan. Pierre had the house built a few years ago. The mortgage takes about a quarter of his salary, which averages between 8000 and 10 000 francs (US$1050-1300) a month depending on sales.

Pierre spends a lot of time on the road, driving to see clients. Then he often thinks about his wife Françoise and how they met. He had just finished his compulsory year in the army and was working in his father's small grocery, in a poor district of Perpignan. Françoise lived with her mother and would often come into the shop. The couple fell in love and they married in St Jean's Cathedral in Perpignan. Their children, Alain (14) and Anne-Marie (12), go to a private Catholic school. Pierre hopes Alain will go to university, and perhaps become a doctor or a lawyer.

After seeing his clients, Pierre returns home at around noon for a tasty lunch of fried aubergines and fillet of sole prepared by Françoise. The children eat at school, so Pierre can watch the 1 pm news on television and have a nap in peace before setting off again at around 2 pm.

He generally does not set out for home until 7 or 8 in the evening. Beforehand, he often likes to go to the local café to have a couple of glasses of Pernod and a chat. Conversations are frequently about politics. Pierre, a centre-rightist, does not hold strong political beliefs but he likes exchanging ideas with his friends.

When he comes home, he changes into casual clothes before sitting down to dinner with the family. The children talk about their day at school and after the meal they watch television, usually a film, but sometimes variety shows, political debates or quizzes. At about 11 pm, Pierre goes to bed and reads a few pages of a thriller before falling asleep. His life may not be as adventurous as the hero's, but he enjoys it.

Peru See p. 510

Perugia ***Italy*** Beautifully preserved medieval city set in the Apennines about 135 km (85 miles) north of Rome. Perugia was a centre of Renaissance painting, and was the home of Pietro Perugino (1445-1523), teacher of Raphael (1483-1520), and Bernardino Pinturicchio (1454-1513). Works by these painters are scattered through the city's churches and public buildings. The Palazzo dei Priori (1243-1443) is one of the best medieval palaces in Italy and home of the Umbrian National Gallery. The richness of Perugia's cultural life continues today, with two universities – one for foreigners – and many educational and artistic institutions.

The economy of the city, which is the capital of Umbria, is founded upon tourism and chocolate making, with a few textile and furniture factories.

Population 142 200

Map Italy Dc

pervious rock A PERMEABLE ROCK that has no pores. It lets liquids pass through it via cracks and fissures, joints and bedding planes. Limestone and chalk are common examples of pervious rocks.

Pesaro ***Italy*** Adriatic port and holiday resort about 130 km (80 miles) east of Florence. The operatic composer Gioacchino Rossini (1792-1868) was born there. The town's main industries are tourism, agricultural machinery and pottery.

Population 90 500

Map Italy Dc

Pescadores (P'eng-hu Lieh-tao) ***Taiwan*** Group of about 64 islands in Taiwan Strait, some 45 km (28 miles) off the west coast of Taiwan. Low-lying and mainly formed from coral, the islands have a total land area of 127 km^2 (49 sq miles). Only 24 of the islands are inhabited, and most of the people are fishermen or farmers. The main town, Ma-kung (population 60 000), was a Japanese naval base during the Second World War; it lies on P'eng-hu, the largest island.

Population 130 000

Map Taiwan Ac

Pescara ***Italy*** Tourist resort and seaport on the Adriatic coast about 155 km (97 miles) east and slightly north of Rome. Its industries include fishing, shipbuilding, machinery, textiles and furniture.

Population 131 700

Map Italy Ec

Peshawar ***Pakistan*** Historic town that guards the foot of the Khyber Pass in North-West Frontier Province. It has Buddhist shrines – one of which contained relics of the Buddha himself until 1909, when they were sent to Burma – and Mogul, Sikh and British buildings. Its bazaars have traded for thousands of years, dealing in goods brought down the pass from central Asia by camel caravans.

Population 555 000

Map Pakistan Ca

Petch ***Yugoslavia*** See PEC

Peru

ONCE THE CENTRE OF THE RICH AND MIGHTY INCA EMPIRE, PERU IS NOW STRUGGLING AGAINST POVERTY AND TORN BY INTERNAL VIOLENCE

Just over 500 years ago, the greatest Indian civilisation in South America – that of the sun-worshipping Incas – had its heartland in Peru. The well-organised Inca Empire, rich in gold and silver, stretched north into Ecuador and south into Chile. The spectacular remains of the Inca cities at CUZCO and MACHU PICCHU are among the continent's most amazing monuments.

But Peru today also has its ugly side. It is scarred by poverty and violence, the victims resulting from an unequal social structure and the struggle to survive in shanty towns round the cities or win a meagre living from the soil in the countryside.

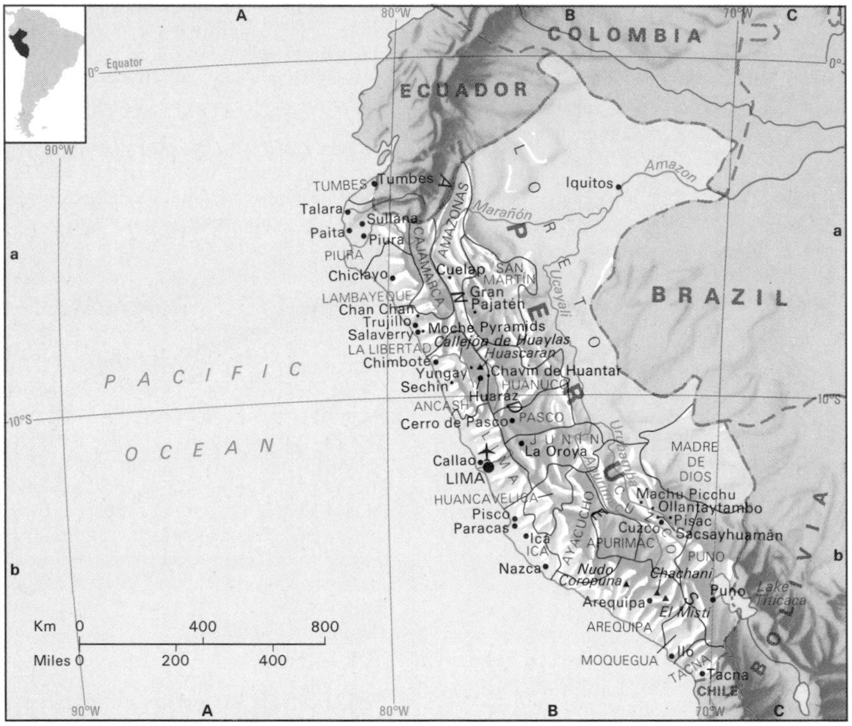

THREE LANDS

The climate and geography of the country have helped to preserve Peru's ancient treasures – and have also, because they are obstacles to prosperity, helped to bring about today's troubles. There are three distinct regions west to east: the coast, the high sierra of the Andes and, over the mountains, the tropical jungle.

The narrow coastal belt west of the Andes is mainly desert, created by the influence of the cold Peruvian Current (once called the Humboldt Current) flowing north from Antarctica along the Pacific coast. The current cools the air, preventing the formation of rain. The dry air has helped to preserve ruins, and has saved the relics of Indian civilisations thousands of years older than the Incas.

The heights of the Andes, where the Incas made their last stand against the Spaniards, are wet mist-soaked solitudes where the vegetation varies with altitude, from dense forest to mountain scrub.

To the east of the Andes is the Montaña region, the Amazon lowland covered in *selvas* (tropical forests). Archaeologists are still finding ancient settlements in this jungle, buried in the fast-growing foliage.

The combination of desert, mountain and jungle limits the areas that can be lived in or cultivated. The Incas grew crops on hillside terraces and irrigated areas. Subsistence farming still continues in the mountains, but today most large-scale agriculture is in the oases and fertile, irrigated river valleys that cut across the coastal desert. Sugar and cotton are the main export products.

The Peruvian Current has in the past brought great wealth from the sea, in the form of vast shoals of anchovies. Coastal settlements developed around industries based on fish canning and the production of fishmeal. In recent years the shoals have become depleted and the government imposed a fishing ban in 1983 to allow stocks to recover.

At their normal levels, the anchovies attract huge flocks of sea birds, and during part of the 19th century the Peruvian economy was based mainly on the huge, thick layers of seabird droppings (*guano*) in the desert which were dug and exported as fertiliser.

Nature is not always so bountiful to the Peruvians. Occasionally, a warm marine current known as EL NINO reverses the cooling effect of the Peruvian Current. Then the anchovies disappear, and storms and floods on land destroy crops, bringing hunger.

In the mountains, large areas are unfarmable, and some farmers who grow food on the slopes are faced by the problem of soil erosion. Furthermore, the zone is subject to earthquakes, which have repeatedly brought destruction and grief to the mountain people. In 1970 an earthquake dislodged a glacier which buried the town of Yungay and 18 000 of its people.

The mysterious, densely forested Montaña region covers some 60 per cent of the country yet has only 5 per cent of its people, and some parts can still be reached only by river. The boom in wild rubber in the early 20th century gave rise to a few towns such as IQUITOS, which boasts a town hall designed by the same Gustav Eiffel who built the Eiffel Tower; the materials for the building were imported from France to this improbable location by a rubber baron.

Today's source of wealth is oil – and some exploitation is beginning to intrude into an area previously the preserve of missionaries working with Indian tribes. New oil discoveries are needed; present reserves are likely to be exhausted by the end of the century.

TWO NATIONS

Two million people in Peru speak only the Indian languages of Quechua or Aimara, and there are some 50 Indian tribes throughout the country. These people have shared very little in what limited benefits economic growth brought Peru, and live in conditions of great hardship.

In the countryside many peasants scratch a meagre living from tiny patches of land. And in the towns they sell food, vegetables and clothing in markets which are often picturesque but desperately poor. The diet and general health of these people is equally poor; 30 per cent of children die before they are one year old. Ninety-nine per cent of the rural population and 60 per cent of town dwellers have no running water or drainage.

The majority of the people of the capital city of LIMA, for example, live in shanty-town settlements called *barriadas* which they illegally built on government land, and are only gradually being supplied with water and sanitation.

The combined effect of the difficulties of climate, terrain and natural disasters is widespread poverty which has resulted in civil discontent. This reached new heights in the early 1980s, especially where the rural poor have had the opportunity of an education.

A movement which began in the University of AYACUCHO, in the middle of an impoverished mountain region, is based on Chinese Maoist doctrine, and is called the Shining Path (*Sendero Luminoso*). The *Senderistas* guerrillas are bitterly opposed to the government, and the inhabitants of the mountain villages are caught between the government's security forces and the Senderistas.

This rebellion is the greatest threat to the government, which is headed by a democrati-

▲ **MARKET DAY The open-air Indian market in the hill-town of Chincheros, in south-central Peru, attracts both natives and tourists. The wares on offer range from wine jugs to colourful Indian shawls.**

cally elected president. Fernando Belaúnde Terry was ousted by a military coup in 1968, but re-elected in 1980. Alan García Pérez succeeded him in July 1985, and introduced widespread reforms, including a price freeze, reduced budget and a campaign against the country's drug runners.

Before the Inca Empire was established in the 15th century, several great Indian civilisations flourished. Over a period of 1400 years these rose in turn – the Paracas, the Nazcas, the Mochicas and the Chimus, until finally the highland Inca tribe absorbed the remains of them into its culture.

The Inca civilisation originated in the Cuzco region and Cuzco itself became the empire's capital and the centre of the nation's religion, which was based on sun worship. The empire, for all its splendour, survived for only 100 years before it fell to the Spanish conquerors in 1532. Parts of the Inca city walls of Cuzco survive today, but much of the Inca capital was lost beneath the Baroque buildings of the Spanish colonisers.

The empire was rich in gold and silver, but treated these metals as purely ornamental, for the Incas used no money. Taxes were paid in the form of labour, and the product of the land was divided between court, administration, priests and the people. But stories of gold brought the Spanish to Peru in 1532.

CONQUERED BY 200 MEN

When the Spanish invaded under Francisco Pizarro, the Inca nation was split by an internal crisis. The last great emperor, Huayna Cápac, had divided his territory between two sons, who established rival capitals at Cuzco and QUITO, and went to war. Both sides hoped for help from the Spanish in their own fight, but were betrayed by the conquerors and the empire fell to a force of only 200 men.

But one isolated Inca citadel remained undiscovered: Machu Picchu, straddling a high mountain in the Andes above the Urubamba river, about 50 km (30 miles) north of Cuzco. Here the Inca Virgins of the Sun, the temple priestesses, took refuge from the violence of the Spaniards. The city lay undisturbed for four centuries.

Pizarro needed a base from which to exploit Peru's treasures, so Lima, the capital, was founded. To handle exports to Spain, the port of Callao was developed on the coast, 13 km (8 miles) to the north-west.

Lima is said to be a different world from the rest of Peru, but it might be truer to say that every part of Peru is different from the rest. On the ALTIPLANO near the Bolivian border, for instance, the bowler-hatted Indians herd llamas and alpacas, make clothes from the wool of these creatures, eat their meat and even keep warm by using their dried dung for fuel. Straddling the border with Bolivia, the highest navigable lake in the world, Lake TITICACA, has been inhabited for centuries by Indians living on reedy islands.

PERU AT A GLANCE
Area 1 285 215 km² (496 222 sq miles)
Population 20 000 000
Capital Lima
Government Parliamentary republic
Currency Sol = 100 centavos
Languages Spanish, Quechua, Aimara
Religion Christian (Roman Catholic)
Climate Temperate on coast; cooler in Andes, tropical in selvas. Average temperature in Lima ranges from 13-19°C (55-66°F) in August to 19-28°C (66-82°F) in February
Main primary products Cotton, sugar cane, coffee, rice, potatoes, fruit, vegetables, livestock, fish; crude oil, copper, silver, zinc, lead, iron ore, tungsten
Major industries Agriculture, mining, oil production and refining, fish processing, textiles, cement, leather goods, plastics, chemicals, metal refining, fishing
Main exports Petroleum, copper, sugar, iron ore, silver, cotton, zinc, coffee, lead
Annual income per head (US$) 1010
Population growth (per thous/yr) 24
Life expectancy (yrs) Male 56 **Female** 60

Petén ***Guatemala*** Most northerly department of the country, between Mexico and Belize in the south of the Yucatán peninsula. It covers about 36 000 km² (13 900 sq miles) and is mainly jungle and grasslands dotted with lakes. The region's Indian farmers practise slash-and-burn agriculture, moving from one patch of jungle to another every few years as their crops exhaust the thin soil.

Flores, the main town of the region (population 19 700), is on an island in the Petén Itzá lake, about 260 km (160 miles) north of the capital, Guatemala City. In the jungle nearby are extensive Mayan ruins dating from before the 10th century AD.

Population 194 200
Map Guatemala Aa

Peterborough ***United Kingdom*** Cathedral city in Cambridgeshire, 120 km (75 miles) north of London. The cathedral, built largely in the 12th and 13th centuries, is one of the finest Norman buildings in Britain. The modern city has agricultural industries and is a residential overspill for London. Much of the capital was built from bricks produced near Peterborough.

Population 88 000
Map United Kingdom Ed

Petra ***Jordan*** Ruined rock city and the country's main tourist attraction, situated 98 km (61 miles) south-west of Karak. It has been called 'the rose-red city' after a line in the poem *Petra* by the Victorian writer the Reverend John Burgon (1845-88), because of the effect of sunrise and sunset on its pinkish-red sandstone.

Petra was founded about 1000 BC by the Edomites, said to be the descendants of Esau, the son of Isaac and Rebecca. In about 300 BC the Edomites were driven out by the Naboteans, who made Petra their capital. But all that remains of it are the rock faces with numerous dwellings cut into them. The Street of Façades – great crags with buildings cut into them – contains the striking Treasury of Khazneh. There are many other cave dwellings and a Roman theatre.

Petra is ringed by high mountains and is approached through the Siq, a steep gorge over 3 km (2 miles) long, whose sheer sides rise to 100 m (about 330 ft). On a hill overlooking the city is the Tomb of Aaron, the elder brother of Moses and high priest of the Hebrews.

Map Jordan Ab

Petrikau ***Poland*** See PIOTRKOW TRYBUNALSKI

Petrograd ***USSR*** See LENINGRAD

Petrokov ***Poland*** See PIOTRKOW TRYBUNALSKI

petroleum, crude oil A natural, thick, yellow-to-black liquid hydrocarbon mixture found beneath the earth's surface. The name petroleum is derived from the Greek, meaning 'rock oil'. See OIL AND NATURAL GAS (p. 488)

Petropavlovsk ***USSR*** Timber town on the Trans-Siberian Railway in northern Kazakhstan, 265 km (165 miles) west of Omsk. Its industries now include food, clothing and canned goods as well as timber and wood pulp.

Population 222 000
Map USSR Hc

Petropavlovsk-Kamchatskiy ***USSR*** Pacific port and naval base used by Russian whalers and fishing boats on the south-east coast of the Kamchatka peninsula. Its population has trebled since the 1960s with the expansion of 'factory-ship' canning, and the servicing and repair of Soviet naval vessels.

Population 241 000
Map USSR Qc

Petrovgrad ***Yugoslavia*** See ZRENJANIN

Petrozavodsk ***USSR*** Capital of Karelia, lying on the western shore of Lake Onega. It was founded in 1703 as an ironworking centre, and has metalworking, ship-repairing and timber-based industries.

Population 247 000
Map USSR Eb

Pforzheim ***West Germany*** City about 30 km (19 miles) north-west of Stuttgart. It is the centre of West Germany's watchmaking and jewellery trade and has a museum of jewellery.

Population 105 000
Map West Germany Cd

phacolith, phacolite Body of intrusive IGNEOUS rock shaped like an upturned saucer and lying in the up-fold of beds or layers of rock. It formed by injection of hot molten rock either during the folding or after it.

Phaistos ***Greece*** Large excavated Minoan palace on a hill, 60 km (37 miles) south of Heraklion in Crete. It was built about 1700 BC on the site of an earlier palace. Parts of both have been uncovered. The Phaistos Disc, a pottery disc stamped on both sides with hieroglyphics as yet undeciphered, was found in the new palace and is in a museum in Heraklion.

Map Greece Dd

Phayao ***Thailand*** Rice-growing province and town in the north, about 630 km (390 miles) north of Bangkok. The originally 13th-century, largely wooden town lies beside Lake Khwan Phayao, which covers an area of 25 km² (10 sq miles).

Population (town) 141 700; (province) 477 900
Map Thailand Bb

phenocrysts Name given to crystals set among a mass of smaller crystals in an IGNEOUS rock. An example is porphyry. The two sizes result from the rock solidifying in two stages. A period of slow cooling produces the larger crystals, followed by a phase of fast cooling to produce the ground mass of smaller crystals.

Philadelphia ***USA*** City and port in Pennsylvania, about 160 km (100 miles) up the Delaware river from the Atlantic Ocean. It was settled in 1681 by colonists sent by the English Quaker William Penn, who laid out the city. The centre retains a flavour of the old city and its buildings are protected. The American Declaration of Independence was signed there in 1776. The city has engineering and shipbuilding industries and is an educational and arts centre, with three universities.

Population (city) 1 646 700; (metropolitan area) 4 768 400
Map United States Kc

Philae ***Egypt*** Island in the Nile river just below the Aswan High Dam, about 700 km (430 miles) south of Cairo. The island was dedicated to the fertility goddess Isis from 500 BC to about AD 380. Its ancient temples were covered by a reservoir when the old Aswan Dam was built in 1907. They were revealed again in 1970, when the new Aswan High Dam held back the river farther upstream. The Temple of Isis has since been reconstructed on a nearby island, Agilkia.

Map Egypt Cc

Philippines See p. 514

Philipsburg ***Netherlands Antilles*** See ST MARTIN

Phitsanulok ***Thailand*** Agricultural province and prosperous market town 355 km (220 miles) north of the capital, Bangkok. The province's farmers mostly grow rice, maize and beans.

Population (town) 215 900; (province) 729 350
Map Thailand Bb

Phlegrean Fields (Campi Flegrei) ***Italy*** Volcanic area to the west of Naples formed by collapsed cones and craters, still partially active. It contains the crater lake of Averno, the legendary entrance to Hell, and Mount Nuovo, an ash cone 140 m (460 ft) high created by an eruption in 1538.

Map Italy Ed

Phnom Malai ***Cambodia*** Hilly region north of the CARDAMOM MOUNTAINS in western Cambodia. It became the main refuge for the Khmer Rouge guerrillas after their overthrow from government by the Vietnamese invasion in 1978.

Map Cambodia Ab

Phnom Penh ***Cambodia*** National capital at the confluence of the Tonle Sap and Mekong rivers. It has been the capital almost continuously since 1434, but recent wars have scarred the city badly and it has only partly recovered from forced depopulation imposed by the Communist Pol Pot regime in 1975. Electricity and water supplies are sporadic. A large number of Vietnamese military and civilians have moved into the city.

Population 500 000
Map Cambodia Ab

Phoenicia ***Middle East*** See LEBANON; SYRIA

Phoenix ***USA*** Capital of Arizona. It lies in the centre of the state, about 200 km (125 miles) north of the Mexican border, and is a centre for tourism and the aluminium and aircraft industries. The city is also a focus of education and welfare facilities for the American Indians.

Population (city) 853 300; (metropolitan area) 1 714 800
Map United States Dd

Phoenix Islands ***Kiribati*** Group of scattered small islands to the east of the Gilberts. Some of them, including Kanton and Enderbury, were claimed by the USA until it signed a treaty of friendship with Kiribati in 1979, abandoning its claim so long as they are not used by another country for military purposes.

Map Pacific Ocean Ec

Phou Bia ***Laos*** Mountain which rises to 2820 m (9252 ft) immediately south of the Plain of Jars. It is the country's highest peak.

Map Laos Ab

Phuket ***Thailand*** The country's largest island covering 801 km² (309 sq miles), set in the Andaman Sea 675 km (420 miles) south and slightly west of the capital, Bangkok. It is joined to the mainland by a causeway and is the centre of a group of 37 beautiful islands.

Phuket mushroomed as a holiday centre after the bay of Phangnga nearby was chosen as a setting for the 1974 James Bond film *The Man With The Golden Gun*. It also has tin and rubber industries, but in 1986 there were protest riots against industrial developments that might damage the tourist trade. The island's main town is also called Phuket.

Population (town) 86 800; (island) 146 400

Map Thailand Ad

Piacenza ***Italy*** City on the Po river about 60 km (35 miles) south-east of Milan. The centre preserves its Roman chequerboard plan, and there are about 5 km (3 miles) of 16th-century walled fortifications still standing. The city's economy is based on livestock, textiles, engineering and chemicals.

Population 106 700

Map Italy Bb

Piatra Neamt ***Romania*** Tourist centre on the Bistrita river in the north-east of the country, 120 km (75 miles) south of the border with the USSR. The city contains a late 15th-century church and belfry, and the Carpathian Natural History Museum. Nearby is the 17th-century Agapia Monastery, the 16th-century Bistrita Monastery, the spectacular Ceahlau mountains, Lake Izvoru Muntelui, Lacu Rosu, and the Bicaz gorge. Piatra Neamt contains a papermill, oil refinery and chemicals, food processing and textile plants.

Population 92 300

Map Romania Ba

Piauí ***Brazil*** See TERESINA

Picardy (Picardie) ***France*** Northern region of 19 399 km² (7490 sq miles) between Paris and Flanders. It is composed of sweeping chalk plateaus drained by the Oise and Somme rivers. Parts were badly damaged by shelling during the First World War, and some villages and towns have since been completely rebuilt. Large mechanised farms now work the region's loamy soils to grow wheat and sugar beet. AMIENS is the largest town.

Population 1 740 000

Map France Eb

Pichincha ***Ecuador*** North-central province stretching across the large fertile Quito basin and into the jungle of the north-west lowlands around Santo Domingo. Mount Pichincha (4794 m, 15 728 ft) which overlooks QUITO, the provincial and national capital, was the scene in May 1822 of the decisive battle in which the Spaniards were defeated by forces led by the South American liberator Antonio José de Sucre.

Population 1 370 000

Map Ecuador Bb

Pico Spanish and Portuguese name for 'mountain' or 'peak'. For names of geographical features beginning 'Pico', see the main part of the name – for example NEBLINA, PICO DA.

Pidurutalagala ***Sri Lanka*** See NUWARA ELIYA

piedmont Formed or lying at the foot of a mountain or mountain range, for example a piedmont glacier. The name is derived from Piedmont province at the foot of the Alps in northern Italy.

Piedmont (Piemonte) ***Italy*** Region covering 25 399 km² (9807 sq miles) in the north-west. Its capital is TURIN and other major towns include ASTI, Alessandria, Cuneo, Novara and Vercelli. Piedmont is divided into three by the great sweep of the Alps that all but encircles the region; the flat plain of the upper Po valley where most of the towns stand; and by the Monferrato and Langhe hills.

Piedmont, through its statesman Camillo Cavour (1810-61), was the driving force behind the unification of Italy, which was largely completed in 1860. Turin was the nation's first capital, losing out later to Florence and then Rome. The region is rich in farmland, iron mines and sawmills.

Population 4 403 000

Map Italy Ab

Pieniny ***Czechoslovakia/Poland*** See BESKIDY

Pierre ***USA*** Capital of South Dakota, on the Missouri river in the centre of the state. It stands opposite the fur trading post of Fort Pierre founded in 1817.

Population 12 400

Map United States Fb

Pietermaritzburg ***South Africa*** Capital of Natal Province, named after two Voortrekker (Afrikaner settler) leaders, Piet Retief and Gerrit Maritz. It lies about 70 km (45 miles) north-west of Durban, and was founded in 1839 after the Zulus were defeated at the Battle of Blood River. The city has a Voortrekker Museum (built in 1838), Fort Napier (1843), a military encampment named after the Cape governor, Sir George Napier, and a campus of Natal University (1909).

Population 187 200

Map South Africa Db

Pietersburg ***South Africa*** Market town and mining centre on the Cape-to-Cairo road in north Transvaal, about 300 km (186 miles) north-east of Johannesburg. In 1900 it was briefly the seat of the governments of both the Transvaal and Orange Free State republics.

Population 25 500

Map South Africa Ca

Pigg's Peak ***Swaziland*** Village, about 60 km (37 miles) north of Mbabane, named after a prospector, William Pigg, who found gold nearby in 1884. Today the mine is defunct, but the settlement is the centre of a timber-producing and tourist area where the fine scenery includes several magnificent waterfalls, among which are the Malolotsha Falls on the Malolotsha river.

Population 3200

Map Swaziland Aa

▼ IN TRADITIONAL STYLE Houses on stilts and long boats line the shore at a quiet fishing village on the island of Phuket in Thailand.

Philippines

SEVEN THOUSAND ISLANDS WHERE TYPHOONS BLOW AND VOLCANOES ERUPT, THE PHILIPPINES IS A LAND OF PASSION AND BEAUTY – A PIECE OF LATIN AMERICA IN THE PACIFIC

The fury of the typhoons that hit this beautiful island chain in the western Pacific, and the destructive power of its active volcanoes are matched by the passions of the people in both religion and politics.

The Philippines is the only predominantly Christian country in Asia. At Easter, Roman Catholic penitents simulate the crucifixion; yet when Pope Paul visited the capital, MANILA, in 1970, an attempt was made to assassinate him. And so too in politics: Benigno Aquino, the chief rival of former President Marcos, was shot as he stepped off his plane at Manila airport on his return from exile in 1983.

Named after Philip II of Spain (1527-98), the 7107 islands of the Philippine archipelago stretch 1800 km (1120 miles) between Taiwan and Borneo. Together they have a coastline longer than that of the USA. Spain colonised the Philippines in 1565 and ruled there until the Americans took over in 1898. When the Americans left in 1946, a cynic said that the Filipinos had spent 'four hundred years in a convent followed by fifty years in Hollywood'. The Filipinos are the least oriental of the orientals; the republic has been called 'a piece of Latin America in the Pacific'.

In 1941, the Philippines was invaded by the Japanese, then liberated by the Americans in 1945. The greatest sea battle in history was fought off the island of LEYTE in 1944, when American naval forces destroyed the might of the Japanese fleet.

Healing the scars of war and re-establishing economic and political life were complicated by a rebellion in central LUZON by the Communist-led Hukbalahap guerrillas (Huks). The Filipino name means 'People's Anti-Japanese Army' – a description of its activities during the occupation. After the war, the Huks openly declared their Communist orientation and launched an armed revolt against the government, seeking sweeping rural land reforms (which attracted many peasants to their ranks) and political power. Military pressure, progress on a fairer distribution of agricultural land and economic reforms led to the collapse of the revolt in 1954, when its leader Luis Taruc surrendered. The need for further land reforms continued, and the Huks became active again in the 1960s. They surrendered in 1972, but a breakaway Maoist group – the New People's Army – has continued terrorist activity.

The total land area of the Philippines is 300 000 km² (115 830 sq miles) – about the size of Italy – but the islands are scattered over an expanse three times as great. They form four main groups. To the north are Luzon, on which stands the sprawling city of Manila, and MINDORO. In the centre are the VISAYAN ISLANDS, including SAMAR, Leyte CEBU, NEGROS and PANAY. In the south are MINDANAO and the SULU ARCHIPELAGO, while isolated in the south-west is PALAWAN.

NATURE'S VIOLENCE

Most of the seas around the Philippines are shallow and easily navigable. The archipelago is dotted with small ports and anchorages. These are usually on the western coasts and interior seaboards, sheltered from the prevailing east winds which blow from the Pacific. From June to December there is a risk of typhoons – in some years there are 200 or more. In 1984 two typhoons with winds of up to 298 km/h (185 mph) sank 11 ships and caused over 1600 deaths.

The central and western parts of the islands are partly sheltered from the worst of the Pacific storms by mountains on the eastern coasts. On Luzon, the SIERRA MADRE mountains form the eastern barrier, while Mindanao is protected by the Diuata range. On both islands there are more mountains farther west; Mindanao also has the Philippines' highest peak, Mount Apo (2954 m, 9690 ft).

Through several of the smaller islands, such as Cebu and Palawan, run hilly backbones flanked by narrow coastal plains, with little land for settlement. The main lowland areas are the central plain and the CAGAYAN valley on Luzon, and the valleys of the Agusan and Mindanao rivers in Mindanao.

The Philippines were created by a disturbance of the earth's crust between 50 and 60 million years ago. There are more than 30 active volcanoes and some have erupted violently in recent times. In 1984 the MAYON volcano on Luzon hurled boulders as big as cars down its slopes as terrified villagers fled their land. The eruption sent a cloud of ash 10 km (6 miles) high over southern Luzon. Earthquakes are also common. One in east Mindanao in 1976 sent huge waves (tsunami) inland and 8000 people were reported dead or missing.

Off the east coast of the Philippines, the movement of the earth's crust has created one of the deepest ocean trenches in the world, plunging to 10 540 m (34 578 ft or over 6½ miles) near Mindanao.

The islands have three seasons: rainy from June to October, cool from November to February and hot from March to May. In the lowlands the climate is hot and humid – the average daily temperature is 28°C (82°F) – but in the highlands temperatures can drop to 16°C (61°F). Manila has 2080 mm (82 in) of rain a year, western Luzon 1000 mm (39 in) and north-east Mindanao 3800 mm (150 in).

RESOURCES AND PEOPLE

More than 40 per cent of the Philippines is covered by tropical rain forest containing many hardwoods including teak. The rain forest of Mindanao is one of the most productive in the world, and timber is an important foreign exchange earner. But too much cutting and too little replanting has taken its toll. Several of the central Visayan Islands are suffering from erosion, and much of the previously forested island of MASBATE is now rolling tropical grassland of little value. The country's minerals include chrome, copper, gold and iron ore.

Of the 7107 islands in the archipelago, only some 400 are inhabited. The first known settlers, 25 000 years ago, were the negritos, or pygmies, who are believed to have arrived by a land bridge from mainland Asia that existed at the time. This primitive group still hunt with spears in the upland jungles of Negros, Samar and Leyte, remote from modern life. Between 2000 and 3000 years ago, the Ifugaos arrived, probably from Taiwan, and began building rice terraces which survive today on the island of Luzon. The terraces, chiselled out of mountain sides and then filled with soil and gravel, depend on a complex system of sluices and canals to water the rice.

The next wave of immigrants were Malays from Indonesia in the 3rd century BC. They were the ancestors of the main present population, and the local administrative unit is still called the *barangay* after the small sailing vessels that carried the Malays to the islands. In this mountainous archipelago, life was concentrated in the coastal lowlands. Each new wave of settlers fished the rivers and grew rice along the banks, pushing earlier inhabitants back into the hills; this is one reason why the rice terraces became so extensive.

The Portuguese explorer Ferdinand Magellan, on a Spanish round-the-world expedition via South America, brought the first Europeans to the islands in 1521. Magellan was killed in a clash between rival native groups, but in 1565 Miguel Lopez de Legaspi formally claimed the Philippines for Spain.

SPANISH AND AMERICAN HERITAGE

In the northern two-thirds of the Philippines, the Roman Catholic religion which the Spanish invaders brought is practised with fervour. Catholics make up 83 per cent of the population, and Protestants 9 per cent. Many Protestants are members of the Nationalistic Philippine Independent Church. The extreme south was never subdued by the Spanish, and about 5 per cent of the total population – mainly in western Mindanao and the Sulus – are Muslims. Among primitive farmers and fishermen living in remote parts animism is still practised.

Under Spanish occupation, the coconut palms which are a feature of almost every village were cultivated commercially for their copra; it can be seen drying on the roadside throughout the archipelago. In areas such as the ILOCOS region of Luzon, tobacco cultivation was introduced in the 18th century, and many smallholders grow little else. Galleons plied between Manila and South America, Chinese junks sailed in with silk, jade and

▶ NATURE TAMED Rice terraces begun more than 2000 years ago climb the steep hillsides of northern Luzon. It is estimated that if all the terraces were laid end to end they would reach halfway round the world.

spices, and a prosperous Chinese merchant community grew up.

In some parts of the new Spanish colony, much of the old village organisation was retained. But in the more central areas, such as the fertile central plain of Luzon and the island of Negros, sweeping changes took place. Large estates were given by the Spanish administration to Spanish settlers and religious orders. The Filipinos had to be content with small tenanted holdings; the size of these is a major social problem in rural areas today, and has brought about the demands for land reform. Sugar cane and pineapples are extensively grown, but the slump in the world sugar market has caused extensive hardship on Negros.

continued, and in 1936 a transitional semi-independent Commonwealth of the Philippines was formed. In the Second World War Filipinos fought with Americans against the Japanese invaders, and in 1946 full independence came.

Under American influence the urban Filipinos developed an assertiveness and informality unusual in South-east Asia. Close ties with the USA – including several military bases – remain, and English is widely used. Today there are two official languages: Pilipino, based on Tagalog (the main language of central Luzon), and English. Spanish is still commonly spoken, together with nearly 80 regional languages and dialects.

Filipino men wear sports shirts and slacks

Most small towns in the Philippines resemble their Spanish equivalents, with a central square which has the church on one side and secular buildings on the other. In education, the Dominican and Franciscan friars have left many institutions of learning; one Filipino in four goes on to higher education and the adult literacy rate is 90 per cent.

In 1898 the four-month Spanish-American War began over Spanish oppression in Cuba. An American squadron destroyed the Spanish fleet in Manila Bay and Manila was occupied with the help of Filipinos, who were already in revolt against Spanish rule. But Filipino hopes of immediate independence were dashed: the USA 'bought' the islands from Spain for US$20 million. Nationalist agitation

by day, and a *barong tagalog* (a loose, almost transparent shirt worn outside the trousers) for formal wear. Filipinas wear light summer frocks, and *ternos* (formal dresses with butterfly sleeves) on special occasions.

PROBLEMS OF INDEPENDENCE

Since independence the Philippines has faced many problems reminiscent of Latin-America: inflation, government debt, high unemployment, social inequalities, crime and corruption. Social unrest and austerity increased as the country's industrial expansion in textiles, chemicals and food processing slowed. Ambitious projects such as finding alternative sources of energy to imported oil have caused increasing debt. The hill peoples object to long-established homelands being flooded for hydroelectricity. The Muslims resent control from the Christian north, and especially the migration of land-hungry northerners to parts of Mindanao.

For 12 years Muslims of the Moro National Liberation Front waged guerrilla warfare in the south, demanding autonomy. Some 60 000 died in the mid 1970s, but the rebellion has faltered through internal dissension and lack of outside support. However, another long-standing threat to security remains: the Maoist New People's Army (NPA) now operates in three-quarters of the country's provinces. The NPA's 12 000 troops have attacked army camps, convoys and towns, and in 1984 they killed nearly 3000 people, including 800 soldiers and 70 policemen. Many observers feel that growing Communist insurgency in the Philippines was fuelled by the repressiveness of government forces under Marcos.

Ferdinand Marcos, aided by his wife Imelda, was the dominant political figure in the Philippines for two decades after his election in 1965. He imposed martial law from 1972 to 1981, and was re-elected in 1981. His chief rival, Benigno Aquino, was killed in 1983. However, in the elections of 1986, Corazon (Cory) Aquino – widow of the assassinated Benigno – made sweeping gains. Both Marcos and Aquino claimed victory, but after allegations of widespread election fraud, pressure from overseas and the defection of most of the armed forces to the Aquino cause, the ailing Marcos and his family fled to exile (and reportedly a vast fortune) in the USA.

Cory Aquino was inaugurated president amid a wave of popular rejoicing. The new government released many political prisoners and set about the enormous tasks of reconciliation, land and political reforms, and the rebuilding of the country's shaky economy. No one believed it would be an easy job, especially when the NPA ignored appeals to surrender. Political instability still lurks beneath the surface charm, colour and sunshine of the Philippines.

PHILIPPINES AT A GLANCE

Area 300 000 km² (115 830 sq miles)

Population 58 115 000

Capital Manila

Government Republic

Currency Peso = 100 centavos

Languages Pilipino, English, Spanish and local dialects

Religions Christian (92%), Muslim (5%)

Climate Tropical; average temperature all year round in Manila ranges from 21°C (70°F) to 34°C (93°F)

Main primary products Rice, maize, coconuts, sugar cane, abaca (Manila hemp), rubber, tobacco, pineapples, bananas, coffee, timber, fish; copper, chrome, gold, iron, nickel, coal

Major industries Agriculture, food processing, textiles, chemicals, forestry, fishing, mining

Main exports Electrical goods, clothing, metal ores, coconut oil, sugar, fruit and vegetables, timber, Manila hemp

Annual income per head (US$) 696

Population growth (per thous/yr) 23

Life expectancy (yrs) Male 61 **Female** 65

Pigs, Bay of (Bahia de Cochinos) ***Cuba*** Bay on the south coast, about 175 km (110 miles) south-east of Havana. It was the scene of an attempted invasion by exiled anti-Castro Cubans, trained and equipped by the American CIA, on April 17, 1961. The 1600 invaders were quickly overcome and many were killed or captured.

Map Cuba Aa

Pikes Peak ***USA*** Mountain rising to 4301 m (14 109 ft) in the Rocky Mountains in Colorado. It is one of the state's main skiing areas. Gold used to be mined nearby.

Map United States Ec

Pilbara, The ***Australia*** Iron-rich area in Western Australia, centred on the Hamersley Range some 1200 km (750 miles) north of Perth. It was inhabited only by Aborigines until the first iron ore mines were opened in the 1960s. Marble Bar, 160 km (100 miles) south-east of the main outlet, Port Hedland, is one of the hottest places in the country. Temperatures there often soar to 49°C (120°F).

Map Australia Bc

Pílion (Pelion) ***Greece*** Heavily forested mountain range rising abruptly on the east coast of THESSALY, and stretching in a peninsula into the Aegean Sea north of Euboea. Mount Pílion itself – the main peak in the range – rises to 1548 m (5078 ft). In legend, it was the land of Centaurs and its foothills, washed by abundant streams and with mild winters, was a wild garden of herbs. The district is good for camping and walking, dotted with charming villages of timber-framed houses, such as Portariá, Makrinítsa, Zagorá, Miléai, Khania and Tsangaráda. There are ski slopes and, on the Aegean coastline, several holiday resorts, including Makryrakhi and Ayios Yiannis.

Map Greece Cb

pillow lava Form of lava that has solidified underwater, consisting of a series of closely fitting pillow-shaped, or globular, masses usually about 500-1000 mm (20-40 in) across.

Pílos (Navarino) ***Greece*** Harbour town on Navarino Bay, on the south-west coast of the Peloponnese. In AD 1827 at the naval battle of Navarino, a combined French, English and Russian fleet defeated the Turks. The remains of 58 vessels sunk during the battle can be seen in calm weather under the water. Nearby, parts of the palace of the Homeric King Nestor have been excavated.

Map Greece Bc

Pilsen ***Czechoslovakia*** See PLZEN

Pinar del Río ***Cuba*** Capital of Pinar del Río province, 160 km (100 miles) south-west of Havana. It manufactures tobacco products, pharmaceuticals and furniture.

Population (city) 150 000; (province) 650 000

Map Cuba Aa

Pindhos ***Greece*** Mountainous backbone of Greece, a massive range extending from Albania in the north and falling away into the Gulf of Corinth. Its peaks include Smólikas (2637 m, 8651 ft), Grámmos (2503 m, 8210 ft), Gióna (2510 m, 8235 ft) and the sacred Parnassus (2457 m, 8061 ft) with its twin summits, one dedicated to Apollo, the other to Bacchus. The range is mainly limestone with a gently rolling plateau, and has a pass – at the village of Metsovon near Katara. Deep ravines are clothed in forests of beech, oak, chestnut and fir. Five major rivers of Greece rise in the range: the AKHELOOS, the Arakhthos and the Aoos, which drain into the Ionian Sea; and the Aliákmon and Piniós, which empty into the Aegean Sea.

Map Greece Bb

Pines, Isle of ***New Caledonia*** Beautiful tourist resort island, area 134 km² (52 sq miles), about 45 km (28 miles) off the south-eastern tip of New Caledonia. The 30-45 m (100-150 ft) 'pines', relatives of the monkey-puzzle tree, are unique to the area.

Population 1300

Map Pacific Ocean Dd

pingo Dome-like mound of earth found in Arctic regions, enclosing a core of ice, where ground water is trapped by permafrost. The water freezes and expands, forcing up the mounds like earth blisters. Pingos are up to 90 m (295 ft) high and 800 m (0.5 mile) across. Some have craters with circular ponds, where the ice has melted and caused collapse.

P'ing-tung ***Taiwan*** Industrial and commercial city in the south of the island, 22 km (14 miles) east of Kao-hsiung. It lies in a region that produces rice, sugar, bananas and sisal, and has food processing industries.

Population 300 000

Map Taiwan Bc

Pinsk Marshes ***USSR*** See POLES'YE

Piotrkow Trybunalski (Petrikau; Petrokov) ***Poland*** Industrial city 45 km (28 miles) south of the city of Lodz. Because of its central position in the country, it was the seat, between 1578 and 1792, of Poland's highest court, the Crown Tribunal – from which it gets its name. Today it makes textiles, glass, timber and machinery.

Population 77 200

Map Poland Cc

Pipestone National Monument ***USA*** Prairie area of 114 hectares (282 acres) in south-west Minnesota, about 10 km (6 miles) from the South Dakota border. It has prehistoric and recent pipestone quarries where Indians dig the material for their ceremonial peace pipes.

Map United States Gb

Piraeus ***Greece*** The country's largest port, on a promontory beside Athens. It has plants for engineering, cement, chemicals, textiles, oil refining and shipbuilding.

Piraeus has been the port of Athens since 482 BC, the main harbour being a natural basin surrounded by land on three sides. There are two smaller harbours, Pasalimani, known in ancient times as Zea, and where ancient boat sheds can be seen under the water, and Microlimano. A museum on Pasalimani Quay traces the history of the port's shipping.

Population 196 400

Map Greece Cc

Piran (Pirano) ***Yugoslavia*** Fortified Venetian-built port 32 km (20 miles) south-west of the Italian border. A lovely town of steep, narrow streets and red-roofed houses around a 14th-century cathedral and bell tower, it has a charming piazza named after the violinist and composer Tartini, born there in 1692. One of the fine palaces on the square, built by a rich merchant from Venice for his mistress, carries the inscription '*Lassa pur dir*' ('Let them talk'). The merchant is said to have ordered the inscription as a defiant riposte to townsfolk who disapproved of his amorous visits from across the Adriatic.

Population 5500

Map Yugoslavia Ab

Pisa ***Italy*** City on the Arno river about 70 km (44 miles) west of Florence. It was once a Roman port but is now 10 km (6 miles) from the sea because of silting in the estuary.

Its renowned bell tower, 55 m (180 ft) high and leaning at an angle of 5°30', is climbed by 294 steps. It was started in 1173 and completed over 200 years later; in recent years work has been carried out to prevent any further settling of its foundations. The summit, used by the scientist Galileo Galilei (1564-1642) to conduct gravity experiments, has fine views over the city. The Civic Museum has a fascinating display of early Pisan sculptures and paintings. There are many Romanesque and Gothic churches and richly ornamented 16th-century palaces such as the Palazzo dei Cavalieri.

Pisa's industries are based on glass, engineering, clothing and pharmaceuticals.

Population 104 300

Map Italy Cc

Pisa (Piso), Lake ***Liberia*** Shallow salt-water lake covering nearly 600 km² (230 miles) about 70 km (some 45 miles) north-west of Monrovia, the capital. Hills covered by thick tropical forest descend to the wide sandy beach along the lake shore, and extend along the two peninsulas that form the lake's narrow outlet to the sea. Fishing villages line the shores, but the lake is also becoming popular with tourists, who are drawn by its calm waters and the wildlife on the island of Massating – which includes monkeys, boars, porcupines, jackals and a wide variety of birds.

Map Liberia Aa

Pisac ***Peru*** Magnificent ruins of an Inca fortress in the Urubamba valley 32 km (20 miles) north of Cuzco. They consist of extensive terraces and towers. The Temples of the Sun and Moon face a large burial ground.

Map Peru Bb

Pisco ***Peru*** Coastal town, founded in 1640, which gives its name to the national brandy. It stands in Ica department and is made up of two districts. Pisco Puebla has Spanish colonial-style houses and a statue of the liberator General José San Martín, who landed with his troops on September 7, 1820 at nearby Paracas Bay. Pisco Puerto is the commercial port and centre of the fishmeal, cotton-seed oil and cotton textile industries.

Population 82 300

Map Peru Bb

Piso, Lake ***Liberia*** See PISA, LAKE

Pistoia ***Italy*** City about 32 km (20 miles) north-west of Florence. It has a fine range of art treasures, including works of the early sculptors of Pisa and Como and of the outstanding Renaissance figures such as Verrocchio (1435-88). The Piazza del Duomo is one of Italy's finest medieval squares. The magnificent cathedral, midway in style between Pisan and Florentine Romanesque, is dominated by a large 13th-century bell tower.

The economy of the city is based on vegetables and fruit, railway industries, textiles, shoes, beds, mattresses and furniture.

Population 91 600

Map Italy Cc

Pitcairn Islands ***Pacific Ocean*** Group of four small islands about halfway between New Zealand and Panama, comprising Pitcairn itself – with an area of 4.5 km^2 (1.75 sq miles) – and the uninhabited islands of Henderson, Ducie and Oeno. It is administered by the British High Commission in New Zealand. Pitcairn is renowned for its association with Fletcher Christian and eight other mutineers from HMS *Bounty*, who settled there with a group of Tahitian women in 1790, and whose descendants still live there.

Population 50

Map Pacific Ocean Gd

Pitch Lake ***Trinidad and Tobago*** Natural deposit of bitumen (pitch), extending over 46 hectares (114 acres) in the south-west of Trinidad, near San Fernando. The lake, which is hard around the edges and viscous towards the centre, is thought to be formed by the seepage of crude oil, which on evaporation has left a residue of bitumen. The supply, which is replenished from underground, has yielded millions of tonnes since its discovery in the 16th century, and is the world's main source of natural pitch. Fossils of prehistoric animals have been found preserved in the pitch.

Map Trinidad Aa

pitch, plunge Angle made by the axis of a fold in geological strata with the horizontal.

Pitesti ***Romania*** Industrial city on the Arges river, 105 km (65 miles) north-west of Bucharest. Oil was discovered there in the early 1960s, and since then Pitesti – a former market town – has rapidly expanded. It is also the country's main automobile manufacturing centre. Pitesti's other products include electric motors, chemicals, textiles and shoes.

Population 135 000

Map Romania Ab

Pitons ***St Lucia*** Twin conical mountains, Gros Piton (798 m, 2619 ft) and Petit Piton (750 m, 2460 ft), rising sheer from the sea near the village of Soufrière on the south-west coast. They are the laval plugs, now exposed by erosion, of volcanoes which erupted about 15 000 years ago; they are one of the most dramatic sights in the Caribbean. Soufrière is named after another nearby volcano, which is surrounded by numerous hot sulphur springs.

Map Caribbean Cc

Pittsburgh ***USA*** Industrial city in western Pennsylvania, about 175 km (110 miles) from Lake Erie. It dates from the mid-18th century and lies in a coal-mining area. By the 19th century it was a flourishing iron-making town, and it became known as Steel City under the influence of the industrialist and philanthropist Andrew Carnegie (1835-1919). Other industries include engineering, glass making, electrical engineering and food processing.

Population (city) 402 600; (metropolitan area) 2 172 800

Map United States Kb

▼ INHOSPITABLE COAST A long boat nears a jetty on the precipitous, volcanic island of Pitcairn, near Adamstown, the island's only village.

Piura ***Peru*** City and department on the border with Ecuador. Rice and high-grade cotton are grown under irrigation on the broad coastal plain. The city was founded in 1532 by the Spanish conquistadores and is the birthplace of Miguel Grau, hero of the War of the Pacific with Chile (1879-84). To the north lie the oil fields of Lobitos and Negritos, and to the south the shifting sands of the Sechura Desert.

Population (department) 1 125 900; (city) 186 500

Map Peru Aa

placer deposit Surface material deposited by a river, ice, the wind or waves and containing valuable minerals accumulated by weathering. The minerals are heavy, resistant minerals such as gold, platinum, cassiterite (tin ore) and diamonds.

plain An extensive low-lying area of land of gentle variations in height.

Planina ***Yugoslavia*** See POSTOJNA

Plassey ***India*** Historic village in West Bengal, about 150 km (93 miles) north of Calcutta. In June 1757 it was the scene of an epic battle between 3000 British troops commanded by Robert Clive (later Baron Clive of Plassey) and 50 000 Indians under the Nawab of Bengal. Clive's subsequent victory transformed the British from a mere trading company into the ruling force in India.

Map India Dc

Plate, River ***South America*** See LA PLATA

plate margin The edge of one of the rigid, adjoining plates that make up the earth's lithosphere. See PLATE TECTONICS opposite.

plate tectonics The theory that the earth's crust is composed of a series of rigid plates and that movement of these plates in relation to one another is responsible for the major structural features of the earth's surface. See opposite.

plateau Elevated, and comparatively level, expanse of land.

Platte ***USA*** River in Nebraska, formed by the confluence of the North and South Platte rivers which rise in Wyoming and Colorado. The Platte flows about 500 km (310 miles) eastwards into the Missouri. The Oregon Trail, used in the 19th century by settlers on their way westwards, followed the Platte and North Platte valleys.

Map United States Gb

Plenty, Bay of ***New Zealand*** Wide inlet on the north-east coast of North Island, about 200 km (125 miles) south-east of Auckland. The region around the bay produces citrus fruits, dairy products and sheep. Timber is the main industry and is exported from the port of Mount Maunganui near TAURANGA. There is a large wood-processing factory on the shores of the bay at Whakatane.

The bay was named by the explorer Captain James Cook in 1769. It was here that friendly Maoris supplied him with food and water.

Length 260 km (160 miles)

Map New Zealand Fb

PLATE TECTONICS

Most geophysicists now agree that the earth's outer crust, the lithosphere, is made up of at least 15 separate but adjacent plates which constantly move in relation to one another. Where plate margins adjoin, massive forces are unleashed which manifest themselves in three different ways, the products of which are ocean ridges, ocean trenches and transform faults. The study of this movement is known as plate tectonics.

Ocean ridges occur at the constructive plate margins where molten rock wells up from the earth's interior as the plates move apart, and cools to make new crust. Oceanic trenches mark destructive plate margins or subduction zones, where the edges of some plates slide down into the earth's hot interior as two plates collide. Along transform faults, plates slide past each other, and crust is neither created nor destroyed, as along the San Andreas Fault in California. However, most transform faults occur at the ocean ridges, cutting across the ridges at right angles to the plate margins.

All types of plate margin are characterised by active volcanoes or earthquakes or a combination of both. As a result, all plates are continuously changing size and shape, growing along constructive plate margins and being consumed along destructive margins.

WHEN CONTINENTS COLLIDE

1
When continental blocks move together, ocean plate between is destroyed as it slides beneath the continental plate to form a trench. Earthquakes, volcanoes and mountain-building occur on the land.

2
Sediments are scraped off the ocean floor by the continental plate and build up thickly. Sediments on the continental margin are also compressed and raised. Oceanic plate continues to be consumed.

3
When the ocean floor has been completely consumed, the landmasses collide. Accumulated sediments are compressed, folded and uplifted to form mountain ranges. Older volcanic ranges are also compressed and raised.

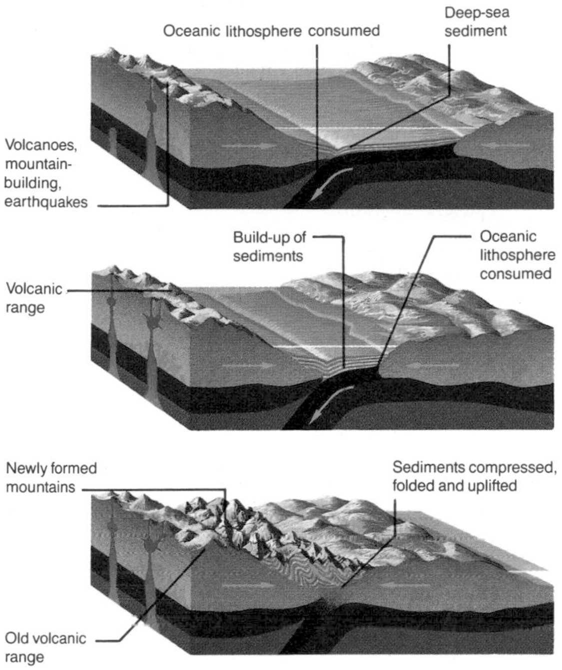

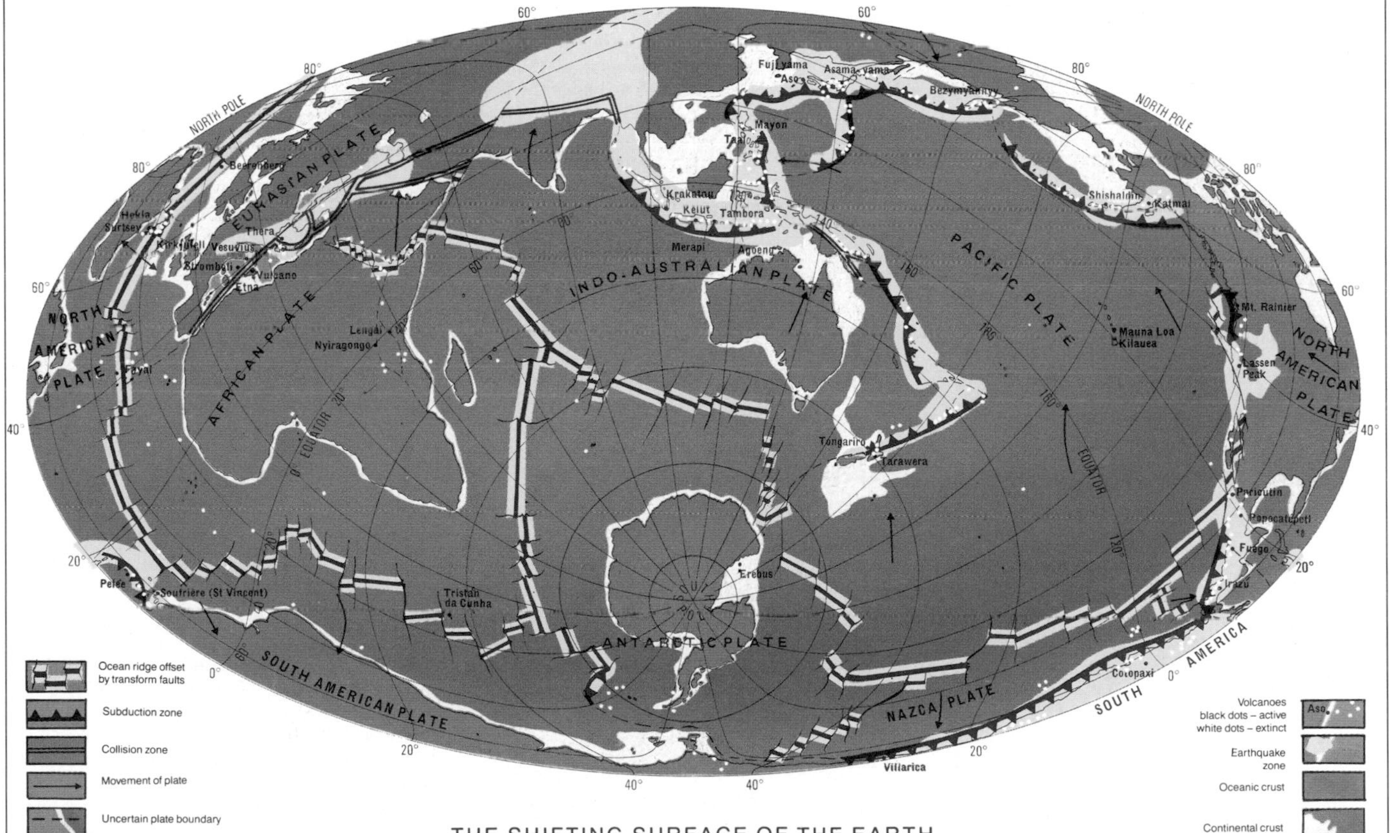

THE SHIFTING SURFACE OF THE EARTH

The mosaic of lithospheric plates that makes up the earth's crust floats on semi-molten rock – the asthenosphere. Each plate moves at about 12.5 mm (0.5 in) each year, but what causes the plates to move is not yet known. Some scientists believe that convection currents in the asthenosphere may be responsible. It is visualised that pancake-like convection cells rise beneath the ocean ridges and spread outwards with a return current coming back only a few hundred kilometres below the upper current. Another theory suggests that gravity is the driving force, pulling the denser lithospheric plate into the less dense asthenosphere at the subduction zones. Alternatively, new material injected at the ocean ridges forces the plates apart.

Pleven ***Bulgaria*** Capital of Pleven province, 137 km (85 miles) north-east of Sofia. A major market, industrial and service centre for the fertile Danubian Platform, it produces textiles, machinery, cement and ceramics. In 1877, Pleven fell to Russian troops after a 143 day siege during a Russo-Romanian invasion which forced the Turks to recognise Bulgarian independence. There is an impressive mausoleum dedicated to those who died in the battle, and a museum depicting the siege.
Population 140 000
Map Bulgaria Bb

Pliocene Fifth, and last, epoch of the Tertiary period in the Cenozoic era of the earth's time scale. See GEOLOGICAL TIMESCALE

Pliska ***Bulgaria*** See VARNA

Plitvice ***Yugoslavia*** Thickly forested national park in Croatia some 20 km (12 miles) north-west of the town of Bihac. It has a string of 16 lakes – linked by cascades and waterfalls, one of them 76 m (249 ft) high – and is noted for its crayfish.
Map Yugoslavia Bb

Plock (Plozk) ***Poland*** City about 95 km (60 miles) north-west of the capital, Warsaw. It was founded in the 10th century and still has a 12th-century cathedral. Today the city's postwar factories make foodstuffs and farm implements, and there is an oil refining and petrochemicals complex fuelled by Russian oil. Each June, Plock hosts the National Festival of Folklore and Folk Art, and a fair where traditional wooden carvings, ceramics and embroideries are sold.
Population 112 400
Map Poland Cb

Ploiesti ***Romania*** Major oil city in the Carpathian foothills, about 60 km (35 miles) north of Bucharest. Oil production and refining began there in 1856. The industry expanded rapidly this century and in the Second World War was so crucial for Germany's war machine that the Allies subjected the city to heavy air attacks. Intensive exploitation which continued after the war left the fields much depleted. But the installations were rebuilt, and today refineries and petrochemical works light up the sky at night.
Population 207 000
Map Romania Bb

Plovdiv ***Bulgaria*** Capital of Plovdiv province, 140 km (87 miles) south-east of Sofia and situated on the Maritsa river. It is a market town in a region producing rice, fruit, wine and tobacco. Industries include machinery, electrical goods, textiles, footwear and food. There is an agricultural college, and the ruins of a 13th-century fortress. Plovdiv was founded in the 1st century BC as Trimontium and became the capital of Thrace in 46 BC. It was frequently sacked during the Middle Ages, fell to Turkey in the 14th century, and became Bulgarian in 1885. It is now the country's second largest city after the capital, Sofia. Plovdiv has impressive Roman remains, including a theatre. Old Plovdiv has charming 19th-century houses.
Population 373 000
Map Bulgaria Bb

Plozk ***Poland*** See PLOCK

plug See NECK

pluton, plutonic rock Large mass of IGNEOUS rock formed beneath the surface of the earth by the consolidation of magma.

Plymouth ***Montserrat*** Chief port and capital, on the south-west coast. Georgian houses on the main street were built of stone brought from England as ship's ballast.
Population 3200
Map Caribbean Cb

Plymouth ***United Kingdom*** Port and naval base some 300 km (190 miles) south-west of London in the English county of Devon. It was one of the bases from which the English fleet set out to defeat the Spanish Armada in 1588, and was the final departure point for the voyage of the Pilgrim Fathers to North America in 1620. There has been a naval dockyard at neighbouring Devonport on the estuary of the River Tamar since 1691.

The city of Plymouth was badly damaged by Second World War bombing, but parts of the old port survived, including the statue of the naval hero Sir Francis Drake (1540-96), whose home was at nearby Buckland.
Population 255 000
Map United Kingdom Ce

Plymouth ***USA*** Town in Massachusetts about 55 km (35 miles) south and slightly east of Boston. It was the site of the first permanent European settlement in New England – the Plymouth Colony, set up by the Pilgrim Fathers in 1620. They sailed from England in the *Mayflower* to escape religious persecution. The town is now a major tourist attraction, with several museums in 17th-century houses.
Population 37 100

Plzen (Pilsen) ***Czechoslovakia*** Regional capital of western Bohemia and an industrial city 84 km (52 miles) south-west of Prague. The Old Town has a 14th-century cathedral and 16th-century town hall. It expanded after 1850 with heavy industries producing Skoda armaments, locomotives, turbines and vehicles. Plzen's older products include Pilsner lager beers, which originated there in 1842.
Population 174 100
Map Czechoslovakia Ab

Po ***Italy*** Longest river in Italy, flowing 652 km (405 miles) from the Cottian Alps to its delta on the Adriatic south of Venice. It is navigable for much of its length and has been an important trade route since Roman times. The risk of flooding has meant that few towns except Turin and Piacenza stand on its banks, but its fertile plains are densely settled.
Map Italy Bb

Pobedy, Pik (Tomur Feng) ***USSR/China*** Mountain in the Tien Shan range on the Soviet-Chinese border. It is the USSR's second highest peak after Communism Peak, reaching 7439 m (24 406 ft).
Map USSR Jd

Podgorica ***Yugoslavia*** See TITOGRAD

Podgoritsa ***Yugoslavia*** See TITOGRAD

Podhale ***Poland*** Southern region around the upper valley of the DUNAJEC river. It is a popular holiday area. ZAKOPANE is the main resort.
Map Poland Cd

P'ohang ***South Korea*** Industrial city on the east coast, about 120 km (75 miles) north-east of the port of Pusan. It is the site of the country's largest iron and steel works, built on land reclaimed from the sea.
Population 201 200
Map South Korea Dd

Pohjois-Karjala ***Finland*** See KARELIA

Pohnpei (Ponape) ***Federated States of Micronesia*** State consisting of one main island, Pohnpei itself, and eight outlying atolls. The national capital Kolonia (population 5500) is on the north coast of Pohnpei island, which covers about 300 km^2 (116 sq miles). In the south-west are the ruins of Nan Madol, a group of stone buildings erected on man-made islands more than 600 years ago.
Population 23 000
Map Pacific Ocean Cb

point Promontory or cape projecting into the sea.

point bar Low, curving sandy or gravelly spit that forms on the inside of a river meander when the main-stream channel shifts towards the outer bank of the bend.

Point Lisas ***Trinidad and Tobago*** Industrial complex on the west coast of Trinidad, about 10 km (6 miles) north of San Fernando. Lying on former mangrove and sugar-cane land, the complex has been developed to make use of the abundant supply of oil and natural gas for the production of petrochemicals, fertilisers and steel. It has a deepwater harbour, a plant for producing liquid oxygen and nitrogen, and the eastern Caribbean's largest power station.
Map Trinidad Aa

Pointe-à-Pitre ***Guadeloupe*** Main commercial centre and port of Guadeloupe standing beside a large and protected bay in the south-west of Grande Terre island. Together with its suburb of Abymes, it forms the largest town in the French overseas department of Guadeloupe.
Population (Pointe-à-Pitre) 23 005; (Abymes) 53 100
Map Caribbean Cb

Pointe-Noire ***Congo*** The country's main seaport, situated on the Atlantic coast at the end of the Congo-Océan railway from Brazzaville, 370 km (230 miles) away. The port was opened in 1939, five years after completion of the railway. The city then grew rapidly and was capital of the Middle Congo region of French Equatorial Africa from 1950 to 1958.

Its exports include cotton, ivory, hardwoods, various minerals, including some – such as manganese – from Gabon, palm products and rubber. Manufacturing is increasing and the city has a refinery to process oil from coastal oil fields. It also has an international airport. Besides its own trade, the port handles the foreign trade of Gabon, Chad and the Central African Republic. About 20 km (12 miles) north-west of Pointe-Noire is the former port of Loango, where Congo's first Roman Catholic mission was founded in 1883.

Population 200 000

Map Congo Ab

Poitiers ***France*** Capital of Vienne department, about 95 km (60 miles) south of Tours. In 1356, during the Hundred Years' War, Edward the Black Prince and his English longbowmen defeated the French there and captured the French king, John II.

Poitiers has several medieval churches, the remains of Roman amphitheatres and a university founded in the 15th century. It lies in a wine, wheat and livestock producing area.

Population 107 700

Map France Dc

Poitou-Charentes ***France*** Largely rural region, mostly to the south and west of the city of POITIERS. The first French dairy-farming cooperatives were established there in the 1880s. The region's main products are butter, cheese, cereals and Cognac brandy. Poitiers and LA ROCHELLE are the chief cities.

Population 1 568 000

Map France Cc

Pojezierze Mazurskie ***Poland*** See MASURIA

Pokhara ***Nepal*** Town of wood and plastered houses in the centre of the country, 140 km (87 miles) north-west of Kathmandu. It is set in a valley bottom at the relatively low altitude for Nepal of about 1000 m (nearly 3300 ft). The area's climate is subtropical, and rice, bananas and citrus fruits are grown.

The town is dominated to the north by the bulk of the Annapurna massif, and is used as a base by tourists trekking on pony or on foot into the Himalayas.

Population 20 600

Map Nepal Aa

Pola ***Yugoslavia*** See PULA

Poland See p. 522

Poland, Greater (Wielkopolska) ***Poland*** Historical heartland of the Polish state, where in the 10th century Mieszko I, a minor prince, began to unite the tribes which later called themselves Poles. The region lies west of the city of Lodz, south of the Vistula and Notec rivers, and north and east of the Oder river. Its flat-bottomed, fertile valleys produce cereals, sugar beet and timber. The Wielkopolski National Park south-west of Poznan, the regional capital, preserves 100 km² (39 sq miles) of low, forested hills and lakes, and is popular with tourists.

Map Poland Bb

Poland, Little (Malopolska) ***Poland*** The region of southern and south-eastern Poland, called 'Little' to distinguish it from Greater Poland. It includes the city of CRACOW, the Carpathian mountains; also the Cracow Jura upland, the HOLY CROSS MOUNTAINS and the Lublin Plateau.

Map Poland Dc

polar air mass Cold air which has originated in middle latitudes, either over an ocean (polar maritime), or over a continental interior (polar continental). See AIR MASS

polar front A frontal zone in the North Pacific and North Atlantic oceans, along which polar maritime and tropical maritime air meet, in which depressions are formed. See AIR MASS

polar wandering Movement of the GEOMAGNETIC POLES away from, towards or around the GEOGRAPHIC POLES.

polder An area of low-lying reclaimed land, especially in the Netherlands, that is protected by a DYKE.

▼ WOMAN'S WORK A farmer's wife in a mountain village near Pokhara turns precious millet to dry after the harvest in late November.

Poland

THROUGH CENTURIES OF OPPRESSION, THE POLISH SPIRIT HAS FLOWERED UNDAUNTED; THE FLOWER'S LATEST BLOOMING IS THE FREE TRADE UNION, 'SOLIDARITY'

Like a granite reef in a stormy ocean for most of its thousand-year history, Poland has been battered, overwhelmed and not infrequently submerged. Yet always it appears again in the sunlight, sturdy, unbroken and unchanged. Its beginning was in the 10th century, when the Slav peoples who lived between the Vistula and Oder rivers were united under a single Christian ruler, Mieszko I. Thereafter, Poland's borders contracted or expanded according to the whims of powerful neighbours and the efficiency of successive monarchs. Its great age began in the 14th century, in the reign of Casimir the Great, who made Poland into a major European power, and continued until the 1680s, when King John Sobieski drove the Turks from the gates of Vienna and saved Christian Europe from the Muslim onslaught.

But Poland's weakness has always been the vulnerability of its borders. Situated on the wide North European Plain, its frontiers are open to invaders on almost every side. The Mongols came from the east, and the Teutonic Knights from the west, followed by Russians and Germans down the centuries to our own time. The gentle, sandy dunes of the Baltic coast, too, were difficult to defend against Swedish invaders, and only in the south, where the CARPATHIAN MOUNTAINS climb up to almost 2500 m (8200 ft), is there any degree of natural protection.

In the second half of the 18th century Poland became progressively weaker and was partitioned between Prussia, Russia and Austria. The last Polish king, Stanislaus August Poniatowski, abdicated in 1795. From then until 1918, Poland virtually disappeared from the map.

But throughout the years of occupation, the fires of Polish freedom still burned; flaring up occasionally into open rebellion. The most desperate of these were the risings of 1830 and 1863 against the Russians. The reprisals that followed led to the Great Emigration, which began in 1831. Leading national figures such as the composer Frédéric Chopin, the poet Adam Mickiewicz and Marie Curie, discoverer of radium, gravitated to Paris, while poorer folk left in droves to work in the coal mines of the Ruhr and northern France. Later, in ever-increasing numbers, they departed for the USA; today, some 10 million people of Polish descent live abroad.

In 1918, at the end of the First World War, Poland once again became a state, but its

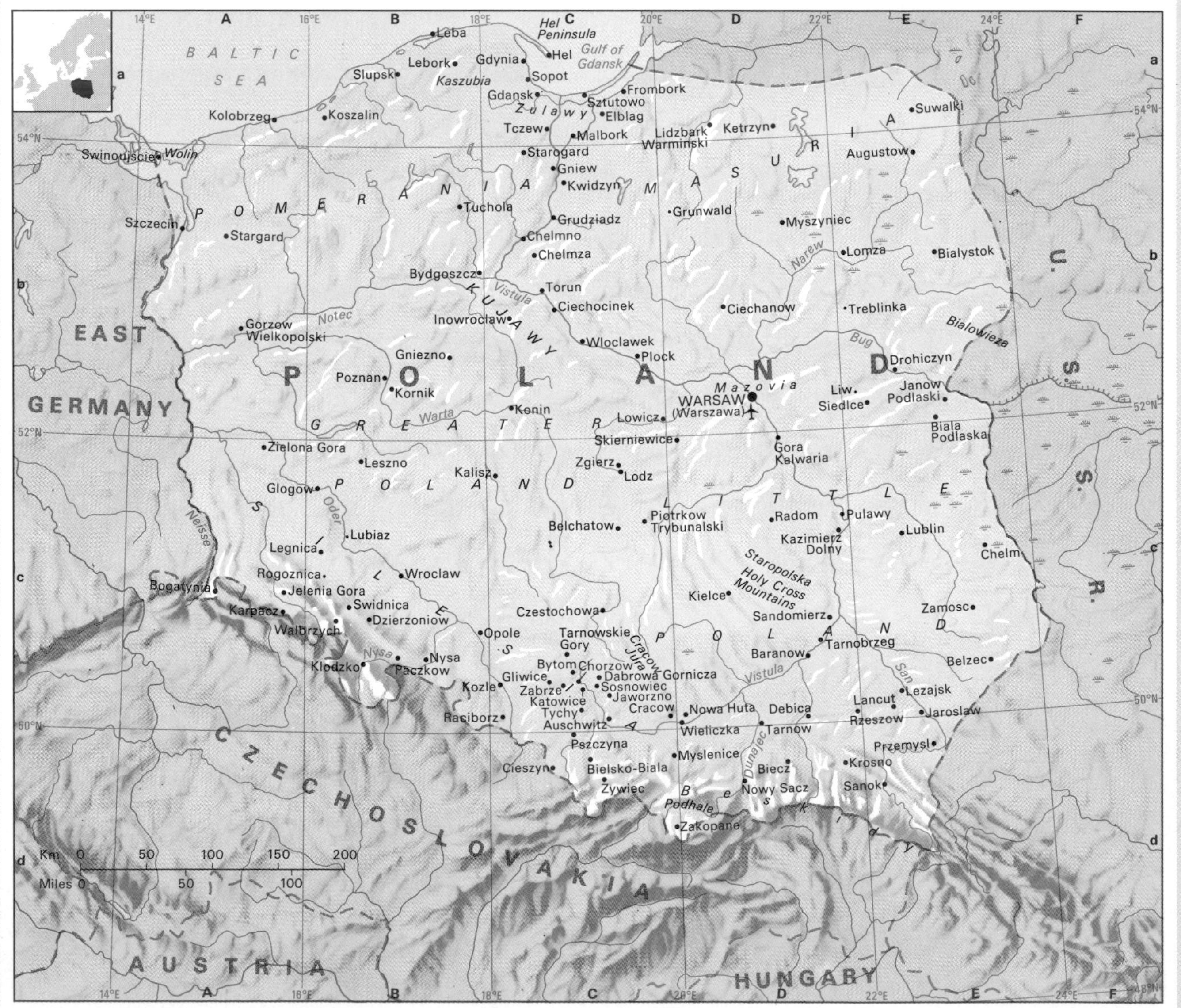

independence was short-lived. Two decades later, Hitler's armies crossed the frontier in the act that precipitated the Second World War. In the years that followed, Poland suffered more perhaps than any other nation. Large areas of the country were devastated, and 6 million Polish citizens, including 3 million Jews, were exterminated. Most died in notorious Nazi death camps such as AUSCHWITZ (Oswiecim), near Cracow. In addition, 14 000 Polish officers, taken prisoner by the Russians, were executed in the KATYN forest, though on this occasion, few historians doubt, it was probably the Russians, and not the Germans, who were responsible. Once again, the Poles were caught between two fires, and once again they fought back with incredible bravery. The rising of the Warsaw Ghetto and later the fight of the Polish Home Army against impossible odds stand among the greatest legends of the war. In other parts of the world, soldiers, sailors and airmen of the Free Polish Forces fought gallantly to bring about an Allied victory – in the Battle of Britain, at Tobruk, Alamein, and Cassino, and in dozens of other campaigns.

But it availed them little. In 1944, while the Home Army fought to the death through the ruins and sewers of Warsaw, the Russians paused on the far bank of the Vistula. While German forces crushed the resistance, the Russians were laying the foundations of Communist Poland. The Soviet-supported Polish United Workers' Party, known simply as 'The Party', rules Poland to this day.

CITIES REBORN

Whatever their persuasion, the new rulers had some fearful problems to overcome – to restore a land laid waste, and to feed and house a population that was virtually destitute. The United Nations Relief and Rehabilitation Association (UNRRA) and war reparation funds levied from the Germans provided some relief. But under Soviet pressure, and in common with all other Eastern Bloc countries, Poland refused Marshall Aid from the USA and looked to its own resources, plus some help from the USSR, for recovery. Ancient cities, such as the capital WARSAW and GDANSK (formerly Danzig), rose again, lovingly remodelled, so far as was possible, upon their former images. Some things, of course, could never be replaced; where the Warsaw Ghetto once stood, for example, there is now only a dark monument of bronze and marble. Fortunately, the Germans were unable to destroy CRACOW before they were driven out by the Russians, and the city's medieval, Renaissance and Baroque buildings are unscathed.

Taken all round, it might have been imagined that Poland was at last emerging into the sunlight; but once again, the storm clouds were gathering. The new rulers were busy remodelling Poland according to the Soviet pattern. The nationalisation of agriculture and industry was coupled with grandiose investment projects and wasteful management of the economy. The cost of living soared, the political preferences of the population were ignored and the secret police became all-powerful. Coupled with this was the new government's anticlerical attitude, deeply offensive to a devout Roman Catholic nation. Peasants, students and workers rioted in protest against Communist and Soviet control in 1956, 1968, again in 1970 when the police fired upon the shipyard workers of Gdansk, and yet again six years later. Concessions won on these occasions usually gave way to further repressions. In 1980, there were widespread strikes, which gave birth to Solidarity, the first free – that is, not state-controlled – trade union in postwar Eastern Europe. But the period of freedom generated by Solidarity lasted only 16 months. The new prime minister and Party boss, General Jaruzelski, reacted strongly. He declared martial law, ordered mass arrests and banned the union. Martial law was repealed in 1983 and most of the prisoners released, but relations between workers and government remain uneasy. In 1983, Lech Walesa, leader of Solidarity, was awarded the Nobel Peace Prize as a salute to the brave stand made by the union.

Between 1939 and 1946, Poland's population shrank from 35 million to less than 24 million. Many of those lost were war casualties, but part of the decline was due to one of those boundary changes only too familiar to the Poles. This one, settled by the major Allies, was drastic, in that the country's area was reduced from 388 600 km² (150 038 sq miles) to 312 683 km² (120 727 sq miles). The land was lost to the USSR, though at the same time, Poland's border was pushed west to the Oder-Neisse line in what had been East Prussia. In this way, Poland regained some of its former territories and 700 km (435 miles) of coastline including the port of Gdansk. German settlers fled these territories, but several million former Polish citizens who now found themselves in the USSR were not so fortunate, in that unless they could prove Polish birth, they had to stay. These changes, coupled with war deaths, changed Poland's multinational population structure. The land that before the war had contained a mixture of Poles, Jews, Germans, Lithuanians, Byelorussians and Ukrainians was now almost exclusively Polish.

▲ CULTURE CAKE Warsaw's Palace of Culture and Science, erected in the 1950s, towers above snow-covered streets and parkland. The vast 'wedding cake' skyscraper houses theatres, museums and sporting facilities – an official symbol of the new Poland.

Native Poles emigrating from the territories lost to the USSR moved to the cities or to the lands of the west and north regained from Germany, This, and a general drift from the countryside to the towns, has brought about a further shift in the Polish kaleidoscope. Though the rural population has remained steady at 15 million, the population as a whole has increased from 24 million in 1946 to 37 million in 1986. The increase was therefore entirely among townspeople, whose numbers grew from 9 to 22 million, reflecting the enormous upsurge in Polish industrialisation since the war.

The country has made good use of its living space, both old and new. From 1919-39, the only coast it possessed was the 140 km (86 miles) strip of the 'Polish Corridor', containing GDYNIA, its only major port. The additional 560 km (350 miles) that it gained in 1945 included not only the major ports of Gdansk and SZCZECIN, but numerous smaller harbours as well. Upon these outlets, Poland based a thriving merchant marine, a deep-sea fishing fleet, ferry links with Sweden and East

Germany and a liner service to Montreal. Along the Baltic coast too, seaside resorts have been established, though these have lost some of their popularity of late, due to industrial pollution.

The Polish papermills take their raw material from the vast forests stretching inland from the coast; these also support a thriving furniture industry, whose low-cost, modernistic products are widely exported.

Northern Poland has poor soils and a cool climate. Agricultural production is limited, even though nationalisation of former German estates provided ready-made large farms suitable for mechanisation and scientific management. By contrast, west-central and south-west Poland have fertile soils and a better climate; these are the nation's 'breadbasket' and 'sugar bowl', producing wheat and sugar beet. These regions also have a dense conglomeration of towns linked by railways, and considerable natural resources such as coal, Europe's largest copper deposits, salt and water power.

Central and eastern Poland, the 'old' 1918-39 areas remaining within the present boundaries, are more rural, with landscapes rather like those of the eastern shires of England. Urban development is concentrated mainly in Warsaw and LODZ, though provincial cities such as BIALYSTOK, KIELCE, LUBLIN, RADOM and RZESZOW have grown rapidly.

The farms of central and eastern Poland are the chief producers of the national 'staples' – rye, potatoes, cabbage – and of specialities like tobacco in the south-east, and fruits and flowers around Warsaw. Much of the land is still divided into family farms worked in the traditional way, but they are more productive than the state-owned agricultural combines. Raising productivity is a slow and painful process and food shortages, especially of meat, are a common complaint. There are many problems – a 50 per cent rise in the Polish population since 1946 for example, and pressures to export food to the USSR in exchange for oil, and to the West to pay for machinery and technology.

During the 1960s and 1970s, the government made particular efforts to build up industry in these predominantly agricultural regions. There are food, textile and shoe factories and plants to process local resources such as sulphur, gas and coal, and convert imported Soviet oil into petrochemicals. Skill-intensive industries such as pharmaceuticals, electronics and engineering have been established around Warsaw and Lodz.

Poland's most valuable acquisition in 1945 was the Silesian coalfield, united within Polish borders at last, after having been variously shared over two centuries or so with Austria and Prussia. Silesia contains the richest coal reserves in Europe, and the Poles made good use of them, and of the other mineral resources in the area. From 1945 to 1979, coal output rose from 47 million to 200 million tonnes per year, and steel from 1 to 20 million tonnes. Besides exporting coal, steel, lead and zinc to Hungary, Romania, East Germany, Yugoslavia and other Eastern Bloc countries, the newly awakened Silesia has fuelled the large-scale growth of Polish shipbuilding, heavy engineering and lively automobile and chemical industries.

Altogether, Poland's industrial progress has been dazzling, though purchased at a price – some of the worst environmental despoliation and air pollution in Europe.

The nation's path since 1945 has been rough and uphill. The cities, with their ballooning populations, were hardly able to cope, even after reconstruction. Improvements were slow in coming and consumer goods even slower. As the queues outside almost every shop grew longer with the years rather than shorter, and expenditure on the military and other government departments increased, it occurred to many of the workers that they had been striving for Soviet, not Polish, aspirations.

THE PEOPLE REBEL

These were the resentments that led to the founding of Solidarity in the Gdansk shipyards, and to the union's subsequent and rapid spread throughout the country. It began in the late 1970s with falling standards of living. Escalating government debts to Western banks led to the steady disappearance of consumer goods – household appliances, cars, cigarettes, canned foods, imported drinks and so on – from the shops, though many were obtainable at government-run stores for US dollars or other Western currencies. Poland's black market in foreign currency – the most blatant in Eastern Europe – exploded, and the value of the zloty dropped from 150 to the US dollar in the 1970s to 800 in the early 1980s. Workers and farmers watched the purchasing power of their labour dwindle away, and resentfully contrasted their lot with that of the Party elite, state officials and media people. But the issue was not only wages and shortages: the workers wanted to gain some influence over their daily lives and the government of their country. The situation remains much the same today, and the workers' grievances are not allayed by the strong-arm methods which often meet their protests.

To be fair, the government is all too aware that if it does not set its own house in order, then the task might be taken over by a large and impatient neighbour – Russia. Also, part at least of its financial trouble is due to its considerable expenditure on expanding and improving medical, educational and training facilities for the postwar generations. State encouragement too is given to the arts, to such composers as Lutoslawski and Penderecki, and to film directors such as Wajda and Zanussi. Wajda's brilliant war trilogy – *A Generation, Canal, Ashes and Diamonds* – achieved international fame.

But Poland's greatest unifying influence is, as it has always been, the Roman Catholic Church. Whatever its political affiliations, this is one of the few countries in Europe where church services on Sundays are filled to overflowing, and Christmas, Shrove Tuesday, Easter, Whitsun and saints' days are still celebrated with ancient fervour. Though its roots are deep in the villages, the Church's influence on towns and cities is just as great. Though church marriages are not recognised as legal, Polish couples seal the civil contract with the religious service. Each year on December 4, the miners of Silesia celebrate *Barburka*, the day of St Barbara, patron saint of miners. On special feast days, as many as half a million people converge on CZESTOCHOWA, a city overlooked by the Jasna Gora (Hill of Light) monastery, a place of international pilgrimage that contains a famous Byzantine painting of the Black Madonna.

Over the years the authorities have tried, unsuccessfully, to limit the influence of the Church. In the 1950s, the then primate was imprisoned, as were many of his priests. Until recently church building was restricted and Mass could not be broadcast. That this is not so today is largely due to the continuing influence of Solidarity. Many Poles consider Roman Catholicism to be their own faith, while Communism is seen as an alien creed. Every Pole rejoiced with fierce pride when Cardinal Wojtyla of Cracow was chosen as Pope in 1978, for the act seemed to confirm the nation's oneness with the Church. It was an ancient emotion. Down the centuries, through partition and foreign oppression, it was the Church that kept the fires of Polish nationalism aglow. Its priests went to concentration camps and death rather than retract their message of spiritual and national faith. The last martyr in such a cause was Father Jerzy Popieluszko, a supporter of Solidarity who, in 1984, was kidnapped, beaten, strangled and thrown into a reservoir near Warsaw. Some 250 000 mourners appeared at his funeral, and riots all over the country forced the authorities to bring four members of the security police to trial for his murder. The lengthy sentences awarded quietened the riots, but the trial did little to convince the population that high-ranking state and party officials were not involved.

The Poles say of themselves that their communism is like the radish, red on the surface, but white within – not a bad summing-up of so proud and tenacious a nation.

POLAND AT A GLANCE
Area 312 683 km^2 (120 727 sq miles)
Population 37 600 000
Capital Warsaw
Government Communist republic
Currency Zloty = 100 groszy
Language Polish
Religions Christian (95% Roman Catholic, 1% Polish Orthodox, 0.25% Lutheran)
Climate Continental; average temperature in Warsaw ranges from −5 to 0°C (23-32°F) in January to 15-24°C (59-75°F) in July
Main primary products Cereals, sugar beet, oilseed, potatoes, livestock, fish, timber; coal, sulphur, copper, zinc, lead, iron ore
Major industries Machinery, iron and steel, mining, chemicals, shipbuilding, food processing, agriculture, petroleum refining, fishing, forestry
Main exports Machinery, coal, foodstuffs, chemicals, non-ferrous metals, ships and boats, motor vehicles, clothing, iron and steel
Annual income per head (US$) 1800
Population growth (per thous/yr) 10
Life expectancy (yrs) Male 70 **Female** 75

poles See GEOGRAPHIC POLES; GEOMAGNETIC POLES

Poles'ye (Pinsk Marshes; Pripyat' Marshes) ***USSR*** Europe's biggest marsh and peat bog. It lies in Belorussia and the Ukraine, along the Pripyat' river about 800 km (500 miles) south-west of Moscow. Almost uninhabited, densely forested, and frozen in winter, it was divided between the USSR and Poland from 1920 to 1939, and taken over by the Russians in 1945.
Map USSR Dc

Polonnaruwa ***Sri Lanka*** Ruined city in the east of the country, 170 km (105 miles) north-east of Colombo. It was the Sinhalese capital in the 11th and 12th centuries. The city and the surrounding plain were irrigated by a dammed lake and vast waterway system. The lake, Minneriya Giritale, was originally built in the 3rd century, was restored in 1903, and is still in use. The city itself, once protected by three concentric walls, had many beautifully carved and frescoed palaces, temples and shrines. The walls of one temple still rise 22 m (72 ft) around a statue of the Buddha 14 m (46 ft) high.
Map Sri Lanka Bb

Polotsk ***USSR*** Market and industrial town in Belorussia, about 550 km (340 miles) west of Moscow. First mentioned in Russian chronicles in AD 862 – making it one of the country's oldest towns – its industries today include oil refining, timber processing, textiles, flax, leather goods and beer.
Population 78 000
Map USSR Dc

Polyarnyy ***USSR*** See MURMANSK

Polynesia ***Pacific Ocean*** Largest of the three main regions of the Pacific Islands (with Melanesia and Micronesia) which are based on ethnic differences. Polynesians are tall, handsome, golden-skinned people, generally with straight or wavy black hair. They include the Maoris of New Zealand.

Their ancestors probably came from parts of Indonesia more than 3500 years ago; they were fine sailors, and made long voyages in huge canoes guided by winds, wave patterns and stars. They settled in TONGA about 1140 BC and SAMOA by 1000-500 BC. They later spread to the SOCIETY ISLANDS (AD 200), the MARQUESAS and other parts of present-day FRENCH POLYNESIA, and by AD 500 had reached as far east as EASTER ISLAND and north to HAWAII. Within another 500 years they had reached the COOK ISLANDS and NEW ZEALAND. Other Polynesian islands include TUVALU, the TOKELAUS and NIUE – the name Polynesia (meaning 'many islands') is an apt one. Despite this huge area and long periods of separation, all Polynesian languages are very closely related.

The ancient Polynesians developed a strict caste system with hereditary chiefs. They worshipped many gods and built large temples on platforms called *marae*, where they made offerings and sometimes human sacrifices. Many sacred objects and places were *tabu* (forbidden) – hence our word 'taboo'.
Population (Polynesians only) 780 000
Map Pacific Ocean Dd, Fb

▲ **HAUNTING STONES On Easter Island in Polynesia, some of the fallen 'living faces', 4-8 m (13-26 ft) tall, have been re-erected. They were carved to give the islanders the protection of the ancestors they represent.**

Pomerania (Pomorze; Pommern) ***Poland*** North-western region on the Baltic coast, between the Oder and Vistula rivers. It is a low lakeland region popular with holidaymakers.
Map Poland Ab

Pomorie ***Bulgaria*** See NESEBUR

Pomorze ***Poland*** See POMERANIA

Pompeii ***Italy*** Ancient ruined city near the foot of the volcanic Mount VESUVIUS, about 24 km (15 miles) south-east of Naples. Pompeii was destroyed by a cloud of choking ash and lava bombs when Vesuvius erupted in AD 79. The ash has grotesquely preserved the postures and even the facial expressions of fleeing or sheltering citizens at the moment of death.

More than 60 per cent of this huge Roman city has been excavated to date. There are several kilometres of sunken streets fronted by blank walls of houses. Many of the more important individual exhibits have been removed to museums. The Forum is justly famous for its beautiful setting against the backdrop of Vesuvius. Other highlights are the Amphitheatre, the Large and Small Theatres, the Basilica, the House of Loreius Tiburtinus – a reconstruction of a Pompeian courtyard garden – the House of the Vettii with its painted walls and, outside the main city, the Villa of Mysteries, which contains the finest of all ancient paintings, including scenes of religious celebrations.

There are lesser relics, too, that bring Pompeii's tragedy down to a human scale, such as graffiti scratched on the walls bearing political slogans and notices for public amusements.
Map Italy Ed

Ponape ***Federated States of Micronesia*** See POHNPEI

Ponce ***Puerto Rico*** Seaport and major city of the south coast, 75 km (47 miles) south-west of San Juan. It is linked to San Juan by the Las Americas Expressway, opened in 1975. It exports sugar. Industries include cement, textiles, footwear, paper products, rum distilling and oil refining. There is a university (1948).
Population 190 900
Map Caribbean Bb

Pondicherry (Puduchcheri) ***India*** Port and territory on the Bay of Bengal about 125 km (77 miles) south of Madras. It was founded by the French East India Company in 1683, and later became the capital of French India. It was permanently under French control from 1816 to 1954. Today, it exports groundnuts (peanuts) and its industries include cotton textiles.
Population 251 400
Map India Ce

Ponta Delgada ***Portugal*** Capital and main port of the Azores, the group of islands 1290 km (800 miles) off the west coast of Portugal. Ponta Delgada is on the island of Sõ Miguel.
Population 21 200
Map Atlantic Ocean

Pontevedra ***Spain*** Port near the border with northern Portugal, trading in grain and fruit, and with shipbuilding and fishing industries. A 12 arched Roman bridge spans the Lerez river on which the town stands.

The farming and fishing province beside the town is also called Pontevedra.
Population (town) 65 200; (province) 859 900
Map Spain Aa

Pontine Marshes ***Italy*** See LAZIO

Poole ***United Kingdom*** See BOURNEMOUTH

Poona (Pune) ***India*** City about 120 km (74 miles) east and slightly south of Bombay. It stands on top of an escarpment of the Western GHATS and was the summer capital of Bombay state during the British Raj. It has many beautiful palaces and Hindu temples, and some fine public gardens and government buildings. It came under British rule in 1818, and since India's independence in 1947 has developed as an industrial centre. Its chief products include munitions, pharmaceuticals, machinery, paper, soap and textiles.

Population 1 203 400
Map India Bd

Popayán ***Colombia*** Colonial town at an altitude of 1536 m (5039 ft), about 375 km (230 miles) south of Cali. It was founded in 1537 by the Spanish conquistador Sebastián Belalcázar, who was searching for Eldorado. Colonists set up sugar plantations in the damp valley below and made their homes in the cool, tree-lined tranquillity of the town. It became a religious town and centre of learning with splendid, Spanish-style churches and monasteries. In 1983, an earthquake did shattering damage, leaving only one church intact. Restoration work has begun.

Population 134 000
Map Colombia Bb

Popocatépetl ***Mexico*** Volcano, 65 km (40 miles) south-east of MEXICO CITY, joined by a ridge to its twin Ixtaccíhuatl. It contains a crater about 1 km (0.5 mile) in circumference and 75 m (250 ft) deep. Its name is Indian for 'smoking mountain', and although it has been dormant since 1802, it occasionally emits vast clouds of smoke. The higher of the two peaks, 5452 m (17 887 ft), Popocatépetl is easier to climb, and has three marked trails.

According to legend, Popocatépetl was an Aztec warrior who was falsely reported killed in battle. His lover, Ixtaccíhuatl, died of a broken heart at the news. Popocatépetl returned to find her dead, and watched over her until he too died and they were reunited in heaven.

Porbandar ***India*** Town on the Arabian Sea about 440 km (270 miles) north-west of Bombay. It was the birthplace of the nationalist leader Mahatma Gandhi (1869-1948). Its manufactures include textiles and cement.

Population 133 300
Map India Ac

Pordenone ***Italy*** Town about 65 km (40 miles) north-east of Venice. The older part of the town contains a wealth of Gothic, Renaissance and Baroque buildings including the Palazzo Comunale (1291-1365) and the late-Gothic cathedral. Pordenone is Italy's leading producer of washing machines and refrigerators.

Population 51 800
Map Italy Db

pores See POROUS ROCK

Pori (Björneborg) ***Finland*** Port in the south-west, on the Gulf of Bothnia about 250 km (145 miles) north-west of the capital, Helsinki. It has metallurgical, softwood and food-processing industries, with an outport at Mäntyluoto.

Population 79 100
Map Finland Bc

porous rock Rock with pores – cavities between its mineral grains – which will soak up water or other liquids like a sponge. Sandstone is both a porous and PERMEABLE ROCK. Clay is porous, but it is not permeable.

porphyry IGNEOUS rock in which large crystals, known as phenocrysts, are scattered through smaller crystals of a different composition.

Port Antonio ***Jamaica*** Seaport and one of the earliest tourist resorts in Jamaica, lying on the north-east coast, 40 km (25 miles) north-east of Kingston. The first hotel was built by the United Fruit Company at the turn of the century, and the port is still a major shipping port for bananas.

Population 12 500
Map Jamaica Ba

Port Arthur ***Australia*** Former penal settlement in Tasmania about 50 km (30 miles) south-east of Hobart. It was built in 1830 and abandoned in 1877, after the shipping of convicts from Britain ceased. But some buildings remain, including a church designed and built by convicts. The settlement is a popular tourist centre.

Map Australia Hg

Port Arthur ***China*** See LUSHUN

Port Blair ***India*** Seaport on the south-east coast of South Andaman Island and the capital of the ANDAMAN AND NICOBAR ISLANDS. It was founded in 1789 by the British mariner Captain A. Blair, and became a penal colony in 1858. During the Second World War it was occupied by the Japanese, who used it as a naval base.

Today it exports copra, timber, rubber and coconuts.

Population 49 600
Map India Ee

Port Bouet ***Ivory Coast*** Deep-water port beside the capital, ABIDJAN, in the Ebrie Lagoon. It is open to the sea through the VRIDI CANAL. The first wharf was built at Port Bouet after a pilot canal was cut between 1905 and 1907, but the canal silted up and the port was abandoned. The wharf was rebuilt in 1932, and expanded rapidly after 1950, when the deep-water canal was opened.

Map Ivory Coast Bb

Port Cros, Ile de ***France*** See HYERES, ILES D'

Port Elizabeth ***South Africa*** Seaport in east Cape Province, which developed rapidly after it was linked by rail to the diamond town of Kimberley in 1873. Its industries today include car assembly, food processing, tyre manufacturing and woodworking. Tourists are drawn to its sandy beaches. The main non-white townships near the city are Kwazakele and Zwide.

Population 585 400
Map South Africa Cc

Port Etienne ***Mauritania*** See NOUADHIBOU

Port Gentil ***Gabon*** Capital of Ogooué-Maritime province, and Gabon's chief port, situated on an island between two mouths of the OGOOUE river. It was founded in the late 19th century in order to export okoumé, a valuable hardwood shipped downriver from Ndjolé and Lambaréné. Today it exports refined petroleum products from an installation built in 1957.

Population 108 000
Map Gabon Ab

Port Harcourt ***Nigeria*** Capital of Rivers state and the country's second port after the capital, Lagos. It stands on firm land beside the Bonny river on the east side of the Niger river delta. The deep-water port was begun in 1912 and connected by rail to the city of Enugu in 1916. Its exports include coal from Enugu, palm oil (used in margarine and cooking fats) and groundnuts. Port Harcourt is also the centre for the oil fields of the eastern delta, and a thriving industrial centre producing refined mineral oil, aluminium sheet, bottles, tyres and paints.

Population 288 900
Map Nigeria Bc

Port Hedland ***Australia*** Port in Western Australia, about 1300 km (810 miles) north of Perth, and the state's fastest-growing town. It handles iron ore from mines in the Pilbara region, exporting it mainly to Japan.

Population 12 900
Map Australia Bc

Port Jackson Inlet on the east coast of Australia, also called Sydney Harbour. It is one of the world's finest anchorages and is spanned by the elegant Sydney Harbour Bridge. The inlet is a drowned river valley with many coves and bays and is navigable by the largest vessels, although at weekends it is alive with smaller leisure sailing craft. The harbour contains Sydney Cove where Captain Arthur Phillip established Australia's first convict settlement on January 26, 1788 – now the centre of SYDNEY.

Area 57 km^2 (22 sq miles)
Width 2.4 km (1.5 miles) at mouth

Port Kembla ***Australia*** Port in New South Wales, 88 km (55 miles) south of Sydney. It is part of the Wollongong industrial area, and has the country's biggest steel works and factories producing tin plate, copper and chemicals.

Population 8000
Map Australia Ie

Port Láirge ***Ireland*** See WATERFORD

Port Louis ***Mauritius*** Capital, largest town and chief port of the island. It stands on the north-west coast, sheltered from the prevailing trade winds, at the head of a deep inlet which provides an excellent harbour. Founded in 1736 by the French and named after Louis XV, the town is dominated by an early 19th-century British-built fortress, the Citadel. The island's main commercial centre, Port Louis exports the sugar crop and has factories producing various goods, including clothing.

Population 160 000

Port Moresby ***Papua New Guinea*** Port, university town and capital of Papua New Guinea, standing at the south-east end of the Gulf of Papua. It exports rubber, coconut products, coffee and timber, and has food processing and light manufacturing industries. It has a relatively dry climate but is rather isolated from

much of the country, relying mainly on sea and air transport. During the Second World War it was an Allied base and was severely damaged by Japanese bombing.

Population (metropolitan area) 126 000

Map Papua New Guinea Ba

Port of Spain ***Trinidad and Tobago*** Main seaport and national capital, on the north-west coast of Trinidad. After Kingston in Jamaica it is the largest city in the English-speaking West Indies, with a continuous line of suburbs stretching eastwards as far as Arima, 25 km (16 miles) to the east. It produces rum, beer, plastics, timber and textiles. There is a National Museum, an Art Gallery, and a 27 hectare (65 acre) Botanic Garden dating from 1820. On the western edge of the city, on a 335 m (1100 ft) hill, stands the restored Fort George, from which there are splendid views.

Population (city) 62 700; (metropolitan area) 443 000

Map Trinidad Aa

Port Phillip Bay Large inlet of BASS STRAIT on the coast of southern VICTORIA, Australia, with MELBOURNE at its head. It was claimed for Britain in February 1802 by a young explorer, Lieutenant John Murray. He named it after Captain Arthur Phillip, founder of the first Australian convict settlement at Sydney Cove.

Dimensions 48 km (30 miles) long, 40 km (25 miles) wide

Map Australia Gf

Port Pirie ***Australia*** Port in South Australia, on Spencer Gulf about 200 km (120 miles) north of Adelaide. It is a lead-smelting centre using ore transported from BROKEN HILL, and exports lead and wheat.

Population 14 700

Map Australia Fe

Port Royal ***Jamaica*** Small fishing village at the western end of the Palisadoes (the sandspit enclosing KINGSTON harbour). Its fortified harbour was once a haven for pirates as well as the British naval headquarters in the Caribbean. Much of the town fell into the sea during an earthquake and subsequent tidal wave in 1692. Parts of it were rebuilt. St Peter's Church, dating from 1725, stands on the site of the first cathedral in the New World, built by the Spanish in 1523. The old Naval Hospital (1819) is now an archaeological museum. There is also a maritime museum at Fort Charles, which was founded in 1655 and expanded to become the strongest fort in the Caribbean.

Map Jamaica Bb

Port Said ***Egypt*** City and port at the Mediterranean entrance to the Suez Canal, about 170 km (105 miles) north-east of Cairo. It was opened as a port in 1868, the year before the Suez Canal was opened. It quickly became the world's largest coaling station, and the headquarters of the canal operations and maintenance. In 1956, Egypt nationalised the canal, and Port Said was badly damaged in an Anglo-French military action which failed to win back control of the canal. During the fighting, a huge bronze statue of Ferdinand de Lesseps, builder of the canal, was destroyed.

Port Said has chemical, tobacco and cotton industries, and is a free trade zone.

Population 342 000

Map Egypt Cb

▲ SMOKING MOUNTAIN The smouldering fury of snow-capped Popocatépetl volcano can be seen from the national capital, Mexico City.

Port Stanley ***South Atlantic*** See FALKLAND ISLANDS

Port Sudan ***Sudan*** The country's largest port, built early in the 20th century on the Red Sea coast north-east of the capital, Khartoum. It replaced the ancient Arab port of Suakin, which had been abandoned because it was choked with coral. The port has modern docks and exports cotton, sesame seeds, gum arabic, hides and skins.

Population 205 000

Map Sudan Bb

Portage la Prairie ***Canada*** Town in south-west Manitoba near the Assiniboine river. Founded as a trading post in 1738, it is a railway centre and exports grain and livestock.

Population 13 100

Map Canada Fc

Portalegre ***Portugal*** Weavers' town about 160 km (100 miles) north-east of the capital, Lisbon, at the foot of the Serra de São Mamede. The town, noted in the 16th century for its tapestries, later became important for its silk industry, which was established in the 17th century.

Population 14 800

Map Portugal Cc

Port-au-Prince ***Haiti*** Principal port and capital city, on the south-east shore of the Gulf of Gonâve. The city, founded by the French in 1749, has been destroyed by fire many times, and by two earthquakes (1751 and 1770). French colonial buildings of the 19th century dominate the town's architecture, though the cathedral is 18th century. Among modern buildings are the university (1944) and the technical institute (1962). Haitian history is depicted in the National Pantheon Museum, and the art and culture of the nation in the Museum of Haitian Art. The city refines sugar, flour and cottonseed oil and manufactures textiles.

Population 888 000

Map Caribbean Ab

Portimão ***Portugal*** Fishing port, boat-building and fish-canning town in the Algarve, about 25 km (16 miles) east of Cabo de São Vicente, the country's south-western tip. The neighbouring resort of Praia da Rocha, which is rich in unusual rock formations carved by the sea from the limestone cliffs, lies at the mouth of the Arade river, guarded by two medieval forts.

Population 19 600

Map Portugal Bd

Portland ***USA*** **1.** Atlantic port in Maine at the head of the attractive Casco Bay. It is a shipbuilding centre and important oil port. It is also a centre of the timber, chemicals and textile industries, a fishing port and a tourist resort.

2. Pacific port in Oregon on the Willamette river near its junction with the Columbia, about 100 km (60 miles) inland from the ocean. It is a timber, fishing and shipbuilding centre and a university city.

Population **1.** (city) 61 800; (metropolitan area) 210 200. **2.** (city) 365 900; (metropolitan area) 1 340 900

Map United States (1) Lb; (2) Ba

Porto ***Portugal*** See OPORTO

Porto, Golfe de ***France*** See CORSICA

Porto Alegre ***Brazil*** Port and capital of the southernmost state, Rio Grande do Sul, which covers 282 184 km^2 (108 920 sq miles) beside the Uruguayan border. Its main products are food, textiles, chemicals and leather. Exports include timber, rice, wheat, meat, hides, wool and wine. The rural areas of the state are home to gaucho cowboys who, like their Argentinian and Uruguayan counterparts, wear ponchos and baggy trousers.

Population (city) 1 126 000; (state) 7 800 000

Map Brazil Ce

Porto Novo ***Benin*** Port and administrative capital of Benin, on Ouémé lagoon, which is open to the sea at Lagos, Nigeria. Porto Novo also has an outlet to the sea via Lake Nokoué and the port of Cotonou, but this route is subject to silting. The town's influence has recently declined, as its road and rail links with the interior are inferior to those of Cotonou. As a result, many government departments and embassies have moved to Cotonou.

Porto Novo was the seat of African kings from the early 17th century. It is an attractive town, with narrow streets lined by the red earth walls of merchants' homes built in the traditional style of the Yoruba tribe. Many of the larger buildings date from the early 19th century, when the Portuguese settled there and made it a centre of their trade in slaves. The town's ethnic museum has displays devoted to the history of its African kings.

Population 209 000

Map Benin Bb

Porto Velho ***Brazil*** Market town and capital of the western Amazonian state of Rondônia, about 750 km (465 miles) south-west of the city of Manaus. The state covers 243 044 km^2 (93 814 sq miles), and its main products are timber, rubber, livestock, cocoa, coffee, gold and tin.

Population (town) 135 000; (state) 688 000

Map Brazil Bb

Portoviejo ***Ecuador*** Capital of Manabí province on the east bank of the Portoviejo river, 135 km (85 miles) north-west of Guayaquil. It is an important commercial centre with vegetable oil and cotton processing plants. Irrigation schemes have been established in the river valley after a series of severe droughts.

Population 167 000

Map Ecuador Ab

Portsmouth ***United Kingdom*** Port in the southern English county of Hampshire, 100 km (62 miles) from London, on the Spithead channel which separates the Isle of Wight from the mainland. It was granted a charter in 1194 by Richard I (the Lion Heart), and established as a naval dockyard in 1540 by Henry VII – a role it has retained ever since. It has both Anglican and Roman Catholic cathedrals, an old town of quaint, narrow streets and a resort area, Southsea, with 3 km (2 miles) of beaches.

Portsmouth's principal tourist attractions, however, are two ships: HMS *Victory*, on which Horatio, Lord Nelson died in 1805 at the Battle of Trafalgar, and Henry VIII's great ship the *Mary Rose*, which sank off Portsmouth in 1545; it was raised in 1982, to be restored and put on display. The novelist Charles Dickens (1812-70) was born in the city.

Population 192 000

Map United Kingdom Ee

Portugal See p. 530

Port-Vila ***Vanuatu*** Capital and main port of Vanuatu, on the island of Éfaté in the centre of the group; also known as Vila. It has a good picturesque natural harbour, from which manganese ore – mined at Forari in the east of the island – used to be exported to Japan; mining stopped in 1978.

Population 17 500

Porvoo (Borgå) ***Finland*** Picturesque cathedral town founded in 1346 on the south

A DAY IN THE LIFE OF A BRAZILIAN COLONIST

This is not the first morning that Floriano has looked at his small field in despair. He stands in front of his small house, built of trees he cleared from the land, and can hardly summon the energy to start weeding.

He can remember when he was full of hope, listening to the speech of the governor of the newly created state of Rondônia, as he offered farmers free land in the western rain forest. At the time, Floriano lived in the state of Matto Grosso, where he was a share-cropper, a tenant-farmer who handed over a large part of his crops as rent to his landlord. Deciding to chance his luck, he – like tens of thousands of other peasants and their families – followed the new road into the forest.

He had been given his land grant of 50 hectares (124 acres), and had begun to clear the trees enthusiastically. But since then, he had faced disappointment. Whatever he tried to grow failed. And all around him the forest seemed to shout its success. 'Why,' he asked himself, 'if giant trees can grow here in such profusion, can't I get tiny maize plants to survive?'

Wearily, Floriano begins his task of weeding. He clambers over the charred remains of tree trunks he felled and burnt, as they lie crazily across his field like huge used matchsticks.

Floriano searches out each maize seedling from among the fallen timber. He was not able to plant in proper rows and there is no easy water supply for the seedlings. Floriano reflects bitterly on the need for water in the midst of a rain forest, with the nearest stream more than a kilometre away.

Today is an important day in Floriano's struggle to beat the forest. He spent most of his savings in the nearby town on some coffee plants in one last bid for success. If this crop fails, he and his family will have to abandon this patch of land and return to work someone else's once again.

Floriano's wife and two sons help him to make holes in the earth for the coffee plants, working with a hoe and with their bare hands. They untie the sacking from the roots of the plants, and spread the roots gently across the base of each hole. The soil is replaced, leaving a little hollow to gather water.

At the end of the day Floriano and his family survey the result of their labour – hardly able to make out the plants among the forest debris. Perhaps a prayer will ensure that the work will not have been in vain.

Not far away another family is leaving their land, defeated. No one told them, just as no one told Floriano, that rain-forest soils without the dead remains of the forest are infertile.

coast, 45 km (28 miles) east of Helsinki. It is where the Russian tsar Alexander I formally became Grand Duke of Finland in 1809 when Sweden ceded Finland to Russia.
Population 19 500
Map Finland Cc

Posen ***Poland*** See POZNAN

Positano ***Italy*** Picture-postcard resort about 30 km (19 miles) south-east of Naples on the Gulf of Salerno. It was once an ancient fishing village, but its spectacular position among mountains rising to 1500 m (5000 ft) has proved an irresistible tourist lure.
Population 3600

Postojna (Postumia; Adelsberg) ***Yugoslavia*** Attractive Slovenian town in the Julian Alps, 40 km (25 miles) south-west of Ljubljana. Nearby the KRAS limestone is pitted by 18 km (11 miles) of spectacular, stalactite-hung caves, halls and tunnels carved by the Pivka river. Graffiti on the walls show that the caves were known as early as 1213. Today, an electric railway, 2 km (1 mile) long, takes visitors to the start of a conducted underground tour through grottoes known as the Ballroom and Paradise. Highlight of the caverns is an enormous vaulted chamber called the Concert Hall – where the Italian conductor Arturo Toscanini (1867-1957) once held a full orchestral concert. The chamber, which covers 2800 m² (3350 sq yds) or the equivalent of ten tennis courts, can hold 1000 visitors at a time.

Planina is another caving centre nearby.
Population 5000
Map Yugoslavia Bb

Potchefstroom ***South Africa*** The oldest town and first capital of the South African Republic – now the province of Transvaal. It lies about 105 km (65 miles) south-west of Johannesburg, and was founded by Voortrekkers (Afrikaner settlers) in 1838. It is the commercial centre for the fertile Mooi river valley.
Population 56 600
Map South Africa Cb

Potenza ***Italy*** Market town about 130 km (80 miles) east of Naples, 819 m (2690 ft) up in the Lucanian Apennines. It is the capital of the impoverished Basilicata region, but has several new state-aided industries in food, clothing and construction.
Population 65 200
Map Italy Ed

pothole 1. Hole in the rocky bed of a stream, formed by the grinding effect of pebbles as they are whirled around by eddies. **2.** A popular name for a SINKHOLE.

Potosí ***Bolivia*** City and department bordering Chile and Argentina. It is best known for the Cerro Potosí, the hill in which vast deposits of silver were discovered by the Spanish.

The city, founded in 1546 by the Spaniards Juan de Villarroel and Diego Centeno about 80 km (50 miles) south of Sucre, stands at the foot of the hill at an altitude of 3978 m (13 050 ft). By 1650 Potosí had become the largest city in the Americas, with some 160 000 inhabitants. It still retains much of its colonial heritage, with some splendid 17th and 18th-century Baroque churches and public buildings such as the Mint, the Casa Real de Moneda, rebuilt in 1759. Although the silver deposits have long been exhausted, the mines still produce tin, copper and lead.
Population (department) 824 000; (city) 103 200
Map Bolivia (department) Bc; (town) Bb

A DAY IN THE LIFE OF A BOLIVIAN PEASANT

About 60 per cent of Bolivia's 6 million people are Indians. Among these, the largest single group is the Quechua who live high up in the mountainous Altiplano area of south-eastern Bolivia. These were the original inhabitants of Bolivia, who were forced up to the plateau by the arrival of the Spaniards.

Mariano Guallpa, his wife Augustina and their five children are Quechuas. They live in a one-room house made of sun-baked earth near the town of Potosí, almost 4000 m (over 13 000 ft) above sea level. Although Potosí is an active tin-mining centre, this has little impact on the daily life of Mariano, who ekes out a living as a subsistence farmer on the poor, thin soil.

Mariano owns a few sheep and cattle, but he follows the Quechua tradition of communal land ownership and farms the surrounding area with the other people of his village. Potatoes, maize and *quinoa*, a kind of millet, are the main crops, and form the staple diet.

Mariano's 13-year-old son, Tito, usually helps out in the fields, where the work is hard because of the lack of oxygen at such an altitude and because of the absence of any but the simplest agricultural tools. Augustina often lends a hand, though much of her time is spent at home looking after the younger children and cooking or weaving. She weaves sheep, llama and alpaca wool on a loom handed down from her mother, into brilliant striped cloth. Her dyes, by-products of Bolivia's mining industry, are bought from merchants or at the market in Potosí. They can use some of Augustina's cloth to trade with neighbouring tribes, or with fellow merchants. Some is made up into the cloaks or ponchos that Mariano wears over a thick woollen shirt and trousers – together with a domed black hat, embroidered in bright colours, which comes down over his ears like the helmets once worn by the Spanish conquistadores.

Like others of his tribe, Mariano speaks Spanish, the official language of Bolivia, as well as the native Quechua language, and he has brought his children up to be bilingual as well. But there are no formal schools for the Guallpa children to attend, and like Mariano, they are unable to read or write.

Though he is nominally a Roman Catholic, Mariano's faith in practice owes much to ancient myths and to him the mountains, lakes and caves are the homes of spirits which can harm or help him in his everyday life.

In the evenings he turns to his most prized possession – a small transistor radio – which brings the news of the distant capital, La Paz, cracklingly to life in the remote windy uplands.

Potsdam ***East Germany*** Port on the Havel river 25 km (16 miles) south-west of Berlin. It has many fine palaces and was the summer retreat of the Prussian king Frederick the Great (1712-86). Allied leaders met there in July and August 1945 to discuss the administration of Germany in the aftermath of the Second World War. Today it is a centre of education with academies of political science, law, finance, film, agriculture and medicine. Its factories produce locomotives, textiles and foodstuffs.
Population 137 700
Map East Germany Cb

Potteries, The ***United Kingdom*** Industrial area in the English Midland county of Staffordshire. It is centred on the towns of STOKE-ON-TRENT, Tunstall, Burslem, Hanley and Longton, and was the setting for many of the novels of Arnold Bennett (1867-1931). Pottery has been made there since Roman times. The modern industry, based on local coal – used to fire the kilns – and canal transport, dates from the mid-18th century; its porcelain manufacturers include Wedgwood, Spode and Copeland. Newcastle-under-Lyme, a little to the west, produces bricks and tiles.
Map United Kingdom Dd

Poubara Falls ***Gabon*** Falls on the upper Ogooué river, harnessed since 1975 to produce electricity for the fast-developing mining and industrial region near FRANCEVILLE.

Poverty Bay ***New Zealand*** Bay on the east coast of NORTH ISLAND, about 320 km (200 miles) south-east of Auckland, where the British explorer Captain James Cook made his first landing in New Zealand in October 1769. After a skirmish with Maoris, three or four of whom were killed, Cook named it Poverty Bay 'because it afforded no one thing that we wanted'. The seaside resort of GISBORNE lies at the head of the bay.
Dimensions 10 km (6.2 miles) long, 6 km (3.7 miles) wide
Map New Zealand Fc

Powys ***United Kingdom*** Sparsely populated county of central Wales, created from the old counties of Brecon (Brecknock), Radnor and Montgomery and named after an ancient Celtic principality. It covers 5077 km² (1960 sq miles) and stretches from the English border almost to Cardigan Bay. It is mostly hill country, with the Brecon Beacons and Black Mountains rising to 886 m (2907 ft) and 811 m (2661 ft) respectively in the south, and the Plynlimon ridge – source of the Severn and Wye rivers – to 752 m (2467 ft) in the west. The Brecon Beacons are a major tourist attraction and much of the area is used for sheep farming. The county town is Llandrindod Wells.
Population 111 000
Map United Kingdom Dd

Portugal

THE POOREST COUNTRY IN WESTERN EUROPE HAS KNOWN GLORIOUS DAYS OF EXPLORATION AND EMPIRE

The word Portugal usually brings to mind sunny, sandy beaches and luxury hotels, or the port wine, which has been produced for 400 years in the DOURO valley. In many ways the country, its people and its landscapes remain largely unknown.

Today, Portugal is the poorest country in Western Europe, but it had the first great European overseas empire. Prince Henry the Navigator inspired the great Portuguese discoveries in West Africa and beyond in the 15th century. Vasco da Gama sailed round Africa to reach India in 1498 and the Portuguese were the first Europeans to reach China and Japan by sea. By 1550 they had claimed Brazil. Today, 200 million people around the world speak Portuguese.

Portugal has modern cities and ancient lifestyles. The hustle and bustle of the capital, LISBON, is a world removed from the oxcarts creaking their way up hillside tracks in the MINHO region.

In the north-east are the gorse-clad, steep-sided valleys and mountains of Tras-os-Montes, cold and damp in winter, and still the haunt of wolves and wild boars.

The coastal plain of the north-west, with its woodlands of pine and eucalyptus, has a patchwork landscape of small fields of maize and beans, surrounded by arbours of vines. The grapes are used to make the sparkling white *vinho verde* wines.

The Douro river is the artery of port wine, the route down which countless barrels have been shipped. It runs from the steep slopes of the upper valley between the Spanish border and Péso da Régua, to the lodges at Vila Nova de Gaia, opposite OPORTO.

Moving south, across the MONDEGO river, are the high granite peaks and glacial valleys of the SERRA DA ESTRELA, Portugal's winter sports area. Farther south one reaches lands increasingly hotter and drier in summer: the vast expanses of the ALENTEJO, with its wheat fields and cork plantations.

High-walled towns such as EVORA, ESTREMOZ, ELVAS and Marvao stand on the skyline like guardians watching over the toil of generations of peasant farmers.

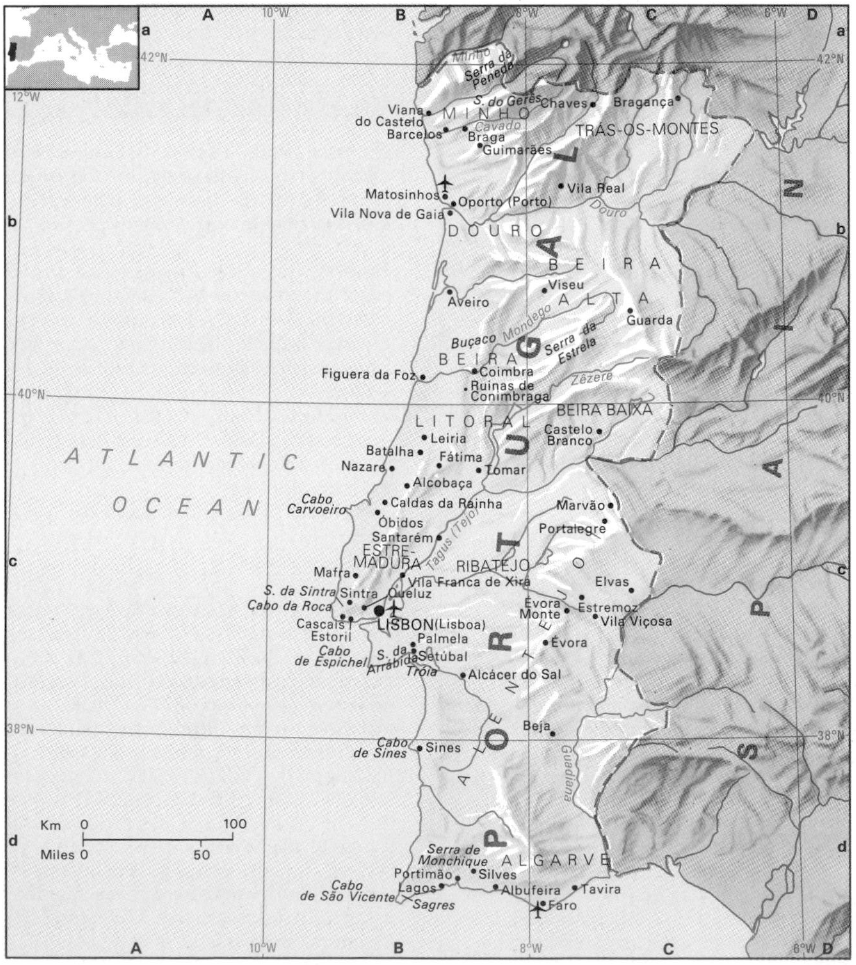

In the west, at the mouth of the TAGUS river, lies Lisbon, built on seven hills, surrounded by wealthy suburbs and the tourist haunts of the Serra da Arrabida and the Serra de Sintra.

The Alentejo continues to the hinterland of the ALGARVE, with its beautiful groves of almond, fig and olive trees. Here the climate is warm and equable even in winter. Along the southern coast of the Algarve are the old fishing harbours of PORTIMAO, FARO and Tavira – centres of an extensive tourist region along the sandy beaches washed by the Atlantic Ocean. The fishing fleets still operate, landing catches of sardines and tuna.

POLITICAL INSTABILITY

A republic since 1910, the country was ruled from 1932 to 1968 by a single prime minister, Antonio de Oliveira Salazar. A right-wing dictator, Salazar, though at first sympathetic to Germany, kept Portugal stable and out of the Second World War. During the war, spies from both sides haunted Lisbon.

From the mid-1950s the country became more and more divided politically: the north remained strongly religious and conservative, but in the south a large landless peasantry began to express discontent.

When Salazar was incapacitated by a stroke and was succeeded by Marcello Caetano in 1968, the costs of colonial wars against nationalist movements in Angola, Portuguese Guinea (now Guinea-Bissau) and Mozambique had begun to place a heavy burden on the country, both in manpower and finance. In 1974 the Armed Forces Movement staged a bloodless coup and a national hero, the monocled General Antonio de Spinola, briefly became president. The African colonies were freed but political confusion followed.

Since 1974 there have been 16 governments, mainly left wing and Socialist alliances. In the 1985 elections, though, the Social Democrats replaced the Socialists in power.

Under Salazar's nationalist policies, foreign investment was low, and political uncertainty following the 1974 coup has continued to hinder foreign investment. Subsequently Portugal's low labour costs attracted some investment, but this fell away as real wages fell by 27 per cent in the years 1983-5, giving rise to strikes and economic turmoil. However, the trading benefits expected from membership of the European Economic Community, which Portugal joined in 1986, are beginning to attract foreign investors.

Portugal's industry is focused mainly in the districts of Lisbon and SETUBAL in the centre-west, and Oporto, AVEIRO and Braga in the north. Around Lisbon, metallurgy and engineering are the dominant industries, whereas in the north textiles take first place.

Industry is mainly labour intensive, but increasing investment is leading to the modernisation of some sectors, such as petrochemicals and shipbuilding. Food processing is also of importance, and no one can visit coastal towns such as Matosinhos or Peniche without noticing the all-pervading smell of fish canning.

Although about 27 per cent of Portugal's work force is employed in agriculture, the

▲ VALLEY OF WINE Port wine grapes grow on terraces above the Douro valley in north-east Portugal. The wine is shipped down the Douro to lodges where it matures for up to 20 years.

country has to import nearly two-thirds of its food. There is a continuing, large balance of payments deficit.

Farms are small in the north – often less than 2 hectares (5 acres) – and large in the south. After 1974 most of the large southern estates were turned into farm cooperatives, but falls in grain production led to the government leasing land back to individual farmers. About two-fifths of the 600 cooperatives were broken up.

The only forestry and agricultural products which Portugal exports in any amount are vegetables.

The small northern farms are old-fashioned; oxen are often used rather than tractors. The farmers grow maize, beans, potatoes and a variety of vegetables, mainly for their own use. Most of the cattle are stall fed, and a common sight in the evenings is to see children taking their family's two or three cows to the collective milking parlour in the village.

An older son may be working abroad, probably in France, spending his holidays back in his home village, perhaps with his foreign car, and using much of his income to build a new house.

The women, often dressed in black, may spend the morning washing clothes at the communal tubs by the stream. After lunch, the men meet in the village café to drink coffee and watch the latest soap opera on television.

The whole rhythm of town life is different. Cars blare their horns in traffic jams; people in brightly coloured clothes view the latest fashions in the shops; men in dark suits go briskly about their work. And yet, in all but the largest towns, the cafés and bars remain very much the preserve of men. Traditions are changing, but the visitor will be surprised at the continuance of old customs and ways of life.

Since 1960, tourism has played an important role in Portugal's economy. In 1984, for example, 9 million tourists visited the country bringing in millions in foreign exchange.

Much of the tourism is based on the coastal resorts, but Portugal also has dramatic inland landscapes and a rich cultural heritage. Much of the pleasure in visiting Portugal lies in exploring underdeveloped parts of the country and staying in state hotels called *pousadas*. These are often buildings of historical interest, such as the castles at Obidos and Palmela.

Beautiful countryside can be found in the National Park of Peneda-Gerês and the Serra da Estrela Natural Park. There are many old castles, such as those at Elvas, Evoramonte and Monsaraz.

Splendid churches and cathedrals are a feature of the towns and cities. The most eye-catching architectural style is the early 16th-century Manueline, in which dramatic use is made of curved and twisted columns, ribs and corbels. One of the best examples is the famous window at the Convent of Christ at Tomar. Other fine religious architecture is found in the Gothic cathedral at Évora and the 15th-century monastery at Batalha.

Portugal's 860 km (530 miles) of coastline and its rivers and streams offer excellent fishing. There are fine golf courses at ESTORIL; Penina; between Portimao and Lagos; Vilamoura and Vale de Lobo, near Faro.

PORTUGAL AT A GLANCE
Area 92 082 km² (35 553 sq miles)
Population 10 100 000
Capital Lisbon
Government Parliamentary republic
Currency Escudo = 100 centavos
Language Portuguese
Religion Christian (96% Roman Catholic)
Climate Hot dry summers, warm moist winters; average temperature in Lisbon ranges from 8-14°C (46-57°F) in January to 17-28°C (63-82°F) in August
Main primary products Cereals, olives, timber, rice, grapes, citrus fruits and vegetables, cork, fish; copper, salt, iron, tungsten, tin
Major industries Agriculture, textiles, machinery, food processing, wood products, chemicals, wine, fishing, mining, tourism
Main exports Textiles, machinery, petroleum products, chemicals, cork, wine, sardines
Annual income per head (US$) 2010
Population growth (per thous/yr) 5
Life expectancy (yrs) Male 70 **Female** 75

Poyang Hu ***China*** One of the country's largest freshwater lakes, joined to the Chang Jiang river. Poyang Hu once covered more than 5000 km^2 (1900 sq miles) but is now only some 3600 km^2 (1400 sq miles) because of silt accumulation and land reclamation. It is impossible to measure its area precisely as the difference between flood and low water levels can be as much as 7.5 m (25 ft).

Map China He

Poza Rica ***Mexico*** Boom town in VERACRUZ state where oil was found in 1930. It now provides 35 per cent of Mexico's oil, most of which is taken elsewhere for refining. Poza Rica has only one small refinery.

Population 567 000

Poznan (Posen) ***Poland*** Historic city 269 km (167 miles) west of the capital, Warsaw. It was founded in the 10th century on an island in the Warta river, and became Poland's first bishopric and the focus of Greater Poland, the heartland of the Polish state. The city, halfway between Gdansk and Prague, and Berlin and Warsaw, thrived as a market and crafts centre.

The medieval St John's Fair – held on June 24 – became, after 1922, the International Trade Fair, now a showcase for trade between the Communist Bloc and the West. Poznan's products include textiles, foodstuffs, agricultural equipment, marine and railway engines, machine tools, electrical goods, rubber and paper.

Population 571 000
Map Poland Bb

Pozzuoli ***Italy*** Coastal town about 12 km (8 miles) west of Naples. It has extensive Roman remains including the amphitheatre of Flavius – one of the largest Roman theatres in Italy. Pozzuoli's economy is based upon fishing, iron and steel.

Population 71 100
Map Italy Ed

Prague (Praha) ***Czechoslovakia*** The country's capital and largest city, also capital of the Czech Socialist Republic. It straddles the Vltava river in the heart of Bohemia, and (like Rome) covers seven hills. It was founded in the 9th century, when the Premysl family of princes established themselves at Vysehrad, a high rock beside the river in the south of the modern city.

Wenceslas, or Vaclav (about 907-29), an early Christian member of the family, was murdered by his heathen brother and later became the patron saint of Czechoslovakia. The 19th-century English carol *Good King Wenceslas* is loosely based on legends about the saint. The massive Palace of Culture (built in 1981) stands near Vysehrad, and across the river is Bertranka, the house (now a museum) where the Austrian composer Wolfgang Mozart (1756-91) stayed during his many visits to Prague.

The Old Town, on the east bank of the river, preserves the Gothic Tyn Church; the 14th-century town hall, with its astronomical clock; Charles University, founded in 1348; and the Jewish ghetto, dating from the 10th century. The Charles Bridge (1357), embellished with 18th and 19th-century statues, links the Old Town with the Lesser Town. This maze of winding streets, with ornate craftsmen's signs, taverns, and palaces, including the Wallenstein Palace, was founded in 1257.

Hradcany hill rises above, crowned by the royal castle and St Vitus Cathedral (1344-1929), the present home of the Bohemian crown jewels. In 1618, during the Thirty Years' War, an event known as 'the Defenestration of Prague' occurred in Hradcany: two deputies of the (Catholic) Holy Roman Emperor were thrown out of a window in the royal castle by angry Bohemian Protestants. Since 1918 the castle has been the official home of Czechoslovak presidents.

The New Town to the south-east of the Old dates mostly from the 19th century, although its ground plan was laid out by the Emperor Charles IV in 1348. It includes the elegant boulevard known as Wenceslas Square.

Chemicals, engineering, electrical, food, clothing, film and pharmaceutical industries have been established in Prague. Its post-1948 outer suburbs, with massive apartment blocks, are now linked to the centre by an underground railway system, which was expanded in the 1980s.

Population 1 235 000
Map Czechoslovakia Ba

Praia ***Cape Verde*** Port and capital of CAPE VERDE. It stands on the main island of São Tiago in the Atlantic Ocean, about 600 km (370 miles) west of the coast of Senegal (Africa). Praia exports sugar cane, castor beans, coffee and bananas.

Population 4055

prairie Flat, temperate grassland in the North American interior east of the Rocky Mountains, similar to steppe. Most of the prairies are now ploughed for cereal and fodder crops. The term is used for similar areas elsewhere.

Praslin ***Seychelles*** Second largest island of the Seychelles group, situated in the western Indian Ocean, 43 km (27 miles) north-east of MAHE, and covering 18 km^2 (7 sq miles). A tourist spot, it is renowned for its sandy beaches and for the coco-de-mer palms of the Valée de Mai.

Population 8000

Prato ***Italy*** Industrial town about 16 km (10 miles) north-west of Florence. Wool textiles, especially those produced from wool rags, are the main industry.

Population 161 700
Map Italy Cc

Precambrian period Oldest and largest division of geological time covering the period from the formation of the earth up to about 570 million years ago. See GEOLOGICAL TIMESCALE

precipitation Moisture that is deposited from the atmosphere in the form of dew, frost, rain, sleet, hail or snow.

Presov ***Czechoslovakia*** Slovak industrial town 36 km (22 miles) north of Kosice. Its factories produce machinery, foodstuffs and textiles. The Gothic old town has an 18th-century Greek Catholic (Uniate) cathedral, and Presov is the centre of Slovakia's Ukrainian community, some 45 000 strong.

Population 78 200
Map Czechoslovakia Eb

Prespa, Lake (Prispansko Jezero) ***Albania/Greece/Yugoslavia*** Lake covering 274 km^2 (106 sq miles) in Macedonia, now divided between the three countries. Up to 54 m (177 ft) deep, it lies cradled in mountains at an altitude of 853 m (2799 ft). It drains by underground streams to Lake Ohrid to the north-west. Popular with tourists, it is noted for its carp, the fine sandy beaches along its eastern shore, and for the pelican and cormorant colonies on the lake island of Golem Grad.

Map Yugoslavia Ed

Pressburg ***Czechoslovakia*** See BRATISLAVA

pressure See ATMOSPHERIC PRESSURE

pressure gradient, barometric gradient Rate at which the ATMOSPHERIC PRESSURE changes on the earth's surface, as indicated by the spacing of the isobars on a weather chart. If the isobars are close together, it indicates a considerable difference of pressure between two points on the earth's surface and a steep gradient, resulting in strong winds. If the isobars are far apart the gradient and the winds are slight.

pressure ridge Ridge of floating ice formed as two ice floes push against each other. A similar pressure ridge may also form on the surface of a glacier where the flow is impeded by the shape of the valley, or at the junction with a tributary glacier.

Preston ***United Kingdom*** County town of Lancashire, standing at the head of the Ribble estuary 45 km (28 miles) north-west of Manchester. It received a charter in 1179, making it one of the oldest boroughs in England, and has had a Member of Parliament since the 13th century. It was the site of a decisive battle of the English Civil War in 1648, when the army of Oliver Cromwell defeated 20 000 Scottish supporters of Charles I. Today Preston is an industrial centre producing aircraft, trucks and buses, textiles, chemicals and machinery. It is the centre of a new town planned to house 285 000 people by the mid-1990s.

Population 87 000
Map United Kingdom Dd

Pretoria ***South Africa*** Administrative capital of South Africa 48 km (30 miles) north of Johannesburg. It is often listed as the national capital, but in fact South Africa has three capitals – the other two are Cape Town (the seat of the legislature, where Parliament meets) and Bloemfontein (the judicial capital and site of the nation's Supreme Court). Pretoria is a stately city, with avenues lined with jacaranda trees. It contains the imposing Union Buildings (government offices) and the Voortrekker Monument, inaugurated in 1949 to commemorate the pioneering trek to the country's interior by Afrikaner settlers in the early 19th century.

Pretoria was founded in 1855 and named after the Voortrekker Andries Pretorius. It became capital of the Transvaal in 1860 and administrative capital of South Africa in 1910. The city's products include cars, steel, chemicals and cement. The main black townships nearby are Mamelodi and Atteridgeville.

The world's largest diamond, the 3106 carat Cullinan, was found at the Premier Diamond

Mine, 40 km (25 miles) to the east. Pieces of the diamond – named after Sir Thomas Cullinan, who discovered the site in 1902 – are now the largest gems in the British Crown Jewels.
Population 739 000
Map South Africa Cb

prevailing wind Wind that blows most frequently from a specific direction in a certain place. For example, the south-westerly winds are the prevailing winds over much of western Britain. The prevailing wind is not necessarily the dominant wind.

Prince Albert ***Canada*** **1.** Town on the North Saskatchewan river, 145 km (90 miles) north-east of SASKATOON. It was founded in 1866 as a Presbyterian mission to the Cree Indians. The town is a centre for the surrounding grain and cattle area, and manufactures timber products. **2.** National park, covering 3875 km² (1496 sq miles), 55 km (34 miles) north of the town. It contains coniferous forest (black and white spruce), parkland and prairie grassland, and hundreds of lakes.
Population (town) 31 380
Map Canada Ec

Prince Edward Island ***Canada*** Island in the Gulf of St Lawrence, and Canada's smallest province, covering 5660 km² (2185 sq miles). Separated from the mainland by the Northumberland Strait, the crescent-shaped island consists of red-sandstone lowlands with a maximum height of 142 m (466 ft). Two-thirds of the area is cultivated, growing potatoes, vegetables and tobacco. Manufacturing is mainly connected with food processing. Tourism is important. Among Prince Edward Island's attractions are deep-sea fishing and horseracing. Prince Edward Island National Park, with dunes, salt marshes and fine bathing beaches, covers 18 km² (7 sq miles). CHARLOTTETOWN is the island's capital and main town.
Population 123 000
Map Canada Id

Prince George ***Canada*** Town in central British Columbia at the junction of the Fraser and Nechako rivers. It is the commercial and industrial centre for a timber, mining and agricultural area. The town is an important road and rail junction.
Population 67 600
Map Canada Cc

Prince Rupert ***Canada*** Important port on the northern coast of British Columbia. It is a tourist centre and a fishing port, and ships grain, ores and timber.
Population 16 200
Map Canada Bc

Príncipe ***São Tomé and Príncipe*** Second largest island of the republic, covering an area of 100 km² (39 sq miles). It lies in the Gulf of Guinea about 200 km (125 miles) west of Equatorial Guinea. Volcanic, tropical and largely forested, Príncipe rises to 948 m (3110 ft) at Pico de Príncipe. The chief town is Santo Antonio. Cocoa, coconuts and coffee are grown on plantations and form the main exports.
Population 5000
Map Gabon Aa

Pripyat' Marshes ***USSR*** See POLES'YE

Pristina ***Yugoslavia*** Capital of the autonomous province of Kosovo in Serbia, 80 km (50 miles) north of Skopje. It was a capital of the medieval Serbian Empire and lies in the plain known as Kosovo Polje, or the 'Field of Blackbirds' – where in 1389 the Turks crushingly defeated the Serbs. The town stagnated under the Turks and became part of Yugoslavia's most backward region, but since 1950 it has become a modern administrative, educational and communications centre, with cotton spinning, metal and electrical industries. Most of the townsfolk are ethnic Albanians, and tension between the two peoples runs high in the town. The lovely 14th-century monastery of Gracanica, with cupolas, Byzantine dome and vivid frescoes dating from the 14th to 16th centuries, lies 10 km (6 miles) to the south-east.
Population 216 000
Map Yugoslavia Ec

Prizren ***Yugoslavia*** Market town at the mouth of a valley on the northern flank of the Sar Planina mountains, some 65 km (40 miles) north-west of Skopje. The old Turkish town in the valley has fine painted houses, mosques, bazaar and 16th-century Turkish baths. There are Orthodox churches and monasteries on the forested slopes above, while the post-war town with modern food processing, tobacco and cotton-spinning factories lies in the plain. The weekly market is a good time to see a wealth of traditional costumes. The ski resort of Brezovica lies above the town.
Population 134 500
Map Yugoslavia Ec

Prokletije ***Albania/Yugoslavia*** See ALBANIAN ALPS

Prokop'yevsk ***USSR*** Coal-mining city in southern Siberia, about 270 km (170 miles) south-east of Novosibirsk. Prokop'yevsk's mines have coal seams up to 40 m (130 ft) thick; but despite modern machinery, some pits have been closed by the government. As a result, it is one of the few Siberian cities with a decreasing population.
Population 274 000
Map USSR Jc

promontory High ridge of land or rock jutting out into a sea or other expanse of water, or into low-lying land.

Proterozoic era The more recent Precambrian era, when various primitive plant and animal life forms evolved from single-celled organisms. See GEOLOGICAL TIMESCALE

Provence ***France*** Historical Mediterranean region of 26 135 km² (10 091 sq miles) between the Rhône river and the Italian border. Settled by the Romans, it became a kingdom in AD 855, and part of France in 1481. Its main cities today are MARSEILLES, NICE and TOULON, but the historic capital was AIX-EN-PROVENCE. The Provençal language had a thriving literature in the Middle Ages and was revived in the 19th century by a group of writers including the poet Frédéric Mistral (1830-1914). The scenery and brilliant light of Provence inspired many painters of the Impressionist school. The modern metropolitan region of Provence-Côte d'Azur covers 31 400 km² (12 125 sq miles) and is administered from Marseilles.
Population (Provence-Côte d'Azur) 3 860 100
Map France Fe

Providence ***USA*** Port and state capital of Rhode Island, at the head of Narragansett Bay about 65 km (40 miles) south-west of Boston. Brown University was founded there in 1770. The traditional textile industry has been outpaced by jewellery making, metalworking and engineering.
Population (city) 154 100; (metropolitan area) 1 095 000
Map United States Lb

Providenciales ***Turks and Caicos Islands*** Most developed tourist island in the group. An international airport was opened in 1984 to serve a newly built holiday resort village there.
Population 1000
Map Caribbean Aa

Provincetown ***USA*** Tourist resort at the northern tip of Cape Cod, Massachusetts. It was the first landing place of the Pilgrim Fathers in 1620 before they founded the PLYMOUTH Colony on the other side of Cape Cod Bay.
Population 3500

Prussia ***Northern Europe*** Former state of north Germany whose capital was BERLIN. Its heartland was Brandenburg, but it eventually covered 295 000 km² (114 000 sq miles) from the Rhine far along the Baltic coast into present-day Poland and USSR, and included the industrial cities of the RUHR.

Prussia was formed as an independent duchy in the 17th century. It became the dominant power of northern Germany and led to the formation of the *Zollverein* (customs union of German states) in the 19th century. The centre of German militarism, it played a leading role in the unification of Germany at the end of the Franco-Prussian War (1870-1). Its king, Wilhelm I, became the first kaiser (emperor) of Germany in 1871, and its leading statesman, Prince Otto von Bismarck, chancellor. Prussia lost its independence under the Nazis and was dissolved by the 1945 POTSDAM Conference.

Prut ***Romania/USSR*** River forming Romania's eastern frontier with the USSR. It rises in the south-west UKRAINE and flows 902 km (560 miles) south to join the Danube 200 km (124 miles) north-east of Bucharest.
Map Romania Ba

Przemysl (Peremyshl) ***Poland*** City in the south-east about 10 km (6 miles) from the Soviet border. It was founded as a fortress beside the San river, and flourished from 981 as a market on the trade route from central Europe to the Black Sea. The city is now part of a huge 'dry port' marshalling complex, where freight is switched between the Polish standard-gauge and Soviet broad-gauge railways.
Population 64 100
Map Poland Ed

Pskov ***USSR*** City 250 km (155 miles) south of Leningrad. It is one of Russia's oldest cities, being mentioned in AD 903. In 1240 it was occupied by Teutonic Knights and became a significant trade power; it was annexed by Moscow in 1510. It then withstood 26 sieges from Poles, Germans and Swedes. In 1917 it was the scene of the abdication of the last tsar, Nicholas II (1868-1918). He and his family were shot by Bolsheviks the following year.

During the Second World War, Pskov was occupied by German troops (1941-4) and was badly damaged. Its most notable buildings and monuments were later painstakingly restored. They include the kremlin (citadel), the remains of the Old Town and the silver-domed 12th-century Trinity Cathedral. It is the centre of a flax-growing area and produces linen.

Population 189 000
Map USSR Dc

Puch'on ***South Korea*** City in the industrial region stretching between Seoul, the capital, and the port of Inch'on.

Population 221 500
Map South Korea Cd

puddingstone See CONGLOMERATE

Puebla (Puebla de Zaragoza) ***Mexico*** Capital of Puebla state, 120 km (75 miles) south-east of MEXICO CITY. It was founded in 1532 by Spaniards from the Talavera region of Spain, who brought their local tile-making skills to adorn Puebla's colonial red-brick houses. The city has 360 churches and the largest church in Mexico, Santa Domingo, built in the 16th century and ornamented with gold leaf. The city is also famous for its wickerwork and pottery and for objects made of local onyx. Recently heavy industry has poured into Puebla, which now has a major car factory. The ancient Toltec site of CHOLULA is 8 km (5 miles) to the west.

Puebla state is mountainous and forested. Industries have overflowed from the national capital into Puebla, which is also important for mining, textiles, sugar cane, cotton, cereals, tobacco, rice, fruit, vegetables and chillies.

Population (city) 835 000; (state) 3 500 000
Map Mexico Cc

Pueblo ***USA*** Industrial city in the south-central Colorado, about 140 km (87 miles) north of the New Mexican border. It specialises in steel making, and is also the centre of an irrigated farming area in the Arkansas river valley.

Population (city) 100 000; (metropolitan area) 124 900
Map United States Fc

Puerto Rico

BIG BUSINESS COMES TO THE RESCUE IN A U.S. OUTPOST WHERE THE FARMERS COULD NOT COPE

The people of Puerto Rico are United States citizens who speak Spanish and live 1600 km (1000 miles) from the US mainland. Spain lost this group of Caribbean islands after the Spanish American War in 1898. Puerto Ricans were given US citizenship in 1917, but they still cannot vote in American elections unless they live on the mainland. There is some pressure for independence, but in 1952 the people voted to become a self-governing commonwealth in association with the United States.

This Spanish-speaking, Roman Catholic, West Indian country, with its people mostly of Spanish or African descent, has more culturally in common with the neighbouring Dominican Republic, Cuba and mainland Central America, than with the nation to which it belongs.

Puerto Rico (Spanish for 'Rich Port') is the main island of the territory, which also includes the two small islands of Vieques and CULEBRA, and a fringe of uninhabited smaller islands and islets. Puerto Rico island has a backbone of mountains, the Cordillera Central, which reach 1338 m (4390 ft) at the peak of Cerro de Punta. Agriculture has cleared most of the island's natural forest, but trees in the Sierra de Luquillo in the north-east are protected in a national park. Cooling winds temper the tropical heat.

AGRICULTURE AND INDUSTRY

Dairy farming is the most important agricultural activity. Sugar is the main crop, followed by tobacco, pineapples, bananas and coconuts, but the whole agricultural sector has been overtaken by industry in recent years.

The change was forced by a rapid growth in population; it is still increasing at 1.7 per cent per year. The need to support more people put pressure on farmland, and slopes were cleared that should have been left forested. This caused serious soil erosion which reduced agricultural output, and eventually farming could not provide enough food and export income for the population, many of whom emigrated to the United States.

The US government encouraged the growth of industry to give Puerto Rico another source of income. Tax reliefs and cheap labour encouraged American businessmen to invest there, and manufacturing has taken over from agriculture as the main export earner. Products include textiles, clothing, electrical and electronic goods, plastics and chemicals, made mainly by subsidiaries of big US companies.

Puerto Rico attracts many tourists to its mountain scenery, excellent beaches, and game fishing. There are many relics of Spanish colonial days, including 16th and 17th-century houses in the old town of SAN JUAN, the capital and chief port, and two great 16th-century clifftop forts which dominate the harbour of San Juan Bay.

PUERTO RICO AT A GLANCE

Map Caribbean Bb
Area 8897 km² (3435 sq miles)
Population 3 345 000
Capital San Juan
Government Self-governing commonwealth in association with the USA
Currency US dollar = 100 cents
Languages Spanish, English
Religion Christian (92% Roman Catholic, 5% Protestant)
Climate Tropical maritime; average temperature ranges from 17°C (63°F) to 36°C (97°F)
Main primary products Sugar cane, livestock (dairy cattle, pigs, goats, poultry), tobacco, pineapples, bananas, coconuts, fish
Major industries Oil refining and petrochemicals, clothing, agriculture, sugar refining, electrical equipment, machinery, fishing, food processing
Main exports Chemicals, petroleum products, clothing, machinery, fish, tobacco products, sugar
Annual income per head (US$) 3900
Population growth (per thous/yr) 17
Life expectancy (yrs) Male 71 **Female** 77

Puerto Cortés ***Honduras*** The country's principal port, in the north-west corner. It exports mostly bananas, and has a large oil refinery.

Population 62 250
Map Honduras Aa

Puerto La Cruz ***Venezuela*** Oil port about 280 km (175 miles) east of Caracas. Once a fishing hamlet, it now has two oil refineries.

Population 80 000
Map Venezuela Ba

Puerto Montt ***Chile*** Seaport and capital of Los Lagos region, on the northern edge of the Gulf of Ancud 900 km (560 miles) south of Santiago. It is a centre for agriculture, fishing and tourism, and is the terminus of the southern railway. The city was founded by German immigrants in 1853, and was devastated by an earthquake in 1960.

Population 110 500
Map Chile Ad

Puerto Presidente Stroessner ***Paraguay*** Tourist town named after General Alfredo Stroessner, who has ruled Paraguay since 1954. It is on the Brazilian border about 320 km (200 miles) east of the capital, Asunción. It has prospered as a low-tax shopping centre and casino resort for foreign tourists and as the base for the nearby ITAIPU hydroelectric scheme.

Population 50 000
Map Paraguay Bb

Puerto Vallarta ***Mexico*** Fastest growing beach resort in Central America, on the Bay of Flags (Bahía de Banderas), Jalisco state. Despite the luxury hotels, it conveys the charm of a fishing village, with red-tiled, white-walled houses, scattered along the Cuale river.

Population 70 000
Map Mexico Bb

Puglia (Apulia) ***Italy*** Region covering 19 250 km² (7500 sq miles), taking in the heel of the boot of Italy. It is known as the 'kingdom of drought and stone' because of its hot, dry climate and the limestone which forms the backbone of the Murge Hills, seamed with gorges and caves. Grapevines are the main crop in the region. Olives, almonds, wheat, tobacco and

vegetables are also grown. There are engineering and food-processing plants at the regional capital, BARI, petrochemicals at BRINDISI and steel making at TARANTO.
Population 3 848 000
Map Italy Ed

Pukaskwa ***Canada*** National park protecting the wilderness of northern Ontario and its wildlife. It covers 1885 km² (728 sq miles), south of Marathon on the north shore of Lake Superior, and is accessible only by boat.
Map Canada Gd

Pula (Pulj; Pola) ***Yugoslavia*** Modern resort, naval base, major shipbuilding centre and chief town of Istra, the Adriatic peninsula just south of the Italian border. The Romans knew the town in the 2nd century BC as Piestas Julia. Ruins of its monumental amphitheatre, built and enlarged between 30 BC and AD 72 to hold more than 20 000 spectators, survive. Its walls rise 32 m (105 ft) above 72 arches. Just north-west of Pula lie the Brijuni (Brioni), a group of small islands where Yugoslavia's former leader Marshal Tito (1892-1980) had a villa.
Population 47 200
Map Yugoslavia Ab

pumice, pumice stone Lightweight, acid volcanic rock full of vesicles (cavities that once held hot gases). It is often used as an abrasive.

puna High, dry plateau at 3000-5000 m (9800-16 400 ft) between the West and East Cordilleras of the Andes in Peru and Bolivia, with a sparse covering of coarse grasses and drought-resistant shrubs.

Punakha ***Bhutan*** A former capital, standing beside the Sankosh river at 1200 m (3950 ft), some 24 km (15 miles) north-east of the present capital, Thimphu. Its large monastery, built in 1637 in the temperate valley, is one of the best examples of Bhutanese architecture.
Map Bhutan Aa

Puncak Jaya ***Indonesia*** The nation's highest mountain, rising to 5030 m (16 500 ft). It lies in Irian Jaya on the island of New Guinea.
Map Indonesia Ic

Pune ***India*** See POONA

Punjab ***India*** North-western agricultural state covering 50 362 km² (19 440 sq miles). It was part of the pre-independence Indian province of Punjab, which in 1947 was divided between Pakistan and India. A new capital for the Indian state of Punjab was built at CHANDIGARH. In 1966 the state was split into three, into a Sikh-dominated Punjab in the west, and what are now the Hindu-dominated states of Haryana and Himachal Pradesh. Punjab, with its rolling expanse of well-irrigated wheatland, is known as the 'Granary of India'.

The name comes from Sanskrit words meaning 'five rivers', referring to the Jhelum, Chenab, Ravi, Beas and Sutlej, which all cross the historic Punjab region; all but the Beas cross the border into the Pakistani Punjab.
Population 16 789 000
Map India Bb

Punjab ***Pakistan*** Province covering 205 344 km² (79 283 sq miles), comprising the larger western part of the British Indian province of Punjab. Economically and politically it is the most important region in the country. It relies on irrigated agriculture using water from the great rivers that cross it – the Indus, Jhelum, Chenab, Ravi and Sutlej. The region produces wheat, cotton and rice. The provincial capital, LAHORE, was once dominated by Sikhs, many of whom still want to unite with the Indian Punjab to form an independent Sikh state.
Population 47 292 000
Map Pakistan Cb

Puno ***Peru*** City and department on the border with Bolivia. The highlands are home for Indian herdsmen who depend on the pastures to support their sheep, llamas and alpacas. The city,

▼ MAJESTIC RETREAT Monks from Bhutan's capital, Thimphu, spend the winter at the dzong (fortified Buddhist monastery) at Punakha.

on the north-west shore of Lake Titicaca – the world's highest large lake at 3812 m (12 506 ft) – is renowned for its fiestas and its music. The town of Pucará is famous for its pre-Inca remains, the pre-Columbian tall burial towers called *chullpas* at Sillustani, and the floating islands of the Urus people on Lake Titicaca.
Population (department) 890 300; (city) 67 600
Map Peru Bb

Punta Arenas *Chile* Capital of Magallanes region and a major port, on the north shore of the Strait of Magellan. It is the world's southernmost city, and the centre of the sheep-farming industry. It exports meat and wool.
Population 70 000
Map Chile Ae

Punta del Este *Uruguay* Resort on a peninsula 139 km (86 miles) east of the capital, Montevideo, at the mouth of the River Plate. Beaches either side of the peninsula offer the breakers of the Atlantic or the shelter of the estuary. Inland are pine and eucalyptus woods near the colonial town of Maldonado, and offshore is Isla de Lobos, a protected island reserve inhabited by some half a million seals.
Population 10 000
Map Uruguay Bb

Puntarenas *Costa Rica* Fishing port and tourist resort on the Pacific coast 80 km (50 miles) west of the capital, San José. The coastal province of the same name produces bananas, cocoa and palm oil.
Population (port) 48 000; (province) 291 000
Map Costa Rica Ba

Pusan *South Korea* The country's second largest city after the capital, Seoul. Lying on the south-east tip of South Korea, it is also the country's leading port. Pusan was developed during Japanese colonial rule (1905-45), and its products now include machinery, textiles and ships. In addition, tourists are drawn to the nearby beach resort of Haeundae and the hot springs of Tongnae. The city's United Nations cemetery contains the graves of soldiers who fell in the Korean War (1950-3).
Population 3 160 000
Map Korea De

Pustynia Bledowska *Poland* See BLEDOW DESERT

puszta Temperate, almost treeless grassland in the plains of Hungary.

Putumayo *Colombia/Ecuador/Peru/Brazil* River rising in the Andes of south-west Colombia near the Ecuadorean border and running south-east for about 1900 km (1180 miles). It forms part of the Colombian border with Ecuador and almost all of that with Peru before crossing into Brazil – where it is known as the Içá – to join the Amazon.
Map Colombia Bc

Puys, Chaine des *France* Mountains of volcanic origin in the AUVERGNE. About 70 summits rise above the main plateau of the Massif Central (800-1000 m, 2625-3280 ft), the highest being the Puy de Sancy (1885 m, 6184 ft).
Map France Ed

Pyatigorsk *USSR* Town and health resort in the northern Caucasus Mountains, 140 km (85 miles) south-east of Stavropol'. Pyatigorsk has numerous sanatoriums, whose mineral waters are used to treat ailments such as rheumatism and arthritis. The Russian poet and writer Mikhail Lermontov (1814-41) was killed in a duel in a glade just outside Pyatigorsk, and the town has a museum devoted to him.
Population 117 000
Map USSR Fd

Pyeongyang *North Korea* See PYONGYANG

Pyinmana *Burma* Town in central Burma, about halfway between Mandalay and Rangoon. Surrounded by a major sugar-cane growing area, it has a new agricultural university which opened in 1975.
Population 172 700
Map Burma Cc

Pyongyang (Pyeongyang) *North Korea* National capital beside the Taedong river, 87 km (54 miles) from its estuary on Korea Bay. Chinese colonists founded the city, then called Lolang, in 108 BC. Renamed Pyongyang, it became capital of an independent kingdom, Koguryo (AD 427-668). After the Japanese occupation of Korea (1905-45), the city became the capital of Communist North Korea in 1948. It was heavily bombed during the Korean War (1950-3).

The new city built since then has broad, tree-lined avenues with many parks and open spaces. There is little motor traffic. Tourists, closely supervised, are shown the Museum of the Korean Revolution, the Tombs of the Revolutionary Heroes, and the 60 m (200 ft) high Triumphal Arch, built in marble and granite.

Pyongyang produces more than a quarter of the country's industrial output – mainly iron and steel, machinery, railway locomotives, rolling stock and textiles.
Population 1 700 000
Map Korea Bc

Pyrenees (Pirineos; Pyrénées) *Western Europe* Mountain chain extending about 440 km (275 miles) from the Bay of Biscay to the Mediterranean, and forming a natural boundary between France and Spain. Within the eastern part of the mountains lies the tiny independent state of Andorra, whose official language is Catalan – a tongue also spoken in north-east Spain and the adjoining area of France. The shape of the hills was largely carved by Ice Age glaciers, but only 34 km² (13 sq miles) of permanent snowfield remain – all at heights above 3000 m (9840 ft). The highest point is Pico d'Aneto (3404 m, 11 170 ft) in the central Pyrenees on the Spanish side of the border.
Map France Ce

pyrite (iron pyrites) Brassy yellow mineral sulphide of iron. It is an important source of sulphur and is also used to produce sulphur dioxide for the manufacture of sulphuric acid. Its metallic yellow lustre has led to it being mistaken for gold, hence its other name – fool's gold.

pyroclastic rock Rock formed from fragments of volcanic material blown into the air by a volcanic explosion. On landing, the fragments become consolidated over a wide area round the volcano. See VOLCANO

pyroxene Any of a group of complex silicate minerals found in basic and ultrabasic IGNEOUS ROCK and some METAMORPHIC ROCK. Pyroxenes include augite (calcium magnesium iron and aluminium silicate) and JADEITE.

Qaanaaq *Greenland* See THULE

Qalat al Bahrain *Bahrain* Historic site 5 km (3 miles) west of Al Manamah. The remains consist of a series of cities – the earliest is 5000 years old – which were part of the Bronze Age state of Dilmun, as Bahrain was then known. Excavations have revealed the remains of palaces, broad streets and a harbour. About 6 km (4 miles) west along the coast of Barbar are the remains of three temples, the oldest dating from 2500 BC. A third site of the same period lies 2 km (just over 1 mile) west of Barbar at Ad Diraz, where traces of village buildings have been discovered. The Dilmun people were buried in single-chamber grave mounds, 100 000 of which lie in north-west Bahrain Island.
Map Bahrain Ba

Qam'do *China* Geographical region of about 270 000 km² (104 250 sq miles) covering eastern Xizang (Tibet), north-west Yunnan and west Sichuan provinces. It is a remote, sparsely populated region of parallel mountain ranges, separated from one another by the deep gorges of the upper Nu Jiang (Salween), Lancang Jiang (Mekong) and Chang Jiang rivers. There is some cultivation of maize, rice and winter wheat.
Map China Ee

Qatif *Saudi Arabia* Coastal oasis on The Gulf, situated 15 km (9 miles) north of Dammam. Numerous underground springs irrigate the surrounding land and for centuries the oasis has been a centre of trade, fishing and agriculture. Qatif is rapidly being modernised, but is still a maze of narrow lanes and old houses linked by overhead bridges.
Population 95 000
Map Saudi Arabia Bb

Qattara Depression *Egypt* Huge, sunken area of the Western Desert midway between Cairo and the Libyan border. It is about 320 km (200 miles) long and 120 km (75 miles) wide, and in parts is 134 m (440 ft) below sea level. It consists almost entirely of soft sand, brackish ponds and salt marshes. During the Second World War, it was a vital, impassable flank to the British defences at the Battle of El Alamein.
Map Egypt Ab

Qazvin (Kasvin) *Iran* Market town founded in the 4th century, about 150 km (95 miles) north-west of the capital, Tehran. It manufactures carpets, textiles and soap.

In the Elburz Mountains above Qazvin is the ruined citadel of Alamut, one of the strongholds of the Assassins a fanatical Muslim sect founded in about 1090 to destroy their religious enemies. They were exterminated in 1264 by the Mongol leader Hulagu, grandson of Genghis Khan.
Population 244 300
Map Iran Aa

Qatar

IMMIGRANTS OUTNUMBER THE NATIVES IN THIS TINY ARAB KINGDOM WHICH IS INVESTING OIL WEALTH TO PROTECT ITS FUTURE

The little emirate of Qatar, halfway along the western coast of The GULF, has a classic rags-to-riches story. Before 1949, when the first oil was exported, it was a tribal society which existed on pearl diving, fishing and camel breeding. In 1982 it had the second highest income per head (the equivalent of US$16 000) in the world after the United Arab Emirates (US$23 000).

The Arabs of this stony desert land have a proverb: 'Hitch your camel first and then depend on God.' They have made it one of the most progressive states in the Arab world, although traditional Muslim values have been respected. The oil money has been used to build up social services, with free education and health care, and to develop heavy industry so that prosperity will not vanish overnight when the oil runs out.

Qatar consists of a peninsula and a few small islands, which cover an area of only 11 437 km² (4416 sq miles). The peninsula thrusts out into The Gulf from the Saudi Arabian coast like a fat thumb.

The summers are long, hot and uncomfortably humid, with temperatures of up to 49°C (120°F); a north-west wind, the *shamal*, helps to cool the land. The winters are mild, with rain in the north. Otherwise it is an arid country, and most fresh water comes either from natural springs and wells or (increasingly today) from desalination plants producing fresh water from sea water. Sheep, goats and some cattle are herded, and the country is noted for the quality of the camels that are bred there – a lucrative activity for the Bedouins.

Arab writers first mentioned Qatar in the 10th century. The people today are descended from migratory tribes who arrived in the 1730s. Qatar was invaded by the Persians in 1783 and later became a dependency of its neighbour BAHRAIN. In 1868 it came under the influence of the British, who installed the present ruling family. The Turks took nominal control at the end of the 19th century. In 1916 Qatar became a British protectorate and it achieved independence in 1971.

OIL AND GAS RESERVES

The discovery and exploitation of oil and natural gas have transformed modern Qatar. The oil field at DUKHAN has an expected life of 40 years and the big offshore North Field has some of the largest reserves of natural gas outside the Communist countries. The new range of heavy industry includes iron and steel, cement, fertiliser and petrochemical plants. The fishing industry has been modernised and frozen shrimps are a substantial export. Dates are also exported.

These modern developments have attracted large numbers of immigrants to work in Qatar, so that they now outnumber the native population. Only about 100 000 of the country's people are Qataris. The rest come from other Arab countries and from India and Pakistan; the white-collar workers include Egyptians, Palestinians, Lebanese and Syrians. It is now a highly urbanised society; 180 000 people live in DOHA, the capital.

In 1947 there was only one hospital and one resident doctor in Qatar. Today there is a wide variety of social services. A home ownership scheme offers every citizen a free plot of land and an interest-free loan.

The emir, Sheik Al Thani, is head of state and prime minister, ruling with the help of an appointed Consultative Council of 30 members. There are no elections. Despite all the economic changes, Qatar remains fundamentally an Islamic society (most Qataris are Sunni Muslims), in which traditional Arab dress is worn, women are enjoined to modesty, and floggings are still carried out for some crimes – though not, a government spokesman has claimed, with great severity. The government is watchful over the divisions between its own people and the more numerous immigrants.

QATAR AT A GLANCE
Area 11 437 km² (4416 sq miles)
Population 311 000
Capital Doha
Government Absolute monarchy
Currency Riyal = 100 dirhams
Languages Arabic (official), English
Religion Muslim
Climate Hot and dry; mild in winter. Summer temperatures reach 49°C (120°F)
Main primary products Livestock, fodder, vegetables, fruit, fish; oil and natural gas
Major industries Oil and natural gas production and refining, cement, iron and steel, chemicals, agriculture, fishing
Main exports Oil and natural gas, frozen shrimps, dates, steel, chemicals
Annual income per head (US$) 16 000
Population growth (per thous/yr) 34
Life expectancy (yrs) Male 56 **Female** 60

Qena ***Egypt*** Market town on the east bank of the Nile about 450 km (280 miles) south of the capital, Cairo. It lies on an ancient route from northern Egypt to the sacred Muslim city of Mecca, in Saudi Arabia. Pottery is made in the town.
Population 93 700
Map Egypt Cc

Qeqertarsuaq ***Greenland*** See DISKO

Qesaria ***Israel*** See CAESAREA

Qilian Shan ***China*** Mountain range rising from the extreme north-east of the Xizang Gaoyuan (Tibetan Plateau). Its towering snow-capped peaks exceed 4000 m (13 000 ft), the highest reaching 6346 m (20 819 ft).
Map China Ec

Qin Ling ***China*** Mountain range stretching 1500 km (900 miles) across central China and rising to 4205 m (13 474 ft). It forms a watershed between the Huang He and Chang Jiang river basins. North of the range, rainfall is light and winter temperatures fall below freezing. To the south, rainfall is abundant and, because of the sheltering effect of the mountains, the winters are relatively mild.
Map China Fd

Qingdao ***China*** City of Shandong (Shantung) province on the Yellow Sea (Huang Hai). It was a German treaty port from 1898 to 1914 and is now Shandong's biggest city. German influence is kept alive in the Qingdao brewery. Other industries include the manufacture of locomotives and railway rolling stock. Treaty ports were cities where Western powers enjoyed privileges obtained through treaties with China, such as immunity for Western nationals from trial under Chinese law.
Population 1 300 000
Map China Id

Qinghai ***China*** Province in the west covering 720 000 km² (280 000 sq miles). It takes its name from a salt lake and includes the Qilian Shan, the Qaidam Pendi (Tsaidam Basin) and part of the Plateau of Tibet. The climate is one of extremes. Winter temperatures fall far below freezing and rainfall is sparse and unreliable. The summers are hot and dry. Cultivated land is confined largely to the valley of the Huang He river, which rises in the highlands of south-east Qinghai. Elsewhere, the breeding of yaks, sheep and horses provides the mainstay of the economy. The province is rich in coal, oil, soda, borax, potash and bromine.
Population 3 720 000
Map China Dd

Qinghai Hu ***China*** Salt lake of Qinghai province, 3000 m (10 000 ft) above sea level. It covers 5957 km² (2300 sq miles).
Map China Ed

Qiqihar ***China*** City of HEILONGJIANG province, 1000 km (620 miles) north-east of Beijing (Peking). It is a centre for the manufacture of locomotives, railway rolling stock and heavy machinery.
Population 1 000 000
Map China Ib

Qom (Qum) ***Iran*** Holy city, about 145 km (90 miles) south and slightly west of the capital, Tehran. Fatima, sister of a Shiite religious leader, Imam Reza, died there in 816, and the 17th-century sanctuary built over her tomb is now a place of pilgrimage for Shiite Muslims. Qom is also the home of Ayatollah Khomeini, the Shiite leader whose followers ousted the last shah and established an Islamic Republic in Iran in 1979. Footwear, rugs and pottery are made in the city, which is also the centre of a grain and cotton region.

Population 424 100

Map Iran Ba

Qornet es Saouda (Qurnet es-Sauda) ***Lebanon*** The country's highest mountain at 3087 m (10 128 ft). It lies in the north of the Lebanon Mountains 30 km (18 miles) south-east of Tripoli. For much of the year, its peak is snow-covered.

Map Lebanon Ba

Quan Phu Quoc ***Vietnam*** Island in the Gulf of Thailand, about 40 km (25 miles) off the south-west coast. It is vital to Vietnam's claims to fishing and exploration rights in the gulf.

Map Vietnam Ac

Quang Yen ***Vietnam*** Coal town about 100 km (60 miles) east of the capital, Hanoi. The surrounding coalfield was developed by the French in the late 19th and early 20th centuries, and by the mid-1970s was producing up to 6 million tonnes a year of high-quality anthracite. But expansion has faltered because of the emigration of skilled Chinese miners after the Chinese invasion of Vietnam in 1979.

Map Vietnam Ba

quartz Hard, crystalline form of silica commonly occurring as colourless, transparent crystals in the form of hexagonal columns surmounted by pyramids. There are many coloured varieties, such as rose, milky and smoky quartz. Amethyst is a purple variety and citrine is yellow. These and other coloured quartzes are prized as semiprecious stones. Quartz crystals are also widely used in electronics.

quartzite Metamorphic rock, consisting almost entirely of silica, formed by the recrystallisation of sandstone.

Quaternary Second of the two periods in the Cenozoic era of the earth's time scale. See GEOLOGICAL TIMESCALE

Quatre-Cantons, Lac des ***Switzerland*** See LUCERNE, LAKE

Quebec ***Canada*** **1.** Largest province, covering 1 358 000 km² (524 300 sq miles) in the east between Hudson Bay and Labrador. The people are mostly French speaking. The capital is Quebec city and the largest city is Montreal. The province includes three geographic regions: the CANADIAN SHIELD in the north; the ST LAWRENCE Lowland Valley, where most of the people live; and the Appalachian region, composed of undulating plateaus which rise to more than 1000 m (3280 ft) near the US border.

The Shield is densely forested and has mines producing copper, iron, zinc, silver and gold. Asbestos is found in the south. The Lowland farms yield sugar beet, tobacco, market garden and dairy products. Abundant hydroelectricity has led to the development of metallurgical industries. Manufactured goods include transport equipment, textiles and clothing.

2. Capital of the province of Quebec and Canada's oldest city, founded in 1608 by the French explorer Samuel de Champlain. It stands on the St Lawrence river at its junction with the Charles river. Quebec was taken from the French by British troops under General James Wolfe in the Battle of the Plains of Abraham (1759). It is the only walled city in Canada, and still retains much of its old French atmosphere. The upper town, built on cliffs, has narrow streets and ancient houses, and contains the main public buildings. The lower town has an excellent harbour and contains the business and industrial areas. Quebec has cathedrals, universities, the Church of Notre Dame des Victoires, and the Chateau Frontenac hotel.

Population (province) 6 438 000; (city) 576 075

Map Canada Hd

Queen Charlotte Islands ***Canada*** Group of islands 160 km (100 miles) off the west coast of British Columbia. The main islands are Graham (6452 km², 2491 sq miles) and Moresby (2567 km², 991 sq miles). Most of the islands are covered by forests. Coal, copper and gold have been mined, but timber and fishing are more important. Most of the inhabitants are Haida Indians, a tribe that survives only on these and on Prince of Wales Island off Alaska.

Population 5620

Map Canada Bc

Queen Charlotte Sound Inlet of the COOK STRAIT in the north-eastern corner of New Zealand's SOUTH ISLAND. The British explorer Captain James Cook, who used it as a base, wrote that 'it is a collection of some of the finest harbours in the world'.

Length 40 km (25 miles)

Map New Zealand Ed

Queen Charlotte Strait A scenic, sheltered waterway running between the north-eastern coast of VANCOUVER ISLAND and the Canadian mainland. It is part of the inland water route (the Inside Passage) from SEATTLE in WASHINGTON state, to ALASKA.

Dimensions 97 km (60 miles) long, 26 km (16 miles) wide

Map Canada Cc

Queensland ***Australia*** Most north-easterly state of Australia, covering 1 272 200 km² (591 200 sq miles). The mountains of the Great Dividing Range split the state in two: the well-soaked coastal plain, very hot and monsoonal in the north; and the arid lands covering much of the interior. The Great Barrier Reef lies off the east coast. There are huge deposits of coal, bauxite (aluminium ore), copper, lead, silver and zinc, while its primary agricultural products are cattle, sugar cane, pineapples, bananas and cereals. More than 40 per cent of the population live in the state capital, BRISBANE, the site of a penal settlement founded in 1824. Queensland became a separate colony in 1859.

Population 2 488 000

Map Australia Gc

A DAY IN THE LIFE OF A QUEENSLAND CANE FARMER

The season has gone from bad to worse. Roy Sorensen stands on the verandah of his house, sipping a cup of tea, watching the downpour turn his fields into red quagmires, and wondering what else can go wrong.

For a Queensland sugar-cane farmer like Roy, Mother Nature must not only be provident, she must also have a sense of timing – and this year her timing has been all wrong. In the summer months of January and February when he wanted the rain for his new cane's growth, the skies stayed clear.

Now it is September. Early this morning the long-awaited rain finally arrived, too late to help the cane grow but right on time to ensure that the harvest will have to be postponed, perhaps for days. And for every day that Roy watches the giant mechanical harvesters standing idle in the muddy fields waiting for the ground to dry out so they can begin cutting, he also watches hundreds of dollars going down the drain.

In normal circumstances, the losses from a harvesting delay would have been bearable – one of the ever-present risks of producing sugar cane in the Bundaberg district, where the subtropical weather patterns are notoriously fickle. But a steep decline in the world market price of sugar has already made life financially tough, for growers and millers alike, who are trying to weather the slump and hoping things will improve.

Now Roy is beginning to wonder whether he might be forced to join the increasing number of farmers – friends and neighbours some of them – who have either left their land for jobs in town or ploughed in their cane to turn their land over to the production of tomatoes, beans and other vegetable crops.

Roy was born in the main bedroom of the farmhouse and married Betty Rehbein, from an adjoining farm. He is 59 now and has put enough aside during the many years when the rain came on time and sugar prices were high to guarantee a comfortable retirement for himself and Betty. But as for his son, Geoff – well, it had always been taken for granted that he would be the fifth Sorensen to grow cane here. Now that is not such a certainty any more.

The life of a cane farmer is all Geoff knows and, despite the long hours of hard work in the fields, it is a good life. The financial rewards – until a few years ago at least – were substantial and reasonably assured. Sugar is still the backbone of coastal Queensland's agricultural economy, and the industry is proud of its reputation as the most efficient in the world – in both growing and milling the cane.

Queenstown ***New Zealand*** Tourist resort on Lake Wakatipu, about 160 km (100 miles) north-west of Dunedin on South Island. It looks across to the 2286 m (7500 ft) mountains of the Remarkables range. One of the country's finest ski resorts – Coronet Peak – lies to the north.
Population 3370
Map New Zealand Bf

Quelimane (Quilimane; Kilimane; Kilmain) ***Mozambique*** Seaport on the mouth of the Quelimane river, about 300 km (185 miles) north-east of the port of Beira. Its coconut plantation, one of the world's largest, covers 202 km² (78 sq miles). A notorious slave port in the 18th and 19th centuries, Quelimane now exports copra (a source of coconut oil), cotton, maize, sisal (a fibre used for ropes), sugar, tea and tobacco. It is the capital and only major port of Zambézia province, which covers 105 008 km² (40 533 sq miles).
Population (port) 71 800; (province) 2 600 200
Map Mozambique Ab

Quelpart ***South Korea*** See CHEJU DO

Quemoy ***Taiwan*** See CHIN-MEN TAO

Quercy ***France*** Former province, 5215 km² (2013 sq miles) in area, on the south-west of the Massif Central. Its limestone plateaus are cut by steep-sided valleys. The main activities are arable farming in the wider valleys, vine growing on the sunny slopes, and grazing and forestry on the plateaus. Cahors is the main town.
Map France Dd

Querétaro ***Mexico*** **1.** Small state on Mexico's central plateau. In the north, the mountain slopes are used for cattle grazing, and cotton, maize, coffee and beans are grown in the fertile valleys. The lower-lying south has many industries encouraged by a government plan to establish industry outside the Federal District around Mexico City. Gemstones, such as agates, opals, amethysts and topaz, are mined.

2. Capital of Querétaro state, 175 km (110 miles) north-west of Mexico City. It was founded by the Otomí and Chichimeca Indians in 1445 and was taken by the Spanish in 1532. It produces cotton textiles and pottery, and processes crops from the surrounding area. Its cotton factories are among the most important in Mexico, and new industries, encouraged by the government, have increased its population five-fold in the last 20 years. Emperor Maximilian was tried and shot here in 1867. The city has a magnificent 18th-century aqueduct, a 16th-century cathedral, and 55 churches.
Population (state) 802 000; (city) 167 000
Map Mexico Bb

Quetta ***Pakistan*** Capital of Baluchistan province, about 80 km (50 miles) from the Afghan border. The name means 'fort' and the town, tucked in the arid hills between the Bolan and Khojak passes, was founded to control the route between the Indus valley and the Afghan city of Kandahar. The British built a military college in Quetta which is still used by the Pakistani army. The town was destroyed by an earthquake in 1935, but has since been rebuilt.
Population 285 000
Map Pakistan Cb

Quetzaltenango ***Guatemala*** Second largest city in the country after Guatemala City, and about 110 km (70 miles) west of the capital, in the heart of a rich agricultural area. It is 2335 m (7660 ft) above sea level on the slopes of Santa Maria volcano (3768 m, 12 362 ft), which destroyed it in 1902. The city's name means 'Place of the Quetzal', after the rare jungle bird worshipped by the Mayas. Quetzal is also the name of the national currency. The city has two universities, and textile and brewing industries.
Population 96 100
Map Guatemala Ab

Quezon ***Philippines*** Province occupying the narrow isthmus that connects the Bicol peninsula with the rest of LUZON; it includes the isolated Bondoc peninsula. There are plans to develop Infanta, on the Pacific coast in the north-east, as a port and overflow city for MANILA, but most of Quezon remains remarkably undeveloped for an area so close to the capital. This is largely because much of it is mountainous and forested; part is a national park. The main agricultural product is coconuts.
Population 1 116 000

Quezon City ***Philippines*** Administrative capital of the country from 1948 to 1976, but now merely a major suburb of Metro MANILA. It contains many embassies and the huge University of the Philippines.
Population 1 165 000
Map Philippines Bc

quicksand Bed of loose, round-grained sand saturated with water. The water acts as a lubricant and the round grains reduce friction, causing the mixture to behave as a liquid and swallow up any heavy object coming in contact with its surface.

Quilimane ***Mozambique*** See QUELIMANE

Quimper ***France*** Capital of Finistère department at the end of the Brittany peninsula. It manufactures pottery, known as Quimper or Brittany ware. The other main industries are sardine fishing and tourism. In July a festival of Breton folklore takes place, with bagpipes, folk dancing and music.
Population 60 200
Map France Ab

Quintana Roo ***Mexico*** Mexico's newest state, created in 1974 on the north-east coast of the Yucatán peninsula, an area of swampy tropical lowland and white beaches. The capital, Chetumal, a free port, lies in the extreme south near the Belize border. But the real centre of Quintana Roo is the purpose-built Caribbean resort of CANCUN in the north.
Population 256 000
Map Mexico Dc

Quito ***Ecuador*** National capital and capital of Pichincha province, lying almost on the Equator 2820 m (9250 ft) up in a narrow valley surrounded by mountains. It was founded in 1534 by the Spanish conquistador Sebastián de Belalcázar on the site of an Indian town of the same name. In 1822 the South American liberator Antonio José de Sucre defeated the Spanish on Pichincha volcano just to the north-west, and gained Ecuador's independence. The cathedral in the central square contains Sucre's tomb.

One of the country's two major industrial

▼ IDYLLIC LAKELAND Queenstown nestles beside Lake Wakatipu at the foot of Queenstown Hill – a superb viewpoint. Sheep graze the green hills dotted with lakes and farmsteads.

centres (the other being Guayaquil), Quito manufactures textiles, clothing, footwear and pharmaceuticals. It also produces handicrafts from leather, wood, gold and silver. Its churches and old mansions contain some of the finest art treasures in South America. Among them are the church and monastery of San Francisco and the Jesuit church of La Compañia. There are two universities and an international airport.

Population 1 110 000
Map Ecuador Bb

Qum ***Iran*** See QOM

Qwaqwa ***South Africa*** Self-governing black state covering 482 km² (186 sq miles). It is a former Bantu homeland once called Witsieshoek, bordering north-east Lesotho. It is occupied by the South Sotho people. The capital is Phuthaditjhaba.

Population 305 000
Map South Africa Cb

Rab (Arbe) ***Yugoslavia*** Resort island of 93 km² (36 sq miles) in the Adriatic off the north Croatian coast, and noted for its marble quarries and silk industry. The town of the same name on the south coast is a former Venetian fortified settlement dating from the 15th century; it has a fine cathedral, several palaces built by wealthy Venetians and four Romanesque bell towers.

Map Yugoslavia Bb

Rabat ***Malta*** Town 10 km (6 miles) west of the capital, Valletta. In Roman times, Rabat and the adjoining smaller Mdina formed the capital city, then called Melita, where St Paul landed after his shipwreck (Acts 28:1). He is supposed to have lived in a grotto adjoining the church named after him. This was the first Maltese church to have been built on the grand scale following the prosperity brought by the Knights of St John, who made the island their headquarters in 1529. Many other churches in Rabat are built over or within the early Christian catacombs.

Population 12 000
Map Malta Db

Rabat ***Morocco*** National capital of the country, Rabat stands on the site of a 10th-century stronghold, on the Atlantic at the mouth of the Bou Regreg river. Founded in the 12th century, it remained small and unimportant until 1913 when the French made it the capital in place of FES. The old city is surrounded by 12th-century battlements, and has many souks (bazaars) offering a range of handicrafts. It also has a 12th-century mosque, and the Oulayas Kasbah, a citadel which looks out over the river to the ancient corsair town of SALE. The new city is spacious with many good hotels and restaurants; there are fine beaches close by. When the king is in residence at the royal palace there is an impressive procession when he goes to the mosque to pray. Outside the walls is Chella, the site of a former Roman town, which the Arabs turned into a cemetery in the 14th century. An ancient ruined mosque stands over the tombs, which are beautifully carved in Arabic script and surrounded by gardens.

Population 520 000
Map Morocco Ba

Rabaul ***Papua New Guinea*** Port and capital of East New Britain province. It was founded in 1910 at the northern tip of the island of New Britain as a German colonial headquarters, but was destroyed by Allied bombing after the Japanese occupied it during the Second World War. It is now the country's biggest town off the New Guinea mainland, and its industries include the manufacture of furniture and building materials. Coconut products and cocoa are the main exports.

Two active volcanoes overlook the town, and its magnificent harbour is a sea-filled crater. A violent eruption in 1937 caused Rabaul to be evacuated, and experts warned of a repeat performance in 1984; people fled but gradually returned when no eruption came.

Population 30 000
Map Pacific Ocean Ca

race **1.** Strong, swift flow of seawater through a restricted channel, produced by marked tidal differences at either end of the channel. **2.** Channel created by tapping a stream and leading the water to drive the wheel of a water mill.

Raciborz (Ratibor) ***Poland*** Industrial city near the Czech border, 65 km (40 miles) south-west of the city of Katowice. It was the capital of an independent Silesian duchy from 1172 to 1336, and is now a busy rail junction, with factories producing metals, electrical goods and foodstuffs.

Population 58 800
Map Poland Cc

radiation fog Fog formed inland by radiation cooling of the earth's surface, which in turn cools the air near the ground to below dew point. Favourable conditions for the formation of radiation fog are a long night with clear skies, moist air and light winds.

radiogenic heat Heat produced in the earth's CRUST and upper MANTLE by radioactive decay of minerals.

radiolarian ooze Sediment of red clay and organic remains, primarily skeletons of small organisms called Radiolaria, found on the ocean bed in deep tropical seas, particularly the mid-Pacific and Indian oceans. See OOZE

Radnor ***United Kingdom*** See POWYS

Radom ***Poland*** Industrial city, 98 km (61 miles) south of the capital, Warsaw. It flourished in the 14th and 15th centuries as a cloth-making centre at the crossroads of two major trade routes – Greater Poland to Silesia, and Lithuania to the Ukraine. Today Radom is still an important manufacturing centre, lying on a railway junction, with engineering and machine-building industries. It also makes fruit and vegetable products, glass, textiles, chemicals, electrical and leather goods.

Population 201 100
Map Poland Dc

rag, ragstone Hard, coarse, rubbly sandstone, such as the Kentish Rag, widely used as a stone for building.

Ragang ***Philippines*** See LANAO

Ragusa ***Italy*** Town in the south-east corner of Sicily. Its economy is traditionally agricultural, but there are now oil, asphalt and chemical plants.

Population 67 300
Map Italy Ef

Ragusa ***Yugoslavia*** See DUBROVNIK

rain forest See TEMPERATE RAIN FOREST; TROPICAL RAIN FOREST

rain shadow Leeside of a mountain or hill range where precipitation is low, such as east of the Rockies in North America, or eastern Britain.

rain spell A period of 15 consecutive days each with a rainfall of at least 0.2 mm (0.01 in).

rainbow An arc of colours – red, orange, yellow, green, blue, indigo, violet – appearing in the sky opposite the sun. It is seen when sunlight falls on rain and is caused by the reflection and refraction of the light in the water droplets which act as prisms. Sometimes some of the sunlight falling on the water droplets is reflected twice and a second rainbow is formed, less distinct than the first and with the colours in the reverse order. In the primary rainbow, red is on the outer side of the arc and violet on the inside.

rainfall Total quantity of water deposited on a given area in a given time, as measured on a rain gauge. It may be condensed or precipitated as rain, snow, hail, dew, hoar frost, rime or sleet. (See also chart under WEATHER.)

raised beach Wave-cut platform backed by old cliffs that has been raised above the shoreline by earth movements, or left behind by a fall in sea level.

Rajang ***Malaysia*** The country's longest river. It flows 564 km (350 miles) from the hills of central Borneo westwards through Sarawak's swampy forest land into the South China Sea west of the town of Sibu.

Map Malaysia Db

Rajasthan ***India*** North-western state covering 342 239 km² (132 104 sq miles) from the Aravalli Range to the Pakistani border in the Thar desert. Its name comes from the Rajput tribes who settled there in the 9th century. They founded various noble lineages and the area contained many princely states during the British Raj. The state is noted for its arts and crafts, particularly its paintings, many of which are miniatures in Mogul style and others are painted on silk.

Much of the state is arid, and contains one of the world's longest irrigation canals which is being extended in an attempt to make the desert arable. The Rajasthan Canal flows for 216 km (134 miles) across Punjab without distributing water, then has a working length of 469 km (291 miles), branches of 837 km (520 miles), and a total length of minor distributaries of 5474 km (3400 miles). The state capital is the city of Jaipur.

Population 34 261 900
Map India Bb

Rajshahi ***Bangladesh*** Chief town of the Rajshahi district, which covers 9456 km² (3651 sq miles) in western Bangladesh. It stands beside the Ganges river and is home of a new university and centre of the silk industry.

Population (district) 5 270 000; (town) 254 000
Map Bangladesh Bb

Raleigh ***USA*** State capital of North Carolina. It lies in the centre of the state, about 185 km (115 miles) from the Atlantic Ocean. The city was founded in 1792 and named in honour of the English adventurer Sir Walter Raleigh, who had helped open up the New World two centuries earlier. It stands in the heart of tobacco-growing country.

Population (city) 169 300; (metropolitan area) 609 300
Map United States Kc

Ramat Gan ***Israel*** Industrial town to the east of Tel Aviv, of which it is a suburb. Founded in 1921, it has a diamond exchange and produces textiles, processed foods and furniture.

Population 116 500
Map Israel Aa

Ramillies ***Belgium*** Village about 50 km (30 miles) south-east of the capital, Brussels, where in 1706 the British under the Duke of Marlborough defeated the French, led by François de Neufville, Duc de Villeroi (1644-1730), during the War of the Spanish Succession.

Population 4200
Map Belgium Ba

Rancagua ***Chile*** Capital of Libertador region, lying in a fertile agricultural area in the Central Valley, about 80 km (50 miles) south of Santiago. It was the scene in 1814 of an important battle between the Spaniards and troops led by the country's liberator, Bernardo O'Higgins.

Population 143 000
Map Chile Ac

Rance ***France*** River, 100 km (60 miles) long, in northern Brittany. It drains into the Gulf of Saint-Malo via a 19 km (12 mile) long estuary. The world's first tidal power station was completed on the estuary in 1968. A 750 m (2460 ft) long dam was built across the river to take advantage of the unusually large rise and fall in the estuary's tides. Special turbines in the dam can tap the power of both the incoming and outgoing water.

Map France Bb

Ranchi ***India*** Industrial town and former summer capital of Bihar state, on the Chota Nagpur plateau about 320 km (200 miles) west and slightly north of the north-eastern city of Calcutta. Its heavy engineering products include machine tools, and it has a university.

Population 502 800
Map India Dc

rand In South Africa, a ridge of hills or an extended area of high ground, often covered with scrub.

Rand, The ***South Africa*** See WITWATERSRAND

Randfontein ***South Africa*** Westernmost town on the Witwatersrand. It is a gold-mining town, founded in 1890, and once part of the town of Krugersdorp. It became a separate municipality in 1929.

Population 49 000
Map South Africa Cb

Randstad ***Netherlands*** See HOLLAND

range 1. An extended group of mountains or hills. **2.** In America, an extensive area of land where animals, usually cattle, roam in search of food.

Rangoon ***Burma*** Port and national capital beside the Rangoon river, just east of the Irrawaddy river delta; it is also the centre of a 10 171 km² (3927 sq mile) division of the same name.

The city was developed as a port to replace Syriam by the 18th-century Burmese ruler King Alaungpaya, and became the administrative centre of Lower Burma in 1852 and the capital of government and the main outlet for rice exports from the Irrawaddy delta. Since independence in 1948, however, Burma's prosperity has dwindled and with it the money available to modernise the port and maintain the city's buildings. A handful of mosques and Hindu temples remain in the northern part of the city – legacies of the Indians who handled much of the administration under British rule – but the Indians themselves have mostly left since 1948.

The commercial hub of the city today is the Bogyoke Aung San market (named after Aung San, a prime minister under the British who was assassinated just before independence). As well as food and clothes, intricately painted lacquerware is on sale, along with gold and silver, and rubies and jade from Burmese mines. Near the Shwe Dagon pagoda are the Royal Lakes, on whose shores are the villas of the country's political and military elite.

Population (city) 2 459 000; (division) 3 974 000
Map Burma Cc

▲ TRANQUIL CITY A ferryman crosses a quiet inlet in Rangoon. The far-flung city began as a fishing village in ancient times, and its development as a port after 1755 and later as a busy capital have hardly left their mark.

in 1886. Its name may come from the Burmese word *yangun*, meaning 'end of strife'. Some scholars, though, believe that the name is linked with Dagon, the god whose enormous 99 m (325 ft) high Shwe Dagon pagoda dominates the city. The pagoda – a funnel-shaped solid brick spire built in the 6th century and now the centre of a marble-floored Buddhist temple complex – is covered with 8868 thin sheets of gold. At modern prices each sheet is worth about US$6500.

The city was redeveloped with a grid layout after Britain colonised southern Burma in 1824, and became a spacious colonial capital, the seat

Ranong ***Thailand*** Province and town on the Andaman Sea. Both lie beside the southern tip of Burma, about 475 km (295 miles) south-west of the capital, Bangkok. The province's main industry is tin mining, and many of the mine-workers are Burmese immigrants.

Population (town) 50 100; (province) 96 600
Map Thailand Ad

Rapa Nui ***Pacific Ocean*** See EASTER ISLAND

Rapallo ***Italy*** Seaport and tourist resort about 30 km (20 miles) south-east of Genoa. It developed from a fishing village to a resort in the late 19th century. It produces olive oil and wine.

Population 29 300
Map Italy Bb

rapids Fast-flowing rocky section of a river, usually caused by a steepening of its bed.

Rarotonga ***Cook Islands*** Largest and most important island of the group, covering 67 km^2 (26 sq miles) and lying about 2750 km (1700 miles) north-east of New Zealand. It is mountainous, rising to several sharp peaks; Te Manga (652 m, 2139 ft) is the highest. Fruit and vegetable crops are grown on a narrow coastal plain; the main industry is a fruit and fruit juice canning factory. Avarua, the chief town and capital of the Cook Islands, is on the north coast; it retains much of the atmosphere of a 19th-century South Seas trading post. The single-storey Cook Islands parliament building is on the outskirts, near the airport; by law, no building on Rarotonga may be higher than a coconut palm.
Population 9500
Map Pacific Ocean Fd

Ras al Khaymah ***United Arab Emirates*** The most north-easterly of the small emirates, covering 1036 km^2 (400 sq miles) on the west coast of the MUSANDAM peninsula, next to the Strait of Hormuz. Its main industries are fishing and agriculture, and it provides about half of the United Arab Emirates' supply of fresh food. The growing town of Ras al Khaymah is divided in two by a lagoon.
Population (emirate) 83 000; (town) 5300
Map United Arab Emirates Ba

Ras Dashen ***Ethiopia*** Highest peak in the country, rising to 4620 m (15 160 ft) in the Semien Mountains of the north.
Map Ethiopia Aa

Ras Lanuf ***Libya*** One of the new oil terminals built in the 1960s on the Mediterranean coast to handle oil from fields to the south and east, which are connected to it by pipelines. A new town and industrial centre are under construction. It is situated on the Gulf of Sirte.
Map Libya Ba

Ras Tanura ***Saudi Arabia*** Port and major oil terminal on The Gulf, situated 40 km (25 miles) north of Dammam. Ras (Cape) Tanura was built on a peninsula shortly before the Second World War and is the site of the country's first oil refinery, which came on stream in 1946. It ships oil from the Dharan, Abqaiq and Qatif fields; there is also a gas processing plant.
Map Saudi Arabia Cb

Rashid (Rosetta) ***Egypt*** Town and port on the west bank of the Rashid (Rosetta), the main western river of the Nile delta, about 55 km (35 miles) north-east of the port of Alexandria. In 1799, one of Napoleon Bonaparte's soldiers, digging a trench near the town, found a black basalt slab about 1 m^2 (1 sq yard). The slab, now known as the Rosetta Stone, was taken to Britain in 1801, and is now in the British Museum in London. It is inscribed with a decree of Ptolemy V of 196 BC in Greek, Egyptian hieroglyphics and demotic characters, and in 1822 the brilliant French Egyptologist Jean François Champollion used it as the key to decipher the hieroglyphics.
Population 43 000
Map Egypt Bb

Rasht (Resht) ***Iran*** Industrial city near the Caspian Sea, about 250 km (156 miles) north-west of the capital, Tehran. Its chief manufactures are textiles, hosiery and carpets. The surrounding province of Gilan, many of whose inhabitants are Armenian, produces silk, rice and tea.
Population (province) 1 581 900; (city) 259 700
Map Iran Aa

Rastenburg ***Poland*** See KETRZYN

Ratibor ***Poland*** See RACIBORZ

Ratisbon ***West Germany*** See REGENSBURG

Ratnapura ***Sri Lanka*** Gem-mining town on the south-west flank of the central mountains, 65 km (40 miles) south-east of Colombo. All around the town are deep pits, where soil is excavated, washed and sieved for sapphires, rubies and cat's-eyes. There is a gem museum in the town, which is overlooked by a large statue of the Buddha.
Population 37 400
Map Sri Lanka Bb

Rauma (Raumo) ***Finland*** Town founded in 1442 on the Gulf of Bothnia, about 215 km (135 miles) north-west of the capital, Helsinki. Its industrial plants include shipyards which specialise in making large riverboats for export to the USSR. The older part of the town still has fine old frame and timber dwellings lining the streets.
Population 30 800
Map Finland Bc

Ravenna ***Italy*** Historic city about 115 km (70 miles) south of Venice, connected to the Adriatic Sea by a canal 10 km (6 miles) long. It was home of the Roman Adriatic fleet and served as capital of the Western Roman Empire from AD 402, after the fall of Rome. It was the Ostragoth capital in the 5th and 6th centuries, later the Byzantine capital of Italy.

Ravenna is noted for its rich array of Roman and Byzantine remains, including several well-preserved churches founded by the Romans in the 5th century, many containing fine mosaics. There is also the tomb of the Ostragoth king Theodoric (455-526), which stands in a ten-sided stone building roofed with a 250 tonne block of solid stone. About the beginning of the 6th century the banker Julianus Argentarius built the baptistry and church of San Vitale which, with its superb mosaics, is Ravenna's most renowned monument.

The poet Dante died in Ravenna in 1321 at the age of 56, and his remains now lie in a small 18th-century temple with a Dante museum nearby.

Industry came to Ravenna in the 1950s when natural gas was discovered in the area. A new canal port has been built and there are huge refining and petrochemicals complexes.
Population 136 500
Map Italy Db

Ravensbrück ***East Germany*** Site of a Nazi concentration camp for women, about 80 km (50 miles) north of the capital, East Berlin.
Map East Germany Cb

ravine Narrow, steep-sided depression, smaller than a valley but larger than a cleft or gully.

Rawalpindi ***Pakistan*** Ancient town in the northern Punjab on a trans-Indian highway known in the days of the British Raj as the Grand Trunk Road. The town was the military headquarters for British control of the North-West Frontier Province and is now the headquarters of the Pakistani army. It has some engineering industry.
Population 928 000
Map Pakistan Da

Rayong ***Thailand*** Province and fishing port on the Gulf of Thailand, about 145 km (90 miles) south-east of the capital, Bangkok. The town has developed rapidly since 1980 as a gas processing centre for offshore gas fields. And the province's farmers largely specialise in growing cassava – a tuberous food plant.
Population (town) 129 100; (province) 410 600
Map Thailand Bc

Ré, Ile de ***France*** Flat island, 28 km by 5 km (17 miles by 3 miles), just west of La Rochelle on the Bay of Biscay. It is noted for its beaches, oyster beds, vineyards and market gardens.
Population 11 400
Map France Cc

reach Unbroken expanse of water, especially on a river or canal.

Reading ***United Kingdom*** County town of Berkshire, England, on the River Thames 60 km (37 miles) west of London. It has a university and a wide range of industries, including electronics and engineering. The town's prison was made famous in *The Ballad of Reading Gaol* by the Irish writer Oscar Wilde, composed in 1898 after he had served two years' hard labour there for homosexual offences.
Population 139 000
Map United Kingdom Ee

Recent An alternative name for HOLOCENE period. See GEOLOGICAL TIMESCALE

Rechna Doab ***Pakistan*** Region between the Ravi and Chenab rivers in the Punjab. Its population grew rapidly after the building of irrigation canals in the 19th century, and it remains the core of the country's agriculture, growing wheat, cotton and rice.
Map Pakistan Db

Recife ***Brazil*** Capital of Pernambuco state near the country's north-eastern tip. It is the nearest of South America's major cities to Europe and Africa. The city, which is also a tourist resort, is crisscrossed by canals and the channels of the Capiberibe river.

The surrounding state, which covers 98 281 km^2 (37 936 sq miles), was one of the earliest areas settled by the Portuguese. The original Atlantic rain forest of the coast around Recife provided the brazilwood tree which gave its name to the country. Once the trees were felled for their highly prized red dye, brazilin. The land is ideal for sugar cane, now the region's chief crop.
Population (city) 1 205 000; (state) 6 662 000
Map Brazil Eb

red earth Tropical soil, produced by intensive chemical weathering under conditions of high temperatures and humidity, and markedly seasonal rainfall. The result is a loamy mixture of clay and quartz coloured by iron compounds.

red mud Mud derived from erosion of the land surface, and found on the continental slope near the mouth of certain rivers. The colour is due to ferric oxide.

Red River ***Canada/USA*** River, 877 km (545 miles) long, which rises in the USA in Lake Traverse on the South Dakota-Minnesota border and flows north into Lake Winnipeg in south Manitoba. Its valley, first settled in 1812, is one of Canada's best farming areas.
Map Canada Fd

Red River (Song Hong; Yuan Jiang) ***China/Vietnam*** River that rises in the Chinese province of Yunnan and flows for about 1200 km (750 miles) past the Vietnamese capital, Hanoi, and into the Gulf of Tongking through a delta up to 80 km (50 miles) wide. The river carries into the delta thousands of tonnes of the red soil that gives it its name, flooding surrounding flatlands with a rich silt. In parts, it runs between banked-up levees with the water level as much as 12 m (40 ft) above the plain. This makes irrigation simple because gravity takes the water to the fields. However, flooding, when it happens, is severe.
Map Vietnam Aa

Red Sea Branch of the INDIAN OCEAN between north-east Africa and the Arabian Peninsula, occupying part of the GREAT RIFT VALLEY system. It owes its name to a species of alga which imparts a red bloom to the water. OCEAN FLOOR SPREADING is occurring along a valley in the centre of the sea and this is pushing the Arabian plate north-eastwards. At the present rate of expansion, the Red Sea will be as wide as the Atlantic in 200 million years.

The Red Sea is a hot and humid region. Water temperatures in summer reach 29°C (84°F). In 1963 rich deposits of zinc, copper and other minerals were discovered on the seabed in the north and central parts and these are likely to be exploited in the near future.

The sea was a major shipping lane in antiquity, but its importance declined after the discovery of a sea route to India around southern Africa. However, after the opening of the SUEZ CANAL in 1869, it once again became a major transit route.
Area 438 000 km² (169 112 sq miles)
Greatest depth 2246 m (7370 ft)
Map Saudi Arabia Ab

Redwood National Park ***USA*** Forest covering 424 km² (164 sq miles) along the Pacific coast in the extreme north of California between Eureka and Crescent City. The park was created in 1968 to preserve one of the few remaining areas of giant redwood trees.
Map United States Bb

reef Strip or ridge of rocks, sand or soil rising from the seabed and extending to just below the surface of the water, and sometimes exposed at low tide. See also CORAL REEF

re-entrant Marked indentation into an upland area or into a steep slope.

reg Flat, stony desert, especially in the Sahara, where sheets of gravel and pebbles cover the surface.

Regensburg (Ratisbon) ***West Germany*** River port and tourist centre on the Danube about 110 km (68 miles) north-east of Munich. It stands as a living museum of German architecture from the Middle Ages to the 20th century, with more than 20 Gothic or Romanesque churches and a 13th-century cathedral. The first German parliament, or Diet, met in the town hall in 1663, and Diets were held there until 1806, when Napoleon, the French emperor, abolished them. It was while storming the walls of the town in 1809 that Napoleon suffered the second war wound of his career – a spent musket ball bruised his right foot.
Population 129 000
Map West Germany Ed

Reggio di Calabria ***Italy*** Seaport and tourist resort at the end of the toe of Italy, founded in the 8th century BC. It was rebuilt after being destroyed by an earthquake in 1908. Its industries include olive oil, pasta and fruit canning.
Population 177 700
Map Italy Ee

Reggio nell'Emilia ***Italy*** Town about 60 km (37 miles) north-west of Bologna. It was a Roman settlement and the Roman Via Emilia (the road from Piacenza to Rimini, built in 187 BC) runs through the heart of the city. Reggio is the birthplace of the green, white and red Italian flag, first flown in the Palazzo Comunale in 1797. Its industries include electrical equipment, cement and canned meat.
Population 130 300
Map Italy Cb

Regina ***Canada*** Capital of SASKATCHEWAN and centre of one of the world's largest wheat growing plains. The city has been extensively landscaped and the provincial government buildings are set in 920 hectares (2270 acres) of lakes and woods. Regina was founded in 1882 on the Canadian Pacific Railway as it progressed westwards. It was headquarters of the Northwest Mounted Police from 1892 to 1920 when it became the western headquarters of the renamed Royal Canadian Mounted Police. Today it has industries processing oil and gas, and making steel, chemicals, electrical equipment and dairy products.
Population 164 000
Map Canada Ec

regolith Layer of loose material, including soil, sand, rock fragments, volcanic ash and glacial DRIFT, which covers the solid bedrock, and which is subject to weathering.

Rehovot (Rehovoth) ***Israel*** Town on the Judaean plain 21 km (13 miles) south-east of Tel Aviv. It is the centre of an area that grows citrus fruit and produces metal goods, glass, pharmaceuticals, fruit juice and dairy products. The tomb of Chaim Weizmann, Israel's first president (1949-52), and the Weizmann Institute of Science are at Rehovot, as well as the Faculty of Agriculture of the Hebrew University.
Population 70 000
Map Israel Ab

Reichenbach ***Poland*** See DZIERZONIOW

Reichenberg ***Czechoslovakia*** See LIBEREC

Reims ***France*** See RHEIMS

Reinga, Cape ***New Zealand*** Headland at the northern tip of North Island. It is the place from where, according to Maori folklore, the souls of the dead depart on their journey back to their legendary Pacific homeland, Hawaiki.
Map New Zealand Da

rejuvenation Renewed down-cutting of a river caused by a lowering of sea level or an uplift of the land through which the river flows.

relative humidity See HUMIDITY

relief The shape of the earth's surface – the variations in its height and steepness of slopes.

Renfrew ***United Kingdom*** See STRATHCLYDE

Rennes ***France*** Industrial city and capital of Ille-et-Vilaine department, about 100 km (60 miles) north of Nantes. It was severely damaged by fire in 1720 and rebuilt with spacious squares and elegant houses. The 17th-century parliament house was restored. Its industries, several of which have moved to the city from Paris, include car assembly, electrical goods, clothing and chemicals.
Population 241 300
Map France Cb

Reno ***USA*** Gambling resort in western Nevada, about 16 km (10 miles) from the California state border. It is a tourist centre with facilities for quick and easy marriages and divorces, and is a base for exploring the surrounding mountains and lakes.
Population (city) 105 600; (metropolitan area) 211 500
Map United States Cc

Resht ***Iran*** See RASHT

Réthimnon ***Greece*** Town on the north coast of Crete 60 km (37 miles) from Heraklion. Many buildings, including the loggia of the old governor's palace and the church of San Francesco, remain from its days as a Venetian provincial capital. The monastery of Arcadi lies 23 km (14 miles) to the south. In 1866, several hundred Cretans who had fled there to escape invading Turks, blew themselves up with explosives rather than be taken prisoner.
Population 17 700
Map Greece Dd

Réunion ***Indian Ocean*** East of Madagascar, in the south-west of the Indian Ocean, lies a small piece of France. The island of Réunion, formerly Bourbon, has been a French overseas department (administered on the same basis as departments in mainland France) since 1946, although it was first colonised in 1638 and belonged for a while – 1810-15 – to the British. Control of Réunion is expensive, and some argue that it brings France little reward; the inhabitants, too, are divided about the potential advantages of political independence. The present arrangement at least gives the young people

▲ **OLD TOWN, YOUNG WINES The lovely town of Bacharach, founded before 900, stands in the Rhine gorge below Stahleck Castle. The vineyards around are noted for their pungent, full-bodied wines, best drunk while they are young.**

access to jobs in France; unemployment is a growing problem in Réunion.

The island covers 2512 km² (970 sq miles). It is volcanic like its nearest neighbour, MAURITIUS, and has one active volcano and nine dormant cones. There are two mountain masses with peaks above 3000 m (10 000 ft), providing spectacular scenery. They are divided by a depression and skirted by coastal plains.

As in Mauritius, the economy is based on sugar, which is grown on nearly 70 per cent of the cultivated land. This and its products, molasses and rum, are the main exports through the chief port and capital city of SAINT-DENIS. Other farm products are vanilla, plants grown for perfume oil, tea, maize and potatoes. Livestock is raised on the high pastures.

Tourists are attracted by the tropical climate and superb beaches with their offshore coral reefs, but the remoteness of the island limits the tourism potential. About a quarter of the population is of French origin. Others are a mixture of Indian descendants of the workers once brought to farm the plantations (who retain their Hindu culture), Malays, Indo-Chinese, and the Creole descendants of French settlers and African slaves.

Population 530 000

Map Indian Ocean Bd

Reutlingen ***West Germany*** Textile-making city amid the vineyards of the Neckar valley, about 30 km (19 miles) south of Stuttgart. Parts of its medieval walls and gates, and a 13th-century church, remain.

Population 96 300

Map West Germany Cd

Reval (Revel) ***USSR*** See TALLINN

Reykjanes ***Iceland*** Peninsula in the south-west of the island. The town of KEFLAVIK is on the peninsula's northern side. Hot springs on the peninsula are used to provide domestic heating.

Map Iceland Bb

Reykjavík ***Iceland*** Capital and principal port in the south-west of the island on Faxaflói bay. According to tradition, the city was founded by Ingolfr Arnarsson, a Viking settler, in AD 874. A cathedral was built in the 18th century and the city became the seat of government in 1843. A university was founded in 1911.

Industries include fishing, fish processing, metal working, clothing manufacture and printing. Nearby there are cement and fertiliser plants. The port handles about 70 per cent of Iceland's trade. Its airport is at KEFLAVIK.

Many of the city's buildings are heated by water piped from nearby volcanic hot springs. An open-air swimming pool in the city is heated so effectively by this method that it remains in use all year round. The steam heat has also given the city its name: Reykjavík means 'smoking bay'.

Population 87 300

Map Iceland Bb

Reynosa ***Mexico*** Dusty town on the Río Bravo (Rio Grande) in the eastern state of Tamaulipas on the border with Texas. Founded by the Spanish in 1749, it burgeoned in the 20th century with the discovery of oil and natural gas. Now the Gulf Plain around Reynosa produces much of Mexico's gas. Reynosa has become a major centre for commerce, textiles, agriculture and stock breeding.

Population 347 000

Map Mexico Cb

Rharb (Gharb) ***Morocco*** Fertile plain north of Rabat extending for about 80 km (50 miles) along the Atlantic coast and about 110 km (70 miles) inland. Crossed by the Oued (river) Sebou, it is an important agricultural region producing wheat and citrus fruits. During French colonial times the land was greatly improved by irrigating dry areas, draining swampland and introducing modern agricultural methods. Rharb is also a source of phosphates, rock salt and petroleum.

Map Morocco Ba

Rheims (Reims) ***France*** Historic city on the Marne river, and centre of the Champagne wine industry. It is famed for its massive 13th to 14th-century cathedral, longer and higher than the cathedral of Notre-Dame in Paris, where Kings of France were crowned until 1830. In 1429, Joan of Arc, then 17 years old, led an army of 12 000 to Rheims to have the Dauphin crowned as Charles VII.

Most of the town was destroyed in the First World War, and the cathedral was not completely restored until 1938. The city's technical college contains a room which was used as the Second World War headquarters of General Dwight Eisenhower (1890-1969), the Supreme Allied Commander. The formal German surrender was signed in the room on May 7, 1945.

Population 204 000
Map France Fb

Rhine (Rhein, Rhin, Rijn) ***Western Europe*** Major European river, 1320 km (825 miles) long. It rises in the Swiss Alps, and follows a mountain course as far as Lake CONSTANCE. Its waters traverse the lake and plunge over the Rhine Falls westwards to BASLE. There the river abruptly turns to the north, and enters the Rhine rift valley. The rift, let down between parallel faults, has a broad, flat floor, with the VOSGES mountains of France rising on the west side and the BLACK FOREST massif on the east. The rift continues for 300 km (nearly 190 miles); at its end, in Hessen, the Rhine turns west, then north-west, to cut through the Middle German upland in its famous gorge. It emerges at BONN into the North German Plain, flows down past COLOGNE and the RUHR and, just after reaching the Dutch frontier, divides into three: the Waal, the most southerly and the main channel, which reaches the sea via Nijmegen; the Lek, which runs farther north, past Arnhem; and the IJssel, which flows into IJsselmeer.

The Rhine today is fully navigable up to Basle. Large-scale work was done between 1817 and 1876 to shorten the winding course, remove obstacles and rapids, reduce flooding and reclaim farmland. Since the Second World War, the greatest changes have been made by the French, between Strasbourg and Basle, with the building of the Grand Canal d'Alsace.

Freight transport on the Rhine – which exceeds 185 million tonnes a year – dwarfs that carried on all other West European waterways. The heaviest movements occur between the Ruhr industrial region and the Dutch port of Rotterdam. Iron ore, coal and oil products are principal cargoes, but so are motor cars and building stone. Inland from Rotterdam, the largest ports on the river are DUISBURG and LUDWIGSHAFEN-MANNHEIM.

To the tourist, the Rhine evokes vineyards and legends. The vineyards are in the rift valley and on the steep sides of the gorge. The legends are everywhere. The Lorelei is the best example: with a marvellous view over the Rhine gorge, this jutting rock is named after the legendary Siren whose singing lured sailors to their doom.

Unfortunately, the Rhine is heavily polluted all the way to the sea. After Dutch protests international efforts have been made to cut down the pollution levels, with some success.

Map West Germany Cd

Rhineland-Palatinate (Rheinland-Pfalz) ***West Germany*** State stretching along the west bank of the Rhine south of the capital, Bonn. It was ruled from the 10th to the 19th century by the Counts Palatinate of the Holy Roman Empire. Its wooded hills include the wine-growing region of the lower Moselle. MAINZ is the capital.

Population 3 600 000
Map West Germany Bc

rhinn In Scotland, a rugged ridge, such as the Rhinns of Kells.

Rhode Island ***USA*** The country's smallest state, covering 3144 km² (1214 sq miles) on the Atlantic coast between Connecticut and Massachusetts. It was founded in 1636 by nonconformist settlers and was one of the original 13 states. Agriculture centres on potatoes, maize, dairy produce and eggs, and the principal manufactures are machinery, textiles and metals. Fishing and tourism thrive along the coast.

The state's capital is PROVIDENCE, and it contains the yachting centre of NEWPORT on the island of Rhode Island.

Population 968 000
Map United States Lb

Rhodes (Ródhos) ***Greece*** Largest of the DODECANESE islands (1399 km², 540 sq miles) in the eastern Aegean and the most developed of the Greek island tourist centres. Rhodes was settled in about 1400 BC and became a successful maritime trading nation, forming the first rules of the sea, the Rhodian Laws, in about 900 BC. In 305 BC, after beating off an attack by Macedonians, the islanders erected one of the seven wonders of the world, a 32 m (105 ft) high statue of the sun god Helios at Rhodes harbour. The Colossus of Rhodes, as it was called, crumbled in an earthquake in 224 BC.

The Knights of St John, a multinational European military society who had their headquarters in Rhodes from the 13th to the 16th century, made the island a place of pilgrimage before moving to Malta. The Italianate Grand Master's Palace and the Gothic Governor's Palace still stand near the Street of the Knights. Today Rhodes has many modern hotels. It produces wine, and a festival takes place at Rodini between July and September.

Population 88 500
Map Greece Ec

Rhodope Mountains (Rodopi Planina) ***Bulgaria*** Forested mountain system, extending 290 km (180 miles) south-east from the Maritsa river to the Greek border. During the Roman Empire the Rhodopes marked the boundary between Thrace and Macedonia.

Map Bulgaria Ac

Rhône ***Switzerland/France*** River, 812 km (505 miles) long, 522 km (324 miles) of which are in France. It rises in the Swiss Alps, flows through Lake Geneva and enters France through the Jura mountains. It then flows south and west to Lyons where it is joined by the Saône, and on south to the Mediterranean. South of Lyons, 12 major projects built since the 1930s tap the river for hydroelectric power and irrigation, and tame it for navigation.

Map France Fd

Rhône-Alpes ***France*** Large region of eastern France, covering 43 698 km² (16 872 sq miles). It consists of the eastern fringe of the Massif Central, the valleys of the Rhône and Saône rivers, and the northern Alps. Economically it is the most developed region in the country after the area around Paris. The main city is LYONS.

Population 5 016 000
Map France Fd

rhyolite Fine-grained, acidic igneous rock consisting largely of quartz, feldspar and mica. It is the extrusive equivalent of granite.

ria Long, narrow branching inlet caused by drowning of a narrow, steep-sided river valley through a rise in sea level. The River Fal in Cornwall is typical. Examples are also found in south-west Ireland and north-west Spain. Unlike a FIORD, a ria deepens towards the sea.

Riau ***Indonesia*** Province covering the east of Sumatra opposite Singapore. It is the centre of the nation's oil industry.

Population 2 168 500
Map Indonesia Bb

Ribe ***Denmark*** One of Denmark's most picturesque market towns, in south-western JUTLAND, where Denmark's first church was built in 854. The church is gone, but the Romanesque cathedral, dating from about 1130, is one of the earliest brick-built structures in the country. Beside the small medieval port, now virtually silted up, are grassy mounds that mark the site of the Riberhus, a medieval royal castle. Ribe has Denmark's oldest provincial museum.

Population 8000
Map Denmark Bb

Riccione ***Italy*** See RIMINI

Richard Toll ***Senegal*** Market town in the Senegal river delta in the north-west. The town is named after Monsieur Richard, a French horticulturalist who founded an experimental garden (*toll* means 'garden' in the Wolo language) in 1830. Just outside the town lie the remains of a mansion built by Baron Roger, governor from 1822 to 1827.

Lying to the east of Richard Toll is an integrated sugar complex which has done much to revive the local economy. It is being expanded to produce 12 500 tonnes annually. The Delta Irrigation Scheme was opened at the head of the delta in 1947, in an attempt to increase the area's rice production and reduce imports.

Map Senegal Aa

Richard's Bay ***South Africa*** Resort and developing industrial town on a large lagoon on the Natal coast, about 160 km (100 miles) north-east of Durban. A harbour, opened in 1976, is linked to the Witwatersrand by rail and by an oil pipeline, and coal and other minerals are exported.

Population 11 800
Map South Africa Db

Richmond ***Australia*** **1.** Town in Tasmania 18 km (11 miles) north-east of Hobart. It was founded in 1824 and has many historic buildings, including St John's Church (1836-7), Australia's oldest Catholic church. **2.** Town adjacent to WINDSOR in New South Wales.

Population (1) 600; (2) 16 000
Map (1) Australia Hg

Richmond ***USA*** State capital of Virginia. It stands on the James river, about 100 km (60 miles) west of Chesapeake Bay, and is a centre for tobacco, paper making, publishing and engineering. The city, founded in 1737, was the capital of the Confederacy during the Civil War (1861-5).

Population (city) 219 100; (metropolitan area) 796 100
Map United States Kc

Richter scale Logarithmic scale ranging from 1 to 10, devised in 1935 by the American seismologist Charles Richter, and used for measuring the magnitude of earthquakes.

Rideau Canal ***Canada*** Canal 203 km (126 miles) long which links the capital, Ottawa, with Kingston, Ontario. It was built between 1826 and 1832 as a British military route secure from American attack. The canal is now used by pleasure craft.
Map Canada Hd

ridge Long, narrow, steep-sided upland.

ridge of high pressure Elongated extension of high atmospheric pressure between two areas of low pressure.

Riding Mountain ***Canada*** National park covering 2976 km² (1149 sq miles) in west Manitoba. It is an area of forested highland with several lakes and a wildlife sanctuary for bison, elks, moose, wolves, black bears and deer.
Map Canada Ec

Riesengebirge ***Czechoslovakia/Poland*** See SUDETY

Rif Mountains ***Morocco*** Coastal mountain range in the north-east of the country, extending 320 km (200 miles) from CEUTA to MELILLA, and separated from the Middle ATLAS MOUNTAINS by the TAZA Gap. The area, which is heavily eroded, with deep valleys and rugged peaks, is difficult to penetrate and has been the retreat of Berber tribesmen threatened by invaders from the 7th century to modern times. In the east, the great bulge of the mountains reaches a height of 1800 m (5900 ft); in the centre the highest point is Tidirhine, 2456 m (8058 ft).
Map Morocco Ba

rift valley Long, narrow depression in the earth's surface formed when the land sinks between two fairly parallel faults, a structural formation known as a graben. One of the most spectacular is the GREAT RIFT VALLEY, extending from the valley of the River Jordan southwards through East Africa.

Riga ***USSR*** Baltic port and capital of the Soviet republic of Latvia, 290 km (180 miles) north of the border with Poland. It was founded in 1190 and became a major trading centre in the 13th century. Riga was the capital of independent Latvia from 1918 until the Germans took the city in 1941. The Russians captured it in 1944.

The medieval old town is a charming jumble of streets, parks, river banks and a canal; its most imposing buildings are the early 13th-century Doma Cathedral, and Pils Castle, built between 1328 and 1515. Riga's main industries include marine engineering, shipbuilding, chemicals, glass, cement, textiles, electrical machinery and electronics.
Population 875 000
Map USSR Dc

Rijeka (Fiume) ***Yugoslavia*** Croatian seaport at the mouth of the Rijeka river, 129 km (80 miles) south-west of Zagreb. It was a focus of international dispute during the First World War because of its mixed Italian and Yugoslav population, and was the Free City of Fiume from 1920 to 1939. The city became Yugoslav in 1945, and grew rapidly as the country's leading port and shipbuilding centre. Its position on Bakar Bay, its good road and rail communications, and plentiful timber and hydroelectric power from the Gorski Kotar mountains to the east, have fostered engineering, paper, textile, food and oil-refining industries. Rijeka is also a tourist centre for the Adriatic islands, including Krk, and for mainland resorts such as Opatija.
Population 193 000
Map Yugoslavia Bb

Rijn ***Western Europe*** See RHINE

Rikuchu Kaigan ***Japan*** National park, 124 km² (48 sq miles) in area, on the wild Pacific coast of northern Honshu island, about 460 km (285 miles) north-east of the capital, Tokyo, with spectacular cliffs and white beaches. The region has been repeatedly devastated by tsunamis – giant waves generated by underwater earthquakes. One such wave, the Great Sanriku Tsunami of 1896, killed 23 000 people.
Map Japan Dc

rime Accumulation of ice crystals on the windward side of exposed objects, such as trees or telegraph poles. It is formed when water droplets in a fog or mist are driven by a slight wind and freeze on contact with the exposed objects.

Rimini ***Italy*** Seaside resort on the Adriatic Riviera about 155 km (95 miles) south of Venice. Medieval Rimini is preserved in the old town centre. The Malatesta Temple is a huge Renaissance mausoleum, and the town also has the oldest Roman triumphal arch still standing in Italy; it was built in 27 BC.

About 10 km (6 miles) south-east is the resort of Riccione (population 31 300), which has annual film, drama and art festivals.
Population 129 500
Map Italy Db

Rimnicu Vilcea ***Romania*** City on the River Olt, 150 km (93 miles) north-west of Bucharest. Rimnicu Vilcea stands at the neck of a wooded valley, and produces timber, leather and fur goods.
Population 78 000
Map Romania Aa

Río Bravo ***USA/Mexico*** See RIO GRANDE

Rio de Janeiro ***Brazil*** Metropolis, port and the national capital from 1763 to 1960, when the purpose-built city of Brasília took over. Rio de Janeiro – Rio for short – is one of the most dramatically sited cities in the world. The Portuguese, who first discovered the spot on New Year's Day 1502, named it 'January River' because they believed, wrongly, that the entrance to the huge Guanabara Bay – on which Rio now stands – was the mouth of a great river. The bay is dotted with rocky, palm-covered islands and around it are steeply domed mountains, some, like the celebrated Sugar Loaf, reaching to the sea. Another domed peak – Corcovado (meaning 'Hunchback'), which rises to 710 m (2330 ft) – is topped by a giant statue of Christ with arms outstretched.

The city itself is crowded onto a narrow shelf of land around the bay. Nowhere are the sharp contrasts of Brazilian wealth and poverty more apparent. The sleek beach-front suburbs of Copacabana and Ipanema – whose names have become familiar the world over through popular songs – have as their backdrop, wherever the slopes are too precipitous for normal housing, the huge *favelas*, or shanty towns, of the poor. Rio also has magnificent museums, beautiful parks and, in the Maracãna stadium, the world's largest soccer arena, with space for 205 000 people.

It is a commercial, financial and industrial centre and a major tourist attraction. Its main industries include clothing, furniture, chemicals, shipbuilding, glass, tobacco products and processed food.
Population 5 094 000
Map Brazil Dd

Rio Grande ***Guinea/Guinea-Bissau*** See CORUBAL

Rio Grande (Río Bravo) ***USA/Mexico*** River, 3078 km (1885 miles) long, rising in the San Juan Mountains in south-west Colorado, USA. It flows south-east, then south through central New Mexico, then meanders south-east forming the border between Texas, USA and Mexico. It finally empties into the Gulf of Mexico between Brownsville, USA and Matamoros, Mexico. Brownsville is the chief port. The river is dammed in several places for irrigation, flood control and regulation of flow. It is unnavigable, except near its mouth. In Texas it includes the canyons of the Big Bend National Park. In Mexico it is known as the Río Bravo. Illegal Mexican immigrants have entered the USA by swimming the Rio Grande, earning the nickname of 'Wetbacks'.
Map Mexico Bb

Río Grande de Santiago ***Mexico*** See LERMA

Rio Grande do Norte ***Brazil*** See NATAL

Rio Grande do Sul ***Brazil*** See PORTO ALEGRE

Río Muni ***Equatorial Guinea*** Mainland part of the country, covering 26 017 km² (10 043 sq miles), at one time called Mbini. Timber, coffee and cocoa are the main products. The dominant people are the Bantu-speaking Fang.
Population 241 000
Map Equatorial Guinea Bb

Riobamba ***Ecuador*** Capital of Chimborazo province, lying at a height of 2700 m (8860 ft) on a flat plain between the Chimborazo and Sangay volcanoes. Its attractive colonial buildings include the Convento de la Concepción, which houses the Museum of Sacred Art. The original town, founded in 1534 on the site of Cajabamba 25 km (15 miles) to the south-east, was destroyed in 1797 by a landslide which killed more than half its 9000 inhabitants. Riobamba manufactures textiles, carpets and footwear.
Population 150 000
Map Ecuador Bb

Rioja, La ***Spain*** Autonomous area south of the Basque Country. It produces Spain's best

A DAY IN THE LIFE OF A MEXICAN GARAGE OWNER

When a Chevrolet with US licence plates pulls up to the 'Frontera Garage' in Nuevo Laredo at 11 am, garage-owner Luis Alvarez says a silent prayer of thanks to the picture of the Virgin Mary above the doorway.

Business has been bad the last week and he has only sold a couple of tyres and fixed two engines – half a normal week's work. And an American customer means dollars, not inflated pesos. This customer – like most who come to this border-town mechanic's shop – is a *Chicano*, or Mexican-American, crossing the Rio Grande from Texas for a few hours to buy cheaper clothing, food, and in this case a new fan belt.

Luis, like many of Nuevo Laredo's 200 000 residents, stakes his livelihood on the thousands of tourists, day-shoppers and migrant workers who cross the border. He and his cousin Pepe bought the property for US$5000 five years ago. Since the economic crisis caused by the tailing off of the oil boom, along with high interest on Mexico's foreign debt, Luis's income has been eroded by several peso devaluations and inflation.

Fortunately his wife Lupe earns US$75 a week working part-time in a supermarket. And Nuevo Laredo lies in the 26 km (16 mile) wide duty-free zone, which means Luis can import US car parts without paying import taxes.

Luis hopes his 14-year-old son Carlos will attend engineering school and that one of his three young daughters will marry a man who could take over the running of the garage. Then he could retire at 60.

Luis's musings are interrupted by a honking car. A blonde American couple in a Toyota have their radiator replaced for US$20 and ask directions to Mexico City. Luis practises the English he learned while working for four years in California as a mechanic and night-watchman.

By the time the car leaves, the temperature has climbed to 32°C (90°F) and Luis realises it is 2 pm; time for lunch.

The three-hour lunch break is the only period Luis devotes entirely to his family, which includes Lupe's mother. Today the women serve consommé, rice, refried kidney beans (*refritos*) and white flour tortillas, or flat pancake-like bread.

Luis reopens the garage at 5 pm and works on repairing the axle of the Buick of his best friend, Paco. When Paco collects the car three hours later he reminds Luis that several friends are waiting at the El Presidente *cantina* (bar) down the road. There, the men end the day with jokes, gossip and argument as they drink tequila with lime and salt before finally making their way home.

table wines, as well as cereals, and manufactures textiles. Lead, copper and coal are mined. The province's capital is LOGRONO.

Population 254 000

Map Spain Da

rip, rip current, rip tide Turbulence and agitation in the sea caused either by the meeting of tidal streams or a tidal stream suddenly entering shallow water, or by the return of water piled up on the shore by strong waves.

Ripon ***United Kingdom*** Cathedral city in the English county of North Yorkshire, 35 km (22 miles) north-west of York. The cathedral was built between the 13th and 16th centuries, but has an earlier Saxon crypt. In the marketplace, a civic official known as the Wakeman blows his horn at 9 o'clock each evening, a custom going back more than 1000 years, to announce the setting of the watch for the night. The spectacular ruins of Fountains Abbey, the greatest Cistercian remains in Britain, lie to the south-west of the city. It was founded in 1132 and grew rich and splendid on the proceeds of the Cistercians' skill as sheep farmers. It was dissolved by Henry VIII in 1539 and gradually fell into decay.

Population 12 000

Map United Kingdom Ec

river Large natural stream of fresh water flowing in a definite channel towards a sea, lake or other body of water. See NATURE'S WATERY HIGHWAYS (below); HOW WATER SHAPES THE LAND (p. 548).

river terrace Stretch of level land covered with ALLUVIUM and lying along a river valley above the level of its present FLOOD PLAIN. A river terrace develops when the river cuts through the old flood plain into a deeper valley. Each pair of terraces in a paired series marks the level of a former flood plain.

Riyadh ***Saudi Arabia*** Capital and commercial centre of the country, Riyadh is situated almost in the centre of Saudi Arabia in an oasis some 680 km (about 420 miles) west of Medina. Its old walls were pulled down in the 1950s to make way for new buildings, and it is now a city of sharp contrasts. The poor live in central Riyadh and the better-off in the modern suburbs. In the centre there are traditional souks (bazaars) dealing in gold, silver and ornamental coffee pots. Looming over the city, the tall, modern water tower which has become a landmark, dwarfs the still-surviving clay forts.

Until recently, Riyadh was one of the most isolated and little known cities in the world. Extremely puritanical, it is the centre of Wahhabism, the Muslim sect founded in the 18th century by Abd-al Wahhab and noted for its strict observance of the original words of the Koran. Due to its position high in the central NEJD, Riyadh has a pleasant and healthy climate. There is an international airport and also railway links with DAMMAM on The Gulf. The city contains the king's palace, numerous mosques and Saudi Arabia's first museum – the Museum of Archaeology and Ethnology.

Population 300 000

Map Saudi Arabia Bb

Rizal ***Philippines*** Province of central LUZON bordering on Laguna de BAY. Those parts that had become heavily built up and industrialised (including Caloocan, Makati and Quezon city) were made part of Metro MANILA in 1976, leaving the rest as a mainly farming and fishing area.

Population 3 754 000

NATURE'S WATERY HIGHWAYS

Although the world's longest river, the Nile, has an official length of 6695 km (4160 miles) it is actually somewhat shorter than that. This is because it lost part of its watercourse with the creation of Lake Nasser behind the Aswan High Dam in Egypt in 1971. However, geographers do not include this when they measure the river's length.

THE WORLD'S LONGEST RIVERS

River	Location	Length (km/miles)
Nile	Africa	6695/4160
Amazon	South America	6440/4000
Chang Jiang (Yangtze)	Asia	6380/3965
Mississippi-Missouri	North America	6019/3740
Ob'-Irtysh	Asia	5570/3460
Zaire (Congo)	Africa	4670/2900
Huang He (Hwang Ho)	Asia	4670/2900
Amur (Heilong Jiang)	Asia	4510/2800
Lena	USSR	4270/2650
Mackenzie-Peace	Canada	4240/2635
Paraná	South America	4200/2610
Mekong	Asia	4185/2600
Niger	Africa	4170/2590
Yenisey	USSR	4130/2565
Murray-Darling	Australia	3750/2330
Volga	USSR	3688/2292

HOW WATER SHAPES THE LAND

Water is the true genius of landscape, that paints and moulds the basic rough rock of earth's crust. It carves valleys, spreads plains, brings seed, nurtures growing things; after aeons of erosion, it reduces the loftiest peaks down to gentle hills. Even then, its work is not finished, for the destruction of the mountains changes climate and vegetation, and the hills may transform into rich, arable plains. The transition is slow, but eternal, as may be seen by tracing the path of any of the world's great rivers. A typical example is born high on a mountain range, of melted ice and snow. Trickles and streams infiltrate and tear the surrounding rock, then unite into a torrent that grinds pebbles and fragments into sediments which are carried into the quieter reaches of the river below. There, some of the eroded material is deposited as gravel bars and as embankments at river bends, while lighter sediments may be spread by floods to create wide flats of rich, alluvial soil. Throughout its course, the river picks up and lays down material, until it reaches its mouth where the bulk of the sediments are dumped, sometimes creating a delta. The remaining debris is carried out to sea, where it eventually evolves into sedimentary rock.

THE HYDROLOGICAL CYCLE

Virtually all the water in the world and in its atmosphere has existed for at least 3000 million years; almost no 'new' water at all has been created in that time. About 97 per cent of it is in the oceans, and a further 2 per cent is locked into the polar ice caps. The remaining 1 per cent is the mainstay of all life on earth, and is driven by the sun in an eternal cycle of evaporation from the sea, precipitation, run-off from the land, and return to the sea via the rivers. Precipitated water is replaced by evaporation every 12 days.

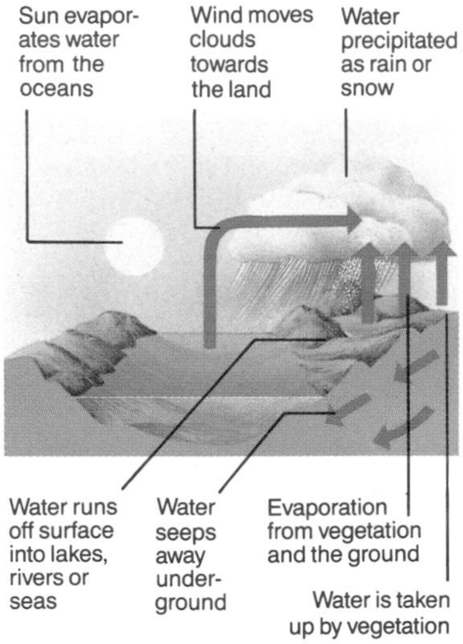

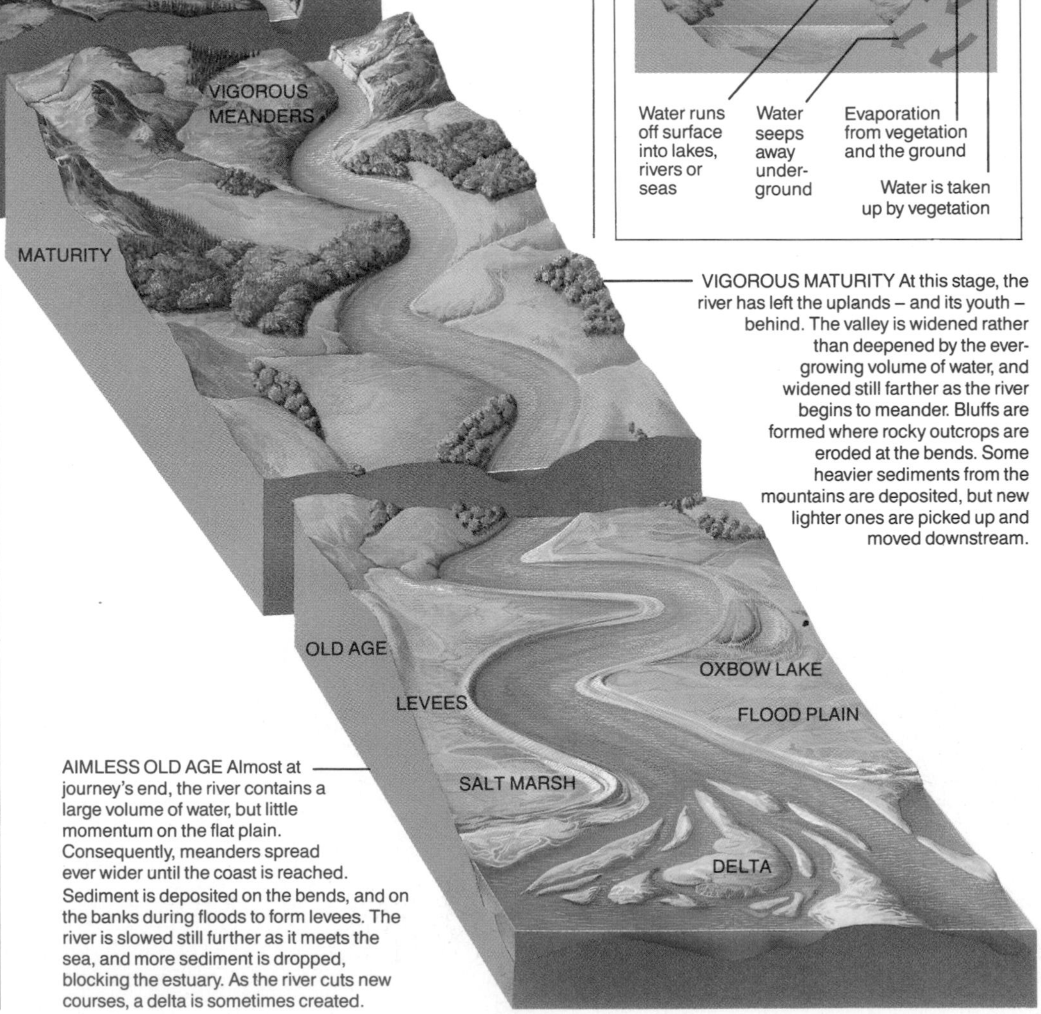

RIVER IN YOUTH Glacier melt-water forms a lake which is drained by a river. This cuts deep gorges in its upper reaches, except on hard outcrops, over which it flows in white rapids or waterfalls.

VIGOROUS MATURITY At this stage, the river has left the uplands – and its youth – behind. The valley is widened rather than deepened by the ever-growing volume of water, and widened still farther as the river begins to meander. Bluffs are formed where rocky outcrops are eroded at the bends. Some heavier sediments from the mountains are deposited, but new lighter ones are picked up and moved downstream.

AIMLESS OLD AGE Almost at journey's end, the river contains a large volume of water, but little momentum on the flat plain. Consequently, meanders spread ever wider until the coast is reached. Sediment is deposited on the bends, and on the banks during floods to form levees. The river is slowed still further as it meets the sea, and more sediment is dropped, blocking the estuary. As the river cuts new courses, a delta is sometimes created.

OXBOW ABANDONED

Meanders or bends in a river's lower course occasionally become so pronounced that the river cuts through the land at the neck of the bend. The stream follows the new course, and the meander becomes a stagnant 'oxbow' lake.

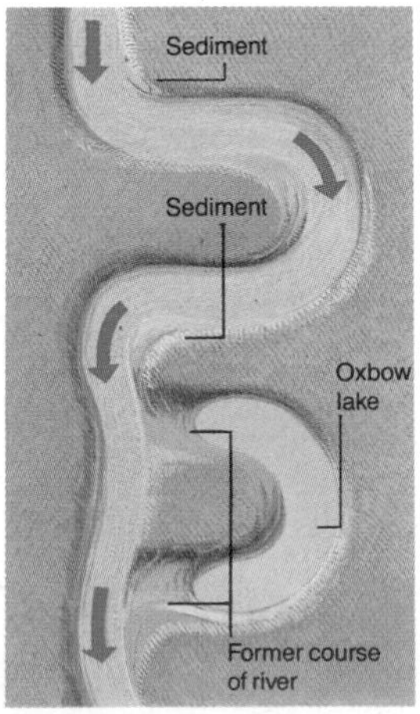

Rjukan ***Norway*** Town in south Norway, 120 km (75 miles) west of Oslo, beside the 104 m (341 ft) high waterfall known as Rjukanfoss. It is the site of the original chemical plant of Norsk Hydro, which made nitrate fertilisers from 1906. One chemical plant – used during the wartime German occupation to make 'heavy water', vital for atomic bombs – was destroyed by Norwegian commandos in 1943. Since the war, the area's chemical industry has largely shifted to the towns of Notodden, about 60 km (40 miles) to the south-east, and Herøya.
Population 5000
Map Norway Cd

Road Town ***British Virgin Islands*** Capital and main port of the Caribbean colony, on the island of TORTOLA. It is a port of call for cruise ships.
Population 3000
Map Caribbean Cb

Roanoke ***USA*** City in the Appalachian Mountains of southern Virginia, about 80 km (50 miles) north of the border with North Carolina. It is a manufacturing centre producing textiles, clothing and electrical equipment, and stands near the Blue Ridge Parkway, a 750 km (466 mile) scenic road through the mountains.
Population (city) 100 700; (metropolitan area) 222 700
Map United States Kc

▼ **REALM OF MISTS Several frost sculpted peaks rise to more than 3050 m (10 000 ft) among the rivers of ice and lakes of Glacier National Park in the Rocky Mountains of Montana.**

Roaring Forties Sailors' name for the region between latitudes about 40° and 60° South, where the prevailing westerly winds blow unimpeded and with considerable force across the oceans. Depressions bring rainy but generally mild weather to this zone of squally, stormy seas – notoriously rough off Cape HORN.

Roca, Cabo da ***Portugal*** Cape about 35 km (22 miles) west of the capital, Lisbon. It ends in a cliff 150 m (490 ft) high. The cape is the westernmost point of continental Europe.
Map Portugal Bc

Rochdale ***United Kingdom*** English town 16 km (10 miles) north of Manchester that was formerly a major centre of the Lancashire cotton industry. It is now part of Greater Manchester, and cotton's importance has dwindled. It was the birthplace of the Cooperative Movement in 1844 and of the popular music-hall singer Gracie Fields (1898-1979).
Population 93 000
Map United Kingdom Dd

Rochester ***United Kingdom*** See MEDWAY TOWNS

Rochester ***USA*** City about 10 km (6 miles) south of Lake Ontario in north-west New York state. It stands in the heart of America's fruit and market-garden belt, and is also the home of the world's largest photographic manufacturer, Eastman Kodak.
Population (city) 242 600; (metropolitan area) 989 000
Map United States Kb

rock crystal Transparent colourless QUARTZ.

rock cycle See THE GEOLOGICAL CYCLE: HOW ROCKS ARE FORMED AND BROKEN DOWN, p. 550.

rock flour Powdered rock produced when rocks are ground together – for example, along the faces of a moving fault, or during movement of a glacier when pieces of rock carried by the ice abrade each other or the underlying rock.

rock salt See HALITE

Rockall ***North Atlantic*** Small uninhabited island of rock about 400 km (250 miles) north-west of Ireland. It is claimed by the UK.
Map Europe Bc

Rockhampton ***Australia*** City on the Fitzroy river in Queensland, lying on the Tropic of Capricorn 520 km (325 miles) north-west of Brisbane. It is the centre of a coal mining and agricultural region. Rockhampton is known as the nation's beef capital, and has two of the country's largest meat processing plants.
Population 50 100
Map Australia Ic

Rocky Mountain National Park ***USA*** Area of 1049 km² (405 sq miles) in the Rocky Mountains north-west of Denver in Colorado. It contains Ice-Age features, alpine valleys and tundra, and has more than 100 peaks above 3000 m (9840 ft).
Map United States Eb

Rocky Mountains (Rockies) ***North America*** Major mountain system of the continent, extending some 4800 km (3000 miles) from the Mexican border north-westwards through the USA and Canada into the *(continued on p. 551)*

THE GEOLOGICAL CYCLE: HOW ROCKS ARE FORMED AND BROKEN DOWN

The surface of the earth is forever changing, subtly but inexorably, in a cycle that takes tens of millions of years to complete before beginning all over again. All of the planet's most powerful forces are concerned in the cycle which begins, or at least recommences, with magma, material that lies beneath the surface crust in a near-molten state and which becomes molten when pressure is released. Then the magma is expelled through fissures and volcanoes to solidify as IGNEOUS ROCKS. These, hard as they are, when exposed to weathering by wind, rain and frost over many years, are broken down into minute fragments that are washed down by streams and rivers as sediments to the sea. In the past 150 million years, for example, the Mississippi has deposited a 12 km (7 mile) thick layer of sediments onto the floor of the Gulf of Mexico. With the pressure of overlying sediments, deeper layers harden into SEDIMENTARY ROCKS. Some are so deep that the pressure from above and thermal energy from below change them into METAMORPHIC ROCKS.

Where crustal plates of the earth (see PLATE TECTONICS) collide, rocks may be pushed up into fold mountains, and these are subjected to weathering. Deep within the mountain core, pressure and heat may turn the rocks to magma, which intrudes into the rocks above and may reach the surface. At ocean trenches, igneous rocks produced at oceanic ridges are returned to the earth's interior. These also replenish the magma supply – and the cycle can begin again.

Weathering occurs in three ways – chemically, mechanically and biologically. In chemical weathering the oxygen and carbon dioxide contained in rain and the humic acid in the soil attack rocks, especially in warm climates, dissolving their components.

In mechanical weathering, the rocks are split by frost, or by the varying expansion and contraction of their components during temperature changes. In biological weathering, sometimes regarded as a type of mechanical weathering, rocks are broken down by the roots of plants, or by the action of animals.

Weathered fragments, transported by ice, wind and water, are ground smaller still against other stones. Ice, wind and water, armed with rock fragments, wear away more of the earth's land surface producing more debris. All the debris is carried down to the lowlands and seas. The combined processes of weathering and transport of its products are known as erosion.

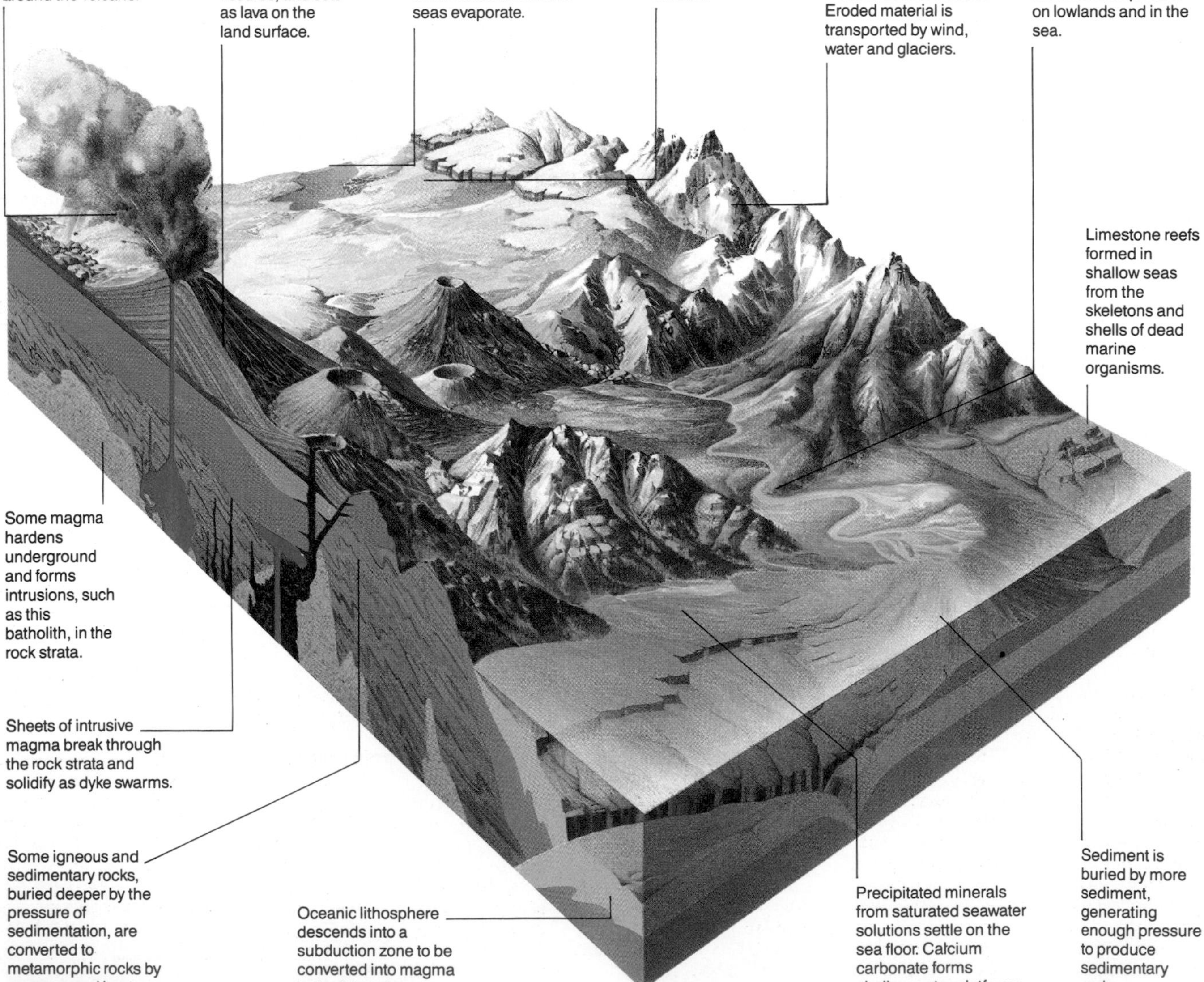

Pyroclastic material ejected during an eruption is deposited around the volcano.
Magma flows to the surface out of volcanoes and fissures, and sets as lava on the land surface.
Some sediments (evaporites) are chemically deposited when lakes and inland seas evaporate.
Wind-borne sand particles form sedimentary beds on land.
Igneous, metamorphic and sedimentary rock on the surface is weathered and eroded. Eroded material is transported by wind, water and glaciers.
Weathered and eroded sediment and dissolved minerals are carried by rivers to be deposited on lowlands and in the sea.
Limestone reefs formed in shallow seas from the skeletons and shells of dead marine organisms.
Some magma hardens underground and forms intrusions, such as this batholith, in the rock strata.
Sheets of intrusive magma break through the rock strata and solidify as dyke swarms.
Some igneous and sedimentary rocks, buried deeper by the pressure of sedimentation, are converted to metamorphic rocks by pressure and heat.
Oceanic lithosphere descends into a subduction zone to be converted into magma in the lithosphere.
Precipitated minerals from saturated seawater solutions settle on the sea floor. Calcium carbonate forms shallow-water platforms.
Sediment is buried by more sediment, generating enough pressure to produce sedimentary rocks.

US state of Alaska. It approaches the Pacific coast in the north, but for much of its extent is more than 1000 km (620 miles) inland. It rises to more than 4000 m (13 132 ft) in the US states of Wyoming, Colorado and New Mexico, and Mount Elbert (4399 m, 14 431 ft) in Colorado is the highest peak.

The Rockies are renowned for their spectacular scenery, and they contain several national parks. They have rich natural resources, including vast forests and minerals including gold, silver, lead, zinc and copper.

Map Canada Cc; United States Da

Ródhos ***Greece*** See RHODES

Rodnei Mountains ***Romania*** Rugged mountain area in the north of the country, rising to 2305 m (7564 ft) above dense forests. It is noted for its many unspoilt villages which are rich in folklore. Other tourist attractions include the Izvorul Tausoarelor Cave.

Map Romania Aa

Rodopi Planina ***Bulgaria*** See RHODOPE MOUNTAINS

Rodrigues Island ***Mauritius*** Volcanic island, surrounded by fringing coral reefs, in the Indian Ocean, 600 km (370 miles) east of Mauritius. Covering 104 km² (40 sq miles), it has a predominantly Creole farming population. The main town is Port Mathurin. Rodrigues is in the cyclone zone of the Indian Ocean, and is liable to be struck by cyclones (hurricanes) in the summer and autumn months.

Population 25 000
Map Indian Ocean Cc

Romania See p. 552

Rome (Roma) ***Italy*** National capital and the country's largest city, standing on the river Tiber 25 km (16 miles) from the west coast. Rome – the so-called 'Eternal City' – has grown twelvefold since 1870 when it was made capital – an explosive rate of growth which largely accounts for two of Rome's greatest problems: lack of planning and chaotic traffic.

It was founded – traditionally by Romulus, the mythical son of Mars, the Roman god of war – in 753 BC and was built on seven hills and surrounded by the Aurelian Wall. Rome is now more than 10 km (6 miles) wide, not counting illegal shanty towns that have sprung up around it. Its character and architectural splendour derive from two widely separated periods: the first three centuries of the Roman Empire (AD 27-300) and the three centuries of the Renaissance and the Baroque (1450-1750).

Although many of the remains have been plundered, the grandeur of the Imperial city was such that probably never in history have so many magnificent public buildings been assembled in one place. After the fall of the Roman Empire there followed a thousand years of neglect and devastation, tempered only by the building of some early Christian churches, and the vast Roman catacombs – an underground system of passages and tombs later used for refuge by persecuted Christians.

In the 15th century the popes brought the full vigour of the high Renaissance into the city. The notable features of this period are the golden-

▲ **ETERNAL CITY The majestic, Baroque colonnade designed by Gianlorenzo Bernini (1598-1680) all but encloses St Peter's Piazza in Rome. It is a supreme example of the enduring fusion of Christian inspiration with classical themes in the city's art.**

brown façades of the plasterwork of many buildings, and the beautifully sculpted monuments of Santa Maria del Popolo. The influence of Michelangelo (1475-1564), painter, sculptor and architect, was crucial. His architectural genius is well illustrated in the Farnese Palace.

The 19th century saw rapid expansion of the city and the break-up of many of the large estates, although the Villa Borghese and many of the major archaeological features such as the Forum and the Caracalla Baths survived. The dictator Benito Mussolini (1883-1945) tried to re-create many of the glorious views of ancient Rome by clearing away old quarters such as the Borgo, and laying out new triumphal avenues such as the Via della Conciliazione, which leads to St Peter's Square, and the 1932 Via dei Fori Imperiali, which runs from Piazza Venezia to the Colosseum.

Rome is not an industrial city. Apart from the construction industry and the industries automatically part of any large city – such as food, clothing and furniture – there is no manufacturing tradition. Above all, Rome is a city of bureaucrats. There are 250 000 of them working for the national, regional, provincial and commune governments and for international organisations based there. It is a major centre of learning – Rome University has 100 000 students – and of publishing, libraries and museums. Its Cinecittà is the largest film studio in Europe, and helps to provide the fashionable Via Veneto with a flow of celebrities.

Places of interest include The VATICAN; the Piazza Campidoglio, designed by Michelangelo in 1536 and containing a 2nd-century AD bronze statue of Emperor Marcus Aurelius; the Spanish Steps (1725) and the Trinità dei Monti church (1495-1585); the 17th-century Villa Borghese, enlarged in the 18th and 19th centuries; the Via Appia Antica, the main Roman road to southern Italy, opened in the 3rd century BC; the Colosseum (72-80 AD); the Foro Romano (the Roman Forum or square) and the Palatine Hill; the Pantheon, the only building of the ancient world to have survived virtually complete; the Arches of Titus (AD 81) and Constantine (AD 315); and numerous other churches and monuments, including San Giovanni in Laterano, the so-called cathedral of Rome, rebuilt in 1649, and the Trevi Fountain (1732-62), into which people throw a coin to ensure their return one day to the city.

Population 2 831 300
Map Italy Dd

Roncesvalles ***Spain*** West Pyrenean village about 35 km (21 miles) north-east of Pamplona. It is celebrated in literature as the scene in AD 778 of an Arab victory over Charlemagne, king of the Franks, and the death of his commander, Roland, which inspired the epic poem *Chanson de Roland*.

Population 90
Map Spain Ea

Ronda ***Spain*** Historic town, about 65 km (40 miles) west of the southern Mediterranean beach resort of Málaga. Set on a steep-sided plateau and divided by a deep gorge – the Tajo – the town is one of Spain's prime tourist attractions. The old town is Moorish, the 'new' town dates from the 15th century and the bullring is claimed to be the country's oldest.

Population 31 400
Map Spain Cd

Rondane ***Norway*** Mountainous area covering 575 km² (222 sq miles) in south-eastern Norway. It was made Norway's first national park in 1962. It is the setting for the legends of Peer Gynt, which were adapted by the Norwegian dramatist Henrik Ibsen and used for his play *Peer Gynt*, written in 1867. The area also contains remains of early hunting pits and Iron Age sites for smelting bog ore – an iron ore found in marshy areas.

Map Norway Cc

Rondônia ***Brazil*** See PORTO VELHO

Romania

LATINS IN A SEA OF SLAVS, THE ROMANIANS HAVE TRIED TO SECURE SOME AUTONOMY WITHIN THE SOVIET BLOC, BUT THEY NOW FACE ECONOMIC PROBLEMS

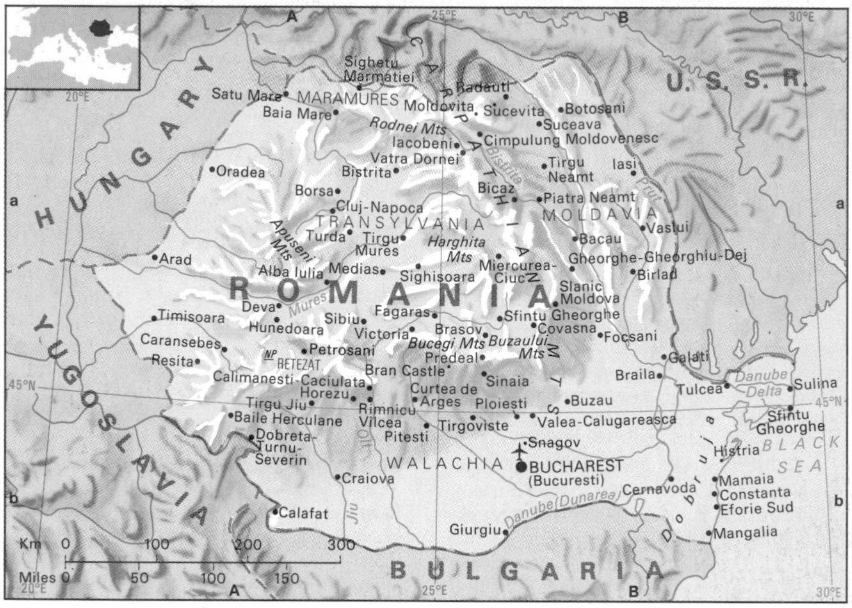

The Romanians are a Latin people who have tenaciously retained their language and customs despite being surrounded by Hungarians and Slav-speaking Bulgars, Yugoslavs and Ukrainians. In turn they fought against Hungarian, Turkish and Hapsburg occupation. Heroic leaders resisted Turkish onslaughts for five centuries; one of them, Vlad Tepes (1456-75), known as the Impaler, was immortalised later as Dracula, the vampire count of the famous horror story.

Romania became an independent state in 1878, fought against Germany in the First World War and with Germany in the Second. Though it joined the Allies in 1944 its postwar fate was decided by the Soviet advance into central Europe: it became a Communist state.

On the map Romania resembles a catherine wheel. In the south-east, the 'fuse' comprises the reed jungle, river channels and wildlife paradise of the DANUBE delta, and the dry rolling farmlands of DOBRUJA separating the river from the BLACK SEA coast. Inland is a ring of rich agricultural plains, flat in the south (WALACHIA) and west (Banat) and hilly in the east (MOLDAVIA).

These plains enclose the sweeping arc of the forested CARPATHIAN MOUNTAINS that rise to 2548 m (8360 ft). On the south-western Yugoslav frontier, the Danube breaches the mountains in the magnificent 160 km (100 mile) IRON GATES gorge. The circular 'core' of Romania is TRANSYLVANIA, within the Carpathian arc.

The old Romanian culture survives most strongly in MARAMURES, Moldavia and Wallachia, where there are hardly any minority groups. Colourful traditional costumes abound in town and country, even among young people. Brightly painted houses of timber or stone have verandahs under roofs of wooden tiles or thatch. Medieval monasteries and churches exhibit strong Byzantine influence. Religion survives under Communism: some 80 per cent of the people are Christians of the Romanian Orthodox Church.

Romania's main minorities, the Hungarians with 8 per cent of the population and Germans (2 per cent), live in Transylvania and Banat. The Hungarians, having lost some degree of self-government and other privileges granted in the 1950s, claim that they are unfairly treated by the Romanians.

The country faces growing economic problems after a period of spectacular growth. Rising overseas debt has led to import restrictions, reducing Romania's debt to the West but causing serious shortages of food and consumer goods. Sugar, flour and cooking oil are rationed; there are power shortages and queues for meat and petrol. One major setback is the drying up of the PLOIESTI oil fields, the country's most valuable resource. Having developed petrochemical and related industries in the 1960s, Romania must now import more oil each year.

For more than two decades Romania pushed through five-year development plans concentrating on investment in heavy industry at the expense of consumer goods. From 1960 to 1970 it had the world's third fastest growing industrial production, from 1970 to 1976 the third fastest growth in agricultural output. The state owns virtually everything except peasant farmland and housing.

The larger estates became state agricultural enterprises (now 17 per cent of farmland) and peasant holdings were forcibly grouped into cooperatives (75 per cent). A substantial tractor and farm machinery industry was developed, and mechanisation combined with larger farms and lack of incentives resulted in a dramatic fall in the number of people employed in agriculture.

Between 1950 and 1984 the proportion of the total labour force engaged in farming fell from 75 per cent to 30 per cent. Horse-drawn carts lit by flickering oil lamps used to jingle through the dusk at harvest time, carrying farm workers home by the dozen. Today there are processions of huge, single-driver combine harvesters, with headlamps ablaze.

The area devoted to maize and wheat has fallen from 86 per cent to 64 per cent, yet output has almost trebled, mainly as a result of mechanisation and wider use of fertilisers.

Industry depends on electricity generated by coal-fired power stations and the rivers flowing from the Carpathians. The hydroelectric dams at Bicaz and CURTEA DE ARGES are impressive, but they are dwarfed by the huge Iron Gates station jointly developed by Romania and Yugoslavia. Yet overambitious planning has plunged the country into a serious energy crisis. Coal and iron ore support steel factories at Resita and HUNEDOARA; the rich mineral deposits in the APUSENI and Maramures mountains supply copper, lead, zinc, gold, silver and bauxite refining industries.

Natural gas in Transylvania and the dwindling supplies of Ploiesti petroleum feed nearby chemical plants. Sawmills and furniture and papermaking factories are scattered in the Carpathians. Textile, engineering and electrical goods factories are in Bucharest and the main towns, while BRASOV makes trucks, tractors and oil drilling equipment, CRAIOVA supplies railway vehicles and PITESTI cars.

ROMANIA AT A GLANCE

Area 237 500 km² (91 700 sq miles)

Population 22 890 000

Capital Bucharest

Government One-party Communist republic

Currency Leu = 100 bani

Language Romanian

Religion Christian (80% Romanian Orthodox, 5% Roman Catholic, 4% Protestant, 3% Greek Catholic)

Climate Continental; average temperature in Bucharest ranges from −2°C (28°F) in winter to 21°C (70°F) in summer

Main primary products Cereals, potatoes, sugar, fruit and vegetables, grapes, livestock, timber, fish; oil and natural gas, coal, lignite, iron ore, salt, copper, lead, gold, silver, uranium

Major industries Agriculture, iron and steel, mining, forestry, chemicals, shipbuilding, machinery, motor vehicles, textiles, fertilisers, coke, food processing

Main exports Machinery, transport equipment, petroleum products, foodstuffs, chemicals, clothing

Annual income per head (US$) 1988

Population growth (per thous/yr) 5

Life expectancy (yrs) Male 69 **Female** 74

Rongelap ***Marshall Islands*** Atoll in the northern Marshalls 160 km (100 miles) east of Bikini. It was badly contaminated by nuclear fallout after a US hydrogen bomb test at Bikini in 1954. The population was evacuated two days later and showed symptoms of radiation sickness. Despite contamination remaining, they were allowed to return in 1957.
Population 235
Map Pacific Ocean Db

Rønne ***Denmark*** Medieval seaport and main town of BORNHOLM island. It has ferries to Copenhagen and the Swedish port of Ystad.
Population 12 400
Map Denmark Cb

Roraima ***South America*** See GUIANA HIGHLANDS

Rorke's Drift ***South Africa*** Mission station on the Buffalo river in northern Natal, about 170 km (105 miles) north of Durban. On January 22-23, 1879, 139 men of the 24th Regiment (the South Wales Borderers) and the Royal Engineers held the mission against 4000 Zulu warriors fresh from their triumph at ISANDHLWANA, winning a total of 11 Victoria Crosses.
Map South Africa Db

Røros ***Norway*** Town in south Norway, 110 km (68 miles) south-east of Trondheim. It was the country's chief source of copper until the 1950s, when the ore ran out. Industrial archaeologists have converted the town into a virtual museum piece with a large tourist industry, and it has become a major winter ski resort.
Population 5460
Map Norway Cc

Ros Comain ***Ireland*** See ROSCOMMON

Ros Láir ***Ireland*** See ROSSLARE

Rosa, Monte ***Italy/Switzerland*** Mountain rising to 4634 m (15 203ft) on the Italian-Swiss border, east of the Matterhorn. The Swiss tourist resort of ZERMATT lies to the north-west.
Map Italy Ab

Rosario ***Argentina*** River port on the Paraná about 280 km (175 miles) north-west of Buenos Aires. Founded in 1689, it is now the main outlet for grain, meat and timber from farms in the valley of the river. Its industrial products include steel and agricultural machinery.
Population 935 500
Map Argentina Cb

Roscommon (Ros Comain) ***Ireland*** Market town about 145 km (90 miles) north-west of Dublin. It is the county town of the surrounding county of the same name which covers 2462 km² (950 sq miles).
Population (county) 54 500; (town) 1730
Map Ireland Bb

Roseau ***Dominica*** Capital, situated on the south-west coast near the mouth of the Roseau (or Queen's) river. Formerly Charlotte Town, it was burned by the French in 1805. Its port, an open roadstead, exports tropical produce.
Population 17 000
Map Caribbean Cb

Rosetta ***Egypt*** See RASHID

Roskilde ***Denmark*** Town on the Danish island of ZEALAND to the west of Copenhagen. It was one of Denmark's major towns from the 10th century into the Middle Ages, and was the scene of the signing of the Peace of Roskilde in 1658 between Denmark and Sweden. Its Gothic cathedral, dating from 1170, has been the burial place of Danish kings for some 600 years. There is a university (1972), and a museum with the remains of six Viking ships from a nearby fiord.
Population 48 300
Map Denmark Cb

Ross and Cromarty ***United Kingdom*** See HIGHLAND; WESTERN ISLES

Ross Ice Shelf ***Antarctica*** The world's largest body of floating ice, lying at the head of the Ross Sea on the Pacific coast of Antarctica. It is estimated to be as big as France. The ice increases in thickness from about 400 m (1300 ft) in the north to more than 750 m (2460 ft) in the south. It moves slowly seawards at about 1000 m (3280 ft) a year. Huge icebergs break off, leaving ice cliffs up to 70 m (230 ft) high.
Map Antarctica d

Ross Island ***Antarctica*** Island about 75 km (47 miles) long in the Ross Sea beside McMurdo Sound. It is dominated by two volcanic peaks – Mount EREBUS (3794 m, 12 447 ft), which is active, and Mount Terror. Both were named after ships of Sir James Ross, their discoverer. The American McMurdo and the New Zealand Scott research bases are on the island.
Map Antarctica Ic

Ross Sea Deep inlet in the coast of Antarctica, discovered in 1841 by the British explorer Sir James Clark Ross. The inlet was the departure point for the shortest route across Antarctica from the sea to the South Pole.
Map Antarctica Hc

Rossing ***Namibia*** Uranium mine, founded in 1971 and employing 3000 workers. It lies about 200 km (125 miles) west of Windhoek.
Map Namibia Ab

Rossiya ***USSR*** See RUSSIAN SOVIET FEDERATED SOCIALIST REPUBLIC

Rosslare (Ros Láir) ***Ireland*** Resort on the east coast, 121 km (75 miles) south of Dublin. The harbour, 8 km (5 miles) south-east of the town, is a ferry terminal, and is the major arrival port for visitors from Fishguard in south-west Wales and Le Havre, France.
Population 590
Map Ireland Cb

Rostock ***East Germany*** The country's chief port, on the Baltic Sea about 200 km (120 miles) north-west of the capital, East Berlin. It is also a shipbuilding and marine engineering centre. Other products include electrical equipment, machinery and textiles.
Population 242 000
Map East Germany Ca

Rostov (Rostov-Yaroslavskiy) ***USSR*** One of the oldest towns in Russia, first mentioned in AD 862 and formerly called Rostov-Velikiy; it lies 195 km (120 miles) north-east of Moscow. In the Middle Ages it grew into a flourishing centre for handmade enamelware and linen goods – in which it still specialises. Rostov has a magnificent lakeside kremlin (citadel), a 13th-century cathedral and some fine 17th-century churches.
Population 31 000
Map USSR Ec

Rostov-na-Donu (Rostov-on-Don) ***USSR*** Industrial city and port on the Don river near the north-east tip of the Sea of Azov. Founded as a customs post in 1750, when Turkey still ruled the coast of the Sea of Azov 50 km (30 miles) downstream, Rostov now produces ball bearings, road-building machinery, electrical equipment, chemicals, barges, agricultural machinery, glass, wine and cigarettes.
Population 983 000
Map USSR Ed

Rothenburg ob der Tauber ***West Germany*** Walled medieval town on the River Tauber about 65 km (40 miles) west of Nuremberg. Guard towers and gates survive from the Middle Ages, and inside are galleried and timbered houses, Gothic and Renaissance churches and flagstoned streets. Tourists crowd it in summer.
Population 11 300
Map West Germany Dd

Rotherham ***United Kingdom*** Industrial town on the River Don 10 km (6 miles) north-east of Sheffield in the English county of South Yorkshire. It has been a market centre since Saxon times, but grew with the 19th-century development of coal mining and the spread of steel making from Sheffield.
Population 82 000
Map United Kingdom Ed

Rotorua ***New Zealand*** Tourist city and health resort about 160 km (100 miles) south-east of Auckland. It lies beside Lake Rotorua in North Island, in an area steaming with hot springs, mud pools and geysers. Some of the volcanically warmed springs are used to heat nearby buildings and homes in winter, and to power refrigeration units in summer. Pohutu, the most powerful geyser, sometimes shoots water 30 m (100 ft) into the sky.

Whakarewarewa, on the edge of the city, is a Maori village whose inhabitants use the hot springs for washing, swimming and cooking.
Population 48 300
Map New Zealand Fc

Rotterdam ***Netherlands*** The country's largest city and the world's busiest port, handling a greater tonnage of cargo each year than any other and capable of handling more than 300 ships simultaneously. It was founded in the 13th century as a fishing village where the Rotte river flowed into the Rhine-Maas delta, about 60 km (35 miles) south-west of Amsterdam.

The Rhine is the 'Main Street' of Europe, and after the NIEUW WATERWEG gave the city deep-water access to the North Sea in 1872, ocean cargo vessels were quick to take advantage of Rotterdam, where river and ocean traffic could meet. The port has expanded all the way to the sea; docks and oil refineries line the waterway, and EUROPOORT stands at its seaward end. The building problems were immense in an area where most land is below sea level and made of delta mud.

Rotterdam's city centre, wiped out by German bombs in May 1940, has been rebuilt, with new residential and shopping areas, conference and concert facilities, and several museums, including one devoted to maritime history, and the Boymans, with a splendid art collection. The docks, warehouses, shipyards and plants processing chocolate, margarine, coffee and tobacco, provide most of the jobs in the city.

Population (city) 558 800; (Greater Rotterdam) 1 024 700
Map Netherlands Bb

Roubaix ***France*** See LILLE-ROUBAIX-TOURCOING

Rouen ***France*** Port and capital of Seine-Maritime department, on the Seine 112 km (70 miles) north-west of Paris. The French national heroine Joan of Arc, captured by the Burgundians and sold to the English, was tried as a heretic and sorceress and burned at the stake there in 1431. Rouen was the medieval capital of Normandy, and has many medieval buildings, including a splendid 13th-century cathedral (restored after damage in the Second World War) and the Abbey of St Ouen, where Joan of Arc was imprisoned. It is now one of the country's largest ports, exporting manufactured goods, and importing coal and oil.

Population 385 800
Map France Db

Roussillon ***France*** Former province covering some 2000 km^2 (about 770 sq miles) beside the Mediterranean and the Spanish border, now part of LANGUEDOC-ROUSSILLON. Irrigated areas on the plain produce wheat, fruit, vines and vegetables. PERPIGNAN is the main town.

Map France Ee

Rouyn-Noranda ***Canada*** Centre of a gold and copper-mining area in western Quebec, 386 km (240 miles) north-east of Ottawa. Other industries include dairy products and timber.

Population 28 650
Map Canada Hd

Rovaniemi ***Finland*** City often called the capital of Finnish Lappland. It lies 160 km (100 miles) north of the port of Oulu on the Gulf of Bothnia, at the broad confluence of the Kemi and Ounas rivers. It was virtually destroyed during the Second World War and reconstructed on the basis of a plan by the Finnish architect Alvar Aalto. It is now a commercial and tourist centre.

Population 30 300
Map Finland Cb

Rovinj (Rovigno) ***Yugoslavia*** International resort on the Istran coast about 55 km (35 miles) from the Italian border. The old town, a pastel-painted Venetian settlement of narrow, winding alleys and colourful piazzas, lies on a promontory surrounded by pine-clad islands.

Map Yugoslavia Ab

Roxburgh ***United Kingdom*** See BORDERS

Roznava ***Czechoslovakia*** Medieval gold-mining town in eastern Slovakia, 13 km (8 miles) from the Hungarian border. Its main square stands above the former mines. The town is a base for tourists visiting the Slovak Ore mountains to the north, and the caves and gorges of the Slovak Karst (Slovensky Kras), which extends into Hungary. Iron ore is mined nearby.

Population 19 500
Map Czechoslovakia Eb

RSFSR ***USSR*** See RUSSIAN SOVIET FEDERATED SOCIALIST REPUBLIC

Ruapehu, Mount ***New Zealand*** Active volcano about 230 km (140 miles) north-east of the capital, Wellington, in Tongariro National Park. At 2797 m (9176 ft), it is the highest point on North Island; it is popular with hikers in summer and skiers in winter. Near the summit is Crater Lake, its cloudy green water strongly acidic and warmed by volcanic steam.

On Christmas Eve 1953, an eruption sent a torrent of mud, water and rocks from the lake smashing down the volcano's flanks and thundering through a river gorge. The torrent swept away a railway bridge in its path minutes before the Wellington-Auckland express was due to pass over it. In the darkness the speeding train plunged into the gorge, killing 151 people.

Map New Zealand Ec

Rub al-Khali ***Saudi Arabia/South Yemen*** Desert known as 'The Empty Quarter', covering some 650 000 km^2 (251 000 sq miles) on the Arabian peninsula. It extends from Nejd into South Yemen, and from Yemen to Oman. One of the world's most arid and hostile deserts, Rub al-Khali has yet to be fully explored. It was first crossed by a European in 1930, when the British traveller and writer Bertram Thomas followed the one well-watered route, used by the Bedouins, from the west coast north-east across the central sands to the Qatar peninsula. The Shaybah oil field lies in the north-east of the Rub al-Khali, some 200 km (125 miles) due south of ABU DHABI city.

Map Saudi Arabia Bc

Rudolf, Lake ***Kenya*** See TURKANA, LAKE

Rugby ***United Kingdom*** Town in the English Midlands county of Warwickshire, 45 km (28 miles) east of Birmingham. It has a number of engineering plants and is the home of the public school where the game of rugby football originated in 1823, when a soccer player picked up the ball and ran with it.

Population 60 000
Map United Kingdom Ed

Rügen ***East Germany*** The country's largest Baltic island, covering 927 km^2 (358 sq miles). It is a popular summer resort, fringed by chalk cliffs and dotted with beechwoods, sandy coves, fishing villages and prehistoric remains.

Population 85 000
Map East Germany Ca

Ruhr ***West Germany*** River in the north-west flowing 235 km (146 miles) west to the Rhine at Duisburg. Centred on its lower valley is Western Europe's largest coalfield and industrial region, with mines and steel and engineering plants stretching from Dortmund in the east to Duisburg. The Ruhr region is known as 'the Forge of Germany', and the mining of top-grade coking coal began there in the 1850s. But since the 1970s, the world recession and a cut in the number of mines from 145 to 25 have forced the eastern Ruhr to diversify into lighter and computer-based industries. Petrochemical industries flourish in the west, based on the Rhine ports and pipelines, and soft coal is produced west of the Rhine.

Population 10 000 000
Map West Germany Bc

Ruhuna (Yala) ***Sri Lanka*** The country's second largest national park after Wilpattu. It is 980 km^2 (378 sq miles) in area and lies near the island's south-east coast, some 180 km (110 miles) from Colombo. Many species of animals, including elephants, leopards, bears, buffaloes, deer and wild boars roam freely over its thorny scrub and open jungle. The nearby Kumana bird sanctuary is a refuge for ibis, pelicans, parrots, peacocks and storks.

Rumaila ***Iraq*** One of the largest oil fields in the world, about 50 km (30 miles) south-west of The Gulf port of Basra. It started production in 1972. The oil field has led to the creation of the nearby industrial towns of Az Zubayr, Shu'aiba and Umm Qasr.

Map Iraq Cc

run In America, a fast-flowing stream or brook, such as Bull Run, Virginia.

run off Amount of water, originating as rain or snow, that reaches streams and rivers, and flows away to the sea. It consists of the water which flows off the surface, as well as some of the water that sinks into the ground and later reaches the streams.

runnel, swale Long narrow hollow on a beach between two ridges running parallel to the coastline which are exposed at low tide.

Runnymede ***United Kingdom*** Meadow on the south bank of the River Thames, about 8 km (5 miles) south-east of Windsor. The English King John sealed Magna Carta – the Great Charter setting out the rights of citizens to justice and security – there on June 15, 1215. There are three memorials on the site: to Magna Carta; to the 20 000 Commonwealth airmen who died in the Second World War with no known grave; and to the American President John F. Kennedy, who was assassinated in Dallas, Texas, in 1963.

Map United Kingdom Ee

Ruse ***Bulgaria*** River port and capital of Ruse province, on the Danube 244 km (152 miles) north-east of Sofia. It is near the only bridge which links Bulgaria and Romania across the Danube. Founded as a Roman fortress in the 2nd century AD, it expanded after 1878 and is now a major industrial centre with shipbuilding and tanning industries. It manufactures rubber goods, agricultural machinery, electrical equipment and plastics. A steamer service runs to the port of Vidin, 280 km (175 miles) upstream.

Population 181 200
Map Bulgaria Bb

Rwanda

ONE OF THE MOST DENSELY POPULATED RURAL NATIONS ON EARTH, WHERE ISOLATION HINDERS ECONOMIC DEVELOPMENT

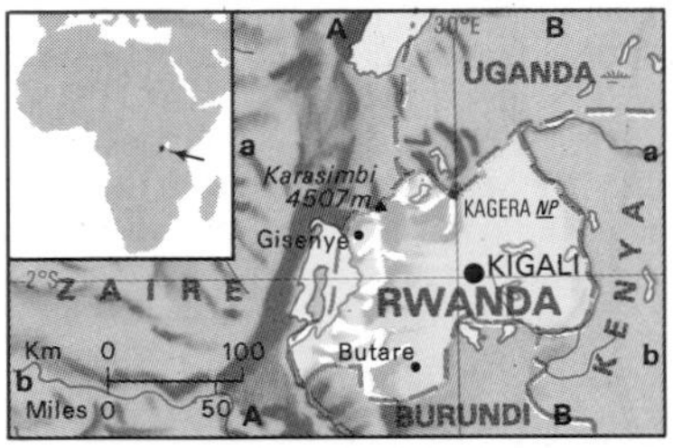

In the heart of Africa, the little known republic of Rwanda is one of the world's most densely populated rural countries. It is almost wholly composed of steep hills, whose slopes are intensively cultivated. Rwanda and the neighbouring country of Burundi are almost identical twins. Until 1962 they were linked as Ruanda-Urundi and were administered by Belgium.

Rwanda has a central spine of highlands from which streams flow west to the Zaire river, and east to the Nile. In the north are active volcanoes rising to about 4500 m (14 760 ft). The land falls gently eastwards towards Tanzania, and more sharply westwards to Lake KIVU, which forms most of the boundary with Zaire. Soils are not very fertile, and rainfall is less than might be expected in such mountainous country – ranging from 750 to 1750 mm (30-70 in) per year – but the rain is more reliable than in much of Africa and serious droughts are rare.

An estimated 6.5 million population gives a density of 250 people per km² (640 per sq mile), which is very high for a country where 96 per cent of the people are in rural areas. At independence in 1962 there were hardly any towns; the country had been governed by the Belgians from BUJUMBURA in Urundi. The settlement of KIGALI, selected as the Rwandan capital, had a mere 5000 inhabitants. A rapid influx has increased Kigali's population to about 170 000, but there are only three other towns, and they each have fewer than 35 000 inhabitants.

Most Rwandans live not even in villages, but in homes dispersed throughout the countryside, generally family units of three or four huts built of mud or adobe (mud bricks). Rwanda is one of the world's least developed countries and agriculture is mainly for subsistence. Most families farm tiny plots producing just enough bananas, cassava roots, sweet potatoes and pulses to feed themselves. A small minority grow coffee for sale, and this accounts for over half of Rwanda's total exports, while some rice is grown commercially on valley floors. Many families have a few goats and chickens, and there are some cattle, but these are kept mainly by the richer Tutsi tribesmen in the east.

The Tutsi are a small minority with less than 15 per cent of the population, against 85 per cent for the Hutu. There are also a small number of pygmy Twa people. In the past, the Tutsi formed a ruling aristocracy, but the Hutu had gained political control by the time the country became independent.

Rwanda's economic development has been hindered by isolation from world trade routes. The main access to the sea is by road through Uganda and Kenya, and transport costs are extremely high. Rwanda's main contacts with the outside world are through aid from the European Economic Community, and through Christian missions which help with education and health.

RWANDA AT A GLANCE
Area 26 338 km² (10 169 sq miles)
Population 6 480 000
Capital Kigali
Government Military-ruled republic
Currency Rwanda franc = 100 centimes
Languages Kinyarwanda, French, Kiswahili
Religions Christian (65%), tribal religions (25%), Muslim (10%)
Climate Tropical modified by altitude; average annual temperature in Kigali is 19°C (66°F)
Main primary products Sweet potatoes, cassava, potatoes, beans and peas, sorghum, bananas, groundnuts, coffee, tea, pyrethrum, cattle, sheep, goats, pigs, timber; tin, tungsten
Major industries Agriculture, mining, forestry, processing hides and skins, food processing
Main exports Coffee, metal ores, tea, hides and skins, pyrethrum
Annual income per head (US$) 250
Population growth (per thous/yr) 37
Life expectancy (yrs) Male 44 **Female** 47

Rushmore, Mount ***USA*** Mountain rising to 1707 m (5600 ft) in the Black Hills of South Dakota, and site of the Mount Rushmore National Memorial, covering 13 km² (5 sq miles). The memorial features giant carvings of the heads of four presidents of the United States, up to 18 m (60 ft) tall, cut out of the face of the mountain. The heads are of Washington, Jefferson, Lincoln and Theodore Roosevelt. They were carved between 1927 and 1941.
Map United States Fb

Russia (Rossiya) ***USSR*** Name loosely applied to the UNION OF SOVIET SOCIALIST REPUBLICS (USSR) and, more correctly, to its largest constituent republic, the RUSSIAN SOVIET FEDERATED SOCIALIST REPUBLIC (RSFSR). It is also used for the pre-revolutionary Russian Empire.

Russian Soviet Federated Socialist Republic (Rossiya; RSFSR; Russian Federation) ***USSR*** Largest of the 15 Soviet republics, with an area of 17 075 400 km² (6 592 800 sq miles) – almost as large as the USA and Canada combined. It stretches from the Baltic Sea to the Pacific Ocean, from the Arctic Ocean to the borders of China and Mongolia, forming some 75 per cent of the entire country. It contains more than half the nation's people, with over 100 nationalities and languages – though Russians are in a large majority. Administratively, it is divided into 16 autonomous republics and a number of regions and territories.

Until the Bolshevik Revolution of 1917, the republic formed most of the tsarist Russian Empire. Its main cities include the capital MOSCOW, LENINGRAD and GOR'KIY. It produces 80 per cent of the USSR's timber and paper, about 65 per cent of the country's coal, steel and electricity, 50 per cent of the fertilisers and between 40 and 50 per cent of the textiles, clothing, shoes and electrical appliances. Its other main products include wheat, livestock and motor vehicles. In all, these represent almost 60 per cent of the Soviet Union's entire industrial production and half of its agricultural output.
Population 142 117 000
Map USSR Fb

Rustenburg ***South Africa*** Mining town about 100 km (60 miles) north-west of Pretoria. The world's largest platinum mine is nearby, and chrome and nickel are also mined.
Population 30 400
Map South Africa Cb

Rutanzige, Lake ***Uganda/Zaire*** See EDWARD, LAKE

Rutland ***United Kingdom*** See LEICESTER

Ruwenzori ***Uganda/Zaire*** Mountain range between lakes Edward and Albert, thought to be the legendary Mountains of the Moon, where the ancient Greeks believed the Nile rivers rose. The highest point is Margherita (5110 m, 16 765 ft), one of the peaks of Mount Stanley (Mount Ngaliema). Dense forests skirt the mountains, but above 1650 m (5410 ft) the forests give way to successive bands of mountain grassland, bamboo thicket, tree heathers, giant senecios, and finally tundra, with snowfields and glaciers at the top.
Map Zaire Ba

Ryazan' ***USSR*** Industrial city on the Oka river, 175 km (110 miles) south-east of Moscow. Ryazan' was founded in 1095 but destroyed by the Tatars in 1237. It was rebuilt and in the late 18th century became a grain-milling centre. The city began refining oil in the early 1960s, and its other chief products now include electrical appliances, machine tools, chemicals, leather goods and farm equipment.
Population 488 000
Map USSR Ec

Rybinsk Dam ***USSR*** Large hydroelectric dam on the Volga river, 270 km (170 miles) north of Moscow. Completed in 1941, it flooded the nearby lowland to create a vast reservoir covering 4550 km² (1760 sq miles). The inhabitants of hundreds of drowned villages were rehoused in the neighbouring city of Rybinsk (now ANDROPOV).
Map USSR Ec

Rysy, Mount ***Poland*** See TATRA MOUNTAINS

Ryukyu Islands (Nansei-shoto) ***Japan*** Chain stretching some 1200 km (750 miles) south-west from Kyushu island to Taiwan. It includes the OKINAWA group. About 150 000 people died there – most on Okinawa – in fighting during the American invasion of Japan in 1945. The USA retains military bases on some islands.
Population 1 366 600
Map Japan Gf

Rzeszow ***Poland*** Modern industrial city 153 km (95 miles) east of the city of Cracow. The medieval town at its heart was frequently looted – by Tatars, Swedes and Austrians – but a fine 17th-century Cistercian church survives. Since 1945, new factories have been set up there, making railway equipment, machinery, chemicals and clothing – and the city's population has more than quadrupled.
Population 134 500
Map Poland Dc

Saada (Sa'dah) ***Yemen*** Town and road junction 190 km (118 miles) north-west of SAN'A. It is connected to the capital by a road built with Chinese aid and opened in 1977. A 79 km (49 mile) long stretch of new road connected Saada with Zahran in Saudi Arabia.
Population 4500
Map Yemen Aa

Saale ***West Germany/East Germany*** River rising just across the border in West Germany and flowing 426 km (265 miles) into the Elbe river north of Leipzig. Its upper course is dammed for reservoirs and hydroelectric power.
Map East Germany Bc

Saarbrücken (Sarrebruck) ***West Germany*** Coal-mining and industrial city and capital of Saarland. It stands on the Saar river near the French border, about 155 km (95 miles) south-west of Frankfurt am Main. Both French and German are spoken. Steel was the main industry, but the world steel recession since the 1970s has seriously affected production.
Population 189 000
Map West Germany Bd

Saarland ***West Germany*** State on the French border opposite Lorraine. Since the 17th century the region has been fought over by France and Germany, but in 1954 the people voted to become part of West Germany – and the integration was completed in 1957. It is a hilly, forested region of coalfields, heavy industry and vineyards. SAARBRUCKEN is the capital.
Population 1 100 000
Map West Germany Bd

Sabac ***Yugoslavia*** Major north Serbian market town and river port on the Sava river about 65 km (40 miles) west of Belgrade. It clusters around the ruins of a massive 15th-century fortress built by Sultan Mehmet II as the base for the Turkish conquest of Croatia.
Population 119 700
Map Yugoslavia Db

Sabah ***Malaysia*** Most easterly state of Malaysia, occupying about 73 700 km² (28 450 sq miles) on the island of Borneo. The state was known as North Borneo until it joined Malaysia in 1963. Rapid growth of its hardwood industry since then has made it the richest state in the country. There is concern, though, that logging may upset the ecology of the hilly interior. In parts, the cleared jungle has been replaced by commercial tree crops such as rubber and oil palm. Other parts are being protected; Kinabalu National Park, for instance, has been set up on the slopes of Mount Kinabalu – at 4094 m (13 432 ft) Malaysia's highest peak.

Copper and offshore oil have been found, and surveys suggest that Sabah may have considerable mineral reserves.
Population 1 034 000
Map Malaysia Ea

Sabrahta (Sabrata) ***Libya*** Ancient city with foundations dating back to the 5th century BC. It is 67 km (42 miles) west of Tripoli, and once stood at the end of the caravan route bringing gold, slaves, ivory and ostrich feathers from the south. As a Roman city it was of importance, exporting olive oil. A Graeco-Roman theatre facing the sea has been restored and seats 5000. Other remains include baths, temples, a basilica, forum, shops and houses with mosaics.
Population 14 500
Map Libya Ba

Sacramento ***USA*** **1.** State capital of California, about 120 km (75 miles) north-east of San Francisco. It grew on a ranch where gold was discovered in the 1840s. Foodstuffs, transport equipment, weapons and defence systems are made. **2.** California's longest river, rising near Mount Shasta in the north of the state and flowing some 560 km (350 miles) southwards through the Sacramento Valley (the northern section of the Central Valley) into Suisun Bay, an extension of San Francisco Bay. Many towns along the valley were established during the California gold rush of 1848-9.
Population (city) 304 100; (metropolitan area) 1 219 600
Map (city) United States Bc; (river) United States Bb

Sacsayhuamán ***Peru*** Inca fortress near Cuzco – its name means 'coloured falcons'. There are three massive zigzag walls containing 21 bastions, made of enormous stones weighing up to 300 tonnes. The walls flank a parade ground and an Inca throne carved out of solid rock. The fortress was razed to the ground by the Spaniards after they had defeated the Inca Manco Capac in the 16th century. They used the stones for their own buildings in Cuzco.
Map Peru Bb

saddle Saddle-shaped depression in the crest of a ridge.

Safad (Zefat) ***Israel*** Hill town, 12 km (7 miles) north-west of the Sea of Galilee. It was a stronghold of resistance against the Romans in the 2nd century, and in the 16th century became a centre of Jewish mysticism. The remains of a Crusader castle built in the 12th century survive today and, in the Arab quarter, there is an artist's colony. As well as being a summer resort, Safad is a base for skiing on Mount Hermon.
Population 17 200
Map Israel Ba

Safajah ***Egypt*** See BUR SAFAGA

Saffaniyah ***Saudi Arabia*** The world's largest offshore oil field, situated in The Gulf about 230 km (143 miles) north-west of Dammam. It is twice the size of the Forties field in the North Sea and can produce up to 1 million barrels of oil a day.
Map Saudi Arabia Bb

Safi ***Morocco*** Atlantic port 220 km (140 miles) south-west of Casablanca that handles phosphates mined at Youssoufia, 100 km (60 miles) to the east. It is also the world's biggest exporter of sardines, handling over 75 000 tonnes each year. Although it is a growing industrial town, with a chemical plant that processes the phosphate rock, it retains its medina (old Muslim residential quarter), an old Portuguese fortress, a palace and potters' workshops.
Population 198 000
Map Morocco Ba

Safid Rud ***Iran*** Name given to the lower course of the Qezel Owzan river, whose source is near Sanundaj. The whole course runs for 800 km (500 miles) and flows into the Caspian Sea. Its delta, to the east of the city of RASHT, is a rice-growing area.
Map Iran Aa

Sagaing ***Burma*** Town and capital of Sagaing division, which stretches far to the north, covering 94 626 km² (36 535 sq miles) between the Irrawaddy and Chindwin rivers. It produces wheat, cotton and beans. Standing beside the Irrawaddy river opposite the city of Mandalay, the town is overlooked by pagoda-lined hills.
Population (town) 190 000; (division) 3 856 000
Map Burma Bb

sagebrush Scrub vegetation in the semi-desert regions of the south-western United States and on the Mexican plateaus.

Sagres ***Portugal*** Small port on the windswept Cabo de São Vicente peninsula at the country's south-western tip. It was there in about 1416 that Henry the Navigator, the Portuguese prince and patron of explorers, founded a school of navigation, where Vasco da Gama's voyages of discovery to Africa and India were planned.
Map Portugal Bd

Sagvar ***Hungary*** See SIOFOK

Sahara ***Africa*** The world's largest desert, occupying nearly one-third of the African continent. It covers 9 065 000 km² (3 500 000 sq miles), and stretches from the Atlantic Ocean to the Red Sea; it is bordered by the Mediterranean in the north and by the SAHEL belt lying to the south.

Most of the Sahara is a high plateau some 460 m (1500 ft) above sea level; its lowest point is the Qattara Depression in Egypt which falls to 133 m (436 ft) below sea level. Only about one-ninth of the desert is sandy – notably the Libyan Desert, Egypt's Western Desert and, in Algeria, the Great Eastern Erg and Great Western Erg.
Map Africa Cb

Saharan Atlas ***Algeria*** See ATLAS MOUNTAINS

Sahel ***Africa*** Semiarid region stretching from Senegal to Sudan and forming a belt dividing the Sahara desert from the tropical forests. The terrain is mostly SAVANNAH, and there is some cropland watered by the Niger and Senegal rivers. Drought in this region has caused widespread famine in recent years.
Map Africa Cc

Sahul Shelf ***Timor Sea*** The undersea extension of the continental platform off north-west Australia. The shelf is at relatively shallow depth and is a potential source of petroleum which has already led to dispute in the area south of Timor.
Map Australia Ca

Saida ***Algeria*** Administrative centre in the Atlas Mountains, 200 km (124 miles) south of Mostaganem on the main western rail and road route across the mountains to the SAHARA. It stands in the middle of an important agricultural region producing mainly wheat, where the country's 'agricultural revolution' – a plan to make Algeria self-sufficient in food – was launched in the 1970s.
Population 60 000
Map Algeria Ba

Saigon ***Vietnam*** See HO CHI MINH CITY

Saimaa ***Finland*** 1. Finland's biggest lake, covering 4400 km² (1700 sq miles) in the south-east of the country. Its shoreline extends for a total of about 15 000 km (9320 sq miles). 2. Canal completed in 1856 in Karelia between the lake and Vyborg Bay, now in the USSR at the head of the Gulf of Finland. The canal is 42.5 km (26.5 miles) long, including the passage through several lakes along its route.
Map Finland Dc

▼ SAHARAN SEA Sands stretch towards the Atlas Mountains – part of the *Djezira el Maghreb* ('the Western Isle') between the desert and Mediterranean.

St Albans ***United Kingdom*** English cathedral city in Hertfordshire, some 32 km (20 miles) north of London. It was founded nearly 2000 years ago as Verulamium and was one of the largest towns of Roman Britain. Remains of the Roman baths, walls, theatre and a street are preserved near the city centre. The cathedral, 168 m (550 ft) long, dates from the 11th century, with 13th and 14th-century additions. It stands on a hill where Alban, a Roman soldier, was executed in AD 209 for harbouring a Christian priest; he became Britain's first saint and gave his name to the Saxon town that replaced Verulamium.
Population 54 000
Map United Kingdom Ee

St Barthélémy ***Guadeloupe*** Small island, commonly called St Barts, situated 210 km (130 miles) north-west of the main islands of Guadeloupe, of which it is a dependency. The inhabitants are descended from Norman immigrants, and speak in a 17th-century Normandy dialect. The economy depends upon fishing and tourism, and the harbour of Gustavia, the capital, is a favourite anchorage for yachtsmen.
Population 3000
Map Caribbean Cb

St Bernard Passes ***France/Italy/Switzerland*** Two important Alpine passes used since Roman times.

The Grand St Bernard, 25 km (16 miles) east of Mont Blanc, is 2473 m (8113 ft) high and joins the Valle d'Aosta in Italy with Martigny and the Rhone valley in Switzerland. The modern road dates from 1905 but is bypassed by a tunnel opened in 1963.

The Little St Bernard, 2199 m (7178 ft) high, is the pass through which, some scholars believe, the Carthaginian general Hannibal and his elephants invaded Italy in AD 218. It is about 18 km (11 miles) south of Mont Blanc, joining the Valle d'Aosta with the Isère valley and Chambéry in France. The present road was built in 1871.
Map Italy Ab

Saint Catharines ***Canada*** Industrial centre in a fruit-growing region at the outlet of the WELLAND CANAL in southern Ontario. Industries include shipbuilding, food processing, automobile parts and electrical equipment.
Population 124 000
Map Canada Hd

St Croix ***US Virgin Islands*** Largest of the group, covering 218 km² (84 sq miles). It contains the territory's only extensive area of lowland. The island is the site of the huge Amerada Hess oil refinery, an alumina processing plant and factories which distil rum and make a variety of goods, especially watches, from imported materials. The two small towns of Christiansted and Fredriksted contain some attractive Danish-style buildings.
Population 50 000
Map Caribbean Eb

Saint-Denis ***France*** Industrial suburb of northern Paris. It manufactures chemicals, plastics, diesel engines, leather and fireworks. The abbey of Saint Denis, founded in 626, was the home of Peter Abelard, a 12th-century French philosopher whose tragic love affair with one of his pupils, Héloise, has become part of romantic legend. Abelard became a monk and Héloise a nun. They are both now buried in a single sepulchre in the Père Lachaise cemetery in Paris.
Population 91 300
Map France Eb

Saint-Denis ***Réunion*** Capital of the island, lying on the north coast and backed by volcanic peaks in the interior. It has attractive 18th-century French and Creole buildings in the centre, but modern development is spreading west to the island's main shipping port, Le Port, and east to the airport.
Population 110 000

Saint-Emilion ***France*** Fortified town near Libourne in the lower Dordogne valley. It is surrounded by vineyards that produce the red wine named after the town.
Population 3000
Map France Cd

Saint-Étienne ***France*** Industrial city about 50 km (30 miles) south-west of Lyons. With coal mines, metalworking and textile factories, it was the cradle of the industrial revolution in France in the early 19th century. Coal mining decreased during the 1970s and ceased in 1981. Its dominant industries now include precision engineering, clothing and plastics.
Population 319 500
Map France Fd

St Eustatius ***Netherlands Antilles*** Small island, also known as Statia, in the Caribbean Leeward chain, consisting of two volcanoes joined by an area of lowland. There is a little farming and pumice mining and some tourism, but the island is largely undeveloped. An oil terminal has been opened recently.

The island was settled by the French in 1625, and became Dutch in 1632. For much of the 18th century it was a rich duty-free trading centre. Indeed, it was the first foreign government to recognise the fledgling United States and was an important supply source for the revolutionaries. Incensed by this, Britain invaded and captured the island in 1781, sacking warehouses and homes. The invaders continued to fly the Dutch flag and captured several American and enemy merchant ships by this ruse.

The capital, Oranjestad, stands 90 m (295 ft) above the sea on cliffs on the south-west coast. Also known as Upper Town, it contains the remains of a 17th-century Dutch fort.

Population 1500
Map Caribbean Cb

St Gall (Sankt Gallen; St Gallen) ***Switzerland*** Cotton-manufacturing town, about 65 km (40 miles) east of Zürich. It also makes glass and metal goods. Most of the people in the town and the surrounding canton of the same name are German speaking.

Population (town) 76 000; (canton) 404 000
Map Switzerland Ba

St George's ***Grenada*** Port and capital city of the island, lying on the south-west coast. It is built around a horseshoe-shaped bay, the brightly painted houses rising up the wooded hillsides. Point Salines airport is about 6 km (3.5 miles) south of the town.

Population 30 800
Map Caribbean Eb

St Gotthard (St Gothard) ***Switzerland*** Pass through the central Alps about 55 km (35 miles) south and slightly east of Lucerne, near the Italian border. A 15 km (9 mile) long tunnel, built 1872-80, carries a railway line under the pass. And in 1980 a 16 km (10 mile) long road tunnel replaced the old winding road over the top, which is usually open from June to mid-October.

Map Switzerland Ba

St Helena ***South Atlantic*** Mountainous island, 1930 km (1200 miles) west of Africa, where Britain held the exiled Napoleon Bonaparte for six years until he died in 1821. The island is only 122 km² (47 sq miles) in area. About 1500 people live in the capital and port, Jamestown. The climate is mild, with enough rain to grow hemp, vegetables and sweet potatoes; cattle are raised, too. St Helena is a British colony and ASCENSION ISLAND and TRISTAN DA CUNHA are its dependencies.

Population 5500
Map Atlantic Ocean Ee

St Helens, Mount ***USA*** Volcano 2549 m (8364 ft) high in the Cascade Range in western Washington state. It erupted in 1980, destroying forest up to 20 km (12 miles) away, killing 65 people and losing some 400 m (1312 ft) in height. The mountain lies within the 445 km² (172 sq mile) Mount St Helens National Monument established in 1982.

Map United States Ba

Saint John ***Canada*** Seaport city on the Bay of Fundy at the mouth of the Saint John river in southern New Brunswick. The French had a trading post there in 1630, but the first major settlement was made in 1783 by a group of New Englanders. In 1785 it received Canada's first city charter. Saint John has an excellent ice-free harbour and large dry docks. Industries include shipbuilding, papermaking and oil refining. Tourist attractions include the 18th-century court house and loyalist cemetery and the Reversing Falls rapids.

Population 114 000
Map Canada Id

Saint John ***USA/Canada*** River, 673 km (418 miles) long, which rises in Maine, USA, and forms part of the US-Canadian border before flowing across New Brunswick into the Bay of FUNDY at the port of SAINT JOHN. It is navigable for small vessels as far as FREDERICTON, New Brunswick. At Grand Falls, on the US border, the river drops 23 m (75 ft) in a great cataract and at Reversing Falls rapids, near the mouth of the river, the flow is reversed by high tides surging up the Bay of Fundy.

Map Canada Id

St John ***US Virgin Islands*** Smallest of the territory's three main islands, covering 52 km² (20 sq miles). An expensive but uncrowded tourist resort, about two-thirds of the island is national parkland and it has the USA's only underwater National Park.

Population 2400
Map Caribbean Eb

St John's ***Antigua and Barbuda*** Capital and chief port of Antigua. The town is laid out in a formal grid plan and contains some attractive 18th and 19th-century buildings, which survived a serious earthquake in 1843.

Population 30 000
Map Caribbean Cb

St John's ***Canada*** Capital of NEWFOUNDLAND and the easternmost city of Canada. It is a major fishing port and one of the oldest settlements in North America; it was colonised by the British under Sir Humphrey Gilbert in 1583. The city extends across a hillside overlooking the harbour. Fires in the 19th century destroyed much of the old town, but some of the old wooden buildings remain. Alcock and Brown made the first non-stop transatlantic air flight from there in 1919, and Marconi received the first transatlantic wireless message at the Cabot Tower on Signal Hill in 1901. Industries include fish processing, shipbuilding, iron foundries, textiles and paper.

Population 155 000
Map Canada Jd

St Lawrence ***Canada/USA*** River, 1197 km (744 miles) long, that flows out of Lake Ontario through Quebec and Ontario into the Gulf of St Lawrence. It is the main trade artery of Canada and a major source of hydroelectricity. It forms part of the boundary between Canada and the USA, and is open to US shipping. Shoals, rapids and islands (the Thousand Islands) slow navigation, but the scenery is a magnet for tourists.

Map Canada Id

St Lawrence, Gulf of Major waterway and fishing ground in eastern Canada, the outlet to the ATLANTIC OCEAN for ships using the ST LAWRENCE SEAWAY. It was discovered by the French explorer Jacques Cartier in 1535. Fogs and drifting ice are hazards from mid-December to mid-April.

Area 259 000 km² (10 000 sq miles)
Map Canada Id

St Lawrence Seaway and Great Lakes Waterway ***Canada/USA*** A system of canals, dams and locks linking the Great Lakes through the ST LAWRENCE river with the Atlantic Ocean. The system, some 3800 km (2350 miles) long, was opened in 1959. It permits ocean-going vessels up to 221 m (725 ft) long, 23 m (75 ft) wide and with up to 7.9 m (26 ft) draught to sail from Montreal, Quebec, to the west end of Lake SUPERIOR. The system provides the water that drives hydroelectric plants in ONTARIO and NEW YORK State.

Map Canada Hd

St Louis ***Senegal*** Seaport and regional capital of the north, founded by the French in 1659 on an island in the mouth of the Senegal river. It was the first French settlement in West Africa. In the 17th and 18th centuries, the French, Dutch, Portuguese and British grew rich on the trade in slaves and gum arabic, but St Louis was superseded by the deep-water port at Dakar as the mouth of the river silted up. It was the capital of French West Africa from 1895 to 1902, however, and remained Senegal's capital until 1958.

Population 95 000
Map Senegal Aa

St Louis ***USA*** City in eastern Missouri, on the Mississippi river near its confluence with the Missouri river. In the 19th century it was a key post for traders and settlers heading west, and it remains a busy trading city. Industries include car, aircraft and clothing manufacture, and brewing. Its suburbs extend across the river into Illinois.

Population (city) 429 300; (metropolitan area) 2 398 400
Map United States Hc

St Maarten ***Netherlands Antilles*** See ST MARTIN

Saint-Malo ***France*** Fishing and ferry port in northern Brittany, 65 km (40 miles) north-west of Rennes. It was founded in the 6th century as a monastic settlement, which was later fortified. Today it is a popular tourist resort with sandy beaches nearby. Only the medieval ramparts and some granite houses survived destruction in 1944 during the Second World War. But the town has been rebuilt in the ancient style.

Population 47 300
Map France Cb

St Martin ***Guadeloupe/Netherlands Antilles*** Island in the Leeward group just south of Anguilla. The northern section is administered by France as part of Guadeloupe, which lies 240 km (150 miles) to the south-west. The southern section, known as St Maarten, is part of the Netherlands Antilles, whose capital is on Curaçao, 900 km (560 miles) to the south-west. There

St Kitts and Nevis

THE 'MOTHER COLONY OF THE WEST INDIES', WHERE FRENCH AND BRITISH BATTLED FOR POSSESSION TWO CENTURIES AGO

The islands of St Christopher (usually called St Kitts) and Nevis lie in the LEEWARD ISLANDS group in the eastern Caribbean. They were discovered by Christopher Columbus in 1493 but there were no settlers until the British came to St Kitts in 1623 and Nevis in 1628. St Kitts became a base for the British colonisation of neighbouring islands and is proud of its title: 'Mother Colony of the West Indies.'

For a time St Kitts was settled simultaneously by the French and British, but in 1783 it became solely British. Administered for many years with ANGUILLA, St Kitts and Nevis became an independent Commonwealth country in 1983 with Elizabeth II as head of state. The inhabitants of both islands are descended from African slaves and are English-speaking.

St Kitts consists of three young but extinct volcanoes which are linked by a narrow sandy isthmus to much lower volcanic remains in the south. Mount Liamuiga (formerly Mount Misery) at 1156 m (3792 ft) is the island's highest point. Around most of the island the gentle slopes which rise from the coast are covered with fertile soil on which sugar cane is grown; sugar is the chief export, and since 1975 the industry has been nationalised. Market gardening and livestock rearing are being expanded on the steeper lands above the cane fields.

There are some light industries which use imported materials to make electronic equipment, footwear and batik (dyed cloth) for export. Older industries continue – sugar processing, brewing, distilling and bottling.

The white sands and relatively dry climate of southern St Kitts led to a major tourist development at Frigate Bay, and the government is anxious to build up this money-spinner. Many of the older buildings of historical interest in BASSETERRE, the capital and deep-water port, have been carefully restored. The partly constructed fort of Brimstone Hill, built on a cliff top in the 17th and 18th centuries, was the scene of battles between the British and French. It is now being restored and provided with a museum.

Nevis lies 3 km (2 miles) south-east of St Kitts, with which it has daily links by boat and light aircraft. The island is an extinct volcano which rises to 985 m (3232 ft) at Nevis Peak. The cloud-covered peak reminded Columbus of snow-capped mountains in Spain. The Spanish word for snow is *nieves*, and this gave the island its name.

Farming has declined on Nevis and the government is trying to increase food production. Small factories process locally grown copra and sea-island cotton seed into oil. There is a small boat-building industry. Tourism is the main source of income, thanks to the fine beaches. Visitors stay in cottages in the grounds of old plantation houses. CHARLESTOWN is the island capital.

ST KITTS AND NEVIS AT A GLANCE

Map Caribbean Cb
Area 261 km² (101 sq miles)
Population 43 900
Capital Basseterre
Government Parliamentary monarchy
Currency East Caribbean dollar = 100 cents
Language English
Religion Christian (75% Protestant)
Climate Tropical; no recognised rainy season. Average temperature is between 24°C (75°F) and 27°C (81°F) all year
Main primary products Sugar, cotton, coconuts, vegetables, livestock
Major industries Agriculture, textiles, electronic equipment, sugar processing, tourism
Main exports Sugar, clothing, electronic equipment, footwear
Annual income per head (US$) 1200
Population growth Declining
Life expectancy (yrs) Male 69 **Female** 72

are no customs or other barriers between the two sections; the French capital of Marigot and the Dutch Philipsburg – both with good harbours – are only 10 km (6 miles) apart. At one time, the production of salt by evaporating sea water was the main source of income, but today tourism is the major industry. Most of the population are of African origin.
Population 24 000
Map Caribbean Cb

St Moritz ***Switzerland*** Fashionable resort and spa in the south-east, about 15 km (9 miles) north of the border with Italy. It has thermal baths and is noted for its winter sports and summer mountaineering.

It is the site of the Cresta Run – 1212 m (3977 ft) long, with a drop of 157 m (515 ft) – for bobsleighs. In 1986, Nico Barachi of Switzerland set a record for the course of 51.31 seconds – an average speed of 85 km/h (53 mph). St Moritz was the site of the Winter Olympics in 1928 and 1948.
Population 7800
Map Switzerland Ba

Saint-Nazaire ***France*** Shipbuilding town at the mouth of the Loire river, developed in the 19th century as a deep-water port for the city of Nantes, 50 km (30 miles) to the east. During the Second World War it was an important German submarine base for U-boats attacking Allied merchant shipping convoys on their way across the Atlantic. The British bombed the base several times. Today, there is a viewing terrace over the remains of the submarine pens.
Population 131 800
Map France Bc

St Lucia

OIL TANKERS AND CRUISE LINERS BRING REVENUE TO A CARIBBEAN ISLAND SWAPPED A DOZEN TIMES BETWEEN BRITAIN AND FRANCE

One of the WINDWARD ISLANDS in the eastern Caribbean, St Lucia was settled briefly by the British in 1605. Over the next 200 years, control passed some 10 to 14 times between them and the French before it was finally captured by the British in 1803. English is the official language, but many of the people, mostly descendants of African slaves, speak a French patois. St Lucia became a fully independent member of the Commonwealth, with the Queen as head of state, in 1979.

St Lucia is an island of extinct volcanoes. Mount Gimie is the highest peak at 950 m (3117 ft). To the south lies the horseshoe-shaped Qualibou which contains 18 lava domes and seven craters. In the west are the impressive peaks of the PITONS – Gros Piton and Petit Piton – which rise directly from the sea to more than 750 m (2460 ft).

Bananas are the main crop and chief export, but production is often affected by hurricanes, drought and disease. Coconuts are the second crop. The free port at Vieux Fort in the south has attracted industries whose products include electrical components for export.

Grande Cul de Sac Bay, to the south of the capital, CASTRIES, is one of the deepest and most modern tanker ports in the Americas for oil transshipment. The beautiful Castries harbour is a regular calling point for cruise liners.

ST LUCIA AT A GLANCE

Map Caribbean Cc
Area 619 km² (239 sq miles)
Population 123 000
Capital Castries
Government Parliamentary monarchy
Currency East Caribbean dollar = 100 cents
Languages English, French patois
Religion Christian (90% Roman Catholic, 10% Protestant)
Climate Tropical; dry season January to April. Average temperature is between 25°C (77°F) and 28°C (82°F) all year
Main primary products Bananas, cocoa, coconuts, vegetables, fruit, spices
Major industries Food processing, rum distilling, clothing, electrical components, fishing, tourism
Main exports Bananas, cardboard boxes, clothing, cocoa, coconut products
Annual income per head (US$) 1020
Population growth (per thous/yr) 11
Life expectancy (yrs) Male 65 **Female** 71

St Paul ***USA*** Twin city of Minneapolis and the state capital of Minnesota. It was founded in 1841 at the navigable limit of the Mississippi river. The city is a centre for tourism and has a popular winter carnival.
Population (city) 265 900; (metropolitan area) 2 230 900
Map United States Hb

St Petersburg ***USSR*** See LENINGRAD

St Pierre ***Martinique*** See PELEE, MONT

St Pölten ***Austria*** Industrial town 56 km (35 miles) west of Vienna. In the Middle Ages it was a centre for the leather and textile industries, and in the 1850s became one of the country's earliest industrial towns. Its other manufactures include turbines, paper, machinery, furniture and rolling stock.
Population 51 200
Map Austria Da

Saint-Raphaël ***France*** See SAINT-TROPEZ

St Thomas ***US Virgin Islands*** Main tourist island of the territory, covering 83 km² (32 sq miles). Its fine harbour, beside which stands the capital Charlotte Amalie, was long the haunt of pirates but now attracts cruise ships and luxury yachts.
Population 44 000
Map Caribbean Eb

Saint-Tropez ***France*** Fashionable tourist resort on the Mediterranean, 100 km (60 miles) east of Marseilles, with excellent beaches and yacht marinas. It has been popular with artists since the 1890s – mainly because of its fine, clear light. The resort of Saint-Raphaël is just to the north.
Population 6300
Map France Ge

Saint Vincent, Cape ***Portugal*** See SAO VICENTE, CABO DE

St Vincent, Gulf Resort-lined inlet in South Australia. ADELAIDE stands on its eastern shore and Kangaroo Island at its mouth. The south-western outlet to the INDIAN OCEAN, Investigator Strait, was named after the *Investigator*, the ship used by the British navigator Matthew Flinders, who discovered the gulf in 1802 during his circumnavigation of Australia.
Dimensions 145 km (90 miles) long, 72 km (45 miles) wide
Map Australia Fe

Saintes ***France*** Industrial town on the Charente river, 60 km (37 miles) south-east of La Rochelle. It produces iron goods, pottery, tiles and brandy. Known as Mediolanium in ancient times, it contains the remains of a Roman amphitheatre. There is also a fine 15th to 16th-century church.
Population 27 500
Map France Cd

Saintes, Iles des ***Guadeloupe*** Group of eight tiny volcanic islands situated 10 km (6 miles) south of Basse Terre. The only inhabited islands are Terre de Haut and Terre de Bas. Terre de Haut's fishermen are descended from 17th-century settlers from northern France. The fishermen of Terre de Bas are mostly black.
Population 2900
Map Caribbean Cb

Saintonge ***France*** Historic province on the Bay of Biscay, north of the Gironde estuary. It was a fief of Aquitaine and part of England from 1154 to 1375. The chief town is SAINTES.
Map France Cd

Saipan ***Northern Marianas*** Largest and most populous island of the territory, covering 122 km² (47 sq miles) and lying 200 km (125 miles) north-east of Guam. It has high cliffs except on the western side, where the capital, Susupe (population 8000), is situated. The capture of Saipan by US forces in June 1944, after a bloody battle, was a major step towards the recapture of the Philippines and the defeat of Japan in the Second World War.
Population 17 000
Map Pacific Ocean Cb

Sakai ***Japan*** Port and industrial city on south Honshu island, just south of Osaka. It was a

St Vincent and the Grenadines

CAPTAIN BLIGH OF THE 'BOUNTY' TOOK BREADFRUIT TREES TO ST VINCENT TO PROVIDE FOOD FOR SUGAR SLAVES

In the beautiful botanic gardens of KINGSTOWN, capital of St Vincent in the WINDWARD ISLANDS, grows a breadfruit tree descended from one brought from Tahiti by Captain William Bligh in 1792. It was Bligh's second attempt; the first ended in 1789 with the famous mutiny aboard HMS *Bounty*. The aim was to provide a source of food for the African slaves working on the West Indian sugar plantations. This plan failed – the slaves preferred other foods such as bananas and plantains – but St Vincent is still sometimes called the 'Tahiti of the Caribbean', partly because of its productive soil and luxuriant vegetation.

Christopher Columbus may or may not have spotted the island in 1498 – there is some controversy – but he was certainly not the first person to see it. The Ciboney Indians are believed to have lived there from about 4300 BC, followed by the Arawaks and, in the 16th century, the Caribs. Only in the 17th century did Europeans – English and French – seriously attempt to settle. The Caribs resisted them, but after the last Carib War (1795-7) most of the Indians were deported to the Bay Islands, off Honduras. Meanwhile, control switched many times between Britain and France, the island being finally ceded to Britain in 1783.

With the Grenadine islands of Bequia, MUSTIQUE, Canouan, Mayreau and Union to the south, St Vincent became an independent country with the Queen as head of state in 1979. Two-thirds of the population are descended from African slaves and most of the rest are of mixed race. Some descendants of the Carib Indians live on the slopes of Mount SOUFRIERE in the north of St Vincent. English is the main language.

A chain of volcanoes makes up the backbone of St Vincent. Soufrière, 1234 m (4048 ft) high, is one of the two active volcanoes in the eastern Caribbean (the other is MONT PELEE on Martinique). Its last violent eruption was in 1979. In 1812 it killed 56 people, and in 1902, more than 1000 died from breathing the hot dust, from burns or as victims of falling stones.

St Vincent is one of the poorest islands in the eastern Caribbean and unemployment is high. Farming is the main occupation and bananas for the United Kingdom are the main export. The island is the world's leading producer of arrowroot starch (used to make medicines, fine flour and paper for computers). Some copra and coconut oil are exported, but sugar is now grown only for local use. Much of the island is forested.

There is little manufacturing; the few factories process farm products and make cement and furniture. A new flour mill serves all the Windward Islands. The government is trying to develop tourism to provide employment, and cruise ships call at the deep-water harbour of Kingstown.

The seas around the tiny Grenadine islands and the many bays are among the best boating waters in the world. Luxury tourism and yachting have made the inhabitants of Bequia and Mustique more prosperous than those on the main island of St Vincent.

ST VINCENT AT A GLANCE
Map Caribbean Cc
Area 389 km² (150 sq miles)
Population 103 000
Capital Kingstown
Government Parliamentary monarchy
Currency East Caribbean dollar = 100 cents
Languages English, French patois
Religion Christian (75% Protestant, 13% Roman Catholic)
Climate Tropical; dry season January to May. Average temperature is between 26°C (79°F) and 28°C (82°F) all year
Main primary products Bananas, arrowroot, eddoes (taro), coconuts, spices
Major industries Agriculture, food processing, rum distilling, flour milling, cement manufacture, forestry
Main exports Bananas, arrowroot starch, eddoes (taro), copra
Annual income per head (US$) 740
Population growth (per thous/yr) 14
Life expectancy (yrs) Male 58 **Female** 59

leading city from the 14th to the beginning of the 17th century – independent, with its own militia, and trading as far afield as the Philippines. The city's economy collapsed – until its modern revival as an oil and steel centre – when Japan outlawed most links with the outside world in the 17th century.
Population 818 400
Map Japan Cd

Sakhalin ***USSR*** Island in the Soviet Far East, separated from the Japanese island of Hokkaido to the south by the 40 km (25 mile) wide La Perouse Strait. Long, narrow and mountainous, with thick forests in the north, Sakhalin covers an area of about 76 400 km² (29 500 sq miles). It was originally settled by Russian fishermen in 1853 and was made a convict colony. The southern part was taken over by Japan in 1905. Russia took it back in 1945, following the defeat of Japan in the Second World War.

The inhabitants' main occupations are fishing (mostly for crabs, herring and cod), canning, coal mining, oil drilling, paper making and hunting (mostly for sables, foxes and bears). The main town and capital is Yuzhno Sakhalinsk (formerly Toyohara).
Population 660 000
Map USSR Pd

Sakhira (La Skhirra) ***Tunisia*** Port on the Gulf of Gabès, 100 km (about 60 miles) south-west of Sfax. It is the oil terminal for the pipeline from the Algerian Edjele oil field and from Tunisia's El Borma field.
Map Tunisia Ba

Salalah ***Oman*** Port and capital of DHOFAR province, situated 100 km (62 miles) east of the border with South Yemen. Salalah is a modern city set among coconut groves and banana plantations. Parts of the old city remain, including the former residence of the sultan, some traditional Arab houses and an old mosque. In addition to its own harbour, Salalah is served by the new port of Raysut, 20 km (12 miles) to the east. It has an airport built in 1977.
Map Oman Ab

Salamanca ***Spain*** University town about 170 km (105 miles) north-west of Madrid. The whole town has been declared a national monument because of its magnificent 17th and 18th-century buildings. Its industries include tanning, brewing and flour milling. It is surrounded by an agricultural province of the same name.
Population (town) 167 100; (province) 368 100
Map Spain Cb

Salamis ***Cyprus*** Site of a ruined Greek city, founded in 1180 BC, on the east coast. It was the scene of a naval victory for the Greeks in 306 BC when Demetrius I defeated Ptolemy I of Egypt. It has the remains of an amphitheatre, a temple of Zeus and a forum (main square).
Map Cyprus Ab

Salamís ***Greece*** Island in the Saronic Gulf, 13 km (8 miles) west of Piraeus. In the narrow strait between the island and the coast of Attica the Athenians destroyed a Persian fleet in 480 BC. It is now a major Greek naval base.
Population 28 600
Map Greece Cc

Salang Tunnel ***Afghanistan*** Tunnel through mountains 100 km (62 miles) north of Kabul. Opened in 1964, it cuts the distance from Kabul to the northern plains by 200 km (125 miles). Lying 3360 m (11 030 ft) above sea level, the tunnel is 2676 m (8780 ft) long and 8 m (26 ft) wide. In the winter, it is closed at night as the temperature can drop to −20°C (−4°F).
Map Afghanistan Ba

Salaverry ***Peru*** Modern port 8 km (5 miles) south of Trujillo. It exports copper from Quiruvilca 120 km (75 miles) inland, and sugar from the valleys and plains of La Libertad.
Population 5250
Map Peru Ba

Saldanha ***South Africa*** Town on Saldanha Bay about 110 km (70 miles) north-west of Cape Town, used as a naval base in the Second World War. In the 1970s it was redeveloped as a port to export iron ore from Sishen in northern Cape Province. A modern plant in the town extracts salt, potassium and bromine from seawater.
Population 27 500 (including nearby town of Vredenburg)
Map South Africa Ac

Salé ***Morocco*** Atlantic seaport and suburb of RABAT, to which it is linked by bridge and ferries across the Bou Regreg river. Founded on the site of an ancient Roman port, it was a major commercial centre from the Middle Ages. It has a 12th-century mosque, a Koranic college dating from 1341, a monumental gateway (Bab Mrisa) and many examples of 13th and 14th-century Moroccan architecture. Its markets specialise in basketwork, jewellery and embroidery, and there are streets for cabinet makers, stone carvers, blacksmiths and brass workers. In the 17th century it was the home of the Barbary pirates, known as the Sallee Rovers.
Population 290 000
Map Morocco Ba

Salem ***USA*** **1.** City in Massachusetts, about 20 km (12 miles) north-east of Boston. It is one of the oldest settlements in the country, founded in 1626. The city became notorious after a series of witchcraft trials in 1692, which ended in the execution of 20 townsfolk, mostly women. It later became a prosperous port, and today it is also an industrial centre and tourist resort. Salem still has several 17th-century buildings, including the House of Seven Gables, immortalised in the novel of the same name by Nathaniel Hawthorne (1804-64).

2. State capital of Oregon, situated on the Willamette river about 70 km (45 miles) south and slightly west of Portland. It is the centre of a rich farming region, and has numerous food-processing plants and a paper mill. It was founded in 1840 by Methodist missionaries, and has a university founded two years later.
Population 1. (city) 38 600; (metropolitan area) 259 100. **2.** (city) 90 300; (metropolitan area) 255 200
Map United States (1) Lb; (2) Bb

Salentina ***Italy*** Area forming the heel of Italy. Vines and tobacco are the main crops. The towns, especially LECCE, are noted for their ornate Baroque architecture.
Map Italy Fd

Salerno ***Italy*** Seaport about 45 km (28 miles) south-east of Naples. It was renowned in the Middle Ages as a centre of medical science and has a fine 11th-century cathedral. Its industries include cement, machinery and textiles.
Population 155 900
Map Italy Ed

Salford ***United Kingdom*** See MANCHESTER

salina In the south-western USA, a SALT PAN or a marsh, spring, pond or lake containing salt water.

Salinas Valley ***USA*** Fertile lettuce and fruit-producing region between two parallel ridges of the Coast Ranges in California. The Salinas river runs 240 km (150 miles) north-west into Monterey Bay south of San Francisco.
Map United States Bc

salinity Measure of the total quantity of salts dissolved in water, especially seawater, expressed in parts per thousand by mass. Seawater's average is 35 parts per thousand.

Salisbury ***United Kingdom*** English cathedral city in Wiltshire, 32 km (20 miles) north-west of Southampton. It is best known for its beautiful cathedral, begun in 1220, whose slender spire – at 123 m (404 ft) the tallest in Britain – was completed in 1334. It replaced an earlier cathedral 3 km (2 miles) to the north, a hilltop site now known as Old Sarum – the Norman name of the town.
Population 40 000
Map United Kingdom Ee

Salisbury ***Zimbabwe*** See HARARE

Saloniki (Thessaloniki) ***Greece*** The country's second largest city after Athens, and Macedonia's chief port, lying on the Thermaic Gulf. It is a university town, a NATO base and a major industrial centre with oil refineries, engineering and textile plants. The city was founded in 315 BC. Little remains of the original buildings, but Roman ruins – the arch and palace of the 4th-century Roman emperor Galerius – and Byzantine ramparts are still visible, and there is an archaeological museum containing some of the treasures of ancient Macedonia, including most of the finds from the tomb of King Philip (383-336 BC) at Vergina. The city stages an International Trade Fair each September.
Population 706 200
Map Greece Ca

Salpausselkä ***Finland*** A terminal moraine – a line of hills formed from debris left by retreating ice sheets during the Ice Age. The hills extend across southern Finland from the province of Karelia in the south-east to the Hanko peninsula in the south-west of the country.
Map Finland Cc

Salt ***Jordan*** Small town and administrative centre, situated 29 km (18 miles) north-west of Amman set in fertile hill country. Salt has been inhabited since 3000 BC. It has the remains of some Roman tombs and the mosaic of an early Byzantine church.
Population 35 000
Map Jordan Aa

salt dome, salt plug DIAPIR, or rock dome, with a core of salt. The rock surrounding a salt dome is heavily folded and faulted and may act as a reservoir for oil and gas. The salt in a dome can be up to 1000 m (3300 ft) deep.

salt flat Wide, level stretch of country that has very salty soil or a surface covering of salt, representing the bed of a dried-up salt lake.

Salt Lake City *USA* State capital of Utah and the headquarters of the Mormon Church. The site, about 22 km (14 miles) east of the Great Salt Lake, was chosen in 1847 by Brigham Young, who had led the Mormons more than 1800 km (1100 miles) from Illinois and Missouri, across the Rockies to their new home. It was laid out around the Mormon Temple and Tabernacle with wide tree-lined avenues and numerous parks, and is now a manufacturing, cultural, commercial and tourist centre.
Population (city) 164 800; (metropolitan area) 1 025 300
Map United States Db

salt marsh Marsh alongside a low-lying shore that is frequently flooded by the sea.

salt pan Basin containing an evaporating salt lake, bordered by a deposit of salt.

Salta *Argentina* City about 1280 km (795 miles) north-west of Buenos Aires. Founded in 1582, it has suffered a number of earthquakes, although some of its colonial buildings – among the best in the country – have survived. It stands at an altitude of 1190 m (3900 ft) just east of the Andes in a region that is rich in Inca and other pre-Columbian archaeological sites. Its main industries are tourism, mining of iron, zinc, lead, tin and silver, and cattle raising.
Population 260 300
Map Argentina Ca

Saltillo *Mexico* Capital of the northern state of Coahuila, and a thriving manufacturing and commercial town making machinery, silverware, pottery and textiles. It is known for its *sarapes* – brightly coloured woollen overgarments. Just outside Saltillo is the Angostura battlefield, scene in 1847 of one of the bloodiest battles of the Mexican-American War.
Population 321 800
Map Mexico Bb

salting Slightly higher area of a SALT MARSH where there is grass and little bare mud.

Salto *Uruguay* Second largest city in the country after the capital, Montevideo. It is on the Argentinian border about 430 km (270 miles) north-west of Montevideo. It is a river port, in a citrus fruit growing area and makes wine, processed meat and boats.
Population 80 000
Map Uruguay Ab

Salut, Iles du *Guyane* Atlantic islands with a grim past. At one time the whole of Guyane was a French penal colony. The most infamous of the prisons were these islands – and they include Devil's Island, which became notorious. Captain Alfred Dreyfus was sent there in 1894 after being court-martialled and wrongly sentenced as a German spy. The 20th-century writer and former inmate Henri Charrière wrote about his term there in *Papillon*.

The rocky islands lie 16 km (10 miles) offshore. Palm-covered Devil's Island measures only about 1200 by 400 m (3940 by 1300 ft). The islands were used as prisons from 1852 to 1953. Some of the prison buildings still stand; the former wardens' mess hall is now a hotel.
Map Guyane Ba

▼ HIGH FORTRESS Hohensalzburg towers over Salzburg's Old City. The 11th-century castle on the tree-girt Mönchsberg was a home of the prince-bishops who ruled the city for more than 1000 years from 798.

Salvador *Brazil* Port and capital of Bahia state about 1200 km (745 miles) north-east of Rio de Janeiro. Traditionally, it was the first European landfall in Brazil – by the Portuguese admiral Pedro Alvares Cabral in 1500 – and the country's first capital. Much of the Alta or Upper City – on a bluff overlooking the bay – is a national monument; it contains more than a hundred 17th and 18th-century churches and civic buildings.

Bahia – once the name of the city and still the name of the state – means 'bay' and refers to the city's huge natural harbour. The port's main exports are cigars and other tobacco goods, petrochemicals, coffee, hides and vegetable oils. The main product of the state, which covers 561 026 km² (216 556 sq miles), is cocoa.
Population 1 507 000; (state) 9 470 000
Map Brazil Ec

Salween (Mae Nam Khong; Nu Jiang) *China/Burma/Thailand* River, some 2900 km (1800 miles) long, rising in eastern Tibet in China. It flows east in a deep valley, then south into eastern Burma, across the hilly Shan and Kayah states. It continues southwards, forming the Burma-Thai border for about 125 km (80 miles), and then crosses lowlands to the Gulf of Martaban at the Burmese city of Moulemin. Rapids, shoals and other dangers confine navigation to the section below Kayah state.
Map Burma Cb; China Ee

Salzburg *Austria* **1.** Mountainous central state renowned for its Alpine spas and resorts, which attract thousands of tourists each year. At one time an independent state ruled by prince-archbishops, Salzburg became part of Austria in 1816, after the Napoleonic Wars. It is noted for its salt deposits and its iron, copper and gold mines. **2.** Capital of Salzburg state, on the Salzach river, 255 km (160 miles) west of Vienna. It was the birthplace in 1756 of the composer Wolfgang Amadeus Mozart, and a festival is held in his honour each August. Mozart was born in the old town, and his house is now a major tourist attraction.

The city, which lies between two high and craggy hills, was built by successive bishops in the 16th and 17th centuries. Its narrow streets include the Getreidegasse, lined with six-storey buildings, each bearing a wrought-iron shop sign depicting its wares. The city was badly bombed during the Second World War, and almost half of it was rebuilt. Among its older buildings are the 17th-century cathedral and an 11th-century castle.
Population (province) 441 840; (city) 140 000
Map Austria Cb

Salzkammergut *Austria* Mountain and lake region in the eastern Alps, crossed by the Traun river. Once important for salt production, it is now a tourist region of spas with Bad Ischl as its chief centre. There are 27 lakes, of which the largest is the Attersee. The highest peak is Dachstein (2998 m, 9836 ft).
Map Austria Cb

Samanala ***Sri Lanka*** See ADAM'S PEAK

Samar ***Philippines*** Third-largest island of the Philippines, covering 13 080 km² (5050 sq miles). Most easterly of the VISAYAN ISLANDS, it is exposed to Pacific typhoons. The island consists mainly of a rugged dissected plateau, with dense forests and few roads except along the west coast. It has few mineral resources other than copper mined at Bagacay, and the main occupation is subsistence rice farming.

The main towns include Calbayog, which has an airport, and Catbalogan, a seaport, both on the west coast. Tacloban on LEYTE (which is linked to Samar by a bridge) serves as its main trading port.

Population 1 100 000
Map Philippines Cc

Samaria ***Israel/Jordan*** Region of ancient Palestine extending from the Mediterranean to the Jordan river, and lying south of Galilee and north of Judaea. It was made a division of the province of Judaea by the Roman Emperor Augustus in AD 6. Its main city was also called Samaria, the ruins of which now lie in the part of Jordan occupied by Israel in 1967. The city, founded in 887 BC, was the holy city of the Samaritans.

Map Jordan Aa

Samarkand ***USSR*** Ancient city in Uzbekistan, about 250 km (155 miles) north of the Afghan border. In 329 BC, when the Macedonian conqueror Alexander the Great captured it, it was known as Maracanda – and it was then already an established city. In the 7th century AD it became a way station on the Silk Road from China to Europe, and it was visited by the Venetian explorer Marco Polo in the 13th century.

In the 14th century it became the capital of the Tatar warlord Tamerlane (1336-1405). Parts of his tomb, the Gur Amir, still stand – and the dome upon it, copied from a Syrian mosque at Damascus which Tamerlane burnt down in 1402, inspired an international style. The bulging dome of the tomb was copied for the onion-shaped domes of the Moscow Kremlin. The style was also taken south by Baber, one of Tamerlane's descendants, who founded the Mogul empire of India. It was one of Baber's dynasty, Shah Jahan, who in 1648 used the same style of dome to crown his own wife's tomb – the Taj Mahal.

Samarkand became a Russian city in 1868, and modern blocks of flats now surround the mosaic-filled Muslim buildings of the city centre. The city's factories produce cotton and silk textiles, carpets, leather, shoes, tea, tobacco, canned fruit and wine.

Population 515 000
Map USSR He

Samarra ***Iraq*** Holy city on the Tigris river, about 100 km (60 miles) north-west of the capital, Baghdad. Built about AD 836, it became a capital city under the Abbasid caliphs who ruled the region. Its 17th-century, copper-domed mosque is revered by Shiite Muslims, and its ruined Great Mosque, built about AD 852, was once the largest in the world.

Population 62 000
Map Iraq Bb

Samoa, American ***Pacific Ocean*** Group of five islands and two atolls to the east of WESTERN SAMOA and about 2600 km (1600 miles) north-east of New Zealand. They have been American territory since 1900, but there is little pressure for reunion with Western Samoa, whose standard of living is much lower and whose chiefs would outrank those of American Samoa in a combined country.

American Samoa has an elected legislature and elected executive governor. The US government gives substantial grants, although the US Navy base at PAGO PAGO, the capital, closed in 1951. The USA also buys almost all the territory's exports, mainly canned tuna fish. Taro, coconuts and bananas are grown for local consumption, although most of the islands (including TUTUILA, the main one) are hilly and densely forested. The total land area is 197 km² (76 sq miles).

Population 36 000
Map Pacific Ocean Ec

Sámos ***Greece*** Wooded island in the eastern Aegean Sea only 2 km (1.2 miles) from the Asia coast of Turkey. It was the birthplace of the mathematician and philosopher Pythagoras (about 582-507 BC), the man who gave his name to the theorem still taught in schools all over the world. The island's main town and port, Vathí, was built in the 19th century.

Population 31 600
Map Greece Ec

Samothrace (Samothráki) ***Greece*** One of the northern Aegean islands, 40 km (25 miles) from the coast of Thrace. In Homeric legend the sea god Poseidon watched the battle for Troy from the island's Mount Fengari (1600 m, 5250 ft). A marble sculpture of the winged goddess of victory, now in the museum of the Louvre in Paris, was found during excavations of the Sanctuary of Great Gods near the island's port of Kamariotissa.

Population 2900
Map Greece Da

Samsun ***Turkey*** Black Sea port about 420 km (260 miles) north-east of the capital, Ankara. It is the capital of the province of the same name. In May 1919 Kemal Ataturk, founder of modern Turkey, launched the revolution in the area to rid Turkey of foreign control.

The port's main cargo today is tobacco grown in the fertile and heavily populated province, which covers 9579 km² (3698 sq miles).

Population (town) 280 100; (province) 1 123 100
Map Turkey Ba

Samut Prakan ***Thailand*** Industrial province and town, both of which have become a southern suburb of the capital, Bangkok. The main industries are steel, glass, electrical goods, textiles and pharmaceuticals.

Population (town) 247 900; (province) 646 800
Map Thailand Bc

San Agustín ***Colombia*** Town 300 km (190 miles) south-west of Bogotá. It lies beside the San Agustín National Park, which contains many remarkable stone carvings of a pre-Columbian farming society. Hundreds of rough-hewn figures of men, animals and birds line the

Western Samoa

TRADITIONAL LIFE THRIVES IN A TWICE-COLONISED NATION AGAIN UNDER A NATIVE KING

When the Pacific islands of Western Samoa became independent in 1962, it was the second time that they had achieved freedom in 73 years. The mountainous tropical islands had been squabbled over by the Dutch, British, Germans and Americans since their discovery by Europeans in the 18th century. In 1889 they were given independence under King Malietoa Laupepa, but after the king died in 1898, the islands came under German control, by agreement with USA and Great Britain.

English influence had already been well established by the arrival in 1830 of missionaries, who quickly converted the people to Christianity. New Zealand annexed the German-ruled islands in 1919 and administered them until 1962.

The islands lie in the Polynesian sector of the Pacific, about 720 km (450 miles) north-east of Fiji. The two main islands are SAVAI'I and UPOLU. Savai'i is largely covered with volcanic peaks and lava plateaus; Upolu has two-thirds of the population and the capital, APIA.

According to legend, Samoa was the homeland from which the Polynesians set out to colonise other parts of the Pacific, and the traditional lifestyle still thrives. Most people live in *fale* (traditional open-sided thatched houses) in coastal villages. Many tourists visit the grave of Robert Louis Stevenson and his home, Vailimá, now the official home of King Malietoa Tanumafili II.

WESTERN SAMOA AT A GLANCE
Map Pacific Ocean Ec
Area 2831 km² (1093 sq miles)
Population 164 000
Capital Apia
Government Parliamentary monarchy
Currency Tala = 100 sene
Languages Samoan, English
Religion Christian
Climate Tropical; average temperature in Apia is between 28 and 29°C (82 and 84°F) all year
Main primary products Bananas, coconuts, cocoa, taro, timber, fish
Major industries Agriculture, food processing, forestry, fishing, tourism
Main exports Copra, coconut oil, taro, cocoa, timber, bananas
Annual income per head (US$) 650
Population growth (per thous/yr) 9
Life expectancy (yrs) **Male** 61 **Female** 65

forest paths and small streams of the Valley of Statues. The earliest of these date from about AD 600. Archaeologists have discovered little about the people, other than that they lived in bamboo huts and grew crops. Many of the stone animals guard elaborate tombs. Pottery, gems and necklaces found in some of the tombs can be seen at the museum in the town.

Population 16 850

Map Colombia Bb

San Andreas Fault ***North America*** See CALIFORNIA; FAULT; SAN FRANCISCO

San Antonio ***Chile*** Seaside resort and port 113 km (70 miles) west of Santiago. It handles the output from El Teniente, one of the world's largest underground copper mines.

Population 60 800

Map Chile Ac

San Antonio ***USA*** Industrial city and market centre in southern Texas, midway between Houston and the Mexican border. There are several military bases, and the San Antonio Missions National Historical Park preserves four 18th-century Spanish missions in the city. Another mission, the Alamo, was made into a fort in 1793, and 187 Texans – including the folk hero Davy Crockett – held out there against the Mexican army from February 23 to March 6, 1836, during the Texas Revolution; they were slaughtered to a man.

Population (city) 842 800; (metropolitan area) 1 188 500

Map United States Ge

San Bernardino ***USA*** Industrial city in California, about 90 km (55 miles) east of Los Angeles, forming part of the Riverside metropolitan area. It was founded by Mormons in 1851 between the San Bernardino and San Gabriel Mountains. It is also a fast growing residential centre in a citrus fruit producing area.

Population (city) 130 400; (metropolitan area) 1 810 900

Map United States Cd

San Cristóbal ***Venezuela*** Capital of Táchira state, high in the Andes, and about 650 km (400 miles) south-west of the capital, Caracas. It was founded by the Spanish in 1561 and retains much of its old colonial appearance.

Population 288 000

Map Venezuela Ab

San Diego ***USA*** Southernmost port in California, about 20 km (12 miles) north of the Mexican border. It was founded as a Spanish mission in 1769 and has become a naval base, tuna fishing centre and an industrial city specialising in aircraft and defence equipment.

Population (city) 960 500; (metropolitan area) 2 063 900

Map United States Cd

San Fernando ***Trinidad and Tobago*** Port and second largest city (after Port of Spain) in Trinidad, on the west coast 40 km (25 miles) south of Port of Spain. It is an important market and administrative centre serving the southern sugar-growing districts.

Population 40 000

Map Trinidad Aa

San Francisco ***USA*** City in California on a hilly neck of land between San Francisco Bay and the Pacific Ocean, and the centre of one of the USA's major conurbations, including OAKLAND and SAN JOSE. It grew around a Spanish mission founded in 1776 and enjoyed a boom during the gold rush of 1848-9. Much of the city was destroyed by earthquake and fire in 1906 – it lies on the San Andreas Fault – and has been rebuilt and extended onto land reclaimed from the bay. Some of the hills have been contoured to ease building, but it is renowned for the cable-car trams that climb its steeply sloping streets. San Francisco is one of the country's main ports and is a major financial and cultural centre, and has several universities. It suffers from fogs but remains a mecca for tourists, particularly to its Chinatown.

Population (city) 712 800; (metropolitan area) 1 541 900; (conurbation) 5 684 600

Map United States Bc

San Francisco Bay ***USA*** Largest bay on the coast of California. It is about 95 km (60 miles) long and up to 20 km (12 miles) wide. The bay is fed by the Sacramento and San Joaquin rivers, and opens to the Pacific Ocean through the Golden Gate, spanned by the Golden Gate Bridge – a suspension bridge spanning 1280 m (4200 ft) completed in 1937.

Map United States Bc

San Gimignano ***Italy*** Beautifully preserved medieval town about 40 km (25 miles) south-west of Florence. It is set amongst the vineyards and gentle hills of southern Tuscany. Its skyline is dominated by palace and church towers.

Population 7400

Map Italy Cc

▼ SILICON VALLEY The electronics centre of San Jose in California stretches into the distance like a piece of giant computer circuitry.

San Joaquin ***USA*** River much used for irrigation in California. It rises in the Sierra Nevada mountains some 300 km (185 miles) east of San Francisco and flows about 560 km (350 miles) north-west through a rich citrus fruit and wine producing region to the Sacramento river near San Francisco Bay.

Map United States Bc

San José ***Costa Rica*** National capital founded by Spanish settlers from the nearby city of Cartago in 1738. It replaced Cartago as capital in 1823 because it has a pleasant climate all year round – average temperatures range from 14-27°C (57-81°F) – thanks to its altitude of 1170 m (3840 ft). It also has an attractive mix of Spanish and modern buildings, with wide avenues and a number of parks. Its main industries are clothing manufacture, food processing, textiles and pharmaceuticals.

San José province produces much of Costa Rica's sugar cane and coffee on its rich, deep volcanic soil. It also has numerous dairy farms.

Population (city) 249 000; (province) 893 300

Map Costa Rica Bb

San Jose ***USA*** City about 20 km (12 miles) south-east of San Francisco Bay in California. It was once renowned for its wine grapes and other fruits but is now the main town of 'Silicon Valley', a centre of the computer industry. The city's population has trebled since 1960.

Population (city) 686 200; (metropolitan area) 1 371 500

Map United States Bc

San Juan ***Argentina*** Wine-making city about 1000 km (620 miles) west and slightly north of Buenos Aires. It is the capital of the province of the same name. Founded in 1562, the city was flattened by an earthquake in 1944 but has since been largely rebuilt. The province extends into the Andes to the Chilean border.

Population (city) 290 500; (province) 466 000

Map Argentina Cb

A DAY IN THE LIFE OF A CALIFORNIAN COMPUTER TECHNOLOGIST

Roger Gerski sinks back into his armchair, nightcap in hand, as his wife Anne waves goodbye to their dinner guests. It has been a long day.

At 6.30 he was up and jogging round the block – at his favourite time of day when the air is refreshingly crisp, and the sun just filters through the leaves of the garden palms before the scorching heat of the day. He was back for a swim, a shower and a breakfast of grapefruit, muesli, fresh orange juice and decaffeinated coffee by the pool. As he left for the office, where he works as a software designer for a computer company, the kids were only just getting up for school. He had hardly seen them today, and wanted to talk to them. Especially to Peter, for whom he has high hopes; already he is a computer whizz at school and there is plenty of scope for a career in his father's footsteps.

At 7.30 Roger backed his car out of the garage. His office is on El Camino Road, on the northern outskirts of San José, to the south of San Francisco. He slipped through the quiet streets of suburban Palo Alto and filtered onto the busy freeway.

It was a tricky morning. The company was launching a new computer software product that Roger had designed, and he had to attend a marketing meeting. Marketing was not really his scene, but as a company director he had to keep tabs on the 'machine' – the network of sales and marketing staff hyping and selling their products – and carry the can if anything went wrong. The most successful entrepreneurs in new high-tech industries such as his need a mixture of skills, some of which he fears, when he is honest with himself, he does not have. Behind those tinted-glass walls in the air-conditioned office world, life is competitive and precarious.

He rushed his lunch, snatching a hamburger at the McDonald's drive-through. Roger's afternoon was spent on paperwork, except for half-an-hour devoted to new thoughts on his latest software project – which he regarded as the only real work of the day. He got home just in time to welcome his dinner guests – an old friend and his boss from a rival transnational software company, along with their wives. They had vaguely talked of a new job for Roger; more or less the same work but with an even higher salary than the US$100 000 a year he earns now.

Roger wonders for a minute, before he heads for bed, if the money would be worth it; he suspects the job might only be a faster way to get an ulcer.

San Juan ***Puerto Rico*** Seaport and capital city, situated near the eastern end of the north coast. It exports sugar, tobacco, pineapples, bananas, oranges and cacao, manufactures clothing, cigars and cigarettes and refines sugar. The chief industry is tourism. The School of Tropical Medicine, part of the University of Puerto Rico, is situated there; the rest of the university is at Rio Piedras, 11 km (7 miles) away. The oldest part of the city, founded by the Spaniards in 1521, is built on a small island connected to the mainland by a causeway and bridges.

Population 435 000

Map Caribbean Bb

San Luis Potosí ***Mexico*** **1.** Industrial and agricultural state in north-central Mexico which is rich in gold and silver. The territory was inhabited before the Spanish Conquest by the semi-nomadic Chichimeca Indians; they were displaced by Spanish speculators seeking gold and silver.

2. Capital of San Luis Potosí state. The site of the present city, an elegant colonial mining town of pink sandstone buildings, had been inhabited by the Chichimeca Indians for three centuries before the Spanish Conquest, but it was formally founded by the Spaniards in 1592. It was named after the Franciscan mission of San Luis, founded there in 1590, and the Bolivian mining town of Potosí. Lead, silver, zinc and iron are still mined there, and flour, beer and leather goods manufactured.

Population (state) 1 800 000; (city) 407 000

Map Mexico Bb

San Marino ***San Marino*** The towers and battlements of this tiny, picturesque capital of Europe's smallest republic are perched more than 610 m (2000 ft) high on the craggy western slope of Monte Titano. Triple walls guard it, only one road enters it, and the narrow winding streets are mostly closed to vehicles. The city is on the site of a hermitage built by a stonemason named Marino about AD 300: his bones now rest in the 16th-century Basilica. The Rocca Fortress, with three linked towers on the three peaks of Monte Titano, offers magnificent views – including one across the plain below to Rimini, 23 km (14 miles) to the north-east, on the Adriatic coast.

Several art museums, six churches, a Gothic Government House and other fine public buildings also draw tourists to the prosperous city of 4500 or so people. Industries include silk and ceramics manufacture – and selling postage stamps, which are popular with collectors.

Population 4500

Map Italy Dc

San Miguel ***El Salvador*** Manufacturing town about 110 km (68 miles) east of the capital, San Salvador, producing textiles, pharmaceuticals and plastics. It was founded in 1530 and has an 18th-century cathedral.

The surrounding department of the same name grows coffee, sisal and cotton and has gold and silver mines. San Miguel volcano, overlooking the town, is 2130 m (6990 ft) high. It last erupted in 1976.

Population (town) 161 200; (department) 440 000

Map El Salvador Ba

San Marino

EUROPE'S SMALLEST REPUBLIC, SURROUNDED BY ITALY BUT INDEPENDENT FOR CENTURIES, STILL HAS A MEDIEVAL FLAVOUR

The smallest European independent republic, lies in the eastern foothills of the Apennines. It has wooded mountains, pastureland, citadels and medieval villages, clustered around the peaks of Monte Titano, which rises to 743 m (2438 ft).

Monte Titano was settled about AD 300 by Marino, the stonemason saint after whom the republic is named. The country became an independent commune in the 13th century and has remained so. It is governed by a Great and General council of 60 elected members; coalitions have ruled since 1957, when 12 years of Communist control ended. Most citizens between 16 and 55 can be called on by the local militia in an emergency, but the police are hired from Italy. San Marino entered a customs union with Italy in 1862.

The republic draws some 3.5 million tourists a year – more than 150 to each inhabitant. One attraction is the capital, the tiny fortified town of SAN MARINO.

Much of the republic's revenue comes from the sale of stamps, postcards, souvenirs and duty-free liquor. A substantial contribution to the economy is now also made by light manufacturing industries, and there is a long tradition of pottery making and silk weaving. The majority of Sanmarinese, however, still work on the land and in forestry.

Despite the inevitable influence of Italy, San Marino still values its separate administration and identity.

SAN MARINO AT A GLANCE

Map Italy Dc

Area 61 km² (24 sq miles)

Population 23 000

Capital San Marino

Government Parliamentary republic

Currency Italian lira

Language Italian

Religion Christian (95% Roman Catholic)

Climate Mediterranean

Main primary products Wheat, olives, vines, dairy cattle; building stone

Major industries Agriculture, tourism, food processing, textiles, quarrying, ceramics, forestry

Main exports Postage stamps, wine, textiles, furniture, stone, ceramics

Annual income per head (US$) 6000

Population growth (per thous/yr) 2

Life expectancy (yrs) Male 70 **Female** 77

San Miguel de Allende (Allende) ***Mexico*** Charming colonial town of narrow cobbled streets, and houses with hand-carved wooden doors, in Guanajuato state, central Mexico. Once important for silver mining, it is now a meeting place for artists, actors, writers and intellectuals.

Population 50 000

San Miguel de Tucumán ***Argentina*** Provincial capital of Tucumán about 1090 km (680 miles) north-west of Buenos Aires. It is the centre of Argentina's sugar-cane industry. Because of its age – it was founded by the Spanish in 1565 – it attracts tourists exploring Argentina's colonial past.

Population (city) 497 000; (province) 973 000
Map Argentina Ca

San Pedro Sula ***Honduras*** North-western town near the Guatemalan border. It is the country's second largest city. Good, paved roads to the capital, Tegucigalpa, and to Caribbean ports, and direct flights to Tegucigalpa, Miami and New Orleans, combined with new industries – such as small steel rolling mills, textiles and clothing, plastic ware, food processing, zinc roofing, cement and furniture – make it Central America's fastest-growing town.

Population 398 000
Map Honduras Aa

San Remo ***Italy*** Seaport and resort about 115 km (70 miles) south-west of Genoa and 20 km (12 miles) from the French border. It abounds in sedate villas and subtropical plants. The medieval town was developed as a resort by aristocratic British tourists in the 19th century. It is the region's principal market for flowers, which are grown on protected terraces either side of the town.

Population 62 100
Map Italy Ac

San Salvador ***Bahamas*** Small island in the east central group of the archipelago, of historical interest as the first land on which Columbus set foot in the New World on October 12, 1492. The day is now a national holiday. There are three monuments marking the event because the exact landing spot is disputed.

Population 850
Map Bahamas Cb

San Salvador ***El Salvador*** National capital and largest city, producing one third of the country's industrial output. Its industries include textiles, food processing, pharmaceuticals, plastics and dairy produce, though the basis of its wealth is coffee, grown on the volcanic soils of the nearby mountains.

The city was founded by the Spaniard Diego de Alvarado in 1525, but has been rebuilt several times as a result of earthquakes. The modern city is laid out in the form of a cross, with low buildings designed to withstand new tremors.

The San Salvador volcano (1960 m, 6430 ft high) overlooks the city. The summit crater is 15 km (9 miles) across and in places 1000 m (3280 ft) deep. Its wooded inner slopes sweep down to a small, black cone – the remnant of its last eruption in 1917.

Population 884 100
Map El Salvador Ba

San Salvador de Jujuy ***Argentina*** Ancient city and capital of Jujuy province, about 1255 km (780 miles) north-west of Buenos Aires. Founded in 1565, the city lies at an altitude of 1260 m (4130 ft), and is surrounded by the windswept Altiplano, or 'High Plateau'. Before the Spanish Conquest, the whole area was part of the Inca Empire. There are reserves of iron, zinc, lead, tin and silver which are being exploited, but agriculture is limited by the altitude. Nearby is the Indian town of Humahuaca, standing at 2940 m (9650 ft), and the Valle Grande National Park.

Population (city) 124 500; (province) 410 000
Map Argentina Ca

San Sebastián ***Spain*** Seaport and industrial city on the Bay of Biscay near the French border, and capital of the province of Guipúzcoa. Its industries include tourism, fishing, chemicals, cement and metal goods.

Population (city) 175 600
Map Spain Ea

San'a ***Yemen*** Capital and leading city situated in the middle of the main plateau, 2286 m (7500 ft) above sea level. Surrounded by high, thick stone walls containing eight gates, San'a is said by the Arabs to be the world's oldest city and the birthplace of the Arab nation. Its founder is claimed to be Shem, the eldest son of Noah, and it was first called Medinet Sam, or 'City of Shem'. The name San'a comes from the period of Abyssinian rule in the 6th century AD and means 'fortified'.

San'a has long been a centre of handicraft industries, such as weaving and jewellery making; and in recent years larger industries have been set up. These include a textile plant, iron foundry, plants to make building fixtures and fittings, tools and water pumps. New water and electricity projects are also complete.

Places of interest include the Medina (the old city and Arab quarter with the most important of the many mosques); the citadel (Al Qasr); the ruined fort of al-Birash; and the Mutawakil, which formerly contained the palace of the imams and is now the Dar el Saada museum.

San'a is connected by good roads to HODEIDA on the coast; SAADA in the north; and TAIZ in the south. There is a university (founded in 1970).

Population 210 000
Map Yemen Aa

Sanchi ***India*** Village in the north, about 590 km (365 miles) south of Delhi. It is the site of India's biggest Buddhist temple complex, which lay hidden under the jungle until uncovered in the early 19th century. Sanchi prospered under the great Indian emperor Asoka (273-232 BC), when the Great Stupa (a stone cairn, more than 30 m (100 ft) high) was built to house holy relics, and it thrived for 1000 years. Like SARNATH, it was abandoned, but the monks' living quarters and many stupas survive.

Map India Bc

sand Loose, rounded particles of disintegrated rock, especially quartz. Fine sand ranges from 0.02 to 0.2 mm (0.0008 to 0.008 in) in diameter. Coarse sand is between 0.2 and 2 mm (0.008 and 0.08 in) in diameter. Sand is finer than gravel and coarser than silt. If the particles are angular, it is called grit.

Sandakan ***Malaysia*** Port in east Sabah on the north-east coast of the island of Borneo. It handles most of the exports of the state's profitable hardwood industry.

Population 113 500
Map Malaysia Ea

sandbank Submerged ridge of sand in a sea or river, formed by currents and often exposed at low tide.

sandstone Variously coloured SEDIMENTARY ROCK consisting of compressed or cemented particles of sand. Among the most important sandstone deposits in Europe are the Old Red Sandstone, laid down in the DEVONIAN period, and the New Red Sandstone of the PERMIAN and TRIASSIC.

Sandwich Islands ***Pacific Ocean*** See HAWAII

Sandwip ***Bangladesh*** An exposed island off the south-east coast in the delta of the Ganges, north-west of Chittagong. Many thousands of people died there in a cyclone in 1970 and again in 1985 when 20 000 died in coastal areas.

Map Bangladesh Cc

Sankt Gallen ***Switzerland*** See ST GALL

Santa Ana ***El Salvador*** The country's second largest city after the capital, San Salvador. It has textile and food-processing industries, and coffee is grown in the surrounding department of the same name. The city retains some colonial architecture, including a neo-Gothic cathedral.

Santa Ana volcano, the country's highest mountain at 2365 m (7760 ft), broods near the city above Lake Coatepeque, which is fed by hot springs warmed by the volcano.

Population (city) 208 300; (department) 568 000
Map El Salvador Ba

Santa Anna Hot, dry, föhn-like wind blowing from the north and north-east, descending from the Sierra Nevada across the deserts of southern California.

Santa Barbara ***USA*** Attractive resort and residential city on the California coast about 135 km (85 miles) north-west of Los Angeles. It grew up around a beautiful Spanish mission established in 1786, and has electronics and aerospace development industries. There are offshore oil wells, one of which leaked in 1969, causing serious beach and harbour pollution.

Population (city) 76 900; (metropolitan area) 322 800
Map United States Cd

Santa Catarina ***Brazil*** See FLORIANOPOLIS

Santa Clara ***Cuba*** Capital of Villa Clara province, on the plateau 255 km (160 miles) south-east of Havana. Founded by the Spanish in 1689, it has become an important route and commercial centre in an agricultural region producing sugar, tobacco and coffee. Industries include tobacco manufacture and leather processing. Since the 1959 revolution, its university, founded in 1949, has become a major educational centre for the island.

Population (city) 175 000 (province) 775 000
Map Cuba Ba

Santa Cruz ***Bolivia*** City and the largest and potentially the richest department in the country, covering the central and southern areas of the Oriente eastern lowlands. It is Bolivia's major centre of new colonisation from the Andean region. The central area grows a wide variety of crops, especially sugar cane, rice, coffee and cotton. Oil fields have been developed at Caranda and Colpa, and natural gas at Río Grande. Large iron ore and maganese deposits have been discovered in the south-east.

The city, founded in 1557 by the Spaniard Nuflo de Chávez, originally on a site 322 km (200 miles) east of the present location, is the fastest-growing in the country.

Population (department) 944 000; (city) 377 000
Map Bolivia Bb

Santa Cruz de la Palma ***Spain*** Chief town of La Palma island in the CANARY ISLANDS.

Population 16 700
Map Morocco Ab

Santa Cruz de Tenerife ***Spain*** Seaport on Tenerife Island, in the CANARY ISLANDS, and also the name of one of the island's provinces. The town exports bananas, tomatoes and potatoes.

Population (town) 190 800; (province) 688 300
Map Morocco Ab

Santa Fe ***USA*** Capital of New Mexico, in the north of the state about 145 km (90 miles) south of the Colorado border. It was settled by Indians 700 years ago and by Spanish colonists in 1610; it became a key trade centre on the route south and west before the railways. The nuclear weapons research centre of Los Alamos, where the first atomic bomb was made, is 40 km (25 miles) to the north-west.

Population (city) 52 300; (metropolitan area) 100 500
Map United States Ec

Santa Isabel ***Equatorial Guinea*** See MALABO

Santa Marta ***Colombia*** Port resort and capital of Magdalena department, on the Caribbean coast about 100 km (60 miles) east of Barranquilla. Tourists are attracted to its long, sandy beaches and warm climate, and to the Tairona Nature Park, 35 km (22 miles) to the east, an unspoilt stretch of wooded coastline.

Santa Marta's deep-water dock exports bananas, and coal from reserves discovered in the nearby Guajira Peninsula. Colombia's liberator, the Venezuelan-born Simón Bolívar, died penniless near Santa Marta in 1830, aged 47.

Population 303 800
Map Colombia Ba

Santander ***Colombia*** North-east Andean department about 310 km (190 miles) north of Bogotá. The capital is BUCARAMANGA. Most of Colombia's oil reserves lie to the west of the department. The country's largest refinery is at Barrancabermeja beside the Magdalena River.

Population 1 353 000
Map Colombia Bb

Santander ***Spain*** Seaport, industrial city and resort on the Bay of Biscay. Capital of the province of Cantabria. It exports minerals, wine and wheat and its industries include shipbuilding, oil refining, tanning, chemicals, cables and machinery. At Altamira, 27 km (17 miles) to the south-west, are caves containing Stone Age paintings between 13 000 and 20 000 years old. Like the caves at Lascaux in France, entrance is now restricted.

Population 180 300
Map Spain Da

Santiago ***Chile*** Capital and main industrial centre of the country, in the Central Valley about 90 km (55 miles) from the coast. It was founded by the Spanish Conquistador Pedro de Valdivia in 1541 at Santa Lucía Hill, now a city park. The city's main street is the Avenida O'Higgins (the Almeda), which is 100 m (330 ft) wide and more than 3 km (2 miles) long. It contains statues of Chile's liberator Bernardo O'Higgins and José de San Martín, and the monument commemorating the Battle of Concepcíon in 1879.

Near the main square, the Plaza de Armas, is a museum in what was formerly the Casa Colorado, built in 1769 for the Spanish governor, and the cathedral built in 1558. A cable car runs 228 m (750 ft) to the top of the highest vantage point in the city, San Cristóbal Hill, which is crowned by an enormous statue of the Virgin Mary. Industries include food processing, textiles, pharmaceuticals, chemicals, clothing and leather goods.

Population 4 132 000
Map Chile Ac

Santiago de Compostela ***Spain*** Pilgrimage and university city in the north-west, 56 km (35 miles) south of Corunna. Its cathedral, built between the 11th and 13th centuries, contains the remains of the Apostle St James the Great, discovered nearby in AD 813. The town is named after the saint, who is known in Spanish as *Sant' Yago*. In the Middle Ages, Santiago was the most popular place of pilgrimage in Christendom, after Jerusalem.

Population 93 700
Map Spain Aa

Santiago de Cuba ***Cuba*** Seaport and capital of Santiago de Cuba province, on the south coast 145 km (90 miles) from the western tip of the island. Founded in 1514, it was the island's capital until 1589, and is now the second largest city after Havana. It was the scene, in July 1953, of Fidel Castro's first revolutionary strike against the Batista government – a raid on an army barracks – which resulted in his capture. He later left the island under an amnesty in 1955 and returned a year later to launch a new uprising. The port ships iron, manganese and copper ore, sugar, rum and tobacco. It has textile mills, distilleries and an oil refinery.

Population (city) 345 000; (province) 922 000
Map Cuba Ba

Santiniketan ***India*** See SHANTI NIKETAN

Santo ***Vanuatu*** See ESPIRITU SANTO

Santo Domingo ***Dominican Republic*** Main port and capital city, lying on the south coast. It was founded in 1496 by Bartholomew, the brother of Christopher Columbus, and is the oldest continuous European settlement in the Americas. The cultural and commercial centre of the state, it has two universities and an early 16th-century cathedral containing the tomb where Christopher Columbus was buried before his remains were taken to Seville Cathedral in Spain. Formerly Ciudad Trujillo, it was renamed in 1961 after President Rafael Trujillo was assassinated. Sugar is exported, and tourism is a principal industry.

Population 1 313 000
Map Caribbean Db

Santorini (Thíra) ***Greece*** Island of the CYCLADES group, north of Crete. It is thought by some archaeologists to be the site of the lost and legendary land of Atlantis. A volcanic eruption in about 1500 BC sank a crater in one side of the island covering 84 km² (32 sq miles). Today the remains of the crater's ring wall form an arc of almost vertical cliffs which rise some 300 m (980 ft) from the sea.

Excavations since 1967 at Akrotiri on the island have revealed evidence of a flourishing civilisation before the eruption. The main present-day town, Thíra, stands spectacularly on the volcanic clifftop.

Population 7100
Map Greece Dc

Santos ***Brazil*** The country's largest port, about 60 km (38 miles) south-east of the city of São Paulo. More than 40 per cent of Brazil's imports, and half of its exports, pass through Santos. It also has oil refineries and chemical plants.

Population 417 000
Map Brazil Dd

São Francisco ***Brazil*** River rising in the hills west of the city of Belo Horizonte, and flowing some 2900 km (1800 miles) into the Atlantic about halfway between the cities of Salvador and Recife. Hydroelectric dams along its length provide electricity for the whole region. The dams also provide irrigation waters, allowing the cultivation of crops such as grapes in an area of perpetual sunshine but little rain.

Map Brazil Dc

São Miguel ***Portugal*** Largest island (770 km², 298 sq miles) in the Azores group, in the Atlantic, about 1290 km (800 miles) off the west coast of Portugal. It has been the scene of numerous volcanic eruptions since the 15th century, and the landscape is pitted with volcanic cones and lake-filled craters. The soil is extremely fertile and maize, pineapples, vines, oranges, figs and tea are all grown on the island. The capital is Ponta Delgada.

Population 131 900

São Paulo ***Brazil*** Industrial city and capital of São Paulo state, about 360 km (225 miles) south-west of Rio de Janeiro and almost exactly on the Tropic of Capricorn. It vies with Mexico City as the fastest growing and largest conurbation in the world. It is also the nation's industrial powerhouse, with 60 per cent of the country's industry concentrated in the state.

The state, covering 247 900 km² (95 690 sq miles), is about the size of Britain. It was transformed by the coffee boom of the late 19th century which brought massive immigration: 1 million people from Italy, 500 000 from Portugal and Spain, and 250 000 from Japan. This magnetic attraction continues today, with considerable internal migration, especially from the poor north-east of Brazil. Apart from its industrial pre-eminence, the state is Brazil's leading producer of sugar – with its associated sugar-alcohol fuel programme – and of livestock.

Population (city) 8 500 000; (metropolitan area) 16 000 000; (state) 25 000 000

Map Brazil Dd

São Tomé ***São Tomé and Príncipe*** National capital and main port, on an inlet on the north-eastern coast of São Tomé island. It was founded in the late 15th century by Portuguese colonists. It still retains many old colonial buildings, but lacks tourist facilities, such as hotels – a shortcoming the government plans to remedy. Its main exports are cocoa and coffee.

Population 25 000

Map Gabon Aa

São Vicente, Cabo de (Cape St Vincent) ***Portugal*** Cape at Portugal's south-western tip, ending in cliffs 75 m (245 ft) high. In 1797 an English fleet under Sir John Jervis (later Lord St Vincent) defeated a Spanish fleet off the cape.

Map Portugal Bd

Saône ***France*** River, about 480 km (300 miles) long in east-central France. It rises in the Vosges mountains and flows south past the vineyards of Burgundy to merge with the Rhône at Lyons.

Map France Fc

Sapporo ***Japan*** Capital of Hokkaido, in the south-west of the island. The city was founded in 1871 as a springboard for the colonisation of the island. Now a thoroughly modern city with a gridiron street plan, it hosted the 1972 Winter Olympics. Its nearby ski slopes and hot-spring resorts make it a winter sports paradise.

Population 1 543 000

Map Japan Db

Saqqara ***Egypt*** Village alongside an ancient necropolis and site of the 'Step' Pyramid about 23 km (14 miles) south and slightly west of Cairo. In February 1986 the tomb of Maya, treasurer to the boy king Tutankhamun, who died in about 1352 BC, was found there. The discovery was made by two archaeologists, Dr Geoffrey Martin of University College, London, and Dr Jacobus Van Dijk of the Leiden Museum in the Netherlands. They had spent ten years excavating the tomb – that of Maya and his wife Merit.

Map Egypt Bb

Sar Planina ***Yugoslavia*** Mountain range separating the Serbian autonomous province of Kosovo from Macedonia. It rises to 2702 m (8865 ft) at Titov Vrh. A region of dense forests, lakes and pastures, it is also a fine skiing area, served by the cities of Tetovo and Prizren.

Map Yugoslavia Ec

Saragossa ***Spain*** See ZARAGOZA

Sarajevo ***Yugoslavia*** Capital of Bosnia Herzegovina, lying on the Miljacka river, 200 km (about 125 miles) south-west of Belgrade. Footprints on the pavement near Princip's Bridge (Principov Most) mark the spot where a 19-year-old Bosnian nationalist, Gavrilo Princip, stood on June 28, 1914. Bosnia was then under Austrian rule, and on that day Princip assassinated Archduke Franz Ferdinand, heir to the throne of the Austro-Hungarian Empire, as he passed the spot in an open car – an act that unleashed the First World War.

The city has some 80 mosques, and the old Turkish quarter below the city's citadel contains some of the finest Ottoman architecture in Yugoslavia. It has a fine post-war road, rail and air links, research centres and a university, and many factories producing metal goods, machinery, porcelain, electrical equipment, tobacco, textiles, carpets and processed foods. Tourism is booming. The Bjelasnica mountains, the Bosnian 'rooftops', offer superb skiing. The Winter Olympics were held there in 1984.

Population 448 500

Map Yugoslavia Dc

São Tomé and Principe

A GROUP OF SMALL ISLANDS, LYING OFF AFRICA'S WEST COAST, WHERE COCOA IS VITAL TO THE NATION'S ECONOMY

The volcanic islands of São Tomé and Principe are Africa's second smallest independent nation after the Seychelles. The Portuguese discovered the islands in 1471 and ruled there from 1740 until independence in 1975, and a colonial plantation economy – the main crop is cocoa – still dominates life today. Most food is imported, and the country relies on foreign aid.

São Tomé and Principe is mainly composed of the two islands in its title. São Tomé covers 845 km² (326 sq miles). Its highest point is a peak of 2024 m (6640 ft); there are many lower summits, all extinct volcanic cones. Sparkling streams rush down steep hillsides and surge through rain forests on their way to the hot, humid coastlands. PRINCIPE is another craggy island, reaching 948 m (3110 ft). In addition, there are two islets, Pedras Tinhosas and Rolas.

The native islanders, known as *filhos da terra* (Portuguese for 'sons of the land'), are mostly descendants of imported slaves, contract labourers and European settlers. About 11 per cent of the work force work in small manufacturing industries, commerce and services, and about 70 per cent on the land. A high proportion of the plantation workers are itinerant, and live on the plantations, now state-owned.

SÃO TOMÉ AT A GLANCE

Map Gabon Aa

Area 964 km² (372 sq miles)

Population 89 000

Capital São Tomé

Government One-party republic

Currency Dobra = 100 centimes

Languages Portuguese, Creole and local African languages

Religions Christian (Roman Catholic), animist

Climate Warm and humid; average temperature in São Tomé ranges from 21°C (70°F) to 31°C (88°F)

Main primary products Cocoa, coconuts, palm kernels, bananas, coffee, timber

Major industries Agriculture, food processing, timber products

Main exports Cocoa, copra, palm kernels and nuts

Annual income per head (US$) 410

Population growth (per thous/yr) 8

Life expectancy (yrs) Male 62 **Female** 62

Saransk ***USSR*** Industrial town 240 km (150 miles) south of Gor'kiy. Saransk is capital of Mordovia, an autonomous republic of the RSFSR covering 26 200 km² (10 100 sq miles).

Population (republic) 971 000: (city) 301 000

Map USSR Fc

Saratov ***USSR*** Industrial city and river port in the south-west, sprawling for about 50 km (30 miles) along the west bank of the Volga river, 320 km (200 miles) north of Volgograd. Saratov was founded on a neighbouring site in 1590, and it grew in the 1870s after the building of a railway from Moscow, 715 km (445 miles) to the north-west. Set amid rich farmland, its main products are tractors, farm machinery, fertilisers, flour and processed food.

Population 894 000

Map USSR Fc

Sarawak ***Malaysia*** The country's largest state covering 125 205 km² (48 342 sq miles) and lying south-west of Sabah on the island of Borneo. From 1841 until the Japanese invasion in the Second World War, it was ruled by the 'White Rajahs' – descendants of Sir James Brooke, to whom the territory was ceded by the Sultan of Brunei – and was a British protectorate from 1881. It became part of Malaysia in 1963, triggering a three-year dispute with Indonesia.

Sarawak's forbidding highland interior, bordering Indonesia, and its swampy plain have limited human settlements to the river deltas and valleys. The Iban people – mostly slash-and-burn farmers who live in communal long-houses – and the Sea Dayak people, coastal fishermen related to the Iban, make up most of the population. Until the 1980s, Sarawak's people lived as they had lived for centuries. But offshore oil discoveries and a spreading road network are now challenging their way of life.

Population 1 323 000

Map Malaysia Db

Sardinia (Sardegna) ***Italy*** Second largest island in the Mediterranean after Sicily, covering 24 089 km² (9301 sq miles). It is sparsely populated because of its rugged landscape and isolation. There are few towns apart from CAGLIARI, the capital, and the provincial capitals of Nuoro, Oristano and SASSARI. Much of it is covered by granite mountains, culminating in the Gennargentu massif which rises to 1834 m (6016 ft). The main lowland, the Campidano, runs between Cagliari and Oristano.

Sardinia's economy is mainly pastoral, based on sheep and goats. Cork oaks grow in the north-west and wheat and vines are grown in the east. Petrochemical industries have been developed at Porto Torres and Cagliari since the Second World War. Coal, lead, zinc and iron mining has a long tradition in the south-west, but this activity is now much reduced as reserves dwindle. Tourism flourishes at ALGHERO and along the COSTA SMERALDA. Sardinians, particularly the shepherds, have a legendary tradition of hospitality, but this is also the land of *vendetta* (blood-feud) and kidnapping for ransom.

During the Bronze Age (4000-2000 BC) the native shepherd people built megalithic fortress-villages of squat, cone-shaped towers. The remains of more than 6000 of these *nuraghi* still dot the landscape.

Population 1 633 400
Map Italy Bd

Sargasso Sea Region of comparatively calm water in the north ATLANTIC OCEAN between the North Equatorial Current, the Gulf Stream and the Canaries Current. It contains many patches of floating sargassum weed, from which it gets its name, and is the breeding place of European and North American eels. But because there is little movement of water, the Sargasso is short of nutrients and plankton, so it contains comparatively few large marine animals.

Sailors once believed that the Sargasso was a graveyard for wrecked ships brought there by gods or devils. Many thought that ships trapped in the weed would forever sail in circles, manned by the ghosts of dead sailors. In reality, there is not enough weed to trap the smallest ship.

Area About 5 200 000 km² (2 007 720 sq miles)
Map Atlantic Ocean Bc

Sarh ***Chad*** Trading and industrial town on the Chari river about 500 km (310 miles) south-east of Ndjamena, near the border with the Central African Republic (CAR). It lies on a busy route used for the export of cattle to Bangui in the CAR. The area around Sarh is one of the richest and most densely populated parts of Chad, where good-quality cotton is produced. It is also one of the country's main industrial centres, with cotton textile mills, an abattoir and tannery, and light industries.

Population 65 000
Map Chad Ac

Sark ***English Channel*** See CHANNEL ISLANDS

Sarnath ***India*** Site just north of the city of VARANASI where the Buddha (Siddhartha Gautama, about 563-483 BC), the founder of Buddhism, preached his first sermon. A monastery was founded there, and its ruins survive, with those of Buddhist shrines, including one built by the great Indian emperor Asoka (273-232 BC). The site was deserted following the decline of Buddhism in India from the 6th century AD, partly because of the revival of popular Hinduism, and later because Muslim invaders desecrated it. Some of the sculptures are in the local museum. A modern temple contains Japanese Buddhist paintings.

Map India Cc

Sarnia ***Canada*** Lake port on the Ontario shore of the St Clair river at the southern end of Lake Huron. It is an oil refining and petrochemicals centre, and is connected to Port Huron, across the US border, by tunnel and bridge.

Population 50 900
Map Canada Gd

Sarospatak ***Hungary*** See BORSOD

Sarrebruck ***West Germany*** See SAARBRUCKEN

sarsen Large sandstone boulder, dating from the EOCENE or OLIGOCENE epochs, found mainly on the chalk lands of southern England. Sarsens were used in the construction of the outer circle of STONEHENGE.

Saskatchewan ***Canada*** **1.** Central prairie province covering 651 900 km² (251 700 sq miles) that is the country's biggest wheat producer. The northern third forms part of the CANADIAN SHIELD; the rest is plains sloping from 1000 m (3280 ft) above sea level in the west, down to 500 m (1640 ft) in the east. It is a rural province with a relatively small population and many farms of 850 hectares (2100 acres) or more. The territory was mainly settled in this century, after the railways came. Mineral resources which are being exploited include coal, oil, gas, zinc and copper. The provincial capital is REGINA. **2.** River, 1930 km (1200 miles) long, which flows from the Rocky Mountains into Lake Winnipeg. Its upper part is divided into the North and South Saskatchewan rivers – which join near Prince Albert. The river provides irrigation and hydroelectric power.

Population (province) 968 000
Map Canada Ec

Saskatoon ***Canada*** City on the South Saskatchewan river, 240 km (150 miles) north-west of Regina. It is a major service centre for the surrounding wheat-growing area, and has industries which mill flour, refine oil, pack meat and make dairy products. It also has art galleries, museums and a university. In 1882 a temperance colony from Ontario settled on the east bank of the river; the following year the railway arrived and put its station on the west bank to start a new settlement. The two communities were amalgamated in 1906.

Population 154 000
Map Canada Ec

Sasolburg ***South Africa*** Town in the Orange Free State, about 80 km (50 miles) south of Johannesburg. It is the site of South Africa's first oil-from-coal plant, set up in 1950. Its name comes from SASOL, the government-run South African Coal, Oil and Gas Corporation.

Population 26 000
Map South Africa Cb

Sassandra ***Ivory Coast*** River rising in the Guinea Highlands, and flowing 563 km (350 miles) to the sea at the town of Sassandra, 65 km (40 miles) along the coast from the port of San Pedro. A hydroelectric dam at Buyo was completed in 1980.

Map Ivory Coast Ab

Sassari ***Italy*** Second town of Sardinia after the capital, Cagliari, and about 170 km (105 miles) to the north-west of it. Its university was founded by the Jesuits in the 16th century. It produces a wide range of foods, including pasta, olive oil, wine and cheese.

Population 119 800
Map Italy Bd

▼ BLACK AND WHITE The lovely Romanesque Church of Santa Trinita de Saccargia stands on the outskirts of Sassari, a city of dazzling white limestone surrounded by olive groves.

Satpura Range ***India*** East-west range of hills rising to 1325 m (4347 ft) in central India. It lies between the trough-like valleys of the westward flowing Tapti and Narmada rivers, and with the Vindhya Range to the north marks the northern end of the Deccan plateau.
Map India Bc

Sattahip ***Thailand*** Port on the east coast of the Gulf of Thailand. It has been developed since 1980 as a deep-water alternative harbour to Bangkok, 130 km (80 miles) to the north-west.
Population 83 600
Map Thailand Bc

Satu Mare ***Romania*** Manufacturing town set in rich farmland in the north-west corner of the country, 10 km (6 miles) east of the border with Hungary. Satu Mare was founded at least 750 years ago beside the Somes river by boatmen taking salt down to central Romania. Its products include mining and transport equipment and textiles.
Population 113 600
Map Romania Aa

saturation State reached when air is holding as much water vapour as is possible at a given temperature.

saturation zone See WATER TABLE

Sau ***Yugoslavia*** See SAVA

Saudi Arabia See p. 572

Sault Sainte Marie ***Canada*** Port on the north bank of St Mary's river between Lake Superior and Lake Huron in Ontario. Manufacturing includes chemicals, lumber, pulp, paper and steel, and it is a centre for tourist activities. Opposite, on the south bank, is Sault Sainte Marie, Michigan. Five ship canals, four in the USA and one in Canada, connect the lakes.
Population 82 700
Map Canada Gd

Sava (Sau; Szava) ***Yugoslavia*** The country's longest river, flowing 945 km (587 miles) in the northern part of the country. It has two sources in Slovenia – in the Karawanke Alps above Planica (near the Austrian border) and Lake Bohinj in the Julian Alps – and flows south-east to meander sluggishly across the Croatia-Slavonia plains to join the Danube at Belgrade.
Map Yugoslavia Db

Savai'i ***Western Samoa*** Largest island of Western Samoa, lying to the west of Upolu and covering 1820 km² (702 sq miles). The volcanic peak of Silisili is 1858 m (6096 ft) high, and much of the north-east of the island is covered by black lava flows. After eruptions in the early 20th century, many people migrated to Upolu. Savai'i has been inhabited by Polynesians for 3000 years, and has several ancient temples.
Population 49 000
Map Pacific Ocean Ec

savannah Tropical grassland with tall grasses and scattered trees and bushes, covering large areas of Africa, South America and northern Australia. See CAMPOS

Savannah ***USA*** Chief port in Georgia, near the mouth of the Savannah river. The attractive and historic city, popular with tourists, was founded in 1733 and is laid out in a series of squares leading to Forsyth Park. The city's industries include oil refining, paper products, cotton and wood processing.
Population (city) 145 400; (metropolitan area) 232 900
Map United States Jd

Savannakhet ***Laos*** Town and rice-growing province in southern Laos. The town is on the Thai border about 210 km (130 miles) north-west of the town of PAKSE.
Population (province) 543 600; (town) 51 000
Map Laos Ab

Savoie ***France*** Former independent duchy covering 10 416 km² (4022 sq miles) in the northern Alps. It became French territory in 1860, and now forms two departments – Savoie and Haute-Savoie. Hydroelectric power and winter tourism are important to its economy. CHAMBERY is the main town.
Population 310 000
Map France Gd

Savonlinna ***Finland*** Town, founded in the 17th century, on an island in Saimaa lake in south-east Finland. It is dominated by the ruined medieval castle of Olavinlinna, one of the largest in northern Europe. It is a summer resort and the setting of an opera festival each July.
Population 28 600
Map Finland Dc

Sayaboury ***Laos*** Province and town in north-western Laos between the Mekong river and the Thai border. Once part of Thailand, the province was annexed by Laos's French colonial rulers in 1909. A hilly and underpopulated area with only narrow lowlands along the Mekong river, it has recently been the scene of border disputes between Thailand and Laos.
Population (province) 223 600; (town) 14 000
Map Laos Ab

Sbeitla ***Tunisia*** Small town on an important road junction on the eastern edge of the Tell Atlas, close to the El Douleb oil field and 30 km (19 miles) north-east of Kasserine. It was the site of the 2nd-century Roman town of Sufetula, whose impressive remains include a triumphal arch, forum, baths and three finely preserved temples built in red sandstone. They are believed to be dedicated to the Roman god Jupiter, and the goddesses Juno and Minerva.
Population 12 100
Map Tunisia Aa

Scafell Pike ***United Kingdom*** See LAKE DISTRICT

scalded flat In Australia, a plain whose soil is of little use because of a high salt content.

Scapa Flow ***United Kingdom*** Natural anchorage enclosed by the Scottish ORKNEY ISLANDS. It was a base for British fleets in both World Wars. In 1919 the defeated German navy scuttled 71 of its ships in Scapa Flow, where they had been interned. In 1939 a German submarine slipped past the British defences and sank the British battleship *Royal Oak*. Churchill Barrier, a series of concrete causeways built between islands, blocks the eastern approaches.
Area 130 km² (50 sq miles)
Map United Kingdom Gj

scar Steep, bare rock face or crag. The name is chiefly used in northern England to indicate a limestone cliff.

Scarborough ***Trinidad and Tobago*** Main port and capital of Tobago, on the south-east coast. Remains of a French fort stand on a 130 m (430 ft) hill above the town.
Population 3000
Map Trinidad Aa

Scarborough ***United Kingdom*** English seaside resort and port in North Yorkshire, 58 km (36 miles) north-east of York. It developed as a spa in the mid-17th century. It has a ruined Roman signal station and Norman castle, and is a popular angling and conference centre.
Population 36 700
Map United Kingdom Ec

scarp See ESCARPMENT

scarth Bare rock face, especially in the English Lake District.

Scebeli ***Ethiopia/Somalia*** See SHABEELLE

Schaffhausen ***Switzerland*** Watchmaking town about 40 km (25 miles) north of Zürich. Its other industries include textiles and tourism. The main tourist attraction is the nearby Falls of the Rhine. German is the principal language of the townspeople and the inhabitants of the surrounding canton of the same name.
Population (town) 34 300; (canton) 70 000
Map Switzerland Ba

Schelde (Scheldt; Escaut) ***France/Belgium/Netherlands*** River which rises in north-east France and flows for 435 km (270 miles) through the Belgian cities of Tournai, Ghent and Antwerp. It used to empty into the North Sea through two estuaries, both in the Netherlands, but dyking has cut off the Oosterschelde (East Schelde), leaving the Westerschelde (West Schelde) as the only outlet.
Map Belgium Aa

schist Any of a variety of medium-to-coarse grained metamorphic rocks composed of parallel, often wavy and flaky, bands. The rock breaks easily along the bands, and is usually named after its dominant mineral, for example, mica schist.

Schlesien ***Poland*** See SILESIA

Schleswig-Holstein ***West Germany*** Northernmost state of the country occupying the southern part of the Danish peninsula. Two duchies were united in 1773 to form a province, which became part of Prussia in 1864. It is agricultural country with dairy farming, sugar beet, wheat and vegetables. Its only industries lie around the ports of Kiel, the capital, and Lübeck, both on the Baltic coast.
Population 2 614 000
Map West Germany Da

▼ **ATLANTIC ISLES Temperatures rarely fall below 7° C (45° F) on the Isles of Scilly — England's westernmost point, warmed by the North Atlantic Drift.**

Schönbrunn ***Austria*** Old district of Vienna noted for its palace, which was built as a summer residence for the Hapsburg emperors. The Schönbrunn Palace and gardens were redesigned in the 18th century for the Empress Maria-Theresa in an effort to outshine those at Versailles, near Paris. The palace, with its 1200 rooms, is open to the public in the summer.

Schottegat ***Netherlands Antilles*** See CURACAO

Schwarzwald ***West Germany*** See BLACK FOREST

Scilly, Isles of ***United Kingdom*** Group of islands some 40 km (25 miles) south-west of Land's End in England. The largest island is St Mary's, followed by St Martin's and Tresco. St Mary's is served by ship and helicopter from Penzance in Cornwall. The mild climate is ideal for spring flowers and early vegetables, which are exported to the mainland, and it also attracts many tourists. Palms and other exotic plants flourish in Tresco's Abbey Gardens.

Population 2000

Map United Kingdom Bf

scoria, cinders, slag Rough, angular fragments of rapidly cooled lava containing cavities caused by bursts of escaping gas. Scoria is darker and more cindery than pumice.

Scotland ***United Kingdom*** Northern constituent country of the UK, covering some 78 762 km² (30 410 sq miles). The kingdom of Scotland – peopled by native Picts and immigrant Scots from Ireland, and Scandinavians – was formed in the 9th century AD. King James VI became James I of England in 1603, and the kingdoms joined in 1707. Scotland's capital is EDINBURGH and its chief industrial cities are GLASGOW, ABERDEEN and DUNDEE.

Population 5 035 000

Map United Kingdom Cb

Scott Base ***Antarctica*** Headquarters of the New Zealand Antarctic research programme, situated on Ross Island on the shores of McMurdo Sound. It was founded in 1957.

Map Antarctica Jc

scree, talus Loose, usually angular fragments of rock debris which have fallen and collected at the foot of a weathered steep slope or cliff.

scrub Vegetation consisting of stunted trees and low-growing shrubs with occasional taller trees, found in regions where the soil is poor and rainfall is too scanty to maintain forest growth.

scud Ragged mass of very low cloud driven by a strong wind, below the main rain clouds.

Scutari ***Albania*** See SHKODER

Scutari ***Turkey*** See USKUDAR

sea A part or branch of an ocean, especially if partly surrounded by land, or a large inland stretch of salt water. For geographical features whose name begins 'Sea of', see under main part of name.

sea breeze Breeze blowing during the afternoon and evening, from the sea towards an area of low atmospheric pressure over the land which has heated up more quickly than the sea, particularly in tropical regions. It is the opposite of a LAND BREEZE.

sea level See MEAN SEA LEVEL

seamount Isolated volcanic peak rising from the ocean floor.

season One of the periods into which the climatic year is divided according to variations in temperature, duration of daylight and rainfall. In latitudes outside the tropics, a seasonal rhythm of temperature is imposed by the large differences in the angle of elevation of the sun that occur as the earth follows its orbital path around the sun. In the 'summer', the angles are at their highest and the days longest, which combine to give the highest temperatures of the year. The converse applies in 'winter'. The transitional seasons are spring and autumn. In tropical areas which are always hot, the rainfall regime is the factor that distinguishes the seasons from one another (see MONSOON).

Seattle ***USA*** Port in Washington state which handles much trade with the Far East and Alaska. It stands on Puget Sound, about 160 km (100 miles) south of the Canadian border, and is the main financial centre for the Pacific North-West region. The city is the home of the Boeing aircraft company and also has flourishing timber and tourist industries.

Population (city) 490 000; (metropolitan area) 1 677 000

Map United States Ba

Sebenico ***Yugoslavia*** See SIBENIK

Sechin ***Peru*** One of the most important pre-Columbian ruins, standing on the coast about 365 km (nearly 225 miles) north of the capital, Lima. There are many adobe and stone buildings decorated with carvings of warriors.

Map Peru Ba

secondary depression Small concentrated area of low atmospheric pressure on the margins of main DEPRESSION.

Secunda ***South Africa*** Town in east Transvaal, 64 km (40 miles) east of Johannesburg, where the second and third oil-from-coal plants were set up by the government in the 1970s and 1980s. The first such plant was in SASOLBURG.

Population 15 000

Map South Africa Cb

Secunderabad (Sikandarabad) ***India*** Twin city just north of HYDERABAD in southern India. Its cantonment (permanent military base) was one of Britain's largest in India. Here in 1898, while in the Indian Medical Service, the British scientist Sir Ronald Ross (1857-1932) discovered that the malaria parasite is carried by mosquitoes. He won a Nobel prize for the work.

Population 136 000

Map India Cd

Sédhiou ***Senegal*** Market town on the Casamance river, some 45 km (28 miles) from the Guinea-Bissau border. The town lies at the heart of a farming area producing rice and bananas and, to a lesser extent, groundnuts, maize, tomatoes, vegetables and tropical fruits.

Population 150 000

Map Senegal Ab

sediment Material comprising weathered particles of rocks, or particles of chemical or organic origin deposited by wind, water or glacial ice. Sediment is the raw material of SEDIMENTARY ROCKS.

Saudi Arabia

DESPITE CHANGES BROUGHT ABOUT BY ENORMOUS OIL WEALTH, SAUDI ARABIA REMAINS A STRICT MUSLIM SOCIETY AND THE FOCAL POINT OF THE ISLAMIC FAITH

The largest Arab country in area and the birthplace of the Prophet Muhammad and of the Islamic faith, the huge, sandy waste of Saudi Arabia contains more oil than any other nation on earth. Its reserves are estimated at some 165 000 million barrels – a quarter of the world's supply. With the quadrupling of oil prices following the 1973 Arab-Israeli War, enormous wealth poured into the desert kingdom – more than US$100 000 million a year at its peak. A nation of desert nomads became one of the richest in the world.

Using its oil revenues, Saudi Arabia launched a massive development programme, so that suddenly its towns were bustling with foreigners: contractors, immigrant workers and business representatives from all over the world. Towns and ports were transformed into major commercial centres with modern office blocks, factories and housing estates. RIYADH, the capital, JEDDAH, the main port on the RED SEA, and DAMMAM, the oil centre on The GULF, were among the major centres of change. Yet despite radical changes it remains a strict Muslim society, with laws based on the Koran.

THE LAND AND ITS PEOPLE

Saudi Arabia, with an area of 2 150 000 km² (830 000 sq miles), covers most of the great Arabian peninsula. It is thinly populated, with an estimated 11.5 million people, and it contains some of the hottest, bleakest desert in the world: temperatures often rise to 45°C (113°F). In the north the NEFUD desert stretches into Jordan and Syria. In the west, the narrow, humid coastal plain along the Red Sea is backed by steep mountains, reaching 2600 m (8530 ft) in the HEJAZ to the north and 3133 m (10 278 ft) at Jebel Abha in the ASIR highlands.

Beyond the mountains, the land slopes gradually down to the Ad Dahna plain, about 240 km (150 miles) wide, bordering The Gulf. Fierce winds and sandstorms sweep the deserts and rocky regions in the centre of the country. In the south-east is the forbidding desert of the RUB AL KHALI ('Empty Quarter') – it is waterless, uninhabited and almost featureless. Large areas of the interior are still unexplored and the southern boundaries with Yemen, South Yemen and Oman remain vague.

The people of the peninsula are the descendants of the original Arabs who swept out of Arabia across the Middle East and North Africa in two centuries of conquest after Muhammad proclaimed the new faith of Islam in the 7th century. On the Red Sea coast, the descendants of African slaves have intermingled with the Arabs. But today, because of rapid industrialisation, there are 1.5 million immigrants in addition to 10 million Saudis. Yemenis doing mainly unskilled work number about a million, and other workers include Palestinians, Jordanians, Sudanese, Indians, Syrians and Egyptians. At the managerial and technical levels, Americans, Britons and other Europeans do tours of duty.

THE DESERT KINGDOM

The Ottoman Turks held sway over the area for four centuries. Then, in the First World War, they were ousted by the Arabs in a daring guerrilla campaign organised by a British colonel, T.E. Lawrence (Lawrence of Arabia). But the creator of modern Saudi Arabia was Ibn Saud, who in tribal warfare between 1902 and 1924 united the four main regions of NEJD, AL-HASA, the Hejaz and Asir. In 1932, as King Abdul Aziz Ibn Saud, he proclaimed the new kingdom of Saudi Arabia.

The king died in 1953 and the present ruler, King Fahd Ibn Abdul Aziz, is the fourth of his sons to govern the country. It is an absolute monarchy; the king is also prime minister and owns all the undistributed land in the country. There are no elections and no political parties. There are about 5000 princes in the royal family and many important government offices are in their hands. Enlightened help for the people has gone hand in hand with royal extravagance: possibly 10 per cent of the oil wealth has gone to the royal family.

King Fahd, however, has said that he intends to create a *majlis-al-shura* (consultative assembly). At first its members will be appointed. Later, half of them will be elected through a system of provincial assemblies.

DEVELOPMENTS BASED ON OIL

Oil was first discovered in the 1930s, but it made little difference to the country until well after the Second World War; pilgrims visiting the holy cities of MECCA and MEDINA remained a major source of income. Then from 1973 Saudi Arabia led OPEC (the Organisation of Petroleum Exporting Countries) in drastically cutting oil production and pushing up its

price; it also nationalised the oil reserves owned by foreign oil companies. But since 1980 it has had a more moderating influence on world markets; it boosted production to stabilise price increases when the Iran-Iraq War again disrupted supplies, and it led in cutting output when overproduction and a fall in demand in the West pushed the price of oil down again.

However, by 1985 the Saudis had tired of bearing the brunt of production cuts, of seeing their market share and income dwindle while other OPEC nations exceeded agreed quotas and sold as much oil as they could. Saudi Arabia again stepped up output, with the result that world oil prices tumbled to below US$10 per barrel in early 1986.

The oil fields that are the source of nearly all the country's wealth lie in the eastern region, near and under The Gulf. GHAWAR is the largest oil field in the world. It is 241 km (150 miles) long and 35 km (22 miles) wide and can produce 5 million barrels a day, more than twice the total production of Britain's North Sea oil fields. There are 46 other oil fields, although with falling demand only one-third are at present in production.

Development plans using the oil riches began with housing, schools and water supplies. At one stage three schools a day were being opened, and there are now 2 million students and seven universities. Education is free but illiteracy still remains high – about 60 per cent. Free land and interest-free loans are provided for housing, and there is now a surplus of apartments in Riyadh and Jeddah.

In the last ten years, 1750 factories have been built. Less than 1 per cent of the land is suitable for agriculture, but lavish subsidies increased grain production from 3000 tonnes to 1 300 000 tonnes a year. Large artificially watered market gardens sheltered beneath huge plastic sheets have been built, and at one time there was even a plan to tow Arctic icebergs into The Gulf to provide fresh water.

The 1980-5 development plan has cost about US$250 000 million. It includes two new industrial complexes at Yanbu on the Red Sea and Jubail on The Gulf. These will have five oil refineries, eight petrochemical plants, a steel mill, desalination plants to distil fresh water from seawater, power stations and support industries. The country also has vast gas resources. However, future industrial development is likely to be curtailed due to the fall in oil prices and government spending.

▲ ISLAM'S HOLIEST PLACE Pilgrims throng around the *Kaaba*, in the centre of Mecca. The building contains the sacred Black Stone, said to have fallen from Paradise with Adam.

GUARDIANS OF ISLAM

The Saudis see themselves as the guardians of Islam and have one of the strictest Muslim societies in the world. In AD 622 the Prophet Muhammad, driven out of his birthplace, Mecca, began his teaching in the city of Medina. Within a hundred years the Arabs had spread Islam as far as Spain in the west and India in the east. Then in 750 the centre of Islamic power moved to BAGHDAD and Arabia lapsed into obscurity. But in the 18th century a new prophet arose – Muhammad

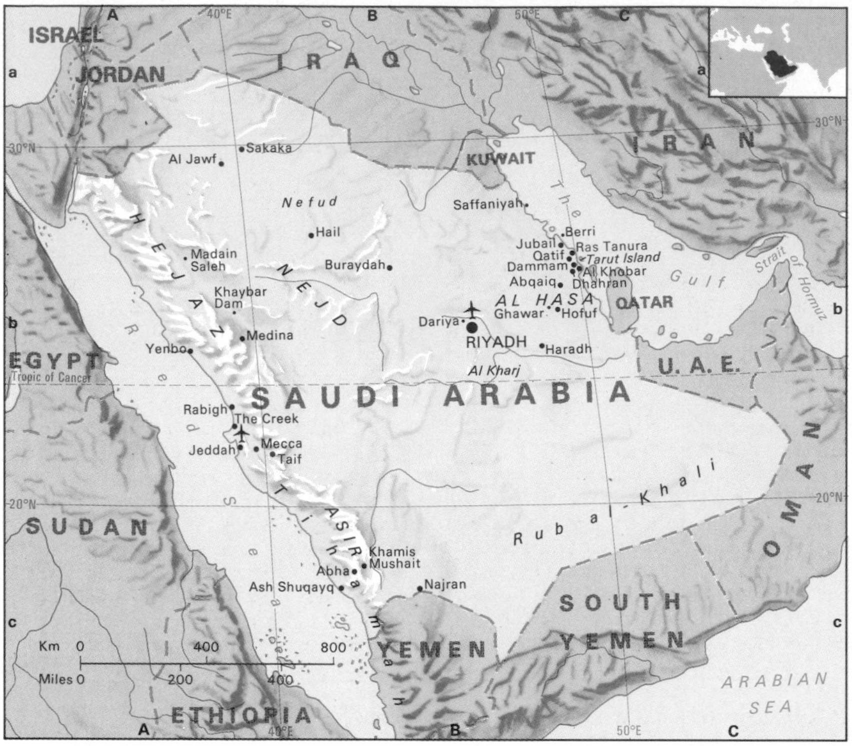

Abdul Wahhab. He preached an austere Islamic doctrine; his followers became members of the Wahhabi sect and this is the basis of religion and law in Saudi Arabia today.

The possession of alcohol, drugs and pornographic material (which may include fashion magazines) is strictly prohibited, even for foreigners. Detection of alcohol on a person's breath may result in up to two months' imprisonment and a lashing. Driving under the influence of alcohol may result in six months' jail, a fine and lashes.

Drug offences carry up to 15 years' imprisonment and even someone with prescribed drugs may be detained while the drug is analysed.

Theft is punished by imprisonment, and repeated theft by amputation of the right hand. Murder, rape, adultery (by a married person) and renouncing the Islamic faith carry the death penalty. This is carried out in public after Friday prayers – by stoning for adultery, and beheading in most other cases. Up to 60 people may be kept in a large cell awaiting trial, and 200 after conviction; they pay for their food and sleep on the floor.

Oil wealth, a growing number of immigrants, Muslim fanaticism elsewhere, and demands for change from the young all pose problems for Saudi Arabia's rulers and their conservative, male-dominated society.

Women still go veiled in public, may not drive cars, and can work only in all-female establishments. Less than 1 per cent are employed. Girls receive primary education but in higher education their numbers drop. At universities they are allowed to watch lectures by men only on closed-circuit television. Money, however, has enabled some Saudi women to travel, especially to the USA and Europe, and inevitably there has been questioning of the restrictions.

Saudi cuisine has widened as a result of immigration from other Arab countries, but meat and rice form the staple dish. When entertaining friends, mutton, chicken, goat or camel will be cooked; the meat may be stuffed with rice, nuts and herbs and will be placed on a large bed of rice. The guests eat from the communal dish with the fingers of the right hand. Soups and salads may be flavoured with chives, coriander or dill. A variety of fish is available in coastal regions; *shami* or *samul* bread is eaten with every meal. Dates are a traditional sweet, but there is also *muhallabiyah*, a milk pudding, and the sticky *kneifah* or *bakhlava*, a kind of pastry filled with honey and nuts. Ground coffee spiced with cardamom seeds completes the meal.

FOR THE VISITOR

Bedouin hospitality is legendary. Among the nomads a stranger had only to drink water, coffee or tea with a tribe or kiss their tent to secure hospitality and protection for three-and-a-third days without question. Today's Bedouins may search for pasture in a fast car, drive their herds to new areas in trucks, or bring up water by road tanker.

Camels and asses, herded by their nomadic owners, are a familiar sight. The wildlife includes leopards, wolves, foxes, hyenas, the rare oryx (a large antelope) and three types of gazelle. Baboons are found in the mountains; sand cats, ratels (honey badgers), hyraxes (the 'coney' referred to in the Bible), and a variety of snakes in the desert. Migrant birds from Europe rest along the Red Sea coast, and falcons are highly prized and trained for hunting. The rich underwater life of the Red Sea has been made famous by the French film maker Jacques Cousteau.

Although a desert country, Saudi Arabia is not without vegetation. Wild flowers appear in the Asir mountains in the spring and TAIF is famous for its roses and pomegranates. Date palms grow in oases and the higher regions support acacias, junipers and tamarisk trees.

The country is developing a network of new roads, and most important places are connected by air. The capital, Riyadh, on an oasis about 400 km (250 miles) inland from The Gulf, is the focal point for trade and desert travel; it can be cold in winter and night temperatures may fall below freezing point. Its buildings are a mixture of the classic Arabic style and large modern structures. In fact, some startling modern architecture has appeared in many of the developing cities.

A lavish patron of the 20th-century phenomenon known as 'industrial sculpture' is Jeddah, Saudi Arabia's commercial and diplomatic capital. Colossal monuments in concrete, stainless steel, aluminium and other metals stand in the city's squares and on the waterfront. The walled city, on the Red Sea, is also the entrance for the 2 million pilgrims a year who come to visit Mecca and Medina.

The holy city of Mecca is 70 km (44 miles) from Jeddah in a narrow valley overlooked by hills and castles. It contains the Great Mosque, the *Haram*, which encloses a small, sacred building, the *Kaaba* (House of God), where pilgrims kiss the venerated Black Stone, a meteorite that fell from the sky in the forgotten past. Medina, 180 km (112 miles) from the Red Sea, is surrounded by double walls with nine gates. Its mosque contains the tombs of Muhammad, his daughter Fatima and the Caliph Omar, and there is an Islamic University. Muslims try to visit Mecca and Medina at least once in their lifetime, but non-Muslims are banned. The Hejaz railway, with its fort stations dating from the First World War, is a curiosity for railway enthusiasts. Perhaps the Asir mountains (the Arabia Felix of the Romans) are the most beautiful area.

SAUDI ARABIA AT A GLANCE
Area 2 150 000 km² (830 000 sq miles)
Population 11 520 000
Capital Riyadh
Government Absolute monarchy
Currency Rial = 20 quirsh = 100 hallalas
Languages Arabic (official), English
Religion Muslim (85% Sunni, 15% Shiite)
Climate Hot and dry, mild in winter; temperature in Riyadh ranges from 8-21°C (46-70°F) in January to 26-42°C (79-108°F) in July
Main primary products Cattle, goats, sheep, poultry, alfalfa, dates, grapes, water melons, wheat, sorghum; crude oil and natural gas
Major industries Crude oil and natural gas production and refining, cement, chemicals, fertilisers, steel
Main exports Crude oil and refined products, natural gas
Annual income per head (US$) 13 400
Population growth (per thous/yr) 30
Life expectancy (yrs) Male 53 **Female** 56

sedimentary rock Type of rock formed by the consolidation of particles weathered from older rock, or from particles of chemical or organic origin. Rocks composed of hardened sediments or particles abrased from mountains and hills by wind and weather are called clastic or detrital rocks; examples are sandstone, shale and conglomerates. Organic sedimentary rocks, such as limestone, are made up of the shells and other remains of uncountable long-dead sea creatures, while chemical sedimentary rocks, like gypsum and rock salt, had their beginnings in naturally occurring chemical solutions.

Sedlez ***Poland*** See SIEDLCE

Sedom ***Israel*** Town, 6 km (4 miles) from the south end of the Dead Sea. Established in 1934 by the Palestine Potash Company of Kaliya, it is now the headquarters of chemical and fertiliser industries using salt and minerals extracted from the Dead Sea.

The Biblical town of Sodom (from which Sedom takes its name), which was destroyed with Gomorrah, is believed to lie buried beneath the waters of the Dead Sea south of the El Lisan peninsula.

Map Israel Bb

Ségou ***Mali*** Town on the Niger River, some 250 km (155 miles) downstream from the capital, Bamako. It was the capital of the Bambara civilisation from 1660 to 1861. Today it is a stopping place on the route to the towns of Mopti and Gao and a market for fish, cattle, hides, salt and cotton.

The headquarters of the Office du Niger are there and some of its attractive office buildings are built in traditional style in spacious grounds. The organisation, which is concerned with improving agriculture in the Niger Delta, is one of the town's main employers.

Population 65 000
Map Mali Bb

Segovia ***Spain*** Ancient fortified town 68 km (42 miles) north-west of Madrid, whose Roman aqueduct still supplies water. Its industries include flour milling, tanning and pottery, while the surrounding province of the same name grows cereals and raises sheep.

Population (town) 53 400; (province) 149 300
Map Spain Cb

seiche An oscillation of the water in lakes, inland seas, bays and channels, resembling a tide, caused by SEISMIC WAVES or atmospheric disturbances such as pressure changes and wind. The rise and fall may be a few centimetres or several metres.

seif dune Ridge of sand, often many kilometres long, crossing a desert parallel to the direction of the prevailing wind. The steep-sided dune may form when a line of BARCHANS joins up. Its height and width are increased by cross-winds.

Seine ***France*** River of northern France 775 km (482 miles) long. It rises on the Plateau de Langres, near Dijon, and flows north-west through Paris. The stretch between Paris and its mouth, at Le Havre, is navigable.

Map France Db

seismic energy Energy released by an EARTHQUAKE.

seismic focus Centre of an earthquake, below the earth's surface, from which the SEISMIC WAVES are emitted.

seismic waves The vibrations emitted by an EARTHQUAKE or man-made explosion.

seismograph Instrument for detecting and measuring SEISMIC WAVES.

seistan Strong, northerly wind, sometimes exceeding 110 km/h (nearly 70 mph), blowing in the Sistan (Seistan) region of eastern Iran during the four months of summer. Sometimes known as 'the wind of 120 days'.

Sekondi ***Ghana*** Coastal town about 170 km (105 miles) west of the capital, Accra. It was the country's first railway terminus, serving a hinterland rich in gold. It also handled most of Ghana's import and export trade until 1923, when it was superseded by adjoining Takoradi. The two were united in 1963; Sekondi remains a residential and trade centre. Sekondi-Takoradi is the capital of Ghana's Western Region.

Population (Sekondi-Takoradi) 254 500
Map Ghana Ac

Selangor ***Malaysia*** The country's most developed state, on the west coast of Peninsular Malaysia surrounding the capital, Kuala Lumpur. Selangor, which covers 7962 km^2 (3974 sq miles), produces much of Malaysia's rubber and 40 per cent of the nation's industrial output. Its capital is Shah Alam.

Population 1 517 500
Map Malaysia Bb

Selenge ***Mongolia/USSR*** River rising in north-west Mongolia and flowing 2120 km (1317 miles) north-east into Lake Baykal in the Soviet Union. Most of Mongolia's arable land is concentrated in the lower Selenge Valley.

Map Mongolia Cb

Selkirk ***United Kingdom*** See BORDERS

Selous Game Reserve ***Tanzania*** Africa's largest wildlife reserve, covering some 45 000 km^2 (17 375 sq miles) and containing more than a million large animals. It lies beside the Rufiji river about 200 km (125 miles) south-west of Dar es Salaam.

Map Tanzania Ba

selva Commonly, the equatorial forest of the Amazon basin in South America. The term is also applied to similar forests in other regions.

Semarang ***Indonesia*** Textile city and port on the north coast of Java. It has a large Chinese community.

Population 503 200
Map Indonesia Dd

Semendria ***Yugoslavia*** See SMEDEREVO

semidesert Semiarid region between the true desert and grassland, characterised by scrubby, thorny plants, coarse grass and bare patches of sand or stony waste.

Semipalatinsk ***USSR*** Industrial city and port on the Irtysh river in eastern Kazakhstan, 685 km (425 miles) south-east of Omsk. It has one of the USSR's largest meat processing plants, and its other manufactures include textiles, leather and processed food.

Population 307 000
Map USSR Jc

Semmering Pass ***Austria*** Pass carrying road and rail routes south-west from Vienna to the Mur and Mürz valleys and the town of Bruck-an-der-Mur, 135 km (84 miles) south-west of the capital. The village of Semmering, a popular winter sports resort, lies near the 985 m (3232 ft) high summit of the pass.

Map Austria Db

Sendai ***Japan*** City near the east coast of Honshu island 300 km (185 miles) north and slightly east of the capital, Tokyo. It is the home of many central government offices and a dozen universities and colleges. Industrially it specialises in oil refining and foodstuffs. The old town was severely damaged in the Second World War, but has since been rebuilt with tree-lined boulevards appropriate to its nickname, the 'Capital of the Forests'.

Population 700 200
Map Japan Dc

Senegal See p. 576

Senegal ***Guinea/Mali/Mauritania/Senegal*** West African river, which is a total of about 1790 km (1110 miles) long. It rises as the Bafing river in the Fouta Djallon uplands of central Guinea, and flows north into Mali, where it is joined by the Bakoy to form the Senegal proper. The enlarged river marks the Senegal-Mauritania border, and reaches the Atlantic at St Louis, Senegal.

The Senegal, which keeps flowing all year round, was a highway for French exploration of West Africa. It is still much used for local transport, despite shifting sandbanks and an erratic flow – a shallow stream in the dry season (October-May) and, usually, flooding in the wet season (May-October). However, river control schemes in Senegal and Mali, due to be completed by 1990, will even out the flow and reduce the dependence of farmers on the region's erratic rains.

Map Senegal Ba

Sensuntepeque ***El Salvador*** See CABANAS

Seoul (Soul) ***South Korea*** Capital city and teeming metropolis, on the Han river 60 km (37 miles) upstream from the Yellow Sea.

Seoul became the capital of unified Korea under the Yi dynasty in AD 1394. Royal palaces were built in the north of the city below Mount Pukak, and remnants of defensive walls survive. In 1948, after Japanese colonial rule (1905-45), it became the capital of the Republic of Korea. During the Korean War (1950-3) the city was occupied by invading Communist armies from the north, and severely bombed. Its population shrank to 50 000.

Reconstruction began in 1953, and since the 1960s the city's expansion has been dynamic, fuelled by wholesale migration from the countryside. Today, the city accounts for almost half

Senegal

WITH A BUSTLING EUROPEANISED CAPITAL ON THE COAST, AND TRADITIONAL RURAL AREAS IN THE INTERIOR, SENEGAL HAS UNITED THE DEVELOPED WEST AND DEVELOPING AFRICA

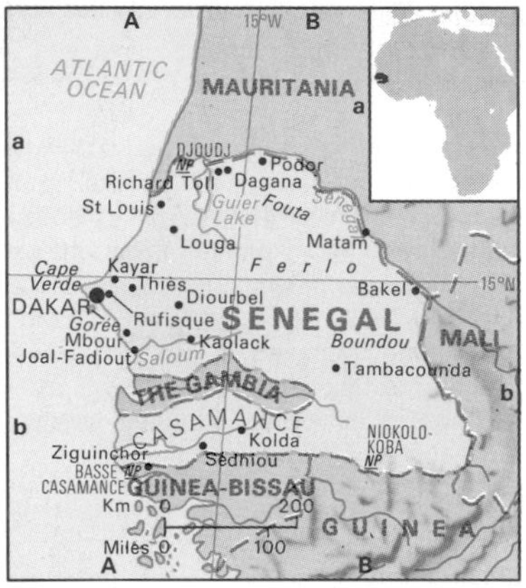

As France's first and most favoured colony in West Africa, Senegal is still very French in character, although it has been independent since 1960. It was the only colony where French citizenship was granted to Africans and where conscious efforts were made to educate Africans to become 'black Frenchmen'. The largest ethnic group in Senegal is the Wolof people (about one-third of the population), traditionally farmers on the savannah, but who now include wealthy, educated townspeople who control Senegalese economic and political life.

The economic and political nerve centre of Senegal is the capital, DAKAR, on the volcanic CAPE VERDE peninsula which juts into the Atlantic and is the most westerly point in Africa.

Most of Senegal consists of low plains, covered with sand or dry savannah of the SAHARA and SAHEL. The lowest, coolest and wettest part is the south-west, which is the most densely populated.

The rural areas are inhabited and farmed by over 80 per cent of the Senegalese. In the villages, the huts of extended families (parents, children, and other close relatives living together) are enclosed in fenced compounds. Work on the land for the whole family begins in June or July when the rains bring relief from the oppressive heat, and almost overnight the flat, brown countryside turns green. The men go to the fields early and return late; the women help there, as well as doing domestic chores. Tribal customs determine whether women are allowed to work in the millet fields, the main source of food for the family. However, the women work with the crops for sale – weeding, spreading fertiliser and harvesting. The main cash crop is groundnuts; Senegal is one of the world's leading producers of groundnut oil. The women also have their own plots of vegetables, and if they are near rivers which can be used to flood the fields they grow rice too. The vegetables and rice supplement a meagre diet which is based on millet. Surpluses are sold to bring the women an income – giving them some independence in this Islamic society.

About 90 per cent of Senegalese are Muslims. Indeed, the name of the country comes from the Zenaga Berbers of Mauritania, who invaded the region in the 11th century, and converted the people to Islam. However, the strictest rules of the faith are not always observed. South of the Sahara, for instance, women do not customarily wear the veil.

Cattle, goats and sheep are kept by most families, but it is bad taste to ask a Senegalese farmer how large his herds or flocks are. These are the symbols of his wealth and prestige, their number known by his neighbours but never admitted by the owner.

Hard times have come to Senegal with the drought that has afflicted the Sahel across the whole of West Africa. Grain stores are low and food does not always last the year; many go hungry for two or three months before the October harvest. More irrigation is needed to make the food supply secure against the erratic rains. And agricultural techniques in the dry areas need to be improved. Today, the country depends on food imports and international aid.

Phosphate is mined near Thies and petroleum products, fertilisers and textiles are Senegal's main manufactures. A film industry based in the capital is the centre of black Africa's film-making.

The government is stable and the country is one of the few multiparty democracies in Africa. In 1982, Senegal joined The Gambia in the Senegambia Confederation.

SENEGAL AT A GLANCE

Area 196 192 km² (75 750 sq miles)
Population 6 971 200
Capital Dakar
Government Parliamentary republic
Currency CFA franc = 100 centimes
Languages French (official), African languages – mainly Wolof
Religions Muslim (90%), Christian (6%), tribal religions (3%)
Climate Tropical; average temperature in Dakar ranges from 18-26°C (64-79°F) in January to 24-32°C (75-90°F) in September
Main primary products Rice, millet, sorghum, maize, groundnuts, fish, timber; phosphates
Major industries Agriculture, petroleum refining, phosphate fertiliser manufacture, textiles, beverages, fishing, forestry, cement and processed food
Main exports Petroleum products, fish, phosphate fertiliser, cotton fabrics, groundnut oil
Annual income per head (US$) 410
Population growth (per thous/yr) 32
Life expectancy (yrs) Male 43 **Female** 45

of South Korea's industrial output. High-rise blocks tower over the business centre; in the suburbs, houses are crammed together; and massive traffic congestion blocks the river bridges daily.

Little of the city's past survives. A few art treasures are, however, held in the National Museum, and the Toksugung Palace offers a glimpse of Yi architecture.

Population 8 364 000
Map South Korea Cd

Sepik ***Papua New Guinea*** River rising in the Victor Emanuel Range and flowing north and east for 1200 km (750 miles) into the Bismarck Sea. It is navigable for about 475 km (300 miles) from its mouth. People of the region are renowned for their woodcarving.

Map Papua New Guinea Ba

Sept Iles (Seven Isles) ***Canada*** Port city on the St Lawrence river in Quebec, founded as a trading post in 1651. The port handles iron ore from mines in Labrador, 560 km (350 miles) to the north.

Population 29 300
Map Canada Ic

Sequoia National Park ***USA*** Area covering 1631 km² (630 sq miles) in the Sierra Nevada mountains of California, about 260 km (160 miles) north of Los Angeles. It has groves of giant sequoia trees, the biggest of which – 'General Sherman' – is about 82 m (269 ft) tall. The park also contains Mount Whitney (4418 m, 14 495 ft), the highest peak in the country outside Alaska. The park was founded in 1890.

Map United States Cc

sérac Ice pinnacle between intersecting crevasses in an ice fall on a glacier.

Seraing ***Belgium*** Industrial town 5 km (3 miles) south-west of the city of Liège. It is the centre of the country's steel and heavy engineering industry, and headquarters of the steel-making and shipbuilding firm of Cockerill, whose English founder John Cockerill built a foundry and machine factory at Seraing in 1817.

Population 63 000
Map Belgium Ba

Seram (Ceram) ***Indonesia*** Mountainous island of 17 148 km² (6621 sq miles) in the Maluku (Molucca) group between Sulawesi and New Guinea. Sago palms are grown, and timber is exported.

Map Indonesia Gc

Seram Sea (Ceram Sea) An arm of the PACIFIC OCEAN in Eastern Indonesia which was formerly called Pitt Passage. It lies between Halmahera and western New Guinea to the north and Seram island to the south.

Map Indonesia Gc

Serbia (Srbija) ***Yugoslavia*** Largest and most populous of the country's republics, lying in the east of Yugoslavia, and bordered by Albania, Bulgaria, Romania and Hungary. It covers 88 361 km² (34 107 sq miles) and comprises three administrative areas: KOSOVO, VOJVODINA and Serbia Proper (55 968 km², 21 604 sq miles). The republic consists of rolling hill

A DAY IN THE LIFE OF A TANZANIAN GAME WARDEN

The first pale light of day throws weak shadows from acacia trees across the tall grasslands of the Serengeti, and the morning chorus of insects and birds begins. This wakes Joseph in his one-roomed banda, and the light shines through its small window, onto the whitewashed concrete walls. Joseph, opens his eyes, blinks at the corrugated ceiling, and then focuses on the room's one chair and beside it the box containing his belongings. His home is far away, near Orusha, but a game warden has to live where his job is.

Joseph thinks of his family, in a new village created by the government's *ujamaa* collectivisation programme which moved black people from their scattered plots and farms to newly established villages where they work communal land. Now, with his new job, Joseph sends money to help them.

Dressed in his khaki uniform with warden's insignia, he eats a quick breakfast of maize meal and walks to the chief warden's office for the day's briefing. As he arrives, a spotter plane calls in to report a wounded elephant some 40 km (25 miles) away. Detailed to go in search of it, Joseph and a fellow warden collect rifles and set out with several helpers in two Land Rovers. The game park, and especially the elephants, are continually threatened by poachers; tusk ivory fetches high prices on the black market.

Leaving the dirt road at the last possible moment, the cars cross the bush. The party is subdued; wounded elephants can be dangerous and one could overturn a Land Rover with ease. Joseph spots the elephant, still standing, but motionless, with two poacher's poisoned arrows in its side. Knowing it is too late to save the animal, Joseph shoots it.

The poachers were probably disturbed by the early morning arrival of the spotter plane and may still be in the area. Joseph radios for help because poachers have killed wardens in the past. Meanwhile, the second Land Rover party gives chase. Waiting for reinforcements, Joseph begins the task of removing the elephant's tusks and hide. This means other poachers will not profit from them, and the carcass will provide good meat to supplement the diet of the wardens and helpers.

The radio crackles into life with the news that two poachers have been found and captured relatively easily. They will be handed over to the police.

It takes Joseph and his helpers all day to finish their task with the elephant. They return to headquarters under a glorious sunset, and, as darkness falls, Joseph thinks he would rather be here than anywhere else in the world.

country rising from the Sava-Danube plains through densely populated, arable and orchard country southwards to mountains – including Zlatibor and Kopaonik which are cut by the Morava rivers.

The Serbs, a South Slav people, settled the area in the 7th century. It became the leading Slav kingdom under Stefan Dusan in the 14th century, but was crushingly defeated by the Turks at the Battle of Kosovo near Pristina in 1389. Most of Serbia was part of the Turkish Ottoman Empire from 1459, but revolts led by Serbian nationalists in the early 19th century resulted eventually in complete independence in 1878. The capital has gradually moved north – from the city of Novi Pazar in the Middle Ages, and under the Turks, to Kraljevo, to Kragujevac in the 19th century, and finally to Belgrade in 1918.

Outside Belgrade, only Nis and Kragujevac are major industrial centres, manufacturing being largely dispersed among valley towns where factories process local cereals, fruits, timber, clay, coal, copper, lead, zinc and antimony, and make textiles, carpets, machinery and metal goods, mostly powered by hydroelectricity.

There are many attractive spas in central Serbia, including Kursumlija, Mataruska, Niska and Vrnjacka. The area is also noted for its fine monasteries, including Zica, Studenica, Gradac, Stubovi and Gracanica.

Population (republic) 9 314 000; (Serbia proper) 5 694 000
Map Yugoslavia Ec

Seremban ***Malaysia*** Capital of the state of Negeri Sembilan, about 65 km (40 miles) south-east of the national capital, Kuala Lumpur. It is the centre for the shrinking tin-mining industry of the state.
Population 136 300
Map Malaysia Bb

Serengeti ***Tanzania*** Savannah plain in the north-east of the country stretching from Lake Eyasi northwards to the border with Kenya. It contains many buffaloes, elephants, giraffes, lions, leopards and more than 2 million gazelles, wildebeests and zebras. The Serengeti National Park – one of Africa's most spectacular game reserves – occupies 14 500 km² (5600 sq miles) of the plain.
Map Tanzania Ba

Seria ***Brunei*** Major oil-producing town about 70 km (44 miles) south-west of Bandar Seri Begawan. Oil was first discovered there in 1929 and, although the field (which is half onshore and half offshore) is now almost exhausted, the town remains the administrative and service centre for the Sultanate's oil industry.
Population 30 000
Map Brunei Ab

series Subdivision of a system that represents the rocks formed during an epoch. For example, Carboniferous Limestone is a series within the Carboniferous system, and Chalk is a series within the Cretaceous system.

Seringapatnam ***India*** See SHRIRANGAPATTANA

serir Stony desert of the central and eastern Sahara within Egypt and Libya which is covered with sheets of angular gravel. It is similar to the REG of the western Sahara.

▼ ELEPHANT COUNTRY The permanent snows of Oldeani (3188 m, 10 459 ft) rise above the torrid grasslands of Tanzania's Serengeti plain.

Serowe ***Botswana*** Chief village of the Bamangwato people, 140 km (90 miles) south-west of Francistown. It was founded in 1902 in a fertile area around a tree-covered hill, which became the burial ground for the Bamangwato royal family.

Population 24 000

Map Botswana Cb

serpentine Dark green or brownish rock, often variegated or mottled like the skin of a snake, composed mainly of minerals of the serpentine group (hydrated magnesium silicates). The rock is a source of magnesium and used as a decorative stone in building. Asbestos is a fibrous mineral of the group.

Serra da Estrela ***Portugal*** Range of mountains about 120 km (75 miles) south-east of the city of Oporto. It is the highest range in mainland Portugal. Estrela, the main summit, is 1991 m (6532 ft) high, with a skiing centre nearby at Penhas da Saúde.

Map Portugal Cb

Serra do Mar ***Brazil*** Mountain range running along the south-east coast between the cities of Rio de Janeiro and Porto Alegre. Historically a considerable obstacle to the settlement of the interior, the range reaches its highest point at Serra dos Orgãos, a 2200 m (7215 ft) peak overlooking Rio.

Map Brazil Dd

Sétif ***Algeria*** Market town 112 km (70 miles) west of Constantine on the road and rail route to Algiers. Lying 1067 m (3500 ft) above sea level in the Kabylie Mountains, it is the administrative centre for the district and serves an important agricultural region dealing mainly in grain and cattle. Sétif was founded by the Romans in the 1st century AD, and remains from that period can still be seen. The town's chief industry is flour milling.

Population 195 000

Map Algeria Ba

Seto Naikai ***Japan*** See INLAND SEA

Setúbal ***Portugal*** City and port about 30 km (20 miles) south-east of the capital, Lisbon. Its industries include cement, chemicals, shipbuilding and food processing. The area around Setúbal produces salt, fish, oranges and wine.

Population 76 800

Map Portugal Bc

Sevastopol' ***USSR*** Black Sea port on the south-west coast of the Crimea. It has been the main base of the Russian Black Sea fleet since the early 19th century, and has large dockyards and arsenals. It is also a spa and has engineering, food processing and clothing factories.

Sevastopol' has twice been besieged: in 1855 during the Crimean War between Russia and the forces of England, France and Turkey (when it stood firm); and in 1942 during the Second World War, when the Germans took possession. In 1944 the Russians recaptured Sevastopol', and there are some 400 monuments in the city commemorating its turbulent past. The coastal village of Balaklava – the scene of the disastrous charge of the Light Brigade in 1854, during the Crimean War, when 247 out of a force of 673 British cavalrymen rode to their deaths – is 15 km (9 miles) to the south-east.

Population 335 000

Map USSR Ed

Seven Pagodas ***India*** See MAHABALIPURAM

Severn ***United Kingdom*** The country's longest river, rising on the slopes of Plynlimon in central Wales and flowing 350 km (220 miles) into England and out into its estuary below Gloucester. The funnel shape of the estuary constricts the incoming tide and produces a BORE. The Severn is rich in salmon.

Map United Kingdom Dd

Severnaya Zemlya (North Land) ***USSR*** Group of ice-covered islands in the Arctic Ocean off Cape Chelyuskin, the most northerly point of Asia. The islands, which cover some 37 000 km^2 (14 300 sq miles), contain weather, radar and research stations.

Map USSR Ka

Seville (Sevilla) ***Spain*** Port and industrial city near the country's southern tip. Some of Europe's highest temperatures – up to 48°C (118°F) – have been recorded there. It was the birthplace of the Spanish artists Murillo (1617-82) and Velazquez (1599-1660). It once had the monopoly of Spain's fabulously wealthy New World trade, but now exports fruit (particularly oranges) and wine from its fertile province.

The Old City contains the world's largest Gothic building – a cathedral begun in 1402; the Alcázar fortress and palace started by the Moors on the bank of the Guadalquivir, and the fascinating old Jewish quarter of Santa Cruz where Murillo lived.

Population (city) 653 800; (province) 1 477 400

Map Spain Bd

Sèvres ***France*** Suburb of south-western Paris near Versailles. It is the site of the renowned state-owned porcelain factory which was transferred there by Louis XV in 1756. Previously the factory had been in the Parisian suburb of Vincennes.

Population 20 300

Map France Eb

Sewa ***Sierra Leone*** Central river flowing some 320 km (200 miles) to the Atlantic. Diamonds are found in its valley gravels.

Map Sierra Leone Ab

Sfax ***Tunisia*** The country's second city and a major port and industrial centre, situated on the northern shore of the Gulf of Gabès. It grew from two ancient Phoenician settlements and became an early trading centre. However, Sfax was of little importance before the 12th century when the Sicilians took it.

There is a huge old medina (Muslim residential area) and numerous souks (markets) selling metalwork, woodwork and embroidery. The Grand Mosque, with its finely decorated, three-tiered minaret, dates from AD 849. The city's industries include chemicals, leather, food processing, olive oil, soap and fishing, especially for sponges and octopus. Together with Gabès it is the main port for the export of phosphates.

Population 232 000

Map Tunisia Ba

Seychelles

AN ISLAND PARADISE IN POLITICAL TURMOIL, BUT WAS IT THE ORIGINAL GARDEN OF EDEN?

The Garden of Eden lay in the Seychelles, according to some islanders. Indeed, it is in many ways an island paradise, with a superb climate and idyllic palm-fringed beaches. Its isolation – 1200 km (750 miles) from the coast of Africa – has given rise to unique species of plants and animals, including rare birds and, on ALDABRA, giant tortoises.

The two main islands, MAHÉ and PRASLIN, are partly granite and reach heights of about 900 m (2950 ft). Granite is also the bedrock of 30 more of the 100 or so islands; the rest are coral, low-lying and largely uninhabited.

Ninety per cent of the people live on Mahé, site of the capital, VICTORIA. The population is descended from French settlers, who came in 1770, and African slaves from MAURITIUS. The islanders' diet is mainly coconut, imported rice and fish. The fishing industry is expanding, but tourism accounts for 90 per cent of foreign exchange earnings.

Politically and socially, however, the Seychelles are a long way from paradise. The islands became a republic in 1976 after 162 years of British rule. The first president, playboy James Mancham, was replaced by Albert René after a 1977 coup. Mancham still leads opposition to the government, which has survived several attempted coups. Society is divided between the landowners (mostly white) and landless (usually mixed race).

The Russians have gained a foothold in the Seychelles, and Soviet-made missiles have been installed as part of the islands' defence system.

SEYCHELLES AT A GLANCE

Map Indian Ocean Bc

Area 453 km^2 (175 sq miles)

Population 66 000

Capital Victoria (on Mahé)

Government One-party republic

Currency Seychelles rupee = 100 cents

Languages Creole, French, English

Religion Christian (90% Roman Catholic, 8% Anglican)

Climate Tropical; average temperature 26°C (79°F); heavy rainfall

Main primary products Fish, fruit, vegetables, coconuts

Major industries Tourism, fishing, brewing, cigarette manufacture

Main exports Fish, copra, cinnamon bark, fertiliser (guano)

Annual income per head (US$) 2300

Population growth (per thous/yr) 9

Life expectancy (yrs) Male 65 **Female** 71

's-Gravenhage ***Netherlands*** See HAGUE, THE

Sha Tin ***Hong Kong*** City in the NEW TERRITORIES on the south-eastern side of a sea inlet known as the Tolo Harbour. One of eight new towns built to relieve congestion in Hong Kong's built-up areas, Sha Tin consists largely of soaring apartment blocks built on land reclaimed from the sea. The city also contains the campus of the Chinese University of Hong Kong, and is well equipped with recreational facilities, including the racecourse of the Royal Hong Kong Jockey Club and the Jubilee Sports Centre.
Population 300 000
Map Hong Kong Ca

Shaanxi ***China*** Province in the north covering 190 000 km² (73 000 sq miles) of the central part of the LOESSLANDS OF CHINA. It contains the Wei river valley, the heartland of Chinese civilisation. The great Huang He river forms the eastern boundary of Shaanxi and the southern boundary follows the crests of the Qin Ling. The dusty loess plateau is prone to drought but produces wheat and millet. Farmers used to live in caves dug in the loess, with chimney holes rising into the fields above. The area is rich in coal reserves, much of them undeveloped, and has valuable deposits of molybdenum and mercury. The capital is XI'AN.
Population 28 070 000
Map China Fd

Shaba ***Zaire*** Administrative region covering 496 965 km² (191 828 sq miles) in the south-east, and formerly called Katanga. It lies on a plateau between 900 and 1900 m (2953 and 5900 ft) high. The region has great mineral wealth, including copper, cobalt, manganese, platinum silver, uranium and zinc. The area includes two national parks: Kundelungu and Upemba. The plateau, which is mostly wooded savannah, with mountains in the east, is drained by headstreams of the Zaire river. Cattle are reared in the south and west, and crops produced on plantations include citrus fruits. The capital is Lubumbashi.

When Zaire (then the Congo-Kinshasa) became independent in 1960, Katanga seceded and remained separate until 1962, when its rebellion was quelled with the aid of United Nations troops. The mining industry, which before independence was controlled by European companies, is now substantially state-owned.
Population 3 823 000
Map Zaire Bb

Shabeelle (Shebelle; Scebeli) ***Ethiopia/Somalia*** River rising in central Ethiopia and flowing roughly southwards for more than 1700 km (1055 miles). It comes within 30 km (18 miles) of the Indian Ocean near the capital, Mogadishu, then turns to join the Jubba river 350 km (220 miles) farther south-west. The Shabeelle and the Jubba are the only two rivers in Somalia that never dry up, and the water of both is used for irrigation.
Map Somalia Ab

shake-hole See SINKHOLE

shale Fine-grained sedimentary rock composed mainly of clay minerals, with well-defined narrow strata, which readily split apart.

▼ FESTIVAL OF REPENTANCE Hooded, bare-foot penitents follow carvings of the Stations of the Cross through the streets of Seville during Holy Week.

Shan ***Burma*** State covering 155 801 km² (60 155 sq miles) – much of it a hilly, forested plateau – in the eastern part of the country bordering China, Laos and Thailand. It is named after the Shan people, the largest ethnic minority in Burma, who are closely related to the Thais. The state capital is TAUNGGYI.

The plateau rises to between 600 m and 1000 m (2000 and 3300 ft). Rice, potatoes and onions are the main food crops of the Shans, but hill-dwellers in the east of the state, including the Wa and Palaung peoples – formerly headhunters – grow opium in regions largely beyond the control of the Burmese government. The crop is sold to opium warlords operating in the area known as the GOLDEN TRIANGLE.
Population 3 719 000
Map Burma Cb

Shandong ***China*** Province in the north covering 150 000 km² (580 000 sq miles) on the coast. It takes in much of the North China Plain. The Huang He river reaches the sea in the north of the province. Shandong's climate is more humid and temperate than the northern interior of China, but farming is often hampered by build-ups of salt in the soil due to evaporation of moisture during the summer. Wheat, maize, cotton and groundnuts are grown, and Shandong has long been famous for its fine quality silk cloth – known in English as shantung. The Shengli oil field, close to the estuary of the Huang He, is the second largest in China after DAQING. The capital is Jinan.
Population 72 310 000
Map China Hd

Shanghai ***China*** China's most populous city and greatest concentration of manufacturing industry. It stands in a 6185 km² (2400 sq mile) municipality on the densely settled and fertile delta of the Chang Jiang river. Shanghai grew rapidly after the arrival of the Western imperial powers in China in the 19th century. Foreign merchants and diplomats enjoyed rights of residence in the city, which became notorious for the stark contrast which it presented between European prosperity and Chinese poverty. By the 1930s, Shanghai had developed into one of the great seaports of Asia, with opulent banks and office buildings flanking the harbour on the Huangpu river. Today, its fortunes are based more on manufacturing than on overseas trade. Its industries include shipbuilding, iron and steel, chemicals and textiles.
Population 11 860 000
Map China Ie

Shannon (An Tsionainn) ***Ireland*** Longest river in the British Isles, 386 km (240 miles) long including its 90 km (56 mile) estuary. It rises in north-west County Cavan near the Northern Ireland border, and runs south and west through Loughs Ree and Derg. It enters its estuary at Limerick and empties into the Atlantic. It is a salmon-fishing river, and at Ardnacrusha in County Clare it powers a hydro-electric scheme built in the 1920s.
Map Ireland Bb

Shanti Niketan (Santiniketan) ***India*** Rural retreat and university in West Bengal, about 130 km (80 miles) north and slightly west of Calcutta. It was founded at the turn of the century by Sir Rabindranath Tagore (1861-1941), India's best known poet and leading educator, and a Nobel Prize winner, to help regenerate India's villages and to widen international understanding with the aid of a multi-cultural university.
Map India Dc

Shantou ***China*** City and port on the coast of Guangdong province. It is one of several areas designated as 'special economic zones' where overseas investment is encouraged.
Map China Hf

Shanxi ***China*** Province in the north of the country, reaching into the LOESSLANDS OF CHINA. It is a fertile area covering 150 000 km² (58 000 sq miles), but the soil erodes easily and needs careful management. Rainfall is light and variable, and Shanxi is vulnerable to drought. Maize, wheat and cotton are grown. The province is also China's leading coal producer. Coal has been mined in Shanxi for centuries, but enormous reserves remain to be exploited. The capital is Taiyuan.
Population 24 472 000
Map China Gd

Shaoshan ***China*** Village in Hunan province, 104 km (65 miles) south-west of Changsha. It is the birthplace of Mao Zedong, leader of Communist China from 1949 to 1976. Mao's family house and the nearby museum attract thousands of Chinese visitors every year.
Map China Ge

Shari ***Equatorial Africa*** See CHARI

Sharjah ***United Arab Emirates*** Fourth largest of the seven emirates, covering 159 000 km² (61 000 sq miles). Its main town, also called Sharjah, lies on The Gulf, 13 km (8 miles) north-east of Dubai. As it has little oil, it concentrates on commerce and encourages new companies to establish themselves in its territory. Manufactures include paint, plastic pipes and cement. There is a container service at Port Khaled. The town has impressive new buildings, including the post office and the Sharjah New Souk (bazaar), as well as new port facilities. There is also an old, covered souk, and a few traditional Arab houses remain.
Population (emirate) 184 000; (town) 126 000
Map United Arab Emirates Ba

Shark Bay Inlet in Western Australia about 670 km (416 miles) north of PERTH. Dirk Hartog Island, at the mouth of the bay, was visited in 1616 by the Dutch navigator Dirk Hartog, who left behind an engraved pewter plaque to mark his visit. The plaque was rediscovered more than 100 years later.
Dimensions 240 km (150 miles) long, 100 km (62 miles) wide
Map Australia Ad

Shatt al Arab ***Iran/Iraq*** Waterway formed by the union of the Euphrates and the Tigris rivers at the Iraqi city of Al Qurnah, about 400 km (250 miles) south-east of Baghdad. From Al Qurnah, it flows south-east for 170 km (105 miles) to The Gulf. The lower course forms the disputed Iran-Iraq border.
Map Iraq Cc

Shayib el Banat ***Egypt*** Highest mountain – at 2187 m (7175 ft) – in the Eastern Highland which runs the length of Egypt's Red Sea coast and stretches from the Gulf of Suez to the Sudanese border.
Map Egypt Cc

shear The movement of one body of rock relative to another along a fault or thrust plane when two forces act parallel with each other.

Sheffield ***United Kingdom*** City in the English county of South Yorkshire, 55 km (35 miles) east of Manchester. Its name has for centuries been synonymous with cutlery, steel and Sheffield plate, a fusion of silver and copper. Its industries are fuelled by the coalfields surrounding the city. The guild of craftsmen known as the Cutler's Company was founded in the 17th century under James I. Sheffield has a cathedral dating from the 15th century and a university founded in 1905.
Population 545 000
Map United Kingdom Ed

Shenandoah ***USA*** River in northern Virginia which flows some 90 km (55 miles) north-east to the Potomac at Harpers Ferry. Its valley, between the Allegheny and Blue Ridge mountains, includes the Shenandoah National Park, and was the scene of several Civil War engagements. The anti-slavery hero of the song 'John Brown's Body', with 18 men, captured the Federal arsenal at Harpers Ferry in 1859 and was hanged for it.
Map United States Kc

Shenyang ***China*** Capital of Liaoning province. It is the largest city in the north-east and stands at the centre of the railway network of DONGBEI (Manchuria). It produces machine tools, metals, textiles and aircraft.
Population 4 000 000
Map China Ic

Shenzhen ***China*** City in Guangdong province just across the border from Hong Kong. It is the largest of the country's 'special economic zones' set up to attract investment from overseas.
Population 350 000
Map China Hf

's-Hertogenbosch (Den Bosch; Bois-le-Duc) ***Netherlands*** Capital of the province of Noord-Brabant, some 80 km (50 miles) south and slightly east of Amsterdam. Its name (meaning 'duke's wood') probably derives from Henry I, Duke of Brabant, who in the 12th century had a hunting lodge nearby. The town's cathedral of St John, built in the 11th-16th centuries, is the country's largest Catholic church.
Population (city) 89 500; (Greater 's-Hertogenbosch) 186 600
Map Netherlands Bb

Shetland ***United Kingdom*** Group of more than 100 islands over 160 km (100 miles) north-east of the Scottish mainland. They were settled by Norsemen in the 9th century and became Scottish only in 1469. The festival of Up-Helly-Aa, each January, when a longship is ceremonially burnt, recalls the Viking connection. The islanders relied on fishing and sheep before the discovery of North Sea oil, but the oil industry is now one of the major employers, notably at the Sullom Voe terminal. The capital, Lerwick, is a ferry port linking the islands to Aberdeen and Norway.
Population 28 000
Map United Kingdom Gg

Shewa ***Ethiopia*** Region (85 500 km², 33 003 sq miles) around the capital, Addis Ababa. It has formed the political and economic heartland of the country for more than 100 years. The main products include barley, maize, sorghum, skins and hides. There is no large-scale industry, though there are diverse small industries in Addis Ababa and other main towns.
Population 7 000 000
Map Ethiopia Ab

shield See CRATON

Shijiazhuang ***China*** Capital of Hebei province 250 km (155 miles) south of Beijing (Peking). It is a centre for the manufacture of woollen textiles.
Population 973 000
Map China Hc

Shikoku ***Japan*** Smallest of the country's four main islands – the others are Honshu, Hokkaido and Kyushu. It has a rugged, mountainous core flanked by small alluvial plains. There is a slow, traditional flavour to life on the island, though new bridge links may bring more industrialisation. The north coast is dry and warm, while the south is exposed to wet summer winds allowing two crops of rice to be grown each year. Tomatoes, cucumbers and melons are the main winter crops.
Population 4 227 200
Map Japan Bd

Shillong ***India*** Capital of Meghalaya state, about 500 km (310 miles) north-east of Calcutta. It is nearly 1500 m (4930 ft) up in the Khasi Hills, with a golf course and racetrack, and was a popular hill resort for British colonial families of Bengal.
Population 174 700
Map India Ec

Shimizu ***Japan*** See SHIZUOKA

Shimonoseki ***Japan*** Deep-sea fishing port and city on the south-west tip of Honshu island. An April festival there commemorates the sea battle of Dannoura in 1185 in which the fleet of the Taira clan was destroyed and the seven-year old Emperor Antoku drowned.
Population 261 700
Map Japan Bd

Shipka (Sipka) Pass ***Bulgaria*** Mountain pass (1326 m, 4350 ft) in the Balkan Mountains between Gabrovo on the north and Kazanluk on the south. It was the scene of major battles in 1877 during the Russo-Turkish war for Bulgarian independence.
Map Bulgaria Bb

Shiraz ***Iran*** Provincial capital of Fars, about 700 km (435 miles) south and slightly east of the national capital, Tehran. Shiraz, which may have been founded before the 6th century BC, was by the 12th century AD the literary capital of Persia. It is the birthplace of the poets Saadi (1184-1292) and Hafiz (1300-88), both of whom are also buried there. From 1750 to 1786 it was the capital of Persia.

Celebrated for many centuries for its wine and silk rugs, Shiraz today is a regional metropolis with an airport and a university. Its products include textiles, carpets, sugar, fertilisers and cement, and it is the commercial centre of a farming region producing cereals, sugar beet and grapes.
Population 801 000
Map Iran Bb

Shire ***Malawi/Mozambique*** River flowing about 400 km (250 miles) south from Lake Malawi to join the River Zambezi in Mozambique. The upper course, through the shallow Lake Malombe, has a gentle gradient. By contrast, the middle course below Matope drops 383 m (1256 ft) in 80 km (50 miles). Here the river descends through scenic gorges in a series of cataracts, called the Murchison Cataracts by the missionary and explorer David Livingstone after a president of the Royal Geographical Society. There are hydroelectric power stations at the Nkula Falls and, 8 km (5 miles) downstream, at the Tedzana Falls. The cataracts end north of Chikwawa and the river then flows across a broad swampy plain.

Map Malawi Ac

Shire Highlands ***Malawi*** Fertile plateau in Southern province rising to 1700 m (5600 ft). It stretches about 80 km (50 miles) from south of Blantyre to Zomba, and was the chief area of European settlement in the colonial period. The main products are tea, tobacco and tung oil (used in waterproof paints and varnishes).

Map Malawi Bc

Shizuoka ***Japan*** City on central Honshu island, about 145 km (90 miles) south-west of the capital, Tokyo. With the nearby city of Shimizu (population 242 200) on Suruga Bay, it makes an extensive industrial zone producing tea, canned fruit, petrochemicals and ships.

Population 468 400

Map Japan Cc

Shkodër (Scutari) ***Albania*** Quiet, largely undeveloped riverside town in a beautiful setting near the lily-covered Lake Shkodër and backed by an arc of the Albanian Alps. It is overlooked by the ruins of a 14th-century Venetian citadel – a legacy of the days when the port was regularly visited by traders from Italy. Now its river is silted up and shallow and there is no railway to replace it. The inadequate transport facilities and the town's closeness to Yugoslavia have deterred greater development, other than a few light industries, since Albania's closed frontier policy has deprived Shkodër of its natural hinterland over the border.

Population 69 700

Map Albania Ba

Shkodër, Lake (Lake Scutari; Lake Skadar; Skadarsko Jezero; Liqen i Shkodrës) ***Albania/Yugoslavia*** The largest Balkan lake, covering 391 km² (143 sq miles) astride the Albanian-Montenegrin border. Bare limestone mountains descend abruptly into the lake, where summer mirages are common. Water chestnuts, a local delicacy, and fish, including carp, are plentiful. Virpazar ('whirlpool bazaar'), in Yugoslavia, and Shkodër, in Albania, are the main towns beside the lake. Outside Crnojevica – known as the 'Venice of Montenegro' – near the lake's northern shore, are the remains of Obod monastery, where in 1493 the first books in the Cyrillic alphabet were printed.

Map Yugoslavia Dc

shoal Ridge of sand, mud or pebbles just below the surface of the sea or a river, creating a hazard to navigation.

Shonaland ***Zimbabwe*** See MASHONALAND

shore Area between the lowest low tide mark and the highest point reached by storm waves.

shott, chott Shallow temporary salt lake in a hot desert, or the hollow in which such a lake lies. The term is used especially in North Africa in the areas bordering the Atlas Mountains.

shoulder Gentle slope above a steep-sided valley in a mountainous area.

Shreveport ***USA*** Industrial city at the navigable limit of the Red River in north-west Louisiana. It is an oil refining and cotton textile centre.

Population (city) 220 000; (metropolitan area) 360 900

Map United States Hd

Shrewsbury ***United Kingdom*** County town of Shropshire, England. It stands on a wide loop of the River Severn 65 km (40 miles) north-west of Birmingham, and was an English stronghold through centuries of conflict with the Welsh. It has many fine half-timbered buildings, one of the handful of round churches in Britain and a public school, whose pupils included the naturalist Charles Darwin (1809-82), exponent of the theory of evolution by natural selection.

Population 60 000

Map United Kingdom Dd

Shrirangapattana (Seringapatnam) ***India*** Fortress town in the south, on the Kaveri river about 400 km (250 miles) west and slightly south of the east-coast city of Madras. It was the capital of Tipu (the 'Tiger'), Sultan of Mysore. A successful siege by the British resulted in Tipu's death in 1799, and the establishment of British power in southern India. The city's breached walls survive.

Map India Be

Shropshire ***United Kingdom*** County of west-central England, bordering Wales. It lies just west of the industrial West Midlands and stretches from the fringes of the industrial north to the placid countryside around the historic town of Ludlow in the south – a total area of 3490 km² (1347 sq miles). There are many hills, such as The Wrekin (407 m, 1335 ft), which is topped by a 2000-year-old fort and from which, it is said, 17 counties can be seen on a clear day. The county's fertile valley of the River Severn provides rich agricultural land.

Shropshire was a cradle of Britain's Industrial Revolution. The world's first iron bridge was built at Iron-Bridge in 1778, from girders produced in a foundry at Coalbrookdale nearby. The new town of Telford has attracted some of Britain's most promising high-technology industries. The county town is SHREWSBURY.

Population 390 000

Map United Kingdom Dd

Shumen (Kolarovgrad) ***Bulgaria*** City and capital of Shumen province, 80 km (50 miles) west of Varna. It trades in grain and wine, and manufactures farm machinery, beer, furniture and enamelware. A strategically important stronghold, it surrendered to the Russians in June 1878 during the Russo-Turkish war for Bulgarian independence.

About 10 km (6 miles) to the east is the ruined ancient city of Madara whose remains date from the 5th century. Carved in the nearby cliffs is the figure known as the Horseman of Madara, which dates from the 8th century.

Population 104 800

Map Bulgaria Cb

Shush ***Iran*** See SUSA

Shuwaikh ***Kuwait*** The country's largest port, lying in the western suburbs of Kuwait city. About 40 per cent of the goods which pass through it are re-exports – the result of the country's expanding manufacturing sector. There is a special industrial zone, with land at a nominal price to attract factory development. Power and water plants have already been established. Shuwaikh is one of the campuses of the University of Kuwait, which has over 18 000 students. It takes some 60 per cent of its students from around The Gulf and other Arab states. Expansion of the university, including the building of a sports complex at Shuwaikh, is due to be completed by 1990.

Population 9100

Map Kuwait Ab

Shwebo ***Burma*** Town and former national capital during the reign of the warrior King Alaungpaya (1752-63), who is buried there. Today the town – which lies about 75 km (45 miles) north-west of Mandalay – is a market for cotton and beans grown in irrigated fields along the Mu river.

Population 167 000

Map Burma Bb

sial General name given to the rocks of the earth's upper continental CRUST that are rich in silica and alumina. The word is formed from the first two letters of these two minerals.

Sialkot ***Pakistan*** Garrison town in the Punjab about 100 km (60 miles) north of Lahore. It was the scene of a bloody insurrection (which was put down) during the 1857 Indian Mutiny against British rule. The town now manufactures sports goods, particularly hockey sticks and squash and tennis rackets.

Population 296 000

Map Pakistan Da

Siam See THAILAND

Sibenik (Sebenico) ***Yugoslavia*** Seaport and fishing and naval base at the mouth of the Krk river in Dalmatia, 50 km (about 30 miles) north-west of Split. The heavily fortified old town, with one of Dalmatia's finest cathedrals (built 1431-1536), and local coves and beaches, are popular with tourists. Divers obtain corals and sponges around the nearby islands of Zlarin and Krapanj.

Population 29 600

Map Yugoslavia Bc

Siberia (Sibir) ***USSR*** Vast north Asian region lying between the Ural mountains and the Pacific, covering 13 800 000 km² (5 300 000 sq miles), mostly in the RSFSR. It is rich in timber, metals and wildlife, and the steppes of western

Siberia are fertile agricultural land if irrigated. But the extremely harsh climate and the inhospitable land in the north and east – where tundra, swamps and swampy forests are frozen, snow-covered or flooded for much of the year – have greatly restricted settlement. Even so, the region now has a population of some 40 million.

Originally peopled largely by Tatars, most of Siberia was conquered by Russian Cossacks between 1582 and 1633. It was colonised by settlers in search of furs, timber, gold and silver, and political prisoners and criminals were exiled there and forced to work in the camps and mines. During the 19th century about 1 million exiles (many of them from the Baltic region) were sent to Siberia, and many more followed after the Russian Revolution – especially in the 1930s and 1940s, during the rule of Joseph Stalin (1879-1953).

The TRANS-SIBERIAN RAILWAY, built between 1891 and 1905, linked Siberia with Russia proper – and opened up the area. From the mid-1920s, development spread eastwards, with the growth of large industrial complexes using local resources – such as coal, zinc, lead, copper and iron ore, and hydroelectric power. This led to the growth of some of the region's chief cities – including IRKUTSK, KRASNOYARSK, NOVOSIBIRSK and OMSK. Since 1945 discoveries of huge gas and oil reserves have boosted the region's industrial development.

Population 39 857 000

Map USSR Hb

Sibiu ***Romania*** Major industrial and tourist city in southern TRANSYLVANIA, 210 km (130 miles) north-west of Bucharest. Originally a Roman colony, Sibiu was settled by Germans in the 12th century. Some ethnic Germans still live there and much of its architecture is Germanic. Its 14th-century walls and fortifications are still standing, and they surround and are surrounded by narrow streets, small squares, and red-roofed houses with dormer windows.

The old town radiates from Republic Square, near which is the 18th-century Brukenthal Museum, which houses a fine collection of Romanian paintings – as well as canvases by the Flemish artists Peter Rubens (1577-1640) and Anthony Van Dyck (1599-1641). Sibiu's industrial products include footwear, textiles, building materials, machinery and chemicals.

Population 161 000

Map Romania Aa

Sichuan (Szechwan) ***China*** The nation's most populous province, lying in the west of the country. It covers 570 000 km² (220 000 sq miles) of the Red Basin, a vast region surrounded almost entirely by high mountains. The climate is humid and temperate and its soils are unusually fertile. Rice, wheat, millet, sugar cane and citrus fruits are grown. Sichuan is noted for its spicy cuisine. The Wolong nature reserve west of the basin is the last refuge of the giant panda. There are spectacular gorges in the east created by the Chang Jiang river as it cuts down through the mountains.

Population 97 740 000

Map China Ee

Sicily (Sicilia) ***Italy*** The Mediterranean's largest and most populous island, covering 25 708 km² (9926 sq miles). Its capital and largest city is PALERMO; other major towns and cities include the provincial capitals of AGRIGENTO, CATANIA, ENNA, MESSINA, RAGUSA, SYRACUSE and TRAPANI. A mountain chain crosses the north between Palermo and Messina. Mount ETNA, at 3323 m (10 902 ft) Europe's largest active volcano, towers above the east.

The central and southern parts of the island are a jumble of open hill country, with widespread sulphur outcrops. Agriculture in this inland area is poor and backward, but in the south-east corner, along the north coast and around the lower slopes of Etna, where Sicily's lemons grow, it is rich and flourishing. Industry is limited to recently established oil refining, petrochemical and car-assembly plants. Tourism focuses on the major archaeological sites and on resorts such as TAORMINA and Cefalù.

The island was named after the ancient Siculi tribe and was colonised in the 7th and 8th centuries BC by the Greeks, who left the remains of several splendid city-states, notably at Agrigento, GELA and Syracuse. The Romans conquered Sicily, exploiting it for its wine, olive oil, wheat and sulphur. Later it became the centre of an Islamic Empire in the central Mediterranean, and Palermo was the great Arab capital. Arab and Byzantine influences are found in the characteristic Norman architecture of the 11th and 12th centuries, best seen in the splendid cathedrals of Palermo, Cefalù and Monreale. Catalan Gothic influences were introduced by Spain in the 15th century when the island was an impoverished viceregal colony.

Unification with Italy in 1860 brought few benefits to the island; indeed, the bitter joke about the Italian boot kicking Sicily dates from this time. Sicilian society is highly stratified, and the criminal Mafia organisation, which the Italian authorities are trying to eradicate, has spread from Sicily to many other countries – particularly the USA, to which many Sicilians have emigrated during this century.

Population 5 065 000

Map Italy Df

A DAY IN THE LIFE OF A SICILIAN DOCTOR

Maria Ferro, 30, graduate in medicine from the university of Catania, is, unusually for a Sicilian, blonde and blue-eyed – which means that she is a descendant of the Normans who came to the island around AD 1000.

Jobs for doctors are scarce in Italy; there are simply too many of them (one for every 230 inhabitants in 1985 – more than twice the number of doctors in the UK or US) and incomes are relatively low – 12-24 million lire (US$7150-14 300) a year. Maria does not have a full-time job. But every day she drives her Lancia to the hospital in Catania to practise on a voluntary basis. Sometimes she earns a little as a stand-in for other doctors. She makes 29 000 lire (US$17) for every private patient she sees and is also paid for the days she works in a health centre when the regular doctor is away. But money is not a major worry; she lives with her parents, who own a shoe shop and can support their only child.

Maria has a knack for gaining people's confidence, and for providing psychological and emotional comfort for those of her patients – and there are many in Sicily – who go to the doctor as much for emotional help as for the diagnosis and cure of physical illnesses. With her male patients, there is sometimes a certain shyness to be overcome: it is still unusual for Sicilian women to have a profession.

Maria's fiancé, Giovanni d'Antona, whom she met at the hospital, has a salaried position as an assistant doctor. They will be married in a few months' time – in a quiet Catholic ceremony attended only by members of the family.

Marriage will not mean the end of work for Maria: she wants at least two children, but intends to have a hospital career as well. Like Giovanni, she would like to reach at least the rank of *primario*, the head of a specialised department. The prospects for both of them are poor.

If Maria has children, she is entitled to six months' paid maternity leave. If she has to work nights Giovanni will mind the children. Her parents will look after them during the day.

Maria snatches a quick lunch in the hospital cafeteria; her main meal of the day will be in the evening, at home with her parents. Her mother mostly cooks lots of pasta, in all different ways, or *schiaccata*, Maria's favourite, a kind of Sicilian pizza with onion and cheese, or rice and meat dishes. Tomorrow Maria is going out to eat with Giovanni and will dress up: her wardrobe consists of stylish casual clothes for work, and very elegant dresses for evening. Afterwards, they may go for an evening stroll, or *passegiata*, along the seafront.

Sidi Barrani ***Egypt*** Coastal village about 400 km (250 miles) west of Alexandria. During the Second World War, it was captured by the Italians in September 1940, by the British in December 1940, by the Germans in 1941, and finally recaptured by the British during Erwin Rommel's retreat in November 1942.

Map Egypt Ab

Sidi Bel Abbès ***Algeria*** Walled town 80 km (50 miles) south of ORAN. It grew up around a French camp built in 1843 and was the headquarters of the French Foreign Legion until Algerian independence in 1962. The town is the administrative centre of the district as well as an important road and rail crossroads. Situated in a major agricultural region, it trades in cereals, wine, olives, livestock, tobacco and esparto grass. Local manufactures include flour, cement and furniture.

Population 158 000

Map Algeria Aa

Sidon (Saida) ***Lebanon*** Port and third city of Lebanon after Beirut and Tripoli, situated 40 km (25 miles) south-west of the capital. It is the Mediterranean terminus of the pipeline from the Saudi Arabian oil fields and has an oil refinery. The city grows and exports citrus fruit – and its residential areas have spread beyond the citrus orchards and banana groves which formerly marked its boundaries. Recently, the city's business and social life has been disrupted by civil disturbances and war.

The site – on a promontory facing an island – was inhabited as early as 4000 BC and later became one of the Phoenicians' main cities. It was noted for the manufacture of glass and purple dyes. From the 16th to 19th centuries it flourished under the Ottoman Turks as the chief port for Damascus. Decline set in with the expulsion of French traders by the Ottoman Turks in 1791 and an earthquake in 1837.

Places of interest include the 13th-century Crusader castle on the promontory, and the harbour with Roman remains; the Great Mosque, whose walls date from the 13th century; the Castle of St Louis built by Crusaders on the acropolis; and the citadel, to the south of which is Murex Hill, a mound of crushed shells of the marine snail *Murex trunculus* which provided the purple dye known as Tyrian purple made and traded by the ancient Phoenicians.

Population 24 700
Map Lebanon Bb

Sidra, Gulf of ***Libya*** See SIRTE.

Siedlce (Syedlets; Sedlez) ***Poland*** Engineering town 85 km (53 miles) east of Warsaw.
Population 61 300
Map Poland Eb

Siegen ***West Germany*** Iron and steel town on the river Sieg about 75 km (45 miles) east and slightly south of Cologne. The artist Peter Paul Rubens (1577-1640) was born in the town while his parents were in exile from religious persecution of Protestants in Antwerp.
Population 112 000
Map West Germany Cc

Siegerland ***West Germany*** Hilly region east of the capital, Bonn, where the river Sieg rises. It was once a major source of the country's iron, when charcoal from the nearby forests was used for smelting. With the discovery of coking coal in the Ruhr in the 1850s, its industries declined but steel is still made at Siegen. Siegerland is a largely forested holiday area.
Map West Germany Bc

Siem Reap ***Cambodia*** Western province, north of Tonle Sap lake. It is the site of the ANGKOR civilisation. The town of the same name, in the province, has an airport, which was built in the 1960s to encourage tourism.
Population (province) 313 000; (town) 20 000
Map Cambodia Ab

Siena ***Italy*** Superb medieval Tuscan town built on red earth hills about 50 km (30 miles) south of Florence. It gives its name to burnt siena – a warm reddish-brown pigment made from the earth and used in painting.

Siena was an independent city-state from the 12th-16th centuries and evolved its own brilliant culture which made it one of Europe's major capitals of Gothic art. The 13th and 14th centuries also saw the development of a distinctively Siennese school of sculptors, of whom the greatest was Jacopo della Quercia (1374-1483).

Sierre Leone

A HOME FOR RESCUED SLAVES IN THE 19th CENTURY, SIERRA LEONE HAS A POPULATION NOW SHACKLED BY POVERTY

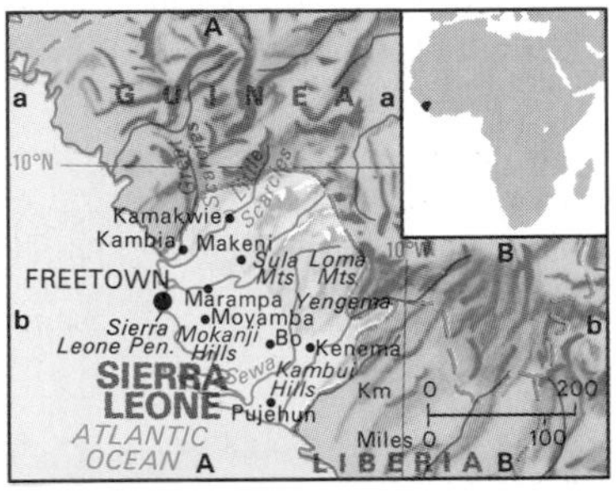

The creation of Sierra Leone's capital, FREETOWN, in 1792 as a refuge for liberated slaves marked the beginning of the end of the Atlantic slave trade – a major milestone in Africa's history. From then through the 19th century the British navy intercepted slave ships in the Atlantic, released their West African prisoners and brought them to settle here.

The liberated slaves acquired the cultural ideals of the British, and became one of the most highly educated and influential elites in West Africa. But conflict arose with the native tribes of Sierra Leone, predominantly the Mende in the south and the Temne in the north who together make up 55 per cent of the population. They resented the superior attitude of this elite. The power of the Creoles (the descendants of the liberated slaves originating from many African countries) has diminished, since independence in 1961, in favour of political parties dominated by one or other of the two main indigenous groups. Since 1978 there has been a one-party government under the Temne-backed All People's Congress.

The name of Sierra Leone ('Lion Mountain') was conferred on the country in about 1462 by Portuguese sailors – either because of the lion-like shape of one of the forested mountains of the peninsula, where Freetown stands, or, some say, because they mistook the thunder which accompanies the frequent torrential rains of this coast for the roaring of lions in the mountains. The rains fall between May and October, and swell the dense mangrove swamp forests and flood the grasslands along the coast.

Rice, the main food of Sierra Leoneans, is grown in the swamplands, which are cleared for cultivation by the local people. The task is dangerous because of leeches.

Freetown Bay – one of Africa's finest natural harbours – is lined with beautiful, palm-fringed beaches which attract tourists. But life for Sierra Leoneans is not easy. In the capital, and in the smaller inland towns such as BO and KENEMA, unemployment is extremely high despite industries such as timber production. The people have to pay for education as well as for medical treatment. In towns such as Bo the hospitals are merely huts where patients supply all their own blankets and food, buy their own medicines, and even pay for kerosene for the hospital generators. About 80 per cent of adults are illiterate and disease is rife.

Around the towns of Bo, Kenema and Moyamba, the coastal swamps give way to tropical forest, in which farmers make a living from small plantations of coffee, cocoa and oil palm. Under the trees, or on nearby plots, yams, cocoyams, cassava, vegetables and ginger are grown; along with rice, these comprise a monotonous diet for the locals. If life for the urban Sierra Leoneans is not easy, in the villages it is yet more precarious. Almost everything depends on the harvest. There is no running water, no electricity, no sanitation, no emergency health care and little motor transport.

Beyond the wide forested coastal plains the land rises to a mountainous plateau in the east, which is an extension of the GUINEA HIGHLANDS. Here much of the forest has been haphazardly cleared for agriculture, although the government has tried to stem this by creating large reserves. On the drier plateaus and mountains of the far north the forest gives way to open savannah, where millet and sorghum are grown for food and groundnuts for cash.

Although the majority of the people make a living from farming, mining provides most of Sierra Leone's export revenue. Diamonds are panned from the rivers Moa and Sewa, and there are extensive iron ore deposits in the Sula mountains, as well as bauxite (aluminium ore), rutile (titanium ore) and some gold.

But the output of the mines is declining, and prices for export crops have fallen. Manufacturing is backward, and the country's balance of payments is deteriorating. Large foreign loans, a flourishing black market and smuggling add to its problems.

SIERRA LEONE AT A GLANCE

Area 71 740 km² (27 691 sq miles)
Population 3 980 000
Capital Freetown
Government Parliamentary republic
Currency Leone = 100 cents
Languages English (official), Krio, Mende, Temne and other local languages
Religions Animist, Muslim, Christian
Climate Tropical; average temperature in Freetown: 23-31°C (73-88°F)
Main primary products Rice, cassava, groundnuts, coffee, cocoa, timber, palm nuts, fish; diamonds, bauxite, iron ore
Major industries Agriculture, mining, fishing, forestry
Main exports Diamonds, bauxite, coffee, cocoa, iron ore, palm kernels
Annual income per head (US$) 280
Population growth (per thous/yr) 26
Life expectancy (yrs) Male 46 **Female** 49

Siena's university was founded in 1247, and its medieval heritage is also evident in its popular festivals such as the *Palio*, a thrilling city-centre horse race held twice each summer between teams in costume representing the city's medieval quarters.

The shell-shaped Piazza del Campo is one of the finest medieval squares in Europe. It is completely surrounded by ancient houses and fronted by the Palazzo Pubblico (1297-1340), with its graceful tower, the 102 m (335 ft) Torre del Mangia. The palace is now Siena's town hall and civic museum, and has a fine collection of local art. The vast Gothic cathedral dates from the 13th century.

Population 60 500; (province) 254 200

Map Italy Cc

sierra Rugged range of mountains having a serrated or irregular profile, particularly in Spain and Spanish America.

Sierra de Guadarrama *Spain* Forested mountain range about 180 km (110 miles) in length, to the north-west of Madrid. The highest point is Pico Peñalara (2430 m, 7970 ft).

Map Spain Cb

Sierra Leone See p. 583

Sierra Leone Peninsula (Freetown Peninsula) *Sierra Leone* Forested and steep-sided peninsula some 40 km (25 miles) long and 16-20 km (10-12 miles) wide. It rises to 888 m (2913 ft) at Picket Hill. The capital, Freetown, lies on the peninsula's north side.

Map Sierra Leone Ab

Sierra Madre *Philippines* Mountain range extending more than 500 km (310 miles) from the northern tip of LUZON island down its eastern flank to Quezon province. It reaches 1850 m (6068 ft) at its highest point. Descending steeply to the Pacific Ocean, the range bears the brunt of the trade winds and typhoons, and is among the wettest places in the islands.

Map Philippines Bb

Sierra Madre del Sur *Mexico* Mountain range stretching southwards from the Río Balsas along the Pacific coasts of Guerrero and Oaxaca states to the isthmus of Tehuantepec. It reaches a height of 3850 m (12 630 ft) in central Oaxaca.

Map Mexico Cc

Sierra Madre Occidental *Mexico* Rugged mountain range stretching for 1610 km (1000 miles) along the western side of Mexico, parallel with the Pacific coast. It has canyons up to 2000 m (6560 ft) deep and is the source of several rivers, including the Lerma and Yaqui. There are several peaks above 3000 m (10 000 ft).

Map Mexico Bb

Sierra Madre Oriental *Mexico* Mountain range in eastern Mexico stretching for about 1130 km (700 miles), roughly parallel with the Gulf of Mexico. The highest point is the volcano, Citlaltépetl (5700 m, 18 700 ft), which is also the highest point in Mexico. The headwaters of the Papaloapán and Panuco rivers rise in the eastern Sierras.

Map Mexico Bb

Sierra Maestra *Cuba* Mountain range in the extreme south of the island rising to 2005 m (6578 ft) at Pico Turquino. The steep, densely wooded slopes at the western end of the range provided an inaccessible base from which Fidel Castro, who had been exiled in 1955, began the Communist uprising in 1956.

Map Cuba Bb

Sierra Nevada *Spain* Mountain range extending about 95 km (60 miles) near the southern city of Granada. Mulhacén (3482 m, 11 424 ft) tops the range and is mainland Spain's highest peak. (The highest peak in the entire country is Pico de TEIDE in the Canary Islands.)

Map Spain Dd

Sierra Nevada *USA* Mountain range running through the east of the state of California and rising to 4418 m (14 495 ft) at Mount Whitney. It is the largest range west of the Rockies.

Map United States Bc

Sierra Nevada de Santa Marta *Colombia* The highest range of mountains in the country, rearing out of the Caribbean coastal plain north of the main Andes range. It reaches 5800 m (19 029 ft) in the twin peaks of Cristóbal Colón (Christopher Colombus) and Simón Bolívar.

Map Colombia Ba

Sighetu Marmatiei *Romania* Border town in the north-west of the country, overlooking the river Tisza and Russia on the river's far bank. The area to the south of the town is a centre of traditional peasant culture, with unspoilt villages, beautiful wooden churches, people in national costumes, and woodcarving and embroidery workshops.

Population 42 000

Map Romania Aa

Sighisoara *Romania* Town in the geographic heart of the country, 230 km (145 miles) north-west of Bucharest. It was the birthplace about 1430 of the bloodthirsty prince Vlad Tepes, upon whom the fictional Count Dracula is thought to have been based. Founded by German settlers in the late 13th century, Sighisoara was built around a hilltop castle. The citadel is the best preserved of its kind in Transylvania. There are several medieval towers along the city wall, the most notable being the 14th-century clock tower. A 13th to 14th-century church dominates the old town.

Population 33 100

Map Romania Aa

Sigiriya *Sri Lanka* Spectacular palace fortress on a 180 m (600 ft) high granite block in central Sri Lanka, 150 km (93 miles) north-east of Colombo. It was built by King Kasyapa (477-495), and its 2 hectare (5 acre) summit is reached by a staircase that passes between the paws and through the throat of a colossal carved lion. Fifth-century frescoes depicting some 20 voluptuous half-naked girls survive in a grotto in the side of the rock.

Map Sri Lanka Bb

Sikandarabad *India* See SECUNDERABAD

Sikasso *Mali* Town about 280 km (175 miles) south-east of the capital, Bamako, near the borders with Ivory Coast and Burkina. It stands in the part of the country that has escaped the ravages of drought and is an important centre for cotton. It is also the clearing-house for trade – both legal and illegal – with the neighbouring countries.

Population 60 000

Map Mali Bb

▼ CHALLENGING PEAK Mount Whitney in the Sierra Nevada of California provides one of the longest routes for rock-climbers in the world.

Sikkim ***India*** State covering 7096 km² (2739 sq miles) on the Chinese border between Nepal and Bhutan, astride one of the main Tibetan-Indian trade routes. The valleys of the Tista river and its tributaries stretching up into the Himalayas occupy most of the state, and its people call it *Denjong* (meaning 'Valley of Rice'). In 1890 the British made the kingdom of Sikkim a protectorate, which passed to India at independence in 1947. The monarchy was abolished and it became India's 22nd state in 1975. Gangtok is the capital.

Population 316 400

Map India Db

Siklós ***Hungary*** See PECS

Silesia (Slezsko; Schlesien; Slask) ***Czechoslovakia/East Germany/Poland*** Region in east central Europe which now lies mostly in Poland, and includes the upper valley of the Oder river.

Upper Silesia contains Poland's industrial core – a congested 19th-century agglomeration of 12 cities grouped around Katowice, with a total of 2.2 million inhabitants. The area mines 200 million tonnes of coal a year, and is Europe's largest producer outside the USSR. It also supplies electricity, steel, lead, zinc, heavy metal goods and chemicals.

Lower Silesia – north-west of Upper Silesia – is a warmer, fertile area with some forests, and produces wheat, sugar beet and oilseeds. Copper is mined and refined around the cities of Legnica and Glogow.

Map Poland Ac

silica Crystalline compound of silicon and oxygen, which are the two most abundant elements in the earth's CRUST and MANTLE. Silica takes many forms, occurring as minerals such as quartz, agate and flint, as a common constituent of sedimentary rocks such as sandstone, or in combination with other compounds to produce the rock-forming minerals called silicates.

silicates Most abundant group of minerals found in rocks. They are composed of silica in complex chemical combinations with oxides of aluminium, potassium, calcium, sodium, magnesium and iron, as in amphiboles, feldspars and pyroxenes.

Silicon Valley ***USA*** See SAN JOSE

Silistra ***Bulgaria*** River port and capital of Silistra province, on the Danube 112 km (70 miles) north-east of Ruse. It ships grain, and manufactures bricks, furniture and cotton goods. Lake Sreburna nature reserve is 20 km (12 miles) east of the town.

Population 58 300

Map Bulgaria Ca

Siljan ***Sweden*** Lake covering 355 km² (137 sq miles) in southern-central Sweden, about 250 km (155 miles) north-west of Stockholm. Its shores are lined with tourist resorts, including Leksand, Mora and Rättvik. Some 10 000 contestants take part in a long-distance cross-country ski tournament held to the north of the lake each winter.

Map Sweden Bc

Silk Road ***China*** Ancient route leading from north China through central Asia to Bukhara, Samarkand and ultimately to Europe. The road, on which Chinese silk was transported west as early as Roman times, leaves the city of Yumen ('Jade Gate') in Gansu and skirts the Gobi and Taklimakan deserts. The 13th-century Venetian explorer Marco Polo journeyed to China on this well-used trail. Modern roads, partly metalled, follow the line of the old Silk Road.

Silkeborg ***Denmark*** Health resort in the hill and lake district of east-central JUTLAND. In the town museum is the mummified head of a 2200-year-old man, known as Tollund man, which was found in 1950 in a Jutland peat bog.

Population 26 100

Map Denmark Ba

sill Sheet-like body of intrusive igneous rock formed by the injection of hot molten rock between the beds or layers of existing rock. Sills vary in thickness from a few centimetres to hundreds of metres. One of the best known is the Great Whin Sill that runs across northern England and southern Scotland.

silt Deposit of fine particles, between 0.002 and 0.02 mm (0.00008 and 0.0007 in) in diameter, laid down by rivers and streams. Silt is coarser than clay but finer than sand. The term also applies to particles of this size in soils, whether waterborne or windborne.

Silurian Third of the six periods in the Palaeozoic era of the earth's time scale. See GEOLOGICAL TIMESCALE

sima General name given to the rocks of the earth's oceanic CRUST that are rich in silica and magnesia. The word is formed from the first two letters of these two minerals.

Simferopol' ***USSR*** City and capital of the Crimea, set in a fruit-growing area 55 km (35 miles) north-east of Sevastopol'. It was founded in 1784 on the site of a Tatar town destroyed in battles between the Russians and the Crimean Tatars. Nearby are the sites of the ancient settlements of Ak-Mechet (once the home of Tatar tribesmen) and Neapolis (occupied in the 3rd century BC by the fierce Scythian people). Today Simferopol's industries include machinery, tools, textiles, perfume, cigarettes and processed food.

Population 328 000

Map USSR Ed

Simla ***India*** Hill resort and capital of Himachal Pradesh state, about 265 km (165 miles) north of Delhi. It stands on wooded Himalayan ridges some 2000-2430 m (6600-8000 ft) above sea level, where temperatures rarely exceed 25°C (80°F). The British began the resort in about 1820 and made it the summer capital of British India. The town has a bizarre profusion of Victorian styles of architecture.

Population 80 200

Map India Bb

Simon's Town ***South Africa*** Naval base and holiday resort on False Bay, 32 km (20 miles) south of Cape Town. A Dutch naval base from 1741, it became the base of the British South Atlantic Naval Squadron in 1814 and was known as the Gibraltar of the South, because of its strategic position. The base was transferred to the South African navy in 1957. Britain gave up its remaining rights to use the base in 1975 in disapproval of apartheid.

Population 6300

Map South Africa Ac

simoom, simoon A swirling, intensely hot, suffocating, sand-laden wind blowing across the northern Sahara desert in summer. It is the result of intense local heating and convection.

Simplon Pass ***Switzerland*** Alpine pass, 2009 m (6590 ft) high, on the route between Brig, Switzerland, and Domodossola, Italy. The road, built in 1800-7 by Napoleon, has been less used since the opening of the 20 km (12.5 mile) long Simplon Tunnel in 1905.

Simpson Desert (Arunta Desert) ***Australia*** Uninhabited arid region covering about 130 000 km² (50 000 sq miles) in the Northern Territory, South Australia and Queensland. The region was entered by the explorer Charles Sturt in 1845, but was not crossed until 1939, when Cecil Thomas Madigan explored it and named it after the Australian geographer A.A. Simpson.

Map Australia Fc

Sinai ***Egypt*** Peninsula, mostly desert and mountains, stretching from the Mediterranean Sea between the Suez Canal and the Israeli border to the Red Sea. At the southern end, a mass of sharp peaks culminate in Gebel Katherina 2637 m (8651 ft), Egypt's highest mountain. It is the land link between Africa and Asia, and has been used by traders and armies since the time of the pharaohs. Since the state of Israel was set up in 1948, it has been a focus of conflict between Arabs and Jews. It was occupied by Israel in 1967, but returned to Egypt in 1982 under a treaty signed by the two countries in 1979. Oil, manganese and iron have been found in the barren El Tih plateau of the south.

Map Egypt Cb

Sinaloa ***Mexico*** Large coastal state in the north-west crossed in the south by the Tropic of Cancer. Once the territory of the nomadic Chibcha Indians of Colombia, it is now a region of silver mining and settled farming – of livestock, cereals, sugar cane, rice, tomatoes, cotton and groundnuts. Shrimp fishing and tourism, around the resort of Mazatlán, are also economically important.

Population 2 000 000

Map Mexico Bb

Sind ***Pakistan*** Province on the lower reaches of the Indus, covering 140 914 km² (54 407 sq miles). It is mostly a scorching desert plain, relying on irrigation for crops (mainly cotton). Water supplies depend on how much is taken upstream in the Punjab, and how much is stored in dams such as Mangla and Tarbela on the edge of the Himalayas. The province was annexed from Afghanistan by the British in the 19th century. The main cities are KARACHI, the capital and Pakistan's largest city, and HYDERABAD.

Population 19 029 000

Map Pakistan Cb

Singapore

TINY SINGAPORE'S WELL-ORGANISED STATE HAS CREATED PROSPERITY AND HARMONY – BUT AT THE EXPENSE OF SOME SOCIAL FREEDOM

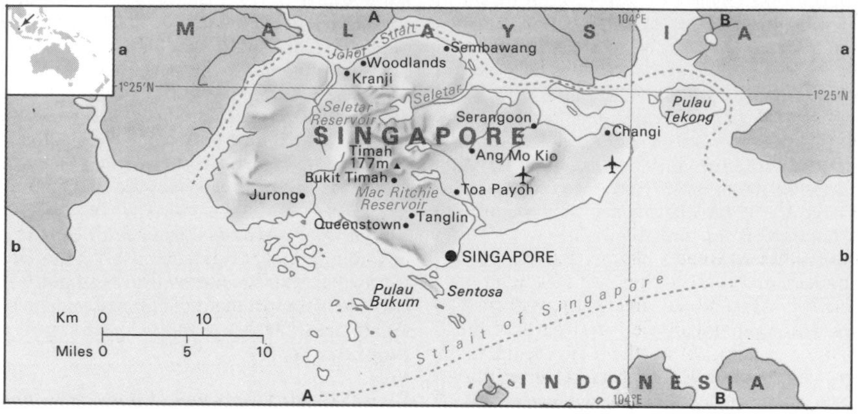

Densely populated Singapore – there are almost 4200 people per square kilometre (10 880 per sq mile) – is one of the world's smallest yet most successful countries. It comprises 60 islands at the foot of Malaysia just north of the Equator, but most of the 2.6 million population live on the main island of Singapore.

Once a great British naval base, Singapore is now a sleek, efficient international business centre at the crossroads of Asia, and one of the world's busiest ports. On any day there are some 600 ships in port, and every ten minutes a ship enters or leaves. Its people are the richest in South-east Asia after Brunei – income per head is US$5600. Unemployment, bad housing and illiteracy have been almost wiped out.

The architect of Singapore's 'miracle' is the austere Lee Kuan Yew, premier since 1959, whose People's Action Party holds an overwhelming majority of the seats of Parliament. The price is a strictly regulated society; the prize is the cleanest, most efficient, most corruption-free state in South-east Asia.

Jaywalking, litter offences and smoking in public places are punished with fines equivalent to about US$225. Islanders are admonished not to chew gum (to save public cleaning bills), not to waste water when cleaning their teeth, and not to have more than two children.

There is a 175 per cent import tax on cars to reduce traffic jams. In addition, cars carrying less than four people are forbidden to enter the centre of SINGAPORE city during peak traffic hours without a special pass.

LEASED FROM THE SULTAN

Singapore (the name means 'lion city') was part of a Sumatran trading empire in the 13th century. It was destroyed by the Javanese in 1376 and was being reclaimed by the jungle when Sir Stamford Raffles, of the British East India Company, arrived in 1819. Raffles leased it from the Sultan of Johor, and seven years later it became part of the new colony of the Straits Settlements.

Singapore has few resources of its own – it grows very little of its own food and even has to get water from Malaysia, to which it is linked by a causeway. Its success is due to its position and its people. It dominates the narrow waterways which lead through South-east Asia to link the Indian and Pacific Oceans, and has a natural deep-water anchorage, Keppel Harbour. Under British control, it quickly attracted Chinese and Malay merchants; with the development of Malaya at the beginning of the 20th century it became one of the world's leading ports for the export of tin and rubber.

As Singapore grew, large numbers of Chinese arrived from the south China coast to work in the port and form the link between European enterprises and Eastern communities. Today 77 per cent of the population is Chinese, though much of the old Chinese city has gone because of redevelopment. About 15 per cent of the people are Malays and 6 per cent Indians or Pakistanis. There are four languages – Malay, Chinese (mainly Mandarin), Tamil and English.

From 1924 to 1938 the British built their biggest Far East naval base at Sembawang, and Singapore became an important marine servicing centre. The base was claimed to be an impregnable fortress, but its guns were pointing the wrong way. The island was fortified against attack from the sea but, in the Second World War, the Japanese attacked through Malaya and the British garrison quickly surrendered. After the war, Singapore experienced Communist terrorism that spread from Malaya before it became self-governing in 1959. It joined briefly in the Federation of Malaysia in 1963, but became an independent republic in 1965.

In the last 20 years Singapore has become a busy metropolis of soaring glass and concrete offices, attracting banks, finance houses and industrial companies. Orchard Road is a vibrant commercial centre of hotels and duty-free shops. There are excellent telecommunications and airline services, and the airport at CHANGI is the best in South-east Asia.

Many of the sights and sounds of the old city are gone, including most of the Chinese shophouses offering gold and silverware and other goods. The street vendors selling noodles, satay (kebabs), rice and dumplings have been moved to modern hygienic stalls.

New industrial and housing developments are springing up all over the islands, and land is being reclaimed from the sea to add to the area available for development.

The former British naval dockyard has been transformed to service international shipping and the oil-exploration industry. Many of the ships use facilities such as the container ports at Tanjong Pagar and Pasir Panjang or the oil terminals on several offshore islands.

Most people work in industry on the outer edges of the main island. The JURONG industrial estate, for instance, has 1900 plants engaged in shipbuilding or manufacturing a vast range of goods. To house the workers, vast new towns of high-rise flats have been built, complete with shopping and leisure complexes. Four out of five people live in these government flats, which have been described as 'filing cabinets in the sky'.

The government sees itself as the guardian of the people's interests in the mould of the Chinese imperial past; rules and regulations abound. One controversial move was to create 'brighter babies'. Mothers under 30 with poor education and one or two children were offered about US$4500 towards an apartment if they agreed to be sterilised. And unmarried professionals, with university degrees or civil service backgrounds, were offered incentives to settle down and have children.

For more than a quarter of a century, Lee Kuan Yew presided over a steady increase in material wealth and racial harmony. The people seemed prepared to pay the price in social freedom – Lee's party won 77 of the 79 parliamentary seats in the 1985 elections. But in 1986, Lee announced his intention to retire from the premiership in 1988, and the economy showed some decline due to the world recession.

SINGAPORE AT A GLANCE
Area 620 km² (239 sq miles)
Population 2 590 000
Capital Singapore
Government Parliamentary republic
Currency Singapore dollar = 100 cents
Languages Malay, Chinese, Tamil, English
Religions Buddhist, Muslim, Hindu, Christian, Taoist, Confucianist
Climate Equatorial; copious rain throughout the year. Average daily temperature is 24-32°C (75-90°F) all year
Main primary products Rubber, coconuts, fruit, vegetables, livestock, fish
Major industries Oil refining, oil-drilling equipment, chemicals, textiles, processed foods, printing, shipbuilding and repair, electronic equipment, international trade
Main exports Petroleum products, machinery, rubber, electrical and electronic equipment, chemicals, vegetable oils
Annual income per head (US$) 5600
Population growth (per thous/yr) 12
Life expectancy (yrs) Male 70 **Female** 74

Singapore ***Singapore*** **1.** Island off the southern tip of the Malay peninsula which forms the main part of the Republic of Singapore. The island covers 570 km² (220 sq miles) and is separated from the mainland by the Johor Strait, less than 1 km (0.6 mile) wide and spanned by a causeway. The city of Singapore lies on the island's southern shore.

2. The country's capital city and a major international port, on the Strait of Singapore on the south side of the main island. It is home to more than 90 per cent of the people in the country. Orchard Road is the heart of the city's commercial and tourism area, with huge international hotels and tax-free shopping complexes. Kalang is one of the oldest parts of the port area – dating from the 19th century – where the Kalang river has been widened into a tidal basin, now a marina. The western suburb of Queenstown, and People's Park in the heart of the old city, are two of the first housing estates built to rehouse people from the overcrowded Chinese city centre. ANG MO KIO and JURONG are more recent new towns.

Population (city) 2 308 300
Map Singapore Ab

Singapore, Strait of Important shipping lane running between the south coast of Singapore island and the Indonesian Riau Islands. It links the Strait of MALACCA to the South CHINA SEA.

Dimensions 105 km (65 miles) long, 16 km (10 miles) wide
Map Singapore Ab

Singora ***Thailand*** See SONGKHLA

Sinj ***Yugoslavia*** See SPLIT

sinkhole, sink Natural depression in a limestone (KARST) land surface communicating with a subterranean passage into which surface water disappears (a swallow-hole) or once disappeared. Sinkholes are formed by solution below the soil, or the collapse of a cavern roof (a shake-hole). They are popularly known as potholes.

Sinop ***Turkey*** Black Sea port about 300 km (186 miles) north-east of the capital, Ankara. Founded in the 8th century BC by colonists from the Greek Ionian city of Miletus, it was the birthplace of the Greek philosopher Diogenes (about 400-325 BC), who taught the virtues of the simple life. Today the port handles mostly tobacco, fruit and timber.

Population 25 000
Map Turkey Ba

sinter Crust of porous SILICA, deposited around a hot spring or geyser when its water is oversaturated with silica. Sinters occur round many geysers in Iceland and in the Yellowstone Park in the USA. See also Travertine

Sintra ***Portugal*** Town about 25 km (15 miles) west of the capital, Lisbon, in the Serra da Sintra. It was the favourite summer resort of Portuguese royalty and has two palaces and a ruined Moorish castle. Its scenery was celebrated by the English poet Lord Byron (1799-1824) in his poem *Childe Harold.*

Population 20 200
Map Portugal Bc

Siófok ***Hungary*** The largest holiday resort on the south shore of Lake Balaton, 100 km (62 miles) south-west of Budapest. It is the lake's most modern tourist centre, with a complex of hotels, sanatoria, holiday homes and campsites behind a fine sandy beach. Sagvar, 9 km (6 miles) to the south-east, has the remains of the Roman fort of Tricciani, dating from the 3rd century AD. Several resorts lie to the south-west along the lake: Zamardi, a picturesque village with traditional cottages and old wine cellars; Balatonendred, noted for lace making; Szantod-Rev, at the narrowest part of the lake, with a ferry to Tihany; and Fonyod, an older, fashionable resort amid picturesque hills.

Population 22 000
Map Hungary Ab

Sion (Sitten) ***Switzerland*** Tourist town about 80 km (50 miles) south of the capital, Berne. It has a pilgrimage church of the 9th to 13th centuries, a Gothic cathedral with a 9th-century clock tower, a 17th-century town hall, and two medieval castles built to guard trade routes along the Rhône Valley. The town is the capital of the wine-producing and mainly French speaking canton of Valais (Wallis).

Population (town) 25 000; (canton) 232 800
Map Switzerland Aa

Sioux City ***USA*** Regional centre on the Missouri river in western Iowa, near the South Dakota and Nebraska state borders. It deals in livestock and grain, and has meat-packing factories.

Population (city) 81 800; (metropolitan area) 118 200
Map United States Gb

Sioux Falls ***USA*** Industrial city in south-east South Dakota, near the Minnesota and Iowa state borders. It markets livestock and is a meat-packing centre.

Population (city) 87 600; (metropolitan area) 118 100
Map United States Gb

Sipka Pass ***Bulgaria*** See SHIPKA PASS

Siracusa ***Italy*** See SYRACUSE

sirocco, scirocco Hot, south or south-east wind of North Africa, southern Italy, Sicily and the other Mediterranean islands, originating in the Sahara as a dry, dusty wind, but becoming moist and increasingly oppressive as it passes over the Mediterranean. Its dust gives rise to 'blood rains' which occasionally reach northern Europe.

Síros (Syra) ***Greece*** Island in the centre of the CYCLADES group in the western Aegean Sea. It is a tourist centre also producing textiles and leather goods.

Population 19 700
Map Greece Dc

Sirte (Sidra), Gulf of ***Libya*** Arm of the MEDITERRANEAN SEA between the towns of MISRATAH and BENGHAZI. It is bordered mainly by desert and oil terminals connected by pipelines to inland oil fields.

Width 440 km (273 miles)
Map Libya Ba

Sitra ***Bahrain*** Small island just off the north-east coast of Bahrain Island; a causeway links the two. It has an oil refinery, Bahrain's main power station, and a desalination plant which turns 90 million litres (20 million gallons) of seawater into fresh water every day.

Population 23 000
Map Bahrain Ba

Sitten ***Switzerland*** See SION

Sittwe ***Burma*** See AKYAB

Sivas ***Turkey*** Industrial town about 440 km (275 miles) east of the capital, Ankara. A former Roman city, it stood around the point where caravan routes from Persia and Baghdad joined on their way to Europe. It is now a market centre making cement, carpets and textiles, as well as dealing in agricultural produce.

In the 14th century the town was sacked by the Tatar warlord Tamerlane. He is said to have promised the city elders that not a drop of the defenders' blood would be shed if they surrendered. Tamerlane kept his promise to the letter: 4000 Armenian soldiers in the city were buried alive; Christians were strangled or drowned; and children were trampled to death by his Mongol cavalry.

Population 197 300
Map Turkey Bb

Siwa ***Egypt*** Oasis town in northern Egypt between the Libyan border and the Qattara Depression, about 470 km (293 miles) south-west of Alexandria. Siwa lies on an ancient North African caravan route and grows dates, olives and grapes. The uninhabited houses of the old abandoned caravan village stand on a craggy outcrop near modern Siwa. It was the site of the oracle of a Libyan god, Jupiter Ammon, and is said to have been visited by the legendary Greek hero Heracles.

Population 3600
Map Egypt Ab

Sjælland ***Denmark*** See ZEALAND

Skadar, Lake (Skadarsko Jezero) ***Albania/Yugoslavia*** See SHKODER, LAKE

Skagen ***Denmark*** Northernmost town in Jutland and one of the country's two chief fishing ports (the other is Esbjerg). It was first recorded in 1299. To its north lies The Skaw, a dangerous sandy spit at the point where the waters of the Skagerrak and the Kattegat meet at the tip of Denmark. Skagen attracted a colony of artists and authors at the end of the 19th century. It is now a popular tourist resort.

Population 11 700
Map Denmark Ba

Skagerrak Channel between Norway and Denmark linking the NORTH SEA to the KATTEGAT, ORESUND and the BALTIC SEA. Navigable throughout the year, the water is deepest near the Norwegian coast, but shallow near JUTLAND where the shore is lined with sandbanks. Ferry services connect a number of ports in the two countries.

Dimensions 240 km (150 miles) long, 130-145 km (80-90 miles) wide
Map Sweden Ad

Skåne ***Sweden*** The most south-westerly province in the country, covering 11 280 km² (4355 sq miles). Its chalk rocks form the basis of a thriving cement industry, and its fertile plains are rich in grain and root crops – particularly sugar beet. It contains the ports of MALMO, HELSINGBORG and TRELLEBORG, and the university city of LUND.

Population 1 028 800
Map Sweden Be

Skeleton Coast ***Namibia*** Stretch of arid, often fog-bound coast and shifting sands in the north-west. It is littered with shipwrecks that have taken place there and gets its name from the human and whale bones that are periodically uncovered by the sands along with the wrecks. The 16 000 km² (6178 sq miles) Skeleton Coast National Park protects this eerie region.

Map Namibia Aa

Skellefteå ***Sweden*** Port on the Gulf of Bothnia, about 812 km (505 miles) north-east of Stockholm. It lies at the mouth of the Skellefte river, along which logs are transported from forests inland. As well as timber, Skellefteå exports mineral products such as copper from the nearby Boliden mining area.

Population 74 200
Map Sweden Db

skerry Low, rocky island, chiefly in Scandinavia and northern Scotland.

Skíathos ***Greece*** Most westerly of the Sporades islands in the Aegean Sea. It is a pine-forested resort with some fine beaches, including Koukounaries west of the harbour. The home of the Greek novelist Alexandros Papadiamandis (1851-1911) stands near the port.

Population 4200
Map Greece Cb

Skien ***Norway*** Central town in an industrial area about 100 km (60 miles) south-west of Oslo. The conurbation includes the smaller towns of Porsgrunn and Brevik, and has a combined population of 100 000. Skien, which has had an iron industry since the 16th century, became in 1885 the first town in Norway to install electric lighting.

Population 28 200
Map Norway Cd

Skikda (Philippeville) ***Algeria*** Seaport on the Mediterranean coast, 80 km (50 miles) west of 'Annaba. Once an ancient Phoenician trading centre and then a Roman settlement, it is now chiefly important as an oil and gas terminal at the end of pipelines from the Sahara. It also manufactures petrochemicals and plastics, and has a large natural gas liquefaction plant. The modern town was founded by the French in 1838, and is today the administrative capital of the district. Other exports include agricultural products, wood, esparto grass, iron ore and marble.

Population 146 000
Map Algeria Ba

Skopje (Skoplje; Üsküb) ***Yugoslavia*** Capital of the Macedonian republic, on the Axios (Vardar) river 320 km (about 200 miles) south of Belgrade. It grew as a major communications hub on the Belgrade-Aegean and Dubrovnik-Constantinople routes, and is an industrial centre producing textiles, tobacco, steel, ceramics, glass, chemicals and metalwares.

In the summer of 1963 much of Skopje was destroyed by an earthquake, which left seven in ten of its people homeless. International aid helped the city's rebirth, and its modern buildings are now designed to withstand earthquakes of up to magnitude 10 on the Richter scale. Much of the Turkish old town, parts dating from 1392, survived the 1963 earthquake or was rebuilt. It contains some fine mosques, including that of Mustapha Pasha, bazaars, *hans* (caravanserais), streets of artisans, and the small 17th-century Orthodox Church of the Holy Saviour (Sveti Spas), which has a magnificent carved wooden screen. Muslim and traditional Macedonian costumes are common around the city.

Population 506 500
Map Yugoslavia Ed

BUILDINGS THAT SCRAPE THE SKY

The world's first metal-frame skyscraper, the ten-storey Home Insurance Building (52 m, 171 ft), was erected in Chicago in 1884-5. It was then the tallest office block in the world – a title now held by the Sears Tower, also in Chicago. However, the world's tallest structure – made of galvanised steel and completed in 1974 – is the Warszawa Radio Mast near the city of Plock in Poland.

SOME OF THE WORLD'S TALLEST STRUCTURES

Name	Location	Height (m/ft)
Warszawa Radio Mast	Plock, Poland	646/2120
CN Tower	Toronto, Canada	553/1814
Sears Tower	Chicago, USA	443/1454
World Trade Center	New York, USA	411/1350
Empire State Building	New York, USA	381/1250
John Hancock Center	Chicago, USA	343/1127
Centrepoint Tower	Sydney, Australia	305/1000
Eiffel Tower	Paris, France	300/984

sky cover, cloud cover Amount of sky covered or obscured by cloud, measured in Britain on a scale of 0 (cloudless) to 8 (entirely covered). Internationally it is measured in tenths of the sky covered.

Skye ***United Kingdom*** Largest island of the Inner Hebrides off the west coast of Scotland. It covers 1417 km² (547 sq miles) and measures about 80 km (50 miles) north to south, yet its many inlets mean that no point is more than 10 km (6 miles) from the sea. It is covered by rock, moorland and bog, and is used mainly for sheep farming. The Cuillin Hills, which rise to 993 m (3258 ft), are generally considered to provide the finest climbing in the British Isles. The only major town is the fishing port of Portree on the east coast. After his defeat by the English at the Battle of Culloden in 1746, Charles Stuart ('Bonnie Prince Charlie') crossed to Skye disguised as a woman, an episode commemorated by the *Skye Boat Song*.

Population 8000
Map United Kingdom Bb

skyscrapers See BUILDINGS THAT SCRAPE THE SKY (above)

slash Wet or swampy ground in the south and south-east United States.

Slask ***Poland*** See SILESIA

slate Fine-grained METAMORPHIC ROCK formed by the effects of heat and pressure on shale. Slate splits easily into thin, smooth-surfaced layers, long used for roofing.

Slavkov (Austerlitz) ***Czechoslovakia*** Small town in southern Moravia 19 km (12 miles) east of Brno. It is the site of the Battle of Austerlitz, at which on December 2, 1805, Napoleon and his French army of 70 000 men routed an allied army of 86 000 Russians and Austrians under the Russian general Kutuzov. The allies lost 18 500 men, Napoleon only 900. A museum and chapel mark the battlefield.

Population 6320
Map Czechoslovakia Cb

Slavonia (Slavonija) ***Yugoslavia*** The part of Croatia that lies south-east of Zagreb, between the Drava and Sava rivers. It is an area of rolling farmland, orchards, forested plains and hills, including the lovely Papuk (953 m; 3127 ft) west of the city of Osijek. Except for Osijek and Bosanski Brod, the towns are small and largely agricultural marketplaces.

Slavonia is rich in folklore, with many traditional events, as at Djakovo (north-east of Bosanski Brod), where a festival in July features folk dances, song, costume, crafts (especially embroidery) and wedding coaches.

Map Yugoslavia Cb

Slezsko ***Poland*** See SILESIA

slickenside Polished and grooved rock surface caused by one rock mass sliding past another under pressure, as along a fault.

Sliema ***Malta*** The main resort and biggest town on the island, lying just across Marsamxett Harbour from the capital, Valletta. It is totally different in character from the 16th-century Baroque harmony of the capital. Hotels, modern buildings and a vigorous nightlife are

the main attractions. It stands on the sea and has a yacht marina but no beach.
Population 20 200
Map Malta Eb

Sligo (Sligeach) ***Ireland*** Market town 217 km (135 miles) north-west of Dublin. The poet William Butler Yeats (1865-1939), some of whose work describes the surrounding countryside, spent much of his childhood at his grandfather's home in Sligo. He is buried at Drumcliff, 6 km (4 miles) to the north. Just south-east of the town lies Lough Gill, with Yeats's 'Lake Isle of Inisfree'. Sligo's port now caters mainly for tourists. The town is the county town of Sligo which covers 1796 km² (693 sq miles).
Population (county) 55 400; (town) 17 200
Map Ireland Ba

Sliven ***Bulgaria*** Capital of Sliven province, on the main road and rail route 95 km (60 miles) west of Burgas. Its chief industries are silk, woollen textiles and woodwork. Bulgaria's first factory – a textile mill – was set up there in 1834.
Population (town) 102 000
Map Bulgaria Cb

Slovakia (Slovenska Socialisticka Republika; Slovak Socialist Republic) ***Czechoslovakia*** The eastern of the country's two constituent republics – the other is the Czech Socialist Republic. Slovakia covers 49 032 km² (18 931 sq miles). Carpathian ranges, popular with tourists, cover most of it, including the Tatra, Fatra and Slovak Ore mountains and, in the west, the White and Little Carpathians. The region's chief products are wine, timber, potatoes, sugar beet, cereals, livestock, iron ore and salt. BRATISLAVA is the capital.
Population 5 013 000
Map Czechoslovakia Db

Slovenia (Slovenija) ***Yugoslavia*** Prosperous republic lying in the extreme north-west of the country and covering 20 251 km² (7817 sq miles). It has a flourishing transit trade between landlocked central Europe and the Adriatic, where the port of Koper near the Italian border provides Slovenia with a toehold on the sea. The republic has considerable mineral resources, including coal, lignite, mercury, lead and zinc. Major industrial centres include the capital Ljubljana, Maribor, Celje and Brezice. These lie in fertile basins, which, with the densely populated valleys of the Sava and Drava rivers, produce maize, wheat, sugar beet, potatoes, cattle and pigs. Many smaller towns have textile and timber industries using local materials and water power.

Tourism contributes a large share of Slovenia's prosperity. The beautiful KARAWANKE and JULIAN ALPS in the north give way southwards and eastwards to forested hills and the rugged limestone KRAS containing spectacular cave systems, such as those near POSTOJNA. The vine-growing area of north-east Slovenia includes wine-producing centres such as Ljutomer.
Population 1 584 400
Map Yugoslavia Ba

Slunchev Bryag ***Bulgaria*** See NESEBUR

Slupsk (Stolp) ***Poland*** Industrial city about 100 km (60 miles) west of the Baltic port of Gdansk. Once a fortified market town and a member of the Hanseatic League of mostly north German towns, it now makes foodstuffs, timber and metal goods.
Population 90 600
Map Poland Ba

Småland ***Sweden*** Southern province surrounding Lake Vättern, south-west of Stockholm, covering 29 322 km² (10 163 sq miles). In the 19th century it was one of the country's poorest farming areas, and many of its inhabitants emigrated to North America. Subsequently it became renowned for its innovative industrialists and thriving small manufacturing towns.
Population 698 833
Map Sweden Bd

Smederevo (Semendria) ***Yugoslavia*** Serbian city on a bluff overlooking the Danube 40 km (25 miles) south-east of Belgrade. Remains of a magnificent moated fortress, with 20 towers and inner citadel and palace, survive. About a kilometre around its walls, the city was built in 1429-30 and was Serbia's capital from then until 1459, when the fortress fell to the Turks, who held the city until 1809. Smederevo has long been the centre of a vine-growing area producing fine red and rosé wines, and is also a steel and heavy engineering centre.
Population 107 400
Map Yugoslavia Eb

smog Thick, yellow, polluted fog over a built-up area where smoke particles promote condensation and sulphur dioxide contributes to the acridity. Since the introduction of pollution control laws in many industrialised countries, there has been a considerable decrease in this atmospheric killer, which has been responsible for many deaths from bronchitis and pneumonia. The term is coined from the words smoke and fog.

Smolensk ***USSR*** Industrial city on the Dnieper river, 370 km (230 miles) west and slightly south of Moscow. First mentioned in the 1st century AD, it was a way station on the trade route from Byzantium (the site of modern Istanbul, Turkey) to the Baltic region. Much of the old town was burnt or blown up during the invasion of Russia by Napoleon in 1812, and by the Germans in the Second World War. But there are still some fine 12th-century churches and an impressive 17th to 18th-century cathedral. Smolensk's industries include engineering, timber, glass, food and textiles.
Population 326 000
Map USSR Ec

Smyrna ***Turkey*** See IZMIR

Snæfellsjökull ***Iceland*** Volcanic peak on a peninsula of the same name in the west of the country. The peak is 1446 m (4744 ft) high, and is celebrated as the setting for Jules Verne's book *Journey to the Centre of the Earth*.
Map Iceland Cb

Snake ***USA*** River rising in Yellowstone National Park in north-west Wyoming and flowing 1670 km (1038 miles) west then north to the Columbia river in southern Washington state. It has been harnessed for irrigation and hydroelectric power.
Map United States Cb

Sniardwy, Lake ***Poland*** See MASURIA

snow Precipitation formed when water vapour condenses at a temperature below freezing point, and passes directly from the gaseous to the solid state in the form of ice crystals. These unite and produce soft, feathery flakes – hexagonal structures occurring in myriads of different patterns – as they fall.

snow line Lower level of a snow-covered area, such as the snowcap of a mountain. It is higher in summer than in winter. The *permanent* snow line is the level above which some snow remains throughout the year, in spite of summer melting, rising from sea level in polar regions to over 5000 m (16 400 ft) in the tropics.

Snowdonia ***United Kingdom*** Region of mountains in north-west Wales that contains Britain's highest peak outside Scotland: Snowdon (1085 m, 3560 ft). There are four other peaks – the lowest just 999 m (3279 ft) high – separated by passes. It is ideal climbing and walking country, though dangerous in bad weather, but there is also a rack-and-pinion railway to the summit of Snowdon from the village of Llanberis, which lies just north of the mountain.
Map United Kingdom Cd

Snowy ***Australia*** River rising in the Snowy Mountains, near Mount Kosciusko in south-eastern New South Wales, and flowing about 435 km (270 miles) generally south into Victoria, to enter the Pacific Ocean about 130 km (80 miles) west of the border. Its head waters fall quickly through deep gorges. The Snowy and its tributary the Eucumbene have been dammed as part of the Snowy Mountains hydroelectric scheme; their waters are tunnelled through the mountains to eight power stations. It also irrigates farmland on the much drier region to the north-west of the mountains.
Map Australia Hf

Snowy Mountains ***Australia*** Mountains straddling the border of New South Wales and Victoria, forming the southern end of the GREAT DIVIDING RANGE. They include Mount Kosciusko, at 2230 m (7316 ft) the country's highest.
Map Australia Hf

Sobat ***Ethiopia/Sudan*** Major tributary of the Nile, about 750 km (460 miles) long, rising in Ethiopia and joining the White Nile near the town of Malakal in Sudan. Several headstreams feed it. When in flood (November-December), the Sobat's discharge carries a whitish sediment that gives the White Nile its name.
Map Sudan Bc

Sochi ***USSR*** Black Sea port, spa and tourist resort 270 km (170 miles) north-west of the Turkish border. Its medicinal springs, at Matsesta 5 km (3 miles) inland, have become increasingly popular since the 1960s – and Sochi's population has trebled in that time. Over 50 sanatoriums cater for visitors to the springs.
Population 307 000
Map USSR Ed

Society Islands ***French Polynesia*** Group of mainly mountainous islands in the centre of the French overseas territory. They consist of two main groups: the Windward Islands, including TAHITI and neighbouring Moorea, and the Leeward Islands, about 200 km (125 miles) to the north-west, which include Raiatea, Bora-Bora and several others. The main product is copra.
Population 142 000
Map Pacific Ocean Fc

Socotra ***South Yemen*** Island in the Indian Ocean lying 240 km (150 miles) north-east of Raas Caseyr (Somalia). It is mostly a barren plateau rising to 1433 m (4700 ft). But the small coastal plains and valleys produce dates, myrrh, frankincense and aloes – and are the home of cattle and goats. The island's main town is Tamrida. Socotra became a British protectorate in 1866, and in 1967 it chose to join the new, independent state of South Yemen.
Population 12 000
Map South Yemen Bb

Södermanland ***Sweden*** Province covering 8388 km² (3239 sq miles) just to the west of Stockholm. It was one of the areas around which the Swedish nation was formed in the 11th century from territory controlled by the main Germanic tribes, the Svears and Gothars.
Population 973 000
Map Sweden Cd

Södertälje ***Sweden*** Industrial town about 30 km (19 miles) south-west of Stockholm. It stands on the 3 km (2 mile) long Södertälje Canal which links Lake Mälaren to the Baltic. A former Viking trading post, it now manufactures pharmaceuticals, machinery and cars.
Population 79 800
Map Sweden Cd

Sodom ***Israel*** See SEDOM

Soenda Strait See SUNDA STRAIT

Sofala ***Mozambique*** Province of 68 018 km² (26 255 sq miles) on the country's central coast. About 160 km (100 miles) north of the capital, Beira, is Mozambique's leading wildlife reserve, the Gorongosa National Park. The scenery varies from grassland to forest, from savannah to marsh and river, and the animals include waterbucks, zebras, hippos, elephants, lions and buffaloes. The province is largely a broad, coastal plain drained by the Zambezi and Save rivers. Its people include the cattle-rearing Shona-Karanga tribe. Sugar and cotton are the leading cash crops.
Population 1 107 800
Map Mozambique Ab

Sofia (Sofiya) ***Bulgaria*** Capital of Bulgaria and of Sofia province, 65 km (40 miles) from the Yugoslav border. Situated on the railway at the crossroads of the Belgrade-Istanbul and Bucharest-Skopje road routes, it is the country's principal transportation centre. The chief industries are metallurgy, engineering, chemicals, textiles and food processing. There is an opera house, a national art gallery housed in the former Royal Palace, several museums, a 19th-century cathedral and a university. The city is famous for its hot mineral springs.

Founded by the Romans with the name of Serdica in the 2nd century AD, it was a favourite residence of the 4th century Roman emperor, Constantine the Great. Roman remains include large sections of Serdica's walls, a portion of paved street, and carvings and other pieces of stonework. Among the few buildings remaining from the Turkish occupation of the country is the 16th-century Banya Bashi mosque, with fine interior decorations. The modern city was severely damaged by Allied bombing during the Second World War, and has been largely rebuilt since 1945.
Population 1 093 800
Map Bulgaria Ab

softwood Timber obtained from coniferous trees – primarily pine, larch, fir and spruce – often grown in huge plantations. Not all conifers produce soft wood; yew is a conifer and technically a softwood, but its timber is harder than many hardwoods. Softwood timber is used for construction work and joinery, and for making paper pulp and cellulose.

Sokodé ***Togo*** The country's second largest town after Lomé, the capital. It lies about 320 km (200 miles) north of Lomé, and is a market town for agricultural products such as rubber, tobacco, groundnuts and kapok – soft fibres from the kapok tree which are used in upholstery. Sokodé is also a base for hunters, who shoot deer, porcupines, jackals and wild boars for food.
Population 40 000
Map Togo Bb

Sokoto ***Nigeria*** The country's second largest state (after Borno), covering 102 535 km² (39 589 sq miles) in the extreme north-west. Its capital is the town of Sokoto, about 760 km (475 miles) north and slightly east of the national capital, Lagos. The town thrived at the end of a Saharan caravan route by the 12th century. In 1803 it was captured by the Fulani people who established the powerful Sokoto Sultanate, and the town is still the spiritual home of the Fulani, with a mixture of mud houses in Islamic style, modern buildings and mosques. Sokoto state is arid, but the Bakolori Dam on the Sokoto river will irrigate 30 000 hectares (74 130 acres) for growing sugar cane and cereals, including rice.
Population (town) 144 300; (state) 4 539 000
Map Nigeria Ba

solano East or south-east, hot oppressive wind that often brings rain to south-east Spain and the Strait of Gibraltar in summer.

Solent, The Channel which forms the western part of the seaway between Hampshire and the Isle of WIGHT, in southern England. It leads to Southampton Water and SPITHEAD in the east. At its western end are The Needles – white pillars of chalk beneath the westernmost point of the Isle of Wight – an unmistakable landmark for ships approaching SOUTHAMPTON or PORTSMOUTH from the west.
Dimensions 24 km (15 miles) long, 6-8 km (3.7-5 miles) wide
Map United Kingdom Ee

Soleure ***Switzerland*** See SOLOTHURN

Solomon Islands

WHERE FIERCE SECOND WORLD WAR BATTLES RAGED, JAPANESE ARE NOW COMMERCIAL PARTNERS

The Solomon Islands were named after the Biblical king by Alvaro de Mendana, a Spanish conquistador who discovered them on a voyage from Peru in 1568. But he found no gold, let alone the fabled King Solomon's mines.

The nation – an independent member of the Commonwealth since 1978, having been a British protectorate since the 1890s – consists of six large islands and a myriad of smaller ones in the Pacific, about 1600 km (1000 miles) north-east of the Queensland coast.

The climate is hot and wet, and prone to typhoons. One in 1986 killed over 100 people and made another 90 000 homeless. The land is mountainous and forested. The trees cover the scars of the battles in 1942-3 between the Japanese and Americans, especially on the main island of GUADALCANAL, where nearly 24 000 Japanese died in seven months of bloody fighting for possession of an airfield. Today, the Japanese are partners in developing the Solomons' fishing industry, and the islands' luxuriant tropical vegetation has been partly cleared for oil and coconut palm plantations and cattle raising.

However, for most Solomon Islanders, who are Melanesians, the way of life has not changed for centuries. They grow fruit and vegetables and catch fish – but are no longer cannibals.

SOLOMON ISLANDS AT A GLANCE
Map Pacific Ocean Dc
Area 29 785 km² (11 500 sq miles)
Population 280 000
Capital Honiara
Government Parliamentary monarchy
Currency Solomon Is dollar = 100 cents
Languages English (official), Pidgin, 87 tribal languages
Religions Christian (34% Anglican, 17% South Sea Evangelical, 19% Roman Catholic), tribal religions
Climate Equatorial; temperature in Honiara averages 27°C (81°F) all year
Main primary products Sweet potatoes, rice, taro, yams, coconuts, livestock, palm kernels, timber, fish; gold, bauxite
Major industries Agriculture, fishing, forestry
Main exports Fish, timber, copra
Annual income per head (US$) 620
Population growth (per thous/yr) 37
Life expectancy (yrs) Male 54 **Female** 54

solfatara Volcanic vent that gently emits sulphurous vapours, usually associated with the approaching extinction of volcanic activity. It is named after a low volcano in the Campi Flegrei (Phlegraean Fields) near Naples, Italy.

Solingen ***West Germany*** Cutlery-making town 20 km (12 miles) east of Düsseldorf.
Population 162 000
Map West Germany Bc

Solomon Sea Arm of the PACIFIC OCEAN, between south-eastern New Guinea, New Britain and the Solomon Islands, containing two basins – the Solomon Basin in the south and the New Britain Basin in the north. The northern basin contains the deep New Britain Trench (9140 m, 29 988 ft), where the Indo-Australian plate descends beneath the Pacific plate (see PLATE TECTONICS). The Solomon Sea was the scene of several US-Japanese naval engagements in the Second World War.
Area 720 000 km^2 (288 000 sq miles)
Greatest depth 9140 m (29 988 ft)
Map Pacific Ocean Cc

Solothurn (Soleure) ***Switzerland*** Town popular with tourists, about 30 km (19 miles) north of the capital, Berne. Originally a Roman settlement, it is now a market and watchmaking centre. Many of its buildings date from the 16th to 18th centuries and it has a 19th-century Baroque cathedral. It is the capital of the German-speaking canton with the same name.
Population (town) 17 900; (canton) 218 900
Map Switzerland Aa

Solway Firth Funnel-shaped arm of the IRISH SEA, between Cumbria, England, and the Scottish region of Dumfries and Galloway, known for its tides which 'sweep in as fast as a galloping horse' according to locals. It has been suggested that the strong tides could be harnessed to produce electricity. Fishing is important, especially for salmon.
Width 32 km (20 miles) (at mouth), 3 km (2 miles) (at north-east end)
Length 64 km (40 miles)
Map United Kingdom Dc

Somerset ***United Kingdom*** County in the south-west of England reduced to 3458 km^2 (1335 sq miles) after losing some of its territory to neighbouring Avon. It is an area of rich farmland, fringed by the Mendip Hills in the north-east and Exmoor in the west. It is famous for cider and for its firm yellow cheese – named after Cheddar, a limestone gorge in the Mendips pitted with caverns. The medieval city of Wells has a magnificent cathedral dating from the 12th century, while 10 km (6 miles) to the south-west is the town of GLASTONBURY. The county town is Taunton.
Population 440 000
Map United Kingdom De

Somme ***France*** River, 245 km (152 miles) long, in northern France. It rises near Saint-Quentin, about 130 km (80 miles) north-east of Paris, and flows through Amiens to enter the English Channel between Boulogne and Dieppe. The Battle of the Somme – between July 1 and November 18, 1916, during the First World War – was one of the bloodiest engagements in history, with a total of 600 000 Allied and 450 000 German deaths. It was also the first battle in which tanks were used.
Map France Eb

Somalia

A NOMADIC NATION WITH A RICH CULTURE AND A POOR LIFESTYLE, WHERE A MARXIST REGIME IS SUPPORTED BY THE UNITED STATES

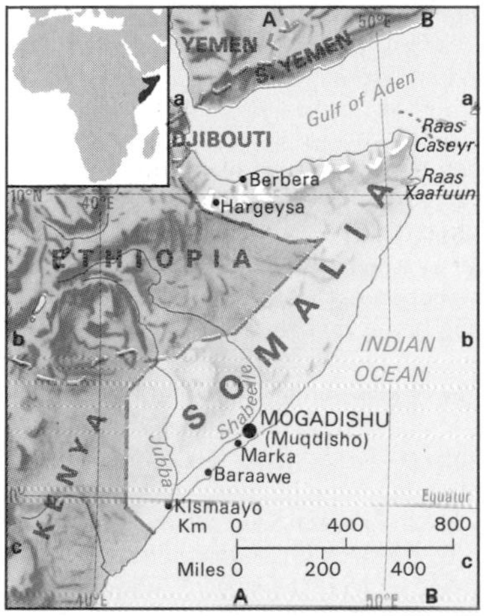

The rich cultural heritage of Somalia contrasts with a poverty that is extreme even by African standards. Life expectancy is a mere 39 years, and the infant mortality rate is 145 per 1000 live births. Most Somalis are nomads with few possessions other than their sheep, goats and camels; when drought kills their animals they very quickly face starvation as in 1974-5, and since 1981.

The country is arid, and most of it is low plateaus with scrub vegetation. There are mountains near the northern coast, and two rivers, the JUBBA and the SHABEELLE, which are used to irrigate crops. The settled population is concentrated in the mountains and river valleys, and in a few coastal towns.

Over 70 per cent of exports consist of live animals, meat, hides and skins, sold mainly to Arab countries. A few large-scale banana plantations by the rivers provide the second export. Priority in irrigated areas is now being given to basic food crops, to reduce dependence on aid and imported food.

Somalia is rare among African states, in that it was a united nation long before it became an independent republic in 1960. Its tribes began merging into a cohesive Islamic nation centuries ago, and it developed trading links with the Arab world and far across the Indian Ocean. In colonial days Britain occupied the north (1884) and Italy the south (1905), and these two territories were joined on independence.

The years since then have been turbulent, for one of Somalia's goals has been the reunification of all Somali peoples, which would involve the annexation of parts of KENYA, ETHIOPIA and DJIBOUTI. Conflict has been most bitter with Ethiopia, which has resisted a Somali invasion of the OGADEN region where the population is mainly Somali. Conflict and drought have led to a massive exodus of refugees from the Ogaden into Somalia.

During the early 1970s Somalia looked to Russia as its chief overseas ally, but Russia began to support Ethiopia and this changed Somalia's position. It now tries to combine Marxist ideology with reliance on military and economic aid from the United States.

Somalia's main strength lies in its culture. Among the most positive recent developments is agreement on a written form for the Somali language. Previously there was only oral drama and poetry, extending back over a thousand years; written information was in Arabic, Italian or English.

The country is changing: young people are tending to forsake the nomadic life and look for jobs in the towns, or in oil-rich Arab states, and there are government schemes to settle refugees and nomads in farming and fishing communities. But Somalia's fortunes depend almost entirely on the length and severity of the drought.

SOMALIA AT A GLANCE
Area 637 657 km^2 (246 199 sq miles)
Population 7 822 900
Capital Mogadishu
Government One-party republic
Currency Somali shilling
Language Somali
Religion Muslim
Climate Hot and dry. Average temperature in Mogadishu ranges from 23°C (73°F) to 32°C (90°F)
Main primary products Sheep, goats, camels, bananas, sugar cane, cotton, maize, millet
Major industries Stock-rearing and small-scale textiles, sugar refining, flour milling
Main exports Meat, hides and skins, livestock, bananas
Annual income per head (US$) 220
Population growth (per thous/yr) 30
Life expectancy (yrs) Male 39 **Female** 39

Søndre Strømfjord (Kangerlussuaq) ***Greenland*** Fiord 160 km (100 miles) long in west Greenland, and site of an American military base established in 1944. It is the centre for air travel between Greenland and Denmark.
Map Arctic Cc

Song Da ***China/Vietnam*** See BLACK RIVER

Song Hong ***China/Vietnam*** See RED RIVER

Songhua Jiang ***China*** River 1840 km (1150 miles) long in the DONGBEI region (Manchuria). It is a major tributary of the Amur and is navigable upstream as far as the city of Jilin.
Map China Ib

Songkhla (Singora) ***Thailand*** South-eastern

province and beach resort near the Malaysian border. The town, which is being developed as a deep-water port, lies at the entrance to a large lake. Rice and rubber are the main crops in the surrounding province.

Population (town) 74 700; (province) 969 200
Map Thailand Bd

Sonora ***Mexico*** State in the north-west bordering the United States. Its wooded inland mountains yield gold, silver, mercury and iron, and its fertile valleys and coastal plains are used to grow vines, sesame, cotton, cereals, alfalfa and flax, and to rear cattle. The state capital, Hermosillo, mixes colonial with modern buildings, including a new university. The state's main port is Guaymas, on the Gulf of California.

Population 1 600 000
Map Mexico Ab

Sonsonate ***El Salvador*** City in the traditional heartland of the country, about 60 km (37 miles) west of the capital, San Salvador. The surrounding department, also called Sonsonate, is dotted with Indian villages and is the nation's main cattle-raising area. Sugar, cotton, rice, tropical fruits and tobacco are grown around the city, which has a fine colonial cathedral. The cathedral's domes are covered in white porcelain. In Easter week there are flower-strewn processions through the city's streets.

Population (city) 53 000; (department) 240 000
Map El Salvador Ba

Sopoćani ***Yugoslavia*** See NOVI PAZAR

Sopron ***Hungary*** Town in the north-west of the country between Lake Fertö (Neusiedler See) and the Austrian border, only 5 km (3 miles) away. It is largely Baroque, and its notable buildings include: the City Tower, originally built in the 12th century; a Benedictine church begun in 1280; a former Dominican church dating from the 18th century; the Arany Sas pharmacy, a going concern since 1724; a museum devoted to the life of Franz Liszt (1811-86), the Hungarian composer and virtuoso pianist who was born at nearby Dobojan; and a former palace of the aristocratic Esterházy family. The town produces Soproni Kekfrancos, a fine red wine, textiles, preserved fruit and sugar. The Hungarian statesman Count Istvan Szechenyi, a moderate leader of the Hungarian nationalists in the 1830s and 1840s, lived in a fine Baroque mansion at Nagycenk, 13 km (8 miles) south-east of Sopron, and his house is now a memorial museum.

Population 55 000
Map Hungary Ab

Sorrento ***Italy*** Resort town on the southern shore of the Bay of Naples. It has been used as a holiday place since Roman times. There are several fine medieval churches and palaces. Sorrento is noted for its lace, inlaid wood, nuts, wines and citrus fruits.

Population 17 600
Map Italy Ed

Sosnowiec ***Poland*** Upper Silesian industrial city 6 km (4 miles) east of the city of Katowice. It produces coal, electricity and steel.

Population 252 000
Map Poland Cc

Soufrière ***St Vincent*** Active volcano 1234 m (4048 ft) high in the north of the island. During a violent eruption in 1979, the crater lake disappeared and a lava dome rose in its place. Some 150 000 tonnes of fine ash from the eruption fell on Barbados, 160 km (100 miles) away, and ash destroyed one-third of St Vincent's banana crop.

Map Caribbean Cc

Soul ***South Korea*** See SEOUL

sound **1.** Narrow passage of water, generally wider than a strait, connecting two larger bodies of water. **2.** A channel between the mainland and an island. **3.** A coastal inlet.

Soúnion, Cape ***Greece*** Cape on the south-eastern corner of Attica 60 km (37 miles) south-east of Athens. A white Doric temple to the sea god Poseidon – dating from 444 BC – stands on its precipitous tip, with panoramic views over the Aegean Sea.

Map Greece Dc

Sour ***Lebanon*** See TYRE

Sourou ***West Africa*** See BLACK VOLTA

Sousse ***Tunisia*** Seaport and one of Tunisia's three major tourist areas, built around an old walled city on the Gulf of Hammamet on the east coast. Sousse lies on the site of ancient Hadrumetum, an important Carthaginian city which Hannibal, the Carthaginian general, used as a base in his campaign against Rome during the Second Punic War (218-201 BC). It became an important Arab city during the 9th and 11th centuries and was taken by the Sicilians in the 12th century. It was recaptured by Arabs in the 13th century.

The old walled town, built on the side of a hill facing the sea, contains an ancient medina (Muslim residential area) and kasbah (citadel). Among the historic buildings of note are the 9th-century Ribat fortress, with a high tower and seven bastions. The local museum has a fine collection of Roman, Byzantine and Christian mosaics. Just outside the town are a series of 2nd to 4th-century Christian catacombs containing some 10 000 burials. Apart from tourism the main industries are textiles, engineering and automobile assembly.

Population 83 500
Map Tunisia Ba

South Africa See p. 594

South Australia ***Australia*** State covering almost 984 380 km² (380 070 sq miles) in the central south. Founded in 1836, it was the fourth British colony established in Australia.

Only the south-east has significant rainfall; inland and westwards, farmland gives way to scrub and finally desert that covers two-thirds of the state. More than 60 per cent of the population live in the state capital, ADELAIDE. There are rich deposits of iron ore, uranium, silver, lead and gold. Wheat, livestock and, along the Murray river, fruit are the agricultural mainstays. Grapes are produced for wine in the BAROSSA VALLEY.

Population 1 347 000
Map Australia Ed

South Carolina ***USA*** South-eastern state covering 80 432 km² (31 055 sq miles) between the Blue Ridge Mountains and the Atlantic Ocean, to the east of Georgia. It was first permanently settled in 1670 and was one of the original 13 states. Agriculture concentrates on soya beans, tobacco and cattle, while the chief manufactures are textiles, chemicals and machinery. Tourism is the second largest money earner after textiles. The state capital is COLUMBIA and CHARLESTON is the main port.

Population 3 347 000
Map United States Jd

South China Sea See CHINA SEA

South Dakota ***USA*** State covering 199 552 km² (77 047 sq miles) of the Great Plains. The Black Hills rise to 2207 m (7242 ft) in the south-west and the state is bisected by the Missouri river. South Dakota is an agricultural region, and cattle, wheat, maize and pigs are the mainstays. Gold is mined in the Black Hills and manufacturing is mainly concerned with meat processing, industrial machinery, timber and wood products. PIERRE is the state capital.

Population 708 000
Map United States Fb

South Georgia ***South Atlantic*** Mountainous, barren island about 1300 km (800 miles) east of the Falkland Islands, of which it is a dependency. It is 160 km (100 miles) long and 30 km (19 miles) wide, and has an area of 3750 km² (1450 sq miles). It rises to a peak at Mount Paget, 2934 m (9626 ft) high. The British explorer Sir Ernest Shackleton died on the island in 1922 and is buried there. It used to be a base for whaling and sealing, and now has a British Antarctic research base.

Map Antarctica Ba

South Island ***New Zealand*** Larger of the country's two principal islands, covering 151 484 km² (58 488 sq miles). Mountains – the Southern Alps – extend for most of its length, but it also has pastureland for some 32.5 million sheep – almost half the country's total. The main cities are CHRISTCHURCH, DUNEDIN, INVERCARGILL, NELSON and TIMARU.

Population 852 700
Map New Zealand Cd

South Magnetic Pole ***Antarctica*** The point on the earth's surface at the southern end of the axis of the earth's magnetic field. It is the point to which the southern end of a magnetic compass points. Its precise location changes by about 10-15 km (6-9 miles) a year; in 1987, its estimated position was 65.1°S, 139.5°E – about 90 miles off the coast of Adélie Land.

Map Antarctica Lb

South Orkney Islands ***South Atlantic*** Barren, uninhabited group of islands in the South Atlantic Ocean, about 1360 km (850 miles) north-east of the Antarctic Peninsula. They are part of the British Antarctic Territory, in an area claimed also by Argentina, and were once used by whalers and sealers. The United Kingdom maintains a scientific base there (Signy on Coronation Island), as does Argentina (Orcadas on Laurie Island).

Map Antarctica Cb

South Polar Plateau ***Antarctica*** Desolate, ice-covered core of Antarctica. The ice has an average thickness of some 2100 m (6900 ft, or over 1.25 miles).
Map Antarctica d

South Pole ***Antarctica*** Southern end of the earth's axis, lying about 480 km (300 miles) south of the Ross Ice Shelf. The southern geographic pole lies 2992 m (9816 ft) above sea level, where the ice is more than 2500 m (8200 ft) thick. It was first reached by the Norwegian explorer Roald Amundsen on December 14, 1911, and then by the British explorer Robert Scott 34 days later. Today it is the site of the US AMUNDSEN-SCOTT research station.
Map Antarctica d

South Sandwich Islands ***South Atlantic*** Uninhabited group of barren, volcanic islands in the South Atlantic Ocean north of the Weddell Sea. They are a dependency of the Falkland Islands – 1100 km (685 miles) to the north – and cover 340 km² (130 sq miles).
Map Antarctica Ba

South Shetland Islands ***Antarctica*** Uninhabited, barren, largely volcanic islands lying north of the Antarctic Peninsula. Formerly a sealing and whaling centre, they are now a site for scientific surveys. They are part of the British Antarctic Territory in an area claimed also by Chile and Argentina.
Map Antarctica Dd

Southampton ***United Kingdom*** Port city at the head of Southampton Water on the south coast of England, 110 km (68 miles) south-west of London. It is the largest city in the county of Hampshire, and has been a port since Roman times. Some parts of its medieval walls remain. The rise of the modern city dates from the 19th-century development – by a railway company – of the docks that made it, in the era of the great transatlantic liners, the principal British port for passengers to North America and South Africa. As liners declined in the 1960s, oil refineries and container traffic took over.
Population 206 000
Map United Kingdom Ee

Southend-on-Sea ***United Kingdom*** See ESSEX

southerly buster Strong, dry, cold wind bringing low temperatures to New South Wales, Australia, when a mass of polar air pulled northwards behind a depression bursts into warmer areas. It corresponds to the pampero of Argentina and Uruguay.

Southern Ocean See ANTARCTIC OCEAN

Southland ***New Zealand*** Southernmost region of South Island covering 29 624 km² (11 437 sq miles). It is mainly a sheep-farming region, but the town of Bluff has the country's only aluminium smelter. Invercargill is the main town.
Population 108 200
Map New Zealand Af

Southport ***United Kingdom*** Seaside town in the English county of Merseyside, about 28 km (17 miles) north of Liverpool. Founded in the late 18th century as an elegant residential and holiday resort, it still maintains this exclusive reputation in contrast to nearby Blackpool.
Population 90 000
Map United Kingdom Dd

Sovetskaya Gavan' ***USSR*** Port and naval base on a fine natural harbour on the Tatar Strait, opposite Sakhalin in the Soviet Far East. Completion in 1984 of the Baikal-Amur railway line, linked to the Trans-Siberian Railway, greatly boosted its role in the shipping of freight between Japan and Europe.
Population 80 000
Map USSR Pd

Soviet Union See UNION OF SOVIET SOCIALIST REPUBLICS

Soweto ***South Africa*** Sprawling collection of black townships, mainly of government-built homes, in the extreme south of Johannesburg. Its name is an abbreviation of South-west Townships, and it developed as a residential area for blacks employed in white areas. Riots erupted in June 1976 over a government ruling (later dropped) that the Afrikaans language should be used in schools. In a week of rioting 176 people died, and the riots spread to other townships, killing 575 in all. Soweto has remained a centre of resistance to apartheid.
Population 829 400
Map South Africa Ch

Spa ***Belgium*** Resort 32 km (20 miles) south-east of the city of Liège. It gave its name to all resorts which, like the town, have mineral or medicinal springs.
Population 10 000
Map Belgium Ba

Spain See p. 598

Spalato ***Yugoslavia*** See SPLIT

Spanish Town ***Jamaica*** Town in the south-east, 20 km (12 miles) west of KINGSTON. Founded by Diego Columbus (the son of the explorer Christopher Columbus) about 1520-6, it was the national capital until superseded by Kingston in 1872. It contains several historic buildings, including St Catherine's Cathedral (1714), the Court House, Eagle House (17th century) and the ruined Old King's House (1762) which was the official residence of Jamaica's governors until 1870. The town now processes sugar, coffee, cocoa and fruit.
Population 50 000
Map Jamaica Bb

Sparta ***Greece*** Town in the south-east Peloponnese at the foot of Taíyetos mountain. Sparta rose to power in the 7th century BC as a ruthless, militaristic city state. It fought beside Athens against the Persians at the Battles of Thermopylae and Salamis in 480 BC, then against Athens in the Peloponnesian War (431-404 BC). It fell later under Roman rule and was sacked by Visigoths in AD 395. The modern town was founded in 1834 near the ancient ruins. Olives and oranges are the main products.
Population 14 400
Map Greece Cc

▼ APPALLING REFUGE Men of the British polar expedition led by Sir Ernest Shackleton spent 105 days on Elephant Island in the South Shetland group in 1916, after the *Endurance* sank.

South Africa

FIVE MILLION WHITE PEOPLE RULE 23 MILLION BLACKS IN A RICH AND BEAUTIFUL COUNTRY WHICH KEEPS THE REST OF THE WORLD ON TENTERHOOKS

The rich and beautiful land of South Africa is out on a limb both literally and metaphorically: geographically, it occupies a swathe of territory at the foot of the African continent; politically it is isolated from much of the world because of its government's insistence on maintaining white supremacy through *apartheid*, the separate development of racial groups.

South Africa and its satellite, NAMIBIA (illegally occupied by South Africa, according to a United Nations resolution), are the only African countries still ruled by whites, and its racial policies are opposed with varying intensities by the countries across its borders.

To the south there are more helpful neighbours – the Indian and Atlantic oceans which wash the coastline east and west and give the country great strategic importance. Much of the oil from the Middle East still travels to Europe and the USA around the foot of Africa, in tankers that are too big for the shorter route through the SUEZ CANAL.

To add to its influence on world peace and world oil, South Africa has some of the world's most important mineral wealth, from gold, diamonds, uranium and silver to coal and natural gas, as well as asbestos, chrome, copper, iron, magnesium, nickel and platinum.

South Africa is too important for other nations to shrug their shoulders about events there. World powers are always on tenterhooks about the country's domestic policies and political stability, but any outside opposition to the government is tempered by recognition of South Africa's strategic and economic importance. One of the most effective ways in which the outside world has made clear its views on apartheid has been by a boycott of South African sport and sportsmen.

THE SHAPE OF THE COUNTRY

The country occupies a huge, oval, saucer-shaped plateau surrounded by a belt of land, 55 to 240 km (35 to 150 miles) wide, which drops in steps to the sea. The plateau averages 1200 m (3950 ft) above sea level, and the rim of the saucer, the GREAT ESCARPMENT, rises in the east to 3482 m (11 425 ft), just over the Lesotho border in the DRAKENSBERG, or Dragon Mountains.

The country is divided into four provinces: CAPE PROVINCE runs across the south; NATAL stretches from the Cape up the east coast to Mozambique; the ORANGE FREE STATE occupies much of the central plateau, and TRANSVAAL lies in the north-east, bordering Botswana, Zimbabwe and Mozambique. At the junction of the Cape, Natal and Orange Free State provinces is the mountainous kingdom of Lesotho, which is landlocked inside, and economically dependent on, South Africa, though strongly opposed to apartheid.

Much of South Africa's northern border is formed by Rudyard Kipling's 'great, gray-green, greasy LIMPOPO river'. The ORANGE is the longest river, running westwards for 2090 km (1299 miles) to the Atlantic, and together with its tributaries drains most of the plateau. A number of other rivers flow from the escarpment to the sea around the coast, but the country as a whole is short of water.

Most of the rain falls in the summer months from November to April in the subtropical strip of the east coast, but in the winter months from May to October in the southern and south-western Cape Province. In some areas, particularly on the east coast, annual rainfall can be as high as 1500 mm (60 in). North of the Cape's mountain ranges and west of the Drakensberg the country lies in a rain shadow with less than 500 mm (20 in) a year, and rainfall decreases steadily towards the NAMIB and KALAHARI deserts in the west.

Temperatures are fairly uniform over most of the country, with a mean annual temperature of 17°C (63°F). The tropical Mozambique Current, flowing down the eastern seaboard, keeps the coastal temperatures several degrees higher than those of the west coast, which is cooled by the Benguela Current flowing from the Antarctic.

VARIED REGIONS AND CULTURES

Despite the overall pattern, there are great regional differences in climate. In winter, while the south-western Cape is experiencing wet weather with temperatures of about 12°C (54°F), the centre of the country has dry conditions when temperatures can fall to −12°C (10°F). In summer on the plateau, green-black thunderclouds bring torrential storms.

Vegetation too is varied. Along the Cape coast, where the climate is Mediterranean, a scrub known as fynbos includes an extraordinary variety of species, particularly of ericas (heathers) and proteas (evergreen shrubs with large flower heads). These plants are astonishingly resistant to drought – and to fire. The giant protea is the South African national emblem. Pockets of temperate evergreen forests are found along this coast.

North of the Cape mountains is the semi-arid KAROO region, flat and sparsely tufted with grasses and dotted with water tanks, wind pumps, groups of sheep and lone, table-topped geological formations. By day it bakes under a harsh heat.

The temperate grasslands (VELD) of the Orange Free State have mostly given way to vast maize fields, north of which is the thorn bush of the Transvaal. The north-east Transvaal was once covered with tropical thorn forests, but these have been cut down to provide shoring (pit props) for the tunnels of gold and coal mines. The forests have been replaced with pine and eucalyptus plantations. From the craggy Drakensberg down to the eastern coast is the VALLEY OF A THOUSAND HILLS, in which evergreen subtropical bush grows.

The source of South Africa's political difficulties lies in the fact that its 23 million blacks are politically dominated by 5 million whites. About 25 per cent of the black people are Zulus; the rest are divided into eight different groups. Their languages have a common base, but their individual cultures differ and many are determined to preserve their identities. This is a cause of frequent and bloody clashes.

South Africa's racial policy separates people not only into blacks and whites, but also into coloured people (about 3 million) and Asians (1 million). Coloured people have diverse origins dating back to 1652, when the Dutch East India Company established a garrison at what is now CAPE TOWN. The coloured class was created by intermarriage between Dutchmen who left the army and became farmers, native Khoikhoi (Hottentots) and San (Bushmen), Huguenots who settled on the land after fleeing from religious persecution in France, and slaves imported as farm labour from Malaya and Madagascar.

The Asian population is mainly descended from Indians brought to work the sugar plantations around 1860.

South Africa's white population is a mixture of the original Dutch and French settlers, who called themselves Afrikaners or Boers (a Dutch word for farmers), English and Germans who joined them later, and other European influxes in modern times. They are split between the Afrikaans-speaking groups, who comprise 65 per cent of the white population, and the English-speaking groups. Most Boers belong to the Dutch Reformed Church, a staunch supporter of white supremacy.

There is hostility between the Boers and the English-speaking whites which stems from the British occupation of the Cape in 1795, and their subsequent treatment of the Boers. In the early 1700s the Dutch colonists had expanded eastwards from Cape Town, staking their farms so far apart that a farmer could not see the smoke from his neighbour's chimney. They spread as far as the Fish River, where they clashed with the Xhosa people, who were being pushed southward by the Zulus. Over the next 100 years, the Boers and the Xhosas fought nine wars on the banks of the river.

By the 1830s the British administration was making itself felt, not least in denying the farmers black slaves.

THE GREAT TREK NORTH

The Boers found British rule oppressive and objected strongly to the enforced changes. This resentment, and the stalemate reached with the Xhosa at the Fish River, led the Boers to decide on a mass exodus to the north, in search of a new Promised Land.

It was known as the Great Trek, and in the years 1835-8 it took 10 000 Boers and their ox wagons across vast tracts of land dominated by marauding Zulu warriors. There were several battles with Zulus, and the Zulu King Dingaan slaughtered a party of 60 Boers under Piet Retief.

In 1838, a Zulu army attacked the Boers in their heavily fortified camp on the Ncome river. The Boers could muster only 470 soldiers, and they faced more than 12 000 Zulu

warriors. Faced with huge military superiority, the Boers resorted to prayer, promising that if they were delivered from these 'forces of evil' they would forever revere the day. In the battle, 3 000 Zulus were killed, trapped in a ditch surrounding the Boer camp and mown down by musket fire. The river ran red with the blood of the dead Zulus, and the battle became known as 'Blood River'; the Boers had only three men wounded. The anniversary of the battle, December 16, is still celebrated by the modern Boers as the Day of the Vow (or Dingaan's Day).

In 1841, the Boer republic of Natalia (now Natal) was formed in former Zulu territory but was annexed by Britain two years later. Two more republics free of British rule were established – the Orange Free State and the South African Republic (now Transvaal). British interest in the territories was aroused by the discovery of diamonds and gold; the British annexed the diamond fields and then the whole of the Transvaal.

Independence was restored to the Transvaal following the first Anglo-Boer War (1880-1), but British eyes still coveted the riches that lay beneath the veld and, in 1899, the second Anglo-Boer War broke out. Following a disastrous series of military defeats, British Imperial troops finally managed to reach PRETORIA, capital of the South African Republic. But the Boers resorted to a guerrilla campaign in which small raiding parties (*kommandos*) scored a series of victories. The British destroyed large areas of farmland and moved Boer families into concentration camps to deny the kommandos any bases from which to operate. About 26 000 Boers – mainly women and children – died in epidemics which swept through the camps.

The Boers were defeated after a three-year war. The British tried to meet the problem of Boer nationalism by forming a self-governing union of the South African states in 1910, but the successive governments of the union were increasingly dominated by the Boers. There was some support for Britain among the Boers, notably from Jan Smuts, a successful Boer military leader before he became a politician and a statesman; he was prime minister 1919-24 and 1939-48.

ORIGINS OF APARTHEID

Smuts and similar political moderates were scorned by Boer extremists, who promoted the nationalism that is a main root of apartheid. Most of the apartheid laws were introduced by the government of Daniel Malan and his National Party, who came to power in 1948. Malan retired in 1954, but the party still governs the country, which left the Commonwealth of Nations and became a republic in 1961.

In the 1970s and 1980s, South Africa faced increasing international opposition to apartheid. In 1985 the government abolished the

▲ FAIREST CAPE Table Mountain forms an unforgettable backdrop to Cape Town – as majestic a sight as when Sir Francis Drake described the Cape peninsula as 'the fairest cape we saw in the whole circumference of the earth' in the 16th century.

Mixed Marriages and Immorality Acts which banned marriage and sexual relations between different races.

The Group Areas Act ensures that white, black, coloured and Asian people live separately in the cities, the blacks and coloureds in 'townships'. However, the blacks have been gradually forced into segregated areas away from the cities since the Natives Land Act of 1913. Consolidation since then has resulted in ten different ethnic areas in which blacks are to have local political rights – the 'homelands'. These areas, comprising 13 per cent of the total area of the country, are occupied by 12 million blacks.

Blacks are not allowed to live outside these areas, in the cities for example, unless they fulfil certain conditions, which effectively mean that they must either have lived in the area since birth or have worked there continuously for the same employer for ten years or more. However, most blacks seeking work outside the homelands are employed only on one-year renewable contracts – the 'contract labour' system – to prevent them attaining residential rights. This 'influx control' is enforced by the Pass Laws, which require all blacks to carry passbooks detailing their legal status. However, in 1986 a new policy was announced: the replacement of passbooks by identity cards. These will be issued to all South Africans over the age of 16, and while they will indicate the holder's ethnic group, failure to carry them will incur no penalty.

The homelands are fragmented, forming a broken crescent around South Africa's eastern and northern borders, and overcrowding has ravaged the land. The KWAZULU homeland allotted to the Zulus is composed of nine separate areas in Natal.

TRANSKEI and CISKEI, which are occupied by the Xhosa, VENDA, which is occupied by the people of that name, and BOPHUTHATSWANA (the Tswana people) have already been given independent status as separate countries, but there is international pressure against recognising this status. The homelands were created to give recognition to different cultures, and also to form a labour pool from which South Africa could draw workers. But it is estimated that over 3.5 million blacks have been moved against their wishes from other parts of South Africa into the homelands.

Blacks do not have the right to vote in general elections, although some of those living in townships elect their own local councils. The denial of direct voting power is a source of bitter discontent and rebounds on the black councillors, who are criticised for collaborating with the whites. The main black political organisation, the African National Congress, is banned.

The other non-white groups have had a direct say in government since 1984, when South Africa established a three-chambered parliament. The House of Assembly is composed of 178 white members chosen by white voters, the House of Representatives contains 85 coloured members chosen by coloured voters, and the House of Delegates has 45 Asian members elected by Asian voters.

These houses legislate on their 'own' affairs, such as education and health, and try to agree jointly on 'general' matters such as defence and foreign affairs. If they cannot agree, the final decision is taken by the white (and government) dominated President's Council.

WEALTH OF MINERALS

South Africa's extraordinary mineral wealth overshadows all its other natural resources, and mining accounts for more than 14 per cent of the total wealth produced by the country each year (its gross domestic product).

The first mineral discovery was diamonds in Griqualand in 1867; more were found along the VAAL river and buried deep in clay at KIMBERLEY, where the Big Hole mine is the largest hole ever dug by hand. The hole covers an area of 15.5 hectares (38 acres) and is 365 m (1200 ft) deep. The first big coal mines were opened in the 1870s in the Transvaal. Coal was first used for producing steam power and electricity in the gold mines, but is now more important as a source of oil. Coal is turned into oil using a process called Sasol.

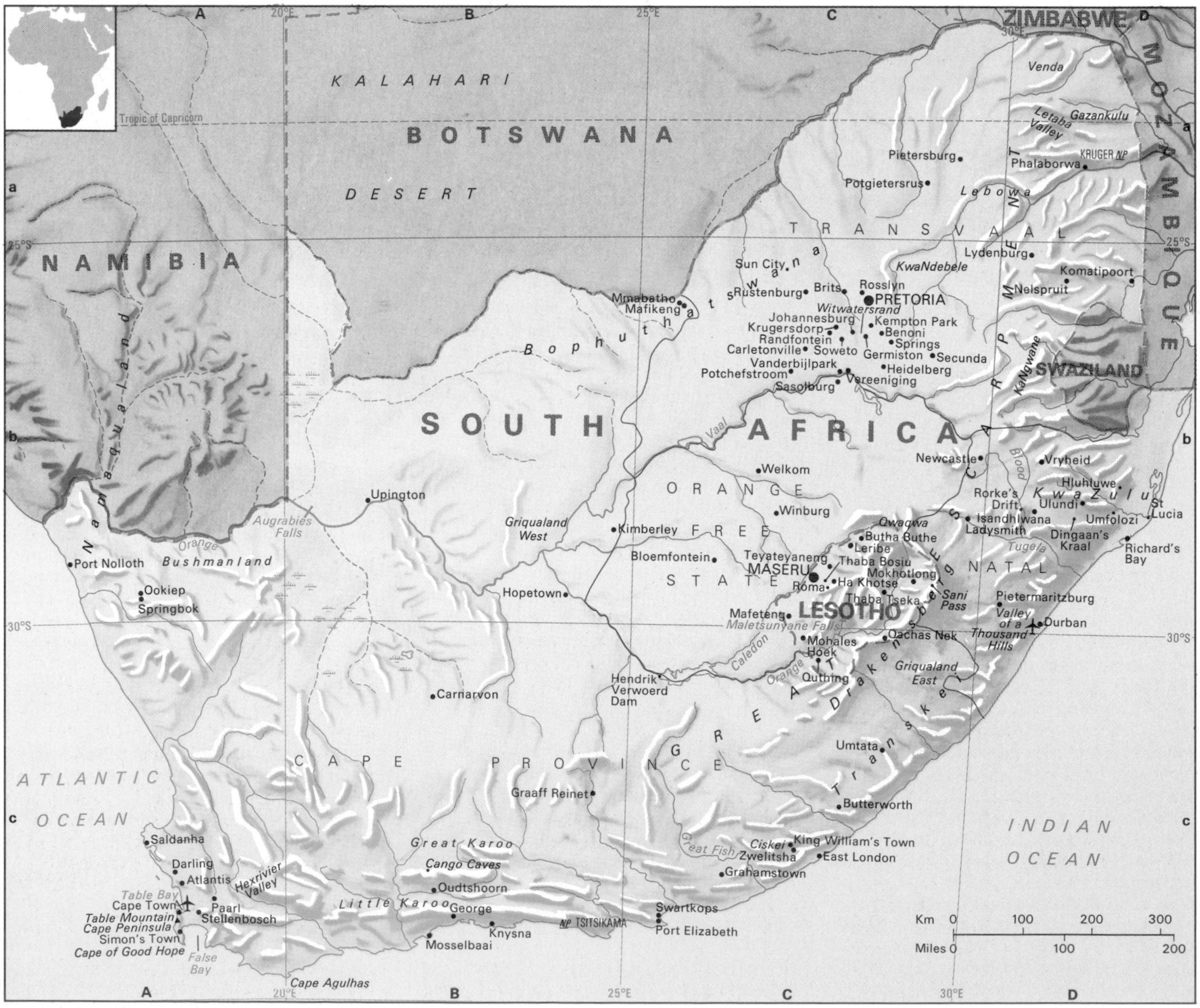

In 1886 the world's richest gold-bearing ore was discovered under a ridge known as the WITWATERSRAND. This was the birthplace of JOHANNESBURG, known in Zulu as *Egoli* (City of Gold), and now South Africa's biggest city. A wide range of important minerals appeared over the next 70 years. The last was a vast reservoir of natural gas found under the Atlantic Ocean, 370 km (130 miles) out from the mouth of the Orange river in the 1970s.

South Africa is the most economically developed country in Africa, but it is mainly dependent on mining. Poor world markets for diamonds and gold, combined with several years of severe drought in the country, resulted in severe recession in the mid-1980s with high inflation (16 per cent and still rising in 1986) and unemployment.

South Africa is working hard to develop non-mineral exports. Blacks are being given more technical education to create a larger skilled work force, and manufacturing now provides 25 per cent of the gross domestic product.

Some restrictions on black businesses have been removed and financial encouragement is being given to black entrepreneurs. The need for black labour and enterprise is increasing black power, some of it organised into trade unions.

Agriculture provides less than 10 per cent of the gross domestic product, but it produces enough food for the country's needs and for export – Outspan oranges, Cape apples and grapes and South African sherry are familiar items in the shops of many countries. Without irrigation, much of the territory is suited only to livestock farming, but a huge scheme called the Orange River Project is bringing large tracts of arid land into cultivation, as well as producing hydroelectric power.

Commercial activity is helped by an extensive road and railway system linking the major cities and the ports. There are three international airports – Jan Smuts, outside Johannesburg, D.F. Malan, outside Cape Town, and Louis Botha, outside Durban.

THE COST OF SECURITY

South Africa needs a healthy export market, not least to maintain its heavy spending on security. Money is needed to maintain a 46 000-strong police force and a standing army of 67 000, with another 130 000 reserves who undergo regular training. Half of the police force is black. The army is made up of white conscripts serving terms of national service, plus white and non-white volunteers.

The economic decline of the mid-1980s hit blacks the hardest, and the worst effects were felt in the black townships and homelands, where an annual 150 deaths per 1000 population were recorded, mainly due to malnutrition. Black unrest grew, bringing demonstrations against price rises and rent increases. School boycotts and protest marches led to clashes with the police which escalated into riots. Hundreds lost their lives between September 1984 and June 1968 when the government announced a 'state of emergency'. This, among other things, banned all meetings, introduced reporting restrictions and provided for detention without trial. Calls for trade sanctions were made around the world, particularly in the black African states.

These ugly events contrast tragically with a beautiful backcloth; South Africa has some of the world's most spectacular scenery, especially along the coasts and in the mountains. The southern coast has superb hiking walks such as the Otter Trail and the Outeniqua Trail. The coast of the eastern Cape, known as the Wild Coast, is famous for fantastic rock formations carved by the sea. Farther up the east coast, the city of DURBAN offers some of the best surfing beaches in the world; residents drive to work with surfboards on their roofracks so that they can go surfing in their lunch hour. From Durban, tourists travel inland to the Drakensberg mountains, where dramatic basalt formations bear paintings by prehistoric San.

CITY AND COUNTRYSIDE

The most distinctive architecture is in the Cape Province, where Cape-Dutch buildings have whitewashed façades with gables curling above Georgian oak-framed sash windows, and tiled *stoeps*, or verandahs.

Cape Town, the legislative capital of the country and seat of parliament, curls in the lap of TABLE MOUNTAIN, which is often draped with a cloud of mist known as the Tablecloth. This attractive city contains the Kirstenbosch Gardens, which stretch up the slopes of Table Mountain, and the Malay quarter, with its distinctive architecture.

The administrative capital of Pretoria is dominated by the dour redstone, civic architecture of Union Buildings, on a hillside overlooking the city. Pretoria is famous for the display of its jacaranda trees from October to November, which purples the streets and pavements with fallen blossom. Just outside Pretoria is the solid bastion of Afrikanerdom – the Voortrekker Monument – which commemorates the achievement of the pioneers who made the Great Trek.

To the south is Johannesburg, still retaining some of its early brashness as a gold-strike town. Tourists can visit a gold mine and be roused by a Zulu mine dance, or shoot up the 201 m (660 ft) Carlton Tower in a high-speed lift for a panoramic view of a city which throbs with commerce and high living.

Native communities provide picturesque attractions for tourists. In the Transvaal, villages of the Ndebele people dot the landscape, the walls of their huts painted with striking geometric designs.

Game conservation is an important concern. The largest reserve is the KRUGER PARK, 20 700 km² (8000 sq miles), on the border of Mozambique, which has leopards, lions, zebras, elephants and rhinoceroses. Other reserves include the UMFOLOZI, where the white rhino has been saved from extinction and is now shipped to reserves around the world. There is wildlife outside the reserves: in parts of the Drakensberg and in hill country near Pretoria, the grunts of leopards punctuate the night as they hunt prey.

The climate encourages the outdoor life, the preferences being summed up in this radio commercial of the 1970s: '*Braaivleis* (barbecue), rugby, sunny skies and Chevrolet' (today the typical family car is more likely to be a Japanese Toyota than an American Chevrolet).

Many whites live in large, detached houses with landscaped gardens and swimming pools, but there are also poor white slum areas. Equally, there are some large, attractive homes in the black, coloured and Indian areas. The government is giving financial assistance to blacks to build their homes. The luxurious lifestyle of many whites is not without its attendant problems, and the white community has some of the world's highest figures for alcoholism, coronary disease and divorce.

While the whites want to protect their lifestyle, most of them realise that black unrest makes political reform necessary in order to maintain stable government. In addition to revoking the Mixed Marriages Act, the government has recently granted blacks freehold rights to their properties in townships, and President Botha has promised that blacks living in townships will not be shipped back to homelands.

One of the most emotive issues between black and white is the continued imprisonment of Nelson Mandela, leader of the banned African National Congress (ANC), who was sentenced to life imprisonment in 1964 for treason. Mandela was offered his freedom in 1984, when he was 65 years old. He refused to leave prison because the government would not recognise the ANC as a legitimate group.

President Botha has raised the possibility of giving the blacks more political rights, but he has said that he will never agree to a unitary state where everyone has an equal vote.

SOUTH AFRICA AT A GLANCE

Area 1 221 042 km² (471 445 sq miles) including black homelands

Population 33 340 000 including black homelands

Capital Pretoria (administrative)

Government Parliamentary republic

Currency Rand = 100 cents

Languages Afrikaans and English (official), Bantu languages especially Zulu and Xhosa

Religions Christian (70%), Bantu churches (26%), Hindu (2%), Muslim (1%), Jewish (1%)

Climate Temperate; warm and sunny. Average temperature in Cape Town ranges from 9-17°C (48-63°F) in July to 16-27°C (61-81°F) in February

Main primary products Cereals, sugar, grapes, livestock, potatoes, citrus fruits, apples, pineapples, tobacco, cotton, timber, fish; gold, coal, iron ore, diamonds, copper, manganese, limestone, chrome, silver, nickel, phosphates, asbestos, tin, zinc, vermiculite

Major industries Iron and steel, mining, food processing, motor vehicles, mineral refining, machinery, chemicals, petroleum refining, agriculture, tobacco products, clothing, paper, textiles, forestry, fishing

Main exports Gold, food, coal, iron and steel, diamonds, gold coins, metal ores

Annual income per head (US$) 1870

Population growth (per thous/yr) 27

Life expectancy (yrs) Male 61 **Female** 65

Spain

EUROPE'S FAVOURITE TOURIST DESTINATION, WHICH HAS STILL NOT LOST THE MYSTERY AND ROMANCE OF ITS ONCE GLORIOUS HISTORY

Each year, 9 per cent of all international travellers – some 30 million pale-faces – go to Spain in search of the sun. The figure may or may not include the movements of the tens of thousands of older people living in Spain who have discovered that pensions go farther there than in their native lands. Housing is cheaper, almost everything costs less than anywhere else in Europe and, provided a suitable settlement is chosen, there is hardly any need to learn Spanish at all.

However, relatively few visitors, residents or tourists, venture away from the golf courses, tower blocks and hastily built urbanisations on the Mediterranean coast, to discover one of the most fascinating countries in Europe – a country that would take several lifetimes to know properly, for each region is different, physically and culturally, from every other, and the whole is very different from the rest of Europe.

One reason is isolation. Sealed off from the main body of the Continent by the PYRENEES, some of whose peaks rise to 3400 m (over 11 000 ft), Spain itself is a mountainous country, much of which is a great plateau slashed across by valleys and gorges. The western Pyrenees are part of the BASQUE REGION, which in Spain extends westwards along the north coast through NAVARRA. Here the mountains of CANTABRIA begin, and stretch along the Bay of Biscay to GALICIA, the corner of Spain which lies on the Atlantic coast north of Portugal. ANDALUCIA in southern Spain also has mountains, the SIERRA NEVADA, and an Atlantic coast, squeezed between Gibraltar and Portugal.

But Spain's longest shoreline is the one that borders the Mediterranean, from Gibraltar to the frontiers of France. This is the coast whose hot dry summers and mild winters lure the visitors. Inland the story is rather different – baking searing heat in summer, and savage cold in winter – though it is there, away from the resorts, that the true Spain is revealed. There too are the most splendid evidences of its golden past.

The greatest days of Christian Spain followed almost immediately upon the expulsion of the Moors of North Africa from GRANADA, their last European stronghold, and the unification of the country under King Ferdinand of ARAGON and Queen Isabella of CASTILE. It was these far-sighted monarchs who dispatched Columbus on his momentous journeys to the New World, then recouped their investments a million times over. Beginning with the conquest and sacking of Mexico by Hernán Cortés in 1521, and of Peru by

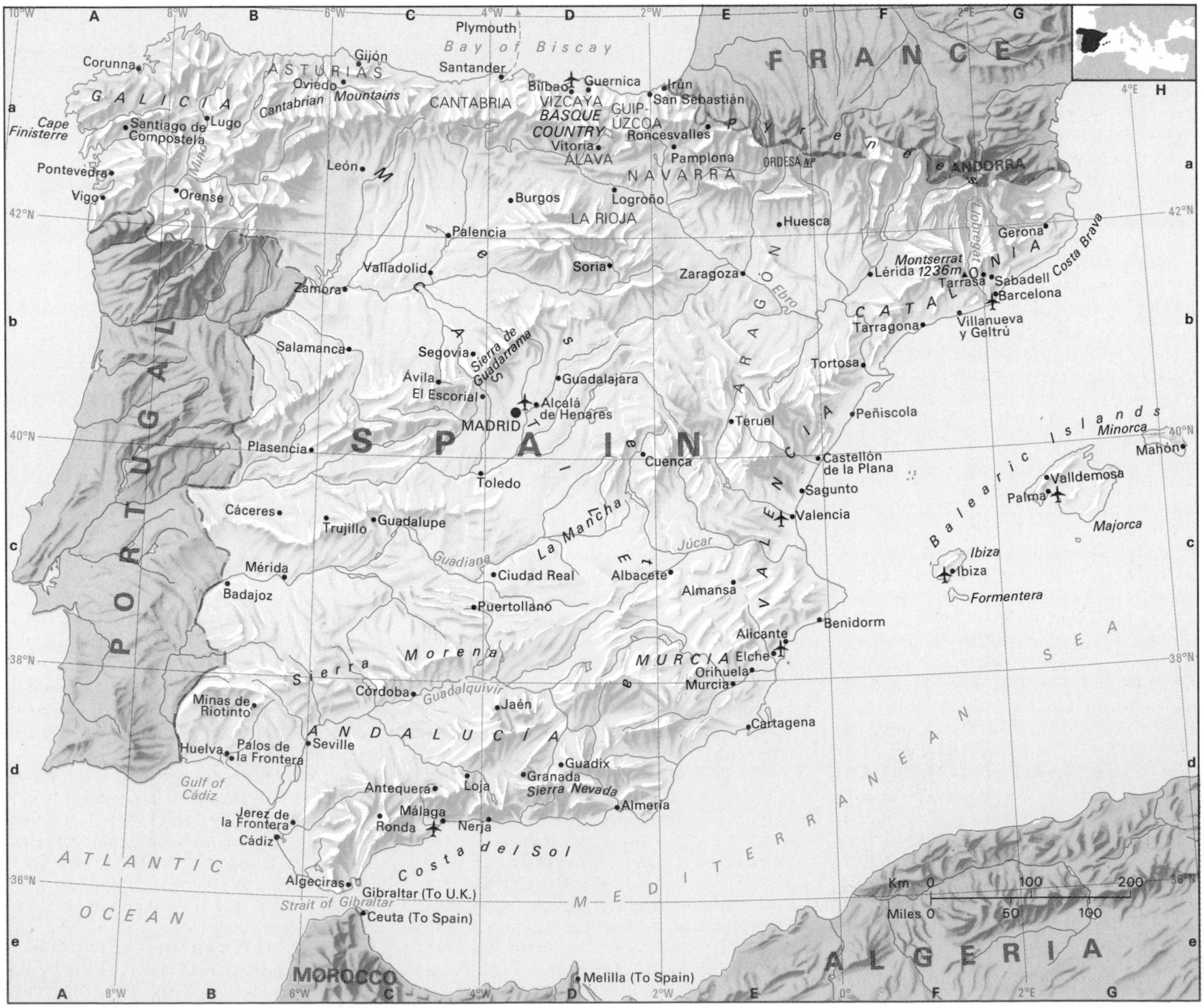

Francisco Pizarro in 1533, riches hitherto undreamt of poured into Spain.

In the 17th and 18th centuries, the power of Spain declined. It lost Gibraltar to the British during the War of the Spanish Succession (1702-13), and when the homeland was occupied by Napoleon's troops from 1808 to 1814, most of its colonies took advantage of the situation to gain their independence.

Spain's story in the first half of the 20th century is not a happy one. It was backward industrially, poverty was rife and after the First World War – in which Spain was neutral – anarchists, trade unionists and Basque nationalists united in violent clashes against the authorities. With the backing of the king, Alfonso XIII, a military dictatorship under General Primo de Rivera was established in 1923. But the required economic reforms were beyond him, and he resigned in 1930; elections were held, a republic was declared and the king driven into exile. In 1936, pressure from the Popular Front – left-wing Republicans, Socialists, Communists and anarchists – caused the government to totter. Meanwhile, the right-wing General Francisco Franco took advantage of the situation to make his own bid for power, and landed an army of crack Moroccan troops at CADIZ.

So began the Spanish Civil War. The opponents were the Loyalists – a mingling of the old Republic and the Popular Front – and the Nationalists, a blending of the regular army, conservatives and much of the Church. The Loyalists were supported by International Brigades of left-wing idealists from all over the world. The Nationalists' allies were thinly disguised and magnificently equipped regular army and air-force units lent by Hitler and Mussolini. Victory went to the Nationalists' big battalions, and in 1939 Franco became Chief of State, virtually dictator, of Spain.

Franco's regime was conservative in the extreme, and repressive, brooking no opposition. Though sympathetic to the Axis cause, he kept Spain out of the Second World War, with the effect that his was by far the longest lived of the European fascist dictatorships. His principal supporters were the industrialists and landowners, the Church hierarchy and the army, making for a moralistic and traditional society in which divorce was illegal and crime was severely punished. A fairly muted opposition consisted of urban workers who objected to state-controlled trade unions, young, socially conscious priests, and Basque nationalists, but was kept firmly in its place by the paramilitary Guardia Civil.

It was a long time before Spain began to recover from the ravages of the Civil War. Already backward agriculturally and industrially, there was much ground to be made up if the country were to regain its place among the European nations. In the 1950s, the tourist trade burgeoned. At the same time the National Institute for Industry was set up. Foreign capital, attracted by cheap labour, began to flow in and new industry developed. Spain's largest car manufacturer, Seat, was established in 1950 with the participation of the Italian company, Fiat. In 1951, the Ensidesa steelworks was established in northern Spain with British collaboration. In the 1960s industry continued to grow with help and investment from the United States – in exchange for naval and air bases. Mechanisation was stepped up on the land, at least on the larger farms and estates. The peasant subsistence farmers carried on much as before.

The rapid growth of tourism opened up the country to outside influences and changed social attitudes and tastes, particularly among the young, but it was not until Franco's death, in 1975, that true liberalisation could begin. In 1947, Franco had promised to restore the monarchy, and in accordance with his wishes, Juan Carlos, grandson of Alfonso XIII, succeeded him and took the oath as head of state and King of Spain. No doubt Franco intended that his successor would continue to maintain a government of the far right, but no sooner was he in power than the young king began to institute reforms unheard of for decades. Political prisoners were freed, opposition parties were permitted (as was divorce), censorship was relaxed, and in 1977 the first general election for 41 years was held. Then, in 1983, the King's firmness defeated a coup attempt by right-wing army officers – and Juan Carlos's democratic credentials were confirmed.

Since Franco, urban Spain at least has become more cosmopolitan – and less distinctively Spanish – with crowds of commuters, fast-food chains and fashionably dressed youngsters. Signs of a post-Franco freedom are everywhere: political graffiti, satirical journalism, and topless bikinis on the beaches. The new Spanish society also has its problems, notably high unemployment and a rising crime rate among the young.

But beneath this modern veneer the old Spain lives on. The flamenco guitar holds its own against its electric counterpart, there are more *tapa* bars selling the traditional snacks of anchovies and spiced sausage than hamburger counters, and a top *torero* (bullfighter) is as big a hero as any footballer. Of course, these are generalisations. To this day, Spain is more regionally conscious, probably, than any other nation in Europe.

Most of the central part of Spain consists of a great plateau, the MESETA Central (Central Tableland). It rises to about 600 m (1970 ft), and is partly covered with a low scrub of thyme, rosemary and lavender – the very scent of Spain. The capital, MADRID, lies in the centre of the plateau, at the heart of Castile. A diagonal line across the plateau from the south-west to north-east more or less divides Old Castile (Castilla y León) – to the north – from New Castile (Castilla-La Mancha). Two ranges of mountains follow the diagonal: the Sierra de Gredos, rising to 2592 m (8504 ft), and the slightly lower Sierra de Guadarrama. Lying north of Madrid, the mountains are loved by Madrilenos and tourists alike for their clear air and beautiful scenery. The Meseta is largely arid; its main agricultural activities are sheep rearing on the mountain slopes, and the growing of wheat and saffron which stretch in vast fields across the wide landscape. The area was more productive in the days of the Moors who irrigated it; later, it was impoverished by medieval sheep grazing. In the extreme south of the Meseta, however, vines are grown in the valley of the GUADALQUIVIR, which cuts Castile off from southern Spain, dividing the plateau from the mountains of the Sierra Nevada. A similar river valley borders Castile to the

▲ CITY OF STEEL Toledo's 16th-century Alcazar fortress stands above the Tajo (Tagus) river. The city was famous for its fine steel swords during the Moorish occupation.

north; there, the EBRO separates the plateau from the Cantabrian mountains (Cordillera Cantábrica) that outline the northern shore.

Geographically and historically, Castile is the heart of Spain. Its dialect, Castilian Spanish, is the official language of the nation. As the country's industrial and commercial centre, focused on Madrid, it is in some ways one of the most modern parts of Spain.

All Castile proclaims a fervent Catholicism in a splendid array of churches and monasteries. There, too, are fairy-tale castles, especially in the province of SEGOVIA.

The Meseta Central is mostly hot and dry; contrary to the song, the rain falls mainly on BILBAO in the north. The wet north-west, though cool, is nevertheless warmed by the North Atlantic Drift in the Bay of Biscay. This part of Spain could be – and constantly expresses a wish to be – another country. Along the steep north coast of Navarra, into the Pyrenees, and over into the south-west corner of France, is the land of the Basques. Its inhabitants, who call themselves Eskualdunak, have lived here possibly since the Stone Age, and speak a language (Euskera) that may be the oldest in the world. The Basques defended themselves against the Moors in the 8th and 9th centuries when most of Spain capitulated.

Basque customs and costume have survived along with the language. Most Basque costume is now only worn on ceremonial occasions, with the exception of the traditional black beret, which has been adopted by armies – and artists – all over the world.

The Basques' chief characteristic is a yearning for separate nationhood, especially in Spain, where there are 2 million of them. Another quarter of a million live in the three Basque departments of France, but there the Euskera-speaking proportion is much smaller. The French Basques are mostly farmers and fishermen, but the Spanish Basques include a vociferous, city-dwelling intelligentsia, whose 'capital' is Bilbao. Among these people a movement has grown up which seeks to create an independent Basque socialist state. The Basques enjoyed some rights of self-government between the 9th and 19th centuries, but restrictions multiplied until, under Franco, all separatist tendencies were suppressed – even the Basque language was banned in schools. The militant wing of the Basque separatists is the ETA, which has instituted a campaign of murders and bombings to drive its point home. The Basque country was given a degree

▲ **HEART OF THE NATION The magnificent 12th-century castle at Alarcón stands on the arid Meseta plateau of Castile, about 165 km (100 miles) south-east of the capital, Madrid. The castle is now an exclusive hotel.**

of autonomy in 1979, but this has not placated the extremists.

In the north-east CATALONIA, too, has its separatist movements. The region also bridges the Pyrenees, between France and Spain, from French Perpignan to the Ebro delta, and has its own language, Catalan, which resembles Provençal, the old tongue of southern France.

Catalonia and the BALEARIC ISLANDS in the Mediterranean – MAJORCA, MINORCA and IBIZA – share a common history and tongue. Catalonian seafarers have been famous since the Middle Ages, and their capital, BARCELONA, is still a great seaport. Textile manufacture, a traditional industry, also thrives, while vineyards, olive groves and orchards provide Catalonia's main agricultural wealth.

Spain's principal tourist area is the COSTA BRAVA, running south from the French-Spanish border to just north-east of Barcelona. Much of the coast is built up. But it is still striking with its red-brown headlands, cliffs, and pine forests that reach down to sandy beaches. South of Barcelona is the Costa Dorada where the old Roman city of TARRAGONA remains unspoilt.

Farther south still, the pine forests give way to orange groves, and Catalonia to VALENCIA. The port of Valencia is Spain's third largest city, after Madrid and Barcelona, and is surrounded by steelworks and cement factories.

In the southern part of Valencia begins the Costa Blanca – the White Coast, a name derived from its clear white light. There is the specially created resort town of BENIDORM. There, too, begins the warm, sunny south.

The south coast provides most foreigners' images of Spain, although in fact it is the area in which the Moorish heritage is strongest of all. This is Andalucia, the land of flamenco dancers, of gypsies and toreadors, a land of romance and blood.

In SEVILLE, where magnificence is commonplace, the Moorish tower (La Giralda) is still outstanding, even though its neighbour is the largest and most opulent cathedral in Spain, lavishly ornamented with Baroque flamboyance, while within, paintings, precious metals and jewels glow about the tomb of Columbus.

The conjunction of Moorish and Baroque is not unusual in Andalucia: CORDOBA has a splendid, simple mosque, so beautiful that even the victorious Christians forbore to destroy it. Instead, on the orders of Charles V, a Baroque cathedral was built inside.

The famous staccato flamenco is the dance of the Andalucian gypsies. The women's costume for the dance – tight, flounced skirts, high combs and lace mantillas – many foreigners take to be the traditional dress of all Spain. Gypsies still live in Andalucia; in the Moorish quarter of Granada, for instance, where some of them live picturesquely, but quite comfortably, in caves bristling with TV aerials. The old quarters of Andalucian towns are perfection; blinding-white with lime wash, the air jasmine-scented and the courtyards hung with vines and birdcages, brightly splashed with ceramics and geraniums, and cooled by fountains. Seated by outdoor tables, the Andalucians take their evening sherry, whose English name is derived from the wine-producing town of JEREZ DE LA FRONTERA, and watch the world go by.

No wonder Andalucia appeals to tourists, even if for the most part they seldom leave the gravelly beaches of the sunny coast – the Costa del Sol – and its miles-long backdrop of hotels, villas, golf courses and high-rise apartments that run from MALAGA to the fashionable resorts of Marbella and San Pedro de Alcántara. There, the descendants of the Moors have begun a new invasion, as is apparent from a splendid new mosque and the palaces of Arab royalty.

Galicia is the north-west corner of Spain. Like the provinces which lie to the west of Castile, near the Portuguese border, it is one of the least visited areas of the country. Until the discovery of America, Galicia was known as Finisterre – the end of the earth. It is this very isolation that has preserved the Spanishness of the north-west; this region held out against the Moors for longer than the rest of the country, and the mountain fastnesses of ASTURIAS, on the north coast, the Moors never conquered at all.

The countryside of Galicia and Asturias is lush and green, rather like the landscape of southern Ireland, and ideal for growing cereals. The west coast of Galicia, on the Atlantic, is deeply indented with *rias*, submerged river valleys. The region's riches are principally agricultural, with a valuable addition of timber from its pine and eucalyptus forests. Galicia is also a holy place. Throughout the Middle Ages, millions of pilgrims from all over Europe journeyed to the town of SANTIAGO DE COMPOSTELA and its cathedral, in which are kept the relics of St James the Apostle.

An appreciation of Spain's past is vital to an understanding of the country. Yet it is clear that a new Spain is emerging, and one that is very much part of the modern age. Spain today seeks to become part of Europe in more ways than imitating its social habits. It joined the European Economic Community in January 1986, and the same year the country (though Socialist-led) voted in a referendum to remain a member of NATO.

SPAIN AT A GLANCE

Area 504 782 km² (194 897 sq miles)

Population 38 820 000

Capital Madrid

Government Parliamentary monarchy

Currency Peseta = 100 centimos

Languages Castilian Spanish (official), Catalan, Galician, Basque

Religion Christian (predominantly Roman Catholic; about 90 000 Protestants)

Climate Mediterranean in the southern and eastern coastlands, temperate elsewhere. Average temperature in Madrid ranges from 1-8°C (34-46°F) in January to 17-31°C (63-88°F) in July

Main primary products Cereals, fruit, vegetables, sugar cane, grapes, tobacco, cotton, almonds, olives, timber, cork, fish; coal, lignite, iron, lead, copper, tin, zinc, mercury, tungsten

Major industries Agriculture, food processing, wine making, yarns and textiles, iron and steel, mining, petroleum refining, chemicals, engineering, transport equipment, forestry and timber products, fishing, cement manufacture, metal processing

Main exports Machinery, motor vehicles, fruit and vegetables, wine, olive oil, iron and steel, chemicals, petroleum products, small metal manufactures, shoes, textiles

Annual income per head (US$) 4200

Population growth (per thous/yr) 5

Life expectancy (yrs) Male 72 **Female** 76

Speightstown ***Barbados*** Small fishing port on the north-west coast. After Bridgetown, it is the island's most important town – and in the 18th century was a major sugar port. Some of its old colonial buildings are still standing. Close to the town is the new Heywoods Beach tourist development, a luxury complex built to provide jobs in northern Barbados and to relieve the tourist pressure on the south.

Population 3000

Map Barbados Ba

Spencer Gulf Finest harbour in South Australia, between Eyre and Yorke peninsulas. It was explored in 1802 by the British navigator Matthew Flinders, who named it after the 2nd Earl Spencer (1758-1834), the First Lord of the Admiralty who had approved Flinders' voyage.

Dimensions 320 km (200 miles) long, 130 km (80 miles) wide

Map Australia Fe

Speyer ***West Germany*** City in the Rhine valley about 20 km (13 miles) south of Mannheim with one of the largest Romanesque cathedrals in Europe. The building was started in the 11th century and eight Holy Roman emperors are buried in it. In 1529 a protest in the town by princes and local officials against the Roman Catholic Church's treatment of the reforming priest Martin Luther (1483-1546) gave the reform movement its name: Protestantism.

Population 44 000

Map West Germany Cd

Sphinx ***Egypt*** See EL GIZA

Spice Islands ***Indonesia*** See MALUKU

spit Narrow ridge of sand or shingle built up by longshore drift and stretching from the shore out to sea or across a bay. If it extends right across the mouth of a bay it is called a bar.

Spithead Deep channel off the entrance to PORTSMOUTH Harbour, between the Isle of WIGHT and the English mainland. It is a busy shipping lane that copes with naval traffic as well as supertankers, bulk carriers and ferries.

Map United Kingdom Ee

Spitsbergen ***Arctic Ocean*** Island group in the Svalbard archipelago in the Arctic Ocean, 580 km (360 miles) north of Norway; also the name of the main island in the group, formerly called West Spitsbergen. Spitsbergen island, which covers 39 000 km² (15 060 sq miles), has extensive national parks and nature reserves, as well as both Norwegian and Russian coal-mining camps. These are a legacy of the mining rights given to all 40 signatories of the 1920 convention that recognised Norway's sovereignty over the islands. The principal Norwegian settlement is at Longyearbyen, named after an American, J. M. Longyear, who in 1905 was the first to mine coal there. Norwegian production is about 500 000 tonnes a year.

Population (Norwegian) 500; (Russian) 1500

Map Arctic Rb

spitskop, spitzkop In South Africa, a hill with a sharply pointed top, caused by the presence of hard rock, such as dolerite, which has resisted erosion.

Split (Spljet; Spalato) ***Yugoslavia*** Largest Dalmatian city, 170 km (106 miles) south-west of Sarajevo. It is a major Adriatic port, fishing and naval base, cultural centre, and an industrial area producing ships, cement, plastics, timber, wine and foodstuffs. The city is also an important tourist resort serving Dalmatia, and the old city is an architectural treasure set on a promontory below a ridge which rises to 1330 m (4363 ft). The city grew up around the retirement home of the Roman emperor, Diocletian, who was born nearby in AD 245. He retired in 305 to the magnificent palace he had built, and was buried there in 313. Afterwards the fortified palace – a 185 by 220 m (about 600 by 720 ft) rectangular complex of luxury apartments, public halls, temples, garrison quarters and catacombs – became a factory producing Roman uniforms, and the nucleus of a city. Early in the 7th century, after the nearby Roman town of Salona (Solin) was sacked by the nomadic Avars from central Asia, refugees turned the palace into a fortified city with a warren of narrow streets, and Diocletian's mausoleum became the city's cathedral. The city prospered under the Byzantines (812-1089) and Venetians (1420-1797), and spread beyond the walls, with graceful campanile (bell tower) and fine buildings. Split has several ancient churches, including Sv Nikola and Gospa od Zvonika, fine museums and a gallery devoted to the works of the Yugoslav sculptor Ivan Mestrovic (1883-1962), which is housed in the sculptor's own villa; a number of Mestrovic's works adorn the city.

Klis, inland from Salona, has the remains of a magnificent 5th-century hilltop fortress. The ancient textile town of Sinj, 30 km (19 miles) north-east of Split, is famous for the Sinjska Alka – a medieval tournament held annually on August 15 to commemorate a battle in 1715, when the townsmen, though heavily outnumbered, soundly defeated the Turks.

Population 236 000

Map Yugoslavia Cc

▼ SQUATTERS' PALACE Diocletian's vast palace (right) overlooks the harbour at Split. In the 7th century, the disused building was taken over by refugees fleeing Barbarian invaders.

Spokane ***USA*** City in eastern Washington state, about 25 km (16 miles) from the Idaho border. It is a market and railway centre, a tourist base and has many timber mills.

Population (city) 173 300; (metropolitan area) 352 900

Map United States Ca

Spoleto ***Italy*** Town about 90 km (55 miles) north of Rome. It is noted for its Roman theatre, Arch of Drusus, a vast aqueduct built by the Lombard dukes in the 10th century and a Gothic cathedral consecrated by Pope Innocent III in 1198. Its industries include leather and textiles. An international music and drama festival is held there each summer.

Population 38 000

Map Italy Dc

Spratly Islands ***South China Sea*** Scattered group of reefs midway between Vietnam and north Borneo. Japan seized them for a submarine base during the Second World War, but they are now claimed by a number of countries, including China, Malaysia, the Philippines and Vietnam.

Map Vietnam Cd

spring Natural fountain or flow of water from the ground. A spring is formed where the water table intersects the surface of the ground. Its position is related to the structure of the local rocks, the level of the water table and the shape of the land. For example, a spring is formed when water sinks through a porous rock layer until it meets the saturated rock above an impervious layer. Water within the saturated rock will flow laterally to the spring. Spring water may be hot or cold, hard or soft, depending on its place of origin underground. See GEYSER; HOT SPRING; MINERAL SPRING

spring tide Tide of highest range which occurs twice a month about the time of the new moon and the full moon, when the sun, moon and earth are approximately aligned. See TIDES

Springbok ***South Africa*** Capital of Namaqualand in north-west Cape Province, founded in 1862 about 485 km (300 miles) north of Cape Town. It is the centre of a copper-mining and sheep-rearing region. An expedition led by the Cape's governor, Simon van der Stel, sank a shaft there in 1685, but mining began only in the 19th century.

Population 8200

Map South Africa Ab

Springfield ***USA*** **1.** State capital of Illinois, about 280 km (175 miles) south-west of Chicago in the middle of a rich agricultural and coal-mining area. President Abraham Lincoln (1809-65) lived in Springfield for many years and is buried there.

2. City in Massachusetts, on the Connecticut river about 130 km (80 miles) west and slightly south of Boston. It was founded as a trading post in 1636 and was the site of a national arsenal from 1777 to 1968. Manufactures today include firearms, electrical equipment, plastics and chemicals. The game of basketball was invented there in 1891.

Population 1. (city) 101 600; (metropolitan area) 190 100. **2.** (city) 150 300; (metropolitan area) 515 900

Map United States **1.** Ic; **2.** Lb

Springs ***South Africa*** Mining town on the East Rand, founded in 1885 to supply coal to Johannesburg, 40 km (25 miles) to the west. Gold was discovered in 1908, and later uranium.

Population 78 700

Map South Africa Cb

spruit In South Africa, a small river or rivulet.

spur A prominent ridge projecting from the mountainside, mountain range or valley side.

squall Sudden, violent, shortlived burst of wind, often accompanied by a shower. See LINE SQUALL

Srbija ***Yugoslavia*** See SERBIA

Sri Jayawardhanapura ***Sri Lanka*** See KOTTE

Sri Lanka See p. 604

Sri Padastanaya ***Sri Lanka*** See ADAM'S PEAK

Srinagar ***India*** North-western city, in the Vale of Kashmir, about 645 km (400 miles) north and slightly west of Delhi. It is cradled by mountains at the end of Lake DAL, which has floating gardens and traditional houseboats. Srinagar is the capital of Jammu and Kashmir state, and is a busy tourist centre, with palaces, mosques, museums and a university. The vast variety of craft goods available in the bazaar include carpets of silk and wool, furs, jewellery and semiprecious stones, and carved wooden furniture.

Population 606 000

Map India Ba

stack Pillar of rock rising from the sea, which has been isolated from the land through erosion caused by wave action.

Stafford ***United Kingdom*** Town and county in the Midlands of England, taking in The POTTERIES and the moorlands that fringe the Peak District National Park. The county, which covers 2716 km² (1049 sq miles), lies north and north-west of the industrial West Midlands and is centred on the peaceful valley of the River Trent and the town of Lichfield with its 13th-century cathedral. Burton upon Trent, in the east of the county, is the centre of the British brewing industry.

The town of Stafford itself, about 32 km (20 miles) north and slightly west of Birmingham, dates back to Anglo-Saxon times and was the birthplace in 1593 of the angler and author Izaak Walton. Today it is a busy industrial town specialising in electrical engineering. South-east of the town is the 120 km² (46 sq mile) heath known as Cannock Chase; in medieval times this was a royal hunting forest.

Population (county) 1 018 000; (town) 56 000

Map United Kingdom Dd

stalactite Icicle-shaped deposit of calcite (calcium carbonate) hanging from the roof of a cave.

stalagmite Icicle-shaped mass, similar to a stalactite, but growing upwards from the floor of a cave.

Stalin Peak ***USSR*** See COMMUNISM PEAK

Stalinabad ***USSR*** See DUSHANBE

Stalingrad ***USSR*** See VOLGOGRAD

Stalino ***USSR*** See DONETSK

Stalinsk ***USSR*** See NOVOKUZNETSK

Stanley ***Hong Kong*** Fishing village on a peninsula on the southern coast of HONG KONG ISLAND. Although there are many shops selling goods to tourists, Stanley remains less affected by modernisation than most other settlements on Hong Kong Island, and is a convenient base from which to explore the coves and beaches along the south-eastern coast of the island.

Population 15 000

Map Hong Kong Cb

Stanley, Mount ***Uganda/Zaire*** See RUWENZORI

Stanley Falls ***Zaire*** See BOYOMA FALLS

Stanley Pool ***Zaire/Congo*** See ZAIRE

Stanleyville ***Zaire*** See KISANGANI

Stara Planina ***Bulgaria*** See BALKAN MOUNTAINS

Stara Zagora ***Bulgaria*** Capital of Stara Zagora province, on the south slopes of the Balkan Mountains, 80 km (50 miles) north-east of Plovdiv. It is an agricultural trading centre dealing in wheat, wine and attar of roses. Its chief industries are brewing, distilling, tanning, textiles, fertilisers and flour milling. In 1877 it was the scene of a Turkish victory over the Russians, during the Russo-Romanian invasion which resulted in the formation of the Principality of Bulgaria (1878).

Population 145 000

Map Bulgaria Bb

Stargard (Stargard in Pommern) ***Poland*** Medieval fortress town in Pomerania, 32 km (20 miles) south-east of the port of Szczecin. Much of its 13th-century walls and the fine Renaissance town hall survive. The town now produces metals, foodstuffs, soaps and detergents.

Population 63 300

Map Poland Ab

Starnberger See ***West Germany*** Bavarian lake 15 km (9 miles) south of Munich. It is 20 km (13 miles) long. The tourist town of Starnberg lies at the northern end. In 1886, the Bavarian King, Ludwig II, confined in a chateau on the lakeshore because of insanity, drowned his doctor and himself while out for a walk.

Map West Germany De

Staropolska ***Poland*** The country's oldest heavy industrial area, about 130 km (80 miles) south of the capital, Warsaw. The Romans began the smelting of ores there, and the first charcoal-iron furnace began production in 1598. Today, foundries and works producing 'Star' trucks continue the metal-working tradition at the area's main town, Ostrowiec.

Map Poland Dc

Statia ***Netherlands Antilles*** See ST EUSTATIUS

Stava ***Italy*** See TRENTINO ALTO ADIGE

Stavanger ***Norway*** City in the extreme south-west of the country on the Stavanger Fjord. It is one of Norway's largest ports, now focusing on the construction of drilling rigs and tankers for liquefied petroleum gas. Because it is near Norway's North Sea oil and natural gas, it has become the headquarters of the Norwegian Oil Board, and of many supply and charter firms. Founded in 1125, when building started on its cathedral, Stavanger also has a theatre and several museums.

Population 92 000

Map Norway Bd

Stavropol' ***USSR*** **1.** Town in the northern foothills of the Caucasus Mountains, about 235 km (145 miles) east of Krasnodar. It was founded in 1777 as a fort, and became a trading centre on the main Russia-Georgia road. Its products today include canned meat, fruit juice, wine and leather goods. **2.** Former town on the Volga river; see TOL'YATTI.

Population 287 000

Map USSR Fd

Sri Lanka

RENOWNED FOR ITS TEA, ITS SCENERY AND ITS BUDDHIST MONUMENTS, THIS BEAUTIFUL ISLAND IS TODAY TORN BY CIVIL CONFLICT

Shaped like a teardrop that has fallen from the cheek of India, the lush tropical island of Sri Lanka is today marred by a civil conflict that might well make India weep. Over the centuries many people have traded with Sri Lanka – the name means 'Resplendent Land' – colonised it or settled there; the Romans called it Taprobane, the Arab merchants Serendip, the Portuguese Ceilas and the British Ceylon. It was from Serendip that the word 'serendipity' – the making of happy discoveries by accident – was coined by the English writer Horace Walpole in 1754.

Today's strife results from divisions between two of the oldest races on the island – the Tamils and the Sinhalese.

The first inhabitants of Sri Lanka, the Veddas, lived there from about 3000 BC. They were conquered around 600 BC by the Sinhalese, who came by sea from north India. The Sinhalese were probably Aryans, since their language is of Aryan origin. They settled mostly in the dry areas of the east and centre of the island, establishing great civilisations which relied on elaborate irrigation systems, with dams and water tanks. Their capital city was ANURADHAPURA, and their religion was, and for most Sinhalese still is, Buddhism.

FOOTPRINT ON A MOUNTAIN

Buddhists say that Buddha left his footprint at the top of a mountain in the south central highlands called ADAM'S PEAK (2243 m, 7358 ft). Hindus claim that the 'footprint', 1.6 m (5 ft) long, was made by the god Siva, and Muslims attribute it to Adam. The spot is a point of pilgrimage for all these religions.

All over the island there are Buddhist temples and statues, attended by saffron-robed monks, who also conduct weddings or lead funeral processions, banging on their drums of mourning.

The Veddas dwindled in number, and only a few small groups survive in the remote interior of Sri Lanka, but another people were to become the lasting enemy of the Sinhalese.

In the 11th century, Hindu Tamils invaded from south India, a mere 29 km (18 miles) away across the Palk Strait. They drove the Buddhist Sinhalese southwards, forcing them to establish a new kingdom in the mountainous interior, around the city of KANDY. The city still contains temples dating from the 11th century when it was the Buddhist capital. The Tamils are still mostly Hindu, although some today are Christian.

Spice-seeking Arab merchants and Portuguese and Dutch colonisers were the next people to arrive in Sri Lanka. Then in 1796 the British took the island, although the

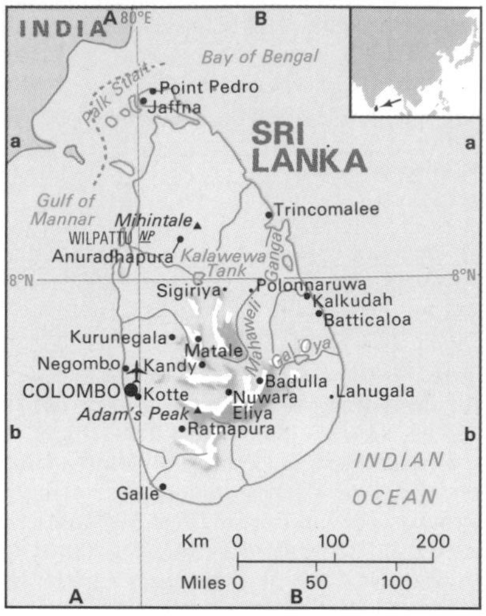

Kandy area held out as an independent Sinhalese kingdom until 1815. The British developed coffee plantations in the mountains, but the crops were devastated by disease in the 1870s and replaced by the Ceylon tea, now bringing in more than half Sri Lanka's export earnings. Rubber and coconut plantations were established on the coast – and in the 19th century more Indian Tamils were brought in to work on the plantations. This influx of people changed the distribution of the population; the wet south-west became heavily settled and the city of COLOMBO became the new capital. The drier lands of the east mainly reverted to jungle.

The descendants of the Tamil immigrants now form about 21 per cent of the population, living mainly in the north around JAFFNA, and down the east coast. In the central mountains Tamil girls still pluck the tender top leaves from the tea bushes, and live in harmony with their Sinhalese neighbours, but in the north trouble between Tamils and Sinhalese has led to martial law and a shoot-on-sight curfew which empties the streets of Jaffna at dusk.

The problems that caused the conflict have surfaced since independence from Britain in 1948. Under British rule, the Tamils, although a minority, held the majority of responsible white-collar jobs. Ceylon became the Republic of Sri Lanka in 1972, and since then the Sinhalese majority has reclaimed the top jobs. Both communities are well-educated – Sri Lanka has the highest literacy rate (over 80 per cent) of any Asian country except Japan – and the Sinhalese feel that they have simply redressed a numerical imbalance. The Tamils argue that they are being discriminated against in higher education and employment.

As unemployment among educated classes has risen in Sri Lanka, what began as a non-violent campaign against discrimination has developed into a violent guerrilla fight for an independent homeland for Tamils, to be called Eelam, in the northern and eastern provinces. The government regards the separatist movement as a Communist threat, and the mainly Sinhalese army has taken on the Tamil guerrillas – sometimes acting very violently against ordinary Tamil citizens in retaliation for terrorist attacks on soldiers. Street fighting in Colombo in 1983 left 400 dead – most of them Tamils. There are fears that India, which has a large Tamil population in its southernmost province – many of them refugees from the island – could become involved in the conflict.

The elected Sri Lankan government which has to deal with this trouble is headed by the right-wing J.R. Jayawardene. He followed the left-wing Bandaranaike family who had dominated Sri Lankan politics since 1956; the world's first woman prime minister was Mrs Sirimavo Bandaranaike, who held office from 1960 to 1965 and from 1970 to 1977.

At independence in 1948, the population had one of the highest living standards in Asia; income per head was twice as high as that in India. However, one problem continued to face the island's economy. Although overwhelmingly agricultural, the island has been unable to grow enough food for its people, because farms outside the tea, rubber and coconut plantations have always been small-scale. Large imports of rice have been necessary, paid for and heavily subsidised by the all-important exports of tea, rubber, coconut products and gemstones.

LITERACY AND HEALTH

Since 1948, successive governments have continued to provide free education for all, which has given the country its excellent literacy record. They have also given food subsidies to poorer people, and health care is good. These factors, however, have led to a large rise in population (running until recently at a rate of 2 per cent a year – now, with family planning campaigns, down to 1.8 per cent). The population growth combined with a heavy fall in the export prices of coconut fibre (coir), rubber and tea has resulted in a standard of living that is little above India's.

The government is now trying hard to encourage development and growth. An ambitious scheme to dam the Ganga river at MAHAWELI will provide irrigation and hydro-electric power – needed if industry is to develop. It has been partly financed by foreign aid. Costs have soared since the project was planned, and it has drained money from the economy. It is hoped that one-tenth of the population will eventually live on the newly irrigated land.

The irrigated region will be used to make the country self-sufficient in rice and may even provide a surplus for export. Small farmers already grow rice on the tropical lowlands which make up 80 per cent of the island, but mainly to feed their families.

The export trade of the plantations and gem mines is the mainstay of the capital and port of Colombo, a city of 600 000 people. The population of Sri Lanka is very young (60 per cent under 25), and Colombo swarms with poor urchins who hawk plastic combs, polish shoes, and work as houseboys and market bearers.

Attempts are being made to increase revenue from tourism. It is a beautiful island and is rich in wildlife. Animals include elephants,

▲ A KING'S BEAST A giant lion, hewn from granite, fronts the magnificent palace fortress of Sigiriya, in Sri Lanka's Central Province. In the 5th century it was the citadel of King Kasyapa.

leopards, bears, water buffaloes and wild boars. There are two major national parks and a refuge for orphan elephants.

But the economy is heavily dependent on the world tea markets – and what Western tea-drinkers are prepared to pay.

SRI LANKA AT A GLANCE

Area 65 610 km² (25 332 sq miles)

Population 16 500 000

Capital Colombo

Government Parliamentary republic

Currency Sri Lankan rupee = 100 cents

Languages Sinhala (official), Tamil, English

Religions Buddhist (69%), Hindu (15.5%), Muslim (7.6%), Christian (7.5%)

Climate Tropical; temperatures range from 23-31°C (73-88°F) in Colombo and from 14-24°C (57-75°F) in the highlands

Main primary products Rice, coconuts, tea, rubber, cassava, fruit, spices, timber, fish; gemstones and semiprecious stones (sapphire, ruby, beryl, topaz, spinel, garnet, moonstone), iron ore, graphite

Major industries Agriculture, textiles, mining, forestry, fishing, oil refining

Main exports Tea, rubber, coconut products, textiles, gemstones, petroleum products

Annual income per head (US$) 282

Population growth (per thous/yr) 18

Life expectancy (yrs) Male 67 **Female** 71

Stein am Rhein ***Switzerland*** One of the country's best preserved medieval towns, with its castle, churches, marketplace and town gates all intact. It also has some old timbered houses with elaborately painted façades. It stands beside the western tip of Lake Constance, near the West German border.

Population 2700

Map Switzerland Ba

Stellenbosch ***South Africa*** The second oldest European settlement after Cape Town, 40 km (25 miles) to the west. Stellenbosch was founded in 1679 and is the centre for a rich fruit and wine-producing district. Six prime ministers – Field-Marshal Jan Smuts, General J.B.M. Hertzog, Dr Daniel Malan, Johannes Strijdom, Hendrik Verwoerd and John Vorster – were students at the university there.

Population 37 700

Map South Africa Ac

Stelvio National Park ***Italy*** Alpine park covering 1316 km² (508 sq miles) in a remote area near the Swiss frontier, about 150 km (95 miles) north-east of Milan. It has a number of peaks over 3500 m (11 500 ft) high, including Ortles (3905 m, 12 812 ft), and also takes in Italy's biggest glacier, the dei Forni, which is 5 km (3 miles) long.

Stelvio was founded in 1935 and is the home of deer, chamois, lynx and bears, and of rare subalpine forest, tundra moss, lichen and wild grasslands.

Map Italy Ca

Stelvio Pass ***Italy*** Alpine pass between Italy and Switzerland about 150 km (95 miles) north-east of Milan. It was developed in 1825 to provide the Austrian rulers with an easier route to Milan. It rises to 2758 m (9048 ft) and leads through the spectacular scenery of the STELVIO NATIONAL PARK.

Map Italy Ca

Stendal ***East Germany*** Former weaving town and major rail junction about 110 km (68 miles) west of the capital, East Berlin. It is rich in Gothic churches.

Population 46 300

Map East Germany Bb

steppe Mid-latitude grassland consisting of level, mostly treeless, plains extending across Eurasia from the UKRAINE to MANCHURIA.

Stereá Ellás ***Greece*** Mountainous central region south of Thessaly and Epirus, bounded by the Aegean Sea in the east and the Ionian Sea in the west. It includes the capital ATHENS, and the classical centres of DELPHI, Orehomenos and THEBES. Stereá Ellás (24 500 km², 9572 sq miles) contains some of Europe's largest reserves of bauxite, the main ore of aluminium. Its main industries are agriculture, fishing, tourism, textiles, aluminium, engineering and chemicals.

Population 4 127 200

Map Greece Cb

Stettin ***Poland*** See SZCZECIN

Stewart Island ***New Zealand*** Largest of the country's offshore islands, covering 1746 km² (674 sq miles), off the southern tip of South Island. It is a haven for some of the country's rarest birds, such as the kakapo, a type of owl-like, nocturnal parrot which lives on the ground.

Population 600

Map New Zealand Ag

Steyr ***Austria*** Attractive old industrial city on the Enns river, 28 km (17 miles) south of Linz. The old city dates from the 13th century and has many fine houses, particularly in the main square. Today its manufactures include lorries, tractors, mopeds, iron goods and sporting guns.

Population 38 970

Map Austria Da

Stirling ***United Kingdom*** City of Central Scotland, about 35 km (22 miles) north-east of Glasgow. The rocky outcrop above the city, which guards the valley of the River Forth, has been fortified probably since Roman times. The present castle dates back to the 11th century and is a smaller version of Edinburgh's. To the south of the city, a monument commemorates the nearby Battle of Bannockburn in 1314, when the Scots under Robert Bruce defeated the English and won independence from Edward II. To the north of Stirling is its university, founded in 1967. The former county of Stirling was divided in the 1970s between CENTRAL REGION and STRATHCLYDE.

Population 39 000

Map United Kingdom Db

stock See BATHOLITHS

Stockholm ***Sweden*** The Swedish capital, beautifully set on some 20 islands and peninsulas beside the Baltic Sea, grew up in the mid-13th century around a stockaded island fortress. Its name comes from the Swedish words *stock*, meaning 'pole', and *holm*, meaning 'island'. Today the historical heart of the city, now linked to nearby islands, is known as 'the town between the bridges'. It contains the Houses of Parliament, the royal palace and Riddarholm Church, the burial place of Sweden's kings. Many of the medieval streets are now pedestrian precincts, crammed with antique shops, and the 17th-century burghers' houses have been turned into offices and private homes.

To the east of the old town is a group of islands whose buildings look across to waterfronts where motor boats, yachts and the occasional schooner are moored. On Island Djuvgåden the capital's maritime past also lives on in a museum housing the *Wasa*, a 60 m (200 ft) long warship that sank on her maiden voyage in Stockholm harbour in August 1628 and was raised to the surface again in 1961.

Over to the north, much of the 19th-century shopping and residential district has been replaced by the high-rise Högtorget quarter. The city's garden suburbs and newer districts are linked by underground railway to central Stockholm. The open-air museum – a living reminder of Sweden's rural past – was a pioneer museum of its type. In total contrast is the royal summer palace of Drottningholm.

The modern city is Sweden's largest manufacturing centre, producing machinery, chemicals, textiles, rubber and beer – and it also has flourishing electrical, engineering and printing industries. There is a frequent ferry service to Finland, and the main international airport is at Arlanda, 40 km (25 miles) north of the city. Each December the Nobel prizes – founded by Alfred Nobel (1833-96), the Swedish chemist and pacifist who invented dynamite and gelignite – are awarded in Stockholm, except for the Peace Prize which is awarded in Oslo, Norway.

Population 1 435 500

Map Sweden Cd

▼ MYSTERIOUS STONES A storm sweeps over Stonehenge, a lonely sentinel on Salisbury Plain. To preserve the monument, access is now restricted.

Stockport ***United Kingdom*** Town on the south-eastern outskirts of Manchester. Originally a market town, it developed as a cotton-spinning centre in the 19th century. Today its main industries are in engineering. A railway viaduct over 30 m (100 ft) high, with 22 arches, spans the valley of the River Mersey, which runs through the town.

Population 139 000

Map United Kingdom Dd

Stockton ***USA*** Industrial city and port in California, on the San Joaquin river about 100 km (60 miles) east and slightly north of San Francisco. It grew during the gold rush of 1848-9, and produces agricultural equipment, timber and foodstuffs.

Population (city) 171 700; (metropolitan area) 398 600

Map United States Bc

Stockton-on-Tees ***United Kingdom*** Town on the north bank of the River Tees in the English county of Cleveland, 50 km (30 miles) south of Newcastle upon Tyne. Originally a quiet market town – with the widest main street in Britain – it developed as an industrial centre after the opening in 1825 of the world's first public railway, the Stockton and Darlington. This brought coal from mines inland to wharves on the Tees. Today there are huge chemical works in the northern suburb of Billingham.

Population 173 000

Map United Kingdom Ec

Stoke-on-Trent ***United Kingdom*** Principal city of The Potteries in the central English county of Staffordshire. It is about 65 km (40 miles) north of Birmingham and is the home of some of the world's largest china and porcelain firms, including Wedgwood and Spode.

Population 250 000

Map United Kingdom Dd

Stolp ***Poland*** See SLUPSK

Stonehenge ***United Kingdom*** Prehistoric stone monument near Amesbury in the English county of Wiltshire, about 120 km (75 miles) west and slightly south of London. It was built between about 2750 and 1300 BC, possibly as a religious or ceremonial centre, and the remains are made up of some 150 boulders – the tallest 6.7 m (22 ft) high and weighing about 45 tonnes – in two main circles. There are traces of other earth and stone circles.

Although Stonehenge was reputedly used by Celtic Druids as a temple for sun worship, the Druids did not arrive in Britain until the 3rd century BC – at least 1000 years after Stonehenge was completed. Many of the stones were transported more than 320 km (200 miles) from quarries in Wales.

Map United Kingdom Ee

Store Bælt ***Denmark*** See GREAT BELT

storm **1.** Violent atmospheric disturbance with strong winds accompanied by rain, snow or other precipitation, and often by thunder and lightning. **2.** A strong wind rated Force 10 and Force 11 (violent storm) on the BEAUFORT SCALE, with speeds of 88-101 km/h (55-63 mph) and 102-119 km/h (64-73 mph) respectively.

storm beach Accumulation of coarse material such as shingle and cobbles formed above the foreshore during a period of powerful storm waves at the highest tides.

storm surge Rapid rise of sea level above predicted tidal levels when water is piled up against a coast by powerful onshore winds.

strait, straits Narrow passage of water, narrower than a sound, joining two larger bodies of water. For geographical features whose name begins 'Strait', see main part of name.

Stralsund ***East Germany*** Port on the Baltic Sea about 185 km (115 miles) north of the capital, East Berlin. It has many medieval buildings and is a growing fishing, shipbuilding, manufacturing and tourist town.

Population 75 400

Map East Germany Ca

Strasbourg ***France*** Industrial city and port near where the Ill river flows into the Rhine on the German border almost due east of Paris. It is the capital of Alsace, and has many canals. The city's Gothic cathedral – dating from the 11th to 15th centuries – is built of red sandstone from the VOSGES, and many of the city's 17th-century houses have carved timbered exteriors. The Council of Europe and the European Parliament of the Common Market hold their meetings in the city.

Population 378 500

Map France Gb

strata See STRATUM

Stratford ***Canada*** Town 160 km (100 miles) west of Toronto which is the home of Canada's famous summer Stratford Shakespearean Festival, started in 1953. It is an important railway engineering centre.

Population 26 300

Map Canada Gd

Stratford-upon-Avon ***United Kingdom*** English town in the county of Warwickshire, 40 km (25 miles) south and slightly east of Birmingham. It was the home town of the playwright and poet William Shakespeare (1564-1616). His birthplace, his beautiful half-timbered house, the school he probably attended and the family home of his wife, Anne Hathaway (at nearby Shottery), have been preserved. The Memorial Theatre overlooking the River Avon runs a full programme of his 38 plays. The town has given its name to many other Stratfords throughout the English-speaking world, for example in Ontario, Canada.

Population 22 000

Map United Kingdom Ed

strath In Scotland, a steep-sided, broad, flat-floored valley that is wider than a glen.

Strathclyde ***United Kingdom*** Administrative region in the west of Scotland, named after the valley of the River Clyde and centred on the city of GLASGOW. It covers 13 856 km^2 (5350 sq miles) and takes in the old counties of Ayrshire, Bute, Dunbartonshire, Lanarkshire and Renfrewshire, as well as parts of Stirlingshire and Argyllshire, so that it extends from within 50 km (30 miles) of the English border to the southern Highlands and islands.

Most of the north is mountain and moorland, but the south is heavily industrialised. There are engineering industries in towns such as Kilmarnock. The port of Hunterston imports iron ore for the only surviving Scottish steel plant, at Ravenscraig near Motherwell. Hamilton and Coatbridge developed as iron towns in the 19th century, providing the materials and machines that, for a century or more until the 1960s, supplied the world with ships and railway locomotives.

Population 2 373 000

Map United Kingdom Cc

stratocumulus Low cloud formation of large, soft grey rolls frequently covering the whole sky, usually associated with dry, dull weather.

stratopause See ATMOSPHERE

stratosphere See ATMOSPHERE

stratum Distinct bed or layer of sedimentary rock. The plural – strata – applies to several beds. One stratum is usually called a bed.

stratus Low cloud formation of white foggy sheets, often bringing drizzle.

strike Direction of a horizontal line in the bedding planes of inclined strata. It lies at right angles to the direction of DIP.

Stromboli ***Italy*** Active volcano-island in the EOLIAN ISLANDS to the north of Sicily. Its summit is 926 m (3038 ft) high. The volcano's nightly fireworks are spectacular, though rarely dangerous, but it erupted strongly in 1966.

Population 400

Map Italy Ee

Stuttgart ***West Germany*** State capital of Baden-Württemberg, about 150 km (93 miles) south of Frankfurt am Main. In the 10th century, a German duke, Liutolf, set up a *stutgarten*, or stud farm, in this natural basin at the north of the Swabian Jura. The city takes its name from Liutolf's stud and keeps a horse in its coat of arms today. It also lives up to its name as a *garten* (the German for 'garden') – more than half of the land in the city centre is filled with parks and woods.

Stuttgart's port is on the Neckar river, north-east of the city, and industrial plants line the river's banks. Among them is the Daimler-Benz factory, the world's oldest car plant, founded in 1890, which has a museum and the original workshop of Gottlieb Daimler (1834-1900). Other products include electronic goods, machinery, precision instruments and textiles. The city is also an educational, cultural and publishing centre.

Population 600 000

Map West Germany Cd

Styria (Steiermark) ***Austria*** Forested, mountainous *land* (state) in the south-east, crossed by the rivers Mur, Mürz and Enns. It is noted for its mining and metal industries. The capital is GRAZ.

Population 1 187 500

Map Austria Db

subduction zone Long narrow zone along which an oceanic plate is moving down into the earth's MANTLE. It is also known as a Benioff zone. See PLATE TECTONICS

Subotica (Subotitsa; Szabadka; Maria-Teresiopel) ***Yugoslavia*** Second largest city in the Serbian autonomous province of Vojvodina after Novi Sad, lying 160 km (about 100 miles) north-west of Belgrade. It is a centre for the fertile Backa Plains which stretch south to the Danube; its factories produce heavy machinery, farm equipment, foodstuffs, furniture and chemicals. The city lies less than 30 km (19 miles) from the Hungarian border, and its people are largely ethnic Hungarians. It has many ornate buildings in the Hungarian style.

The resort of Palić on the shore of Lake Palić, surrounded by acacias, vineyards and orchards, lies 8 km (5 miles) to the east.

Population 154 600

Map Yugoslavia Da

subsidence Sinking of part of the earth's surface relative to its surroundings due either to local factors, such as mining activities, or to major processes affecting the earth's crust, such as the formation of a RIFT VALLEY.

subtropical climate Warm climate in low, temperate latitudes, with distinct seasonal rhythm. There are two types: dry subtropical, known as Mediterranean climate, which has long, dry summers and is usually found on the west side of continents; and wet subtropical, with higher rainfall and wet summers, found on the east side of continents.

Sucre ***Bolivia*** City and the seat of the judiciary, Sucre is the country's original capital. It was founded in 1538 as Charcas by the Spaniard Pedro de Anzures and renamed in 1826 after Antonio José de Sucre, the Venezuelan who freed parts of Peru and Bolivia from Spanish rule. It has some fine colonial buildings, including the legislative palace where Bolivia's independence was declared in August 1825.

Population 79 900

Map Bolivia Bb

Sudan See p. 610

Sudbury ***Canada*** Mining town 370 km (230 miles) north-west of Toronto. It is the centre of one of the world's richest nickel-mining areas and also has copper and platinum mines.

Population 150 000

Map Canada Gd

sudd Floating mass of compact vegetation that often obstructs navigation on the White NILE, consisting chiefly of plants from swamps lying alongside the river. The mass may sometimes form a barrier over 30 km (19 miles) long and up to 5 m (16 ft) thick.

Sudd ***Sudan*** Swampland in the south covering some 129 500 km² (50 000 sq miles) – about the size of England. It disperses and allows to evaporate much of the waters of the White Nile, though its effects will be reduced by the Jonglei Canal which will drain some of the swamp and provide some irrigation.

Map Sudan Ac

Sudetenland ***Central Europe*** See SUDETY

Sudety (Sudetes; Sudeten) ***Czechoslovakia/Poland*** Series of mountain ranges between Czechoslovakia and Poland. The Riesengebirge (Karkonosze; Krkonose; Giant Mountains) in the west are the highest. They reach 1602 m (5257 ft) at Mount Sniezka (Sneska), on the border, and are a noted skiing area.

The Czech Sudety were part of Austria-Hungary until 1919 and had a significant proportion of German-speaking inhabitants. This 'Sudetenland' was handed over to Nazi Germany by the Munich agreement of 1938. However, together with similar areas of north-west and south-west Czechoslovakia – also called Sudetenland – it was returned to Czechoslovakia in 1945, and its German-speaking people were later expelled.

Map Czechoslovakia Ba

Suez Canal ***Egypt*** Ship canal built by the French engineer Ferdinand de Lesseps to link the Mediterranean with the Red Sea. It is 173 km (107 miles) long, running from Port Said on the Mediterranean coast through Lake Timsah and the Bitter Lakes to the port of Suez. The canal was opened in November 1869, and it shortened the journey to India by more than 11 000 km (7000 miles).

It was run by a French company, which Britain took over in 1875. Britain remained a major shareholder until 1956. Then the Egyptian president, Gamal Abdul Nasser nationalised it, provoking the Suez conflict with Britain, France and Israel, during which the Egyptians blocked it with scuttled ships. The canal was blocked again in 1967 during the Six-Day War with Israel, but was reopened in June 1975. More than 20 000 vessels a year now pass through the canal, which has been enlarged to take large tankers. It is entirely controlled by Egypt, and Israeli shipping is allowed through.

Map Egypt Cb

Suez (El Suweis) ***Egypt*** Town at the southern end of the Suez Canal, for which it is a key control point. Suez has a fuelling station for ships, two oil refineries and manufactures fertilisers. To the south are long stretches of sandy beaches.

Population 195 000

Map Egypt Cb

Suez, Gulf of Horn-like north-western arm of the RED SEA. It is an important shipping lane because of the SUEZ CANAL which links it to the MEDITERRANEAN SEA. It also contains offshore oil wells. This branch of the GREAT RIFT VALLEY system is bordered by Egypt's Eastern Desert in the west and SINAI in the east.

Dimensions 290 km (180 miles) long, 32 km (20 miles) wide

Map Egypt Cb

Suffolk ***United Kingdom*** County of eastern England covering 3800 km² (1467 sq miles) and forming part of East Anglia. The south-east has the estuaries of the Stour, Orwell and Deben, the county town of IPSWICH and the ports of Harwich and Felixstowe. In the west are the sandy heathlands of the area known as the Breckland, north of BURY ST EDMUNDS, and the grasslands of the horse-racing town of Newmarket. In between is a gentle agricultural land immortalised by the paintings of the Suffolk artists John Constable (1776-1837) and Thomas Gainsborough (1727-88).

Population 619 000

Map United Kingdom Fd

Sukhothai ***Thailand*** Agricultural province and market town about 385 km (240 miles) north of the capital, Bangkok. The old town was the country's first capital, flourishing in the 13th and 14th centuries, and has been restored as a historical park containing the buildings of that period. The main crops are rice, mung beans, soya beans and cotton.

Population (town) 11 500; (province) 562 800

Map Thailand Ab

sukhovey Hot dry wind, blowing mainly from the east during summer in southern European USSR and Kazakhstan.

Sukhumi ***USSR*** Port and tourist resort near the eastern end of the Black Sea, below the southern foothills of the Caucasus Mountains. With its warm, tideless water, sandy beaches and subtropical climate, Sukhumi is one of the area's most popular holiday places. It also produces leather goods, tobacco, canned fruit and wine. It was settled by the ancient Greeks and Romans (and called Dioscurias); it is now capital of the Abkhazsk Republic, an autonomous republic of Georgia.

Population 124 000

Map USSR Fd

Sulawesi (Celebes) ***Indonesia*** Curious-shaped island of 179 370 km² (69 255 sq miles) between Borneo and New Guinea. It is largely mountainous, but has some good farmland, producing rice, cassava, beans and maize.

Population 10 409 600

Map Indonesia Ec

Sulu Archipelago ***Philippines*** Group of some 400 named islands (and many more unnamed) covering 982 km² (379 sq miles) between MINDANAO and BORNEO, and forming the southern boundary of the SULU SEA. It consists of two main island chains – the northern Pangutaran group, mainly coral reefs, and the southerly main group of larger, mostly volcanic islands, including BASILAN and JOLO. It is a distinctive region whose people – the Moros – make a living mainly by fishing and raising such crops as rice, cassava and coconuts for local use.

The islands formally became part of the Philippines only in 1940. For long the home of notorious pirates, they were ruled by the Sultans of Jolo until the late 19th century, then coming under shaky Spanish and later American control. The population remains staunchly Muslim, and the islands are a centre of resistance to the Manila government.

Population 360 600

Map Philippines Bd

Sulu Sea (Sea of Mindoro) Lying between the south-western Philippines and Borneo, the sea was once dominated by Moro (Philippine Muslim) pirates. Moro boats with their vividly coloured sails are still much in evidence, but fishing has replaced piracy.

Area 260 000 km² (100 386 sq miles)

Map Philippines Ad

Sumatra (Sumatera) ***Indonesia*** Island covering 473 607 km² (182 860 sq miles) off the west coast of Peninsular Malaysia. It is the centre of Indonesia's oil industry and also produces rubber and timber for export.
Population 28 016 200
Map Indonesia Ab

Sumba ***Indonesia*** Island covering 11 153 km² (4306 sq miles) in the Lesser Sundas group, south of Flores. It produces maize, tobacco, rice, coconuts and fruit.
Population 251 100
Map Indonesia Ed

Sumbawa ***Indonesia*** Island 15 448 km² (5965 sq miles) in the Lesser Sundas group, west of Flores. Its highest volcano, Tambora, at 2821 m (9258 ft), is the remnant of a much larger volcano which devastated the island in 1815, killing 50 000 people and causing 35 000 to emigrate. Many tropical products are grown, including rice, sweet potatoes and soya beans.
Population 195 600
Map Indonesia Ed

Sumer ***Middle East*** Ancient kingdom in the region of Mesopotamia founded around 5000 BC. The Sumerians are credited with inventing the earliest form of writing about 3500 BC. It consisted of cuneiform (wedge-shaped) symbols and pictures inscribed on damp clay tablets which were then dried. Some 50 000 such tablets, depicting the Sumerian way of life, have been discovered at Nippur, the greatest city of Sumer about 100 km (60 miles) south of Babylon. Sumer was conquered in about 1950 BC by the Babylonians; the area is today part of Iraq.

Sun City ***South Africa*** Modern holiday complex, in the black state of Bophuthatswana, 150 km (93 miles) north-west of Johannesburg. It has a gambling casino – which would not be permitted in white South Africa. There is also a wildlife sanctuary, sporting facilities and a large entertainment centre.
Map South Africa Cb

Sun Moon Lake (Jih-yüeh T'an) ***Taiwan*** Largest lake in Taiwan, covering 8 km² (3 sq miles) in the central mountain ranges some 45 km (28 miles) south-east of T'ai-chung. A temple on its shore houses remains of Hsuan Chuang, a Tang dynasty monk largely responsible for introducing Buddhism to China. The lake, which is a major tourist attraction, supplies hydroelectric power from a plant built by the Japanese in the 1930s.
Map Taiwan Bc

Sunch'on (Suncheon) ***South Korea*** Railway junction and market town, in a vegetable-producing area about 145 km (90 miles) west and slightly south of the port of Pusan. It is the tourist gateway to the holiday beaches and islands of the south-western coast.
Population 108 000
Map South Korea Ce

Sunda Islands ***Indonesia*** Collective name for islands of the Malay archipelago. The Greater Sunda Islands consist of JAVA, SUMATRA, BORNEO, SULAWESI and adjacent islands. They cover 1 333 677 km² (514 953 sq miles). The Lesser Sunda Islands (NUSA TENGGARA) comprise the chain of islands east from Bali to include Alor and Timor but not Wetar. Their total area is 76 249 km² (29 441 sq miles).
Population (Greater Sundas) 115 000 000; (Lesser Sundas) 8 500 000
Map Indonesia Cc

▲ **INTERNATIONAL ARTERY Oil tankers make their way through the Suez Canal. The waterway can now take all but the largest supertankers.**

Sunda Shelf ***South China Sea*** Submarine extension of the South-east Asian mainland which underlies much of the South China Sea. It is one of the widest areas of continental shelf in the world, and its relatively shallow waters make it an attractive place to search for oil. It is also potentially a rich source of fish.
Map Asia Ih

Sunda (Soenda) Strait Strait between JAVA and SUMATRA, Indonesia. It contains many small volcanic islands, including KRAKATAU which erupted in 1883 with such force that its explosions were heard 4700 km (2930 miles) away. Explosions of another kind occurred in 1942, when the US fleet suffered heavy losses in the strait during the Battle of the JAVA SEA.
Width 26-110 km (16-68 miles)
Depth 27-183 m (90-600 ft)
Map Indonesia Bd

Sundarbans ***Bangladesh*** Dense mangrove swamp and jungle of the Ganges delta. The area teems with wildlife, including snakes, deer and crocodiles, but it is plagued by leeches and mosquitoes. There is some development of a timber industry, but much of the jungle is now a wildlife sanctuary which is one of the last refuges of the Bengal tiger.
Map Bangladesh Bd

Sunderland ***United Kingdom*** English industrial town at the mouth of the River Wear, 16 km (10 miles) south-east of Newcastle upon Tyne in the county of Tyne and Wear. It thrived in the 19th century on coal mining and shipbuilding – at one point it was the greatest shipbuilding centre in the world – but these have declined and been only partly replaced by other forms of engineering. St Peter's Church dates from AD 674, but was partly rebuilt in the 1870s.
Population 200 000
Map United Kingdom Ec

Sundsvall ***Sweden*** Timber port on the Gulf of Bothnia, about 406 km (250 miles) north of Stockholm. It manufactures and exports wood pulp and paper.
Population 93 200
Map Sweden Cc

Sungei Kolok ***Malaysia/Thailand*** River rising close to the Thailand border in the northern Malaysian state of Kelantan. For much of its length it forms the border between Malaysia and Thailand. It turns north-west into Thailand before joining the South China Sea just north of the border.
Map Malaysia Ba

Sunset Crater National Monument ***USA*** Volcanic cinder cone in northern Arizona, about 180 km (110 miles) south of the Utah border. It was formed by an eruption in about 1065, and the colour of the upper part is reminiscent of a sunset.
Map United States Dc

Sunyani ***Ghana*** Largest town and capital of Brong-Ahafo region, about 320 km (200 miles) north-west of the capital, Accra. It is also a busy market for the cocoa and kola nuts intensively grown in the surrounding area.
Population 25 000
Map Ghana Ab

Sudan

A LAND OF GREAT POTENTIAL THAT COULD GROW FOOD ENOUGH FOR MANY OF ITS NEIGHBOURS, YET MILLIONS FACE STARVATION

The republic of Sudan is the largest country in Africa, and also has some of its biggest problems – drought, famine, economic crisis, political instability and civil war. The famine has grown steadily worse as annual rainfall has decreased since the mid-1970s. In addition, a million refugees have poured in from Ethiopia and Chad, fleeing from war and famine. The crisis-ridden economy suffers from a lack of investment and an inability to earn foreign currency.

The fragile political scene, held together by the dictatorial President Jaafar Nimeiri – there were 12 attempted coups from the time he seized power in 1969 – faced new uncertainties when a successful army coup ousted him in April 1985. In addition to these problems, civil war between the north and south has been on and off the boil since 1955, caused by differences in race and religion.

Sudan is a melting pot of races. The people of the north are Arab and Muslim; their affinities tend to be with the Middle East. The people of the south are black Africans – some Christian though mostly pagan – speaking their own tribal languages. Southern tribes include the Dinka, Nuer (some of the tallest people on earth, with many men reaching more than 2 m – over 6.5 ft), Shilluk, Bari and Azande. Each has its own customs and beliefs, music, dances and handicrafts.

Arabic is the main language of Sudan and English is often understood, but altogether 115 languages are spoken. About a quarter of the people live in towns; however, Sudan still has a long way to go in developing its social services. About one child in every eight dies before the age of one. Half the children go to primary school but only 16 per cent to secondary school. The adult literacy rate is 32 per cent.

Sudan occupies much of the Upper NILE basin, from the foothills of the East African Highlands to the SAHARA. The river flows 3475 km (2160 miles) from the Uganda border to Wadi Halfa on the border with Egypt. The White Nile drains the East African Highlands with their year-round rain, the Blue Nile drains the mountains of Ethiopia where rain comes only in summer, causing the great annual flood so important to Egypt. The rivers join at KHARTOUM.

In the north the Nile winds through the Libyan and Nubian deserts, extensions of the Sahara, its palm-fringed valley a narrow strip of habitable land. Towards the RED SEA in the east is broken hill country where irregular rain gives scant pasture for sheep and camels of the Beja nomads. Across central Sudan stretches the savannah that produces most of the country's food; 250-750 mm (10-30 in) of rain normally falls in summer, while east of the White Nile there is large-scale irrigation. In the far west the DARFUR Highlands rise to 3071 m (10 073 ft) at Jebel Gimbala.

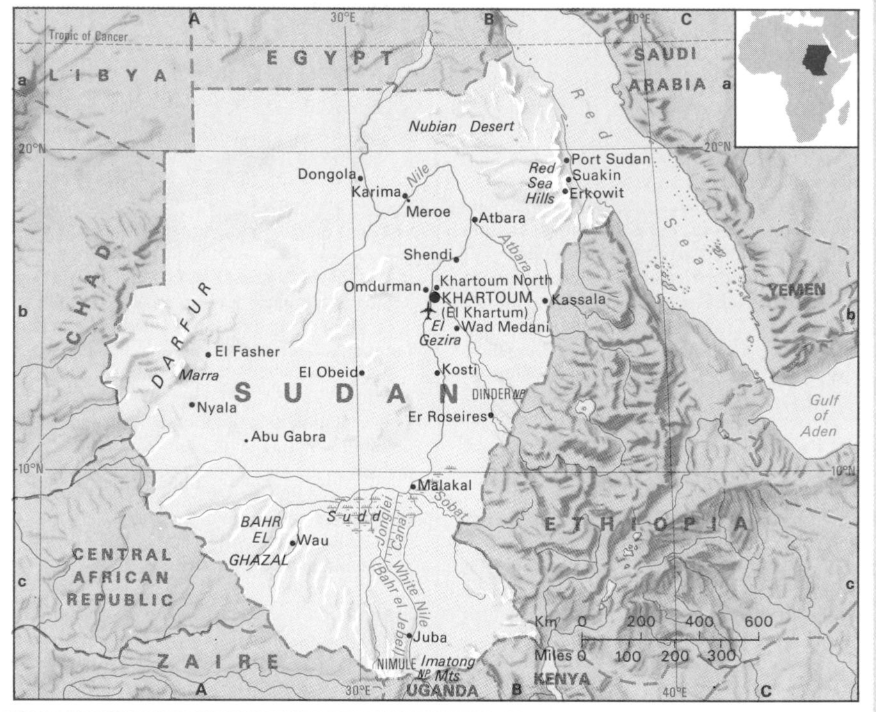

In southern Sudan, the flat, well-watered clay plains are flooded in the summer wet season, while areas of permanent swamp, papyrus beds and floating vegetation block navigable channels in the SUDD region. In the dry season grass fires sweep the area, so few trees grow.

Sudan is a country of tropical heat. Summer temperatures range from 27°C (81°F) in the north to 46°C (115°F) in the south, winter temperatures from 16°C (61°F) to 27°C (81°F).

The economy is largely dependent on agriculture: subsistence farming accounts for 80 per cent of production. Many northern people are nomads whose way of life has changed little in generations. Until the recent droughts, which caused large-scale famine and death, the country had 19.5 million cattle, 15 million sheep and 12.9 million goats.

Commercial farming has expanded with several large-scale projects. Cotton is produced under the Rahad and long-established EL GEZIRA cooperative schemes, and cotton and cotton products normally account for about two-thirds of exports. The similar KENANA scheme aims to make Sudan an exporter of sugar. In the south the JONGLEI CANAL, 320 km (200 miles) long, will drain much of the vast Sudd swamp. There is also a plan to irrigate the triangle of land at the junction of the White and Blue Niles, which has very rich soil.

But a constant problem has been to find foreign exchange to pay for the farm machinery, spare parts, fertilisers and oil that these huge developments need. An ambitious scheme in the 1970s envisaged using Arab oil money to develop Sudan as a 'breadbasket' for the Middle East, but the petrodollars never materialised.

Sudan is the world's greatest source of gum arabic, used in medicines and inks, and also produces sesame and groundnuts. Oil and natural gas have been discovered, and development of a large oil field near Bentiu in the Sudd would provide all the country's needs. But its exploitation has been delayed by guerrilla warfare in the area.

A small manufacturing sector produces building materials, cigarettes, petroleum products, sugar and textiles. Many skilled Sudanese work in The Gulf states. The money they send home helps the economy, but their absence is felt in a country which urgently needs skills for its own development.

In ancient times two civilisations flourished in Sudan: Nubia and Kush. In the 6th century three Sudanese kingdoms were converted to Christianity, and they resisted the spread of Islam from Egypt until the 14th century. In 1821 armies sent by Muhammad Ali of Egypt occupied much of the north and developed Sudan's trade in ivory and slaves. Ismail Pasha tried to extend Egypt's influence farther south and appointed the British General Gordon governor-general of the Sudan (1874-9).

In 1881 the Mahdi (divine guide) Muhammad Ahmed led a revolution to reform Islam and drive out foreigners from Sudan. Gordon went back with orders to evacuate the Egyptian army, but was killed when Khartoum fell in 1885 after a ten-month siege – two days before a relief force arrived. In 1898 the army of the Khalifa, the Mahdi's successor, was defeated at OMDURMAN by an Anglo-Egyptian army under General Kitchener. The two countries established a condominium over Sudan, but in practice it was run by the British.

In 1955 civil war broke out as independence approached. The black African south tried to break away from the Arab north. Independence in 1956 led to political instability. A multi-party democracy was followed by a military coup in 1958 and a return to civilian rule in 1964. In 1969 Colonel Nimeiri led a second coup, and three years later ended the civil war by giving regional autonomy to the three southern provinces. The war had lasted 17 years and cost nearly 500 000 lives.

Nimeiri established a leftist government and nationalised industries and banks. In 1973 he made Sudan a one-party state. In 1977, however, he expelled his Russian military advisers and began to turn to the West for aid. By the mid-1980s, Sudan was receiving more than US$700 million of foreign aid each year – more than any other African country except Egypt – but the country was near bankruptcy.

Successive years of drought have hit the country, especially the north-west and north-east, making large tracts of land parched and spoiled; rivers have run dry. No green foliage grows, and the carcasses of animals lie out in the sun. Refugees have converged on Khartoum, living in makeshift settlements outside the city.

To make the problem worse, a million refugees have crossed the border, mostly from starving Ethiopia, fleeing war between the military regime and the Tigre People's Liberation Front. About 150 000 have come from Chad, where both France and Libya have intervened in a civil war. Nomads from eastern Chad have moved into north-west Sudan in search of pasture.

Horrifying scenes of famine brought international aid organisations to the rescue, but they were hampered by poor administration, transport difficulties and fuel shortages.

In 1983 Nimeiri tried to consolidate his support among fundamentalist Muslims by introducing Sharia law, with its harsh punishments of whippings, maimings and hangings. Alcohol was banned; thieves were maimed; couples unable to prove they were married were lashed. A 76-year-old political moderate, Mahmoud Muhammad Taha, was hanged in public for criticising the new law.

It also sparked rebellion in the south by non-Muslims. The Sudanese People's Liberation Army sank a Nile steamer, killing hundreds and halting river traffic; attacked foreign installations; mined roads and railways. By 1986 they controlled much of the south.

The first democratic elections for 18 years, held in April 1986, brought a coalition of northerners led by Sadiq al Mahdi to power. He soon began talks with the rebels, but faced enormous political and economic problems. He also inherited an uneasy relationship with Sudan's two powerful northern neighbours, Egypt and Libya.

SUDAN AT A GLANCE
Area 2 505 813 km² (967 500 sq miles)
Population 22 350 000
Capital Khartoum
Government Republic
Currency Sudanese pound = 100 piastres = 1000 milliėmes
Languages Arabic, English, local languages especially Nubian
Religions Muslim (73%), tribal (18%), Christian (9%)
Climate Tropical; average temperature in Khartoum ranges from 15-32°C (59-90°F) in January to 26-41°C (79-106°F) in June
Main primary products Cotton, dates, groundnuts, sesame seed, gum arabic, sorghum, wheat, beans, livestock; crude oil and natural gas, chromium
Major industries Agriculture, oil production and refining, food processing, cement
Main exports Cotton, sesame seed, groundnuts, cereals, gum arabic, livestock, petroleum products
Annual income per head (US$) 390
Population growth (per thous/yr) 27
Life expectancy (yrs) Male 45 **Female** 48

▼ WEEKLY CELEBRATION The Dinka gather near the southern capital of Juba for a huge tribal dance every Friday. They and their herds migrate from villages in the savannah to riverside pastures in the dry season.

Superior, Lake ***Canada/USA*** Largest and highest of the Great Lakes and the world's largest sheet of fresh water, covering about 82 400 km² (31 800 sq miles). The Soo (or Sault Sainte Marie) Canals enable ships to overcome the 6 m (20 ft) difference in water level between Superior and Lake HURON.
Map United States Ia

Suphan Buri ***Thailand*** Market town set in a prosperous agricultural province of the same name, about 80 km (50 miles) north-west of the capital, Bangkok. The province's farmers grow rice and sugar cane.
Population (town) 134 500; (province) 761 500
Map Thailand Bc

Surabaya ***Indonesia*** The nation's second largest city after the capital, Jakarta, and the provincial capital of East Java. It is an industrial port with shipyards, textile mills and a fishing fleet, and exports sugar, rubber, tobacco and other agricultural products.
Population 2 470 000
Map Indonesia Dd

Surat ***India*** West-coast port at the mouth of the Tapti river, about 260 km (160 miles) north of Bombay. It was India's chief port under early Mogul emperors (1550-1650), and once was the main western port of British India, even surpassing Bombay for a time in the value of goods handled. It still produces traditional gold and silver threadwork and hand-woven cottons.
Population 913 800
Map India Bc

Surat Thani (Ban Don) ***Thailand*** Coastal province and town about 560 km (350 miles) south of the capital, Bangkok, developing as an agricultural area. The nearby island of Ko Samui is a growing tourist resort. The province's major crop is the oil palm – from which palm oil, used in margarine, cooking fats and soap, is made.
Population (town) 35 200; (province) 669 600
Map Thailand Ad

Surin ***Thailand*** Province and town about 335 km (208 miles) north-east of the capital, Bangkok. The province – which lies along the Cambodian border – is the site each November of a spectacular round-up of wild elephants. At the round-up, the elephants – which are later tamed and trained to transport logs in the teak forests – are paraded, raced and pitted against 70 to 100 humans at a time in tugs-of-war. The elephants usually win. Many of the province's farmers are Khmers from across the border. Rice is their main crop.
Population (town) 33 000; (province) 1 149 900
Map Thailand Bc

Surrey ***United Kingdom*** English county of 1655 km² (639 sq miles) on the southern border of London, now partly absorbed by residential overspill from the capital. However, rural conservation laws have protected most of the North DOWNS, the chalk hills that cross the county. The county town, Guildford, 45 km (28 miles) south-west of London, has a modern cathedral and is the seat of the University of Surrey.
Population 1 012 000
Map United Kingdom Ee

Surinam

A NATION OF MIXED CULTURES BROUGHT TOGETHER BY DUTCH COLONISERS IN SOUTH AMERICA

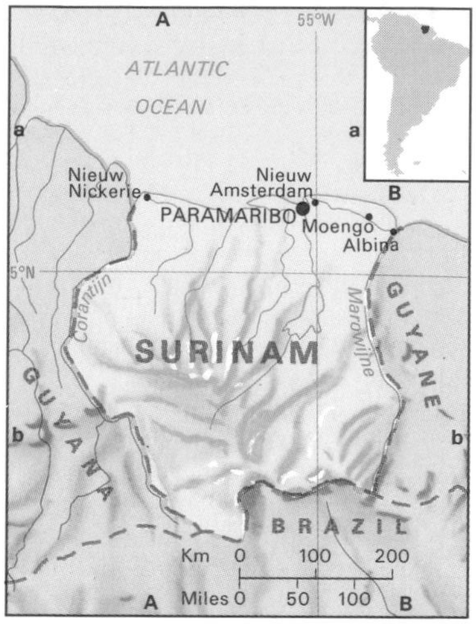

In 1667, the Dutch gave a North American settlement to the British in exchange for a territory on the northern coast of South America. The settlement, then called New Amsterdam, became New York. The territory became Dutch Guiana and is now the independent state of Surinam.

The Dutch had occupied the territory in the 16th century, but European wars led to treaties which alternately took away and returned the territory to them. The country became an official Dutch colony in 1814 and independence was declared in 1975. The Dutch brought slaves from Africa in the 17th century and, after the abolition of slavery in 1863, indentured labourers from India, Java and China to work on the sugar plantations.

The population is now a mixture of Creole descendants of African slaves, who make up most of the 180 000 people in the capital city of PARAMARIBO; Indians, Chinese and Indonesians, who mostly farm rice and sugar on the coastal plains; Europeans (who also live mainly near the coast); and native Amerindians, small numbers of whom live mainly in the dense jungle and savannahs of the interior which rise towards the Guiana Highlands in the south.

The Amerindian tribes have been nudged deeper into the jungles by settlements of 'Bush Negroes' who are the descendants of escaped slaves, but all the races live in harmony. Between one-third and one-half of the entire population emigrated, mainly to the Netherlands, during the 1970s.

The economy is dependent on bauxite (aluminium ore), which comprises 80 per cent of exports. Surinam has under-exploited resources of oil and timber, and great potential for agriculture and tourism. But the economy desperately needs an influx of money. Aid from the USA and Netherlands was cut off after a military junta seized power in 1980 and subsequently executed opposition leaders.

SURINAM AT A GLANCE

Area 163 265 km² (63 037 sq miles)
Population 384 000
Capital Paramaribo
Government Socialist-military state
Currency Surinam guilder = 100 cents
Languages Dutch (official), English, Sranang Tongo (Surinamese or Taki-Taki), Hindi, Javanese, Chinese, Spanish
Religions Christian (22% Roman Catholic, 15% Moravian), Hindu (26%), Muslim (19%)
Climate Tropical; temperature in Paramaribo averages 26-27°C (79-81°F) all year
Main primary products Rice, sugar, timber, bananas, citrus fruits, coconuts, coffee, cocoa, some cattle; bauxite
Major industries Agriculture, bauxite mining, alumina refining, aluminium smelting, forestry, timber processing
Main exports Bauxite, alumina and aluminium, rice, timber and its products
Annual income per head (US$) 2500
Population growth (per thous/yr) 18
Life expectancy (yrs) Male 67 **Female** 71

Surtsey ***Iceland*** Volcanic island which rose from the sea off the south coast of Iceland, some 20 km (12 miles) from the WESTMAN ISLANDS, in November 1963. Within a few months the island – named after Surtr, the god of fire in Norse mythology, was 150 m (490 ft) high and covered about 2.5 km² (1 sq mile). And within months after the eruptions grumbled to a halt, plants – whose seeds were carried in by wind and birds – were colonising the new land.
Map Iceland Bb

Susa (Shush) ***Iran*** Archaeological site on the Susiana Plain about 100 km (60 miles) from the Iraqi border. Clay tablets and pottery dating from before 3500 BC have been found there. For centuries the capital of the Elamite kingdom, it was destroyed in the 7th century BC by the Assyrian king Assurbanipal. Cyrus the Great (ruled 559-530 BC) rebuilt it as his first capital before he conquered all Persia. It was destroyed again by the Macedonian conqueror Alexander the Great (356-323 BC) and again in the 4th century AD. It was eventually abandoned towards the end of the 14th century, following the Mongol invasion and devastation by Tamerlane. Four *tells*, or mounds, have been excavated since the mid-19th century, uncovering citadels, palaces and other buildings.
Map Iran Aa

Susquehanna ***USA*** River rising in central New York state and flowing 715 km (444 miles) southwards into Chesapeake Bay about 50 km (30 miles) north-east of Baltimore.
Map United States Kb

Sussex ***United Kingdom*** Twin counties of southern England – East and West Sussex – on the English Channel. Their combined area is 3811 km² (1471 sq miles). The hills of the South DOWNS were once covered by forests, which provided timber for shipbuilding and charcoal for a primitive iron industry. Today the downs are mostly open pastureland. The coast is lined with resorts, of which the largest is BRIGHTON. Inland the chief towns are CHICHESTER (county town of West Sussex), Horsham and Crawley, which has grown with the development of nearby Gatwick as London's second airport. Lewes, with its Norman castle, is the county town of East Sussex, but the county's most famous town is HASTINGS, near the site of the battle in 1066 that established Norman rule in England. Herstmonceux Castle became the home of the Royal Greenwich Observatory in the 1950s, but there are plans to move the observatory to Cambridge University in the 1990s.
Population 1 361 000
Map United Kingdom Ee

Susupe ***Northern Marianas*** See SAIPAN

Sutherland ***United Kingdom*** See HIGHLAND

Sutjeska ***Yugoslavia*** National park west of the upper Drina river in Bosnia Herzegovina. It includes the lovely mountain of Maglić, which reaches 2387 m (7831 ft) near the Montenegrin border, rare virgin forests around Perucica, and the magnificent Sutjeska gorge west of Maglić. A striking white marble memorial complex at Tjentiste commemorates the month-long Battle of Sutjeska in 1943 when 19 700 partisans were surrounded by 127 000 heavily armed Nazi troops. The partisans escaped eventually, but 7356 of them were killed and Yugoslavia's subsequent leader Marshal Tito (1892-1980), then the partisan commander, was wounded.
Map Yugoslavia Dc

Sutlej (Satluj) ***China/India/Pakistan*** River which rises in the Himalayas of south-west Tibet in China, and flows some 1370 km (850 miles) west into India, then south-west through the Punjab to the Chenab river in Pakistan. It is dammed at BHAKRA. The Sutlej is part of the Indus river system, and its headwaters were allocated to India following a treaty between India and Pakistan in 1960. Pakistan has had to build new canals and dams to keep the irrigation schemes of the lower Sutlej valley alive.
Map India Bb; Pakistan Db

Suva ***Fiji*** Capital and major port of Fiji, on the south-east coast of the island of Viti Levu. Cargo and cruise ships use its fine natural harbour. It is the home of the University of the South Pacific, founded in 1968. A museum displays old Fijian war canoes and weapons. Tourism and copra and food processing are the main industries.
Population (city) 74 000; (metropolitan area) 133 000

Suwon (Suweon) ***South Korea*** City founded in the 18th century, 27 km (17 miles) south of Seoul. It produces textiles and chemicals.
Population 310 500
Map South Korea Cd

Suzdal' ***USSR*** One of the major principalities of Russia in the Middle Ages, now a small city and tourist centre 190 km (118 miles) north-east of Moscow. It dates from the 11th century, and among its architectural treasures are a 12th to 13th-century kremlin (citadel), several fortified monasteries, a 13th-century cathedral, several churches and a market square.
Population 10 000
Map USSR Fc

Suzhou ***China*** City of Jiangsu province south of the Chang Jiang delta. Dating from the 6th century AD, it is an ancient and attractive city crisscrossed by canals, sometimes called the Venice of China. It is famous for its fine silk embroidery.
Population 900 000
Map China Ie

Svalbard ***Norway*** Arctic Ocean archipelago covering 62 049 km² (23 958 sq miles) due north of mainland Norway. The name means 'the cold coast'. Discovered by Vikings in 1194, and then rediscovered by the Dutch navigator William Barents in 1596, Svalbard became a centre for international whaling in the 17th century. Norwegian sovereignty was recognised internationally in 1920 and became effective in 1925. The islands are now demilitarised, but signatories of the 1920 convention have the right to mining concessions – which is why there is a Russian coal-mining operation on the main island, SPITSBERGEN.
Population 3500
Map Arctic Rb

Sverdlovsk ***USSR*** Industrial city on the eastern fringe of the Ural mountains, about 1400 km (870 miles) east of Moscow. It was founded as an iron-making centre in 1721 by the emperor Peter the Great (1672-1725), and flourished after the building of the Great Siberian Highway (1783) and the Trans-Siberian Railway (1878). The last of the Russian tsars, Nicholas II (1868-1918), and his family were imprisoned and shot by the Bolsheviks in Sverdlovsk (then known as Ekaterinburg or Yekaterinburg) in 1918, the year after the Russian Revolution. Today it is the largest city of the Ural region; its industries include heavy machinery, electrical equipment, copper, chemicals and ball bearings.
Population 1 288 000
Map USSR Hc

Swabia ***West Germany*** Former province of south-west Germany south of the city of Stuttgart. It is now part of Baden-Württemberg and south-western Bavaria.
Map West Germany Cd

Swakopmund ***Namibia*** Resort at the mouth of the Swakop river about 260 km (160 miles) west of Windhoek. Once Namibia's chief port, it was superseded by Walvis Bay 32 km (20 miles) to the south. Many workers at the nearby Rossing uranium mine live there.
Population 13 500
Map Namibia Ab

swale See RUNNEL

swallow-hole See SINKHOLE

swamp A permanently waterlogged lowland region and its vegetation – often reeds. Swamps can form when a lake basin fills and the surface is so flat that the run-off of rainwater is very slow. In some permanently waterlogged areas, certain highly adapted trees can grow – for example, swamp cypresses in the Florida Everglades and mangroves on tropical coasts.

▼ 'IMPERIAL RIVER' The Southern Grand Canal, 400 km (250 miles) long, glides through Suzhou. It joins the Da Yunhe (Grand Canal) with the city of Hangzhou, and was built by the Zhou emperors in the 4th century BC.

Swaziland

ALMOST SURROUNDED BY SOUTH AFRICA, SWAZILAND HAS BECOME INCREASINGLY DEPENDENT POLITICALLY AND FINANCIALLY ON HER GIANT NEIGHBOUR

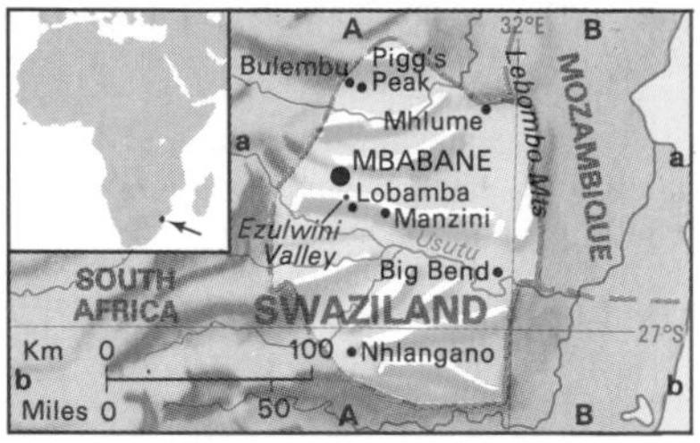

One of the three monarchies remaining in Africa, Swaziland is faced by an unusual mix of political and economical problems, as the country tries to balance the old and the new.

Surrounded on three sides by South Africa and, to the north, by Mozambique, this landlocked hilly enclave is well-watered, has a pleasant climate, and contains splendid scenery. The land drops in three great steps from west to east. The first, the highveld (over 900 m, about 3000 ft), consists of rugged mountains clothed in forestry plantations and deeply cut by streams and waterfalls. The middleveld (400-850 m, 1310-2790 ft), planted with orange groves and pineapple fields, falls away to the gently undulating, park-like eastern lowlands of the lowveld (150-300 m, 490-980 ft), where sugar-cane plantations flourish on the irrigated lands. The extreme eastern boundary is formed by the LEBOMBO MOUNTAINS (up to 825 m, about 2700 ft), indented by gorges and dividing Swaziland from Mozambique and the Indian Ocean.

Swaziland, which gained independence from Britain in 1968, has been an absolute monarchy since 1973, ruled by a king called the *Ngwenyama* (Lion). The death in 1982 of King Sobhuza II, who had reigned for 61 years, brought into the open the struggle between traditional and modern elements. For a time, the nominal head of the country was the Queen Regent, known as the Ndlovukazi (Great She-elephant) – one of the late king's 100 wives. However, she was no more than a figurehead – behind her lay the Liqoqo, the supreme tribal council of senior princes and chiefs. Then, in April 1986, the throne was taken by her son, the former Crown Prince Makhosetive – aged 18 and a pupil at a British public school. He immediately dispensed with the Liqoqo.

Just over half the land is held in trust for the 691 000-strong nation by the king, and it supports peasant farmers who produce chiefly maize and cattle to feed themselves and grow a few crops to sell for cash, such as cotton and tobacco. The rest is worked by commercial companies, who grow all the main export crops – sugar (the most valuable), timber, and citrus fruits – and mine the country's mineral resources, principally asbestos in the north-west and, until 1980, iron ore (now exhausted but in the past a major export). Manufactures include fertilisers, textiles, leather and tableware. Tourism also thrives, most tourists being South Africans attracted by the relaxed atmosphere and the spas and casinos.

Swaziland is overshadowed by the Republic of South Africa, with which it is in customs union. South Africa – whose president, P.W. Botha, attended the young king's enthronement – backs the Swazi traditionalist elite.

Cooperation with South Africa includes a non-agression pact signed in February 1982, which inhibits Swazis from sheltering South African political refugees, and the opening of a South African trade mission in 1984 – the first new South African diplomatic mission in a black African country for 19 years. Many Swazis still have hopes of a land deal with South Africa whereby parts of the tribal homelands adjoining Swaziland – Kangwane and Ingwavuma, with nearly 1 million Swazi people – will be ceded to them; these are some of the lands lost to European settlers in the 19th century. But the modern faction in the government and fellow nations of the Organisation of African Unity are highly critical of such a deal, which they say would allow South Africa to set aside its responsibilities to South African-born citizens and so would make Swaziland an ally of *apartheid*.

SWAZILAND AT A GLANCE

Area 17 400 km² (6705 sq miles)

Population 691 000

Capital Mbabane

Government Absolute monarchy

Currency Lilangeni = 100 cents

Languages English, siSwati (Swazi)

Religions Christian (70%), traditional religions (30%)

Climate Subtropical moderated by altitude; average temperature in Mbabane ranges from 6-17°C (43-66°F) in June to 15-25°C (59-77°F) in January/February

Main primary products Maize, cotton, rice, sugar, citrus fruits, tobacco, cattle, timber; asbestos, coal

Major industries Agriculture, forestry, mining, timber processing, chemicals, tourism

Main exports Sugar, wood pulp, chemicals, asbestos, citrus fruits

Annual income per head (US$) 740

Population growth (per thous/yr) 30

Life expectancy (yrs) Male 51 **Female** 56

Swansea (Abertawe) ***United Kingdom*** Port in the Welsh county of West Glamorgan and the second city of Wales after Cardiff, 55 km (35 miles) to the east. It stands at the mouth of the River Tawe and rose to prominence in the 19th century with the export of coal and the growth of industries such as tin plating in the Tawe Valley. The city has been heavily rebuilt since it was bombed in the Second World War. The poet Dylan Thomas (1914-53) was born and grew up in Swansea. The splendid scenery and beaches of the Gower peninsula lie just to the south-west.

Population 168 000

Map United Kingdom De

swash Rush of seawater up a beach following the breaking of a wave.

Swat ***Pakistan*** Former tribal state ruled by the Wali of Swat. It covers 7511 km² (2900 sq miles) in the mountains north-east of Peshawar. The region was self-governing until 1960. The family of the Wali has maintained opposition to many Pakistani governments.

Population 900 000

Map Pakistan Da

Sweden See p. 616

swell Regular undulating movement of waves out in the open sea, with no breaking and with considerable distance between successive crests.

Switzerland See p. 620

Sydney ***Australia*** Largest, brashest and oldest city in the country and the state capital of New South Wales. It is centred on the magnificent harbour, which was named Port Jackson (but never explored) by the English navigator James Cook in 1770, and now spreads north, west and south over 1735 km² (670 sq miles).

The city was settled in 1788 by 1487 people – 759 of them convicts – who had sailed in a fleet of 11 ships from Britain. Its historic buildings include St James's Church and Hyde Park Barracks, both designed by the convict architect Francis Greenway in the early 19th century. But its most spectacular features are more modern: the Centrepoint Tower, complete with revolving restaurant, at 305 m (1000 ft) the tallest building in the country; the shell-roofed Opera House, opened in 1973; and the Sydney Harbour Bridge, opened in 1932. The bridge, which carries two railway tracks, eight lanes of roadway and a cycle path and footway across a span of 503 m (1650 ft), is the world's widest long-span bridge. It links the city centre, south of the harbour, with the rapidly expanding second commercial centre of North Sydney, formerly a residential suburb.

The city draws thousands of tourists, who flock to its famous beaches, such as Bondi. Sydney is also Australia's busiest port, has three universities, and since the Second World War has developed an easy-going cosmopolitan flavour, with large numbers of immigrants from Britain, Italy, Greece, the Lebanon, South-east Asia and many other places.

Population 3 332 600

Map Australia Ie

Sydney ***Canada*** Port on Cape Breton Island, Nova Scotia, and a coal, iron and steel centre. Industries include mining equipment, marine and offshore engineering and electronics. It was recently designated a free trade zone.

Population 29 440

Map Canada Id

Syedlets ***Poland*** See SIEDLCE

A DAY IN THE LIFE OF AN AUSTRALIAN BUS DRIVER

In the early 1950s, Jim Arnold was among the thousands of young Britons who seized the chance of a new life 'down under' in Australia. Jim, then 21, had taken some savings and invested £10 (US$28 – a week's wages in those days) in the price of his assisted sea passage, and was full of high hopes.

The reality of Australia was disappointing at first. There was stiff competition for the best jobs and Jim tried a series of trades, including working on a building site, driving a long-distance truck and being a cook on a cattle station. Then he met Anne Carter at a barbecue in Sydney. They were engaged within a few months, and he settled down in his present job – driving a Sydney bus.

His wife's great-great-grandfather was a convict transported from Britain to Australia in the 1840s, a heritage of which she is quite proud. He was convicted for stealing a handkerchief.

The Arnolds live in a three-bedroomed bungalow in the eastern Sydney suburb of Bondi. Many of their neighbours are also immigrants. Jim became a naturalised Australian citizen 15 years ago; now his parents are dead he has no desire to live in – or even visit – Britain again.

The climate and relaxed lifestyle have won Jim over completely to his new country. With his salary of 17 000 Australian dollars (US$11 670), his overtime pay, and Anne's earnings from working part-time in a shop, they can live comfortably. Their son Greg studied engineering, and is now an engineer on oil-exploration projects in Queensland.

Jim's bus route begins at Watson's Bay, an affluent suburb with a splendid view up Sydney Harbour to the skyscrapers of the city centre. The 11 km (7 mile) route ends there – at Circular Quay, a terminal for buses, trains and ferries next to the dock where Jim himself first landed in Australia.

Jim would like to move out to more spacious surroundings – but not if it means going too far from the beach; unless he's on a very early shift, Jim takes a swim before breakfast at Bondi. He hopes to buy a house near one of the superb beaches that stretch for 25 km (16 miles) north of the harbour to the wealthy suburb of Palm Beach, and settle into a retirement of fishing, swimming and basking in the sun.

On Saturday evenings Jim and Anne often go to the South Sydney Leagues club, a huge social centre, for a meal, to watch the cabaret and put a few dollars in the 'pokies' – the poker machines or one-armed bandits. Jim, like most Australians, likes to gamble. 'We'd move to Palm Beach tomorrow if we won the lottery', he says.

Sylhet ***Bangladesh*** Town and district in eastern Bangladesh near the Indian border south of Shillong. The district, which covers 12 718 km² (4910 sq miles), is the home of Bangladesh's tea industry.

Population (district) 5 656 000; (town) 166 800

Map Bangladesh Cb

Sylte ***West Germany*** See FRIESIAN ISLANDS

syncline A downfold in the BEDROCK, in which rocks incline together from opposite sides.

Syra ***Greece*** See SIROS

Syracuse ***USA*** City in central New York state, about 230 km (145 miles) east of Lake Erie, to which it is linked by canal. There are salt springs nearby and the city was once a leading salt producer. Its manufactures now include electronics, electrical equipment, chinaware and pharmaceuticals.

Population (city) 164 200; (metropolitan area) 650 500

Map United States Kb

Syracuse (Siracusa) ***Italy*** Seaport on the east coast of Sicily about 50 km (30 miles) from its south-eastern tip. It was founded by Corinthian Greeks in 734 BC and grew rapidly to become the most notable city of the Hellenic Age outside Greece itself. At its peak it probably had more inhabitants – about 200 000 – than any other city in the world. Greek Syracuse covered not only the island of Ortygia – its centre today – but also large areas of the mainland. The mathematician Archimedes was born there in 287 BC. He was killed by the Romans when they seized the city in 212 BC.

Ortygia is the site of the extraordinary cathedral, moulded around a Greek temple and with an imposing Baroque façade; the Castello Maniace (about 1239); the Palazzo Bellomo (13th and 15th centuries); the Arethusa fountain; and the National Museum, which contains some of the finest Greek exhibits outside Athens. The greatest wonders of ancient Syracuse on the mainland include the Roman amphitheatre, the Greek theatre – the largest in existence – the *latomie* or ancient quarries, now planted with beautiful gardens, and the catacombs of San Giovanni. The Castello Euralio, constructed by the tyrant Dionysius II between 402 and 397 BC for defence against the Carthaginians, is 9 km (5.5 miles) to the north.

Apart from nearby petrochemical plants, tourism is the main industry, and the port exports locally produced wine, olive oil and citrus fruits.

Population 119 200

Map Italy Ef

Syrdar'ya ***USSR*** Central Asian river, 2860 km (1780 miles) long. Known at its source, in the Tien Shan mountains near the Chinese border, as the Naryn, it flows west between the cities of Tashkent and Samarkand, then north-west to the Aral Sea. Along the way it is dammed to provide electricity below Begovat, a steel town near Leninabad in the Soviet republic of Uzbekistan, and tapped to irrigate parts of the so-called Hungry Steppe north and north-west of Tashkent. The river was known to the ancient Greeks as the Iaxartes.

Map USSR Hd

▼ SEA, SUN AND SAND Serious surfers, sunbathers and those who just enjoy relaxing throng the beach at Bondi, a suburb of Sydney on the Pacific.

Sweden

A COUNTRY WHERE SOPHISTICATED TECHNOLOGY AND EFFICIENT INDUSTRY EXIST SIDE BY SIDE WITH THE WILD BEAUTY OF GREAT FORESTS AND MYRIAD LAKES

Now deep into its second century without war, with a living standard higher than that of the United States, and with one of the most extensive social welfare programmes on earth, Sweden is in many respects the envy of Europe. But earthly paradises do not come cheaply; the Swedes carry the world's heaviest tax burden, and their welfare services cost more than 70 per cent of the national budget. Some Swedes feel that social featherbedding has led to over-government, and are uneasily aware of a blandness and uniformity in the world about them. But no one could deny that there is a great deal of comfort too.

Sweden is a large, handsome, northerly country whose southernmost tip is on roughly the same latitude as the border between England and Scotland. Its northernmost tip, some 1500 km (1000 miles) away, lies well within the Arctic Circle. Much of the south is flat, much of the north mountainous, and the coastline is a bewildering scatter of 20 000 and more islands and islets. But to summon up a general picture, most of Sweden is an undulating plain, whose chief elements are coniferous forest and water, dark lakes scoured out of granite during the Ice Age. On the margins of the picture, however, there are some startling variations, such as beech and oak forests in the south while the northernmost latitudes are treeless. Moving north, the coniferous forests gradually give way to tundra-like landscape, which in summer at least is a colourful wilderness of bog, lichens and mosses illuminated by cotton grass, wild flowers and banks of cloudberries, bilberries and wild strawberries.

On the whole, the climate is extreme. Summers can bring heatwaves, but they are brief. Winters in the north are seven months long and in the south, three; to all areas they bring deep snow and ice.

FORESTS AND LAKES

About half the country is covered by forest; spruce and pine, with an admixture of birch – the national tree – being the usual pattern. In the past, Swedish woodlands were endangered by overcutting for charcoal, tar, timber and fuel. But today they are among the best maintained in the world, with more trees being planted annually than are cut – a policy that is enhanced by draining and planting some of the extensive marshlands. Huge areas of wilderness still remain, however, where flocks of migrant birds move in and out with the swing of the seasons and elks, roe deer and reindeer roam. In addition, an estimated 250 000 domesticated reindeer migrate between the fells in summer and the forests in winter. With such enormous populations, there is grave danger of overgrazing and serious damage to the forests. Consequently, tens of thousands of deer are culled annually.

In Sweden, land and water are everywhere intermingled. Off the coasts there are archipelagoes, great and small – the most extensive and varied being in the Stockholm area. The bald, storm-lashed archipelago of the west coast presents a striking contrast, while the shoreline of the Gulf of Bothnia, still rising from the compression of the Ice Age, produces new islands, skerries and shoals every two or three generations. There are islands too in the lakes of the interior, some of which – including VANERN and VATTERN – are among Europe's largest. The rivers that drain them have wild falls, torrents and rapids. The most spectacular rivers are found in Norrland, the northern two-thirds of the country. They issue from high-level lakes along the Norwegian border and race down parallel valleys to the Gulf of Bothnia, plunging over majestic waterfalls, most of which have been harnessed to produce electricity. The fish in the lakes, rivers and surrounding seas include such delicacies as North Sea and Baltic herring, salmon, lobsters, crabs and river trout.

FARMS AND MINES

Fewer than 8.5 million people live in this large country. The southern third has always been the most populous, while vast areas have been and remain thinly occupied. Most Swedish farmers are owner-occupiers, with the family providing the labour force; produce is sold through cooperatives. Dairy farming predominates, even though all cattle have to be housed during the winter. About 7 per cent of Sweden is cultivated, with the emphasis on fodder crops, grain and sugar beet. Southern farms are larger than those in the north, but northern forests, intensively worked for timber products, are far greater in extent than the woodlands of the south. Despite all this activity, only some 5 per cent of the population is engaged in farming, forestry and fisheries.

Forestry is now a separate industry, though traditionally it was a winter occupation for farmers. So too was mining, particularly around Bergslagen in southern Sweden. There, iron ore was extracted, smelted with charcoal and forged with the aid of water power. The introduction of new processes in the mid-19th century rendered many small foundries obsolete and iron and steel production were concentrated instead upon a few towns. Sweden is now one of the world's leading producers of iron ore, most extracted from the LAPPLAND mines of Küranavaara, Luossavaara and Svappavaara. Lapland ores are processed at LULEA in the north, while southern Sweden's principal steel plant is at Sandviken. Mines around Boliden on the Bothnian coast produce zinc, silver, gold and copper, though Europe's longest-worked copper mine at Falun expired, like the small foundries, in the 19th century.

The 19th-century industrial boom that transformed the steel industry also dragged wood processing from tiny plants powered by upriver waterfalls to large factories sited on navigable waterways. These large-scale enterprises produce a wide range of papers, wallboards, laminates, prefabricated units, chemicals and, of course, matches. The story of the match industry is told in the museum at JÖNKÖPING.

Sweden has very few deposits of coal or oil, so its industrial development depended very largely on the development of hydroelectric techniques. The demand for energy has increased phenomenally – a demand not least due to the entirely electrified state railway system. The trouble is that most of the principal sources of hydroelectric power are in the north of the country, while the greatest demand is in the south. Faced with this problem, Swedish engineers led the world in developing methods of transmitting electricity. Later, the installation of nuclear energy plants and imports of cheap oil provided the basis for a final phase in energy expansion; then public opinion swung against nuclear reactors at just about the same time as the mid-1970s oil crisis. Sweden has therefore had to find alternative sources of energy – imported coal from Germany and Poland, natural gas from Denmark and, potentially, from Norway.

INDUSTRIAL EXPERTISE

Sweden came rather late to the industrial field. But having adopted British and German methods of steel production in the late 19th century, together with mechanical and chemical processes for paper and pulp production, native Swedish inventiveness began to assert itself, though often in matters surprisingly warlike for such a peace-loving nation. Alfred Nobel invented dynamite and established an industrial empire before making amends with his prizes, while Bofors, a munitions factory in central Sweden, was known the world over for the light automatic cannon it produced. More praiseworthy, perhaps, was Sweden's impact upon the design of furniture and household utensils – clean, unfussy lines in glassware, porcelain, textiles and kitchenware.

Heavy industry also benefited from the well-coordinated energy programme, with electrical and engineering firms and their products acquiring world stature. Major shipbuilding yards grew up at the great ports of GOTHENBURG, MALMO and Uddevalla. The Husqvarna company pioneered sewing machines and motorcycles in the early 20th century, while the leap to international fame in automobile design was achieved in not much more than a generation. The success of the industry is apparent in the immense assembly plants of Volvo, Saab and Scania-Vabis. The Swedish aircraft industry too has gained a name for itself, especially for its single-seater fighters. As in most countries, specialised industries have become associated with particular places – ESKILSTUNA is best known for cutlery; Orrefors for glass; VASTERAS for electrical components; and BORAS for textiles. The nation's craftsmen have built a reputation for skilled workmanship, though this might seem to be somewhat endangered by the new Swedish expertise of robotics.

Since the Swedish home market is limited,

▲ THE BIG COUNTRY The sun glints on a tumbling river in Jamtland, in central Sweden. This is a crisp, clear-eyed 'away from it all' region of lakes, rivers, mountains, forests – and very few people.

large-scale industrial developments have only been possible by looking to the outside world. There, many major Swedish concerns have established subsidiary plants, some of which employ larger labour forces than the parent company at home. The country also sells its skills abroad – banking and insurance, statistical advisory services and engineering. Swedish engineers are particularly active in the Third World, building dams, hydroelectric power plants, high voltage transmission systems, timber processing plants, and silos for grain storage, as well as airfields, hospitals and public buildings. Not infrequently, the revenue derived from such projects exceeds that generated by Swedish companies at home.

Sweden's late arrival on the industrial scene had one distinct advantage; its towns were able to avoid the horrors of the slums and satanic mills that beset the first generation of manufacturing towns in other parts of Europe. Many Swedish towns are of ancient origin, and usually have at least one or two historic buildings, but most down the years have been devastated by fire, the plague of all timber-built settlements. When they were reconstructed, the Swedish love of order manifested itself in gridiron street plans. The width of streets, the size of residential lots and of public spaces were and are strictly regulated. In the larger cities, apartment blocks and tower blocks predominate; save in the suburbs, small terraced or detached privately owned houses are less common. A considerable number of towns have only a few thousand inhabitants and centre around a single enterprise – a mine, a softwood processing plant, a specialised factory. In the north, where rural life is harder, the drift to the towns continues, and from northern towns to those of the south. Unemployment is consistently higher in the north, and many 'northerners' feel that they are living in an underprivileged territory. Heavy investment in communications, subsidies for industry, the establishment of growth centres and salary adjustments ease, but do not cure, the situation.

A SEASONAL LIFESTYLE

To some extent, the problem is one of a small population in a very large country. On the other hand, it is the abundance of space that enables most Swedes to enjoy their particular lifestyle, and to create, for example, such great national parks as the 1940 km^2 (750 sq mile) Sareks, the largest such park in Europe. Just about every family has a car; many have a motor or sailing boat as well, and everywhere there is easy access to the countryside. Many families have a second home, whether a substantial villa or a prefabricated unit put together by a do-it-yourself buff. If winterised – that is, with water, fuel and sewage pipes buried beneath the reach of frost – these leisure homes can also be used for autumn hunting and winter skiing.

Life is geared very much to the seasons. Swedes make the most of the short summer, with school holidays running from early June until late August; during this period, business activities start and finish earlier in the day. The National Day, June 6, is a major summer occasion; but Midsummer Eve, when every community erects a tall cross decked with greenery, flowers and ribbons, and throws an all-night party, is summer's climax. As the nights lengthen, there are August crayfish parties to herald summer's close. Gloom begins to descend in November and deepens until the midwinter solstice, when south Sweden has about six hours of daylight, and north Sweden not much more than three. In Arctic Sweden, street lighting is never extinguished during the winter months and snow ploughs, snow blowers, defrosting machines and icebreakers stand by, constantly ready for action.

Each year a brave challenge is thrown into the face of darkness – the midwinter festivals of light. On the fourth Sunday before Christmas, Advent Sunday, candelabra burn in every window and Christmas trees are already aglow in town squares. On December 13, St Lucia's Day, girls put on crowns of lighted candles, representing St Lucia, the patron saint of light, and boys wear hats decorated with stars. The festive drink is *glögg*, mulled wine aflame with brandy – Christmas gingerbreads are baked, and tables are decorated with the red-capped trolls and straw goat of Yuletide. The festive fare is the Christmas cold table – the most extravagant of the *smörgåsbord.*

Temperatures drop to their lowest about the end of February, though March can be radiant with sunshine. Schools have skiing holidays, while skiing competitions, local and national, are held as the days lengthen. Meanwhile, the students begin to look forward to May Day Eve – Walpurgis Night – when they put on their white caps and everybody celebrates the arrival of spring with bonfires, dance and song.

Although Sweden is a byword for neutrality, she has had her share of military adventure. The Protestant king, Gustavus Adolphus, was the greatest commander of the Thirty Years' War (1618–48). But the country has not fought a war since 1814.

On the whole, Sweden is about as near to a model state as the 20th century is likely to see. It is technically advanced, efficiently organised, affluent, egalitarian, politically neutral. In practice it is not without its problems and paradoxes. Some Swedes object to the standardisation, state interference in private life, and high level of taxation that are necessary adjuncts to the welfare state.

Others fear that standards cannot be maintained and that the generous donation of 1 per cent of GNP that Sweden contributes to international aid may also have to be cut back. There are those too who are disturbed about possible racial tensions. A post Second World War influx of guest workers and refugees from Finland, the Baltic States, Italy, Greece and Yugoslavia, has had the result that one in ten of Swedish citizens have either been born abroad, or born of immigrant parents. A large majority clearly supports Sweden's neutral stance, though it is an armed neutrality with a large defence budget, military conscription and remunerative arms industry. Perhaps only in religious matters have the Swedes ceased to conform. No more than a small minority are still active churchgoers, though most citizens happily pay the tithes asked by the state church as an aid towards maintaining its splendid buildings. Nevertheless, the stern Lutheran ethic is firmly entrenched in the Swedish soul. From it stem the national virtues of discipline and hard work, and the value placed upon achievement.

The even tenor of Swedish politics was shaken severely in February 1986, when the Prime Minister, Olof Palme, was assassinated in a Stockholm street.

SWEDEN AT A GLANCE

Area 449 793 km² (173 665 sq miles)

Population 8 340 000

Capital Stockholm

Government Parliamentary monarchy

Currency Krona = 100 öre

Language Swedish

Religion Christian (94% Evangelical Lutheran)

Climate Cool temperate with short, hot summers and long, cold winters. Average temperature in Stockholm ranges from −5 to −1°C (23-30°F) in February to 14-22°C (57-72°F) in July

Main primary products Dairy products, cereals, potatoes, sugar beet, rapeseed, timber; iron ore, copper, lead, zinc

Major industries Engineering and electrical goods, motor vehicles, mining, timber products including wood pulp and paper, furniture

Main exports Engineering and electrical goods, timber, timber products, paper and wood pulp, motor vehicles, chemicals, iron and steel, petroleum products

Annual income per head (US$) 10 350

Population growth (per thous/yr) Static

Life expectancy (yrs) Male 75 **Female** 79

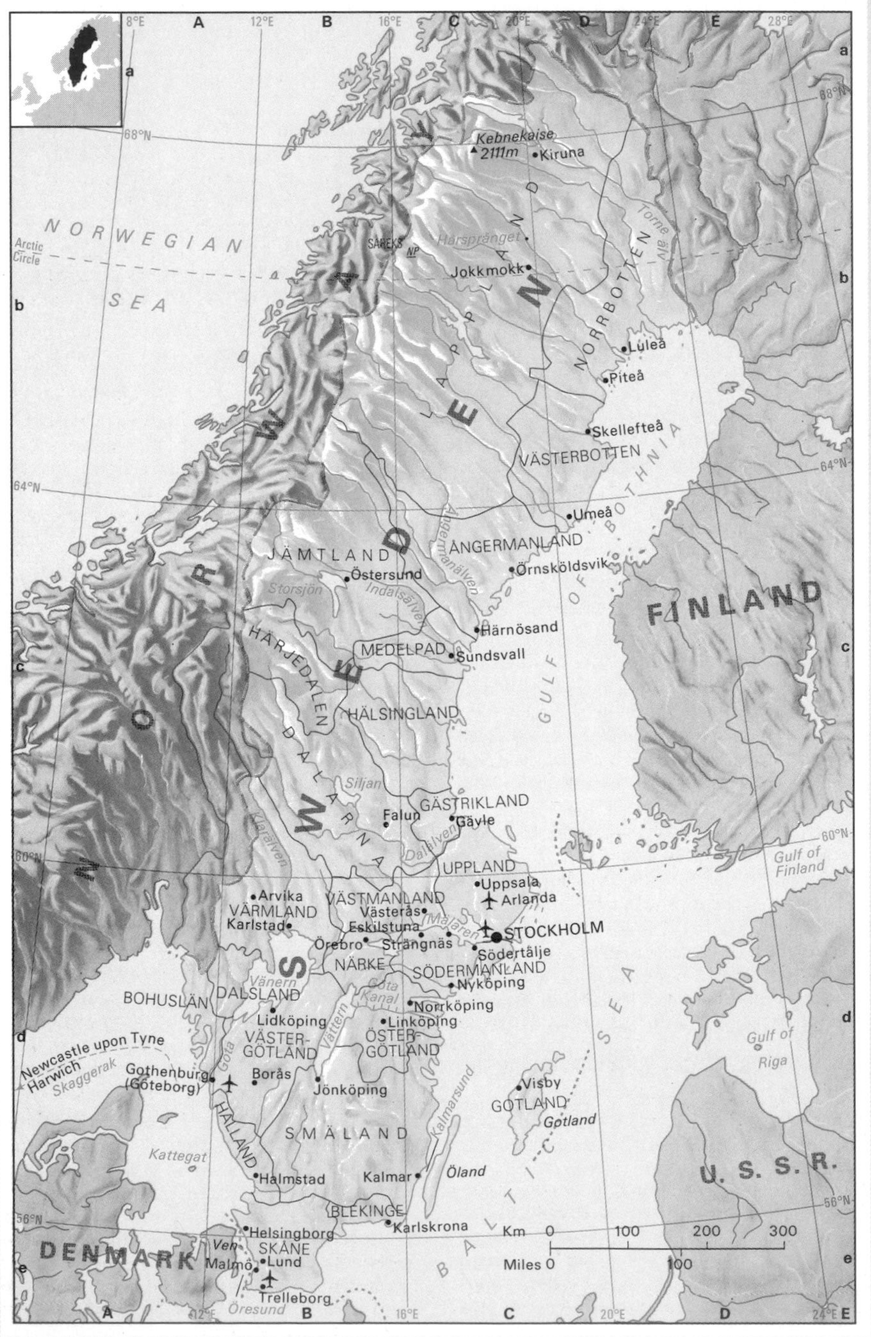

Syria See p. 622

Syriam ***Burma*** Oil town on the southern bank of the Rangoon river, opposite the capital. Formerly a Portuguese trading post, it is now the site of the country's main oil refinery.

Population 102 000

Map Burma Cc

system Succession of rocks formed during a geological period such as the Cambrian or the Triassic.

Szabadka ***Yugoslavia*** See SUBOTICA

Szantod-Rev ***Hungary*** See SIOFOK

Szava ***Yugoslavia*** See SAVA

Szczecin (Stettin) ***Poland*** Port in the north-west, on the Oder river 55 km (34 miles) from the open sea. It rivals Gdansk as Poland's leading Baltic port and shipbuilding centre. Between 1720 and 1945 it was the main port for the German city of Berlin. Today some of its cargoes are destined for East Germany – about 10 km (6 miles) to the west.

Population 390 200

Map Poland Ab

Szechwan ***China*** See SICHUAN

Szeged ***Hungary*** River port on the Tisza, some 10 km (6 miles) from the Yugoslav border, and the economic and cultural focus of the southern Great ALFOLD (Plain). The old town, extensively damaged by river floods in 1879, was rebuilt with concentric boulevards linked by avenues radiating from the inner city. The heart of the city is dominated by a neo-Romanesque cathedral (1912-29), one of the country's largest churches. An open-air festival of the performing arts has been held in the cathedral square virtually every July-August since 1913. Szeged has two universities, several scientific institutes and a theological college.

Long famous for its fiery paprika spice, fine restaurants and food-processing factories, Szeged now also has chemical plants, which use local and Romanian natural gas, and engineering and electrical works. Ujszeged (New Szeged), a resort across the Tisza, is noted for water sports.

Population 176 300

Map Hungary Bb

Székesfehérvár ***Hungary*** Chief city of the central county of Fejér, 65 km (40 miles) south-west of Budapest on the motorway to Lake Balaton. The kings of Hungary were crowned in the city from 1000 to 1527, and several are buried there. Székesfehérvár is a beautiful city with a fine Baroque cathedral (1758-78), churches, palaces and houses set among attractive parks and gardens. Long a market and commercial centre for a wine and fruit-producing district in the Bakony foothills, it has expanded rapidly since 1960 with the building of factories making aluminium, Ikarus buses, and electrical and telecommunciations equipment. Lake Velence, some 8 km (5 miles) to the east, has summer resorts and a wildfowl refuge.

Population 110 000

Map Hungary Ab

Szekszárd ***Hungary*** Chief town of the central county of Tolna, 137 km (85 miles) south of Budapest on the road to Pecs. Over the ruins of an 11th-century castle stands the old County Hall, an impressive palace built in the early 19th century. It lies in a region producing Szekszardi red wine, and close to the Gemenc wildlife reserve.

Population 37 000

Map Hungary Ab

Szolnok ***Hungary*** City on the Tisza river at the heart of the Great ALFOLD (Plain) 89 km (58 miles) south-east of Budapest. It has a fine Baroque church, but is best known for its water sports and river regattas. The city is the administrative centre of Szolnok county and a busy communications and commercial centre at the heart of a prosperous farming area. It has food processing and fertiliser factories.

Population 78 600

Map Hungary Bb

Szombathely ***Hungary*** City in the west of the country 18 km (11 miles) east of the Austrian border. As Savaria it was a capital of the Roman province of Pannonia, which covered the central Danube plains. The city has two large squares, with superb Baroque houses, cathedral and bishop's palace. It is the chief city of Vas county, a prosperous wine-producing area. Major new industries include chemical plants using local oil and natural gas as raw materials. The grounds of the ruined 2nd-century Roman temple of Isis to the south are used for the Savaria Festival of music each summer. Jak, 11 km (7 miles) south of Szombathely, has an outstanding 13th-century triple-aisled romanesque basilica restored in 1896.

Population 86 100

Map Hungary Ab

Sztalinvaros ***Hungary*** See DUNAUJVAROS

Taal ***Philippines*** See BATANGAS

Taban Bogdo Ula ***Mongolia*** See ALTAI

Tabasco ***Mexico*** Swampy and forested state covering 25 267 km² (9756 sq miles) in the south-east, on the Gulf of Campeche. Rains are exceptionally heavy and vast acres are often flooded by swollen rivers. Once only a source of timber, dyes, coffee, sugar and bananas, this jungle state was opened up for oil exploration in the late 1960s. Now Tabasco produces much of Mexico's oil, which is refined in the state capital, VILLAHERMOSA.

When the jungle was cleared for the oil industry, huge carved stone heads were uncovered. They were made by the Olmec Indians who settled in Tabasco about 1200 BC, and who revered the head as a symbol of power.

Population 1 200 000

Map Mexico Cc

Tabgha ***Israel*** Traditional site of Christ's miracle of the Feeding of the 5000, about 10 km (6 miles) north of Tiberias, on the west shore of the Sea of Galilee. The 6th-7th century Church of the Multiplication has a Byzantine mosaic depicting a basket containing loaves and two fishes.

Map Israel Ba

Table Bay Inlet in south-western Africa, overlooked by TABLE MOUNTAIN, discovered around the end of the 15th century by Portuguese navigators seeking a sea route to India. In 1652 the Dutch founded a settlement on the bay, Kaapstad (CAPE TOWN), which became the cradle of modern South Africa.

Dimensions 5 km (3 miles) long, 10 km (6.2 miles) wide

Map South Africa Ac

Table Mountain ***South Africa*** Flat-topped sandstone mountain reaching 1086 m (3563 ft) and overlooking Cape Town. It is visible from ships some 150 km (95 miles) away, and is often seen swathed by its 'tablecloth' – a layer of cloud that rolls over the top when a south-east wind blows, and falls down the precipitous northern side. A cable railway, built in 1929, runs to the top.

Map South Africa Ac

Tábor ***Czechoslovakia*** Town in Bohemia 76 km (47 miles) south of Prague. It was founded in 1420 by followers of John Huss (about 1369-1415), the Czech religious reformer who lived for several years at nearby Kozí Hrádek castle. The town was the Hussite capital during the wars in which Huss's followers sought to conquer Bohemia, and the 15th-century town hall is now a Hussite museum.

Population 33 800

Map Czechoslovakia Bb

Tabor, Mount ***Israel*** Mountain about 10 km (6 miles) east of Nazareth, overlooking the Jezreel Valley. It is the traditional site of the miracle of the Transfiguration, when Christ's face 'shone like the sun and his garments became white as light' (Matthew 17:2). At the summit, 588 m (1929 ft) high, there is a Franciscan church and hospice and a Greek Orthodox church. There are also numerous remains of Crusader fortifications.

Map Israel Ba

Tabora ***Tanzania*** Regional capital about 340 km (210 miles) east of Lake Tanganyika. It was founded by the Arabs in 1820, and trades in millet, groundnuts (peanuts) and cotton.

Population 100 000

Map Tanzania Ba

Tabriz ***Iran*** Provincial capital of eastern Azarbaijan, about 550 km (350 miles) north-west of the national capital, Tehran. Set among high mountains, Tabriz possibly dates back to Assyrian times and is known to have existed in the days of the Macedonian conqueror Alexander the Great (356-323 BC). Earthquakes have destroyed it several times. Today it is a commercial and industrial city, manufacturing textiles, carpets, leather and soap. Farms in the surrounding province produce almonds and dried fruit.

Population 853 000

Map Iran Aa

Ta-chia ***Taiwan*** River of west-central Taiwan. It rises in the northern part of the central mountains and flows westwards about 140 km (90 miles) to the Taiwan Strait. The fast-flowing waters of its upper reaches have been harnessed for hydroelectric power.

Map Taiwan Bb

Switzerland

INDEPENDENCE AND NEUTRALITY, THE CORNERSTONES OF THE NATION'S POLICY, HAVE BROUGHT PEACE AND PROSPERITY TO THE SWISS

To the Swiss, it seems that everybody else wants to be Swiss too. Not permanently, perhaps: but long enough to take advantage of the benefits provided by the country. The Italians want to work there. The French and many others want to keep their money there. The Americans want to set up international institutes there. And the British want their daughters to learn French there. Meanwhile the Swiss work steadily at creating for themselves one of the world's highest standards of living. National income per person is US$14 066, and the gross national product is increasing by 2.3 per cent annually. Though primarily an industrial economy, with machinery, chemicals, watches, instruments and textiles among the main exports, Switzerland has huge earnings from international finance and tourism. Switzerland has become a leading centre for international banking and insurance. There are some 1750 banks and financial institutions. Apart from the country's economic stability, the great attraction for foreign customers is Swiss banking secrecy.

Being Swiss is a less definite state of affairs than being, say, French or Italian. For Switzerland is a confederation of 23 cantons, or sovereign states (three of which are divided into 'half-cantons'), and people are citizens first and foremost of their canton. There is even a word for this cantonal loyalty – *Kantönligeist*. It is the canton which issues a residence permit or runs a university, 'Swiss' refers to a ski team, to a railway system; above all (and this is probably the greatest unifying force in the nation) to an army – indeed, one of the world's most remarkable armies, in which every Swiss male serves actively or on the reserve until he is 50 (55 for officers). So as to be prepared, every Swiss soldier keeps all his equipment, arms and ammunition at home.

Nor do the Swiss intend to be unprepared for nuclear war. Underground nuclear shelters will be provided for the entire population by AD 2000.

There is no Swiss language by which to identify Swiss people. There are three major ones in use: the most widespread is German (73.5 per cent of the population), followed by French in the west (20 per cent) and Italian in the south (4.5 per cent). There is even a fourth official language: it is called Romansch and is spoken by 1 per cent, mainly in the ENGADIN. Most Swiss people will respond politely if addressed in any of the three main languages, but they will talk to one another in a quite different language – Swiss German.

Such are the attractions of living in Switzerland that nearly 15 per cent of the population are foreigners. A further 100 000 workers cross the frontier each day, mainly from Italy, to work in Switzerland. In addition, another 100 000 seasonal workers are admitted for a few months at a time, mainly to work in hotels and cafés. To bring in his family, however, a worker must have at least an annual work permit, and it is the holders of these who, after long probation, can eventually become Swiss. In 1984, they numbered 956 000 out of a population of 6.4 million. Yet the national unemployment rate has for years been below 1 per cent.

Switzerland began in 1291 as a defensive alliance between three cantons – Unterwalden, Uri and Schwyz (from which the country takes its name). The other 20 joined the league over succeeding centuries, to form what became in 1848 a federal republic. GENEVA, VALAIS and NEUCHATEL joined only in 1815; Jura, a new canton, was established in 1979.

The system of government is federal: the capital is BERNE and it houses the assembly, the council of states and a seven-man federal council, one member of which acts as president of the Swiss Confederation (the nation's official name) each year. But the cantons cling to their individual rights, and only such central functions as control of the army, the railways or the postal service are left to the federal government.

The nation was born out of resistance to Austria and its Hapsburg rulers: it fought off various takeover bids by Savoy (1559) and France (during the Napoleonic Wars) and survived unscathed the boast by Germany's Adolf Hitler that, after his armies had conquered Europe, he would take Switzerland with the Berlin fire brigade.

Independence and neutrality – guaranteed in perpetuity by international treaty in 1815 – are the essentials of Swiss policy. To defend that neutrality the Swiss have a formidable army, although Swiss neutrality has been far too useful to other nations for any country, however aggressive, to find it worth violating. For example, Switzerland or Swiss-based agencies have acted as go-betweens in the exchange of prisoners-of-war, the release of hostages, and in times of war to protect and help military and civilian victims.

Probably no other nation has made quite such a business out of neutrality as the Swiss. Ever since 1864, when the International Red Cross was initiated there, Switzerland has become the natural place for international organisations to establish their headquarters – there are now more than 150 of them. One of these, the ineffective League of Nations (1919), provided Geneva with a splendid white elephant of a building which, after the Second World War was filled with branches and offices of the League's successor organisations – many of them associated with the United Nations. Other international bodies in Geneva include the World Council of Churches and World Wildlife Fund.

THE FACE OF THE LAND

To many people, Switzerland is the country of the ALPS, though not all of it is mountainous. Northern Switzerland, like neighbouring regions of eastern France and south-west Germany, is a land of hills and woods but also of cities and industries. BASLE is world famous for pharmaceuticals and ZÜRICH and its suburbs for electrical engineering and machinery. It is non-alpine Switzerland that produces the cheese, the chocolate, the clocks and the watches for which the country is renowned.

The Alps occupy the southern half of the country. They form two main east-west chains, divided by the straight line of the upper valleys of the RHONE and RHINE. The northern chain is wholly in Switzerland, and contains such peaks as the EIGER (3970m/13 025 ft) and JUNGFRAU (4158 m/13 642 ft). The southern chain contains the highest summits, but these Switzerland shares with its neighbours – the highest peak of Mont BLANC (4807 m/15 771 ft) lies just in France, and the MATTERHORN (4478 m/14 690 ft) and the Dufour Spitze (4634 m/15 203 ft) – the highest peak of Monte ROSA – lie on the Italian border.

In alpine Switzerland, snowfields feed numerous glaciers, of which the largest is the ALETSCH (23.6 km/15 miles long), and winter sports traffic turns small towns like DAVOS and ZERMATT into large ones during the season. The high alpine passes – FURKA, GRAND ST BERNARD, ST GOTTHARD and SIMPLON – play a major role in the life of the region.

The Alps also influence Swiss life through their impact on climate. They divide the Mediterranean world from the central European. Going south, to emerge from the Gotthard or Simplon rail tunnels is to enter a new environment. Italian Switzerland is a land of lakes and subtropical vegetation, distinctive in architecture as well as in speech. Going north there is less sunshine, and also an uncomfortable hot, dry south wind – the Föhn – blows periodically.

There are many regional variations in climate. But generally the mountain air is clear and clean – a factor that has resulted in Switzerland attracting invalids from all over the world to its hospitals and clinics.

North of the Alps lies the Swiss lowland with its many lakes, its agriculture (here are the homes of Gruyère and Emmenthal cheese) and the bulk of the Swiss population. The region which lies between Basle, Zurich and WINTERTHUR is heavily industrialised.

The third and smallest section of Switzerland is formed by the mountains of the JURA, which it shares with France. This region lies north-west of Lakes Geneva and Neuchâtel and comprises a number of steep, forested ridges running from south-west to north-east and reaching 1400-1600 m (about 4600-5250 ft). In the heart of the Jura lies CHAUX-DE-FONDS, centre of Swiss watchmaking.

In overcoming the problem of linking a nation together in the face of the many physical obstacles, the Swiss have produced a brilliantly engineered and superbly integrated transport system. A visitor may arrive on a train, and then change to a lake steamer, cable car or post bus: everything links up with everything else. Special techniques employed include spiral tunnels, which corkscrew up through mountains like spiral staircases.

The hotel and catering trades employ 180 000 workers, looking after the needs of tourists by the million.

SWITZERLAND AT A GLANCE
Area 41 293 km² (15 943 sq miles)
Population 6 510 000
Capital Berne
Government Federal republic
Currency Swiss franc, or franken = 100 centimes or rappen
Languages German, French, Italian, Romansch
Religion Christian (48% Roman Catholic, 44% Protestant)
Climate Warm summers, cold winters; average temperature in Zurich ranges from −3 to 2°C (27-36°F) in January to 13-24°C (55-75°F) in July
Main primary products Wheat, barley, potatoes, grapes, apples, livestock, timber; salt, building stone
Major industries Tourism, machinery, chemicals, pharmaceuticals, banking and insurance, clock and watchmaking, instruments, textiles and yarns, forestry, paper and wood pulp, cement, iron and steel
Main exports Machinery, chemicals, pharmaceuticals, precious metals and jewellery, clocks and watches, instruments, textiles, food
Annual income per head (US$) 14 066
Population growth 2
Life expectancy (yrs) Male 74 **Female** 78

▼ HAPPY VALLEY Prosperity beams like the sun on Lauterbrunnen, cosily enfolded in its valley in the Bernese Oberland. The Breithorn (3782 m, 12 408 ft), in the background, is only 11 km (7 miles) from the mighty Jungfrau, Mönch and Eiger.

Syria

A MUSLIM LAND WHERE CHRISTIAN CRUSADER CASTLES DOT THE LANDSCAPE AND ARAB NATIONALISM IS A MAJOR EXPORT

Armed with Russian missiles and revolutionary fervour, Syria holds the key to any lasting settlement between Israel and her Arab neighbours. Though Egypt has made its peace with Israel, Syria has not. Its troops, who lost the commanding GOLAN HEIGHTS in the Six Day War with Israel in 1967, now occupy parts of Lebanon. Here, Syria has manipulated the rival political and religious factions to ensure the virtual destruction of the Palestine Liberation Organisation (PLO); like Israel, Syria does not want a revolutionary Palestinian state as a neighbour.

People have lived in the capital, DAMASCUS, in unbroken succession since 2500 BC; it is the oldest continuously inhabited city in the world. From the coast the Phoenicians (9th-4th centuries BC) set out to trade in their ships, built from the cedars of Lebanon. St Paul was converted to Christianity on his way to Damascus and his refuge, Hanania's House, still stands off the Via Recta, the Street called Strait.

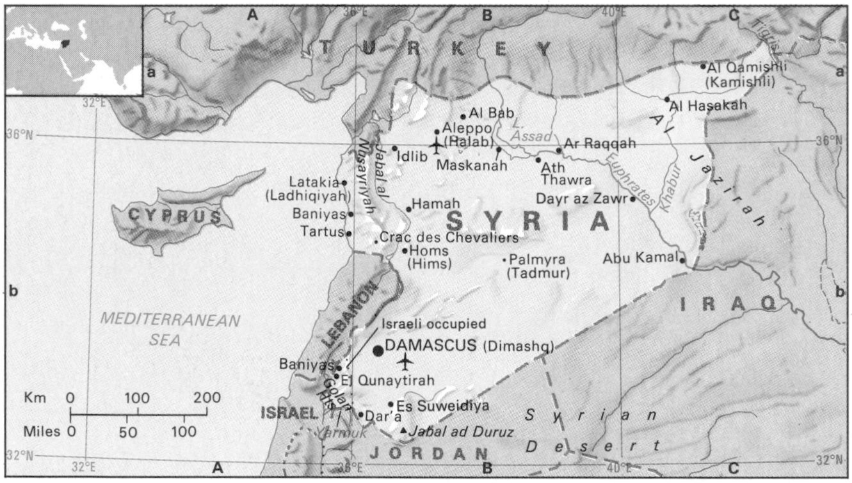

Today Syria is one of the most nationalist and radical of Arab states. Agriculture still flourishes in the Syrian part of the rich FERTILE CRESCENT formed by the ORONTES and EUPHRATES rivers, but now there is oil (though not a lot) and Arab nationalism to export.

THE LAND AND ITS PEOPLE

Much of Syria is mountainous. Behind the narrow, fertile coastal plain, the mountain range of JABAL AL NUSAYRIYAH, scored by deep valleys, drops eastwards to the GREAT RIFT VALLEY which continues to the Red Sea and into Africa. Farther south, the ANTI-LEBANON range rises to Mount HERMON (2814 m, 9232 ft). Most of the eastern part of the country is desert or semidesert, a stony and inhospitable land.

The coast has warm dry summers and cool wet winters; inland, the climate becomes hot and there is little rain. In the west, average temperatures range from 8°C (46°F) in January to 32°C (90°F) in July, but in the eastern desert temperatures can soar to 46°C (115°F). In complete contrast, snow falls on the mountains in winter.

Vegetation varies from Mediterranean scrub – tamarisks, camel thorns and acacias – to coniferous forests in the highlands; much of the country behind the coastal range is steppeland that turns green after the sparse rains encourage grasses to grow. Wildlife includes gazelles, wolves and wild cats; eagles, buzzards, kites and falcons are found in the mountains.

Ninety per cent of Syria's people are Arab, but the country presents an intriguing mixture of communities, languages, religions and customs. Some of the minorities can trace their histories back long before the coming of the Arabs in the 7th century.

The Kurds of the north make up 6 per cent of the population, and have their own language, culture and national dress. There are Armenians in ALEPPO, Circassians (mountain people like the Kurds) in the south-west, Assyrians from Iraq, Turkmans from central Asia, and the nomadic Bedouins of the desert. Many Arab Palestinians live in camps around Damascus, and there is a small Jewish community in the capital.

Nearly all Syrians are Muslims. The orthodox Sunni Muslims outnumber the Shiites (the sect that holds sway in Iran) by about seven to one, but it is the small Alawite sect, an offshoot of Shiism, that controls the government. Though President Assad is Iran's ally, he tolerates no Islamic militancy: alcohol is freely available and Western dress is common among men and women in the cities. In the south, the secretive Druse sect practise a religion containing both Muslim and Christian elements. There are about a dozen Christian communities, the strongest being the Greek Orthodox.

Half Syria's work force get their living from agriculture; 31 per cent of the land is arable and another 45 per cent provides some form of pasture, though this becomes increasingly poor on the edges of the desert. Agriculture produces one-fifth of Syria's income. Cotton, barley, wheat, tobacco, fruit and vegetables are grown; sheep, goats and cattle are raised; mules and camels are beasts of burden. The Euphrates dam at Al-Thaura, built with the help of Russian technology, now feeds a vast reservoir, Lake ASSAD, which has enabled large-scale irrigation and also supplies 70 per cent of the country's electricity.

MINERAL WEALTH

Syria's oil reserves are small by Middle East standards – about 1520 million barrels. Production is only one-sixteenth of neighbouring Iraq's, but it is enough to make the country self-sufficient and provide three-quarters of the nation's export earnings. Most of the oil comes from fields round Karachuk in the north-west, while the Al Jozirah region in the north-east is yielding natural gas – reserves are nearly 20 000 million m^3 (700 000 million cu ft). Other minerals found include limestone, phosphates (around Al Shargiya and at Khneifis), salt, manganese, iron, gypsum and asphalt.

In the last 20 years Syria has diversified and greatly expanded industries such as textiles, leather, chemicals and cement with the help of its oil revenues. Most of the industrial growth has taken place in the major towns, such as Damascus, HOMS, HAMAH and Aleppo. These have drawn people from all over the country. Indeed, half the population are now town dwellers and there has been a rapid growth of slums and shanty towns, especially in Damascus and Aleppo.

Nearly half the population is under 15, so there is great pressure on schools; Syria's population increase (4 per cent a year) is one of the world's highest. Education is free and one-third of university students are women. The adult literacy rate is 50 per cent. Health facilities have improved in recent years: there is now one doctor to every 2300 people, a figure comparable with several other Middle East states but lower than the developed Western countries. The average life expectancy is 65 years, but 6 per cent of children still die in their first year.

BATTLEGROUND OF THE CRUSADES

Syria has been a trading crossroads since ancient times. The Phoenicians were great seafaring traders and overland trade routes extended to Arabia, India and China; even today oil pipelines across Syria follow ancient trading routes. Syrian traders are found throughout the Middle East and in parts of Africa, where the word 'Syrian' is a synonym for 'merchant'.

After the Phoenicians, Syria came in turn under the Greek, Roman and Byzantine empires. Then, three years after the death in AD 632 of the founder of the Muslim faith, the Prophet Muhammad, two Arab armies took Damascus. It became the most important political and military base in an Islamic empire ruled by descendants of Umayya, the Prophet's great-great uncle. From 661 until 750, 14 successive caliphs of the Umayyad dynasty held sway in Damascus.

In 1095 Christians in Europe launched the first of their crusades. Crusader knights, sworn to recover the Christian holy places from the Islamic infidels, landed in Asia and in 1099 took Jerusalem. For two centuries, wars and battles sparked off intermittently, during which period the Crusaders built a defensive chain of superb castles along the coast, many of which are still standing.

The Syrians fought back; in 1114 Damascus beat off an attack and Sultan Nur ad-Din (1146-74) captured many of the Christians' northern strongholds. In 1169 he took Cairo. His son Salah ad-Din (the famous Saladin) recaptured Jerusalem in 1187. Five years later peace came and Saladin returned to Damascus in triumph. But he had taken no share of the spoils and died in poverty; his friends had to borrow to pay for his funeral.

For two centuries Syria was united with Egypt under the rule of the Turkish Mamelukes, then, in 1516, it was absorbed into the Turkish Ottoman Empire. At that time, Syria included the territory that is now Lebanon. When Turkey sided with Germany in the First World War, the Allies invaded Syria, and after the war the Levant (Syria and Lebanon) became a French mandate. In 1926 Lebanon became a separate state, taking with it the important ports of BEIRUT and TRIPOLI. Twenty years later, Syria became independent.

LEADING THE ARAB REVIVAL

Since then, Syria has played a leading role in the Arab revival that has been so marked a feature of Middle East politics. In 1958 it joined Egypt to form the United Arab Republic, but broke away after a coup in 1961. Other coups followed until Hafez al Assad came to power in 1970.

Under President Assad, Syria has achieved reasonable stability despite war with Israel, a bitter feud with Iraq, hostility towards Jordan (for her more pragmatic policies towards Israel) and intervention in Lebanon. Although defence expenditure is heavy (65 per cent of the Syrian budget), the country seems able to afford it.

Syria has always been hostile to Israel. The Six Day War was followed by another war in 1973. The government sent 20 000 troops into Lebanon as a 'peace-keeping force' in 1976; they remain in a commanding position there. Syria's pro-Soviet stance has strained relations with the United States, especially after attacks on the US embassy and marines in Lebanon in 1983 were attributed to Syrian extremists. Plans for union with Libya, announced in 1980, have not taken effect.

The country is governed by a people's council of 195 members. The ruling National Progressive Front, a coalition of the Ba'ath and four other parties, is opposed by the National Alliance of the Syrian People, which represents 19 parties.

Syria has fostered tourism and has the hotels and other facilities to make visitors welcome. In a good year, 2 million people may come to enjoy the climate, the scenery, the historic sites and the modern places of interest.

FOR THE VISITOR

Damascus is mentioned in early Egyptian records, and legend says it was the town of Shem, son of Noah. The walled city lies in an oasis where the Barada river descends like a torrent from the Anti-Lebanon mountains. Today, with a population of 2 million, it is a blend of the old and the new, with the Great Mosque, the Citadel and Saladin's tomb to see. The site of the Umayyad Mosque has been a place of worship for at least 3000 years, serving Aramaeans, Romans, Byzantines and Arabs in turn. The present building was erected as a Christian church for the Romans under Emperor Theodosius I in AD 379, but was rebuilt as a mosque in 705 after Arabs had taken Damascus.

Aleppo, with a population of 750 000, was a key town on trade routes for thousands of years. It has a citadel, nearly 200 minarets, a 13th-century royal palace and Syria's biggest *souk* (market).

Tourist villages are being developed on the Mediterranean at Ras Al-Basit, near the port of LATAKIA. Along the coast are some of the great Crusader castles: the CRAC DES CHEVALIERS, on a hilltop near Homs, is a classic example. There are mountain resorts at Zebadani and Bloudan, near Damascus. In the desert are the remains of the city of PALMYRA.

Syrian cuisine includes many *mezza*, or hors d'oeuvres. *Burgol* (boiled, crushed wheat like semolina) is a staple food; often it is rolled into balls which are stuffed with a mixture of minced meat, onions and nuts. Minced meat dishes include *kibbe mechwiye* (with wheat) and *kafta antakiya* (with parsley and lemon). *Shish taouk* is chicken on a skewer with mushrooms or truffles, and *chawarma* consists of large roasted pieces of mutton.

▲ ONCE PROUD CITY The Grand Arch stands amid the ruins of Palmyra. The city grew in an oasis on the edge of the Syrian desert, and prospered from trade under the Romans. However, the conquerors destroyed Palmyra in AD 273 after its queen, Zenobia, defied them.

SYRIA AT A GLANCE
Area 185 180 km² (71 498 sq miles)
Population 11 000 000
Capital Damascus
Government Socialist republic
Currency Syrian pound
Language Arabic
Religions Muslim (90%), Christian (8%)
Climate Mediterranean on the coast; hot and dry inland, cold winters in highlands. Average temperature in Damascus ranges from 0-12°C (32-54°F) in January to 18-37°C (65-99°F) in August
Main primary products Sheep, goats, cattle, cotton, fruit, potatoes, sugar, wheat, barley, vegetables; crude oil, natural gas, crude asphalt, phosphates, salt, manganese
Major industries Agriculture, oil and gas production and refining, textiles (wool and cotton), cement, flour, soap, leather goods, glass, metal goods
Main exports Oil products, natural gas, cotton, fruit, vegetables, barley, wool, wheat
Annual income per head (US$) 1800
Population growth (per thous/yr) 40
Life expectancy (yrs) Male 64 **Female** 67

Tacloban ***Philippines*** See LEYTE

Tacna ***Peru*** City and department on the border with Chile. Its arid lands are now being irrigated to produce crops of grapes and olives. The city lies in the shadow of the Campo de la Alianza, the site of a decisive battle between Peru and Chile during the War of the Pacific (1879-84), and was under Chilean rule from then until 1929. Its fountain and cathedral were designed by the French engineer Gustave Eiffel (1832-1923), who built the Eiffel Tower in Paris.
Population (department) 143 100; (city) 97 200
Map Peru Bb

Tacoma ***USA*** Port in Washington state. It lies on Puget Sound about 40 km (25 miles) south of Seattle, and has timber and nonferrous ore processing industries.
Population (city) 159 400; (metropolitan area) 515 800
Map United States Ba

Tadzhikistan ***USSR*** Central Asian constituent republic of the USSR near the Afghan and Chinese borders. It is dominated by the snow-capped Pamir mountains; only 7 per cent of the republic's 143 100 km² (55 250 sq mile) land area is below 1000 m (3280 ft) and more than half is over 3000 m (9850 ft). The lowland area is now irrigated, mainly in the FERGANA and AMUDAR'YA valleys, and used to grow cotton, mulberry trees (for silkworms), fruit, wheat and vegetables. Hydroelectricity from tributaries of the Amudar'ya powers cotton and silk mills, food factories, mines and smelting works. The republic is rich in minerals such as coal, zinc, lead and molybdenum, and oil and gas. Its chief towns are the capital DUSHANBE, LENINABAD and Ura-Tyube.

Tadzhikistan was invaded by the Persians in the 6th-7th centuries BC, and subsequently by the Greeks (under Alexander the Great), Arabs, Tatars and Mongols. In the 16th century it was ruled by Bukhara, and was taken over by Russia in the late 19th century. Its people are predominantly Muslim.
Population 4 365 000
Map USSR Ie

Taegu ***South Korea*** The country's third largest city – after the capital, Seoul, and the port of Pusan. It lies about 95 km (60 miles) north and slightly west of Pusan. Taegu is the country's chief textile centre, making silk, cotton, woollen and synthetic fabrics. In 1950, during the Korean War, the advance of the North Korean armies was checked just to its north.
Population 1 604 900
Map Korea De

Taejon (Daejeon) ***South Korea*** Modern city about 150 km (95 miles) south of the capital, Seoul. Destroyed during the Korean War (1950-3), it has been rebuilt as a manufacturing centre for textiles, ceramics and furniture.
Population 651 800
Map Korea Cd

tafelberg See TAFELKOP

tafelkop In South Africa, a mountain with a flat top. Large tafelkops are known as tafelbergs.

Tafilalt (Tafilet) ***Morocco*** Largest Saharan oasis in Morocco, comprising 1375 km² (530 sq miles) of fortified villages and groves of date palms stretching for 50 km (30 miles) along the Ziz valley near Erfoud. In AD 757 the Berbers established an independent kingdom here, with its capital at Sijilmassa on the caravan route from the NIGER river to the Mediterranean. Sijilmassa was destroyed by Arab invaders in 1363, refounded in the 17th century by Emir Moulay Ismail (from whom the present Moroccan king descends), then devastated by nomadic tribesmen in 1818. The town of Rissani stands near the ruins of the city. Tafilalt is noted for its dates, which it exports worldwide.
Population 70 000
Map Morocco Ba

Taganrog ***USSR*** Seaport and industrial town on the Sea of Azov, east of the Crimea. It ships grain and has a steel industry, shipyards and commercial fisheries. Founded as a fortress and naval base in 1698 by Tsar Peter the Great, it was occupied by the Turks in 1730-69. The Russian playwright Anton Chekhov (1860-1904) was born there; his birthplace is now a museum.
Population 289 000
Map USSR Ed

Tagaytay City ***Philippines*** See CAVITE

Tagus (Tajo; Tejo) ***Spain/Portugal*** Longest river in the Iberian peninsula, rising in eastern Spain and flowing 1001 km (626 miles) south-westwards through the Portuguese capital, Lisbon, to the Atlantic.
Map Portugal Bc

Tahiti ***French Polynesia*** Largest island of the territory, covering 1042 km² (402 sq miles), and main island of the Society group. It is mountainous, reaching 2241 m (7352 ft) at Orohena, and covered in dense vegetation and tropical flowers. Coral reefs fringe the island, which was the original South Seas paradise island. Its beauty has been immortalised by the painter Paul Gauguin (who lived there 1895-1901) and such writers as Pierre Loti and Herman Melville. It is a popular tourist destination, and also produces coconuts, vanilla and tropical fruits. The capital is PAPEETE. The smaller but equally beautiful island of Moorea lies 16 km (10 miles) to the west.
Population 96 000
Map Pacific Ocean Fc

Tahoe, Lake ***USA*** Lake covering 500 km² (193 sq miles) astride the Nevada-California state border about 250 km (155 miles) north-east of San Francisco. It is a popular tourist area near the gambling centre of Reno.
Map United States Bc

Tai Po ***Hong Kong*** City in the NEW TERRITORIES, stretching westward from a sea inlet known as the Tolo Harbour. Originally a small market centre for the surrounding region, it has been developed, much of it on reclaimed land, since 1973 as one of the new towns aimed at relieving population pressure in the central built-up areas of the colony.
Population 116 000
Map Hong Kong Ca

Tai Shan ***China*** Mountain in Shandong province 85 km (52 miles) south of Jinan city in the Shandong Peninsula. It is the highest peak in the peninsula at 1545 m (5100 ft) and is one of the five sacred mountains of the Taoist religion.
Map China Hd

T'ai-chung (Taizhong) ***Taiwan*** City and port, about 125 km (80 miles) south-west of T'ai-pei. Developed as a planned settlement during the Japanese colonial period (1895-1945), it has emerged as an expanding industrial city since the mid-1960s. Construction of its port, on the coast 20 km (12 miles) west of the main built-up area, began in 1971; although incomplete, it handles a high volume of cargo. T'ai-chung has an export processing zone where foreign companies manufacture products ranging from electronics and optical goods to plastics.
Population 671 000
Map Taiwan Bb

Taif ***Saudi Arabia*** Mountain resort situated 60 km (37 miles) south-east of Mecca. Some fine old Arab houses remain near the souk, which sells traditional Arab clothes made locally. Its streets are lined with pepper trees. Set 1700 m (5577 ft) above sea level, the roads leading up to Taif provide spectacular views of brown, wind-sculpted rocks overlooking fertile green valleys, in which grow figs, peaches, apricots, pomegranates, almonds and vegetables. Taif itself is noted for its roses.
Population 25 000
Map Saudi Arabia Bb

taiga Coniferous forest region of northern Europe bordered on the north by the tundra and on the south by deciduous forest or steppes. The term is also used for similar forest in North America.

T'ai-lu-ko Gorge ***Taiwan*** See HUA-LIEN

T'ai-nan ***Taiwan*** City of south-western Taiwan, about 50 km (30 miles) north of Kaohsiung. It is Taiwan's oldest city and dates from the late 16th century. The site was occupied briefly by the Dutch in the early 17th century, and some of their fortifications survive. The city was the capital of Taiwan until superseded by T'ai-pei in 1885. Although now industrial, T'ai-nan retains many buildings of great historical interest. The Confucian Temple, over 300 years old, is one of the best examples of traditional Chinese architecture on the island. The Koxinga Shrine commemorates the general who expelled the Dutch in 1662.
Population 629 000
Map Taiwan Bc

T'ai-pei ***Taiwan*** Capital and largest city of Taiwan, lying near the northern end of the island on the east bank of the Tan-shui river at its confluence with the Keelung. The capital since 1885, it grew rapidly during the Japanese colonial administration of Taiwan (1895-1945), but even more dramatically after the Second World War. Its population expanded from 335 000 to nearly 2.5 million in 40 years, and it is now the centre of a wide metropolitan area.

A busy, bustling city of broad avenues, T'ai-pei contains government offices and parliament buildings as well as the headquarters of many of

Taiwan's industrial and commercial companies. Industries include engineering, textiles, food, chemicals and printing.

Little of old T'ai-pei remains, but the nightly market, with open-air foodstalls, survives. The most imposing buildings are recent structures, many combining uneasily the traditional Chinese and 20th-century Western architectural styles. More historic structures include the late 19th-century city wall and gates, and the ornate and colourful Lungshan Temple, some 250 years old but largely rebuilt after the Second World War. The National Palace Museum, about 8 km (5 miles) from the centre, in the Wai-shuang-hsi hills, houses a spectacular collection of Chinese art treasures – some 250 000 items – many removed from mainland China on the eve of the Communist revolution in 1949. They include carved jade, enamel and lacquer ware, porcelain, bronze items, paintings, tapestries and books. T'ai-pei has several parks, a zoo and a botanical garden containing some 700 species of trees, shrubs and palms.

Population (city) 2 498 800; (metropolitan area) 4 000 000

Map Taiwan Bb

Taiwan Strait Separating Taiwan ideologically as well as geographically from the Chinese mainland, the strait is subject to typhoons which sweep in from the central PACIFIC OCEAN. Its former name was the Formosa Strait.

Narrowest width 160 km (100 miles)

Depth About 70 m (230 ft)

Map Taiwan Ab

Taiyuan *China* Capital of Shanxi province 400 km (250 miles) south-west of Beijing (Peking). It is a centre for the manufacture of iron and steel, heavy machinery and chemicals.

Map China Gd

Taiz *Yemen* The country's second largest city situated 80 km (55 miles) north-east of Mocha. A hill resort with panoramic views, Taiz lies at the foot of Saber mountains. It has government offices and several mosques, of which the most important are the Ashrafiyah Mosque and the Mudhafar Mosque. There is also an international airport.

Taiz is the centre of a coffee-growing area, and its industries include cotton weaving, tanning, jewellery making and plastics. New water and electricity projects have been set up.

Population 123 000

Map Yemen Ab

Taj Mahal *India* See AGRA

Takamatsu *Japan* City in north-east Shikoku island, badly damaged by Second World War bombing and now a ferry port and industrial centre. Its Ritsurin park has a graceful bridge, tranquil ponds and a forest backdrop.

Population 327 000

Map Japan Bd

Takasaki *Japan* City in central Honshu island, about 100 km (60 miles) north-west of the national capital, Tokyo. It is a gateway to the Japan Alps and is noted for *daruma*, red roly-poly dolls said to bring good fortune.

Population 231 800

Map Japan Cc

Taiwan

AN ISLAND REFUGE FOR GENERAL CHIANG KAI-SHEK'S CHINESE NATIONALISTS IN 1949, TAIWAN HAS BECOME ONE OF ASIA'S LEADING TRADING NATIONS

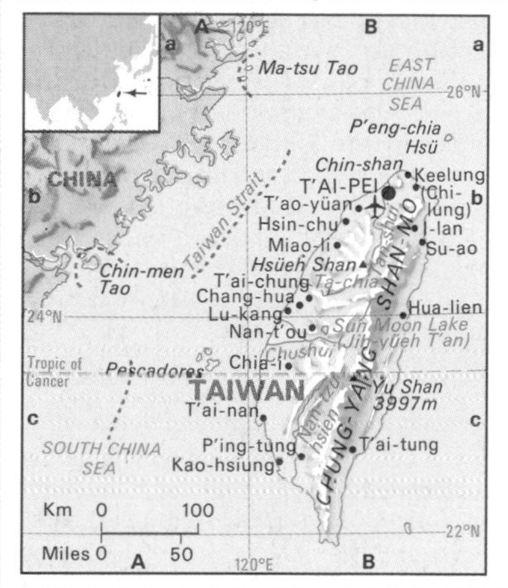

Taiwan is a political anomaly: a major trading country whose very existence few governments officially recognise. Virtually none today supports its claim to be the sole legitimate Republic of China; most prefer to recognise the Communist People's Republic, ruled from BEIJING (Peking) since 1949. Yet Taiwan – or Formosa, as the island was once known – cannot be ignored. In the last 20 years, it has emerged as one of the world's most dynamic newly industrialised nations.

Sixteenth-century Portuguese navigators called Taiwan *Ilha Formosa* – the beautiful island – and despite modern cities, factories and motorways, Taiwan still contains spectacular landscapes.

Lying about 160 km (100 miles) off the south-east coast of China, the island is predominantly mountainous, with many majestic peaks; the tallest is YU SHAN. Cultivation is largely restricted to the western and northern coastal plains, which also contain most of Taiwan's population and industry. Apart from the main island, which is about 375 km (235 miles) long and 145 km (90 miles) wide, the country includes the PESCADORES islands and CHIN-MEN TAO and MA-TSU TAO island groups.

Taiwan lies astride the Tropic of Cancer, and the climate is warm and humid for most of the year. Winters are mild, average lowland temperatures rarely falling below 10°C (50°F); summers can be oppressive, July temperatures in T'AI-PEI, the capital, often exceeding 30°C (86°F). But a wide range of crops – tea, sugar cane, bananas and especially rice – thrive in the high temperatures, abundant rainfall and fertile soil.

On the other hand, fuel and mineral resources are meagre. Domestic coal and hydroelectric power are not sufficient to meet industrial needs, so oil and other raw materials must be imported. Forty per cent of Taiwan's electricity now comes from nuclear power.

Japan annexed the island from the Chinese Empire in 1895, following the Sino-Japanese War. After the Second World War, Taiwan reverted to China, then controlled by the right-wing Nationalist Party. Routed by Mao Zedong's Communist forces, some 2 million Nationalists, under their leader Chiang Kai-shek, fled from the mainland to Taiwan in 1949 and set up a government in T'ai-pei.

With American help, the government started to develop the island's economy from its agricultural base, beginning with a successful land reform programme and industrial expansion. The first aim was to replace imports, but from the mid-1960s the emphasis switched to promoting exports. Taiwan's export-processing zones are amongst the most successful of their kind in the world, accommodating domestic and overseas companies. Exports, including electronic goods, clothing and footwear, today drive the Taiwanese economy forward, helping to ensure a spectacular rise in living standards.

Overshadowing the future of this prosperous and politically stable (if undemocratic) island is the question of its relationship with China. Two major blows were its expulsion from the United Nations (in favour of mainland China) in 1971, and the switch of American diplomatic recognition from T'ai-pei to Beijing in 1979. China has repeatedly invited Taiwan to rejoin the fold, even offering to preserve the island's economy and way of life, but the T'ai-pei government's response so far has been contemptuous and hostile.

TAIWAN AT A GLANCE

Area 36 174 km² (13 967 sq miles)

Population 19 650 000

Capital T'ai-pei

Government One-party republic

Currency New Taiwan dollar = 100 cents

Language Chinese (Mandarin – official, Hokkien, Hakka)

Religions Buddhist, Confucian, Taoist, Christian

Climate Subtropical maritime; in T'ai-pei temperatures range from 12-18°C (54-64°F) in February to 24-33°C (75-91°F) in July.

Main primary products Rice, sweet potatoes, sugar cane, bananas, citrus fruits, pineapples, pigs, poultry, fish, timber; coal, sulphur

Major industries Electronic and electrical goods, textiles, chemicals, fertilisers, paper, cement, glass, plastics, iron and steel, mining, agriculture, forestry, fishing

Main exports Electrical and electronics equipment, clothing, textiles, food, footwear, toys and sporting goods

Annual income per head (US$) 2351

Population growth (per thous/yr) 15

Life expectancy (yrs) Male 72 **Female** 72

Taklimakan ***China*** The country's largest desert, covering more than 327 000 km² (126 000 sq miles) of western China. In the language of the Uygurs, the main ethnic group in and around the desert, Taklimakan means 'no escape after entry' – an apt description of the desert's immense trackless wastes. Some 85 per cent of the desert consists of drifting sand dunes, some as high as 300 m (1000 ft).

Map China Bc

Takoradi ***Ghana*** Town just west of Sekondi, 170 km (105 miles) west of the capital, Accra. It grew only after the country's first deep-water harbour was built there in 1923, cutting the cost of exports. The port was improved in 1953 and is now fully mechanised, handling about 2 million tonnes of cargo a year, mainly timber, cocoa and mineral exports. Takoradi has a thriving industrial area producing cocoa butter and powder, plywood and veneer, cigarettes, hardware and cement. It was linked with Sekondi as a city in 1963, and Sekondi-Takoradi is the capital of Ghana's Western Region.

The nearby village of Nkroful was the birthplace of the independent country's first president, Dr Kwame Nkrumah, who held power from 1957 until he was deposed in a coup in 1966. He is also buried there.

Population (Sekondi-Takoradi) 254 500

Map Ghana Ac

talc Fine-grained, white, grey, brown or pale green mineral, magnesium silicate, having a soft, soapy texture. It is used in talcum powder and as an electrical insulator. It has a hardness of 1 on the MOHS SCALE – the lowest value.

Talca ***Chile*** Province and city in the Maule region of the Central Valley. It is a major wheat growing, wine producing and stock-farming area. The city, which is also capital of the region, lies 240 km (150 miles) south of Santiago. It was founded in 1692, and twice destroyed by earthquakes. Its industries include flour and paper mills and leather and shoe factories.

Population (province) 262 200; (city) 138 000

Map Chile Ac

Tallahassee ***USA*** Capital of Florida, in the north of the state between the Gulf of Mexico and the border with Georgia. Tourism, chemicals, paper and timber products are the economic mainstays.

Population (city) 112 300; (metropolitan area) 207 600

Map United States Jd

Tallinn ***USSR*** Baltic Sea port and capital of Estonia, on the Gulf of Finland opposite Helsinki; formerly Reval or Revel. Its industries include shipbuilding, machinery, electrical equipment, textiles, paper and chemicals. It began as a fortified trading post in the 10th century and was captured by the Danes in 1219. It was subsequently occupied by Germans, Swedes and, from 1710 to 1918, Russians. During the Second World War it was badly damaged by German bombing, and has since been substantially rebuilt. There is a medieval city wall, a 13th-century cathedral and remains of a 13th to 16th-century citadel.

Population 458 000

Map USSR Dc

talus See SCREE

Tamale ***Ghana*** Largest town in the Volta river basin, about 440 km (275 miles) north of the capital, Accra. It is the capital of Ghana's Northern Region, with its own airport, and is a market for rice, groundnuts and cotton. However, the land is increasingly threatened by soil erosion.

Population 219 200

Map Ghana Ab

Tamanrasset ***Algeria*** Market town and administrative centre, 1400 m (about 4600 ft) above sea level, on the edge of the Hoggar mountain region in the deep south. It is the key starting place for desert travel to AGADEZ in Niger and GAO in Mali. It is also the capital and meeting place of the Tuaregs (the desert nomads), who wander in the Hoggar mountains. In recent years Tamanrasset has become quite a tourist spot – the red, mud houses in the old native quarter are very attractive – and an airport links the town with Algiers, to which it is also linked by a 1370 km (850 mile) highway.

Population 10 500

Map Algeria Bb

Tamatave ***Madagascar*** See TOAMASINA

Tambov ***USSR*** Town on the plain of the Oka and Don rivers, 420 km (260 miles) south-east of Moscow. Its industries include engineering, chemicals and food processing. Tambov was founded as a fort in 1636 to defend Moscow against the Tatars.

Population 290 000

Map USSR Fc

Tamil Nadu ***India*** Rural state in the south-east, formed from the Tamil-speaking southern 130 058 km² (50 202 sq miles) of the former state of Madras. The main commercial crop of the coastal plain, where myriad 'tanks' (small artificial lakes) provide water for irrigation, is rice. Inland, the Deccan plateau produce cotton, tobacco and millet, and the Nilgiri Hills, tea and rubber. MADRAS is the state capital.

Population 48 408 100

Map India Be

Tammerfors ***Finland*** See TAMPERE

Tampa ***USA*** Port on the west coast of Florida at the head of Tampa Bay. It produces large numbers of cigars, is the centre for a nearby early vegetable growing area and exports phosphates (used in fertiliser manufacture).

Population (city) 275 500; (metropolitan area) 1 810 900

Map United States Je

Tampere (Tammerfors) ***Finland*** Second largest city in Finland after the capital, Helsinki. Tampere was founded in 1779 in the south-west of the country, about 165 km (105 miles) north and slightly west of Helsinki. It lies beside a waterfall between Näsi and Pyhä lakes. The waterfall provided energy for a cotton mill established by a Scot, James Finlayson, in the 1820s. The city grew up around the mill to become Finland's leading textile centre. Its industries now also include locomotive and rolling stock production, engineering, timber processing, plastics, footwear and printing. The city has an open-air theatre with a revolving auditorium and several museums and art galleries.

Population 167 300

Map Finland Bc

Tampico ***Mexico*** Oil-refining city in the state of Tamaulipas, near the mouth of the River Pánuco on the Gulf of Mexico. Summers there are hot and humid, and torrential rains in late September often cause floods. Despite its trail of refineries and storage tanks along the river, Tampico is a popular resort.

Population 401 000

Map Mexico Cb

Tana ***Ethiopia*** The country's largest lake (2849 km², 1100 sq miles), in the highlands about 350 km (215 miles) north and slightly west of the capital, Addis Ababa. It is the source of the Abay, or Blue Nile. More than 60 rivers flow into the lake, which is nowhere more than 15 m (50 ft) deep.

Map Ethiopia Aa

Tana ***Kenya*** The country's longest river, flowing 800 km (500 miles) from the Aberdare Mountains south-eastwards into the Indian Ocean about 90 km (55 miles) north-east of Malindi. The upper section is used to generate electricity via a series of dams built in the 1960s and 1970s. In the lower, coastal region, irrigation supports cultivation of bananas, cassava, maize and rice.

Map Kenya Cb

Tananarive ***Madagascar*** See ANTANANARIVO

Tanga ***Tanzania*** Regional capital and port on the Indian Ocean, about 50 km (30 miles) south of the Kenyan border. It exports sisal, coffee and copra, and has a fertiliser industry.

Population 150 000

Map Tanzania Ba

Tanganyika, Lake ***Burundi/Tanzania/Zaire/Zambia*** Africa's second largest lake (after Lake Victoria), covering 32 900 km² (12 700 sq miles) in the western arm of the East African section of the GREAT RIFT VALLEY. With depths up to about 1480 m (4755 ft), it is the world's second deepest freshwater lake after Lake Baikal in the USSR.

The first Europeans to reach the lake were the British explorers Richard Burton and John Hanning Speke in 1858. In 1871 Ujiji on the eastern shore was the meeting place of Dr David Livingstone and Henry Morton Stanley, who together finally determined that Lake Tanganyika was not a source of the Nile. In fact, the lake has only one outlet, the Lukuga river. There are two main feeders, the Ruzizi and Malagarasi rivers.

The lake is navigable and has fine beaches, fishing and wildlife. The chief ports are Bujumbura (Burundi), Kalémié (Zaire) and Kigoma (Tanzania).

Map Zaire Bb; Tanzania Ba

Tangier (Tanger) ***Morocco*** Seaport at the south-western end of the Strait of GIBRALTAR. Situated on the closest part of the African landmass to Europe, it has been the gateway between the two continents for centuries.

Greeks, Phoenicians and Romans all built cities on the site. The Roman city, Tingis, fell successively to the Vandals (in the 5th century), Byzantines (6th century) and Arabs (8th century). In 711 it became the springboard for the Arab invasion of Spain. Later occupants of Tangier included the Portuguese (1471-1580; 1656-62), Spanish (1580-1656), and English (1662-84), before the city again became part of Arab Morocco.

The centre of Tangier is the busy Grand Socco market, alongside which is the old city dominated by its kasbah (citadel) containing the Dar al Makhzen (the Sultan's Palace), the apartments of which have been transformed into a museum of Moroccan arts, displayed region by region. The Museum of Moroccan Antiquities lies in the Dar Shorfa Palace. Also in the old city is the 17th-century Great Mosque built by Moulay Ismail (1672-1727), the ruler who pacified Morocco's warring tribes and drove out the European interlopers. The city has a distinctly international flavour with numerous banks, a famous casino and an exciting nightlife.

To the east and west of Tangier lie Cape Malabata and Cape Spartel, both with lighthouses. Between them are some 22 km (14 miles) of fine beaches and rocky bays with views across the straits to Spain.

Population 267 000
Map Morocco Ba

Tangshan ***China*** Coal-mining city of Hebei province 150 km (95 miles) east of Beijing (Peking). It was devastated in 1976 by an earthquake measuring 7.8 on the Richter scale; 500 000 people were killed. Tangshan's mines have now been repaired and the city, with more than 200 new factories, is rebuilt.

Population 1 400 000
Map China Hc

Tanjore ***India*** See THANJAVUR

Tanna ***Vanuatu*** Island in the south of the group whose active volcano, Yasur, is a popular tourist attraction; its crater continuously issues smoke, steam and lava fragments. Tanna is a stronghold of 'cargo cult' (see MELANESIA).

Population 15 600
Map Pacific Ocean Dc

Tannenberg ***Poland*** See GRUNWALD

Tanta ***Egypt*** Town in the centre of the Nile delta, 83 km (51 miles) north of Cairo and midway between the Dumyat (Damietta) and Rashid (Rosetta), the river's two main branches. Tanta processes cotton and wool produced on farms in the delta.

Population 285 000
Map Egypt Bb

Tanzania See p. 628

Taolanaro ***Madagascar*** Indian Ocean port in the extreme south-east, formerly called Fort Dauphin then Faradofay. It exports beans, cattle, timber and rice. The first French settlement on the site was founded in 1643. Its governor was Etienne de Flacourt, whose *History of the Great Island of Madagascar* (1658) remains a major source of information about the island.

Population 13 000
Map Madagascar Ab

Taormina ***Italy*** Well preserved medieval town and resort on Sicily's east coast, about 60 km (37 miles) from its north-east tip. The old town's position, 250 m (820 ft) high, has protected it from tourist development, and hotel building is restricted on the beaches below. There are many strikingly attractive buildings from the later Middle Ages that are ornamented with lava and different coloured stones. The view of Mount ETNA from the Graeco-Roman theatre is the most breathtaking in Sicily.

Population 10 300
Map Italy Ef

T'ao-yüan ***Taiwan*** City some 20 km (12 miles) west of T'ai-pei, on Taiwan's northern plain. It is at the centre of a prosperous agricultural region producing rice and tea. Nearby is Chiang Kai-shek International Airport, opened in 1979, Taiwan's principal airport.

Population 100 000
Map Taiwan Bb

Tapiola ***Finland*** See ESPOO

Tapti (Tapi) ***India*** River rising on the northern Deccan in central India and flowing 724 km (450 miles) west to the Gulf of Khambhat at the city of Surat. Its flow varies greatly with the seasons and a massive irrigation scheme is under construction.

Map India Bc

tar pit Region where natural BITUMEN has accumulated and is exposed at the surface, such as at Rancho La Brea in California.

tar sand Sandstone in which crude oil has been trapped and the lighter components have evaporated, leaving behind a residue of asphalt or bitumen that fills the pores or holes in the rock. An example is the tar sands of the Athabasca valley in northern Canada. See also OIL AND GAS

Tara ***Yugoslavia*** Montenegrin river, which unites with the Piva to form the Drina. The Tara rises near the Albanian border, and flows generally north-west, cutting a spectacular gorge below Mojkovac. East of the town of Zabljac it is spanned by a 365 m (1197 ft) long bridge, 150 m (about 490 ft) above the churning river.

Map Yugoslavia Dc

Tara Planina ***Yugoslavia*** See VISEGRAD

Tarai (Terai) ***Nepal*** The southern lowlands of Nepal, containing one-third of the country's population – over 5 million people. The area is humid and thick with jungle and elephant grass. Growing up to 4.5 m (15 ft) high, the grass is almost impenetrable – and even elephants have difficulty in forcing their way through it. Much of it has been cleared for the cultivation of rice, jute and tobacco. Nepal's main industrial centre is at BIRATNAGAR, in the south-eastern Tarai near the border with India.

Map Nepal Ba

Taranaki ***New Zealand*** District covering 9720 km² (3753 sq miles) in North Island, some 160 km (100 miles) north and slightly west of the capital, Wellington. It has some of the richest dairy pastures – around the volcanic cone of Mount Egmont (called Taranaki by the Maoris) – and many dairy processing factories producing butter and cheese. The inland hill country is mostly given over to sheep. It is also an oil and gas-producing centre.

Population 105 200
Map New Zealand Dc

Taranto ***Italy*** Seaport and naval base on the 'instep' of Italy, about 250 km (155 miles) south-east of Naples. It was founded by the Spartans in 706 BC, but there are few ancient remains. The city's medieval core is on a rectangular island connected to the mainland by two bridges which separate the inner and outer harbours. This tightly packed district is the site of a castle built by the Spaniard Ferdinand of Aragon in 1480, a cathedral, originally 10th-11th century but with a 1713 Baroque façade, and the 14th-century Gothic church of San Domenico. Modern Taranto sprawls onto the mainland where many new industries, including one of Europe's largest steelworks, have been established.

Population 244 600
Map Italy Fd

Taranto, Gulf of Arm of the IONIAN SEA occupying the 'instep' of Italy's 'boot'. The main port, Taranto, sheltered much of Italy's fleet during the Second World War – though British carrier-borne aircraft inflicted much damage on November 11-12, 1940. Today it is a NATO base. The gulf yields shellfish.

Dimensions 140 km (90 miles) long and wide
Map Italy Fd

Tarawa ***Kiribati*** Most important atoll of the group, with nearly 40 per cent of Kiribati's population and containing its administrative centres. It lies towards the north-western end of the Gilbert Islands chain. It was the site of a fierce Second World War battle when, in November 1943, the first major American amphibious landing took the heavily fortified Japanese positions. In four days of fighting, only 17 Japanese troops survived out of 4700; there were 3300 American casualties.

Population 22 500
Map Pacific Ocean Db

Tarbela Dam ***Pakistan*** Largest dam in the country, built in the 1970s on the Indus river about 50 km (30 miles) north-west of Islamabad to provide irrigation water and hydroelectric power for large areas of the Indus valley. It is 148 m (485 ft) high, and contains 121 000 000 m³ (158 000 000 cubic yards) of rock and earth. Its projected life is less than 80 years; after that it will be irretrievably choked with silt.

Map Pakistan Da

Tarbes ***France*** Capital of Hautes-Pyrénées department, about 120 km (75 miles) west and slightly south of Toulouse. Developed by the Romans, Tarbes was for a time occupied by the Saracens in the 8th century. It still retains echoes of the Arab period in the form of stud farms, which have won an international reputation since the early 19th century for the horses they have bred from English and Arabian stock.

Population 81 000
Map France De

Tanzania

THOUGH IT HAS MADE LITTLE PROGRESS IN ECONOMIC DEVELOPMENT, TANZANIA HAS SUCCEEDED IN BUILDING A STABLE SOCIALIST STATE

With some 800 km (about 500 miles) of virtually untouched palm-fringed coast bordering the warm seas of the Indian Ocean, Tanzania abounds in scenic superlatives. Mount KILIMANJARO is Africa's highest mountain and Lake TANGANYIKA the continent's deepest and longest freshwater lake. Lake VICTORIA is the world's second largest lake and the volcano Ngorongoro contains its second largest crater. Two of the largest rivers in the world – the NILE and the ZAIRE (Congo) – have their head-waters in Tanzania.

In the 1950s, the country – then known as Tanganyika – was very much the Cinderella of British colonial territories, especially by comparison with its East African neighbours Kenya and Uganda. But by the 1970s it had become one of the best-known post-colonial African countries, largely because of the personality of Julius Nyerere (who resigned from the presidency in 1985) and his innovative approach to the social and political problems of the country. His policies have been hailed as a shining example to all poor countries – and bitterly criticised as total failures. The truth lies somewhere between. Incomes have hardly risen in real terms, and the state of the national economy is critical. On the other hand, social welfare has improved greatly, and far more has been achieved in nation-building than in most parts of Africa.

Tanzania is entirely a colonial creation. Occupied by the Germans in the 1880s, the area became British-administered Tanganyika after the First World War. Independence was obtained with little opposition in 1961, but Tanzania was not formed until 1964, when the small state of ZANZIBAR joined the vastly larger mainland territory. The name 'Tanzania' is itself a fusion of 'Tanganyika' and 'Zanzibar'.

Mainland Tanzania consists mostly of plateaus, broken by mountainous areas and the East African section of the GREAT RIFT VALLEY. The south-eastern plateau, covered by dry grasslands, rises behind the narrow, reef-fringed coastal plain, and is bordered on the west by the southern highlands. Between them and the Kilimanjaro volcanic region in the north-east is the Masai Steppe, a plateau covered by grass and thorn bush. Mount Kilimanjaro itself rises to 5895 m (19 340 ft). Although only 3 degrees south of the Equator, its peak is snow-capped all year.

Between the eastern and western arms of the Rift Valley is the vast interior plateau, about 1200 m (4000 ft) above sea level. It is dry, infertile and covered by poor grassland. In the north it descends to the more fertile Lake Victoria basin. Tanzania's western border follows the western arm of the Rift Valley.

Height makes the land much cooler than it would otherwise be so close to the Equator. The mountains are temperate and the central areas warm and dry. Only the coast is very hot, with temperatures averaging 25-35°C (77-95°F) all the year. It is dry almost everywhere, but none of the country is desert. Most areas have just enough rainfall for agriculture. In complete contrast, the islands of Zanzibar and PEMBA are hot and humid.

The country's 22.4 million people are widely scattered, with the main population concentrations occurring along the borders and on the narrow coastal plain. The heartland is relatively empty. Of the 120 different tribes – most of which are of Bantu origin and Swahili-speaking – the largest is the Sukuma (over 1 million). Its people are cotton growers on the lowlands near Lake Victoria. The Chagga tribe, living in the green and fertile foothills of Kilimanjaro, has grown relatively prosperous from coffee growing. On the Masai Steppe, some of the nomadic Masai people still herd their humpbacked cattle. However, the enclosure of their land for government projects and the death of their cattle through drought have led many to seek work in the towns.

Among the other tribes are the Nyamwezis, who established the early trade routes into the interior; the Makondes, who excel in sculpture and are noted for their tribal ceremonies involving stilt-walking; and the Wahehes, a warrior tribe who strongly resisted the Germans in the 1880s.

In addition, the population contains a large amount of Arab blood. Over the past 1000 years, Arabs from The Gulf countries have settled and intermarried with the local people. There are immigrants from the Indian subcontinent also, who have settled during the last 300 years and now number about 50 000. Europeans account for about 20 000 of the population. On Zanzibar, the people are a mixture of Shirazis (descendants of early Persian settlers), Arabs and Comorans (natives of the Comoros group of islands lying between Tanzania and the northern tip of Madagascar).

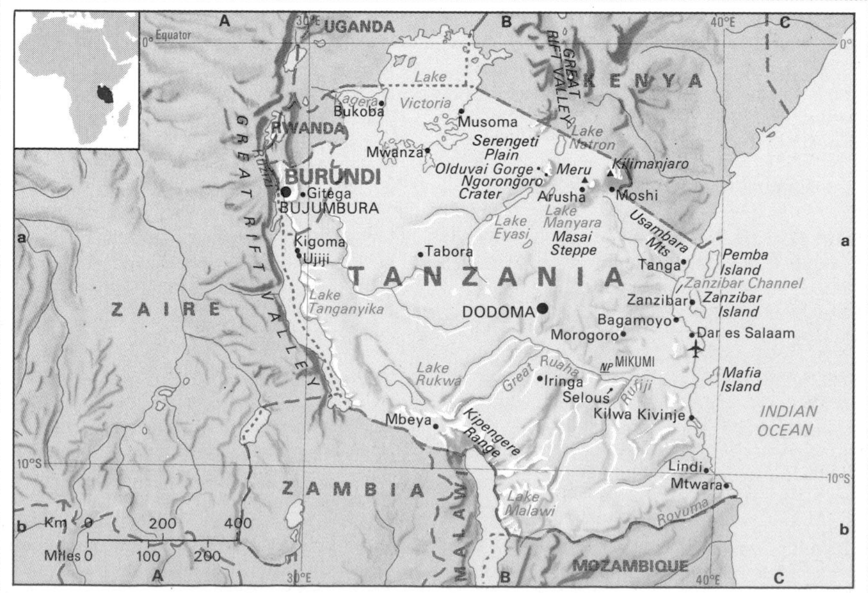

THE GROUNDNUTS DISASTER

Despite the harsh environment, 80 per cent of Tanzanians make a living from the land. Productivity is low, however, and there is no surplus from the crops they grow – principally maize – after they have fed themselves. Only a few farmers have a regular income from cash crops such as cotton and coffee. Food has to be imported for the urban population.

Large-scale farming, using machinery and a large labour force, was tried in the colonial period. One such operation – the Groundnut Scheme – failed hopelessly, partly because of inadequate rain but also because of inadequate machinery. Another – sisal estates on the coast – provided useful export earnings for a time, but demand for sisal fibre (used for making rope) has fallen since the development of synthetic substitutes in the 1960s.

The islands of Zanzibar and neighbouring Pemba, with higher and more reliable rainfall, are more successful agriculturally than the mainland. The western half of Zanzibar in particular has very fertile soil, where citrus fruits, rice, maize, plantains and cassava are grown. But of greater importance are the coconut and clove plantations. Copra and coconut fibre are leading exports, and the two islands are the world's leading suppliers of cloves.

With the exception of one diamond mine, Tanzania's mineral resources are either too low grade or too isolated to be worth working. Manufacturing is limited.

Tanzania began as one of the poorest countries, and so it remains today. In 1967, however, there were high hopes for changing the position. In the Arusha Declaration, President Nyerere stated his government's objectives for the country's future economic, social and political development. They included the raising of incomes through increased agricultural output and through some industrial growth, improved welfare and education for everyone, and a fair distribution of wealth. There was to be no elite social group of the rich, of the educated, of the politically powerful, or of the town dwellers. And Tanzania was to avoid aligning with either the capitalist West or the communist East.

To achieve these objectives, two main courses of action were taken. One was the nationalisation of a large part of the country's economy, from sisal estates to banks, from mining to tourism. The other, more radical, move was the rearrangement of Tanzania's communities. Based on the principle of *ujamaa*, a Swahili word meaning 'familyhood', 13 million people were moved from far-flung rural settlements and concentrated in new villages. The new communities were to manage themselves on democratic lines, and were to be supplied with everything necessary for communal agriculture – such as fertilisers, high quality seeds and irrigation. The scheme would also provide health care and education to people previously too scattered to have received them. At first people joined the new villages willingly, but later coercion was used.

In the event, resources and the resettled farmers both fell short of the government's hopes. Some people preferred to cultivate their own plots, rather than take part in communal schemes. Corruption plagued the system, despite government measures to stamp it out, and funds were 'misdirected'. Gradually people drifted back to the lands they had left; overall, the upheaval of resettlement led to a decline in agricultural productivity from which the country has not recovered. Today, there are still food shortages.

But the plan had its successes, particularly in the area of social welfare. Primary school education is universal and free, and adult literacy has improved from 28 per cent in 1967 to 79 per cent in 1985.

The nationalisation programme was less successful. Most factories are working far below capacity and some have closed down for lack of spares or essential imported raw materials, for which no foreign exchange is available. The range of goods in the shops is limited, especially when compared with what can be bought in Tanzania's more prosperous neighbour Kenya.

Under Nyerere, Tanzania avoided political alignment with the great powers, but its financial dependence on the rich world is greater than ever. Some US$3000 million in aid from the West, in particular loans from Canada and Western Europe, poured into Tanzania in the 1970s – more aid per head of population than for any other African country. But many of Tanzania's economic problems seem to have arisen because Nyerere overestimated the country's capacity to change and to grow, given its limited finances and skills. Furthermore, drought and increases in the price of imported oil have not helped. Oil imports now consume over half Tanzania's export earnings.

Nyerere did, however, lead Tanzania into the forefront of African affairs. It was he who deposed President Idi Amin from Uganda in 1979, after Amin decided in 1978 to annex a part of Tanzania. But the invasion of Uganda was another financial drain on Tanzania's limited resources – it cost an estimated US$500 million.

Tanzania has all the necessary ingredients for successful tourism. One of the country's greatest attractions is its abundant wildlife, including elephants, rhinoceroses, lions, leopards, buffaloes, hippopotamuses, zebras, giraffes, cheetahs, elands, impalas, wildebeests, warthogs and monkeys. There are 17 national parks and game reserves, occupying some 100 000 km² (nearly 40 000 sq miles). Among the most celebrated of these is the SERENGETI Plain in the north, a vast expanse dotted with rocky outcrops, acacia bushes, forest and small rivers, all overlooked by the Kilimanjaro range. The highest concentration of wildlife is found in the NGORONGORO CRATER, east of the Serengeti, and the greatest concentration of elephants in the world is in Manyara Park in east central Tanzania. The SELOUS GAME RESERVE in southern Tanzania, covering 50 000 km² (19 000 sq miles), is the largest game reserve in the world.

Tanzania's wonders also include archaeological sites such as OLDUVAI GORGE on the western edge of the Serengeti Plain, where the remains of early man – some 2 million years old – were discovered in 1964 by the British archaeologist and anthropologist Louis Leakey, assisted by his wife and sons. Further evidence of primitive man includes the 30 000-year-old Stone Age rock paintings of wild animals and hunters around Kondoa, east of the Masai Steppe.

Reminders of the country's early settlement by Arab traders can be found along the coast. Remains of mosques and palaces dating from as early as the 12th century abound here, as well as relics of the slave trade.

There are monuments to the early days of European exploration. At BAGAMOYO, a port 60 km (37 miles) north-west of DAR ES SALAAM, there is a memorial to the British explorers Sir Richard Burton (1821-90) and John Speke (1827-64); the house where the explorer and journalist Sir Henry Morton Stanley (1841-1904) lived; and the chapel where the body of the Scottish missionary David Livingstone (1813-73) lay before being taken to Westminster Abbey for burial.

▲ BELOW KILIMANJARO A thin band of snow outlines and defines the shape of Mount Kilimanjaro, a dark blue mass against the blue African sky. In the foreground is one of Tanzania's smaller game reserves, Amboseli.

TANZANIA AT A GLANCE

Area 945 050 km² (364 886 sq miles)
Population 22 430 000
Capital Dar es Salaam effectively, though due to be replaced by Dodoma
Government One-party socialist republic
Currency Tanzanian shilling = 100 cents
Languages Swahili (official), English
Religions Christianity (40%), Islam (23%), tribal religions (23%), Hinduism (4%)
Climate Hot and humid coast; dry inland with temperatures governed by altitude. Average temperature in Dar es Salaam ranges from 19°C (66°F) to 31°C (88°F)
Main primary products Maize, cotton, coffee, sisal, cloves, coconuts, tobacco, cassava, beans; diamonds
Major industries Agriculture, food processing, textiles, cement, oil refining
Main exports Cotton, coffee, cloves, coconut products, diamonds, sisal
Annual income per head (US$) 270
Population growth (per thous/yr) 32
Life expectancy (yrs) Male 50 **Female** 53

Tarija ***Bolivia*** Department and city in the south, on the borders of Argentina and Paraguay. It has flourishing vineyards and orchards and is rich in oil, particularly around Sanandita and Bermejo near the Argentine border.

The city, founded in 1574 by the Spaniard Luis de Fuentes, is noted for its colourful niño (child) processions held in September and for a flower festival which is held each October during the Southern Hemisphere spring.

Population (department) 246 700; (city) 54 000
Map Bolivia Bc

Tarim Pendi ***China*** Basin covering 530 000 km² (205 000 sq miles) in the west, bounded by the massive mountains of Altun Shan, Karakoram, and Tien Shan. The Taklimakan desert lies at the heart of the basin. Wheat, maize, some rice and millet, cotton and fruit are grown in oases on its rim. There are oil reserves, but they have not yet been exploited because of the area's remoteness.

Map China Bc

Tarn ***France*** River in the south-west, 375 km (233 miles) long. It rises on Mont Lozère about 70 km (45 miles) north-west of Nîmes, and flows west to join the Garonne river near Montauban. In its upper reaches, the Tarn has cut spectacularly beautiful canyons, known as Les Gorges du Tarn.

Map France De

Tarnopol ***USSR*** See TERNOPOL

Tarnow ***Poland*** Industrial city 72 km (45 miles) east of the city of Cracow. Its products include nitrates, textiles, timber and machinery.

Population 111 000
Map Poland Dc

Tarnowskie Gory (Tarnowitz) ***Poland*** The cradle of Upper Silesia's industry, 26 km (16 miles) north-west of the city of Katowice. Its mines were producing lead and zinc by the 13th century, the first smelter was established 300 years later, and Poland's oldest mining school opened there in 1803. A square of buildings dating from the 16th to 18th centuries lies at the heart of the old town, and the city makes boilers, mining machinery and safety equipment, clothing, chemicals and meat products.

Population 65 300
Map Poland Cc

Taroudannt ***Morocco*** Ancient town with 17th-century battlements lying about 80 km (50 miles) east of Agadir in the fertile Sous valley, with the High and Anti ATLAS MOUNTAINS rising north and south. Founded in the 11th century, the town rose to prominence in the late 16th century as the capital of the country under the Saadid dynasty. Its past glories have faded away and now Taroudannt is no more than a sleepy, sun-baked market town.

Population 17 100
Map Morocco Ba

Tarragona ***Spain*** Ancient city on the Mediterranean about 95 km (60 miles) south-west of the port of Barcelona. Pontius Pilate was once its governor and it has many Roman and medieval remains. It exports wine and manufactures chemical products and electrical equipment.

The surrounding province of Tarragona produces wine and fruit, and has deposits of lead, copper, silver and marble.

Population (town) 111 700; (province) 516 100
Map Spain Fb

Tarsus ***Turkey*** Agricultural market town about 380 km (235 miles) south-east of the capital, Ankara. It is the birthplace of St Paul, the 1st-century Christian apostle.

Population 160 200
Map Turkey Bb

Tartu ***USSR*** Estonian university town about 275 km (170 miles) south-west of Leningrad; formerly Yur'ev. The university – the second oldest in the USSR – was founded by the Swedish king Gustav II Adolf in 1632, when Estonia was ruled by Sweden and Tartu was called Dorpat. The town manufactures machinery, footwear, food and timber products.

Population 110 000
Map USSR Dc

Tartus ***Syria*** Coastal fishing port and agricultural centre, 64 km (40 miles) south of Latakia. Crops grown include cereals, fruit, olives and cotton. In the Middle Ages Tartus was one of the arrival points for pilgrims making their way to the Holy Land. It became a major supply port for the Crusaders, and its defence was entrusted to the Knights Templar – an order founded in 1119 to protect Christian pilgrims. Tartus fell to the Muslims in 1291. The medieval Cathedral of Notre Dame de Tortose is now a museum.

The port is the terminal for the pipeline from Karachuk, the country's first oil field (1956). Industrial development is also taking place.

Population 348 000
Map Syria Ab

Tashkent ***USSR*** Capital of Uzbekistan, about 725 km (450 miles) south-east of the Aral Sea. It is the centre of an irrigated area producing cotton, rice, fruit, and tobacco. It is also the fourth largest city of the USSR, a major industrial centre producing machinery, cotton and silk textiles, electrical equipment, chemicals, furniture and foodstuffs.

In 1966 it was hit by an earthquake in which 300 000 people were made homeless; it was rebuilt with an exciting modern centre.

Population 1 986 000
Map USSR Hd

Tasman Bay Inlet on the north coast of SOUTH ISLAND, New Zealand, where, in 1770, Captain James Cook recorded evidence of Maori cannibalism. After landing, he saw a Maori with the bone of a human forearm picked clean of flesh. His journal recorded: 'They gave us to understand that but a few days before they had taken, killed, and eaten a boat's crew of their enemies or strangers.' The town of NELSON lies on the bay's south-eastern shore.

Dimensions 72 km (45 miles) (east-west), 48 km (30 miles) (north-south)
Map New Zealand Dd

Tasman Sea Arm of the South PACIFIC OCEAN, lying in the ROARING FORTIES between south-eastern Australia, Tasmania and New Zealand. It was discovered in 1642 by the Dutch navigator Abel Tasman, after whom it was named in 1890. It is often stormy.

Area 2 300 000 km² (888 030 sq miles)
Width 1600 km (1000 miles) between Australia and New Zealand
Maximum depth More than 5200 m (17 060 ft)
Map Pacific Ocean Cd

Tasmania ***Australia*** Island state off the south-east coast, covering 68 330 km² (26 380 sq

▼ **WILDERNESS COUNTRY Mount Ruby (771 m, 2529 ft) towers over Port Davey and the forested ranges of Tasmania's Southwest National Park.**

miles). The capital is HOBART. Tasmania was discovered in 1642 by the Dutch navigator Abel Tasman. It was originally called Van Diemen's Land after the governor-general of the Dutch East Indies, on whose order Tasman sailed to Australia. The British took over the island in 1803 and used it as a penal settlement. It became a separate colony in 1825 and was renamed Tasmania in 1855. Tasmania lies in the moist, temperate zone, and produces apples, potatoes, other vegetables, sheep and cattle. Forestry and paper making are major industries, as is the mining of tungsten, tin, copper, silver, lead and zinc. The heavy rainfall and mountainous terrain enables hydroelectric plants to supply 88 per cent of Tasmania's electricity.

Population 434 700
Map Australia Hg

Tassili N'Ajjer ***Algeria*** Sandstone plateau 2000 km (about 1240 miles) south-east of Algiers, resembling a moon landscape of great beauty with gigantic canyons, deep gorges, petrified forests and sandstone features fashioned by erosion. The highest point of the Tassili is 2254 m (about 7400 ft) above sea level. The region is one of the richest in prehistoric art in the world. There are thousands of fine rock paintings, dating back to around 6000 BC, which provide a record of the life of the area when it was a grassland area supporting animals such as elephants and giraffes. At the foot of the Tassili in the east is the oasis town of Djanet. This is the main centre for the plateau.

Map Algeria Bb

Tatar Republic ***USSR*** Autonomous republic of the RSFSR around the Volga river centred on the city of KAZAN', its capital. It covers an area of 68 000 km² (26 250 sq miles) and is populated mainly by Muslim Tatars – descendants of the Mongol horde and the peoples it conquered when it galloped across Russia in the 13th century and held the land in thrall for 250 years. These now form the second largest group after the Uzbeks of non-Slavic peoples in the USSR. The inhabitants' main occupations today are agriculture and dairy farming. There are oil and gas fields, and the region's industries include engineering, tanning, timber and chemicals. Apart from Kazan', the chief towns are Bugul'ma and Zelenodol'sk.

Population 3 494 000
Map USSR Gc

Tatra Mountains (Tatry) ***Czechoslovakia/Poland*** Three ranges of the central Carpathians. Along the border lie the Western Tatras (Zapadné Tatry; Tatry Zachodnie) and, to the east, the High Tatras (Vysoké Tatry; Tatra Wysoke). The Western Tatras reach 2250 m (7382 ft) at Mount Bystrá in Czechoslovakia. The High Tatras rise to 2663 m (8737 ft) at Gerlach Peak (Gerlachovsky Stít), which is the highest point in Czechoslovakia and in the Carpathians, and to 2499 m (8199 ft) at Rysy on the border, one of whose peaks is Poland's highest mountain. Tourist resorts include Zakopane and Podhale in Poland, and Strbske Pleso and Tatranská Polianká in Czechoslovakia. The Low Tatras (Nízké Tatry) to the south, within Slovakia, rise to 2043 m (6703 ft) at Dumbier peak and include the lovely Demanova Valley.

Map Czechoslovakia Db

Tatung ***China*** See DATONG

Tauern Mountains ***Austria*** See HOHE TAUERN

Taunggyi ***Burma*** Tourist resort and capital of Shan state, near Inle lake 175 km (110 miles) south-west of Mandalay. It is a centre of the Shan handicraft industry.

Population 149 000
Map Burma Cb

Taunus ***West Germany*** Range of mainly forested hills north-west of Frankfurt am Main. The hills, which rise to 800 m (2625 ft), are mostly flat-topped with steep-sided valleys.

Map West Germany Cc

Taupo, Lake ***New Zealand*** The largest lake in the country, rich in trout. It covers 606 km² (234 sq miles) on North Island, midway between Auckland and the capital, Wellington. The lakeside town of Taupo (population 15 400) is used by tourists as a base for exploring the Volcanic Plateau in summer and for trout fishing throughout the year.

Map New Zealand Ec

Tauranga ***New Zealand*** Market town on North Island, about 160 km (100 miles) south-east of Auckland on the Bay of Plenty. It is the centre of a prosperous fruit and farming area. The nearby port of Mount Maunganui exports dairy produce, meat and timber.

Population 53 000
Map New Zealand Fb

Taurus Mountains ***Turkey*** Mountain range running east to west that separates the Anatolian plateau from the Mediterranean Sea. The western end from the Gulf of Antalya has peaks rising to more than 3000 m (9840 ft), the highest of which is Mount Erciyas Dagi at 3916 m (12 848 ft). The Cilician Gates, a pass in the eastern Taurus, has been used for centuries by armies and traders on the highway from Europe to Syria and Persia.

Map Turkey Bb

Tavastehus ***Finland*** See HAMEENLINNA

Taveuni ***Fiji*** See VANUA LEVU

Taxco ***Mexico*** Delightful hillside town in GUERRERO state, where the twin-spired Santa Prisca church towers over red-tiled roofs and small plazas with fountains and flowers. Silver mines, some worked as early as 1521, made Taxco prosperous, and today the town's narrow, winding streets are crowded with shops selling finely worked jewellery and silverware. Taxco has been declared a national monument to restrict any new building which might spoil its colonial character.

Population 60 000
Map Mexico Cc

Taxila ***Pakistan*** Town with the excavated remains of three ancient cities covering 25 km² (10 sq miles) about 36 km (22 miles) north-west of Rawalpindi. The first city, Bhir Mound, dates from about the 7th century BC and was the capital of the kingdom of Gandhara, which became a province of the Persian Empire of Darius I. It was then overrun by Alexander I in 326 BC and the Indo-Greek site – the second city – flourished as Sirkap under a succession of rulers, the last of which were the Parthians. The city then fell to the Kushans in the 1st century AD, who founded the third city on the site, Sirsukh. It became a famous seat of learning throughout the Buddhist world. The domestic buildings were multi-storeyed and there were several Buddhist monuments. Eventually, in the 5th century, it was sacked by the Huns. By the 7th century, the city was ruined. A museum in Taxila displays artefacts found at the site.

Population 38 000
Map Pakistan Da

Tayside ***United Kingdom*** Administrative region of east-central Scotland, taking its name from the River Tay. It covers 7668 km² (2961 sq miles) and includes the former counties of Angus, Kinross and most of Perthshire. The population is concentrated in the cities of DUNDEE and PERTH. The south has some of Scotland's finest agricultural land, producing soft fruits, potatoes and livestock. The north rises to the Highlands and draws tourists to its mountains and ski slopes.

Population 395 000
Map United Kingdom Db

Taza ***Morocco*** Eighth-century Arab town on an ancient site, situated 88 km (55 miles) east of Fès on the Taza Gap, a pass that separates the RIF MOUNTAINS from the Middle ATLAS MOUNTAINS. In the 12th century its strategic position, 550 m (1800 ft) above sea level, led the Almohad conquerors to construct its fortifications. Today it manufactures footwear and building materials, and is famous for its carpets. There is an interesting merdesa (ancient religious college) and a mosque with a fine Hispano-Moorish bronze lantern.

Population 56 000
Map Morocco Ba

Tbilisi ***USSR*** Capital of the Soviet republic of Georgia in the Caucasus Mountains near the Turkish border; formerly Tiflis. It manufactures machine tools, electrical equipment, locomotives, chemicals and petroleum products.

Founded in AD 455 or 458, when the capital of the Georgian Kingdom was transferred from Mtskheta, it was for centuries an important centre on trade routes between Europe and Asia. Captured in turn by Persians, Byzantines, Arabs, Mongols and Turks, it came under Russian rule in 1801.

Tbilisi is one of the oldest remaining cities in the world. Many of the older parts are built on the slopes of Mount Mtatsminda. They include medieval buildings and courtyards, a 6th-century cathedral (rebuilt in the 16th century), a basilica dating from the 6th and 7th centuries and a 13th-century castle.

Gori, 65 km (40 miles) to the north-west of the city, was the birthplace of Joseph Stalin (1879-1953), the Soviet leader from 1924 to 1953. Mtskheta, the Georgian capital in the 2nd-5th centuries, lies 20 km (12 miles) north of Tbilisi. The tiny town has several old churches, including an 11th-century cathedral which incorporates a 6th-century church and contains the tombs of Georgian kings.

Population 1 140 000
Map USSR Fd

Tczew (Dirschau) ***Poland*** Vistula river port 32 km (20 miles) south-east of the Baltic port of Gdansk. Its 890 m (2920 ft) long bridge, which was built across the river in 1857, was at the time Europe's longest iron bridge.
Population 49 900
Map Poland Ca

Tébessa (Theveste) ***Algeria*** Road and rail junction near the Tunisian border and the administrative capital of the district, which is an important phosphate mining region. It also manufactures some of the best carpets in Algeria. Founded by the Romans in AD 71, the town has some of the finest Roman remains in Africa, including a basilica, an arch and an enormous circus. It was one of the first towns in Africa to adopt Christianity.

Tébessa was destroyed by the Vandals in the 5th century AD, but restored by the Byzantines in the 6th century, and the modern city lies within Byzantine walls. It was strategically important to the Allies in February 1943, when it was captured from the Germans and defended against one of the last counter-thrusts in the North African campaign made by the German general, Rommel.
Population 68 000
Map Algeria Ba

tectonics See PLATE TECTONICS

Tegucigalpa ***Honduras*** National capital since 1880, when the president of the time moved from Comayagua because the citizens there resented his marriage to an Indian.

Founded as a gold-mining camp in the mid-16th century, the city still has some fine Spanish colonial architecture, including the Church of St Francis, completed in 1592 and now rebuilt. It produces textiles, chemicals and processed food, mostly for domestic consumption.
Population 534 000
Map Honduras Ab

Tehran (Teheran) ***Iran*** National capital about 100 km (60 miles) south of the Caspian Sea at the foot of the Elburz Mountains. Agha Muhammad Shah, founder of the Qajar dynasty (1794-1925), made Tehran his capital in 1794, and during the 19th century it was enlarged and developed by his descendants. However, although the southern part of the city retains its 18th-century character, the centre of Tehran was rebuilt by Reza Shah (ruled 1926-41), who changed the face of the capital by demolishing the old fortifications and adopting a geometrical layout with broad avenues, open parks and modern buildings.

In 1943 the Tehran Conference, the first wartime meeting between the Allied leaders Winston Churchill, Franklin Roosevelt and Joseph Stalin, was held in the city. Today Tehran is one of the largest cities in the Middle East. Its industries include chemicals, textiles, car assembling, tanning and glass making.

Places of interest include the Pahlavi Palace; the Majlis (parliament building); the 19th-century King's Mosque; the Archaeological Museum, which includes items from the ruined city of Persepolis; the huge bazaar; and the Sepahsalar Mosque, with its eight minarets.
Population 6 000 000
Map Iran Ba

Tehuacán ***Mexico*** Spa town in the south-eastern state of Puebla, with a mild, pleasant climate, and mineral springs where visitors bathe. Its mineral water is bottled and sold throughout Mexico.
Population 112 000

Teide, Pico de ***Spain*** Volcanic mountain 3718 m (12 198 ft) high on Tenerife in the Canary Islands. It is the highest point on Spanish territory and in the Atlantic.
Map Morocco Ab

Tejo ***Portugal*** See TAGUS

Tel Aviv-Jaffa (Tel Aviv-Yafo) ***Israel*** Capital of Tel Aviv district, on the Mediterranean coast about 50 km (30 miles) west of Jerusalem. Founded in 1909, Tel Aviv was the country's capital until 1950 and is still the site of foreign embassies, since Jerusalem is not recognised as Israel's capital by most foreign countries. It is the nation's second largest city after Jerusalem and the main industrial centre. Its products include chemicals, textiles, metal goods and processed foods.

In 1950 the neighbouring area of Jaffa (Yafo) – the Biblical port of Joppa – became part of Tel Aviv. The Old City has been rebuilt and now has a colourful artists' colony, a marina and cafés, shops and restaurants.

Among the city's museums are the Museum of the Diaspora, telling the story of the Jewish people in exile from AD 70 up to the founding of the State of Israel in 1948, the Municipal Museum and the Helena Rubinstein Pavilion (both art museums), and the Ha'aretz Museum complex, which contains exhibitions of science and technology, glass, folklore, coins and archaeology.

The city's airport is at Lod (Lydda), about 15 km (10 miles) to the south-east.
Population 323 400
Map Israel Aa

Telemark ***Norway*** Mountain and lake region in south Norway, west of Oslo. Its highest point is Gausta (1883 m, 6178 ft). The 'telemark', a type of turn made while cross-country skiing, was perfected in the area.
Map Norway Bd

Tell Atlas ***Algeria*** See ATLAS MOUNTAINS

Tema ***Ghana*** The country's chief port since 1961, about 30 km (19 miles) east of the capital, Accra. It has a well-planned modern town with attractive residential areas and excellent hotels – including one on the Greenwich Meridian. There is an industrial area near the railway, with an aluminium smelter, an aluminium utensil works, an oil refinery and textile mills. The port specialises in handling cocoa. The AKOSOMBO Dam across the Volta provides electricity for the factories.
Population 54 600
Map Ghana Ab

temperate climate An ambiguous term sometimes used loosely to define a climate without extremes. Correctly, however, the term denotes climates of temperate (mid) latitudes, while 'equable' defines more accurately climates with a lack of extremes. Many temperate-latitude climates are far from equable. See COOL TEMPERATE CLIMATES: SEASON

temperate rain forest Dense forest of tall trees – often up to 30 m (100 ft) tall, but lower and less dense than in TROPICAL RAIN FORESTS – found in areas with an equable climate and heavy rainfall, such as north-western North America, southern Chile, the west coast of New Zealand, south-east Australia and southern Japan. The trees often have leathery leaves and form a dense canopy; ferns and other plants needing little light grow on the forest floor.

temperature See chart under WEATHER

Temuco ***Chile*** Capital of Araucanía region, 610 km (380 miles) south of Santiago. It was founded in 1881 and is the centre of a rich agricultural area producing cereals, fruit and timber. A peace treaty with the warring Araucanian Mapuche Indians was signed in 1881 under a tree on the nearby peak of Cerro Nielol.
Population 186 000
Map Chile Ac

Tenasserim ***Burma*** Hilly division, covering 43 344 km² (16 735 sq miles) in the extreme south of the country, stretching into the narrow isthmus which forms the northern part of the Malay peninsula. The south-west monsoon – which brings 4000 mm (158 in) of rainfall to the area each year – occurs between May and October. The scattered population work largely on rubber and fruit plantations, or fish along the coast. Tavoy (population 102 000) is the capital. Other towns include MERGUI.
Population 918 000
Map Burma Cd

Tenerife ***Spain*** Largest of the CANARY ISLANDS at 2059 km² (795 sq miles).
Population 500 000
Map Morocco Ab

Tennant Creek ***Australia*** Gold, silver and copper mining centre in the Northern Territory, 450 km (280 miles) north of Alice Springs.
Population 3000
Map Australia Eb

Tennessee ***USA*** **1.** South-central state covering 109 412 km² (42 244 sq miles) between the Mississippi river and the Appalachian Mountains. The Tennessee and Cumberland rivers have both been extensively dammed for flood and navigation control. Cattle, tobacco, dairy goods and soya beans are the main farm products. Coal and zinc are the chief minerals, and chemicals, electric equipment and foodstuffs the principal manufactures. NASHVILLE is the state capital and MEMPHIS the largest city.

2. River rising in the Appalachian Mountains of North Carolina and flowing 1050 km (652 miles) south-west, then west and north through Alabama, Tennessee and Kentucky into the Ohio river, a tributary of the Mississippi. In 1933 the Federal government set up the Tennessee Valley Authority (TVA) to provide flood control and navigation improvement on the Tennessee and lower Mississippi rivers. It also aimed to improve the living conditions in a depressed area by encouraging better agricultural and forestry practices and reducing soil

erosion, and by providing hydroelectric power for domestic and industrial use. The TVA covers some 103 500 km² (39 950 sq miles) in seven states; its headquarters are at Knoxville.
Population (state) 4 762 000
Map (state) United States Ic; (river) United States Id

Teotihuacán ***Mexico*** Ruined capital of the Teotihuacán culture, which extended its influence over central Mexico between the 2nd and 6th centuries AD. Known as the 'City of the Gods', the site consists of 18 km² (7 sq miles) of palaces, temples and courtyards and is dominated by the Pyramid of the Sun (64 m, 210 ft high) and the Pyramid of the Moon (30 m, 98 ft high). The pyramids are connected by the broad Street of the Dead. The site's most famous building, the Temple of Quetzalcóatl, is adorned with elaborate stone carvings of the feathered serpent and the rain god, Tláloc.

Teplice ***Czechoslovakia*** Bohemia's oldest spa, which grew around a monastery built next to the spring in 1158. It lies north-west of Prague, 11 km (7 miles) from the East German border.

The small town of Duchcov (Dux) to the south-west has made fine porcelain for more than 300 years, including the type known as Royal Dux. The Italian adventurer and rake Giovanni Giacomo Casanova (1725-98) retired to Dux Castle and died there at the age of 73.
Population 53 500
Map Czechoslovakia Aa

Tequendama Falls ***Colombia*** See BOGOTÁ

terai Area of marshy jungle along the foothills of the Himalayas, found chiefly in Uttar Pradesh state in India and in Nepal. See TARAI

Terengganu (Trengganu) ***Malaysia*** East coast state peopled predominantly by Malays. It covers 12 955 km² (5002 sq miles) and is mostly hilly jungle with a densely populated coastal region. Most of the people are rice farmers, but offshore oil is bringing new prosperity and new roads have boosted coastal tourism.

On the beaches north of the town of Kuantan from May to September, leatherback turtles, some of them more than 2 m (6.5 ft) long and weighing almost a tonne, waddle ashore at night to lay eggs in the sand.
Population 542 300
Map Malaysia Bb

Teresina ***Brazil*** Market town and regional capital about 800 km (500 miles) west of the country's north-eastern tip. It is reputed to be the country's hottest district capital – with an average temperature of 30°C (86°F).

The surrounding state of Piauí, covering 250 934 km² (96 886 sq miles), is the poorest in the country. Its main products are vegetable oils, tropical fruits and livestock.
Population (city) 379 000; (state) 2 140 000
Map Brazil Db

Terni ***Italy*** City about 75 km (47 miles) north of Rome. It is one of central Italy's most important industrial towns, specialising in steel, engineering, chemicals and hydroelectricity.
Population 111 400
Map Italy Dc

Ternopol' ***USSR*** West Ukrainian industrial town, formerly Tarnopol, 120 km (75 miles) east and slightly south of L'vov. Its products include concrete, leather goods, footwear and canned foods.
Population 178 000
Map USSR Dd

terrace See RIVER TERRACE

Terschelling ***Netherlands*** See FRIESIAN ISLANDS

Tertiary First of the two periods in the Cenozoic era. See GEOLOGICAL TIMESCALE

Teschen ***Poland*** See CIESZYN

Tesin ***Poland*** See CIESZYN

Tessin ***Switzerland*** See TICINO

Tete ***Mozambique*** Capital of Tete province (100 724 km²; 38 879 sq miles), on the Zambezi river. It is the trade centre for west-central Mozambique, exporting cattle, cotton, hides and skins, and could become a major industrial centre based on power from the CABORA BASSA DAM, coal from Moatize and other local minerals.
Population (town) 39 000; (province) 864 200
Map Mozambique Ab

Tethys Sea which lay between the ancient continents of Laurasia and Gondwanaland. See CONTINENTAL DRIFT

Tétouan ***Morocco*** The capital of Tétouan province, 64 km (40 miles) south-east of Tangier and 10 km (6 miles) from the Mediterranean. It was founded in the 14th century and was the capital of former Spanish Morocco. There is a fine medina (old Muslim residential quarter) and a museum with antiquities from the ancient site of Lixus (see LARACHE). Manufactures include textiles, tiles, leather goods and soap.
Population 200 000
Map Morocco Ba

Tetovo ***Yugoslavia*** Macedonian city 40 km (25 miles) west of Skopje. An old Turkish settlement, it has colourfully painted houses and mosques, and textile and carpet industries. It is a centre for the ski resorts of the Sar Planina mountains.
Population 162 400
Map Yugoslavia Ec

Teutoburger Wald (Teutoburg Forest) ***West Germany*** Chalky ridge of wooded hills in the north-west. It is an extension of the highlands south of the North German Plain, and lies in an arc from Osnabrück to Paderborn, where it rises to 468 m (1535 ft).

In AD 9, Arminius (or Hermann in modern German), a local tribal chief, ambushed the main Roman expeditionary force in the forest and virtually annihilated three legions. The disgraced Roman general, Varus, committed suicide, and the Romans never again made any great effort to conquer the tribal territories to the east of the Rhine.
Map West Germany Cb

▼ SHORT-LIVED CELEBRATION This grandiose edifice stands at the heart of Tehran. It was opened as a museum called the Shahyad Monument by Iran's last shah in 1971 to mark 2500 years continuity of successive Persian and Iranian royal states.

Tevere ***Italy*** See TIBER

Texas ***USA*** Second largest state after Alaska, covering 692 408 km² (267 339 sq miles) between the Rio Grande (which forms the Mexican border), the Red River and the Gulf of Mexico. It was first settled in 1682, and was for long ruled by Spain and later Mexico until it won independence as the Republic of Texas in 1836. It became a state of the USA in 1845, but its boundary with Mexico was finally agreed only after the Mexican-American War (1846-8).

The climate ranges from subtropical in the south-east to continental in the interior, and in places irrigation is necessary. Cattle, cotton and wheat are the main farm products and Texas has about half the nation's cotton-growing area. It is also the main oil and gas producer and there are rich coal reserves. The principal manufactures are chemicals, transport and space equipment, and clothing. The state capital is AUSTIN and the largest cities HOUSTON, DALLAS and SAN ANTONIO.

Population 16 370 000
Map United States Fd

Texel ***Netherlands*** See FRIESIAN ISLANDS

Thaba Bosiu ***Lesotho*** Steep-sided natural fortress occupied by the Basotho King Moshoeshoe I in 1824 as a refuge for his people during the conflict between African, Boer and British interests. It stands 19 km (12 miles) east of the capital, Maseru, and contains the grave of Moshoeshoe I, who died in 1870.

Map South Africa Cb

Thai Binh ***Vietnam*** Province with an area of 1344 km² (519 sq miles), about 80 km (50 miles) east of the capital, Hanoi. It is one of the most densely populated rural areas of South-east Asia, with 1284 persons per km² (3326 per sq mile). Rice is cropped twice a year from its fertile plain.

Population 1 726 400
Map Vietnam Ba

Thai Nguyen ***Vietnam*** Steel-producing town about 50 km (30 miles) north of the capital, Hanoi. The integrated steel plant, built during French rule and fuelled by the nearby Quang Yen coalfield, has been modernised with Russian and Chinese aid.

Population 158 200
Map Vietnam Ba

Thailand See p. 636

Thailand, Gulf of Arm of the South CHINA SEA between the Malay Peninsula, Thailand, Cambodia and the south-western tip of Vietnam. Its shallow coastal waters are fishing grounds, and the west and north coasts, in Malaysia and Thailand, are lined with beautiful resorts. Some offshore oil is produced.

Dimensions 720 km (450 miles) long, 480-560 km (300-350 miles) wide
Map Thailand Bc

Thames ***United Kingdom*** English river, 335 km (210 miles) long, on which London stands. It rises in the Cotswold Hills in Gloucestershire, and flows generally eastwards past the city of Oxford, through the Chiltern Hills, and through London to the North Sea. The river has been an important trade and transport route since prehistoric times, with many large houses and estates on its banks, including seven past or present royal residences – at Windsor, Hampton Court, Richmond, Kew, Westminster, the Tower of London and Greenwich.

Ocean-going shipping used to sail up to the Pool of London in the heart of the city, or to docks a little downstream, but today most ships berth at modern cargo ports, such as Tilbury, on the lower reaches. The Thames used to be liable to flood at exceptionally high water, but an antiflood barrier – completed at Woolwich in 1984 – now protects the capital. Above London, boating has become a popular tourist attraction, and many riverside towns, notably Henley, hold regattas in summer.

Map United Kingdom Ee

Thames, Firth of Inlet in the NORTH ISLAND of New Zealand, forming the south-eastern arm of HAURAKI GULF. Captain James Cook explored it in 1769 and named the river that flows into it the River Thames. The river is now known by its Maori name, Waihou, but the name Thames is used for the Firth and for the town on its shores.

The small industrial town of Thames (population 6500) is about 65 km (40 miles) south-east of Auckland. Its industries include iron and steel foundries, a car assembly plant and a factory producing prefabricated houses.

Dimensions 35 km (20 miles) long, 16 km (10 miles) wide
Map New Zealand Eb

Thane (Thana) ***India*** Town at the head of the Bombay estuary, and now a north-eastern suburb of Bombay city. It has much of western India's new industries, including chemicals and automobiles, and has grown rapidly since 1960. It also refines oil from the offshore 'Bombay High' field.

Population 1 486 200
Map India Bd

Thanjavur (Tanjore) ***India*** Silk-producing city in the Kaveri river delta in the south-east, about 290 km (180 miles) south and slightly west of the city of Madras. It was the capital of several empires, including that of the Cholas in the 10th century. The city has a Great Fort and palace, and its 11th-century temple of the Hindu god Siva has adjoining libraries of Sanskrit and Tamil manuscripts.

Population 184 000
Map India Ce

Thar (Indian Desert) ***India/Pakistan*** Area of sand desert and rocky hills covering about 259 000 km² (100 000 sq miles) astride the India-Pakistan border. It has up to 510 mm (20 inches) of rain a year, but this is erratic, and temperatures reach 50°C (122°F) in May and June. In places the underground water is salty and useless for irrigation, although scant crops of wheat are grown in some hollows. The desert scrub and grassy dunes support herds of cattle and goats, and camels provide transport. Several cities, including Jodhpur and Bikaner, lie in the Thar, and the massive Rajasthan Canal project will bring new agriculture to the area.

Map India Ab

A DAY IN THE LIFE OF A NEW ZEALAND SHEEP FARMER

Geoff Williams wakes late after the excitement of the previous day. It is Sunday and he did not get home yesterday until well after midnight from the races at the nearby small country town of Paeroa near Thames on North Island. His neighbour, Brian Macinroe, had a horse running which won its race and the celebrations went on through the afternoon and evening.

Geoff looks at his sleeping wife, Jeannette, as he rises to make a cup of tea. The morning peace and quiet makes him aware of his worries over the farm and his family. Jeannette is pregnant. The thought of another child to bring up and pay for is a troublesome one. The price of wool and meat from his 4000 sheep has been declining and the future looks uncertain. He already has three children. His two sons Dick and Jack are at boarding school not far away in Auckland, and his daughter Susan is on a home science course at Otago University in Dunedin on South Island.

Some of his friends have turned their sheep farms over to cultivation. But Geoff's land is not good enough. He has seriously thought of selling his 400 hectare (988 acre) hill property and trying his hand at something else, such as running a motel, perhaps at a beach resort in New Zealand, or even across the sea in Queensland, Australia.

His thoughts are broken by the yapping of his sheepdogs outside. Geoff strides out into the open air. He lets his dogs loose, kick-starts his Honda motorcycle, and roars off into the hills behind his house to move a small herd of beef cattle to fresh grazing. He has bought the cattle to fatten and resell.

When Geoff gets back, Jeannette is cooking a big breakfast of bacon, eggs and bubble-and-squeak.

Geoff spends an easy day setting up electric fences in preparation for moving more stock later. He tours his boundary fences, checking where he will have to make repairs, searches out his shotgun for the duck-hunting season, and digs over the vegetable patch. It is autumn, and he has time to catch up on farm maintenance before his busiest times of lambing in the spring and shearing in the summer.

Jeannette delivers a carload of jams and preserves to the local church bazaar and in the afternoon heads for Miranda Beach on the Hauraki Gulf, about half an hour's drive away, where friends keep a boat for water-skiing. This time, Jeannette does not take her turn; she has had to give up the sport until after the baby is born. Geoff joins the group in his car before sunset for a quick swim and a few beers. He and Jeannette stay for a barbecue on the beach.

thaw Period of relatively warm weather when snow and ice melt. Thaw occurs each spring in high latitudes, freeing icebound seas, lakes and rivers.

Thebes ***Egypt*** Ruins of the ancient city of the New Kingdom pharaohs, and capital of Egypt for almost 1000 years from about 1600 BC. It lies beside the Nile about 500 km (310 miles) south and slightly east of the present capital, Cairo. Ancient Thebes was a walled city described by Homer, the Greek poet, as 'hundred-gated Thebes'.

Huge temples to the gods and temple tombs for the pharaohs were built at the city's twin religious sites, Luxor and Karnak, which are linked by an avenue of sphinxes. To stop looting, which had emptied the pyramid resting-places of earlier pharaohs, the Thebans buried their royal dead in great chambers carved out of the sandstone cliffs in the Valley of Kings across the river. The largest of these, the tomb of the teenage king Tutankhamun, who died in about 1352 BC, was discovered in 1922 by the British archaeologist Howard Carter. It was crammed with an astonishing collection of gold furniture, masks and jewelled ornaments, most of which are now in the Cairo Museum.

Thebes declined after it was sacked by the Mesopotamian ruler Assurbanipal and his Assyrian hordes in the 7th century BC. Today it is one of Egypt's main tourist attractions.

Map Egypt Cc

Thebes (Thívai) ***Greece*** Modern city 50 km (31 miles) north-west of Athens on the site of ancient Thebes, the city of seven gates and the legendary capital of Oedipus, the mythical king who unwittingly killed his father and married his mother.

Population 18 700

Map Greece Cb

thermal Local, rising convection current of warm air. It occurs because some land surfaces such as built-up areas and bare sand are heated by the sun much more rapidly than others such as grassland and forest. Thermals assist the ascent of birds and gliders.

thermal equator See HEAT EQUATOR

thermal spring See HOT SPRING

thermocline Temperature gradient in an ocean or a lake, in which there is a marked decrease of temperature with depth. It occurs because the surface water is warmed by the sun and becomes less dense, and therefore remains on the surface. The water below is colder and more dense, and so does not rise.

In oceans, because only a thin layer of the surface is heated, and because there is no large-scale convection mixing, the surface water cools off quickly in winter. Seasonal changes in the relative temperature of the ocean surface and land cause changes in wind patterns, such as the monsoons.

thermosphere See ATMOSPHERE

Thessalia ***Greece*** See THESSALY

Thessaloniki ***Greece*** See SALONIKI

Thessaly (Thessalia) ***Greece*** Rural central region of Greece (12 939 km², 5382 sq miles) bounded by Macedonia in the north, the Aegean Sea in the east and Stereá Ellás in the south. It contains the largest plain of Greece, drained by the Piníos river.

Population 695 700

Map Greece Bb

Thetford Mines ***Canada*** Centre of an asbestos mining area in southern Quebec. The town is the largest producer of asbestos in the Western world. There is skiing near by.

Population 19 965

Map Canada Hd

Thimphu (Thimbu) ***Bhutan*** Capital of the Himalyan kingdom since 1960 and home of its ruler, the 'Dragon King'. The town stands in a broad valley in western Bhutan at a height of 2370 m (7775 ft), and is linked to Assam in India by one of the country's few roads. The *dzong* (monastery-fortress) of Tashi Chho dominates the town; it was built in 1641 but largely reconstructed (without using nails) in the 1960s. It houses government officials and the king's throne room; his palace is on the opposite side of the river. About 8 km (5 miles) from the town, at the entrance to the Thimphu valley, is Bhutan's oldest dzong, Simtokha (1627).

Population 12 000

Map Bhutan Aa

Thionville ***France*** Former iron-mining town on the Moselle river near the southern tip of Luxembourg. It was a favourite residence of Charlemagne in the late 8th century. It grew rapidly in the early 20th century with the Lorraine steel industry, but has been hit in the 1980s by cutbacks in steel.

Population 139 000

Map France Eb

Thíra ***Greece*** See SANTORINI

Thívai ***Greece*** See THEBES

Thon Buri ***Thailand*** Twin city of Bangkok, on the west bank of the Chao Phraya river and, for five years, the national capital after the Burmese sacked the royal capital of AYUTTHAYA in 1767. Traditionally, much of its population – a large part of which is Chinese – lives on boats moored in the river, and many still live a canal-side existence. The city's most famous building is Wat Arun, or Temple of the Dawn, which was built between 1824 and 1851; this Buddhist temple is elaborately decorated with wood-carvings and porcelain.

Population 919 000

Map Thailand Bc

Thorn ***Poland*** See TORUN

thorn forest Dense thickets of thorny scrub growing where the rainfall is too scanty or unreliable for the growth of normal forests. See CAATINGA

Thousand Islands ***Canada/USA*** Group of about 1500 islands and islets extending some 128 km (80 miles) along the St Lawrence river from Gananoque to Brockville, Ontario. Those on the west side, including Wolfe (the largest, covering 127 km², 49 sq miles), Howe and Grenadier, are mostly Canadian. Those on the east, including Grindstone and Wells, are in New York State. Some of the islands are large enough to support villages.

The St Lawrence Islands National Park, covering 405 hectares (1000 acres) includes 18 of the Canadian islands, 80 islets and part of the Ontario coastline. The five-span Thousand Islands International Bridge, opened in 1938, crosses the river between Coelins Landing, New York State, and Ivy Lea, Ontario.

Map Canada Hd

Thrace ***South-east Europe*** Historical region spanning the borders of modern Bulgaria, Greece and Turkey. It stretched at one time from Macedonia to the Black Sea and Sea of Marmora, and from the Aegean to the Danube. Inhabited by various warlike tribes, it became a Roman province in AD 46 and, from 1453, part of the Turkish Ottoman Empire. The present frontiers were established in the 1920s.

The Bulgarian part of Thrace (Trakiyska; Nizina) north of the Rhodope Mountains – centred on PLOVDIV and the EVROS (Maritsa) valley – became known as Eastern Rumelia and was annexed to Bulgaria in 1885. Turkish Thrace corresponds to European Turkey, stretching from the Evros (Meriç) to the Bosporus and Dardanelles straits, an area of 23 973 km² (9256 sq miles). Apart from ISTANBUL, the largest town is EDIRNE. The modern Greek region of Thrace (Thraki; sometimes called Western Thrace to distinguish it from the Turkish Eastern Thrace) covers 8578 km² (3312 sq miles). The largest towns are Komotiní, Xanthí and Alexandroúpolis; there are excavated ruins at Abdera, birthplace of the philosopher Democritus (about 460-370 BC).

In ancient times, Thrace was renowned for its gold and silver mines; today it is mainly agricultural, producing tobacco, wheat, maize, cotton, silk, olive oil and fruits.

Population (Turkish) 4 325 300; (Greek) 345 200

Map Bulgaria Bc, Greece Da, Turkey Aa

Thraki ***Greece*** See THRACE

Three Pagodas Pass ***Burma/Thailand*** Mountain pass linking the south of Burma's Karen state with Thailand – though it is not used, except illegally, for cross-border transit. It carries the road from the Burmese port of Moulmein south-east to Bangkok, the Thai capital. It was the route for the infamous Second World War 'death railway' constructed by Allied prisoners of war for the Japanese. The bulk of the line on the Thai side of the pass has now been abandoned, and the line on the Burmese side, though planned, was never built.

Map Burma Cc

Three Rivers ***Canada*** See TROIS RIVIERES

throw Vertical distance between a rock bed on one side of a FAULT and its displaced continuation on the other side. The distance may vary from a few millimetres to thousands of metres.

thrust Force of compression in the earth's crust, resulting in a low-angle reverse FAULT. The term is also used for such a fault.

Thailand

SUBTLE DIPLOMACY BY ITS KINGS SAVED THE COUNTRY, FORMERLY SIAM, FROM EUROPEAN COLONISATION – NOW WESTERN TOURISTS FLOCK TO THIS ORIENTAL GEM

Like a scintillating jewel that flashes myriads of colours, Thailand sparkles with a special magic unlike any other country of the Orient. Perhaps it is because this country, alone in South-east Asia, has never been a European colony and had its traditions diluted by Western influences; even the word *Thai* means 'free'. Perhaps it is because Thailand's culture – the gilded temples and devotion to the Buddhist religion, the houses on stilts, the sinuous classical dances, the adulation of the royal family – goes hand in hand with a welcome to the West.

But there is another, less-attractive face of Thailand. Its capital, BANGKOK, has a reputation as the world's sex capital, where the 'bar girls' of Patpong Road – often little more than children, many of them sold into prostitution by their families – entertain tourists in seedy establishments once frequented by American servicemen on 'rest and recuperation' from Vietnam.

Some other parents take the barely preferable alternative of sending children to work in 'sweatshop' factories; while the whole family may inhabit a tiny dwelling perched above a filthy canal – regarded as 'picturesque' by tourists, yet in reality a squalid slum. Again largely for economic reasons, for thousands of farmers in the GOLDEN TRIANGLE – the isolated area spanning the borders of Thailand, Burma and Laos – the main crop is the opium poppy, and despite government efforts Thailand remains a major source of heroin.

Yet Thailand has much genuine beauty. It is a tropical country of mountains and jungles, rain forests and emerald green plains, *klongs* (canals) and rice fields, huge statues of the Buddha and miles of unspoilt beaches. Bordered by Burma, Laos, Cambodia and Malaysia, the country is about the size of France. Within it live 53 million people, the main group – the Thais proper – being descended from tribes that came south from China about 1000 years ago. There are minorities of Chinese, Malays and Indians, while in the hills live the Meo, Yao and Karen tribes. Half the population of Bangkok is said to have some Chinese blood, while thousands of refugees from Communist Cambodia live in camps in the south-east.

Most Thais are Buddhists, but it is a Buddhism of their own which borrows from Hinduism and other religions. (For example, the walls of a Buddhist temple may be decorated with scenes from the great Hindu epic the Ramayana.) There are 27 000 *wats* (temples), splendid festivals and a quarter of a million monks, for almost all young Thai men spend time as monks as well as in national service.

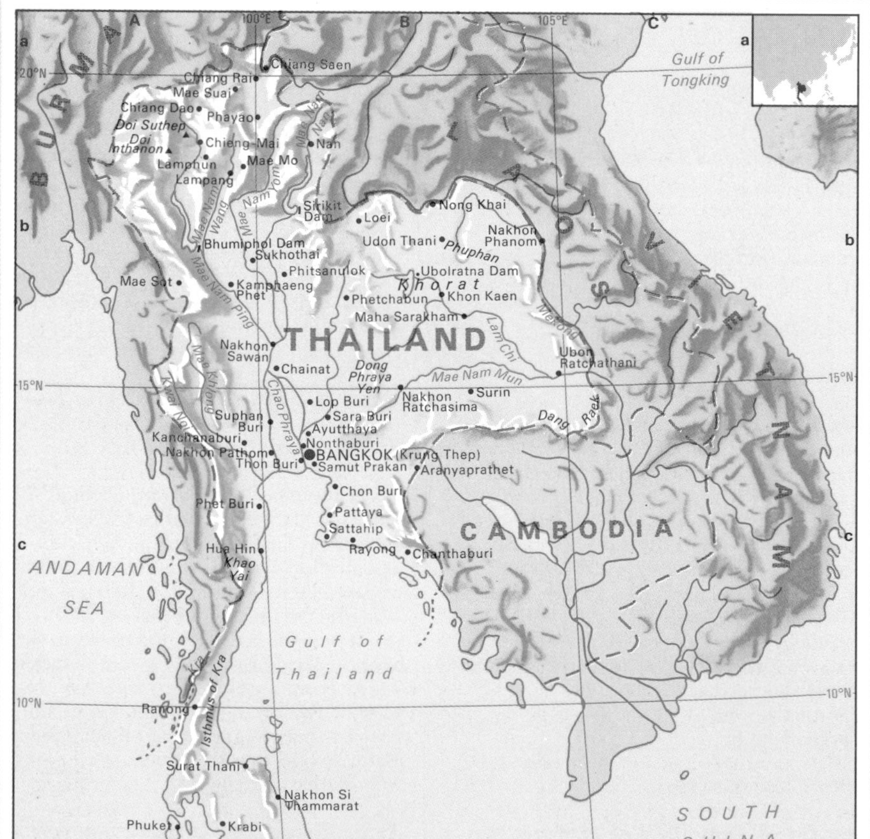

THE HEART OF THE COUNTRY

Bangkok, which the Thais lyrically call Krung Thep, the 'city of angels', is a sprawling metropolis of 5.5 million people, choked by traffic and loud with Western pop music in its modern shopping plazas, yet it retains a flavour of traditional Thailand. In the 300 temples the visitor can sense the inner tranquillity of the Buddhist religion.

The Grand Palace symbolises the reverence which the Thai people feel for King Bhumibol (formally, King Rama IX) and Queen Sirikit. The klongs which traverse the city, and the River CHAO PHRAYA on which it stands, gave Bangkok its unofficial title, the 'Venice of the East'. Many of the canals have been filled in to make roads, but on those that remain stilted houses, some draped with fishing nets, are visited daily by market vendors in boats piled high with fruit and vegetables.

The heartland of the country is the alluvial central plain, north of Bangkok, formed by the River Chao Phraya. Here there are three seasons – cool, hot and monsoon – all of them sweltering. A 'cool' maximum temperature between November and February is 26°C (79°F); a 'hot' maximum between February and May is 36°C (97°F). In the monsoon, the weather is hot, very humid and wet – more than 1500 mm (59 in) of rain may fall between May and October. The central plain contains a vast expanse of paddy fields, for Thailand is the world's leading exporter of rice. The Chao Phraya is crisscrossed by transport and irrigation canals which serve the paddy farmers who live on their banks.

In this rice bowl, 80 km (50 miles) from Bangkok, lies AYUTTHAYA, the old capital of Siam. Once the centre of a large empire, it was sacked in 1767 by the Burmese, who left its proud buildings in ruins. Today it is a quiet provincial town, but the remains of its former grandeur can still be seen.

In the far north of the country, thickly forested hills form a series of north-south ridges, separated by wide flat valleys and rising to 2595 m (8514 ft) at DOI INTHATHON. Here the climate is much cooler than in the south. Teak is hauled to the water by elephants and floated to the timber mills downriver. Rice is grown in the valleys, as are such temperate crops as garlic and onions.

At festival times, many people from the surrounding hills flock into the regional capital of CHIENG MAI, Thailand's second largest city. But for most of the year the hill tribes remain in their small villages on the forested ridges, up to 1000 m (about 3300 ft) above sea level.

The hill tribes can be distinguished from one another by their style of costume, mainly in black and white or red and blue. They practise shifting agriculture, clearing forest and brush and abandoning the land after two or three years. Some tribes, such as the Meo, have taken to illegal opium cultivation, but production in Thailand has fallen from 145 tonnes in 1967 to less than 40 tonnes as the government has opened up the northern territories, making them easier to patrol and regulate.

The north-east, known as the Isan region, is a low undulating sandstone plateau, cut off from the rest of the country by steep slopes. In the dry season from November to April, a red dust is thrown up from the roads. It pervades the atmosphere, settling on trees, on houses, on everything everywhere. The people wear checked cloths, akin to the cowboy bandana, around their heads and mouths. This is the mark of the Isan, or Thai-Lao, whose speech and culture resemble those of the Lao across the border.

If the north-east is dry Thailand, the narrow southern peninsula is wet Thailand. Here, in the upland massifs and coastal plains that lead to the Malaysian border, the landscape is more luxuriant. This is rubber country; Thailand is the world's third largest producer of rubber. Natural rain forest blends with the dark green of the man-made forest of rubber trees, grown mainly on small plots by individual farmers.

Much of the landscape resembles Malaysia, and this is where Thailand's Malay-speaking minority live. Here rice is grown in small plots lined with rubber trees. A wide variety of fruit – rambutans (red and oval, with soft spines), oranges, grapefruit and limes – is also grown. Muslim mosques are to be seen and there are few houses on stilts as elsewhere in Thailand. The Thai speech becomes more clipped and singsong, the sarong replaces the shorter *phakama* of the northern areas, and (as in Malaysia) shadow plays are an important cultural feature.

Thailand did not resist Japanese occupation in the Second World War, and under pressure even declared war on the United States. America refused to accept the declaration, saying it could not regard Thailand as an enemy. Since the war Thailand has taken a consistently pro-Western stance and has achieved stability despite numerous coups and long spells of military rule; the monarchy has been a stabilising force, respected by military and civilian governments alike. Thailand aided America during the Vietnam War, providing bases and airfields, and has received large grants from the USA.

The aftermath of the war still lingers. Thailand has taken in 80 000 refugees from Vietnam (including many 'boat people'), and more than 100 000 Cambodians are housed in camps along the border. Fighting between the Vietnamese in Cambodia and the Khmer Rouge guerrillas has frequently spilled over into Thailand.

THAILAND TODAY

Since 1960 Thailand's economic growth has been rapid. Traditionally the economy was dominated by four major commodities: rice, rubber, tin from the south and teak from the northern hills. In the 1960s new highways were built across the country, opening up vast tracts of land. Farmers in the uplands began to grow maize, cassava, jute and sugar cane. Irrigation schemes and storage dams extended the cultivation season, high-yielding rice increased crops and tractors began to replace the water buffalo.

In 25 years the country's forest area shrank from 60 per cent of the land to 25 per cent, leading to soil erosion in some areas. There are other problems: the Gulf of Thailand is being overfished, tin resources in the ANDAMAN SEA are being drained by hundreds of suction boats, and Bangkok itself has a growing flood problem. The city is sinking a little each year because its heavily used underground water supplies are drying up.

More country people are seeking their fortunes in the towns, only to find that opportunities in industry are limited. However, as Hong Kong and Taiwanese industry produces more sophisticated products, there is scope for Thai firms – whose labour is cheaper – to export a growing proportion of simple manufactured goods, led by textiles and electronics. Natural gas discoveries in the Gulf of Thailand and a small onshore oil field are spawning a small petrochemical industry. Still, for many young people, even the educated, the only opportunities lie in the service sector – above all, in the tourist trade: more than 2 million people visit Thailand each year, some for its more dubious pleasures but most for the exotic attraction of its sights.

▲ GLITTERING DEVOTION A host of colourful Buddhas and shrines fills the compound of the Grand Palace in the old walled city in Bangkok. The palace was the home of the sumptuous court when the country was called Siam, and was forbidden to more lowly mortals.

THAILAND AT A GLANCE
Area 514 820 km² (198 720 sq miles)
Population 53 700 000
Capital Bangkok
Government Military-dominated monarchy
Currency Baht = 100 satang
Language Thai
Religions Buddhist (95%), Muslim (4%)
Climate Tropical, monsoon from May to October. Temperature in Bangkok: 20°C (68°F) in December, 35°C (95°F) in April
Main primary products Rice, rubber, maize, sugar cane, cassava, pineapples, bananas, timber, fish; tin, tungsten, iron, manganese
Major industries Agriculture, processed foods, cement, paper, textiles and clothing, mining, forestry
Main exports Rice, tapioca, rubber, sugar, tin, textiles, maize
Annual income per head (US$) 793
Population growth (per thous/yr) 20
Life expectancy (yrs) Male 61 **Female** 65

Thule ***Greenland*** Northernmost settlement in the country, on the north-west coast 1530 km (950 miles) from the North Pole. It has an airport, a weather station (Dundas) and a large American military base. During the Second World War, all the inhabitants were moved to New Thule (Qaanaaq) at Inglefield Bay.
Population 800
Map Arctic Db

Thunder Bay ***Canada*** Port at the western end of Lake Superior, created in 1970 by the merger of Fort William and Port Arthur. It is also a centre for hunting, fishing, winter sports and water sports.
Population 121 380
Map Canada Gd

thunderhead Term, used chiefly in North America, for the swollen upper part of a cumulonimbus cloud with shining white edges. Thunderheads usually herald storms.

thunderstorm Storm in which intense heating induces air to rise rapidly and form vast CUMULONIMBUS CLOUDS, with heavy rain, lightning, thunder, and sometimes hail.

Thüringer Wald ***East Germany*** Mountains in the south-west, stretching 80 km (50 miles) near the West German border. They rise to 987 m (3238 ft) on the Beerberg, and are popular with skiers.
Map East Germany Bc

Thysville ***Zaire*** See MBANZA-NGUNGU

Tiahuanaco ***Bolivia*** Ruined city near the southern end of Lake TITICACA, thought to date from about 500 to 1000 AD. These pre-Inca ruins are the remnants of one of the oldest civilisations in South America. The major surviving structures on site are the Gate of the Sun, the Acapana pyramid and the Temple of Kalasasaya, covered with hieroglyphics that have still to be deciphered. The main statues are in the fine National Museum, the Museo de Tiahuanaco, in La Paz.
Map Bolivia Bb

Tianjin (Tientsin) ***China*** Third largest city in China after Shanghai and Beijing (Peking). Founded as a garrison town where the Da Yunhe (Grand Canal) joins the Hai river, it grew in the 19th century as a port for the capital 100 km (60 miles) away. From 1860 to 1949 foreigners enjoyed rights of residence in the city and many of its buildings are in the Western style. Tianjin is an important centre on the communications network of northern China and has a wide range of industries, including textiles, steel and chemicals. Most overseas trade is now handled by nearby XINGANG.
Population 7 390 000
Map China Hc

Tiber (Tevere) ***Italy*** Third longest river in the country after the Po and Adige. The Tiber rises near Mount Fumaiolo (1407 m, 4616 ft) to the east of Florence and flows 405 km (252 miles) south through Umbria to Rome and its mouth at Ostia. The towns of Todi, Perugia and Cittá di Castello overlook the Tiber valley.
Map Italy Dc

Tiberias ***Israel*** City and health resort, with hot springs, about 50 km (30 miles) east of Haifa, and 207 m (680 ft) below sea level. It stands on the west shore of the Sea of Galilee (now also known as Lake Tiberias), where Jesus walked on the waters. The city was founded by Herod Antipas in the 1st century AD, and named after the Roman Emperor Tiberius. In the 11th to 16th centuries, it was captured in turn by the Crusaders, the Arabs and the Turks. Tiberias has many ancient synagogues and tombs of Jewish scholars – and is a centre for pilgrims to the Galilee area.

To the south, at Hammat, there is a 1st-century synagogue with a fine mosaic floor depicting the signs of the Zodiac. At the nearby spa of Hammat Gader, Roman antiquities can be seen and there are hot mineral springs which feed a natural pool.
Population 29 500
Map Israel Ba

Tibesti ***Chad*** Chain of extinct volcanic mountains in the Sahara of north-western Chad, in the prefecture of Borkou-Ennedi-Tibesti near the Libyan border. There are dramatic cliffs, curiously shaped rocks and deep ravines, and the highest peak is Emi Koussi (3415 m, 11 204 ft). Fossil finds in the area include the skull of an ape-man called *Chadanthropus*, and ancient rock drawings of wildlife indicate that the climate used to be much wetter than it is now.
Map Chad Aa

Tibet ***China*** See XIZANG AUTONOMOUS REGION

Tibetan Plateau ***China*** See XIZANG GAOYUAN

Ticino (Tessin) ***Switzerland*** The country's only Italian-speaking canton, on the southern slopes of the Alps near the Italian border. It covers an area of 2811 km² (1085 sq miles) and includes Lake Lugano and the northern end of Lake Maggiore.
Population 266 000
Map Switzerland Ba

tidal current Movement of tidal water into and out of a bay, estuary, harbour or other body of water at the flood and ebb tide.

tidal flat Area of sand or mud uncovered at low tide.

tidal wave Incorrect name for a TSUNAMI

tide Daily rise and fall of sea level. See HOW TIDES ARE PRODUCED (opposite)

Tien Shan (Tian Shan) ***China/USSR*** Mountain range on the border of north-western China and the Soviet republic of Kirghizia, extending eastwards across the northern part of China's Xinjiang Uygur Autonomous Region. There are many peaks over 5000 m (16 500 ft); the highest is Pik Pobedy (Tomur Feng; 7439 m, 24 406 ft), on the Russian-Chinese border.
Maps China Ac; USSR Id

Tientsin ***China*** See TIANJIN

Tierra del Fuego ***Argentina/Chile*** Archipelago at the southern tip of South America, separated from the rest of the continent by the Strait of Magellan. It was discovered by the Portuguese explorer Ferdinand Magellan in 1520, when he made passage through the strait and saw fires lit by the native people; the name means 'Land of Fire'. The total land area is some 73 700 km² (28 450 sq miles), two-thirds of which comprises the main island, also called Tierra del Fuego. This is divided between Chile and Argentina; most of the other islands are Chilean. The main Chilean town is Porvenir, the Argentine, Ushuaia – the continent's and the world's most southerly town. The region's economy is based on oil and sheep farming.
Map Argentina Cd

Tiflis ***USSR*** See TBILISI

Tigray (Tigre) ***Ethiopia*** Province (65 700 km², 25 360 sq miles) of the north-east which has been devastated by civil war and drought. The people differ ethnically from the Ethiopians (Amhara). In early 1986, the Tigray People's Liberation Front, resisting central government policies such as the imposition of the Amhara language, was in effective control of most of the land.
Population 3 000 000
Map Ethiopia Aa

Tigris ***Middle East*** Shared by Turkey, Syria and Iraq, the Tigris – 1840 km (1143 miles) long – is one of the major rivers of south-west Asia. It rises in a mountain lake in Turkish Kurdistan, and for a short distance forms the north-east border between Turkey and Syria. It then enters Iraq, passing through Mosul and Baghdad. At Al Qurnah it joins the EUPHRATES to form the SHATT AL ARAB. The Tigris has a number of large tributaries, and since ancient times it has been a valuable means of irrigation – especially in the area known as the FERTILE CRESCENT. It reaches its highest levels in April, following the spring thaw in Turkey, when it can rise as much as 300 mm (12 in) an hour, and produce severe flooding.
Map Middle East Ba

Tihamah ***Saudi Arabia/Yemen*** Coastal plain running parallel to the Red Sea. It varies in width from 15 km (9 miles) to 80 km (50 miles). Agriculture dominates the Saudi Arabian area, with entire families – most are descendants of African slaves – working in the fields. But rainfall is low, there is no oil, and the people are the poorest in the country. In Yemen, the plain is well irrigated and produces vegetables and cotton.
Map Saudi Arabia Bc; Yemen Aa

Tijuana ***Mexico*** Border city at the top of the Baja California peninsula, busy with American tourists who cross into Mexico for a few hours of cheap entertainment and shopping. As a result, Tijuana claims the highest per capita income in Mexico, and even its surrounding shanty towns are gradually being transformed.
Population 535 000
Map Mexico Aa

Tikal ***Guatemala*** Largest of the pre-Columbian Maya Indian cities in Central America. It lies in the Petén jungle of northern Guatemala, about 300 km (185 miles) north and slightly east of the capital, Guatemala City. Forest and

HOW TIDES ARE PRODUCED

The endless cyclical movement in the great waters of the earth – the tides – is powered by the gravitational pull of the moon and, to a lesser extent, by that of the sun. Although the sun is 26 million times larger than the moon, it has about half as much gravitational pull on the earth because it is 390 times farther away from the planet.

In mid-ocean, the difference between high and low tide is only about 0.6 m (2 ft), but in shallow coastal seas and in estuaries the difference is much greater. The Bay of Fundy in eastern Canada, for example, has a tidal range of 15 m (49 ft).

At any time, there are two high tides on the earth – the direct tide on the side facing the moon, and the indirect tide on the opposite side. The direct tide is a bulge produced by the moon pulling the earth's water towards it more powerfully than it pulls the solid earth because the ocean – on the surface of the earth – is nearer to the moon. On the other hand, on the opposite side of the earth farthest from the moon, the solid earth is nearer the moon, and the moon's pull on it is greater than on the water. So, as it were, the earth is drawn away from the water, and a second bulge, the indirect tide, is produced.

Since the moon orbits the earth once every 24 hours 50 minutes, there are two high tides and two low tides at any point in the world's oceans during that period. Local tidal timetables are, however, very much modified by coastal features such as bays, estuaries and promontories that speed or slow the water's movement.

The world's tides could be used to generate huge supplies of electricity, but only limited attempts have been made to harness this form of energy. One system in operation is the 24 megawatt generating station near St Malo in Brittany, France, opened in 1966. There a 750 m (2460 ft) tunnelled barrage blocks the River Rance estuary. When the tide comes in, the sea is fed through turbines in the dam, turning them to generate electricity, and filling the basin beyond. As the tide goes out, water stored in the basin is fed back through the turbines to generate more power.

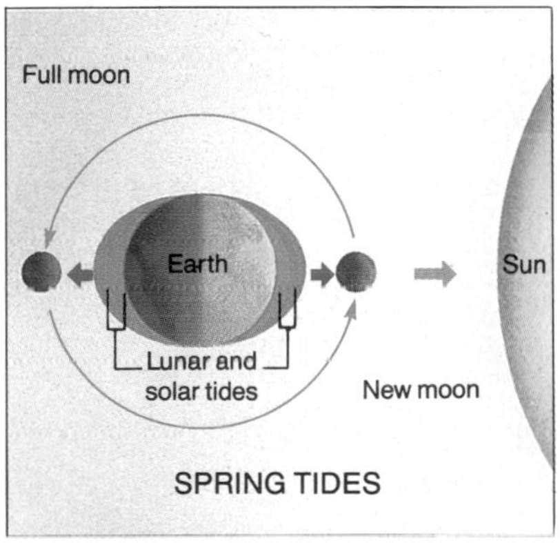

The highest tidal ranges occur at new and full moon when the earth, sun and moon are in a straight line, the sun and moon reinforcing each other's gravitational pull.

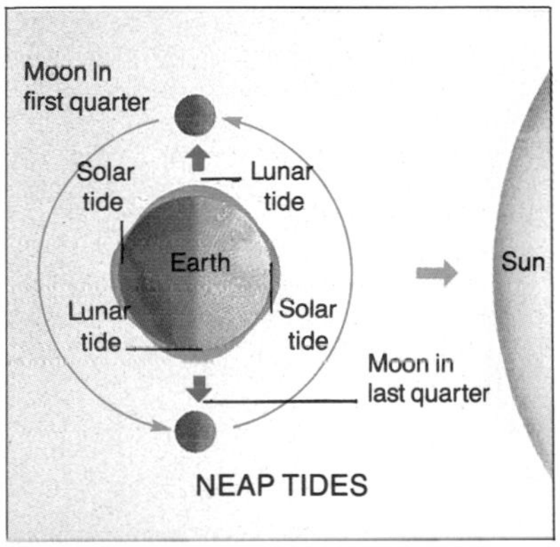

The lowest tidal ranges occur when the moon is in its first and last quarters. At these times, the pull of the sun and moon are at right angles to each other.

undergrowth cover most of what was one of the largest pre-industrial cities in the world, of which 16 km^2 (6 sq miles) have been cleared so far. Beside the great plaza in the city centre stand the tallest temple pyramids in the Americas, some almost 70 m (230 ft) high, with steep steps and carved chambers below the pinnacle sanctuaries.

The Mayas built at Tikal for more than 1100 years, creating in stone ever-taller sacrificial shrines, evolved from the basic design of a wooden native hut. Traces of the oldest buildings have been dated to 600 BC. More than 3000 buildings and 200 stone monuments have been found at the site, but 10 000 earlier buildings are thought to lie beneath them. The largest temples are believed to date from about AD 600 to AD 800. Building ended in about AD 900, when the Mayas seem to have abandoned the city, for reasons which are still a mystery.

Map Guatemala Ba

Tilburg ***Netherlands*** Spacious, industrial town about 90 km (55 miles) south and slightly east of Amsterdam, specialising in woollen textiles. Rebuilt after the Second World War, it has on its outskirts Efteling Park, a Dutch equivalent of the Disneyland pleasure parks in the United States.

Population (town) 153 900; (Greater Tilburg) 220 900

Map Netherlands Bb

till See BOULDER CLAY

Timaru ***New Zealand*** Port and seaside resort on South Island, about 150 km (95 miles) south-west of Christchurch. It exports flour, wool and locally frozen meat.

Population 28 500

Map New Zealand Cf

timberline See TREE-LINE

Timbuktu (Tombouctou) ***Mali*** Town on the edge of the Sahara which, for most Westerners, encapsulates the mystery of the Dark Continent. It was founded by the Tuareg people in the 11th century, and rapidly became a Saharan centre of the caravan trade because of its closeness to the Niger river, which laps at its walls when in flood. In the Middle Ages, its gold and salt were known in Europe and an Italian, Benedetto Dei, visited the city in 1470. Tribal wars from the 16th century sent it into decline. The French explorer René Caillé found it in 1828, and the French occupied the city in 1893.

Timbuktu has for centuries been an important Islamic centre, and had its own religious university and more than 100 schools in the 16th century. Its fortunes have been partly revived as a refuge for nomadic herdsmen forced to abandon their lifestyle because of drought, and as a tourist centre. It is best reached by camel, four-wheel-drive vehicle or air.

Population 20 000

Map Mali Bb

time zones See WHAT TIME IS IT WHERE? (p. 640)

Timgad ***Algeria*** Ruined Roman city in the north-east. It was founded by the Emperor Trajan (c. AD 100) and destroyed by Berbers in the 7th century. Remains of the forum, capitol, theatre, several baths, a splendid Arch of Trajan and well preserved mosaics have been excavated.

Map Algeria Ba

Timisoara ***Romania*** Major industrial city in the south-west of the country, 30 km (19 miles) east of the border with Yugoslavia. It dates from the 13th century and has an impressive 14th-century castle, a beautiful 18th-century town hall and several fine religious buildings. Products include textiles, food, metalwork, rolling stock, machine tools and chemicals.

Population 287 500

Map Romania Aa

Timor ***Indonesia*** Island of 30 775 km^2 (11 883 sq miles) in the Lesser Sundas group, midway between Sulawesi and northern Australia. The eastern half was taken over in 1975 from Portugal, but resistance from pro-independence guerrillas continued into the 1980s. The western part is a cattle-rearing area.

Population 3 085 000

Map Indonesia Fd

Timor Sea Situated between Timor, Indonesia, and north-western Australia, the sea was explored by the Dutch navigator Abel Tasman in 1644 and later by the explorer and buccaneer William Dampier, whom a contemporary described as 'the mildest mannered man that ever scuttled ship or cut a throat'.

Area 450 000 km^2 (173 745 sq miles)

Map Pacific Ocean Bc

Tindouf ***Algeria*** The most westerly town in the country near the borders of Morocco and Mauritania. It is a tourist centre for desert travellers, and an airfield links the town to Algiers. The roads beyond Tindouf into neighbouring countries are very rough going.

Population 7000

Map Algeria Ab

Tinian ***Northern Marianas*** Second largest island of the group, covering 101 km^2 (39 sq miles). It is separated from Saipan, to the north, by a 6 km (4 mile) channel. After it was captured from the Japanese in June 1944, during the Second World War, the Americans built a huge air base from which to bomb Japan. The aircraft that dropped the first atom bombs, on Hiroshima and Nagasaki, in August 1945 took off from there.

Population 800

Map Pacific Ocean Cb

Tínos ***Greece*** Hilly island of the CYCLADES group, south of Andros. It was occupied by the Venetians for 500 years (AD 1207-1712), longer than any other part of Greece. The pilgrimage church of Panayia Evangelistria, in the harbour town, has an icon said to work wonders. The island is famed for its marble.

Population 7800

Map Greece Dc

Tipperary (Tiobraid Árann) ***Ireland*** Market town about 160 km (100 miles) south-west of Dublin. North of the town is the Golden Vale, a lush land with thriving dairy farms. The surrounding county of the same name has an area of 4254 km^2 (1642 sq miles), and Clonmel is the county town.

Population (county) 135 200; (town) 4980

Map Ireland Cb

Tiranë (Tirana) ***Albania*** National capital founded by Turks in the 17th century; it lies in what was originally woodland beneath a limestone ridge at the southern end of a fertile plain. By 1920, when it became the capital, there were 10 000 people in the town. Later, King Zog – the first and only king of independent Albania – hired Italian architects to build a new city centre and government offices around Skanderbeg Square, named after a national hero who fought unsuccessfully against Turkish invasion in the 15th century.

By 1939 the population had risen to 35 000: Axis forces occupied the city during the Second World War, then after the Communists took over in 1946 a big expansion took place under Soviet influence. Twenty new factories were built, to turn out a wide variety of products, from cigarettes to cement. New housing went up, a hydroelectric power station was constructed, and a modern water supply laid on. Rail links connected Tiranë to the country's main port of Durrës and the steel plant at Elbasan, and an airport was built.

Population 230 000

Map Albania Bb

Tirgoviste ***Romania*** City and former capital of the province of WALACHIA, 71 km (44 miles) north-west of Bucharest. It was a trading and market centre in the Middle Ages and has several fine medieval buildings – including the 15th-century Chindia Tower which contains a museum devoted to Vlad Tepes, the notorious 15th-century prince who is said to have been the model for the fictional character Count Dracula.

The nearby 16th-century Dealu Monastery contains the tombs of several Walachian rulers. In its crypt is the head of a Romanian national hero, Michael the Brave (1558-1601), who in 1600 briefly united the provinces of Walachia, TRANSYLVANIA and MOLDAVIA.

Population 74 400

Map Romania Bb

WHAT TIME IS IT WHERE?

Everyone regards midday as being the time when the sun is at its zenith, but due to the earth's rotation, this occurs at different times in different places. Using the sun's position as the arbiter of the clock, actual time varies with every kilometre travelled to east or west, and varies very considerably over great distances. The situation imposes timetable problems, as became apparent when the railways began. In 1884, the International Meridian Conference divided the 360 degrees of longitude into 15 degree time zones, one for each of the 24 hours. As the 0 degree line ran through Greenwich, England, the basic time was established as Greenwich Mean Time (GMT). Time to the east is ahead of GMT, that to the west behind it until, on the other side of the globe along the International Date Line – 180 degrees longitude – time is 12 hours different from Greenwich. Travellers crossing the line from east to west add 24 hours and omit a day. People travelling from west to east subtract 24 hours and repeat a day.

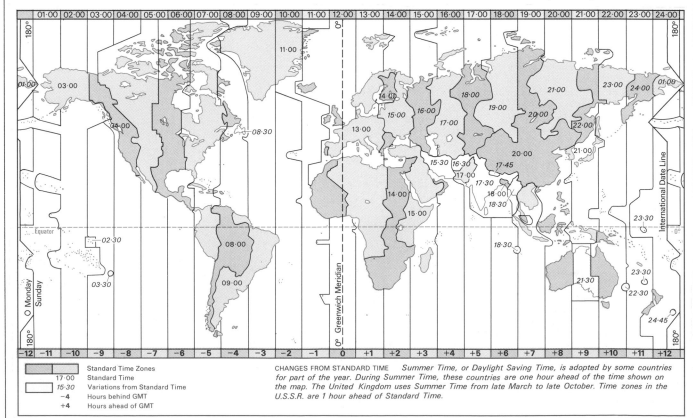

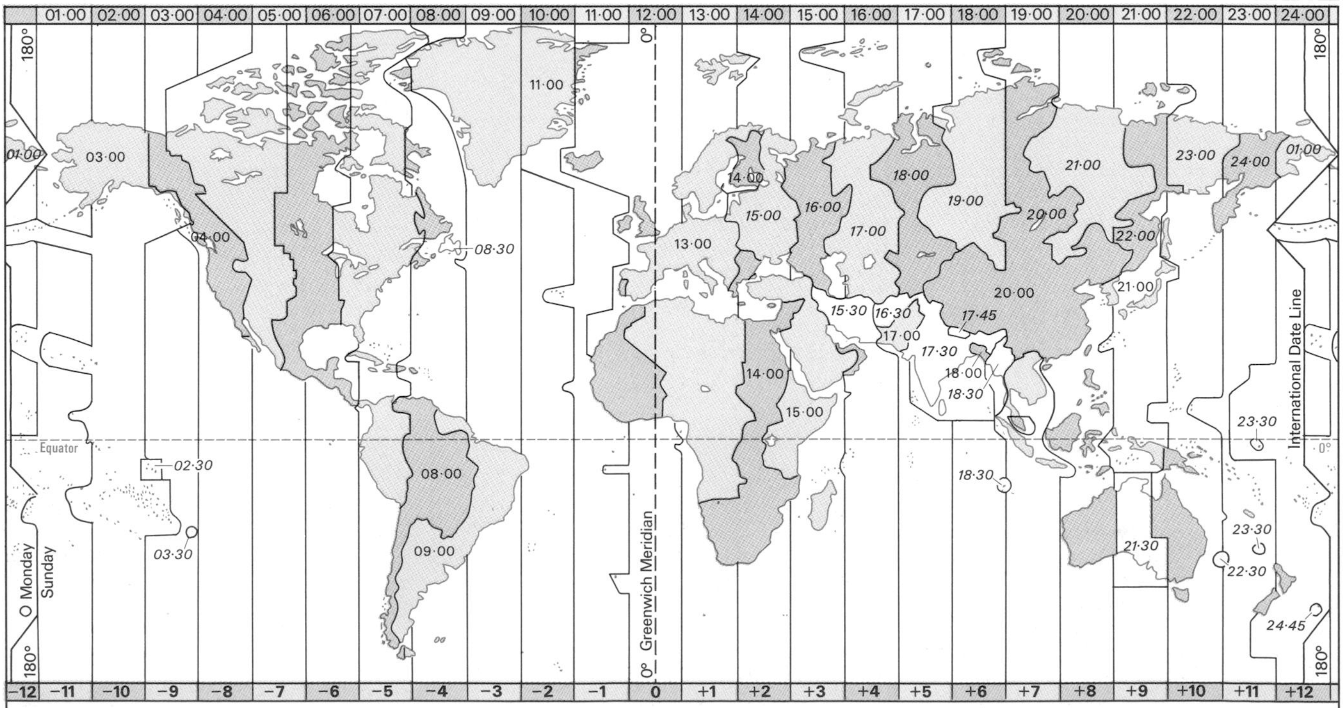

Standard Time Zones
17·00 Standard Time
15·30 Variations from Standard Time
−4 Hours behind GMT
+4 Hours ahead of GMT

CHANGES FROM STANDARD TIME *Summer Time, or Daylight Saving Time, is adopted by some countries for part of the year. During Summer Time, these countries are one hour ahead of the time shown on the map. The United Kingdom uses Summer Time from late March to late October. Time zones in the U.S.S.R. are 1 hour ahead of Standard Time.*

Tirgu Mures ***Romania*** Industrial and cultural centre in northern TRANSYLVANIA, 145 km (90 miles) south of the border with Russia and 250 km (155 miles) east of the Hungarian border. Tirgu Mures has a large Hungarian-speaking population, and the city radiates from a large central square surrounded by a Hungarian-style inner town. Among its historic buildings are a 15th-century fortess, a 15th-century church and several Baroque houses. There is also a Jesuit monastery and two 18th-century Orthodox churches, one of them wooden. The city stands on the Mures river. Its products include metal goods, machinery, food, clothing and chemicals.
Population 138 800
Map Romania Aa

Tiruchchirappalli (Trichinopoly) ***India*** City on the Kaveri river in the south-east, about 305 km (190 miles) south and slightly west of the city of Madras. With the surrounding plains it is dominated by the Rock hill, some 885 m (270 ft) high, whose fort (now ruined) the British and French fought for in the 18th century. The city has several Protestant churches and schools.
Population 609 500
Map India Ce

Tista ***Bangladesh*** River, 310 km (193 miles) long, that flows from Himalayan Sikkim onto the plains, and is liable to sudden and extensive floods. It has built a sandy alluvial fan in the Dinajpur and Rangpur districts, and shifted its course several times during the past century. Any scheme to link the Indian sections of the Brahmaputra and Ganges rivers would have to cross the Tista in Bangladesh, presenting formidable engineering and political problems.
Map Bangladesh Bb

Tisza ***Hungary/USSR/Yugoslavia*** East European river some 960 km (600 miles) long. It rises high in the Carpathians of the Ukraine in Russia, and flows west, then south-west across Hungary and Yugoslavia to join the Danube some 45 km (28 miles) north of the Yugoslav capital, Belgrade. Between Chop on the Soviet border and Szeged near the Yugoslav frontier, the river drops only 104 m (341 ft) in 440 km (273 miles), as it meanders sluggishly across the Great Alföld (Plain).

Much of the river's course in Hungary is lined with levees (banks) built after devastating floods in 1879. There are towns along this part of the river only where natural terraces rise above the flood plain on both banks to provide easy foundations for bridges: at Tokaj, Szolnok, Csongrad and Szeged.
Map Hungary Bb

Titicaca, Lake ***South America*** Largest lake in South America and, at 3812 m (12 506 ft), the highest navigated body of water in the world. It covers 8786 km² (3392 sq miles) and is shared by Bolivia and Peru. The first steamer to ferry passengers across the lake was built in England in 1862, dismantled on the coast of Peru, hauled up the Andes and reassembled on the lake to link the two countries.

The local Uru Indians live on floating 'islands' made out of the lake's totora reeds. Their homes on these huge rafts are made out of the same reeds, as are the boats from which they fish. Ruins of the Inca and pre-Inca Tiahuanaco civilisations stand on the Islands of the Moon and the Sun north of the Copacabana peninsula, and at the ruined city of TIAHUANACO just south of the lake.
Map Bolivia Bb

Titograd ***Yugoslavia*** Capital of Montenegro, 20 km (12 miles) north of Lake Shkodër. It is a new town named in 1946 after the former Yugoslav leader Marshal Tito (1892-1980), and was built after the Second World War beside the scant remains of the largely Turkish town of Podgorica (Podgoritsa). There is a busy airport, a university, and factories making tobacco products and aluminium.
Population 132 300
Map Yugoslavia Dc

Tivoli ***Italy*** Hill town about 25 km (16 miles) east of Rome. There are sumptuous Renaissance villas, such as the Villa d'Este and the Villa Gregoriana, set in magnificent gardens by some spectacular waterfalls. Emperor Hadrian (AD 76-138) built a summer resort there, and its remains, including some statues and mosaics, are still intact.
Population 52 200
Map Italy Dd

Tiwi ***Philippines*** See BICOL

Tizi Ouzou ***Algeria*** Tourist centre 88 km (55 miles) east of Algiers, lying in the Kabylie hill country. It is the administrative capital of the surrounding district and in recent years the town, which is typical of the region, has expanded rapidly. Figs are processed and olive oil is produced. About 10 km (6 miles) north-east, at Draa Ben Khedda, is the country's biggest (and most important) textile factory.
Population 103 000
Map Algeria Ba

Tlemcen (Tlemsen) ***Algeria*** Road and rail junction on the northern edge of the Tell Atlas close to the Moroccan border. It is the capital of the district, exporting blankets, carpets, olive oil, leather and esparto grass through the port of Rashgun. Its manufactures include leatherwork, carpets, brassware, hosiery, footwear and furniture. Originally founded by the Romans, it later became the Moorish capital of the region, until it fell to the Turks in 1553, when it started to decline. In 1842, the French captured it.

Architecturally, it retains its medieval Moorish character, with some of the best preserved buildings of the early Arab conquerors dating from the 12th century onwards. Among the numerous splendid mosques are the Grand Mosque and the Mosque (and museum) of Bel Hassane. Other medieval remains include towers, walls and minarets.
Population 160 000
Map Algeria Aa

Toamasina ***Madagascar*** The republic's chief port, on the Indian Ocean about 200 km (125 miles) north-east of the capital, Antananarivo. The city, formerly called Tamatave, was devastated by a hurricane in 1927. It exports cloves, coffee, hardwood, meat and vanilla, and processes textiles, leather and tobacco.
Population 77 000
Map Madagascar Aa

Toba ***Indonesia*** Volcanic crater lake in northern Sumatra. The oval lake, formed after a massive eruption blew the top off a mountain some 60 000 years ago, covers some 1160 km² (450 sq miles) and is in places 500 m (1640 ft) deep. Popular with tourists for its peaceful tropical setting, the lake is also the home of the Batak people, thought to be descended from 14th-century Chinese settlers.
Map Indonesia Ab

Tobago ***Trinidad and Tobago*** Smaller and more northerly of the two Caribbean islands constituting the Republic of Trinidad and Tobago. It lies about 32 km (20 miles) off the north-eastern tip of Trinidad and was supposedly the home of Robinson Crusoe in Daniel Defoe's book. It produces cocoa, coconuts and bananas, but agriculture has declined in recent years as many inhabitants have emigrated to Trinidad to find better paid work. Tourism is the main industry. SCARBOROUGH, on the south-east coast, is Tobago's capital and main port.
Population 39 500
Map Trinidad Aa

Tobruk (Tubruq) ***Libya*** Town and port on the north-east Mediterranean coast, 120 km (75 miles) from the Egyptian border. It has an airfield and an excellent harbour. It was occupied by the Italians in 1912 and was the scene of much fighting, including a siege of the British in the town, during the Second World War.
Population 59 000
Map Libya Ca

Tocorpuri ***Bolivia*** Permanently snow-capped mountain of 5833 m (19 137 ft) in the Western Cordillera range on the border with Chile.
Map Bolivia Bc

Togliatti ***USSR*** See TOL'YATTI

Togo See p. 642

Togo, Lake ***Togo*** Former estuary of the Chio river, now forming the country's largest lagoon. It covers 78 km² (30 sq miles) between the capital, Lomé, and Aného, and is a rich source of fish and phosphate. The phosphate, with reserves of 100 million tonnes, is Togo's chief export. It vies for third place with South Africa as the continent's largest producer of phosphate – after Morocco and Tunisia.
Map Togo Bb

Tokaj ***Hungary*** Town in the north-east of the country 48 km (30 miles) east of the city of Miskolc. It stands above the Tisza river on sunny volcanic slopes and has been Hungary's most famous wine-producing centre for more than 700 years. Tokay, a golden dessert wine, is sold in numerous wine cellars in the old town's narrow streets.
Population 4700
Map Hungary Ba

Tokelau ***Pacific Ocean*** New Zealand dependency consisting of three atolls – Atafu, Nukunono and Fakaofu – about 480 km (300 miles) north of Western Samoa. The total land area is 12 km² (4.6 sq miles). Some copra is exported.
Population 1600
Map Pacific Ocean Ec

Togo

TOGO SAYS IT IN FRENCH – BUT WAS ONCE GERMAN; ITS MILITARY RULERS ARE COMMITTED TO PEACEFUL PROSPERITY

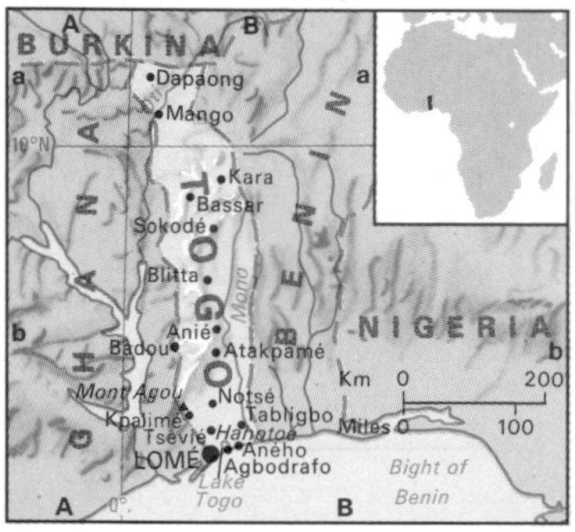

The West African nation of Togo is unusual in being a French-speaking country that was once a German colony. Moreover, its geographical diversity is far greater than its tiny area and long, narrow shape would suggest.

Grassy plains in the north and south are separated by the Togo Highlands, which run from south-west to north-east, rising to nearly 1000 m (about 3280 ft). High plateaus, particularly in the more southerly ranges, are heavily forested with teak, mahogany and bamboo.

The original colony of Togoland included part of present-day Ghana and was a German protectorate from 1884 until 1914. The country then came under British and French mandates, and in 1922 the League of Nations divided it up – the large, eastern region being administered by France. This region became the independent nation of Togo in 1960.

The main ethnic group is the Ewe tribe, concentrated in the south, a traditionally better-developed region. They regard themselves as the intelligentsia, and usually hold the important administrative posts. However, the president, General Eyadéma, is a Kabré, the dominant ethnic group of the north who are noted for their farming skills.

AGRICULTURE AND INDUSTRY

Traditionally, agriculture has been the mainstay of Togo's economy and it still involves over 80 per cent of the population. In the north, millet, sorghum and beans are subsistence crops and groundnuts are the main cash crop. In the highlands and the south, yams, cassava, maize, oil palms, vegetables and rice are grown for local consumption – while coffee, cocoa and cotton provide cash. Agriculture used to be the main export earner, but this gradually declined. However, a 'Green Revolution' launched in 1977 has almost made the country self-sufficient in basic foods.

The decline in cash crops is partly due to years of drought, and partly due to neglect. The drought has taken its toll of trees, but there has also been a failure to replant at regular intervals and to guard against disease – particularly in the cocoa plantations. Replanting schemes are being financed by the World Bank to overcome problems of ageing, disease and decay in tree crops, but it will be some years before any benefits are reaped from this operation.

Minerals, especially phosphates, are now the main export earners. In 1974, Togo launched major developments, based on the income from phosphates, but by 1978 the bottom had fallen out of the market and Togo's sales could not finance its development schemes. The government had to borrow from abroad and is still doing so. Phosphate prices have recovered somewhat, but Togo's exports are still suffering from the recession in its major markets in Western Europe.

DECLINING PORT TRAFFIC

Several large development projects have run into trouble. The port and capital, LOME, for example, expanded to handle increased mineral exports and transit trade for landlocked nations of the interior (Burkina, Mali and Niger), is handling about 12 per cent less traffic than before expansion. An oil refinery built at Lomé to process crude oil from Nigeria and Gabon is currently in mothballs, because Nigeria has failed to maintain promised supplies.

Heavy investment has been made in tourism, with facilities being developed in all the major centres of the country. But tourism, too, is waiting for an upturn in the world economy.

Politically, the military regime under General Eyadéma has been stable since 1967, the central policy being that peace and stability are needed to ensure prosperity. Togo is a founder-member of the Economic Community for West African States (ECOWAS), and is a meeting place for leaders of conflicting groups in tropical Africa.

TOGO AT A GLANCE
Area 56 785 km² (21 925 sq miles)
Population 3 110 000
Capital Lomé
Government One-party republic
Currency CFA franc = 100 centimes
Languages French (official), Ewe, Kabra and other local languages
Religions Tribal (50%), Christian (27%), Muslim (12%)
Climate Tropical; average annual temperature in Lomé 27°C (81°F)
Main primary products Cassava, maize, millet, oil palms, groundnuts, cotton, cocoa, coffee, livestock; phosphates
Major industries Mining, cement, agriculture, textiles, food processing
Main exports Phosphates, cocoa, coffee
Annual income per head (US$) 280
Population growth (per thous/yr) 31
Life expectancy (yrs) Male 46 **Female** 50

Tokushima ***Japan*** Virtually destroyed by Allied bombing in the Second World War, this city on the east coast of Shikoku island is now a thriving commercial and industrial centre specialising in chemicals.

Population 257 900

Map Japan Bd

Tokyo ***Japan*** National capital and one of the world's largest cities – prosperous, vast, overcrowded, vibrant, and a melting pot of Japanese and Western civilisation. Powerful ministries, corporation headquarters, the top universities, national theatres and museums, and a thousand shrines are concentrated there. The city has more telephones than the entire continent of Africa. More than a quarter of Japan's 121 million people live within 50 km (31 miles) of the walls and moats of the Imperial Palace at its heart. Land reclaimed from Tokyo Bay sustains an industrial juggernaut whose products range from books and electrical goods to heavy machinery.

Originally founded in 1457 with the name of Edo, it became the headquarters of Ieyasu, the first of the Tokugawa clan of shoguns, or warlords. Japan's ruling shogun in the early 17th century, Ieyasu kept an eye on fractious provincial warlords – and drained their purses – by requiring them to build and maintain elaborate residences in his capital. The city was renamed Tokyo (meaning 'Eastern capital') in 1868 after the last of the Tokugawa shoguns was deposed by Emperor Meiji. It twice rose from its own ruins. In 1923 it was devastated by an earthquake which killed about 100 000 people, and in 1945 it was severely damaged in the firestorms caused by Allied bombing.

The modern capital has 26 satellite cities and is itself divided into distinct districts – including Marunouchi, Japan's financial and business heart; Kasumigaseki, the government quarter containing the marble Diet (parliament) building; and the neon-lit Ginza shopping and nightlife centre. Prices in hotels, restaurants and places of entertainment make it one of the most expensive cities in the world for the visitor – especially for business entertainment. Public transport is fast and frequent though crowded in rush hours. Farther from the metropolitan core, railway stations have become major shopping and entertainment complexes in their own right, at Shibuya, Ikebukuro and Shinjuku.

Among the city's main Buddhist temples is Sensoji, where a tiny golden image of Kwannon, revered as the Buddhist goddess of compassion, is enshrined. The statue is said to have been found by fishermen in the Sumida river on which the city stands in AD 628. Tokyo also has Japan's most spectacular festivals – including the Dezome Shiki, or Firemen's Parade, in which acrobats perform hair-raising stunts atop tall bamboo ladders.

Population (city) 8 353 700; (Greater Tokyo) 11 670 600

Map Japan Cc

Tolbukhin ***Bulgaria*** City and capital of Tolbukhin province on the Varna-Constanţa railway 40 km (25 miles) north of Varna. Formerly called Dobrich, it was renamed in 1944 after the Soviet Marshal Tolbukhin who liberated it from the occupying Nazis. A market town for the fertile Dobrudja region, it processes foods and makes farm machinery. Following collectivisation after the Second World War it now has the country's biggest state farm and agricultural-industrial complex.

Population 102 300

Map Bulgaria Cb

Toledo ***Spain*** Historic city about 65 km (40 miles) south-west of Madrid. Perched on a river-flanked granite hill, Toledo has a wealth of fine Moorish and medieval Christian architecture and Roman remains. It is the seat of the Primate of Spain and – appropriately for a town whose swords were the world's finest – manufactures firearms. The Cretan painter El Greco (1501-1614) spent most of his life at Toledo, where he painted his masterpieces. The surrounding province – also called Toledo – grows cereals and raises sheep and goats.

Population (city) 60 100; (province) 471 800

Map Spain Cc

Toledo ***USA*** City in Ohio, on the south-western tip of Lake Erie. It is one of the country's main inland coal ports and also handles grain and timber. Industries include oil refining, food processing, glass making and vehicle manufacture – it is the home of the Jeep.

Population (city) 343 900; (metropolitan area) 610 800

Map United States Jb

Toluca ***Mexico*** The capital of the state of MEXICO, 56 km (35 miles) south-west of Mexico city. It is a thriving commercial and industrial city, making cars and industrial machinery. It is also the centre of an agricultural region which produces dairy products, sausages and dried meats, as well as pottery.

Population 357 000

Map Mexico Cc

Tol'yatti (Togliatti) ***USSR*** Town on the Volga river 65 km (40 miles) north-west of Kuybyshev. The former town of Stavropol' was drowned in the 1960s by the Kuybyshev reservoir, created when the Volga was dammed nearby. A new town was built beside the new shoreline and named Tol'yatti after Palmiro Togliatti (1893-1964), leader of the Italian Communist Party, who died while on a visit to the USSR. The new town is an industrial centre with ship-repairing, engineering, motor car, chemical and food-processing industries.

Population 576 000

Map USSR Fc

tombolo Sand or shingle bar that links an island with the mainland, such as Chesil Beach in Dorset, England, joining the Isle of Portland to the mainland west of the town of Weymouth.

Tombouctou ***Mali*** See TIMBUKTU

Tomsk ***USSR*** Siberian town on the Ob' river about 200 km (125 miles) north-east of Novosibirsk. Founded as a fort in 1604, it developed as a regional capital after the discovery of gold nearby in 1824. Today it is an industrial town producing electrical equipment, machinery, timber, paints, dyes and rubber goods. It is also an educational centre.

Population 467 000

Map USSR Jc

Tonga

THE PACIFIC'S FRIENDLY ISLANDS, IN NAME AND IN NATURE, HEADED BY THE WORLD'S HEAVIEST KING

When the coronation of Elizabeth II of England was televised around the world in 1953, one of the guests who emerged as a popular personality was Tonga's Queen Salote, who smiled broadly from an open carriage in pouring rain, underlining the name Friendly Islands that Captain James Cook gave Tonga in the 1770s. She died in 1965, and her son Taufa'ahau Tupou IV now rules a Tonga that ceased to be a British protectorate in 1970. His brother, Prince Tu'ipelehake, is prime minister.

Tongans are fat as well as friendly. The king weighed some 194 kg (440 lb, or 31½ stones) in 1980, making him the world's weightiest monarch. Women on one island average 1.6 m (5 ft 4 in) tall and weigh 74 kg (163 lb, 11.65 stones)

Hereditary monarchs have ruled Tonga for a thousand years. The royal palace stands in Tonga's capital, NUKU'ALOFA, on its largest island, Tongatapu. More than half the population of Polynesians live on this island, although the nation includes another 35 inhabited and 120 uninhabited islands and islets.

The government owns all land. Males over 16 are entitled to rent a town plot and a 3 hectare (8 acre) allotment for growing food, but population growth – over half is under 20 years old – has made land scarce. Most Tongans live on their own produce and fish from the sea. Bananas and coconuts are grown for export. There are plans to build a regional oil distribution depot.

TONGA AT A GLANCE

Map Pacific Ocean Ed

Area 747 km² (288 sq miles)

Population 109 000

Capital Nuku'alofa

Government Constitutional monarchy

Currency Pa'anga = 100 seniti

Languages Tongan (official), English

Religion Christian (75% Protestant, 19% Roman Catholic)

Climate Subtropical; average temperature in Nuku'alofa ranges from 18-25°C (64-77°F) in July to 23-29°C (73-84°F) in February

Main primary products Coconuts, bananas, sweet potatoes, cassava, taro, citrus fruits, vanilla, fish

Major industry Coconut processing

Main exports Copra, coconut oil, vanilla, watermelons, taro, bananas

Annual income per head (US$) 660

Population growth (per thous/yr) 19

Life expectancy (yrs) Male 60 **Female** 61

Tonga Trench See KERMADEC-TONGA TRENCH

Tongariro, Mount ***New Zealand*** Volcano rising to 1968 m (6457 ft) in North Island's Tongariro National Park, about 230 km (140 miles) north and slightly east of the capital, Wellington. It has several volcanic lakes and hot springs. The national park, which today covers 767 km² (296 sq miles), was originally founded in 1894 after the land around the volcano was presented to the nation by a Maori chief, Te Heuheu.

Map New Zealand Ec

Tongatapu ***Tonga*** Main island of the kingdom, lying in the south; its name means 'sacred Tonga' or 'sacred garden'. It is a coral island with a land area of 256 km² (99 sq miles), largely covered by coconut plantations. Over half of Tonga's people live there. NUKU'ALOFA, the national capital, is on the north coast.

Population 62 500

Map Pacific Ocean Ed

Tongking (Tonkin), Gulf of Arm of the South CHINA SEA between north-eastern Vietnam, mainland China and the island of Hainan. In August 1964 two US destroyers reported an attack by North Vietnamese torpedo boats in the gulf. President Lyndon B. Johnson responded by ordering air attacks on North Vietnam, and the US Congress's Tongking Gulf Resolution led to an increase in American involvement in the Vietnamese conflict.

Dimensions 480 km (300 miles) long, 240 km (150 miles) wide

Map Vietnam Ba

Tonkin ***Vietnam*** Name of the northern MIEN BAC region in French colonial times.

Tonle Sap ***Cambodia*** Freshwater lake at the centre of the Cambodian lowland. Once a branch of the sea, the Tonle Sap (Great Lake) has been formed by the silting up of the Mekong river. It acts as a natural regulator of the Mekong's flow. For most of the year its waters drain via the Tonle Sap river into the Mekong. At the height of the Mekong floods in October, however, the flow is reversed and the lake swells from 2600 km² (1000 sq miles) to some 10 400 km² (4000 sq miles). This annual spreading of fertile silt makes the land around the lake Cambodia's richest rice-growing region. The lake is also rich in fish.

Map Cambodia Ab

Tønsberg ***Norway*** The country's oldest town, founded in about AD 870 under King Harald Fairhair. Tønsberg was a whaling port until the curtailment of Norway's commercial whaling over the last 30 years, and is still a timber port near the mouth of Oslofjord, about 70 km (45 miles) south-west of Oslo.

Population 36 800

Map Norway Cd

Toowoomba ***Australia*** City in Queensland, 110 km (70 miles) west of Brisbane on the edge of the Great Dividing Range and the DARLING DOWNS. It is a market centre for the surrounding farming and grazing lands.

Population 64 000

Map Australia Id

topaz Semi-precious stone that can be colourless, blue, yellow or pink, found in IGNEOUS ROCK and QUARTZ veins. It is composed of hydrous aluminium fluosilicate.

Topeka ***USA*** State capital of Kansas, on the Kansas river about 90 km (55 miles) west of Kansas City. It is a railway repair centre and has metal processing, printing and agriculture-based industries.
Population (city) 118 900; (metropolitan area) 159 000
Map United States Gc

tor Isolated mass of weathered rock, usually granite, on or near the summit of a large rounded hill. Tor is a term local to south-west England, but similar features are present in many parts of the world.

Torbay ***United Kingdom*** Popular English tourist centre on Tor Bay, on the south coast of the county of Devon, 45 km (28 miles) east of Plymouth. The district's main resorts are Torquay, Paignton and Brixham, a fishing port. The area is famous for its mild climate, subtropical trees, water sports and fine beaches.
Population 114 000
Map United Kingdom De

Torino ***Italy*** See TURIN

tornado Twisting column of fast-moving air, usually accompanied by a funnel-shaped downward extension of a cumulonimbus cloud, whirling destructively at speeds of up to 480 km/h (300 mph). A tornado occurs at a cold front when humid, warm air suddenly releases a large amount of water as it cools. The energy liberated by the condensing water drives the tornado, producing a fast spinning column of cloudy air that leaves a trail of destruction as it moves, swirling up quite large objects – even cars and horses have been carried aloft. Over sea, a tornado produces a waterspout. Over dusty land, it produces a whirlwind of dust.

Tornadoes are short-lived – perhaps only an hour or two – and are usually no more than about 100 m (330 ft) in diameter. They commonly occur in the USA, in the Great Plains and the south-eastern part of the country.

Torne ***Sweden/Finland*** River marking the border between Sweden and Finland at the head of the Gulf of Bothnia. It is 510 km (317 miles) long and is noted for its salmon fisheries. Its Finnish name is Tornionjoki.
Map Sweden Db

Toronto ***Canada*** Canada's largest city and capital of Ontario, on the north shore of Lake Ontario. It is the main industrial and commercial centre of Canada, and an important port, shipping grain, meat and livestock. Its industries include the manufacture of electrical equipment, iron and steel, farm machinery and aircraft. The business area is close to the lake, and is dominated by many tall buildings including the tallest freestanding structure in the world. This is the 553 m (1814 ft) CN tower, which has a revolving restaurant at 347 m (1140 ft). The Canadian National Exhibition is held each year at Exhibition Park, and the Royal Ontario Museum houses a famous collection of Chinese art. The University of Toronto (1827) is Canada's largest university.

Toronto occupies the site of an old French trading post, Fort Rouille, founded in 1749. The British bought the site from local Indians in 1787 and Toronto (then called York) was chosen as the capital of Upper Canada. The city has a rich and varied population mix, including people of Italian, Portuguese, Greek, Chinese and Polish descent.
Population 2 999 000
Map Canada Hd

Torquay ***United Kingdom*** See TORBAY

Torremolinos ***Spain*** See MALAGA

Torrens, Lake ***Australia*** Lake covering 5775 km² (2230 sq miles) west of the Flinders Range in South Australia, about 400 km (250 miles) north and slightly west of Adelaide. It is normally a saline mud flat, but has been known to overflow after heavy rains.
Map Australia Fe

Torreón ***Mexico*** City in the state of Coahuila, some 800 km (500 miles) north-west of Mexico City. It is one of the main textile centres of the Laguna cotton-growing district. It also has breweries, flour mills and chemical factories.
Population 438 000
Map Mexico Bb

Torres Strait Islands ***Australia*** Group of about 70 islands in Torres Strait, off the far northern tip of Queensland, of which 30 or so are inhabited. The main centre is Thursday Island, 16 km (10 miles) off the mainland and covering 100 km² (38 sq miles). Most of the population are Torres Strait Islanders, a Melanesian group more closely related to the people of Papua New Guinea, to the north, than to Australian Aborigines.

The shallow strait links the ARAFURA and CORAL seas. Its discovery in 1606 by the Spanish navigator Luis Vaez de Torres proved that New Guinea was an island and not part of a southern continent. It is little used by shipping today because of its dangerous shallows and reefs. The strait has beds of pearl-shell and bêche-de-mer, but world demand for their commodities is low and many of the islanders are consequently unemployed. Many of them now live in mainland Queensland.
Population 6600
Map Australia Ga; Papua New Guinea Ba

Tórshavn ***Denmark*** Administrative centre and principal port of the FAEROE ISLANDS on the island of Streymoy. It is the islands' main educational centre and a Faroese Academy of Higher Learning was established there in 1965. There is an airport on the island of Vágar, some 30 km (20 miles) from Tórshavn.
Population 14 200

Tortola ***British Virgin Islands*** Main island of the Caribbean colony, covering 54 km² (21 sq miles). It is mountainous, reaching 521 m (1710 ft) at Mount Sage, but fertile farmland grows sugar cane, sweet potatoes, avocados, bananas and other crops.
Population 9300
Map Caribbean Eb

Torun (Thorn) ***Poland*** Industrial city 177 km (110 miles) north-west of the capital, Warsaw. The Polish astronomer Nicolaus Copernicus (1473-1543) was born there. Torun still has much of its medieval walls, including the Leaning Tower, which is 35 m (115 ft) high and 1.4 m (4.5 ft) out of true. Its industries include chemicals, timber, engineering and printing.
Population 165 100
Map Poland Cb

Toscana ***Italy*** See TUSCANY

Touggourt (Tuggurt) ***Algeria*** Oasis town in the north-east of the country at the end of the railway, 210 km (about 130 miles) south-west of 'Annaba. It is in the Souf region of dunes and is an important centre of communications for the oil fields to the south. The Touggourt oasis produces large crops of dates.
Population 45 500
Map Algeria Ba

Toulon ***France*** The nation's most important naval base, occupying a sheltered Mediterranean harbour about 50 km (30 miles) east and slightly south of Marseilles. Much of the French fleet was scuttled there in 1942 to prevent it falling into German hands after the fall of France. Shipbuilding and ship repairing are now the major industries.
Population 418 600
Map France Fe

Toulouse ***France*** Capital of Haute Garonne department, on the Garonne river about 90 km (56 miles) from the Spanish border. Its characteristic red-brick houses give it its nickname of *la ville rose* ('the pink town'). A centre of the French aviation industry, it was the birthplace of the French Concorde aircraft and the European Airbus.
Population 550 500
Map France De

Touraine ***France*** Area in the Loire valley, centred on the city of TOURS and known as the Garden of France. It is famous for rich crops of wheat and vegetables, and for its wines, produced at such villages as Vouvray, Touraine and Bourgueil.
Map France Dc

Tourane ***Vietnam*** See DA NANG

Tourcoing ***France*** See LILLE-ROUBAIX-TOURCOING

Tournai (Tournay; Doornik) ***Belgium*** Medieval city 50 km (31 miles) south of the city of Ghent. Its centre retains the circular outline of the medieval town, but the walls have been replaced by boulevards. It produces carpets, embroidery and leather goods.
Population 70 000
Map Belgium Aa

Tours ***France*** Capital of Indre-et-Loire department. It lies on a triangle of land between the Loire and Cher rivers about 210 km (130 miles) south-west of Paris. Despite extensive damage during the Second World War, it possesses many historic buildings such as the 12th to 16th-century cathedral and two churches built during

the same period. Engineering, electronics, wine-making and tourism are its main industries.
Population 268 650
Map France Dc

Townsville ***Australia*** Port in Queensland, about 1125 km (700 miles) north-west of Brisbane. It is the administrative, commercial and industrial capital of northern Queensland and home of the James Cook University. It exports meat, wool and minerals from the interior and sugar and timber from the fertile coastal region. Its industries include meat packing, sawmilling and copper refining. Townsville is also a base for exploring the Great Barrier Reef and nearby Magnetic Island.
Population 86 100
Map Australia Hb

Toyama ***Japan*** Industrial city near the north coast of Honshu island, about 255 km (158 miles) north-west of the capital, Tokyo. Once famous for patent medicines and silk spinning, it is now noted for its expertise in robotics and pharmaceuticals.
Population 314 100
Map Japan Cc

Trá Lí ***Ireland*** See TRALEE

Trâblous ***Lebanon*** See TRIPOLI

Trabzon (Trebizond) ***Turkey*** Port near the south-east corner of the Black Sea. It is the capital of a province of the same name. The town's 13th-century Aya Sofia Church has some of Turkey's finest surviving Byzantine frescoes. The province covers an area of 4685 km² (1809 sq miles) and produces beans, hazelnuts, fruits and tobacco.
Population (town) 156 000; (province) 817 500
Map Turkey Ba

trade winds Winds blowing from the subtropical high-pressure belts to the equatorial low-pressure belts. They blow from the north-east in the Northern Hemisphere and from the south-east in the Southern Hemisphere. See WEATHER AND WIND SYSTEMS

Trafalgar, Cape ***Spain*** The country's most south-westerly point, some 60 km (37 miles) west of Gibraltar. The waters off the cape were the scene of a decisive British naval victory over a French and Spanish fleet in 1805, during the Napoleonic Wars, during which the British admiral, Lord Nelson, was killed.

Tralee (Trá Lí) ***Ireland*** County town of Kerry, 91 km (57 miles) north-west of the southern city of Cork. It is a tourist centre, and an industrial centre linked to the sea by canal. The song *Rose of Tralee*, written in the town by William Mulchinock (1820-64), has led to an August festival at which a Kerry girl is chosen as Rose for the year. Sir Roger Casement (1864-1916), an Irish-born British consular official, was captured at McKenna's Fort, an earthwork on Tralee Bay in 1916, when he landed there after attempting to gain German help for the Irish nationalists during the First World War. He was executed for treason in London.
Population 16 500
Map Ireland Bb

▲ **WRITHING DEVIL A waterspout cavorts off Key West, Florida. It formed when the downward extension of cloud in a tornado joined up with a cone of spray thrown up by the stormy sea.**

tramontana Cold north wind sweeping down from the mountains in Italy and the Mediterranean region.

Tran Ninh Plateau ***Laos*** Upland plateau in the centre of the country, consisting of limestone and sandstone hills that were once covered with tropical monsoon rain forest. Shifting cultivation by the hill people has left the plateau covered by tropical grasslands, with scattered oaks and pines along the stream courses. Unworked deposits of alluvial gold, antimony, copper, lead, zinc and silver exist there. The inner core of the plateau is known as the Plain of JARS.
Map Laos Ab

Trans-Amazonian Highway ***Brazil*** Network of roads extending for about 4800 km (3000 miles) across north-central Brazil. It runs from Cruzeiro do Sul near the Peruvian border to Recife on the Atlantic coast. Construction of the highway opened up the Amazon River Basin to new settlement and allowed the tapping of its vast and largely unexploited mineral wealth – including iron, copper, tin and petroleum – but also threatened its ecological and social balance.

Transantarctic Mountains ***Antarctica*** Vast mountain range dividing the Antarctic into its eastern and western regions. The mountains extend more than 3200 km (2000 miles) from Victoria Land to the Weddell Sea, and reach a height of more than 4500 m (14 750 ft) in the Queen Maud Mountains. The range is cut by numerous glaciers, including the BEARDMORE.
Map Antarctica d

Trans-Canada Highway ***Canada*** The world's longest national road. It runs for a total of about 8000 km (5000 miles) from Victoria (British Columbia) on the Pacific coast to North Sydney in Nova Scotia, and across the island of Newfoundland from Port Aux Basques to St John's on the Atlantic coast. It was completed in 1962.

Transkei ***South Africa*** Former Bantu homeland, in east Cape Province, covering 43 798 km² (16 910 sq miles). It was declared an independent republic in 1976, but is recognised only by South Africa, on which it is economically dependent. It is peopled mostly by Xhosa farmers growing maize and rearing goats, sheep and cattle for mohair, wool and beef. The coast has a growing tourist trade, particularly near the border with Natal, and the whole wild coast is rated high for its scenery and beaches. The capital is Umtata.
Population 3 298 000
Map South Africa Cc

Trans-Siberian Railway ***USSR*** The world's longest railway line, stretching some 9300 km (5780 miles) from Moscow to the Pacific coast ports of Vladivostok and Nakhodka. The main Siberian section was built between 1891 and 1905, with simultaneous construction east from the Urals city of Chelyabinsk and west from Vladivostok. It marked the turning point in Siberia's development, opening up vast areas to exploitation, settlement and industrialisation.

Through trains now travel four to seven times a week via Sverdlovsk, Novosibirsk, Irkutsk and Khabarovsk, where foreign travellers change for the Nakhodka train. (Vladivostok is a closed military city.) The journey takes eight days, with 92 stops. There are also trains between Moscow and Beijing (Peking) via Ulan Bator, Mongolia. A new 3200 km (2000 mile) extension, the Baikal-Amur-Magistral line, completed in 1984, passes north of Lake Baikal to Komsomol'sk-na-Amure and the port of Sovetskaya Gavan', opening up more of Siberia.

Transvaal ***South Africa*** Northern province which includes South Africa's main industrial region, the Pretoria-Witwatersrand-Vaal Triangle (PWV area). Boer (Afrikaner) farmers settled the area – which covers 262 499 km² (101 351 sq miles), including Gazankulu, KaNgwane, KwaNdebele and Lebowa – from the mid-1830s, and in 1852 Britain recognised their right to rule beyond the Vaal river. The Transvaal became the South African Republic in 1857, but Britain annexed it in 1877. Self-government was restored in 1881 but, after the Anglo-Boer War (1899-1902), Transvaal became a British colony. In 1910 it became a province of South Africa, with its capital at Pretoria.

Population 8 950 500

Map South Africa Ca

Transylvania ***Romania*** Region in the centre and north of the country, covering 55 166 km² (21 300 sq miles). It consists largely of a triangular plateau some 400-650 m (1310-2130 ft) high. It is enclosed on the north, south and east by the CARPATHIANS, and is cut by the Mures river. Transylvania was once part of the Roman province of Dacia. In the Middle Ages it was conquered first by the Hungarians, then by the Turks. It was incorporated into the Hapsburg Empire in 1799 and became part of Hungary in 1867. After the First World War it was ceded to Romania.

Map Romania Aa

Trápani ***Italy*** Port near Sicily's western end. Its closeness to Africa, 230 km (140 miles) away, made it an important port and naval settlement under Carthaginian, Roman, Arab and Spanish rule. Its products include tuna fish, macaroni and wine.

Population 73 300

Map Italy De

Trasimeno ***Italy*** Lake covering 128 km² (49 sq miles) about 16 km (10 miles) west of Perugia. Three small islands lie in the lake, which reaches a maximum depth of only 6 m (20 ft). The main towns, Castiglione and Passignano, are developing as popular bathing resorts. It was the scene of the Carthaginian general Hannibal's crushing defeat of the Romans in 217 BC.

Map Italy Dc

Travancore ***India*** Former state ruled by Muslim princes in the south-west. Following Indian independence in 1947, it merged with neighbouring Cochin to form a state which was renamed KERALA in 1956.

Map India Be

Trebizond ***Turkey*** See TRABZON

Treblinka ***Poland*** Nazi concentration camp near the town of Malkinia, about 100 km (60 miles) north-east of the capital, Warsaw. The camp is preserved as a museum and memorial to the 750 000 people – mostly Jews and Polish partisans – who died there between July 1942 and November 1943.

Map Poland Eb

tree-line Altitude above which trees do not grow. It is not a uniform line since it depends on local as well as general conditions of relief, climate and soil, but in general it rises with increasing temperatures from sea level in the tundra to 4000 m (13 000 ft) in East Africa.

Trelleborg ***Sweden*** Seaport on the south-west coast about 30 km (18 miles) south of Malmö. It was founded in the 13th century and is now the main ferry port link with Germany.

Population 34 100

Map Sweden Be

tremor Minor earthquake with a low level of intensity.

trench See OCEAN TRENCH

Trengganu ***Malaysia*** See TERENGGANU

Trent ***United Kingdom*** River of central England, rising in the hills of Staffordshire and flowing 270 km (167 miles) through the cities of Stoke-on-Trent and Nottingham into the HUMBER. Its main use today is to provide cooling water for power stations on its banks.

Map United Kingdom Ed

Trentino-Alto Adige ***Italy*** Mountainous region covering 13 613 km² (5256 sq miles) in the Alps to the north of Verona. It was once Austrian territory and was incorporated into Italy in 1919. Many people, particularly in Alto Adige (formerly South Tyrol) in the north, still speak German. The region's highest peak is Ortles at 3905 m (12 812 ft) near the Swiss border. At the heart of the region are the valleys of the ADIGE and Isarco rivers, which contain the provincial capitals of TRENTO and BOLZANO and the main railway and motorway from Verona to the BRENNER PASS into Austria.

Hydroelectricity is the chief natural resource of the region, which is also Italy's leading producer of timber and apples. In July 1985 a burst dam attached to a fluorite mine at Stava near the town of Tesero caused a flood of water and mud that killed over 200 people.

Population 877 800

Map Italy Ca

Trento (Trient) ***Italy*** Capital of the Trentino-Alto Adige region, about 120 km (75 miles) north-west of Venice. It stands on the road to the BRENNER PASS into Austria. Trento's oldest quarter is between the 12th-13th century cathedral and the imposing Castello del Buonconsiglio, originally 13th century but with later additions. The economy of the city is based on tourism, electrical products, chemicals, pottery and silk.

Population 100 000

Map Italy Ca

Trenton ***USA*** State capital of New Jersey, standing at the limit of navigation on the Delaware river, about 40 km (25 miles) north-east of Philadelphia. Trenton was first settled in the late 17th century and has a wide range of industries, including steel, textiles and rubber.

Population (city) 92 100; (metropolitan area) 313 800

Map United States Lb

Trèves ***West Germany*** See TRIER

▼ LONELY OUTPOST Cottagers on Tristan da Cunha, whose first settlers came from St Helena in the 19th century. The islanders are proud of their curious 19th-century form of spoken English.

Treviso ***Italy*** Old fortified town about 25 km (16 miles) north of Venice. It has an 11th-century cathedral and a 13th-century palace. Its main industries are textiles, ceramics and lamp-making.

Population 85 800
Map Italy Db

Triassic First of the three periods in the Mesozoic era of the earth's time scale. See GEOLOGICAL TIMESCALE

Trichinopoly ***India*** See TIRUCHCHIRAPPALLI

Trient ***Italy*** See TRENTO

Trier (Trèves) ***West Germany*** City on the Moselle river, 10 km (6 miles) from the Luxembourg border. It is the trading centre for the Moselle-Saar-Rüwer wine-growing area and beneath its streets are kilometres of cellars, where more than 30 000 million litres (6600 million gallons) of wine are stored.

Trier is one of West Germany's oldest towns. Julius Caesar conquered it and by about 15 BC it was a Roman city, later to become a summer residence of the emperors. Baths, a basilica and Emperor Trajan's amphitheatre survive, as well as the Porta Negra, one of the fortified gates.

Karl Marx (1818-1883), the founder of Communism, was born in the city.

Population 94 600
Map West Germany Bd

Trieste ***Italy*** Seaport at the head of the Adriatic Sea, just west of the Yugoslav border. It started life as the Roman city of Tergeste, and a theatre and a triumphal arch remain from those days. The cathedral church of San Giusto and the adjacent castle date from the Middle Ages. From 1945 to 1954 it was occupied by Yugoslavia. Trieste is much used as a shopping centre by Yugoslavs, and most of its port trade is funnelled through to Austria and eastern Europe. Its chief industries include steel, oil and tourism.

Population 251 400
Map Italy Db

Triglav ***Yugoslavia*** See JULIAN ALPS

Trim (Baile Átha Troim) ***Ireland*** Town on the River Boyne, 40 km (24 miles) north-west of Dublin. It has Ireland's largest Anglo-Norman castle: a 13th-century fortress covering more than 1.2 hectares (3 acres), built on the site of an earlier castle visited by King John. Parliaments were held at the castle in the 15th century.

Population 2340
Map Ireland Cb

Trincomalee (Trinkomali) ***Sri Lanka*** Port in the north-east, some 230 km (140 miles) from Colombo. Its well-preserved fort, built on a peninsula by the Portuguese in the 17th century, guards Trincomalee Bay. The British admiral Lord Nelson (1758-1805) described this as the finest harbour in the world. Trincomalee was a British naval base until 1957, when the Sri Lankan navy took it over. North of the town is one of the island's finest beaches, 30 km (19 miles) long, centred on the resort of Nilaveli.

Population 44 900
Map Sri Lanka Ba

Trinidad and Tobago See p. 648

Trinkomali ***Sri Lanka*** See TRINCOMALEE

Tripoli (Trâblous) ***Lebanon*** Port and the country's second city, situated 72 km (45 miles) north-east of the capital, Beirut. Tripoli was founded by the Phoenicians, probably in the 7th century BC, and was the capital of the colony of Tripolis – from the Greek *tripolis*, meaning 'with three cities'. The cities forming the colony were Tripoli itself, Sabrata and Leptis Magna. Despite its ancient pedigree, most of the city's monuments are no older than the 14th and 15th centuries.

It is the terminus for the oil pipeline from Iraq. Its chief manufactures are textiles, cement and soap, and its main exports are citrus fruit and tobacco. It also trades in silk.

Places of interest include the 12th-century Crusader castle of St Gilles; the 13th-century Great Mosque; Teinal Mosque and the Mosque of Emir Wartawi, both 14th century.

Population 175 000
Map Lebanon Aa

Tripoli (Tarabulus) ***Libya*** Capital and principal city of Libya on the north-west coast, 200 km (124 miles) from the Tunisian border. It is a major port with an international airport. Industries include fishing and fish processing, tobacco, textiles, bricks, salt and leather goods.

The city was founded by the Phoenicians and developed by the Romans as one of the three main cities of the province (with LEPTIS MAGNA and SABRAHTA), but there are few ancient remains except the Arch of Marcus Aurelius in the old city, which is Moorish in character. The Spanish castle by the harbour dates from the 16th century and houses a museum which covers prehistory, archaeology, ethnology, ancient carved inscriptions and natural history.

The old city has many narrow alleyways and souks (markets) for weavers, coppersmiths, goldsmiths and leatherworkers. There are several mosques containing rich art treasures and some marvellous 18th-century gardens. The modern city owes much to the architecture of the Italian period and has a fine seafront and eucalyptus-lined avenues. It was heavily bombed during the Second World War, but has expanded massively since the 1969 revolution.

Population 620 000
Map Libya Ba

Tripolitania ***Libya*** Smallest of Libya's three historical regions but the most productive and containing 75 per cent of the country's population. Agriculture is highly developed, with groundnuts, castor oil, tobacco, and fruit and vegetables as the major crops. Tripolitania's Mediterranean coastal plain, 10 km (6 miles) wide, is the country's most fertile region. During the 2nd century AD it supplied one-third of ancient Rome's grain imports.

Map Libya Ba

Tristan da Cunha ***South Atlantic*** Lonely Tristan island, only 104 km² (40 sq miles) in area, lies halfway between South Africa and South America. It is formed almost entirely by a large, 2060 m (6760 ft) volcano, which unexpectedly erupted in 1961. The population was evacuated to Britain, but returned in 1963. Their main crop is potatoes, and they also raise cattle, sheep and pigs; fishing is good. The island is British – part of the St Helena Dependencies – and is one of a group of four. The others are uninhabited.

Population about 325
Map Atlantic Ocean Ef

Trivandrum ***India*** Seaport and capital of Kerala state, on the south-west coast, about 90 km (55 miles) north-west of the mainland's southernmost point. It has an 18th-century fort, museums and a university, and there are good beaches nearby and hill resorts inland.

Population 520 100
Map India Be

Trnava ***Czechoslovakia*** Slovak market and industrial town 44 km (27 miles) north-east of Bratislava. It has many Gothic and Baroque churches and a university founded in 1635, and makes washing machines, hardware and foodstuffs. The country's first nuclear power station opened at nearby Jaslovske Bohunice in 1972.

Population 67 600
Map Czechoslovakia Cb

Trogir ***Yugoslavia*** One of Dalmatia's loveliest towns, lying on a small island 20 km (12 miles) west of Split. It is linked to the mainland by a bridge. At one end of the island stands a Venetian campanile (bell tower) and a Romanesque-Gothic cathedral which has an elaborately carved door. Limestone houses, churches, monasteries and palaces lie between them and the Kastel-Kamerlengo, a ruined 15th-century fortress, at the island's other end. Between Trogir and Split there are the remains of seven medieval castles, from which the 'Riviera of the Seven Castles' or 'Kastela Riviera' gets its name.

Population 6200
Map Yugoslavia Cc

Trois Rivières (Three Rivers) ***Canada*** Industrial town on the north bank of the St Lawrence river in southern QUEBEC. Abundant hydroelectricity has fostered industrial development, which includes the manufacture of paper, iron, textiles and electrical equipment.

Population 111 500
Map Canada Hd

Trojmiasto ***Poland*** See GDANSK

Tromsø ***Norway*** Chief town of north Norway, on a small island some 340 km (210 miles) north of the Arctic Circle. It is the country's biggest fishing centre, especially for herring, and is the supply port for Arctic expeditions and trading. Tromsø has a cathedral and the world's most northerly university, founded in 1968. Its museum has a Lapp research centre and there is an institute for the study of the *aurora borealis* (northern lights). In 1944 the German battleship *Tirpitz* was sunk by British planes off Tromsø.

Population 47 300
Map Norway Ea

Trondheim ***Norway*** Third largest city of Norway after Oslo and Bergen, and the nation's medieval capital. Trondheim is a port in central Norway on the south shore of Trondheimsfjord. The town was founded in AD 977 and first named Nidaros. The 12th-century cathedral is

Trinidad and Tobago

COLOURFUL AND EXCITING, TRINIDAD IS A LAND OF CARNIVAL, CALYPSO AND CRICKET

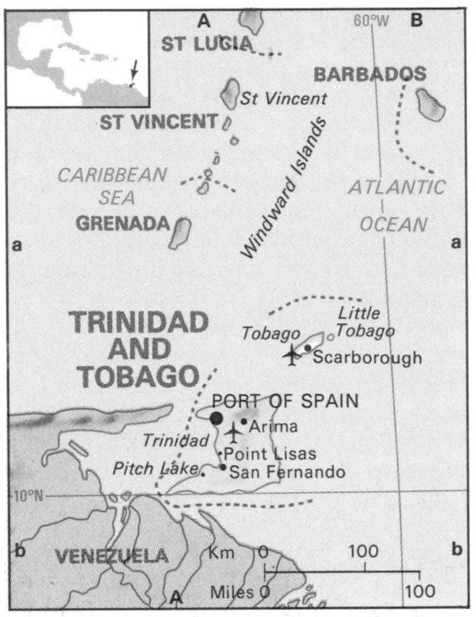

Land of calypso, carnival and steel bands, tropical Trinidad and its tranquil neighbour Tobago lie off the ORINOCO river delta of north-eastern Venezuela. The islands, 31 km (19 miles) apart, have secluded palm-fringed beaches and bays sheltered by coral reefs. They form the third largest Commonwealth country in the West Indies (after the Bahamas and Jamaica).

They were discovered in 1498 by Christopher Columbus, who claimed them for Spain. Trinidad was settled by the Spanish from about 1532 and they retained it until the island was captured by the British in 1797. TOBAGO changed hands several times before it was finally ceded to Britain by France in 1814. The islands were linked to form a single administrative unit in 1889. They gained full independence from Britain in 1962, and in 1976 Trinidad and Tobago was declared a republic.

The people, largely English speaking and keen cricketers, are a complex racial and religious mixture; about 43 per cent are descended from African slaves, 40 per cent are of Asian origin, and the rest are of mixed or European descent. The Asians, both Hindu and Muslim, came to Trinidad from India as indentured labourers to work on the sugar estates after slavery was abolished.

The islands are the most southerly in the Lesser Antilles. Trinidad is crossed by three upland ranges: the mountainous Northern Range, a continuation of the Coastal Range of Venezuela, reaches 940 m (3085 ft) at Mount Aripo. Two undulating plains separate the much lower Central and Southern Ranges.

The climate is tropical. Temperatures vary little throughout the year, but a rainy season lasts from June to December. Trinidad has extensive forests, parts of which have been made into nature reserves and national parks.

Income from oil has provided Trinidad's people with a standard of living unmatched by other Commonwealth countries in the West Indies. It is one of the oldest oil-producing countries in the world. The first well was drilled in 1867, only eight years after the first American oil well, in Pennsylvania. Commercial production from fields in the south-east and south-west began in 1909. In the 1950s an offshore field near the south-west coast was put on stream, and in the 1970s an oil and gas field south-east of the island began production.

Output is small by international standards but oil provides 90 per cent of Trinidad's exports and most of its revenue.

There are two large refineries, and the government has encouraged the development of industries to make use of the oil and natural gas produced, and also imported oil. On the new POINT LISAS industrial estate there are chemical, fertiliser, iron and steel plants, and a natural-gas power station.

Near La Brea in the south of Trinidad is the amazing PITCH LAKE, the world's main source of natural asphalt. Crude oil oozing through porous rocks from an underground reservoir formed the lake, which is 87 m (285 ft) deep in the middle and about 1.5 km (nearly a mile) around the edge.

Agriculture has declined in importance in Trinidad, and food now accounts for about 10 per cent of the country's imports. Sugar is the main crop and is grown mostly on large estates in the west of the island. Cocoa is the second crop, cultivated chiefly by small farmers.

Tobago lives almost entirely on tourism. Visitors are attracted by its relaxed, friendly atmosphere, the hilly scenery, and the fine sandy beaches. The capital is SCARBOROUGH, which consists mainly of hotels and guesthouses. Much of Tobago's agricultural land has been abandoned as farmers have sought a better living in tourism or in the oil industry in Trinidad. As a result, the main crops of cocoa and coconuts have declined considerably.

TRINIDAD AND TOBAGO AT A GLANCE

Area 5128 km² (1980 sq miles)	
Population 1 200 000	
Capital Port of Spain	
Government Parliamentary republic	
Currency Trinidad and Tobago dollar = 100 cents	
Languages English, Hindi, French, Spanish	
Religions Christian (35% Roman Catholic, 15% Anglican), Hindu (25%), Muslim (6%)	
Climate Tropical; average temperature on the coast is 24-26°C (75-79°F) all year	
Main primary products Sugar, cocoa, coffee, rice, citrus fruits, bananas; oil and natural gas	
Major industries Oil and gas production and refining, steel, fertilisers, agriculture, food processing, cement, paints, plastics, tourism	
Main exports Crude oil, petroleum products, chemicals, sugar, cocoa, coffee, fruit	
Annual income per head (US$) 5200	
Population growth (per thous/yr) 15	
Life expectancy (yrs) Male 70 **Female** 74	

built over the tomb of King Olaf II, which is now a shrine. Olaf ruled Norway from 1015 to 1028, died fighting to regain his throne at the Battle of Stiklestad in 1030, and was canonised because of his campaign to convert his people to Christianity. He is now recognised as Norway's patron saint. Until 1906 Norway's kings were crowned in the cathedral, which was much restored in the 19th and 20th centuries. The city's fortifications date from the 17th century. Trondheim also contains Stiftsgården, the largest wooden building (about 1150 m², 12 380 sq ft) in Norway, built in 1774 and now used as a royal residence. The city is a commercial and shipping centre, and manufactures wood products, electronic equipment and clothing.

Population 134 700
Map Norway Cc

Troödos ***Cyprus*** Range running for 60 km (37 miles) across the south of the island. The highest point – and that of the island itself – is Mount Olympus at 1951 m (6401 ft). The peak is also known as Mount Troödos. Named after the Greek mountain which is the legendary home of the ancient Greek gods, Olympus is now the site of an early-warning radar scanner.

The mountains contain copper and asbestos, many Byzantine monasteries and the rare mouflon (wild mountain sheep). Many of the tourist resorts in the hills are winter sports centres with excellent ski slopes.

Map Cyprus Ab

tropical air An air mass that has originated in low latitudes, either over the ocean (tropical maritime), or over a continental interior (tropical continental).

tropical climates Low-latitude climates characterised by high temperatures throughout the year. There are two types: **1.** Tropical maritime with no dry season, though with a summer maximum of rain associated with the arrival of the intertropical convergence zone. This climate covers east coast regions and islands in the tropics (see TROPICAL RAIN FOREST and EQUATORIAL CLIMATE). **2.** Tropical continental with a distinct winter dry season. This is found on the west side of continents and in their interiors, corresponding with the SAVANNAH areas.

tropical cyclone Very low pressure area, 80 to 160 km (50-100 miles) in radius, that originates over the sea in tropical regions and is frequently marked by circulating winds of hurricane strength. At the centre of the circulating winds is a calm – the eye of the storm. Wind systems of this type are called hurricanes in the West Indies and North America, cyclones in Australia and India, and typhoons in east Asia. They can cause great damage when they move inland, though they are generally short-lived over land.

tropical rain forest Dense, evergreen forest of prolific growth, dominated by tall, hardwood trees which are often festooned with trailing lianas. The tree crowns form a canopy beneath which are layers of smaller trees, shrubs and other plants. The large trees are also host to plants called epiphytes which grow on them. Many orchids are epiphytes. The forest has no seasonal rhythm, hence its evergreen appearance, as each tree – or even branch – buds, flowers and sheds its leaves independently of the other vegetation.

This forest is typical of areas of hot, humid, tropical maritime climates, such as the Amazon basin, Zaire basin, and the coastal lowlands of South-east Asia, Central America and West Africa. Also known as Selva. See also TEMPERATE RAIN FOREST

tropics Two lines of latitude 23.5°N and 23.5°S of the Equator, which mark the limits of the zone within which the noon sun is overhead at some time of the year. The northern limit (Tropic of Cancer) is reached on June 21 and the southern limit (Tropic of Capricorn) on December 21 – the solstices.

tropopause The boundary between the troposphere and the stratosphere in the earth's atmosphere. It can be seen from high-flying jet aircraft, especially over oceans, appearing as a hazy layer containing trails of cirrus cloud.

troposphere See ATMOSPHERE

trough See TRENCH

trough of low pressure A narrow area of low pressure between two areas of higher pressure, often an extension of a DEPRESSION.

Troy (Ilium) ***Turkey*** Ruined city at the southern end of the Dardanelles strait, about 270 km (165 miles) south-west of Istanbul. Excavations in the late 19th century have revealed nine levels of occupation dating back to the Bronze Age and extending for more than 4000 years up to the Roman period. The largest remains are the House of Pillars and the city's fortifications and towers.

The story of the legendary Trojan War – which may have been based on a real war fought in the 13th century BC – is told in the *Iliad*, the epic poem by the Greek poet Homer.

Map Turkey Ab

Troyes ***France*** City on the Seine river 140 km (90 miles) south-east of Paris. It gave its name to the troy system of measuring weights, probably first used at medieval trade fairs there and still used for weighing gems and precious metals. Textiles, knitwear and electrical goods are now the city's main industries.

Population 127 200

Map France Fb

Trujillo ***Peru*** Capital of La Libertad department. It was founded in 1534 by Spanish colonists, and refounded the next year by the Spanish adventurer Francisco Pizarro who named it after his birthplace. It is a bustling modern city, but retains much of its colonial heritage, including the large cathedral – twice rebuilt after earthquakes in the 17th and 18th centuries – the monastery of San Carmen and the churches of La Compania and San Francisco.

Population 750 000

Map Peru Ba

▼ STEAMY HEAT Swathes of mist rise to the tree-top canopy in Surinam's tropical rain forest. It covers three-quarters of the country.

Truk ***Federated States of Micronesia*** Most populous state of the island federation, with more than half the total population. It consists of a number of atolls with nearly 300 small islands in all; the administrative centre is Moen island in Truk atoll. Truk was a Japanese fortress and major naval base in the Second World War and was heavily bombed in 1944. In its lagoon are many sunken Japanese ships.
Population 38 000
Map Pacific Ocean Cb

Truong Son (Annamite Chain) ***Vietnam*** See MIEN TRUNG

Trust Territory of the Pacific Islands ***Pacific Ocean*** United Nations trusteeship established in 1947, administered by the USA and consisting of four territories. Two of them – the MARSHALL ISLANDS and the Federated States of MICRONESIA – had in the mid-1980s signed independence agreements allowing the Americans to retain defence rights in order to exclude the military presence of other nations. The path of the third territory, BELAU, to independence has been roughened by its ban on nuclear materials from its territory. The fourth – the NORTHERN MARIANAS – decided to keep closer American links by becoming a Commonwealth of the USA. All the agreements have to be ratified by the United Nations. The trusteeship was created after the Second World War defeat of Japan, which from 1922 administered the islands under a League of Nations mandate.
Map Pacific Ocean Cb

Tsaratanana Massif ***Madagascar*** Highlands in the north which include the island's highest peak, Maromokotro (2876 m, 9436 ft).
Map Madagascar Aa

Tsavo National Park ***Kenya*** Wildlife reserve covering 21 000 km² (8108 sq miles) of thorny scrubland either side of the main road and railway between Mombasa and Nairobi. It was the setting for Colonel Patterson's book *The Man-eaters of Tsavo*. Luxurious tourist lodges near waterholes such as Mzima Springs, 40 km (25 miles) south-west of the road, offer close-up views of the reserve's large population of lions and elephants.
Map Kenya Cb

Tselinograd ***USSR*** City in Kazakhstan 190 km (120 miles) north-west of Karaganda; formerly called Akmalinsk. Originally a market town for cattle in semiarid steppe country, it became a key base in the 1950s and 1960s for a Russian project to open up the so-called Virgin Lands of Siberia for settlers and agriculture. Its industries now include meat packing, farm machinery and fertilisers.
Population 256 000
Map USSR Ic

Tshikapa ***Zaire*** Town on the Kasai river, about 580 km (360 miles) south-east of the capital, Kinshasa. It is the centre of Zaire's gem diamond industry. Diamonds were first discovered there in 1907.
Population 100 000
Map Zaire Bb

Tsien Tang Kiang ***China*** See FUCHUN JIANG

Tsuen Wan ***Hong Kong*** Most populous of the eight new towns established by Hong Kong in the NEW TERRITORIES to the north-west of KOWLOON. Founded in 1963, it is more self-sufficient than the other new towns and contains many factories which provide local employment opportunities. An industrial city of skyscraper apartment blocks and broad looping highways, it is connected to Kowloon and VICTORIA by the underground Mass Transit Railway. Tsuen Wan's container terminal, at Kwai Chung, is one of the largest in the world.
Population 700 000
Map Hong Kong Ca

Tsukuba ***Japan*** A 'science city', 60 km (37 miles) north-east of Tokyo, begun in the late 1960s and used as the site of the 1985 Expo industrial exhibition. It has been designed to house research and educational institutions relocated from Tokyo.
Population 150 100
Map Japan Dc

tsunami Very large ocean wave caused by an underwater EARTHQUAKE or volcanic eruption. Tsunami – the word is Japanese for 'overflowing wave' – increase in size as they approach the coast, and surge inland over coastal areas, causing much destruction. The largest recorded was one 85 m (278 ft) high off Ishigaki Island in the Ryukyu group of islands of south Japan in 1971. A tsunami 37 m (120 ft) high which hit the coasts of Java and Sumatra after the Krakatoa eruption of 1883 drowned 36 000 people.

Tsuruoka ***Japan*** Old town near the west coast of northern Honshu island, about 340 km (210 miles) from Tokyo. The Chidokan, once a school for the children of samurai warriors, is now a national historic property.
Population 100 200
Map Japan Cc

Tsushima ***Japan*** Island in Tsushima Strait, between Kyushu island and South Korea. The Russian fleet was annihilated by the Japanese under Admiral Togo near the island in 1905, a battle which established Japan as a world power.
Population 48 900
Map Japan Ad

Tuamotu ***French Polynesia*** Archipelago of about 80 islands, mostly low-lying coral atolls, stretching over more than 1250 km (800 miles) of the Pacific to the east and north-east of Tahiti. The islands produce pearls and pearl-shell (especially at Hikueru), coconuts, breadfruit and pandanus (screw pines, whose leaves are used for thatch and baskets). Mururoa atoll is used for French nuclear tests; Rangiroa is the largest atoll. *Kon Tiki* – a raft built by the Norwegian explorer Thor Heyerdahl to support his theory that the Polynesian peoples could have migrated from South America – landed on Raroia atoll at the end of its voyage in 1947.
Population 11 200
Map Pacific Ocean Fc

Tübingen ***West Germany*** Medieval university town on the Neckar river about 35 km (22 miles) south of Stuttgart. The university was founded in 1477, and among its students have been the philosopher Georg Hegel (1770-1831) and the astronomer Johannes Kepler (1571-1630).
Population 75 000
Map West Germany Cd

Tubruq ***Libya*** See TOBRUK

Tubuai (Austral) Islands ***French Polynesia*** A 1300 km (800 mile) chain of volcanic islands, atolls and reefs about 600 km (375 miles) south of Tahiti. Rurutu and Tubuai are the main islands.
Population 6300
Map Pacific Ocean Fd

Tucson ***USA*** City in southern Arizona, about 100 km (60 miles) north of the Mexican border. It was founded in 1776 in a desert valley surrounded by mountains, and is the seat of the state university. The city (whose name is pronounced 'Tooson') attracts visitors with its dry, sunny climate. It has a number of high-technology industries and is a service centre for the surrounding mining and agricultural area.
Population (city) 365 400; (metropolitan area) 594 800
Map United States Dd

tuff Rock formed from ash ejected from a volcano and fused together on the ground. The particles are mostly less than 2 mm (0.08 in) across.

Tugela ***South Africa*** River rising in south Natal near the Lesotho border in the Drakensberg range, and flowing 500 km (310 miles) east into the Indian Ocean about 75 km (45 miles) north-east of Durban. It hurtles over waterfalls and through narrow gorges for much of its course. The Tugela Falls, a series of five falls with a total drop of 948 m (3110 ft), include a single, uninterrupted fall of 411 m (1350 ft), which is Africa's highest waterfall.
Map South Africa Db

Tula ***Mexico*** Archaeological site in the central state of Hidalgo, which is believed to be Tollán, the ancient capital of the Toltec civilisation which flourished in the 10th to 13th centuries AD. The site is dominated by the Atlantes – giant carved columns of black basalt which stand 5 m (15 ft) high on the central pyramid. The columns once supported the roof of the priests' temple, and represent Toltec warriors. Tula also contains a well preserved frieze of jaguars, eagles and coyotes, and a reclining statue of the god 'Chac-Mool'.
Map Mexico Cb

Tula ***USSR*** City 190 km (120 miles) south of Moscow, founded in 1146. The site of Tsar Peter the Great's arms factory in the 18th century, it is now an industrial town producing weapons, sewing machines, musical instruments, machinery and samovars (Russian tea urns). Chemicals are produced from local deposits of coal, and iron and steel from local ore.

There is a Museum of the History of Arms (built 1724), and opposite this is the 1530 kremlin (fortress).
Population 529 000
Map USSR Ec

Tulcea ***Romania*** Main inland port of the Danube Delta, standing 70 km (43 miles) west of the Black Sea. Tulcea is built on tree-clad slopes overlooking a graceful bend of the river. It is a halfway mark for seagoing cargo ships sailing to and from the upriver port of Braila. Tulcea was noted as a port in the time of the Persian king Darius the Great (about 549-486 BC). Today it is also a manufacturing centre, producing food, timber and building materials. Its Museum of the Danube Delta houses many examples of the plants and wildlife found in the area's nature sanctuary and game reserve.

Population 70 200

Map Romania Ba

Tulsa ***USA*** City in north-east Oklahoma, on the Arkansas river about 160 km (100 miles) north-east of Oklahoma City. Tulsa stands in the heart of an oil field and is a refining centre. It also makes aircraft and metal goods.

Population (city) 374 500; (metropolitan area) 725 600

Map United States Gc

Tumbes ***Peru*** City and department on the border with Ecuador. It is a major oil-producing region. The Spanish adventurer Francisco Pizarro landed on the coast in 1532 near the present city, which is surrounded by a fertile oasis growing tropical and subtropical crops such as rice, bananas and maize.

Population (department) 103 800; (city) 47 900

Map Peru Aa

tundra Area between the perpetual snow and ice of the Arctic regions and the tree-line of northern latitudes, having a permanently frozen subsoil. Tundra has only low-growing vegetation such as lichens, mosses, dwarf shrubs and stunted birch trees.

Tungabhadra ***India*** River in southern India, about 640 km (400 miles) long. It is formed from the Tunga and Varada which rise in the Western Ghats, and flows north-east to the Krishna river. The Tungabhadra Dam, 2441 m (8008 ft) long and 49 m (162 ft) high, is the world's biggest masonry dam.

Map India Bd

Tunguska ***USSR*** Two rivers, tributaries of the Yenisey, in central Siberia. The Stony (Podkamennaya) Tunguska is 1550 km (960 miles) long; the Lower (Nizhnyaya) Tunguska is 2690 km (1670 miles) long.

In June 1908 the Tunguska plateau was the scene of a huge explosion caused by a fireball – possibly a large meteorite that crashed to earth. It made a crater 1.5 km (almost 1 mile) wide and about 190 m (625 ft) deep; trees were destroyed up to 32 km (20 miles) away from the point of impact.

Map USSR Kb

Tunis ***Tunisia*** Capital of the country and the political, administrative, commercial and industrial centre. It is built on the west bank of a shallow lake, Lac de Tunis. An artificial harbour is linked by a deep-water channel to the Bay of Tunis where the new port at La Goulette serves the city and receives goods which are then distributed throughout the country.

Tunis lies near the site of ancient CARTHAGE and existed as a small town during the Carthaginian Empire. It remained unimportant until the 7th-century Muslim conquest, and then in the 9th century it became the administrative capital of Tunisia (replacing KAIROUAN). It consists of two adjacent districts, widely different in character – the Muslim old town (the medina) and the modern city.

The main entrance to the old, walled medina is through the Bab el-Bahar Gate. Inside is an impressive collection of medieval Islamic buildings – there are more than 700 ancient monuments including mosques, mausoleums, palaces and madrassas (Islamic places of learning) which have been turned into apartment houses. The medina is a place of winding alleys with *souks* (markets) selling their wares in vaulted and ceramic-tiled *zawias* (courtyards). Various streets specialise in particular crafts – textiles, brass and copperwork, gold and silver jewellery, leatherware, perfumes and spices.

Other places of interest include the 9th-century Great Mosque of Zitouna; the Lapidary Museum of Sidi Bou Krissan (in the garden of the 11th-century el-Ksar Mosque) containing a collection of Arab tombstone inscriptions; the 18th-century Palace of Dar Hussein, housing the National Institute of Archaeology and Arts; and the Museum of Islamic Art.

About 5 km (3 miles) north of the city is the Bardo Museum, which houses one of the finest collections of Roman, Carthaginian and Islamic art in the world. Just north of the port of La Goulette lie the ruins of Carthage.

Population (Greater Tunis) 1 180 877

Map Tunisia Ba

Tunisia See p. 652

▲ RUINED METROPOLIS The capital of the powerful Toltecs near Tula (Mexico) covered more than 8 km² (3 sq miles) and was home to over 20 000 people.

Tunja ***Colombia*** Capital of Boyacá department, in the eastern Andean mountains about 135 km (85 miles) north-east of Bogotá. It was founded by the Spanish in 1539, on the site of a Chibcha Indian capital. The decisive battle in the fight for Colombia's independence was fought in 1819 by revolutionary troops under Simón Bolívar at Boyacá Bridge, 16 km (10 miles) south of the city. A throne of the Zipa, the Chibcha chief, is carved into a rock face just outside the city.

Population 113 000

Map Colombia Bb

tunnels See UNDER THE SEA BY BULLET TRAIN (p. 653)

Turin (Torino) ***Italy*** Second city of northern Italy after Milan and capital of Italy between 1861 and 1865. It stands on the Po river about 125 km (78 miles) south-west of Milan. The 20th-century growth of Turin has been fuelled by the growth of the Fiat car empire, founded there by Giovanni Agnelli in 1899.

The heart of Turin, with its Roman grid street layout, is well-ordered and picturesque, with fine squares and tree-shaded boulevards. Apart from a Roman gateway, some relics of the medieval citadel and its Romanesque cathedral, Turin's important buildings date almost entirely from the 17th and 18th centuries, when it was a ducal capital. Its link with the former Duchy of Savoy gives it an architectural tone more French than Italian.

Turin's most sacred possession, kept in the Santa Sindone chapel of the cathedral, is a piece of cloth measuring 4.1 by 1.1 m (13.5 by 3.5 ft) that bears the faint image in negative of a man. Many Christians believe that this is the shroud in which Christ's body was wrapped after his Crucifixion. The shroud, which has been the subject of many investigations, was last put on display in 1978. *(continued on p. 653)*

Tunisia

ONE OF THE MOST EUROPEANISED PARTS OF THE ARAB WORLD, TUNISIA IS A PRO-WESTERN AND PROGRESSIVE ISLAMIC SOCIETY

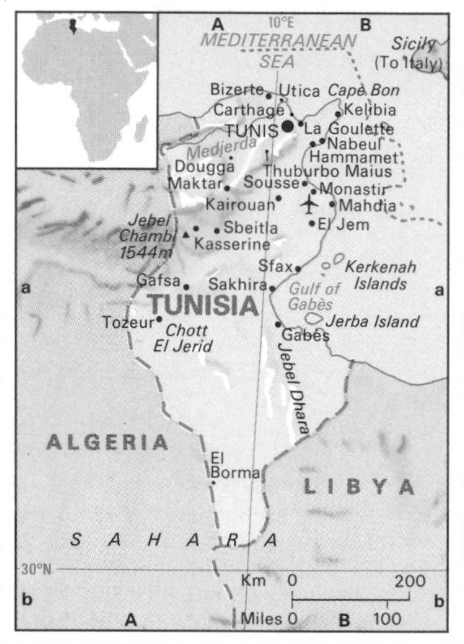

Phoenician traders first sailed to Tunisia's jasmine-scented coast in 1200 BC from Syria and Lebanon. In 814 the Phoenicians founded CARTHAGE, the seat of one of the greatest and richest ancient empires until the Romans razed it in 146 BC. Vandals, Byzantines, Arabs, Turks and the French have invaded over the centuries. Now over 2 million tourists invade every year, drawn by the beautiful beaches, the sunshine and the relics of great civilisations.

Tourism is one of the mainstays of the modern economy. Oil, produced in the SAHARA near the Algerian border, and phosphates are also great revenue-earners and these three sources of wealth have underpinned the traditional agricultural economy.

Tunisia was dominated by the French from 1881 until independence in 1956. A year later the country overthrew its Bey (King) Sidi Lamine, to become a republic under President Habib Bourguiba. He has led Tunisia ever since, managing to stay out of serious international trouble. Relations with Colonel Gaddafi's Libya to the east and south have been uneasy since Bourguiba thwarted plans to unite Tunisia with Libya in 1974. But peace is maintained with large and powerful Algeria to the west, and Tunisia is remarkable among Arab countries for its stability and moderation.

Northern Tunisia consists of plains, hills and valleys, and is divided south-west to north-east by ranges which extend from the eastern end of the ATLAS MOUNTAINS. The hills are forested with evergreen oaks, including the cork tree whose outer bark is stripped for cork. The Cape Bon peninsula east of Tunis is the country's market garden and is fragrant with orange and lemon groves, mimosas, magnolias, jasmine and roses. Farther south along the east coast are the orchards and olive groves around the resort of SOUSSE, and vast olive groves around the port of SFAX; olive oil brings 15 per cent of Tunisia's export earnings – half as much as petroleum revenue. Off the coast are small islands, including the popular palm-covered resort island of JERBA in the hot south.

Inland, the Dorsale 'backbone' mountains separate the Tell coastal zone from the central plains before the land drops to the *chott* country. Chotts are white salt pans, sometimes lower than sea level, which turn to quagmires after rain but are cracked and dry in summer, and glisten with a strange mirage effect. Around the chotts are fruitful oases – with luxury hotels. Tourist excursions run from the hotels into the Sahara in the southern half of Tunisia.

Most of the people of Tunisia are Arabic-speaking Muslims. Arabic is the official language but French is widely spoken, and three of the five main Tunisian newspapers are in French. More than half the population live in towns, which are mainly along the east coast from BIZERTE to Sfax, and especially in the capital, TUNIS. There are still some Europeans in the cities, particularly the descendants of French and Italian settlers. But most settlers left at the time of independence, which followed guerrilla warfare in the 1950s. Greeks, Spaniards and Sicilians still make a home here. Tunisia also has one of North Africa's most substantial Jewish communities around Tunis and Jerba, as well as a few Berber tribesmen who live by farming in the hills of the centre and south. Black descendants of slaves inhabit the oases of the desert.

INDUSTRIES NEW AND OLD

Tunisia's varied resources have given its people a relatively good standard of living. There is a huge gulf between rich and poor still, but some of the government's projects have been to the general good. Health facilities have been improved in recent years, and Tunisia boasts one of the best educational systems in Africa, free all the way from primary level through to university.

Industry is growing, making use of iron ore, lead and zinc deposits, and substantial offshore reserves of natural gas, as well as oil and phosphates. The government has encouraged foreign industrial investment with tax concessions and the availability of cheap labour. More traditional industries process agricultural products – olives, wheat, barley, almonds, fruit, vegetables and dates – as well as making good wine.

But it is the ancient crafts that give Tunisia much of its colourful reputation. Women sitting on low cushioned seats weave beautiful carpets on hand looms, or embroider cloth with traditional patterns of flowers and leaves in brilliant colours. Ceramic tiles are patterned with elaborate interwoven floral shapes, in the familiar Islamic style which decorates the interiors of mosques. It was these tile patterns that the Victorian painter William Morris transferred to his renowned wallpapers, bringing the exotic Arab world into British homes. Pottery, lacework, copperwork and leather crafts also contribute to the bright array of goods for sale in the *souks* (markets) which fill the narrow alleys of the *medinas* (old Muslim quarters) of the towns.

It is in the medinas that the mosques are found. Although Muslim, Tunisia is not a fanatical country, and since independence it has liberated its women from the veil. However, the 9th-century Grand Mosque at KAIROUAN is one of the holiest places for Muslims in the world; to visit Kairouan seven times is an act of piety equal to a visit to MECCA. Tunisia is rich in Islamic architecture, with fine domed and minareted mosques in Tunis – the Zeitouna – Sousse and Sfax, as well as Kairouan.

MODERN UNCERTAINTIES

Tunisia also has exceptional Roman remains. Its countryside is dotted with temples, baths, amphitheatres, forums and colosseums – at DOUGGA, EL JEM, Maktar and UTICA, for example. The legends of Carthage draw the tourists there, although, lying as it does in a suburb of Tunis, the site may be disappointing. Such was the hatred of the Romans for the civilisation they destroyed at Carthage, that they levelled the city, burnt the ground for seven days, and salted the earth to make it infertile. What remains are the ruins of the city the Romans built in its place.

In today's Tunisia, the prosperity that has been enjoyed is suffering; mainly because of the fluctuations in international markets, unemployment is about 20 per cent and there are rising taxes and inflation. Discontent with the results of this struggling economy even led briefly to riots. And squabbling between political parties threatens the stability which has characterised Tunisian politics in the past.

TUNISIA AT A GLANCE
Area 163 610 km² (63 170 sq miles)
Population 7 530 000
Capital Tunis
Government Parliamentary republic
Currency Dinar = 1000 millimes
Languages Arabic (official), French
Religions Muslim (99%), Jewish and Christian minorities
Climate Temperate on the coast, very dry inland. Average temperature in Tunis ranges from 6-14°C (43-57°F) in January to 21-33°C (70-91°F) in August
Main primary products Wheat, barley, olives, tomatoes, dates, timber, citrus fruits, livestock, fish, wine; crude oil and natural gas, phosphates, salt, iron, lead
Major industries Mining, tourism, agriculture, oil refining, textiles, phosphate processing, cement, steel, food processing
Main exports Crude oil and refined products, phosphates, textiles, chemicals, food
Annual income per head (US$) 1100
Population growth (per thous/yr) 24
Life expectancy (yrs) Male 59 **Female** 63

UNDER THE SEA BY BULLET TRAIN

The world's first successful underwater transport tunnel – the Thames Tunnel – was built in London between 1825 and 1843 and runs for 1253 m (4111 ft) between Rotherhithe and Wapping. It is still used as a railway tunnel. By contrast, the world's longest railway tunnel runs for 54 km (33.5 miles) and links the Japanese island of Hokkaido with the island of Honshu. It will be opened in March 1988 after more than 24 years' work, and passengers may travel through it on the Shinkansen, or Bullet Train, though its final use has still to be decided.

THE WORLD'S LONGEST TRANSPORT TUNNELS

Tunnel	Location	Length (km/miles)	Opened
Seikan (underwater rail)	Japan	54/33.5	1988
Moscow Metro	USSR	31/19 (Medvedkovo–Belyaevo)	1979
London Underground	UK	28/17 (East Finchley-Morden)	1939
Daishimizu (rail)	Japan	22/13.6	1982
Simplon II (rail)	Switzerland-Italy	20/12.4	1922
Shin-Kanmon (underwater rail)	Japan	19/11.8	1975
Great Appenine (rail)	Italy	18.5/11.5	1934
St Gotthard (road)	Switzerland	16/10	1980
Rokko (rail)	Japan	16/10	1972
Henderson (rail)	USA	15.8/9.8	1975

Modern Turin is one of Italy's leading centres of publishing, art, learning and fashion. Its major industries, apart from motor vehicles, include chemicals, textiles, metal goods, rubber, plastic, pharmaceuticals and television and radio sets.

Population 1 103 500
Map Italy Ab

Turkana, Lake ***Kenya/Ethiopia*** Lake extending more than 240 km (150 miles) north to south, most of it in Kenya. Its maximum width is less than 55 km (35 miles) since it is hemmed in on the east and west by the walls of the GREAT RIFT VALLEY. The lake was formerly known as Lake Rudolf, and is often called 'The Jade Sea' because of its colour. On its barren shores, tools and bones of man-like creatures 2.5 million years old have been found.

Water flows into the lake from the Omo river in the north and the Turkwel and Kerio rivers in the south, but there is no outlet. The lake's level is maintained by evaporation in the fierce heat – a process that leaves the water salty. Fish caught commercially in the lake are sent some 450 km (280 miles) south to the Kenyan capital, Nairobi. Turkana is also the name of a large district west of the lake, and of the tribe of nomadic herdsmen who live there.

Map Kenya Ca

Turkey See p. 654

Turkmenistan ***USSR*** Constituent republic of the USSR to the east of the Caspian Sea, beside the borders of Iran and Afghanistan; its area is 488 100 km² (188 450 sq miles). The west and central parts (over 80 per cent of the region) consist of the KARA KUM desert. The east, which is bordered by the Amudar'ya river, is a plateau. It is the driest of the Soviet republics, and the population is concentrated in irrigated oases along the region's rivers. Their chief occupations are growing cereals, fruit and cotton, and rearing Karakul sheep. Silk, oil, natural gas and sulphur are also produced. At Krasnovodsk, on the Caspian Sea, water is desalinated by atomic power. The capital ASHKHABAD, beside the Iranian border, is irrigated by the Kara Kum Canal.

The area has been settled by nomadic peoples since ancient times, and was on the caravan routes linking Asia with Europe. It was under Persian rule in the 6th-4th centuries BC, and was subsequently invaded by Alexander the Great, by the Arabs and by the Tatars and Mongols. It was ceded to Russia in the late 19th century, but retains its own language and Muslim religion.

Population 3 118 000
Map USSR Ge

Turks and Caicos Islands ***West Indies*** British colony consisting of about 30 small and scrub-covered islands south-east of the BAHAMAS. They cover a total of 430 km² (166 sq miles). They were discovered by Christopher Columbus in 1492. For 300 years the main industry on the islands was the production of salt by evaporating sea water – an occupation that began after salt-rakers from BERMUDA settled there in 1678.

Most of the islanders today are descendants of African slaves brought from the USA in the 18th century and freed in the mid-19th century. Fishing – particularly for lobster and conch – became an important export industry after demand for salt declined in the early 1960s, but since then tourism has taken over as the main employer. Indeed, direct flights have been introduced from Miami to the new (1984) international airport on the main tourist island of Providenciales. In addition, the islands' status as a tax haven has encouraged many offshore financial companies to open offices there. However, scandal was brewing in the mid-1980s with allegations of organised crime operations in the islands.

Only six of the islands in the group are inhabited – among them Grand Turk, the site of the capital, COCKBURN TOWN.

Population 7400
Map Caribbean Aa

Turku (Åbo) ***Finland*** The country's third largest city – after Helsinki and Tampere – and a principal seaport, on the Gulf of Bothnia, 155 km (95 miles) west and slightly north of Helsinki. Founded in 1229, it was the capital of Finland until 1812. It was also the site of Finland's first university – founded in 1640 – which was transferred to Helsinki after Turku was mostly destroyed by fire in 1827. Today there are two universities – Turun Yliopisto (Finnish-speaking) and the Swedish-speaking Åbo Akademi. There is a restored 13th-century castle, and a late-medieval cathedral built in the German Gothic style.

The city's industries include shipyards, engineering, food processing, textiles and clothing, tobacco processing and cement. The city celebrates its own history on Turku Day, held in mid-September each year.

Population 161 400
Map Finland Bc

Turpan Hami ***China*** Basin of western China at the eastern end of the TIEN SHAN range, covering 50 000 km² (19 305 sq miles). It is renowned for its summer heat, when the temperature can rise above 47°C (117°F). The centre of the basin lies 154 m (500 ft) below sea level, the lowest part of the country. Hardly any rain falls in the basin, but fruit, wheat and cotton are grown with the aid of irrigation.

Map China Cc

Tuscaloosa ***USA*** City in west-central Alabama, about 75 km (45 miles) south-east of Birmingham. It has been the seat of the University of Alabama since 1831, and its industries are based on locally produced coal, iron, cotton and timber.

Population (city) 73 200; (metropolitan area) 138 900
Map United States Id

Tuscany (Toscana) ***Italy*** Central region of 22 992 km² (8877 sq miles) of which FLORENCE is the capital. It is the heartland of the nation's medieval history and culture, with such historic towns and cities as LUCCA, PISA and SIENA, and the Tuscan dialect is the foundation of modern Italian. *(continued on p. 656)*

Turkey

ONLY HALF A CENTURY AGO A BACKWARD MUSLIM COUNTRY, TURKEY HAS NOW EMERGED AS A MODERN WESTERN STATE

Turkey forms the bridge between Europe and Asia, with land on both continents. It guards the sea passage between the Black Sea and the Mediterranean, which runs through the BOSPORUS, the Sea of MARMARA and the DARDANELLES, scene of First World War slaughter. For 1000 years Turkey was the hub of the Byzantine Empire; for nearly 500 years it was at the centre of the Ottoman Empire; and today it forms the south-eastern flank of the NATO alliance.

The Turks, who once controlled a quarter of Europe, have a reputation as a race of warriors, and violence lies near the surface in domestic politics. Three times since 1960 the military have taken over the government. The last time, in 1980, some 20-30 people a day were being killed by political extremists on the left and the right before the army moved in. In 1983 three political parties were allowed to contest an election, and the Motherland Party came to power with plans to reduce long-standing state controls.

THE FACE OF THE LAND

Modern Turkey has an area of about 780 000 km^2 (301 000 sq miles) and a population of about 50 million – it is the most populous country in the Middle East. It consists of THRACE, on the European mainland, and the much larger area of ANATOLIA in Asia. European Turkey is fertile agricultural land with a Mediterranean climate. Anatolia is shaped like a saucer, with a mountainous rim and a central plateau which averages 750 m (2460 ft) high. The Pontus Mountains in the north and the TAURUS MOUNTAINS in the south converge eastwards, rising to 5165 m (16 945 ft) at Mount ARARAT. This mountain is the legendary resting place of Noah's Ark, but expeditions to the site have found no concrete evidence of the Bible story.

The weather ranges from Mediterranean warmth on the coasts to the hot summers and bitterly cold winters of the central plains and the snows of the high mountains. Average daily temperatures in the capital, ANKARA, vary from −4°C (25°F) in January to a summer maximum of 31°C (88°F). There are occasional destructive earthquakes.

About a quarter of Turkey is covered by forests, nearly all of which are state-owned. These are mainly coniferous, but also include deciduous beech, poplar, walnut and oak. The north is the most densely wooded area, and there are subtropical forests by the BLACK SEA. In the south and west, large areas are covered by thick Mediterranean scrub.

▲ DOMES OF GLORY With the mosque of Bayezid II in the foreground, the mystery and beauty of Istanbul unfold. Within the city walls are byways with names like the Elephant's Path and the Street of the Chicken That Could Not Fly.

Animal life includes wild boars, bears, lynx and leopards. Wolves can be a menace in the mountains, and cause havoc among flocks of sheep when the snows drive them to lower ground. Each village has 'wolf dogs' to protect its sheep; they are smaller than sheepdogs but more powerful, and trained to attack wolves. Walkers can also be at risk in winter – hungry wolves may attack a lone traveller who is unwise enough to venture between remote villages. Migrant birds abound in the Bosporus in autumn, heading south in winter.

RICH MIXTURE OF PEOPLE

The first Turks came from the central steppes of Asia, but over the centuries the population has become a rich blend of peoples and cultures. They are of Mediterranean stock in the west, and Armenians from the CASPIAN area in the east; while in the south-east there are several million Kurds, ethnically close to the Iranians. Ninety per cent of the people speak Turkish, a language of many dialects which was changed from Arabic script to Roman alphabet in 1928. Most are Muslims.

The country has a substantial industrial sector, good agricultural land and mineral resources, including large deposits of chrome. Industry is mainly around ISTANBUL and Ankara, and in the south-east near ADANA and ISKENDERUN. Manufactures include iron, steel, textiles, motor vehicles, petrochemicals, processed food and Turkey's famous carpets, tobacco, meerschaum pipes and pottery. The two great rivers of the Middle East, the TIGRIS and EUPHRATES, both rise in Turkey, and hydroelectric plants supply half of the country's energy needs.

Encouraged by the government, farmers have achieved the best rate of agricultural production in the Middle East outside Israel. Despite its dryness, the Anatolian plateau is the granary of Turkey. The main crops are cereals, rice, cotton, fruit and tobacco, and most of the country's sheep, goats and cattle are raised there.

Mechanisation and an increasing population have forced many peasants to seek jobs in the overcrowded cities. Others have emigrated. About 1.5 million Turks, for instance, work in West Germany, while some 175 000 are in the oil-rich Middle Eastern states. Their remittances home help the country's continual struggle to pay for its imports.

HOW EMPIRES ROSE AND FELL

The first settlers came to Anatolia in about 7000 BC, and the Hittites, warriors from central Asia, founded an empire there about 1800 BC. In turn the Persians, Macedonians and Romans came; Constantine the Great moved the eastern Roman capital to Byzantium (now Istanbul) in AD 330. He renamed the city Constantinople. When the empire split, the eastern half continued as the Byzantine Empire. The Ottomans, a Turkish tribe, took Constantinople in 1453, and in the 16th century their Muslim empire stretched from the Danube to The Gulf, from the Crimea to Morocco. But then a slow decline began.

In the First World War Turkey sided with Germany and defeated an Allied invasion force at GALLIPOLI on the Dardanelles, but its armies were beaten back in the Middle East. The Ottoman Empire came to an end, but nationalists led by Mustafa Kemal, a successful general, rejected the proposed peace, which favoured Turkey's old rival Greece. They set up a national assembly, deposed the Sultan and drove out the Greeks.

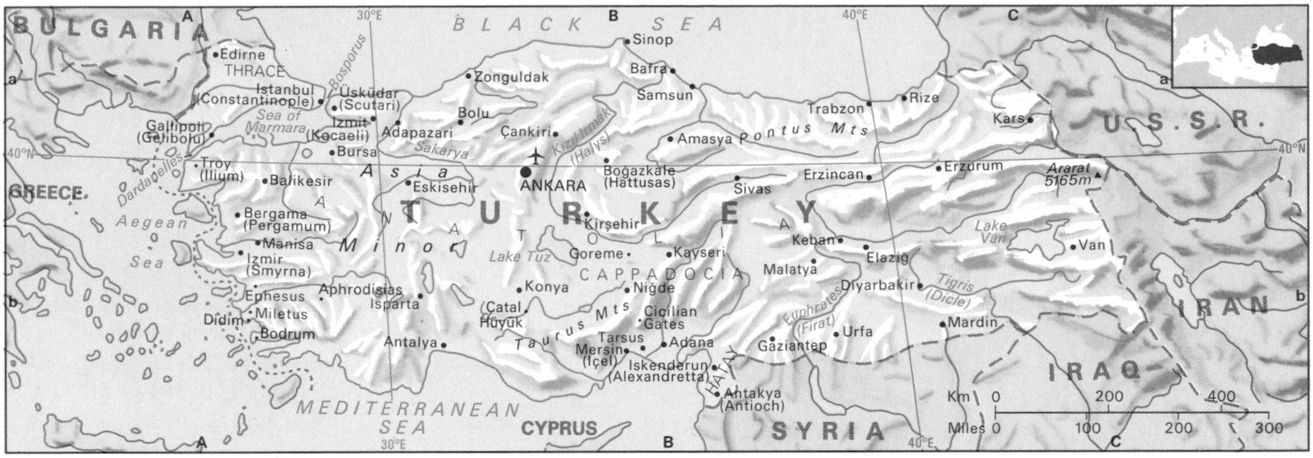

Kemal, who later took the surname Ataturk, became president of the republic in 1923 and ruled virtually as a dictator until 1938. He was the father of modern Turkey, bringing a backward Muslim state into the modern world. To Westernise his country, he separated religion and the state, abolished polygamy, banned men from wearing the fez (the brimless felt cap of Muslims) and discouraged women from using the veil. He sent everyone under 40 to classes to learn a new Latin alphabet for the Turkish language. He decreed that everyone should have a surname.

In the Second World War Turkey remained neutral until February 1945, when it declared war on Germany. Turks fought alongside Americans in the Korean War, and the USA still maintains military bases in Turkey. In 1974, after a Greek-inspired coup in CYPRUS, the Turks invaded the island, claiming to protect the Turkish minority; they continue to hold the northern half with about 20 000 troops. Other points of dispute also irritate Turkish-Greek relations, including claims to minerals beneath the AEGEAN SEA.

In Turkey itself, military leaders keep a close watch over the latest civilian government. After the 1980 coup, many leading figures were banned from politics and, according to Amnesty International, some 20 000 people were held as political prisoners – some being tortured. The Press was censored. Between 1980 and 1983 Turkish exports doubled, but inflation rose to more than 100 per cent and is still high.

LIFE OF THE PEOPLE

The country needs increased prosperity to improve the lot of many people who live in poverty and discontent while the professional and middle classes enjoy a comfortable lifestyle. Urban life is centred on the three largest, most industrialised and Westernised cities: Istanbul, Ankara and IZMIR.

Life for most Turks remains simple. The small café in town or village is the place where men meet to drink tea and talk, read the papers or play *tavla* (backgammon). The café reflects a male-dominated Islamic society. There have nevertheless been women cabinet ministers, and in the cities women take part in professional activities and are becoming more influential in public life. But generally far fewer jobs are open to them. Even a waitress is uncommon. In the country women remain subservient to men. Women are tolerated in mosques, for example, but traditionally are expected to say their prayers at home. The custom of arranged marriages and payment of dowries still persists.

Turkish food is distinctive – a happy blend of the culinary traditions of central Asian pastoral people and the Mediterranean influence. The fact that the country produces all its own food is reflected in the freshness of its vegetables, fruit, meat and fish. The coffee is world-famous but has become increasingly expensive. The wine is often excellent.

The national sports are soccer, wrestling, horse riding and archery. A Turkish form of wrestling is *yagli gures*, greased wrestling. The contestants, who wear long tight leather breeches, pour olive oil over themselves to make holds more difficult.

FOR THE VISITOR

For the adventurous tourist, Turkey offers high mountains and rugged scenery to the north, south and east of the Anatolian Plateau, as well as the spectacular landscapes of the plateau itself. For the historically inclined, the plateau has ancient Neolithic remains (ÇATAL HÜYÜK), Hittite (BOGAZKALE), Greek (TROY) and Roman (ANTAKYA and KAYSERI), as well as fine examples of Byzantine art, and mosques and museums of the Ottoman Empire in many of the cities. For relaxation, it offers the beautiful Mediterranean and Aegean coasts.

The old city of Istanbul, destination of the famous Orient Express train from Paris, was built on seven hills around the Golden Horn, an inlet of the Bosporus. With its Asian section, it has a population exceeding 3 million. Under its glistening domes and minarets, there is continual bustle and movement: from shoppers and street sellers, from vehicles rumbling along its cobbled streets, from the coming and going of the ferries.

Topkapi was the great palace of the Ottoman sultans and is now a museum. The basilica of Saint Sophia, also a museum, was built by Constantine the Great and renovated by Emperor Justinian: its dome rises 50 m (164 ft) high. The covered bazaar has a different street for each trade. A graceful suspension bridge, built in the 1970s, spans the Bosporus, linking Europe with Asia.

Ankara, Turkey's second largest city, had a population of only 75 000 when Kemal Ataturk made it his capital in the 1920s; now it has over 2 million. It is very much his creation, and contains many modern buildings, many trees, the Ataturk Museum and the Museum of Anatolian Civilisations with examples of Hittite art going back to 6000 BC. Izmir (Smyrna), probable birthplace of the Greek poet Homer, is an important port and a starting point for excursions to the ancient Greek towns of Pergamum (BERGAMA) and EPHESUS.

TARSUS, near the Mediterranean coast, was the birthplace of St Paul. The Aegean coast is rich in Greek remains, including the site of Troy near Canakkale. The Black Sea coast has such towns as TRABZON (the ancient Trebizond) with Greek and Byzantine remains.

TURKEY AT A GLANCE

Area 779 452 km² (300 946 sq miles)

Population 52 340 000

Capital Ankara

Government Parliamentary republic

Currency Lira = 100 kurus (piastres)

Languages Turkish (official), Arabic, Greek, Circassian, Armenian, Yiddish, Kurdish

Religion Muslim (98%)

Climate Mediterranean on the coast, continental inland; average temperature in Ankara ranges from −4 to 4°C (25-39°F) in January to 15-31°C (59-88°F) in August

Main primary products Wheat, barley, maize, cotton, sugar beet, pulses, cattle, sheep, goats, tobacco, fruit (olives, nuts, figs), vegetables, fish; coal, lignite, iron, crude oil, chrome, boron, copper

Major industries Agriculture, steel, textiles, tobacco and food processing, oil refining, chemicals, paper, fishing, mining

Main exports Textiles, cotton, nuts, fruit, tobacco, cereals, pulses

Annual income per head (US$) 1100

Population growth (per thous/yr) 21

Life expectancy (yrs) Male 60 **Female** 64

The Apennine mountains form the eastern border of Tuscany. At its heart is the ARNO river valley, where most of its main towns stand. The bare hills of Siena and the Maremma lie to the south. The terraced hillsides produce olives and vines for Chianti wine in this predominantly farming area. There are important producers of steel at Piombino, clothing in AREZZO province, motor scooters at Pontedera, marble at CARRARA and wool at PRATO. Tourism is a vital part of the economy.
Population 3 578 000
Map Italy Cc

Tutuila ***American Samoa*** Main island of the territory, covering 135 km² (52 sq miles). It is mountainous, rising to 653 m (2142 ft) at Mount Matafao. The beautiful natural harbour of PAGO PAGO almost divides the island in two.
Population 30 000
Map Pacific Ocean Ec

Tuva Republic ***USSR*** Mineral-rich autonomous republic of the RSFSR on the border of north-west Mongolia covering 170 500 km² (65 800 sq miles). The people, formerly nomadic herdsmen, now live mostly in farming villages. The republic's capital is Kyzyl (population 71 000), which has leather, timber and food industries. Tuva's rich deposits of gold, cobalt, asbestos, iron ore, salt and coal have so far been largely unexploited because of the region's remoteness.
Population 276 000
Map USSR Kc

Tuxtla Gutiérrez ***Mexico*** Busy modern capital of the state of CHIAPAS, which replaced the less accessible San Cristóbal de las Casas as the seat of government in 1892. A distribution centre for the surrounding area's coffee and tobacco crops, Tuxtla also has markets selling varnished gourds, leatherwork, and the gold jewellery worn by the Zoque Indian women.
Population 166 500
Map Mexico Cc

Tuzla (Dolnja Tuzla) ***Yugoslavia*** Town in north-east Bosnia, 80 km (50 miles) north-east of Sarajevo. Its salt mines have been exploited since before the Middle Ages, and it has developed into a major chemicals centre.
Population 121 700
Map Yugoslavia Db

Tver ***USSR*** See KALININ

Twante ***Burma*** Canal linking the capital, Rangoon, with the rice lands of the Irrawaddy delta. It is the most important of a network of transport canals running through the delta.
Map Burma Cc

Tychy ***Poland*** Industrial town, 16 km (10 miles) south of the city of Katowice. It was once a small, sleepy town – its name means 'quiet' – but has grown since 1951 as a residential district for people working in Katowice, and since 1970 as a production centre for Polski Fiat cars.
Population 178 100
Map Poland Cc

Tyne and Wear ***United Kingdom*** County of north-east England named after the two rivers that cross it. It covers 540 km² (208 sq miles), and was formed in 1974 from parts of the counties of Northumberland and Durham. It is centred on the city of NEWCASTLE UPON TYNE and includes the declining shipbuilding and industrial towns of Jarrow, Wallsend, Tynemouth, South Shields and SUNDERLAND. During the 1930s and again in the 1980s it was one of the parts of the country worst hit by unemployment, reaching 20 per cent in 1985.
Population 1 145 000
Map United Kingdom Ec

Tuvalu

A GROUP OF TINY PACIFIC ISLANDS WHERE THE MAIN EXPORT IS POSTAGE STAMPS

The nine coral atolls of Tuvalu have a total area of only 26 km² (10 sq miles); none of them rises more than 4.5 m (15 ft) out of the South Pacific. The main island and capital, FUNAFUTI, covers a mere 2.8 km² (1.1 sq mile) and has a population of 2800. The other islands are scattered in a chain 600 km (375 miles) long, about 960 km (600 miles) north of Fiji.

In 1974, the Polynesian inhabitants of Tuvalu, then the Ellice Islands, voted to separate from the Gilbert Islands (now Kiribati, whose people are Micronesians), with which they had formed a joint British colony since 1916. Since 1978 Tuvalu has been independent within the Commonwealth, with Elizabeth II as head of state.

The atolls' natural vegetation is coconut palms, breadfruit trees and palmlike pandanus trees, with large buttress roots above ground and crowns of narrow leaves. The coconut provides the main agricultural export of copra and fishing licences also bring foreign exchange, but most export revenue comes from the sale of elaborate postage stamps to philatelists.

There are about 8100 islanders, mostly living in thatched huts beside lagoons and producing their own food in gardens and by fishing. The population was once much higher, but between 1850 and 1875 the islands were raided by 'blackbirders' – Europeans who took most of the inhabitants, willingly or as slaves, to work the plantations of other Pacific islands. They reduced a population of 20 000 to a mere 3000. Another European invasion occurred in the Second World War, when the USA used the islands as air and naval bases.

TUVALU AT A GLANCE
Map Pacific Ocean Dc
Area 26 km² (10 sq miles)
Population 8100
Capital Funafuti
Government Parliamentary monarchy
Currency Tuvaluan and Australian dollar = 100 cents
Languages Tuvaluan, English
Religion Christian (mainly Protestant)
Climate Tropical; average temperature is about 25-32°C (77-90°F) all year
Main primary products Coconuts, fruit, vegetables, fish
Major industry Fishing
Main exports Postage stamps, copra
Annual income per head (US$) 560
Population growth (per thous/yr) 17
Life expectancy (yrs) Male 57 **Female** 59

typhoon Small, intense tropical cyclone in the China Sea and western Pacific, accompanied by winds of great force (160 km/h, 100 mph, or more), torrential rain and thunderstorms.

Tyre (Sour) ***Lebanon*** Small seaport 65 km (40 miles) south of the capital, Beirut. It was fought over and badly damaged during the Israeli invasion of southern Lebanon. Founded in 1500 BC, it was the Phoenicians' main trading centre. Gradually, Tyre's trade spread throughout the Mediterranean world and reached the Atlantic coast of Morocco – where its merchants established colonies – and Tunisia, where they founded the city of Carthage. Tyre was famed for its prosperity and for its glass and purple dye (Tyrian purple).

The Phoenician alphabet – adapted by the Greeks about 1000 BC – was attributed to Cadmus of Tyre, and his sister Europa gave her name to the continent of Europe. The Babylonian King Nebuchadnezzar (died about 562 BC) besieged Tyre without success for 13 years; and in the 4th century BC, Alexander the Great overcame and sacked the city.

Tyre became important again in Roman times, and today most of the remains are Roman. They include a monumental archway, an aqueduct, baths, a theatre, a necropolis and a large hippodrome.
Population 14 000
Map Lebanon Ab

Tyrol (Tirol) ***Austria*** Most mountainous of the country's states, with the Bavarian Alps in the northern border and the Ötztaler Alps in the south-central part. It is crossed by the Inn and Lech rivers. The capital, INNSBRUCK, is linked with Bolzano (Italy) by the BRENNER PASS. The main industry is tourism, and other occupations include dairy farming, forestry and wine making. South Tyrol was ceded to Italy in 1919 (see TRENTINO-ALTO ADIGE).
Population 586 200
Map Austria Bb

Tyrone ***United Kingdom*** Largest of Northern Ireland's six counties, covering 3136 km² (1211 sq miles). It is a predominantly rural area, growing potatoes, turnips and oats. It lies west of Lough Neagh and rises to 683 m (2240 ft) in the Sperrin Mountains. The county was dominated by the O'Neills, kings of Ulster, until the end of the 16th century when Hugh O'Neill was forced to flee to mainland Europe to escape the English. The main towns are Omagh and Strabane.
Population 159 300
Map United Kingdom Bc

Tyrrhenian Sea Deepest basin in the western MEDITERRANEAN SEA, bordered by the west coast of mainland Italy, Sicily, Sardinia and Corsica. It includes the volcanic EOLIAN ISLANDS in the south-east; resort islands, such as CAPRI and ISCHIA; and the mountainous island of ELBA in the north. Named after Tyrrhenus, legendary founder of the Etruscan kingdom, it is generally placid in summer, though electric storms are common. In winter, depressions moving in from the Atlantic and a northerly wind, the GREGALE, sweeping down from the Apennines, can cause choppy seas. The SIROCCO, a wind blowing from the Sahara in spring and early summer, can bring hot and sticky weather.

Area 155 400 km² (60 000 sq miles)
Greatest depth nearly 3660 m (12 000 ft)
Map Italy Cd

Tyumen' ***USSR*** Western Siberian town about 300 km (185 miles) east of Sverdlovsk. Founded in 1586, it was the first Russian town to be settled east of the Urals. It is the centre of a rich oil and natural gas region, and its industries today include boat-building, engineering, timber, tanning, oil refining and petrochemicals.

Population 411 000
Map USSR Hc

UAE See UNITED ARAB EMIRATES

Ubangi ***Equatorial Africa*** One of the largest tributaries of the Zaire river (also known as the Congo). It is formed by the meeting of the UELE and BOMU rivers and flows 1060 km (660 miles) west then south into the Zaire in Congo. It handles much of the Central African Republic's foreign trade.

Map Central African Republic Bb

Ubud ***Indonesia*** Village on Bali. It is famous for its artistic traditions and contains studios and galleries of several European and Balinese painters who have depicted the Balinese way of life on canvas. The Puri Lakisan Palace of Paintings contains a fine collection of Balinese paintings and sculptures from the 1930s to the present day. There is also a school of traditional music and dancing where the visitor can see the finest performances on the island.

Ucayali ***Peru*** River formed by the meeting of the Apurímac and Urubamba about 360 km (225 miles) east of the capital, Lima. It flows north through Pucallpa, the centre of thriving gas, oil and timber industries, to join the Marañón, forming the Amazon. The Ucayali is 1928 km (1198 miles) long and is navigable by boats of up to 3000 tonnes for 1600 km (995 miles). The department of the same name covers 129 658 km² (50 061 sq miles) of eastern Peru.

Population (department) 200 700
Map Peru Ba

Udagamandalam (Ootacamund) ***India*** Health resort and tourist centre at 2268 m (7440 ft) in the NILGIRI HILLS. It was the colonial summer capital of the former state of Madras, and is still known as the 'Queen of Hill Stations' – a fundamentally Victorian town, where the (English) Club still flourishes, strictly observing rules of dress.

Population 63 000
Map India Be

Udaipur ***India*** Walled city on the edge of the Thar desert, some 625 km (390 miles) north and slightly east of Bombay. Its many palaces and Hindu temples make it popular with tourists. The 16th-century maharajah's palace of glistening white stone stands in an artificial lake at the foot of low hills.

Population 232 600
Map India Bc

Udi Plateau (Donga Ridge) ***Nigeria*** Upland in the south-east, on which the town of ENUGU stands. Its north-eastern edge is an escarpment which, in the east, rises more than 457 m (1500 ft) above the Cross River plains. The plateau's sandstone ridges run south-westwards from the escarpment, then east towards Cameroon.

Map Nigeria Bb

Udine ***Italy*** City at the foothills of the Alps about 60 km (37 miles) north-west of Trieste. It has a 13th-century cathedral with an unfinished 15th-century bell tower, a 15th-century town hall (restored after a fire in 1876) and a 16th-century castle. Its economy is based on chemicals, textiles, leather, iron goods and the casting of bells.

Population 101 200
Map Italy Da

Uele ***Zaire*** River about 1200 km (750 miles) long. It rises north-west of Lake Albert and flows west to the Bomu river to become the Ubangi river. The Uele was discovered in 1870 by the German botanist Georg August Schweinfurth, who was exploring the Nile-Zaire divide.

Map Zaire Ba

Ufa ***USSR*** Capital of Bashkiria, lying on the western fringe of the Ural mountains about 360 km (225 miles) south of Perm'. Founded in 1574 as a fort to guard trade routes across the Urals, it grew as an industrial centre from the late 19th century and especially after the exploitation of the Volga-Urals oil fields in the 1950s. Its industries include machinery, chemicals, timber products, electrical equipment, telecommunications and office equipment.

The satelite town of Chernikovsk, 32 km (20 miles) north-east, has several oil refineries and petrochemical works.

Population 1 048 000
Map USSR Gc

Uganda See p. 658

Uíbh Fhailí ***Ireland*** See OFFALY

Ujiji ***Tanzania*** See KIGOMA

Ujjain ***India*** Centre of early Hindu civilisation in northern India, about 565 km (350 miles) north-east of Bombay. It is one of the seven Hindu holy cities where, every 12 years at a Kumbha mela (religious festival and fair), pilgrims bathe in the Sipra river. Ujjain has mosques dating from its time under Muslim rule (1235-1750), temples and palaces, and the ruins of an observatory built by the astronomer Jai Singh, Maharajah of Jaipur (1699-1743).

Population 282 200
Map India Bc

Ujszeged ***Hungary*** See SZEGED

Ujung Pandang ***Indonesia*** Port and largest city on Sulawesi. Formerly called Makassar, the city stands in the extreme south-west of the island and is an important trading centre. Ferries link it to the rest of Indonesia.

Population 709 100
Map Indonesia Ed

Ujvidek ***Yugoslavia*** See NOVI SAD

UK See UNITED KINGDOM

Ukraine (Ukraina) ***USSR*** Large, fertile and densely populated constituent republic of the USSR consisting mostly of steppes – vast, treeless plains. The Ukraine extends from Belorussia in the north to the Black Sea in the south, and lies along the borders of Poland, Czechoslovakia, Hungary, Romania and Moldavia. It covers an area of 603 700 km² (233 100 sq miles), making it the third largest of the republics after the RSFSR and Kazakhstan. But it is second only to the Russian Federation in population and in industrial and agricultural output.

The fertile uplands of VOLHYNIA form the western Ukraine. The northern part of the republic is dominated by the lowlands of the DNIEPER and DONETS rivers, which include marshes as well as good arable land; here the capital, KIEV, stands. The plains west of KIROVOGRAD are rich in fertile black earth called *chernozem*; those to the east, while less fertile, still support many farming communities. Much of this land was contaminated by radioactivity after the 1986 Chernobyl' accident (see KIEV). Spread through these eastern steppes are some of the USSR's major industrial cities, including DONETSK, DNEPROPETROVSK, KHAR'KOV and KRIVOY ROG.

In the far south are the extremely dry lowlands bordering the Sea of Azov and the Black Sea. The main port in this area is ODESSA, while farther south again is the CRIMEA, one of the Soviet Union's favourite tourist areas.

The republic's main industrial products are iron and steel, coal, agricultural machinery and chemicals. There are also many food-processing plants, handling its agricultural output, which includes wheat, barley, rye, maize, sugar beet, sunflower seeds, cotton, potatoes and other vegetables, meat and milk.

The Ukraine, or 'Little Russia', was first settled by Ukrainians (Ruthenians) – an East Slav group – in the 6th and 7th centuries AD, and Kiev was one of Russia's main principalities until conquered by the Golden Horde (Tatar warriors) in the 13th century. The Ukraine became part of Lithuania in the 14th century and was divided between Russia and Poland in the 17th century before being annexed entirely by Russia in 1793. It declared its independence as a republic in 1918 but became part of the USSR in 1922, although part of it was controlled by Poland from 1919 to 1939. The region was the scene of fierce fighting in the First and Second World Wars. The Ukraine and Belorussia are the only two Soviet republics with a separate vote in the United Nations, apart from that of the USSR as a whole.

The Ukrainians are the country's second largest ethnic group after the Russians. They retain a colourful folk culture and their own language.

Population 50 667 000
Map USSR Ec

Uganda

THE 'PEARL OF AFRICA' IS NOW A WAR-TORN LAND, REDUCED TO A STATE OF ECONOMIC COLLAPSE BY A TYRANNICAL RULER

Long ago, in colonial days, Winston Churchill described Uganda as 'the pearl of Africa'. Then, the description was not unmerited. Though its only access to the sea was by road and rail through Kenya to Mombasa, Uganda was relatively prosperous. It was a British Protectorate, not a colony, and was encouraged to manage its own affairs and develop its own resources.

For the most part, Uganda was and is a richly fertile land, well watered and with a kindly climate.

Much of the north and centre is a high plateau, 1000-1400 m (3280-4590 ft) above sea level, that lies between the east and west arms of the GREAT RIFT VALLEY system. It is a wide savannah sea, nurturing one of the last great wildlife communities of the African grasslands. To the west are the RUWENZORI mountains, whose spiky peaks, reaching up to 5110 m (16 765 ft), are capped with snow and perpetually wreathed in drifting veils of mist. Below, in the Bushenyi region, plantations clothe the foothills. The south-eastern border of Uganda runs through the basin of Lake VICTORIA. The lowlands around the lake were once forested, but have now mostly been cleared for cultivation. From the lake, the Victoria NILE runs north-west through central Uganda, into Lake KYOGA and onward to found the main course of the Nile itself.

Lakes and rivers between them, married to the elevation of the land, keep Uganda fairly cool despite its position athwart the Equator. The chief exception is the KARAMOJA district in the north-east, where rainfall is uncertain and the terrain semi-desert. The people are nomadic herdsmen, living only a step from disaster, as was apparent from the drought of 1979-81, after which 30 000 of them died.

Uganda gained its independence in October 1962, an occasion greeted with celebrations and high hopes. Of all former colonial possessions in Africa, it had the highest percentage of Western-educated people, and was well supplied with hospitals, schools and centres of higher learning. Yet within a decade and a half, optimism crumbled into a tragic shambles – a ruined economy, internal strife and a frightened population.

The seeds of disaster were sown long before independence. Uganda was a portmanteau state of half-a-dozen kingdoms and tribal regions. The heart of the protectorate was the powerful and ancient kingdom of Buganda, with the weaker kingdoms of Bunyoro, Toro and Ankole lying to the west. To the north and east are tribal lands, including those of the Langi and Acholi, while far to the south-west is the Kigezi district. Culturally, linguistically and racially, the people living at one end of Uganda have about as much in common with people living at the other as the Irish with the Chinese.

As the largest ethnic group in Uganda – about 18 per cent of the population – as well as being the most prosperous, the best educated and the possessors of the capital KAMPALA, it would have been thought that the Bagandans of Buganda would have dominated the new nation in the same way as the Kikuyu dominated Kenya. As it turned out, the Bagandan leaders had little interest in the republic of Uganda and preferred instead to concentrate upon tribal affairs. Consequently, the reins of national power were taken by the leaders of the less powerful tribes; the first prime minister, Milton Obote, for example, was a Lango from the north.

THE YEARS OF TERROR

In 1966, Buganda attempted to secede from the union, which Obote used as an excuse to abolish the constitution and set up a central government with himself as president. Without Bagandan support, however, his reign was uneasy, and during his absence at the Commonwealth Prime Ministers' Conference in 1971, he was deposed by the Minister of Defence, General Idi Amin. The general's move may have been accelerated by a court of inquiry which was investigating the loss of US$5 million from defence ministry funds.

So began the sordid reign of Idi Amin. At first he was acclaimed as a strong man, a man of the people, and welcomed by the British government too, who were becoming alarmed at Obote's nationalising of British assets. Ugandan cheers at least swiftly died away as Amin quashed all political activity and despatched his military death squads to smell out opposition. Chronically short of the cash necessary to maintain his repressive rule, in 1972 he expelled all Asians with British passports – virtually the entire business community – and seized their property. The businesses were handed over to Ugandans who had no experience in running them, and the economy sagged still further. Amin then proceeded to 'eliminate' all those who offered the slightest challenge, real or imaginary, to his rule, and was prepared to condone, if not encourage, repeated atrocities by his soldiers.

Many of the atrocities were the settling of old scores, arising out of ancient tribal jealousies and hatreds. It is variously estimated that, during his regime, between 100 000 and 300 000 people were brutally slain. Amin's ruthlessness became notorious throughout the world. His downfall came at last in 1979, when he decided to annex a portion of Tanzania. Tanzania's president, Julius Nyerere, instructed his forces not only to remove Amin's invaders but to remove Amin from Uganda. Amin fled first to Libya and then to Saudi Arabia.

He left behind a country in a desperate state. The economy was shattered, law and order had broken down, and there was a deep mistrust between the country's many factions.

The two presidents who succeeded Amin in 1979 were speedily deposed. In 1980 a national election brought victory to the Ugandan People's Congress party led by the returned Milton Obote. But there were those who claimed that the election was rigged. One of these dissidents was Yoweri Museveni, who had been foreign minister in one of the short-lived post-Amin governments. He took to the bush with his National Resistance Army, formed out of units of the old Uganda National Liberation Army that had helped the Tanzanians to oust Amin.

At the same time, trouble broke out in the Karamoja district of the north-east where armed cattle rustlers clashed with government troops. All in all, probably well over 200 000 people have died in the troubles since Amin's departure.

As always, inter-tribal rivalries were at the root of the problem. Amin was a Muslim of the northern Kakwa group, and so favoured Uganda's Muslims, some 10 per cent of the population. Obote promoted Langi soldiers of his own tribe – in order to give them an advantage in an army which, since the days of the colonial King's African Rifles, had been mostly recruited among the Acholi. In Obote's time, fighting frequently broke out around the capital between Acholi and Langi factions within the army. Then, in 1985, the army ousted Obote in a coup supported by Ugandan exiles, and the army commander, Lieutenant-General Tito Okello, established a military caretaker government and became head of state. He invited Museveni and the NRA to throw in their lot with the new regime. An

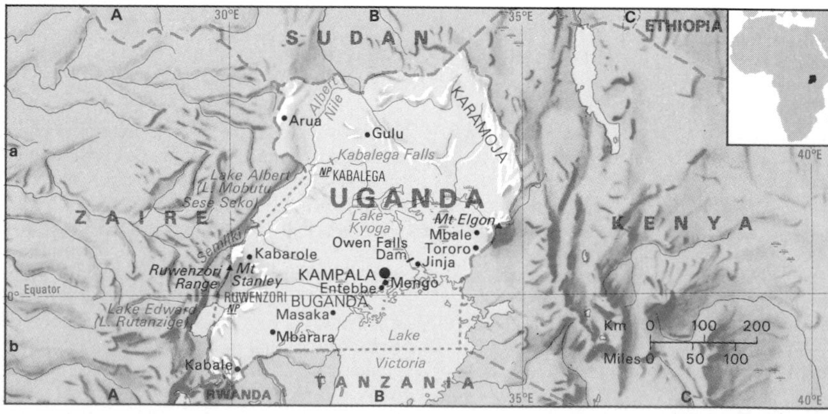

▲ MURCHISON FALLS In a land of waterfalls, this is one of the most stupendous. The waters of the Victoria Nile seem to boil as they plunge 122 m (400 ft) in three thunderous cascades, of which Murchison is the first.

accommodation between the two sides came to an end early in 1986 when NRA guerrillas drove the army out of Kampala and Mr Museveni was sworn in as president.

Nevertheless, parts of Uganda have been hardly touched by the horrors of the past two decades. Oppressive governments in Kampala could not directly influence all aspects of people's lives throughout the country. Only 10 per cent of the people live in towns, and the country dwellers are widely scattered. In remote areas, life has hardly changed.

It is fortunate that most Ugandans still grow their own food, staving off widespread famine despite the breakdown of the economy. Family ties are strong, and even townsfolk are often able to obtain food from rural kin. Not that a great variety of foodstuffs is called for. In the south, the staple is a green banana called *matoke*, while in the north it is millet and sorghum.

Cotton and coffee were Uganda's main cash crops in colonial times. Coffee became, and still is, Uganda's most important export. During the 1970s production seemed to decline sharply, but in fact large quantities of coffee were being smuggled into Kenya and Zaire. A sharp increase in coffee prices in the late 1970s helped keep Amin in power when bankruptcy seemed likely to topple him. Coffee survives because it is a perennial plant that continues to produce whether it is tended or not. Cotton, on the other hand, must be newly planted each year, and shortages of seed, labour and pesticides resulting from the general breakdown have considerably reduced the output. Many northern farmers, who once grew cotton, have reverted to subsistence farming, though the coffee growers of the south still remain relatively prosperous. But what remains of a cash economy in Uganda is largely dependent on the black market.

Attempts are now being made to restore enterprises which were under way at the time of independence in the 1960s. These include the tea plantations of the far west, two large sugar plantations near Lake Victoria, a copper mine (also in the west) and new manufacturing industries based in Kampala and near the Owen Falls Dam and power station at JINJA.

A hopeful start. But a few new factories do not make a nation, and to most of Uganda's 15 million inhabitants, tribal and family ties are considerably more binding than attachment to a country that has given them little. Efforts are now being made to restore a sense of national identity, and especially to make it meaningful to the 50 per cent of the population who are still children. The slaughter of the 1970s was on such a scale that it affected the population growth rate, and in some areas violent death is a commonplace still. Even so, births greatly exceed deaths, violent or otherwise and, as in most parts of tropical Africa, the total population can be expected to double over the next 25 years. If a viable political structure can be established, Uganda may again be considered one of the more fortunate African countries.

UGANDA AT A GLANCE
Area 236 580 km² (91 344 sq miles)
Population 15 200 000
Capital Kampala
Government Republic
Currency Uganda shilling = 100 cents
Languages English (official), Swahili, Bantu and Luganda languages
Religions Christian (65%), Muslim (10%)
Climate Equatorial, tropical; cooler in mountain areas. Temperature ranges from 15°C (59°F) to 26°C (79°F)
Main primary products Coffee, cotton, tea, sugar, livestock, millet, maize, bananas, sorghum, yams, timber; copper, phosphates
Major industries Agriculture, food processing, textiles, cement, copper processing, motor vehicles, metal processing, mining
Main exports Coffee, cotton, tea
Annual income per head (US$) 180
Population growth (per thous/yr) 32
Life expectancy (yrs) Male 46 **Female** 49

Union of Soviet Socialist Republics

RUSSIA'S GREAT REVOLUTION HAS CHANGED THE WORLD, BUT A HARSH REGIME STILL HAS SECRET POLICE AND POLITICAL PRISONERS

Part in Europe and part in Asia, the Soviet Union is the world's largest state, covering nearly one-sixth of the earth's land surface. It stretches from MURMANSK near the Finnish border to VLADIVOSTOK on the Sea of Japan, and from the BLACK SEA to well north of the ARCTIC CIRCLE. Its climate ranges from vast frozen wastes to subtropical deserts. The Soviet Union has Europe's highest mountain, ELBRUS (5642 m, 18 481 ft), and its longest river, the VOLGA (3688 km, 2293 miles). The country also has the world's deepest lake, BAIKAL, which plunges to 1620 m (5315 ft).

This is the temple of Communism, the socialist doctrine born in the minds of 19th-century philosophers such as Karl Marx, which became political reality in 1917 when the Bolshevik leaders Vladimir Lenin (1870-1924) and Leon Trotsky (1879-1940) took over a war-weary nation and bent it to their will. The Tsar and the royal family were executed, the well-to-do were hounded, imprisoned or killed, and the state became all powerful. Lenin died in 1924, and Trotsky, having lost a power struggle, was finally liquidated in Mexico in 1940. Backed by a reign of terror in which millions died, Joseph Stalin (1879-1953) reigned supreme for 30 years.

Cynically signing a pact with Fascist Germany in 1939, Stalin seized half of Poland, attacked Finland and took over the Baltic states, only to face a German invasion which cost 20 million Russian lives. Allies in war, the victorious Americans and Russians soon fell out, and so began the Cold War which has lasted to this day. The once backward Soviet Union became a nuclear power with vast military strength, its influence stretching from Communist satellites in Europe to newly independent African nations and Fidel Castro's Cuba.

The Union of Soviet Socialist Republics comprises 15 republics. They are dominated by the Russian republic, which stretches for

► RED SQUARE Visitors watch the changing of the guard outside the mausoleum containing Lenin's body. The 16th-century Cathedral of St Basil the Blessed (or Pokrovsky Cathedral), a unique combination of nine chapels, is now a museum.

10 000 km (6250 miles) east to west – a quarter of the way round the world. It contains nearly three-quarters of Soviet territory – 17 075 400 km² (6 592 812 sq miles) – and 140 580 000 people – half the country's total. The other republics form an arc around the Soviets' western and southern borders. They are the Baltic republics (ESTONIA, LATVIA and LITHUANIA); three Slav republics (BELORUSSIA, MOLDAVIA and the UKRAINE); three Transcaucasian republics (GEORGIA, ARMENIA and AZERBAIDZHAN); KAZAKHSTAN, and four Central Asian republics (TURKMENISTAN, UZBEKISTAN, TADZHIKISTAN and KIRGHIZIA).

The borders of the Soviet Union exceed 65 000 km (40 600 miles) in length. Seas that freeze for up to eight months of the year in the north and east, and the rugged mountains in the south, provide natural defences for many of the frontiers. But there is a critical 2000 km (1250 mile) stretch of lowland on the border with China, in the south-east. And the vast plains lying west of the URAL mountains which divide Europe from Asia are easily crossed by land or by the river systems of the DNIEPER and Volga. Between the 7th and 16th centuries there were invasions by Vikings from the north; Bulgarians and Turks from the south-west; Mongols and Tatars from the east; and Poles and Lithuanians from the west. Napoleon reached MOSCOW in 1812, but was forced into a humiliating retreat by the bitter Russian winter.

In turn the Russians took territory from the invaders. The tsars Ivan III (1462-1505) and Ivan IV 'The Terrible' (1533-84) pushed out the borders as they ruled from their Kremlin fortress in Moscow. From 1703 Peter the Great (1682-1725), Catherine the Great (1762-96) and Alexander I (1801-25), directed the country's expansion from the imperial capital of St Petersburg (now LENINGRAD).

Russia's historic openness to attack is the root of her apparently paranoid concern for strong military defence and centralised govern-

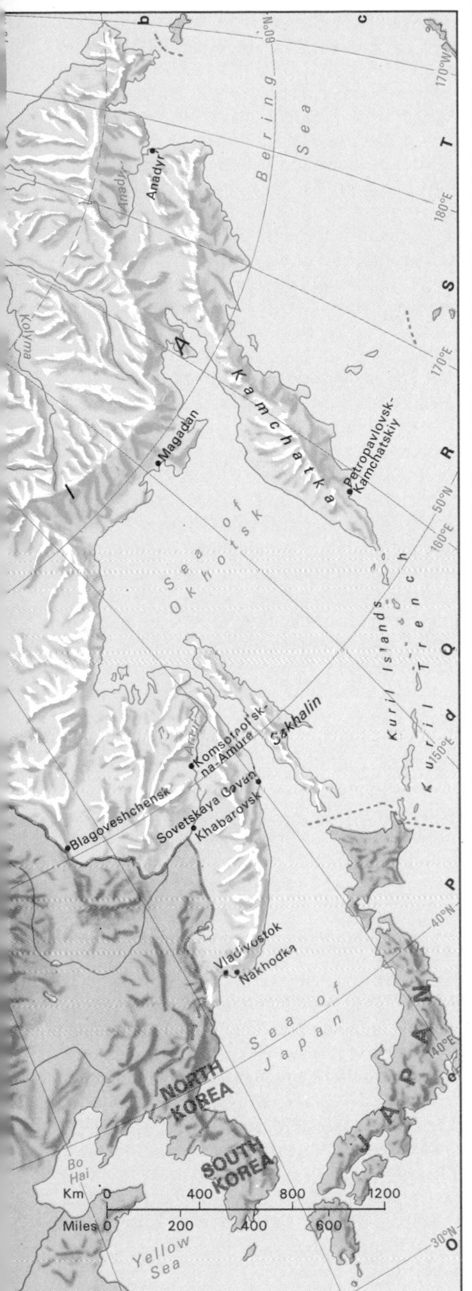

ment. In the time of the tsars, controls were already deeply rooted in the Russian Empire. The tsars' rule was authoritarian, mostly reactionary and intolerant of opposition. Serfdom was not abolished until 1861. Peter the Great, who brought much West European civilisation to Russia, sent the first political prisoners to Siberia, in 1710.

Following an uprising in 1825, the numbers exiled to Siberia or to south-eastern deserts swelled. They included Moscow University professors, writers like Fyodor Dostoyevsky, and later, revolutionaries like Lenin. Paul I (1796-1801) and Nicholas I (1825-55) built up a secret police force and introduced censorship, with bans on foreign books dealing with reform or the plight of oppressed European peasantry. Restrictions on movement prevented intellectuals like the poet Alexandr Pushkin (1799-1837) from travelling abroad. People were already being isolated from the outside world.

Many people made the revolution possible, but two names stand out: Vladimir Ilich Ulyanov (Lenin) and Lev D. Bronstein (Trotsky). Both were Marxists, intellectuals, visionaries, orators and organisers. Following Lenin's death in 1924, Joseph Dzhugashvili (Stalin) used ruthless purges, executions and labour camps to maintain a position of absolute, unquestioned power as Secretary-General of the Communist Party. Terror, fear of the knock on the door by secret agents in the night, achieved social obedience. It enabled Stalin to push through gigantic plans for industrialisation, collective farms, canal and railway building as part of successive five-year plans from 1928 onwards.

There is still great indifference to human rights. The US State Department reported in 1983 that 4 million people were held in forced labour camps in various parts of the Soviet Union.

AGRICULTURE AND INDUSTRY

Between 1922 and 1985 the Soviet population has doubled to more than 280 million. Farm output has quadrupled and industrial production has increased 140-fold. There has been commitment to technical modernisation, military supremacy, improved welfare and self-reliance. The Soviet Union has become the world's leading producer of petroleum, coal, hydroelectric power, iron ore, steel, timber and some nonferrous metals. Railways have been improved, enormous frontier wildernesses tamed, and 1050 new towns built. Fertilisers and farm machinery are produced for the long underprivileged people of extensive arable regions; woollen cloth and shoes are made to equip people adequately for the severe winters.

To do this, the USSR has had to accommodate the biggest rural-to-urban migration in the developed world this century. The 1926 census showed 25 million people (18 per cent of the population) living in towns. Only three cities (the capital Moscow, Leningrad and KIEV) had more than 1 million inhabitants. Today 181 million (66 per cent) live in towns, and 26 cities were over the million mark in 1984.

Rural Russia has become the urbanised Soviet Union. Yet it lags behind the USA, Japan and West Europe in consumer goods and technology. The sharpest contrasts for Western visitors are the queues in shops (where the abacus is still used for counting) and the congestion on public transport. Shortages are facts of everyday life. In the West, goods are chasing money; in the USSR the reverse is true. Wages and salaries are low, but rent, heating, lighting, cinemas, theatres and public transport are cheap, and education and health care are free.

A US supermarket may have 52 brands of coffee. A housewife's choice depends on her need, budget and taste. A working Soviet woman's choice is between no coffee at all or one brand, if she knows when and where it will be delivered and can be there at the right time at the head of the queue. Soviet people swop information about such deliveries among family, friends and workmates or, better still, 'connections' who can 'short-circuit' the queues and supply the goods and services either at a premium or on a barter basis.

All this explains why the *babushka* (grandmother) holds such an esteemed place in the crowded city family flat. She can mind children and queue in the shops in daytime, allowing the young mother to go to work.

In public transport, Moscow is the fare-setter for Soviet cities: its flat-rate fares are unchanged since 1934.

Food supplies have been a persistent problem. From 1928 to 1940, 21 million peasant holdings were amalgamated into 203 000 collective farms and 4200 state farms. The process was resisted by peasants who slaughtered livestock for their own use, so greatly reducing milk and meat production, and siphoned off cereals and other crops between the farm gate and state collection point. In addition, the war damaged the western agricultural areas.

Food shortages were so great that when Nikita Khrushchev (1894-1971) became the Russian leader in 1953 he launched the Virgin Lands scheme to grow crops in Kazakhstan and the ALTAY territory in southern Siberia. At the same time, production of fertilisers was increased to raise yields in the fertile Ukraine and elsewhere. The campaign was only partly successful and though mechanisation and fertilisers have substantially raised yields in western areas, consumption has increased because of population growth and better diets. When summer is cool and wet, as in several recent years, the USSR has to spend scarce foreign currency or divert more cars, cameras, watches, refrigerators and other manufactured goods from the home market to pay for US and Canadian wheat, Argentine beef or EEC butter.

A LAND OF MANY LANGUAGES

The average Muscovite shopping in Moscow's most famous stores, GUM (State department store) and *Detskii Mir* (Children's World), may understand the languages spoken by only two out of every three fellow shoppers. Command of the Russian language varies enormously, for the Soviet Union is a multinational and multiracial society with 102 nationalities speaking more than 100 different languages. The Russians play the leading role in cultural, economic and political life.

Other Slav peoples of the Indo-European language group who mostly inhabit the western USSR are Ukrainians (50 million), Belorussians (9.5 million) and Poles (1.2 million). There are Lithuanians and Latvians in the Baltic regions, Moldavians in the south-west, and Armenians and Tadzhiks in the south.

Altaic peoples of Turkic origin live mostly between the southern Urals, the CASPIAN Sea and the PAMIRS. They include Uzbeks, Tatars, Kazakhs, Azerbaidzhanians, Chuvashes, Turkmens and Kirghizians. Caucasian peoples between the Black and Caspian seas make up a third group, including Georgians, Dagestani and Chechen. Uralians comprise the Mordovians, Udmurtians and the Maris in the central Volga-Ural area, Estonians along the Baltic. The Uralians speak Finno-Ugrian languages akin to Hungarian and Finnish.

THE ANCIENT CIVILISATIONS

There were several non-Russian civilisations in what is now the USSR. Persians came to the Caucasus and Central Asia 3000 years ago. In the 7th-6th century BC, the Khorezm state produced crops using elaborate irrigation systems in what later became the Turkmen and Uzbek deserts, straddling the AMUDAR'YA river south of the ARAL SEA. Trading posts grew up along the Silk Road, the caravan route between the Eastern Mediterranean and China, at places like Maracanda (now SAMARKAND) founded in 379 BC. Some of these became fine Greek cities under Alexander the Great (356-323 BC): Alexandria Oxus (Termez) and Ai-Khanum along the upper Amudar'ya, and Alexandria Eschate (now LENINABAD) and Cyropolis along the Syr-Darya river in the fertile Fergana Valley. The Armenian highlands between the Black and Caspian seas became the cradle of the Iron Age before 1500 BC. The town of ECHMIADZIN, near YEREVAN, boasts the oldest standing building in the Soviet Union, a shrine built in AD 303.

Ancient civilisations were largely destroyed by plundering invaders: Arabs in the 7th century and Mongols in the 13th century. Tamerlane, the Mongol conqueror, created a new empire in central Asia after 1360, based on trade along the Silk Road, crafts and slaves, and poured its wealth into the magnificent Islamic buildings of the capital, Samarkand. But decline set in from the 16th century as internal feuding fragmented the empire and the caravanners lost their business to maritime traders from Portugal and Spain, who shipped goods via the Americas or the Cape of Good Hope. Within the Russian Empire, the Caucasus and central Asia became bulwarks against invaders by providing men for armies. They were also the source of raw materials such as cotton and wool for Moscow's industries, luxury goods and drinks for the court at St Petersburg (Leningrad), and country estates for the rich. Middle Eastern cultural and religious influences and Russian colonialism meant that these peoples entered 20th-century Soviet society without experiencing the developments of the 19th century.

Non-Slav families have many children: Caucasians average 28 births per 1000 population per year, and central Asians 40. The people, especially in the higher mountains, are long-lived and the southern non-Slav republics increased their share of the population from 14 per cent in 1940 to 25 per cent in 1984.

Westerners often view the rapid increase in non-Slav peoples as a potential source of ethnic tension within the USSR, and a threat to Russian dominance – especially given the world awakening of Muslims from Morocco to Pakistan. Russians claim that Caucasians and central Asians enjoy higher living standards than average Muslims of other countries, and that they will help to solve Soviet labour shortages. But Kirghizians, Tadzhiks and Uzbeks are reluctant to leave their homes, and employing them locally in industry would use scarce water resources.

The Soviet system believes in equality between men and women. Caucasian and central Asian women became more emancipated after the revolution: this is symbolised in the leading Azerbaidzhanian city of BAKU by a monument of a woman throwing off her *yashmak*, or veil. Nowadays, 40 per cent of the deputies in the Azerbaidzhan supreme soviet are women, compared with 35 per cent in the Russian republic.

But despite increased provision of crèches, kindergartens and canteens, attitudes in the south are that the woman's place is in the home. Few of the men do housework or heavy labour, and women are absent from cafés, restaurants and the shaded tea-drinking gardens (*chaykhana*) in Uzbekistan. Women in the big cities may not be much freer: in addition to housework, shopping and child-minding, they hold jobs to which they must often commute for up to an hour each way.

THE CENTRAL ASIAN REGION

Transcaucasia and Central Asia excel in high-value farm products. Shepherds manage large flocks of sheep on high mountain pastures in summer, descending to the valleys in winter. The sheep produce wool, milk and cheese, and the desert vegetation, though sparse, supports millions of karakul and astrakhan sheep with coarse, curly fleeces. Glaciers and winter snows feed rivers which, with the new KARA KUM canal connecting the Amudar'ya with the Caspian Sea, irrigate parched valleys and deserts. Cotton, tobacco, apricots, peaches, nuts, lemons, oranges, pomegranates, tea, early flowers and vegetables are grown. Armenians and Georgians are the main producers (and heavy drinkers) of Soviet wines and brandies, but the legacy of Islam and the dry heat make tea the main beverage.

People still dress in their national costumes daily in the towns and villages. Uzbek and Tadzhik women wear thin headscarves (*shalwars*) and long colourful tunics: men put on square dark-coloured skullcaps embroidered with white patterns. Turkmens protect themselves from the sun with shaggy fur hats and quilted robes.

The southern cities are unlike those elsewhere: colourful, exotic, individual. Narrow twisting streets, workshops, bazaars and state-owned shops mingle in old city centres with medieval churches, in Transcaucasia, and mosques in Central Asia. Public buildings since 1960 often incorporate local architectural styles, like the 'filigree in concrete' on the walls of buildings in TASHKENT. Many cities are oases in arid surroundings. Central TBILISI and towns created after 1922 like FRUNZE, ALMA-ATA and Shevchenko have broad, tree-lined boulevards, parks, gardens and fountains to give shade or cool the air in summer when temperatures exceed 30-40°C (86-104°F). Earthquakes, hardly known in the Soviet Union except in east Siberia, razed ASHKHABAD (Turkmenistan) in 1948 and TASHKENT (Uzbekistan) in 1966, and both were rebuilt.

Here oil and natural gas are key resources; much is piped to the Moscow region and cities in the Urals. With hydroelectricity from mountain rivers and varied (but not large) metal ore deposits, they support local industries in line with the Soviet policy of developing backward regions. Soviet Central Asia has become one of the main manufacturing regions in the USSR, producing chemicals, metals, cotton textiles, clothing, carpets, shoes, food, machinery and electrical appliances. Geologists predict that the deserts cover enormous mineral reserves.

Kazakhstan, to the north, is larger than all the EEC countries put together. From the Caspian shores in the west, the land gradually rises eastwards across 3000 km (1875 miles) of semiarid steppe and plateau to 4000 m (13 120 ft) on the Chinese frontier. Historically an area of perpetual migration, it became home in the 15th century for nomadic Kazakh horsemen, grazing huge herds on vast pastures. It remained like that, except in the north where farm settlement began after 1700.

Development started in earnest after the Nazi occupation of the Ukraine in 1941. Heavy industry, with big steel, nonferrous metals, chemicals and engineering complexes using local coal, metals and non-metallic minerals, is found in KARAGANDA in the east and Aktyubinsk in the west. In the 1950s, an area of desolate grey virgin steppe larger than Britain was ploughed up, sown and developed with towns, roads and railways centring on TSELINOGRAD (formerly Akmolinsk) to supply scarce cereals. Though much land is still cultivated, some areas quickly became dustbowls. Since Yuri Gagarin became the first man in space in 1961, BAYKONYR in central Kazakhstan has been developed as the Soviet launching site for spacecraft. Here the skies are cloudless for 300 days a year.

The Kazakhs have adjusted from nomadic living in portable felt tents (*yurts*) to settled ways in urban flats. Shortages of labour and skills in factories and on farms in virgin lands brought a large-scale influx of Russians and Ukrainians who were attracted by better salaries, and now comprise two-thirds of the population.

THE BALTIC REPUBLICS

The Baltic republics of Estonia, Latvia and Lithuania in the north-west are in another world. Sand-dune coasts once rich in amber, a fossil resin used for jewellery, give way inland to low wooded hills, marsh, lakes and meandering rivers. Livestock graze in green meadows. Baltic Sea air brings changeable mild weather, cloud and rain to temper biting winter Arctic or Siberian winds, or hot dry summer breezes. Ice-free ports permit maritime trade, shipbuilding, fishing and naval activity throughout the year.

The Baltic peoples escaped Mongol-Tatar occupation and have a strong Western culture. Their brief independence (1918-40) from Russia has encouraged a nationalism which erupts now and then into open discontent. People use the Latin alphabet and literacy was high even before 1917 (the government now claims a literacy level of 100 per cent throughout the USSR). The main Baltic cities, RIGA in Latvia and TALLINN in Estonia, have medieval Teutonic fortifications and Gothic buildings. In VILNIUS the Italian Renaissance style expresses Lithuania's strong Polish and Catholic connections. The city, with more than half a million people, has 500 colourful cafés and restaurants, offering a wide variety of food and decor. Baltic republic governments put people first, providing sanatoriums and health centres, winning prizes for new housing estates, and introducing the first Soviet pedestrian precincts.

Indeed, the Baltic people enjoy the highest Soviet living standards. This is based on people's skills, precision and hard work in making machinery, tools, electrical and electronic products, and traditional craft items. Farmers achieve high yields from drained marshland.

SHARED LANGUAGE AND RELIGION

The vast Russian republic as distinct from the other 14 republics within the Soviet Union – stretches east to the Pacific Ocean. The Slav republics of Belorussia (the size of England and Wales), the Ukraine (larger than France) and Moldavia separate the Baltic from the regions of Transcaucasia, Kazakhstan and Central Asia. They share similar languages and the Orthodox religion.

A broadly triangular zone, from Leningrad to ODESSA in the west, tapering to KRASNOYARSK in central Siberia, is the country's economic heart. This gently undulating plain is one of the world's great agricultural regions, managed in huge state and collective farms, sown with cereals, fodder and sugar beet. The west and north are moister, greener and forested: the east and south, drier steppe.

Large industrial complexes produce coal and steel (in the DONETS BASIN, mostly in the Ukraine, and the Kuzbas in west Siberia), oil and salt (the Volga region), nonferrous metals and iron ores (the Urals). Timberworking and paper making are scattered along the north, food processing in the south. Post-war development enabled planners to divert factories from Moscow, Leningrad and the Donets Basin, where labour was scarce, west into Belorussia, east to the Volga and Siberia, or into southern rural areas between L'VOV and SARATOV.

MODERN TOWNS, OLD VILLAGES

New towns of repetitive high-rise flats were built after 1950 on virgin land to serve modern industries. Some, like TOL'YATTI (cars) and BREZHNEV, formerly Naberezhnyye Chelny (the world's largest truck plant), already have more than 400 000 people. They take people from overcrowded cities like Moscow, Kiev, GOR'KIY, SVERDLOVSK, OMSK and NOVOSIBIRSK.

Modern apartments have been built in country areas since 1960 to move farm families from villages into 'agro-towns' where collective and state farms have merged into fewer, larger units. But many areas are still backward. Numerous villages of simple log cabins are served by poor tracks and lack piped water: supplies are drawn by bucket from wells and carried on shoulder yokes. Such areas are reservoirs of labour to swell city populations. Despite having the densest rural population in the Soviet Union, Belorussia, the Ukraine and the Russian republic today house 60-72 per cent of their people in cities, compared with 21-34 per cent in 1940.

The Ukrainians retain strong national characteristics. Donets Basin miners, with their racing pigeons and Baptist inclinations, are a breed apart from Belorussians with their hunting and farming customs like *Dozhinki* (the harvest festival). Folklore, songs, dances and food vary greatly between southern, KRASNODAR and Ural Russians. Minorities must learn Russian as the main language.

The vast, thinly populated and wilderness regions of Siberia and the Far East are a rich storehouse of fuel, energy, minerals and timber. They are difficult and costly to develop

▼ REMOTE VALLEY The Varzab river cuts through the Pamirs in Soviet Central Asia. The word Pamir means 'valley at the foot of a mountain peak', and the range includes Communism Peak (7495 m, 24 590 ft), the highest Soviet peak.

and to live in because the weather is so severe. One-third is lowland marsh west of the YENISEY river: two-thirds consists of north-sloping plateaus and magnificent mountains east of the river. In the northern *tundra*, inside the Arctic Circle, moss and lichen are under snow for up to nine months a year. In summer much of the region is a quagmire with permanently frozen subsoil. This is the land of Chukchi Chukotskiy Eskimos who fish, trap and herd reindeer, and the adopted home of metal workers (since 1934) and coal miners (since 1940). Southwards, stunted larch gives way to the *taiga* primeval forest zone, waterlogged and impassable except by boat along the rivers OB and IRTYSH in west Siberia. Wells drilled between TYUMEN and Surgut since 1965 now yield most of Soviet oil, and in the east Siberian plateaus forest covers huge mineral deposits, including massive coal reserves. Western Siberia has vast natural gas reserves and a gas pipeline runs from Siberia to the Czechoslovak border and into Western Europe.

Arctic air lowers winter temperatures in the region, known as the 'pole of cold', to −78°C (−108°F) at OYMYAKON and a January average of −44°C (−47°F) at YAKUTSK, while BRATSK in the southern taiga has only 120 frost-free days a year. Modern flats, shops, offices and factories in towns like Yakutsk are supported on concrete stilts to allow freezing air to circulate underneath and so preserve the permafrost. Otherwise the ground would melt and buildings would sink into the mud. At temperatures of −45°C (−49°F) or lower, truck engines and diesel motors must be left running day and night in winter. Yet summer brings heat, enabling people in Yakutsk to raise cattle in clearings in the taiga.

The new BAM (Baikal-Amur-Magistral) railway, extending 3200 km (2000 miles), from Taishet on the Trans-Siberian railway to the Pacific port of Sovetskaya Gavan', will make south-east Siberia, with its timber, coal, copper, iron ore, asbestos and other minerals, one of the world's major development areas. Previously, industry was concentrated between Krasnoyarsk, BRATSK and IRKUTSK: it was based on power supplies from the world's largest hydroelectric power stations on the Yenisey and Angara rivers, and on KANSK coal. But since 1975, 60 boom towns have emerged on the BAM route beyond Bratsk. For decades, settlement was restricted to land alongside the Trans-Siberian railway. Beyond Lake Baikal and the Yablonovyy range, this line descends the Amur and Ussuri valleys to Vladivostok through a cool, moist area with fertile soils watered by monsoon rains. Wheat, maize, sugar beet, soya beans, apples, grapes, vegetables and rice are cultivated. These are processed in cities like BLAGOVESHCHENSK, KHABAROVSK and KOMSOMOL'SK-NA-AMURE, where factories also manufacture equipment for the fishing, naval and trading port complex of Vladivostok-Nakhodka.

PROBLEMS OF THE RUSSIAN JEWS

Prohibited for years after the Second World War, Jewish emigration became a reality in 1972, but only for Jews wishing to join their relatives in Israel – a state not recognised by the USSR. Atheistic Soviet governments view Israel as an American-supported base for an international religious (Zionist) movement, and every practising Soviet Jew as its potential agent. Considered as 'alien nationals' despite their Soviet birth, the Jews do not enjoy equal rights. Teaching of Hebrew is illegal and punishable by harsh sentences.

The Russians might logically allow all Jews (2 million out of a population of 280.7 million) to emigrate. Permanent exit visas were issued to a record 52 000 in 1979, but by 1984 emigration had dwindled to less than three people a day. Under détente, allowing Jews to emigrate was one US condition for supplying Western technology to the USSR. After the Soviet invasion of AFGHANISTAN détente broke down: arms spending increased, the USA put embargoes on exports of 'high-tech' electronics from NATO to Warsaw Pact countries, and there was tit-for-tat boycotting of the 1980 Moscow and 1984 Los Angeles Olympics. These factors, and the Israeli occupation of LEBANON, encouraged Soviet governments to block Jewish emigration, to create a queue of thousands of *refuseniks* (people refused exit visas), as a bargaining counter. The shift in world power created after 1974 by the Arab OPEC states and Islamic militancy encouraged Russian anti-Semitism.

Stalin's policy of concentrating Jews in their own autonomous region in a remote area on the Sino-Soviet frontier failed: its population declined from 180 000 in 1959 to 18 000 in 1979. Jews hold many top professional, academic, scientific and artistic posts in leading Russian cities. Yet once Jews become refuseniks they often also become dissidents, although most of the government's leading critics are Russians or Ukrainians, like Alexander Solzhenitsyn, Vladimir Bukovsky and Andrei Tarkovsky.

Instability in Soviet leadership during the waning years of Leonid Brezhnev (1906-82) and the caretaker governments of Yuri Andropov (1914-84) and Konstantin Chernenko (1911-85) made sure that the KGB secret police would crack down on dissidents and refuseniks. It was reported in 1983 that political arrests were running at 20 a month, and that over the previous eight years about 200 people had been detained in psychiatric clinics for political reasons.

THE RUSSIAN WORLD TODAY

Since Stalin's time, every Russian has had to carry an internal passport registering his home and work place. Official permission must be obtained to travel to other Soviet cities for more than three days. Stalin died in 1953, and within three years his 'personality cult' was denounced by the new leader, Nikita Khrushchev. In 1961 Stalin's embalmed body was removed from Red Square in Moscow, where it had lain in state beside Lenin's. It was the final disgrace.

Khrushchev quelled a Hungarian uprising in 1956, upset the Chinese with his 'revisionism' and brought nuclear war close by shipping missiles to CUBA in 1962. Ousted in 1964, he was succeeded as Secretary-General of the Communist Party by Leonid Brezhnev, who ordered troops into CZECHOSLOVAKIA in 1968 when Dubcek's regime became too 'liberal'.

Moves towards détente by America and Russia began to founder in 1979 when Soviet troops entered Afghanistan to bolster a weak Communist regime. Brezhnev died in 1982 and was briefly followed by Andropov and Chernenko, both elderly. But in 1985 the rising star of Communism was Mikhail Gorbachev, elected Secretary-General of the Communist Party at 54. With the Soviet defence budget reaching US$18 000 million a year, talks with America on reducing the number of missiles and avoiding a possible space war slowly got under way. Meanwhile, along the Chinese border, 30 Soviet armoured divisions stand alert against possible trouble.

Contrasting attitudes and patterns of behaviour are evidence of a significant generation gap which has developed between Russians born before 1953 and those born since. Those under 30 never knew Stalinist terror and have lived, especially in the 1960s, under less harsh conditions. Incentives have begun to replace directives, for example in labour migration to Siberia. Nor have most young men experienced the trauma of war. There have been gradual improvements in living conditions: better food, moves to new city flats for those in old, substandard and overcrowded homes as the supply of new housing trebled after 1958. The supply of television sets, refrigerators, hi-fi units, furniture, clothes and shoes all improved – until the breakdown of détente led to rises in military expenditure. But young people are constantly reminded in books, films and monuments of 20th-century threats to their socialist state.

USSR AT A GLANCE
Area 22 402 200 km² (8 649 538 sq miles)
Population 280 700 000
Capital Moscow
Government Communist federal republic
Currency Rouble = 100 kopeks
Language Russian (official) and more than 100 others
Religions Christian (25%) (22% Russian Orthodox, 2% Protestant, 1% Roman Catholic); Muslim (11%), Jewish (1%)
Climate Continental; arctic in the north. Average temperature in Moscow ranges from −16 to −9°C (3-16°F) in January to 13-23°C (55-73°F) in July
Main primary products Cereals, livestock, fruit and vegetables, timber, fish; oil and natural gas, coal, lignite, iron ore, diamonds, bauxite, copper, lead, zinc, salt, uranium
Major industries Iron and steel, cement, transport equipment, engineering, armaments, electronic equipment, chemicals, fertilisers, oil and gas processing, fishing, shipbuilding, mining and mineral refining, agriculture, forestry and timber processing
Main exports Oil and refined products, natural gas, machinery, iron and steel, transport equipment, timber, coal, textile yarns and fabrics, chemicals, metal ores
Annual income per head (US$) 3400
Population growth (per thous/yr) 8
Life expectancy (yrs) Male 70 **Female** 75

Ulan Bator (Ulaanbaatar) ***Mongolia*** Capital city, 1300 m (4265 ft) up on the Mongolian plateau on the Tuul river, a tributary of the Selenge.

The city dates back to the 17th century and the founding of the Temple of the Living Buddha. The settlement which grew around the temple became an important base for caravans crossing the Gobi Desert. By the 19th century, when it was known as Urga, it was one of the country's few substantial towns. The name was changed to Ulan Bator – 'Red Hero' – in 1924 after the Communist revolution in Mongolia.

Ulan Bator is a bleak city of wide streets and modern buildings, many of them reminiscent of Stalin's Russia. Scores of high-rise apartment blocks have been built, especially since the early 1970s, and most buildings have steam central heating – essential in the bitterly cold winters. Yurts – the traditional portable felt homes of the nomadic herdsmen – are still commonplace in the outskirts, but today they are supplied with electricity.

The city's State Central Museum has numerous prehistoric remains unearthed from the Gobi Desert, including 20 cm (8 in) long dinosaur eggs and the remains of a giant rhinoceros. Other cultural attractions include the art museum, and the state circus, opera and ballet. There is also a university.

The main industries are meat packing and processing, flour milling, printing, textiles, and the manufacture of building materials and prefabricated houses.

Population 435 000

Map Mongolia Db

Ulan-Ude ***USSR*** Capital of the Buryat Republic, a railway city 235 km (145 miles) east of Irkutsk, formerly Verkhneudinsk. Ulan-Ude was founded as a fort in 1648 and came to prominence in 1900 as a junction on the Trans Siberian Railway. A branch line to the Mongolian capital, Ulan Bator, 445 km (275 miles) to the south, was opened in 1949. Ulan-Ude's products include locomotives and rolling stock, glass, processed food and timber.

Population 329 000

Map USSR Lc

Ulcinj ***Yugoslavia*** See BAR

Uleåborg ***Finland*** See OULU

Ullswater ***United Kingdom*** Second largest lake (after Windermere) in the LAKE DISTRICT of north-west England. It is nearly 13 km (8 miles) long and is a popular tourist attraction.

Map United Kingdom Dc

Ulm ***West Germany*** Industrial city on the Danube river about 70 km (43 miles) south-east of Stuttgart making vehicles, electrical goods, textiles, clothing and food products. The lovely Old City, founded before 800, has a 14th-century town hall and a cathedral with the highest spire in the world (161 m, 528 ft). It was the birthplace of the physicist Albert Einstein (1879-1955), creator of the Theory of Relativity.

Population 98 700

Map West Germany Cd

Ulsan ***South Korea*** Industrial port on the south-east coast about 50 km (30 miles) from Pusan. It has oil refineries and petrochemical plants, and makes motor vehicles and ships.

Population 418 300

Map South Korea De

Ulster (Cuige Ulaidh) ***Ireland/United Kingdom*** Most northerly of Ireland's four ancient provinces. Ulster was colonised by the British in the 16th and 17th centuries. In 1922, six of Ulster's nine counties (Antrim, Armagh, Fermanagh, Down, Londonderry and Tyrone) remained within the United Kingdom as the province of Northern Ireland; Donegal, Monaghan and Cavan became part of what is now the Republic of Ireland. Some people of Northern Ireland call their province Ulster.

Population (Northern Ireland) 482 000; (Republic) 230 200

Map United Kingdom Bc

Ulundi ***South Africa*** Capital of the black state of KwaZulu about 160 km (100 miles) north of Durban. It was founded in 1873 by the Zulu king Cetewayo and it was here on July 4, 1879, that the final battle of the Anglo-Zulu War took place. A memorial commemorates the fallen.

Map South Africa Db

Ul'yanovsk ***USSR*** City on the Volga river, 170 km (105 miles) south of Kazan'. Set on the top of a hill, Ul'yanovsk is a major stage on the Volga riverboat route. It was formerly called Simbirsk, and was renamed in 1924 in honour of the revolutionary leader Vladimir Ilich Lenin (1870-1924), who was born there as Vladimir Ilich Ulyanov. His birthplace is now a memorial museum. The city's industries include engineering, machinery, electronics, textiles, food processing, vodka and beer.

Population 524 000

Map USSR Fc

Umbria ***Italy*** Landlocked central region covering 8456 km² (3265 sq miles) in the Apennine mountains to the north of Rome. Its green hills and carefully cultivated valleys have changed little over the centuries. There is a dense scatter of medieval hill towns, including ASSISI, GUBBIO, SPOLETO, ORVIETO, Todi and Città di Castello. Farming is the main occupation. There is industry in the two provincial capitals – steel at TERNI, confectionery at PERUGIA.

Population 816 000

Map Italy Dc

Umeå ***Sweden*** Port and timber-processing town on the Gulf of Bothnia, 676 km (420 miles) north-east of Stockholm. It has Sweden's northernmost university, and is linked by ferry with the Finnish city of Vaasa.

Population 85 110

Map Sweden Dc

Umfolozi Game Reserve ***South Africa*** Wildlife sanctuary in the black state of KwaZulu, about 270 km (170 miles) north of Durban. It was set up in 1897 to protect both the black and the rare white rhinoceros – which gets its name not from its colour but because it has a wide snout. The Afrikaans word for wide is *weit*. Antelope, buffalo, cheetah, giraffe, leopard, lion, wart hog and zebra also roam its 480 km² (185 sq miles).

Map South Africa Db

Umm al Qaywayn ***United Arab Emirates*** One of the seven emirates. It stretches for 24 km (15 miles) along The Gulf and covers 518 km² (200 sq miles). The sheikdom has large unexploited reserves of natural gas.

The chief town, also called Umm al Qaywayn, has a small cargo port. Industries include cement manufacture (the mainstay of the emirate's economy), an asbestos factory and some small plants producing agricultural fertilisers and chemicals. Fine examples of old Arab architecture contrast with high-rise office and apartment blocks. An old fort serves as the police headquarters. Offshore are the Umm al Qaywayn islands, with 18th-century watchtowers and a wide variety of wildlife.

Population (emirate) 14 000; (town) 2900

Map United Arab Emirates Ba

Umm Said (Musay'id) ***Qatar*** Important industrial town situated 40 km (25 miles) south of Doha. It contains Qatar's main oil terminal, and has a number of heavy industries – including steel, fertilisers and petrochemicals.

Population 7000

Map Qatar Ab

Umtali ***Zimbabwe*** See MUTARE

Umtata ***South Africa*** Capital of the black state of Transkei, about 285 km (175 miles) south-west of Durban. A market town settled by Europeans in 1860, it was transferred to Transkei when it was made an independent black state in 1976. Construction of government buildings has led to an economic boom. It has the University of Transkei, dating from 1977.

Population 45 000

Map South Africa Cc

unconformity Break in a sequence of SEDIMENTARY ROCKS, representing an interval when no sediments were deposited. The older rocks may be tilted and eroded before deposition resumes. Another type of unconformity occurs when sediments are laid down on top of IGNEOUS ROCKS such as a lava flow.

undertow An undercurrent moving down a beach, caused by a back flow of water piled up on the beach by a breaking wave.

Union of Soviet Socialist Republics (USSR) See p. 660

United Arab Emirates (UAE) See p. 668

United Kingdom (UK) See p. 670

United Provinces (of Agra and Oudh) ***India*** See UTTAR PRADESH

United States of America (USA) See p. 676

Unzen-Amakusa ***Japan*** National park, 257 km² (99 sq miles) in area, in western Kyushu island. The park includes the Shimabara peninsula, where the Christian armies of Kyushu were finally defeated by the Tokugawa shogunate in 1638. After the battle, a few Christians took refuge on the nearby Amakusa Islands, where their descendants kept their faith alive in secret for 250 years until the Tokugawa era ended.

Map Japan Bd

United Arab Emirates

PIRACY AND PEARLING, SLAVES AND SANDALWOOD HAVE LONG SINCE GIVEN WAY TO OIL AND ALUMINIUM, STEEL AND CEMENT AS SOURCES OF WEALTH

The coast of the United Arab Emirates, 600 km (375 miles) long, was once known as the Pirate Coast. Armed with fearsome curved daggers (*kunjahs*) and scimitars (*qattaras*), Arab corsairs preyed on European ships trading in THE GULF. Their ruined forts and watchtowers can still be seen along the shores. Piracy ended in the 19th century after Britain imposed a truce, and for this reason the area became known as the Trucial Coast. Today, the sigh of the trade winds in the rigging and the fierce cries of battle have given way to the sound of the engines of oil tankers making their way along The Gulf.

SEVEN SHEIKDOMS

The UAE is a federation of seven sheikdoms formed in 1971 when Britain left the region: ABU DHABI, DUBAI, SHARJAH, RAS AL KHAYMAH, AL FUJAYRAH, UMM AL QAYWAYN and AJMAN. Since 1962, oil production has brought undreamt-of riches to the Emirates, which once traded in spices, pearls, sandalwood and slaves. Abu Dhabi, the largest of the group, is the richest state in the world with an average yearly income for every man, woman and child equivalent to US$26 000. Dubai, the second largest, is also a major oil producer.

The country, which covers 83 600 km^2 (32 300 sq miles), has a short coastline east of the MUSANDAM Peninsula, on the Gulf of OMAN, as well as that on The Gulf. The land is mainly flat, sandy desert but to the north, on the peninsula, the HAJAR MOUNTAINS rise to 2081 m (6826 ft). Temperatures in the hot, humid summer can reach 49°C (120°F) in July and August, but from October to mid-May the weather is delightful, with warm, sunny days and pleasantly cool evenings. Scant rain is brought mainly by winter storms. The emirate of Ras al Khaymah, the coastal plain of Al Fujayrah and the oases are fertile, but most of the remaining land is desert.

As many as 80 per cent of the Emirates' estimated 1 380 000 population are expatriates, mainly Pakistanis, Indians and other east Asians, and only 15 per cent of the workforce are UAE nationals; the development financed by oil wealth has drawn immigrants like a magnet. Men outnumber women two to one, so up to half the men now marry girls from outside The Gulf – Egyptian and Indian women are popular choices.

The seven states abolished their separate ministries in 1974 and the country is ruled by a council of ministers. The Sheik of Abu Dhabi is president, the Sheik of Dubai vice-president and prime minister. The country is Muslim, strict but not fanatical; for example, alcohol is allowed for non-Muslims.

INTO THE MODERN WORLD

Since the discovery of oil and gas, the UAE has used its oil wealth to diversify the economy at a staggering rate, finance social improvements and invest overseas. Abu Dhabi and Dubai are the main industrial centres. Among the ancient forts along the coast there now stand an oil refinery, gas liquefaction and fertiliser plants, a steel rolling mill and aluminium smelter, and factories making cement and other building materials. The bustling construction industry has turned tiny fishing ports into modern oil terminals and harbours for bulk carriers and other vessels.

In the towns, new office blocks have replaced many of the traditional Arab buildings. Dubai's 39 storey International Trade Centre is The Gulf's tallest building, and four modern airports cater for international flights. Where once health and educational facilities were limited there are now hospitals and clinics, schools and universities. Fine modern roads now link the sheikdoms.

Irrigation projects, based on giant desalination plants producing fresh water from sea water, have extended agriculture beyond the oases. Fields of wheat, tobacco and alfalfa lie among the sand dunes while thriving market gardens provide tomatoes, aubergines and melons for the home market. Date palms abound in the oases and once found many uses. Dates provided the staple diet, while the trunks of the trees were used for building, the leaf stems for *barasti* (light, portable homes), and the leaves for basket weaving.

In the past, fish were one of the few sources of protein and fishing is still an important industry. The clear Gulf waters are rich in prawns, tuna, mackerel, anchovies and sardines, and for sport there are sailfish, sharks and marlin.

The food is international and the visitor can find anything from the traditional *kabsa* (a whole sheep stuffed with rice, spices and almonds) to fish and chips. The *shwarma* consists of pieces of lamb roasted on a spit and served with tomatoes, parsley and other herbs, and chick pea sauce. In the *souks* (markets) you can buy exotic foods, spices, traditional silver jewellery, rugs and carpets, hi-fi and electronic gadgets and gold ornaments. In fact much of Dubai's wealth once came from gold trading – and smuggling.

Piracy has given way to the excitement of camel racing, falconry, wrestling and even ice skating on artificial rinks. Camel races usually take place soon after the cool desert dawn, either at established tracks or at impromptu meeting places in the desert. There is no betting, but cheering spectators follow the animals by truck until the winners emerge in a cloud of sand and dust. Falcons are trained for hunting – though the bustard, a common prey, is becoming scarce. As a complete contrast to this ancient sport, football has been imported; it is played and watched with passion – just another step towards the Western way of life for these ancient sheikdoms.

UNITED ARAB EMIRATES AT A GLANCE

Area	83 600 km^2 (32 300 sq miles)
Population	1 380 000
Capital	Abu Dhabi
Government	Non-elected federal council
Currency	dirham = 100 fils
Languages	Arabic, English
Religion	Muslim
Climate	Hot, mild in winter; average temperature in Sharjah ranges from 12-23°C (54-73°F) in January to 28-38°C (82-100°F) in August
Main primary products	Goats, sheep, camels, cattle, fruit, vegetables, fish and shellfish; oil and natural gas
Major industries	Oil production, natural gas liquefaction, steel milling, cement, aluminium smelting, fishing, fish processing, fishmeal
Main exports	Crude oil, natural gas, shrimps, prawns, fishmeal
Annual income per head (US$)	23 000
Population growth (per thous/yr)	44
Life expectancy (yrs)	**Male** 61 **Female** 65

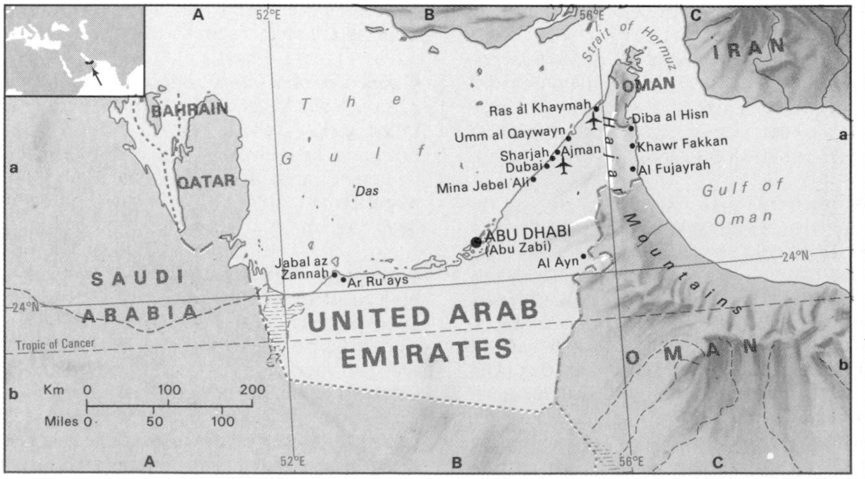

Upemba National Park ***Zaire*** Reserve covering 11 730 km² (4529 sq miles) in central Shaba region west of Lake Mweru. It contains swampland, with crocodiles and hippopotamuses in the west. To the south and east there is savannah country with animals such as antelopes, buffaloes, elands and zebras. The park was established in 1939.
Map Zaire Bb

Upington ***South Africa*** Market town on the Orange River in north-west Cape Province, about 475 km (295 miles) from the Atlantic coast. It was founded as a mission station in 1875. It is now a road and rail junction on the line to Namibia, in a prosperous, irrigated farming region.
Population 25 900
Map South Africa Bb

Upolu ***Western Samoa*** Second largest but most populous island of the Pacific nation, covering 1100 km² (425 sq miles). It is mountainous, rising to 1113 m (3652 ft) at Mount Fito; some parts have rain forests and high waterfalls. Coconuts are the main crop. The capital, APIA, is on the north coast.
Population 133 000
Map Pacific Ocean Ec

Upper Austria (Oberösterreich) ***Austria*** Hilly, well-forested, northern state noted for its agriculture. Industry is concentrated at LINZ, the capital, STEYR and Ranshofen.
Population 1 270 500
Map Austria Ca

Upper Egypt ***Egypt*** See LOWER EGYPT

Uppland ***Sweden*** Province covering 12 673 km² (4893 sq miles) in the east of the country, just north of Stockholm. The iron-mining district of Dannemora is on its northern edge. UPPSALA is the provincial capital.
Population 1 128 400
Map Sweden Cd

Uppsala ***Sweden*** Medieval city 68 km (42 miles) north of Stockholm. An important cultural centre, it has the country's oldest university (founded in 1477), which houses manuscripts going back to the 6th century. There is a 13th to 15th century cathedral, where Sweden's kings were once crowned, and a 16th-century castle. Three kilometres (2 miles) to the north is Old Uppsala, with the impressive prehistoric burial mounds of legendary kings. Uppsala produces machinery, building materials, pharmaceuticals and food products.
Population 154 900
Map Sweden Cd

up-valley wind See ANABATIC WIND

Ur ***Iraq*** Archaeological site of the ancient Sumerian town that was the home of the Old Testament patriarch, Abraham. It lies about 310 km (190 miles) south-east of the capital, Baghdad. The city, founded between 6000 and 5000 BC, fell into ruin in the 4th century BC after the course of the Euphrates changed and river trade by-passed the city. The city figures in the Bible as Ur of the Chaldees (Chaldeans).
Map Iraq Cc

Urals ***USSR*** North-south mountain range forming the geographical boundary between Europe and Asia. They run for 2000 km (1240 miles) from the Kara Sea in the north of Russia to the Aral Sea and the Kazakh steppes in the south. The Urals have an average height of about 1000 m (3500 ft) and are divided into three contrasting sections.

The Northern Urals are the narrowest and highest, reaching a peak in Mount Narodnaya (1894 m, 6214 ft). The mineral-rich Middle Urals consist mainly of a low, densely forested plateau forming the historic gateway to Siberia. They contain the Ural Industrial Region, an industrial and mining area possessing many minerals – including iron, bauxite (from which aluminium is made), copper, asbestos, chrome, platinum and gold. The region also has large oil fields. The Southern Urals, made up of rich pastureland, contain the source of the Ural river, which flows south-eastwards for 2525 km (1570 miles) into the Caspian Sea.
Map USSR Gc

Ural'sk ***USSR*** Market town on the Ural river in Kazakhstan, 470 km (290 miles) north of the Caspian Sea. It trades mainly in cattle and grain, and its chief products are processed food and leather goods.
Population 188 000
Map USSR Gc

Urbino ***Italy*** Tourist town about 110 km (70 miles) east of Florence. Its narrow streets, cheek-by-jowl buildings and lack of modern development make Urbino seem like a time capsule of the Middle Ages. Raphael, one of the greatest of the Renaissance painters and one of the architects of St Peter's in Rome, was born there in 1483. Urbino's pride is its vast, 15th-century ducal palace – which now houses the Marche National Gallery, containing the finest picture collection in the region, covering many Italian schools of painting. The town's university dates from 1506.
Population 15 900
Map Italy Dc

Uri ***Switzerland*** German-speaking canton in central Switzerland said to be the home of the legendary Swiss folk hero William Tell, who opposed Austrian rule. It was one of the three cantons of the infant Swiss state formed in 1291. Uri, which covers an area of 1076 km² (415 sq miles), consists largely of mountains, forests and glaciers. The Canton's arms is a bull's head, and the name Uri probably comes from *urochs*, meaning 'wild bull'.

The main industry is forestry, and the capital is the small town of Altdorf (population 8000) – where there is a statue of William Tell on the spot where he was supposed to have shot the apple from his son's head.
Population 34 000
Map Switzerland Ba

Urmia ***Iran*** See ORUMIYEH

Uruguay See p. 683

Uruguay ***South America*** River which gives its name to the nation of Uruguay. Its source is in the mountains of southern Brazil close to the Atlantic. The river marks part of both the Brazil-Argentina border and the frontier between Argentina and Uruguay. After 1600 km (990 miles), it joins the Paraná river about 75 km (47 miles) north of the Argentinian capital, Buenos Aires, to form the River Plate. The Uruguay is navigable for 350 km (220 miles) below the city of Salto. Above Salto, it is dammed to generate hydroelectricity.
Map Uruguay Ab

Urümqi (Urumchi) ***China*** Capital of the Xinjiang Uygur Autonomous Region in the far west of the country. It is the biggest administrative centre for western China and contains steel-making, engineering and textile factories.
Population 1 200 000
Map China Cc

US Virgin Islands ***Caribbean*** See VIRGIN ISLANDS

USA See UNITED STATES OF AMERICA

U-shaped valley Trough-like VALLEY which has been deepened, widened and straightened by the passage of a glacier.

Üsküb ***Yugoslavia*** See SKOPJE

Üsküdar (Scutari) ***Turkey*** Residential suburb of ISTANBUL on the eastern side of the Bosporus. It was the site of the hospital run by the British nurse Florence Nightingale during the Crimean War (1853–6).
Map Turkey Aa

USSR See UNION OF SOVIET SOCIALIST REPUBLICS

Ussuri ***USSR/China*** See WUSUL JIANG

Ustí nad Labem ***Czechoslovakia*** Regional capital of north-west Bohemia and industrial city north-west of Prague and 16 km (10 miles) from the East German border. It produces chemicals, textiles, foodstuffs and glass, and is also a tourist centre for the area to the north-east known as Bohemian Switzerland.
Population 90 200
Map Czechoslovakia Ba

Ustinov (Izhevsk) ***USSR*** Industrial city and capital of the Udmurt Republic, near the western edge of the Ural mountains, 225 km (140 miles) south-west of Perm'. It was founded in 1760 and has been making steel products – including weapons and armaments – since then. Its other main products are machine tools, motor vehicles, pianos and furniture.
Population 603 000
Map USSR Gc

Usumacinta ***Mexico*** One of Mexico's principal rivers, about 1000 km (620 miles) long. It rises as the Chixoy in north-west Guatemala where it flows east then north-west through Chiapas state, forming the boundary between Mexico and Guatemala. It then meanders over the Tabasco lowlands, and continues north-west to empty into the Grijalva near its mouth in north Tabasco. The river is navigable for 500 km (310 miles); plans are under way to develop the river's fertile basin for agriculture.
Map Mexico Cc

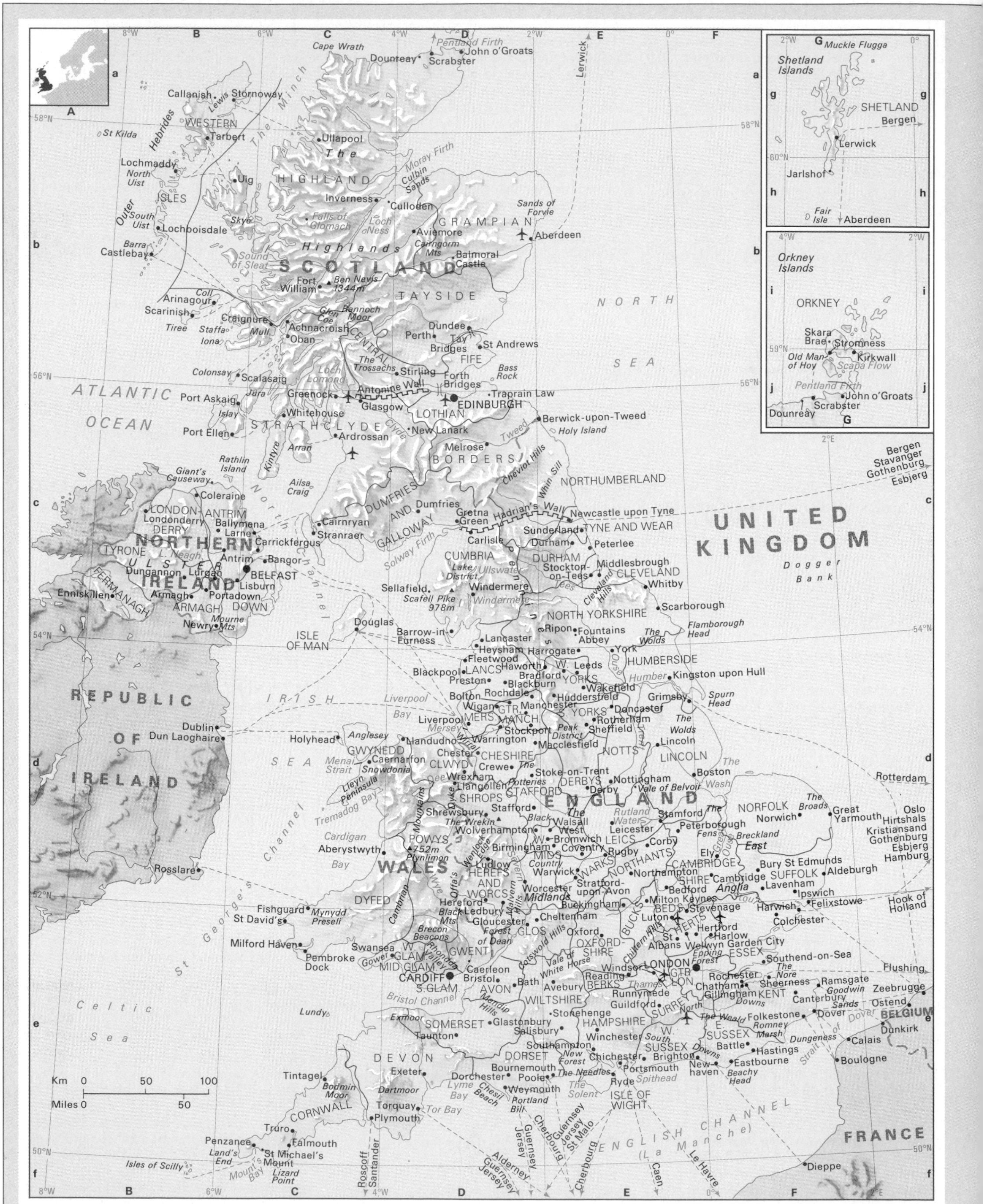

UNITED KINGDOM
SCOTLAND
ENGLAND
WALES
NORTHERN IRELAND
REPUBLIC OF IRELAND
ISLE OF MAN
FRANCE
BELGIUM
ATLANTIC OCEAN
NORTH SEA
IRISH SEA
Celtic Sea
ENGLISH CHANNEL (La Manche)
St George's Channel
North Channel
The Minch
Cape Wrath
Dounreay
Pentland Firth
John o'Groats
Scrabster
Lerwick
Callanish
Lewis
Stornoway
WESTERN ISLES
Outer Hebrides
St Kilda
Tarbert
Ullapool
The Highlands
Lochmaddy
North Uist
Uig
Skye
South Uist
Lochboisdale
Barra
Castlebay
Inverness
Culloden
Moray Firth
Culbin Sands
Falls of Glomach
Loch Ness
Aviemore
Cairngorm Mts
GRAMPIAN
Sands of Forvie
Aberdeen
Balmoral Castle
Sound of Sleat
Fort William
Ben Nevis 1344m
Coll
Arinagour
Scarinish
Tiree
Staffa
Iona
Mull
Craignure
Achnacroish
Oban
Glen Coe
Bannoch Moor
TAYSIDE
Dundee
Perth
Tay Bridges
St Andrews
FIFE
CENTRAL
The Trossachs
Loch Lomond
Stirling
Forth Bridges
Bass Rock
Colonsay
Scalasaig
Jura
Port Askaig
Islay
Port Ellen
Greenock
Antonine Wall
Glasgow
EDINBURGH
Traprain Law
LOTHIAN
Whitehouse
STRATHCLYDE
Clyde
New Lanark
Berwick-upon-Tweed
Holy Island
Tweed
Ardrossan
Arran
Kintyre
Melrose
BORDERS
Cheviot Hills
Rathlin Island
Giant's Causeway
Ailsa Craig
Whin Sill
NORTHUMBERLAND
Coleraine
LONDONDERRY
Londonderry
ANTRIM
Ballymena
Larne
Carrickfergus
Cairnryan
Stranraer
DUMFRIES AND GALLOWAY
Dumfries
Gretna Green
Hadrian's Wall
Newcastle upon Tyne
TYNE AND WEAR
Sunderland
Durham
Peterlee
TYRONE
L. Neagh
ULSTER
Antrim
Bangor
BELFAST
Solway Firth
Carlisle
CUMBRIA
DURHAM
Middlesbrough
Stockton-on-Tees
CLEVELAND
Dungannon
Lurgan
Lisburn
FERMANAGH
Enniskillen
Armagh
Portadown
ARMAGH
DOWN
Mourne Mts
Newry
Lake District
Ullswater
Sellafield
Scafell Pike 978m
Windermere
Cleveland Hills
Whitby
Scarborough
NORTH YORKSHIRE
Flamborough Head
Douglas
Barrow-in-Furness
Lancaster
Ripon
Fountains Abbey
The Wolds
Heysham
Harrogate
York
Fleetwood
Blackpool
LANCS
Haworth
W. YORKS
Leeds
Bradford
HUMBERSIDE
Preston
Blackburn
Wakefield
Humber
Kingston upon Hull
Bolton
Rochdale
Huddersfield
Grimsby
Spurn Head
Liverpool Bay
Wigan
GTR. MANCH.
Manchester
Doncaster
MERS.
Liverpool
Mersey
S. YORKS
Rotherham
Stockport
Sheffield
Peak District
Dublin
Dun Laoghaire
Holyhead
Anglesey
Llandudno
Warrington
Macclesfield
NOTTS
Lincoln
GWYNEDD
Caernarfon
Menai Strait
Snowdonia
Chester
CHESHIRE
LINCOLN
Crewe
CLWYD
Stoke-on-Trent
Nottingham
Boston
The Wash
Rotterdam
Lleyn Peninsula
Wrexham
Llangollen
The Potteries
DERBYS
Derby
Vale of Belvoir
Tremadog Bay
SHROPS
STAFFORD
Stafford
Rutland Water
Stamford
NORFOLK
The Broads
Norwich
Great Yarmouth
Oslo
Hirtshals
Kristiansand
Gothenburg
Esbjerg
Hamburg
Shrewsbury
The Wrekin
The Black Country
Walsall
West Bromwich
Leicester
LEICS
Peterborough
Fens
Breckland
East Anglia
Cardigan Bay
Wolverhampton
Birmingham
Coventry
Rugby
Corby
Ely
Aberystwyth
POWYS
Plynlimon 752m
Wenlock Edge
Ludlow
W. MIDS
Warwick
WARKS
NORTHANTS
CAMBRIDGE
Bury St Edmunds
SUFFOLK
Aldeburgh
HEREFS AND WORCS
Worcester
Stratford-upon-Avon
Northampton
CAMBRIDGESHIRE
Cambridge
Lavenham
Ipswich
Rosslare
Cambrian Mountains
Offa's Dyke
Malvern Hills
Midlands
Bedford
BEDS
DYFED
Hereford
Ledbury
Buckingham
Milton Keynes
Harwich
Felixstowe
Hook of Holland
Fishguard
Mynydd Preseli
St David's
Black Mts
Gloucester
Cheltenham
BUCKS
Stevenage
Colchester
Brecon Beacons
Forest of Dean
GLOS
Cotswold Hills
Oxford
OXFORDSHIRE
Luton
St Albans
HERTS
Hertford
Harlow
Welwyn Garden City
Milford Haven
Pembroke Dock
Swansea
Gower
W. GLAM
Rhondda Valley
GWENT
Chiltern Hills
Epping Forest
ESSEX
Southend-on-Sea
MID GLAM
CARDIFF
S.GLAM
Caerleon
Bristol
Vale of White Horse
Reading
BERKS
Windsor
LONDON
GTR LON
Thames
Rochester
Chatham
Gillingham
The Nore
Sheerness
Ramsgate
Flushing
Bath
Avebury
Runnymede
KENT
Goodwin Sands
Zeebrugge
Bristol Channel
AVON
WILTSHIRE
Guildford
North Downs
Canterbury
Ostend
Lundy
Mendip Hills
Stonehenge
SURREY
The Weald
Folkestone
Dover
Exmoor
SOMERSET
Glastonbury
Salisbury
HAMPSHIRE
W. SUSSEX
E. SUSSEX
Romney Marsh
Strait of Dover
Dunkirk
Taunton
Winchester
South Downs
Battle
Dungeness
Calais
Southampton
Chichester
Brighton
Hastings
DEVON
DORSET
New Forest
Portsmouth
Newhaven
Eastbourne
Beachy Head
Boulogne
Exeter
Bournemouth
Dorchester
Poole
The Needles
Ryde
Spithead
Tintagel
Bodmin Moor
Dartmoor
Lyme Bay
Chesil Beach
Weymouth
The Solent
Portland Bill
ISLE OF WIGHT
Torquay
Tor Bay
Plymouth
CORNWALL
Truro
Falmouth
Penzance
Land's End
St Michael's Mount
Mount's Bay
Lizard Point
Isles of Scilly
Roscoff
Santander
Cherbourg
Jersey
Guernsey
Alderney
St Malo
Caen
Le Havre
Dieppe
Dogger Bank
Bergen
Stavanger
Gothenburg
Esbjerg
Km 0 50 100
Miles 0 50
Shetland Islands
Muckle Flugga
SHETLAND
Jarlshof
Fair Isle
Orkney Islands
ORKNEY
Skara Brae
Stromness
Kirkwall
Old Man of Hoy
Scapa Flow

United Kingdom

LOSS OF EMPIRE LEAVES BRITAIN STILL SEEKING A NEW WORLD ROLE – AND A NEW SOURCE OF WEALTH WHEN THE NORTH SEA OIL RUNS DRY

Two World Wars and the power politics of the 20th century have wrought drastic changes in the small, island state that once held sway over the greatest empire the world has seen, that unfurled its flag in every quarter of the globe. Within the lifespan of a single generation, the term British Empire has been almost forgotten, though its ghosts haunt television and cinema screens with flickering images of durbars and red-coated regiments charging to tunes of glory. This loss of empire has left the United Kingdom of Great Britain and Northern Ireland still seeking a new world role – a search that is proving prolonged and painful for a nation that has given the world so much.

To many of the great nations she once ruled, and to many others, Britain left her greatest legacies of all – the poetic and expressive language in which Shakespeare wrote; and the principles of parliamentary democracy, hard won over many centuries and preserved at appalling cost in the two great struggles for survival which all but bled her to death.

Indeed, if the First World War left her seriously weakened, the Second left Britain shattered. Dominions and colonies were demanding – sometimes forcibly – the freedom and independence they, too, had fought for. Britain, her financial and economic resources devastated, had little choice but to start dismantling an empire that was becoming an increasingly insupportable burden.

Besides, the mother country herself had embarked upon an expensive social revolution of her own – erecting the cradle-to-grave umbrella of the Welfare State. She could not afford that, either. But somehow, with much belt-tightening, it was done and, like so many things British, became an example for others to learn from and adopt.

Setting an example, 'doing the right thing', ranks high among the qualities valued by many Britons. In general they are proud of their country and its record – which has led many former subject nations to model their governments and institutions upon Britain's. Indoctrinated from childhood with ideals of sportsmanship, 'fair play' and social responsibility, they can also exhibit a dogged determination to get their own way, and a sublime faith in their ability to do so. Inventive and practical, they often succeed – by ingenuity or, as they put it, 'muddling through'.

Slow to anger, courageous and cunning when roused, they have disposed of many adversaries who have taken their penchant for modesty, understatement and readiness to negotiate as signs of weakness. As recently as 1982, in a brief, bloody war fought 8000 miles from home, they destroyed a numerically superior, well-equipped Argentinian army which had occupied one of their few remaining colonies, the Falkland Islands.

Apart from the fighting qualities of the men engaged, a notable factor in this victory was the success of Britain's vertical takeoff Harrier jet fighter, a versatile and highly manoeuvrable aircraft. It was a typical product of British inventiveness, and the admiring American armed forces have bought hundreds. But not all such inventions are successfully exploited in Britain. From electronics to pharmaceuticals, aeronautical engineering to medical technology, too many British scientists have had to look overseas for the backing to exploit their discoveries – or have watched while other countries have made their ideas reality.

One cause has been a lack of technical expertise at the very top of many British firms; another a feeling that marketing and salesmanship are somehow distasteful – anything made in Britain was automatically the best, and any sensible person ought to know that.

Disillusionment came as postwar shortages in some products and prewar work practices reduced Britain's capacity to produce at competitive prices, and high-quality – often superior – products began to flow from overseas industries employing more advanced equipment and work methods. Notably successful in penetrating Britain's markets, both at home and overseas, were her former adversaries, Germany and Japan.

Britain is fighting back, but with over 3 million unemployed and once-great industries in decline, all her resources of brainpower, energy, sheer grit and will to win are needed if she is to resume a truly effective role in world affairs.

DIVIDED NATION

There are encouraging signs of recovery. Prosperity has returned already to a large area of Britain – a huge triangle stretching across southern England between CAMBRIDGE, BRISTOL and BRIGHTON. Here are expanding new towns, such as Milton Keynes, Harlow, Stevenage and Crawley, and resuscitated older ones like Swindon. From HAMPSHIRE to the valleys of the CHILTERNS, villages have doubled in population. The electronics industry, much of it defence-related, is among the growth leaders of a new, high technology industrial revolution. However, the decline of the older 'smokestack' industries – coal, iron and steel, textiles, heavy engineering and so on – has deeply affected Scotland, Wales, THE MIDLANDS and the north of England, creating major unemployment areas and draining away manpower to the more affluent south-east.

This was a factor in an upsurge in regionalism in the 1970s, when nationalist political parties in Scotland and Wales gained sufficient strength to make themselves felt in the General Election of 1974. Devolution, the dispersal of political power from the central parliament in Westminster to proposed new assemblies in Scotland and Wales, became a dominant topic of political discussion. The issue was finally put to Scottish and Welsh voters by a referendum in 1979 – with the result that schemes for regional parliaments were abandoned.

The new realignment of industry and workers is, in fact, a fairly modest reversal of the huge population shifts generated by the mightiest social and economic upheaval of all – the Industrial Revolution that made the nation an empire and changed the world itself. British ingenuity was what sparked it; coal was what fuelled it – coal by the billion tonnes won from seams scattered through England, Scotland and Wales.

The revolution created immense wealth and prestige for the country. It financed Britain's expansionist policies abroad and the military muscle to safeguard them; and the wealth of empire poured back to the mother country to complete the cycle. It also created poverty, squalor and misery as men, women and children were drawn into the insatiable maws of mines and factories. Wages were often at starvation level, working hours long and housing frequently congested – miles of brick terraces in streets blackened by the smoke that now enveloped the ever expanding industrial towns and cities. With a few notable exceptions, the new factory owners and entrepreneurs paid scant attention to the cost in human terms of products stamped 'British Made'.

Reform was slow, despite the efforts of a few philanthropically minded industrialists to improve the lives of their employees. Mill owners such as Sir Richard Arkwright, in the early years of the revolution, and Sir Titus Salt, in its Victorian heyday, were far ahead of their times. Arkwright (1732-92) and his son built two cotton mills in the Derbyshire village of Cromford – and lavished money on the village, building a church, a school and houses for themselves and their employees.

Titus Salt (1803-76) was a BRADFORD wool merchant's son, and by the 1840s owned four mills in the city. But he became sickened by the filth, overcrowding and pollution there, and by the conditions endured by working people. On a beautiful, green-field site outside Bradford, he built not only an extraordinary new mill, but a whole village for its workers, incorporating the newest ideas in planning, engineering and hygiene. He paid good pensions 36 years before government ones were introduced, and built a school for 750 pupils, an exquisite church, almshouses – even a boathouse and dining room on the river, with boats for hire. The village, Saltaire, remains almost as he built it.

The paternalism of Salt and Arkwright, deeply religious men like many others of their time, sprang from a belief that the possession of wealth imposed a moral duty to improve the lot of the workers who had helped to create it. It was an attitude increasingly affordable (although by no means universal) as Britain rapidly became the wealthiest nation on earth. And though aggressively pursuing the expansion and consolidation of empire, the British developed a genuinely held conviction that they were bringing order and civilisation to the territories they had explored or conquered.

At the pinnacle of imperial power, in the closing years of the 19th century and the first decades of the 20th, this was truly, as the slogans of the time proclaimed, 'an empire on which the sun never sets'. Huge areas of the world atlas were coloured pink, denoting a British possession; a quarter of the world's population owed allegiance to the British Crown. Britain saw herself as presiding over a brotherhood of nations, an attitude that found ultimate expression after the First World War, when the dominions secured equality in the comity of nations by their admission to the League of Nations.

The first real crack in the edifice appeared only three years later – on Britain's own doorstep. For most of the 19th century and up to 1922, the United Kingdom embraced the territories of four historic peoples – the English, Scots, Welsh and Irish. Then, after centuries of bitter and often bloody strife, 26 of IRELAND's 32 counties broke away to form the Irish Free State. By 1931 the Statute of Westminster was acknowledging the self-governing status of Australia, Canada, New Zealand and South Africa as 'freely associated members' of the Commonwealth. However, Britain's territories still included India, and lands in east and west Africa, the Far East, Central America and the Mediterranean.

NEW ALLEGIANCES

The Second World War (1939-45), in which beleaguered Britain fought Nazi Germany alone for 18 months, saw the USA and USSR emerge as the new 'super powers'. The whole pattern of international politics was being upended, and after India and Pakistan gained independence in 1947, Britain's influence on world events declined steeply. By 1949 the term 'British' was being dropped from the Commonwealth's title – but the monarch remained at its head. Nations which had won independence even as republics could remain members, or rejoin, if they acknowledged the Crown as head of the Commonwealth. Burma and Ireland had opted out – but the six counties forming Northern Ireland remained firmly British, and are in continuous and violent conflict with those in both the north and south who wish them to join as one nation.

South Africa was forced to withdraw in 1961 over its racial policies, and Pakistan quit in 1972 when East Pakistan was recognised as Bangladesh. But 49 members remain, and their leaders find plenty of mutual concern to discuss at two-yearly meetings.

The Commonwealth may survive, but Britain has in fact surrendered political control of all but 17 small dependencies: Anguilla; Bermuda; British Antarctic Territory; British Indian Ocean Territory; British Virgin Islands; Cayman Islands; Falkland Islands; Falkland Islands Dependencies; Gibraltar; Hong Kong; Montserrat; Pitcairn, Ducie, Henderson and Oeno; St Helena; St Helena Dependencies (Ascension, Tristan da Cunha); the Isle of Man; Channel Islands; and the Turks and Caicos Islands. Hong Kong is due to return to Chinese rule in 1997, when Britain's 99 year lease runs out.

New naval and military roles more in keeping with Britain's reduced circumstances were evolved with the formation in 1949 of the NATO military alliance. Britain was a founder-member and has allowed American airfields, missile-launching sites, submarine bases and radar tracking stations to be established on her soil. Nevertheless, Britain's own numerically small armed forces (about 325 000 strong) retain their own formidable defence and strike capabilities, armed with nuclear submarines, advanced aircraft and high-technology tanks, missiles, guns and communications equipment. There are 55 000 of them in the British Army of the Rhine.

Commitment to Europe in political and economic terms did not come until 1972, when Britain finally joined the European Community. Founded 20 years earlier by France, West Germany, Belgium, Luxembourg, the Netherlands and Italy, it was conceived as a new power bloc – a vehicle for cooperation between members at many levels. In fact it is composed of three organisations – the European Economic Community (EEC or Common Market), European Coal and Steel Community (ECSC) and European Atomic Energy Community (Euratom). It is now the world's largest trading bloc, accounting for about one-third of all international trade. Britain's entry was achieved only after overcoming some bitter opposition abroad – notably from France – and equally bitter wrangles among her own politicians. Indeed, the internal arguments raged on even after Britain joined, culminating in a national referendum in 1975 – the only one it has ever held – when those wanting to quit the Common Market were substantially out-voted.

Reluctance to recognise any authority outside its own shores is perhaps understandable in a country which proudly proclaims itself ruled by a constitutional monarch and governed by the 'Mother of Parliaments' – a very model of enlightened democracy. Britain has no written constitution and Parliament is the country's supreme legislative body. It has three elements – the Queen, the House of Lords and the House of Commons. The Lords is composed of hereditary and lifetime peers of the realm, the two archbishops and 24 senior bishops of the Church of England, and the 'Law Lords' (Lords of Appeal, senior judges). The 650 Members of the Commons (MPs) are elected by universal adult suffrage – British and Commonwealth citizens aged 18 or over who are resident in the country are eligible to vote in a secret ballot. Irish citizens living in Britain may also vote. Voting is not, however, compulsory, and 73 per cent of the 42 million eligible people voted in the elections of 1983.

Britain is governed in the name of the Crown, but the Queen meets her Lords and Commons only on symbolic occasions such as the ceremonial State Opening of Parliament. In practice, procedures ensure that legislative power is firmly in the hands of the Commons, although in principle both Houses must approve the measure proposed. The Queen follows the advice of the Prime Minister – the leader of the political group (party) able to command a majority vote in the Commons, and which forms the Government. As a member of the European Community, Britain also sends 81 elected representatives (MEPs) to sit in the European Parliament, and recognises certain types of legislation passed there.

As it happens, about 70 per cent of the EEC budget goes to subsidise farm and food prices. And about the same percentage of Britain's total land area is farmed. Of a total of 243 000 farms and holdings, about half provide full-time work for only one person, while 31 000 larger farms account for half the total agricultural output, which topped US$16 billion in 1983 – around 2.3 per cent of the Gross

▲ NORTHERN LIGHTS Edinburgh – called the 'Athens of the North' – is the capital of Scotland, which has been ruled by the same monarch as England and Wales since 1603, when Scottish King James VI became James I of England.

National Product (GNP). Fewer than 700 000 people work in agriculture, less than 3 per cent of the national workforce.

Some of their products helped to swell the EEC's colossal agricultural surpluses, in butter, beef, grain and milk powder. EEC support policies have encouraged large-scale crop-growing and, to accommodate ever larger machines, thousands of kilometres of hedges, ditches and ancient boundaries have been bulldozed and deep-ploughed into oblivion. Some modern farms, fully mechanised to eliminate the farm worker, have begun to resemble factories – an unlovely cluster of steel and concrete buildings, silage towers and sheds.

Perhaps fortunately, relatively little of Britain's landscape lends itself to large-scale operations of this nature, and mounting public concern about its environmental and ecological consequences is beginning to exert political pressure. Indeed, much of Britain's farming remains on more traditional lines. In the mountainous, rain-swept west and north, and their green-pastured, river-laced borderlands, pastoral farming predominates, with sheep grazing the hillsides and cattle browsing the lush meadows, as they have done for 1000 years and more. The lowlands of the south and east, sheltered from the rain of an Atlantic climate by the western highlands, have always favoured crop growing. Wheat, barley, potatoes and vegetables are harvested from soils which are richer and deeper than those of highland Britain.

Across the scarped uplands and clay vales of southern England the imprint of nearly 2000 years of cultivation can still be seen. Chalk downs bear the marks, however faint, of terraces (lynchets) farmed by early Iron Age people, as on Salisbury Plain and near the Uffington White Horse in Oxfordshire. Near Northleigh in the same county, a Romano-British village is surrounded by relics of the ridges and furrows of medieval communal cultivation.

ERAS OF CHANGE

Many ancient sites such as these have been affected by radical changes imposed upon large areas of the countryside between 1350 and 1600. The Black Death (bubonic plague) from 1348 onwards wiped out more than one-third of the population within a short space of time. Then landlords began to enclose the land and evict villagers to make way for a huge expansion of sheep farming to satisfy a soaring demand for wool from British and continental cloth manufacturers, and to benefit from rising meat prices. Over little more than 100 years, at least 3000 villages and hamlets were deserted, as countless strips of medieval open-field cultivation were turned over to pasturage for sheep. Some never knew a plough again, and so retain the grassed-over marks of their former use.

A legacy of the great wealth created can be seen in the magnificent 'wool' churches built in towns and villages by pious merchants, manufacturers and farmers – particularly in the COTSWOLDS, where the warm tones of the mellowed local stone suffuse towns such as Chipping Campden and villages like Bibury.

Later, in the 18th and 19th centuries, landlords and clan chieftains swept tenant farmers and crofters from large areas of the Scottish HIGHLANDS, again to make way for sheep and deer. These wholesale evictions, often enforced by destroying crofts and cottages, became infamous as the 'Clearances', and led to large-scale emigration to North America, especially Canada. Much of the Scottish Highlands remains virtually empty to this day.

A landslide victory by the Labour Party over the Conservatives in the post-Second World War election of 1945 heralded another upsurge of social change in Britain: the establishment of the Welfare State and the nationalisation of the Bank of England and basic industries – coal, gas and electricity, road, rail and air transport, and finally steel production. Between 1942 and 1945, as final victory in the war approached, a series of planning commissions had been set up to formulate domestic policies and plans for reconstruction after the war. The most far-reaching report was that prepared by Sir William Beveridge, which proposed the unification of social benefits to cover the whole population, from cradle to grave. From it sprang the National Health Service, providing free medical care for all, National Insurance, financed by employers, workers and the state; and National Assistance, which provided care for those not qualifying for insurance benefits, or in danger of becoming destitute. Family Allowances – cash allowances for children – had already been introduced before Labour gained power.

This revolutionary legislation was pushed through at a time when Britain's overseas debts were high. Factories, homes and other buildings worth US$6800 million had been destroyed, one-third of the merchant fleet sunk and 60 per cent of export trade lost. A period of austerity and hardship followed: wartime food rationing continued until 1954; clothes were rationed until 1949. Problems were exacerbated by bleak events abroad – the 'Cold War' between Russia, her East European satellite countries and the Western powers; and the Korean War, in which British troops fought in the United Nations forces.

▲ CENTRE OF REVOLUTION
Arkwright's cotton mill at Cromford, Derbyshire – one of many 18th and 19th-century factories that housed the new technology of the Industrial Revolution and generated the wealth that built a huge empire.

Labour barely scraped in with a majority of 6 in the 1950 General Election. In October 1951, the Prime Minister, Clement Attlee, called another General Election and Labour was defeated by the Conservatives. Labour did not regain power for 13 years.

But despite continuing difficulties, Britain began to prosper. In 1957, an ebullient Conservative Prime Minister, Harold Macmillan (later Lord Stockton), was moved to remark that 'most of our people have never had it so good' – a phrase that returned to haunt him in less-prosperous times soon to come. But in the 1960s 'Swinging Britain' became a world trendsetter in design and the expanding pop music and entertainment industries. A growing laissez-faire attitude to social behaviour heralded the so-called Permissive Society.

But the textile industries of LANCASHIRE and YORKSHIRE were in massive decline, as was shipbuilding on the Tyne, CLYDE and MERSEYSIDE, and footwear manufacture in Northamptonshire and Leicestershire. The energy crisis of the early 1970s and continuing recession in the 1980s merely accelerated the trend. Steel mills closed and even the car industry – centred on the WEST MIDLANDS,

▲ **BLACK SAVIOUR A production platform pumps oil from the North Sea's Brent field – one of the world's largest offshore fields. Oil contributed 6 per cent of the UK's gross domestic product in 1982, helping to maintain living standards in a stagnant economy.**

ESSEX, the Clyde Valley and Merseyside, and highly successful in the 1950s and 1960s – ran into serious trouble.

By 1973 the Welfare State had an already unacceptable half million unemployed; within a decade the total had exceeded 3 million.

However, responsibility lay not only with the recession and competition from younger and more efficient industries abroad: the very structure of British industry had been undergoing profound change. In 1909 half the country's manufactures were produced by 2000 or more firms; by 1970 only 140 firms were turning out the same proportion of a much larger quantity of goods. The demise of the small firm and the rise of the large corporation – or the multinational, with its roots outside the country – was all but complete.

Moreover, the decline in manufacturing brought a shift in emphasis to service industries, and by 1983 13.2 million of the 20.8 million Britons in full employment were in these industries. They included a massive 4.6 million in public administration, health and education. The National Health Service alone employed nearly 1.3 million people – the biggest single employer in Europe and among the world's top ten. Only 5.5 million people were employed in manufacturing – a drop of 2.2 million in ten years.

On top of this industrial imbalance, the population is virtually static and ageing. The average family has barely the theoretical 2.1 children needed to replenish a population of 56 million. The population actually fell between 1974 and 1982, and despite calculations that it should reach 58 million by the year 2001, the real prospect is of 'zero growth'. A tradition of immigrants being outnumbered by emigrants also continues – 2.27 million left in 1973-82, while 1.8 million arrived. But improving health standards and medical advances are allowing more Britons to live longer – to pose further long-term problems of providing for more state pensions.

THE ISLAND RACE

If the current social and economic outlook is giving many Britons that old beleaguered feeling again, it is a situation they have, as an insular people, long endured. In fact, about 7500 years ago Britain was finally cut off from the rest of Europe's landmass by the Strait of DOVER. This happened when the last cold phase of the Ice Age waned, and waters locked in the great ice sheets returned to the oceans. The North Sea submerged the forest-and-marsh lowland which had once joined Britain to the Continent. However, this did not stop migration. Neolithic settlers, who first arrived about 6000 years ago, are known from their burial mounds. They settled at places like AVEBURY and STONEHENGE which were developed by later Bronze Age peoples into grand megalithic circles. Late in the Bronze Age (2200-650 BC) and through the early Iron Age (650 BC-AD 43) came the Celts and other tribes who built the large earthen encampments that still crown many uplands. Their languages survive in the Welsh and Gaelic still used in western areas.

Then – following Julius Caesar's raids of 55 and 54 BC – came the Roman invasion of AD 43, and an occupation lasting over three centuries. The Romans established order, built a road network and laid out the first real towns – places such as CHESTER and CHICHESTER. As the Romans withdrew, their empire crumbling, Saxons, Danes and Vikings arrived between the 5th and 11th centuries. Saxon speech gave rise to the English language; Danes and Vikings added their elements. Many of their place names survive in recognisable form – including DERBY, previously known to Saxons as Northworthy. Finally, in 1066, came Duke William and his Normans, to conquer and then merge with the indigenous population, bringing new facets to the language – and even more new place names.

The Normans were the last conquerors of England, but it was more than 700 years before the United Kingdom came about. The Anglo-Norman conquest of Wales began in the 12th century, the last principality (Gwynedd) falling in 1282, and Wales was brought under English law by the Acts of Union of 1536 and 1542. The Kingdom of Scotland, formed in the 9th century when the Picts joined the Scots, resisted successfully. But James VI of Scotland, a great-great-grandson of Henry VII of England, succeeded to the English throne in 1603, and ruled both kingdoms. The Scottish parliament voted to unite with England and Wales in 1707. The first Anglo-Normans reached IRELAND in about 1170. English control gradually spread over the whole island, and the Act of Union enforced in 1801 created the United Kingdom.

The green and fertile land inherited by the Normans was not only beautiful but contained rich mineral wealth which was barely tapped until the Industrial Revolution. Iron ore was fairly widespread, and lead, tin, copper, silver and a little gold were mined in CORNWALL, northern England and Wales. But it was coal that largely provided the power for 'the workshop of the world'.

But the 20th century brought the discovery of new energy sources – oil, natural gas and nuclear reaction. In the 1950s coal still provided 95 per cent of the nation's energy needs. By 1979 that proportion was down to 36 per cent, and oil and gas were providing 59 per cent. Gas was discovered in the North Sea in 1965, and oil four years later, and by 1984, 25 oil fields were on stream. The tax revenues generated helped compensate for the slump in manufacturing, but the oil and gas reserves are expected to begin running down in the 1990s.

The decline of manufacturing industry has been particularly hard on Britain, one of the most urbanised countries in Europe, where over three-quarters of the population live in towns and cities. Skills generated by a century and more of hard-won, years-long apprenticeships have been made worthless as demands change and computer-controlled machines turn out finished products. As in farming, but on a far larger scale, skilled workers have become redundant in many manufacturing industries – especially major ones such as textiles, engineering and shipbuilding. The only remaining 'native' mass-production car manufacturer survived on massive Government subsidies and close links with a Japanese company. Now the Japanese are building their own cars in Britain.

YOUNG AND OLD

Leisure is becoming an increasing preoccupation in a country where, apart from the unemployed – whose leisure is enforced – numbers of people are accepting redundancy payments and virtual retirement, or early retirement pensions. More people of pensionable age are living to enjoy their retirement, and the number of centenarians increased from 271 in 1951 to 2410 in 1981.

There have, however, been significant changes over recent years in many of England's inner cities. Here, despite the national drop in population, there are growing numbers of teenagers and young adults – a factor in recent civil disturbances and a pattern that could persist into the 1990s. Many of the younger children have been born to parents of Pakistani or New Commonwealth (that is, excluding Canada, Australia and New Zealand) origins, of whom only 40 per cent were themselves born in Britain.

In inner LONDON alone the number of children 4 years old or less increased by 21 per cent between 1981 and 1984, and was thought to be due to a high birthrate among the capital's Asian communities. In 1984 more than half the children born in four London boroughs were to mothers from overseas – 64 per cent in one borough, Brent. Outside the capital the figures for Slough were 41 per cent, Leicester 36 per cent and BIRMINGHAM 30 per cent. Even in declining areas such as Accrington, Burnley and Nelson, in Lancashire, with large Asian communities, birthrates were high.

In general, the more entrepreneurial Asians have prospered, while people of African and West Indian origins have found life difficult in an era of low employment prospects, and during the 1980s their frustrations erupted in serious rioting in London, LIVERPOOL, Birmingham and Bristol.

However, rioting has become almost a way of life among a violent – and largely white – minority. Soccer hooligans have driven millions away from the grandstands and terraces of Britain's traditional winter game. Yet the British almost invented modern spectator sports. The Football Association was formed in England in 1853; lawn tennis was first played in England in 1873-4 and the first Wimbledon championships were held in 1877. Golf is generally accepted as having originated in Scotland, and rugby began at RUGBY School, in Warwickshire, in the 19th century. Squash originated at Harrow School in 1850, and the first organised track and field athletics of modern times took place in London in 1850. The origins of cricket – most English of games and the great national summer sport – are obscure, but it is probably 400 years old.

Cricket continues to draw good crowds, but rugby union is losing spectators. Tennis, golf, horse racing, motor racing and athletics remain popular, and television coverage has made darts and snooker major interests. Sports such as jogging and distance running, surfing, sailing, hang-gliding and climbing are also increasingly popular. Leisure pursuits of a less athletic nature are influenced by television watching, in which the British indulge themselves for an average 21 hours a week. Listening to the radio, records or tapes is popular, along with working on home improvements, gardening, visiting the countryside and museums, and walking the numerous public footpaths.

These paths – some of them rights of way and drovers' roads trodden since prehistoric times – are jealously guarded by numerous conservation societies and organisations such as the National Trust, which has acquired large areas of land and scores of fine historic houses and buildings for public benefit. The efforts of the Trust have taken on a wider significance with the growth of tourism, which has become a major industry, employing more than 1.3 million persons in 1985. Fifteen million visitors came to Britain in that year.

For tourists, the radical changes which have taken place in the country are obscured by a history so long and rich in incident and character that it has imposed itself upon almost every aspect of both landscape and townscape. Great cathedrals such as those at CANTERBURY, DURHAM, ELY and LINCOLN, raised almost 1000 years ago, divert their eyes from the present. The church music sung in them is a basic ingredient of English culture, celebrated in the Three Choirs Festival, held in turn at GLOUCESTER, WORCESTER and HEREFORD cathedrals. Other festivals of the arts have become established since the Second World War – notably the great international event held in EDINBURGH every autumn, not to mention others in YORK, HARROGATE, BATH and Cheltenham. The literary trail, too, has its powerful associations – the LAKE DISTRICT with sonnets by Wordsworth; the wild, romantic EXMOOR of Lorna Doone, and Burns's Ayrshire countryside; the Brontës' Yorkshire, Thomas Hardy's DORSET and Dylan Thomas's magical west Wales.

But for most tourists, Britain's greatest attraction is the capital itself, London; while many travel on to view the ancient university seats of Oxford and Cambridge, historic cities like York and Chester, timbered towns like Ledbury, and resorts like Brighton and BLACKPOOL. Most of all, perhaps, they travel to STRATFORD-UPON-AVON, birthplace – and burial place – of William Shakespeare, whose matchless plays and poems may well have etched upon hearts and minds images of Britain less fleeting than those cast by television.

THE UNITED KINGDOM AT A GLANCE
Area 244 119 km² (94 255 sq miles)
Population 56 400 000
Capital London
Government Parliamentary monarchy
Currency Pound sterling = 100 pence
Languages English, Gaelic, Welsh
Religions Christian (55% Protestant, 10% Roman Catholic), Muslim (2%), Jewish, Hindu and Sikh minorities
Climate Temperate; average London temperature ranges from 2-6°C (36-43°F) in January to 13-22°C (55-72°F) in July
Main primary products Wheat, barley, potatoes, sugar beet, fruit and vegetables, fish; oil and natural gas, coal
Major industries Agriculture, oil and gas extraction and refining, coal mining, machinery and transport equipment, iron and steel, metals, food processing, paper and paper products, textiles, chemicals, clothing, light industry, finance and business services
Main exports Agricultural and industrial machinery, crude oil and petroleum products, chemicals, transport equipment, vehicles, aircraft, electrical goods, iron and steel, non-ferrous metals, textiles, food products
Annual income per head (US$) 7550
Population growth (per thous/yr) 1
Life expectancy (yrs) Male 71 **Female** 77

United States of America

TWO CENTURIES AND MORE AGO, THE UNITED STATES WAS BORN AMID SOME OF THE NOBLEST CONCEPTS MANKIND HAS PRODUCED; ALL IN ALL, THE NATION HAS BEEN TRUE TO ITS WORDS

This land is your land;/This land is my land./From California to the New York island,/From the redwood forest to the Gulf Stream waters,/This land was made for you and me.' So begins folk singer Woody Guthrie's interpretation of the American Dream, a vision that has beguiled millions since first it was given wings by the Declaration of Independence, drafted in a room over a stable in PHILADELPHIA in the baking summer of 1776. It is a vision not without criticism during the intervening couple of hundred years or so, but its essence is indestructible, and remains one of mankind's noblest concepts.

But what is the essence of America? Many Americans, the most communicative people on earth, can only tell you what it is not.

It is not NEW YORK with its Aladdin's Cave shops, stone and glass canyons that alternately broil and freeze, and its ghettos. Nor is it LOS ANGELES – 'several dozen suburbs in search of a city.' – or the ravishing coast of CALIFORNIA. SAN FRANCISCO is too quirky; NEW ORLEANS, though it gave its music – jazz – to the world, is too colonial; while NEVADA, where LAS VEGAS and RENO blaze neon signs to the desert stars, is too brash. No one quite knows what to make of FLORIDA, with its emerald and sapphire seas, its dripping swamps, space shuttles, Disney World, cheap resorts, rich mansions and hotels, and sky-high crime rate.

Neither is it CHICAGO, nicknamed 'The Windy City', with its Hollywood movie image of gangsters, underworld violence and political chicanery – an image of the past now, for today's city offers enough art, music and theatre to make it America's cultural runner-up to New York. Nor is it mighty TEXAS, land of oil and cattle barons – or so it might be imagined from the TV soap-opera image of the citizens of Dallas. Texas, annexed by congress in 1845 after nine years as an independent republic, has a distinctive character that serves as a reminder that the South once wanted to be its own country. A quarter bigger than France, and beset with a love of all things gigantic, Texas prompted novelist John Steinbeck to call it 'a state of mind', recalling when most of the region minded being states.

The Old South is only now carving a new identity out of what was left by the Civil War of the last century. Not even NEW ENGLAND, where the United States began, can be considered as typically American. The creeper-clad universities (hence Ivy League), the small

ports, the 17th-century brick or white weather-boarded churches and villages seem instead a little wistful for the Old World – though at the same time declaring the traditional Yankee virtues of hard work and thrift.

Nevertheless, Americans will often confess to two positive if contradictory images. The first is of a small town that so many city dwellers feel is waiting for them somewhere, complete with main street, courthouse, church, war memorial, barber shop, drug store, pretty houses surrounded by lawns, maybe a Civil War cannon, general store and, for convenience's sake, a supermarket and a fast-food restaurant facing a vast car park. It is likely to be somewhere in the MIDWEST and is an America that is devoutly desired.

Another vision that grips Americans is the sheer magnitude of their country, its great distances that impart a sense of freedom to take the breath away. For anyone who doubts it, there is Interstate 80, a double ribbon of highway that runs for nearly 4700 km (about 2900 miles) from NEW JERSEY to San Francisco on the Pacific coast with never a traffic light on the way. Or the older US1 that stretches from the Canadian border to Florida; or the cornlands of the Midwest like an inverted

golden sky; or the huge sweep of the high western plains; or the mighty MISSISSIPPI river that drains half a continent on its 3779 km (2348 mile) journey from MINNESOTA to the Gulf of MEXICO.

So vast a land is the United States that dawn comes eight hours later to its citizens who live on the westernmost islands of ALASKA than to those in eastern MAINE; even California is three hours behind the east coast. In the north there are Americans to whom the possibility of frostbite is an annual hazard, while in the south many of their fellow citizens grow palms and hedgerows of hibiscus.

In between these outriders stretches the great continent, embracing almost every kind of climate and terrain that the planet has to offer: polar tundra in northern Alaska and hot, sandy deserts in southern California. OREGON and WASHINGTON have vast coniferous forests, Florida has mangrove swamps, and in the central and western states there are seemingly endless grasslands where the buffalo used to roam. There are ranges of snow-capped peaks such as the lovely, majestic ROCKY MOUNTAINS and the SIERRA NEVADA.

The climate is as varied as the features and vegetation. Around the Gulf of Mexico it is humid and subtropical, while Arctic Alaska is intensely cold and arid. The western coast's rainfall decreases from north to south; much of it falls on the coastal ranges, resulting in deserts, such as the MOJAVE, farther inland. In direct contrast there is coastal California's Mediterranean climate, the cool, maritime region around New England and the continental extremes of climate of the central plains and western plateaus.

In the brief 200-odd years of the nation's existence, most of these domains have been exploited to the full, largely to support an ever-growing population, but also for the production of exports. The Atlantic coastlands are market gardens for the eastern cities; the ploughed grasslands to the south-west of the GREAT LAKES produce maize – 'corn' to the Americans – and soya beans to feed the continent and support huge numbers of cattle and pigs. Farther west there is the Wheat Belt, and then more cattle and sheep roaming the open ranges that run up to the Rocky Mountains. The northern Pacific coast produces about half the nation's timber requirements as well as apples and berry fruits, while California's Central Valley grows soft and citrus fruits, grapes and rice. Florida and the south-eastern states also produce citrus fruits, cotton and rice, and add to them sugar cane, peanuts (groundnuts) and tobacco. Among the agricultural products in which the United States leads are maize and soya beans – about half and two-thirds respectively of the world's total

▲ GLITTERING DREAM Or neon nightmare? Downtown Las Vegas – self-proclaimed Entertainment Capital of the World – gleams with the wealth attracted (and generally lost) to its gambling tables and floorshows. It is one version of the American Dream.

crops – tomatoes one-seventh, oranges a quarter, peaches one-fifth and nuts (excluding groundnuts) about a quarter of the world's annual harvest. The United States is also the foremost producer of industrial timber, poultry, beef and cheese and is by far the biggest exporter of wheat, though second in actual production to the USSR.

Despite this amazing output, the USA is not primarily an agricultural nation. A century ago, about 15 per cent of the population worked on the land. Now no more than 1.5 per cent do so. The high yields produced by such a small work force are due to the 20th-century American genius for technology, which also turned most of the rest of the population into urban workers.

Raw materials are plentiful, for the continent was as amply provided with minerals as it was with the means of growing things. The United States leads the world in the production of copper, gypsum, kaolin, mica, salt, phosphates and sulphur, and is close to being the premier producer of lead and pig iron. Another American 'first' is aluminium, though bauxite, the mineral from which it is made, is largely imported from countries that do not possess the cheap energy resources required in the refining process.

Harnessed energy is one of the USA's greatest strengths. Its rivers and dams produce hydroelectric power in abundance, and it is the world's leading producer of coal. Though its oil reserves are only the fourth largest in the world – after Saudi Arabia, the USSR and Mexico – it is second only to the USSR as an oil producer, with 8.9 million barrels a day against the USSR's 11.9 million. Owing to its ownership pattern, however, the USA's production comes from five times as many wells as the USSR's, and from 191 refineries compared with 38; the USA has, in fact, the largest refining capacity in the world. It has the third largest reserves of natural gas, and is the foremost producer of uranium, supplying 80 nuclear power stations, with more to come on line if environmental objections are resolved. Produced by whatever means, the USA generates and consumes about a quarter of all the electricity generated in the world.

High consumption rates cause problems. Despite its huge resources, the USA must still import oil, and was consequently badly hit by the oil crisis which began in 1973. This, together with overseas industrial competition and changes in trade, has given the USA some bad years for employment. At one point in 1982, the unemployment rate reached almost 11 per cent; in 1983, it was 9.6 per cent. The 1985 figure was 7.2 per cent. Heavy industries have suffered particularly, as have the poorer working populations of the northern inner cities. But American finance is lighter on its feet than that of many European countries, and it began to advance the development of the new, high-technology industries – such as those in California's Silicon Valley – in other parts of the west and the Deep South. In the early 1980s, many northern workers followed suit, with scarcely a backward glance, to settle in this so-called Sun Belt.

Always a trendsetter in industrial and sociological affairs, America seems to be heading for a 'post-industrial' future in which manufacture is based on automation and its people live in smaller, more widespread communities. Already, more than three-quarters of its work force are engaged in communications, the wholesale and retail trades, health care, banking and finance, government, catering, and in leisure and other services.

CREATING A NATION

The continent that was named after a Florentine merchant, Amerigo Vespucci, was first colonised by people who crossed the Bering Sea from Asia via a land bridge about 25 000 years ago. They were the ancestors of the 'Indians' of South and North America and of the Inuit (Eskimos). North America was sighted and briefly settled by Norsemen in the 10th and 11th centuries. Christopher Columbus never saw the northern mainland, but John Cabot did in 1497. Thereafter, it became host to a multiplicity of peoples.

In 1565, the Spanish founded St Augustine in Florida, the first stable European settlement in North America; a century later, the French began exploring the lower Mississippi valley, naming the entire area LOUISIANA after King Louis XIV. But it was the British, the Dutch and, later, the Germans who were the most enthusiastic colonisers of the North American territories. Some fled poverty or religious persecution, others came in search of fortune or adventure. JAMESTOWN, the first permanent British colony, was founded in VIRGINIA in 1607, and 13 years later the *Mayflower* landed the Pilgrim Fathers at PLYMOUTH in MASSACHUSETTS. With them they brought the Puritan work ethic that still underlies American industriousness and ambition today. The southern colonies were found to be ideal for growing cotton, tobacco and other warm-climate crops, and from 1619 African slaves were brought in increasing numbers to work on the plantations, founding today's population of black Americans.

The colonists' steady determination to control their own destiny led, in 1773, to the celebrated BOSTON Tea Party, in which a band of patriots, dressed as Red Indians, threw 342 chests of the British East India Company's tea into Boston harbour – a double protest against attempted British monopolies and British taxation. This act was one of the opening moves of the War of Independence which played out on a larger scale what the Tea Party had implied: the former colonists were now and forever American. On July 4, 1776, the 13 American colonies, through the Declaration of Independence, pronounced themselves to be the United States of America. After a long and hard-fought conflict, the infant republic drove the British into capitulation at YORKTOWN in 1781.

The 1781 Articles of Confederation, which bound the 13 states to one another, provided the steppingstone for the 1787 Constitution and, in 1791, for the first ten amendments to it, known as the Bill of Rights. The Constitution established the federal republic with a system of checks and balances so that no arm of government could hold sway over the others, and divided power between three branches of authority: the executive (the government), the legislature (the law-makers), and the judiciary (the courts). In 1789, George Washington became the first Chief Executive, or President.

With further amendments from time to time, the Constitution has proved remarkably far-sighted and durable. Elected every fourth year, the American President is undoubtedly the most powerful man in the country, able to initiate and veto legislation and – perhaps more crucially – to guide the national mood. He is also commander in chief of the armed forces. His power can never be absolute, however, since it is restrained by the other two branches of government: the legislative Congress, consisting of the Senate and the House of Representatives, and the judicial Supreme Court, whose nine justices are appointed for life and can overturn any executive or legislative ruling that it finds contrary to the spirit of the Constitution.

It took nearly 180 years for the original 13 states to become 50, in which time the area comprising the USA grew from 2 301 692 to 9 372 614 km² (888 685 to 3 618 772 sq miles). The most important steps were the purchase of the Mississippi basin from France in 1803 – the Louisiana Purchase – and the acquisition of the south-west (including California) after the Mexican War (1846-8).

Throughout the 19th century, the USA expanded westwards as pioneers tamed new lands that were eventually admitted to the Union as fully fledged states. This expansion led to conflicts between homesteaders (who were granted free land under the 1862 Homestead Act), cattlemen (who claimed the right to graze their stock freely on the open range) and the Indians (who had been there long before either of the others, depended on the free-ranging wild buffalo and who died or were forced into reservations).

The earliest westward settlers crossed the country in wagon trains, but the completion of the first transcontinental railway line in 1869 caused a great surge of growth in the west, and between 1870 and 1900 the country's population almost doubled, to 76 million. Founded on the unalienable principle that 'All men are created equal', the United States has prided itself in offering asylum to emigrés for more than two centuries.

At the entrance to New York harbour stands France's tribute to America, the torch-bearing Statue of Liberty, on whose base are inscribed the words of Emma Lazarus, a wealthy young New Yorker who, horrified at Russian oppression of the Jews, became an ardent exponent of the American Dream: 'Give me your tired, your poor/Your huddled masses yearning to breathe free . . . I lift my lamp beside the golden door.' The words summarised the hope that brought millions of European immigrants to America in the decades around the turn of the century.

With no money to travel farther, many of the newcomers stayed in the east, but by no means all. CLEVELAND in OHIO has more people of Hungarian descent than any city outside Budapest, and Chicago almost as many people with Polish names as Warsaw.

▲ GLEAM OF HOPE 'Miss Liberty' has welcomed arrivals to New York – millionaires on luxury liners and impoverished emigrés alike – since 1886; she was scrubbed and restored for her hundredth birthday. Behind rise the twin World Trade Center towers.

Many immigrants, Irish and Italian especially, settled in Boston. South-west of New York lie the middle-Atlantic states – New Jersey, PENNSYLVANIA, MARYLAND, DELAWARE and the District of Columbia. The last of these, which contains WASHINGTON DC, is not actually a state at all, but a 'territory' carved out of neighbouring states in the 1790s to provide a site for the nation's capital.

South and west of the capital lie the former

slave-owning states that were the backbone of the breakaway Confederacy during the Civil War of 1861-5 – MISSISSIPPI, Louisiana, ALABAMA, Florida, GEORGIA, NORTH and SOUTH CAROLINA and Virginia. The western 40 counties of Virginia, a land of small farmers who had no interest in keeping slaves, voted against secession from the Union and became the new state of WEST VIRGINIA in 1861.

The secession and the Civil War that followed were not based primarily upon the issue of slavery, but upon States' Rights – the doctrine that states may control their own destinies regardless of federal consent. In this case, the slave-owning states refused to accept the federal curb on the expansion of slavery into the new western territories, and seceded from the Union. The federal states regarded this move as unconstitutional and rebellious, and so went to war. The conflict was one of the bloodiest in history, leaving half a million men dead – one in six of those taking part.

The Civil War smashed the economy of the South, and parts of some states – Mississippi, TENNESSEE and Virginia among them – are still desperately poverty-stricken. Other parts rose from the ashes – literally in ATLANTA's case – and took on successful new commercial and industrial roles to offset the destruction of the plantation economy. The entire area is immensely popular with tourists, since everywhere is propagated a nostalgic flavour of Dixie, the 'Ole Plantation' and mint juleps. For music buffs there are jazz bands in New Orleans, blues and rock 'n' roll in MEMPHIS and country and western in NASHVILLE, while LOUISVILLE offers the Kentucky Derby.

The most recent partners in the Union, Alaska and HAWAII, were admitted as the 49th and 50th states in 1959. Alaska was wrested from the Indians by the Russians around 1800 and bought for the USA in 1867 by Secretary of State William H. Seward for $7.2 million. Though dubbed 'Seward's Icebox' or 'Seward's Folly' at the time, it was one of the greatest bargains in history, for it is exceedingly rich in timber and minerals, including gold, petroleum and coal.

Although over 3700 km (2300 miles) west of San Francisco, Hawaii is in no doubt that it is American – though as late as 1893 its 130

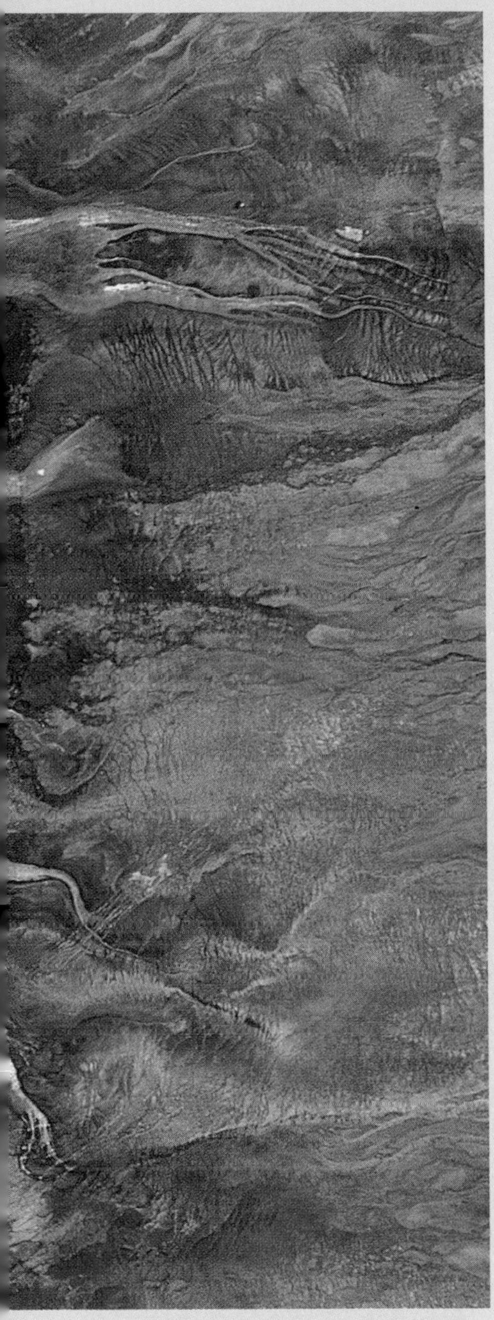

◄ **NATURAL WONDER Multicoloured algae and bacteria flourish in the mineral-rich hot waters of Grand Prismatic Spring in Yellowstone National Park, the oldest of hundreds of national parks and other sites of which Americans are justifiably proud.**

or so islands formed a Polynesian kingdom. Originally charted as the Sandwich Islands, they were discovered by Captain Cook in 1778; he was slain there by natives the following year. Since then, Hawaii has been more hospitable to mariners and other wanderers, so that today its population is a polyglot mixture of Polynesian, Asian and European.

DEMOCRACY IN ACTION

The vital thing to comprehend about the country is inherent in its name – the *United States* of America. Each state is a separate entity with jealously guarded rights and powers, though united to every other state by acknowledging the Constitution as 'the supreme law of the land'. Each state has irretrievably delegated to the Union the basic powers of waging war, conducting foreign policy, levying federal taxes and regulating foreign and interstate commerce. Though federal power has tended to increase during the 20th century, many essential powers still remain with the states. Chief among these is the right to levy state taxes and to make laws governing such matters as private property, business, public health, works and education – provided such laws are not contrary to the letter or spirit of the Constitution.

The United States is a land of many governments, for in addition to federal and state, there are also city, county, town and village governments and several other bodies with local powers, including school boards. Americans are by no means overawed by authority, however, and more than the citizens of most nations, perhaps, are prepared to challenge their rules in the courts. Of the 241 842 civil cases filed in district – that is federal – courts in 1983, for instance, more than half were actions brought by private citizens aggrieved at some aspect of labour, social security, tax, patent, antitrust or civil rights laws.

The same courts that hear civil actions also try criminal cases – and in recent years a sharply rising crimes rate has kept them busy. In 1960, there were 3 384 200 crimes reported nationwide; in 1983, 12 070 213. The 1983 figures include 1 237 979 violent crimes and 18 673 murders, as against 9110 murders committed in 1960. The high murder rate is frequently blamed on the easy availability of handguns in many states, which in turn is often defended as a Constitutional right. The Second Amendment actually says that 'A well-regulated Militia being necessary to the security of a free State, the right of the people to keep and bear arms shall not be infringed' – which does not quite imply that lethal weapons should be available to everyone.

Thirty-seven states retain the death penalty, but no executions were carried out between 1968 and 1976. In 1977, a convicted murderer in Utah demanded the right to be executed, and, after a lengthy legal battle, he was shot by firing squad. This established a precedent for the execution of other murderers.

Drugs are probably the greatest single cause for the rise in the crime rate – crimes carried out by addicts in order to obtain money to buy drugs, and by racketeers endeavouring to gain control of the trade. It is highly lucrative. In 1984, police smashed a ring run by the Sicilian Mafia that had smuggled $1650 million worth of heroin into the USA and sold it through a chain of pizza restaurants.

Each state has its own educational scheme, though all have a basic programme that provides 12 years' free and compulsory education for every child, plus kindergarten. In addition, most states offer further years of voluntary and free attendance at public (state) school up to the age of 21. Patterns of education vary considerably, but two of the most usual are: six years or grades of elementary school, three years of junior high school and three years of senior high school; or eight years of elementary school followed by four of high school.

Generally, pupils may leave at 16 (one or two years older in some states), but the system is geared to encourage them to continue to 17 or 18. Promotion from one grade to the next is earned as much by good progress throughout the year as by end-of-year exams. All American public schools are co-educational and racially integrated by federal law.

A college (university) education is highly regarded by Americans, and is the gateway to promotion in many fields. Apart from state establishments – including community colleges (which offer two-year business and technical courses) and junior colleges (which prepare less-well-qualified students for transfer to a full four-year university course) – there are many famous independent universities, such as Harvard, Yale and Princeton. Altogether, there is no shortage of centres of higher education; in the vicinity of Boston alone there are 24 four-year universities, including Harvard, the Massachusetts Institute of Technology and Boston University. In the academic year beginning in autumn 1984, 27.1 per cent of Americans aged between 18 and 24 enrolled for higher education.

The success of the education system is illustrated by the USA's record of innovation, as shown by the success of its technology-based industries and by the fact that Americans have won more than twice as many Nobel prizes as the people of any other nation. That only eight were in the literature category belies America's achievements in the arts: from music and literature to architecture, painting and sculpture, the USA is a powerhouse.

HEALTH AND WELFARE

In recent years, Americans have become much more weight, diet and exercise conscious. The number of smokers in one of the world's leading tobacco-producing countries has noticeably declined, and a great many people take exercise – jogging, cycling or following the workouts prescribed on TV. As a consequence, the figures for diseases of the heart and circulation have fallen significantly – 40 per cent fewer fatal strokes since 1972, for instance – a blessing to the average person's health and to the average purse.

Medical services in the USA are extremely expensive; having a baby, for example, costs between $2000 and $3000. To lessen the blow, many people are covered by insurance schemes, whose premiums are mostly paid by employers. Few of these schemes pay more than 80 per cent of a medical bill and a large number of Americans have no medical insurance at all. To these, a doctor's bill can be ruinous. Assistance is available to the poor through Medicaid and to the aged and handicapped through Medicare. Some hospitals have a limited number of charity beds, and all will take emergency cases regardless of ability to pay. Dentistry too is expensive, but many companies include dental care in the insurance package they offer employees.

One factor that has pushed up costs in recent years has been patients' increasing tendency to sue their doctors if anything goes wrong with treatment. Courts have awarded such huge damages that doctors' own professional idemnity insurance premiums have skyrocketed – and they have increased their fees as a result. Some doctors even said that they could no longer afford to remain in practice, and the federal government stepped in with moves to limit court awards.

The social security system was founded during the great economic depression of the 1930s as part of President Franklin D. Roosevelt's 'New Deal', and was greatly extended in the 1960s. It incorporates a federal system of medical and supplementary income benefits for the aged, the disabled and dependants. It also provides various welfare programmes for the unemployed. Apart from basic cash allowances, beneficiaries may also be eligible for food stamps, Medicaid and subsidised housing. In 1984, according to the US Census Bureau, 33.7 million Americans were living below the poverty level – then defined as $10 609 per annum for a family of four or $5278 for a single person. However, burgeoning federal budget deficits in the 1980s threatened levels of aid.

LIVING IN AMERICA

Americans have mastered, like no other nation, the techniques of living in a vast country. Their road system – over 6 million km (3.8 million miles) of turnpikes and freeways (highways with and without tollgates), rural and urban roads – carries about 160 million vehicles and is America's prime mover both of freight and of passengers making shortish journeys – to work, to the next town, to the drive-in bank or movie theatre, or to the supermarket with its huge car park and loading bays.

There is an excellent and cheap nationwide bus service, but it tends to be used mostly by students, the poor and by foreign tourists who want to get close to the country. Americans, businessmen especially, making long journeys tend to use the internal airline networks which link not only cities but quite small towns.

Railways are not what they were in the prewar heyday of such famous transcontinental trains as the *Super Chief*, the *California Zephyr* and the *Royal Palm*. Few people nowadays travel long-distance by rail, though many, especially around Chicago and New York, use commuter services to get them to work. Nevertheless, in the last few years, there have been serious stirrings on the part of the railway companies to recapture the nation's interest with Sunshine Specials and excursion trains with glass-domed observation cars to such picturesque regions as the Rockies.

Due to the immense distances involved and the amount of 'junk' mail carried, the postal service is poor, so Americans in a hurry make use of the overnight and express delivery services operated by such private companies as Flying Tigers. They also make great use of the telephone; according to the statistics, there is almost one telephone for every man, woman and child in the country.

Distance too is the reason why there are so few national newspapers, though of course there are many local ones of national and even international repute. Many Americans will take the *New York Times* wherever they live, and the *Wall Street Journal* is required reading for all businessmen. Racy tabloids like the *National Inquirer* and *Star* are on sale in every supermarket, as is the first true national daily, *USA Today*, which is printed in colour in regional centres from electronically circulated copy. Also stacked in the supermarket racks is a staggering array of magazines covering a multitude of interests.

Television is the true dispenser of news – about six hours of it a day, local, national and international. Almost all of America's 83 million homes have at least one TV set, and almost all programmes are financed by advertising, except on 'educational' channels which have no advertising and are paid for by public subscription and government grants. As well as this Public Broadcasting Service, there are three main national networks – ABC, CBS and NBC – and more than 700 local stations. Then there are some 5000 cable networks serving about 36 million subscribers, so that a city dweller may have dozens of different channels to choose from. He can pick channels that play nothing but serious music, or jazz, or sport, or old Westerns, or news or any other subject he chooses, including religion – a diversity repeated on AM and FM radio. In the nation that contains Hollywood and Broadway, entertainment is big business, but it is also capable of immense variety – from banal soap operas to the best of serious and popular drama – making the USA one of the world's greatest cultural exporters.

The religion-only channels are not surprising in a country where, according to the latest figures (1983), 139 603 059 people – 59.6 per cent of the population – are church members. Yet there is no official religion in the United States. This is not a matter of omission, but of deliberate policy, as laid down in the First Amendment to the Constitution: 'Congress shall make no law respecting an establishment of religion, or prohibiting the free exercise thereof . . .' It is an edict that is firmly upheld. In 1984, a federal judge ruled that town officials of Birmingham, MICHIGAN, were acting unconstitutionally in displaying a Nativity scene on a public site at Christmas (though this was later overruled), and in the same year the Supreme Court ordered that an Alabama law permitting teachers to lead prayers in public schools should be abolished.

INTO THE MODERN WORLD

By the turn of the 19th century, after only 120 years of independence, the USA had transformed itself from a pioneer colonial economy into an exporter of industrial goods. In less than a century since then it has become the richest and most powerful state the world has ever known – and yet it is still not entirely sure of its role in international affairs.

'America cannot be an ostrich with its head in the sand', declared President Woodrow Wilson in 1917, calling for American involvement in the First World War. Americans took some persuading. Isolationism had been official policy – known as the Monroe Doctrine after James Monroe, President in 1817-25 – and Americans were too busy building their promised land to concern themselves much with foreign affairs, and particularly foreign wars.

The feeling reasserted itself after the First World War – reinforced by the domestic problems of the 1930s depression – and it took Japan's attack on PEARL HARBOR in December 1941 to tip the balance in favour of involvement in the Second. From that moment, however, the USA has, like it or not, been a major world power in the political as well as the economic sense, a fact emphasised when it dropped the first atomic bombs on Japan in August 1945.

Although the USA and USSR were victorious allies in 1945, it was an alliance of convenience that soon turned to enmity. The two giants confronted each other economically, in the arms race, in space, through propaganda, and on numerous secondary battlefields – notably Korea (1950-3) and Vietnam (1964-73). Defeat in Vietnam led to a resurgence of isolationist sentiments, but continuing confrontation with Communism and the growth of international terrorism, culminating in US action against Libya in 1986, made it clear that America was once again being held accountable for the principles – 'life, liberty and the pursuit of happiness' – on which it was founded by the Declaration of Independence on July 4, 1776.

USA AT A GLANCE

Area 9 372 614 km² (3 618 772 sq miles)

Population 241 000 000

Capital Washington DC

Government Federal republic

Currency Dollar = 100 cents

Language English; many others spoken

Religions Christian (32% Protestant, 22% Roman Catholic, 2% Eastern Orthodox), Jewish (3%), Muslim (1%)

Climate Mainly temperate; variations include subtropical in the south, Mediterranean-type in southern California. Mainly hot summers, cold winters in north. Average temperature in Washington DC ranges from −3 to 6°C (27-43°F) in January to 20-31°C (68-88°F) in July

Main primary products Cereals, soya beans, cotton, tobacco, potatoes, citrus fruits, sugar cane, oilseeds, vegetables and soft fruits, timber, livestock, fish; coal, oil and natural gas, iron ore, copper, lead, zinc, gold, silver, molybdenum

Major industries Iron and steel, chemicals, motor vehicles, aircraft, telecommunications equipment, computers, electronics, textiles, forestry, paper and pulp, mining, fishing

Main exports Machinery, electrical and electronic goods, chemicals, cereals, motor vehicles, aircraft, soya beans, coal, instruments, petroleum products, small metal manufactures, textiles, tobacco, fruit and vegetables

Annual income per head (US$) 11 695

Population growth (per thous/yr) 9

Life expectancy (yrs) Male 70 **Female** 77

Uruguay

ONCE A MODEL OF DEMOCRACY AND PROSPERITY, THIS COUNTRY OF UNDULATING PASTURES AND EXTENSIVE BEACHES HAS RECENTLY KNOWN DARK TIMES

Uruguay, one of the continent's smallest countries, was once dubbed 'the Switzerland of South America' because of the republic's contentment, prosperity and neutrality. For most of the 20th century it has had a democratic government, a generous welfare state and a high standard of living. However, in the past few decades Uruguay has come to experience all the evils of troubled Latin American nations: a failing economy, strikes, urban guerrillas, military rule, hyperinflation and abuses of human rights.

Prosperity was the result of Uruguay's rich terrain and mild climate. Ninety per cent of the land is suitable for agriculture, but only 10 per cent is cultivated. The rest of the rolling country is grazed by vast herds of cattle and flocks of sheep.

In 1903, a socially minded president, José Batlle y Ordoñez, took over a country wounded by civil war, repeated invasions – before independence from Spain in 1828 – and a war with Paraguay. Under him booming exports led to affluence and much immigration, mainly from Spain and Italy. Exports financed a democracy and the continent's first welfare state.

But the heyday lasted only until mid-century, when a recession began. The welfare state may have been too generous. Revenues from exports of meat, wool and hides began to decline from 1945, and from 1960 there were few years when the economy grew. During the 1960s inflation rose to 50 per cent a year, living standards fell and Uruguayans could no longer afford the imports they had become accustomed to in days of prosperity.

Their discontent led to widespread strikes and an urban guerrilla movement called the *Tupamaros*, recruited from the young of the MONTEVIDEO middle class. This left-wing movement with anarchist views was responsible, among other things, for kidnapping politicians and industrialists for ransom. The government's inability to stop the Tupamaros led to the military taking power in 1973 and launching a harsh campaign of political arrests and torture.

Uruguay earned a new reputation as the country with the most political prisoners per head of population. In 1976 there were 4700 political detainees. The military government failed to reverse the economic decline, and the world recession aggravated it. By 1984 inflation was rampant, unemployment was running at over 15 per cent, and the foreign debt had risen dramatically to US$5500 million.

The military, faced by almost universal opposition, washed their hands of the problems. Amid great celebrations, 2 million Uruguayans went to the polls in 1984 to end 11 years of military rule. But when the military handed over to President Julio Sanguinetti, leader of the liberal Colorado (Red) party, a few months later there were veiled threats of another coup should the civilian government lose control of the country.

Uruguay has been compared to ancient Athens, because the country is really a city-state – one major city with a huge ranch of countryside attached to it. Of a population of nearly 3 million, half live in the capital city of Montevideo.

The country's 175 000 farmers feed the 1.5 million Montevideans and tend vineyards, orchards, rice fields and groves of olives and citrus fruits. The rest of the country is the domain of the *gauchos* (cowboys), who work on the huge *estancias* (ranches).

Uruguay has no gold or silver, and its native Indians, the Charruas, had no treasures to attract invaders. Today, in the absence of oil and gas, other energy sources are all-important, notably the hydroelectric power stations at Palmar and Salto Grande.

URUGUAY AT A GLANCE

Area 176 215 km² (68 037 sq miles)
Population 2 950 000
Capital Montevideo
Government Parliamentary republic
Currency New peso = 100 centésimos
Language Spanish
Religion Christian (66% Roman Catholic)
Climate Temperate; average temperature in Montevideo ranges from 6-14 °C (43-57°F) in July to 17-28°C (63-82°F) in January
Main primary products Cattle, sheep, wheat, rice, maize; amethysts, topaz, marble
Major industries Agriculture, meat processing, tanning and leather goods, wool and textiles, cement, fishing, tourism, oil refining
Main exports Canned meat, meat, wool, hides and leather goods, fish, rice
Annual income per head (US$) 3043
Population growth (per thous/yr) 3
Life expectancy (yrs) Male 69 **Female** 74

Usutu (Usuthu) ***Southern Africa*** River 216 km (135 miles) long. It rises in Transvaal in South Africa and flows east through central Swaziland into Mozambique through a scenic gorge in the Lebombo Mountains. In Mozambique it becomes a tributary of the Maputo river (called the Pongola in South Africa), reaching the sea in Maputo (formerly Delagoa) Bay.

The Usutu is Swaziland's most important river, and provides water for irrigation and hydroelectric power. Its name, meaning 'dark brown', refers to its muddy appearance.

Map Swaziland Aa

Utah ***USA*** Western state covering 219 932 km² (84 916 sq miles) that was acquired from Mexico in 1848 at the end of the Mexican-American War, having been settled by the Mormons from 1847. Clashes between the Mormon Church authorities and the US government – particularly over the Mormons' belief in polygamy – delayed the achievement of statehood until 1896.

Utah is mountainous, rising to 4123 m (13 528 ft) on the edge of the Rocky Mountains in the north east, and has vast stretches of desert. The GREAT SALT LAKE is the largest of the state's many lakes. Cattle, dairy goods, fodder and wheat are the main farm produce. Copper, coal and oil are mined, and machinery, metals and transport equipment are the principal manufactures. SALT LAKE CITY is the state capital and Mormon headquarters.

Population 1 645 000
Map United States Dc

Utica ***Tunisia*** Site of an ancient Phoenician city and port lying near the mouth of the Medjerda river and 40 km (25 miles) north-west of Tunis. It is claimed to be the oldest Phoenician settlement on the North African coast and was founded in 1100 BC. Following the destruction of Carthage by the Romans it rose to become the capital of the Roman province of Africa. For a time it was commanded by the Roman statesman and general Cato the Younger, a bitter opponent of Julius Caesar. It became famous as the scene of Cato's suicide after the Battle of Thapsus in 46 BC. Although once a port, the site is now 10 km (6 miles) inland, due to silting up of the river mouth. The remains include a Phoenician necropolis of the 6th century BC, Roman baths and a museum containing a fine collection of Roman mosaic floors.

Map Tunisia Ba

Utica ***USA*** City in central New York state about 300 km (185 miles) east of Lake Erie, to which it is linked by canal. Its wide range of manufactures includes car and aircraft parts and dairy products. The city is a tourist gateway to the Adirondack Mountains and the Thousand Islands – some 1500 islets in the St Lawrence River where it leaves Lake Ontario.

Population (city) 72 900; (metropolitan area) 321 400
Map United States Kb

Utrecht ***Netherlands*** City 42 km (26 miles) south and slightly east of Amsterdam, on the site of a Roman fortress commanding a crossing of a former course of the Rhine. Its first bishop was St Willibord (658-739), the Northumbrian 'Evangeliser of the Friesians', and under its bishops it prospered as a medieval cloth-producing centre. In 1579, the Union of Utrecht united the seven northern Netherlands provinces against Spanish rule, and was the foundation of the later Kingdom of the Netherlands.

The old city includes the 14th-century tower of St Martin's Cathedral, at 113 m (370 ft) the tallest church tower in the country, a 15th-century Catholic cathedral, a house built in 1523 for the only Dutch pope, Adrian VI (1522-3), who was born in Utrecht, and the university founded in 1636. The city has the Netherlands mint, the National Museum of Christianity and the National archives. It also is the centre of the country's rail network, and has a railway museum.

Utrecht is also the capital and largest town of the country's smallest province, Utrecht, covering 1396 km² (539 sq miles). Its main products are dairy goods, poultry, fruit and vegetables, and it has many fine country houses founded by rich Amsterdam merchants between 1650 and 1720.

Population (city) 231 800; (Greater Utrecht) 498 900; (province) 886 000
Map Netherlands Ba

Utsunomiya ***Japan*** City in east Honshu, about 95 km (60 miles) north of the capital, Tokyo. Its modern industrial parks have attracted a variety of firms from the metropolitan area around the capital. Cars, office equipment and microchips are made there.

Population 405 400
Map Japan Cc

Uttar Pradesh ***India*** The country's most populous state, covering 294 411 km² (113 643 sq miles) in the north. It includes part of the Himalayas, with Nanda Devi (7817 m, 25 645 ft), and much of the Ganges, Yamuna and Ghaghara river plains. The irrigated western plains adjoin the industrial area around the national capital, Delhi. They contrast with the wetter, much more densely populated east, where poverty is rife. The state's products include rice, wheat, sugar, cotton textiles and leather goods. Formerly known as the United Provinces (of Agra and Oudh), it was the heartland of early Hindu civilisation. It has many of the holiest Hindu cities, including Varanasi and Allahabad, and is the core of the Hindi-speaking area. Lucknow is the state capital.

Population 110 862 000
Map India Cb

Uzbekistan ***USSR*** Central Asian constituent republic of the USSR between Kazakhstan and Turkmenistan, centred on the Aral Sea. It consists mainly of deserts and plains, and covers 447 700 km² (172 850 sq miles). It is named after Khan Uzbek (1312-42), one of the leaders of the Golden Horde – the Mongol army that swept across central Asia and eastern Europe in the 13th century and established its own state in the area.

The republic contains several contrasting zones – including the fertile, irrigated FERGANA region; the mountain ridges of the TIEN SHAN, among which stand the cities of TASHKENT, the capital, and SAMARKAND; the Kyzyl Kum desert, which is 480 km (300 miles) long and is rich in oil and gas; the Turan lowland of steppes and deserts; the oases and irrigated areas of the lower Amudar'ya river; and the stony desert of the Ustyurt plateau.

Modern irrigation systems installed since the mid-1920s have established Uzbekistan as the USSR's main source of cotton. The republic also rears Karakul lambs for wool and meat. Its main industrial products are coal, chemicals, copper, agricultural machinery and textiles.

The region was invaded by Alexander the Great's troops in the 4th century BC and by Arabs in the 7th century AD; the latter introduced Islam, which remains the predominant religion. Powerful Islamic Khanates were established in the 16th to 18th centuries, but they were ceded to Russia in the late 19th century.

Population 17 500 000
Map USSR Gd

Vaal ***South Africa*** River rising in south-east Transvaal near the Swazi border. It flows about 1120 km (695 miles) west, forming much of the Transvaal's border with the Orange Free State, and it joins the Orange River about 112 km (70 miles) south-west of Kimberley. With its tributaries the Vaal is the chief source of water for the Witwatersrand, Pretoria and south Transvaal. It is tapped for urban water supplies at the Vaal Dam, 29 km (18 miles) upstream from the town of Vereeniging, and for irrigation at the Bloemhof Dam and at the Vaal Hartz irrigation scheme, 80 km (50 miles) north of Kimberley.

Map South Africa Cb

Vaasa (Vasa) ***Finland*** Seaport founded in 1606, about 370 km (230 miles) north-west of the capital, Helsinki. It was destroyed by fire in 1852 and rebuilt in 1860 on its present site on the Gulf of Bothnia.

Vaasa's main industries are ship repairing, flour milling, sawmilling, sugar refining, and the making of machinery, textiles and soap. It exports timber and wood products. There is a ferry service from the town to Umeå in Sweden.

Population 54 400
Map Finland Bc

Vadodara (Baroda) ***India*** Industrial city about 300 km (186 miles) north of Bombay. Its main products are jewellery, textiles, chemicals, metal goods, tobacco, wool and cotton. Vadodara's Markapura palace is noted for its deep and ornately carved wells.

Population 744 900
Map India Bc

Vaduz ***Liechtenstein*** Capital of the tiny principality, on the Swiss-Austrian border about 80 km (50 miles) east of the Swiss city of Zürich. Once a small market town, it is now an international finance centre, due to its liberal taxation laws. It is overlooked by the 18th-century castle of its princes and has a modern art gallery and museum.

Population 5000
Map Switzerland Ba

Váh ***Czechoslovakia*** Longest river of Slovakia, 378 km (235 miles) long. It rises in the Tatra Mountains and joins the Little Danube river at the town of Kolarovo, near the Hungarian border. It is dammed for hydroelectric power at several places.

Map Czechoslovakia Db

Valais ***Switzerland*** See SION

Valdai Hills ***USSR*** Range of hills and plateaus about halfway between Moscow and Leningrad. With an average height of 180-300 m (600-1000 ft), the hills rise to a peak of 343 m (1125 ft) – the highest point in the European Russian Plain. The Valdai are well forested, with many lakes and peat bogs, from which flow the Dnieper, West Dvina, Volga and Volkhov.

Map USSR Ec

Valdivia ***Chile*** Province in the south-central region of Los Lagos. It has many picturesque lakes which provide excellent fishing, such as Panguipulli, Calafquén and Riñihue. The capital city, Valdivia, lying 800 km (500 miles) south of Santiago, was founded in 1552 by the Spanish conquistador Pedro de Valdivia. It was severely damaged by an earthquake and tidal wave in 1960. It is surrounded by fertile agricultural land and has several food-processing factories and leather, paper and timber industries.

Population (province) 329 800; (city) 116 000
Map Chile Ac

vale A broad valley.

Valence ***France*** Capital of Drôme department, on the Rhône river 90 km (55 miles) south of Lyons. It is a market centre for fruit and vegetables, and has textile and food industries.

Population 108 300
Map France Fd

Valencia ***Spain*** Seaside city and port on the central Mediterranean coast with many medieval and Renaissance buildings and one of Spain's largest bullrings, seating 18 000. It has numerous industries, including shipbuilding, railway engineering and chemicals. And it exports wine and fruit from the fertile province and wider region of the same name, whose *huertas* – irrigated lands – yield more than one crop a year. It was the capital of a Moorish kingdom, captured for Spain by the Christian hero El Cid in 1094, retaken by the Moors in 1099 and finally secured for Spain in 1238. In the Civil War (1936-9), it was the last Republican city to fall to General Franco's Nationalists – in March 1939.

Population (city) 751 700; (province) 2 066 400
Map Spain Ec

Valencia ***Venezuela*** Capital of Carabobo state, about 100 km (60 miles) west of the national capital, Caracas. Founded in 1555, it is one of the country's largest cities. Its factories produce cars, processed foods, beer, animal fodder and glassware. Like its Spanish namesake, Valencia is noted for its oranges.

Population 540 000
Map Venezuela Ba

Valenciennes ***France*** Industrial town on the Escaut river, 10 km (6 miles) from the Belgian frontier, south-west of Lille. It was once famous for its lace, but its chief industry now is the manufacture of metal goods.

Population 350 600
Map France Ea

Valladolid ***Spain*** Industrial city about 160 km (100 miles) north-west of Madrid, the main industries being railway engineering, motor-car manufacturing, tanning, flour milling, brewing, textiles and chemicals. It was the seat of the Spanish royal court for brief periods in the 16th and 17th centuries, and the explorer Christopher Columbus died there in 1506. Notable buildings include several royal palaces, the massive 15th-century Monastery of San Pablo and College of San Gregorio (now a national museum of sculpture), the late Renaissance cathedral and numerous fine churches.

The province of the same name produces cereals, fruit and wine.

Population (city) 330 200; (province) 489 600
Map Spain Cb

Valle d'Aosta ***Italy*** French-speaking Alpine region covering 3262 km² (1259 sq miles) in the north-west, centred on the valley of the same name and the regional capital, AOSTA. It is linked to Switzerland and France by the Grand and Little ST BERNARD PASSES and to France by the Mont BLANC tunnel.

Population 113 500
Map Italy Ab

Valle del Cauca ***Colombia*** Department in Colombia dominated by the Cauca river, whose western valley gives the department its name. Most of the country's sugar cane is grown in the department. CALI is the capital.

Population 2 965 000
Map Colombia Bb

Valletta ***Malta*** National capital, standing on a ridge-like peninsula on the east coast. It hangs over the Grand Harbour to the south and Marsamxett Harbour to the north. The city was named after Jean Parisot de la Vallette, Grand Master of the Order of the Knights of St John, who masterminded the city's reconstruction after the island had beaten off the Turks in the Great Siege of 1565.

Valletta remains a beautiful, if occasionally dilapidated, Baroque city. It is heavily fortified with huge stone walls meeting at the castle of St Elmo. A chequerboard of streets pays scant regard to the hilly lie of the land, and many of them end in steep stairways plunging to the quaysides. The main street is Kingsway, which runs parallel to Strait Street, unflatteringly called the 'Gut'. Strait Street was formerly the only street where knights could duel and later became a 'red light' district of bars, cheap restaurants and brothels, frequented by visiting sailors.

Valletta has many fine churches and palaces, notably St John's Cathedral and the *auberges* or knights' lodgings. The Auberge de Castille, grandly remodelled in the 18th century, is now the prime minister's office. Grand Harbour has several dry docks, and can handle vessels of up to 300 000 deadweight tonnes.

Population 14 100
Map Malta Eb

valley Region of low-lying land surrounded by higher land, usually a long, narrow depression with a regular slope from one end to the other. Valleys may be carved out by a stream or river, or formed by FAULT activity, later modified by running water. A young valley is narrow with steep sides; an old valley may be very wide, with only a shallow slope and a broad FLOOD PLAIN. Valleys are also eroded, or profoundly modified by GLACIERS, creating a characteristic U-profile. Valleys formed solely by stream erosion are V-shaped until they widen with age.

Valley of a Thousand Hills ***South Africa*** Broad valley dotted with rolling hills and Zulu villages between Durban and Pietermaritzburg. It is a popular tourist attraction.

Map South Africa Db

Valona ***Albania*** See VLORE

Valparaíso ***Chile*** Capital of Aconcagua region, and the country's chief port, 100 km (60 miles) west of Santiago. It handles the bulk of Chile's imports, and its industries include foundries, ship repairing, chemicals, textiles and leather goods. The business area stands on land reclaimed from the sea, while the residential parts are a maze of streets winding up the hills around the bay. Founded in 1536, the city has been subject to frequent earthquakes, as in 1906 when it was devastated.

Population 267 000
Map Chile Ac

Vals Bay ***South Africa*** See FALSE BAY

Van ***Turkey*** Lakeside town about 1000 km (620 miles) east of the capital, Ankara. The town markets grain, fruit and hides. It stands on the eastern shore of Lake Van, Turkey's largest lake, which covers 3675 km² (1419 sq miles). The main products of the surrounding province, which is also called Van and which has an area of 19 069 km² (7363 sq miles), are coal, rye and apples.

Population (town) 46 800; (province) 573 800
Map Turkey Cb

Van Allen belt See ATMOSPHERE

Van Diemen's Land ***Australia*** See TASMANIA

Vancouver ***Canada*** Third largest city in Canada and the country's chief Pacific port. Vancouver, on an inlet off the Strait of Georgia opposite Vancouver Island, is the commercial and industrial centre of British Columbia, with fish and other food processing plants, oil refineries and electronics factories. The original town, facing the inner harbour, has grown northwards across the Burrard Inlet, southwards towards Roberts Bank (where new port facilities have been opened) and eastwards to Burnaby and Port Moody. Vancouver has an excellent year-round harbour and is the western terminus of trans-Canadian railways, roads and airways. The city is famous for its parks: Stanley Park (364 hectares, 900 acres) has specimens of native trees and a zoo. Vancouver's mild climate and picturesque setting, with harbour and mountain views, make it popular with tourists.

Population 1 268 000
Map Canada Cd

Vancouver Island ***Canada*** The largest island on the Pacific seaboard (31 285 km², 12 079 sq miles), separated from the British Columbia mainland by the Strait of Georgia. There are several fine harbours and the interior is rugged and forested, rising to 2200 m (7219 ft) at Golden Hinde. People live mainly on the eastern and southern sides. Created a British Crown Colony in 1849, the island became part of British Columbia in 1866. Coal, iron ore and copper are mined, and timber, fishing and tour-

▼ GEM OF THE PACIFIC Greater Vancouver lies around its magnificent harbour. The city grew from a small loggers' settlement after the arrival of the first transcontinental train from Montreal in May 1887.

ism are all sources of income. VICTORIA is the main city.

Map Canada Cd

Vanderbijlpark ***South Africa*** Industrial town on the Vaal river, about 65 km (40 miles) south of Johannesburg. It grew around an iron and steel plant built in the 1940s.

Population 61 200

Map South Africa Cb

Vänern ***Sweden*** The country's largest lake, 145 km (90 miles) long and stretching across 5586 km^2 (2157 sq miles). Set in the south-west of the country, it drains by way of the GOTA river into the Kattegat strait between Sweden and Denmark.

Map Sweden Bd

Vanua Levu ***Fiji*** The country's second largest island, lying 65 km (40 miles) north-east of Viti Levu, the main island, and covering 5556 km^2 (2145 sq miles). Like Viti Levu, it is drier on the western side than to the east, where there are extensive coconut plantations. Sugar is grown in the north around the main town, Labasa (population 12 000), which has a large sugar mill. The island has an extensive coral reef around it.

Off the south-east tip is Taveuni, Fiji's third largest island; it has a central volcanic spine and is planted extensively with coconuts.

Population 108 000

Map Pacific Ocean Dc

vapour trail See CONDENSATION TRAIL

Varanasi (Benares) ***India*** Ancient and holy Hindu city on the river Ganges. It stands in the north-east of the country about 230 km (142 miles) south of the Nepalese border. According to Hindus, those who die in the city automatically go straight to heaven – and so many Hindus come to spend their last days in Varanasi. It is also the annual destination of more than a million pilgrims.

The city is a maze of narrow lanes and hundreds of temples. Along the arc of the river are the bathing ghats, or slopes, where the deceased are cremated, so that their ashes may be strewn in the sacred waters.

The city manufactures silks, chemicals, brassware, lacquered toys and jewellery. It has a Hindu university, founded in 1915, on the Central Hindu College which was founded in 1896.

Population 797 200

Map India Cc

Varangerfjord ***Norway*** One of the most northerly Norwegian fiords – 68 km (42 miles) long – and the one nearest the Soviet border. Kirkenes, the terminal port for Norway's coastal shipping, is on an inlet on the south side of the fiord near the Finnish border, and is the site of the vast Sydvaranger iron mines. The Varanger peninsula, with the port of Vadsø, is on the north side of the fiord.

Map Norway Ha

Varazdin (Varasd; Warasdin) ***Yugoslavia*** Ancient Slav town in Croatia, lying on the Drava river 61 km (38 miles) north-east of Zagreb. A busy textile and commercial centre, it dates mostly from the 17th century, with fine palaces, churches and Baroque houses around a chateau-like fortress. There are medicinal springs at Krapina to the south-west.

Population 34 600

Map Yugoslavia Ca

Vanuatu

VOLCANIC ISLANDS WHERE LUSH FORESTS ALIVE WITH TROPICAL BIRDS RING THE NEW AND PROFITABLE CATTLE RANCHES

The New Hebrides became the independent republic of Vanuatu in 1980, so ending nearly a century of joint British and French rule over this chain of western Pacific islands lying south-east of the Solomons and 1750 km (1100 miles) east of Australia. The Portuguese were the first Europeans to visit the islands, in 1606, and they were charted by the British explorer Captain James Cook in 1774. The islands proved valuable as a major Allied base during the Second World War.

The republic comprises about 80 islands and islets. Two islands, ESPIRITU SANTO and TANNA, made a bid for separate independence when the new republic was founded, and troops from nearby newly independent Papua New Guinea had to suppress the rebellion when the British and French failed to do so. The main islands are ringed with coral reefs. Five of them – including Tanna, Ambrym and Lopevi – have active volcanoes, and earth tremors are frequent. Birds abound in lush forests.

The indigenous people are Melanesians, a dark-skinned people. There are a few immigrants from Europe, China, Vietnam and other Pacific islands. Since manganese mining on Éfaté (the island on which the capital PORT-VILA stands) stopped in 1978, the main exports have been copra, coconut oil, beef and fish. Cattle ranching is being developed with scientific help from overseas.

VANUATU AT A GLANCE

Map Pacific Ocean Dc

Area 14 760 km^2 (5700 sq miles)

Population 138 000

Capital Port-Vila

Government Parliamentary republic

Currency Vatu

Languages Bislama (Pidgin), English, French, about 40 tribal languages

Religions Christian (53% Protestant, 12% Roman Catholic), tribal religions

Climate Tropical, with heavy rainfall; average temperature in Port-Vila ranges from 19-26°C (66-79°F) in August to 24-31°C (75-88°F) in February

Main primary products Coconuts, cocoa, coffee, yams, taro, cassava, bananas, beef cattle, timber, fish

Major industries Copra processing, fishing, food processing, forestry

Main exports Copra, fish, beef

Annual income per head (US$) 510

Population growth (per thous/yr) 27

Life expectancy (yrs) Male 58 **Female** 60

Vardar ***Greece/Yugoslavia*** See AXIOS

Varese ***Italy*** Town standing in the Alpine foothills near Lake Varese, about 50 km (30 miles) north-west of Milan. It is noted for its sanctuary, the Sacro Monte. Its economy is based mainly on engineering, textiles, furniture, footwear and leather goods.

Population 89 200

Map Italy Bb

Varna ***Bulgaria*** Major seaport and capital of Varna province, on the Black Sea 90 km (56 miles) north of Burgas. It is also a seaside resort, with sandy beaches. It trades in grain, livestock and canned fish. Industries include shipbuilding, engineering, textiles, furniture and foodstuffs. It is a railway terminus and has a hydrofoil service to the seaside resorts of Albena, Balchik and Druzhba. There are a number of museums, and an international Summer Music Festival is held each year in July.

Founded in the 6th century BC as Odessus, Varna came under Ottoman Turkish rule in 1391 and was ceded to Bulgaria in 1878.

The ruined 7th-century city of Pliska, Bulgaria's first capital, is set amid cornfields 40 km (25 miles) to the west of Varna. Pliska is being reconstructed as a monument to Bulgarian nationalism.

Population 295 300

Map Bulgaria Cb

varve Pair of distinctive bands of sand and silt deposited in one year in meltwater lakes near the margins of glaciers or ice sheets. The coarser material settles first during the spring melting; the finer material, held in suspension, later in the year. By counting the varves, it is possible to estimate the time since the last Ice Age.

Vasa ***Finland*** See VAASA

Västerås ***Sweden*** Manufacturing city and inland port on the north shore of Lake Mälaren, 110 km (68 miles) west of Stockholm. Founded in the 12th century, it has a medieval castle, a 13th-century cathedral, and the country's oldest secondary school, established in 1623. The introduction in the 16th century of hereditary succession to the Swedish throne – earlier monarchs had been elected – resulted from a conference held in the city in 1544. It is the home of several of Sweden's largest electrical and engineering concerns.

Population 117 700

Map Sweden Cd

Vatnajökull ***Iceland*** The largest ice sheet in the Northern Hemisphere outside Greenland. Vatnajökull is in south-east Iceland and parts of it are more than 2000 m (6560 ft) above sea level. In all, the ice sheet covers an area of 8538 km^2 (3296 sq miles), and in places the ice is nearly 1000 m (3280 ft) thick.

Heat from volcanoes beneath the ice sheet has created large lakes and rivers under the

Vatican City State

THE SEAT OF GOVERNMENT OF THE ROMAN CATHOLIC CHURCH, THE WORLD'S SMALLEST INDEPENDENT STATE, LIES IN THE HEART OF ROME

On a low hill on the west of the Tiber in the city of Rome lies the world's smallest independent state, the headquarters of the Roman Catholic Church. The Vatican State consists of a walled city within Rome which embraces the Vatican Palace, the Papal Gardens, St Peter's Square and St Peter's Basilica – the mother church for the world's 700 million Roman Catholics. In addition, the Vatican State includes 12 other buildings in and outside Rome.

The popes have lived here since the 5th century – apart from the period between 1309 and 1378 when the papacy moved to Avignon. They enjoyed great political power after the foundation of the Holy Roman Empire in AD 800, and the papacy once held states extending for 44 000 km² (17 000 sq miles) through central and northern Italy, with 3.7 million inhabitants. But when Italy was unified in 1860-70 these states were incorporated into the new nation under Italy's first king, Vittorio Emmanuel. The popes, who refused to recognise the new arrangement, considered themselves prisoners in the Vatican. In 1929 Mussolini negotiated the Lateran Treaty with the Church, which recognised the independence of the Vatican City in return for the papacy recognising the kingdom of Italy under the House of Savoy.

Today the state has its own police, newspaper, coinage, stamps and railway and radio stations. Among the main tourist attractions are the frescoes of the Sistine Chapel painted by Michelangelo Buonarroti (1475-1564), who was the architect of the great Renaissance church of St Peter – the largest church in the world. He also designed the colourful uniforms of the Swiss Guard, the Pope's personal bodyguard, recruited in the Roman Catholic cantons of Switzerland.

VATICAN AT A GLANCE

Map Italy Dd
Area 0.44 km² (0.17 sq mile)
Population 1000
Capital Vatican City
Government Papal Commission
Currency Vatican City lira = 100 centesimi; Italian currency also used
Languages Latin (official), Italian
Religion Christian (Roman Catholic)
Climate Mediterranean

ice – giving it its name, which means 'Water Glacier'. About once every ten years, one of these buried volcanoes erupts, releasing what Icelanders call a *jökullaup*: a catastrophic torrent of water, ice chunks, boulders and other debris which can surge across the countryside at speeds of up to 100 km/h (60 mph), sweeping away everything in its path.
Map Iceland Cb

Vättern ***Sweden*** The country's second largest lake after Vänern. It covers 1912 km² (738 sq miles) and lies about 220 km (135 miles) south-west of Stockholm. It is linked to the Baltic Sea by the GOTA river.
Map Sweden Bd

Vaud ***Switzerland*** See LAUSANNE

veering Clockwise change of direction of the wind, for example from north-east through east to south-east. It is the opposite of backing.

Veglia ***Yugoslavia*** See KRK

vein Thin sheet of crystalline minerals – often metallic ores – which has filled a fissure or joint in a rock.

Vejle ***Denmark*** Market town and small industrial centre at the head of Vejle Fjord in south-eastern Jutland. In medieval times, it was an important way station on the cattle export route to north Germany. Nearby is the village of Billund (population about 4000), where the construction toy Lego was invented by a local carpenter, Ole Kirk Christiansen, in 1954. Billund is also the home of Legoland, a 10 hectare (25 acre) park containing model villages, cars, ships and planes built entirely with Lego bricks – more than 30 million in all.
Population 48 600
Map Denmark Bb

veld Area of open country in South Africa, especially in the Transvaal. It is often divided into HIGHVELD (above 1500 m, nearly 5000 ft), MIDDLEVELD (900-1500 m, about 3000-5000 ft), and LOWVELD (below 900 m, 3000 ft). The highveld consists of rolling, treeless grassland. The middleveld and lowveld are mainly scrub.

Velence, Lake ***Hungary*** See SZEKESFEHERVAR

Velgam Vihare ***Sri Lanka*** Ancient temple about 16 km (10 miles) west of Trincomalee. It was built as a Buddhist shrine at about the beginning of the Christian era, and was restored in the 11th century by Hindu Chola kings. It is one of the world's most interesting amalgams of Buddhist and Hindu art and architecture.

Veliki Beckerek ***Yugoslavia*** See ZRENJANIN

Veliko Turnovo ***Bulgaria*** City and capital of Veliko Turnovo province on the Yantra river, 88 km (55 miles) south-east of Pleven. It processes meat and makes beer, furniture and textiles. It was the capital of the Bulgarian Empire from 1186 until it came under Turkish rule in 1394. In 1908 the independent kingdom of Bulgaria was proclaimed there.
Population 65 100
Map Bulgaria Bb

Vellore ***India*** Fortified south-eastern city about 115 km (70 miles) west of the city of Madras, where a savage mutiny of Indian troops against their British commanders took place in 1806. It has a temple devoted to the Hindu god Siva, and a noted medical college.
Population 247 000
Map India Ce

Venda ***South Africa*** Two land areas in north-east Transvaal, covering a total of 6500 km² (2510 sq miles). Formerly a Bantu homeland, Venda was declared independent in 1979, but it is recognised only by South Africa. The population is mostly involved in farming and forestry. The capital is Thohoyandou.
Population 435 000
Map South Africa Da

Vendée ***France*** Department of 6720 km² (2595 sq miles) in western France between Nantes and La Rochelle. A rebellion there against the Revolutionary government was crushed by Robespierre in 1793 although localised uprisings continued until 1796. The main occupations today are farming and tourism.
Population 483 000
Map France Cc

Veneto ***Italy*** Region covering 18 337 km² (7080 sq miles) of which VENICE is the capital. Part of the PO river plain and the western Italian Alps, including part of the DOLOMITES, lie within its borders, as do the artistic cities of PADUA, TREVISO, VERONA and VICENZA.

Until the 1960s it was a region of rural poverty and overpopulation. Now agriculture is prospering, with rich harvests of apples, pears and peaches, vines (Valpolicella, Bardolino and Soave wines), sugar beet and vegetables, and a thriving livestock sector. Industry – including oil refining and petrochemicals – is based on the Venetian lagoon.

Tourism is booming, especially in Venice, the Adriatic coast resorts such as Lido di Jésolo, the Alps and the Dolomites.
Population 4 366 100
Map Italy Cb

Venezia ***Italy*** See VENICE

Venezuela See p. 688

Venice (Venezia) ***Italy*** Seaport and one of the world's great architectural and cultural treasures, built on a large group of islands in the Lagoon of Venice, near the northern end of the Adriatic Sea. It was founded in the 5th century AD by Romans fleeing from the Lombard invaders of northern Italy, and became a republic in the 9th century. For the next 500 years it was the most powerful European maritime state, trading with East and West and developing its own industries, particularly textiles, lace, glass and shipbuilding. Venice today is regional capital of the Veneto, but it has virtually no industry left; this is now based across the lagoon on the mainland at Mestre and Porto Marghera, which have large petrochemical plants.

The original builders set Venice a few centimetres above high-tide levels on foundations of wooden piles driven into the mud of the lagoon. Silt-bearing rivers were diverted away from the lagoon while the (*continued on p. 690*)

Venezuela

UNCONTROLLED SPENDING AND A HUGE INTERNATIONAL DEBT HAVE LEFT VENEZUELA WITH MASSIVE ECONOMIC PROBLEMS

When Venezuela began to produce oil in 1922, it was transformed from an economic backwater into the richest country in South America. Cities boomed, motorways were built, and Venezuela made a belated leap into the 20th century. But the poor were left behind. Successive governments failed to make sure that everyone shared in the country's prosperity, and even today, more than 60 years later, poverty is as much in evidence as wealth.

The country, which forms the northernmost crest of South America, is the size of France and East and West Germany together. The Caribbean coastline is the most densely populated and developed strip of Venezuela, but behind this stretches a vast – and, in many places, little-explored – interior. To the north-west of the country, a spur of the snow-capped ANDES mountains, the Cordillera de Mérida, runs south-west to north-east. It continues in the central highlands – covered by tropical forest – which lie behind the capital CARACAS. In the centre of the country, north of the great ORINOCO river (which cuts Venezuela in two), lie undulating plains, the *llanos*, where huge herds of humpbacked zebu cattle are reared. Farther south and east are the GUIANA HIGHLANDS containing some of the oldest mountains in South America; this largely unexplored area occupies nearly half the country. It is the region that inspired Sir Arthur Conan Doyle's novel *The Lost World*, with its huge isolated plateaus and outcrops of solid sandstone rising suddenly from the rolling savannah. The extreme south, lying on the rim of the AMAZON basin, is covered with dense tropical jungle scattered with small frontier towns, Protestant missions and communities of Indian tribes.

OIL WEALTH AND INDIANS

The precious oil fields, capable of producing 2.4 million barrels a day, lie in the north-west, around Lake MARACAIBO – now a forest of 10 000 oil derricks. Strictly speaking, the lake is a river estuary, forming a vast but slow-moving pool, which flows past the narrow straits on which stands MARACAIBO, the oil capital, to empty its sluggish waters into the Caribbean. The metropolis of Maracaibo has a humid climate, and a population that is a brash mixture of oil executives, cattlemen and competitive merchants – and the gentle nomadic Guajiro Indians from the Guajira Peninsula to the west.

The Guajira Peninsula is semidesert, with wild brushland and lagoons crowded with flamingos. It is divided between Venezuela and Colombia, but the Guajiros regard themselves as a nation and so do not take account of this frontier. They rear cattle and herd goats, and although they have a veneer of Spanish Catholicism, their way of life follows its own traditions. The men wear a headband of feathers to go visiting; the women are always dressed in long, flowing, black or patterned cotton mantles.

South-west of Maracaibo lives an Indian tribe of very different character: the Motilones, a warlike people who violently resisted European colonisation, retaliating when their villages were burnt and firing arrows at geologists who came to explore for oil – sometimes with fatal results. Only recently have they been pacified, and now they have become part of the tourist scene – for instance, at the Tucuco Mission near Machiques.

North of Maracaibo city, on the Gulf of Venezuela, live the Paraujano Indians, in huts on stilts. It was these houses that gave Venezuela its name. When the Italian explorer and mapmaker Amerigo Vespucci first saw the country in 1499, the year after Columbus had claimed it for Spain, he thought the waterways that divided the homes looking out over water resembled his own Venice, and called the land 'Little Venice', or, in Spanish, 'Venezuela'.

The Cordillera de Mérida rises to 5007 m (16 427 ft) at Pico Bolivar and separates the oil fields around Maracaibo from the rest of Venezuela. The world's longest (12 km, 7.5 miles) and highest (4765 m, 15 633 ft) cable-car ride runs up from the colonial, coffee-producing town of MERIDA to Pico Espejo.

East from Maracaibo stretch 1700 km (1100 miles) of beaches, from remote coves with underwater wonderlands of coral reefs just offshore, past busy tourist centres near Caracas to the island maze of the Orinoco delta where the hunting and gathering Guarauno Indians also live in houses on stilts. A favourite resort is MARGARITA Island, a four-hour ferry ride out into the Caribbean from CUMANÁ. It is named after and noted for its pearls (*margarita* is Spanish for 'pearl').

Caracas, stretching along a narrow strip between mountains and shore, is the focus for the country's culture, industry and commerce, except for oil, which is based in Maracaibo.

A vast community of poor live in a maze of brick and corrugated iron boxes they built themselves on the steep hillsides around the city. As many people live in these *barrios* as in the city itself, where most of the population live in flats. Single family houses are affordable only by the very wealthy – or are built in makeshift fashion by the very poor.

Appropriately enough for the birthplace of the great liberator Simón Bolívar, Venezuela's government is democratic, though from 1821 to 1958 military rule was virtually uninterrupted. It survived the wave of military coups that swept the rest of South America in the 1960s and 1970s – an achievement of which the Venezuelans are justly proud. When oil was first discovered, the bloodthirsty Juan Gómez was dictator; he ruled intermittently from 1908 to his death in 1935. He is now referred to as *El Benemérito*, 'The Deserving', in ironic comment on his greed and ruthlessness.

The military stayed in power until 1945; then, after a brief democratic respite, they returned in 1948. The last military dictator, Major Marcos Jiménez, was overthrown in 1958, and democratic government returned; it has continued to the present day. Indeed, voting is compulsory in Venezuela resulting in a 90 per cent turnout in elections. But in spite of democracy there is still a wide gulf between rich and poor.

Much of the problem revolves around Venezuela's oil wealth. For a long time the industry was controlled by multinational companies, most of whom employed foreigners. Many of the profits were siphoned off to the parent companies – a fact that has left Venezuelans suspicious of big foreign companies, particularly American ones. Determined to control its own resources and the profit from them, Venezuela nationalised the industry in 1976.

An economic boom followed and many projects, such as agricultural land development, roads, railways and huge apartment blocks, were financed by the government. However, administrative inefficiency and uncontrolled spending by the government produced a financial crisis. This was further compounded by an economic recession when world oil demand fell, with a resultant fall in oil revenue. Dependence on oil has been considerable – 75 per cent of all government revenues come from it, though only 3 per cent of the population are employed in the industry – an important reason why national wealth has not been spread among the people.

With the oil recession, Venezuela built up a huge national debt to international banks – reaching US$34 000 million in the mid-1980s – amassed largely because imports far exceed exports. Food imports, for example, far exceed what they need be, as the country could be much closer to self-sufficiency than it is. Oil revenue was so easy to come by in the past that other sources of prosperity have not been nurtured; for example, mineral resources such as iron ore, bauxite and coal have only recently begun to be exploited.

Although it has troubles, Venezuela's stability as a democracy gives it a good atmosphere in which to sort them out. With Colombia, Mexico and Panama, it is a member of the Contadora Group of Latin American countries, and has a voice which is listened to in the continent's affairs. It is only neighbouring Guyana with whom relations are poor – the two countries have a long-standing dispute over the border.

POTENTIAL OF THE LLANOS

The region that is seen as the potential source of future prosperity is the great expanse of plain, the llanos, between the populated northern coastline and the southern jungle. This is the region that makes the country already mostly self-sufficient in meat; it is Venezuela's Wild West, where cowboys, called *llaneros*, spend their lives herding cattle across the trackless plains. It is they who dance Venezuela's national dance, the *joropo*, with its fancy footwork accompanied by a handheld harp. To show off their skill at rodeos, which are occasions for great barbecues, or *asados*, they dress in simple white linen suits, with military collars, and wear stetsons and boots. Days are spent in the saddle and nights in a hammock; the llaneros's greatest pride is their skill and toughness, their greatest fears jaguars and disease-carrying vampire bats, both of which threaten their cattle.

The llaneros are of part Indian descent. Unlike some of the other countries of South America, Venezuela has no evidence of great pre-Columbian Indian civilisations, but Indian influence is still felt in its culture, religion and folklore. A blend of Roman Catholicism and paganism is practised in the hills between Maracaibo and Caracas, and even the capital has its statue of the pagan fertility goddess María Lionza.

But the Venezuelans are not only a mixture of European and Indian. During colonial days traders brought black slaves to the Caribbean coast and sold them for coffee and tobacco. Many of today's population have some African ancestry.

The communities that have managed to stay out of this racial melting pot are the Indian tribes of the remote jungle high up the Orinoco river, whose culture has been the same since the Stone Age. Some are threatened by such possible economic developments as exploiting the Orinoco still further to generate hydroelectric power (HEP). One HEP complex has already been built at the confluence of the Orinoco and Caroní rivers, where iron and steel and aluminium production have led recently to the rapid growth of a completely planned new city, CIUDAD GUAYANA. Even this idealistically designed experiment has not escaped its fringe of homemade barrios, as the rural population comes in from the country to cling to the outskirts.

These poor migrants are not the first people to seek wealth in Venezuela. From the European colonisers seeking gold in the 16th century to geologists prospecting for oil, this seems to be the theme of the country's history. Sir Walter Raleigh twice sailed up the Orinoco in search of the fabled golden city of El Dorado. As late as 1935 an American pilot-adventurer, Jimmy Angel, flew a small plane into the Guiana Highlands in search of gold. Instead he found a wonder that immortalised him: tumbling down from the plateau of Auyan Tepuí, the 'Devil's Mountain', was a waterfall 20 times higher than Niagara, to which he gave his name. Today small planes follow Jimmy's route to bring tourists to see the thundering trail of the Angel Fall plunging 980 m (3215 ft) down a sheer cliff face into the green jungle below.

With vast areas of uninhabited jungle and open plains, Venezuela is the most sparsely populated country in South America. Its 18.34 million people are mostly concentrated in a few limited areas. But both the busy and the remote regions share a passion for festivities, which sweep the country on frequent religious holidays. Forgetting poverty and troubles, every town and village commemorates its own patron saint with a festival, often a whole week of fairs, bullfights and cockfights, of drinking and dancing in the streets, of games and races. It is in the uninhibited revelry that the culturally rich mixture that is Venezuela's character becomes most apparent.

▲ TROPICAL HIGHLANDS The Cordillera Mérida, a spur of the Andes mountains, runs diagonally across the north-west of Venezuela. Small Indian settlements snuggle in the foothills, and are surrounded by scrubland and green, fertile hills and valleys.

VENEZUELA AT A GLANCE
Area 912 050 km² (352 143 sq miles)
Population 18 340 000
Capital Caracas
Government Parliamentary republic
Currency Bolívar = 100 céntimos
Language Spanish
Religion Christian (95% Roman Catholic, 2% Protestant)
Climate Tropical; humid lowlands; cooling progressively with altitude. Temperature at Caracas averages 24-27°C (75-81°F) all year. Los Nevados (village in Andes) constantly cold: 14°C (57°F). Rainy season May to November
Main primary products Livestock, coffee, cocoa, maize, rice, sugar, cotton, tobacco, fruit, vegetables; petroleum, iron ore, diamonds, manganese, bauxite (aluminium ore), gold, coal
Major industries Steel, transport equipment, ships, petroleum products, chemicals, food processing, textiles, cement, aluminium
Main exports Crude oil and products
Annual income per head (US$) 4260
Population growth (per thous/yr) 30
Life expectancy (yrs) Male 66 **Female** 70

seaward defences across the sandbars were strengthened from the 14th century onwards culminating in the great sea-works of the 18th century. The causeway carrying the road and railway to the main island was finished in 1846.

The city plan is one of bridged islets – more than 100 of them. Each islet forms a community with its own patron saint, church, central square, palace and noble family. The most splendid buildings, such as the Doge's Palace – the most spectacular secular Gothic building in the world – are set around the Piazza San Marco and along the Grand Canal. But there are hundreds of other palaces and dozens of churches.

The problem of the sinking of Venice is at last being solved. The main cause of the sinking is the pumping of water from the subsoil, allowing this to compact. Since disastrous floods in 1966, this has been banned, and a new system of coastal defences across the entrances to the lagoon will prevent the surging high tides sweeping into the city. A much longer-term problem, however, is how to renovate decaying houses and check the exodus of the population to better housing and job opportunities on the mainland.

Venice is turning more and more into a treasure-house city to be enjoyed chiefly by tourists. It was the birthplace of many famous artists, including Giovanni Bellini, Giorgione and Titian in the 15th century, Tintoretto and Veronese in the 16th, and the Tiepolo family in the 17th and 18th centuries. Many of their works are displayed in the 14th to 15th-century Doge's Palace, the Accademia, the city's churches, and its famous 'schools' or guild halls. The city was also immortalised in the canvases of another Venetian painter, Canaletto (1697-1768), perhaps the best known being his *Fête on the Grand Canal*.

Its historic attractions include the Piazza San Marco, with St Mark's Cathedral (11th-15th centuries); the Grand Canal, fronted by more than 100 palaces such as the Ca d'Oro (1420-40); the Rialto Bridge; the fishing island of Burano; the island of Torcello with its 7th to 11th-century cathedral; and the Lido di Venezia, the city's elegant beach resort.

Population (city) 100 600; (commune) 332 800
Map Italy Db

Veracruz *Mexico* **1.** The country's most productive agricultural state, stretching for 1600 km (1000 miles) along the south-east coast of the Gulf of Mexico, and inland to the mountains of the SIERRA MADRE ORIENTAL. Rice and bananas are grown in the swampy southern lowlands, and coffee and citrus fruits on the temperate mountain slopes. Maize feeds the people as well as earning them an income. The more arid areas in the north are used for grazing cattle. Rich in oil, Veracruz is also industrially important, with factories and refineries.

2. Mexico's principal port, founded by Hernando Cortes in 1519. It stands at the mid-point of the Gulf state of Veracruz. It seems both Spanish and Afro-Caribbean in character since it was a port of entry for a quarter of a million black slaves in the 16th century. A lively, sunny resort with beaches, arcades and pavement cafés, its pace of life quickens at carnival time in the spring. Veracruz is also important industrially, principally in petrochemicals.

Population (state) 1 200 000; (port) 332 000
Map (state) Mexico Cb; (port) Mexico Cc

Verapaz *Guatemala* Agricultural lowland region between the southern fringe of the Petén rain forest and the northern slopes of the Guatemalan central highlands, about 130 km (80 miles) north of the capital, Guatemala City. It consists of two departments, Alta (High) and Baja (Low) Verapaz. It was known as the Land of War by Spanish conquistadores in the 16th century because of the fierce resistance they met from the local population. But when the Spanish priest Bartolomé de las Casas finally converted the Quiché Indian chiefs to Christianity in 1524 the name was changed to Verapaz – meaning 'True Peace'.

The region is the home of the country's rare national flower, the White Nun orchid, and of the rare quetzal, the sacred bird of the Mayas. Verapaz produces coffee, oil, fruits, sugar and maize. The Languin caves, 80 km (50 miles) north of Guatemala City, are estimated to extend for 400 km (250 miles), with caverns as much as 50 m (165 ft) high.

Population (Alta) 457 600; (Baja) 164 500
Map Guatemala Aa

Verde, Cape (Cap Vert) *Senegal* Atlantic peninsula between the Senegal and Gambia rivers. It formed when a volcanic island became linked to the mainland by sediments. Dakar, the national capital, is on the peninsula's south side, and the western tip, known as Almadies Point, is mainland Africa's most westerly point. The peninsula's coastal road, known as the Route de la Corniche, follows the shore past fine sandy beaches, Dakar's colourful Soumbédioune market, the city's museum and university, the Mamelles – two extinct volcanoes – and the airport at Yoff.

Map Senegal Ab

Verdun *France* Small industrial town in eastern France, 58 km (36 miles) west of Metz. During the First World War the nearby fortifications were the scene of one of the longest and fiercest battles on the Western Front. The battle lasted from February to December, 1916, and cost a total of 350 000 French lives and 330 000 Germans. Huge war cemeteries now mark the battlefield.

Population 24 100
Map France Fb

Vereeniging *South Africa* Industrial town in south Transvaal on the Vaal river, about 55 km (35 miles) south of Johannesburg. The Treaty of Vereeniging ending the Anglo-Boer War (1899-1902) was negotiated there. Its main industries today are coal mining and steel.

Population 60 700
Map South Africa Cb

Verkhoyansk *USSR* Tin and gold-mining centre in the Yakut Republic of north-east Siberia, on the Yana river 120 km (75 miles) north of the Arctic Circle. The surrounding mountains, which rise to more than 2300 m (7550 ft), trap cold Arctic air in the river valley, causing winters of extreme severity; the lowest recorded temperature is −67.5°C (−89.5°F). Oymyakon, in the mountains 630 km (390 miles) south-east, is even colder, with the Northern Hemisphere record of −72°C (−97.6°F).

Population 1600
Map USSR Ob

Vermont *USA* North-eastern state in New England, sometimes called the Green Mountain State after the forested range that runs across it from north to south and rises to 1339 m (4393 ft) in the north. The state covers 24 887 km^2 (9609 sq miles) between the Canadian border and Massachusetts and between New York state and New Hampshire. It was first settled in the 1720s, and became a state in 1791 – the first to do so after the original 13. Cattle, maple syrup and apples are the main farm products, and minerals include marble, slate and granite. Vermont is popular with tourists, particularly for skiing and its autumn colours. MONTPELIER is the state capital.

Population 535 000
Map United States Lb

Verona *Italy* City about 100 km (60 miles) west of Venice. It is based on the Roman grid street plan and contains some fine Roman remains, including a 22 000-seat arena, a theatre carved into the hillside overlooking the Adige river and the Porta dei Borsari gateway. Many Romanesque churches were built in the 11th-13th centuries, some of them unusually striped with red and white stone. The most striking example is the splendid church of San Zeno, flanked on one side by a massive tower and on the other by a slender bell tower.

Verona reached its peak in the 14th century under the Della Scala dynasty; the family's palace, now the Prefecture, still survives. The fine Castelvecchio, now the art museum, overlooks the fortified Scala Bridge, which was rebuilt after its wartime destruction in 1945. Perhaps the most remarkable Della Scala memorials are the so-called Arches of the Scala, *Arche Scaligere*, a series of flamboyantly pinnacled Gothic tombs in the grounds of the church of Santa Maria Antica. The city was the birthplace of the artist Paolo Veronese (1528-88), some of whose pictures hang in the church of San Giorgio Maggiore.

The city's chief industries include textiles, leather goods, chemicals, paper, printing and wine. It is also a busy grain market and tourist centre.

Population 260 200
Map Italy Cb

Versailles *France* Formerly a town, now a south-western suburb of Paris and the site of the magnificent 17th-century Palace of Versailles, built for Louis XIV, who became known as the *Roi Soleil* ('Sun King') for the brilliance of his court. Elegant gardens were laid out in 100 hectares (250 acres) of grounds around the palace, and a new town was built nearby for the palace's servants, soldiers and traders. The palace was ransacked in 1793 during the French Revolution, but was restored after the First World War. The Treaty of Versailles, which officially ended that war, was signed in the palace's Hall of Mirrors on June 28, 1919. Each year almost 2 million people visit the palace, which is now a museum.

Population 95 300
Map France Eb

Vert, Cap *Senegal* See VERDE, CAPE

vesicle Small cavity in volcanic rock formed by a bubble of gas trapped in the lava.

▲ **SUN FOR THE SUN KING In summer, palm and orange trees grace Versailles' Orangery – a courtyard sun-trap at the south end of the great palace leading down to the Swiss Lake (far left).**

Vesterålen ***Norway*** Island group covering 2400 km² (927 sq miles) in north Norway just north-east of the Lofoten Islands. The main port and capital is Harstad.

Population 34 000
Map Norway Da

Vestmannaeyjar ***Iceland*** See WESTMAN ISLANDS

Vesuvius ***Italy*** Volcano about 15 km (9 miles) east of Naples. Its most violent eruption was in AD 79, when it destroyed the Roman towns of POMPEII and HERCULANEUM. There were major eruptions in 1631, 1794 and 1906, and it erupted again in 1944. The road to the 1277 m (4190 ft) summit snakes through fertile vineyards and orchards and then across a black lava desert. A cableway goes up the final 500 m (1650 ft).

Map Italy Ed

Veszprém ***Hungary*** Administrative centre of Veszprém county on the north side of Lake Balaton. The hilly city, 105 km (65 miles) south-west of Budapest, grew on the southern flanks of the Bakony hills around a medieval castle perched on a limestone crag some 30 m (100 ft) high. It has a heavily rebuilt cathedral begun in the 11th century which contains fine collections of vestments and plate; a Baroque bishop's palace, and a 13th-century chapel with superb wall paintings.

New suburbs in the east house workers in engineering and electronics industries. The Herend porcelain factory (with museum) is 10 km (6 miles) west of the city.

Population 63 000
Map Hungary Ab

Viangchan ***Laos*** See VIENTIANE

Viareggio ***Italy*** Seaport and tourist resort about 80 km (50 miles) west of Florence. The English poet Percy Bysshe Shelley (1792-1822) was drowned off the coast there.

Population 58 100
Map Italy Cc

Viborg ***Denmark*** Town and capital of Viborg county, in north-central Jutland. It was a medieval bishopric and has a heavily restored Romanesque cathedral. It was once Jutland's largest town (now overtaken by others). It lies in the middle of the Jutland heaths and is the headquarters of the Danish Heath Society, responsible for the afforestation, reclamation and improvement in the area.

Population 25 500
Map Denmark Ba

Vicenza ***Italy*** City about 60 km (40 miles) west and slightly north of Venice. Vicenza's most famous resident was the architect Andrea Palladio (1508-80), originator of the Palladian style that became fashionable in Europe, and especially in Britain, in the mid-18th century. He gave the city some of his finest works, including the Basilica, the Olympic Theatre, the Palazzo Chiericati, the Capitaniato Loggia and his own house, the Casa del Palladio. More of his elegant villas are scattered around the nearby countryside.

The city's industries include iron and steel, chemicals, machinery, food processing, textiles, furniture and glass.

Population 113 900
Map Italy Cb

Vichy ***France*** Spa and health resort about 120 km (75 miles) north-west of Lyons. Its nine springs have been used since Roman times and they now provide the raw material for a thriving mineral water industry. The factory where the water is bottled is open to the public. In 1940, during the Second World War, the town was made capital of the unoccupied southern zone of France, and seat of the government under Marshal Henri Philippe Pétain (1856-1951). The government was disbanded in 1942 when Germany occupied the whole of France.

Population 64 400
Map France Ec

Vicksburg ***USA*** City at the junction of the Yazoo and Mississippi rivers in western Mississippi state. Its National Military Park has preserved the Civil War fortifications of the 1862-3 siege, which culminated in the North winning control of the Mississippi river.

Population 26 200
Map United States Hd

Victoria ***Australia*** State covering 227 600 km² (88 875 sq miles) in the extreme south-east. Formerly part of New South Wales, it became a separate colony in 1851. The state capital, MELBOURNE, lies on the coastal lowlands, backed by the Great Dividing Range. Beyond are the arable and grazing country of the inland plains. Sheep, cattle and wheat are the state's agricultural mainstays.

Gold was discovered in 1851 at BALLARAT and BENDIGO, starting a rush, but today brown coal is the chief mineral, although gas and oil deposits are now being exploited. After New South Wales, it is the country's second most important industrial state. Nearly two-thirds of the population live in Melbourne.

Population 4 053 400
Map Australia Gf

Victoria ***Canada*** Port city and capital of BRITISH COLUMBIA, at the southern end of Vancouver Island. It is a business and tourist centre, the Pacific headquarters of the Canadian navy and the base for a deep-sea fishing fleet. There is also a university. Industries include timber processing, shipbuilding and papermaking.

The city's many parks and gardens, its notable Victorian architecture, and its mild climate have made it popular with tourists. It is linked by ferries to VANCOUVER on the mainland and to SEATTLE in the USA.

Population 233 480
Map Canada Cd

Victoria ***Hong Kong*** Seaport city and capital of the British Crown Colony of Hong Kong, on the north-west coast of HONG KONG ISLAND overlooking Victoria Harbour. One of the most prosperous cities in the Far East, Victoria began as a settlement on land reclaimed from the sea after Britain acquired Hong Kong Island in 1842. Its fortunes have been inextricably linked with Victoria Harbour, one of the greatest natural harbours in the world, which is used by more than 10 000 ocean-going ships every year as well as myriads of smaller craft engaged in local trade. Victoria is connected to KOWLOON, on the other side of the harbour, by ferries, a railway and a cross-harbour road tunnel.

Central Victoria contains the island's business section, while towards Kennedy Town in the west and Causeway Bay in the east, densely populated tenements are mixed with offices and factories. Several streets run parallel with the shoreline, each representing a successive stage of land reclamation. Behind the waterfront zone, narrow streets hemmed in by high tenements run up the steep slope of VICTORIA PEAK. These 'ladder streets', as they are sometimes called, are filled with dozens of curio stores, tiny pavement workshops and crowded open-air markets.

Higher up the slopes of the mountain are the colony's Botanic Gardens, Government House and Hong Kong University, all of which have superb views of the bustling city and harbour lying below.

Population 1 026 900
Map Hong Kong Cb

Victoria *Malta* Capital of the charming island of Gozo. It is usually called by its older name of Rabat by the locals. The citadel hill, known as Gran Castello, was the centre of the Norman town and now houses the Law Courts, the Gozo Museum, the cathedral and the Old Bishop's Palace. The modern town is built round the shady main square, it-Tokk. Old Town, a tangled knot of balconied streets, lies to the south and west. It has many goldsmiths' workshops.

Population 5410
Map Malta Ca

Victoria *Seychelles* Capital, chief port and commercial and tourist centre of the Seychelles. It stands on the north-east coast of MAHE, with a background of mountains.

Population 25 000

Victoria, Lake *Kenya/Tanzania/Uganda* The world's second largest sheet of fresh water after Lake Superior in North America. It covers 69 500 km² (26 800 sq miles) – about the size of the Republic of Ireland – but is nowhere deeper than 80 m (265 ft). It is one of the main sources of the Nile, the lake's only outlet, which flows out of the lake beside the Ugandan town of Jinja on the north shore.

The lake – discovered and named in 1858 by the British explorer John Speke – has provided hydroelectric power since completion of the Owen Falls Dam just down the Nile in 1954. The dam drowned the falls, some 20 m (65 ft) high. The lake's main value, however, to the people who live around its shores is as a source of fish. Shipping services link the towns and countries beside the lake.

Map Uganda Bb

Victoria Falls *Zambia/Zimbabwe* One of the world's great waterfalls, plunging 108 m (354 ft) on the Zambezi river near the north-west tip of Zimbabwe. The first European to see them, in 1855, was the Scottish missionary and explorer David Livingstone. The local name for the falls, *Mosi-oa-Toenja*, means 'the Smoke that Thunders', referring to the clouds of spray that soar up to 500 m (1600 ft) above the chasm into which the river tumbles.

Map Zimbabwe Ba

Victoria Peak *Hong Kong* Mountain (551 m, 1808 ft) in the north-west of HONG KONG ISLAND. It is the highest point on the island, and was the last place to be surrendered to the Japanese by the British in the fall of Hong Kong on Christmas Day 1941. From the summit, there are fine views over VICTORIA, KOWLOON and, on a clear day, the NEW TERRITORIES.

Map Hong Kong Cb

Vidarba *India* See BIDAR

Vidin *Bulgaria* River port and capital of Vidin province, on the Danube river near the Yugoslav border 280 km (175 miles) west of Ruse, to which there is a connecting steamer service along the river. It trades in wine, grain and olives. A Shakespeare festival is held each July in the medieval citadel of Baba Vida, built on the site of the Roman town of Bononia.

Population 61 200
Map Bulgaria Ab

▼ A CITY REBORN The fine dome of the Karlskirche towers over Vienna's magnificent Baroque centre, built after the devastating Turkish siege of 1683.

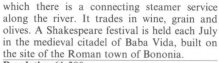

Vienna (Wien) *Austria* Once the capital of the mighty Austro-Hungarian Empire (1867-1918), Vienna now rules over Austria alone and is the capital that is almost too large for its country. Set mostly on the right bank of the Danube in north-east Austria, Vienna is also a separate state and the capital of LOWER AUSTRIA. Divided into 23 districts, the city is grouped in two sweeping circles around the historic Inner City.

The modern city grew after the removal, in the late 1850s, of the ramparts guarding the Inner City and the building of the 4 km (2.5 mile) long Ringstrasse boulevard along their site. Most of the city's main and most imposing buildings are lined along the Ringstrasse. They include the handsome Opera House; the Hofburg, the old imperial palace (another stands in formerly suburban SCHONBRUNN); the neo-Gothic Rathaus (City Hall); the Greek-style Houses of Parliament; the Burgtheater (City Theatre) and the Musikverein, home of the Vienna Philharmonic Orchestra. In the centre of the city is the Cathedral of St Stephen, Austria's finest Gothic church, built in the 14th and 15th centuries.

For centuries, Vienna has been a meeting-place for East and West and it now has the headquarters of several world agencies, including the Organisation of Petroleum Exporting Countries and the UN International Atomic Agency. Recently, Vienna has expanded along the main route to the south. Large new shopping complexes have gone up, rivalling the busy shopping street of Mariahilferstrasse and the traditional shops found in Kärntner Strasse and Am Graben.

Although Vienna is noted for its industries – which include clothing, electronics, paper, chemicals and beer – it is one of the world's great tourist centres. People flock there to savour the atmosphere of a city which began life as a Celtic settlement, was later a Roman camp, and from 1273 to 1918 was the home of the powerful Hapsburg dynasty. At the end of the First World War it became the capital of Austria, and in March 1938 the Nazis entered the city and Austria became part of Germany. Because of this, Vienna was severely bombed by the Allies in the Second World War.

In 1945 it was split into four zones which were occupied until 1955 by Britain, the USA, USSR and France. It was against this background that the English writer Graham Greene set his script for the thriller film *The Third Man*, complete with its catchy theme tune played on the zither. Indeed, Vienna is a city of music, the home of the waltz and the birthplace in 1825 of Johann Strauss the Younger, who wrote more than 400 waltzes, including *The Blue Danube* and *Tales from the Vienna Woods*. Musicians' Square, in the Central Cemetery, is the burial place of Beethoven, Brahms and Schubert, whose music is regularly heard in the city.

Population 1 531 000
Map Austria Ea

Vientiane (Viangchan) ***Laos*** Northern province and the nation's capital city. The city is on the north bank of the Mekong river opposite Thailand. The city is about the size of a French provincial town and, before the Communist takeover in 1975, it had much of the flavour of France, with pavement cafés, vendors selling French patisseries, and even its own smaller Arc de Triomphe monument, modelled loosely on the arch in Paris. The city also contains several Buddhist temples and shrines including the That Luang pagoda, which is said to contain relics of the Buddha. The province of Vientiane is mostly a low-lying plain which is partly irrigated by the Nam Ngum river and is the country's most densely populated area.

Population (province) 377 400; (city) 177 000
Map Laos Ab

Vierwaldstätter See ***Switzerland*** See LUCERNE, LAKE

Vietnam See p. 694

Vigo ***Spain*** Atlantic port and resort about 15 km (10 miles) north of the Portuguese border. It has fish processing, shipbuilding, oil refining and distilling industries, and manufactures chemicals, soap and cement.

Population 258 700
Map Spain Aa

Viipuri ***USSR*** See VYBORG

Vijayanagar ***India*** Ruined capital of the great Hindu empire of the same name in southern India. It lies at Hampi, about 540 km (335 miles) south-east of Bombay. The empire, founded in 1336, fought off Muslims from the north for 200 years, and was finally destroyed by them in 1565. The ruins, a massive fort and temple complex, include palaces and the elephant stables of the royal retinue.

Map India Bd

Vijayapura ***India*** See BIJAPUR

Vijayawada (Bezwada) ***India*** Town in the south-east at the head of the Krishna river delta, about 385 km (240 miles) north and slightly east of the city of Madras. It is the focus of an ancient system of irrigation canals which are still in use.

Population 543 000
Map India Cd

Vijosë (Aóös Potamós) ***Albania/Greece*** Southern Albania's largest river, 269 km (167 miles) long. It rises in the Greek Pindhos mountains and flows north-west into the Adriatic, north of VLORE. Its valley provides a comparatively easy route for a recently constructed international road from Yannina in Greece to Korçe, Vlorë and Durrës in Albania.

Map Albania Cb

Vila ***Vanuatu*** See PORT-VILA

Vila Pery ***Mozambique*** See CHIMOIO

Vila Viçosa ***Portugal*** Town about 150 km (95 miles) east of the capital, Lisbon. From the 15th century it was the seat of the Dukes of Braganza. There is a 13th-century castle, modified in the 17th century. The ducal palace, dating from the 16th century, contains personal possessions of the royal family.

Population 4400
Map Portugal Cc

Villa Cisneros ***Western Sahara*** See AD DAKHLA

Villahermosa ***Mexico*** Oil town on the Grijalva river and capital of the state of TABASCO. Until 1915 the town was known as San Juan Bautista, and its original charm as an agricultural centre was destroyed by the arrival of the Nationalised Mexican Petroleum Industry (PEMEX) in the late 1960s.

Now ablaze with the burning chimneys of the oil-processing plants, Villahermosa is noisy, crowded with migrant oil workers and, despite its wealth, shabby. Ironically, its name means 'beautiful town'.

Population 250 900
Map Mexico Cc

Villány ***Hungary*** See PECS

Villmanstrand ***Finland*** See LAPPEENRANTA

Vilnius (Vilna; Wilno) ***USSR*** Capital of Lithuania, 165 km (103 miles) north-west of Minsk. Vilnius was founded in the 13th century and was the capital of independent Lithuania from 1323 to 1795. Its extensive Old Town surrounds the ruins of a 14th-century castle, and there is also a mosque, a synagogue and almost 40 churches.

The city's university is the oldest in the USSR; it was first founded in 1579, but was closed by the Russian government in 1832 because of the 'revolutionary activities' of the students. It was reopened by the Polish government, which ruled the city from 1919 until 1939 – when it was occupied by Russian troops.

Vilnius now manufactures farm machinery, machine tools, electrical products, textiles, chemicals and processed food.

Population 536 000
Map USSR Dc

Viña del Mar ***Chile*** Picturesque city and popular seaside resort, 9 km (5.5 miles) north of Valparaíso. The president's summer palace and the Quinta Vergara, scene each February of an international song festival, are among the splendid buildings that stand on its palm-fringed avenues and plazas.

Population 300 000
Map Chile Ac

Vinnitsa ***USSR*** City in the Ukraine, set among rich farmland 190 km (120 miles) south-west of Kiev. The surrounding area is noted for its pottery, embroidery, woven goods and carpets. Vinnitsa has the USSR's largest sugar refineries, and it manufactures fertilisers, farm machinery, shoes and clothes.

Population 360 000
Map USSR Dd

Vinson Massif ***Antarctica*** Highest peak in the Antarctic, rising to 5140 m (16 863 ft). It is part of the Ellsworth Mountains at the southern end of the Antarctic Peninsula.

Map Antarctica Ec

Virgin Islands, British ***West Indies*** British Crown Colony in the Caribbean consisting of some 40 islands and cays in the LEEWARD ISLANDS, to the east of the US Virgin Islands. Their total area is 130 km^2 (50 sq miles). Only 16 of the British Virgin Islands are inhabited, among them TORTOLA (the biggest), Virgin Gorda, Anegada and Jost Van Dyke. Christopher Columbus discovered the islands in 1493, and British settlers took control from the original Dutch colonists in 1666. There is some farming but tourism has become the chief employer since the mid-1960s, aided by the islands' position close to the United States and to the sheltered waters which provide some of the best sailing in the world. About two-thirds of the visitors stay on yachts. The colony operates a strict conservation policy designed to preserve the beauty and diversity of the islands' coral reefs. The capital, ROAD TOWN, is on Tortola, which is also the main tourist island.

Population 12 000
Map Caribbean Cb

Virgin Islands, US ***West Indies*** American dependency consisting of three main islands – ST THOMAS, ST CROIX and ST JOHN – and more than 60 mainly uninhabited cays and small islands in the Caribbean. They lie in the LEEWARD ISLANDS to the east of Puerto Rico; the main islands total 342 km^2 (132 sq miles). They were discovered by Christopher Columbus in 1493 and settled by the Dutch, British and French in turn before they were acquired by Denmark. They were then bought in 1917 by the USA for US$25 million. The islands are of strategic importance to the United States because they command the deep Anegada Passage from the Atlantic Ocean to the Caribbean Sea and its approach to the Panama Canal.

The inhabitants of the Virgin Islands were made American citizens in 1927, but constitutionally the islands are an 'unincorporated territory'. Like those of the smaller neighbouring British Virgin Islands, the US islanders are mostly descendants of African slaves, brought from the USA in the 18th century when the islands' economy depended on sugar.

Geologically, both the British and the US islands are the peaks of a volcanic ridge running eastwards below sea level from Puerto Rico. Agriculture is limited by the islands' low and unreliable rainfall. There is some commercial fishing, but game fishing is more important.

Tourism is by far the most important industry. Cruise ships bring large numbers of visitors, and there are also airborne day-trippers who take advantage of the duty-free shopping on St Thomas. Other islands are less busy and offer a quiet holiday with clear seas and good sailing. The local population is too small to cope with the tourist influx, and as a result, about one in three of the present residents of St Thomas are Puerto Ricans or citizens of other Caribbean islands.

Watches, textiles and pharmaceuticals are also manufactured on the island with imported materials, and rum is a valuable export. St Croix has one of the world's largest oil refineries, and an alumina plant which processes African bauxite for refining into aluminium in the United States.

Population 100 000
Map Caribbean Cb

Vietnam

TRIUMPHANT BUT POVERTY-STRICKEN, VIETNAM'S COMMUNISTS HAVE MORE BATTLES TO FIGHT

In this South-east Asian country, formerly the separate republics of North and South Vietnam, the United States for the first time lost a war. Its forces fought for eight years against the Viet Cong, a Communist guerrilla movement supported by the North Vietnamese government. The Americans dropped 7 million tonnes of bombs, sprayed defoliant chemicals on 16 200 km^2 (6250 sq miles) to strip the land of vegetation, and sacrificed 57 000 lives. But the most modern machinery of death could not prevail against the North Vietnamese guerrillas.

About 2 800 000 Americans served in Vietnam during 1965-73. They went in to preserve pro-Western South Vietnam's independence from the Communist North. The theory was that, if South Vietnam fell, other South-east Asian states would follow, like falling dominoes. In the slaughter that followed, 2 million Vietnamese – capitalist and Communist – died, and as the weary years dragged on, America lost the will to win. After a temporary truce and two years after American forces withdrew, South Vietnam finally surrendered on April 30, 1975. Vietnam was formally unified under Communist control in 1976.

Whether for economic or political reasons, hundreds of thousands of Vietnamese fled the country in small boats after the Communist takeover, mainly in 1978 and 1979. Many of these 'boat people' died in storms or at the hands of pirates, but the survivors posed serious refugee problems to neighbouring countries before, gradually, they were resettled in places as far apart as Australia, Canada, France and the United States.

ENDLESS WAR

Vietnam has known war for almost half a century. In 1940, when it was a French colony, the Japanese invaded the country. They were defeated in 1945, and nationalists led by Ho Chi Minh took to arms to prevent the restoration of French rule. The French withdrew in 1954 after the Battle of DIEN BIEN PHU, but international agreements left the country divided along the 17th parallel (17°N), and the Communist North began a 20 year struggle against South Vietnam, aimed at reunification. By 1965 there was full-scale war.

Ever since 1975, Vietnamese forces have been involved in fighting in neighbouring CAMBODIA, where they dislodged the Communist Khmer Rouge regime of Pol Pot and installed their own government. On Vietnam's northern border, China (which supports the Khmer Rouge) launched diversionary attacks in support of its Cambodian allies.

This embattled history emphasises two important characteristics of modern Vietnam: first, the all-pervasive presence of the military in the country, for ever since 1975 Vietnam has maintained a standing army of 1 200 000 which is used in all aspects of the country's life; second, the extraordinary patience and resilience of a whole generation of Vietnamese people, who have never seen real peace, in the face of the country's desperate problems.

Vietnam is one of the poorest countries in Asia. For most of its 62 million people, life remains a struggle, despite the fact that armed conflict is now on or beyond the country's borders. Its continuing involvement in Cambodia still drains its resources; in 1980, when the world's attention was drawn to the plight of the starving Cambodian people, Vietnam itself was suffering from food shortages that brought reports of widespread malnutrition among the young.

Many other basic goods remain in short supply, from petrol to drugs, paper to cement. Much of the country's industry – largely based in the north – was damaged by American bombing and Chinese incursions in 1979.

Only in the cities of former South Vietnam are consumer goods more widely available, particularly in the former capital Saigon, now renamed HO CHI MINH CITY after North Vietnam's great nationalist leader. Some goods were left by the Americans, but most are sent by the thousands of Vietnamese who fled the country to the families they left behind.

Vietnam today is very much a tale of two cities. In the south, Ho Chi Minh City still retains something of the atmosphere of pre-1975 times. Then, swollen with refugees from the countryside, Saigon was a brash and noisy place, full of the trappings of Western civilisation, a playground for large numbers of American troops. The city made its living from the artificial economy created by the war. Today the only foreigners are a few Western journalists and visitors from the Eastern bloc. But the young people of Ho Chi Minh City can still be seen in the roadside cafés listening to popular Western music; the moped, introduced by the Americans, is still the typical means of transport; and street stalls sell a range of goods from Thai cigarettes to Japanese cassette recorders.

By contrast, HANOI, in the north, is drab and strait-laced, with little of the life of its southern counterpart. Few black market goods are available and bicycles rather than mopeds provide transport. The Lao Dong, the Vietnamese Workers' (that is, Communist) Party which controls Vietnam, is everywhere.

Vietnam is also a tale of two rivers. The country, which lies on the eastern coast of Indo-China looking out over the South CHINA SEA, is shaped like a dumbbell, with a narrow central area linking the broader plains to the north and south. The plains are centred on

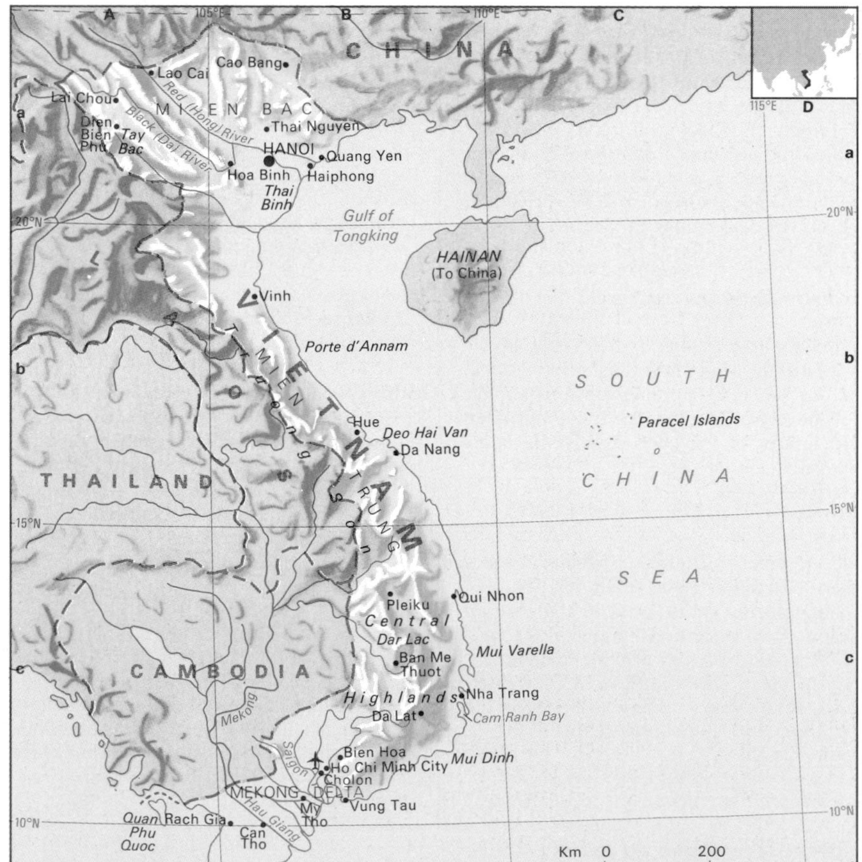

the valleys of the RED and MEKONG rivers. The Red River (or Song Hong) is largely confined to Vietnam, but the Mekong starts in China and flows through Burma and Laos, along the border of Thailand and through Cambodia before reaching the sea south of Ho Chi Minh City. The two river plains are separated by a narrow, hilly central zone, which has always made communications between north and south a problem – a significant factor in Vietnamese history.

This narrow 'handle' of the dumbbell – once known as Annam and crossed by the 17th parallel – divides Vietnam not only geographically, but also culturally. The true Vietnamese are a Mongoloid people originating from GUANGDONG province of southern China. They settled in the Red River valley (the area subsequently called TONKIN) over 2000 years ago at a time when the south was dominated by the Cham and Khmer peoples. It was only in the 17th century that the Vietnamese began to move southwards in any numbers, absorbing the Chams and pushing back the Khmers to the TONLE SAP basin to the west. Even then, the migration was limited before the arrival of the French in the middle of the 19th century. By 1867 the south had become the French colony of COCHIN CHINA, and by 1884 Tonkin and Annam were French protectorates.

CONTRASTING RICELANDS

The valley of the Red River is the true heartland of Vietnam. Perhaps the resilience of the north Vietnamese over the past 40 years has something to do with their long experience of this wild and dangerous river which takes the name Red from the heavy load of silt it carries, far out of proportion to its length. Brought down from the hills of north-west Vietnam and China, this silt is redeposited as the river changes gradient in the northern plains of Vietnam. Over the years it has built up the bed of the Red River and its tributaries to 12 m (40 ft) above the level of the surrounding plain. Throughout history, the Vietnamese have faced the danger of flooding. The French, in particular, attempted to strengthen the river's levees or dykes and raise them to a safe level, but breaches and floods occur regularly, with disastrous consequences for the surrounding farmland. Their vulnerability to bombing during the war can be imagined.

However, the waters of the river and its chief tributaries are easily led off for use in agriculture, and in many parts of north Vietnam two crops of rice are grown each year. The higher areas are cultivated at the peak of the monsoon from May to November, while low-lying lands yield a winter crop after the floods have subsided. Tonkin is heavily populated, with over 1100 people per km² (2850 per sq mile) in the province of Thai Binh, and modern intensive rice-growing techniques imported from China are used. Intensive cultivation is also found in the coastal lowlands in central Vietnam. Here smaller streams which fall steeply from the hills of the ANNAMITE CHAIN provide irrigation water for the plains. In some places five crops of rice can be grown in two years.

After North Vietnam became independent in 1954, agriculture in this area was organised into cooperatives. Groups of villages held and worked their land in common to fulfil production targets set by the central government.

This system seemed to work in the war years, when food production in the north increased rapidly. But after the war the farmers were reluctant to produce more food without incentives. Since 1979 farm prices have been raised and the cooperatives have turned over land to individual families on a contract basis. A share of the produce is given to the state and the rest can be sold.

▲ RED REMEMBRANCE On Lenin's birthday in the Vietnamese capital Hanoi, workers pass portraits of the grim-faced Russian leader. One sports a 1975 vintage Western suit – rarer in the strait-laced north than in freewheeling Saigon, now Ho Chi Minh City.

The southern delta of the Mekong shares some features with the north, notably the importance of waterborne transport, but the landscape is very different. When the Mekong reaches southern Vietnam it is a wide and majestic river that has already begun to separate into different branches. Much of the land through which it passes is swampy, and sea water intrudes when the river's flow is at its lowest. The higher parts of the delta were drained by the French, and much of the new land was turned into large estates which produced rice for sale to the north and for export. The tradition of commercial farming has remained, and since 1975 it has proved difficult to impose cooperatives on this area; over half is still privately farmed.

Since 1975, the government has tried to reduce overcrowding in the north by resettling people in the central hill areas, disrupting the way of life of the local tribes. Despite opposition the plan has gone ahead, with the military helping to clear the difficult and malarial terrain. Such development needs outside help, and thus foreign exchange to pay for it. Eastern Europe is the only source of such aid since Vietnam's intervention in Cambodia. The Vietnamese do not welcome their role as a provider of tropical raw materials to the Communist Bloc, but it seems a necessity if this overcrowded country is to recover from its long history of warfare and build a new life for its long-suffering people.

VIETNAM AT A GLANCE
Area 329 565 km² (127 245 sq miles)
Population 61 950 000
Capital Hanoi
Government Communist republic
Currency Dong = 10 hao = 100 xu
Languages Vietnamese, Chinese, French, English
Religions Buddhist (55%), Christian (8%)
Climate Tropical; temperatures vary much with seasons only in the north. Average temperature in Hanoi ranges from 17°C (63°F) in January to 29°C (84°F) in June
Main primary products Rice, cassava, sweet potatoes, bananas, maize, pineapples, coffee, tea, rubber, tobacco, pigs, cattle, timber, fish; coal, anthracite, lignite, iron, chrome, manganese, titanium, gold, phosphates, limestone, salt
Major industries Agriculture, mining, food processing, textiles, cement, cotton and silk manufacture, forestry, paper
Main exports Coal, clothing, rubber, tea, coffee
Annual income per head (US$) 150
Population growth (per thous/yr) 24
Life expectancy (yrs) Male 54 **Female** 58

Virginia ***USA*** State covering 105 716 km² (40 817 sq miles) mainly west of Chesapeake Bay on the Atlantic coast. It reaches into the Appalachians, and includes parts of the Blue Ridge and Allegheny mountains, which rise to 1746 m (5729 ft). Dairy goods, tobacco and chickens are the chief farm products, and coal is the main mineral. Manufactures include tobacco and food products, and chemicals. In 1606 Virginia was the first American colony to be chartered from England. It was settled the following year, and became one of the original 13 states. In 1789 it ceded territory for the new federal capital of Washington, and during the Civil War (1861-5) the western part remained loyal to the Union while the rest of Virginia joined the Confederacy; as a result the separate state of WEST VIRGINIA was formed in 1863.

Virginia is sometimes known as the Mother of Presidents, for eight presidents – including the first, George Washington (1732-99) – were born there. The attractive scenery and coastline and the state's many historic associations make it popular with tourists. RICHMOND is the state capital and NORFOLK the largest city.

Population 5 706 000

Map United States Kc

Virunga National Park ***Zaire*** Reserve covering some 7600 km² (2934 sq miles) on the Ugandan and Rwandan borders. It includes the former Albert National Park, the Ruwenzori range, Lake Edward and the Parcs des Volcans, incorporating the volcanic Virunga mounts, two of whose peaks (NYIRAGONGO and Nyamlagira) are active. Pygmies live in parts of the park which shelters abundant wildlife, including elephants, giraffes, hippopotamuses, leopards, lions and okapis.

Map Zaire Bb

Visayan Islands (Visayas) ***Philippines*** Island group forming the central part of the Philippine archipelago. The main islands are BOHOL, CEBU, LEYTE, MASBATE, NEGROS, PANAY and SAMAR. Ferdinand Magellan's first landfall after crossing the Pacific Ocean in 1521 was on the small island of Homonhon, which lies in Leyte Gulf south of Samar.

Map Philippines Bc

Visby ***Sweden*** Main port and seaside resort on the Baltic island of Gotland. In the Middle Ages it was one of the richest trading centres in northern Europe, controlling much of the Baltic and minting its own money. Its 3 km (2 miles) of medieval walls are studded with more than 40 massive towers. Within them, the old city has many churches and a 12th to 13th-century cathedral.

Population 19 300

Map Sweden Cd

Visegrád ***Hungary*** Historic village with fewer than 3000 inhabitants overlooking the great Danube bend, where it turns south 43 km (27 miles) north of Budapest. It was once a thriving medieval town and royal residence, and is now popular with tourists for its fine views, ruined 14th-century citadel, and below it, the restored sections of a magnificent 350 room palace built by the Hungarian King Matthias Corvinus in the 15th century.

Map Hungary Ab

Visegrad ***Yugoslavia*** Town at the lower end of the Drina river gorge in eastern Bosnia Herzegovina, 72 km (45 miles) east of Sarajevo. Its graceful white bridge, built by the Turks in 1577, has ten arches and is 180 m (591 ft) long. The town has timber and machinery factories, and is a tourist centre, especially for visitors to the lake behind the hydroelectric dam downstream, and for holidaymakers heading for the Tara Planina, a mountain resort area to the north-east.

Map Yugoslavia Dc

Viseu ***Portugal*** City about 75 km (45 miles) south-east of the city of Oporto, in the centre of the Dão wine region. The old, mainly medieval town, which was founded in Roman times, has attractive narrow streets. Its cathedral was built between the 16th and 18th centuries.

Population 21 000

Map Portugal Cb

Vishakhapatnam (Vizagapatam) ***India*** Sugar-exporting port on the east coast, about 610 km (380 miles) north-east of the city of Madras. It has India's most important shipyards, an oil refinery and is a naval base.

Population 603 600

Map India Cd

Visla ***Poland*** See VISTULA

Viso, Monte (Monviso) ***Italy*** Highest peak of the Cottian Alps rising to 3841 m (12 602 ft) near the French-Italian border about 65 km (40 miles) south-west of Turin. The PO river rises on its eastern slopes.

Map Italy Ab

Vistula (Wisla; Visla; Weichsel) ***Poland*** The country's longest river (1047 km, 651 miles). It rises near the Czech border south-west of the city of Cracow and flows roughly north and east through Cracow and Warsaw. Near its mouth, a canal takes the river directly into the Gulf of Gdansk, bypassing the Vistula lagoon and Soviet territory.

Map Poland Cb, Dc

Vitebsk ***USSR*** Belorussian city on the West Dvina river about 470 km (290 miles) west of Moscow. Vitebsk was a trading centre in the Middle Ages and was later occupied by Lithuania, Poland and finally Russia in 1772. Its industries today include machine tools, radios, hosiery, furnishing and processed food.

Population 330 000

Map USSR Ec

Viterbo ***Italy*** Well preserved medieval town about 65 km (40 miles) north and slightly west of Rome. Among its splendours are the 12th-century Gothic cathedral, the 13th-century episcopal palace and town hall, and the 15th-century Farnese palace. There are a number of beautiful churches dating from the 9th to the 16th centuries. Its manufactures include furniture, textiles and pottery.

Population 57 800

Map Italy Dc

Viti Levu ***Fiji*** Main island of Fiji, covering 10 429 km² (4027 sq miles). It is mountainous, rising to 1324 m (4344 ft) at Tomaniivi (Mount Victoria). On the eastern side, facing the prevailing trade winds, very heavy rain falls, and there are dense rain forests. The western side is drier but still receives substantial rainfall; this is the main sugar-growing region, centred on LAUTOKA. Fiji's capital, SUVA, is on the south-east coast. There are many resorts along the Coral Coast west of Suva, on the low-lying islands off the west coast and in the vicinity of the international airport at Nadi.

Population 500 000

Map Pacific Ocean Dc

Vitória ***Brazil*** State capital and port about 400 km (250 miles) north-east of Rio de Janeiro. The port, which is on a beautiful offshore island, exports coffee, along with timber and iron ore from the mountains of Minas Gerais.

Espírito Santo, the agricultural state of which Vitória is capital, covers 45 597 km (17 605 sq miles). Its main products are coffee and tropical hardwoods.

Population (city) 208 000; (state) 2 012 000

Map Brazil Dd

Vitoria ***Spain*** Basque city, capital of Álava province, about 60 km (35 miles) south-east of the Bay of Biscay port of Bilbao. It manufactures bicycles, furniture and agricultural machinery, and has sugar refining, tanning and distilling industries. Founded in 1181, it has a 12th-century fortress-cathedral and a still unfinished New Cathedral. The British, under Sir Arthur Wellesley and helped by Spanish auxiliaries, defeated the French nearby during the Peninsular War in 1813.

Population 192 800

Map Spain Da

Vlaanderen ***Belgium*** See FLANDERS

Vladimir ***USSR*** Tourist city 160 km (100 miles) east of Moscow. Visitors flock there to see its fine kremlin (citadel), its imposing Golden Gate, built in 1158-64 as a fortified entrance on the road from Moscow, and its two 12th-century cathedrals, St Dmitri and Uspenski Sobor, which were restored in the 19th century. Vladimir is also a prosperous manufacturing city with chemical, engineering, electrical and textile industries.

Population 326 000

Map USSR Fc

Vladivostok ***USSR*** Largest port on the Soviet Pacific coast, a naval base and shipyard only 50 km (30 miles) east of the Chinese border and 135 km (85 miles) from North Korea. Its products include machinery, timber, oil refining, beer, tea and flour. The city stretches around a sheltered harbour on the Sea of Japan. It was founded in 1860 and became the base for the Russian Pacific Fleet in 1872. It also has fishing and whaling fleets. In 1903 it became the terminus of the Trans-Siberian Railway; however, it is a restricted military area and foreigners must use nearby NAKHODKA.

Population 590 000

Map USSR Od

Vlissingen ***Netherlands*** See FLUSHING

vloer In South Africa, a salty clay flat on an arid plateau.

Vlorë (Valona) ***Albania*** Port in southern Albania. Like DURRES, the country's chief port, Vlorë has been occupied and influenced by a series of foreign masters, including the Venetians and Turks. The main town developed amid olive groves, about 3 km (2 miles) inland from a small fishing village. After the Communist takeover in the 1940s, it was expanded to serve Albania's industrial development. Bitumen, oil, rice and cotton flowed out to be traded for imports of Soviet, and later Chinese, equipment.

Population 66 400

Map Albania Bb

Vlotslavsk ***Poland*** See WLOCLAWEK

Vltava (Moldau) ***Czechoslovakia*** Bohemian river, 433 km (269 miles) long. It rises south-west of Prague, near the West German border, and flows through the city to join the Elbe. It is dammed for hydroelectric power at six places.

Map Czechoslovakia Bb

Vojvodina (Voyvodina) ***Yugoslavia*** Autonomous province of Serbia, covering 21 506 km^2 (8301 sq miles) and lying north of the Sava river. It is drained by the Danube and Tisza rivers. Apart from the hills of the Fruska Gora, the volcanic heights of Vrsac Planina and the sandy hills of Deliblatska Pescara between Vrsac and Belgrade, it is flat and extremely fertile. The province is Yugoslavia's granary and sugar bowl, and a major producer of vegetables, sunflower seed, soya beans and meat.

Only about half of the people in Vojvodina are Serbs. The province was formerly part of Hungary and ethnic Hungarians now account for 19 per cent of its inhabitants, the remaining 27 per cent being largely Croat, Slovak or Romanian.

The province's major towns include Novi Sad, the capital, Subotica, Senta, Zrenjanin and Vrsac, renowned for its September Grape Harvest and Folk Festival. The province is crossed by the 750 km (466 mile) long Veliki Backu-Vrsaki Canal, which creates a short cut from the Danube near the Hungarian border to the Tisza river at the town of Becej and on south-eastwards to the Danube just above the Djerdap (Iron Gates) gorge.

Population 2 035 000

Map Yugoslavia Db

Volcanic Plateau ***New Zealand*** Area of volcanic mountains and hot springs around Lake Taupo in the centre of North Island. The most active volcano is Mount Ngauruhoe at 2291 m (7516 ft). In the 1920s and 1930s, 172 000 hectares (425 000 acres) of the plateau were planted with radiata pine trees – today the Kaingaroa Forest created by the planting is the basis of a flourishing timber industry. Pastoral farming also thrives in the area.

Map New Zealand Ec

volcano A fissure at a weak point in the earth's crust through which lava – molten rock from the depths – ash and gases reach the surface. The term – derived from Vulcan, the Roman god of fire – is applied both to the fissure and to the conical mountain of lava and other debris that forms about it. Volcanoes are classified as active, dormant or extinct. It is estimated that there are some 500 active volcanoes on earth, but the other two categories are uncertain at best. A volcano may remain dormant for thousands of years – a short time in geological terms – and then, without warning, erupt violently. See HOW A VOLCANO IS CREATED (p. 698)

Volcano Islands ***Japan*** See IWO JIMA

Volga ***USSR*** Europe's longest river and a trade route between Europe and Asia since the Middle Ages. It rises in the Valdai Hills about 300 km (185 miles) north-west of Moscow and runs south-eastwards for 3688 km (2292 miles) – through Kalinin, Yaroslavl', Gor'kiy, Kazan', Ul'yanovsk, Kuybyshev, Saratov and Volgograd – into the Caspian Sea. It is navigable for most of its length, and was used by the tsar Ivan the Terrible in 1552-6 to ship supplies and reinforcements to his troops as he expanded his empire along the river against fierce Tatar opposition.

Since 1930 dams have been built at several points on the Volga to generate power and irrigate nearby steppes, and to provide road and rail crossings. Canals built in the 1950s and 1960s link the Volga to the Don river and to the Baltic and White seas, enabling ocean-going ships to sail into the USSR's industrial heartland in the Donets Basin.

Map USSR Ec, Fc

Volgograd ***USSR*** Major port on the Volga river at the eastern end of the Volga-Don Canal, about 440 km (275 miles) from the Caspian Sea. It was founded in 1589 as Tsaritsyn, a fort designed to protect trade along the river from marauding Cossacks. In 1925 it was renamed Stalingrad in honour of the Soviet leader Joseph Stalin (1879-1953); in the 1918-20 civil war he had organised the defence of the city for the Bolsheviks against the White Army, which included the former forces of Tsar Nicholas II (1868-1918).

In the Second World War – from August 1942 to February 1943 – it was the scene of fierce fighting as the Russians gradually recaptured the city from the Germans, who had occupied parts of it after a siege lasting for more than two months. During the Battle of Stalingrad some 350 000 German troops were cut off and mostly killed – and the city was virtually destroyed. It was rebuilt after the war, and there is now a large park beside the Volga with monuments to the Russians who died in the battle.

The city was renamed Volgograd in 1961, five years after Stalin's rule was officially denounced at the 1956 Communist Party Congress. Its industries today include oil refining, petrochemicals, steel, shipbuilding, agricultural machinery, building materials, fibres, textiles, shoes and processed food.

Population 989 000

Map USSR Fd

Volhynia ***USSR*** Fertile, wooded region in the western Ukraine. Its rolling hills stretch west to Poland, which ruled the region in the 16th century. Previously it had been an independent principality, then from 1340 part of Lithuania. Volhynia became part of Russia in 1793-6, and from the early 1920s it was shared by the USSR and Poland. In 1939 the Russians seized the

▼ VOLCANIC PERFECTION Winter snows cover Mount Ngauruhoe, the dramatically symmetrical cone of an active volcano in Tongariro National Park on the Volcanic Plateau.

Polish sector and have retained it ever since. Volhynia's main town is L'VOV.

Map USSR Dc

Vólos ***Greece*** City, port and industrial centre in eastern THESSALY. The city, founded in the 14th century, is also a centre for tourists exploring the nearby ancient sites of Iolkos – legendary port from which Jason and the Argonauts set out – and Demetrias, and lies close to Mount PILION. Vólos has cement and steel works and produces tobacco and textiles.

Population 107 400

Map Greece Cb

Volta ***West Africa*** River with three headstreams, which rise in Burkina and flow southwards into Ghana. There they merge in Lake Volta, which drains to the sea some 110 km (70 miles) east of the capital, Accra. The Black Volta, which forms part of the Burkina-Ghana and Ghana-Ivory Coast borders, is the main headstream, and it is 1520 km (944 miles) from its source to the sea. The White Volta, which receives the Red Volta as a tributary, flows some 885 km (550 miles) into the lake. Only a short section, between lake and sea, is known by the name Volta. The Volta system is Ghana's main waterway; its basin covers nearly three-quarters of the country.

Map Ghana Ab

Volta, Lake ***Ghana*** One of the world's largest artificial lakes. It lies behind the AKOSOMBO Dam on the Volta river, and covers 8480 km^2 (3251 sq miles) – or 3.5 per cent of Ghana's entire area.

The dam, the lake, a modern aluminium smelter, and a bauxite (aluminium ore) quarry and processing plant make up West Africa's largest development scheme. Some 76 000 people had to be resettled when the lake formed after completion of the dam in 1966. The lake provides irrigation water for crops and livestock. It has opened up transport in an area where communications were particularly poor and is a rich source of fish. However, there are now fears that bilharzia, river blindness, malaria and other waterborne diseases will increase.

Map Ghana Bb

Volubilis ***Morocco*** Ruined capital of the ancient Roman province of Mauretania, lying about 30 km (19 miles) north of Meknès. The 1800-year-old remains are 3 km (2 miles) from MOULAY IDRISS. The most notable Roman remains in Morocco, they include a capitol, forum, basilica, baths, triumphal arch, several houses – many with intact mosaics – a bakery and 2nd-century walls. The remains of many oil mills are a reminder that the area was once rich in olive groves.

Map Morocco Ba

Vorarlberg ***Austria*** The country's smallest and most westerly state, noted for its breathtaking Alpine scenery. It covers 2601 km^2 (1004 sq miles) centred on the ARLBERG PASS. The state's prosperity comes from tourism (there is fine skiing), dairy farming and textiles. Its capital is the resort of BREGENZ on the shore of Lake Constance.

Population 307 900

Map Austria Ab

HOW A VOLCANO IS CREATED

According to the theory of PLATE TECTONICS, the earth's crust or outer shell consists of a number of rigid plates which float upon the magma – the molten rock of the planet's interior. Where one plate collides with or slides beneath another, the crust may be weak enough to permit the magma – squeezed by immense pressures – to force its way up to the surface. Thus a volcano is created.

In many cases this is a symmetrical, conical mountain, such as Japan's Mount FUJI or Kenya's KILIMANJARO. These were formed by a single central vent pouring forth a mixture of lava – magma become runny through the release of pressure – and hot ash, dust and rock fragments called pyroclasts. As this material was forced up the vent, it rolled down the sides of the cone in all directions, building it ever higher but still symmetrical in shape.

If volcanic activity ceases for a time, the lava in the vent hardens into a plug. Renewed subterranean pressure may force this plug out, resulting in a cataclysmic explosion. In other cases the lava forces new exits or lateral intrusions through the strata of the mountainside. A violent eruption may create a huge empty chamber where once there was magma; the subsequent collapse of the volcano's peak can create a vast crater called a caldera – which often fills with water to form a lake, or even a bay if on the shore.

The behaviour of a particular volcano depends on such factors as the nature and structure of the crust in the vicinity and on how fluid the lava is. This last depends in turn on its chemical composition; generally the higher the proportion of oxygen and silicon in a lava – called a pasty lava – the less runny it is and the more violent an eruption is likely to be. All eruptions are unpredictable, but they fall into several major categories, named after typical examples.

The Plinian type, for example, is caused by an explosion of gases in the magma chamber deep below the volcano. This creates a chimney through which a column of red-hot ash shoots as through a gun barrel, rising many kilometres into the air before falling and smothering the entire countryside downwind. Such was the fate of Pompeii and Herculaneum when VESUVIUS erupted in AD 79, killing the Roman naturalist Pliny the Elder – whose name nevertheless lives on in recognition.

Most violent of all is the Peléan, after Mont PELEE on the island of Martinique. There, in 1902, a previously slowly rising volcanic plug gave way in a series of violent explosions, sending a Plinian column of ash far up into the stratosphere, accompanied by fountains of lava and rocky debris from the disintegrated plug. At the same time, a cloud of superheated steam and cinders rolled down the mountainside, killing virtually everyone in the nearby town of St Pierre. The explosion that rent the island of KRAKATAU in 1883 was Peléan, too, but there the damage was due to nine huge waves caused by the subsequent collapse of the volcanic island into its own magma chamber.

Not all eruptions are so violent. The Strombolian, for example – named after Mount STROMBOLI in Sicily – is generally noisy and spectacular, with lava bombs shooting up a little way and falling back again, but does little damage. The Hawaiian type is usually a slow outpouring of very fluid basic lava from a wide, low crater. And volcanoes bring benefits as well as destruction – particularly to soil fertility.

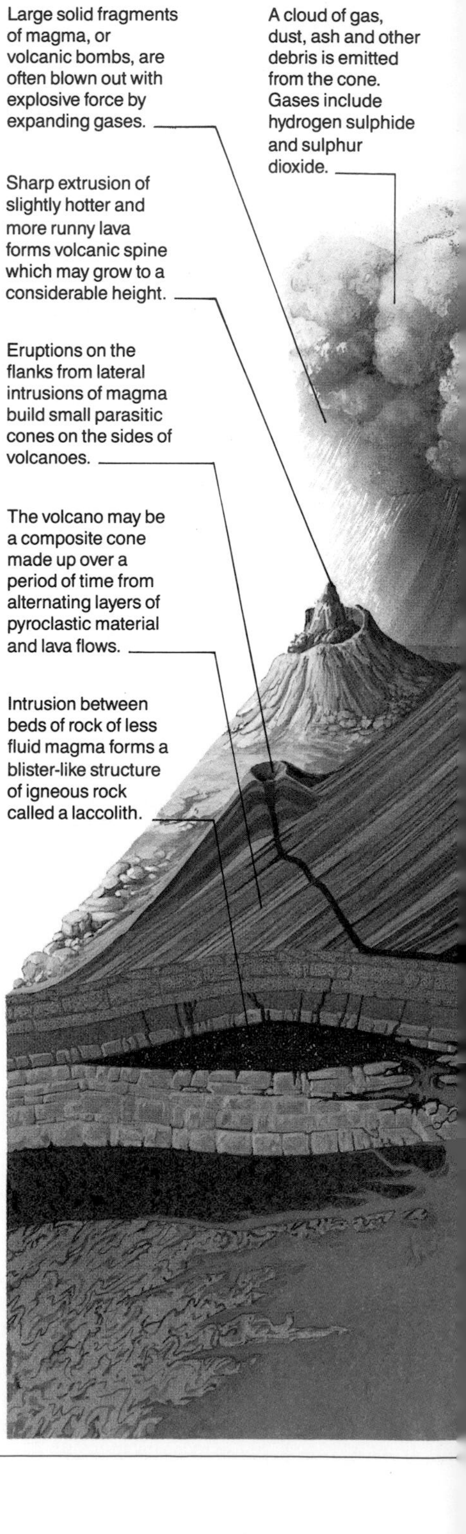

Large solid fragments of magma, or volcanic bombs, are often blown out with explosive force by expanding gases.
A cloud of gas, dust, ash and other debris is emitted from the cone. Gases include hydrogen sulphide and sulphur dioxide.
Sharp extrusion of slightly hotter and more runny lava forms volcanic spine which may grow to a considerable height.
Eruptions on the flanks from lateral intrusions of magma build small parasitic cones on the sides of volcanoes.
The volcano may be a composite cone made up over a period of time from alternating layers of pyroclastic material and lava flows.
Intrusion between beds of rock of less fluid magma forms a blister-like structure of igneous rock called a laccolith.

Plinian Violently explosive eruptions are named after the Roman writer Pliny, who died when Vesuvius erupted in AD 79.

Peléan A deadly avalanche of fiery gases and ash characterised the Mont Pelée eruption of 1902, which killed 30 000.

Strombolian Any explosion is of limited force and usually is accompanied by all-engulfing streams of white-hot lava.

Hawaiian Seldom explosive – rather a slow-spreading seepage of lava from a wide crater that creates a gently sloping cone.

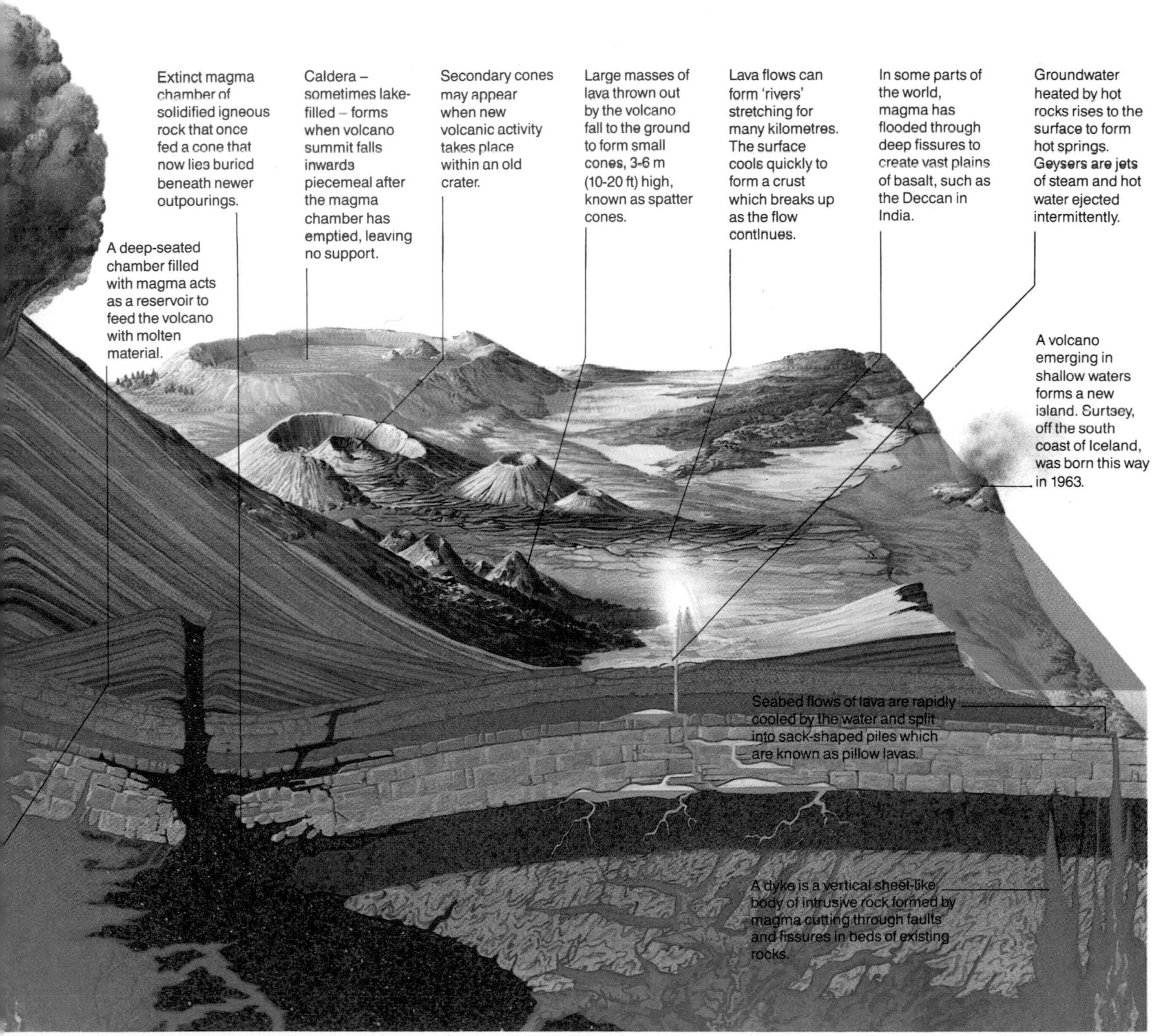

Extinct magma chamber of solidified igneous rock that once fed a cone that now lies buried beneath newer outpourings.
Caldera – sometimes lake-filled – forms when volcano summit falls inwards piecemeal after the magma chamber has emptied, leaving no support.
Secondary cones may appear when new volcanic activity takes place within an old crater.
Large masses of lava thrown out by the volcano fall to the ground to form small cones, 3-6 m (10-20 ft) high, known as spatter cones.
Lava flows can form 'rivers' stretching for many kilometres. The surface cools quickly to form a crust which breaks up as the flow continues.
In some parts of the world, magma has flooded through deep fissures to create vast plains of basalt, such as the Deccan in India.
Groundwater heated by hot rocks rises to the surface to form hot springs. Geysers are jets of steam and hot water ejected intermittently.
A deep-seated chamber filled with magma acts as a reservoir to feed the volcano with molten material.
A volcano emerging in shallow waters forms a new island. Surtsey, off the south coast of Iceland, was born this way in 1963.
Seabed flows of lava are rapidly cooled by the water and split into sack-shaped piles which are known as pillow lavas.
A dyke is a vertical sheet-like body of intrusive rock formed by magma cutting through faults and fissures in beds of existing rocks.

Voronezh ***USSR*** City on the Voronezh river 450 km (280 miles) south of Moscow. It was founded as a fortress in the late 16th century. Today it is a largely industrial city making machinery, electrical products, chemicals, cigarettes and processed foods.
Population 841 000
Map USSR Ec

Voroshilovgrad ***USSR*** City in the Ukraine, about 275 km (170 miles) south-east of Khar'kov. It lies at the heart of the Donets coal-mining basin, and was known as Lugansk before 1935 and again in 1958-70. The city was founded in 1795, when an iron foundry making arms for the Russian Black Sea fleet was established there. It later specialised in railway equipment, and by 1941 produced half of the country's locomotives. It is now an expanding engineering and research centre, and also produces textiles, foodstuffs and chemicals.
Population 491 000
Map USSR Ed

Vosges ***France*** Mountains along the Rhine valley in eastern France. They are separated from the Jura mountains to the south by the Belfort Gap. The Vosges' highest point is the Ballon de Guebwiller, at 1424 m (4671 ft).
Map France Gb

Voss ***Norway*** Lakeside resort, popular in both summer and winter, on the Oslo-Bergen railway about 70 km (45 miles) north-east of Bergen. It has a small medieval church.
Population 14 000
Map Norway Fa

Vostok Base ***Antarctica*** Russian research station on the South Polar Plateau some 1200 km (750 miles) from the South Pole. It holds the record for the coldest temperature recorded on earth, −89.6°C (−129.3°F), in July 1983.
Map Antarctica Mc

Voyvodina ***Yugoslavia*** See VOJVODINA

Vratsa ***Bulgaria*** Capital of Vratsa province, 56 km (35 miles) north of Sofia. Its chief industries are silk, textiles, chemicals, cement and ceramics.
Population (town) 66 500; (province) 290 000
Map Bulgaria Ab

Vridi Canal ***Ivory Coast*** Deep-water channel, 3 km (nearly 2 miles) long and 14 m (45 ft) deep, cut through a sandbar and linking the port of Abidjan, PORT BOUET, with the open sea. It was opened in 1950.
Map Ivory Coast Bb

Vriesland ***Netherlands*** See FRIESLAND

Vryheid ***South Africa*** Market town in north Natal, whose name is the Afrikaans word for 'freedom'. It lies about 235 km (145 miles) north and slightly west of Durban, and was founded by Afrikaners in 1884 as the capital of the New Republic, one of several small states set up on the Transvaal borders. The New Republic was absorbed by the South African Republic (Transvaal) in 1888 and became part of Natal in 1903.
Population 11 300
Map South Africa Db

V-shaped valley Narrow, steep-sided valley formed by stream erosion and mass movement, commonly indicating youthfulness, as opposed to the broader valleys with flood plains of more mature rivers or to the U-shaped valleys formed by glacial erosion.

Vulcano ***Italy*** Volcanic holiday island covering 21 km² (8 sq miles) in the EOLIAN ISLANDS to the north of Sicily. It has three volcanoes: Vulcano Vecchio, rising to 500 m (1640 ft) and now extinct; Gran Cratere, which looms 391 m (1283 ft) over the harbour and last erupted in 1888; and Vulcanello (123 m, 404 ft), to the north. These volcanoes have created thermally heated seawater and curative, bubbling mudbaths.
Population 430
Map Italy Ee

Vyborg (Viipuri) ***USSR*** Fishing port 120 km (75 miles) north-west of Leningrad and only 30 km (18 miles) from the Finnish border. It is a gateway into Finland (of which it was part from 1918 to 1940) by road, rail and canal. Apart from fishing, its main industries are machinery, shipbuilding, timber and furniture.
Population 79 000
Map USSR Db

'W' National Park ***West Africa*** Huge wildlife reserve covering some 10 230 km² (3950 sq miles) of savannah on the borders of Benin, Burkina and Niger. It gets its name because the Niger river forms a huge W shape as it flows through. It contains a vast range of animals and birds, but is less developed than Benin's adjoining Pendjari National Park. Tracks in the 'W' National Park are open only from January to June (when they have dried out following the end of the wet season); motorists must take adequate supplies of petrol with them.
Map Benin Ba; Niger Ab

Waal ***Netherlands*** See RHINE

Wad Medani ***Sudan*** Market town about 160 km (100 miles) south-east of the capital, Khartoum. It is the centre of the El Gezira cotton-growing region. It has increased in importance since the Gezira irrigation scheme was established and is the headquarters of the research division of the Ministry of Agriculture.
Population 153 000
Map Sudan Bb

Waddenzee Shallow inlet of the NORTH SEA between the West FRIESIAN ISLANDS and the Netherlands mainland used as a wildlife refuge, a recreational area and as a nursery for fish and shellfish. It was once part of the ZUIDER ZEE, a former lake which became a saltwater gulf when it was flooded by the sea in the 13th century. A dyke completed in 1932 divided the Zuider Zee into two parts: the northern Waddenzee, which was left to the sea, and the southern IJSSELMEER, which has been largely reclaimed.
Area 10 000 km² (3861 sq miles)
Map Netherlands Ba

wadi In Arabic countries, a steep-sided valley, gully or riverbed of desert terrain that usually remains dry, but sometimes contains a short-lived torrent after heavy rain.

Wadi el Natrun ***Egypt*** Salt-marsh valley about 100 km (60 miles) south-west of Alexandria. Most of the valley, which contains extensive deposits of sodium carbonate (used in glass-making, ceramics, detergents and soap), lies below sea level.
Map Egypt Bb

Wagga Wagga ***Australia*** Prosperous agricultural and educational centre on the Murrumbidgee river in New South Wales, 380 km (240 miles) south-west of Sydney.
Population 36 800
Map Australia Hf

Waikato ***New Zealand*** Longest river in the country, rising on the eastern slopes of Mount Ruapehu in the centre of North Island and flowing 425 km (264 miles) over rapids and through gorges to the Tasman Sea about 40 km (25 miles) south of Auckland. It has been harnessed for hydroelectric power, with eight dams along its upper course.

The Huka Falls, 4 km (2.5 miles) from Lake Taupo – where the river is squeezed into a natural millrace and bursts over a 12 m (39 ft) drop like the jet from a giant hosepipe – is a popular tourist attraction.
Map New Zealand Eb

Waipou Forest ***New Zealand*** See NORTHLAND

Wairarapa ***New Zealand*** Sheep and dairy-farming area covering 9800 km² (3784 sq miles) on North Island. The main market town is Masterton, about 80 km (50 miles) north-east of the capital, Wellington. The fertile lowlands were among the first to be settled by Europeans, in the 1840s.
Population 39 600
Map New Zealand Ed

Waitaki ***New Zealand*** South Island river rising in the Southern Alps and flowing 153 km (95 miles) into the Pacific, 130 km (80 miles) north of the east coast port of Dunedin. Much of the water has been harnessed for hydroelectric power. The dam, lake and power station at Benmore, completed in 1965, is the largest of the many schemes along the river. Much of the electricity is sent to North Island through a submarine cable across Cook Strait.
Map New Zealand Cf

Waitangi ***New Zealand*** Historic North Island site on the Bay of Islands, about 200 km (125 miles) north and slightly west of Auckland. A group of Maori chiefs signed the Treaty of Waitangi there in 1840, ceding sovereignty to the British, and in return were guaranteed their land and other possessions. Today the site has a museum, a Maori meeting house and a Maori war canoe.
Map New Zealand Ea

Waitemata Harbour ***New Zealand*** See MANUKAU HARBOUR

Waitomo Caves ***New Zealand*** North Island cave system discovered in 1887 about 200 km (125 miles) south of Auckland and now visited by more than 250 000 tourists a year. The most spectacular chamber is Glow-worm Grotto –

which is lit by the eerie blue light of millions of glow-worms. The worms – larvae of a type of gnat – hang sticky threads from the chamber's ceiling to trap small flying insects.

Map New Zealand Ec

Wakatipu ***New Zealand*** S-shaped lake in a glacier-carved valley on South Island about 160 km (100 miles) north-west of the port of Dunedin. It covers an area of 293 km^2 (113 sq miles) and is a major tourist attraction.

Map New Zealand Bf

Wakayama ***Japan*** City on the south coast of Honshu island, about 60 km (35 miles) south-west of the city of Osaka. Its 16th-century castle commands a view over the Kii Channel between Honshu and the island of Shikoku. Its industries include a steelworks.

Population 401 400

Map Japan Cd

Wake Island ***Pacific Ocean*** Atoll about 1100 km (700 miles) north of Kwajalein in the Marshall Islands. American territory since 1899, it was a relay station on the transpacific telegraph, a submarine and air base (held by the Japanese in 1941-4), and a refuelling stop for military and commercial transpacific flights before the advent of long-range jets. It remains a US Air Force base.

Map Pacific Ocean Db

Wakefield ***United Kingdom*** Cathedral city in West Yorkshire, 12 km (7 miles) south of Leeds. It was the area's chief town in the Middle Ages, but with the industrialisation of the wool trade at the end of the 18th century, factories in Leeds and Bradford quickly took over. The cathedral dates from the 14th century.

Population 61 000

Map United Kingdom Ed

Walachia ***Romania*** Region and former principality between the southern Carpathians, to the north, and the River Danube to the east. Covering an area of 76 586 km^2 (29 570 sq miles), it consists mainly of plains, river valleys and farmland. The national capital, BUCHAREST, is in the centre of the plain, with roads leading to every corner of the region.

Walachia was united with Moldavia to form the new nation of Romania in 1862, but it still retains its own folklore, and many of its rural homes retain a characteristic regional style, with deep verandahs and wooden-tiled roofs.

Map Romania Ab

Walbrzych (Waldenburg) ***Poland*** Coal, iron and steel town 68 km (42 miles) south-west of the city of Wroclaw. Hochberg Castle – built near the town between the 14th and 16th centuries – was used as a retreat by the Nazi leader Adolf Hitler (1889-1945).

Population 130 000

Map Poland Bc

Walcheren ***Netherlands*** Former island in the south-west, at the mouth of the Schelde river. It covers 212 km^2 (82 sq miles) and is joined to Zuidbeveland island by polder (reclaimed land). Middelburg and the port of Flushing are the main towns. Most of Walcheren lies below sea level, and severe flooding occurred when its dykes were breached by military action in 1944 and by the sea in 1953. The Delta Plan for land reclamation, begun in 1956, is designed to minimise the danger.

Map Netherlands Ab

▲ RUGGED FASTNESS Two lovely lakes, Llyn Idwal and Llyn Ogwen beyond, lie below the Devil's Kitchen, a craggy basin on the flanks of Glyder Fawr (999 m, 3278 ft) in Snowdonia, Wales.

Waldenburg ***Poland*** See WALBRZYCH

Wales ***United Kingdom*** Western constituent principality of the kingdom. (The sovereign's oldest son is generally Prince of Wales.) It covers some 20 761 km^2 (8016 sq miles) and was united with England in 1536-42. Today there is a strong feeling of nationalism, and the Welsh language is known by about a quarter of the people. CARDIFF is the capital and Newport and SWANSEA the main industrial cities.

Population 2 749 600

Map United Kingdom Dd

Wallace's Line See LOMBOK

Wallis and Futuna ***Pacific Ocean*** French overseas territory consisting of three main islands and many islets in the South Pacific to the north-east of Fiji. The Hoorn group, about 240 km (150 miles) from Fiji, consists of Futuna (64 km^2, 25 sq miles) and uninhabited Alofi (51 km^2, 20 sq miles); both are mountainous, Futuna reaching 765 m (2510 ft). Wallis (Uvéa) island, 160 km (100 miles) farther north-east, covers 159 km^2 (61 sq miles) and is hilly, with a coral reef. Wallis has 60 per cent of the population; the capital, Mata-Utu, is on its east coast. The people, who are Polynesians, grow coconuts, taro, yams and bananas, and raise pigs. Little is exported, and many of the people have emigrated to New Caledonia or Vanuatu.

Population 13 500

Map Pacific Ocean Ec

Wallsend ***United Kingdom*** See HADRIAN'S WALL; NEWCASTLE UPON TYNE

Walsall ***United Kingdom*** Industrial and market town in the English metropolitan county of West Midlands, about 12 km (8 miles) north-west of Birmingham. It has numerous engineering and other industries.

Population 179 000

Map United Kingdom Ed

Walvis Bay ***Namibia/South Africa*** Namibia's chief port, about 260 km (160 miles) west and slightly south of the capital, Windhock. It was given its name, meaning Whale Bay, by whalers who frequented it in the 19th century. Walvis Bay was annexed by Britain in 1878 and was transferred to Cape Colony in 1884. It became part of South Africa in 1910 and, although administered as part of Namibia from 1922, it is still regarded as South African. The port serves the livestock and mining areas of the interior and is a fishing centre.

Population 23 500

Map Namibia Ab

Wanganui ***New Zealand*** North Island river rising on Mount Ruapehu and flowing 288 km (180 miles) into the Tasman Sea at the port of Wanganui (population 39 600), about 160 km (100 miles) north of the capital, Wellington.

Map New Zealand Ec

Wangaratta ***Australia*** See BEECHWORTH

Wankie ***Zimbabwe*** See HWANGE

Warasdin ***Yugoslavia*** See VARAZDIN

warm front Boundary plane between the warm and cold air in a depression, at which the mass of warm air is rising above the cold air which it is overtaking. As the front moves, warm air replaces cold air on the ground, hence its name. Low nimbostratus cloud usually produces a belt of rain along the front.

warm sector Area of warmer air in the central part of an atmospheric DEPRESSION, bounded by a WARM FRONT and a COLD FRONT.

Warri *Nigeria* Port in the west of the Niger river delta, about 270 km (170 miles) east and slightly south of the capital, Lagos. It dates back at least to the 15th century when it was visited by Portuguese missionaries. Later, it was a base for Portuguese and Dutch slave traders. Today the town has an oil refinery and imports drilling equipment. It exports groundnuts and cotton products, which are brought down the Benue river from northern Nigeria, Cameroon and Chad, and cocoa, palm oil (used in margarine and cooking fats), rubber and timber from southern Nigeria.

Population 88 900

Map Nigeria Bb

Warrington *United Kingdom* Beer brewing and industrial town in Cheshire, in north-west England. It stands on the River Mersey and the Manchester Ship Canal midway between Manchester and Liverpool.

Population 136 000

Map United Kingdom Dd

Warsaw (Warszawa) *Poland* The country's capital and largest city. Despite a history reaching back to the Middle Ages, it is also one of Poland's most modern cities, for when the Russians liberated it in January 1945, less than one in ten of the buildings was fit for use, and 800 000 of its inhabitants – nearly two in three of the city's prewar population – were dead or deported.

In a massive rebuilding project the snugly walled, medieval and Renaissance Stare Miasto (Old Town) was painstakingly re-created. Its narrow streets are now for pedestrians only, and teem with cafés and bars, echoing the lively commercial and crafts centre that existed before 1596, when it was made the national capital in place of Cracow. Along the route towards Cracow spread the 17th and 18th-century expressions of Warsaw's new status: the Royal Palace (reopened in the 1980s); town palaces of the nobles, including the Radziwill and Potocki families; centres of art and learning, which now form the city's university; churches and convents; recreational areas, including Lazienki Park; and – 10 km (6 miles) from the centre – Wilanow, the royal summer residence, which was built for the Polish king Jan Sobieski in the 1680s. The opening of the Vienna-Warsaw railway in 1845 unleashed industrial development: to the west of Muranow, where Jewish immigrants settled, and to the east across the Vistula river in Praga.

By 1939 there were 1 290 000 Warsovians, 35 per cent of them Jewish, and nearly 30 per cent Yiddish-speaking. In 1940, the occupying Germans confined 450 000 Jews to a walled ghetto – on rations providing only 184 calories a day – to await transfer to death camps. Survivors were removed and the ghetto razed following an abortive Jewish rising in 1943. The following summer, the Germans began a systematic destruction of the rest of the city after a rising led by the Polish Resistance failed while the Russian Red Army was poised just outside the city.

Rebuilding has created a spacious new city outside the Old Town. Arterial roads and a motorway have replaced the former maze of streets. The highways stretch out to modern high-rise suburbs, with plants producing steel, cars, cement, tractors and electronic goods.

Among Warsaw's many powerfully evocative monuments are the 'Warsaw *Nike*' (opposite the Grand Theatre), which commemorates the city's war dead, and the simple memorial to the heroes of the ghetto at Muranow.

Population 1 641 000

Map Poland Db

Wartburg *East Germany* Historic 11th-century castle in the extreme south-west. It is perched on a hill 172 m (564 ft) above the town of Eisenach and is accessible only on foot or by donkey. It has a magnificent Romanesque palace and half-timbered houses.

The religious reformer Martin Luther was sheltered there in 1521-2 while he translated the Bible from Latin into German. The castle was used as a setting by the composer Richard Wagner (1813-83) in his opera *Tannhäuser*.

Map East Germany Bc

Warwick *United Kingdom* Town and county covering 1981 km^2 (765 sq miles) in the Midlands of England. Farming and tourism are the main sources of income. There is an imposing medieval castle in Warwick town – 30 km (20 miles) south-west of Birmingham – and the remains of another castle are in nearby Kenilworth. Leamington Spa was a fashionable 18th-century watering place. But the county's biggest attraction is STRATFORD-UPON-AVON, the birthplace and home of the poet and dramatist William Shakespeare (1564-1616).

Population (county) 478 000; (town) 22 000

Map United Kingdom Ed

wash Fine material, particularly alluvium, found especially in the dry beds of streams that flow intermittently, and on desert surfaces where flowing water is rare. The term is also used for fine material moved down the slope of a hill or mountain by rain, especially where there is little vegetation.

Wash, The *United Kingdom* Shallow inlet of the NORTH SEA between the counties of Lincolnshire and Norfolk on the east coast of England. The Great Ouse, Nene, Welland and Witham rivers empty into the bay through the surrounding Fenland. Much land has been reclaimed from The Wash by drainage and dyking projects carried out by surrounding landowners since Roman times. Today The Wash is 35 km (20 miles) long and about 25 km (15 miles) wide.

Map United Kingdom Fd

Washington *USA* State covering 176 617 km^2 (68 192 sq miles) in the north-west, bordering Canada. It is traversed from north to south by the Coast and Cascade ranges; the latter rises to 4392 m (14 410 ft) at Mount Rainier. Wheat, cattle and apples are the main farm produce, and the region is rich in fish, especially salmon. Industries include aerospace, timber processing and food processing. OLYMPIA is the state capital and SEATTLE the chief town and port.

Population 4 409 000

Map United States Ba

Washington DC *USA* The national capital, lying in the District of Columbia – Federal territory now covering 179 km^2 (69 sq miles) astride the Potomac river between Maryland and Virginia, about 200 km (125 miles) from the Atlantic coast. The territory was ceded by Maryland and Virginia in 1788-9, and the new capital, the world's first purpose-built seat of government, was laid out on the east side of the river by the French engineer Major Pierre L'Enfant. Congress (the Federal legislature) moved there from Philadelphia in 1800.

The city, which incorporates the once-separate GEORGETOWN and now spreads beyond its original boundaries into the adjoining states, has a spacious centre, with the Congress buildings on Capitol Hill at one end of The Mall, and the Washington Monument and Lincoln Memorial at the other. The Mall is flanked by public buildings, including the Smithsonian Institution, the National Gallery of Art and government offices. To the north lies the White House, official residence of the president. Not far away, in contrast, are some crowded slums. On the west bank of the Potomac stands the Pentagon, the USA's military headquarters.

The civil service is Washington's main employer, but there are also many business, financial and foreign institutions, including embassies. There are more than 700 parks, a cathedral, five universities and many cultural establishments including the John F. Kennedy Center for the Performing Arts, and two international airports. The city is a popular tourist attraction, with some 16 million visitors a year.

Population (city) 622 800; (metropolitan area) 3 429 400

Map United States Kc

water meadow Meadow on the flood plain of a river that is irrigated and fertilised with silt by the periodic flooding of the river. In addition to meadows subject to natural flooding, from late medieval times artificial water meadows were created in Europe by farmers digging channels controlled by sluices. The running water did not often freeze, so grain continued to grow giving early feed for livestock.

water table, saturation level Boundary between the aeration zone – the region of soil and rock in which air and water mix – and the saturation zone, a region of porous rock in which all the pores are filled with water (see GROUNDWATER). The water table is uneven, depending on the nature of the pores or holes in the rock; it also varies according to the amount of rainfall, rising in wet weather and falling in dry weather. SPRINGS may appear on the surface at the junction of the porous rock with an underlying layer of impermeable rock. The rise and fall of the water table affects the flow from springs as well as the level of water in wells and artesian supplies.

waterfall Point where a river plunges more or less vertically over a steeply sloping rock face or an overhang. It may result from a layer or a DYKE of hard, wear-resistant rock lying across softer, more easily eroded strata; or from the arrival of a river at a coastal cliff, the edge of a plateau, a fault-line ESCARPMENT or the mouth of a HANGING VALLEY. See MIGHTY CATARACTS AND SPARKLING CASCADES (opposite)

Waterford (Port Láirge) ***Ireland*** City and port 135 km (84 miles) south and slightly west of the capital, Dublin. Waterford was originally settled by Danish invaders, but they were driven out by the Anglo-Normans in 1170. The quays of its busy port line the River Suir. Waterford manufactures food, metal and engineering products, pharmaceuticals and furniture, but is best known for its light opera festival held each September and its crystal glass. Its glass factory is the largest in the world. The city has two cathedrals, two 13th-century monasteries and a tower, built by the Danes in 1003. The tower is now a museum. County Waterford covers 1837 km² (702 sq miles), and its county town is Dungarvan.

Population 38 500

Map Ireland Cb

Waterloo ***Belgium*** Battlefield town 16 km (10 miles) south of Brussels. It lies near the scene of the great battle on June 18, 1815, when the French emperor Napoleon was defeated by the British under the Duke of Wellington, with the crucial help of the Prussians. The battlefield still draws tourists to the town, which is now a suburb of the capital. There is a museum in Wellington's former headquarters, and several monuments in the village church.

Population 18 000

Map Belgium Ba

watershed The divide between two different river systems. Watersheds may be marked by steep ridges, but may also run across plateau surfaces and even lakes and bogs.

waterspout TORNADO or similar whirlwind occurring over water and resulting in a whirling column of sea, spray and mist.

Wau ***Sudan*** Provincial capital of Bahr el Ghazal and major cattle market. It lies about 1050 km (630 miles) south-west of the capital, Khartoum, to which it has a rail link.

Population 58 100

Map Sudan Ac

Wave Rock ***Australia*** Natural rock formation and tourist attraction in Western Australia, near the town of Hyden, 300 km (186 miles) east and slightly south of the capital, Perth. The granite rock, carved by wind and rainwater over tens of millions of years, is 15 m (50 ft) high and is part of a 2 km (1 mile) long chain of outcrops known collectively as Hyden Rock. It looks like a huge wave about to break over the dry, flat country surrounding it.

Map Australia Be

wave-cut platform Horizontal rock surface at the base of sea cliffs formed by wave action, which cuts a notch in the cliff base and then cuts back the cliffs themselves.

Waza National Park ***Cameroon*** Nature reserve covering 170 000 hectares (420 000 acres) in the far north, on the Nigerian border. It is considered one of the best game parks in West Africa, and its forest and grassland is home for a wide range of wildlife.

Map Cameroon Ba

Weald, The ***United Kingdom*** See DOWNS, THE

MIGHTY CATARACTS AND SPARKLING CASCADES

Waterfalls are among nature's most awe-inspiring phenomena, whether tall and relatively slim or lower but broad like Niagara and Victoria Falls. With the flooding of the Guaíra Falls on the Alto Paraná of South America in 1982, the mightiest of all in terms of annual water flow are the Boyoma Falls of Zaire.

THE WORLD'S HIGHEST WATERFALLS

Waterfall	Location	Height (m/ft)
Angel	Venezuela	980/3215
Yosemite (total drop)	USA	739/2425
Mardalsfoss	Norway	655/2150
Utigård	Norway	600/1970
Sutherland	New Zealand	580/1904
Gavarnie	France	422/1384
Tugela	South Africa	411/1350
Krimml	Austria	381/1250

BONE DRY AND SOAKING WET

The driest places on earth are the Dry Valleys in the cold desert landscapes near McMurdo Sound, opposite Ross Island in Antarctica. Scientists estimate that no rain has fallen there for the last 2 million years. Another of the world's driest places – Death Valley in California – is also one of the hottest, with temperatures reaching 56.6°C (134°F), and one of the lowest, 86 m (282 ft) below sea level.

The greatest recorded rainfall in the world occurred in the Indian village of Cherrapunji in July 1891, when more than 11 836 mm (466 in) of rain fell during the month. With an average annual rainfall of 10 874 mm (428 in), it is the wettest inhabited place on earth. But the wettest known location of all is Mount Waialeale in Hawaii, with an average rainfall of 12 344 mm (486 in).

THE WORLD'S DRIEST PLACES

Place	Location	Rainfall (mm/in per year)
Dry Valleys	Antarctica	No rain for 2 million years
Death Valley	USA	3/0.1
Arica Desert	Chile	3/0.1
Gobi Desert	Central Asia	5/0.2
Sahara (parts)	North Africa	25/1
Lake Eyre Basin	Australia	101-152/4-6

AND SOME OF ITS WETTEST

Mount Waialeale	Hawaii	12 344/486
Cherrapunji	India	10 874/428
Mount Cameroon	Cameroon	10 160/400
Sprinkling Tarn	Cumbria, England	6528/257
North-west Washington State	USA	2997/118

weather The day-to-day condition of the atmosphere at any place, with respect to the various elements, such as pressure, temperature, sunshine, wind, clouds, fog and precipitation. See WEATHER AND WIND SYSTEMS (p. 704); for weather records, see BONE DRY AND SOAKING WET (above) and FROM BOILING HOT TO FREEZING COLD (p. 705)

weathering Processes of disintegration and decomposition by which rocks at or near the earth's surface are broken down into loose fragments. Weathering provides the materials with which the agents of EROSION work.

Weddell Sea Part of the South ATLANTIC OCEAN between Coats Land and the ANTARCTIC PENINSULA. The British sealer and explorer James Weddell discovered it in 1823. In 1915, Sir Ernest Shackleton's *Endurance* became trapped in this ice. After ten months, he abandoned the ship, recording: 'It was a sickening sensation to feel the decks breaking up under one's feet, the great beams bending and then snapping with a noise like gunfire.' Shackleton and his crew reached safety after an epic journey over the ice and across the ocean by small boats.

Area 800 000 km² (308 880 sq miles)

Map Antarctica Cb

WEATHER AND WIND SYSTEMS

The fundamental cause of all weather is the temperature difference between the poles and the Equator. The Equator receives more heat from the sun than it loses into space: the poles lose more heat to space than they receive from the sun. The redistribution of heat from the Equator towards the poles is the action that produces all weather, modified in its operation by spin of the earth. This spin tends to 'thicken' the atmosphere along the Equator and, other things being equal, would cause high pressure at the Equator. But other things are not equal.

The intense heat at the Equator causes hot, moisture-laden air to rise and creates a low pressure zone – the Intertropical Convergence Zone (ITCZ); the high pressure belt that should exist is divided into two and displaced to the tropics. The resultant pattern of pressure belts gives rise to wind systems. Winds blow towards the Equator and towards the mid-latitudinal ($23\frac{1}{2}°$-$66\frac{1}{2}°$ in Northern and Southern Hemispheres) low pressure systems from the high pressure belts at the tropics. Winds are also directed to the mid-latitudinal regions from the polar high pressure regions.

The spinning of the earth deflects the winds, for example, 'pulling' the trade winds of the tropics to the west. In the middle latitudes the spin gives rise to the steady westerlies that have such a marked effect upon the weather of Europe, North America and Australia.

This is the pattern at ground level but high aloft some counter-currents operate. Circulation over the globe, ultimately transporting heat towards the poles, consists of a number of cells – the so-called Hadley Cells.

If the earth were a smooth sphere, covered only by water or flat landscape, then the movement of winds would be constant and unchanging, and all weather patterns predictable. However, air currents are diverted and distorted by mountains, valleys, ice caps and deserts – even the heat radiated by cities. A major influence is the contrast between land and sea, which affects maritime climates at all latitudes. Land warms and cools more quickly than water, so that it is hotter than the sea in summer and colder in winter. This has a profound seasonal effect on wind and weather for the same reasons as for the daily shift that takes place on coasts. There, in the day, cool breezes off the sea move in to replace the hot air rising off the land; at night, as the sea retains its warmth and the land cools, the system goes into reverse. In middle latitudes, north and south of the tropics, the weather is governed by CYCLONES (depressions) and ANTICYCLONES. These are whirlpools of air set in motion by the rotation of the earth, which occur at low level where warm air from mid-latitudes meets the polar air from the polar high. They are much influenced by topographical and other features, and are therefore all the less predictable in their effect.

ANATOMY OF A HURRICANE

Hurricanes originate over tropical seas where the water surface has been heated to over 27°C-(80°F). An area of intense low pressure may form, and is thrown into a spin by the rotation of the earth. Winds bearing water vapour from the warm sea are dragged up into the spiral to create towering banks of cloud. Energy is released in torrential rains and winds of up to 300 km/h (190 mph).

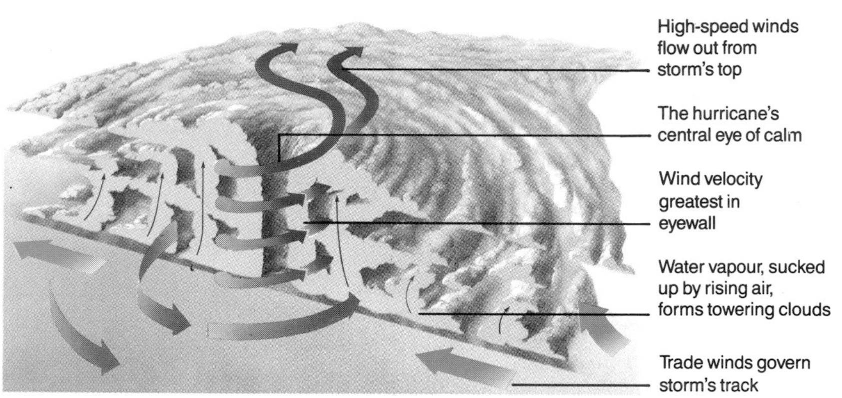

WHERE THE WIND BLOWS

Major circulation systems of rising and falling air, Hadley Cells, lie either side of the Equator. Winds within the cells are curved by the 'drag' of the earth's spin. The rotation rate in upper latitudes disrupts the cells, which otherwise would stretch as two great loops to the poles.

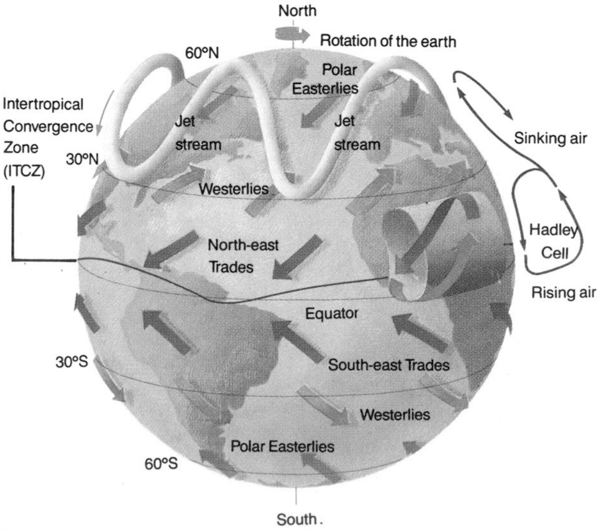

DEVELOPMENT OF A DEPRESSION

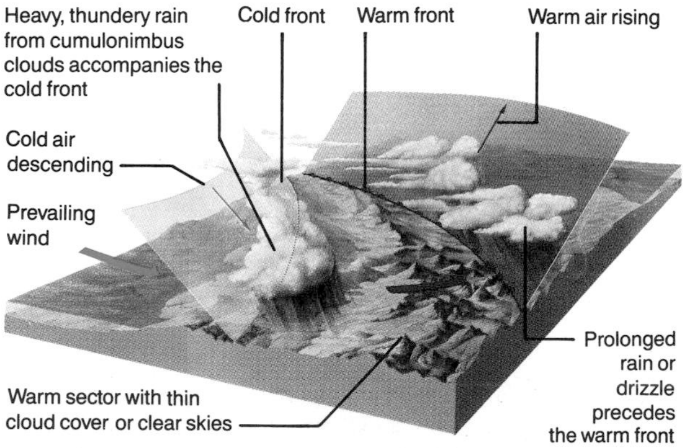

A depression is a low-pressure region into which whirlpools of air spiral. Cold polar air and warm tropical air converge in the depression and, as the depression moves, warm and cold air masses are forced against each other to form fronts. The warm front's passage is marked by drizzle and rain from dark, cloudy skies, the cold front's by heavy rain from towering cumulonimbus or thunder clouds.

FROM BOILING HOT TO FREEZING COLD

The highest shade temperature in the world – 58°C (136.4°F) – was recorded in the Libyan village of Al' Aziziyah, about 40 km (25 miles) south of Tripoli, one day in 1922. However, the world's consistently hottest place is Death Valley in California, where in 1917 maximum temperatures of more than 48°C (118°F) were recorded on 43 days in succession.

The 'award' for the world's coldest recorded temperatures has been won three times running by the Russian Antarctic research station, Vostok Base. At recordings made between August 1958 and July 1983 the extremely low temperatures at the base dropped progressively from −87.4°C (−125.3°F) to −88.3°C (−126.9°F) to −89.6°C (−129.3°F).

wedge of high pressure See RIDGE OF HIGH PRESSURE

Wei He ***China*** River rising in Shaanxi province and flowing 800 km (500 miles) into the Huang He (Yellow River). Its fertile valley in the LOESSLANDS OF CHINA is the cradle of Chinese civilisation and contains many important archaeological sites. These include Lantien, site of the discovery of Lantien Man, a skull 500 000 years old; Banpo, site of a Neolithic (5000 BC) village of the Yangshao culture; Xi'an, capital of China under the Qin dynasty (221-206 BC), and the site of Chang'an, capital during the Han dynasty (206 BC-AD 220) and the Tang dynasty (AD 618-906).

Map China Fd

Weichsel ***Poland*** See VISTULA

Weimar ***East Germany*** City about 80 km (50 miles) south-west of Leipzig. The poets Johann Wolfgang von Goethe and Friedrich von Schiller lived there and their fame made it the cultural centre of 18th-century Germany. The composer J.S. Bach was court organist and concert master to the Elector of the duchy of Saxe-Weimar from 1708-17, while Franz Liszt was musical director there in the mid-19th century. Today it is an exquisitely restored city of half-timbered houses, painted stone buildings with shuttered windows – including Schiller's birthplace – castles and Baroque palaces. In 1919 the German National Assembly met there to draw up a constitution, and the town gave its name to the republic which survived until Adolf Hitler came to power in 1933.

Population 64 000

Map East Germany Bc

Weipa ***Australia*** Mining settlement in the extreme north of Queensland on the western coast of Cape York Peninsula. It has one of the world's largest deposits of bauxite.

Population 2400

Map Australia Ga

Welland Canal ***Canada*** Waterway linking Lake Erie with Lake Ontario. The first canal, with 25 locks, was built between 1824 and 1829 to enable ships to avoid the 98 m (320 ft) drop of NIAGARA FALLS. The latest, with eight locks, built between 1913 and 1932 and improved in 1973, now carries ships up to 221 m (725 ft) long with up to 7.9 m (26 ft) draught.

Map Canada Hd

Wellington ***New Zealand*** National capital and port in the extreme south-west of North Island. Built on a steep and often windswept slope around a deep, sheltered harbour, it exports dairy produce, wool, meat and hides, and manufactures motor vehicles, footwear, machinery, metal goods, textiles and chemicals.

The city was founded in 1840 by the New Zealand Company, a British organisation which promoted the colonisation of the country. Wellington was named after the British military hero and prime minister, the Duke of Wellington (1769-1852), who gave the company his support. It succeeded Auckland as the capital in 1865. Many of its outstanding buildings date from the Victorian period; they include the Houses of Parliament (1877), the Victoria University (1897), St Paul's Church, the cathedral (1866) now a well-preserved historic monument, and the Thistle Inn (1860). Its industries are concentrated in the nearby Hutt Valley and the town of Porirua.

Population 342 500

Map New Zealand Ed

Welo ***Ethiopia*** Province of 79 000 km² (30 500 sq miles) about 160 km (100 miles) north of Addis Ababa, devastated by famine in 1984-5 after five successive years of drought.

Population 3 000 000

Map Ethiopia Aa

Weser ***West Germany*** River formed by the confluence of the Fulda and Werra rivers about 25 km (15 miles) north-east of Kassel. It flows about 320 km (200 miles) northwards to the North Sea at Bremerhaven.

Map West Germany Cb

West Bank ***Israeli-occupied Jordan*** Area situated west of the River Jordan and the Dead Sea, seized by Israel during the Six-Day War of 1967. Covering 5858 km² (2262 sq miles), it contains some of Jordan's most fertile regions. East Jerusalem is the only part formally annexed by Israel, but the rest has been increasingly integrated into the Israeli economy. However, West Bank agriculture has stagnated since 1967 and there is little industrial development.

Under the Israelis, a number of controversial Jewish settlements have been set up; and some 111 600 Jews now live on the West Bank. The Israelis regard it as part of the traditional 'Land of Israel'. However, there is continuing Arab-Israeli conflict over the area. The Rabat summit of Arab heads-of-state in October 1974 declared that the Palestinian Liberation Organisation (PLO) was the sole legitimate representative of the Palestinian people (with the inference that the West Bank would eventually become an independent Palestinian Arab state). This strongly undermines Jordan's right to negotiate the return of the West Bank, though King Hussein of Jordan has not disclaimed sovereignty over the area.

Map Jordan Aa

West Bengal ***India*** The western part of the former Indian state of Bengal, which was divided at Indian independence in 1947. (East Bengal became East Pakistan, now Bangladesh.) West Bengal covers 88 752 km² (34 258 sq miles), mostly in the Ganges river plain and delta, and has most of India's Bengali-speaking people. It produces rice, jute, wheat and coal. Factories make steel, cars, aluminium and fertilisers, mainly around Calcutta, the capital, and in the Damodar river valley.

Population 54 580 600

Map India Dc

West Bromwich ***United Kingdom*** Major industrial centre on the western outskirts of Birmingham, in the English county of West Midlands. Coal has been mined since the late 18th century, and industry – including, today, oil refining and chemicals – soon followed.

Population 155 000

Map United Kingdom Ed

West Indies ***North Atlantic*** Group of islands to the east of Central America. They consist of three main groups: The Greater Antilles, including CUBA, HISPANIOLA, JAMAICA and PUERTO RICO; the Lesser Antilles, including BARBADOS, TRINIDAD AND TOBAGO, and the LEEWARD, WINDWARD and VIRGIN ISLANDS.

West Midlands ***United Kingdom*** Metropolitan county in central England, covering 899 km² (347 sq miles) and taking in the industrial area known as the Black Country. It includes the cities of BIRMINGHAM, COVENTRY and WOLVERHAMPTON, and is the centre of Britain's engineering and car industries.

Population 2 658 000

Map United Kingdom Dd

West Point ***USA*** Site of the US Military Academy, standing on the Hudson river about 72 km (45 miles) north of New York City. The academy, founded in 1802, was built on the site of a defence post established in 1778 during the War of Independence.

Map United States Lb

West Rand ***South Africa*** Part of the Witwatersrand west of Johannesburg. It includes the gold-mining towns of Roodepoort, Krugersdorp and Randfontein.

West Virginia ***USA*** State covering 62 629 km² (24 181 sq miles) between the Allegheny Mountains and the upper Ohio river. Much of it is mountainous and farming is limited, cattle, chickens and apples being the main products. The state is the USA's leading producer of bituminous coal, and also produces oil and gas; these fuel its chemicals, metal processing and glass industries.

It was originally part of Virginia, but its people opposed slavery and remained loyal to the Union during the Civil War (1861-5); they broke away to form a separate state in 1863. Harpers Ferry National Monument at the junction of the Shenandoah and Potomac rivers in the north-east corner of the state commemorates John Brown's rising against slavery in 1859. CHARLESTON is the state capital.

Population 1 936 000

Map United States Jc

westerlies Winds that blow from the subtropical high-pressure systems to the temperate low-pressure systems, from 35 to 65 degrees north and south of the Equator. They blow from the south-west in the Northern Hemisphere and from the north-west in the Southern Hemisphere.

Western Australia ***Australia*** Largest state in the country, covering 2 525 500 km² (975 100 sq miles) – an area twice the size of South Africa. It was the second British colony in Australia, being founded in 1829. More than 60 per cent of the population live in the state capital, PERTH. Much of the land is flat and arid but the PILBARA and the KIMBERLEY PLATEAU in the north-west are highland regions; the former is extremely rich in minerals, notably iron. Farming relies on sheep, but the 'Mediterranean' south-west produces wheat and fruit, and in the far north attempts are being made to grow irrigated rice and cotton.
Population 1 300 000
Map Australia Bc

Western Desert ***Egypt*** See SAHARA

Western Isles ***United Kingdom*** The Inner and Outer Hebrides off the west coast of Scotland, totalling 2901 km² (1120 sq miles). The Outer Hebrides stretch more than 200 km (120 miles) from the Butt of Lewis in the north, through North and South Uist to Berneray in the south. Lewis is the largest island, and has the only substantial town, Stornoway. The Inner Hebrides, nearer the Scottish mainland, include SKYE, MULL, Jura and Islay. Sheep farming, production of woollen goods, whisky distilling and fishing are the principal occupations. The islands are linked with each other and with the mainland ports of Oban and Ullapool by ferry.
Population 32 000
Map United Kingdom Ba

▼ FANTASTIC UNDERWORLD Shapely stalactites and stalagmites give an eerie beauty to a limestone cave at Yallingup near Western Australia's south-western coast.

Western Province (Barotseland) ***Zambia*** Region of 126 386 km² (48 798 sq miles) bordering Angola. Its provincial capital, Mongu, is about 550 km (340 miles) west of the national capital, Lusaka. The state is the home of the Lozi people, who live mostly on the savannah-covered floodplain of the Zambezi river and fish and grow maize, millet and pulses on small subsistence farms. They also rear cattle, sheep and goats.
Population 488 000
Map Zambia Bb

Western Samoa See SAMOA, WESTERN

Westland ***New Zealand*** South Island district stretching along the coast west of the Southern Alps. It covers 15 415 km² (5952 sq miles) and was a gold-mining area in the 19th century. Today its industries are farming, logging and coal mining. The magnificent mountains, glaciers, lakes, forests and the coastline make it a popular tourist area. The main towns are Greymouth and Hokitika.
Population 34 200
Map New Zealand Bc

Westman Islands (Vestmannaeyjar) ***Iceland*** Group of 15 islands – covering in total about 16 km² (6 sq miles) – about 10 km (6 miles) off the south coast of Iceland. The islands' main port, on the island of Heimaey, was partly destroyed by a volcanic eruption in 1973. However, the town has since been completely rebuilt and the fishing harbour improved.
Population 5500
Map Iceland Bb

Westmeath (An Iarmhí) ***Ireland*** Cattle-raising county covering 1763 km² (681 sq miles) of flatlands and lakes in central Ireland, about 50 km (30 miles) north-west of Dublin. The county town is Mullingar.
Population 61 500
Map Ireland Cb

Westminster ***United Kingdom*** See LONDON

Westmorland ***United Kingdom*** See CUMBRIA

Westphalia ***West Germany*** See NORTH RHINE-WESTPHALIA

wet spell Period of at least 15 consecutive days during which the daily rainfall is 1 mm (0.04 in) or more.

Wewak ***Papua New Guinea*** Capital of East Sepik province, on the north coast of New Guinea west of the SEPIK river mouth. It is a port handling exports of coffee, cocoa and rubber, and is the transit point for air travel to Irian Jaya, the Indonesian part of New Guinea. Tourism is increasing. At Cape Wom, just west of the town, Japanese forces in New Guinea formally surrendered on September 13, 1945.
Population 20 000
Map Papua New Guinea Ba

Wexford (Loch Garman) ***Ireland*** Market town, county town and port, 116 km (72 miles) south of Dublin. Wexford became a stronghold during the Anglo-Norman invasion of 1169. Its importance as a port has declined as the river Slaney has silted up, but industry has been developed, including a car plant.

The county of the same name, covering 2351 km² (908 sq miles), consists of an arable plain, backed in the north-west by the Blackstairs Mountains, which rise to 796 m (2610 ft).
Population (county) 99 000; (town) 11 900
Map Ireland Cb

Whangarei ***New Zealand*** Main city of Northland, situated on a fine natural harbour 135 km (85 miles) north of Auckland. It is an industrial and regional market centre.
Population 40 200

whirlwind Column of rotating air centred on a small, local area of low atmospheric pressure. It is less violent than a TORNADO.

White Russia ***USSR*** See BELORUSSIA

White Sands National Monument ***USA*** Area of hills of glistening white gypsum and dunes up to 15 m (50 ft) high in New Mexico, about 100 km (60 miles) north of El Paso, Texas. The monument covers 586 km² (226 sq miles). Some 65 km (40 miles) farther north is the site where the first atomic bomb was exploded on July 16, 1945.
Map United States Ed

A DAY IN THE LIFE OF A CANADIAN 'MOUNTIE'

This chilly Saturday in February is John Pickwood's favourite day of the year. Today the 15 000 inhabitants of the little mountain-fringed town of Whitehorse are celebrating the Sourdough Rendezvous – a commemoration of the Klondike gold rush of the 1890s.

The Sourdough Rendezvous, named after the pancakes the grizzled miners used to eat, captures for John Pickwood all the romance of living in Canada's frozen northern wastes and of being a Mountie – which is, after all, like living a legend. The name 'Mountie' is, for John, something of a misnomer. He has no horse. The Royal Canadian Mounted Police are a national force which polices all Canada's towns and most cities (only the half dozen big cities have a City Force). Most today patrol by car.

John, 25, and his wife June, 24, have no regrets about leaving their home in Vancouver to fly up north. They go back twice a year to see friends and family, but find the bustling Pacific port almost intimidating. In the Yukon, June has happily settled into teaching at the state-run high school – and John revels in the job of being a Mountie in a land of rugged individualists.

This morning he strides down the ice-covered main street to keep an eye on the traffic; trucks are piling in and the streets are crowded with revellers in the dress of the 1890s.

In the afternoon he goes down to the frozen Yukon river to cast a supervisory eye over the dog-sled races. The huskies' breath clouds the icy air, and drivers gather in huddles to discuss form. John finds out who is the favourite for today's 6 km (4 mile) dash down the river.

The night is exceptionally busy. The bars are full to overflowing – and the beer mugs too. After the long winter, everyone is eagerly looking forward to the arrival of spring. Arrests, as it turns out, are few, and the fun is light-hearted. John's main responsibility is to watch that no drunks pass out in the snow, where the cold would soon kill them. Temperatures in winter can be as low as −57°C (−70°F), although it is not so cold as that tonight.

John and June are, in a sense, exceptions to many of the inhabitants of the Yukon: they have managed to stay together since they moved here. The area has a high proportion of divorces and separations, partly because of a new 'gold rush' Yukon – the oil boom which has brought in young strangers with money to burn and has changed the face of the towns and disrupted their societies. It is an unsettling time for everyone.

But John and June are happy in each other's company – and on Sourdough Rendezvous day they would not change places with anyone in the world.

Western Sahara

A MINERAL-RICH DESERT LAND THAT BROUGHT NEIGHBOUR NATIONS INTO BITTER RIVALRY

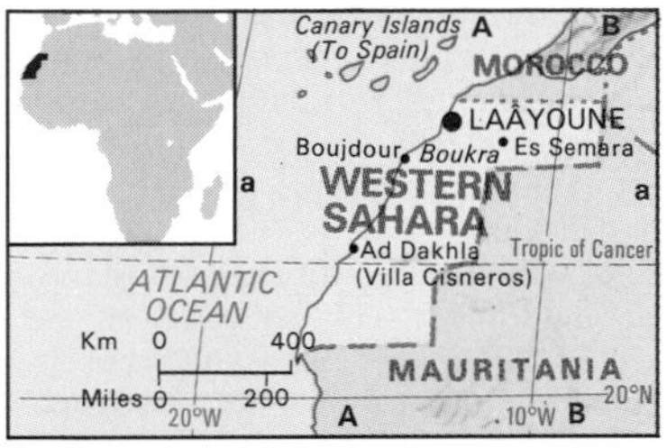

This troubled territory of phosphates, desert and nomads has split the Organisation of African Unity (OAU) from top to bottom. There has been warfare in the land since 1976 involving the Sahrawis (the indigenous people of the area) and the Moroccans, who have long claimed it.

The country, covering 266 770 km² (103 000 sq miles) on the north-west coast of Africa, consists entirely of desert – mainly stony but with some sandy areas. There are a few scattered oases and a little poor pastureland. Annual rainfall is less than 50 mm (2 in).

Most of the 300 000-400 000 Sahrawis are a mixture of Arab and Berber, and almost all are Muslims. The area was colonised by Spain from 1884 and declared a Spanish province in 1960, but like the rest of Africa was affected by growing nationalism. In 1973 the Sahrawi liberation movement, the Polisario Front, demanded independence. Then the problems started.

PEACEFUL INVASION

Morocco has historical claims to the territory as part of 'Greater Morocco', and in 1975 King Hassan sent 350 000 unarmed Moroccans on a peaceful 'green march' into Western Sahara in an attempt to force Spain to hand it over. Apart from historical claims, Morocco's interest lay in Western Sahara's deposits of phosphate rock – the world's largest. With Morocco's own considerable output of phosphate – raw material for fertiliser – these deposits would give the Moroccans a dominant position in world markets.

The peaceful occupation was successful and in 1976 Spain withdrew from the colony, agreeing that it should be divided between Morocco and Mauritania, Morocco taking the northern 162 000 km² (about 62 000 sq miles). The Polisario, however, declared an independent Saharan Arab Democratic Republic which was supported by Algeria and (for a time) Libya. The Polisario found themselves fighting on two fronts, but in 1979 Mauritania gave up its claim, and its territory was occupied by Morocco.

Some of the more conservative African states supported Morocco's claim, others backed the Polisario. At its summit meeting, in 1984, the OAU agreed that the Polisario representatives should take their seat. As a result, Morocco left the OAU in protest.

By then the war had been going on for eight years and Morocco was largely in control of most of the northern section, including the chief town, LAAYOUNE, and the phosphate deposits at BOUKRA. They had built defensive fortifications across the country to restrict the guerrillas' movements to the inhospitable stony desert to the south but guerrilla incursions into the north continued. Some 125 000 Sahrawi refugees were living in camps in south-western Algeria.

The Moroccans have already invested heavily in the part of Western Sahara they control, improving transport, fisheries, water supplies and education. Phosphate exports could finance much more development, both economic and social. A referendum may be held on the territory's future, but with a large nomadic population this could be difficult to conduct.

WESTERN SAHARA AT A GLANCE

Area 266 770 km² (103 000 sq miles)
Population 300 000-400 000 (including refugees in Algeria)
Chief town Laayoune
Government Disputed territory occupied by Morocco and Polisario guerrillas
Currency Moroccan dirham = 100 centimes
Languages Arabic, Berber, Spanish
Religion Muslim
Climate Hot and dry
Main primary products Phosphates, iron ore, fish
Major industries Phosphate mining, fishing
Main exports Phosphates
Annual income per head (US$) 800
Population growth (per thous/yr) 18

White Sea Almost enclosed inlet of the BARENTS SEA off the USSR which largely freezes over in winter, although icebreakers keep the main ports open. It was used during the Second World War to get essential convoys with supplies into the USSR. The sea is rich in fish. Timber is a major export from its chief port ARCHANGEL (Arkhangelsk). The northern DVINA and Onega rivers flow into the sea.
Area 95 000 km² (36 680 sq miles)
Map USSR Eb

Whitehorse ***Canada*** Capital of Yukon Territory, at the upper limit of navigation on the Yukon river, 84 km (52 miles) north of the border with British Columbia. It is the terminus of the White Pass and Yukon Railway to Skagway, 90 km (55 miles) south in the Alaska 'panhandle', and is just off the Alaska Highway. It was an important staging post during the Klondike gold rush in the 1890s.
Population 14 800
Map Canada Bb

whiteout Polar weather condition caused by a heavy cloud cover over the snow, in which the light coming from above is approximately equal to the light reflected from below. Whiteouts are characterised by the absence of shadow, the invisibility of the horizon and the discernibility of only very dark objects.

Whitney, Mount ***USA*** Highest peak in the country outside Alaska. It rises to 4418 m (14 495 ft) in the Sequoia National Park in California, about 280 km (175 miles) north of Los Angeles.
Map United States Cc

Whyalla ***Australia*** Major port and steel-making city in South Australia on Spencer Gulf, 225 km (140 miles) north-west of Adelaide. It exports iron ore from open-cast mines inland.
Population 29 900
Map Australia Fe

Whydah ***Benin*** See OUIDAH

Wichita ***USA*** City straddling the Arkansas river in southern Kansas. It is a centre for aircraft production, oil-field equipment and other engineering and a grain and meat market.
Population (city) 283 500; (metropolitan area) 428 600
Map United States Gc

Wicklow (Cill Mhantáin) ***Ireland*** Resort, market town and county town 45 km (28 miles) south of Dublin. The county of the same name covers 2025 km² (782 sq miles), and includes the Wicklow Mountains – rising to 926 m (3038 ft) – and part of Ireland's central plain. John Halpin, captain of the *Great Eastern*, the ship that laid the first one-piece transatlantic telegraph cable (in 1866), was born in the town.
Population (county) 87 000; (town) 5000
Map Ireland Cb

Wielkopolska ***Poland*** See POLAND, GREATER

Wiener Neustadt ***Austria*** Industrial city 40 km (25 miles) south of Vienna. Most of it was destroyed in the Second World War, when only 18 of its 4200 buildings were left undamaged. The Holy Roman Emperor Maximilian I (1459-1519) was born in the city, and is buried there in the Church of St George, though the memorial tomb to the emperor lies in INNSBRUCK. The city's products include cars, aircraft, rolling stock and textiles.
Population 35 000
Map Austria Eb

Wienerwald ***Austria*** Forested hills called the Vienna Woods, to the west and north-west of the capital. The villages of the region's eastern foothills are set among vineyards and have become commuter areas for Vienna. The new wines can be tasted in the courtyard bars of converted farmhouses. The hills are an extension of the eastern Austrian Alps.
Map Austria Da

Wiesbaden ***West Germany*** Spa town and state capital of Hessen beside the Rhine 32 km (20 miles) west of Frankfurt am Main. Rheumatics sufferers have visited its hot saline springs since Roman times. Wiesbaden is a centre of the German publishing and film industries, has the national library of West Germany and is the site of a large American military base.
Population 272 600
Map West Germany Cc

Wigan ***United Kingdom*** English town formerly in Lancashire, now in Greater Manchester, about 25 km (16 miles) north-east of Liverpool. There was a Roman fort on the site and it received its royal charter in 1246, but Wigan grew to industrial prominence with the 19th-century development of local coal mines – the last of which closed in 1967 – and the cotton textile industry (which is also in decline). New industries include engineering, food processing, plastics and retail distribution.
Population 80 000
Map United Kingdom Dd

Wight, Isle of ***United Kingdom*** Island county covering 381 km² (147 sq miles) off the south coast of England, and separated from the mainland by the Spithead and Solent channels. It is about 36 km (22 miles) long from the town of Bembridge in the east to The NEEDLES chalk stacks in the west, and is a world-famous yachting centre, with a regatta at Cowes each August. In the 19th century, Queen Victoria and the poet Alfred, Lord Tennyson had homes on the island, and the queen died there in 1901. It has remained a popular tourist and retirement area. Newport is the county town.
Population 120 000
Map United Kingdom Ee

Wigtown ***United Kingdom*** See DUMFRIES AND GALLOWAY

▼ STORMY PAST The peaceful church of Fort Amsterdam, built by the Dutch at Willemstad from 1634, still has a cannonball embedded in its wall – a gift from English attackers in 1804.

Wilhelm, Mount ***Papua New Guinea*** The country's highest peak, rising to 4509 m (14 793 ft) in the Bismarck Range.
Map Papua New Guinea Ba

Wilhelmshaven ***West Germany*** North Sea port about 65 km (40 miles) north-west of Bremen. It was founded in the 19th century as a naval base. Since the Second World War it has become an oil-importing town, with a pipeline link to the industrial Ruhr. It makes petrochemicals and precision instruments.
Population 98 200
Map West Germany Cb

Willemstad ***Netherlands Antilles*** Seaport and capital of CURACAO and of the Netherlands Antilles, situated on the Schottegat on the south-west coast of Curaçao. It is one of the largest ports in the world in terms of the tonnage it can accommodate – nearly 100 million tonnes of shipping at a time – and has the largest dry dock in the Americas. It is also a major oil-refining centre, and handles crude oil from the Lake Maracaibo oil fields of Venezuela.
Population 50 000
Map Caribbean Bc

Williamsburg ***USA*** Historic city between the James and York rivers in Virginia, about 180 km (110 miles) south of the national capital, Washington. It was the state capital for nearly 80 years after 1699. The College of William and Mary, founded in 1693, is the second oldest such institution in the country after Harvard.

Tourists flock to Colonial Williamsburg, a huge restoration project occupying the centre of Williamsburg where the guides wear period costume. The project has more than 400 buildings renovated and re-created in their original settings. With JAMESTOWN and YORKTOWN it lies in the 38 km² (15 sq mile) Colonial National Historical Park.
Population 11 000
Map United States Kc

Wilmington ***USA*** Industrial and port city on the west bank of the Delaware river, in Delaware state, some 40 km (25 miles) south-west of Philadelphia. It is one of the world's largest centres of the chemical industry and the home of the du Pont multinational chemical firm.
Population (city) 69 700; (metropolitan area) 539 700
Map United States Kc

Wiltshire ***United Kingdom*** County of central southern England covering 3481 km² (1344 sq miles). It contains two of Britain's oldest landmarks – STONEHENGE and AVEBURY stone circles – and the army training grounds of Salisbury Plain. Most of the county is agricultural, but Swindon – a former railway town whose locomotive workshops are now closed – has a number of computer and light industrial companies. The beautiful cathedral of SALISBURY has the tallest spire in England at 123 m (404 ft). The county town is Trowbridge.
Population 540 000
Map United Kingdom De

Winburg ***South Africa*** Oldest town in the Orange Free State, about 105 km (65 miles) north-east of Bloemfontein. Founded by Voortrekkers (Afrikaner settlers) in 1836, it is now a market town.
Population 7000
Map South Africa Cb

Winchester ***United Kingdom*** Cathedral city in the English county of Hampshire, about 80 km (50 miles) south-west of London. It was the capital of Saxon kingdom of Wessex before the 7th century. There are remains of the Roman city of Venta Belgarum, and of a Norman castle, but Winchester's best-known landmark today is its cathedral. It replaced a Saxon building, and was begun in 1079 but only completed three centuries later, with further additions until 1525. Around the lawns of the cathedral close are the ruins of two bishop's palaces and several buildings dating from before 1500. They include parts of Winchester College – one of the most famous of England's public schools, founded in 1382 by William of Wykeham, Bishop of Winchester.

A round table said to have belonged to the legendary King Arthur hangs in the hall (the only remaining part) of the Norman castle; but the table dates only from about 1150, some six centuries after Arthur died. It is another king – Alfred the Great (849-99), called the founder of the English nation – who is chiefly commemorated in this historic city. It was his capital and his main base in his struggle against the Danish invasion.
Population 31 000
Map United Kingdom Ee

wind Current of air on the earth's surface induced by difference in atmospheric pressure, varying in strength from light wind to hurricane. Winds are identified by the direction from which they come, though many have local names. See PRESSURE GRADIENT and BEAUFORT SCALE

wind deflection Difference in the direction of the wind from that expected from the pressure gradient, caused by the rotation of the earth. The deflection is negligible at the Equator and greatest at the poles. In the Northern Hemisphere, winds are deflected to the right, causing them to circulate anticlockwise around low-pressure systems and clockwise around highs. The opposites apply in the Southern Hemisphere. The deflection is apparent, rather than real, since it is the earth that moves and not the wind that changes direction.

Windermere ***United Kingdom*** Largest lake in England and a town in the LAKE DISTRICT of north-west England, some 110 km (70 miles) north of Liverpool. The narrow lake is 16 km (10 miles) long, with many islands, and is a popular water sports and tourist attraction.
Map United Kingdom Dc

Windhoek ***Namibia*** National capital and industrial city. It was founded as the military headquarters of German South West Africa in 1890 and still has many German buildings, including the Alte Fest (Old Fort) and the German Evangelical Lutheran church. German festivals are still held in the city, and local Herero women wear dresses modelled on those worn by the wives of 19th-century missionaries. Industries include meat canning, brewing, manufacture of bone meal, and processing of 'Persian lamb' pelts from karakul lambs.
Population 105 000
Map Namibia Ab

Windsor ***Australia*** With its twin town Richmond (population 16 000), Windsor is one of Australia's oldest settlements. Both were founded about 1810 on rich farming land beside the Hawkesbury river about 50 km (30 miles) north-west of Sydney. They contain many fine colonial buildings.
Population 15 400

Windsor ***Canada*** Canada's most southerly city, on the Detroit river in south-west Ontario. It is linked by bridge and tunnel with the city of Detroit, across the US border. It makes cars, machine tools and chemicals, has salt mines and distilleries, and is the centre of a farming region.
Population 246 110
Map Canada Gd

Windsor ***United Kingdom*** Royal English town on the River Thames in Berkshire, 35 km (22 miles) west of London. There has been a castle at Windsor since the Norman Conquest in 1066. The base of the great round tower dates from the 12th century and St George's Chapel from the 16th. Most of the remainder of the huge castle – which covers more than 5 hectares (12 acres) – was built in the 19th century. It is one of the homes of the British royal family, and the burial place of several kings and queens. The castle and the 1940 hectare (4795 acre) Great Park beside it enclose the town of Windsor on the east and south sides. Across the Thames is Eton, whose public school was founded in 1440 by Henry VI.
Population 31 000
Map United Kingdom Ee

Windward Islands ***Caribbean*** See LEEWARD ISLANDS

Windward Islands ***French Polynesia*** See SOCIETY ISLANDS

Winnipeg ***Canada*** **1.** Capital of Manitoba, lying at the junction of the Assiniboine and Red rivers. First settled as a trading post in 1738, it became a city in 1873 with a population of fewer than 4000. It is now one of the largest cities of the prairie provinces and has one of the world's biggest wheat markets. It has two universities, the Royal Winnipeg Ballet and a symphony orchestra. In addition to grain elevators and flour mills, the city has meat packing and food processing plants, clothing factories and railway workshops.

2. Lake covering 24 500 km² (9460 sq miles) in southern Manitoba; it drains via the Nelson and other rivers into HUDSON BAY. Like Lake Manitoba to the west, it formed part of an important transport route before the railways.
Population 585 000
Map Canada Fc

Winterthur ***Switzerland*** Industrial town about 20 km (12 miles) north-east of the city of Zürich. Its manufactures include locomotives and textiles. It was founded in 1175 and has a 16th-century late Gothic church.
Population 90 000
Map Switzerland Ba

Wirral ***United Kingdom*** Peninsula on the Irish Sea in north-west England between the estuaries of the Mersey and Dee rivers. It is divided between the counties of Cheshire and Merseyside. The industrial and port conurbation of Birkenhead and Wallasey smothers the tip. The town of Chester is at the peninsula's southern end.
Map United Kingdom Dd

Wisconsin ***USA*** Midwestern state covering 145 439 km² (56 154 sq miles) between the Great Lakes and the Mississippi river. Dairy farming is the main agricultural activity and the state also produces beef, maize, pigs and fruit. Machinery, processed food and transport equipment are the principal manufactures, and tourism is expanding. MADISON is the capital and MILWAUKEE the largest city.
Population 4 775 000
Map United States Hb

Wisla ***Poland*** See VISTULA

Wittenberg ***East Germany*** Industrial town midway between Leipzig and the capital, East Berlin, which was the cradle of the Reformation. Plaques on the doors of the Schlosskirche (castle church) mark where in 1517 the priest and reformer Martin Luther nailed his 95 theses against the Roman Catholic Church. Luther is buried in the town.
Population 54 200
Map East Germany Cc

Witwatersrand (The Rand) ***South Africa*** Gold-mining and industrial area in south Transvaal; its name means 'ridge of white water'. This ridge, between about 1500 and 1800 m (4900 and 5900 ft) high, extends about 80 km (50 miles) from Randfontein in the west, through Johannesburg to Springs in the east.

Mining began in 1886 near Johannesburg after a prospector named George Harrison reported that he had found a 'payable gold field'. Harrison later sold his claim for just £10 (about US$50), leaving fortunes to be made by later miners. Since the 1920s the East and West Rand have become increasingly important as the Central Rand mines have been worked out. The South African currency, the rand, gets its name from the region.

Map South Africa Cb

Wloclawek (Vlotslavsk) ***Poland*** Paper-making city about 140 km (86 miles) north-west of the capital, Warsaw. It was founded as a fortress in the 11th century, and has a 14th-century cathedral. The city also makes nitrate fertilisers.

Population 113 400

Map Poland Cb

Wolds, The ***United Kingdom*** Series of rounded chalk hills in eastern England on both sides of the Humber estuary in Lincolnshire, Yorkshire and Humberside. They are mostly under 180 m (600 ft) high, and provide good grazing and arable land.

Map United Kingdom Ed

Wolf Creek ***Australia*** Site of a meteorite crater measuring 840 m (2755 ft) across and 90 m (295 ft) deep. It lies near the town of Halls Creek in the Kimberleys of Western Australia about 160 km (100 miles) from the state's border with the Northern Territory. The perfectly preserved circular crater – which is littered with glassy green balls of rock formed by the impact – is thought to be about 6000 years old.

Map Australia Db

Wollomombi Falls ***Australia*** Said to be the highest waterfall in the country, plunging 357 m (1171 ft) near Armidale in the New England region of New South Wales, about 370 km (230 miles) north of Sydney.

Map Australia Ie

Wollongong ***Australia*** Port and major industrial city in New South Wales, 80 km (50 miles) south of Sydney. It is a coal-mining centre, and the urban area includes PORT KEMBLA, where Australia's biggest steelworks is situated.

Population 235 000

Map Australia Ie

Wolverhampton ***United Kingdom*** Town in the West Midlands of England, the so-called Black Country, some 22 km (14 miles) north-west of Birmingham. It began as a settlement around a 10th-century monastery and grew, in the 19th century, into a coal mining and manufacturing town, noted for its locks and safes.

Population 255 000

Map United Kingdom Dd

Wonga Wongue National Park ***Gabon*** Nature reserve covering 3800 km² (1462 sq miles) north of Ogooué river in western Gabon. It contains antelopes (including the rare situtonga), buffaloes, crocodiles, elephants, hippopotamuses and leopards.The unusual Bam Bam amphitheatres in the park are beautiful semicircular basins eroded by torrential rain into the sides of hills that rise about 100 m (328 ft) above the surrounding plain. The scenery is reminiscent of the BADLANDS of the USA.

Map Gabon Ab

Wonsan ***North Korea*** East coast industrial city and port, 140 km (87 miles) east of Pyongyang. It was heavily bombed during the Korean War (1950-3) but has since been rebuilt. It now has chemicals and oil-refining plants, as well as shipbuilding yards and factories making railway rolling stock.

Population 215 000

Map Korea Cc

Woodlands-Kranji ***Singapore*** The largest new-town development on Singapore Island, begun in 1971 on the north side, near the southern end of the causeway that carries the road, railway and water pipeline links with the Malaysian mainland. It is planned to accommodate up to 300 000 people.

The Japanese invasion force landed at the largely undefended site in December 1941, during the Second World War.

Population 68 600

Map Singapore Aa

Woomera ***Australia*** Town in South Australia, in a semiarid region 160 km (100 miles) north-west of Port Augusta, from which water is piped to it. It was established in 1947 to house workers at the nearby missile testing site, which was operating in the 1950s and 1960s. The town gets its name from an Aboriginal word for a spear-thrower – a notched wooden stick used by hunters to extend the range of their weapons. Some 17 km (11 miles) south of Woomera is the Joint Defence Space Communications Station, operated jointly by the US and Australian Departments of Defence.

Population 4100

Map Australia Fe

Worcester ***United Kingdom*** City and former county of western England, the latter now amalgamated with its neighbour Herefordshire. The city stands on the River Severn about 45 km (28 miles) south-west of Birmingham. Its cathedral was built between the 11th and 14th centuries and shares the annual Three Choirs Festival with its counterparts in Hereford and Gloucester. Royal Worcester porcelain and Worcestershire sauce are its best known products.

Population 76 000

Map United Kingdom Dd

Worcester ***USA*** City in Massachusetts about 60 km (38 miles) west of Boston. It was mainly a textile centre, but now its varied products include chemicals and precision instruments. It is the seat of Clark University.

Population (city) 159 800; (metropolitan area) 404 700

Map United States Lb

Worms ***West Germany*** Cathedral city on the Rhine about 20 km (12 miles) north and slightly west of the city of Mannheim. At Worms in 1521, Martin Luther (1483-1546), a priest already excommunicated by the Pope, appeared before the Imperial Diet, or assembly, and repeated his religious defiance to the Holy Roman Emperor Charles V, taking the Protestants another step towards their break with Rome. Worms is in a grape-growing region which produces high-quality white wine.

Population 72 900

Map West Germany Cd

Wrath, Cape ***United Kingdom*** North-western tip of mainland Scotland, in Highland region. Nearby cliffs, 260 m (850 ft) high, are the tallest in Britain.

Map United Kingdom Ca

Wrexham ***United Kingdom*** The main industrial and commercial centre of North Wales, lying in the county of Clwyd 40 km (25 miles) south of Liverpool. Its 19th-century prosperity was based on coal mining, steel and brickmaking, but as these declined new industries such as chemicals and textiles have developed. Elihu Yale (1648-1721), whose donations helped to found Yale University in the American state of Connecticut, is buried in Wrexham churchyard.

Population 40 000

Map United Kingdom Dd

Wroclaw (Breslau) ***Poland*** Industrial city 306 km (190 miles) south-west of the capital, Warsaw. Founded in the 10th century astride several tributaries of the Oder river, it prospered as a trading town. Under Austrian rule (1526-1741), craft industries developed, and the city's many Renaissance and Baroque houses, churches and the Jesuit College (now the university) date from that time.

Wroclaw became an industrial centre under Prussian-German occupation (1741-1945). It now produces electric railway locomotives, wagons, machinery, electrical appliances, chemicals, foodstuffs, paper, timber and artificial silk. Some 100 bridges span the 90 km (56 miles) of canals and rivers, giving the city a Venetian air.

Population 631 300

Map Poland Bc

Wuhan (Hankow) ***China*** Capital of Hubei province. It stands at the confluence of the Han Shui and Chang Jiang rivers and is made up of the cities of Wuchang, Hankow and Hanyang. In 1911, Wuhan was the scene of the Nationalist unrest which led to the overthrow of the Manchu emperors, the last of imperial China.

The Qing dynasty (1644-1911) was established by Manchu invaders from their power base in Manchuria (now called Dongbei). They seized northern China in 1644 and extended their rule over the whole country. In the 19th century the Chinese government became increasingly corrupt and incapable of resisting the domination of China by Western powers and Japan. The indemnity which China had to pay to the powers after the Boxer Rebellion of 1900 provoked strong anti-foreign sentiment among the Chinese. In 1911 the Chinese court tried to purchase a Chinese-owned railway (probably to sell it to foreigners). This sparked off a rebellion. Chinese troops mutinied in Wuhan in September 1911 and the republic was declared on January 1, 1912. Today it is one of the most important concentrations of manufacturing in central China, with iron and steel, cement, paper-making, and cotton textile factories.

Population 3 832 500

Map China He

Wuppertal ***West Germany*** Textile manufacturing city about 25 km (15 miles) east of Düsseldorf. An overhead tramway runs through the city and along the Wupper river.

Population 386 000

Map West Germany Bc

▲ **BISHOPS' SEAT Medieval Marienburg fortress, home of the ruling bishops of Würzburg for 500 years, overlooks the city and its vineyards – noted for exuberant, fine white wines.**

Würzburg ***West Germany*** University city about 100 km (60 miles) east and slightly south of Frankfurt am Main. It was the seat of prince-bishops, who from 1650 created a Baroque city with cathedral, bridges, palace and gardens. The city was badly damaged in the Second World War, but has been restored.
Population 127 000
Map West Germany Cd

Wusul Jiang (Ussuri) ***China/USSR*** River rising in the USSR and flowing 800 km (500 miles) into the Heilong Jiang (AMUR) river. It forms part of the boundary between China and the Soviet Union where clashes between Chinese and Russian forces occurred in the mid-1960s.
Map China Kb

Wuxi ***China*** City of Jiangsu province on the north shore of Tai Hu lake, 120 km (75 miles) west of Shanghai. Its major industries are cotton textiles and silk reeling – the drawing out of raw silk threads from silkworm cocoons to be wound on bobbins.
Population 1 000 000
Map China Ie

Wyoming ***USA*** Western state covering 253 597 km² (97 914 sq miles). It is the ninth largest state in area but is the second smallest only to Alaska in population numbers and density. The Rocky Mountains in the west and south rise to 4208 m (13 804 ft) at Gannett Peak, and plains more than 1000 m (3300 ft) above sea level stretch north-eastwards to the Black Hills on the South Dakota border. Irrigation is necessary for some crops because of the semiarid climate, and cattle, wheat and sugar beet are the main farm products. Wyoming has extensive deposits of coal, oil and natural gas, and its fine scenery – including that of YELLOWSTONE NATIONAL PARK – increasingly attracts tourists. CHEYENNE is the capital.
Population 509 000
Map United States Eb

Xai-Xai ***Mozambique*** Seaport on the mouth of the Limpopo river, in southern Mozambique, about 230 km (145 miles) from the South African border. It was formerly called João Belo. The port is the centre of a productive farming area and exports cotton, maize, rice and sugar.

Xai-Xai is the capital of Gaza province, which covers 75 709 km² (29 224 sq miles).
Population (province) 1 030 500
Map Mozambique Ac

Xauen ***Morocco*** See CHAOUEN

xenolith Rock fragment different from the IGNEOUS ROCK mass in which it occurs.

Xi Jiang ***China*** Longest river in the south, rising in the mountains of Yunnan province. It flows eastwards for 2197 km (1370 miles) to the Zhu Jiang delta on the South China Sea. The river is an important transport artery and its water is used extensively for irrigation.
Map China Gf

Xiamen (Amoy) ***China*** Port of Fujian province, opposite Taiwan. It was the first Chinese port used by Portuguese, Dutch and English traders, from the 16th century, and became a treaty port (where foreigners could live and trade) in 1842. It is now being developed as one of China's 'special economic zones' – areas set up to attract overseas investment.
Map China Hf

Xi'an ***China*** Capital of Shaanxi province. It is an ancient city in the Wei He valley and, under the name of Chang'an, was the capital of imperial China during the Tang dynasty (AD 618-906). Among the many archaeological sites around Xi'an is the vast tumulus tomb at Mount Li of Qin Shi Huangdi (ruled 221-210 BC), China's first emperor and founder of the Qin or Ch'in dynasty, from which China gets its name. The tumulus, discovered in 1974, contains a buried army of some 7000 life-sized terracotta figures in three pits. A fourth has still to be fully excavated. Today Xi'an is an important centre for textile and electrical manufactures.
Population 2 020 000
Map China Gd

Xiang Jiang ***China*** River in south-central China, flowing 1150 km (719 miles) north through Hunan province to Dongting lake and the Chang Jiang.
Map China Ge

Xigaze ***China*** Town in the Xizang Autonomous Region (Tibet), lying in the Yarlung Zangbo (Brahmaputra) valley 230 km (140 miles) south-west of Lhasa. It is the site of the Tashi-Lhunpo Buddhist monastery. Before the Chinese moved on Xizang in 1950 the monastery had 3000 monks. It still survives, though with fewer monks and without their spiritual leader, the Panchen Lama, whose authority was second only to that of the Dalai Lama.
Population 40 000
Map China Ce

Xingang ***China*** Port in the north of the country at the mouth of the Hsi He river, 45 km (28 miles) east of Tianjin. It has been developed since the 1950s as the main harbour for Tianjin, and is now one of the largest ports in China.
Map China Hc

Xining ***China*** Capital of Qinghai province, about 200 km (125 miles) north-west of LANZHOU. It is a centre for the manufacture of iron and steel, farm machinery and fertilisers.
Population 860 000
Map China Ed

Xinjiang (Sinkiang) Uygur Autonomous Region ***China*** Region covering 1 600 000 km² (620 000 sq miles) of western China, historically known as Chinese Turkestan or Dzungaria. For centuries, nomadic herding was the mainstay of the region's economy, and Uygur, Kazakh, Kirghiz and Tadzhik herdsmen still graze their animals on the vast pastures.

Since the mid-1950s, at first with Soviet assistance, the great mineral wealth of the deserts has begun to be exploited. There is oil beneath the Junggar and Tarim basins, along with other minerals, including coal and metal ores. A railway connects the region with Gansu province and northern China, but the successful development of Xinjiang's mineral riches will depend on further improvements in communications.
Population 12 560 000
Map China Bc

Xishuangbanna ***China*** Region in Yunnan province famous for its rubber, shellac (from resin secreted by the lac insect), quinine derived from the cinchona bark, and rare plants.
Map China Ef

Xizang Autonomous Region (Tibet) ***China*** One of the most remote regions in the world. It lies north of the great Himalaya mountain range

▲ **RUSSIAN RIVIERA Yalta has as much sunshine as Nice in the south of France. The playwright Anton Chekhov (1860-1904) was a school teacher in the Crimean resort, and is commemorated by a museum (his villa) and a theatre.**

and encompasses the vast Xizang Gaoyuan (Tibetan Plateau), the upper valley of the Yarlung Zangbo (BRAHMAPUTRA) river and part of the QAM'DO region. Much of the 1 200 000 km² (460 000 sq mile) plateau is uninhabited. The small population is concentrated mainly in the Yarlung Zangbo valley, where barley, rye and wheat are grown. In the Qam'do area, tea is produced. Almost half of the population herds yaks, sheep and cattle.

Tibet was formerly a Buddhist kingdom, ruled from Lhasa by a priestly aristocracy headed by the Dalai Lama, the spiritual leader of Tibetan Buddhists. In earlier centuries China intermittently extended its rule over Tibet, which was absorbed into China during the Yüan dynasty (1279-1368) and again during the Qing dynasty; in the latter period Chinese armies occupied Lhasa in 1717 and Tibet was thereafter incorporated within the Chinese Empire until the revolution of 1911. In 1913 the Dalai Lama declared his country independent. China invaded in 1950, and the Dalai Lama fled to India after an uprising in 1959.

During the Cultural Revolution (1966-76), many of Tibet's monasteries and shrines were destroyed and practising Buddhists were persecuted. Since 1976, China has adopted a more moderate policy towards Tibet. It is China's poorest province and strong tensions remain between the 120 000 ethnic Chinese and the 1 710 000 native Tibetans.

Population 1 830 000
Map China Ce

Xizang Gaoyuan (Tibetan Plateau) ***China*** An immense expanse of uninhabited highland, covering roughly 1 760 000 km² (679 500 sq miles) and 4000 m (15 000 ft) above sea level, the biggest and highest plateau in the world. The climate is extremely harsh – even at the height of summer, temperatures rarely rise above freezing. It is rich in mineral resources, but these have not yet been exploited because of its remoteness.
Map China Cd

Xuzhou ***China*** Strategic city in north-west Jiangsu, 625 km (390 miles) south and slightly east of Beijing. It is a transport hub and centre of a rich coal mining area. Its products include machinery and textiles.
Population 700 000

Yakut Republic ***USSR*** Autonomous republic of the RSFSR in east-central Siberia, covering an area of 3 103 200 km² (1 198 150 sq miles) – making it almost as large as the whole of Western Europe. The republic consists largely of barren, sub-Arctic plains, hill country and marshy pine forests. The climate is appallingly severe, and the towns of Oymyakon and VERKHOYANSK are two of the coldest places on earth, with winter temperatures plummeting to below −60°C (−76°F). The main agricultural occupations are herding – often of reindeer – fishing and hunting. Gold, tin and coal are mined.

The republic's capital, Yakutsk – a name sometimes used for the whole republic – lies 480 km (300 miles) south of the Arctic Circle. It was founded in 1632 as a fort beside the 10 km (6 mile) wide LENA river. The city is now a fur-trading centre and its main industries are leather tanning, brick-making and timber milling.
Population (republic) 965 000; (city) 175 000
Map USSR Nb

Yakutsk ***USSR*** See YAKUT REPUBLIC

Yala ***Sri Lanka*** See RUHUNA

Yalta ***USSR*** Tourist and health resort in the Crimea, 50 km (30 miles) east of Sevastopol'. Stretched along a sheltered bay, Yalta is bedecked with flowering shrubs and plants. It has attracted holidaymakers and the ailing since the early 19th century and today has numerous sanatoriums.

In February 1945 the small resort of Livadiya, 3 km (2 miles) south of Yalta, was the scene of the historic Yalta conference to plan the final stage of the Second World War and the division of power in postwar Europe. The conference took place in the Grand Palace, built mostly of marble in 1911 by Russia's last tsar, Nicholas II (1868-1918), and was attended by the British Prime Minister, Winston Churchill (1874-1965), the President of the United States, Franklin D. Roosevelt (1882-1945), and the Soviet leader, Joseph Stalin (1879-1953).
Population 85 000
Map USSR Ed

Yalu ***China/North Korea*** River flowing 795 km (500 miles) along the countries' border. It is a source of hydroelectric power.
Map China Ic

Yamagata ***Japan*** City in north Honshu island, about 320 km (200 miles) north and slightly east of the capital, Tokyo. It now specialises in foodstuffs and electrical goods.
Population 245 200
Map Japan Dc

Yambol ***Bulgaria*** River port and capital of Yambol province, on the Tundzha river 72 km (45 miles) east of Stara Zagora. It ships wool and wine. Industries include metal goods, textiles and tanning.
Population 88 700
Map Bulgaria Cb

Yamoussoukro ***Ivory Coast*** Birthplace of Felix Houphouët-Boigny, the country's president since independence in 1960 from French colonial rule. The town is in the centre of the country, 220 km (137 miles) north-west of ABIDJAN. In 1983 the government announced plans to make Yamoussoukro the country's new capital in place of Abidjan. Work started on new administrative buildings there, and the transfer of government departments began.
Population 45 000
Map Ivory Coast Ab

Yamuna (Jumna) ***India*** Northern river, 1376 km (855 miles) long. It rises in the Himalayas in north-west Uttar Pradesh state, and flows generally south past Delhi, then south-east to the Ganges at Allahabad.
Map India Cb

Yan'an ***China*** Town in Shanxi province. The Chinese Communists set up their headquarters there in 1937 after their Long March through the mountains from the south. Yan'an's revolutionary museum commemorates the feat.
Map China Gd

Yangshuo ***China*** Town in Guangxi-Zhuang Autonomous Region famous for its spectacular mountain scenery. It is the terminus for tourist boats down the Gui Jiang from Guilin.
Map China Gf

Yangtze Kiang ***China*** See CHANG JIANG

Yankari Game Reserve ***Nigeria*** Wildlife park on the Geji river about 860 km (535 miles) east and slightly north of Lagos. It covers 2224 km² (858 sq miles) of wooded savannah.
Map Nigeria Cb

Yannina (Ioannina) ***Greece*** Main town of the region of EPIRUS, on a promontory in Lake Ioannínon, at an altitude of 520 m (1708 ft) and overlooked by the Pindhos mountain range. The town was built by Normans in the 11th century. It fell to the Serbs (1345), then to the Turks (1430-1913). Yannina has a university, a Norman fortress, a mosque and many monasteries. Local industries include farming, tobacco, cereals, fine filigree silverwork and jewellery crafts. A literature and art festival is held in August.
Population 44 800
Map Greece Bb

Yaoundé ***Cameroon*** Capital of the country. It is set among beautiful hills, overlooked by Mont Febé and on the edge of dense jungle. Western influences are strong in its luxury hotels, restaurants, nightclubs and sporting facilities, yet it is only minutes from a traditional African rural environment. There is a university, and there are plans to build an international airport and to improve the road to the port of Douala some 200 km (125 miles) to the west.
Population 500 000
Map Cameroon Bb

Yap ***Federated States of Micronesia*** Most westerly of the FSM's four states, consisting of Yap island, its three adjoining islands and some 130 outliers. Yap retains much of its strong traditional culture, including stone money with 'coins' up to 4 m (13 ft) across.
Population 8200
Map Pacific Ocean Bb

Yaracuy ***Venezuela*** North-western state, covering an area of 7096 km² (2741 sq miles). Bauxite (an aluminium ore) and copper are mined, but its economy is predominantly agricultural. Cattle, sugar and citrus fruits are the main products. Its capital is San Felipe.
Population 350 000
Map Venezuela Ba

yardang Sharp desert ridge up to 6 m (20 ft) high and separated from its parallel neighbour by a furrow. It results from the scouring effect of sand-laden winds on weakly consolidated sediments.

Yaren ***Nauru*** District in the south of the island containing the parliament and government buildings and also the airport.
Population 400

Yarlung Zangbo ***China*** See BRAHMAPUTRA

Yarmuk ***Syria/Jordan*** River rising in the Jabal ad Duruz region and flowing west to the Sea of Galilee. For part of its 80 km (50 mile) long course it forms the boundary between the two countries. It is an important source of irrigation water for both.
Maps Jordan Aa; Syria Ab

Yaroslav ***Poland*** See JAROSLAW

Yaroslavl' ***USSR*** Industrial city and river port on the Volga river, 240 km (150 miles) north-east of Moscow. Founded in the early 11th century by Prince Yaroslav I, it became a market and crafts town – as well as a staging post on the North Russian fur trade routes. What was then the country's largest textile mill was built there in 1772 and still produces linen. The main products today, though, are petrochemicals, plastics, dyes, synthetic rubber, tyres, diesel engines, trucks, shoes and food. There is a fortified 12th-century monastery, a 16th-century cathedral and an 18th-century theatre – the oldest in the USSR.
Population 623 000
Map USSR Ec

Yazd (Yezd) ***Iran*** Textile town about 520 km (325 miles) south-east of the capital, Tehran, on the edge of the Dasht-e-Kavir desert. It produces woollen and silk fabrics and carpets. The Zoroastrian religion, founded in Persia in the 6th century BC, is still practised in the surrounding area.
Population 193 300
Map Iran Ba

Yellow River ***China*** See HUANG HE

Yellow Sea (Huang Hai, Hwang Hai) Arm of the PACIFIC OCEAN between China and Korea. It is named after the yellow loess or fine-grained silt deposited by the HUANG HE and other rivers. At the present rates of deposition, the sea will silt up in about 24 000 years time unless a rising sea level compensates. Shifting sandbanks and fogs are today's navigational hazards.
Dimensions 640 km (400 miles) long and wide
Map China Id

Yellowknife ***Canada*** Capital of the NORTH-WEST TERRITORIES, on the Yellowknife river, at the north-west end of Great Slave Lake. It developed after gold was discovered in the area in 1934.
Population 9500
Map Canada Db

Yellowstone National Park ***USA*** The country's oldest national park, established in 1872, and its largest outside Alaska, covering 8991 km² (3472 sq miles) of the Rocky Mountains. It spreads over parts of three states: north-western Wyoming, southern Montana and eastern Idaho. It has some 200 geysers, about 10 000 hot springs and mud pots, and spectacular alpine scenery. There are canyons and waterfalls (dropping 33 and 94 m, 109 and 308 ft) on the Yellowstone river and much wilderness country. The Old Faithful geyser erupts regularly about every 67 minutes.
Map United States Db

Yemen See p. 714

Yemen, South See p. 716

Yenangyaung ***Burma*** Town in central Burma, about 200 km (125 miles) south-west of Mandalay. From 1886 to 1970 it was Burma's most important oil town – its name means 'smelly water creek'. But the oil fields became exhausted and the town is now a marketplace for the surrounding agricultural region.
Population 128 000
Map Burma Bb

Yenbo (Yanbu) ***Saudi Arabia*** Port on the Red Sea, situated 160 km (99 miles) west of Medina – for which it is the main entry place for pilgrims. It is being modernised and developed as an industrial centre. It lies at the western end of an east-west oil pipeline, which will supply the raw materials for refineries and petrochemical plants. Until its modernisation, the port was mainly concerned with the export of dates. Its occupations also include fishing, boat-building and salt-panning.
Population 45 000
Map Saudi Arabia Ab

Yendi ***Ghana*** Town about 440 km (275 miles) north of the capital, Accra. It developed from the 17th century as a trading centre for savannah products from the north and products from the forests of the south. However, Yendi is now overshadowed by Tamale, some 85 km (53 miles) to the west.
Population 25 000
Map Ghana Ab

Yengema ***Sierra Leone*** The country's main diamond area, just east of the Nimini Hills and near the Guinea border. The diamonds, which are mostly of industrial quality, not suitable for jewellery, are found in river gravels.
Map Sierra Leone Ab

Yenisey ***USSR*** River rising near the north-west border of Mongolia and flowing northwards through Siberia for 4130 km (2565 miles) into the Kara Sea. Its tributaries include the Angara and the Stony and Lower Tunguska. It is navigable for most of its length, despite being frozen over in parts for much of the year. It was settled by Cossacks in the early 17th century, and the main city is Krasnoyarsk.
Map USSR Jb, Kc

Yerevan ***USSR*** Capital of Armenia, only 12 km (7.5 miles) from the Turkish border. It stands on the site of a fort built in 783 BC, amid orchards and vineyards. The city's main products are chemicals, synthetic rubber, tyres, copper, aluminium, machinery, electrical appliances, clocks, food, wine and brandy.
Population 1 114 000
Map USSR Fd

Yerushalayim ***Israel*** See JERUSALEM

Yevpatoriya (Evpatoriya) ***USSR*** Seaport and resort on the west coast of the Crimea, 70 km (43 miles) north of Sevastopol'. Russia took Yevpatoriya from the Turks in 1783, and in more recent times it has grown into a popular holiday place. Lying back from its beaches is the old quarter – a jumble of twisting alleyways, merchants' quarters, craft shops and a mosque. In contrast, the new port has a gleaming array of fishing quays, sanatoriums, hotels, parks and gardens. Just to the east of the town is Lake Saki, whose medicinal mud is used to treat arthritis and nervous disorders.
Population 102 000
Map USSR Ed

Yemen

TEXANS STRIKE OIL IN THE LAST REMNANT OF ANCIENT ARABIA WHERE THE QUEEN OF SHEBA REIGNED OVER A PROSPEROUS NATION

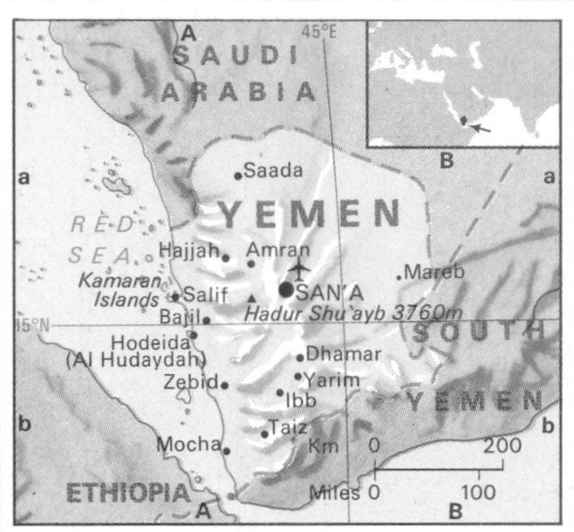

This land of rugged mountains, green hills and trackless desert at the heel of the Arabian peninsula has a history that goes back to Biblical times. It is said that the capital, SAN'A, was founded by Noah's eldest son, Shem, and that Noah's body lies buried in the mountains beyond. MARIB, farther east, was the Queen of Sheba's capital, from where, rich with the profits of trade with India, she set off on her visit to King Solomon nearly 3000 years ago.

The ancient city of San'a, 2210 m (7250 ft) above sea level, still has an atmosphere redolent of *The Arabian Nights*. Despite modern development, within the old city wall the visitor will find bazaars, mud-brick palaces, and white-domed mosques.

The country was extremely backward until recent years, and remains poor by Arabian standards. It was almost entirely dependent on agriculture until a Texan company struck oil in 1984. The reserves have been estimated at a modest 300 million barrels, but production should soon be enough to supply Yemen's needs, with some to spare.

Yemen is the most mountainous, best watered and most fertile land in the Arabian peninsula. It is also strategically important: with its Marxist neighbour, SOUTH YEMEN, it guards the entrance to the route through the RED SEA to the SUEZ CANAL.

The Red Sea coastal plain is hot and humid: summer temperatures at HODEIDA, the main port, reach 40°C (104°F). Behind the plain the land rises steeply, and the country's highest peak, Hadur Shu'ayb, is 3760 m (12 336 ft) above sea level. Although there are oases in the semidesert of the lower slopes, most vegetation grows on the hills above 1000 m (3280 ft), to which the summer monsoon brings about 900 mm (35 in) of rain each year. Just above 1000 m, scrub of acacias, euphorbias and tamarisks flourish. Above 2300 m (7545 ft) there are extensive juniper forests, and woodland covers about 8 per cent of the country. Winters in the hills are cool, and snow falls on the mountains. The north-east of the country is part of the Arabian desert.

The climate enables three-quarters of the people to live by agriculture. More than one-third of the land provides pasture for the largest herds of sheep, goats and cattle in the Arabian peninsula.

Cultivation, mainly at heights of 1400 to 2300 m (4590 to 7545 ft), includes crops of cereals, citrus fruit, grapes and vegetables, and cotton and coffee for export. The shrub *qat* (whose leaves have a narcotic effect when chewed) is a highly profitable export.

Life for the peasants and nomads is hard and simple. Many villages lack clean water; illiteracy (only 21 per cent of the population can read and write) and infant mortality (190 per thousand live births) are high.

In early times, Yemen was a major trading area. The ancient kingdom of Sheba flourished here from about 750 to 115 BC. The country came under the Turks in 1517, but their control was spasmodic. Yemen was an independent kingdom from 1918 until 1962, when the army proclaimed a republic. This was followed by a bitter civil war in the 1960s and a border war with South Yemen in the 1970s. In 1979 the two sides agreed to a truce.

The government has taken care to remain friendly with both the Eastern and Western blocs. Saudi Arabia props up the country's economy with generous aid, arms are supplied by both America and Russia, and the Chinese and South Koreans are building roads. Ironically there is a serious shortage of skilled workers to develop the country – some 1.5 million Yemenis work elsewhere in the Arab world, but the money they send home is an important part of the economy.

YEMEN AT A GLANCE
Area 195 000 km² (75 300 sq miles)
Population 6 220 000 excluding expatriates
Capital San'a
Government Military-ruled republic
Currency Yemeni rial = 100 fils
Languages Arabic (official), English, some Russian
Religion Muslim
Climate Hot and humid on the coast; cooler, with summer rain in the hills. Temperatures at San'a average 14°C (57°F) in January; 22°C (71°F) in July
Main primary products Millet, sorghum, coffee, qat, cotton, fruit; crude oil
Major industries Agriculture, handicrafts, cotton, textiles, leather goods
Main exports Qat, coffee, cotton, sugar
Annual income per head (US$) 610
Population growth (per thous/yr) 27
Life expectancy (yrs) Male 44 **Female** 48

► STANDING TALL Like a watchful eagle in its eyrie, an ancient Yemeni village near Manakhah stands above the road from San'a, the capital, to the port of Hodeida. The houses, built like fortresses, lie amid ancient cultivation terraces.

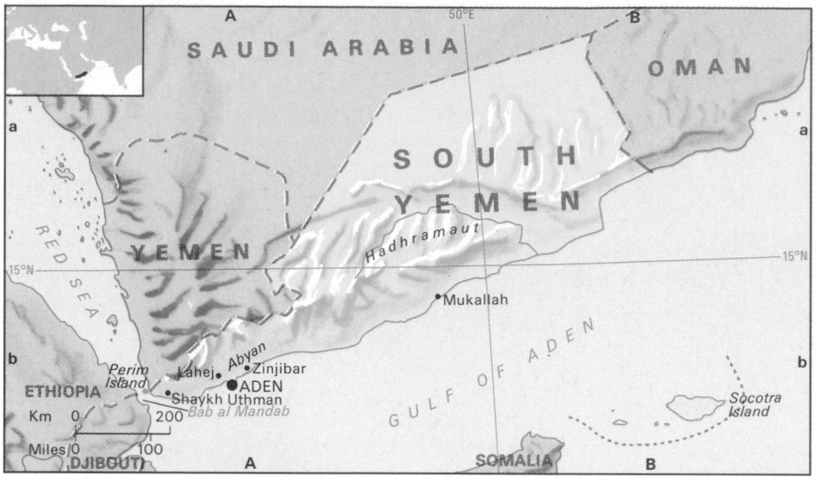

South Yemen

AMONG THE POOREST OF THE ARAB STATES, THE PEOPLE'S DEMOCRATIC REPUBLIC OF YEMEN IS SEEKING CLOSER TIES WITH CAPITALISM

The Marxist state of South Yemen was born in 1967 out of the Federation of South Arabia after a bloody three-sided struggle between rival national movements and the British garrison in ADEN.

South Yemen is a mountainous desert land. Agriculture is the main occupation, though only 1 per cent of the land is fertile – mainly the HADHRAMAUT valley; the only industry of any size is the oil refinery at Aden which processes imported oil.

The coastal plain along the Gulf of Aden is narrow and arid; summer temperatures reach 38°C (100°F) with high humidity, and vegetation is sparse. Inland the mountains, cut by *wadis* (rocky ravines) which flood in the summer rains, rise to 2500 m (8200 ft) on the border with YEMEN (the Yemen Arab Republic). In the north-east the country merges into the Empty Quarter of SAUDI ARABIA, and goes for years without rain.

Britain, recognising the value of Aden as a port controlling the southern entrance to the RED SEA, annexed it in 1839. Aden grew in size and importance after the SUEZ CANAL opened in 1869, but the 20-odd sultanates and sheikdoms of the interior and coast remained undeveloped. Britain signed treaties with their rulers between 1882 and 1914, establishing them as protectorates under the governor of Aden. In 1963, despite local opposition, Aden was incorporated with the other states in the Federation of South Arabia. Britain promised this independence by 1968, but constitutional talks failed.

A guerrilla campaign against the British and the local rulers was mounted by the National Liberation Front (NLF) and the rival Front for the Liberation of South Yemen (FLOSY); the NLF emerged as the dominant group. When the British were forced to withdraw in 1967, the NLF declared independence and its Marxist wing took control. The sheiks fled or were deposed. Private owners were allowed 8 hectares (20 acres) of irrigated land, or 16 hectares (40 acres) without irrigation; the rest was redistributed. Banks, insurance and trading companies and the refinery were nationalised.

The Yemeni Socialist Party (including the NLF) is the only legal political party. The country is governed by a head of state who is also prime minister, and a supreme people's council of 111 members, elected by committees in each village. Welfare spending has been concentrated on education, which is free at primary level, but rapid population growth has slowed expansion.

Agricultural modernisation schemes have begun with United Nations, Arab and East European funds, and a new fishing port is being built near Nishtun. However, development is slow and to attract funds the government has now offered to allow privately owned businesses to operate.

SOUTH YEMEN AT A GLANCE
Area 287 682 km² (111 074 sq miles)
Population 2 280 000
Capital Aden
Government One-party socialist republic
Currency Yemeni dinar = 1000 fils
Languages Arabic (official), English
Religion Muslim
Climate Hot and dry; temperatures in Aden range from 22-28°C (72-82°F) in January to 29-37°C (84-99°F) in July
Main primary products Cotton, coffee, sorghum, millet, sesame, wheat, qat, fish
Major industries Agriculture, oil refining, textiles, fishing, fish processing
Main exports Refined oil, processed fish, cotton, coffee
Annual income per head (US$) 380
Population growth (per thous/yr) 29
Life expectancy (yrs) Male 44 **Female** 47

Yezd ***Iran*** See YAZD

Yin Shan ***China*** Range running 1200 km (750 miles) west-to-east through Nei Mongol. The peaks, between 2000 and 25 000 m high (6560 to 8200 ft), mark the southern edge of the Mongolian plateau.
Map China Gc

Yinchuan ***China*** Capital of the Ningxia-Hui Autonomous Region. A city on the HUANG HE (Yellow River) close to the GREAT WALL of China, it is now a centre for processing hides, wool and tobacco.
Population 635 800
Map China Fc

Yixing ***China*** City of Jiangsu province, 55 km (34 miles) south-west of Wuxi on the west side of Tai lake. It is famous for its pottery and stoneware and for the spectacular Shan Juan cave, 15 km (10 miles) south-west of the city. The cave extends for 5000 m² (1.24 acres). Its spectacular stalactite and stalagmite formations are a popular tourist attraction.
Population 30 000
Map China He

Yogyakarta (Jogjakarta) ***Indonesia*** City and former sultanate of central Java. It was the centre of the Indonesian revolution against Dutch rule in 1825-30 and of the fight for independence in 1945-9. It is the focus of Javanese culture, and its Kraton (Sultan's Palace) keeps alive court traditions and preserves valuable artefacts. The city's craftsmen produce a range of handicrafts such as batik, shadow puppets and silverware. There are many fine temples nearby, notably BOROBUDUR.
Population 400 000
Map Indonesia Dd

Yoho ***Canada*** National park covering 1313 km² (507 sq miles) in eastern British Columbia. It comprises glaciers, mountains and alpine meadows and has a profusion of wildlife and flowers. The town of Field is the park's headquarters and tourist centre.
Map Canada Dc

Yokkaichi ***Japan*** Industrial port in south Honshu island, on the Ise Bay about 105 km (65 miles) east of Osaka. In the 1960s its petrochemical plants were responsible for 'Yokkaichi disease', a form of asthma brought on by high levels of sulphur dioxide in the air.
Population 263 000
Map Japan Cd

Yokohama ***Japan*** The nation's leading port and second largest city after neighbouring Tokyo. A fishing port until the mid-19th century, Yokohama, on Tokyo Bay in south-east Honshu island, became the capital's deep-water port. It was rebuilt after a disastrous earthquake in 1923 and again after devastating Allied bombing during the Second World War. Industry includes oil refining and the manufacture of electrical machinery and automative products. Its factories, many on land reclaimed from the bay, contrast with its nightlife, which is enlivened by Japan's most colourful Chinatown.
Population 2 992 600
Map Japan Cc

York ***United Kingdom*** Historic English city on the River Ouse in North Yorkshire. It was founded by the Romans as Eboracum in AD 71, and Constantine the Great was proclaimed emperor by his men in the city in 306. The Angles made it their capital and it became an archbishopric in 634. The Vikings burnt it and then made it their capital of Jorvik from 867.

But the city's chief claim to fame is its magnificent cathedral, or minster, completed in 1470 and the largest medieval church in northern Europe. Its beautiful 15th-century stained-glass east window is one of the biggest in the world and another, dating from about 1150, is probably the oldest in Britain; its lantern tower is the tallest in England at 60 m (198 ft). The roof was badly damaged by fire in 1984, and needed extensive repairs.

The medieval city walls, built on Roman foundations, are 5 km (3 miles) around. There is a modern university and many museums, including the National Railway Museum and Viking Museum. York's products include chocolates and other confectionery.

Population 103 000

Map United Kingdom Ed

York, Cape ***Australia*** Northernmost point on the Australian mainland, lying in Queensland across Torres Strait from Papua New Guinea. The peninsula on which it stands, Cape York Peninsula, extends north for about 750 km (470 miles) between the Coral Sea to the east and the Gulf of Carpentaria to the west. It is covered by tropical forest and grassland to the west. During the summer 'wet', rivers become torrents, making overland travel impossible. On the west coast of the peninsula is the bauxite mining and exporting centre of WEIPA.

Map Australia Ga

Yorkshire ***United Kingdom*** Former county of northern England. The ancient county, centred on the city of YORK, was Britain's largest, covering 15 859 km² (6123 sq miles), and was divided into three Ridings – named from an Old English word meaning 'a third'. These were reorganised in 1974 into several new areas whose boundaries correspond only here and there with those of the old Ridings. Parts were cut off to form sections of the new counties of Cleveland and Humberside.

North Yorkshire (8317 km², 3211 sq miles) is the largest English county, stretching across the country from the east coast to within 20 km (12 miles) of the west. Much of it is moorland. The North York Moors (or Yorkshire Moors), with some of Britain's loveliest and most open landscapes, lie in the east and the Pennine hills – cut by the Yorkshire Dales – in the west. Between them is the fertile Vale of York, an area of rich farmland. Tourism is concentrated on the seaside resort of SCARBOROUGH, the spa town of HARROGATE and the Pennine hills. Northallerton is the county town.

South Yorkshire (1560 km², 602 sq miles) is the centre of British coal mining and metalworking. The principal city is SHEFFIELD on the River Don, which then flows past ROTHERHAM, a Saxon foundation overwhelmed by industry in the 19th century, and the railway town of DONCASTER. North of Sheffield lies BARNSLEY, whose name, perhaps more than that of any other British town, is synonymous with coal. Yet even in this heavily industrialised county there are wide areas of open moorland.

West Yorkshire (2039 km², 787 sq miles) is focused on the cities of LEEDS, BRADFORD and WAKEFIELD, and on a number of other large towns such as HUDDERSFIELD and HALIFAX which grew with the rise of the woollen industry in the 19th century. These towns, with their huge mills and their vast town halls, reflected the civic pride of Yorkshire men and women. But today many of the mills and the factories have closed down or disappeared – and so has much of the employment.

Population 4 058 000

Map United Kingdom Ec

Yorktown ***USA*** Historic town on the York river in south-east Virginia, about 190 km (120 miles) south and slightly east of the national capital, Washington. It was settled in 1631 and was an important tobacco port by 1750. During the War of Independence, it was besieged and captured from the British by George Washington (who later became the country's first president) on October 19, 1781 – the final defeat that won America its freedom. It was again besieged, by Union forces, in the Civil War; Confederate troops relinquished it on May 4, 1862. It lies in the Colonial National Historical Park.

Map United States Kc

Yorubaland ***Nigeria*** Traditional homeland of the Yoruba people. It covers some 90 700 km² (35 000 sq miles) on the coast, south-west of the Niger river. It is fertile and densely populated, and produces cocoa, yams, tapioca and kola nuts (chewed as a mild stimulant). The area includes the ancient towns of Ife and Oyo, Lagos, the national capital, and Ibadan.

Map Nigeria Ab

Yosemite National Park ***USA*** Spectacular mountainous area of 3083 km² (1190 sq miles) about 225 km (140 miles) east of San Francisco in the Sierra Nevada range of California. It has deep gorges and ravines, towering granite cliffs and mountains – some over 3960 m (13 000 ft) high – many waterfalls and giant sequoia trees. The Yosemite Falls tumble a total of 739 m (2425 ft).

Map United States Cc

▲ LIVING HISTORY York's city walls enclose the magnificent minster and fine buildings which chart the city's history – and that of England too.

Youghal (Eochaill) ***Ireland*** Port and resort town, 42 km (26 miles) east of Cork. According to legend, Sir Walter Raleigh, the English explorer, who was mayor of Youghal in 1588-9, planted the first potatoes in Ireland at Myrtle Grove in the town.

Population 5870

Map Ireland Cc

Youth, Isle of (Isla de la Juventud) ***Cuba*** Island in the Caribbean Sea, 48 km (30 miles) off the south-west coast. Formerly the Isle of Pines, it was renamed in 1978 in recognition of the Cuban young people's contribution to its agricultural development. Once notorious for its large prison, it is now an important growing region for citrus fruit. The chief town is Nueva Gerona on the north coast.

Population 61 000

Map Cuba Aa

Ypres (Ieper) ***Belgium*** Battlefield town in Flanders, about 65 km (40 miles) west of the city of Ghent. In the 13th and 14th centuries it was one of Europe's largest cloth-making cities, with a population of 200 000. Later it fell into decline, only to find a different kind of fame after being flattened during three great battles in the First World War. Cemeteries now surround the rebuilt town – known to British troops, who struggled to pronounce its name, as 'Wipers' – and the Menin Gate is a memorial to 58 000 Allied soldiers who died on the Ypres front and have no known grave.

Population 22 000

Map Belgium Aa

Yssel ***Netherlands*** See IJSSEL

Ysselmeer ***Netherlands*** See IJSSELMEER

Yü Shan (Hsin-kao Shan) ***Taiwan*** Highest mountain peak in Taiwan and in the maritime fringes of east Asia, reaching 3997 m (13 113 ft). It stands almost on the Tropic of Cancer, in the island's central mountains 180 km (111 miles) south of T'ai-pei. The vegetation on its flanks is extremely varied, ranging from subtropical camphor trees on the lower slopes to subalpine plants nearer the summit.

Map Taiwan Bc

Yuan Jiang ***China*** See RED RIVER

Yucatán ***Mexico*** State on the Yucatán peninsula at the south-east end of Mexico. It is a low-lying jungle-covered region, and was the home of an advanced Mayan culture before the 10th century AD. It includes the archaeological sites of CHICHEN ITZA, Kabah and Uxmal. MERIDA, Yucatán's capital, is an important tourist centre, but the state's economy depends principally on the henequen plant, which yields fibres used to make ropes, nets and bedding.

Population 1 100 000

Map Mexico Db

Yucatan Channel (Yucatan Strait) Channel between the YUCATAN peninsula of Mexico and Cuba, linking the CARIBBEAN SEA to the Gulf of MEXICO. Part of the westward-moving Equatorial Current flows through the channel from the Caribbean into the Gulf of Mexico. It emerges from the gulf through the Straits of Florida as the GULF STREAM.

Width 217 km (135 miles)

Map Mexico Db

Yugoslavia See p. 720

Yukon Territory ***Canada*** Vast area of mountain ranges, forests and plateaus in the extreme north-west of Canada covering 483 450 km² (186 660 sq miles). It contains Canada's highest peak, Mount Logan (5951 m, 19 524 ft). The Yukon is a land of great beauty with high mountains and sweeping valleys cut by great rivers. Summers are hot and winters cold, but the climate is more temperate than that of the adjacent Northwest Territories.

The 1896 gold strike on a tributary of the Klondike river led to the development of DAWSON. Today gold, silver, lead and other minerals are mined. Hydroelectric power has been developed, there is a forestry industry and tourism is growing. Water and road links are used, but most travel is by air.

Population 23 150

Map Canada Bb

Yumen ***China*** City in Gansu province, once the eastern terminus of the SILK ROAD. It is now a centre for oil production. The Yumen oil field is one of the largest in China.

Map China Ec

Yungas ***Bolivia*** Name for the area of hillside forests, below about 2500 m (8200 ft). The tangled and mossy forests, often wreathed in cloud, mark the transition from the treeless Andean highlands of Bolivia to the lush lowland Amazon jungle to the east. The best developed are north-east of the capital, La Paz, at Coroico and around Cochabamba – where rich alluvial soil washed down from the Andes supports a wide range of crops such as coffee, cacao, sugar cane, maize, bananas, oranges and other citrus fruits, and coca.

Map Bolivia Bb

Yungui Plateau ***China*** Plateau of roughly 160 000 km² (61 800 sq miles) covering western Guizhou and eastern Yunnan provinces in southern China. It lies between 1000 and 2000 m (3300 and 6600 ft) above sea level. Much of the plateau has a cool, spring-like climate all year round.

Map China Fe

Yunnan ***China*** Province in the south-west of the country, covering 390 000 km² (150 000 sq miles) and bordering on Vietnam, Laos and Burma. It is a mountainous area, embracing the western half of the Yungui Plateau and the great ranges and canyons of the southern Qam'do region. While the plateau surface enjoys a mild, temperate climate, the south is subtropical and tropical. The mountains contain some bizarre scenery – including a 'forest' of tree-shaped 30 m (100 ft) high limestone crags, pitted with caves. They are also a botanist's paradise, with more than 15 000 plant species. The best agricultural land is around the city of Kunming, where rice, wheat and maize are grown. Pu'er, in the south, is noted for its black tea. Yunnan contains many ethnic groups related to the peoples of South-east Asia.

Population 31 350 000

Map China Ef

Zabljak ***Yugoslavia*** See DURMITOR

Zabrze (Hindenburg) ***Poland*** Industrial city 24 km (15 miles) west of the city of Katowice. It produces coal, steel, coke, chemicals and machinery, and is linked to the Oder river by the Klodnica Canal.

Population 211 300

Map Poland Cc

Zacatecas ***Mexico*** **1.** Mineral-rich state on the central plateau, crossed by the Sierra Madre Occidental mountains. It is one of the best cattle-raising areas in the country. **2.** Capital of Zacatecas state, 600 km (375 miles) north-west of Mexico City. Its name is Aztec for 'Place of the Grasses', but Zacatecas made its fortune from silver. The world's largest silver mine, which processes about 10 000 tonnes of ore a day, lies just outside the city at Real de Angeles. The city has several magnificent Spanish Baroque churches, including the Jesuit Church of Santo Domingo, and picturesque pink stone houses.

Population (state) 1 200 000; (city) 71 000

Map Mexico Bb

Zadar ***Yugoslavia*** Ancient capital of Dalmatia, 116 km (72 miles) north-west of the port of Split. The town, on a promontory above the Adriatic, contains several beautiful buildings: the 13th-century Church of Sv Franje (St Francis); the 9th-century, circular, castle-like St Donat's Church, built in the remains of a Roman forum; a Renaissance Bishop's palace; and the Romanesque Cathedral of Sv Stosija (St Anastasia) – Dalmatia's largest church. The cathedral's nearby bell tower, begun in 1452, was finished by the English architect Sir Thomas Jackson, in 1892.

The Baroque Church of Sv Simun (St Simon) contains the saint's extravagant reliquary. This was made from 250 kg (551 lb) of silver for the Hungarian Queen Elizabeth in 1380, and is now supported by bronze angels cast from Turkish cannon. Modern Zadar is a major port, shipbuilding centre, naval base, industrial city and tourist resort.

From the 9th to 14th centuries, Nin, 18 km (11 miles) to the north, was the first capital of the Croatian kings and bishops. Its Sveti Kriz (Holy Cross) Church dates from the 7th to 9th centuries and is Yugoslavia's oldest intact church.

Population 116 200

Map Yugoslavia Bb

Zagazig ***Egypt*** Rail centre and canal junction on the Nile delta waterway system, about 65 km (40 miles) north-east of Cairo. It trades in cotton and grain grown on delta farms.

Population 203 000

Map Egypt Bb

Zagorje ***Yugoslavia*** Beautiful region of rolling hills topped by Baroque churches, woods and castles, lying in north-west Croatia between Zagreb and the Drava river. It has more than 100 castles, including that at Trakoscan. The area is densely populated, largely with small textile towns.

Map Yugoslavia Ba

Zagorsk ***USSR*** Town and place of pilgrimage, 60 km (37 miles) north-east of Moscow. It has a fortified 14th-century monastery, Trinity-St Sergius, and a seminary founded in 1742 which is now the country's main training college for monks. The monastery complex contains the 15th-century Trinity Cathedral and the 16th-century Cathedral of the Assumption. The monks' traditional craft was toy-making, and there is a toy museum crammed with examples of their work.

The modern town of Zagorsk has engineering and chemical industries and produces electrical goods.

Population 111 000

Map USSR Ec

Zagreb (Zagrab; Agram) ***Yugoslavia*** Capital of the republic of Croatia and the country's second largest city after Belgrade, 118 km (73 miles) south-east of Ljubljana. Each spring and autumn it hosts a trade fair, and each July a folklore festival. The city grew as twin towns on a hill overlooking the Sava river plain. Gradec (or Gornji Grad, meaning 'small town'), once a royal town, was fortified against Tatar attacks in the 13th century, and some of its medieval walls and gates survive today. There are several fine Gothic and Baroque churches and palaces. Kaptol, a religious foundation lower down the hill, is dominated by the filigree spires of the cathedral and the fortified bishop's palace.

The pastel-painted buildings of Gradec and Kaptol contrast with the grey stone of the 19th-century hub of the city, the Donji Grad, or 'lower town', and since 1960 the focus has moved again, across the Sava, with new commercial and administrative offices, the university, a fair site, shopping centres, and high-rise flats and hotels.

Zagreb has several museums, including the Yugoslav Academy of Art and Science and an archaeological museum. Good communications make it a fine tourist centre, with the Zagorje hills, the Plitvice lakes, and ski resorts on Sljeme (1035 m; 3396 ft) within easy reach. The city's manufactures include electrical machinery and appliances, chemicals, cement, pharmaceuticals, engineering equipment, clothing, leather, food and printing.

Population 1 174 500
Map Yugoslavia Bb

Zagros Mountains ***Turkey/Iran*** Range nearly 1600 km (1000 miles) long, stretching south-east from eastern Turkey along The Gulf to the Strait of Hormuz. The range consists of parallel chains of mountains cut by rivers and deep gorges, with large areas of forest. The highest peak is Zard Kuh, at 4548 m (14 918 ft), west of the city of Esfahan.

Map Iran Aa

Zaire See p. 722

Zaire (Congo) ***Zaire/Congo*** Africa's second longest river after the Nile. It is some 4670 km (2900 miles) long, and its drainage basin of 3 626 000 km^2 (1 400 000 sq miles) and its flow are second only to those of the Amazon.

The Zaire's mouth was discovered by a Portuguese mariner, Diogo Cão, in 1482, but its course was not explored until the British adventurer Sir Henry Morton Stanley sailed down the Lualaba river to the Boyoma Falls, and on to Malebo Pool in 1876-7.

In the 16th century, the river was called the Zaire, a corruption of several words in local dialects meaning 'river'. It was renamed Rio Congo in the 17th century after the Kongo people in its lower course; but the Zairean government reintroduced the original name in 1971.

The chief headstream is the LUALABA river. The upper course contains rapids, waterfalls and lakes. The middle course starts at Kisangani, and here the channel is broad and the volume of water is increased by such large tributaries as the Ubangi and Kasai. About 560 km (350 miles) from the sea, the river widens into Malebo Pool (formerly Stanley Pool), on which stand the cities of Kinshasa and Brazzaville. Below Malebo Pool, the river passes through the unnavigable Livingstone Falls before reaching the head of its estuary at Matadi. More than 12 000 km (7500 miles) of the Zaire and its tributaries are navigable – and the rivers have an estimated one-sixth of the world's total hydroelectric potential.

Map Zaire Ab

Zákinthos (Zante) ***Greece*** One of the most southerly of the IONIAN ISLANDS off the coast of the Peloponnese. It is a tourist resort of gardens and flowers with a rugged coast. Zákinthos was occupied by the Venetians from AD 1487 to 1797, and a Venetian castle stands on the site of the ancient Acropolis (citadel). The island's main business now is tourism, but it also produces melons, currants, white wine and olives. A medieval drama festival is held each August.

Population 30 000
Map Greece Bc

Zakopane ***Poland*** The country's main tourist resort, with more than 2.5 million visitors a year. It lies at 800 m (2625 ft) in the Carpathian mountains, 84 km (52 miles) south of the city of Cracow. Zakopane is surrounded by magnificent scenery with marked trails for walkers and climbers, and good skiing facilities, including a cable car to Kasprowy Wierch, which rises to 1985 m (6514 ft) on the Czech border.

The Gorale (highlanders) wear regional costumes as a tourist attraction and keep alive folk crafts and customs. And the town and the hillsides around it are dotted with traditional mountain chalets, with carved gables and steeply pitched roofs.

Population 29 700
Map Poland Cd

Zamardi ***Hungary*** See SIOFOK

Zambezi ***Southern Africa*** Longest river of southern Africa, flowing some 2700 km (1700 miles) south and east from north-western Zambia to the Indian Ocean. It crosses part of western Angola, forms a small part of the Zambian border with Namibia and all of that with Botswana and Zimbabwe, and crosses central Mozambique. It is dammed for hydroelectricity, forming huge lakes, at KARIBA – some 450 km (280 miles) downstream from the mighty VICTORIA FALLS – and at CABORA BASSA in Mozambique.

Map Zambia Cb; Zimbabwe Ba

Zambézia ***Mozambique*** See QUELIMANE

Zambia See p. 724

Zamboanga ***Philippines*** Chief trading port of the southern Philippines, at the tip of the Zamboanga peninsula in western MINDANAO. Originally a fort built to protect Christian settlers against the local Muslim population, it was rebuilt early this century.

It is known as the 'city of flowers' and has a colourful market. It exports timber, copra, coconut oil, Manila hemp, rubber and other crops produced in the region, while *kumpits* (motorboats) carry barter goods to and from islands of the SULU ARCHIPELAGO and ports in BORNEO. Smaller craft called *vintas* are the homes of the Badjaos, or sea gypsies.

Population (provinces) 1 500 000; (city) 250 000
Map Philippines Bd

Zamora ***Spain*** Town about 210 km (130 miles) north-west of Madrid whose many churches of the 12th and 13th centuries make it a living museum of Romanesque architecture. It trades in cereals and wine and has flourmilling, cement, textiles and soap industries.

The surrounding province of the same name is a sheep-rearing area.

Population (town) 59 800; (province) 224 400
Map Spain Cb

▼ DESOLATE ROAD The Dempster Highway linking Dawson and Inuvik makes its way through the Richardson Mountains of the northern Yukon – a bleak, forbidding land in winter.

Yugoslavia

A SPECIAL BRAND OF COMMUNISM HAS BEEN DEVELOPED FROM A BALKAN MELTING POT OF PEOPLES AND CULTURES

Prewar Yugoslavia was a poverty-stricken agricultural country where peasants scratched a miserable living from the land. In the KRAS mountain region, for instance, peasants had spent centuries creating arable fields on barren ground.

In 1945, about 42 000 peasant families like these were resettled in state farms by the new Communist government. The farms were created on fertile northern plains by taking more than 11 per cent of the agricultural land. It was the most symbolic act of a Communist state which was determined to make the ideology work. Yugoslavia was the first European country outside Russia to introduce a Soviet-style five-year plan. All industry, trade, transport and finance was nationalised, and from 1947 ambitious projects for industry were directed from the capital, BELGRADE.

As industry developed and towns grew, millions of people left rural homes and lifestyles to live in high-rise city flats and work in factories and offices. They now enjoy educational, medical and recreational facilities undreamt of before the Second World War. But all is not as well as it was. In recent years inflation has hit 80 per cent and foreign debts exceed US$20 000 million. Living standards have declined and the reputation of Marshal Tito – architect of Yugoslavia's postwar progress and its autocratic leader until his death in 1980 – has been demoted.

Despite a revolution in lifestyle since 1945, the Yugoslavs still preserve much of their rich folklore and traditions, including picturesque regional costumes, peasant dances like the *kolo*, and gypsy-like music. Nor is life drab or regimented in the new Communist state.

Tourists flock to the country, bringing large amounts of foreign currency that helps to reduce trade deficits. The visitors come mainly from Western and Eastern Europe, most of them making for the coasts and islands of ISTRA and DALMATIA for the Mediterranean climate, unspoilt beaches and rocky coves. Inland, mountains and high plateaus cover three-quarters of the country.

Until 1918 Yugoslavia was a patchwork of small states that made the BALKANS a byword for political fragmentation. Yet out of this melting pot of races, cultures and conflicts emerged today's united nation of 23 million people. They include 8.3 million Serbs, 4.6 million Croats, 1.9 million Slovenes, 1.5 million Macedonians and 600 000 Montenegrins.

The Slovenes, who settled the north-west Alpine areas in the 6th century, never enjoyed sovereignty. Bavarian rule after about 740 gave way to Austrian control from about 1380 to 1918. The Slovenes kept their language and culture but they absorbed German attitudes.

The Croat state, consolidated from 925 in an arc comprising lowland Pannonia and Adriatic Dalmatia, lasted only until 1102. It then came under 800 years of Hungarian (later Hapsburg) control. Foreign rule swung across the south-east regions like a pendulum, bringing the Macedonians successively under the control of Greece, Bulgaria, Serbia and, for more than 500 years, Turkey.

In the fertile rolling country of the MORAVA and IBAR river systems in the east, the Serbs developed a state, later an empire, from the 13th century. Though subjected to Ottoman rule in the 15th century, they regained their sovereignty in 1815 to form the kernel of modern Yugoslavia. Only the Montenegrins, inhabiting the barren limestone Karst mountains around Lovcen, were never subjugated.

Centuries of comparative freedom enabled the Serbs and Montenegrins to nurture a common Serbian language. The Slovenes, Croats and Macedonians preserved their distinctive languages, often under severe foreign pressure. Serbo-Croat, recognised as the main language today in Yugoslavia, is two separate tongues having many common elements but differing in alphabet, pronunciation and everyday words. Despite 30 years of compulsory education, younger Serbs and Croats still have difficulty in reading each other's writing.

The Roman Catholic Slovenes and Croats in the west use Latin script; the Orthodox Christian Serbs, Montenegrins and Macedonians in the east write in Cyrillic. There is also a Muslim minority of about 2 million, mainly in BOSNIA HERZEGOVINA where the religion was fostered by Ottoman rule after 1500.

THE BULLET THAT CHANGED HISTORY

The most famous city in Yugoslavia is probably SARAJEVO, where an assassin's bullet sparked the First World War in 1914. The city was then the capital of Bosnia-Herzegovina in the Austro-Hungarian Hapsburg Empire. The assassin was a Bosnian-Serb nationalist, and his victim was Archduke Franz Ferdinand, heir to the Austro-Hungarian throne. Austria-Hungary declared war on Serbia, and other European powers joined the conflict.

Yugoslavia was created after the war by the union of six territories: Bosnia Herzegovina, SLOVENIA, CROATIA, SERBIA, MONTENEGRO and MACEDONIA.

Marshal Tito's Communist regime started as a partisan uprising against German invaders in the Second World War. He led guerrilla forces who fought against the Germans and against a rival group of partisans, the *Chetniks*, led by Draza Mihajlovic. By the time the country was freed in 1944, Tito had become a popular hero and he proclaimed himself head of an independent Yugoslavia.

Equal rights were proclaimed for all citizens, irrespective of nationality, to defuse explosive ethnic rivalries, and this approach, and the personal dominance of Tito until he died in 1980, produced a united nation which resisted all foreign influences.

Tito led a sweeping programme to develop agriculture, industry and tourism, and open the country to the outside world. Yugoslavia's progress and Tito's personality, coupled with an alliance with Albania and growing influence in Greece, alarmed the Russians, who saw a threat to their command over Eastern Europe.

From 1948 to 1955 Russian and East European governments blockaded Yugoslavia. They hoped that by cutting off supplies such as Soviet and Czech machinery and Polish coal they would wreck the economy and bring down the Tito regime.

The blockade united the Yugoslavs as never before, leading them to rethink Marxism and seek their own road to socialism. Under Tito they introduced daring experiments. They gave workers responsibility in management, delegated public administration to regional authorities and communes, and in international relations encouraged non-alignment of Third World countries.

From 1948, Western aid helped Yugoslavia to develop hydroelectricity, timber, copper, aluminium and steel industries in the central mountains and the eastern region. Engineering and electrical industries were set up around the old cities of LJUBLJANA, ZAGREB, Sarajevo, Belgrade and SKOPJE, and shipbuilding was developed in PULA, RIJEKA and SPLIT.

When Soviet-Yugoslav relations were restored in 1955, Russian delegations were astonished at the progress that had been made by Yugoslav industry. Workers were producing far more than their counterparts in other Communist countries – products including consumer goods for home and export.

But times changed, and by the late 1970s industries were in trouble, facing restrictions on sales to West European and Communist markets, and stiff Asian competition in the Third World. Since 1979 unemployment has risen to nearly 15 per cent, partly due to rapid population growth (50 per cent since 1946). Regional inequalities have aggravated the problem. Decentralisation enhanced the prosperity of the more developed west and north (Slovenia, Croatia and VOJVODINA), which are nearer the prosperous markets of Austria, Italy and West Germany. It retarded job creation in the backward centre and south-east (especially Bosnia, Macedonia and KOSOVO), where population growth is fastest.

Meanwhile there have been second thoughts on agricultural policy. Attempts to consolidate more peasant holdings into collective farms were stopped in 1952. Most food supplies (about 80 per cent of maize, meat, milk, fruit, vegetables, tobacco and hops) still come from peasant farms.

Policies to encourage cooperation between state and private farms have not worked well because of the regional concentration of state farms and peasant suspicions. Since 1960 a drift to the towns has added to the problem.

Yugoslavia is a complicated hotch-potch of peoples, cultures and territories which seems to have been simplified into a fairly happy Communist state. That cracks are beginning to appear in this edifice is hardly surprising when, as Tito once stated: 'I am a leader of one country which has two alphabets, three languages, four religions and five nationalities living in six republics surrounded by seven neighbours, a country in which live eight national minorities.' Tito, in fact, understated his case: there are at least 18 ethnic minorities in modern Yugoslavia.

▲ FREE CITY The fortified port of Dubrovnik (formerly Ragusa) survived as a free city state for more than 600 years, until the early 19th century. Its leaders – 'rectors' elected for one month only – lived in the palace to the left of the domed cathedral.

YUGOSLAVIA AT A GLANCE
Area 255 804 km² (98 766 sq miles)
Population 23 000 000
Capital Belgrade
Government One-party Communist republic
Currency Yugoslav dinar = 100 paras
Languages Serbo-Croat, Slovene, Macedonian; some Hungarian, Albanian
Religions Christian (35% Serbian Orthodox, 26% Roman Catholic), Muslim (9%)
Climate Mediterranean on coast, temperate inland. Average temperature in Belgrade ranges from −3 to 3°C (27-37°F) in January to 17-28°C (63-82°F) in July
Main primary products Wheat, barley, maize, potatoes, sugar, tobacco, fruit and vegetables, livestock, timber; bauxite, lignite, zinc, lead, oil and natural gas, iron, coal, copper
Major industries Steel, cement, chemicals, food processing, fertilisers, textiles, forestry and timber products, mining
Main exports Chemicals, machinery, food, clothing, footwear, motor vehicles, textile yarns and fabrics, timber, livestock, non-ferrous metals
Annual income per head (US$) 2500
Population growth (per thous/yr) 7
Life expectancy (yrs) Male 69 **Female** 72

Zaire

A VAST LAND OF EQUATORIAL RAIN FORESTS AND SAVANNAHS IN THE HEART OF AFRICA, DRAINED BY ONE OF THE WORLD'S GREAT RIVERS

Zaire lies in the heart of what was 'Darkest Africa' to 19th-century Europeans, a country of nightmarish horror according to Joseph Conrad's 1899 novel, *Heart of Darkness*. Zaire is vast, nearly 77 times the size of Belgium, which ruled the country as the Belgian Congo from 1908 to 1960. But it has less than 34 million people (little more than three times Belgium's population) and contains large tracts of virtually uninhabited equatorial rain forest.

Europeans penetrated this forest in the 19th century by way of the great Congo river (or the ZAIRE). A narrow corridor, following the lower course of the river, reaches westwards from the capital KINSHASA to the Atlantic Ocean, where Zaire has a short coastline of 40 km (25 miles) and a port on the Zaire estuary at MATADI.

The Zaire and its tributaries, though interrupted in places by waterfalls and rapids, were essential lines of communication in the 19th century and are still important today. Although slow, river transport is cheap and rivers remain a vital part of the transport network, forming pathways through, in Conrad's words, 'the towering multitude of trees . . . the immense matted jungle'.

For instance, one of the chief outlets for copper and other metals from the main mining province of SHABA (formerly Katanga) is a railway to ILEBO and then river transport down the KASAI and Zaire rivers to Kinshasa. From Kinshasa, goods are carried by rail to the coast, bypassing the LIVINGSTONE FALLS.

Central Zaire, which occupies a shallow depression in the central African plateau, has an equatorial climate, with abundant rain and high temperatures throughout the year. In the north and particularly in the high plateau country in the south, there are marked dry winter seasons. There, luxuriant rain forest merges into wooded savannah. But nowhere is the contrast in vegetation as great as in the highlands, such as the RUWENZORI mountains, that overlook the great lakes in the GREAT RIFT VALLEY on the eastern border.

Zaire's rain forests cover 55 per cent of the country and contain valuable hardwoods including mahogany and ebony. Ecologists hope that exploitation, now confined to under 2 per cent of the forest area, will not be expanded.

In 1980 agriculture, forestry and fishing (mainly in inland waters) employed 75 per cent of the workforce (compared with 13 per cent in manufacturing and 12 per cent in service industries). Yet cultivated land makes up less than 3 per cent of the country and grazing land – limited to those savannah regions which tsetse fly does not afflict – just over 10 per cent.

The main food eaten by the rural people and grown in most parts of Zaire is cassava, a starchy root. Other foods include plantains and various root crops in wet areas, rice in flooded valleys, and beans, groundnuts, guinea corn and maize in drier places. Many people are subsistence farmers who practise shifting cultivation, clearing land for crops and moving on when the soil has become exhausted. Women do most of the farmwork.

Commercial agriculture produces export crops, especially coffee, palm oil and rubber, and also cocoa, cotton and tea. It was the Europeans who established the plantations, where these cash crops are grown, during the colonial days. Since 1970 the Africanisation of plantations and the lack of skilled local managers – the Europeans left without making any provisions for a takeover of management – means they produce less than they did. Cash crops are still important to the economy, though it is Zaire's massive mineral and hydroelectric resources that make it potentially one of Africa's richest countries. Zaire leads the world as a cobalt producer and ranks second in diamonds (mainly industrial diamonds) and sixth or seventh in copper. Copper is the main export, accounting for nearly half the country's total value of exports. Next in importance are cobalt, coffee and diamonds.

Apart from southern Kasai, where most of the diamonds are mined, the chief mineral region is the huge industrial complex in Shaba. It includes KOLWEZI, Likasi, Kipushi and LUBUMBASHI, and is second only in Africa to South Africa's WITWATERSRAND. The area produces cadmium, cobalt, copper, germanium, silver, zinc and other metals, using power from hydroelectric stations on the Lufira and Lualaba rivers. The largest hydroelectric project is at INGA DAM, not far from the Zaire estuary, downriver from fast-growing Kinshasa. But overall hydroelectric output is only about 20 per cent of capacity. There is some offshore oil production.

About 100 000 pygmies live in the forests, descendants of Zaire's first inhabitants. They are divided into several groups, including the Mbuti of the north-eastern ITURI FOREST, who still live by hunting and gathering.

The Mbuti, who call themselves *bamiki ba'ndura* (children of the forest), live in small bands, or 'families', of up to 30 households. Members of a band regard each other as kin, though they are linked by their struggle to make a living rather than common blood. The Mbuti have maintained their identity largely by supplying nearby villagers with the products they want from the forest.

The pygmies are one of the many ethnic and language groups in Zaire. More than 200 languages and dialects are spoken, so French remains the common (and official) language. About two-thirds of the people speak languages which belong to the Bantu family. The largest clusters of Bantu people include the Kongo in the west, the Mongo in the centre, the Luba in the south-centre and the Lunda in the south. There are also Sudanese groups, including the Azande and Mangbetu, in the north; Nilotes, including Alur and Lugbara, in the north-east; and a few Hamites, including the Tutsi in the KIVU region.

Several groups developed advanced societies before the arrival of Europeans. For example, when Portuguese navigators first reached

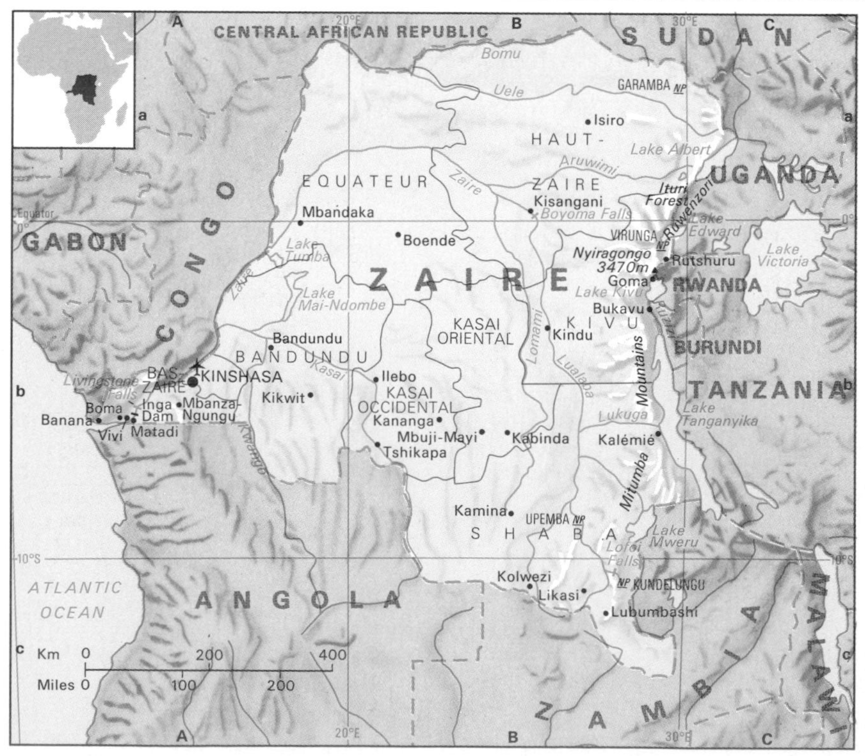

the Zaire river estuary in 1482, they made contact with the kingdom of Kongo, a powerful nation ruled from the royal capital Mbanza, some 250 km (155 miles) inland. The Kongo had elaborate legal and political institutions and the system was financed by taxes and tributes. In the 16th century, the Portuguese traded with the Kongo for slaves and drastically depleted the population, eventually undermining the kingdom in the 17th century.

Most rural people live in villages of 300 or so. Standards of living are low, diets are poor and limited, and malnutrition and disease take a heavy toll of life. Misfortune is often regarded by the people as a consequence of witchcraft, and sorcery is often their first, and modern medicine their last, resort when they fall ill.

COLONIAL DAYS

European influence in the interior did not begin until the late 19th century, after the explorations of Henry Morton Stanley in 1874-7. His later expeditions were financed by King Leopold II of Belgium who decided to claim the Congo Free State (as it was known) as his personal property. Leopold granted rights to exploit the land to Belgians of his choice – who sometimes abused their position. Exploitation of the natives was exposed by the British consul, Roger Casement, and Belgium was forced to take control of the Congo from Leopold. Casement was knighted in 1911 for his services, but five years later he was executed as a traitor for plotting an Irish uprising against Britain. Belgium's policies, though humanitarian, were similar to those of Leopold, in that private concessionaires were granted rights to collect ivory and forest products, and large companies were encouraged to establish plantations and mines.

This style of colonial development was successful in creating living standards that, by the 1950s, were among the highest in tropical Africa. But no great effort was made to prepare the country for independence – higher education, for instance, became available to black Africans only after 1954. Political parties, first permitted in the 1950s, were generally tribally or regionally based and many wanted independence from the rest.

The Belgians lost control of events and mounting unrest culminated in nationalist riots in 1959. Independence came on June 30, 1960, but within a few days there was anarchy in the newly named Republic of the Congo. The army of black soldiers rebelled against its white officers, there were outbreaks of communal fighting and rebellions in the provinces of Kasai, Stanleyville and Katanga. In Katanga (now Shaba), the rebel government, backed by European business interests, seceded from the rest of the country and ran its own affairs until the breakaway was ended by United Nations troops in 1963.

The two leading political figures at independence were Patrice Lumumba and Joseph Kasavubu, leaders of rival nationalist factions who became respectively prime minister and president. A power struggle ended with Lumumba's assassination in 1961.

In 1965 General Joseph Désiré Mobutu seized power. He has ruled ever since, putting down conspiracies and invasions by exile armies into Shaba. Since 1971 Zaire has been a one-party nation: as leader of the *Mouvement Populaire de la Révolution* (MPR) President Mobutu is automatically head of state.

In 1971 President Mobutu introduced his policy of 'authenticity', an attempt to return the country to its African origins, especially by changing many European names, both personal and geographical, to Zairean ones. The Congo became Zaire and citizens were required to drop their (European) Christian names. Mobutu changed his own name to Mobutu Sese Seko, dropping 'Joseph Désiré'.

▲ LIFE AND DEATH A painted native dancer pretends that a knife is passing through his head in this traditional ceremony used for initiations and funerals. The dancers come from the Bandundu region in south-west Zaire.

Political stability brought about by strong government (though the country has an imperfect record on human rights) led to an economic revival. But a stark economic crisis, caused by ill-conceived nationalisation programmes in all branches of the economy, rising import costs and falling copper prices, corruption and excessive borrowing, hit Zaire in the mid-1970s. In the late 1970s, the once healthy trade balance went into deficit. Aid from the United States, Belgium and the International Monetary Fund, tied to strict conditions, reduced Zaire's massive debt repayments and, by the mid-1980s, the rapid economic decline had been halted.

Tourism is not a major industry – there were about 19 000 visitors in 1976. However, for adventurous travellers, Zaire contains some of the world's most unspoilt places, with magnificent wildlife in its national parks and reserves (which make up 15 per cent of the country). Zaire has more elephants than any country in Africa. But more than 8000 are killed by poachers every year and the population is falling rapidly. In 1979 there were 371 000; now there are about 150 000. Zaire's tourist attractions include scenic mountains and lakes in the east, thundering waterfalls, including the majestic 340 m (1113 ft) Lofoi falls, and tranquil rivers flowing through dense forests.

ZAIRE AT A GLANCE
Area 2 344 885 km² (905 360 sq miles)
Population 33 940 000
Capital Kinshasa
Government One-party republic
Currency Zaire = 100 makuta
Languages French (official), Lingala, Swahili, Kikongo, Tshiluba
Religions Tribal (50%), Christian (45%), Muslim (1%)
Climate Tropical; equatorial in centre; average temperature in Kinshasa ranges from 18°C (64°F) to 32°C (90°F)
Main primary products Cassava, plantains, maize, sugar cane, groundnuts, bananas, palm oil and kernels, coffee, rubber, cotton, cocoa, tea, timber; copper, oil and natural gas, cobalt, zinc, tin, diamonds, gold
Major industries Agriculture, foodstuffs, oil refining, textiles, clothing, mining, forestry
Main exports Copper, cobalt, coffee, diamonds
Annual income per head (US$) 160
Population growth (per thous/yr) 29
Life expectancy (yrs) Male 48 **Female** 52

Zambia

RELIANCE ON COPPER EXPORTS HAS LEFT ZAMBIA EXPOSED ON A FALLING MARKET; NOW SHE MUST DEVELOP HER VAST AGRICULTURAL POTENTIAL TO SURVIVE

A single product, copper, has governed the development of Zambia's economy, dictated where people live, and determined where the main roads and railways run. The mineral, a basic raw material used in industries worldwide, provides about 95 per cent of Zambia's export income and over half of all government revenue. Indeed Zambia is second only to the USA in reserves of the ore, and is one of the largest producers in the world.

Since independence from Britain in 1964 (before which the country was known as Northern Rhodesia), the mines have been nationalised, with the result that any changes in world copper prices directly affect Zambia's whole economy. In addition, two other factors influence the well-being of the country's copper industry: an efficient system for transporting the mineral out of the country (landlocked Zambia relies on railways to get its copper to the coast for export); and cheap and secure sources of power for mining and processing the ores.

At the time of independence, Zambia's major rail outlets passed through Southern Rhodesia (now Zimbabwe) to ports in Mozambique and South Africa, or through Zaire to LOBITO in Angola. Since then, however, Zambia has been forced to reorganise its export links because of troubles with its neighbours.

THE CLOSED BORDER

The illegal Unilateral Declaration of Independence (UDI) by Southern Rhodesia in 1965 and the ensuing internal war, along with international sanctions against Southern Rhodesia, led to Zambia's closing of the mutual border. Lengthy liberation wars in Angola (1961-74) and Mozambique (1964-74), followed by continued fighting by anti-government forces since their independence, also hit Zambia's outlets through these countries.

Large quantities of electrical power, the second factor affecting the copper industry, are needed by Zambia's mines. Before independence, when the country was Northern Rhodesia, it became a key partner in the 1953-63 Federation of Rhodesia and Nyasaland (now Malawi). The huge copper revenues helped to build the vital KARIBA Dam on the Zambezi, which provided power for industry in both Northern and Southern Rhodesia. Unfortunately for Zambia, the hydroelectric power station was on the Southern Rhodesian side of the Zambezi, which forms the border between the two countries. During the UDI period Southern Rhodesia was in a position to put pressure on Zambia against joining in sanctions, by threatening to cut off its power. At great expense, Zambia built a power station on its side of the river, and another – now the country's biggest – on the KAFUE.

Faced by these problems in the 1960s and 1970s, Zambia was forced to turn north for a solution. Here lay Tanzania, the only neighbouring state that was then in any way similar – newly independent and black-ruled. In 1968, an oil pipeline was completed, running south from DAR ES SALAAM in Tanzania to Ndola on Zambia's Copperbelt in the north of the country. The links have since been strengthened with the completion of a railway in 1975, known as the Tazara (or Tanzam), to connect Zambia's rail system with Dar es Salaam, and so to ship out the copper. The Tazara was built with Chinese finance, under Chinese direction and partly with Chinese labour. The Chinese involvement was less a reflection of Zambia's political leanings than the fact that the West rejected the project as an uneconomic investment. Indeed, port congestion at Dar es Salaam and operating problems on the railway reduced the potential of Tazara to release Zambia from its transport dependence on the south. In 1978, after severe economic hardship, Zambia was forced to re-establish rail links through Southern Rhodesia (Zimbabwe).

Though independence wars are now over in Angola, Mozambique and Zimbabwe, the situation has not got better for Zambia's copper exports. The Angolan and Mozambique railways are often cut by rebels in both countries. The accumulated debts of the 1960s and 1970s, including that for the Tazara railway, are a huge drain on Zambia's foreign exchange and its ability to import. Medium and long-term debt is estimated at around US$4000 million. These problems, combined with a depressed market for copper – world prices have halved in real terms since 1964 – have led to enormous financial problems. In the 1980s Zambia has rescheduled its debt repayments, which account for 40 to 50 per cent of the country's earnings from exports of goods and services, and the International Monetary Fund imposed tight controls on expenditure.

Most of Zambia is a high plateau, 900-1500 m (about 3000-5000 ft) above sea level. Bordering it to the south is the Zambezi river, and in the south-west the sands of the desolate KALAHARI desert. The country has a number of other large rivers, including the Luangwa, and some large lakes – the largest is Lake BANGWEULU, but parts of lakes MWERU and TANGANYIKA are also in Zambia.

The country lies in the tropics but the height of the land reduces temperatures slightly. From May to August it is dry and cool, with temperatures averaging 9-23°C (48-73°F). September to November is hot and dry with temperatures of 18-31°C (64-88°F). The hot, rainy season from December to April is only slightly cooler. The natural vegetation is mainly savannah, and ranges from dry grassland in the south to more luxuriant grassland with scattered trees and bushes, and some woodland, in the north.

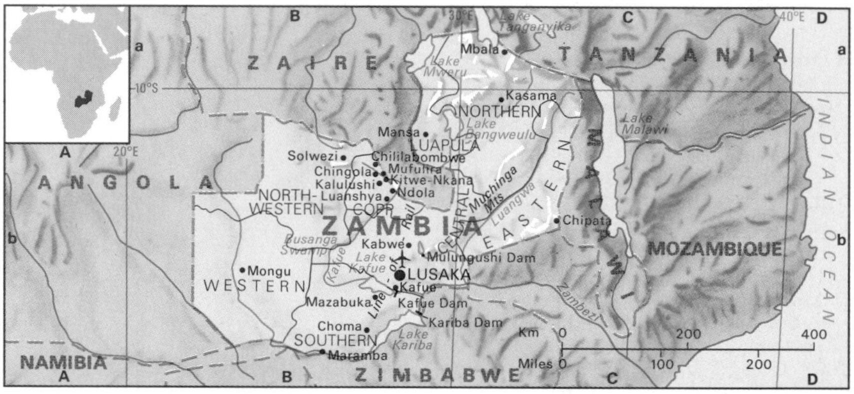

The country contains a great variety of wildlife and there are large game parks on the Luangwa and Kafue rivers with antelopes, Cape buffaloes, elephants, leopards, lions, giraffes, hippopotamuses, rhinoceroses and zebras. With nature parks, huge open spaces and scenic attractions such as the VICTORIA FALLS on the Zambezi, the tourist industry is developing.

THE RURAL COMMUNITY

Zambia's population of nearly 7 million contains over 72 different tribal groups, each with its own language. Although the population is growing fast – 3 per cent per annum – there is no great pressure on land, since the country covers some 750 000 km² (about 290 000 sq miles), over three times the size of the United Kingdom.

The majority of Zambians are peasant farmers living in scattered villages. Large commercial farms also exist – particularly along the railway route between the Copperbelt and the Zimbabwe border. Many of them are still owned by Europeans or large international companies, such as Lonrho. However, the bulk of the country's agricultural output is grown by the peasants – maize is the staple crop – who also eat most of it. Their farming methods are primitive; hand hoes are used for cultivation, and bush-fallow techniques (by which much of the land is left unused and allowed to grow wild at any one time) are widespread. Women, as in most of Black Africa, do much of the work.

Despite the number of people working on the land, and despite the extensive water resources and vast tracts of fertile land, food shortages are constant, malnutrition is common, and both contribute to an infant mortality rate of 105 per 1000 live births. The government has been forced to turn to expensive food imports to feed its people but, because of its huge debts, these are drastically limited.

In the ten years since the decline in the price of copper, Zambia has done little to use its huge agricultural potential as a substitute foreign exchange earner. There are no major irrigation schemes; transport systems and marketing facilities for isolated areas are inadequate; pricing policies are unattractive and produce marketing boards are incompetent. As a result there has been little incentive for the farmers to grow a surplus for sale; and huge numbers of people have left the countryside to seek new opportunities in the towns.

However, the government is planning to halt the slide. Scattered communities have been brought together in larger village groups to make it more economical to provide services – health clinics, primary schools, cooperative marketing schemes, and better roads.

THE DRIFT TO THE TOWNS

The movement of people from the countryside to the towns has produced one of the most urbanised societies in Africa. Today, more than 40 per cent of Zambia's population live in the cities. In particular, the string of Copperbelt towns (among them NDOLA, CHINGOLA, KITWE-NKANA, Chilila-Bombwe and Luanshya) form one of the largest urban concentrations in tropical Africa, though the capital, LUSAKA, to the south of the mining area, remains Zambia's largest city with well over half a million inhabitants.

Many of the rural migrants have been unable to find paid jobs and have therefore set up in a variety of self-employed jobs – from bicycle repairs to hawking, beer brewing and tailoring. The increasing populations in the cities have also caused housing shortages and many people now build their own homes in settlements around the towns. The government has started a programme to improve many of these areas – for example, George in Lusaka – by providing services such as water, electricity and shops. Sites have also been laid out, with services laid on, where people can build in a more organised fashion.

Many townspeople still retain strong links with their rural homelands. It is still common for some migrant workers who have lived in towns for many years to retire to their tribal villages, where the system of communal landholding allows them access to a plot of land to farm. Other townspeople – the children of migrants, born in the cities – regard themselves as permanent city dwellers and have lost their ethnic or old regional affiliations. For example, most of the original migrants to the Copperbelt were from Bemba-speaking regions in the north-west; now their descendants largely regard themselves as Copperbelters.

One significant effect of independence has been the steady takeover by Zambians of what were previously regarded as White jobs. During the colonial period, the mining companies employed single men as a ready source of unskilled, casual labour. But as mining techniques advanced, it was in the interest of the mines to take on a more permanent, skilled African labour force. Thus the settled population of the copper towns began to grow. Certain jobs, however, were reserved for Whites and fiercely guarded by the White trade unions, so that it was not until the 1950s that any concessions were made. After independence, the process of Africanisation of skilled jobs was speeded up and the number of non-Zambian employees in the mines has fallen from over 7000 to about 1750 today, while Zambians have taken the key positions. This has created a well paid, well educated Zambian elite, many of whom have adopted a European lifestyle.

Ruling over all – it is a one-party republic – is Kenneth Kaunda, who has been head of state since independence. A much-respected national figure, he is one of the key leaders of the so-called Front Line States, which played a major diplomatic role in bringing about independence in Zimbabwe and now oppose apartheid in South Africa. His concern for his people has coloured much of the country's social and economic policies, with the result that Zambia is an inextricable mix of capitalism and socialism.

Zambia, along with eight other African countries, is a member of the Southern Africa Development Coordination Conference (SADCC), which is committed to rehabilitating and improving the railways and ports which have been the cause of so many of Zambia's economic problems. If this succeeds – and financial support from the West is critical – the country's economic burden may be lightened. However, the poor market prospects for copper, a resource that will run out eventually, make it imperative for Zambia to press on with developing her vast agricultural potential.

▲ RIVER OF SPLENDOUR The blue ribbon of the Luangwa winds through the browns and greens of the Zambian landscape. The Luangwa, 806 km (500 miles) long, joins the Zambezi on the Mozambique border. On its way it flows through two great national parks, each rich in wildlife and abounding in magnificent scenery.

ZAMBIA AT A GLANCE
Area 752 620 km² (290 586 sq miles)
Population 6 990 000
Capital Lusaka
Government One-party republic
Currency Kwacha = 100 ngwee
Languages English (official), Bemba, Nyanja, Lozi, Tonga and other local languages
Religions Christian (65%), tribal (30%)
Climate Tropical; cool in the highlands. Average temperature in Lusaka ranges from 9-23°C (48-73°F) in July to 18-31°C (64-88°F) in October
Main primary products Maize, cassava, sugar, tobacco, cotton, groundnuts, beef cattle; copper, cobalt, zinc, lead, coal
Major industries Agriculture, mining, brewing, chemicals, cement
Main exports Copper, cobalt, zinc, lead
Annual income per head (US$) 430
Population growth (per thous/yr) 32
Life expectancy (yrs) Male 49 **Female** 52

Zante ***Greece*** See ZAKINTHOS

Zanzibar ***Tanzania*** Island in the Indian Ocean about 50 km (30 miles) north of Dar es Salaam. Zanzibar was visited by Arabs and became a Portuguese trading place in the early 16th century. By 1700, Arabs had made Zanzibar a thriving centre of the East African ivory and slave trades. It was made a British protectorate in 1890 – together with the island of PEMBA 60 km (35 miles) to the north-east. The two islands became an independent country for four months in 1963-4. They then joined mainland Tanganyika to form the republic of Tanzania. Zanzibar covers 1660 km² (640 sq miles), and its chief products are cloves and copra.

The Zanzibar Channel, a branch of the Indian Ocean, 31 km (19 miles) wide, separates Zanzibar island from the mainland.

The port of Zanzibar on the west coast of Zanzibar island is the main town. In the 19th century it was used as a base by European explorers and missionaries heading for the African interior.

Population (island) 330 000; (town) 140 000
Map Tanzania Ba

Zaozan ***Japan*** Mountainous area in northern Honshu, about 260 km (160 miles) north of Tokyo. It rises to 1841 m (6040 ft) and is popular with visitors because of its volcanic scenery and, in winter, its skiing and curious *juhyo* – 'snow monsters' formed by ice and snow draped over the mountain pines.

Map Japan Dc

Zaporozh'ye ***USSR*** City on the Dnieper river in the Ukraine, 250 km (155 miles) south of Khar'kov. The nearby Dnieper hydroelectric dam powers a huge industrial complex which now makes iron and steel, aluminium, vehicles, machinery and chemicals.

Population 844 000
Map USSR Ed

Zaragoza (Saragossa) ***Spain*** University city about 275 km (170 miles) north-east of Madrid, with engineering, sugar refining, flourmilling and wine-making industries.

Its province of the same name supports some agriculture – sugar cane, cereals, vineyards and olive groves – despite low rainfall and extreme temperatures.

Population (city) 590 800; (province) 842 400
Map Spain Eb

Zaria (Zegzeg) ***Nigeria*** Historic walled town, about 690 km (430 miles) north-east of the capital, Lagos. It was founded in the 14th century as the capital of the Hausa Kingdom of Zazzan, and was a centre of the slave trade. Its ancient monuments include the beautiful Friday Mosque and the Emir's palace. Today, it is a large-scale producer of cigarettes, palm oil (used in margarine and cooking fats), cotton textiles, printed matter, cosmetics and bicycles. Traditional weaving, dyeing and basket-making also play their part.

Population 267 300
Map Nigeria Ba

Zarqa ***Jordan*** Country's second largest city and an important industrial centre, situated 20 km (12 miles) north-east of Amman. Zarqa developed from a 19th-century village and is now the administrative centre for the Zarqa region – which includes an extensive phosphate mining area. The town is on the Zarqa river, which flows into the River Jordan and provides much of the water that irrigates the Jordan Valley. The King Talal Dam was built on the Zarqa between 1972 and 1978.

Population 270 000
Map Jordan Ba

Zealand (Sjælland) ***Denmark*** The country's largest island (7014 km², 2708 sq miles), between the GREAT BELT channel and the Öresund, the main entrance to the Baltic Sea. COPENHAGEN, the Danish capital, is on the island's eastern coast. It has a gently rolling landscape, humped with ridges left by retreating Ice Age ice sheets. It also embraces Denmark's largest lake, Arresø (41 km², 16 sq miles).

The island is intensively cultivated, with 10 per cent of the land covered by well-managed woodlands. It is also the most urbanised part of Denmark. There are numerous towns and settlements: from Kalundborg in the north-west, with its five-towered fortified church, to the old port of Køge in the east, with its richly timbered houses; from ELSINORE with Hamlet's castle in the north-east to Korsør, with its naval base, in the west. Baronial homes such as Løvenborg, Borreby Castle, and the royal hunting seat at Frederiksborg Castle, HILLEROD, recall Denmark's feudal past.

Many parts of Zealand have fine beaches, and there are large numbers of summer cottages. Zealand is also the site of 40 per cent of Denmark's industries, most of which are clustered near the capital.

Population 1 855 500
Map Denmark Bb

Zeebrugge ***Belgium*** Ferry and cargo port linking the city of Bruges to the North Sea by a 13 km (8 mile) long canal. The waterway was opened in 1903 to provide deep-water access to the city. Its harbour – used by German submarines in the First World War – was the scene of a daring naval raid by the British on April 23, 1918, which resulted in the destruction of the mole and the blocking of the port.

Population 2500
Map Belgium Aa

Zeeland ***Netherlands*** South-western province consisting of the southern islands of the Maas-Schelde delta and the mainland area south of the Schelde estuary. Most of its 2745 km² (1060 sq miles) are below sea level and protected by dykes built under the Delta Plan for land reclamation, begun in 1956. Zeeland produces cereals, potatoes, flax and fruit, and Middelburg is its capital.

Population 354 900
Map Netherlands Ab

Zegzeg ***Nigeria*** See ZARIA

Zelazowa Wola ***Poland*** Village on the edge of the Kampinos forest 53 km (32 miles) west of the capital, Warsaw. The Polish composer and pianist Frédéric Chopin (1810-49) was born in the village's small manor house. The house is now a museum, where some of the world's greatest pianists give recitals in summer.

Zelle ***West Germany*** See CELLE

Zemlya Frantsa-Iosifa (Franz Josef Land) ***USSR*** Group of many small Arctic islands north of Novaya Zemlya. They are the most northerly pieces of land in the Eastern Hemisphere and have a total area of about 20 700 km² (8000 sq miles) – most of it ice-covered, with a few pockets of lichen. They were discovered by Austrian explorers in 1873 and claimed in the 1920s by the Russians – who established research stations there.

Map Arctic Ocean a

Zenica ***Yugoslavia*** Industrial town in Bosnia Herzegovina, 65 km (40 miles) north-west of Sarajevo. It is Yugoslavia's chief coal, steel and paper-producing centre. To the west lies the ancient town of Travnik.

Population 132 700
Map Yugoslavia Cb

zeolite Any of a group of complex silicate minerals usually containing sodium, potassium, calcium or barium, or a combination of these elements, used chiefly as filters or as water softeners.

Zermatt ***Switzerland*** Resort village in the south-west, about 10 km (6 miles) north of the Italian border. It is ringed by mountains rising to 4000 m (13 100 ft) or more – most notably the MATTERHORN. Zermatt, which lies at an altitude of 1616 m (5302 ft) and is accessible only by rail, is a centre for winter sports, mountaineering and summer walks.

Population 3200
Map Switzerland Aa

Zetland ***United Kingdom*** See SHETLAND

zeuge Isolated flat mass of hard rock standing on a narrower pillar of softer underlying rock in a desert, or surviving as a relic of former climatic conditions. Zeugen are produced by sand-laden winds eroding the softer rock and undercutting the harder mass. They stand up to about 30m (100 ft) high.

Zhangjiakou ***China*** City in Hebei province. It was once a junction of caravan routes on the dry northern plateau of Hebei. Today it serves the nearby Xuanhua iron-ore field.

Population 630 000
Map China Hc

Zhanjiang ***China*** Industrial city in Guangdong province 400 km (250 miles) west of Guangzhou (Canton). It is a centre for the manufacture of chemical fertilisers and is a base for offshore oil exploration in the South China Sea.

Map China Gf

Zhdanov ***USSR*** Port on the north shore of the Sea of Azov; formerly Mariupol. It was founded in 1779 on the site of an ancient Greek colony. Its main industries are iron and steel, engineering and chemicals, and it exports coal, grain and salt.

Population 520 000
Map USSR Ed

Zhejiang ***China*** Province covering 100 000 km² (38 500 sq miles) south of the delta of

the Chang Jiang (Yangtze River). Hills and mountains make up more than 70 per cent of the province, and cultivated land is restricted to valleys and the narrow coastal plains. Rice, wheat, tea, maize, sugar cane and tangerines are grown in the subtropical province. It is also a leading producer of silk. Dense forests of pine, spruce and bamboo cover the mountainsides in the interior, and they are exploited to serve a flourishing timber and paper industry.
Population 37 920 000
Map China He

Zhengzhou ***China*** Capital of Henan province, south of the Huang He (Yellow River) on the North China Plain. It is the main railway junction of north China and a centre for the manufacture of textiles and food products.
Population 1 290 000
Map China Gd

Zhenjiang ***China*** Ancient city in Jiangsu province, on the south bank of the Chang Jiang river. It manufactures machinery, chemicals and paper. The beautiful wooded hills in the southern suburbs are dotted with Buddhist temples, some of which date back to the 6th century.
Population 270 000
Map China Hd

Zhitomir ***USSR*** Ukrainian city 130 km (80 miles) west of Kiev. It is the centre of the USSR's largest hop-growing and beer-brewing region. It has large grain mills, timber mills and sugar refineries – and manufactures machinery, furniture and clothing.
Population 270 000
Map USSR Dc

Zhob ***Pakistan*** River valley, town and district in north-east Baluchistan. Chromite (chromium ore) is mined there for export. The valley is rich in wild flowers and fruit orchards.
Population (town) 33 000; (district) 120 000
Map Pakistan Cb

Zhu Jiang (Pearl River) ***China*** River formed by the triple confluence of the Xi Jiang, Bei Jiang and Dong Jiang rivers near Guangzhou (Canton). It flows 176 km (110 miles) into a broad delta – one of the richest lowlands in southern China and the most densely populated part of Guangdong province. The British and Portuguese enclaves of HONG KONG and MACAU lie on either side of the main estuary.
Map China Gf

Zibo ***China*** City of Shandong province on the Shandong Peninsula about 370 km (230 miles) south-east of Beijing (Peking). It is a centre for the manufacture of glass, aluminium and petrochemicals.
Population 2 000 000
Map China Hd

Zielona Gora (Grünberg) ***Poland*** Industrial city 145 km (90 miles) north-west of the city of Wroclaw. It was producing fine cloth and wine in the Middle Ages, and still holds an annual wine festival in early autumn after the grape harvest. The city also has metal, clothing, linen, food and printing industries.
Population 107 800
Map Poland Ac

Ziguinchor ***Senegal*** River port and regional capital just north of the Guinea-Bissau border. It was founded by the Portuguese in the 16th century on the Casamance river, some 70 km (43 miles) from the Atlantic. But it now has more of a French flavour, with elegant buildings and broad, palm-lined streets built during the French colonial period (1840-1960).

From the town, tourists can visit the beach of Cap Skirring, or go on trips into the surrounding tropical forests, or ride a boat upriver past mangrove swamps and villages of the region's traditional conical huts.
Population 72 700
Map Senegal Ab

Zihuatanejo ***Mexico*** Picturesque fishing port and increasingly popular beach resort on the Pacific coast of Guerrero state. New developments, including an international airport, have not spoilt the town's natural beauty. Neighbouring Ixtapa, by contrast, is a purpose-built resort specialising in water sports.
Population 50 000
Map Mexico Bc

Zikhron Ya'aqov ***Israel*** Village in the Sharon Valley, about 25 km (15 miles) south of Haifa. The village was founded by the Rothschild family in 1882 and is the centre of a region producing grapes and wine.
Population 5000
Map Israel Aa

Zilina ***Czechoslovakia*** Industrial town in Slovakia 74 km (46 miles) south-east of Ostrava. It makes chemicals, timber products, textiles and foodstuffs. The Old Town includes a Baroque market square and several theatres and museums.
Population 87 800
Map Czechoslovakia Db

Zillertal Alps ***Austria*** Part of the Eastern Alps between the Brenner Pass and the Hohe Tauern range, on the border with Italy. They rise to 3509 m (11 513 ft) at the Hochfeiler. There are several popular ski resorts, such as Mayrhofen, in the valley of the Zill river.
Map Austria Bb

Zimbabwe See p. 728

Zinder ***Niger*** Walled market town and former national capital, about 750 km (465 miles) east of the present capital, Niamey, and at the southern end of a road that stretches northwards across the Sahara to the Algerian capital, Algiers. Its wares include groundnuts (peanuts), skins, hides, leather goods and blankets.

Zinder, replaced as capital in 1926, is dominated by Muslims who have close family and trade links with the city of KANO in northern Nigeria. It has a dense maze of alleyways, a former sultan's palace of 1860, a mosque, and a fort, dating from the French occupation.
Population 60 000
Map Niger Ab

Zion, Mount ***Israel*** See JERUSALEM

Zion National Park ***USA*** Area of 593 km² (229 sq miles) in south-west Utah. It contains the sheer canyon of the Virgin river, mesas and colourful rock formations.
Map United States Dc

▼ MAGICAL CONSULTING ROOMS The motley paraphernalia of his trade adorn a witch doctor's premises on Zanzibar. Many islanders of African descent consult such practitioners to find the source of their misfortunes and for an insight into the future.

Zimbabwe

A BEAUTIFUL LAND WITH A TURBULENT HISTORY – RICH IN MINERALS AND TEEMING WITH WILDLIFE

Everything about Zimbabwe is dramatic, from its spectacular landscapes to its violent politics, from its teeming wildlife to its turbulent history. In the north the ZAMBEZI river flings itself over the mile-wide VICTORIA FALLS and flows through a man-made lake nearly 300 km (186 miles) long; in the south the 'great, grey-green, greasy LIMPOPO river' as Kipling described it marks the border with South Africa. Mountain ranges ridge the east, peaking in 2592 m (8504 ft) Inyangani; and to the west lies superb safari country, forest and savannah full of lions, elephants, buffaloes and giraffes.

Then there is the gold that attracted the English imperialist Cecil Rhodes to the territory which was named Rhodesia after him. He brought white colonial rule after a bloody war. Less than 100 years later, another war ravaged the country before blacks ruled again and called it Zimbabwe.

There never was gold in the quantities hoped for by Rhodes, already rich from his South African mining operations. But Zimbabwe, although a minor gold producer, is rich in other minerals and has a great wealth of farmland. It is also one of Africa's major manufacturing nations.

Rhodes was one of the larger-than-life men whose names punctuate Zimbabwe's history, from the canny Matabele King Mzilikazi and his son Lobengula, to the present prime minister, the scholarly Robert Mugabe. They include Ian Smith, last of the white rulers, who defied the British Government and made a unilateral declaration of independence (UDI) in 1965. Not to mention the Scottish missionary-explorer David Livingstone, the first white man to reach the Zambezi, in 1851, and to see the Victoria Falls, four years later.

INDEPENDENCE UNDER BLACK RULE

Many of Zimbabwe's 225 000 whites left after the country became truly independent under black rule in 1980, but Mugabe, needing their skills and know-how, has tried hard to win their confidence and stop the exodus. About 150 000 remain – many of them farmers, including Ian Smith – and they have 20 reserved seats in parliament. From 1987 onwards this arrangement can be changed if 70 MPs and two-thirds of senators approve. The whites continue to enjoy a high standard of living, with African servants.

But 96 per cent of the 8.9 million population is black: 80 per cent of them belong to the Shona people (known also as the Mashona), who have been in Zimbabwe since the 11th or 12th centuries; the other 20 per cent are Ndebele (known also as Matabele), descendants of Mzilikazi's warrior tribe who broke away from the Zulu nation in the 1830s.

Broadly speaking, Mugabe's Zimbabwe African National Union (ZANU) represents the Shona majority and Joshua Nkomo's Zimbabwe African People's Union (ZAPU) the Ndebele. The parties were born as guerrilla movements which fought and brought down the Smith regime.

The newly independent country took its name from GREAT ZIMBABWE, a site of massive, stone ruins near Masvingo; they are believed to date from a Shona civilisation that dominated the region from the 12th to the 19th centuries. The Shona were mainly farmers.

In the 1830s the invading Ndebele, who lived largely on cattle, the spoils of raiding parties and tributes from those they had conquered, established a community in what is now BULAWAYO. Their descendants still live in the surrounding region, called MATABELELAND (or Ndebeleland). The Shona are in the north, around the modern capital HARARE (formerly Salisbury) in the region known as MASHONALAND or Shonaland.

Almost all this scenic country is more than 300 m (1000 ft) above sea level. A great central plateau, called the highveld, is 1200-1500 m (4000-5000 ft) high and occupies 52 000 km² (about 20 080 sq miles). Massive granite outcrops called *kopjes* dot the landscape, but this is a fertile savannah area, with temperatures averaging a pleasant 20°C (68°F) and an annual rainfall at 700-900 mm (28-35 in), most of it in summer storms.

The highveld runs roughly from north-east to south-west, and on either side the land falls away to the middleveld, again fine plateau country 600-1200 m (about 2000-4000 ft) high. Below 600 m is the lowveld, a narrow strip in the Zambezi valley and a broader band between the Limpopo and Sabi rivers.

Tobacco is the most important crop, accounting for about 20 per cent of export earnings. The biggest food crop is maize, grown mainly for internal consumption, followed by other grains, tea, coffee and sugar. Cotton is another significant crop.

A geological feature called the GREAT DYKE runs for 515 km (about 320 miles) across the highveld from north-west to south-east, and forms the low ridge on which Harare stands. In and around the Dyke are rich deposits of gold, chrome, nickel and other minerals.

The modern history of Zimbabwe started in the 1860s when emissaries from Cecil Rhodes persuaded King Lobengula to grant them mining rights. Rhodes decided that Lobengula's concession amounted to a licence to grab the whole country. He proceeded to do so, forming the British South Africa Company and sending 'pioneer columns' to invade the territory.

The outraged Ndebele went to war against the company in 1893, and a bloody conflict ended in the defeat and subsequent death of Lobengula. In 1896-7 the Ndebele rebelled again and were joined by the Shona, who had retained lands in the north. It was a final, desperate resistance to colonialism that ended with British regular troops being called in.

The territory so won became the colony of Southern Rhodesia, and in 1930 a Land Apportionment Act limited the areas in which Africans could own land, creating, in effect, native reserves. Whites ended up owning almost all the best land.

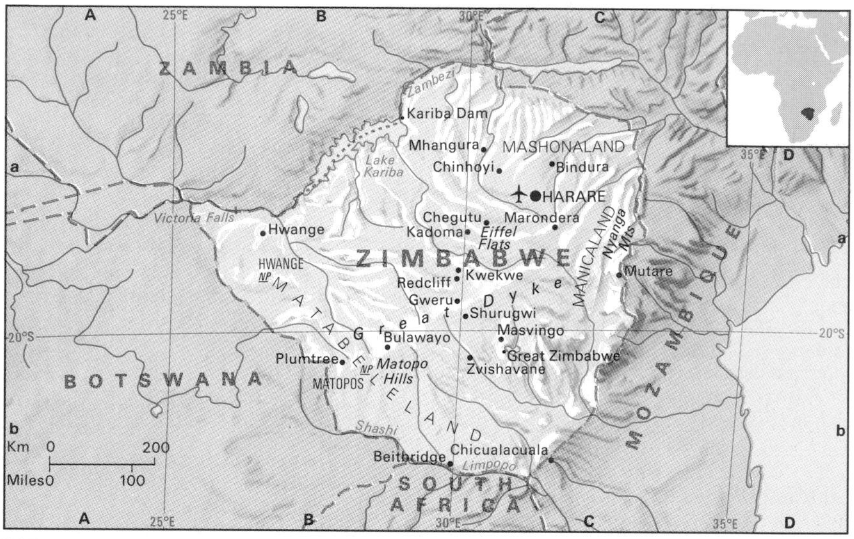

In 1953 Southern Rhodesia became part of a federation with Northern Rhodesia (now ZAMBIA) and Nyasaland (now MALAWI). But this broke up in 1963 primarily through black opposition to white rule in Southern Rhodesia. Zambia and Malawi both achieved independence with black governments, but the whites in Southern Rhodesia would not accept a similar solution. In 1965, led by Ian Smith and his Rhodesian Front Party, Southern Rhodesia rebelled against Britain and declared itself independent as Rhodesia.

Britain cut off trade and financial links with the breakaway colony. The United Nations followed with international sanctions.

Neighbouring South Africa and Mozambique continued to allow vital supplies into Rhodesia, which already had a strong industrial base and was manufacturing a range of

▶ TEMPLE OF MYSTERY This enclosure, known as the Temple, is part of the mysterious ruins of Great Zimbabwe, near Masvingo. It was built as a king's residence by the Shona Bantu-speaking people, who settled the area from about AD 300.

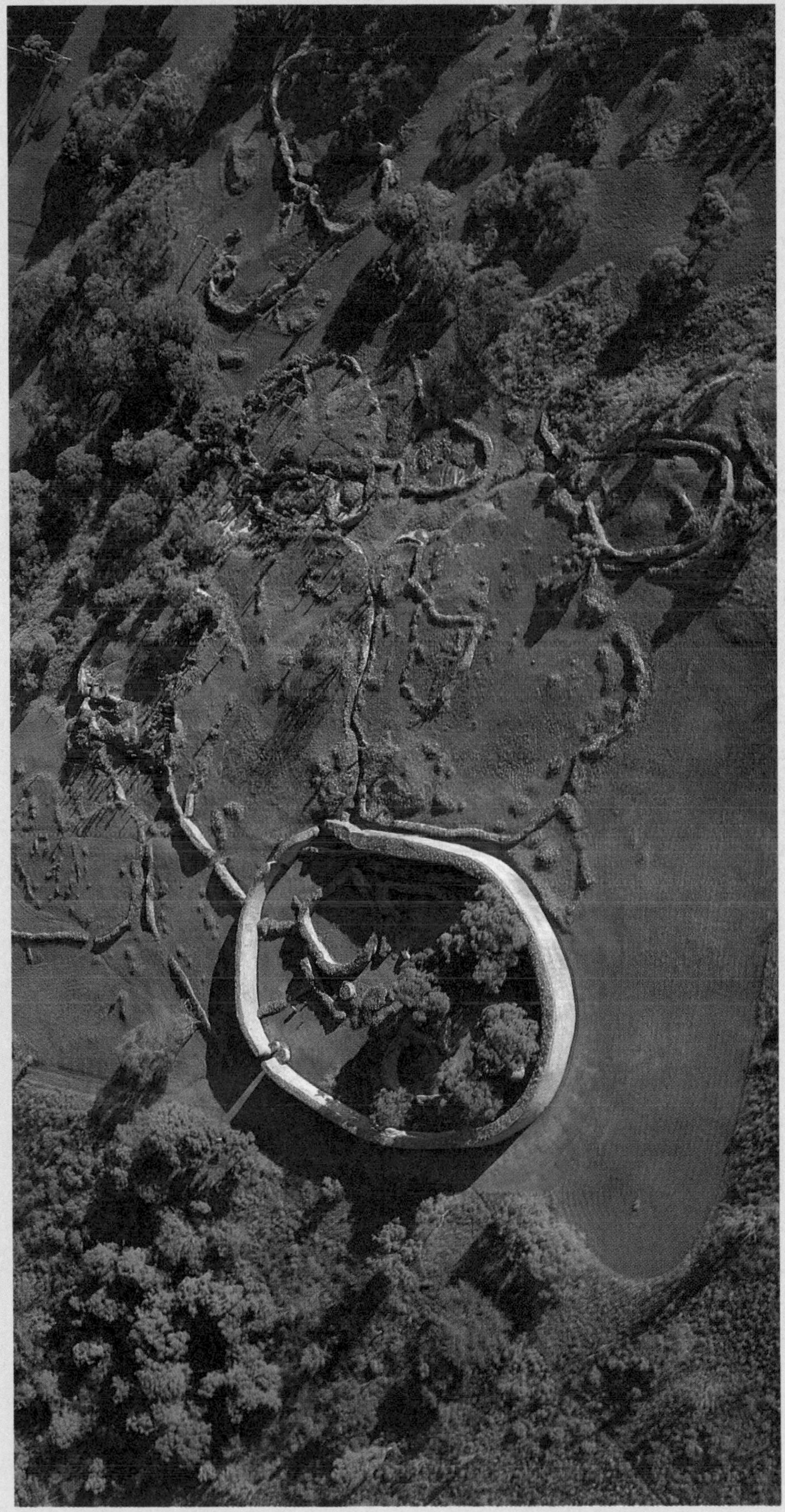

substitutes for blocked imports. In 1970, Smith declared a republic, and guerrilla action against him escalated rapidly into civil war.

Eventually the combined effects of the war, sanctions and diplomatic pressure persuaded Smith to make way for black majority rule. At a conference in London in 1979 a new constitution was thrashed out. Elections followed in 1980, Zimbabwe was born, and it rejoined the Commonwealth.

The years since have seen a post-independence boom fade away, and deepening rifts between ZANU and ZAPU. Unemployment and tension over land ownership are mounting problems among a population which is growing by 3.3 per cent annually. Britain and the European Economic Community (EEC) are helping to finance the settlement of black peasants on land bought from white farmers. China is partnering the government in mining, manufacturing and agricultural ventures aimed at improving Zimbabwe's exports.

Tourism is a major growth industry, which attracted more than half a million visitors in 1985-6. This beautiful country, with its pleasant climate, has much to offer them. In addition to the Victoria Falls and Great Zimbabwe, there are the glorious Eastern Highlands, laced with trout streams and home to leopards and eagles. Lake KARIBA, in the north, is 5000 km² (nearly 2000 sq miles) of blue, island-studded water offering fishing, sailing and game-rich shores. In the MATOPO HILLS, near Bulawayo, there are 2000 km² (about 800 sq miles) of grotesque, wind-sculptured, granite hills, with caves painted by ancient peoples. Then there are national park and wildlife reserves covering a staggering 44 700 km² (about 17 260 sq miles), or 11 per cent of the whole country.

ZIMBABWE AT A GLANCE
Area 390 600 km² (150 810 sq miles)
Population 8 950 000
Capital Harare (formerly called Salisbury)
Government Parliamentary republic
Currency Zimbabwe dollar = 100 cents
Languages English (official), Shona, Ndebele
Religions Christian (55%), tribal (40%), Muslim (1%)
Climate Subtropical; average temperature in Harare ranges from 7-21°C (45-70°F) in June/July to 16-27°C (61-81°F) in November
Main primary products Maize, tobacco, wheat, sugar, millet, cotton; gold, asbestos, coal, iron, chrome, nickel, copper, tin
Major industries Agriculture, cigarettes, fertilisers, cement, coke, iron and steel, textiles, mining, tourism
Main exports Tobacco, foodstuffs, ferrochrome, gold, asbestos, cotton
Annual income per head (US$) 860
Population growth (per thous/yr) 33
Life expectancy (yrs) Male 53 **Female** 57

Zipaquirá ***Colombia*** Town about 50 km (30 miles) north of Bogotá. It has a huge rock salt mine, said to be able to supply the world's needs for 100 years. Parts of the mine are open to the public. One section was shaped in 1954 into a vast underground cathedral, with a salt block as its altar.

Population 45 000
Map Colombia Bb

Ziz ***Morocco*** See ERFOUD

Zlatibor ***Yugoslavia*** Thickly forested mountain massif rising to 1496 m (4908 ft) in west Serbia some 140 km (about 85 miles) south-west of Belgrade. In autumn 1941, Yugoslavia's former Communist leader Marshal Tito (1892-1980) declared the first free partisan state in Titovo Uzice, a town in the Djetinia valley to the north-west.

The massif has an exceptionally clear atmosphere and 2000 hours of sunshine a year, and is a popular summer and winter resort.

Map Yugoslavia Dc

Zlatoust ***USSR*** City in the Ural mountains 200 km (125 miles) south of Sverdlovsk. It grew out of iron and copper works set up near local mines in 1754, and specialises in making cutlery, swords and engraved and stainless steels, and also machine tools, clocks and watches.

Population 204 000
Map USSR Gc

Zomba ***Malawi*** Capital of Nyasaland 1885-1964 and of independent Malawi 1964-75, when it was displaced by LILONGWE, 240 km (150 miles) to the north-west. It was originally the base from which the slave trade was destroyed in the 19th century. It is now the centre of a fertile farming region which produces coffee, cotton, tobacco and tung oil, used in making waterproof paints and varnishes.

Population 49 400
Map Malawi Bc

Zouerate ***Mauritania*** Mining town about 650 km (405 miles) north-east of Nouakchott, connected by rail to the port of Nouadhibou. Its iron ore provides some 60 per cent of the country's total exports.

Population 22 000
Map Mauritania Ba

Zrenjanin (Petrovgrad; Veliki Beckerek; Nagybecskerek) ***Yugoslavia*** One of the largest cities in the Serbian autonomous province of Vojvodina. It lies on the Becej river – a tributary of the Tisza – 48 km (about 30 miles) north-east of the province's capital, Novi Sad. A market, transport and industrial centre, it produces farm machinery, beer, spirits, sugar and processed foods.

Population 139 300
Map Yugoslavia Eb

Zug ***Switzerland*** Town and resort on the north-east shore of Lake Zug, about 20 km (12 miles) south of Zürich. It is capital of the canton of the same name. The town has a 16th-century town hall and a clock tower dating from 1480. It makes textiles and electrical goods.

Population (town) 23 000; (canton) 79 000
Map Switzerland Ba

Zugspitze ***West Germany*** The country's highest mountain at 2963 m (9721 ft). It is in the Bavarian Alps beside the Austrian border, 90 km (55 miles) south-west of Munich.

Map West Germany De

Zuiderzee ***Netherlands*** See IJSSELMEER

Zulawy ***Poland*** Fertile farming region (2515 km², 970 sq miles) in the delta of the Vistula river to the east of the Baltic port of Gdansk. Wheat, sugar beet and oilseed rape are grown, Friesian and Holstein cattle graze the fenland pastures, Dutch-type windmills survive, and half-timbered churches and houses abound. The Teutonic Knights began draining the area in the 14th century, and imported settlers from the Netherlands and German Friesland to farm it.

Map Poland Ca

Zululand ***South Africa*** See KWAZULU

Zürich ***Switzerland*** The country's largest city standing on a lake of the same name, about 20 km (12 miles) south of the German border. The site was a lakeside settlement even before the Romans arrived in 58 BC. Today Zürich is Switzerland's main financial and commercial centre, housing numerous insurance and international banking headquarters. The city, which is popular with tourists, has a Romanesque cathedral and a 12th-century town hall.

Swiss Protestantism was founded in Zürich by the religious reformer Ulrich Zwingli (1484-1531). The city embraced the new faith in 1523 after Zwingli – while a priest at the cathedral – violently denounced Roman Catholicism. He was killed in 1531 in an attack on Zürich by the forces of the Catholic cantons.

Population 422 700
Map Switzerland Ba

Zutphen ***Netherlands*** Industrial town on the IJssel river, about 95 km (60 miles) east and slightly south of Amsterdam. The English poet and soldier Sir Philip Sidney was fatally wounded there in 1586, when the town was besieged by the Spanish during the 80 Years' War (1568-1648). The town makes paper, leather, soap and glue.

Population 31 500
Map Netherlands Ca

Zwelitsha ***South Africa*** Town in the black state of Ciskei, in eastern Cape Province, near King William's Town about 50 km (30 miles) north-west of the port of East London. It became the seat of government for Ciskei in 1981 while a new capital was being built at Bisho. Zwelitsha means 'new era'.

Population 47 000
Map South Africa Cc

Zwickau ***East Germany*** Industrial city about 65 km (40 miles) south of the city of Leipzig. The Romantic composer Robert Schumann (1810-56) was born there. Its manufactures include porcelain, dyes, pharmaceuticals, textiles and, most importantly, cars.

Population 120 100
Map East Germany Cc

Zwolle ***Netherlands*** Capital of the province of Overijssel, some 75 km (45 miles) east and slightly north of Amsterdam. Within its canal-encircled medieval centre stands the marketplace, town hall and a 15th-century church. Its manufactures include ships, iron and chemicals.

Population 86 400
Map Netherlands Ca

▼ SECRETIVE CAPITAL The Limmat glides into the lake near Zürich's lofty twin-towered cathedral. The country's banking capital – noted for its confidentiality – is also its industrial heart. The Bahnhofstrasse – the 'Fifth Avenue of Switzerland' – is a shopper's paradise for luxury goods.

FLAGS OF THE WORLD

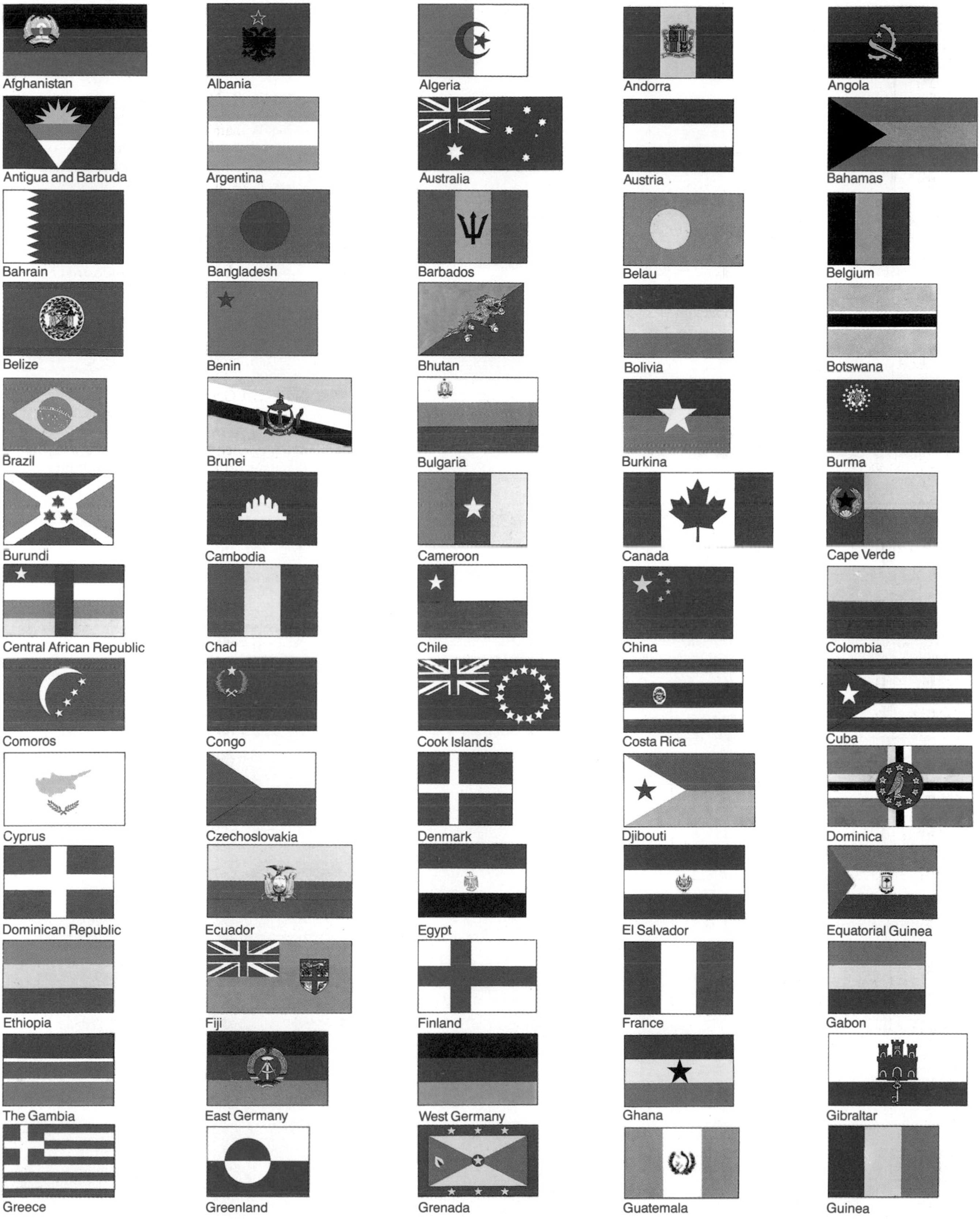
Afghanistan
Albania
Algeria
Andorra
Angola
Antigua and Barbuda
Argentina
Australia
Austria
Bahamas
Bahrain
Bangladesh
Barbados
Belau
Belgium
Belize
Benin
Bhutan
Bolivia
Botswana
Brazil
Brunei
Bulgaria
Burkina
Burma
Burundi
Cambodia
Cameroon
Canada
Cape Verde
Central African Republic
Chad
Chile
China
Colombia
Comoros
Congo
Cook Islands
Costa Rica
Cuba
Cyprus
Czechoslovakia
Denmark
Djibouti
Dominica
Dominican Republic
Ecuador
Egypt
El Salvador
Equatorial Guinea
Ethiopia
Fiji
Finland
France
Gabon
The Gambia
East Germany
West Germany
Ghana
Gibraltar
Greece
Greenland
Grenada
Guatemala
Guinea

FLAGS OF THE WORLD

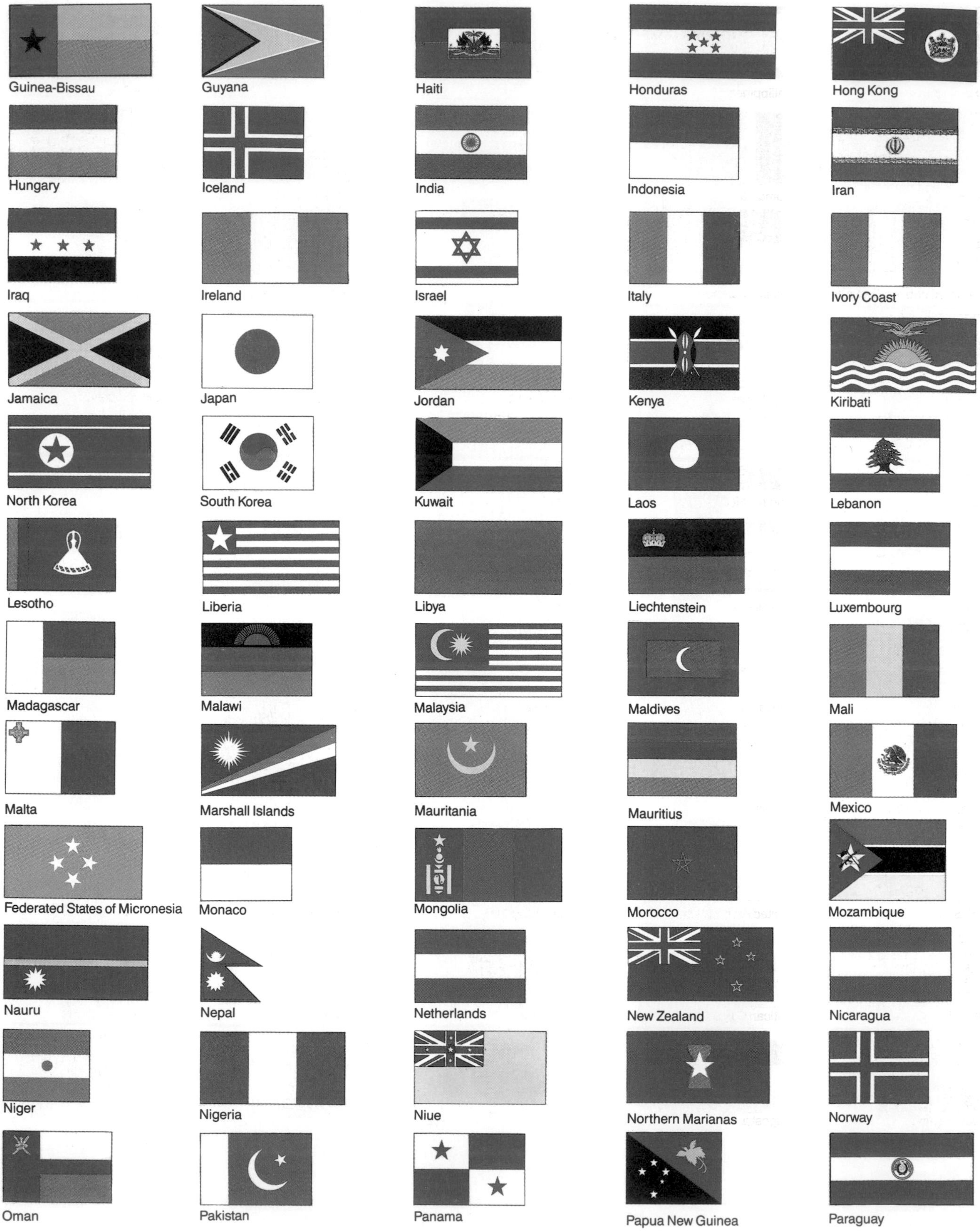
Guinea-Bissau
Guyana
Haiti
Honduras
Hong Kong
Hungary
Iceland
India
Indonesia
Iran
Iraq
Ireland
Israel
Italy
Ivory Coast
Jamaica
Japan
Jordan
Kenya
Kiribati
North Korea
South Korea
Kuwait
Laos
Lebanon
Lesotho
Liberia
Libya
Liechtenstein
Luxembourg
Madagascar
Malawi
Malaysia
Maldives
Mali
Malta
Marshall Islands
Mauritania
Mauritius
Mexico
Federated States of Micronesia
Monaco
Mongolia
Morocco
Mozambique
Nauru
Nepal
Netherlands
New Zealand
Nicaragua
Niger
Nigeria
Niue
Northern Marianas
Norway
Oman
Pakistan
Panama
Papua New Guinea
Paraguay

FLAGS OF THE WORLD

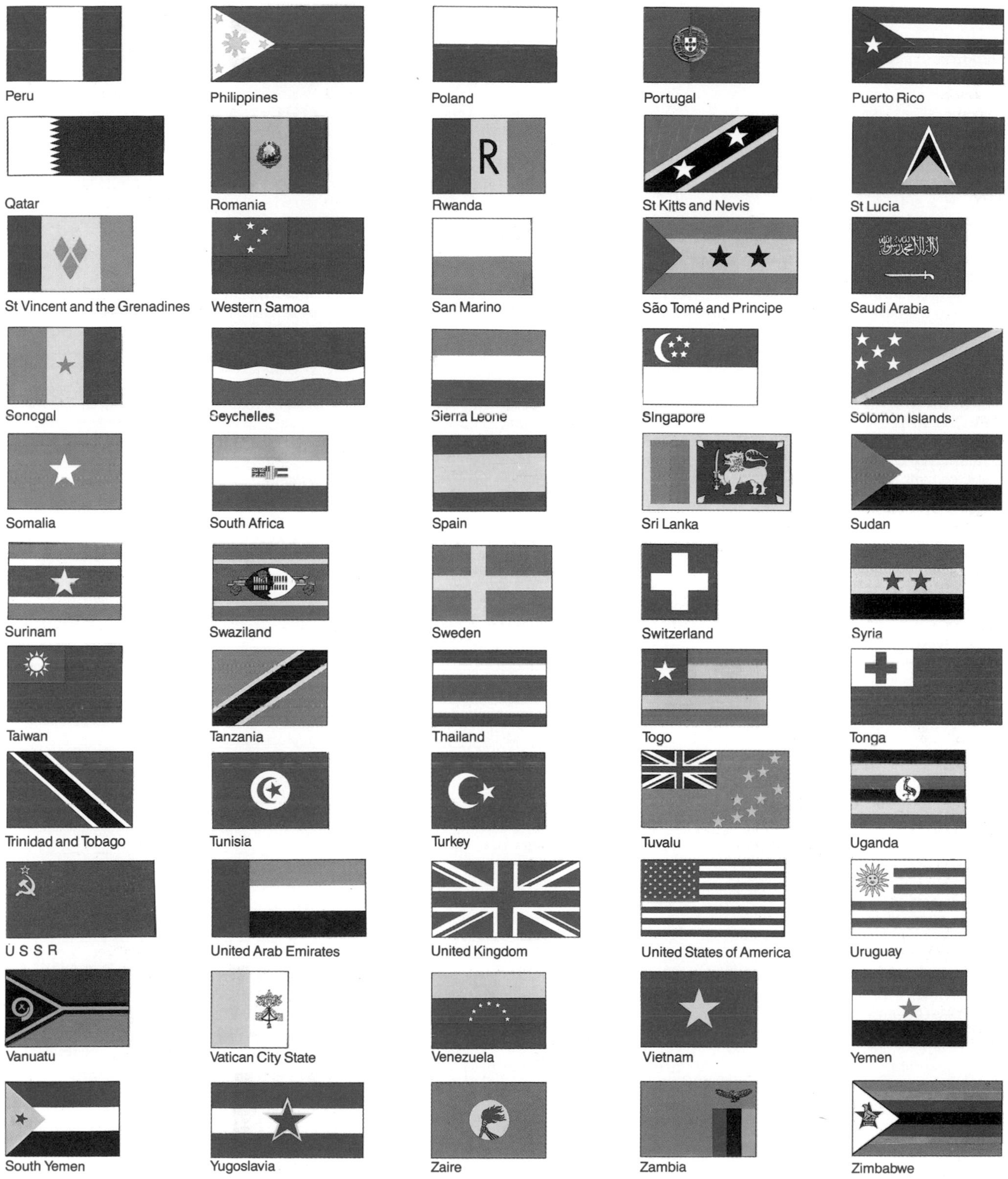
Peru
Philippines
Poland
Portugal
Puerto Rico
Qatar
Romania
R
Rwanda
St Kitts and Nevis
St Lucia
St Vincent and the Grenadines
Western Samoa
San Marino
São Tomé and Principe
Saudi Arabia
Senegal
Seychelles
Sierra Leone
Singapore
Solomon Islands
Somalia
South Africa
Spain
Sri Lanka
Sudan
Surinam
Swaziland
Sweden
Switzerland
Syria
Taiwan
Tanzania
Thailand
Togo
Tonga
Trinidad and Tobago
Tunisia
Turkey
Tuvalu
Uganda
U S S R
United Arab Emirates
United Kingdom
United States of America
Uruguay
Vanuatu
Vatican City State
Venezuela
Vietnam
Yemen
South Yemen
Yugoslavia
Zaire
Zambia
Zimbabwe

International Organisations and Alliances

There are dozens of international alliances and agencies devoted to promoting international cooperation of one kind or another. Some are global, others regional; some are much more active and effective than others. Among the more important and vigorous are the following.

Association of South East Asian Nations (ASEAN)

Founded in 1967 by a group of non-Communist nations of the region in order to aid economic, cultural and social development. It was recently active in trying to obtain a Vietnamese withdrawal from Cambodia. Members are Brunei, Indonesia, Malaysia, the Philippines, Singapore and Thailand. Its headquarters are in Jakarta, Indonesia.

Caribbean Community (CARICOM)

Founded by four nations in 1973 and expanded the next year by taking in members of the Caribbean Free Trade Association (CARIFTA). It fosters economic cooperation (through the Caribbean Common Market) as well as aiming to coordinate foreign policy and social development. Antigua and Barbuda, Bahamas (not a member of the Caribbean Common Market), Barbados, Belize, Dominica, Grenada, Guyana, Jamaica, Montserrat, St Kitts and Nevis, St Lucia, St Vincent, and Trinidad and Tobago are members. Its headquarters are in Georgetown, Guyana.

Colombo Plan

An economic development programme for South and South-east Asia and the Pacific region. It began in 1950 with seven members, and now has 26: Afghanistan, Australia*, Bangladesh, Bhutan, Burma, Cambodia, Canada*, Fiji, India, Indonesia, Iran, Japan*, Laos, Malaysia, Maldives, Nepal, New Zealand*, Pakistan, Papua New Guinea, the Philippines, Singapore, South Korea, Sri Lanka, Thailand, the United Kingdom*, and the USA*. The six countries marked with an asterisk are donor members who, between them, provided development aid under the Plan exceeding US$3750 million in 1984. Its headquarters are in Colombo, Sri Lanka.

The Commonwealth

One of the loosest but most far-reaching international organisations, with no formal charter but (as of 1986) spanning 49 nations and a quarter of the world's population – all at one time ruled by or a protectorate of Britain. Originally defined in 1926 as a group of 'autonomous communities within the British Empire', what ties member states today is recognition of Queen Elizabeth II as head of the Commonwealth. More than half the members are republics, five have their own monarchs and in only 18 is the Queen also head of state. Two nations have left the Commonwealth – South Africa (1961) and Pakistan (1972) – while several others did not join on becoming independent.

Commonwealth heads of government meet regularly (in different capitals) to discuss common political and economic concerns, and there is a permanent secretariat in London. The Commonwealth Games are held every four years; the 1986 venue was Edinburgh.

Council for Mutual Economic Assistance (CMEA, COMECON)

Main economic organisation of the Eastern Bloc, founded in 1949 by Bulgaria, Hungary, Poland, Romania and the USSR. Other members are Cuba, East Germany, Mongolia and Vietnam; Yugoslavia is a part-member, while Albania joined in 1949 but left in 1961. COMECON's headquarters are in Moscow.

European (EC) Community

Powerful western European body that has, since 1967, combined three formerly quite separate communities: the European Coal and Steel Community (the first, founded in 1952), the European Atomic Energy Community (Euratom, concerned with civilian nuclear power) and – best known of the three – the European Economic Community (EEC or Common Market); Euratom and the EEC were founded in 1958 following the signing of the Treaty of Rome in 1957. The EEC forms a customs union to abolish trade barriers between members, operates the Common Agricultural Policy (which regulates community farm prices) and the European Monetary System (to stabilise exchange rates), and ensures the free movement of goods, labour and capital between members.

The original six EC members – Belgium, France, Italy, Luxembourg, the Netherlands and West Germany – were joined in 1973 by Britain, Denmark and Ireland; Greece joined in 1981, and Spain and Portugal in 1986. Greenland, which became a member as a dependency of Denmark, withdrew in 1985. The Community's main headquarters are in Brussels, where the European Commission – its executive – is based. Other Community bodies include the Council of Ministers (of member states); the European Parliament, which meets in Luxembourg and Strasbourg and is mainly consultative; and the European Court of Justice, also in Luxembourg. The Community's main source of income is a proportion of value-added taxes collected in member countries, and its main expenditure is on the Common Agricultural Policy.

European Free Trade Association (EFTA)

Formed as a tariff-free area by seven countries in 1960, following the failure of Britain's first attempt to join the Common Market, this is now reduced to five as Denmark, Portugal and the UK have left to join the European Community. There remain original members Austria, Norway, Sweden and Switzerland, plus latecomer Iceland; Finland is an associate member. Following agreements between EFTA and the EEC, there is now virtually free trade in industrial goods between members of the two bodies.

League of Arab States (Arab League)

Regional, primarily political, grouping founded in 1945 by seven Arab nations, and now expanded to 21 stretching from Mauritania and Morocco to Iraq and Oman. The Palestine Liberation Organisation is a member, but Egypt's membership was suspended in 1979 after it signed a treaty with Israel. The League's secretariat is in Tunis.

North Atlantic Treaty Organisation (NATO)

Main Western military alliance, founded in 1949 by Belgium, Canada, Denmark, France, Iceland, Italy, Luxembourg, the Netherlands, Norway, Portugal, the UK and USA. Subsequently, West Germany, Greece, Turkey and Spain also joined. NATO headquarters are in Brussels, and its military command – which in an emergency would assume control over members' armed forces except those of Iceland (which has none) and France (liaison status only) – is also in Belgium, near Mons.

Organisation for Economic Cooperation and Development (OECD)

Formed in 1961 from members of the Organisation for European Economic Cooperation (OEEC) to promote world trade and economic development. Its members include all the main western European nations (including neutral Austria, Sweden and Switzerland) plus Australia, Canada, Japan, New Zealand and the USA; Yugoslavia has a special status. The organisation is based in Paris.

Organisation of African Unity (OAU)

Founded by 32 nations at a conference in Addis Ababa, Ethiopia (where its headquarters remain), this now has some 50 member states throughout the continent. Its aims are to promote African unity and development, and to eliminate colonialism.

Organisation of American States (OAS)

Founded in 1948 but tracing its origins back to the first International Conference of American States in 1890, this now has over 30 members in North, Central and South America and the West Indies. Its original emphasis was on regional security and independence, but it has increasingly promoted economic and social development. Its secretariat is in Washington DC.

Organisation of Petroleum Exporting Countries (OPEC)

Founded in 1960 to promote the interests of oil producers, it was highly effective in raising crude oil prices in the 1970s and early 1980s. But its influence waned with the oil glut of the mid-1980s and the failure to control the output of major oil-producing non-members such as

Mexico, Norway, the UK and USA. Its headquarters are in Vienna, but over half its 13 members are in the Middle East.

South Pacific Forum

Regional political and economic grouping of 13 independent states: Australia, Cook Islands, Fiji, Kiribati, Nauru, New Zealand, Niue, Papua New Guinea, Solomon Islands, Tonga, Tuvalu, Vanuatu and Western Samoa. It has vocally opposed French nuclear testing in the Pacific and continued French rule in New Caledonia. Its headquarters are in Suva, Fiji.

The United Nations (UN)

Founded in 1945 as the successor to the League of Nations to foster international peace, security and cooperation. Most countries are members – 161 in all in 1986, including two of the constituent republics of the USSR, Belorussia and the Ukraine, as well as the USSR itself. Notable exceptions are the two Koreas, Switzerland (on account of its neutrality) and Taiwan (replaced by mainland China in 1971). All members have a seat in the General Assembly, which meets annually, but only five (China, France, the USSR, UK and USA) are permanent members of the Security Council, which functions continuously and is responsible for maintaining international peace and security; an additional ten members serve on the Security Council for two-year terms.

The main UN headquarters are in New York, but the International Court of Justice, another UN body, is based in The Hague. There are also numerous specialised agencies, but not all UN members participate in all of these. They include the Food and Agriculture Organisation (FAO), based in Rome, General Agreement on Tariffs and Trade (GATT; Geneva), International Atomic Energy Agency (IAEA; Vienna), International Labour Organisation (ILO; Geneva), International Monetary Fund (IMF; Washington DC), United Nations Educational, Scientific and Cultural Organisation (UNESCO; Paris) and World Health Organisation (WHO; Geneva).

Warsaw Pact

The Eastern Bloc's equivalent to NATO, founded in 1955 and also known as the Eastern European Mutual Assistance Treaty. Its headquarters are in Moscow and its members are Bulgaria, Czechoslovakia, East Germany, Hungary, Poland, Romania and the USSR. Yugoslavia is not a member and Albania ceased to participate in 1962.

Acknowledgments

Artwork on pages 484 (bottom), 519 (bottom) was originally commissioned for *The Reader's Digest Great World Atlas*. All other artwork was commissioned for this book.

The photographs in this book came from the sources listed below. Those commissioned by Reader's Digest appear in *italics*.

Front cover TL, J. Jackson/Robert Harding Picture Library; BR, M. Cator/Impact Photos. Back cover TL, G. Tortoli/Zefa; TR, G. Gerster/John Hillelson Agency; BR, J. Langley/Aspect Picture Library. Inside, Maurice & Katia Krafft. 1 Michael Freeman. 2 B. Glinn/Magnum/John Hillelson Agency. 5 TL, D. Turner/Colorific!; TR, Preman Sotomayor; B, V. Englebert/Susan Griggs Agency. 6-7 I. Morath/Magnum/John Hillelson Agency. 7 J. Yates/Susan Griggs Agency. 15 Borrel, Campion/Gamma/Frank Spooner Pictures. 16-17 R. & S. Michaud/John Hillelson Agency. 19 J. Gardey/Robert Harding Picture Library. 20-21 S. Kraserman/Bruce Coleman Ltd. 24 Robert Estall Photographs. 27 G. Tortoli/Zefa. 29 Michael Freeman. 36-37 B. Gerard/Hutchison Picture Library. 45 P. Frey/The Image Bank. 50 N. Westwater/Robert Harding Picture Library. 55 U. Seer/The Image Bank. 58 G. Ricatto/Zefa. 59 A. Woolfitt/Susan Griggs Agency. 61 H. Munzig/Susan Griggs Agency. 67 B. Barbey/Magnum/John Hillelson Agency. 71 R. Perry/Impact Photos. 73 Patrick Eagar. 75 H. Sykes/Impact Photos. 81 K. Benser/Zefa. 85 W. Filser/The Image Bank. 90 F. Breig/Zefa. 91 J. Cleare/Mountain Camera. 92 R. Smith/Tony Stone Worldwide. 95 T. Ives/Susan Griggs Agency. 96 B. Barbey/Magnum/John Hillelson Agency. 101 W. Rawlings/Robert Harding Picture Library. 104-5 M. Vautier/vdn Picture Library. 108 B. Barbey/Magnum/John Hillelson Agency. 111 P. Vauthey/Sygma/John Hillelson Agency. 117 M. Freeman/Bruce Coleman Ltd. 119 F. Damm/Zefa. 122 P. Griffiths/Magnum/John Hillelson Agency. 126-7 A. Farquar/Daily Telegraph Colour Library. 128 T. Fincher/Colorific! 135 Tom Hanley. 141 A. Woolfitt/Susan Griggs Agency. 143 A. le Garsmeur/Impact Photos. 145 M. Koene/Colorific! 147 B. Barbey/Magnum/John Hillelson Agency. 148-9 B. Barbey/Magnum/John Hillelson Agency. 152 E. Arnold/Magnum/John Hillelson Agency. 157 A. Evrard/Susan Griggs Agency. 158-9 N. Tomalin/Bruce Coleman Ltd. 163 D. Beatty/Susan Griggs Agency. 166 R. Woldendorp/Susan Griggs Agency. 169 J. Gascoigne/Robert Harding Picture Library. 172-3 S. Meiselas/Magnum/John Hillelson Agency. 175 B. Seed/John Hillelson Agency. 177 L. Woodhead/Hutchison Picture Library. 178 J. Davis/Robert Harding Picture Library. 181 *Malcolm Aird*. 182 E. Young/Robert Harding Picture Library. 185 A. Durand/Robert Harding Picture Library. 195 A. le Garsmeur/Impact Photos. 201 M. Beebe/The Image Bank. 204-5 K. Langley/Aspect Picture Library. 209 A. Deane/Bruce Coleman Ltd. 211 B. Campbell/John Hillelson Agency. 213 G. Gerster/John Hillelson Agency. 215 D. Burnett/Contact Press Images/Colorific! 217 D. Barrault/Robert Harding Picture Library. 221 D. Muscroft/Falklands Pictorial. 227 H. Wiesner/Zefa. 229 G. Mangold/The Image Bank. 230 G. Davis/Colorific! 233 Robert Estall Photographs. 234 H. Munzig/Susan Griggs Agency. 237 L. Arepi/The Image Bank. 238 T. Okuda/Aspect Picture Library. 245 H. Uthoff/The Image Bank. 247 M. Fogden/Bruce Coleman Ltd. 249 I. Berry/Magnum/John Hillelson Agency. 251 Hilmar/Zefa. 252 H. Verhufen/Focus. 255 Robert Estall Photographs. 261 P. Carmichael/Aspect Picture Library. 263 H. Hanser/The Image Bank. 265 J. Stage/The Image Bank. 267 A. Woolfitt/Susan Griggs Agency. 269 S. & D. Cavannagh/Robert Harding Picture Library. 270 W. Ferchland/Zefa. 275 Michael Freeman. 276 D. Reed/Impact Photos. 277 P. & C. Leimbach/Robert Harding Picture Library. 284 C. Joyce/Impact Photos. 289 C. Pillitz/Impact Photos. 293 G. Gerster/John Hillelson Agency. 295 Photri/Zefa. 297 UWF/Zefa. 300 K. de Francke/Bilderberg. 302 J. Poncar/Bruce Coleman Ltd. 304 R. & S. Michaud/John Hillelson Agency. 308 M. Vautier/vdn Picture Library. 315 D. Burnett/Contact Press Images/Colorific! 317 H. Sykes/Impact Photos. 319 P. Carmichael/Aspect Picture Library. 320-1 C. Molyneux/Bruce Coleman Ltd. 323 A. Havlicek/Zefa. 326-7 Rainbird/Robert Harding Picture Library. 328 P. Carmichael/Aspect Picture Library. 331 J. Lister/Hutchison Picture Library. 334 E. Bleicher/Zefa. 337 K. Kurita/Gamma/Frank Spooner Pictures. 339 R. Magruder/The Image Bank. 341 A. Woolfitt/Susan Griggs Agency. 343 S. Sassoon/Robert Harding Picture Library. 344 C. Jopp/Robert Harding Picture Library. 348 J. Cleare/Mountain Camera. 352 B. O'Connor/Robert Harding Picture Library. 354-5 R. Gilmor/Bruce Coleman Ltd. 357 P. Baker/Photobank. 358 R. Harding/Robert Harding Picture Library. 364 Sally & Richard Greenhill. 368 Sally & Richard Greenhill. 375 C. Jones/Impact Photos. 379 D. Simpson/Hutchison Picture Library. 383 P. Chauvelot/Sygma/John Hillelson Agency. 384 I. Yeomans/Susan Griggs Agency. 389 P. Tweedie/Sunday Times, London. 391 M. Cator/Impact Photos. 394 C. Jones/Impact Photos. 398 J. Langley/Aspect Picture Library. 401 R. Smith/Tony Stone Worldwide. 403 G. Cubitt/Bruce Coleman Ltd. 404 R. Harding/Robert Harding Picture Library. 409 M. Vautier/vdn Picture Library. 411 I. Griffiths/Robert Harding Picture Library.412 A. Woolfitt/Susan Griggs Agency. 413 D. Houston/Bruce Coleman Ltd. 414 James Davis Photography. 417 M. Andrews/Susan Griggs Agency. 422 R-N. Guidicelli/Hutchison Picture Library. 425 Zefa. 427 B. Glinn/Magnum/John Hillelson Agency. 429 Hoagland/Liaison/Frank Spooner Pictures. 432 B. Davis/Aspect Picture Library. 435 G. Lincoln/Aspect Picture Library. 436 Stephanie Colasanti. 439 J. Highet/Hutchison Picture Library. 440 G. Zimbel/Colorific! 443 J. Jackson/Robert Harding Picture Library. 444 M. Jacot/Susan Griggs Agency. 445 D. Levenson/Colorific! 447 C. Bonnington/Bruce Coleman Ltd. 449 S. Meiselas/Magnum/John Hillelson Agency. 453 M. Reardon/Tony Stone Worldwide. 459 D. Bayes/Aspect Picture Library. 460-1 M. Cator/Impact Photos. 463 A. Hutchison/Hutchison Picture Library. 466 K. Benser/Zefa. 469 G. Hunter/Bruce Coleman Ltd. 475 G. Gerster/John Hillelson Agency. 481 R. Lomas/Zefa. 485 Tony Stone Worldwide. 486 T. Nebbia/Aspect Picture Library. 489 R. Smith/Zefa. 491 R. Ellis/Robert Harding Picture Library. 492 J. Langley/Aspect Picture Library. 498 G. Gerster/John Hillelson Agency. 501 J-G. Jules/John Hillelson Agency. 505 F. Mayer/John Hillelson Agency. 508 Maroon/Zefa. 511 M. Koene/Colorific! 513 Cameraman/Zefa. 515 B. Coates/Bruce Coleman Ltd. 518 M. Berge/Bruce Coleman Ltd. 521 J. Cleare/Mountain Camera. 523 A. le Garsmeur/Impact Photos. 525 G. Gerster/John Hillelson Agency. 527 J. Margesson/Robert Harding Picture Library. 531 Stephanie Maze/Woodfin Camp, Inc. 535 A. le Garsmeur/Impact Photos. 539 M. Collier/Robert Harding Picture Library. 541 C. Jones/Impact Photos. 544 G. Mather/Robert Harding Picture Library. 549 N. Devore/Bruce Coleman Ltd. 551 Starfoto/Zefa. 557 Tony Stone Worldwide. 562 T. Okuda/Aspect Picture Library. 564 J. Belog/Black Star/Colorific! 569 G. Tortoli/Colorific! 571 P. Toler/Impact Photos. 572-3 A. Hander/Daily Telegraph Colour Library. 577 J. Silvestris/Frank Lane Agency. 579 L. Taylor/Hutchison Picture Library. 584 M. Shrimpton/Tony Stone Worldwide. 593 L. Myers/Bruce Coleman Ltd. 595 Terence McNally. 599 A. Woolfitt/Susan Griggs Agency. 600-1 A. Woolfitt/Susan Griggs Agency. 602 C. Weckler/The Image Bank. 605 R. Singh/John Hillelson Agency. 606 J. Glover/Tony Stone Worldwide. 609 J. Langley/Aspect Picture Library. 611 S. Errington/Hutchison Picture Library. 613 M. Yamashita/Colorific! 615 B. Davis/Aspect Picture Library. 617 W. Ferchland/Zefa. 621 M. St. Maur Sheil/Susan Griggs Agency. 623 M. Hackforth-Jones/Robert Harding Picture Library. 629 J. Rowan/Aspect Picture Library. 630 J-P. Ferrero/Ardea, London. 633 J. Fry/Bruce Coleman Ltd. 637 D. Hiser/The Image Bank. 645 R. Chesher/Planet Earth Pictures. 646 P. Steyn/Ardea, London. 649 F. Lanting/Bruce Coleman Ltd. 651 Michael Holford. 654 R. & S. Michaud/John Hillelson Agency. 659 I. Berry/Magnum/John Hillelson Agency. 660-1 J. Bryson/The Image Bank. 665 M. Andrews/Susan Griggs Agency. 672 Patrick Thurston. 673 Clive Coote. 674 M. St. Maur Sheil/Susan Griggs Agency. 678-9 Michael Freeman. 680 Michael Freeman. 681 J. Langley/Aspect Picture Library. 685 J. Korman/Bruce Coleman Ltd. 689 P. Dickerson/Susan Griggs Agency. 691 A. Burman/Aspect Picture Library. 692 J. Bulmer/Susan Griggs Agency. 695 Nakamura/Sygma/John Hillelson Agency. 697 W. Ruth/Bruce Coleman Ltd. 701 *Patrick Thurston*. 706 A. Williams/Planet Earth Pictures. 708 M. Freeman/Bruce

Coleman Ltd. 711 C. Folgmann/Zefa. 712 I. Bradshaw/Colorific! 714 J. Jackson/Robert Harding Picture Library. 717 A. Woolfitt/Susan Griggs Agency. 719 S. Kraserman/Bruce Coleman Ltd. 721 L. Hellbig/Zefa. 723 P. Maitre/Gamma/Frank Spooner Pictures. 725 R. Campbell/Bruce Coleman Ltd. 727 Bullaty/Lomeo/The Image Bank. 729 G. Gerster/John Hillelson Agency. 730 P. Justitz/Zefa. Endpaper, G. Gerster/John Hillelson Agency.

The publishers acknowledge their indebtedness to the following books, which were consulted for reference.

Africa: A Natural History by Leslie Brown (Hamish Hamilton); *Africa on a Shoestring* by Geoff Crowther (Lonely Planet, Australia); *Albania* (New Albania Magazine, Tiranë); *All-Asia Guide* (Far Eastern Economic Review, Hong Kong); *The Alps* (National Geographic Society, USA); *Amnesty International Report* (annual); *The Atlas of Mankind* (Mitchell Beazley); *The Atlas of the Universe* (Mitchell Beazley and George Philip); *Australia – A Travel Survival Kit* by Tony Wheeler (Lonely Planet, Australia). *Baedeker's Germany*; *Bali and Lombok – A Travel Survival Kit* by Mary Covernton and Tony Wheeler (Lonely Planet, Australia); *Berlitz Guides* to various parts of the world; *British Columbia* by G. E. Mortimore (Collins, Toronto). *Chambers's Biographical Dictionary* (Chambers); *Chambers's World Gazetteer* edited by T. C. Collocott and J. O. Thorne (Chambers); *Collins Atlas of the World*; *The Companion Guide to Jugoslavia* by J. A. Cuddon (Collins, London; Prentice-Hall, USA); *Continents Adrift: Readings from Scientific American* (W. H. Freeman). *A Dictionary of Geography* by F. J. Monkhouse (Edward Arnold). *Eastern Europe: A Geography of the Comecon Countries* by Roy E. H. Mellor (Macmillan Press); *Encyclopedia Americana*; *Encyclopedia Britannica* and its *Books of the Year*; *An Encyclopedia of World History* edited by William L. Langer (Harrap); *Explore Australia* (George Philip and O'Neil, Australia). *Facts About Germany* (Lexikothek Verlag, Gütersloh); *Fielding's Travel Guide to Europe* (Fielding, New York); *Fodor's Guides* to various parts of the world (Traveltex, New York; Hodder and Stoughton, London). *Gazetteer of the British Isles* (Bartholomew); *The Geographical Digest* (George Philip; annual); *A Glossary of Geographical Terms* edited by Sir Dudley Stamp and Audrey N. Clark (Longman); *The Great Cities* series, including *Dublin* by Brendan Lehane (Time-Life Books); *The Great Geographical Atlas* (Mitchell Beazley); *Guide to East Africa* by Nina Casimati (Travelaid, London; Hippocrene Books, New York); *The Guinness Book of Records* (Guinness Superlatives; annual). *Hong Kong, Macau and Canton* by Carol Clewlow (Lonely Planet, Australia). *Indonesia Handbook* by Bill Dalton (Moon Publications, California); *Insight Guides* to various countries, notably *Malaysia* and *Philippines* (APA Productions, Hong Kong; Harrap, London; Lansdowne, Australia; Prentice-Hall, USA); *Iraq: A Tourist Guide* (State Organisation for Tourism, Baghdad); *Isles of the South Pacific* (National Geographic Society, USA). *Korea Guide* by Edward B. Adams (Seoul International Publishing Co). *Library of Nations* series, notably *Arabian Peninsula, Australia, China, Germany, The Soviet Union* and *The United States* (Time-Life Books); *Life Nature Library* series, notably *South America* (Time-Life Books). *Man's Religions* by John B. Moss (Macmillan, New York); *Meteorological Glossary* (HMSO for the Meteorological Office); *Michael's Guide to South America* by Michael Shichor (Inbal Travel Information, Tel Aviv); *The Milepost: All-the-North Travel Guide* (Alaska Northwest Publishing Co, Anchorage); *Monuments of Civilization* series, notably *India* by Maurizio Taddei (Cassell and Reader's Digest). *National Geographic Atlas of the World* (National Geographic Society, USA); *The New Columbia Encyclopedia* (Columbia University Press); *The New Larousse Encyclopedia of the Earth* by Leon Bertin (Hamlyn); *North Sea Fields: Facts and Figures* (Shell UK). *Pacific Islands Yearbook* edited by Stuart Inder (Pacific Publications, Sydney); *The Penguin Dictionary of Geography* by W. G. Moore (Penguin Books); *The Penguin Dictionary of Geology* by D. G. A. Whitten with J. R. V. Brooks (Penguin Books); *Penguin Travel Guides* to various parts of the world; *Philips' Pocket Guide to the World* by Bernard Stonehouse (George Philip); *The Physical Earth* (Mitchell Beazley Joy of Knowledge Library); *Planet Earth* series, including *Continents in Collision* and *Volcanoes* (Time-Life Books); *Poland* by Marc Heine (B. T. Batsford); *Post Guides* to various parts of East and South-east Asia (South China Morning Post, Hong Kong). *Rand McNally Cosmopolitan World Atlas* (Rand McNally, USA); *The Rough Guide to Yugoslavia* by Martin Dunford and Jack Holland (Routledge and Kegan Paul). *Saudi Arabia: A MEED Practical Guide* (Middle East Economic Digest); *The Shell Guide to Europe* edited by Diana Petry (Michael Joseph); *South America and Central America: A Natural History* by Jean Dorst (Hamish Hamilton); *The South American Handbook* edited by John Brooks (Trade and Travel Publications; annual); *South Pacific Handbook* by David Stanley (Moon Publications, California); *Spirit of Asia* by Michael Macintyre (BBC Publications); *The Statesman's Yearbook* edited by John Paxton (Macmillan Press); *Statistical Yearbook* and various other UN publications (United Nations); *Switzerland* by François Jeanneret, Walter Imber and Franz auf der Maur (Kümmerly and Frey, Berne); *Syria Today* by Jean Hureau (Editions J. A., Paris). *The Times Atlas of China* (Times Books); *The Times Atlas of the Oceans* (Times Books); *The Times Atlas of the World* (Times Books); *The Times World Index Gazetteer* (Times Publishing Co); *Travel Guide to the People's Republic of China* by Ruth Lor Malloy (William Morrow; New York); *Traveller's Guide to West Africa* (IC Magazines). *United Nations List of National Parks and Protected Areas* (IUCN, Switzerland and UK). *Volcanoes* (HMSO for the Institute of Geological Sciences). *Webster's New Geographical Dictionary* (Merriam-Webster); *Whitaker's Almanack* (J. Whitaker; annual); *The World Factbook* (Central Intelligence Agency; annual); *The World in Figures* (The Economist); *World Tables* (Johns Hopkins University Press for the World Bank); *The World's Great Religions* by the editorial staff of Life (Collins); *The World's Wild Places* series (Time-Life Books); and official guides, surveys and statistical information published by numerous national tourist and information offices and national and regional governments.

Various editions of the following newspapers, magazines, periodicals and journals were also consulted: *The Economist*; *The Financial Times*; *Geo*; *The Geographical Journal* (Royal Geographical Society); *The Geographical Magazine*; *Keesing's Contemporary Archives* (Longman); *National Geographic*; *Newsweek International*; *The Observer* and its colour magazine; *South*; The Sunday Times (London) and its colour magazine; *Telegraph Sunday Magazine*; *Time* (Atlantic Edition); *The Times* (London).

Also of great value were a number of books published by regional Reader's Digest offices and their associated companies, including the following: *America the Beautiful* (USA); *Antarctica: Great Stories from the Frozen Continent* (Australia); *Atlas of Australia* (Australia); *Atlas of Canada* (Canada); *Atlas of Southern Africa* (South Africa); *Book of British Towns* (Drive Publications, UK); *Great World Atlas* (UK); *Illustrated Guide to Britain* (Drive Publications, UK); *Illustrated Guide to Britain's Coast* (Drive Publications, UK); *Library of Modern Knowledge* (UK); *Natural Wonders of the World* (France/USA); *Reader's Digest Almanac and Yearbook* (USA); *Reader's Digest Book of the Great Barrier Reef* (Australia); *Southern Africa: Land of Beauty and Splendour* (South Africa); *Trésors de France* (France); *Vie et Paysages des Montagnes de France* (France); *Wild Australia* (Australia); *Wild New Zealand* (Australia).

Typesetter: Sprint Productions Ltd, London
Separator: Colourscan Co. Pte. Ltd, Singapore
Paper: C. Townsend Hook Paper Co. Ltd, Snodland
Cloth: BN International UK, London
Printing: HunterPrint Group Plc, Peterlee
Binding: Hazell Watson & Viney Ltd, Aylesbury
Cartography: Cartographic Services (Cirencester) Ltd

40-042-1

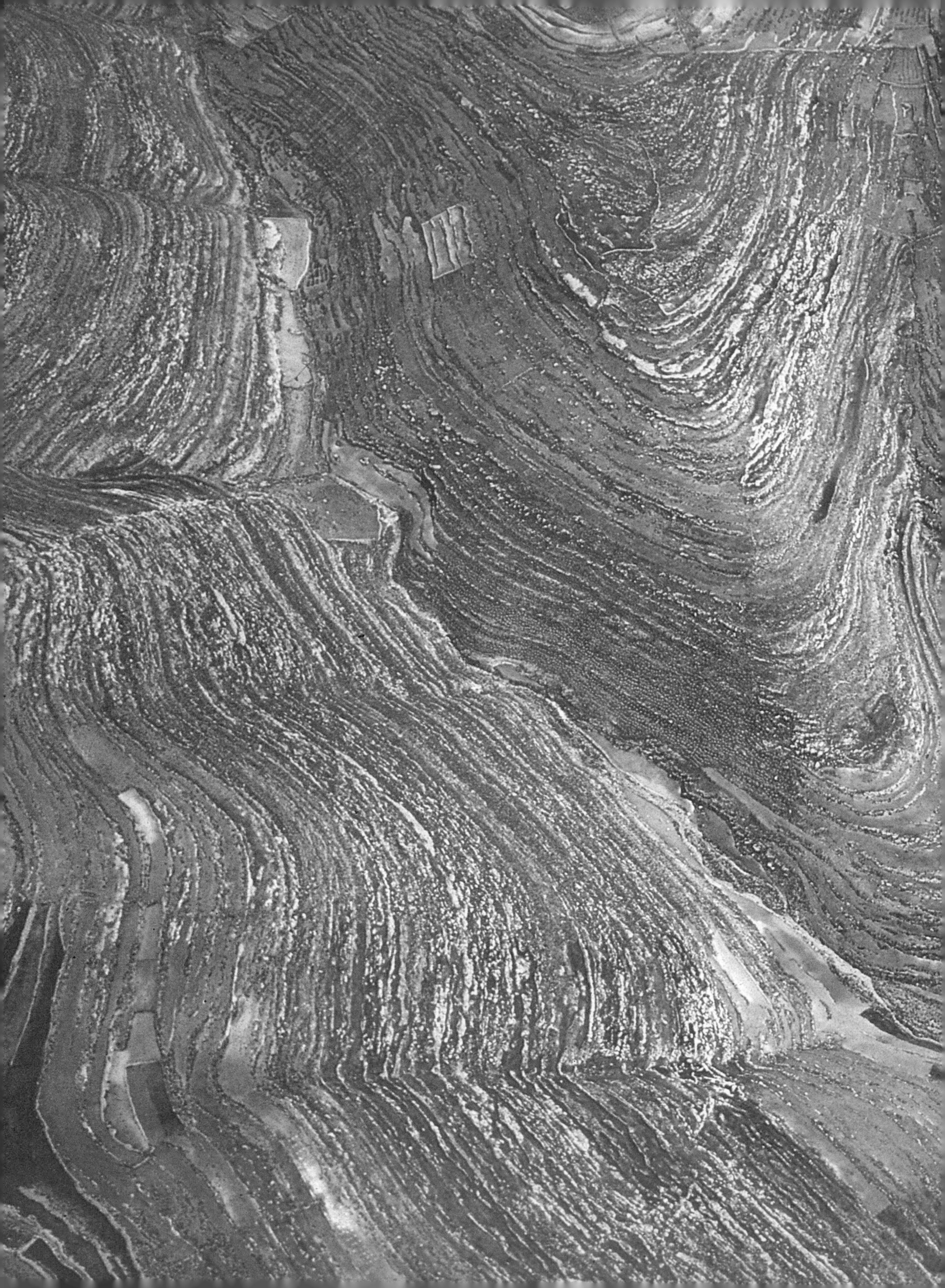